10th International Conference on Trends in Welding Research and 9th International Welding Symposium of Japan Welding Society (9WS)

Tokyo, Japan
11 – 14 October 2016

Volume 1 of 2

ISBN: 978-1-5108-4403-2

Printed from e-media with permission by:

Curran Associates, Inc.
57 Morehouse Lane
Red Hook, NY 12571

Some format issues inherent in the e-media version may also appear in this print version.

Printed by Curran Associates, Inc. (2017)

For permission requests, please contact American Welding Society
at the address below.

American Welding Society
8669 NW 36 Street, #130
Miami, Florida 33166-6672
USA

Phone: 800-443-9353
305-443-9353

www.aws.org

Additional copies of this publication are available from:

Curran Associates, Inc.
57 Morehouse Lane
Red Hook, NY 12571 USA
Phone: 845-758-0400
Fax: 845-758-2633
Email: curran@proceedings.com
Web: www.proceedings.com

Proceedings

10th International Conference on

Trends in Welding Research

&

9th International Welding Symposium of Japan Welding Society (9WS)

October 11 - 14, 2016

Hitotsubashi Hall,
National Center of Sciences Building
Tokyo, Japan

Organizing Society
American Welding Society (AWS)
Japan Welding Society (JWS)

AWS JWS

http://trends2016.org

Program

October 11th (Tuesday)

9:00-10:00 **Registration**

Main Hall

10:00-10:10 **Opening**
Opening addresses by Toshihiko Koseki, Stan A. David and Fumiyoshi Minami

10:10-10:50 **Keynote Lecture 1**
Session chair: Toshihiko Koseki

[S1] Arc welding 1
Session chair: Masaya Shigeta

Room 1

[S2] Weldability of high strength steel
Session chair: Kenji Oi

Program - October 11th (Tuesday)

Room 2

[S3] Laser welding 1
Session chair: Hiroto Yamaoka

Room 3

[S4] Solid-state bonding 1
Session chair: Kazuhiro Ito

Room 4

[S5] Residual stress 1
Session chair: Zhili Feng

Program - October 11th (Tuesday)

Main Hall

14:00-14:40 **Keynote Lecture 2**
Session chair: Toshihiko Koseki

15:00-17:30 **9WS Special Session: Activities of Japan Welding Society, Trends of Welding Research Activities in Japan, Trends of Welding Engineering Activities in Japan**
Fumiyoshi Minami, Manabu Tanaka, Kenji Shinozaki, Satoru Asai, Kenji Oi, Mitsuyoshi Nakatani

Room 1

[S6] Microstructure of steel weld
Session chair: Norbert Enzinger

Room 2

[S7] Laser welding 2
Session chair: Satoshi Igi

Program - October 11th (Tuesday)

Room 3

[S8] Solid-state bonding 2

Session chair: Yansong Zhang

Room 4

[S9] Residual stress 2

Session chair: Michael Andreassen

October 12th (Wednesday)

9:00-9:30 **Registration**

Main Hall

Room 1

Room 2

Room 3

[S13] Solid-state bonding 3

Session chair: Katsuyoshi Kondoh

Room 4

[S14] FSW process

Session chair: Yuri Hovanski

Program - October 12th (Wednesday)

Main Hall

Room 1

Room 2

Room 3

[S18] Dissimilar and carbon materials joining

Session chair: Shoichi Nambu

Room 4

[S19] FSW modeling

Session chair: Satish Kailas

October 13th (Thursday)

9:00-9:30 **Registration**

Main Hall

9:30-10:10 **Keynote Lecture 5**
Session chair: Stan A. David

[S20] Arc welding 4
Session chair: Yoshinori Hirata

Room 1

[S21] Welding of stailess steel
Session chair: Kazuyoshi Saida

Room 2

[S22] Computational fracture modeling
Session chair: Mitsuru Ohata

Room 3

[S23] Additive manufacturing 1

Session chair: Soshu Kirihara

Room 4

[S24] Steel FSW

Session chair: Hidetoshi Fujii

Program - October 13th (Thursday)

Main Hall

[S25] Arc welding 5

Session chair: Anthony Murphy

Room 1

[S26] Welding of stainless steel & Corrosion

Session chair: Kazuhiro Ogawa

Room 2

[S27] Computational welding mechanics

Session chair: Junya Inoue

14:10-14:30 **[S27-3] Prediction of Residual Stresses Distribution in External Attachment Welding Applications to Evaluate the Need of PWHT** 515
Rashed Alhajri, Stephen Liu

14:30-14:50 **[S27-4] Numerical Study on Influence of Specimen Preparation Procedure upon Fatigue Strength of Weld Joints** 521
Hidekazu Murakawa

14:50-15:10 **[S27-5] Numerical Investigation of Residual Stresses in Ti-6Al-4V Linear Friction Welds** 522
Clement Buhr, Paul Andrew Colegrove, Anthony McAndrew

Room 3

[S28] Additive manufacturing 2

Session chair: Wei Zhang

13:30-13:50 Invited **[S28-1] Machines and Equipment for Large Scale Wire + Arc Additive Manufacture** 526
Paul Andrew Colegrove, Anthony McAndrew, Pavel Kurzynski, Filomeno Martina, Stewart Williams, Jialuo Ding, Stephen Porter, Johnny v. Zwaag, Stephen Newman, Fangda Xu

13:50-14:10 **[S28-2] Effects of process parameters on defects in Ti-6Al-4V specimen built with EBM** 530
Masato Yamada, Yuuki Kozue, Masashi Mouri, Koji Nezaki

14:10-14:30 **[S28-3] Microstructure Control of Alloy 718 Additively Manufactured by Selective Electron-Beam Melting** 534
Yuichiro Koizumi, Shi-Hai Sun, Xiao Ding, Kenta Yamanaka, Kenta Aoyagi, Akihiko Chiba

14:30-14:50 **[S28-4] Stereolithographic Additive Manufacturing of Geometrically Modulated Metal and Ceramic Components** 535
Soshu Kirihara

14:50-15:10 **[S28-5] Thermal management in manufacture of thin-walled components produced by arc-based additive manufacturing** 536
Jayprakash P Sharma, Suresh Kumar Goak, Surya S Kumar, Kazuyuki Kohama, Abhay Sharma, Kazuhiro Ito

Room 4

[S29] Dissimilar FSW & FSSW

Session chair: Pedro Vilaca

13:30-13:50 Invited **[S29-1] Solid-State Joining of Dissimilar Materials with Friction Stir Scribe** 540
Yuri Hovanski, Piyush Upadhyay, Leonard Fifield, Saumyadeep Jana

13:50-14:10 Invited **[S29-2] Joining Aluminum Alloys to High Strength Steels by Friction Spot Welding** 544
Uceu F.H. Suhuddin, Vanessa F. Fischer, Jorge F. dos Santos

14:10-14:30 **[S29-3] Dissimilar Joining between Al and Mg alloys with FSW** 545
Masatoshi Enomoto, Shigeki Kakiuchi, Masafumi Araki, Kazuhiro Nakata

14:30-14:50 **[S29-4] Microstructural Characterization and Mechanical Properties of Fiction Spot Welded Aluminum Alloy AA6181-Ti6Al4V Dissimilar Joint** 549
Uceu F.H. Suhuddin, Jorge F. dos Santos, Athos H Plaine, Nelson G Alcântara

14:50-15:10 **[S29-5] Improving thermal stability of fine grain structure in weld nugget via in-situ intermetallic-particle formation during friction stir welding** 553
Cunli Liu, Muyang Deng, Ji Chen, Lijie Guo, Aidang Shan, Ke Chen

Program - October 13th (Thursday)

Main Hall

[S30] Arc welding 6
Session chair: Hiroyuki Shimizu

Room 1

[S31] Welding for automotive
Session chair: Yongbing Li

Room 2

[S32] Simulation for weld microstructure
Session chair: Dieter Siegele

Program - October 13th (Thursday)

Room 3

[S33] Surface modification

Session chair: Toshitaka Ikeshoji

Room 4

[S34] Monitoring and inspection

Session chair: Tetsuo Era

October 14th (Friday)

9:00-9:30 **Registration**

Main Hall

9:30-10:10 **Keynote Lecture 6**
Session chair: Toshihiko Koseki

[S35] Simulation for welding process 1
Session chair: Fumikazu Miyasaka

Room 1

[S36] Cold cracking in weld
Session chair: Tadashi Kasuya

Room 2

[S37] Automation
Session chair: Satoru Asai

Room 3

[S38] Nano/micro joining

Session chair: Tomokazu Sano

Room 4

[S39] FSW metallurgy in Al alloys

Session chair: Egoiz Aldanondo

Program - October 14th (Friday)

Main Hall

[S40] Simulation for welding process 2

Session chair: Masaya Shigeta

Room 1

[S41] Hot cracking in weld 1

Session chair: Hiroaki Mori

Room 2

[S42] Welding consumable

Session chair: Hiroyuki Shimizu

Room 3

[S43] Brazing and soldering

Session chair: Jaepil Jung

Room 4

[S44] FSW metallurgy

Session chair: Yutaka Sato

Program - October 14th (Friday)

Main Hall

[S45] Simulation for welding process 3
Session chair: Shinichi Tashiro

Room 1

[S46] Hot cracking in weld 2
Session chair: Kenji Shinozaki

Room 2

[S47] Pipeline
Session chair: Satoshi Igi

Program - October 12th (Wednesday)

Poster

16:10-18:00

Fundamental and applications of welding and joining process

Weld properties and performance

Program - October 12th (Wednesday)

October 11th
(Tuesday)

Advancement of Science and Technology in GMAW Processes - from Understanding to Prediction of Welding Phenomena -

Yoshinori HIRATA, Prof. Dr. Eng.
Osaka University, Institute for Academic Initiatives-1st Division, Suita, Osaka 565-0871, Japan
e-mail: hirata@mapse.eng.osaka-u.ac.jp, phone: +81-6-6879-4242

Abstract

Recent progress of visualization and simulation technology in gas shielded arc welding process has been making possible to measure and analyze welding phenomena with high resolution in both of time and space. Quantitative understandings result in evolution of welding process with more reliability and higher productivity. Arc plasma model provides heat source features at the surface of base metal depending on welding variables such as arc current and joint geometry. Metal transfer mode in GMAW can be predicted by using unified model considering interaction among arc plasma, molten droplet and weld pool.

Introduction

GMAW process is widely used in various production sectors because it provides high deposition rate and/or it is suitable for automatic/robotic welding. These years, high performance welding machine has been available, which enables fast and precise control of welding current and wire feeding. But GMAW processes have been still complex phenomena in which the behavior of gas metal arc with metal transfer are influenced by many variables: material, joint geometry, welding position, welding current and so on. Recent progress of visualization and simulation technology in welding process has been making possible to observe, measure and analyze welding phenomena with high resolution in both of time and space.

In 1941, Rosenthal found the analytical solution of quasi-state temperature distribution of the plate by moving heat source derived from heat conduction equation [1]. After World War II, electronic computer was developed and advanced year by year. Together with progress in both of hardware performance and numerical analysis technique, it was applied to various scientific and technological calculations. In welding technology, FEM analysis was spread into the welding mechanics field and stress distribution in weld joint was numerically analyzed in order to ensure the safety of welding structure in 1970s. FDM analysis of heat conduction equation was widely used to obtain temperature distribution of the base plate and predict weld penetration shape with using various heat source models.

With the aid of FDM analysis, Ohji quantitatively showed surface profile of weld pool in fillet welding and position welding through numerically solving Laplace equation on the basis of surface/interface tension theory [2]. Heiple and Roper found occurrence of penetration shape change induced by Marangoni flow which is caused by difference of surface tension depending on minor elements in stainless steel [3]. It promoted research on molten metal flow in the weld pool. Yokoya struggled to obtain the penetration shape of TIG arc spot welding considering convective heat transport of molten metal flow driven by surface tension, electro-magnetic force, shear stress of plasma flow and buoyancy [4]. Development of 3D numerical model in 1980s provided great help to understand heat and metal flow in weld pool with moving heat source [5][6]. David and DebRoy pointed out the importance of prediction of micro-structure and properties of weld region which determine the joint performance [7]. Numerical result of temperature and flow field of TIG arc plasma by Ushio contributed to understand heat generation in arc plasma, relationship between current path and plasma temperature, formation of flow field in electro-magnetic fluid and so on [8]. Numerical model of bridging transfer was developed by Hirata, et al. [9] and made clear the electro-magnetic pinch effect of short-circuiting current on breakup of liquid metal bridge.

These 10 years, remarkable progress of computer technology has been promoting numerical simulations of welding phenomena. As for deeply understanding the arc welding phenomena of electric arc, metal transfer and weld pool, various theoretical models and simulations have been proposed and discussed all over the world. Tanaka, et al. showed that penetration shape of TIG arc spot is more exactly predicted considering metal vapor from anode pool surface [10]. Fan, et al. numerically described temperature and flow field of MIG arc plasma with metal transfer [11]. But before around 2005, majority of numerical model related with welding phenomena was 2D axially symmetrical model [12]-[18]. Then numerical analysis have been almost conducted by 3D model which can treat asymmetric phenomena [19]-[24]. Recently, it is orienting to unified 3D numerical model considering interaction among arc plasma, weld pool and metal drop for making clear the complex welding phenomena.

Arc plasma model and heat source features

Recently, the research object on arc plasma model has been mainly oriented to obtain the solution of thermal MHD fluid under specific boundary condition on the assumption of local thermal equilibrium (LTE). In the calculation, shielding gas in welding process is used as plasma gas and temperature field is dynamically kept by balance of Joule's heating by electric current of a few hundred ampere and heat loss of conduction, convection and radiation. As temperature of arc plasma is distributed from room temperature to over 20,000K in the space range of around 10 mm, temperature dependence of the physical properties of the shielding gas must be considered. For example, as for electrical conductivity, at room temperature gas is almost electrical insulator but beyond around 4,000K it acts as electrical conductor. Currently modeling potential has been gradually going up for predicting arc plasma behavior in the wide range of welding variables. Welding phenomena include not only physical phenomena of mass, momentum and energy transport but various chemical reactions. But when we look at the heat source model widely used in commercial simulation software, arc plasma model makes possible to give heat source features on the physical base. Here are described heat source features derived from arc plasma model.

Figure 1 shows temperature and velocity field of TIG arc plasma calculated for current of 200A and arc length of 3mm [23]. As seen, the peak temperature of 21,800K is located close to the electrode tip. Arc plasma higher than 10,000K is conical shape as shown in *Fig. 1(a)*. Just above the anode surface, maximum temperature is 15,200 K at the center of TIG arc. The area higher than 10,000K is circular shape of about 8 mm in diameter as shown in *Fig.1 (b)*. And plasma flows from cathode electrode toward the anode plate, and then it flows along the anode surface. This plasma flow causes arc pressure and drag force at the weld pool surface.

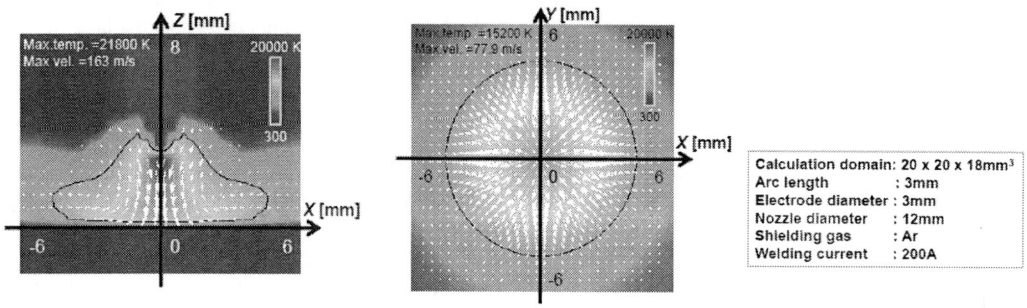

(a) Cross section *(b) Above the anode surface*

Fig.1 Temperature and velocity distribution of TIG arc on flat anode.

From calculation result, heat input distribution at the anode surface can be obtained. Heat input from TIG arc into base metal is mainly provided by heat conduction and current carrying electrons but exactly it is given by the following equation:

$$q_A = j(V_A + V_T + V_W) + q_c + q_r \qquad (1)$$

where, j : current density [A/m^2], V_A : anode fall voltage [V], $V_T (= 5k_B T / 2e)$: kinetic energy of electron, V_W : work function of anode metal [V], $q_c\left(= -\lambda \cdot \partial T / \partial z|_{\text{anode surface}}\right)$: heat conduction from arc plasma [W/m^2], q_r : heat input by radiation [W/m^2], k_B : Boltzmann constant [1.38×10^{-23} J / K], T : plasma temperature [K] and e : electron charge [1.6×10^{-19} C]. *Figure 2(a)(b)* shows temperature distribution just above the anode surface of the flat plate and current density distribution of electron into anode surface, respectively. From temperature distribution, we can obtain the temperature gradient at the anode surface, and then heat input conducted from arc plasma. Then, heat input carried by electron is provided by multiplying current density j by electron condensation potential $(V_T + V_W)$. As a result, we can obtain the heat input distribution at the anode surface as shown in *Fig.2(c)*. As seen, heat input calculated shows Gaussian like distribution, of which the peak value reaches an order of magnitude 10^8 W/m^2 and the size is about 6 mm in diameter. As for the flat anode, it is reasonable to describe heat input distribution approximated by Gaussian distribution.

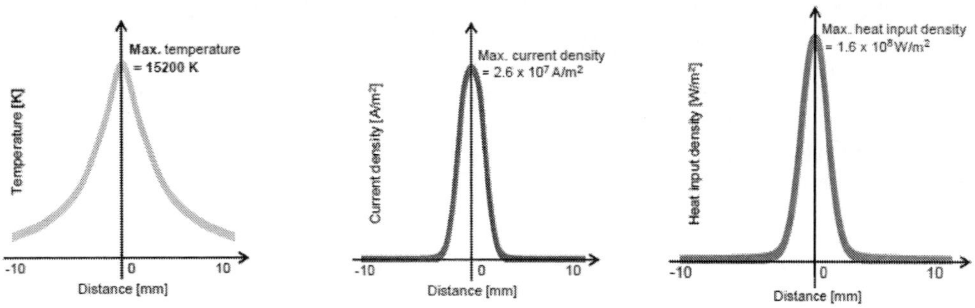

(a) Temperature distribution above anode surface *(b) Current density distribution* *(c) Heat input distribution*

Fig. 2 Heat input distribution of TIG arc derived from current density and plasma temperature above anode surface.

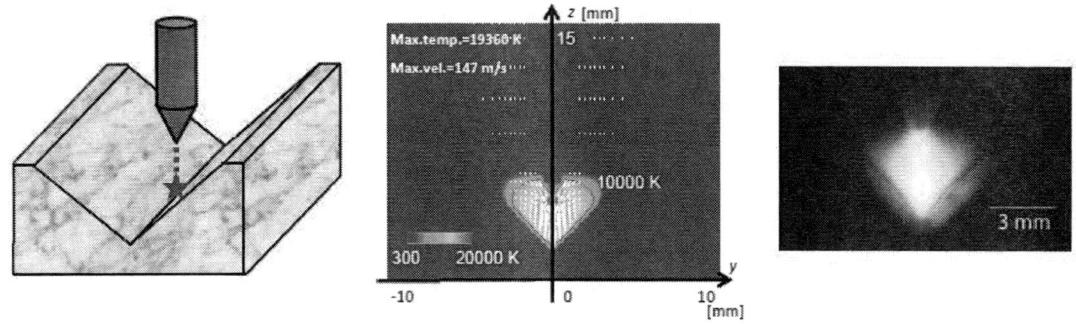

(a) TIG arc in V-grooved anode. *(b) Temperature and flow field of TIG arc plasma in V-groove calculated.* *(c) TIG arc appearance in V-groove.*

Fig.3 Temperature distribution and appearance of TIG arc in V-grooved anode.

In the case of TIG arc on V-grooved anode as shown in *Fig.3 (a)*, heat input distribution is numerically investigated. Base plate or anode of electric arc is one of boundary conditions. Accordingly arc plasma shape must be varied with the anode shape. As seen in *Fig.3 (b)*, temperature distribution of arc plasma calculated is much changed compared with that on flat anode. The calculation result agrees well with TIG arc appearance as shown in *Fig.3 (c)*. Next, heat input distribution on the grooved anode is calculated with using equation (1). As shown in *Fig.4*, two peaks of heat input exist at the both groove surfaces. This is mainly caused by current path in the arc plasma, which branches out into two paths toward both anode surfaces. As arc plasma is electrical conductor, current path and heat input distribution are much influenced by configuration of both electrodes.

As mentioned above, arc plasma shape and current path dominates heat source characteristics. Welding arc shape is varied with welding process and welding variables: current, arc length, shielding gas, electrode, etc. In addition, it is changed by joint geometry, groove shape and welding position. So, practically it is necessary to establish the procedure for obtaining heat input distribution according to specific welding process and welding variable. Then, those research activities will surely contribute to remarkable improvement in reliability and validity of numerical simulation with using available commercial software.

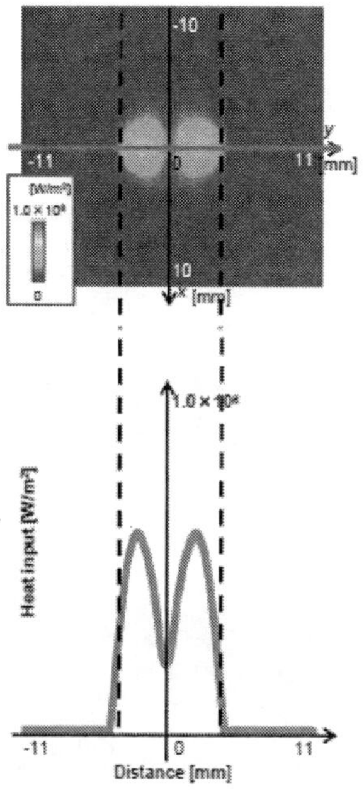

Fig.4 Heat input distribution in V-grooved anode.

High performance control of GMAW processes and numerical simulation

The digital controlled welding machine provides not only supplying very fast and precise current/voltage but wire feeding digitally controlled, which was considered as a weak point in GMAW processes for long years. Then, GMAW processes have been becoming high quality welding process giving extremely stable arc, which is equivalent to TIG welding arc, and simultaneously high productivity welding process giving high deposition rate. But in order to ensure the stability of the GMA equivalent to TIG arc, it would be required to prepare and maintain software for welding phenomena control providing the optimum combination of current, voltage and wire feeding. This means that information of dynamic behavior of arc, electrode wire and weld pool must link with control variables of digital welding machine.

As well known, in short-circuiting welding process, research and development have been carried out to realize stable welding with spatter free by means of adjustment and/or control of only welding current together with progression of power source using inverter-circuit in not only Japan but Western countries. But in practical applications it has not been satisfied because of restriction of the range of welding condition such as current, shielding gas and so on. The presentation and publication of CSC process in Aluminum MIG welding by Huismann notified that control of wire feeding speed and direction was very effective for stabilizing short-circuiting process and spatter reduction [27]. After then, welding machine makers developed to establish simultaneous control of current and wire feeding for practical fabrications [28]. *Fig.5* shows numerical simulation of bridging transfer to be controlled by wire feeding. *Fig.5 (a)* shows time-chart of welding current, wire melting speed and wire feed speed in wire feeding controlled method. *Fig.5 (b)* shows time-change of liquid metal profile calculated. As seen in the phase 1, electrode wire is melting and molten drop grows up at the wire tip by arc heating. In the phase 2, the arc current switches to low level value, then wire melting hardly occur. On the other hand, the wire feeding continues, then molten drop is approaching the surface of weld pool. In the phase 3, just after the contact of the drop with weld pool, the liquid metal flows into the pool by the motive force of capillary pressure. Then, the liquid bridge is forming between wire tip and weld pool, and subsequently breaking up by pulling up the wire tip together with wire-retracting motion. This shows the breakup of the liquid bridge is adjustable mechanically by precise movement of wire feeding in addition to current modulation. So, effective use of numerical simulation will be connected with suitable control variables corresponding to various requirements of fabrication.

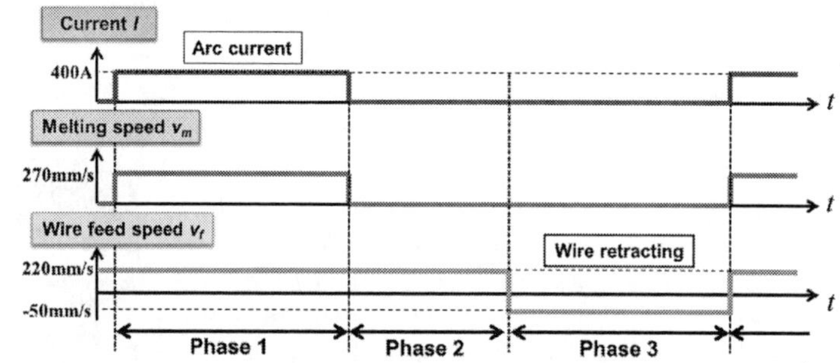

(a) Time-chart of welding current, wire melting speed and wire feed speed

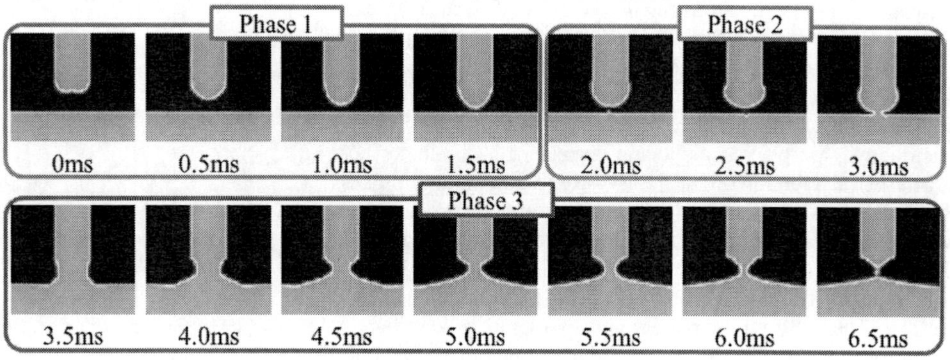

(b) Time-change of liquid metal profile calculated

Fig.5 Time-change of liquid metal profile in bridging transfer process by wire feed speed control method.

Figure 6 shows the numerical simulation results of metal transfer in GMAW with shielding gas of pure argon gas and mixture gas of Ar+18%CO_2. The arc current is set to be 260 A with use of 1.2 mm in dia. mild steel wire. The left-hand sides of the figures show the metal temperature, and the right-hand sides show the gas temperature. *Fig.6 (a)* shows the calculation of metal transfer in Ar gas shielded GMAW. The droplets detaching from the wire tip are smaller than the wire diameter. This means that numerical simulation agrees well with experiment in which spray transfer occurs. On the other hand, when the mixture gas is used, the droplet becomes larger than the wire diameter as shown in *Fig.6 (b)*. In the case of the mixture gas, globular transfer occurs in the lower current range than 280 A. Even if the arc currents are set to be the same, metal transfer mode depends on the shielding gas. As seen, the temperature in the core region of arc plasma is around 8000 K which is lower than that of the peripheral region. The numerical simulation also demonstrates temperature distribution of the Ar arc plasma which was spectroscopically measured [29]. The calculation results show the maximum temperature of the droplets is around 3000 K. Then, metal vapor is emitted from the surface of molten drop. Electric conductivity of arc plasma increases by mixing with a small amount of metal vapor because of low ionization potential. And radiation loss of metal vapor is higher than that of argon gas plasma. So, the temperature of the core plasma becomes lower than that of the peripheral plasma.

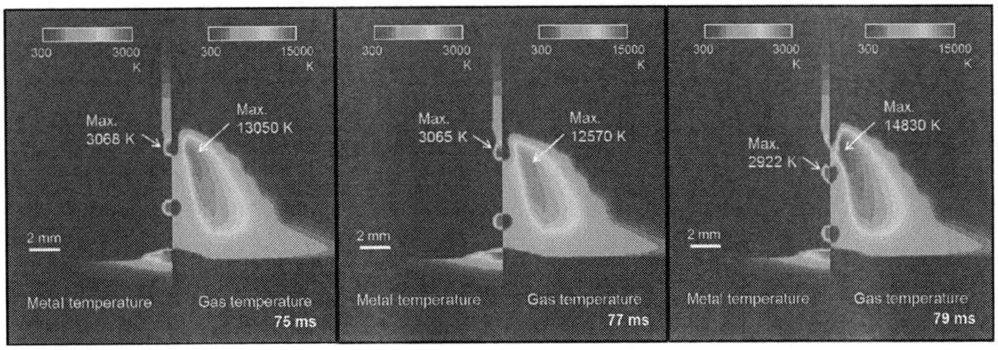

(a) Ar gas shielded

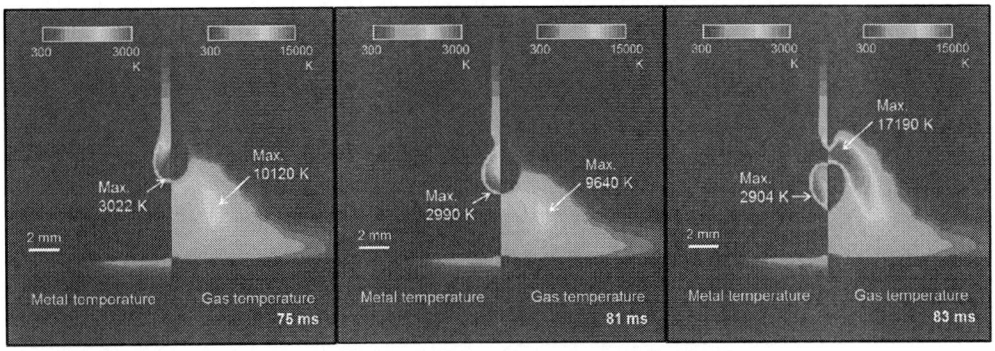

(b) Ar+18%CO_2 mixture gas shielded

Fig.6 Influence of shielding gas on plasma and metal temperature (arc current: 260 A)

Conclusions

Recent progress of visualization and simulation technology in gas shielded arc welding process has been making possible to measure and analyze welding phenomena with high resolution in both of time and space. Change from qualitative understanding to quantitative understanding result in evolution of welding process with more reliability and higher productivity. These years, high performance digital camera have been spread into industries and applied to practical fabrication: detection of electrode position and groove shape, monitoring arc/weld pool during welding operation, etc. These struggles inside industries will be more actively carried out by increase of available explicit knowledge.

Temperature and electron density of arc plasma are directly connected with the electrical conductivity, and then the current path in the arc discharge is governed by the characteristics of arc plasma. Metal droplet and/or molten pool surface reach considerably high temperature. Accordingly, metal vapor is emitted from the liquid metal surface and diffused into the plasma. Then, it greatly affects heat input mode and/or current path to base material and electrode wire.

While, in order to verify the numerical model and/or more exact set up of the boundary condition, it is necessary to conduct higher accurate measurement of temperature and fluid flow of weld pool and molten drop. The verification of existing numerical model leads to develop more advanced model. Then, the prediction based on the refined model is repeatedly compared with the experiment, and then it will clarify the mechanism of the arc welding phenomena. Consequently, it will be more connected for the development of the high-performance simulation software. It is also expected to predict phenomena of the wire arc additive manufacturing (WAAM) which is modern process but originated from GMAW process.

References

[1] D.Rosenthal: Mathematical Theory of Heat Distribution During Welding and Cutting, Weld. J., 1941, Vol.20, No.5, pp.220s-234s.
[2] T.Ohji: Analytical study on surface profile of weld bead and stability of molten pool based on capillary theory, 1978, doctoral thesis (in Japanese).
[3] C.R.Heiple and J.R.Roper: Mechanism for minor element effect on GTA fusion zone geometry, Weld. J. 1982, Vol.61, No.4, 97s-102s.
[4] S.Yokoya: Heat and metal flow mechanism in arc spot welding, 1989, doctoral thesis (in Japanese)
[5] T.Zacharia, A.H.Eraslan and D.K.Aidun: Modeling of Non-Autogenous Welding, Weld. J., 1988, Vol.67, No.1, 18s.
[6] T.Zacharia, A.H.Eraslan and D.K.Aidun: Modeling of Autogenous Welding, Weld. J., 1988, Vol.67, No.3, 53s
[7] S.A.David and T.DebRoy: Current Issues and Problems in Welding Science, Science, 1992, Vol.257, pp.497-502.
[8] M.Ushio, F.Matsuda: Numerical model of TIG arc plasma, Quarterly J. JWS, 1988, Vol.6, No.1, pp.91-98 (in Japanese).
[9] H.Maruo, Y.Hirata, N.Goto: Effect of electromagnetic pinch force on bridging transfer of pendant drop, Quarterly J. JWS, 1992, Vol.10, No.2, pp.43-50 (in Japanese).
[10] K. Yamamoto, M. Tanaka, S. Tashiro, K. Nakata and A. B. Murphy: Metal Vapor Behavior in GTA Welding of a Stainless Steel Considering the Marangoni Effect, Trans. on Electrical and Electronic Engineering, 2009, 4, pp. 497-503.
[11] H. G. Fan and R. Kovaceivic: A unified model of transport phenomena in gas metal arc welding including electrode, arc plasma and molten pool, J. Physics D: Applied Physics, 2004, 37, pp. 2531-2544
[12] M. Tanaka, S. Tashiro and J. J. Lowke: Predictions of weld formation using gas tungsten arcs for various arc lengths from unified arc-electrode model, STWJ, 2007, 12, pp. 2-9.
[13] J. Hu and H. L. Tsai: Metal Transfer and Arc Plasma in Gas Metal Arc Welding, J. Heat Transfer, 2007, 1028, 129, pp. 1025-1035.
[14] F. Lu, X. Tang, H. Yu and S. Yao: Numerical simulation on interaction between TIG welding arc and weld pool, Comp. Material Science, 2006, 35, pp. 458-465.
[15] J. Hu, H. Guo and H. L. Tsai: Weld pool dynamics and the formation of ripples in 3D gas metal arc welding, Inter. J. Heat and Mass Transfer, 2008, 51, pp. 2537-2552.
[16] J. Hu and H. L. Tsai: Modeling of transport phenomena in 3D GMAW of thick metals with V groove, J. Physics D: Applied Physics, 2008, 41, 065202.
[17] Z. H. Rao, J. Zhou, S. M. Liao and H. L. Tsai: Three-dimensional modeling of transport phenomena and their effect on the formation of ripples in gas metal arc welding, J. of Applied Physics, 2010, 107, 054905.
[18] W. Dong, S. Lu, D. Li and Y. Li: GTAW liquid pool convections and the weld shape variations under helium gas shielding, Inter. J. of Heat and Mass Transfer, 2011, 54, pp.1420-1431.
[19] P. Freton, J. J. Gonzalez, and A. Gleizes: Comparison between a two- and a three-dimensional arc plasma configuration, J. Physics D: Applied Physics, 2000, 33, pp. 2442-2452.
[20] G. Speckhofer and H. S. Schmidt: Experimental and Theoretical Investigation of High-Pressure Arcs - Part II: The Magnetically Deflected Arc (Three-Dimensional Modeling), IEEE Trans. on plasma science, 1996, 24, 4, pp. 1239-1248.

[21] G. Xu, J. Hu, H. L. Tsai: Three-dimensional modeling of the plasma arc in arc welding, J. of Applied Physics, 2008, 104, 103301.
[22] M. Schnick, G. Wilhelm, M. Lohse, U. Fussel and A. B. Murphy: Three-dimensional modeling of arc behavior and gas shield quality in tandem gas-metal arc welding using anti-phase pulse synchronization, J. Physics D: Applied Physics, 2011, 44, 185205.
[23] Y. Ogino, Y. Hirata and K.Nomura: Numerical analysis of the heat source characteristics of a two-electrode TIG arc, J. Physics D: Applied Physics, 2011, 44, 215202.
[24] A. B. Murphy: A self-consistent three-dimensional model of the arc, electrode and weld pool in gas-metal arc welding, J. Physics D: Applied Physics, 2011, 44, 194009.
[25] M. Hertel, A. Spille-Kohoff, U. Fuessel and M. Schnick: Numerical simulation of droplet detachment in pulsed gas-metal arc welding including the influence of metal vapour, J. Physics D: Applied Physics, 2013, 46, 224003.
[26] Y. Ogino, Y. Hirata and A. B. Murphy: Numerical simulation of GMAW process using Ar and an Ar-CO_2 gas mixture, Welding in the World, 60 (2016), pp. 345-353.
[27] G.Huismann: Introduction of a new MIG process - Advantages and possibilities, IIW Doc.212-952-99(1999).
[28] K.Himmelbauer: THE CMT-PROCESS - A REVOLUTION IN WELDING TECHNOLOGY, IIW Doc.XII-1 845-05(2005)
[29] M.E.Rouffet, M.Wendt, G.Goett, R.Kozakov, H.Schoepp, K.D.Weltmann and D.Uhrlandt: Spectroscopic investigation of the high-current phase of a pulsed GMAW process, J. Physics D: Applied Physics, 2010, 43, 434003.

Experimental investigations on single-sided CMT welding of hybrid aluminum-steel blanks

Zahra Silvayeh, Rudolf Vallant, Christof Sommitsch
Graz University of Technology, Institute of Materials Science and Welding, Graz, Austria
zahra.silvayeh@tugraz.at, rudolf.vallant@tugraz.at, christof.sommitsch@tugraz.at

Bruno Götzinger, Werner Karner
Magna Steyr Engineering Austria AG & Co KG and Magna Steyr Fahrzeugtechnik AG & Co KG, Graz, Austria
bruno.goetzinger@magna.com, werner.karner2@magna.com

Abstract

Single-sided Cold Metal Transfer welding of aluminum alloy EN AW 6014 T4 and zinc-coated dual phase steel HCT450X + ZE75/75 was investigated. These materials are typically used for automotive lightweight application, *e.g.* in multi-material car bodies. The microstructures of the weld seam and of the intermetallic layer between the steel sheet and the weld seam were characterized by conventional optical light microscopy and by scanning electron microscopy with energy-dispersive X-ray spectroscopy. The results of the experimental investigations indicate that the intermetallic layer has tongue/tooth-like morphology of less than 20 μm thickness, which is acceptable to meet the requirements regarding the mechanical properties. However, hot cracks which deteriorate the formability of the metal blanks in subsequent processing steps may form during the welding process. In order to reduce the hot cracking tendency different welding process parameters and filler materials were tested.

Introduction

Al alloys become increasingly important in the modern automotive industry, since their comparatively low density offers high weight-saving potentials. However, steels still dominate many applications in car body manufacturing. In order to benefit from the specific advantages of each of these materials, dissimilar joining of Al and steel is the key process for obtaining weight-optimized car bodies [1]. Two categories of welding processes are basically distinguished: fusion welding and solid state welding. Particularly fusion welding of Al and steels causes serious difficulties, because these materials exhibit significant differences in their chemical and physical properties and because brittle intermetallic phases may form [2],[3]. In order to overcome these problems solid state welding processes of reduced heat input can be applied, taking into account that their efficiency and flexibility are limited. An opportunity for joining Al to steel by combining the effects of both fusion as well as solid state welding offers the Cold Metal Transfer (CMT) welding process [4]. In comparison to conventional gas metal arc (GMA) welding processes the heat input of the CMT process is significantly reduced. Though the temperature is sufficient to melt the Al base and filler materials, the steel base material remains solid. Thus, dissimilar joining is achieved by the combination of Al welding and steel brazing. Fundamental investigations reveal that CMT welding is a stable process of high reproducibility particularly for joining thin sheets in the automotive industry [5],[6]. However, microstructure defects which deteriorate the mechanical properties and the formability of the welded Al-steel blanks in subsequent processing steps may form. The probably most serious defects occurring in welding of Al are pores and hot cracks. Gas porosity formation is mainly attributed to the distinct difference of hydrogen solubility in liquid and solid Al. During solidification, *i.e.* with decreasing temperature, hydrogen is expelled from the solid into the surrounding liquid where hydrogen bubbles may form beyond a certain level of content [7]. The weld porosity is influenced by four major factors: the initial hydrogen contents of the base and filler materials, the rate of hydrogen absorption, the volume of gas kept in solid solution in the weld and the rate at which gas bubbles escape [8]. An additional reason for the formation of pores exists particularly in arc welding of Zn-coated steel. Zn which evaporates at high arc temperatures from the galvanized surface of the steel may be entrapped into the liquid weld pool where it induces a large number of pores. Turbulent flows and fast solidification of the weld pool may enhance porosity formation [9]. Hot cracking is determined by a combination of metallurgical, thermal and mechanical factors. Hot cracks form in the weld bead at temperature regimes close to solidus, if tensile stresses induced by solidification shrinkage, thermal contraction or weldment restraints, expand the dendritic network. Hot cracks occur typically at grain boundaries where microsegregations accumulate, *e.g.* at the centerline of welds [7]. In the present work the influence of different CMT welding process parameters (filler material, torch type and torch-to-workpiece distance) on the tendencies of hot cracking and porosity formation inside of the weld seam are experimentally investigated, since reducing these serious welding defects is crucial for obtaining technologically sound weldments of high quality.

Experimental procedure

Single-sided CMT welding of Al alloy EN AW 6014 T4 (sheet thickness 1.15 mm) with galvanized dual phase steel HCT450X (sheet thickness 0.80 mm, Zn layer thickness 7.5 μm) was investigated. The dimensions of the metal sheets to be joined were approximately 250 mm x 150 mm. The sheets were clamped in horizontal butt joint position without overlapping. The vertical distance between the welding torch and the workpiece, d, was 6 mm or 12 mm, respectively. Six different Al-based filler materials designated with code letters A, B, C, D, E and F, were tested. Fillers A-D were commercial wires typically used in industrial welding applications, whereas fillers E and F were test wires especially produced for the experiments. The diameter of each wire was 1.2 mm. The welds were produced with a KUKA KR 30-2 welding robot. Two different torch types, the Fronius CMT Braze+ torch and the Fronius Robacta500 torch, were used. Compared to the standard nozzle of the Robacta500 torch the narrow conical nozzle of the CMT Braze+ torch increases the emerging speed of the shielding gas. This constricts the arc, improves cooling and finally enables faster welding [4]. All experiments were performed with a constant welding speed of 0.4 m/min using Argon with a flow rate of 12 l/min as shielding gas. Based on the welding speed, v [m/min], the mean welding current, I [A], and the mean welding voltage, U [V], the nominal heat input, E [J/mm], is calculated as $E = 0.06UI / v$. The welding process parameters considered in the present study are given in Table 1 and the welding process configuration is shown in Figure 1.

Table 1: Welding process parameters (w is the wire feed rate)

Sample	Filler material	Filler code	Torch type	d [mm]	w [m/min]	U [V]	I [A]	E [J/mm]
05	EN AW 5183	A	CMT Braze+	6	3.7	8.0	69	83
06	EN AW 5183	A	CMT Braze+	12	3.7	8.0	69	83
07	EN AW 5183	A	Robacta500	12	4.1	8.2	74	91
09	EN AW 5087	B	CMT Braze+	6	4.7	9.3	68	95
10	EN AW 5087	B	CMT Braze+	12	3.4	7.9	66	78
08	EN AW 5087	B	Robacta500	12	3.9	8.1	71	86
12	EN AW 6082	C	CMT Braze+	6	3.6	8.0	68	82
11	EN AW 6082	C	CMT Braze+	12	3.8	8.1	70	85
13	EN AW 6082	C	Robacta500	12	3.9	8.1	71	86
14	Al-3Si-1Mn	D	CMT Braze+	6	3.7	8.0	69	83
15	Al-3Si-1Mn	D	CMT Braze+	12	3.7	8.0	69	83
16	Al-3Si-1Mn	D	Robacta500	12	3.9	8.1	71	86
17	Al-4Mg-0.6Sc	E	CMT Braze+	6	3.7	8.0	69	83
18	Al-4Mg-0.6Sc	E	CMT Braze+	12	3.9	8.1	71	86
19	Al-4Mg-0.6Sc	E	Robacta500	12	4.1	8.2	74	91
20	Al-0.3Mg-0.5Sc-0.4Zr	F	CMT Braze+	6	4.0	8.1	73	89

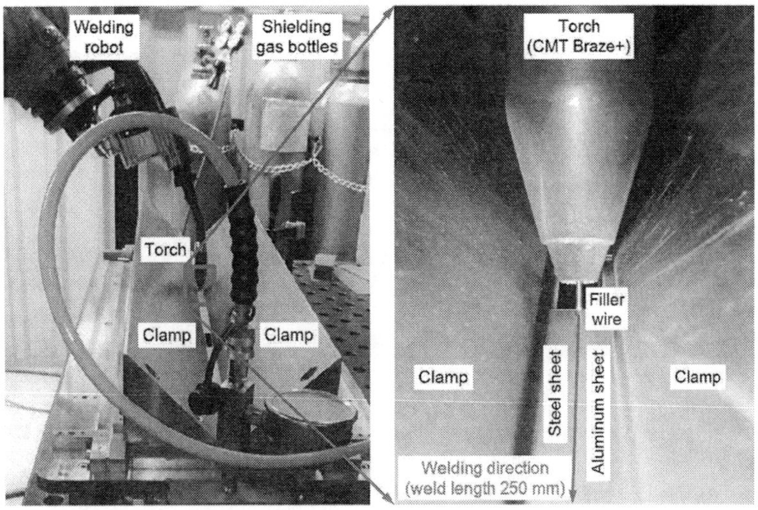

Figure 1: Configuration of the single-sided CMT welding process

Results and discussion

The macroscopic examination reveals that the metal blanks are fully penetrated by the weld seams, though the weld seams are produced with the single-sided CMT process where the torch is located above the horizontally clamped sheets to be joined, as depicted in Figure 1. The following discussion focuses exemplarily on sample 20, which shows the most promising microstructure almost free from pores and cracks. The quality of the weld surfaces is similar in all experiments, but a closer look reveals that the surfaces of the welds produced with 6 mm distance between the torch and the workpiece are generally smoother than the surfaces of the welds produced with 12 mm distance. This seems to be independent from the filler material or from the torch type used. Exemplarily, Figure 2 shows the top side (a) and the bottom side (b) of the weld seam of sample 20. Obviously, at both sides flat and spatter-free weld surfaces without visible cracks are achieved. Close to the weld seam the Zn coating of the steel sheet is almost unaffected by the welding arc, which is favorable for the corrosion resistance of the joint.

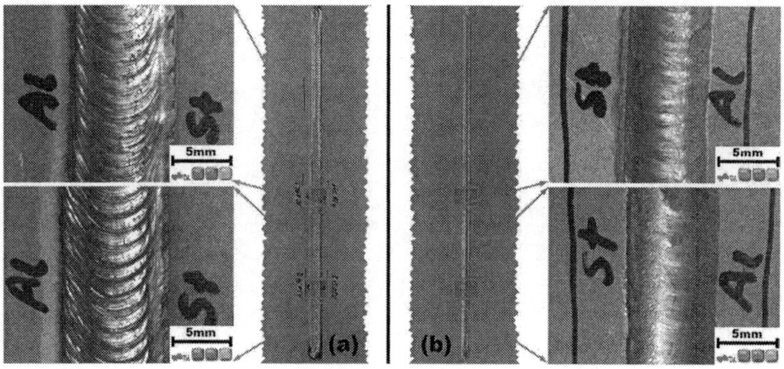

Figure 2: Top side (a) and bottom side (b) of the weld seam of sample 20

The microstructure of the weld seam cross-section of sample 20 is shown in Figure 3(a). It demonstrates that the steel remained solid during CMT welding. Bonding between the weld seam and the steel sheet is just achieved by the intermetallic layer. The micrograph also demonstrates that the weld contains several defects like gas pores and small hot cracks. Lacks of fusion are not observed for the weld produced with filler material F. In contrast, partial lack of fusion is observed between the steel sheet and the weld seam for filler materials A and B. However, the weld seam contains comparatively large pores at the cutting edge of the steel sheet, as marked with the red arrow in Figure 3(a). Since this type of porosity occurs for each of the filler materials tested, it seems to be specific for the single-sided CMT process where the cutting edge of the steel sheet is not galvanized and therefore poorly wettable for the molten filler material. The weld seams produced with filler material E contain a large number of almost uniformly distributed pores. This type of porosity is presumably attributed to the high hydrogen content of the filler wire due to insufficient melt treatment before producing the wire, *i.e.* due to ineffective melt degassing.

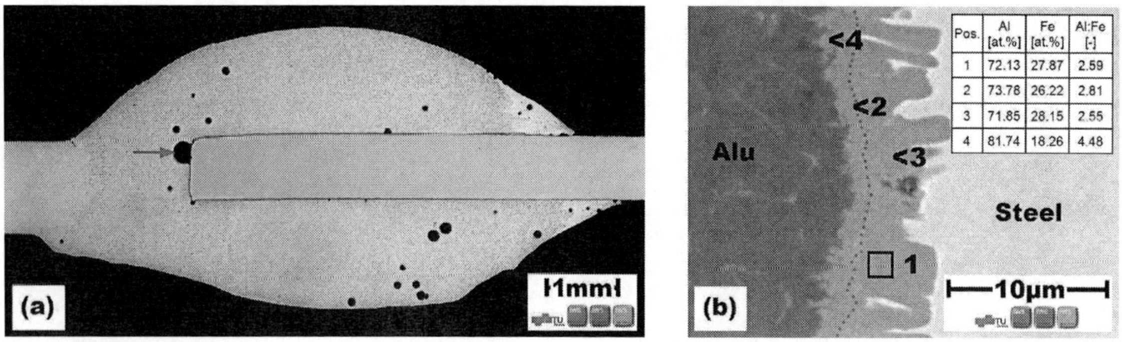

Pos.	Al [at.%]	Fe [at.%]	Al:Fe [-]
1	72.13	27.87	2.59
2	73.78	26.22	2.81
3	71.85	28.15	2.55
4	81.74	18.26	4.48

Figure 3: Microstructure of the weld seam cross-section (a) and positions of the EDX scans (b) of sample 20. Since position 4 is located close to the Al-rich weld a comparatively high Al concentration is detected there.

The chemical composition of the intermetallic layer between the steel sheet and the weld seam was analyzed with SEM-EDX. The positions of the EDX spot scans are visualized in Figure 3(b). The ratios between the detected Al and Fe concentrations indicate that the intermetallic layer consists of the Al_5Fe_2 phase close to the steel surface and of the Al_3Fe phase which penetrates the weld. Their interfaces are tongue/tooth-like (between Al_5Fe_2 and the steel sheet) or highly serrated (between Al_3Fe and the weld) which is in accordance to the literature [6],[10]-[12]. During solidification firstly Al_5Fe_2 forms presumably by chemical reaction and secondly Al_3Fe forms by diffusion of Fe towards the weld with the already formed Al_5Fe_2 acting as nucleation site [12]. The growth of both phases is then mainly controlled by interdiffusion [11]. In Figure 3(b) the boundary between the Al_5Fe_2 and the Al_3Fe sub-layers is approximated with a dashed line. Obviously, the thickness of the intermetallic layer shows spatial variations. For example, the layer of sample 20 has a local thickness maximum of ~ 20 μm and a mean thickness of ~ 10 μm. The values measured at the other samples are in the same order of magnitude.

Conclusions

This experimental work investigates the influence of different process parameters, *i.e.* the filler material, the torch type and the torch-to-workpiece distance, on the tendencies of hot cracking and porosity formation in Cold Metal Transfer (CMT) welding of hybrid aluminum-steel blanks. The conclusions are summarized as follows:

- Thin sheets of aluminum alloy EN AW 6014 T4 and galvanized dual phase steel HCT450X + ZE75/75 are successfully butt welded with the single-sided CMT process. The blanks are fully penetrated by the welds.
- The smoothest weld surfaces are achieved with the Fronius CMT Braze+ torch when the vertical distance between the torch and the workpiece is 6 mm.
- The tendency of hot cracking depends on the chemical composition of the filler material. Particularly the weld seam produced with filler material Al-0.3Mg-0.5Sc-0.4Zr is almost free from cracks.
- However, the weld seams contain comparatively large pores at the cutting edge of the steel sheet which seems to be specific for the single-sided CMT process where the cutting edge is not galvanized.
- An intermetallic layer which consists of Al_5Fe_2 and Al_3Fe sub-layers is formed between the weld and the steel sheet for each of the filler materials tested. This layer has a local thickness maximum of ~ 20 μm.

In order to confirm the results of the microstructure investigations discussed in this study, testing the mechanical properties of the weldments (*e.g.* strength and deformability) is required which is an ongoing work.

Acknowledgment

The authors would like to thank Mrs. Raphaela Mühleder for her experimental assistance. The present work was financially supported by the Austrian Research Promotion Agency (FFG).

References

[1] U. Dilthey and L. Stein: "Multimaterial car body design: challenge for welding and joining", *Sci. Technol. Weld. Joi.*, Vol. 11, No. 2 (2006), pp. 135-142.

[2] M. Mazar Atabaki, M. Nikodinovski, P. Chenier, J. Ma, M. Harooni and R. Kovacevic: "Welding of Aluminum Alloys to Steels: An Overview", *J. Manuf. Sci. Prod.*, Vol. 14, No. 2 (2014), pp. 59-78.

[3] N. Jank, H. Staufer and J. Bruckner: "Schweißverbindungen von Stahl mit Aluminium - eine Perspektive für die Zukunft", *BHM*, Vol. 153, No. 5 (2008), pp. 189-192.

[4] Fronius International GmbH: http://www.fronius.com/

[5] J. Feng, H. Zhang and P. He: "The CMT short-circuiting metal transfer process and its use in thin aluminium sheets welding", *Mater. Design*, Vol. 30 (2009), pp. 1850-1852.

[6] H.T. Zhang, J.C. Feng, P. He, B.B. Zhang, J.M. Chen and L. Wang: "The arc characteristics and metal transfer behaviour of cold metal transfer and its use in joining aluminium to zinc-coated steel", *Mat. Sci. Eng. A*, Vol. A499 (2009), pp. 111-113.

[7] J.A. Dantzig and M. Rappaz: *Solidification*, EPFL Press: Lausanne (2009), pp. 461-548.

[8] R.A. Woods: "Porosity and Hydrogen Absorption in Aluminum Welds", *Weld. Res. Supp.*, Vol. 53, No. 3 (1974), pp. 97-108.

[9] Y. Su, X. Hua, Y. Wu: "Quantitative characterization of porosity in Fe-Al dissimilar materials lap joint made by gas metal arc welding with different current modes", *J. Mater. Process. Tech.*, Vol. 214 (2014), pp. 81-86.

[10] A. Shyam, S. Suwas and S. Bhargava: "Microstructural Features of Iron Aluminides Formed by the Reaction between Solid Iron and Liquid Aluminium", *Pract. Metallogr.*, Vol. 34 (1997), pp. 264-277.

[11] K. Bouché, F. Barbier and A. Coulet: "Intermetallic compound layer growth between solid iron and molten aluminium", *Mat. Sci. Eng. A*, Vol. A249 (1998), pp. 167-175.

[12] L. Agudo, D. Eyidi, C.H. Schmaranzer, E. Arenholz, N. Jank, J. Bruckner and A.R. Pyzalla: "Intermetallic Fe_xAl_y-phases in a steel/Al-alloy fusion weld", *J. Mater. Sci.*, Vol. 42, No. 12 (2007), pp. 4205-4214.

Weld pool formation mechanism in plasma keyhole arc welding

Nguyen Van Anh[1,*]*, Shinichi Tashiro*[1] *and Manabu Tanaka*[1]
1) Joining and Welding Research Institute, Osaka University, Japan
11-1 Mihogaoka, Ibaraki, Osaka 576-0047, Japan
**) anh.nguyenvan@jwri.osaka-u.ac.jp*

Abstract
The purpose of this investigation is to clarify the formation mechanism of weld pool and keyhole in plasma keyhole arc welding process. In order to clarify this issue, the convective pattern of weld pool was observed. In order to observe the convective pattern inside weld pool, an X-ray transmission observation system was applied. This observation system consists two sets of X-ray power sources, high speed video cameras and image intensifiers. In order to measure the velocity field and acceleration of convective pattern, the tungsten particle with diameter of 0.5 mm was employed as a tracer. By using this system, the three–dimensional structure of convective pattern can be discussed. The results showed that there were two circular convective patterns appearing in the weld pool. As for the main domination of shielding gas flow along the top surface and marangoni force in upward direction, a rotating convective pattern in clockwise direction appeared near the top surface of weld pool. In contrast, because of main domination of shear force in downward direction and marangoni force in outward direction, a rotating convective pattern in counterclockwise direction appeared near the bottom surface of weld pool.

Introduction
Plasma Arc Welding (PAW) is one of the fusion welding processes of high penetration and low distortion over conventional Gas Tungsten Arc Welding. In comparison with Laser Beam Welding and Electron Beam Welding processes, PAW is more cost effective and more tolerant of joint penetration, even though its energy is less dense. Consequently, it has found wide applications in industry such as: aerospace, shipbuilding, structure steel, etc.. Such as LBW and EBW, PAW can be operated in Plasma Keyhole Arc Welding (PKAW) mode [1]. *Fig. 1* depicts the schematic illustration of PKAW. In PKAW, the molten flow by arc near the keyhole is transported backward through both sides of the keyhole and therefore produces a weld pool. It is required to fill the void behind the keyhole with the molten flow for maintaining welding stability. There are two necessary conditions for welding stability: the molten flow on both sides must bridge the gap formed at the rear part by the passage of keyhole and, in addition, the molten flow is supported by the surface tension acting on the back surface of molten pool.

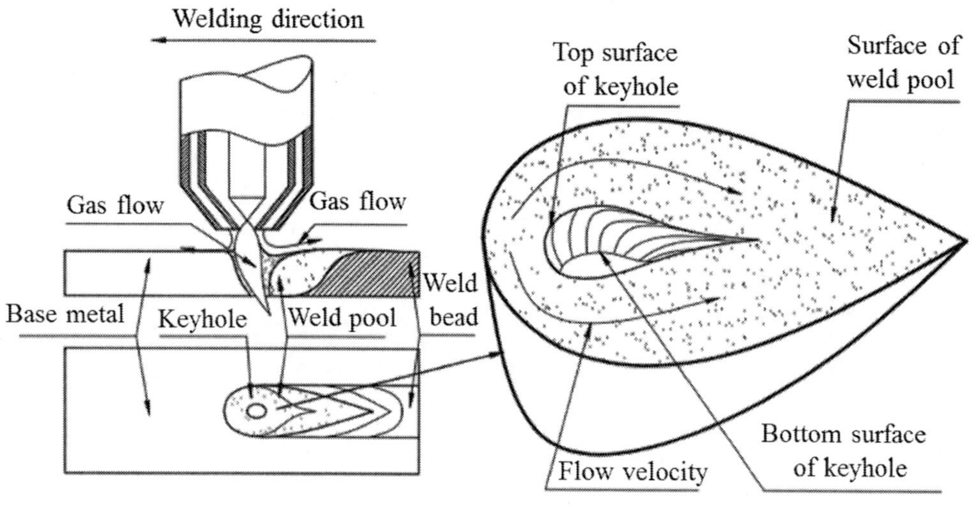

Fig. 1 Schematic illustration of PKAW process

To improve this welding stability, it is essential to study the keyhole formation mechanism by considering the behavior of the convective pattern in weld pool. In order to understand and improve the behavior of the convection flow, it is necessary to research mechanisms involving transportation of the convective pattern in weld pool in further details. The behavior described above is a result of a balance of forces acting on the surface and inside PKAW weld pool. As indicated in a previous paper [2], there were four principal forces driving the welding convection flow of welding arc processes: (i) Drag force (Shear force) on the liquid surface is produced by plasma flow through keyhole and plasma flow on the weld pool surface; (ii) Buoyance force is created due to the temperature difference within the weld pool; (iii) Lorentz force is generated due to the self-magnetic field of arc current within the weld pool and (iv) Marangoni force is caused by the temperature gradient on the surface of weld pool. These forces, which depend mainly on the welding conditions used such as: welding current, welding voltage, shielding gas, welding speed, base metal, etc. Recently, in order to can be obtained good welding quality, PKAW was investigated. However, these researches focused mainly upon to control the keyhole and the weld pool surface or control welding robot.

For purpose improving the quality of welds, Zhang et al. [3,4] have attempted to observe the influence of welding speed and welding current to the change of the keyhole diameter and the weld pool surface using an Insulated Gate Bipolar Transistor (IGBT) power module. By using this module, welding current was controlled in pulse form to decrease the brightness of main arc. During the time of weak brightness of main arc, the keyhole and the weld pool surface were clearly visualized by High-Speed Video Camera (HSVC).

Wu et al. [5,6] purposed to archive good quality of weld joints through controlling the keyhole stability. In order to do that, a welding current control unit for PKAW process was developed. During welding process, the keyhole and the weld pool surface were captured the image, as well measured the temperature from the bottom side of base metal. The main results showed the relationship of welding current and keyhole geometry parameters.

In order to achieve sound weld bead, Yamane et al. [7,8] investigated mainly developing algorithms for controlling feedback of welding robot through processing the weld pool images to track the weld-line based on the boundary between base metal and weld pool. During welding process, however, it was difficult to clearly distinguish the boundary due to too high brightness of main arc. To correctly control, the pulsed current wave form was used to cut off main arc for a very short time.

As above aforementioned, the behavior of convective pattern of weld pool depends mainly on various driving forces. However, its behavior and the keyhole formation mechanism have been not fully understood. As for overcoming this issue, present investigation focused on clarifying the influence of main dominated forces on the keyhole formation mechanism and convective pattern of weld pool. In order to discuss this issue, the convective pattern of weld pool was investigated through observing the motion of small tungsten particles inside the weld pool.

Experimental method

Welding equipment using in this research consisted of a transfer-type plasma arc welding torch (100WH, Nippon Steel Welding & Engineering Co.,Ltd.) and a welding power (NW-300ASR, Nippon Steel Welding & Engineering Co.,Ltd.). The base metal was stainless steel SUS304 with dimensions of 300x100x4 mm. The observed equipment utilizing in present investigation was an X-ray observed system. An experimental set up of this observed system is presented in *Fig. 2*. It consisted two sets of X-ray power sources, HSVCs and image intensifiers. The X-ray 1 was setup from the upper side of the base metal. Its power source was 1 mmA of tube current and 225 kV of tube voltage. Meanwhile, the X-ray 2 was set up from the lower side of the base metal. Its power source was 3.5 mA of tube current and 230 kV of tube voltage. The tilt angle of each of two incident X-rays (from X-ray 1 source and X-ray 2 source) and the horizontal plane of base metal was fixed at 30^0. The incident X-rays cut each other at the center point of the keyhole on the neutral plane of base metal. In this setting, the X-ray transmission images were recorded by two HSVCs at the same frame rate of 1000 fps (frame per second). The trigger signals of both of two HSVCs were controlled in the synchronization. The back shielding gas box put on the welding jig. The shielding gas flow was blown into back shielding gas box for protecting the backside of welding joint without the negative influence of air through two small holes. In order to observe the behavior of convective pattern, the tungsten particle with diameter of 0.5 mm was employed. Using tracer software, afterwards, features of three-dimensional structure, the convective pattern, velocity field and acceleration of the weld pool were discussed. Other welding conditions consisted of welding current: 120 A; welding speed: 0.3 cm/s; shielding gas and plasma gas: pure Ar; shielding gas flow rate: 7.5 l/min; back shielding gas flow rate: 10 l/min, plasma gas flow rate: 1.7 l/min and arc length: 5 mm.

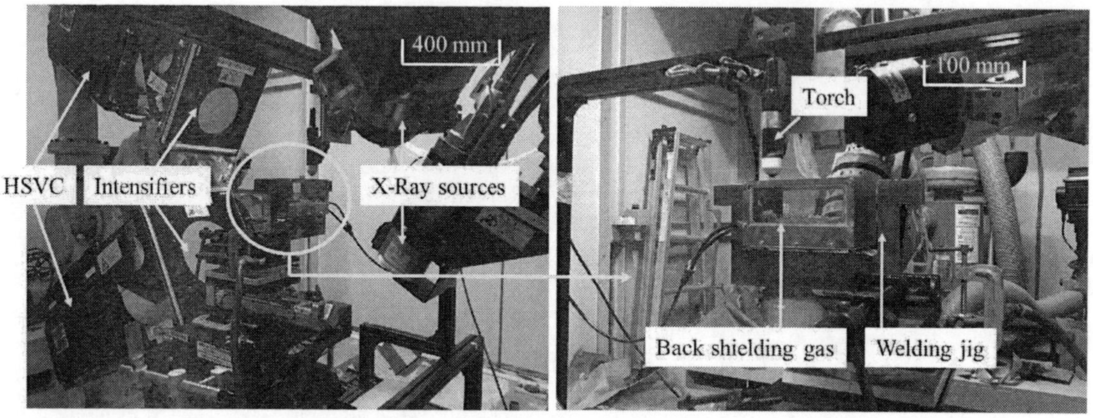

Fig. 2 Photograph of experimental set up of X-ray observation

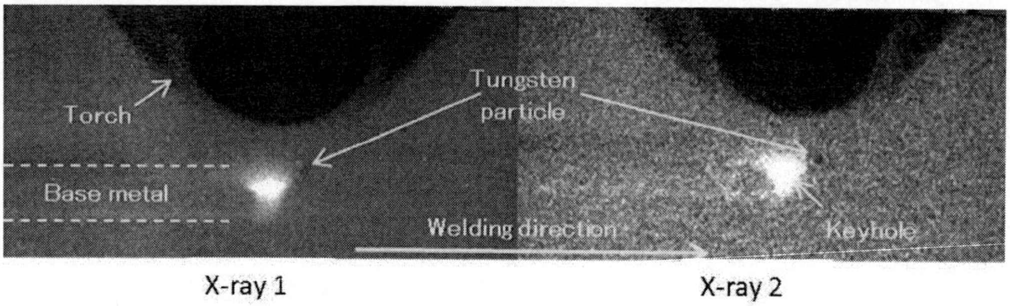

Fig. 3 two typical images of X-ray observed system

Fig.3 showed two typical images of X-ray observed system. In this figure, two images were captured at same time during welding before the tungsten particle entering in the weld pool. It may seem that the image acquired of X-ray 1 was darker than that of X-ray 2. This was because the power of X-ray 1 was higher in comparison with the power of X-ray 2. In addition, the resolution of HSVC in X-ray 1 was higher than the solution of HSVC in X-ray 2. Hence, the image obtained of X-ray 1 was finer in comparison with that of X-ray 2.

Results and discussion

As indicated in *Fig. 4*, three-dimensional structure of the convective pattern inside weld pool was drawn using tracer software through the motion of the tungsten particle inside weld pool. There were two trends in the movement of tungsten particle. Since the tungsten particle belong to near the top surface, it was accelerated in upward direction near the keyhole before moving to rear part of weld pool. Afterwards the tungsten particle was in sinking tendency to bottom surface before moving back to near the keyhole and practicing in a new motion cycle. Whilst, since the tungsten particle belong to near the bottom surface it trended in downward direction near the keyhole before moving to the rear part of weld pool around the bottom surface. Afterwards, it moved back to near the keyhole before entering in a new motion cycle. This phenomenon was repeated several times in each experiment. In addition, it is also measured that the maximum velocity reached 0.40 m/s on the top and 0.30 m/s on the bottom surface near the keyhole. The average velocity of the tungsten particle was 0.12 m/s for the convective pattern near the top surface and 0.10 m/s for the convective pattern near the bottom surface.

As of observed phenomena and measured results above, it can be explained through considering the activation of main dominated forces in weld pool. Due to the main driving of marangoni force in upward direction and gas flow in outward direction, the convective pattern near the top surface was in clockwise direction. Whilst, because of the main domination of shear force in downward direction and the maragoni force in outward direction the convective pattern near the bottom surface was in counterclockwise direction. On the other hand, in a previous paper [9], the simulated results shown that there were two convective patterns in weld pool near the keyhole. Therefore, results in this paper matched well with results of numerical calculation.

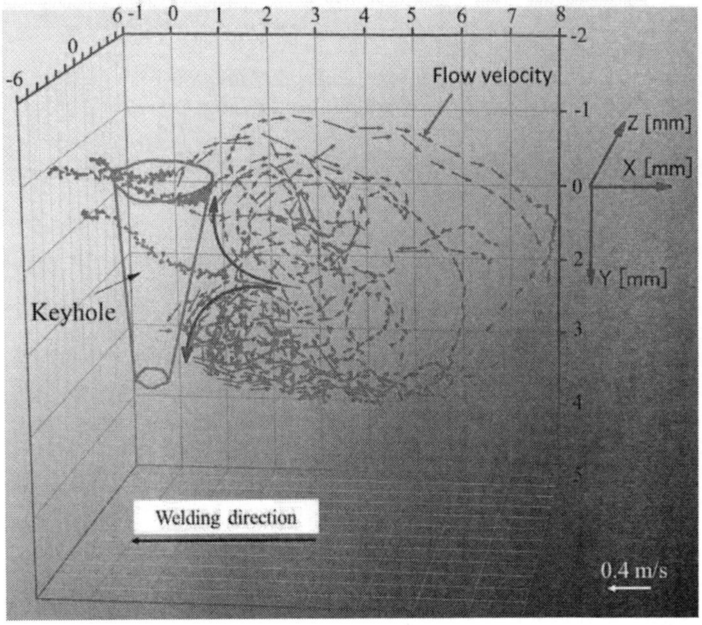

Fig. 4 Convective pattern inside weld pool

Conclusions

The convective pattern inside weld pool was completely observed. Several main results can be drawn as following:

- There were two circular convective patterns appearing inside weld pool.
- The convective pattern near the top surface was in upward and outward direction.
- The convective pattern near the bottom surface was in downward and outward direction.
- The velocity of convective pattern near the top surface was stronger than that of convective pattern near the bottom surface.
- The convective patterns of weld pool were estimated in matching well with results of numerical simulation.

References

[1] H G fan and R Kovacevic: "Keyhole formation and collapse in plasma arc welding," *J. Phys. D:Appl. Phys.*, 32 (1999), pp. 2902-2909.

[2] M. Tanaka and J J Lowke: "Predictions of weld pool profiles using plasma physics," *J. Phys. D: Appl. Phys.*, 40 (2007), pp. 1-23.

[3] Y. M. Zhang and S. B. Zhang: "Observation of the keyhole during plasma arc welding," *Welding journal*, (1999), pp. 53-58.

[4] W Lu, Y M, Zhang and John Emmerson: "Sensing of weld pool surface usingnon-transferred plasma charge sensor," *Meas. Sci. Technol.*, 15 (2004), pp. 991-999.

[5] Z. M. Liu, C. S. Wu and J. Chen: "Sensing dynamic keyhole behaviors in controlled-pulse keyholing plasma arc welding," *Welding journal*, 92 (2013), pp. 381-389.

[6] Wu CS, Jia CB and Chen MA: "A control system for keyhole plasma arc welding of stainless steel with medium thickness," *Welding journal*, 89 (2010), pp. 225-231.

[7] Satoshi Yamane, Hikaru Suzuki, Jiro Toma, Takahiro Godo, Kazumichi Hosoya, Tooru Nakajima and Hikaru Yamamoto: "Imaging processing of the weld pool in robot plasma welding system," *Japanese welding society*, 31 (2013), 44-47.

[8] S. Yamane, T. Godo, K. Hosoya, T. Nakajima and H. Yamamoto: "Detecting and tracking of welding line in visual plasma robotic welding," *Quarterly Journal of the Japan welding society (in Japanese)*, 31 (2013), pp. 175-180.

[9] Shinichi Tashiro, Manabu Tanaka, Tomoyuki Ueyama and Tetsuo Era: "Numerical simulation of keyhole welding of aluminum thick plate with plasma arc," *Japanese welding Society*, 84 (2010), pp. 119-124.

Development of Two Piece Type Contact Tip for MIG/MAG Welding

Muneharu Kutsuna*, Takaaki Murakami**
** Advanced Laser Technology Research Center Co.,Ltd., Anjo-shi, Aichi-ken, Japan*
Altrec-kutsuna@nifty.com, +81-566-91-2281,81-566-91-2282)
***Taikoh Engineering Lim. Benten-jima, Maisaka Nishi-ku,Hamamatsu-shi, Japan*
t-murakami@muf.sakura.ne.jp, +81-53-592-5350

Abstract

One piece type contact tip is now used as a contact tip for gas shielded arc welding. The life time should be improved for automatic or robotic welding systems. In the present work, two piece type contact tip was developed to improve the life time and arc stability. 24 hours MAG welding operation (arc time =12hours) using the two piece type contact tip for wire of 1.2mm in dia. shows 0.034～0.057mm wear of tip groove.

The tip cover and contact tip unit are also developed for most of the arc welding torch in Japan. Comparing with the life time of a conventional contact tip, the life time of developed one is 10 ～20 times longer than that of conventional one. Indeed, the application of the tip to motorcycle industry, the total life time was 384hours at the welding current of 210A for the filler wire of 1,2mm in diameter. So, 13 tips were enough for one welding robot in a year, comparing with 264 conventional tips.

1. Introduction

1) Wear of Tip groove

In gas metal arc welding (GMAW), an electric arc is established between a continuously fed consumable electrode (filler wire) and the work piece. Welding current is supply from power source to the work piece through a contact tip and the electrode in the welding torch. So, the contact tip is a key component of a welding torch for stable GMAW process. The contact tip has two main functions. One of them is to feed the current to the work piece through the electrode which is fed through the tip. Another one is to introduce welding arc and electrode tip to the correct position during welding to get stable welding process, especially in robotic welding. ,

The life of contact tip is important in automatic welding, because the changing time of contact tip reduce the arc time and increases the welding cost. The contact tip should have long life time and lower price. Then, several research are reported on a conventional contact tip (one piece type). For instance, contact tip wear was studied. If severe wear is occurred on the contact tip as shown in Figure 1, The robotic welding operation can not continue. conventional contact tip should be changed in each several hours. Then, two piece type contact tip has been developed to solve the problems. One of the characteristics is long life time and another one is stable arc phenomena without any troubles.

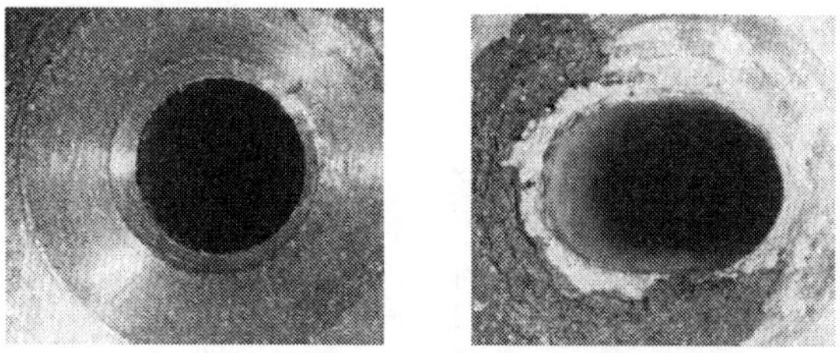

(a) Unused contact tip *(b) Tip wear after hours*
Fig.1 The wear of contact tip groove during welding [1]

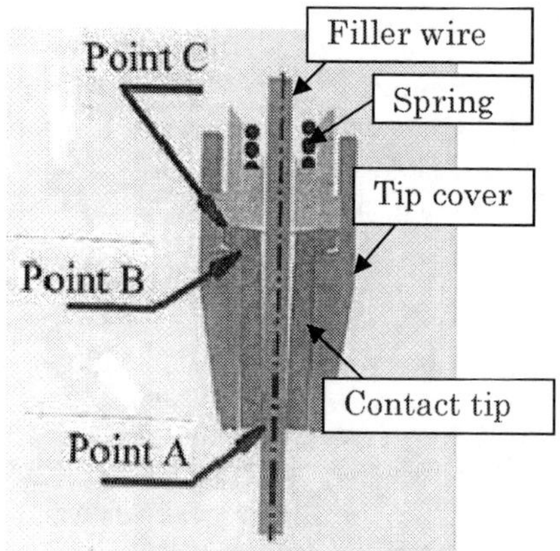

Fig.2 Current supply mechanism of new contact tip

2) Current supply mechanism of two piece type contact tip

To get a smooth/stable supplying of current from contact tip to electrode surface, low stress was applied to the root of contact tip (point C) using a small spring in tip base as shown in Fig.2. Therefore, another end (point A) of contact tip can be rotated around point B. Then, the point A of contact tip is pushed to the surface of filler wire during welding. When a bended filler wire is supplied, the gap of each contact tip can be expanded. The applied stress to the root of contact tip is about 150N. Welding current is always supplied at the point A during welding. Joule heating of wire is occured from point A to tip of wire. Therefore, the temperature of contact tip at point A is not higher than that of conventional tip in which Joule heating occurred easily. Then the wear of tip is not severe, comparing with that of the conventional tip.

3) development of two piece type contact tip

The weight of developed contact tip (two pieces) is about 3.2g. That of conventional one is 17g as shown in Figure 3. The material is copper alloys. The manufacturing process is different from that of conventional one. In the case of two piece type contact tip, Roll forming, stamping and forge cutting are used for the production. High production rate can be obtained using these forming processes. Fig.4 shows a type of developed contact tip and contact tip unit for GMAW process using an electrode wire of 1.2mm in diameter.

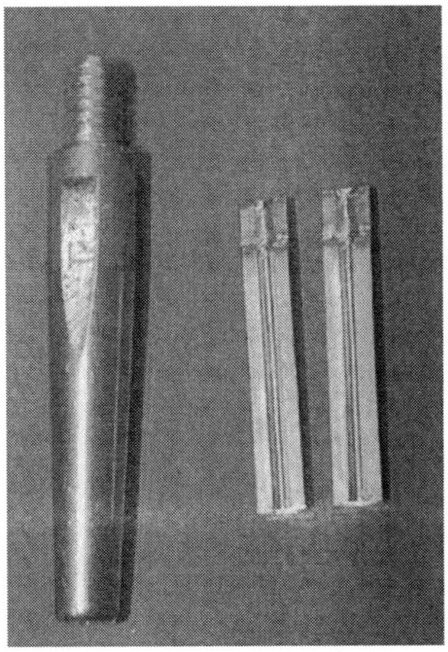

Fig.3 Contact tips (Left: conventional, Right; new one)

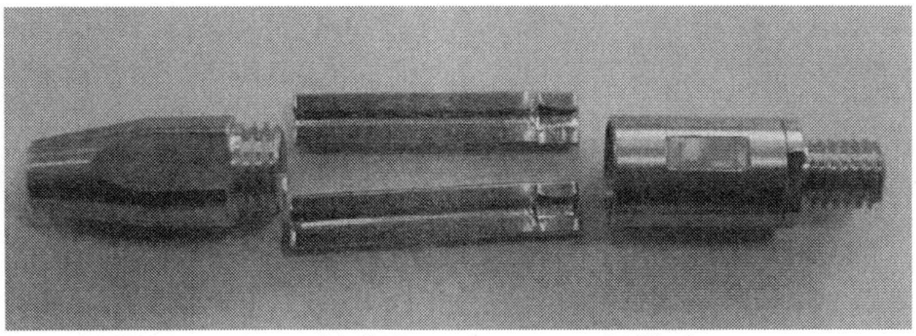

(a) Tip cover (b) Contact tip (c) Tip base

Fig. 4 Two piece type contact tip and tip units for GMAW

3) Advantages of two piece type contact tip

The two piece type contact tip has the following features;

1) Excellent current transferability:
2) The life time of contact tip is 10 ~20 times longer than that of conventional tip.
3) Higher arc stability without feeding hesitation : Resulting in less spatter and good arc starts
4) Good linear adjustment of filler wire to arc location

2. Experimental procedure

1) Tip wear test in CO2 and MAG welding

The size of screw of conventional contact tip is the same as a developed tip unit. So, It is easy to change the contact tip from a conventional tip to the developed tip unit as shown in Fig.5..

Mild steel SS400 plates were welded under the following;

Filler wire : MG-50 (1.2mm in diameter)
Welding current : 280A ~ 307A.
Arc voltage : 32.7V ~32.9V
Welding speed : 600mm/min
Shielding gas : 100%CO2 or 80%Ar +20%CO2

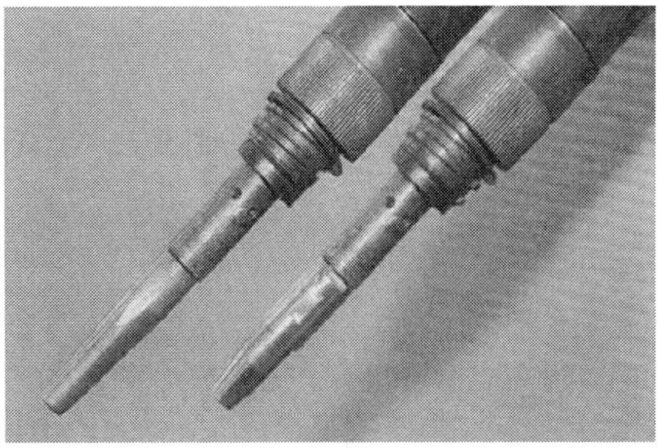

Fig.5 Contact tip unit with a ceramic guide (Left: Conventional, Right; Two piece type)

Fig.6 shows the robotic welding operation. Arc time of each test piece was 6 minutes and weld length was 3.6m each as shown in Fig.7. 120 test pieces were welded in 24 hours. The arc time was 12 hours. Tip wear was measured by a laser microscope after each 2 hours of arc time. Nozzle spatter was measured after each 18minutes or 36minutes of arc time.

Fig.6 Welding test for contact tip using a robot (Wire extension : 17mm-19m)

Fig.7 Bead appearance of welded plates

3. Experimental results

1) Wear test of contact tip

Fig.8 shows the results of groove depth measurement after each two hours of arc time in MAG Process. Measurement error is about ±15μm. Wear occurs mainly tip of the contact tip (line a). Total (Tip No.6 and No.10) groove wear was 37μm after 12 hours of arc time. In the case of conventional tip, the groove wear was 263μm after 2 hours as shown in Fig.9. It was 39 times higher than that of two piece type contact tip. In case of CO2 arc welding, Total groove wear was 57μm after 12 hours of arc time. Amount of Spatter on nozzle and contact tip was measured each 36 minutes of operation. it was .0187g/min.

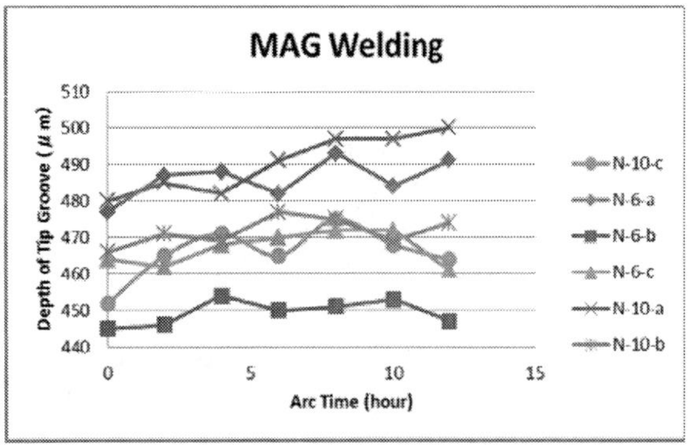

Fig. 8 Measured groove depth of contact tip in MAG process using a two piece type of contact tip (Tip No.6 and No.10 a) Tip of contact tip b) middle point, c) Root of contact tip)

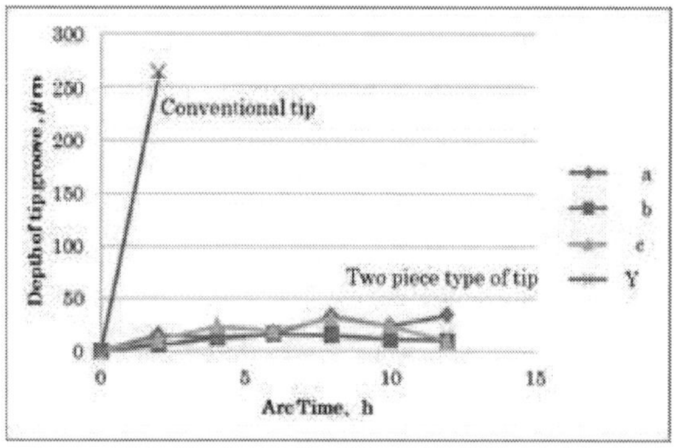

Fig.9 Comparison of new tip wear to conventional tip

Fig.10 shows the photos of welding nozzles. Spatter on the nozzle and tip are a small amount. Color changing of filler wire occurred from the tip of contact tip to tip of wire as shown in Figure 11. This means Joule heating of wire starts from the tip of contact tip. The temperature of wire is not high at the tip of contact tip.

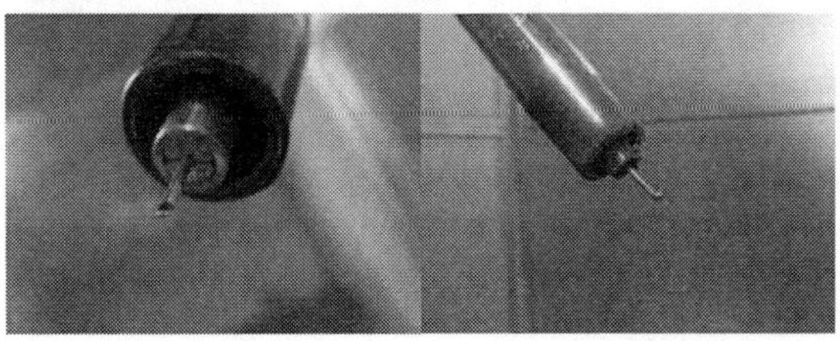

(a) Before welding (b) After 18 minutes

Fig.10 Photos of Welding nozzles

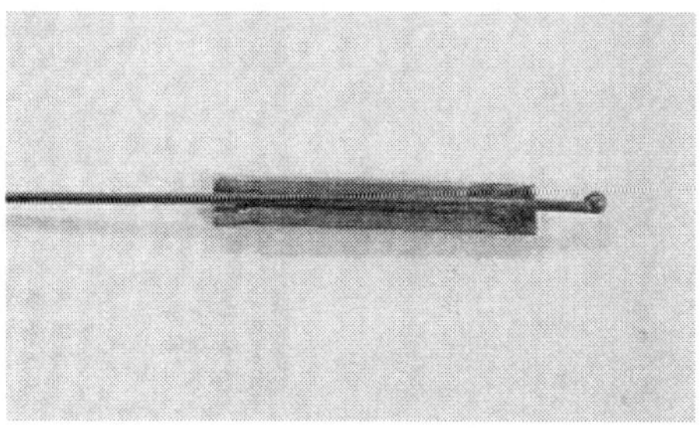

Fig.11 Appearance of filler wire and contact tip after 4 hours welding operation

2) Arc stability test in CO2 and MAG welding

For evaluation of arc stability using a two piece type contact tip, CO2 arc welding and pulsed MAG welding were conducted using a conventional and a Two piece type contact tip. DS1A filler wire of 1.2 mm in diameter was used under the following welding condition;

welding current : 208 A for pulsed MAG, 230A for CO2 arc Welding.
arc voltage : 27V ~ 30 V
travel speed : 50 cm/min

Spatter on the nozzle using a developed contact tip is more less than that by a conventional one. In addition, the characteristics of arc starts was checked. The developed contact tip showed good arc starts. In addition, the welding current and arc voltage are recorded during pulsed MAG welding using a memory recorder. The results are shown in Fig.12 and 13.

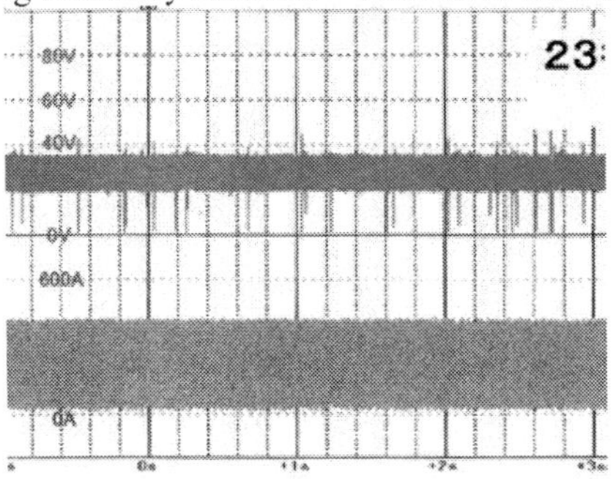

Fig.12 Welding current and arc voltage during welding using a developed contact tip in pulsed MAG

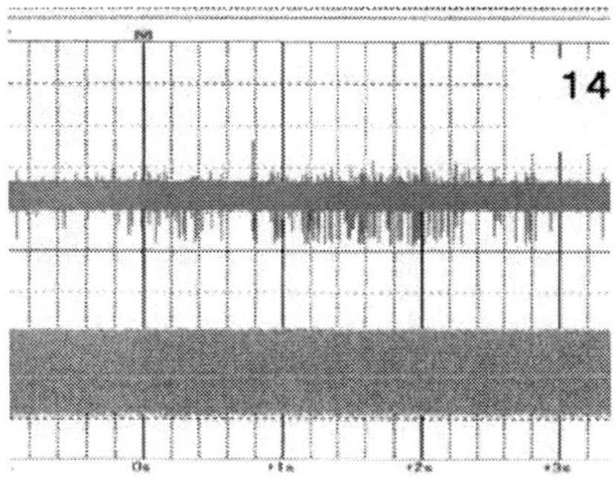

Fig.13 Welding current and arc voltage during welding using a conventional contact tip in pulsed MAG

Welding current changes using contact tip in both welding process are small comparing with that using a conventional one. This means that arc stability using a developed contact tip is higher than that using a conventional one.

3) Application test in a production line

In production line using a welding robot, Contact tip is often changed after each 4 hours or 8 hours of welding operation. Two piece type contact tip was used to verify the performance for the motor cycle part production line in Philippine. The welding procedure are as follows;

Welding process : MAG process (75%Ar+25%CO2),
Filler wire diameter : 1.2 mm indiameter (Solid wire)
Welding current : 210A
Arc voltage : 23V

Results of welding test shows the follows;

■Two piece type contact showed its long life. Cleaning time:48hours(3 days), Changing (life)time:384
In case of conventional tip, no cleaning time and changing (life) time was 16hours.
■Two piece type contact showed good bead appearance, low spatter and low chance of welding defect.
In case of conventional tip,occurrence of NG weld bead, more welding defect, more spstter and miswelding.
■Total consumption of contact tip was 13 pieces in a year using a two piece type contact tip.
. In case of conventional tip, total tip was 264 pieces in a year.

4. Conclution

Two piece type contact tip was developed. It has the following features;

1) Excellent current transferability
2) Long life of contact tip
3) Higher arc stability without feeding hesitation
4) Good linear adjustment of filler wire to arc location
5) Small size and low weight of copper tip

These characteristics are shown in the welding tests comparing with a conventional contact tip.
The life time of a developed one is 10 ~20 times longer than that of conventional one.

References

[1] J. Villafuerte, "Understanding Contact Tip Longevity for Gas Metal Arc Welding," *Welding Journal*, vol.78, pp. 29-35, Dec. 1999.

Advantage of Narrow Gap Welding Technique in Dissimilar GMA Weld Joint of Austenitic Stainless Steel to Micro-alloyed steel

Ramkishor Anant , P.K.Ghosh
Department of Metallurgical and Materials Engineering, Indian Institute of Technology, Roorkee, India
ram_meta@yahoo.co.in

Abstract
Dissimilar metal joints between thick section of austenitic stainless steel and high strength micro-alloyed steel are widely used in high-temperature applications like power stations and petrochemical plants. In this paper the development and advanced study of narrow gap technique to produce sound weld joint of thick section of austenitic stainless steel and micro-alloyed steel with developed mechanical and metallurgical properties has been discussed. Weld joints are prepared by P-GMA welding process by using multi-pass single seam per layer deposition technique which Significantly reduces the residual stresses in dissimilar weld joints as compare to conventional weld joint. Narrow gap weld joint of dissimilar metal produced by single seam per layer deposition is very effective procedure to produce sound weld joints and this study indicated that the narrow gap weld joint is useful in reducing the residual stresses in the weldment where it is one of the main factors which cause failures in dissimilar weld joints.

Introduction
Dissimilar metal welding (DMW) of thick wall austenitic stainless steel (ASS) and high strength low-alloy steel (HSLA) plates are used in defense, power generation, chemical, petrochemical, and nuclear industries where, the temperature elevates up to a range of the order of 350-450°C [1, 2]. The adoption of this combination has been justified based on both the technical and economical reasons.

The dissimilar metal welding (DMW) of thick sections of the ASS and HSLA is critical. The difficulties primarily include the problems largely associated with difference in coefficient of thermal expansion (CTE) and thermal conductivity respectively affecting the undesirable development and distribution of residual stresses, development of undesirable weld chemistry due to dilution, metallurgical incompatibility primarily with respect to the formation of undesirable phases in the weld and HAZ and the segregation of high and low melting phases due to chemical mismatch [3-5].

Fusion welding as one of the most commonly used process; it is also largely used for the joining of dissimilar metals. For versatile application in general weld fabrication of thick sections of ASS and HSLA, there may be several choices of fusion welding processes, such as shielded metal arc, gas tungsten arc, gas metal arc and submerged arc welding. Every process is having different weld thermal cycle depending upon rate of weld deposition, welding parameters and shielding environment. The amount of weld deposition which also considerably influences the severity of thermal characteristics of a weld can be considerably reduced by using narrow gap welding technique[7]. The primary aims and objectives of this investigation are to characterize the weld joints with respect to its service requirements so that a further improvement in welding process, procedure and parameter can be proposed for production of better weld joint of HSLA steel and stainless steel plates. In case of the solution emerged out as the use of P-GMAW process and narrow gap welding technique are the superior one over the use of SMAW process with the conventional groove weld, suitable welding procedure specification (WPS) shall be recommended for consideration of their applications.

The P-GMAW can also facilitate the application of narrow gap gas metal arc welding [6,7] which may be further effective for lowering of thermal severity and its adverse influences on the dissimilar weld joint of thick sections of ASS and HSLA. However, the narrow gap P-GMAW process often produces lack of fusion in the groove wall due to low angle of attack (arc strike), low heat input and small molten weld pool. Thus, in order to produce minimum residual stresses in a defect free dissimilar weld joint with improved or comparable mechanical, fracture mechanics and corrosion properties in comparison to those either of the base materials, of widely different physical properties the design of appropriate narrow groove and selection of pulse parameters is very much important. This may be achieved by developing a clear understanding over the influence of narrow groove design and pulse parameters on amount of weld deposition, weld thermal cycle and microstructure of the multipass weld being largely influenced by the solidification behaviour [7] of the P-GMA weld deposit in the joint of dissimilar materials.

Experiment

Several weld joints shall be prepared to optimize the joint design and WPS for fabrication of sound narrow gap weld with desired superiority over the weld joint having conventional weld groove. The optimization of WPS with respect to the narrow groove design shall be primarily made in reference to angle of attack to the groove wall, heat input, welding parameters, control of effective shielding gas flow and sequential control of positioning weld bead deposition to reduce distortion of weld joint.

The preparation of narrow gap dissimilar weld joints of the thick dissimilar plates shall be carried out by using newly designed weld grooves and conventional welding torch head fitted with narrow nozzle (designed by us).

Fig.1 Schematic diagram of conventional V-groove and narrow groove.

Fig.2 Schematic diagram of weld metal deposition procedure by P-GMAW process.

Fig. (1 and 2) Showing the design of conventional (28mm) and narrow weld groove (8mm) and its filling procedure which is called single bead multi layer deposition and single bead single layer deposition technique. Fig 2 clearly shown the number of passes as well as mass of deposition in narrow groove is very less as compared to conventional weld groove.

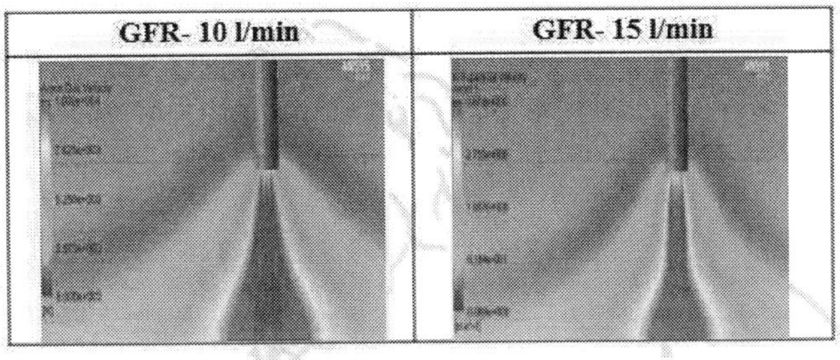

Fig.3 Effect of gas flow rate on arc length and diameter inside the narrow groove.

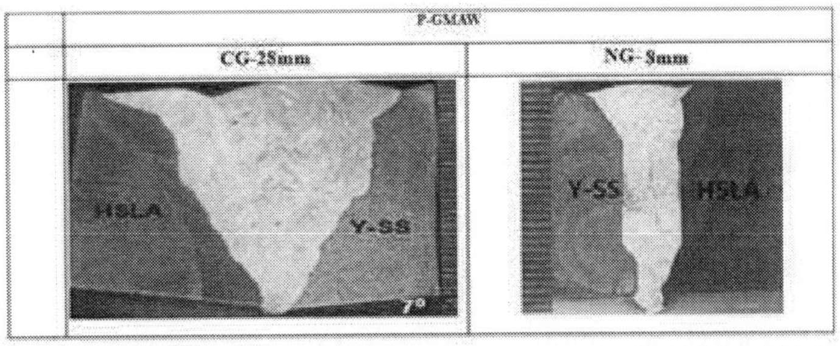

Fig.4 Typical macrograph of dissimilar conventional & narrow gap P-GMAW weld joints of HSLA and Y-SS steel.

Due to huge changes in design of conventional torch nozzle study and optimization of gas flow rate inside the narrow groove is needed. The simulation study of gas flow dynamics inside the narrow groove has been done through ANSYS 14.5 software. Fig 3 shows the gas flow behaviour inside the narrow groove which is clear that the arc length and diameter with the gas flow rate. Gas flow rate has been optimized with minimum acceptable turbulence to produce stable arc inside the narrow groove and by using this newly designed torch nozzle, weld joints are successfully prepared (Fig 4). Fig (5 and 6) clearly shows that the microstructure of HAZ and fusion zone of dissimilar conventional and narrow groove sound weld joint. Fig 6 also shows that the sound weld without lack of side wall fusion on both side. Microstructure of narrow groove weld joint shows that the very less effect on the grain size in HAZ zone of both side of groove wall due to low mass of deposition and heat.

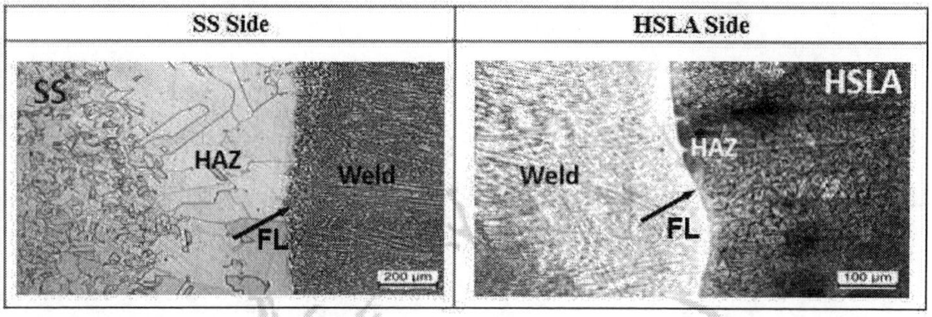

Fig.5 Typical microstructure of HAZ adjacent to FL on both sides of conventional weld joint of γ-SS and HSLA steel at 9.81kJ heat input.

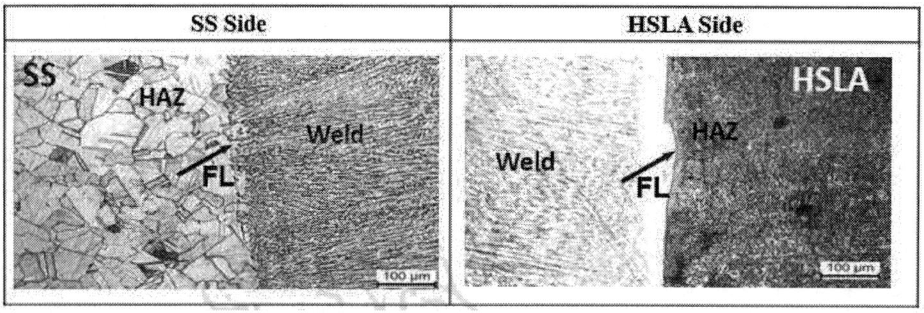

Fig.6 Typical microstructure of HAZ adjacent to FL on both sides of weld joint of γ-SS and HSLA steel 9.81kJ heat input.

Observations and Conclusions

- The newly developed narrow torch nozzle device perform satisfactorily during single bead per layer deposition to produce narrow gap weld joint of thick plates with stable arc and proper gas shielding and without disturbing the welding operations due to excessive heating and burning of the nozzle head.
- Gas flow rate is optimized for narrow groove weld joint i.e. 10-12 l/min to produce stable arc.
- Single bead per layer deposited narrow gap P-GMA weld joints significantly reduce the transverse shrinkage by about 55-65% than that of the conventional groove P-GMA weld.
- Single bead per layer deposited narrow gap P-GMA lowers the deflection of weld joint by about 75-80% than that of the conventional groove weld joint.
- Single bead per layer deposited narrow gap P-GMA lowers the residual stresses in weld centre, fusion and HAZ zone of weld joint by about 45-57% than that of the conventional groove weld joint
- Single bead per layer deposited narrow gap P-GMA weld produces comparatively finer microstructure and reduces the width of heat affected zone (HAZ).

Acknowledgment

The authors would like to express their gratitude to Defence Metallurgical Research Laboratory (DMRL-DRDO), Hyderabad, India for financial support to carry out this work.

References

[1] P.K. Ghosh and Vivek: "Weldbonding of stainless steel", *ISIJ Int.*, Vol. 43, 1, (2003), pp. 85- 94.
[2] P.K. Ghosh and Pawan Kumar Arora: "Control of residual stresses using narrow gap technique in welding of structural steel", *Ind. Weld. J.*, Vol. 36, 1 (2003), pp. 9-15.
[3] P.K. Ghosh et.al, "Improvement of weld characteristics by application of narrow gap technique in SMA welding of 304LN stainless steel pipe", *Procd. Seminar on Advances in Welding Technology (Weld Tech – 2003)*, IIT Kharagpur, March 14-15, 2003.
[4] P.K. Ghosh, "Health monitoring of steel bridge components by NDT coupled with fracture mechanics concepts", *Procd. National workshop on "Bridge instrumentation for health monitoring",* IIT Roorkee,India, May 2003.
[5] Sandeep Bansal, S.K. Nath, P.K. Ghosh, S. Ray, J. Chattopadhyay and H.S. Kushwaha, "Measurement of stretched zone width (SZW) of fracture surface of carbon steel pipes for evaluation of initiation fracture toughness", *Procd. XIII National Conf. of Indian Soc. of Mech. Engineers (ISME-2003)*, IIT Roorkee, December 30-31, 2003.
[6] K. Devakumaran, M. Ravi Reddy and P.K. Ghosh, "Experimental investigation on the transverse shrinkage stress and distortion generated in butt welded joints", *Procd. Int. Symp. of Research Studies on Mater. Sc. & Engg.*, IIT Madras, 20-22 December 2004.
[7] Agarwal, B.P. and P.K. Ghosh, "Thermal modeling of multipass Narrow gap pulsed current GMA Welding by single seam per layer deposition", *Materials and Manufacturing Processes*, (2010), pp.1251-1268

Application of Martensite Tempering Metallurgy and Prediction Models to the Welding of Advanced High Strength Steels

Elliot Biro[1,2*], and Y. Norman Zhou[2]

[1]ArcelorMittal Global R&D, 1390 Burlington St. E, Box 2460, Hamilton, Ont., Canada, L8P 2K5

[2]Mechanical and Mechatronics Engineering Department, University of Waterloo, 200 University Ave. W., Waterloo, Ont., Canada, N2L 3G1

elliot.biro@arcelormittal.com

nzhou@uwaterloo.ca

Abstract

Martensite tempering has been long studied due to its importance for industrial applications. How martensite decomposes is well understood under normal heat treatment conditions; however, because of the very short time-scale during welding, multiphase microstructure of advanced high strength steels (AHSS), and prestrain from forming operations before welding, it is not clear how to apply classical tempering theory to welding applications. Understanding the tempering process during welding and rapid post-welding heat treatments is necessary to design weld process parameters and predicting post-welded joint properties. This work studies how tempering occurs during welding of AHSS from a fundamental level, showing how physical properties change during tempering and how prestrain, multiphase structures and chemical composition affect this process. Finally, it discusses how knowledge of tempering is integral to designing both weldments and welding processes in AHSS.

Introduction

Understanding tempering is important to welding processes. Tempering occurs in the sub-critical heat-affected zone (SCHAZ) during welding, and is sometimes used to improve post-welded properties. Tempering is a well-studied process in martensitic steel under normal heat treatment conditions and occurs in four stages [1]. In stage 1 (70-100°C), C segregates to dislocations and ε carbides form in martensite with more than 0.2 wt% C. In stage 2 (200-350°C), retained austenite decomposes to ferrite and cementite. In stage 3 (200-450°C), segregated C and ε carbides transform to cementite. Finally, at higher temperatures, in stage 4, cementite coarsens and secondary carbides form. During welding, other issues complicate the tempering process, such as: high heating and cooling rates, various chemical compositions, base material prestrain, and multi-phase microstructures. This review examines these issues and discusses how they affect the tempering process. Finally, topics that need future study to weld new automotive steels are discussed.

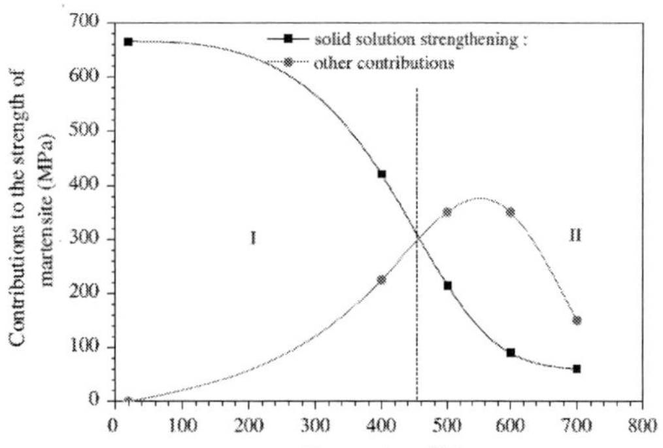

Figure 1: Contributions of solid solution and precipitation strengthening to the strength of rapidly tempered martentsite [3].

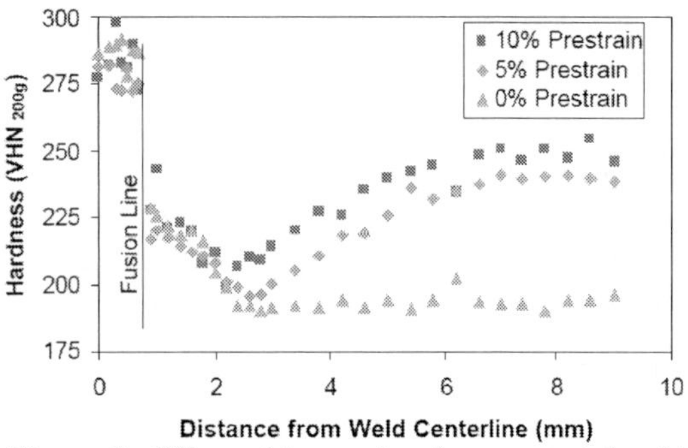

Figure 2: Effect of increasing base material cold work on the hardness profile of gas metal arc welded DP600 [4].

Microstructural Changes during Welding

In welding, tempering occurs at very short time scales due to high temperatures experienced in the SCHAZ near the Ac_1 isotherm (600-700°C). Here, tempering is effectively a two-step process. First carbon leaves solid solution and forms cementite (stage 3 tempering), then these carbides grow by Ostwald ripening (stage 4 tempering) [2][3]. The first step is very rapid, occurring on the order of 10^{-5}-10^{-3} s [2]. The second step is much slower, continuing throughout the rest of the time the weld is at elevated temperatures.

Martensite hardness changes throughout the tempering process. When cemenetite forms, solution strengthening of the structure decreases, as C comes out of solution with the martensite. This step is responsible for about 27% of the decrease in hardness [2]. However, change in solution strengthening cannot explain the total drop in hardness. Cementite is able to strengthen the structure by precipitation strengthening. When cementite is small and densely distributed, its strength contribution is high. However, as the cemenite coarsens, its strength contribution will drop as the particle density drops (see *Fig. 1*), leading to an overall decrease in material hardness.

Influence of Prestrain on Tempering

During automotive manufacture, welds are seldom made on as-received parts. More often, the material has some degree of prestraining due to forming operations. It has been shown that prestraining increases material hardness due to cold work, and during tempering softening increases because cold work is removed during recrystallization at tempering temperatures, increasing the difference between the tempered and the untempered hardness. However, weld samples that have been cold worked before tempering have a higher tempered hardness than samples that are tempered in their as-received state. Biro and Lee showed that when various dual-phase (DP) steels with 600 MPa ultimate tensile strength (UTS) were welded, the minimum heat-affected zone (HAZ) hardness increased with increasing prestrain [4] (see *Fig. 2*). When this was repeated by Kapustka *et al.*, it found that this phenomenon was even more pronounced on transformation induced plasticity (TRIP) steels [5].

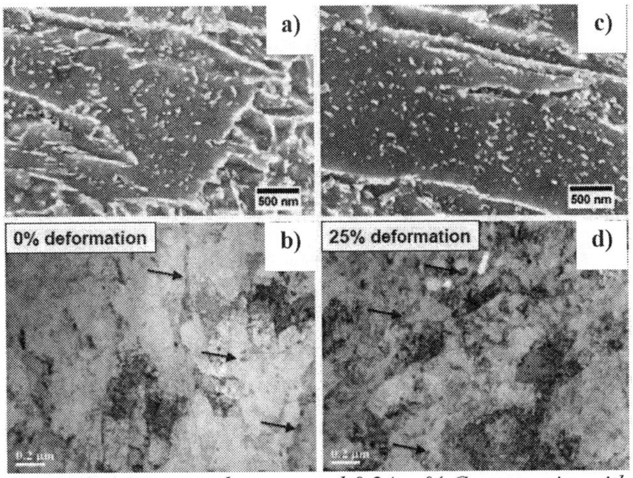

Figure 3: Mircrographs tempered 0.24 wt% C martensite with no prestrain as seen under a) SEM and b) TEM and with 25% prestrain as seen under c) SEM and b) TEM [6].

Saha et al. carried out Gleeble rapid tempering experiments and thermodynamic simulations on as-produced and prestrained martensite [6]. They found that the increased dislocation density (due to prestraining) decreased the driving force required to form cementite. This resulted in smaller, more densely distributed cementite particles in the tempered martensite (see *Fig.3*). As the mean free distance between cementite in prestrained tempered martensite was smaller than in as-produced tempered martensite, the contribution of precipitation strengthening increased in the prestrained materials, leading the observed increased minimum hardness values [4][5]. Furthermore, as it is known that when TRIP steels are strained they have a higher dislocation density than DP steels with similar UTS [7], it is expected that the cementite density will be higher in tempered prestrained TRIP steels, than in in similarly prestrained DP steels. The increased cementite density would result in the tempered TRIP steel having a higher hardness than the DP steel, explaining the observations of Kapustka *et al.* [5].

Tempering in Multi-Phase Structures

As it is projected that the majority (>80%) of the AHSS used in automotive steel structures will be multiphase DP and TRIP steels, not martensitic or press hardened steels with fully martensite structures [8], it is important to understand how tempering of the martensitic phases affects the strength of the material.

Dual-Phase Steels

DP steels are an AHSS with microstructure composed of martensite and sometimes bainite within a ferrite matrix. In the simplest case (ferrite/martensite structure), the hardness of the base material can be can be easily modelled using a rule of mixtures approach, understanding C partitioning between the ferrite and martensite phase [9]. As it has been shown that when these steels are tempered, only the hardness of the martensite decreases, and the hardness of the ferrite does not change [10], it would be expected that predicting the tempered hardness of DP steels could also be calculated using the rule of mixtures, if the change in martensite hardness is known. However, this approach under-predicts the tempered hardness [9]. There are many theories of why this occurs, such as: dislocations in the ferrite adjacent to the tempered martensite grains increasing flow stress in the tempered structure, high martensite Mn content due to partitioning during the inter-critical anneal slowing tempering kinetics, strain hardening during tempering due to the density difference between martensite and ferrite, and strain hardening due to differences in the coefficient of thermal expansion of martensite and ferrite. However, it is most likely that the high dislocation density in the ferrite grains adjacent to the martensite allow for a finer cementite structure to nucleate. This is consistent with tempering work on prestrained martensite [6] and the observation that the difference between the tempered martensite hardness and predictions by rule of mixtures increases with martensite C content [9], which increases local dislocation density in the DP steel.

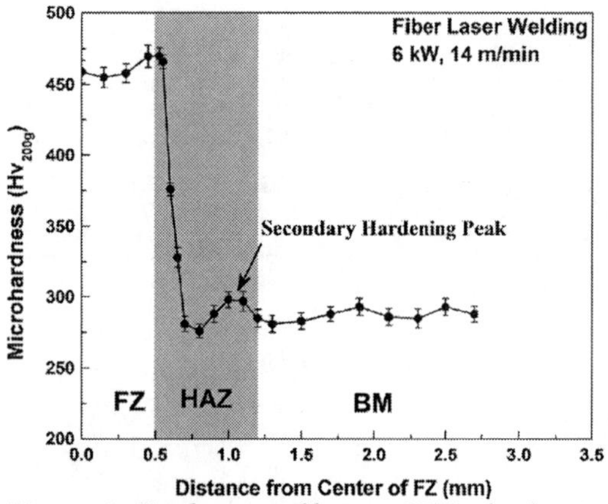

Figure 4: Hardness profile in a LBW of a bainite containing DP780 laser beam weld [11].

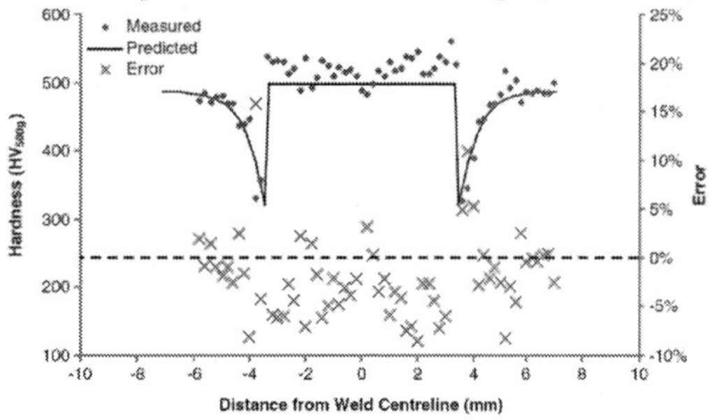

Figure 5: Comparison between actual and predicted hardness profile of a RSW in 0.21wt% C martensite [16].

Multi-Phase Structures Containing Bainite

Bainite in DP steels further complicates the tempering process. When these multi-phase structures are welded, it was seen that an area of the SCHAZ increases in hardness above the base material hardness (see *Fig. 4*). This was investigated using Gleeble rapid tempering, where it was seen that a peak hardness was reached when the sample was tempered for 500°C for 0.5 s [11]. When examined using transmission electron microscopy, it was seen that that samples exhibiting this hardness peak, the Fe_3C and TiC had dissolved from within the bainite structure, forming M_4C_3 and M_2C carbides, which led to secondary hardening.

Tempering Kinetics during Welding

As tempering is a diffusion driven process, it is important to understand the role of tempering kinetics on the change in properties. One of the largest contributing factors to tempering kinetics is the chemical composition of martensite. Three major compositional effects have been seen which affect either carbide nucleation or size. Increasing martensite C content will increase tempering rate [12]. This increase is due to increased C diffusion rate during the nucleation of carbides from the supersaturated martensite. The addition of carbide formers (i.e. Cr, Mo, V), will retard softening [12][13]. It has been shown that during rapid tempering, carbide formers create complex carbides such as: M_3C_2, M_7C_3, and $M_{23}C_6$ which are smaller than the cementite particles [13][14]. Finally, Si and Mn alloying additions will slow cementite growth due to solute partitioning [15]. Again, this will strengthen the tempered martensite due to precipitation strengthening. However, it should be noted that Si additions were seen to reduce cementite diameters at tempering times as low as 30 s at all tempering temperatures, whereas Mn only retarded tempering at temperatures above 500°C and was most effective at longer (>300 s) tempering times. So, it is believed that during welding, Mn will play a limited role in affecting tempering kinetics.

Once the influence of alloying elements on tempering kinetics are known, then the tempering process can be modelled. Several studies have shown that tempering may be modelled using the Johnson-Mehl-Avrami-Kolmogorov (JMAK) equation. However, as diffusion during tempering is heavily dependent on chemical composition and diffusion path, and changes in physical properties are non-linearly correlated to carbide size, literature values for JMAK parameters cannot be used [2]. Instead, JMAK parameters for individual steels may be found by carrying out rapid tempering experiments. Once found, the JMAK equation can be applied to all of the local temperature histories though the weld zone to predict the post-weld properties [16] (see *Fig. 5*).

Applying Tempering to Welding of AHSS

Tempering is encountered in two areas of welding AHSS. Tempering is seen in the SCHAZ (HAZ softening) where it is an outcome of the welding process. Secondly, resistance spot weld (RSW) nuggets may be deliberately tempered using a tempering cycle to lower their hardness [14]. During welding of martensite containing AHSS, HAZ softening occurs, locally reducing the strength of the weldment. During vehicle design this must be accounted for to accurately predict joint strength. Determining HAZ softening is currently done experimentally, however as modelling of the softening process improves, it may also be modelled. It is interesting to note, HAZ softening may sometimes be beneficial. In AHSS with a tendency to form highly martensitic RSW nuggets, HAZ softening will cause button pullout by promoting failure in the HAZ. Without HAZ softening, interfacial failure of the nugget will occur [17]. Again, to improve RSW properties, spot weld procedures in highly hardenable materials are sometimes designed as a two pulse cycle. The first pulse creates the weld nugget and the second tempers it after partial cooling, improving its toughness. Currently these welding cycles are designed experimentally, but they will also be modelled as understanding of tempering kinetics is improved.

Future Work

Although understanding of tempering in welds has greatly improved over the last several years, there is still much that is unknown. Currently, work is needed complete the tempering model, as it does not account for Si, which is well known to retard tempering, or microalloying elements such as Ti and V, which have unknown effects on rapid tempering. Furthermore, there is little understanding on the effects that tempering has on steels with retained austenite. Although there have been some studies done investigating HAZ softening in TRIP steels, there has not been detailed work to understand the changes in material structure after tempering which will determine post-weld mechanical properties in forming of tailor welded blanks or in vehicle crashes. Understanding these effects will become ever more important as TRIP, TWIP and Q&P steels become a larger mix of the body-in-white.

Acknowledgements

The authors would like to acknowledge ArcelorMittal for granting permission to publish this work and financial support from the Natural Sciences and Engineering Research Council of Canada.

References

[1] M. Jung, S-J. Lee and Y-K Lee, "Tempering Kinetics of S45C Martensitic Steel", *Solid State Phenom*, Vol. 118 (2006), pp. 375-380.

[2] E. Biro, J.R. McDermid, S. Vignier, and Y.N. Zhou, "Decoupling of the Softening Process During Rapid Tempering of a Martensitic Steel", Mater Sci Eng A, Vol. 615 (2014), pp. 395-404.

[3] V. Massardier, M. Gouné, D. Fabregue, A. Selouane, T. Douillard, and O. Bouaziz, "Evolution of Microstructure and Strength During the Ultra-Fast Tempering of Fe-Mn-C Martensitic Steels", J Mater Sci, Vol. 49 (2014), pp. 7782-7796.

[4] E. Biro and A. Lee, "Welding Properties of Various DP600 Chemistries", *Sheet Metal Welding Conference XI*, Sterling Heights, MI, May 11-14, 2004, paper 5-2.

[5] N. Kapustka, C. Conrardy, S. Babu, and C. Albright, "Effect of GMAW Process and Material Conditions on DP 780 and TRIP 780 Welds", *Weld J*, Vol. 87 No. 6 (2008), pp. 235-s – 148s.

[6] D.C. Saha, "Kinetics of Martensite Tempering: Thermodynamic and Crystallographic Analysis", Internal Presentation, University of Waterloo, Dec 8, 2015.

[7] S. Sangal, N.C. Goel, And K. Tangri, "A Theoretical Model for the Flow Behavior of Commercial Dual-Phase Steels Containing Metastable Retained Austenite: Part II. Calculation of Flow Curves" *Metall Trans A*, Vol. 16A (1985), pp. 2023-2029.

[8] Advanced High-Strength Steels Application Guidelines, Version 5.0, ed. S. Keeler and M. Kimchi, WorldAutoSteel (Brussels, 2014), pp. 1-15.

[9] S. Vignier, E. Biro, and M. Hervé, "Tempering of Dual-Phase Steels Modelling and Application to Welding : HAZ-HARD Model", ArcelorMittal Internal Report 2012 17729 RDMA (2012).

[10] V.H. Baltazar Hernandez, S.K. Panda, M.L. Kuntz, Y. Zhou, "Nanoindentation and Microstructure Analysis of Resistance Spot Welded Dual Phase Steel", *Mater Lett*, Vol. 64 (2010), pp. 207-210.

[11] D.C. Saha, S.S. Nayak, E. Biro, A.P. Gerlich, and Y. Zhou, "Mechanism of Secondary Hardening in Rapid Tempering of Dual-Phase Steel", *Metall Mater Trans A*, Vol. 46, No. 13 (2014), pp. 6153-6162.

[12] E. Biro, J.R. McDermid, J.D. Embury, and Y. Zhou, "Softening Kinetics in the Subcritical Heat-Affected Zone of Dual-Phase Steel Welds", *Metall Mater Trans A*, Vol. 41A, No. 9 (2010), pp. 2348-2356.

[13] J.B. Lee, N. Kang, J.T. Park, S-T. Ahn, Y-D. Park, I-D. Choi, K-R. Kim, and K-M. Cho, "Kinetics of Carbide Formation for. Quenching and Tempering Steels During High-Frequency Induction Heat Treatment", *Mater Chem Phys*, Vol. 129 (2011), pp. 365-370.

[14] X. Liao, X. Wang, Z. Guo, M. Wang, Y. Wu, Y. Rong, "Microstructures in a Resistance Spot Welded High Strength Dual Phase Steel", *Mater Charact*, Vol. 61 (2010), pp. 341-346.

[15] G. Miyamoto, J.C. Oh, K. Hono, T. Furuhara, T. Maki, "Effect of Partitioning of Mn and Si on the Growth Kinetics of Cementite in Tempered Fe–0.6 mass% C Martensite", *Acta Metall*, Vol. 55 (2007), pp. 5027–5038.

[16] S. Vignier, E. Biro, and M. Hervé, "Predicting the Hardness Profile Across Resistance Spot Welds in Martenistic Steels", *Weld World*, Vol. 58, No. 3 (2014), pp. 297-305.

[17] Y. Okita, V.H. Balatazar Hernandez, S.S. Nayak, and Y. Zhou, "Effect of HAZ-Softening on the Failure Mode of Resistance Spot-Welded Dual-Phase Steel", *Sheet Metal Welding Conference XIV*, Livonia, MI, May 11-14, 2010, paper 1-4.

Weldability of Direct-Quenched Steel with a Yield Stress of 960 MPa

Juho Taavitsainen, Pedro Vilaça
Department of Mechanical Engineering, School of Engineering, Aalto University, Espoo, Finland
juho.taavitsainen@aalto.fi, pedro.vilaca@aalto.fi

David Porter
Materials Eng. and Production Technology, Faculty of Technology, University of Oulu, Oulu, Finland
david.porter@oulu.fi

Pasi Suikkanen
SSAB Europe, POB 93, 92101 Raahe, Finland
pasi.suikkanen@ssab.com

Abstract
The increasingly stringent environmental regulations have raised interest in the use of higher-strength steel, which can enable the design and fabrication of lighter, high-performance structures. Recent innovations in steel processing have made it possible to produce lean alloyed steels with exceptional mechanical properties. As welding is one of the most important manufacturing processes used in industry, the ability to exploit the favourable properties of ultra-high-strength steels is dependent on their weldability. In this work, a commercial, direct-quenched, bainitic-martensitic ultra-high-strength steel with a specified minimum yield stress of 960 MPa, has been studied. The plate thickness was 10 mm and Y-groove butt joints were MAG welded with two passes. The feasibility of using Gleeble specimens to represent specific key-zones of the HAZ was also evaluated. Thus, thermally simulated samples were Charpy impact tested in order to determine the impact transition temperatures of the principal subzones of the heat-affected zone. Hardness measurements and microstructural analyses were used to evaluate the correspondence between the simulated and real heat-affected zones. The relation between the mechanical and metallurgical results on real welds and Gleeble specimens are relevant. A model relating prior austenite grain size and hardness for the CGHAZ was developed and is ready for further exploitation. With low heat input, the 28J transition temperature obtained for the ICCGHAZ was -75.8 °C, revealing a significant improvement from the -29.2 °C obtained for the CGHAZ.

Introduction
Steels are used worldwide and are the most important engineering material for construction and fabrication. Nevertheless the ultra-high strength steels (UHSS), i.e. for tensile strengths exceeding 780 MPa are relatively new to the materials world when compared to the mild steels [1]. As welding is one of the most important manufacturing processes used in industry, the ability to exploit the favourable properties of special steels is dependent on their weldability. Thus, from a manufacturing point of view, the weldability of these steels is of critical importance and represents nowadays a major limitation in the widespread use of the UHSS [2]. Furthermore, the alloying and processing of the steel are the critical factors influencing the microstructures that develop in the heat-affected zone (HAZ) of the welded joints, and hence, the weldability [3]. As a result, there is a strong need to understand the link between the HAZ microstructure and the resulting mechanical properties.

This work addresses the weldability of a modern UHSS. In particular, HAZ properties of a commercial low-alloyed, direct-quenched UHSS [4] with a yield stress of 960 MPa, are investigated. The objective is to investigate the microstructure and mechanical properties that develop in the various subzones of the HAZ in two-run butt welds produced by gas metal arc welding (GMAW). Physical simulation, using the Gleeble thermomechanical simulator, is employed to reproduce samples that can be used for mechanical testing, including impact tests. The purpose of using physical, or thermal, simulation is to enable the reproduction of the HAZ microstructures in quantities that allow characterization of the desired microstructure, which would be difficult, or as in case of impact testing, impossible with samples extracted from the real welds due to the presence of sharp microstructural gradients. Therefore, physical simulation is used for determination of impact transition temperatures. Hardness measurements and microstructural analyses were used to evaluate the correspondence between the simulated and real HAZ. A model relating prior austenite grain size and hardness for the CGHAZ was developed and is evaluated.

Experimental Conditions

Specimens for optical micrography were polished down to 1 μm diamond paste and etched with 2 % nital solution. The optical micrographs were taken from specimens using Nikon Epiphot 200 inverted metallurgical microscope. The microscope was equipped with Nikon DS-U1 digital lenses. For SEM-EBSD, Zeiss Ultra 55 field emission electron microscope (FE-SEM) was used, equipped with Nordlys II EBSD detector by Oxford Instruments. Sample preparation for EBSD was conducted ensuring a deformation-free surface to obtain a sufficiently high diffraction pattern quality.
To measure both prior austenite grain sizes and effective grain sizes on a fine scale, an automated grain size measurement script in Python was implemented according to the mean linear intercept method described in standard ASTM E 1382. To capture and account for any directionality in grain sizes, the measurements are thus done horizontally, vertically and diagonally in rotation increments of 45°. In addition to single-field measurements, a variant of the code capable of measuring grain size gradients along the horizontal direction of a stitched panorama grain boundary image was developed. This was used in determination of the prior austenite grain boundary gradient in the CGHAZ.
Hardness and charpy impact tests were conducted for welded and Gleeble samples. The Gleeble tests were performed using Gleeble 3800 thermomechanical simulator. Gleeble simulations were made to enable a throughout investigation of the three main zones in the HAZ of the welds, namely: Coarse Grain HAZ (CGHAZ); Intercritical HAZ (ICHAZ) and Intercritically reheated CGHAZ (ICCGHAZ). HV1 measurements were performed using a Buehler MMT-7 digital microhardness tester. Charpy impact test were conducted according to standard ISO 148-1. Due to constraints placed by plate thickness and Gleeble test specimens, a reduced-section, i.e. sub-size specimen was used [mm]: 55 x 10 x 5, and test results are fitted using the modified fitting algorithm [5].

The *base material* was an industrial thermomechanically processed and direct-quenched S960 grade steel, with bainitic–martensitic microstructure. The plate thickness was 10 mm. The composition was [wt. %]: 0.092 C; 0.187 Si; 1.10 Mn; 0.01 P; 0.0012 S; 0.033 Al; 0.002 Nb; 0.011 V; 0.014 Cu; 1.14 Cr; 0.398 Ni; 0.0054 N; 0.183 Mo; 0.023 Ti; 0.0017 Ca; 0.0022 B; 0.016 Co. Average grain size was 1.58±1.38 mm (measured from EBSD grain boundary maps using mean linear intercept method and 15° misorientation criterion). The estimated volume fractions of martensite and bainite were 25 % and 75 %, respectively, and in the EBSD maps, no retained austenite was found. Tensile tests were performed according to EN 10149-2 for standard grade S960MC. In longitudinal/rolling direction (RD): yield strength 970 MPa; tensile strength 1123 MPa; elongation 11.6 %. In transverse direction: yield strength 1069 MPa; tensile strength 1192 MPa; elongation 9.8 %. Charpy-V impact toughness tests were performed to get reference properties at low temperatures. Average Cv were as follows for longitudinal direction: 66 J@-20 °C; 54 J@-40 °C; 27 J@-60 °C; and for Longitudinal direction: 109 J@-20 °C; 88 J@-40 °C; 65 J@-60 °C. Thus, in general the requirement of 27 J specified in standard EN 10149-1was clearly overmatched, and only at -60 °C, for the transverse direction, the value is at the limit.

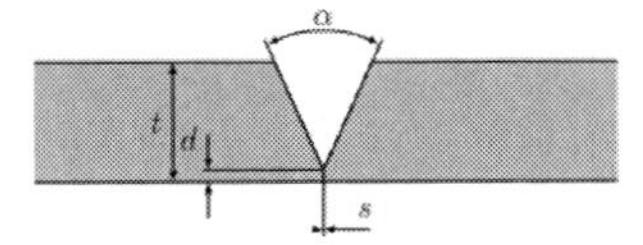

	t (mm)	d (mm)	s (mm)	α (°)
Weld A	10	1	0	45
Weld C	10	1	0	50

a)

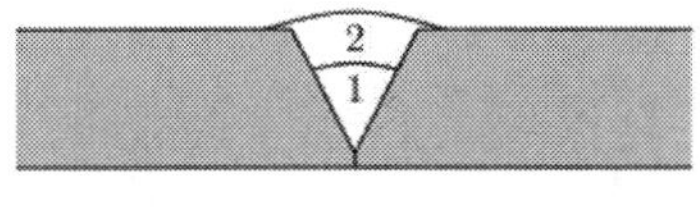

	Run	Polarity	I [A]	E [V]	v_f [m/min]	v [cm/min]	EI/v [kJ/mm]
Weld A	1	DC+	316	26.6	10.6	62.8	0.8
	2	DC+	291	26.5	9.8	57.7	0.8
Weld C	1	DC+	315	26.9	9.9	62.8	0.81
	2	DC+	316	28.8	10.6	36.2	1.51

b)

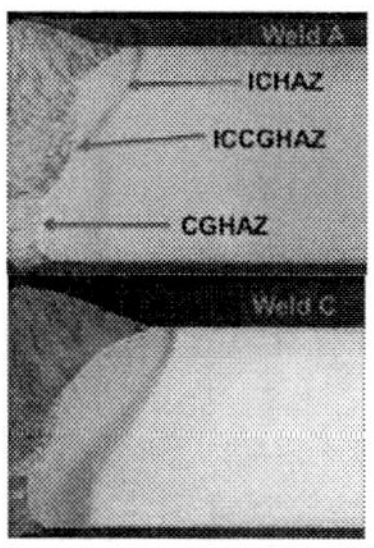

c)

Figure 1: Weld joint design, sequence of weld passes and parameters: a) Weld joint design; b) Welding sequence and parameters; c) Macrograph of the welds with identification of the investigated zones at the HAZ.

The *welding tests* were conducted according to standard ISO 15614-1. The plate dimensions [mm] before welding were 1000 (RD and welding direction) x 200. Two MAG weld samples were produced, designated by "A" and "C" as represented in *Fig. 1*. Both are 2-passes/runs, V-groove butt joint welds, implemented with no air-gap and in flat position without backing. The edge preparation was done by machining. The joint dimensions and weld parameters are depicted in *Fig. 1a* and *1b*. The consumable electrode was Union X 96 with composition [wt. %]: 0.12 C; 0.8 Si; 1.90 Mn; 0.45 Cr; 2.35 Ni; 0.55 Mo (EN 12534). The shielding gas was Ar + 8 % CO2 + 0.03 % NO. No post-weld treatments were applied to welds. As can be seen by comparing the process parameters in *Fig. 1*b, the welds differ only in the second-run parameters. Second pass of low HI weld "A" corresponds to an estimated cooling time, $t_{8/5}$=4s, and second pass of high HI weld "C" corresponds to an estimated cooling time, $t_{8/5}$=15s.

Analysis of the Results

The hardness measured in the cross section, near the face and root of the 2-passes welds is presented in *Fig. 2*. In general the hardness at the face overmatches 10 to 30 HV the hardness at the root for both weld conditions. The minimum hardness occurs about 3 mm from the fusion line with an undermatching of about 100 HV for weld "A" (low HI) and 130 HV for weld "C" (high HI). The undermatching in the weld metal is about 100 HV for weld "C", and 50 HV for weld "A", although at the face, due to highest cooling rate for the 2nd pass with low HI, the hardness of weld metal nearly matches the base material. The extension of the HAZ with hardness undermatching induced by the weld is about 7 mm from fusion line (FL) for the low HI weld "A" and about 12 mm for the high HI weld "C".

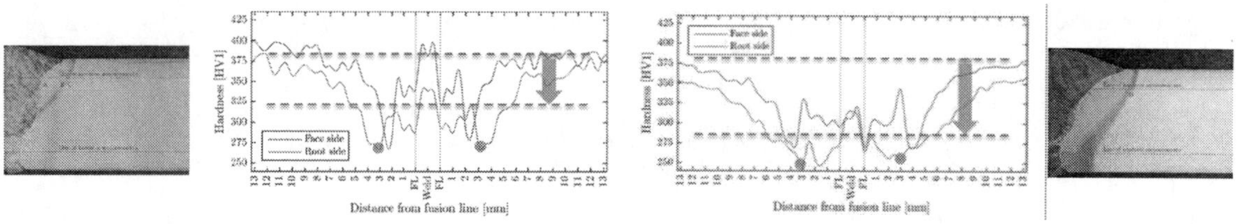

a) b)

Figure 2: Hardness distribution at the face and root sides: a) Low HI weld "A"; a) High HI weld "C".

The *Fig. 3* presents two microstructures of CGHAZ and ICHAZ of low HI weld and the Charpy-V impact toughness of all the Gleeble simulated samples. Concerning the influence of the HI weld conditions: the increase in HI increased the toughness of the CGHAZ; In opposite effect, the increase in HI decreased the toughness in the ICHAZ and ICCGHAZ. With low heat input, the 28J transition temperature obtained for the ICCGHAZ was -75.8 °C, revealing a significant improvement from the -29.2 °C obtained for the CGHAZ. The Fig. 4 complements the microstructural analysis with the EBSDinverse pole figures of all the different zones and conditions. The CGHAZ microstructure is similar in both samples, with microstructural features reminiscent of bainite and martensite in a relatively coarse prior austenite grain structure. Both ICHAZ shows no texture effects and the one simulating the high HI weld shows the lower grain size. The extent of reaustenitization in ICCGHAZ seems to be significantly lower than what it is observed in the ICHAZ.

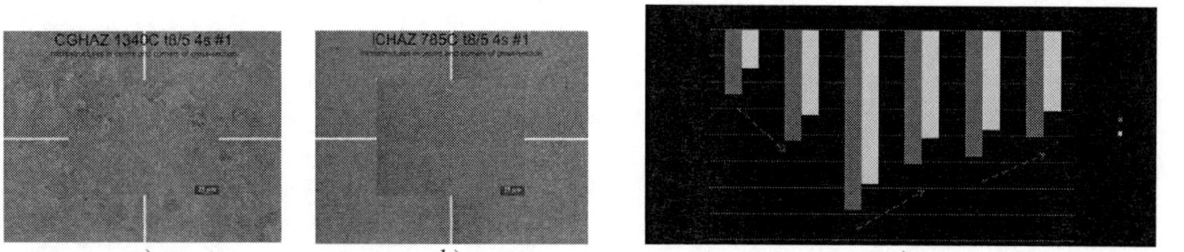

a) b) c)

Figure 3: Results from the Gleeble simulation samples of CGHAZ, ICHAZ, and ICCGHAZ for low HI weld "A" and high HI weld "C": a) and b) Microstructure of CGHAZ and ICHAZ of low HI weld "A"; c) Charpy-V impact toughness.

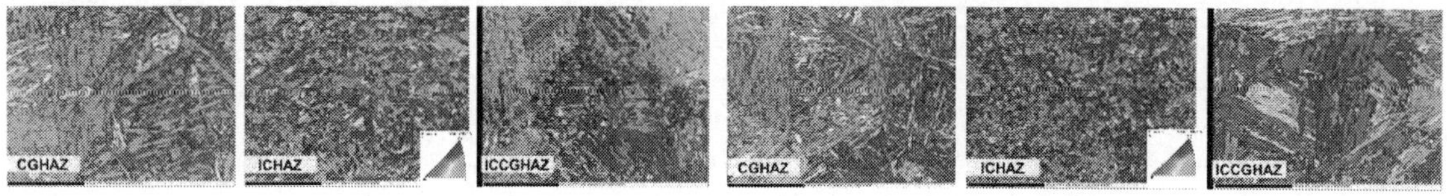

a) b)

Figure 4: SEM/EBSD characterization of Gleeble simulation samples: a) Low HI weld "A"; b) High HI weld "C".

To investigate the effect of the prior austenite grain size on the hardness of CGHAZ the prior austenite grain size (*Fig. 4*) is correlated with the hardness in *Fig. 5*. A linear regression model is fitted to the data, using the linear least squares method. From the results it can be seen that there is a weak but statistically significant ($p < 0.05$) decreasing trend in the hardness. The findings are in line with literature [6], anyway the existence of a Hall-Petch type relationship between prior austenite grain size in CGHAZ and hardness was tested but it is weak, mainly for the high HI weld "C".

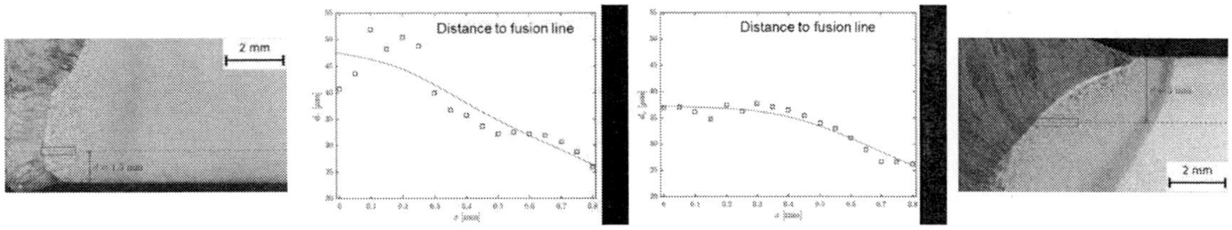

a) b)

Figure 4: Size of prior austenite grain size in CGHAZ of real welds: a) Low HI weld "A"; b) High HI weld "C".

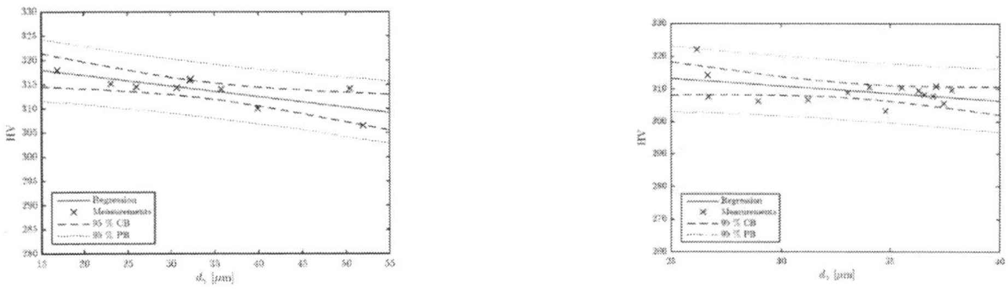

a) b)

Figure 5: Effect of prior austenite grain size, d_γ, on the hardness of CGHAZ: a) Low HI weld "A"; b) High HI weld "C". Note: CB = Confidence bound: uncertainty in slope of regression line; PB = Prediction bound: uncertainty in a single prediction.

Conclusions

Along the weld zone, the hardness near the face (second weld pass) is higher than the hardness near the root for both welds. Minimum hardness occurs about 3 mm from the fusion line with an undermatching of about 100 HV for low HI weld "A" and 130 HV for high HI weld "C". The toughness increases with the increase of HI for the CGHAZ. However, an increase in HI decreased the toughness of the ICHAZ and the ICCGHAZ. The effect on toughness of the ICHAZ thermal cycle is preponderant over the previous CGHAZ. With low HI, the T28J transition temperature obtained for the ICCGHAZ was -75.8 °C, revealing a significant improvement from the -29.2 °C obtained for the CGHAZ. A relationship between the prior austenite grain size in CGHAZ and hardness was tested for both welds, but not found.

References

[1] H. Bhadeshia et al: Steels: microstructure and properties Elsevier Science (2011).
[2] P. Kah, et al: "Welding of ultra-high strength steels". Advanced Materials Research, Vol. 849 (2014) pp. 357–365.
[3] K. Easterling: Introduction to the Physical Metallurgy of Welding, Butterworth-Heinemann (Oxford, 2nd Edition 1992), pp. 126-190.
[4] D. A. Porter: "Developments in hot-rolled high-strength structural steels", *In Nordic Welding Conference*, Tampere, Finland, November, 2006.
[5] K. Wallin. Fracture toughness of engineering materials: estimation and application. EMAS Publishing (Warrington, 2011), pp. 321-372.
[6] S. Morito, et al:. "Effect of block size on the strength of lath martensite in low carbon steels". Materials Science and Engineering: A, Vol. 438 (2006), pp. 237–240.

Mechanical Properties of Ultra High-Strength Steel Welded Joints

Mikko Peltonen, Hannu Hänninen
Aalto University School of Engineering, Espoo, Finland
mikko.j.peltonen@aalto.fi
hannu.e.hanninen@aalto.fi

Akseli Kukkonen
ABB Oy, Helsinki,Finland
akseli.kukkonen@abb.com

Sami Nummela
SSAB Europe Oy, Hämeenlinna, Finland
sami.nummela@ssab.com

Abstract

The objective of the study was to compare conventional welding methods and their suitability for high-strength steels as alternative choice for laser or laser-hybrid welding. The joining methods were metal-active gas welding (MAG), plasma arc welding (PAW) and submerged arc welding (SAW). Test welds were evaluated by metallography, tensile tests, impact tests, fatigue tests and hardness profiles. Also, the effects of residual stresses of base material on the welding distortion in butt-welded joints were studied. Test materials were ferritic-bainitic Optim 700 MC Plus and bainitic-martensitic Optim 900 QC steels. The studies showed that the conventional welding methods can produce high quality welds if the possibilities and limitations of each welding method are recognized, and single-pass butt welds can be produced up to material thickness of 8 mm. Mechanical properties of the weldments are controlled by the heat input, cooling time $\Delta t_{8/5}$ and consumable

Introduction

The demand for lighter structures in industry has greatly increased over the years. Applications can be found in the field of transportation such as material handling and containers, lifting such as mobile cranes, and construction engineering. Novel high-strength steels are extremely sensitive to welding, especially when dealing with high heat input processes. Attention must be paid to the welding procedure to minimize the heat input without losing the properties in the heat-affected zone (HAZ). The manufacturer needs to develop, employ, and control the necessary equipment and techniques to reduce the effects of the welding process that promote distortion. Alternative welding methods for laser beam welding, laser-hybrid welding and friction stir welding such as MAG, PAW and SAW are used to study the limitations and possibilities for single-pass butt welds. For using the benefits of the mechanical properties, pros and cons of each welding method can be categorized and the most suitable welding process for each application can be chosen.

Test materials and test procedure

In order to select the most suitable welding method for different applications the requirements of the structure have to be known. Conventional welding methods have their own limitations when dealing with AHSSs. The objective was to determine the parameter window for each welding method and material thickness. In addition, the scope was to define whether or not air gap and backing are needed. Parameter window defines limits where the welding can take place and the weld quality is acceptable. The goal was to weld a single-pass butt joint without an air gap and backing. The following welding equipment were used in MAG welding: Kemppi FastMig Pulse 450, FastMig MXF 65, in PAW: Esab LTG 400, PW3000, PTW 300, CW3000 and in SAW: Esab LAE 1250, A6 SFD 1, PEG 1.

Two different strength level steels were studied which met the requirements of standard EN 10149-2. First material was Optim 700 MC Plus which is manufactured using thermomechanical rolling. In the scope of the studies, material thickness 6 and 8 mm was used. The second material was thermomechanically rolled and quenched Optim 900 QC with material thickness 4, 6 and 8 mm. Under normal conditions preheating in welding is not necessary because of the low carbon content, however heat input and cooling time should be restricted in case the goal is to achieve base material-like properties. Dimensions for the test plates were 200 x 600 mm. Mechanical properties are presented in Table 1.

Table 1 Mechanical properties of the test materials.

Material	Yield Strength $R_{p0,2}$ [MPa]	Tensile Strength R_m [MPa]	Elongation A_5 [%]	Impact Energy KV [J]
Optim 700 MC Plus	≥ 700	750 - 950	≥ 13	≥ 40 / -60 °C
Optim 900 QC	≥ 900	930 - 1200	≥ 8	≥ 27 / -40 °C

The goal was to weld a single-pass butt joint without an air gap or backing. Groove geometry of the joint was determined in preliminary welding tests which addressed that 4 mm plates are the only ones that can be welded without air gap or backing using MAG. As the thickness of the plate increases to 6 and 8 mm, air gap of 1 and 2 mm is needed. Backing is also needed in 8 mm plates with 2 mm air gap. Using keyhole PAW all the plates can be welded without air gap or backing. Furthermore, air gap is not needed in SAW, only backing support is necessary to control the penetration.

Mechanical properties of the base material and weld joints were determined transverse to rolling direction using tensile testing according to SFS-EN ISO 4136 and 15614-1. Microhardness profiles were measured from mid-thickness of the plate. Impact testing was carried out according to SFS-EN ISO 148-1 and SFS-EN ISO 9016. Based on the test material thickness, standard 10 x 10 x 55 mm impact test specimens could not be used. Fatigue tests were only made to Optim 900 QC 4 mm plates according to ASTM E 466-96. Cooling time, $\Delta t_{8/5}$, was determined for each weld both by calculations and measurements using thermocouples. Microstructures were evaluated from the welds and different zones of the HAZ. Welding parameters are presented in Table 2.

Table 2 Welding parameters.

	Material	Consumable	Welding Speed [mm/min]	Current [A]	Voltage [V]	Q [kJ/mm]	$\Delta t_{8/5}$ [s]
MAG	Optim 700 MC Plus - 6 mm	OK AristoRod 69	900	338	29,1	0,56	9,2
	Optim 700 MC Plus - 6 mm	Tubrod 14.03	900	336	31,0	0,59	13,5
	Optim 900 QC - 6 mm	OK AristoRod 89	900	319	26,3	0,48	9,0
	Optim 900 QC - 6 mm	Coreweld 89	900	347	28,3	0,56	9,6
PAW	Optim 700 MC Plus - 6 mm	OK AristoRod 69	230	232	27,5	1,00	23,6
	Optim 700 MC Plus - 6 mm	Tubrod 14.03	220	232	27,8	1,06	24,8
	Optim 900 QC - 6 mm	OK AristoRod 89	230	231	27,5	0,99	24,8
	Optim 900 QC - 6 mm	Coreweld 89	230	230	27,5	0,99	24,0
SAW	Optim 700 MC Plus - 6 mm	OK Autrod 13.43 + OK Flux 10.62	1100	701	27,7	1,06	38,8
	Optim 700 MC Plus - 6 mm	OK Tubrod 15.72S + OK Flux 10.62	850	701	28,5	1,41	43,6
	Optim 900 QC - 6 mm	OK Autrod 13.43 + OK Flux 10.62	1050	703	27,7	1,11	39,0
	Optim 900 QC - 6 mm	OK Tubrod 15.72S + OK Flux 10.62	900	701	28,5	1,33	42,2

Figure 1a) presents the microstructures together with EBSD maps. The microstructure of Optim 700 MC Plus consists of bainite and ferrite. The volume fraction of bainite is higher than that of ferrite. Based on microstructure analysis of the samples, the volume fraction of bainite is evaluated to be approximately 70 % and ferrite is roughly 30 %. In Figure 1b) Optim 900 QC consists of martensite and bainite. Based on microstructure analysis, the volume fraction of bainite is approximately 70 % and martensite is about 25 %. Ferrite and retained austenite was also observed [1].

Consumables for welding

Commercially available consumables were selected for each base material based on steel manufacturer recommendations. Both solid and cored wires were used and the yield strength varied between 700 - 930 MPa. The diameter of the filler wire was 1,2 mm for MAG and PAW, excluding SAW where the wire diameter was 3,0 mm. Consumables are more alloyed than the base material, due to dilution. Higher alloying content is used to prevent the effects of dilution, when the weld metal (WM) can be markedly undermatching. Following filler wires were used: OK Aristorod 69, OK Tubrod 14.03, OK AristoRod 89, Coreweld 89, OK Autrod 13.43 and OK Tubrod 15.27S.

Residual stresses and welding distortions

Residual stress measurements had an important role in the studies. Due to steel manufacturing, the material experiences high amounts of deformation and the temperature differences are causing internal stress development. The goal was to determine if these stresses have an effect on the distortions after the welding. Residual stress measurements were performed in transverse and longitudinal direction using X-ray diffraction. All the X-ray diffraction results were verified using hole drilling measurements [2].

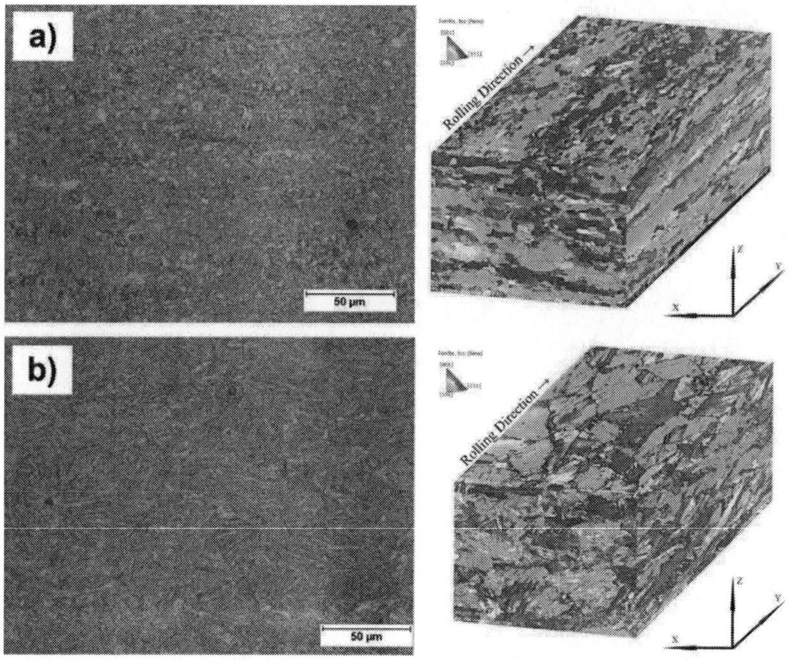

Fig. 1 Microstructures of the base materials with EBSD maps: a) Optim 700 MC Plus and b) Optim 900 QC. The maps are combined from three different samples of different orientations.

The distortions caused by the welding were evaluated in transverse direction based on angular distortion. In the longitudinal direction shrinkage and bending distortions were evaluated. To minimize the residual stresses and welding distortions, the plates were not clamped into the welding jig, and hence all the deformation happened freely. As a reference for the residual stress measurements, stress relieved plates were used.

Results and discussion

Tensile tests were made for every material thickness. Due to large number of the samples, only the results of the 6 mm plates are presented. The weld cross-sections of the 6 mm Optim 900 QC plates are presented in Figure 2, where the joint geometry and HAZ can be seen. Tensile test results for Optim 700 MC Plus are compared between each welding method (Figure 3). The results do not have any significant difference between each other, although the heat input and cooling times in PAW and SAW are higher compared to MAG. Microhardness results indicate that the softened HAZ is wider as the heat input is increased, which can be seen especially in PAW and SAW welds. This does not have a remarkable effect on tensile strength. The lowest heat input and the fastest cooling time can be achieved using MAG welding.

In Figure 3 tensile test results are presented for Optim 900 QC. Novel high-strength steels are extremely sensitive to welding, especially when dealing with high heat input processes. Therefore none of the used welding methods could not achieve base material-like properties. Due to high and slow temperature cycle, microstructure of the base material is changed greatly, so that the undesired microstructures become more dominant. Consequently, small differences in the heat input can have drastic effect on the tensile strength.

Impact tests were done for all material thicknesses and the test results are corrected to correspond the impact test results of standard full-size specimen. Due to size factor, Charpy-V impact testing was not carried out for the 4 mm plates. Welded samples were tested within temperature range from -60 to 60 °C and the notch was located in weld metal, fusion line and fusion line +1 mm. Figure 4 presents impact energies for 6 mm Optim 700 MC Plus for each welding method. At the test temperature of -20 °C MAG and SAW welded samples behaved in the most favorable way in all zones. Based on the studies, the microstructure in the samples favors acicular ferrite. It can be justified that increased heat input has a direct influence on the impact toughness, which can be seen from MAG and PAW samples where the same filler wire was used. The PAW samples have low impact energies through the whole test procedure, especially in the HAZ. This may be caused by the high heat input which increases grain growth. In addition, the dilution was great because air gap was not present in PAW tests.

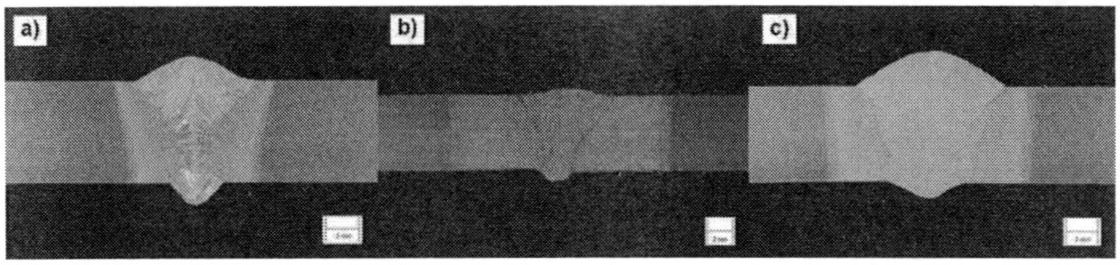

Fig. 2 Macro-sections of welded Optim 900 QC samples: a) MAG, b) PAW and c) SAW.

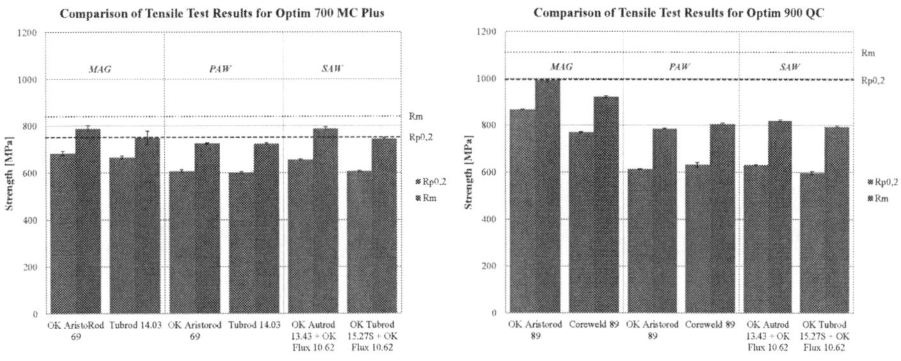

Fig. 3 Comparison of the tensile test results between each welding method for 6 mm plates. Base material yield and tensile strength are marked on the chart.

Test results of the 6 mm Optim 900 QC show that the cored wires can produce higher impact energies than solid wires, although more welding energy is needed to make a single-pass weld. This can be seen from MAG and SAW results in Figure 4 which favored acicular ferrite formation. In SAW, the cored wire was more alloyed than the solid wire which compensated the high heat input and caused the good results in WM. Higher content of Ni produced desirable microstructure. Even at FL and FL+1 the impact energies are very good, even though the heat input and cooling times were increased. The lack of air gap decreases impact toughness values in PAW which can be caused mainly by the dilution rate and high heat input combined with increased cooling time. Grain boundary ferrite, Widmannstätten ferrite, carbides and polygonal ferrite were seen in all tested zones, which can explain decreased impact toughness in PAW.

Fatigue tests were carried out for the 4 mm Optim 900 QC plates and results were compared between each welding method. The actual weld geometry was taken into account together with the influence of misalignments caused by the welding when the structural hot-spot stress approach was applied. The geometry of the weld bead has an important role in terms of the structural fatigue life, and hence the actual hot-spot stresses next to the weld joint were determined [3]. Fatigue life results are presented in Figure 5. PAW specimens have the highest fatigue strength. One of the advantages of the PAW is the smooth and low angle toe of weld which makes crack initiation more difficult and hot-spot stress lower. SAW has also very good fatigue strength, although undercutting was observed. All the MAG welds had the lowest fatigue strength. The main reasons for the low fatigue strength are the MAG weld joint geometry and high angle weld toe which can be improved using smaller size filler wire, air gap and mechanical treatment for the weld toe.

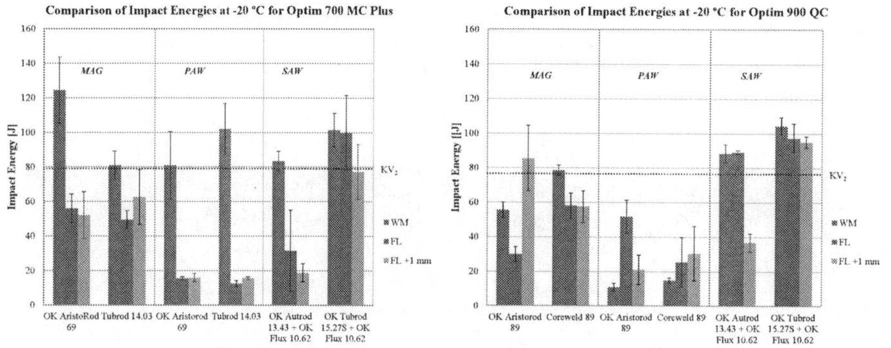

Fig. 4 Comparison of impact energies between each welding method for 6 mm plates. Base material impact energy level is marked on the chart.

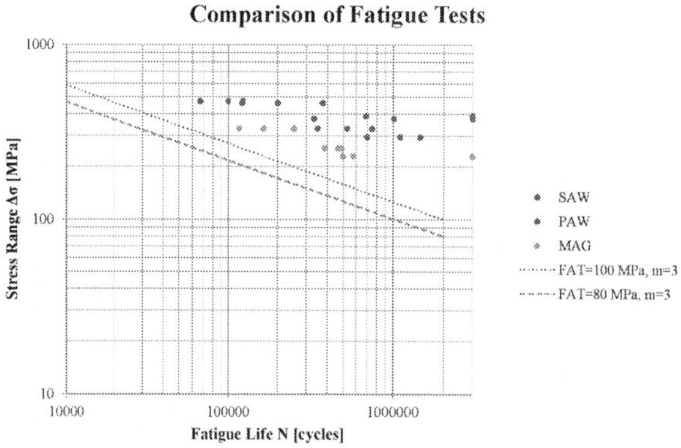

Fig. 5 Comparison of fatigue test results using hot-spot stress approach between each welding method for 4 mm Optim 900 QC specimens [3].

Residual stresses before welding on the surface of base materials are different on different sides of the plate and the stress level is dependent on the material thickness and microstructure. Results show that the residual stresses of the base material had little effect on the welding distortion. Whereas, the effect of residual stresses of the base material before welding on residual stresses after welding is considerable. In Figure 6 are shown residual stresses before and after welding in 6 mm Optim 900 QC plates. Used heat input is directly proportional to the residual stresses, that is, welding distortions are also large as the heat input is increased. For angular distortions, welding speed is a significant factor together with the material thickness.

Conclusions

The studies showed that the conventional welding methods can produce high quality welds, if the limitations of each welding method are recognized. MAG welding is the most suitable for all test material thicknesses. It combines good penetration and low heat input. The downside of the process is the fact that only 4 mm plates can be welded without air gap and backing, which restricts the use. PAW welding is suitable if air gap and backing are not accepted, but the effects of dilution rate and high heat input must be taken into consideration. In addition, the material thickness is restricted to 6 mm. The microstructure of SAW samples is similar with PAW samples, but the used SAW consumables favor acicular ferrite in WM, which improves the impact toughness. The effect of heat input can be seen in CGHAZ, where the difference in size is evident when MAG and SAW samples are compared.

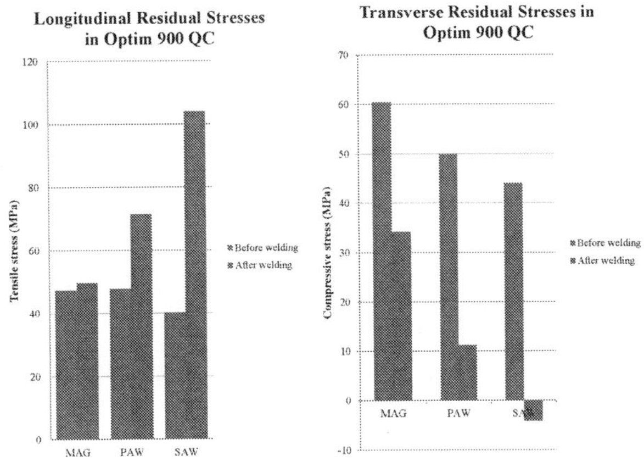

Fig. 6 Comparison of changes in longitudinal and transverse residual stresses on face side by using different welding methods.

The angular misalignments have a significant effect on the fatigue strength when nominal stress approach is used. The structural hot-spot stress approach shows that all the results are above the fatigue class FAT100. The highest fatigue strength can be achieved using PAW. One reason for the good fatigue behavior is the smooth weld geometry, especially the low angle weld toe.

Residual stresses are higher on the other side of the plate in both longitudinal and transverse direction. In butt-welded plates, the final stress field is the sum function including initial residual stresses and mechanical properties of the material, heat input and welding distortions. Bending distortion is higher in Optim 900 QC than in Optim 700 MC Plus and is directly proportional to the heat input. It is also directly proportional to the thickness of the welded material. Tests show that angular distortion is directly proportional to heat input of welding process per welding speed (Q/v). Residual stresses before welding do not seem to affect angular distortion.

Acknowledgements

The research was a part of the FIMECC (Finnish Metals and Engineering Competence Cluster) program: Light and efficient solutions strategic platform research (LIGHT) – Novel manufacturing technologies.

References

[1] M. Peltonen, Weldability of high-strength steels using conventional welding methods, 2014, Aalto University, Master´s Thesis.

[2] A. Kukkonen, Residual stresses and distortions in welded ultra-high-strength steels, 2014, Aalto University, Master´s Thesis.

[3] H. Remes, M. Peltonen, T. Seppänen, A. Kukkonen, S. Liinalampi, I. Lillemäe, P. Lehto, H. Hänninen, J. Romanoff, S. Nummela, Fatigue strength of welded extra high-strength and thin steel plates based on nominal and structural hot-spot stress approach, 2014, MARSTRUCT2015.

Observation of the precursor phenomenon of the brittle crack initiation in auroformed Bainitic steel

Shohei Asako [1], Tomoya Kawabata [1], Shuji Aihara [1], Shintaro Kimura [2], Kiyoshi Kagehira [3]

[1] *the University of Tokyo, Japan*
[2] *Mitsubishi Heavy Industries, Ltd, Japan*
[3] *Primetals Technologies Japan, Japan*

In recent years, one of the high tensile steels ,bainite, have got much attention in steel structural fields. because it has good balance of strength, ductility and toughness.We must consider not only the strength but also the brittle fracture toughness when we choose structural materials for some structures. Although it is well known that the brittle fracture toughness of the fine bainitic structure manufactured using modern TMCP is much superior to ordinary strength steels, we still have to improve its brittle fracture toughness so as to use it under the severer conditions. The theoretical mechanism of brittle fracture initiation of bainite is still not clear because of its complexity of its structure. According to previous research including the observation of the fracture surfaces or cross sections of fractured specimens, the prior phenomenon of brittle fracture may occur at either the boundary area between MA and matrix or inside of MA. However, the details of the prior event for the macroscopic fracture mechanism are still unknown. The purpose of this study is clarifying the precursor phenomenon and estimate the quantitative criterion of brittle fracture initiation of bainite steels, which have fine microstructures by observing the process directly by the in-situ observation methods.

Development of Laser Welding for Advanced High Strength Steels Directly into the Bainitic Range

Lima, M.S.F.*, Liu, S.
Colorado School of Mines, Golden, CO, USA
E-mail: sliu@mines.edu, phone: 1 (303) 273-3796
** permanent address: Photonics Division, Institute for Advanced Studies, Sao Jose dos Campos, SP, Brazil*

Abstract
Laser welding of Advanced High Strength Steels (AHSS) produces an almost fully martensitic fusion zone, which extends to some portion of the heat-affected zones. Subsequent tempering or direct hot stamping increases the direct processing costs. In addition, the manipulation of a hard and brittle joint could produce stress concentrators or part breakage. The present study proposes an innovative method of laser welding at high temperatures, where the rapid solidification and austempering are conducted together in a same equipment. The AHSS parts are heated to a given austempering temperature and kept for a given time to ensure an isothermal condition, after which laser welding is performed. The welded part is then maintained at the austempering temperature for a time sufficient to develop a bainitic microstructure, and then cooled down to the ambient temperature. In order to test the applicability of the method, a 22MnB5 hot press AHSS was welded in ambient and high temperature conditions. Ordinary ambient temperature welds using a fiber laser of 800 W power and 3 m/min speed produced an almost fully martensitic seam. Using the same laser parameters, but after holding at 445°C for10 minutes, a bainite plus retained austenite microstructure was generated. This methodology is a viable method to produce welds directly in the bainitic range as long as the laser and heating sources are conjugated.

Introduction
Advanced high strength steels (AHSS) are a class of materials particularly suitable for the automobile body structures, where low weight, reliability and safety are required [1]. AHSS comprise many steel designations such as dual phase (DP), transformation induced plasticity (TRIP) and martensitic steels. It is a common industrial practice that one or more AHSS are combined together in a same part in order to fulfill the requirements of the car designer, such as stiffness, toughness and/or impact absorption. These steels have been joined with the use of lasers, producing the tailored blanks, and then send to cold or hot pressing. On one hand, cold pressing is the most economical way, because tools and dies are only lubricated by water and emulsions, and the cycle takes only few seconds. On the other hand, hot pressing is a more sophisticated process where the blank enters a furnace, is kept at austenitizing temperatures for a given time, and is then press-hardened using a cold tool. In this case, the final microstructure is mostly martensitic, with very high hardness and negligible ductility, which is suited for energy absorbing structures such as the car pillars. A particularly good AHSS for hot pressing is the grade 22MnB5, which is the subject of the current study. The problem of joining one or more parts of 22MnB5 steel is the intrinsic brittleness of the fusion zone (FZ) and the heat-affected zone (HAZ) due to the high content of martensite. The joint is so brittle that automatic mechanical manipulation could break it during press loading. Additionally, many hot press blanks of 22MnB5 steel are coated with an aluminum-based alloy in order to resist the extensive oxidizing environment during heating. If not properly removed, this coating induces further embrittlement of FZ due to the appearance of Fe-Al intermetallic compounds [2].

Many authors indicated that the formation of bainite instead of martensite in the FZ of AHSS could greatly increase the part toughness [3]. However, few authors could propose a methodology where the bainite could be formed directly during welding. For example, the classical austempering treatment comprises austenitizing the blanks in temperature above A3 and then quench them into a convenient medium, such as molten salt, at a temperature just above the martensite start (Ms). The resulting microstructure is bainite and retained austenite, which combines good mechanical strength and excellent formability.

The objective of this work is to propose, for the first time, a procedure comprising both laser welding and austempering in the same equipment. The potential benefits of the technique are huge and could redefine the weldability of materials, possibility of cold forming of difficult-to-weld materials, reduction of deformation and residual stresses, and possibility to form new metastable phases. In the present context, the methodology was planned to produce bainite in the fusion zone of a 22MnB5 weld, as well as to characterize the welded parts from the microstructural and mechanical behavior points of view.

Material and Methodology
The studied material was furnished by AK Steel Corporation, OH, U.S.A. in the form of sheets measuring 105 mm

x 44 mm x 1 mm. The steel classification is 22MnB5, under the trade name of ULTRALUME PHS (Press Hardenable Steel) [4]. The steel composition was Fe-0.22C-1.18Mn-0.33Si-0.27Cu-0.054Al -0.204Cr (wt.%) and its surface was coated by a 20 μm thick Al-9%Si alloy. The interface between the coating and the steel exhibited a reaction layer of approximately 6 μm. The base material is composed of pearlite and ferrite, with pearlite accounting for about 44% of the volume fraction of the base material.

The laser used in this study was an IPG Photonics fiber laser with an output power of 1000 W, model YLR-1000, equipped with a 100 μm inner diameter optical fiber for beam delivery. The beam parameter product (BPP) was 7.5 mm.mrad. The laser head motion was carried out by step motors mounted on a CNC table and controlled by Mach3 CNC software. The CNC software also controls the laser on/off and laser power.

The heating of the samples was realized using an induction radio frequency (RF) power source by Ameritherm Inc.. The induction coil is a pancake-type with four complete turns and positioned below the sample to be heated. Figure 1a presents the experimental setup for the high temperature welds. An alumina bridge was fabricated in order to keep a given clearance between the coil and the sample and to avoid short-circuits. Figure 1b presents the actual set-up viewed from the front for better visualization of the coil and the sample positions.

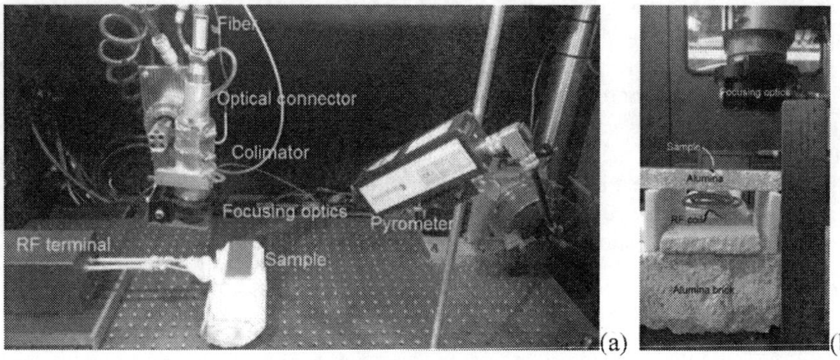

(a) (b)

Figure 1. (a) Set-up for high temperature laser experiments. (b) Front-side view of the experimental set-up.

The accuracy of the infrared pyrometer measurements was assured by a comparison with thermocouples soldered to some samples. The differences between the thermocouple and the pyrometer measured temperatures were ± 1°C. According to these measurements, the optical emissivity of the substrate was set at 48% for all experiments.

The fixed process parameters were: weld speed (v) 50 mm/s (3 m/min); laser power (P) 800 W and focus on the upper surface. The focus on the upper surface ensured a minimum spot diameter of 0.1 mm. Considering both laser power and spot diameter, the intensity was 1×10^{7} W/cm^2 and the laser heat input was 16 J/mm.

In agreement with our *a priori* calculations, the holding temperature T* was set as 455°C by using a RF current of 135 A. The "AT20" condition belongs to the ambient temperature welds and the "HT455" condition is the high temperature weld. Even a small variation in the sample position could influence the high frequency coupling and thus the actual T* of HT455 for each sample could change between 440°C and 469°C. Each experimental condition was repeated four times.

The procedure for the high temperature condition HT455 was as follows:

a) Both the pyrometer recording and induction current were started at the same time;
b) The time (t1) during the heating time was 10 minutes for all the samples;
c) The laser weld is realized in the middle of the sample, as bead-on-plate;
d) The sample, already laser processed, is maintained at T* (~455°C) for 10 minutes (t2);
e) After the total time (t1+t2) of 20 minutes the induced current is turned off and the sample was cooled down to the ambient temperature over the alumina plate.

A graphic representation of the high temperature welding is presented in Figure 2.

Optical microscopy (OM) examination was performed on samples prepared following standard metallographic procedures, polishing of the transverse cross sections of the samples to 6 and 1 μm (diamond suspension), followed by etching with a 2% Nital solution (2% HNO_3 in ethanol). The images were recorded by a camera connected to an inverted metallurgical microscope LECO/Olympus model PMG3.

Scanning electron microscopy (SEM) examination was carried out using a QUANTA600 FESEM and a JEOL JSM 7000F, equipped with an EDAX Octane Plus, for the chemical analyses.

The thermal simulation was carried out using Sysweld Software© [5]. Sysweld is a finite element modeling (FEM) software specially designed for welding and heat treatment of metals and alloys. For the current purpose a mesh refined around the laser path was designed with a lateral resolution of about 0.001 mm. The exact 22MnB5 material properties are not available in the Sysweld database, therefore TRIP750 material properties were used instead. The model dimensions and welding procedure followed those described earlier in the experimental design, starting with the material at ambient temperature and heating it up to T*. The software produces many valuable outputs, but those of interest here are the time-temperature evolution in different regions of the sheet metal. The thermodynamic data are obtained using TCFE6-steel database of the ThermoCalc© software [6].

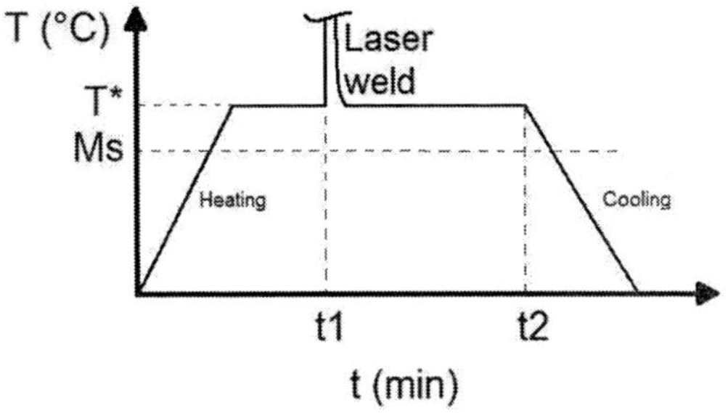

Figure 2. A schematic evolution of temperature as a function of time for the experiments.

Computational Results

The use of computational methods before the experimental development is of utmost importance to define the process parameters instead of the trial-and-error approach. Here, two theories are important: defining the onset of the phase formations and kinetics and determining the cooling curves.

Time-temperature-transformations

In bainite, nucleation corresponds to a point where the slow, thermally activated migration of glissile partial dislocations gives way to rapid, breakaway dissociation [3]. It is possible to observe two sets of transformation units, the first consisting of very fine embryo platelets below the size of the operational nucleus, and the second the size corresponding to the rapid growth to the final size. There are several theories of bainite nucleation and growth such as the models proposed by Bhadeshia [7], Rees and Bhadeshia [8] and Singh [9], all from the University of Cambridge. According to Santofimia et al. [10], the Singh model [9] predicts better the bainite fraction as a function of temperature for an alloy (Steel A – Fe-0.31C-0.25Si-1.22Mn) whose composition is similar to the current 22MnB5 steel. Considering the similarity, the time-temperature-transformation (TTT) diagrams for each alloy was generated by a software called MAP_STEEL_MUCG83 developed by the same research team [11]. MUCG83 was used for modeling of the thermodynamics and kinetics of solid state transformations in steels and calculates Widmanstätten, Bainite and Martensite start temperatures. Figure 3 presents the curves for Widmanstätten ferrite start (F), Bainite start (B) and martensite start (Ms) temperatures as a function of time for the current alloy. The Widmanstätten start temperature range is between 600 and 760 °C. The nucleation-limited bainite start temperature is 480 °C. The martensite start temperature is estimated as 412°C, which corroborates the value obtained by Karbasian and Tekaya [12].

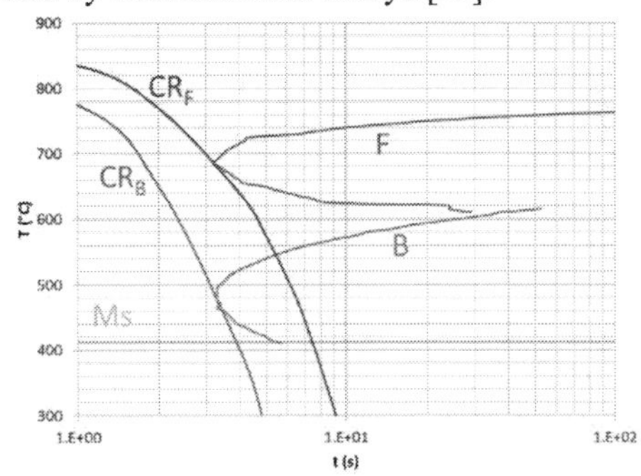

Figure 3. Calculated time-temperature-transformation curves for the current alloy. (F) Widmanstätten ferrite start, (B) Bainite start, (Ms) martensite start.

Considering the TTT curves in Figure 3 and an initial austenitizing temperature of 900°C, the critical cooling rates for Widmanstätten Ferrite (CR_F) and Bainite (CR_B) are 65°C/s and 15°C/s, respectively. When the CR is below CR_F a primary diffusion-controlled ferrite appears. When the CR is higher than CR_B, martensite is the main reaction product. Finally, between CR_F and CR_B, bainite will growth from the primary austenite. These values are only estimates, since the TTT curves will be displaced during continuous cooling conditions. However, it is easy to see that the any trial for locating the weld cooling rate between CR_F and CR_B is going to be a difficult task.

Welding thermal evolution and cooling curves

The original TTT curves in Figure 3 were plotted together with the FEM cooling curves for each experimental condition (AT20 and HT455). Figure 4 presents the results for the AT20 condition. The isotherms Ms and A3 and the curves for bainite (B) and ferrite (F) are plotted together with the calculated cooling curves from the weld centerline - CL (y0) to the HAZ (y0. 610), where the number at the right of "y" is the distance from CL. As can be seen, a fully martensitic structure is expected from 0 to 0.366 mm from CL. Partial austenitization is expected between 0.488 mm and 0.610 mm from CL, since these curves peak temperatures are below A3. These last points are considered as part of the partially transformed region (PTR) and may be composed by the original base material ferrite plus fresh martensite.

The results are quite different for T*=455°C (HT455), Figure 5. The austenitization is complete from CL to y0.610 mm, but the decomposition product will be bainite instead of martensite. This behavior is exactly the aim of the present work.

Since the formation of bainite is of central importance here, it could be of cardinal importance to have a real measure of the temperature. Figure 5 presents the results of a pyrometer temperature measurement (Py – dashed line). As can be seen, the Py curve is very close to some part of the estimated cooling curves. However, some key facts must be pointed out: Firstly, the pyrometer maximum temperature is 700°C and thus there is no possible correlation between experimental and calculated T(t) above this threshold. Secondly, the spot size of the pyrometer is about 1 mm, i.e. the temperature is an average between y-0.5 to y+0.5 from CL. This latter point is not a serious issue because it is expected that the cooling rates between y0 and y0. 488 (Figure 5) are about the same. One useful result from the pyrometer measurements shown in Figure 5 is the latter part of the cooling; Simulation did not provide cooling rate data for the end of the austempering, i.e. after 600 s. For HT455, the cooling rate was 2.9°C/s from T* to 300°C. The relatively small cooling rate is due to the mass of alumina under the sample (Figure 1b).

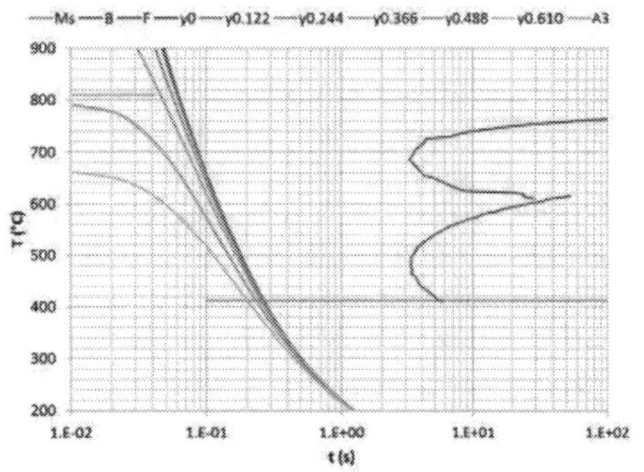

Figure 4. Calculated TTT curves for the current alloy together with calculated cooling curves for AT20 condition. The number after "y" indicates the distance in millimeters from the centerline.

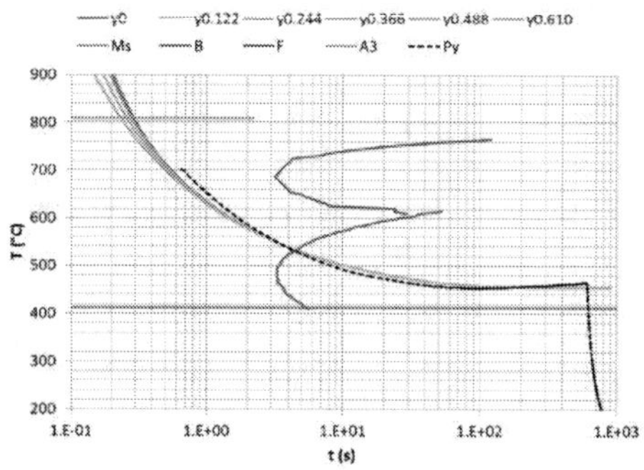

Figure 5. Calculated TTT curves for the current alloy together with calculated cooling curves for HT455 condition and the temperature profile obtained by the Pyrometer (Py). The number after "y" indicates the distance in millimeters from the centerline.

Microstructural Results and Discussion

Ambient temperature welds – AT20

The welds produced at ambient temperature, AT20, presents the typical laser keyhole shape, as shown in Figure 6a. The width of the fusion zone (FZ) depends on the depth z, being 1 mm, 0.4 mm and 0.6 mm at the top, middle and bottom of the weld, respectively. The extension of the heat-affected zone (HAZ) measured 0.48 mm in the middle of the sheet. The microstructure observed in the FZ is presented in Figure 6b. As can be seen, the as-welded microstructure is composed of primary dendrites of austenite and interdendritic ferrite. The end-of-solidification ferrite morphology is sometimes referred to as skeletal-type [13]. The dendrites fill with dark needles of the martensitic after transformation. Figure 6c is a back scattered electron image of the AT20 - FZ. The dendritic structure and the intradendritic martensite needle are visible. Using an automatic particle counting software, it could be estimated that the interdendritic ferrite accounts for about 20% vol. near to the center of the weld.

Figure 7 presents the EDS chemical analyses of the intradendritic and interdendritic regions, such as those presented in Figure 6c. Each region was measured six times. The carbon composition could be disregarded because, although present a low standard deviation of the current samples, the peak was too close to the beginning of the spectra and subject to a large variation depending on the analysis method. From Figure 7, it could be observed a measurable amount of ferrite stabilizing elements, such as Al, Si and Cr in the interdendritic spaces. This segregation explains the appearance of a light-colored zone within dendrites in Figure 6b. Another fact of relevance is the high Al content in the FZ as a result of the coating dilution. Therefore, the coating might not have been appropriately removed from the surface prior to welding and the TTT curve (Figure 3) of ferrite (F) may be displaced to the left (shorter times) [11].

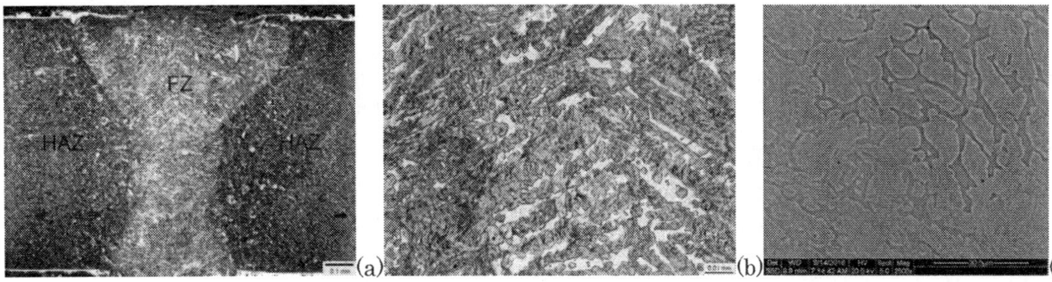

Figure 6. (a) Macrostructure of the sample AT. FZ and HAZ mean fusion zone and heat-affected zone (OM).(b) The microstructure of the FZ in the sample AT20. Optical microscopy (OM).(c) The microstructure of the FZ in the sample AT20 (SEM).

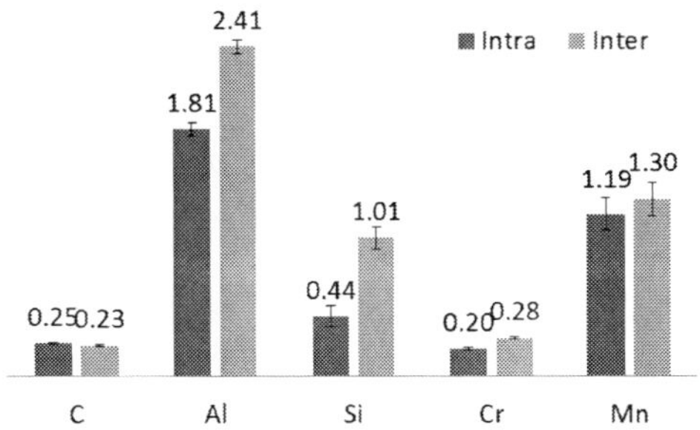

Figure 7. Chemical analysis by EDS of the intradendritic and interdendritic regions of the AT20 weld.

High temperature welds - HT455

Figure 8 presents an optical image of the sample in the HT455condition. The width of the fusion zone (FZ) depends on the relative weld depth, being 0.9 mm, 0.4 mm and 0.6 mm at the top, middle and bottom of the weld. The extension of the evident HAZ measured around 0.67 mm in the middle of the sheet (Figure 8a). The observed microstructure of the FZ for the sample HT455 is presented in Figure 8b andFigure 8c. Figure 8b shows a low magnification of a region in the middle of the FZ, showing a typical bainitic (B) microconstituent and residual austenite (A) phase. Figure 8c presents a higher magnification image where the ferritic leaves and polygons are more visible. According to the phase analyses, the total amount of bainite is 66% and the average bainite platelets thickness is 0.4 μm.

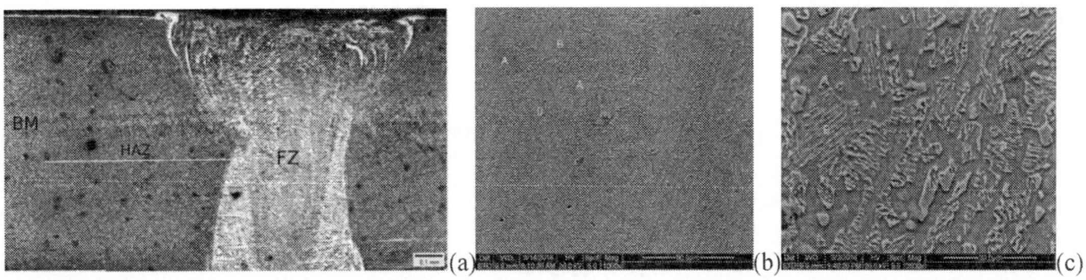

Figure 8. (a) Macrostructure of the sample HT455. (b) and (c) SEM micrographs of the FZ in HT455. The letters A and B indicate austenite and bainite, respectively.

Although the high content of aluminum in the FZ, the Widmanstätten ferrite did not appear as a primary phase. The reason is the rapid quenching of the laser weld, because less severe cooling rates might produce reconstructive reactions (F in Figure 3). According to Christian [14], reconstructive, or civilian, reactions occur by thermally activated random jumps across the interphase boundaries. Consequently, the current method should be effective only for high energy density sources such as laser or electron beams.

Conclusions

A combination of laser welding and induction heating was developed to produce 22MnB5 steel welds directly in the bainitic range. The methodology for such process was to preheat the sample to a temperature T* above the martensite start temperature and holding during 10 minutes before the laser weld. After laser welding, the temperature T* was held for another 10 minutes in order to enter into the bainite range.

The as-welded ambient temperature welds, designated as AT20, presented a microstructure composed of primary austenite dendrites that transformed to martensite during cooling, and interdendritic ferrite. The interdendritic region is richer in ferrite stabilizing elements such as Al, Si and Cr.

After high-temperature welding, the samples changed the microstructure to a typical bainite plus austenite grain. The sample with T*=455°C, HT455, presents around 66% of bainite in an austenite matrix.

The proposed methodology successfully attained the objective of this work to produce laser welds directly in the bainitic range, without the need of further heat treatments.

Acknowledgments

One of the authors (Lima) thanks the financial support from the *Fundação de Amparo à Pesquisa do Estado de São Paulo*, FAPESP, by means of a post-doctoral scholarship, project number 2014/26930-7. The authors also acknowledge the testing material received from AK Steel corporation.

References

[1] Dolfi J (chairman). Auto/Steel Partnership Joining Project: Advanced high strength steel (AHSS) weld performance study for autobody structural components. Report; 2006.

[2] C. Kim, M. J. Kang and Y. D. Park: "Laser welding of Al-Si coated hot stamping steel", *Procedia Engineering*, Vol.10 (2011), pp. 2226–2231

[3] H.K.D.H. Bhadeshia: Bainite in Steels, 2nd Edition, Institute of Materials (London, 2001), 478p.

[4] AK Steel. ULTRALUME®-Aluminized Type 1 Press Hardenable Boron Steel. Brochure, 2012, 6p.

[5] ESI Group. Sysweld. Avaliable at https://www.esi-group.com/software-solutions/virtual-manufacturing/welding-assembly. Acessed May19th, 2016.

[6] J.O. Andersson *et al*: "Thermo-Calc and DICTRA, Computational tools for materials science", *Calphad*, Vol. 26 (2002), pp. 273-312.

[7] H.K.D.H. Bhadeshia: "Bainite: Overall Transformation Kinetics", *Journal de Physique IV*, Vol. 43 (1982), pp. 443–448.

[8] G.I. Rees and H.K.D.H. Bhadeshia: "Bainite Transformation Kinetics: Part I Modified Model", *Mater. Sci. Technol.*, Vol. 8 (1992), pp. 985–993.

[9] S.B. Singh: Phase Transformations from Deformed Austenite, Ph.D. Thesis, University of Cambridge, (Cambridge, 1998) 222p.

[10] M.J. Santofimia *et al*: "Evaluation of Displacive Models for Bainite Transformation Kinetics in Steels". *Materials Transactions*, Vol. 47, No. 6 (2006), pp. 1492-1500.

[11] MAP Program MAP_STEEL_MUCG83: Avaliable at http://www.msm.cam.ac.uk/map/steel/programs/mucg83.html. Acessed Jan 9th, 2016.

[12] H. Karbasian and A.E. Tekkaya: "A review on hot stamping", J. Mater. Proc. Tech., Vol 210 (2010) pp. 2103–2118.

[13] M. Xia, *et al*:. "Fusion Zone Microstructure Evolution of Al-Alloyed TRIP Steel in Diode Laser Welding", *Materials Transactions*, Vol. 49, No. 4 (2008), pp. 746-753.

[14] J.W. Christian: The Physical Properties of Martensite and Bainite. The Metals Society (London, 1965), pp. 1-19.

Effects of Angle of Incidence or Focal Position upon Spatter Reduction on Basis of Three-Dimensional X-Ray Transmission In-Situ Observation of Laser Welding Phenomena

Yousuke KAWAHITO, Kouki NAKADA, Yousuke UEMURA, Masami MIZUTANI, Seiji KATAYAMA
Joining and Welding Research Institute, Osaka University,11-1 Mihogaoka, Ibaraki, Osaka, Japan
kawahito@jwri.osaka-u.ac.jp

Hiroshi KAWAKAMI
Graduate School Engineering, Mie University, 1577 Kurimamachiya-Cho, Tsu, Mie, Japan
kawakami@mach.mie-u.ac.jp

Koji NISHIMOTO Koji
National Institute of Technology, Anan College, 265 Aoki Minobayashi, Anan, Tokushima, Japan
knishi@anan-nct.ac.jp

Abstract

The objective of this research is to reveal effects of angel of incidence or focal position upon spatter generation and reduction in melt-run welding of a SUS304 stainless steel plate with a 6-kW power laser beam on basis of a high-speed video camera and a three-dimensional X-ray transmission in-situ observation apparatus. The high-speed video images at 150 mm/s in welding speed show the convex molten-pool surface behind a keyhole inlet was elongated and scattered as large spatters over 0.1 mm in size. According to the X-ray images, a melt flew to the elongated part at approximately 2.3 m/s in velocity, which caused the rapid increase of the spatters in size. A 2-mm inner defocused position or a 20-degrees angle of advance for an incident laser beam decreased number of spatter by half or one third owing to circulating melt flow behind the keyhole inlet. It was revealed that the large spatter reduction was effective to deceleration or change in direction of the melt flows behind the keyhole inlet, which led to not only suppressing formation of the elongated surface but also improving the frequency that the elongated went back to the molten pool.

Introduction

Welding is a fundamental metal-joining technology, and better understanding of the melt behavior in molten pool is required to realize high-quality, defect-free joints. Typical defects encountered in laser welding include porosity (cavities formed upon solidification of molten metal), underfill (dents in the nugget surface), and spatter (expulsion of molten metal). The movement of molten metal is deeply related to all of them, and thus it is particularly important to deeply understand how the melt flows within the pool. For example, in partial penetration welding of stainless steel at a welding speed of 17 mm/s with a 6-kW high-power beam, sporadic formation of bubbles is observed at the leading edge of the keyhole, which are transported by melt flow swirl around the leading edge of the keyhole and become trapped by the solidification front.

The mechanism of porosity defect generation has been clarified through two-dimensional X-ray transmission observation[1,2]. On the other hand, the melt flow at a welding speed of 100 mm/s differs from that at a low welding speed. It is reported that the melt flow from the inlet of the keyhole to the rear is observed clearly, and the generation of bubbles which cause porosity is not observed. A numerical analysis of three-dimensional melt flow has been reported[3], but no three-dimensional measurement has been reported. Thus, the relationship between melt flow and welding defects has not been elucidated sufficiently.

In this study, we focused on the relationship between melt flows and spatter formation in melt-run welding of a SUS304 stainless steel plate with a 6-kW power laser beam and investigated effects of angle of incidence and focal position upon spatter reduction with a high-speed video camera and a three-dimensional X-ray transmission in-situ observation apparatus.

Materials and experimental procedures

The test material was a SUS304 (austenitic) stainless steel plate as base material and a disk laser oscillator (consecutive oscillation type, max. power: 16 kW) was used as the welding device. **Figure 1** is a schematic illustration of the experimental setup.

The beam was transmitted from the oscillator through a fiber having a core diameter φ of 0.2 mm. It was then focused by a lens (focal distance: 280 mm) to produce a beam profile as shown in **Fig. 2**. The beam spot diameter, φ was 0.3 mm, and peak power density was 200 kW/mm^2. The focal position was fixed. In this arrangement, melt-run welding was performed at welding speeds of 50, 100, 150, and 250 mm/s at a 6-kW power setting. Subsequently, focal position of -4, -2, 0, 2 mm or angle of incidence of -20, -10 ,0 10, 10 degrees were changed.

A real-time X-ray transmission imaging system was used to observe three-dimensional melt flow within molten pool. The system consists of a micro-focused X-ray beam (max. resolution: 4 μm, resolution at time of experiment: 70 μm, max. voltage: 230 kV, max. current: 1 mA, max. stainless steel plate thickness photographable at 1,000 fps: 5 mm) and a mini-focused X-ray beam (resolution: 0.4 mm, maximum voltage: 225 kV, maximum electric current: 3.5 mA, max. stainless steel plate thickness photographable at 1,000 fps: 10 mm). The X-rays can be observed by means of a fluorescent image intensifier (afterglow time: 0.1 ms) and a high-speed camera (10,000 fps). From the image data measured at two different angles simultaneously, the three-dimensional velocity of the melt flow can be measured by using parallax. In addition, the test metals used for X-ray transmission observation are T-shaped to prevent omission. The X-ray penetration was 5 mm at a welding speed of 25 mm/s, 3 mm at 50 mm/s, and 1.5 mm at 100 mm/s or above. Small (diameter: 0.5 mm) tungsten carbide balls were embedded into the test materials before welding as tracers for the visualization of melt flow.

Moreover, behaviors of molten pool surface and laser-induced plume were observed at 10,000 fps by using a high-speed video camera.

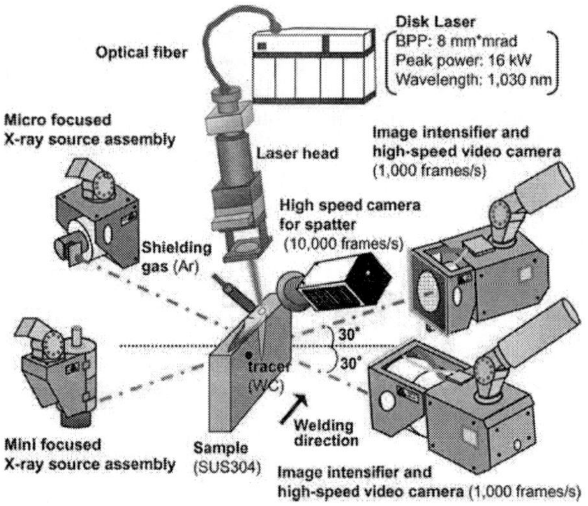

Fig.1: Schematic illustration of experimental set-up of high-speed video camera and three-dimensional X-ray transmission system for observation of welding phenomena with high power laser.

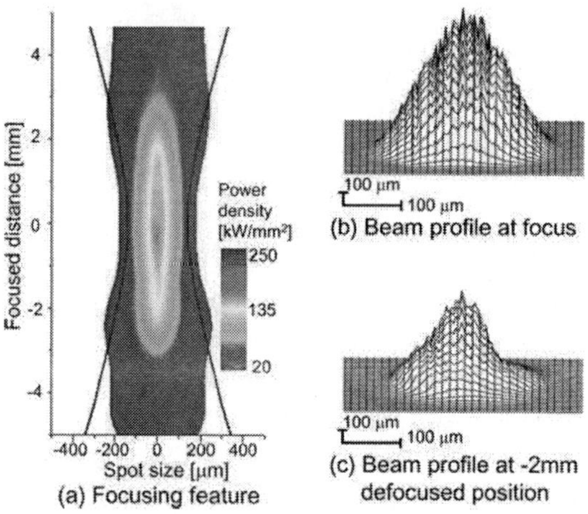

Fig. 2: Optical feature of incident laser beam.

Three-dimensional observation of melt flows: spatter generation

Increase of welding speed generated spatters according to the obtained weld beads as shown in **Fig. 3**. To investigate the melt flow under the conditions that led to the spatter generation upon deep penetration welding at a welding speed of 150 mm/s, three-dimensional X-ray transmission observation was performed using 17 tungsten carbide balls embedded 1 mm, 2 mm, and 3 mm beneath the plate surface. **Fig. 4** shows the results of the investigation. There were two kinds of melt flows, one being in the upper part and another in the lower part of molten pool at the rear of keyhole. The melt flow was fast at the keyhole inlet and keyhole tip. The average speed of overall melt flow in the molten pool was 0.49 m/s. The melt flew to the elongated part of the molten pool surface behind a keyhole inlet at approximately 2.3 m/s in velocity, which caused the rapid increase of the spatters in size. It is also noted that the melt flow speed at the keyhole increases with increasing welding speed.

The high-speed video images at 150 mm/s in welding speed show the convex molten-pool surface behind a keyhole inlet was elongated and scattered as large spatters over 0.1 mm in size as observed in **Fig. 5**.

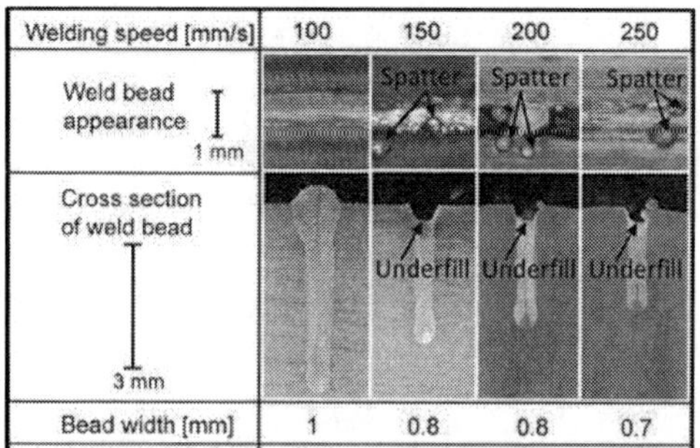

Fig.3: Photos of surface appearances and cross sections of SUS 304 weld beads obtained.

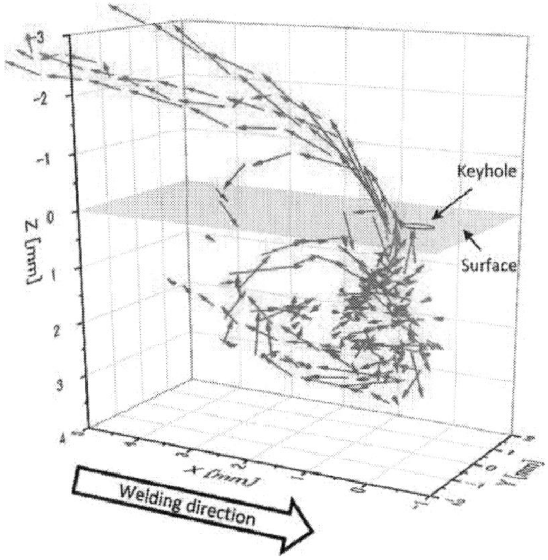

Fig. 4: Display of three-dimensional melt flows calculated from 17 tracers during laser welding at focal point.

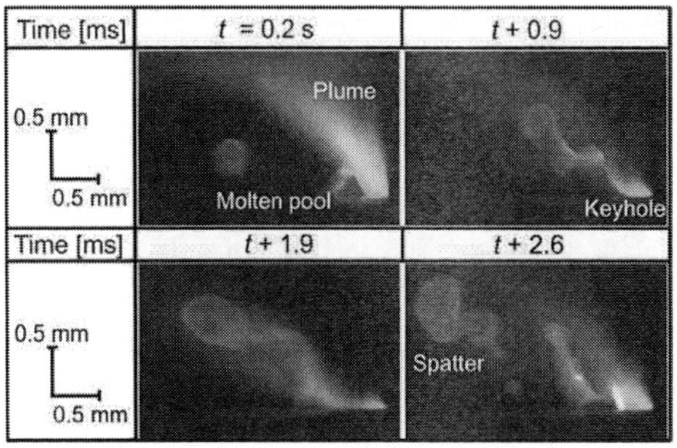

Fig. 5: Typical high-speed video pictures of spatter formation over 0.1 mm in size.

Effects of angle of incidence or focal position upon spatter reduction

A 2-mm inner defocused position decreased number of spatters by half in comparison with that at focus position and zero angle owing to circulating melt flow behind the keyhole inlet as demonstrated in **Fig. 6**. On the other hand, a 20-degrees angle of advance for an incident laser beam also reduced number of spatters by one third due to circulating melt flow behind the keyhole inlet as indicated in **Fig.7**.

It was revealed that the spatter reduction was effective to deceleration or change in direction of the melt flows behind the keyhole inlet, which led to not only suppressing formation of the elongated surface which became a spatter but also improving the frequency that the elongated surface or spatters went back to the molten pool.

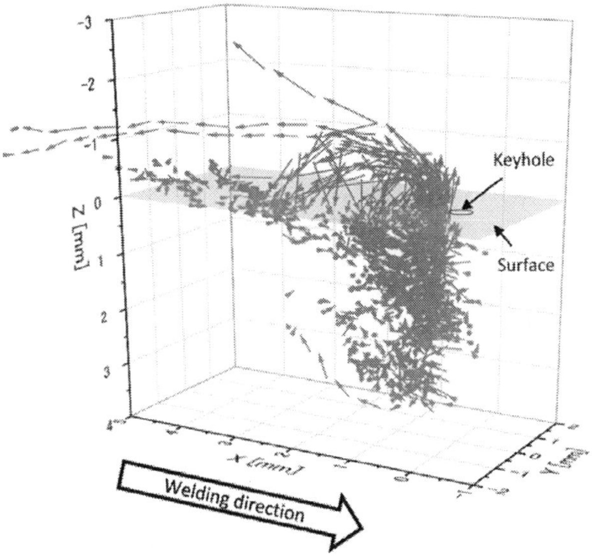

Fig. 6: Display of three-dimensional melt flows calculated from 80 tracers during laser welding at -2 mm defocused position.

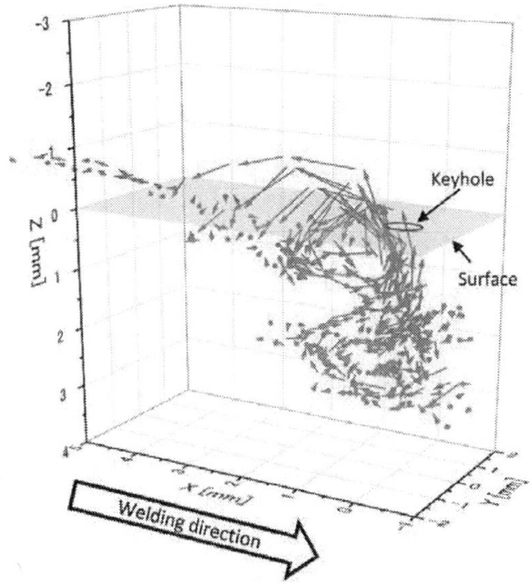

Fig. 7: Display of three-dimensional melt flows calculated from 20 tracers during laser melt-run welding at 20 incident degrees.

Conclusions

In laser welding, there were two kinds of melt flows, one being in the upper part and another in the lower part of molten pool at the rear of keyhole. The melt flow was fast at the keyhole inlet and keyhole tip, and heat transfer in the melt flow governed energy transfer in molten pool. Furthermore, the speed and the direction of the melt flows were sensitive to welding speed, focal position or angel of incidence. In spatter generation, the melt flow accelerated from the inside of keyhole and flowed straightly into the keyhole inlet. Proper angel of incidence or focal position produced the circulated melt flow in the upper part of the molten pool, which led to spatter reduction.

Acknowledgments

This work was partly supported by a Grant-in-aid for Scientific Research (Kakenhi No. 25289240).

References

[1]. K. Kinoshita, *et al*: "Phenomena Elucidation in High-Power Fiber Laser Welding of Stainless Steel", Quarterly Journal of The Japan Welding Society, Vol.25, No.1 (2007), pp.18-23. (in Japanese)

[2] Y. Kawahito, *et al*: " High quality welding of stainless steel with 10 kW high power fibre laser", Science and Technology of Welding & Joining, Vol.14, No.4(2009), pp.288-294.

[3] A. Otto, *et al*: "Numerical Simulations - A Versatile Approach for Better Understanding Dynamics in Laser Material Processing" Physics Procedia, Vol.12, Part A(2011), pp.11-20.

Effects of Vacuum Environment on Laser Welding

Kosuke Watanabe, Yuto Nishioka, Yu Murakami, Naoyuki Matsumoto, Fumio Matsuzaka
IHI Corporation, 1 Shin-Nakahara-cho, Isogo-ku, Yokohama-shi, Kanagawa, Japan
kosuke_watanabe@ihi.co.jp, +81-45-759-2812

Abstract

There are obvious advantages in welding using the vacuum environment, such as lower porosity and increasing the penetration depth of welded joints. The mainstream of its process is covering whole work pieces with a large chamber during welding. Recently, various studies have been conducted. We suggest the new concept of laser welding process under vacuum conditions. This method is applied with principle of Venturi effect using jet stream and realizes a local area vacuuming that sufficient for welding without a large chamber. Advantages of this method are preventing the dump of laser energy from pollution of protective windows using gas flow that causes of welding spatter and fume, and also providing the vacuum environment by using a same source. This study presents a detail of evaluation in the case of applying to Ni/Fe dissimilar welded joint. By increasing gas flow rate, amount of pressure around weld zone and the number of porosity in the weld metal became lower, the penetration depth of welded joint was increased. It was confirmed that the welding process using the local vacuuming method with gas flow realizes the low porosity and deep penetration welds. The paper overviews the volume and distribution of porosity evaluated by using microfocus x-ray computed tomography and cross sections of welded joints through microscopic examination.

Introduction

Laser welding in vacuum environment is a technique which is known as realizes such as high penetration depth welding and reducing porosity in weld metal [1]. Recently, various studies have been conducted using a large chamber and a large capacity vacuum pump to generate vacuum environment during welding [1-2]. Simplifying mechanism of this technique will be significant for using this technique in mass production or large structure construction. In the present study, penetration depth, amount and distribution of porosity in the Ni/Fe dissimilar welded joint using a new concept of laser welding process under vacuum conditions applied with principle of Venturi effect using jet stream that sufficient for welding without a large chamber and a vacuum pump were evaluated to clarify the performance of the local vacuum welding system.

Experimental Procedure

The round shaped Ni alloy and Fe alloy components were welded using an in-house developed local vacuum laser welding system, which is shown in Figure 1. This laser welding system is comprises a casing with an ejector that arranged near the welding zone, a rotator and a fiber laser system. The welding was conducted under three levels of vacuum conditions using a N_2 gas flow control. During laser welding, the phenomena of weld zone were recorded using a CCD camera. After welding, penetration depth, amount and distribution of porosity in the Ni/Fe dissimilar welded joint were evaluated by metallographic examination and using microfocus x-ray computed tomography (as shown in Figure 2). The specimens were prepared for metallographic examination using automatic polishing equipment. The polished surface was etched chemically with aqua regia reagent and microstructure was performed using an optical microscope.

Results and Discussion

By increasing gas flow rate of ejector, amount of pressure around weld zone became lower. As shown in Figure 3, low ambient pressure enables high penetration depth. These trends were consistent with the findings that have been reported by using a vacuum chamber equipped with a vacuum pump [1-2]. Figure 4 shows the phenomena around weld zone during laser welding and the microfocus x-ray computed tomography images of weld metal. As a result of comparison of the visualized surface of weld zone between 0.88 atm and 0.16 atm pressure condition, it was confirmed that the former (0.88 atm) was covered with spatter and plume. A large amount of porosity in the weld metal was visible.

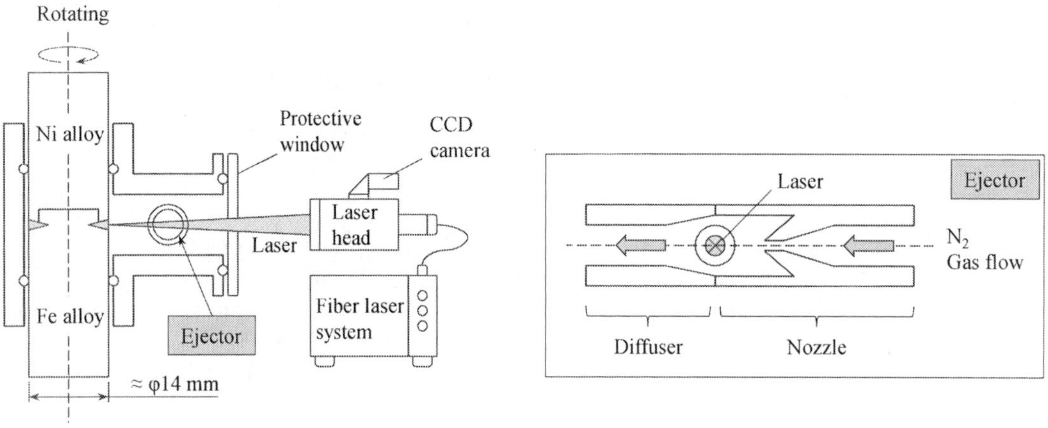

Fig. 1 Schematic showing the local vacuum welding system.

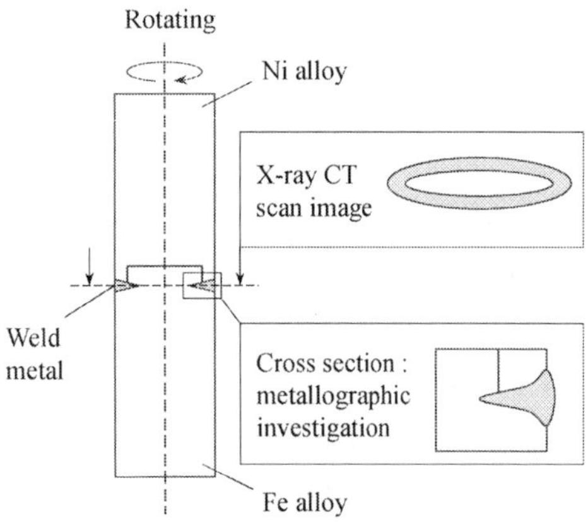

Fig. 2 Schematic showing the evaluation of weld joint.

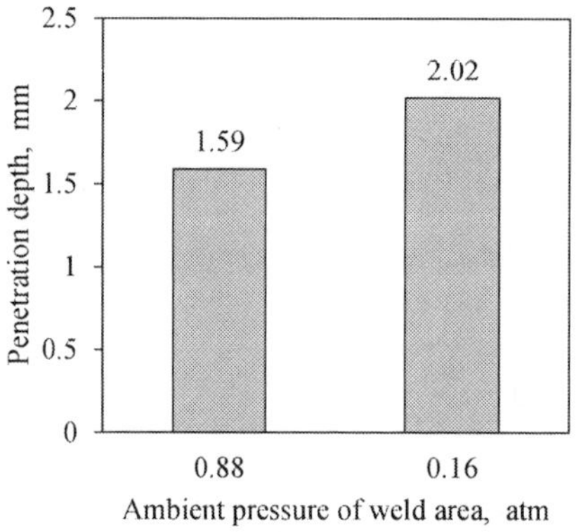

Fig. 3 Relation between ambient pressure of weld area and penetration depth.

On the other hand, spatter and plume that covered weld zone during welding under 0.16 atm pressure condition was reduced. The number of porosity in the weld metal became lower in spite of partial penetration welding. The spatter and plume that generated during laser welding are one of the inhibitor known as causes of attenuation and scattering of laser [1-6]. As mentioned above, the vacuum environment removed spatter and plume, which inhibiting factors of laser. Therefore, it was considered that the penetration depth increased and porosity of weld metal was reduced. It can be inferred that drastically reduced porosity was the result of effect of stability of keyhole. Also, Figure 4 shows that the route of spatter bended left to right direction by N_2 gas flow from ejector. The gas flow performs not only creating the vacuum environment but also blowing away the inhibiting factors of laser. It is considered that the gas flow using the ejector is great advantage for removing inhibiting factor near the condensing spot of laser during welding.

Figure 5 shows the metallographic cross section and the microfocus x-ray computed tomography images of weld joint that welded under 0.02 atm pressure condition after adjusting the local vacuum welding system. Sharp shaped weld metal and minimizing the porosity were realized.

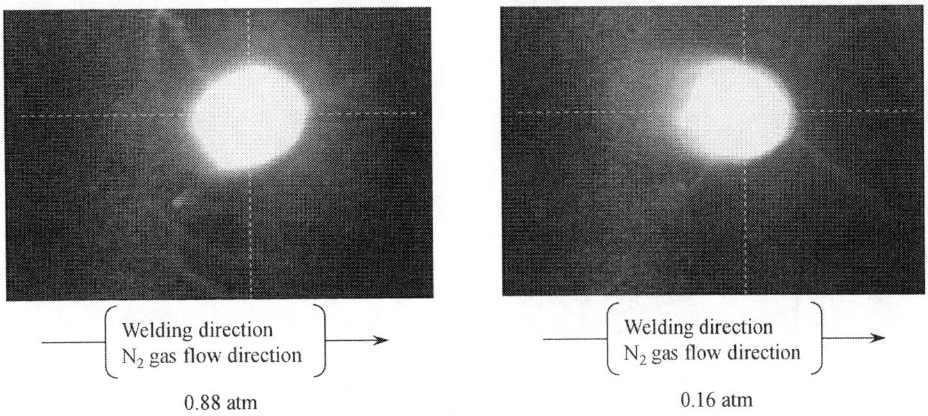

a) CCD camera images of weld zone during welding

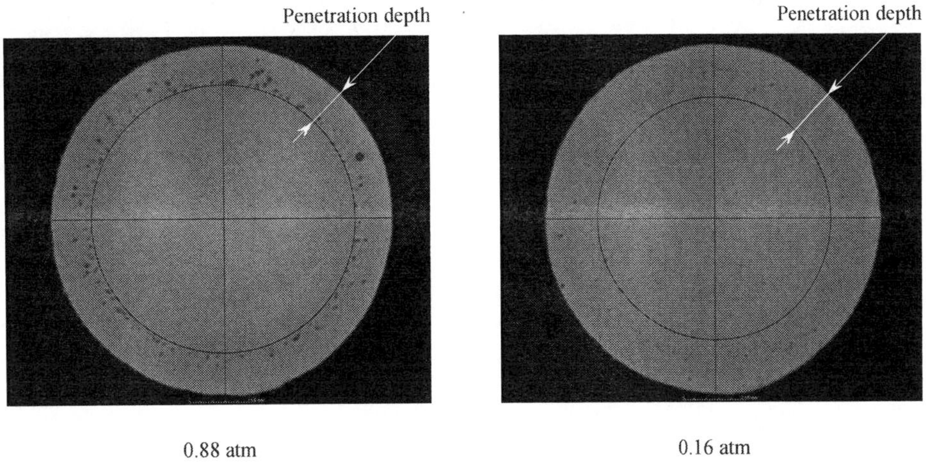

b) Microfocus x-ray computed tomography images of weld metal (Dots indicates porosity)

Fig. 4 a) CCD camera images of weld zone during welding and
b) Microfocus x-ray computed tomography images of weld metal.

Conclusions

- There is a correlation between amount of pressure, porosity in the weld metal, and penetration depth using an in-house developed local vacuum laser welding system.
- By increasing gas flow rate of ejector, spatter and plume that covered weld zone during welding became reduced.
- A weld joint that welded under 0.02 atm pressure condition has sharp shaped weld metal and very few porosity.

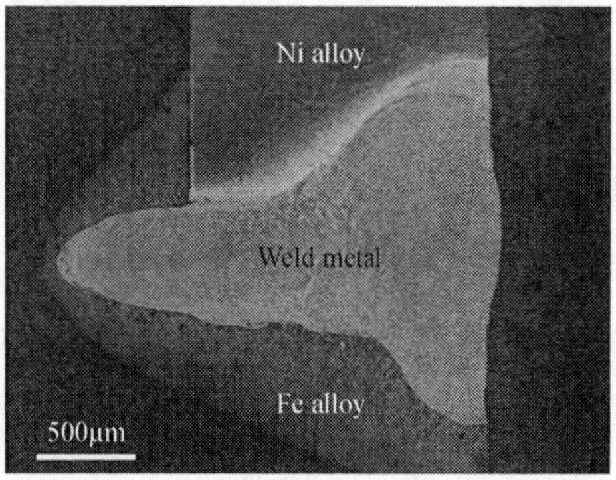

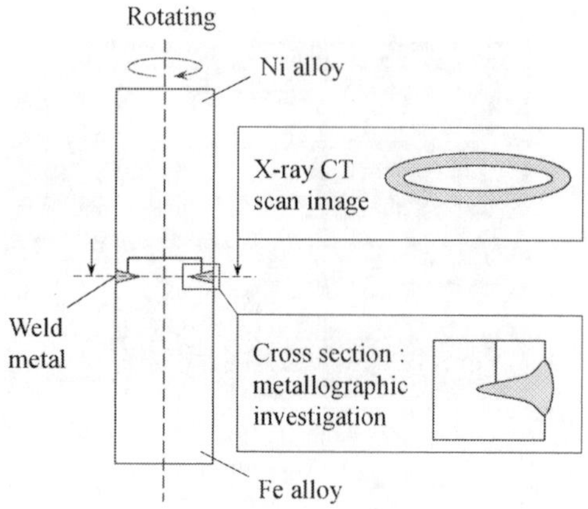

a) Optical microscopy image of the weld zone

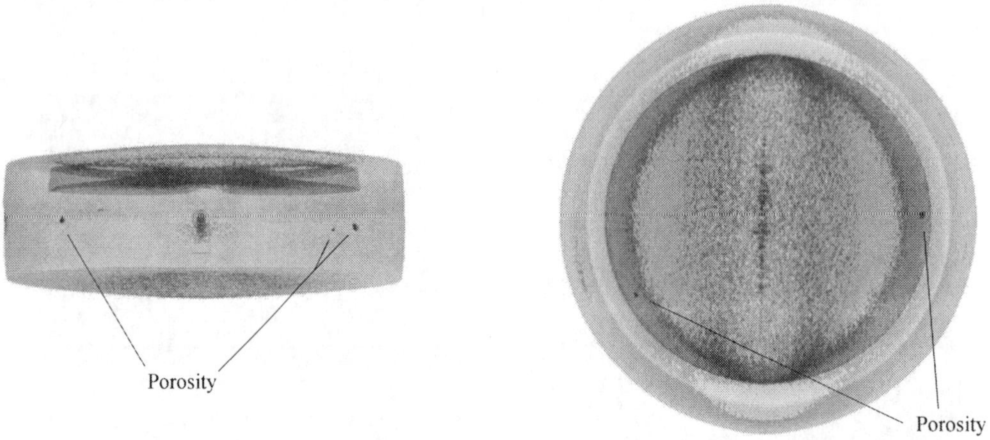

b) Microfocus x-ray computed tomography images of weld metal (Dots indicates porosity)

Fig. 5 a) Optical microscopy image of the weld zone and
b) Microfocus x-ray computed tomography images of weld metal, welded under 0.02 atm pressure condition.

References

[1] Y. Abe, *et al*: "Effect of Reduced Pressure Atmosphere on Partial Penetration Welds of Stainless Steel and Aluminum Alloy with High Power and High Brightness Laser," *Q. J. Jpn. Weld. Soc.*, Vol. 31, No. 1 (2013), pp. 48-55.

[2] S. Katayama and Y. Kawahito: JP. Patent 5,234,471

[3] S. Katayama, *et al*: "Visualization of Welding Phenomena with High Speed Video Camera and X-Ray Transmission Real-Time Observation System," *J. High Temp. Soc.*, Vol. 33, No. 3 (2007), pp. 118-127.

[4] Y. Kawahito, *et al*: "Defect Formation Mechanism and Reduction Procedure in 10 kW High Power Fiber Laser Welding of Stainless Steel," *Q. J. Jpn. Weld. Soc.*, Vol. 26, No. 3 (2008), pp. 203-209.

[5] S. Katayama: "Laser Welding," *J. Jpn. Weld. Soc.*, Vol. 78, No. 2, (2009), pp. 124-138.

[6] S. Katayama: "Recent Progress in Laser Welding Technology," *J. Vac. Soc. Jpn.*, Vol. 55, No. 11, (2012), pp. 471-480.

Clarification of plume along laser axis above specimen

Masami Mizutani, Yousuke Kawahito, Seiji Katayama
Joining and Welding Research Institute, Osaka University, 11-1 Mihogaoka, Ibaraki, Osaka, Japan
mizutani@jwri.osaka-u.ac.jp, +81-6-6879-8692

Abstract

In order to clarify the effect of plume along laser axis on scattering and absorption which deteriorate laser power and its density to be reached onto specimen during laser welding, the space where plume forms between the laser processing head and the specimen was observed. The welding process by the processing laser (5 kW fiber laser: 1070 nm) with high power probe laser (0.5-10 kW disk laser: 1030 nm) irradiated from lateral direction was observed with synchronous high-speed video recording of color image, 1030 nm and 1070 nm band-limited monochrome images. As the observation results, laser scattering by ultrafine particles could be visualized by band-limited monochrome images. As emission of visual light along the high power probe laser was observed, it can be said that ultrafine particles were reheated or re-excited by the high power probe laser. Plume was categorized as primary plume ejected from keyhole and secondary plume which consists of the matter reheated or re-exited from ultrafine particles by processing laser along laser trajectory. Laser scattering distribution where stronger scattering at the outer boundary of laser axis and hollow cavity of scattering at the center of laser axis formed were observed. The existence of hollow cavity in the processing laser axis was also confirmed as the scattering of the high power probe laser was unseen there. Percentage graph of the relationship between laser irradiation power and laser extinction using the probe laser showed that the laser extinction decreases with an increase in laser power. From above results, it can be said that laser extinction is little, but the thermal lensing effect by the media at the space between the laser processing head and the specimen is not negligible.

Introduction

The effect of the plume on the laser propagation during laser welding has been disputed[1-4]. It has been reported recently that in a process of fiber laser remote welding, the state of air, plume and ultrafine particles in the space above specimen greatly affects welding bead formation. The report shows that the laser seemed to be fluctuated and defocused on the specimen during welding without an air flow to eliminate the media above specimen. And the phenomenon was interpreted as the result of the laser deflection due to the media possessing temperature gradient by a theoretical calculation and the subsequent experiments. Experiments such as shadow graph observation suggested that the reason why the laser irradiated area was fluctuated around the weld pool during welding is because of the effect of formation of refractive index gradient due to unsteady temperature distribution above specimen caused by the updraft of the hot air due to welding on laser trajectory. But another reason why laser focal position was shifted to far position from the origin, i.e. thermal lensing effect by the media above specimen is still uncertain. Furthermore, there must be scattering and absorption by plume to hinder the laser propagation onto the specimen.

In this study, in order to clarify the effect of plume along laser axis on scattering and absorption which deteriorate laser power and its density to be reached onto specimen during laser welding, the space where plume forms between the laser processing head and the specimen was observed. And the relationship between laser irradiation power and laser extinction was investigated using high power probe laser (0.5-10 kW disk laser: 1030 nm). The welding process with probe laser was observed with synchronous high-speed video recording of color image, 1030 nm and 1070 nm band-limited monochrome images

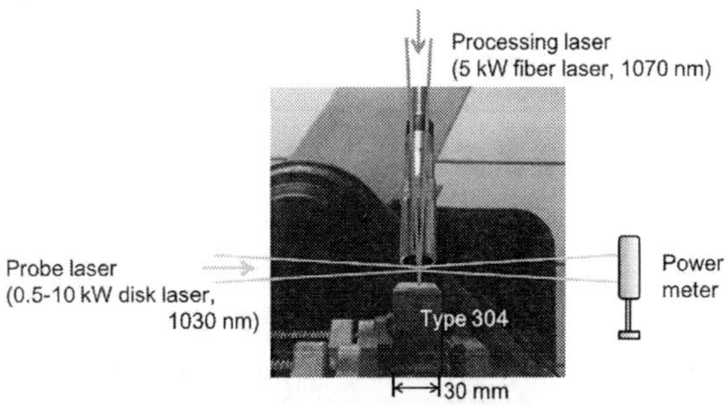

Fig. 1 Experimental configuration, showing probe laser and processing laser direction schematically.

Experimental methodology

Fig. 1 shows experimental configuration. A processing laser of 5 kW fiber laser was irradiated onto the surface of the specimen, type 304 at the welding

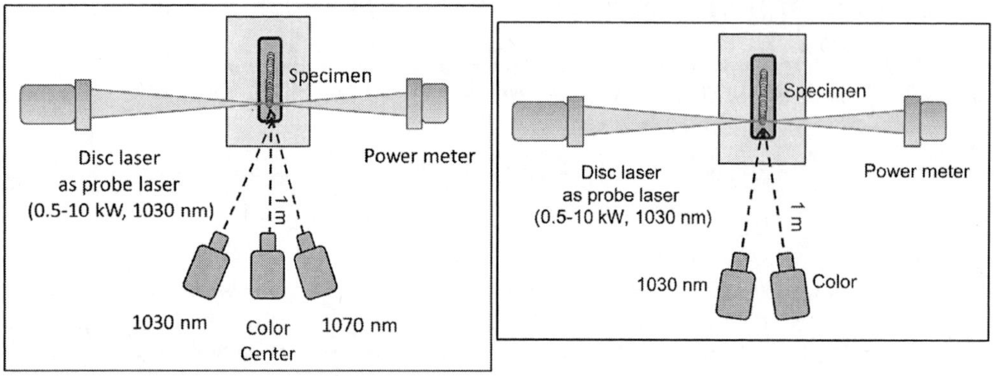

Fig. 2 Overhead views of high-speed video camera. setup.

speed of 0.3 or 0.6 m/min. During laser welding, a probe laser of 0.5 to 10 kW disk laser was being irradiated laterally from one side to another where a power meter was set to measure laser power after passing through the media above specimen. The setup of the high-speed videos is shown in **Fig. 2** as overhead views, schematically. The welding process with probe laser was observed with synchronous high-speed video recording of color image, 1030 nm and 1070 nm band-limited monochrome images. **Fig. 3** shows in detail, side view of set up arrangement with the position (x, h) through which probe laser is targeted. Shielding gas was Ar of 30L.

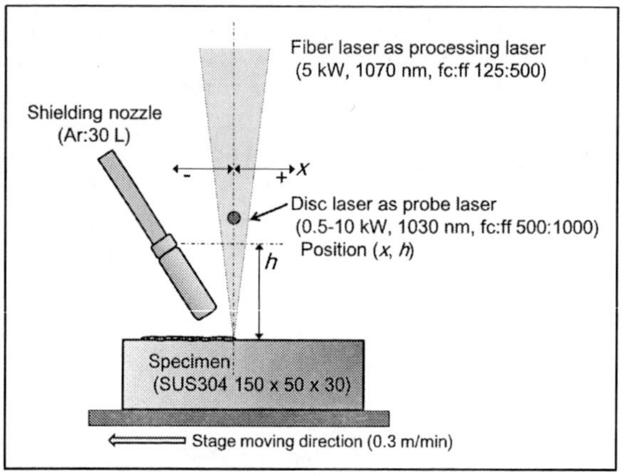

Fig. 3 Side view of setup arrangement of specimen, shielding nozzle and laser, showing .probe laser position (x, h).

Distribution of scattering of probe laser and plume ejection from the specimen

Fig. 4 shows simultaneous pictures of monochrome images with 1030-nm bandpass filter and color images at representative time and their averaged images from 1 to 6 s. monochrome images indicate there must be scattering along 10-kW probe laser propagation. And the linear emission along the probe laser in color image also shows that matters above specimen during laser welding can be reheated or re-excited. Furthermore, from monochrome images, just on the processing laser axis the scattering is unseen, even in the averaged image. This may mean the inside of laser axis with the certain laser power density is hollow in terms of scattering.

Therefore, the power dependency of scattering and emission was observed as shown on **Fig. 5**. The pictures are averaged images from 1 to 6 s. Every 1030-nm image shows hollow distribution at the processing laser axis. The liner scattering along probe laser propagation is stronger from 1000 to 5000 W. From color images, it can be said that the increase of liner emission is with an increase of laser power.

At the same time as the power dependency was observed, the laser extinction was measured by a laser power meter. The probe laser started to be irradiated toward the power meter 60 s prior to the irradiation of the processing laser and then the initial value of the laser power without media above specimen was measured as A. And during laser welding, the laser power passing through the media was measured as B. **Fig. 6** shows the relationship between the probe laser power and its extinction as the actual power, A-B and the percent, (A-B)/A x 100. This percentage graph shows that the laser extinction decreases with an increase in laser power. This result can be interpreted as in the case of a laser welding with a certain high power laser, laser extinction is little because the inside of high power laser axis is hollow in terms of scattering.

Fig. 7 shows a schematic illustration as a brief summary of the media distribution above the specimen during a high power laser welding. At the laser axis inside, the ultrafine particles are reheated or re-exited to atoms or plasma resulting in less scattering compared from the ultrafine particles outside of the laser axis. Finally it can

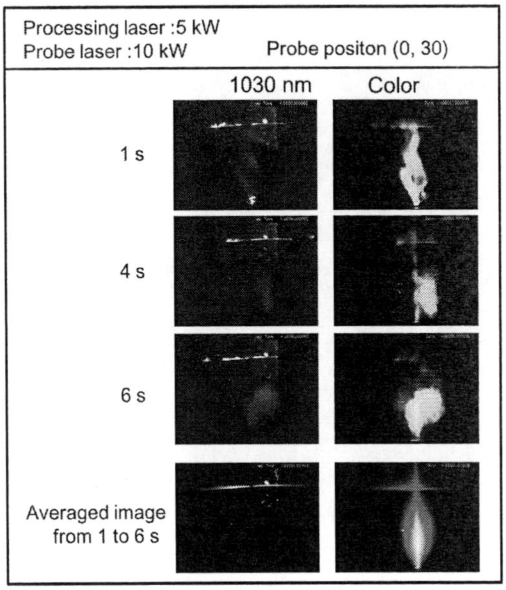

Fig. 4 Example of simultaneous pictures of 1030-nm bandpass and color taken from high-speed observation, showing time dependent and averaged images.

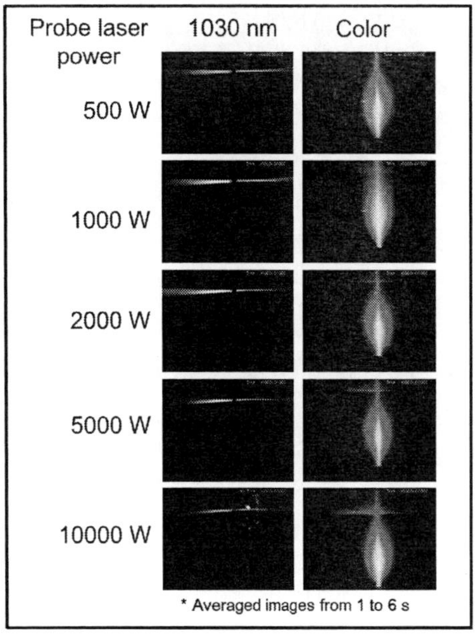

Fig. 5 Power dependency of scattering and emission of probe laser shown as averaged images from 1 to 6 s..

be said that the laser extinction by scattering is little but the this distribution may form temperature distribution where the temperature is high at the center and low at the fringe of laser axis, which can induce the focal position far, i.e. the thermal lensing effect.

Conclusions

Representative results obtained are as follows;

1) Laser scattering by ultrafine particles could be visualized.
2) As emission of visual light along the high power probe laser was observed, it can be said that ultrafine particles were re-excited or reheated by the high power probe laser.
3) Plume was categorized as primary plume ejected from keyhole and secondary plume which consists of the matter reheated or re-exited from ultrafine particles by processing laser along laser trajectory.
4) The existence of hollow cavity in the processing laser axis was also confirmed as the scattering of the high power probe laser was unseen there.
5) Percentage graph of the relationship between laser irradiation power and laser extinction using the probe laser showed that the laser extinction decreases with an increase in laser power.
6) From laser scattering distribution where stronger scattering at the outer boundary of laser axis and empty scattering at the center of laser axis were observed, the temperature was thought to be higher at the center of laser axis than at the outer boundary because less scattering at the center may mean the matter consists of atoms or molecules whose temperature is higher than ultrafine particles. This may cause the thermal lensing effect in the space above specimen.

From above results, it can be said that laser extinction is little, but the thermal lensing effect by the media at the space between the laser processing head and the specimen is not negligible.

References

[1] M. Mizutani, S Katayama:"Numerical Calculation of Laser Beam Path Influenced by High Temperature Gas Above Specimen during Laser Welding" ICALEO 2009 Paper #1403

[2] S. Oiwa, Y. Kawahito, M. Mizutani, S. Katayama:"Effect of atmosphere above specimen on welding results during remote welding" J. Laser Appl. 23, 022007 (2011)

[3] P.Yu. Shcheglov, S.A. Uspenskiy, A.V. Gumenyuk, V.N. Petrovskiy, M. Rethmeier, V.M.

Yermachenko:"Plume attenuation of laser radiation during high power fiber laser welding" Laser Phys. Lett. 8, No. 6, 475–480 (2011)
[4] P. Yu. Shcheglov, A. V. Gumenyuk, I. B. Gornushkin, M. Rethmeier, V. N. Petrovskiy:"Vapor–plasma plume investigation during high-power fiber laser welding" Laser Phys. 23 (2013) 016001 (7pp)

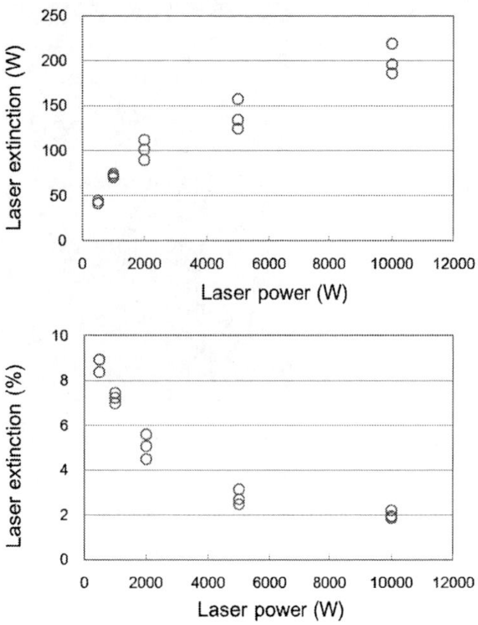

Fig. 6 Relationship between laser power and laser extinction, showing comparison between actual power (W) and percent (%).

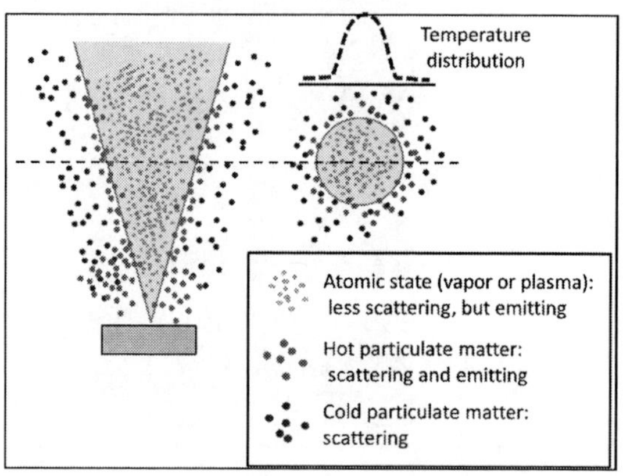

Fig. 7 Schematic illustration to explain structure of media above specimen during laser welding, showing distribution of ultrafine particulate matter surrounding laser axis and vapor or plasma.in laser propagating area.

Room Temperature Bonding of Solid Substrates using Thin Metal Films (Atomic Diffusion Bonding)

T. Shimatsu and M. Uomoto
FRIS, Tohoku University, Sendai, Japan
shimatsu@riec.tohoku.ac.jp, +81-22-217-5494

Abstract
Room-temperature bonding using thin metal films (atomic diffusion bonding) was applied to bonding mirror-polished solid substrates: metals to ceramics and metals to metals. The technical potential and current status were reviewed. We demonstrated fabrication of a viewing port for ultra-high vacuum chambers by bonding a vacuum flange made of Al alloy with a sapphire plate. Bonding of Al alloys (such as A1050) with Fe, Cu, and stainless alloys was achieved with bonding strength greater than the creep strengths of the aluminum alloys.

Introduction
Atomic diffusion bonding of two flat surfaces with thin metal films [1–3] is a promising process to produce solid substrate bonding at room temperature. This process was originally proposed for semiconductor or ceramic wafer bonding to fabricate new thin film devices. Now it is actually used in mass production. In this process, thin metal films are sputter-deposited on two flat surfaces. Bonding of the two metal films on the wafers is accomplished subsequently in vacuum or in air, depending on the metal film materials used for bonding. Any mirror-polished wafer can be bonded using thin metal films without loading pressure. High surface energies of metal films and a large atomic diffusion coefficient at the grain boundaries and film surfaces enable bonding at room temperature. We applied this bonding method to bonding mirror-polished solid substrates: bonding of metals to metals and metals to ceramics can be achieved using this process, although loading pressure is necessary for bonding. As described herein, we review the technical potential and current status of atomic diffusion bonding for bonding of substrates: metals to ceramics and metals to metals.

Bonding procedure
Two processes are used for atomic diffusion bonding. One is bonding in vacuum: after metal films are fabricated on two flat substrates surfaces using sputter deposition, the two films are bonded on the substrates in vacuum (Fig. 1(A)). Another bonding process is bonding in air. Bonding of the two films on the substrates can be done in air after sputter film deposition in vacuum [2] (Fig. 1(B)). This process is limited to use with Au and Au alloy films, but it is convenient. Two solid substrates that can be bonded in air can also be bonded in vacuum. This review specifically assesses the bonding process in air to examine the technical potential of the bonding technique.

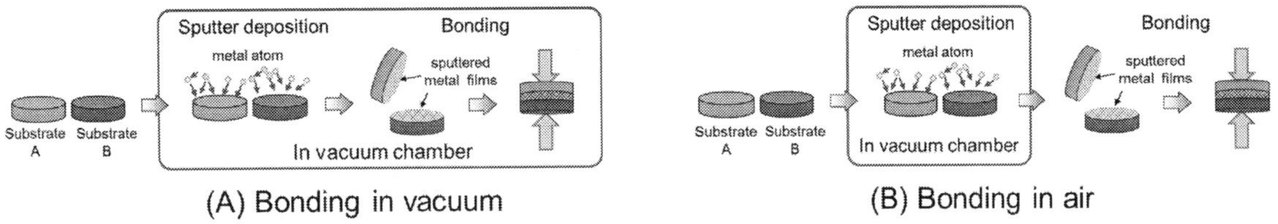

Fig. 1. Atomic diffusion bonding processes: (A) bonding in vacuum and (b) bonding in air.

Bonding of ceramics and metals
As a representative examples, Figure 2(A) portrays an image of a viewing port for ultra-high vacuum chambers, fabricated by bonding an ICF114 vacuum flange made of Al alloy to a sapphire plate [4]. Au films (50 nm thickness) with Ti (5 nm) underlayers were deposited on the mirror polished surfaces on both sides. Then bonding was conducted with a loading force. Figure 2(B) shows a TEM cross-section image for the bonded interface. The bonded Au–Au films show no interface corresponding to the original film surface, indicating that recrystallization occurred at the original surfaces of the bonded Au-Au films at room temperature, as is true also for wafer to wafer bonding [3]. Actually no gas-leak was observed in the leak test using He gas, as shown in Figure 2(C).

Bonding of metals and metals

Any mirror-polished metal substrate can be bonded using this method. Bonding of Al alloys with Fe, Cu, and stainless alloys has been demonstrated [5]. Our experimentally obtained results show that the bonding strength of Al alloys (such as A1050) with stainless alloys is greater than the creep strengths of the aluminum alloys. As one examples, Figure 3 (A1) portrays an Al alloy substrate (A5052) and a stainless alloy substrate (SUS316L) bonded using Au(50 nm) films with Ta (5 nm) underlayers on each side. Panel (A2) shows the bonded interface observed using scanning acoustic tomography (SAT). Panels (B1) and (B2) respectively show a bonded A5052 substrate and a Cu substrate (C1020). The 8-mm-thick substrate sizes are 14 mm × 14 mm. One of the 14 mm × 14 mm surfaces was mirror polished and used for bonding. No vacancy was observed in the SAT images shown in (A2) and (B2). Share tests of these samples revealed extremely high shear strength: greater than 51 MPa for bonded A5052 and SUS316L, and 44 MPa for bonded A5052 and C1020.

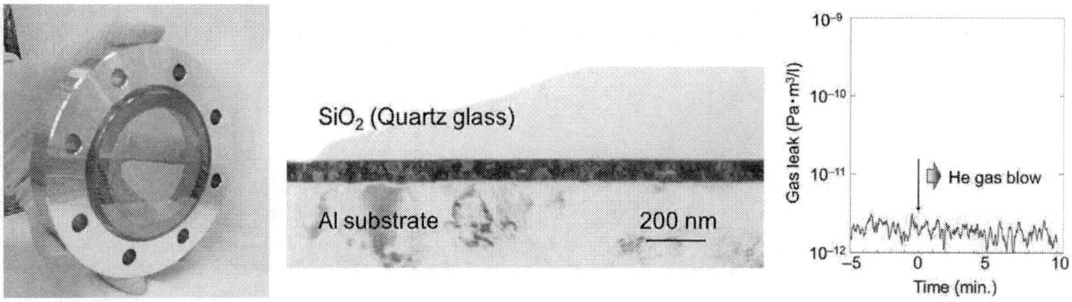

(A) (B) (C)

Fig. 2. (A) A viewing port for ultra-high vacuum chambers fabricated by bonding an ICF114 vacuum flange made of Al alloy with a sapphire plate. (B) TEM cross-section image for the bonded interface. (C) Result of the leak test using He gas.

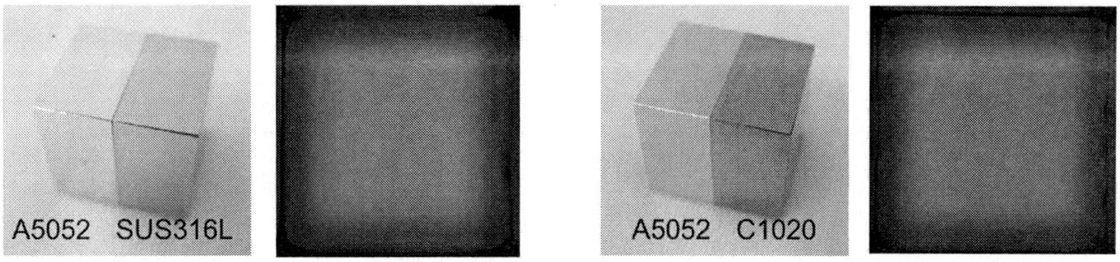

(A1) (A2) (B1) (B2)

Fig. 3. (A) Image of A5052 and SUS316L substrates bonded using Au (50 nm) films with Ta (5 nm) underlayers on each side. (A2) SAT image of the bonded interface. (B1) and (B2) are images showing bonded A5052 and C1020 substrates.

Conclusions

This bonding process holds great potential for the fabrication of new machinery components and new devices although more intensive efforts must be undertaken to enhance the bonding strength further.

References

[1] T. Shimatsu and M. Uomoto, *J. Vac. Sci. Technol. B*, vol. 28 (2010), pp. 706–714.
[2] T. Shimatsu and M. Uomoto, *ECS Transactions*, vol. 33, no. 4 (2010), pp. 61–72.
[3] T. Shimatsu, M. Uomoto and H. Kon, *ECS Transactions*, vol. 64, no. 5 (2014), pp. 317–328.
[4] T. Shimatsu and M. Uomoto, *Ceramics Japan*, vol. 51, no.2 (2016) pp. 91–93.
[5] M. Uomoto and T. Shimatsu, *Proceedings of 30th Spring Meeting of Japan Institute of Electronics Packaging* (2016) pp. 47-48.

Atomic behavior and mechanism of metal/metal solid-state bonding at low temperature

Kittipan Pongmorakot*, Shoichi Nambu, Yasushi Shibuta and Toshihiko Koseki
Department of Materials Engineering, The University of Tokyo, 7-3-1 Hongo, Bunkyo, Tokyo 113-8656 Japan
Email: kittipan.p@metall.t.u-tokyo.ac.jp

Abstract

Similar and dissimilar constituents with clean surface were hot-pressed and isothermally held in vacuum atmosphere at relatively low temperatures, approximately $0.5T_m$, and mechanical properties along with the microstructure were investigated. In case of IF/IF steel constituent, the evolution of the interfacial strength was found to comprise of two stages. The interfacial strength sharply increases after the compression in the first stage, meanwhile the interfacial strength increases at slower rate during the second stage. TEM observation indicates that the as-compressed interface are not extensively coherent at all, resulting in reduced interfacial strength. These incoherency continuously becomes coherent during the isothermal holding in the first stage through short-range atomic rearrangement, leading to strength increase without long-range diffusion. Together with similar constituent solid-state bonding, steel/Mo and steel/Ni constituent solid-state bonding, which do not contain reaction product, were also investigated. The interfacial strength test results reveal similar behavior with IF/IF steel with separated 2 stages, although the bonding temperature and isothermal holding range is different.

Introduction

Solid-state bonding is a welding process, which induces the coalescence of based materials below the constituents' melting temperature. It has been widely utilized for several advanced applications, because solid-state bonding offers several superiorities over the traditional welding process such as the absence of hazardous heat-affected zone (HAZ). Although solid-state bonding has been studied for decades, the researches were focused on the physical evolution and the kinetics of the mass transfer at the bonding region. To clarify the dominant mechanism of the evolution of interfacial strength, Derby and Wallach [1,2] and Takahashi [3] proposed that the atoms was transported from voids' surface to the voids' tip by long-range diffusion, leading to the increase in contact area and subsequent interfacial strength. Also, several other theories were proposed for the mechanism: energy barrier [4], oxide film [5] and etc. According to these proposals, high temperature, severe compression or long bonding time are required to attain decent interfacial strength. However, bonding at high temperature potentially causes rapid kinetics for evolution of interfacial strength. As a result, mechanism in very small scale, which is limited in very short time, might possibly be overlooked.

In the previous study [6], the evolution of interfacial toughness of surface-activated bonding at relatively low temperature between dissimilar constituents, type 304 stainless steel and pure nickel, was found to consist of two stages with much faster evolution in the first stage than in the second stage. However, the dominant mechanism for the faster evolution in the first stage is still not clarified. Thus, this study aims to investigate the mechanism for evolution of interfacial strength in the first stage in the solid-state bonding of similar and dissimilar constituents, which do not contain reaction products.

Experimental procedures

For similar constituent solid-state bonding, the ultralow-carbon IF steel, whose chemical composition is listed in *Table 1*, was employed to the both sides of the bonding. One side of the bonding surface was prepared by grinding and polishing to have 3 nm root-mean-square (rms) roughness and 300 nm average wavelength. The other side was grinded to achieve 30 nm rms roughness and 1000 nm average wavelength, measured by scanning probe microscopy (SPM). In case of dissimilar constituents, Mo, which does not contain reaction product with steel at high temperature [7], or Ni and steel were employed to both sides of specimen. The surface of steel was prepared identically to the first side of similar constituent bonding, while the surfaces of Ni and Mo specimens possess 30 nm rms roughness and 1000 nm average wavelength.

A pair of surface-prepared specimens were stacked and followed by a thermo-mechanical process, as shown in *Fig. 1*. After being heated up to the targeted bonding temperature of IF/IF steel, steel/Mo and steel/Ni, which were varied from 600 to 700 °C, both sides of specimens were momentarily compressed with 5% compressive strain, and then isothermally held for a given time. Subsequently, the bonded specimens were machined to a plate shape with 1.5 mm thickness. The interfacial strength was measured at room temperature by tensile test perpendicular to the bonding interface with crosshead speed of 5.56×10^{-6} $m\cdot s^{-1}$. The microstructure of the bonding interface were observed by scanning electron microscopy (SEM) and transmission electron microscopy (TEM).

Table 1: Chemical composition of the ultralow-carbon interstitial-free steel (mass%).

C	Si	Mn	Ti	Al	Fe
0.001	0.008	0.20	0.038	0.021	bal.

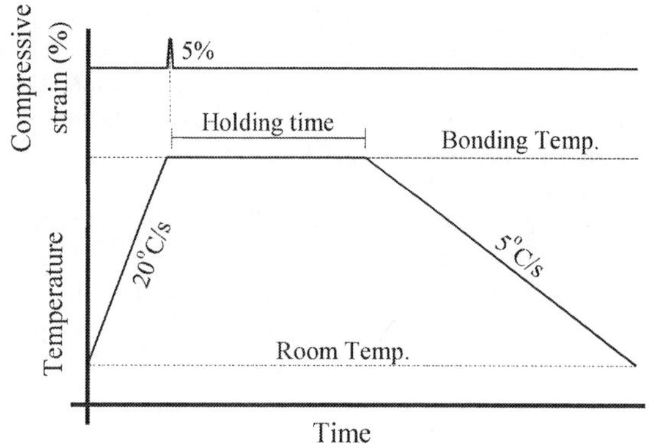

Figure 1: Thermo-mechanical process used in the experiment of similar and similar constituents.

Molecular Dynamics (MD) simulation was performed to study the atomic behavior at the bcc-bcc interface with Finnis-Sinclair (FS) potential, whose details was illustrated by Shibuta [8]. The number of atoms, which is 60272 atoms, the simulation volume and the temperature were controlled to be constant during time step, 5 fs. The simulated specimen comprises of upper lattice, whose [010] and [001] directions are parallel to the *y* and *z* axis, and the lower lattice, which was tilted by 30 ° around the *x* axis relatively to the upper lattice. Then, both lattices were stacked and bonded by thermo-mechanical process, shown in *Fig. 2(b)*.

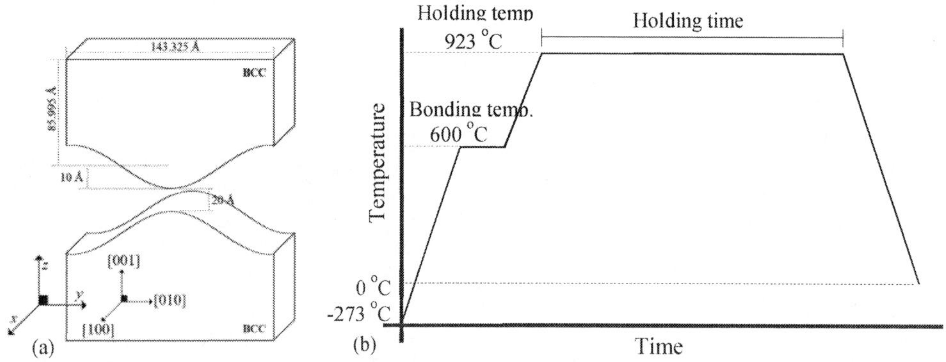

Figure 2: (a) Simulated bonding specimen and (b) Thermo-mechnical process used in the simulation

Results and discussion

Fig. 3 shows the evolution of interfacial strength with isothermal holding time on logarithmic scale of IF/IF steel interface after the compression. From *Fig. 3*, the gradients the relationship appear not to depend on the bonding temperature. Thus, the result clearly reveals two stages, which is distinguished by the difference of the gradient along the isothermal holding time. The evolution of interfacial strength of the first stage is more rapid, meanwhile the evolution of interfacial strength of the second stage becomes gradual. In case of the first stage, the gradient is found to be approximately 1.3, which is probably related to interface reaction. On the contrary, the gradient of the second stage, which is approximately 0.4, reveals that the diffusion dominates the evolution.

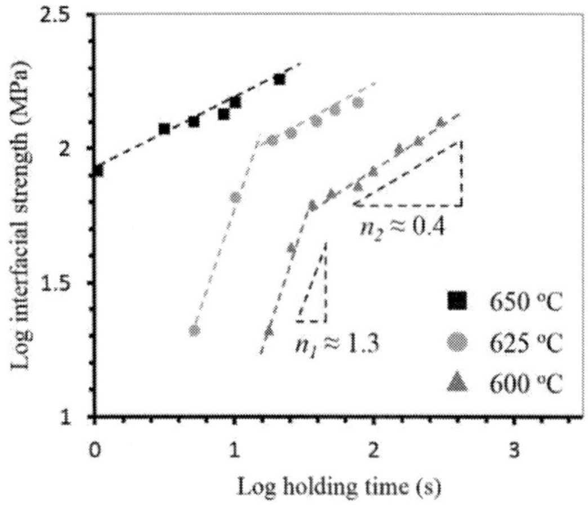

Figure 3: Relationship between isothermal holding time and interfacial strength on logarithmic scale of IF/IF steel interface.

The interface of as-compressed specimen was examined by Fourier transformed TEM lattice image analysis. Also, the interface of the one at the end of the first stage was observed. The observation of as-compressed interface revealed a large area of discontinuity between both sides of bonding lattices. Then, the interface became more coherent during the first stage, potentially leading to significant increase in interfacial strength during the first stage. From the MD simulation, the potential energy of the as-compressed interface is high and it appears to decrease rapidly at the early stage of holding time, then gradually decrease along the holding time. The interface in as-compressed conditions involves several distorted small grains and also the disordered layer, introducing discontinuity between upper and lower lattices. *Fig. 4* shows the atomic rearrangements at the interface at the beginning of the first stage and after the first stage. The lattices of the upper and lower lattices appeared to become more coherent along the holding time by the atomic short-range movement without extensive long-range diffusion.

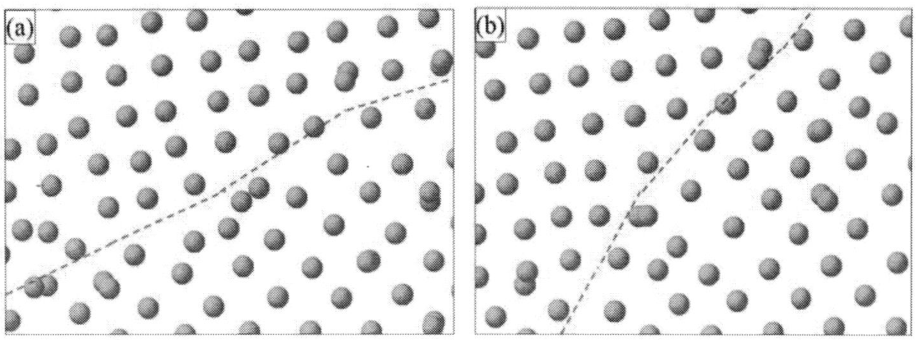

Figure 4: Atomic arrangements of the interfaces (a) at beginning of the first stage with incoherency and (b) after the first stage with more coherency, where red dashed lines are boundaries.

In case of dissimilar constituent solid-state bonding, the evolution of interfacial strength of the interface without reaction product shows similar behavior with the IF/IF steel interface. The evolutions of interfacial strength of both IF steel/Mo and IF steel/Ni comprise of two stages, distinguished by the difference of the gradients during the isothermal holding time, as shown in *Fig. 5(a)* and *5(b)*, respectively. According to the result of similar and dissimilar constituents with the 5% constant compressive strain, the gradient of the first stage appears to be independent to the bonding constituents with IF steel. Thus, the same mechanism dominates the evolution of both similar and dissimilar constituents. However, the evolution of interfacial strength of dissimilar constituents bonding obviously requires higher bonding temperature and longer isothermal holding time than that of similar constituent.

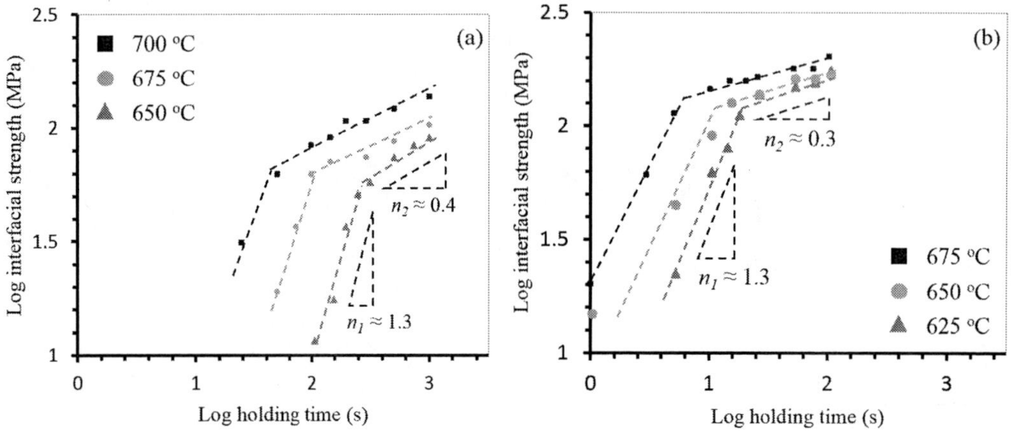

Figure 5: Relationship between isothermal holding time and interfacial strength on logarithmic scale of (a) steel/Mo and (b) steel/Ni interface.

Conclusions

The evolution of microstructure and interfacial strength at the similar and dissimilar constituents (without reaction product) interface bonded by the solid-state bonding at relatively low temperature involves two stages. The first stage starts immediately after the compression with rapid increase in interfacial strength, then the strength gradually increase in the second stage. The evolution of interfacial strength in the first stage is achieved by the atomic rearrangement near the interface.

References

[1] B. Derby and E.R. Wallach: "Theoretical model for diffusion bonding," Metal Science, Vol. 16, No. 1 (1982), pp. 49-56.

[2] B. Derby and E.R. Wallach: "Diffusion bonding: development of theoretical model," Metal Science, Vol. 18, No. 9 (1984), pp. 427-431.

[3] Y. Takahashi and K. Inoue: "Recent void shrinkage models and their applicability to diffusion bonding," Materials Science and Technology, Vol. 8, No. 11 (1992), pp. 953-964.

[4] H.A. Mohamed and J. Washburn: "Mechanism of solid state pressure welding," Welding Journal, Vol. 54, No. 9 (1975), pp. 302-310.

[5] K.I. Johnson and D.V. Keller Jr: "Effect of contamination on the adhesion of metallic couples in ultra-high vacuum," Journal of Applied Physics, Vol. 38, No. 4 (1967), pp. 1896-1904.

[6] S. Nambu, M. Michiuchi, J. Inoue and T. Koseki, Kinzoku (Materials Science and Technology), Vol. 80 (2010), pp. 289-293. (In japanese)

[7] F. Hatakeyama, K. Suganuma, and T. Okamoto: "Solid-state bonding of alumina to austenitic stainless steel," *Journal of materials science,* Vol. 21, No. 7 (1986), pp. 2455-2461.

[8] Y. Shibuta, S. Takamoto, and T. Suzuki: "A molecular dynamics study of the energy and structure of the symmetric tilt boundary of iron," *ISIJ international,* Vol. 48, No. 11 (2008), pp. 1582-1591.

Magnetic Pulse Welding of Dissimilar Metals in Tube-to-Tube Configuration

J. Lueg-Althoff*, S. Gies, A. E. Tekkaya
Institute of Forming Technology and Lightweight Construction, TU Dortmund, Germany
**Corresponding author: Joern.Lueg-Althoff@iul.tu-dortmund.de, phone: +49 231 755 6922*
J. Bellmann, E. Beyer
Institute of Manufacturing Technology, TU Dresden, Germany and Fraunhofer IWS, Dresden, Germany
S. Schulze
Fraunhofer IWS, Dresden, Germany

Abstract
Magnetic Pulse Welding (MPW) by electromagnetic compression is a solid-state welding technique favorable for the creation of strong metallurgical bonds between both similar and dissimilar metals. The process is based on the plastic deformation of one of the joining partners and the subsequent high velocity collision with the other joining partner. Due to the absence of external heat, brittle intermetallic phases and other heat induced defects are largely avoided which makes the process especially suitable for multi-material joints applied in lightweight design. However, the stationary part must be able to withstand the harsh impact without inadmissibly permanent deformation. This is why mostly solid cylinders are used as parent parts, limiting the lightweight potential. The paper focuses on the MPW of aluminum (EN AW-6060) to steel (C45) tubes with and without supporting mandrel inside of the parent part. Experimental studies were performed in which the wall thickness of the parent steel part was reduced successively. The deformation behavior of both the flyer and parent parts was recorded during the experiments by a periscope-style photonic Doppler velocimetry (PDV) setup. It was found that defect-free MPW of aluminum tubes on steel tubes without supporting mandrel is generally possible but strongly depends on the characteristics of the electromagnetic pulse due to the initiation of elastic waves that may deteriorate the weld quality. Critical values for the parent wall thickness were deduced for various pulse shapes and joining partner geometries and the feasibility of supporting elements was investigated.

Introduction
The implementation of lightweight concepts in automotive or aerospace industries demands the joining of hollow structures made out of various materials and combinations of them. Especially the welding of dissimilar metals like aluminum and steel is challenging. Magnetic pulse welding (MPW) is a solid-state impact welding process patented by Lysenko et al. in 1970 [1], which meets these demands. In this process, a flat or tubular joining partner (the "flyer") is accelerated by Lorentz forces created by the discharge of a capacitor via a tool coil (**Fig. 1b**). The flyer is deformed plastically and, after overcoming a standoff of few millimeters, collides with the "parent" part, the second joining partner. Under correct conditions with respect to radial impact velocities $v_{i,r}$, collision angles β, and impact pressures p_i (Fig. 1a), a metallic bond is formed without excessive heating of the parts. Brittle intermetallic phases and other heat-induced defects can thus largely be avoided. The physics of MPW is closely related to other impact welding processes like explosion welding [2], laser impact welding [3] and vaporizing foil actuator welding [4].

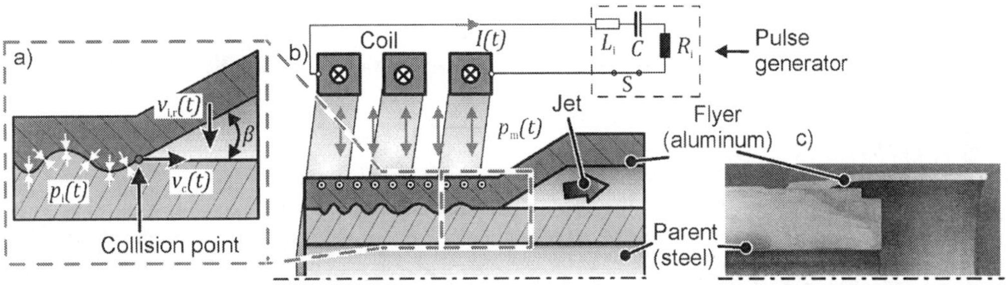

Figure 1: Principle of magnetic pulse welding. a) key collision parameters, b) setup, c) example joint

Numerous applications of MPW comprise hollow components like pipes, filters, and capsules [5]. Besides a proper weld quality, these parts have another requirement: the integrity of the inner cross-section after the MPW. This integrity can be affected due to the impact of the flyer on the parent resulting in unwanted permanent reduction of the inner diameter. In order to prevent this, either the walls of the parent part can be stiffened or additional supporting mandrels can be used. The problem with the latter is a limited removability and reusability. Additional unwanted phenomena observed in MPW of thin-walled parts are delamination and fracture because of an unfavorable interaction of reflected stress waves with the weld zone. Ben-Artzy et al. [6] studied the influence of the free path length inside of the parent on the interfacial wave formation. However, resilient and transferable boundary values are missing which enable an MPW user to decide if a supporting mandrel is necessary or not. First results of the authors imply that the weld formation itself is largely unaffected by the deformation of the parent part [7]. Good welds between two tubes with the same wall thickness of 2 mm were obtained. The objective of this paper is to expand those investigations by looking at the influence of the characteristics of the electromagnetic pulse as well as providing background information on the chronology of the process phases impact, weld formation and deformation of the parent.

Experimental setup and measurement technology

Welding trials with flyer tubes made of aluminum EN AW-6060 and steel C45 parent parts were performed on two different pulse generators, which exhibit distinctly different pulse characteristics with respect to the frequency and amplitude of the discharge current. The parameters of the two machines, a Maxwell Magneform 7000 and a Bmax MPW 50/25, are summarized in **Table 1**. For all setups, energy values were chosen which were adequate for welding with solid cylindrical parents [8].

Table 1: Characteristics of the deployed pulse generators

	Maxwell Magneform 7000 (Bank 2+3)	Bmax MPW 50/25
Maximum charging energy	12 kJ	50 kJ
Maximum charging voltage	8.16 kV	25 kV
Capacitance	362 μF	160 μF
Inductance	78 nH	58 nH
Resistance	5.4 mΩ	2.0 mΩ
Discharge frequency	Approx. 7 kHz	Approx. 21 kHz

At the Maxwell machine, an eight-turn compression coil was used in combination with a fieldshaper with an inner diameter of 41 mm and a concentration zone of 10 mm. The Bmax machine was connected to a directly acting one-turn coil with an inner diameter of 42 mm and a concentration zone of 10 mm. The geometrical parameters of the experiments are depicted in **Fig. 2**. The aluminum flyer tubes were in a heat-treated condition with a measured yield strength of 60 MPa, the steel parent part in normalized state. For selected experiments, supporting cylindrical mandrels made of aluminum EN AW-6060 and steel C45 were applied inside of the parent tubes. During the experiments, the discharge current was measured using a Rogowski current probe CWT 3000 B from PEM Ltd. The velocities of both the flyer during the radial compression (v_f) and the parent during and after impact (v_p) were recorded inline by a photonic Doppler velocimetry system, as described in [7]. The ratio of the maximum parent velocity $v_{p,max}$ and the maximum flyer velocity $v_{f,max}$ is used as an indicator for the parent deformation. The analysis of the welding result was done in instrumented mechanical tests applying peel and tensile force on a Zwick Z250 universal testing machine. A 3D digitizer ATOS of GOM mbH was used for the digitization of the welded specimens in order to analyze a potential permanent deformation of the parent part.

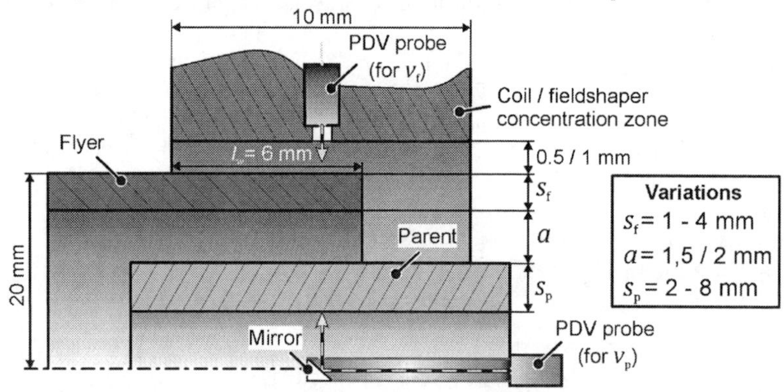

Figure 2: Experimental setup and variations of geometrical parameters

Influence of the wall thicknesses of flyer and parent

It was found, that defect-free welds between flyer tubes with a wall thickness s_f = 1.5 or 2 mm and parent part wall thicknesses s_p down to 2 mm could be produced with the Maxwell pulse generator. **Fig. 3a** shows that the ratio of $v_{p,max}$ and $v_{f,max}$ is around 10 % for parent parts with a wall thickness larger than s_p = 3 mm. It increases to values higher than 30 % for s_p = 1 mm. With smaller s_p, the difference between the two flyer thicknesses s_f increases, with a tendency to higher ratios for s_f = 1.5 mm. Moreover, the scans in Fig. 3a visualize that for a parent wall thickness of 5 mm, no plastic deformation of the parent part can be observed. A permanent deformation of the parent part starts at a parent wall thickness s_p of about 3 mm. For further reduced wall thicknesses, the deformation gets more intense, but successful welding can be obtained nevertheless. Fig. 3b shows that the velocity ratio for MPW experiments with the Bmax system is in the same range as with the Maxwell machine, at similar energies. The absolute velocity values however differed. Moreover, the increasing influence of the flyer wall thickness on the velocity ratio is shown.

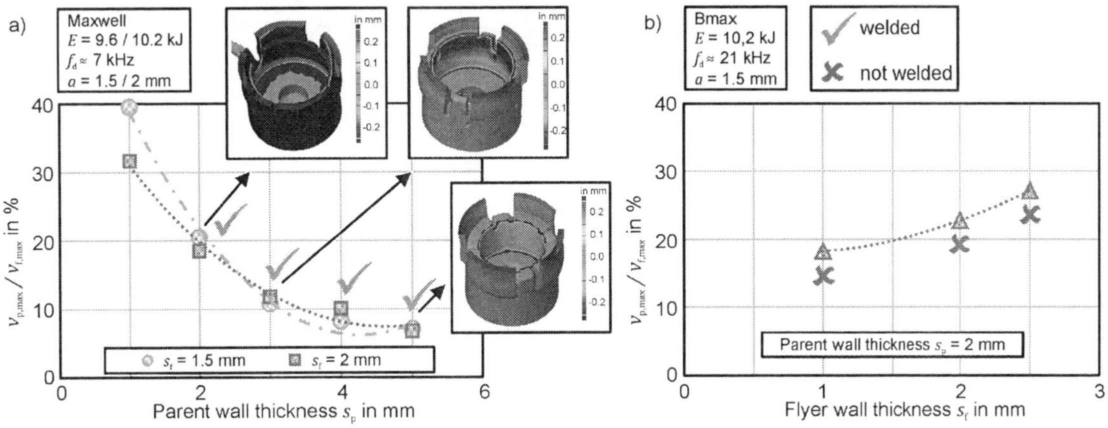

Figure 3: a) Ratio of maximum flyer and parent velocities during impact for experiments at the Maxwell machine and related 3D-scans of parts showing the deviation of the inner parent contour from a perfect cylinder, b) the same ratio for the Bmax system

By comparing the PDV curves of experiments with thick-walled and thin-walled parent tubes, indications regarding the chronology of the sub-processes weld formation and parent deformation can be found, see **Fig. 4**. For both thick- and thin-walled parent parts, the acceleration and impact phases have comparable durations. The impact phase is supposed to be approx. equal to the welding duration. At least that is the time when the jetting phenomenon takes place. For the experiments with thick-walled inner partner, no parent velocity and thus no deformation after the impact can be measured, which is in agreement with the 3D-scans of those specimens (Fig. 3). However, the PDV-curve for a thin-walled parent (Fig. 4b) shows that the velocity does not return to zero immediately after the impact but it decreases rather slowly. That phase is supposed to be the phase of the deformation of the thin-walled parent, which obviously takes more than three times the impact process itself.

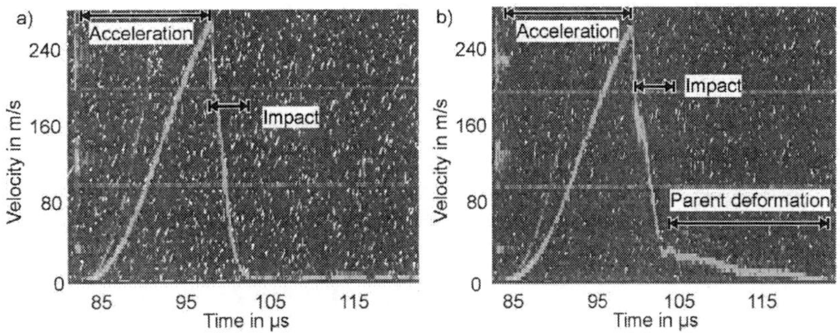

Figure 4: a) Spectrogram of flyer deformation in MPW with a parent with 5 mm wall thickness, b) spectrogram of flyer deformation in MPW with a parent with 2 mm wall thickness

Influence of the pulse characteristics

In contrast to the previous results, with the Bmax system, which features an approx. three times higher discharge frequency f_d compared to the Maxwell-System, no defect free welds could be produced with flyer tubes with

s_f= 2 mm and hollow parent parts with wall thicknesses s_p between 6 and 2 mm. Even with a wall thickness of approx. 11 mm, the weld failed in the peel test. The reason for this is attributed to a different behavior of elastic waves that travel through the parent part and are reflected from the free surface leading to an unfavorable interaction with the weld formation. Consequently, introducing a steel rod with an outer diameter matching the inner diameter of the parent part, lead to successful welds.

Influence of supporting mandrels

The experiments summarized in Fig. 3b were repeated with supporting mandrels made of aluminum EN AW-6060 or C45 steel, respectively, inserted into the parent tubes. **Fig. 5a** shows selected parts before the MPW process. For all combinations, successful welds could be obtained, which was not possible without supporting mandrels. This supports the hypothesis mentioned above regarding the important role of elastic stress waves interfering with the weld seam. A drawback of using supporting mandrels is that those steel and aluminum mandrels must be manufactured specifically for every combination of parent diameter and wall thickness. Moreover, it can be difficult to remove them after the process, since they get jammed inside the parent, see Fig. 5b. A repeated use is thus limited.

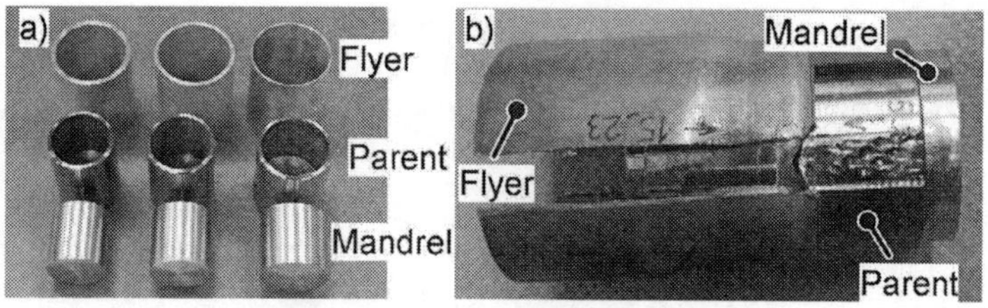

Figure 5: a) Parts before the MPW process, b) jammed mandrel inside the parent part

Conclusions

An experimental study of MPW with thin-walled inner joining partners was presented. It can be concluded:

- The electromagnetic pulse and especially the discharge frequency has a strong influence on the result of the welding process with hollow parents.
- With low discharge frequencies, welding of two tubes with similar thickness is feasible. For higher discharge frequencies, solid supporting mandrels or devices for dispersion of elastic stress waves need to be used.
- The monitoring of tube-to-tube MPW with PDV technology and subsequent 3D scanning is feasible.
- The chronology of the MPW process with thin-walled parents can be enlightened by the help of the PDV technology.

Future investigations comprise the development of mandrels that prevent the parent deformation and are easy to remove. Furthermore, a numerical analysis of the process is planned to support the hypotheses regarding the temporal course of the process.

Acknowledgments

The presented results originate from research carried out within the subproject A1 of the priority program 1640 ("joining by plastic deformation") funded by the German Research Foundation (DFG). The financial support is greatly acknowledged.

References

[1] D.N. Lysenko, et al.: U.S. Patent 3,520,049.
[2] B. Crossland: Explosive welding of metals and its application. Clarendon Press (Oxford, UK, 1982).
[3] H. Wang, et al.: "Laser impact welding: Design of apparatus and parametric optimization", *J Manuf Proc*, Vol. 19 (2015), pp. 118–124.
[4] A. Vivek, et al.: "Vaporizing foil actuator: A tool for collision welding", *J Mater Proc Technol*, Vol. 213, No. 12 (2013), pp. 2304–2311.
[5] V. Shribman: "Magnetic pulse welding for dissimilar and similar materials", *Proc 3rd Int Conf High Speed Forming* (ICHSF 2008), pp. 13–22.
[6] A. Ben-Artzy et al.: "Wave formation mechanism in magnetic pulse welding", *Int J Impact Eng*, Vol. 37, No. 4 (2010), pp. 397–404.
[7] J. Lueg-Althoff et al.: "Influence of the wall thicknesses on the joint quality during magnetic pulse welding in tube-to-tube configuration", *Proc 7th Int Conf High Speed Forming* (ICHSF 2016), pp. 259–268.
[8] J. Bellmann et al.: "Effects of surface coatings on the joint formation during magnetic pulse welding in tube-to-cylinder configuration", *Proc 7th Int Conf High Speed Forming* (ICHSF 2016), pp. 279–288.

Magnetic Pulse Welding: Joining within Microseconds – High Strength Forever

J. Bellmann*, E. Beyer
Institute of Manufacturing Technology, TU Dresden and Fraunhofer IWS Dresden, Germany
**Corresponding author: Joerg.bellmann@tu-dresden.de, phone +49 351 83391 3716*

J. Lueg-Althoff, S. Gies, A. E. Tekkaya
Institute of Forming Technology and Lightweight Construction, TU Dortmund, Germany

G. Kirchhoff, S. Schulze
Fraunhofer IWS Dresden, Germany

Abstract
The commercial availability of Magnetic Pulse Welding (MPW) equipment is opening up new prospects for the technology within industrial line production. To establish MPW as a clean and safe solid state joining technology for cylindrical parts as well as sheets, comparisons to conventional fusion welding technologies are needed. The focus of this paper is on MPW-samples made of aluminum tubes (EN AW-6060) on steel or aluminum rods (C45 or EN AW-6082). They were tested in a cyclic torsion test with up to two million cycles to evaluate their fatigue strength and to compare them with the fusion welded aluminum parts of the same geometry. Even dissimilar MPW parts show an excellent performance due to the absence of critical intermetallic phases and allow for weight reduction in the powertrain. Furthermore, this work describes a possibility for energy reduction in the MPW process that allows for the downsizing of pulse generators.

Introduction
Magnetic pulse welding (MPW) is a solid state joining technology for metals [1] that is often compared with explosive welding [2], whereas a direct comparison to conventional fusion welding technologies have not been found in literature. From an end user's perspective this is of great interest in order to choose the optimal joining process for production. The focus of this study is to compare MPW and other processes on the same geometry, but with process-wise optimized joint geometry. MPW for example requires a standoff distance between flyer tube and parent part in order to accelerate the flyer through the strong magnetic pressure generated by a working coil as shown in Fig. 1. The controlled collision between two parts under a certain angle β and radial impact speeds $v_{i,r}$ is used to create a surface cleaning jet and finally the joint [2]. New publications also support the thesis of a very small melting zone [3], for example due to thermal activation of the compressed medium in the welding gap [4]. Thus intermetallic phases are reduced to an uncritical minimum making MPW favorable for dissimilar metal joining [5]. Hybrid metal joints are a promising solution to ensure lightweight parts with high strength, good electrical conductivity and gas tight joining zones. Driveshafts are one example for the class of structures mentioned first [6].

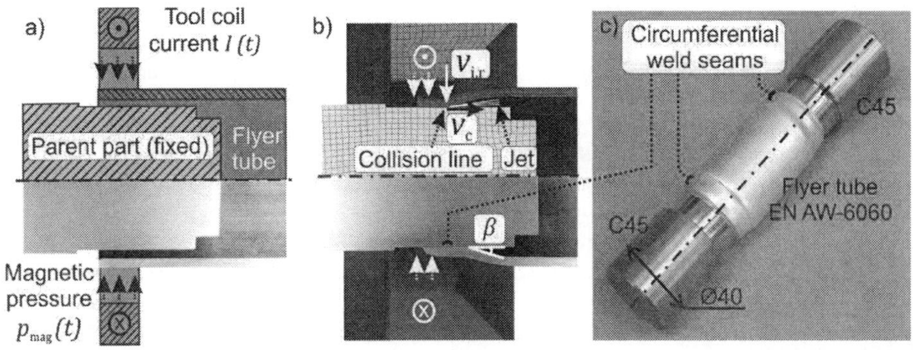

Fig. 1 MPW of tubular parts: a) magnetic pressure acts on the flyer tube's outer surface, b) time and place depended collision of accelerated flyer tube with inner parent part, c) sample with two dissimilar metal joints

In order to establish MPW as a fast and clean process in production [7], it has to be efficient as well. The MPW process can be called efficient, if the weld seam is generated with a minimum charging energy. This correlates with a minimum radial impact velocity $v_{i,r}$, if all other parameters like the electrical and mechanical properties of the flyer and the parent part as well as the welding setup are fixed. For tube welding, the minimum impact energy might be too high and deform the inner part inadmissibly [8]. In a previous publication the application of certain surface coatings during MPW were presented as another possibility for energy reduction [9].

Production of magnetic pulse welded parts

The proper welding parameters collision angle and velocity depend on various factors, for example on the part geometry and material, coil design and pulse characteristic. MPW of aluminum flyers (EN AW-6060 T4, 2 mm thickness) on steel parents (C45, normalized) are investigated in [9] and serve here as a base for the process adjustment. Finally, an optimized experimental setup designed by Bmax was used in this study with a welding gap of 1.5 mm and a slightly increased charging energy of 13.5 kJ to take the higher strength of the T66 condition into account. Weld lengths of more than two times the flyer thickness and waves with approx. 10 µm height appeared at the welding interface as shown in Fig. 2b and c. Another MPW configuration was tested to weld EN AW-6060 and EN AW-6082. Even if not optimized in term of energy, a configuration at 23 kJ was found with a different pulse generator (discharging frequency of 33.5 kHz instead of 20 kHz). Nevertheless, the maximum current amplitude was similar of around 740 kA. As a consequence of the higher impact velocity the deformation of the parent parts increased significantly (see Fig. 3c) with a maximum wave height of 140 µm. The welding time was less than 0.1 ms, so almost instantaneous. The weld qualities for both material combinations were secured by peel tests over the complete circumferences.

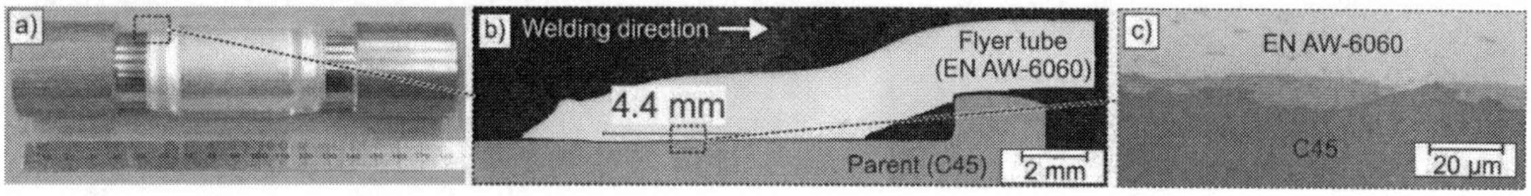

Fig. 2 a) MPW-joint between aluminum and steel, b) cross section of the welded zone with 4.4 mm weld length, c) wavy interface between aluminum and steel

For comparison reasons, two conventional fusion welding processes were used for joining the same parts; a laser beam welding process (LBW) with laser power of 1.7 kW and a welding time of 3.9 s and an electron beam welding process (EBW) with 0.83 kW and a welding time of 6.3 s. At first glance, the laser beam (6.6 kJ) and the electron beam processes (5.2 kJ) consume less energy than MPW. However, due to their lower efficiency the total energy consumption is typically three times higher for laser systems and two times higher for electron beam systems, respectively. Additionally, MPW also offers other possibilities for energy reduction. As presented in [9], an electroplated nickel layer with a thickness of 5 µm on the parent part doubles the weld length. Further experiments were conducted at the lower boundary of the welding window. It was found that the application of a nickel layer decreases the minimum collision velocity between EN AW-6060 and C45 from 350 to 230 m/s and the charging energy from 5.8 to 2.9 kJ with all other parameters being constant. The reduction of the charging energy does not only save energy but also allows for choosing a smaller pulse generator with lower investment costs for the specific joining task.

Torsion testing of welded parts

The material's strength in the magnetic pulse welded zone is normally similar or higher than the weaker base material due to hardening effects [10]. Thus components for torque transmission like driveshafts, are a promising application for magnetic pulse welded parts [6]. The main load cases for those tubular parts are static and cyclic torsion. Therefore torsion tests with simplified magnetic pulse welded driveshafts (Fig. 1c) and fusion welded samples (Fig. 3d) were conducted using a servo-hydraulic torsional testing machine from Inova. The axial force in the samples was set to zero during all tests. Each part consisted of two massive rods as parent parts (Ø40 x 70 mm) that were connected by a tube (Ø40 x 2 x 70 mm, EN AW-6060, T66). The length of the tube was reduced compared to real driveshafts to avoid buckling and assure the ability to test the weld seam. The minimum sample length was limited by the mounting ability of the rods' ends in the clamping system of the testing machine. The necessary welding gap for the flyer acceleration was machined in the rods' outer contour (Fig. 2b). Table 1 lists all tested samples configurations including the fusion welded parts used as a benchmark. For static testing, a constant angular velocity of 10 °/minute was applied while the torque was measured. Cyclic testing was performed using an alternating torque (stress ratio = -1) with constant amplitude and constant frequency of 20 Hz. The initial torque amplitude was incrementally reduced after each test, if the sample failed before 2 million stress cycles. If the sample passed, the torque amplitude was increased, respectively. With this step test method the torque amplitude for 50% probability to default was evaluated. A direct comparison of weld seam strength is difficult to obtain due to technology related differences in the geometrical seam configurations (see Table 1).

Here it is beneficial that an identical global sample geometry was chosen for all three welding technologies: a comparison is possible based on their overall torque performance.

Table 1: Sample configuration for torsion test, tube material EN AW-6060 T66

Material		Joining		No. of samples	
Rod	Tube	Technology	Joint geometry	Static	Cyclic
Al (EN AW-6082)	Al (EN AW-6060)	LBW	Butt	3	6
Al (EN AW-6082)	Al (EN AW-6060)	EBW	Butt	3	6
Al (EN AW-6082)	Al (EN AW-6060)	MPW	Overlap	3	6
St (1.0503, C45)	Al (EN AW-6060)	MPW	Overlap	3	6

Fig. 3a shows an increased static load capability until failure of the MPW samples compared to the fusion welded samples. All LBW and EBW samples failed due to a circumferential rupture next to or in the welded zone as a consequence of heat affects in the seam; see Fig. 3b. The welded zone for the MPW samples is twice as large due to the overlap geometry but without a heat affected zone. Thus the base material of the tube failed either by a circumferential rupture next to the weld or by buckling, see Fig. 3c and e, respectively. Cyclic tests revealed almost equivalent maximum torque amplitudes for all three welding technologies. Cracks appeared longitudinally in the tubes as well as radially *in* the fusion welded seams (see Fig. 3d) and radial, with some *distance*, to the MPW seam as shown in Fig. 3f. The crack propagation was not recorded during the experiments. Thus the assumption of a crack initiation in the tube and propagation towards the weld seam has to be verified in further experiments.

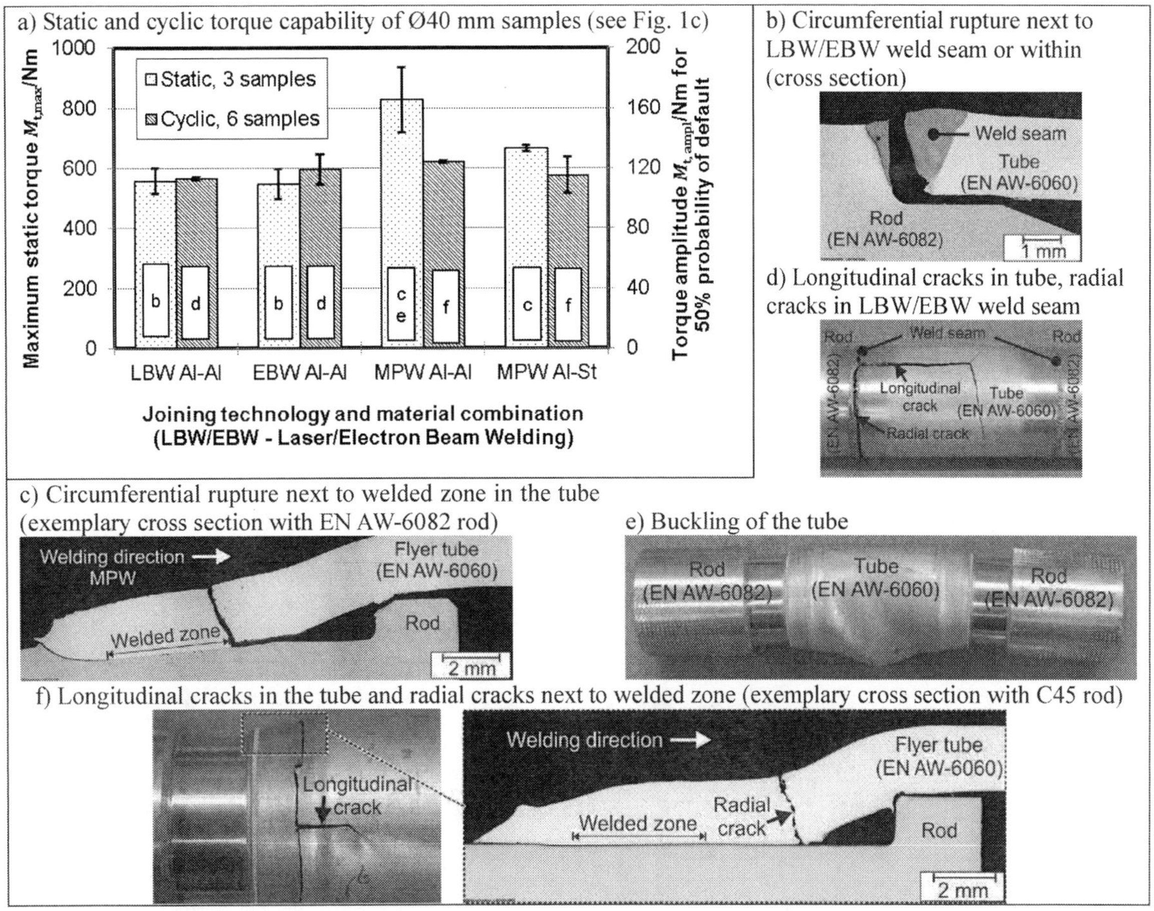

Fig. 3 a) Static and cyclic torque capability with similar and dissimilar material combinations and different welding technologies b)-f) indication of failure mode

The enlargement of the sample number would have been also necessary for statistical cross checking. Nevertheless it can be stated that, in contrast to fusion welding technologies, MPW offers also the possibility of dissimilar metal joining. This is shown for the combination of steel and aluminum in Fig. 2c. Due to the low heat input no critical intermetallic phases are formed and thus the applicability of lightweight mixed material parts joined by MPW is proven in Fig. 3a.

Conclusions

Within this study, the production process of magnetic pulse welded samples was characterized and compared with fusion welded parts of the same geometry, but with process-wise optimized joint geometries. Samples generated with MPW, LBW and EBW were tested under static and dynamic torsion load and directly compared with each other. Three statements regarding MPW can be formulated:

I. Compared to fusion welded parts the capability for static torsion loads is higher due to the process specific overlap welding configuration and the absence of heat affected zones. Fatigue tests show similar maximum load amplitudes for all three investigated processes, probably due to crack initiation in the base material of the tube.
II. In contrast to fusion based processes, MPW can join dissimilar metals like steel and aluminum enabling lightweight design with excellent load bearing capabilities.
III. Electroplated nickel coatings on the parent part allow for energy reduction during MPW of aluminum on steel and thus downsizing of pulse generators.

For future investigations it is planned to analyze the crack propagation in detail. Furthermore, an economical comparison of the welding technologies will be carried out, including the weight saving potentials of dissimilar MPW joints.

Acknowledgments

The authors would like to thank "Bmax" (Toulouse, France) for supplying the sample's material and preparing the magnetic pulse welded parts for the torsion tests. Furthermore the authors would like to thank the "University of Applied Science Dresden" (Dresden, Germany) for producing the EBW samples. The study on the coating's influence was part of the subproject A1 within the priority program 1640 ("joining by plastic deformation") funded by the German Research Foundation (DFG). The financial support is greatly acknowledged.

References

[1] D. Lysenko, V. Ermolaev, and A. Dudin, "Methods of pressure welding": U.S. Patent 3,520,049
[2] J. Verstraete, W. de Waele, and K. Faes, "Magnetic Pulse Welding: Lessons to be learned from explosive welding," *International Journal of Sustainable Construction and Design (2011)*, pp. 458–464.
[3] S. Sharafiev, M. F.-X. Wagner, C. Pabst, and P. Groche, "Microstructural characterisation of interfaces in magnetic pulse welded aluminum/aluminum joints," *Tagungsband zum 18. Werkstofftechnischen Kolloquium*, 2016, pp. 294–298.
[4] M. Koschlig, M. Veehmayer, and D. Raabe, "Production of Steel-Light Metal Compounds with Explosive Metal Cladding," *Proceedings of the 3rd International Conference on High Speed Forming*, 2008, pp. 23–32.
[5] G. Göbel, J. Kaspar, T. Herrmannsdörfer, B. Brenner, and E. Beyer, "Insights into intermetallic phases on pulse welded dissimilar metal joints," *Proceedings of the 4th International Conference on High Speed Forming*, 2010, pp. 127–136.
[6] Bmax, "Magnetic Pulse Welding - The ultimate solution for driveshaft manufacturers." Whitepaper. Accessed on: Mar. 02 2016.
[7] A. Kapil and A. Sharma, "Magnetic Pulse Welding: An efficient and environmentally friendly multi-material joining technique," *Journal of Cleaner Production*, no. 100 (2015), pp. 35–58.
[8] J. Lueg-Althoff et al, "Influence of the Wall Thicknesses on the Joint Quality during Magnetic Pulse Welding in Tube-to-Tube Configuration," *Proceedings of the 7th International Conference on High Speed Forming*, 2016, pp. 259–268.
[9] J. Bellmann et al, "Effects of Surface Coatings on the Joint Formation During Magnetic Pulse Welding in Tube-to-Cylinder Configuration," *Proceedings of the 7th International Conference on High Speed Forming*, 2016, pp. 279–288.
[10] S. D. Kore et al, "Application of electromagnetic impact technique for welding copper-to-stainless steel sheets," *International Journal of Advanced Manufacturing Technology*, vol. 54 (2011), pp. 949-955.

Characterization of residual stresses of a Tekken test block

Miguel Yescas [1], John Bouchard [2], Denis Pollier [1]

[1] *AREVA NP, France*
[2] *The Open University, England*

AREVA NP in France is looking at different ways to demonstrate that low preheating temperatures applied during Temperbead welding repairs are possible without compromising the integrity of repaired nuclear components, particularly when these repairs are carried out using the automatic GTAW process with a nickel base filler metal. One way of doing this, is by using one or more of the cold cracking severity test currently available to the industry. The implant test and the Tekken test have been considered in the first instance, however, preliminary results have oriented the project to continue with the Tekken test since it seems to be better adapted to our needs. In addition this test is being used worldwide and it is described by European and international standards. In order to demonstrate the conservativeness of the Tekken tests in terms of self-constraint compared with residual stress levels of actual welding geometries, residual stresses have been successfully characterized on a Tekken test block. Two different techniques have been applied, namely Neutron diffraction and the Contour method. The next step of this project is to carry out finite element numerical simulations in order to compare experimental residual stresses with numerical simulations. This paper deals with the residual stress measurements. The measured stresses are of a magnitude that would be expected given consideration of the material yield strength.

Comparison of Welding Residual Stresses of Hybrid Laser-Arc Welding and Submerged Arc Welding in Offshore Steel Structures

Michael Joachim Andreassen
Technical University of Denmark, Department of Civil Engineering, DK-2800 Kgs. Lyngby, Denmark
mican@byg.dtu.dk, +(45) 45 25 18 14

Zhenzhen Yu, Stephen Liu
Colorado School of Mines, Department of Metallurgical and Materials Engineering, Golden, CO 80401, USA
zyu@mines.edu, +(1) 303 273-3798, sliu@mines.edu,+(1) 303 273-3796

Martha Patricia Guerrero-Mata
Universidad Autonoma de Nuevo Leon, Facultad de Ingenieria Mecanica y Electrica, San Nicolas de los Garza, N.L. 66455, Mexico
martha.guerreromt@uanl.edu.mx, +(52) 81 8329 4020

Abstract
In the offshore industry, welding-induced distortion and tensile residual stresses have become a major concern in relation to the structural integrity of a welded structure. Particularly, the continuous increase in size of welded plates and joints needs special attention concerning welding induced residual stresses. These stresses have a negative impact on the integrity of the welded joint as they promote distortion, reduce fatigue life, and contribute to corrosion cracking and premature failure in the weld components.
This paper deals with the influence and impact of welding method on the welding induced residual stresses. It is also investigated whether the assumption of residual stresses up to yield strength magnitude are present in welded structures as stated in the design guidelines. The fatigue strength for welded joints is based on this assumption.
The two welding methods investigated are hybrid laser-arc welding (HLAW) and submerged arc welding (SAW). Both welding methods are applied for a full penetration butt-weld of 10 mm thick plates made of thermomechanically hot-rolled, low-carbon, fine-grain S355ML grade steel used in offshore steel structures.
The welding residual stress state is investigated by means of computational welding mechanics, experiments, and in accordance with existing production procedures to determine the real distribution and magnitude of the residual stresses. The experimental validation of the FE simulations includes temperature and hole-drilling measurements.

1 Introduction
Supersize monopiles used for offshore wind turbine support structures are being developed and have already been installed on several offshore wind farms. The demand for fabrication of monopiles with large wall thicknesses is high. However, the implications on design and especially the joining of cans by circumferential welding with these wall thicknesses are not yet fully understood. Particularly, the welding induced residual stresses need special attention. The stresses have a negative impact on the integrity of the welded joint as they promote distortion, reduce fatigue life, and contribute to corrosion cracking and premature failure in the weld components. Research within the field of welding residual stresses is important as the size of structures and welds are rapidly increasing. The steel fabricators raise the demands, as deeper foundations are needed due to design requirements.
This paper deals with the influence and impact of welding method on the welding induced residual stresses. In this relation, it is also investigated whether the assumption that residual stresses up to yield strength magnitude are present in welded structures as stated in the design guidelines, [1,2,3]. The fatigue strength for welded joints is based on this assumption. The two welding methods investigated are submerged arc-welding (SAW) and hybrid laser-arc welding (HLAW). Both welding methods are applied for a full penetration butt-weld of 10 mm thick plates produced of thermomechanically hot-rolled, low-carbon, fine-grain S355ML grade steel.
By investigating the welding residual stresses through numerical simulations, experiments and in accordance with existing production procedures, an optimization of the fatigue design is expected, leading to a more efficient and improved design. In this context, this research is expected to benefit the offshore industry by leading to an improved design which consequently can be included in present norms and standards. Further, it will optimize the correlation between numerical simulations and physical models.
3D FE models are very CPU time consuming. Thus, this paper deals with a steel plate thickness of 10 mm which makes it easier to adjust and optimize the FE models. The experiences gained from this model will then form the basis of the modelling technique utilized for the thicker plates used for the construction of monopiles.

2 Material Tests and Results

The material used for the experiments is thermomechanically hot-rolled, low-carbon, fine-grain structural steel of grade S355ML in accordance with the European Standard, [4]. This steel grade is widely used for offshore structures and possesses excellent weldability. It has a lower carbon content than grade S355NL steel (normalized and rolled) which lowers the cold cracking tendency and increases the toughness in the HAZ. The increased performance against cold cracking makes it possible to reduce the preheating, which leads to cost reduction in terms of reduced fabrication time and energy savings. The chemical composition of grade S355ML steel is shown in Table 1.

Table. 1: Chemical composition (in wt. pct.) of S355ML steel.

C %	Si %	Mn %	P %	S %	Al %	Cr %	Ni %	Mo %	Cu %	V %	Nb %	Ti %	N %	B %	Cae %
.082	.332	1.37	.01	.0005	.045	.024	.012	.004	.016	.003	.027	.014	.0052	.0002	.32

Experiments include the determination of material properties such as elastic modulus, yield stress and ultimate strength. Thus, 25 room and high temperature monotonic tensile tests have been performed in the temperature range from 20°C to 1000°C. The tests followed the European tensile testing standards for metallic material, [5,6]. Fig. 1 shows the tensile test setup.

Fig. 1: High temperature monotonic tensile test setup.

An MTS 312.31, 250 kN, universal testing machine was used. Displacement and load were logged by the testing machine and output was a stress-strain curve. An Instron, type 2620-601, extensometer with a range of 50 mm was used on the specimens to measure the displacement over an initial gauge distance, which was used to determine the modulus of elasticity.

Table. 2: Mechanical properties of S355ML steel.

Temp. (°C)	20	100	200	300	400	500	600	700	800	1000
E (MPa) x 10^5	2.18	2.03	2.15	1.86	1.06	1.54	0.91	0.53	0.39	0.14
f_y (MPa)	404	380	366	325	244	210	132	104	34	22
f_u (MPa)	601	535	556	597	521	372	179	121	46	39

Table 2 shows the mean values of elastic modulus, yield stress and ultimate strength given in the temperature range from 20°C to 1000°C. It is seen how the elastic modulus, yield stress and ultimate strength decreases as the temperature increases.

3 Welding Experiments

The following three subsections contain a brief description of the welding experiments. The first subsection concerns the test weld specimens while the second and third subsections concern the submerged arc welding process and the hybrid laser-arc welding process, respectively.

3.1 Test Weld Specimens

The geometry of the welding application comprises a full penetration, one-sided butt-weld joining two identical plates, each with a dimension of 480 × 240 × 10 mm. The perpendicular edges have been prepared by water jet cutting and the weld grooves by flame cutting. The weld grooves are different for the two welding methods and will thus be described in the following two subsections. Fig. 2 shows an isometric view of the test plate geometry.

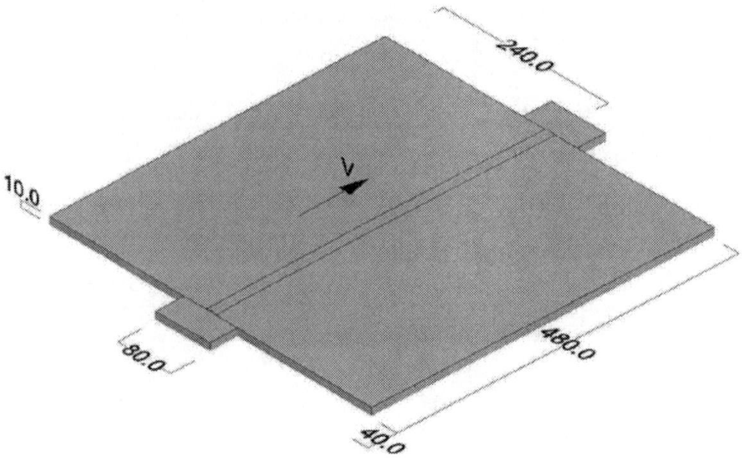

Fig. 2: Isometric view of geometry.

The two main plates have been constrained (tack-welded) by a run-on and a run-off plate of 80 × 40 × 10 mm at each end of the weld plate.

3.2 Submerged Arc Welding Experiments

The Submerged Arc Welding (SAW) butt-weld consists of two weld-passes; a root-pass and a cover-pass, which have an interpass time of 940 seconds. The weld geometry comprises a bezel angle of 35°, a nose height of 3.0 mm and a nose opening of 0.0 mm. An underlying bench supports the plates in the vertical downward direction.

Table 3: Welding procedure specifications for SAW.

Pass	AC/DC	Efficiency	Wire (mm)	Current (A)	Voltage (V)	Travel Speed (cm/min)/(mm/s)	Heat input (kJ/cm)
SAW 1	DC+	0.98	4	535	27.3	36/6	24
SAW 2	DC+	0.98	4	535	27.3	36/6	24

Table 4: Properties of filler wire and flux powder.

	C (%)	Si (%)	Mn (%)	P (%)	S (%)	Cu (%)	f_y (MPa)	f_u (MPa)	El. (%)	Impact Energy (J)	T(°C)
OK Autrod 12.22/L	0.1	0.2	1.0	0.018	0.018	0.25	–	–	–	–	
With OK flux 10.71/72	0.08	0.5	1.5	–	–	–	440	530	30	50	−40

Table 3 presents the welding procedure specification and the filler wire and flux powder properties are given in Table 4.

3.3 Hybrid Laser-Arc Welding Experiments

The Hybrid Laser-Arc Welding (HLAW) consists of two passes; first, a Metal Active Gas (MAG) weld pass from the bottom side of the plates followed by a Hybrid Laser-Arc pass from the top side. The MAG pass is performed in order to reduce the risk of laser beam penetration between the two plates. The weld geometry comprises a bezel angle of 15°, a nose height of 5.0 mm and a nose opening of 0.0 mm. The groove angle is only used for the MAG pass. Consequently, the HLAW pass is performed from the opposite side without any groove angle. An underlying bench supports the plates in the vertical downward direction.

Table 5: Welding procedure specifications for the root pass (MAG) and cover pass (HLAW).

Pass	Travel Speed (m/min)	Wire Diameter (mm)	Stick out (mm)	MAG Current (A)	MAG Voltage (V)	Laser (kW)	Weld Speed (mm/min)
1 (MAG)	9.5	1.2	25	310	33.5	–	650
2 (HLAW)	9.5	1.2	25	310	25	8	1000

Table 6: Chemical composition (in wt. pct.) and mechanical properties of the wire OK Autrod 12.51 and weld.

	C (%)	Si (%)	Mn (%)	P (%)	S (%)	f_y (MPa)	f_u (MPa)	A4–A5 (%)	Z	Impact Energy (J)	T (°C)
Min	0.06	0.80	1.40	–	–	–	–	–	–	–	–
Max	0.14	1.00	1.60	0.025	0.025	–	–	–	–	–	–
As welded	0.10	0.72	1.11	0.013	0.012	470	560	26	68	130, 90, 70	20, 20, 30

Table 5 provides the welding procedure specifications and Table 6 lists the chemical composition and mechanical properties of the weld.

4 FE Model

Experiments normally form the basis for developing welding procedures contained in a Welding Procedure Specification (WPS). This methodology also implies that the basis for WPS is observations of e.g. joint integrity, absence of defects, microstructure and mechanical testing [7]. With increasing computational power over the last decades, one of the significant tools applied for additional design considerations is the finite element analysis. FEA is rarely used in the process of developing welding procedures; however, it is expected that simulations can be used to complement the experimental procedures for preparing a WPS since the resulting residual stress and distortion can then be considered when comparing different welding procedures. Furthermore, simulations are useful in designing the manufacturing process as well as the manufactured component itself. A coupled 3D FE model simulates the present welding process and analyses the residual stress state. The 3D model includes the transient heat transfer along with the material properties at different temperatures. It was decided to consider the coupling of the heat transfer and the stress analysis as one-way coupled, in order to reduce the computational costs considerably. The commercial FE program Abaqus models the welding process using a computer capacity of 8 CPUs and 128 GB RAM. Fig. 3 lists the model geometry and the used mesh configuration.

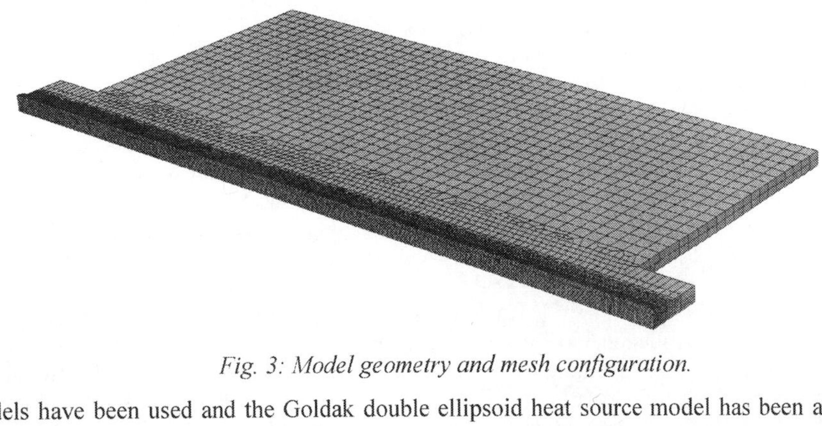

Fig. 3: Model geometry and mesh configuration.

Isotropic models have been used and the Goldak double ellipsoid heat source model has been applied by using the Dflux subroutine.

5 Temperature Distribution

Temperature measurements during welding monitor in-situ the temperature evolution in the plates. The measurements verify the FE model and guide the mapping of the residual stresses in the hole drilling experiments. Fig. 4 shows the heat flow pattern from the FE models for both the submerged arc welding and the hybrid laser-arc welding. It is noticed that the SAW has a wider temperature distribution than the HLAW for both the transverse and the longitudinal direction, and that the maximum temperature is highest for the HLAW.

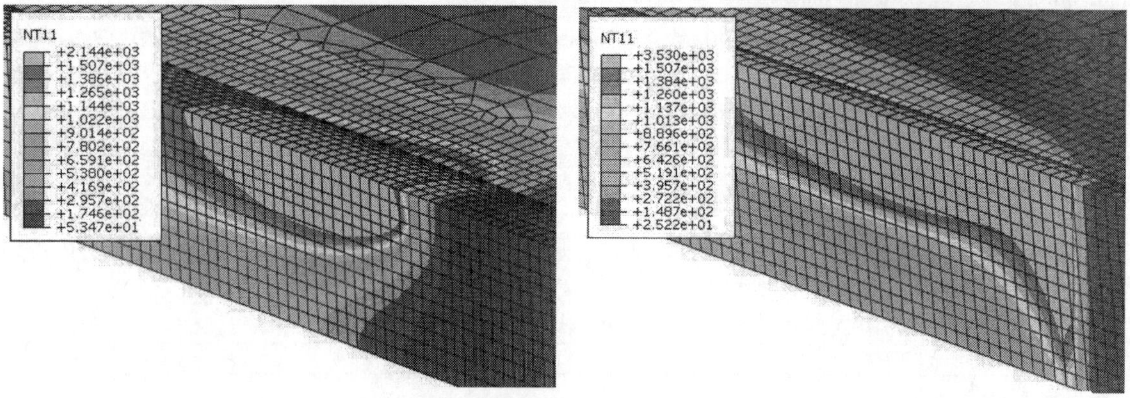

Fig. 4: Heat flow pattern calculated with the FE models for submerged arc welding (left) and hybrid laser-arc welding (right). All temp. in ℃.

In the analysis of the thermal gradients from the welding procedure, it is desired to have measurements as close as possible to the weld pool due to buildup of residual stresses, which are highly influenced by the peak temperatures caused by the transient heat source. The distance between the thermocouples is therefore shortest close to the weld. The position of the thermocouples is in the middle of the plate corresponding to 5 mm below the surface. The thermocouples are type "K" MI with a probe-diameter of 1.5 mm and a length of 150 mm. A HP, model 34970A, data logger acquired the thermocouple data. The temperature measurements have an interval of 1 second.
For the SAW temperature measurements the first pass reached a maximum temperature of 598°C and the second pass a maximum temperature of 679°C at a distance of 10 mm from the center of the weld. The second pass was expected to have a higher maximum temperature since it started at an initial temperature of 75°C compared with the room temperature for the first pass.
For the HLAW experiments, the MAG welding pass reached a maximum temperature of 324°C and for the HLAW the maximum temperature was 395°C, both at a distance of 10 mm from the center of the weld.
In relation to the temperature distribution, the SAW weld is expected to result in a wider region of welding residual stresses compared to the HLAW because of the wider temperature distribution reached.

6 Residual Stress Distribution

Hole-drilling measurements were performed to map the stress distribution as a function of the distance away from the welds produced with the two different welding methods. The hole-drilling measurements as well as the temperature measurements validated the FE model. Prior to the hole-drilling measurements the surface was prepared by grinding to remove the mill scale for proper bonding of the strain gauges. Micro-measurements RS-200 hole-drilling equipment and software as well as standardized hole-drilling strain gauge rosettes E837 type A, 031RE and 062RE, with a maximum stress data depth of 0.5 mm and 1.0 mm, respectively, was used. As the hole-drilling method measures the residual stresses close to the surface, the comparison between the experimental and FE model predictions is limited with respect to the vertical distance from the surface. Moreover, it is not possible to measure the stresses in the weld by using the hole-drilling method, as the strain gauges require a flat and smooth surface. It is anticipated that such a weld surface would also change the residual stresses in the weld. These concerns means that the maximum residual stresses measured by the FE model and the hole drilling method cannot be directly compared. However, it is possible to measure and compare the stresses close to the surface at a horizontal distance of 4 mm from the weld fusion line and further away. Fig. 5 compares the maximum longitudinal residual stress distribution for both welding methods. The distribution is from the plate centerline cross section. The weld beat width is 10 mm for the SAW and 5 mm for the HLAW. In relation to the mentioned temperature distribution, which are widest for the SAW, hence it is expected that the SAW will introduce a wider region and a greater amount of welding residual stresses compared to the HLAW.

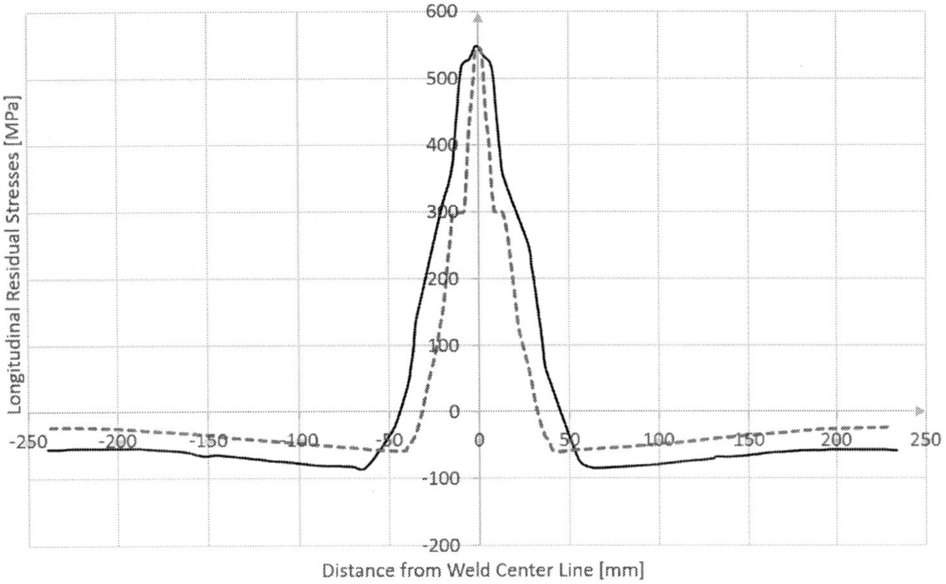

Fig. 5: Longitudinal welding residual stresses for SAW (solid line) and HLAW (dotted line).

Fig. 5 depicts a wider stress distribution for the SAW than for the HLAW. However, both welding methods result in approximately the same maximum longitudinal residual stress value. The maximum FEM longitudinal welding residual stresses calculated in the middle of the cross section are 565 MPa and 562 MPa for SAW and HLAW, respectively. Thus, the maximum residual stresses calculated are approximately the same for the SAW and HLAW; however, the distribution of the stresses is different. The distribution is much wider for the SAW than for the HLAW thus making the total amount of residual stresses larger for the SAW.

7 Conclusions

Two different butt-welds obtained by SAW and HLAW of 10 mm thick steel plates have been analyzed by experiments, FE models and in accordance with existing production procedures. The use of a coupled 3D FE model to simulate both welding processes has shown to be sufficient to map the residual stress distribution. The Goldak double ellipsoid distributed heat source was applied via Dflux subroutine on the 3D model. The developed 3D FE models show good agreement with the experimental measurements of the weldments. The SAW method has a wider distribution of the residual stresses than the HLAW method as well as higher maximum compression residual stresses. However, both welding methods results in approximately the same maximum tensile residual stress value. As the maximum longitudinal stresses are higher than the yield strength for both welding methods an underestimation is found when assuming residual stresses up to yield strength magnitude in welded structures.

Acknowledgments

The authors are grateful to Lindoe Welding Technology for the support provided developing the HLAW testing.

References

[1] Eurocode 3: "Design of steel structures - Part 1-9: Fatigue". Danish Standard, 2005.

[2] Det Norske Veritas, DNV-OS-J101, Design of Offshore Wind Turbine Structures. DNV, 2010.

[3] J. Krebs, *et al*: Influence of Welding Residual Stresses on Fatigue Design of Welded Joints and Components: IIW Doc. IIW-1805-07; 2007.

[4] DS/EN 10025-4-2004, Hot rolled products of structural steels – Part 4: Technical delivery conditions for thermomechanical rolled weldable fine grain structural steels, 2004.

[5] DS-EN ISO 6892-1-2013 - Metallic material - Tensile testing Part 1: Method of test at room temperature.

[6] DS-EN ISO 6892-2-2013 - Metallic material - Tensile testing Part 2: Method of test at elevated temperature.

[7] L.E. Lindgren: "Finite Element Modeling and Simulation of Welding - Part 1: Increased Complexity", Journal of Thermal Stresses, Vol. 24, 2001.

Welding Induced Distortion Prediction in a Double-Skin Aluminum-Alloy Structure

Yu-Ping Yang [1], Guoqing Gou [2], Qimeng Zhu [2], Hui Chen [2]

[1] *EWI, USA*

[2] *Southwest Jiaotong University, China*

Double-skin structural aluminum-alloy plate was developed for reducing the structure weight and maintaining the structure stiffness, which has been used in the fabrication of large welded structures such as a ship and a transportation vehicle. Typically, the double-skin plate is narrow (0.5m) and long (20m). Often, several plates are butt welded together using gas metal arc welding (GMAW) process with top passes and bottom passes forming a structural wall. Out-of-plane distortion is often observed after welding because the top welds and the bottom welds do not balance each other. Flame straightening has to be used to correct the out-of-plane distortion, which brings additional manufacturing cost and time. An efficient distortion prediction method was developed and validated using experimental data to predict distortion on the double-skin aluminum structure. The prediction methods included a local model and a global model. The local model was analyzed with a thermal-elastic-plastic method to predict plastic strains by inputting welding parameters. The global model was analyzed with an elastic analysis to predict distortion by mapping the predicted plastic strains. The local model were calibrated and validated with experimental measured data including temperature history, weld macrographs, hardness, residual stress, and deformation. The global-model predicted distortion had a good agreement with the measured one in production parts. The developed distortion prediction method was used to evaluate potential distortion mitigation methods including pre-cambering, weld balancing by adjusting the volume of top and bottom welds, and replacing GMAW using friction stir welding (FSW) process. Analysis results showed that all three distortion mitigation methods were effective to reduce the distortion, but distortion was not eliminated from the welded structure. A combined method such as pre-cambering and FSW can achieve distortion-free fabrications during welding the double-skin aluminum-alloy structure.

Control of Weld Residual Stress in a Thin Steel Plate through Low Transformation Temperature Welding Consumables

Zhenzhen Yu [1], Zhifen Wang [1], Stephen Liu [1], Zhili Feng [2], Joe Scott [3]

[1] *Colorado School of Mines, USA*
[2] *Oak Ridge National Laboratory, USA*
[3] *Devasco International, USA*

Low transformation temperature welding (LTTW) consumables are characterized by low martensite start temperature and large fraction of martensite forming in the weld. It can efficiently reduce the tensile residual stress because the volume expansion associated with the martensitic transformation compensates for the thermal contraction during cooling. In this work, neutron diffraction technique was used to obtain the residual stress distributions along three principal directions in 2mm thick A36 steel plates welded with a newly-developed LTTW and conventional ER70S-3 wires, respectively. Middle thickness line mapping along transverse direction on the thin plates showed that the maximum residual stress in heat affected zone along longitudinal direction was reduced from 330MPa to 240MPa by using the LTTW wire. EBSD scan and X-ray diffraction measurement confirmed that 89% volume fraction of martensite was present in the LTTW weld. Distortion pattern caused by the LTTW wire was opposite from that of ER70S-3 wire. Fatigue property of lap joint panels was almost doubled in comparison with ER70S-3 wire. Finite element modeling of the welding procedures was performed to reveal the microstructure evolution and provide a more complete analysis of the residual stress profile.

Application of ICME to Weld Process Innovations and Residual Stress Management

Zhili Feng

Oak Ridge National Laboratory, USA

Integrated computational materials engineering (ICME) is an emerging discipline that aims to integrate computational materials science tools into a holistic system that can accelerate materials development, transform the engineering design optimization process, and unify design and manufacturing. With the advancement in computational modeling methodology in various aspects of materials joining processes and availability of high-performance computers, ICME has become an increasingly important tool in today's materials joining research and development, and technology innovation. In this presentation, the applications of ICME to advance welding technologies are discussed. An integrated weld process-property-performance modeling framework is described first, as well as the experimental work for model development and validation including in-situ neutron/synchrotron experiments. Several examples of modeling applications are presented next, for proactive weld residual stresses for nuclear power plant life-cycle management, design and engineering of innovative welding processes to repair irradiated materials, and improvement of the weld fatigue life for automotive applications. Perspectives to integrate welding modeling into the overall manufacturing process and structure performance optimizations will be provided.

Effects of Titanium and Oxygen to Maximize Intragranular (Acicular) Ferrite Volume Fraction in GMA welds

Yuji Kisaka, Adrian P. Gerlich
University of Waterloo, 200 University Avenue West, Waterloo, Ontario, N2L 3G1, Canada
ykisaka@uwaterloo.ca

Abstract

Maximizing intragranular (acicular) ferrite volume fraction in weld metals is a typical strategy to satisfy mechanical property requirements, considering this offers high yield strength, high fracture toughness, and low hardness. In order to achieve this, titanium oxides are usually introduced as nucleation sites for intragranular ferrite, and most studies of gas metal arc welding consumables have considered titanium content less than 0.3 wt. %. A secondary factor that has seldom studied is the role of varying CO_2 content in Ar/CO_2 mixed shielding gas, especially when simultaneously varying the titanium content in weld metals. In the present study, a relationship between titanium content and intragranular ferrite volume fraction was investigated by changing weld metal titanium content using two welding wires and adding a thin pure titanium wire for in-situ alloying. Furthermore, the influence of oxygen content was evaluated by using various Ar/CO_2 ratios of shielding gas. The results revealed that volume fraction of intragranular ferrite was maximized when titanium content is 0.03 to 0.04 wt. % and oxygen content is 0.025 to 0.028 wt. %, respectively. Finally, in order to evaluate the product of these elements, the oxides within the intragranular ferrite were characterized by transmission electron microscopy. Ti_2O_3 participates incorporating MnS were observed and a manganese depression zone (MDZ) was found at the outer surface of the oxides, which supports a theory that the MDZ promotes acicular ferrite nucleation. Based on the chemistry and quantitative microscopy of microstructures, the ideal balance between titanium and oxygen in the weld metals is proposed to maximize intragranular ferrite volume fraction.

Introduction

Typically, weld metals must satisfy multiple mechanical properties such as strength overmatching, high fracture toughness, and low hardness, even though these may potentially be at odds with each other. On the other hand, it has widely recognized that intragranular (acicular) ferrite offers fine, homogeneous grain structures, and low hardness, which can achieve most strength and toughness requirements. Hence, promoting intragranular ferrite (IGF) has been a common target when designing weld metals for thicker steel joints. Nowadays, there are many welding consumables for high strength steels, which can promote IGF microstructure, and these normally incorporate titanium oxides which constitute strong IGF nucleation sites, whereas the detailed mechanisms for this remain controversial [1-6]. Many researchers have investigated the effects of both titanium and oxygen content in the weld metals respectively; however, most papers have focused on the case of titanium content being less than 0.3 wt.% [2-4]. In contrast, few papers mention the effects of chemical content on IGF volume fraction when titanium content exceeds 0.5 wt.%, and one suggested a method to control titanium contents in the weld metals by adding a thin pure titanium wire for in-situ alloying during welding [5,6]. This indicates that IGF volume fraction in the weld metals could be promoted by the in-situ alloying method; however, the combined effects between titanium and oxygen are not mentioned. Thus, there remains a need to consider both titanium and oxygen in order to determine how to maximize IGF volume fraction in weld metals.

In the present study, several weld metals containing varying IGF volume fractions will be produced by adding a thin pure titanium wire to the molten pool as an in-situ alloying technique, while also employing various ratios of premixed Ar/CO_2 shielding gases. This provides comparison of various IGF volume fractions in the welds to the titanium and oxygen content. Consequently, an ideal balance of titanium and oxygen content in the weld metals is suggested in order to maximize IGF volume fraction.

Experimental procedure

The specimen list for this work is shown in Table 1. All weld joints were produced by a robotic GMA welding machine for butt-welding, a U-shape groove, and in the flat position. The base material consisted of 1/8 circumferential sections of API 5L X80 grade pipe of 24" diameter and 15.1mm thickness. Welding consumables included two welding wires with different chemical compositions and pure thin titanium (CP-Ti) wire with a diameter of 0.2 mm. Also, three kinds of premixed shielding gas with different Ar/CO_2 ratio were used, with CO_2 contents of 50, 30, and 15%. Six layers of multi-pass welding were performed using following welding condition; 290-320A, 24.5-28.0V, and 120cm/min for the root pass, 280-310A, 24.5-28.0V, and 70cm/min for hot and filler pass, and 210-230A, 21.5-25.0V, and 30cm/min for cap pass. Preheat and interpass temperature was 150°C. Weld samples designated as AA-series were produced by adding a thin titanium wire for alloying in-situ during welding excluding cap pass, this wire was fixed with resistance spot welding every few centimeters to avoid being blown away. On the other hand, the titanium wire was not added for specimens designated as B series, since wire B included titanium already.

After welding, microstructures were obtained from the weld cross section at a location 40mm away from the crater for all welds, and observed at 200X and 500X magnifications with an optical microscope after polishing and etching with 5% Nital solution. For consistency, the observed position which detailed characterization and quantitative microscopy was performed was at the bead center of the as-deposited weld metal in the hot pass. Then microstructures were distinguished as grain boundary ferrite, intragranular ferrite (which was predominantly acicular ferrite, but could also include Widmanstätten ferrite + polygonal ferrite + bainitic ferrite), and bainite (upper and lower bainite). A point counting method was applied based on the definition proposed by Thewlis [7], in which a grid was drawn at a 10 micrometer pitch on micrographs taken at 200X magnification, and the number of intersections for each microstructure were counted. Then, the volume fraction values were determined by averaging the point counts from five micrographs to avoid bias and scatter. Moreover, the actual chemical composition of each weld metal were measured by inductively coupled plasma spectroscopy, with the results are also shown in Table 1. Transmission electron microscope (TEM) observation was performed to characterize the titanium oxide precipitates which are usually considered as IGF nucleation points, and the distribution of elements around and on the precipitates was also investigated with energy dispersive x-ray spectroscopy (EDX) during TEM.

Table 1 Specimen list and chemical composition in the weld metals

Specimen ID	Welding wire	Shielding gas	Chemical composition in weld metals (wt.%)						
			C	Si	Mn	Al	Ti	O	N
AA-15	Wire A +Ti wire	80%Ar-15%CO_2	0.084	0.57	1.79	<0.01	0.06	0.018	0.0032
AA-30	Wire A +Ti wire	70%Ar-30%CO_2	0.082	0.53	1.72	<0.01	0.04	0.026	0.0036
AA-50	Wire A +Ti wire	50%Ar-50%CO_2	0.079	0.47	1.63	<0.01	0.03	0.027	0.0036
B-15	Wire B	80%Ar-15%CO_2	0.056	0.54	1.68	<0.01	0.04	0.022	0.0032
B-30	Wire B	70%Ar-30%CO_2	0.055	0.50	1.66	<0.01	0.03	0.031	0.0039
B-50	Wire B	50%Ar-50%CO_2	0.057	0.45	1.17	<0.01	0.03	0.034	0.0033
A-30	Wire A	70%Ar-30%CO_2	0.082	0.48	1.69	<0.01	0.01	0.029	0.0032

Results

The optical micrographs in weld metals for the AA series and B series are shown in Fig. 1. In all cases, the IGF microconstituent is the most dominant; however, the AA-15 weld metal which was produced with wire A, thin titanium wire, and 85%Ar-15%CO_2 shielding gas includes a much higher volume fraction of bainite with some aligned carbide, and B-50 which is produced with wire B and 50%Ar-50%CO_2 shielding gas exhibits a higher amount of grain boundary ferrite and bainitic ferrite compared to others.

The relationship between titanium content and IGF volume fraction in the weld metals (Fig. 2a) and oxygen content and IGF volume fraction (Fig. 2b) are demonstrated as Fig. 2. The IGF volume fraction is maximized when the titanium content in weld metal is from 0.03 to 0.04 wt.% and oxygen content in weld metal is from 0.025 to 0.028 wt.%, respectively.

A titanium oxide precipitate observed by TEM along with the chemical distribution around/on the particle measured using EDX are shown in Fig. 3. It appears that the precipitate has a core/shell structure, where a titanium oxide is at the center and another compound based on manganese sulfide is coating the particle surface. In addition, it can be noted that a manganese depletion zone appears around the apparent manganese sulfide compound.

Discussion

As illustrated in Fig. 2a and b, the IGF volume fraction indicates a peak regarding titanium and oxygen content, respectively; thus, the combined effect of these elements should be considered to maximize the IGF volume fraction. Considering this critical range of composition, the ratio of titanium and oxygen may shed light into the nucleation mechanism for IGF. There have already been other studies mentioning the IGF nucleation mechanism, for example the work of Yamamoto et al., which showed that MnS and TiN around Ti_2O_3 particles are effective in IGF nucleation. MnS in particular has been shown can easily adhere to Ti_2O_3 and generate an MDZ around itself [3]. The formation of the MDZ has been shown to increase the Ar_3 transformation temperature locally, and this promotes nucleation of ferrite [3]. Additionally, when TiN is present close to this zone, formation of IGF is possible since TiN has a low lattice misfit with ferrite, reducing the barrier for nucleation [3]. Although these results have been obtained from weld metals or steels containing less than 0.3 wt.% titanium content, similar results have been also obtained when titanium concentration is higher, which is mentioned by Seo et al [5]. They found Ti_2O_3 particles with a MDZ in a weld metal involving 410 ppm of titanium, whereas the presence of TiN was not pointed out. The present work could not confirm the crystallographic structure of the titanium oxides; however, the evidence indicates that the titanium oxide is Ti_2O_3 based on measured chemistry.

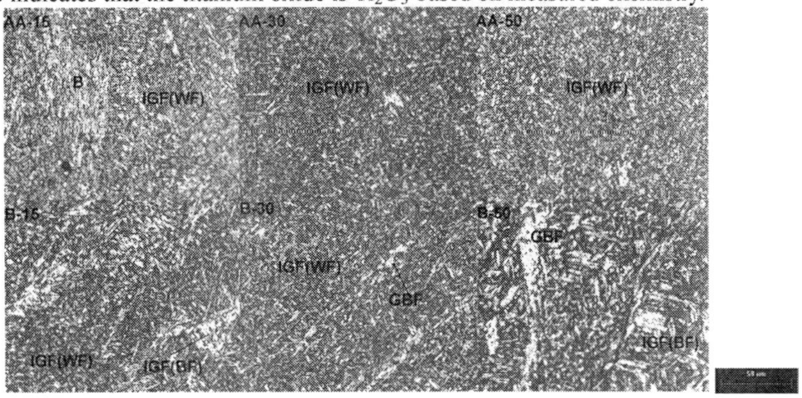

Fig.1 Optical micrographs of weld metals for the AA series and B series

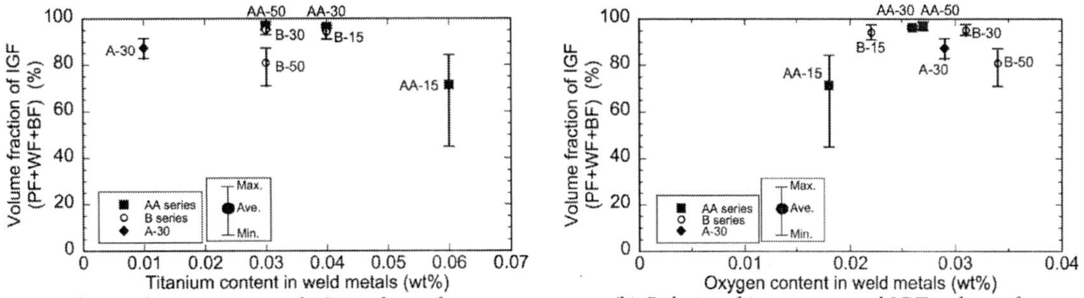

(a) Relationship titanium and IGF volume fraction *(b) Relationship oxygen and IGF volume fraction*

Fig.2 Relationship between titanium, oxygen content and IGF volume fraction in the weld metals

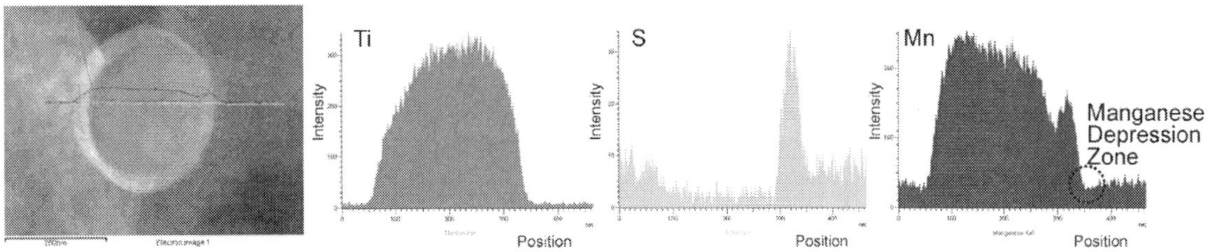

Fig.3 TEM photograph and EDX results for a precipitate appearing in weld metals

It is expected that the most effective balance of titanium and oxygen contents to maximize IGF volume fraction in weld metals should correspond to neither a stoichiometric surplus nor deficit of either element. For instance, if excess oxygen remains once reaction with titanium is complete, the overall silicon and manganese contents will be decreased since oxygen should react with these next and generate compounds such as SiO_2 and MnO (which are mostly discharged from the weld metal as slag). Also, it is known that compounds such as $MnOSiO_2$ and Al_2O_3MnO act as IGF nucleation sites as well as titanium oxides. Therefore, reducing silicon and manganese should restrict IGF nucleation. Moreover, the result investigated by Terashima et al. indicates a grain boundary ferrite increase with increasing oxygen content in weld metals [4]. Hence, it is expected that surplus oxygen lead to a decrease in IGF. For support of this, the specimen B-50 which has low titanium and high oxygen content exhibits much more grain boundary ferrite as shown in Fig. 2. In contrast, if excess titanium remains in the molten weld metal, TiN will be generated. Although TiN provides a preferred ferrite nucleation site, it will also increase bainite microstructures since it is known that TiN distributes typically at the austentite grain boundaries, which promotes grain boundary nucleation. This is supported by microstructures observed in specimen AA-15, which has a high titanium and low oxygen content, and comprises more bainite structures than other specimens.

The relationship between the ratio of molar number between titanium and oxygen when their ratio is 2:3, and

IGF volume fraction in the weld metals is shown in Fig. 4, including the data reported by Seo et al. for comparison [6]. The ideal ratio of molar number between titanium and oxygen, $R_{Ti2/O3}$, for nucleation of Ti_2O_3 is given by:

$$R_{Ti_2/O_3} = \frac{Ti/2Ti_m}{O/3O_m}$$

Where;

Ti = Titanium content in the weld metals (wt.%), Tim = Atomic weight for Titanium (= 47.867)

O = Oxygen content in the weld metals (wt.%), Om = Atomic weight for Oxygen (= 15.999)

When this $R_{Ti2/O3}$ value is one, this represents the condition when both titanium and oxygen are in the ideal ratio to generate Ti_2O_3. The figure demonstrates that the curves have a peak when the ratio is approximately one, ie: at the ideal 2:3 ratio of Ti:O, which agrees with the hypothesis. The difference between the two curves can be attributed to a difference in either/both silicon or/and manganese content in the weld metals (silicon and manganese content in the present study are 0.45-0.5 wt.% and 1.57-1.79 wt.% respectively; however, while the work of Seo et al. involved silicon and manganese values of 0.20-0.23 wt.% and 1.28-1.31 wt.% respectively). Also, minor differences in the definition of IGF volume fraction may be noted (where the present work defines IGF as the sum of Widmanstätten ferrite, polygonal ferrite, and bainitic ferrite). Although this ratio to maximize IGF fraction in the weld metals cannot be universally applied because it does not consider other elements which can affect IGF nucleation (such as aluminum), it can be a useful guideline to determine titanium and oxygen balance in similar weld metals.

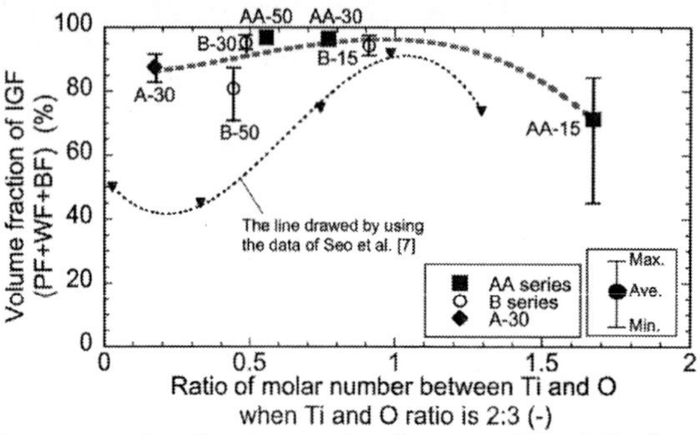

Fig.4 Relationship between ratio of molar number between Ti and O when Ti and O ratio is 2:3

Conclusions

The present study investigated the ideal balance of titanium and oxygen content in the weld metals in order to maximize IGF volume fraction. The results are summarized as follows;

1. Titanium and oxygen content in the weld metals could be controlled by in-situ alloying using thin titanium wire additions and employing varying Ar/CO_2 ratio shielding gas in order to generate various weld joints with different chemistries and microstructures.
2. It was proposed that IGF volume fraction is maximized when titanium and oxygen content neither surplus nor deficit stoichiometrically, in the target ratio of 2:3 to generate Ti_2O_3.

Acknowledgement

The authors are grateful to Nippon Steel & Sumikin Engineering Co., Ltd (Japan), TransCanada Pipelines, and the Natural Sciences and Engineering Research Council (NSERC) of Canada for supporting this study.

References

[1] T. Koseki: "A Review on Inclusion-Assisted Microstructure Control in C-Mn and Low-Alloy Steel Welds." *Welding in the World,* Vol.49, No.5-6 (2005), pp.22-28.

[2] T. Yamada, et al.: "Relation between inclusion surface and acicular ferrite in low carbon low alloy steel weld. " *ISIJ international,* Vol.*49*, No.7 (2009), pp.1059-1062.

[3] K. Yamamoto, et al.: "Effect of Boron on Intra-granular Ferrite Formation in Ti-Oxide Bearing Steels." *ISIJ international* Vol. 36, No.1 (1996), pp.80-86.

[4] S. Terashima, and H. K. D. H. Bhadeshia: "Changes in toughness at low oxygen concentrations in steel weld metals." *Science and Technology of Welding and Joining,* Vol.11, No.5 (2006), pp.509-516.

[5] J.S. Seo, et al.: "Effect of Ti addition on weld microstructure and inclusion characteristics of bainitic GMA welds." *ISIJ international* Vol. 53, No.5 (2013), pp.880-886.

[6] J.S. Seo, et al.: "Effect of titanium content on weld microstructure and mechanical properties of bainitic GMA welds." *Welding in the World* Vol. 58, No.6 (2014), pp.893-901.

[7] G. Thewlis: "Classification and quantification of microstructures in steels." *Materials Science and technology* Vol. 20, No.2 (2004), pp.143-160.

In-situ investigation of solidification behaviour of an Al-containing Transformation Induced Plasticity steel melt pool

G. Agarwal, M.J.M. Hermans and I.M. Richardson

Department of Materials Science and Engineering, Faculty of 3mE, Delft University of Technology, Mekelweg 2, 2628CD, Delft, The Netherlands

g.agarwal@tudelft.nl, m.j.m.hermans@tudelft.nl and i.m.richardson@tudelft.nl, phone +31 15 2786 036, +31 15 2782 286 and +31 15 2785 086

M. Amirthalingam

Department of Metallurgical and Materials Engineering, Indian Institute of Technology Madras, Chennai, 600036, India

murugaiyan@iitm.ac.in, phone +91 44 2257 4784

Abstract

In-situ investigation of solidification of a TRIP steel melt pool was carried out using a high temperature laser scanning confocal microscope. Disk shaped thin specimens were heated in an infrared furnace by a 1.5 kW halogen lamp. A liquid melt pool was formed at the center of the specimen while the outer rim remained solid. A cooling rate of 5 K min^{-1} was applied to study the solidification of the liquid pool. Solidification starts with the formation of delta ferrite; subsequently, the peritectic reaction occurs in which austenite starts forming at the interface of the delta ferrite and the liquid. It is found that the peritectic transition is incomplete and delta ferrite stabilises at the newly formed austenite grain boundaries at relatively low temperatures. Incomplete peritectic transition could be due to the partitioning of Al from the melt pool to the solidified delta-ferrite.

Introduction

Advanced high strength steels (AHSS) are rapidly being implemented in automotive applications [1]. AHSS are evolving, with increases in strength and formability, helping the automotive industry to provide safe, economic and environmentally efficient transportation. One of the steels belonging to the AHSS family are the TRIP steels, these steels rely on the transformation of austenite into martensite during deformation to achieve increased ductility and strength. In the 1980s this effect was demonstrated in low alloyed steels containing 0.2 C, 1-2 Mn and 1-2 Si (all in wt. %) [2–4]. A typical microstructure of TRIP steel consists of 40-60 wt. % allotriomorphic ferrite, 20-30 wt. % bainite and 10-20 wt. % retained austenite. Silicon, in these steels inhibits the formation of cementite, making more carbon available to stabalise the austenite. However, increase in the Si content beyond (0.5 wt. %) reduces the galvanisability by forming silicon oxide [5]. Both Al and P, also retard the cementite formation and can partially substitute Si. Recently, Chatterjee *et al.*[6] and Yi *et al.*[7] have designed steels containing higher Al contents in which the delta-ferrite remains stable at the end of solidification. Amirthalingam *et al.* found delta-ferrite at the fusion boundaries of welded TRIP steel [8]. This could be advantageous as a fully martensitic microstructure can be avoided in the heat affected zone.

The attributes of the peritectic transition like the solidification temperature range, phase stability, segregation of alloying elements *etc.* affect the weld metal strength during the final stages of solidification and in particular it's susceptibility to solidification cracking. In this study, *in-situ* obervation of the peritectic transition during the solidification of an Al (1.1 wt. %) containing TRIP steel was made using a high temperature laser scanning confocal micrcoscope (HTLSCM).

This technique is frequently used to study *in-situ* solidification events like peritectic transformations and solid-state transformations, details of which can be found in the literature [9][10].

Experimental

A commercial TRIP steel sheet with composition 0.19 C, 1.63 Mn, 1.1 Al, 0.35 Si, 0.089 P and 0.019 Cr (all in wt. %) was examined in this study. *Fig.1* shows the equilibrium phase fraction of this steel calculated using Thermo-Calc software with TCFE6 database. Circular disk specimens with a diameter of 10 mm and a thickness between 200 µm - 250 µm were prepared using electro-discharge machining. Each specimen was placed in an alumina crucible, which in turn was held in a platinum holder. Specimens were placed at the upper focal point of a gold plated ellipsoidal cavity in an infrared furnace beneath a quartz view port (see *Fig.2*) under an ultra-high purity inert atmosphere, >99.9999% Ar. A B-type thermocouple wire was welded to the platinum holder. A 1.5 kW halogen lamp located at the lower focal point in the cavity heats the specimen by radiation. The power input to the halogen lamp is controlled by an Omron ES100P digital PID controller, which in turn was connected to the thermocouple at the crucible holder for a feedback signal. The temperature measured by the thermocouple incorporated in the crucible holder was recorded while simultaneously the images were taken at a rate of 30 frames per second. A melt pool with a diameter of 4 mm was obtained at the centre of the specimen while the outer rim remained solid. During cooling at a rate of 5 K min^{-1} until 1423 K, solidification progresses in a concentric manner.

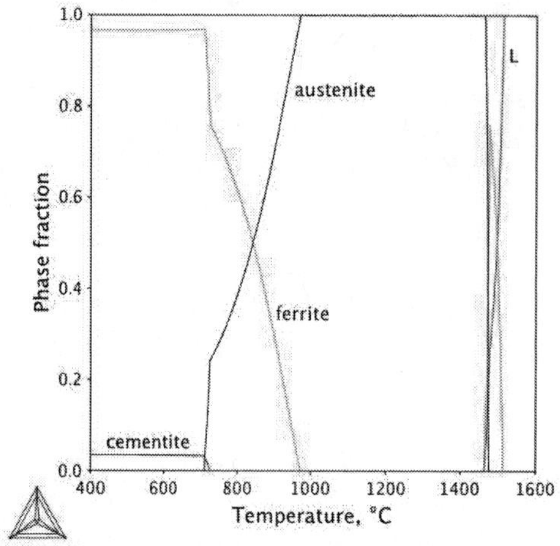

Figure 1: Equilibrium phase fraction

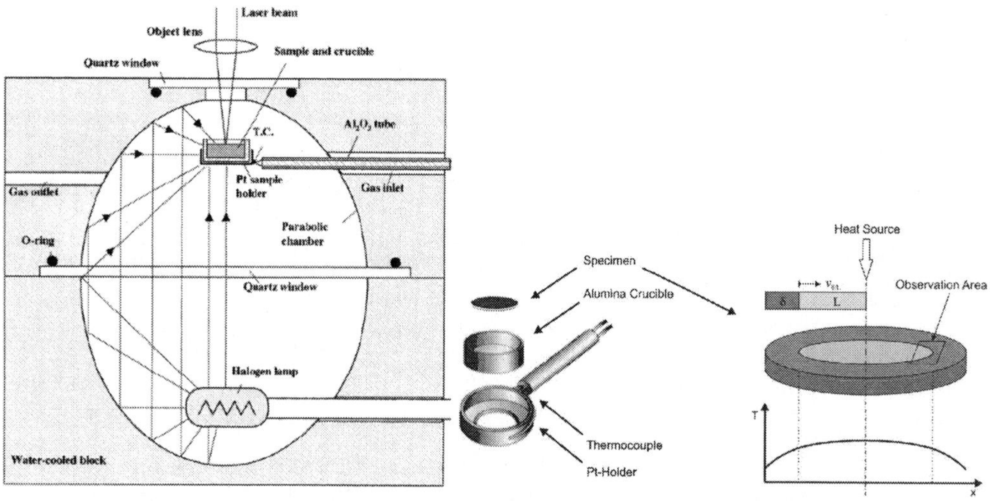

(a) Schematic of the infrared furnace

(b) Crucible holder and specimen geometry

Figure 2: Experimental setup[11]

Results and discussion

Peritectic transition

The sequence of solidification events is shown in *Fig.3*. Solidification begins with the formation of the delta-ferrite (δ) phase at the interface of the melt pool and the outer solid rim (*Fig.3b*). Due to the low cooling rate (5 K min^{-1}), the solidification front was mostly observed to remain planar. On further cooling, peritectic reaction occurs and austenite (γ) forms at the liquid (L)-δ interface (*Fig.3c*). At this stage all the three phases (L, γ and δ) are in contact and the liquid and δ phases react to form γ. This reaction is rapid and currently believed to be diffusion controlled [11]. As the reaction progresses, a thin film of γ phase separates the liquid and the δ phase. Subsequently, further formation of the γ phase occurs in the liquid and δ phases, growing concurrently and γ thus forming planar L/γ and δ/γ interfaces. Additionally, γ was observed to form both at the grain boundaries and intragranular regions of the δ phase (*Fig.3d*). The L/γ interface propagates on an average at a faster rate (0.9 µm s^{-1}) than the δ/γ interface (0.25 µm s^{-1}). Upon further cooling, the austenite transforms to a combination of ferrite and cementite. It is found that even at low temperatures, the δ to γ phase

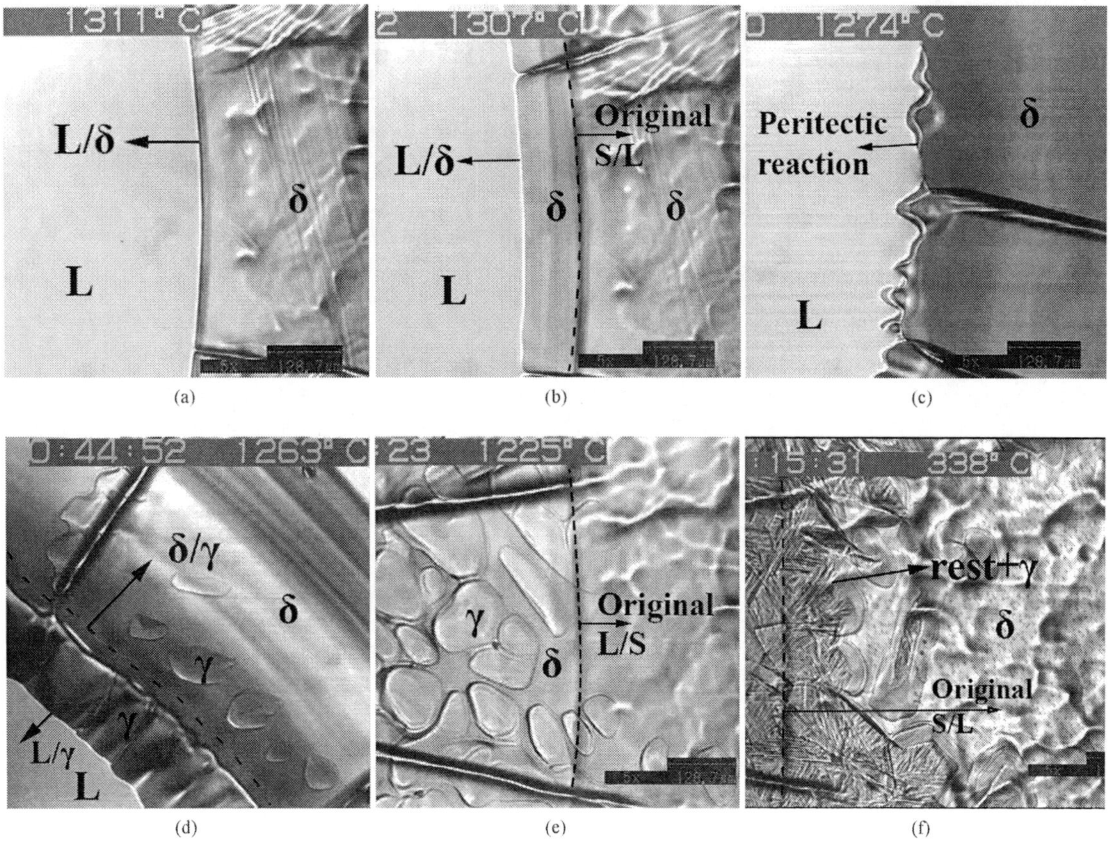

Figure 3: (a) Melt pool surrounded with the δ phase, (b) Solidification starts, δ phase forms at the L/δ interface, (c) Peritectic reaction occurs, γ forms at the L/δ interface, (d) Subsequently, γ forms in the L and δ phase. L/γ interface propagates faster than the δ/γ interface, (e) γ continues to grow in the δ matrix. δ to γ phase transformation is incomplete. L/δ fusion boundary is also indicated, (f) δ phase remains in a metastable state at low temperature. The temperature indicated in the images is measured at the periphery of the platinum holder and is not the temperature of the specimen.

reaction is incomplete and delta ferrite exits as a metastable state (*Fig.3f*). The retention of the δ phase at low temperatures does not conform with the equilibrium phase fraction calculation. In general, during solidification the solute atoms are partitioned from the solid into the liquid phase. However, Al has a higher solubility in δ phase compared to the liquid as seen in *Fig.4* [12]. This promotes the direct solidification of γ from the liquid phase since the diffusion of solutes from δ phase is not required [13]. The argument is supported from our observations as γ grows into the liquid phase at a faster rate than the decomposition of the δ phase, which can explain the retention of δ phase at low temperatures.

Conclusion

The peritectic phase transition of an Al containing TRIP steel has been studied *in-situ*. δ phase was observed to exist at low temperatures in contrast to the equilibrium phase fraction calculation. The direct solidification of γ in the liquid phase was found to occur faster than the δ phase decomposition. This could be due to the depletion of Al in the liquid phase during solidification.

Acknowledgements

This research was carried out under project number F22.8.13485a in the framework of the Partnership Program of the Materials innovation institute M2i (www.m2i.nl) and the Foundation for Fundamental Research on Matter (FOM) (www.fom.nl), which is part of the Netherlands Organisation for Scientific Research (www.nwo.nl). The authors would like to thank Professor R. Dippenaar and Dr. S.C. Moon for access to instrumentation and assistance with the HTLSCM experiments at the University of Wollongong, Australia.

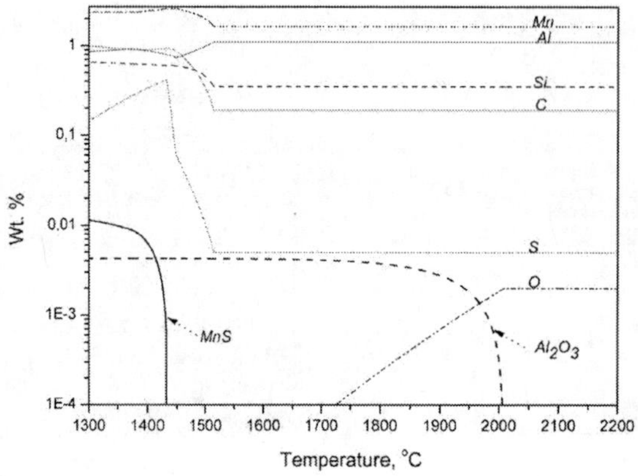

Figure 4: Equilibrium element content in the melt pool of TRIP steel, indicating that once the solidification starts, Al starts partitioning from the melt pool to the solidified δ phase [12].

References

[1] Steel is the solution to achieve 54.5 MPG by 2025" http://www.autosteel.org/research/growth-of-ahss.aspx.

[2] O. Matsumura, Y. Sakuma, and H. Takechi. Enhancement of elongation by retained austenite in intercritical annealed 0.4C-1.5Si-0.8Mn steel. *Transactions of the Iron and Steel Institute of Japan*, 27(7):570–579, 1987.

[3] O. Matsumura, Y. Sakuma, and H. Takechi. Trip and its kinetic aspects in austempered 0.4C-1.5Si-0.8Mn steel. *Scripta Metallurgica*, 21(10):1301–1306, 1987.

[4] H. Takechi, O. Matsumura, and Y. Sakuma. Japan Kokai Tokyo Koho Japan Patent, 62, 188, 729 1987.

[5] J. Mahieu, J. Maki, B.C. De Cooman, and S. Claessens. Phase transformation and mechanical properties of Si-free CMnAl transformation-induced plasticity-aided steel. *Metallurgical and Materials Transactions A: Physical Metallurgy and Materials Science*, 33(8):2573–2580, 2002.

[6] S. Chatterjee, M. Murugananth, and H.K.D.H. Bhadeshia. δ TRIP steel. *Materials Science and Technology*, 23(7):819–827, 2007.

[7] H.L. Yi, K.Y. Lee, and H.K.D.H. Bhadeshia. Extraordinary ductility in al-bearing δ-TRIP steel. *Proceedings of the Royal Society A: Mathematical, Physical and Engineering Sciences*, 467(2125):234–243, 2011.

[8] M. Amirthalingam, M. Hermans, and I. Richardson. Microstructural development during welding of siliconand aluminum-based transformation-induced plasticity steels-inclusion and elemental partitioning analysis. *Metallurgical and Materials Transactions A: Physical Metallurgy and Materials Science*, 40(4):901–909, 2009.

[9] H. Shibata, Y. Arai, M. Suzuki, and T. Emi. Kinetics of peritectic reaction and transformation in Fe-C alloys. *Metallurgical and Materials Transactions B: Process Metallurgy and Materials Processing Science*, 31(5):981–991, 2000.

[10] M. Reid, D. Phelan, and R. Dippenaar. Concentric solidification for high temperature laser scanning confocal microscopy. *ISIJ International*, 44(3):565–572, 2004.

[11] D. Phelan, M. Reid, and R. Dippenaar. Kinetics of the peritectic phase transformation: In-situ measurements and phase field modeling. *Metallurgical and Materials Transactions A*, 37(3):985–994, 2006.

[12] M. Amirthalingam. *Microstructural development during welding of TRIP steels*. PhD thesis, TU Delft, 2010.

[13] Y. J. Choi. Non-equilibrium solidification of δ-TRIP steel. Master thesis, Graduate institute of ferrous technology, Pohang university of science and technology, 2011.

Analysis of ferrite nucleation on single crystal oxide to understand acicular ferrite formation in steel welds

S. Nambu, R. Ikeda, J. Inoue, T. Koseki
Department of Materials Engineering, The University of Tokyo, Tokyo, JAPAN
nambu@metall.t.u-tokyo.ac.jp

Abstract

To obtain finer microstructures in heat affected zone of steel welds, it is well known that B1-type compounds are effective as a nucleating agent for the austenite-to-ferrite transformation in steel. In this study, the nucleation behavior of ferrite from a B1-type compound was analyzed using single crystal Fe alloy particles formed on low-index planes of single crystal MgO substrate. Thin film of pure iron or Fe-Ni alloys with various Ni concentrations was deposited on the MgO substrate, then heated to obtain spheroidized, submicron single crystal austenite particles, and cooled down to cause the ferrite transformation in the particles. It was confirmed by EBSD that each particle was single crystal at both austenite and ferrite stages, implying that only one nucleation of ferrite occurred at the austenite/MgO interface. It was shown that ferrites having Baker-Nutting (B-N) orientation relationship with MgO form preferably on (100) MgO. It was also revealed by combining XRD measurement that the number fraction of ferrites having B-N orientation relationship with the (100) MgO decreased as the transformation temperature decreased due to the increase of the cooling rate or Ni concentration in alloy.

Introduction

In heat affected zone (HAZ) of steel welds, the degradation of toughness is a huge problem due to the coarsening of the brittle phase nucleated from the grain boundary of coarse-grained austenite. In order to prevent the degradation of toughness in HAZ, the grain refinement by intragranular ferrite (IGF) which nucleates from compound particles dispersed in an austenite grain has been applied. B1-type compounds are effective as a nucleating agent for transformation of IGF because they have good crystal orientation relationship with ferrite (α), called Backer-Nutting (B-N) orientation relationship ($\{001\}_{B1}//\{001\}_{\alpha}$, $<100>_{B1}//<110>_{\alpha}$). Thus many studies have been conducted to clarify the nucleation mechanism of IGF [1-3]. In the previous studies, however, it is very difficult to identify the nucleation site of IGF because there exists many possible nucleation sites at the compound in steels, the possibility that the interested IGF secondary-nucleates from the adjacent IGF, and so on. To solve these problems, we have to focus only on the austenite-to-ferrite transformation at the interface between austenite and compound interested by excluding the nucleation sites such as austenite grain boundary and austenite/ferrite interface. In our study, the fundamental investigation of hetero-nucleation behavior of IGF from a compound was attempted by the following method (Fig. 1); many single crystal austenite particles with submicron size are formed on a low-index plane of single crystal B1-type compound substrate, and then the transformation behavior from single crystal austenite to single crystal ferrite at the interface of compound is observed. By investigating many single crystal particles on the substrate statistically, the method is considered to be useful for quantitating the effect of compound on the hetero-nucleation of IGF that is a stochastic phenomenon. In this paper, the effect of composition and transformation temperature of Fe alloys on the crystal orientation relationship between ferrite/compound and the selection in orientation relationship was investigated using Fe-Ni alloys with various Ni concentrations to change the transformation temperature.

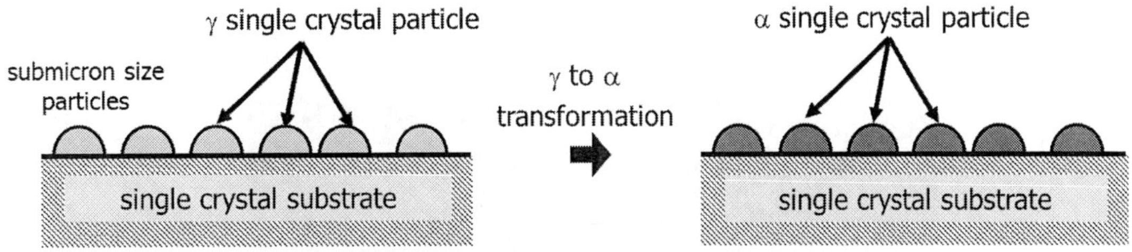

Fig. 1 Schematic image of the method in this study.

Experimental Procedures

Single crystal MgO substrates having low-index planes which are mirror polished were prepared as a B1-type compound. Thin film of pure Fe or Fe-Ni alloys with various Ni concentrations was deposited using electron beam on (100) or (110) plane of the MgO substrate. The thickness of thin film was about 30 nm. After the deposition, the concentrations of Ni in Fe-Ni alloys were investigated by EPMA. The schematic of the procedure is shown in Fig. 2. Firstly, the deposited sample is heat up to 1100 °C at heating rate of 50 °C/s in Ar-3%H_2 gas atmosphere to obtain many single crystal γ particles with spheroidized shape and submicron size on the MgO substrate. Then, the transformation from γ to α is caused by cooling down with various cooling rate. During the cooling process, the transformation start temperature was measured by XRD. After the heat treatment, the crystallographic analysis by SEM/EBSD was also conducted for the obtained single crystal α particles, and the crystal orientation relationship between α/MgO substrate was evaluated quantitatively using 200-400 α particles.

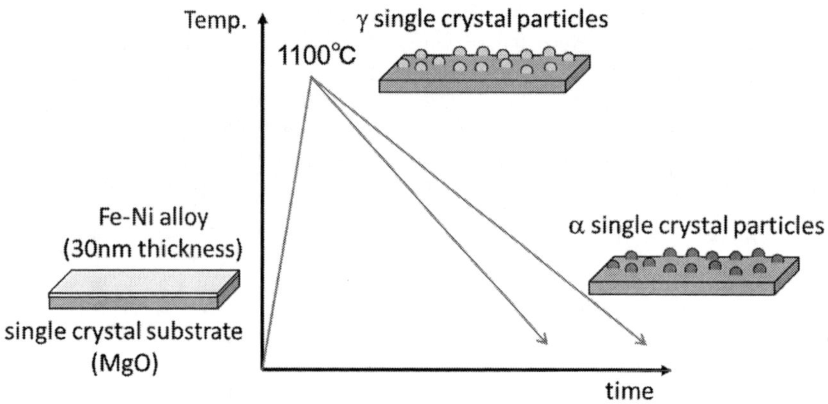

Fig. 2 Schematic image of experimental procedure.

Results and discussion

Ni concentrations of Fe-Ni alloy films deposited on MgO substrates were obtained as 3wt% and 8wt%, so hereafter the samples having these Ni concentrations are called 3Ni and 8Ni, respectively. Figure 3 shows the SEM observation result and its crystal orientation analysis by EBSD for α particles after the heat treatment. As shown in Fig. 3a, many submicron sized particles can be formed by the heat treatment and the average grain size is about 0.5 μm. The corresponding inverse pole figure (IPF) map shown in Fig. 3b suggests that these obtained particles are single crystal α. The sample that pure Ni was deposited on MgO substrate was also prepared in order to confirm the crystal orientation and size of single crystal austenite particles because pure Ni has fcc structure during the present process and similar lattice constant with γ-Fe. The result after heat treatment shows that γ particles have submicron grain size and are single crystal randomly oriented to MgO substrate. Consequently, it was confirmed that each particle was single crystal at both γ and α stages, implying that only one nucleation of α occurred at the γ/MgO interface.

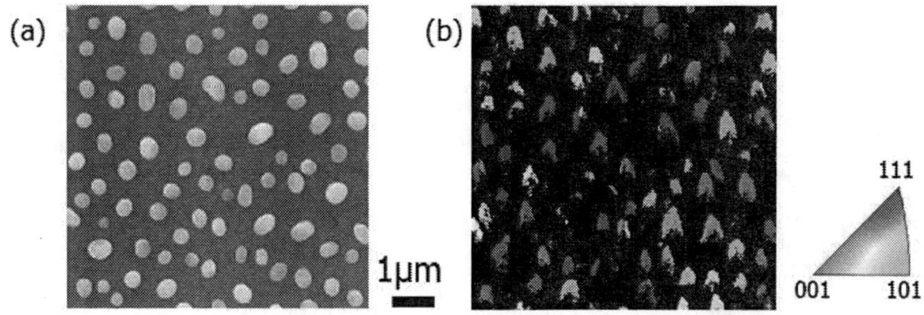

Fig. 3 Obtained ferrite particles on MgO substrate (a) observed by SEM and (b) analyzed by EBSD.

The crystal orientations of obtained α particles for each sample after heat treatment were plotted in the (001) pole figures as shown in Fig. 4. There is clear difference in the crystal orientation relationships between α/MgO owing to the Ni concentration in Fe-Ni alloy, the cooling rate, and the plane of MgO substrate. The specific orientation relationships found in the experimental results are also plotted in the simulated pole figures: B-N orientation relationship, OR1 ($\{001\}_{B1}//\{221\}_{\alpha}$, $<001>_{B1}//<011>_{\alpha}$), and OR2 ($\{011\}_{B1}//\{211\}_{\alpha}$, $<001>_{B1}//<011>_{\alpha}$) are shown in Fig. 5. It is noted that [010] and [001] directions of MgO parallel to RD and TD, respectively in case of (100) MgO sample, and [001] and [1-10] directions of MgO parallel to RD and TD, respectively in case of (110) MgO sample. The crystal orientation analysis for α nucleated on (100) MgO substrate demonstrates that B-N orientation relationship is preferably selected and OR1 also selected secondly. By increasing of Ni concentration in Fe-Ni alloy, the fraction of these two orientation relationships becomes small. The decrease of the fraction is also found by increasing of cooling rate. The XRD measurement during heat treatment suggests that the transformation temperature also decreases as the increment of Ni concentration and cooling rate. In the case of α nucleated on (110) MgO substrate, OR2 is preferably selected, and the fraction of OR2 becomes small with increasing of Ni concentration and cooling rate as well as (100) MgO substrate case. By comparing the plane of MgO substrate, the fraction of α having specific orientation relationships with (100) MgO is larger than that with (110) MgO. The transformation start temperature of (100) MgO sample is higher than that of (110) MgO sample. This comparison suggests that α/(100) MgO interface is effective for the nucleation of IGF rather than α/(110) MgO interface.

The crystal orientation relationship which shows the best integrity between α/compound is selected preferentially in case of the transformation at relatively high temperature, while α having the crystal orientation relationship which is relatively good integrity also nucleates as the transformation temperature decreases. Furthermore when the transformation temperature decreases, it was indicated that Fe/compound interface acts as an inert interface and α having various crystal orientation relationships between α/compound nucleates.

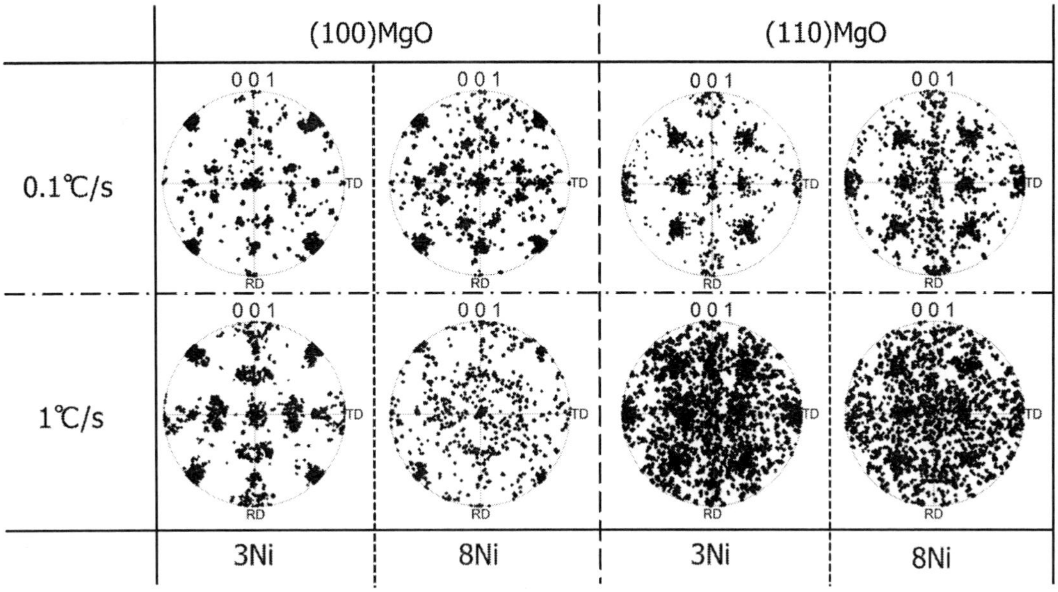

Fig. 4 The obtained (001) pole figures of crystal orientations for single crystal α particles on MgO substrate under various conditions; cooling rate, Ni concentration in Fe-Ni alloy, and plane of MgO substrate.

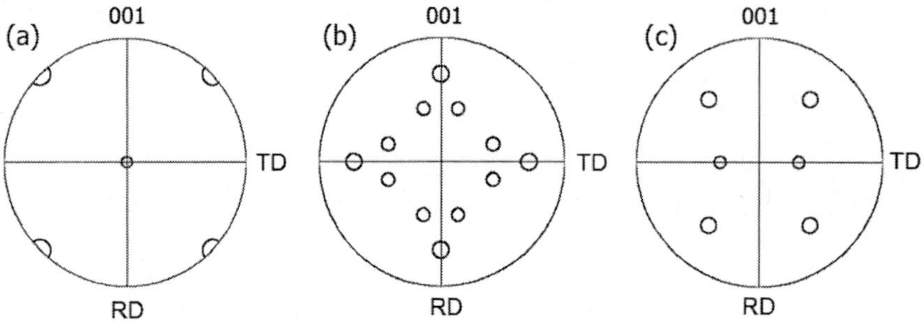

Fig. 5 The simulated (001) pole figures of α for (a) B-N orientation relationship, (b) OR1, $\{001\}_{B1}//\{221\}_{\alpha}$, $<001>_{B1}//<011>_{\alpha}$, and (c) OR2, $\{011\}_{B1}//\{211\}_{\alpha}$, $<001>_{B1}//<011>_{\alpha}$

Conclusions

The nucleation behavior of ferrite from a B1-type compound was analyzed using single crystal Fe alloy particles formed on low-index planes of single crystal MgO substrate. The following conclusions are obtained.

1. Baker-Nutting orientation relationship is the most preferable when ferrites nucleate on (100) MgO substrate.
2. The fraction of the specific orientation relationships having good integrity between ferrite and MgO becomes small with decrease in the transformation temperature due to increase in cooling rate or Ni concentration in Fe-Ni alloy.
3. It is indicated that ferrite/(100) MgO interface is better nucleation agent than ferrite/(110) MgO interface because the transformation start temperature of (100) MgO sample is higher than that of (110) MgO sample.

References

[1] T. Koseki, *et al*: "Inclusion assisted microstructure control in C-Mn and low alloy steel welds," *Mater. Sci. Technol.*, 21 (2005), pp. 867-879.

[2] K. Kasai *et al*: "Transformation behavior of ferrite at steel/B1 compounds interface," *Tetsu-to-Hagane*, 96 (2010) pp. 123-128.

[3] H. Nako *et al*: "Crystal orientation relationships between acicular ferrite, oxide, and the austenite matrix," *ISIJ Int.*, 54 (2014), pp. 1690-1696.

On the effective grain size of steel weld

Hidenori Terasaki, Kohei Hayashi, Yu Miyahara
Kumamoto University, Kumamoto, Japan
terasaki@mech.kumamoto-u.ac.jp, +81-96-342-3857

Mitsuru Ohata
Osaka University, Suita, Osaka, Japan
ohata@mapse.eng.osaka-u.ac.jp, +81-6-6879-7545

Koji Moriguchi, Yusaku Tomio, Kotaro Hayashi
Nippon Steel & Sumitomo Metal Corporation, Tokyo, Japan
moriguchi.5ym.kohji@jp.nssmc.com

Abstract

Effective grain size of steel weld is important property and it is related to heat-input during welding process. Understanding the formation process of effective grain is important to secure the safe weld. In the present study, quantification method of effective grain size using electron backscatter diffraction pattern (EBSD) method was introduced. Effective grain size of weld metal, simulated-coarse grained heat-affected zone (CGHAZ) and Intercritically Reheated Coarse-grain Heat Effected Zone (ICCGHAZ) were visualized using EBSD and it was discussed from viewpoint of toughness of steel weld.

Furthermore, effect of phase transformation temperature on the effective grain size was discussed from viewpoint of variant pairing under the Kurdjumov-Sachs orientation relationships.

Introduction

It is well-known that heat input of welding governs transition temperature of steel weld. Higher heat input cause higher transition temperature.[1] Then maximum heat input should be regulated in welding procedure sheet (WPS). In a high tensile strength steel such as 780MPa grade, the transition temperature highly depends on the heat input of welding. The concept of "effective" grain size is used to rationalize the relation between transition temperature and heat input of welding.

According to recent progress of crystallographic analysis of steel [2,3], the effective grain size can be visualized by using electron back scattering diffraction pattern (EBSD) method [4-8]. The effect of transformation temperature on the effective grain size was well-visualized in simulated heat-affected zone of bainite steel. [5-8]

In the present study, effective grain size of weld metal of HSLA steel and intercritically reheated coarse-grain heat affected zone (ICCGHAZ) are visualized. Furthermore, the effect of phase transformation temperature on the effective grain size was discussed from viewpoint of variant pairing [9] under the Kurdjumov-Sachs orientation relationships, by visualizing the variant pair in simulated CGHAZ of bainite steel.

Experimental method

The detail of EBSD measurement method was descried in elsewhere. [5-8]

The definition of variant (V) pair in Kurdjumov-Sachs orientation relationships (KSOR) was according to the method by Morito *et al* [9]. Their method [9] take into account the shifted KSOR and the experimental misorientation angle and axis were defined. For example, variant pair of V1-V2 pair was defined as a rotation in angle of 68 degree and axis of <011>. In the present study, boundary length of each variant pair was derived and divided by observed area. Then the unit of boundary density is expressed with 'μm^{-1}'.

Boundary length of variant pair was measured using EBSD data of simulated CGHAZ180 [5] and CGHAZ25 [6]. CGHAZ180 means as it takes 25 and 180 seconds to go from 1073 K to 773 K (800 °C to 500°C) in the cooling cycle, as described in previous paper [6]. The chemical composition of used steel was Fe-0.1C-0.3Si-0.25Cu-1.0Ni-0.4Cr-0.4Mo-0.039V (mass%). The peak temperature of CGHAZ25 and 180 was 1350 °C. Simulated ICCGHAZ specimen was prepared in the present study and the first peak temperature was 1400°C and the second peak temperature was 815°C.

Results and Discussions

Boundary density of variant pair in CGHAZ25[6] and CGHAZ180[5] condition was summarized in Table 1. It is clear that boundary density of V1-V2 drastically increased from 0.017 to 0.439. Furthermore, boundary density of V1-V3/V5 and V1-V6 also increased. On the other hand, boundary density of V1-V4 slightly decreased. The increasing of V1-V2, V1-V3/V5 and V1-V6 pair supplies high angle boundaries and refine the effective grain size. Morito *et al* [2] showed all 6 variants showed the lowest shape strain. Lower transformation temperature demand self-accommodation of transformation strain and all 6 variant became active as described above. It is the one of the reasons that transition temperature of high tensile strength steel is low showing small effective grain size, in low heat input condition.

Table 1: Boundary density of variant pair in CGHAZ25 and CGHAZ180 condition.

Variant Pair	Boudary density CGHAZ25 (μm^{-1})	Boudary density CGHAZ180 (μm^{-1})
V1-V2	0.439	0.017
V1-V3/V5	0.163	0.011
V1-V6	0.042	0.003
V1-V4	0.083	0.138

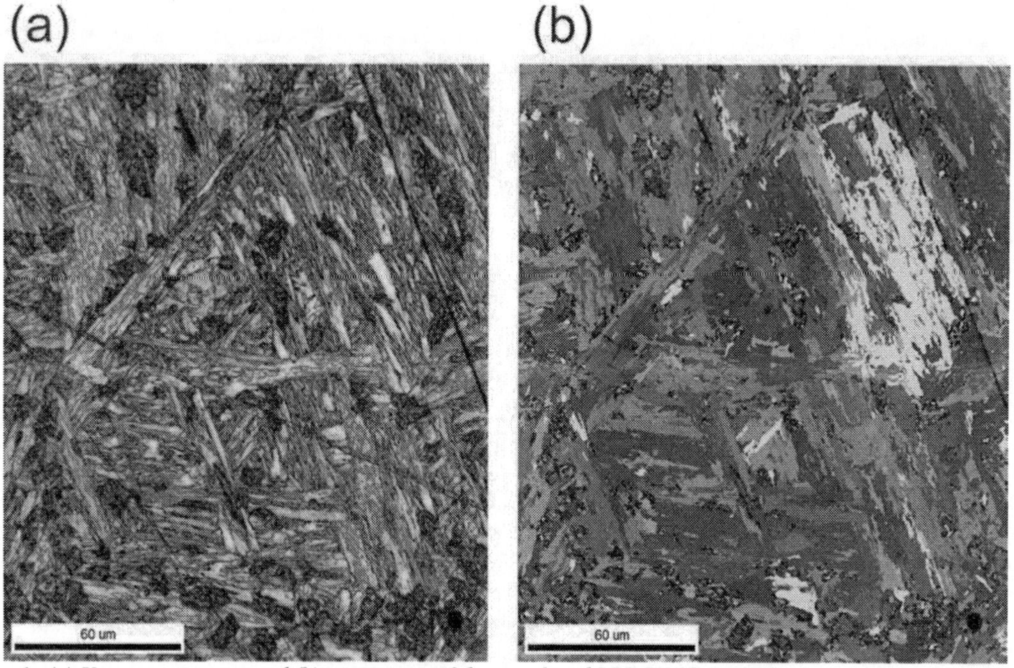

Figure 1: (a) Variant pair map and (b) grain map of the simulated ICCGHAZ specimen

Figure 1 shows (a) variant pair map and (b) grain map of the simulated ICCGHAZ specimen. The simulated ICCGHAZ contain much V1-V2, V1-V3/5 and V1-V6 pair boundaries reflecting the high cooling rate of 15C/s at first cycle. The grain map was formed by setting the grain size of 15 degrees and it represented the effective grain. It is clear that there is correlation between density of V1-V2, V1-V3/5 and V1-V6 pair boundaries and effective grain size.

Conclusions

Effect of phase transformation temperature on the effective grain size was discussed from viewpoint of variant pairing under the Kurdjumov-Sachs orientation relationships. Majority of boundary density of variant pair was transited from V1-V4 pair to V1-V2 pair as weld heat-input decreased. The fraction of boundary density of variant pair well rationalizes the size of effective grain size of steel weld. The accommodation of transformation strain governs variant pair distributions and it determine the effective grain size in steel weld.

References

[1] A. Lambert-Perlade, A.F. Gourgues, J. Besson, T. Sturel, and A. Pineau: "Mechanisms and modeling of cleavage fracture in simulated heat-affected zone microstructures of a high-strength low alloy steel," *Metall. Mater. Trans. A*, Vol. 35A, No. 13 (2004), pp. 1039-1053.

[2] S. Morito, H. Tanaka, R. Konishi, T. Furuhara and T. Maki: "The morphology and crystallography of lath martensite in Fe-C alloys," *Acta. Mater.*, Vol. 51, No. 6 (2003), pp. 1789-1799.

[3] N. Takayama, G. Miyamoto and T. Furuhara: "Effects of transformation temperature on variant pairing of bainitic ferrite in low carbon steel," *Acta. Mater.*, Vol. 60, No. 5 (2012), pp. 2387-2396.

[4] A.F. Gourgues: "Electron backscatter diffraction and cracking" *Mater. Sci. Technol.*, Vol. 18, No. (2002) pp. 119–133.

[5] H. Terasaki and Y. Komizo: "Morphology and Crystallography of Bainite Transformation in a Single Prior-Austenite Grain of Low-Carbon Steel," *Metall. Mater. Trans. A*, Vol. 44A, No. 6 (2013), pp. 2683-2689.

[6] H. Terasaki, Y. Shinotome, A. Takada, Y. Komizo, K. Moriguchi and Y. Tomio: "Visualization and Analysis of Variant Grouping in Continuously Cooled Low-Carbon Steel Welds," *Metall. Mater. Trans. A*, Vol. 45, No. 8 (2014), pp. 3554-3559.

[7] H. Terasaki, Y. Shinotome, Y. Komizo, M. Ohata, K. Moriguchi and Y. Tomio: "Effect of Close-Packed Plane Boundaries in a Bain Zone on the Crack Path in Simulated Coarse-Grained HAZ of Bainitic Steel," *Metall. Mater. Trans. A*, Vol. 46, No. 5 (2015), pp. 2035-2039.

[8] H. Terasaki, Y. Miyahara, M. Ohata, K. Moriguchi, Y. Tomio and K. Hayashi: "Visualization of Microstructural Factor Resisting the Cleavage-Crack Propagation in the Simulated Heat-Affected Zone of Bainitic Steel," *Metall. Mater. Trans. A*, Vol. 46, No. 12 (2015), pp. 5489-5493.

[9] S. Morito, A.H. Pham, T. Hayashi and T. Ohba: "Block Boundary Analyses to Identify Martensite and Bainite," *Materials Today: Proceedings*, Vol. 2, No. 3 (2015), pp. S913-S916.

Spot Joining of AA6061-T6 Aluminum Alloy to Galvanized DP590 Steel by Cold Metal Transfer Arc Welding

HaiYang Lei, YongBing Li*, Zhongqin Lin
State Key Laboratory of Mechanical System and Vibration & Shanghai Key Laboratory of Digital Manufacture for Thin-walled Structures, Shanghai Jiao Tong University, Shanghai, P.R. China.
E-mail: yongbinglee@sjtu.edu.cn, Tel.: 86-21-34206305
Blair E. Carlson
Manufacturing Systems Research Lab, General Motors Research & Development Center, Warren, MI, USA
E-mail: blair.carlson@gm.com

Abstract
In order to save energy and reduce emission, lightweight materials like aluminum alloys were introduced into steel car body, which requires the development of effective joining processes. In this paper, an arc spot joining process was proposed to join aluminum alloy AA6061-T6 to hot dip galvanized (HDG) advanced high strength steel (AHSS) DP590 based on cold metal transfer (CMT) welding process. Three different welding modes, which were direct welding (DW) mode, plug welding (PW) mode, and edge plug welding (EPW) mode, were proposed and investigated in comparison. Results showed that EPW mode could homogenize the welding heat input into the weld pool, and thus produce the welds with best mechanical performance for optimal joint morphology and smooth IMC layer among the three welding modes.

Introduction
The application of lightweight materials in the automobile body assembly is the trend nowadays to save energy and reduce CO_2 emissions, which drives the hybrid use of aluminum and steel. Resistance spot welding (RSW) is the most popular welding process and fusion brazing the aluminum to steel is preferred [1]. However, the thickness of the hard and brittle Fe_xAl_y intermetallic compound (IMC) is hard to control [2-4] and large electric current is required. Friction stir welding (FSW) is a solid state joining process with significantly lower thermal effect [5], but the process hole left in the joint and high wear rate limit its application. Self-piercing riveting (SPR) is a double-sided, mechanical joining process with no thermal effect which was widely used in joining aluminum to steel [6]. However, for riveting ultra-high strength steel (UHSS) to low ductility materials, the SPR process faces great challenges [7].

Cold metal transfer (CMT) process is a modified arc welding process with low heat input which has produced an acceptable joint quality in seam welding of aluminum to galvanized steel [8]. However, seam welding distortion is inevitable due to the different thermal expansion coefficient of aluminum and steel. In this study, to CMT spot welding aluminum alloys and galvanized DP590 steel, three spot welding modes, i.e., direct welding (DW), plug welding (PW), and edge plug welding mode (EPW), were proposed and investigated systematically.

Experimental Procedure
The materials used in this study were 1 mm thick AA6061-T6 aluminum alloy containing 0.8%~1.2% Mg and

0.4%~0.8% Si and 1.2 mm HDG (Hot Dip Galvanized) DP590 steel. 4043 Al-Si welding wire containing ≤5% Si with a diameter of 1.2 mm was adopted as the filler metal. Fronius Transpuls Synergic 3200 CMT welding system integrated with an industrial robot was set up for the arc spot welding process, as shown in Fig. 1(a). Shielding gas for the arc spot welding process was 99.999% pure Argon with a flow rate of 20 L/min. Three different welding modes, i.e., direct welding (hereafter DW for short), plug welding mode (hereafter PW for short) and Edge plug welding mode (hereafter EPW for short) were proposed in this study, as presented in Fig. 1(b). A welding time of 0.8 s was used for all three modes and a circling speed of 32.6 mm/s was set for EPW mode. The control of the wfs (from 4.0 to 7.2 m/min), welding current and voltage are integrated together in the CMT welding power source

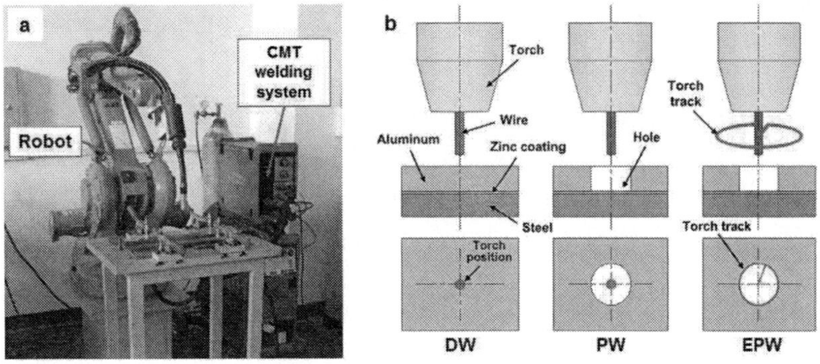

Fig. 1 (a) CMT robotic welding system, (b) illustration of the three welding modes.

Results and Discussion

The effect of wire feed speed, wfs, on the joint strength of lap shear testing for the three welding modes was studied, as presented in Fig. 2. For DW mode at low wfs, the joint strength is very unstable and the fracture mode is partial interface fracture. With the increase of the wfs over 6.6 m/min, the joint strength increased as a whole and fractured entirely from the interface. For PW mode, the joint strength is also unstable and the fracture mode was full interfacial fracture. The joints at EPW mode with homogenized welding heat input present best performance. The strongest joint of EPW mode could reach 4.0 kN at the wfs of 5.6 m/min, and the fracture mode was button pullout. The minimum average strength could reach 3kN at the wfs of 4.2 m/min and partial thickness fracture was obtained. At larger welding heat input, the joint strength decreased and the joint fractured from the interface like DW mode.

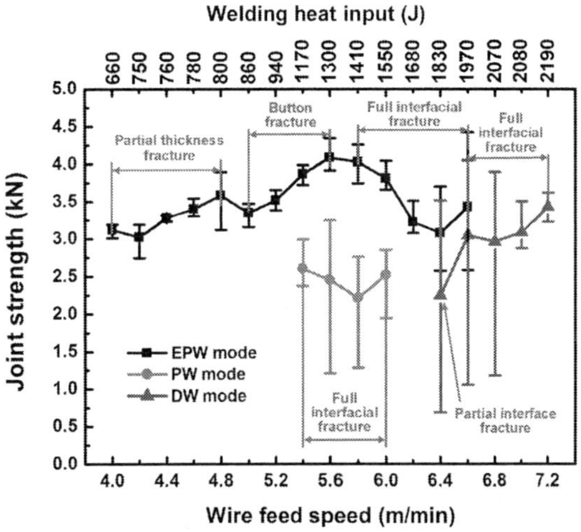

Fig. 2 Effects of welding modes and wire feed speeds on joint strength [9].

Fig. 3 presents the typical weld appearance, cross-section and microstructure of CMT welded 1 mm thick AA6061-T6 aluminum alloy and 1.2 mm thick HDG DP590 steel joints with DW, PW and EPW modes, respectively. As shown in Fig. 3(a), for DW mode at the wfs of 6.4m/min, the diameter of the joint was the biggest and serious welding defects such as partial tearing and incomplete brazing could be found in the weld due to the largest welding heat input. The weld diameter of PW mode at the wfs of 5.4 m/min was the smallest, whereas the defect-free EPW mode weld was larger than PW mode since more aluminum base metal melted during the circling welding around the hole, as shown in Figs. 3(b) and 3(c).

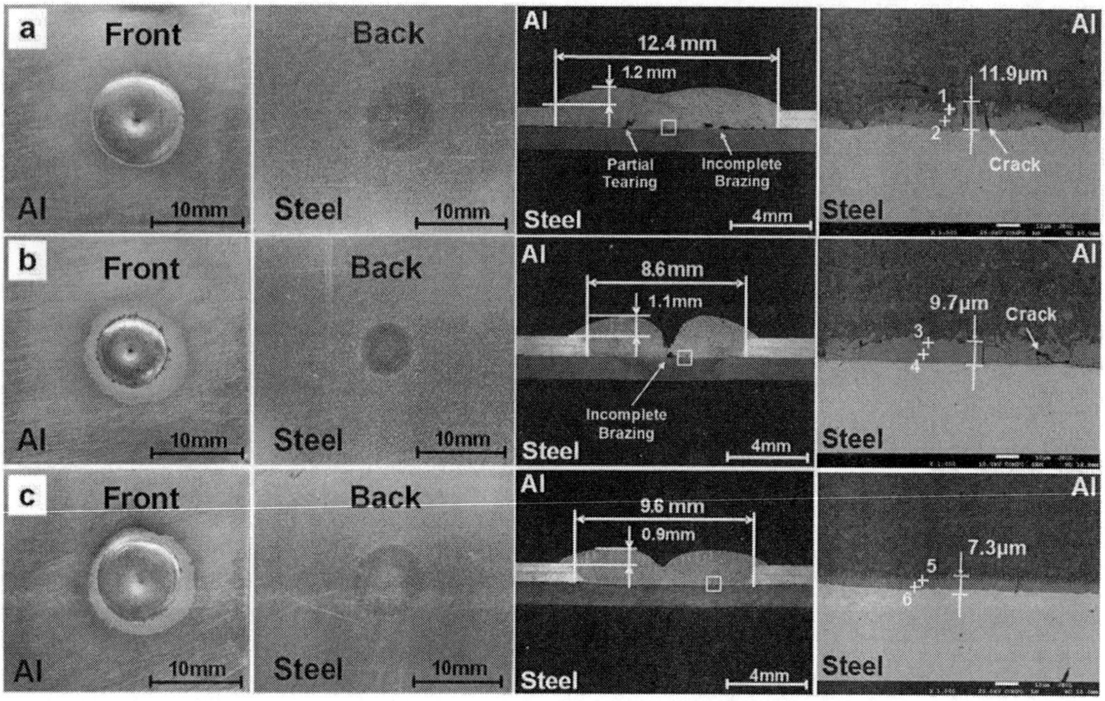

Fig.3 Macro morphology, cross-section and microstructure of the joints with three welding modes at different wfs: (a) DW mode, 6.4 m/min, (b) PW mode, 5.4 m/min, (c) EPW mode, 5.6 m/min.

The thickest part of the IMC layers under each welding mode within the box area were observed by SEM, as shown in Fig. 3. For the mode of DW and PW, refer to Figs. 3(a) and 3(b), the IMC layer was much thicker due to the concentrated heat input in the weld center, which also caused the serious crack in the IMC layer. For EPW mode, refer to Fig. 3(c), the IMC layer is relatively thin and uniform in thickness.

EDS analysis was conducted to further identify the chemical composition of different positions of the IMC layer for each welding mode with the reference to Fig.3. For DW and PW mode, the layer consisted of two layers. The phase on the top of the layer, i.e., positions 1 (75.84% Al, 21.84% Fe, 2.32% Si and 0% Zn) and 3 (76.08% Al, 16.60% Fe, 3.27% Si and 0.05% Zn), was $FeAl_3$ with higher aluminum content, whereas the phase at the bottom layer, i.e., positions 2 (46.06% Al, 50.45% Fe, 2.96% Si and 0.53% Zn) and 4 (62.37% Al, 34.83% Fe, 2.59% Si and 0.21% Zn), near the steel side was Fe_2Al_5. For the mode of EPW, only Fe_2Al_5 formed in the thin IMC layer, i.e., positions 5 (62.58 % Al, 33.33% Fe, 3.70% Si and 0.39% Zn) and 6 (50.13% Al, 46.69% Fe, 2.42% Si and 0.76% Zn), moreover, the content of zinc in the IMC layer increased from the top to the bottom. Very little zinc was found in the IMC layer of DW and PW mode weld due to the evaporation of zinc coating on the steel surface under the extensive welding heat input acting in the weld center.

Conclusions

(1) Welding with DW and PW mode caused large heat input and severe welding defects like incomplete brazing and uneven IMC layer along the faying surface, and thus produced unstable and brittle joints. Welding with EPW mode could homogenize the heat input into the hole, produced a uniform and thin intermetallic compound, and thus resulted in joints of highest strength.

(2) The best joint strength, 4.1 kN, could be obtained at a wfs of 5.6 m/min (heat input of 1300 J) which correlated to a maximum IMC peak layer thickness of 7.3 μm.

Acknowledgments

The authors would like to acknowledge the supports of GM Research and Development Center, National Natural Science Foundation of China (Grant Nos. U1564204, 51322504, 51275300 and 51421092), and the Program of Introducing Talents of Discipline to Universities (Grant No. B06012).

References

[1] Ranfeng Qiu, Chihiro Iwamoto, Shinobu Satonaka: "Interfacial microstructure and strength of steel/aluminum alloy joints welded by resistance spot welding with cover plate," *Journal of Materials Processing Technology*, Vol.209 (2009), pp. 4186–4193.

[2] Metals Handbook, Vol.63, 10th Edition, ASM International, (1992).

[3] A. Bouayad, Ch. Gerometta, A. Belkebir, A. Ambari: "Kinetic interactions between solid iron and molted aluminum," *Materials Science and Engineering A*, Vol. 363 (2003), pp. 53-61.

[4] E. Schubert, M. Klassen, I. Zerner, C. Walz, G. Sepold: "Light-weight structures produced by laser beam joining for future applications in automobile and aerospace industry," *Journal of Materials Processing Technology*, Vol.115 (2001), pp. 2-8

[5] Takehiko Watanabe, Hirofumi Takayama, Atsushi Yanagisawa: "Joining of aluminum alloy to steel by friction stir welding," *Journal of Materials Processing Technology*, Vol.178 (2006), pp. 342–349.

[6] R. Cacko, Czyżewski, A Kocańda: "Initial optimization of self-piercing riveting process by means of FEM," *Steel Grips*, Vol. 2 (2004), pp. 307–310.

[7] M Lou, YB Li, YT Li, GL Chen: "Behavior and Quality Evaluation of Electroplastic Self-Piercing Riveting of Aluminum Alloy and Advanced High Strength Steel," *Transactions of the ASME, Journal of Manufacturing Science and Engineering*, Vol.2, No.135 (2013), pp. 1-9.

[8] H.T.Zhang, J.C.Feng and P.He: "Interfacial phenomena of cold metal transfer (CMT) welding of zinc coated steel and wrought aluminum," *Materials Science and Technology*, Vol.24 (2008), pp. 1346-1349

[9] HaiYang Lei, YongBing Li, ZhongQin Lin, Blair E Carlson: "Cold Metal Transfer Spot Joining Of AA6061-T6 To Galvanized DP590 Under Different Modes," *Transactions of the ASME, Journal of Manufacturing Science and Engineering*, Vol.137, No.5 (2015), pp. 051028-051028-10

[10] HaiYang Lei, YongBing Li, Blair E Carlson, ZhongQin Lin: "Microstructure and Mechanical Performance of Cold Metal Transfer Spot Joints of AA6061-T6 to Galvanized DP590 Using Edge Plug Welding Mode,," *Transactions of the ASME, Journal of Manufacturing Science and Engineering*, Vol.138, No.7 (2016), pp. 071009-071009-13

Changes in weld metal microstructure and hardness due to addition of boron trioxide in flux during submerged arc welding on modified 9Cr-1Mo (P91) steel.

Joydeep roy[1], Ram Naresh Rai[2] and Subhash Chandra Saha[3]

Assistant professor, Department of Mechanical Engineering, National Institute Technology Agartala [1], Tripura, India, deeproymech@gmail.com
Associate professor, Department of Production Engineering, National Institute Technology Agartala[2], Tripura, India, nareshray@yahoo.co.in
Professor, Department of Mechanical Engineering, National Institute Technology Agartala[3], Tripura, India, subashchandrasaha@yahoo.in

Keywords: P91 steel, Submerged Arc Welding, hardness, boron trioxide

Abstract:

The present study investigates the influence of boron on microstructure and hardness of P91 steel weldment in submerged arc welding process. Boron trioxide powder was mixed with fresh commercial flux in different ratio (2.5-12.5%) to manipulate the weld metal chemical composition. Welding parameters were kept constant. Optical emission spectroscopy results showed that wt% of B varies from 0.00 to 0.0068 with addition of boron trioxide into welding flux. Weld metal microstructure without addition of Boron, reveals the presence of tempered martensite and ferrite phase, whereas B added microstructures highlighted the dominating nature of tempered martensite. SEM analysis also showed that the dispersed carbides especially $M_{23}C_6$ were increased with addition of Boron. A macrohardness profile has been drawn across both columnar and refined regions on a transverse section of P91 weld metal in the as-post weld heat treatment (PWHT) condition. It was found that columnar region exhibits a much higher hardness than the refined region. The hardness values in the columnar regions vary from 259 to 343 HV2, whilst the hardness values in the refined regions decreases gradually in WM and become steady at the base material. Improvement of hardness has positive effect on the creep strength which is the main objective of this material.

Introduction:

The key factors in protecting the global environment are the energy security and lower carbon dioxide emissions. In modern fossil-fired power-plant, the application of grade 91 steels allows higher operating temperatures and pressures, and therefore higher efficiencies [1]. During the development of steels for advance power plant applications at 650°C, it is shown that 9Cr based steels are superior compared to 12Cr steels due to their good high temperature mechanical properties and weldability [2]. The modified 9Cr 1Mo steel known as **P91/T91** has been successfully accepted world-wide since the late 80's for header and steam piping in power industries due to its excellent elevated temperature strength and creep behavior [3].

In order to achieve excellent high temperature strength, many kinds of strengthening elements have been added to the conventional Cr-Mo steels. In P91, the stable chromium carbide in the precipitation sequence is $M_{23}C_6$. The creep resistant properties of P91 are attributable to the pinning of dislocations and grain boundaries by these secondary phase precipitates [4]. Ti addition combined with C and N promotes the precipitation of Ti-MX particles. Ti-MX particles have an extremely high stability and thus a high potential for strengthening of martensitic/ferritic steels. Ti precipitates showed a very slow coarsening rate [5]. In many of the 9-12%Cr steels, 0.005-0.01% boron is added to improve the creep strength. It has been found that boron can segregate to the interface between $M_{23}C_6$ particles and the matrix to decrease the rate at which the carbide can coarsen, thus stabilizing the microstructure, since $M_{23}C_6$ particles help in pinning the sub grain boundaries [6]. It is also found that boron assists in the nucleation of VN particles by the mechanism of latent creep resistance [7]. Most of the boron atoms are in solution. They can combine with vacancies to reduce the self-diffusion coefficient of ferrite matrix. They can also form a Cottrell atmosphere around dislocations and thus increase creep deformation resistance [8]. The above study revealed that very little research works have been done on various aspects of boron on P91 steels. So the aim of the present investigation is to study the influence of boron trioxide additions in the fluxes on the microstructure and hardness in submerged arc welds.

Experimental Details:

Square butt welding was performed in a submerged arc welding machine [Model- MAESTRO 1200 (F)] available in NIT Agartala, India workshop on 10mm thick Modified 9Cr-1Mo (P91) steel plate. B_2O_3 metal powder was mechanically mixed with the commercial fluxes in amounts of 2.5%, 5%, 7.5%, 10% and 12.5%. Consumables for P91 steel is supplied by D&H Scheron electrodes pvt. Ltd. Electrode Wire AWS.SFA.5.23 EB9 having diameter is 3.15mm Flux used for welding is max flux SAF12. Welding process parameters remain same for all the welding conditions viz. Wire feed rate 200 mm/min, stickout 28mm, traverse speed 0.45m/min, Voltage 34V and Current 625A. Welding was performed on six sets of P91 steel plates and the average value was taken for the discussion.

The pre heat temperature of 200-250°C is maintained throughout the welding. A PWHT is performed at 760°C for 3 hours then cooled into room temperature in the furnace. For metallographic study the welded specimens were polished followed by standard metallographic procedure. Polished specimens were etched with a mixture of Viella's solution to reveal the various zones of weldments. Grain size and bead geometry parameters were measured by using an optical microscope [Leica DMI 3000M] attached with an image analyzing software. Linear intercept method has been used for the grain size measurement according to ASTM E112. Scanning electron microscopy [SEM] was carried out for the microstructural analysis. The macrohardness tests were carried out in a Matsuzawa MMT-X7B Vickers macrohardness machine throughout the weldment with a constant load of 2 kg and dwell time 10 s.

Results and Discussion:

Changes in Weld metal Chemical composition:

The weld metal chemical composition analysis was conducted by optical emission spectroscopy and the results are provided in table 1. The weld metal chemical composition results show that the values for each element are fairly uniform, except the boron, which varied systematically from 0.00 to 0.0068%.

Table 1 Weld metal chemical composition with various B_2O_3 (wt%).

Element	C	Mn	Si	P	S	Cr	B	V	Mo
	Nb	Al	N						
Parent Metal	0.105	0.432	0.12	0.014	0.0008	8.975	0.00	0.202	0.855
	0.075	0.013	0.0036						
B-0	0.096	0.55	0.36	0.014	0.0071	9.82	0.00	0.201	0.96
	0.077	0.018	0.0056						
B-2.5	0.092	0.52	0.349	0.013	0.0072	9.59	0.0018	0.196	0.97
	0.074	0.016	0.0059						
B-5	0.092	0.53	0.323	0.016	0.0077	9.92	0.0024	0.198	0.96
	0.077	0.014	0.0057						
B-7.5	0.093	0.57	0.34	0.012	0.007	9.7	0.0036	0.195	0.95
	0.076	0.012	0.0055						
B-10	0.09	0.52	0.323	0.015	0.0076	9.92	0.0049	0.201	0.96
	0.078	0.016	0.0057						
B-12.5	0.098	0.498	0.332	0.016	0.0068	9.62	0.0068	0.20	0.99
	0.079	0.018	0.0058						

Microstructure of P91 steel weld metal enriched by B_2O_3:

The optical micrograph of the weld metal produced with matching electrode is shown in figure 1. In the as-welded condition, P91 electrodes produce a predominantly untempered martensitic structure in which the solidification substructures were not seen. Microstructural evidence for martensitic transformation for PWHT above the *AC*1 temperature is available after etching the specimens with the Viella's regent, which reveals the fresh martensite and ferrite in white, tempered martensite and bainite in a dark color [4]. Weld metal microstructure consists of 55.08% black phase and 44.45% white phase. The average grain size of black phase and white phase are 9.3 μm and 8.6 μm.

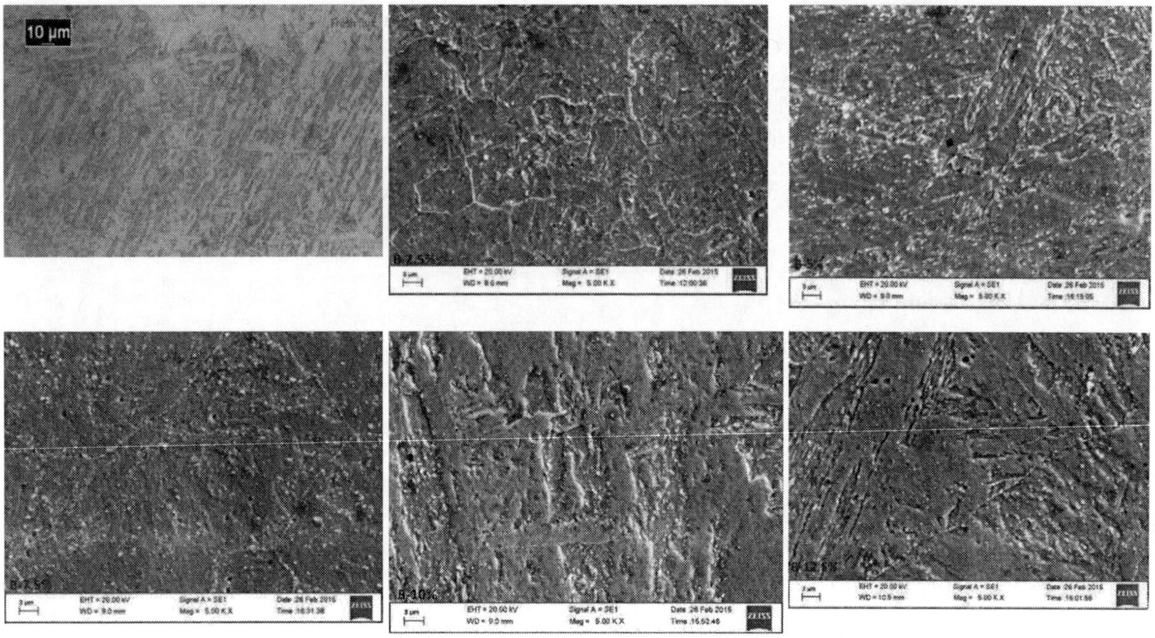

Fig. 1 Optical Micrograph of WM without B_2O_3 (fresh flux) and SEM micrographs of WM (2.5-12.5%) with various B_2O_3.

SEM micrographs show finely dispersed precipitates in all steel weld metals. Precipitates are mainly aligned along the prior austenite grain and the martensitic lath boundaries. Some finer precipitates can be observed inside the martensitic laths. SEM micrographs with different % of B_2O_3 addition reveal the changes in precipitation quantity and formation of carbides as shown in figure 1. It has been seen that with the addition of boron, precipitation percentage has been increases. Boron can only improve the creep behavior if it is either enriched in other precipitates or dissolved in the matrix. Boron dissolved in $M_{23}C_6$ carbides is said to retard their coarsening and, therefore, improve the long-term stability of the microstructure during creep exposure [6]. The $M_{23}C_6$ carbides are distributed mainly along prior austenite grain boundaries and along martensite lath, block and packet boundaries inside the grain [7].

Hardness profile with various B_2O_3%:

A macrohardness profile across both columnar and refined regions on a transverse section of P91 weld metal in the as-PWHT condition is shown in figure 2. It can be seen that columnar region exhibits a much higher hardness than the refined region. Hardness profile shows that boron has a significant effect on the WM hardness, which shows the hardenability effect of boron. Maximum hardness is found for the maximum B_2O_3%. Hardness of HAZ does not showing any appreciable improvement with boron content. The hardness values in the columnar regions range from 258 to 342 HV2 with the B_2O_3%, whilst the hardness values in the refined regions decreases gradually and became steady at the base material. In the CGHAZ the hardness value varies from 250 to 340 HV. After that in

FGHAZ hardness is decreasing drastically and varies in between 190 to 250 HV with respect to B_2O_3. ICHAZ is showing the minimum hardness in HAZ, where the hardness varies within 190-200 HV.

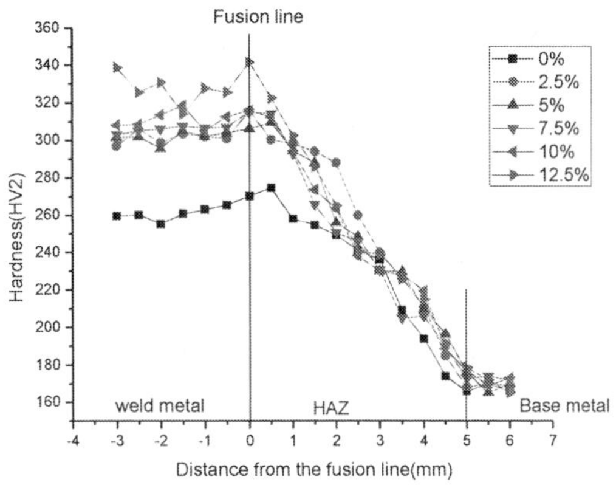

Conclusion:

Addition of boron trioxide in to the flux is a useful method to change the weld metal chemical composition. Weld metal chemical composition shows that boron varies in the range of 0.00-0.0068%. SEM micrographs reveal that carbide ($M_{23}C_6$) precipitates percentage increases with increase of B content which could be beneficial for the long term stability under elevated temperature. Hardness value of weldment significantly increases with respect to the B addition which has positive effect on the improvement of high temperature application.

References:

1. Agyakwa, Pearl. "Creep and microstructural development in P91 weldments at elevated temperature." PhD diss., University of Nottingham, 2004.
2. Vaillant, J. C., B. Vandenberghe, B. Hahn, H. Heuser, and C. Jochum. "T/P23, 24, 911 and 92: New grades for advanced coal-fired power plants—Properties and experience." International Journal of Pressure Vessels and Piping 85, no. 1: 38-46, 2008.
3. Haarmann, K., J. C. Vaillant, W. Bendick, and A. Arbab. "The T91/P91 Book."Vallourec & Mannesmann Tubes 1999.
4. Zhang, Yan. "Changes in microstructure and mechanical properties of P91 weld metal during creep." PhD diss., University of Nottingham, 2009.
5. Rojas, D., J. Garcia, O. Prat, G. Sauthoff, and A. R. Kaysser-Pyzalla. "9% Cr heat resistant steels: Alloy design, microstructure evolution and creep response at 650° C." Materials Science and Engineering: A 528, no. 15: 5164-5176, 2011.
6. Das C. R. "Influence of boron on microstrcture and mechanical properties of modified 9cr-1mo steel weldments" Ph. D thesis, IIT Madras, 2011.
7. Horiuchi, Toshiaki, Masaaki Igarashi, and Fujio Abe. "Improved utilization of added B in 9Cr heat-resistant steels containing W." ISIJ international 42, no. Suppl: S67-S71, 2002.
8. Abe, Fujio, Masaki Taneike, and Kota Sawada. "Alloy design of creep resistant 9Cr steel using a dispersion of nano-sized carbonitrides." International Journal of Pressure Vessels and Piping 84, no. 1: 3-12, 2007.

Development of Vertical Welding Technology for Thick Steel Plate using Hot-wire Laser Welding Method

Motomichi Yamamoto, Kenji Shinozaki *; Hiroshima University, Higashi-Hiroshima, Hiroshima, Japan*
motoyama@hiroshima-u.ac.jp
Hiroshi Yajima *; Yajima Material Integrity Laboratory, Ngasaki, Nagasaki, Japan*
Tsutomu Fukui *; Nippon Kaiji Kyokai, Chiyoda, Tokyo, Japan*
Shin Nakayama *; Mitsubishi Heavy Industries, Ltd., Ngasaki, Nagasaki, Japan*
Kenji Kamita *; Imabari Shipbuilding Co. Ltd., Marugame, Kagawa, Japan*
Shinji Kodama *; Nippon Steel & Sumitomo Metal Corporation, Futtsu, Chiba, Japan*
Hiroshi Watanabe *; Mitsubishi Hitachi Power Systems Engineering Co., Ltd., Yokohama, Kanagawa, Japan*

Abstract

The objective of this research was the development of the single-pass vertical welding process for thick steel plates using the hot-wire laser welding method. A laser diode with a large rectangular beam spot and hot-wire feeding system were employed in the proposed process. Welding phenomena of the proposed process were investigated using stationary and weaving laser irradiation with small and large groove widths. The result showed that the single-pass vertical welding for thick steel plates with a low heat input and low dilution could be achieved by using the proposed process. It became also clear that there was the critical energy density for each welding speed to begin fusion of the groove surface. Moreover, the energy density affected largely on fusion of the groove surface compared with the welding speed in the proposed process. Weaving laser irradiation made it possible to maintain the high energy density and to fit the large gap variation.

Introduction

In recent years, larger container ships have been built for more efficient maritime transportation, and consequently thicker steel plates have been used for its structure [1]. An electrogas arc welding (EGW) process is generally used to join heavy-thick steel plates on the vertical direction in ship building, especially has been applied for a hatch side coaming [2] in the part of a container ship structure. However, the EGW process causes very high heat input, and it affects seriously on toughness deterioration of a welded joint due to grain coarsening and widening in a heat affected zone (HAZ) [3]. Therefore, the novel vertical welding process with a lower heat input is strongly demanded to join heavy-thick steel plates having high strength and toughness.

The hot-wire laser welding process for various types of joints and materials has been developed, and it has achieved a low heat input, low dilution, high efficiency and high quality joint simultaneously [4]. The hot-wire method allows the high deposition rate independently from the main heat source since a filler wire is heated up to its melting point only by Joule heating, in other words, the main heat source can be used only for base metal melting.

The present research proposes the novel vertical welding process for thick steel plates using a laser diode as a heat source in combination with the hot-wire method in order to achieve much lower heat input compared with the EGW process. This welding process with a low heat input should prevent grain coarsening and widening HAZ width, and keep toughness of a welded joint. In addition, this process needs high efficiency by single-pass welding for thick steel plates in the vertical direction as with the EGW.

Figure 1 shows the schematic illustration of the proposed welding process. In this process, a high-power laser diode is used as a main heat source and the hot-wire method is used for efficient deposition. A large rectangular spot shape of a laser beam, which fits a groove width (gap) and plate thickness, is employed. The laser is irradiated continuously from the above the joint into the groove to create and keep a molten pool during welding. A reflected laser on a molten pool surface is utilized for melting groove surfaces just in front of molten pool efficiently. Filler wires are fed from both sides of the groove, and filler wires are heated up to its melting point by Joule heating using the hot-wire system before entering the molten pool. Jigs and a moving way are almost the same as these of an EGW process. It can be expected from the aforementioned features that the proposed hot-wire laser welding method realizes single-pass vertical welding for a heavy joint with a low heat input, low dilution, high efficiency and reduction of a laser power.

The welding phenomena of this process were studied and effects of primary parameters on welding phenomena were investigated first. A stationary laser beam without weaving and a specimen having a relatively small gap were used for basic investigations. Then a weaving laser beam and a specimen having a relatively large gap were employed in order to investigate the effect of a laser irradiating condition on welding phenomenon. In addition, the potential of the proposed process for a practical use was shown from this investigation.

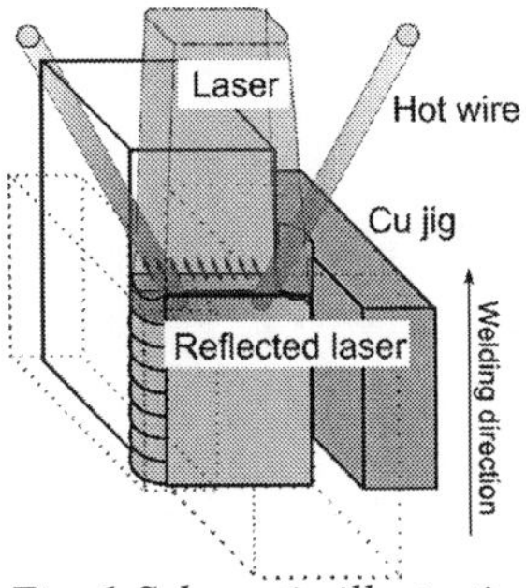

Fig. 1 Schematic illustration of proposed vertical welding process using hot-wire laser welding method.

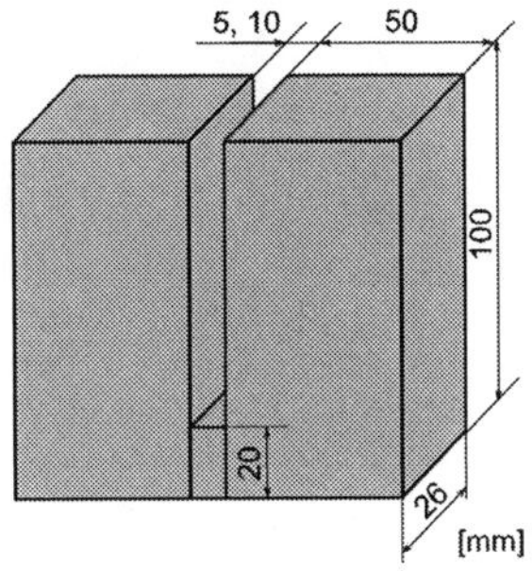

Fig. 2 Specimen dimension used for hot-wire laser welding.

Table 1 Welding conditions.

Gap, mm	5	10
Laser irradiation method	Stationary	Weaving (15 Hz)
Laser power, kW	4 ~ 6	6
Spot size, mm	3.5^{W} x 26^{T}	2^{W} x 26^{T}
Defocus amount, mm	50	80
Welding speed, m/h	1.0, 2.0	1.0, 2.0
Wire feed speed, m/min	0.54, 1.12	1.24, 2.48
Wire current, A	51 ~ 113	91, 124
Wire feeding angle, degree	45	45
Wire feeding position, mm	1 ~ 4	0
Shielding gas (Argon), LPM	1	20

Materials and Experimental procedure

KE47 steel plates and filler wires of JIS Z 3325 YGL2-6A (AP) with its diameter of 1.6 mm were used in this research. The dimension of a specimen was 50 (width) x 100 (height) x 26 mm (thickness) as shown in Fig. 2.

Table 1 shows the welding conditions. For the basic investigation, the stationary laser irradiation without weaving and a specimen having a relatively small gap (5 mm) were used. The laser power and welding speed were changed as 4 ~ 6 kW, 1 and 2 m/h respectively in order to obtain effects of primary parameters on welding phenomena. The rectangular laser spot size of 3.5 x 26 mm and irradiation condition were fixed. The hot-wire feeding speed was adjusted to the welding speed, and the wire current was set to heat a filler wire tip up to just under its melting point appropriately. Two hot-wires were inserted from both sides of the groove. Argon gas was used for shielding.

In the next investigation, weaving laser irradiation and a specimen having a relatively large gap (10 mm) were used. The weaving system used in this research can sweep a large rectangular laser spot having a long length fitting a plate thickness and a narrower width than a gap parallel to the gap width direction. The weaving amplitude can be adjusted to a variable gap and the weaving frequency can be also changed. This weaving system provides the gap flexibility and maintains a high power density. The laser spot with a width of 2 mm and length of 26 mm at a just focus position was used under the laser power of 6 kW. The weaving amplitude was controlled to fit to a 10 mm gap, and the weaving frequency was set as 15 Hz. The welding speed was changed 1 and 2 m/h. The other welding conditions are same as those using stationary laser irradiation.

Result and discussion

In-situ observations using a high-speed camera indicated stable formation of a molten pool and stable feed of filler wires during welding. It became clear that the proposed hot-wire laser welding method could stably form a large molten pool filling a groove constantly using a laser diode with a large rectangular spot as a heat source.

Figure 3 shows macro cross-sections of welded joints at the thickness center in the vertical direction, and Fig. 4 shows the lack of fusion ratio and the width of HAZ measured on vertical sections under various welding conditions in the cases of the 5 mm gap. With increasing in the laser power up to 6 kW, the width of weld bead and HAZ increase, nevertheless, the width of HAZ is narrower than that of the conventional EGW process (15~20 mm). It was cleared that the single-pass vertical welding for thick steel plates with a low heat input and low dilution could be achieved by using the proposed process.

When the relatively low laser power of 4 or 4.5 kW was applied, only a fairly top region on the groove surface was melted, then lack of fusion occurred under some conditions. On the other hand, when the relatively high laser power of 5 or 6 kW was applied, the groove surface was sufficiently melted, then the lack of fusion decreased and sound weld bead could be obtained.

In contrast with the effect of laser power, and welding speed on melting of base metal and the lack of fusion ratio is very small, as shown in Fig. 3 and 4. It can be implied from these results that the laser power density affects largely on fusion of the groove surface and sound weld bead formation compared with the welding speed.

Figure 5 shows the relationship between the energy density (laser power / laser irradiating area) and the penetration width of the groove in each welding speed using stationary laser irradiation and the 5 mm gap. It is clear that there is the critical energy density, which is about 25 and 35 W/mm2 for each welding speed of 1 and 2 m/h, to begin fusion of the groove surface. Moreover the penetration width increases linearly with increasing in the energy density on both welding speeds. It can be considered that the energy density is the primary factor to obtain sound fusion in this process since the critical energy density does not double even if the welding speed doubles.

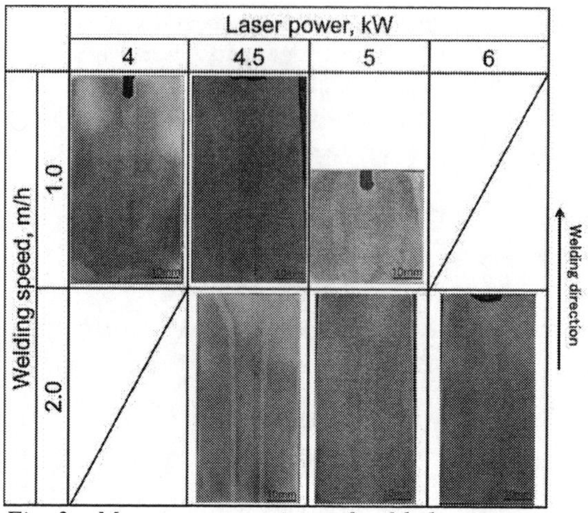

Fig. 3 Macro cross-sections of welded specimens in case of 5 mm gap in case of 5 mm gap with stationary laser irradiation.

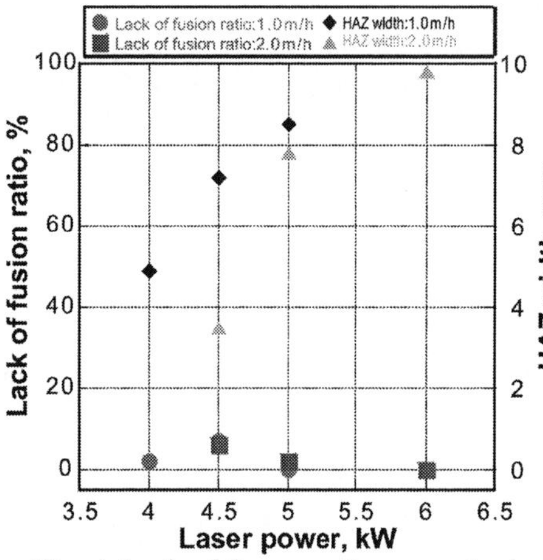

Fig. 4 Lack of fusion ratio on vertical sections and HAZ width in case of 5 mm gap with stationary laser irradiation

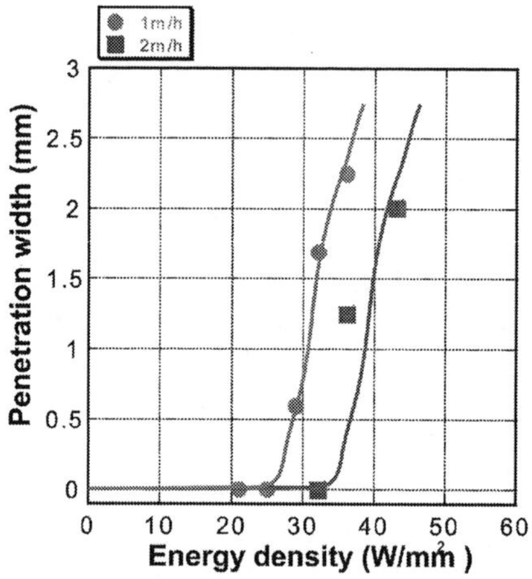

Fig. 5 Relationship between energy density, welding speed and penetration width of weld metal.

It was found from in-situ observation using a high-speed camera that weaving laser irradiation achieved stable formation of a molten pool and stable feed of filler wires during welding when the gap became 10 mm under the condition of 6 kW laser power.

Figure 6 shows macro cross-sections of welded joints on vertical direction at the thickness center and horizontal direction under the welding speed of 1 and 2 m/h in the cases of the 10 mm gap using weaving laser irradiation. Observation of the cross- section in the vertical direction shows stable fusion of the groove surface and sound beads on both welding speeds. It can be also observed from the cross-section in the horizontal direction that although the lack of fusion was seen at both groove edges, the small and uniform penetration width along the groove was achieved.

Figure 7 shows the lack of fusion ratio and the bead width on horizontal sections. The bead width created on both welding speeds is just over 10 mm which is the initial gap, in other words, the dilution ratio is very small. The lack of fusion is found only at groove edges and its percentage to the full thickness is about 20 % on both welding speeds, while perfect fusion is achieved on about 80 % middle region of the groove in the thickness direction. The optimum energy distribution at the groove edge must be considered to obtain adequate fusion and the sound bead at this region. It can be also seen that the welding speed does not significantly affect on the bead width and lack of fusion ratio.

These results indicate that the weaving laser irradiation with the narrow rectangular laser spot make it possible to form the stable molten pool and create suitable fusion at the groove surface on a large gap since this method can keep the high laser power density.

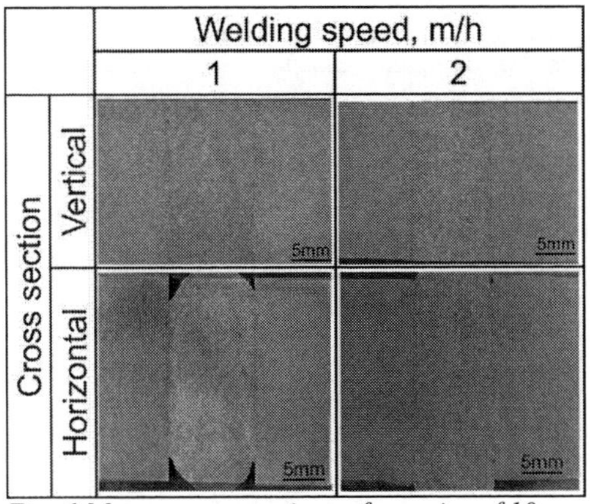

Fig. 6 Macro cross-sections of gap size of 10 mm using weaving laser beam irradiation.

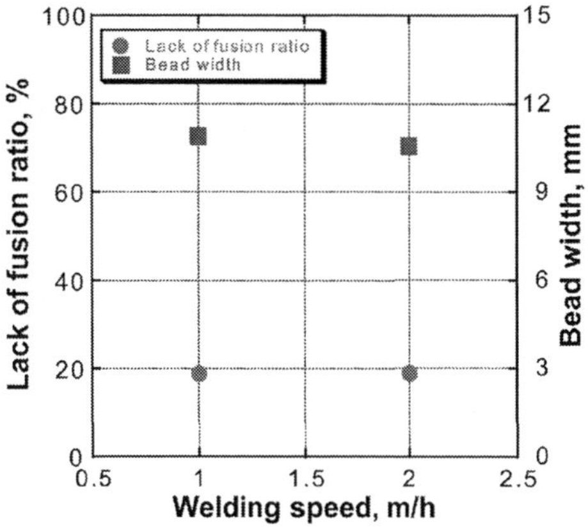

Fig. 7 Lack of fusion ratio on horizontal sections and bead width in case of 10 mm gap.

Conclusions

In this research, the novel single-pass vertical welding process for thick steel plates using the hot-wire laser welding method was proposed and its melting phenomena were investigated. The conclusions are as follows.

(1) Stable formation of a molten pool and stable feed of filler wires during single-pass vertical welding on a joint of 26 mm thick steel plates were achieved using the laser diode with a large rectangular beam spot and the hot-wire system.
(2) The energy density is the principal parameter to obtain adequate fusion and a sound weld bead, and it affects largely on melting phenomena compared with a welding speed.
(3) Stable formation of a molten pool and sound fusion were achieved by using the laser weaving irradiation method with a narrow rectangular laser spot even when a large gap was employed. The combination of a long and narrow rectangle laser spot fitting to a plate thickness and the laser weaving irradiation achieves the large gap tolerance.
(4) Although the energy distribution in the thickness direction must be optimized to obtain the sound bead at bead surfaces, it can be said that the proposed hot-wire laser welding process has the potential to be used as the single-pass vertical welding process for thick steel plates with low heat input and low dilution.

Acknowledgments

This research was supported by Nippon Kaiji Kyokai in the "Joint R&D with Industries and Academic Partners" research program.

References

Number literature references in the order in which they appear in the manuscript, like [1]. The following format must be used.

[1] S. Okano, Y. Obanya, K. Abe, K. Hosoi: Latest Technical Trends in Steel Plate and Action in Kobe Steel, Kobe Steel Eng. Reports, 58-1 (2008), 2-7.
[2] T. Yokura: Practical Lecture - A Class in Joining & Welding - Shipbuilding (Process & Construction), J. JWS, 81-6 (2012), 43-50.
[3] Y. Funatsu, J. Otani, K. Hirota, T. Matsumoto, H. Yajima: Development of Higher Toughness YP47 (460N/mm2) Class Steel Plate for Large Container Ships, Proc. 2010 ISOPE (2010), 102-106.
[4] R. Phaonaim, K. Shinozaki, M. Yamamoto, K. Kadoi, S. Tsuchiya, A. Nishijima: Development of a highly efficient hot-wire laser hybrid process for narrow-gap welding - welding phenomena and their adequate conditions, Welding in the World, 57 (2013), 607-613.

A Preliminary Study of Thick Steel Plate Welding with 100kW Power Fiber Laser

Hiroyuki SUMI, Yasushi KITANI, Kenji OI
JFE Steel Corporation, Chiba, Japan
h-sumi@jfe-steel.co.jp, phone:+81-44-262-2418, fax:+81-44-262-2117
y-kitani@jfe-steel.co.jp, phone:+81-44-262-2418, fax:+81-44-262-2117
ke-ooi@jfe-steel.co.jp, phone:+81-44-262-2052, fax:+81-44-262-4155

Daichi SUMIMORI, Haruki MIZUTANI, Shingo ITO
NADEX PRODUCTS Co.,Ltd , Fukui, Japan
daichi.sumimori@nadex-p.jp, phone:+81-770-25-2266, fax:+81-770-25-2267
haruki.mizutani@nadex-p.jp, phone:+81-770-25-2266, fax:+81-770-25-2267
shingo.itou@nadex-p.jp, phone:+81-770-25-2266, fax:+81-770-25-2267

Yousuke KAWAHITO, Seiji KATAYAMA
Osaka University, Osaka, Japan
kawahito@jwri.osaka-u.ac.jp, phone:+81-6-6879-8652, fax:+81-6-6879-8642
katayama@jwri.osaka-u.ac.jp, phone:+81-6-6879-8652, fax:+81-6-6879-8642

Abstract

High power laser welding is greatly expected as a high-efficiency welding process of thick steel plate because it is capable of deep penetration welding with high speed and low heat input. Recently, 100kW power fiber laser has been developed and commercialized as for industrial applications. However, experimental research reports of thick steel plate welding using such a high power laser aren't available. In this study, the high power laser welding properties of low-carbon thick steel plate with a thickness of 50mm were investigated by using 100kW power fiber laser. Meltrun laser welding tests were carried out under the conditions of laser power range from 10 to 100kW, welding speed range from 0.5 to 4m/min and the laser beam spot diameter of 1mm. Welding properties were evaluated by observation of cross-section macrostructures of the welds. Relations between the welding conditions and penetration shapes of welds were made clear. The results revealed that penetration depth of 100kW laser power reached more than 50mm at welding speed 1m/min and 32mm at welding speed of 4m/min. In the case of welding speed 1m/min or more, weld beads had underfill because of spatter generation. The sound welding conditions without spatter generation and underfill were laser power range below 75kW and welding speed range below 0.75m/min. Furthermore, the cross-sectional area of weld metal was proportional to heat input (laser power / welding speed) in all welding conditions.

Introduction

Laser welding is capable of deep penetration welding with high speed by using a high energy density heat source. Therefore, it is greatly expected as a high-efficiency welding process of thick steel plate. Recent years, the improvement in the performance of high power laser oscillator has been progressed remarkably, and 100kW power fiber laser has been developed and commercialized for industrial applications [1]-[3]. However, experimental research reports of thick steel plate welding using such a high power laser have not been available yet.

In this paper, the high power laser welding properties of low-carbon thick steel plate were investigated by using 100kW power fiber laser.

Experimental procedures

In this research, TS490MPa grade steel plate with thickness of 50mm was used. Table 1 shows the chemical compositions of the steel plate.

Figure 1 shows the schematic illustration of experimental setup of 100kW laser welding used in this research. The laser oscillator was a continuous wave fiber laser device with a maximum laser power of 100kW (YLS-100000, manufactured by IPG). The welding head equipped the parabolic mirror with focal length 1000mm. The core diameter of process fiber was 0.5mm and the laser beam focus diameter was 1mm.

The meltrun laser welding conditions are shown in Table 2. The laser welding tests were carried out under the conditions of laser power range from 10 to 100kW, welding speed range from 0.5 to 4m/min. The beam incident angle of was forehand angle 80deg. The focal position was just on the surface of the steel plate. After laser welding, the welding properties were evaluated by observation of cross-section macrostructures of the welds. The sampling position of observation specimens are shown in Figure 2.

Table 1: Chemical compositions of steel plate used.

(mass%)

Material	Thickness	C	Si	Mn	P	S
TS490MPa grade steel plate	50mm	0.15	0.33	1.46	0.024	0.005

Table 2: Laser welding conditions.

Laser power	10 ~ 100 kW
Welding speed	0.5 ~ 4 m/min
Focal position	±0 mm
Beam incident angle	80 deg. (Forehand angle)

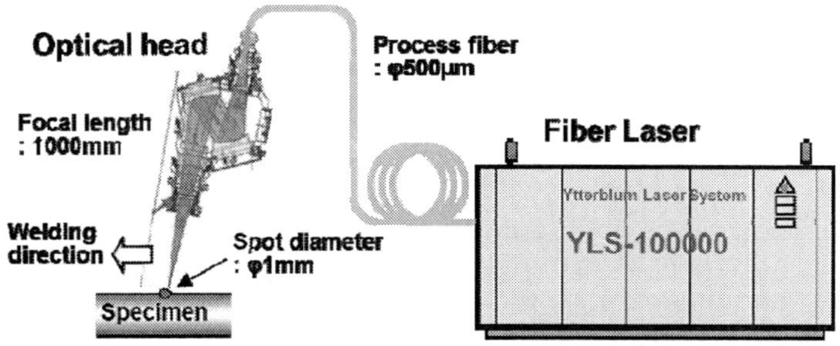

Fig.1: Schematic illustration of experimental setup of 100kW laser welding.

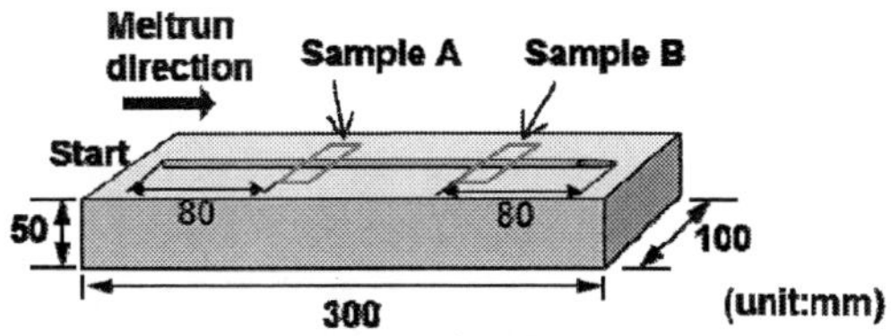

Fig.2: Sampling position of specimens for the cross-section observation.

Results and discussions

Figure 3 shows examples of cross-section macrostructures. Penetration depth became deeper with lower welding speed and higher laser power. The penetration depth of 100kW laser power reached more than 50mm at welding speed 1m/min. The penetration shape of welding speed 0.5m/min was an inverted triangle shape which indicated wider penetration width at upper part of weld metal. In contrast, the penetration shape of welding speed 1m/min or more was an elongated shape which indicated almost same penetration width at upper and bottom part of weld. In the case of high power laser welding with high speed, the evaporation on the molten pool was intense and a large amount of the spatter was generated and resulted in the underfill formation on the weld bead. In this research, the sound welding conditions without spatter generation and underfill were laser power range below 75kW and welding speed range below 0.75m/min.

Figure 4 shows the relationship between the welding speed and the penetration depth at various laser power. The penetration depth of 100kW reached more than 50mm at welding speed 1m/min and 32mm at welding speed of 4m/min.

In order to clarify the influence of welding conditions on the penetration shapes, the heat input of laser welding was defined as the ratio of laser power / welding speed. Figure 5 shows the relationship between the heat input and the cross-sectional area of weld metal. The cross-sectional area of weld metal was proportional to the heat input in all welding conditions. In other words, this shows the fact that the cross-sectional area of weld metal is governed by the heat input regardless of the laser power.

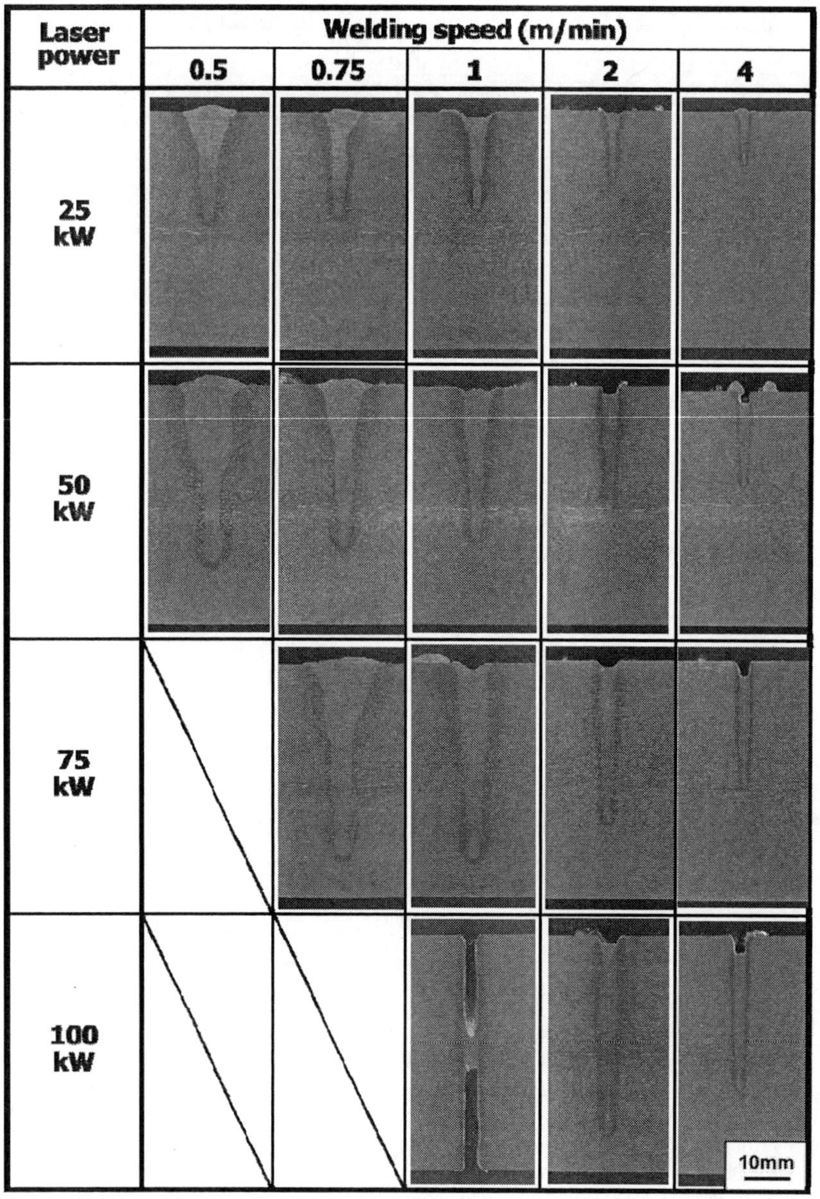

Fig.3: Cross-section macrostructures of laser weld bead, showing effects of laser power and welding speed on the penetration.

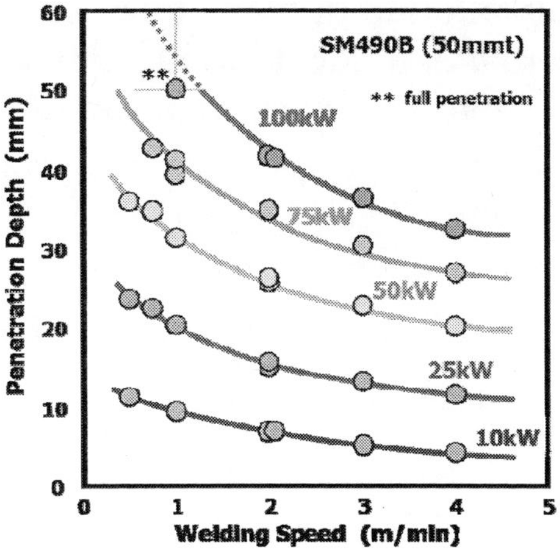

Fig.4: Relationship between welding speed and penetration depth at various laser power.

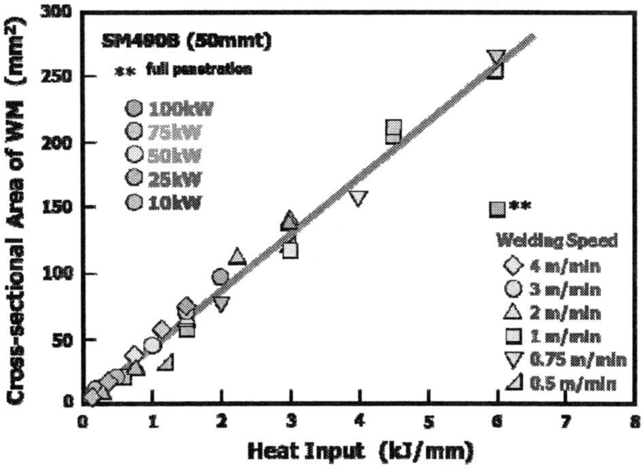

Fig.5: Relationship between heat input and cross-sectional area of weld metal.

Conclusions

This paper described the welding property of low-carbon thick steel plate welding with 100kW power fiber laser. The results of this research are summarized as follows.

1) The penetration depth of 100kW laser power reached more than 50mm at welding speed 1m/min and 32mm at welding speed of 4m/min.
2) In the case of welding speed of less than 1m/min, the penetration width became wider at upper part of weld and in the case of welding speed 1m/min or more, the penetration width was almost same at upper and bottom part of weld.
3) The cross-sectional area of weld metal was proportional to heat input in all welding conditions.

References

[1] S. Katayama: "Latest Development Trend of Laser Welding Technology", *Journal of JWS*, Vol. 80, No. 7 (2011), pp. 593-601. (*in Japanese*)

[2] S. Ito: *Yousetu-Gizyutu*, No.1 (2015), pp. 78-84. (*in Japanese*)

[3] S. Katayama, et al: "Possibility of Weldability at 100kW High Power Laser" *Preprints of the National Meeting of JWS*, 96 (2015), pp. 162-163. (*in Japanese*)

Application to duplex stainless steel of narrow gap oscillation laser welding

○ABE Yohei, NAKATANI Mitsuyoshi
Osaka University (Hitachi Zosen Corporation), 11-1, Mihogaoka, Ibaraki, Osaka, 567-0047, Japan
Email: y_abe@hitachizosen.co.jp
YAMAZAKI Yosuke, HIOKI Yukio
HITACHI ZOSEN CORPORATION, 2-11, Funamachi 2cho-me,Taisho-ku, Osaka, 551-0022, Japan

Abstract

Duplex stainless steel has excellent corrosion resistance and high strength, and it is used in the various structures including the chemical plant. But the welding deformation by arc welding is large because of its low heat conductivity and correction process takes much time. The narrow gap welding can reduce the groove section area drastically compared to other welding process. And it achieves the low heat input and very small deformation of weld joint. In our group, the narrow gap oscillation laser welding using high power laser as heat resource has been studied. Additionally, in order to prevent the lack of fusion which becomes a problem when the narrow gap welding, the automatic system that follows the fluctuating center of groove and groove width has been also developed. This system has the mechanism which recognizes the groove shape by processing the image obtained by CCD camera mounted coaxially with laser beam axis and controls the center of laser beam oscillation and the laser beam oscillation width. In this study, the automatic system and the welding process are established in order to apply the narrow gap oscillation laser welding to the butt weld of duplex stainless steel product.

1. Introduction

Duplex stainless steel has excellent corrosion resistance and high strength, and it is used in various structures including chemical plants. However, it has lower heat conductivity than carbon steel. As a result, weld deformation is significant with arc welding, requiring much time to correct the distortion. A process which can reduce weld deformation is narrow gap welding. The narrow gap welding process can reduce the groove section area drastically compared to other welding processes, achieving a weld joint with low heat input and resulting in minimal distortion. Especially in recent years, narrow gap oscillation laser welding has been studied[1-3]. This process uses a high-power laser as its heat source and oscillates it rapidly. Setting the control of the center of laser beam oscillation and the laser beam oscillation width to adjust the groove's position and width can be expected to prevent the lack of fusion on the groove side wall often occurring in narrow gap arc welding. This report establishes the narrow gap oscillation laser welding process for plate joint welding of duplex stainless steel, establishes a laser multi-layer welding system in order to apply to actual equipment production.

2. Welding duplex stainless steel

2.1 Weld target

A sample chemical plant component of duplex stainless steel is shown in **Fig. 1**. The groove shape and pass sequence is shown in **Fig. 2.** The figure shows a disk-shaped component with a diameter of 2500 mm and a thickness of 60 mm. This component is made of two semicircular pieces joined together by plate joint welding.

The groove is a double U groove of 30 mm per side, produced using GTAW. The number of passes is 37 on the BP (backing pass) side and 38 on the FP (finishing pass) side for a total of 75, a full penetration weld. When plate joint welding was produced using GTAW, weld distortion had, in the end, caused the component to bend in approximately 10 mm at the weld site, approximately 0.9 degrees when converted to angular deformation (**Fig. 3**). Much time is required to correct this deformation. This report has as its objective the reduction of time used to correct deformation and the time required for production by applying the narrow gap oscillation laser welding process to reduce the amount of deformation, and establishes the multi-layer welding system.

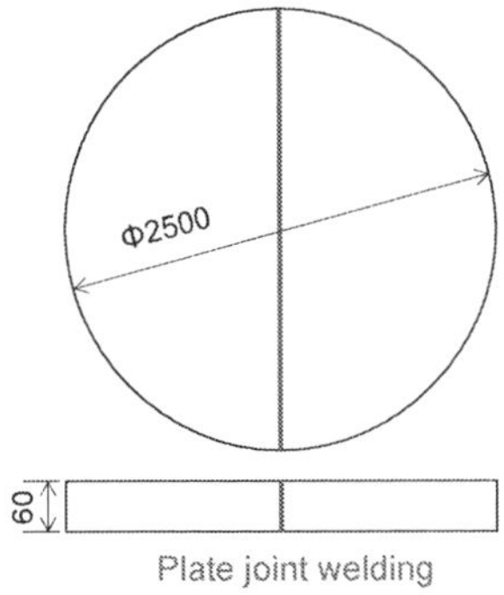

Fig.1 Weld Target

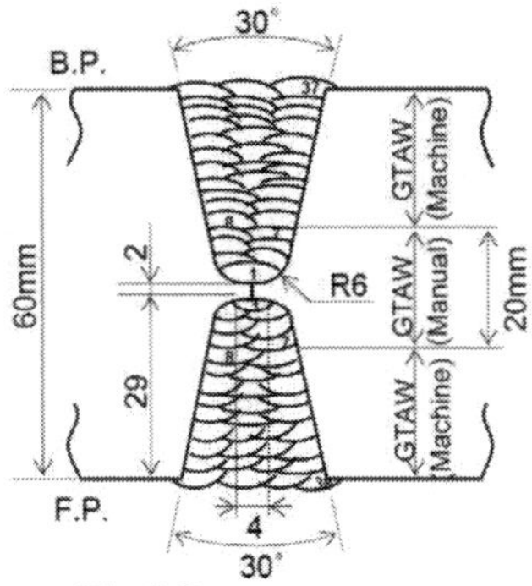

Fig.2 Pass sequence

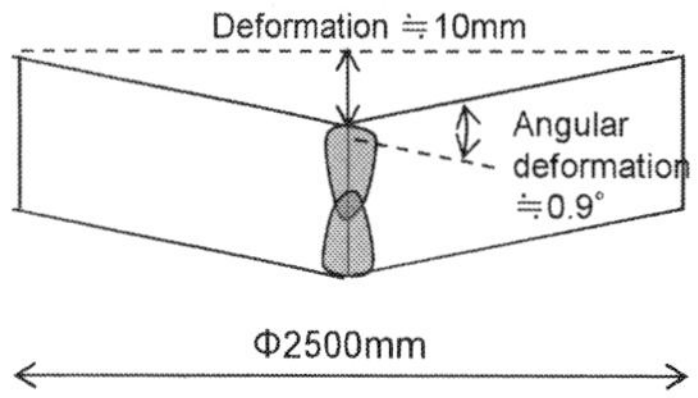

Fig. 3 Deformation using with GTAW

2.2 Production of test piece

The test piece is a plate with a thickness of 60 mm with strongback attached to reproduce actual equipment constraint. Its double U groove has a root face of 6 mm, a groove radius of 1.5 mm, and a groove angle of 6°. Narrow gap oscillation laser welding was performed with the welding condition shown in **Table 1**. The groove shape and pass sequence is shown in **Fig. 4**. The procedure is as follows: finish welding on the BP side, turn over, reattach the strongback, weld the FP side. The welding wire is one for GTAW (with a diameter of 1.2 mm), and the shield gas is N_2.

Table 1 Welding condition

	Root face	Backing pass		Finishing pass	
Pass	1	2~4	5~10	11~17	18~19
Laser power [W]	5000	5000	5000	6000	6000
Welding speed [mm/min]	500	300	300	300	300
Defocused distance [mm]	+10	+50	+10	+50	+10
Wire feeding speed [m/min]	—	3	3	3	3

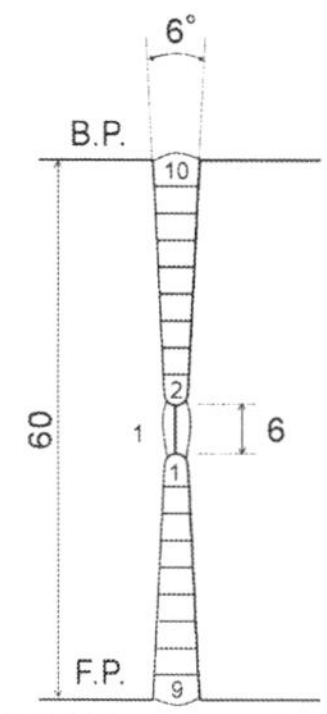

Fig.4 Welding Pass Sequence

The experiment was performed using the welding system shown in **Fig. 5**. A laser head equipped for oscillation and an automatic gap sensing system are built into this apparatus. The resulting mechanism is that the center of laser beam oscillation and the laser beam oscillation width are automatically adjusted. The oscillation width of the laser was set at 0.5 mm less than the detected groove width, and the oscillation frequency was 20 Hz. The result of the macrostructure examination of the weld cross section is shown in **Fig. 6**. Multi-layer welding was completed with a total of 19 passes (10 on the BP side and 9 on the FP side). However, the lack of fusion was confirmed in the first layers on both the BP and FP sides. This is thought to be the result of insufficient laser heat input causing incomplete melting of the groove and the fed wire in the area of the starting layers. Other zones resulted in good weld with no defects.

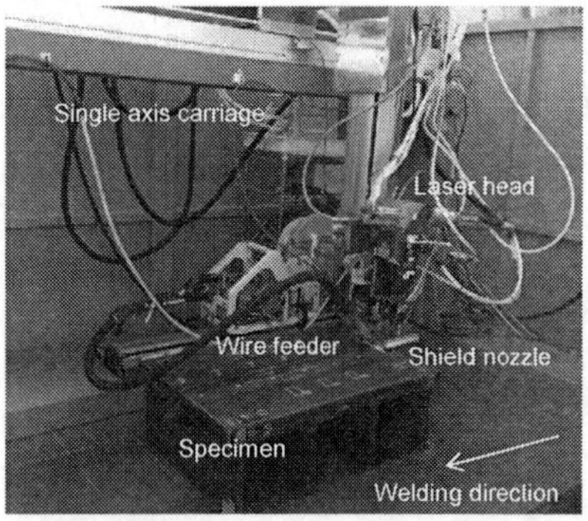

Fig.5 Overview of welding system

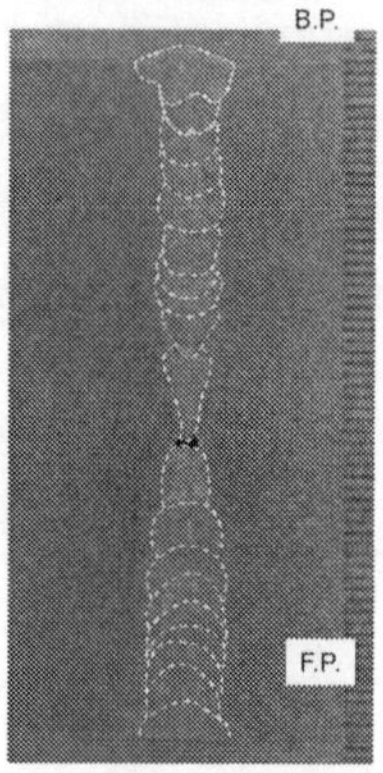

Fig.6 Cross section produced with LBW

2.3 Consideration of reduction of cost and deformation

The amount of angular deformation after each welding pass and production process is shown in **Fig. 7**. The angular deformation increased as welding on the BP side progressed, and the amount was 0.8° when the BP side was completed and the strongback was reattached. Deformation in the opposite direction resulted as the FB side was welded, culminating in a final angular deformation of 0.2°. The deformation was reduced to approximately 1/4 of amount resulting from GTAW (0.9°). Next, the welding time was calculated for the same duplex stainless steel plate with a thickness of 60 mm when a diameter of 2500 mm is joined with plate joint welding when produced by GTAW and when produced by narrow gap oscillation laser welding. The result is shown in **Fig. 8**. The welding time required for GTAW was calculated using standard condition based on welding speed and pass sequence. Compared with the approximately 31 hours required for multi-layer welding of a 60 mm plate by GTAW, narrow gap oscillation laser welding requires only 2.6 hours, reducing execution time to less than 10%. Additionally, GTAW requires turning over the piece five times. Laser welding requires this only once or twice, greatly reducing the time required for turnover as well.

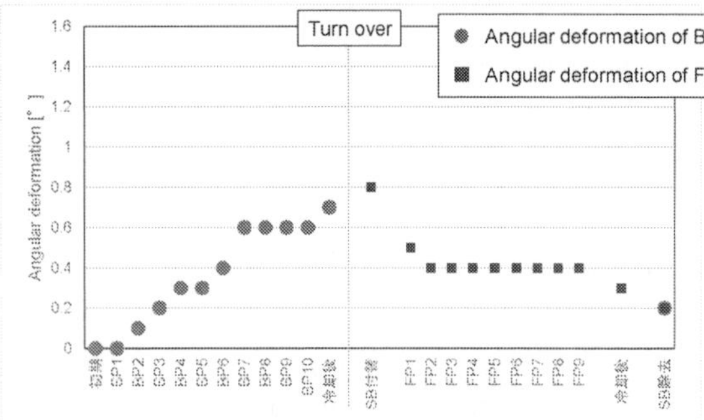

Fig.7 Angular deformation at each pass

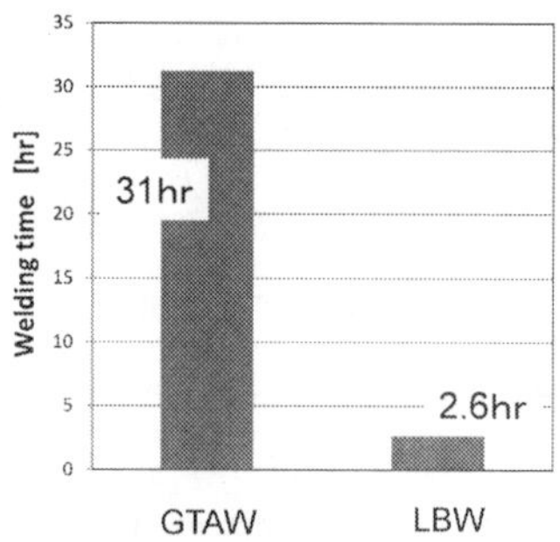

Fig.8 Comparison of welding time

3. Establishment of laser multi-layer welding system

The appearance of a laser multi-layer welding system in order to apply to actual equipment production is shown in **Fig. 9**. This system is capable of (1) oscillating the laser beam, (2) gap sensing system ;recognizing the joint shape and automatically adjusting, (3) z-axis direction automatic adjusting system ;automatically maintaining a constant distance between the weld surface and the laser head, and (4) feeding the wire. The (2) gap sensing system first uses a CCD camera with a filter to take a coaxial image of the area where the laser is irradiated during welding, and sends it to the control PC. The sent image is then processed to calculate the groove center and the groove width. Finally, the determined groove center position and the amount of displacement of the oscillation center position of the current oscillation are sent to the uniaxial slider driver to control the change in groove center position, and the determined groove width is sent to the laser oscillation controller to control the oscillation width. The time elapsed between the receipt of the image and the execution of the control is approximately 75 ms. This system repeats this sequence of actions from the start to the finish of welding to perform in-process adjustment to match the ever-changing groove. The (3) z-axis direction adjusting system uses a laser range finder to measure the distance between the laser head and the weld surface along the full length of the weld line before starting to weld, and controls to maintain a constant distance between the laser head and the weld surface during welding. The distance between the laser head and the weld target of this system is set based on a distance of 300 mm, the focal point of the laser head, and is adjustable in the range of df+10–30 mm.

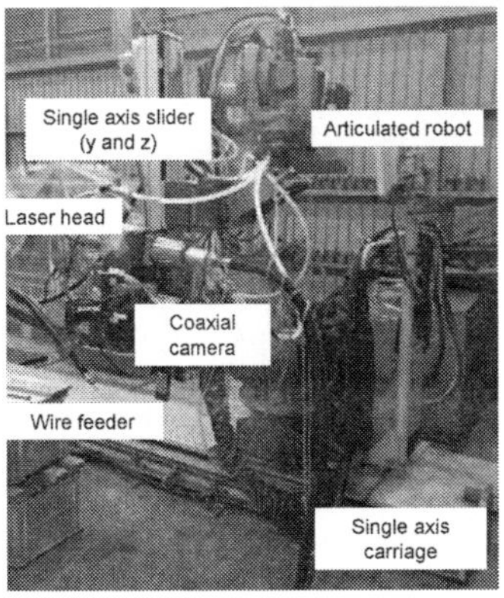

(a) Overview of system

(b) Front view

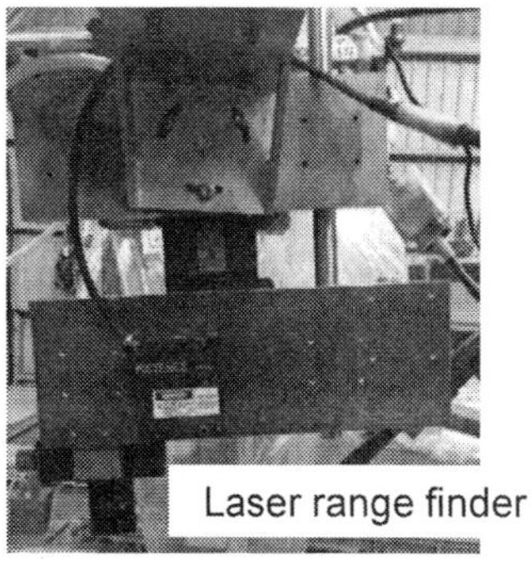

(c) Back view

Fig.9 Laser multi-layer welding system

4. Conclusion

•Confirmed that laser multi-layer welding reduces angular deformation to approximately 1/4 and welding time to under 10% compared to conventional GTAW.

•Established a laser multi-layer welding system in order to apply to actual equipment production.

References

[1] T.Tsukamoto et al, "Laser narrow gap welding of thick carbon steels using high brightness laser with beam oscillation", ICALEO2011, #305, 141-146.

[2] K.Shinozaki et al, "Behavior and prevention of lack of fusion and solidification cracking of modified 9Cr-Mo steel during hot wire laser welding with a narrow gap", proc. Of the 77th Laser Materials Processing Conference (2012), 21-26 (in Japanese).

[3]Y.Yamazaki et al, "Development of Narrow Gap Multi-layer Process using Oscillation Laser Beam", Quarterly Journal of the Japan Welding Society, vol.32(2014) No.4, 258-266

Influence of controlled atmosphere of argon in fiber laser welding of austenitic stainless steel

Yadaiah Nirsanametla [1], Sohini Chowdhury [1], Swarup Bag [2], Christ Prakash Paul [3], Muralidhar Manapuram [1]

[1] *Department of Mechanical Engineering, North Eastern Regional Institute of Science and Technology (NERIST), Nirjuli, Itanagar 791109, Arunachal Pradesh, India*

[2] *Department of Mechanical Engineering, Indian Institute of Technology Guwahati, Guwahati 781039, Assam, India*

[3] *Laser Material Processing Division, Raja Ramanna Centre for Advanced Technology, Indore, 452013, Madhya Pradesh, India*

In recent past, the fiber laser become a significant fusion welding heat source as compared to other laser sources due to its outstanding advantages such as high beam quality and efficiency; reliability, compact size and lower running and maintenance cost. The quality of a weld joint is major aspect for the assessment of service life and selection of the process parameters that determines the weld joint quality. Weld joint quality depends on several input process parameters and protection of molten weld pool. Moreover, the gases such as oxygen, hydrogen and nitrogen in atmosphere may become part of final weld joint that severely affect the weld joint quality and weld metal properties. Hence, the welding atmosphere and protection of weld pool plays a noticeable role on the final quality of the weld joint. Therefore, in the present work, the authors have examined an effect of controlled atmosphere of argon in fiber laser welds. Also, experiments are carried out in open atmosphere to study the characteristic difference between two different atmosphere welds. A mixed mode beam power of a 2-kW ytterbium fiber laser has been used to conduct the experiments. The experimental investigation specifies that the weld depth of penetration and aspect ratio are higher in case of argon atmosphere welds as compared to open atmosphere welds. Both the microstructures of heat affected zone (HAZ) and fusion zone (FZ) at atmospheric conditions are analyzed. It is also obvious from the experimental results in case of welds in argon atmosphere that the top surface profile is better and weld surface is very clean as compared to open atmospheric condition. Keywords: Fiber Laser, Fusion Welding, Controlled Atmosphere, Argon, Microstructure, Weld Bead Dimensions, Aspect Ratio.

Development of high power direct diode laser equipped remote laser welding robot system "LAPRISS" and welding process

Atsuhiro Kawamoto, Yasushi Mukai, Shigeki Yonemori,
Hitoshi Nishimura, Junji Fujiwara, Tatsuyuki Nakagawa
Panasonic Welding Systems Co., Ltd. Toyonaka, Osaka Japan
Email kawamoto.atsuhiro@jp.panasonic.com, phone 06-6866-8566,

1. Abstract

In the welding industry, high productivity has been required to reduce the manufacturing cost. The demand of faster welding speed and improving welding tolerance have been becoming more and more strongly. Due to increasing requirement of CO_2 emission reduction to prevent global warming, especially in the automotive industry, improving fuel efficiency has also been requested. As a large effect is expected, weight saving of car itself is requested. So multi material of the use of lightweight materials in right parts has been accelerating.

For these, many pressed parts are used in automotive, so if gap is generated between the welding parts or the aiming point shift, burn through generate at laser welding. Since the post-processing is required, it becomes a factor of productivity reduction.

In addition, regarding lightweight of car itself, multi material is being advanced for the introduction of lightweight materials such as aluminum, resin, CFRP, instead of steel. For this, dissimilar materials joining is required.

In this study, using laser robot welding system called LAPRISS [1] (Fig.1), we will introduce welding process to improve welding tolerance against gap or aiming point shift in high speed welding which is strong point of laser welding, and new dissimilar materials joining process for multi material.

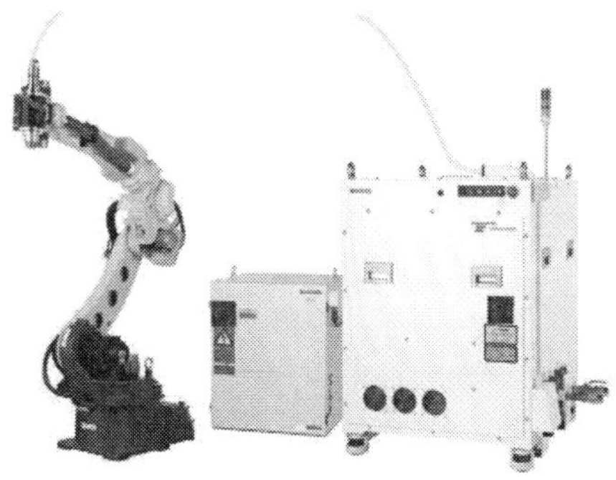

Fig.1 LAPRISS

2. Welding equipment

This consideration was evaluated with laser robot welding system called "LAPRISS" that integrates all elements required for the LASER processing robot such as oscillator, head, etc. that can be operated by one controller. LAPRISS is equipped with a trepanning head of the small size and light weight. This head makes it possible to draw a welding pattern of circle shape with a high accuracy. A plurality of parallel plates are arranged inside this head. We shift the exit position itself of the beam by driving the parallel plate with AC servo motor. Due to this configuration, it is possible to weld patterns such as circle, robot itself remains stopped.

3. Consideration of expanding welding tolerance

The introduction of laser welding makes it possible to increase the productivity by high speed welding. But In the manufacturing site, burn through occurs against gap between welding parts and aiming point shift, so production efficiency improvement is disturbed.

For these challenges, 2 items of gap and aiming point shift were considered.

3-1-1. Gap tolerance

First consideration is the gap tolerance. Two steels of lap joint that were 0.8mm thickness where there was a gap were welded. Two welding process that were straight and Laser Spin were welded. While drawing a circle in the trepanning head, robot is moving to the weld line direction. (Fig.2)

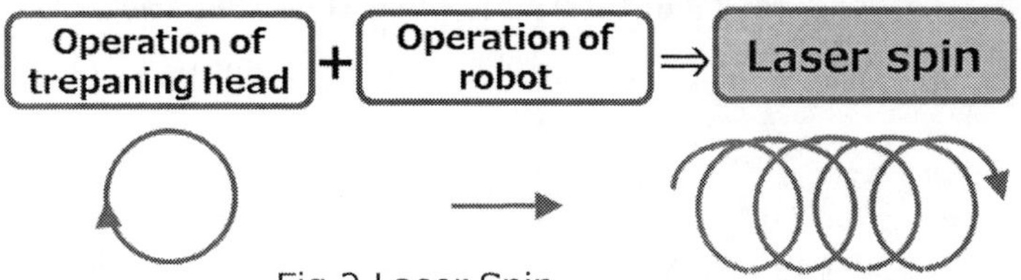

Fig.2 Laser Spin

3-1-2. Results

In the straight process, it was possible to weld without burn through in the case of up to gap 0.3mm, but burn through occurred in the case of over gap 0.4mm. On the other hand, in Laser Spin process, it was possible to weld without burn through even in the case of gap 0.5mm. Laser Spin process showed gap tolerance about 1.7times higher than straight process. (Fig.3, Fig.4)

	welding speed : 2mpm	
Gap(mm)	Straight (2.5kW)	Laser spin (3.4kW φ1.0)
0.0	○	○
0.1	○	○
0.2	○	○
0.3	○	○
0.4	×	○
0.5	×	○
0.6	×	×
	×··burn through	

Fig.3 Gap tolerance

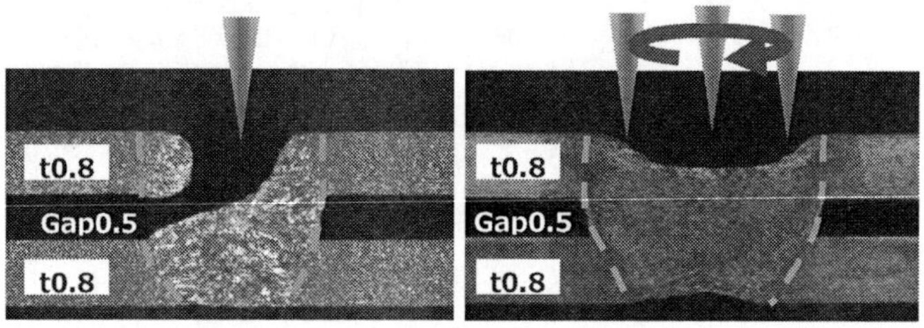

Fig.4 Macro section

3-2-1. Aiming point tolerance

Second consideration is the aiming point tolerance. Two steels of butt joint that were 0.8mm thickness with 0.2mm gap were welded at 1.25kW. Two welding process that were straight and Laser Spin were welded.

3-2-2. Results

In the straight process, it was possible to weld without burn through in the case of up to aiming point shifted 0.1mm in the case of gap 0.2mm. But in Laser Spin process, it showed aiming point tolerance 9 times higher than straight process. (Fig.5)

Position shift (mm)	Straight	Laser Spin (Spin φ1.0)
gap	0.2mm	0.2mm
+0.7	–	×
+0.6	–	○
+0.5	–	○
+0.4	–	○
+0.3	–	○
+0.2	–	○
+0.1	×	○
±0.0	○	○
-0.1	○	○
-0.2	×	○
-0.3	–	○
-0.4	–	×
-0.5	–	–
-0.6	–	–
-0.7	–	–
Aiming point tolerance	0.1mm (-0.1~+0.0)	0.9mm (-0.3~+0.6)

Fig.5 Aiming point tolerance

4. Consideration of dissimilar material joining

In weight saving of car itself, multi material has been proceeding for introduction of lightweight materials such as aluminum alloy, resin, CFRP instead of steel. For this, the joining process such as adhesive or bolt fastening is used so far. However adhesive reduces productivity due to take a long time until hardening and is not easy to weld narrow range due to wetting and spreading on a joint part. On the other hand, in bolt fastening it takes a long time to fasten and prevents downsizing and lightweight of parts due to enlargement of bolts and nuts for enhancing fastening force.

For these challenges, we considered Laser Rivet process that shortens operating time without increasing weight of a car body.

4-1. Principle of Laser Rivet

Principle of Laser Rivet is to sandwich dissimilar material between same metals and weld with laser. (Fig.6) It fixes and joins by compressed force due to solidification shrinkage of welded part. Weld quality is stable because of same metal welding. For this reason, it is a new dissimilar materials joining process that is possible to join any dissimilar materials.

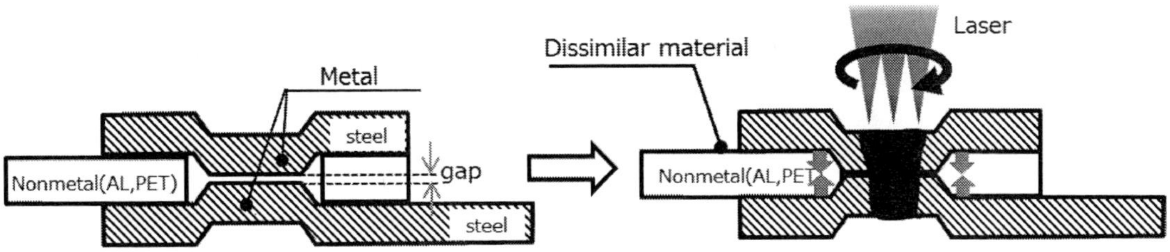

Fig.6 Principle of Laser Rivet

4-2-1. Consideration of gap shrinkage

Key point of Laser Rivet is sandwich force from upper and lower metals. That force is generated by solidification shrinkage of welded part. For this we measured amount of solidification shrinkage after welding the case of changing the gap amount. Before welding we lapped over two steels that were 1.2mm thickness with gap A (Fig.7) using shim plate. After welding we measured the gap B(Fig.7), and calculated the shrinkage (A-B). Fig.7 is the result in the case of changing the gap.

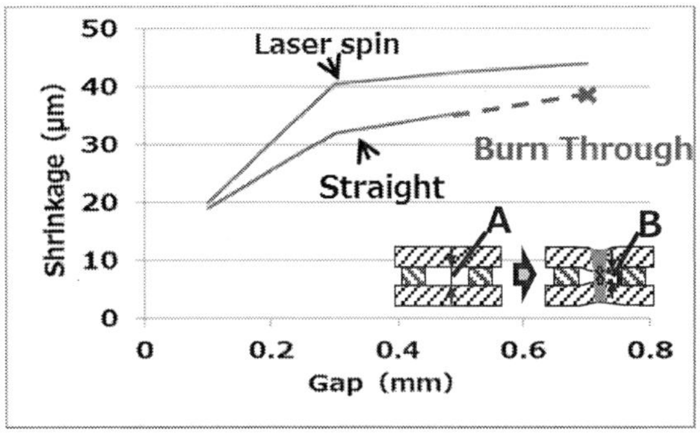

Fig.7 Gap shrinkage

Two welding process that were straight and Laser Spin were implemented. In case of straight process, laser power is 3.3kW, welding speed is 2.5mpm. In case of Laser Spin process, laser power is 3.8kW, welding speed is 1.5mpm, spin diameter is φ2.0.

4-2-2.Result

Since Laser Spin process was enable to widen bead width and increase molten metal against straight process, gap tolerance was improved. With the gap increase, solidification shrinkage was increased. (Fig.7)

4-3-1. Comparison of tensile shear strength and weight

Joining of A5052 2mm and mild steel 1.6mm thickness was compared by bolt fastening and Laser Rivet. Tensile shear strength and weight were measured. Hole diameter of dissimilar material was set based on diameter of bolt. Hole diameter of both process were same. The gap of Laser Rivet was

0.3mm.

4-3-2.Result

In the case of same hole diameter, tensile shear strength of both process were equal. With the hole diameter increase, that strength increased and all samples were fractured at a base material.(Fig.8) Regarding weight, Laser Rivet is enable to save weight because there is no nut that is essential for bolt fastening. With the hole diameter increased and tensile shear strength increased, diameter and thickness of bolt and nut increase, as a result weight gain. Regarding Laser Rivet, in spite of hole diameter increase, since only the diameter of the element is a little bigger, it is possible to suppress the increase in weight. (Fig.9)

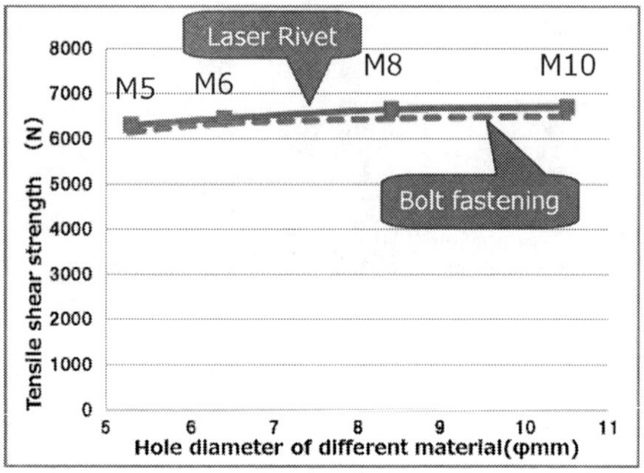

Fig.8 Tensile shear strength

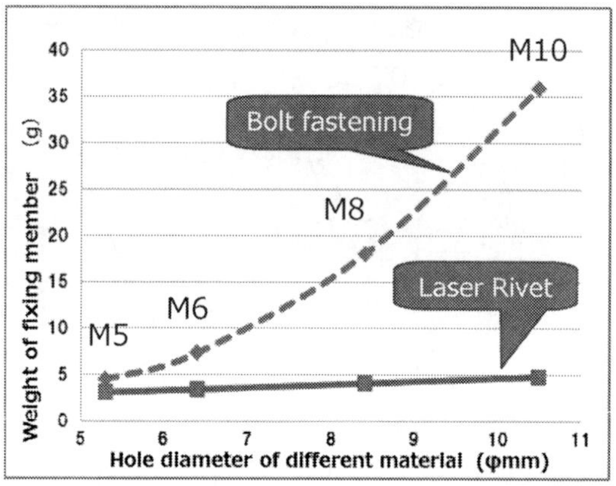

Fig.9 Weight of fixng member

5. Conclusions

Study on improving welding tolerance for high speed welding at laser welding production site and dissimilar materials joining for multi material was carried out.

(Regarding tolerance of gap and aiming point shift)

- Laser Spin process is wider tolerance than straight process.

(Regarding Laser Rivet)

- With the gap increased, amount of solidification shrinkage in case of Laser Spin process was more than straight process, sandwich and fixing force of Laser Spin increased. Laser Spin is enable to improve gap tolerance and strengthen the fixing force of Laser Rivet.
- Laser Rivet process which does not need nut is enable to save weight against bolt fastening. With the strength increase due to hole diameter increase of dissimilar material, weight saving effect of Laser Rivet is increased. From the above, Laser Rivet is new dissimilar material joining process which is enable to shorten operating time and downsize and save weight against bolt fastening.

References

[1] Junji Fujiwara, "Development of remote laser welding robot system "LAPRISS" to enable innovative manufacturing" : Welding Technology, Nov (2015),pp53-61

Effects of Laser Brazing Condition and Filler Metal Composition on Dissimilar Lap Fillet Joint Formation of Austenitic Stainless Steel and Pure Titanium

Masao Hadano, Yasuo Murai
Shinko Welding Service Co., Ltd. (Kobe Steel Group) , Fujisawa, Japan.
hadano.masao@kobelco.com, +81-466-20-3225
Taizo Kobashi
Kobe Steel, Ltd., Fujisawa, Japan.
kobashi.taizo-1@kobelco.com, +81-466-20-3400
Kunika Ueno, Katunori Kanno
Tosei Electrobeam Co., Ltd., Tokyo, Japan
k-kanno@tosei.co.jp, +81-42-556-0611
Kazuhiro Nakata, Tetuo Suga
Osaka University, Osaka, Japan
suga@jwri.osaka-u.ac.jp, +81-6-6879-8647

Abstract

Stainless steel features high corrosion resistance but may undergo intergranular corrosion in some specific environments especially in seawater. To cope with this problem, it can be considered to a useful technique to apply titanium material selectively to the part exposed to such specific environments. However, if austenitic stainless steel and titanium are fused together, Fe and Ti create brittle intermetallic compounds, which make it difficult to configure highly reliable joint with suitable welding parameters [1],[2]. The authors have studied the laser brazing process to minimize the fusing of the base metals to join the lap joint of stainless steel and titanium.

2. Experimental procedures

2.1 Testing materials

The testing base metals used were austenitic stainless steel plates of SUS304 (JIS G 4305) and commercially pure titanium plates of TP340 (JIS H 4600), both of which had the dimensions of 2-mm thick. × 80-mm width × 200-mm length. The testing filler metals used were the commercially available 1.2-mm dia. wires of five different types of Cu-Si, Cu, Cu-Ni, Ni-Cu, and Ni. The typical chemistries of the filler metals are shown in Table 1.

Table 1 Chemical compositions of filler metals

Filler metal		Chemical composition (wt. %)					
Type	JIS	Cu	Si	Mn	P	Fe	Ni
Cu-Si	YCuSi A	96.2	2.37	1.2	<0.01	—	—
Cu	YCu	98.8	0.29	0.38	<0.01	—	—
Cu-Ni	YCuNi-1	bal.	0.03	0.92	0.002	1.01	10.1
Ni-Cu	YNiCu-7	bal.	0.22	2.79	0.007	0.24	65.6
Ni	YNi-1	0.01	0.01	0.15	0.001	0.06	96.3

2.2 Laser brazing process

The semiconductor laser used in this study was the four-wave mixing type with the wavelengths of 910, 940, 980, and 1030 nm, featuring a maximum laser power of 4 kW and a process fiber diameter of 1 mm. The beam spot diameter was kept constant at 3.0 mm. The filler metal was fed to the lap fillet joint from the front of the traveling direction. An argon shielding gas was fed at 20 L/min coaxially with the filler metal.

In the examination of the brazing conditions with a commercially available Cu-Si type filler metal, the laser power was kept constant at 2 kW, and laser-beam irradiation angle, the shift of beam aiming position, brazing speed, and filler-metal feed speed were varied. The laser irradiation conditions are shown in Fig. 1, and the brazing conditions are shown in Table 2.

Secondly, with the brazing condition that resulted in an acceptable weldment without crack in the use of Cu-Si type filler metal, the type of filler metal was changed to examine its effect on the characteristics of the brazed joint. The brazing conditions are shown in Table 3.

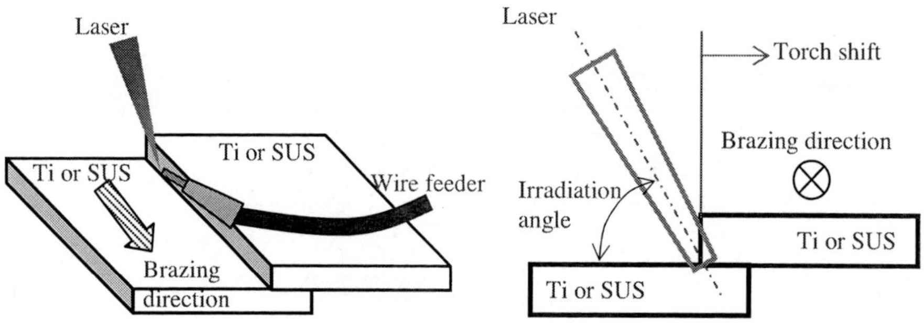

Fig.1 Laser irradiation condition

Table 2 Brazing test conditions for optimizing the dissimilar laser brazing process of Ti and SUS

<table>
<tr><th rowspan="2">No.</th><th rowspan="2">Joint geometry</th><th colspan="2">Filler metal</th><th rowspan="2">Irradiation angle (°)</th><th rowspan="2">Brazing speed (m/min)</th><th rowspan="2">Filler metal feed speed (m/min)</th><th rowspan="2">Torch shift (mm)</th></tr>
<tr><th>Type</th><th>JIS</th></tr>
<tr><td>1</td><td>Up Ti ⁄Under SUS</td><td rowspan="10">Cu-Si</td><td rowspan="10">YCuSiA</td><td rowspan="2">80</td><td rowspan="4">0.7</td><td rowspan="6">1.1</td><td rowspan="8">0.2</td></tr>
<tr><td>2</td><td rowspan="9">Up SUS ⁄Under Ti</td></tr>
<tr><td>3</td><td>70</td></tr>
<tr><td>4</td><td>60</td></tr>
<tr><td>5</td><td rowspan="6">80</td><td>0.8</td></tr>
<tr><td>6</td><td>0.6</td></tr>
<tr><td>7</td><td rowspan="4">0.7</td><td>1.3</td></tr>
<tr><td>8</td><td>0.9</td></tr>
<tr><td>9</td><td rowspan="2">1.1</td><td>0</td></tr>
<tr><td>10</td><td>0.4</td></tr>
</table>

Table 3 Brazing conditions for evaluation of filler metals experiment

<table>
<tr><th rowspan="2">No.</th><th colspan="2">Filler metal</th><th rowspan="2">Irradiation angle (°)</th><th rowspan="2">Brazing speed (m/min)</th><th rowspan="2">Filler metal feed speed (m/min)</th><th rowspan="2">Torch shift (mm)</th></tr>
<tr><th>Type</th><th>Standard JIS</th></tr>
<tr><td>2</td><td>Cu-Si</td><td>YCuSiA</td><td rowspan="5">80</td><td rowspan="5">0.7</td><td rowspan="5">1.1</td><td rowspan="5">0.2</td></tr>
<tr><td>11</td><td>Cu</td><td>YCu</td></tr>
<tr><td>12</td><td>Cu-Ni</td><td>YCuNi-1</td></tr>
<tr><td>13</td><td>Ni-Cu</td><td>YNiCu-7</td></tr>
<tr><td>14</td><td>Ni</td><td>YNi-1</td></tr>
</table>

2.3 Nondestructive and destructive tests

Penetrant test, macroscopic test, Tensile shear test, and EPMA analysis were carried out to evaluate the braze joints.

3. Test results and discussions

3.1 Effect of joint arrangement on brazed joint characteristics

The result of the penetrant test on the brazed joint surface is shown in Fig. 2. The titanium/stainless steel brazed joint (No.1) exhibited an indication of crack. In contrast to this, on the stainless steel/titanium brazed joint (No.2), no indication of crack was observed as shown in this figure. From the above-mentioned results, the stainless steel/titanium joint can be considered appropriate for the top plate/bottom plate material combination

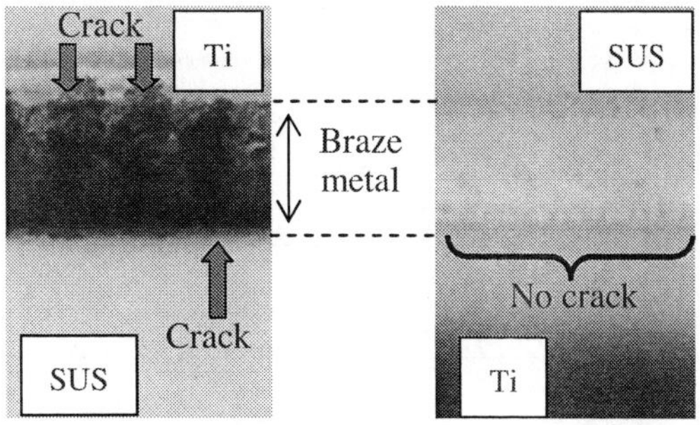

(a) Titanium/SUS(No.1) *(b) SUS/Titanium(No.2)*
Fig. 2 Results of penetrant testing of different joint arrangements

3. 2 Effects of brazing conditions on joint strength

Fig.3 shows the effect of representative brazing parameter on the tensile-shear fracture force. As the beam irradiation angle and brazing speed increased, the tensile-shear fracture force tended to increase.

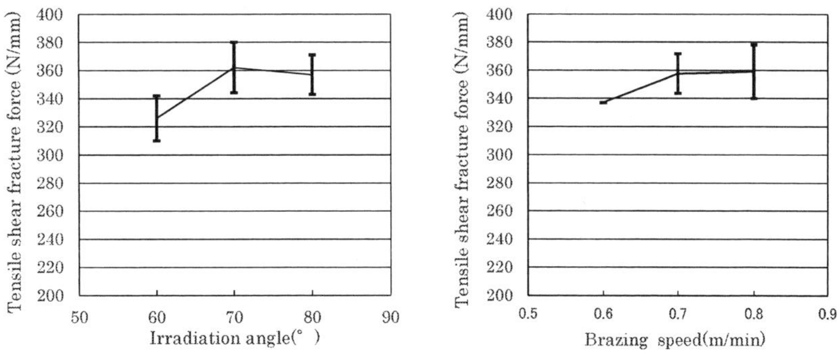

(a) Irradiation angle *(b) Brazing speed*
Fig.3 Effects of brazing parameters on the tensile shear fracture force of each brazed joint
(Filler metal: Cu-Si, Laser power: 2kW const.)

The results of SEM observation and EPMA analysis of the joint made with brazing condition No.2 are shown in Fig. 4. The brazed joint and SUS side consists of the mixture of stainless steel and copper alloy structures. On the other hand, in the vicinity of the interface on the titanium side, the compound phases of Ti_2Cu, TiCu, and Ti_2Cu_3 are generated in this order in the width of approx. 50 μm, expanding from the area close to the titanium-side interface towards the braze metal.

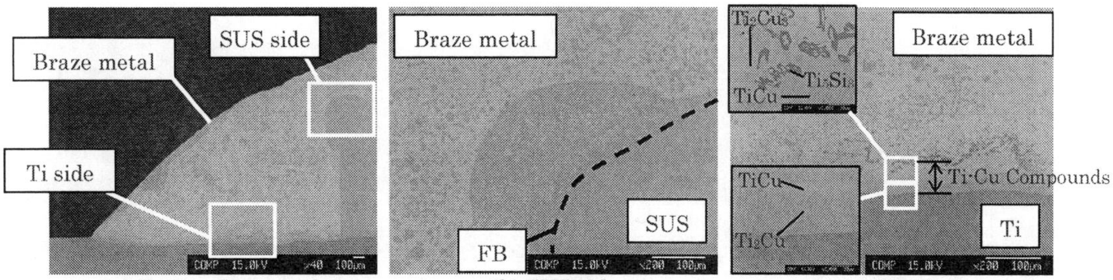

(a) Entire photograph *(b) SUS side* *(c) Ti side*
Fig.4 SEM microstructure showing the joint cross-section of SUS/ Titanium(No.2)

Fig.5 shows the tensile shear stress as a function of titanium-side penetration area. Clearly, as the titanium-side penetration area increased, the tensile shear stress tended to decrease. With an increase in the titanium-side penetration area, the amount of interfused titanium in the brazed joint may have increased, thereby increasing the quantity of Ti-Cu system compound. That is, the increase in the amount of Ti-Cu system compound may have caused the decrease in the tensile shear stress of the brazed joint.

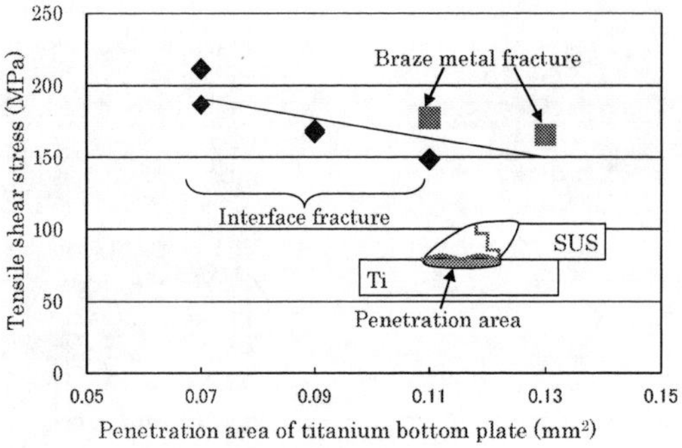

Fig.5 Relation between penetration area and tensile shear stress of the brazed joint(Cu-Si filler metal)

3.3 Examination of filler metals

Fig.6 shows the results of tensile shear test of SUS/ Titanium brazed joint for different filler metals. The tensile shear stress of the joints made with Cu-Si type, Cu type, and Cu-Ni type filler metals exceeded the minimum strength of 140 MPa specified by the JIS standard for the titanium clad steel. Furthermore, all the brazed joints made with Ni-Cu type and Ni type filler metals contained cracks, and thus the tensile shear test could not be executed.

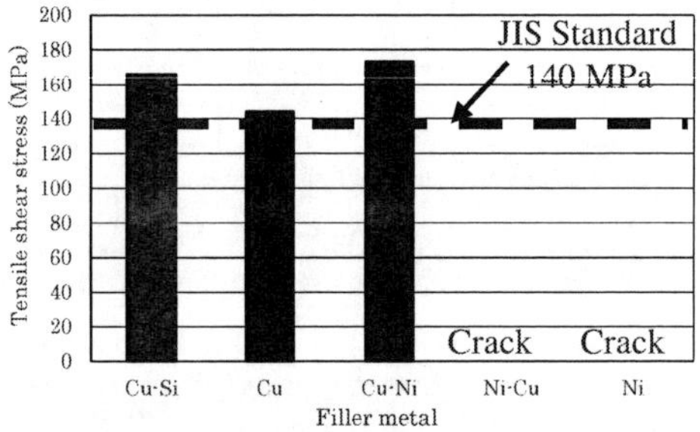

Fig.6 Tensile shear stress of Titanium/SUS brazed joint for different filler metals

4. Conclusions

(1) From the results of examining the material arrangement of the lap joint by using Cu-Si type filler metal, the joint arrangement of stainless steel (top plate)/titanium (bottom plate) was more suitable than that of titanium (top plate)/stainless steel (bottom plate).

(2) The tensile-shear fracture force can be affected by the amount of Ti-Cu system compound. For increasing the tensile-shear fracture force of the brazed joint, it is important to lessen the penetration area in the titanium plate.

(3) With Cu-Si type, Cu type, and Cu-Ni type filler metals, no crack was observed in the optimized brazing conditions, and the brazed joints satisfied the tensile-shear fracture strength specified by the JIS standard for the titanium clad steel.

References

[1] H. Terashi, *et al*: "Direct lining method of Ti on steel structure. Part 3," *Materials and Processes*, Vol.18, No.3 (2005), pp.534.

[2] Z.Hiraga, *et al*: "YAG Laser Welding of Pure Titanium to Stainless Steel," *Preprints of the National Meeting of Japan Welding Society*, Vol.67 (2000), pp.250-251.

Rotary Friction Welding – Advanced Control Techniques

Daniel C. Adams
Manufacturing Technology, Inc.,(MTI) South Bend, Indiana USA
Email: adams.daniel@mtiwelding.com, Phone: +1.574.233.9490

Abstract
Friction welding dates back to the early 1950's when companies commercialized rotary fiction welding for joining a wide range of metallic products. Friction welding is a solid-state welding process that generates heat through mechanical friction between workpieces in relative motion to one another, with the addition of an axial force to plastically displace and bond the materials. Because no melting occurs, the benefits associated with friction welding are significant and have been well documented.

A new rotational friction welding machine (Tri-Mode) used for this research combines the widely established processes known as 'continuous drive' and 'inertia' systems. The limitations and advantages of the two previously known systems are summarized and a description of the new control modes will be presented, together with weld data plots to explain it in practical terms.

The Tri-mode uses a high performance Electric Spindle Drive Motor directly coupled to the Spindle. New advanced control techniques have been developed using this capability combined with advanced control algorithms.

Torque Modulation is a control technique that is used for Upset Control. This technique uses the ability to close the loop to accurately repeat a previously defined weld Upset profile. Torque Modulation for Upset Control can be extended to Length Control by adding Dynamic Profile Modification where the Upset profile is mathematically scaled to target a new target Upset.

Orientation is a control technique that enables the two parts being welded to be radially aligned after welding. The controller stores a previously created weld profile that maps the Spindle Position over the weld time. MTI uses this profile to close the loop on Spindle Position enabling accurate control of the final Spindle Position.
These advanced control techniques and the Tri-Mode Welder make previously difficult to weld finished parts a reality.

Introduction
The Tri-Mode welding machine made it possible to develop the advanced control techniques that combine the advantages of both the standard Inertia Friction Welder and Direct Drive systems. Before describing the advance control techniques developed with the Tri-mode and the associated advantages the processes and associated limitations will be discussed in some detail.

The main difference between a standard Inertia Friction Welder and the Tri-Mode Friction Welder is that the Tri-Mode machine can execute the three main types of Rotary Friction Welding described below, while an Inertia Friction Welder can only execute the Inertia Friction Welding cycle.

<u>Direct Drive Friction Welding</u> – The Direct Drive Friction Welding (DDFW) cycle uses a directly coupled electric motor to provide the energy needed for welding by using relative motion required to create frictional heat. The DDFW cycle typically consists of rotating at a constant RPM to provide a constant energy source. Varying the axial load influences the torque at the weld interface and can control the rate of energy conversion. After Upsetting a desired amount of material from the work-pieces being welded (or a designated amount of time), the electric motor is typically decelerated quickly to stop rotation prior to applying full forge load, which will result in additional material upsetting (and Upset variation or final part length). This reduces the torque input to the work-pieces before the weld is completed which can affect the strength of the joint.
Limitations: Typical DDFW require a large spindle drive motor (when compared to IFW) not only to obtain optimum weld quality but to overcome the initial torque requirements at part contact. It is not uncommon for DDFW systems to start out with low axial loads in 1st or 2nd Friction Phase in order to avoid stalling the electric

motor. This decreases the rate at which energy can be converted in the weld. As the coefficient of friction decreases with the increasing temperature of the work-pieces, the axial load may be increased.

Inertia Friction Welding – The Inertia Friction Welding (IFW) cycle uses flywheels to store kinetic energy needed for welding. The Flywheels and the rotating work-piece are accelerated up to a predetermined speed and then allowed to coast, storing a predetermined amount of kinetic energy which is converted to heat by the relative motion and applied axial load. This energy conversion causes the rotation of the flywheels and attached work-piece to slow; when all of the kinetic energy is converted, the work-piece rotation has stopped, completing the weld cycle. This results in a very large torque peak at the end of the weld resulting in additional hot working of the material at the weld interface.
Advantages: IFW is recognized as the preferred process based on the transformation of the mechanical energy to thermal energy and is faster in IFW than the DDFW process. The faster transformation produces a joint with a smaller heat affected zone, which can translate to improved weld quality in some applications. The IFW cycle is the standard for Aerospace applications because the geometries and materials of aerospace components require high torque input to create frictional heat, and this torque input is easily scaled to larger parts using flywheels.

Hybrid Friction Welding – The Hybrid Friction Welding cycle combines the constant energy input phase of a DDFW cycle with the stored energy in the Flywheels of the IFW cycle to create a different weld cycle that can be used to optimize challenging part or material combinations. During the constant energy phase, the motor is driving and providing the welding energy. This issue of motor stalling at part contact is mitigated by the fact that there are flywheels attached, reducing the size of the spindle motor required. The end of the weld transitions to the flywheel stored energy conversion phase by disengaging the electric motor (commanding 0 torque) allowing the weld to be completed as if it were an IFW cycle.
Advantages: This process does not have the initial torque limitations of a DDFW cycle, but it can use the constant energy input phase. A Hybrid weld cycle typically ends with the flywheel aspect of an IFW including the significant torque peak at the end of the cycle for additional hot working of the materials.

Advanced Control Techniques

By introducing the advanced control techniques it is possible to overcome the limitations and leverage the advantages associated with the two standard rotary friction welding processes.

Pressure Modulation - is an Upset control technique that relies on the forge force (hydraulic pressure of the forge actuator) adjustments during welding to close the loop on Upset. Forge Force controls the rate of energy consumption, so increasing the forge force will increase the rate of energy consumption resulting in a slight increase in the rate of Upset. This will lead to an increase in total Upset. The opposite effect occurs when Forge Force is reduced resulting in a lower total Upset for the weld. The challenge with Pressure Modulation is change in Forge Load also causes changes in Tooling Compression that cannot be differentiated from weld Upset without linear transducer feedback in the tooling to have direct measurement of the position of the part back-ups.

Torque Modulation - is an alternative advanced control technique for Upset control. This technique uses the ability to close the loop to accurately repeat a previously defined weld Upset profile from a weld made with no Upset control. This profile of Upset versus time is used to compare against future welds where the error signal for the closed loop control compares the Upset versus time results of the profile to the actual weld. This error signal is converted to a Torque command in the spindle drive that either adds or removes energy form the flywheels in order to increase or decrease the Upset needed to reproduce the Upset in the profile weld. Torque Modulation for Upset Control can be extended to Length Control by adding Dynamic Profile Modification, which is described in detail in the next section. Torque Modulation is a much more efficient way of controlling Upset compared to Pressure Modulation because it increases and decreases the energy input directly (instead of relying on rate of energy conversion). This technique does not require a special back-up sensor for upset measurement, and it is not subject to tooling deflection variations. In addition, Pressure Modulation can only make a few adjustments during a weld due to the slow reaction of the hydraulic system and the slow reaction of the weld Upset to changes in Forge Load. The spindle motor can react hundreds of times faster and the energy adjusts impact weld Upset immediately.

The *(Fig.1)* shows a Weld Data File of a Torque Modulated weld on a production machine showing a weld made in closed-loop. This deceleration curve (red line) resembles a true IFW weld, and the black line shows the torque feedback of the motor making adjustments to the Energy Input in order to manipulate Upset (yellow line) in closed loop control. From the Torque trace you can see the large number of very fine (less than 1% available torque) adjustments that are made to control Upset. This was a long weld taking about 6.5 seconds from start of thrust to zero speed.

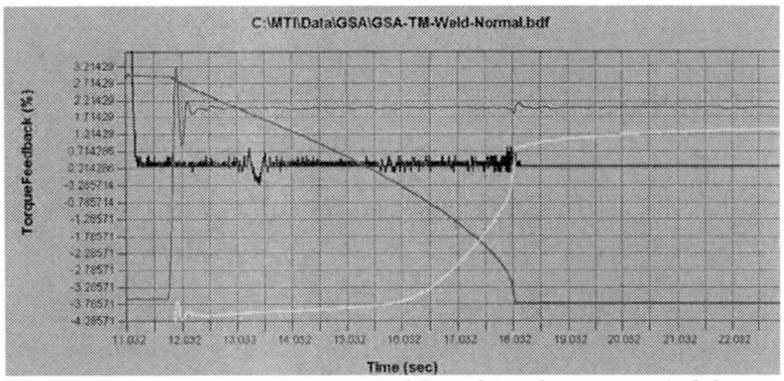

Figure 1-*MTI's weld graph showing an Inertia Friction Weld made with Torque Modulation*

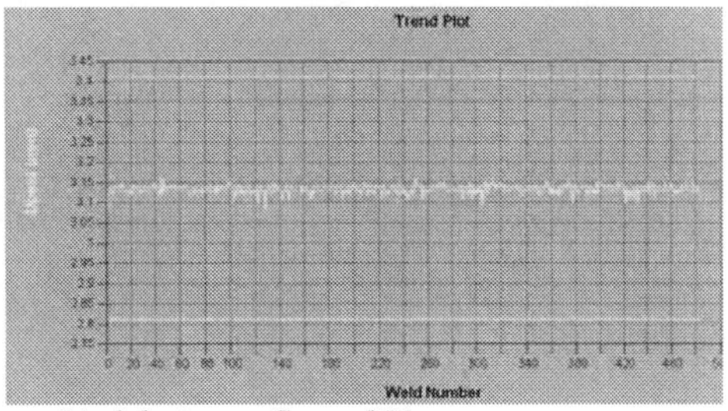

Figure 2: *Un-welded Diesel Piston that used Torque Modulation to Control Upset.*
Figure 3: *Trend Plot showing almost 500 Torque Modulated Welds with a total range of approx.. 0.07mm variation*

An example of Torque modulation is shown in (*Fig. 2 & 3)*, this diesel engine piston upset tolerance was .07 mm vs. a standard IFW upset tolerance of 0.44mm. Holding a tighter tolerance results in a significant reduction in pre-weld blank material and post machining time.

Dynamic Profile Modification - is an extension to Torque Modulation (or Pressure Modulation), that allows the user to target an Upset (different than the Upset of the profile weld) or target a Welded Length (if incoming part lengths are known and the target Upset needs to be calculated). This patented technique mathematically adjusts the profile (same profile used in Torque Modulation) to increase or decrease the desired Upset to achieve the new target. Performance and repeatability of Torque Modulated welds using Dynamic Profile Modification can achieve target upset better than a total range of .05 mm.

Torque Modulated Welds with Dynamic Profile Modification has been demonstrated in a production environment. This application consisted of a machined flange being welded to a stamping (*Fig. 4)*. The incoming part stack-up variations exceeded 0.3mm. Normal welding tolerances would have been guaranteed to achieve 0.5mm length variations, but the customer's requirement was to improve incoming part tolerances and to achieve 0.15mm finished length after welding (*Fig. 5*).

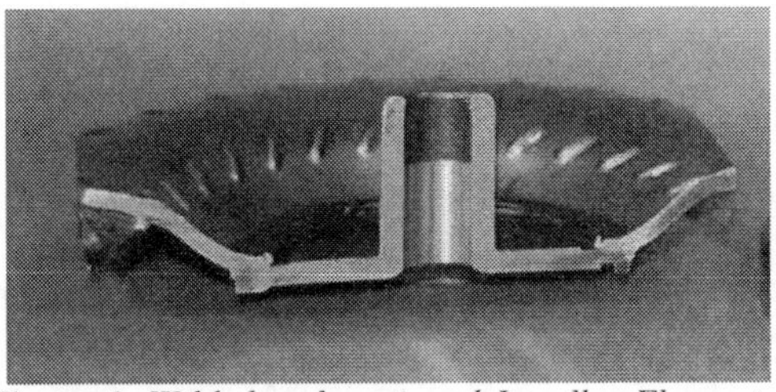

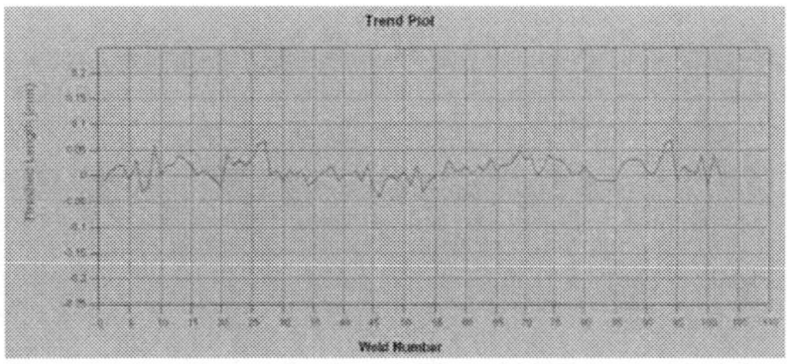

Figure 4: *Welded and sectioned Impeller Flange to Stamping* ***Figure 5***: *Trend Plot: 100 Torque Modulated Welds with Dynamic Profile Modification resulting in total range of 0.08mm of Welded Length variation.*

Weld Orientation - is an advanced control technique that allows the two parts to be radially aligned after welding. The controller stores a previously created weld profile that maps the Spindle Position over the weld time. MTI uses this profile to close the loop on Spindle Position enabling accurate control of the final Spindle radial Position. This desired location is typically repeatable to better than ±0.5°, but it is application specific. The weld Orientation technique can be applied to the DDFW weld cycle and Hybrid cycle; however, most accurate results will be based upon the true IFW cycle.

(Fig. 6 and 7) show an aluminum drive shaft with the yolk welds on either end oriented to each other and a similar weld plot. This application was a thin walled aluminum tube weld that oriented to less than 0.5 degrees of target position. This is a true IFW weld, and the black line shows the torque feedback of the motor making adjustments to the Spindle Position in closed-loop control. From the Torque trace it appears that most (but not all) of the effort was done with braking. You can see that this weld takes less than 0.7 seconds from start of thrust to zero speed.

This application also required a very high surface velocity due to the material and geometry. This is a further challenge to weld Orientation because the of the spindle drive motor requirement. In this case the weld RPM was about 3 times higher than the base speed of the motor, so the majority of the weld is completed prior to being in an RPM where the motor has full torque capability.

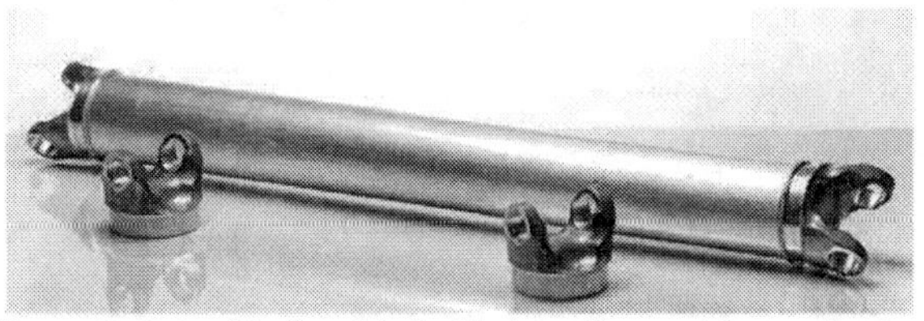

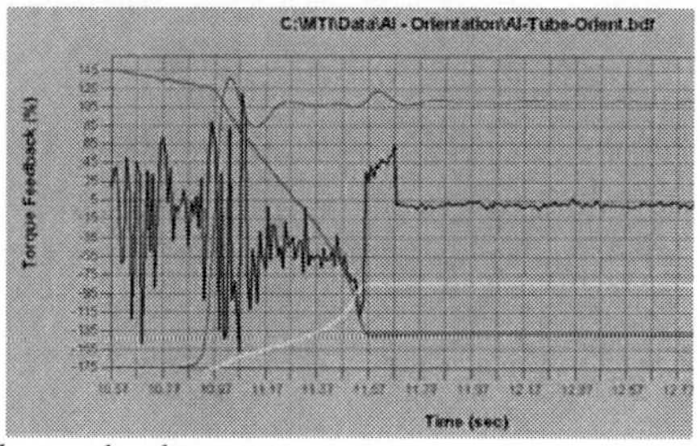

Figure 6: *Drive Shaft welded with the yolks on either end oriented to each other.*
Figure 7: *Weld Plot of an Orientation Control of IFW of a larger diameter aluminum tube*

Conclusion

While Rotary Friction Welding (RFW) dates back to the early 1950's it is still viewed as a superior joining technology for a wide range of metallic products. The advanced control techniques applied to RFW offer improved length and alignment control. Torque Modulation and Torque Modulated Welds with Dynamic Profile Modification are two different advanced techniques that improve the Upset/Length repeatability. While the results are dependent upon the incoming part tolerances, these techniques allow improved results that approach maching tolerances. Weld orientation is also application specific, but the MTI technique is a significant improvement over historical capabilities. These improved part tolerances will be critical to many new and existing near net shape component applications. Holding tighter tolerances in any of the three rotary friction welding cycles results in a significant reduction in pre-weld blank material and post machining time, hence lowering costs while providing a superior weld.

The Tri-Mode rotational friction welding machine was critical to the development of these new control modes. This system will be instrumental in the optimization and implementation of these control modes on new applications. The improved performance and repeatability associated with these advance control techniques makes producing stronger, less expensive, near net shape parts a reality.

Rotary Friction Welding of Large Molybdenum Tubes

M. Stütz, N. Enzinger
Institute of Materials Science and Welding, Graz University of Technology, Austria
markus.stuetz@tugraz.at, norbert.enzinger@tugraz.at

J. Wagner, N. Reheis, H. Kestler
Plansee SE, Reutte, Austria
josef.wagner@plansee.com, nikolaus.reheis@plansee.com, heinrich.kestler@plansee.com

E. Raiser
Klaus Raiser GmbH & Co. KG, Eberdingen, Germany
elmar@raiser.de

Abstract

In the scope of this work, a test series was carried out on a heavy-duty continuous drive Rotary Friction Welding machine, joining commercially pure molybdenum tube samples of cross sections larger than 4,000 mm². The experiments show that the process behaves differently compared to steel tubes of the same geometry. The friction phase and the plasticization process of the pure refractory metal molybdenum require special investigation. Moreover, suitable welding parameters have to be defined. By using custom process monitoring and adapted machine control, it was possible to achieve successful welds. Subsequent penetration and ultrasonic tests, as well as metallography, revealed sound joints. Nevertheless, a further investigation for a better understanding of the behavior of Mo during friction welding is necessary.

Introduction

Molybdenum is a refractory metal with enhanced performance at high temperatures. *Table 1* lists properties, which are considered the most important for friction welding, of Mo and a low-carbon steel for comparison. Molybdenum is frequently used in technical applications, where e.g. excellent high-temperature strength, high thermal conductivity are crucial requirements. Among others, Mo-tubes are used as sputter targets for producing TFT-monitors and solar panels [1]. The production of tubes longer than as-sintered is important for processing route in this industry. Hence, welding is of major interest.

Table 1: Selected physical properties of molybdenum [2] *and a low-carbon steel AISI 1000 series* [3] *at room temperature*

	Molybdenum (99.96%)	Low-carbon steel AISI 1000 series
Crystal structure	bcc	bcc, fcc
Density ρ (g/cm³)	10.22	7.85
Melting temperature T (°C)	2,623	1,430
Recrystallization temperature T_{Rxx} (°C)	1,000-1,200	600-700
Coefficient of thermal expansion α (10^{-6}/K)	4.8	12.0
Specific thermal capacity c (J/g K)	0.254	0.511
Thermal conductivity λ (W/m K)	138	55
Thermal diffusivity $\lambda/(c{\cdot}\rho)$ (10^{-6} m²/s)	53	14

Friction welding is a solid state welding process, where the required heat is generated by relative movement between two parts. The heat generation is accounted to friction and large plastic deformation of the joining partners. However, in literature different theories about the actual cause of heat generation are discussed [4]. The maximum temperature reached at the friction interface is not clear either, but it is agreed that liquefaction and subsequent re-solidification do not occur. Therefore, the joining process is less prone to typical welding defects observed in fusion welding (e.g. pores). High temperatures and large plastic deformation cause dynamic recrystallization which results in a finely grained structure. The process for common materials like steels typically is stable within large parameter windows and suitable for wide commercial application without deep knowledge of the underlying physics.

Friction welding of pure molybdenum and its alloys was investigated in some publications [5–7], however, process parameters, their scalability to larger cross sections, and the parameter influences were not addressed.

Experiments

For this work, test series were carried out on a heavy-duty continuous drive rotary friction welding (RFW) machine (max. forge force 1,250 kN), welding tubes of steel to steel and Mo to Mo. To gain a sound insight of the process, custom measuring equipment was installed and the following process parameters were recorded: axial force, upset, spindle speed, and motor load.

Based on data from literature [5] and preliminary experiments [7], the required process parameters were estimated. Literature states that with increasing high-temperature strength, the friction pressure and thus energy input has to be raised. Moreover, it is recommended to increase both the relative speed and the friction pressure with increasing high-temperature strength. [8]. For the current experiments, a tube format with 4,400 mm^2 (Ø130x10 mm) was used. The utilized RFW machine was able to provide up to 150 kW driving power. Hence, the available power to cross section ratio was 34 W/mm^2.

Against the theory of requiring more friction pressure, the first welding trials of molybdenum with large cross sections revealed that the heat generation was sufficient at lower pressure values. Furthermore, the good heat conductivity in combination with the low thermal capacity of molybdenum (*Table 1*) lead to a wide heat affected zone and significant cooling of the welding zone. Therefore, the tube specimens were pre-heated in a furnace to 450°C, to 1) reduce the temperature gradient and thus reduce the heat flux away from the weld zone, 2) enhance the plasticization process at shorter welding times leading to a reduced heat input necessary, and 3) also to stay above the ductile-brittle transition temperature.

The final process parameters for welding Mo were defined to a friction stage, where the friction pressure was gradually increased from 10 to 15 MPa to prevent an overload of the RFW machine at the initial contact. The forging pressure was defined at 152 MPa. For the steel sample, a standard program was used, where the friction pressure was increased to 42 MPa after 2 s friction time. The welding speed was set to 900 min^{-1} to achieve a maximum possible relative velocity between the welding specimens. In most welds the motor of the welding machine could not maintain the desired welding speed. The friction time was set to 5 s, additionally, custom machine control was programed to initiate the forging stage if the spindle speed dropped below a certain limit value. With this setup, the time between spindle halt and forging stage was minimized. Finally, the forging pressure was applied and maintained for 10 s.

Results and Discussion

Fig. 1a and *b* show the measured course of friction pressure and upset of steel and molybdenum, respectively. It is observed that a similar friction pressure and process time for steel and Mo specimens resulted in substantially different upset courses. The AISI 1000 steel specimens showed a total upset of 5 mm – most of the upset occurred in the forging stage. The increase of friction pressure after 2 s friction time had a minor influence on the upset, a slight increase in upset is observed. At the present low pressure level during friction, steel specimens only reached a phase of slow plastic flow, which is also visible by the low upset rate of 0.1 mm/s.

On the contrary, molybdenum upset vigorously fast at an even lower pressure level between 10 and 15 MPa, as observed in *Fig. 1a* and *b*. The total upset measured for Mo was 26 mm, and 70% of the upset occurred during the friction phase. The upset rate derived from the measured upset was 7 mm/s.

In *Fig. 1a*, the spindle speed during the welding process of steel and Mo is compared. First, the spindle was accelerated to welding speed, second the weld pieces were brought in contact at t = 0 s. This is indicated by the slight drop in speed (*Fig. 1c*). In the case of Mo it was observed that upon friction contact, the welding machine operated at maximum power, while the speed gradually decreased and came to a total halt.

The comparison of the thermal diffusivities (see *Table 1*) shows that Mo possesses a four times higher value than steel. Therefore, the temperature is distributed very fast over a wide area, causing material softening of a large volume, which in the present case consumed the entire provided drive power on the one hand and lead to a faster shortening on the other hand.

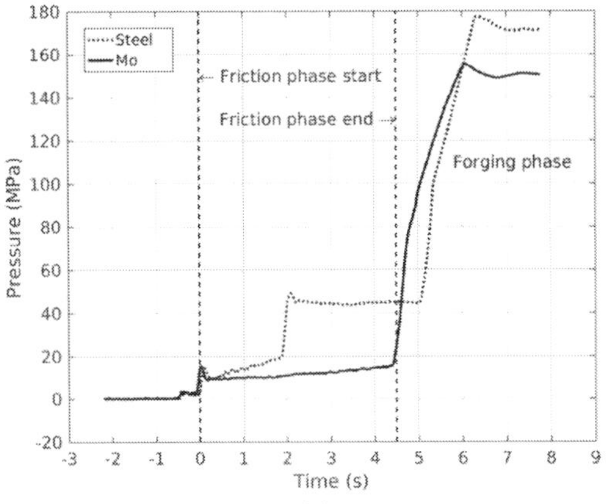

(a)

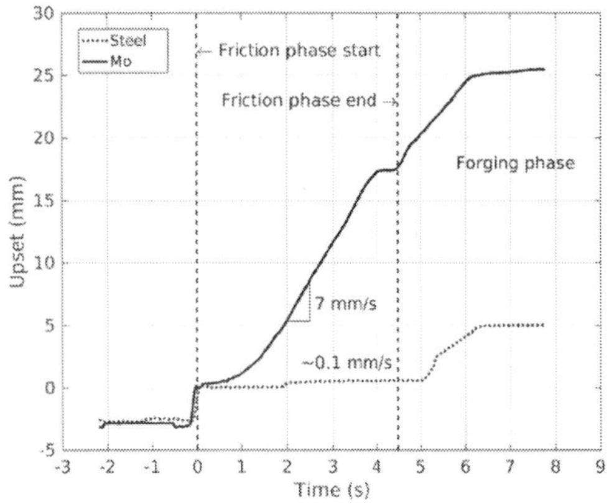

(b)

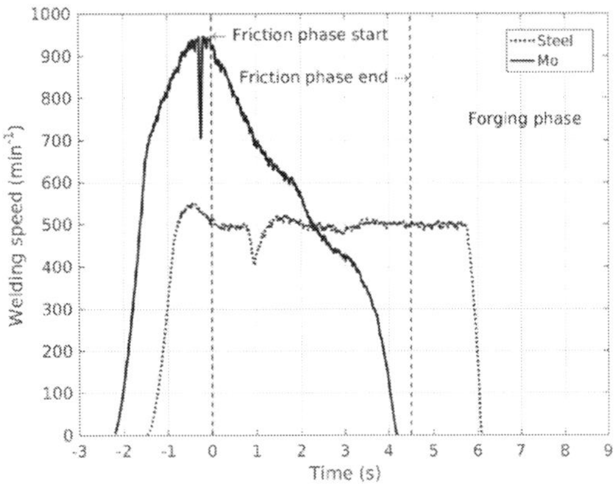

(c)

Fig. 1: Measured parameters for steel and Mo of (a) axial pressure, (b) upset, and (c) welding speed over time

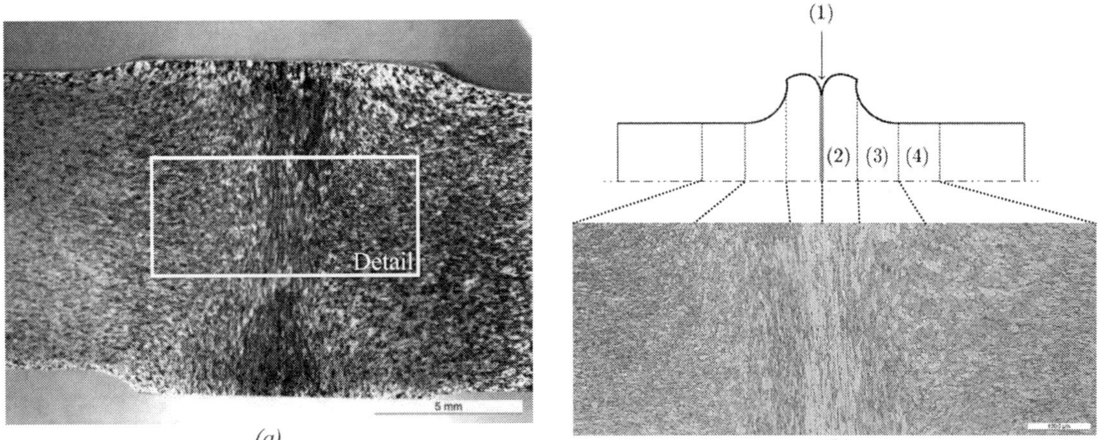

(a) *(b)*

Fig. 2: (a) Micrograph of a successful molybdenum weld and
(b) detail of the weld with (1) welding plane, (2) fully plasticized zone, (3) partly plasticized zone, and (4) undeformed zone [4]

The etched micrograph in *Fig. 2a* shows the microstructure of a molybdenum friction weld. No macroscopic or microscopic defects are observed in the investigated cross sections. The detail in *Fig. 2b* shows dynamically recrystallized and recovered grains, which exhibit large plastic strains. The orientation is radially in the vicinity of the contact plane, which is characteristic for rotary friction welds [4]. In the ultrasonic tests some fusion defects were observed. These defects were not observed in every weld and they were accounted to limited power and insufficient clamping which resulted in an alignment deviation on the machine. The clamping was realized by tight fitted, face sided keys.

Conclusions

In this contribution, the attempt to friction weld molybdenum tubes with a large cross section is presented and compared to friction welded steel specimens of the same dimensions. Preliminary test series were carried out and the machine control was accustomed to optimize the welding process. The following was revealed:

1. It was shown that it is possible to successfully friction weld commercially pure molybdenum tubes.
2. The best welding parameters were low friction force, high power input, high speed, and short friction time.
3. For steel the upset was 5 mm in total, which is low regarding the dimension of the tubes. Contrary, the molybdenum tubes upset accounted to 26 mm and a high upset rate of up to 7 mm/s was observed.
4. Pure molybdenum showed a different plasticization behavior than steel. Experimentally, this was observed in the much more sudden and steeper increase of the upset, as well as the increased power demand, in spite of a reduced friction pressure than for welding steel.
5. The large plasticized zone, fast upset, and high drive power requirement suggests that an even more rapid energy input is required to oppose the fast heat dissipation and sudden plasticization of the molybdenum parts.
6. The pre-heating of the Mo specimens aided the welding process as it lowered the temperature gradient through the welding partners. On the contrary, it increased the size of the heat affected zone, with potential negative effects on the weld quality.

To conclude, the general guidelines of friction welding are well established and proved for steel. Regarding the refractory metal molybdenum, a different approach for parameter choice is required as the different heat generation and plasticization processes alter the requirements substantially. A systematic and more basic study of molybdenum during friction welding is currently in progress to further develop the rotary friction welding process of molybdenum and its alloys.

Acknowledgments

The K-Project Network of Excellence for Metal JOINing is fostered in the frame of COMET – Competence Centers for Excellent Technologies by BMWFW, BMVIT, FFG, Land Oberösterreich, Land Steiermark, Land Tirol, and SFG. The program COMET is handled by FFG.

References

[1] A. F. Shields, *Applications of Mo Metal and Its Alloys*. International Molybdenum Association, 2013.
[2] "Info-Booklet Molybdenum Material Properties and Applications," *PLANSEE SE*, 2013.
[3] "Overview of materials for AISI 1000 Series Steel," *Matweb Website*, 2016. [Online]. Available: www.matweb.com. [Accessed: 21-Jun-2016].
[4] M. Maalekian, "Friction welding – critical assessment of literature," *Sci. Technol. Weld. Join.*, vol. 12, no. 8, pp. 738–759, Nov. 2007.
[5] A. Ambroziak, "Friction welding of molybdenum to molybdenum and to other metals," *Int. J. Refract. Met. Hard Mater.*, vol. 29, no. 4, pp. 462–469, Jul. 2011.
[6] B. Tabernig and N. Reheis, "Joining of molybdenum and its application," *Int. J. Refract. Met. Hard Mater.*, vol. 28, no. 6, pp. 728–733, Nov. 2010.
[7] D. de Pretis, "Rotational Friction Welding of Titanium-Zirconium-Molybdenum (TZM)," Graz University of Technology, 2014.
[8] "Merkblatt DVS 2909-1 Reibschweißen von metallischen Werkstoffen - Rotationsreibschweißen: Verfahren, Begriffe, Werkstoffe." Deutscher Verband für Schweißen und verwandte Verfahren e. V., 2009.

Importance of flow stress based on microstructure evolution in modelling Inertia Friction Welding

F. Di Simone, J.W. Brooks, H.C. Basoalto
School of Metallurgy and Materials, University of Birmingham, Birmingham, United Kingdom
fxd313@bham.ac.uk
J.Brooks@bham.ac.uk
H.Basoalto@bham.ac.uk

S. Bray
Rolls-Royce plc., Derby DE24 8BJ, United Kingdom
Simon.Bray@Rolls-Royce.com

Abstract
Inertia Friction Welding (IFW) is a solid state joining process widely used for Ni- based superalloys disks because of the high quality of joint that can be obtained. The self-cleaning nature of the process ensures a clean weld zone that allows good performance of the joint during operation. However to define the welding parameters required to obtain a good joint it is beneficial to model the process and see how changes of the key process variables can influence the weld zone microstructure prior to conducting costly experimental trials. Any successful model requires a definition of the flow behaviour of the alloy to be welded and although flow stress curves can be obtained from experimental trials this is problematic for the conditions relevant to IFW where Thermo-Mechanically Affected Zone (TMAZ) is characterized by high values of strain rate and temperature, so elevated that it is difficult to obtain the flow response experimentally. Furthermore the use of interpolated experimental data in simulations can give rise to numerical instability. It is therefore necessary to use a mathematical model to predict the flow stress over the wide range of values required. The current models for high volume fraction Ni-based superalloy show a lack of sensitivity at high strain rate values, however, they do not take into account the microstructural state of the material so that, for example, the models do not capture the dissolution of particles, evolution of the mobile dislocations or the dynamic recrystallization that actually occur during welding. The present work focuses on the prediction of the γ' particle dispersion across the weld and the evolution of grain size during the process.

Introduction
The new generation of high volume fraction gamma prime (γ') strengthened nickel-based superalloys is used in high pressure compressor and turbine disks thanks to their ability to maintain their mechanical properties at elevated temperatures. The chemical composition and the heat treatments have been carefully selected to obtain a high volume fraction of strengthening γ' particles, frequently present in a tri-modal distribution, with a homogeneous grain size. Such alloys are not easy to weld, such that enhanced micro-cracking has been found when solidification occurs after welding [1]. Inertia Friction Welding (IFW) is the solid-state joining process of choice for Ni-based superalloy rotor assemblies as it minimises the occurrence of melting during joining. This type of joining process consists of bringing an axis-symmetric and rotating workpiece in contact with a second stationary workpiece under axial load. If the right key process variables (KPVs) are chosen, such as the rotating speed and the applied pressure, the frictional heating is enough to soften the two interfaces without introducing melting thus allowing the pieces to be joined. The challenge in using this technique is that the interfaces can reach such high temperatures and strain rates that the microstructure in the welding zone is heavily modified and recrystallization and γ' dissolution can occur potentially causing a lack of integrity in the joint. If the microstructure changes, the flow stress behaviour changes as well and it is, therefore, important to capture how the microstructure evolves during IFW. In this work the evolution of γ' particles and grain size have been predicted in a form suitable for incorporation into a new set of constitutive equations that capture the microstructure evolution during IFW simulation.

Model setup
In order to simulate IFW process, the Finite Element Method (FEM) has been used. The model has been created with DEFORM v11.02 package using torsional 2.5D elements. It is a rigid-plastic model with a fine mesh (≈0.1 mm) at the interface graded into larger elements at the end of workpieces (≈1.5 mm) as shown in

figure 1a. This was done because the evolution of γ' happens in the TMAZ of the joint and far away from this region the temperatures do not reach high enough values to cause changes in microstructure.

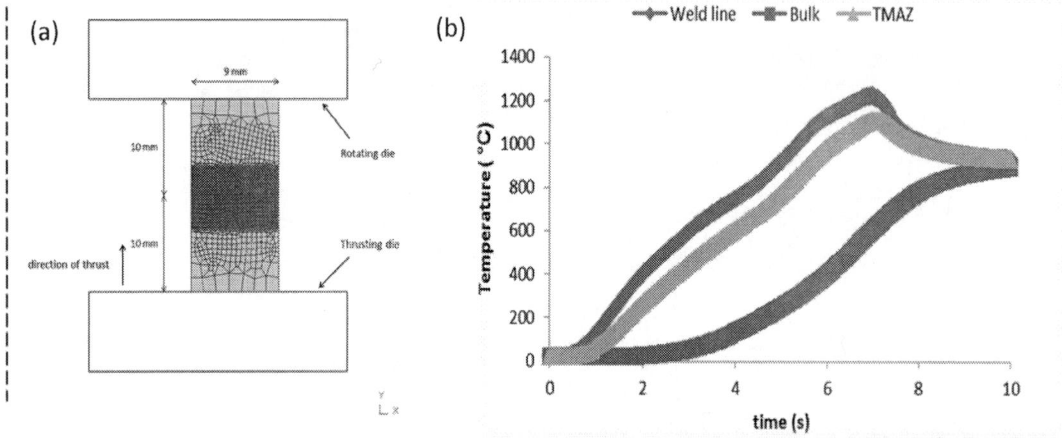

Fig.1 (a) Initial geometry of the two plastic workpieces with meshes; (b) predicted thermal fields

The simulation consists of four parts with 2 identical cylindrical workpieces attached to a rotating die and a thrusting die respectively. In order to ensure that the energy from the rotating die and the axial load were applied in a correct way the dies have a stick boundary condition with the workpieces while at the interface of the components a pure and constant frictional shear has been applied. The welding parameters, such as initial angular velocity and axial load, are the same as used in the experimental trials. The model has been carried out with a time-step of 0.001 seconds, until all the energy stored by the flywheel is depleted and the rotating die comes to a halt. A hybrid friction model was calibrated to obtain agreement with experimental rundown data to provide thermal profiles as close as possible to the actual welding trial.

Results

Three locations from the top workpiece were selected to extract temperature and strain rate evolution data from the welding model to provide the boundary conditions required for the microstructure prediction codes. These locations were on the weld line, on the TMAZ and one in the bulk of the component. Once calculated the γ' and grain size evolution outputs were used in the flow stress constitutive equations to predict the flow behaviour and allow comparison with the existing flow stress data.

γ' evolution during IFW

A mean field approach was chosen to predict the γ' evolution using the approach of M.J. Anderson et al. [2] which follows the research done by Svoboda et al. [3]. This consists of the application of the thermodynamic extremum principle to calculate the particle growth rate and atomic attachment rate solely as a function of alloy composition. Using this approach it is possible to capture particle coarsening and nucleation as well as the kinetics of a statistically representative number of particles. The model also provides a description of the evolution of the multi-modal particle size distributions.

The evolution the tri-modal γ' precipitate distributions have been predicted from the thermal profiles taken from the IFW model and Figure 2 represents the final distribution of γ' at the end of the weld simulation. The weld interface (purple curve) reaches such a high temperature (above the γ'phase dissolution temperature T_{solvus}) that one can see the dissolution of both primary and secondary γ' and the subsequent nucleation of re-precipitated γ' during the cooling phase. In the TMAZ (blue curve) temperatures are not high enough to dissolve completely the primary γ' but there is partial dissolution and shrinkage of secondary γ'. As anticipated, in the bulk of the component (green curve), there are no changes from the initial condition.

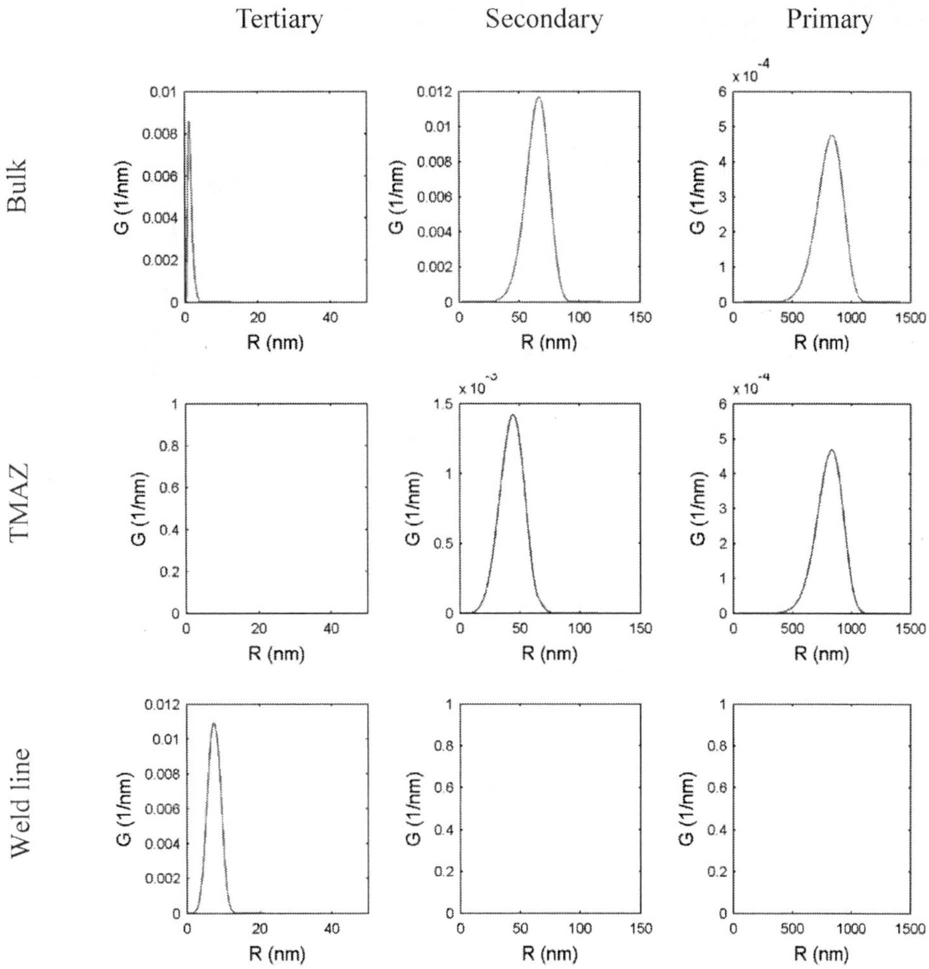

Fig.2 γ' distribution at the end of IFW simulation. Graphs are particles distribution vs particles radius.

Dynamic Recrystallisation (DRX) during IFW

During IFW the zone adjacent to the weld interface is affected by both high temperatures and strain rates and this can lead to grain refinement and to associated changes in the flow behaviour. In order to predict the DRX a code has been adapted for high volume fraction γ' Ni-based superalloys that is based on the work of Shen et al. [4]. The grain size is temperature and strain rate dependent and by calculating the rate of recrystallization it is possible to determine the volume fraction of recrystallized grains. Temperature and strain rate profiles have been taken from the IFW model from the same locations as shown in the previous section and input into the DRX model to obtain the grain size evolution. Figure 3 shows the evolution of the grain size distribution (volume fraction of grains with the same size) that changes when dynamic recrystallisation occurs. The green bars are the initial grain size distribution (the initial size is 6 μm, or 11 ASTM) while the yellow/orange bars represent the grain size distribution when the grain size has a higher value in ASTM (which means finer grains). From figure 3a it is apparent that at the weld line, which is affected by higher temperatures and strain rates, the grain size starts to change after ~6s and, as the welding process proceeds, new finer grains begin to appear such that when the welding is finished there is complete recrystallisation at the interface with a grain size of ~1.8 μm (or 15 ASTM). In the TMAZ (figure 3b) there is a different situation. Here the temperatures and strain rates are high enough to start the DRX process (at around 7s) but are not sufficient to reach complete recrystallisation at the end of the welding, such that at the end of the joining process there are new grains (~1.8 μm) in the TMAZ but with a volume fraction of only ~0.2. The bulk of the workpiece (figure 3c) does not reach the temperatures or strain rates required to initiate DRX and consequently the grain size distribution remains the same throughout welding.

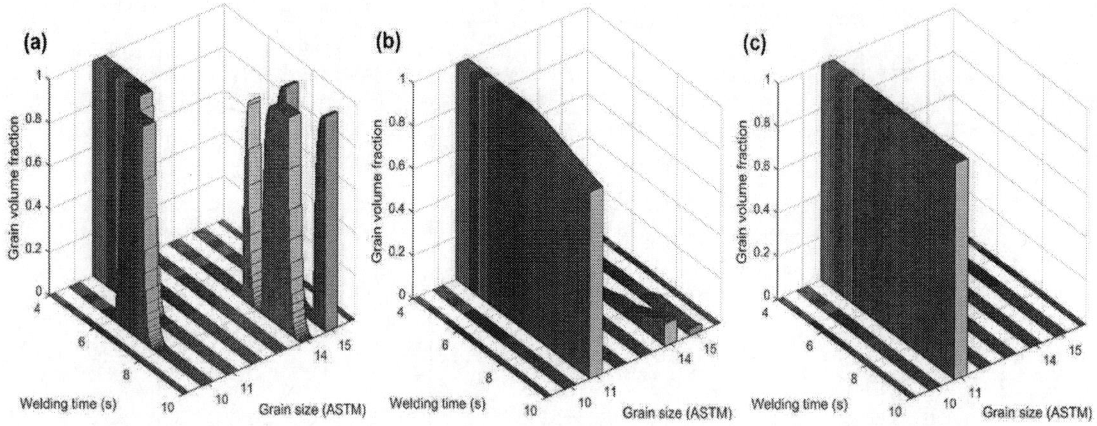

Fig.3 Evolution of the grain size during IFW model (a) in the weld line (b) in the TMAZ (c) in the bulk

Flow constitutive equations

A physics-based constitutive description of plastic flow has been developed that explicitly includes aspects of the precipitation behaviour and recrystallisation [5]. A number of micro-mechanisms has been included in the modelling approach: (i) back stress partitioning between plastically hard and soft phases (precipitates and grains); (ii) dislocation evolution; and (iii) grain size evolution.

Conclusion

A mean field model and a dynamic recrystallisation model have been used in order to capture the evolution of γ' particles and grain size respectively during an Inertia Friction Welding process. The outputs from these models have been inserted into new constitutive equations to provide improved predictions of material behaviour and it will be used in the future IFW simulations.

Acknowledgments

The authors thank Dr. R. Turner for his support in the initial FEM setup stage and Dr. M.J. Anderson for his support and assistance in the mean field code. The research project is financially supported by EPSRC and Rolls-Royce plc. One of the author (FDS) is grateful for the Armourers and Brasiers Award.

References

[1] Ojo, O.; Richards, N. & Chaturvedi, M. "Contribution of constitutional liquation of gamma prime precipitate to weld HAZ cracking of cast Inconel 738 superalloy", Scripta Materialia, 2004, 50, 641-646.
[2] M. Anderson,; A. Rowe, J. Wells, & H. Basoalto, "Application of a multi-component mean field model to the coarsening behaviour of a nickel-based superalloy", Acta materelia, 2016.
[3] J. Svoboda F.D. Fischer, P. F. & Kozeschnik, E. "Modelling of kinetics in multi-component multi-phase systems with spherical precipitates, I: Theory", Materials Science and Engineering A, 2004.
[4] G. Shen, S.L. Semiatin, R. Shivpuri, "Modeling Microstructural Development during the Forging of Waspaloy", Metallurgical and materials transaction A, vol 26A, July 1995, pp 1795-1803.
[5] H.C. Basoalto, "A micromechaniscs description of high temperature deformation of 2-phase alloys", Internal Report PR-2015-002, 2005.

Relative motion and growth of bonded area in ultrasonic bonding of 6061 aluminum alloy

Takuya Chiyozawa[1,a], *,Takumi Kamimura*[1,b] *Tomohiro Sasaki*[1,c],*Jonas Aditya Pramudita*[1,d]
[1]*8050 Ikarashi 2-nocho, Nishi-ku, Niigata 950-2181, Japan*

[a]*f15b129j@mail.cc.niigata-u.ac.jp, bf16b089a@mail.cc.niigata-u.ac.jp*
[c]*tomodx@eng.niigata-u.ac.jp,* [d]*jpramudita@eng.niigata-u.ac.jp,*

Abstract

Ultrasonic bonding process of 6061 aluminum alloy was discussed focusing on growth of bonded area at the bonding interface and relative motion involving bonding metals and a bond tool. In the bonding of 6061 aluminum alloy, the relative motion mainly occurred at the tool edge/bonding metal interface, which had no significant effect on growth of bonded area. In order to control the relative motion behavior, interfacial state was changed by partially anodizing on the bonding surface. The relative motion between specimens was enhanced because anodized region inhibited the bond formation, while initial bonds locally formed on non-anodized region developed. This caused significant plastic deformation at the bonding interface owing to shear stress concentration to the non-anodized region.

1. Introduction

Ultrasonic bonding is widely used in automotive industry for joining thin and soft metals such as pure aluminum, copper and nickel. With the development of high power transducer, the application range is extended to joining alloys. The bond process is basically achieved through removal of surface impurity by interfacial friction under normal force, leading to formation of initial bonds. The initial bond subsequently grows with its shear deformation. The relative motion between bonding metals induced by the bond tools has a critical role on these phenomena. In addition, since the relative motion is hindered with the growth of initial bond, the relative motion often occurs between the bonding metal and the bond tool which applies vibration and the normal force. This makes the ultrasonic bonding process be complicated. Thus, it is essential to understand the correlation of the microscopic phenomena and the relative motion behavior. The aim of the present study is to investigate how the initial bond develops by the relative motion. The interfacial state is controlled by adding surface impurity including anodized oxide film or lubricant, in an attempt to control the relative motion and enhance the growth of initial bond. The effect of interfacial state on the bond microstructure is discussed comparing with the relative motion behavior obtained high speed camera analysis.

2. Experimental method

AA6061 aluminum alloy sheets (AA6061-T6) with 0.5 mm thick and 0.8 mm thick were cut into specimens of 50 mm in length × 11 mm in width for upper specimen, and 50 mm in length × 11.5 mm in width for lower specimen. The longitudinal direction of specimens was parallel to the rolling direction. Anodizing was partially conducted on the bonding surface of lower specimen as shown in *Fig. 1*. After masking process by photolithography, the bonding surface was anodized in a solution of 8% oxalic acid with a voltage of 40V for 5 minutes. Then, resist pattern was removed by acetone. A lateral-drive ultrasonic welder with a power output of 2400 W and a vibratory frequency of 15 kHz was used for the experiment. The vibration amplitude under no-load condition was approximately 53 μm (peak to peak). As shown in *Fig. 2*, the lower specimen was clamped on an anvil with fixtures in the vibration direction. The specimens were jointed with 10 mm × 10 mm bonding tips which has knurled edge pattern (*Fig. 2*). Vibration was applied for 0-1000 ms under a constant normal force of 1470 N. Motion of the specimens and the weld tool during the bonding process were observed using a high-speed camera (100,000 frames/s). The observation area (24 × 128 pixels) included the tool edge (edge) and the specimens (*Fig. 2*). Subsets of this area (10 × 20 pixels) were set for the edge and the specimens. The image comprised of gray pixels with 256 bits was expanded by bi-linear interpolation. The resolution of the interpolated image corresponded to approximately 2.0 μm/pixel. Displacements in the vibration direction (x axis) of the edge and the specimen during vibration were measured using a digital image correlation method in the same manner as a previous study [1]. The data point of displacement (x-axis) for every time interval of 1 ms was subjected to Fourier transformation, and the spectrum amplitude was computed. Joint strength was evaluated by tensile test using t-bent specimen. The maximum load achieved in the tensile test was defined as joint strength in this study.

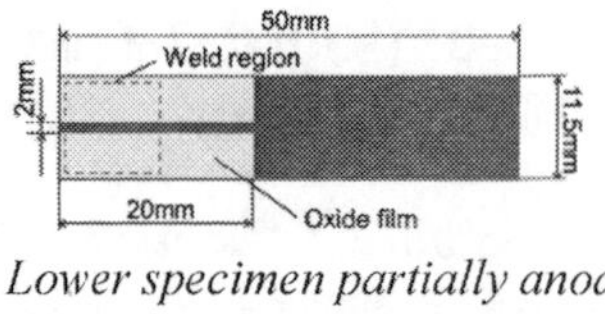

Fig. 1 Lower specimen partially anodized

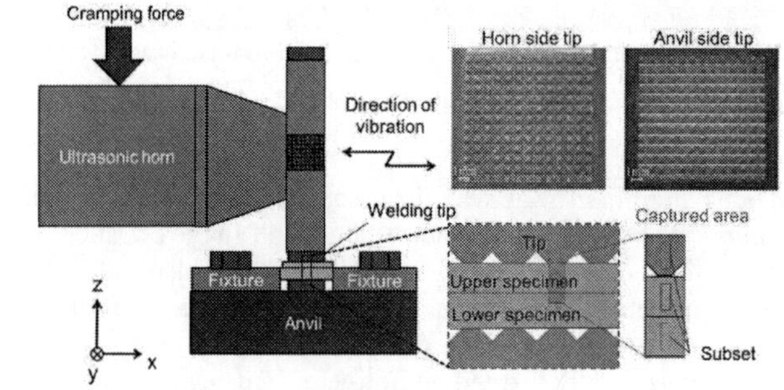

Fig. 2 Schematic of the ultrasonic bonding process in this study.

3. Result and discussion

3.1 Joint strength and fracture surface

Fig. 3 shows relation between joint strength and bonding time. The maximum load in the tensile test of the joint bonded using the non-anodized specimen (untreated joint) increased up to approximately 60 N, and it saturated. The joints showed interfacial fracture in all the bonding time conditions. On the other hand, the maximum load of the partially anodized joint increased up to 220 N with increasing the bonding time. *Fig. 4* shows fracture mode of the partially anodized joint. At the earlier bonding time of 400 ms, the fracture mode changed from the interfacial fracture to base metal fracture in which the fractured upper specimen remained on the non-anodized region of lower specimen as shown in *Fig. 4 (e)*. *Fig. 5* shows SEM micrograph of the fracture surface. It was observed that white spots which imply the formation of initial bonds were locally formed in the non-anodized region, and each of these expanded with increasing welding time from 200ms to 400ms as shown in *Figs. 5 (a)* and *(b)*. In the upper part of *Fig. 5 (b)*, the remained base metal of upper specimen was confirmed on the lower specimen, indicating the growth of initial bond. When joining the untreated specimens, the expansion of white spots was less observed despite of increasing bonding time from 200 ms to 600 ms as shown in *Figs. 5 (e)* and *(f)*. These findings suggest that the partially anodizing on the surface enhanced the growth of initial bonds locally formed in the non-anodized region, leading to increasing of joint strength.

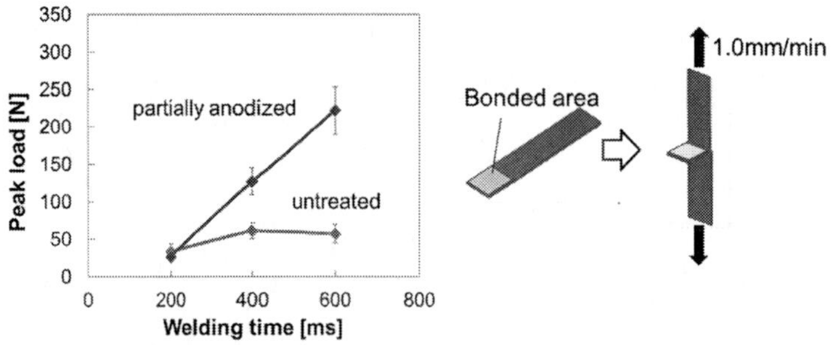

Fig. 3 Relation between joint strength and bonding time. The error bars indicate the scatter in measured values.

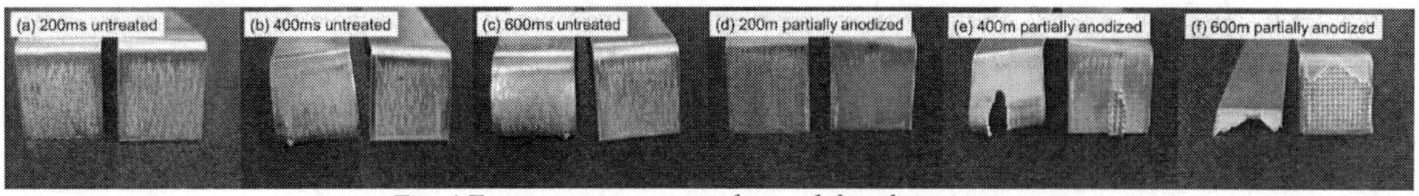

Fig. 4 Fracture appearance for each bonding time.

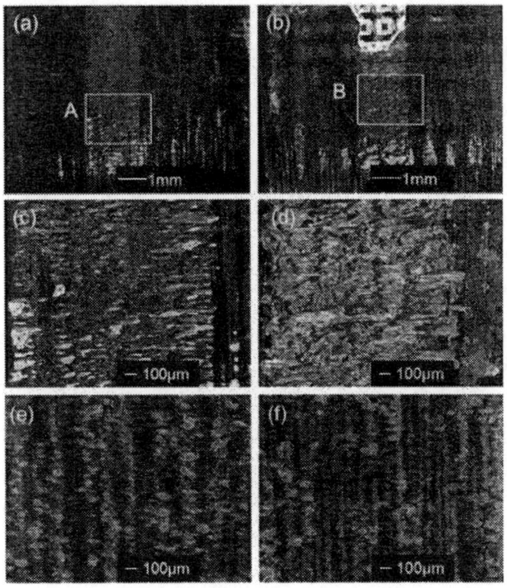

Fig. 5 SEM images of the fracture surface of AA6061 joint bonded using partially anodized specimen at the bonding time of (a)200 ms, (b)400 ms. (c) and (d) are higher magnification images of region A and B. respectively. (e)and (f) show the fracture surface of untreated specimens at 200 ms and 600 ms .

3.2 Effect of oxide film pattern to the relative motion behavior

Fig. 6 shows amplitudes of the tool edge, the upper specimen and the lower specimen plotted as a function of bonding time. When the untreated specimen was bonded (*Fig. 7(a)*), the amplitude of upper specimen decreased rapidly for the bonding time of approximately 50ms, then, it kept almost same amplitude as the lower specimen. It can be said that relative motion between the tool edge and the upper specimen predominantly occurred compared to the relative motion between the bonding materials. On the other hand, when bonding the partially anodized specimen (*Fig. 7(b)*), the upper specimen's amplitude kept larger amplitude than that of the lower specimen until the bonding time approximately 800ms. This result indicates that the anodized region enhanced the relative motion at the bonding interface, *i.e.*, the thick oxide film by the anodizing interrupted the formation of initial bonds.

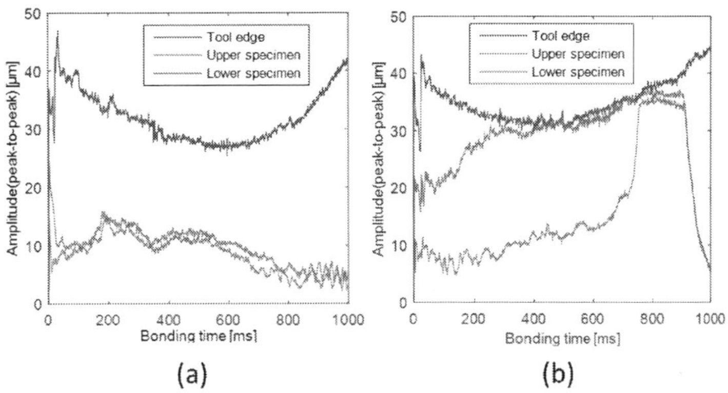

Fig. 6 Relation between amplitude and bonding time (a) untreated (b) partially anodized.

3.3 Weld microstructure

Fig. 7 shows cross-sectional micrograph of bonded joints at the bonding time of 1000 ms. When jointing untreated specimen, bonding interface was observed clearly as shown in *Fig. 7 (a)*, and few initial-bonds were confirmed in an enlarged micrograph of the interface (*Fig. 7 (c)*). Our previous work has been reported that the relative motion between tool edge and upper specimen causes the plastic flow involving the bonding interface, resulting in increasing the joint strength [1]. When jointing AA6061 alloys, the large relative motion between edge and upper specimen was observed (*Fig. 6 (a)*). However, as shown in *Fig. 7 (a)*, the plastic flow did not reach to the bonding interface. This is probably due to higher flow stress of AA6061. On the other hand, *Figs. 7 (b)* and *(d)* show significant plastic deformation at the center of bonding region which corresponds to the non-anodized region. At the center of weld region, the bonding interface was invisible, indicating occurrence of severe plastic flow. In contrast, outside of the bonding region, the interface was observed clearly owing to the thick anodized oxide film which inhibits the bond formation. The effect of partial anodizing can be concluded as follows. Friction resistance at the bonding interface was macroscopically reduced by the anodized surface via reduction of initial bond formation. This fact causes the large relative motion between materials. In other word, the vibration energy applied by the bond tool effectively propagated into the bonding interface. At the same time, the shear stress induced by the vibration was concentrated to the initial bonds partially formed on the non-anodized surface, leading to the plastic flow and increase of bond strength.

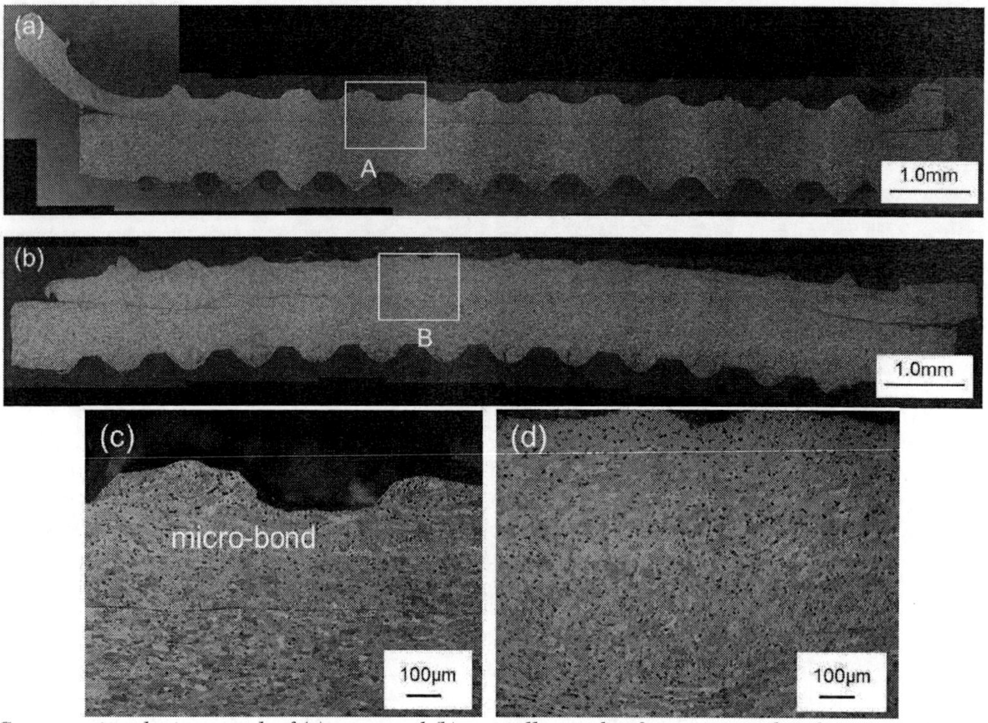

Fig. 7 Cross-sectional micrograph of (a) untreated (b) partially anodized specimen at bonding time of 1000ms. (c) and (d)are higher magnification images of regions A and B, respectively.

Conclusions

1. Joint using the partially anodized specimen showed higher joint strength than that using untreated specimen at longer bonding time of 400ms and 600ms. The fracture mode changed interfacial fracture to base metal fracture and fractured upper specimen remained on the non-anodized region of lower specimen at the welding time of 400ms.
2. Significant plastic deformation was observed at the center of bonding area which corresponds to the non-anodized region when joining partially anodized specimen. This resulted in the higher joint strength when joining partially anodized specimen.
3. Partially anodizing of the surface of specimen enhanced relative motion between specimens owing to inhibition of formation of initial bond by thick oxide film. This relative motion leaded to growth of the locally bonded area at the center of bonding area.

References

[1] T. Sasaki, T. Watanabe, T. Hosokawa and A. Yanagisawa: 'Analysis for relative motion in ultrasonic welding of aluminum sheet', Science and Technology of Welding and Joining, 18(2013), 19-24.

Plastic deformation behavior within AZ31B Mg alloy joints welded by ultrasonic spot welding

Yuichi Higashi
National Institute of Technology, Kagoshima College, Kirishima, Kagoshima, Japan
higashi@kagoshima-ct.ac.jp, +81-995-42-9103

Chihiro Iwamoto
Ibaraki University, Hitachi, Ibaraki, Japan
chihiro.iwamoto.77@vc.ibaraki.ac.jp, +81-294-38-5061

Abstract
In this study we examined the microstructure within an AZ31B Mg alloy joint which was welded using ultrasonic spot welding. And the relationship between the plastic deformation behavior within the joints and the weldability was evaluated. After welding, the weld diameter and the weld strength increased with the welding energy. The fracture modes of welded specimens transformed in to a pull-out fracture mode at higher welding energies. The oxide layers were observed at weld interface, and the plastic deformation behavior within the joints was revealed by analyzing the behavior of the oxide layers. At conditions which exhibited the pull-out fracture mode, the oxide layers were curved violently in the slope region of the joints. Consequently, the enhancement of the plastic deformation in the slope region contributed to the expansion of the weld diameter with an increase in the welding energy. And then, it was considered that the weld strength increased with the welding energy and the fracture modes transformed to the pull-out fracture mode.

Introduction
The Mg alloys, which are the lightest utility metals, are expected to serve as lightweight structural components in the automotive and aircraft industries [1], [2] and [3]. Use of the Mg alloys has been significantly increasing [4], since they have the best strength-to-weight ratio among all structural metals [1], [5] and [6]. However, Mg alloys are known to have intense anisotropy, low ductility and relatively poor workability at room temperature [7], [8] and [9], because its deformation has a strong tendency to slip on the basal plane [10].

In contrast, ultrasonic spot welding has been used in various industries such as electrical, electronics and especially in automotive industry [11], [12] and [13]. The friction by ultrasonic vibrations during the ultrasonic spot welding process breaks oxide layers on the faying surfaces, thus allowing solid-state bonding at contacts between clean metal surfaces [14] and [15]. Accordingly, ultrasonic spot welding is an attractive method because bonding is quickly obtained without extra heat in the air [16]. And then, it is the solid-state bonding technology in which energy efficiency is good for welding the structural components of the thin plates at the spot [11], [12], [17] and [18]. Furthermore, the application to the structural components had been expected from the point of view of energy conservation, because ultrasonic spot welding can be conducted with significantly less energy compared with resistance spot welding [11] and [18].

Our study applied ultrasonic spot welding to weld a commercial Mg alloy AZ31B. For the AZ31B alloys, the weld strength distribution and the microstructure evolution of ultrasonic spot welded joints as a function of the welding energy has been indicated [4] and [19]. In addition, since the plastic deformation behavior within the ultrasonic spot welded joints that depended on the macro shape of the welding tip has an effect on the weldability [16], to clarify the relationship between the mechanical properties and the microstructure of the joints would enhance the understanding of the ultrasonic spot welding technology to weld the AZ31B alloys. In the present study, the plastic deformation behavior within the joints as a function of the welding energy was evaluated, and then the relationship between the mechanical properties and the microstructure of the joints was discussed.

Materials and experimental procedure
As materials, AZ31B plates were machined in the strip from as-received rolled sheets in the direction parallel to that of rolling. The size of the specimens was 70 mm in length, 20 mm in width, and 0.76 mm in thickness. The faying surfaces of the specimens were ground using 2000 waterproof abrasive paper.

Fig. 1 (a) shows a schematic illustration of ultrasonic spot welding. Ultrasonic spot welding was conducted using an ultrasonic welder (Toshiba Mitsubishi-Electric Industrial Systems Corporation) and an oscillator (Sonomac Japan INC.). The constant power, the clamping force and the vibration frequency of the machine were 1200 W, 1100 N and 20 kHz, respectively. The specimens were welded at energy levels from 600 to 3500 J at the constant power. Fig. 1 (b) shows a welding tip used in the present experiment. The welding tip has a flat region of the initial contact surface of Φ5.0 mm which contacts with the specimens and 0.5 mm pitch knurling was machined on the surface. In addition, the welding tip has a slope region outside thereof. The shape of the joints transferred the shape of the welding tip. All samples for the mechanical properties investigation of the joints were welded in a lap joint with 50 mm overlap, as shown in Fig. 1 (c). The weld strength was investigated by a lap shear tensile test with a crosshead speed of 5 mm/min. All samples for the microstructure observation were also welded in the same manner. The cross-section of the joints was cut, as shown in Fig. 1 (d), and the microstructure within the joints was observed with a scanning electron microscope (SEM: JEOL JSM-6010LA).

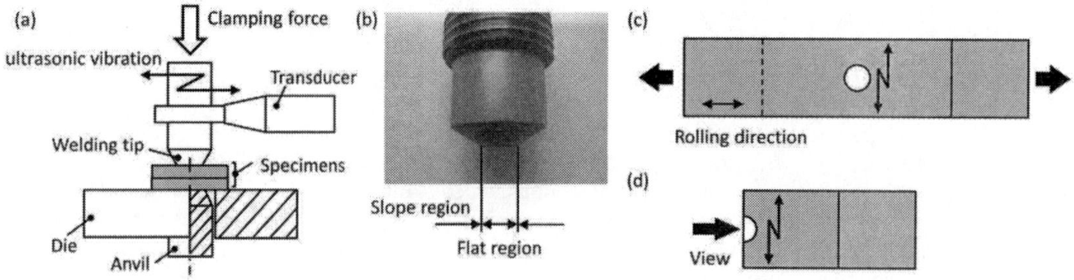

Fig. 1. (a) Schematic illustration of ultrasonic spot welding. (b) Welding tip used in the present experiment. (c) Schematic layout of specimens. (d) View of the microstructure observation samples.

Results and discussion

Fig. 2 (a) depicts the dependence of the weld strength and the weld diameter as a function of the welding energy. The weld strength increased with the welding energy. The peak was obtained at a welding energy of 2300 J and the value was approximately 2350 N. The weld diameter also significantly expanded with the welding energy and plateaued at roughly 9.0 mm from the welding energy of approximately 2000 J. Fig. 2 (b) and (c) show the fracture modes of samples welded at the welding energies of 1000 and 1100 J, respectively. An interfacial fracture mode was exhibited with a welding energy of 1000 J. By contrast, a pull-out fracture mode was exhibited with a welding energy of 1100 J and the fracture modes transformed from the interfacial fracture mode to the pull-out fracture mode at the welding energies of 1100 J and more. The present study puts its focus on the fracture mode transformation, and then the microstructure within the joints was investigated.

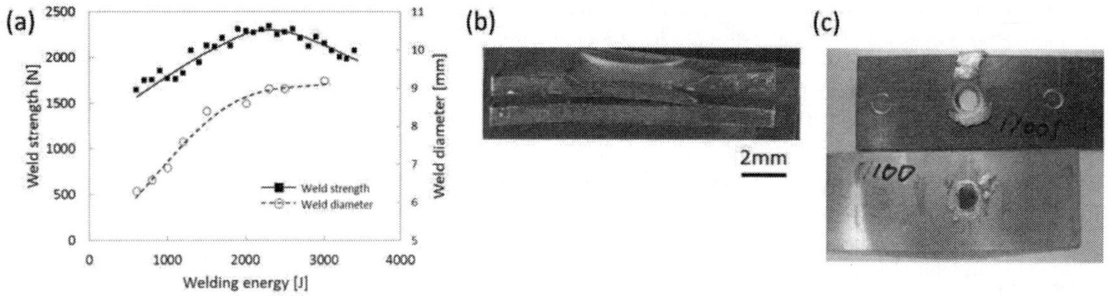

Fig. 2. (a) Relationship between the weld strength and weld diameter as a function of the welding energy. Fracture modes at different energy levels; (b) 1000 J, (c) 1100 J.

In order to examine the plastic deformation behavior within the joints, the microstructure observation was carried out. Results of observation of the microstructure, white substances were observed at the weld interface, as shown in Fig. 3 (a). To clarify the identity of the white substances, it was analyzed using EDX. Fig. 3 (b) shows the result of the elemental mapping. Mg and O were exhibited by blue and red, respectively. The white substances region was exhibited by the red. Thus the substances were oxide layers which were present on the faying surface of the initial material, and then the oxide layers were remained at the weld interface without being removed. And then, the plastic deformation behavior within the joints was revealed by analyzing the behavior of the oxide layers.

Fig. 4 (a), (b), (c) and (d) show SEM images in the slope region at the welding energies of 800, 1000, 1200 and 1400 J, respectively. As a guide indicating the contact surface of the original two plates, the oxide layers were depicted in white lines. From these results, the intensity of the plastic deformation in the slope region was found. At first, the oxide layers were continuously observed at a welding energy of 800 J and were flat (Fig. 4 (a)). The oxide layers were also observed continuously at a welding energy of 1000 J and indicated a slightly wavy form (Fig. 4 (b)). By contrast, the behavior of the oxide layers significantly changed at conditions which exhibited the pull-out fracture mode. The oxide layers indicated a violently curved form at a welding energy of 1200 J (Fig. 4 (c)) and the form further intensified at a welding energy of 1400 J (Fig. 4 (d)). And then, the oxide layers were fractionally observed, and welded areas were also observed. Hence, the plastic deformation was enhanced with the welding energy in the slope region, and then it was found that the intense plastic deformation occurred at the conditions which exhibited the pull-out fracture mode. In the case of ultrasonic spot welding, the plastic deformation behavior within the joints has an effect on the weldability [16], [17] and [18]. And then, the

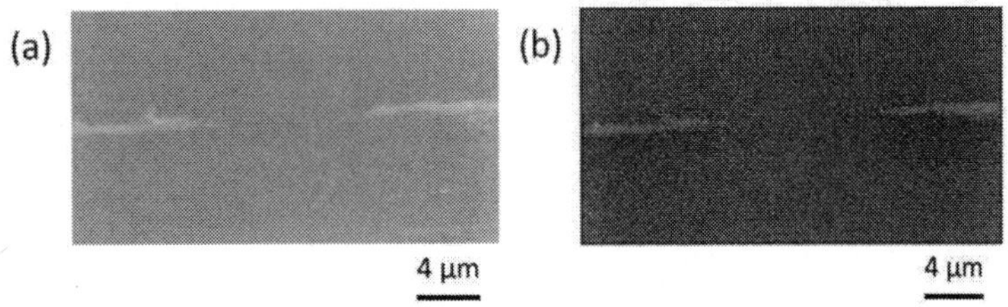

Fig. 3. (a) SEM image of the enlarged oxide layers. (b) The element mapping of oxide layers using EDX.

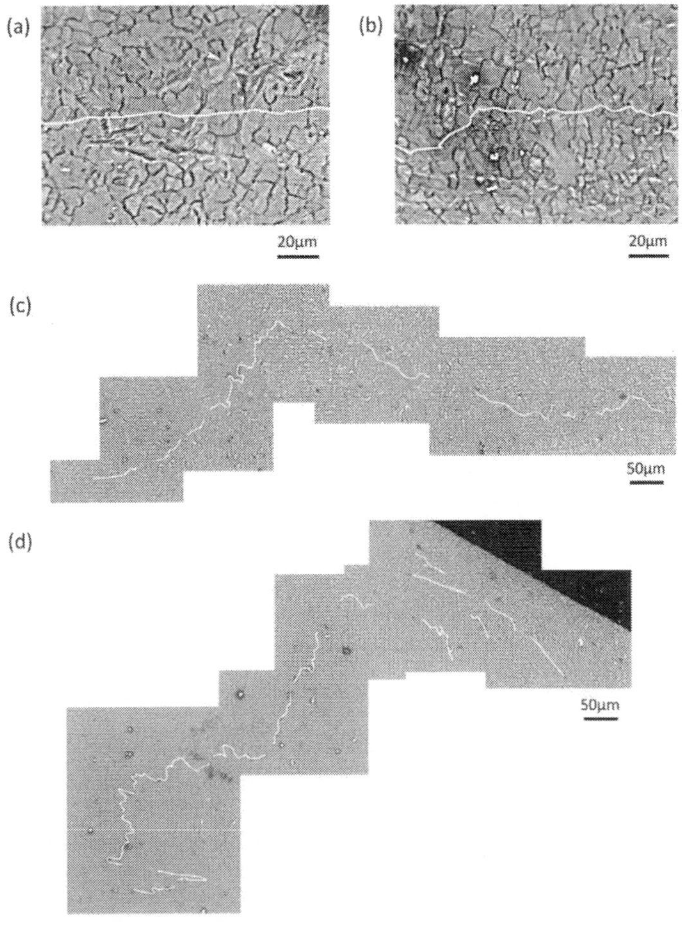

Fig. 4. The oxide layers form within the joints at different energy levels; (a) 800 J, (b) 1000 J, (c) 1200 J, (d) 1400 J.

enhancement of the plastic deformation contributes to the expansion of the weld diameter [16] and [17]. In the AZ31B joints which were welded using the welding tip having the slope region, the enhancement of the plastic deformation in the slope region affected to the expansion of the weld diameter. And then it was considered that the enhancement of the plastic deformation contributed to an increase in the weld strength and the transformation of the fracture modes from the interfacial fracture mode to the pull-out fracture mode.

Conclusions

The weld strength of ultrasonic spot welded AZ31B joints increased with the welding energy, and the fracture modes transformed to the pull-out fracture mode from the welding energy of 1100 J. Furthermore, the intense plastic deformation occurred in the slope region at the welding conditions which exhibited the pull-out fracture mode. The enhancement of the plastic deformation in the slope region contributed to the expansion of the weld diameter and it was revealed that the macro shape of the welding tip affected the weldability.

References

[1] L.M. Zhao and Z.D. Zhang: "Effect of Zn alloy interlayer on interface microstructure and strength of diffusion-bonded Mg–Al joints," Scr. Mater., Vol. 58 (2008), pp. 283–286.

[2] A.A. Luo: "Recent magnesium alloy development for automotive powertrain applications," Mater. Sci. Forum, Vol. 419–422 (2003), pp. 57–66.

[3] X. Cao and M. Jahazi: "Effect of welding speed on the quality of friction stir welded butt joints of a magnesium alloy," Mater. Des., Vol. 30 (2009), pp. 2033–2042.

[4] V.K. Patel, S.D. Bhole and D.L.Chen: "Influence of ultrasonic spot welding on microstructure in a magnesium alloy," Scr. Mater., Vol. 65 (2011), pp. 911-914.

[5] M. Pareek, A. Polar, F. Rumiche and J.E. Indacochea: "Metallurgical evaluation of AZ31B-H24 magnesium alloy friction stir welds," J. Mater. Eng. Perform., Vol. 16(5) (2007), pp. 655-662.

[6] M. Harooni, B. Carlson and R. Kovacevic: "Dual-beam laser welding of AZ31 magnesium alloy in zero-gap lap joint configuration," Opt. Laser Tech., Vol. 56 (2014), pp. 247-255.

[7] Z. Yu and H. Choo: "Influence of twinning on the grain refinement during high-temperature deformation in a magnesium alloy," Scr. Mater., Vol. 64 (2011), pp. 434-437.

[8] H. Watanabe, T. Mukai, M. Kohzu, S. Tanabe and K. Higashi: "Effect of temperature and grain size on the dominant diffusion process for superplastic flow in an AZ61 magnesium alloy," Acta Mater., Vol. 47 (1999), pp. 3753-3758.

[9] A. Yamashita, Z. Horita and T.G. Langdon: "Improving the mechanical properties of magnesium and a magnesium ally through severe plastic deformation," Mater. Sci. Eng. A, Vol. 300 (2001), pp. 142-147.

[10] J. Koike and T. Miyamura: "Microscopic mechanisms of plastic deformation in polycrystalline magnesium alloys," J. JILM, Vol. 54, No. 11 (2004), pp. 460-464 (in Japanese).

[11] E.T. Hetrick, J.R. Baer, W. Zhu, L.V. Reatherford, A.J. Grima, D.J. Scholl, D.E. Wilkosz, S. Fatima and S.M. Ward: "Ultrasonic metal welding process robustness in aluminum automotive body construction applications," Weld. J, Vol. 88 (2009), pp. 149-158.

[12] M. Shakil, N.H. Tariq, M. Ahmad, M.A. Choudhary, J.I. Akhter and S.S. Babu: "Effect of ultrasonic welding parameters on microstructure and mechanical properties of dissimilar joints," Mater. Des., Vol. 55 (2014), pp. 263-273.

[13] E.A. Nappiras: "Ultrasonic welding of metals," Ultrasonics, Vol. 3 (1965), pp. 128-135.

[14] E. Mariani and E. Ghassemieh: "Microstructure evolution of 6061 O Al alloy during ultrasonic consolidation: An insight from electron backscatter diffraction," Acta Mater., Vol. 58 (2010), pp. 2492-2503.

[15] C.Y. Kong, R.C. Soar and P.M. Dickens: "Optimum process parameters for ultrasonic consolidation of 3003 aluminium," J. Mater. Process. Tech., Vol. 146 (2004), pp. 181-187.

[16] Y. Higashi, C. Iwamoto and Y. Kawamura: "Microstructure evolution and mechanical properties of extruded $Mg_{96}Zn_2Y_2$ alloy joints with ultrasonic spot welding," Mater. Sci. Eng. A, Vol. 651 (2016), pp. 925-934.

[17] U.I. Chang and J. Frisch: "On Optimization of Some Parameters in Ultrasonic Metal Welding," Weld. J. Vol. 53 (1974), pp. 24-35.

[18] D. Bakavos and P.B. Prangnell: "Mechanisms of joint and microstructure formation in high power ultrasonic spot welding 6111 aluminium automotive sheet," Mater. Sci. Eng. A, Vol. 527 (2010), pp. 6320-6334.

[19] V.K. Patel, S.D.Bhole and D.L.Chen: "Ultrasonic spot welded AZ31 magnesium alloy: Microstructure texture and lap shear strength," Mater. Sci. Eng. A, Vol. 569 (2013), pp. 78-85.

Development of bonding interface between steel and Ni by ultrasonic welding

J.Y. Lin, K. Seto, S. Nambu and T. Koseki
Department of Materials Engineering, School of Engineering, The University of Tokyo, Tokyo, Japan
jhe_yu_lin@metall.t.u-tokyo.ac.jp, nambu@metall.t.u-tokyo.ac.jp, koseki@material.t.u-tokyo.ac.jp

Abstract
Ultrasonic welding is a solid-state welding technique that can bond materials at relatively low temperature and pressure. In this study, steel and Ni with thicknesses of 1 mm were successfully bonded by ultrasonic welding, and the development of bonding interface between steel and Ni was investigated. The bonding strength obtained by lap shear test increases with welding time, and the fraction of bonding area observed by SEM also increases with welding time. The sequence of the bonding behavior was investigated by the SEM observation and EBSD analysis of cross-section at bonding interface. It was revealed that abrasion is caused by the oscillation to form small particles consisting of steel and Ni, and that the particles are grown and subsequently flattened with time. The bonding is achieved by the flattened particles spread along the bonding interface without any voids.

Introduction
For weight reduction of car body in the automobile industry, combining widely-used steel with other materials is a reasonable method, and thus choosing proper welding technique is crucial to bond dissimilar materials. However, conventional welding techniques conducted at high temperature and pressure could have side effects, for example, coarsening grains near the weld, or diffusion of solute elements in base materials, to degrade the mechanical properties of the weld and base materials. To prevent the above effects on the weld, ultrasonic welding is obtaining much attention because it is able to bond materials at relatively low temperature and pressure. Ultrasonic welding is a solid-state bonding technique of which the principle can be described as follows: a clamping force is vertically pressed to workpieces to keep them in contact; a high frequency vibration is applied to workpieces by a horn; and heat is generated at the contact interface by friction to weld the workpieces. It has advantages, for example, short welding period (around a second in usual), less energy consumption, possible in air instead of high vacuum environment, and minimizing heat damage. Based on the above, ultrasonic welding is a potential technique that could potentially be widely used to bond dissimilar materials.

Until recently, numerous studies for bonding soft metals, for example, aluminum alloys, or CFRP materials [1-2], are conducted to show that higher bonding strength is obtained by ultrasonic welding in few seconds. Regarding interfacial phenomena, broken surface oxide layers from base materials [3], mechanical interlocking caused from plastic deformation or plastic flow [4], and formation of intermetallic compounds at the bonding interface [5], have been reported. According to the above, various interfacial phenomena are reported in the investigation of ultrasonic welding, but the understanding of bonding strength evolution mechanism is still unclear. To clarify the bonding strength evolution mechanism, the development of bonding interface between dissimilar metals is necessary to be studied. In the present study, Ni and steel were bonded by ultrasonic welding, and the development of bonding interface between Ni and steel was investigated.

Experimental procedure
Interstitial-free (IF) steel (Fe-0.001wt%C-0.008wt%Si-0.16wt%Mn) and pure Ni sheet with the thickness of 1 mm were prepared. IF steel and Ni sheets were firstly cut to the dimension of 15mm×7.5mm. Contact surfaces of IF steel and Ni were mirror-polished by Al_2O_3 with the particle size of 1μm. After mirror-polished step, metal sheets had the average surface roughness (Ra) of about 10nm. In the setup of ultrasonic welding, Ni sheet was positioned on IF steel sheet. A clamping force was then applied by a horn in the normal direction to the bonding interface to keep Ni and IF steel in contact. Experimental parameters for ultrasonic welding are shown in Table 1, which indicate clamping force, horn size, frequency and amplitude of ultrasonic vibration, and welding time. In the present study, 0.1s of duration was required to reach to the maximum amplitude for the horn, and hereafter the welding time was started to be counted. To obtain bonding strength, lap shear test was conducted to each welded specimen. The lap shear direction was parallel to the bonding interface, and perpendicular to the oscillation direction. Fracture surface after lap shear test and interfacial microstructure were investigated by SEM observation and EDS compositional analysis, and the microstructural analysis of cross section at the bonding interface were also examined by EBSD.

Table 1. Experimental parameters established for bonding experiments by ultrasonic welding.

Clamping Force	Horn size	Frequency	Amplitude	Welding time
2000N	10×10 mm	20kHz	41μm	0 – 5.0s

Results and discussion

Figure 1 shows the result of the lap shear test for Ni/IF steel joints obtained from ultrasonic welding for various welding time. According to *Figure 1*, bonding strength increases with the welding time. When the welding time is longer than 1.75s, the bonding strength reached to 25MPa, and the failure at substrate was occurred. Based on the above, Ni and IF steel were successfully bonded, and strong bonding interface was obtained by ultrasonic welding.

Figure 2 shows SEM micrographs of fracture surface of IF steel after the lap shear test, in which the welding time is from 0.05s to 1.5s. Also, the arrow above *Figure 2* indicates the direction of oscillation. White area with island shape was observed in SEM micrograph. These white areas on fracture surface are considered as bonding area, because dimple-like fracture was observed at the area, and the compositional analysis for bonding area by EDS shows that Ni was detected on the fracture surface of IF steel, and Fe was detected on the fracture surface of Ni vice versa. Further, bonding area was increased, and each island-shape bonding area were also grown with the welding time. It can be suggested that Ni and IF steel were contacted under the clamping force, and were endured abrasion caused by the oscillation, to form bonding interface during ultrasonic welding.

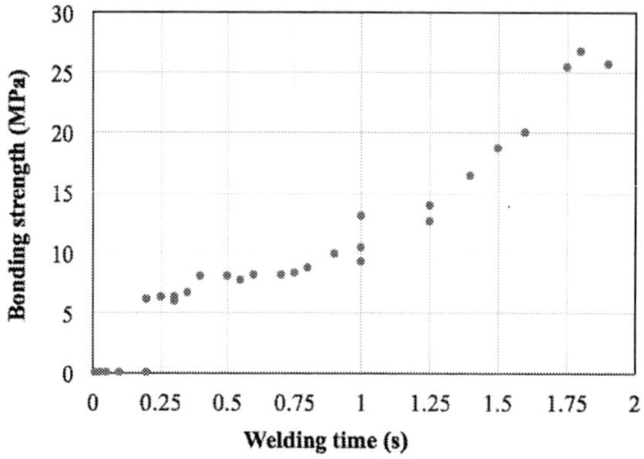

Figure 1. The relationship between the welding time and bonding strength of Ni and IF steel.

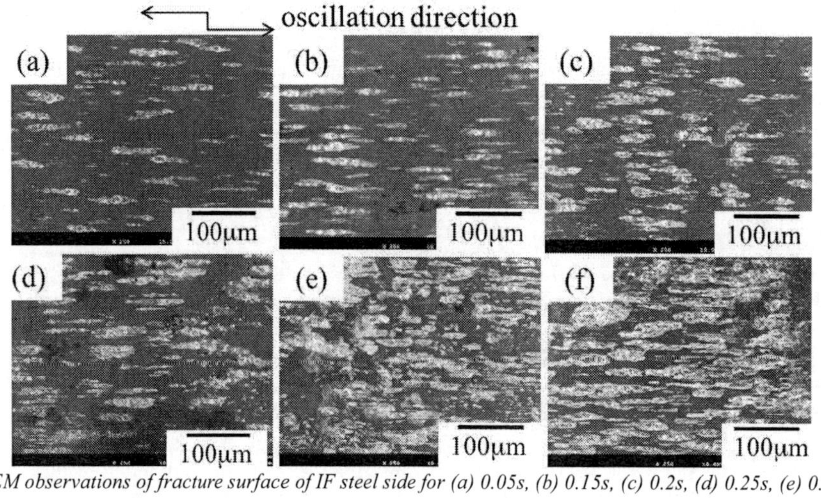

Figure 2. SEM observations of fracture surface of IF steel side for (a) 0.05s, (b) 0.15s, (c) 0.2s, (d) 0.25s, (e) 0.75s, (f) 1.5s.

Figure 3 shows the relationship between bonding strength and fraction of bonding area. At the beginning of welding, it is seen that bonding strength of the welded specimens were too weak, and it is thus unable to be measured while fraction of bonding area was less than 0.04 and the welding was less than 0.15s. When fraction of bonding area reached to 0.04, bonding strength emerged. As the welding time reached to 1.5s, the fraction of bonding area was 0.49 and the bonding strength was 19MPa. SEM micrographs for development of bonding area are shown in *Figure 4*. A small scratch initially was formed from contact area of Ni and IF steel under the oscillation of ultrasonic welding, as shown in *Figure 4-(a)*. *Figures 4-(b) to 4-(d)* show the development of bonding area following from small scratch, wherein bonding area in *Figure 4-(d)* shows dimple fracture with ductility. According to the above, it is suggested that at the beginning of the welding, Ni and IF steel after ultrasonic welding were without bonding strength due to low fraction of bonding area. It is suggested that as the welding time increases, bonding area grew sequentially (as shown in *Figures 2 and 4*), and subsequently formed good bonding.

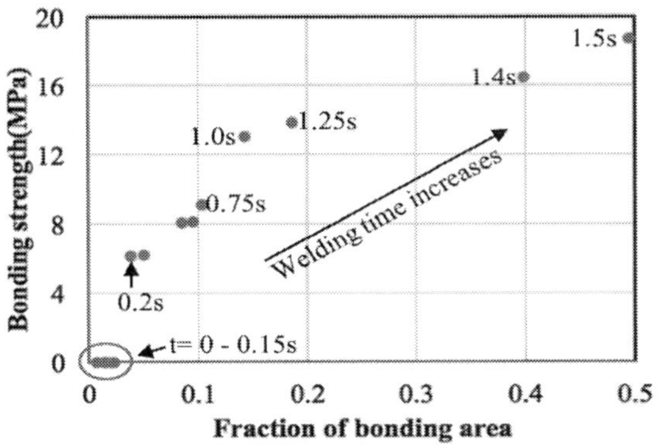

Figure 3: The relationships between the welding time, bonding strength, and bonding area ratio.

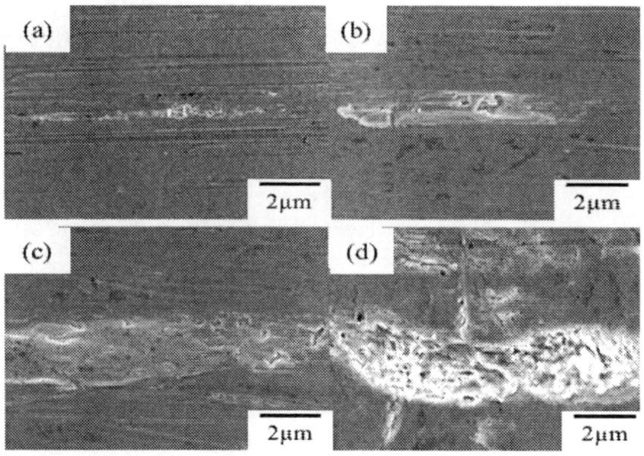

Figure 4. SEM micrographs for the development of bonding area, were taken from fracture surfaces of IF steel.

Further, SEM observation and EBSD analysis of cross section at the bonding interface was conducted to investigate the development of bonding area, as shown in *Figures 5 and 6*. At the beginning of welding, the asperity of IF steel surface is seen (*Figure 5-(a)*), and a slight amount of Ni was detected on the bonding surface of IF steel by EBSD (*Figure 6-(a)*). Then, as shown in *Figure 5-(b)*, particle having spherical shape were observed. In later stage of welding, flattened particle was observed, and examined by EBSD that it is consisted of Ni and IF steel phase (*Figures 5-(c) and 6-(b)*). Subsequently, the flatten particle were spread along the bonding interface to achieve the bonding of Ni and IF steel with the decreasing of voids, as shown in *Figure 5-(d)*.

It is suggested that at the bonding interface, Ni and IF steel were started to scratch, attach, and shave to each other repeatedly due to abrasion caused from the oscillation. Debris peeled from Ni and IF steel were further mixed and clustered mutually, and thus sphere particles was grown. In later stage of welding, particles, formed from debris peeled from Ni and IF steel, were flattened, and expanded to the bonding interface. Subsequently, bonding area was formed at the bonding interface and thus the weld with high strength is achieved.

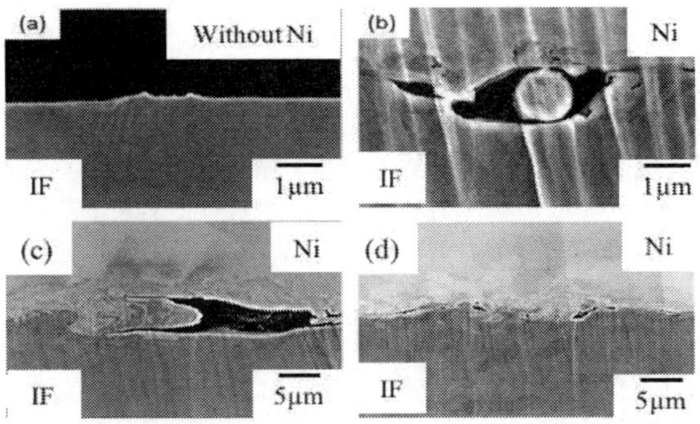

Figure 5. SEM micrographs of the cross section for the development in bonded area.

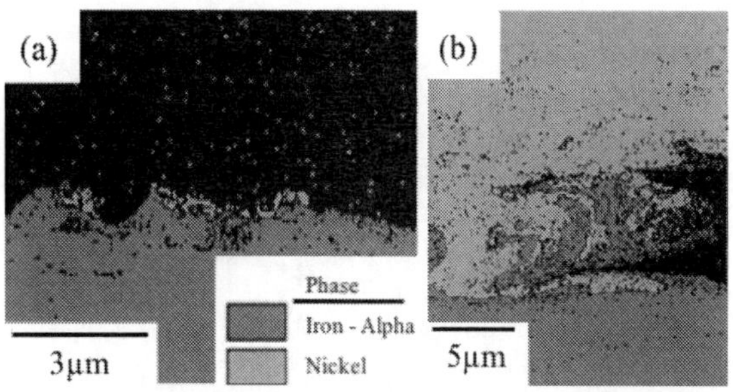

Figure 6. EBSD phase map for: (a) Fig.5-(a); and (b) Fig.5-(c).

According to above discussion, bonding strength evolution mechanism of Ni and IF steel of ultrasonic welding is schematically summarized in *Figure 7*. Firstly, the abrasion is caused from oscillation and tiny particles formed at contact area of the bonding interface (the red circle of *Figure 7(a)*). Then, tiny particles are grown to larger particles by mixing and clustering materials peeled from Ni and IF steel. In the meantime, new tiny particles are continuously formed from contact points in which Ni and IF steel shave to each other (the red circle of *Figure 7-(b)*). Under clamping force, larger particles are flattened to spread along the bonding interface (*Figure 7-(c)*). Subsequently, voids between Ni and IF steel are vanished, and intermixed bonding interface with high strength is achieved (*Figure 7-(d)*).

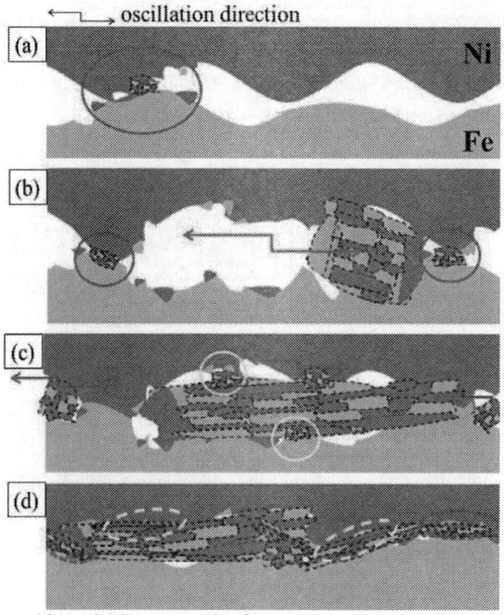

Figure 7. The mechanism of ultrasonic welding: (a) Tiny particles form at contact point; (b) Particles become larger with peeled debris from Ni and IF steel, and particles sequentially form at new contact points; (c) The large particles are flattened; and (d) Intermixed interface is achieved.

Conclusions

To investigate the evolution mechanism of bonding strength during ultrasonic welding, Ni and IF steel were used and successfully bonded by ultrasonic welding. Both of the bonding strength obtained by lap shear test and the fraction of bonding area increase with the welding time. The investigation of the bonding interface development was revealed that at the beginning of ultrasonic welding, the abrasion is caused by the oscillation to form small particles consisting of Ni and IF steel. The small particles are firstly form at contact point of the bonding interface, and are gradually grown to become larger particles by clustering and mixing debris peeled from Ni and IF steel. Subsequently, larger particles are flattened to spread along the bonding interface without any voids, and thus intermixed bonding interface with high strength is obtained.

References

[1] G.G. Harman, *et al*: "The Ultrasonic Welding Mechanism as Applied to Aluminum-and Gold-Wire Bonding in Microelectronics," Parts, Hybrids, and Packaging, IEEE Transactions on, Vol. 13, No. 4 (1977), pp. 406-412.

[2] F. Balle, *et al*: "Ultrasonic Spot Welding of Aluminum Sheet/Carbon Fiber Reinforced Polymer–Joints," Materialwissenschaft und Werkstofftechnik, Vol. 38, No. 11 (2007), pp. 934-938.

[3] C. Born, *et al*: "Ultrasonically Welded Aluminium Foams/Sheet Metal–Joints," Advanced Engineering Materials, Vol.8, No.9 (2006), pp. 816-820.

[4] D. Bakavos, *et al*: "Mechanisms of Joint and Microstructure Formation in High Power Ultrasonic Spot Welding 6111 Aluminium Automotive Sheet," Materials Science and Engineering, Vol. 527, No.23 (2010), pp. 6320-6334.

[5] A. Panteli, and J. Robson, *et al*: "Modelling Intermetallic Phase Formation in Dissimilar Metal Ultrasonic Welding of Aluminium and Magnesium Alloys," Science and Technology of Welding and Joining Vol. 17, No. 6 (2012), pp. 447-453.

Optimal hull penetrations positions and frame spacing leading to minimal welding distortion

Duncan Camilleri

University of Malta, Malta

Hull penetration and openings are governed by standards that ensure the structural integrity of the assembly. However, a number of stiffeners are usually welded to the panel, giving rise to welding residual stresses and distortion. This study identifies the optimal position and shape of penetrations relative to two stiffener spacing configurations, leading to the least welding distortion. A numerical parametric study is adopted where an uncoupled thermo-elastoplastic technique based on the inherent strain method is employed. The results have shown that the final distortion is highly dependent on the buckling mode shape predicted by the numerical models that in turn is influenced by the structural stiffness and support position of the assembly. When the predicted out-of-plane distortion follows a similar buckling mode the results have shown that a stiffener spacing of 500mm leads to less distortion when compared to 700mm stiffener spacing. Centrally positioning penetrations results in symmetric panel stiffness, leading to near symmetric out-of-plane distortion and ultimately minimal distortion, when compared to positioning penetrations next to the stiffener. The difference in the predicted distortion for square and circular penetrations and the case studies considered, is relative small.

Adaption of heat control concepts for welding stress optimization in high-strength steel components

Dirk Schroepfer, Arne Kromm, Thomas Kannengiesser, Michael Rhode
Bundesanstalt für Materialforschung und –prüfung (BAM), 9.4 Division - Weld Mechanics,
Unter den Eichen 87, 12205 Berlin, GERMANY
Tel.: +49 (0) 30 8104 4376, E-mail: dirk.schroepfer@bam.de

Abstract

High-strength structural steels from 690 MPa are increasingly applied at present. Major reasons are lightweight design trends and potential cost reductions. The structural design of welds and their manufacturing become more challenging with increasing material strength. By reasons of a higher yield ratio of these steels, the development of high residual stresses has to be avoided, since they are detrimental to the components safety and performance. Local restraint stresses and welding loads due to external shrinkage restraints occur. Frequently, this leads to critical tensile residual stresses in the weld and HAZ. In this study, influences of welding process parameters and restraint conditions on the residual stress state in welded components of high-strength steels were investigated. Multilayer GMAW tests under free shrinkage and experiments under well-defined restraints in special in-house developed testing facilities were accomplished. The tests permitted analyses of the resulting local residual stresses measured by means of X-ray diffraction and global reaction stress build-up while welding and cooling. Significant effects were found for heat control, seam configuration and restraint condition. Besides high restraints, elevated preheating and interpass temperatures lead to increased welding stresses. An adaption of welding parameters considering heat control, weld run sequence and seam configuration proved to be beneficial.

Introduction

An increasing portfolio of high-strength base and filler materials is provided by the steel producers. A sustainable and economical application of high-strength structural steels from 690 MPa, e.g. in mobile crane industry, depends on the load-bearing capacity and safety of their welds, since welding is the major joining technology [1]. The challenge regarding the welding process increases according to the material strength due to the requirements for the mechanical properties of the weld and for the component safety. This is due to a reduced heat control working range regarding $\Delta t_{8/5}$-cooling times (from 800 °C to 500 °C) to provide an adequate microstructure and, above all, a sufficient hydrogen effusion to avoid cold cracking. Various empirical approaches based on carbon equivalents and weld tests are available to evaluate a proper heat control. In addition, high tensile residual stresses are disadvantageous regarding cold cracking resistance, strength and service life of welded components [3]. In component welds, high restraints, quantified by the restraint intensity (e.g. R_{Fy}) concept [4], are common; see *Fig. 1*. These increase residual stresses due to superimposed reaction stresses. Thus, the residual stress state of welded components is subjected to a complex interaction of welding process, applied material and restraint of the weld, observed in numerous experimental and numerical analyses [5–8]. Currently, the relationship between residual stresses levels and heat control of welded components is only qualitatively understood. Assumptions of residual stresses at yield strength level lead to conservative designs of today's structures. Especially for high-strength steel component welds further works concerning residual stress prediction and reduction are absent; a better understanding of the effects regarding *Fig. 1* is needed.

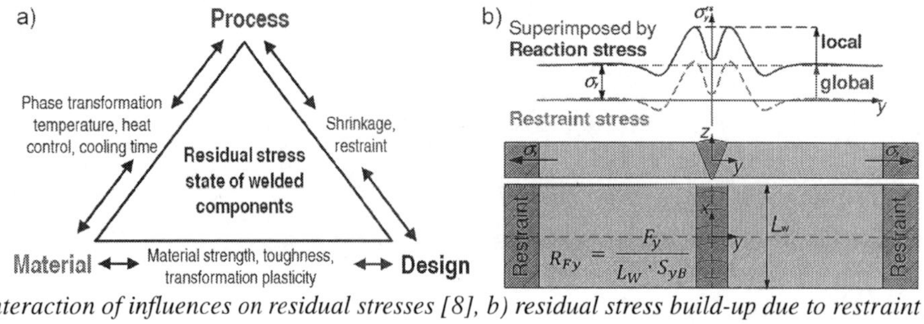

Fig. 1. a) Interaction of influences on residual stresses [8], b) residual stress build-up due to restraint [3,6].

Experimental

In this research the restraint intensity in weld transverse direction R_{Fy} were experimentally and analytically determined at real components and transferred to special in-house developed testing facilities. Automated multilayer GMAW were applied in V-groove butt joints in free shrinking and restrained specimens; see *Fig. 2* and *Table 1*.

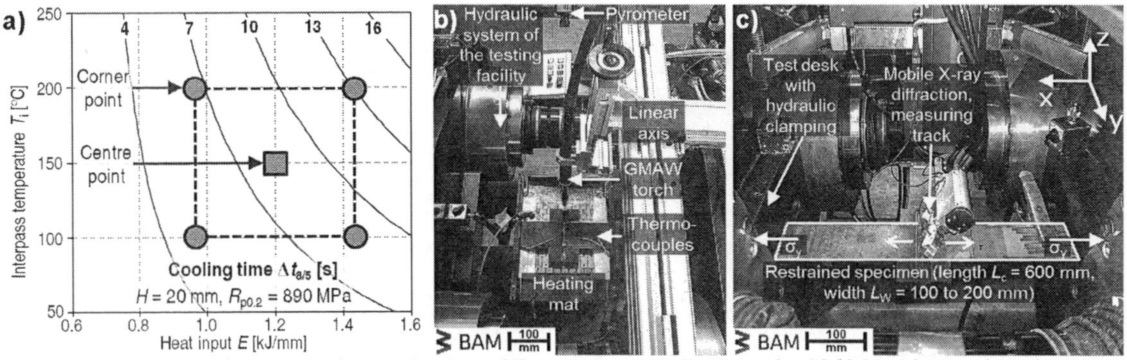

Fig. 2. a) Variation of heat control using Design of Experiments (DoE, Example: S960QL, 20 mm *plate thickness) over a contour plot of the statistical evaluation of the detmermined $\Delta t_{8/5}$-cooling times; Overview of b) weld test setup in the 2-*MN*-testing facility and c) residual stress analysis in the testing facility at a clamped specimen.*

Table 1. Welding parameters and variation of design, welding process and material by means of DoE.

Weld preparation	Welding current	Welding voltage	Welding speed	Wire feed speed	Welding process
V groove, 45°	220 to 290 A	22 to 29 V	280 to 550 mm/min	6.0 to 8.7 m/min	Conv.[*]
V groove, 30°	320 A	29 V	400 to 420 mm/min	11 m/min	Mod.[*]

Varied design parameters		Varied heat control and welding process		Varied material, BM/FM	
Groove angle [°]	30 and 45	Working temperature $T_{p/i}$ [°C]	50 to 200	S690QL	G69[b]
Plate thickness [mm]	8 and 20	Heat input [kJ/mm]	0.6 to 1.6	S960QL	G89[c]
Restraint intensity [kN/(mm·mm)]	0[a] to 4[b]	Welding process[*]	Conv./Mod.	S960QL	G89, root G46[d]

[*]Welding process of the power source: Conv.: transitional arc, Mod.: modified spray arc; [a] Free shrinking; EN ISO 16834-A: [b]G 69 6 M21 Mn4Ni1.5CrMo, [c]G 89 6 M21 Mn4Ni2CrMo; [d]EN ISO 14341-A-G 46 4M21 4Si1.

Table 2. Chemical composition (FES, Fe balanced) and mechanical properties of the 20 mm *thick test materials.*

Element in %	C	Si	Mn	Cr	Mo	Ni	V	Nb	Ti
Base material (S690QL, EN 10025-6)	0.14	0.71	1.15	0.03	0.17	0.07	0.009	0.005	0.010
Base material (S960QL, EN 10025-6)	0.16	0.32	1.24	0.20	0.61	0.05	0.040	0.015	0.003

Property	$R_{p0.2}$ in MPa	R_m in MPa	A_5 in %	Av in J at -40 °C	HV10
S690QL	768	821	19	198	270
S960QL	1035	1050	17	96	332

Specimens were tack welded and clamped into the test setups. A synchronous restraint and analysis of reaction forces during welding and cooling was achieved; see *Fig. 2b*. Thermocouples were attached 10 mm adjacent to the weld, to monitor the working temperatures T_p and T_i. $\Delta t_{8/5}$-cooling times were determined by a two colour pyrometer (350 °C to 1300 °C) at the weld layer surface. The influence factors according *Fig. 1* were varied, in particular heat control, welding parameters and seam configuration as well as the restraint intensity for two different high-strength steel grades in 8 mm and 20 mm thick plates; see *Table 1*. Modified welding parameters enabled the welding of 30° V-groove joints with a less amount of weld runs. High-strength quenched and tempered fine-grained structural steels S690QL and S960QL were welded with matching high-strength solid filler wire; see *Table 2*. All parameters were variated by means of DoE, for an adequate statistical evaluation of the results. *Fig. 2a* exemplarily shows the parameter matrix consisting of centre and corner point tests for heat control variation of 20 mm thick S960QL welds over a contour plot of the regression analysis of the determined $\Delta t_{8/5}$-cooling times. Centre point tests were repeated to provide an adequate statistical basis for the results. The heat control range was selected according to $\Delta t_{8/5}$-cooling time ranges specified by steel producers. Mobile X-ray diffraction was used for local stress analyses in the weld seam area of the restrained and unrestrained specimens; see *Fig. 2c*.

Results

The contour plots of the regression of the $\Delta t_{8/5}$-cooling times in *Fig. 2a* shows the significant effect of heat control. Significant effects of steel grade and plate thickness were also observed. All specimens welded with centre point parameters regarding heat control were subjected to a testing of mechanical and technological properties. In these weld tests, $\Delta t_{8/5}$-cooling times were obtained, which are typically applied in high-strength steel fabrication: 6 s to 8 s for S960QL and 12 s to 15 s for S690QL. All properties of the welds fulfilled the specified requirements.

Global reaction stresses

Fig. 3a exemplarily shows the reaction force $F_y(t)$, temperature $T(t)$ and bending moment $M_x(t)$ for a weld in the 2-MN-testing facility with S690QL. While preheating of the tack welded specimen low compressive forces are initiated and starting the root weld, the reaction force is $F_y = -15$ kN with a bending moment more or less zero. During root welding, the already solidified inserted weld metal produces transversal shrinking forces and an increasing bending moment. Both increase further to a first maximum of $F_y = 62$ kN and $M_x = 0.5$ kNm during cooling to $T_i = 150$ °C. Forces and moments decrease while the second layer is welded due to a local heat input combined with stress relief. Subsequent cooling to T_i leads to an increase of the reaction force and bending moment, the next weld run to a reduction. This evolution was detected for each weld sequence. The amplitudes of the reaction force and bending moment rise with the amount of welded layers as a result of the welding heat input. Subsequent cooling to ambient temperature after completion of the weld leads to maxima of $M_x = 1.95$ kNm and $F_{y,end} = 535$ kN. *Fig. 3b* shows a comparison of the reaction stress build-up $\sigma_y(t)$ of weld tests with S690QL for different interpass temperatures. $\sigma_y(t)$ was determined using $F_y(t)$ and the actual height of the load bearing cross section $H_{layer}(t)$. As the two graphs exhibit the same tendency, an increase of the total welding time and higher reaction stresses due to the lower interpass temperature are obvious. However, cooling to ambient temperature leads to 19 % higher end reaction stresses $\sigma_{y,end}$ due to an increased heat quantity within the specimen, welded with higher interpass temperature. The same tendency is obvious for the bending moment due to restraint angular distortion; see *Fig. 3c*. The longer cooling phases between each layer cause a lower temperature profile across the specimen, which initiates a 60 % lower bending moment reaching ambient temperature.

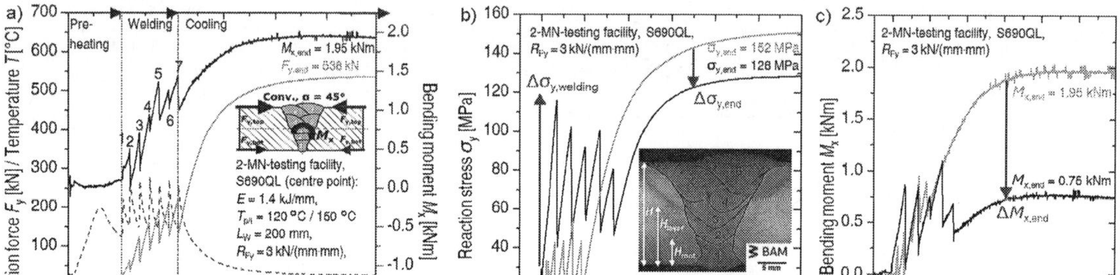

Fig. 3. a) Reaction force $F_y(t)$, temperature $T(t)$ and bending moment $M_x(t)$ for 2-MN-weld test; comparison of reaction stresses (b) and bending moments (c) for two different interpass temperatures at S690QL.

Due to the bending moments, bending stresses σ_{Mx} occur in the welds; see *Fig. 4a*. The total reaction stress at the top of the weld is the result of normal reaction stresses σ_y and the bending stresses σ_{Mx}. The regression analysis of the determined total reaction stresses of all restrained weld tests is shown by contour plots in *Fig. 4b to d*. A major effect on the total reaction stresses is obvious for the weld run amount needed to produce the weld. The amount of weld runs depends on either the weld seam volume (groove angle) or the deposition rate while welding (heat input).The effect of interpass temperature and restraint is significant as well. The application of different base and filler materials also affects the total stress build-up due to diverse phase transformation behaviours.

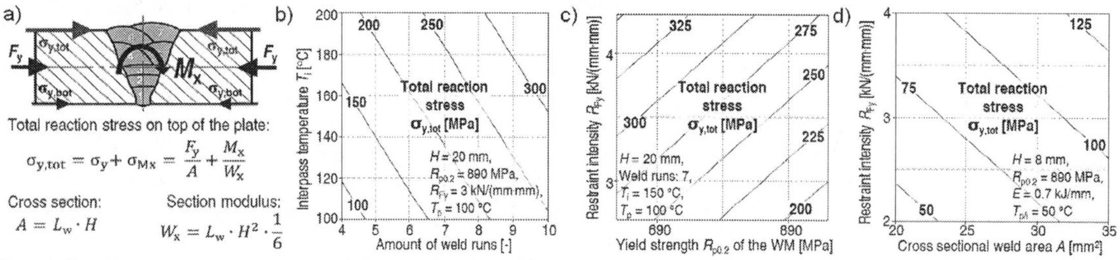

Fig. 4. Total reaction stress $\sigma_{y,tot}$: a) Determination, b) vs. interpass temperature and amount of weld runs needed to finish the weld, c) vs. restraint intensity and steel grade, d) vs. restraint intensity and seam configuration.

Local welding stresses

Fig. 5 presents typical transverse residual stress distributions for steels with undergoing phase transformation according to common concepts [9]. For conventional welds at a 45° V-groove a high effect reveals for the restraint condition on the residual stresses in the HAZ; see *Fig. 5a*. The residual stresses in the weld metal are moderately increased due to the restraint. The modified welds with 30° groove angle show lower residual stresses in the HAZ and a decreased influence of the restraint on the stresses in the HAZ, which correlates with the measured total reaction stresses; see *Fig. 5b*. Local residual stress analyses mainly support the global results, e.g. the significant effect of restraint intensity and seam configuration on the residual stress level in the HAZ. *Fig. 5c* shows the distinct decrease of local residual stresses for a lower amount of necessary weld runs due to the modified weld seam.

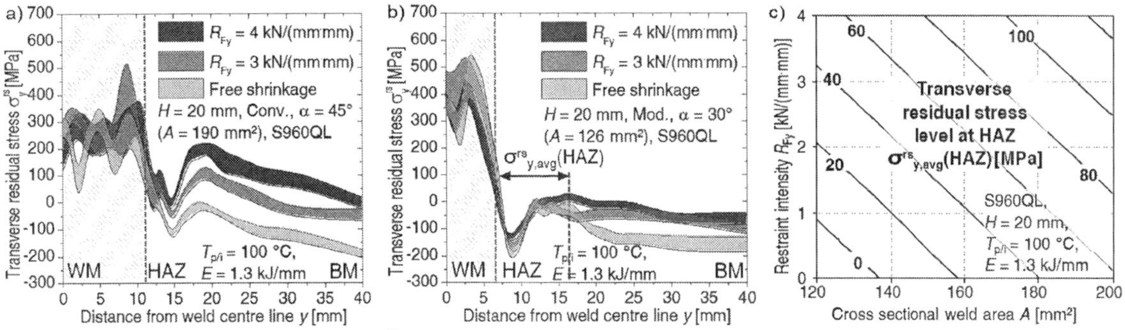

Fig. 5. Transverse residual stresses. $σ^{rs}_y(y)$ *welded and measured at different restraints: a) Conv., 7 weld runs, b) Mod., 4 weld runs; c) stress level in the HAZ* $σ^{rs}_{y,avg}$(HAZ) *vs. restraint inensity and seam configuration.*

Conclusions

This work gives an example of the research focused on the adjusting of heat control to optimize stresses due to welding in high-strength steel components. High restraint intensities strongly elevate welding stresses. It revealed that lower working temperatures lead to significantly decreased global loads in restrained welds. The consideration of bending moments due to a restrained angular distortion shows predominantly reduced total stresses at welds with a less amount of weld runs. This is obtainable either with narrow weld seams, weldable with modified processes, or with higher deposition rates of filler wire. As a rule, higher deposition rates increase the overall heat input. From this point of view, a lower working temperature and a higher heat input, in order to adjust the cooling times, are beneficial, especially at narrow seam configurations. This applies similarly for the local transverse residual stresses particularly in the HAZ. Higher strength filler wires used in filler beads cause less shrinking forces and bending moments due to diverse phase transformation behaviour; see *Fig. 4*. In contrast, under-matching filler metals are advantageous to prevent critical stresses in the root weld. Based on these studies, further advanced heat control concepts for the required weld properties as well as improved residual stress states are achievable.

Acknowledgments

The studies were funded by the AIF-projects IGF-Nr. 17267 N/ Fosta P922 and IGF-Nr. 17978 N/ Fosta P1011. Sincere thanks are given for this support and to the companies, actively involved in the project boards.

References

[1] H.-P. Guenther: Use and Application of High-Performance Steels for Steel Structures", IABSE (Zürich, 2005).
[2] P. Wongpanya et al.: "Ways to Reduce the Cold Cracking Risk in High Strength Structural Steel Welds", *Proc: Int. Conf. Int. Inst. Weld.*, Johannisburg, 2008.
[3] J. Lu: "Prestress Engineering of Structural Material: A Global Design Approach to the Residual Stress Problem", in: Handb. Residual Stress Deform. Steel, 2002, pp. 11-26.
[4] K. Satoh et al.: "Japanese Studies on Structural Restraint Severity in Relation to Weld Cracking", *Weld. World.* 15 (1977), pp. 155-189.
[5] D. Schroepfer, T. Kannengiesser: "Correlating Welding Reaction Stresses and Weld Process Conditions for High-strength Steel S960QL", *Weld. World.* 58 (2014) pp. 423-432.
[6] D. Schroepfer et al.: "Improving Welding Stresses by Filler Metal and Heat Control Selection in Component-related Butt Joints of High-strength Steel", *Weld. World.* 59 (2015), pp. 455-464.
[7] D. Schroepfer, T. Kannengiesser: "Stress build-up in HSLA Steel Welds due to Material Behaviour", *J. Mater. Process. Technol.* 227 (2016), pp. 49-58.
[8] T. Lausch et al.: "Multi-Axial Load Analysis of Thick-Walled Component Welds made of 13CrMoV9-10", *J. Mater. Process. Technol.* 213 (2013), pp. 1234-1240.
[9] T. Nitschke-Pagel, H. Wohlfahrt: "The Generation of Residual Stresses due to Joining Processes" in: Residual Stress. - Meas. Calc. Eval. (1991), pp. 121-133.

Residual Stress Tensor Distributions around Welded Bond of Stainless Steel

Shun-Ichiro Tanaka
NICHe, Tohoku University, Sendai, Japan
sitanaka@tagen.tohoku.ac.jp

Takehiko Watanabe
Niigata University

Abstract
Residual stress distribution in welded stainless steel was measured by 3 dimensional analysis, such as X-ray two dimensional (2D) method, around welded part where three axis stress states remained. Normal stress values perpendicular to the bead increased from 100 MPa in compression to 200 MPa in tensile in HAZ and reached maximum 250 MPa at the interface of HAZ and bead. Normal stress along the bead reached 300 MPa at the center of bead. Stress in depth direction was as high as 50 MPa at the interface. Principal stress in welded stainless steel, only maximum principal stress, $\sigma 1$, in both phases showed maximum 400 MPa at the interface and inside the bead which predict crack initiation point comparable to yield strength of the steel.

Introduction
Residual stress in welded steel has been evaluated by various experimental and simulation methods since serious accident occurred via stresses such as a corrosion cracking. The former experimental method widely ranged, for examples using strain gauge, hole drilling and X-ray or neutron diffraction. Although stress values in ceramics are easy to measure in a confined region [1], those in metals are limited to wider area because of large grains grown in the cooling process. Conventional measuring method using X-ray has been a $\sin^2\psi$ method, although it is based on plane stress hypothesis and depth profile could not be evaluated. A micro-area X-ray two dimensional (2D) method [2] was newly developed to obtain stress tensor utilizing curved Debye ring by two-dimensional detector. The diameter of measured area ranges 0.1-1 mm.
In this paper, recent outputs on measurement of stress tensor distribution around welded interface of stainless steel will be introduced.

Experimental
Plate specimen was Gas Tungsten arc (TIG) welded SUS440/SUS304 using SUS390 as a filler rod where these phases were ferrite for SUS440 and austenite for SUS304, and welding conditions were 300 mm/min for speed and 75A for welding current.
Three axis residual stress distributions around butt joint of dissimilar stainless steels were measured as a tensor by the X-ray two dimensional (2D) method equipped with two-dimensional detector. X-ray conditions were 38 kV, 16 mA Cr-Kα collimated to ϕ0.5 mm with diffraction angles of 2θ=156.0 deg for αFe(211) and 128.5 deg for γFe(220).
Stress tensors across and along welded part were obtained to check the reliability of joint which contained normal, shear stress, principal stresses, Euler angles and stress ellipsoid.

Residual stress tensor distribution

Normal stress distribution across welded joint
In SUS304 part, normal stress values perpendicular to the bead increased from 100 MPa in compression to 200 MPa in tensile in heat affected zone (HAZ) and reached maximum 250 MPa at the interface of HAZ and bead. Normal stress along the bead reached 300 MPa at the center of bead. Those in SUS440 showed similar tendency. Stress in depth direction was as high as 50 MPa at the interface. Residual stress tensor distributions of welded joint in sub-mm-scale area should be rechecked including their depth profile.

Maximum principal stress distribution across welded joint

Principal stress can be converted from normal and shear stress essential to discuss the crack initiation point in brittle materials. Although discussion should be done with equivalent stress in welded stainless steel, only maximum principal stress, σ1, were shown in Fig. 1. σ1 of both phases showed maximum 400 MPa at the interface and inside the bead which predict crack initiation point. This value coincides with yield strength of the steel easy to collapse the joint.

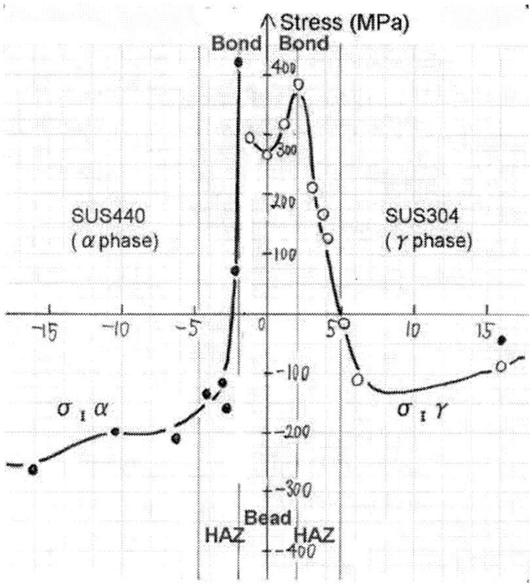

Fig.1 Distribution of maximum principal stresses, σ 1, across TIG weld line of SUS440/SUS304. σ 1 was obtained for α and γ phase composing stainless steel showing tensile maximum in HAZ and bead.

Conclusions

1. Normal stress values perpendicular to the bead increased from 100 MPa in compression to 200 MPa in tensile in HAZ and reached maximum 250 MPa at the interface of HAZ and bead. Normal stress along the bead reached 300 MPa at the center of bead. Stress in depth direction was as high as 50 MPa at the interface.

2. Principal stress in welded stainless steel, only maximum principal stress, σ1, in both phases showed maximum 400 MPa at the interface and inside the bead which predict crack initiation point. This value coincides with yield strength of the steel easy to collapse the joint.

3. Stresses should be evaluated by 3 dimensional analysis, such as X-ray 2D method, around weld region where three axis stress states remained.

References

[1] Shun-Ichiro Tanaka and Yumiko Takahashi: "Effects of X-Ray Beam Collimation on the Measurement of Residual Stress Distribution in a Si_3N_4/Steel Joint", *ISIJ International*, **30** (1990) 1086-1091
[2] Bob B. He: Two-Dimensional X-Ray Diffraction, John Wiley & Sons (2009)

Residual stresses and distortion of electron beam dissimilar welded structural intermetallics Ti_2AlNb alloy and conventional titanium alloy

Yue Zhao [1], Yanjun Li [1], Aiping Wu [1], Ruican Zhu [2], Guoqing Wang [3]

[1] *Department of Mechanical Engineering, Tsinghua University, P.R. China*
[2] *Capital Aerospace Machinery Company, Beijing, P.R. China*
[3] *China Academy of Launch Vehicle Technology, Beijing, P.R. China*

Ti_2AlNb alloy, as a Ti-Al series structural intermetallics with amount of niobium, has high specific strength and good creep resistance at the temperature range of 600 to 800 degree centigrade. It has been attracted much attentions as the potential aerospace intermediate-temperature structural material. Dissimilar welding of Ti_2AlNb alloy and conventional titanium alloy is often necessary for its applications. Electron beam welding is a suitable method to weld Ti_2AlNb alloy and titanium alloy, however, the welding residual stresses and distortion characteristics and their effect factors were not clear yet. In the present study, the residual stresses and distortion of electron beam dissimilar welded Ti-22Al-25Nb / TA15 alloy and Ti-22Al-25Nb / TC4 alloy were investigated, the influence of restraint conditions, welding parameters, molten pool shape on the residual stresses and distortion were clarified. The results indicated that the welding distortion is dominated by out-of-plane deformation which could be presented by transverse angular deflection and longitudinal deflection. A smaller distortion could be acquired in the conditions of fully restraint, I-shaped molten pool, lower welding speed and lower electric current with the same heat input. The residual stresses are mainly influenced by the molten pool shape, I-shaped molten pool leads to higher residual stresses. The Ti-22Al-25Nb / TC4 joint has smaller distortion and lower residual stresses, compared with Ti-22Al-25Nb / TA15 joint.

Improving the Weld Procedures for Repairing Aged Heat Resistant Nickel-Base HP-Mod Alloy Reformer Tubes

Kiran Rampat,
Industrial Plant Services Ltd., Point Lisas, Couva, Trinidad and Tobago
Kiran.rampat@gmail.com

Abstract

Repair welding of service-exposed, aged Heat Resistant Nickel-Base alloys is notorious for being extremely difficult. There is a high propensity for cracking in the heat affected zone of the weld which manifests during or immediately after the welding process. Industry experience and current literatures converge upon one conjecture- that the success of the weld is based largely on optimistic 'trial and error'.

Prior Failure Analysis of characteristic weld cracks observed upon welding revealed the root cause to be the synergistic action of high Heat Input of the welding process and the residual stress which lead to a localized Creep Embrittled HAZ which eventually failed due to brittle intergranular fracture. Three weld parameters were identified for the reduction of heat input and residual stress; Groove design, Welding Pattern and Interpass Temperature. The joint was upgraded from a Double-angle V-groove to a U-groove, use of a Stringer Pattern instead of Weaving and a higher Interpass Temperature.

Mechanical testing revealed that the modified procedure was superior. Void Counts of the HAZ showed a reduction in void percentage from 5.31% (Original Procedure) to 0.74% (New Procedure) near the tube's ID and a reduction from 13.3% (Original Procedure) to 3.40% (New Procedure) at OD areas. Comparative Hardness Testing of the Original and New Weld Procedures' weld joints concurred with the creep void analysis. Hardness profiles of the New Procedure showed higher values compared to the Original Procedure suggesting superior tensile strength of and less fissuring.

The success of the new weld procedure was further authenticated by tensile testing. The 0.2% Yield Strength of the New Procedure was found to be 383.18N/mm^2 versus 339.70N/mm^2 (Original Procedure) reflecting an improvement of 43.48N/mm2. UTS results also reflected an increase from an average of 475.10N/mm^2 (Original Procedure) to a superior 492.55N/mm^2 (New Procedure).

Introduction

To date, there is no single panacea to prevent the cracking of the aged HP-Mod (*25Cr-35Ni-1.5Nb*) alloy in its Heat Affected Zone (HAZ) upon attempted welding. The cause of cracking lies with the cast microstructure and its changes in service due to aging.

Prior to aging, welding is straightforward. In the 'As-cast' form, the microstructure consists of an austenitic matrix with fine dispersions of chromium-rich $M_{23}C_6$ carbides and Niobium-rich MC carbides [1]. However, at elevated temperatures, typically between 590-650°C secondary carbides precipitate into the inter-dendritic spaces. This carbide precipitation gives HP alloys their superior creep rupture properties [2]. As the reformer tube temperature is increased beyond 650 up to 970°C carbides begin to coalesce. In the fine intergranular form, $M_{23}C_6$ precipitation is beneficial for enhancing creep resistance. However, in the coarsened and coalesced state, they spread grain boundaries apart and reduce ductility.

There are some practices currently used to improve the weldability of these components. These include Solution Annealing, Buttering, and Shot Peening. However, even when used cracking still occurs. This work identifies and shows the results of testing of three additional weld parameters to improve the weldability of aged HP-Mod.

Results of Failure Analysis of Cracks with Typical Weld Procedure

Several test joints (using aged tube stock material) were prepared and welded. Visual inspection revealed typical characteristic cracking in the HAZ starting at the Tube OD on the toe of the weld. These cracks had been extracted for visual, microstructural and SEM analysis. Based on the metallurgical failure analysis performed then, it was determined that the root cause was a creep-like degradation of the microstructure caused by the high heat in the HAZ during the welding process itself coupled with residual stress thereby leading to brittle intergranular fracture.

Development of New Weld Procedure

Analysis of various weld parameters revealed that the main three parameters which would have the largest impact on reducing the cracking phenomena due to the Creep Embrittlement and Residual Strain Mechanism are:

1. Interpass Temperature- The Interpass temperature was set to a range of (200°C – 230°C)
2. Groove Design- Upgrade from Double-Angled V-groove to a U-groove
3. Welding Pattern- Upgrade from Weave Pattern to Stringer.

Interpass Temperature

Interpass temperatures need to be high enough to ensure a slow cooling rate which in turn reduces residual stress. Contrastingly, interpass temperatures that are too high are associated with a reduction in tensile properties and notch toughness [3]. The original procedure used an Interpass temperature of 100°C maximum. Work by Petro in 2011 tested the welding of Nickel-Base alloys with higher interpass temperatures up to 500F (260°C). It was revealed that there was no significant loss in tensile properties. Also, there were no phase changes in the microstructure nor was any cracking evident in the HAZ or the weld metal [4]. In 2001 Hasegawa experimented with the use of Preheat and Interpass temperatures as high as 400°C. It was found that there could be benefits in preventing HAZ cracking with a slight risk of solidification cracking in the weld metal [5].

Groove Design

The type of groove profile affects both residual stress and heat input due to the amount of weld metal required to be deposited. Moreover, the profile of the groove edges determine the plane of the solidification front of the weld metal and the partially melted zone. This in turn affects the area on which shrinkage forces act thereby increasing or reducing contraction stresses.

In comparison to V-grooves and double-angled V-grooves, joint designs incorporating J-grooves and U-grooves are more effective at distributing shrinkage and restraint forces. Not only is the stress applied on a larger surface area than in V-grooves but also, the fusion line follows a divergent curved surface. As such, shrinkage forces with each pass do not have overlapping vectors. The superiority of U-grooves over V-grooves has been shown to reduce cracking in highly restrained joints (those with high residual stress) [6].

A new U-groove was designed in accordance with standard recommended U-groove guideline data from the International Institute of Welding's Recommended Guidelines [7]. The recommended modification to the Groove angle is shown below in Figure 1 where the U-groove profile is overlaid onto the old Double Angle V-groove design for comparison.

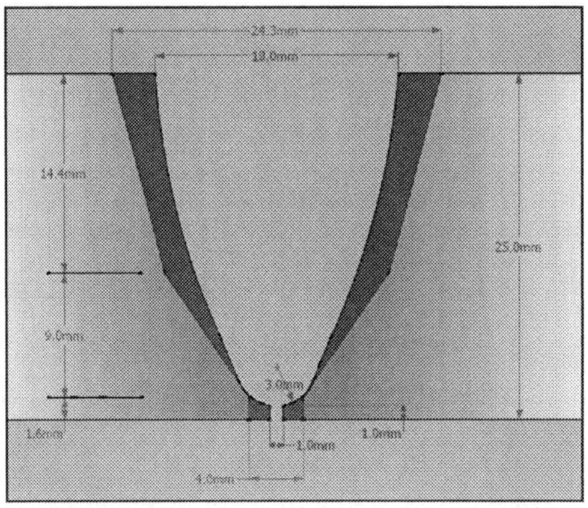

Fig. 1: Comparison between New Weld Procedure U-Groove Design (Blue) and Old Weld Procedure Groove Design (Red)

Weld Pattern

The original weld procedure specified a weave pattern weld bead. In weave beads, the welder moves the arc and the puddle of molten metal in a side-to-side motion across the joint during its forward motion. Stringer beads on the other hand involves either dragging/pulling (electrode pointing back) or pushing (electrode pointing forward) the torch along the weld joint without (or minimal) side-to-side movement. Since Stringer beads has a much faster Travel Speed, the overall heat input is reduced.

Comparing Original Weld Procedure with Modified Weld Procedure

Welds were prepared using the procedure with modified weld parameters specified. These were tested and compared to that of the original procedure.

Optical Microscopy

With the new procedure, the microstructure there was a massive improvement when compared to the original. No macro-cracking, micro-cracking or fissuring was observed in the HAZ upon visual and microstructural inspection. Several isolated voids were observed, however these were scattered and not connected nor aligned. Refer to Figures 2 and 3.

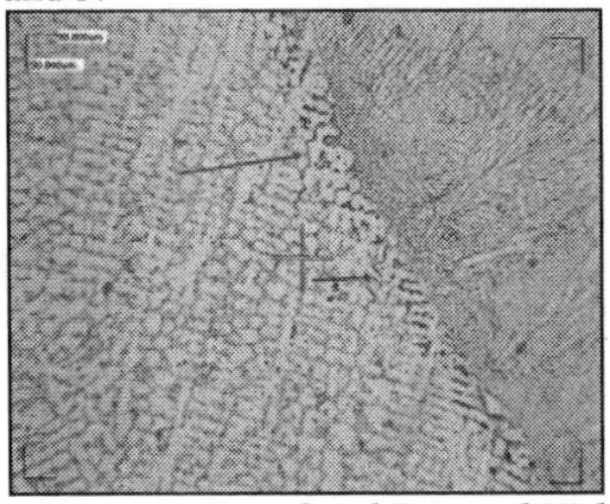

Figure 2: Sample of Original Weld Procedure: Parallel Fissuring at Fusion Line (50X)

Figure 3: Sample of New Weld Procedure: No fissuring observed.

Void & Fissure Count Analysis

The light optical micrographs of the stock material, original weld procedure and the modified weld procedure were observed and compared. Due to the root cause of the HAZ cracking having been determined to be the synergistic phenomena between Residual Stress and Creep Embrittlement, focus was attributed to comparative void and fissuring assessment. Table 1 below shows the comparative Void Counts observed for the Stock Material, UBM and HAZ for both the Original and New Weld Procedures.

Table 1: Comparison of Surface Areas of Voids & Fissures for Stock Material, Original Procedure and the New Procedure

	Percentage Voids & Fissures (% Surface Area)				
		Original Weld Procedure		New Weld Procedure	
	Stock Material	Unaffected Base Metal	HAZ	Unaffected Base Metal	HAZ
ID	0.6	3.09	5.31	0.36	0.74
OD	0.6	3.44	13.3	2.8	3.40

Hardness Testing Analysis

Hardness tests were carried out on the stock material, the original weld procedure joint and the new weld procedure joint. In the Heat Affected Zone of the weld, there was a sharp drop in the hardness values obtained using the old procedure. Micro-hardness testing revealed that hardness values dropped sharply adjacent to voids, fissures and cracks. Contrastingly, carbide precipitation increases hardness. Refer to Figure 4 which shows the comparative average Hardness values for both the Original and Modified Procedures.

In the HAZ, there is a sharp drop in the hardness of the weld of the original procedure. The drop in the original procedure is quite sharp. This is expected of the creep embrittlement phenomena.
With the new weld procedure however, the hardness transition is much more gradual. This indicates that the reduced heat input (U-joint, Stringer Bead) and slower cooling rate (Higher Interpass temperature) was successful in improving and alleviating the creep embrittlement phenomena.

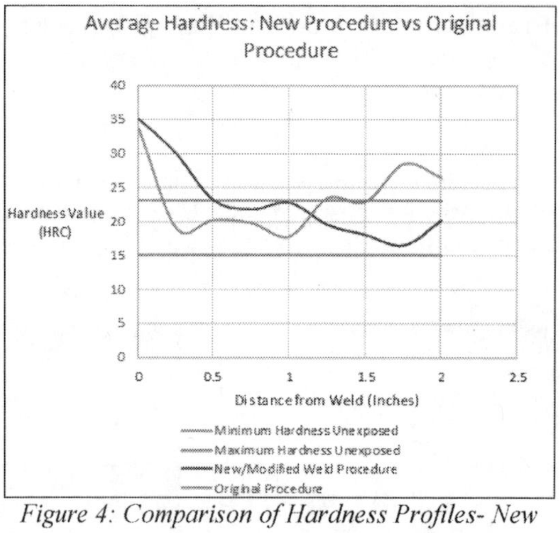

Figure 4: Comparison of Hardness Profiles- New Procedure vs Original Procedure

Tensile Tests

Tensile testing was carried out in accordance with specifications set in ASTM A297 and ASTM A370-14. Testing was done on the original weld procedure joint as well as the new weld procedure joint. Ultimate Tensile Strengths and 0.2% Yield Strengths were calculated from the base results and compared in Table 2 below. The yield strength of the new weld procedure samples was calculated to be 383.18N/mm^2. This was 43.48N/mm^2 higher than that of the original weld procedure average of 339.70N/mm^2 reflecting an improvement of 12.8%. A comparison of the average UTS showed an increase from 475.10N/mm^2 with the Original Procedure to 493.55N/mm^2 of the New Weld Procedure. This reflects an improvement of 18.45N/mm^2 or 3.9%.

Table 2: Summarized Calculated Data from the Result of Tensile Testing of the Original Weld Procedure and New Weld Procedure Joints

Mechanical Property (N/mm^2)	Original Weld Procedure Specimen				New Weld Procedure Specimen			
	O2	O3	O4	Avg.	N1	N2	N3	Avg.
0.2% Yield Str.	342.42	350.57	326.11	339.70	383.18	383.18	383.18	383.18
Ult. Tensile Stress	501.10	465.20	459.01	475.10	495.75	482.68	502.22	493.55

Conclusion

It was proven via Hardness, Tensile, and Microstructural Analysis that the weldability of aged HP-Mod Reformer tubes can be improved significantly by adjusting Interpass Temperature, Groove Design and Weld Pattern as specified. However, further work needs to be done to optimize these and other parameters e.g. Welding Sequence, Solution Annealing Temperature, and even use of Pulsed Welding etc. for optimum specifications.

References

[1] Lamb, S., and J.E. Bringas. 1999. Practical Handbook of Stainless Steels & Nickel Alloys. 1st Edition ed: ASM International
[2] Blair, Malcom, and Thomas L. Stevens. 2004. Steel Castings Handbook. Edited by B. Linskey. Vol. Supplement 9- High Alloy Data Sheets, Heat Series: Steel Founders' Society of America.
[3] Funderburk, R.S. 1998. The Importance of Interpass Temperature. Welding Innovation Quarterly Vol. 15 (No. 1).
[4] Petro, J.S. 2011. Effect of Interpass Temperature on the Structure and Properties of Multipass Weldments in High Performance Nickel Alloys, Department of Mechanical Engineering, Colorado State University, Fort Collins, Colorado.

[5] Hasegawa, Katsunobu. 2001. Repair Welding and Metallurgy of HP-Modified Alloy after Long Term Operation. Paper read at Metal 2001, May 2001, at Czech Republic.
[6] Dixon, B.F., R.H. Phillips, and J.C. Ritter. 1983. Solidification Cracking: Causes, Remedies and the Transvarestraint Test. Paper read at 31st Annual Conference of the Australian Welding Institute, 16-21 October, at Sydney.
[7] Welding, The International Institute of. 1965. Recommended Welded Connections for Pipework. In Section 1: Arc Welded End to End Pipe Butt and Sleeve Joints. London: The Institute of Welding.

Weld Residual Stresses in Dissimilar Metal Welds - Modeling and Measurement

Zhili Feng, Xinghua Yu, Jian Chen, Stan David

Oak Ridge National Laboratory, USA

Weld residual stresses (WRS) in dissimilar metal welds (DMWs) are a critical factor causing performance degradation of welded structures in a number of important industries and applications. For example, it is as an important driver for primary water stress corrosion cracking in nuclear power plant safety-related components. Recent studies have shown that weld residual stress prediction using today's DMW residual stress models strongly depends on the strain-hardening constitutive model chosen. The commonly used strain-hardening models (isotropic, kinematic, and mixed) are all time-independent and thereby inadequate to account for the time-dependent (viscous) plastic deformation at the elevated temperatures experienced during welding. For materials with profound strain-hardening, such as stainless steels and nickel-based alloys that are widely used the DMWs, the degree of strain hardening can be highly dependent on the recovery and recrystallization processes. These processes are in turn a strong function of temperature, time, and deformation rate. Recently, we have developed a new strain-hardening constitutive model - the so called dynamic strain-hardening constitutive model - that can adequately describe the strain hardening behavior of DMWs. The application of such a model has resulted in improved weld residual stress prediction. In this talk, I will highlight our recent progress, including: (1) The development of a special high-temperature measurement technique to experimentally quantify the dynamic strain-hardening behavior; (2) The development of the dynamic strain hardening constitutive model based on the considerations of the kinetics of the recovery and recrystallization of flow stress; (3) The implementation of the dynamic strain hardening relation in weld residual stress modeling; and (4) comments on the measurement techniques for measurement of DMW residual stresses.

Experimental Study of Surface Strain during Friction Stir Welding

Krishnapada Sinha [1], Tanmoy Medhi [2], Barnik Saha Roy [3] and Subhash Chandra Saha [4]
[1]Research Scholar, Department of Mechanical Engineering, National Institute Technology Agartala, Tripura, India, krishnapadasinha@gmail.com
[2] Research Scholar, Department of Mechanical Engineering, National Institute Technology Agartala, Tripura, India, tanmoymedhi@gmail.com
[3]Assistant Professor, Department of Mechanical Engineering, National Institute Technology Agartala, Tripura, India, barnik2707.bs@gmail.com
[4]Professor, Department of Mechanical Engineering, National Institute Technology Agartala, Tripura, India, subashchandrasaha@yahoo.in

Abstract
Friction stir welding (FSW) is a solid-state welding technique which requires rigid clamping of plates, unlike the fusion welding. This restricts the expansion and contraction of the material and significant residual stresses are developed although no melting takes place. In this study, FSW of AA6061-T6 plates is carried out to measure the longitudinal and transverse surface microstrain in the advancing and retreating side of the weldment with the help of electrical resistance strain gauges. The study of surface microstrain and the effect of welding parameters i.e. tool rotational speed and traverse speed on the peak microstrain is also done. The result shows that the welding parameter have much effect on the surface strain during friction stir welding.

Introduction
Friction stir welding (FSW) is an innovative solid state welding process invented by TWI, England in the year 1991[1]. During FSW due to the different temperature profiles across the weld joint, the material undergoes deformation and strain is developed both inside and outside the weld bead. A theoretical study of the strain rates and strain components within the stir zone (SZ) and thermo-mechanically affected zone (TMAZ) by using a coupled three-dimensional viscoplastic flow and heat transfer model [2] was done where it was seen that the longitudinal strains are tensile in front of the tool and compressive behind the tool while the transverse strains are compressive in front of the tool and becomes slightly tensile behind the tool. A computation of the equivalent strains in FSW using solid mechanics [3] was done where the maximum values of equivalent strain varied from 6 to 133 for different welding conditions. The study of strain distribution in FSW weld by FEM model [4]; it was observed that when the rotation speed increases, the temperature of the weld plate increases and hence the strain increases. The strain reduces as the distance from the weld zone increases. The asymmetry of the welding process was also explained.

Though several papers have been published in the study of strains within the weld bead but experimental study of surface strains in the weldment surface is yet to be discussed elaborately and the variation of surface strain with welding parameters are not seen experimentally to understand the behavior of the metal at the time of welding with extensive holding pressure from the holding devices. In this study, the electrical resistance strain gauges are used to measure the transient microstrains in AA6061-T6 plates. A dynamic strain measuring system is used to observe the variation of microstrain with time in the advancing and retreating side along the longitudinal and transverse direction. Experiments are performed by varying the tool rotational speed and the tool traverse speed. The results obtained are used to understand the effect of welding parameters on deformation of the material and the formation of surface strain on the weldment.

Experimental details
FSW butt welds were carried out in nine samples. Each sample had two plates to be joined and each plate having a dimension of 110mm x 65mm x 6.3 mm. The material used for the experiment is AA6061-T6 panels. Temperature compensated strain gauges are used in the experiment. In one sample four strain gauges are installed in longitudinal and transverse directions as shown in Figure 1. A strain gauge in the sample plate is connected with a strain gauge of the dummy plate of same layout as sample plate to complete the Wheatstone half bridge circuit. A threaded pin tool is used for welding having pin diameter of 6mm, pin length of 5mm and shoulder diameter of 13mm.

The samples are welded by varying tool traverse speed and tool rotational speed. The downward force and the tilt angle are kept constant during the process. Each of the 9 samples is welded with a given tool traverse speed and tool rotational speed. The first three samples (samples 1, 2, and 3) are welded at 900rpm, the next three samples (samples 4, 5, and 6) are welded at 1200rpm and last three samples (samples7, 8, and 9) are welded at 1500rpm at varying welding speed of 1mm/s, 2mm/s and 3mm/s.

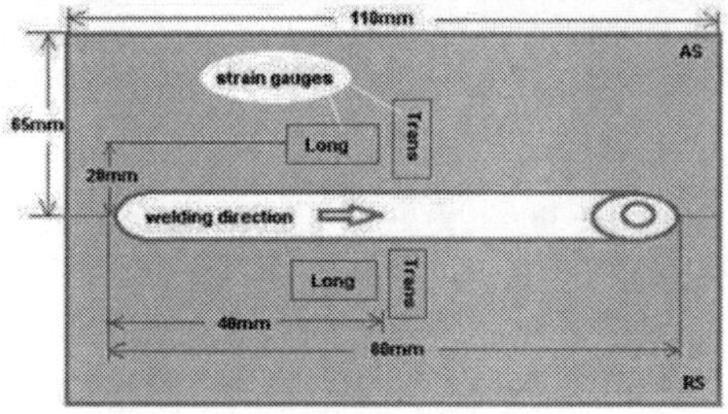

Figure 1: Sample description

Materials

AA6061-T6 aluminium plates are used for the experiment. The heat-treatable 6000 group of aluminium alloys usually serves for automotive panels and aircraft fuselages, because of being strengthened after forming. It is also used to manufacture a shell of rockets, aircraft cabins, train wagons, and pipes [5].

The strain is measured by using a self-temperature compensating Ni-Cr alloy foil strain gauge (ZFLA-6-350-11) attached to the upper surface of the plate. This is a high temperature strain gauge having the backing of polyimide that is applicable for metals and ceramics. The strain limit of the strain gauge is 1%(10000 X 10-6 strain) operational within a temperature range of -20~ +300°C. This train gauge has the same coefficient of thermal expansion as that of the base material.

Result

Based on parametric combination as discussed above, FSW butt welds were made with the placement of the strain gauges as shown in Figure 1. During the experiment, few strain gauges were damaged for welding parameters 900 rpm & 3mm/sec and 1200 rpm & 3 mm/sec.

Transient longitudinal microstrain curves as a function of time

The variation of transient longitudinal microstrains welded at a constant traverse speed of 1mm/s and variable tool rotational speeds are shown in Figure 2 (a,b,c). In Figure 2(a), during welding when the tool is near to the strain gauges, a sharp increase in the strain values is observed due to the increase in temperature which results in expansion of the material. The surface microstrains in the AS and RS are negative which indicate that the contraction of the material takes place but this contraction is constrained by the adjacent high-temperature material resulting from tensile surface stresses. The surface microstrains in Fig. 2(b) and Fig. 2(c) are positive in AS and negative in RS indicating compressive and tensile surface stresses respectively. The increase in peak microstrains with an increase in rpm is due to the increase in heat input. Similar trends are observed in a variation of longitudinal microstrain experiment was done at a welding speed of 2mm/s and 3mm/s and variable rotational speed respectively. Lower rotational speed produced lower peak microstrains compared to higher rotational speed due to the increase in heat input with an increase in rpm. Also, the microstrains in the AS are higher than the RS which shows the asymmetry of the process. The surface microstrains in all the cases are negative which may indicate the development of tensile residual stresses. The longitudinal microstrains at 900rpm and 3mm/s of sample 3 cannot be obtained due to the damage of the strain gauges. A similar process of stress formation was also observed in GMAW of thin walled aluminium alloy [6].

Similar changes are observed when microstrains in the transverse direction are measured at a welding speed of 1mm/s, 2mm/s and 3mm/s and variable rotational speed respectively. Higher microstrains in the AS compared to RS shows the asymmetry of the process.

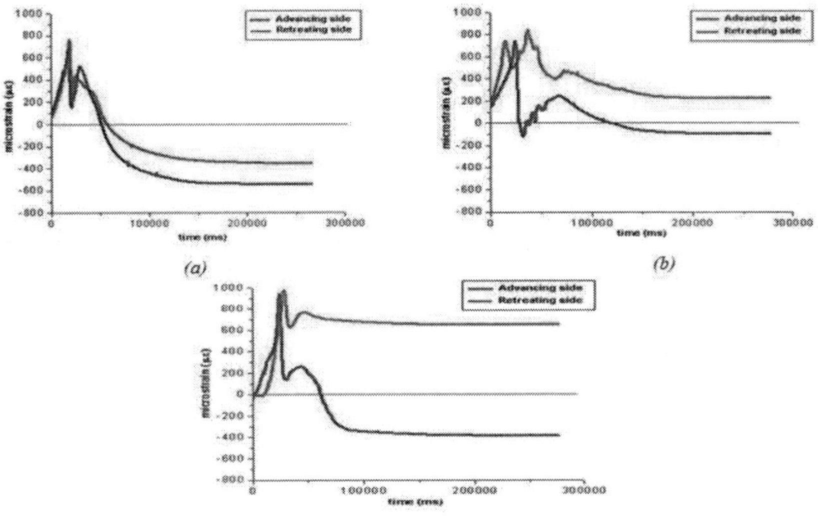

(c)

Figure 2: Transient longitudinal microstrain in AS and RS (a) 900 rpm and 1mm/s, (b) 1200 rpm and 1mm/s, (c) 1500 rpm and 1mm/s

Effect of tool rotational speed on peak transient microstarins

The effects of tool rotational speed on peak longitudinal microstrains at traverse speeds 1mm/s and 2mm/s are shown in Figure 3(a) and Figure 3(b) respectively. The peak longitudinal microstrains increase with an increase in tool rotational speed due to the increase in heat input both in the AS and RS. It can also be seen that at 1mm/s the difference between the peak microstrains in the AS and RS is low since the heat input is high as the welding speed is very low. As a result, the strains in the RS are also significantly high. A greater difference of peak microstrains are observed between the AS and the RS due to higher welding speed of 2mm/s and the heat input is lower thus reducing the strains in the RS. It can also be observed that the peak microstrains are higher at 1mm/s than at 2mm/s due to the higher heat input at lower welding speed.

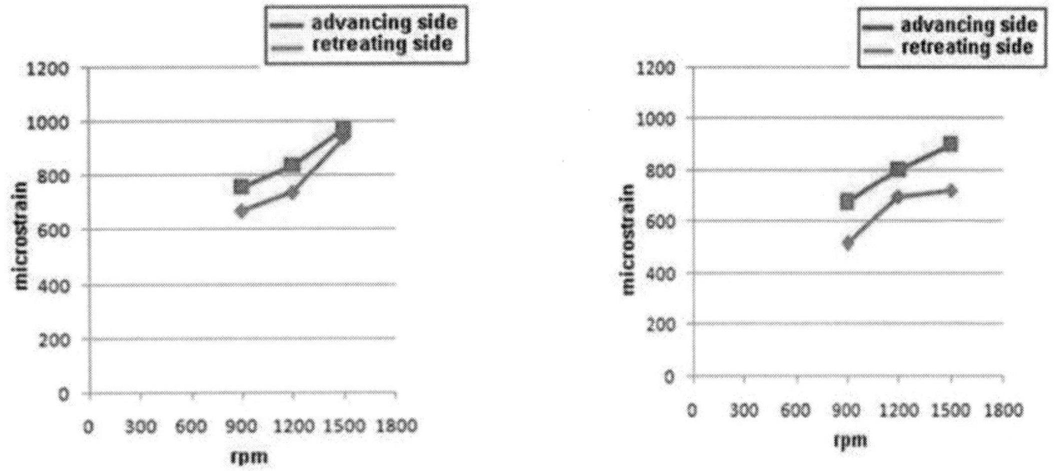

(a) (b)

Figure 3: Effect of tool rotational speed on peak longitudinal microstrain at (a) 1mm/s, (b) 2mm/s

Similar effects of tool rotational speed on peak transverse microstrains at traverse speeds 1mm/s and 2mm/s are observed.

Effect of traverse speed on peak transient microstrains

The effects of traverse speed on peak longitudinal microstrains at 1200 rpm and 1500 rpm are shown in Figure 4(a) and Figure 4(b)respectively. It can be seen that the peak longitudinal microstrains decrease with increase in welding speed both in the AS and RS. Also, the peak microstrains are high at 1mm/s, due to higher heat input but are low at 3mm/s due to lower heat input both at 1200rpm and 1500rpm. Again, the peak microstrains are higher in the AS than the RS which shows the asymmetry of the welding process. Additionally, it can be seen that the peak microstrains are higher at 1500rpm than at 1200rpm due to higher heat input at 1500rpm.

Similar effects of traverse speed on peak transverse microstrains at 1200 rpm and 1500 rpm are observed.

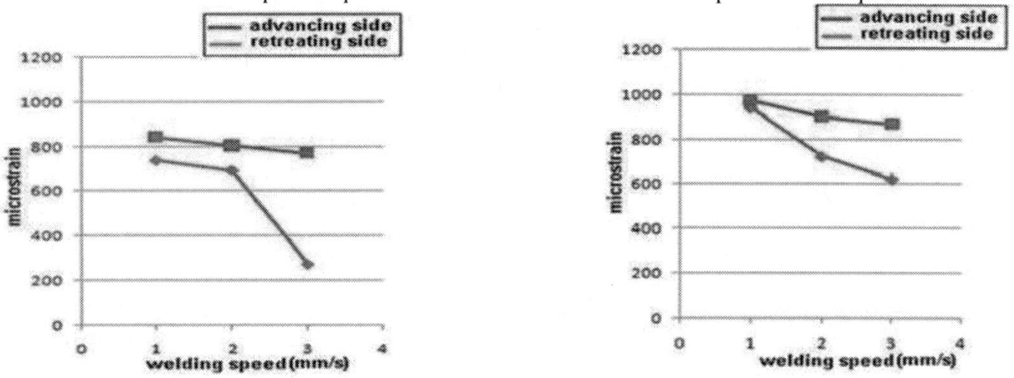

(a) (b)

Figure 4: Effect of traverse speed on peak longitudinal microstrain at (a) 1200rpm, (b) 1500rpm

Conclusion

The strain in the advancing side is higher than the retreating side both in longitudinal and transverse directions. The significant difference is observed between the longitudinal and transverse microstrains. The peak longitudinal and transverse microstrains increase with an increase in tool rotational speed both in advancing and retreating side. The peak longitudinal microstrains decrease with increase in traverse speed. The peak longitudinal and transverse microstrains decrease with increase in traverse speed. The surface microstrains are negative in the longitudinal direction indicating tensile residual stresses and the surface microstrains in the transverse direction are low positive/negative indicating low compressive/tensile residual stresses. The proposed method for the study of strains is feasible but with some restrictions.

References

[1] Thomas, W.M., Nicholas, E.D., Needham, J.C.,Murch, M.G.,Temple-Smith, P., and Dawes, C.J., 1995. Friction Welding. The Welding Institute(Cambridge, GB), Vol. 5, 460, 317.

[2] A. Arora, *et al*: "Strains and strain rates during friction stir welding" Scripta Materialia, 61(2009), 863–866.

[3] G. Buffa, *et al*: "A continuum based FEM model for friction stir welding –model development" Materials Science and Engineering: A, 419 (2006), 389–396.

[4] R. Jain, *et al*: "Finite Element Simulation of Temperature and Strain Distribution in Al2024 Aluminum Alloy by Friction Stir Welding" (2014) 5th International and 26thAll India Manufacturing Technology, Design and Research Conference (AIMTDR 2014), IIT Guwahati, Assam, India.

[5] L. Cao: "Analysis on the Surface Stresses in the Aluminum Alloy Friction Stir Welded Joints" Advanced Materials Research, 291-294(2011), 958-963.

[6] S.K. Chionopoulos, *et al*: "From transient to permanent phenomena during GMAW of a thin-walled marine aluminium alloy" International Journal of Materials Science and Applications, 3(2) (2014)58-69.

October 12th
(Wednesday)

Science and Technology of Welding and Joining at Micro- and Nano-scale

Y. Norman Zhou
University of Waterloo, Waterloo, Ontario, Canada
nzhou@uwaterloo.ca, (519)888-4567 ext. 36095

Abstract
Welding and joining is a crucial part of manufacturing from macro- to micro- and nano-scale, providing mechanical support and integration, electrical connection, optical coupling and environmental protection. This presentation looks at recent innovation and challenges in microjoining and nanojoining. Specifically, case studies are used to illustrate how fundamental concepts, established through research and practice in macroscopic welding and joining, can be applied to help developing joining processes at micro- and nano-scale, and even new materials processing technology in general. On the other hand, examples are provided to show how research work on microjoining and nanojoining has fundamentally expanded our understanding of the science and technology of welding and joining, facilitating exciting new applications.

Introduction
Similar to "macro-scale" welding and joining for heavy manufacturing, effective microjoining has been an integral part of light manufacturing for many decades in microelectronics, medical, and other industries [1]. Because of ever advancing miniaturization, nanojoining has recently become a new frontier of advanced materials joining research [2]. Industrial scale production of nanoscale devices and systems will not be possible, mirroring the experience both at macro- and micro-scales, without successful nanojoining. Normally, researchers and engineers from "macro-scale" welding and joining, and those from micro-/nano-joining belong to different industrial communities. They seldom communicate with each other even from the same organizations. This presentation will provide a number of case studies to illustrate that lessons learned through research and practice in "macro-scale" welding and joining can be used to help developing joining processes at micro- and nano-scale, and also that research work on microjoining and nanojoining can fundamentally expand our understanding of the science and technology of welding and joining.

Two solid surfaces, if perfectly clean and atomically flat, will join to each other if brought into intimate contact, because of interatomic forces. The reduction of the Gibbs free energy of the system by replacing two surfaces with an interface depends on how well atoms pair across the interface [1]. Since most engineering surfaces are characterized as rough and contaminated (with oxides, etc.) various joining operations involve essentially different techniques applying energy, usually heat and/or pressure to overcome these barriers to joining, i.e., surface oxide and roughness, and interfacial lattice mismatching [1]. Of course, successful welding and joining in practice would involve many more process and metallurgical issues and require knowledge from many branches and fields of science and engineering. The following will cover just a few examples of these fundamental and practical aspects, focusing on the similarity and differences between welding and joining at different scales.

Surface oxide removal
The Ellingham diagram is often used to understand how and why surface oxides could be reduced, especially in brazing and soldering and solid-state bonding processes. This is still the case in joining of metal substrates using Ag nanoparticles but in a slightly twisted way [3]. Nanoparticles are used to develop low (even room) temperature joining processes, e.g., as an alternative for Pb-free soldering. During nanofabrication, metallic nanoparticles (such as of Ag, Cu and Au) are generally coated by thin organic shells (e.g., PVP) to prevent agglomeration. During joining, this organic shell will be removed/decomposed to allow sintering between nanoparticles and bonding of nanoparticles to the substrates. Interestingly, the carbon generated from the decomposition of the organic shell can act as reducing agent to remove the oxide on the substrate surfaces. However this only works for those substrate metals (Ag and Cu) with Gibbs free energy lines that lie above those of carbon in the Ellingham diagram and the Ni joints are weak because the oxide reduction by carbon becomes difficult when the bonding temperature drops below 550-700 K as shown in Figure 1 [3]. On the other hand, the Al and Ti free energy lines are at the bottom of the Ellingham diagram (not shown in Figure 1) so bondability of Al and Ti using organic shelled Ag NPs is very poor.

Lattice matching in cold welding of Ag nanowires
While epitaxial solidification provides near perfect lattice matching at the interface between base and molten

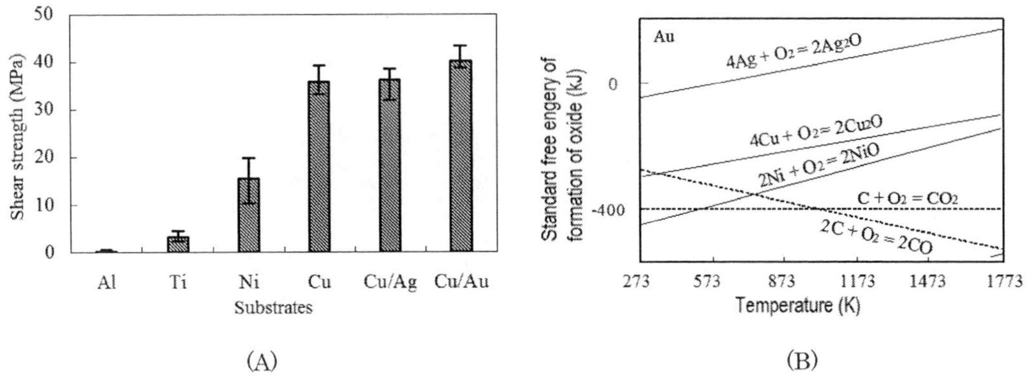

Figure 1: (A) Strength of metal joints bonded using Ag NPs; (B) Top part of Elligham diagram (modified from ref [3])

material in liquid state joining processes, deformation is generally needed to realign the atomic structure at the faying interface in solid-state joining processes [4], which seems still the case at nano-scale. For example, in a study where a nano-indenter was used to weld Ag nanowires at room temperature, the molecular dynamic simulation (Figure 2) showed that larger deformation was needed to weld pentagonal- or five-twinned nanowires because of the large crystallographic orientation mismatching as compared to single crystal nanowires [5].

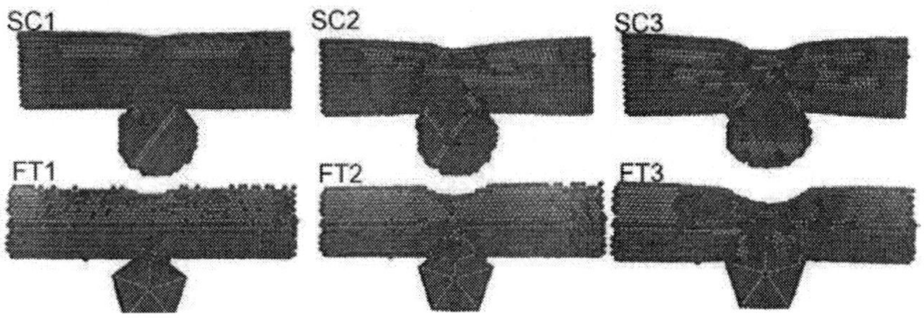

Figure 2: Atomic structures of welded nanwires at different stages of deformation by molecular dynamic simulation with SC1, SC2 and SC3 for single crystal nanowires, and FT1, FT2 and FT3 for five-twinned nanowires. The atoms are colored by red for fcc, pink for hcp and blue for other crystalline structures [5].

Covalent bond formation between carbon nanotubes

This case study will look at how direct joining could occur between two crossing single-walled carbon nanotube (CNTs). Since the CNTs are in ordered and hence stable hexagonal structure, simply stacking them across each other cannot combine two surfaces into one interface because of the energy barrier. Terrones et al. [6] applied a focused electron beam to induce structural defects, "vacancies" and "interstitials", which resulted in dangling bonds at the crossing point of two CNTs (Figure 3A). Inter-nanotube junctions formed by cross-linking of those dangling carbon bonds at 800°C into heptagonal or octagonal structures (Figure 3B). The molecular dynamics simulation clearly indicated that lowering of the energy of the system was the driving force for forming these junctions, which could be used in developing multiterminal electronic devices [6]. This process is obviously different from what has been experienced at macro- and micro-scale but may be grouped as solid-state bonding according to the traditional classification of welding and joining since no melting occurs.

Active elements in joining of ceramics to metals

It is well known that active elements, such as Ti, V, Nb, Zr, are needed in fillers for brazing/soldering or interlayers for solid-state bonding of ceramics to metals because those active elements would cause chemical reactions leading to joints between them [7]. This is the case when ultrasonic nanowelding was used to join CNTs to Ti electrodes for the development of carbon nanotube-based photovoltaic cells [8]. Low Ohmic contact was achieved between the Ti electrodes and CNTs only when applied ultrasonic energy was high enough to form covalent C-Ti bonds (at high power: 0.21 W in Figure 4). The difference here between macro-scale and nano-/micro-scale is that joining of ceramics to metals at macro-scale requires good joint strength while desirable joint electrical characteristics are the goal at nano-/micro-scale.

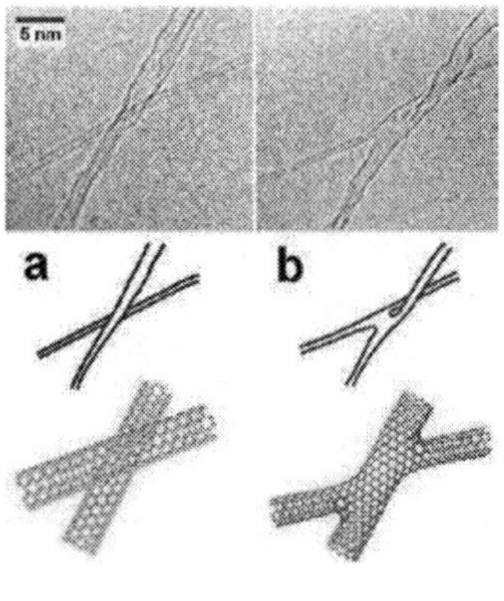

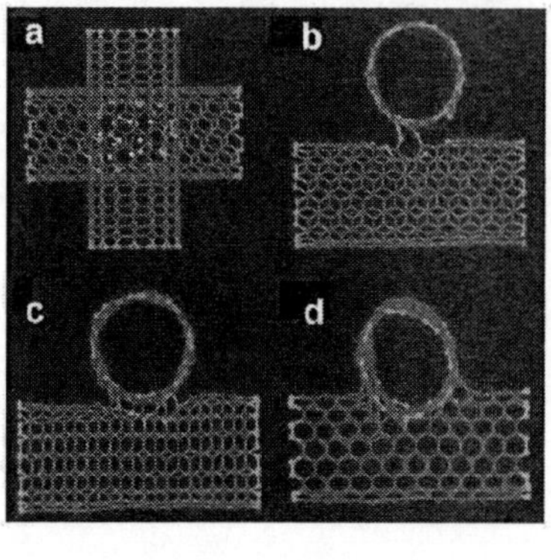

(A) (B)

Figure 3: (A) TEM micrographs, skeches and molecular dynamic models of two crossing single-walled CNTs (a) before and (b) after joining with heptagonal rings indicated in red. (B) Molecular dynamics simulated sequences of merging between two CNTs: (a) The simulation starts with randomly created 20 vacancies in the lattices in the localized neighboring region (top view); (b) After 10 ps, two links between the two defective carbon structures are formed via carbon chains (side view). (c) After 100 ps, the connection between the two tubes is established, although some sp3-carbon atoms (red) and dangling bonds (green) are still remaining. (d) After 220 ps, constructed junction contains six heptagons, one octagon, one pentagon, and two dangling bonds [6].

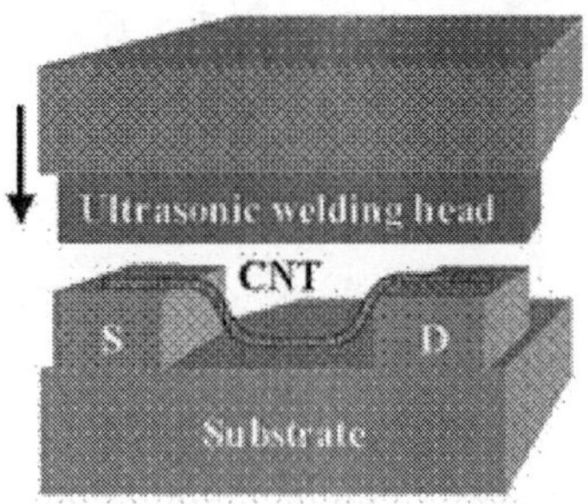

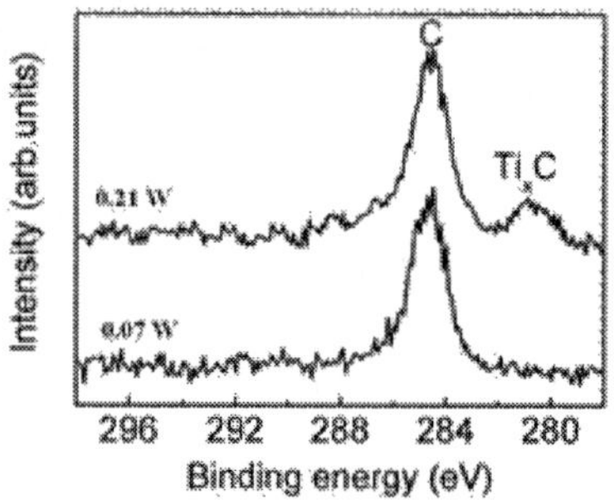

(A) (B)

Figure 4: (A) Setup for ultrasonic welding with a clamping force (indicated by the arrow) to press the welding head against the canbon nanotube (CNT), and Ti electrodes (S and D); (B) XPS spectra of the weld zone at ultrasonic power of 0.07 W and 0.21 W [8].

Technology innovation based on laser vaporization

There is a great amount of research work published on laser-material interaction in laser welding, e.g., vaporization in fusion zone as shown in Figure 5A [9]. This understanding of why and how laser irradiation affects fusion zone vaporization is very useful for welding researchers and engineers to develop and optimize welding parameters to minimize the vaporization and hence its harmful influence on base metal microstructure and mechanical properties. But interestingly this knowledge has been also used in the development of an innovative laser processing technology to embed additional memory and/or pseudoelasticity in NiTi products, in which laser beam is used to produce local preferential Ni vaporization, and hence to alter the Ni/Ti ratio and martensite phase transformation temperature [10]. But laser parameters still need to be tightly controlled so no expulsion will occur as shown in Figure 5B [11].

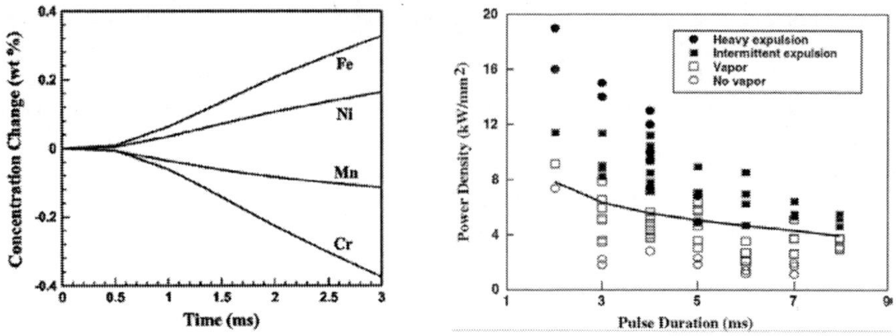

(A) (B)

Figure 5: (A) Elemental change due to vaporization [9]; (B) Process window showing vaporization and expulsion [11] in laser welding of SS304.

Integrated micro-sensors in ultrasonic wire bonding

Ultrasonic wire bonding is the dominant method for electrical interconnection at the IC level with trillions of wire bonds made annually. To fit into a tiny space surrounding a typical bond (joint) at 70 μm in diameter (Figure 6), strain gages need obviously to be greatly shrunk to provide precise measurement of the stresses in the substrates surrounding the bond. In fact, micro-scale sensors are not only much smaller in scale, but are also integrated with custom made chips, which can provide accurate in situ signals of forces acting on the brittle substrate during bonding. For example, piezo-resistive microsensors were fabricated using the source/drain diffusion zones of commercial CMOS processes to get force signals (Figure 6B).

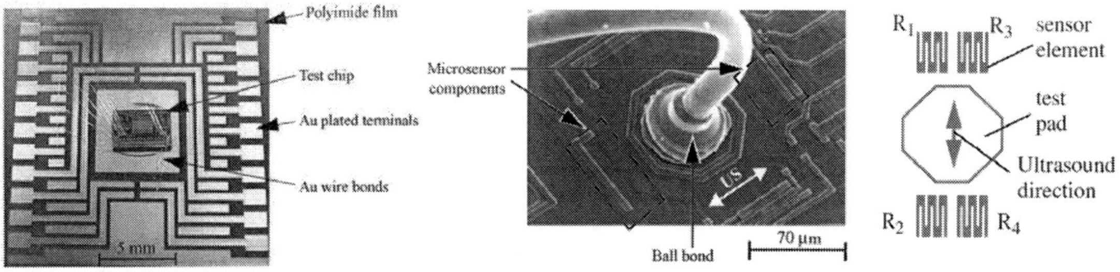

(A) (B) (C)

Figure 6: (A) Microsensor chip on substrate; (B) A ball bond on the microsensor test pad; (C) Illustration of piezo-resistive force sensor element shape and location (not to scale) [12].

Concluding remarks

While mcirojoining has been an integral part of manufacturing for many decades, microjoining continues to face challenges because of ever advancing miniaturization in microelectronics, medical, aerospace, and defense industries. On the other hand, after many years of significant progress in nanoscience and nanotechnology, there is an emerging need to develop effective joining technology for commercially successful assembly of nanoscale devices and systems.

However, similar to welding and joining at "macroscopic scale", micro- and nano-joining aims to establish interatomic bonding between the faying surfaces of the building blocks and to ensure the assembled structures and systems perform desired functions under the service conditions. The engineering challenges remain almost the same, i.e., to remove surface barriers (surface contamination and lattice mismatching, etc.) using different energy sources, which requires precision manipulation of workpieces/building blocks, precision control of energy sources, precision monitoring and control of the processes, and assurance of product functionality and reliability. For these reasons, fundamental concepts and engineering know-how, established through research and practice in "macroscopic" welding and joining, have been inspirations for developing joining processes at micro- and nano-scale. On the other hand, due to the differences in scales and also materials to be joined, microjoining, and especially nanojoining, have their own scientific and engineering issues to deal with. Advances in micro/nano-joining research and development have fundamentally expanded and will continue to strengthen our understanding of the science and technology of welding and joining, facilitating exciting new applications.

Acknowledgments

Funding support from the National Science and Engineering Research Council of Canada and Canada Research Chairs is acknowledged.

References

[1] Y. Zhou (editor), Microjoining and Nanojoining, Woodhead Publishing (Cambridge, 2008).

[2] Y. Zhou and A. Hu: "From Microjoining to Nanojoining", Open Surf. Sci. J., Vol. 3 (2011), pp. 32-41.

[3] A. Hirose and K.F. Kobayashi: "Bonding using Nanopartciles", in Microjoining and Nanojoining (ed. Y. Zhou), Woodhead Publishing (Cambridge, 2008), pp. 250-268.

[4] J.E. Gould, "Mechanisms of Solid-state Bonding Processes", in Microjoining and Nanojoining (ed. Y. Zhou), Woodhead Publishing (Cambridge, 2008), pp. 3-24.

[5] L. Liu, D. Shen, G. Zou, P. Peng, Y. Zhou, "Cold welding of Ag nanowires by large plastic deformation", Scripta Mater., Vol. 114 (2016), pp. 112-116

[6] M. Terrones, F. Banhart, N. Grobert, J.C. Charlier, H. Terrones, and P.M. Ajayan: "Molecular Junctions by Joining Single-Walled Carbon Nanotubes", Phys. Rev. Lett., Vol. 89 (2002), pp. 075505-1 to 075505-4

[7] A. Wu, "Ceramic/metal Bonding", in Microjoining and Nanojoining (ed. Y. Zhou), Woodhead Publishing (Cambridge, 2008), pp. 758-785.

[8] C. Chen, L. Yan, E. Kong and Y. Zhang, "Ultrasonic nanowelding of carbon nanotubes to metal electrodes" Nanotechnology, Vol. 17 (2006), pp. 2192-97

[9] X. He, T. DebRoy and P.W. Fuerschbach: "Composition change of stainless steel during microjoining with short laser pulse", J. Appl. Phys., Vol. 96, No. 8 (2004), pp. 4547-4555.

[10] M.I. Khan, A. Pequegnat and Y. Zhou, "Multiple Memory Shape Memory Alloys", Adv. Eng. Mater., Vol 15, No. 5 (2013), pp. 386-393

[11] X. He, J.T. Norris, P.W. Fuerschbach and T DebRoy: "Liquid metal expulsion during laser spot welding of 304 stainless steel", J. Phys. D: Appl. Phys., Vol. 39 (2006), pp. 525-534

[12] A. Shah, M. Mayer, Y. Zhou, S.J. Hong, J.T. Moon: "In situ Ultrasonic Force Signals during Low-Temperature Thermosonic Copper Wire Bonding", Microelectron. Eng., Vol. 85 (2008), pp. 1851-1857

Visualization of Metal Transfer and Weld Pool Phenomena in SAW

V. Sengupta[a,c], D. Havrylov[a,b], P. F. Mendez[a,d]
[a]Canadian Centre for Welding and Joining, Edmonton, Canada [b]The E.O. Paton Electric Welding Institute, Kiev, Ukraine. [c]vsengupt@ualberta.ca, [d]pmendez@ualberta.ca

Abstract

High speed, high quality videos of metal transfer in SAW have been produced with an innovative technique consisting of inserting a thin-gage steel tunnel along the welding path. The tunnel locally disrupts the flux cover, but the images obtained seem to be representative of the actual mode of metal transfer in unperturbed welds. The possible contamination of the weld cavity with atmospheric gases was tested using a neon marker, and no external gases were detected. Analysis of the electrical signal did not show a significant impact of the presence of the tunnel. Analysis of the weld cross sections show a small but measurable effect of the tunnel, possibly due to the presence of occasional stray arcs between the electrode and the tunnel. A thorough literature search was carried out to compare the observations with non-disruptive X-ray imaging techniques. For comparable conditions, the same clearly identifiable features are seen in X-ray and optical imaging both at high and low current levels. The observations at high current (above 600 A) revealed a new metal transfer method likely governed by an electromagnetic kink instability. At lower currents the metal transfer resembles a very chaotic globular transfer. A transition from recirculating flows to gouging penetration is seen both in the videos and in the increased penetration measured in the cross sections. With the increasing amount of evidence supporting this experimental technique, a whole new world of possibilities is available for the design of advanced waveforms in SAW aiming at higher quality, higher productivity, and novel of fluxes and alloys.

Introduction

The physical phenomenon in submerged arc welding (SAW) takes place under a flux bed. The phenomena like metal transfer and weld pool dynamics even though of utmost importance in determining the quality of the weldment have not been studied thoroughly over the years due to lack of visibility of the arc. Researchers [2-4] investigated the process using X-rays and established the presence of an arc. Franz [5] inserted a ceramic tube in the flux bed and used a high-speed camera to record the metal transfer at 3000 f/s and was able to obtain good images at 500 A.

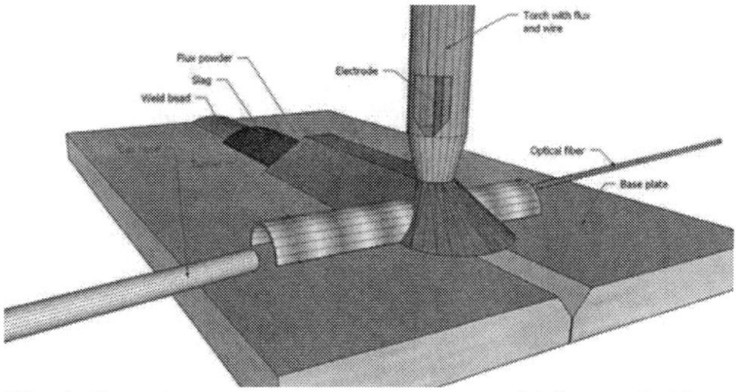

Fig. 1: Experimental arrangement to create high-speed videos of metal transfer in SAW[1]

A technique has been developed to observe the physical phenomena like metal transfer and weld pool movements in SAW. A tunnel made of thin gage steel is placed in the welding path perpendicular to the welding direction. The tunnel provides a brief opening of the cavity; a high-speed camera is utilized to capture the phenomenon going on in the cavity during this brief opening at 10,000 f/s. The welding machine used was Lincoln AC/DC. 1000; the flux used was Lincolnweld 980 (basicity index 0.6); the wire used was Lincolnweld L-50 wire (3.2 mm diameter). The high-speed camera used was a Phantom V210 at a frame rate of 10,000 fps. A 180 mm lens was used with extension tubes to accomplish the desired magnification. A single 850 nm long pass filter was used. Spectra were collected using an Ocean Optics HR4000CG spectrometer.

High-speed video of metal transfer and weld pool phenomena in SAW

Three high-speed videos of metal transfer and weld pool phenomena in SAW are uploaded as supporting online material [8]. Table 1.shows the parameters used to obtain the videos.

Video SOM2 corresponds to Experiment 5 (500 A, DCEP), and Fig 2, summarizes frames 800 to 1800 every 100 frames. The mode of metal transfer is free flight, as it is also appreciated in Fig. 2. A visual count of the number of droplet detachments yields an approximate frequency of detachment of 9 Hz, leading to a droplet diamet.er of 3.6 mm. Video SOM2 shows an irregular shaped droplet with chaotic explosions. Video SOM3 corresponds to

Experiment 12 (500 A, AC). The most interesting feature of this video is the movement of the cathode spot in the EN cycle. Frame 2582 (during an EN cycle) capture a cathode spot, which looks like a very bright area at the bottom left surface of the droplet. After starting, this cathode spot and moves higher and to the right, pushing and deforming the droplet as it moves. The droplet in the EP cycle is similar to the one observed in video SOM2. A detachment count results in a detachment frequency of 13 Hz and droplet diameter of 3.3 mm. Video SOM4 and Fig. 3 corresponds to Experiment 10 (1000 A, DCEP). The most important feature of this video is that the arc is "buried," meaning that the weld pool surface is so depressed that the arc happens almost completely out of sight below the original surface of the workpiece. The mode of met.al transfer seems to be unstable and based on the electromagnetic kink instability, somewhat similar to rotating transfer, except that in this video the kink does not seem to rotate. Frames 7912-8010 show an episode of ejection of molten metal in the form of a kink instability. Metal flies sideways and upwards at high speeds.

Table 1: Parameters of experiments corresponding to uploaded videos

Experiment	Video	Target Current A	Target Voltage V	Polarity	Average WFS m/min	Average WFS ipm	Travel speed m/min	Travel speed ipm	Gas
5	SOM2.	500	30	DCEP	1.61	63.4	0.457	18	CO_2
12	SOM 1, SOM3	500	30	AC	1.79	70.5	0.457	18	CO_2
10	SOM4	1000	38	DCEP	4.47	176.0	0.711	28	CO_2

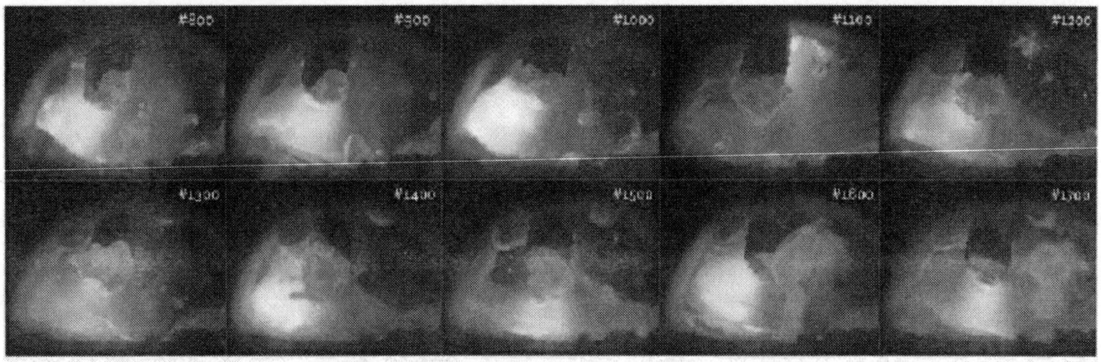

Fig. 2: Metal transfer in SAW during Experiment 5 (500 A, 30 V, DCEP). The figure shows frames 800 to 1700 at 100 frame intervals (total time elapsed is 0.09 s). A is a detachment event is observed between frames 1600 and 1700. For a sense of scale, consider that the wire diamet.er is 0.125 in (3.2. mm).

Atmosphere in the weld cavity

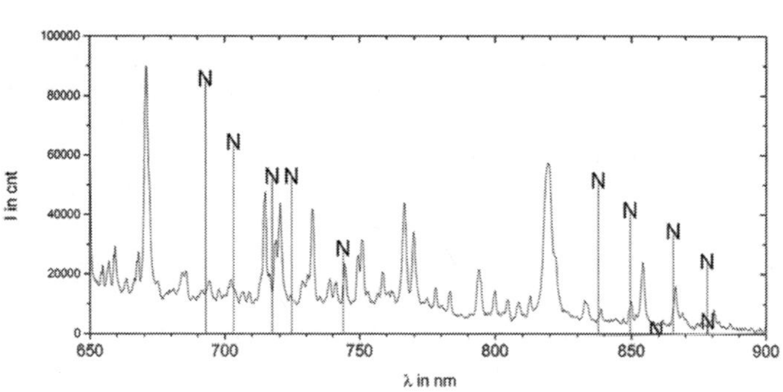

Fig. 4: Spectrum of Experiment 15. Vertical lines indicated as "N" represent the location of spectral lines to be expected from Ne

During the experiments a strong ejection of gas is observed from both ends of the tunnel; this pressure release gas should prevent any contamination of the cavity from atmospheric gases. Spectroscopic investigation was carried out to check this hypothesis. Ne was injected into the weld cavity from one side; if any external gas entered the tunnel, it would carry Ne and would be detected easily with the spectrometer due to its distinctive peaks. Fig. 4 shows the spectral lines of Ne (represented by 'N') for the test of atmosphere, realized with the same nominal parameters of Experiment 5. The absence of any 'Ne' peaks indicates that there is no contamination of the cavity from atmospheric gases.

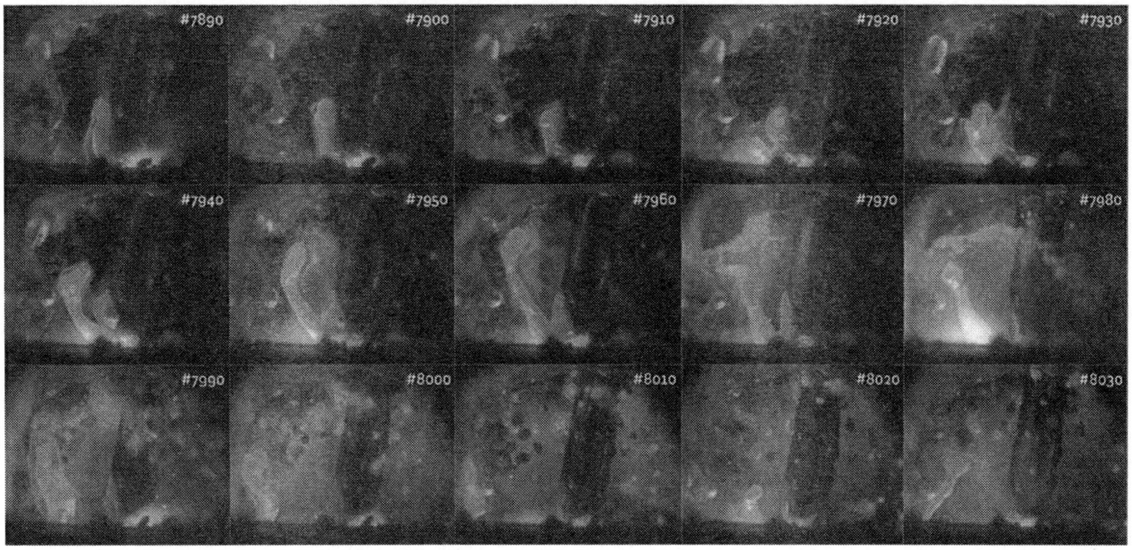

Fig. 3: Metal transfer in SAW during Experiment 10 (1000 A, 38 V, DCEP). The figure shows frames 7890 to 8030 at 10 frame intervals (total time elapsed is 0.014 s). The arc is "buried" and there is metal ejection resembling a kink instability.

Analysis of weld cross sections

Cross sections corresponding to the experiments 5, 12, 15 are compared before, during and after the tunnel. Fig. 5 and Table 2 shows the comparison. Analysis of the cross sections shows a small, but measurable difference in weld width and penetration under the tunnel possibly due to the stray arc between the electrode and the tunnel.

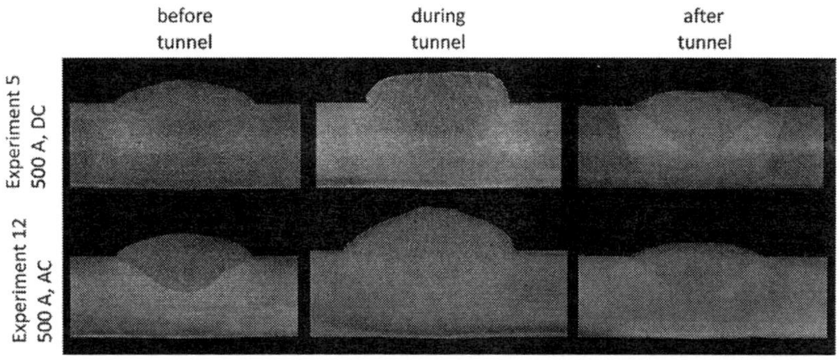

Fig. 5: Cross sections before, during, and after the tunnel. All photos have the same magnification. For reference of scale, the average plate thickness is 9.45 mm.

Comparison of results with other researchers

The images of metal transfer obtained around 500 A, among different researchers [4-7] show a large droplet at the tip of the electrode with a diameter similar to the electrode diameter. A comparison of images at high current (800 A) was made with the unperturbed X-ray images by Ref. [3]. The images show a similar type of metal transfer based on electromagnetic kink instability. The similarity in the metal transfer confirms that the new technique is not altering the process much and can open new doors to revolutionize the research in SAW.

Effect of Current and Fluxes on Metal Transfer

Additional videos at 10,000 f/s were taken for welds between 500 A and 1000 A for both DCEP and AC polarities. At 500 A, metal transfer shows an irregular shaped droplet with chaotic, non-axial metal transfer in both DC and AC polarities. Between 600 A and 1000 A, a tapering electrode tip ejecting a molten tail through a mechanism resembling an electromagnetic kink instability was observed in both DC and AC polarities. The frequency of droplet detachment increases significantly with current.

The effect of five fluxes was tested at 500 A. The fluxes used were Lincolnweld 760 flux (Basicity Index (B.I.) = 0.8), ES200 (an electroslag welding flux with 48% CaF_2), Lincolnweld 980 flux (B.I. = 0.6), Lincoln 880 M flux (B.I. = 3.3) and an experimental flux with 85% CaF_2. The mode of metal transfer under all the fluxes was found to be comparable. In all case except the experimental flux with 85% CaF2 the metal transfer appeared chaotic with frequent explosions. For the fluxes with 48% and 85% CaF2 content, the observable arc length (arc length above the original plate surface) was shorter compared to the other fluxes.

Conclusions

4 high-speed videos of high quality are presented to visualize the metal transfer and weld pool phenomena in SAW; obtained by a technique consisting of inserting a thin-gage tunnel along the welding path. At low current (500 A) a chaotic globular transfer is observed. At high current (beyond 600 A) a new metal transfer method likely governed by an electromagnetic kink instability is observed. The possible contamination of the weld cavity with atmospheric gasses was tested using a Ne and it was concluded that there is no contamination from the external gas. Analysis of the electrical signal did not show a significance of the presence of the tunnel. Analysis of the weld cross sections shows a small but measurable change before and during the tunnel. For comparable conditions, the same clearly identifiable features are seen in X-ray and optical imaging both at low and high current levels. A transition from recirculating flows to gouging penetration is seen both in the videos and in the increased penetration measured in the cross sections.

Table 2: Measurements of cross sections of Fig. 5. Accuracy of measurement is of the order of 0.1 mm

Experiment	Location relative to tunnel	Thickness of plate (mm)	Bead width (mm)	Penetration (mm)	Reinforcement (mm)
5 (500 A, DC)	before	9.441	16.64	5.54	2.64
	during	9.449	16.94	4.15	3.42
	after	9.442	17.68	6.35	1.89
12 (500 A, AC)	before	9.463	16.86	4.39	2.84
	during	9.452	20.40	3.54	4.88
	after	9.454	16.35	5.05	1.91
10 (1000 A, DC)	before	18.857	11.18	14.17	5.70

Acknowledgements

The authors would like to thank Lincoln Electric for supplying the welding equipment and consumables; Praxair for neon used; Nairn Barnes and Gentry Wood for assistance with experiments; Ying Wang for assistance with data analysis; and NSERC for partial funding of this project.

References

[1] Mendez, P.F., Goett, G., and Guest, S.D., "High-Speed Video of Metal Transfer in Submerged Arc Welding." *Welding Journal*, 2015. 94 (10): p. 325s-332s.

[2] N.G. Ostapenko and B.I. Medovar: "X-ray analysis of submerged arc zone (Russian)", *Avtogennoe Delo,* No. 11 (1947), pp. 16–20.

[3] P.G. Grebelnik, "X-ray study of the process of automatic submerged arc welding (Russian)", *Avtomatich. Svarka,* No. 6 (1950), pp. 18–29.

[4] I.K. Pokhodnya, "A method of investigating the processes of melting and transfer of electrode metal during welding", *Automatic Welding*, Vol. 17, No. 2 (1964), pp.1–10.

[5] U. Franz. "Operations in the cavern of submerged arc welding Part I (German)", *Schweisstechnik*, Vol. 15, No. 4, pp. 145–150, 1965.

[6] K. Akahide et al. "Correlation between arc phenomena and welding parameters in submerged-arc welding", *Journal of the Japan Welding Society,* Vol. 50, No. 5 (1981), pp. 520–524.

[7] Uwe Reisgen, et al. "Diagnostic process analysis of the submerged arc by welding high-strength fine-grained structural steels," *68th IIW Annual Assembly and International Conference,* Helsinki, Finland, July 2015, *Doc. No. 212-2205-15.*

[8] http://www.ualberta.ca/~ccwj/Publications/WJ_HSV_SAW/
SOM1: ccwj_000015.001.mp4, SOM2: ccwj_000015.002.mp4, SOM3: ccwj_000015.003.mp4, SOM4: ccwj_000015.004.mp4

Study on In-situ Measurement of Weld Pool Configuration Using Laser Ultrasonic Technique

S. Asai , K.Nomura*,T.Matsuda*, Y.Mine*, T. Hoshi** and A.Sugawara***
**Graduate school of Engineering, Osaka University, Japan*
**e-mail:asai@mapse.eng.osaka-u.ac.jp Tell:+81-6-6879-7547*

*** Toshiba Corporation, Yokohama, Japan*
**e-mail: takeshi.hoshi@toshiba.co.jp Tell:+81-45-770-2307*

Abstract
A new method for in-situ monitoring of welding based on laser ultrasonic technique is investigated. When a laser pulse is irradiated onto a surface of base metal, an ultrasonic pulse is generated due to ablation of material from surface. The ultrasonic waves are detected with a laser interferometer as micro displacement on the surface and the propagation time of the ultrasonic pulse is measured. This technique is non–contacting ultrasonic technique and then can be applied to detect the weld surface of the high temperature for in-situ monitoring weld quality during welding. Firstly, the laser ultrasonic technique is tried to use to measure the weld pool configuration during welding. A prototype measuring system using laser ultrasonic technique was manufactured. This system is applied to measure the artificial defect of 1.0mm dia. hole and the weld pool of TIG. Consequently, it is found that the laser ultrasonic signals are detected with good signal-to- noise ratio under welding and the weld pool configurations are estimated. Also, it is recognized that the accuracy and the measuring area.is required to improve.

Introduction
Laser ultrasonic technique that uses lasers as an alternate use of conventional contact transducers has been established [1]. Thermal expansion or ablation induced by short pulse laser irradiation to target surface are used as transmitters. As receivers, laser Doppler vibration measurement is used. This principle is shown in Fig.1.This combination was originally limited for non-contact measurement in laboratory due to low sensitivity. A long pulsed laser with enhanced peak power and an interferometer applicable to rough surface have brought laser ultrasonic technique into industrial application. The advantages of laser ultrasonic technique is applicability to inspection of high temperature target with contactless. Authors have been proposing in-process quality control system of welding [2]. In-process quality control system of welding means conducting welding and inspection simultaneously with monitoring. By using in-process quality control system of welding, defects can be detected at early time during welding. Especially, it is effective for the welding of the wall thickness over 100mm and the mass production. Authors developed the in-process quality control system using laser ultrasonic technique for welding of 150mm wall thickness with the preheat temperature at over 200 degree C and this system applied to the welded steam turbine rotor in production [3]. Also, if the weld pool configuration can be measured during welding, welding parameters can be controlled with feedback for the assurance of weld quality. The applicability of laser ultrasonic technique for measuring the weld penetration depth was investigated and it was effective to apply [4].
In this study, in-situ measurement system of weld pool configuration using laser ultrasonic technique developed will be explained and the preliminary test results using the prototype system will be also shown.

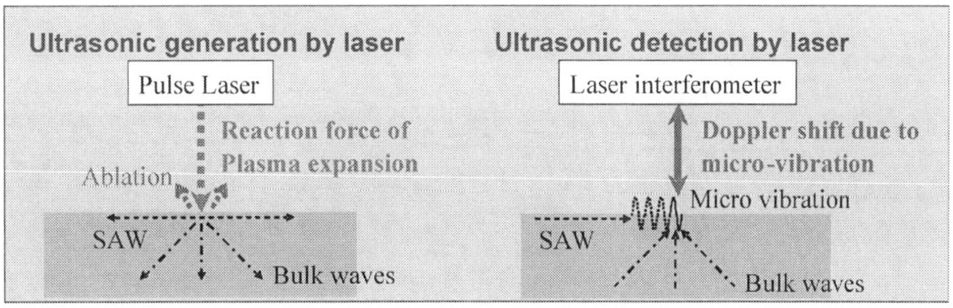

Fig.1 Principle of laser ultrasonic technique

In-situ measurement system of weld pool configuration

Fig.2 shows the constitution of the prototype in-situ measurement system for weld pool configuration using laser ultrasonic technique produced. The generation laser was irradiated on the backside surface and the ultrasonic wave propagated was detected at the same surface by detecting laser with a laser interferometer. Giant pulsed Nd:YAG laser with wave length of 1064nm was used as a generation laser that generates ultrasonic waves. The pulse width was 8ns and the average energy was 90mJ/pulse. The generation laser can be moved to the 0.1mm pitch on the surface by the optical mirror. The detection laser of Nd.YAG with wave length of 532nm was transmitted by optical fibers and irradiated using optical probe. Also, the detection laser can be detected at the 3 points of ±7mm from the center. Obtained signals were recorded and used for signal processing by a personal computer. The image of weld pool configuration was displayed by the visualization system.

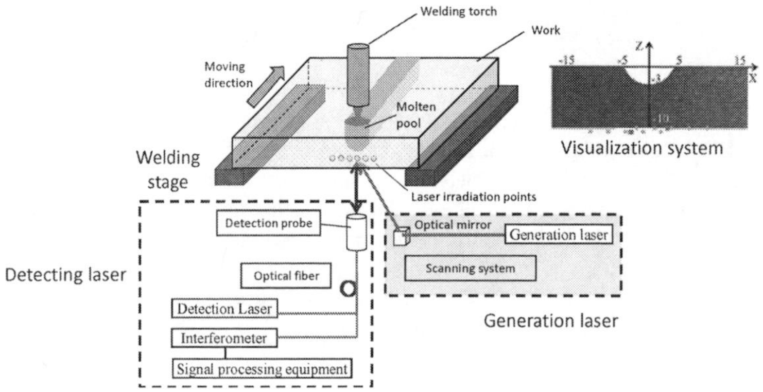

Fig.2 Schematic constitution of weld pool configuration measurement system using laser ultrasonic technique

Measuring method

The generation laser was irradiated at the backside surface and ultrasonic waves were generated. The propagated ultrasonic waves reflected at the interface of weld pool and propagated to the backside surface. The distance from the backside surface to the weld pool interface can be calculated by measuring the propagation time with detecting laser. The example of the detected signals is shown in Fig.3. To visualize the image, surface wave that may become a noise was removed by the calculation of the arrangement between generation laser and detection laser. Also, the synthesis aperture focusing technique (SAFT) was adopted for improving the signal to noise ratio of laser ultrasonic technique [5].

It is effective to detect at the multi points in order to measure the whole of weld pool configuration. In this system, the whole of weld pool configuration can be clarified by combining the images obtained from the 3 detected points as shown in Fig.4.

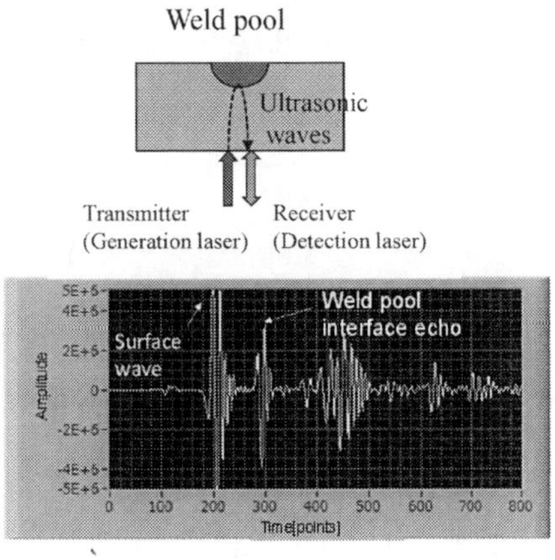

Fig.3 Example of detected signals

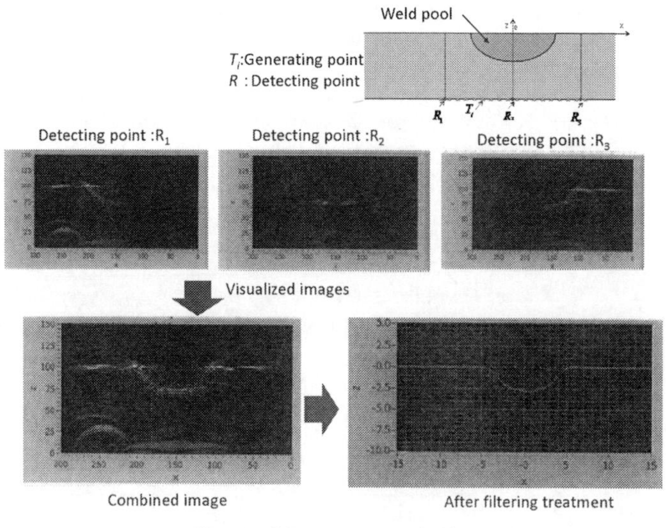

Fig.4 Measuring method

Also, the temperature dependency of ultrasonic velocity have to be considered in the high temperature field as like the weld pool. Fig.5 shows the effect of the temperature distribution on the configuration of weld pool. The temperature distribution used is shown in Fig.6. In the case of using the ultrasonic velocity with temperature distribution, the depth of weld pool was 0.9mm shifted upward by the calculation compared with the case of the ultrasonic velocity without temperature distribution. From this result, the calculation of the propagation time using ultrasonic velocity with the dependency of temperature was adopted in this system.

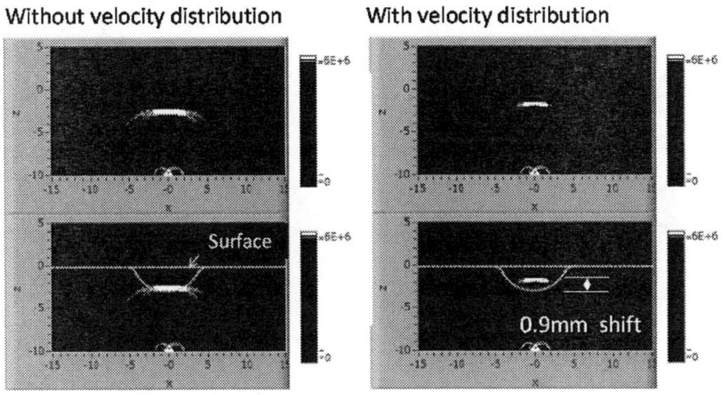

Fig.5 The effect of temperature distribution on the weld pool depth

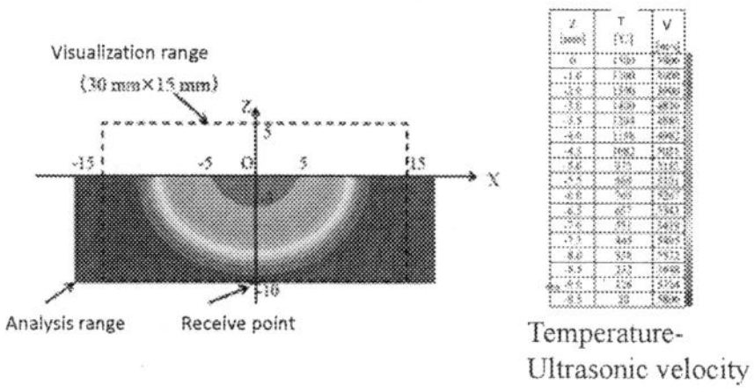

Fig.6 Temperature distribution used to calculate propagation time

Test results of detecting the artificial defect

This weld pool configuration measurement system was applied to the test measuring the artificial defect which had the hole of 1.0mm dia. made by machining in the flat plate. Fig.7 shows the configuration of test plate. Austenitic stainless steel (JIS SUS304) was used as base metal of test plate. The combined image of weld pol configuration and intensity distribution of ultrasonic wave on the z direction are shown in Fig.8. Table1 shows the test results. In the test results, the distance from the surface to the bottom of the hole was 5.9mm measured by this system compared with 5.9mm of actual value. Also, diameter of hole was 1.1mm measured by this system compared with 1.0mm of actual value. The signal to noise ratio was 4.73.In the case of the 3.0mm hole, the result was similar. It was confirmed that this system had the sufficient detecting accuracy.

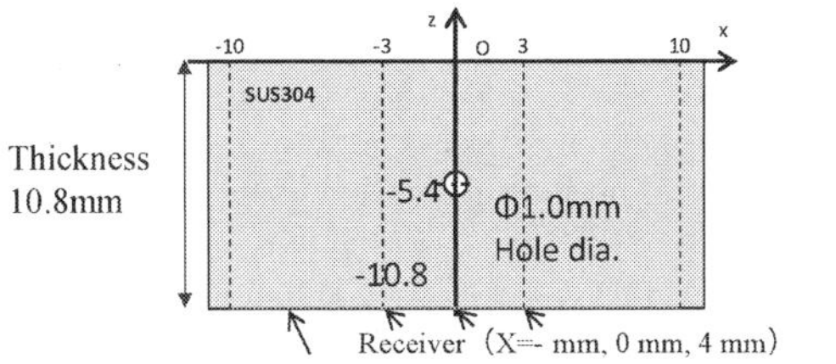

Fig.7 Test plate configuration with the artificial defect

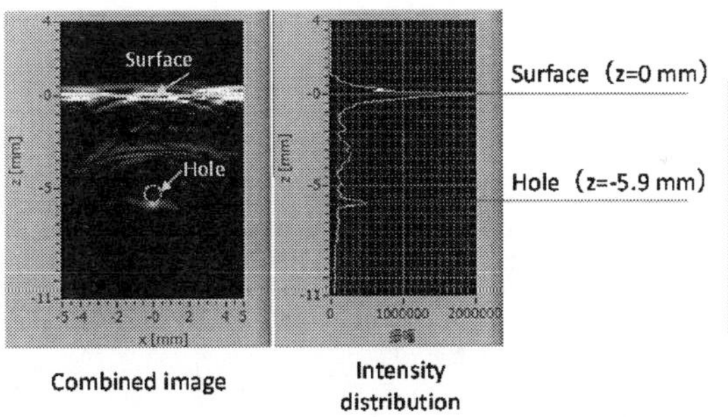

Result

	Actual value	Measured value
Depth	z=-5.9 mm	z=-5.9 mm
Diameter	Φ1.0 mm	Φ1.1 mm
SNratio		4.7

Fig.8 Test results

Test results of detecting weld pool configuration

The weld pool configuration of TIG arc was measured by using this system. The thickness of test plate used was 10.8mm. Base metal was the Austenitic stainless steel (JIS SUS304). TIG welding was carried out without traveling as like spot welding. Welding current was 100A and welding time was 3min.Fig.9 shows the bead appearance and sectional macro photograph. The penetration depth was 1.75mm and bead width was 9.85mm by the sectional macro photograph. In the measurement of the weld pool configuration, the propagation time of ultrasonic wave was calculated by using the ultrasonic velocity with the temperature distribution assumed as shown in Fig.10. Fig.11 shows the combined image as measured and the image of the weld pool configuration after the signal processing. In the test results, the weld pool depth was 0.8mm and the weld pool width was 8.0mm, which were different from the measured values by sectional macro photograph. It is recognized that the measurement accuracy is required to improve by the optimization of the temperature distribution on the weld area and. the measuring method

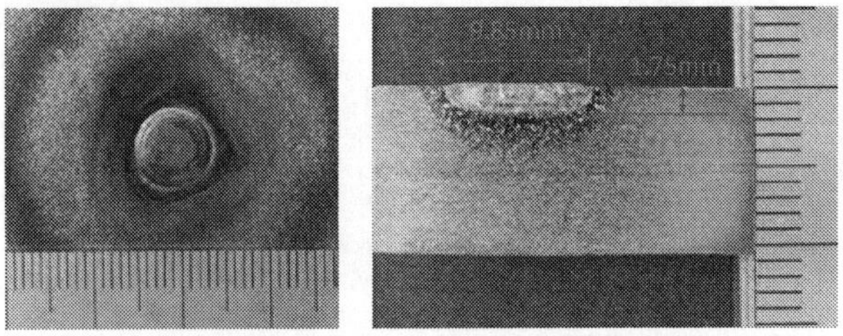

Bead appearance Sectional macro photo.

Fig.9 Bead appearance and sectional macro photograph

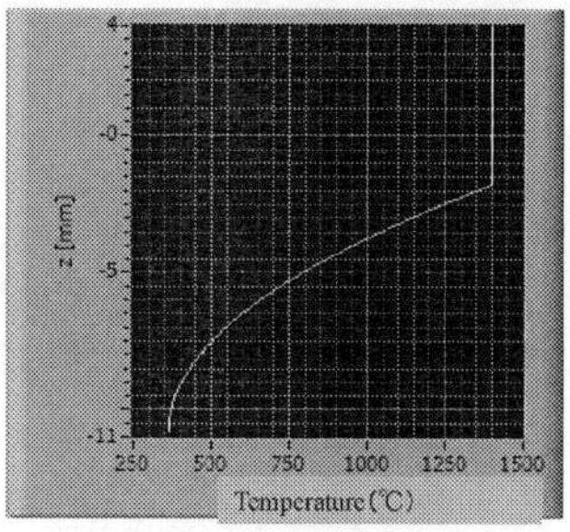

Fig.10 Temperature distribution

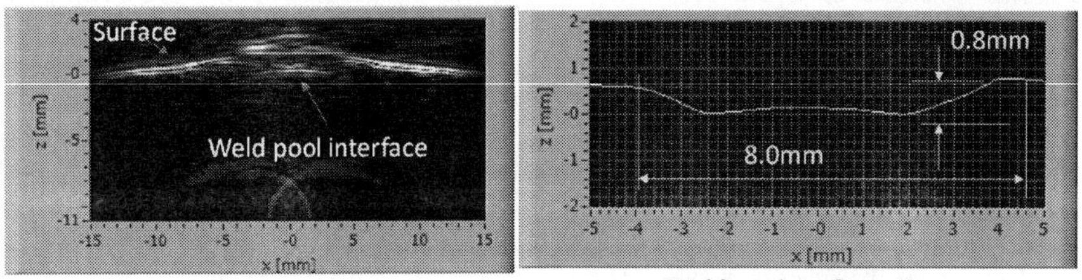

Measured image

Weld pool configuration after signal processing

Fig.11 Weld pool configuration of TIG arc spot measured

Conclusions

The prototype in-situ measuring system of weld pool configuration using laser ultrasonic technique was developed. This system is applied to measure the artificial defect of 1.0mm dia. hole and the weld pool of TIG. Consequently, it is found that the laser ultrasonic signals are detected with good signal-to- noise ratio under welding and the weld pool configurations are estimated. Also, it is recognized that the measuring accuracy is required to improve by the optimization of the temperature distribution on the weld area.

Acknowledgments

This work was supported by Council for Science, Technology and Innovation(CSTI), Cross-ministerial Strategic Innovation Promotion Program (SIP), " Development of Simulation Techniques for Performance Assurance of Weld Joints" (Funding agency:JST).

References

[1] C. B. Scruby and L. E. Drain: "Laser Ultrasonic Techniques and Applications", Adam Hilger, Bristol, 1990.
[2] Y.Fujita, et al : "Development of a Welding Monitoring System for In-process Quality Control of Thick Walled Pipe", *Welding in the World*, Vol.56, No,11-12, ,2012
[3] S.Asai,et al:" Development of In-process Quality Control System for Welding of Steam Turbine Rotor", *Proc. of THE THERMAL AND NUCLEAR POWER GENERATION CONVENTION*,2013
[4] M.Ochiai et al:"In-process Identification of Molten Pool Configuration", *J.Japan Weld.Soc.*,Vol.17,No.3(1999)
[5] S.Yamamoto, et al :"Defect Detection in Thick Weld Structure Using Welding In-Process Laser Ultrasonic Testing System" ,*Materials Transactions*, Vol. 55, No.7 (2014) pp. 998-1002.

Reconsideration over the Reliability of Arc Sensor

Katsuya KUGAI, Nobuhiro NAKAMURA
KINDAI University Technical College, Nabari-shi, Mie Pref., Japan
kugai@ktc.ac.jp, phone+81-595-41-0111, fax +81-595-62-1320

Abstract

Arc sensor is a conventional technique toward the compensation of deviation through the arc welding which causes from the heat distortion or unevenness of work piece. But, at present, only less than 5% of arc welding robot has arc sensing function. It is hard to use in actual welding line because of occurring of miss sensing. Many studies have been performed about arc sensing [1][2][3], but many of them study about the performance of sensing. So, we began to reconsider over arc sensing in the meaning of reliability. We will apply theoretical equations so as to separate the signal and the noise toward unsettled arc phenomena. We already researched the characteristics of arc voltage toward arc length. When arc length becomes long, the fluctuation of arc voltage increases. To estimate arc length from arc voltage data, we adopt histogram. By averaging around the mode, we can avoid the undesirable error from outlier data. As a next step, this paper explain about the effect of considerable disturbance toward arc length estimation using arc voltage data, and examine about signal processing method so as to reduce the estimation error.

Introduction

Arc sensor detects the aiming deviation through the arc phenomena. For example, arc sensor measures arc voltage or welding current or wire feeding speed and so on. However, the arc phenomena include many physical elements such as the effect of electromagnetism, heat distribution and convection in a molten pool. Under the combination of these physical elements, many indefinite phenomena will occurs. Indeterminacy of phenomena will leads some miss detection of aiming deviation. So, in this research, we tried to separate the indefinition of each physical phenomena. At first, to investigate the characteristics of arc phenomena only, we made TIG welding on copper plate. This means no molten pool, and non-consumable electrode [4]. In addition, to make the indefinite situation, we used the round tip tungsten electrode and oil applied copper plate V groove, so as to compare the arc voltage fluctuation toward sharp tip electrode and clean copper plate.

Experiment Apparatus

TIG torch is attached to the welding robot, and the robot moves along the weld line of V groove with weaving motion. Welding robot is FD-H5 of DAIHEN, welding power supply is AVP-300 of DAIHEN and data logger is MR8847 of HIOKI. For measuring weaving position, we used the laser distance sensor ZX0-LD50A61 of OMRON. Overview of experimental apparatus is shown in *Fig.1*.

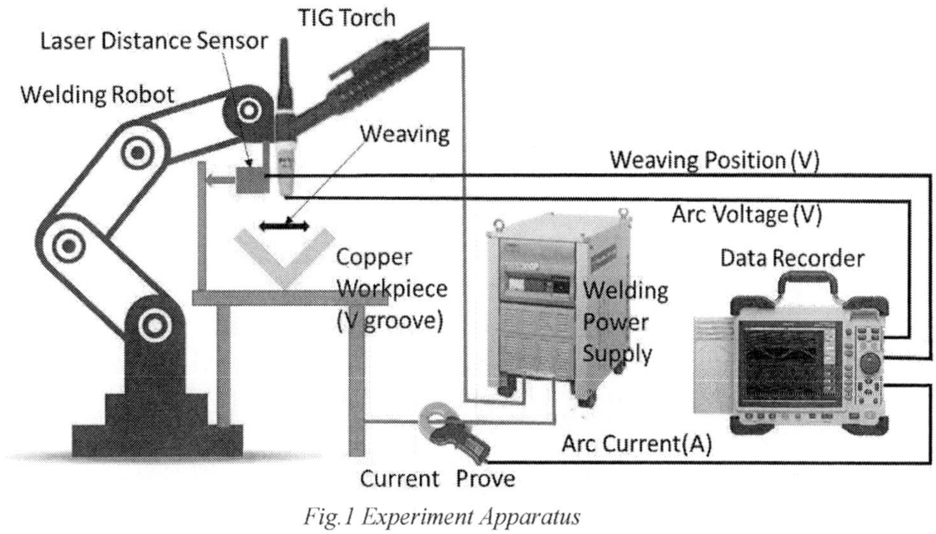

Fig.1 Experiment Apparatus

In this experiment, no molten pool will appear. But for future experiment on steel plate which has molten pool, welding conditions are tuned under actual TIG welding on steel plate. We tuned the welding condition to get proper bead width for 3.2t steel plate, biggest weaving width, and no welding defect. So, the welding conditions becomes as shown in *Table 1*. And welding bead on steel V groove becomes shown in *Fig.2*.[5]

Table 1. Welding Conditions

Item	Value
Welding Current	130A
Arc Length	3.5mm at groove center
Weaving Width	±0.7mm
Weaving Frequency	2Hz
Welding Speed	13cm/min

Fig.2 TIG welding bead under Table.1 conditions

Characteristics of the arc sensing in TIG welding

By using Table.1 welding conditions, we made TIG welding on the copper V groove. That means no molten pool. Typical arc voltage waveform toward weaving position became like *Fig.3*.

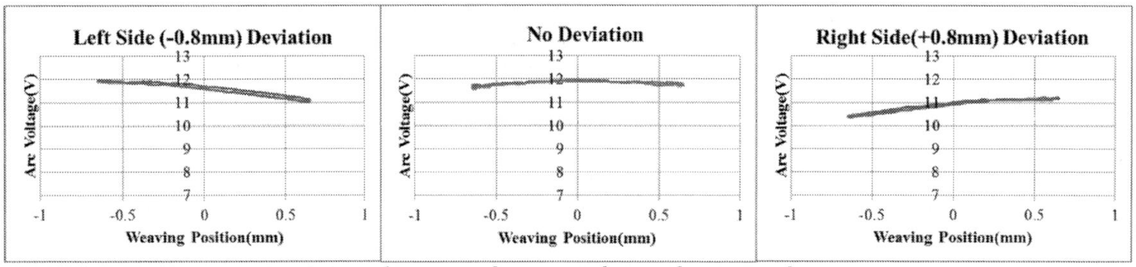

Fig.3 Arc voltage waveform according to the aiming deviation

Then we calculated the "Voltage difference" between left side and right side, so as to correspond with aiming deviation. When the average voltage during weaving left side is expressed as V_L, and during weaving right side is expressed as V_R. Then "Voltage difference V_d" is determined as the equation below.

$$V_d = V_L - V_R \qquad -(1)$$

Fig.4 shows the relationship between the aiming deviation and V_d under ideal condition of TIG welding on the copper V groove. The word "ideal" means the tip of tungsten electrode is sharp, the surface of copper plate is clean and welding current flows symmetry. So we can find the coefficient of determination is $R^2 = 0.9006$ and this shows the aiming deviation can be estimated by V_d in sufficient precision [6]. In former experience, we know that over 0.5 of coefficient of determination brings the practical seam tracking function by arc sensing.

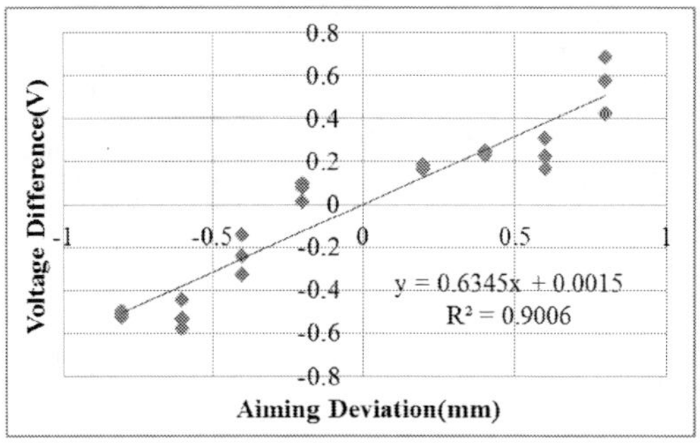

Fig.4 The relationship between aiming deviation and V_d

Influence of considerable unstable factor

As for the TIG welding in copper V groove, we prepare the two unstable factor. The first is round tip tungsten electrode. And the second is oil applied copper surface. The result of experiments are shown in *Fig.5*.

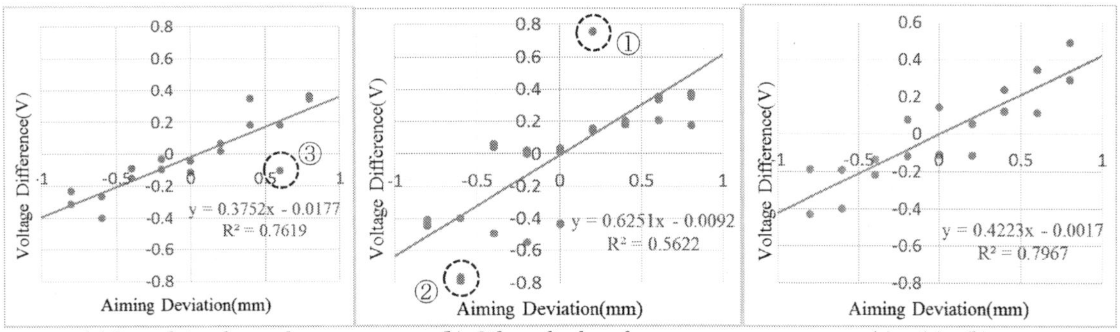

(a)Round tip electrode *(b) Oil applied surface* *(c) (a)+(b)*

Fug.5 The relationship between aiming deviation and V_d under the considerable unstable factor

By looking these results, we can find some outlier data from the approximate line. For example, at the point ① of *Fig.5(a)*, we found the arc from the electrode does not go straight down, but goes to the far side plate as shown in *Fig.6(b)*. Because of the round tip electrode, the arc cannot concentrate and becomes wide spread state. So, we think the arc will easily moves to the left or the right side. In this way, the arc sensor will mistake to detect the aiming deviation. To avoid this mistake, during the sensing process, sensor has to remember the past sensing data (V_d). If the last V_d value changes the certain level toward the mean of past value, it can be judged as the outlier and will be ignored. This ignoring algorism will work well under the condition that the welding line does not have discrete point or steep curve. When arc sensor can find the outlier, for example, arc sensor can ignore the data ① in *Fig.5(a)*, the coefficient of determination becomes better from $R^2 = 0.7619$ to $R^2 = 0.8855$ as shown in *Fig.7*.

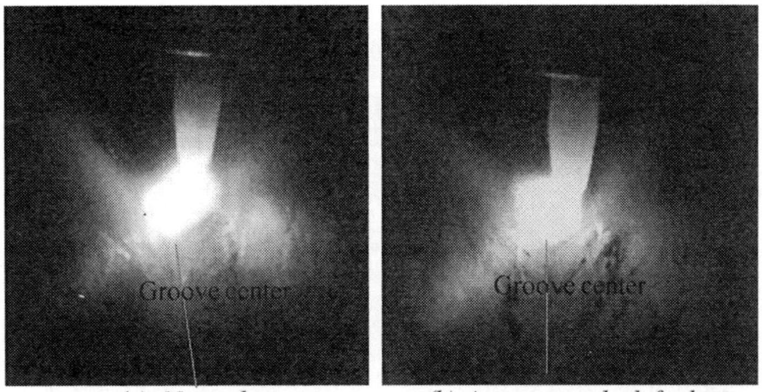

(a) Normal state *(b) Arc goes to the left plate*

Fig.6 Arc shape from the electrode to the copper V groove

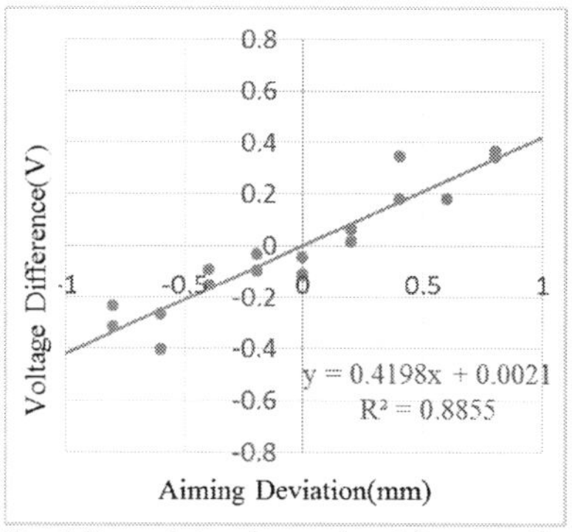

Fig.7 Round tip electrode without outlier

As for oil applied surface, we can find the fluctuation of arc voltage. This phenomenon is shown in *Fig.8*. The vapor of evaporated oil will bother the uniformity of sealed gas. Because of the arc voltage is depend on the physical characteristics of sealed gas, even if the length between the tip of electrode and V groove center will not change, the arc voltage will be fluctuate. By detecting this arc voltage fluctuation, for example to use a threshold level, we can ignore the data ② and ③ of *Fig.5(b)*. Then the coefficient of determination becomes better from $R^2 = 0.5622$ to $R^2 = 0.8671$ as shown in *Fig.9*. This ignoring algorism will also work well under the condition that the welding line does not have discrete point or steep curve.

By using these two ignoring algorithm, R^2 value of arc sensing becomes better like *Fig.7* and *Fig.9* even if some indefinite situations are exist. From *Fig.7* and *Fig.9*, we can find the R^2 value becomes close to ideal state *Fig.4*. But during the sensing, some data will be ignored and not used for sensed value. So to use enough number of V_d toward the seam tracking by arc sensing, high frequency weaving is desired.

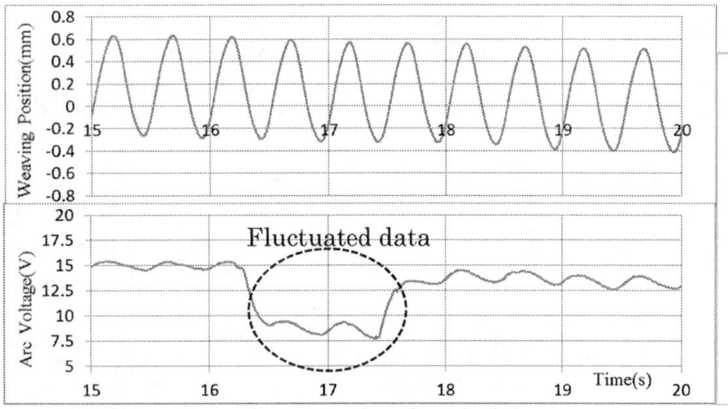

Fig.8 Arc voltage fluctuation by oil applied V groove

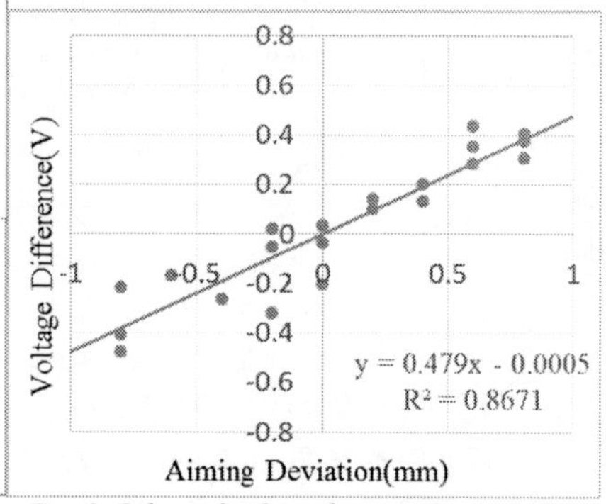

Fig.9 Oil applied surface without outlier

Conclusions

Our final target is CO2 welding because that is most popular in Japan. The characteristics of CO2 welding is consuming electrode and short circuit droplet transfer. As the first step to the final target, this paper reports about the effect of arc phenomena toward arc sensing independently. Therefore we used TIG welding toward copper V groove. In this experiment, the electrode does not consume and workpiece does not melt. At first, we made an experiment with sharp tip tungsten electrode and clean surface copper plate. For arc sensing algorism, we used the voltage difference between weaving left side average and right side average as the sensed value. The relationship between this sensed value and aiming deviation has linear characteristic and its coefficient of determination became 0.9006. This value means the detection accuracy of arc sensing toward aiming deviation is enough high to execute the seam tracking.

Then we added the two unstable factor. One is round tip electrode and another is oil applied workpiece. As the result of these effect, some miss detections are occurred in sensed value. So we searched the cause of miss detection, and found the abnormal state in the arc shape and the arc voltage value by measuring video data and arc voltage data. Then we proposed two ignoring algorisms for abnormal state. By using these algorisms, we can get the close result as no unstable factor case. That means the coefficient of determination becomes 0.8855 or 0.8671 that is close to 0.9006.

But these two ignoring algorism are not all-around. They have the limitation that the welding line does not have discrete point or steep curve. And because of ignoring some sensed value, to have enough data number for seam tracking, weaving frequency is desired to be high.

From now, we will advance to research about the effect of molten pool. So, TIG welding on the steel V grove will be tried.

Acknowledgments

To proceeding this study, Tooru Migimatsu who is the technical officer of our college prepared many items for experiment. We greatly appreciate for his effort.

References

[1] S.Kodama, *et al*: "Process Modeling of Short Circuiting GMA Welding and Its Application to Arc Sensor Control", Technical report of Shin-nittetsu co.ltd, No.385, p.64-68, 2006
[2] M.Ushio, *et al*: "Dynamic characteristic analysis of an arc sensor in GMAW welding", The 111th arc physical committee of Japan welding society,1993
[3] K.Kugai, *et al*:"An approach by simulation model about the absolute detection in arc sensing", The 135th arc physical committee of Japan welding society,1999
[4] K.Kugai, H.Kawamura: "Re-study about the reliability of arc sensor", The 93th collection of the Japan Welding Society national tournament lecture outline, p.138-139, 2013
[5] K.Kugai, T.Kitabatake: "Re-study about the reliability of arc sensor", The 95th collection of the Japan Welding Society national tournament lecture outline, p.176-177, 2014
[6] K,Kugai, N.Nakamura: "", Japan Welding Society collected papers, 4th number for the 33rd reel,p.309-316,2015

Infrared Visual Monitoring and Seam Deviation Detection for Swing Arc Narrow Gap Welding

*Jie Zhu, Lei Zhang, Na Su, Guoxiang Xu, Jiayou Wang**

Jiangsu Provincial Key Laboratory of Advanced Welding Technology,

School of Materials Science and Engineering, Jiangsu University of Science and Technology,

2 Mengxi Road, Zhenjiang, Jiangsu 212003, P. R. China.

**Corresponding author: E-mail: jywang@just.edu.cn*

Abstract

The present work uses an infrared high-speed camera with optical filters to capture welding image as the arc weaves to the swing midpoint for reducing arc light disturbance, and then sends the image into a computer for displaying welding region in real time and extracting weld deviation by image processing. The proposed deviation detection approach can rapidly obtain a deviation value by calculating the distance between welding wire center and groove center through one frame of welding image. Experimental results show that the developed monitoring system can acquire steadily and display clearly infrared welding images with low arc radiation and welding fume interferences; this swing midpoint image detection algorithm can perform twice deviation extractions in one period of arc swing, thus providing higher detection efficiency. As the planning weld deviation linearly varies within ±0.5 mm, the maximum detection error of this approach is 0.052mm, which thus demonstrates the effectiveness of the proposed approach.

Introduction

To improve the welding efficiency of high strength steel and thick-wall structures, narrow gap gas metal arc (NG-GMA) welding processes have been utilized in modern equipment manufacturing industry [1]. For single-wire NG-GMA welding, the research hotspot is primarily focused on improving sidewall penetration of narrow groove. To date, several typical single-wire NG-GMA processes have been developed, such as wave wire, rotation arc and swing arc approaches [2, 3]. Owing to the parameters controllability and good directivity of swing arc, swing arc (or weaving arc) NG-GMA welding is becoming more and more attractive [1, 3-5].

However, similarly to the other narrow gap welding processes, the swing arc NG-GMA welding area is difficultly observed due to deep and narrow groove. Furthermore, the swing trajectory midpoint of arc usually deviates from the centre of groove to be welded, because of welding thermal distortion, groove preparation error and clamping error, thus limiting its application. So, it is urgent to develop some effective welding process monitoring and weld tracking detection techniques by suitable sensors for swing arc NG-GMA welding. Among numerous sensors used for welding, passive visual sensors [6] are the most acceptable due to their superiority on rich information, non-contact, high accuracy and strong reliability, and very suitable for narrow gap welding.

To raise further the environmental adaptability and welding quality of swing arc NG-GMA process, this study develops an infrared visual monitoring system, and proposes a new weld deviation detection approach. The detection accuracy and efficiency of this approach are then investigated by swing arc NG-GMA weld deviation detection experiments.

Visual Monitoring and Weld Deviation Detection System

Fig. 1 shows a schematic diagram of the infrared visual monitoring and weld deviation detection system developed. It mainly consists of swing arc NG-GMA welding torch, infrared camera with optical filters, and industrial computer. A straightened electrode wire is fed to pass through the hollow axis motor and the inner hole of bending conductive rod, and finally stretches out from the contact tip. The motor directly turns the bending rod and the contact tip, and then to oscillate circularly the arc around itself axis of the welding torch at several hertz frequencies. The position and frequency of arc swing are detected by a photoelectric switch which is installed on the axis of motor. During welding, the torch moves at the speed of V_w.

The infrared camera with optical filters is focused on welding region from the front of welding groove, the capturing angle of which is about 30 degrees to horizontal plane. According to the arc position signal P_{arc} output by the photoelectric switch, this camera captures the regional image of swing arc NG-GMA welding, and then sends the image into the industrial computer for displaying in real time and extracting weld deviation by image processing.

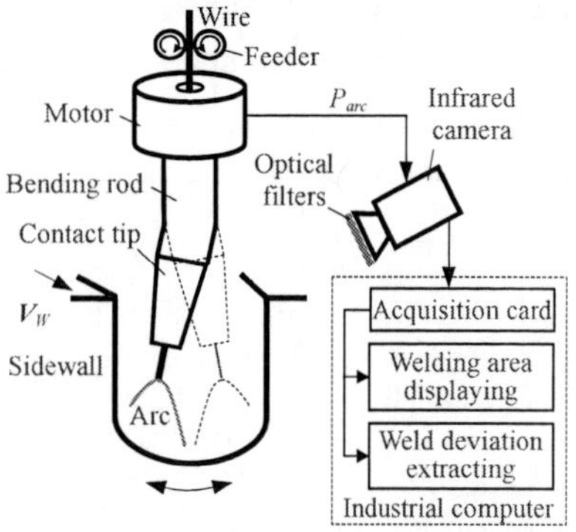

Fig. 1 Schematic diagram of visual monitoring and weld deviation detection system for swing arc narrow gap welding

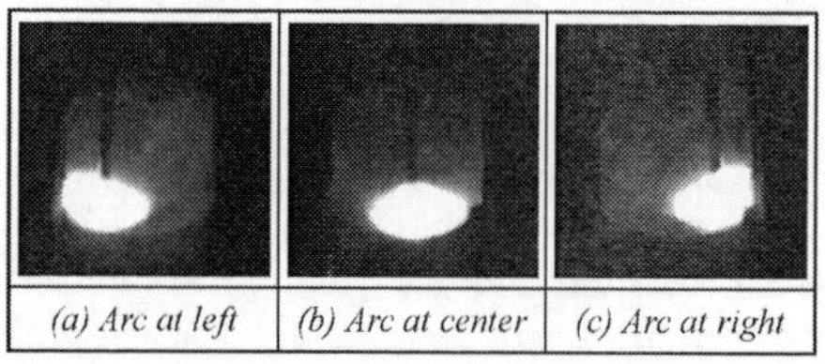

Fig. 2 Infrared images of swing arc narrow gap welding

Fig. 2 gives three frames of swing arc NG-GMA infrared welding images captured in an arc swing period. Here, the welding parameters were: ~280 A for *d.c.* arc current, 29.5 V for average arc voltage, 18 mm for torch standoff height and 20.3 cm/min for welding speed; 1.2 mm for the diameter of solid wire, 25 L/min for the flow rate of Ar-20%CO_2 shielding gas, and 14mm for groove gap width; 2.5 Hz for swing frequency, 0.1 s for at-sidewall staying time, 68° for swing angle and 8.19° for bending angle of conductive rod; 970 nm for center wavelength of infrared filter with 20 nm width, 30% for pass rate of neutral density filter, 0.2 ms for exposure time and *f*/16 for camera aperture.

In **Fig. 2**, the arc, groove sidewalls, wire, pool, spatter and slag are clearly visible. For welding operators, the relative position of arc to groove sidewalls can be distinctly observed from theses images with low arc radiation and less fume disturbances. The images in **Fig. 2(a)** and **(c)** were captured as the arc swung to stay at sidewall, in which the arc was respectively closest to the left and the right sidewalls of groove and more arc heat was put into the groove sidewalls for enough sidewall penetration. However, the groove edges were disturbed by violent arc light and welding spatters, which will bring troubles on the detection of weld deviation.

Weld Deviation Detection Approach

Fig. 2(b) shows an infrared welding image captured as the arc wove to its swing midpoint. Since the arc was far away from two sidewalls of groove, the effect of arc light disturbance on each groove edge reduced. Accordingly, a weld deviation detection approach based on arc swing midpoint image is developed, as shown in **Fig. 3**.

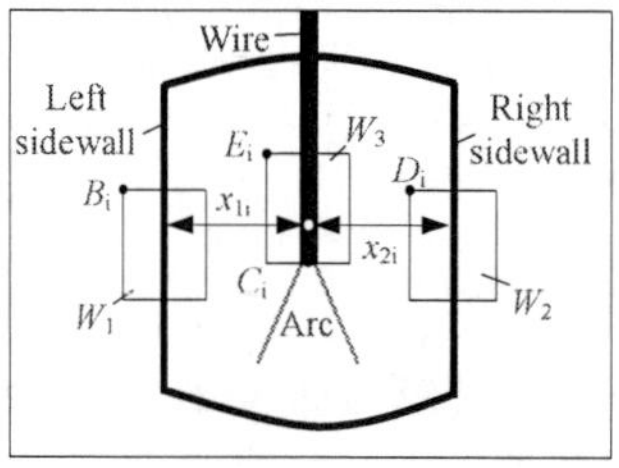

Fig. 3 Principle of seam deviation detection

When the arc weaves to the midpoint of swing trail, the photoelectric switch outputs the arc position signal P_{arc}. The infrared camera immediately captures one frame of welding image, and then sends it into the industrial computer. Subsequently, the highest point of arc C_i is detected by global threshold from welding image. Then, three moving region-of-interest (ROI) windows W_1, W_2 and W_3 are respectively created for left and right sidewalls of groove and welding wire by adaptively locating three points B_i, D_i and E_i. After using the Otsu threshold for two ROI window images of groove sidewalls and the adaptive threshold for ROI window image of welding wire, the groove edges of sidewalls and the center of welding wire are extracted by Canny edge operator respectively. Finally, the relative position of welding wire to groove left sidewall x_{1i} can be obtained by

$$x_{1i} = x_{Wi} - x_{Li} \tag{1}$$

where x_{Wi} and x_{Li} are the relative horizontal positions of welding wire center and groove left sidewall to the left edge of the i-th global image. Similarly, the relative position of welding wire to groove right sidewall x_{2i} can be wrote as

$$x_{2i} = x_{Ri} - x_{Wi} \tag{2}$$

where x_{Ri} is the relative horizontal position of groove right sidewall to the left edge of the i-th global image. When x_{1i} and x_{2i} are detected out from the welding image, the weld deviation value Δx_i which indicates the deviated degree of welding torch center from groove center during the unidirectional weaving movement of arc can be calculated by

$$\Delta x_i = \tfrac{1}{2}(x_{2i} - x_{1i}) = \tfrac{1}{2}(x_{Li} + x_{Ri}) - x_{Wi} \tag{3}$$

If Δx_i is larger than zero, welding torch deviates toward left sidewall of groove; if Δx_i is smaller than zero, welding torch approaches to the right sidewall of groove. As the arc passes twice through the swing midpoint in one swing period, two sequent frames of welding images are captured, from which another weld deviation value $\Delta x_{(i+1)}$ can be acquired similarly. So, this approach has high detection efficiency.

Weld Deviation Detection Experiments

To investigate the effectiveness of the weld deviation detection approach proposed, a number of swing arc narrow gap GMA welding experiments were carried out at flat position. Welding conditions were the same as the experimental parameters for **Fig.2**. Furthermore, the weld deviation varying within ±0.5 mm was planned on a testpiece of 30 mm thickness and 150 mm length.

ROI Windows Location

Fig. 4 shows the image processing procedure of ROI windows location. **Fig. 4(a)** shows the welding image captured when the arc wove to the swing midpoint. **Fig. 4(b)** gives arc region extracted from **Fig. 4(a)** by global threshold. By detecting the highest point of arc, three ROI windows were adaptively created for the welding wire and two groove sidewalls respectively, as shown in **Fig. 4(c)**.

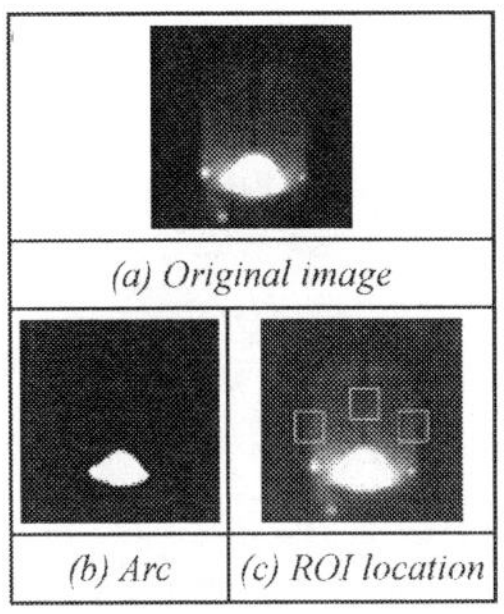

Fig. 4 Image processing procedure of ROI windows location

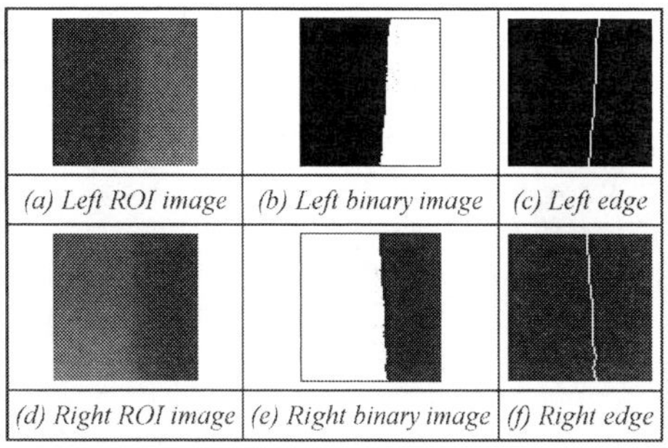

Fig. 5 Image processing procedure of groove edge extraction

Groove Edge Extraction

Fig. 5 gives the image processing procedure of groove edge extraction. **Fig. 5(a)** and **(d)** are the ROI windows images of groove sidewalls intercepted from **Fig. 4(c)**. **Fig. 5(b)** and **(e)** show two binary images of groove sidewalls acquired by using a combined filter and Otsu threshold. **Fig. 5(c)** and **(f)** indicate two groove edges extracted by Canny operator. Finally, the positions of two groove edges were detected out by seeking the straightest section of groove edges.

Wire Center Extraction

Fig. 6 shows the image processing procedure of wire center extraction. **Fig. 6(a)** is the ROI window image of welding wire intercepted from **Fig. 4(c)**. Similarly, the binary image of welding wire was obtained after the combination filtering and adaptive threshold, as shown in **Fig. 6(b)**. The edges of welding wire were extracted by Canny operator, as shown in **Fig. 6(c)**, but the left edge of wire was clearly disturbed by arc light. As a result, according to average position of two outermost wire edges, the wire center was detected as indicated in **Fig. 6(d)**, where the effect of arc light disturbance was effectively reduced.

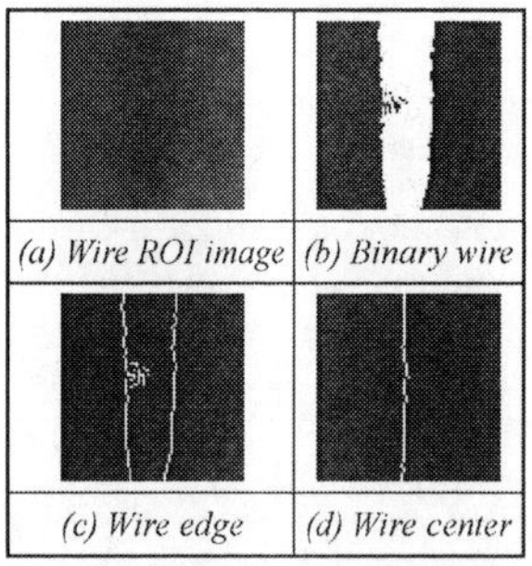

***Fig. 6** Image processing procedure of welding wire center extraction*

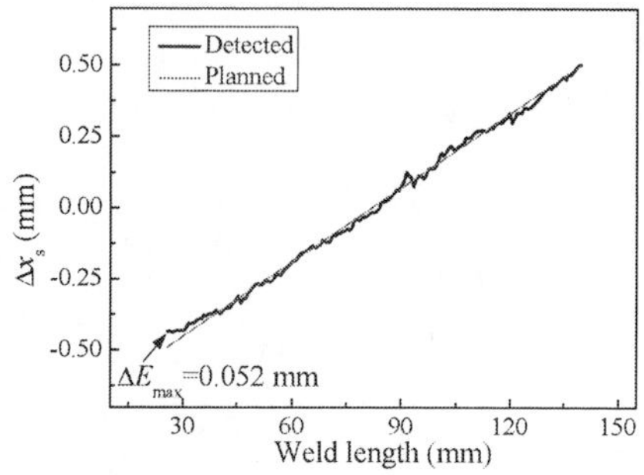

***Fig. 7** Weld deviation detection results*

Varying Deviation Detection Experiment

Fig. 7 shows the weld deviation detection results, where Δx_s denotes average filtering value of weld deviation detection values Δx_j ($j=i$, ..., (i-4)) extracted from five late swing periods. As the planned weld deviation linearly varies within ±0.5 mm, the maximum detection error ΔE_{max} is 0.052 mm. From **Fig. 7**, it can be seen that detection accuracy of weld deviation is quite high, which thus demonstrates the effectiveness of the proposed approach.

Conclusions

The infrared visual monitoring system developed for swing arc NG-GMA welding can steadily capture and display clearly welding image with low arc light and less fume interferences. Weld deviation detection approach based on arc swing midpoint image detection is effective under the disturbance of violent arc light, and led to the high efficiency and accuracy of detection.

Acknowledgements

This work is financially sponsored by the National Natural Science Foundation of China (Grant No. 51475218), and the Specialized Research Fund for the Doctoral Program of Higher Education (Grant No. 20133220110001).

References

[1] M. Murayama, *et al*: "Narrow gap gas metal arc (GMA) welding technologies", *JFE Technol. Rep.*, No. 20 (2015), pp. 147-153.

[2] J. Y. Wang, *et al*: "Novel rotation arc system for narrow gap MAG welding", *Sci. Technol. Weld. Join.*, Vol. 12, No. 6 (2007), pp. 505-507.

[3] J. Y. Wang, *et al*: "A swing arc system for narrow gap GMA welding", *ISIJ Int.*, Vol. 52, No. 1 (2012), pp. 110-114.

[4] J. Y. Wang, *et al*: "Development of swing arc narrow gap vertical welding process", *ISIJ Int.*, Vol. 55, No. 5 (2015), pp. 1076-1082.

[5] J. Y. Wang, *et al*: "Effect of arc swing parameters on narrow gap vertical GMA weld formation", *ISIJ Int.*, Vol. 56, No. 5 (2016), pp. 844-850.

[6] W. H. Li, *et al*: "A modified welding image feature extraction algorithm for rotating arc narrow gap MAG welding", *Ind. Robot*, Vol. 42, No. 3 (2015), pp. 222-227.

Effect of RTE treatment on toughness of HSS weld metal

Horst Cerjak, Norbert Enzinger
IWS TU Graz, Graz, Austria
Email: horst.cerjak@tugraz.at, norbert.enzinger@tugraz.at

Ozan Caliskanoglu, Gunter Figner
Stirtec GmbH, Premstätten, Austria
Email: ozan.caliskanoglu@gmail.com; gunter.figner@stirtec.at

Milan Pudar
Magna Steyr Fahrzeugtechnik AG & CO KG, Graz, Austria
Email: pudmilan@yahoo.com

Francisca Mendez-Martin
Department Physical Metallurgy and Materials Testing, Montanuniversität Leoben, Austria
Email: mendez@unileoben.ac.at

Maria Domankova
Institute of Materials Science, Slovak University of Technology, Trnava, Slovakia
Email: maria.domankova@stuba.sk

Abstract

Weld deposits from advanced high-strength steel constructions may suffer from severe embrittlement reactions caused by non-qualified PWHT procedures. This effect, if not detected, can lead to catastrophic brittle failure of the component. A temper embrittlement (TE) reaction in the weld deposit was detected as root cause of this effect. By qualified application of reversible temper embrittlement (RTE) treatment, characterized by a short-time heating at about 600°C, the toughness of the embrittled weld material can be significantly recovered. Basic investigations were performed to explain the embrittlement as well as the de–embrittlement effect.

Introduction

Storage and pumping storage hydro plants play an important role in hydropower generation [1]. Severe unexpected embrittlement has been detected in weld metal of a completed penstock of a pumped-storage hydropower plant just before commissioning of the plant [2]. The penstock was manufactured from plates of quenched and tempered high strength steel S690 according to DIN EN 100025-6. Submerged arc welding with filler materials according to AWS 23.F11A4-tc-F5 were applied. A post weld heat treatment (PWHT) at 570°C 2h was applied by using an electrical heating device. In difference to the common practice of the usual 50°C/h, the heating and cooling speed applied was 15°C/h.

The chemical analysis of the weld deposit is (in wt.-%): C: 0.057%, Si: 0.241%, Mn: 1.43%, Cr: 0.526%, Mo: 0.386%, Ni: 2.78%, P: 0.012%, S: 0.0049%. The mechanical properties of the weld deposit are matching compared to that of the specified plate material properties. Samples removed from original welds revealed Charpy V values down to 8J at testing temperature –20°C and therefore declared as very questionable to assure a safe service behavior of the plant. For this reason, a solution has been sought in order to develop applicable procedures for recovering the toughness of the brittle weld deposit in the inclined shaft. Details can be found in [2].

Causes of embrittlement

The analysis of the manufacturing process showed that acceptable toughness values have been reported during the respective welding procedure qualification (WPS) tests. However, it was very conspicuous that slow cooling rates of 15°C/h were applied during the described PWHT procedure in contrast to the specified 50°C/h. Subsequent laboratory experiments on representative test welds confirmed that slow heating and cooling rates have a strong influence on the weld materials toughness. This behavior of the test weld material is illustrated in Fig. 1.

The investigations additionally confirmed that only the cooling rate had a major impact on the materials embrittlement. The heating rate and the holding time at peak temperature, on the other hand, did not play a significant role. This result suggested that the respective effect was temper embrittlement.

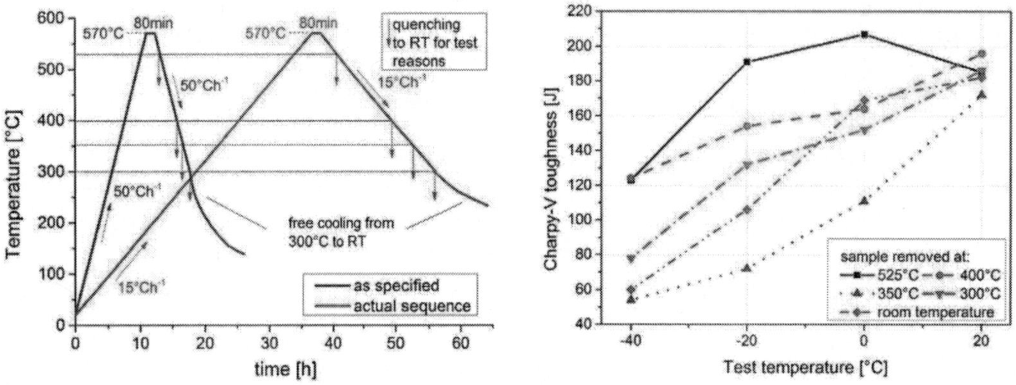

Figure 1: a) PWHT sequence specified and actual applied (left) b) Charpy-V toughness of samples according to 1a) (right)

Additional investigations have been performed by application of step cooling procedures, which have been developed for the investigation of the temper embrittlement susceptibility of steels for heavy forgings [3]. The obtained results revealed that the weld deposit investigated is very sensitive to temper embrittlement (Fig. 2). This phenomenon has already been identified to occur on heavy forgings and is described in detail in literature [4, 5]. Furthermore, various investigations have already been performed for the weld material as well [6].

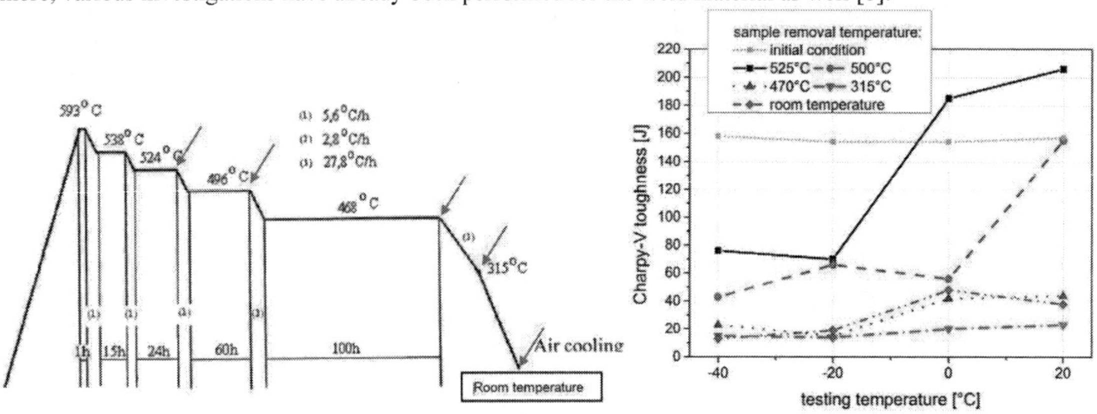

Figure 2: a) Step cooling according to [3] (left) b) Charpy-V Toughness of samples removed according to Fig. 2a (right)

Examinations performed by scanning electron microscopy (SEM) on broken Charpy impact specimens, which have been removed from the original welds of the penstock, as well as taken from the laboratory investigations described in Fig.1 and 2, showed high proportions of intergranular fracture [2]. The results of these investigations clearly revealed that a temper embrittlement effect can be considered as cause of the occurred embrittlement of the weld material.

Reversible temper embrittlement (RTE) treatment

The effect of temper embrittlement has been well known for decades in the literature and it can be reversed by short time heating to a peak temperature of around 600°C followed by a subsequent fast cooling (reversible temper embrittlement, RTE-treatment) [7; 8]. Based on this knowledge, samples of the brittle weld materials have been removed from the penstock and were exposed to a short-time heat treatment in order to investigate its influence on the materials toughness behavior. The results of these experiments as a function of peak temperature and holding time are presented in Fig. 3a. The positive impact on the recovery of the Charpy-V toughness deduced the expected RTE effect [7; 8] revealing clear ductile fracture proportions of Charpy-V samples [2].

Systematic investigations using the Gleeble simulator have been performed to quantify appropriate parameters for the practical application of the RTE-treatment in the existing inclined shaft. The respective results are illustrated in Fig 3b. It can thus be derived that a short-time heat treatment between 550°C and 750°C of the brittle weld metal investigated causes the toughness to become recovered.

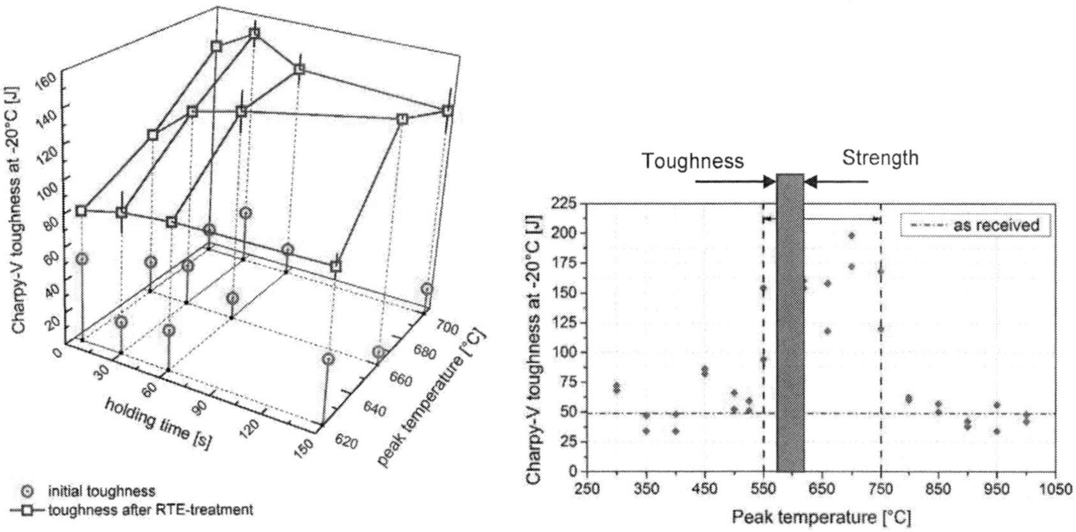

Figure 3: a) Increase of Charpy-V toughness at -20°C after short time RTE-treatment (left). b) Influence of Gleeble RTE treated Charpy-V samples as a function of peak temperature (heating rate: 15°C/s, holding time: 15s, cooling time t8/5) (right)

Root causes of effects of RTE treatment

The surprising observation of a respectable recovering of toughness of high-strength weld materials by RTE treatment by short time application of temperature around 600°C for seconds is undoubtedly from high interest. A number of investigations have been started to understand the mechanism of that effect which also includes the interest on the root causes of the effect of temper embrittlement TE. Investigations have been performed to explain the mechanism of RTE treatment concentrated on the investigation of the comparison of defined material conditions: TE and TE RTE treated.

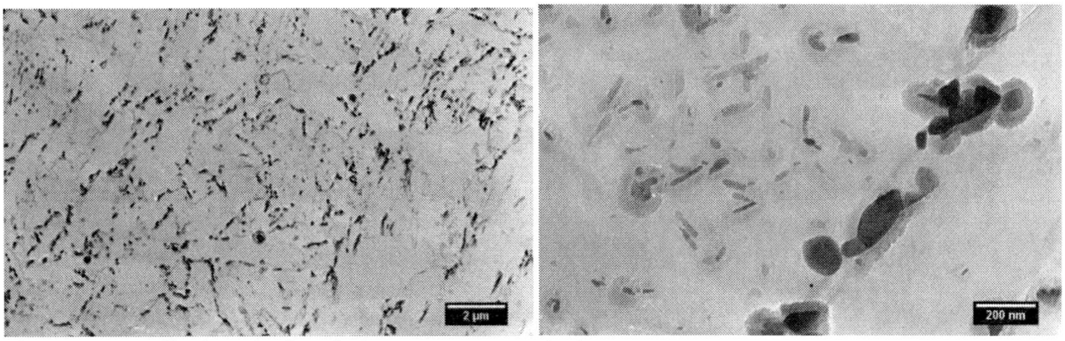

Figure 4: TEM replica investigation of microstructure as received

By reviewing the extensive accessible literature about the root causes of TE, the most authors come to the conclusion that grain boundary segregation of residual elements is responsible for the brittle intercrystalline fracture behavior [4; 5]. Under consideration of the fact that a short heat treatment of a few seconds at 600°C cause surprisingly an improvement of the Charpy V toughness, the microstructural changes are from primary interest.

Metallographic investigations:

Metallographic investigations using LOM and SEM confirmed the existence of a number of precipitations which can be expected in this type of alloying system.

TEM investigations on replicas as well as on thin foils revealed that larger precipitates are mainly concentrated on grain boundaries, while smaller precipitates are concentrated mainly inside the grains, Fig. 4, By quantitative evaluation of TEM results, a limited increase of interparticle distance (IPD) could be observed, Fig. 5. Typical differences of dislocation structures could not be detected. Differences observed in microstructures of TE embrittled and RTE treated are limited and could not fully explain the recovery effect caused by the RTE-treatment.

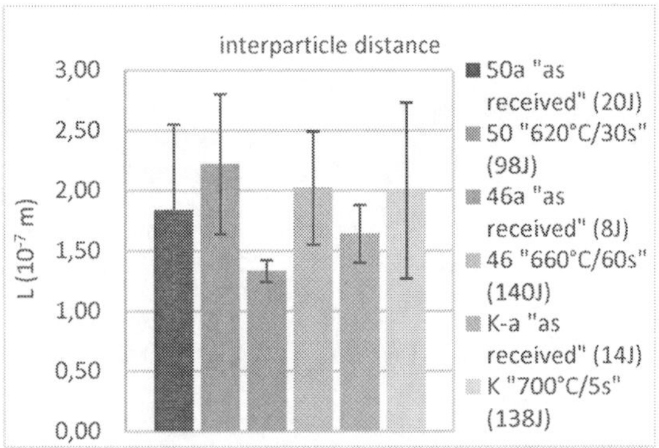

Figure 5: Average interparticle distance measured by TEM of as-received and RTE treated samples.

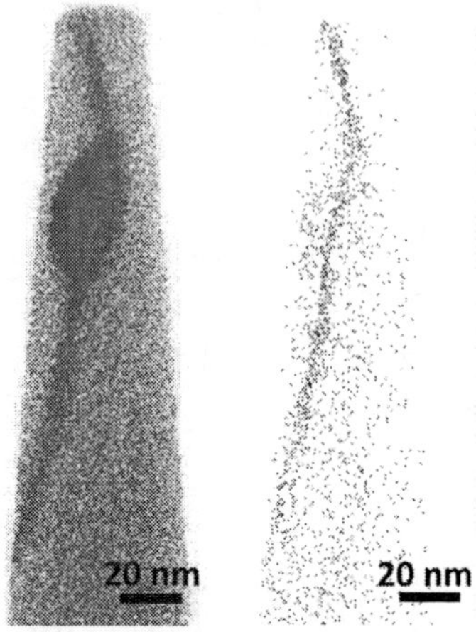

Figure 6: Atom probe investigation at specimen 8J. Blue color represent Cr atoms and pink color C atoms distributed in a $M_{23}C_6$ located in a grain boundary (left). P distribution at the grain boundary (right).

Atom probe investigations:
The main aim of limited atom probe investigation performed was to investigate an expected possible concentration of doping elements on grain boundaries: In Fig 6, results of this investigation are shown. The results reveal the presence of $M_{23}C_6$ carbides as well as segregation of Cr, C and P in the TE condition. Further investigations will be performed.

Conclusions

Severe embrittlement of weld metal in advanced high strength steel welds of a hydropower penstock has been detected. The embrittlement was caused by unqualified PWHT characterized by extremely low cooling rate of 15°C/h from holding temperature at 570°C. Temper embrittlement has been found as root cause for the embrittlement. Reversible temper embrittlement (RTE) treatment, a short time heat treatment at around 600°C, was found as method to recover the toughness of the embrittled welds in the penstock. Electrical mat heating was selected and qualified to be applied in the shaft [2].
Basic investigations focused on the scientific explanation of the surprising positive effect of RTE treatment on the recovery of the toughness of temper embrittled weld deposit have been performed by application of different advanced test methods. However, we had to learn, that even for the effect of temper embrittlement, since decades under intensive investigation by the related community, many fundamental questions remain still open.
Nevertheless, the results of the industrial applications of RTE treatment performed are doubtless very successful and reproducible.

References

[1] Cerjak, H., N. Enzinger u. M. Pudar: Development, Experiences and Qualification of Steel Grades for Hydro Power Conduits. Wasserwirtschaft – Hydropower (2015), S. 109/13.
[2] Cerjak, H., O. Caliskanoglu, N. Enzinger, G. Figner, M. Pudar: Versprödung von hochfestem Schweißgut beim Spannungsarmglühen. Schweissen und Schneiden (3/2016), S. 128-132.
[3] Erwin, W. E., u. J. G. Kerr: The Use of Quenched and Tempered 2 1/4Cr-1Mo Steel for Thick Wall Reactor Vessels in Petroleum Refinery Processes – An Interpretive Review of 25 Years of Research and Application. Weld. Res. Counc. Bull. (1982).
[4] Zabil'skii, V. V. "Temper embrittlement of structural alloy steels (review)." Metal Science and Heat Treatment 29.1 (1987): 32-42.
[5] Hickey, J. J., and J. H. Bulloch. "The role of reverse temper embrittlement on some low and high temperature crack extension processes in low carbon, low alloy steels: A review." International journal of pressure vessels and piping 49.3 (1992): 339-386.
[6] J. Grosse-Wördemann u. S. Dittrich: Prevention of Temper Embrittlement in 2 1/4 Cr-1 Mo Weld Metal by Metallurgical Actions. Wdg. J. 62 (1983), H. 1, S. 123-s/28-s
[7] Houdremont, E.: Handbuch der Sonderstahlkunde. Dritte Auflage (1956)
[8] Stein, D. F.: Reversible Temper Embrittlement. Annu. Rev. Mater. Sci. 7 (1977), S. 123

Development of High Strength Steel Plate with superior low temperature toughness for large heat input welding

Koichi NAKASHIMA, Kazukuni HASE, Yoshiaki Murakami, Taiki ETO, Hideaki FUKAI
JFE Steel Corporation, Kawasaki, Kanagawa, Japan
koi-nakashima@jfe-steel.co.jp

Abstract

High strength steel plate with superior low temperature toughness for large heat input welding has been successfully developed by applying an advanced micro-alloying design. Optimization of dispersion of martensite-austenite constituent (MA) was applied to developed YP390MPa grade F steel. Suppression of MA formation in the coarse-grained HAZ and promotion of MA formation in the fine-grained HAZ were achieved simultaneously in order to improve the HAZ toughness and the strength of welded joint. The developed steel was produced in an actual plate mill, and showed the satisfactory property of base metal and welded joint of large heat input welding.

Introduction

In recent years, upscaling of various types of welded structures, including ships, buildings, industrial plants, etc., has heightened the need for high strength, heavy-thick steel plates. In the shipbuilding field, grade E steel plate (Base metal toughness: −40°C specification, Joint toughness: −20°C specification) are generally used in merchant ships. For this application, heavy thickness E class steel plates with thicknesses exceeding 50 mm have been developed in both the YP390 MPa class and the YP460 MPa class (YP: Yield point)[1−5]. On the other hand, the operating environment becomes severe because the northern sea route is developed and the gas or oil fields are shifting to the cold area. Therefore, the steel plate with superior low temperature toughness is required to satisfy these demands.

Steels up to YP390 MPa class have been developed and applied practically as grade F hull structural steels for low temperature specifications (Base metal toughness: −60°C specification, Joint toughness: −40°C specification), but in order to guarantee the low temperature toughness of the heat affected zone (HAZ), multilayer welding by gas metal arc welding (GMAW) is applied. From the viewpoint of improving welding efficiency and reducing construction time, large heat input welding such as electro gas arc welding (EGW) is preferred to weld these heavy-thick plates. However, in large heat input welding, a decrease in toughness[6,7] occurs due to coarsening of the HAZ microstructure, while a decrease in strength of welded joint[8] occurs due to enlargement of the softened HAZ region. Although the hardening of softened HAZ is effective for increase in strength of welded joint, it leads to the decrease in HAZ toughness. For this reason, practical application of YP390MPa grade F steel for large heat input welding had been difficult. In this research, YP390MPa grade F steel for EGW which satisfies both high strength of the base metal and low temperature toughness of welded joint has been successfully developed by applying advanced micro-alloying design. This paper introduces the basic alloy design concept of the developed steel and the properties of the steel produced in the actual mill.

Alloy Design of Developed Steel

When steels are welded, a HAZ forms, in which the microstructure that was adjusted in TMCP process changes and various types of microstructure are formed depending on its thermal cycle in HAZ. A region near the fusion line, grain coarsening occurs and toughness is decreased (Fig. 1 : coarse-grained HAZ(CGHAZ)), while a region heated just above Ac_3 temperature is softened (Fig. 1 : fine-grained HAZ(FGHAZ)) and this leads to the decrease in strength of welded joint. In large heat input welding, the time heated at higher temperature below the melting point is longer and the deterioration of these properties is emphasized. In order to obtain a high strength steel having excellent low temperature toughness in large heat input welding, it is essential to satisfy both HAZ toughness and suppression of HAZ softening. To achieve this, the authors focused on martensite-austenite constituent (MA), which is a hard second phase, and optimized the MA dispersion in the respective regions of the HAZ.

Figure 2 shows the concept of HAZ microstructure control. To improve the toughness of the CGHAZ, JFE Steel developed the "JFE EWEL™" technology[1,2], and established a technology in which the growth of γ grains at high temperature is restrained by optimization of TiN dispersion, and the HAZ microstructure is refined by promoting the $\gamma \rightarrow \alpha$ transformation within prior γ grains by addition of B and Ca. In addition to the above-mentioned matrix microstructure control, second phase microstructure control was also intended by focusing on MA in this development. It has been reported that MA formation in the CGHAZ depends on the Si content and the amount of MA is reduced by reducing Si in multilayer welding[9]. The results in this study confirmed that the similar MA reduction effect also occurs in 1-pass large heat input welding. However, if the MA in the HAZ is simply reduced, this will leads to a decrease in the hardness of the softened HAZ. Therefore, the reduction of MA in CGHAZ and the increase of MA in the FGHAZ should be achieved simultaneously in the same plate. For the aimed microstructure, optimized alloy design was investigated and it was confirmed that optimization of Mn-Cu-Ni balance is effective for good combination of HAZ toughness and strength of the welded joint. Figure 3 shows examples of HAZ microstructures in the CGHAZ and FGHAZ. The aimed HAZ microstructures have been obtained in the developed steel. In the conventional steel, the C-enriched phase in the FGHAZ has undergone pearlite decomposition, whereas, in the developed steel, it has become MA, contributing to an increase in the hardness of the FGHAZ. In the coarse-grained region, ferrite forms on prior γ grain boundaries, and the bainite transformation then proceeds in the γ grains. In contrast to this, because diffusional transformation of ferrite is predominant in the fine-grained region, it is thought that the alloying elements are readily distributed to γ, which promotes the formation of MA.

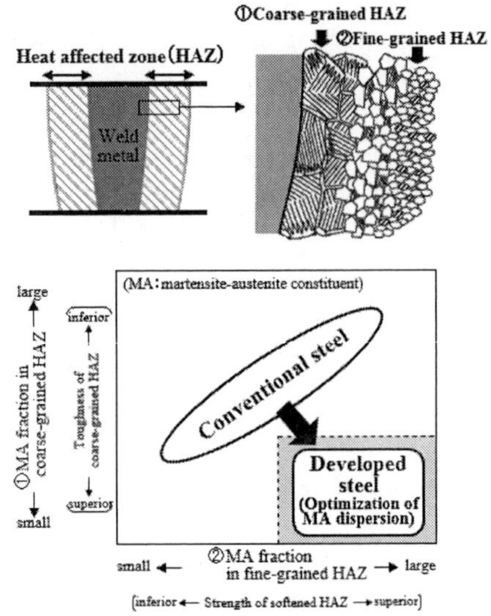

Fig.1 Concept diagram of MA control for developed steel.

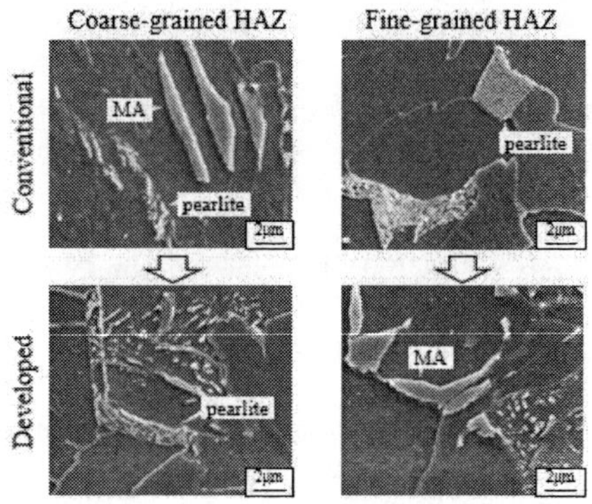

Fig.3 HAZ microstructure of developed steel.

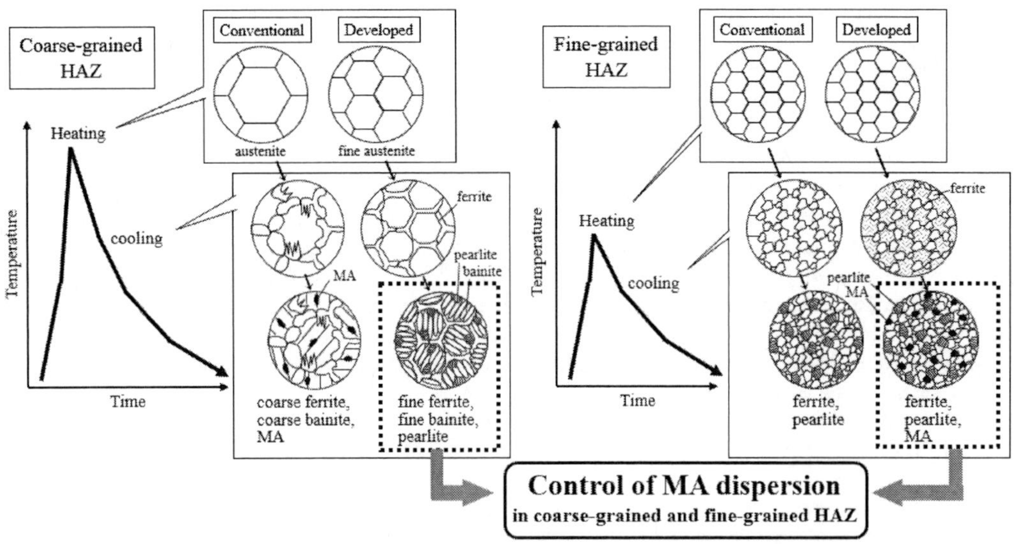

Fig.2 Concept of HAZ microstructure control.

Property of Developed Steel

Mechanical property and microstructure of base metal

Based on the concept of HAZ microstructure control, a YP390MPa grade F steel for large heat input welding was manufactured by an actual plate mill. Table 1 shows the chemical composition of the developed steel. The contents of the various added alloying elements were optimized in order to satisfy both strength of welded joint and low temperature toughness. Applying the above-mentioned JFE EWEL™ technology, the contents of Ti and N were optimized and Ca and B were added. The content of Si was also reduced to suppress formation of MA in the CGHAZ. In addition, Cu and Ni were also added, resulting in high strength of welded joint. Low Pcm design was adopted to enable preheating-free welding. Under the optimum rolling and cooling conditions in TMCP process, plate with a thickness of 50 mm were produced. The microstructure of the developed steel is shown in fig. 4, and the results of the tensile test and Charpy impact test are shown in Table 2. By applying the optimum TMCP process, it was possible to obtain a fine microstructure, and the properties of the base material satisfied the target values for the strength and toughness of YP390MPa grade F steel. With a y-groove weld cracking test and a maximum hardness test, it was also confirmed that the developed steel has excellent weldability as a result of low Pcm design.

Property of large heat input welded joint

In order to evaluate the property of large heat input welding joint, 1-pass electrogas arc welding (EGW) joints were prepared under the conditions shown in Table 3. Examples of the macrostructure and the microstructure in the vicinity of the fusion line are shown in fig. 5. The tensile strength of welded joint is shown in Table 4. The welded joint have adequate strength exceeding the required property (510 MPa). The results of a test of the Charpy impact properties are shown in fig. 6. High absorbed energy was obtained at all notch positions, and it was possible to satisfy the property requirements for Charpy toughness in the large heat input welding joints of YP390MPa grade F steel, which had been difficult with the conventional technology.

Table 1 Chemical compositions of developed steel.

C	Si	Mn	Nb	Ti	Others	*Ceq*[*1]	*Pcm*[*2]
0.05	0.06	1.56	0.01	0.01	Cu, Ni, Ca, B, etc.	0.38	0.16

[*1] *Ceq*=C+Mn/6+(Cr+Mo+V)/5+(Cu+Ni)/15
[*2] *Pcm*=C+Si/30+Mn/20+Cu/20+Ni/60+Cr/20+Mo/15+V/10+5B

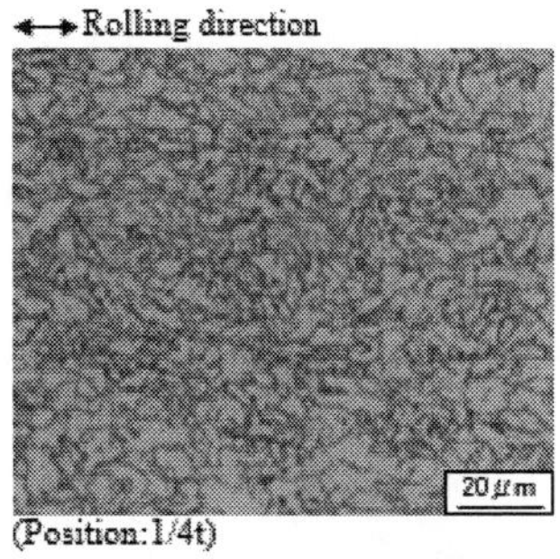

Fig.4 Microstructure of developed steel.

Table 2 Mechanical properties of developed steel.

Position	YP (MPa)	TS (MPa)	El (%)	$vE_{-60℃}$ (J)
1/4t	476	567	26	334
1/2t	422	518	28	262

Tensile test specimen: NK U14A, T-direction
Charpy impact test specimen: NK U4, L-direction
YP: Yield point TS: Tensile strength
El: Elongation vE: Absorbed energy

Table 3 Welding conditions for EGW.

Thickness (mm)	Welding method	Edge preparation	Welding consumable	Pass	Current (A)	Voltage (V)	Speed (cm/min)	Heat input (kJ/cm)
50	EGW	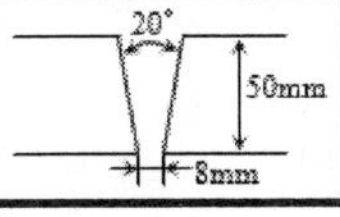	DW-S1LG (1.6mmφ) KL-4 (Kobe Steel, Ltd.)	1	410	42	3.5	302

EGW: Electrogas arc welding

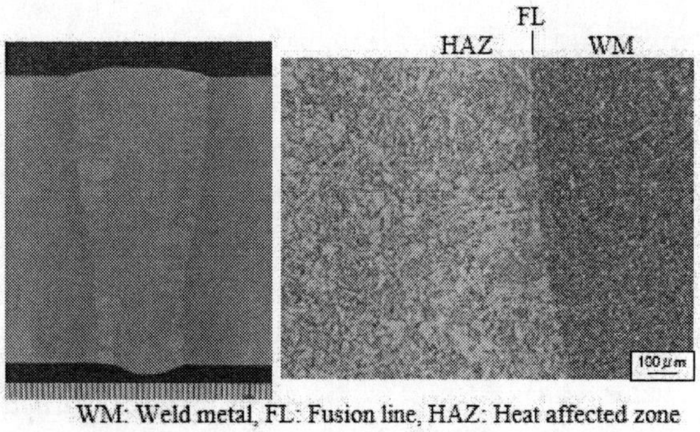

Fig.5 Macrostructure and microstructure at fusion line of EGW welded joint

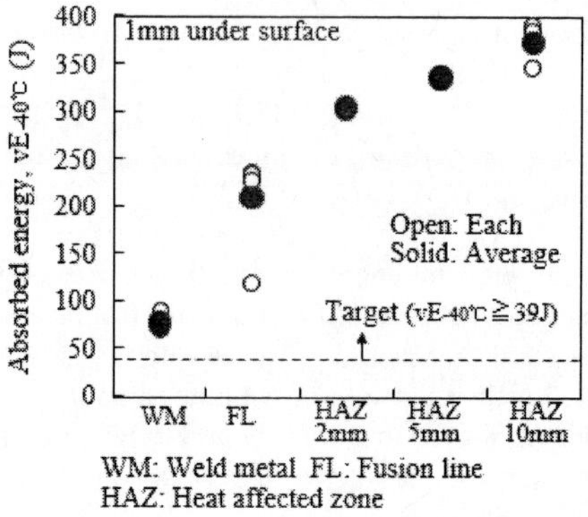

Fig.6 Charpy impact properties of EGW welded joint.

Table 4 Tensile test results of EGW welded joint

Test piece	TS (MPa)	Fracture positions
NKU2A	549, 545	Base metal
NKU2B	524, 517	Base metal

TS: Tensile strength

Conclusions

The YP390MPa grade F steel for large heat input welding was developed by applying advanced microalloying technology. Conventionally, multipass welding with a small heat input had been used with grade F steels in order to guarantee the low temperature toughness of the heat affected zone (HAZ).

HAZ microstructure was optimized by controlling of MA dispersion. Suppression of MA formation in the CGHAZ and promotion of MA formation in the FGHAZ were achieved at the large heat input welding simultaneously. The advanced technology has made it possible to develop a high strength, heavy thickness steel plate that displays excellent low temperature toughness and is also suitable for high efficiency 1-pass large heat input welding.

References

[1] S. Suzuki, *et al*: JFE Technical Report, No. 5 (2005), pp. 24–29.
[2] K. Ichimiya, *et al:* JFE Technical Report, No. 11 (2008), pp. 7–12.
[3] K. Abe, *et al*: Kobe Steel Engineering Reports, Vol. 55, No. 2 (2005), pp. 26–29.
[4] M. Kaneko, *et al*: Kobe Steel Engineering Reports, Vol. 58, No. 1 (2008), pp. 39–41.
[5] M. Minagawa, *et al*: Nippon Steel Technical Report, No. 90 (2004), pp. 7–10.
[6] T. Koseki: NMS-ISIJ 191 · 192, (2007), pp. 25–45.
[7] S. Tsuyama: NMS-ISIJ 191 · 192, (2007), pp. 47–65.
[8] K. Satoh *et al*: Journal of the Japan Welding Society, Vol. 40, (1971), pp. 885–900.
[9] O. Tanigawa et al: Kawasaki Steel Giho, Vol. 25, No. 1 (1993), pp. 13–19.

Improvement of notch toughness after PWHT for HSLA weld metals with rutile-type flux-cored wire

Yoshihiko Kitagawa, Mana Takawa, Shuji Sasakura and Kazuyuki Suenaga
Welding Business, Kobe Steel, Ltd., Fujisawa, Kanagawa, Japan
kitagawa.yoshihiko@kobelco.com, +81-466-20-3271

Masahiro Inomoto, Hidenori Nako and Yoshitomi Okazaki
Materials Research Laboratory, Kobe Steel, Ltd., Kobe, Hyogo, Japan
inomoto.masahiro@kobelco.com, +81-78-992-5503

Abstract
To improve the notch toughness after postweld heat treatment (PWHT) of weld metal for HT610 class steel by the flux-cored arc welding process, the microstructure and morphology of the carbide have been investigated in detail after PWHT at 620°C. As a result, the size of the cementite precipitated at the prior-austenite grain boundary (PAGB) has been clarified to be the dominant factor of the notch toughness of welds; this is supposedly because the precipitates act as the origin of the intragranular fracture. It was effective for the refinement of the PAGB cementite to control the carbon content and the carbide formative elements such as chromium and molybdenum. Based on these results, an optimal chemical composition of weld metal for HT610 class steel was obtained, which offers an excellent balance of strength and low-temperature notch toughness at –40°C.

Introduction
In the construction of structures such as spherical tanks and pressure vessels, the weldment is subjected, after welding was finished, to postweld heat treatment (PWHT) for reducing the residual stresses induced by welding and for improving the fracture toughness and fatigue properties of the weld. For these structures, there is a tendency to be built with a larger size and to be operated at a higher pressure in tandem with the recent growing in energy demand, and thus the steel materials used are strengthened increasingly. In association with the strengthening of steel material, the welding consumables for such applications are also desired to possess higher strengths; however, for the flux-cored wires (FCWs) of rutile type for HT610-or-higher-class steel materials, there was the problem of remarkable decrease in the impact toughness of weld metal after PWHT.

The embrittlement of weld metal by PWHT was considered attributable, for one cause, to the precipitation hardening by the formation of carbides of impurities such as niobium and vanadium [1]. However, for the welding consumables for HT610-or-higher-class steel materials, only the reduction of these impurities was not enough to obtain the weld metal with sufficient impact toughness after PWHT, and hence the toughness was needed to be improved further by the control of microstructure.

In the present study, the rutile type FCWs for HT610 class steel were prepared to produce the weld metals with different chemical compositions for studying the relationship between impact toughness and microstructure after PWHT. Furthermore the effects of individual chemical elements on the microstructure of weld metal were investigated in order to clarify the most appropriate chemical composition of weld metal for better impact toughness after PWHT.

Experimental procedures
The FCWs with various chemical compositions for high-strength low-alloy (HSLA) steel (HT610 class) were experimentally produced and tested in the multi-layer MAG welding with a shielding gas of 80%Ar-20%CO_2. The testing steel plates for multi-layer welding were prepared with the dimensions of 20-mm thickness, 20-degree groove angle, and 16-mm root gap. In this welding procedure, the average heat input was 1.2KJ/mm, preheating temperature was 90-110°C, and interpass temperature was 140-160°C. After welding was finished, the weldment was subjected to PWHT at 620°C for 8 hours, and tensile test specimens and Charpy impact test specimens were machined from the center of the weld metal for evaluation of strength and toughness. In addition, the microstructure of weld metal was observed by optical microscope and transmission electron microscope (TEM).

Table 1 Chemical composition range of weld metal investigated in this study (mass%)

C	Si	Mn	Ni	Cr	Mo	Ti	B	Fe
0.04-0.06	0.1-0.3	1.3-2.0	0.9-1.0	0.02-0.76	0.3-0.5	0.05	≦0.003	bal.

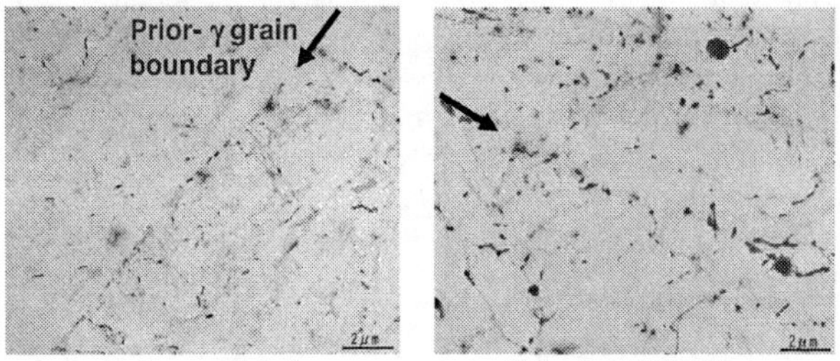

(a) as welded *(b) after PWHT*

Fig. 1 TEM micrographs of weld metal

Results and discussions

Table 1 shows the chemical composition range of the weld metals made with the experimentally produced FCWs in the present study. Fig. 1 shows the TEM-observed micrographs obtained by the extraction replica method with the weld metals (nominal chemistry: 0.05C-0.31Si-1.85Mn-1.0Ni-0.43Mo-0.02 Cr-0.05Ti-0.003B-Fe) in the as-welded condition and after PWHT. With these weld metals, Charpy impact test was conducted at –40°C and, the resultant absorbed energy was 90J in the as-welded condition but decreased remarkably to 35J after PWHT. The microstructure after PWHT exhibits the inter- and intra-granular carbides coarsened by PWHT. Particularly in the prior-austenite grain boundary (PAGB) area, coarse carbides are precipitated continuously, which suggest the possibility of acting as the initiation site of fracture in the impact testing, thereby reducing the notch toughness.

The absorbed energies at –40°C for the weld metals in the range of chemical composition shown in Table 1 are plotted (Fig. 2) as a function of the size of the PAGB carbide. The carbide size was obtained in such a way that the weld metal specimen made by the extraction replica method was observed by TEM to measure the individual carbide areas for carbides precipitated along PAGB, from which the circle-equivalent diameters were calculated, and then their average value was obtained. In this figure, a clear relationship can be observed between the absorbed energy and the carbide size, from which it has been clarified that with an increase in the carbide size, the absorbed energy decreases. Therefore, in this research, the effects of the major carbide forming elements such as carbon, chromium, and molybdenum on carbide formation and impact toughness were studied.

Fig. 3 shows the relation between the carbon content of weld metal and the size of the PAGB carbide after PWHT. The chemical composition of weld metal was 0.17Si-1.32Mn-0.90Ni-0.41Cr-0.20Mo-0.05Ti-0.003B-Fe other than carbon. As shown in the figure, the carbide size tends to become larger with an increase in the carbon content. From this, it can be considered that an increase in the carbon content simply contributes to an increase in the size of carbide.

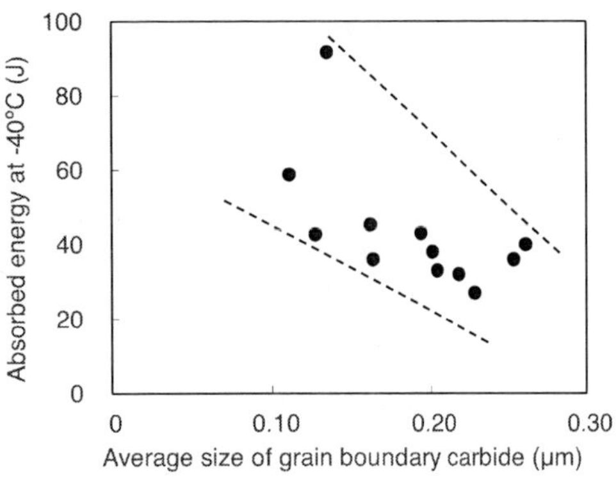

Fig. 2 Relationship between average size of PAGB carbide and absorbed energy at -40 °C of weld metal

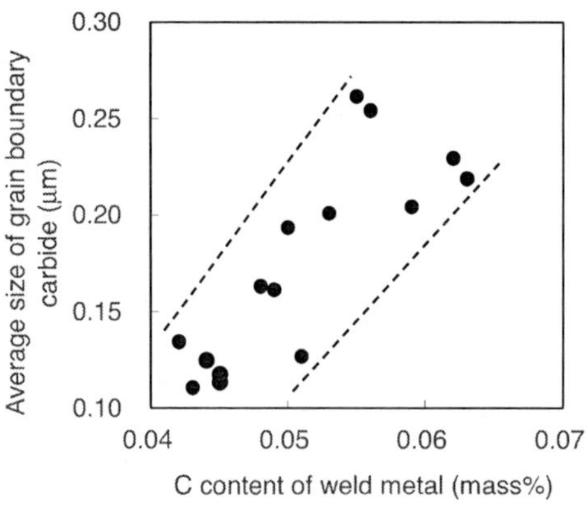

Fig. 3 Relationship between C content and average size of PAGB carbide of weld metal

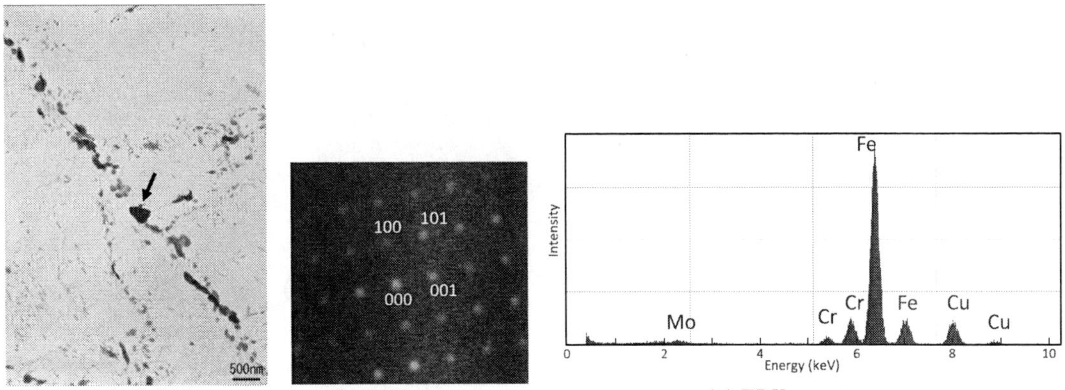

(a) Bright field image *(b) SAD pattern* *(c) EDX spectrum*

Fig. 4 TEM observation of PAGB carbide (SAD pattern (b) and EDX spectrum (c) were obtained from the carbide indicated by an arrow in the bright field image (a))

Fig. 4 shows the TEM observation results of the carbides precipitated around the PAGB of the 0.05%C weld metal after PWHT. For the coarse carbide precipitated along PAGB, iron and chromium were detected by EDX analysis and identified to be cementite ($(Fe,Cr)_3C$) by selected area diffraction (SAD) pattern analysis. In contrast, for the many intragranular carbides precipitated with comparatively small sizes, molybdenum was detected; hence, these carbides could be deemed to be Mo_2C. From these findings, the coarsening of carbide observed after PWHT can be considered attributable mainly to the growth of cementite. In order to suppress the growth of cementite, it is desirable to decrease the carbon content; however, excessive reduction in the carbon content lessens the hardenability of weld metal and may cause the formation of coarse bainite in a multi-layer weld metal, particularly in its reheated coarse grain zone. The optimum carbon content, therefore, was judged to be about 0.05%.

Fig. 5 shows the relation between the chromium content of weld metal and the size of the PAGB carbide after PWHT. Although the plots are varied due to the varieties of carbon content, the size of carbide is smallest at around 0.2%Cr. As shown above in Fig. 4, chromium was detected by EDX analysis of the cementite. These results implied that chromium was dissolved to form a solid solution in the cementite; hence, it could be considered that the Ostwald ripening of the cementite was controlled by chromium (a substitutional element with small diffusion rate) to prevent the cementite from coarsening.

Fig. 6 shows the relation between the absorbed energy at –40°C after PWHT and the molybdenum content of the weld metal. With an increase in the molybdenum content, the notch toughness tends to decrease, and this tendency becomes remarkable over 0.4%Mo. From the results shown above in Fig. 4, molybdenum probably precipitated as an intragranular carbide of Mo_2C in the weld metal after PWHT. The precipitation of Mo_2C may have suppressed the growth of the PAGB cementite; however, the embrittlement by the precipitation hardening may have affected more strongly, resulting in the reduction in the notch toughness. Hence, in order for the weld metal to suppress the embrittlement and assure the tensile strength equivalent to HT610 class steel, the molybdenum content should likely be controlled to be around 0.3%.

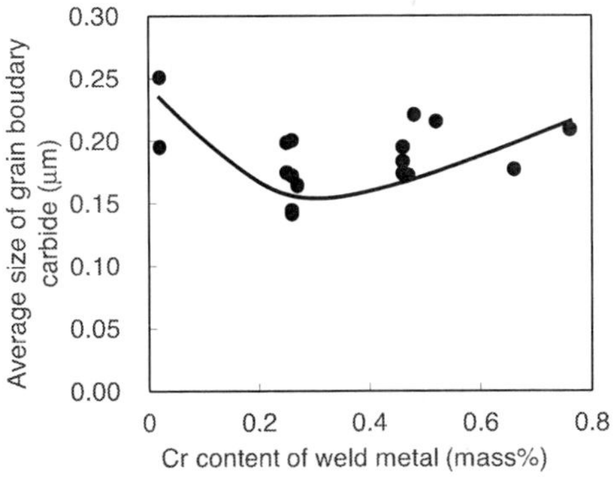

Fig. 5 Relationship between Cr content and average size of PAGB carbide of weld metal

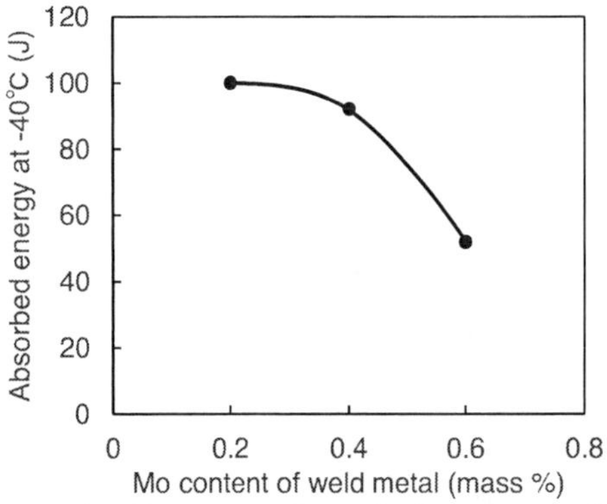

Fig. 6 Relationship between Mo content and absorbed energy at -40°C of weld metal

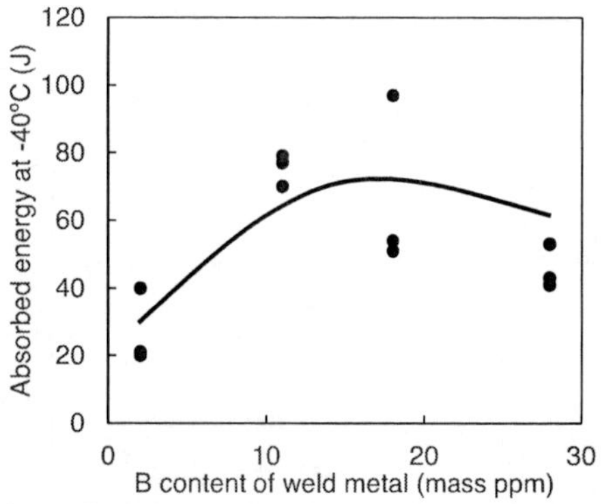

Fig. 7 Relationship between B content and absorbed energy at -40°C of weld metal

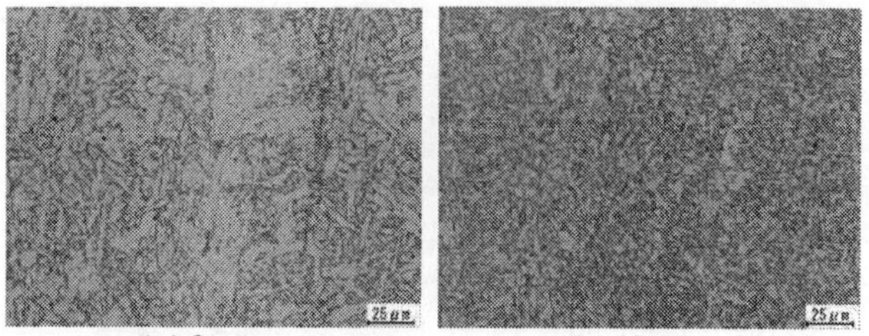

(a) 2 ppm *(b) 18 ppm*

Fig. 8 Optical micrographs of weld metal with different B content

Table 2 Chemical composition and mechanical property of weld metal

Chemical composition (mass%)									TS (MPa)		vE-40°C (J)	
C	Si	Mn	Ni	Cr	Mo	Ti	B	Fe	as welded	PWHT	as welded	PWHT
0.05	0.15	1.40	0.95	0.40	0.35	0.05	0.002	bal.	681	653	111	92

Lastly, the optimum content of boron in the weld metal was examined. Fig. 7 shows the relation between the absorbed energy at –40°C after PWHT and the boron content in the weld metal. In general, the formation of coarse ferrite or bainite at grain boundaries can probably be suppressed by the boron segregation at PAGB. Fig. 8 shows the optical microstructures of the weld metals that contain 2 ppm B and 18 ppm B, respectively; the weld metal with 2 ppm B exhibits the coarse bainite formed at the grain boundaries, the notch toughness of which was resulted in a low level. Thus, the optimum boron content was deemed to be 10～30 ppm (0.001~0.003%).

From the examination results mentioned above, the authors have obtained the optimum chemical composition of weld metal and its mechanical properties as shown in Table 2. The tensile strength of 653MPa after PWHT at 620°C for 8 hours is high enough for HT610 class steel. For the impact absorbed energy of the weld metal at –40°C, no noticeable reduction by PWHT is observed; this good toughness has been obtained by the control of the size of grain boundary carbide.

Conclusions

For the weld metals of the FCWs for high-strength low-alloy steel (HT610 class), the relationship between the microstructure and notch toughness after PWHT was studied to examine the appropriate chemical composition of weld metal for better notch toughness. The experimental results obtained in the present study are summarized in the following.

1. As the result of the research on the relationship between the size of the PAGB carbide and the notch toughness of the weld metal after PWHT, the particular correlation was observed between both variables, and the reduction in the notch toughness by PWHT was considered attributable mainly to the coarsening of carbide (cementite).
2. The effects of alloying elements on the size of the grain boundary carbide after PWHT were researched; consequently, it was found that the coarsening of carbide was promoted by carbon and suppressed by chromium.
3. The weld metal with the optimum chemical composition obtained through the examination was verified to fulfill the tensile strength specified for HT610 class steel and exhibit excellent notch toughness after PWHT.

References

[1] T. Suga, K. Ikemoto, M. Konishi, T. Kurokawa and K. Hosoi: "Toughness of Weld Metal by MAG Welding Flux-Cored Wire for Low Temperature Service Steel", IIW Doc. XII-1492-97 (1997).

Relationship between κ-carbide behavior and HAZ softening in age hardened Fe-Mn-Al-C lightweight steel

Seonghoon Jeong, Bongyoon Kim, Changhee Lee
Division of Materials Science & Engineering Hanyang University, Seoul, Republic of Korea
sunghoonid@hotmail.com, gain-yoon@hanmail.net, chlee@hanyang.ac.kr, +82-2-2220-0388

Joonoh Moon, Seong-Jun Park
Korea Institute of Materials Science, Changwon, Republic of Korea
mjo99@kims.re.kr, hyega@kims.re.kr, +82-55-280-3431

Abstract

Strengthening for safety and lightening for increasing energy efficiency have been a major issue in automobile industry, and light weight vehicle became more important especially these days because cultural restriction is strengthened. Although many parts of vehicle are substituted with other materials, steel is still used the most. For this reason, interest in lightweight steel is Increasing steadily. Weld heat affect zone shows various microstructures and properties based on κ-carbide precipitation behavior because of high temperature exposure near the fusion line. In this study, the relationship between κ-carbide precipitation and HAZ softening for an age hardened 'Fe-Mn-Al-C' lightweight steel was analyzed. Gleeble simulator was utilized for HAZ simulation, and mechanical test and microstructure analysis were performed for understanding it.

Introduction

Various researches related to high strength steel have been progressed, since strength is generally considered as very important mechanical property. [1-3] However, steel only with fine mechanical properties cannot secure competitiveness because market steadily requires more functional, advanced properties these days. In addition, steel should be eco-friend in order to satisfy strict cultural regulation of carbon dioxide emission. [4] Thus, automobile industry constantly tries to develop lightweight steel with high strength assuring enhanced fuel efficiency and safety. With regard to this, Fe-Mn-Al alloy is one of the most noticeable materials due to its high strength and low density resulting from lattice expansion and density decrease by added Al elements. Also, by Mn and C elements addition, austenite based Fe-Mn-Al-C alloy can be successfully manufactured with low density and improved mechanical properties. High ductility of Fe-Mn-Al-C came from micro band induced plasticity (MBIP) due to its high stacking fault energy. [5,6] On the other hand, the high strength and hardness of lightweight steel mainly originates from nano scale $(Fe, Mn)_3AlC$ κ-carbide that is precipitated under proper composition and temperature condition. [4,7-9] κ-carbide can be formed easily in lightweight steel, even during casting or homogenization involving quenching process. For these reasons, Fe-Mn-Al-C steel has been studied actively and still possesses high potential to be investigated. Meanwhile, the most important thing is that welding process is always accompanied to manufacture products in various industrial fields such as shipbuilding, aviation, automobile and construction. Accordingly, there are numerous researches about microstructural evolution and mechanical properties of Fe-Mn-Al-C lightweight steel according to different chemical composition or age hardening conditions. [10,11] However, in spite of many studies were conducted, there is only few studies about properties of welded region. Since welded parts experience dramatic heating and cooling thermal cycles, unexpected microstructural change can be occurred in the fusion zone and heat affected zone (HAZ), and such parts show different mechanical properties compared to base metal due to several reasons. Hence, in this research HAZ properties of age hardened Fe-Mn-Al-C in relation with κ-carbide precipitates were investigated.

Experimental procedures

The composition of designed alloy used in this research was Fe – 30 wt.%, Mn – 12 wt.%, Al – 0.9 wt.% C. Original ingots of base metal were homogenized for 2 hours at 1150 °C, and then, were hot-rolled into 13 mm plate. After hot-rolling, solution heat treated for 2 hours at 1050 °C and water quenched. Age hardening was conducted at 550 °C in a commercial furnace under air atmosphere. For HAZ simulation, Gleeble thermo-mechanical simulator (Gleeble 1500, DSI, USA) was used. Hardness was measured with micro-indentation tester (HMV-2, SHIMADZU, Japan) and microstructures were analyzed using TEM (JEM-2100F, JEOL, Japan), prepared in a solution of 5% perchloric acid solved in methanol at -40 °C.

Results and discussion

Fig. 1 shows hardness change in accordance with aging time at 550 °C. Since κ-carbide in Fe-Mn-Al-C lightweight steel is usually well-known for vigorous precipitation around relevant temperature according to preceding researches. [7,8,12]

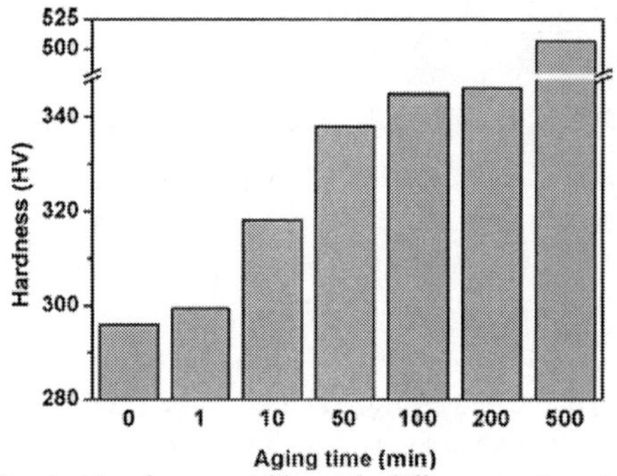

Fig. 1: Hardness profile with different aging time

As shown in Fig. 1, even little time of aging resulted in certain increase of hardness by κ-carbide precipitation. In the case of 100 min aged specimen, hardness increased 50 HV compared to the base metal which is showing only 295 HV. However, hardening by κ-carbide precipitation saturated after 100 min. When the specimen is aged much longer than saturation point (i.e. 500 min), hardness increased once more. This second stage of hardening is related to formation of β-Mn phase which is very hard but severely brittle. [13-15] Hence, the aging for 100 min was selected as peak aging time with best efficiency.

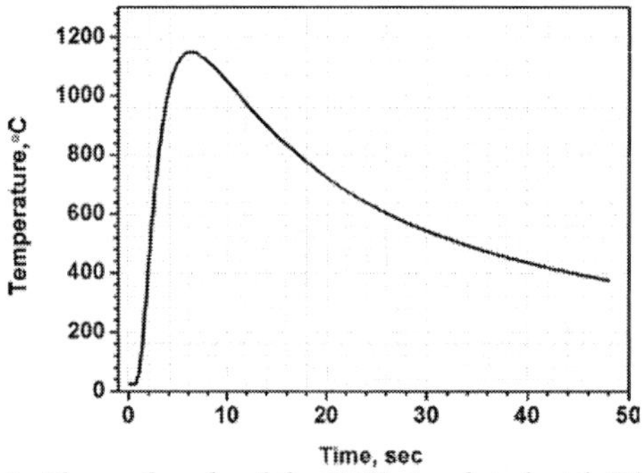

Fig. 2: Thermal cycle of the HAZ simulated with Gleeble

The thermal cycle for HAZ simulation was calculated based on Rosenthal's equation. Heat input was 30kJ / cm and the peak temperature was 1150 °C. A schematic HAZ thermal cycle is illustrated in Fig. 2. The peak temperature should dissolve κ-carbide because κ-carbide dissolution temperature was found to around 900 °C. [9]

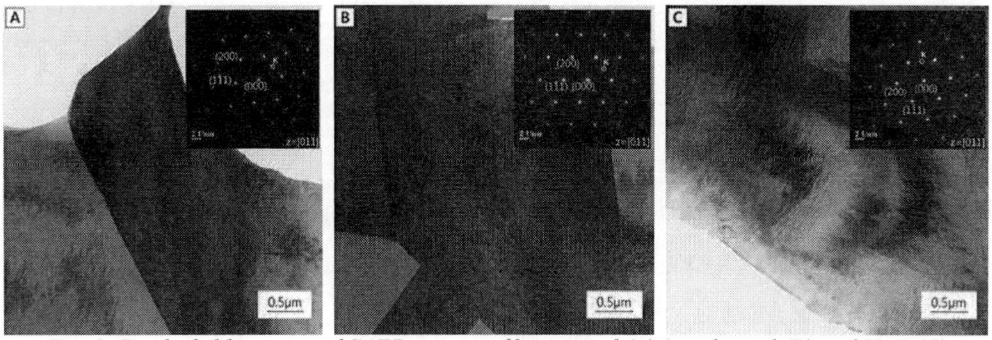

Fig. 3: Bright field image amd SAED pattern of base metal (A), aged metal (B) and HAZ (C)

Representative bright field TEM images and SAED patterns of base metal, aged metal and HAZ are shown in Fig.

3. Basically, differentiation among microstructures is almost impossible through these bright field images. This is because κ-carbide precipitates have high coherency with their austenite matrix and also the size of κ-carbide precipitates is extremely small to observe with general way. [4] Fortunately, κ-carbide evidence can be found in SAED pattern even in base metal. Therefore, for quantitative measuring κ-carbide, dark field TEM image was used.

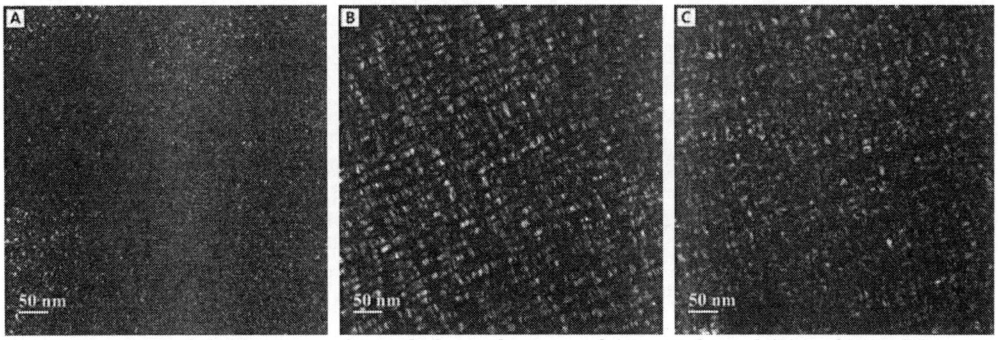

Fig. 4: Dark field images of κ-carbides on base matal (A), aged metal (B) and HAZ (C)

TEM dark field images of specimens are shown in Fig. 4. It can be figured out that κ-carbide precipitates in base metal and HAZ were small and ambiguous, compared to aged metal. To clarify the effect of κ-carbide particle size and amount on hardening, the mean κ-carbide particle size in all specimens was measured in statistical way. Average κ-carbide particle size of base metal was 3.5 nm. Aged metal showed 7.3 nm κ-carbide particles, twice larger than that in the base metal. HAZ had 6.0 nm κ-carbide particles, 22% smaller than those of Aged metal.

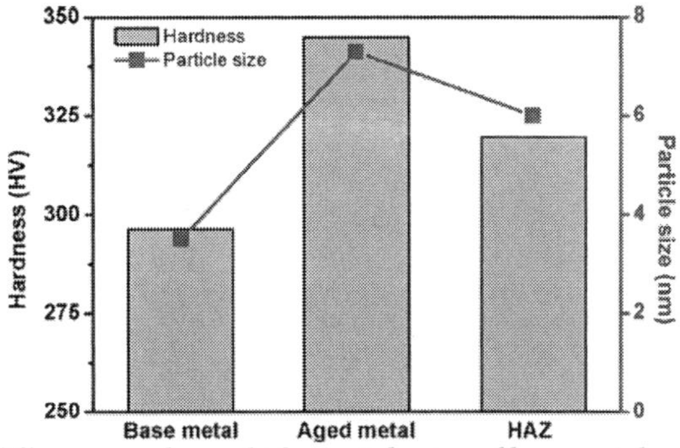

Fig. 5: Hardness difference and κ-carbide particle size of base metal, aged metal and HAZ

Fig. 5 shows the relationship between hardness and mean κ-carbide particle size among specimens. Hardness of HAZ simulated specimen was 320 HV. According to this result, it seems clear that mean κ-carbide diameter of specimen increased, directly proportional to its hardness value. Meanwhile, HAZ specimen still has higher hardness and bigger κ-carbide relative to the base metal. This is because, κ-carbide was dissolved during heating while HAZ simulation, but formed again during the cooling stage. Here, as mentioned above, base metal was water quenched after hot-rolled, but HAZ specimen was air-cooled after peak temperature, which has more time for κ-carbide precipitation.

Conclusions

In this study, relationship among κ-carbide precipitation, heat treatment such as aging and HAZ simulation and hardness of Fe–30Mn–12Al–0.9C were explored. Proper heat treatment led to the improvement of mechanical properties by κ-carbide precipitation. In this study, 100 min aging at 550 °C was set as peak aging condition according to hardness profile, and 1150 °C was selected for HAZ specimen based on Rosenthal's equation. It was hard to determine κ-carbide precipitation only with general observation because of their coherency with very low misfit strain and extremely small size (< 10 nm), but their particle size can be figured out with TEM dark field image. Average particle size increased in the aged metal as compared to the base metal, but decreased in the HAZ during high temperature thermal cycle. HAZ still has bigger κ-carbide precipitates, compared to the base metal.

This is because base metal was water quenched but HAZ was air-cooled during HAZ simulation, which had more time for κ-carbide precipitation. The shape of κ-carbides in base metal and HAZ were ambiguous relative to the aged metal. Hardness among these specimens showed same tendency with particle size change.

Acknowledgments
This research is supported by Materials and Components Technology Development Program (10048157) funded by the Ministry of Trade, Industry and Energy (MOTIE, Republic of Korea).

References
[1] O. Grässel, L. Krüger, G. Frommeyer and L. Meyer:"High strength Fe–Mn–(Al, Si) TRIP/TWIP steels development—properties—application," International Journal of Plasticity, vol. 16, no. 10, pp. 1391-1409,

[2] J. Moon, S. Kim, J. Lee, B. Hwang, C. G. Lee and C. Lee:"Effect of Cu and B addition on tempering behavior in the weld CGHAZ of high strength low alloy plate steel," Materials Science and Engineering: A, vol. 497, no. 1, pp. 153-159,

[3] N. Zhong, X. Wang, L. Wang and Y. Rong:"Enhancement of the mechanical properties of a Nb-microalloyed advanced high-strength steel treated by quenching–partitioning–tempering process," Materials Science and Engineering: A, vol. 506, no. 1, pp. 111-116,

[4] H. Kim, D.-W. Suh and N. J. Kim:"Fe–Al–Mn–C lightweight structural alloys: a review on the microstructures and mechanical properties," Science and Technology of Advanced Materials, vol. 14, no. 1, pp. 014205,

[5] J. D. Yoo and K.-T. Park:"Microband-induced plasticity in a high Mn–Al–C light steel," Materials Science and Engineering: A, vol. 496, no. 1, pp. 417-424,

[6] J. D. Yoo, S. W. Hwang and K.-T. Park:"Factors influencing the tensile behavior of a Fe–28Mn–9Al–0.8 C steel," Materials Science and Engineering: A, vol. 508, no. 1, pp. 234-240,

[7] K. Sato, K. Tagawa and Y. Inoue:"Age hardening of an Fe-30Mn-9Al-0.9 C alloy by spinodal decomposition," Scripta metallurgica, vol. 22, no. 6, pp. 899-902,

[8] W. Song, W. Zhang, J. von Appen, R. Dronskowski and W. Bleck:"κ-Phase Formation in Fe–Mn–Al–C Austenitic Steels," steel research international, vol., no., pp.,

[9] M. Li, H. Chang, P. Kao and D. Gan:"The effect of Mn and Al contents on the solvus of κ phase in austenitic Fe-Mn-Al-C alloys," Materials chemistry and physics, vol. 59, no. 1, pp. 96-99,

[10] J. Moon and S.-J. Park:"Microstructure and Mechanical Property in the Weld Heat-affected Zone of V-added Austenitic Fe-Mn-Al-C Low Density Steels," Journal of Welding and Joining, vol. 33, no. 5, pp. 31-34,

[11] I. Gutiérrez-Urrutia and D. Raabe:"High strength and ductile low density austenitic FeMnAlC steels: Simplex and alloys strengthened by nanoscale ordered carbides," Materials Science and Technology, vol. 30, no. 9, pp. 1099-1104,

[12] I. Gutierrez-Urrutia and D. Raabe:"High strength and ductile low density austenitic FeMnAlC steels: Simplex and alloys strengthened by nanoscale ordered carbides," Materials Science and Technology, vol. 30, no. 9, pp. 1099-1104,

[13] H.-H. Huang and T.-H. Chuang:"Erosion-and wear-corrosion behavior of Fe–Mn–Al alloys in NaCl solution," Materials Science and Engineering: A, vol. 292, no. 1, pp. 90-95,

[14] T. Liu and C. Wu:"β □ Mn structure in an Fe □ Al □ Mn □ Cr alloy," Scripta metallurgica, vol. 23, no. 7, pp. 1087-1092,

[15] I. Kalashnikov, A. Shalkevich, O. Acselrad and L. Pereira:"Chemical composition optimization for austenitic steels of the Fe-Mn-Al-C system," Journal of materials engineering and performance, vol. 9, no. 6, pp. 597-602,

A Study on the Development of Welding Consumables Applied to P91 Steel for Thermal Power Plants

Yasutaka BANNO*, Hideaki TAKAUCHI*, Shigenobu NANBA, Hidenori NAKO****

* Technical Center, Welding Business, Kobe Steel, Ltd. Kanagawa, Japan, ** Materials Research Laboratory, Kobe Steel Ltd., Hyogo, Japan

e-mail: banno.yasutaka@kobelco.com, phone : +81-466-20-3287, fax : +81-466-20-3289

Abstract

It is performed to investigate microstructure, A_{c1} temperature and creep rupture time of P91 deposited metals. Reductions of Mn and Ni contents promoted the formation of retained δ-ferrite. In contrast, it was revealed that addition of 0.4-0.8 % Co was effective to suppress the δ-ferrite formation. Moreover, as the Mn+Ni content decreased, the A_{c1} and A_{e1} temperatures rose. And an addition of 0.4-0.8 % Co had no effect on them. Furthermore, the reduction of the Ni content lengthened the creep rupture time.

1. Introduction

P91 steel has already been used in a number of fossil-fuel thermal power boilers due to its excellent properties of oxidation resistance, linear expansion coefficient and high temperature strength. The international standards relating to welding consumables for P91 steel, such as ASME (American Society of Mechanical Engineering) and AWS (American Welding Society) have been changed over the last decade. Key changes are the requirements for restriction of Mn+Ni content and PWHT (Post Weld Heat Treatment) temperature, which is closely related to Mn and Ni contents. P91 steel welded joints are subject to PWHT under suitable conditions to form tempered martensite microstructures with a large number of fine carbo-nitride precipitates in order to secure mechanical properties such as creep rupture strength. However, the mechanical properties of these weld metals could be unstable if the PWHT temperature exceeds their A_{c1} transformation temperature, because fresh martensite microstructure could form [1]. Therefore, the upper limits of the Mn+Ni content and PWHT temperature have been revised continuously in ASME and AWS. For instance, the upper limit of PWHT temperature is specified as 775 °C in case of filler metal having a Mn+Ni content of 1.0-1.2 %, and 800 °C in case of less than or equal to 1.0 % respectively in the ASME B&PV Code Section I:2015.

Nevertheless, although the welding consumables, satisfying the requirement of Mn+Ni<1.0 % or 1.2 % are demanded, the influence of Mn+Ni contents on the properties of P91 deposited metals still remain to be clarified. In the present study, therefore, the authors have evaluated the effects of Mn and Ni contents on the properties of P91 deposited metals.

2. Experimental procedures

2.1. Specimens

Shielded Metal Arc Welding (SMAW) deposited metals were prepared with current of 150 A DCEP (Direct Current Electrode Positive) polarity, voltage of 24 V and an average welding speed of 10 cm/min. The linear heat input was calculated at 22 kJ/cm. 6 trial covered electrodes with different levels of Mn, Ni, Cr and Co contents were used. The chemical compositions of deposited metals are shown in *Table 1*. The welding groove configuration and pass-sequences are shown in *Fig. 1*. All specimens except for A_{c1} temperature measurement and some TEM observation were carried out PWHT at 760 °C for 4 hours.

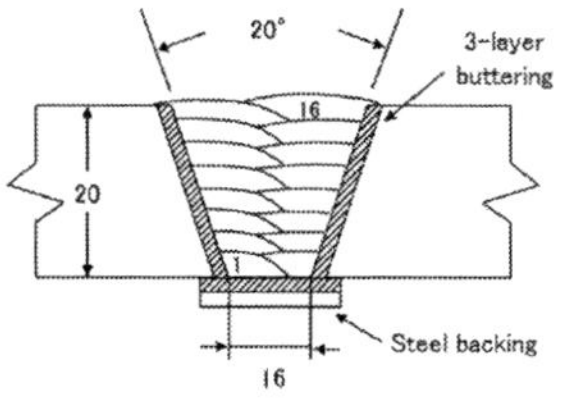

Fig. 1 Welding groove configuration and pass Sequences for SMAW (mm)

2.2. A_{c1} transformation temperature measurement

A_{c1} temperature is defined as a temperature at which the α / γ reverse transformation starts during heating. In this study, the measurements of A_{c1} temperatures were carried out using cylindrical specimens 12 mm in height and 8 mm in diameter taken longitudinally from the center of deposited metals. A_{c1} temperature was measured by detecting dimensional change in the diameters of the specimen using a high-accuracy LED apparatus during a induction-heating process in a vacuum of 10^{-3} MPa. The heating rate was set to 5 °C/second up to 600 °C, then changed to a lower rate of 5 °C/min in a range between 600-1050 °C, where reverse transformation is expected to occur. The validity of this method was sufficiently verified by a preliminary study [1].

2.3. Creep rupture test

Creep rupture specimens have 6 mm diameter and 30 mm gauge length and they were taken from the middle of deposited metals longitudinally. Creep rupture tests were conducted at 650 °C and applied 100 MPa.

2.4. Microstructure evaluation

The microstructure was observed by optical microscopy in order to know the influence of chemical composition on remaining δ-ferrite. In addition, for some deposited metals, the configurations of the precipitates and lath microstructures in before and after creep rupture testing were observed using a Transmission Electron Microscope (TEM, JEOL JEM-2010F operated at 200 kV).

3. Results and discussion

3.1. Optical microstructure

Optical microstructures of the deposited metals after PWHT are shown in *Fig. 2*. Martensitic microstructure was observed over the entire surface of the code (a) deposited metal. However, retained polygonal δ-ferrite grains were formed in codes (b) and (c) which have lower Ni content than code (a). In particular, when the Ni content was reduced below 1.0 % as in code (c), obvious amounts of δ-ferrite were observed. On the other hand, by reducing the Cr content from 9.0 % to 8.5 %, smaller amount of δ-ferrite was observed in code (d) compared to code (b). In this study, δ-ferrite could be observed only in the as-casted zone of the last bead which was not re-heated by following passes.

According to previous researches, carried out on P91 and 12Cr steels, long term creep rupture properties are

Table 1 Chemical composition of deposited metals (mass%)

Code	C	Si	Mn	Ni	Cr	Mo	V	Co	Nb	N	Mn+Ni
(a)	0.10	0.3	0.9	0.5	9.0	1.0	0.24	<0.02	0.04	0.03	1.4
(b)	0.10	0.3	0.9	0.3	9.0	1.1	0.24	<0.02	0.04	0.03	1.2
(c)	0.10	0.3	0.5	0.3	9.0	1.1	0.26	<0.02	0.04	0.02	0.8
(d)	0.09	0.3	0.7	0.4	8.5	1.0	0.26	<0.02	0.03	0.03	1.1
(e)	0.09	0.3	0.7	0.4	8.5	1.0	0.26	0.4	0.03	0.03	1.1
(f)	0.09	0.2	0.7	0.4	8.4	1.0	0.26	0.8	0.03	0.03	1.1

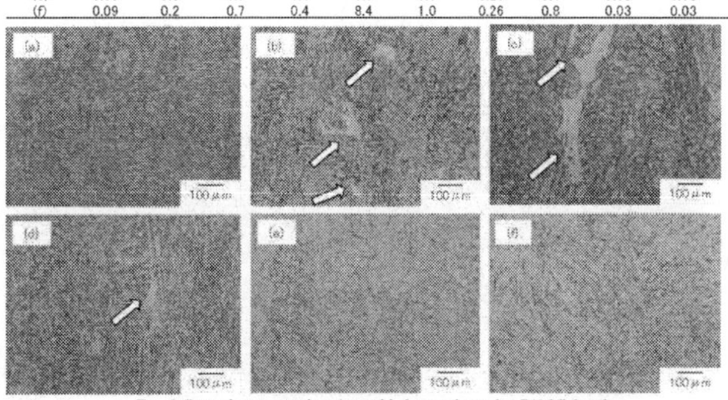

Fig. 2 Optical micrographs of as welded zones heated at 760 °C for 4h.

degraded by the presence of δ-ferrite [2-3]. In addition, the EPRI (Electric Power Research Institute) notes that all Gr. 91 products should exhibit a fully tempered martensite microstructure, which is free of δ-ferrite [4]. In our study, the influence of Co on δ-ferrite was evaluated as well. Generally, Co is not added as an alloying element for either P91 steel or welding consumables. However, in high Cr ferritic heat resistant steel, it may be added to suppress the formation of δ-ferrite [5]. In the AWS A5.5/5.5M: 2014 Specification for Low-alloy Steel Electrodes for Shielded Metal Arc Welding, the addition of Co in the B91 classification is not restricted but required to be reported if Co is added intentionally or Co content is larger than 0.20 %. In this study, δ-ferrite was not observed both in codes (e) and (f) deposited metals containing 0.4 % and 0.8 % of Co respectively, albeit high Mn+Ni content 1.2 %. It was confirmed that Co addition is effective to inhibit formation of δ-ferrite in the P91 deposited metals.

3.2. A_{c1} transformation temperature

Fig. 3 shows the relationship between the contents of Mn+Ni and A_{c1} (measured) and A_{e1} (calculated by Thermo-Calc. Ver. S / Data base: TCFE_7) temperatures of deposited metals with various Co content. As is well known, with the increase of the Mn+Ni content, the A_{c1} and A_{e1} temperatures became lower in the deposited metals as well. The A_{c1} temperatures estimated in this study were 790-800 °C with Mn+Ni=1.2 %, and 810-820 °C with Mn+Ni=1.0 %. These results support the validity of the maximum PWHT temperatures specified in ASME Sec. I

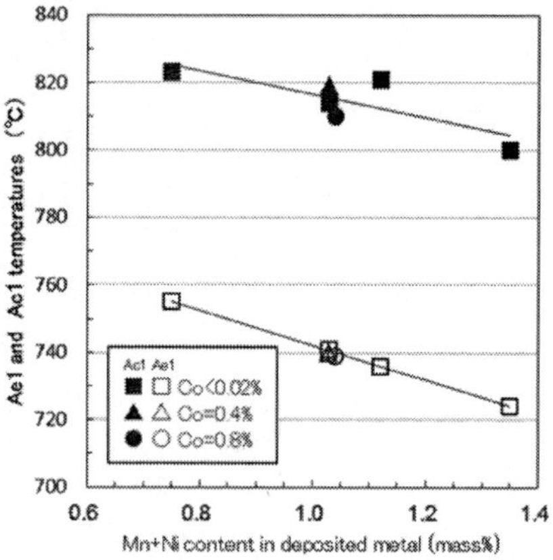

Fig. 3 Relationship between Mn+Ni, Co contents and A_{c1}, A_{e1} temperatures of deposited metals.

In this study, the influences of Co on A_{c1} and A_{e1} temperatures were evaluated as well. Zhang *et al.* demonstrated that additions of Ni, Mn and Co reduced the A_{c1} temperature of P92 deposited metals, though Co had only a 40 % effect compared to the other 2 elements [6]. On the other hand, in *Fig. 3,* it can be seen that the A_{c1} temperatures of the deposited metals containing 0.4 % and 0.8 % Co were plotted in varying ranges of other experimental results. Furthermore, the A_{e1} temperatures of the deposited metals containing 0.4 % and 0.8 % Co did not differ from those of the deposited metals containing less than 0.02 % Co. It is concluded that the Co addition of up to 0.8 % did not have a significant effect on A_{c1} and A_{e1} temperatures in the P91 deposited metals. *Fig. 4* shows the binary phase diagrams of the Fe-Mn, Fe-Ni and Fe-Co systems [7]. Focusing on the γ-region in each diagram, Mn, Ni and Co are effective to raise the A_{e3} temperature. Meanwhile, only Mn and Ni lower the A_{c1} temperature and Co has no effect on it. This peculiar feature of Co can support our experimental results, which Co is effective to suppress the δ-ferrite without lowering the A_{c1} temperature.

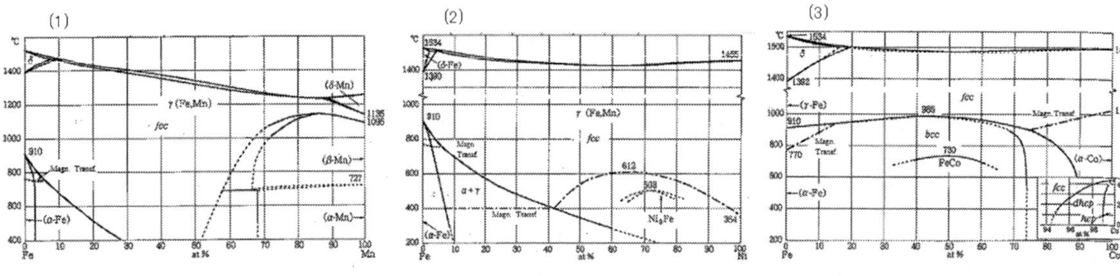

Fig. 4 Binary phase diagrams of (1)Fe-Mn, (2)Fe-Ni, (3)Fe-Co systems[7].

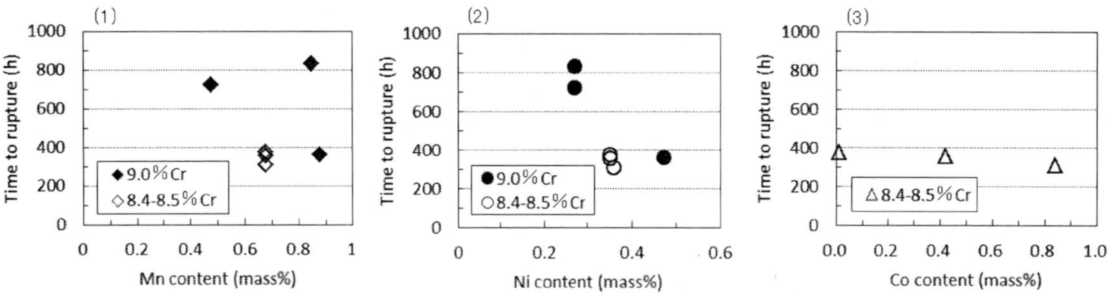

Fig. 5 Relationships between (1)Mn, (2)Ni, (3)Co contents and time to rupture of deposited metals heated at 760 °C for 4h.

3.3. Creep rupture property and microstructural change during creep testing

3.3.1. Influences of Mn, Ni and Co on creep rupture time

Fig. 5 shows the results of creep rupture tests as a function of Mn, Ni and Co contents. Since all the specimens for these tests were taken longitudinally from the center of deposited metal, their microstructures were martensitic without any δ-ferrite. Hence, the influence of δ-ferrite on the creep rupture time was not involved here. It could be seen clearly that rupture time became longer with a decrease in Ni content. There was no clear correlation between rupture time and either Mn or Co contents.

3.3.2. Carbide precipitation

Fig. 6 shows the replica TEM images of the deposited metals before creep testing. Code (b) and (c) deposited metals contain lower Ni and lower both Mn and Ni respectively compared to code (a). Submicron-size carbides were observed in all deposited metals. *Fig. 7* shows a replica TEM image of code (c) deposited metal at higher magnification. Fine carbides of nano-scale were observed (indicated by thick arrow) in addition to the submicron-size carbides, which are observed in *Fig. 6*. The EDS (Energy Dispersive x-ray Spectroscopy) spectra taken from points (a) P1 and (b) P2 (thin arrows) are shown in *Fig. 8*. In both points, there were strong peaks of Cr. Thus, these submicron-size carbides were considered to be $M_{23}C_6$. According to equilibrium calculation by Thermo-Calc software, the stable carbides are $M_{23}C_6$ (where M=Cr, Fe) and MC (where M=V, Nb) in the code (c). The fine carbides of nano-scale, therefore, were presumed to be MC. The number and size of $M_{23}C_6$ particles were almost the same in all deposited metals shown in *Fig. 6*, albeit the variations of Mn and Ni contents.

The carbides in the deposited metals after creep testing are shown in *Fig. 9*, where "1" and "2" indicate the difference of magnification. Comparing *Fig. 9* with *Fig. 6*, the coarsening of $M_{23}C_6$ took place and the number of $M_{23}C_6$ was decreased. On the other hand, the coarsening of MC was not observed. There was only slight a difference in size and number of $M_{23}C_6$ particles among codes (a), (b) and (c). The number of MC particle in codes (b) and (c) deposited metals containing lower amounts of Ni clearly increased compared to code (a). There was no influence of the Mn content on the number of MC particles.

3.3.3. Lath microstructure

Fig. 10 shows the bright field TEM images of deposited metals before creep testing. Here, elongated lath microstructures are observed. In the deposited metals (b) and (c) with lower Ni content, the width of martensite lath is smaller than in code (a). On the other hand, comparing code (b) with (c), the difference of lath width was negligible small. This result implies the influence of Mn content on lath width was not large.

Fig. 11 shows the bright field TEM images of deposited metals (a), (b) and (c) after creep testing. Equiaxed-subgrain was observed instead of lath microstructures, which had existed before creep testing. The increase of the subgrain size associated with the progress of creep deformation has been detected as reported in previous study [5]. In comparison with code (a), codes (b) and (c) had smaller subgrain. There was no clear correlation between Mn content and subgrain size.

The boundary of these recovered lath microstructures were composed of rearranged dislocations as shown in *Fig. 12-(1)*. Some of dislocation edges were in contact with $M_{23}C_6$ and lath boundaries, or trapped by MC (*Fig. 12-(2)*). These facts could suggest that $M_{23}C_6$, MC and lath boundaries acted as obstacles to dislocation motion in creep testing.

3.3.4. The influence of Ni

Creep rupture time became longer with a decrease of Ni content, as shown in *Fig. 5*. A remarkable smaller lath microstructure of low Ni deposited metal after creep testing was considered to be a direct reason for the longer rupture time [8]. On the creep test condition of high temperature and applied stress, lath recovery involves the reduction in dislocation density and the rearrangement of dislocations [8]. Thus, it is presumed that the inhibition of dislocation motion down was caused by the reduction of the Ni content. In this section,

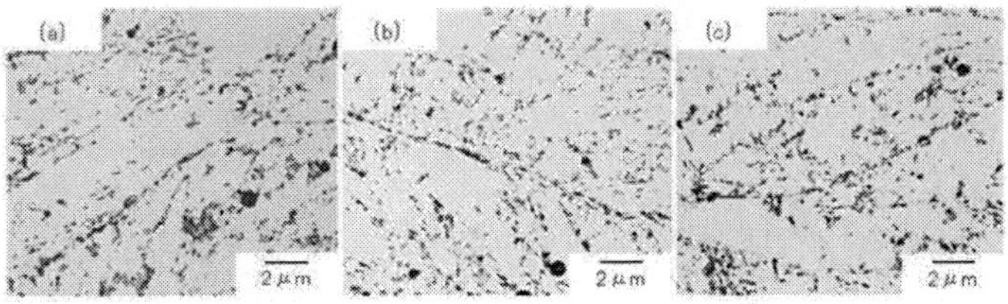

Fig. 6 Replica TEM images of the deposited metals heated at 760 °C for 4h.

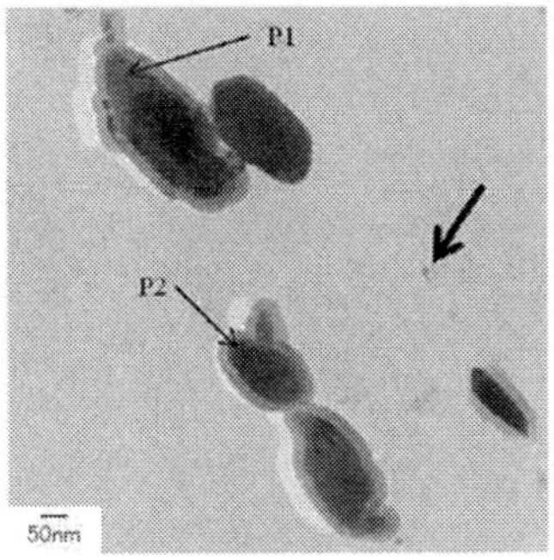

Fig. 7 Replica TEM image of the deposited metal (C) heated at 760 °C for 4h.

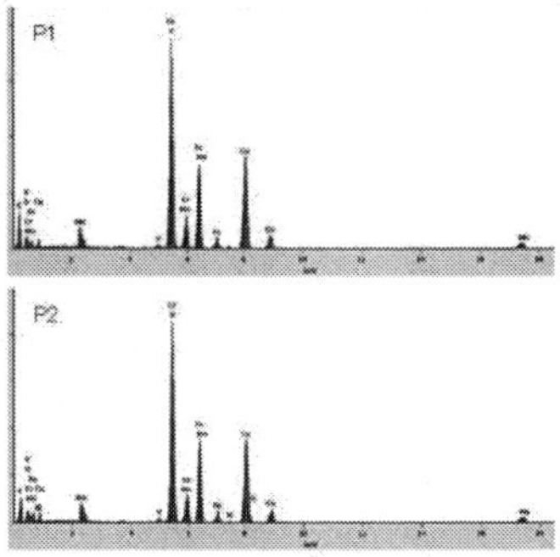

Fig. 8 EDS spectra from points P1 and P2 indicated in Fig. 7.

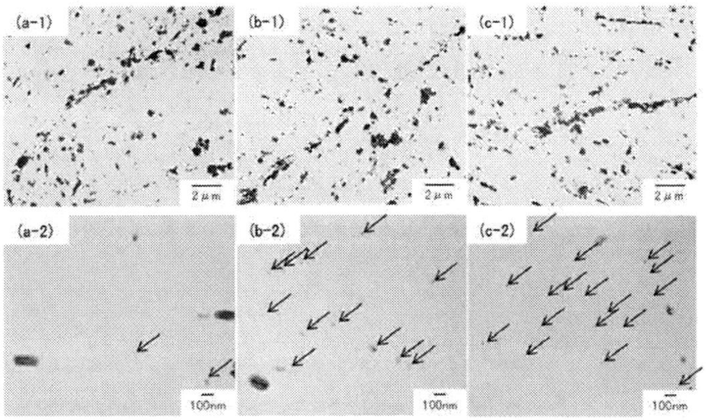

Fig. 9 Replica TEM images of the deposited metals after creep testing (arrows indicate MC particles).

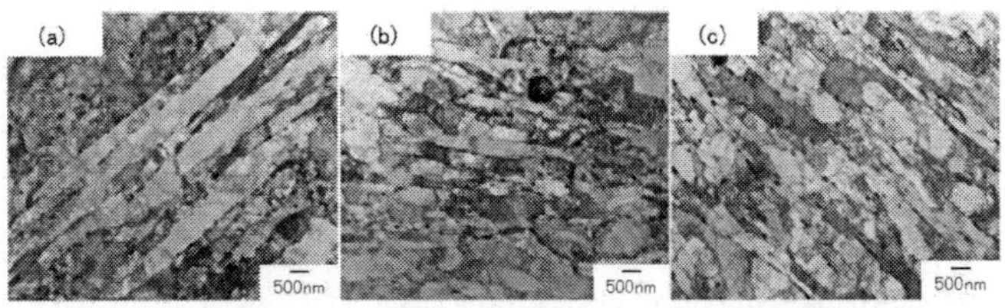

Fig. 10 Bright field TEM images of the deposited metals heated at 760 °C for 4h.

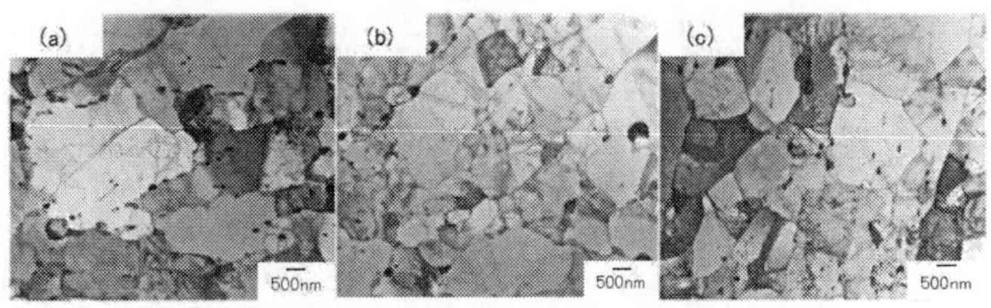

Fig. 11 Bright field TEM images of the deposited metals after creep testing.

therefore, we would like to consider the mechanism of slower dislocation motion through the reduction of the Ni content.
First of all, one of the main obstacles in dislocation mobility is the pinning effect by carbide particles. As mentioned above, the size and number of $M_{23}C_6$ did not differ among each deposited metal. Conversely, it should be noted that a great number of fine MC particles were observed in low-Ni deposited metal after creep testing. The recovery and lath growth was considered to be delayed because MC particles acted as obstacles to dislocation motion. Therefore, the question is why MC particles become more stable with a decrease of Ni content. The following three mechanisms can be stated in general:

A) Delay of the MC / Z-phase transition.
B) The inhibition of Ostwald ripening in MC.
C) A decrease of dislocation mobility.

A) In P91 steel, it is recognized that the precipitation of the Z-phase occurs as a consumption of MC particles in creep testing, resulting in the degradation of pinning effect of MC for dislocation motion [5]. The decrease in creep rupture strength of Grade 91 steel by Ni addition has been reported by Kimura *et al.*[3].In his study, it is considered that this was due to the nucleation and growth of the Z-phase, which was promoted by Ni addition. However, the formation of Z-phase was observed in long-term creep rupture tests with test time of over a thousand hours. In this study, which dealt with short-term creep rupture tests, the formation of Z-phase was not observed.

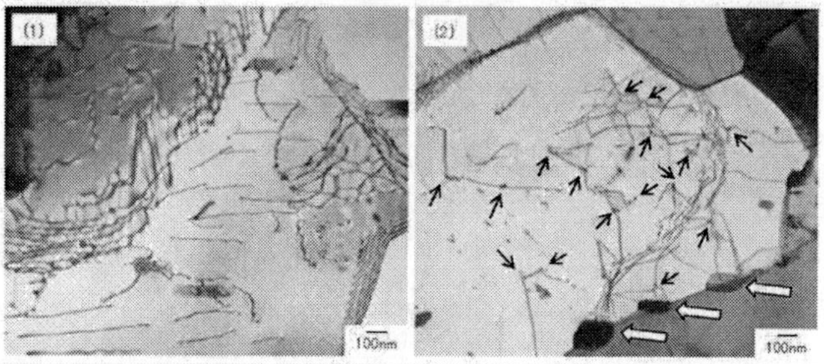

Fig. 12 M23C6 particles (white arrows) and MC particles (black arrows) observed in the bright field TEM images of the deposited metal (a) after creep testing.

B) Since MC contains Nb and V, the Ostwald ripening of MC particle was considered to be controlled by the diffusion of Nb, which has the lowest diffusion coefficient at the testing temperature of 650°C [9]. It is known that the reduction of Ni content decreases diffusion coefficient of high Cr ferritic steel [5]. It is probable that decrease of Ni content lowered diffusion coefficient of Nb, resulting in the inhibition of the Ostwald ripening of MC. In this case, reduction of Ni content should be also effective to inhibit the Ostwald ripening of $M_{23}C_6$. However, there was little difference in size or number of $M_{23}C_6$ among the deposited metals containing various amount of Ni. Hence, it is considered that change in diffusion coefficient of alloy elements by Ni reduction is not large enough to affect the inhibition of the Ostwald ripening of MC.

C) The study for the mechanism of toughness improvement by Ni addition was carried out by Maeno *et al.*, who concluded that the improvement of toughness was caused by the increase of dislocation mobility together with Ni at low temperature below -40 °C [10]. Also, it is probable that the decrease of Ni content lowered the dislocation mobility and delayed the recovery of lath microstructures. The reason that the number of MC particles after creep testing increased by lowering of Ni content can be considered as follows:

As shown in *Fig. 12*, a dislocation is pinned by several MC particles during creep testing. Therefore, it is presumed that the dislocation acted as a diffusion path of alloy elements among MC particles, resulting in the acceleration of the Ostwald ripening. In addition, it is supposed that a decrease of dislocation mobility in a low-Ni deposited metal causes a reduction of number of dislocations pinned by MC particles. Based on this assumption, it is believed that the Ostwald ripening of MC particles by pipe diffusion through dislocations was suppressed in the deposited metals with lower Ni content, which causes larger number density of MC particles after creep testing. If this hypothesis would be true, the inhibition of the recovery of lath microstructures was considered to be accomplished synergistically by the decrease of dislocations mobility due to the large number density of MC particles.

4. Conclusions

In the P91 deposited metal,

1) A reductions of Mn and Ni contents promoted the formation of retained δ-ferrite. The addition of 0.4 and 0.8 % Co was effective to suppress δ-ferrite formation.
2) As the Mn+Ni content decreased, the A_{c1} and A_{e1} temperatures rose. An addition of 0.4 or 0.8 % Co had little effect on the A_{c1} and A_{e1} temperatures.
3) The reduction of the Ni content lengthened the creep rupture time by stabilizing the MC particles and inhibiting the recovery of lath microstructures.

References

[1] L. Chen *et al.*: Effect of PWHT temperature and mechanical properties of High-Cr ferritic heat-resistant steel weld metal, Weld in the world, 56 (2012).
[2] F. Abe: Structural Alloys for Power Plants, Development of creep-resistant steels and alloys, Woodhead Publishing (2014), 277.
[3] K. Kimura *et al.*: Influence of Chemical Composition and Heat Treatment on Long-term Creep Strength of Grade 91 Steel, Procedia Engineering, 55 (2013), 2.
[4] Electric Power Research Institute: Guidelines and Specifications for High-Reliability Fossil Power Plants 2nd Edition 3002006390 Final report, (2015).
[5] K. Maruyama *et al.*: Strengthening Mechanisms of Creep Resistant Tempered Martensitic Steel, ISIJ International, 41-6 (2001), 641.
[6] Z. Zhang *et al.*: Properties of T/P92 Steel Weld Metals for Ultra Super Critical Power Plant, OMMI, 6-1 (2009).
[7] S. Nagasaki *et al.*: Binary Phase Diagram, AGNE Gijutsu Center (2001).
[8] K. Sawada *et al.*: Residual Creep Life Assessment by Change of Martensitic Lath Structure in Modified 9Cr-1Mo Steels, Tetsu-to-Hagané, 84-8 (1998), 580.
[9] The Japan Institute of Metals and Materials: Metal Data Book, Maruzen Publishing, (2004).
[10] K. Maeno *et al.*: Changein Dislocation Mobility with Ni content in Ferritic Steels and Its Effect on Brittle-to-Ductile Transition, Tetsu-to-Hagané, 98-12 (2012), 667.

Magnetically Assisted Resistance Spot Welding of Dual Phase Steels

Yongbing Li
State Key Laboratory of Mechanical System and Vibration & Shanghai Key Laboratory of Digital Manufacture for Thin-walled Structures, Shanghai Jiao Tong University, Shanghai, P.R. China;
Email: yongbinglee@sjtu.edu.cn
Stan A. David
Oak Ridge National Laboratory, Oak Ridge, Tennessee, USA.
Email: standavid@charter.net

Abstract
AHSSs present a big challenge to resistance spot welding (RSW) widely applied in automotive industries because the rapid heating and cooling procedures during RSW produce hardened weld microstructures, which lower the ductility and fatigue properties of welded joints and raise the probability of interfacial failure under external loads. Magnetically assisted resistance spot welding (MA-RSW) process has been proposed to improve the weldability of dual phase steels of different grade. In this paper, magnetically assisted resistance spot welding (MA-RSW) process was reviewed. The effect of the external magnetic field on weld geometry, microstructures, performance and softening zone were analyzed systematically to reveal the advantages of MA-RSW process in welding AHSS.

Introduction
Dual-phase (DP) AHSSs have been substantially used in automotive industry for their excellent performance. In DP steels, carbon, manganese and silicon are the main adding elements. Carbon has a strengthening effect on martensite, Manganese acts as a solid solution strengthener, and silicon promotes ferrite formation. As a result, the base metal microstructure of DP steels consists of soft ductile ferrite with hard martensite/bainite pockets along the grain boundaries. The amount of martensite/bainite controls the strength of the DP steels. This type of structure is produced by intercritical annealing followed by critical cooling that produces pockets of dispersed martensite/bainite. The combination of hard martensite phase in a soft ferrite matrix gives DP steels a higher strength and ductility.[1] However, the local heat input associated with resistance spot welding (RSW) can greatly increase the risk of cracks, element segregation, and interfacial failure, and thus reduce the mechanical performance of the welded structure. Postweld tempering has been used in production to diminish the hardening rate of the weld and thus reduce the risks, but it adds to cycle time and energy consumption. Since the 1960s, an externally applied magnetic field has been used in arc welding to refine grains in welding aluminum alloys and stainless steels [2] and enable high-speed welding. [3] Since 1993, the magnetic assistance technique was introduced to resistance spot welding of stainless steels and found that it could eliminate porosities in fusion zone, increase weld size, and thus improve weld fatigue life and weld strength.[4-5] Since 2009, the authors proposed the magnetically assisted resistance spot welding (MA-RSW) process, and applied it to the DP steels. [6-10]

In this paper, the progresses in MA-RSW are reviewed in light of the challenges that arise in welding DP steels. The MA-RSW experimental setup, the role of the external magnetic field and its effects on weld geometry, microstructure, and performance will be reviewed to assess the ability of the MA-RSW process to improve the weldability of DP steels.

Experimental setup
A schematic of the MA-RSW process is shown in Fig. 1. Considering the symmetry of weld nugget formation, a pair of ring-shaped permanent magnets was installed symmetrically about the faying surface of the steel sheets. As shown in Fig. 1(a), the south poles of the two magnets were designed to repel each other so that the vertical components of the upper and lower magnetic fields cancel each other near the faying surface of the workpieces while the horizontal components superimpose and strengthen each other. Thus, the external magnetic field on the faying surface is radially oriented, which is perpendicular to the circumferential magnetic field induced by the welding current, as shown in Fig. 1(b). Therefore, the resultant magnetic force inside the weld nugget is composed of two components. One is the radially induced magnetic force; the other is the circumferential external magnetic force, as shown in Fig. 1(c). Based on above setup, it can be expected that during the MA-RSW process, the molten metal driven by the resultant magnetic force inside the weld nugget will flow both radially and circumferentially without interference with each other.

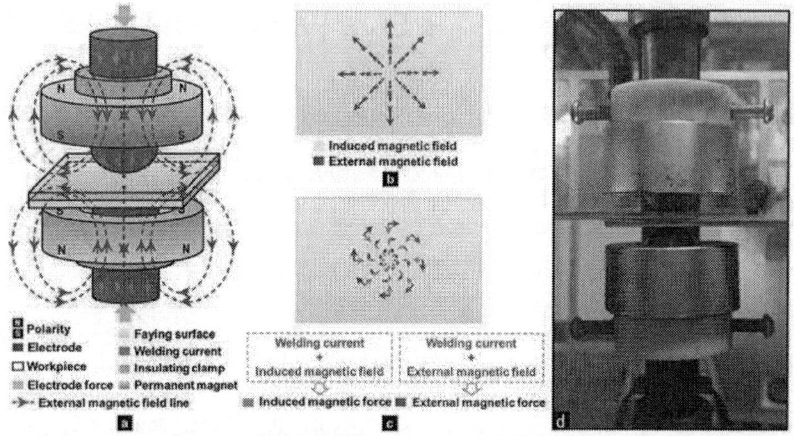

Fig. 1 Principle and experimental setup of MA-RSW process. (a) MA-RSW process principle, (b) top view of the magnetic field on the faying surface, (c) top view of the magnetic force field on the faying surface, (d) arrangement of the magnets with reference to the sheets and electrodes.[10]

The RSW system used in this study consists of a Fanuc R2000-Ib210f robot, a Medar 5000s MFDC welding controller, and an Obara C-type servo gun. Two identical dome-shaped electrodes with 5.0 mm tip diameters were used. A pair of 20 mm thick ring-shaped NdFeB permanent magnets with an inner diameter of 20 mm and an outer diameter of 40 mm was used for MA-RSW. The remanence of the magnet is 1.22~1.25 T, the maximum energy product is 287~310 kJ/m^3, and the intrinsic coercive field is greater than 907 kA/m. Distance from the South pole of a magnet to the tip of the nearest electrode was defined as working distance, which was fixed at 3mm for all the welds. The radial external magnetic flux density measured by a gauss meter was found to increase monotonically from the center to the edge of the faying surface with a maximum 0.06 T at the edge.

Results and discussion

Effect on weld geometry and microstructure

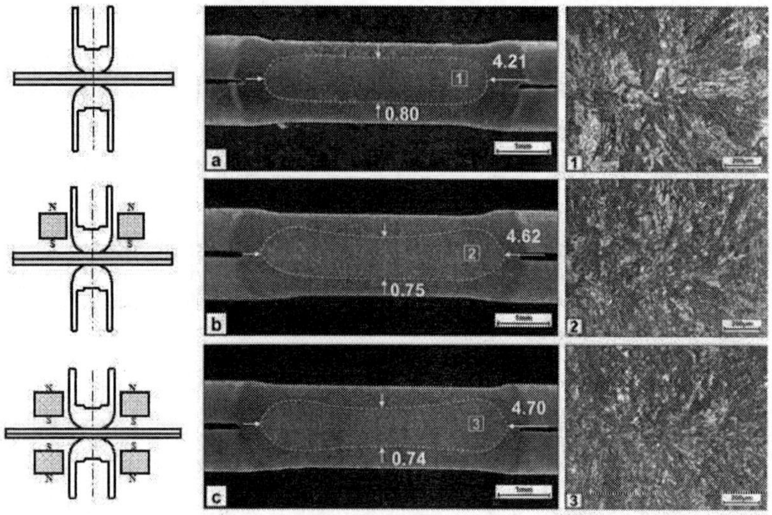

Fig. 2 Weld nugget geometry and microstructures. (a) Traditional weld, (b) MA-RSW using a single magnet, and (c) MA-RSW using a pair of magnets. Process parameters: welding current, 6.0 kA; welding time, 150 ms.[9]

Figure 2 shows the microstructure of the welds made with 0.8 mm thick DP980 sheets.9 Compared with the traditional welds, the nugget diameters obtained under a single magnet and a pair of permanent magnets was increased by 9.7% and 11.6%, respectively. In addition, the two types of MA-RSW nuggets were dog-bone shaped, in which the edges were thicker than the middles [Figs. 2(b) and (c)], in contrast with the ellipsoidal nugget of the traditional weld [Fig. 2(a)]. The differences in nugget shape can be explained by the effect of the

Lorentz force on the flow mode of the molten metal. In the MA-RSW process, the molten metal will flow both radially in symmetric plane because of the induced magnetic field and circumferentially because of the external magnetic field. As a result of the centrifugal flow, more heat will be brought from the center of the nugget to its edge, and thus the thickness of the nugget is decreased in the middle portion but the width of the nugget is increased. The stirring effect near the edge of the faying surface is much stronger than that at the center of the nugget, so a large pressure gradient forms around the edge along the width direction. Because the molten metal is constrained by the surrounding unmelted solid metal, it moves upward and downward, and thus the nugget will thicken around the edge area and result in the dog-bone shape shown in Fig. 2(c). However, the nugget under a single permanent magnet is not symmetrical along the faying surface because of the weak magnetic field on the bottom sheet. Differences in MA modes not only change the nugget shape, but also affect the microstructures within the nugget. As shown in Fig. 2, the macro-solidification directions in the two MA-RSW nuggets are both less obvious compared with that of the traditional weld. Moreover, under a pair of permanent magnets, the oriented growth of the dendrites is less directional, and the faying surface is less visible.

Effect on Weld Performance

Figure 3 showed the tensile-shear and cross-tension test results on the traditional RSW and on MA-RSW welds. The two types of welds were both button-pullout fracture. The MA-RSW welds performed better than the traditional welds in terms of failure load and displacement at the failure. The tensile-shear force (TSF) of the MA-RSW weld was 17.7 kN, about 12.0% higher than that of the traditional welds. The corresponding total displacement at the failure point of the EMS-RSW welds was 3.93 mm, about 15.9% longer than that of the traditional welds. Results of the cross-tension testing were consistent with those of the tensile-shear testing, exhibiting the button-pullout fracture mode for both types of welds. The cross-tension force (CTF) and the corresponding total displacement at failure of the MA-RSW welds were, respectively, 27.4% higher and 8.7% longer than those of the traditional welds. The results suggest that the MA-RSW method could improve weld strength and enhance weld ductility, which could be explained exactly by the wider weld nugget and the refined grains in the MA-RSW process. Figure 4 showed the tensile-shear test fracture surfaces at the plastic shear zone of the fracture faces. For the conventional weld, as shown in Fig. 4(a), both small dimples and large quasicleavage fracture were found in the microfractograph. However, for the weld made with MA-RSW, as shown in Fig. 3(b), only dimples were found in the fracture. Moreover, the dimples were significantly stretched, indicating large plastic deformation prior to fracture.

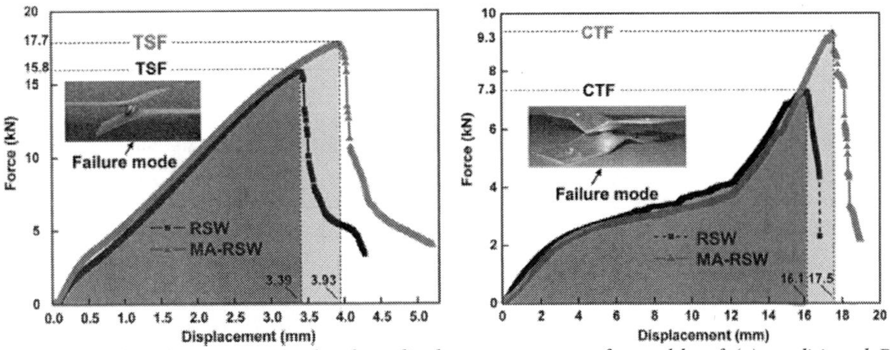

Fig. 3 Typical tensile-shear/cross-tension load vs displacement curves for welds of (a) traditional RSW and (b) MA-RSW of 1.25 mm DP780 at 7.8 kA welding current.[7]

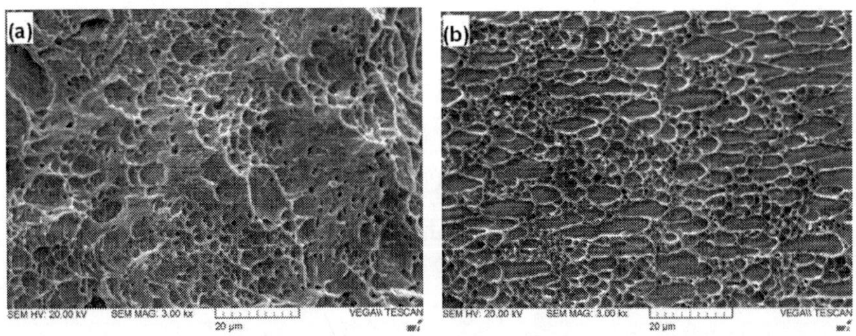

Fig. 4 Morphology of plastic shear zone for welds of (a) traditional RSW and (b) MA-RSW of 1.8mm thick DP780 at 8.0kA welding current.[6]

Figure 5 showed the fatigue life of both types of the welds tested using a SHIMADZU electro-hydraulic servo fatigue testing machine in tension-tension mode with a load ratio (R) of 0.1 and a frequency of 25 Hz. The overall trends for the two curves are similar, gradually extending with the decrease of the maximum load. However, the fatigue lives of the MA-RSW welds are always longer than those of the traditional welds under the same maximum load, especially in high cycle conditions under relatively low loads. When the maximum load was set to 2.0 kN, the endurance limit of the EMS-RSW welds was nearly 1.1×10^7, whereas it was only about 1.4×10^6 for the traditional welds. Obviously EMS notably improved the fatigue performance of RSW welds, especially in low-load (high-cycle) conditions. The improvement was attributed primarily to the better ductility of refined grains and the increased nugget diameter from the external magnetic field.

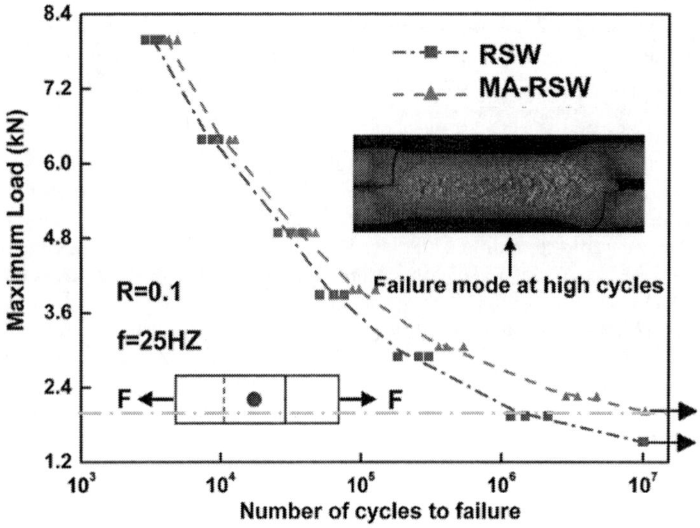

Fig. 5 Fatigue testing results for welds of traditional RSW and MA-RSW of 1.25 mm DP780 at 7.8 kA.[7]

Conclusions

The MA-RSW process was demonstrated capable of improving the resistance spot weldability of DP steels by increasing nugget size using the enhanced flow of the molten metal in the heating stage and by homogenizing and refining microstructures using the inertia flow of the molten metal in the cooling stage. Compared with the traditional welds, MA welds were generally dog-bone-shaped (i.e., the nugget edges were thicker than the middle), and exhibited higher tensile-shear strength and ductility, stronger energy absorption capacity and prolonged fatigue life.

Acknowledgments

The authors would like to acknowledge the supports of National Natural Science Foundation of China (Grant Nos. U1564204, 51322504, 51275300 and 51421092), and the Program of Introducing Talents of Discipline to Universities (Grant No. B06012).

References

[1] B. Yan, M. Gallagher, Material Science and Technology, 2006, 2, 97-101.
[2] G.K. Hicken and C.E. Jackson. Welding Journal, 1966, 45(8), 513s-518s.
[3] L. Wang, C. S. Wu and J. Q. Gao. Science and Technology of Welding and Joining, 2016, 21(2), 131-139.
[4] V. A. Popov: Weld. Int., 1993, 7(11), 905–7.
[5] Y. Watanabe, T. Takeda and H. Sato: ISIJ Int., 2006, 46(9), 1292-96.
[6] Y. B. Li, Q. Shen, Z. Q. Lin and S. J. H: Science and Technology of Welding and Joining, 2011, 16(5), 465-69.
[7] Q. Shen, Y. B. Li, Z. Q. Lin and G. L. Chen: ASME Transactions, Journal of Manufacturing Science and Engineering, 2011, 133(5), 051001-1-7.
[8] Q. Shen, Y. B. Li, Z. Q. Lin and G. L. Chen: IEEE Transactions on Magnetics, 2011, 47(10), 4116–19.
[9] Y. B. Li, Y. T. Li, Q. Shen and Z. Q. Lin: Welding Journal, 2013, 92(4) 124-32.
[10] Y. B. Li, D. L. Li, S. A. David, Y. C. Lim, and Z. Feng, Science and Technology of Welding and Joining, 2016, 21(7), 1-9

Improvement of Fatigue Strength of Resistance Spot Welded Joints in High Strength Steel Sheets by Shot Blast Processing

Hiroki Fujimoto, Hideki Ueda
Nippon Steel & Sumitomo Metal Corporation, Chiba, Japan
fujimoto.fh2.hiroki@jp.nssmc.com, ueda.896.hideki@jp.nssmc.com

Rintaro Ueji, Hidetoshi Fujii
Joining and Welding Research Institute, Osaka University, Osaka, Japan
ueji@jwri.osaka-u.ac.jp, fujii@jwri.osaka-u.ac.jp

Abstract
This paper discusses the effect of shot blast processing on the fatigue strength of resistance spot welded joints in 980 MPa steel sheets. The fatigue limit load of shot blasted spot welded joints increased to approximately twice that of non-blasted specimens. While high compressive residual stress was conferred to the outer surface of the shot blasted steel sheets, there was little variation in the residual stress on the side of the overlapped surface. Shot blasting on the outer sheet surface increased the initiation life of fatigue cracks that occurred on the overlapped surface and reduced the propagation speed of the cracks that grow from the overlapped surface toward the outer surface. The initiation and propagation of fatigue cracks were affected in the region where compressive residual stress by shot blasting was not conferred.

Introduction
Resistance spot welding is a major welding method used for assembling automobiles. To achieve lighter automotive bodies and higher safety during collisions, the application of high strength steel sheets in automotive bodies has increased in recent years. However, the fatigue strength of spot welded joints in steel sheets has been reported to be generally identical irrespective of the tensile strength of the steel sheets[1][2]. Also, the fatigue strength of the spot welded joints deteriorates when thinner steel sheets are used[3]. To enable the use of thinner high strength steel sheets for parts that require high fatigue resistance, the fatigue strength of spot welded joints needs to be improved.

Shot blasting is used to improve the fatigue strength of machine components such as gears and shafts. In this case, fatigue strength is considered to be improved by the direct effect of compressive residual stress by shot blasting to lower the average stress. Recently, we used shot blasting to develop a new method that can improve the fatigue strength of spot welded joints even without directly affecting the initiation points of fatigue cracks. This method can contribute to improving the fatigue properties of spot welded joints, especially for the parts used in the chassis of automobiles where high fatigue properties are required.

This study demonstrates that the fatigue life of spot welded joints in high strength steel sheets can be improved by shot blast processing. In addition, the phenomenon of improved fatigue properties in cases where fatigue cracks initiate and propagate from an overlapped surface was investigated.

Experimental Method
The steel sheets used in this study were cold rolled steel sheets with a tensile grade strength of 980 MPa and a thickness of 1.2 mm. The shape and dimensions of the fatigue specimens are shown in Fig. 1. Figure 2 shows the shot blasting method. The shot blasting was performed using a pneumatic shot blasting machine that blasted steel beads of 0.3 mm diameter on both sides of the spot welded specimens at an air pressure of 0.35 MPa and a blasting distance of approximately 150 mm. Shot blasted spot welded specimens and non-shot blasted specimens were prepared, and their fatigue strengths were evaluated in accordance with JIS Z3138 (Fatigue Testing Method of Spot Welded Joints). Testing was conducted under load control at a load ratio of $R = 0.1$. When the displacement of the fatigue specimens reached a value of 1 mm from the initial displacement, the specimen was identified as a failure and the number of cycles to failure was recorded.

In addition, the residual stress distribution around the spot welded joints was measured. To measure the residual stress of the overlapped surface at which the fatigue cracks initiate, this study adopted the method shown in Fig. 3. The test specimen for measuring the residual stress was made up of three parts. One part had a pre-drilled hole, and another part is a small piece of steel sheet that fits in the hole. Before spot welding was performed, the steel sheet with a plugged hole was overlapped on another steel sheet for measuring the residual stress. For specimens without shot blasting, the residual stress on the overlapped surface and outer surface of the steel sheets was measured after removing the steel sheet used to plug the hole. For the specimens with shot blasting, the steel sheet used to plug the hole was removed after shot blasting. Then, the residual stress in the radial direction was measured by X-ray diffraction. In addition, the residual stress in the depth direction was investigated by polishing the surface with electrolyte polishing at a position 10 mm away from the nugget center.

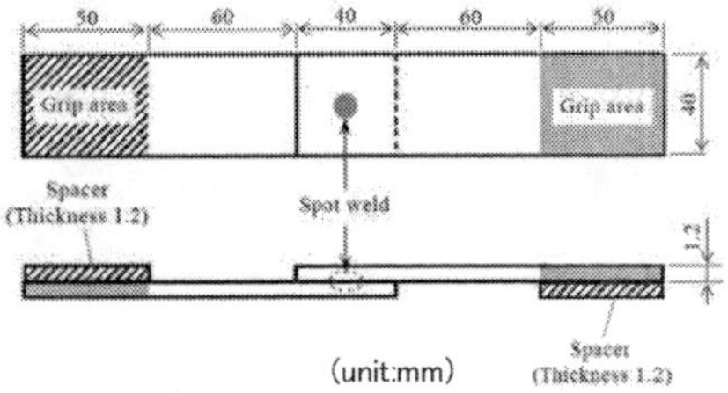

Fig. 1 Shape and dimensions of the fatigue test specimen

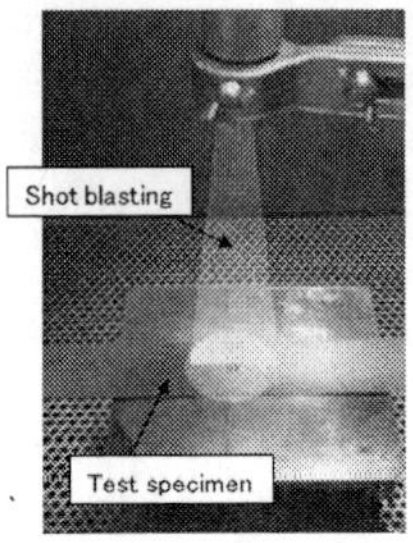

Fig. 2 Shot blasting method

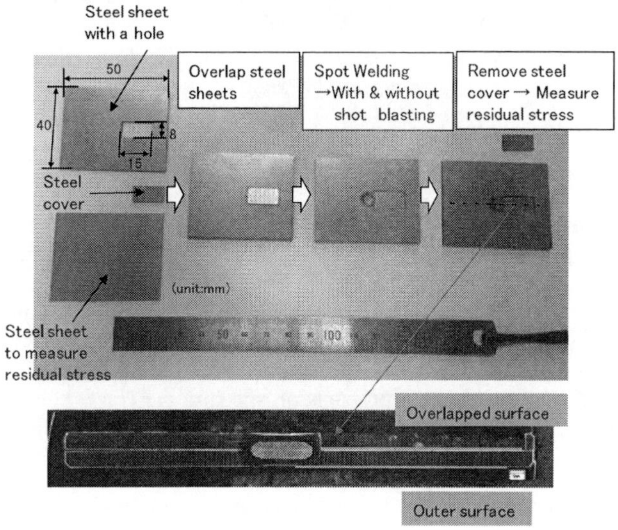

Fig. 3 Procedure for preparing the specimen used to measure residual stress by X-ray

Results and Discussion

Effects of shot blasting on the S-N curves of the spot welded joints

The fatigue test results of the spot welded joints with and without shot blasting are shown in Fig. 4. The shot blasted spot welded joints had longer life cycles to failure, and the fatigue limit load was approximately twice that of the non-shot blasted joints. This result clearly indicates that the fatigue properties of the spot welds were improved by the shot blast processing.

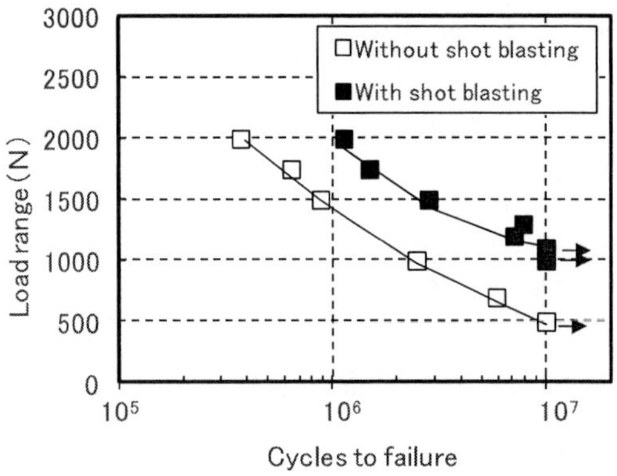

Fig. 4 Fatigue test results of spot welds with and without shot blasting

Effects of shot blasting on the residual stress of the spot welded joints

Figure 5 (a) shows the residual stress distribution on the outer surface of the spot welded joints. The residual stress in the spot welded joints without shot blasting is high tensile stress at the nugget center that decreases with increasing distance from the center. The tensile stress condition was observed approximately 3 mm away from the nugget center, and this distance coincides well with the distance where the fatigue cracks propagated. In the spot welded joints with shot blasting, a compressive residual stress of around 500 MPa was identified at all of the measurement points. The residual stress distributions on the overlapped surface are shown in Fig. 5 (b). The shot blasting did not significantly affect the residual stress on the overlapped surface. The joints mainly showed tensile residual stress, and their orders were similar to each other at a position approximately 3 mm away from the nugget center. It should be noted that this area is in the vicinity of the initiation point of the fatigue cracks. This indicates that while a large compressive residual stress can be introduced on the outer surface of the steel sheets by the shot blasting, the residual stress of the overlapped surface, which is the initiation point of fatigue cracks, is not impacted by the shot blasting and remains at the same level compared to the specimens without shot blasting. Figure 6 shows the residual stress along the radius direction that was measured in the depth direction from the outer surface of the steel sheet 10 mm away from the nugget center. A tensile residual stress of approximately 50 MPa was identified in the joints without shot blasting. On the other hand, the joints with shot blasting had a compressive residual stress of approximately 500 MPa to a depth of approximately 0.1 mm from the surface, and the magnitude of the compressive stress decreased as the depth increased. When the measurement location reached 0.25 mm in depth, the residual stress had the same value as the specimens without shot blasting. It is generally considered that tensile residual stress to balance the compressive residual stress by shot blasting is sometimes generated, but this phenomenon was not clearly observed. This might be due to the slight deformation of the specimens, which results in relief of the tensile residual stress.

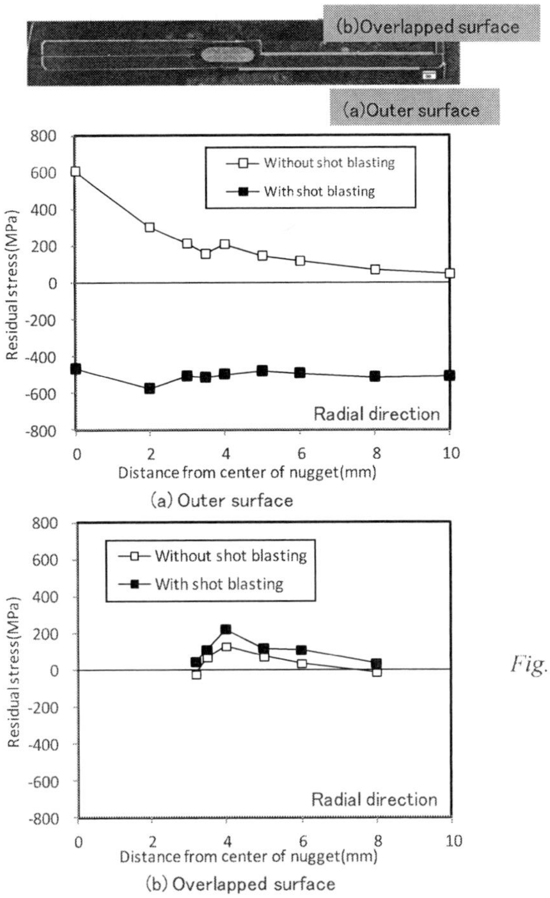

Fig. 5 Residual stress distributions around the spot welds

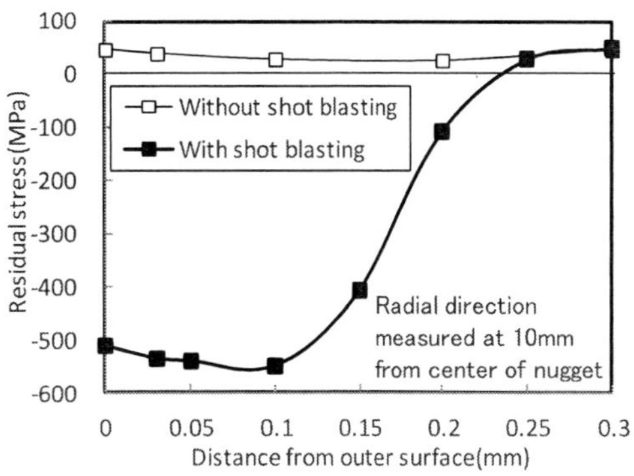

Fig. 6 Residual stress distributions in the thickness direction from outer surface

Effects of shot blasting on the initiation/propagation of fatigue cracks from the overlapped surface

To clarify the effect of shot blasting on the initiation/propagation behavior of fatigue cracks, the lengths of the fatigue cracks that occurred during the fatigue tests were measured. Figure 7 shows cross sections of a specimen loaded in $4x10^4$ cycles. Small fatigue cracks were identified in the plate thickness direction from the notch tip of the corona bond in the joints without shot blasting, and in contrast, no fatigue cracks were identified in the shot-blasted joints. The lengths of the fatigue cracks are shown in Fig. 8. Since fatigue cracks were observed in the two right and left positions of the nugget, the lengths of both were plotted. Figure 8 shows that the initiation of fatigue cracks was delayed in the joints with shot blasting compared to those without shot blasting. In addition, the logarithmic scale in the horizontal direction of this graph indicates that the joints with shot blasting had slower rates of crack propagation, as plotted on the right side, compared to those without shot blasting. Interestingly, both the crack initiation and propagation were delayed in the region where compressive residual stress is not conferred.

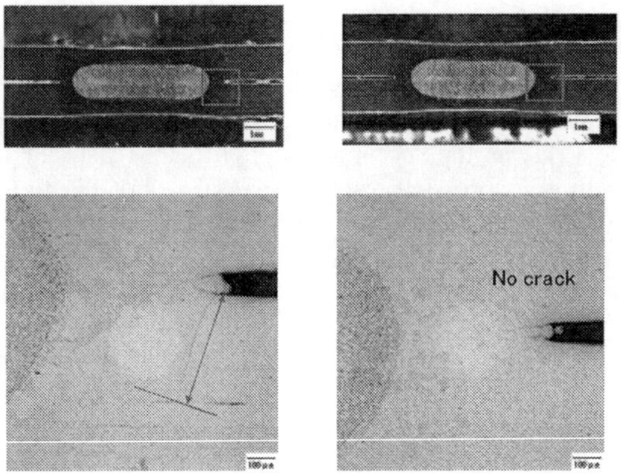

(a) Without shot blasting *(b) With shot blasting*

Fig. 7 Comparison of the fatigue crack behavior of spot welds with and without shot blasting (Load range: 1.75 kN, $4x10^4$ cycles*)*

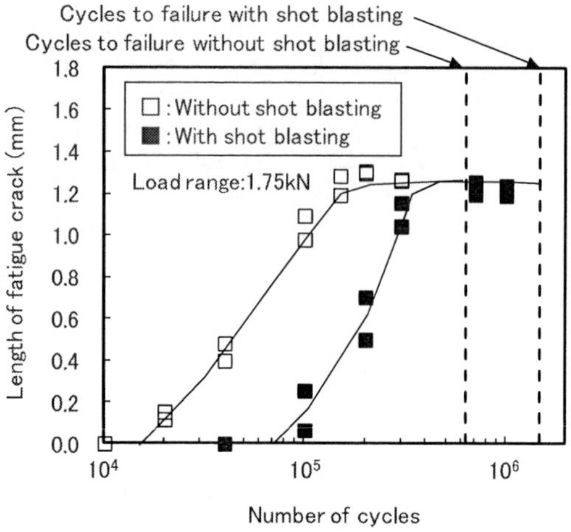

Fig. 8 Effect of shot blasting on the fatigue crack length of the spot welds

Conclusions

This study demonstrated that shot blast processing can improve the fatigue strength of spot welded joints using high strength steel sheets with a tensile grade strength of 980 MPa. The phenomenon by which the fatigue properties are improved from the standpoint of residual stress was also investigated. The obtained results are as follows:

1) Shot blast processing after spot welding improved the fatigue strength of the spot welded joints.
2) While high compressive residual stress was conferred on the outer surface of the shot blasted steel sheets, there was little variation in the residual stress on the side of the overlapped surface.
3) Shot blasting on the outer surface of the steel sheets delayed both the initiation of fatigue cracks, which occurred from the non-shot blasted overlapped face, and their propagation toward the outer surface.
4) The crack initiation and propagation were delayed in the region where compressive residual stress is not conferred by the shot blasting.

References

[1] R.Iyengar, M. Amaya, J. Bonnen, K. Citrin, H. T. Kang, S. Laxman, A. Khosrovaneh, N. Schillaci and H. S. Shih: Great Design in Steel , American Iron & Steel Institute, Michigan(2008)
[2] E. Nakayama, O. Kazuo, M. Miyahara, M. Yoshida, K. Fukui and H.Fujimoto: SAE Technical Paper 2003-01-2830(2003)
[3] M. Ono, M. Nagae, M. Kabasawa and J. Tanaka: Tetsu-to-Hagane 70(1984), S585. (in Japanese)

Strength of Resistance Spot Welded Lap Joints under Combined Loads due to New Hardened Zone using Laser Heating

Takanori KITAMURA, Tetsuya AKIYAMA

Kyushu Institute of Technology, Fukuoka, Japan
kitamura.takanori599@mail.kyutech.jp , +81-93-884-3383

Abstract

We have studied the strengthening of spot-welded lap joints due to new hardened zone (HAZ). The factors needed for strengthening were determined by a previously proposed equation to predict the strength. Using these factors, the strengthening of spot-welded lap joints was examined. First, in order to increase the size of the hardened zone, the edge of the nugget was heated by laser. Next, tensile testing under combined loads was conducted by using spot-welded lap joints that were heated by laser. It was found that the strength of resistance spot welded joint with laser-heated part was increased under combined loads. And, it was shown by experiments that there was a difference in the effect of the laser heating on strengthening with the different load angles.

Introduction

About the strength of resistance spot welded joints, the tensile shear strength and cross tension strength have been studied. Recently the tensile shear strength and cross tension strength have been studied to investigate the strength of resistance spot welded joints, , it has been proposed that several equations can predict the strength [1-3]. Based on these strength prediction equations, the authors focused on fracture diameter D (mm) which is commonly included in the proposed prediction equations [4,5]. And, the authors strengthen the joint by hardening the specimens in the vicinity of nugget edge by laser heating. It is clear that the combined loads of the tensile shear load and cross tension load were already applied in the real environment. Lee [6] has been studied spot welded joints strength under combined loads, however there are very few studies on strengthening of the joint under combined loads. This study was carried out to examine the strength under combined loads of the resistance spot welding joints which were heated in the vicinity of nugget edge by laser.

A method to strengthen the resistance spot welding joints using laser heating

A schematic diagram of a tensile shear test is shown in Fig.1 (a). A schematic diagram of a cross tension test is shown in Fig.1 (b). Maximum load obtained in the tensile shear test is the tensile shear strength (*TSS*). On the other hand, maximum load obtained in the cross tension test is the cross tension strength (*CTS*).

Prediction equation of the joint strength has been proposed by many researchers [1-3]. In the case of the fracture occurred outside of the nugget, the tensile shear strength (*TSS*) prediction equation is shown in eq. (1).

$$TSS = h \times \sigma_{BM}^{B} \times (1 + 0.0059El) \times 1.95 \times D \cdots (1)$$

In addition, the cross tension strength (*CTS*) prediction equation is shown in eq. (2).

$$CTS = h \cdot \pi \cdot D \cdot (\beta \quad \sigma_{BM}^{B}) \cdot \sin\theta \cdots\cdots (2)$$

where h (mm): Plate thickness, σ_{BM}^{B} (N/mm^2): Tensile strength of base metal, El (%): Elongation of base metal, D (mm) : Fracture diameter, θ (deg.) : Deformation angle of the fracture part, β : Constant value

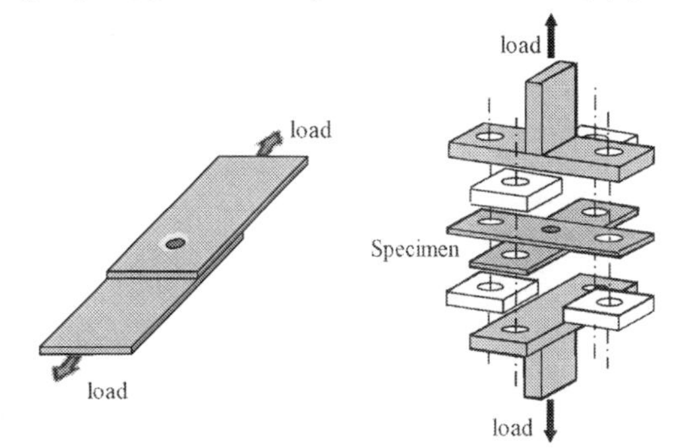

(a) Tensile shear strength test (JIS Z 3136) (b) Cross tension strength test (JIS Z 3137)

Fig. 1 Tensile strength test for spot welded lap joints.

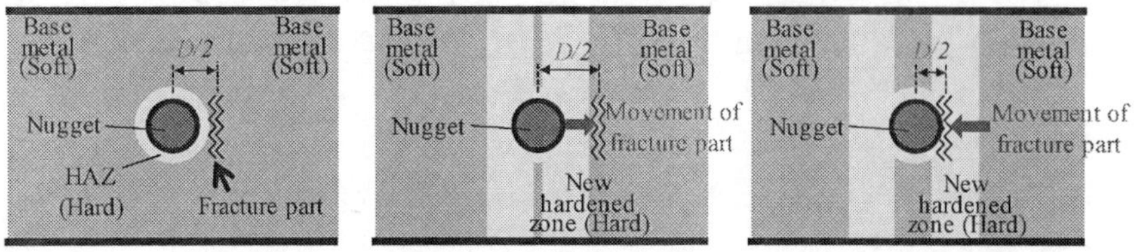

(a) The case of spot welding only. (b) The case of hardening by laser heating near the nugget edge. (c) The case of hardening by laser heating far from the nugget edge.

Fig. 2 Relationship between fracture part and Vickers hardness.

The authors proposed (provided)a method to increase the tensile shear strength and cross tension strength based on strength of prediction equation (1), (2). The Proposed method is as follows.

The diagram shown in Fig.2 explains the transition fracture location in the case of generating the hardened zone by laser heating. Fracture position of spot welding only is shown in Fig.2 (a). Fracture position of hardening by laser heating near the nugget edge is shown in Fig.2 (b). Fracture position of hardening by laser heating far from the nugget edge is shown in Fig.2 (c). The joint strength and the relationship of the distance from the nugget center to fracture position are shown in Fig.3.

In Fig.3, ○, △marks of the plots represent the relationship between $D/2$ and maximum load P_{max} of conventional spot welded specimen and of laser heating specimen, respectively. As shown in Fig.2 (b), after generating a hardened zone by laser heating, the distance of fracture position and the nugget edge is significantly extended compared with Fig.2 (a). Then, since the fracture position moves away along the thick broken line in Fig.3, the joint strength becomes higher. As the fracture position moves further, the strength becomes stronger than that of the nugget edge. And the fracture position is shifted to the nugget edge as shown in Fig.2 (c). Therefore, as the thick solid line arrow in Fig.3 shows, the fracture position is shifted to the nugget edge. In other words, if a hardened zone near the nugget edge is generated, it is possible to improve the joint strength.

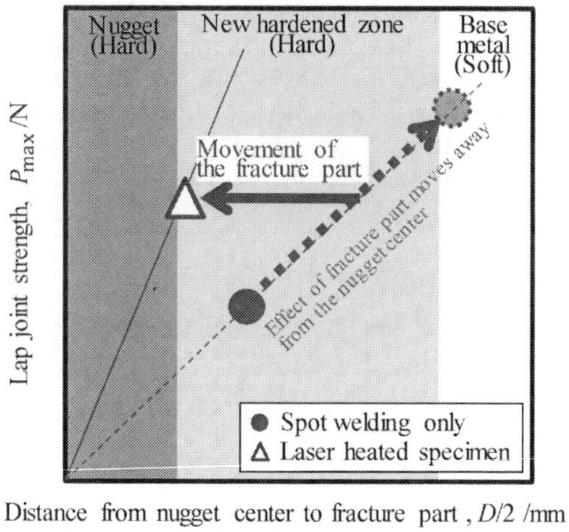

Fig. 3 Concept behind the strengthening of spot welded lap joint specimens.

Static strength test of resistance spot welded joints under combined loads

Test material and laser heating method

Test material were hot-rolled steel sheet SPHC of thickness h = 1.55mm. The chemical composition was shown in Table 1. A schematic diagram of a laser heating to generate the hardened zone was shown in Fig.4. The position deviated from the copper block edge 5mm was heated by laser. The Nd-YAG was used as the heat source. Laser heating conditions were the laser power P = 1500W, defocus distance d_f = 24mm, travelling speed of torch v = 500mm / min.

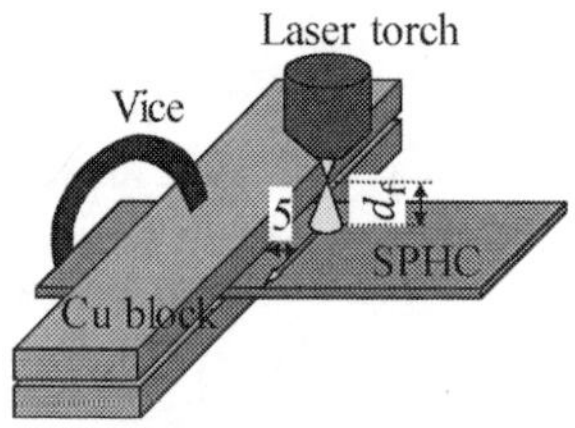

Fig. 4 Specimen and jig for generating a hardened zone by laser heating.

Table 1 Chemical composition for SPHC

Unit:mass%

	h	C	Si	Mn	P	S	Cu	Ni	Cr	Mo	V	Al
SPHC	1.55	0.03	0.0	0.215	0.011	0.006	0.03	0.03	0.04	0.01	0.002	0.049

Tensile test method under combined loads

A schematic diagram of the test specimen was shown in Fig.5. The size of the steel plates was the width 50 mm, length 150 mm. After laser heating, the two steel plates were bended and then welded by resistance spot welder. Welding conditions were the current 11600A, the welding time 14cycles (welding power source frequency 60Hz) and the pressure 3.6kN. DR-shaped copper electrode (electrode tip diameter 6mm, electrode tip curvature radius 40mm) was used. Tensile testing machine used in this experiment was a static tensile testing machine AG-20kNNIST (manufactured by Shimadzu Corporation). The tensile speed was 0.5mm / min. A test jig which was used under a combined load is shown in Fig.6 (a). A photograph of the test piece is shown in Fig.6 (b). A schematic diagram of attaching the test piece to the jig is shown in Fig.6 (c). As shown in Fig.6(c), load angel was defined.

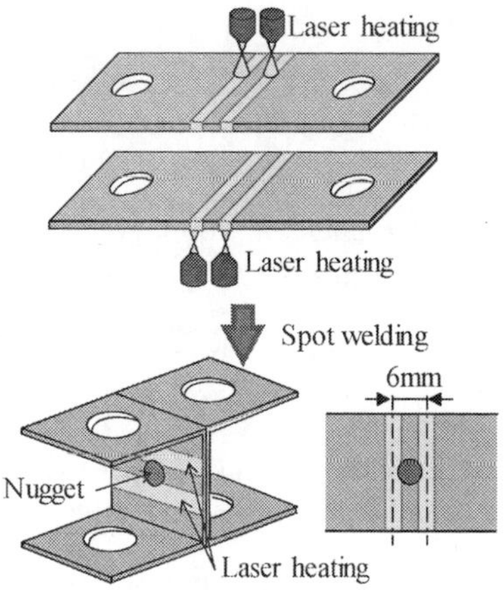

Fig. 5 Specimen for lap joint strength test with new hardened zone by laser heating.

Experimental results and discussion

Experimental results

The photographs of the test piece, the jig in the tensile testing machine, and the enlarged photographs of joint part just before the fracture are shown in Fig.7. The photographs of the load angle of 30 degree and 82.5 degree are shown in Fig.7 (a), (b), respectively.

Deformation of the test piece of 30 degree load angle was larger than that of 82.5 degree load angle. Relation between tensile load – displacement between chucks is shown in Fig.8.

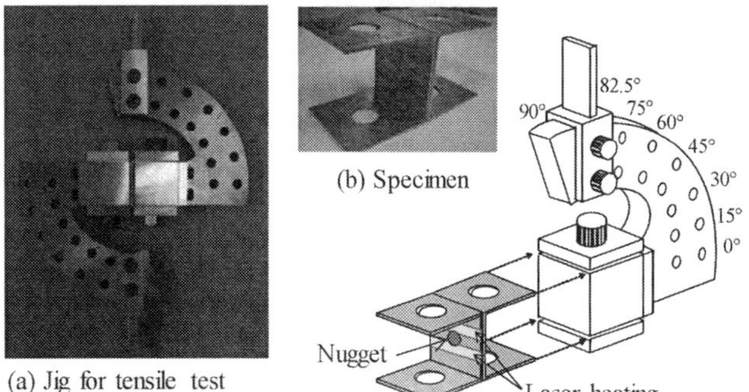

(a) Jig for tensile test under mixed load (b) Specimen (c) Scheme of jig and specimen

Fig. 6 Jig and specimen for tensile test under combined loads.

The thin line shows the experimental results of the test pieces in only spot welding experiment, and the thick line shows the experimental results of the laser heating test piece. The solid line shows the case of the 30 degree load angle, and the broken line shows the case of the 82.5 degree load angle. And the peak value of the load is the joint strength. As shown in Fig.8, in the case 82.5 degree load angle, the joint strength of the test specimens in only spot welding experiment is approximately equal to the joint strength of the laser-heated specimens. However, in the case of the 30 degree load angle, the joint strength of the laser-heated specimen is stronger than the specimen in only spot welding experiment. The relationship of joint strength and load angle is shown in Fig.9. ● and ○ marks show the experimental results of the test piece that fractured far from the nugget edge. ▲ and △ marks show the experimental results of the test piece that fractured at the nugget edge. □ shows the experimental results of test piece which was fractured partly like as the plug shape. In addition, the solid marks are the experimental results of the specimens in spot welding only experiment, and the open marks are the experimental results of laser-heated specimens. As the load angle increases from 0 degree to 90 degree, the joint strength becomes lower. And the same tendency was also shown in the experiment studies of Lee et al [6]. As shown in Fig.9, at the small load angle (0 ~ 30 degree), joint strength is improved by generating a hardened zone using laser heating. On the other hand, at the large load angle (45 ~ 90 degree), joint strength is not improved even if the specimens were heated by laser.

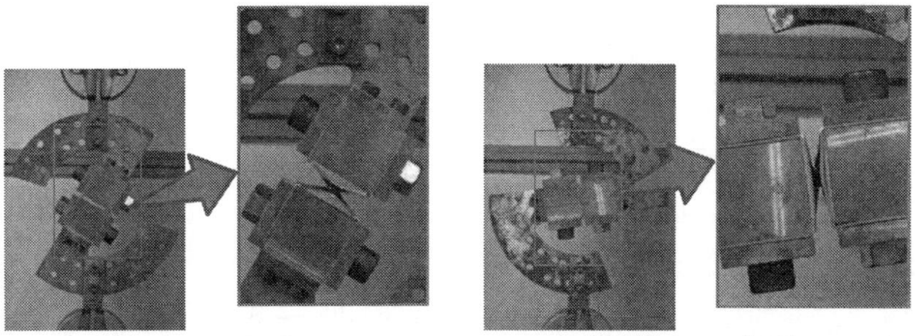

(a) 30 degree direction (b) 82.5 degree direction

Fig. 7 Tensile test under combined loads.

Discussion

For the case of the 30 degree load angle and the 82.5 degree load angle, the experimental results of the joint strength are shown in Fig.10. As shown in Fig.10, at a small load angle (30 degree), the joint strength of the test piece after laser heating is improved compared with the joint strength of the test specimens in only spot welding experiment. If the fracture point is at the base metal (137Hv), the strength is increased along the dashed line by expansion of the hardened zone. However, the joint strength reaches approximately P_{max}=10000N, and exceed the strength of the nugget edge. Then the fracture position is shifted to the nugget edge. As a result, the joint strength was improved. On the other hand, as shown in Fig.10, in a large load angle (82.5 degree), the joint strength of specimens in spot welding only experiment and the strength at nugget edge are almost the same. So, the joint strength is not improved even if the specimens were heated by laser.

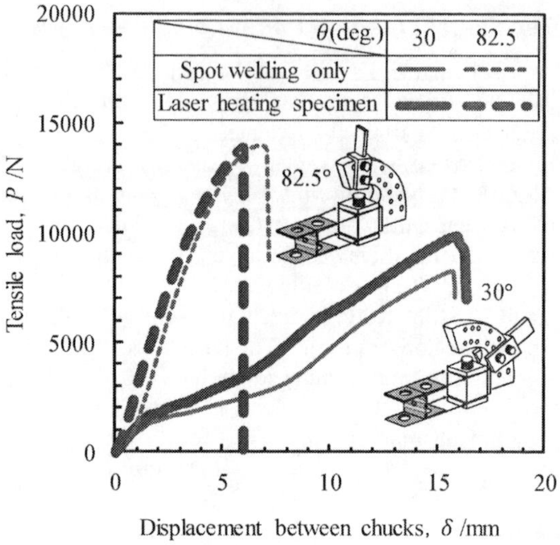

Fig. 8 Tensile load – displacement between chucks diagram.

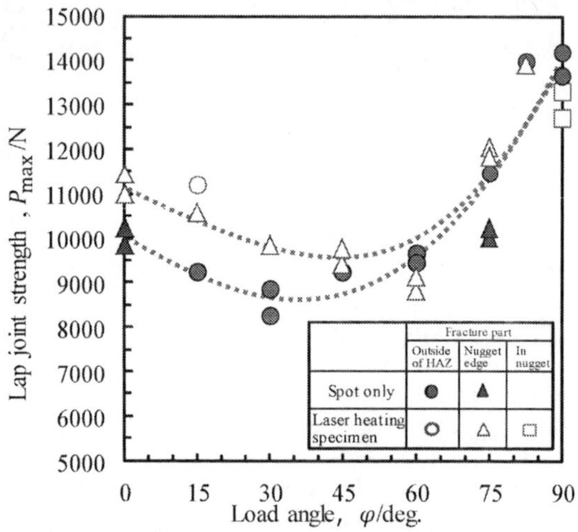

Fig. 9 Relations of lap joint strength and load angle.

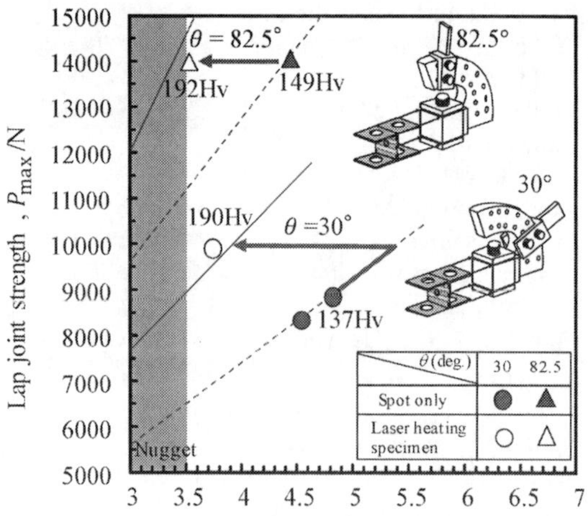

Fig. 10 Relation between lap joint strength and fracture radius D/2.

Conclusions

1. By hardening in the vicinity of nugget edge using laser heating, resistance spot welded joints under a complex load were enhanced.

2. It showed experimentally that there is a difference in the effectiveness of the laser heating on the degree of strengthening by the load angle.

References

[1] M.Kabasawa, Y.Funakawa, K.Ogawa and M.Tamura: Estimation of Tensile Shear Strength of Spot Welded Joint of Steel Sheets, Quarterly Journal of the JWS, 14,4(1996), 754-761. (in Japanese)

[2] Y.Funakawa, M.Kabasawa : Estimation of Cross Tension Strength of Spot Welded Joint, Preprints of the National Meeting of JWS, 54 (1994), 256-257. (in Japanese)

[3] Y.Funakawa, M.Kabasawa : Estimation of Cross Tension Strength of Spot Welded Joint, Preprints of the National Meeting of JWS, 54 (1994), 258-259. (in Japanese)

[4] T.Kitamura, T.Akiyama, Y.Masuda : Strengthening of Spot Welded Lap Joints by New Hardened Zone, Quarterly Journal of the JWS, 31,2(2013), 89-95. (in Japanese)

[5] T.Kitamura, T.Akiyama, M.Asada : Enhancement of Cross-Tension Strength of Resistance Spot-Welded Lap Joints by Laser Heating, J. Japan Inst. Met. Mater, 79, 10 (2015), 518-522 (in Japanese)

[6] Y. Lee, T. Wehner, M. Lu, T. Morrissett, E.Pakalnins : Ultimate Strength of Resistance Spot Welds Subjected to Combined Tension and Shear, Journal of Testing and Evaluation, JTEVA, 26, 3 (1988), 213-219.

Techniques for improving spot weld joint strength of high tensile strength steel sheets

Chisato Wakabayashi, Yasunobu Miyazaki, Seiji Furusako and Masanori Yasuyama
Nippon Steel & Sumitomo Metal Corporation, Futtsu city, Chiba, Japan
Wakabayashi.e93.chisato@jp.nssmc.com, +81-70-3514-8073

Abstract
Spot weld is widely used in automotive assembly. The joint strength with high tensile strength steel sheets has a tendency to decrease. This is due to the high tensile strength steel sheets being composed of high weight fraction carbon and other additional elements. Therefore, the spot weld including Heat Affected Zone (HAZ) loses its toughness. This study proposes techniques to improve the spot weld joint strength with the post-heating. There are two effects based on the structure of weld parts. One is the reduction of segregation in the nugget. This reduction of segregation increases the toughness of the nugget. As a result, fracture morphologies improve and the joint strengths increase. The other effect is expected to strengthen the toughness of the HAZ. Compared to the conventional spot welding and other methods, the amount of carbide precipitation increases with the longer post heating. Consequently, auto-tempering is promoted, resulting in stronger joint strength.

Introduction
To improve the fuel consumption of automobiles, high tensile strength steel sheets are extensively used. They offer the advantages of improved crashworthiness without increasing the body weight, thereby meeting the strong pressure requirements [1, 2]. Furthermore, high tensile strength steel sheets can achieve a reduction in body weight. One of the critical problems associated with the usage of high strength steels in automotive body components is the resistance spot weldability. The tensile strength of spot weld in the peel direction, or the cross tension strength decreases as the base metal strength increases over 780MPa[3]. Therefore, to increase the usability of high strength steel, it is important to circumvent the above-mentioned limitations.
The joint strength of the welds obtained in this process is usually evaluated by performing tensile shear and cross tension tests. Joint strength increases with the increase in the tensile strength of sheets. However, the cross tension strength (CTS) does not increase or decrease in many cases, especially for strength grades higher than 780 MPa or 980 MPa. Thus far, several methods have been proposed to overcome this problem. One of the most feasible approaches is to enlarge the nugget diameter. The joint strength increases with the increase in the nugget diameter [4]. This strengthening could be attributed to the fact that the increase in nugget diameter can reduce the stress concentration on the nugget edge, thereby delaying the crack generation and propagation. On the other hand, tempering after nugget formation by controlling the current pattern has long been used as another solution to this problem [5]. Typically, spot welded parts of high tensile strength sheets are normally brittle martensite, which is particularly remarkable in high-carbon steel. In order to obtain tougher welded parts, the welded part is cooled below the martensite start temperature and re-heated again for tempering during the spot welding process. These two processes pose great difficulty from the practical usage viewpoint. Of the two strategies, the approach of enlarging the nugget diameter requires a larger welded area and larger weld flange. This, in turn, would contradict the reduction in car body weight. On the other hand, the tempering method takes much longer, limiting its practical feasibility for use in car body assembly. In recent years, a much more pragmatic method has been proposed in the literature [6]. This method involves short time cooling via current-off. Although solidifying segregation is reduced during spot welding, the continuous use of lower current for a shorter time reduces the formation of segregation peaks. For this second current duration, around 0.05- 0.1sec is recommended in the literature. According to this process, the nugget edge becomes tougher. Moreover, another study has reported a different approach, wherein a pulsed current is applied after nugget formation [7]. These two methods do not demand a long time for the welding process. Therefore, it is considered to be a much more viable option for introduction to the car assembly line. Nevertheless, in the near future, the demand for much higher tensile strength steels will intensify. Hence, the importance of the techniques to improve weldability will increase.

Experimental
In this study, a high tensile strength steel sheet with a thickness of 1.2 mm was used. The tensile strength of base metal is 1180MPa in steel. This sheet is Fe-C-Mn alloy. The chemical compositions, Ceq (Ceq=C+1/6Mn+1/5(Cr+Mo+V)+1/15(Ni+Cu) (mass%)) and P are shown in Table 1.

Table 1: Chemical composition and Ceq of steel sheet

C[mass%]	Mn[mass%]	P[mass%]	Ceq
0.21	2.9	0.010	0.69

Welding was performed using a single-phase AC spot welding machine with a servo gun. Fig. 1 shows the schematic of the welding current pattern. In the typical process, post-heating time t_2 was varied from 0 to 1.4 s. The first step of this current history I_1 is to make the nugget. The targeted nugget size of $5\sqrt{t}$ (=5.5mm) was obtained by changing the I_1 value. As the second step, after a short cooling (current-off) period, the post-heating current I_2 was maintained at 95% of I_1. Hold time was fixed to 0.1s for all samples. Specimens were prepared under the above-mentioned welding conditions. And the Cross Tension Strength (CTS) was measured based on JIS Z 3137. In addition, the microstructure was observed by an Electron Probe Micro Analyzer (EPMA) and Scanning Electron Microscope (SEM).

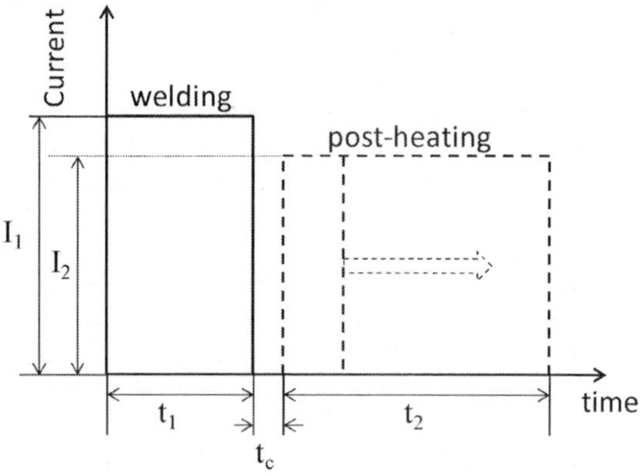

Fig. 1: Schematic image of current pattern

Results and Discussions

The CTS of the weld samples with varied post-heating times were examined and plotted in Fig. 2. For each welding condition, two or three samples were tested, and all of the data is shown in Fig. 2. Obviously post heating improves CTS. It increases gradually. For example, with post heating 1.0s, CTS increases 50%. Hereafter, to consider the effect of post heating, the microstructures of nuggets and heat affected zones (HAZ) of the weld were observed.

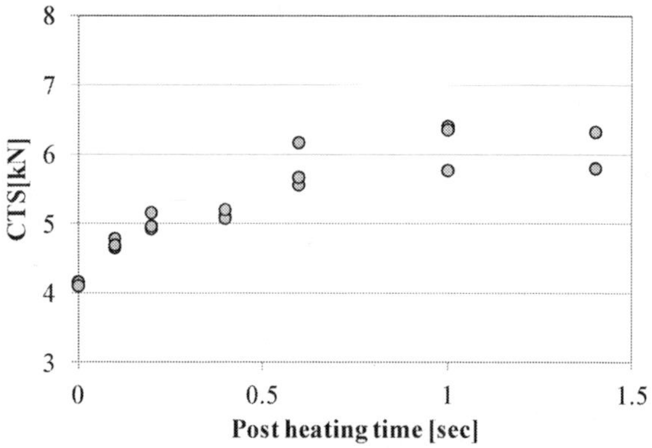

Fig. 2: Post heating time t_2 and CTS

Segregation Reduction

To clarify the effect of post heating on solidifying segregation reduction, EPMA mapping was conducted for a larger area and is shown in Fig.3. The post heating time t_2 is considered to be one of the major influences on the segregation reduction effect. Therefore, this report specifically focused on post heating duration. In previous reports [6, 7], short post-heating (0.1s) and the pulse current after nugget formation have an effect on segregation reduction. This effect was also observed in this experiment. Furthermore, it was confirmed that the lessening effect is larger as the post heating time was longer.

In addition, the fracture morphologies of CTS tested samples are shown in Fig.3. Without post heating, samples fractured in the nugget (Partlial plug fracture). In contrast, with post heating 0.1s, samples fractured outside the nugget (Plug fracture). Therefore, 0.1s post heating is sufficient to prevent cracking in the nugget. Longer post heating samples after testing CTS were plug fracture. Thus the post heating improves CTS.

Fig.3: EPMA map of phosphorous

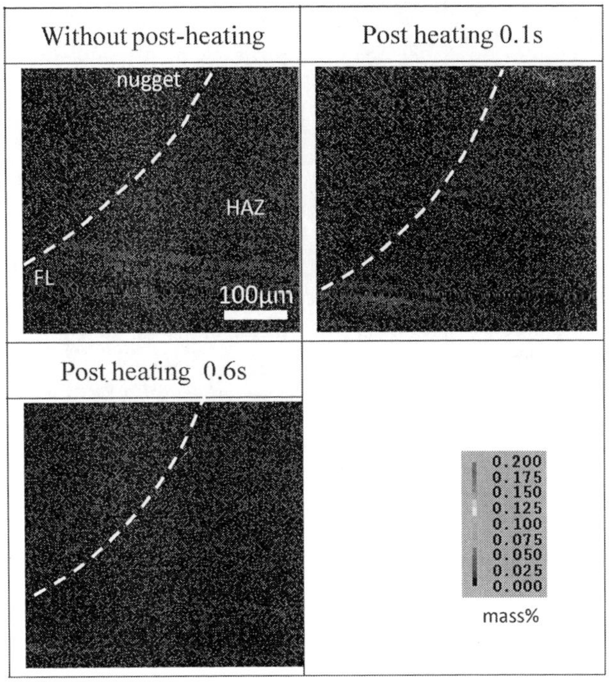

Fig.4 : Cross section of post CTS measurement

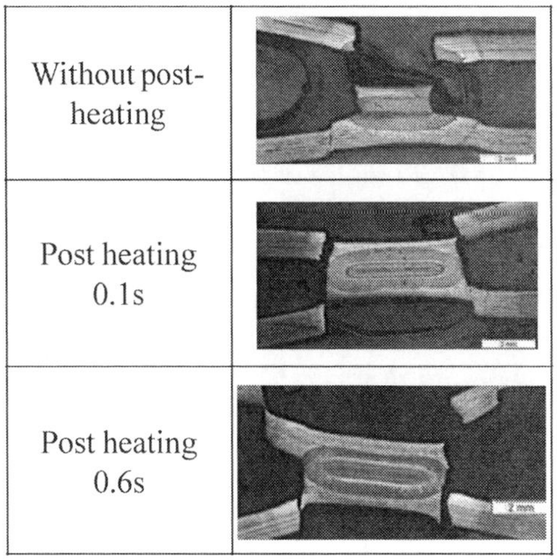

Fig. 5 : SEM images of HAZ

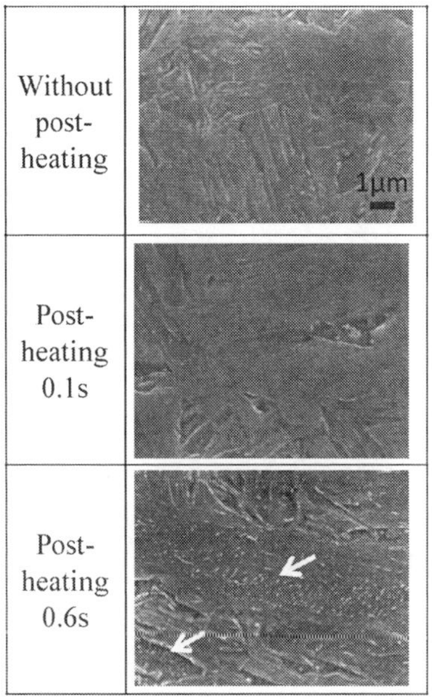

The Effect of Auto-tempering of HAZ

As described above, segregation reduction inside of the nugget can avoid the crack propagation into the nugget and the plug fracture can be achieved. Therefore, the difference with or without post heating can be understood. Nevertheless, the duration of the post heating effect on CTS is still unclear. According to Fig. 2 longer post heating gives higher CTS. Both samples after measurement of CTS were plug fracture, but the crack pass of HAZ seems to differ. The short post heating sample's cracking path is near the fusion line compared to the longer one. Hence, the HAZ region was observed by SEM. The observed location was 100μm from the nugget edge, 200μm from the interface of the sheet along its vertical direction. In Fig.5 SEM images are shown, respectively. Inside the blocks indicated by the white arrows several precipitations were confirmed. The precipitations are seen only in the 0.6sec post heating sample. This precipitation is carbide that has precipitated during cooling, so-called auto-tempering. Generally tempered martensite is tougher than martensite without tempering [8]. This microstructure difference is considered to be the cause of the difference in the behavior of CTS that is improved by prolonging post heating duration.

Conclusions

This report described post heating merits on the spot weld of high tensile strength steel sheets. The post heating increases the joint strength of the spot weld. The time it took for these phenomena to occur was shorter than 1.0s. Therefore, these techniques seem to be used practically.

The post heating effects were divided into two. One is solidifying segregation reduction, the other is accelerating the tempering of HAZ. The first phenomena segregation reduction is achieved with short post heating, for example 0.1s. In this report, phosphorous concentration change was described. The effect is more superior with longer post heating. The other is the improvement of the HAZ toughness. According to the prolongation of post heat, the amount of precipitation of carbide increases at HAZ.

References

[1] M. Takahashi: "Development of High Strength Steels for Automobiles," Nippon Steel Technical Report, No.88, July 2003

[2] A. Satoh: Koukyodokouhan no mondaiten to sonotaioh, Journal of the JSTP, Vol. 46(2005), No. 534, pp. 548-551 (in Japanese).

[3] H. Oikawa, G. Murayama, T. Sakiyama, Y. Takabayashi and T. Ishikawa, "Resistance Spot Weldability of High Strength Steel (HSS) Sheets for Automobile," Nippon Steel Technical Report, No.385, 2006, 36-41

[4] M. Pouranvari, H. R. Asgari, S. M. Mosavizadch, P. H. Marashi, M. Goodarzi, Science and Technology of Welding and Joining, Vol.12, No. 3, 2007, pp. 217 – 225

[5] WES 7301-1986, p11

[6] F. Watanabe, S. Furusako, H. Hamatani, Y. Miyazaki and T. Nose, Mathematical Modelling of Weld Phenomena, Vol.**10**. 2013, 653-667

[7] C. Sawanishi, T. Ogura, K. Taniguchi, R. Ikeda, K. Oi, K. Yasuda, and A. Hirose, Science and Technology of Welding and Joining 2014, Vol. **19**, pp. 52-59

[8] G. Krauss, "Deformation and Fracture in Martensitic Carbon Steels Tempered at Low Temperatures," Metallurgical and Mat. Trans. B, Vol.32B, 2001, pp. 205-221

Effect of knurl pattern of sonotrode tip on stress distribution and deformation behavior in Al-Cu ultrasonic welding

Yansong Zhang, Kunkun Chen

Shanghai Jiao Tong University , P.R.China

Ultrasonic welding (USW) is a rapid joining process during which high frequency ultrasonic energy is used to produce a solid-state bond between two pieces of metals. Specimens, which was placed between sonotrode and anvil, would get into intimate contact under the constant clamping force. Bond comes into formation at the faying interfaces of specimens. In USW process, stress distribution and plastic deformation would be mainly affected by the knurl pattern of sonotrode tip that plays significant role for the joint formation process It is important to determine the optimized knurl pattern of sonotrode tip for USW process. In this study, Finite element modelling (FEM), which incorporated of the coupled thermal-mechanical fields and high-strain-rate deformation of metallic materials, has been developed to reveal the influence of sonotrode geometry in Al-Cu USW process. Contact pressure, stress distribution and deformation behavior along defined route at contact interface were modeled. The optimized knurl pattern of sonotrode tip was determined by FEM and experimental verification. The results showed that the serrated knurl pattern of sonotrode tip would greatly affect the in-process variables (including stress, strain and displacement) at the contact surfaces of specimens. The localized concentration of plastic strain caused by the specific design of serrated tip knurl pattern, turned out to be the major cause for joint formation. At the initial period of USW process, the average in-process variables at the contact surface grow quickly. Then the growth rate decreases and finally nearly comes into a saturation status. Bond mainly occurs in the initial period. As for the later period, the primary function is only to keep plastic flow of materials.

Weld formation in ultrasonic welding of Al alloy to Cu

Hiromichi T. Fujii, Hiroki Endo, Yutaka S. Sato, Hiroyuki Kokawa
Tohoku University, Sendai, Japan
fujii@material.tohoku.ac.jp, +81-22-795-7353

Abstract
The microstructures of ultrasonic dissimilar weld between Al alloy and Cu were systematically examined to understand the underlying weld formation. Dissimilar welds between Al alloy and Cu were prepared by an ultrasonic spot welding technique for the welding time ranging from 0.1 s to 0.6 s. The ultrasonic welds between anodized Al alloy and Cu were also prepared in order to make it easier to observe the oxide layer behavior at the weld interface. The microstructure around the weld interface consisted of fine and equiaxed grains through recrystallization in the Al alloy during USW. The metallurgical weld between Al and Cu developed with high-strain-rate shear deformation and deformation heating around the weld interface.

Introduction
Ultrasonic metal welding technique has shown great promise for the application to electronic industry. This is because dissimilar welds between Al alloy and Cu can be achieved by a lower energy input, a shorter welding time and thinner wokpieces than other conventional welding techniques. The weld formation in ultrasonic welding (USW) is achieved under normal clamping force with ultrasonic vibration. Ultrasonic vibrations typically of 20 – 100 kHz frequency are locally applied parallel to the workpieces through an ultrasonic horn. Since it is a solid state process, some metallurgical phenomena, such as plastic deformation, deformation heating, plastic flow and dispersion of surface oxide layer, are considered to be key factors in sound weld formation. In USW, the normal force and ultrasonic energy will introduce local strains that disperse oxides, generate local heating through deformation and friction, and creates the necessary nascent surfaces. However, there is little academic understanding on these phenomena in USW of Al alloy to Cu despite its engineering significance.

Several studies have been performed thus far to characterize ultrasonic welds between Al alloy and Cu [1-5]. Zhao *et al.* observed the voids, swirl like structure and intermetallic compound layers around the weld interface in the USW of Al alloy 6061 to pure Cu [2]. They concluded the swirls lead to a mechanical interlocking between the materials and increase the weld strength. Balasundaram *et al.* clarified the effects of zinc interlayer on ultrasonic welding of Al alloy 5754 to pure Cu [3]. Welds with a Zn interlayer formed a eutectic structure of Al and Al_2Cu intermetallic compound around the interface. Consequently, the lap shear tensile strength of the welds was found to be drastically improved. Additionally, many studies [6-17] have shown that the welding conditions exert significant influences on bond characteristics between many different combinations of dissimilar metals. However, there are few systematic studies that reveals the nature of ultrasonic weld formation between Al alloy and Cu. The aim of the current work was to improve understanding of the fundamental phenomena involved in weld formation in the USW of Al alloy to Cu.

Experimental
This study used sheets of the commercial Al alloy 1050-H24 and Cu sheets with dimensions of $50 \times 10 \times 1.0$ mm^3. The Al alloy 1050-H24 with 5 μm anodized oxide layer was also used to observe the oxide layer behavior around the weld interface during USW. The rolling direction (RD) of the specimens was set perpendicular to the direction of ultrasonic vibration. The shape of the horn tip was a square with 6 mm sides; thus, the welding area was approximately 6×6 mm^2. The surfaces of the ultrasonic horn and anvil were knurled to prevent them from slipping on the specimen surfaces. *Table 1* lists the welding parameters used in this study. During USW, the temperature was measured by using a type-K thermocouple embedded on the interface between the Al alloy 6061-T6 and 304 stainless steel sheets.

To evaluate the weld strength of the specimens, lap shear tensile strength tests were employed at a crosshead speed of 3 mm/min. The tensile direction was set perpendicular to the direction of ultrasonic vibration. Vickers hardness measurements were made on the cross section parallel to the vibrating direction. The microstructures of the welds were characterized with a field emission gun scanning electron microscope (FEG-SEM). Anodized oxide layers around the weld interface were identified by energy-dispersive X-ray spectroscopy (EDS) installed

Table 1: Ultrasonic welding conditions in this study

Material	Thickness, t / mm	Ultrasonic vibration		Normal load, F / N	Welding time, t / s
		Frequency, F / kHz	Amplitude, A / µm		
A1050-H24	1.0	19.15	51	589	0.10 – 0.60
Cu	1.0				

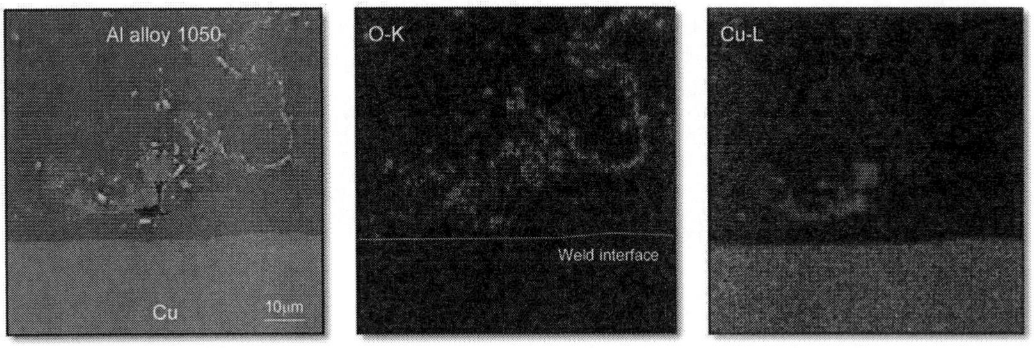

Figure 1 SEM image around the weld interface obtained at 0.4 s. Element maps of O and Cu of the area shown in the SEM image were obtained by EDX.

with the SEM equipment. Electron backscattered diffraction (EBSD) analyses were undertaken to observe the crystallographic microstructure. The electron beam was scanned at step sizes of 0.7µm on the transverse direction– normal direction (TD–ND) plane. In order to understand the behavior of the interfacial reaction at the weld interface during USW, transmission electron microscopy (TEM) was used to observe specific points picked up at the interface by the focused-ion beam (FIB) micro-sampling technique.

Results and discussion

All specimens were successfully welded when the welding time was longer than 0.10 s irrespective of presence of the anodized oxide layer was on the Al alloy surface. To evaluate the weld strength, lap shear strength tests were conducted in every specimen. The lap shear strength increased with the welding time before plateauing at 1.0-1.2 kN. This indicates that the welded area increased with the welding time until it reached to the contact area between ultrasonic horn and Al alloy surface. The fracture mode changed from interface debonding to nugget pull-out when the welding time was longer than 0.35 s. Subsequently, hardness profiles around the weld interface were measured in the Al alloy side. Vickers hardness in the welded area decreased with welding time. This would be attributed to the recrystallization caused by the plastic deformation and the deformation heating during USW. These mechanical behaviors were not changed regardless of the existence of anodized oxide layer. Therefore, the welding mechanism in the USW of anodized Al alloy to Cu would be same as that in the USW of Al alloy to Cu. This suggests the behavior of the anodized oxide layer around the ultrasonic weld interface during USW is almost same as the behavior of the natural oxide film.

Since the fracture and dispersal of the oxide layer must occur to achieve a metallurgical Al–Cu weld during the USW process, the weld interface between anodized Al alloy 1050 and Cu was observed. *Figure 1* shows SEM image and corresponding elemental maps (O and Cu) obtained by EDS around the weld interface. The oxide layer was clearly fractured and dispersed in the Al alloy side and the direct contact between Al and Cu was achieved. In addition, some Cu particles were also dispersed in the Al alloy. As mentioned above, Zhao *et al.* also observed the swirl like structure of Cu in the Al alloy 6061 around the ultrasonic weld interface. These structures would be formed by the plastic flow with high strain rate shear deformation around the weld interface softened by deformation heating.

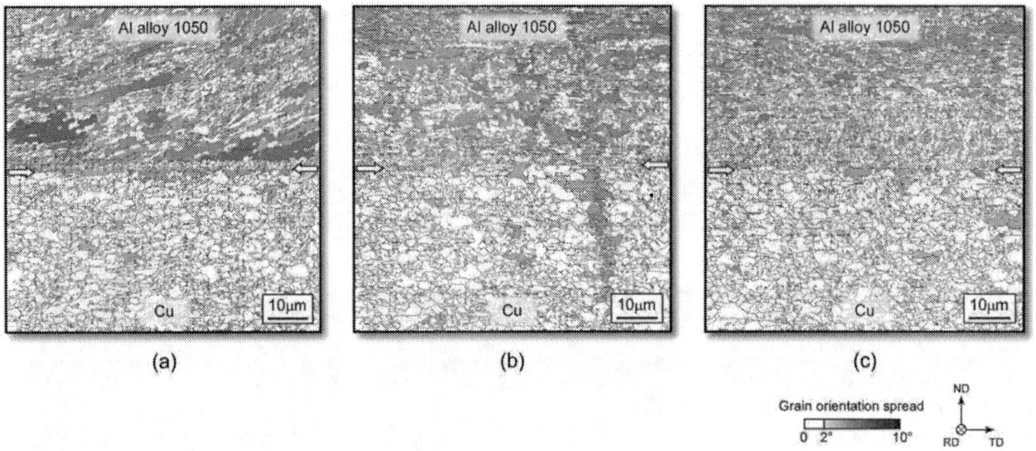

Figure 2 Grain orientation spread maps around the ultrasonic weld interface between Al alloy 1050 and Cu obtained at (a) 0.2 s, (b) 0.3 s and (c) 0.4 s. The white and gray grains correspond to the recrystallized and the non-recrytallized one, respectively.

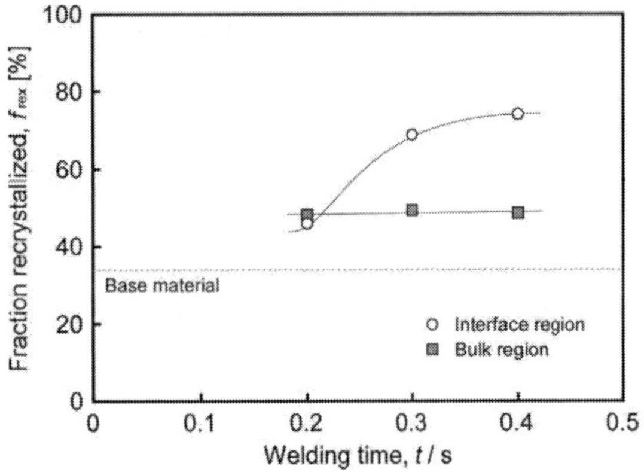

Figure 3 Welding time evolution of the area fraction of recrystallized grains around the interface and the bulk regions. For comparison, the fraction of base material is also shown by broken line.

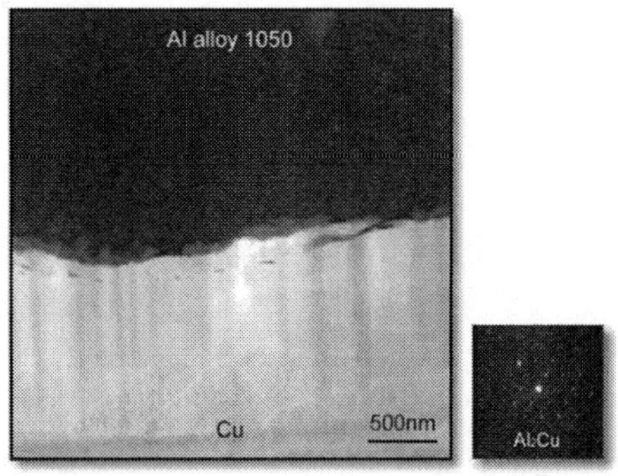

Figure 4 The annular dark field STEM image around the weld interface between Al alloy 1050 and Cu obtained at 0.2 s and selected area diffraction pattern obtained from interfacial reaction layer.

The EBSD analysis was undertaken in order to understand the microstructural characteristics and deformation behavior around the weld interface in more detail. *Figure 2* shows a set of grain orientation spread maps around the weld interface obtained at (a) 0.2 s, (b) 0.3 s and (c) 0.4 s. In this study, the grains with less than 2° orientation spread were defined as the recrystallized grains, which correspond to the white grains in *Fig. 2*. The approximate location of interface between Al alloy and Cu is represented by white colored arrows. As seen in every map, there is a clear microstructural difference between interface and bulk regions in the Al alloy. Many grains in bulk region, which is far from the interface, are elongated along the rolling plane of original sheet and fine grains are seen between elongated grains. The length of major axis of elongated grains is greater than 10 μm and the size of fine grains is about 1 μm. On the other hand, many grains close to the interface are fine and recrystallized. The thickness of interface region, which is mainly composed of fine and recrystallized grains, increased with welding time. Moreover, the area fraction of recrystallization in the interface region also increased with welding time as shown in *Fig. 3*.

Texture analyses were conducted in each region of *Fig. 2* to reveal the deformation behavior during USW. At the early stage of USW (welding time = 0.2 s ~ 0.3 s), {100}<011> texture was formed around the welding plane in the Al alloy. This indicates that the compressive deformation predominantly occurred around the weld interface. Thereafter, the weld formation proceeded to some extent (welding time = 0.4 s), {111}<011> texture was formed around the welding plane. This indicates that the shear deformation predominantly occurred around the weld interface. This would be because the Cu restrained the Al alloy's motion by ultrasonic vibration once the weld was achieved.

TEM observation was made to understand the interfacial reaction at the weld interface. *Figure 4* shows annular dark field STEM image at the weld interface between Al alloy 1050 and Cu obtained at 0.2 s. New reactive layers with 20 ~ 100 nm thickness were also found at some parts of the weld interface as shown in *Fig. 4*. They were identified to be Al_2Cu intermetallic compound phase by the selected area diffraction pattern and the chemical analysis using TEM-EDS technique. The density of dislocations in Al alloy would be significantly high around the weld interface owing to the high-strain-rate shear deformation during USW. This causes the pipe diffusion of Cu to Al around the weld interface which is much more rapid than usual diffusion.

The above results suggest the following sequence for the weld formation during USW of Al alloy 1050 to Cu. At the early stage of welding, the compressive deformation predominantly occurs around the weld interface. At this time, the Al alloy is softened by deformation heating, which causes the plastic flow in the Al alloy. Consequently, the oxide layer is fractured and dispersed in the Al alloy and some Cu particles are also dispersed in the Al alloy. When the weld proceeds to some extent, shear deformation of the Al alloy occurs along the ultrasonic vibrating direction because the motion of Al alloy is restricted by Cu. Finally, the metallurgical weld is achieved between Al and Cu through the Cu diffusion in to the Al alloy.

Conclusions

The weld formation process in the ultrasonic welding of Al alloy 1050 to pure Cu was experimentally clarified by the microstructural characterization. The main results obtained are as follows. The contact with ultrasonic vibration between Al alloy and Cu surfaces would break and disperse the oxide layer at the weld interface. In addition, the broken oxide particles were moved into Al alloy side by the plastic flow in Al alloy caused by the interfacial friction and severe deformation. Consequently, Al–Cu metallic bonding without surface oxide layers was mainly achieved and Al_2Cu intermetallic compound particles were dispersively formed at the weld interface.

Acknowledgments

The authors are grateful to Mr. Akira Honda and Dr. Kosei Kobayashi for technical assistance. This work was supported by a Grant-in-Aid for Young Scientists (B) (24760590) from the Japan Society for the Promotion of Science (JSPS). This support is greatly appreciated.

References

[1] S. Matsuoka and H. Imai: *J. Mater. Process. Technol.*, **209** (2009), pp. 954-960.
[2] Y. Y. Zhao, D. Li and Y. S. Zhang: *Sci Technol. Weld. Join.*, **18** (2013), pp. 354-360.
[3] R. Balasundaram, V.K. Patel, S.D. Bhole and D.L. Chen: *Mater. Sci. Eng. A*, **607** (2014), pp. 277-286.
[4] J.W. Yang, B. Cao, X.C. He and H.S. Luo: *Sci Technol. Weld. Join.*, **19** (2014), pp. 500-504.
[5] X. Wu, T. Liu and W. Cai: *J. Manufact. Process.*, **20** (2015), pp. 321-331.
[6] T. Sanada and H. Mishina: *Wear*, **74** (1981-1982), pp. 263-274.
[7] M. Militzer, W.P. Sun, J.J. Jonas: *Acta Metal. Mater.*, **42** (1994), pp. 133-141.
[8] K.L. Murty, K. Detemple, O. Kanert and J.T.M. Dehosson: *Metal. Mater. Trans. A*, **29** (1998), pp. 153-159.
[9] S. Elangovan, S. Semeer and K. Prakasan: *J. Mater. Process. Technol.*, **209** (2009), pp. 1143-1150.
[10] T. Watanabe, H. Sakuyama, and A. Yanagisawa: *J. Mater. Process. Technol.*, **209** (2009), pp. 5475-5480.
[11] P. Prangnell, F. Haddadi, and Y.C. Chen: *Mater. Sci. Technol.*, **27** (2011), pp. 617-624.
[12] A. Panteli, J. D. Robson, I. Brough, and P.B. Prangnell: *Mater. Sci. Eng. A*, **556** (2012), pp. 31-42.
[13] M. Shakil, N.H. Tariq, M. Ahmad, M.A. Choudhary, J.I. Akhter, and S.S. Babu: *Mater. Design*, **55** (2014), pp. 263-273.
[14] C.Q. Zhang, J.D. Robson, O. Ciuca and P.B. Prangnell: *Mater. Charact.*, **97** (2014), pp. 83-91.
[15] H.T. Fujii, Y. Goto, Y.S. Sato and H. Kokawa: *Mater. Sci. Forum*, **783-786** (2014), 2747-2752.
[16] G. Zhang, Y. Takahashi, Z. Heng, K. Takashima and K. Misawa: *Mater. Trans.*, **56** (2015), pp. 1842-1851.
[17] H.T. Fujii, Y. Goto, Y.S. Sato and H. Kokawa: *Scripta Mater.*, **116** (2016), pp. 135-138.

Configurations of Ultrasonic Complex Vibration Welding Systems and Welding Characteristics Using Two-dimensional Vibration Stress

Jiromaru Tsujino
Kanagawa University, Yokohama, Japan
Asahi EMS Co. Ltd, Yokohama, Japan
tsujino@kanagawa-u.ac.jp

Abstract
Various complex vibration systems with two dimensional vibration locus were developed for improving welding characteristics of ultrasonic metal welding. Ultrasonic complex vibration welding system could be applied effectively for joining same and dissimilar metal and ceramics, and has superior quality compared with conventional welding with linear vibration locus. Welding of dissimilar metal specimens including many metal foils such as aluminum and copper are required for electronic devices and multi-layer fuel cell, battery or EDLC electrodes for electric or hybrid automobile and other various industry fields.

Introduction
Ultrasonic metal welding is used for joining same and dissimilar metal materials and obtained welded part has superior mechanical and electrical characteristics. Welding mechanism is based on atomic force and welded interface has no diffusion and no intermetallic structure.
But, conventional ultrasonic welding method using linear vibration locus has limited welding characteristics. For improving welding characteristics, ultrasonic complex vibration welding method using two-dimensional vibration stress was developed and proved significantly effective for small to large welding specimens.
Ultrasonic complex vibration systems could be configured using multiple vibration systems using same and different frequencies and multiple power amplifiers, and various complex vibration systems up to 1 MHz were developed.
Complex vibration converter using diagonal slits driven by a longitudinal vibration source was proposed. The longitudinal to torsional complex vibration converter vibrates in circular to elliptical vibration locus at a free edge of the converter in the case where vibration phase difference between longitudinal and torsional vibrations is about 90 degrees. The converter was designed using equivalent electrical transmission line method and FEM. Various welding tips could be installed in the free edge part with a connecting bolt.
Welding characteristics of various metal specimens such as aluminum, copper, nickel and aluminum alloy plates and many lapped foils were studied using 20, 27 and 40 kHz complex vibration welding systems. Various welding tips could be installed in the free edge part with a connecting bolt for different purposes.
Using complex vibration, required vibration velocity becomes one-third to quarter compared with conventional welding and weld strength near to material strength was obtained independent of specimen position and direction, and multiple spot or continuous seam welding becomes possible using complex vibration equipment.
As an example by 27 kHz system, 1.0-mm-thick aluminum and copper plates are welded stably from only 1.5 μm (peak-to-zero value) to 2.1 μm with specimen material strength. Stable weld range is wider compared that using conventional linear vibration equipment.
Various dissimilar metal specimens including many metal foils such as aluminum and copper were successfully welded using ultrasonic complex vibration welding systems. These welding are essential for electronic devices and multi-layer fuel cell, battery or EDLC electrodes for electric or hybrid automobile and other various industry fields.

Configurations of ultrasonic complex vibration systems

Ultrasonic complex vibration converter using diagonal slit
Fig. 1 (1) shows the configuration of the 19.5 kHz, 2 kW ultrasonic complex vibration welding equipment with a half-wavelength transverse complex vibration welding tip The ultrasonic welding system consists of a 20 kHz bolt-clamped Langevin type longitudinal transducer (BLT, 50 mm in diameter), a stepped catenoidal horn for enlarging vibration velocity with a supporting flange, a complex vibration converter with diagonal slits. The complex vibration converter is designed using equivalent electrical transmission line method and FEM analysis. The converter is 40 mm in diameter. Twelve diagonal slits were cut directly along the circumference of the converter rod using a spark or milling machine. Longitudinal vibration is partially converted by diagonal slits to torsional vibration. The complex vibration system is driven using a 2 kW frequency auto-tracking and constant

vibration velocity control driving system. The admittance loop is single loop. Quality factor is about 3,000. Circular vibration loci of the complex vibration welding tip are also shown in the Fig. 1 (2). The welding tip vibrates in almost circular vibration locus during welding process.

Complex vibration welding system with a thin and long welding tip.

For welding of narrow and deep area, 27 kHz and 19.5 kHz complex vibration welding systems with a thin and long complex vibration welding tip were developed.

Fig. 2 (1) shows a 19.5 kHz welding system with a 3.0-mm-diameter, 79-mm-length hard metal complex transverse vibration welding tip installed in a half wave-length vibration conical holder by thermal insert. Using a welding tip with a resonant tip holder, resonance frequency of the driving system does not change by installing the welding tip. Transverse distribution along the welding tip measured is shown in Fig. 2 (2). A conical vibration holder has some vibration velocity transform ratio, and also vibration velocity increases by connecting small diameter welding tip. Total vibration transform ratio of the welding tip with a conical holder is about three.

Longitudinal and torsional vibration distributions along the converter and stepped horn

Fig. 3 (1) shows configuration of the converter with thin and long welding tip and longitudinal and torsional vibration distributions along the converter and driving stepped horn. The converter was designed as the converter and horn vibrate 3/4 wave-length longitudinal and 5/4 wave-length torsional vibration modes from the free end of convertor to nodal part of horn. The complex vibration system supported at a nodal part of the horn. Fig, 3 (2) shows a free admittance loop of the complex vibration system.

Complex vibration welding systems with various non-resonant welding tips

Fig. 4 (1) shows 23-mm-wide, 9-mm-height complex vibration welding tip installed at free end of complex vibration converter. This non-resonant welding tip vibrates nearly as a mass but vibrates slightly transversally.

Fig. 4 (2) shows non-resonant hard metal welding tip protruded from the free end of converter for welding narrow area. The welding tip vibrates as a mass. These non-resonant welding tips require precise power factor compensation of the complex vibration system according to these mass weight for obtaining circular or elliptical vibration locus.

(1)

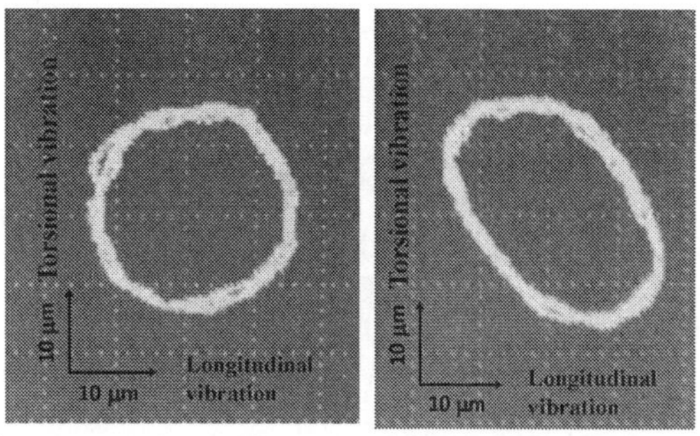

(2) Examples of vibration loci under welding condition

Fig. 1 (1) 19.5 kHz ultrasonic complex vibration welding system using a complex vibration converter with diagonal slits and a half-wavelength complex vibration welding tip. (2) vibration loci under welding process.

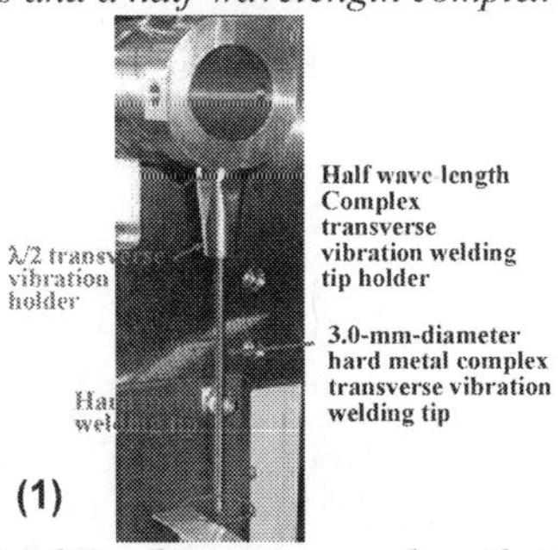

(1)

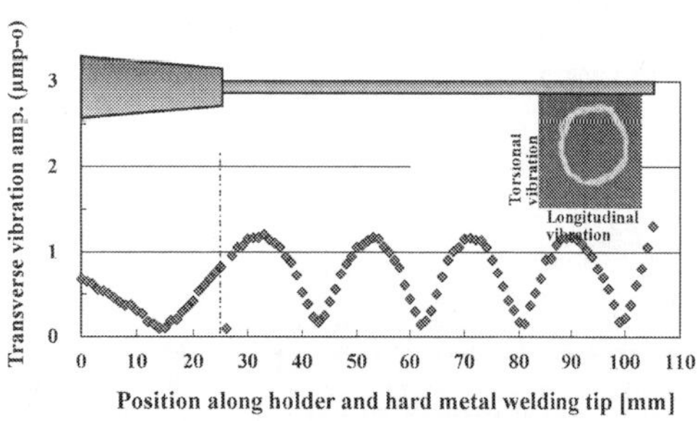

(2)

Fig. 2 (1) 19.5 kHz ultrasonic complex vibration welding system with a 3.0-mm-diameter, 79-mm-length complex vibration welding tip installed in a half wave-length holder. (2) Transverse vibration distribution along a 3.0-mm-diamete welding tip and half wave-length holder.

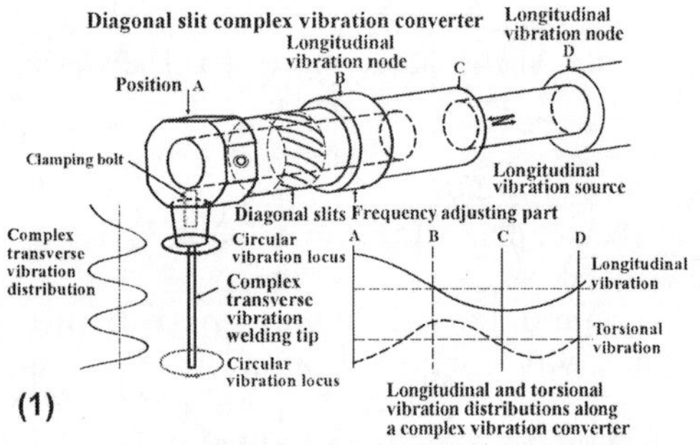

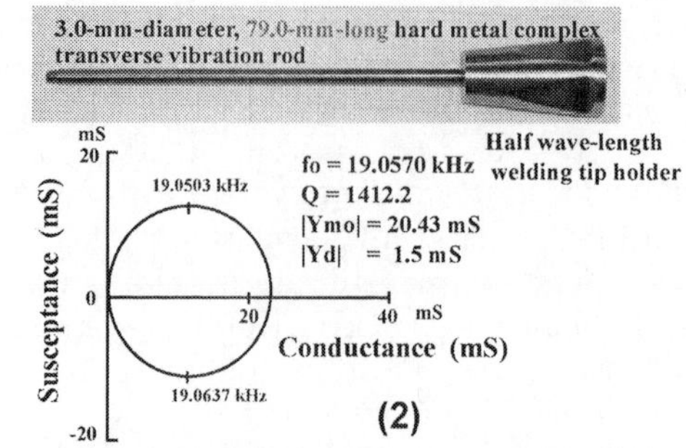

Fig. 3 (1) Longitudinal and torsional vibration distributions along a complex vibration converter and driving stepped horn. (2) Free admittance loop of welding system with a hard metal welding tip.

(1)

(2)

Fig. 4 (1) 19.5 kHz non-resonant 23-mm-wide, 9-mm-height welding tip installed in a complex vibration converter using a connecting bolt. (2) Hard metal welding tip for welding of narrow welding area.

Welding conditions of ultrasonic complex vibration welding systems

Fig. 4 (1) shows welding condition of 1.0-mm-thick aluminum and copper plate using a 27 kHz 10-mm-diameter complex vibration welding tip of circular vibration locus. 1.0-mm-thick aluminum and copper plates are welded stably from only 1.5 μm (peak-to-zero value) to 2.1 μm with specimen material strength. Stable weld range is wider compared that using conventional linear vibration equipment.

Fig. 4 (2) shows welding condition of 0.3-mm-thick aluminum and 1.0-mm-thick copper plate using a 19.5 kHz 3.0-mm-diameter, 79-mm-length hard metal complex vibration welding tip of circular vibration locus. Maximum welding strength over 20 N is obtained from 0.1 s to 0.25 s welding time under 10 μm vibration amplitude and maximum strength is obtained between 0.4 s and 1.0 s. Maximum strength was obtained at wider welding condition compared with a conventional linear vibration welding system.

Using a complex vibration welding tip, required vibration amplitude and static pressure become smaller and stable welding range becomes wider. Furthermore, large and uniform welding strength is obtained independent of welding specimen direction.

(1)

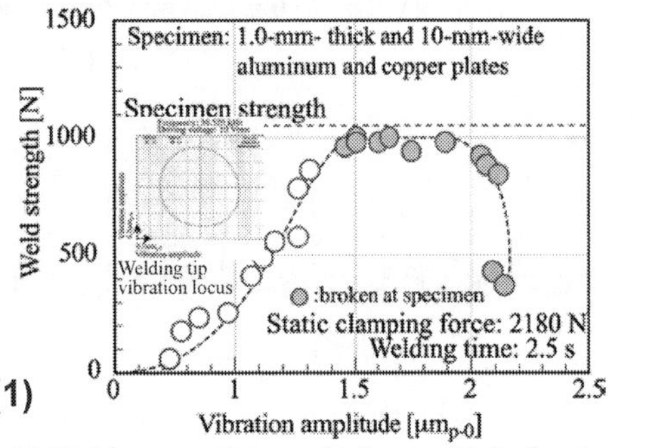

(2)

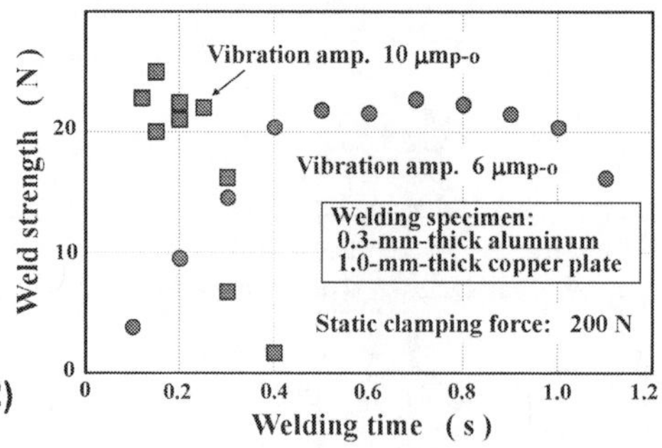

Fig. 4 (1) Welding condition of 1.0-mm-thick aluminum and copper plate using a 27 kHz 10-mm-diameter complex vibration welding tip of circular vibration locus. (2) Welding condition of 0.3-mm-thick aluminum and copper plate using a 19.5 kHz 3.0-mm-diameter hard metal welding tip.

Conditions of lapped foil electrodes and nickel-clad copper terminals

Fig. 5 (1) shows welding tip and anvil side conditions of thirty lapped 0.020-mm-thick aluminum foil electrodes and nickel coated coper terminal welded using 19.5 kHz 23-mm-wide complex vibration welding tip. The specimens were welded successfully without large deformation and damage. Fig. 5 (2) shows conditions of two 0.16-mm-thick, 3.0-mm-wide nickel clad copper terminals and nickel coated steel case welded using 19.5 kHz 3.0-mm-diameter, 79-mm-long hard metal welding tip.
Fig. 6 (1) shows TEM images of cross sections of aluminum-copper and aluminum-nickel plate specimens. No diffusion area and no intermetallic structure and no specific structure are observed in welded interfaces.
Fig. 6 (2) shows successfully welded conditions of two 0.16-mm-thick, 3.0-mm-wide nickel clad copper terminals and deep nickel coated steel case bottom welded using a 19.5 kHz 3.0-mm-diameter, 79-mm-long hard metal complex transverse vibration welding tip.

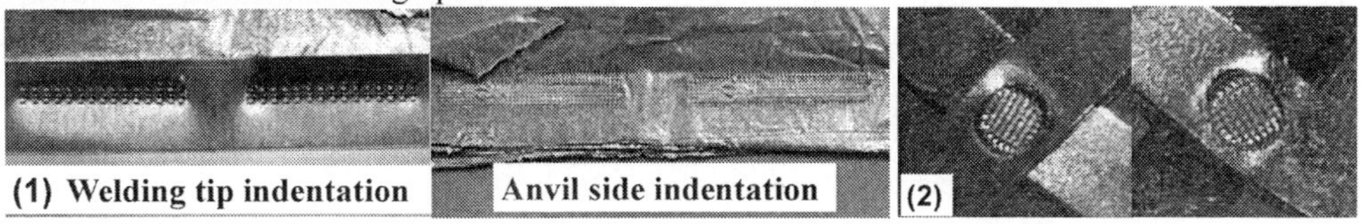

Fig. 5 (1) Conditions of thirty lapped 0.020-mm-thick aluminum foil electrodes and nickel coated coper terminal welded using 19.5 kHz 20-mm-wide complex vibration welding tip. (2) Conditions of two 0.16-mm-thick, 3.0-mm-wide nickel clad copper terminals and nickel coated steel case welded using 19.5 kHz 3.0-mm-diameter, 79-mm-long hard metal welding tip.

(1)

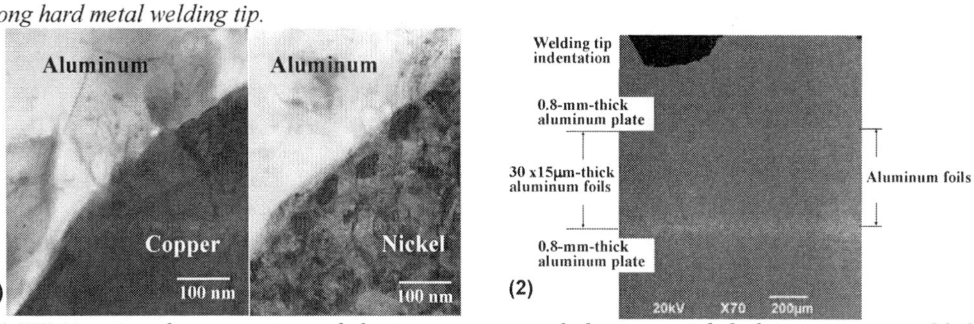

Fig. 6 (1) TEM images of cross sections of aluminum-copper and aluminum-nickel plate specimens welded using an ultrasonic welding system. (2) Cross section of thirty lapped 0.015 mm aluminum foil electrodes and 0.3-mm-thick aluminum terminal plates.

Conclusions

Various complex vibration systems with two dimensional vibration locus were developed for improving welding characteristics of ultrasonic metal welding.
Ultrasonic complex vibration welding system using two dimensional vibration stress has superior welding characteristics and required vibration velocity and clamping force were decreased significantly compared with conventional linear vibration system.
Welding of dissimilar metal specimens including many metal foils such as aluminum and copper which are essential for electronic devices and multi-layer fuel cell, battery or EDLC electrodes for electric or hybrid automobile and other various industry fields.

References

[1] J. Tsujino and E. Sugimoto: Welding characteristics of aluminum, copper, nickel and aluminum alloy with alumina coating using ultrasonic complex vibration welding equipments, Proc. 2009 IEEE Ultrasonics Symposium, pp. 1211-1214 (2009)
[2] J. Tsujino: Ultrasonic welding using complex vibration and high frequency vibration - Various vibration sources and frequency characteristics of ultrasonic welding - (invited), Proc. 2011 IEEE Ultrasonics Symposium, 1910-1919 (2011)
[3] J. Tsujino and E Sugimoto: Ultrasonic welding of electronic parts and devices using a long and thin complex vibration welding tip, Proc. 2014 EEE Ultrasonics Symposium, pp. 947-950 (2014)
[4] J. Tsujino and E Sugimoto: Ultrasonic welding using a long and thin complex transverse vibration welding tip with vibration detector and static pressure controller, Proc. 2015 IEEE Ultrasonics Symposium, (2015), IEEE Xplore.

Effect of welding tip geometry on the mechanical properties of high power ultrasonic spot welded AA6111-TiAl6V4 dissimilar joints

Chaoqun Zhang[1,2]
[1]*Shanghai Key Laboratory of Digital Manufacture for Thin-Walled Structures, School of Mechanical Engineering, Shanghai Jiao Tong University, Shanghai 200240, PR China*
[2]*Materials Science Centre, School of Materials, University of Manchester, Grosvenor Street, Manchester M13 9PL, UK*
chaoqunzhang@sjtu.edu.cn, +86-18321351096

Abstract
Dissimilar joining of aluminum to titanium is problematic due to the large difference in melting point of the two metals and the formation of potentially brittle intermetallic phase on the weld interface. High power ultrasonic spot welding is a low energy input solid-state welding technology that has potential capacity to solve this problem. In this study, AA6111 aluminium alloy and TiAl6V4 titanium alloy were successfully welded by high power ultrasonic spot welding using two different types of welding tips, a rectangular tip and a round tip. Lap shear strength testing and high resolution three-dimensional laser surface profile scanning were applied to examine weld geometry and weld mechanical performance. The effects of welding time on peak load and fracture energy were investigated. The weld interface microstructure was investigated using scanning electron microscopy and transmission electron microscopy. The joints welded by the round tip has a shallower and smoother footprint, larger welded area and a longer fracture path than the rectangular tip, which result in higher lap shear strength and fracture energy.

Introduction
Dissimilar joining of titanium alloy to aluminium alloy could be a promising means to reduce aerospace and automotive structure weight and cost[1]. However, brittle intermetallic phases form easily between aluminium and titanium when they are welded by a fusion welding means, such as laser welding [2-4].
Intermetallic phases formation can be controlled effectively by using solid state welding method, which has a low welding temperature[5]. Friction stir welding, a solid state welding method, has been applied for Al-Ti dissimilar welding, the intermetallic phases formation has been controlled compared with fusion welding [5, 6]. But intermetallic phase has been detected though less [6, 7] and friction stir welding tool can be easily damaged or worn because of the contact between the tool and titanium alloy.
Ultrasonic spot welding (USW), due to its even lower welding temperature and less energy input[8-10], no intermetallic phases has been detected[11]. Besides, there is no direct contact between USW welding tip and titanium, which can prevent welding tip damage and wear.
For solid state welding, tool geometry has always been very important for the microstructure and mechanical performance of welds, since welding energy and plastic deformation were transmitted by welding tool[12-14]. The influence of friction stir welding tool geometry on weld geometry and mechanical properties has been studied a lot due to its high importance [14, 15]. However, only a small number of studies have been done on USW tools as application of USW on thick metals only has been developed in recent years[16, 17].Watanabe et al have studied the effect of two types of welding tip, one is C-tip with a cylindrical contact face and the other one is K-tip with a flat contact face. It has been demonstrated that C-tip welds have better mechanical performance[17].But the effect of tip geometry on the precise weld geometry, stress concentration and the fracture path has not been discussed. In this paper, to understand more deeply about the effect of tip geometry, high precision 3D laser surface profile scanning was applied to investigate the weld geometry, the influence of tip geometry on the precise weld geometry, stress concentration, welded area, the fracture path and mechanical properties are examined.

Body of paper
Experiments
25×75×0.93 mm 6111-T4 aluminium and 25×75×1mm TiAl6V4 titanium sheets were used for present study. The sheets were lap welded with aluminium sheet on the top, using a 2.5kW single reed ultrasonic welder. A 9×6 mm flat serrated sonotrodetip and a 10 mm diameter round sonotrode tip were located and pressed at the centre of a 25mm overlap when ultrasonic spot welding was performed. The mechanical properties of the joints were measured by tensile lap shear tests performed with a crosshead speed of 1 mm/min.
A 3D NanoFoucus µscan SC200 profilometer based on laser scanning was applied to measure the weld geometry precisely. The samples were cross-sectioned perpendicular to the welding direction to examine the weld geometry.

Results and discussion
Figure 1 shows a picture of the top surface of a round tip and a rectangular tip used for ultrasonic spot welding. The round tip has a diameter of 10mm and the size of the rectangular one is 8×6 mm. The area of the round tip is around 1.6 times the area of the rectangular tip. The rectangular one has more teeth and its tooth is sharper and is

shaped like a triangle. The tooth of the round tip is wider and blunter and has a square shape. As a result, the rectangular tip is easier to grip aluminium plate and plunge into it. The cross-sections of two different type welds are shown in Figure 2.

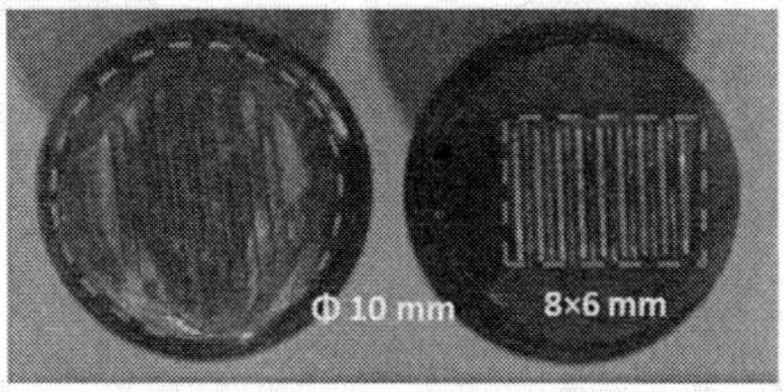

Fig. 1 Size of round and rectangular tips

Fig. 2 Macrograph of cross-section of two different type welds

In general, tip geometry can influence: 1. the depth of welding tip footprint; 2. the extent of stress concentration; 3. the fracture path. Then, the mechanical properties of USWed joints were affected.

(1) The footprint depth

Figure 3a) shows the high resolution profiles of a rectangular tip weld and a round tip weld in the positions labelled by the red lines in Figure 3b). The profiles were collected using a laser profilometer with a resolution of 10 μm. These two welds were joined using the same welding time 1.2 s and other same parameters, except the welding tip. However, the depth of their footprints has a large difference, as shown in Figure 3. In Figure 3c), d1 and d2 stands for the distance between the surface of aluminium plate with the first valley bottom at the edge of the round tip weld (d1) and the rectangular tip weld (d2). Here, d1 is 120 μm and d2 is 455 μm. Since first valley bottom at the edge of welds is the first weak region exposed to stress when load is applied (other valley bottoms are protected by the wave crest regions and the interface bonding), its depth has a strong influence on the load bearing capacity of welds. Due to the large difference between d1 and d2, when other things are equal and fracture occurs at the weak region of aluminium and goes through aluminium, the strength and fracture energy of rectangular tip welds are lower than that of round tip welds, as shown in Figure 4.

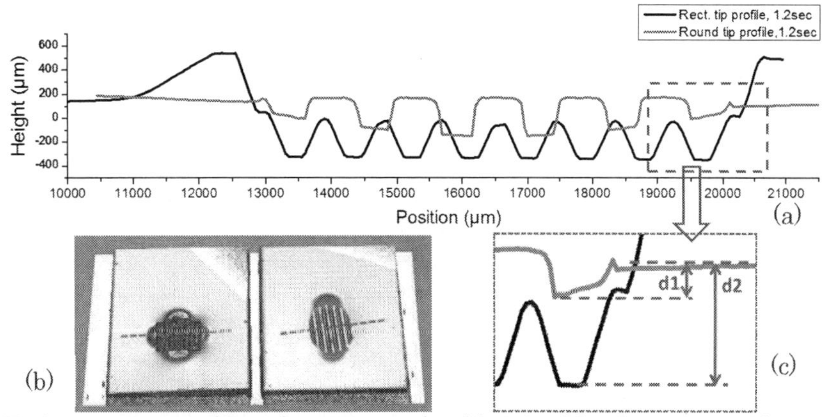

Fig. 3 (a) High precision profiles of two types of welds joined by the same welding time, (b) the positions (labelled by the red lines) of profile curves and (c) the partial enlarged view of the edge region.

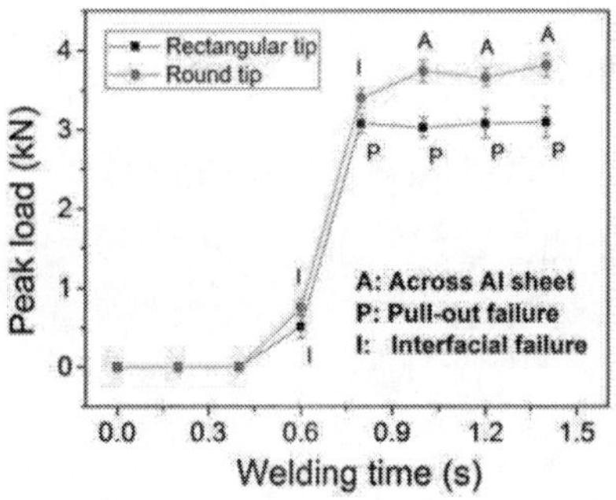

Fig.4 Comparison of peak leads between as-welded rectangular-tip welds and round-tip welds.

(2) The extent of stress concentration

As shown in Figure 3, the significant difference of height between the wave crest at the edge of weld and the first adjacent valley bottom in the rectangular tip weld caused much higher stress concentration than round-tip welds, which has a much smooth transition from welded region to the unwelded base metal. When load was applied, the higher stress concentration led to low shear strength.

Previous research by Watanabe et al also demonstrated that the joint produced by a tip with a geometry similar to the round tip has a higher strength than the joint produced by a tip with a geometry similar to the rectangular tip [17].

(3) The fracture path

As shown in Figure 5, in as-welded condition, for both rectangular-tip welds and round-tip welds, the fracture type is ductile fracture, but the fracture path is different because of the influence of footprints of two different tips. For rectangular-tip welds, the fracture mode is the 'Pull-out' mode, in which fracture path goes through the edges of welded region, which are thin and weak. For round-tip welds the fracture mode is 'Across aluminium fracture' mode, in which fracture path mainly goes through aluminium base metal which is much thicker and stronger than the edges of welded region of rectangular-tip weld. As a result, the fracture of round-tip welds required more energy than that of rectangular-tip weld.

Also, generally the round tip weld has a larger welded area, which is beneficial for the weld properties. Jahn et al. [16] also found tip geometry has important effects on the properties of welds. They found that the welded area increased with the increase of tip size and a trend of increasing peak load with the increase of welding tip size.

In all, the results suggest that a welding tip with smooth edges and a larger area is beneficial for joint mechanical properties.

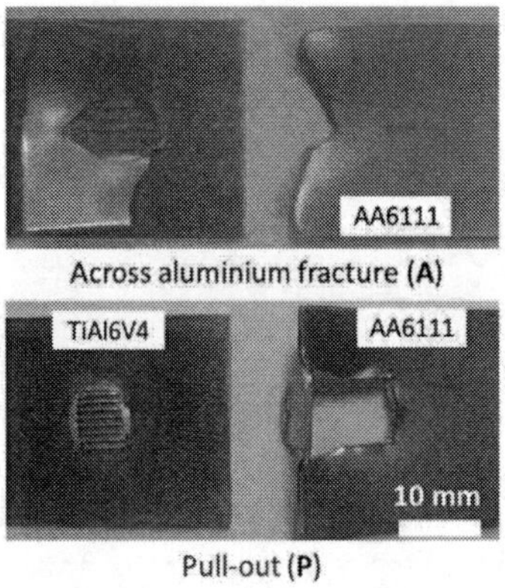

Fig.5 Fracture path of round tip weld (upper) and rectangular tip weld (lower), both in as-welded condition.

Conclusions

The feasibility of two different types of welding tips for welding aluminium alloy AA6111 to titanium alloy was studied. According to the present mechanical property and weld geometry investigation, the following conclusions were reached:

1. It has been shown that 0.93 mm thick AA6111 sheet and 1 mm thick TiAl6V4 sheet can be successfully welded by high power ultrasonic spot welding using both the rectangular tip and the round tip.
2. The footprint of rectangular tip is deeper than that of round tip.
3. The fracture path of as-welded round-tip welds goes through unaffected aluminium; the fracture path of

as-welded rectangular-tip welds goes through the thinned edge of welded region.
4. When other things are equal, the lap shear strength and fracture energy of round tip welds are higher than those of round tip welds.

Acknowledgments

This work was funded by the EPSRC through LATEST2, Light Alloys towards Environmentally Sustainable Transport (EP/G022402/1) and Friction Joining—Low Energy Manufacturing for Hybrid Structures in Fuel Efficient Transport Applications (EP/G022402/1.JLR). The authors acknowledge the China Scholarship Council (CSC, 2011612001) for financial support and would like to thank Constellium and Airbus UK for the provision of materials.

References

[1] Rendigs, K.H., Aluminium structures used in aerospace - Status and prospects, in Aluminium Alloys: Their Physical and Mechanical Properties, Part 4/Supplement, J.H. Driver, et al., Editors. 1997. p. 11-23.
[2] Vaidya, W.V., et al., Structure-property investigations on a laser beam welded dissimilar joint of aluminium AA6056 and titanium Ti6Al4V for aeronautical applications Part I: Local gradients in microstructure, hardness and strength. Materialwissenschaft Und Werkstofftechnik, 2009. 40(8): p. 623-633.
[3] Song, Z.H., et al., Interfacial microstructure and mechanical property of Ti6Al4V/A6061 dissimilar joint by direct laser brazing without filler metal and groove. Materials Science and Engineering a-Structural Materials Properties Microstructure and Processing, 2013. 560: p. 111-120.
[4] Lv, S.X., et al., Influence of Zr addition on TIG welding-brazing of Ti-6Al-4V to Al5A06. Materials Science and Engineering a-Structural Materials Properties Microstructure and Processing, 2013. 568: p. 150-154.
[5] Chen, Y.-h., Q. Ni, and L.-m. Ke, Interface characteristic of friction stir welding lap joints of Ti/Al dissimilar alloys. Transactions of Nonferrous Metals Society of China, 2012. 22(2): p. 299-304.
[6] Bang, K.S., et al., Interfacial Microstructure and Mechanical Properties of Dissimilar Friction Stir Welds between 6061-T6 Aluminum and Ti-6%Al-4%V Alloys. Materials Transactions, 2011. 52(5): p. 974-978.
[7] Chen, Y.C. and K. Nakata, Microstructural characterization and mechanical properties in friction stir welding of aluminum and titanium dissimilar alloys. Materials & Design, 2009. 30(3): p. 469-474.
[8] Prangnell, P., F. Haddadi, and Y.C. Chen, Ultrasonic spot welding of aluminium to steel for automotive applications - microstructure and optimisation. Materials Science and Technology, 2011. 27(3): p. 617-624.
[9] Bakavos, D. and P.B. Prangnell, Mechanisms of joint and microstructure formation in high power ultrasonic spot welding 6111 aluminium automotive sheet. Materials Science and Engineering a-Structural Materials Properties Microstructure and Processing, 2010. 527(23): p. 6320-6334.
[10]Robson, J., A. Panteli, and P.B. Prangnell, Modelling intermetallic phase formation in dissimilar metal ultrasonic welding of aluminium and magnesium alloys. Science and Technology of Welding and Joining, 2012. 17(6): p. 447-453.
[11]C. Q. Zhang, J. D. Robson, O. Ciuca, and P. B. Prangnell, Microstructural Characterization and Mechanical Properties of High Power Ultrasonic Spot Welded Aluminum Alloy AA6111–TiAl6V4 Dissimilar Joints, Materials Characterization, 97 (2014), 83-91.
[12]Shin, H.S., Tool geometry effect on the characteristics of dissimilar friction stir spot welded bulk metallic glass to lightweight alloys. Journal of Alloys and Compounds, 2014. 586: p. S50-S55.
[13]Lin, B.Y. and J.J. Liu, Effect of Tool Geometry on Fatigue Properties and Fracture Behavior of Friction Stir Spot Welded Aluminum Sheets. Journal of the Chinese Society of Mechanical Engineers, 2013. 34(4): p. 277-284.
[14]Tozaki, Y., Y. Uematsu, and K. Tokaji, Effect of tool geometry on microstructure and static strength in friction stir spot welded aluminium alloys. International Journal of Machine Tools & Manufacture, 2007. 47(15): p. 2230-2236.
[15]Chen, Y.C. and K. Nakata, Effect of tool geometry on microstructure and mechanical properties of friction stir lap welded magnesium alloy and steel. Materials & Design, 2009. 30(9): p. 3913-3919.
[16]Jahn, R., R. Cooper, and D. Wilkosz, The Effect of Anvil Geometry and Welding Energy on Microstructures in Ultrasonic Spot Welds of AA6111-T4. Metallurgical and Materials Transactions A, 2007. 38(3): p. 570-583.
[17]Watanabe, T., K. Nishihara, and T. Sasaki. Effect of Weld Tip Geometry on Ultrasonic Welding of Aluminum Alloy. in Materials Science Forum. 2010. Trans Tech Publ.

Fundamental understanding of the mechanism of solid-state weld formation in dissimilar metal welds using ultrasonic additive manufacturing

N.Sridharan*, S.S.Babu
Department of Mechanical Aerospace and Biomedical Engineering, University of Tennessee, Knoxville, TN, USA
* niyanth.sridharan@gmail.com
D.Isheim, D.Seidman
Department of Materials Science and Engineering and Northwestern University Center for Atom-Probe Tomography (NUCAPT), Evanston, IL, USA
P.J.Wolcott, M.J.Dapino
Department of Mechanical and Aerospace Engineering, The Ohio State University, Columbus, OH 43210, USA

Introduction:

Welding metals in the solid state always involves plastic deformation of the metals that need to be welded. Plastic deformation serves to remove the oxide layers and collapse asperities bringing nascent materials in contact [1]. However some underlying fundamental mechanisms are yet to be resolved. One of the areas where a fundamental gap exists is if co deformation is necessary to initiate a solid state weld in a dissimilar metal combination. To address this question attempts were made to weld aluminum to titanium and characterize the interfaces. Al and Ti were chosen primarily for the difference in hardness. In addition to the above objective aluminum and titanium were chosen due to their industrial relevance. Solid state welds are necessary to fabricate Al-Ti welds due to the formation of intermetallic compounds at the interface during fusion welding [2]. UAM is an important additive manufacturing technique that has gained significant importance in the present times to fabricate complex geometries using dissimilar metals for a wide gamut of applications. Though UAM has been used to successfully join Al 3003 and CP Ti [3] there is a fundamental gap in understanding the actual bonding mechanism due to the lack of characterization. Hence to address the exact mechanism of bonding a detailed multiscale characterization coupled with mechanical testing is performed in this work.

Experimental Procedure:

Bilayers of Al-1100 and CP-Ti both 0.005" thick were welded onto an Al-6061-T6 substrate as shown schematically in figure 1(a). The builds were made using a Fabrisonic Sonic layer 4000 9 kW using a weld force of 3500 N, weld speed of 25.4mm/sec, and vibration amplitude of 41.55 μm. During deposition, the substrate preheated to 93.3°C. The samples were then sectioned for multi scale and prepared using standard metallography techniques. Optical microscopy was performed in a Leica DM 750P microscope. EBSD was performed on a JEOL 6500 FEG Scanning Electron Microscope (SEM). Post weld heat treatment was performed at 600°C for 1 hour to improve bond strengths by promoting intermetallic formation [2, 4]. Following multiscale characterization mechanical testing was performed using the pushpin test. A brief description of the test is found elsewhere [5]

Results and Discussions:

Optical microscopy:

A schematic of the bilayer arrangement is shown in Figure 1 (a). In the bilayer arrangement, the Al-1100 foil is in contact with the Al substrate as illustrated in Figure 1 (b) and marked in the micrograph. As shown in the micrograph no intermetallic formation was observed as the interface [6]. The roughness in the micrographs is a result of the sonotrode contacting on the top layer of the Ti. Al being the softer of the two materials flows around these contours and hence the interface experiences significant plastic flow. The effect of the deformation on the crystallographic grain structure needs to be examined to gain a fundamental understanding of the bond formation mechanism. This was realized using EBSD. A detailed description of this technique and the various methods of analysis can be found elsewhere [7].

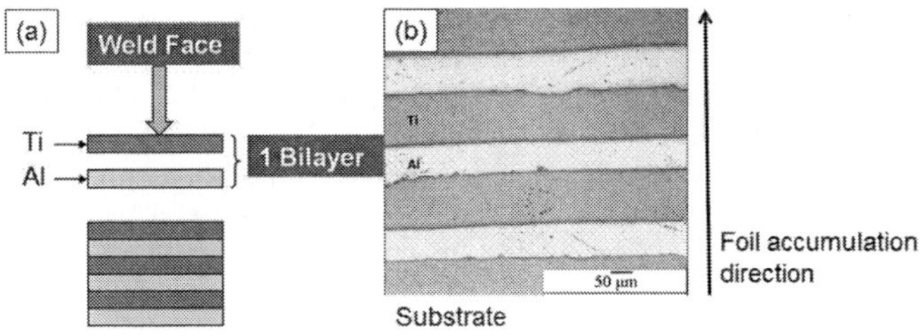

Figure 1: (a) Schematic illustration of the bi layer arrangement used for sample fabrication (b) Optical micrograph of the Al-Ti welded interface. The substrate is at the bottom and the direction of foil accumulation is marked

Mechanism of bond formation:

Figure 3 shows the inverse pole figure of the Al-Ti welds in the as welded condition. Deformation concentrated predominantly only in the softer metal while the harder Ti did not deform. A detailed evolution of the microstructure may offer significant in sight to the bond formation mechanism. To analyze the nature of the deformation and grain refinement mechanisms, the crystallographic texture was analyzed. The crystallographic texture of three layers of Al-1100 is presented in the form of a pole figure in figure 2.

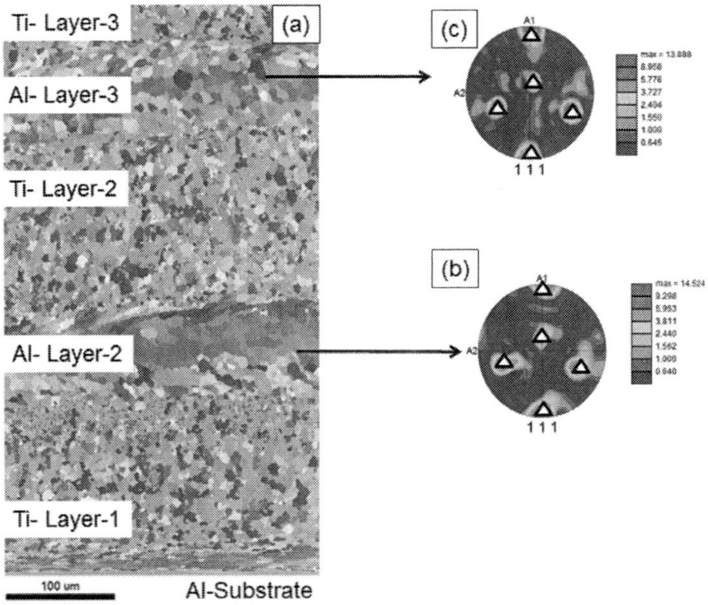

Figure 2: (a) EBSD of the Al-Ti interface. The Al interfaces show extensive grain refinement. (b) Al pole from layer-2 (c) Al pole from layer-3. The filled triangles mark the position of the copper component

The texture changed significantly from a rolling texture present in the original aluminum foils to a copper texture where {112} ||ND and <111>||RD in the as processed condition. The presence of the copper texture in the Al foil after UAM shows a similarity with textures developed after ARB. Due to this similarity in the texture it can be concluded that the underlying mechanism of bond formation is similar. There are two criteria to initiate a solid state weld.

i. Removal of surface oxide
ii. Asperity collapse to bring the metals into intimate contact

In the case of ARB (accumulative roll bonding), bonding is achieved by the plastic deformation. The plastic deformation serves to remove the oxide layers and bring nascent metals in intimate contact to achieve a solid-state bond [8].

Heat treatment and characterization of heat treated sample:

As pointed out previously heat treatment is an effective method to promote interdiffusion and hence increase the extent of intermetallic formation at the interface. In addition heat treatments may lead to alterations in the grain growth mechanisms leading to a change in mechanical properties. EBSD was performed at the interfaces to observe changes in the morphology of the grains at the interface. The results are shown in figure 3(a). Though it would be natural to expect significant grain growth, previous work on Al-3003 similar welds shows that absence of grain growth at the interface during post weld heat treatment [9, 10]. The reason for such behavior has not yet been rationalized. As expected and shown in figure 3(a) (marked with black arrows) the Al grains at the Al-Al welds do not show any grain growth. However surprisingly the grains at the interfaces of dissimilar metal weld interfaces showed significant growth after post weld heat treatment in this case. This could be attributed to the fact that the inherent mechanisms of the bond formation are different for the similar and dissimilar metals resulting in completely different behavior of the interfaces.

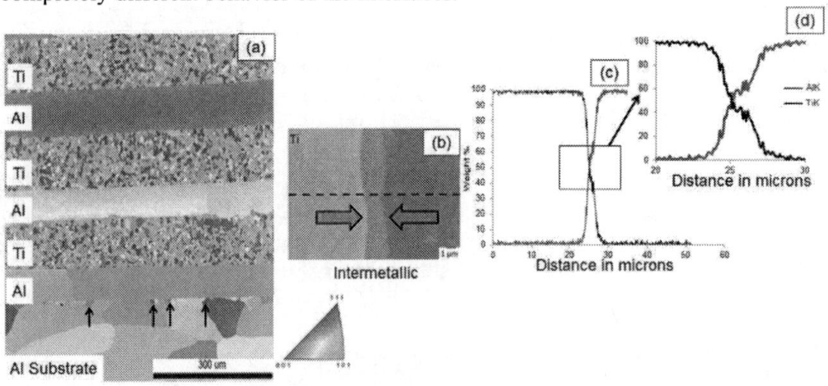

Figure 3: (a) EBSD of the heat treated interface. Note the absence of grain growth at the interface between the Al substrate and the Al-1100 tape interface. These regions are marked using black arrows.

To investigate the extent of interdiffusion and chemistry adjacent to the interfaces zone, a diffusion profile was measured, shown in figure 3. The region where the EDS line scan was performed in shown in figure 3(b). The line scan results show significant interdiffusion and evidences for a line compound/intermetallic formation (~2μm thickness) as shown in figure 3(c).

Mechanical testing and fractography of builds

Mechanical testing was performed on a Gleeble ® the results from mechanical testing are presented in figure 4.

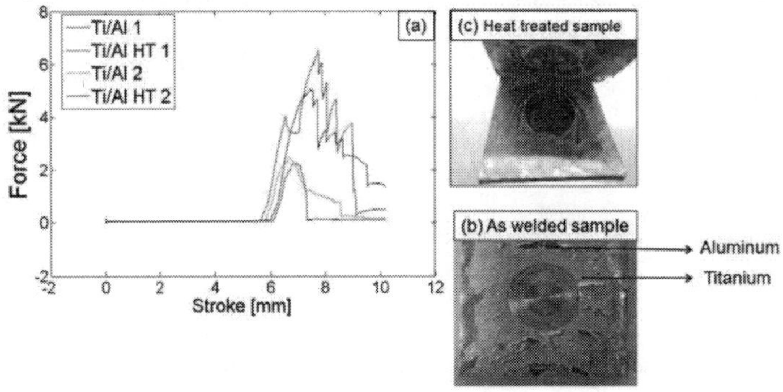

Figure 4: (a) Load vs displacement curves from the push pin tests. Note the significant increase in the area under the curve after heat treatment. (b) Fracture surface of the as welded specimen. The regions marked Al are regions where the Al failed in the foil as against the substrate. (c) The fracture surface of a heat treated specimen.

The area under the curves shows a drastic rise for the heat treated material on comparison to the as welded material. This rise could be attributed to the interdiffusion of Al and Ti and the formation of a thin layer of intermetallic which decreases the surface energy of the interface making the metallurgical bonding more complete [11]. This effect has been observed and documented in friction stir welding [2, 4, 12]. The morphology of the difference in fracture of the surfaces is shown in figure 4(b) and (c). The figures show a transition from a brittle fracture to a more ductile failure mode arising as a result of the heat treatment. The area examined shows a predominantly brittle region but localized ductile regions are observable. Al though the heat treated samples exhibited a drastic increase in strength, the as welded samples also showed evidences for ductility in several localized regions. Once could attribute this localized ductility to the presence of a metallurgical bond in the as welded state itself. If a solid-state bond had formed between the Al and Ti then the examined region should be not show any Ti signals since the Al should be bonded to the Ti and hence peeled away. This is what is observed when the as welded fractographs were examined as shown in figure 4 where Al layers were still present in a largely Ti side. This confirms that the localized ductility in the as welded sample is due to the presence of a metallurgical bond.

Summary and Conclusions:

Using multi scale characterization Al-Ti builds fabricated using 9 kW UAM has been investigated. The key findings can be summarized as follows

i. Plastic deformation is a key step to initiate metallurgical bonding. Metallurgical bonding occurred in the case of Al and Ti with plastic deformation concentrated exclusively in Al alone. In other words co deformation of both the metals are not necessary to initiate a solid state bond
ii. Heat treatment showed significant enhancement in mechanical properties and could be attributed to the interdiffusion of Al and Ti at the interfaces.

Acknowledgments

Funding for this work was provided by the Israeli Ministry of Defense. The authors would like to thank M. Norfolk of Fabrisonic LLC for providing technical support. Special thanks are given to D. Tung of the OSU Welding Engineering Program for his time and help in using the Gleeble 3800.

References

[1] D. Milner, G. Rowe. Fundamentals of solid-phase welding, Metallurgical Reviews 7 (1962) 433-480.

[2] Y.-C. Kim, A. Fuji. Factors dominating joint characteristics in Ti–Al friction welds, Science and Technology of Welding & Joining 7 (2002) 149-154.

[3] J. Obielodan, B. Stucker, E. Martinez, J. Martinez, D. Hernandez, D. Ramirez, L. Murr. Optimization of the shear strengths of ultrasonically consolidated Ti/Al 3003 dual-material structures, Journal of Materials Processing Technology 211 (2011) 988-995.

[4] A. Fuji. In situ observation of interlayer growth during heat treatment of friction weld joint between pure titanium and pure aluminium, Science and Technology of Welding & Joining 7 (2002) 413-416.

[5] C.S. Zhang, A. Deceuster, L. Li. A method for bond strength evaluation for laminated structures with application to ultrasonic consolidation, Journal of materials engineering and performance 18 (2009) 1124-1132.

[6] P. Wolcott, N. Sridharan, S. Babu, A. Miriyev, N. Frage, M. Dapino. Characterisation of Al-Ti dissimilar material joints fabricated using ultrasonic additive manufacturing, Science and Technology of Welding and Joining (2015) 1362171815Y.0000000072.

[7] A.J. Schwartz, M. Kumar, B.L. Adams, D.P. Field. Electron backscatter diffraction in materials science, Springer, 2009.

[8] L. Li, K. Nagai, F. Yin. Progress in cold roll bonding of metals, Science and Technology of Advanced Materials 9 (2008) 023001.

[9] K. Sojiphan. Effects of Very High Power Ultrasonic Additive Manufacturing Process Parameters on Hardness, Microstructure, and Texture of Aluminum 3003-H18 Alloy. The Ohio State University, 2015.

[10] D.E. Schick. Characterization of aluminum 3003 ultrasonic additive manufacturing. The Ohio State University, 2009.

[11] R. Dixon, S. Chen. Fundamentals of Metal and, ASM handbook: Welding, brazing, and soldering 6 (1993) 143.

[12] Y. Chen, K. Nakata. Microstructural characterization and mechanical properties in friction stir welding of aluminum and titanium dissimilar alloys, Materials & Design 30 (2009) 469-474.

[13] W.-B. Lee, M. Schmuecker, U.A. Mercardo, G. Biallas, S.-B. Jung. Interfacial reaction in steel–aluminum joints made by friction stir welding, Scripta Materialia 55 (2006) 355-358.

[14] M. Maalekian. Friction welding–critical assessment of literature, Science and Technology of Welding & Joining 12 (2007) 738-759.

[15] M. Uday, M. Ahmad Fauzi, H. Zuhailawati, A. Ismail. Advances in friction welding process: a review, Science and Technology of Welding & Joining 15 (2010) 534-558.

[16] T. Watanabe, H. Takayama, A. Yanagisawa. Joining of aluminum alloy to steel by friction stir welding, Journal of Materials Processing Technology 178 (2006) 342-349.

[17] P. Xue, B. Xiao, D. Ni, Z. Ma. Enhanced mechanical properties of friction stir welded dissimilar Al–Cu joint by intermetallic compounds, Materials science and engineering: A 527 (2010) 5723-5727.

Effective Friction Stir Welding of Thick Section Aluminium Alloys

Jonathan Martin
TWI Technology Centre (Yorkshire) Ltd, Rotherham, South Yorkshire, UK
Jonathan.martin@twi.co.uk, +44(0)114 269 9046,

Abstract
There are many applications for which joining of thick section aluminium components is required. These include aircraft wing spars, liquefied natural gas (LNG) tanks, armour plate, and railway headstocks. Using conventional fusion welding techniques for such applications can be challenging, as typically:

- Component edge preparation and a shielding gas are necessary;
- Preheat is required;
- The process is difficult to control;
- A multi-pass technique is required, leading to low productivity;
- Welds are prone to porosity and cracking;
- Welding can lead to component distortion;
- Some aerospace alloys, particularly in the 2xxx and 7xxx series, are difficult, or impossible, to fusion weld.

Recent innovations at TWI, using the unique twin headed capabilities of the *PowerStir*^{TM} friction stir welding (FSW) machine, have demonstrated several techniques which have the potential to minimise production cycle times, making joining of thick section aluminium alloys both technically and commercially advantageous. This paper describes these techniques and the work undertaken at TWI to develop them. Process benefits and weld properties are discussed, highlighting the potential benefits including:

- Reduced component processing times through shorter weld times and minimal part handling;
- Improved weld quality through use of reduced and uniform heat input;
- Increased tool life, leading to reduced tooling cost;
- Reduced distortion.

Introduction
There are many applications for which joining of thick section aluminium components is required. These include aircraft wing spars, LNG tanks, armour plate, and railway headstocks. Using conventional fusion welding techniques can be challenging, as typically:

- Component edge preparation and a shielding gas are necessary;
- Preheat is required;
- The process is difficult to control;
- A multi-pass technique is required, leading to low productivity;
- Welds are prone to porosity and cracking;
- Welding can lead to component distortion;
- Some aerospace alloys in the 2xxx and 7xxx series are difficult, or impossible, to fusion weld.

Friction stir welding (FSW) has been the subject of rapid development and application across the world since its invention, by TWI, in 1991 [1], and is now a well-established joining technique for aluminium alloys. The majority of industrial applications involve joining components in the thickness range of 2-8mm. Companies gain many benefits from using this technique; however, the technical and commercial benefits of joining thicker aluminium alloys are less well established. Producing thick FSW section welds places additional demands on the FSW machine/tooling. This leads to slow welding speeds/cycle times and high weld heat input, which can have a negative effect on the weld properties and tool life. Typically, to weld thicker plate, a Weld-Flip-Weld approach is used. This consists of using a FSW tool with half thickness component penetration. After the first weld has been made, it is normally necessary to remove any flash from the weld surface before releasing the component from the fixture. Some plate distortion can occur, so once the partially fabricated plate has been lifted and turned, it must be placed back in the fixture and forced into position prior to completing the second weld. These additional operations (component cleaning, pilot holes, handing, and refixturing), especially when fabricating large and heavy components, can be a very significant portion of the total time required to fabricate the component.

TWI has been developing several novel FSW techniques to minimise cost, production cycles, and weld heat input, making joining thick section aluminium alloys both technically and commercially more advantageous. All these techniques require the unique capabilities of the *PowerStir* FSW machine (Fig. 1) at the TWI Technology Centre. This multi-functional R&D machine has opposed synchronised spindles, capable of running and supporting tooling. Although there are many FSW machines with opposed spindles, it is believed that this is the only machine that can run tooling capable of at least 50mm penetration on each head.

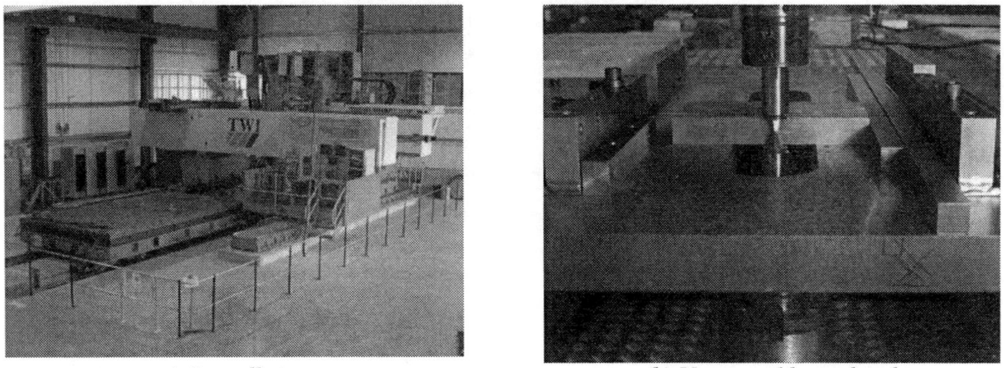

a) Overall view *b) Upper and lower head*

Fig. 1 PowerStir FSW Machine.

Initial trials, when simultaneously welding from either side of 50mm thickness aluminium alloy plates, have shown that the welding speed can be significantly increased when compared with the Weld–Flip–Weld joining technique.

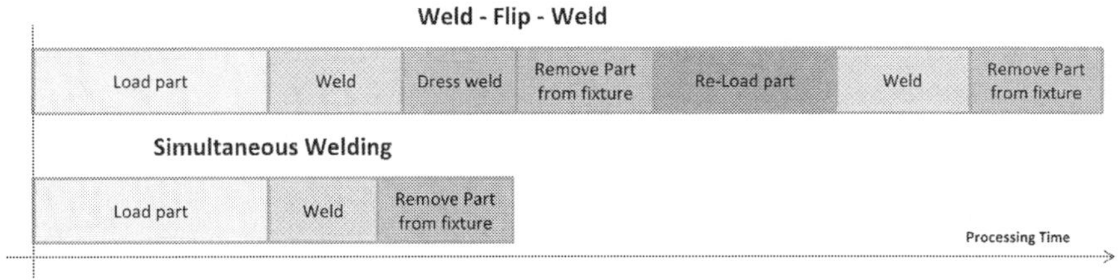

Fig. 2 Comparison of total process time for Weld-Flip-Weld and simultaneous welding.

The effect on the total process cycle is very significant, as shown in Fig 2. Not only is the welding time reduced, but secondary operations such as dressing, drilling a pilot hole, and refixturing a potentially distorted component, are eliminated. These benefits have to be considered against the increased capital cost of ownership of a more capable FSW machine.

One of the main concerns when welding thicker materials is the likelihood of tool breakage. Welds are typically made at a lower traverse speed to minimise the forces on the tool, but this may have a negative effect on weld properties. Welding with smaller tools from either side reduces the risk of tool failure, as bending stresses are reduced; two new techniques are being developed and assessed which can potentially:

- Produce a through-thickness weld in one setup;
- Eliminate the need for a second driven head;
- Reduce stress and deflection on the tool, leading to longer life;
- Operate smaller and cheaper tooling.

The first of these techniques is called Supported Bobbin FSW. The bobbin FSW, or floating bobbin FSW technique, used in these trials can offer the following advantages:

- Eliminate kissing bonds;
- Lower forces on fixture and machine;
- Lower distortion due to more uniform heat input.

However, as the component thickness increases, it becomes increasingly difficult to control the position of the bottom bobbin shoulder. Fig. 3 shows a macrograph taken through a weld made in 40mm thickness AA6082-T6, which shows that the upper and lower weld surfaces are not parallel. There is a narrowing of section on the weld advancing side (right) and a thickening on the retreating side. This phenomenon was termed 'wedging'.

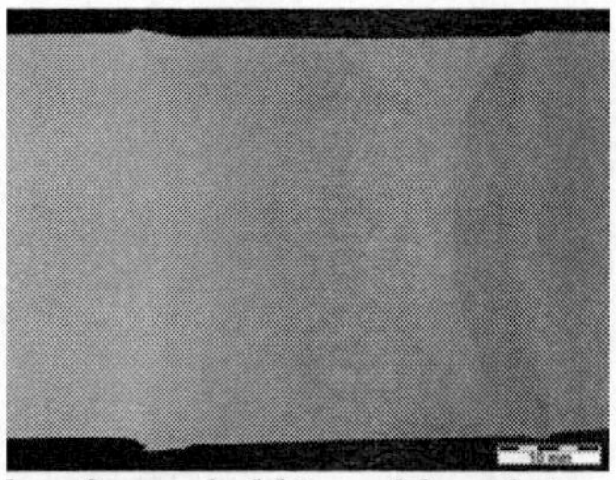

Fig. 3 A macrograph taken from a bobbin weld made in 40mm thick AA 6082-T6.

The probe connecting the upper and lower shoulder of the bobbin is therefore rotating in an abnormal manner during welding, which leads to premature tool failure.

In the Supported Bobbin technique, the bottom shoulder is radially restrained by a bearing in the under head as shown in Fig. 4.

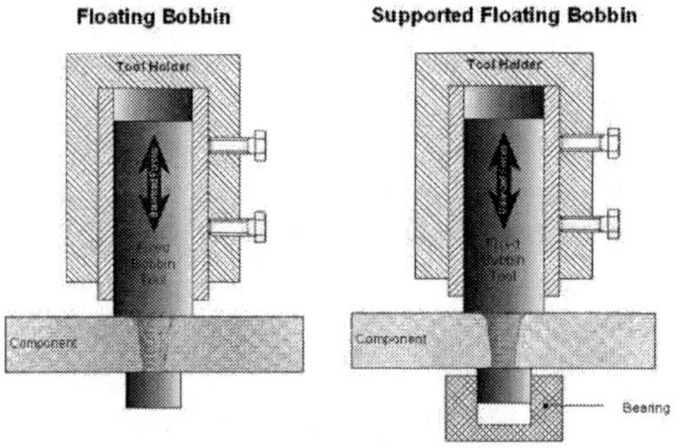

Fig. 4 Schematic of Floating Bobbin FSW and Supported Bobbin FSW.

An example of a weld made using this technique is shown in Fig. 5 (a). This weld was made in 50mm thick AA5083-O, produced using a welding speed of 120mm/min, without tool failure, and the upper and lower weld surfaces were nominally parallel.

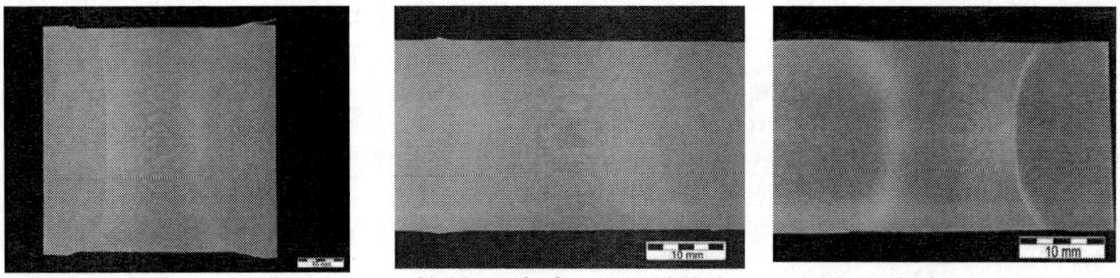

a) 50mm thickness AA5083-O *b) 25mm thickness AA5083-O 25mm probe* *c) 25mm thickness AA5083-O 16mm probe*

Fig. 5 Macrographs taken from a supported bobbin welds.

Further trials were undertaken to assess the performance of smaller sized bobbin tools. Using conventional floating bobbin tools to join 25mm thickness AA5083-O, the maximum welding speed which could be realised was 100mm/min without 'wedging'. Using a bobbin tool design featuring a smaller probe and shoulders, it was possible to increase the tool rotation, and also welding speed to 250mm/min, using the supported bobbin technique. The calculated spindle energy required to make these welds reduced from 4.7kJ/mm for the larger tool to 2.5kJ/mm for the supported smaller tool. Macrographs taken from these welds are shown in Fig. 5 (b) and (c) respectively.

The second technique is called Supported FSW, and is based on the dual-rotation FSW technique. In these trials, the shoulder was kept stationary, and only the probe rotated. A schematic diagram of this technique is shown in Fig. 6 (a).

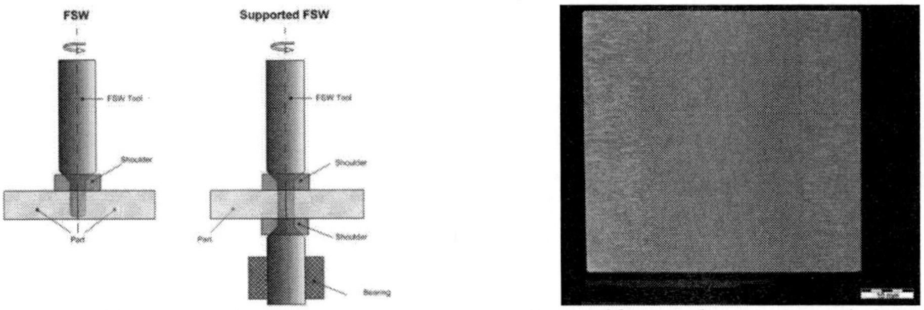

a) Schematic drawing *b) A welding made using Supported FSW in 50mm thickness AA6082-T6*

Fig. 6 Supported FSW.

The basic principle uses two stationary shoulders, top and bottom, with the FSW probe passing through the shoulders and component, and then supported by a bearing in the lower head. It was expected that this technique, similar to the supported bobbin technique, would:

- Increase the weld depth possible by a single weld;
- Use small diameter probes/shoulders, and therefore increase welding speeds;
- Be impossible to get lack of penetration defects.

However, with this technique, as heat is generated only by the probe, a lower weld heat input was expected. This may be beneficial when joining aluminium alloys, which obtain their strength by solution heat treated and aging, as less thermal damage to the material properties would occur. A limited number of welds have been made using this technique, and it has been demonstrated that it is possible to make sound welds as shown in Fig. 6 (b).

Conclusions

Recent innovations at TWI, using the unique twin headed capabilities of the PowerStir FSW machine, have demonstrated several techniques which have the potential to minimise production cycle times, making joining of thick section aluminium alloys both technically and commercially advantageous. These techniques need further development to understand more fully the extents of the process window, weld properties, and limits on tool geometry; however, this work has demonstrated the potential to:

- Reduce component processing times through shorter weld times and minimal part handling;
- Improve weld quality through use of reduced and uniform heat input;
- Increase tool life leading to reduced tooling cost;
- Reduce distortion.

References

[1] Thomas W M, Nicholas E D et al: 'Improvements Relating to Friction Welding', European Patent Specification 0 615 480 B1, December 1991.

A study on the correlation between mechanical properties and axial force of friction stir welded AA6082-T6 joint

Han Li, Gong Zhang, Qing-yu Shi*

State key laboratory of tribology, Department of mechanical engineering, Tsinghua University, Beijing, China

shqy@mail.tsinghua.edu.cn

Abstract

Friction stir welding (FSW) is an important technic for manufacturing high strength light metals. Axial force is introduced as a primary control parameter in FSW process in recent years. Yet, unlike rotating speed and welding speed, researches on the correlation between axial force and joint properties are relatively limited. In this study, FSW butt welds of AA6082-T6 plates are obtained under different axial forces. After welding, tensile tests and root bending are carried out to examine the mechanical properties of joints. It is noticed from the analysis of the experiment results that sufficient axial forces is essential for material joining during FSW, tunnel defect forms when axial force is low, and axial force is also a determinate parameter on root defect generation. Yet, once axial force is sufficient for producing defect free weld, increasing axial force further would slightly reduce the tensile strength but increase the elongation.

Key words: Friction Stir Welding, Axial Force, Mechanical property, Defect

Introduction

Friction stir welding(FSW) is a promising welding technology, as a solid phase process FSW can prevent some major defects caused by fusion process, such as hot crack and porosity. This advantage make FSW especially suitable for welding aluminum and magnesium alloy, and it has been widely used in transportation industry for welding high strength joints[1]. Mechanical properties and internal defects of FSW joints are determined by welding parameters such as rotating speed, welding speed, plunge depth, etc.[2]. In recent years development of robotic FSW introduced force control technic into FSW, making axial force another major control parameter of FSW process. Due to this hardware development, effects of axial force on FSW have come into the research scope. Rajakumarx[3-4] et al. studied the effect of axial force on tensile properties of different aluminum alloys, find that when axial force is tensile strength of joint would firstly increase to a maximum and then decrease, meanwhile elongation of joint would increase monotonically. Upadhyay[5] studied axial force, hardness and material temperature of FSW AA6056, find that axial force affect hardness profile by affecting temperature history. Jayaraman[6] studied cast A356 aluminum FSW, and find tunnel defect generated when axial force was either too low or too high. However, the researches about axial force are still limited comparing the massive studies of rotating speed and welding speed. What's more, force controlled architecture could reduce or eliminate the plunge depth reduction caused by device deformation, thus it helps reducing root defect of FSW joint and raising its mechanical properties, yet experimental study on this topic is still rare. In this paper an attempt is made to understand how axial force affect mechanical properties of AA6082-T6 butt joint, and explores the correlation between axial force and root defect, in this way it would provide some preliminary knowledge on the role of axial force in FSW.

Experiment

In this study 5mm thick plates were butt welded on mild steel anvil. 5 welds are produced under 6000, 8000, 10000, 12000, 14000N axial forces, all other parameters are kept constant in 5 trials. Process parameters except of axial force are shown in Table 1. Former experiments show that welds with good surface formation can be made with axial forces within this range, lower axial force could lead to severe groove and higher axial force may cause the tool shoulder submerged into the workpiece, destabilizing the welding process and destroying the weld. After the joints have been made, longitudinal specimens for tensile and root bend tests are taken from the welded plate. The configuration and size of the specimens are controlled by ISO 4136-2001 and ISO 5173-2000 standards. Then tensile and bend tests are performed under room temperature on a WDW3020 universal testing machine with 0.5mm/min machine head moving speed. Vickers hardness along the half thickness line of joint cross sections is tested with 1mm interval on a MHVD-5MP micro hardness tester. And optical micrographs of joint cross sections perpendicular to the welding direction are taken in a microscope for root defect observation.

Table 1: Process parameters

Pin shape	Pin length(mm)	Pin diameter(mm)	Shoulder diameter(mm)	Rotational speed(rpm)	Welding speed(mm/min)
Threaded cylinder	4.7	5	20	1800	200

Results and discussion

Tensile strengths and elongations obtained from the tensile tests are shown in Figure 1. It could be noticed that when axial force is raised from 6000N both tensile strength and elongation rise rapidly, but when axial force is further raised, tensile strength reaches a maximum at 10000N axial force and then mildly drop down, meanwhile elongation before break keep rising monotonically.

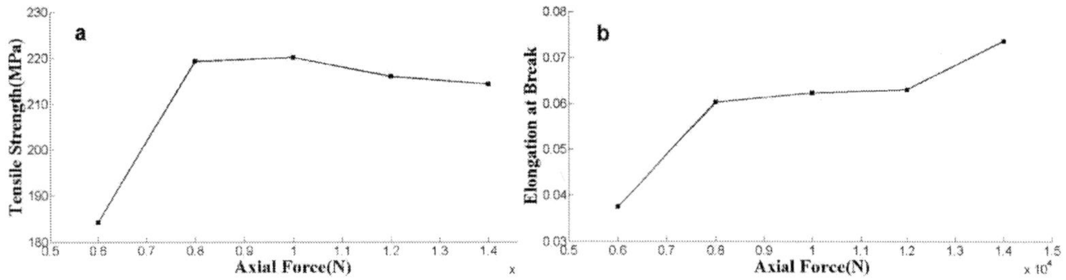

Figure 1: (a)Tensile strength evolution with axial force, (b)Elongation evolution with axial force

The rapid rise of strength and elongation at the beginning could be explained by severe internal defects generated under low axial force. Tunnel defects are found in the joint welded by 6000N axial force, as shown in figure 2, and no obvious defect could be found in joints welded under higher axial force. Internal defect could reduce actual load bearing area of joint and initiate crack under tensile stress, reducing joint mechanical properties. That is the reason why tensile strength and elongation of specimen welded under 6000N are significantly lower than other specimens. This phenomenon indicates sufficient axial force is necessary for good joint integrity.

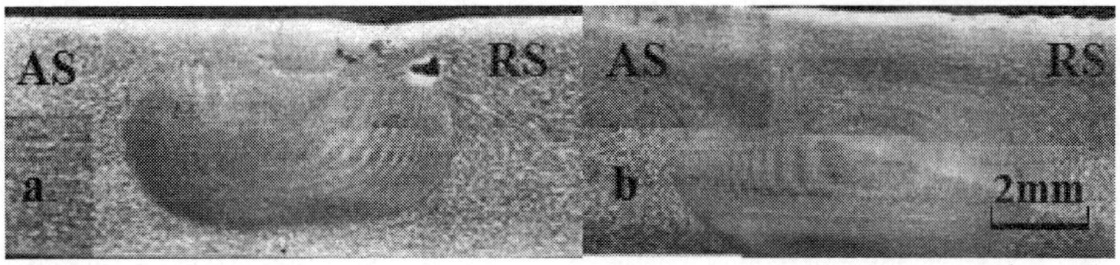

Figure 2: Optical micrographs of joint cross sections perpendicular to the welding derection, (a)joint welded under 6000N axial force,(b)joint welded by 12000N axial force, notations "AS" and "RS" mean advancing side and retreating side

The break locations of the tensile tests and crack status after root bend 180° are shown in table 2, notations "HAZ" means heat affected zone. When axial force is relatively low, the tensile break locations are in the nugget and joints would crack after root bend, giving a significant signal of root defect existence. When axial force is high, the tensile break locations move to RS HAZ and joints are not cracked after root bend. This indicates the root defect is diminished by high level axial force.

Table 2: Tensile test break location and root bend test crack status

Axial force(N)	6000	8000	10000	12000	14000
Tensile break location	nugget	nugget	RS HAZ	RS HAZ	RS HAZ
Crack status after root bend 180°	cracked	cracked	cracked	non cracked	non cracked

After polishing and etching root defect can be observed in microscope. Identifying the root defect length as length between root defect tip and joint back surface, effect of axial force on root defect could be analyzed quantitatively, as shown in figure 3. Root defect scale decreases with increasing axial force, and when axial force increases to 12000N root defect is eliminated. This proves root defect generation can be restrained by high axial force.

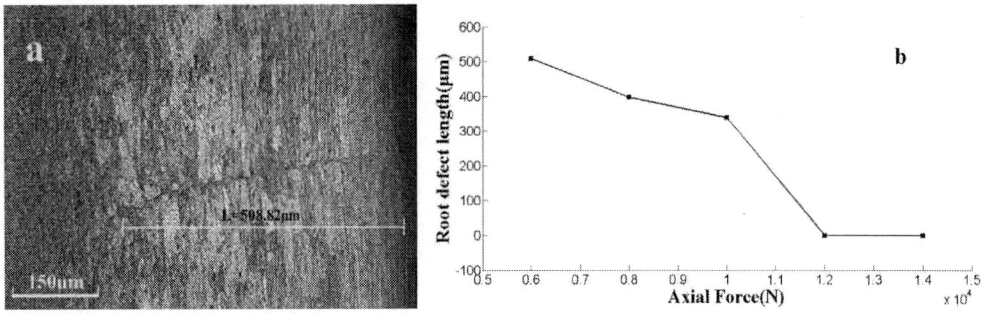

Figure 3: (a)Root defect observed in microscope, (b)Root defect length eolution under rising axial force

Figure 4 shows micro Vickers hardness profiles of the 5 joints. When axial force increases from 6000N to 14000N, hardness profile evolves from "U" shape to "W" shape. The lowest hardness always locates in HAZ, and it continues decreasing with increasing axial force. This indicates HAZ is the weakest part of joint and higher axial force would weaken it further. Since joint's overall strength is restricted by its weakest part, these hardness profiles show that joint's tensile strength should decrease with increasing axial force if defect influence is excluded. It implies the ascending stage of tensile strength evolution in figure 1.a reflect the process of eliminating defects.

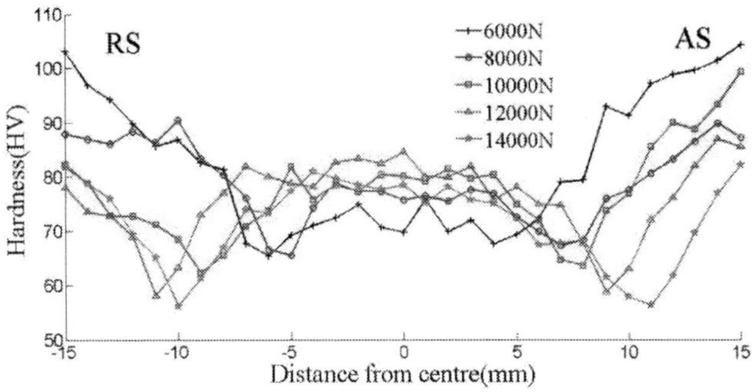

Figure 4: Micro Vickers hardness profiles

Conclusions

1) Axial force could improve integrity of FSW joint, sufficient axial force is necessary for producing defect free joint. Insufficient axial force would cause tunnel defect and root defect, when axial force is raised root defect scale gradually decreases until root defect is eliminated.

2) When axial force is insufficient to avoid internal defects, raising axial force could reduce defects and in turn improving tensile strength. After axial force is sufficient to avoid defects, raising it further would reduce tensile strength. But elongation would increase monotonically when axial force is raised.

References

[1] B.T. Gibsona, et al: "Friction stir welding: Process, automation, and control", Journal of Manufacturing Processes, Vol.16(2014), pp.56-73

[2] Mishra R S, et al: "Friction stir welding and processing", Materials Science and Engineering R, Vol.50(2005), pp.1-78

[3] S. Rajakumar , et al: "Influence of friction stir welding process and tool parameters on strength properties of AA7075-T6 aluminium alloy joints", Materials & Design, Vol.32(2011), pp.535-549

[4] K. Elangovan, et al: "Influences of tool pin profile and axial force on the formation of friction stir processing zone in AA6061 aluminium alloy", International Journal of Advanced Manufacturing Technology, Vol.38(2008), pp.285-295

[5] P. Upadhyay, et al: "Effects of forge axis force and backing plate thermal diffusivity on FSW of AA6056", Materials Science & Engineering A, Vol.558(2012), pp.394-402

[6] M. Jayaraman, et al: "Effect of process parameters on tensile strength of friction stir welded cast A356 aluminium alloy joints", Transactions of Nonferrous Metals Society of China, Vol.23(2013), pp.605-615

Refill Friction Stir Spot Welding for Aerospace application

Shintaro FUKADA, Mitsuo FUJIMOTO, Syuhei YOSHIKAWA, Shunsuke HARUNA
KAWASAKI HEAVY INDUSTRIES, LTD., Corporate Technology Division, Akashi, Hyogo, Japan
fukada_shin@khi.co.jp, +81-78-921-1692

Kenichi KAMIMUKI, Hideki OKADA,
KAWASAKI HEAVY INDUSTRIES, LTD., Aerospace Division Kakamigahara, Gifu, Japan

Abstract

The Friction Stir Spot Welding (FSSW) is an emerging solid-state spot joining method that creates a molecular-level bond between the work-pieces through a thermo-mechanical processing without consuming any filler or foreign material. Especially the refill type of FSSW, which is named as Refill Friction Spot Joining (RFSJ), is a highly anticipated technology for the aircraft manufacturing due to its flush surface finish without a key or exit hole. The most of precursor RFSJ study has demonstrated promising results of joining high-strength aerospace aluminum alloys with bare surface finish. However, in aircraft manufacturing, parts are usually pre-treated before assembly process. A surface treatment, such as chemical conversion or anodic coating, is essential for ensuring the best possible adhesion and properties of the organic coating on the metallic surface. In addition, surface treatments prevent the substrate surfaces from corrosion. Therefore, the presented study investigated the weld ability and mechanical properties of surface-treated aluminum alloys, and discussed further process improvement.

Introduction

The Friction Stir Spot Welding (FSSW) is a solid-state joining technology that can produce a lap spot joint between sheet metals, preferably aluminum alloys, without any filler or foreign materials. During the FSSW process, the rotating tool plunges into the work-pieces and stirs them together to establish a fully-consolidated spot joint. As the solid-to-liquid phase transition is not involved in the FSSW process in general, there is no lack of fusion or material deterioration caused by the phase transition. Kawasaki Heavy Industries has introduced Refill Friction Spot Joining (RFSJ) [1] and a similar refill friction stir spot welding (RFSSW) technique was developed by GKSS-GmbH [2]. The remarkable feature of these methods is that it can make joints with no keyholes, a perfectly flush surface finish.

The aerospace industry employs solid rivets for fastening the primary structures as they have well-established standards and specification. However, the riveting process consists of tedious resource-intensive steps, such as hole-drilling, deburring, fastener insertion, and fastener clinching, which reduce the cost efficiency of the entire assembly process. Furthermore, the rivet head and the flange overlap introduce additional weight to aircraft. Refill FSJ is expected to replace the rivets to achieve the weight reduction, the cost saving and the manufacturing time reduction.

In aerospace industry, most of aluminum parts are surface-treated before assembled; the surface treatment is required to ensure the adhesion of primer paint and corrosion resistance in aerospace industry. Okada et al presented metallurgical and mechanical properties and corrosion resistance of RFSJ joint with chemical conversion coating [3]. There are some kinds of coating, not only chemical conversion coating but also anodic coating. There is not enough discussion for metallurgical and mechanical properties of RFSJ with each kind of coatings. Therefore, this study investigated the joint properties of surface-treated materials, and SEM analysis on anodic coated material.

Process

The RFSJ system uses a weld tool with a retractable probe and a shoulder that are contained inside a cylindrical clamp (*Fig. 1*). RFSJ has several variations, which are including the probe-plunging process and the shoulder-plunging process [4,5]. *Fig. 1* shows the illustrations of the shoulder-plunging process which was used in this study. The shoulder-plunging process is as follows; (1) Clamping and pre-heating, (2) Plunging rotating shoulder into the material, (3) Re-plunge, (4) Tool removal. The shoulder-plunging process can make the nugget larger than that of the probe-plunging process and can produce stronger joints.

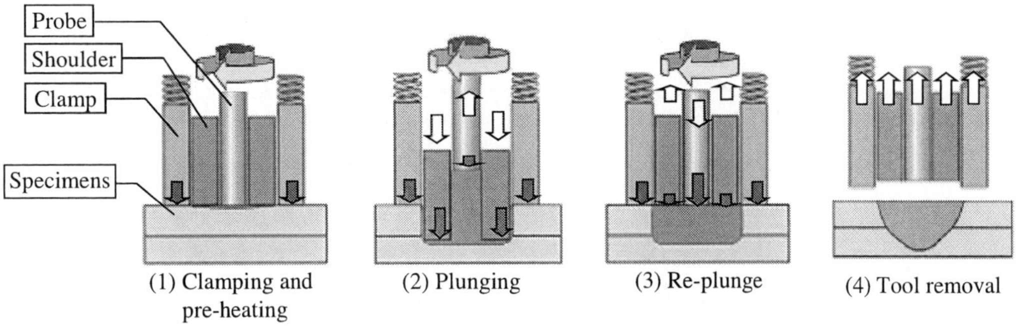

Fig. 1: Schematics of RFSJ processes

Fig. 2 shows RFSJ equipment systems. The welding head has 3 independent servomotors for the load of the joining tool, the rotational speed, and the position of the probe. The maximum Z force is 14,700N and rotation speed is 2160rpm, so that it can join high-strength 7xxx series aluminum alloys up to 4mm of upper-sheet thickness. The end-effector can also be operated with robotic system which Okada et al presented for actual prototype trials [3,6]. The close-up of joining tool is shown in *Fig. 2 (b)*. There is the probe in the center. The shoulder encloses the probe and the clamping system encloses the shoulder. The clamping system grips and holds the work-pieces tightly to prevent burring, displacement and mechanical deformation during the welding process. In addition, the clamping system with springs enables the probe and the shoulder to recede from the work-piece surface. In addition, the high performance coating of the tool prevents work-piece alloys from adhering to the tool surface.

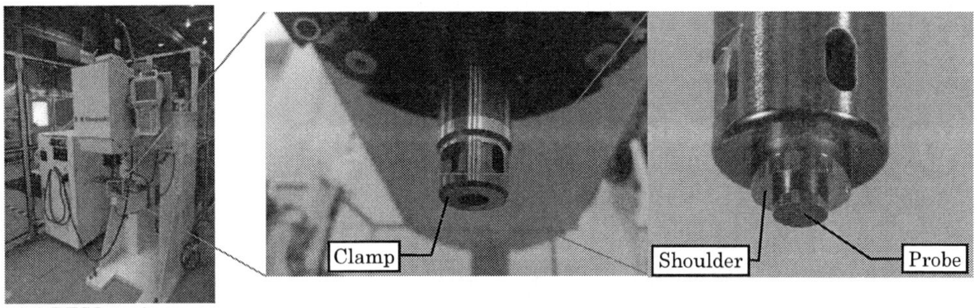

(a) RFSJ equipment *(b)Close-up of RFSJ tool*

Fig. 2: Refill FSJ system

Set-up

In this study, AA7075-T6 t1.016mm (0.040 inch) aluminum alloy sheets and 2 types of surface treatment were used. *Table 1* shows chemical components of sheet material, specification of surface treatments, and illustration of the shape of the specimens. The test coupons were fixed on the fixture with some toggle clamps. 9mm diameter RFSJ tool was applied in this study.

Table 1: Chemical components, specification of surface treatments, and illustration of the shape of the test pieces

Sheet material of specimens										
AA7075-T6 alloy (AMS 4045)										
	Chemical components[Wt.%]									
	Al	Cr	Cu	Fe	Mg	Mn	Si	Ti	Zn	Others
MAX	91.4	0.28	2.0	0.5	2.9	0.3	0.4	0.2	6.1	0.15
MIN	87.1	0.18	1.2	-	2.1	-	-	-	5.1	-

Surface treatments	
Non-coating	-
Chemical conversion coating	MIL-C-5541 Class 1A Chromic acid coating (Alodine #1200)
Anodic coating	MIL-A-8625 F Type 1, Class 1 Chromic acid coating

Shape of specimens

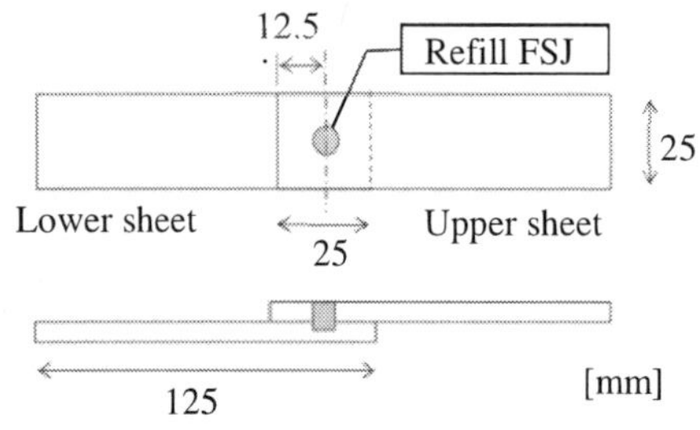

Joint strength

Fig. 3 shows tensile shear test result on non-coated specimens, chemical conversion and anodic coated specimens. The failure mode of all specimens was nugget pull out as shown in *Fig. 3 (a)* in this study. *Fig. 3 (b)* shows tensile shear strength, the strength of joints with chemical conversion coating resulted in almost same strength as non-coated joints. However, the strength of joints with anodic coating was lower than other two conditions.

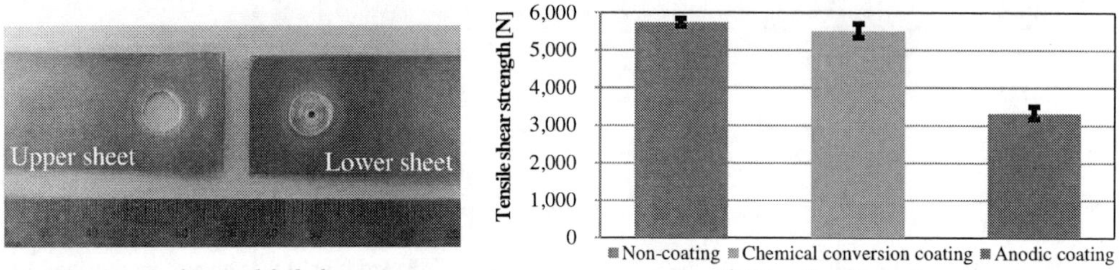

(a) Appirace of typical failed specimen *(b) Tensile shear strength*

Fig. 3: Tensile shear test results

SEM Fractography

Fig. 4 shows SEM analysis on fracture surface with anodic coating. *Fig. 4 (a)* shows over-view of failed area on upper sheet, and *Fig. 4 (b)* shows SEM image of the fractured surface on red boxed area in *Fig.4 (a)*. The fracture surface could be divided into two different texture areas; the dimple texture which indicated ductile fracture was observed in upper side, on the other hand, stripe texture without dimple was observed on interface side. *Fig. 4 (c)* shows magnified Back Scatter Electron (BSE) image on the striped area. Some white particles were observed in this area. Energy Dispersive X-ray (EDX) spectrometry indicated these white particles included Chrome and Oxygen, therefore they were supposed to be anodic coating.

Fig. 5 shows tensile shear tested specimen stopped before failure. As shown in *Fig.5 (a)* and *(b)*, cracks were observed at the edges of stir zone, and these cracks were likely identified as the striped area shown in *Fig. 4*. *Fig. 5 (c)* shows the results of EDX analysis on the red circle of *Fig.5 (b)*. This EDX results indicated Chrome and Oxygen were observed at the crack area, it supported the assumption that the cracks in cross-section fitted in the stripe area in fracture surface.

These analyses suggested that anodic coating remained at the edge of stir zone caused unjoined or weakly-joined interface, it is called as "Hooking", and it should make joint strength lower. Further study or investigation can reveal how this "Hooking" was created and it will contribute to improve joint strength with anodic coated materials.

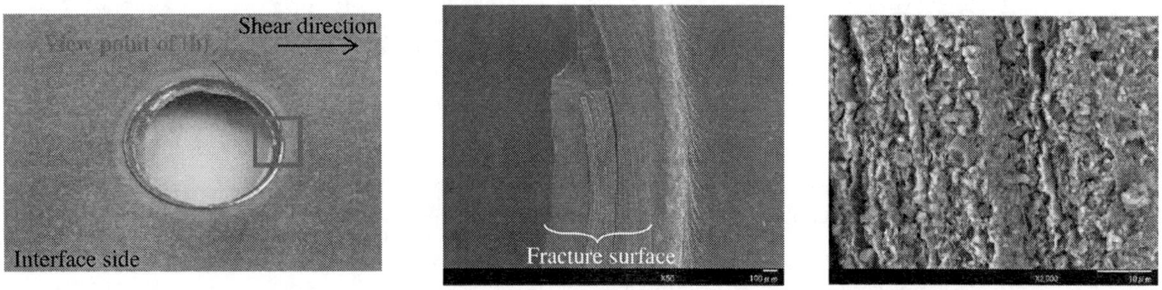

(a) Overview of failed area on upper sheet *(b) SEM image of fracture surface (Redboxed area in (a))* *(c) BSE image on stripe area (Red boxed area in (b))*

Fig. 4: SEM analysis on fracture surface

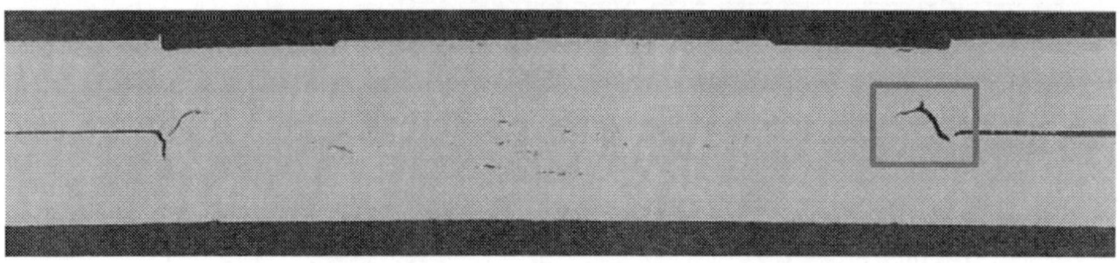

(a) Overview of cross-section before plug fractured

Fig. 5: Cross-section analyses before failure

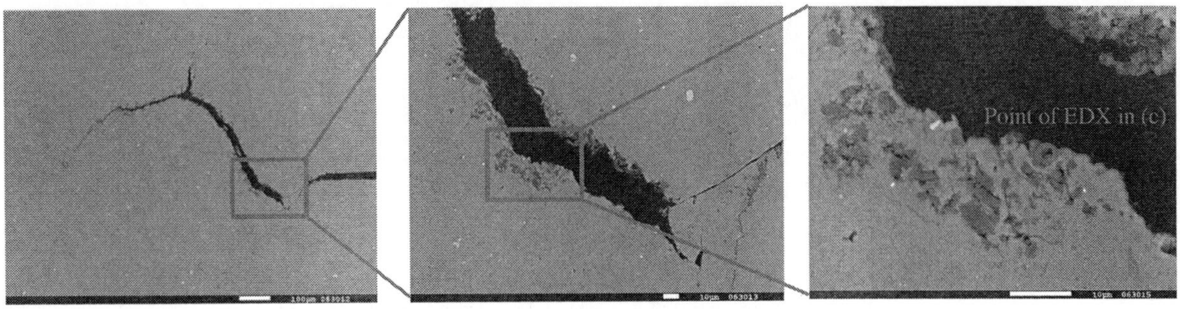

(b) SEM images on the crack (Red boxed area in (a))

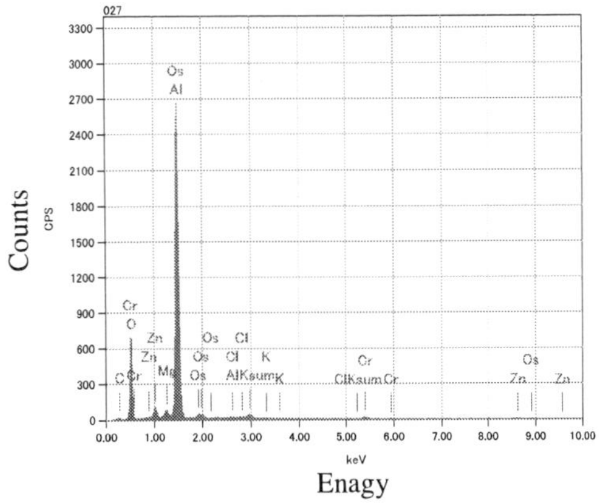

(c) EDX analysis on the crack (Red circulered area in (b))

Fig. 5 Continued

Conclusions

Basic characterization on the joint strength and fracture was conducted in this study. Nugget pull out failure was observed on non-coated, with chemical conversion coating and anodic coating. However, tensile shear strength with anodic coating indicated lower strength than the others. From the SEM fractography on anodic coated specimens, the coating was remained at the edge of stir zone as some particles and they caused “Hooking” which make joint strength lower. Further study or investigation can reveal how this “Hooking” was created and it will contribute to improve joint strength with anodic coated materials.

References

[1] Okada H., Kamimuki K., and Fujimoto M.: “Application Study of Refill FSSW for Aerospace Industry”, *Proceedings of Friction Stir Welding and Processing Symposium VII*, TMS, 2013.

[2] Schilling C., and Dos Santos J.: “METHOD AND DEVICE FOR JOINING AT LEAST TWO ADJOINING WORK PIECES BY FRICTION WELDING” U.S. Patent 6722556.

[3] Okada H., Kamimuki K., Yoshikawa S., Fukada S.: “Refill Friction Spot Joining for Aerospace Application”, *SAE Technical Paper*, 2015-01-2614, 2015, doi:10.4271/2015-01-2614.

[4] Badarinarayan H., Hunt F., and Okamoto K., “Chapter12: Friction Stir Spot Welding,”, Friction Stir Welding and Processing, First Edition, edited by Mishra R. and Mahoney M.W., The ASM International, Novelty, OH, ISBN 978-0-87170-840-3, pp. 235-272, 2007.

[5] Okada H., Kamimuki K., and Fujimoto M., “Friction Spot Joining for Aerospace Industries” *The Japan society for Aeronautical and Space Sciences, 47th Aircraft Symposium*, JSASS-2009-5169, Nov 2009 (in Japanese)

[6] Okada H., Kamimuki K., and Fujimoto M., “Assembly Study of Refill FSSW,” *SAE Int. J. Aerosp.*, 6(1), pp.299-304, 2013, doi: 10.4271/2013-01-2310.

Flame-resistant Mg alloy joint formed by asymmetry double-sided FSW

Mengran Zhou, Yoshiaki Morisada, Hidetoshi Fujii

Joining and Welding Research Institute Osaka University, Osaka, Japan

email: zhoumr@jwri.osaka-u.ac.jp Tel: +0081-06-6879-8643

1. Introduction

Magnesium alloys are potential candidates to replace aluminum alloys in many structural applications owing to their unique properties of low density, high strength/weight ratio, etc. They are also considered as advanced materials in terms of electromagnetic shielding, energy conservation and environmental pollution regulation. However, the oxidation and combustion of magnesium alloys during the melting, forging and heat treatment process significantly detar their application. Fortunately, there have been enormous efforts in developing non-combustive magnesium alloys in recent years [1]. The alloying method by adding elements, such as alkaline, rare earth and their combinations, to magnesium alloys to improve the oxidation and combustion resistance is of great research and application prospects.

In order to broaden the application for this brand new magnesium alloy, an effective and inexpensive joining technique is necessary. Friction stir welding(FSW)[2], a novel solid state joining method invented by The Welding Institute(TWI) in 1991 has proved to be an effective defect-free joining method for this magnesium alloy. Because of the hexagonal close-packed (hcp) structures with few slip systems of the magnesium alloy, which results in a relatively poor formability and ductility, the conventional FSW joint of the magnesium alloy shows a strong anisotropy and lower tensile property in the stir zone which has limited its applications [3]. A previous study [4] indicated that by introducing a more complicated metal flow during the FSW of a magnesium alloy, such as setting an asymmetrical layout of double tools on both the upper and lower sides of the plate, a randomized texture and mechanical property optimized joint can be formed.

However, the effects of the lower tool for the AZX Mg alloy by asymmetry double-sided FSW (DFSW) have not been reported yet. Consequently, the aim of this study is to clarify the mechanical properties of the AZX Mg alloy joints formed by asymmetry DFSW, and the texture evolutions in several regions of the stir zone caused by the change of the lower tool rotation rate and rotation direction.

2. Experiment Procedure

The extruded flame-resistant AZX612 (Al-6% Zn-1% Ca-2% Mg-bal) magnesium alloy sheets were used in this study. The dimensions of the plates are a width of 65mm, a length of 200mm, and a thickness of 3mm. The asymmetry DFSW was conducted using the upper tool with a threat probe and a flat lower tool (schematically shown in Fig.1). The welding parameters are listed in Table 1.Both the upper tool and lower tool were made of JIS-SKD61 tool steel (equal to AISI H-13 steel) and rotated CW during the joining, while only the upper tool was tilted 3 degrees. A schematic illustration of the asymmetry DFSW is shown in Fig.1. In this figure, AS means advancing side, while RS represents the retreating side. A previous study indicated that the AZX Mg alloy can be successfully joined at 600rpm for both the upper and lower tools. In this study, the upper tool rotation rate was set at 600rpm, while the rotation rate of lower tool was changed from 400rpm to 600rpm in order to investigate the effect of lower tool rotation rate on the joint, and upper tool and lower tool rotated reversely at 600rpm in order to investigate the effect of the lower tool rotation direction. The joining speed was set at 500mm/min. A conventional one-sided FSW with only the upper tool was done for comparison.

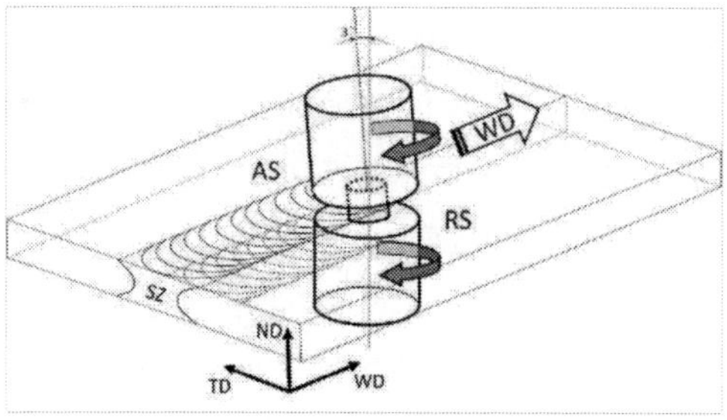

Fig.1 Schematic illustration of asymmetry DFSW with a flat lower tool.

	Upper tool rotation rate	Lower tool rotation rate	Tilt angle	Rotation direction	Joining speed
Asymmetry DFSW	600rpm	600,500,400rpm	3°	CW	500mm/min
Asymmetry DFSW	600rpm	600rpm	3°	Upper CW Lower CCW	500mm/min
One-sided FSW	1000rpm	-	3°	CW	500mm/min

Table.1 Joining parameters.

Optical microscopy and scanning electron microscopy (JEOL JSM-7001FA SEM) equipped with an electron back scattered diffraction (EBSD) system were used to examine the microstructures of the joint formed by the asymmetry DFSW. The optical microstructures were examined at the center of the joints. The EBSD measurements were conducted at the center line of the cross section parallel to the transverse direction (TD) and analyzed by TSL OIM. Samples were prepared using the same mechanical polishing parameters, then electronically polished. The tensile test was conducted at room temperature by SHIMAZU AGS-10kNX at the tensile speed of 1mm/min.

3. Results and Discussion

3.1 Optical Microstructures

Under all the conditions, asymmetry DFSW produced defect-free sound joints which can be observed in the optical microscopy image shown in Fig.2. The optical images of the cross-section indicated that the joints formed by the asymmetry DFSW (b,c,d,e) are completely different from that formed by the conventional one-sided FSW (a) due to the presence of the lower tool. The cross section of the DFSW joint is vertically symmetric and no clear boundary of the RS/BM can be observed. With the decreasing lower tool rotation rate, the shape of the stir zone was slightly changed (b-d). Additionally, the boundary of the SZ/BM on the AS side is distinct, while that on the RS side is not. This indicated that the material flow on the RS side during the asymmetry DFSW is different from that during the conventional one-sided FSW [3][4].

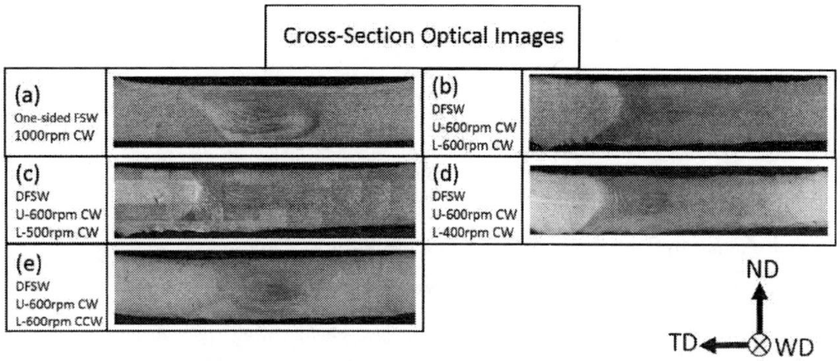

Fig.2 Transverse cross-sectional microstructure of conventional one-sided FSW (a) , asymmetry DFSW(b-e).

4. Microstructure and Texture Evolution

Fig.3 shows the orientation color maps from the EBSD measurement results at: (c) the center of the stir zone (SZ); (b, d) 1mm from the center of the stir zone on both the AS and RS sides (AS1, RS1); and (a, e) 2mm from the center of the stir zone on both the AS and RS sides (AS2, RS2). For comparison, the conventional one-sided FSW joint was examined. The orientation color maps of the conventional one-sided FSW undergoes a change in its color scheme with the change in position. Sato et al. [5] reported these results due to the following combined reasons: (1) relatively simple metal flow during the conventional one-sided FSW and (2) the hexagonal close-packed (hcp) structures with few slip systems of the magnesium alloy. For the asymmetry DFSW, when the upper tool and lower tool were both set at 600rpm, the orientation color maps from all 5 regions in stir zone show a randomized color scheme of several colors.

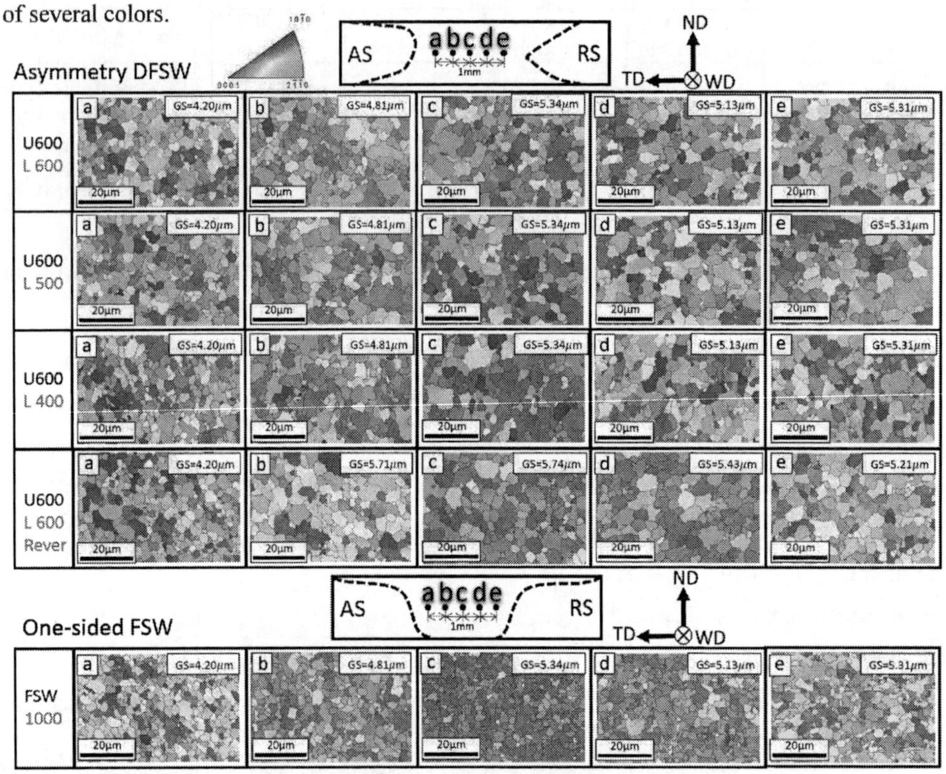

Fig.3 Orientation color maps of several regions in the stir zone

5. Tensile Properties

Fig.4 shows the transverse tensile properties of the joints and BM. The black bar shows the ultra-tensile strength, while the slanted bar indicates the elongation rate. With the decrease in the lower tool rotation rate and reverse lower tool rotation, the tensile strength decreased because the randomization of the crystal orientation decreased. Especially, for the reverse upper and lower tool rotation directions, the areas (c,d) were changed into the color scheme of that for the conventional one-sided FSW. Overall, the U600 L500 joint simultaneously shows a high tensile strength and elongation rate which is considered to be the preferable joining parameter for the flame-resistant Mg alloy.

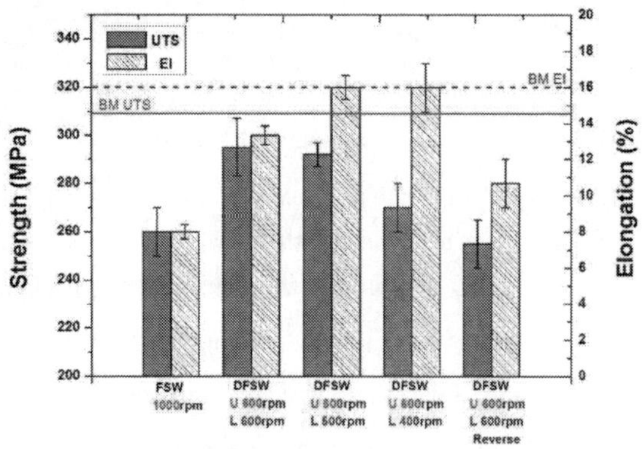

Fig.4 Transverse tensile properties of joints and BM

6. Conclusion

The flame-resistant Mg alloy was successfully joined by the asymmetry DFSW with different lower tool rotation rates. Based on the tensile test and texture measurement, the asymmetry DFSW is believed to be the preferable joining method of the AZX series flame-resistant Mg alloys.

(1). The asymmetry DFSW with a flat lower tool proved to be the preferable joining method than the conventional one-sided FSW for the AZX series flame-resistant Mg alloys because the joints were optimized in both tensile strength and elongation rate.

(2) Due to the asymmetry and asynchrony layout of 2 tools from the upper and lower sides, a more randomized crystallographic orientation in the stir zone caused by a more complicated metal flow was formed. It is considered to be the reason for the optimized tensile property.

7. Acknowledgement

This study was supported by the New Energy and Industrial Technology Development Organization (NEDO) under the "Innovation Structural Materials Project (Future Pioneering Projects)" and a Grant-in-Aid for Science Research from the Japan Society for Promotion of Science.

8. Reference

[1] Yoshihito Kawamura, " Flame-resistant Magnesium Alloys with High Strength", Magnesium Research Center (MRC), Kumamoto University, Japan

[2] Thomas W, Dawes C, Gittos M, Anderws D. TWI Bull 1998;May/June:44.

[3] Suhuddin, U. F. H. R., et al. "Grain structure evolution during friction-stir welding of AZ31 magnesium alloy." Acta Materialia 57.18 (2009): 5406-5418.

[4] Chen, Juan, Rintaro Ueji, and Hidetoshi Fujii. "Double-sided friction-stir welding of magnesium alloy with concave–convex tools for texture control." Materials & Design 76 (2015): 181-189.

[5] Mironov, S., et al. "Microstructure evolution during friction-stir welding of AZ31 magnesium alloy." Acta Materialia 100 (2015): 301-312.

[6] Fonda, R. W., and K. E. Knipling. "Texture development in friction stir welds." Science and Technology of Welding and Joining 16.4 (2011): 288-294.

[7] Xin, Renlong, et al. "Changes in texture and microstructure of friction stir welded Mg alloy during post-rolling and their effects on mechanical properties." Materials Science and Engineering: A 582 (2013): 178-187.

Fundamental role of FSW tool for stir zone formation

Naoki TAKEOKA, Yoshiaki MORISADA, Yufeng SUN, Hidetoshi FUJII
Joining and Welding Research Institute, Osaka University, Ibaraki, Osaka, Japan
fujii@jwri.osaka-u.ac.jp, +81-6-6879-8663

Introduction

Friction Stir Welding (FSW) [1] is a solid state joining process developed by TWI in 1991 and widely used for low-melting-point metals such as aluminium. [2-5] The tool used for FSW consists of a shoulder and a probe. There have been many studies evaluating the effect of the tool shape and the welding conditions on the formation of the stir zone.

Tozaki et al. proposed a no-probe tool with a scroll groove on its shoulder surface for preventing development of the Keyhole[6]. Colegrove and Shercliff developed "Trivex" tools and researched the welding parameters and material flow formed by various tool shapes [7]. Fujii et al. investigated the effect of the probe shape and welding conditions on the defect formation[8]. Sun et al. changed the tool shape and threads, and studied the influence of the tool design on the microstructure [9]. However, the fundamental role of the shoulder and the probe are still not sufficiently understood. It is difficult to evaluate it using the conventional tool because the tool parts form one unit.

In this study, "an adjustable tool" which can separately rotate the shoulder and the probe was used to clarify their fundamental role for forming the stir zone. The adjustable tool can rotate the shoulder part and the probe part separately. Additionally, the adjustable tool can control the heat input and strain rate more precisely.

Experimental procedure

Pure aluminum (A1050) plates of 5 mm×50 mm×300 mm were used as the work pieces. The FSW was conducted using an adjustable tool which can rotate the shoulder part and the probe part separately. Figure 1 shows a schematic illustration of the adjustable tool. The shoulder diameter was 15 mm and the probe length was 4.9 mm. In this study, four types of probes; i.e., a 6, 8 and 10 mm probe diameter with threads and a 6mm probe diameter without threads were used. Each part of the tool was made of tool steel (ISO 4957 40CrMoV5). The rotation rate of the shoulder and the probe were varied from 0 to 1500 rpm and 500 to 1500 rpm, respectively. The welding speed was 400 mm/min and the tool tilt angle was 3 degrees. The FSW was conducted in the position control mode. Cross sections of the FSWed joints were observed by optical microscopy for investigating the shape of the stir zone and existence of defects.

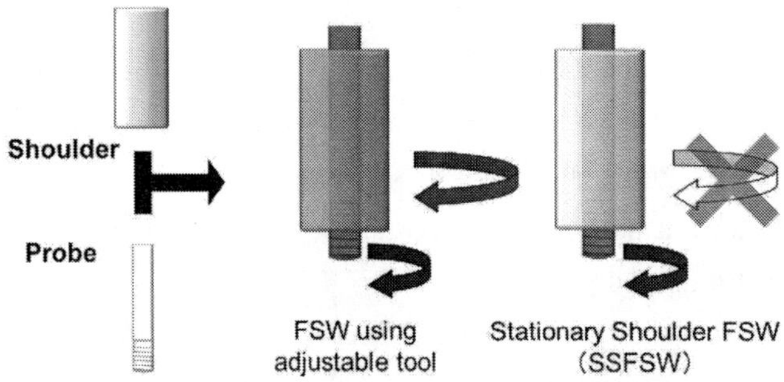

Fig.1 Schematic illustrations of the adjustable tool.

Results and discussion

Process window

Table 1 shows the process window in this study. The circle-mark means that no defects were formed under the welding conditions, while the cross-mark means that the joint had a tunnel defect at the bottom of the stir zone.

At the shoulder rotation rate of 500 rpm, a defect was formed when the probe rotation rate was from 500 to 1250 rpm. On the other hand, there was no defects under the condition when the probe rotation rate was 1500 rpm. It was considered that the defect formation was prevented by increasing the probe rotation rate, even though the shoulder rotation rate was very low. In Stationary Shoulder FSW (SSFSW) [10-11], the defect formation condition was much narrower than that of the FSW with the rotating shoulder of 500 rpm. It seems that a simpler material flow in the SSFSW expanded the optimal FSW conditions.

Table 1 The process window by using adjustable tool.

		Rotation rate of shoulder (rpm)			
		0	500	1000	1500
Rotation rate of probe (rpm)	500	×	×	×	○
	750	×	×	○	○
	1000	○	×	○	○
	1250	○	×	○	○
	1500	○	○	○	○
		SSFSW	FSW using adjustable tool		

Shape of the stir zone

Fig. 2 shows optical microscopic images of the cross sections of the FSWed joints formed by the adjustable tool at the various rotation speeds. Fig. 2 (a) and (b) show that the shoulder-affected zone extended in the depth direction with the increasing shoulder rotation rate from 500 to 1500 rpm. On the other hand, even with the increasing probe rotation rate from 500 to 1500 rpm, Fig. 2 (a) and (c) show that the probe-affected zone was not changed.

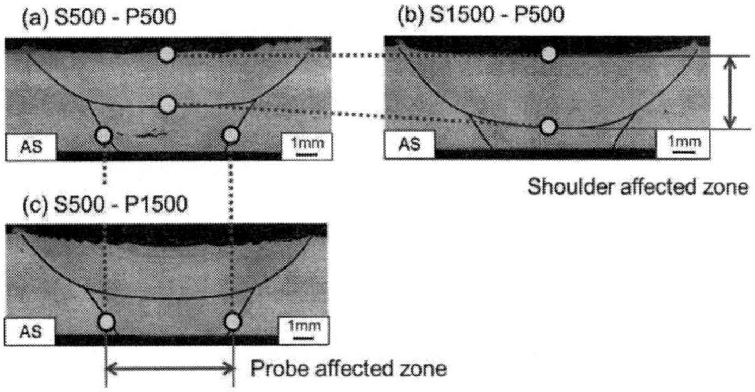

Fig. 2. OM images of the cross sections of the FSWed joints using the adjustable tool at various rotation rates: (a) shoulder rotation rate of 500 rpm / probe rotation rate of 500 rpm, (b) shoulder rotation rate of 1500 rpm / probe rotation rate of 500 rpm and (c) shoulder rotation rate of 500 rpm / probe rotation rate of 1500 rpm.

Width of the stir zone formed by the probe

In order to remove the effect of the shoulder on the stir zone, SSFSW was conducted, and the stir zone formed by only the probe was evaluated. Fig. 3 shows the relationship between the width of the stir zone formed by the probe and the rotation rate of the probe. The measurement location was at the center in the thickness direction (2.5 mm from the bottom). The stirred material was overflowed during the process with the M10 probe at the rotation rates of 500 and 750 rpm. Therefore, these two plots were not validated in this study. For either probe, the width of the stir zone formed by the probe was not changed by the rotating speed of the probe. Moreover, the shape of the stir zone was almost the same with or without threads. These results indicated that the width of the stir zone are not affect by the rotating rate regardless of the tools.

Fig. 4 shows the average width of the stir zone formed by the different diameter probes; the difference between the width of the stir zone and the tool diameter was not changed. These results indicated that the stirring areas formed beside the probe were not affected by the probe diameter, either.

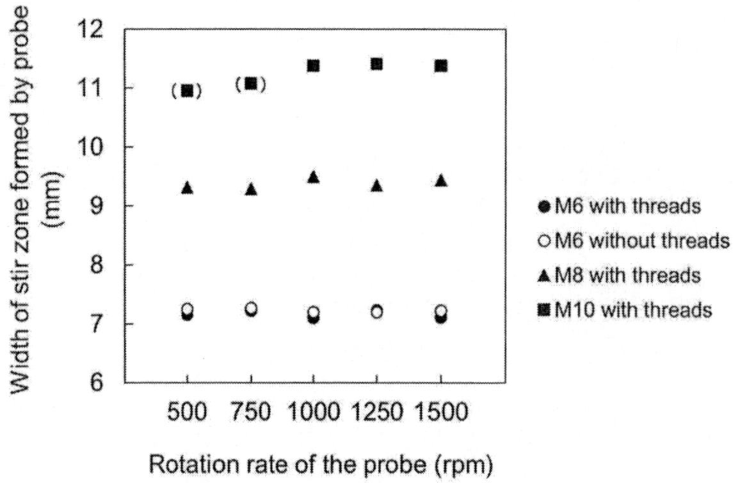

Fig. 3 Relationship between the width of the stir zone formed by the probe and the rotating speed of the probe.

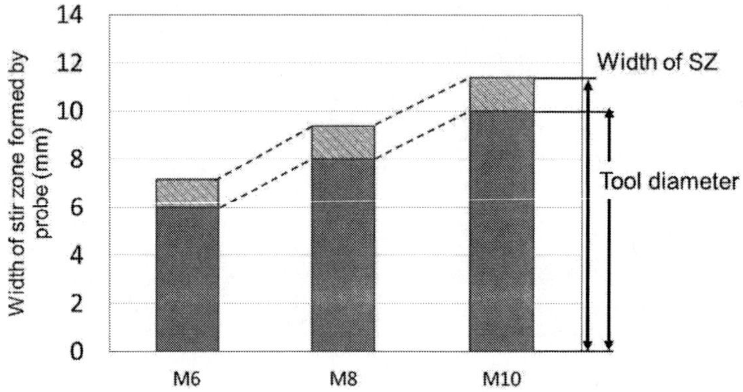

Fig. 4 The stirring area formed by the stationary shoulder FSW.

Conclusion

The FSW was conducted using an adjustable tool which can rotate the shoulder part and the probe part separately. The effect of the shoulder part and the probe part on the formation of the stir zone was studied by changing the rotation speeds of each part of the tool. The obtained results can be summarized as follows.

(1) The shoulder-affected stir zone was extended in the depth direction with the increasing rotation speed of the shoulder.
(2) The shape of the probe-affected stir zone was not changed with the increasing rotation speed of the probe.
(3) The simpler material flow formed by only the probe expanded the optimal FSW condition compared to the FSW with a lower rotation rate of the shoulder than that of the probe.
(4) The stirring area formed by the SSFSW was not changed by the rotating rate, diameters and the threads of the probe.

Acknowledgements

This study was partly supported by the New Energy and Industrial Technology Development Organization (NEDO) under the "Innovation Structural Materials Project (Future Pioneering Projects)" and a Grant-in-Aid for Science Research from the Japan Society for Promotion of Science.

References

[1] W.M. Thomas, E.D. Nicholas, J.C. Needham, M.G. Murch, P. Temple Smith and C.J. Dawes (TWI): Int. Patent Application No. PCT/GB92/02203 and GB Patent Application No. 9125978.9, (1991)

[2] C. G. Rhodes, M. W. ahoney, W. H. Bingel, R. A. Spurling, and C. C. Bampton: "Effects of friction stir welding on microstructure of 7075 aluminum" , *Scripta Mater.*, 36, (1997) pp. 69-75

[3] S. Benavides, Y. Li, L. E. Murr, D. Brown, and J. C. McClure: Low-temperature friction-stir welding of 2024 aluminum" , *Scripta Mater.*, 41, (1999) pp. 809-815.

[4] R. S. Mishra, M.W. Mahoney, S. X. McFadden, N. A. Mara, and A. K. Mukherjee: "High strain rate superplasticity in a friction stir processed 7075 Al alloy" , *Scripta Mater.*, 42, (2000) pp. 163-168

[5] H. J. Liu, H. Fujii, M. Maeda, and K. Nogi: "Tensile properties and fracture locations of friction-stir-welded joints of 2017-T351 aluminum alloy" , *Mater. Proc. Technol.,* 142, (2003) pp. 692-696

[6] Y. Tozaki, Y. Uematsu, and K. Tokaji: "A newly developed tool without probe for friction stir spot welding and its performance" , Journal of Materials Processing Technology, vol. 210, no. 6-7, (2010) pp. 844–851

[7] P.A. Colegrove & H.R. Shercliff: "Development of Trivex friction stir welding tool Part 2 – three-dimensional flow modelling" , Science and Technology of Welding and Joining 9:4, (2004) pp. 352-361.

[8] H.Fujii, L.Cui, M.Maeda and K.Nogi: "Effect of Tool Shape on Mechanical Properties and Microstructure of Friction Stir Welded Aluminum Alloys" , Mater. Sci. Eng. A, 419 (2006) pp. 25-31.

[9] N. Sun, Y. H. Yin, A. P. Gerlich and T. H. North: "Tool design and stir zone grain size in AZ31 friction stir spot welds" , Science and Technology of Welding and Joining, 14-8 (2009) pp. 747-752

[10] Wynne BP, Threadgill PL, Davies PS, Thomas MJ, Ng BS: "Microstructure and texture in static shoulder friction stir welds of Ti-6Al-4V" , Proceedings of the 7th International Friction Stir Welding Symposium, Awaji Island, Japan (2008)

[11] Davies PS, Wynne BP, Rainforth WM, Thomas MJ, Threadgill PL: "Development of microstructure and crystallographic texture during stationary shoulder friction stir welding of Ti-6Al-4V" , Metall Mater Trans A 42 (2011) pp. 2278–2289

Intelligent Welding Robot and Beyond: Learning From Human Welder Behavior

Yukang Liu, YuMing Zhang
Department of Electrical and Computer Engineering and Institute for Sustainable Manufacturing
University of Kentucky, Lexington, KY 40506, USA
ymzhang@engr.uky.edu, +1 859 323-3262

Abstract
Next generation intelligent welding robots which combine the advantages of human welders (intelligence and sensing versatility) and automated welding robots (precision and consistency) could outperform skilled welders in various precision welding tasks. The challenges lie in both the *hardware* that is able to tracking the human welder hand movements and welding process information altogether, as well as the *software* that learns the human welder behavior. To this end, an innovative human-machine welding paradigm, virtualized welding, is proposed that can transfer human intelligence to welding robots. In this new paradigm, a welding robot is augmented with sensors to observe the welding process and performs actual welding, in a welding station. A human welder operates a virtual welding torch and freely adjusts its movement in 3D space based on the visual feedback from the sensors, in a virtual station. The adjustments together with the reconstructed 3D weld pool surfaces can thus be recorded, analyzed, and utilized to form models representing human welder intelligence. To demonstrate the underlying concepts and applications of the virtualized welding system, human welder's adjustments are "selectively" learned and transferred to the welding robot to perform automated Gas Tungsten Arc Welding (GTAW). Experimental results verified the effectiveness of the learned human response models. A foundation is thus established to rapidly extract human intelligence and transfer it into welding robots.

Introduction
Precision welding using Gas Tungsten Arc Welding (GTAW) [1] typically requires skilled welders with years of training and practice to adjust welding parameters based on the observation of the welding process, in particular, the weld pool, which is considered the center of the welding process where complex phenomena originates. These optimal adjustments often lead to desired weld bead geometry. However, such ability will become luxury for many manufacturers due to the increased shortage of skilled welders, increased degree of automation and increased cost for skilled workers [2].

Automated welding robots seem to be a reasonable approach to solve the skilled welder shortage issue. They could be rigorously programmed for certain precision welding tasks. However, this is at the cost of precision joint preparation to minimize variations in order to compensate the lack of the aforementioned adaptation ability. For customized small batch jobs (which is recognized as an important feature of future manufacturing [3]), facilitating such programming and precision joint preparation could become very expensive. It would thus be extremely appealing that next generation welding robots can emulate welder skills or even outperform skilled welders who suffer from various physical limitations including inconsistent concentration, fatigue, and stress, etc.

Developing such intelligent welding robots involves three fundamental challenges: 1) the robot can view the welding process similarly like a human welder – sensing; 2) the robot can make decision on how to respond to the sensed information similarly like a skilled welder – decision making; 3) the robot can make movement adjustments similarly like a human welder – execution. While the third issue appears easy to address given recent advancements in robotics, other two issues deserve further attention. Numerous sensing methods have been proposed to sense GTAW process with successes, including pool oscillation [4], radiography [5], infrared [6, 7], and vision [8-10]. However, only sensing 3D weld pool surface can lead to emulating human welders who adaptively make adjustments based on observation of the weld pool. As such, an innovative method has recently been proposed/developed to monitor the 3D free weld pool surface in GTAW in real-time at the University of Kentucky Welding Research Laboratory (UK-WRL) [11, 12]. The reflection law has led to accurate real-time algorithm to process the reflection image, and to measure and control 3D weld pool surface in real-time [13, 14].

With this unique capability to see similarly as human welders, it is now possible to address the challenges in learning human welder decision making which processes observation of 3D weld pool surface. Preliminary work has been done at the UK-WRL to learn how welders make adjustments on the welding current [15-18]. However, such learning, although meaningful and significant, can only be considered a straightforward application of our unique new capability for weld pool surface measurement but there are no great additional challenges. Torch and sensor are all fixed while work-piece moves; human welder only holds a current regulator to watch the weld pool and adjust the current! In reality, a human welder carries/moves the torch and makes adjustments on torch orientation and travel speed all freely in 3D space.

The key challenge in learning human welder decision making for intelligent welding robot development thus lies in the availability of another unique ability that allows the human welder to perform naturally and freely in 3D space while still being capable of monitoring the inputs (weld pool surface) and outputs (adjustments on welding parameters) of the decision process. The laser and imaging plane, which images the laser reflection, need to move together with the torch with fixed geometrical relationship while the human welder operation in 3D space must still be natural and free. The innovative robotic welding platform – referred to as Virtualized Welding System – proposed/developed in this paper is exactly such a system with all needed capabilities. As will be further detailed below, this system uses a welding robot to carry the torch and sensors to perform the welding and measure the weld pool surface. The human welder holds a virtual torch whose operation is similar as a real torch such that his operation and adjustment are still natural and free in 3D space. The weld pool surface image will be projected to him under the virtual torch on the mock-up work-piece such that he can observe the weld pool surface similarly as during welding. His adjustments will be accurately monitored and then followed by the welding robot. Using such a system, the inputs and outputs of the human decision making process can be accurately monitored/recorded while the operation is still natural and free similarly as during welding.

Hardware - Virtualized Welding System

In this section the developed innovative virtualized welding system [19] is detailed. The system (Fig. 1(a)) consists of two workstations: welding station and virtual station. In the virtual station a human welder can view the monitor or mock up where the weld pool images are displayed, and moves the virtual welding torch (Fig. 1(b)) accordingly as if he/she is right in front of the work-piece. The human welder movement is accurately captured by a Leap sensor, and the obtained 3D coordinates will be sent to the PC. The robot arm is equipped with the welding torch and 3D weld pool sensing system. It receives commands via Ethernet from the PC, including the next robot tool pose (i.e., 3D positions and orientations), and robot tool movement speed. It then executes the command and sends the current robot tool pose back to the PC. The communication delay in this system is less than 30 ms being negligible for GTAW process.

Virtual Station

More specifically, the virtual station employs a mock up with the same dimensions as the actual work-piece which is a pipe in this paper. A projector projects rendered video on its surface as the visual feedback for human welder to view.

Leap sensor is an advanced motion sensor that can accurately capture the movement of the torch when operated by a welder freely in 3D space. It uses two cameras and three infrared LEDs such that it can observe a roughly hemispherical area to a distance of about 1 m to track fingers or similar items to a spatial precision of 0.01 mm [19]. By utilizing a predictive algorithm [19] accurate tracking of human hand movement is achieved by the robot. Fig. 2 shows the visualization of the work-piece/pipe including both eye view of the work piece (Fig. 2(a)) and

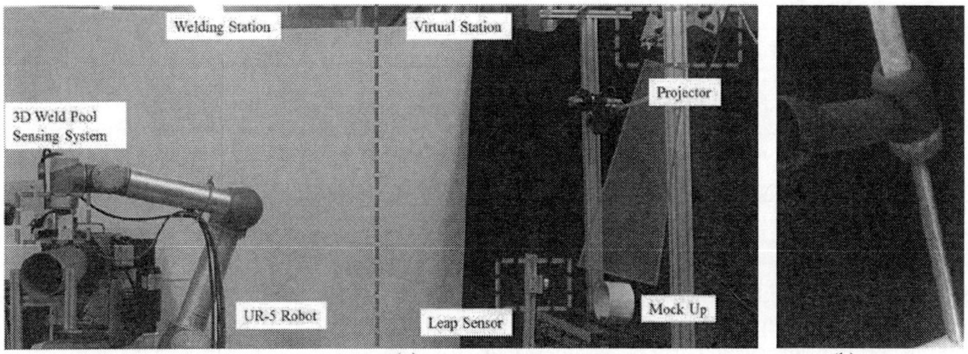

(a) (b)

Fig. 1 Virtualized welding system proposed/developed. (a) System overview (b) Virtual welding torch.

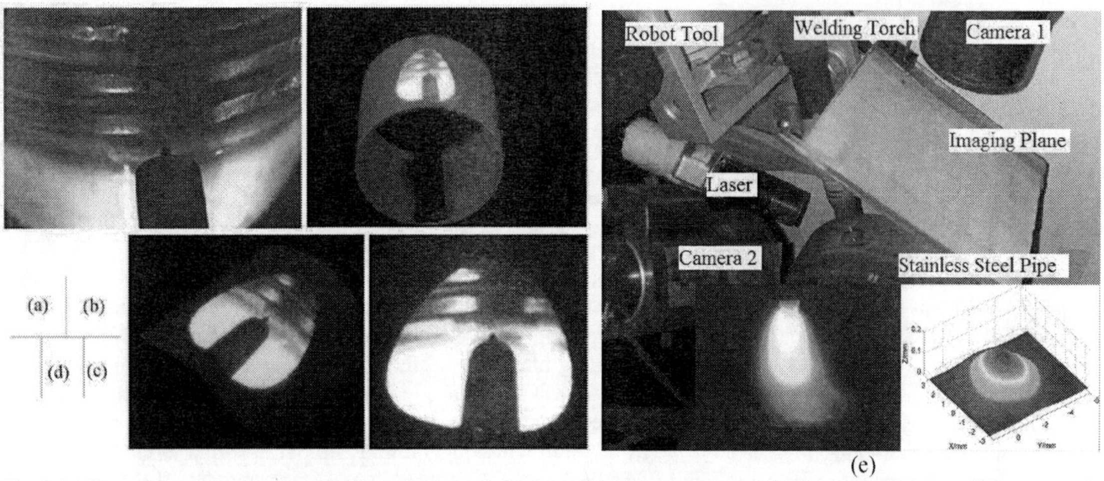

Fig. 2 (a) Eye view of work-piece (b) Virtual view of mock-up (c) Close-up view (d) Mock-up view at a different angle (e) Detailed view of the 3D weld pool sensing system.

virtual view of the mock-up (together with visualized eye view image).

Welding Station

Welding station consists of an industrial welding robot, eye view camera, and a compact 3D weld pool sensing system. The robot utilized is Universal Robot UR-5 with six Degree of Freedom. The accuracy of the robot's repetitions is 0.1 mm. Fig. 2 (e) depicts a detailed view of the 3D weld pool sensing system [19]. Camera 2 (eye view camera) captures the weld pool image and sends it back to the PC (a sample image is shown in Fig. 2(e) lower left). A low power laser (20 mW) projects a 19 by 19 dot matrix pattern to the weld pool surface (a fluctuating liquid surface mirror under strong arc) and its reflection from the pool surface (mirror) is intercepted by an imaging plane and imaged by a CCD camera (Camera 1 in Fig. 2 (e)). 3D weld pool surface can thus be reconstructed in real-time (a sample reconstructed weld pool is shown in Fig. 2(e) lower right).

We envision our virtualized welding being able to operate at three different levels: (a) teleoperated welding: similar to the well-known da Vinci surgical robot, now a welder can remotely operate the welding robot by controlling a virtual welding torch in virtual station [19], as if he/she is right in front of the work piece. This would enable welding in hazardous areas for human operators; (b) autonomous welding: in this level the welding robot is running in a fully automatic way under closed-loop control [20-22]. The control decision is based on the sensor inputs, i.e., 3D weld pool surface. This type of feedback control is fundamentally more advanced than the current welding robots; and (c) cooperative welding: unskilled human welder operates a virtual welding torch determining the welding speed while the welding machine could compensate his/her incorrect movement by adjusting the welding current and other welding parameters. This paper will use this system in its level (a) – teleoperated welding – to collect the welder response data and then in its level (b) – autonomous welding – such that the robot realizes the learned decision making.

Software - Response Rating and Modeling

Automated Response Rating

Because of the intrinsic stochastic nature of human controlled processes, real-time adjustments/responses made by the human welder might not always be ideal/consistent. It is thus important to "filter" "bad response" data and only preserve "correct" adjustments made by the welder.

To better distill such "correct" response, the human welder evaluates the measured data (including the welding parameters, weld pool characteristic parameters) and corresponding back-side bead width to assign a rating (0 to 10, with 10 for the most desirable consistent penetration) in each 5 s interval. Assigning such a rating is an off-line process requiring no real-time operation/control and is thus much less skill demanding from the welder. If all the data points are used to model the human welder response, the resultant model would represent the "average" responses of the human welder which may be less ideal than "good responses" rated by human welder.

The classifier (i.e., response rating system) is defined as:

$$R = f(W, L, C, A) \quad (1)$$

where W, L, C represent the measured weld pool characteristic parameters, and A is the adjustment in welding parameters made by the human welder (e.g., welding speed, arc length, and torch orientations, etc.).

The classifier trained from the experimental data collected from welding using the virtualized welding system and associated human ratings will be used to automatically classify the data from training experiments by automatically assigning data segments with ratings calculated from the trained rating model (1). Selected measurements (e.g., with associated ratings larger than 8) are then utilized for modeling being discussed below.

Welder Response Modeling

The welder response model is expressed as the following correlation:

$$A = g(W, L, C) \quad (2)$$

Linear model may capture this correlation with certain accuracy. However, it can only represent the average correlation over the entire range of the weld pool geometrical parameters.

Actually, as a human inference mechanism, the human welder's response to the 3D weld pool surface is inevitably fuzzy and nonlinear. (The nonlinearity will be discussed in detail by analyzing the implication of an identified linear model, i.e., eq. (3), in the Demonstrative Example.) Such abstract thoughts or concepts in human reasoning, however, are difficult to extract. Neuro-fuzzy approach [17] provides an effective method to determine the parameters in fuzzy models using automated learning techniques developed in neural networks. Adaptive Neuro-Fuzzy Inference System (ANFIS), as a type of neuro-fuzzy model, possesses the advantages of adaptive rule changing capability, fast convergence rate, and does not require extensive experiences about the process to construct the fuzzy rules. As such, from data $\{A, W, L, C, R\}$, model (2) can be obtained. Mathematical details, which have been documented in [21], are not repeated again here.

Demonstrative Examples

To demonstrate the effectiveness of the virtualized welding system (hardware) and response rating based ANFIS response modeling (software), human welder adjustment on the welding speed is first "selectively" learned and transferred to the welding robot to perform automated GTAW [21] as a straightforward yet complete example to demonstrate concepts. In the second example, human welder's full adjustments in 3D space including torch orientation and travel speed are "selectively" learned and the resultant human response models are transferred to the welding robot to perform automated GTAW welding. While example 1 which is relatively straightforward will be detailed to demonstrate the concepts/procedures, the more complex 3D learning in example 2 will be presented more briefly.

Example 1 - Learning Welder Adjustment in Speed

Thirteen dynamic training experiments have been conducted to collect the data needed to model the human arm movements to 3D weld pool surface [21]. Stainless steel pipe is welded using the direct current electrode negative GTAW process. The outer diameter and wall thickness of the pipe are 113.5 and 2.03 mm, respectively. In these experiments the welding current is randomly changed in [40 A, 48 A] resulting in a fluctuating weld pool surface. The welder sees the projected weld pool image and moves the virtual welding torch accordingly. The robot follows the welder's movement and actually performs the welding. Three weld pool characteristic parameters (weld pool width, length, and convexity) are selected as the system inputs, which are considered as the major information sources human welder perceive. Welder's fluctuating arm movement speed (i.e., welding speed) is filtered to suppress the unintentional shaking during the experiments and the filtered speed S is used as the human welder response.

Fig. 3 depicts the measured data from these training experiments. The sampling period is 0.5 s because the welder controls arm movement by observing the weld pool which does not change extremely fast. Fig. 4 shows the assigned rating. It is seen that over 60% of the data points are rated above 8, however about 10% of the data points have been rated below 4. The rating model, i.e., Eq. (1), can be identified using the measurements in Fig. 3 as the inputs and the human rating in Fig. 4 as the output. In particular, linear model has been fitted using standard least squares method [21]. Substantial fitting errors are frequently observed. ANFIS has thus then be utilized to improve the classifier performance. It is apparent that compared to linear model result, the proposed ANFIS model

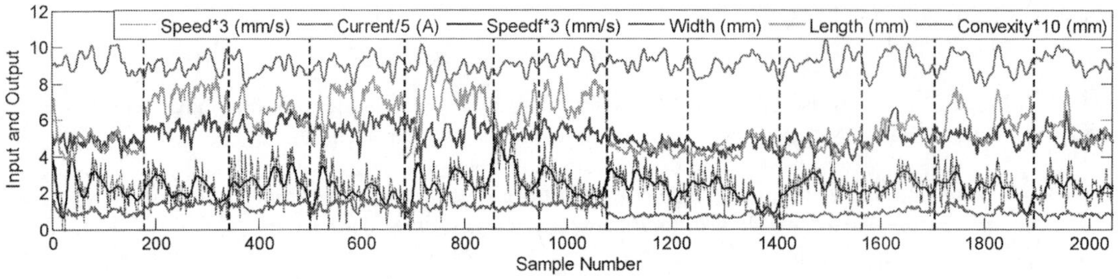

Fig. 3 Measured welding current, weld pool characteristic parameters and human arm movement speed in thirteen dynamic training experiments.

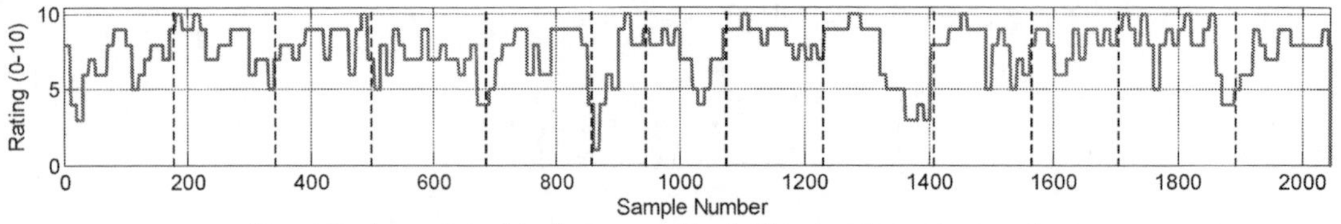

Fig. 4 Rating assigned by human welder in thirteen dynamic experiments.

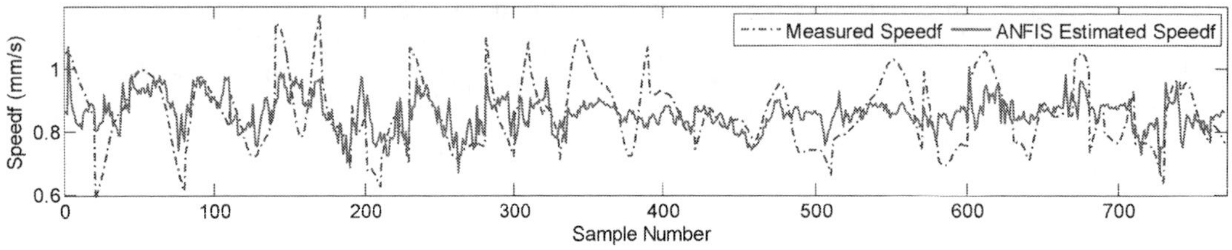

Fig. 5 Supervised ANFIS modeling of the welder response.

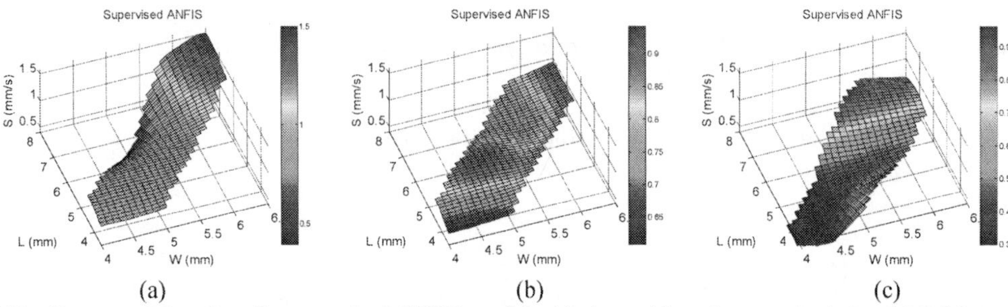

(a) (b) (c)

Fig. 6 Nonlinear model surface for supervised ANFIS model with the weld pool convexity being (a) 0.06 mm, (b) 0.11 mm, (c) 0.16 mm.

provides much better fitting result, and is thus finally utilized to automate the rating process [21], i.e., to obtain model (1). Data with ratings larger than 8 are finally selected to model human welder adjustments.

ANFIS model is thus then employed to better describe this complex nonlinear correlation between the weld pool characteristic parameters and the welder's movement speed [21]. The model is developed based on the classified/selected data and is thus referred to as supervised ANFIS model. The modeling result is shown in Fig. 5. Careful observation of Fig. 5 suggests that the adjustment/fluctuation in the speed are much more accurately fitted by the ANFIS model. (Note: The large variations in the human welder speed caused by operation incosistency/randomness is not modelable and thus would bound the lower limit of the modeling error. Actually the inconsistency should not be learned and modeled to transfer into welding robots. Modeling thus provides a method to further better retrieve the "good response" from the human welder.)

The identified supervised ANFIS model can provide more accurate prediction on the welder's adjustment on the welding speed than the linear model. However, it is much more complicated. For its intended use in welder response modeling and control, this complexity is not a concern because of the strong computation power of

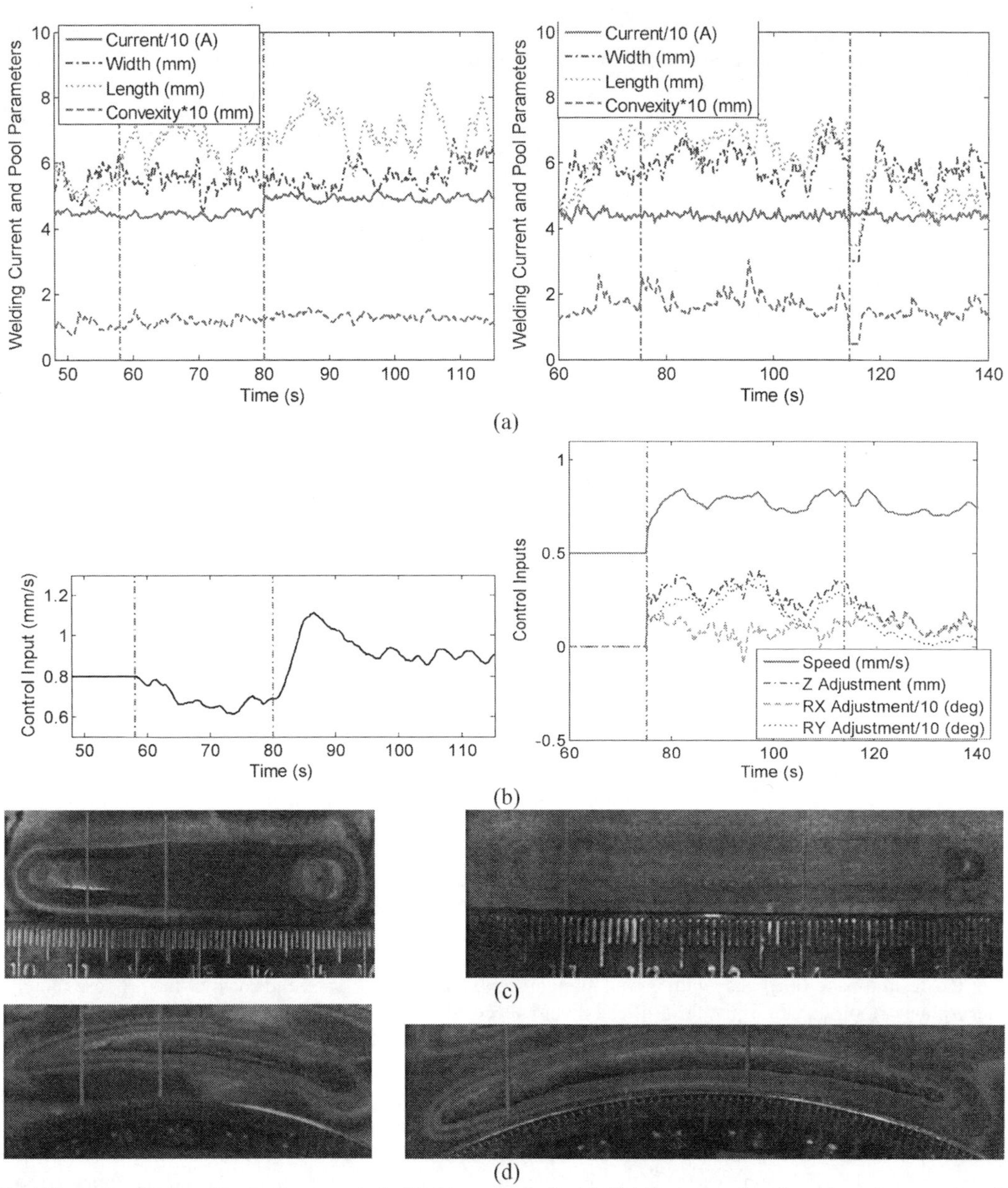

Fig. 7 Experimental results. Left: example 1; Right: example 2. (a) Welding current and weld pool parameters (b) Control input (c) Front-side bead (d) Back-side bead.

today's industrial computers and embedded systems. However, the understanding of its implication is less straightforward than that of the linear model. The model's reasonableness thus needs to be examined against our observations and knowledge on weld pool and welding process since it was identified statistically from experimental data.

Since it is difficult to visualize the speed adjustment with three inputs, Fig. 6 visualizes the nonlinear speed adjustment under different weld pool convexities. (1) When the convexity is small (i.e., 0.06 mm shown in Fig. 6(a)), the speed gradually increases as the pool width and length increases. When the pool is extremely large (for a width of 6.5 mm and length of 8 mm), the welding speed reaches 1.5 mm/s. Actually for small convexity the weld penetration is mainly determined by the weld pool size. A small pool size indicates small amount of metals being melted and thus an insufficient penetration. In this case, the welding speed should be kept small. As the pool size increases, more metals are melted and the welding speed should be increased accordingly. The response generated by the supervised ANFIS model coincides with our understanding about the welding process. (2) For medium weld pool convexity (about 0.11 mm shown in Fig. 6 (b)), supervised ANFIS model remains at 0.75 mm/s to 0.85

mm/s for most of the weld pool size. This is understandable because for medium convexity the weld penetration is larger than that of the small convexity. (3) For large weld pool convexity (Fig. 6(c)), small pool size indicates that no penetration occurs because of the large surface tension force being applied. In this case, the speed is expected to be extremely small to produce enough heat input for full penetration. Supervised ANFIS model is able to generate a welding speed as slow as 0.3 mm/s to compensate the insufficient penetration. As the pool size increases, the supervised ANFIS model outputs an increased welding speed. The speed reaches its maximum (about 0.95 mm/s) for a weld pool width at 5.5 mm and length at about 8 mm. In this case, large weld pool size and relatively large convexity indicates a sufficient penetration, and the welding speed should be relatively large to maintain a consistent penetration status. As such, it can thus be concluded that the supervised ANFIS model generates correct welding speed adjustment for varying weld pool geometry, and can provide more detailed intelligence possessed by human welder than the linear model.

In order to demonstrate the effectiveness of the distilled welder response model as an intelligent controller, automated control experiments have been designed and conducted by the welding robot under different welding currents [21]. In the example shown in Fig. 7 (left), before 58 s it is the open loop period with no closed loop control during which the speed is a pre-set constant which does not necessarily produce the needed penetration for the used nominal current 43 A. As observed from Fig. 7(c) left, the full penetration is not produced. From 58 s to 80 s the nominal welding current is still 43 A but the speed is controlled by the human response model such that the speed is reduced to achieve the full penetration. From 80 s to 115 s the welding current is purposely changed to 48 A (Fig. 7(a) left). However, it is observed that the controller (human welder response model) is able to automatically increase the welding speed accordingly to maintain a consistent penetration (about 4.5 mm back-side bead width shown in Fig. 7(c) left) despite the increased current.

Example 2 - Learning Welder Adjustments in 3D

In this example, human welder's full adjustments in 3D space (including torch orientation, arc length and travel speed) are "selectively" learned and the resultant human response models are transferred to the welding robot to perform automated GTAW welding. The experimental results are depicted in Fig. 7 right. Before 75 s it is the open loop period with no closed loop control during which the welding speed is pre-set at 0.5 mm/s. Other welder adjustments are set to zeros. From 75 s to 113 s the welding parameter adjustments are controlled by the human response models such that the welding speed is increased to about 0.75 mm/s to achieve the desired full penetration. Other welding parameters are also adjusted as depicted in Fig. 7 (b). An artificial error between the actual and measured characteristic parameters of the weld pool surface is then applied to purposely generate an artificial disturbance. At 113 s, the measured weld pool width, length, and convexity are set to 3 mm, 3.5 mm, and 0.05 mm, respectively. As can be seen from Fig. 7 (b) right the welding speed is slightly decreased from 0.8 mm/s to about 0.76 mm/s, and the back-side bead width is slightly increased (Fig. 7 (d) right). Other welding parameters are also adjusted by the controller accordingly. By applying the controller the desired back-side bead width is well maintained at about 3 mm. The robustness of the proposed intelligent controller is thus demonstrated despite the artificial disturbance.

Conclusions

An innovative human-machine welding paradigm, virtualized welding system, has been proposed/developed such that a human welder can freely adjust/operate in 3D space while can still capture the weld pool surface and adjustments. The software foundation has been established to mathematically model his/her decision for the adjustment from the weld pool surface and better model using "good responses". It is found from the demonstrative example that the human welder response/adjustment is fundamentally nonlinear and this intrinsic nonlinearity can be modeled using ANFIS, and enhanced by the proposed response rating system.

The availability of the hardware and software foundations established in this paper provides conditions to conduct future work to study human welders beyond the demonstrative examples presented. For example, it is apparent that this system can also be used to speed up the welder training process. More specifically, response models from unskilled welders being trained can be compared with those from skilled welders to understand their differences.

Acknowledgments

This work is funded by the National Science Foundation under grants CMMI-0927707 "Machine-Human Cooperative Control of Welding Process" and IIS-1208420 "Virtualized Welding: A New Paradigm for Intelligent Welding Robots in Unstructured Environment".

References

[1] R. O'Brien, Ed., Welding Handbook, 8th Edition vol.2-Welding Processes, AWS, 1998.

[2] G. D. Uttrachi, "Welder shortage requires new thinking," *Welding Journal*, vol. 86(1), pp.6, 2007.

[3] A Roadmap for US Robotics: From Internet to Robotics, http://www.us-robotics.us/reports/CCC%20Report.pdf, May 2009.

[4] R. Renwick, R. Richardson, "Experimental investigation of GTA weld pool oscillations," *Welding Journal*, vol. 62 (2), pp. 29s-35s, 1983.

[5] A.C. Guu, S.I. Rokhlin, "Arc weld process control using radiographic sensing," *Materials Evaluation*, vol. 50(11), pp. 1344, 1992.

[6] J. B. Song, D.E. Hardt, "Closed-loop control of weld pool depth using a thermally based depth estimator," *Welding Journal*, vol. 72(10), pp. 471s-478s, 1993.

[7] P. Ghanty, M. Vasudevan, D. P. Mukherjee, etc. "Artificial neural network approach for estimating weld bead width and depth of penetration from infrared thermal image of weld pool," *Science and Technology of Welding and Joining*, vol. 13(4), pp. 395-401, 2008.

[8] K.A. Pietrzak, S.M. Packer, "Vision-based weld pool width control," *Journal of Engineering for Industry-Transactions of ASME*, vol. 116(1), pp. 86-92, 1994.

[9] Y.M. Zhang, R. Kovacevic, "Neuro-fuzzy model based predictive control of weld fusion zone geometry," *IEEE Transactions on Fuzzy Systems*, vol. 6(3), pp. 389-401, 1998.

[10]H. Chen, F. Lv, T. Lin, et.al., "Closed-loop control of robotic arc welding system with full-penetration monitoring,", *Journal of Intelligent and Robotic Systems*, vol. 56(5), pp.565-578, 2009.

[11]W.J. Zhang, Y.K. Liu, X. Wang, Y.M. Zhang, "Characterization of three dimensional weld pool surface in GTAW," *Welding Journal*, vol. 91(7), pp. 195s–203s, 2012.

[12]Y.K. Liu, W.J. Zhang, Y.M. Zhang, "Estimation of weld joint penetration under varying GTA pools," *Welding Journal*, vol. 92(11): 313s-321s, 2013.

[13]Y.K. Liu, Y.M. Zhang, "Model-based predictive control of weld penetration in Gas Tungsten Arc Welding," *IEEE Transactions on Control Systems Technology*, vol. 22(3), pp. 955-966, 2014.

[14]Y.K. Liu, Y.M. Zhang, "Control of 3D weld pool surface," *Control Engineering Practice*, vol. 21(11), pp. 1469-1480, 2013.

[15]Y.K. Liu, Y.M. Zhang, L. Kvidahl, "Skilled human welder intelligence modeling and control: Part I - Modeling," *Welding Journal*, vol. 93, pp. 46s-52s, 2014.

[16]Y.K. Liu, Y.M. Zhang, L. Kvidahl, " Skilled human welder intelligence modeling and control: Part II - analysis and control applications," *Welding Journal*, vol. 93, pp. 162s-170s, 2014.

[17]Y.K. Liu, W. Zhang, Y.M. Zhang, 2014, "Dynamic neuro-fuzzy based human intelligence modeling and control in GTAW", *IEEE Transactions on Automation Science and Engineering*, vol. 12(1), pp. 324-335, 2015.

[18]Y.K. Liu, W.J. Zhang, Y.M. Zhang, "Adaptive Neuro-Fuzzy Inference System (ANFIS) modeling of human welder's response to 3D weld pool surface in GTAW", *Journal of Manufacturing Science and Engineering-Transactions of the ASME*, vol. 135, pp. 0210101-02101011, 2013.

[19]Y.K. Liu, Y.M. Zhang, "Toward welding robot with human knowledge: a remotely-controlled approach," *IEEE Transactions on Automation Science and Engineering*, vol. 12(2), pp. 769-774, 2015.

[20]Y.K. Liu, Z. Shao, Y.M. Zhang, "Learning human welder movement in pipe GTAW: a virtualized welding approach," *Welding Journal*, vol. 93, pp. 388s-398s, 2014.

[21]Y.K. Liu, Y.M. Zhang, "Supervised learning of human welder behaviors for intelligent robotic welding," *IEEE Transactions on Automation Science and Engineering*, accepted for publication, 2015.

[22]Y.K. Liu, Y.M. Zhang, "Iterative local ANFIS based human welder intelligence modeling and control in pipe GTAW process: a data-driven approach," *IEEE/ASME Transactions on Mechatronics*, vol. 20(3), pp. 1079-1088, 2015.

Consideration of Gas Tungsten Arc Behavior during Ignition Stage

Ogawa Yoji
Ogawa Giken
ogawa-giken@me.pikara.ne.jp

Abstract

Ultra-high-speed observation on Gas Tungsten Arc welding was carried out to understand actual reactions at arc ignition stage. The highest video rate to observe arc behavior was 1,300k frames per second. The behavior of arc and electrodes spectra on a center line of Tungsten electrode and arc was also observed by the same ultra-high-speed video system up to 100k frames per second. Actual reactions on electrode surface and arc plasma are very complex by evaporation and oxidation reactions of tungsten and included minor elements. Arc spectrum behavior is rather simple, because space is fulfilled by shielding gas. The surface conditions of tungsten electrode after arcing are quite different by temperature distribution. Each electrodes after arcing were observed by Scanning Electron Microscope system. Normally minor element such as thoria was much reduced on top ends surface of electrodes. Just near the cathode area, there are many dendrite crystal of tungsten were generated. Some breakdown occurred this region. And some millimeters from this area, there are many warm shaped thoria picked up on last arcing duration from inside zone of the electrode. Many breakdown occurs on this area.

When the anode surface was painted by special active flux, some jet like propulsion occurs this part. And the behavior changes by time because of temperature increasing by arc.

Evaluation of emission is important on numerical simulation of GTA process. Emission is given as a function of electrode surface temperature. However, actual reaction on electrode surface is very complex by evaporation and oxidation reaction of tungsten and included thoria or other rare elements. Time variation by erosion is also existed. Temperature effect of Tungsten electrode on electron emission is considered by high-speed visual observation.

Introduction

Cathode material and its shape are strongly affected on plasma distribution and thermal transportation on anode in Gas Tungsten Arc welding. Thorium oxide is major element to add tungsten for the cathode. Mechanical property of cathode is increased with a few percent of thorium oxide in tungsten, and amount of thermal emission increases very effective. This is explained that mono-layer of thorium on tungsten cathode much reduced work function[1,2]. Metal vapor from anode material also plays an important role on plasma distribution and cathode reaction[3,4]. Recent remarkable progress on high-speed imaging technology makes it possible to observe plasma configuration and surface reaction on cathode and anode. This paper shows actual behavior of Gas Tungsten Arc welding process, especially cathode surface reaction on arc ignition stage.

Experimental procedure

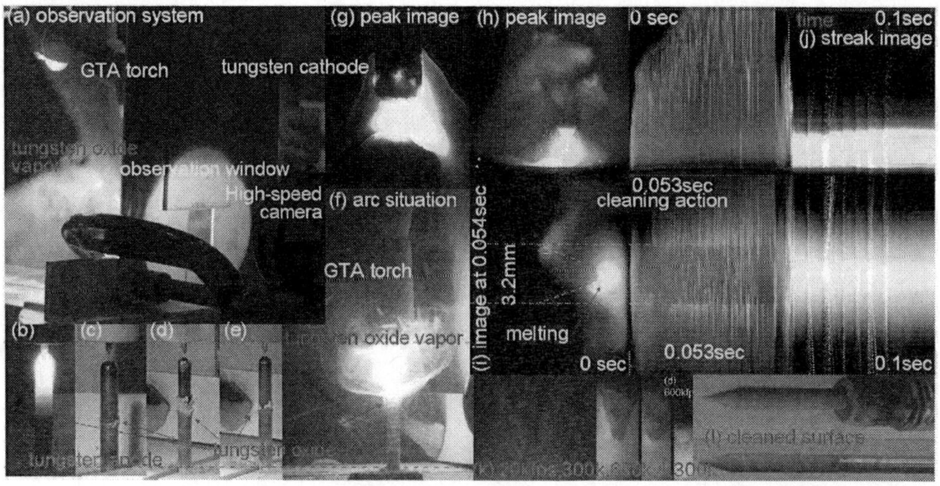

Fig.1 Experimental results for pure tungsten electrode.

Experimental setup is very simple. Gas Tungsten arc generates on material to be used, and whole arc ignition process is captured by high-speed video camera as shown in Fig.1. This is the most simplest case, which arc is generated on a tungsten rod. Various high-speed video cameras were used to capture arc process. The highest capture rate is 130,000,000 frames per second(fps is used later). Spectrum distribution was captured by using of special prism on the top of camera lens. The prism is called as ImSpector. The highest capture rate is 100,000fps.

Experimental results for pure tungsten cathode

Tungsten is excellent material for electrode. However the problem is low boiling temperature of its oxide. Hot tungsten easily reacts surrounded oxygen, and some tungsten oxide evaporates as shown in Fig.1. Some tungsten oxide remains on electrode surface. Another problem is cleaning action in ignition stage. One example is shown in righthand of Fig.1. Cathode spots are running on the cathode surface in the ignition stage for a long time. The period is determined by many conditions such as cathode material, its shape, shielding gas, added voltage, and so on. Melting of the cathode top is essential to provide enough arc current.

Experimental results for thoria added cathode

Results in low high-speed video rate

A few percent of thorium oxides(thoria) are added in tungsten electrode to improve cathode characteristics. Melting point of thoria is higher than pure tungsten. Its work function is lower than tungsten. So, arc ignition becomes very stable as shown in Fig.2. Cleaning action does not appear in this figure, because the video rate is too low. Cone angle of the cathode is very important to produce stable arc ignition. Current passing through the cathode is concentrated by narrowing cross section. So, temperature of cathode top section raises enough to keep electron ignition. But it is a weak point for mechanical and thermal strength, it is easy to broke or exploded..

Transient state of arc ignition finishes in 1 second. However, the electrode temperature profile in stable condition needs more than 10 seconds. Stable arc configuration and cathode temperature profile are affected by cathode material, cathode shape, shielding gas mixture, ambient pressure, and so on. Right hand side of Fig. 2 shows the effect of ambient pressure and shielding gas. Helium was used in this case. Arc is much condensed in high pressure. And only the top of the cathode is heated. On the contrary, arc is spread in low pressure. Very large area of the cathode is in high temperature. This situation causes quite different erosion on the cathode with air added conditions as shown in right of Fig.2. Some content of oxygen in shielding gas makes tungsten oxidation and evaporation, as a result of this reaction, tungsten dendrite forms on cathode as called Rim.

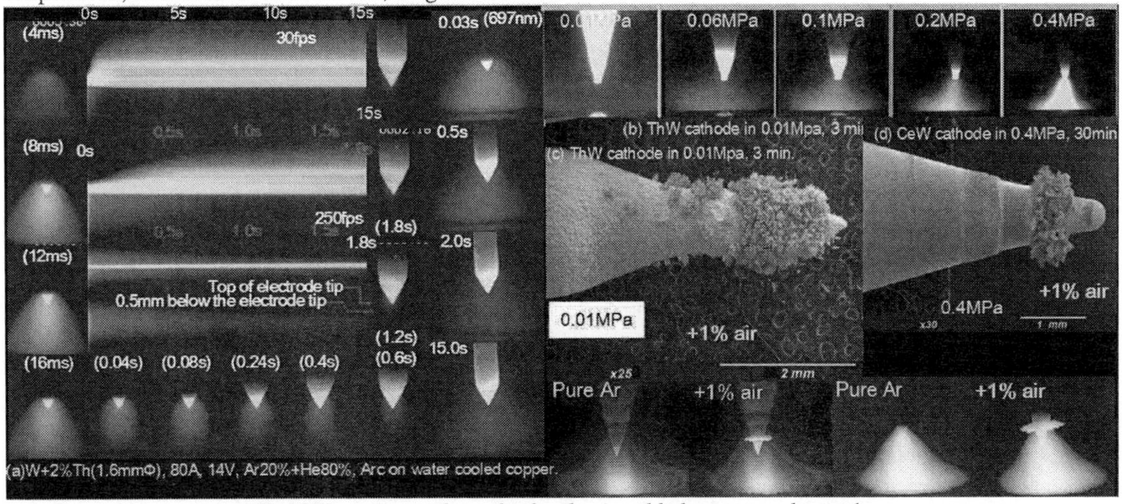

Fig.2 Experimental results for thoria added tungsten electrode.

Figure 3 shows typical condition of the cathode after arc. Thoria is appeared on the surface from inside, because of high temperature during arc welding. Some amount of thoria is reduced by tungsten during hot pressing process to form thoria added electrode. Then thorium can be moved to the surface during arc welding. The tungsten dendrites are formed on surface just apart from arc ignition area. Tungsten oxide evaporates upward position. And tungsten oxide is dissociated by high temperature plasma around the cathode. However plasma temperature and cathode surface temperature are not enough high to maintain tungsten atom in plasma. Some part of tungsten atom hits on the cathode changes to dendrite.

Anode material such as iron and chrome are also evaporated by high temperature arc. Some amount of metal ions hits on the cathode, and attached on the surface when the cathode temperature is low enough as shown in Fig.3.

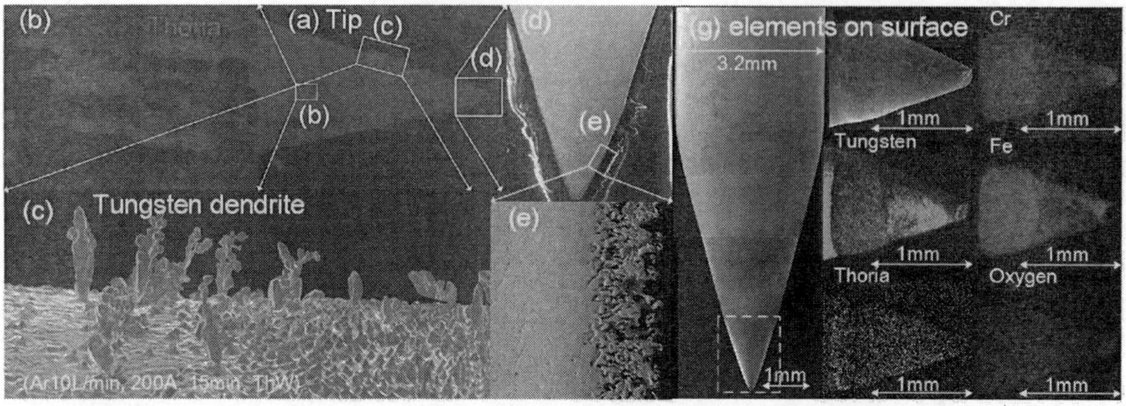

Fig.3 minor elements appeared on cathode after welding.

Results of ultra high-speed video

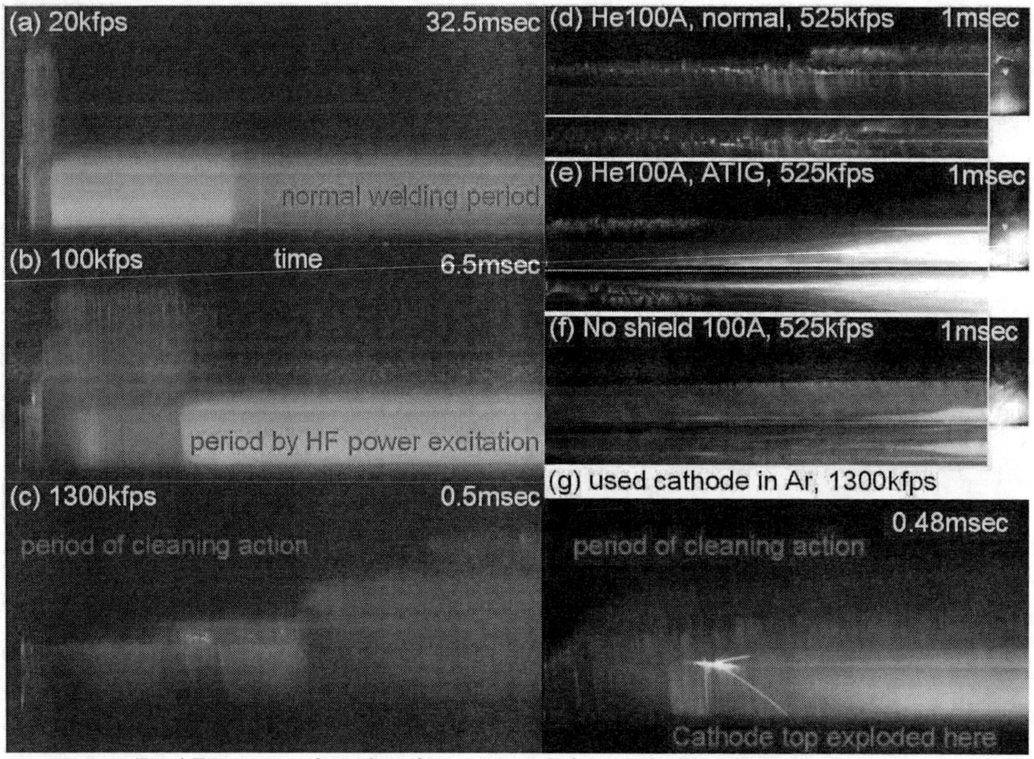

Fig.4 Experimental results od arc ignition behaviors by ultra high-speed video.

Fig.4 shows the cleaning action in arc ignition period for thoria added tungsten cathode which was used welding. Many debris generated on the cathode surface after high temperature heat process as shown in Fig.3. Cleaning action for the debris on the cathode occurs in quite short period in arc ignition stage.

Cleaning action occurs even for clean cathode which is grinded and polished just before the arc ignition, especially in helium shield. However, when the base material contains low heat capacity materials on it, they will easily evaporate and ionize between the cathode and anode. Available current channel between the cathode and the anode is quickly produced in this case, then cleaning action period reduces quite short. Some deteriorated part on the tip is exploded in this ignition stage. They are the reason why electrode loss at ignition stage is high compared to stable condition in case of quite low oxygen content in shielding gas. However, cathode contains some oxygen in it, so erosion of cathode occurs during arc welding even in perfect shielding conditions. Quick erosion occurs in poor shielding condition as explained in Fig.2.

Results of spectrum image by ultra high-speed video

Spectrum distribution is changed time by time, especially in the arc ignition stage. Metal vapor acts important role on plasma condition. Metal vapor from the base metal acts special behavior on the cathode surface reaction as shown in Fig.5 and Fig.6. Spectrum distribution on arc plasma is also much influenced by metal vapor from anode. This paper shows only image pictures for convenience of understanding at a glance. Quantitative analysis shows drastically change of spectrum distribution in plasma. Light emission from recombination of metal ions on hot cathode surface is quite high as shown in Fig.6.

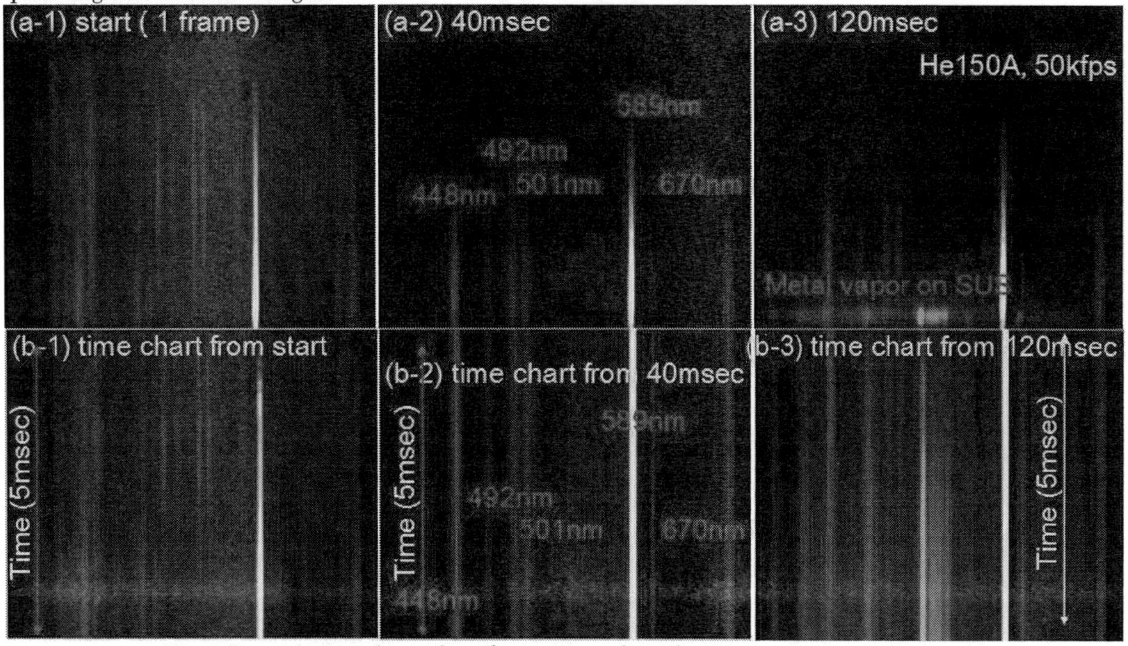

Fig.5 Experimental results of spectrum distribution in arc ignition stage.

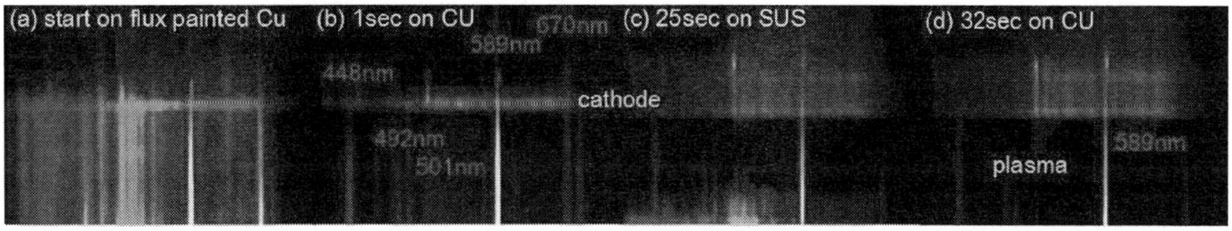

Fig.6 Experimental results of spectrum distribution in different arc conditions during welding.

Conclusions

Cleaning action occurs every kind of cathode material in arc ignition stage, especially debris remained electrode by former welding. When the cathode exposed in air, cleaning action occurs only on the top of cathode, because the temperature at the top increases quite short by oxidation reaction, and metal ion produces enough current path. Plasma contains much metal elements in plasma at early ignition stage, and transient gas excitation is different from stable condition. Much reaction occurs on hot cathode surface metal atoms and ions.

Acknowledgments

The author thanks for high-speed video companies NAC and Photron to prepare excellent high-speed cameras.

References

[1] N.A.Chapin, *et al "Tungsten-Thoria Electrode for Inert Arc Welding, Welding J. Vol30(1951) p.529.*
[2] M.Ushio and F.Matsuda, "Study on Gas Tungsten Arc Electrode, "IIW Doc.212-648(1986).
[3] Y. Ogawa, "High speed imaging technique Part 1 . high speed imaging of arc welding phenomena", Science and Technology of Welding and Joining, Vol.16, No.1(2011), pp.16-26.
[4] Y.Ogawa, "Visual Analysis of Welding Processes", Welding Process(2012),pp.277-304.

Laser Pulse Irradiation Driven Short-circuited Transfer in GMAW

Jun Xiao [1], Shujun Chen [1], Yazhou Jia [1], Yuming Zhang [2]

[1] *Beijing University of Technology, China*
[2] *University of Kentucky, USA*

Previously current-independent free-flight metal transfer was successfully achieved in regular GMAW by applying a pulsed fiber laser to irradiate the droplet. Sufficient laser recoil force is generated to detach the droplet in one-drop-per-pulse mode despite of the welding current amps. However, the metal transfer process can be further modified to controlled short-circuiting and thus the heat input in GMAW process can be minimized. Experiments are conducted by using a CC (constant current) power supply, 0.8 mm ER70S-6 wire, and pure argon shielding gas at 15 L/min flow rate. A single-board computer based controller is established to control the welding current waveform and laser pulse shape. A pulsed fiber laser is used. The metal transfer and weld current/voltage waveforms are recorded synchronously by a high-speed camera and data acquisition set for further offline analysis. Experiments of laser controlled short-circuiting metal transfer are conducted in two modes: 1) laser pulse applied before short-circuiting. In this mode, the arc length is controlled to be relatively short. The laser pulse frequency matches with the welding current. Once the laser pulse is applied, the droplet is actively elongated and short-circuited with the weld pool. The droplet is thus detached from the wire tip by the laser recoil force and surface tension.2) laser pulse applied right after short-circuiting. Similar with STT and CMT, once the short-circuiting occurred and detected, the laser pulse is instantly applied to detach the droplet, and the short-circuiting metal transfer becomes robust and almost without spatters.

Study on separated plasma transferred arc weld surfacing process

Ruiying Zhang, Shujun Chen, Fan Jiang

Beijing University of Technology, China

The separability property of plasma arc has been demonstrated by the authors recently, providing the foundation to expand the application of plasma arc used in plasma transferred arc weld surfacing (PTAW), which is a standard technology used for hard facing and has been applied in many areas. The novel process, namely the separated plasma transferred arc weld surfacing (SPTAW), is proposed. Gr. D steel plates are the base metal and the high chromium iron-base alloy powders (WF372) are the sparing powders. The bead cross-section and hardness of typical joints were tested using confocal laser scanning microscope and micro-hardness tester. The results show that the heat affected zone (HAZ) of the joints made using traditional plasma transferred arc weld surfacing is expanded with the increase of current, powder feed rate and preheating temperature. However, the HAZ of the joints made using SPTAWS is narrowed as the separated current increasing. The average micro-hardness of the spraying layer is HV 711 when the separated current is 40A, which is HV 548 with the total current 130A flowing through the base metal. And the micro-hardness of the HAZ is higher than that of base me

Dissimilar Welding between Ti and Al with Arc Welding

Keita Hatakeyama [1], Masatoshi Enomoto [2]*, Kazuhiro Nakata [3]**

1 Akahoshi Inc., 5-4, Yawatakaigan-diri, Ichihara, Chiba, Japan*
2 Japan Light Metal Welding Association, 4-20, Sakuma-cho, Kanda, Chiyoda-ku, Tokyo,Japan*
3 Osaka Univercity, 11-1, Mihogaoka, Ibaraki, Osaka, Japan*
hatakeyama@akahoshi.co.jp, masatoshi_enomoto@mbe.nifty.com, nakata@jwri.osaka-u.ac.jp

Abstract

Welding of dissimilar metals is hard to complete without any imperfections due to intermetallic compounds formed at interface between base metal and welded metal. Fortunately welding power supply is recently progressing and then it is easy to weld dissimilar metals such as Ti and Al alloy. Welded dissimilar joint of Al alloy and Ti is ruptured at HAZ in Al alloy due to thin IMC formed after welding.

1 Introduction

It is difficult for dissimilar weld to keep the sufficient joint strength due to formation of intermetallic compound (IMC) at the interface between base metal and deposited metal. IMC is also formed after being welded between Al and Ti. Recently, laser welding, FSW and GMAW [1] are applied for this dissimilar welding. However, equipment of laser welding or FSW is very expensive and this is why these process are not currently used. On the contrary, equipment of GTAW is very cheap in comparison with above mentioned two process. So, if GTAW is available for welding dissimilar materials, this process is widely applied for industrial sectors.

In this paper, manual and fully mechanized GTAW are applied for dissimilar welding between A6061-T6 or A6005C and pure Ti and discussed on mechanical property of welded joints.

2 Preliminary Survey

It is confirmed to weld between Al and Ti with manual arc welding. A6005C-T5 and pure Ti JIS TP340 (ASTM Grade 2) are used and dimension of test coupon is 2mm thick, 150mm width and 300mm length. Their chemical compositions are shown in Table 1.

Joint design is I butt and rout gap is 0mm. Manual AC GTAW is applied and shielding gas is Ar 100%. Welding current is 80A.

Surface appearance is very smooth after welding and joint strength is 195N/mm^2. Rapture is occurred at heat affected zone (HAZ) at A6005C and acceptable quality of weld is confirmed. Transverse section of joint is shown in Fig. 1 TP340 is not melted during welding shown in Fig. 1-A. Thickness of IMC at interface between deposited metal and TP340 is about 2~3μm shown in Fig. 1-B. It is confirmed that manual GTAW is available for welding Al and Ti.

Table 1 Chemical compositions of experimental materials

Materials	Chemical compositions (wt%)					
A6005C-T5	Si	Cu	Mn	Mg	Fe	Al
	0.64	0.23	0.01	0.5	0.24	Bal.
TP340 (ASTM Gr.2)	C	H	O	N	Fe	Ti
	0	0.001	0.04	0	0.02	Bal.

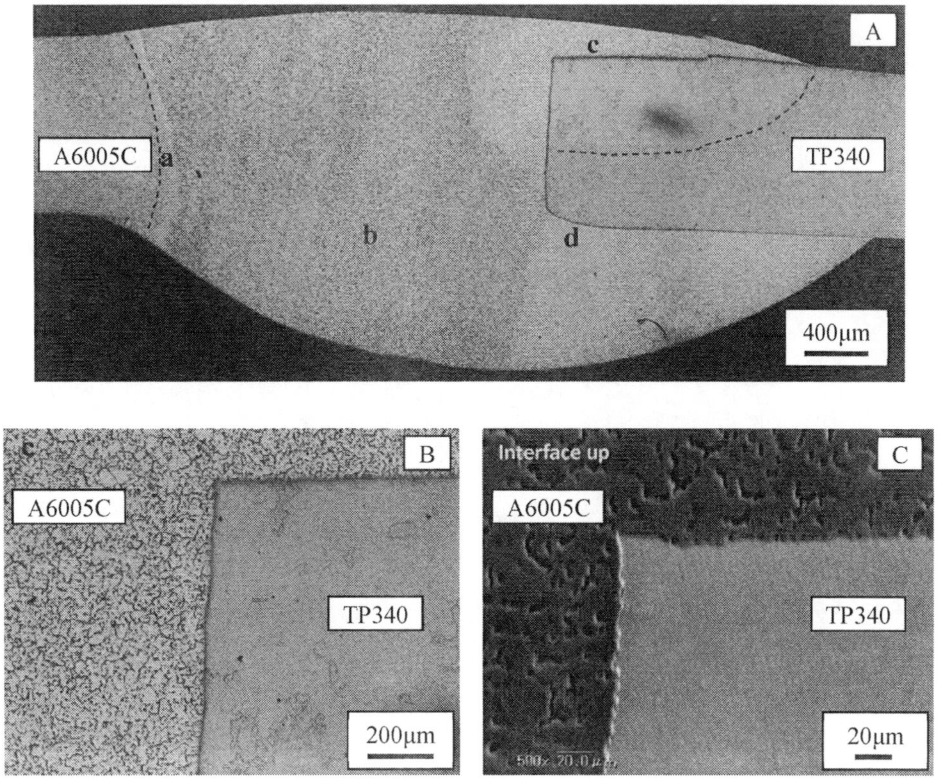

Fig. 1 Macrographs (A, B) and SEM photomicrograph (C) of A6005C/TP340 by manual welding

3 Fully Mechanized GTAW

To establish the dissimilar welding without personal skill, automatic GTAW is applied. A6061-T6 and pure Ti JIS TP270 (ASTM Grade 1) are used and dimension of test coupon is 2mm thick, 150mm width and 300mm length. Their chemical compositions and mechanical properties are shown in Table 2. Joint design is shown in Fig. 2. Welding consumable is A5356-WY 1.6mmφ. Welding parameters are shown in Table 3. This parameters are based on manual GTAW. Visual testing, tensile testing and macroscopic observation on transverse section of welded joint are evaluated without eliminating reinforcement.

Table 2 Chemical compositions and mechanical properties of experimental materials

Materials	Chemical composition (wt%)						Tensile strength (N/mm^2)	Thermal conductivity (W/m·K)
A6061-T6	Si	Cu	Mn	Mg	Fe	Al	333	582~652
	0.58	0.27	0.03	0.697	0.41	Bal.		
TP270 (ASTM Gr.1)	C	H	O	N	Fe	Ti	331	1670
	0	0.001	0.04	0	0.02	Bal.		

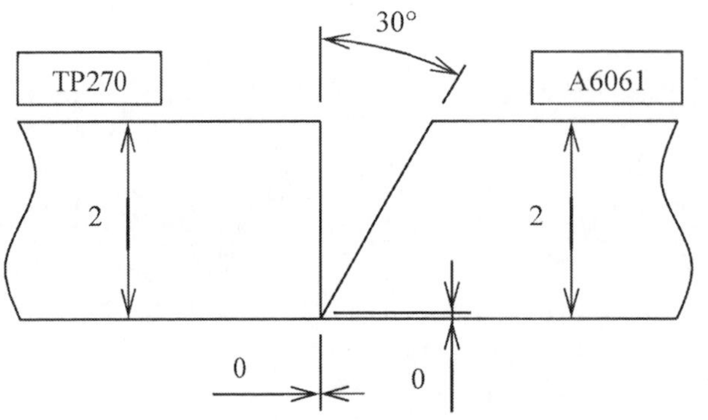

Fig.2 Groove profile (Unit: mm)

Table 3 Welding menu by fully mechanized welding

Base current (A)	45	Welding speed (cm/min)	10,11,13,15,17
Pulse current (A)	105	Wire feed rate (cm/min)	81
Pluse frequency (Hz)	1.1	Welding target position	Center
Toach shield gas flowrate (l/min)		12	

4 Results and discussion

Surface appearance after welding is shown in Fig. 3. Smooth appearance of both top surface and root surface. The high of reinforcement is gradually increasing with decreasing travel speed at constant wire feeding rate. Microscopic observation at transverse section is shown in Fig. 4. Tensile strength of welded joints is shown in Fig. 5. Rapture is occurred at HAZ in A6061 in all tensile specimens. Tensile strength of welded joint is fully satisfied in comparison with that of A6061-O material 145N/mm^2. Tensile strength of welded joint are not changed within experimented travel speed. This is why melting point of Ti is higher than the temperature of melted A5356 and then thickness of IMC is not increasing during welding.

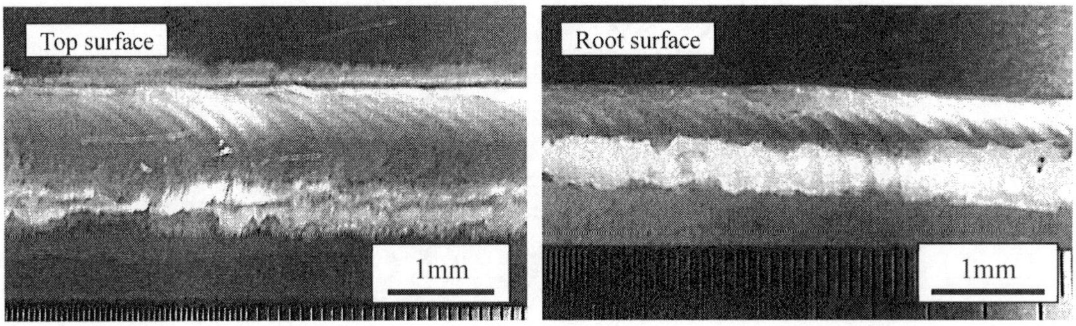

Fig. 3 Surface appearance of A6061/TP270 by fully mechanized welding (Welding speed 15cm/min)

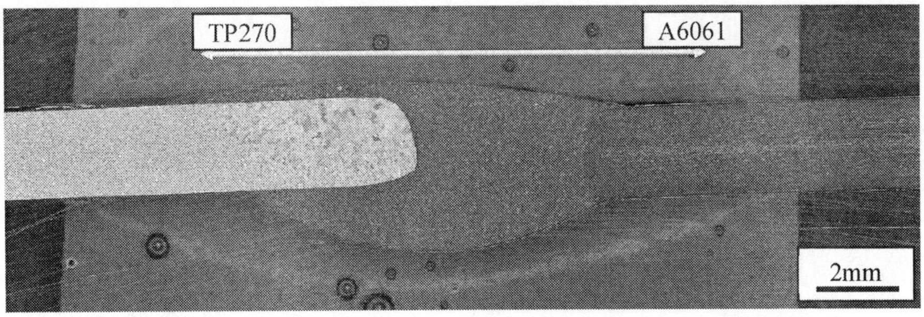

Fig. 4 Macrograph of A6061/TP270 by fully mechanized welding (Welding speed 15cm/min)

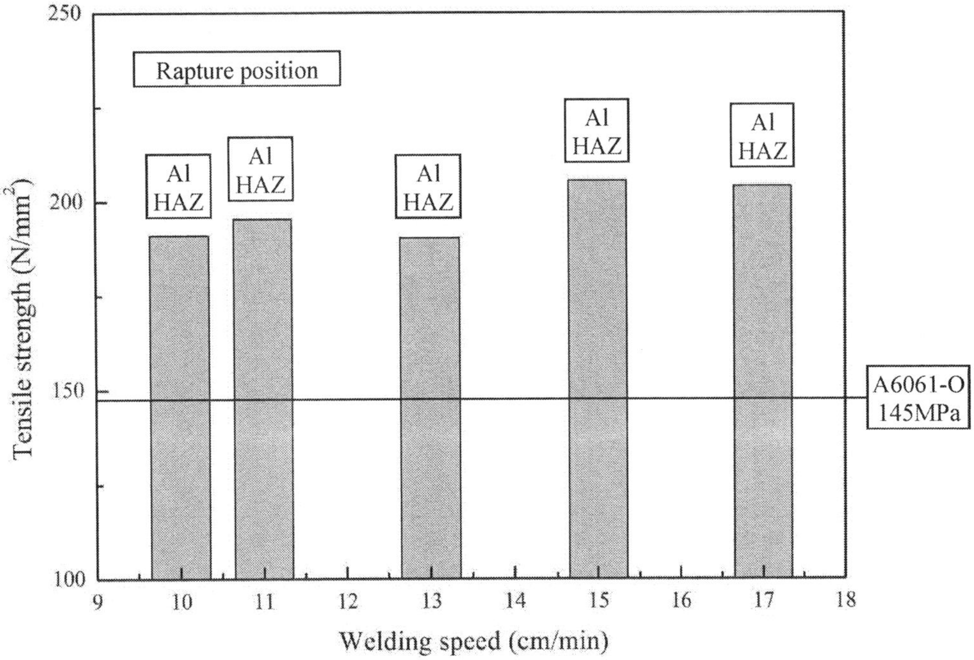

Fig. 5 Tensile strength of A6061/TP270 by fully mechanized welding

5 Conclusions

It is concluded that manual or fully mechanized GTAW is available for dissimilar welding between A6005C of A6061 and pure Ti. This is why melting point of pure Ti is much higher than melted Al consumables and then the thickness of IMC is not increasing during welding. So, the joint strength of welded metal is higher than heat affected zone of Al alloy.

Acknowledgments

We are very grateful to supports by Japan Light Metal Welding Association Dissimilar Joint Committee and The Japan Titanium Society. The authors express heartfelt thanks here.

References

[1] Wei Shouzheng, *et al*: "Microstructure and joining mechanism of Ti/Al dissimilar joint by pulse gas metal arc welding," Int J Adv Manuf Technol 70, Feb 2014, pp. 1137-1142.

Process Stability Assessment of Cold Wire Gas Metal Arc Welding Process in Aluminum Alloy Butt Welding

P. D. C. Assunção
Faculty of Mechanical Engineering, (FEM) Amazon Superior Studies Institute (IESAM). Avenida Governador José Malcher, 1148 - Nazaré. 66060-232.. Belém – PA – Brazil
dangeloassuncao@bol.com.br Phone: +55 (91) 4005-5400

R. A. Ribeiro
Metallic Materials Characterization Laboratory (LCAM), Faculty of Mechanical Engineering (FEM), Federal University of Pará (UFPA). R. Augusto Corrêa, 01 - Guamá. 66075-110. Belém - Pará – Brazil
rafaelaraujo@ufpa.br Phone: +55 (91) 3201-8673

M. Calandrini
Brazilian Navy, Val de Cãs Naval Base (BNVC), Naval Construction Section, Rodovia Arthur Bernardes - Val de Cãs. 66115-300. Belém – PA – Brazil.
calandrini@bnvc.mar.mil.br Phone: +55 (91) 3216-4444

E. M. Braga
Metallic Materials Characterization Laboratory (LCAM), Faculty of Mechanical Engineering (FEM), Federal University of Pará (UFPA). R. Augusto Corrêa, 01 - Guamá. 66075-110. Belém - Pará – Brazil
edbraga@ufpa.br Phone: +55 (91) 3201-8673

Abstract

Nowadays, a common paradigm guides the development of new welding processes, decouple the direct relationship between nominal heat input (kJ/mm) and deposition (kg/h). The new Cold Wire Gas Metal Arc Welding (CW - GMAW) promotes such decoupling since it is characterised by the feeding of an extra cold (non-energized) wire into the welding pool; in order to increase the deposition, theoretically decreasing the transferred heat to the substrate. This work aims to apply the CW - GMAW to butt weld 5mm-thickness AA5052 H-32 plates, used for shipbuilding in local shipyards, in a flat position for two different welding speeds (cm/min), 30 and 40 cm/min. The amount of cold wire feeding is measured as a percentage of the energised wire speed (m/min) which will be maintained constant throughout this work, to know, 14 m/min. The cold wire feeding will be varied from 0%, which characterises the conventional GMAW, to 40%. On one hand, the voltage (V) and current (A) signals were recorded and used to assess the process stability through their coefficients of variation. On another hand, these values were correlated to the average nominal heat input (kJ/mm), and to the cold wire feeding values. The results showed that the welds produced with the introduction of cold wire, for 30 cm/min, feeding generally had a slight decrease in the nominal average heat input to the weld substrate. This was not the same for the 40 cm/min condition.

Introduction

According to Suban and Tusk [1], the stability of a welding process is defined by the arc behaviour, and it is classified as stable when the metal transfer occurs spatterless, uniformly and with constant arc length; for short-circuit transfer the arcing time and short-circuit time should be equal. For spray mode, the time between two metal transferences should be the same. According to the same work [1], one of the simplest forms of assessing arc stability is to record the current and voltage signals with reference to time and analyse them statistically.

This statistical analysis of the current and voltage involves the calculation of coefficients of variation which can be defined as voltage and current standard deviation over the mean current and voltage values [2]. Shinoda et al. [3] propose the use of the coefficient of variation which these authors [3] considered that when the coefficient of variation reached its minimum value the welding parameters that determined this condition are the most stable. Though the coefficient of variation can be used to assess arc stability, Bauné at al. [4] proposed the use of the standard deviation of voltage to measure arc stability due to the relation between voltage and arc length. Also, some researchers used the voltage standard deviation value to determine the different metal transfer modes for GMAW and FCAW [5,6].

Another form of assessing arc stability is through cyclogrammes, these are charts where the voltage is plotted against the current, for short-circuit two distinct regions could be discerned, the first is the short-circuit region characterised by low values of voltage and high values of current. The other zone is called arc burning or arcing area, this is characterised by low values of current and high values of voltage. In cyclogrammes, a welding condition is more

stable if the cyclogramme occupies a smaller area [1].
The main objectives of this work is to demonstrate that welding with the novel process Cold Wire GMA welding (CW-GMAW) is possible and viable from the arc stability point of view and investigate the influence of the welding parameters on arc stability. Another goal of this work is determine for the four conditions of cold wire feeding which is the most stable in comparison with the conventional Gas Metal Arc Welding Process (GMAW) condition.

Experimental Procedure

Two classes of welds were fabricated using CW-GMAW process, the first condition with the welding speed of 30 cm/min and the second condition with a welding speed of 40 cm/min. The welding source used constant voltage (CV) and the material used was a 5 mm thick AA5052 H-32 aluminium alloy, measuring 300 x 150 mm each plate. The joint executed was butt without distance between the plates and the position was flat. During the welding procedures the signals of current and voltage were recorded, with a frequency of 5000 Hz, and after that, they were analysed statistically in order to calculate the voltage and current coefficient of variation and the Carrer criterion [2]. *Fig. 1* shows the welding experimental set-up employed in this work, in it is possible to see the main wire and torch and secondary wire (cold). The welding wire used was the AWS ER5356 for the hot and cold wires with diameters of 1.2 and 1.0 mm respectively.
Table 1 presents the values of the composition of the base metal and of the wires used according to the manufacturers.

Table 1 - Chemical composition (% wt.) of wires used and of the base metal

	Chemical Elements (% wt.)									
Material	**Si**	**Fe**	**Cu**	**Mn**	**Mg**	**Cr**	**Ni**	**Zn**	**Ti**	**Al**
AA5052-H32*	0.25	0.40	0.10	0.10	2.2-2.8	0.15-0.35	-	0.10	-	Balance
ER5356**	0.25	0.40	0.10	0.05-0.20	4.5-5.5	0.05-0.20	-	0.10	0.06-0.20	Balance

*Council, Aluminum Extruders. "International Designations and Chemical Composition Limits for Aluminum Hardeners." (2007).
**According to AWS A5.10/A5.10M:2012. "Welding Consumables – electrode wires and rods for welding of aluminium and aluminium alloys – Classification". America Welding Society (AWS). USA. (2012).

Figure 1 - Experimental set-up used for the AA5052 H-32 butt welds

It is now necessary to define an important parameter to measure the cold wire feeding to the welding pool, this is called R, which is the ratio of the actual Cold Wire Feed Speed (CWFS) in m/min over the main Wire Feed Speed (WFS) in m/min, this is ratio is presented in equation 1.

$$R = \frac{CWFS}{WFS} \times 100 \quad (1)$$

Then, the cold wire feed speed is a percentage of the main wire speed. *Tables 1 and 2* present the welding parameters used in this work along with the nominal heat input of each weldment produced.

Table 2 - Welding parameters for the first set of welds

Shielding gas: 100% Argon with flow of 17 l/min						
Cold wire feed ratio (%)	Electrode Wire Feed Speed (m/min)	Voltage (V)	CTWD (mm)	Welding Speed (cm/min)	Current (A)	Welding Energy (kJ/mm)
0	14	26	22	30	191	0.9932
10	14	26	22	30	189	0.9828
20	14	26	22	30	188	0.9776
30	14	26	22	30	189	0.9776
40	14	26	22	30	188	0.9776

Table 3 - Welding parameters for the second set of welds

Shielding gas: 100% Argon with flow of 17 l/min						
Cold wire feed ratio (%)	Electrode Wire Feed Speed (m/min)	Voltage (V)	CTWD (mm)	Welding Speed (cm/min)	Current (A)	Welding Energy (kJ/mm)
0	14	26	22	40	184	0.9568
10	14	26	22	40	186	0.9672
20	14	26	22	40	188	0.9776
30	14	26	22	40	224	1.1648
40	14	26	22	40	236	1.2272

Results and Discussions

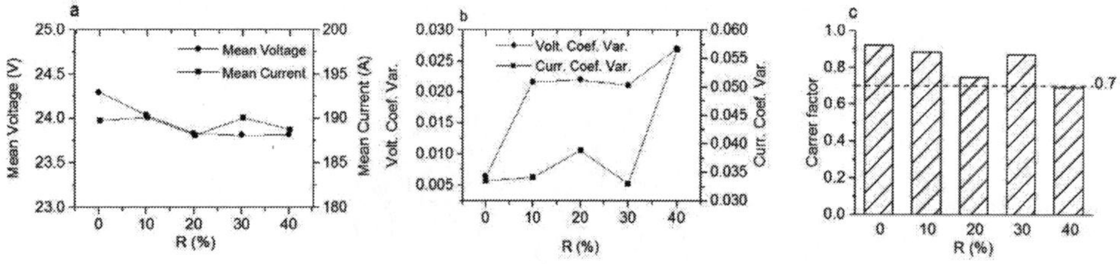

Figure 2 – Arc stability indexes for the welding speed of 30 cm/min.

Fig. 2 shows the arc indexes for the 30cm/min welding speed condition. *Fig. 2a.* shows the mean and voltage plot against the percentage of cold wire injected into the electric arc region. It can be observed that as a measure as the cold wire is introduced into the welding system there is a slight decrease in arc voltage indicating a little decrease in the arc length until the quantity of 20% of cold wire. From this point on, there is practically no change in current intensity. Fig. 2b presents the coefficient of variation of voltage and current as a function of the quantity of cold wire. The general trend observed is that for voltage, the observed increase in coefficient of variation from 0% to 10% is due the increase in voltage standard variation probably because the arc length fluctuation was caused by the introduction of cold wire into the welding system.

The voltage coefficient of variation achieves a plateau between 10% to 30% of cold wire and further increase when the quantity of cold wire reaches 40%. Still in *Fig. 2b.* could be observed that current coefficient of variation almost constant unless for the 20% condition. *Fig 2c.* shows the Carrer coefficient which indicates how stable the welding was processed as a function of cold wire feeding for the constant welding of 30 cm/min. Slazak et al. [2] provided the threshold of more than 0.7 of the Carrer index to indicate welding stability. According to such criterion such the conventional welding (R = 0%) is the most stable condition followed by 10%, 30%, and 20% of cold wire injection. The 40% cold wire injection condition is considered unstable according to the Carrer coefficient.

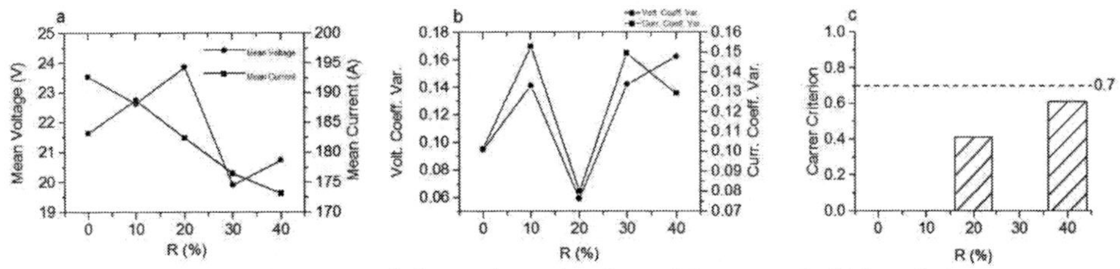

Figure 2 – Arc stability indexes for the welding speed of 30 cm/min.

Fig. 3 presents the stability results for the condition of 40 cm/min of welding speed. *Fig. 2a.* shows the mean voltage and current versus the cold wire injection quantity, R (%). In this figure as in *Fig. 2a* there is the trend for a slight decrease in voltage as the cold wire is put into place following this diminish the voltage increases for R = 20% and abruptly decreases for R = 30% and finally, increasing again for R = 40%. It can be noticed that the general trend of the voltage decreases with the augmenting of cold wire feed speed. The same *Fig. 2a.* shows that the general trend of the voltage is decreased as long as the cold wire injection increases. *Fig. 2b* shows the voltage and current variation versus R (%).
Compared to *Fig. 1b*, *Fig. 2b* shows a more unstable pattern presenting a lesser variation for R=20% condition. Looking into the current signals, recorded during welding for this condition, it can be noticed that many short-circuit occur which leads the current to 0, being this value the minimum found in the current data. This short-circuit leads the Carrer coefficient to 0 for 0%, 10%, and 30% of cold wire feed speed, then unstabilizing the welding conditions. Even for cold wire feeding where such short-circuits did not occur, presented a Carrer criterion lower than 0.7 indicating that these conditions are not stable. It is believed that the combination of 40% of cold wire feed speed and 40 cm/min do not favour a stable metallic transfer because of the high speed the welding pool is moving.

Conclusions

1. The introduction cold wire in welding aluminium butt welds contributed to the slight decrease in the nominal heat input imposed to the plate for the condition of 30cm/min;
2. The introduction of cold wire proved to be beneficial for the arc stability for the welding speed of 30 cm/min for the cold wire feeding of 10%, 20% and 30%. For the welding speed of 40 cm/min, the introduction of cold wire was detrimental to arc stability.
3. The introduction of cold wire influences the arc stability and this stability depends on the welding speed. For the high welding speed condition, the introduction of cold wire led to an instability in the arc;
4. The increasing in welding speed from 30 cm/min to 40 cm/min led to a welding source unstable behaviour leading to minimum values of currents equal to zero.
5. As a response to zero current minimum values for 40 cm/min condition, the calculation of the Carrer criterion showed that this condition is totally unstable.
6. A trend was observed for the 30 cm/min condition, the increase in cold wire feeding leads to a decreasing in current values.
7. The values of voltage and current for the condition of 30 cm/min remain almost constant in the interval of 10% to 30% of cold wire feed speed which characterizes the zone as the most stable.

References

[1] Suban, Marjan, and Janez Tušek. "Methods for the determination of arc stability." Journal of Materials Processing Technology 143 (2003): 430-437.

[2] Slazak, B., J. Slania, T. Węgrzyn, and Abilio P. Silva. "Process stability evaluation of manual metal arc welding using digital signals." In Materials Science Forum, vol. 730, pp. 847-852. Trans Tech Publications, 2013.

[3] Shinoda, T., H. Kaneda, and Y. Takeuchi. "An evaluation of short circuiting arc phenomena in GMA welding." *Weld. Met. Fabr.* 57, no. 10 (1989): 522-525.

[4] Baune, E., C. Bonnet, and S. Liu. "Assessing metal transfer stability and spatter severity in flux cored arc welding." *Science and Technology of Welding & Joining* (2013).59-s.

[5] Liu, S., and T. A. Siewert. "Metal transfer in gas metal arc welding: droplet rate." *Welding Journal* 68, no. 2 (1989): 52-58.

[6] Wang, W., S. Liu, and J. E. Jones. "Flux cored arc welding: arc signals, processing and metal transfer characterization." *Welding Journal-Including Welding Research Supplement* 74, no. 11 (1995): 369s.

Assessment of twin-wire GMAW as a candidate for large scale arc-based Additive manufacturing

Syed Quadir Moinuddin, Manish Kumar, Surya S Kumar, Abhay Sharma

Indian Insttitute of Technology Hyderabad, India

Manufacture of large scale additive manufacturing products needs heavy deposition coupled with sufficient bonding strength between layers. Arc welding is a potential candidate in such scenario, however, thermal management is major challenge. Moreover, the major proportion of heat should be utilised in electrode melting and dilution should be kept lower. In present investigation, twin wire welding, known for higher deposition rate and shallow penetration, is evaluated as a potential candidate for large scale additive manufacturing. A large number of experiments are carried out with change in process conditions including lead and trail wire diameter and their respective currents, welding speed, type of welding current (i.e pulsed and continuous). An index representing suitability of process conditions for additive manufacturing is developed. The index encompasses effects of electrode melting efficiency, weld bead geometry, dilution, and ease and time in producing multi-pass multi-layer components. Representative samples corresponding to different values of the foresaid index are evaluated for their mechanical and metallurgical characteristics. The investigation suggests that the index can be used to evaluate candidacy of an arc welding process under different process conditions for suitability of additive manufacturing. It is observed that process conditions that stabiles fluctuation in arc signals such as lower current at trail wire and higher current density at lead wire, moderate welding speed and pulsed current are more affective in obtaining good additive manufacturing products. A proper weld cooling schedule must be devised in order to minimise interlayer property variation. The results of this investigation offers a frame work of process design for large scale arc-based additive manufacturing through arc welding.

Investigation of Weldability of Aluminium (Al6063T6) using Pulse MIG Welding

Ajai Kumar Pathak[1] (corresponding author), Ravi Pratap Singh Pal[2], Razaullah Khan[3]
NIFFT Hatia, Ranchi, India
[1]akpathak.nifft@gov.in

Abstract
Aluminium is used for various applications, such as, airframes, missile bodies, fuel cells, and satellite components, because of its inherent advantages of high strength-to-weight ratio, high notch toughness at cryogenic temperatures, excellent corrosion resistance, ease in extrusion, and forming. However, weldability of aluminium is not considered good. Hence, it is important to investigate the weldability of aluminium to obtain the proper welding parameters. In this project, the weldability of Al6063T6 was investigated using single pulse and double pulse MIG welding. The welding process parameters like current, voltage and welding speed were optimised practically. Butt joints of 3 mm thick sheet were prepared to investigate percentage elongation, ultimate tensile strength, and microstructure of the welded joint. Butt welded joint of 10 mm thick plate were fabricated for toughness, and hardness tests. The microstructure of the different zones was investigated by using optical microscope. In the fusion zone of single pulse MIG welded joint FeSiAl were observed which increase the hardness of the welded joint and Mg_2Si were observed in greater amount in the double pulse MIG welded joint which decrease the hardness. But in both the cases the hardness of weld metal zone and heat affected zone were found lower than the base metal. The ultimate tensile strength (UTS) and percentage elongation of the welded joint was found lower than the base metal. During tensile test most of the specimens were broken outside the welded joint, hence, it can be concluded that a good welded joint of aluminium can be obtained in proper welding condition.

Keywords: Weldability, Al6063T6, MIG Welding, Mechanical properties, Microstructure

Introduction
Welding of aluminium and its alloys is always considered to be difficult. From the different welding processes, arc welding is most extensively used for aluminium. Basically, aluminium and its alloys can successfully be welded by using conventional GTAW and GMAW process. However, occasionally specialized equipment or techniques, or both, are required due to the inherent unique physical and mechanical characteristics of aluminium and its alloys [1]. This paper deals with GMAW of aluminium and its alloys.

To investigate the weldability alloy 6063 T6 was selected. Alloy 6063, one of the most popular alloys in the 6000 series, provides good extrudability and a high quality surface finish. It is used in standard architectural shapes, custom solid shapes and heat sinks, as well as seamless and structural tube and pipe. In the heat-treated condition, alloy 6063 provides good resistance to general corrosion, including resistance to stress corrosion cracking. Since 6063 is a heat- treatable alloy, strength in its -T6 condition can be reduced in the weld region [2]. Selection of an appropriate filler alloy will depend on the desired weld characteristics.

Experimental work
Extruded sheets of the Al6063T6 of 3 mm and 10 mm thickness were used for the experimental investigations. For welding of sheets filler wire of 4043 was selected as recommended by the Welding Hand Book [3]. Table 1 shows the composition of base metal and filler wire.

Table 1: Composition (wt %) of base metal and filler wire [4]

Element	Base Material Al6063T6	Filler wire 4043
Cu	0.10	0.30
Mg	0.45 - 0.90	0.05
Si	0.20 - 0.60	4.5 – 6.0
Fe	0.35	0.08
Mn	0.10	0.05
Ti	0.10	0.20
Zn	0.10	0.10
Al	Balance	

Preparation of Test Specimen

Plate of 3 mm and 10 mm thick of 6063T6 was cut to the required dimensions using hand hacksaw. The weld joint geometry of the specimens was shown in Fig. 1 [3]. For 3 mm sheet square butt joint and for 10 mm thick single 'V' butt joint was used.

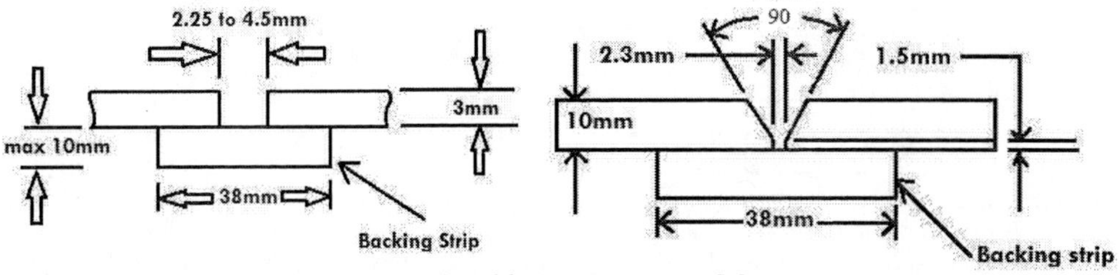

Fig. 1 Weld Joint Geometries [3].

The surface to be joined was cleaned properly before starting welding. Specimens were welded by single pulse and double pulse MIG welding using argon as shielding gas. Welding parameters was recorded in Table 2. Photograph of welded specimen (3 mm and 10 mm thickness) was shown in Fig. 2(a) and 2(b).

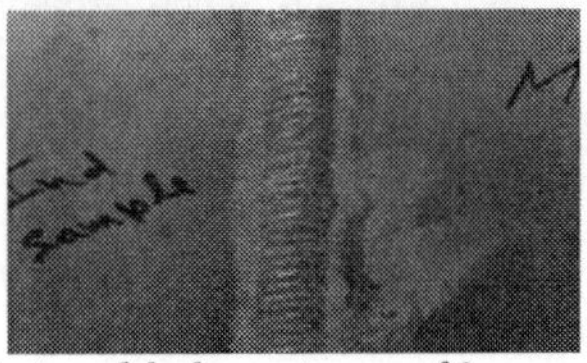

Fig. 2(a) A welded specimen of 3 mm plate

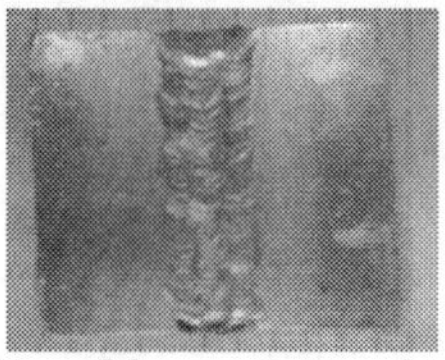

Fig. 2(b) welde specimen of 10 mm plate

Table 2: MIG welding process parameters.

Particular	Single pulsed	Double pulsed
Base Material	6063T6	6063T6
Plate Thickness (mm)	3	3
Electrode	ER4043	ER4043
Electrode diameter (mm)	1.2	1.2
Welding Current (amps)	54.5-61	56-60.5
Welding Voltage (volts)	16.05-17.05	16.25-17.6
Welding Speed (mm/min)	269.52	254.5
Wire feeding speed (mm/min)	2.8-2.9	2.3-3.4
Polarity	DCEP	DCEP

Mechanical testing

Tensile Test: Specimens for tensile test for 3 mm thick plate were prepared as pre ASTM B557 [5]. A total of 18 tensile specimen, six for each, single pulse, double pulse and base metal were prepared. Fig. 3 shows the standard dimension of the tensile test specimen. Fig. 4 shows the specimens of single pulse welded joints. Tensile tests were conducted on Univarsal Testing Machine.

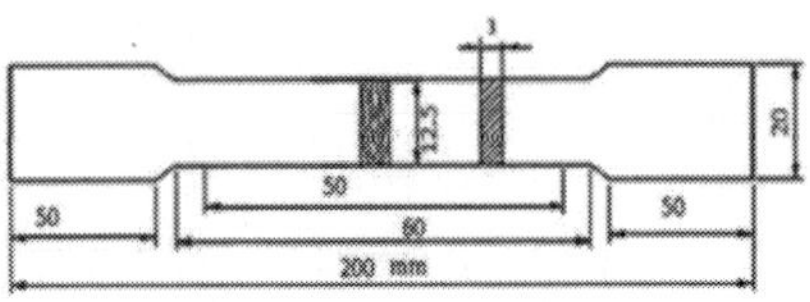

Fig. 3. Dimensions of tensile specimen

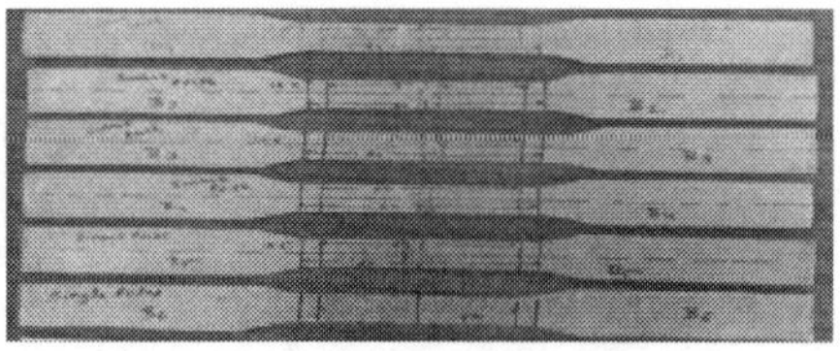

Fig. 4 Tensile specimen of single pulse welded joint

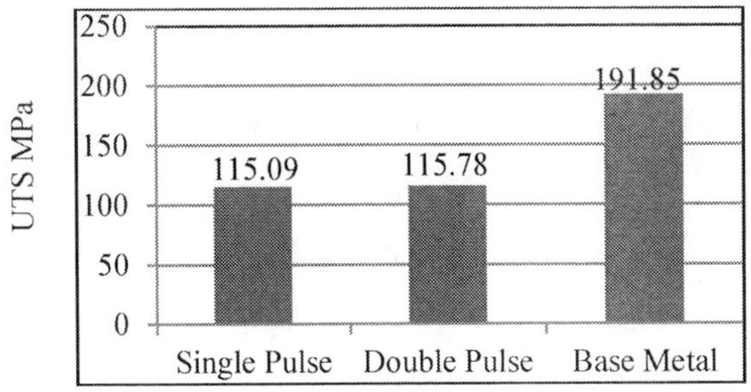

Fig. 5 Comparison of UTS of single pulse welded joint, double pulse welded joint and base metal.

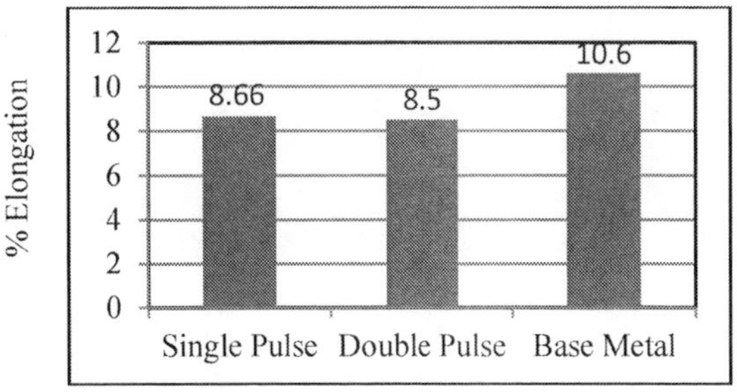

Fig. 6 Comparison of % elongation of single pulse joint, double pulse joint and base metal.

Average value of UTS and percentage elongation of six specimens in each case (base metal, single pulse welded joint and double pulse welded joint) were calculated and compared in Fig. 5 and Fig. 6 respectively. From the figures it can be seen that base metal is having higher value of UTS and percentage elongation than the welded joints. UTS and percentage elongation of welded specimen decreases due to change in heat treatment effect and microstructure due to welding. No significant effect on the value of UTS and percentage elongation were observed due to single pulse and double pulse welding.

Hardness Test

The Rockwell hardness test of the specimen was conducted under a preliminary minor load of 10 kgf. Rockwell hardness number of weld metal zone, heat affected zone and base metal was measured for single pulse and double pulse welded joint. Average of six samples in each case was calculated and compared in Fig. 7 and Fig. 8. It can be seen that hardness value in fusion zone and heat affected zone are less than the hardness of base metal. The hardness value decreases due to change in heat treatment effect and change in composition of material in fusion zone.

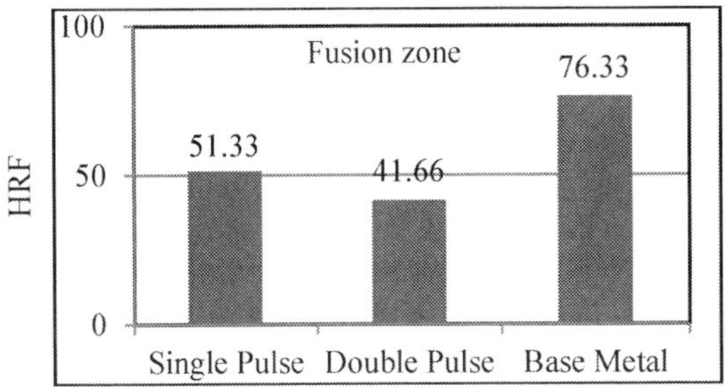

Fig. 7 Comparison of rockwell hardness number at fusion zone.

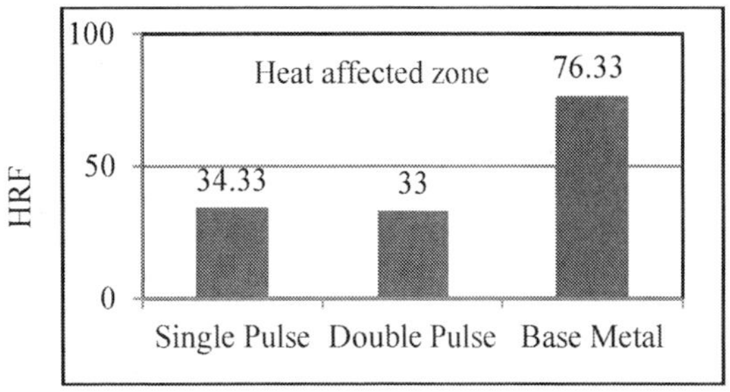

Fig. 8 Comparison of rockwell hardness number at heat affected zone

Metallography

To study the microstructure of the welded joint the specimen were prepared by series of sequential finishing operations. The finished surface was etched for microstructure investigation. For etching of aluminum alloys, mixture of 2 ml HF (15%) + 3 ml HCL (45%) + 15 ml HNO_3 + 190 ml H_2O (25%) and HCL concentration 37% (Tucker's Reagent) was used as an etchant. The investigation of microstructure was done under optical microscope.

Microstructure of 3 mm thick Al6063-T6 plate (base metal) is shown in Fig. 8a. Flowing metal in semi solid condition and hardened portion solid solution have been shown in the microstructure. Elongated grains of aluminium solid solution contain particles of Mg_2Si (black) and FeSiAl. Fig. 8b shows microstructure of HAZ and weld metal zone of single pulse welding. Interdendritic network of aluminium-silicon eutectic (dark) in weld bead, and dark band of Al-Mg_2Si eutectic in the heat-affected zone were observed. Grain size number 7 (48 to 96 grains per square inch) was found in fusion zone. Fig. 8c shows the microstructure of weld metal zone, HAZ and base metal of double pulse welded joint. Presence of interdendritic aluminium-silicon eutectic in small amount in weld metal zone and in large amount in heat affected zone can be seen in Fig. 8c. Grain size number 6 which is 24 to 48 grains per square inch has been found in the weld metal zone.

Fig. 8a Microstructure of base metal Al6063T6.

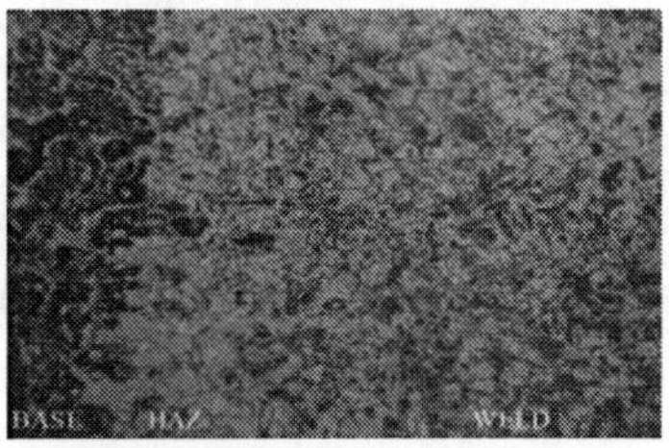

Fig. 8b Microstructure of HAZ + Weld metal zone of single pulse welding.

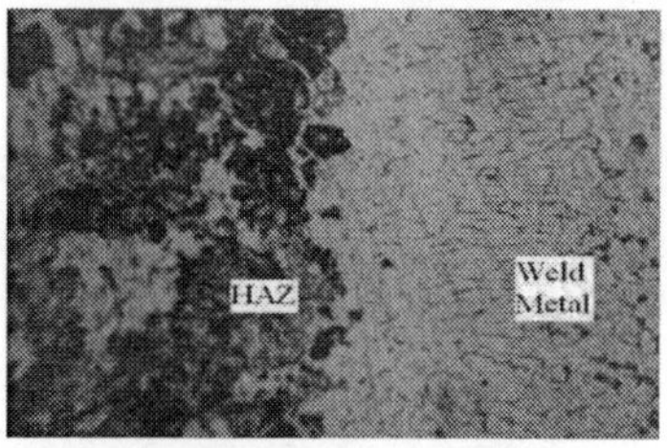

Fig. 8c Microstructure of HAZ + weld metal zone of double pulse welding.

From the photograph we can see that the microstructure is free from under cut, over laps, segregation, pin holes and other discontinuities. Hence defect free and sound welded joint can be obtained in case of aluminium also.

Conclusions

1. Good welded joints of aluminium can be obtained using MIG welding.
2. The UTS and percentage elongation of the welded joints were found to be lower than the base metal. The UTS and percentage elongation reduce due to change in heat treatment condition and microstructure of the weldment.
3. The Rockwell hardness of fusion zone and heat affected zone of single pulse MIG welded joint were found higher than the double pulse MIG welded joint. Higher value of hardness in case of single pulse welding was obtained due to different composition of fusion zone and grain size which were observed from the study of microstructure.

But in both the cases hardness was found lower than the base metal. Reduction of hardness in fusion zone takes place due to change in metal composition and microstructure. Reduction in hardness in heat affected zone is due to change in microstructure.

4. Microstructure of all the welded joints in the weld metal showed variation in grain structure such as fine equiaxed grain at weld centre and columnar grain near the fusion boundary. The grain size of double pulse MIG welding is larger than the single pulse MIG welding. In the single pulse weld microstructure FeSiAl was observed which increase the hardness value. In the double pulse welded microstructure Mg_2S were observed in higher quantity which decreases the hardness value. The grain size number 7 (48 to 96 grains per square inch) for single pulse and number 6 (24 to 48 grains per square inch) for double pulse were found in weld metal zone.
5. Aluminium alloy is inherently difficult to weld and it was experienced during the course of practical work of this project. Lack of fusion is a major problem. Some tensile bars were broken at the welded joint during tensile testing due to lack of proper fusion. Formation of oxide layer and porosity were the other major problem.

References

[1] Arc Welding of Non-Ferrous Metals, 4th edition, published by Kobe Steel, Ltd., 2011, 5-912, Kita-Shinagawa, Shinagawa-Ku, Tokyo, 141-8688 Japan, pp. 1-19.
[2] Understanding extruded aluminium alloys, Alcoa distribution and industrial products, 53 Pottsville Street, Cressona, PA 17929.
[3] Welding Handbook, 7th Edition, Vol. 4, Metal and their Weldability, AWS, pp. 316-388.
[4] ASM hand book V2, Properties and selection: non ferrous alloys and special purpose materials, ASM International.
[5] ASTM B557: Tension testing wrought and cast aluminium and magnesium-alloy products.

Characterization of electron beam and friction stir welds in a novel Al-Mg-Sc-Zr alloy

Johannes Taendl, Cecilia Poletti
Institute of Materials Science and Welding, Graz University of Technology, 8010 Graz, Austria
johannes.taendl@tugraz.at

Abstract

This work investigates the influence of electron beam and friction stir welding on the microstructure and age hardening behavior of a novel AlMg4Sc0.4Zr0.12 alloy. We analyze the welds in as-welded and post-weld heat treated condition for both welding techniques. The post-weld heat treatment (PWHT) is done at 325°C to promote the precipitation of secondary $Al_3(Sc,Zr)$ phases. We study the microstructure evolution by means of light optical microscopy (LOM) and electron backscatter diffraction (EBSD), and correlate the observations to the hardness in each material condition. The fast solidification and cooling rates in the electron beam welded material lead to grain refinement and dissolution of primary $Al_3(Sc,Zr)$ phases, thus increasing the degree of Sc-supersaturation. Therefore, the hardness increase after the post-weld heat treatment is more pronounced in the weld metal due to the presence of a larger number of finely dispersed $Al_3(Sc,Zr)$ precipitates. The mechanical properties of the friction stir welds strongly depend on the heat input given by the process. In case of low heat input, precipitation hardening does not occur during the welding process. Thus, the hardness in the weld zone is at the same level as in the base metal both before and after PWHT. In case of friction stir welds with high heat input, precipitation of $Al_3(Sc,Zr)$ phases occurs during the process at temperatures higher than 325°C, leading to a slight hardness increment in the weld zone. After PWHT, however, the hardness of the base metal is not reached.

Introduction

Al-Mg alloys modified with Sc and Zr possess better mechanical properties than conventional Al-Mg alloys while maintaining beneficial materials characteristics such as the good corrosion resistance. The increased strength level of Al-Mg-Sc-Zr alloys results from the formation of finely dispersed, nano-sized $Al_3(Sc,Zr)$ precipitates that are formed during age hardening treatments [1]. Furthermore, these precipitates stabilize a strain hardened microstructure, thus maintaining the strength of a cold worked sheet by preventing recrystallization [2]. In general, Al-Mg-Sc-(Zr) alloys show good weldability similar to conventional Al-Mg alloys using both fusion [3], [4] and friction stir welding [5], [6]. In case of tungsten inert gas (TIG) welded plates, the hardness in the weld zone decreased due to dissolution of strengthening precipitates [3] but could be increased to the hardness of the base material by applying a post-weld heat treatment leading to re-precipitation of Al_3Sc phases [4]. In friction stir welded material (FSW), on the other hand, the precipitates were stable [5] and the hardness could be maintained to a larger extent [6]. These studies on TIG and FSW demonstrate the suitability of either process, however, all of them were carried out on materials, where the Al_3Sc precipitates were present prior to welding and the Al matrix were free of Sc. Currently, it is aimed to produce Al-Mg-Sc-Zr alloys by rapid solidification to achieve a large degree of Sc-supersaturation after casting in order to promote the formation of the strengthening precipitates in a more targeted way [7], [8]. However, a welded material of such type was not characterized so far.

In this study, we investigate the microstructure and hardness of electron beam and friction stir welds in a novel, rapidly solidified Al-Mg-Sc-Zr alloy produced on industrial belt-casting equipment.

Experimental

The investigated material was an Al-Mg4-Sc0.4-Zr0.12 alloy casted to an 8mm thick, 280mm wide, and 30m long belt using a continuous belt-casting technology. We have already demonstrated that due to the solidification conditions, only 0.13±0.02 wt.% Sc of nominal 0.4 wt.% was in solution after casting, while the remaining Sc was bound in primary and eutectic $Al_3(Sc,Zr)$ phases. Despite the limited degree of Sc-supersaturation, we have shown that secondary $Al_3(Sc,Zr)$ precipitation occurs upon ageing [7], [9] and hot deformation [10].

Electron beam welding (EBW) was carried out using an EBG 45-150 K14 electron beam welding machine (pro-beam). The main parameters were an acceleration voltage of 150kV, a beam current of 10mA, and a welding speed of 10mm/s. A point beam figure was used and the beam was slightly underfocused below the surface of the casted belt.

Friction stir welding (FSW) was performed using an ISTIR FSW machine (MTS). Blind welds were accomplished using a threaded pin tool (Stirtec) with a shoulder diameter of 15mm and a pin length of 5.5mm. The basic processing parameters were a tilt angle of 0° and a tool rotation speed of $1200 min^{-1}$. Two different parameter sets were studied aiming for a variation in heat input. The high heat input weld was carried out using a tool rotation speed of 100mm/min and a vertical down force of 6.5kN, while the low heat input weld was done at 1000mm/min and 32.5kN vertical down force. The temperatures were measured for both welding parameter sets.

For both EBW and FSW, the materials conditions in as-welded (AW) condition were compared to a post-weld heat treated condition. The post-weld heat treatment (PWHT) was carried out at 325°C for 60min to promote the precipitation of secondary $Al_3(Sc,Zr)$ phases. The welds were characterized using several microscopy techniques. Macrosections were etched using Flick's etchant, while microsections studied more in detail by light-optical microscopy (LOM) were electro-chemically etched using Barker's etchant. Electron backscatter diffraction (EBSD) was carried out on a JSM7001F field-emission scanning electron microscope (JEOL) operated at an acceleration voltage of 15kV with a step size of 0.25μm. The temperature was measured during FSW experiments using K-type thermocouples located close to the welds. Hardness testing was done on an automatic M1C010 hardness tester (EMCO-TEST). The used method was HV0.1 Vickers hardness with a load of 0.981N.

Results and Discussion

Fig. 1 (a) shows a macro- and microsection of an electron beam weld with its typical narrow shape and grain size reduction in the weld metal. Fig. 1 (b) and (c) show EBSD images of the base material (BM) and weld metal (WM). The EBW process reduced the average grain size from 30±6μm to 5±2μm and dissolved large primary or eutectic $Al_3(Sc,Zr)$ phases present in BM, as indicated by arrows. The hardness in as-welded condition was at a similar level in BM and WM, in good agreement to other studies reporting on limited strengthening effects of grain size reduction [11], [12]. The hardness increase due to PWHT, on the other hand, was significantly more pronounced inside the weld than in BM. The strength contribution from $Al_3(Sc,Zr)$ precipitates in PWHT condition equaled 23HV0.1 in BM and 50HV0.1 in WM. We attribute this difference to a larger number of precipitates being present in PWHT condition as a result of a higher degree of Sc-supersaturation after welding [9].

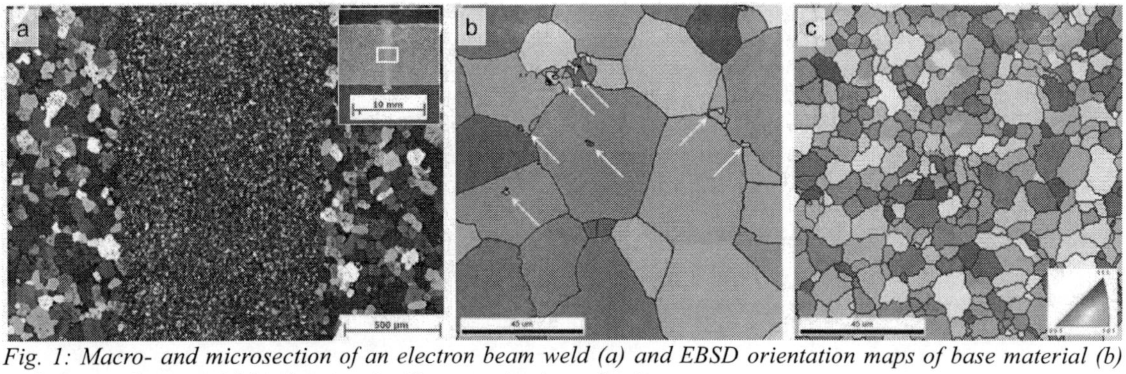

Fig. 1: Macro- and microsection of an electron beam weld (a) and EBSD orientation maps of base material (b) and electron beam weld (c) showing significant grain size reduction.

Fig. 2 shows the top (a) and bottom (b) view of a friction stir welded sample instrumented with 5 thermocouples located at various distances from the weld center on the advancing side of the weld. A pocket was machined in the edge of the sample to position the thermocouples in the center of the casted belt regarding the sample thickness. The final positions of the measurement points with respect to a macrosection of a weld are illustrated in Fig. 2 (c).

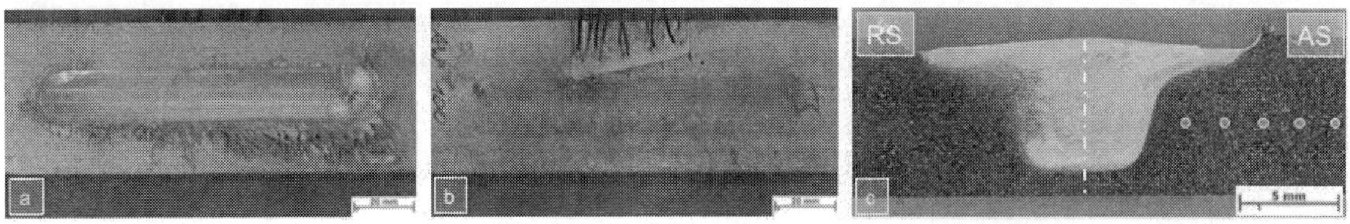

Fig. 2: Top (a) and bottom (b) view of a FSW sample showing the applied thermocouples. Macrosection of a weld and schematic illustration of the temperature measurement points at 5, 7, 9, 11, and 13mm from the center

Fig. 3 shows the measured temperatures as a function of the welding time for different sample positions in high (red) and low heat input (blue) friction stir welds. For each heat input, the temperature decreased with increasing distance from the weld center. Except for the measuring point closest to the weld center, where in both conditions a temperature of around 470°C was measured, the respective temperatures were slightly higher in case of higher heat input. More important, however, was the exposure to elevated temperatures, which was significantly longer for the high heat input welds. Since these temperature values represent measurements close to the thermo-mechanically affected zone (TMAZ), it can be reasonably assumed that the temperatures inside TMAZ and stir zone (SZ) were considerably higher and the exposure to these temperatures was even longer.

Fig. 4 shows hardness patterns across the friction stir welds of low (a) and high heat input (b). The respective dashed line represents the as-welded condition, while the full line corresponds to the PWHT condition. In case of low heat input (Fig. 4 a), the hardness in as-welded condition was at a similar level in the unaffected BM and the different zones inside the weld. After PWHT, the hardness increased in average by 22HV0.1 due to precipitation of $Al_3(Sc,Zr)$ phases leading to a rather constant hardness plateau across the entire weld. In case of high heat input (Fig. 4 b), the hardness in as-welded condition increased with decreasing distance to the weld center with the highest hardening increment of around 12HV0.1 inside the stir-zone. However, after PWHT no further hardening was observed inside the weld and the hardness of the unaffected BM was not reached. Apparently, the strengthening $Al_3(Sc,Zr)$ precipitates were formed already during the welding process at temperatures higher than 325°C. In such a case, the precipitates are coarse and have a low number density [9]. During PWHT at 325°C, therefore, no additional precipitation hardening could be achieved. In contrast, the maximum hardness was lowest compared to the unaffected BM and the low heat input weld.

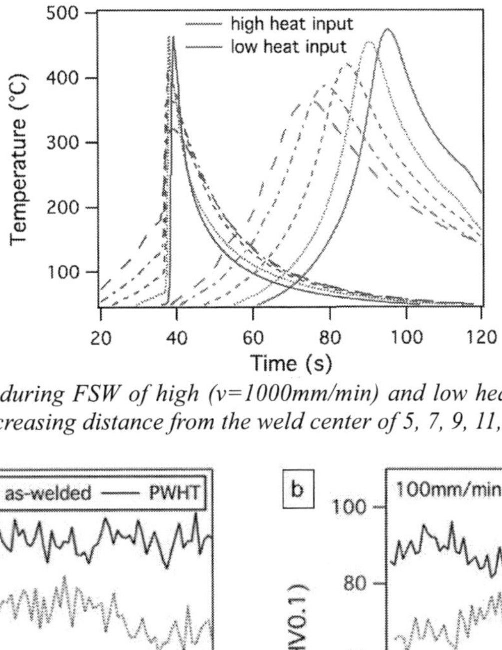

Fig. 3: Temperature evolution during FSW of high (v=1000mm/min) and low heat input (v=100mm/min) welds. Temperature increased with decreasing distance from the weld center of 5, 7, 9, 11, and 13mm.

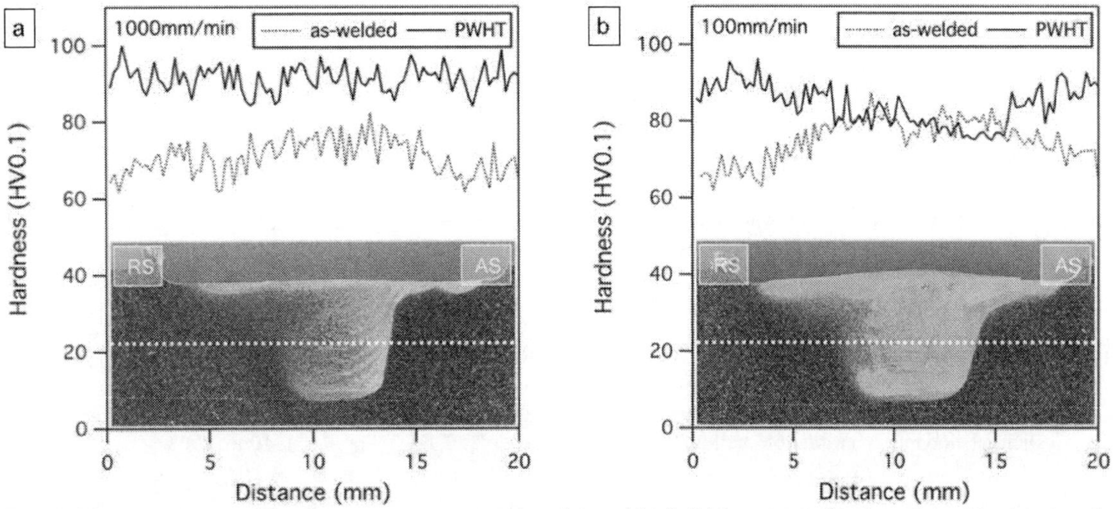

Fig. 4: Macrosections and hardness patterns across low (a) and high (b) heat input firction stir welds showing the influence of the welding parameters and post-weld heat treatment on the $Al_3(Sc,Zr)$ precipitation

Conclusions

In this study, electron beam and friction stir welds in a novel belt-casted Al-Mg4-Sc0.4-Zr.012 alloy were characterized using LOM, EBSD, and hardness testing. The conclusions can be summarized as follows:

1. Electron beam welding leads to significant grain size reduction and dissolution of primary and eutectic $Al_3(Sc,Zr)$ phases, thus increasing the degree of Sc-supersaturation.
2. The hardness in as-welded condition was similar in BM and WM for the EB weld. A post-weld heat treatment at 325°C for 60min increased the hardness both in BM and WM due to precipitation of secondary $Al_3(Sc,Zr)$. The hardening effect was considerably more pronounced inside the welds as a result of the larger degree of Sc-supersaturation prior to ageing.
3. The temperature evolution during friction stir welding significantly depends on the heat input. In case of higher heat input, the weld region was subjected to higher temperatures for considerably longer times compared to the lower heat input weld.
4. For low heat input friction stir welds, the hardness in as-welded condition remained at a similar level as in BM and was uniformly increased to the same hardness level as in BM after applying a PWHT at 325°C for 60min.
5. For high heat input friction stir welds, $Al_3(Sc,Zr)$ precipitation occurred already during the welding process at temperatures higher than 325°C. During PWHT at 325°C for 60min, therefore, the hardness could not be further increased and was finally lower than in BM and all other welds.

Acknowledgments

The authors gratefully acknowledge financial support by the Austrian Research Promotion Agency FFG project 839002 funded within the program TAKE OFF. We want to thank Mr. Andreas Huetter for his valuable support with the friction stir welding experiments.

References

[1] J. Røyset and N. Ryum, "Scandium in aluminium alloys," *Int. Mater. Rev.*, vol. 50, no. 1, pp. 19–44, 2005.

[2] J. Taendl, S. Nambu, A. Orthacker, G. Kothleitner, J. Inoue, T. Koseki, and C. Poletti, "In-situ observation of recrystallization in an AlMgScZr alloy using confocal laser scanning microscopy," *Mater. Charact.*, vol. 108, pp. 137–144, 2015.

[3] J. Zhao, F. Jiang, H. Jian, K. Wen, L. Jiang, and X. Chen, "Comparative investigation of tungsten inert gas and friction stir welding characteristics of Al-Mg-Sc alloy plates," *Mater. Des.*, vol. 31, no. 1, pp. 306–311, 2010.

[4] A. C. Muñoz, G. Rückert, B. Huneau, X. Sauvage, and S. Marya, "Comparison of TIG welded and friction stir welded Al-4.5Mg-0.26Sc alloy," *J. Mater. Process. Technol.*, vol. 197, no. 1–3, pp. 337–343, 2008.

[5] X. Sauvage, A. Dédé, A. C. Muñoz, and B. Huneau, "Precipitate stability and recrystallisation in the weld nuggets of friction stir welded Al-Mg-Si and Al-Mg-Sc alloys," *Mater. Sci. Eng. A*, vol. 491, no. 1–2, pp. 364–371, 2008.

[6] Y. Tao, Z. Zhang, D. R. Ni, D. Wang, B. L. Xiao, and Z. Y. Ma, "Influence of welding parameter on mechanical properties and fracture behavior of friction stir welded Al-Mg-Sc joints," *Mater. Sci. Eng. A*, vol. 612, pp. 236–245, 2014.

[7] J. Taendl, F. Palm, K. Anders, R. Gradinger, and C. Poletti, "Investigation of the precipitation kinetics of a new Al-Mg-Sc-Zr alloy," *Mater. Sci. Forum*, vol. 794–796, pp. 1038–1043, 2014.

[8] S. H. Mousavi Anijdan, D. Kang, N. Singh, and M. Gallerneault, "Precipitation behavior of strip cast Al-Mg-0.4Sc-0.15Zr alloy under single and multiple-stage aging processes," *Mater. Sci. Eng. A*, vol. 640, pp. 275–279, 2015.

[9] J. Taendl, A. Orthacker, H. Amenitsch, G. Kothleitner, and C. Poletti, "Influence of the degree of scandium supersaturation on the precipitation kinetics of rapidly solidified Al-Mg-Sc-Zr alloys," *Acta Mater.*, vol. 117, pp. 43-50, 2016.

[10] J. Taendl, M. Dikovits, and C. Poletti, "Investigation of the hot deformation behavior of an Al-Mg-Sc-Zr alloy under plane strain condition," *Key Eng. Mater.*, vol. 611–612, pp. 76–83, 2014.

[11] S. Costa, H. Puga, J. Barbosa, and A. M. P. Pinto, "The effect of Sc additions on the microstructure and age hardening behaviour of as cast Al-Sc alloys," *Mater. Des.*, vol. 42, pp. 347–352, 2012.

[12] M. S. Kaiser, S. Datta, A. Roychowdhury, and M. K. Banerjee, "Effect of scandium on the microstructure and ageing behaviour of cast Al-6Mg alloy," *Mater. Charact.*, vol. 59, no. 11, pp. 1661–1666, 2008.

Fused-Overlap Zone in Aluminum Arc Welds: Tendency of Formation and Effect on Corrosion

Tyler E. Borchers and Wei Zhang
The Ohio State University, Columbus, Ohio, USA
borchers.72@osu.edu, 1-419-303-3900

Abstract

The present study aims at understanding the tendency to form a fused-overlap zone (FOZ) in Al alloy (AA) arc welds and its effect on corrosion. A wide range of conditions were investigated, including different arc welding processes, joint geometries, welding heat inputs (HIs), and base metal (BM)/filler metal (FM) compositions. The tendency to form a FOZ was found to be "widespread." Intermetallics form along grain boundaries within the FOZ, which are absent in the bulk FZ. With respect to corrosion, these intermetallics can have, depending on alloying elements, a detrimental or neutral effect on the pitting corrosion susceptibility.

Introduction

Because of the demand for strong and lightweight structures, usage of aluminum alloys (AAs) in welded structural applications has been increasing. Flexible in welding parts with complex geometry, fusion welding, especially arc welding, remains a widely used process for welding various series of AAs (e.g., 5xxx and 6xxx). Industry standards such as American Welding Society (AWS) D1.2 Structural Welding Code – Aluminum provide guidelines for safe design and welding of AAs. However, a relatively new failure mode encountered in weldment of AAs is the damage initiated either inside the weld toe (WT) or in the heat-affected zone (HAZ), directly adjacent to the fusion line (FL) [1-2]. This failure mode, reported for multiple base metal (BM) and filler metal (FM) combinations, is typically identified as stress corrosion cracking (SCC) [3-5] or fatigue cracking [6-7]. The crack initiates from local stress concentrations (*i.e.*, corrosion pits or WT geometry). In a previous study of AA 7003 BM welded with AA 5356 FM by the present authors [3], SCC initiation was found to occur at the WT tip, where large intermetallic precipitates existed and were susceptible to pitting corrosion. The large intermetallics, identified as T phase [$(Al,Zn)_{49}Mg_{32}$] by transmission electron microscopy, were only found in the WT tip, indicated as "FOZ" in Fig. 1(a). This local region exhibited an appearance of fusion zone (FZ) flowing/wetting over the BM surface. Enriched with the solutes Mg and Zn, it had a drastically different microstructure than the bulk FZ. It is named the fused-overlap zone (FOZ), since a portion of the FZ overlaps the BM without appreciable penetration, while the FZ/BM interface adequately fuses. Figure 1(b) shows a SCC failure that initiated in the FOZ.

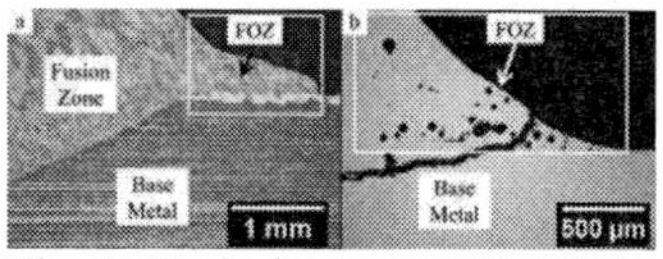

Fig. 1: Optical micrographs of a) location of the FOZ, and b) SCC failure initiation [3]

The solute enrichment, subsequent intermetallic formation, and failure initiations underline the FOZ's significant effect on weld performance. Currently, there is a limited understanding of the FOZ in AA weldments. Particularly, it's not clear whether the FOZ exists only in the specific combination of AA 7003 BM and AA 5356 FM, or whether the FOZ could occur with other alloy combinations. Hence, the present paper is focused on studying the tendency of FOZ formation and, if it does form, the corrosion performance of the FOZ.

Table 1: Key combinations of process/BM/FM studied

Case #	Process	BM	FM
C0	GMAW	7003	5356
C1	GMAW	6061	5356
C2	GMAW	2219	5356
C3	GMAW	1100	5356
C4	GTAW	7003	5356
C5	GTAW	7003	4043

Approach

Different joint geometries, welding heat input (HI), processes, and material combinations were evaluated to study the tendency of FOZ formation, solute enrichment with resulting intermetallics, and effect of intermetallic composition on corrosion. The baseline case was chosen as a weldment of AA 7003 BM and AA 5356 FM using gas metal arc welding (GMAW), for which the FOZ of a lap joint was thoroughly characterized in [3]. In the present paper, three new joint configurations for the baseline case were investigated: bead-on-plate (BOP), fillet, and double-beveled butt joints. HI was varied by changing the wire-feed speed (WFS) from 13 to 5 m/min while keeping all other weld parameters constant. The shielding gas used was a commercially pure Ar. Fronius® cold metal transfer (CMT) power supply with a Motoman® welding robot was utilized to achieve the lowest heat input. Detailed weld parameter information is discussed in [8].

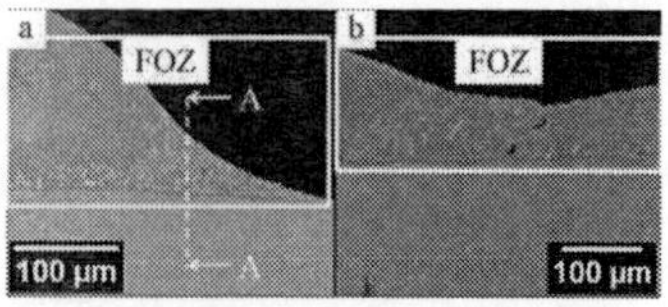

Fig. 2: BSE SEM micrographs of FOZ, where a) transverse weld cross-section; b) longitudinal weld cross-section along A-A marked in (a)

AA 7003 BM predominantly contains alloying additions of Zn and Mg. To study the effect of BM alloying compositions, several other AAs, commonly used in welded structural applications, were tested: AA 6061 (Al, Mg & Si), AA 2219 (Al & Cu), and AA 1100 (Al). The welding tests for those BMs were performed in the BOP joint configuration. The weld parameters were held constant when welding those different AAs, and not optimized for the individual BMs. In addition to GMAW, manual gas tungsten arc welding (GTAW) was used on a Miller Dynasty 280 machine with a welding current of 120 Amperes and a travel speed around 0.2 m/min. Two FMs were used: AA 5356 (Al & Mg) and AA 4043 (Al & Si). Table 1 summarizes the key combinations of welding processes, BMs and FMs investigated in the present paper.

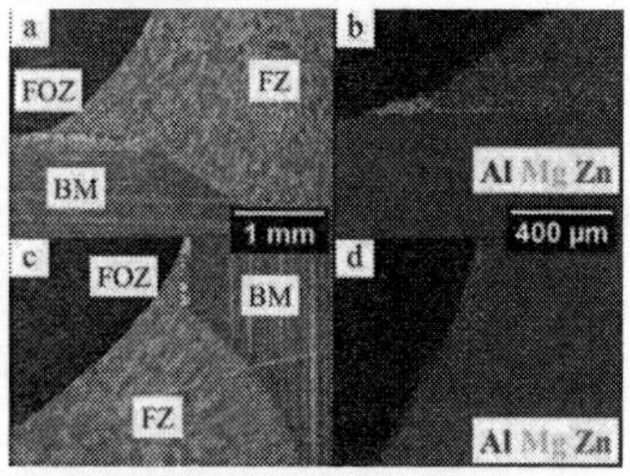

Fig. 3: Optical micrographs of a) horizontal and c) vertical members of fillet weld, where b) and d) are the respective horizontal and vertical EDS maps

It is noted that several welding experiments were also repeated with different welding power supply and robot combinations. Consistent results were obtained, indicating the FOZ phenomenon was not isolated to a single power supply or robot manufacturer. In other words, the effect of welding hardware on FOZ formation is expected to be negligible, and is not investigated further.

A preliminary assessment of corrosion response of different FOZs (if formed) was performed by submerging welded specimens in a 5% NaCl solution for 16 hours. This short testing time was designed to detect differences in pitting severity of FOZs formed under different conditions. To facilitate the analysis of corrosion, prior to immersion, a cross section of each specimen was cold mounted in epoxy and then polished. The polished sample was broken out of the mount and then immersed in the solution. Post immersion, the specimen was cleaned; to observe the extent of pitting, optical microscopy was conducted on the polished surface in the un-etched condition.

Results and Discussion

I. Effect of joint geometry: For BOP welds made on AA 7003 BM with AA 5356 FM by GMAW, a FOZ is observed at both WT tips. A backscatter electron (BSE) image of the FOZ, measured in scanning electron microscope (SEM), is shown in Fig. 2(a), where the islands exhibiting high intensity (i.e., bright regions) correspond to the T phase intermetallics, same as those observed in the lap joint configuration [3]. The image of the opposite WT tip (not depicted) is almost a mirror image of the FOZ depicted in Fig. 2(a). A longitudinal weld cross-section is shown in Fig. 2(b), which reveals that T precipitates persist in the FOZ along the entire length of weld. Finally, it is noted that T precipitates preferentially form near the FOZ/BM interface, as evident from the left side of both images in Fig. 2.

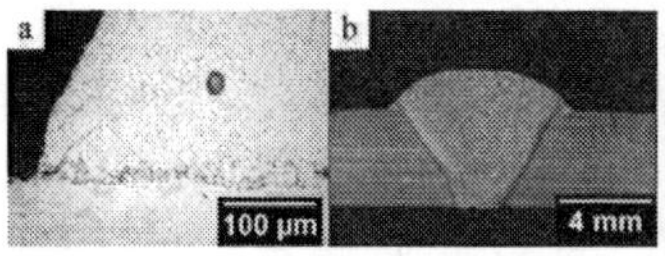

Fig. 4: Optical images showing preferential FOZ etching of a BOP weld with the highest heat input, where a) micrograph of left FOZ, and b) macrograph

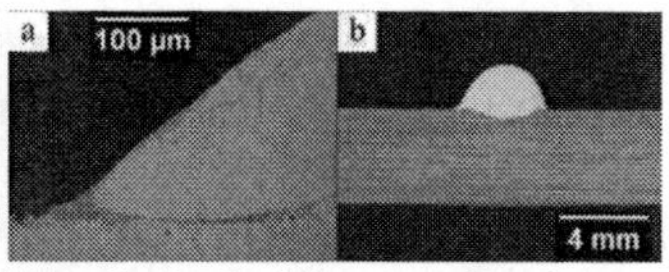

Fig. 5: Optical images showing preferential FOZ etching of a BOP weld with the lowest heat input, where a) micrograph of left FOZ, and b) macrograph

For the fillet joint on AA 7003 BM with AA 5356 FM (i.e., Case # C0 in Table 1), FOZ formation is observed on both horizontal and vertical members (sheets), as shown in Fig. 3. However, it is noted that the vertical FOZ covers 0.8 mm of BM, while the horizontal FOZ covers 1.2 mm. This is evidence that gravity might have a subtle effect on the size of FOZ, but couldn't cease its formation. Additionally, each FOZ is enriched with solutes. Particularly, the FOZ intermetallics possess intense signatures of Mg and Zn, as observed by energy-dispersive x-ray spectroscopy (EDS).

The FOZ microstructure and solute enrichment in the fillet joint are consistent with those observed previously in the lap joint, especially for the lower horizontal member [3]. It is noted that the upper horizontal member in the lap joint experiences FOZ formation intermittently, unlike the consistent appearance of FOZ in the vertical member of the fillet joint. Finally, the welding of a 45° × 45° double-beveled butt joint also results in FOZ formation with local solute enrichment, albeit intermittently (similar to that for the upper horizontal member in the lap joint). Particularly, if the bevel joint is filled perfectly by the FM so that the FZ meets the BM exactly at the top corner of both beveled members, wetting is inhibited and no FOZ forms.

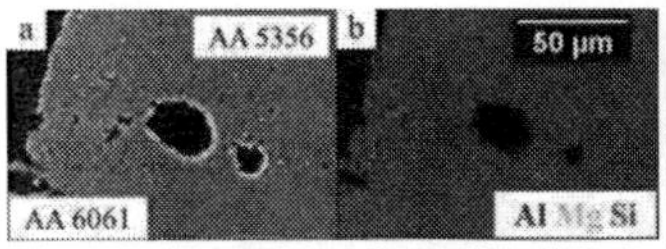

Fig. 6: For C1, a) BSE SEM micrograph of FOZ, and b) associated EDS map

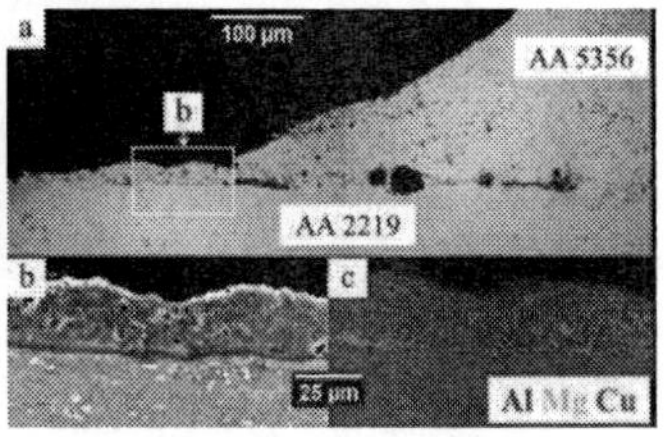

Fig. 7: For C2, a) optical micrograph of FOZ, b) BSE SEM micrograph, and c) associated EDS map for area marked in a)

II. Effect of HI: Regardless of the HI used for GMAW of AA 7003 BM and AA 5356 FM, T phase consistently forms in the FOZ. Fig. 4 shows the case of highest HI in the BOP joint configuration. Each FOZ overlaps the BM by 0.8 mm. As observed in optical microscopy, preferential etching occurs along the FOZ/BM interface, which is due to the likely presence of T phase. Fig. 5 shows the case of the lowest HI, where the left FOZ overlaps the BM by 0.7 mm, while the right overlaps by only 0.6 mm. Again, intermetallic precipitates preferentially exist along the FOZ/BM interface. In summary, similar to the effect of gravity, weld HI subtly affects FOZ size, but cannot cease FOZ formation.

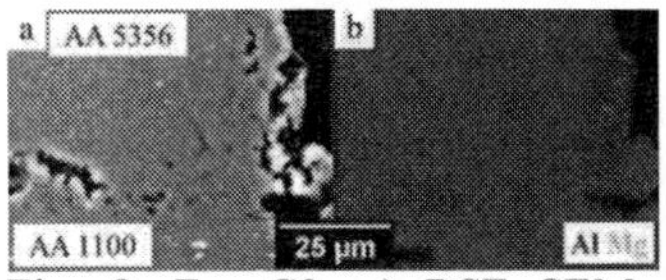

Fig. 8: For C3, a) BSE SEM micrograph of FOZ, and b) associated EDS map

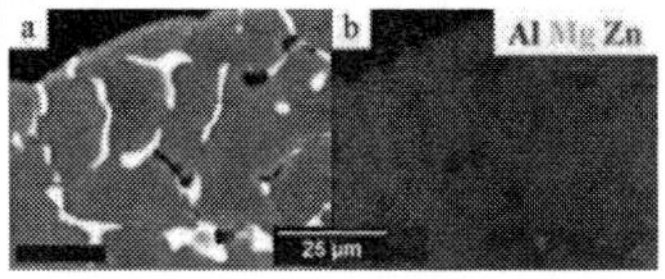

Fig. 9: For C4, a) BSE SEM micrograph of FOZ, and b) associated EDS map

III. Effect of alloying compositions and welding processes: GMAW of different BM/FM combinations (i.e., C1 – C4 listed in Table 1) also results in formation of FOZ containing coarse intermetallics. These FOZs along with their corresponding EDS maps are shown in Figs. 6 to 9 for C1 to C4, respectively. For C1 to C4, where AA 5356 FM is used, the FOZs are all enriched with both Mg (the primary alloying element of FM) and the prominent alloying elements for each BM. In other words, the concentrations of the alloying elements in the FOZs are significantly higher than those in the bulk weld metal. Particularly, C1 (AA 6061 BM) is enriched with Mg and Si (Fig. 6), C2 (AA 2219 BM) with Mg and Cu (Fig. 7), and C3 (AA 1100 BM) with Mg alone (Fig. 8). Based on composition and computational thermodynamic analysis in ThermoCalc™, it is likely that the intermetallics formed from C1, C2, and C3 are Mg_2Si, Cu-rich and Zn-lean T phase, and β phase, respectively. GTAW, C4 in Table 1, resulted in FOZ formation and enrichment with Mg and Zn (Fig. 9), which is similar to that observed in the GMAW of the same BM/FM combination. To further examine the possible phases in the FOZ, Fig. 10 compares the EDS results for the "nominal" intermetallic composition for different cases. The qualitative EDS results show more Zn than Mg in the FOZ precipitates in GTAW (C4) than GMAW (C0), suggesting that some of the precipitates may be η phase ($MgZn_2$). Clearly, FOZs exist for different BMs welded using AA 5356 FM by either GMAW or GTAW. It is noted that additional testing, using other 5xxx series FMs (*i.e.*, AA 5083 and AA 5025), showed a FOZ similarly enriched in solutes like AA 5356 FM welds.

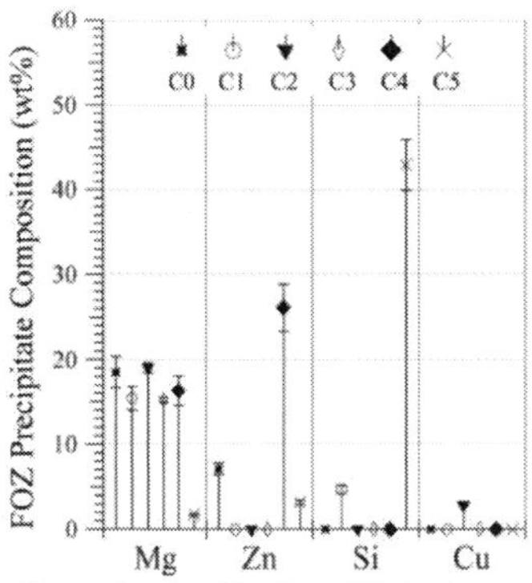

Fig. 10: Qualitative EDS data for the composition of FOZ precipitates for various material/process combinations listed in Table 1

Finally, C5 corresponds to a different FM (i.e., AA 4043 Al-Si alloy) on AA 7003 BM. As shown in Figs. 10 and 11, the intermetallics in the formed FOZ are enriched with Si, Zn, and Mg. Interestingly, the enrichment of Si extends beyond the prominent precipitates enriched with Zn and Mg. Based on ThermoCalc™ results, the Si-rich region is predicted to be a Si-matrix with Zn and Al solubility. Additionally, the prominent precipitates are likely Mg_2Si and a Zn-rich V phase that has solubility for Mg, Si, and Al. In summary, C5 revealed that the enrichment phenomenon is not isolated to Mg-rich (5xxx series) FM.

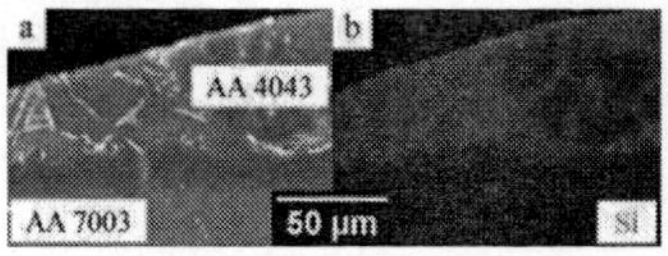

Fig. 11: For C5, a) BSE SEM micrograph of FOZ, and b) associated EDS map

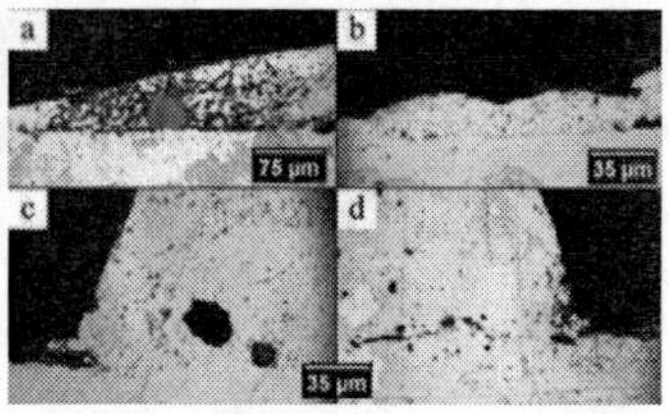

Fig. 12: Optical micrographs of corrosion test specimens, where a) C0, b) C2, c) C1, and d) C3

IV. Corrosion performance: The preliminary corrosion testing results show that C0 (Fig. 12(a)), C3 (Fig. 12(d)), and C4 (not shown) experience pitting of the FOZ intermetallics after 16 hours of immersion. On the other hand, C1 (Fig. 12(c)), C2 (Fig. 12(b)), and C5 (not shown) do not exhibit signs of pitting after the immersion test. Note this doesn't mean that the intermetallics in C1, C2 and C5 would not pit after longer duration of exposure. However, pitting of intermetallics without Si or Cu occurred prior to any pitting of those rich in Si or Cu. One possible explanation is the electrochemistry of different intermetallics. Particularly, β (in C3), T (in C0 and C4), and η (in C4) phases are anodic to the Al-matrix [9], and corrosion of these phases is thus likely severe. The likely phases in C1 and C5 are either cathodic or neutral to the Al-matrix [9]. The lack of pitting corrosion in C2's FOZ, with a likely precipitate of Cu-rich and Zn-lean T phase, requires further investigation in the future.

Conclusions

The FOZ exists as a FZ layer that overlaps the BM by approximately 0.6 to 1.2 mm, and is enriched with primary alloying elements present in both the BM and FM. For the conditions studied in this paper, FOZ formation consistently occurs regardless of joint geometry, welding HI, process (including various welding power supplies and robots), or BM and FM compositions. On the other hand, pitting corrosion resistance of the FOZs does depend on the BM and FM compositions. Such variation in corrosion resistance is likely due to the electrochemistry of precipitates formed in the FOZ. In other words, anodic FOZ intermetallics (those that do not possess Si or Cu) corrode prior to intermetallics that are cathodic or neutral to the Al-matrix.

Although still preliminary, the results reported in this paper indicate that the formation of FOZ can be a prevalent phenomenon in fusion welding of AAs. Given the impact of the FOZ on corrosion and SCC, further FOZ studies are important for improving the quality and performance of welded AA structures for light-weight applications.

Acknowledgments

This research work was supported by Honda R&D Americas (HRA), Inc. through the NSF supported industry/university cooperative research center, Manufacturing and Materials Joining Innovation Center (Ma^2JIC). A special thanks to Alan Seid and Patrick Shafer of HRA, Inc. for their support and direction. Also, thanks to Andrea Peer and Owen Oesterling of The Ohio State University for their assistance with experimentation and microscopy.

References

[1] J. Dabrowski and J. Kish, *Corros.*, vol. 71, pp. 895-907, 2015.
[2] C. Liao, *Corros.*, vol. 49, no. 1, pp. 52-59, January 1993.
[3] T. Borchers, D. McAllister and W. Zhang, *Metall. Mater. Trans. A*, vol. 46A, pp. 1827-1823, 2015.
[4] A. Jha, S. Murty and V. Diwakar, *Eng. Fail. Anal.*, vol. 10, pp. 265-273, 2003.
[5] M. Reboul, B. Dubost and M. Lashermes, *Corros. Sci.*, vol. 25, pp. 999-1018, 1983.
[6] K. Mutombo and M. du Toit, *Inter. J. Fatigue*, vol. 33, pp. 1539-1547, 2011.
[7] A. Kermanidis, A. Zervaki, G. Haidemenopoulos and S. Pantelakis, *J. Mater. Sci.*, vol. 45, pp. 4390-4400, 2010.
[8] T. Borchers, A. Seid, S. Babu, P. Shafer and W. Zhang, *Sci. Technol. Weld. Join.*, vol. 20, pp. 460-467, 2015.
[9] M. Textor, P. Nema, and J. Timm, *Mat.-wiss. u. Werkstofftech.* vol. 26, pp. 318-326, 1995

TIG welding and laser welding of boron alloyed Ti-6Al-4V

***Sakari Tolvanen*[1], *Robert Pederson*[1,2], *Uta Klement*[1]**
[1]*Department of Materials and Manufacturing Technology, Chalmers University of Technology, S-41296 Gothenburg, Sweden, sakari.tolvanen@chalmers.se*
[2]*Department of Engineering Science, University West, S-46186 Trollhättan, Sweden*

Abstract

TIG welds and laser welds have been performed on cast i) Ti-6Al-4V, ii) Ti-6Al-4V+0.06B, and iii) Ti-6Al-4V+0.11B. The effect of boron on the weld microstructure and microhardness is here presented. In the conventional Ti-64 alloy without boron addition large prior β grains exist in the fusion zone and in the heat affected zone, whereas in the alloys with 0.06 wt% and 0.11 wt% boron the prior β grains are found to be approximately 5 times smaller. In the heat affected zones of the boron doped alloys incoherent TiB precipitates are distributed mainly along the prior β grain boundaries, inhibiting grain growth at elevated temperature and being the main reason for the grain refinement. In the fusion zone TiB precipitates are found in stripes at interdendritic regions. Prior β grains are elongated and their size is refined by boron addition. The microstructure in the fusion zones and heat affected zones consist of fine acicular α. TiB precipitates in the fusion zone restricts the length of individual α laths and the size of α colonies. Microhardness measurements across welds reveal increasing hardness with fine acicular microstructure. Boron addition is not found to have significant effect on microhardness in the welds.

Introduction

Within aerospace industry the high specific strength properties of titanium alloys are highly appreciated because this enables the manufacturing of parts and components with lower final part weight. Cost is another just as important parameter for consideration and here includes both the direct material cost as well as costs related with manufacturing of the final part. Recent work [1]–[3] indicated that by adding small amounts of the alloying element boron to a titanium alloy the as cast microstructure becomes significantly finer, actually in the same range as would have been achieved through conventional thermomechanical processing. But in this case without the need of costly additional manufacturing process steps. Similar observations have been reported in additive manufacturing [4] where boron addition has been shown to refine the coarse columnar β grain morphology. Previous work has shown that this slight modification of chemical composition in the titanium alloy – the chemical composition remains within the material specification of the alloy – improves tensile strength, ductility, creep strength as well as fatigue properties [1,4-7]. Fusion welding is a key technology for manufacturing of large structural parts for aerospace applications. Fusion welding involves melting of material. In titanium alloys, the fusion zone in such welds normally consist of large prior β grains. Remembering the previously described findings from adding small amounts of boron to titanium alloys when casting, i.e. melting the material, the current work explores further the influence of boron addition in titanium alloy welds. The titanium alloy investigated in the present work is the α-β alloy Ti-6Al-4V with 0.06 wt% B and 0.11wt% B addition, respectively, and standard grade Ti-6Al-4V material without any B addition as reference. TIG welding and laser beam welding are the two welding processes applied. The same welding process parameters were used when welding the different alloy compositions in order to avoid effects from parameter variations on the weld geometry and microstructure.

Experimental

Induction skull melted rod-like ingots, with dimensions 150 x 120 mm, cast Ti-6Al-4V (hereafter called Ti-64) and with different boron concentrations were obtained from Titanium Castings Ltd, UK for this study. The chemical compositions of the three different alloys were (wt%): a) Ti - 6.16Al - 4.04V - <0.001B - 0.20O - 0.20Fe - 0.010C - 0.002N - <0.001H - <0.001Y; b) Ti - 6.24Al - 4.06V - <0.06B - 0.19O - 0.18Fe - 0.012C - 0.003N - <0.001H - <0.001Y; c) Ti - 6.18Al - 4.02V - <0.011B - 0.24O - 0.19Fe - 0.007C - 0.004N - <0.001H - <0.001Y. Boron was added in the form of TiB_2 particles of size between 1.7-5 mm. After casting, the ingots were hot isostatically pressed (HIP) at 900°C at 100 MPa for 2 hours.

Bead on plate TIG and laser welds were conducted on 8.5 mm thick sections of each alloy composition. For TIG a partial penetration beads and for laser a full penetration beads were obtained. Samples where thereafter cut out from each of these welds to investigate the weld cross sections with regard to geometry, microstructure and microhardness.

Sample preparation was done using conventional metallographic techniques for titanium alloys, involving grinding, polishing, and etching using Kroll's etchant. Microhardness measurements were performed with a Shimadzu HMV-2000 machine using a load of 500 g for 12 seconds. For TIG the hardness profile was acquired starting from the top of the bead continuing through the fusion zone (FZ) and heat affected zone (HAZ) to the base material (BM). For laser welds the measurement was started in the middle of the FZ continuing through the FZ and HAZ to the BM. Each point in the microhardness profiles is the average of three measurements. Prior β grain size in the HAZ was evaluated using average grain intercept method.

Results and discussion

Weld geometry and microstructure

Figure 1 shows the TIG and laser weld cross sections from Ti-64 with and without boron, with the different zones indicated in the micrographs. Laser welds (Fig. 1c) are wider at the top and bottom with a narrow waist in the middle. HAZ is the widest at the top and gets narrower towards the bottom. The width of the laser welds was measured at the top and at the bottom, as indicated in Fig. 1c. The fusion line is clearly visible in the samples with boron but difficult to discern in the Ti-64 welds not doped with boron. The width of the HAZ was determined as the area which consists of transformed β microstructure. The cast microstructure of base material was also heat affected and had coarsened to some extent and can be seen between HAZ and BM in Fig. 1. The FZ in TIG welds is lens shaped (Fig. 1b) and about 1 mm in maximum penetration while the width varied between 4-5 mm. As seen from the results in table 1, the width of the welds seems to decrease in the materials that contain boron, both in TIG and laser welds. It has been reported [8] that boron increases the β transus temperature of the Ti-64 alloy. This could explain the decreased size of the HAZ.

Table 1. Prior β grain size and weld geometries for the different weld cross sections.

	Prior β grain size in HAZ (μm)	Weld geometry (mm) FZ width top	FZ width bottom	HAZ depth/width
TIG Ti-64	>2000	6,2	5,1	9,4
TIG Ti64+0,06B	280±60	4,9	4,5	9,8
TIG Ti64+0,11B	240±50	4,3	4,0	9,1
LBW Ti64	>2000	8,0		10,2
LBW Ti64+0.06B	390±135	7,0		8,9
LBW Ti64+0.11B	240±50	7,0		9,0

The prior β grain size in the FZ and HAZ is strongly affected by the boron addition and decrease significantly when boron is added. This is clearly visible in Fig. 1. The prior β grains in the FZ are elongated, but because of lack of grain boundary α and TiB precipitates, prior β grains are difficult to distinguish in optical micrographs. The size of prior β grains in the FZs was typically in the range of millimeters in the standard grade Ti-64 alloy, and decreased significantly in Ti-64+0.11B alloy (sizes of approximately 200 μm in width and 1 mm in length). Prior β grain size in the HAZs was measured and the results are shown in table 1. In the standard grade Ti-64 alloy the prior β grains could span across the whole HAZ and FZ. In boron containing welds the prior β grains were significantly smaller and the average prior β grain size in the HAZ was here less than 400 μm, in all samples. Higher boron content was found to lead to a finer microstructure. Prior β grain structure was found to be slightly finer in laser welds than in TIG. These finding are in line with previous related work [1] where it was reported that in cast Ti-64 the prior β grain size decreased from approximately 1700 μm (Ti-64) to 200 μm (Ti-64+0.11B) with boron addition. In that work it was also found that in the BM the TiB precipitates were present in the form of needles along the prior β grain boundaries.

The microstructure in the FZ and HAZ in all welds was found to consist of fine acicular α and possibly some martensitic α'. Thin grain boundary α was observed in the FZ and HAZ of standard Ti-64 alloy but was not observed in the welds with boron addition.

Boron addition affects the morphology of α constituents in both the FZ and in the HAZ. In laser welded Ti-64+0.11B the α plates in the FZ (Fig. 2b) are shorter than in the heat affected zone (Fig. 2c and 2d). Similar observation were made in all boron containing welds. TiB precipitates are more evenly distributed in the FZ whereas in HAZ they are mainly located in the prior β grain boundaries. TiB precipitates act as nucleation sites for α plates and they also limit the length of α plates and the size of α colonies. The effect on α morphology and its effect on mechanical properties will be evaluated in future work.

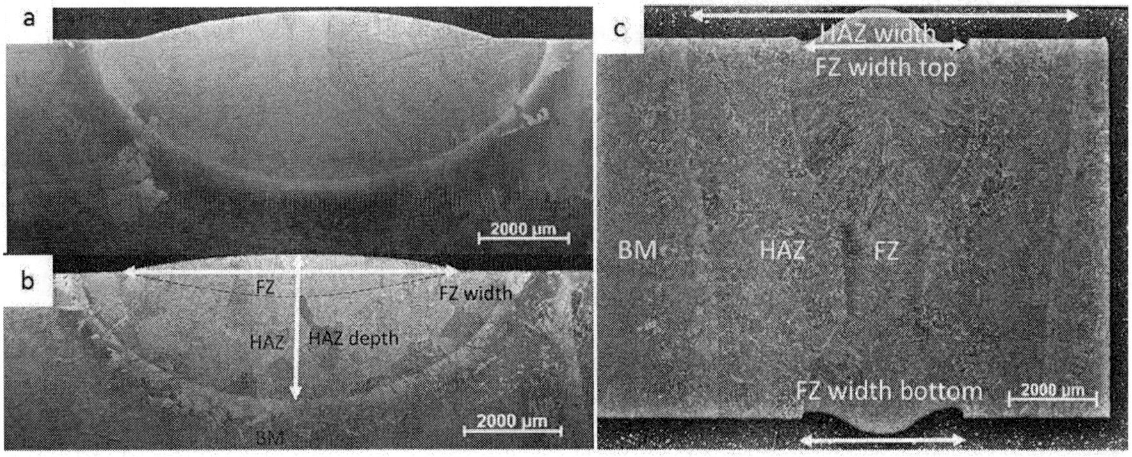

Figure 1. Etched cross sections of a) TIG Ti-64, b) TIG Ti-64+0.06B and c) laser weld Ti-64+0.11B.

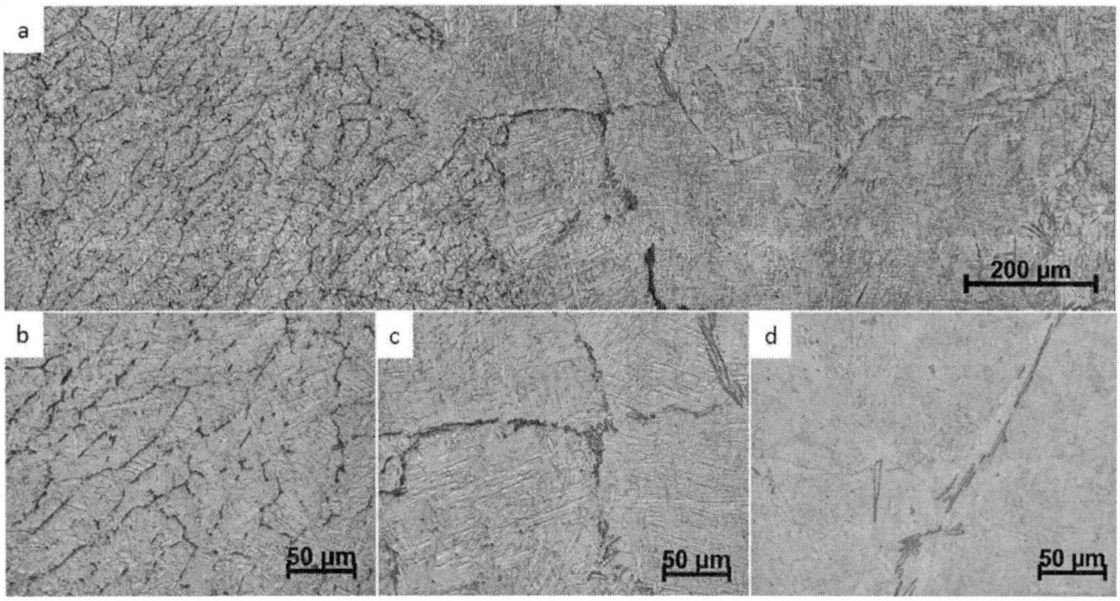

Figure 2. Microstructure of laser weld Ti-64+0.11B. a) FZ and HAZ, b) FZ, c) HAZ close to the fusion line and d) HAZ.

As shown in Fig. 2, the morphology and distribution of TiB precipitates is changed because of the welding process. In the HAZ and some distance away from the FZ line (Fig. 2d), the TiB precipitates seem to be unaffected. They have similar morphology and distribution as the TiB precipitates in the BM. TiB precipitates in the BM and in the HAZ have an average length of approximately 20 µm. Closer to the FZ line but still within the HAZ (Fig. 2c), the TiB precipitates morphology has changed because of the heat from the welding process. Here the TiB precipitates are broken up and their size is about 10 times smaller than in the BM. In the FZ (Fig. 2b) the TiB precipitates are aligned in stripes along the interdendritic regions. The TiB precipitates in these stripes show both round and elongated shapes. The size of the largest TiB precipitates found in the FZ is less than 5 µm.

Further work is needed to explain if the change in the morphology of TiB precipitates is caused by breaking down and spheroidization of the needles or their melting/dissolution to the melt.

Microhardness

In Fig. 3 microhardness profiles are presented. In the TIG welds the measurements starts at the top of the weld, in the FZ, and continuous down across the whole FZ and into the HAZ and BM. In the laser welds, the measurement starts from the middle of the weld and continuous across the FZ and HAZ into the BM.

The hardness increases when going from BM through the HAZ and into the FZ, mainly because of the finer acicular microstructure that exists in the weld zones as compared to the coarser colony microstructure in the BM. No obvious difference in hardness between the two welding process can be found. Adding boron to Ti-64 seems to cause a slight increase in the microhardness in the FZ and in the HAZ but the difference is not significant. This is in line with previous work on related material [1] where boron addition was found to increase microhardness of cast Ti-64.

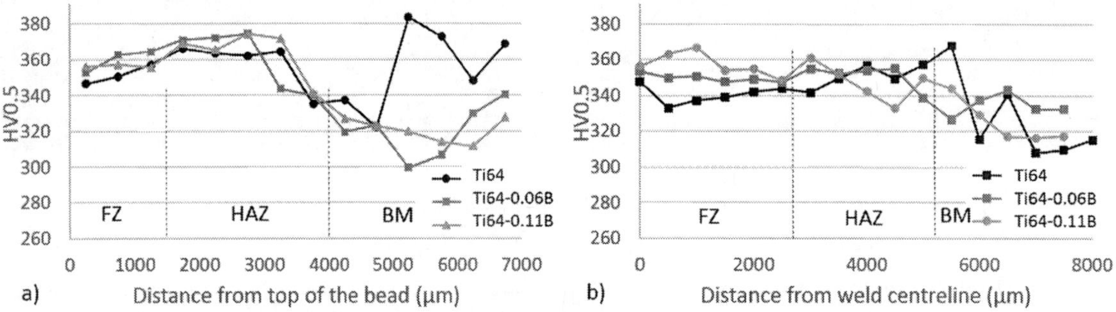

Figure 3. Microhardness profiles across a) TIG welds, and b) laser welds.

Conclusions

Based on the results presented in this work the following conclusions can be made:

- Fusion weld zones in boron doped cast Ti-6Al-4V retains their small prior β grain size after welding.
- TiB precipitates in the prior beta grain boundaries inhibit grain growth in the HAZ.
- Evenly distributed TiB precipitates in the FZ limit the length of individual α plates and the size of the α colonies.
- TIG and laser welding affects the size, morphology and distribution of the TiB precipitates in the weld zone. The smallest TiB precipitates are found in the FZs.
- The microhardness in the TIG welds is slightly higher in the HAZ than in the FZ. The BM shows the lowest hardness. In the laser welds the hardness was found to be similar in the weld zone (i.e. both FZ and HAZ). The BM shows the lowest hardness.

References

[1] R. Pederson, R. Gaddam, and M.-L. Antti, "Microstructure and mechanical behavior of cast Ti-6Al-4V with addition of boron," Central European Journal of Engineering, vol. 2, no. 3, pp. 347–357, 2012.

[2] S. Roy, S. Suwas, S. Tamirisakandala, D. B. Miracle, and R. Srinivasan, "Development of solidification microstructure in boron-modified alloy Ti-6Al-4V-0.1B," Acta Materialia, vol. 59, no. 14, pp. 5494–5510, 2011.

[3] S. Tamirisakandala, R. B. Bhat, J. S. Tiley, and D. B. Miracle, "Grain refinement of cast titanium alloys via trace boron addition," Scripta Materialia, vol. 53, no. 12, pp. 1421–1426, 2005.

[4] M. J. Bermingham, D. Kent, H. Zhan, D. H. StJohn, and M. S. Dargusch, "Controlling the microstructure and properties of wire arc additive manufactured Ti–6Al–4V with trace boron additions," Acta Materialia, vol. 91, pp. 289–303, 2015.

[5] G. Singh, D. V. V Satyanarayana, R. Pederson, R. Datta, and U. Ramamurty, "Enhancement in creep resistance of Ti-6Al-4V alloy due to boron addition," Materials Science and Engineering A, vol. 597, pp. 194–203, 2014.

[6] G. Singh, R. Gaddam, V. Petley, R. Datta, R. Pederson, and U. Ramamurty, "Strain-controlled fatigue in B-modified Ti-6Al-4V alloys," Scripta Materialia, vol. 69, no. 9, pp. 698–701, 2013.

[7] Luan, J. H. et al. Effects of boron on the fracture behavior and ductility of cast Ti-6Al-4V alloys. Scripta Materialia 100, 90–93 (2015).

[8] Tamirisakandala, S. et al. Effect of boron on the beta transus of Ti-6Al-4V alloy. Scripta Materialia 5217–222 (2005).

Weldability Study of Superalloys Waspaloy® and Haynes® 282®

Jonny Jacobsson[1,2], Joel Andersson[3*], Anssi Brederholm[4] and Hannu Hänninen[4]
[1]Department of Materials and Manufacturing Technology, Chalmers University of Technology, Gothenburg, Sweden
[2]Research and Technology Department, GKN Aerospace Engine Systems, Trollhättan, Sweden
[3]Department of Engineering Science, University West, Trollhättan, Sweden
[4]Aalto University School of Engineering, Helsinki, Finland

[3]Email joel.andersson@hv.se, phone +46 (0)739013319*

Abstract
The weldability of Waspaloy® and Haynes® 282® have been evaluated using the Varestraint weldability test. The experiments reveal that Waspaloy® has a higher susceptibility to hot cracking compared to Haynes® 282®. This conclusion is supported by increased total crack length (10 mm or more) for Waspaloy® when compared to Haynes® 282® in Varestraint test. The cracking in Haynes® 282® seems to be connected with a secondary phase which presumably can be associated with Ti-Mo based MC-type carbide observed in the fusion zone. Also, a surrounding segregated area is present near this secondary phase as well as along the grain boundaries. Furthermore, micro-Vickers hardness results revealed more or less the same weld metal hardness (260-280 HV) but a difference in the base metal hardness. The weld metal hardness of Waspaloy® was lower than that of the base metal hardness, while Haynes® 282® had a higher hardness in comparison.

Introduction
Welding of precipitation hardening Ni-based superalloys is of significant importance in the aerospace industry [1]. The aim of this study is therefore to investigate the susceptibility to hot cracking of wrought Haynes® 282® and Waspaloy®, using the Varestraint weldability testing method. Haynes® 282® is a newly developed alloy with limited amount of information available regarding the weldability in the literature [2]. Similarly for Waspaloy®, there is a lack of systematic weldability testing using Varestraint testing method or similar techniques [2] despite the many years that the alloy has been around.

Experimental
Waspaloy® was solution heat treated at 1010 °C for 2 h followed by forced air cooling providing a macro-Vickers hardness of 294HV (1a) and a grain size of ASTM 4 (90 μm), while Haynes® 282® was used in the mill-annealed condition followed by polymer quench providing a macro-Vickers hardness of 203HV (1b) and a grain size of ASTM 5 (64 μm). The macro-Vickers hardness testing on the as-received and as solution heat treated materials was carried out using 10 kg load for 12 s and taken as an average of 10 indents. The micro-Vickers hardness showed similar values, 208HV for Haynes® 282®, but somewhat higher 340HV for Waspaloy®. The micro-Vickers testing was performed on etched surfaces to avoid influence of micro-constituents such as carbides or grain boundaries and an average of 5 indents was taken. Grain size measurements were carried out using the linear intercept procedure. Microstructural preparation was performed according to standard metallographic procedures. Kalling's etchant for Ni-based materials as well as electrolytic etching with oxalic acid were used to reveal the microstructure. The chemical composition (wt. %) of the test materials can be seen in Table 1.

Table 1. Chemical composition (wt. %) of Waspaloy® and Haynes® 282®.

Element wt. %	Ni	Cr	Fe	Co	Mo	Al	Ti	C	P	B	Cu	Mn	S	Si
Waspaloy®	Bal.	19.13	1.13	13.34	4.22	1.36	3.03	0.08	0.004	0.006	0.02	0.02	0.002	0.09
Haynes® 282®	Bal.	19.63	0.35	10.35	8.56	1.41	2.21	0.068	0.002	0.004	-	0.08	0.002	-

The Varestraint testing was performed on plates with a thickness of 3.2 mm. The welding process used in Varestraint testing was gas tungsten arc welding (GTAW). The welding current was set to 85 A together with the welding speed of 2 mm/s and Ar-gas flow rate (18 l/min) for protection. The radii used were 50, 75, 100, 133, 200 and 400 mm imposing ideal augmented strains ranging from 1.1 to 8.6 % (three repetitions were made at each strain level). The stroke rate was set to 16 mm/s.

Results

The total crack length (TCL) in mm versus the augmented strain (%) is presented in Figure 1. Waspaloy® possesses significantly higher hot cracking susceptibility at all strain levels in comparison to that of Haynes® 282®. The difference between the two alloys in terms of TCL is in the range of 10 mm at most strain levels together with the standard deviation of less than 5 mm which supports an advantage for Haynes® 282® in this case.

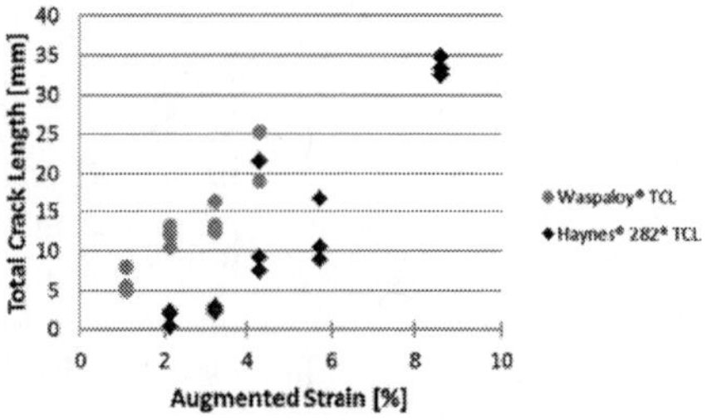

Figure 1. Total crack length (mm) versus augmented strain (%) in Varestraint testing of Waspaloy® and Haynes® 282®.

It was not possible to complete the comparison at the highest strain level due to the limitation of the available test material. However, both materials constituted the same amount of test plates (11 test plates) in the range from 1 to 6 % augmented strain and the additional test plates above 6 % give information about the saturation level. The crack patterns were similar for both Waspaloy® and Haynes® 282®. Irrespective of the strain level, the cracks are located both in the heat affected zone (HAZ) and in the weld fusion zone (FZ). In a small region outside the fusion zone near the tip of a HAZ liquation crack in a Haynes® 282® sample, a secondary phase can be observed, as seen in Figure 2a and 2b. The crack is relatively small and extends almost to, but not into the weld fusion zone, which means that there is not enough liquid network to support backfilling. The secondary phase, possibly forms on a grain boundary and is divided into several small segments which can act as a possible crack path if positioned close enough to each other. A brighter area surrounds the secondary phase and similar regions were observed in the fusion zone. Also, an area close to the crack shows evidences of partially dissolved carbides.

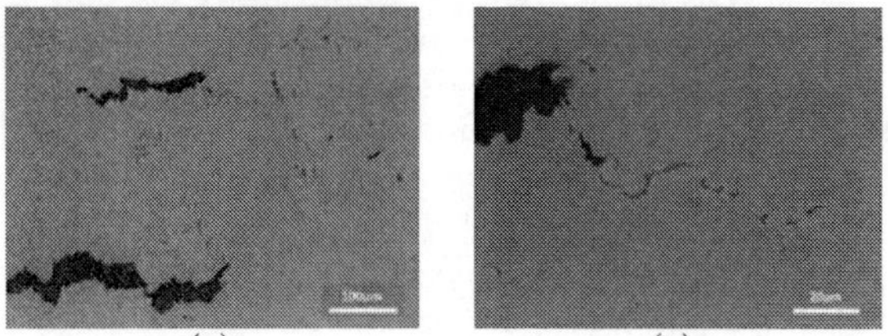

(a) (b)

Figure 2a and 2b. HAZ liquation cracks with a secondary phase in a Haynes® 282® sample (a) together with a closer view (b) where the bright segregated areas surrounding the secondary phase near the crack tip are present.

The micro-Vickers hardness tests were carried out using a 200 g load, and revealed that an overall hardness was generally 100HV higher for Waspaloy®. A similar hardness level of roughly ~250HV is reached for both alloys in the FZ. When comparing the hardness of the centerline in the FZ with that of each base metal, FZ of Waspaloy® has lower hardness than that of the base metal, while FZ of Haynes® 282® has a higher hardness than the base metal.

Discussion

The results from Varestraint tests clearly show that Haynes® 282® has a lower hot cracking susceptibility than that of Waspaloy®. Approximately a 10 mm difference in TCL at all augmented strain levels up to 4.3% was observed between Waspaloy® and Haynes® 282®. Interestingly, strains above 4.3% for Haynes® 282® do not show any tendency for TCL saturation, which suggests that the needed strain level for TCL saturation is above 8.6%. By considering the Varestraint testing data on the as-received Waspaloy®, which were obtained in the previous study (using the same testing parameters) [6] and plotting them together with the results from this study in Figure 3, it can be seen that the solution heat treatment for 2 h at 1010 °C has a very limited effect on the hot cracking susceptibility of Waspaloy®. As seen in the figure, most of the data points for the as-received Waspaloy® material coincide and match an assumed linear extrapolation of cracking results towards higher strain levels for the solution heat treated condition excluding the two data points at the strain level of ~4.5%. The as-received Waspaloy® material came from the same batch of material as the one tested in this study and therefore has the same chemical composition. However, the grain size was ASTM 5.7 (~53 μm) with a macro-Vickers hardness of 238HV [7] which further suggests that neither the specific difference in grain size nor the hardness have any significant influence on the effect of susceptibility to hot cracking. This further supports the advantage of Haynes® 282® in the comparison.

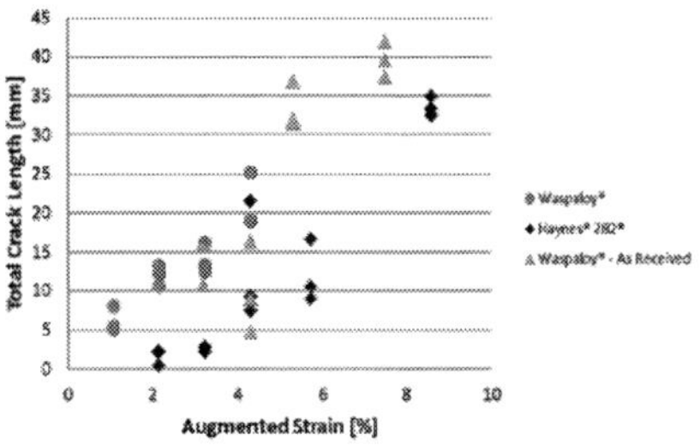

Figure 4. Total crack length (mm) versus augmented strain (%) in Varestraint testing of Waspaloy® and Haynes® 282® as compared to the earlier study of Waspaloy® in the as-received condition [4].

At reduced levels of augmented strain extrapolation to lower strain levels for Waspaloy® leads to zero, while Haynes® 282® seems to be able to sustain a certain level of strain without any hot cracking. However, to validate this, a more detailed study at lower levels of augmented strain is required. The microstructural analyses of the HAZ reveled possible evidence of a secondary phase distributed ahead of the propagating cracks in the Haynes® 282® samples. This secondary phase can be the same as distributed interdendritically in the FZ identified in other studies [3,6] as Ti-Mo based MC-type carbides with an FCC crystal structure. MC carbides were found randomly distributed intergranularly and intragranularly within γ phase with typical size in the range of 2-15 μm. Osoba et al. utilized small angle diffraction pattern SADP technique to detect Mo-B rich particles, which appeared to be M_5B_3-particles, at the grain boundaries as well as GB triple points [3]. The particles seem to be too small when compared with the secondary phase observed in this study. On the other hand the difference in the welding method and the distribution along grain boundaries can explain the appearance. Furthermore, the area around the secondary phase seems to be segregated which can be a prerequisite for the phase transformation. The cracks in both alloys seem to prefer the interdendritic paths in the FZ and along the segregated grain boundaries in the HAZ. For Waspaloy®, the visible area near the cracks does not show any traces of low melting phases, which suggests that a more advanced evaluation at a higher magnification is needed in order to confirm or exclude the presence of MC carbides observed in the earlier study [5].

Waspaloy® had a micro-Vickers hardness of 340HV in the base metal while Haynes® 282® had hardness of 208HV which can primarily be explained by the degree of γ' precipitation hardening. The overall level of hardness was generally higher for Waspaloy® except the similar level of hardness in the FZ, which may mean a higher degree of localized plastic deformation and cracking in the solidifying weld metal. The higher hardness of Waspaloy® leads generally to an increased hot cracking susceptibility, since it may be less capable of coping with the strains which are induced at testing or welding, but if compared to the as-received Waspaloy® material, the difference in hardness seems to be insignificant.

The smaller grain size ASTM 5 (64 μm) of Haynes® 282® compared with ASTM 4 (90 μm) in Waspaloy® may be important for the weldability. In general, it is believed that the larger grains deteriorate the weldability [6], because the grain boundary area decreases and less strain can be accommodated. Considering the as-received Waspaloy® material which possessed the smallest grain size among the three different conditions (ASTM 5.7, 53 μm) it seems that the difference in grain size is insignificant within this specific range and boundary conditions used for testing.

Considering the phosphorous content of 0.004 wt.% in Waspaloy®, which is double compared with that of Haynes® 282® and the boron content of 0.006 wt.% compared with 0.004 wt.%, respectively, the weldability is generally accepted to be worse for Waspaloy® [7-9]. The element segregation of melting point depressants like boron and phosphorous, can reduce the melting point of the grain boundaries.

Conclusions

1. The susceptibility to hot cracking is higher for Waspaloy® than for Haynes® 282® according to the Varestraint testing performed in this study.

2. In Haynes® 282®, HAZ liquation cracking was connected to the presence of an intergranular secondary phase which seems to be similar to what was identified earlier in the FZ, namely Ti-Mo based MC-type FCC carbides.

3. The weld metal hardness was similar (250HV) for both alloys, while in comparison with their respective base metal; Haynes® 282® revealed an increased weld metal hardness as compared to a decreased hardness level of Waspaloy® weld metal.

References

[1] J. Andersson: "Weldability of Precipitation Hardening Superalloys – Influence of Microstructure", PhD thesis, Chalmers University of Technology, Cothenburg, Sweden, 2011.

[2] L. M. Pike: "Development of a fabricable Gamma – Prime (γ') Strengthened Superalloy", Superalloys 2008 eds. R.C. Reed, K.A. Green, P. Caron, T. P. Gabb, M. G. Fahrmann, E. S. Huron, S. A. Woodard, The Minerals, Metals & Materials Society, 2008, pp. 191-200.

[3] L. O. Osoba, "A study on laser weldability improvement of newly developed Haynes 282 superalloy", PhD thesis, University of Manitoba, Winnipeg, Canada, 2012.

[4] J. Andersson, G. Sjöberg, L. Viskari, A. Brederholm, H. Hänninen, and C.S. Knee., " Hot Cracking of Allvac® 718PlusTM, Alloy 718 and Waspaloy at Varestraint testing", Proceedings of the 47th Conference of Metallurgists (COM), ed. by. M. Elboujdaini, Winnipeg, Canada, August, 2008, pp. 401 – 413.

[5] H. Matysiak, M. Zagorska, J. Andersson, A. Balkowiec, R. Cygan, M. Rasinski, M. Pisarek, M. Andrzejczuk, K. Kubiak and K. J. Kurzydlowski, "Microstructure of Haynes® 282® Superalloy after Vacuum Induction Melting and Investment Casting of Thin-Walled Components", Materials, 2013, 6(11), 5016-5037.

[6] J. Andersson, G. Sjöberg, L. Viskari and M. C. Chaturvedi, "Effect of Different Solution Heat Treatments on the Hot Ductility of Superalloys, Part 3 - Waspaloy", Materials Science and Technology, 2013, 29 (1), 43-53.

[7] K. R. Vishwakarma, N. L. Richards, M. C. Chaturvedi: "Microstructural analysis of fusion and heat affected zones in electron beam welded ALLVAC® 718PLUS™ superalloy", Materials Science and Engineering A, Volume 480, Issue 1-2, 15 May 2008, Pages 517-528 .

[8] K. R. Vishwakarma, M. C. Chaturvedi: "Effect of boron and phosphorus on HAZ microfissuring of Allvac 718 Plus superalloy", Materials Science and Technology, Volume 25, Issue 3, March 2009, pp. 351-360.

[9] N. L. Richards: "The effect of grain boundary segregation of boron in cast alloy 718 on HAZ microfissuring - A SIMS analysis", Acta Materialia, Volume 45, Issue 8, pp. 3095-3107.

Mechanisms of Elasto-plastic Contact, Electric Current and Thermal Conduction Triply Coupled Phenomena during Resistance Spot Welding

Tomoyoshi HORIE, Tomoya NIHO, Naoki HAYASHI, Daisuke ISHIHARA
Kyushu Institute of Technology, Iizuka, Fukuoka, Japan
horie@mse.kyutech.ac.jp, +81-948-29-7772

Abstract

Mechanisms of the coupled phenomena during the resistance spot welding are examined using coupled finite element analyses. During the welding process, the electric current conducting just below the electrodes moves outward between the high temperature region and the high contact resistance region accompanied by the growth of contact surface, high Joule heat region and high temperature region. The nugget area is formed and grows under such complicated coupled effect among electric, thermal and contact phenomena. When electric current applied is higher, the conducting current moves outward together with the high temperature region, which leads to larger heat generation on account of the higher electric resistivity in the base metal region, and then leads to larger nugget area. When the electrode force applied is larger, the temperature along the interface becomes lower because of the reduction of current density.

Introduction

Resistance spot welding technique is widely used in industries, where welding conditions are determined based on their longtime production experiences. The phenomena during the welding process, however, are not simple but complicated because of the coupling effect among electric current, heat generation, heat conduction, deformation with contact and contact electric resistance. Many researches have been made for characterization of spot welding by experiments[1] and simulations[2, 3, 4]. Since advanced coupled analysis techniques are needed for such triply coupled analyses, neither the understanding of mechanisms of the coupled phenomena itself nor the application for determining optimum welding conditions considering the effect of the coupled phenomena has not been realized yet. Mechanisms of the coupled phenomena during resistance spot welding are examined using coupled finite element analyses. Effects of applied electric current and electrode force are also examined.

Method of coupled analysis among elasto-plastic, electric current and thermal conduction

Features of the coupled phenomena

The electric-thermal-structural coupled phenomena in resistance spot welding occur from three factors, such as contact and deformation of the electrodes and steel plates, Joule heat by electric current and temperature dependent material properties. The significant feature of the coupled phenomena is its intensity. As for the electric current, influence of contact resistance is large; current concentration or fringing occurs at the contact edge; and temperature dependence of material property is large. As for the heat, temperature range is wide from the room temperature to the melting point; temperature distribution is irregular. In the structure, influence of contact and deformation is large.

Another feature of the coupled phenomena is its complexity arising from the coupling paths. There are coupling paths concerning contact area, current distribution and heat generation as depicted in Fig. 1.
(Path 1) temperature → plate rigidity → contact area → current distribution → heat generation → temperature
(Path 2) temperature → electric resistivity → heat generation → temperature
(Path 2') temperature → electric resistivity → current distribution → heat generation → temperature
There are also six coupling paths concerning contact electrical resistance, which are microscopic and local phenomena.
(Path 3) temperature → electric resistivity → contact resistivity → heat generation → temperature
(Path 4) temperature → yield stress → contact resistivity → heat generation → temperature
(Path 5) temperature → plate rigidity → contact area → contact pressure → contact resistivity → heat generation → temperature
There are also two paths between contact resistivity and heat generation for each path.
contact resistivity → heat generation (Path 3, 4, 5 listed above)
contact resistivity → current distribution → heat generation (Path 3', 4', 5')
Therefore, it is complicated and difficult to predict the influence when temperature distribution or welding parameter of RSW changes.

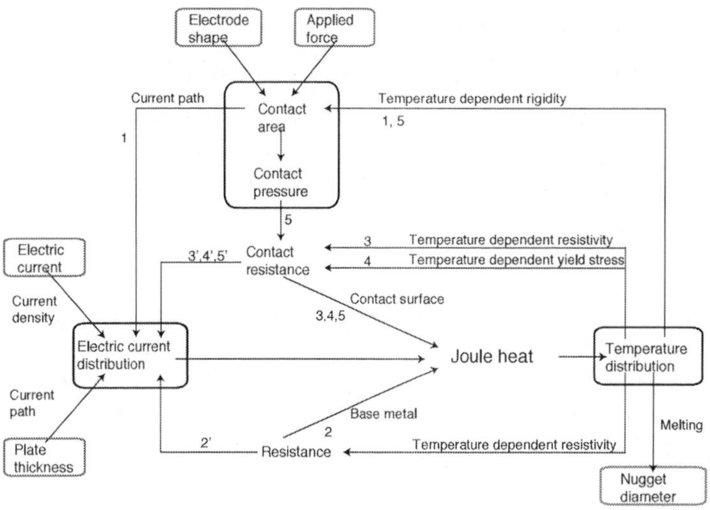

Figure 1: Coupling paths among electric current, heat and structure

Coupled analysis model

In electric current analysis, temperature dependent material properties, pressure and temperature dependent contact resistance[5] are taken into account. In transient thermal analysis, contact heat conduction and Joule heat generation are also considered. Elasto-plasticity, contact, large displacement, finite strain and temperature dependent material properties are also considered in the structural analysis[6].

Coupled analysis of resistive spot welding

Model and analysis conditions

Two steel plates of 1.6 mm in thickness are held by the electrodes with the electrode force of 2.65 kN. The welding current of 9.5 kA is applied for 0.267 s. Axisymmetric coupled finite element analysis was performed.

Analysis results

Electric current and temperature distribution for the early and final stages are shown in Fig. 2. Heat generation at the center and the edge of the plate interface together with the edge of the plate/electrode interface is observed. These high temperature regions are merged afterward, then the nugget area is formed.

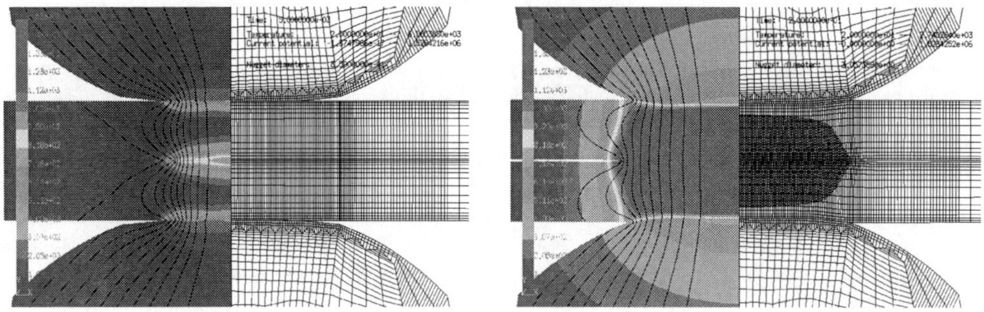

(a) $t = 5.0$ ms (b) $t = 200.0$ ms

Figure 2: Electric current and temperature distribution together with nugget area

A peak of the current density appears at the center of the plate interface, then it moves outward to the edge of the interface as shown in Fig. 3 to avoid high temperature and high resistance area.

Fig. 4 shows the distribution of contact resistance, current density and Joule heat together with the temperature distribution. According to Fig. 4(a), there are three regions on the plate interface. One is the high temperature completely contacted region (region I) which ranges from 0 to 1.5 mm in radius, another is the low pressure and low temperature transition region (region II) ranging from 1.5 to 2.5 mm and the other is the uncontacted region (region III). The current density peak appears on the outer edge of region I. The Joule heat peak appears on the inner edge of region II, whereas the contact resistance peak appears on the outer edge of region II, which corresponds to the end of the interface. Therefore, the temperature is the highest and uniform in region I, and sharply decreases in region II. As time proceeds, all these peaks move outward and come together to the edge of the interface as shown in Fig. 4(b). After the contact resistance heat generation disappears in about 70 or 80 ms, the Joule heat in the RSW process has been shifted completely to the base metal heat generation.

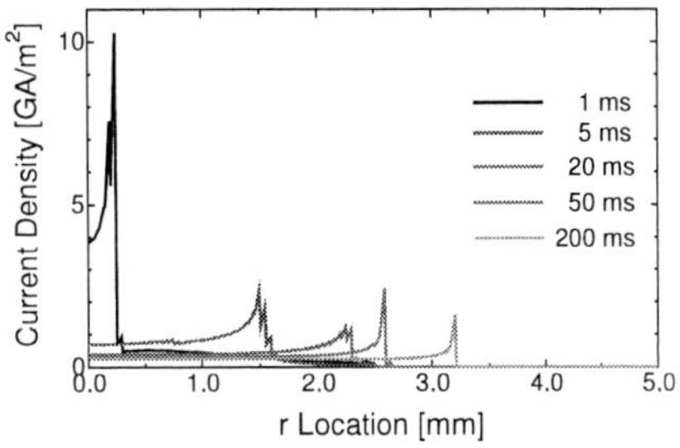

Figure 3: Time history of electric current distribution

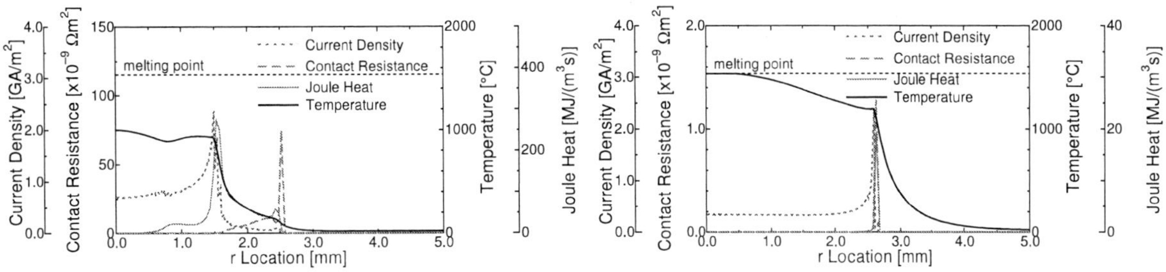

(a) $t = 5.0$ msec (b) $t = 50.0$ msec

Figure 4: Distribution of contact resistance, current density, Joule heat and temperature

Influence of welding conditions

Effect of welding current

Effect of welding current, which is one of the major parameters of RSW, is considered. When the welding current is larger, the total Joule heat becomes larger and the temperature becomes higher according to Path 2. However, there is a possibility that the current density becomes lower because of the larger contact area produced by lower plate rigidity according to Path 1. Then, comparison is made with different level of welding current such as 7.0, 9.5 and 12.0 kA.

Temperature history at the center of the plate interface is shown in Fig. 5. The temperature rises up sharply at about 1 ms but the rise disappears and keeps almost the same temperature for these three cases. It attributes to the disappearance of contact resistance calculated by Bubu's model[5]. Temperature difference after 5 ms is caused by the Joule heat in the base metal. When the welding current is larger, temperature rise becomes earlier and higher.

Effect of electrode force

Welding force is also a major parameter of RSW. When the welding force is larger, contact area on the interface becomes larger, and it may influence the electric current (Path 1). The contact resistance is also affected by the contact pressure (Path 3). Then, effect of welding force is considered using different force of 2.0, 2.65 and 3.0 kN.

Fig. 6 shows the current distribution along the interface at 50 ms. When the applied welding force is larger, electric current density along the plate interface becomes slightly lower, which leads to lower temperature along the interface as shown in Fig. 7. The cause is not the decrease of contact resistance but the reduction of current density due to the increase of contact surface.

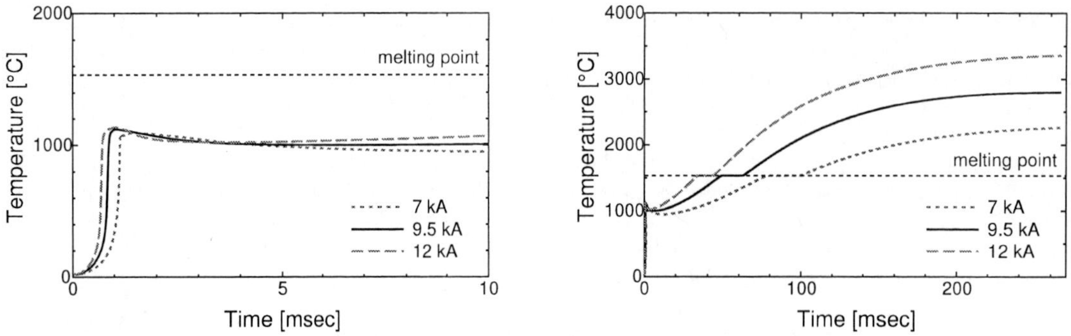

(a) Magnified in time scale (b) Full time scale

Figure 5: Time history of temerature at the center of the plate interface

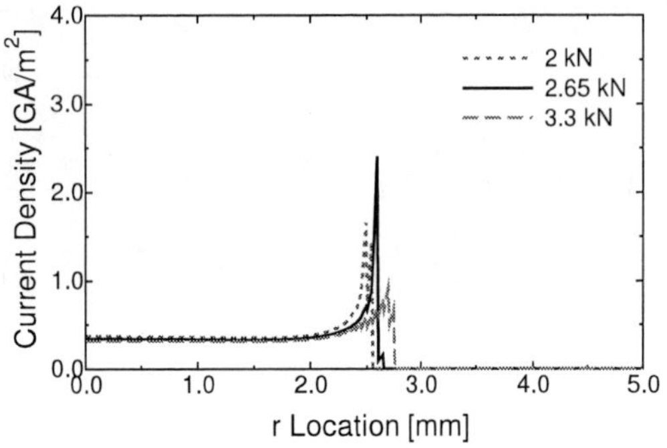

Figure 6: Current density distribution (Time = 50 ms)

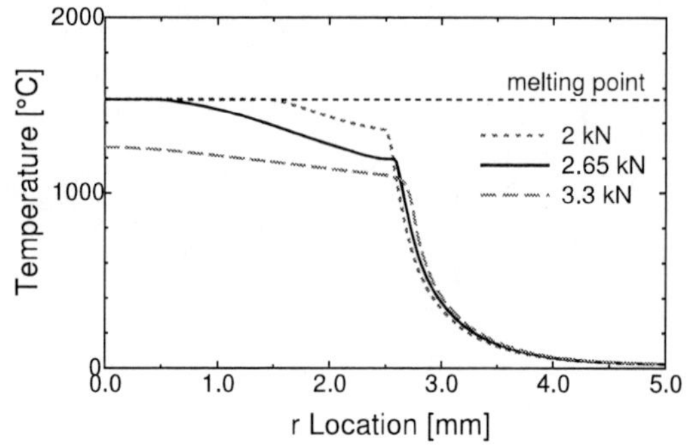

Figure 7: Temperature distribution (Time = 50 ms)

Conclusions

There are three regions on the plate interface, high temperature completely contacted region, low pressure and low temperature transition region and uncontacted region. Current density peak, Joule heat peak and contact resistance peak move outward changing from the contact resistance heating to the base metal heating.

Acknowledgments

This work was supported by JSPS KAKENHI Grant Number 16K05043. The authors wish to express their gratitude for cooperation of TOYOTA MOTOR KYUSHU Inc.

References

[1] D. W. Dickinson: Characterization of spot welding behavior by dynamic electrical parameter monitoring, Welding Journal, **59**–6, (1980), 170s–176s.

[2] Z. Hou, I. Kim, Y. Wang, C. Li, C. Chen: Finite element analysis for the mechanical features of resistance spot welding process, Journal of Materials Processing Technology, **185**–1–3, (2007), 160–165.

[3] N. Ma, H. Murakawa: Numerical and experimental study on nugget formation in resistance spot welding for three pieces of high strength steel sheets, Journal of Materials Processing Technology, **210**–14, (2010), 2045–2052.

[4] J. Shen, Y. Zhang, X. Lai, P.C. Wang: Modeling of Resistance Spot Welding of Multiple Stacks of Steel Sheets, Materials and Design, **32**–2, (2011), 550–560.

[5] S. S. Babu, M. L. Santella, Z. Feng, B. W. Riemer, J. W. Cohron: Empirical model of effects of pressure and temperature on electrical contact resistance of metals, Science and Technology of Welding and Joining, **6**–3 (2001), 126–132.

[6] T. Niho, T. Horie, Y. Morita, D. Ishihara, D. Yamakawa, S. Momii: Elasto-platic contact, electric current and thermal conduction triply coupled analysis model for resistance spot welding, Quarterly Journal of the Japan Welding Society, **33**–3 (2015), 271–282. (in Japanese)

Multiscale Finite Element Multiphysics Analysis of Resistance Spot Welding

Hiroyuki KURAMAE
Osaka Institute of Technology, Osaka, Japan
kuramae@dim.oit.ac.jp, +81-6-6954-4692

Tomoya NIHO, Riku KUSUMOTO, Jyunya YAMADA, Tomoyoshi HORIE
Kyushu Institute of Technology, Fukuoka, Japan
niho@mse.kyutech.ac.jp, +81-948-29-7757

Abstract

Since electrical contact resistance is an important property for resistance spot welding simulation, a multiscale coupled finite element (FE) procedure is newly developed to evaluate the electrical contact resistance accurately. The multiscale analysis consists of macroscopic FE analysis for resistance spot welding and microscopic electrical contact resistance analysis using three-dimensional thermal elasto-plastic contact FE simulation with electrical current analysis.

In this study, a micro-FE model, which corresponds to statistically similar representative volume element (SS-RVE), was constructed based on surface roughness measurement of a steel plate using laser microscope apparatus. The size of SS-RVE in-plane was determined and sampled as $100\times100\mu m^2$ from measured area $284.7\times213.5\mu m^2$ with $0.139\mu m$ interval. A rigid plate is contacted to the micro-FE model with contact pressure and temperature that are obtained by the macro-FE analysis, to calculate electrical contact resistance. The micro-FE analysis was conducted by an elasto-plastic contact FE code including large deformation theory combined with electric current FE code based on the ϕ-method. To compute the electrical contact resistance, electric current analysis is executed using deformed shape of the micro-FE model every time step. A micro-FE analysis is assigned to each of contact FEs on the macro-FE mesh between steel plates to obtain electrical contact resistance. In order to investigate the multiscale FE result, a conventional single-scale FE analysis that is only macroscopic resistance spot welding FE analysis using Bubu's model to evaluate the electrical contact resistance was also performed.

It was confirmed that the nonlinear effects of time history of electrical contact resistance were pronounced for loading paths. Temperature of multiscale analysis was lower than that of single-scale, because electrical contact resistance of the multiscale FE result was lower than that of Babu's model.

Introduction

Electrical contact resistance is an important property in the numerical simulations for resistance spot welding. The electrical contact resistance is strongly dependent on contact pressure and temperature, and microscopic asperities of contact surface on steel plates and electrode. Until now, some mathematical models of electrical multipoint contacts have been developed. In the simple case, the Holm's equation [1] was used for a large number of small equal spots distributed uniformly and densely over a circular area. Greenwood [2] expanded the Holm's equation to a cluster of spots with various sizes and their distributions. Babu *et al.* [3] proposed an empirical model for resistance spot welding simulation by introducing asperity density η, which is the number density of asperities in real contact, into Greenwood's model in order to eliminate parameters of the surface roughness. However, Babu's model is an unconsidered asperity shape, and unexpressive loading path such as contact pressure and temperature on nonlinear behaviour, because this model is a single-valued function of loads. In addition, a lot of experimental data of relationship between electrical contact resistance and contact pressure are required to evaluate the asperity density η.

In this study, three-dimensional multiscale finite element (FE) analysis of electrical contact resistance for spot welding simulation is newly developed. To execute the micro-FE analysis, micro-FE model as a statistically similar representative volume element (SS-RVE) is constructed based on measurement of surface roughness of steel plate. The multiscale spot welding FE analysis is executed and is compared with conventional single-scale FE result using Babu's model.

Method of the multiscale analysis

In this study, two-scale is considered for electrical contact resistance analysis as shown in Fig. 1. One is macroscopic resistance spot welding analysis, which is electric-thermal-structural coupled analysis. The other is a microscopic electrical contact resistance analysis, which is electric-structural FE analysis using temperature dependent material properties. There are contact FEs to evaluate electrical contact resistance on the interface between two steel plates and between the electrode and the steel plate in macro-FE model. Micro-FE analyses are assigned into macro-contact FEs to obtain electrical contact resistance. The initial micro-FE models are all same state, however, applying loads that are contact pressure and temperature to micro-FE analyses are different due to temperature and stress distribution on macro-FE model. Since micro-FE analyses are independent each other, the parallel computing technique is employed by the message passing interface (MPI) library.

Figure 2 shows data flow of the multiscale coupled FE analysis. Elasto-plasticity contact analysis is executed by using MSC. Marc 2012 for both macro- and micro-FE analyses that are incremental form for employing material nonlinear and large deformation analyses. Temperature-dependent material properties such as Young's modulus, yield stress and work hardening rate are considered in the analysis.

In the electric current analysis of micro-scale, the ϕ-method is used to obtain electrical contact resistance of deformed shape. In the macroscopic electric current FE analysis, the T-method is employed to evaluate the Joule heating. The thermal FE analysis computes temperature distribution by Joule heat with contact dependent thermal conduction paths at the macro-scale.

Construction of micro-FE model based on surface roughness measurement

In order to analyze micro electrical contact resistance, micro-FE model as an SS-RVE is constructed based on laser microscope measurement of surface roughness of a 440MPa grade tensile strength steel plate using KEYENCE VK-9700. The resolution was 0.139μm of 284.7×213.5μm^2 area. Figure 3 shows surface morphology and parameters of surface roughness are summarized in Table 1.

To determine the SS-RVE in-plane size, various sizes of quadrate area are sampled from center of the full area size 284.7×213.5μm^2. Frequency of asperity height distribution is compared between the full area size and sampled size. In this study, a correlation analysis was executed for similarity between full area and sampled area. At the size of over 100×100μm^2 area, the coefficient of correlation is saturated as shown in Fig. 4.

The SS-RVE is equally divided into 20-node solid elements with size of 5.0μm in-plane as shown in Fig. 5. On the side face of the micro-FE model, the periodic boundary condition is satisfied. The asperity shape of the top surface is based on laser microscope measurement. The micro-FE model is contact with a rigid flat plate to analyze deformation and electrical contact resistance.

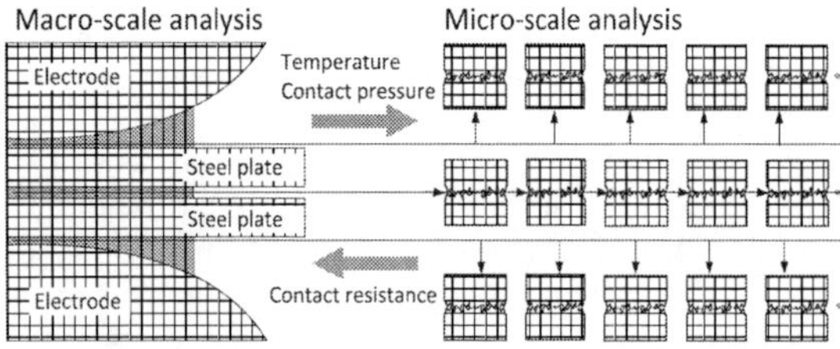

Fig. 1: Assignment of micro-FE analysis into macro-contact FEs.

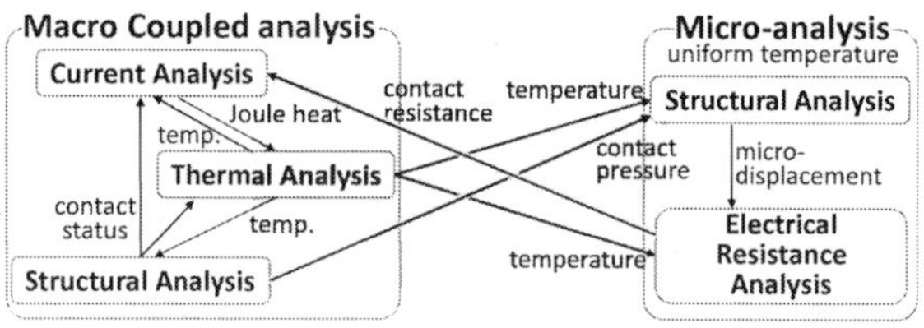

Fig. 2: Schematic diagram of multiscale coupled FE procedure of resistance spot welding.

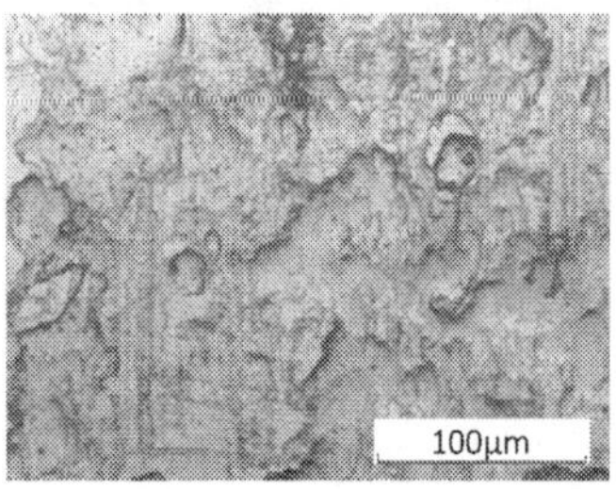

Fig. 3: Surface roughness of steel plate by laser microscope.

Table 1: Measurement results of surface roughness.

maximum roughness profile peak height	*Rp*	10.684μm
maximum roughness profile valley depth	*Rv*	11.701μm
maximum height of roughness profile	*Rz*	22.285μm
arithmetical mean deviation	*Ra*	1.282μm
root mean square deviation	*Rq*	1.575μm
skewness of the roughness profile	*Rsk*	0.1592
kurtosis of the roughness profile	*Rku*	2.5311
roughness profile section height difference	*Rδc*	2.286μm
relative material ratio	*Rmr*	50.00%
material ratio	*Rmr(c)*	60.15%

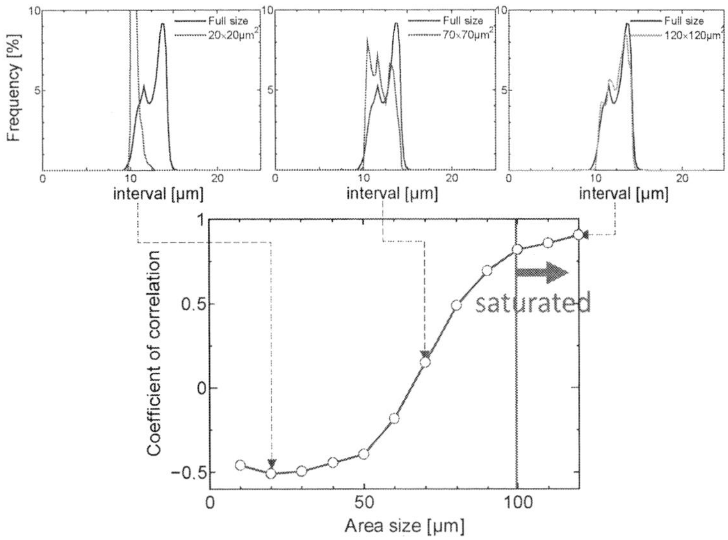

Fig. 4: Correlation analysis result to determine SS-RVE size.

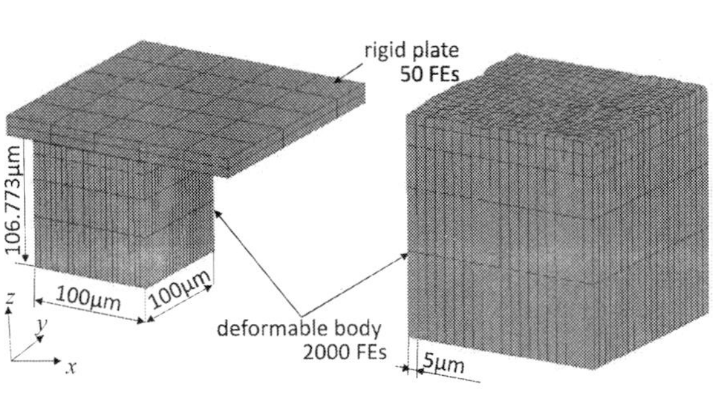

Fig. 5: Micro-FE model based on laser microscope measurement.

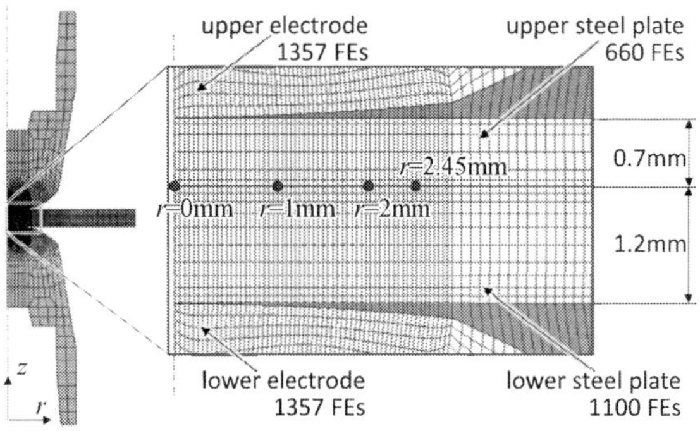

Fig. 6: Macro-FE model.

Figure 6 shows macro-FE model for resistance spot welding FE simulation. It consists of two steel plates and, upper and lower copper electrodes. The macro-FE model is divided using axisymmetric 8-node element. Since there are 60 contact FEs between two steel plates, micro-FE analyses are assigned into the macro-contact FEs to evaluate the electrical contact resistance every time step.

Results and Discussion

Figure 7 shows time history of temperature, contact pressure and electrical contact resistance of multiscale FE results comparing at 4 different points such as r = 0, 1, 2 and 2.45mm. At close to the center which is just below electrode, indicated by black and blue lines, temperature is immediately increased. On the contact edge (r = 2.45mm), indicated by green line, contact pressure and temperature are delay increased.

In order to investigate the multiscale coupled FE results, single-scale coupled FE analysis is also executed using Babu's empirical model [3] for evaluating the electrical contact resistance R_C as follow:

$$R_C = (\rho_1 + \rho_2)\left(\frac{1}{4}\sqrt{\frac{\pi\sigma_{YS}}{\eta P}} + \frac{3\pi}{16\sqrt{\eta}}\right) \tag{1}$$

where ρ_1 and ρ_2 are resistivity of materials, σ_{YS} is yield stress, and asperity density η (m^{-2}) is the number density of asperities in real contact. This model is expressed by only as a single-valued function of the contact pressure P and the temperature T, however, the microscopic shape of the asperities and loading paths of contact pressure and temperature are not considered.

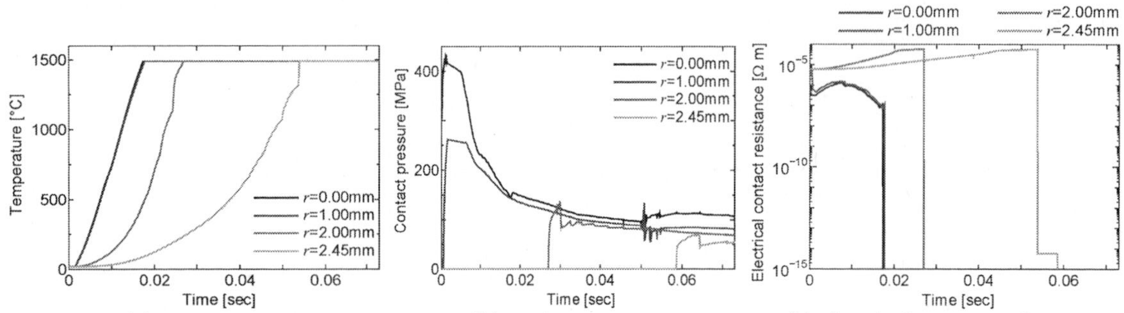

(a) temperature (b) contact pressure (c) electrical contact resistance

Fig. 7: Time history of multiscale FE result.

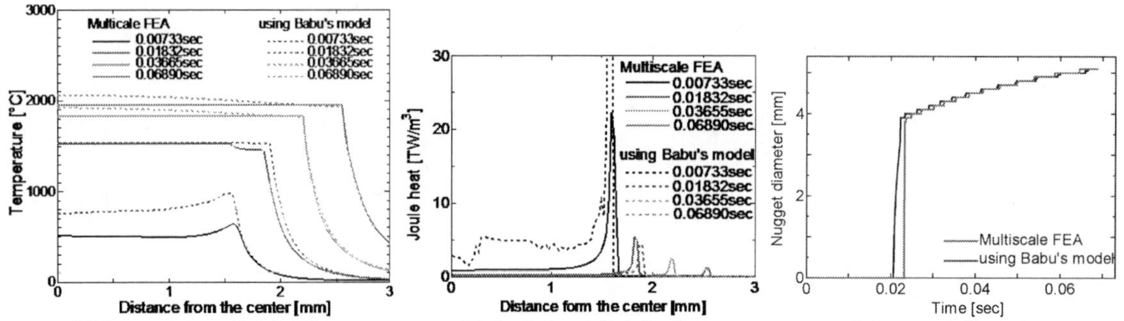

(a) temperature distribution (b) Joule heat distribution (c) nugget diameter

Fig. 8 Comparison of FE results between multiscale FE analysis and single scale FE analysis using Babu's model.

Figure 8 shows comparison of temperature, Joule heat distributions and nugget growth between the multiscale and single-scale results. The temperature of single-scale FE result using Babu's model is higher than that of multiscale one because the Joule heat caused by electrical contact resistance of Babu's model is higher than multiscale FE result as shown in Fig. 8 (b). The nugget growths are different between multiscale and single-scale analyses, however, the final nugget diameters are almost same as shown in Fig. 8 (c) because the electrical contact resistance is eliminated by high temperature.

Conclusions

In this study, multiscale coupled FE analysis of electrical contact resistance is proposed for resistance spot welding simulation. Multiscale coupled FE procedure is developed and micro-FE model based on measurement of surface roughness of steel plate by laser microscope is constructed. Temperature of single-scale analysis is higher than that of multiscale, because electrical contact resistance of the multiscale FE result is lower than that of Babu's model.

References

[1] R. Holm: Electric Contacts, Springer-Verlag, 1958.
[2] J.A. Greenwood: "Constriction Resistance and the Real Area of Contact," *Brit. J. Appl. Phys.*, Vol. 17 (1966), pp. 1621-1632.
[3] S.S. Babu, M.L. Santella, Z. Feng, B.W. Riemer and J.W. Cohron: "Empirical Model of Effects of Pressure and Temperature on Electrical Contact Resistance of Metals," *Sci. Tech. Welding Joining*, Vol. 6, No. 3 (2001), pp. 126-132.

The comparison of transformers designed for resistance spot welding machines

Mariusz Adam Stepien [1], Boguslaw Grzesik [1], Zygmunt Mikno [2]

[1] *Silesian University of Technology, Poland*
[2] *Welding Institute, Poland*

The main objective is the analysis of two transformers that are devoted to resistance spot welding DC machine in order to get information about its suitable parameters, namely, minimized leakage inductances as well as maximized coupling coefficient between windings. The analysis is also aimed at the influence of the design (shape of the cross-section of the magnetic core and also the way the windings are wounded), type of magnetic material and operational frequency on on the power properties of transformer (power loss and power density). Two magnetic materials, ferrite and amorphous one are taken into consideration in the analysis. The analysis is carried out for two frequencies of operation of the transformer, 1 kHz and 10 kHz.

Temperature distribution change by constriction of electric current for resistance welding

Kazuki Wakabayashi, Pasomphone Hemthavy, Shigeki Saito, and Kunio Takahashi*
**corresponding person*
Tokyo Institute of Technology, Tokyo, JAPAN
wakabayashi.k.af@m.titech.ac.jp, (+81)357343689

Abstract
The change in temperature distribution due to constriction of an electric current for resistance welding is investigated. The temperature change is significant for the understanding and optimizing the resistance welding process. Two electrodes are modeled as two semi-infinite body and touch at a point contact. It is assumed that the material constants are uniform and constant. The heat density distribution has a singular point at the periphery of the contact area. The singularity of the heat density distribution is clearly discussed. The change in temperature distribution due to constriction of an electric current is analytically obtained. Types of the temperature distribution are classified into two types according to the maximum temperature. One is the maximum temperature is at the periphery of the contact area. The other is the maximum temperature is at the whole of the contact area. The value of the maximum temperature is derived by the electric current value through the system, elapsed time since the current is passed, material constants and contact radius. The time to melting is obtained by comparing the maximum temperature with melting temperature. Those results are valuable for improvement of the resistance welding process. Several materials are chosen for the materials to discuss. Actual experiment systems are finite system. Then it is important to show applicable range of the theory. A circular system is investigated for the analysis of shape dependence. The BEM(Boundary Element Method) is used for the analysis introducing surface charge density. The surface charge density has singular points. The points are clearly discussed. The result shows the applicable range of the theory.

Introduction
Resistance welding uses heat generation by constriction of an electric current through a contact area. The change in temperature distribution is significant to understand and optimize the resistance process. The potential and current density distribution are analytically obtained for semi-infinite body[1].The heat density distribution is derived from the current density distribution, and the heat density has singular points at the periphery of the contact area. It is necessary to treat properly the singular points for numerical analysis but has not been clearly mentioned[2-3]. Authors analytically show how to treat the singular points[4]. Then, the change in temperature distribution is obtained. The theory can be applied to derive melting condition. In actual systems, the body is finite. Actual experiment systems are finite system. Then it is important to show applicable range of the theory. Cylinder system is typical shape of electrodes. Then, the cylinder system is selected. One of the method to analyze the finite system is BEM(Boundary Element Method)[5].

In this paper, first we discuss the analysis of melting condition for the resistance welding process based on the theory of the change in temperature distribution. Then, the applicable range of the theory for the cylinder system is discussed.

1. Change in temperature distribution by current passing semi-infinite body
Figure 1 shows an axially symmetric semi-infinite conductive system introduced in this research. The dark part represents the contact area with the contact radius a [m]. The origin O is set at the center of the contact area; the axis of symmetry is z axis; radial direction is r axis; and tangential direction of the circle is θ axis. The boundary conditions are set as the potential is Vc [V] at the contact area, the potential equals zero at an infinite distance from the contact area, and the component normal to the surface of the gradient of the potential equals zero at the top surface of the body except the contact area.

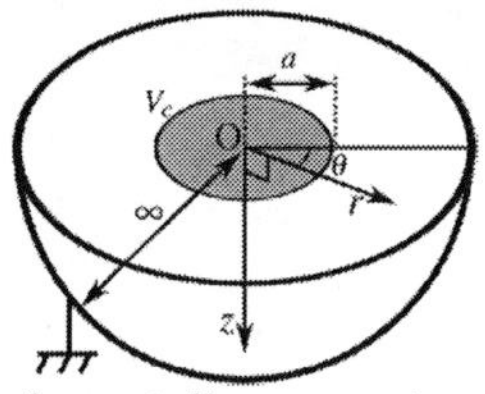

Fig 1: Schematic illustration of an axially symmetric semi-infinite conductive system.

Under an assumption that the magnetic effect is ignored, the potential distribution $V(\boldsymbol{r})$ [V], current density distribution $\boldsymbol{i}(\boldsymbol{r})$ [A/m^2], heat density distribution $q(\boldsymbol{r})$ [J/m^3 s] are derived as equations (1-3)[1,2].

$$\tilde{V}(\boldsymbol{r})=V(\boldsymbol{r})/Vc=\frac{2}{\pi}\arctan\frac{1}{\xi}, \tag{1}$$

$$\tilde{\boldsymbol{i}}(\boldsymbol{r})=\boldsymbol{i}(\boldsymbol{r})/\frac{2Vc}{\pi\rho a}=\frac{\text{grad}\,V(\boldsymbol{r})}{\rho}/\frac{2Vc}{\pi\rho a}=\left(\frac{1}{\sqrt{(1+\xi^2)(\xi^2+\eta^2)}},0,0\right), \tag{2}$$

$$\tilde{q}(\boldsymbol{r})=q(\boldsymbol{r})/\frac{4Vc^2}{\pi^2\rho a^2}=\rho|\boldsymbol{i}(\boldsymbol{r})|^2/\frac{4Vc^2}{\pi^2\rho a^2}=\frac{1}{(1+\xi^2)(\xi^2+\eta^2)}, \tag{3}$$

where $\rho\,[\Omega\,\text{m}]$ is resistivity and $\xi,\eta\,(\xi\in[0,\infty],\eta\in[0,1])$ are components of the spheroidal coordinate system. Under the assumption that the loss of heat from the surface of the body is ignored, the thermal conduction equation is expressed as equation (4).

$$\frac{\partial T(\boldsymbol{r},t)}{\partial t}=D\Delta T(\boldsymbol{r},t)+\frac{q(\boldsymbol{r})}{C\rho_d}. \tag{4}$$

where $T(\boldsymbol{r},t)$ [K] is temperature distribution, $D=\kappa/C\rho_d$ [m^2/s] is the thermal diffusivity, C [J/K kg] is the specific heat, ρ_d [kg/m^3] is the density of mass, κ [W/m K] is the thermal conductivity, and t [s] is time. The time when voltage is applied to the contact area is defined as $t=0$. Under the assumptions that the material constants (specific heat, density of mass and thermal conductivity) are uniform and constant, and that the phase transition is ignored, the change in temperature distribution is obtained as equation (5) from equations (3)(4)[4].

$$T(\boldsymbol{r},t)=T_0+\frac{4Vc^2a}{C\rho_d\rho\pi^2}\int_{t'=0}^{t}\int_{\xi=0}^{\infty}\int_{\eta=0}^{1}\int_{\theta=0}^{2\pi}\frac{\exp\left(-\frac{|\boldsymbol{r}-\boldsymbol{r}'|^2}{4D(t-t')}\right)}{[4\pi D(t-t')]^{3/2}(1+\xi^2)}dt'\,d\xi\,d\eta\,d\theta \tag{5}$$

where T_0 is the initial temperature [K]. $T(\boldsymbol{r},t)$, and t are normalized as equations (6-7).

$$\tilde{T}=(T-T_0)/\frac{Vc^2}{2\rho\kappa}, \tag{6}$$

$$\tilde{t}=t/\frac{a^2}{D}. \tag{7}$$

There are singular points of $\tilde{\boldsymbol{i}}(\boldsymbol{r})$ and $\tilde{q}(\boldsymbol{r})$ at the periphery of the contact area. From equation (6), if $\tilde{q}(\boldsymbol{r})$ has singular points then the temperature change rate with respect to time or the diffusion of heat have an infinite value at the point. When $t>0$, from equation (7), it is clear that temperature has finite value for any point. This means the infinite heat generation is cancelled out by the infinite heat diffusion. When $t=0$, the heat diffusion equals zero and temperature change rate with respect to time has an infinite value. For sufficiently small t, equation (9) is analytically derived.

$$\frac{\partial\tilde{T}}{\partial\tilde{t}}\approx\frac{2}{\pi^{3/2}\sqrt{\tilde{t}}}. \tag{9}$$

Equation (9) shows that temperature change rate with respect to time has an infinite value but temperature, which is the integration with respect to time of equation (9), has finite value.

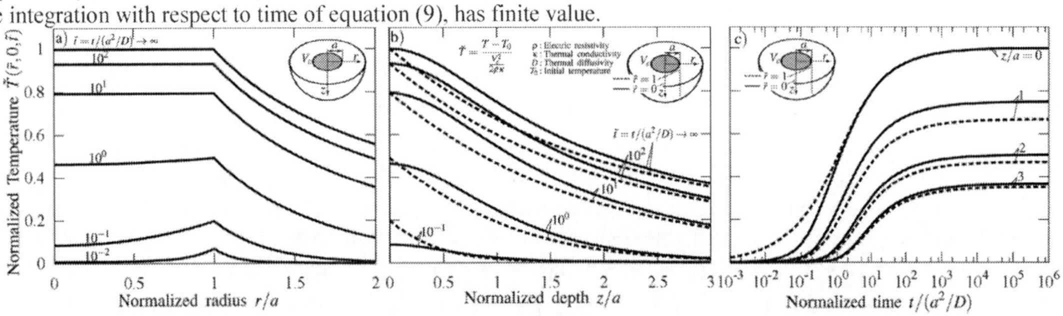

Fig 2: (a)Temeprature distribution change at $z=0$. (b)Temperature distribution change at $r=0,1$. (c)Temeprature change at several points.

Figure 2 shows the change in temperature distribution. It shows that the highest temperature is at the periphery of the contact area until $t \approx 10^1$. After that, temperature inside the contact area is almost same and goes up to $\tilde{T} = 1$.To compare the highest temperature with melting point of selected electrode material, values of potential and time when the melting occurs are derived. Figure 3 shows time to melting with respect to potential by introducing normalized potential $\tilde{V}c$ as equation (10). Table 1 shows that difference of time to melting for several materials.

$$\tilde{V}c = \frac{Vc}{\sqrt{2\rho\kappa(T_m - T_0)}} . \quad (10)$$

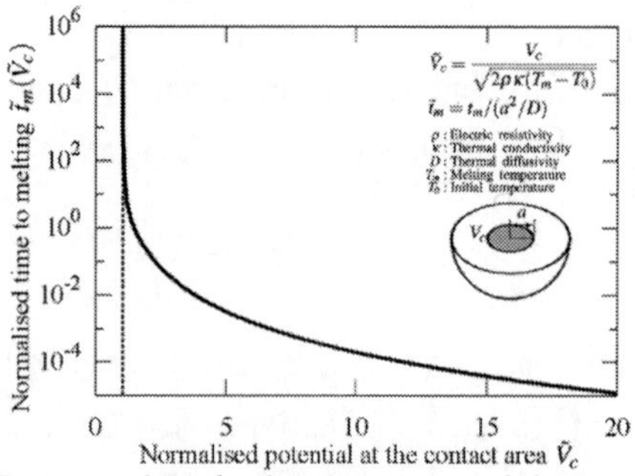

Fig 3:Time to melting by the constriction of electric current.

Table 1: Time to melting for several materials with potential Vc=1[V],contact radius a =0.001[m] and initial temperature T_0=300[K].

Material	Electric resistivity ρ 10^{-8}[Ω m]	Thermal conductivity κ 10^2[W/m K]	Melting temeperature T_m [K]	$\tilde{V}c$ [·]	$\tilde{t}$ 10^{-3}[·]	thermal diffusivity D 10^{-4}[m^2/s]	Time to melting t 10^{-3}[s]
Al	2.64	2.38	1206	9.37	0.255	1.08	0.00235
Au	2.29	3.18	1610	7.24	0.725	1.29	0.00564
Cu	1.73	4.01	1631	7.36	0.675	1.19	0.00565
Sn	0.133	6.68	778	10.9	0.39	0.401	0.00346
NiCr	0.150	0.113	1673	4.64	4.45	0.0299	1.49

2. Potential and Current density distribution for the finite body (Cylinder)

The theory of the change in temperature distribution is under the assumption of semi-infinite body. But the actual experimental setup has finite body. Thus it is important to show the applicable range of the theory for finite body. In this paper, cylinder shape is selected. Figure 4 shows the cylinder body and boundary conditions. BEM (Boundary Element Method) is chosen for the analysis[3]. BEM assumes surface charge distribution at the surface. Potential distribution and current density distribution are calculated by the charge density. The charge density is determined to fulfill the boundary conditions. Then, the potential and current distribution is calculated. At the periphery of the contact area, the current density has singularity in cylinder system as same as in the semi-infinite system. It results in surface charge density also has singularity. Then, charge distribution near the edge of the contact area is assumed as equation (11).

$$\frac{c}{(a^2 - r^2)^{\gamma}} , \quad (11)$$

where $0 < \gamma < 1/2$. Another assumption is that charge at the edge of the cylinder equals to zero. Under those assumptions, potential and current density can be obtained for any set of $(\tilde{l}, \tilde{h})$,where $\tilde{l} = l/a > 1$ and $\tilde{h} = h/a > 0$. If the theory for semi-infinite body is applicable, it is at least necessary that potential and current density distribution around contact area are same as those for semi-infinite body. Here, the amount of current which passes the cylinder system is chosen to evaluate the shape dependence. If the amount of current is same, then the potential and current density distribution around the contact area basically same. Figure 5 shows the amount of current for several sets of $(\tilde{l}, \tilde{h})$. The amount of current is normalized by amount of current for semi-infinite body $I = 4Vca/\rho$, thus $\tilde{I} = 1$ is necessary. From Figure 6, the condition is basically satisfied when $\tilde{l} = \tilde{h}$.

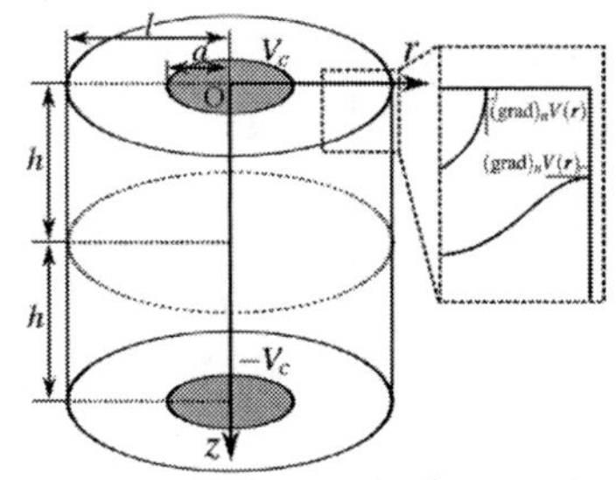

Fig 4: Schematic illustration of a finete cylinder system.

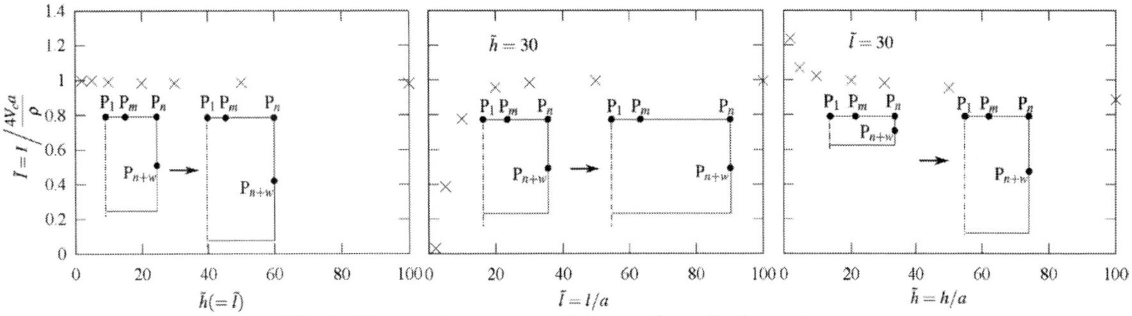

Fig 5: Changes in amount of current for cylinder system..

Conclusions

Melting condition and time to melting by constriction of current are discussed based on change in temperature distribution change revealed by authors work. The potential difference between the contact area and the infinite far surface of the electrode is required. If the potential difference is bigger than the threshold value, time to melting is obtained as Figure 6. Time to melting for several materials in certain condition is also presented.

The theory of the change in temperature distribution is under the assumption that the system is semi-infinite body. But actual systems are finite bodies. Then it is important to reveal the applicable condition of the theory for finite bodies. A finite cylinder system is introduced for the analysis. The analysis method is BEM. Two assumptions are introduced. One is charge distribution near the edge of the contact area is in equation (11) and the other is charge at the edge of the cylinder equals to zero. The results show the range of the shape of the cylinder, where potential and current distribution around the contact area is same as that of semi-infinite system at least.

References

[1] J. A. Greenwood and J. B. P. Williamson, "Electrical conduction in solids II. Theory of temperature-dependent conductors", *Proceedings of the Royal Society of London. Series A, Mathematical and Physical Science*, Vol.246 , No.1244 (1958),pp.13-31.

[2] T. Kubono, "Temperature distribution in the constriction region of electrical contacts carrying the dc-current", *The Transactions of the Institute of Electrical Engineers of Japan A*, Vol.104, No.6(1985) ,pp.47–53.

[3] S. R. Robertson, "A finite element analysis of the thermal behavior of contacts", IEEE Trans.Compon., Hybrids, Manuf. Technol., CHMT-5, No.1(1982), pp.3–10.

[4] K. Wakabayashi et al., "Change in temperature distribution by constriction of electric current through contact area", *Science and Technology of Welding and Joining*, Vol.20, No.4 (2015), pp.337-344.

[5] K. Takahashi et al., "Voltage required to detach an adhered particle by Coulomb interaction for micromanipulation", *Journal of Applied Physics*, Vol.90, No.1 (2001), pp.432-437.

Microstructure due to morphological instability in resistance spot welding

P. S. Wei, T.H. Wu
Department of Mechanical and Electro-Mechanical Engineering
National Sun Yat-Sen University,Kaohsiung,Taiwan, ROC
Email, pswei@mail.nsysu.edu.tw

Abstract
Resistance spot welding is an important technique often used in joining thin workpieces. This presentation theoretically and quantitatively investigates and interprets processes by realistically accounting for transient magneto-fluid mechanics, heat and species transport, and bulk resistance in workpiece, and film and constriction resistances at contact interfaces. Since temperature gradient and solidification rate are found, the computed morphological parameter shows that the effects of electrode geometries on microstructures of the weld nugget can be interpreted, predicted and controlled.

Introduction
In resistance spot welding two metal sheets are compressed between two water-cooled copper electrodes. Current is supplied via the two electrodes to create high joule heating at the faying surface. A molten nugget initiates and grows until current flow is terminated. The joint is completed via solidification due to cooling through electrodes .

Microstructure of the nugget is strongly affected by thermal processes [1-3]. Since the cooling rate is the highest in the direction of electrode, the directions of columnar grains are roughly perpendicular to the welded sheet metal [2-5]. Nugget microstructure of most metals, such as steel, Mo, Ni, Ti alloy and Cu alloy, is a columnar dendritic structure and that of a few metals (i.e. Al alloy) is columnar grains around periphery and equiaxed grains in the center. The nugget of Mg alloy only consists of an equiaxed dendritic structure. It is interesting to know if microstructures can be interpreted by a morphology parameter, defined by the ratio between liquid temperature gradient at the solidification front and solidification rate [2,3]. As the morphological parameter decreases, microstructures or morphologies of the solidification front vary from the planar, columnar, columnar dendrite and equiaxed dendrite. On the other hand, smaller spacings of the primary and secondary dendrite arms due to cooling rates, defined as the product of temperature gradient and solidification rate, occur near the fusion boundary.

In this work, heat transfer or temperature gradient along the liquidus line and solidification rate responsible for microstructure of nugget affected by different geometries of electrodes during resistance spot welding are theoretically and quantitatively investigated. This study uses the self-developed general computer code from previous work [6], accounting for transient magneto-fluid mechanics, heat and species transport, and temperature-dependent bulk resistance in workpiece, and film and constriction resistances at contact interfaces. Microstructures of weld nuggets therefore can be revealed and controlled by optimizing the shape of electrodes.

System model and Analysis
Resistance spot welding, as illustrated in Fig.1, includes electrodes containing a coolant hole, specified by dimensionless face radius (r_e) and truncated length (L_1) of the electrode, and radius ratio between the coolant hole and electrode (r_c).Imposed welding current is AC. The major assumptions are [6]:(1) the mushy zone in workpiece is a porous medium, and (2) properties are allowed to be distinct between phases. With the assumptions, dimensionless continuity, momentum, energy, species and magnetic field intensity equations in workpieces, respectively, become

$$\frac{\partial \rho}{\partial \tau} + \nabla \cdot (\rho \mathbf{V_m}) = 0 \tag{1}$$

$$\frac{\partial \rho u_m}{\partial \tau} + \nabla \cdot (\rho \mathbf{V_m} u_m) = \Pr \nabla \cdot (\rho \nabla u_m) - \frac{\Pr}{Da} \frac{\rho(1-g_\ell)^2}{g_\ell^3} u_m - \frac{\partial p}{\partial z} + \Pr^2 Gr[\theta_0 (T - T_{sol}) + F_0 (f_\ell^\alpha - f_{\ell,Tsol}^\alpha)] - \frac{Lo}{\pi^2} \mu_r H \frac{\partial H}{\partial z} \tag{2}$$

$$\frac{\partial \rho v_m}{\partial \tau} + \nabla \cdot (\rho \mathbf{V}_{\mathbf{m}} v_m) = \Pr \nabla \cdot (\rho \nabla v_m) - \frac{\Pr}{Da} \frac{\rho (1-g_\ell)^2}{g_\ell^3} v_m - \frac{\partial p}{\partial r} - \frac{Lo}{\pi^2} \frac{\mu_r}{r} H \frac{\partial r H}{\partial r} \quad (3)$$

$$\frac{\partial \rho h_m}{\partial \tau} + \nabla \cdot (\rho \mathbf{V}_{\mathbf{m}} h_m) = C \nabla \cdot (K \nabla h_m) + C \nabla \cdot [K \nabla (h_s - h_m)] - \nabla \cdot [\rho (h_\ell - h_m) \mathbf{V}_{\mathbf{m}}] + \frac{Lo \Sigma}{\pi^2} (\frac{1}{\sigma_c} + \frac{1}{\sigma}) [(\frac{1}{r} \frac{\partial r H}{\partial r})^2 + (\frac{\partial H}{\partial z})^2] \quad (4)$$

$$\frac{\partial \rho f_m^\alpha}{\partial \tau} + \nabla \cdot (\rho \mathbf{V}_{\mathbf{m}} f_m^\alpha) = \frac{\Pr}{Sc} \nabla \cdot (\rho f_\ell \nabla f_m^\alpha) + \frac{\Pr}{Sc} \nabla \cdot [\rho f_\ell \nabla (f_\ell^\alpha - f_m^\alpha)] - \nabla \cdot [\rho (f_\ell^\alpha - f_m^\alpha) \mathbf{V}_{\mathbf{m}}] \quad (5)$$

$$\frac{\partial \mu_r H}{\partial \tau} + \nabla \cdot (\mu_r H \mathbf{V}_{\mathbf{m}}) = \Pr_m \nabla \cdot (\frac{1}{\sigma} \nabla H) + \Pr_m [\frac{H}{r} \frac{\partial}{\partial r} (\frac{1}{\sigma}) - \frac{H}{\sigma r^2}] + \frac{\mu_r H v_{\mathrm{m}}}{r} \quad (6)$$

where lengths and time in equations are nondimensionalized by electrode outer radius, and time for thermal diffusion, respectively. The last terms on the right-hand side of Eqs. (2) and (3) stand for Lorentz force, whereas the last term of Eq. (4) combines heat generations at contact surfaces and in bulk workpieces.

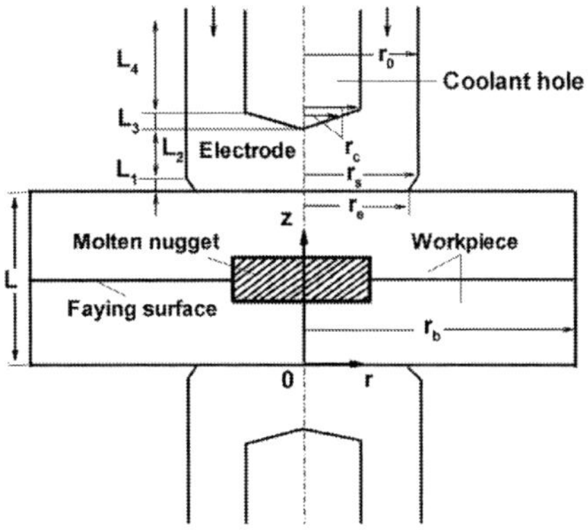

Fig. 1. Sketch for resistance spot welding and coordinate system

Energy and magnetic field intensity equations in the electrode are, respectively,

$$\frac{\partial T_E}{\partial \tau} = \frac{k_E k_s RC}{\rho_E R_E} [\frac{\partial^2 T_E}{\partial \zeta^2} + (\frac{1}{N^2} + M^2) \frac{\partial^2 T_E}{\partial \xi^2} + 2M \frac{\partial^2 T_E}{\partial \xi \partial \zeta} + (\frac{1}{\xi N + r_c} - 2M \frac{dN}{d\zeta}) \frac{1}{N} \frac{\partial T_E}{\partial \xi}] + \frac{Lo \Sigma}{\pi^2 \rho_E R_E \sigma_E} \cdot [(M \frac{\partial H_E}{\partial \xi} + \frac{\partial H_E}{\partial \zeta})^2 + (\frac{H_E}{\xi N + r_c} + \frac{1}{N} \frac{\partial H_E}{\partial \xi})^2] \quad (7)$$

$$\frac{\partial H_E}{\partial \tau} = \Pr_m \eta_E [\frac{\partial^2 H_E}{\partial \zeta^2} + (\frac{1}{N^2} + M^2) \frac{\partial^2 H_E}{\partial \xi^2} + 2M \frac{\partial^2 H_E}{\partial \xi \partial \zeta} + (\frac{1}{\xi N + r_c} - 2M \frac{dN}{d\zeta}) \frac{1}{N} \frac{\partial H_E}{\partial \xi} - \frac{H_E}{(\xi N + r_c)^2}] \quad (8)$$

where spatial coordinates ξ and ζ are immobilization coordinates. Continuities of tangential electric and magnetic field intensities are required at contact surfaces. Electrical contact resistance contains constriction and film resistance. Morphological instability can be simply scaled by considering Gibbs-Thomson equation

$$\tilde{T}_{s\ell} - \tilde{T}_m = \frac{\gamma \tilde{T}_m}{\tilde{\rho} \tilde{h}_{s\ell}} \frac{\partial^2 \eta_{s\ell}}{\partial \tilde{x}^2} + m \tilde{f}_\ell^\alpha \quad (9)$$

Deviation of temperature and concentration profiles from the base state is scaled as $\tilde{T}_{s\ell} - \tilde{T}_m - m\tilde{f}_\ell^\alpha \sim (G_{T\ell} - mG_f)\eta_{s\ell}$, where dimensional liquid temperature and concentration gradients are $G_{T\ell}$ and G_f, respectively. Surface curvature can be scaled as $\partial^2 \eta_{s\ell} / \partial \tilde{x}^2 \sim -\eta_{s\ell} / \lambda_M^2$. Substituting into Eq. (9) gives the wavelength of interface morphology

$$\lambda_M \sim \sqrt{\frac{\gamma \tilde{T}_m}{\tilde{\rho} \tilde{h}_{s\ell} (mG_f - G_{T\ell})}} \quad (10)$$

Morphological instability thus occurs for $mG_f - G_{T\ell} > 0$. Conservation of solute ahead of the solidification front is given by $-\tilde{D}_\ell^\alpha G_f = V(1-k_p)\tilde{f}_\ell^\alpha$. Substituting into Eq. (10) leads to morphology parameter

$$\text{The morphological parameter} \equiv \frac{G_{T\ell}}{V} = \frac{m(1-k_p)\tilde{f}_\ell^\alpha}{\tilde{D}_\ell^\alpha} \qquad (11)$$

Eqs. (1)- (8) are discretized by the control-volume, implicit finite-difference scheme with staggered grids. Using grid systems of 50×53 and 72×75, convergence tolerances for global energy and velocity components and enthalpy, concentration, and magnetic intensity fields were 10^{-2} and 10^{-3}, respectively. This model is confirmed from good agreement with a measured between nugget thickness provided by Gould [5] in welding AISI 1008.

Results and discussion

Values of working parameters were based on properties of typical metals such as iron, copper, and metallurgical properties of a binary alloy of iron and manganese under typical working conditions [6]. Dimensionless time-dependent heat fluxes and corresponding nugget growth in radial and axial directions at the liquidus line on the faying surface and axisymmetric axis for different electrode face radii are shown in Figs. 2(a) and (b), respectively. Dimensionless welding AC is also plotted for a comparison. Heat fluxes and nugget growth in both directions increase rapidly as the molten nugget initiates. The maximum axial temperature gradient at the moment when the power is off can be estimated to be 3×10^5 K/m for steels, agreeing well with the measured results [4]. Heat flux or temperature gradient in axial direction is always greater than that in radial directions, whereas solidification rate in radial direction is always greater than that in axial direction. Therefore, the morphology parameter in radial direction ($q_r/(\partial\delta_r/\partial\tau)$) is always less than that in axial direction ($q_z/(\partial\delta_z/\partial\tau)$).Columnar dendrite structure and equiaxed grains are thus found to occur near the fusion boundary and the centre regions of the nugget, respectively [1]. It also reveals that a decrease in electrode face radius increases the morphology parameter, leading to columnar dendrites. Since cooling rates can be decreased or increased, the effects of electrode face radius on spacings of the primary and secondary dendrite arms are uncertain.

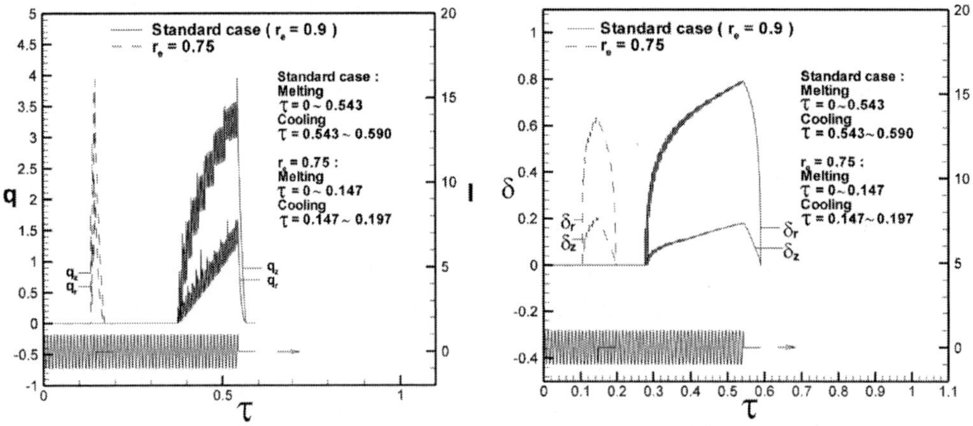

(a) (b)

Fig. 2 Dimensionless time-dependent (a) heat flux and (b) nugget thickness in radial and axial directions for different electrode face radii

Dimensionless heat fluxes and nugget thickness in axial and radial directions decrease with maximum radius of the coolant hole, as shown in Figs 3(a) and (b), respectively. Morphological parameters in axial and radial directions thus can be either increased or decreased. On the other hand, the decrease in the maximum radius of the coolant hole decrease cooling rates, leading to greater spacings of primary and secondary dendrite arms. Figures 4(a) and (b) show that an increase in electrode truncated length decreases heat flux and increases solidification rate, leading to a decrease in the morphology parameter, whereas an increase or decrease in cooling rates.

Conclusions

The conclusions drawn are the following:

1. The morphological parameter in both directions decrease for higher face radius and truncated length of the electrode. Furthermore, morphological parameter in the radial direction near the faying surface is always less than that in the axial direction in the course of solidification.
2. The effects of the location and shapes of the coolant hole away from the electrode face on morphology parameters are uncertain. Smaller spacings of primary and secondary dendrite arms are achieved by increasing maximum radius of the coolant hole.
3. Microstructure of weld nugget can be controlled via designing the shapes of electrode, and coolant hole.

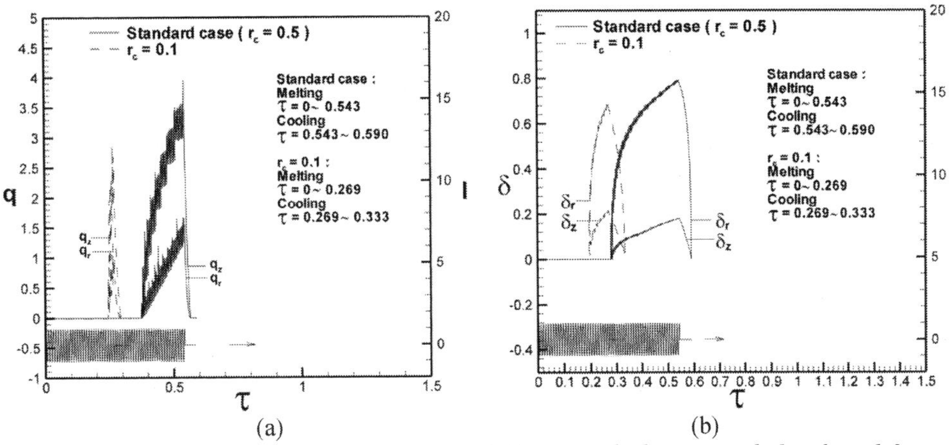

Fig. 3 Dimensionless time-dependent (a) heat flux and (b) nugget thickness in radial and axial directions for different coolant hole radii

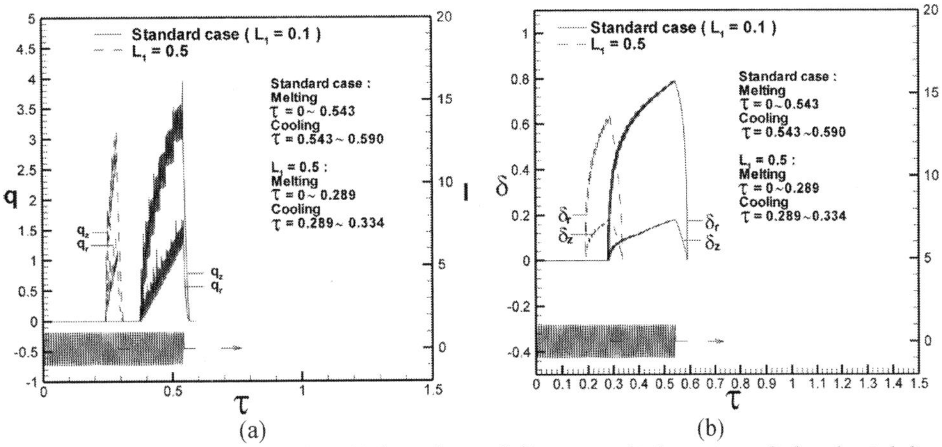

Fig. 4 Dimensionless time-dependent (a) heat flux and (b) nugget thickness in radial and axial directions for different electrode truncation lengths

References

[1] Y. R. Wang et al.: "Microstructure Characteristics of Resistance Spot Welds of AZ31 Mg Alloy," *Sci. Technol. Weld. Join.*, Vol. 11(2006), pp.555-560.

[2] T. DebRoy, S. A. David: "Physical Processes in Fusion Welding," *Rev. Mod. Phys.*,Vol. 67(1995), pp.85-112.

[3] S. Kou, Welding Metallurgy, 2nd edition, Wiley (New York, 2003).

[4] J. E. Gould, and P. H. Chang: "Thermal Modeling of Spot Welds," *J. Metals*, Vol. 38(1986), pp. 39-42.

[5] J. E. Gould: "An Examination of Nugget Development during Spot Welding, Using Both Experimental and Analytical Techniques," *Weld. J.*,Vol. 66 (1987), pp. 1s-10s.

[6] P. S. Wei, T. H. Wu: "Electrode Geometry Effects on Microstructure Determined by Heat Transfer and Solidification Rate during Resistance Spot Welding," Int. J. Heat Mass Transfer, Vol. 79(2014), pp. 408-416.

Direct Bonding of Plastic Materials to Metals Using C=O Double Bond and Its Bonding Mechanism

Katsuyoshi Kondoh, Hisashi Imai and Junko Umeda
Osaka University, Ibaragi, Osaka, Japan
kondoh@jwri.osaka-u.ac.jp, +81-6-6879-4369

Abstract

Direct bonding process by using the conventional hot pressing was applied to prepare the dissimilar materials consisting of the thermoplastic resin and industrial metals such as Ti, Fe, Cu, Al, Mg and their alloys. In employing the pressing temperature at the melting point of plastics, the interfacial bonding shear strength of the dissimilar material using polyamide66 (PA66) was drastically increased. FT-IR, TEM-EDS and XPS analysis clarified that C=O double bonds of the carbonyl group was thermally decomposed and oxygen atoms contributed to diffusion into the metals or oxides formation at the interface. In addition, the chemical reaction of C-H and N-M (M; metal) formation was also effective to the interfacial bonding improvement. On the other hand, when polystyrene (PS) with no functional group including C-O/C=O bond was employed, for example, the PS-Ti dissimilar material showed no bonding strength even then the bonding temperature was close to 300℃.

Introduction

Carbon fiber reinforced plastic (CFRP) is widely employed as engineering materials in the transportation industry such as airplanes, automotives, high-speed trains and marines because of its low density, outstanding corrosion resistance, high specific strength and Young's modulus compared to the conventional industrial metals such as steels and aluminum alloys [1]. In particular, CFRP material, using thermoplastic resins as a matrix, possibly reduces the processing cost due to its excellent formability in pressing and recyclability [2]. From high reliability and enhanced safety points of view, however, the hybrid materials design using both CFRP and metals called as "multi-materials design" is necessary for further lightweight of the components in the transportation industries. In joining the metal sheets and CFRP flat panels, the use of adhesive agents [3,4] or mechanical joints and fastening by bolts or rivets [5,6] is common in the industries because of their ease of assembly. They are, however, not complete in joining due to its insufficient reliability and weight increment of the components. To establish a direct bonding process between CFRP and industrial metals, the fundamental researches using ultrasonic welding [7,8], friction welding [9] and laser welding [10,11] are recently carried out, and successfully achieved a high bonding strength between metal sheets and CFRP or thermoplastic materials by each process. Regarding the bonding mechanism in each process, for example, FTIR analysis on the bonding interface indicated that a physicochemical interaction based on hydrogen bonds originated in the polyamide caused a chemical bonding of the plastic to aluminum materials [11]. In addition, a high gas pressure in gas bubbles generated from the molten plastic material assisted to directly bond the plastic to the metal surface, and resulted in formation of thin oxide layers at the interface [10]. In use of friction welding, the similar mechanism in using lase welding mentioned above was reported [9]. On the other hand, the mechanical interlocking effect at the interface was also necessary for a strong bonding strength in each process [7,9,11]. The dissimilar bonded materials between CFRP or plastic and common metals prepared by the above processes showed an enough bonding stress, and the fractures occurred on the plastic side, not at the bonding interface. The bonding mechanism, however, has not been completely clarified in these studies. In particular, the detail analysis at the local area near the interface is not enough to understand the chemical reaction between the plastic and metals by considering the molecular structural changes of plastic materials at elevated temperature in bonding. This study investigates the effect of C=O double bond of the carbonyl group in polyamide66 on the direct bonding behavior between the plastic and pure titanium plate by using FT-IR, TEM-EDS and XPS, and establishes a chemical bonding model for the dissimilar materials.

Experimental procedure

Five kinds of pure metal plates were employed in this study. The purity of aluminum (Al) was 99.0%, that of titanium (Ti) and iron (Fe) was 99.5%, and both copper (Cu) and magnesium (Mg) had 99.9% purity level. Each metal plate had a mirror-finished surface through various polishing treatments using fine alumina particles and diamond lapping paste compounds. The plates of polyamide66 (PA66), polystyrene (PS) and polyethylene terephthalate (PET) were also prepared for bonding test. The melting point of each plastic was measured by differential thermogravimetric analysis (DTA). The direct bonding process of each plastic to metal plate is schematically illustrated in *Fig. 1 (a)*. The metal plate was heated by contact with a small soldering iron device, and its heat conduction effectively caused a temperature increase of the plastic. The temperature profile at the interface between them was measured by a thermocouple inserted in the interface. When the interfacial temperature reaches the target, the thermocouple is pulled out, and immediately 5 MPa pressure is applied to the metal plate by an upper punch for 8 sec. Examples of the directly bonded materials of CFRP or PA66 to Cu plate are also shown in *Fig.1 (a)*. Transmission electron microscopy with energy dispersive spectroscopy (TEM-EDS) was applied to analyze oxygen atoms behavior at the bonding interface, and Fourier transform infrared spectroscopy (FTIR) and X-ray photoelectron spectroscopy (XPS) analysis was carried out to investigate the molecular structural changes of the plastic materials before and after bonding to the metal plate. A shear bonding strength of the dissimilar material was measured as illustrated in *Fig.1 (b)*.

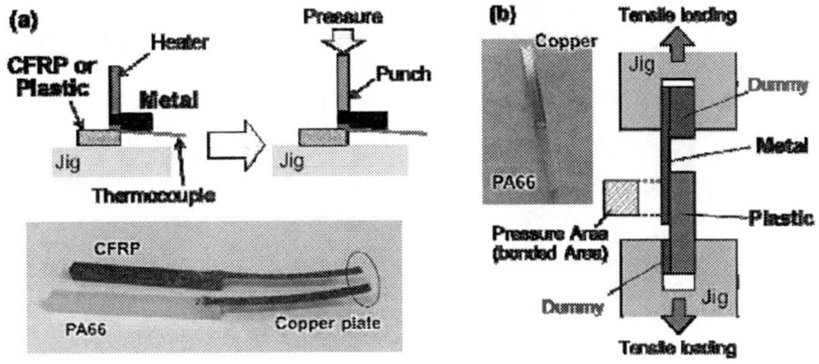

Figure 1: Schematic illustration of direct bonding process between CFRP or plastic and metal plate (a) and shear bonding strength test using 50 kN hydraulic tensile test machine (b).

Results and discussion

Shear bonding strength test results in use of pure Ti plate

Figure 2 shows a dependence of shear bonding strength of the dissimilar bonded materials on the interfacial temperature, where the melting point of PA66, PET and glass transition temperature of PS measured by DTA are 248℃, 242℃ and 98℃, respectively. They are close to the data measured in the previous studies [12,13].

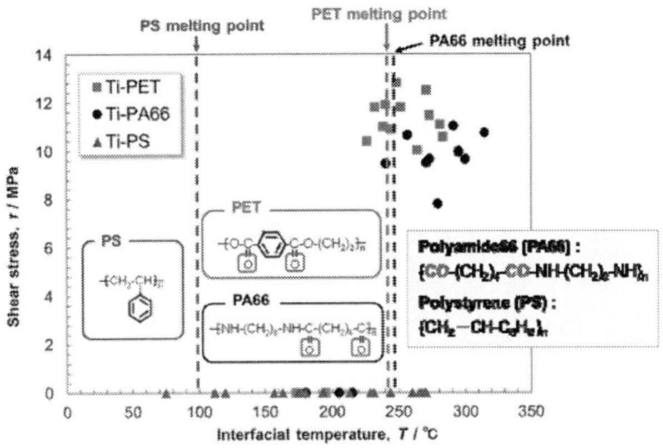

Figure 2: Shear bonding strength of dissimilar materials between pure Ti plate and PA66, PS and PET plastic materials prepared by changing bonding temperature.

In case of the plastic contains a carbonyl group including C=O double bond such as PA66 and PET, their shear bonding strength of both dissimilar materials (Ti-PA66 and Ti-PET) drastically increases when the interfacial temperature is over the melting point of each plastic material. The average of the bonding strength in Ti-PET dissimilar material is a little higher than that of Ti-PA66 bonded material. In addition, the bonded materials easily separate and their shear strength is zero in employing the interfacial temperature in bonding less than each melting point. On the other hand, Ti-PS dissimilar bonded materials have no bonding and joining behavior even the interfacial temperature is over its melting point of 98℃. It is concluded that the carbonyl group with C=O double bond are necessary to directly bond the plastic to metals and the bonding temperature must be controlled over the melting point of the plastic material.

TEM-EDS analysis on outer-most surface of Ti plate at interface of Ti-PA66 material

TEM-EDS analysis in *Fig. 3* indicated oxygen concentration area at the interface of the Ti-PA66 bonded material, and its electron diffraction detects some oxides (TiO_2 and TiO) formation in the layer. By considering an interstitial solubility of oxygen in α-Ti is 33 at.% in Ti-O binary phase diagram [14], the oxides formation suggests many oxygen atoms diffused from PA66 into pure Ti through the bonding interface during hot pressing.

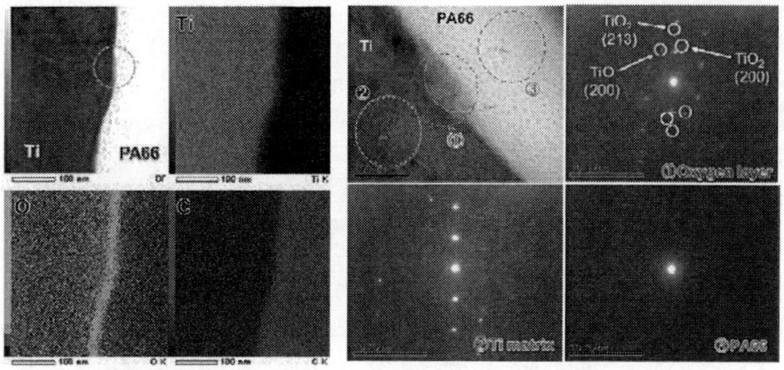

Figure 3: TEM-EDS and electron diffraction analysis on bonding interface of Ti-PA66 bonded material.

FTIR analysis on plastic of Ti-PA66 dissimilar bonded material

As shown in *Fig. 2*, the shear bonding strength test already indicated C=O bond has an important role on the direct bonding of PA66 to Ti plate. First of all, FTIR analysis was conducted to detect C=O spectrum changes in bonding as shown in *Fig. 4*. As mentioned above, the conventional hot pressing process was applied in the preparation of Ti-PA66 dissimilar bonded material. That is, the effects of the thermal factor and pressure in hot pressing on the chemical reaction at the interface should be considered. Therefore, as-received PA66 material, PA66 after heat treatment (HT) at 260℃, PA66 after HT at 260℃ under 5 MPa pressure, and PA66 separated from the bonded material were prepared for FTIR analysis. Each spectrum at wavenumbers 3300 cm^{-1}, 3100 cm^{-1}, 1633 cm^{-1}, 1539 cm^{-1} is amino group (NH_2, stretch), aromatic C-H group (stretch), C=O bond and NH_2 (bending), respectively [15,16]. The specimens (c) and (d) show no change in C=O spectrum compared to the original PA66 (b), that is, the molecular structural change of C=O bond never occurs by pressure and high temperature over its melting point. The specimen (e), however, obviously reveals a significant variation of C=O double bond. The PA66 specimen separated from the dissimilar bonded material shows an increment of wavenumber (κ) compared to the reference specimen (b). That is, it is obvious that the direct bonding behavior between PA66 and pure Ti by hot pressing causes molecular structural changes of PA66 material.

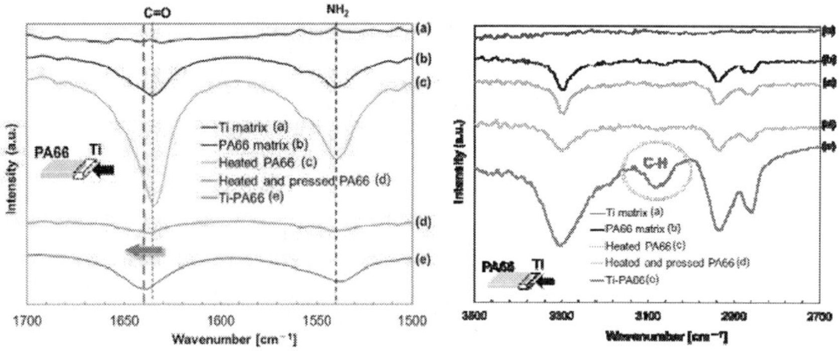

Figure 4: IR spectra of raw pure Ti plate (a), as-received PA66 material (b), PA66 after heat treatment (HT) at 260℃ (c), PA66 after HT at 260℃ under 5 MPa pressure (d), and PA66 of the bonded material(e).

A wavenumber, κ is expressed as a modified Hooke's Law in the below by considering the bond as a spring.

$$\kappa = f/c = 1/(2\pi c)\sqrt{(k/\mu)}, \quad \mu\ (\text{; reduced mass of atoms}) = m_1 m_2/(m_1+m_2) \qquad \text{(Eq.1)}$$

where k ; spring constant for the bond, m ; mass of atom, f ; frequency and c ; velocity of light (=3 x 10^{10} cm/s). When the bonding strength between two atoms (m_1 and m_2) increses, the increment of a spring constant occurs, and results in the increase of a wavenumber (κ value). The commercial PA66 contains two kinds of amides; X-C(=O)-Y and X-C(-O)-Y, and has a resonance structure with both amides [17]. In general the bonding strength of C=O bond is much larger than that of the latter with a covalent bond. As shown in *Fig. 4*, the wavenumber increment in the PA66 of Ti-PA66 bonded material means the number of X-C(=O)-Y relatively becomes much large compared to X-C(-O)-Y amide. It means C=O double bonds in the functional group (carbonyl group) are dominant in the PA66 specimen bonded to pure Ti plate. When a polarity change occurs by the electronegativity due to the bonding of another atom to X or Y atom linked up with C atom in the functional group, the number ratio of C=O relatively increases [17,18]. In considering TEM observation result in *Fig. 3*, Ti atoms would chemically bond to C atoms to form metal complex during hot pressing, and resulted in C=O spectrum variation near the interface of the bonded material as shown above.

Conclusions (Required)

Dissimilar bonded materials between pure Ti and PA66 or PET, having C=O double bonds, showed high shear bonding strength while Ti-PS bonded one had no bonding behavior. C=O bond in the carbonyl group is essential for chemical bonding between plastic and metals by hot press. In employing pure Ti, oxygen atoms thermally separated from C=O bond diffused into Ti specimen, and resulted in oxides formation at the bonding interface.

Acknowledgments

Authors would additionally like to thank Mr. T. Izawa for his technical supports in the experiment.

References

[1] R.F. Gibson: "A review of recent research on mechanics of multifunctional composite materials and structures", *Composites Structures*, Vol. 92, No. 12 (2010), pp. 2793-2810.
[2] K.K. Chawla: Composite Materials, Science and Engineering, Springer (New York, 1998).
[3] L. Tong: "Failure of Adhesive-Bonded Composite Single Lap Joints with Embedded Cracks", *AIAA Journal*, Vol. 36, No. 3 (1998), pp. 448-456.
[4] P.T. Cheuk, L. Tong: "Failure of adhesive bonded composite lap shear joints with embedded precrack", *Composites Science and Technology*, Vol. 62, No. 7-8 (2002), pp.1079-1095.
[5] L.J.H. Smith: "Bonded-bolted composite joints", *Journal of Aircraft*, Vol. 22, No. 11 (1985), pp. 993-1000.
[6] G. Marannano and B. Zuccarello: "Numerical experimental analysis of hybrid double lap aluminum-CFRP joints", *Composites Part B: Engineering*, Vol. 71, No. 15 (2015), pp. 28-39.
[7] F. Balle, G. Wagner, and D. Eifler: "Joining of Aluminum 5754 Alloy to carbon fiber reinforced polymers (CFRP) by ultrasonic welding", *Proc TMS2009 Conf. In Aluminum Alloys: Fabrication, Characterization and Applications II*, San Francisco, CA Feb 2008, pp. 191-196.
[8] F. Balle, G. Wagner, and D. Eifler: "Ultrasonic Metal Welding of Aluminum Sheets to Carbon Fiber Reinforced Thermoplastic Composites", *Advanced Engineering Materials*, Vol. 11, No. 1 (2009), pp. 35-39.
[9] F.C. Liu, J. Liao, K. Kanaka: "Joining of metal to plastic using friction lap welding", *Materials and Design*, Vol. 54 (2014), pp.236-244.
[10] S. Katayama and Y. Kawahito: "Laser direct joining of metal and plastic", *Scripta Materialia*, Vol. 59, No.12 (2008), pp. 1247-1250.
[11] C. Lamberti, T. Solchenbach, P. Plapper and W. Possart: "Laser Assisted Joining of Hybrid Polyamide-Aluminum Structures", *Physics Procedia*, Vol. 56 (2014), pp. 845 – 853.
[12] L.A. Utracki: Polymer Alloys and Blends. Thermodynamics and Rheology, Hanser Pub. (NY, 1989).
[13] M. Evstatiev, S. Petrovich and B. Krasteva: "Calorimetric Studies on PET/PA6 and PA66/PA6 Drawn Blends with Various Thermal Histories", *Bulgarian Journal of Physics*, Vol. 30, No. 1-2, (2003) pp. 70-79.
[14] J.L. Murray and H.A. Wriedt: "The O-Ti (Oxygen-Titanium) System", *Bulletin of Alloy Phase Diagrams*, Vol. 8, No. 2 (1987), pp. 148-165.
[15] R.M. Silverstein, G.C. Bassler, and T.C. Morrill: Spectrometric Identification of Organic Compounds. 4th ed., John Wiley and Sons (New York, 1981).
[16] H. Domininghaus, P. Elsner, P. Eyerer, T. Hirth: Kunststoffe-Eigenschaften und Anwendungen. 8th ed., Springer Verlag (Berlin, 2012).
[17] E.J. Hartwell, R.E. Richards and H.W. Thompson: "The vibration frequency of the carbonyl linkage" *Journal of Chemistry Society*, Vol. 293 (1948), pp. 1436-1441.
[18] R.N. Jones and C. Sandorfy: Chemical Applications of Spectroscopy. Interscience Pub (NY, 1956) 247-580.

Joining of carbon-carbon composites by wetting of noble metal on carbon fiber coated with tungsten trioxide

Masafumi Fukuda, Takahisa Yamazaki

Tokyo Institute of Technology, Japan

Brazed carbon/carbon composites using reactive brazing filler was damaged by infiltration of filler metal into carbon matrix and decomposition of carbon fiber. We attempted high temperature scarf joining of carbon/carbon composites by wetting of noble metal on carbon fiber coated by means of forming the tungsten carbide in vacuum chamber. Tungsten trioxide film and carbon film were coated to silicon carbide fiber to simulate the spreading of molten gold on the nonreactive fiber. Tungsten atoms were deposited on silicon carbide fiber by tungsten filament heating in a vacuum chamber. Tungsten trioxide was formed by oxidation. Oxidation conditions were listed as: holding temperature; 500 ℃, in the air, holding time; 120 minutes. Wetting tests of the gold drop were performed on the coated fiber fixed on graphite plate. Increasing amount of deposited tungsten and raising surface temperature of fiber, solidified gold adhered to coated fiber. It suggested tungsten carbide was formed between carbon and tungsten trioxide. Molten gold can wet on tungsten carbide easier than tungsten trioxide. Tungsten were also deposited on carbon/carbon composites by tungsten filament and carbon/carbon composites was also oxidized to form tungsten trioxide. According to the result of X-ray diffraction pattern, tungsten trioxide and tungsten carbide were formed on the surface of carbon/carbon composites. According to the result of EPMA, gold wetted on the area of existing tungsten trioxide and tungsten carbide. It showed wettability of molten gold on carbon/carbon composites was improved by tungsten trioxide and tungsten carbide. Gold didn't infiltrate into carbon/carbon composites because of the barrier effect of tungsten film and poor wettability of gold on carbon/carbon composites.

The microstructure and mechanical properties of carbon fiber/aluminum composites fabricated by Friction Stir Processing

*Xiong Cao, Gaoqiang Chen, Qu Liu, Gong Zhang, Qing-yu Shi**
State key laboratory of tribology, Department of mechanical engineering, Tsinghua University
shqy@mail.tsinghua.edu.cn

Abstract
Carbon fiber is a low cost engineering material with superior mechanical properties, which has been used as an important reinforcement for the metal-matrix composites to improve their mechanical properties. There are two major challenges in fabricating carbon fiber reinforced aluminum composites: the uniform dispersion of carbon fiber in aluminum matrix and strong bonding at aluminum/carbon fiber interface, which are possible to be solved by direct Friction Stir Processing (FSP). In this study, carbon fiber reinforced Aluminum2024 bulk composites are fabricated by multiple-pass FSP. It is found that the carbon fiber is broken owing to the severe deformation during FSP, and the carbon fiber segments in random orientation with the length of 20 to 40μm are dispersed in the matrix homogeneously. The HREM images demonstrate that sufficient bonding is formed between carbon fiber and aluminum matrix, and no layer is found at the interface between carbon fiber and aluminum matrix. It is showed that the hardness of the composites is significantly higher than the base metal and FSPed matrix metal. The tensile test reveals that the ultimate tensile strength of the composites rise with the increasing of rotation speed. The ultimate tensile strength of the composite increases by 12.2% at the FSP parameter of 1800 rpm and 100mm/min in comparison with the base metal. The strengthening mechanisms are analyzed by using Orowan looping strengthening and load transfer strengthening.

Key words: Friction Stir Processing, aluminum metal-matrix composites, carbon fiber, mechanical property

Introduction
As an ideal reinforcement for light weighted composites, carbon fiber has attracted much attention due to its high specific strength, good flexibility, high elastic modulus and especially the light weight [1, 2]. Many researchers tried to reinforce aluminum with carbon fibers. There are two major challenges in fabricating carbon fiber reinforced aluminum composites: the uniform dispersion of carbon fiber in aluminum matrix and strong bonding at aluminum/carbon fiber interface. Deng et al. [3] reinforced 2024Al with carbon nanotubes using hot pressing sintering, and found the formation of Al_4C_3 during sintering which cause severe degradation of the mechanical properties. Qiurong Yang et al. [4] adopted Cu-coatedwoven carbon fibers to reinforce ZL205A Al. it effectively restrained the interface reaction, and leaded to an improvement of tensile strength. Common fabrication processes bring in large energy input which is likely to lead to the formation of Al_4C_3. Friction Stir Processing has been demonstrated to be an effective method of incorporating the reinforcing particles. In this study, the main objective is to fabricate bulk 2024Al/carbon fiber composites by direct FSP method. Furthermore, the microstructure and strength of the composites is also discussed.

Experimental
2024Al is used in this study with dimensions of 300×100×3mm. T700 long carbon fiber is chosen as the

reinforcements. A groove width 15mm and depth 1mm is machined on the middle of the plates. Carbon fiber is placed in the groove and covered by another aluminum plates without the groove. A five-pass FSP are conducted side by side at the middle of the plates and it repeats for four times. The tool shoulder is 20mm in diameter. A rectangular pyramid pin tool is used with 4.4 mm in length and the bottom of the pin is 7.5mm in diameter.

The microstructure of the composites was observed by an optical microscope and scanning electron microscopy (SEM). Transmission electron microscopy (TEM) was used to reveal the interface of the aluminum and carbon fiber in the composites. The Vickers micro hardness was conducted and the specimens for micro hardness testing were cross-sectioned perpendicular to the FSP direction. The tensile specimens were machined parallel to the FSP direction with a gauge section of 10×5×2mm. Tensile tests were conducted at room temperature at an extension rate of 0.3mm/min. SEM observation was conducted to show the tensile fracture surfaces.

Results and discussion

Microstructure

The optical microscope results show the dispersion of carbon fiber in the matrix metal (*Fig. 1*). After multi-pass FSP, long carbon fiber was broken to segments, and they were dispersed uniformly in the matrix metal. The volume fraction of carbon fiber segments in the composites was estimated to 12% approximately by using the Image-Pro software. *Fig. 2* shows that the length of carbon fiber segments range from 20μm to 40μm. It is shown that almost all the carbon segments keep cylinder shape without being broken but in different orientation (*Fig. 2*b). These indicate that the bulk composites with homogeneously dispersion of carbon fiber are successfully fabricated by FSP.

Fig. 1. Optical microscope images of the aluminum/carbon fiber composites

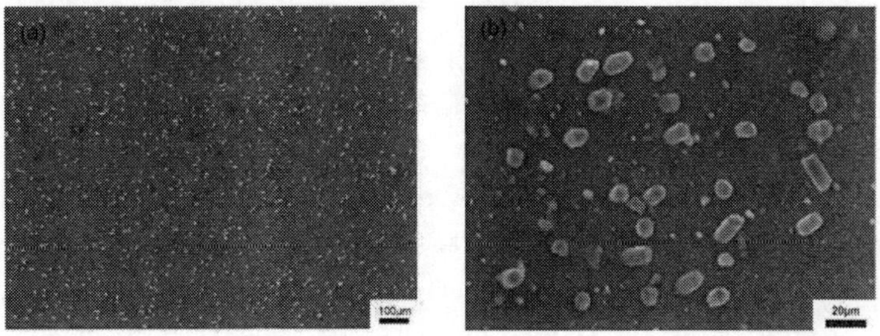

Fig. 2. SEM images showing uniform distribution and different orientation of carbon fiber segments

The TEM images are shown in *Fig. 3*. It reveals that no defects is detected at the interface and no layer is observed between the aluminum and carbon fiber. This result means sufficient bonding is formed between the matrix and the reinforcements, and no layer is found at the interface of carbon fiber and aluminum matrix.

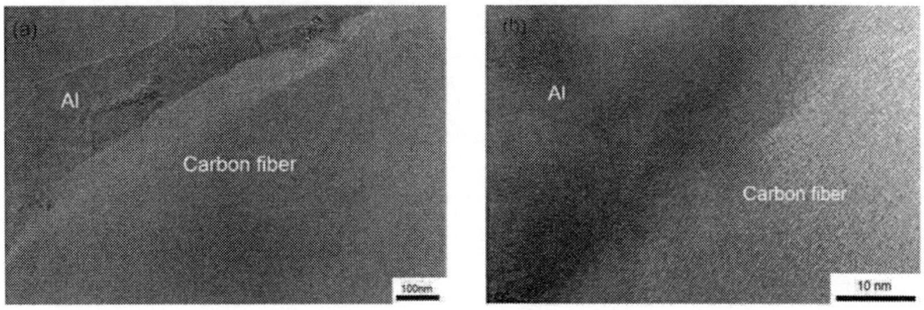

Fig. 3. The interface between aluminum and carbon fiber showing no interfacial layer: (a) TEM and (b) HRTEM

Micro hardness

Fig. 4 shows the average micro hardness of 2024Al, FSPed matrix and carbon fiber reinforced 2024Al composites. The micro hardness of 2024Al is 68.47HV, while it decreased by 17.36% after being FSPed without reinforcements. Adding carbon fiber as reinforcements leads to an apparent increase in micro hardness. In particular, the composites processed at 1000rpm and 50mm/min show the largest improvement in micro hardness, increasing to 101.87HV. It refers to no obvious change with the travelling speed ranging from 50mm/min to 100mm/min. However, increasing the rotation speed results in a decrease in micro hardness. The micro hardness data are averaged of 7 testing results with 1mm as spacing distance between adjacent testing points in the horizontal direction. The low standard deviation reveals to homogeneous distribution of carbon fiber reinforcements.

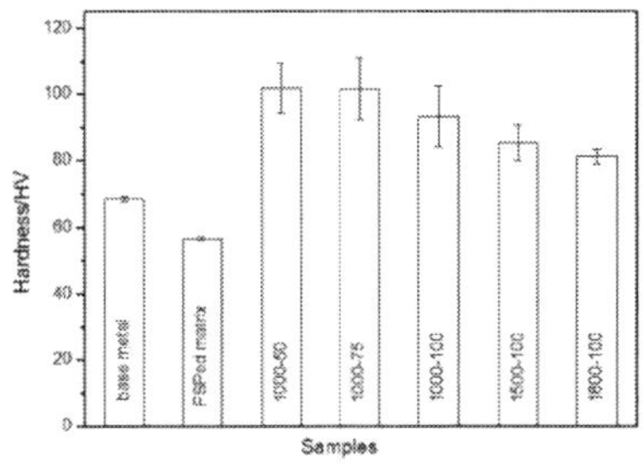

Fig. 4. Micro hardness of 2024Al, FSPed matrix and carbon fiber reinforced 2024Al composites

Tensile properties and fractography

The results of tensile tests of the composites is presented in *Table 1*. The 2024Al shows a good strength property, while the FSP decreases the ultimate tensile strength. The composites FSPed in 1000rpm exhibits similar strengths with the FSPed matrix. However, the UTS of the composites rises rapidly with the increase of rotation speed. It reaches 255.97MPa at 1800rpm and 100mm/min, which is 12.2% higher than that of the base metal. This can be the result of Orowan looping mechanism.

The fractography of 2024Al and the composite is analyzed (*Fig. 5*). The 2024Al represents deep and large dimples on the fracture surface (*Fig. 5*a), whereas the composite shows shallower and smaller dimples (*Fig. 5*b). Furthermore, uniformly dispersed carbon fiber segments are observed in the inners of the dimples. It reveals to that carbon fiber segments are pulled out during tension, which indicates that an applied force is transferred from the matrix to the reinforcement. It is one of the reason that leads to the strength enhancement of the composites FSPed at 1800rpm and 100mm/min.

Table 1: the results of tensile tests

Sample	Parameter of FSP		UTS
	Rotation speed(rpm)	Travelling speed(mm/min)	(MPa)
Base metal			228.23
FSPed matrix	1000	100	206.86
A	1000	100	203.91
B	1500	100	236.82
C	1800	100	255.97

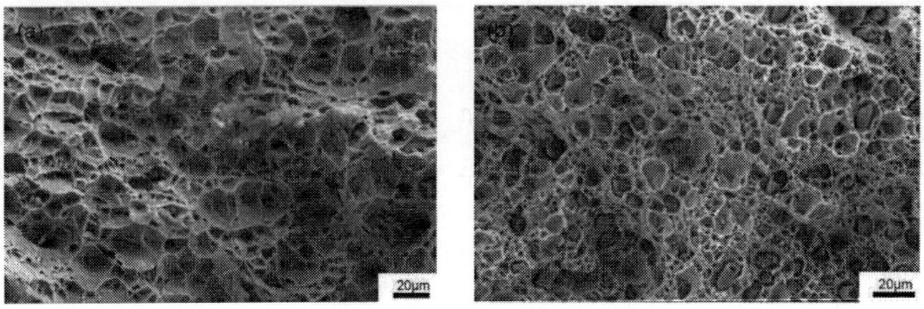

Fig. 5. SEM fractography of (a) 2024Al and (b) the composites FSPed at 1800rpm and 100mm/min

Conclusion

(1) Bulk 2024Al/carbon fiber composites were successfully fabricated by FSP.

(2) Long carbon fibers were broken into segments without breaking cylinder shape, which further dispersed uniformly in the composites with the length of 20-40μm.

(3) The TEM results revealed that sufficient bonding was formed and no layer is found at the interface between carbon fiber and aluminum matrix.

(4) The hardness of composites is obviously higher than the base metal. And the UTS of composites FSPed at 1800rpm and 100mm/min is 12.22% higher than that of 2024Al, which can be attributed to Orowan looping strengthening and load transfer strengthening.

References

[1] Frank E, et al: Carbon fibers: precursors, manufacturing, and properties[J]. Macromolecular materials and engineering, 2012, 297(6): 493-501.

[2] Chand S: Review carbon fibers for composites[J]. Journal of Materials Science, 2000, 35(6): 1303-1313.

[3] Chunfeng D, et al: Chemical stability of carbon nanotubes in the 2024Al matrix[J]. Materials Letters, 2007, 61(3): 904-907.

[4] Yang Q, et al: Fabrication and mechanical properties of Cu-coatedwoven carbon fibers reinforced aluminum alloy composite[J]. Materials & Design, 2014, 57: 442-448.

Characterization of copper to alumina dissimilar joint bonded by continuous drive friction welding

Peng Li, Honggang Dong

School of Materials Science and Engineering, Dalian University of Technology, Dalian 116024, PR China, lipeng2016@dlut.edu.cn, donghg@dlut.edu.cn, +86-411-84706283

Jinglong Li, Chengzongji

Shaanxi Key Laboratory of Friction Welding Technologies, Northwestern Polytechnical University, Xi'an 710072, PR China, lijinglg@nwpu.edu.cn, 631514707@qq.com, +86-29-88460673

Abstract

Dissimilar materials joining of copper and alumina rods was achieved by continuous drive friction welding using 2.5mm thick AA1100 aluminum as interlayer. The effect of friction pressure on the tensile strength of joints were evaluated with universal testing machine, while the interface microstructure and fracture morphologies were analyzed with optical microscope and scanning electron microscope. The microstructure examination shows that microcracks formed at the interface due to thermo-mechanical coupling effects in welding. The tensile test results reveal that as increasing friction time and friction pressure, the tensile strength increased until reaching a peak value of 35MPa, and then decreased. The optimized parameters were determined as friction pressure of 12MPa, friction time of 12s, upsetting pressure of 36MPa and upsetting time of 5s. All samples failed on the aluminum/alumina interface, leaving a pit structure on alumina base. The application of pure aluminum interlayer was essential to obtain sound copper/alumina joints since it could diminish stress concentration during welding.

1 Introduction

Alumina ceramics, which possess the favorable properties of high temperature strength, corrosion resistance, wear resistance, high dielectrics strength, low dielectric loss and stable electrical performance, have been widely applied in electrical industries. In addition, application of copper in electrical industries is also quite prominent due to its excellent conductivities of heat and electric. The hybrid joint between dissimilar alumina and copper combines their advantages and find promising application in the field of vacuum interrupters and vacuum electronic packaging. However, challenges still exist in joining the two parent materials due to large differences in thermal-physical properties, such as linear thermal expansion coefficient and melting point, which lead to high residual stress and decrease the joint strength.

Several welding methods including fusion welding, brazing, and solid-state welding have been attempted to obtain ceramic-metal coupled joints [1-4]. Among these methods, friction welding has gained great attentions due to its solid-state nature and short welding cycle, which can greatly decrease the thermal mismatch between the parent materials. There has been considerable effort directed to friction welding of ceramics and metal couples. Essa et al[5] investigated the friction joining of ceramics to metals and found that it was not possible to friction weld copper to alumina. The reason for this could be that copper could not directly wet the alumina, while the aluminum can wet the alumina well due to the admirable chemical compatibility. Noh et al[6] investigated the friction welding of alumina to mild steel by using aluminum interlayer, and found that it the joint strength is highly dependent on the wettability of aluminum to alumina. The fracture mainly occurred at the aluminum/alumina interface due to weak joining resulted from differences in melting points and higher surface energy compared to aluminum/steel interface. Seli et al[7] also found that aluminum interlayer is essential for friction welding of mild steel to alumina and the fracture occurred between aluminum and alumina. Accordingly, aluminum was chosen as the interlayer in this study.

The aim of this investigation is to fabricate a credible combination between T2 copper and 95% alumina by using continuous drive friction welding. The effect of friction pressure on the interfacial microstructure and tensile strength of the joints were thus investigated. The fracture behavior and welding mechanism of the joints were also discussed.

2 Materials and experimental procedures

Rods of commercially available T2 copper and 95% alumina in the diameter of ϕ14mm were used as base metals. Prior to welding, the faying surfaces were polished with emery paper and ultrasonically cleaned in ethanol and fried in air. Friction welding was carried out using a continuous drive friction welding machine (C40, Changchun CNC Machine Tool Co. Ltd., China), attached with displacement acquisition sensor assembly. The constant welding parameters were set as follows: rotation speed of 1900rpm, friction time of 12s, upset pressure of 36MPa, upset time of 5s, while the friction pressure was selected as the main variable to study its effect on the joint evolution. To this end, the friction pressure varied from 6 to 24MPa in a step of 6MPa. Small level of friction pressure was used in order to diminish the cracking of alumina during the friction welding operations.

Before welding, an AA1100 aluminum rod having the same diameter of ϕ14mm was firstly boned to the copper base, and then it was machined to 2.5mm thickness acting as a buffer interlayer. Thereafter, the conventional friction welding process was conducted. After welding, the axial cross-section of the joint was obtained by using diamond cutting machine and mounted for polishing. The microstructure of the joints was performed by scanning electron microscope (SEM, JSM–6390A). The joint strength was evaluated on a tensile testing machine (Instron 3382) using a crosshead speed of 0.5mm/min at room temperature.

3 Results and Discussion

Fig. 1 shows the appearances of the as-welded joints at different friction pressures. It can be seen that the joint welded at lower pressure (6MPa) has hardly any flash. As the pressure increases, the flash grows bigger in size and better in appearance. However, the flash grows too bigger when the friction pressure reaches 24MPa, thus left less aluminum interlayer in the joint. It is well known that the friction pressure and rotation speed are the main parameters in the welding process due to the heat generation is closely related to these parameters according to the local heat generation equation:

$$dQ = 2\pi p \mu v r dr$$

Where, Q is the heat power, p is the unit pressure, μ is the friction coefficient, v is the linear velocity, r is the local radius. In this study, we do not take the constant velocity (i.e. rotation speed) into consideration. Thus the friction pressure parameter plays the leading role in the joint formation process. It determines the friction torque, heat input intensity, temperature distribution, thickness of the plasticized layer and the axial shortening rate. When the lower pressure of 6MPa was used, the heat generation is relatively low thus closed and plump flash could not form under this condition, leading to the hardly any flash, as shown in *Fig. 1a*. Generally speaking, this is a typical insufficient heating process and will not form a sound joint. The heat generation rapidly increases as the pressure increases thus leading to a higher temperature in the interface, which could be responsible for the larger flash formation due to the lower plastic deformation resistance. However, when the friction pressure reaches 24MPa, the flash presents an oversized and asymmetric feature (as shown in *Fig. 1d*), which is unfavorable for the joint performance.

Fig. 2 shows the corresponding evolvement of the interfacial microstructure of the joints welded at different friction pressures of 12MPa, 18MPa and 24MPa for friction time of 12s. The joint welded at 6MPa was broken during the cross-section metallographic specimen machining process. The thickness of the residual aluminum interlayer rapidly decreases from 0.7mm to 0.2mm, yet the flash size grows larger as the friction pressure increases. Reasons for this could be attributed to that more plastics were formed and extruded from the weld interface, just as the analysis done before. That is, the heat generation grows as the friction pressure increases and leads to more plasticized aluminum interlayer which could be easily deformed to form a collar of flash. The aluminum interlayer left in the joint welded at 24MPa is just 0.2mm, which could not absorb the residual stress cause by the thermo-mechanical coupling effects, cracks was thus formed in the joint interface, as shown in *Fig. 2c*.

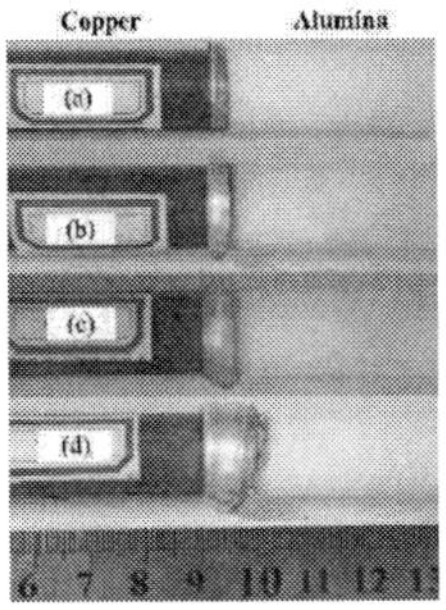

Fig. 1 Appearances of the joints welded at different friction pressures: (a)6MPa, (b)12MPa, (c)18MPa and (d)24MPa.

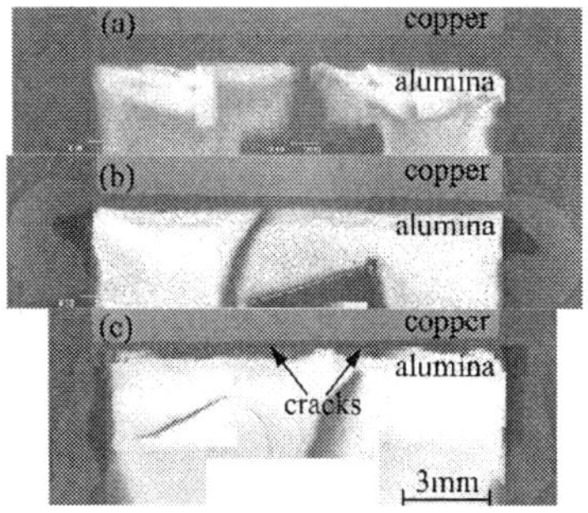

Fig. 2 Interface structures joint welded at different friction pressures: (a)12MPa, (b)18MPa and (c)24MPa.

Fig. 3 depicts the evolution of tensile strength of the joints as a function of friction pressure. The strength rapidly increases with friction pressure till 12MPa, under which the maximum strength of 35MPa was achieved, and then decreases. *Fig. 4* shows the fracture morphologies of the tensile samples. Both of them show the pits and protrusions structure, which indicates that the fracture initials in alumina parent near the interface, finally the joint failed and presents the features as shown in *Fig. 4(a-c)*. The fracture morphologies presented in *Fig. 4d* shows the different feature, i.e., the fracture mainly happens in the joint interface, leaving the broken aluminum interlayer on alumina side. Combined with the interface structure analyzed before, it indicates that thickness of the residual aluminum interlayer plays an important role in determining the joint strength. There exists an optimized value of thickness for

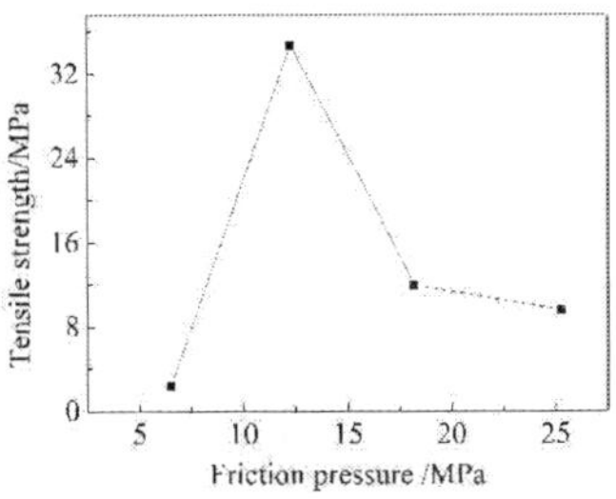

Fig. 3 Tensile strength of the joints as a function of friction pressure

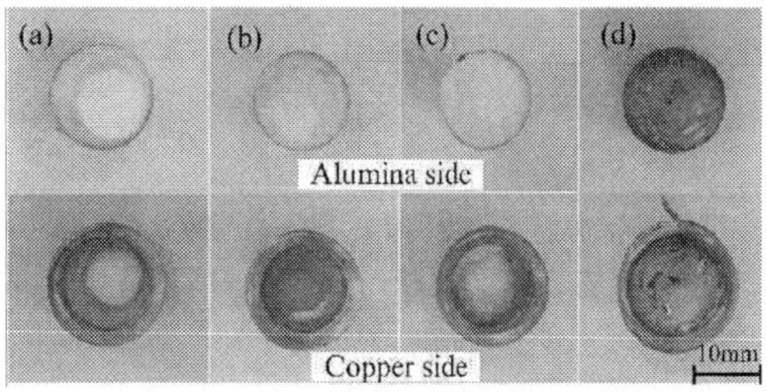

Fig. 4 Fracture morphologies of the joints welded at different friction pressures: (a)6MPa, (b)12MPa, (c)18MPa and (d)24MPa.

the sound alumina/copper joint formation, the corresponding condition is 12MPa friction pressure.

According to the results described above, it was found that the residual aluminum interlayer plays an important role in the joint properties. The joint strength initially increases with increasing of the residual aluminum interlayer thicknesses. The maximum tensile strength was obtained for the joint with 0.7mm thickness aluminum interlayer, while further increase of the thickness decreases the tensile strength. Aluminum actually acts as a buffer interlayer, which could absorb the residual stress generated during the welding process. However, the release effect decreased when the residual thickness was either larger or smaller. That's to say, there exists an optimized aluminum interlayer thickness produced at friction pressure of 12MPa, under which the maximum tensile strength of 35MPa was obtained. However, excessive or lower pressure will induce over or insufficient heating of the joining interface, joints with improper properties were thus produced due to thermal shock or incomplete wettability effect.

4 Conclusions

By using AA1100 aluminum as the buffer interlayer, continuous drive friction welding of T2 copper to 95% alumina has been conducted. The effect of friction pressure on the mechanical and metallurgical properties of the joints were discussed. The results show that the copper/alumina dissimilar joint can be produced by using continuous drive friction welding process. Joint strength as high as 35MPa was achieved at condition of 12MPa friction pressure, beyond which joints with poor strength appear due to thermal shock and incomplete wettability coupled effects. The application of pure aluminum interlayer was essential to obtain sound copper/alumina joints since it could diminish stress concentration during welding.

Acknowledgments

This work was financially supported by the Fundamental Research Funds for the Central Universities (No. DUT16RC(3)009), and State Key Laboratory of Solidification Processing in NWPU (SKLSP201614).

References

[1] K. Suganuma, *et al*: "Joining of ceramics and metals", *Annual Review of Materials Science*, Vol. 18, No. 1(1988), pp. 47-73.

[2] M.B. Uday, *et al*: "Joint properties of friction welded 6061 aluminum alloy/YSZ–alumina composite at low rotational speed", *Materials & Design*, Vol. 59(2014), pp. 76-83.

[3] M.B. Uday, *et al*: "Effect of welding speed on mechanical strength of friction welded joint of YSZ–alumina composite and 6061 aluminum alloy", *Materials Science and Engineering A*, Vol. 528, No. 13(2011), pp. 4753-4760.

[4] M.N.A. Fauzi, *et al*: "Microstructure and mechanical properties of alumina-6061 aluminum alloy joined by friction welding" , *Materials & Design*, Vol. 31, No. 2 (2010), pp. 670-676.

[5] A.A. Essa, *et al*: "The friction joining of ceramics to metals", *Journal of Materials Processing Technology*, Vol. 26, No. 2(1991), pp. 133-140.

[6] M.Z. Noh, *et al*: "Alumina–mild steel friction welded at lower rotational speed", *Journal of materials processing technology*, Vol. 204, No. 1(2008), pp. 279-283.

[7] H. Seli, *et al*: "Characterization and thermal modelling of friction welded alumina–mild steel with the use of Al 1100 interlayer", *Journal of Alloys and Compounds*, Vol. 506, No. 2(2010), pp. 703-709.

On the solid state bonding behavior between the butting surfaces during friction stir welding: A numerical simulation study

Gaoqiang Chen
State Key Laboratory of Tribology, Tsinghua University, Beijing 100084, PRC
Key Laboratory for Advanced Materials Processing Technology, Tsinghua University, Beijing 100084, PRC
Oak Ridge National Laboratory, Oak Ridge, TN 37831, USA
Email: gaoqiang.chen@gmail.com

Zhili Feng
Materials Science and Technology Division,Oak Ridge National Laboratory, Oak Ridge, TN 37831, USA
Email: fengz@ornl.gov

Han Li, Qu Liu, Shuai Zhang, Qingyu Shi (Corresponding author)
State Key Laboratory of Tribology, Tsinghua University, Beijing 100084, PRC
Key Laboratory for Advanced Materials Processing Technology, Tsinghua University, Beijing 100084, PRC
Email: shqy@tsinghua.edu.cn

Abstract

Many microstructural and defective features in friction stir welds are considered to be closely related to the solid state bonding between the original butting surfaces. These features include Zigzag line, kissing bonds, lack of penetration, etc. However, little has been known on the solid state bonding phenomenon that occurs in friction stir welding (FSW) in the previous research. In this paper, we present an investigation on the solid state bonding between the original butting surfaces in FSW. The thermal-mechanical processes, such as heat generation, heat transfer and material plastic flow were numerical simulated by using computational fluid dynamics. The predicted geometry of the deformation zone and Zigzag line was used to valid the numerical simulation. By combining the numerical simulation of the thermal-mechanical processes in FSW with an analytical approach for the solid state bonding, the bonded fraction between the original butting surfaces in FSW was analyzed. It was demonstrated that the material plastic flow plays a critical role in determining the temperature history and pressure history and thus the bonded fraction at the original butting surfaces. It was shown that the thermal plastic flow condition around the pin is favorable to produce sound solid state bonds between metallic surfaces, while bonding difficulties were found to exist in the vicinity of the backing plate. The bonding difficulties in the vicinity of the backing plate were induced by the insufficient interfacial pressure, which is further attributed the inadequate bonding pressure caused by the material flow difficulty.

Introduction

In friction stir welding (FSW), many microstructural and defective features in the welds are considered to be closely related to the solid state bonding (SSB) between the original butting surfaces. These features, such as Zigzag line[1, 2], kissing bonds[3], lack of penetration[4], could have detrimental effects on the mechanical strength and fatigue life [5]. As such, the SSB behavior between the butting surfaces is important issue in the fundamental understanding of FSW. In this paper, we present a computational investigation on the thermal-mechanical processes, such as heat transfer and material flow, and the solid state bonding between the origin butting surfaces in FSW.

Approaches

This investigation is based on our recent developed approaches on studying the thermal-mechanical processes and material bonding behaviors in FSW. The thermal-mechanical processes is analyzed by our numerical simulation based on computational fluid dynamics [6-8]. Based on the simulated thermal-mechanical condition, the interfacial material bonding behaviors are investigated via a differential equation [9], which links the bonded fraction to the bonding conditions.

In the welding experiments, FSW butt joining of two AA2024-T4 sheets was conducted. The dimensions of each workpiece were 145mm×55mm×3mm (length × width × thickness). The welding tool was made by H13 steel. The tool shoulder was 13mm in diameter, and length of the conical pin was 2.4mm. The radius of pin was 2mm at the root and 1.75mm at the tip. A specimen was cut after the welding for examination of the cross-sectional macrostructure. The specimen was grounded, polished and etched with Keller's solution (95ml water, 1.5ml hydrochloric acid, 2.5ml nitric acid, 1ml hydrofluoric acid) for 1min. After that, the macrostructure of specimen was observed by optical microscope.

Thermal-mechanical condition by numerical simulation

The predicted temperature distributions at different welding parameters are shown in Fig 1(a). It could be found that the temperature is relatively high in the vicinity of the tool due to the localized heat generation [6, 10]. A bowl-shaped localized high temperature volume, in which the maximum temperature exceeds 400°C, exists in the vicinity of the tool/workpiece interface. The high temperature volume increases with the tool rotation rate. Fig 1(b) shows the predicted strain rate in the workpiece. It could be found that the size of the deformation zone increases significantly with the tool rotation rate. The size variation of the deformation zone is also supported by the macrostructures in Fig 1(c). In the space between the pin bottom and the backing plate, there is hardly plastic deformation when the welding condition is 500rpm, 40mm/min, while the plastic deformation become more and more significant with the increases of the tool rotation rate.

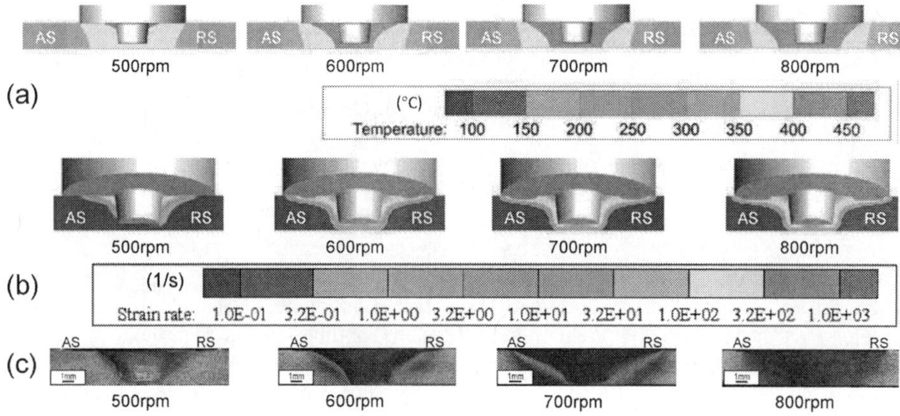

Fig. 1 Distribution of temperature distribution and material flow velocity at different welding parameters. The welding speed is 40mm/min. (a) Temperature; (b) Strain rate; (c) Cross-sectional weld macrostructure.

Geometrical evolution of the original butting interface

As shown in Fig. 2 (a)~(d), the flow paths under the tool are different owing to the different material flow and deformation behaviors at different welding parameters. Due to the differences in the flow paths, the final deposit positions of material are different. Based on the material flow paths, the geometry evolution of the butting interface is reconstructed (Fig 3a). The deposit position of the origin butting interface in the welds is complex. Fig 3(b) shows the distribution of the origin butting interface in the weld. The predicted distribution of the butting interface agrees reasonably well with an experimental result from literature (Fig 3c). By the geometric characterization, the distribution could be divided into three zones, including (I) shoulder zone, (II) pin side zone, and (III) backing plate zone. Furthermore, it could be found that the orientation of the butting surface is alternated significantly when it flows around the welding tool.

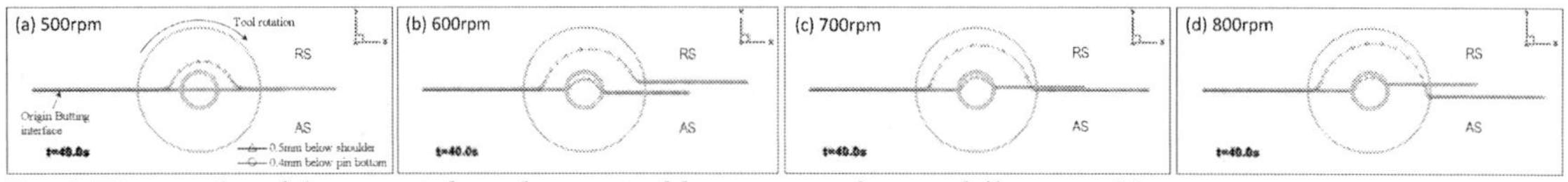

Fig. 2 Flow paths of the materials at the original butting interface at different welding parameters.

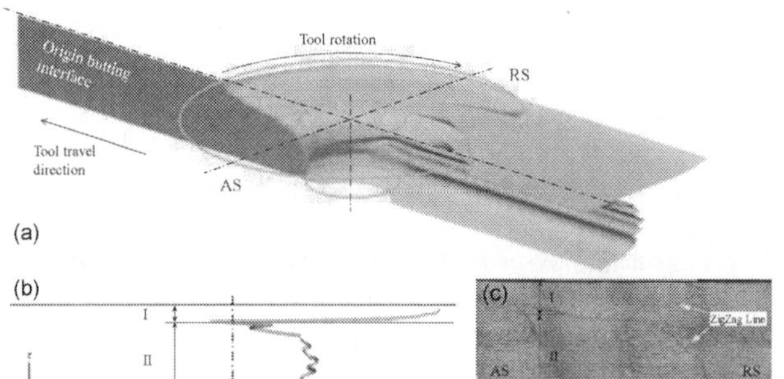

Fig. 3 Predicted geometry of the original butting surfaces. (a) Predicted original butting surfaces during welding; (b) Predicted original butting surfaces in the weld; (c) Original Butting Surfaces in the experiments[11].

Thermal-mechanical condition at the butting interface

Fig 4 shows the thermal-mechanical condition at the butting interface in the welding process. It could be found from Fig 4 (a)~(d) that, although the maximum temperature is similar at positions near the shoulder and near the pin bottom, the stay time at high temperature near the pin bottom is much longer than that near the shoulder. This is because the material velocity is smaller near the pin bottom. Fig 4 (e)~(h) shows the interfacial pressure between the butting surfaces. The interfacial pressure is determined by the axial force and orientation of the butting interface. Before the material enters the deformation zone, the interfacial pressure is 0 as the interface is parallel to the axial load direction. As the orientation variation is caused the plastic deformation, the interfacial pressure varies complexly beneath the tool shoulder. Generally, the interfacial pressure is higher when severe plastic deformation occurs to the material. The peak interface pressure is high near the shoulder, while the peak interface pressure near the pin bottom is lower especially at lower tool rotation rate. From the above discussion, it could be seen that the material plastic flow plays a critical role in determining the thermal-mechanical history in FSW process.

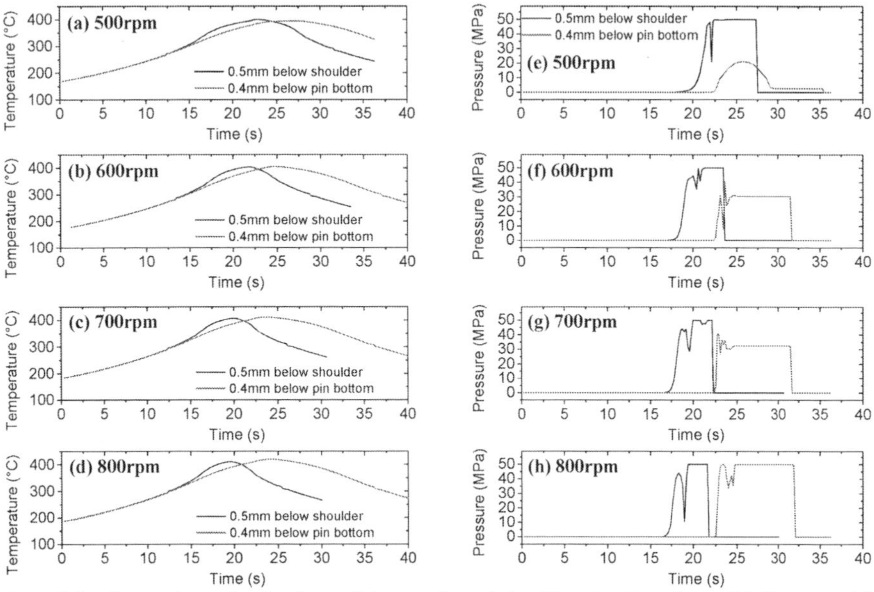

Fig. 4 Evolution of the thermal-mechanical condition at the original butting interface. (a) Pressure history; (b) Temperature history.

Bonded fraction at the original butting interface

Fig 5 shows the predicted bonded fraction at the original butting interface in the weld. It can be found that the thermal plastic flow condition around the pin is favorable to produce sound solid state bonds between metallic surfaces, while bonding difficulties were found to exist in the vicinity of the backing plate. The bonding difficulties in the vicinity of the backing plate were induced by the insufficient interfacial pressure (see Fig 4e~g), which is further attributed the inadequate bonding pressure caused by the material flow difficulty.

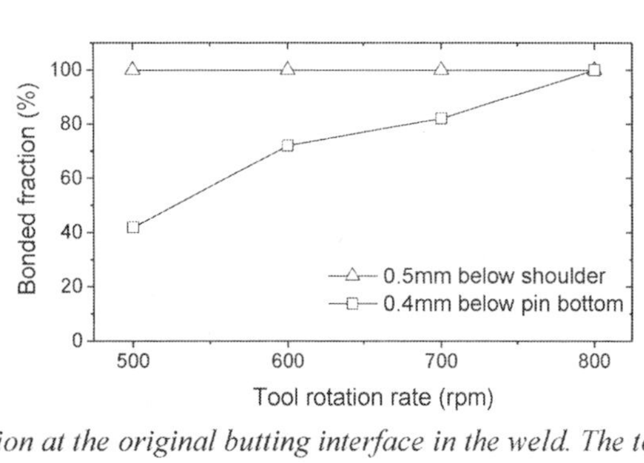

Fig. 5 Predicted bonded fraction at the original butting interface in the weld. The tool travel speed is 40mm/min.

Conclusions

By combining the numerical simulation of the thermal-mechanical processes in FSW with an analytical approach for the solid state bonding, the material bonding behaviors between the initial butting surfaces in FSW was analyzed. The following conclusions could be drawn.

(1) The material plastic flow plays a critical role in determining the temperature history and pressure history experienced by the original butting interface and thus the bonded fraction at the original butting surfaces.

(2) The thermal plastic flow condition around the pin is favorable to produce sound solid state bonds between metallic surfaces, while bonding difficulties were found to exist in the vicinity of the backing plate.

(3) The bonding difficulties in the vicinity of the backing plate were induced by the insufficient interfacial pressure, which is further attributed the inadequate bonding pressure caused by the material flow difficulty.

Acknowledgments

The work was supported by the National Natural Science Foundation of China (grant no. 51375259) and the National Science and Technology Major Project of the Ministry of Science and Technology of China (no. 2012ZX04012-011). Besides, Gaoqiang Chen was supported by the China Scholarship Council (File No. 20130620105) for 2-year study at Oak Ridge National Laboratory (Oak Ridge, TN, USA).

References

[1] Sato YS, Yamashita F, Sugiura Y, Park SHC, Kokawa H. FIB-assisted TEM study of an oxide array in the root of a friction stir welded aluminium alloy. Scripta materialia 2004;50:365-9.

[2] Chen HB, Wang JF, Zhen GD, Chen SB, Lin T. Effects of initial oxide on microstructural and mechanical properties of friction stir welded AA2219 alloy. Materials & Design 2015;86:49-54.

[3] OOSTERKAMP A, DJAPIC OOSTERKAMP L, NORDEIDE A. 'Kissing bond' phenomena in solid-state welds of aluminum alloys. Miami, FL, ETATS-UNIS: American Welding Society; 2004.

[4] Leonard A, Lockyer S. Flaws in friction stir welds. 4th International symposium on friction stir welding, Park City, USA2003.

[5] Zhou C, Yang X, Luan G. Effect of kissing bond on fatigue behavior of friction stir welds on Al 5083 alloy. Journal of Materials Science 2006;41:2771-7.

[6] Chen G-q, Shi Q-y, Li Y-j, Sun Y-j, Dai Q-l, Jia J-y, et al. Computational fluid dynamics studies on heat generation during friction stir welding of aluminum alloy. Computational Materials Science 2013;79:540-6.

[7] Chen G, Shi Q, Fujiya Y, Horie T. Simulation of Metal Flow During Friction Stir Welding Based on the Model of Interactive Force Between Tool and Material. Journal of materials engineering and performance 2014;23:1321-8.

[8] Chen G, Feng Z, Zhu Y, Shi Q. An alternative frictional boundary condition for computational fluid dynamics simulation of friction stir welding. Journal of Materials Engineering and Performance 2016;To be published.

[9] Chen G, Feng Z, Liu L, Shi Q. Analytical modeling of rapid solid state bonding between metal surfaces.To be published.

[10] Su H, Wu CS, Bachmann M, Rethmeier M. Numerical modeling for the effect of pin profiles on thermal and material flow characteristics in friction stir welding. Materials & Design 2015;77:114-25.

[11] Dai QL, Wang XY, Hou ZG, Wu JJ, Shi QY. Microcavities accompanying a zigzag line in friction stir welded A6082-T6 alloy joint. Science and Technology of Welding and Joining 2015;20:68-74.

A Sensitivity Study on Mild Steel Friction Stir Welding Tool Penetration and Depth

Duncan Camilleri*, Daniel Micallef and Clarissa Saliba
Department of Mechanical Engineering
Faculty of Engineering
University of Malta, Msida MSD 2080
Malta.
duncan.camilleri@um.edu.mt, 00356 2340 2527

Abstract

In friction stir welding the material at the joint undergoes a thermo- visco-plastic, mass transportation and shearing phenomena. The amount of material subject to friction, stirring and visco-plastic shearing is dependent on the tool configuration and dimensions, which in turn controls the heat generated at the weld. The tool penetration depth plays an important role in this aspect. Sufficient penetration ensures a sound weld throughout the entire thickness of the workpiece but at the expense of more heat generation and thermal strains. This possibly results in more residual stresses and distortion. On the other hand if the tool depth is too shallow, there will not be sufficient contact between the tool shoulder and workpiece surface, resulting in non-fully penetrated welds. This study investigates the influence of different tool penetration depths on the thermo-mechanically affected zone, thermal strains, residual stresses and distortion developed in friction stir welding of DH36 mild steel. A local-global numerical approach is adopted. Firstly, a local visco-plastic fluid mechanics model is employed to establish the mass transportation, shearing effect and heat generated at the weld. The predicted thermal strains are then applied to a global finite element structural model to identify the welding residual stresses and distortion. The optimal tool penetration depth for a typical mild steel friction stir welded plate, in terms of residual stresses and distortion, is numerically identified.

Introduction

Friction stir welding involves a multi-physics phenomena where the material at the joint undergoes a visco-plastic shearing and stirring to ultimately join assemblies [1]. The process has been applied to a number of advanced materials and most commonly to aluminium alloys. With the advancement in materials and tool technology friction stir welding has also been used to weld mild steel [2], [3]. Unlike conventional welding the material is not liquefied but plasticized to temperatures in the region of 900°C to 1200°C [4]. Nonetheless, the localized heating due to the friction and shearing action developed during the process will inevitably give rise to thermal gradients and consequently thermal strains, residual stresses and possibly distortion [5].

Numerical modelling provides an efficient and cost effective tool that is able to predict the thermal gradients, inherent residual stresses and distortion developed due to welding [7]. A local-global numerical modelling approach can be adopted to first predict the local material shearing, flow, stirring and heat generated due to FSW followed by a structural elasto-plastic analysis that identifies the welding contraction forces and distortion developed. The local material shearing is predicted through a visco-plastic computational fluid dynamics model [4]. Reduced solutions methods based on elasto-platic algorithms are applied to identify the welding contraction forces and angular distortions while numerical FEA are employed to identify the out-of-plane distortion for different plate dimensions [5], [6].

The tool penetration depth determines the quality of the weld. The tool dimensions, profile and penetration depth essentially controls the amount of mass transportation and stirring occurring during FSW. Sufficient penetration ensures a sound weld throughout the entire thickness of the work-piece such that all material residing at the weld line is stirred. If on the other hand the tool penetration depth is too shallow, there will not be sufficient contact between the tool shoulder and work-piece surface, resulting in welds with a surface groove and non-fully penetrated welds. If the penetration depth is too deep, the tool shoulder plunges into the work-piece, creating excessive flash and a concave weld is produced. Over plunging the tool will result in more contact area with the material.

The work done by Toumpis et. al [7] has shown that in cases when a non-fully penetrated weld is achieved the weld properties (in particular fatigue properties) can still be higher than that of the parent material. However there is a fine line on how much non-penetration is acceptable and efforts have to be made to ensure full penetration. In practice this is not always possible due to changes in the original plate thickness and machine set-up. The influence of tool penetration on the generated thermal strains, welding contraction forces, residual stresses and distortion is thus investigated through a series of 4 numerical models simulating constant welding parameters of 200RPM and 175mm/min, subject to similar heat loss boundary conditions, with the tool penetration set to 4 different distances as shown in Figure 1 and Table 1. Two metre long by 0.4m wide 6mm thick DH36 mild steel assemblies are considered.

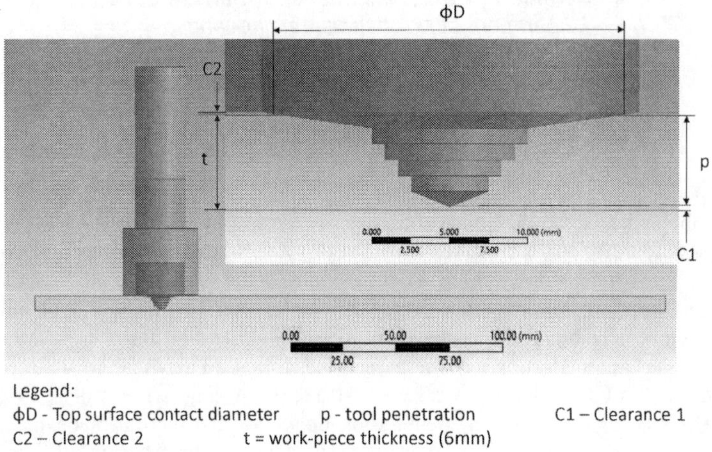

Figure 1: Tool depth and penetration parameters under investigation

Model	p (mm)	φD(mm)	C1(mm)	C2 (mm)
TooldepthLow	5.6	19.16	0.36	0.4
TooldepthNorm1	5.7	20.60	0.26	0.3
TooldepthNorm2	5.8	21.95	0.16	0.2
TooldepthHigh	5.9	23.46	0.06	0.1

Table 1: Tool depth and penetration parameters under investigation

Local thermo-mechanical affected zone (TMAZ) and thermal gradients.

A local visco-plastic fluid mechanics model was used to identify the thermal gradients and thermo-mechanical affected zone. A fixed Eulerian reference frame is adopted such that the frame is attached to the solid tool moving at the predefined rotational and transverse speed. The Navier-Stokes equations are solved in steady state formulation for a non-Newtonian fluid representing the quasi-static thermal and fluid material flow after tool plunge, dwell time and static heat flow. The plate assumed flow stress material constitutive laws based on the Norton-Hoff visco-plastic model [8] and using the material properties identified in [9]. Thermophysical non-linear temperature dependent conductivity and specific established through previous studies on DH36 mild steel [6] were applied. Constant heat loss boundary conditions were applied to all models that were identified through correlation with experimentally measured thermocouple results.

Figure 2 shows the developed temperatures and the TMAZ zone dimensions predicted via the local models for different pin-depth settings. Increasing the tool depth results in more tool work-piece contact area thereby developing more heat by shearing and friction. This will in turn result in higher temperatures as exemplified both in the peak temperatures measured at the TMAZ and further away. For instance at low tool penetration the maximum temperature predicted was 905 as opposed to the 967 in 5.9mm tool penetration. Increasing the tool depth will also extend the TMAZ which ultimately increases the transverse contraction force.

Global welding distortion and residual stresses

Analytical solutions based on the inherent thermal strain developed during FSW are applied to give an indication of the longitudinal contraction force and angular distortion developed due to FSW. These are based on the elasto-plastic analytical approaches identified in [6]. The analytically predicted longitudinal contraction forces and angular distortion for different tool depths is given in Table 2.

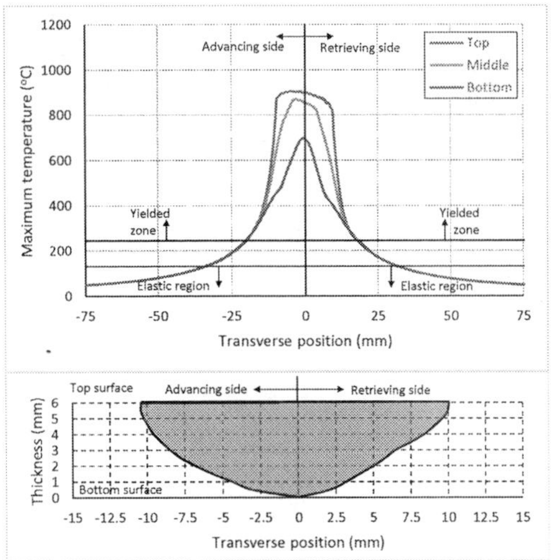

a) TooldepthLow (5.6mm)

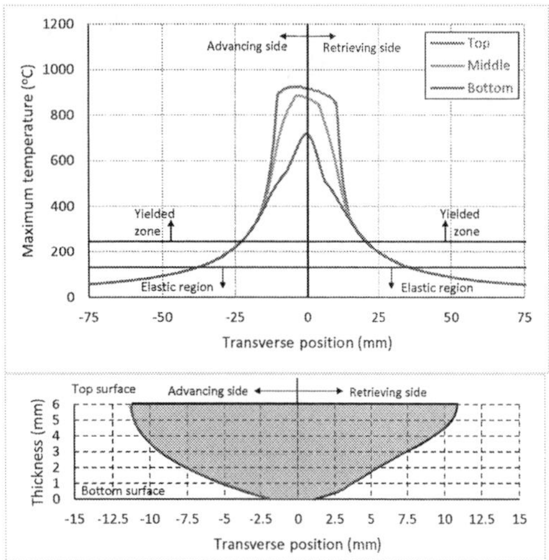

b) TooldepthNorm1 (5.7mm)

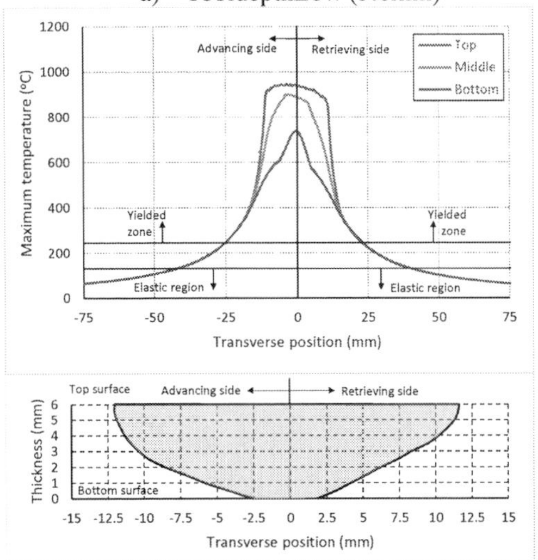

c) TooldepthNorm2 (5.8mm)

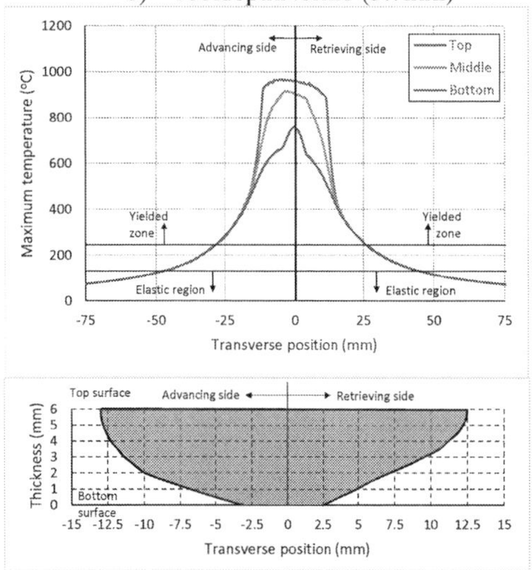

d) TooldepthHigh (5.9mm)

Figure 2: Predicted maximum temperatures and TMAZ subject to different tool depth penetrations

Tool penetration	**Longitudinal contraction force (kN)**	**Angular distortion (°)**
TooldepthLow (5.6mm)	107.24	0.1759
TooldepthNorm1 (5.7mm)	121.84	0.1789
TooldepthNorm2 (5.8 mm)	135.10	0.1833
TooldepthHigh (5.9mm)	152.16	0.1857

Table 2: Analytical predicted longitudinal contraction forces and angular distortion

If the longitudinal force acts uniformly through thickness and about the neutral axis of the plate no out-of-plane angular distortion will develop, however the combination of an increase in TMAZ dimension together with offsetting this force action away from neutral axis, a higher angular distortion is predicted when the tool depth is increased.

Figure 3 and Table 3 show the numerically predicted out-of-plane distortion for constant welding parameters and boundary conditions but different tool depths. A hybrid modelling strategy was adopted [6]. The out-of-plane distortion increases with higher tool depth penetration. This follows the same conclusions attained from the analytical solutions.

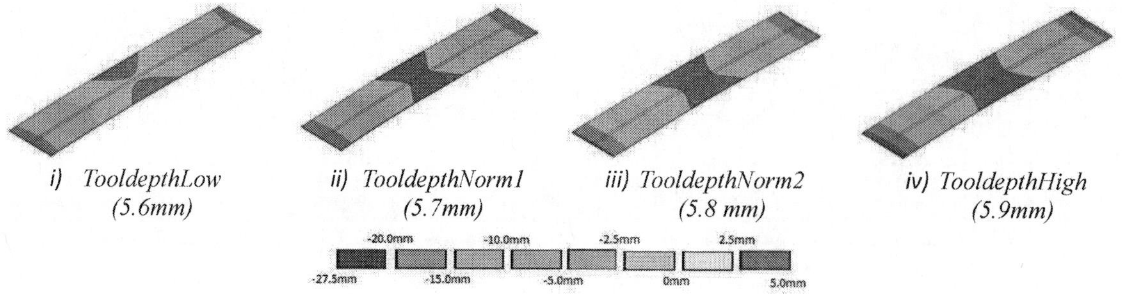

Figure 3: Predicted out-of-plane distortion for different tool depths

Plate dimension	Tool Depth	Out-of-plane distortion				
		Minimum (mm)	*Maximum (mm)*	*Range (mm)*	*Average angular distortion half angle theta (°)*	*Average longitudinal curvature (m^{-1})*
Plate 2m x 0.4m	Low 5.6mm	-18.28	3.54	21.82	0.233	0.04187
	Norm 1 5.7mm	-19.97	3.93	23.92	0.227	0.04606
	Norm 2 5.8mm	-20.88	4.19	25.07	0.205	0.04854
	High 5.9mm	-22.34	4.61	26.95	0.181	0.05244

Table 3: Predicted out-of-plane deformation for different tool depths

Figure 4 shows the numerically predicted residual stresses for constant welding parameters and boundary conditions but subject to different tool depths. As expected the longitudinal residual stresses are more pronounced when the tool depth is increased as a result of higher heat input. Similarly the transverse residual stresses extend over a larger distance but are less in magnitude for higher tool depths. This is an effect of the smaller TMAZ and more localized heating for reduced tool depths.

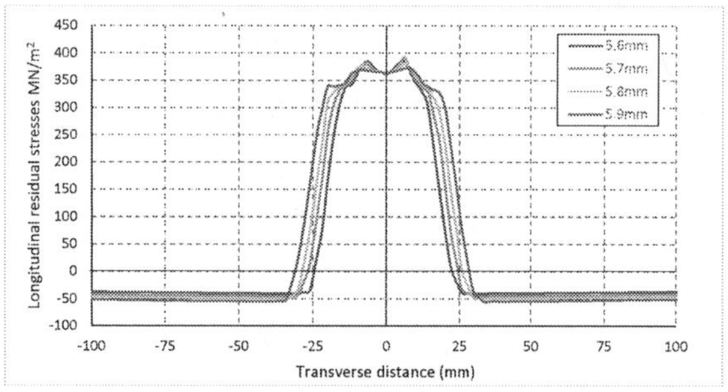

i) Longitudinal residual stresses

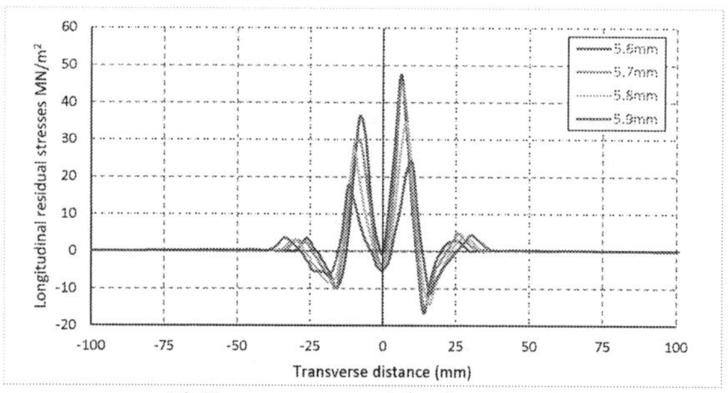

ii) Transverse residual stresses

Figure 4: Predicted longitudinal and transverse residual stresses for slow welding subject to different tool depths.

Conclusions

A numerical and analytical sensitivity analysis on the effect of tool penetration depth in FSW of mild steel assemblies was conducted. The results have shown that increasing the penetration depth gives rise to higher temperatures and consequently more thermal strain, residual stresses and distortion. On the other hand and in the case tool depths set at 5.6mm a non-fully penetrated weld was achieved. A compromise therefore exists on the optimal tool penetration that ensures sufficient material stirring but at the same time reduces the amount of heat developed. In this study it was found that the optimal tool penetration should be set at 5.7mm.

Acknowledgments

This study was funded by the European Commission in Call FP7-SST-2012-RTD-1 under the project titled High Integrity Low Distortion Assembly – HILDA.

References

[1] W.M. Thomas, E.D. Nicholas, J.C. Needham, M.G. Murch, P. Temple-Smith, C.J. Dawes, 'Friction Welding,' International patent application no PCT/GB92/02203and GB patent application n.9125978.8, 1991

[2] A. Toumpis, A. Galloway, S.R. Cater, D.Micallef, D.Camilleri, N.Poletz and L.Arbaoui: "Advances in Friction Stir Welding of Steel: Project HILDA," *Proc of TRA2014 Transport Research Arena*, Paris, France, April 2014.

[3] A. Toumpis, A. Galloway, D.Camilleri and L.Arbaoui: "Recent Developments in Steel Friction Stir welding: Project HILDA," *Proc of ASME 2015 International Mechanical Engineering Congress and Exposition*, Huoston, Texas, USA, Nov 2015.

[4] D. Micallef, D.Camilleri, A. Toumpis, A. Galloway and L.Arbaoui: "Local Heat Generation and Material Flow in Friction Stir Welding of Mild Steel Assemblies," *Journal of Materials Design and Application*, Vol. 230, No.2 (2016), pp.586-602.

[5] D. Camilleri, D. Micallef and P.Mollicone: "Thermal Stresses and Distortion Developed in Mild Steel DH36 Friction Stir-Welded Plates: An Experimental and Numerical Assessment," *Journal of Thermal Stresses*, Vol. 38 (2015), pp.485-508.

[6] T. Gray, D. Camilleri, and N. McPherson: "Control of Welding Distortion in Thin Plate Fabrication: Design support exploiting computational simulation," ISBN: 978-0-85709-047-8 Woodhead Publishing Ltd, 2014, Cambridge U.K.

[7] A. Toumpis, A. Galloway, S. Cater, and N. McPherson, 'Development of a Process Envelope for Friction Stir Welding of DH36 Steel – A Step Change,' Materials and Design, Vol. 62, 2014, pp 64-75.

[8] M. Chuimenti, M. Cervera, C. Agelet de Saracibar, and N. Dialami: "Numerical Modeling of Friction Stir Welding Processes," *Computer Methods in Applied Mechanics and Engineering*, Vol 254, 2013, pp353-369

[9] A. Toumpis, A. Galloway, L. Arbaoui and N. Poletz: "Thermomechanical Deformation Behaviour of DH36 Steel during Friction Stir Welding by Experimental Validation and Modelling," *Science and Technology of Welding and Joining*, Vol 19, no. 8, 2014, pp 653-663.

Prediction of Inherent Strain in Friction Stir Welding by Using A New Coupled Analysis of MPS and FEM

Hisashi SERIZAWA
Joining and Welding Research Institute, Osaka University, Ibaraki, Osaka, Japan
serizawa@jwri.osaka-u.ac.jp, +81-6-6879-8665

Taku HAYAMI, Fumikazu MIYASAKA
Graduate School of Engineering, Osaka University, Suita, Osaka, Japan

Abstract
The inhomogeneous heat density distribution generated near the friction stir welding (FSW) tool could be computed by a moving particle semi-implicit method (MPS) and the heterogeneous heat input distribution of finite element method (FEM) was obtained through the integration method. The transient temperature distributions were calculated by thermal analysis of FEM where the distributions near the tool was replaced by those estimated from MPS. In addition, the elastic-plastic analysis of FEM was conducted using those distributions by assuming the boundary temperature which is as same as the annealing temperature in the welding. The longitudinal inherent strain distributions obtained indicated that the plastic strain distributions near the tool can be predicted by employing the concept of boundary temperature.

Introduction
There has been a few papers regarding the deformation and the residual stress of joint produced by friction stir welding (FSW) [1]. Terasaki *et al.* experimentally studied about the inherent strain produced in FSW [2], which is the origin of welding distortion and residual stress based on the welding mechanics [3]. The authors numerically examined the inherent strain in FSW by using the finite element method (FEM) [4]. However, the effect of plastic flow on the inherent strain has not been studied. In addition, a method using both a moving particle semi-implicit method (MPS) and a finite element method (FEM) has been developed, and the inherent strain can be predicted except for those near the joint interface, which is caused by a simplification in the definition of heat source and a neglect of the plastic flow [5]. In this research, in order to overcome these problems, a new coupling method of MPS and FEM is developed and the applicability of this coupling method is examined.

Model for Analysis
Modeling for Analysis
The model for analysis is the butt joint of aluminum alloy plates (A6063-T5) whose size is 300 mm in length, 60 mm in width and 5 mm in thickness according to the experiment conducted by Terasaki *et al.* as shown in Fig. 1 [2]. FSW tool has a cylindrical shoulder of 15.5 mm in diameter and a probe of 5.5 mm in diameter and the rotational and traveling speed of tool are 1750 rpm and 3.3 mm/s, respectively. The compressive load for tool is 3450 N. In the experiment, a backup plate was attached to the back of aluminum alloy plates using the clamp and the temperature changes at the bottom of aluminum alloy plates were measured by the thermocouples, which were directly attached to the back of aluminum alloy plates by drilling the holes in the backup plate as shown in Fig. 1.

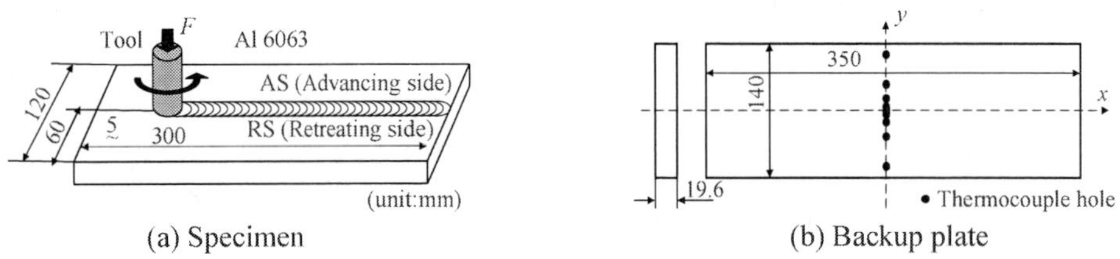

Fig. 1: Schematic illustration of specimen for FSW and position of thermocouple.

MPS Method
In this research, the plastic flow of aluminum alloy near the FSW tool was set to be approximately described by flow of highly viscous fluid. So, the governing equations for plastic flow motion are Navier-Stokes equation. On

the other hand, the value of equivalent flow stress at any temperature and any equivalent strain rate depends on the type of materials. In this study, the equivalent flow stress, σ, is approximated using the following equation, which was proposed by Sheppard *et al.*. [6].

$$\sigma = \frac{1}{\alpha} ln\left(\left(\frac{Z}{A}\right)^{1/n} + \sqrt{\left(\frac{Z}{A}\right)^{2/n} + 1}\right), \quad Z = \dot{\varepsilon} \cdot exp\left(\frac{Q}{RT}\right) \tag{1}$$

Where, R is gas constant and $\dot{\varepsilon}$ is equivalent strain rate. α, A, n, Q are material constants. In this research, α, ln A, n and Q were set as 0.04 (MPa^{-1}), 22.5 (s^{-1}), 5.385, 141 (kJ/mol), respectively according to the report of Sheppard *et al.*. During FSW, heat is generated due to plastic deformation. In this paper, the heat generation per unit volume is assumed to be that 90 % of the work for plastic deformation changes into heat [5].

Modeling for MPS

In order to reveal the temperature distributions near FSW tool precisely, a small volume near the FSW tool was employed for MPS. Figure 2 shows a schematic illustration of the model for MPS. The size of aluminum alloy plate studied was 45 mm in length, 30 mm in width and 5 mm in thickness. Although the diameters of tool shoulder and probe used in the experiment were 15.5 and 5.5 mm, they were assumed to be respectively 15 and 5 mm which are dependent on the initial distance between particles. As shown in Fig. 2, the aluminum alloy plate was described as the highly viscous fluid and its edge was set as the rigid solid. Also, the FSW tool was modeled as the rigid solid. The shapes of shoulder and probe were set as a simple cylindrical though the shape of shoulder has little taper and the probe has thread in usual experiment. So, the vertical velocity which may correspond to the effect of thread was added on the lateral face particles of probe. In actual FSW process, there is the plug-in process of tool before traveling the tool in the plate. However, in this research, the tool was set into the plate at the initial state, where the temperature of whole model was equal to room temperature and the initial velocity was zero, because only the heat generation behavior in traveling the FSW tool is important.

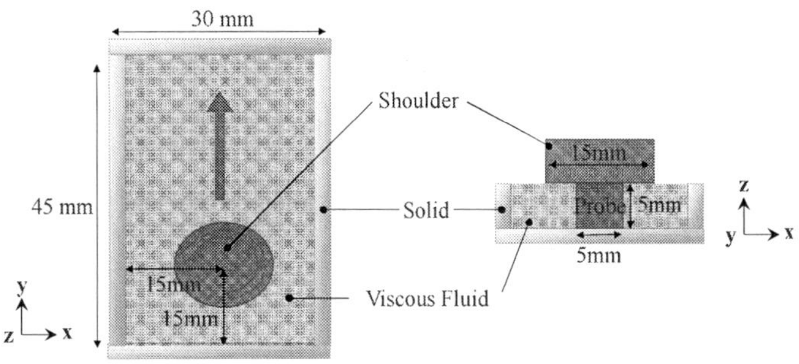

Fig. 2: Schematic illustration of model for MPS method.

Plastic Flow of FSW

In order to define the plastic behavior in MPS, the equivalent stress is computed from the equivalent strain rate $\dot{\varepsilon}$ and is compared with the yield stress of aluminum alloy. From this comparison, it is revealed that the plastic behavior seems to start after the temperature of aluminum alloy decreases to a boundary temperature which is computed to 400 °C.

Coupling Method of MPS and FEM

From our previous studies using both MPS and FEM for FSW process, it was revealed that the transient temperature distributions except for area near the tool could be predicted using a homogeneous volumetric heat source [4,5]. In this research, since a precise and heterogeneous heat density distribution can be obtained as the result of MPS computations, a non-uniform heat input distribution of FEM is calculated from the heterogeneous heat density distribution through the integration method. Then, in order to improve the precision of temperature distributions near the tool, the result of MPS method was superposed on that of FEM. Finally, by using the temperature distributions improved, the inherent strain was computed through the elastic-plastic finite element analysis. In both thermal and mechanical analyses using FEM, the physical properties should be defined to be temperature dependent [4]. In addition, in order to demonstrate the influence of plastic flow during FSW, the equivalent plastic strain at the temperature higher than the boundary temperature is assumed to be zero as same as the annealing process in the welding.

Results and Discussions

Results of MPS

According to our previous studies using MPS method for FSW process, it was revealed that the heat generation area through the stirring of FSW tool can be defined as a circular truncated cone and the average and maximum temperatures of this heat generation area becomes to be almost constant after 2.0 s [4]. Figure 3(a) shows the heat generation distribution on the specimen surface computed after 2.0 s and the difference of heat generation between retreating and advancing sides is obviously demonstrated. Because the computational grids for MPS are much finer than those for FEM, the heat density distribution was computed by superimposing the heat generation distribution of MPS on the nodes of FEM (Fig. 3(b)) and it was employed as the heat input distribution for FEM.

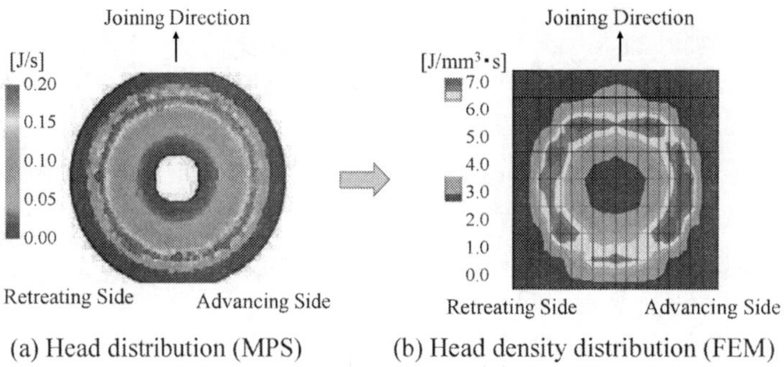

Fig. 3: Heat and heat density distributions on specimen top surface.

By moving the heterogeneous heat density distribution along the joining line, the transient temperature distributions were obtained through the thermal analysis of FEM. Because the temperature distribution near the tool at 2.0 s computed by MPS seems to be quasi-stationary, the temperature at the finite element nodes within 5.0 mm from the center of tool is replaced by the temperature estimated from MPS. The maximum temperature distributions on the central top and bottom surface transverse to the joining line are summarized in Fig. 4, where the experimental results on the bottom surface are also plotted in this figure. From this figure, it is revealed that the coupled result of MPS and FEM shows a smooth curve and the computed result has a very good agreement with the experiment. So, it can be concluded that this coupling method seems to be a good tool for estimating the transient temperature distributions during FSW simply and precisely.

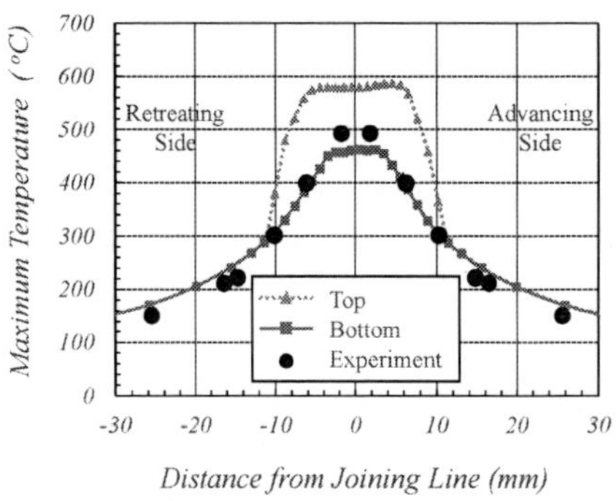

Fig. 4: Maximum temperature distributions on central top and bottom surface transverse to joining line.

Mechanical Analysis of FEM

In order to predict the inherent strain generated at FSW, the elastic-plastic finite element analysis was conducted using the transient temperature distributions obtained by the coupling method of MPS and FEM. The average of longitudinal plastic (inherent) strains computed through the plate thickness are summarized in Fig. 5(a), because the experimental result of Terasaki *et al.* is also the average of the plate thickness. Where the equivalent plastic strain at the temperature higher than 400°C is assumed to be zero. In addition, the calculation result without assuming the boundary temperature is plotted in this figure. From this figure, it is revealed that the numerical result assuming the boundary temperature has a fairly good agreement with the experimental result and the concept of boundary temperature seems to be reasonable for demonstrating the influence of plastic flow on the inherent strain. Then, in order to examine the effect of boundary temperature on the inherent strain distributions, the longitudinal inherent strains are computed by varying the boundary temperature from 300 to 500 °C and the results are summarized in Fig. 5(b). These results suggest that the absolute value of the longitudinal inherent strain near the tool decreased with decreasing the boundary temperature. In addition, it is found that when the boundary temperature is set to be 350 °C, the numerical result has a good agreement with the experiment. Because the plastic flow might enhance the recrystallization in the stir zone and the recovery in thermos-mechanical affected zone, the recrystallization and the recovery might reduce the equivalent plastic strain generated by the plastic flow and then the boundary temperature would be decreased apparently.

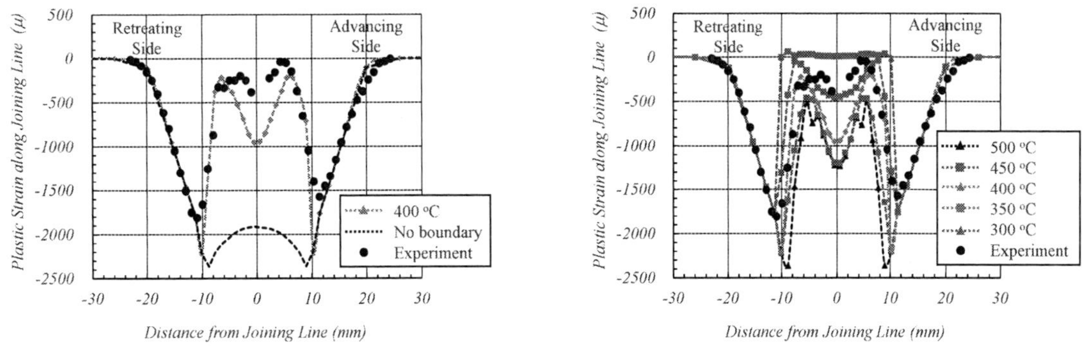

(a) Boundary temperature effect

(b) Effect of boundary temperature

Fig. 5: Distributions of plastic strain along joining line at central view transverse to joining line.

Conclusions

In order to simulate both thermal and mechanical transient behavior in FSW butt joint of two plates simply and precisely, a new coupling method of MPS and FEM was developed where the inhomogeneous heat density distribution for thermal analysis of FEM was obtained from the result of MPS. From the thermal-elastic plastic finite element analyses, it is revealed that the longitudinal inherent strain distribution can be predicted by assuming the boundary temperature which is as same as the annealing temperature in welding.

References

[1] R. Nandan, T. DebRoy and H.K.D.H. Bhadeshia, "Recent Advances in Friction-Stir Welding – Process, Weldment Structure and Properties", *Progress in Materials Science*, Vol. 53 (2008) pp.980-1023.

[2] T. Terasaki and T. Akiyama, "Mechanical Behavior of Joints in FSW : Residual Stress, Inherent Strain and Heat Input Generated By Friction Stir Welding", *Welding in the World*, Vol. 47 (2003) pp.24-31.

[3] Y. Ueda, H. Murakawa and N. Ma, Welding Deformation and Residual Stress Prevention, Butterworth-Heinemann (Massachusetts, 2012).

[4] H. Serizawa, J. Shimazaki and H. Murakawa, "Numerical Study of Factors for Generating Inherent Strain in Friction Stir Welding", *Trends in Welding Research 2012, Proceedings of the 9th International Conference*, Chicago, IL, June 2012, pp.922-929.

[5] H. Serizawa, J. Shimazaki, F. Miyasaka, G. Yoshikawa and H. Murakawa, "Basic Numerical Analysis of Friction Stir Welding with A Combination of MPS and FEM", *Proceedings of the 1st International Joint Symposium on Joining and Welding*, Osaka, Japan, November 2013, pp. 317-324.

[6] T. Sheppard and A. Jackson, "Constitutive Equations for Use in Prediction of Flow Stress During Extrusion of Aluminum Alloys", *Materials Science and Technology*, Vol. 13 (1997) pp.203-209.

A new frictional boundary condition for simulation of friction stir welding by computational fluid dynamics

Gaoqiang Chen
State Key Laboratory of Tribology, Tsinghua University, Beijing 100084, PRC
Key Laboratory for Advanced Materials Processing Technology, Tsinghua University, Beijing 100084, PRC
Oak Ridge National Laboratory, Oak Ridge, TN 37831, USA
Email: gaoqiang.chen@gmail.com

Zhili Feng
Materials Science and Technology Division, Oak Ridge National Laboratory, Oak Ridge, TN 37831, USA
Email: fengz@ornl.gov

Yucan Zhu, Qingyu Shi (Corresponding author)
State Key Laboratory of Tribology, Tsinghua University, Beijing 100084, PRC
Key Laboratory for Advanced Materials Processing Technology, Tsinghua University, Beijing 100084, PRC
Email: shqy@tsinghua.edu.cn

Abstract
Numerical simulation based on computational fluid dynamics (CFD) has become a widely applied approach in optimization and design of friction stir welding (FSW). The frictional boundary condition at the tool/workpiece interface is of critical importance in the prediction of heat generation, temperature and material flow in FSW process. Previous computational modeling studies tended to simplify the material behavior at the interface, generally assuming simple sticking or frictional sliding conditions. In this paper, a new frictional boundary condition is proposed for simulation of FSW by using CFD. The proposed boundary condition is based on the Coulomb friction model and allows the dynamics transition between the sticking condition and sliding condition. We implemented the frictional boundary condition in CFD simulation to analyze the FSW of AA2024 alloy. It is shown that the new boundary condition yields non-uniform distribution of contact state over the tool/workpiece interface. The interfacial sticking state is present over large area at the tool-workpiece interface, while significant interfacial sliding occurs at the shoulder periphery and the lower part of pin side. Due to the interfacial sticking, a rotating flow zone is found under the shoulder, which forms a continuous material flow layer around the pin. The size of the flow zone is smaller than the shoulder diameter, which could be explained by the interfacial sliding at the shoulder periphery.

Introduction
Friction stir welding (FSW) [1, 2] is an advanced materials joining technology. In FSW process, a rotating welding tool is in intimate contact with the workpiece, which generates large interfacial friction between the tool and the workpiece. The interfacial friction leads to significant heat generation, high temperature and significant plastic flow. Such thermal-mechanical processes induced by the interfacial friction play important roles in shaping the microstructure [1] and thus the final in-use properties of the joints. Computational fluid dynamics (CFD) based numerical simulation[3-9] is a powerful approach in quantitative investigation of the friction-induced thermal-mechanical processing condition in FSW for the purpose of fundamental understanding and optimal design.

In the CFD-based simulation for FSW, one of the critical issues is the frictional boundary condition between the welding tool and the workpiece. In the previous simulation, the frictional boundary condition at the tool/workpiece interface could be divided into two categories: (a) velocity-based boundary condition, and (b) shear-stress-based boundary condition. The most important consideration in the frictional boundary condition is the determination of the contact state, i.e. the sliding/sticking state. In a number of the simulation studies [7, 10, 11], the sticking contact state was employed via a velocity-based boundary condition, in which the material was assumed to flow at the same velocity as the welding tool at the tool/workpiece interface. Generally, both the maximum temperature and the deformation zone size were over-predicted using the 'sticking' condition [7]. Nandan et al [8], Atharifar et al [12] and Wang et al [13] adopted a sliding state at the tool/workpiece interface by using the velocity boundary condition, in which a predefined interfacial contact state was required. However, there was no generic way for quantitative description of the interfacial contact state for various welding parameters, tool geometries and materials. In recent CFD simulations, the shear-stress-based boundary condition has been demonstrated to superior to a velocity-based boundary condition in the CFD simulation of FSW [14], as the shear-stress-based boundary condition allows predicting the interfacial contact from the friction model based on the balance of interfacial shear stresses. Liechty et al [14] used a shear-stress-based boundary condition in a simulation for the FSW of plasticine. Chen et al [15] applied the shear-stress-based boundary condition in their model for FSW of AA6061 using a threaded welding tool, which resulted in reasonable prediction of the thermal-mechanical processing condition. Unfortunately, the shear-stress-based boundary condition used in the current simulation [14, 15] was proposed based on the assumption of the sliding state. The presence of 'sticking' state at the tool/workpiece could not be predicted in the current simulation. This limits the application of numerical simulation in optimization and design of the weld process. Therefore, it is of great interest to implement a generic shear stress boundary condition in the CFD simulation for FSW, which is capable to predict the presence of both the sliding state and the sticking state at the tool/workpiece interface.

In this study, we propose a new shear-stress-based frictional boundary condition for the CFD simulation of FSW. In the proposed boundary condition, the transition between the sliding state and sticking state at the tool/workpiece interface is taken into consideration. In this paper, CFD simulation based on the new boundary condition is used to investigate the interfacial contact state and the material flow during the FSW process.

Modeling approaches

In the CFD model, butt joining of two AA2024-T4 sheets by FSW was considered. The dimensions of each workpiece were 145mm×55mm×3mm (length × width × thickness). The welding tool was made by H13 steel. The tool shoulder was 13mm in diameter, and length of the conical pin was 2.4mm. The radius of pin was 2mm at the root and 1.75mm at the tip. In the welding process, the FSW tool rotated at 1600rpm and the workpiece traveled at 20mm/min. No tilt angle was employed in the experiment. Fig 1 illustrates the geometry model in the numerical simulation. The dimensions of the workpiece and the geometry of welding tool are taken as the same as the setup in the experiment. Different boundary conditions are defined for modeling of the quasi-steady-state welding process, as shown in Fig 1. More details of the model are given in reference [16].

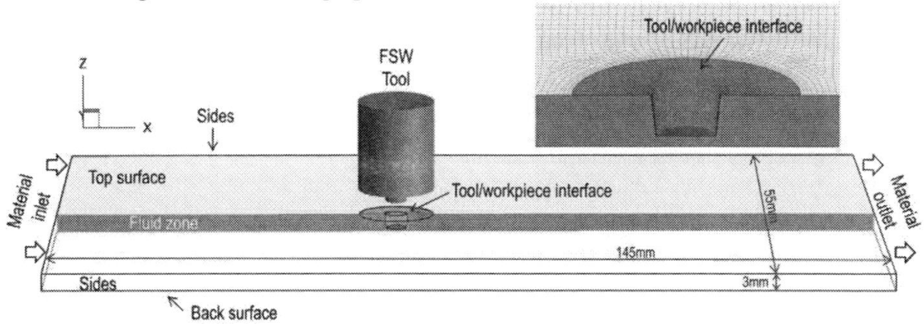

Fig. 1 Geometric model in the simulation.

An 'Inlet-outlet' method is adopted such that the material flows into the 'Fluid zone' from 'Material inlet' at the welding speed and flows out of the domain from the 'Material outlet'. The 'Back surface' and 'Sides' of the work piece are assumed to contact with the fixtures. The material velocity on 'Back surface' and 'Sides' is set to be the same as the welding speed. 'Top surface' is taken as a free surface contacting with air. At the 'Tool/workpiece interface', the frictional shear stress is applied, given as,

$$\tau_f = \mu_f \sigma_n \tag{1}$$

where μ_f is the frictional coefficient, σ_n is the interfacial normal stress (taken as 50MPa[17]). The friction coefficient is defined as temperature dependent [4, 18-20].

Predicted interfacial contact state by numerical simulation

Based on the proposed frictional boundary condition, the contact state at the tool/workpiece interface is provided. Fig 2 shows the predicted workpiece velocity at the tool/workpiece interface. It is shown that the predicted distribution of material velocity is almost symmetric about the tool axis. In order to analyze the interfacial contact state, the velocity of the workpiece at the position shown by dashed line (oabc in Fig 3a) is plotted versus the distance from tool axis in Fig 3b, in comparison with the tool velocity. It could be seen that the contact state is non-uniformly distributed over the tool/workpiece interface in the simulated welding condition. The sticking state is present at most part of the shoulder surface, upper part of the pin side and most part of the pin bottom surface as the interfacial velocity of the workpiece is almost the same as the tool velocity, while significant interfacial sliding occurs at the shoulder periphery and the lower part of pin side.

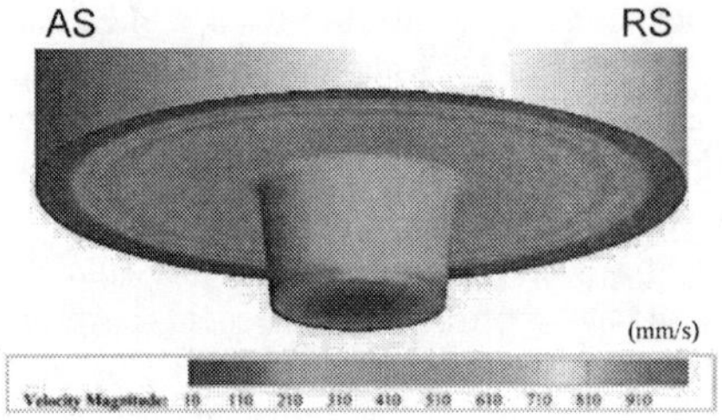

Fig. 2 The predicted velocity of the workpiece at tool/workpiece interface.

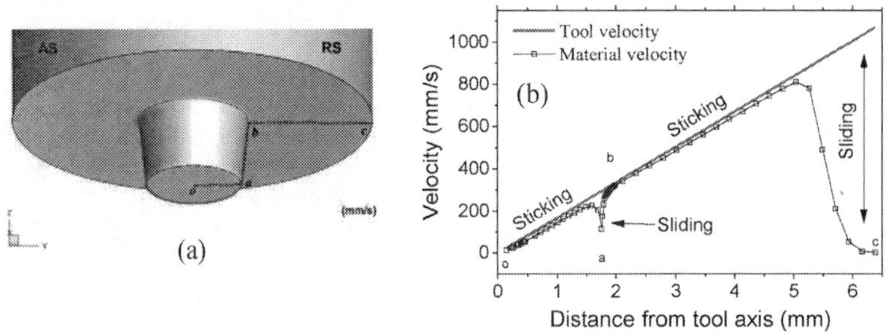

Fig. 3 Predicted contact state at tool/workpiece interface.

Material flow pattern

Fig 4 shows the predicted flow paths in a horizontal plane near the tool shoulder. It is found that a rotating flow zone (RFZ) exists in the vicinity of the shoulder as marked by red dotted lines in Fig 4, which forms a continuous material flow layer around the pin. The presence of such a RFZ under the shoulder is supported by the experiments by Schmidt et al [21] and the recent in-situ observation by Morisada et al [22]. In their studies, a similar RFZ is found to be smaller than the shoulder, in which the material is undergoing fast circular motion. Both the in-situ observation and our simulation result show that the material enters the RFZ in front of the tool and flows around the tool for several circles before depositing behind the tool. As such, the RFZ is a channel for rapid material transportation from the front of the tool to the behind. As shown in Fig 4, the shape of RFZ is almost circulate in the X-Y plane. The diameter of the RFZ is smaller than the shoulder diameter, which is attributed to the interfacial sliding state at the shoulder periphery. It is worth noting that the sliding state at the shoulder periphery may play an important role in the formation of a friction stir weld. As the material is undergoing fast circular motion in the RFZ, excessive loss of the material may be caused due to the centrifugal motion, if the size of the RFZ reaches or exceeds the shoulder diameter with the absence of the interfacial sliding at the shoulder periphery.

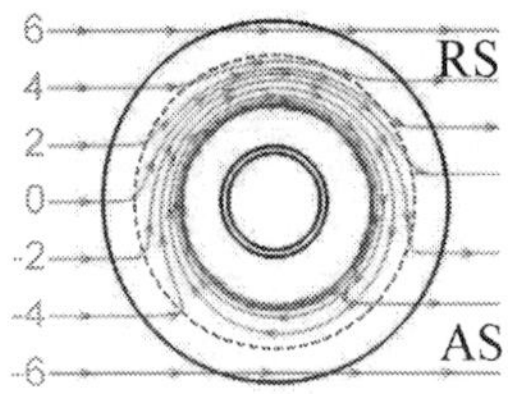

Fig. 4 Material flow paths under the shoulder. The numbers on the left denotes the distances (in mm*) from the welding center line.*

Conclusions

In summary, the conclusions below could be drawn.

(1) A new frictional boundary condition is proposed for simulation of FSW by using CFD. The interfacial contact state at the tool/workpiece interface and material flow in FSW of AA2024 alloy are analyzed.

(2) The new boundary condition yields non-uniform distribution of contact state over the tool/workpiece interface. The interfacial sticking state is present over large area at the tool-workpiece interface, while significant interfacial sliding occurs at the shoulder periphery and the lower part of pin side.

(3) Due to the interfacial sticking, a rotating flow zone is found under the shoulder, which forms a continuous material flow layer around the pin. The size of the flow zone is smaller than the shoulder diameter, which

could be explained by the interfacial sliding at the shoulder periphery.

Acknowledgments

The work was supported by the National Natural Science Foundation of China (grant no. 51375259) and the National Science and Technology Major Project of the Ministry of Science and Technology of China (no. 2012ZX04012-011). Besides, Gaoqiang Chen was supported by the China Scholarship Council (File No. 20130620105) for 2-year study at Oak Ridge National Laboratory (Oak Ridge, TN, USA).

References

[1] Mishra RS, Ma ZY. Friction stir welding and processing. Materials Science and Engineering: R: Reports 2005;50:1-78.
[2] Nandan R, DebRoy T, Bhadeshia H. Recent advances in friction-stir welding–process, weldment structure and properties. Progress in Materials Science 2008;53:980-1023.
[3] Schmidt H, Hattel J. A local model for the thermomechanical conditions in friction stir welding. Modelling and Simulation in Materials Science and Engineering 2005;13:77.
[4] Zhang Z. Comparison of two contact models in the simulation of friction stir welding process. Journal of Materials Science 2008;43:5867-77.
[5] Grujicic M, Arakere G, Yen C-F, Cheeseman BA. Computational Investigation of Hardness Evolution During Friction-Stir Welding of AA5083 and AA2139 Aluminum Alloys. Journal of Materials Engineering and Performance 2010;20:1097-108.
[6] Colegrove PA, Shercliff HR. Experimental and numerical analysis of aluminium alloy 7075-T7351 friction stir welds. Science and Technology of Welding and Joining 2003;8:360-8.
[7] Colegrove PA, Shercliff HR. 3-Dimensional CFD modelling of flow round a threaded friction stir welding tool profile. Journal of Materials Processing Technology 2005;169:320-7.
[8] Nandan R, Roy GG, Lienert TJ, Debroy T. Three-dimensional heat and material flow during friction stir welding of mild steel. Acta Materialia 2007;55:883-95.
[9] Neto DM, Neto P. Numerical modeling of friction stir welding process: a literature review. The International Journal of Advanced Manufacturing Technology 2013;65:115-26.
[10] Yu Z, Zhang W, Choo H, Feng Z. Transient heat and material flow modeling of friction stir processing of magnesium alloy using threaded tool. Metallurgical and Materials Transactions A 2012;43:724-37.
[11] Chen G-q, Shi Q-y, Li Y-j, Sun Y-j, Dai Q-l, Jia J-y, et al. Computational fluid dynamics studies on heat generation during friction stir welding of aluminum alloy. Computational Materials Science 2013;79:540-6.
[12] Atharifar H, Lin D, Kovacevic R. Numerical and experimental investigations on the loads carried by the tool during friction stir welding. Journal of materials engineering and performance 2009;18:339-50.
[13] Wang H, Colegrove PA, dos Santos JF. Numerical investigation of the tool contact condition during friction stir welding of aerospace aluminium alloy. Computational Materials Science 2013;71:101-8.
[14] Liechty BC, Webb BW. Modeling the frictional boundary condition in friction stir welding. International Journal of Machine Tools and Manufacture 2008;48:1474-85.
[15] Chen G, Shi Q, Fujiya Y, Horie T. Simulation of Metal Flow During Friction Stir Welding Based on the Model of Interactive Force Between Tool and Material. Journal of materials engineering and performance 2014;23:1321-8.
[16] Chen G, Feng Z, Zhu Y, Shi Q. An alternative frictional boundary condition for computational fluid dynamics simulation of friction stir welding. Journal of Materials Engineering and Performance 2016;To be published.
[17] Assidi M, Fourment L, Guerdoux S, Nelson T. Friction model for friction stir welding process simulation: Calibrations from welding experiments. International Journal of Machine Tools and Manufacture 2010;50:143-55.
[18] Zhang X, Xiao B, Ma Z. A transient thermal model for friction stir weld. Part I: The model. Metallurgical and Materials Transactions A 2011;42:3218-28.
[19] Kumar K, Kalyan C, Kailas SV, Srivatsan TS. An Investigation of Friction During Friction Stir Welding of Metallic Materials. Materials and Manufacturing Processes 2009;24:438-45.
[20] Li W-Y, Ma T, Li J. Numerical simulation of linear friction welding of titanium alloy: effects of processing parameters. Materials & Design 2010;31:1497-507.
[21] Schmidt HNB, Dickerson T, Hattel JH. Material flow in butt friction stir welds in AA2024-T3. Acta Materialia 2006;54:1199-209.
[22] Morisada Y, Imaizumi T, Fujii H, Matsushita M, Ikeda R. Three-Dimensional Visualization of Material Flow During Friction Stir Welding of Steel and Aluminum. Journal of materials engineering and performance 2014;23:4143-7.

New Development and Testing of FSChanneling Technique: HybridFSC

Heikki Karvinen, Daniel Nordal, Pedro Vilaça
Department of Mechanical Engineering, School of Engineering, Aalto University, 02150 Espoo, Finland
heikki.karvinen@aalto.fi, daniel.nordal@aalto.fi, pedro.vilaca@aalto.fi

Timo Galkin
Nokia, 02610 Espoo, Finland
timo.galkin@nokia-bell-labs.com

Abstract
The Friction Stir Channeling (FSC) is a recent manufacturing technique based on the friction stir processing of a monolithic component, able to produces internal, closed channels in a single step, with any path and constant or continuously modified shape along the path. A new concept for this process: Hybrid Friction Stir Channeling (HFSC), is presented and tested in the production of channels for thermal management of chassis components for electronic application. The new concept includes joining of multiple components and reinforces the high potential of conventional FSC. The thermal management of electronics is a growing challenge as devices become smaller. High operating temperatures reduce the reliability of the devices and increase energy consumption. Therefore, electronic device manufacturers are researching new cooling solutions able to apply air, liquid or a mixture of both. Currently liquid cooling requires multiple operations with complex assembly work using different techniques like milling, drilling and welding, and even including pre-deformed tubes it is difficult to access the heat sources with optimal efficiency. The new HFSC presented in this paper is tested to evaluate the cooling performance, productivity benefits and saving of material when compared with alternative solutions.

A case study is established concerning the conformal cooling of a generic chassis plate made of AA5754-H111 for electronic application with multiple heat source powers and locations. The path for the cooling channel is developed based on FEM thermal analysis, considering the unique geometrical features of the channels. The heat transfer efficiency of the channels produced with the new HFSC is evaluated. The properties of the channels are experimentally assessed via measurement of shape and roughness, hardness distribution in the processed zones and establishment of main metallurgical characteristics resulting from the thermomechanical processing cycle.

Introduction
Internal closed channels are structural features with several high potential applications and usually found in hollow extruded components and heat transfer applications. There are several methods to create this type of channels [1] . The manufacturing method limits the channel design and some methods demand multiple processing stages to produce an internally closed channel. These limitations increase material usage of applications and impair opportunities to optimize structures efficiently. The Friction Stir Channeling (FSC) is a recent manufacturing technique based on the friction stir processing of a monolithic component, able to produces internal, closed channels in a single step, with any path and constant or continuously modified shape along the path. This concept was firstly patented in 2005, by Mishra as a method of manufacturing heat exchangers [2]. Mishra proposed that by reversing the material flow pattern and by selecting the appropriate processing parameters of friction stir processing, it is possible to produce continuous and stable channels inside monolithic metal components. In this first concept, that was tested by Balasubramanian et al. [3], all the material extracted from the metal workpiece is laid on the processed zone bellow the tool shoulder, within a clearance between the tool shoulder and the metal workpiece. A new concept of the original FSC process was patented in 2013 by Vidal and Vilaça [4]. The new concept is based on a distinct material flow, where a controlled amount of viscoplasticised material from the metal workpiece flows out from the processed zone producing the internal channel without any gap between the tool shoulder and the solid metal workpiece. Thus, the material flowing from the interior of the solid metal workpiece is not deposited on the processed surface but directed outside of the processed zone in the form of detachable flash. With this innovation, it is possible to produce channels in a wide range of dimensions by control of the process parameters. These channels are addressed by Vilaça et al. [5] and characterized by Vidal et al. [6-8]. Studies performed by Rashidi et al. [9, 10] reported on an approach where a non-threaded conical probe with tilt

angle and a clearance between the shoulder and the top surface of the metal workpiece are used for extracting material and creating the channel. No new concept is introduced with this last approach.

The conventional FSC, is a flexible technique to manufacture internal closed channels with free path, but intended to be applied in bulk components. A new concept intended to optimize multi-component structures made of similar or dissimilar materials, is introduced in this paper. The new concept is the Hybrid Friction Stir Channeling (HFSC) [11] and it is a new paradigm in the production of internal closed channels with free path.

New Concept: HFSC

The Hybrid Friction Stir Channeling (HFSC) [11] is a new concept to manufacture, in one simultaneous action, tailor-made internal closed channels and a weld joint between multiple components (*Fig. 1a*). The geometrical arrangement of the multiple components in the joint can be designed in an overlap joint, butt joint or a combination of overlapped with butt joint, as represented in *Fig. 1b*. The multiple components can be of similar or dissimilar materials. During the HFSC the material of the multiple components is stirred, activating solid-state joining mechanisms in the processed zone, resulting in the weld. In simultaneous, the geometrical features of the tool extracts part of the stirred material from the processed zone into external flash, resulting in the channel. The HFSC is controllable, repeatable and able to produce channels with a wide range of dimensions and free-path. The channels can have constant or continuously variable cross-section dimensions. The finishing of the channel surfaces is rough, which creates easily turbulent flow at low flow rates allowing a production of efficient heat transfer systems that are operated with low energy. The development of a new HFSC probe allowed for a welded zone to be produced underneath the channel, while the combined effect of the probe and the shoulder creates the channel by flowing the viscoplastic material, out of the processed zone, into the form of detachable flash.

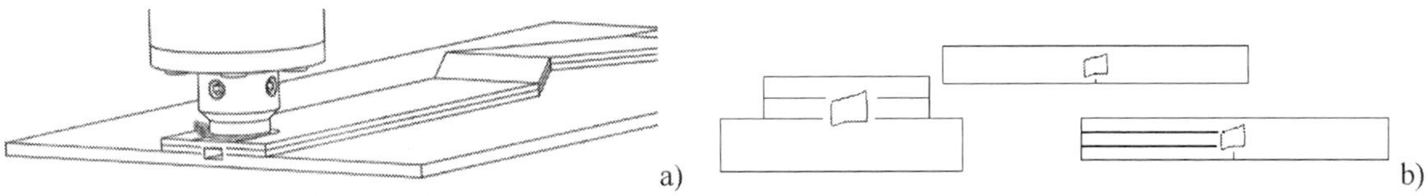

Figure 1: Concept of HFSC: a) The HFSC applied to 3 overlapped components along a non-linear path; b) Schematic representation of alternative joints for HFSC.

Methods and Equipment

The HFSC was implemented with an ESAB LEGIO FSW 5UT machine. The processing parameters for the HFSC were as follows: position control, spindle rotation speed of 350 rpm, travel speed of 70 mm/min, probe length of 10.5 mm and a plunge depth of 9.8 mm. During the HFSC process, a force of about 12 kN was observed. All microscopy samples were prepared by cutting samples from processed material perpendicular to their processing direction. Each sample was mounted in Struers MultiFast phenolic hot mounting resin using a Struers ProntoPress-20 machine set at 180 °C and 25 MPa for 13 min. Each sample was then ground using SiC papers to a 4000 grit finish and polished with a 1 μm diamond compound. Samples were put through hardness tests on a micro-combi tester from CSM Instruments, with a force of 100 mN. A step of 1 mm was used within a grid of 19 x 12 indentations. After processing the harness tests, the samples were repolished and etched with a 10 % hydrofluoric acid for 15 seconds and then optically investigated with a Nikon Epiphot 200 microscope equipped with a Nikon DS-2Mv camera.

The HFSChannels were produced within the processing zone of two overlapped AA5754-H111 plates as can be seen in *Fig. 2b*. The plates were aligned, centered and processed by HFSC along their rolling direction (RD) to ensure uniformity and consistency. The bottom plate has a thickness of 5 mm, with in-plane dimensions [mm] of 500 (RD) x 300, and a hardness of about 1170 MPa. The plate at the top has a thickness of 8 mm, with in-plane dimensions [mm] of 500 (RD) x 80, and a hardness of about 900 MPa.

Characterization of Channel Properties

The 450 mm long channel produced was continuous (*Fig. 2a*) with minor porosity in the roof. Ceiling roughness varies along the length of the channel with minor instances of floor contact. The walls and ceiling have a comparable roughness to conventional FSC. However, the floor of the channel had an increased roughness when compared to conventional FSC. The HFSC probe profile causes the floor to have an increased roughness due to the formation of the weld nugget. One representative sample of the cross-section of the HFSChannel is depicted in *Fig. 2b*. The floor of the channel starts flat but begins to tilt, with the higher edge on the retreating side, as the channel advances. This is caused by the concentrated flow of viscoplastic material in the welded zone at the retreating side, i.e., the flow side. No defects or porosity was found in the welded zone (*Fig. 2b*).

The hardness was relatively consistent throughout the channel, with the hardness field presented in *Fig. 2c*. The hardest points are located within the dynamically recrystallized zones of the weld (bottom of the channel) and roof of channel. The highest hardness values recorded were in the stirred zone of the weld (bottom of the channel) with about 1400 MPa. This represents an increase of strength from the original base material.

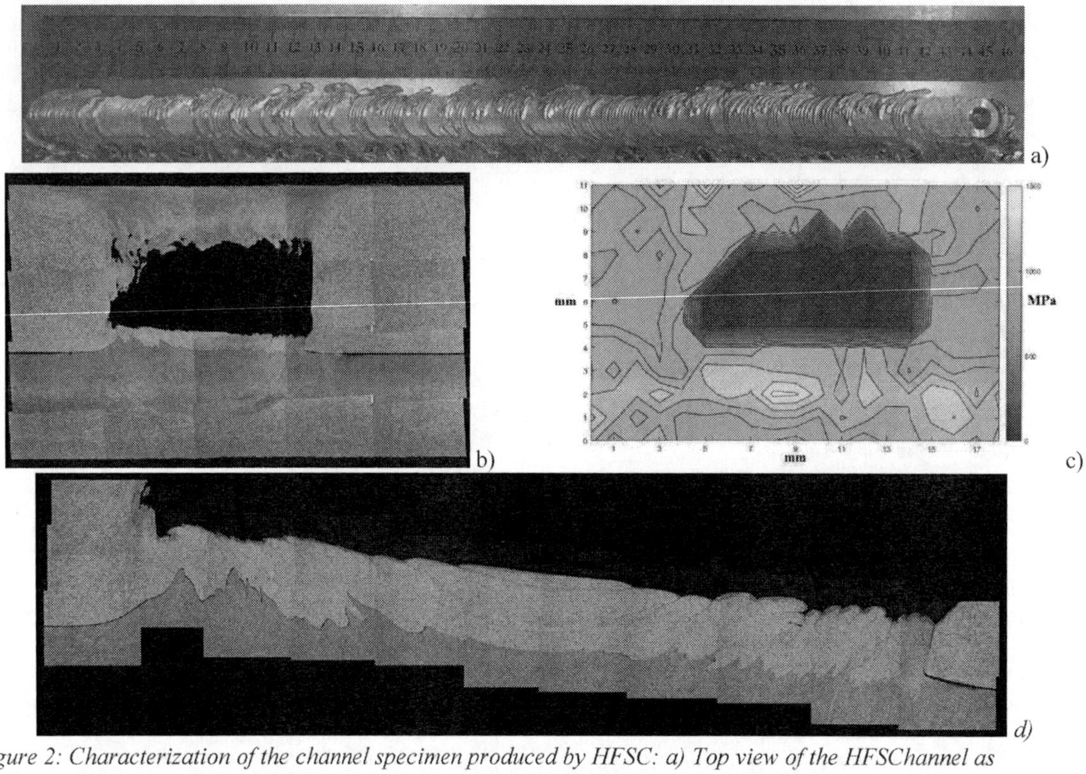

Figure 2: Characterization of the channel specimen produced by HFSC: a) Top view of the HFSChannel as produced; b)Macrograph of a representative cross-section of the channel; c) Hardness field of the cross section; d) Detail of metallurgy of the bottom of the channel emphasizing the material flow and diffusion joining mechanism.

From the analysis of the *Fig. 2d*, it is possible to confirm that there is a successful weld between the two components. Along the interface different levels of diffusion joining mechanism can be identified with higher joining quality towards the advancing side. The retreating side of the weld has poor diffusion between the two plates which is evident from the dark black line in the boundary zone of the two plates. This gap tapers off as the boundary areas approaches the advancing side of the weld. Areas of high flow create finger like propagations that intertwine with each other forming a bond. A crack-like interface exists on the advancing side of the channel floor heading towards the original interface area between the plates.

Establishment of Channel Path

SolidWorks modeling and simulation was performed to develop a good conformal cooling solution and analyze the channel characteristics with regards to material savings and heat transfer efficiency. To simulate a real generic chassis plate for electronic application with multiple heat source powers and locations, eight heat source with different powers were implemented (*Fig. 3a*). The HFSChannel was simulated with same geometry represented in Fig. 2b, with and without roughness. The roughness considered was 100 μm for the bottom and 500 μm for the remaining walls of the channel. The flow considered for cooling media was from 0.5 (*Fig. 3b*) to 6 l/min (*Fig. 3c*). The roughness allows a significant cooling efficiency even for low flows, such as 0.5 l/min, for which the peak temperature in the chassis was about 29.4 °C, while for the 6.0 l/min the peak temperature in the chassis was 24.6 °C.

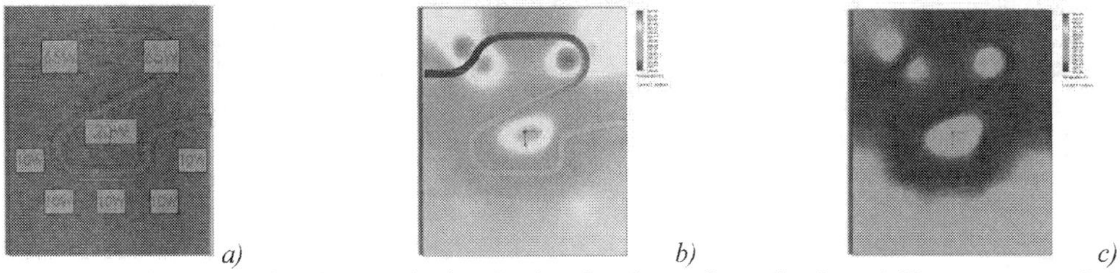

a) b) c)

Figure 3: Modeling of heat transfer of a generic chassis plate for electronic application: *a) Heat sources and channel path; b) Cooling with 0.5 l/min with rough HFSC; c) Cooling with 6.0 l/min with rough HFSC. Note: [T]=K.*

Conclusions

The fundamentals of the new Hybrid Friction Stir Channeling (HFSC) showed to be technologically feasible and it is ready to be explored in the production of internal closed channels with free path. The production of a channel within the joint of two or more plates is one more demonstration of the high industrial potential applications of the friction stir based solutions. In this study, a continuous sound channel resulted from new tool features, designed specifically to manufacture, in one action, closed channels and a weld joint between multiple components. The main geometrical and metallurgical characteristics of the processed zone were established. Although a strong level of diffusion between the two plates was accomplished, this level is not homogeneous and thus represents a challenge for further improvement. The hardness was increased from the original base material values at both the bottom and ceiling of the channel. The modelling and preliminary experimental investigation shown a significant benefit using the HFSC for conformal cooling applications. The development of HFSC continues to bring this new process into a reliable manufacturing process.

References

[1] Miguel Ferraz, et al: "Preliminary Study on Procedures and Channels Characteristics Using Friction Stir Channelling", 65th Annual Assembly, DocIII-1647-12, Denver, USA July 2012.

[2] R.S. Mishra, Patent US 6923362 (2005)

[3] N. Balasubramanian, et al: "Friction Stir Channelling: Characterization of the Channels", Journal of Materials Processing Technology, Vol. 209 (2009), pp. 3696-3704.

[4] P. Vilaça, C. Vidal: Patent PT 105628 (2013).

[5] P. Vilaça, et al: Linear Friction Based Processing Technologies for Aluminum Alloys: Surfacing, Stir Welding and Stir Channeling, Aluminium Alloys-New Trends in Fabrication and Applications, InTech (2012), pp.159-197.

[6] C. Vidal, et al: "Characterisation of fatigue fracture surfaces of friction stir Channeling specimens tested at different temperatures", Engineering Failure Analysis, Vol. 56, No. 10, (2015) pp. 204-215.

[7] C. Vidal, et al: "Fatigue behaviour at elevated temperature of friction stir Channeling solid plates of AA5083-H111 aluminium alloy". International Journal of Fatigue, Vol. 62 (2014) pp. 85-92.

[8] C. Vidal, et al: "Fatigue assessment of friction stir channels", Int. Journal of Fatigue, Vol. 62. (2014), pp. 77-84.

[9] A. Rashidi et al: "Modified Friction Stir Channeling: A Novel Technique for Fabrication of Friction Stir Channel, Applied Mechanics and Materials", Vol. 302, (2013) pp 365-370.

[10] A. Rashidi et al: "Influence of tool pin geometry and moving paths of tool on channel formation mechanism in modified friction stir channeling technique", Adv. Manufacturing Tech., Vol. 80, (2015) pp. 1087–1096.

[11] P. Vilaça, H. Karvinen, Patent FI 20160165 (2016)

October 13th
(Thursday)

Role of Cyclic Phase Transitions in Additive Manufacturing of Metals and Alloys - Lessons Learned from Welding Science

S. S. Babu[1,3], *N. Raghavan*[1,3], *M. Kirka*[2,3], *R. Dehoff*[2,3], *W. Peter*,[3,5] *L. Love*[3,4], *T. R. Watkins*[2] *and J. Turner*[6]

[1]*Department of Mechanical, Aerospace and Biomedical Engineering, Department of Materials Science and Engineering, The University of Tennessee, Knoxville, TN 37996; Email: sbabu@utk.edu*
[2]*Materials Science and Technology Division,* [3]*Manufacturing Demonstration Facility,* [4]*Energy and Transportation Science Division,* [5]*Energy and Environmental Sciences Directorate,*[6]*Computer Science and Mathematics Division, Oak Ridge National Laboratory, Oak Ridge, TN 37831; Email: kirkamm@ornl.gov, dehoffrr@ornl.gov, peterwh@ornl.gov, lovelj@ornl.gov, watkinstr@ornl.gov, turnerja@ornl.gov*

Abstract

Additive manufacturing (also known as 3D printing) of metals is considered to be a disruptive technology, able to produce limited number of high value components with topologically optimized geometries and functionalities. Realization of the above potential for real-world applications is stifled by lack of standard computational design-tools; component certifications, varied starting powder feed stock compositions, methods to probe thermo-mechanical processes, microstructural homogeneity, residual stress, as well as, anisotropic static- and dynamic-material properties. Detailed research of direct energy deposition, laser- and electron-powder bed additive manufacturing demonstrates that the underlying physics of these processes are very similar to welding, except for complex boundary conditions. This paper will review published literature and on-going research with reference to fundamental aspects of heat and mass transfer, solidification under large (10^3 to 10^5 K/m) thermal gradients and (10^{-3} to 10^0 m/s) liquid solid-interface velocities, as well as, solid->solid transformation under repeated thermal excursions. Case studies on model based qualification of Ni- and Ti- alloy builds made by additive manufacturing (AM), based on the above fundamental knowledge will be discussed.

Challenges and Opportunities for AM

In 2005, Prof. Ashby published the hybrid materials [1] concept by which an engineer can access combinations of properties that is not achievable by single material. For example in Fig. 1a, the map of Young's modulus (E) as a function of density (ρ) shows monolithic materials appearing in limited regions with unpopulated areas or "holes" existing for certain combinations. To fill these holes in property space, Ashby proposed four classes of hybrids (i) composite, (ii) sandwich, (iii) lattice and (iv) segments, in which different materials including metals and alloys, ceramics, polymers, glass and elastomers are assembled together in repeating micro-scale configuration within an enclosing macro-level component. This assumes that there exists a manufacturing route that is capable of putting any material in any place within a component. Christodoulou and Venables [2] have argued that these hybrids must include systems with structural load bearing and energy storage, autonomous sensing, multifunctional actuating and electromagnetic functions, as well as, survival and damage tolerant capabilities for defense applications. Although, manufacturing technologies based on metal forming and brazing has shown the feasibility of achieving hybrid materials [3], the need for minimization of cost and number of manufacturing steps for each and every individual constituents has forced us to consider AM as an alternate route [4]. This notion was indeed proven with the building of an integrated (see Fig. 1b) robotic arm [5]. In

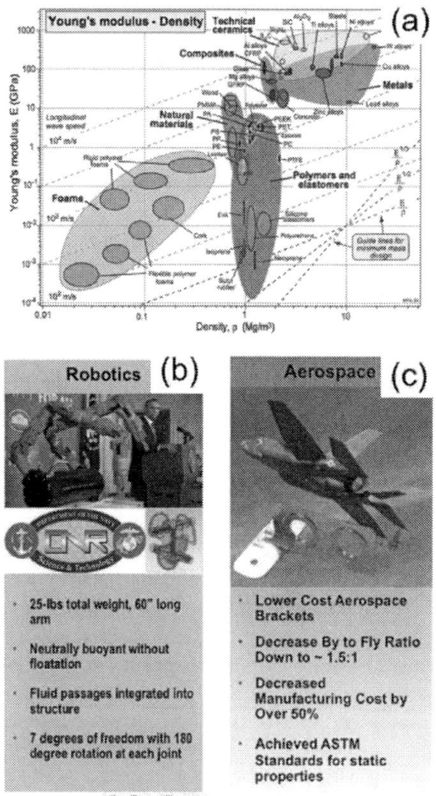

Figure 1: (a) Map of material properties showing holes in properties that cannot be accessed by monolithic materials. (b) An example of robotic arm made using hybrid material concept that incorporates fluid passages and also arrives at improved performance. (c) Another example of AM that leads to reduction in buy-to-fly ratio of aerospace bracket. (After [1,5,6])

addition, additive manufacturing enabled us to produce near-net-shape components, thereby reducing the need for expensive machining. For example, Dehoff et al., [6] produced an aerospace grade Ti-6Al-4V bracket by additive manufacturing with a reduced buy-to-fly ratio. Although, these examples and other literature show the potential of AM, widespread adoption of this technology across industries is still lagging [7].

Hesitancy to adopt AM is related to the current process flow for additive manufacturing [8]. The process flow starts with a computer aided design (CAD) file either with or without topology optimization [9]. The second step leads to materials (powder type and distribution) and process platform selection based on prior experience (e.g. from many commercial vendors like EOS®, Concept-Laser®, etc.). The third step involves trial and error process parameter (e.g. laser power, velocity, hatch distance, scan-length and -modes, gas flow and supports) optimization for single-bead, multi-bead, cube, prismatic sized samples for varied geometries and microstructures [6]. If these samples meet the target properties (e.g., mechanical), the prototype component evaluation is initiated. However, research has shown that the process parameters developed for simple geometries does not translate to the complex due to variations in the thermal signatures at each and every location within a component [10, 11, 12, 13]. As a result, skepticism has set in among engineers by quoting the following limitations of AM: (i) need for process reliability (e.g. spatter formation); (ii) lack of governing requirements (e.g. the need for supports to control distortion), and (iii) need for re-qualification for each and every hardware and software version. We believe that the above limitations can be minimized by leveraging knowledge from welding and joining research [14, 15].

Lessons from Welding Science and Technology

Era of optimization of welding processes based on fundamental udnerstanding of physical processes started with a classic paper by Rosenthal, who developed theory of moving heat sources for welding [16]. With the slow and steady introduction of computational methods and infrastructure, including finite element methods and computational fluid dynamics [15], thermodynamics and kinetics [17], the predictive capability has increased tremendously [14]. These predictive tools developments were aided by detailed *ex-situ* and *in-situ* characterizations of the welding arc, melt-pool formation, fluid flow, defect formation, distortion and microstructure [18]. With these predictive and experimental tools, it is possible to determine the mechanisms for cracking during welding, as well as, degradation of welds during service. For example, Lolla et al., [19] using detailed multi-scale microstructure characterization and thermodynamic modelling, rationalized the overlay failure of valve body (see Fig. 2a). These failures were attributed to the formation brittle ferrite + sigma lamellar microstructure in the dilution region, which is indeed predicted by equilibrium thermodynamic calculations using composition ranges across the lines marked by A and B in the ternary diagram shown in Fig. 2b. Using these results, alloy design methodologies for avoiding such delamination cracks were developed. Many more such examples can be quoted from literature where fundamentals of heat transfer, phase transformations, and thermo-mechanical deformation characteristics have been used to solve industrial problems [20]. The obvious question then is the following: "How can we use the extensive knowledge in welding and joining to address the challenges faced by metal AM processes?".

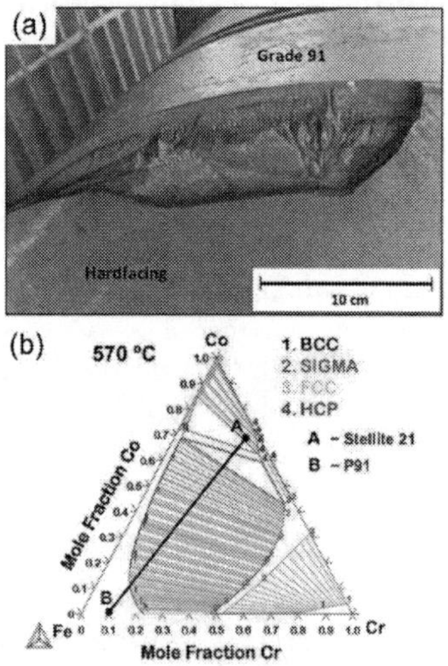

Figure 2(a) Photograph of a gate valve in which an entire section of hardfacing has delaminated. (b) The location of stellite cladding and Fe-Cr substrate composition in Fe-Co-Cr ternary alloy phase diagram, calculated for operating temperature confirming the tendency for the formation of BCC and sigma phases.

AM Research Directions

AM involves of all the steps relevant to traditional manufacturing including design, process selection, parameter control, *in-situ* and *ex-situ* characterization, and qualification, as well as, certification. However, in contrast to traditional manufacturing, for the AM enterprise, the expertise and tools must be available within close proximity and be available within a compressed time for rapid optimization and sustainability.

Design of AM structures

Traditional manufacturing design starts with a concept and dimensions of a product that serves its purpose for targeted cost per product. Most of the CAD tools have incorporated finite element methodologies that allow for engineers to understand the performance of a structure with reference to external loading conditions. However, these tools have been developed with machining (subtractive operation) as the predominant operation. Extension of these CAD tools to AM design often requires agonizing trial-and-error optimization by interacting with other software that perform topology optimization, slicing, scan strategies and support structure design for overhang geometry and thermal management. For example, Bell *et al.* [21] used bi-directional structural optimization within MATLAB software [22]. The penalty function in the topology optimization was updated to minimize the overhang surfaces that were above 45°. With this approach, they were able to design and produce a hybrid thermos-plastic component capable of the same load bearing capacity as bulk geometry, while reducing the weight by 50% (see Figs. 3a and 3b). In this work, these researchers observed strong sensitivity to the starting point, i.e., block of material (see Figs. 3c and 3e). In this research, density of some regions were deliberately changed before topology optimization exercise. As the optimization progressed and completed (see Figs. 3d and 3f), the evolution of geometry was totally different. In this optimization, the mechanical property (modulus and strength) of each voxel is considered to be isotropic and similar. In reality, these properties do vary depending upon the build direction [23]. Design tools are prejudiced to subtractive manufacturing and need to be revised in light of AM's possibilities.

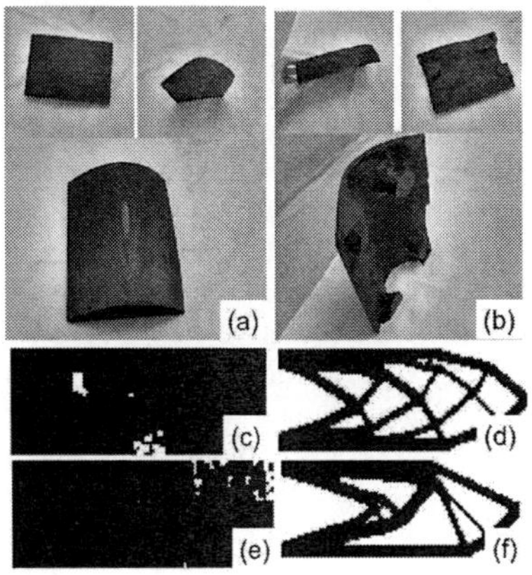

Figure 3: (a) An example of a solid thermo-plastic part before optimization and (b) the same geometrical envelope with 50% reduction in weight through topology optimization. Topology optimization was also found to be very sensitive to the initial starting point of the design operation for the same loading conditions. Different initial shapes of blocks before topology optimization (c) and (e) lead to totally different topologically optimized (d) and (f). geometry, respectively

Process Specifications and Qualifications

All the metal AM systems in the current market go through rapid changes in beam (laser or electron beam) control strategies. These rapid changes are made in response to satisfying a wide range of requirements from users including conformity of geometry, microstructure, static- and dynamic-material properties. Due to lack of any guidance tool, selection of AM process is often done based on prior experience of practitioners [7]. Furthermore, with rapid changes in software and material feed stock, the process parameters (e.g. laser power) and scan sequences (stripe, island or checker board) for the same component geometry made in different time periods cannot meet the stringent standards of reproducibility expected from traditional manufacturing.

This dilemma of using the scattered power sources and consumables have been addressed by welding industry by documenting welding process specifications (WPS) and procedure qualification reports (PQR) through standards (e.g., API (American Petroleum Institute) standards for welding pipelines transferring oil and gas). These reports document the essential variables that allow for faithful reproduction of welds for mission critical applications. However, due to flexibility of AM geometries, the extension of WPS and PQR documentation of the process parameters as a function of geometry becomes harder to implement. Nevertheless, some equipment, like electron beam melting (EBM), do track all process parameters (see Fig. 4) including electron beam current, substrate temperature, layer height, vacuum levels, and powder rake position, as well as, occasional arc trips. Similarly, in certain laser powder bed AM machines, one may be able to analyze the energy transfer to powder bed using QM-Meltpool® modules [24]. Interestingly, some of these machines use random movement of power source (i.e. chess scan strategies) and are often very difficult to document. In this regard, one can use simple video recording and image analyses to extract the laser locations (see Fig. 5).

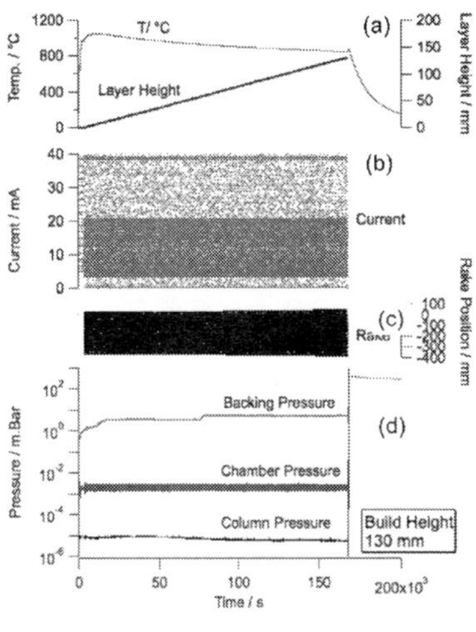

Figure 4: A typical graph of data logged by electron beam powder melting process which could be used for reporting process specification for a given CAD geometry. Some of the graphs (current and rake position) cannot be resolved due to small time resolution (0.1 second) used for monitoring.

In-situ Process Monitoring

Although, AM industry is transitioning towards process quality control strategies mentioned above, it is necessary to understand extent of energy transfer into processing zone for controlling thermal signatures. In this regard, the concept of in-situ monitoring of thermal signatures have been adopted by welding community more than 30 years ealier [25, 26]. Extension of these techniques to understand the defect formation in AM have already started. For example, Dinwiddie *et al.*

[27] have used infrared imaging to monitor the defect formation during EBM. Interestingly, these authors also discuss unique challenges associated with metallization of lenses that may hinder the direct one-to-one comparison of infrared signals to temperature distribution. These results and other literature [28] demonstrate that thermal signatures are quite different at each and every location within a given layer during AM. As a result, an AM component will have severe microstructure and property gradients.

Ex-situ Characterization

Results from *in-situ* characterization of AM have to be interpreted in conjunction with *ex-situ* characterization of defects and microstructures in three dimensions at different length scales. For welded structures, these needs have been addressed by serial sectioning [18] or computer aided X-ray tomography [29], which have also been adopted by AM community. Cordero et al [30] used X-ray tomography to correlate the spatial variation of porosity [see Fig. 6] to thermal signatures. The next step is to understand the crystallographic texture within a AM build, usually achieved through electron back-scattered diffraction (EBSD) imaging [31]. Recently, Dehoff et al [32] have used innovative neutron Bragg-edge tomography methodology to extract three-dimensional distribution of crystallographic texture in nickel base superalloys non-destructively. In contrast, robust measurement of mechanical properties in 3D builds still eludes AM practitioners. Often, these measurements rely on micro-hardness mapping [31] on surfaces of samples prepared for optical microscopy and EBSD. Furthermore, transmission electron microscopy [31] (see Fig. 7) and atom probe tomography are often needed to rationalize the microstructural heterogeneity [18]. Measurements of residual stress was pioneered by welding researchers using neutron and synchrotron beam lines [33, 34]. Recently, Kolbus et al [35] have extended these techniques to measure 3D distribution of residual stresses in identical prismatic geometries processed by EBM (-90 to +90 MPa) and (-500 to +400 MPa) laser powder bed melting processes. These results confirm that the large variations of residual stresses are linked to spatial variation of thermal signatures. Bagg et al [36] have used the similar methodology and observed subtle variations residual stress as a function of scanning strategy (stripes, island and checker board).

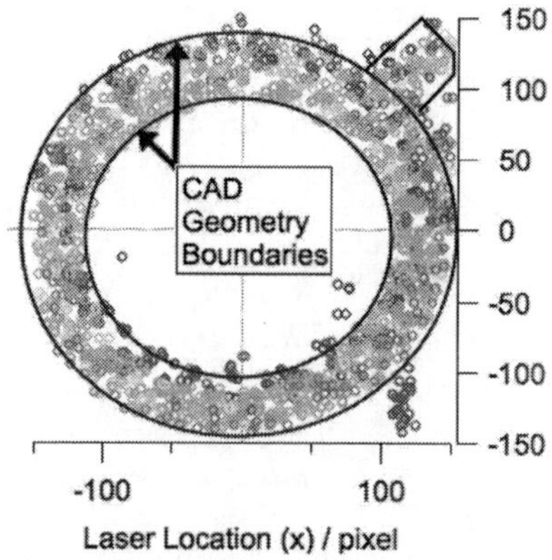

Figure 5: Random locations of square laser interaction region with powder bed extracted from analyses of video. The color scale corresponds to sequence of events (blue-1st; red-1780th) per layer

Figure 6: X-ray tomography of porosity distribution in a cylinder made by EBM process .

Process Modeling

It is clear from above discussions that it is necessary to minimize trial and error optimization of additive manufacturing processes by computational modeling. This approach is similar to the integrated modeling [37] framework proposed for welding by Kirkaldy [38] in early 1990s. The above framework has been implemented within commercial software (e.g. SYSWELD, VrWeld, VFT, etc.) and on-demand (e.g. E-WeldPredictor) Internet tools [39]. This leads to a fundamental question: "Why not adopt extensive tools [40, 41, 42], developed for welding to additive manufacturing?". Often this similarity with welding is questioned by AM practitioners since they do not observe any defects even-though the laser speed is higher than 100 mm/s. At these speeds, the welding process leads to defects such as liquid surface instabilities [43] brought about by the variation of surface tension with temperature as well as different modes of cracking [44]. This is indeed a fair criticism and analyzed further using Fig. 8. Dye, Hunziker and Reed [45], using phenomenological models for defect formation and solidification, rationalized the weldability of Alloy 718 (2-mm thick) sheets. The graph clearly shows that above 5 mm/s the welding will not yield sound components. In contrast, almost all laser powder bed process provides acceptable quality builds at speeds greater than 100 mm/s. However, this discrepancy between AM and Welding literature can be rationalized based on fundamentals of heat transfer [46].

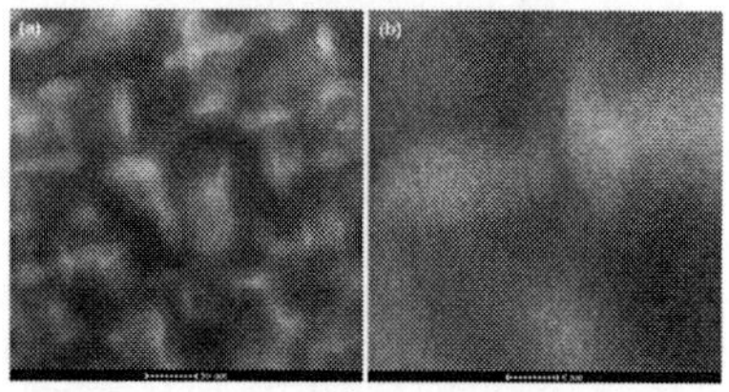
Figure 7: High-resolution electron microscopy (a) low- and (b) high-magnification images of additively manufactured alloy 718 DDM build showing diffused interface close to gamma-double prime precipitates and gamma phase, tentatively correlated to repeated thermal cycling that may induce

With high laser beam velocities (>100 mm/s), the power distribution will change from circular to pseudo-line source due to limited thermal diffusion distance (δ). Thermal diffusion distance is given by $\delta \approx \sqrt{2\alpha\Delta t}$, where α is thermal diffusivity and Δt is the time interval for a given laser scan to return to the same location in a powder bed. If this distance is comparable to the hatch distance or laser beam focal diameter, one can conclude that the laser power distribution is more elongated (or line source) along the scan direction (see Fig. 8 for calculated melt front velocities for laser powder bed AM). Although, we can develop refined and sophisticated methodologies to describe transitions in power distribution modality, it is noteworthy that thermal signatures, defect formatting tendencies, solidification and microstructure evolution should be correlated to melt-front velocities, rather than laser velocity. The systematic increase in laser scan velocity in AM for a given hatch distance, the power distribution will go through transitions from *point* to *line* to *area* to *volume* heat source as a function of geometry being processed. Based on the above argument, by describing fundamental behavior of energy source transition from 1D to 2D to 3D distribution and correlating to *global* and *local* thermal signatures one could control the underlying microstructure in AM components and this concept is evaluated in the next section.

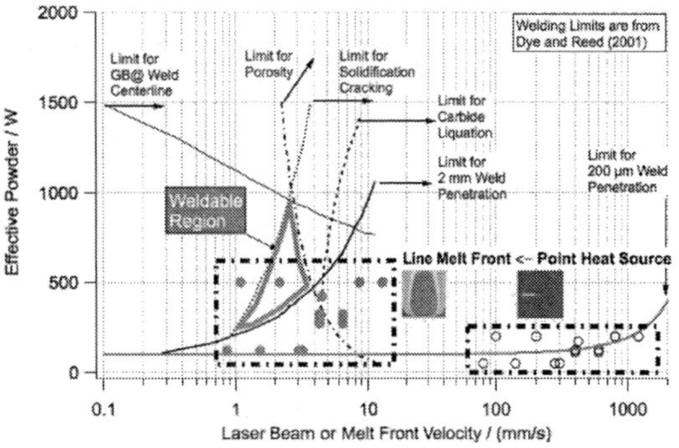

Figure 8: Comparison of weldability map for alloy 718 within (green triangle) with that of experimental data points (unfilled black circles) relevant to SLM processing. By correcting for transition from point to line source (see the inset images) due to interaction with successive scans, the SLM process parameters can be interpreted (red-filled circles) as line heat source with melt-front velocities similar to that of welding.

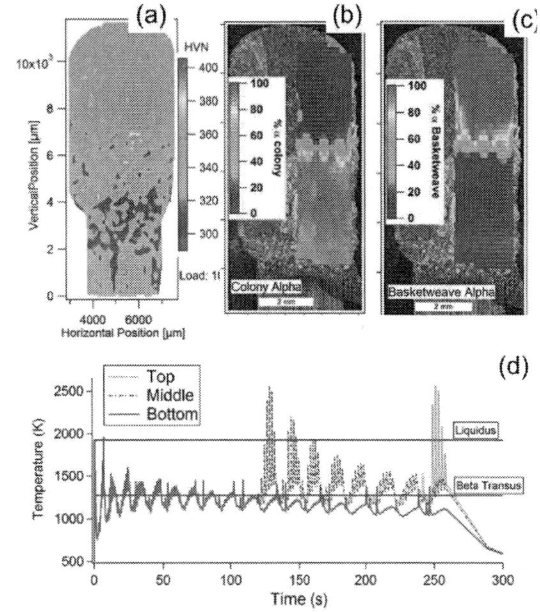

Figure 9: (a) Hardness map across a Ti6Al4V build shows heterogenity as a function of build height. Detailed microtructural measurements show (b) colony microstructure near the bottom and (c) basketweave structure near the top regions of the builds. (d) Predicted thermal signatures at three different locations.

Model Based Qualification Approach

Two published case studies that demonstrate the feasibility of model based qualification of AM components is discussed in this section.

Improvement of Ti6Al4V Fatigue Properties: In this example, poor fatigue properties along z-direction of Ti6Al4V builds, made by a direct energy deposition (DED) process, were correlated to mechanical and microstructural heterogeneity (see Fig. 9). The hardness maps shown in Fig. 9a indicate a soft region (< 340 VHN) close to the substrate (~3000 µm from the bottom) in contrast to hard (>340 VHN) in the top of the builds (~ 7000 µm). Interestingly, these variations correlated to an abrupt transition from predominantly colony microstructure (see Fig. 9b) to predominantly basketweave microstructure (see Fig. 9c). The reasons for this abrupt transition were evaluated by calculating thermal signatures using a finite element model [47]. The results (see Fig. 9d) showed that the top-regions are subjected to *local* thermal gyrations only above the β-transus temperature, while the bottom regions are subjected to *local* thermal gyrations within $\alpha+\beta$ phase fields. With this understanding, processing parameters were designed to maintain the whole build above the β-transus temperature and homogeneous basketweave structure was obtained during final cooling, i.e., imposing *global* thermal signature. With this model based process parameter development [48], and by minimizing the physical defects, fatigue life of Ti6Al4V was improved by two orders of magnitude (1,200 to 374,000 cycles).

Microstructure Control in EBM process: It is noteworthy that interactions of *local* to *global* thermal signatures are characteristic of the electron beam powder melting (EBM) process, where *local* thermal signatures control liquid→solid transitions and *global* thermal cycle control solid state precipitations [49]. For example, Dehoff et al

[32] controlled *local* thermal gradients (G) and liquid-solid interface velocities (R) and demonstrated on-demand control of columnar to equiaxed transition (CET) during processing of alloy 718. Furthermore, solid-state precipitation [δ (delta), γ'(gamma-prime) and γ" (gamma double-prime)] was controlled by *global* slow cooling conditions from build temperature [50]. Based on the above understanding, Raghavan et al [51] developed a high-performance computational (HPC) model (i.e. using TRUCHAS [52]) capable of predicting *local* thermal signatures, as a function of EBM process parameters and geometry. With process parameters derived from this model, grain size and solidification texture for a given geometry (See Fig. 10) were controlled. In contrast, Prabhakar et al., [53] predicted *global* thermal signatures and correlated the same to residual stress and distortion. The above examples provide early stage results that reliable microstructure can be achieved in AM through pragmatic design and control of *global* and *local* thermal signatures for a given class of geometries, a key attribute of model based qualification. Interestingly, these demonstrations fit the emerging materials science paradigm, i.e., integrated computational materials engineering (ICME) [54] , to arrive at microstructure and property control with minimal trial and error experimentation.

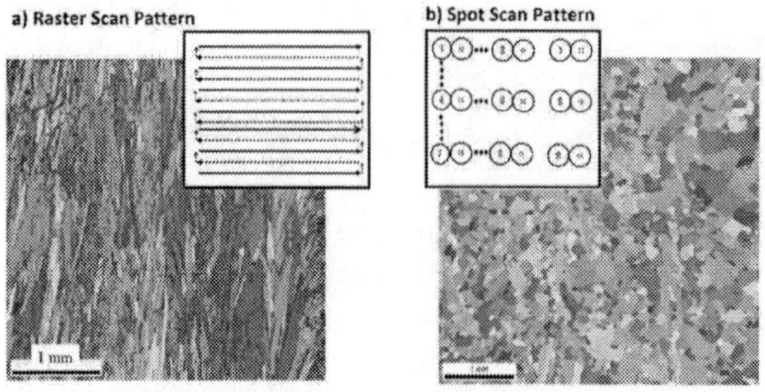

Figure 10: Using a computational heat transfer model and high-performance computing infrastructure, the beam path conditions (raster or spot) that lead to columnar to equiaxed transition in electron beam melting AM were predicted and experimentally validated.

Future Directions

With concurrent developments of computer aided design tools for AM and in materials science of additive manufacturing, it is now possible to imagine a viable pathway to fabricate AM structures with different length scales by leveraging parallel advances made in robotic technology. Recently, ORNL researchers have developed big area additive manufacturing process (BAAM) for thermo-plastic composite structures [5, 55] by integrating these technologies. The extension of underlying technology to low-cost metals (Fe, Ni and Al), referred as m-BAAM, has been proposed. Although, welding based large scale AM has been demonstrated for titanium components [56], the current goals are bolder to push the boundaries of this technology to meet: (i) high deposition rates (100 lbs./h); (ii) low cost [<$10/lbs.] (iii) large components [major axis greater than 6 ft] and (iv) compliance of CAD geometrical and property requirements, simultaneously.

Conclusions

Through detailed analyses of each step in the AM process flow, i.e., design, material feed-stock selection, process selection, process parameter optimization, scanning strategy, in-situ and ex-situ characterization, followed by integrated process modeling, it is possible to accelerate the adoption of additive manufacturing across many industries. This paper has identified the commonality and differences between welding and AM physical processes. The detailed analyses also show that in melt-front velocities are indeed comparable to welding beam velocities, thereby allowing us to extend theories of weld solidification and solid-state phase transformation to AM processes also. This hypothesis was evaluated with two case studies involving Ti-6Al-4V and Inconel 718 alloys. By transitioning from local to global thermal signatures, uniform basketweave microstructure and improved fatigue properties were obtained in direct energy deposition AM process. By careful for control of local thermal signatures, predictable site-specific columnar to equiaxed transition in solidification was achieved in electron beam melting process. These case studies have demonstrated that there is a potential for model based qualification of AM components.

Acknowledgements

Research discussed in this article was sponsored by the US Department of Energy (DOE), Office of Energy Efficiency and Renewable Energy, Advanced Manufacturing Office, under Contract DE-AC05-000R22725 with UT-Battelle, LLC, and by the Laboratory Directed Research and Development Program of Oak Ridge National Laboratory (ORNL), managed by UT-Battelle, LLC. Research at ORNL's High Flux Isotope Reactor was sponsored by the Scientific User Facilities Division, Office of Basic Energy Sciences (BES), and US DOE. Some of the neutron diffraction characterization was support by the Division of Chemical Sciences, Geosciences, and Biosciences, Office of BES, US DOE, under Contract DE-AC05-000R22725 with ORNL, managed and operated by UT-Battelle, LLC. Authors thank Mr. Y. Tian and Mr. K. Makiewicz for giving permission use images from their theses at OSU.

References

1. M. F. Ashby, "Hybrid to fill holes in material property space," *Phil. Magazine,* 2005, Vol. 85, pp. 3235-3257
2. L. Christodoulou and J. D. Venables, "Multifunctional material systems: the first generation," JOM, 2003, Vol. 55, pp. 39-45
3. H. N. G. Wadley, "Cellular metal manufacturing," *Advanced Engineering Materials*, 2002, Vol. 4, pp. 726-733
4. L. Djumas et al., "Enhanced mechanical performance of bio-inspired hybrid structures utilizing topological interlocking geometry," *Scientific Reports*, 2016, doi:10.1038/srep26706
5 C. Holshouser et al., "Out of Bounds Additive Manufacturing, "*Adv. Mater. and Proc.*, 2013, Vol. 171, pp. 15-17
6. R. Dehoff et al., "Case study: Additive manufacturing of aerospace brackets," *Adv. Materials and Processes*,"2013, Vo. 171, pp. 19-22
7. S. S. Babu, L. Love. R Dehoff, W. Peter, T. R. Watkins and S. Pannala, "Additive Manufacturing of Materials: Opportunities and Challenges," *MRS Bulletin*, 2015, Vol. 40, pp. 1154-1161
8. W. Sames et al., "The Metallurgy and Processing Science of Metal Additive Manufacturing," International Materials Reviews, 2016, DOI:10.1080/09506608.2015.1116649
9. L. Korg, A. Tucker, and G. Rollema, "Application of topology sizing and shape optimization methods to optimal design of aircraft components," www.altairproductdesign.com, 2011
10. S. Kelly, Oerlikon-Metco, Personal Communication, 2016
11. S. Moylan,et al., "An additive manufacturing test artifact," *J. Research NIST*, 2014, Vol. 119, pp. 429-459
12. C. K. Sudbrack, Powder-bed fabrication of high temperature Ni-based superalloy LSHR , LANL Workshop, July 20-21, 2015, unpublished research, 2015
13. E. Schwalbach and M. Groeber, "Multi-modal data collection and integration for additive manufacturing." *AAAS Annual Meeting, Integrated Computational Materials Engineering Principles for Additive Manufacturing*, 15 February, 2015
14. S. S. Babu, "Introduction to Integrated Weld Modeling," ASM Handbook, Vol. 22B, Editors D. U. Furrer and S. L. Semiatin, 2010, ASM International.
15. T. Debroy and S. A. David, "Physical processes in fusion welding," *Rev. Mod. Physics*, 1995, Vol. 67, pp. 85-112
16. D. Rosenthal, "Mathematical theory of heat distribution during welding and cutting," *Welding Journal*, 1941, Vol. 20, pp. 220s-234s
17. S. S. Babu, "Thermodynamic and Kinetics Models for Describing Microstructure Evolution during Joining of Advanced Materials," *International Materials Reviews*, 2009, Vol. 54, pp. 333-367
18. J. M. Vitek and S. S. Babu, "Multiscale characterization of weldments," *Science and Technology of Welding and Joining*, 2011, Vol. 16, pp. 3-11
19. T. Lolla, J. Siefert, D. Purdy, S. S. Babu, and D. Gandy, "Delamination of Stellite Hardfacing in Power Plants: A Microstructural Characterization Study," *Science and Technology of Welding and Joining*, 2014, Vol. 19, pp. 476-486
20. Y. Yang et al., "Cracking mitigation during buttering and cladding of a low alloy steel pipe," Proceedings of Trends in Welding Research, Pine Mountain, ASM International, 2008
21. J. R. Bell, H. Rezayat, B. K. Post and S. S. Babu, "Method for minimizing overhang creation in topology optimization for additive manufacturing," Undergraduate Research Project Report, The University of Tennessee, Knoxville, 2016, Personal communication.
22. Y. M. Xie and X. Huang, Evolutionary Topology Optimization of Continuum Structures: Methods and Applications. Jon Wiley and Sons, 2010
23. H. Rezayat, W. Zhao, A. Siriruk, D. Penumadu and S. S. Babu, "Structure-Mechanical Property Relationship in Fused Deposition Modeling [Special Issue on 3D Printing]; *Materials Science and Technology*, 2015, Vol. 31, pp. 895-903
24. T. Craeghs, S. Clijsters, E. Yasa, F. Bechmann, S. Berumen, and J. –P. Kruth, "Determination of Geometrical Factors in Layerwise Laser Melting using Optical Process Monitoring," *Optics and Lasers in Engineering*, 2011, Vol. 49, pp. 1440-1446
25. Y. Ogawa, "Review: High Speed imaging technique Part 1 – high speed imaging of arc welding phenomena," *Sci. Technol. Weld. Join.*, 2011, Vol. 16, pp. 33-43
26. S. Tsukamto, "Review: High Speed imaging technique Part 2 – high speed imaging of power beam welding phenomena," *Sci. Technol. Weld. Join.*, 2011, Vol. 16, pp. 44-55
27. R. B. Dinwiddie et al., "Thermographic In-situ Process Monitoring of the Electron Beam Melting Technology Used in Additive Manufacturing," Thermosense: Thermal Infrared Applications XXXV, 2013, # 87050K
28. W. Sames et al., "The Metallurgy and Processing Science of Metal Additive Manufacturing," International Materials Reviews, 2016, DOI:10.1080/09506608.2015.1116649

29. H. N. B. Schmidt, T. L. Dickerson and J. H. Hattel, "Material flow in butt friction stir welds in AA2024-T3," *Acta Materialia*, 2006, Vol. 54, pp. 1199-1209
30. Z. C. Cordero, R. B. Dinwiddie and R. R. Dehoff, "In-situ observations of Chimney pore formation during Electron beam additive manufacturing," unpublished research, 2016, Oak Ridge National Laboratory, Oak Ridge, TN 37831
31. Y. Tian, et al., "Rationalization of mechanical heterogeneity in IN-718 clads made by direct laser additive manufacturing," *Metallurgical and Materials Transactions A*, 2014, Vol. 45, pp. 4470-4483
32. R. R. Dehoff, et al., "Site-specific control of crystallographic grain orientation through electron beam additive manufacturing," *Materials Science and Technology*, 2015, Vol. 31, pp. 931-938
33. P. J. Withers and H. K. D. H. Bhadeshia, "Overview – Residual Stress Part 1: Measurement Techniques," *Materials Science and Technology*, 2001, Vol. 17, pp. 355-365
34. P. J. Withers and H. K. D. H. Bhadeshia, "Overview – Residual Stress Part 2: Nature and Origins," *Materials Science and Technology*, 2001, Vol. 17, pp. 366-375
35. L. Kolbus, et al., "Comparison of Residual Stress in Inconel 718 Simple Parts Made by Electron Beam Melting and Direct Laser Sintering," *Metall. Mater. Trans. A,* 2015, Vol. 46, Issue 3, pp., 1419-1432
36. S. Bagg et al., "Characterization of Residual Stress and Microstructure in Inconel 718 builds made by Selective Laser Melting with Different Scanning Strategies." Submitted to *Metall. and Mater. Trans. A*, 2016
37. S. S. Babu, "Thermodynamic and Kinetics Models for Describing Microstructure Evolution during Joining of Advanced Materials," *Int. Materials Reviews*, 2009, Vol. 54, pp. 333-367
38. J. S. Kirkaldy: 'Diffusion-controlled phase transformations in steels – theory and applications', *Scand. J. Metall.*, 1991, Vol. 20, 50–61
39. S. S. Babu, "Introduction to Integrated Weld Modeling," ASM Handbook, Vol. 22B, Editors D. U. Furrer and S. L. Semiatin, 2010, ASM International.
40. J. Goldak and M. Asadi, "Challenges in validation of computational weld mechanics code to compute residual stress and distortion in welds," *J. Press. Vess. Tech. – Trans. ASME*, 2014, Vol. 136, #011201
41. Virtual Joining Portal: http://calculations.ewi.org/VJP/
42. W. Zhang, Y. Liu, X. Wang, Y. M. Zhang, "Characterization of Three-dimensional weld pool surface in GTAW," *Welding Journal*, 2012, Vol. 91, pp. 195s-203s
43. Y. Kawahito, M. Mizutani and S. Katayama, "Elucidation of high-power fiber laser welding phenomena of stainless steels and effect of factors on weld geometry," *J. Phys. D. Appl. Phys.*, 2007, Vol. 40, pp. 5854-5859
44. J. M. Vitek, S. A. David., and S. S. Babu "Optimization of Weld Conditions and Alloy Compositions for Welding of Single-Crystal Nickel-Based Superalloys," *Mater. Sci. Forum*, 2007, Vol. 539-543; pp. 3082-3087
45. D. Dye, O. Hunziker, R. C. Reed, "Numerical analysis of the weldability of superalloys," *Acta Materialia*, 2001, Vol. 49, pp. 683-697
46. O. Grong, "Metallurgical modelling of welding," Materials modeling series; 1994, London, The Institute of Materials
47. K. Makiewicz, "Development of Simultaneous Transformation Kinetics Microstructure Model with Application to Laser Metal Deposited Ti-6Al-4V and Alloy 718," Thesis, The Ohio State University, 2013
48. A. Prabhu et al., "Effect of Microstructure and Defects on Fatigue Behavior of LENS deposited Ti-6Al-4V," *Science and Technology of Welding and Joining*, 2015, Vol. 20, pp. 659-669
49. W. Sames et al., "Thermal effects on Microstructural Heterogeneity of Inconel 718 materials fabricated by Electron Beam Melting," *Journal of Materials Research*, 2014, Vol. 29, pp. 1920-1930
50. M. M. Kirka et al., "Microstructure development in electron beam melted Inconel 718 and associated monotonic properties," *JOM*, 2016, Vol. 68, pp. 1012-1020
51. N. Raghavan, et al., "Numerical Modeling of Heat-transfer and the Influence of Process Parameters on Tailoring the Grain Morphology of IN718 in EBAM," *Acta Materialia*, 2016, Vol. 112, pp. 303-314
52. D. B. Kothe et al., "A High Resolution Finite Volume Method for Efficient Parallel Simulation of Casting Processes on Unstructured Meshes," in 8th SIAM Conf. on Parallel Proc. for Scientific Computing, Minneapolis, MN, 1997.
53. P. Prabhakar, et al., "Computational modeling of residual stress formation during the electron beam melting process for Inconel 718," *Additive Manufacturing*, 2015, Vol. 7, pp. 83-91
54. Integrated Computational Materials Engineering: A Transformational Discipline for Improved Competitiveness and National Security," National Research Council, National Academies, 2008
55. L. J. Love, "The importance of carbon fiber to polymer additive manufacturing," *J, Materials Research*, 2014, Vol. 29, pp. 1893-1898
56. J. Ding et al., "Thermo-mechanical analysis of wire and arc additive layer manufacturing process on large multi-layer parts," *Computational Materials Science*, 2011, Vol. 50, pp. 3315-3322

Towards a Complete Model of Arc Welding – Arc Properties, Weld Depth, Thermal History and Residual Stress

Anthony B. Murphy
CSIRO Manufacturing, Lindfield NSW, Australia
tony.murphy@csiro.au

Vu Nguyen
CSIRO Manufacturing, Clayton VIC, Australia
vu.nguyen@csiro.au

Yuqing Feng
CSIRO Mineral Resources, Clayton VIC, Australia
yuqing.feng@csiro.au

David G. Thomas
CSIRO Data61, Clayton VIC, Australia
david.thomas@data61.csiro.au

Abstract

A computational fluid dynamic (CFD) model of MIG welding of aluminium, which includes the arc plasma, the wire and the workpiece in the computational domain self-consistently, has been developed. The model allows the depth and shape of the weld to be predicted for a wide range of welding parameters, without the need for repeated measurements to determine fitting parameters. A graphical user interface (GUI) has been developed, so that the model can be easily used by welding engineers; the model runs on standard desktop or laptop computers. The model also predicts the thermal history at every point in the workpiece, which is the input information that is required for predicting important weld properties such as residual stress and distortion and weld microstructure. Predictions of weld composition are compared with measurements, and the feasibility of coupling the CFD model to a finite element model of residual stress and distortion is demonstrated.

Introduction

Many factors are important in determining the quality of a welded joint. These include the depth of penetration, the residual stress in the welded metal and the resulting distortion, and the microstructure of the fusion zone and heat-affected zone. Different modelling approaches are generally used for predicting each of these properties. Computational fluid dynamic (CFD) models of the flow in the weld pool, and sometimes in the arc plasma as well, are used to predict the depth and shape of penetration. To predict residual stress and distortion requires the application of finite element analysis (FEA), while calculation of microstructure can be performed by approaches such as the phase field method.

There has been little effort to combine these approaches. An opportunity is presented by the fact that models that predict residual stress and distortion and microstructure require the thermal history of the base metal as input data. The thermal history is usually calculated using a heat transfer model in which the arc and weld pool are represented by a simple two- or three-dimensional heat source. But the thermal history is one of the quantities calculated by CFD models.

Here we first describe our CFD model of the workpiece and arc and present some representative results. We then report our efforts to combine the CFD model with an FEA model that predicts residual stress and distortion, using the thermal history and other quantities predicted by the CFD model as input data for the FEA model.

CFD model of the arc and weld pool

The computational domain of the CFD model includes the wire, arc and workpiece. The methods that have been used were presented in previous publications [1]-[3]; we summarize them here briefly. The model solves coupled partial differential equations describing conservation of mass, momentum, energy and charge, together with an equation for the magnetic potential. The equations are those for viscous incompressible flow, modified to treat conducting media and plasma effects. To ensure that the wire, arc and weld pool remain in the centre of the computational domain, the equations are transformed into the frame of reference of the moving wire and arc. They are discretized and solved in three-dimensional Cartesian geometry using a finite volume method incorporating the SIMPLEC algorithm [4],[5]. The sheath regions at the interfaces between the electrodes and arc are not explicitly modelled; instead internal heat and momentum transfer boundary conditions are used.

The droplets transfer mass, energy and momentum to the weld pool. The energy and momentum transfers between the droplets and the arc and weld pool are averaged over the droplet cross-sectional area and over time [2].The transfer of mass by the droplets ensures that the weld pool surface is not flat; the shape of the surface is calculated by minimizing the total surface energy, taking into account the volume of metal transferred, the surface tension and surface curvature of the weld pool, the pressure exerted by the arc and droplets, and buoyancy in the weld pool [1]. Large amounts of metal vapour are formed from the tip of the wire, the droplets and the weld pool; the production and transport of the vapour is calculated self-consistently [3]. The wire and base metal are typically made of different alloys; the mixing of the alloys in the weld pool is calculated using

$$\nabla \cdot (\rho \mathbf{v} Y_d) = -\nabla \cdot (\rho D_d \nabla Y_d) + S_d, \qquad (1)$$

where Y_d is the mass fraction of the droplet alloy, $\mathbf{v}$ the flow velocity, ρ the mass density, D_d the diffusion coefficient and S_d the source term representing the addition of droplet alloy to the weld pool.

A graphical user interface has been developed for the model, allowing it to be run by non-expert users under the Windows 64-bit operating system on a standard desktop or laptop computer.

Results and comparison with measurements

Some results are presented here for a lap fillet weld with the parameters presented in *Table 1*. *Fig. 1(a)* shows the distribution of temperature in a vertical plane through the wire, arc and weld pool. The arc temperature is relatively low, due to the strong cooling caused by the radiation from aluminium vapour, which is calculated to have a concentration of over 80 wt% in the centre of the arc. The molten tip of the wire and the weld pool are clearly apparent in the figure; the liquidus temperature is 875 K. *Fig. 1(b)* shows the mixing of the droplet alloy into the weld pool. The droplets impact the weld pool in the arc attachment region, and drive flow downwards. The droplets mix with the base metal through the recirculating flow patterns that develop, and by diffusion. *Fig. 2* shows that mixing is, however, incomplete, with the predicted mass fraction of the droplet alloy in the fusion zone varying from below 20% to above 50%. The calculated distribution is compared in *Fig. 2* with measurements of the composition of the weld. The silicon concentration in a cross-section of the welded metal was measured using an FEG–EPMA (field-emission gun – electron-probe microanalyser) on a scanning electron microscope. Since the wire and droplets contain silicon, while the base metal does not, the droplet mass fraction and the silicon concentration should correspond. There is good qualitative agreement, with the both the silicon and the droplet alloy having the largest concentration in the lower left of the weld pool, and lowest concentration in the top left.

Table 1: Parameters used in the calculations presented here

Parameter	Value
Arc current (average)	95 A
Shielding gas flow	Argon, 14 L/min
Distance from wire tip to workpiece (along axis of wire)	5 mm (before welding)
Welding speed	0.9 m/min (15 mm/s) in $-y$ direction
Wire diameter	1.2 mm
Wire feed rate	4.32 m/min (72 mm/s)
Droplet detachment frequency	93 Hz
Work and travel angle (from horizontal)	60° , 90°
Wire alloy	AA 4043 (Al + 5.3 wt% Si)
Workpiece thickness (each sheet)	3 mm
Workpiece alloy	AA 5754 (Al + 3.1 wt% Mg)

Thermal history, and residual stress and distortion

As noted in the Introduction, FEA models of residual stress and distortion use the thermal history of the base

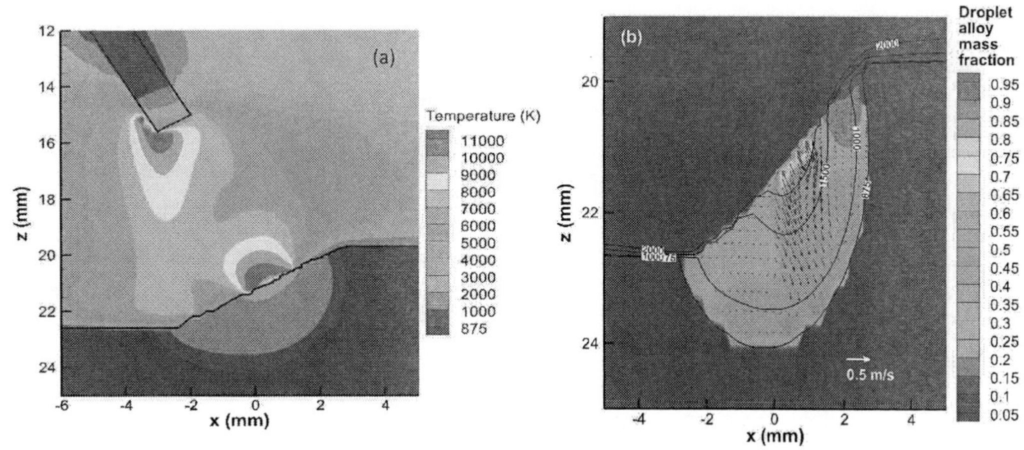

Fig. 1 (a) Temperature in a vertical plane through the wire, arc and workpiece. (b) Mass fraction of the droplet alloy in a cross-section of the weld pool; velocity vectors, and temperature contours for 875, 1000, 1500 and 2000 K are also shown. The welding direction is into the page.

metal as input data. The thermal history is usually calculated using a heat transfer model that represents the arc and weld pool by an approximate two- or three-dimensional heat source, such as a uniform ellipsoidal heat source

$$q = \varepsilon VI \exp\left\{-\left[(x/a)^2 + (y/b)^2 + (z/c)^2\right]\right\}, \tag{2}$$

where V is the arc voltage, I is the arc current, and ε, a, b and c are free parameters respectively representing the efficiency of the heat transfer and the radii of the ellipsoid. The free parameters are determined by comparing measured thermal histories, obtained using thermocouples, with predictions of the heat transfer model [6].

The CFD model calculates temperature and alloy composition throughout the workpiece, and the shape of the reinforcement. It is a simple matter to transform this data into time-dependent quantities using the welding speed. These data can then be input to FEA codes that calculate residual stress and distortion. This replaces the use of an approximate heat source, and the removes the requirement for thermocouple measurements.

We tested this approach by using the thermal histories and reinforcement shape from our CFD model as input to a model of residual stress and distortion, which used the MSC Marc FEA software package. As the weld progresses along the sheet, additional metal elements are added to account for the reinforcement, and the workpiece is re-meshed. The material yielding behaviour was assumed to follow the von Mises criterion, and the Johnson–Cook constitutive model [7] was used to describe the evolution of yield. Because of the difficulty in finding suitable data for the alloys that were used in the CFD model, we instead used thermomechanical data for AA 2219, which is an aluminium alloy containing 6 wt% Cu and 0.3% Mn. The sheets used for the simulation

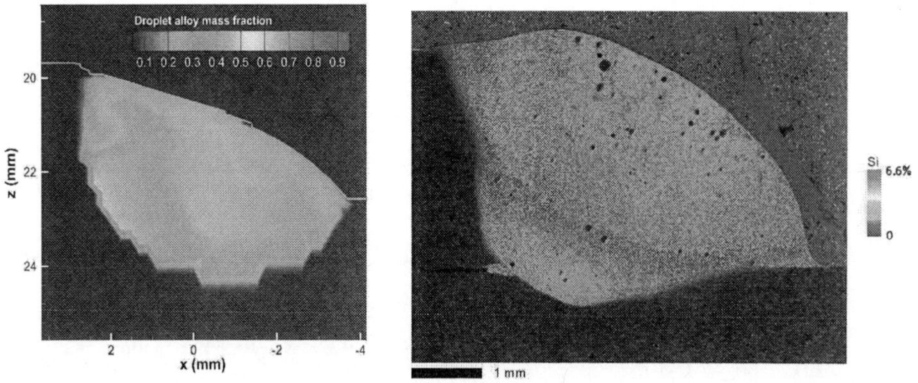

Fig. 2: Calculated droplet alloy mass fraction (left) and measured silicon mass fraction (right) in a vertical weld cross-section. The scale in the top image is from 0% Si (black) to 6.6% Si (red). Note that the x-direction has been reversed compared to Figs. 1 and 2.

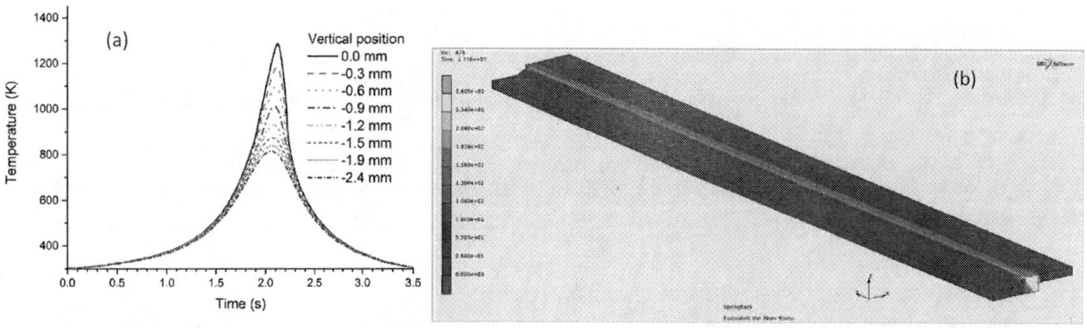

Fig. 3: (a) calculated thermal history at different vertical positions at x = 0 (see Fig. 1); the vertical position is the distance below the top of the lower workpiece sheet before welding. (b) Von Mises stress in the workpiece after welding, cooling and removal of restraints; the scale is from 0 to 260 MPa, in increments of 26 MPa. Welding is from the bottom right to the top left, in the y-direction.

were each 20 mm wide, 300 mm long and 3 mm thick, with an overlap of 5 mm in the case of lap-fillet geometry. The welding speed used was 0.4 m/min, which is significantly slower than that used in the arc model; this means that the time scale in the thermal histories shown in *Fig. 3(a)* was expanded by a factor of 2.25.

Fig. 3(b) shows the predicted von Mises stress in the workpiece after the completion of welding and the removal of restraints. The stress is concentrated around the weld seam. The workpiece is predicted to deform by 1.8 mm in the x-direction, 0.75 mm in the y-direction and 2.1 mm in the z-direction, all relative to the corner closest to the viewer in the figure. Welded sheets of the same length and thickness, but of width 50 mm, made of AA5754 alloy and welded at 0.9 m/min, show a mid-point 1.5±0.5 mm lower than the ends; i.e, a deformation in the z-direction of 3±1 mm. This is in agreement with the predictions; however it is important to note that, because of the differences between the model and experiment, the results should be regarded only as an indication of the usefulness of the approach, rather than a demonstration of its accuracy.

Conclusions

We have demonstrated the feasibility of combining a CFD model of the arc and weld pool with an FEA model of residual stress in the workpiece, through calculating the thermal history and reinforcement shape with the CFD model. This approach means that measurements of the thermal history are not required, unlike in standard approaches that use an approximate representation of the arc and weld pool by a heat source. A similar approach could be applied to models that predict microstructure. This points the way to the development of a complete model of arc welding that requires only the standard benchmarking necessary for any model, rather than detailed thermal history measurements.

References

[1] A.B. Murphy: "A Self-Consistent Three-Dimensional Model of the Arc, Electrode and Weld Pool in Gas–Metal Arc Welding," *J. Phys. D: Appl. Phys.*, Vol. 44, No. 19 (2011), art. 194009.

[2] A.B. Murphy: "Influence of Droplets in Gas–Metal Arc Welding – a New Modelling Approach, and Application to Welding of Aluminium," *Sci. Technol. Weld. Join.*, Vol. 18, No. 1 (2013), pp. 32-37.

[3] A.B. Murphy: "Influence of Metal Vapour on Arc Temperatures in Gas–Metal Arc Welding: Convection versus Radiation," *J. Phys. D: Appl. Phys.*, Vol. 46, No. 22 (2013), art. 224004.

[4] S.V. Patankar: Numerical Heat Transfer and Fluid Flow, Hemisphere (Washington DC, 1980).

[5] J.P. van Doormal and G.D. Raithby: "Enhancements of the SIMPLE Method for Predicting Incompressible Fluid Flows," *Numer. Heat Transfer*, Vol. 7, No. 2 (1984), pp. 147-163.

[6] O. Muránsky, *et al:* "Comprehensive Numerical Analysis of a Three-Pass Bead-in-Slot Weld and its Critical Validation using Neutron and Synchrotron Diffraction Residual Stress Measurements'" *Int. J. Sol. Struct.*, Vol. 49, No. 9 (2012), pp. 1045-1062.

[7] G.R. Johnson and W.H. Cook: "A Constitutive Model and Data for Metals Subjected to Large Strains, High Strain Rates and High Temperatures," *Proc. 7th Int. Symp. on Ballistics*, The Hague, April 1983, pp. 541-547.

Visualization of Metal Evaporation from Tungsten-Based Cathode in Ar-N_2 DC Arc

Manabu Tanaka, Taiki Kondo, Naoki Sakura, and Takayuki Watanabe
Kyushu University, Fukuoka, Japan
watanabe@chem-eng.kyushu-u.ac.jp, +81-92-802-2745

Abstract
The purpose of the present study is to investigate the erosion mechanism of tungsten-based cathode in a Ar-N_2 DC arc at atmospheric pressure. Cathode erosion rates and surface temperatures were measured. In addition, the metal vapor generated from the cathode surface was successfully visualized by a high-speed camera system with appropriate bandpass filters. To understand the effect of N_2 addition on the cathode erosion, the N_2 concentration in the Ar atmosphere was changed from 0 to 70vol%. The Ar-N_2 and Ar-H_2 arcs were compared to clarify the effect of N_2 addition. Obtained results indicated that the increase of N_2 concentration led to drastic increase of the cathode erosion rate, while H_2 addition had negligible effect on the cathode erosion. Measured surface temperature at the cathode tip in Ar-N_2 arc was about 4150 K, which temperature was higher than that of 3600 K in Ar-H_2 arc. This difference can be explained by the different recombination temperatures between N_2 and H_2. Higher recombination temperature of N_2 than the cathode surface temperature led to the cathode heating at cathode surface and/or molten cathode tip. Furthermore, the experimental results from the metal vapor visualization revealed that the tungsten vapor was generated from the periphery of the high temperature cathode spot. These understanding of the electrode phenomena in N_2 arc enable to utilize the N_2 arc in industrial field.

Introduction
Thermionic cathodes are the cathode where thermionic emission is the main electron emission mechanism. These are utilized in arc welding, plasma arc cutting, plasma spraying, high-intensity discharge lamp, nanoparticle synthesis, and so on. All thermionic cathodes erode depending on a specific application. Cathode erosion in the DC arc is of great importance in determining process performance such as welding or cutting characteristics and the electrode lifetime. Therefore, the experimental and theoretical investigation on the cathode erosion have been intensively conducted [1-3].

The DC arc method for nanoparticle synthesis has attracted great interest because it is a promising route to prepare high-purity metal and/or intermetallic nanoparticles in single step. *Figure 1* shows a conceptual diagram of the nanoparticle synthesis process by the DC arc method. First, raw material metal mixture as anode are evaporated from the molten weld pool by arc heating. Then, the vapors of the constituent metals are transported to the low temperature region. As a result, nanoparticles are formed though homogenous nucleation, heterogeneous co-condensation, and particle growth due to collision and coagulation with each particle.

Addition of molecular gas such as H_2 or N_2 into the inert gas arc brings great advances in the nanoparticle production. Effective method for metal nanoparticle production by a "hydrogen plasma-metal" reaction in Ar-H_2 DC arc with high productivity of the order of several g/min at less than 10 kW power input have been reported [4, 5]. Furthermore, Ar-N_2 arc has also attracted many researchers because of its low cost. Mass production of metal nanoparticles via the DC arc method in molecular gases atmosphere enables to enhance capability of thermal plasmas in many industrial processing.

The electrode phenomena of the DC arc in molecular gas atmosphere, however, have rarely been studied. The purpose of the present study is to investigate the cathode erosion mechanism in DC arc with molecular gas. Temperature measurements of the cathode surface were conducted as well as the cathode erosion rate measurements. Furthermore, W vapor visualization by a high-speed camera with appropriate bandpass filters was conducted.

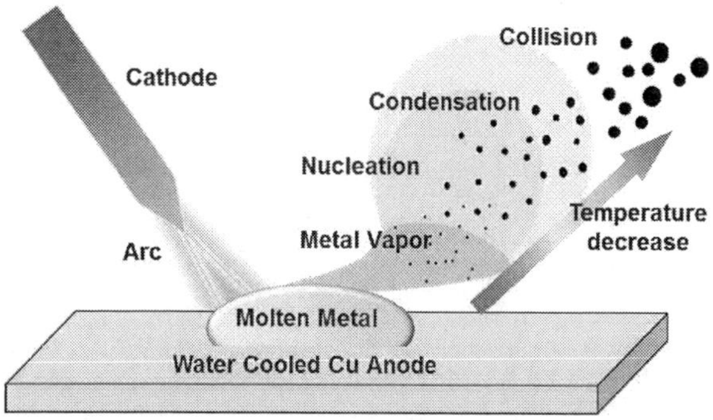

Fig. 1 Conceptual image of the nanoparticles synthesis in the DC arc method.

Experimental Procedure

Experimental Setup of DC arc

Figure 2 shows a schematic illustration of experimental setup. This setup consists of a DC power supply, an arc chamber, a particle collector, and a gas circulator. A 2wt%-thoriated tungsten cathode with 6 mm in diameter and a Mo ingot placed on a water-cooled copper anode are used as electrodes. Ar-N_2 or Ar-H_2 gas mixture was filled in the chamber after several times of the evacuation by the rotary vacuum pump. In order to investigate the effect of molecular gas addition on the electrode phenomena, concentration of N_2 gas was changed from 0 to 70vol% at constant pressure of 1 atm. The concentration of H_2 gas was also changed from 0 to 50vol%. Shield gas of 15 L/min and circulation gas of 150 L/min were fixed. Arc current was changed from 100 to 300 A. Evaporation rate of the cathode was measured by the weight difference of the cathode after 10 min arcing.

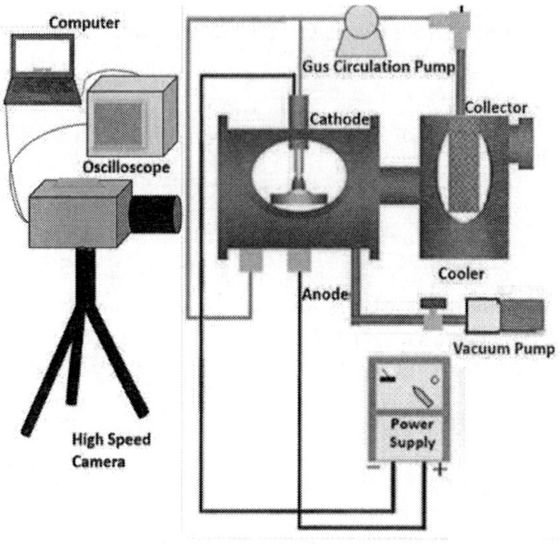

Fig. 2 Schematic image of the experimental setup and measurement system.

Temperature Measurement of Cathode Surface

Cathode surface temperature was measured by the high-speed video camera (FASTCAM SA5, Photron) with appropriate bandpass filters. Conceptual diagram of the measurement system is shown in *Fig. 3*. Two color pyrometry was used to evaluate the two-dimensional temperature distribution of cathode surface. Since the emissivity ε is unknown for many applications, the gray body assumption was applied where it is hypothesized that the surface emissivity is independent of wavelength. Therefore, surface temperature is expressed as follows:

$$T = \frac{C_b(\lambda_1 - \lambda_2)}{\lambda_1\lambda_2}\left[\ln\left(\frac{I_1}{I_2}\right) + 5\ln\left(\frac{\lambda_1}{\lambda_2}\right)\right]^{-1} \quad (1)$$

where, λ_1 and λ_2 are selected wavelengths of two bandpass filters; C_b equals to 1.439×10^{-2} m·K,; and I_1 and I_2 corresponds to radiation intensities at the wavelengths of λ_1 and λ_2, respectively. In this system, the emissions from the arc was divided by an optical splitting mirror. Divided emissions were then passed through the bandpass filters with different wavelengths. Wavelengths of 785 nm ± 2.5 nm and 880 nm ± 5 nm were selected to measure the surface temperature because of negligible emissions from the arc around the above wavelengths region. Considering bandwidths of bandpass filters and the sensitivity of high-speed camera, this pyrometry system was calibrated by a tungsten lamp which was used as a standard light source. The details of this measurement setup can be referred in our previous work [6].

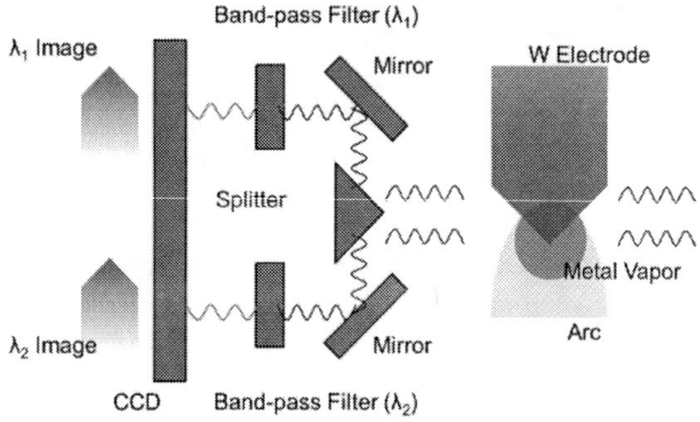

Fig. 3 Conceptual diagram of measurement system for two-color pyrometry and vapor visualization.

High-Speed Visualization of Metal Vapor

Tungsten vapor from 2wt%-thoriated tungsten cathode during the arc operation was visualized by the same system with the above-mentioned temperature measurements as shown in *Fig. 3*. In general, emissions from thermal plasmas include many emissions such as line/continuum spectra from the arc and the continuum spectra from the surface of electrodes and/or inner wall. Hence, it is difficult to observe the only particular emission separating from others. Therefore, the bandpass filter system was applied to visualize the metal vapor in the arc to solve above difficulty. Spectroscopic measurements (iHR550, Horiba Jpbin Yvon) were conducted to determine the appropriate wavelengths of the bandpass filters. A pair of the bandpass filters with the filter wavelengths of 430 nm ± 5 nm for evaporated tungsten vapor and 738 nm ± 3 nm for atomic emission from argon was carefully selected. Relative intensity of tungsten vapor to atomic argon was calculated to evaluate the relative concentration of metal vapor in the arc region.

Results and Discussion

Erosion Rates and Surface Temperature

Figure 4 shows the effect of the molecular gas addition on the cathode erosion rate. An increase of N_2 concentration led to drastic increase of cathode erosion rate. On the other hand, the erosion rate in Ar-H_2 arc was almost same value with that in Ar arc. In order to clarify the reason for this difference among the different molecular gases, the cathode temperature measurements were conducted.

Figure 5 shows the cathode temperature distributions for 50%-N_2 and 50%-H_2 arcs. In the case of the 50%-N_2 arc, cathode tip temperature was higher than melting point of tungsten (3695 K) and cathode tip became molten state. On the other hand, the cathode tip temperature for 50%-H_2 arc was less than the melting point of tungsten. Figure 6 summarizes the effect of molecular gas concentration on the cathode tip temperature. The cathode tip temperatures for 30 to 70vol% of N_2 concentration were similar values of 4150 K. In contrast, the cathode tip temperature for 50%-H_2 arc was about 3600 K.

Above different temperature for different gases can be explained by the different recombination temperatures of the molecular gases. Recombination temperature of nitrogen is 7500 K, which temperature is sufficient higher than the cathode tip temperature. Hence, the nitrogen atoms recombine at cathode surface and/or the molten pool on the cathode tip, resulting in the heating of the cathode tip due to the nitrogen recombination energy. In contrast, recombination temperature of hydrogen is 3500 K, which temperature is comparable with the cathode tip temperature. Therefore, the effect of cathode heating due to the hydrogen recombination energy is relatively lower, resulting in the lower cathode tip temperature for H_2 arc.

Visualization of Metal Vapor in Ar-N_2 Arc

Effect of N_2 addition on the cathode erosion was experimentally clarified in the previous paragraphs. Then, the cathode erosion mechanism in N_2 arc was investigated by the visualization technique of the electrode phenomena.

Figure 7 shows the relative intensity distribution of two synchronized images in the different wavelengths, 430 nm ± 5 nm for W vapor and 738 nm ± 3 nm for atomic Ar. This distribution was observed during 50%-N_2 arc at 300 A of arc current in order to observe the W emission clearly. Three bright regions were observed in the relative intensity distribution. Two of them correspond to the periphery of the arc. According to the spectroscopic measurements, the emissions from W vapor at 427, 429 and 430 nm was observed only from the periphery of the arc. This means that the W vapor existed at the periphery

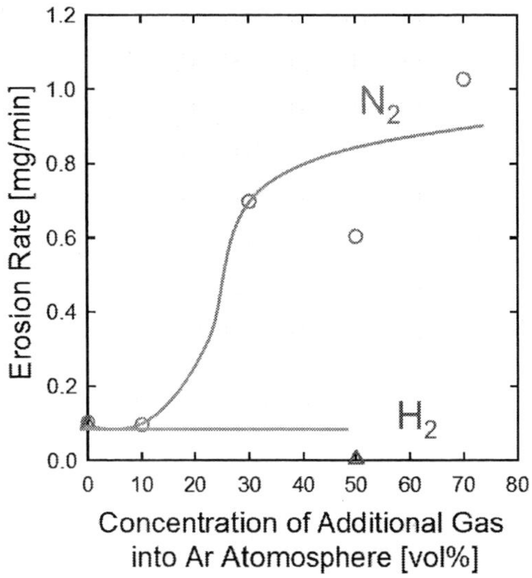

Fig. 4 Effect of molecular gas concentration on cathode erosion rates.

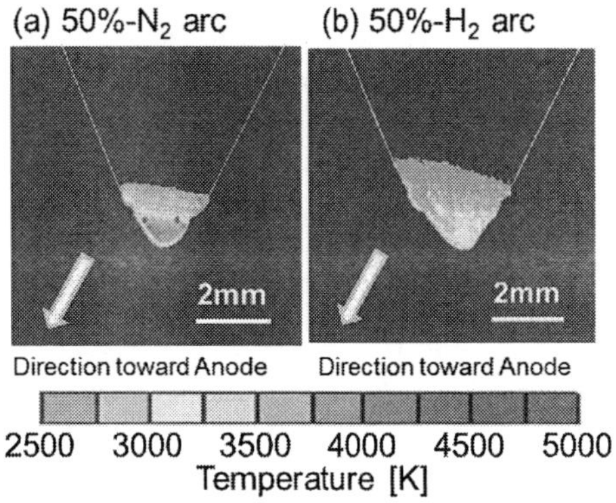

Fig. 5 Cathode temperature distributions for 50%-N_2 (a) and 50%-H_2 arc (b) at fixed arc current of 100 A.

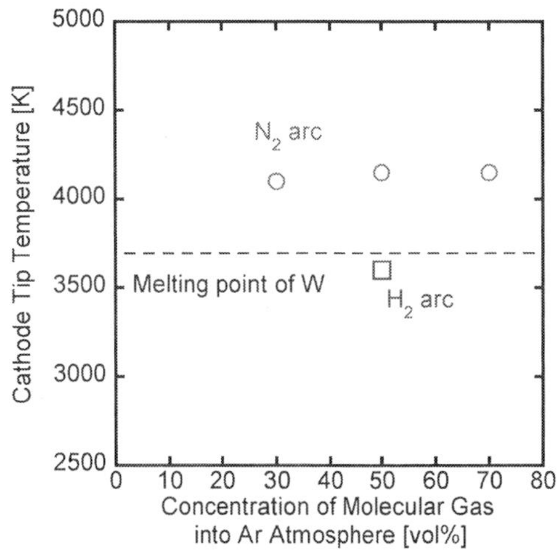

Fig. 6 Effect of molecular gas concentration on cathode surface temperature.

of the arc. Contrastingly, the bright region in the center of the arc corresponds to the emissions from Ar ion at 433 and 435 nm. Although the emission intensities from Ar ion were not negligible, the emissions from Ar ion can be distinguished from the W vapor due to their different existence regions. Consequently, the W vapor from the cathode during Ar-N_2 arc was successfully visualized at the periphery of the arc.

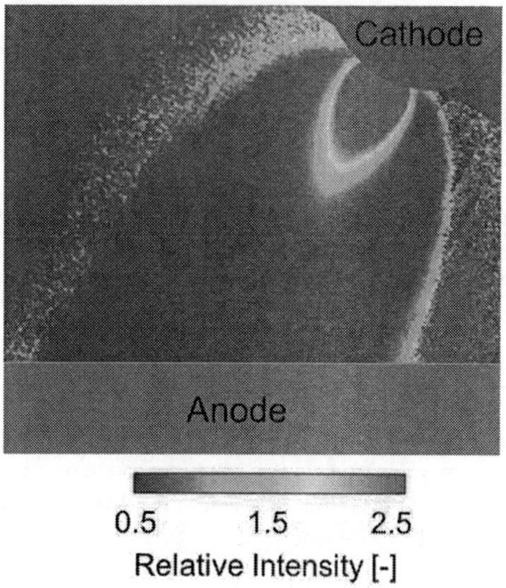

Fig. 7 Relative intensity distribution of arc region filtered at 430 nm to that at 738 nm during 50%-N_2 arc with 300 A of arc current. Image of 430 nm corresponds to line emissions from W vapor and Ar ion. Image of 738 nm corresponds to line emissions from atomic Ar.

The reason for the existence of the W vapors only at the periphery of the arc can be explained by two different mechanisms. First one is discussed based on the present experimental results. Tungsten evaporates from the whole region of the cathode surface where the surface temperature and the corresponding vapor pressure are sufficient high. However, most of W vapor returns back to the cathode surface because the W vapors in the arc center region are ionized soon after their evaporation due to the higher arc temperature than 10000 K. In contrast, the W vapor which evaporates from the fringe of the cathode spot does not return back to the cathode surface because of the lower arc temperature at the periphery of the arc. As a result, the evaporation only from the fringe of the cathode spot has an important role on the cathode evaporation.

Another possible mechanism of cathode erosion is hypothesis based on the formation of intermediate compounds such as tungsten nitride. Evaporation of tungsten nitride with higher vapor pressure than metal tungsten enhances the cathode erosion. These mechanisms are currently under investigation.

Conclusion

Visualization system with the high-speed camera combined with the bandpass filter system was applied to observe the metal vapor from the tungsten-based cathode in Ar-N_2 DC arc. Tungsten vapor evaporated from the cathode surface in the Ar-N_2 DC arc was successfully visualized only at the periphery of the arc. The experimental results revealed that the N_2 addition into the arc led to drastic increase of the cathode erosion. The cathode erosion in Ar-N_2 arc is mainly attributed to the tungsten and/or tungsten nitride evaporation from the periphery of the cathode spot.

References

[1] X. Zhou and J. Heberlein: "An Experimental Investigation of Factors Affecting Arc-Cathode Erosion," *J. Phys. D: Appl. Phys.*, Vol. 31, (1998), pp. 2577-2590.
[2] A.A. Sadek, and M Ushio, *et al.*: "Effect of Rare Earth Metal Oxide Additions to Tungsten Electrodes," *Metallurgical Trans. A*, Vol. 21A, (1990), pp. 3221-3236.
[3] V. Nemchinsky: "Erosion of Thermionic Cathodes in Welding and Plasma Arc Cutting Systems," *IEEE Trans. Plasma Sci.*, Vol. 42, (2014), pp. 199-215.
[4] M. Tanaka and T. Watanabe: "Vaporization Mechanism from Sn-Ag Mixture by Ar-H_2 Arc for Nanoparticle Preparation," *Thin Solid Films*, Vol. 516, (2008), 6645-6649.
[5] M. Tanaka and T. Watanabe: "Mechanism of Enhanced Vaporization from Molten Metal Surface by Argon-Hydrogen Arc Plasma," *Jpn. J. Appl. Phys.*, Vol. 52, (2013), 076201.
[6] M. Tanaka, T. Hashizume, T. Imatsuji, Y. Nawata, and T. Watanabe: "Investigation on Erosion mechanism of Tungsten-Based Electrode in Multiphase AC Arc by High-Speed Visualization of Electrode Phenomena," Jpn. J. Appl. Phys., Vol. 55, (2016), 07LC01.

3D temperature measurement of MIG arc plasma by multidirectional spectroscopic method

Kazufumi Nomura, Kaichiro Yoshii, Kaname Toda, Kohei Mimura, Yoshinori Hirata, Satoru Asai

Osaka University, Japan

Spectroscopic measurements for GMA phenomena have been recently performed. The studies have reported that the metal vapor behavior greatly affects the arc properties. However, they can be applied only to axially symmetric phenomena because of the assumption used for the measurement. GMA welding are normally performed with a travel speed and most of the phenomena become axially asymmetric. Tomographic measurement is required to obtain the internal emission intensity distribution of axially asymmetric GMA phenomena. This study constructed the multidirectional and heterochromatic measurement system which can capture such axially asymmetric plasma. The system was composed of 12 CCD cameras with three types of narrowband interference filters. We used one Ar I line and two Fe I lines. Each line intensity distribution was obtained through the image reconstruction method. Then we converted the line intensities into plasma temperature and iron vapor concentration. For this conversion, we used the relative intensity method, the absolute sensitive calculation, solving Saha equation, LTE assumption. We measured a pulsed MIG welding process and could observed that the temperature of the central area of MIG arc plasma was lower rather than the outer area as shown in recent studies. We found such temperature and iron vapor distributions were drawn against the welding direction. The time variation of iron vapor concentration was corresponding to the current pulse shape. We succeeded to reveal three dimensional temperature distribution and iron vapor concentration.

Radiation and Temperature Distribution of Argon Arc Contaminated with Nitrogen in Pulsed TIG Welding

Toru Iwao*, Hiroki Takahashi*, Yoshifumi Maeda*, Shinji Yamamoto*
**Tokyo City University, Setagaya, Tokyo 158-8557, Japan*
tiwao@tcu.ac.jp

Shinichi Tashiro**, Masaya Shigeta**, Manabu Tanaka**
Joining and Welding Research Institute, Osaka University
*11-1, Mihogaoka, Ibaraki, Osaka, 567-0047, Japan***

Abstract

TIG arc welding is the high-quality joining technology. However, the current has to be small because the cathode melting should be prevented. In this case, the welding pool is shallow, then, the welding defect sometimes occurs. The pulsed TIG arc welding is often used in order to obtain the deep penetration of the welding pool. However, few report has reported the distribution of arc contaminated with nitrogen. It is important to prevent the contamination of nitrogen because the melting depth increases in order to prevent the welding defects. In this paper, the distribution of argon arc contaminated with nitrogen in pulsed TIG welding is elucidated. In addition, the radiation and temperature distribution of argon arc contaminated with nitrogen in pulsed TIG welding are elucidated. The nitrogen concentration, the peclet number and the arc temperature are calculated using the EMTF (Electro-Magnetic Thermal Fluid) simulation at the time of peak current. As a result, the nitrogen concentration into the arc becomes high. The radial peclet number decreased because the arc temperature increases near the cathode. Moreover, the arc shrinks because of an electromagnetic pinch force near the cathode. However, the nitrogen concentration near the anode becomes low. The strong convection outward near the anode is caused by the temperature increment at the axial center.

Introduction

The TIG (Tungsten Inert gas) arc welding is the welding technology which uses the tungsten cathode and the inert gas. It has the characteristics of prevention of melting the electrode, because the shielding gas can separate the inert gas and air at the welding pool. TIG arc welding can prevent the contamination of the impurities to the welding pool, the reliability and quality of welding become very high, and the kinds of joining materials for welding are a lot [1]. Therefore, TIG arc welding has been used for various construction with reliability well. However, TIG arc welding has some defects. Because the cathode does not melt, the high current should not be used. Therefore, the heat transfer of TIG arc welding is smaller than that of MIG and MAG welding. The TIG arc welding can not melt the welding pool deeply. In this case, the welding pool spreads to the radial direction, and the weld defect sometimes occurs. The pulsed arc welding has been used for the improvement of this defect. The pulsed arc welding can control the heat transfer to the anode, the convection and driving force in the welding pool, because the current and temperature of arc can change periodically.

However, these pulsed TIG arc welding has been researched under the conditions of shielding property. Therefore, few report has reported the distribution of arc contaminated with nitrogen in pulsed TIG arc welding. In order to prevent the welding defect, it is necessary to improve the welding pool shape. Moreover, it is necessary to prevent contaminated with nitrogen into the welding pool. For example, the strength of the metal is lowered when the blowhole occurred because the nitrogen is mixed in the weld metal [2]. Moreover, it has been reported that the toughness of the metal deteriorates even if the small amounts of nitrogen is contaminated. The nitrogen contamination of the welding pool is caused by convection and diffusion in the arc, because the thermodynamic transport properties of the arc changes. Thus, the nitrogen contamination in the metal has been reported such as its dependence with the welding conditions [3]. For this reason, it is presumed to affect the contaminated phenomenon of nitrogen, because the arc shape of pulsed arc is changed by time. Therefore, it is important to elucidate the nitrogen concentration of pulsed arc for welding defect prevention of pulsed TIG arc welding.

In this paper, this research elucidates the nitrogen concentration distribution of argon arc contaminated with nitrogen in pulsed TIG welding. In addition, the radiation and temperature distribution of argon arc contaminated with nitrogen in pulsed TIG welding is elucidated.

Calculation method

Fig.1 shows the calculation area of a two-dimensional axisymmetric cylindrical coordinate system is constructed with the electrode and the arc being one system. Table 1 shows the boundary condition. The cathode is thoria tungsten, the anode is SUS304. *Fig.2* shows the current waveform. The current is to be 100 to 200 A and increment of current is 0.1 ms. The atmospheric gas is the nitrogen. The nitrogen contamination ratio is 0 to 100%, the pressure is 0.1 MPa. The gas flow rate is 4 slm. The calculation is LTE, laminar flow, optical thin, metal vapor is not considered, and welding pool is flat. This calculation uses SIMPLER method, based on the finite volume method. This calculation is used by dominant equations with mass conservation equation, momentum conservation equation (axial direction and radial direction), energy conservation equation, current continuity equation, Ohm's law, ampere's law, electrical field. The nitrogen concentration is calculated using *Eqs. (1)* and *(2)*. In this calculation model, the electrode and the arc are regarded as one system to perform calculations. The energy transfer between the arc and the electrode is obtained from the following equations and added as a generation term of energy conservation equation. Income and outgo of energy is calculated at cathode and anode.

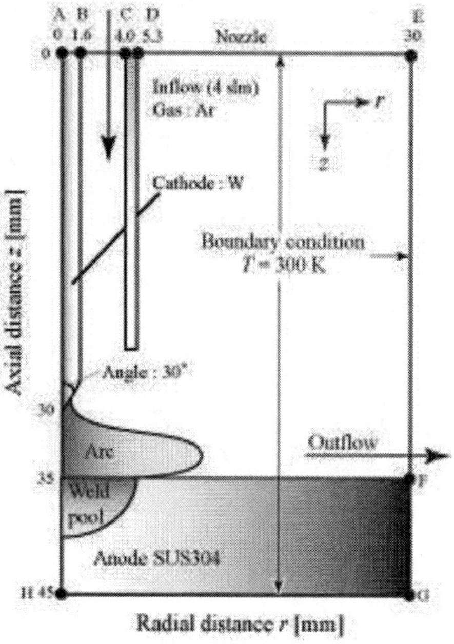

Fig.1: Calculation area.

Convection – diffusion equation

$$\frac{\partial}{\partial t}(\rho C)+\frac{1}{r}\frac{\partial}{\partial r}(r\rho v_r C)+\frac{\partial}{\partial z}(\rho v_z C)=\frac{1}{r}\frac{\partial}{\partial r}\left(r\rho D\frac{\partial C}{\partial r}\right)+\frac{\partial}{\partial z}\left(\rho D\frac{\partial C}{\partial z}\right) \quad (1)$$

Viscosity equation[3)]

$$D=\frac{2\sqrt{2}\left(1/M_1+1/M_2\right)^{0.5}}{\left[\left(\rho_1^2/\beta_1^2\eta_1^2M_1\right)^{0.25}+\left(\rho_2^2/\beta_2^2\eta_2^2M_2\right)^{0.25}\right]} \quad (2)$$

Therein, suffixes 1 and 2 respectively denote Ar and N_2; C is the nitrogen concentration; M is the molar weight; β_1 and β_2 respectively signify the constant number (=1.385).

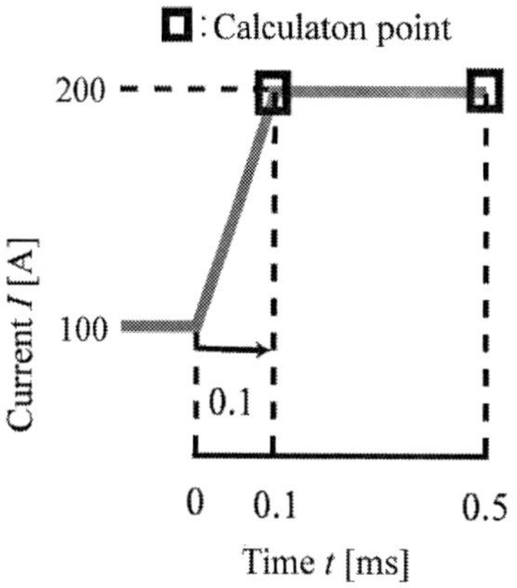

Fig.2: Current wave form.

Results and discussion

Fig.3 shows the distribution of temperature and nitrogen concentration, vector distribution of flow velocity and peclet number. The nitrogen concentration becomes high into the arc at 0.5 ms. The inward force affected by the electromagnetic pinch force and nitrogen are mixed by diffusion and temperature increment.

Fig.4 shows the radial distribution of the radial peclet number in near the cathode (z = 30.0). The convection of inward becomes stronger in the range of 2 - 4 mm at the 0.5 ms. Thus, the increment of nitrogen concentration in the arc means that the nitrogen easily mixed. The change in the peclet number means that the diffusion coefficient and the flow rate is changed.

Fig.5 shows the radial distribution of diffusion coefficient and the radial flow velocity near the cathode (z = 30 mm). The difference between the diffusion coefficient and the radial flow velocity is small. However, the diffusion coefficient is significantly decreased over 3 mm. The temperature of arc near the cathode does not increased, because the arc shrinks affected by the electromagnetic pinch force near the cathode. Thus, the flow velocity of the concentration of nitrogen over 3 mm region is dominant. Therefore, the decrement of the radial peclet number at 0.5 ms is caused by the radial flow velocity of inward increment, because the electromagnetic force increased with increasing the temperature at the axial center over time

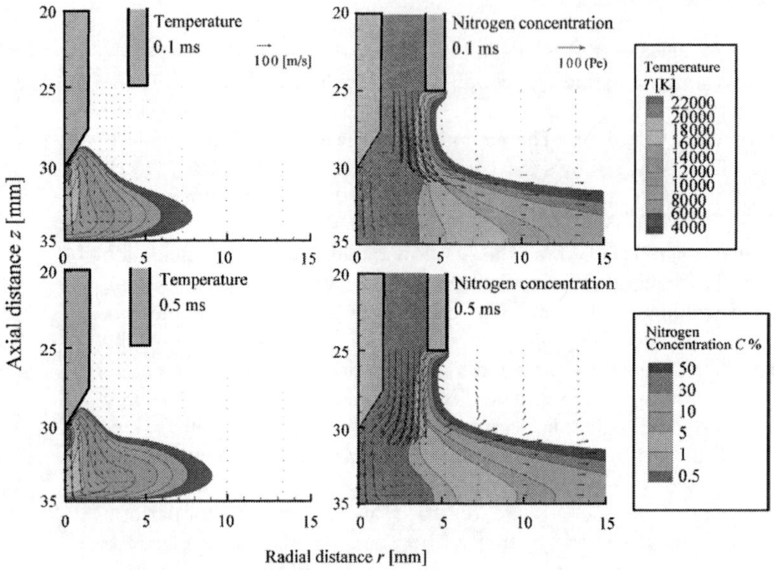

Fig. 3: Distribution of arc temperature, nitrogen concentration and vector distribution of flow velocity and peclet number.

Fig.6 shows the radial distribution of the nitrogen concentration near the welding pool (z = 34.9mm) The nitrogen concentration with time elapsed at the peak time is lower than that at the steady at 200 A. The temperature distribution is constricted. The temperature gradient increased in 0.5 ms because of the temperature at the axial center is high. In this case, the axial flow velocity increased. The axial flow rate increases to the radial direction side when it collided with the anode. Thus, the nitrogen concentration in the axial center becomes almost same as the steady at 200 A. In addition, the diffusion coefficient in the radially decreased, because the arc temperature decreased with the transient response. Therefore, the nitrogen concentration decreased at a whole radial direction, because the convection from the axial center becomes dominant. *Fig.7* shows the radial distribution of peclet number (z = 34.6 mm). The convection is dominant near welding pool.

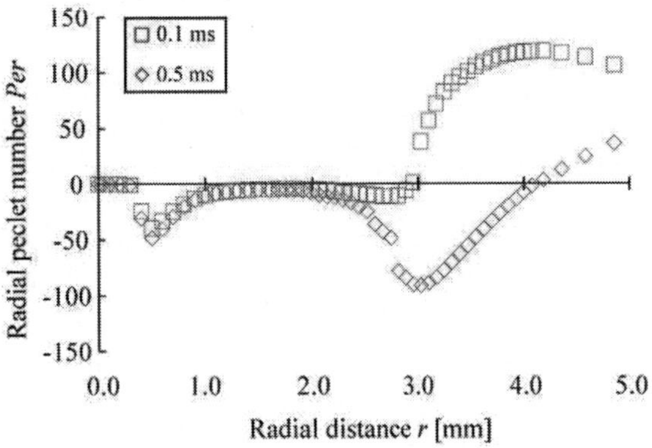

Fig.4: Peclet number with radial distance (z = 30.0 mm)

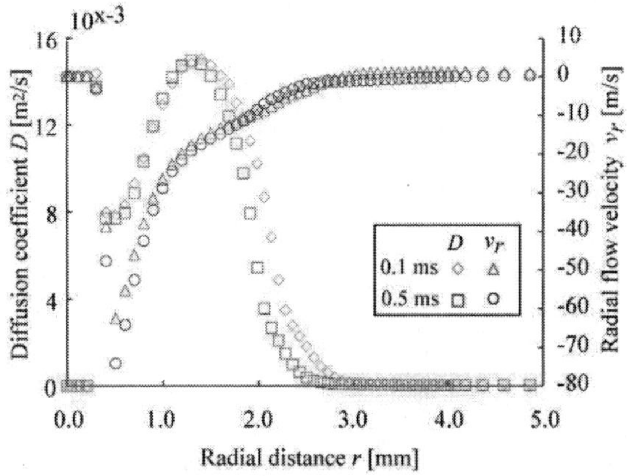

Fig.5: Diffusion coefficient and radial flow velocity (z = 30.0 mm)

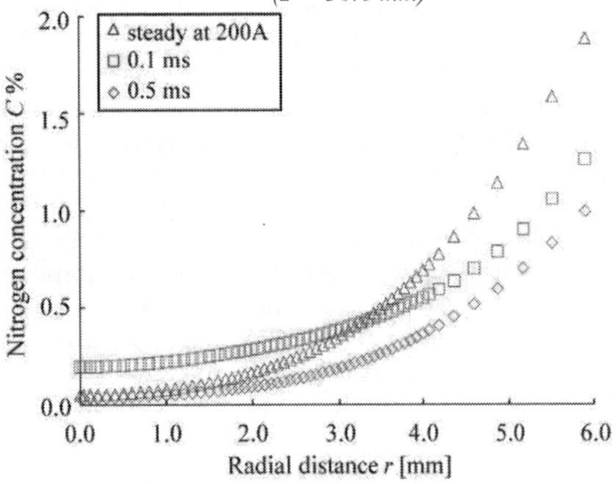

Fig.6: Nitrogen concentration near welding pool (z = 34.9 mm)

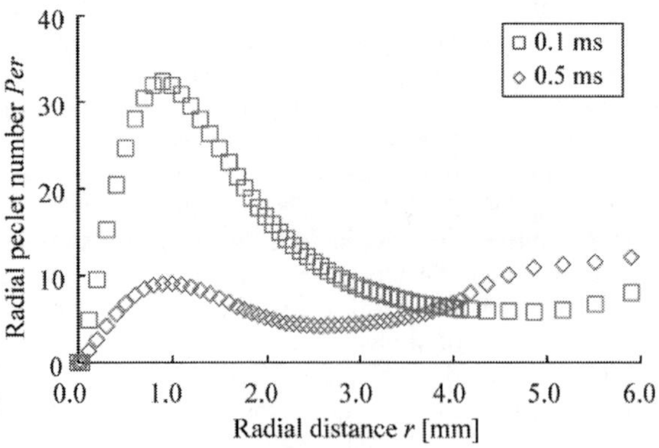

Fig.7: Peclet number (z = 34.6 mm)

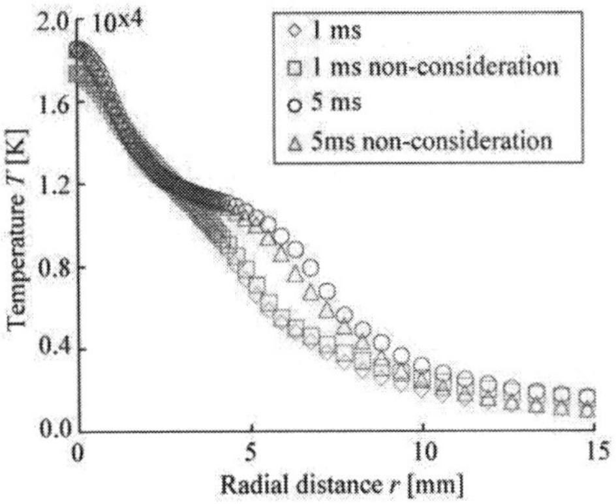

Fig.8: Temperature with radial distance.

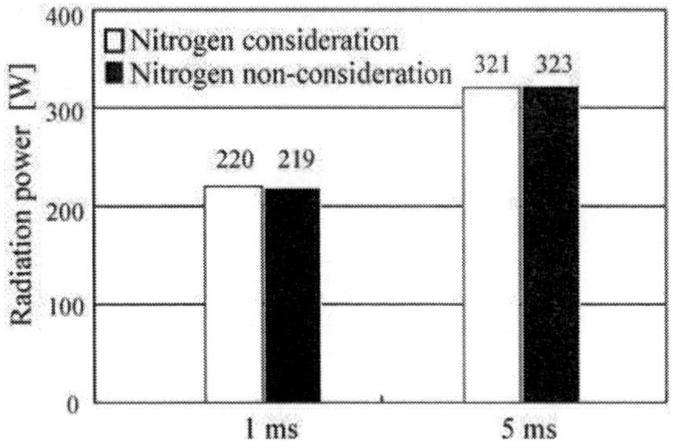

Fig.9: Radiation power at 1 and 5 ms.

Fig.8 shows the temperature with radial distance (z = 32.5 mm). The temperature in the case of nitrogen consideration is lower than that in the case of the nitrogen non-consideration. This was caused at the temperature region of 7500 K that specific heat increased locally. In addition, the temperature at 1ms is lower than that at 5ms. *Fig.9* shows the radiation power a 1 and 5 ms. The radiation loss at 1 ms is lower than that at 5 ms. The temperature decreased because of the transient response of the temperature when the time reached at the peak current. However, the difference in radiation loss is little in the case of nitrogen consideration and non-consideration. The radiation loss depends on the temperature. The following temperature region 12000 K decreased because of the thermal pinch effect, however radiation loss increased in the temperature over 13000 K. Therefore, the radiation loss is almost constant even if the temperature of the periphery part because of the thermal pinch effect in the case of nitrogen consideration and non-consideration.

Conclusions

The nitrogen concentration distribution of argon arc contaminated with nitrogen in the pulsed TIG arc welding by the time is elucidated. The distribution of nitrogen concentration in just after peak current reached and it after the peak time is calculated. The main results are shown below.

1) The radial flow velocity of inward increases with the electromagnetic pinch force, because the temperature of the axial center near the cathode increases. In this case, the nitrogen concentration in the arc increases.
2) The nitrogen concentration near the welding pool decreases with convection dominated by temperature increment.
3) The temperature in the case of nitrogen consideration is lower than that in the case of the nitrogen non-consideration. In addition, the temperature at 1ms is lower than that at 5ms.
4) The radiation loss at 1 ms is lower than that at 5 ms. The temperature decreased because of the transient response of the temperature when the time reached at the peak current.

Therefore, it was calculated in the time after the peak time to obtain a knowledge of the changes in nitrogen concentration in peak current time. These results suggest that the nitrogen concentration decreases near the welding pool affected by increasing the duty ratio. Therefore, it is important to control the temperature using the pulsed parameters.

References

[1] Yoshinori Hirata, "Pulsed arc welding", Journal of the Japan Welding Society, Vol.71, No.5, (2002), pp.389-404,

[2] Tadaaki Yamada, Minoru Kobayashi, "Blowholes on Pulsed MAG Welding" Journal of the Japan Welding Society, Vol.3, No.4, (1985), pp.822-833.

[3] S.KODAMA, K.SUGIURA, S.NAKANISHI, D.SAKAI, Y.TSUJIMURA, M.TANAKA, Anthony B. MURPHY: "Influence of Plasma Characteristicson Nitrogen Mixing into Shielding Gas in Helium Gas Tungsten ArcWelding", QUARTERLY JOURNAL OF THE JAPAN WELDING SOCIETY, Vol. 31, No. 1, (2013), pp.80-87.

Numerical Investigation of Charge Effect on Collective Growth of Iron Nanoparticles Generated in Arc Welding

Shota MINANI [1, a], ***Masaya SHIGETA*** [1, b] ***and Manabu TANAKA*** [1, c]

[1]Joining and Welding Research Institute, Osaka University, Osaka, Japan

[a]s.minami@jwri.osaka-u.ac.jp, [b]shigeta@jwri.osaka-u.ac.jp, [c]tanaka@jwri.osaka-u.ac.jp,

Abstract

Charge effect on collective growth process of iron nanoparticles, which are primary particles of fume generated in arc welding was investigated numerically. The result clearly showed that charging of nanoparticles was a crucial factor for the collective growth process. The model used in this study has several advantages. It can treat particles with multiscale diameters ranging from sub nanometers to a few hundred nanometers. It can also express particle size distributions with various shapes. It can take account of homogeneous nucleation, heterogeneous condensation, coagulation process among particles and particle charge simultaneously. The numerical calculation by this revealed that nuclei generated by homogeneous nucleation grew mainly by heterogeneous condensation. The effect of coagulation was smaller because almost all of the nanoparticles have charges and repelled each other by Coulomb's force. When the model did not take charge effect into account, nanoparticles exhibited larger sizes. The numerical calculation considering charge effect obtained a particle size distribution which showed a good agreement with that obtained by an experiment.

Introduction

Arc welding is one of welding methods, which heats, melts and joins metal materials with arc plasma. Because it is a powerful and effective process, it has been used in various industrial applications of cars, ships, airplanes and bridges. During arc welding melting metals, the high-temperature metal vapor is always generated from the molten metals at the pool surface of the base material, the tip of the welding wire and droplets [1]. The metal vapor is transported with the plasma flow to the outside of the plasma, where, the metal vapor is cooled rapidly. With decrease in temperature, the saturation pressure of the metal vapor decrease, and then the metal vapor pressure exceeds the saturation vapor pressure. This supersaturated state leads to phase change from the vapor to nanoparticles whose diameters range from sub nanometers to a few hundred nanometers. In arc welding processes, smoke released from the arc welding region is often observed. This smoke, which is called "welding fume", is actually aggregates of nanoparticles [2].

Welding fume can cause serious health problems for welder when they inhale it [3]. Reduction of fume is one of important issues in arc welding.

Collective growth of nanoparticles is difficult to observe because it is a process which progresses in microseconds through homogeneous nucleation, heterogeneous condensation and coagulation. Therefore, some recent studies have been based on modelling and numerical calculations. However, the effect of particle charge on the growth process still remains unknown because no models took account of charging processes which could occur around thermal plasma including ions and electrons.

In this study, a model that calculates the collective growth of fume primary particles with charging because of collision with ions and electrons from plasma is developed on the basis of the model [4]. Furthermore, numerical calculation using the developed model is performed to reveal the charge effect on the particle growth under a condition of gas metal arc welding.

Model description

In arc welding, iron vapor is generated from an electrode and a base metal because of high temperature of arc plasma. The iron vapor is transported with a shielding gas flow to the outside of the arc plasma, where the temperature rapidly decreases. Therein, the iron vapor becomes supersaturated, and a large number of nuclei are generated (homogeneous nucleation). Once nucleation occurs, the iron vapor easily condenses on the nuclei (heterogeneous condensation). Through these processes, primary particles of fume are formed. Simultaneously, the particles collide and merge with each other and form larger particles (coagulation). During the growth process, the particles can be charged by collisions with electrons and ions. Because Coulomb force acts between charged particles, the collision frequency is affected considerably. In this study, the net production rate of particles with volume v_k and charge amount z_n can be written as

$$\frac{\Delta N_{k,n}}{\Delta t} = \dot{N}_{k,n}^{(nucl)} + \dot{N}_{k,n}^{(cond)} + \dot{N}_{k,n}^{(coag)}, \quad (1)$$

where $N_{k,n}$ represents number density of particles at node (k, n), $\dot{N}$ means the increment per unit time, superscripts (*nucl*), (*cond*) and (*coag*) stand for the contributions of homogeneous nucleation, heterogeneous condensation and coagulation. Because charge occurs by collisions between particles and ions or electrons, the charging processes are treated at the same regime of heterogeneous condensation in this model. Charge states significantly affect coagulation among particles, whereas the effects are neglected on homogeneous nucleation in this study.

Fig. 1 shows the present computational domain. The computation is performed for the trajectory which starts from the position (x= 11.4 mm, y= 0.4 mm). This streamline starts at the temperature of 2800 K. In the present calculation, the increment of a time step is set to be 0.1 μs. When the temperature reaches T = 300 K, the calculation is finished.

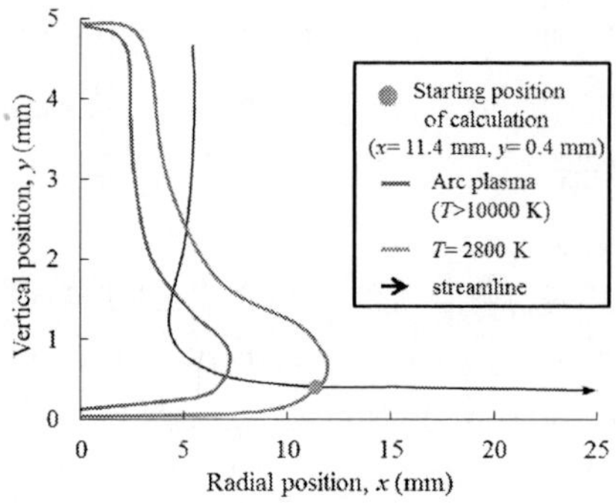

Fig. 1 Computational domain.

Results and discussion

Fig. 2 shows the time evolutions of particle size distributions cooling from 2800 K to 300 K along the streamline. *Fig. 2(a)* and *2(b)* portray numerical results without and with the effect of particle charge, respectively. Regardless of charge, particles smaller than 1 nm increase at early stages from 2465 K to 2307 K. These results indicate that homogeneous nucleation actively takes place in this temperature range. Simultaneously, the particle size distributions expand in the larger size direction increasing the numbers of particles, which means that particles grow up dominantly by heterogeneous condensation. After particles experience 2307 K, small particles start to decrease and the peaks at the larger sizes shift in the larger diameter direction. This mode change indicates that the cause of the dominant growth has changed from condensation to coagulation. Because coagulation is affected by particle charge, the particles exhibit different size distributions at the final stage of 300 K without and with charge effect. With charge effect, small particles do not decrease because repulsive Coulomb force prevents negatively charged particles from coagulating with each other.

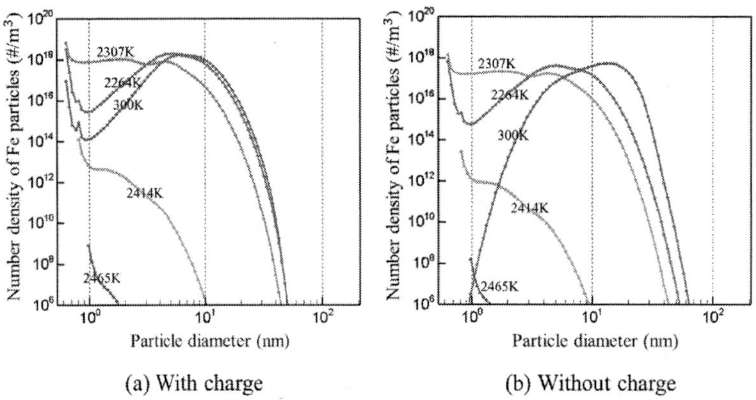

Fig. 2 Time evolutions of size distributions of Fe particles.

Fig. 3 shows the particle size-charge distribution at 300 K. In the figure, d_p and z denote particle diameter and charge amount, respectively. The present calculation reveals that almost all particles generated in a typical condition of MIG welding are negatively charged. Therein, the ratio of non-charged particles is estimated to be only 1.1×10^{-7}.

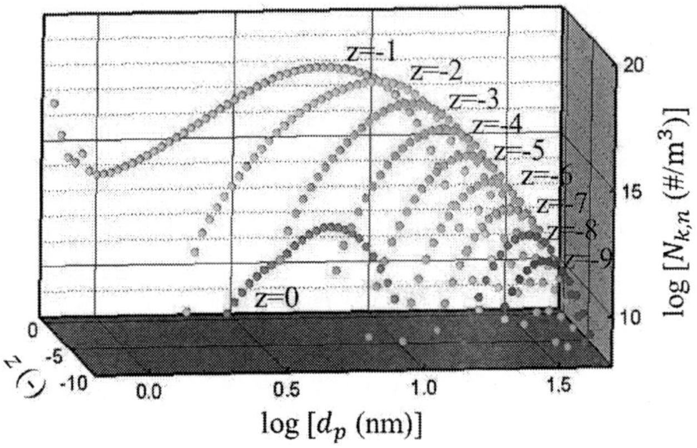

Fig, 3 Particle size-charge distribution at 300 K.

The size distributions of particles finally obtained at 300 K were compared with an available profile of the final product of fume at a MIG welding [2]. The computational result with charge effect agreed with the experimental result, whereas that without charge effect did not.

Conclusions

In this study, a model that calculates the collective growth of fume primary particles with charging due to collisions with ions and electrons from plasma was developed. Using the model, numerical calculation was performed to reveal the charge effect on the particle growth under a condition of MIG welding. In consequence, it was found that almost all particles generated in a typical condition of MIG welding are negatively charged. The effect of particle charge plays an important role on the collective growth process of primary particles of welding fume because the particle charge causes repulsive Coulomb force which prevents the particles from growing by coagulation.

References

[1] H. Terasaki, M. Tanaka and M. Ushio, Effect of Metal Vapor on Electron Temperature in Helium Gas Tungsten Arcs, Metallurgical and Materials Transactions A, Vol. 33, No. 4, (2002), 1183-1188.

[2] B. Berlinger, N. Benker, S. Weinbruch, B.V. L'Vov, M. Ebert, W. Koch, D. G. Ellingsen and Y. Thomassen, Physicochemical characterisation of different welding aerosols, Analytical and Bioanalytical Chemistry, Vol. 399, No. 5, (2011), 1773–1780.

[3] C. Yoshii, Y. Morimoto and M. Kido, Welder's Pneumoconiosis—Pulmonary Disease Caused by Inhalation of Welding Fumes—, Journal of Aerosol of Research, Vol. 20, No. 3, (2005), 238-242 (in Japanese).

[4] M. Shigeta and T. Watanabe, Growth mechanism of silicon-based functional nanoparticles fabricated by inductively coupled thermal plasmas, Journal of Physics D: Applied Physics, Vol. 40, No. 8, (2007), 2407-2419.

Weldability of 22Cr-13Ni-5Mn-2Mo-Nb, V stainless steel containing high nitrogen under high pressure hydrogen gas

Shinnosuke Kurihara, Kana Jotoku, Jun Nakamura, Takahiro Osuki, Tomohiko Omura
Nippon Steel & Sumitomo Metal Corporation, Amagasaki, Hyogo, Japan
Email; kurihara.5th.shinnosuke@jp.nssmc.com Phone; +81-6-7670-5843

Hiroyuki Hirata
Nippon Steel & Sumitomo Metal Corporation, Futtsu, Chiba, Japan

Abstract

22Cr-13Ni-5Mn-2Mo-Nb, V stainless steel containing high nitrogen content with higher strength and similar hydrogen embrittlement compared with Type 316L is rapidly applied for heat exchangers, pipes, tubes and joint etc. to secure the filling efficiency of 70MPa-class hydrogen gas station. The tensile strength of welded joint of 22Cr-13Ni-5Mn-2Mo-Nb, V stainless steel is accomplished over 800MPa which guaranteed minimum tensile strength of base metal by optimizing the activity coefficient of nitrogen in weld metal.

Introduction

Constructions of 70MPa-class high pressure hydrogen gas station have been increased to improve the filling efficiency to fuel cell vehicle on the hydrogen gas station. Type 316L with high Ni-equivalent; Ni+0.65%Cr+0.98%Mo+1.05%Mn+0.35%Si+12.6%C of more than 28.5% is well known as high resistance to hydrogen embrittlement by high pressure hydrogen gas at the temperature range between -10 and -45°C and applicable to pipes, tubes and fittings etc. in the hydrogen gas stations [1].

On the other hand, Type 316L has low minimum tensile strength of 480 MPa at room temperature and is enforced to have thicker wall thickness in the 70MPa-class high pressure hydrogen gas station. It results low filling efficiency by increasing the design pressure of hydrogen gas. Therefore, 22Cr-13Ni-5Mn-2Mo-Nb, V stainless steel containing high nitrogen with higher strength and similar hydrogen embrittlement compared with Type 316L is rapidly applied to secure the filling efficiency of hydrogen gas station. It is important to establish the welding technique of 22Cr-13Ni-5Mn-2Mo-Nb, V stainless steel because of decreasing the tensile strength of welded joint by removing nitrogen from the weld metal during welding. In this study, the effect of nitrogen on the tensile property of weld metal has been investigated.

Experimental procedure

Table 1 shows the nominal composition of test material plate. The 4mm thickness plate was melted in laboratory and manufactured by hot forging and cold rolling. After cold rolling, the plate was solution annealed. The welded joints of this material were produced using commercial filler wires of YS 309MoL, 308H and 309N_2 as shown in *table 2*. *Table 3* and *figure 1* shows the welding condition and bevel configuration.

Tensile test specimens were cut from the welded joint with and without excessive convexity. Microstructure on cross section was observed by optical microscope and δ-ferrite fraction was measured by image analysis. Nitrogen content on the weld metal was measured by gas analysis. In addition, 15mm thickness plate was welded under the heat input range from 0.7 to 1.2kJ/mm, the back shielding gas of pure N_2. The other welding parameters were listed in *table 3* and *figure 1*. Slow Strain Rate Test (SSRT) specimens were cut from the welded joint and the SSRT test was conducted on 90MPa high pressure hydrogen environment at room temperature, -40 and 150°C. The strain rate was 3×10^{-6} s^{-1}.

Table 3 Welding condition

Welding process	Automatic GTAW
Filler wire	1.2mmφ spool
Heat input (kJ/mm)	0.4-0.8
Shielding gas	Pure Ar or Ar+2-8vol%N_2
Back shileding gas	Pure N_2
Preheating	None
Interpass temp.	150°C max.
PWHT	None

Table 1 Nominal composition of test material plate (mass%)

C	Si	Mn	Ni	Cr	Mo	V	Nb	N
0.03	0.4	5.0	13.0	22.0	2.0	0.2	0.2	0.3

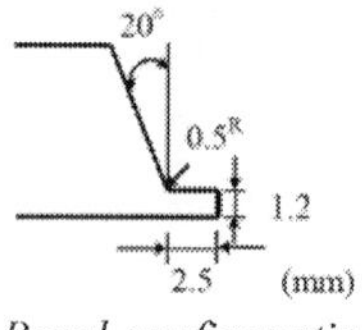

Figure 1 Bevel configuration

Table 2 Nominal composition of materials used. (mass%)

Type	C	Si	Mn	Ni	Cr	Mo	N
YS309MoL	0.02	0.4	2.3	13.7	23.2	2.2	0.09
YS308H	0.06	0.4	2.0	10.2	19.7	-	0.03
YS309N2	0.07	0.2	2.1	9.7	21.4	-	0.15

Results

Properties of welded joint

Figure 2 shows the tensile test results of welded joint using YS 309MoL filler wire. Although the welded joint with the excessive convexity is fractured at the weld metal, the tensile strength is accomplished over 800MPa which guaranteed minimum tensile strength of base metal. The welded joint with the excessive convexity has higher tensile strength by changing the shielding gas from pure Ar to Ar+2vol.%N_2. On the other hand, the tensile strength of welded joint without the excessive convexity is not accomplished over 800MPa despite the compositions of shielding gas. The tensile test results show that an improvement of tensile strength on weld metal is required in order to accomplish over 800MPa tensile strength without the excessive convexity.

Figure 3 shows the effect of temperature on relative reduction of area on welded joint using YS 309MoL filler wire under 70MPa and 90MPa hydrogen gas at room temperature, -40 and 150°C. The relative reduction of area means the ratio of reduction of area under 90MPa hydrogen gas to that under air atmosphere. The relative reduction of area on welded joint shows over 90% and it indicates that the welded joint has excellent resistance to hydrogen embrittlement.

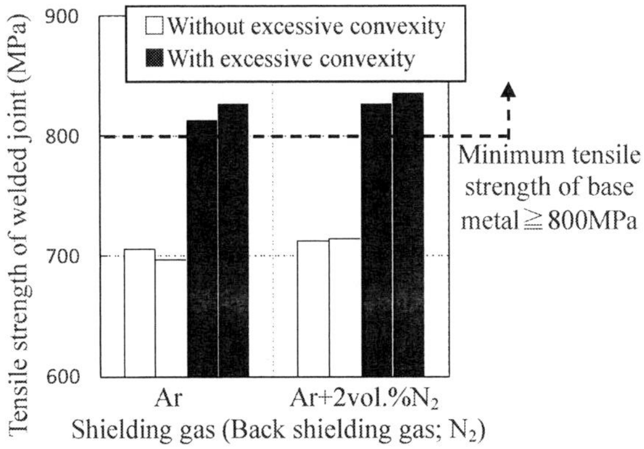

Figure 2 Tensile test results of welded joint with YS 309MoL filler wire.

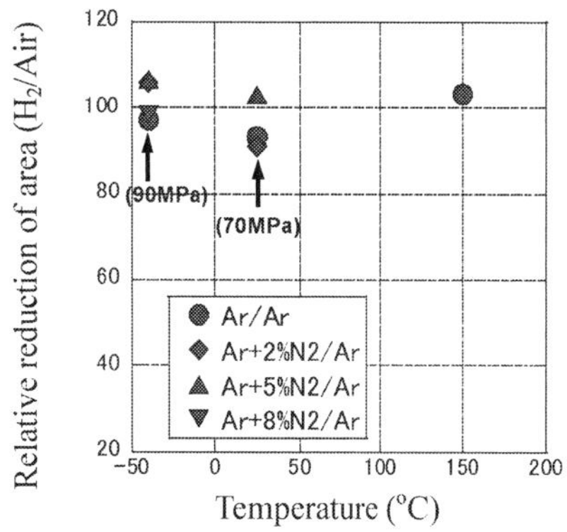

Figure 3 Effect of temperature on relative reduction of area on welded joint with YS 309MoL filler wire.

Improvement of tensile strength on weld metal

The improvement of tensile strength on welded joint is conducted by the increase of N_2 gas content in shielding gas in order to increase nitrogen content as a solution strengthening element of austenitic stainless steel on weld metal. *Figure 4* shows an effect of N_2 gas content in shielding gas on tensile strength of weld metal with YS 309MoL filler wire. The excessive convexity of tensile test specimens was eliminated and they were fractured at the weld metal. Therefore, the tensile strength of welded joint is equal to that of weld metal. The tensile strength is increased until 2vol.%N_2 gas content in shielding gas and saturated over 2vol.%N_2 gas content in despite of the compositions of back shielding gas. *Figure 5* shows an effect of N_2 gas content in shielding gas on nitrogen content in weld metal with YS 309MoL filler wire. As well as the test result of *figure 4*, the nitrogen content in weld metal is increased until 2vol.%N_2 gas content in shielding gas and saturated over 2vol.%N_2 gas content. *Figure 6* shows a relationship between the tensile strength of weld metal and nitrogen content in weld metal with YS 309MoL filler wire. The tensile strength of weld metal has correlation with nitrogen content in weld metal and increases with higher nitrogen content. *Figure 7* shows a relationship between tensile strength of weld metal and nitrogen content in the weld metals using YS 309MoL, 308H and 309N_2 filler wire. The tensile strength of weld metal using YS308N_2 filler wire increases with higher nitrogen content in weld metal as wells as YS309MoL. On the other hand, the tensile strength of weld metal using YS308H filler wire is almost constant until 0.3mass% nitrogen gas content in the shielding gas and is increased over 0.3mass% nitrogen gas content in the shielding gas.

The test results clarify that the increase of nitrogen content in weld metal is effective to improve the tensile strength of weld metal. However, the tensile strength of weld metal by the variation of filler compositions is not accomplished over 800MPa and other approach is required to improve the tensile strength in addition to N_2 gas content in shielding gas.

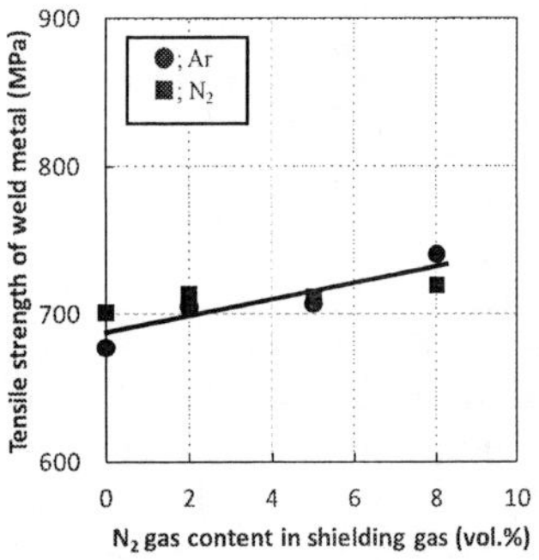

Figure 4 Effect of N_2 gas content in shielding gas on tensile strength of weld metal with YS309MoL filler wire

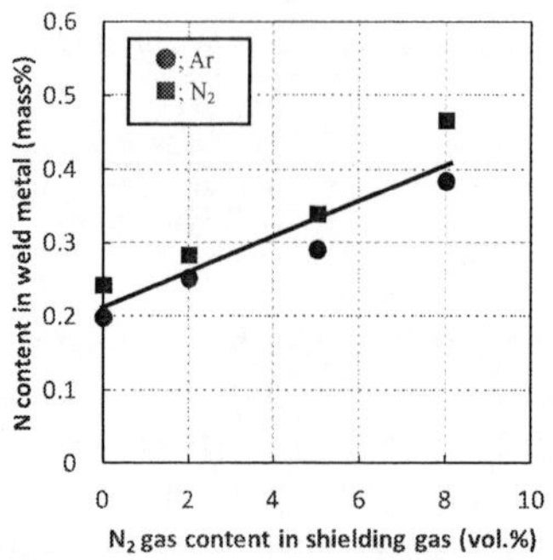

Figure 5 Effect of N_2 gas content in shielding gas on nitrogen content in weld metal with YS309MoL filler wire

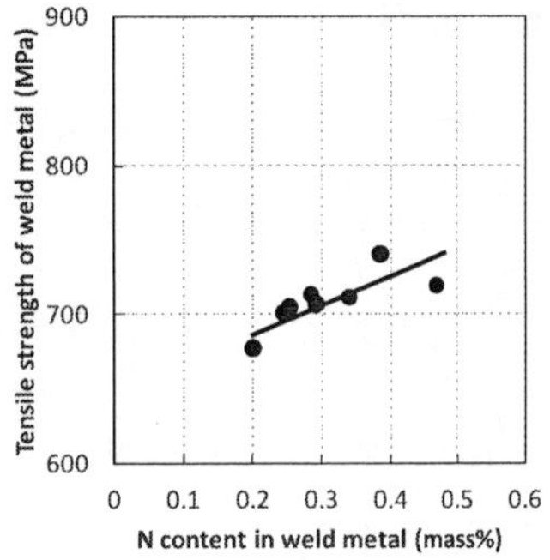

Figure 6 Relationship between tensile strength of weld metal and nitrogen content in weld metal with YS309MoL filler wire

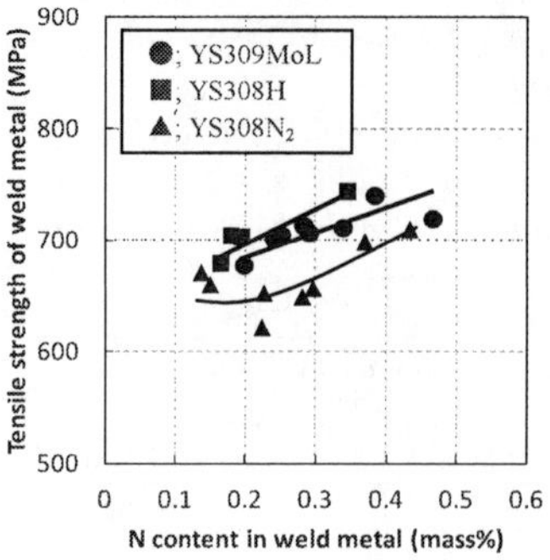

Figure 7 Relationship between tensile strength of weld metal and nitrogen content in several weld metals

Discussion

An optimization of chemical composition in the weld metal is carried out to improve the tensile strength by more increase of nitrogen content in weld metal. Several studies have been reported about nitrogen absorption in weld metal and T. Kuwana et al. have indicated that an influence on element with nitrogen content in weld metal is equivalent to equilibrium nitrogen solubility [2]. The reaction of nitrogen absorption in melting ferrous alloy is shown in Equation (1) and its equilibrium constant K is expressed in Equation (2) [3].

$$N_2(gas) = 2N \quad \text{(in liquid metal)} \qquad (1)$$

$$K = a_N \cdot (P_{N_2})^{-1/2} = f_N \cdot [\%N]_e \cdot (P_{N_2})^{-1/2} \qquad (2)$$

a_N, f_N and P_{N2} indicates nitrogen activity in melting ferrous alloy, activity coefficient of nitrogen in melting alloy and partial pressure of N_2(atm), respectively. The equilibrium constant in melting ferrous alloy under equilibrium condition is also represented in Equation (3) [4]. Nitrogen solubility in weld metal is shown in Equation (4) from Equation (2) and (3).

$$\log K = -518/T - 1.063 \qquad (3)$$

$$\log[\%N] = -518/T - 1.063 - \log f_N + 1/2 \cdot \log\sqrt{P_{N_2}} \qquad (4)$$

When the temperature T and partial pressure P_{N2} is constant, nitrogen solubility in melting ferrous alloy is increased with the decrease of f_N in Equation (4). In the case of weld metal, it is presumed that the relationship between the nitrogen and f_N is equivalent. The fN is calculated from interaction coefficient and concentration of other alloy element in melting ferrous alloy[3].

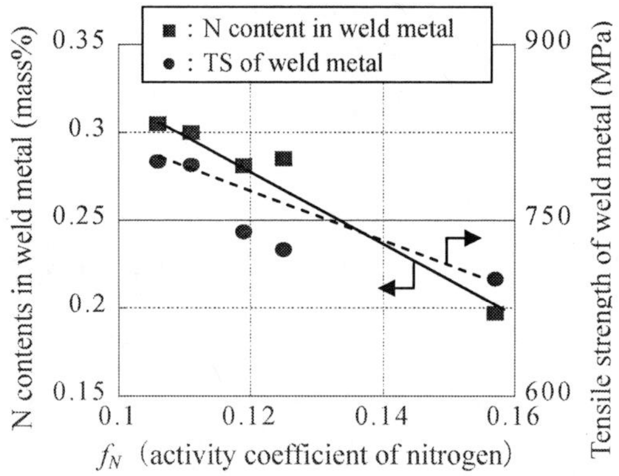

Figure 8 Effect of activity coefficient of nitrogen on nitrogen content in weld metal and tensile strength of weld metal with $YS308N_2$ filler wire.

The effect of activity coefficient f_N of nitrogen on nitrogen content in weld metal and tensile strength of weld metal with $YS308N_2$ was investigated. The test results are shown in *Figure 8*. When the fN is lower, the nitrogen content in weld metal and the tensile strength is higher. Therefore, it is effective to increase the tensile strength of weld metal by decreasing f_N in welding filler wire. On the other hand, the relationship between the nitrogen content in weld metal and the tensile strength of weld metal using YS308H filler wire is different from the other filler wires as shown in *figure 7*. It has been reported that the tensile strength of austenitic stainless steel is influenced by the solution strengthening element, δ-ferrite and twin distance [5]. The factor for improvement of tensile strength of weld metal is both solution strengthening element and δ-ferrite because the twin distance is negligible for the weld metal with solidification microstructure.

Figure 9 shows a microstructure and δ-ferrite of weld metal using several filler wires of YS 309MoL, 308H and $309N_2$. The weld metals with YS 309MoL and $309N_2$ filler wire contain the δ-ferrite fraction. However, the weld metal with YS308H filler wire contains only 0.2% of the δ-ferrite fraction and almost austenitic microstructure when the nitrogen content is increased to 0.22mass%N. The weld metals with YS 309MoL and $309N_2$ filler wire contain the similar nitrogen content and δ-ferrite, and the difference of tensile strength of weld metal is affected by the solution strengthening element except nitrogen. In the case of weld metal with YS308H filler wire, the δ-ferrite solves in the matrix and does not have the effect of increasing tensile strength by the δ-ferrite in increasing the nitrogen content to 0.22mass%. The increasing nitrogen content and solving the δ-ferrite leads to cancel out the effect of the tensile strength. When the nitrogen content in weld metal is over 0.22mass%, it is assumed that the tensile strength is increased by the solution strengthening of nitrogen content because the weld metal is almost austenitic microstructure. As the results, the tensile strength of weld metal with YS308H filler wire has different behavior from that with YS309MoL and $308N_2$ filler wire.

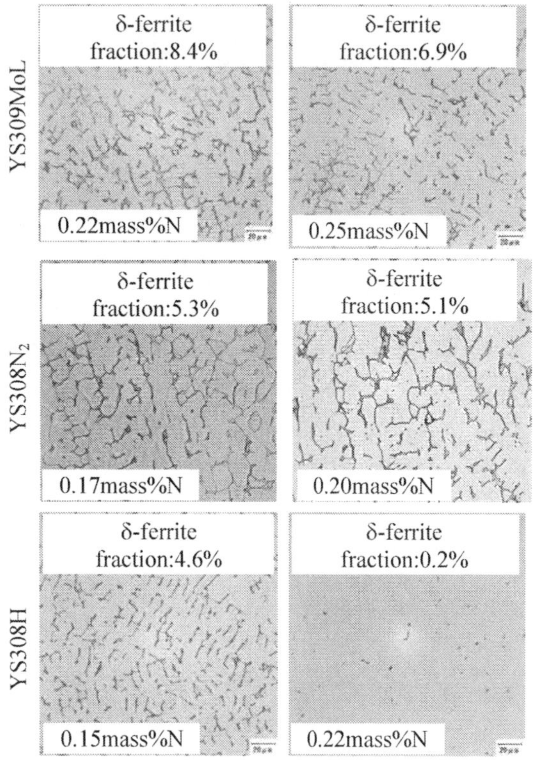

Figure 9 Microstructure and δ-ferrite of weld metal with several filler wires

Therefore, the test results demonstrate that the important factor for the improvement of tensile strength of weld metal is not only utilizing the nitrogen content in weld metal, but also the δ-ferrite and other solution strengthening element.

Conclusions

The effect of nitrogen content on the tensile strength of weld metal of 22Cr-13Ni-5Mn-2Mo-Nb, V stainless steel has been investigated to improve the tensile strength of welded joint.

1) The tensile strength of welded joint for 22Cr-13Ni-5Mn-2Mo-Nb, V stainless steel containing high nitrogen with the excessive convexity accomplishes over 800MPa which guaranteed minimum tensile strength of base metal. In addition, the welded joint has excellent resistance to hydrogen embrittlement.

2) Increasing nitrogen content in weld metal is effective to improve the tensile strength of weld metal. In order to increasing nitrogen content in weld metal, it is effective to add $2vol.\%N_2$ gas content in shielding gas and lower activity coefficient f_N in weld metal.

3) The influence of nitrogen content on tensile strength of weld metal has different behavior depend on the kind of filler wire. It is assumed that the factor for the improvement of tensile strength of weld metal is not only utilizing the nitrogen content in weld metal, but also the δ-ferrite and other solution strengthening element.

References

[1] T.Omura, K.Kobayashi, M.Miyahara and T.Kubo: Hydrogen Embrittlement Properties of Stainless Steels in High Pressure Gaseous Hydrogen Enviroment, Zairyo-to Kankyo, 55-4(2006), 139-145. (in Japanese)

[2] T.Kuwana, H.Kokawa and Y.Nakaya: Nitrogen Absorption of Fe-Cr-Ni Ternary Alloys during Gas Tungsten Arc Welding, Quarterly Journal of the Japan Welding Society, 10-3(1992)403-409. (in Japanese)

[3] H.Kokawa: Nitrogen Absorption and Desorption by Steels during Arc and Laser Welding, Journal of the Japan Welding Society, 72-5(2003),112-121. (in Japanese)

[4] Japan Society for the Promotion of Science, Seiko-Hannou-No-Suisyou-Heikouchi, Nikkan Kogyo Shimbun, ltd., (1984),17. (in Japanese)

[5] F.B. Pickering: Physical Metallurgy and the Design of Steels, Applied Science Pub.,(1978), 231.

Estimation of Mechanical Properties of SUS315J2 Brazed Joint

M.Iwata, M.Ishihara, K.Ohta, and Y.Miyazawa

Tokai University, 4-1-1, Kitakaname, Hiratsuka-city, Kanagawa 259-1292, Japan

E-mail: ymiyazawa @tokai-u.jp

Abstract

Many kinds of stainless steel have been developed, and some of them are added a few alloying elements such as Cu or Si to improve their properties. In our previous study, spreading ability of SUS315J2, which is containing Cu and Si as alloying elements, was investigated. And the fact that paste type Ni-based brazing filler metal has excellent spreading ability on SUS315J2 was observed. The effects of the excellent spreading on the thickness of the brazed layer and the microstructure were considered. Furthermore, the effects on the mechanical properties of the brazed joint were expected due to the thickness of the brazed layer. Therefore, SUS315J2 was brazed with several Ni-based filler metals on both clearances narrow and wide, and the mechanical properties of the brazed joints were investigated in this study. As the results, sound brazed joints were obtained regardless of the clearance. Due to the decreased brazed layer caused by the excellent spreading ability, higher mechanical property of SUS315J2 brazed joint was obtained.

1. Introduction

Many kinds of stainless steel have been developed, and they have been used for many applications ranging from kitchenware to automobiles. A few alloying elements such as Cu or Si are added to improve the properties of stainless steel. However, the influence of the addition of a few alloying elements on the brazing ability has not been investigated still now. SUS315J2 stainless steel, which has good stress corrosion cracking (SCC) resistance and crevice corrosion resistance properties, has been developed [1]. SUS315J2 contains Cu and Si as alloying elements to improve SCC resistance properties [2].

In our previous study, the fact that paste type 613 brazing filler metal has excellent spreading ability on SUS315J2 was observed [3]. And it appears that the presence of Cu and Si changed the specific characteristics of the molten brazing filler metal and the spreading was facilitated. The effect of the excellent spreading on the thickness of the brazed layer and the microstructure was considered. Furthermore, the effect on the mechanical properties of the brazed joint was expected due to the decrease in thickness of the brazed layer. In generally, a thinner brazing layer produces higher strength of a brazed joint, and that was also confirmed in our other study. On the other hand, a filler metal having great wettability is concerned it cannot fill relatively wide clearance.

For these reasons, SUS315J2 was brazed with several brazing filler metals, 613, MBF67 having similar chemical composition to 613, and so on, on two different clearance conditions. And, the mechanical properties of the brazed joints were investigated in this study.

2. Experimental procedure

2.1. Base metal and brazing filler metal

SUS315J2 stainless steel (JIS) was employed as a base material. Chemical composition of the stainless steel was shown in *Table 1.* Chemical compositions of the filler metals used in this study were shown in *Table 2*. MBF20 and BNi-2 were employed as standards. For making difference on the thickness of the brazed layer, brazed joints holding three brazing sheets, or keeping clearance with tungsten wire between the base metals were also prepared.

Table 1 Chemical compositions of SUS315J2 used in this study

SUS No.	Chemical compositions (mass%)									
	Fe	Si	Mn	Cr	Ni	Mo	S	Cu	C	P
315J2	Bal.	2.5-4.0	<2.0	17.0-20.5	11.0-14.0	0.5-1.5	<0.03	0.5-3.5	<0.08	<0.045

Table 2 Chemical compositions and temperature of brazing filler metals used in this study

Filler metals	Chemical compositions (mass%)								Temperature (°C)	
	Ni	Cr	Fe	Si	B	Mo	P	C	Solidus	Liquidus
MBF67	Bal.	25.38	0.08	1.43	0.49	1.09	5.91	0.008	890	970
MBF20	Bal.	7.06	3.43	4.92	3.01	-	0.01	0.003	969	1024
613	Bal.	27.5-31.5	-	3.8-4.2	-	-	5.6-6.4	-	980	1030
BNi-2	Bal.	6.0-8.0	2.5-3.5	4.0-5.0	2.75-3.50	-	-	-	970	1000

2.2. Specimens

Single lap specimens were used to estimate its mechanical properties and observe the cross-sectional microstructure at the brazed joint. The schematic diagram of the specimen was shown in *Fig.1*.

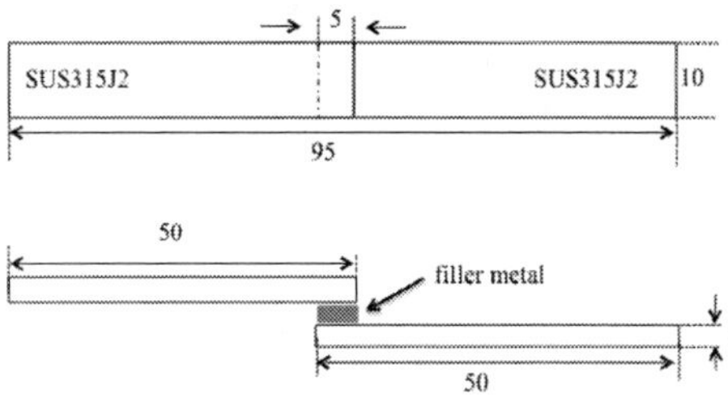

Fig.1 Schematic diagram of the single lap specimen

2.3. Brazing process and estimation method

Brazing was done using a vacuum-type furnace under vacuum atmosphere (5×10^{-3}). The brazing temperatures were 1050℃(for brazing sheets) and 1080℃(for paste type brazing filler metals) taking their liquidus temperatures of the brazing filler metals into account. The brazing time was 5 min. The cross-sectional microstructures at brazed joint, especially focusing on the thickness of the brazed layer, of the single lap specimens were observed with optical microscope (OM). Mechanical properties of the brazed joint were estimated by tensile shear test. Also, elemental distributions were analyzed by using EPMA system.

3. Results and discussion

3.1. Cross-sectional microstructures of the specimens

As shown in *Fig.2*, sound brazed joints were obtained for each specimen. Even the 100μm clearance was kept between the base metals it was filled up enough with the brazing filler metal.

Table 3 shows the change of the thickness of the brazed layer between before and after brazing. The brazed layer with MBF67 showed high decreasing ratio in the thickness. And, the brazed layer with Ni-Cr-Si-P brazing filler metal got thinner significantly. It is considered that the change occurred due to the excellent spreading of these filler metals on SUS315J2.

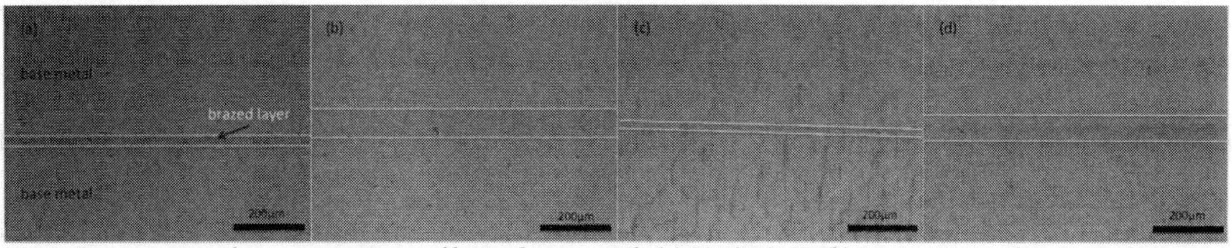

Fig.2 Cross-sectional microstructures of brazed joints with (a) Ni-Cr-Si-P, (b) Ni-Cr-Si-P with 0.1mm clearance, (c) MBF67, and (d) three MBF67 sheets

Table 3 Change in thickness of the brazed layer

Filler metals	Amount of the filler	Thickness of the brazed layer (µm)		Decreasing ratio in the thickness (%)
		Before brazing	After brazing	
MBF67	1 sheet	30	21	30.0
MBF67	3 sheets	90	48.75	45.8
MBF20	1 sheet	38	31.5	17.1
MBF20	3 sheets	114	98.04	14.0
613	a quantum	a little	20	-
613	a quantum	100	94.29	5.7
BNi-2	a quantum	a little	33.16	-

3.2. Tensile test

The results of tensile test were shown in *Fig.3*. All specimens were fractured at brazed layer. The shear strength of brazed join with MBF67 was 1.17 times as much as one with three MBF67 sheets. Also, the shear strength of brazed joint with 613 brazing filler metal was 1.68 times as much as one with 100µm clearance. About brazed joint with MBF20, that strength was 1.15 times as much as one with three sheets. Due to the decrease in the thickness of the brazed layer, relatively high increase in the strength was obtained in the combination having excellent spreading ability.

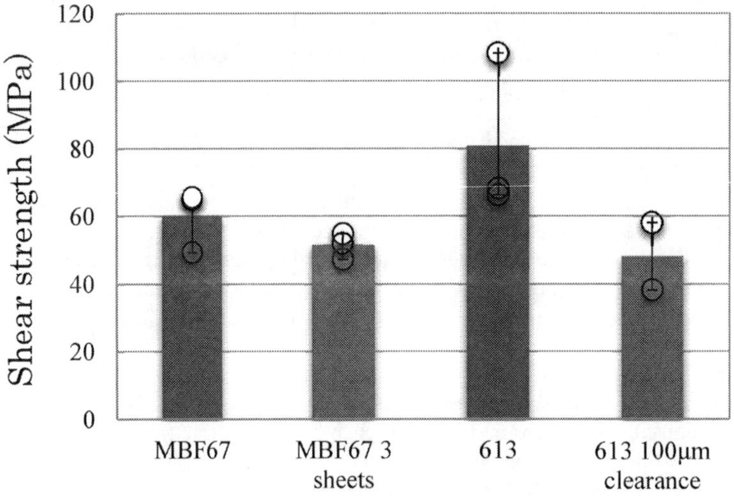

Fig.3 Shear strength of the brazed joints with each filler metals

3.3. EPMA elemental distribution

Fig.4 shows EPMA elemental distributions of the brazed joint with 613 brazing filler metal. The results of the EPMA elemental distributions were typical ones that have been observed in our previous study. There was no significant intermetallic compounds or something in the interface between base metal and brazing filler metal. Cu content is one of the unique characters of SUS315J2, and the diffusion of Cu into brazing filler metal was slightly observed. However, the Cu did not influence significantly on formation of any compounds. So the increase in the shear strength is thought that depends on the thinner brazed layer.

4. Conclusions

The mechanical properties of the SUS315J2 brazed joints with 613 brazing filler metal, and MBF67 on narrow and wide clearances were investigated in this study, and the following results were obtained:

A) Sound SUS315J2 brazed joint with 613 brazing filler metal, and MBF67 was obtained regardless of the clearance.
B) Brazed layer got thinner due to the excellent spreading ability of the brazing filler metals on SUS315J2.
C) The decrease in the thickness of the brazing layer led to increase in the shear strength of the SUS315J2 brazed joint.

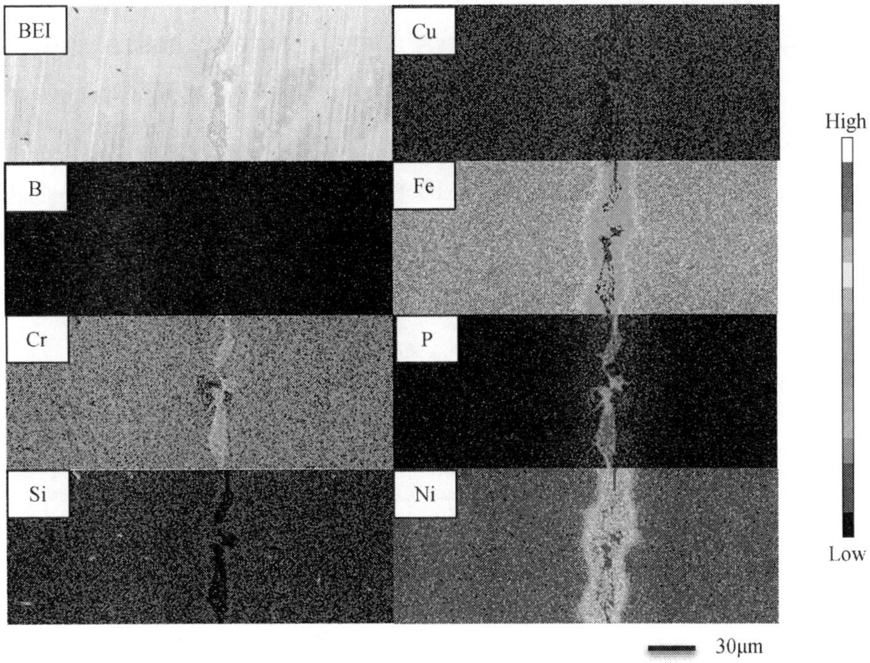

Fig.4 EPMA elemental distribution of brazed joint with 613 brazing filler metal

5. Acknowledgement

I received generous help from my co-worker Yusaku Shinoda in EPMA analysis. I am deeply grateful to him.

5. Reference

[1] Adachi T, Nishikawa M and Yoshii T 1989 Nisshinseikogiho 60 56
[2] Katoh T, Kusaka K and Hisada T 1980 Denkiseiko 51 4 252
[3] Miyazawa Y, Ohta K and Nishiyama A 2014 Effect of stainless steel chemical composition on brazing ability of filler metal, IOP Conference Series: Materials Science and Engineering, Volume 61, conference 1

Influence of welding parameters on the micro resistance spot welding (microRSW) of 316L austenitic stainless steel and titanium alloy

Muhammad Safwan Mohd Mansor [1], Farazila Yusof [1], Mohd Fadzil Jamaludin [2], Tadashi Ariga [3], Mohd Hamdi Abdul Shukor [1]

[1] *Department of Mechanical Engineering, University Malaya, 50603 Kuala Lumpur, Malaysia*
[2] *Advanced Manufacturing and Materials Processing Centre (AMMP Centre), University Malaya, 50603 Kuala Lumpur, Malaysia*
[3] *University Malaya, 50603 Kuala Lumpur, Malaysia*

Micro-resistance spot welding (microRSW) has a very important role as a joining process in various industrial applications including packaging of medical devices, electronic connections, jewellery, etc. For example, in the manufacturing process of miniature or small scale product, it contains a lot of micro parts that needed to be spot welded as this is the most suitable permanent joint technique to be used which involved parts that have small thickness size. The strength and quality of the spot welds are very important to the durability and the safety design of the products. Moreover, the development of dissimilar materials joint will contribute benefits towards the cost, strength and quality. The lack of experience with new materials or combinations of dissimilar materials often results in the use of welding parameters which is not optimal. For this reason, a microRSW process needs the optimum process condition that can afford allowance in parametric values for a good quality welding. In this paper, 316L austenitic stainless steel and titanium alloy grade 5 (Ti-6Al-4V) each with 0.5 mm thickness respectively is used and its tensile shear and peel test were studied. The experimental studies were conducted under varying electrode force, welding current, electrode diameter, and welding time. The settings of welding parameters were determined by using Taguchi experimental design method. The level of importance of the welding parameters on the tensile shear strength is determined by using analysis of variance (ANOVA). The cross-section area of the welded joint is observed by using light optical microscope and scanning electron microscope (SEM). It was found out that welding current and time has dominant effect on the joint characteristics which reflect in the microstructure.

Welding Technology for Metallic Sheathing of Offshore Steel Structures Using Seawater-resistant Stainless Steel Sheet

Kohsuke Inagaki, Fumiaki Kimura, Yasuhiro Kawai, Hirotaka Satoh, Norimitsu Baba

Nippon Steel & Sumikin Engineering Co., Ltd, Futtsu-city, Chiba-pref, Japan
Inagaki.kohsuke.5sc@eng.nssmc.com, +81-80-7754-6771

Abstract
Welding technology for metallic sheathing of offshore steel structures in seawater tidal and splash zones was described. Seawater-resistance stainless steel SUS 312L sheets were lap-welded on carbon steel pipes with this combined welding method - indirect seam welding and plasma welding. The indirect seam welding enables to reduce rapping material thickness (0.4-0.6mm) and fabrication costs. The plasma welding with Inconel 625 filler wire was applies to prevent crevice corrosion at severer environments. A repair welding procedure was also established.

Introduction
It is necessary to provide highly reliable corrosion-preventive treatment in offshore steel structures used in extremely corrosive marine environments such as splash zone and tidal zone. It has been common practice to apply corrosion-preventive organic material to prevent corrosion of structural material in the marine surroundings [1]. The corrosion-preventive coating, however, must be repaired or renewed as required to maintain their long-term durability since it is subject to damage caused by some driftage and age-related deterioration. Because of this, offshore steel structures with the organic coating are hugely expensive to maintain. On the other hand, there are cases where corrosion-preventive metallic sheathing is applied to an offshore steel structure which is required to withstand use for an extremely prolonged periods, for example from 50 to 100 years [2]. When this method is used, the steel structure seldom needs repair and hence, the maintenance cost can be reduced significantly. Nevertheless, the application of metallic sheathing has been limited because the cost of the material and fabrication are higher than those of the conventional corrosion-preventive methods.

From the standpoint of minimizing the lifecycle cost of an offshore structure used in a highly corrosive environment, Nippon Steel & Sumikin Engineering Co., Ltd. has developed and put into practical use metallic sheathing technology using a highly reliable and economic seawater-resistant austenitic stainless steel sheet. This paper describes in detail the welding technology for metallic sheathing that is mentioned in Reference [3], which outlines the metallic sheathing method of offshore steel structures.

Problems in Conventional Welding Methods for Metallic Sheathing
Metallic sheathing is a corrosion-preventive method in which the steel members of a structure are sheathed with a metallic sheet having high strength and excellent corrosion resistance (hereinafter referred to as "covering material") to protect their surfaces from a corrosive environment. This technique is mainly applied to those members in offshore steel structures which are exposed to splash zone and tidal zone. The conventional metallic sheathing method uses a sheet of stainless steel, Monel, cupronickel or some other metal as the covering material, which is applied to carbon steel or low-alloy steel structural members by TIG (Tungsten Inert Gas) welding or shielded metal arc welding. The typical covering material is 1.5 to 2.0 mm thick. From the standpoint of minimizing the cost of covering material and facilitating the application of covering material to structural members, it is desirable that the covering material should be as thin as possible. When the covering material thickness is 1.0 mm or less, thermal deformation of the covering material caused by the welding heat increases and a gap is produced between the covering materials or between the covering material and the structural member. In order to prevent from making the crevice, it is necessary to decrease the pitch of tack welding or adjust the welding speed or wire rod

feeding speed in accordance with the gap size produced during welding. Because of this, welding speed and efficiency are not very good. Thus, if the conventional welding methods are applied to metallic sheathing, any reduction in the covering material thickness, improvement of welding efficiency and the cost cutting of metallic sheathing could hardly be hoped for.

Development of Highly-Efficient, Low-Cost Sheathing Method

The first aim of the present development project, intended for steel pipe members that occupy the greater part of the sheathing area of steel structures, was to establish a new welding technique which permits using much thinner covering materials and welding them more efficiently than conventional welding techniques. Specifically, the ability to weld a covering material 0.4 to 0.6 mm in thickness at a speed of about 1.0 m/min on a stable basis was set as the development target. As the covering materials, seawater-resistant stainless steel SUS312L [4] was used in this study.

Resistance seam welding was investigated for this development stage. In this welding method, as a large current is passed directly to the parts to be joined, the temperatures of the parts are raised by the resistance heat generated by the current so that the parts can be welded together. In particular, the adverse influence of material deformation caused by the welding heat can be avoided since a proper pressure is constantly applied by the electrode during welding. Thus, it is a rational welding method for metallic sheets that are thin and have a large thermal expansion coefficient. Since our new welding technique should be applicable to metallic sheathing of relatively large structures, such as offshore steel structures, we adopted not the commonly-used direct conducting seam welding method in which the welding current is passed from two electrodes to the joint sandwiched between them, but the indirect conducting seam welding method (hereinafter referred to as the "indirect seam welding method") in which a couple of electrodes are applied to the covering material side.

Fig. 1: Appearance of the indirect seam welding

In the indirect seam welding method, it is common practice to provide a back bar of copper, etc. at the back of the sheet to be welded in order to secure a stable current path which runs through the sheet in the thickness direction. In our method, a current path through the steel structural member at the back of the covering material can be obtained. Therefore, it is possible to weld stainless steel sheet of about 0.5 mm in thickness directly to the steel structural member. Fig. 1 shows the appearance of the indirect seam welding equipment developed for practical use. It's notable that high welding speeds up to 1.5 m/min could be attained with this welding method.

Development of Sheathing Method Offering Good Corrosion Resistance

With the sheathing method applying indirect seam welding, it is possible to use a thinner covering material and improve the welding efficiency. However, since the lap-welded structure obtained has crevice between the covering materials, it was considered that the crevice-corrosion resistance of those joints may become a problem depending on the ambient conditions.
With the aim of further improving the corrosion resistance of the welded structure, therefore, we attempted to find a welding method for sheathing which would not produce crevice between the

covering materials and which would still permit using a thinner covering material and offer high welding efficiency. As a result, we came up with a new welding method—the combination of indirect seam welding and arc welding—for sheathing with thin covering material.

With the strip sheathing method using conventional arc welding, the transfer of arc heat to the mild steel at the back of the covering material cannot be expected. In particular, since stainless steel used as the covering material has less heat conductivity than carbon steel, the heat transfer through the covering material is slow. Therefore, the thinner the covering material, the higher the possibility of a burn-through or perforation. On the other hand, with the newly developed sheathing method in which indirect seam welding is followed by arc welding, it is possible to transfer the arc heat to the carbon steel member at the back of the covering material via the indirect seam welded part, thereby preventing the burn-through and perforation even when the covering material is thin. Fig. 2 schematically shows the combination indirect seam plasma arc welding method. By performing indirect seam welding and plasma arc welding at the same time, it is possible to complete the sheathing without causing the welding efficiency to decline.

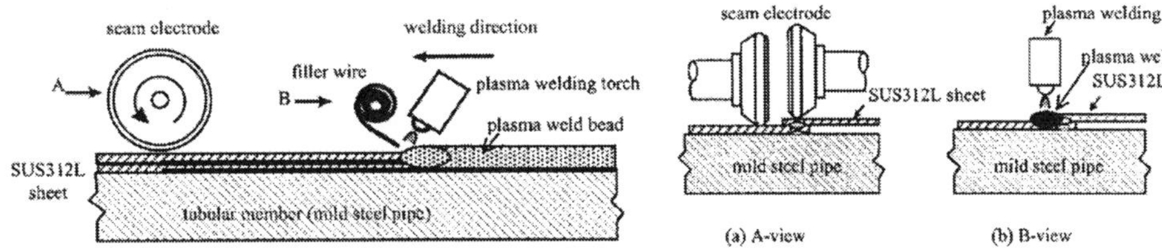

Fig.2 : Schematic illustration of combined welding

Fig. 3 shows the appearance of weld beads and the cross-section of a joint obtained by the combination welding method. A 0.4 mm thick stainless steel sheet used as the covering material was free from burn-through or perforation by the plasma arc welding and produced a good joint without crevice between the covering materials. Fig. 4 shows the distributions of Cr and Mo concentrations in the cross-section of the combination weld, obtained by an X-ray micro-analyzer. It was confirmed that the Cr and Mo concentrations in the weld metal and heat-affected zone remained nearly the same as those in the base metal and hence, there were no problems in terms of corrosion resistance.

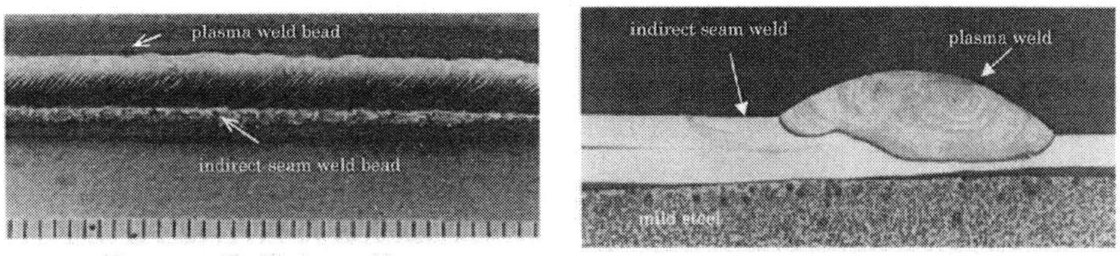

(a) appearance of combined seam weld

(b) cross section of combined seam weld

Fig.3: Appearance of weld bead and cross section

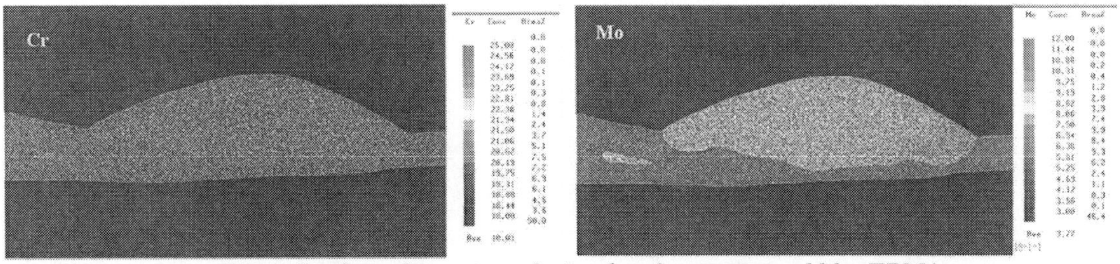

Fig.4: Surface element analysis of a plasma arc weld by EPMA

Repair Welding

The patching method is used for repairing metallic sheathing damaged in service by some driftage, etc. As the patching material, use an SUS 312L sheet 1.0 to 1.5 mm thick. If seawater has entered the room between the covering material and structural member from the damaged part, drill a steam vent 5 mm in diameter in the center of the patching sheet to use the welding heat to expel the moisture inside. Such a steam vent is unnecessary when the part around the defect is dry. Apply lap fillet welding to the part surrounding the defect and block the steam vent by TIG welding. Use Inconel 625 as the welding material. After the repair welding, carry out a liquid penetration test to check for leaks.

Application of the New Welding Method to Offshore Steel Structures

The sheathing with seawater-resisting stainless steel has achieved "low initial cost" and "low maintenance cost". Thus, we have come up with a new corrosion-preventive method that makes it possible to minimize the lifecycle cost of offshore steel structures with prolonged life expectancies. As seen in the example of steel jackets of Tokyo International Airport runway D, this method is now applied to various offshore steel structures (Fig. 5).

Fig.5 : The offshore steel structures

Conclusions

(1) We developed a new metallic sheathing method using indirect seam welding. This technique has made it possible to reduce the covering material thickness to 0.4 to 0.6 mm and improve the welding efficiency.
(2) It has been confirmed that the weld of seawater-resisting stainless steel SUS 312L has sufficient corrosion resistance.
(3) In addition to the above sheathing method, by using the combination indirect seam-plasma arc welding, we developed a new corrosion-resistant sheathing method that produces crevice-less joints.

References

[1] Coastal Development Institute of Technology: Corrosion Prevention of Port and Harbor Steel Structures. Improvement Manual. (1999)
[2] Yamamoto, A., Hase, T., Takahashi, Y.: Titanium. 47(3), 218-226(1999)
[3]Satoh, H.: Nippon Steel Technical Report. (87), 37-42(2003)
[4]Uki, T.: Zairyo-to-Kankyo. 41, 329-340(1992)

A Study on the Microstructure, Hardness, Modeling and Optimization of Laser Welding of Austenitic Stainless Steel

Nikhil Kumar, Asish Bandyopadhyay

Jadavpur University, India

In the present work, Nd:YAG pulsed Laser welding process has been successfully applied and optimized for joining a AISI 304 stainless steel sheet forming butt joints. This particular type of work has various applications for forming the automobile, microelectronics, marine and aerospace parts etc. A mathematical and statistical approach is adopted to study the laser welding of AISI 304 stainless steel. The effect of welding parameters, namely laser power, scanning speed and pulse width on the weld geometry (weld width and weld seam depth) has been investigated using the empirical models developed by response surface methodology (RSM). Weld quality is directly related with the weld geometry. The results of ANOVA indicate that the laser power has most significant effect on the responses. It is observed that average micro-hardness of the weld is higher than the base metal. Higher hardness of the weld is the resultant of grain refinement and δ-ferrite formation in the weld structure. The result suggests that the lower line energy generally produce fine grain structure and improved mechanical properties than the high line energy. The combined effects of input parameters on responses have been analyzed by developed 3-D response surface and contour plots. Finally, multi-objective optimization has been conducted for selecting parameters with complete penetration, minimum seam width and acceptable welding profile. Confirmatory tests have been conducted at optimum parametric conditions to validate the optimization technique.

Numerical Simulation of Brittle Crack Propagation Incorporating Toughness and Residual Stress Distribution in Welds

Tetsuya Namegawa*, Hiroaki Nakai**, Shuji Aihara, Tomoya Kawabata***
The University of Tokyo, Bunkyo, Tokyo, Japan,
*presently, * Nippon Steel & Sumitomo Metal Corp., ** IHI Corp.*
**** aihara@fract.t.u-tokyo.ac.jp, 03-5841-6505*

Hitoshi Yoshinari
National Marine Research Institute, Mitaka, Tokyo, Japan

Abstract
The present paper describes numerical simulation of brittle crack propagation and arrest in steel and welds. The model is based on the local fracture criterion and considers non-straightness of a crack front and irregularity of a crack surface. From static mixed mode stress intensity factors at a crack-tip, stress tensor is calculated, from which cleavage crack orientation is determined. By continuing this process, whole crack surface configuration is calculated. The model was applied to press-notched bend crack arrest tests. Calculated fracture surface morphologies agreed well with experimentally observed ones, including shear-lips, chevron patterns. Then, the model was applied to welds. It was demonstrated that crack propagation path changed, depending on toughness and residual stress distributions.

Introduction
Prevention of brittle fracture is crucially important in welded steel structures including large container ships and cryogenic storage tanks because a brittle crack propagates at high velocity and could bring about fatal damage to the structures. The double integrity concept is often applied for preventing brittle fracture, in which material and structural components are selected and designed so that once initiated brittle crack at welds is arrested in base metal. For realizing this concept, brittle crack arrest toughness of base metal must be evaluated accurately [1] and crack propagation path in welds must be controlled.

To realize the above concept, behavior of brittle crack propagation must be fully understood. To help these purposes, the authors proposed a numerical model to analyze brittle crack propagation in a microscopic scale [2][3]. The model was extended to macroscopic scale to simulate crack propagation path in steel welds, having toughness and residual stress distributions [4][5]. In the present paper, this model was applied to press-notched bend crack arrest toughness test [6]. Formation of shear-lips and chevron patterns, as well as crack arrest length, were analyzed by the model and compared with experiments. Then, the model was applied to steel welds. The model took account of toughness distribution along weld metal, heat-affected zone (HAZ) and base metal, and welding residual stress distribution. Possible effect of the toughness distribution and residual stress on crack path is discussed based on the numerical simulations.

Model
The present model for brittle crack propagation is described in detail in [2]. It is based on the critical stress criterion for cleavage fracture. Material is decomposed into small square elements and a cleavage plane is determined in each element. Macroscopic crack is regarded as aggregate of the cleavage planes. Therefore, the crack surface is irregular and the crack front is non-straight. Mixed mode static stress intensity factors are calculated at a crack front position, from which stress tensor is calculated. Using this stress tensor, a cleavage plane is selected from three planes which are perpendicular with each other in an element by a condition of maximum normal stress σ_n acting on them. This computational process is continued in a stepwise manner and the crack front progresses. In selecting the cleavage plane in each element, critical condition is set,

$$K_n \geq K_{c-local} \quad (1)$$

where, K_n is equivalent stress intensity factor corresponding to σ_n and $K_{c-local}$ is local fracture toughness. If this condition is not met, the crack is arrested. locally. Because the cleavage planes of the neighboring elements are generally twisted, ridges are formed between them. The ridges are assumed to shear-yielded and they reduce local stress intensity factors. Also, remote stress transverse to the crack plane may produce mode-II and III stress intensity factors, due to the crack surface irregularity. These features are taken into account in the model.

The procedure of the cleavage plane selection is performed on the three edges of an element. If the critical condition for an edge is not satisfied, then the edge is "arrested". The crack front is approximated by a continuous function and the stress intensity factor is calculated using this function. Because the calculation of the stress intensity factors is based on the approximate equations and the effect of crack surface irregularity, crack front non-straightness are linearly superposed, it is approximate. Model calculations were conducted for temperature-gradient crack arrest tests by changing $K_{c\text{-local}}$, from which relationship, $K_{c\text{-local}} = 0.44\ K_{ca}$, was established, K_{ca} being calculated from applied stress and arrest crack length [5].

Because stress tri-axiality is lower near the plate surface than at mid-thickness, cleavage fracture is less likely to take place there; shear-lips are formed [6]. To account for this effect, a multiplying factor on $K_{c\text{-local}}$ was introduced, which was assumed to increase near the plate surface.

Experiment

Press-notched bend crack arrest tests were conducted according to WES TR2816 [7][8]. Tested steel plate was normalized high-strength steel. Table 1 shows its mechanical properties. Figure 1 shows specimen configuration. Thickness was reduced to 12.7mm, and 5mm or 10mm deep notch was introduced by pressing a chisel onto the specimen edge. The specimen was cooled by alcohol and dry ice and force was quasi-statically applied until brittle fracture commenced. From the applied force at brittle fracture initiation, P_0, and arrested crack length, a_a, arrest toughness, $K_{ca\text{-PB}}$ was determined. Figure 2 shows dependence of $K_{ca\text{-PB}}$ on test temperature. Figure 3 shows fracture surfaces for selected 4 specimens. Crack arrest positions were clearly identified in the three specimens, while almost complete fracture was evident in specimen N37. Shear-lips were deeper at higher test temperature. Chevron patterns were observed in all the specimens. An interesting feature is that the chevron patterns became weaker at the last half of the fracture surface of specimen N36 and N37. Note that specimen N36 exhibited ductile crack extension from the press-notch root, ductile crack length, 3mm, was included to the initial crack length in calculating $K_{ca\text{-PB}}$ [7].

Table 1 Tested Steel

Thickness t (mm)	Tensile Properties			Charpy V FATT (℃)
	YP (MPa)	TS (MPa)	Elongation (%)	
30	416	541	27	-40 (L)

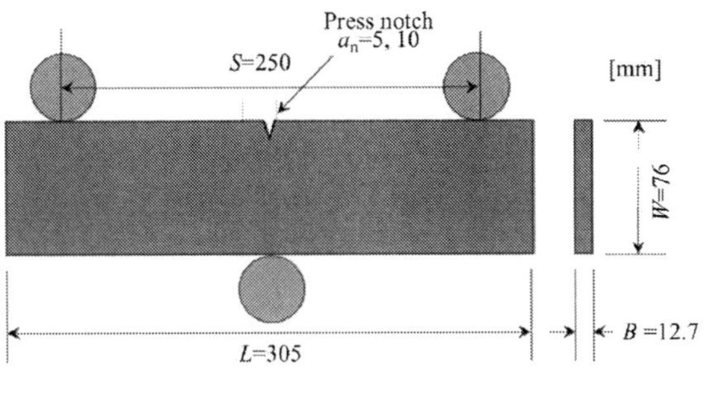

Fig.1 Press-notched bend specimen configuration

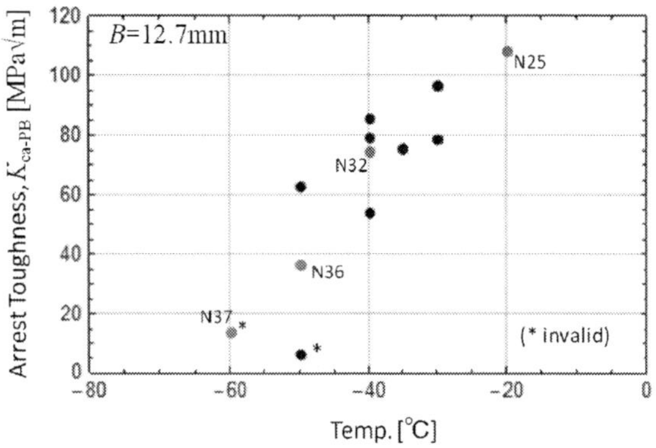

Fig.2 Crack arrest toughness by press-notched bend test

Numerical Simulation of Crack Propagation in Press-Notched Bend Specimens

Brittle crack propagation of the press-notched bend tests was simulated by the present model. Crack arrest toughness, K_{ca}, was assumed according to the empirical formula:

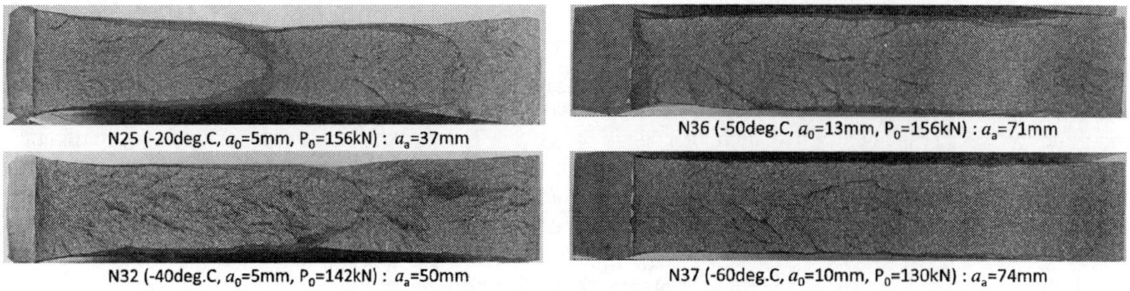

Fig.3 Facture surfaces of test specimens

$$K_{ca} = 480\exp\left[10^{3.488+0.0069vTr_E+0.0055t}\left(0.00327-\frac{1}{T}\right)\right] \tag{2}$$

where, K_{ca} is in [MPa$\sqrt{}$m], vTr_E [℃] is Charpy V absorbed energy transition temperature and assumed equal to FATT (Table 1), t [mm] is plate thickness and T [K] is temperature [5]. Stress intensity factor was calculated from brittle fracture initiation load, P_0, initial crack length, a_0, and crack length, a, using the formula proposed in [7], in which fixed displacement was assumed. Figure 4 shows change of the stress intensity factor for 4 specific specimens in Fig.2 and Fig.3. This was used in the model calculation. Fifty elements of size 0.25mm were aligned in the thickness direction, *i.e.*, 12.5mm thickness. Yield stress for calculating the closure force by the ridges between the cleavage planes was assumed equal to 800MPa, which is higher than YP of the material accounting for the strain-rate effect on yield strength.

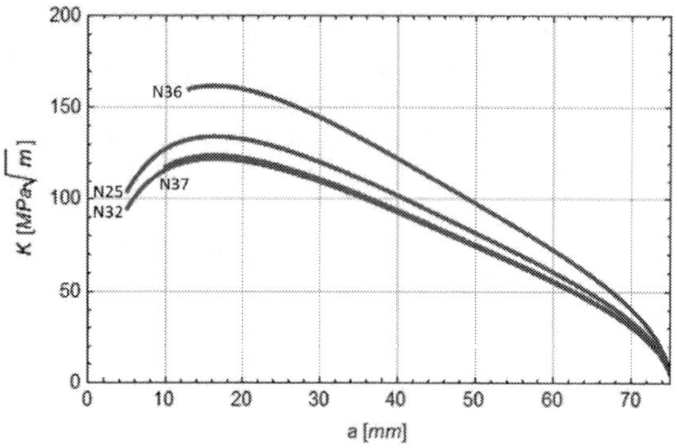

Fig.4 Stress intensity factor of press-notched bend specimen

Figure 5 shows calculated fracture surfaces. The color indicates height. Good agreement is obvious regarding the morphologies of shear-lips and chevron patterns, and arrest crack length with shorter length in specimen N37. Note that the arrested crack front shape changed with temperature: deeper crack tunneling at higher temperature, which agrees with Fig.3. Figure 6 shows crack propagation process before crack arrest in specimen N25. As the crack propagates, ranges of active crack front, where cleavage fracture continues, shrinks and finally all the crack front becomes inactive; the crack is arrested. It is also interesting to note that the chevron patterns decrease at the last half of the fracture surface of specimens N36 and N37, see Fig.3. This feature is reproduced in the simulation, as shown in Fig.5. This phenomenon is presumed different from the temperature-gradient crack arrest test, in which stress intensity factor monotonically increases and chevron patterns increase with crack propagation. Temperature effect, however, should also be considered.

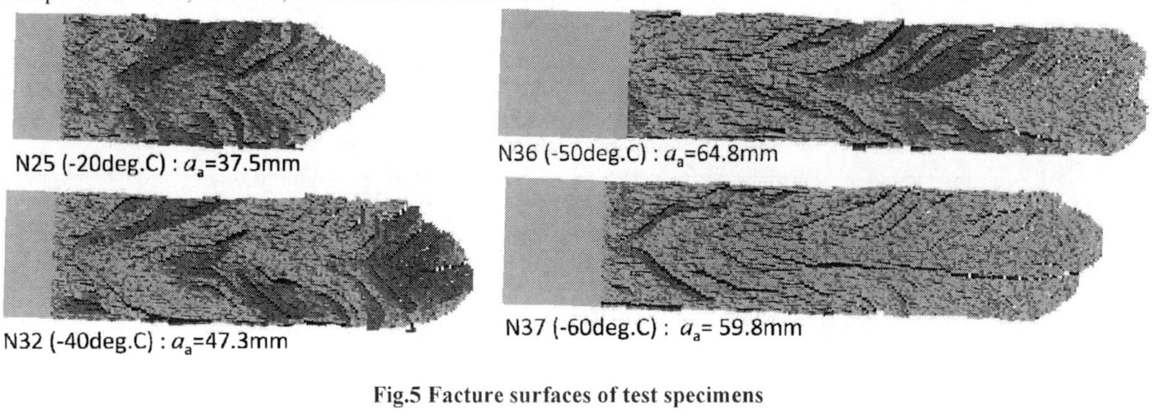

Fig.5 Facture surfaces of test specimens

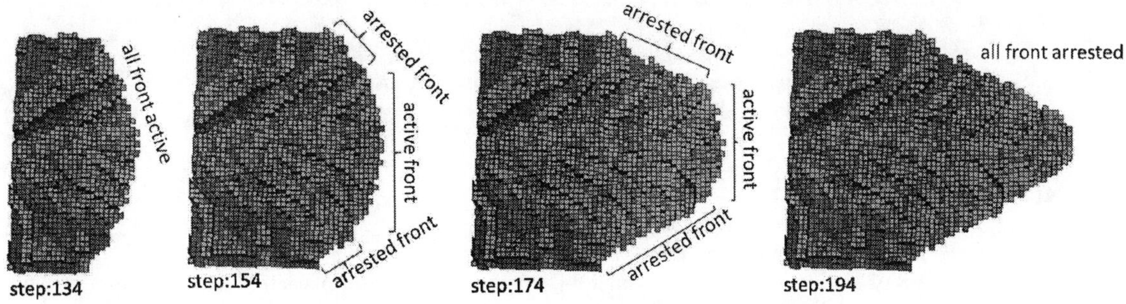

Fig.6 Crack propagation process in specimen N25.

Numerical Simulation of Crack Propagation in Welds

Brittle crack propagation in welds was simulated by the proposed model in special attention to crack path deviation from welds [5]. Welds contain residual stress distributions and toughness distribution at weld metal, HAZ and base metal. Weld groove shapes might affect the crack propagation behaviors, as well. Effect of all these factors were examined by the present model. Detailed results will be shown at presentation.

Conclusions

In the present study, a numerical model was presented to simulate brittle crack propagation and arrest in steels and welds. It was demonstrated that the model calculation agreed well with experiments in terms of arrest crack length and fracture surface morphologies including shear-lips and chevron patterns. Also, the model clarified the effect of toughness distribution and residual stress on crack path deviation in welds. The model might need improvement because it is intrinsically static and based on the liner superposition of approximate stress intensity factor formulas. However, it has a potential to analyze many aspects of brittle crack propagation and arrest behaviors in steels and welds.

Acknowledgments

A part of the present study was conducted at ATE2 committee, Steel division, Japan Welding Engineering Society. The authors express their thanks to the committee members.

References

[1] WES 2815, Test method for brittle crack toughness, K_{ca}, Japan Welding Engineering Society, Jan. 2014.

[2] S. Aihara and Y. Tanaka, "A simulation model for cleavage crack propagation in BCC polycrystalline solids", Acta Materialia, vol. 59 (2011), No.11, pp,4641-4652.

[3] I. Kawata, K. Sugimoto, S. Aihara, "Simulation of cleavage crack propagation in steels with anisotropic crystal orientation", Proc. 20th European Conf. Fracture (ECF20), June 2014, Trondheim, Norway, Vol.3, p.531-536.

[4] H. Nakai, S. Aihara and T. Inoue, "Numerical simulation of cleavage crack path in a welded joint", Proc. 21st Int. Offshore and Polar Eng. (ISOPE2011), Jun. 2011, Hawaii, USA, p.222-229.

[5] H. Nakai, H. Yoshinari, T. Inoue and S. Aihara, "Development of three-dimensional crack path prediction model for brittle fracture in welded joint with residual stress, toughness distribution and various welding groove", Bulletin Japanese Society of Naval and Ocean Engineering, vol. 19 (2014), No.13, pp.123-137.

[6] K. Shibanuma, F. Yanagimoto, T. Namegawa, K. Suzuki and S. Aihara, "Brittle crack propagation / arrest behavior of steel plate – Part I&II", early view, Eng. Fracture Mech., 2016.

[7] WES TS2816, Simplified test method for brittle crack arrest toughness using press-notched bend specimen, Japan Welding Engineering Society, Jan. 2015.

[8] T.Kawabata, T. Namegawa, M. Kaneko, Y. Shiamda, H. Tajika and K. Shibanuma, "Numerical analysis of press-notched bend tests and applicability to simplified method of arrest toughness evaluation", Proc. 25th Int. Offshore and Polar Eng. Conf. (ISOPE2015), June 2015, Hawaii, USA, p.169-176.

On the influence of weld porosity on fatigue strength

Dieter Siegele, Peter Tempel, Majid Farajian
Fraunhofer Institute for Mechanics of Materials IWM, Freiburg, Germany
dieter.siegele@iwm.fraunhofer.de, +49-761-5142116

Abstract

Weld quality could have a significant influence on the fatigue strength of welded components. The correlation between weld imperfections and fatigue strength is defined in different guidelines as IIW-recommendations or the German guideline DVS 0705 in a conservative manner. In order to assess the influence of weld porosity on fatigue strength in a more accurate way, experimental and numerical investigations on welded specimens with internal pores were performed. Experimentally, the specimens were tested under cyclic loading until failure and finally, the fracture surfaces were evaluated with respect to the size and location of the defects and the fracture behavior. Numerically the specimens were simulated including the inner pores and evaluated regarding the number of cycles to crack initiation and crack propagation, respectively.

Besides the conventional approach to fatigue assessment of welded components based on their fatigue classes, the applicability of the fatigue damage parameter according to Fatemi and Socie was examined in order to estimate the fatigue life until the formation of a crack from the initial pore. The remaining life until failure was estimated using fracture mechanics methods in combination with the fatigue crack growth curve da/dN. The results demonstrate that in general the number of cycles until crack formation is larger than the cycles of crack propagation until failure. Therefore, the assumption of the pore as a sharp crack would underestimate the fatigue life significantly. With the combination of both methods, the estimation of crack formation using an appropriate damage parameter and crack extension based on fracture mechanics the fatigue life of welded plates with internal pores can be described in accordance with the experiments.

Introduction

If a welded construction is loaded under fatigue, the rules and standards always recommend the highest weld quality [1]. Therefore, the weld had to be proven with respect to its quality using nondestructive methods as visual proof and ultrasonic techniques. The weld quality consist of several categories in terms of weld imperfections: surface imperfections as weld notches, excessive peaking or cracks and internal imperfections as pores, inclusions or incomplete fusion. For a safe use of a welded construction the imperfections have to be assessed if they can be tolerated or how such imperfections could influence the strength of the component. The assessment groups of weld imperfections and their correlation to fatigue classes are defined in different standards and guidelines as DIN EN ISO 5817 [2], DVS guideline 0705 [3], or IIW recommendations [4]. But in general, these recommendations are based on experimental investigations on specimens or real components welded under normal production conditions. These constructions could contain high tensile residual stresses besides the imperfections that could have reduced the fatigue strength significantly. Also the type of an imperfection and its location in the stress strain field play an important role with respect to the components lifetime. All these factors could influence the assessment of imperfections and its correlation with fatigue classes. Therefore, the different guidelines show some differences in the classification of assessment groups and fatigue classes [4, 5].

As an alternative to traditional methodology for fatigue assessment, computational methods can be employed based on numerically determined stress and strain fields and appropriate fatigue damage models, see e.g. [6], [7], [8]. Such an approach can facilitate to transfer of different types of welded joints, and to estimate fatigue lives of components with different geometry and/or loading conditions. In this paper a methodology for the fatigue assessment of weld imperfections based on finite element calculations will be presented. The procedure concentrates on internal defects as pores but can be transferred also to other weld imperfections. The procedure contains the estimation of lifetime until formation of a crack nucleated from the defect as well as the lifetime until final failure due to crack propagation. The procedure is supported and validated by experiments on welded specimens with internal defects. The proposed methodology allows for the new classification of defects to fatigue classes as well as the assessment of individual weld imperfections.

Method

Fatigue damage models describe the stress strain field in terms of a suitable parameter and correlate them with the Woehler-Curve, i.e. the number of load cycles to failure. Since these fatigue damage parameter (FDP) are based on the local stress strain field they allow for taking into account the effects of the mean stress and stress multiaxiality. In the literature, many FDP are presented and it can be distinguished between the stress based, the strain based, and the energy based parameter that assume the local stresses, strains and energy as the main damage driving parameter, respectively. All parameter have to be evaluated in a critical plane which yields the maximum value of the respective parameter and therefore the highest possibility of failure.

The parameter P_{FS} used in this work has been derived by Fatemi and Socie [8] and is defined as follows:

$$P_{FS} = \frac{\Delta\gamma}{2}\left(1 + k\frac{\sigma_{n,\max}}{\sigma_0}\right) = \frac{\tau'_f}{G}\left(2N_f\right)^{b\gamma} + \gamma'_f\left(2N_f\right)^{c\gamma}$$

It is based on the assumption that the maximal damage occurs in the plane of the highest shear stress amplitude $\Delta\gamma$, whereas the damage will be promoted by the maximum normal stress $\sigma_{n,max}$. In the equation, σ_0 denotes the monotonic yield strength and k is a factor to consider the mean stress sensitivity of the corresponding material. As an alternative to cyclic shear tests also tension-compression tests with different R-ratios could be used to calibrate P_{FS} with N_f.

Within these investigations, round tensile specimens were extracted from the weld material and cyclically loaded at different strain amplitudes. From additional finite element simulations of these fatigue tests the damage parameter P_{FS} was calculated and plotted versus the number of cycles to failure. The left part of figure 1 shows the results of these tests for different values of k. For comparison reasons the damage-life curve of another construction steel 1050 is plotted too. In addition a correlation of P_{FS} with the material hardness proposed by Fatemi [9] is plotted in the diagram. It can be seen that the weld material behaves quite similar as the construction steel and especially, that the weld metal hardness correlation provides a very good approximation of P_{FS}. Therefore, this hardness correlation could be used for the assessment of the specimens with weld imperfections with respect to the damage development e.g. crack formation over the whole stress range.

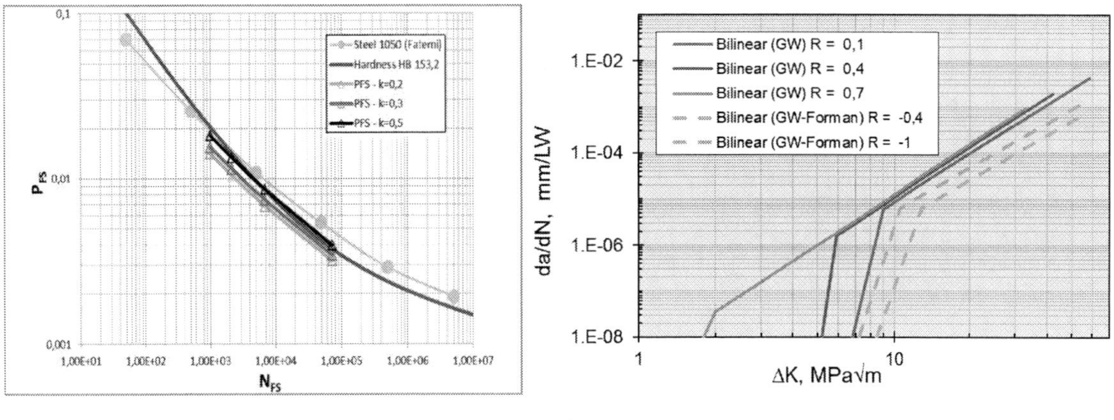

Fig. 1: Damage Parameter P_{FS} vs. number of cycles to failure (left) and crack propagation da/dN vs. range of the stress intensity factor ΔK for the weld material in the construction steel S355J2H

After a crack has been developed at the boundaries of an imperfection its crack propagation could be described with the concept of fracture mechanics in terms of the crack propagation rate da/dN dependent on the range of the stress intensity factor ΔK [10]. The da/dN curves were determined from tests on compact-tension (CT) specimens at different R-rations. The diagram on the right hand side of figure 1 shows these da/dN curves obtained for R-ratios 0.1, 0.4 and 0.7. For the stress ratios -0.5 and -1 the da/dN lines were extrapolated using Forman equation [11]. The dependency of the crack propagation from the R-ratio is clearly visible especially with respect to the threshold value. It should be noted that these threshold values were not fully measured up to the very low crack propagation rate of 10^{-8} but extrapolated from the lowest value. Moreover, the threshold value for R=0.7 was estimated from a specimen with an internal pore at a high superimposed bending stress in order to be able to describe the crack propagation observed in the experiment.

Results

For the specimens with internal imperfections fatigue tests were performed under tension load at R=0.1 and different stress ranges. Regardless of size of the inner pores in the welded specimens, the fatigue failure initiated at the weld toe because of the notch effect. In order to activate the damage initiation at the internal defect and not at the weld toe, the weld reinforcement was removed. Since the specimens were distorted after the welding they experienced a superimposed bending stress under cyclic loading that acts as a mean stress. Depending on the magnitude of distortion angle and the location of the defect in the cross-section the local R-ratio at the pores boundaries could vary over a wide range up to about 0.7. Figure 2 (left) shows the finite element mesh of a typical pore including the stress concentration at the boundaries of a pore. In the analyses, a stress concentration factor up to 3.8 was observed. The right hand part of figure 2 shows the fracture surface of a weld with an included pore, where the crack propagation is clearly visible. Although the pore is located quite in the center, due to the superimposed tensile bending stresses the crack propagates faster to the upper surface, where the specimen finally fails.

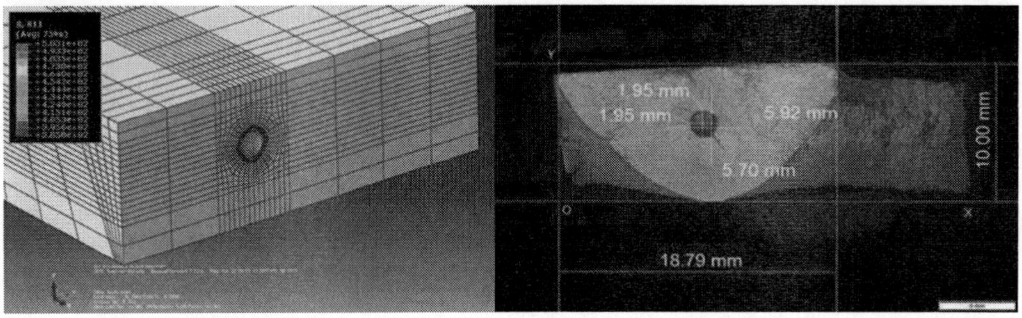

Fig. 2: Finite element model with stress concentration around the pore (left) and fracture surface of a typical specimen with a weld defect.

For all specimens finite element analyses were performed and the number of cycles to crack formation was calculated with the aid of the damage parameter according to Fatemi and Socie. After that, the crack propagation was calculated assuming the pore as a penny shaped crack by using the analytical defect assessment software VERB developed and distributed at Fraunhofer IWM [12].

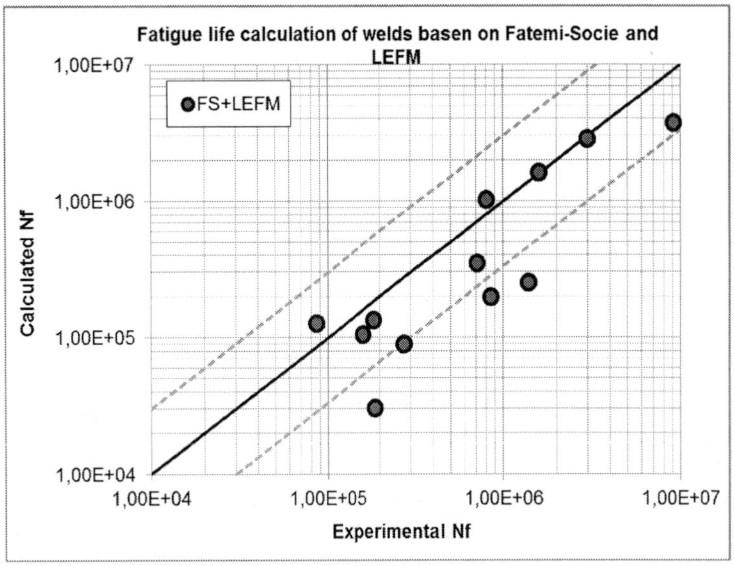

Fig. 3: Lifetime estimation calculated based on crack formation and propagation compared to measurements.

Table 1 shows the calculated number of cycles for crack formation and crack propagation in the specimen in figure 2, compared to the fatigue test. Here and for most other cases the number of cycles to crack initiation is much higher than the crack propagation life (table 1).

The same procedure was used to calculate the total fatigue life of 12 weld specimens with a pore in each of them, based on damage and fracture mechanics (figure 3). It can be concluded that the numerical procedure results in a very good agreement with the test results with the tendency of a conservative prediction which is of importance for practical applications.

Table 1: Calculated and measured cycles to failure for a typical weld with imperfection.

Lifetime Estimation	N_f (NUM)	N_f (EXP)
Crack Formation (P_{FS})	450,000	-
Crack Propagation (da/dN)	146,200	-
Total Lifetime	596,200	706,758

Conclusions

For welds with internal defects fatigue tests were performed and in parallel, a numerical method was established to predict the lifetime of specimens or components containing internal imperfections. The numerical method includes the estimation of crack nucleation at the surface of an imperfection as well as the crack propagation. The crack nucleation is calculated based on elastic-plastic finite element analyses and the fatigue damage parameter according to Fatemi and Socie that can be derived from the local stress-strain field at the internal defect. The parameter includes also some material constants that could be obtained from a limited number of fatigue tests or from a simple analytical correlation with the material hardness. The estimation of crack propagation was performed using the analytical flaw assessment software VERB based on linear fracture mechanics and using crack propagation curves which were measured on CT-specimens.

The analyses of the fatigue tests on specimens with internal defects result in a very good agreement with the measurements in a slightly conservative manner which is of importance for practical applications. The results demonstrate that in general the number of cycles until crack formation exceeds the number of cycles of crack propagation until failure. Therefore, the assumption of the pore as a sharp crack would underestimate the fatigue life significantly. The combination of both methods, the estimation of crack formation using the fatigue damage parameter according to Fatemi and Socie and the crack propagation based on fracture mechanics was demonstrated as an efficient tool to estimate the fatigue life of welded constructions with internal defects.

Acknowledgments

The paper was supported by the German Federal ministry for Economic Affairs and Energy (BMWi) as part of an AiF Project entitled "Influence of the weld quality on fatigue strength". The authors would like to thank the AiF and BMWi for their support.

References

[1] Hobbacher, A., Recommendations for Fatigue Design of welded Joints and Components, Second Edition, Springer Verlag, 2016.

[2] DIN EN ISO 5817, Schweißen - Schmelzschweißverbindungen an Stahl, Nickel, Titan und deren Legierungen (ohne Strahlschweißen) - Bewertungsgruppen von Unregelmäßigkeiten. Beuth Verlag, Berlin 2014.

[3] DVS 0705, Empfehlungen zur Zuordnung von Bewertungsgruppen von Unregelmäßigkeiten von Schmelzschweißverbindungen (außer Strahlschweißen) an Stahl, Nickel, Titan und anderen Legierungen nach DIN EN ISO 5817:2006-10 und deren Vorgängernorm DIN EN 25817:1992-09 (ISO 5817:1992), Düsseldorf 2012.

[4] Hobbacher, A., Kassner, M., On Relation between Fatigue Properties of Welded Joints, Quality Criteria and Groups in ISO 5817, IIW Doc. XIII-2323r1-10

[5] Hobbacher, A., Problems of Effect of Weld Imperfections on Fatigue and their Consideration in Design Codes, Steel Structures, 289-298, 2006.

[6] Smith, R.N., Watson, P., Topper, T.H., A stress-strain parameter for the fatigue of metals. J. Materials 5, 767-778, 1970

[7] Socie, D.F., Multiaxial fatigue damage models. J. Eng. Mater. Tech. 109, 293-298, 1987

[8] Fatemi, A., Socie, D.F.: A critical plane approach to multiaxial fatigue damage including out of phase loading, Fatigue Fract. Eng. Mater. Struct. 11, 149-165, 1988

[9] Roessle, M. L. and Fatemi, A., 2000, "Strain-Controlled Fatigue Properties of Steels and Some Simple Approximations," International Journal of Fatigue, Vol. 22, pp. 495-511.

[10] Paris P. C., Erdogan F., A critical analysis of crack propagation law, Trans ASME D (J Basis Engng) 85, pp. 528-534, 1963.

[11] Forman, R.G., Kearney, V.E., Engle, R.M., Numerical analysis of crack propagation in cyclic loaded structures, Trans. ASME, J. Basic Engineering 89, 459-463, 1967

[12] VERB-Manual, User Guide für das Programm VERB, Fraunhofer IWM, 2016.

Numerical model of ductile damage using the subloading surface model including the stress triaxiality and the Lode angle effects

R. Fincato, S. Tsutsumi
Joining and Welding Research Institute, Osaka University, Osaka, Japan
fincato@jwri.osaka-u.ac.jp, +81-6-6879-8667

Joining and Welding Research Institute, Osaka University, Osaka, Japan
tsutsumi@jwri.osaka-u.ac.jp, +81-6-6879-4887

Abstract
Ductility in metals takes into account the material capability to bear a large amount of plastic deformations before a partial or total degradation of the mechanical properties. This parameter acquired a great importance in structural or components design since its modeling can be used to estimate the material failure under a generic multi-axial stress state. Several authors attempted to give an accurate description of the mechanical properties degradation as a result of the formation of micro-voids inside the media, their growth and finally their coalescence. On the other side it has been experimentally proved that ductile damage is inherently linked with the accumulation of plastic strain, therefore a coupling between damage and elasto-plasticity is necessary for an accurate description of this phenomena. The purpose of this paper is to combine the approach propose by Lemaitre together with the features of an unconventional plasticity model: the Extended Subloading Surface Model, in order to predict material fatigue even for loading conditions below the yield stress.

Introduction
Ductility can be understood as the ability for the material to undergo a certain amount of plastic deformations before the formation of cracks and the consequent material or structural failure. The mechanisms that generate the fracture have been widely investigated in the previous years starting from the pioneering work of Kachanov [1], which led to a series of different constitutive models which aim to describe the mechanical property degradation in the framework of the Continuum Damage Mechanics CDM.
Within the theory of elastoplasticity two main approaches have been developed: the Gurson's void growth model [2] and the Lemaitre's model [3]. In the present paper we followed the second theory describing the damage as an additional internal scalar variable which evolves following the accumulation of the plastic deformations. However, many recent experimental works [4-7] underlined the importance of considering the stress triaxiality as a key parameter in the damage evolution, showing different ductility levels as a function of the notch size in tensile bar tests. Moreover, from the experimental observations carried out by Clausing [8] and the subsequent analyses of Wierzbicki *et al.* [9] and Xue [10], the Lode angle effect has also being recognized as a fundamental aspect influencing the plastic flow and the ductile fracture of metallic materials. An interesting recent paper [11] reports the combination of the stress triaxiality and Lode angle on the ductile fracture for the Q460 construction steel both with numerical and experimental evidence.
The present work adapts the constitutive equations of the Lemaitre's ductile damage model for the unconventional plasticity model Subloading Surface [12] taking into account the effect of the stress triaxiality and the Lode angle for describing the material degradation under different loading conditions.

Constitutive Equations
The subloading surface equations were modified in order to take into account the voids formation that leads to the ductile fracture for metals. A detailed explanation of the plasticity algorithm is not treated in the present paper, the author is therefore referred to [12] for a better comprehension of the model.
The damage is regarded as a scalar internal variable which expresses the loss of material stiffness caused by several processes such as fatigue, void formations or eventually creep, and it is coupled with the plastic variable by modifying the expression of the plastic potential as follows (see *Fig. 1* as reference):

$$\begin{aligned} &f(\hat{\boldsymbol{\sigma}}) = (1-D)F(H),\ \hat{\boldsymbol{\sigma}} = \boldsymbol{\sigma} - \boldsymbol{\alpha} && \textit{normal-yield surface} \\ &f(\overline{\boldsymbol{\sigma}}) = (1-D)RF(H) && \textit{subloading surface} \end{aligned} \tag{1}$$

Here, $\boldsymbol{\sigma}$ is the Cauchy stress, $\boldsymbol{\alpha}$ is the back-stress, F is the isotropic hardening function, H is the isotropic hardening variable, R is the similarity transformation ratio, and $\overline{\boldsymbol{\sigma}}$ and $\overline{\boldsymbol{\alpha}}$ are the conjugate Cauchy stress and conjugate back-stress for the subloading surface, respectively, which are expressed as:

$$\bar{\boldsymbol{\sigma}} = \boldsymbol{\sigma} - \bar{\boldsymbol{\alpha}}, \quad \bar{\boldsymbol{\alpha}} = \mathbf{s} - R\hat{\mathbf{s}}, \quad \tilde{\boldsymbol{\sigma}} = \boldsymbol{\sigma} - \mathbf{s}, \quad \hat{\mathbf{s}} = \mathbf{s} - \boldsymbol{\alpha} \tag{2}$$

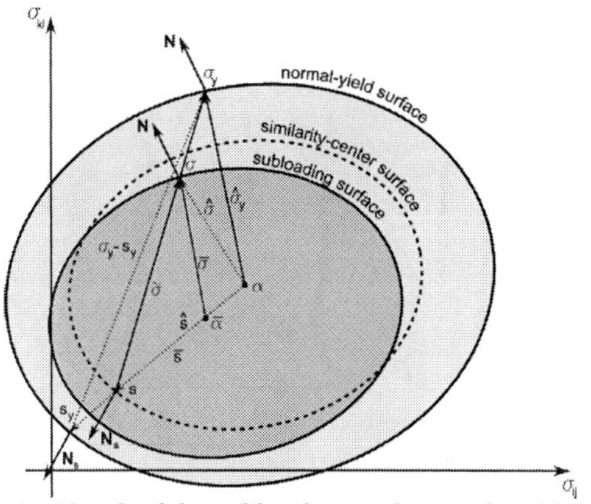

Figure 1 - Sketch of the subloading and normal-yield surface.

The choice of this particular plasticity model is due to the fact that it captures a realistic plastic strain accumulation during cycles under different boundary conditions, therefore it is particularly well suited to be used as a reliable tool for fatigue problems.

Ductile Damage Evolution Law

Following the continuum damage mechanics approach, the damage variable is assumed to be an internal variable that is responsible for the description of the degradation of the mechanical properties arising from microscale imperfections. For sake of simplicity it was assumed that the defects are uniformly distributed in the media allowing us to consider an isotropic scalar variable, whose evolution is derived directly from a dissipative potential as stated by [3].
Briefly the damage evolution law can be written as:

$$\dot{D} = \frac{\lambda^+}{(1-D)}\left(\frac{Y}{s_1}\right)^{s_2}\langle H - s_3\rangle\frac{1}{\mu_\vartheta}; \quad Y = -\left[\frac{f(\bar{\boldsymbol{\sigma}})^2}{6G(1-D)^2} + \frac{\bar{\sigma}_m^2}{2K(1-D)^2}\right]; \tag{3}$$

where λ is the plastic multiplier and superscript + indicates that just the tensile contributions are considered. s_1, s_2, and s_3 are material parameters; s_1, and s_2 affect the energy release rate, Y [3], and s_3 is a threshold for the cumulative plastic strain after which damage begins (the term in the Macaulay brackets is null until H = s3). σ_m is the mean stress and G and K are the shear and bulk moduli, respectively.
As it can be seen in *Eq. (3)* the energy release term Y already takes into account the effect of the stress triaxiality on the damage evolution, whereas the function μ_ϑ is instead responsible to induce a different response depending on the loading condition. Before introducing the Lode angle function, it is important to point out two additional hypotheses [10]:

- The effects of the stress triaxiality and the Lode angle are independent;
- The Lode angle dependence function is symmetric with respect to the plane strain axes ($\chi = 0.5$) and the principal stress axes ($\chi = 0$ or $\chi = 1$);

$$\mu_\vartheta = 4\beta\chi(1-\chi)+1; \qquad \chi = (\sigma_2' - \sigma_3')/(\sigma_1' - \sigma_3') \quad (0 \le \vartheta \le \pi/6 \rightarrow 0 \le \chi \le 1) \tag{4}$$

where ϑ is the Lode angle, χ indicates the type of loading condition as a function of the principal deviatoric stress and β ($0 < \beta \le 1$) is a material parameter which regulates the amount of influence of the Lode angle function on the damage evolution. Looking at *Fig. 2* we can imagine that the effective strain failure surface on the deviatoric plane is a perfect circle if no Lode angle effect is taken into account, meaning that, independently from the loading condition the failure happens at the same value of the effective strain (i.e. cumulative plastic deformation at failure). With the addition of the function in *Eq. (4)* the damage accumulates faster for higher values of the β parameters (*Fig.*3), for instance, the crack will open sooner for a plane strain or pure shear condition ($\chi = 0.5$) compared to a uniaxial extension or compression ($\chi = 1 / \chi = 0$).
It has to be said that experimental evidences [7, 10] showed that the effect of the Lode angle effect is higher for low stress triaxiality regimes and tends to be negligible at high triaxiality, therefore a reduction of the μ_ϑ should

be also evaluated. In this present work however such improvement is not considered and the two factors, stress triaxiality and Lode angle, are assumed as completely independent.

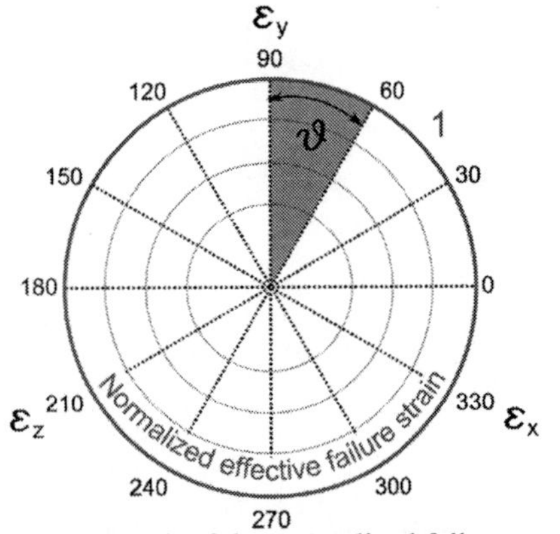

Figure 2 - sketch of the normalized failure strain surface on the deviatoric plane.

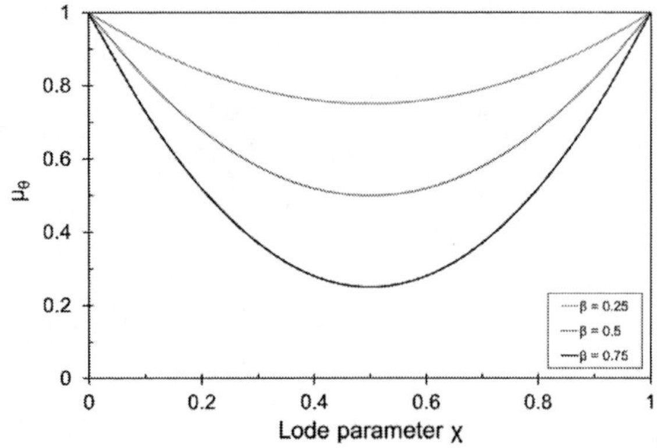

Figure 3 - Lode angle function evolution in one sextant on the deviatoric plane.

Numerical Analyses

The literature, dealing with the experimental research in the investigation of the Lode angle effect on the material failure, is quite scarce and this make very difficult to formulate a quantitative description of the phenomena. However, a qualitative formulation is still possible starting from few experimental results. In particular, Clausing [8] investigated the effective strain ratio between the fracture strain ε_{ps} in a plane strain uniaxial tension case and the one in a axisymmetric uniaxial tension test ε_{ut} against the strain hardening capacity (*Eq. (5)*) for the seven different materials reported in *Table 1*.

$$\left(\sigma_{f,ut}-\sigma_{y,ut}\right)/\left(\varepsilon_{ut}\cdot\sigma_{y,ut}\right) \tag{5}$$

Table 1 – Materials and data from Clausing experiments [8] and numerical strain ratios.

Steel	Yield stress $\sigma_{y,ut}$ [MPa]	Fail. stress $\sigma_{f,ut}$ [MPa]	ε_{ut}	ε_{ps}	Strain hardening	(ε_{ps}/ ε_{ut})	(ε_{ps}/ ε_{ut}) numerical
ABS-C	262.7	428.9	1.04	0.75	0.61	0.72	0.74113
A302-B	371.0	590.2	0.98	0.72	0.60	0.73	/
HY-80	586.1	689.5	1.22	0.73	0.14	0.60	0.59182
HY-130(T)	937.7	986.0	1.06	0.39	0.05	0.37	/
18Ni(180)	1227.3	1289.4	1.00	0.42	0.05	0.42	/
10Ni-Cr-Mo-Co	1254.9	1413.5	1.16	0.36	0.11	0.31	0.32225
18Ni(250)	1710.0	1772.0	0.89	0.15	0.04	0.17	0.193574

For sake of simplicity we limited the analyses to four of the seven materials, implementing the constitutive equations via user subroutine for the commercial code *Abaqus* (v6.14-5). We created two different samples in 2D reproducing an axial symmetry uniaxial tension on a round bar and a uniaxial tensile plane strain condition respectively using 1279 CAX4R and 1100 CPE4R elements. We considered as a condition for the material failure when the damage variable reached the value of 0.6.

A first thing to notice in the Clausing's experimental data is that the value of the effective strain is unanimously smaller in the plane strain compared to the axisymmetric uniaxial cases generating ratios lower than the unity, stressing out the importance of the loading condition in the material failure. Moreover, the ductility seems to decrease qualitatively with a weaker strain hardening. We tried to reproduce the experimental data attributing to β a higher value for the weaker strain hardening materials and a lower β for the stronger ones. The results obtained are in good agreement with the experiments (*Fig.4*), however it has to said that the β coefficients were chosen qualitatively and a precise parameter calibration is particularly difficult for weak strain hardening materials, which show very different effective strain ratios for small variations of the strain hardening.

However, the introduction of a Lode angle function in the damage evolution (*Eq. (3)*) seems to be able to catch the tendency of a different material degradation depending on the loading condition indicating that more intensive experimental investigations are necessary for a more precise characterization.

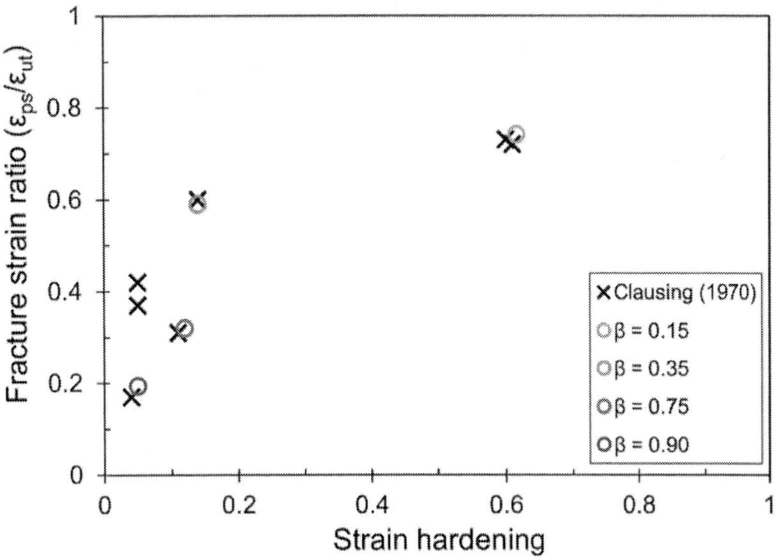

Figure 4 - The ductility ratio of axisymmetric and plane strain tension versus the strain hardening, Clausing experiments vs numerical data with Lode angle function.

Conclusions

This paper presented a first attempt to take into account the Lode angle effect in the ductile damage modeling within the framework of the Continuum Damage Mechanics. Experimental evidence had in fact suggested that the loading condition plays a very important role in regulating the damage variable evolution and that, considerations on the material performances based uniquely on the stress triaxiality are not sufficient to describe exhaustively the material response.

References

[1] L.M. Kachanov: "Time of the Rapture Process under Creep Condition,", *Izv. Akad. Nauk. SSSR, Otd. Tekhn. Nauk.,* Vol. 8 (1958), pp. 26-31.

[2] A.L. Gurson: "Continuum theory of ductile rapture by void nucleation and growth: part I – yield criteria and flow rules for porous ductile media," *J. Eng. Mat. Tech.*, Vol. 99(1977), pp. 2-15.

[3] J. Lemaitre: "A continuous damage mechanics model for ductile fracture," *J. Engng. Mat. Tech.*, Vol. 107(1985), pp. 83-89.

[4] F.A. McClintock: "A criterion for ductile fracture by the growth of Holes", *J. Appl. Mech.*, Vol. 35, No. 2(1968), pp. 363-371.

[5] G.R. Johnson, W.H. Cook: "Fracture characteristics of three metals subjected to various strains, strain rates, temperatures and pressures," *Eng. Fract. Mech.*, Vol. 21, No. 1(1985), pp. 31-48.

[6] Y. Bao, T. Wierzbicki: "On fracture locus in the equivalent strain and stress triaxiality space", *Int. J. Mech. Sci.*, Vol. 46, No. 1(2004), pp. 81-98.

[7] T. Wierzbicki, Y. Bao, Y.W. Lee, Y. Bai: "Calibration and evaluation of seven fracture models," *Int. J. Mech. Sci.*, Vol. 47(2005), pp. 719-743.

[8] D.P. Clausing: "Effect of plastic strain state on ductility and toughness," *Int. J. of Frac. Mech.*, Vol. 6, No. 1(1970), pp. 71-85.

[9] M.L.Wilkins, R.D. Streit., J.E. Reaugh: "Cumulative-strain-damage model of ductile fracture: Simulation and prediction of engineering fracture tests," Technical Report UCRL-53058, Lawrence Livermore National Laboratory, October 1980.

[10]L. Xue: "Damage accumulation and fracture initiation in uncracked ductile solids to triaxial loading," *Int. J. Solids and Structures*, Vol. 44(2007), pp. 5163-5181.

[11]W. Li, F. Liao, T. Zhou, H. Askes: "Ductile fracture of Q460 steel: Effect of stress triaxiality and Lode angle," *J. of Constr. Steel Res.*, Vol. 123(2016), pp. 1-17.

[12]K. Hashiguchi: Elastoplasticity theory. In: F Pfeiffer, P Wriggers (Eds.), Lecture notes in applied and computational mechanics. Springer, Berlin, 42.

Investigation of microstructural factors on fatigue crack initiation behavior using crystal-plasticity finite element method

Takashi Hiraide, Satoshi Igi, Tetsuya Tagawa, Rinsei Ikeda
JFE Steel Corporation, Chiba, Japan
t-hiraide@jfe-steel.co.jp, +81-43-262-2420

Seiichiro Tsutsumi
Osaka University, Osaka, Japan

Abstract

Fatigue crack initiation stage at welded toe is also important as well as crack propagation to achieve an improvement of fatigue life in welded structures. A lot of factors such as residual stress, stress concentration and inhomogeneous microstructure in HAZ (Heat Affected Zone) could affect crack initiation. Therefore, it is difficult to experimentally evaluate an effect of the individual factor. Especially, microstructural effect on fatigue crack initiation behavior has not been fully evaluated. CP-FEM (Crystal-Plasticity Finite Element Method) is expected to describe fatigue crack initiation behavior, because it can express strain localizations caused by inhomogeneous microstructure [1]. In this study, a simulation model using CP-FEM is developed to describe strain localization under cyclic loading. The applicability of the present model to fatigue crack initiation simulation is investigated from the viewpoints of the microstructural effect on crack initiation behavior in welded joint.

Tested material was a simulated weld coarse grain HAZ in low carbon steel. We firstly obtained cyclic stress-strain curves with round bar specimens which were key mechanical properties of the material input into CP-FEM. Microstructural information of individual crystalline grain was also obtained, in which the grain shape and size were quantified by SEM observation and the crystalline orientation was decided by EBSD (Electron Back Scattering Diffraction) measurement. The developed model in the present work can calculate not only macroscopic strain response against applied cyclic stress, but also calculate strain localization in each grain. We focused on accumulated equivalent plastic strain during cyclic loading as an index parameter for fatigue crack initiation.

We conducted fatigue tests with semicircular notched plate specimens in which stress concentration was similar to welded joint. We compared fatigue crack initiation locations in the experiments with the plastic strain localized area calculated by the CP-FEM simulation.

Introduction

It is well known that many factors such as residual stress, stress concentration and an inhomogeneous microstructure in the HAZ (Heat Affected Zone) can affect fatigue fracture of welded joints. Some solutions, for example, hammer peening [2], have been developed to mitigate effects related to stress. For further improvement of fatigue fracture resistance, microstructural design of the base metal and HAZ will be important. However, microstructural effects on fatigue crack initiation behavior have not been fully evaluated due to the difficulty of experimental evaluation. Therefore, an analytical method for evaluating microstructural effects on fatigue crack initiation is required. CP-FEM (Crystal-Plasticity Finite Element Method) is expected to describe fatigue crack initiation behavior, because it can express strain localizations caused by an inhomogeneous microstructure.

In this study, a simulation model using CP-FEM is developed to describe strain localizations under cyclic loading. The applicability of the proposed model to fatigue crack initiation simulation is investigated from the viewpoint of microstructural effects on crack initiation behavior in welded joints.

Tested material and experimental procedure

A test material with a coarse grain HAZ microstructure was prepared by exposing KA36 ship hull structural steel to a simulated welding heat cycle. The heat cycle comprised rapid heating to 1400°C (heating rate: 150°C/s) and cooling by a N_2 gas jet (cooling rate: 40°C/s) after one second at 1400°C. This heat cycle was intended to simulate a weld toe of a transverse non-load-carrying fillet welded joint. The microstructure of the tested material was coarse bainite with a grain size around 100μm. Cyclic stress-strain curves were obtained as key mechanical properties for the material input in CP-FEM. Specimens were exposed to an incremental strain amplitude under strain controlled loading, and the corresponding peak stress for a given strain amplitude was measured.

For fatigue crack initiation tests, a plate specimen with a semicircular notch on a single side edge was prepared. The semicircular notch edge was specially manufactured to identify the internal fatigue damage accompanying surface damage because fatigue crack initiation would be examined by optical observation from the notch edge surface. A cyclic load with a range from 0.18kN to 1.8kN and frequency of 2Hz was applied to the specimen. During the fatigue tests, the side surface around the notch was optically observed, and the test was finished when a fatigue crack was detected. SEM observation and EBSD measurement around the fatigue cracks in the fatigue-damaged specimens were performed, and the obtained information was used in the simulation model described in the following section.

Numerical procedure

Crystallographic information on individual crystal grains of the fatigue specimens was evaluated to establish CP-FEM models. The grain contours were quantified by SEM observation, and the crystalline orientations of individual grains were identified by EBSD measurement. This crystallographic information was reflected in a finite element (FE) model of the specimen. The FE model developed in the present work can calculate not only macroscopic strain response against applied stress, but also strain localizations due to cyclic cumulative deformation in individual grains. We focused on accumulated equivalent plastic strain during cyclic loading as an index parameter for fatigue crack initiation.

CP-FEM calculations were performed with ABAQUS with a user subroutine developed by one of the authors [3]. This model considers the 12 major slip systems of a body-centered cubic lattice. To reduce calculation time, CP-FEM was applied only to the notch root area (0.3 x 1.5mm), and other parts were calculated by conventional elasto-plastic FEM. The CP-FEM area contains about 40 grains. Fig. 1 shows an example of the model with nodes and elements numbers of 79,824 and 65,440, respectively. The minimum element size is set at approximately 10 x 10 x 40μm to express the strain distribution inside a grain. The boundary condition of this model was same as the experiment. Then, the cyclic load range was from 0.18kN to 1.8kN. Local stress and strain response were calculated until 100 cycles, and the numerical plastic strain localization was compared with the fatigue crack initiation sites in the experiments.

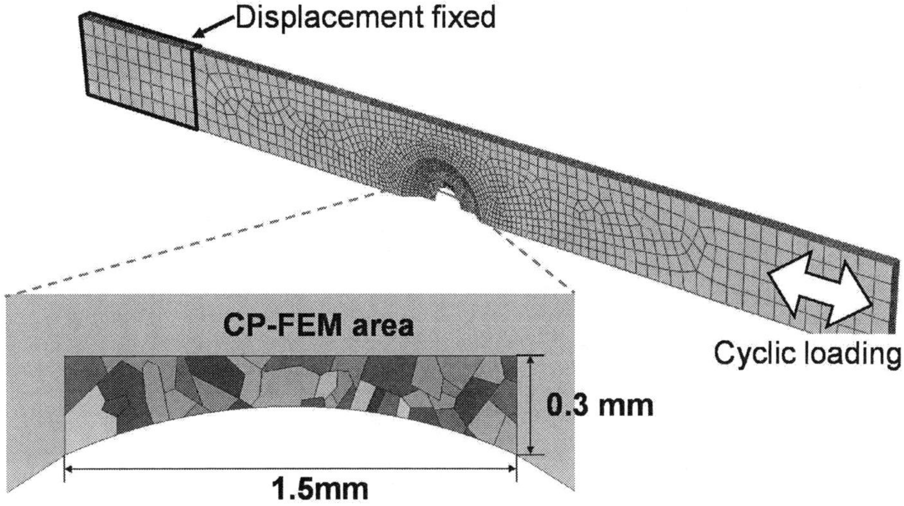

Fig. 1 Finite element model of specimen

Results

Fig. 2 shows contour maps of the equivalent plastic strain around the notch. Equivalent plastic strain is localized in specified grains, and is accumulated with increasing loading cycles.

The accumulations of equivalent plastic strain in several specified grains were compared. Fig. 3(a) shows selected grains in the model of the CP-FEM area, and Fig. 3(b) shows the cumulative behaviors of equivalent plastic strain against the number of cycles. Reflecting the non-homogeneous cyclic plasticity of individual grains, plastic strain is accumulated locally in Grain C immediately after the start of cyclic loading, and reaches about double the strain level of Grain A after 100 cycles. In spite of the partial modelling of the specimen into CP-FEM, the calculation required much CPU time. Although the CP-FEM calculation could not continue to the similar load cycles number to the experiments, the fatigue crack surely initiates at the plastic strain localized area.

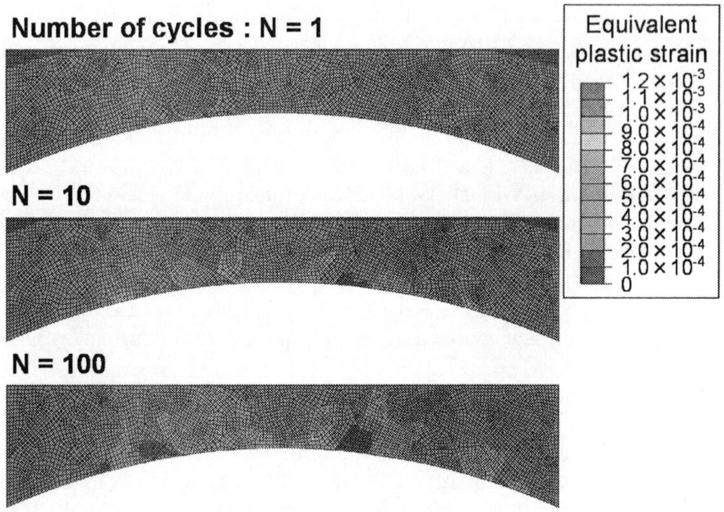

Fig. 2 Contour maps of accumulated equivalent plastic strain around notch

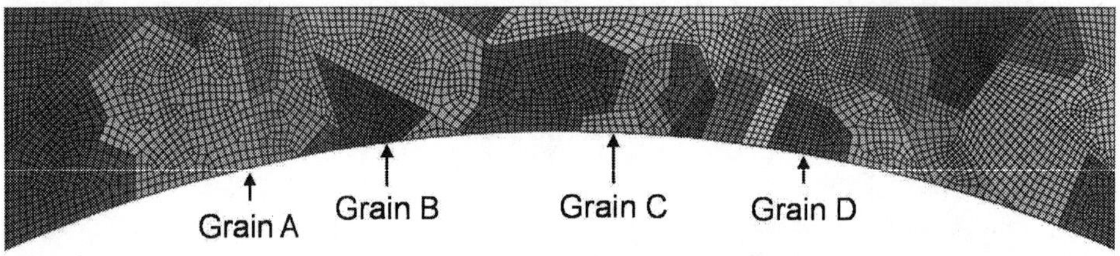

(a) Selected grains in model of CP-FEM area

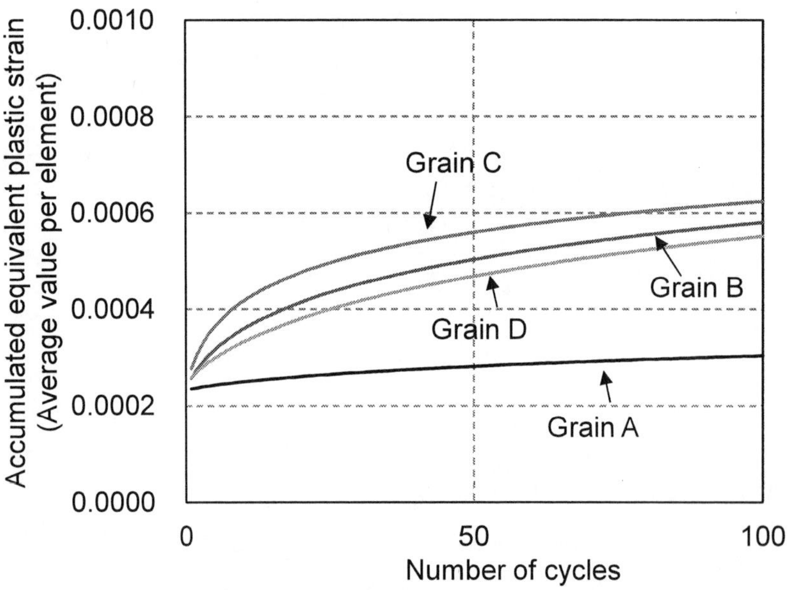

(b) Cumulative behaviors of equivalent plastic strain against number of cycles
Fig. 3 Accumulation of equivalent plastic strain in specified grains

Fig. 4 shows the result of SEM observation after the fatigue test of the specimen. The fatigue crack initiation life in this specimen is 56,640 cycles. The fatigue crack initiation site in Fig. 4 coincides with the grain with the largest accumulated equivalent plastic strain in the CP-FEM shown in Fig. 3. Although the present model calculates accumulated equivalent plastic strain distribution until 100 cycles, the model indicates the most deformed grain, considering crystallographic information such as the crystalline orientation and shape. Since it is known that fatigue crack initiation is related to strain localizations, the fatigue crack initiation site can be predicted as the accumulated equivalent plastic strain peak simulated by the proposed model.

Fig. 4 Fatigue crack initiation site in experiment

Conclusions

1. A simulation model was developed to describe strain localization in CGHAZ under cyclic loading by using CP-FEM (Crystal-Plasticity Finite Element Method).
2. A CP-FEM model with crystallographic information on individual grains of the fatigue test specimen could simulate the equivalent plastic strain localization in a specific grain and its accumulation behavior with increasing loading cycles.
3. The location with the most accumulated plastic strain in the CP-FEM model agreed with the experimental fatigue crack initiation site.

References

[1] F. P. E. Dunne, *et al.*: "Experimental and computational studies of low cycle fatigue crack nucleation in a polycrystal," *International Journal of Plasticity*, Vol. 23, No. 2 (2007), pp. 273-295.

[2] Y. Morikage, *et al.*: "Effect of Hammer Peening Processing on Fatigue Property of Welded Joints after Overload," *Proc. of the ASME 2013 Pressure Vessels & Piping Division Conf.*, Paris, France, July 2013.

[3] Tsutsumi, *et al.*: "Cyclic Softening Behavior under Macroscopically Elastic Stress Condition Predicted by a Crystal Plasticity FE Analysis," Journal of Smart Processing, Vol. 2, No. 3 (2013), pp. 123-127.

Numerical study of influence of mechanical heterogeneity on cyclic plastic energy dissipation of a plate with a center hole

Atsushi Seto
Nippon Steel & Sumitomo Metal Corporation, Amagasaki, Hyogo, Japan
seto.d94.atsushi@jp.nssmc.com, +81-6-7670-5871

Abstract
Relation between nominal stress range and temperature increase at center-hole edge of plates with and without weld bead was studied by combining elasto-plastic stress analysis under cyclic loading with heat conduction analysis of plastic energy dissipation. The temperature increases at the hole edge are relatively small in low nominal stress range, and begin to increase remarkably in higher nominal stress range. The temperature increase at the hole edge of the plate with the weld bead is slightly influenced by the plastic energy dissipation around the edge because of a small plastic deformation around the edge especially in the weld bead.

1. Introduction

Fatigue limits of metallic materials have been studied these days through changes in temperature range and temperature increase obtained by infrared thermographic cameras. The author applied the thermographical measurement to two types of specimens, plate specimens with center holes and those with weld bead to clarify the influence of stress concentration and mechanical heterogeneity [1]. The nominal stress range of the plate specimen without a weld bead, at which the average temperature increase rapidly rose, corresponded to the fatigue limit. However, the nominal stress range of the plate with a weld bead that showed remarkable temperature increase was 10 MPa greater than that of the fatigue limit [1]. The difference in the stress range in the plate with the weld bead was considered to be caused by the difference in the cyclic plastic deformation area. Some research reports that analyze energy dissipation by cyclic plastic deformation and thermal elastic effect, and those for heat conduction analysis for smooth specimens without stress concentration were seen [2,3]. No clear explanations were given to mechanical heterogeneity considering the effect of heat conduction. The influence of mechanical heterogeneity on the cyclic plastic deformation and energy dissipation with heat conduction was studied in this report.

2. Analytical method

Stress analysis model

The shape and size of the plate specimens are displayed in Fig. 1. The plate specimens consist of base metal with and without a weld bead. The finite solid elements of plane-stress with four nodes were selected for the two plate specimens. Only the half parts of these specimens were modeled because of their symmetry to the center line. The total numbers of the elements and nodes are 33084 and 33664, respectively. Elements of 0.08 mm size were used around the center hole edge to enhance the accuracy of the stress and strain in the stress concentration area. The width of the weld bead was 4 mm. The element subdivisions of the plate are displayed in Fig.2 (a) for the overall arrangement and Fig.2 (b) for the detail around the center hole edge.

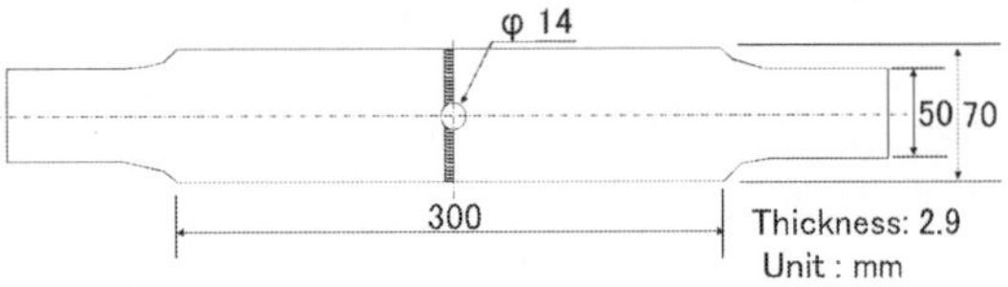

Fig. 1 Plate specimen with a center hole

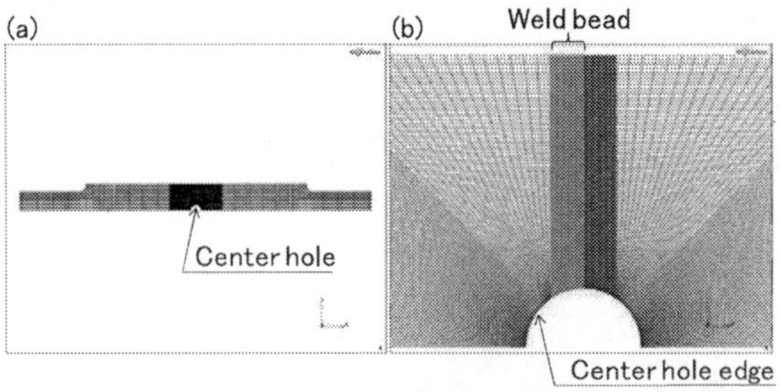

Fig.2 Finite element subdivision of plate with a center hole with weld metal
(a) Overall arrangement of finite elements
(b) Arrangement around center hole

The materials of the stress analysis models consist of a base metal and weld bead. For the base metal, pairs of the cyclic stress and strain ranges were measured at about half a number of cycles of their failure lives through strain-controlled low cycle fatigue tests. The stress-strain relationship under cyclic loading was determined in the form of exponent hardening rule using the data. For the weld bead, the stress-strain relationship of the weld bead was determined so as the stress was 1.5 times of the base metal at the same strain, because the hardness of the weld bead was about 1.5 times of the base metal [1]. Kinematic hardening was applied to both materials as their hardening rule under cyclic loading. The cyclic tensile stress was applied to the plate models with center holes in the longitudinal direction as was the same of the fatigue tests [1]. In the stress analysis, the cyclic stress was repeated up to ten cycles. The difference of the equivalent plastic strain between the maximum and the minimum stress at the tenth cycle were calculated. The software used was MSC-MARC.

Heat conduction analysis model

The plastic deformation behavior around the center hole in cyclic loading is displayed in Fig.3. In the process when the applied nominal stress increases, plastic deformation occurs and continues until the nominal stress reaches the maximum. When the nominal stress starts decreasing, plastic deformation immediately stops. In the next process when the nominal stress decreases, plastic deformation begins again and continues until the nominal stress reaches to the minimum. Therefore, plastic deformation occurs twice in a cycle. The temperature distribution after the time corresponding to a half cycle was analyzed in this study because the time during cyclic plastic deformation was considered to be extremely small compared to that of the half cycle. The initial condition of heat generation was set to energy dissipation by plastic work.

The area of heat conduction analysis was decided as follows. The maximum value of average temperature increase in 0.4 mm square areas along the center hole was calculated because the size of a pixel of the thermographic viewer in the reference [1] was also 0.4 mm. The average temperature increase of the area is generally affected by plastic deformation in and around the area. The area of ±2 mm in the longitudinal direction and 2 mm in the width direction of the plate from the most strain-concentrated point in the hole edge was selected for heat conduction analysis as displayed in Fig.4. As for the plate with the weld bead, average temperature increase was calculated for 0.4 mm square area next to the weld bead, because plastic deformation in the weld bead was smaller than that in the base metal.

The temperature increase generated by plastic deformation was analyzed by heat conduction equation. Plastic work was obtained by integrating equivalent stress by equivalent plastic strain increment. The ratio of converted heat to plastic work was 0.9 [4]. Two-dimensional heat conduction equation for homogeneous isotropic body, i.e. equation (1), was used to calculate the change in temperature increase [5].

$$\frac{\partial^2 T}{\partial x^2} + \frac{\partial^2 T}{\partial y^2} - \frac{1}{k}\frac{\partial T}{\partial t} = 0 \qquad (1)$$

Here, k is thermal diffusivity, obtained by dividing thermal conductivity by specific heat and density. Temperature T (x, y, t) is expressed by equation (2) as a solution of equation (1) when distribution of the initial temperature increase was given in the form of f(x, y) and the hole edge that is almost corresponding to y-axis is regarded as adiabatic boundary [5].

$$T(x,y,t) = 1/(4\pi\,\mathrm{kt})\int_{-2}^{2}\int_{0}^{2} f(x',y')\{e^{-\frac{(x-x')^2+(y-y')^2}{4\pi t}} + e^{-\frac{(x+x')^2+(y-y')^2}{4\pi t}}\}dx'\,dy' \quad (2)$$

Calculation of integrals in equation (2) is conducted by dividing the area by 100 in both the longitudinal direction (x-direction) and the width direction (y-direction).

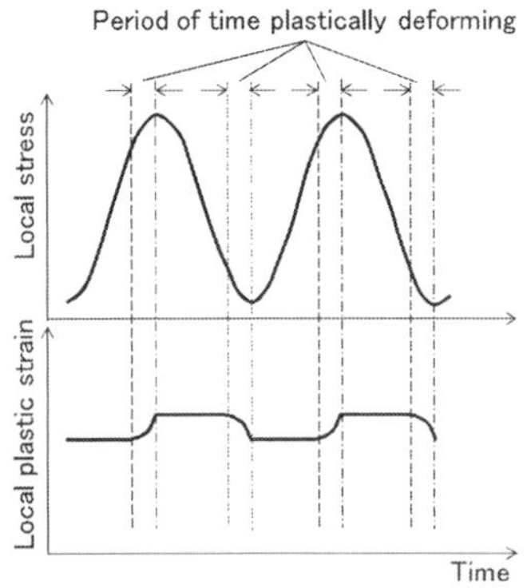

Fig. 3 Period of time plastically deforming

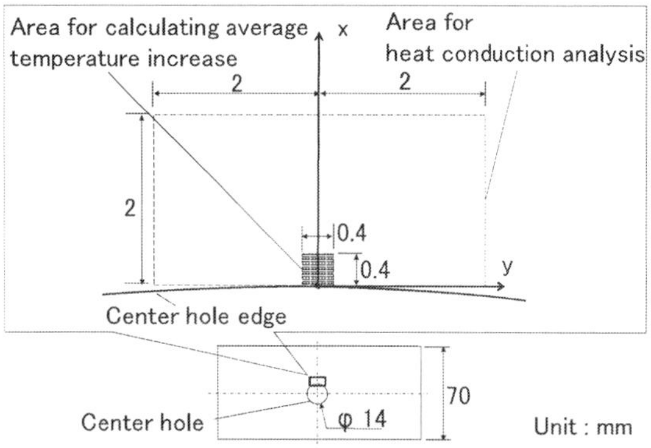

Fig. 4 Area for heat conduction analysis

3. Analytical results

Results of cyclic elasto-plastic stress analysis

First, plastic deformation area by cyclic elasto-plastic stress analysis was investigated. The growth of the plastic deformation area, which is expressed by counter lines of certain equivalent plastic strain, is indicated in Fig.5.

These equivalent plastic strains are obtained at maximum stress of the 10th cycle of three nominal stress ranges that are lower, almost the same, and greater compared to the fatigue limit of the experiments [1]. The equivalent plastic strain of 1.3×10^{-3} and 6.6×10^{-4} were selected for the plate without and with the weld bead, respectively. In the plate without the weld bead, plastic deformation grows along the hole edge symmetrically in Fig.5 (a). At the nominal stress range of 140 MPa, the plastic deformation area expands in the width direction, which is the upper direction in the figure. The plastic deformation area remarkably grows in the width direction at 170 MPa nominal stress range. In the plate with the weld bead, the plastic deformation area is rather small at nominal stress range of 90 MPa in Fig.5 (b). The area grows slightly as the nominal stress range increases to 105 and 120 MPa. And in the weld bead, little expansion of plastic deformation area was recognized. The weld bead is hard to deform because of its high hardness. The weld bead also constrains deformation of the base metal next to the weld bead, and this causes suppression of expansion of the plastic deformation area in the plate.

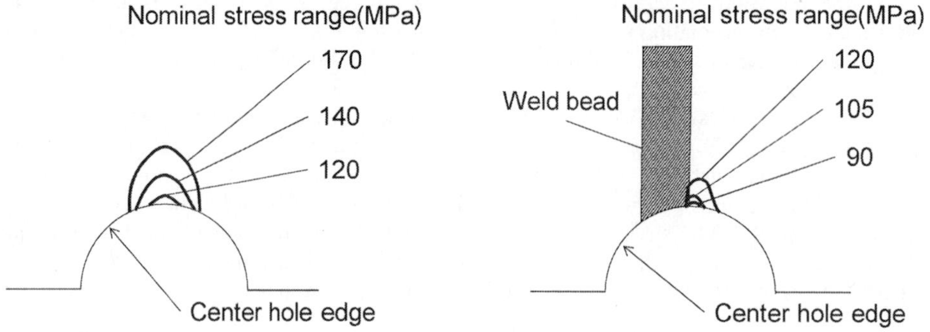

(a) 1.3×10^{-3} equivalent plastic strain of plate without a weld bead *(b) 6.6×10^{-4} equivalent plastic strain of plate with a weld bead*

Fig.5 Contour lines of equivalent plastic strain

As the next step, equivalent plastic strain ranges along the hole edge were calculated and displayed in Fig.6. The transverse of the figures indicates the distance along the edge from the deepest point of the hole. In the plate without the weld bead in Fig.6 (a), the distributions of equivalent plastic strain range are symmetrical to the origin of the axis. The equivalent plastic strain range increases with the increase in the nominal stress range. As a specific phenomenon, the inclinations of the distributions get flat as the nominal stress range increases to 140 and 170 MPa. This tendency of the strain range is the same phenomenon as the extension of the plastic deformation area displayed in Fig.5. In the plate with the weld bead in Fig.6 (b), the distributions of the equivalent plastic strain are asymmetry with respect to the origin of the y-axis. The equivalent plastic strain ranges begin to increase with the increase in the nominal stress range in the right side of the origin, which is the base metal. The equivalent plastic strain range in the weld bead is almost zero at nominal stress range 90 MPa, i.e., lower than the fatigue limit in the experiments. However, occurrence of cyclic plastic deformation is recognized at 105 MPa. The plastic deformation area and its strain range in the weld bead gradually increase with the increase of the nominal stress range.

Results of heat conduction analysis

The influence of the energy dissipation by equivalent plastic strain range on average temperature increase in 0.4 mm square area after half a cycle is displayed in Fig.7. The data in both figures are marked by frequency corresponding to time of the half cycle. Figure 7(a) displays the results of the plate without the weld bead. No remarkable temperature increase is observed in the nominal stress range of 100～110 MPa. The nominal stress ranges larger than 120 MPa show rising of average temperature increase, and their increments increase with the increase in the frequency. The ratio of the temperature increase to the nominal stress range in the figure seems to change at about 120 MPa in the case of 20 Hz. The slope in the nominal stress range larger than 120 MPa is steeper compared to that in the range smaller than 120 MPa. The thresholds of the nominal stress range, which show the change of the slope, are also between 120 and 130 MPa in other frequencies. These thresholds are about 20 MPa smaller than the nominal stress range at which temperature increase starts increasing in experiments [1].

In the plate with the weld bead in Fig.7 (b), the same tendencies are recognized as the average temperature increase rises with the increase in the nominal stress range, and the increments also become large with the increase in the frequency. The threshold of the nominal stress range, which shows the change in the inclination, seems to be about 110 MPa. The threshold values of the plates with the weld bead are also 10 MPa smaller than the nominal stress range obtained in experiments [1].

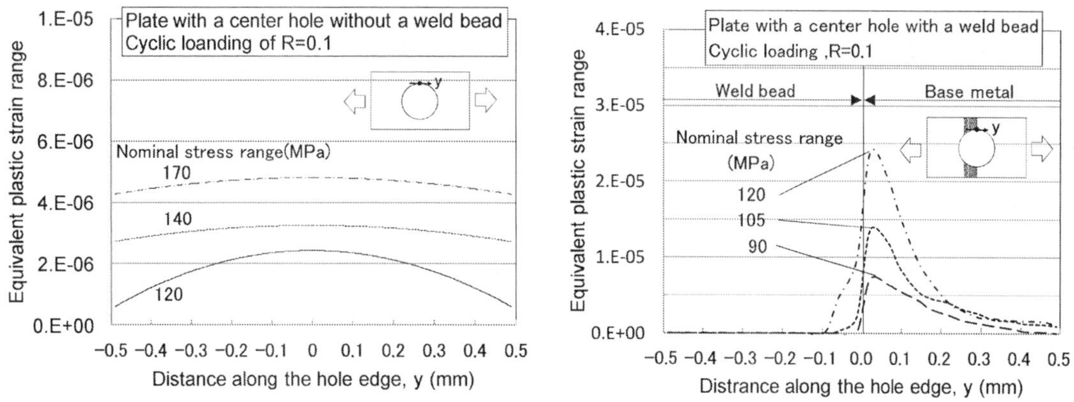

(a) Plate with a center hole without a weld bead *(b) Plate with a center hole with a weld bead*

Fig.6 Distribution of equivalent plastic strain range

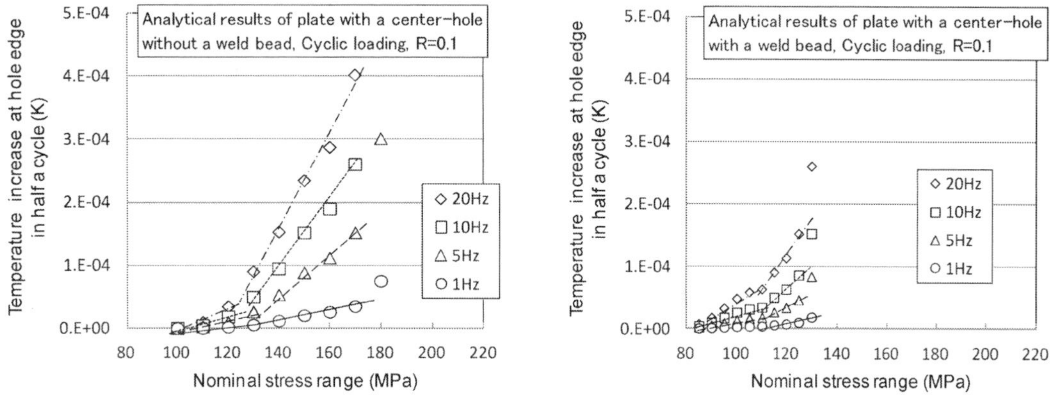

(a) Plate with a center hole without a weld bead *(b) Plate with a center hole with a weld bead*

Fig.7 Temperature increase at hole edge in half a cycle

4. Conclusions

To investigate the effect of the weld bead on the temperature increase by energy dissipation under cyclic loading, elasto-plastic cyclic stress analysis and heat conduction analysis were conducted for plates with center holes with and without a weld bead. The results obtained are described as follows.

(1) In the plate without a weld bead, the distribution of the strain range along the hole edge gets flat with the increase in the nominal stress range. In the plate with a weld bead, little cyclic plastic deformation is recognized at low nominal stress range than that at the fatigue limit.

(2) The increments of temperature increase to the nominal stress range at the deepest point of the hole edge remarkably rise at the range greater than the thresholds in both plates with and without a weld bead.

(3) The temperature increase of the deepest point of the hole edge consists of heat generation of both the deepest point and the vicinity of the point. In the plate with a weld bead, the influence of heat generation around the deepest point is smaller compared to the plate without a weld bead.

References

[1] A. Seto, *et al*: Relationship between Fatigue Limit and Measured Temperature by Thermography in Welded Part, Transactions of Society of Automotive Engineers of Japan, Vol.42, No.4 (2011), pp.935-940

[2] H. A. Ly, *et al*: Numerical Simulation on Rapid Evaluation of Fatigue Limit through Temperature Evolution, Journal of Solid Mechanics and Materials Engineering, 5 (2011), p.459-475

[3] B. Yang, *et al*: Thermographic investigation of the fatigue behavior of reactor pressure vessel steel, A314 (2001), pp.131-139

[4] M. Toyosada, *et al*: Temperature Rise Distribution near a Crack Tip due to Plastic Work under High Loading Rate, Journal of the Society of Naval Architects of Japan, No. 170(1991), pp.651-663

[5] H.S. Carslaw and J.C. Jaeger: Conduction of heat in solids, Oxford University Press (1959), pp.9-56

Meso-scale modeling of formation of balling defect in laser-powder bed fusion additive manufacturing

Yi Li and Wei Zhang
The Ohio State University, Columbus, OH, USA   Email: Zhang.3978@osu.edu

Yousub Lee
Oak Ridge National Laboratory   Email: leey@ornl.gov

Jacob Marchal and Shawn Kelly
Edison Welding Institute   Email: jmarchal@ewi.org

Abstract
In laser-powder bed fusion (L-PBF) additive manufacturing, heat transfer and molten pool fluid flow can significantly influence the as-built quality. In the present study, a meso-scale 3D transient model was utilized to simulate the temperature distribution, fluid flow and surface profile of the molten pool for laser fusion of single tracks on a powder bed. The model took into account the powder particle size distribution, powder packing density, laser power, and scanning speed. Surface tension force and the resulting Marangoni shear stress along with the recoil pressure due to metal vaporization were also considered. For the experiment, two single linear tracks with low and high heat inputs were deposited by L-PBF using AISI 316L powder. High-speed video was used to observe the molten pool region during fusion. The as-built surface topography including balling defect was characterized by optical profilometry. The predicted molten pool shape was consistent with that measured experimentally. The effect of laser power and scanning speed and on the formation of balling defect was discussed.

Introduction
Laser-powder bed fusion (L-PBF) is an additive manufacturing technique for fabrication of net-shape complex structures that are not easily fabricated by traditional subtractive and formative approaches. By fusing powder particles on a power bed in a layer by layer fashion using laser beam, it is able to build a 3D part directly from computer-aided design (CAD) data. L-PBF had been evolving prodigiously over the past decade, but the products manufactured are still prone to the formation of defects, including porosity, cracking and balling defects [1]. These defects can largely impede the productivity and reduce the product quality. Complementary to experimental testing and measurement for quality control, computational modeling has been used to understand the mechanisms of defect deformation. The present study considers AISI 316L, an austenitic stainless steel that is widely used in structural applications for many industries due to its excellent corrosion resistance and mechanical strength [2]. There are several past studies to understand the correlation between the laser processing parameters and the formation of defects especially balling defect in an effort to produce defect-free 316L builds [3, 4]. This study focuses on applying a meso-scale model to better understand the complex physics at the individual particle level and the formation of balling defect. Modeling results are validated using experimental data for L-PBF of 316L.

Modelling Approach
Detailed approach of the meso-scale model is reported in Refs. [5, 6],and only salient features are summarized in the following. The modelling approach comprises of two parts in sequence. The first model utilizes discrete element method (DEM) to simulate the packing information of powder particle on the powder bed. DEM treats individual powder particles as elastic spheres. Each ball has six degrees of freedom: three translational and three rotational velocities in a Cartesian (x,y,z) coordinate system. Mechanical contact interactions between particles and those between particles/container walls are directly considered. In particular, the normal force is calculated using Hertz formula, while the tangential forces are described by Coulomb friction law. In addition to the surface contact forces, the gravitational force, a body force, is also exerted on the particles. The moments resulting from the tangential contact forces are also computed for individual particles. Once the total forces and moments acting on a particle are formulated, the resulting accelerations are then calculated by solving Newton's second law of motion. Integration the accelerations over a time step twice yields the distances traveled by a particle for that time step and thus the particle position at the end of the time step.

The DEM packing model is implemented based on an open source software package Yade [7]. Experimentally measured particle size distribution, defined as a function of cumulative size frequency versus radius, is used to initialize the particle sizes. Packing of one layer with a nominal thickness of 60 μm is simulated. Due to the randomness nature of DEM, the resulting thickness of powder layer varies from 60 to 66 μm. Once the DEM simulation is completed, the powder packing information including particle coordinates and radii is outputted to the second model.

The second model is a 3D transient heat transfer and fluid flow model in the molten pool. Due to the intense computation time necessary for particle-level resolution, a relatively small computational domain is used. The domain has a cross section area of 1000 μm × 270 μm in xy direction, where the laser travels along x direction. Along z direction the bottom layer is the substrate with a height of 90 μm, and the upper layer containing both powder particles and void occupies another 100 μm.

The heat transfer and fluid flow model is developed based on a commercial computational fluid dynamics (CFD) software Flow-3D [8], which solves a set of mass, energy and momentum conservation equations in conjunction with the volume of fluid (VOF) equation by finite difference method. VOF tracks the location of molten pool/void interface. It assigns a scalar (fraction of fluid or *F*) to each cell: zero when a cell is fully void, one when a cell is completely occupied by fluid, and a value between zero and one when a cell is partially filled, indicating a fluid/void interface existing within this cell. The corresponding equations are as follows:

Mass: $$\nabla \cdot \vec{v} = 0$$

Energy: $$\frac{\partial h}{\partial t} + (\vec{v} \cdot \nabla)h = \frac{1}{\rho}(\nabla \cdot \lambda \nabla T)$$

Momentum: $$\frac{\partial \vec{v}}{\partial t} + (\vec{v} \cdot \nabla)\vec{v} = -\frac{1}{\rho}\nabla P + \mu \nabla^2 \vec{v} + \vec{g}[1 - \beta(T - T_m)]$$

VOF: $$\frac{\partial F}{\partial t} + \nabla \cdot (\vec{v}F) = 0$$

where $\vec{v}$ is the fluid velocity, h enthalpy, t time, ρ density, λ thermal conductivity, T temperature, T_m melting temperature, P pressure, μ viscosity, $\vec{g}$ gravitational acceleration, and β coefficient of thermal expansion. The temperature-dependent material properties for AISI 316L are obtained from the literature [9], including density, viscosity, thermal conductivity and specific heat. T_{liq} is 1700 K. Finally, the laser heat input is approximated as a heat flux which has a Gaussian distribution.

The modelling approach combining DEM powder packing and the molten pool models helps provide a relatively comprehensive understanding of the effect of powder packing characteristics and laser processing parameters on the molten pool geometry, surface profile and thermal history.

Experimental Approach

The laser fusion experiment was performed in EWI test bed, a L-PBF additive manufacturing machine. The machine was equipped with several sensors including a high speed video recorder and a local infrared (IR) camera. One layer of AISI 316L powder particles with a height of 60 μm was spread on a machined AISI 316L substrate plate. Two individual linear tracks were deposited using the following two sets of processing parameters: Case 1 with laser power = 200W and scanning speed = 200 mm/s, and Case 2 with laser power = 250W and scanning speed = 1750 mm/s. For these two cases considered, the linear heat input, defined as $H_{in} = \frac{P}{V}$ where P is the laser power and V is the scanning speed, is equal to 1 J/mm for Case 1, which is 7 times as much as that for Case 2 (0.14 J/mm). The laser beam radius was about 60 μm.

After fusion, the unmelted powder was blown off the substrate and the resulting track geometry was first characterized by optical profilometry to measure the surface profile. Afterwards these single tracks were sectioned perpendicular to the laser travelling direction and then grinded and polished. The polished sample was etched electrolytically in a 10% chromic acid and the transverse molten pool geometry was then examined in an optical microscope.

Results and Discussion

Figure 1 illustrates the simulation results by the meso-scale heat transfer and fluid flow model for the two cases. Case 1 (high input 1 J/mm) results in a continuous molten pool shape, as shown in in Figs.1 (a) and (b). On the other hand, a balling defect appears in Case 2 (low heat input 0.14 J/mm), where the molten pool is disconnected into small islands as shown in Figs. 1 (c) and (d). This observation is consistent with the findings of other researches [3, 4], which shows that the higher the scanning speed and/or the lower the laser power, the more likely the formation of balling defect. Another observation is that the molten pool size (width and depth) in Case 1 is significantly larger than that of Case 2, as expected due to the high heat input used in Case 1.

Figures 1 (b) and (d) plot the center longitudinal section of the molten pool where the velocity field is visualized as arrows whose length is proportional to its magnitude. Underneath the laser beam the molten pool is slightly depressed due to the recoil pressure from liquid metal evaporation when the temperature is heated over the boiling temperature by the intense laser heat input. Behind the laser beam the liquid metal flows in the direction opposite to laser scanning (i.e., toward the rear), which is driven by the Marangoni shear stress resulting from the difference in surface tension due to temperature variation on pool surface. The Marangoni shear stress, and to a lesser extent, the recoil pressure, dominate the convective heat transfer and the shape of molten pool and surface profile.

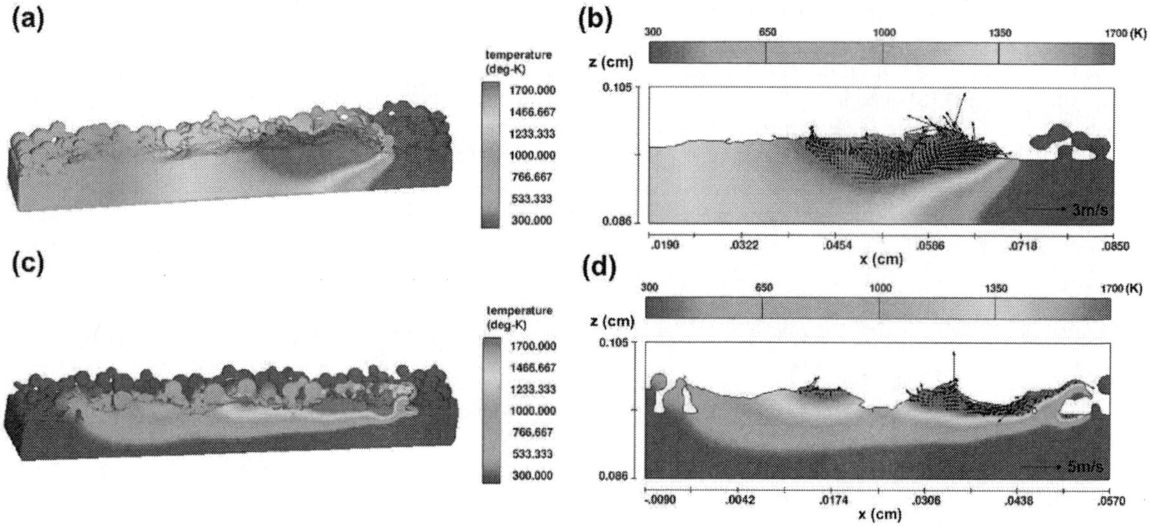

Figure 1. Calculated temperature and velocity fields: (a) & (c) 3D view, and (b) & (d) 2D longitudinal section view along the center XZ plane. (a) & (b) are for Case 1 (high heat input) with laser power = 200 W and scanning speed = 200 mm/s, and (c) & (d) are for Case 2 (low heat input) with power =250 W and speed = 1750 mm/s.

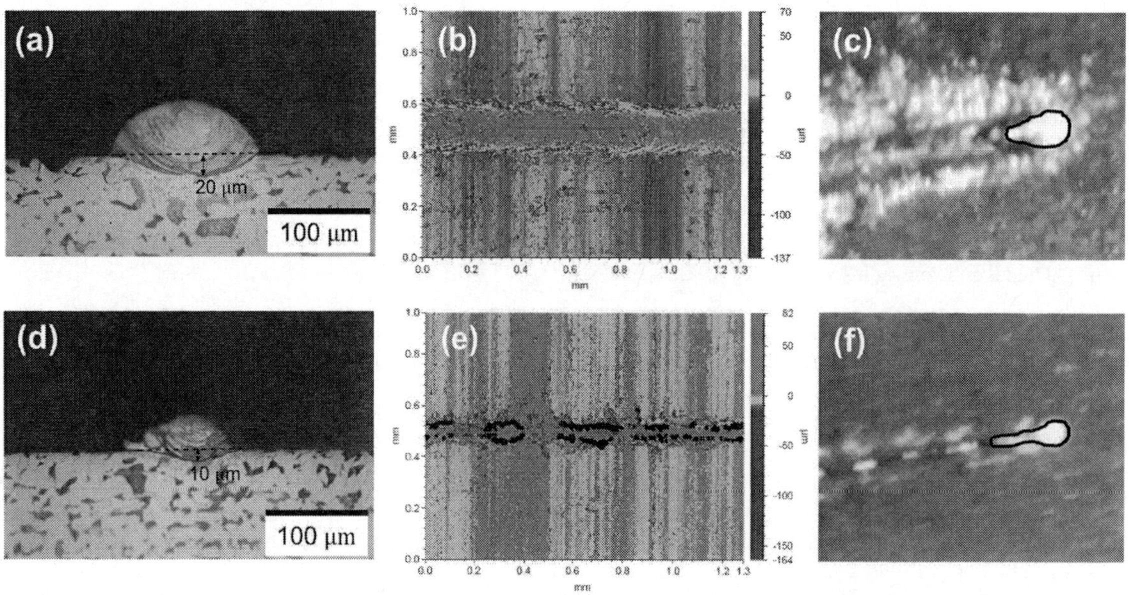

Figure 2. (a) and (d): Micrograph of molten pool transverse geometry (etched electrolytically in a 10% chromic acid), (b) and (e) 2D topographic maps of top surface of fused tracks measured by optical profilometry, and (c) & (f): high speed video snapshot, where black lines trace molten pool boundary. Processing parameters: (a), (b) & (c) are for Case 1 (high heat input), and (d), (e) and (f) are for Case 2 (low heat input).

The corresponding experimental results for Cases 1 and 2 are summarized in Fig. 2. As shown in Figs. 2(a) and (d), it is clearly seen that the molten pool in Case 1 is wider and deeper than that in Case 2, which is consistent with the simulation results. Particularly, both simulation and experimental results show quantitatively that the average penetration depth in Case 1 is about 20-40 μm whereas that in Case 2 is only 10-15 μm. It is noted that there is some small variation in the penetration depth from location to location possibly because of small fluctuation of deposition conditions (e.g., local powder density). The surface profile measured by optical profilometry is plotted in Figs. 2(b) and (e), for Cases 1 and 2, respectively. For Case 1, a continuous track is observed, as indicated by the continuous red band with a maximum height of 50 to 70 μm. The continuous track formed in Case 1 is also predicted by the model, as shown in Fig. 1(a). On the other hand, the surface profile for Case 2 appears irregular and discontinuous. As a matter of fact, there exist narrow regions with sharp change in height, where the light beam is deflected away from the optical lens preventing data from being measured there. Such areas are marked black in Fig. 2(e). It is noted that the vertical strips in the background correspond to the machining traces on the substrate plate before deposition. Such irregular geometry in Fig. 2(e) indicates the formation of balling defect when the heat input is low, consistent with the simulation results shown in Fig. 1(c).

Finally, a snap shot of high speed video recorded during real-time deposition is shown in Figs. 2(c) and (f) for Cases 1 and 2, respectively. For clarity, it is added the black lines that traces the molten pool boundary around laser beam. Case 1 possesses a wider but shorter molten pool shape whose aspect ratio (width/length) is around 0.5. On the other hand, the molten pool in Case 2 is more elongated and narrower. This is consistent with the calculated molten pool shapes shown in Fig. 1. Furthermore, as shown in Fig. 2(f), there are several bright islands separated by dark segments in-between, which likely correspond to the separated molten pool and formation of balling defect.

High-speed IR data was also recorded during real-time deposition. Such data is being analyzed and will be compared to the predicted temperature field in a future study.

Summary and Conclusions

A 3D meso-scale transient simulation of heat transfer and fluid flow is applied to quantitatively understand the interrelation between processing parameters and weld pool characteristics, particularly temperature distribution, fluid flow and surface profile of the molten pool. The model not only is able to resolve individual powder particles but also takes into account surface tension force and recoil pressure due to liquid metal evaporation. The initial powder bed geometry is generated by a DEM based packing simulation and then exported into a 3D fluid dynamics model. For experimental validation, two linear tracks are deposited by L-PBF of AISI 316L powder using high (Case 1: $P = 200$ W and $V = 200$ mm/s) and low (Case 2: $P = 250$ W and $V = 1750$ mm/s) heat inputs. The following conclusions can be drawn:

- The molten pool shape is narrower and more elongated along the beam scanning direction in Case 2 which has a much faster laser scanning speed than Case 1. Moreover, the average penetration depth in Case 2 is shallower (about 15 μm) while the average penetration in Case 1 is about 30 μm.
- Case 1 has a continuous surface profile whereas balling defect occurs in Case 2. This is observed by both simulation results and experimental measurements including optical profilometry and high speed video. The formation of balling defect in Case 2 is in part due to the elongated molten pool which is susceptible to separation under the surface tension force. Hence, overly fast laser scanning speed has a detrimental effect on forming balling defects and subsequently the as-built quality.

Acknowledgements

This material is based upon research supported by, or in part by, the U. S. Office of Naval Research (ONR) under award number N00014-14-1-0688. Helpful discussions with Prof. J.C. Zhao of OSU are also acknowledged.

References

[1] Gu, D. D., et al. International materials reviews, 57.3 (2012): 133-164.
[2] Peckner, D., et al. "Handbook of stainless steels." McGraw-Hill Book Co., New York. 1977.
[3] Antony, Kurian, et al. Journal of Manufacturing Processes 16.3 (2014): 345-355.
[4] Gu, D. D., et al. Materials & Design 30.8 (2009): 2903-2910.
[5] Lee, Y. S., W. Zhang, 26th Solid Free Form Fabrication Symposium, Austin, TX (2015): 1154-1165.
[6] Lee, Y. S., W. Zhang, Additive Manufacturing, in press, doi:10.1016/j.addma.2016.05.003.
[7] The Yade Project, 1st ed., 2010, http://yade-dem.org/doc/.
[8] Flow-3D: Version 11.0.1.2: User Manual, Flow Science, Santa Fe, NM, USA, 2014.
[9] Kim, Choong. No. ANL--75-55. Argonne National Lab., Ill.(USA), 1975.

Modeling Heat Transfer, Fluid Flow and Solidification during

T. DebRoy

The Pennsylvania State University, USA

Numerical modeling of heat transfer and liquid metal flow during additive manufacturing provides the most important factors that affect metallurgical product quality. It solves tens of billions of equations quickly and inexpensively to calculate transient three dimensional temperature and velocity fields, cooling rates, solidification parameters and other variables that determine structure and properties of the metallic products. After the simulations are validated with experimental data, the role of various process variables are examined to understand not just what structure and properties are obtained but why. Furthermore, since additively manufactured metallic parts are susceptible to distortion, lack of fusion defects and compositional changes, the model is applied to examine how these common defects can be alleviated by appropriately selecting process variables based on scientific principles.

Estimation of Melt Pool Size of Selective Laser Melting Process by Numerical Transient Heat Conduction Analysis

Toshi-Taka Ikeshoji, Hideki Kyogoku, Makiko Yonehara

Research Institute of Fundamental Technology for Next Generation, Kindai University ; KURING

ikeshoji.tt@hiro.kindai.ac.jp, +81(82)434-7000

Masahiro Araki, Kazuya Nakamura

Technology Research Association for Additive Manufacturing ; TRAFAM

Abstract

To optimize the process parameters of the powder bed selective laser melting (PB-SLM), a method to predict the melt pool size is proposed. Based on the FEM transient heat conduction analysis, the method involves the conversion of FEM elements from powder layer to bulk material. The modeling of powder layer heat conductivity and specific enthalpy is also discussed. The temperature distribution of multi path laser scanning is compared with the single laser scanning. By comparison of the width and length of melt pool dimensions with experimentally obtained temperature distribution of PB-SLM process for Ni-based alloy, the modeling and the simulation method is accessed.

Introduction

Additive manufacturing (AM) is expected to build 3D objects with free surface which cannot be made by machining processes and precision casting. AM includes several processes *e.g.* the laser deposition and the electron beam melting (EBM). In this research, the selective laser melting for powder bed (PB-SLM) is dealt. PB-SLM is the process compiling the melted powder bed. Each of the melted powder bed layer is the sliced cross section of the building object. Typical thickness of powder bed is very thin, 30 – 50 μm, and the laser spot size is relatively small. These contribute to the higher shape accuracy compared with other AM methods. The deformation during building process, however, appears excessively on the case of Ni-base alloy powder or stainless steel powder[1, 2]. The suppression of deformation during building process requires the estimation of melt pool size. Therefore, in this research, a method to estimate the melt pool size during PB-SLM process is proposed.

Some of the major factors of the building conditions are the laser power p W, the laser spot size d mm, the scanning speed v mm/s, the laser hatching width w mm, and the layer thickness t mm. The four factors of them can be summarized into the energy density, $E = p/(v \times w \times t)$ J/mm^3. But still, it is necessary to search the suitable energy density value. It is idealistic that searching the proper conditions by repeating numerical simulations with variations in those factors. But, in this paper, to describe the simulation method, the major factor values are fixed to those obtained experimentally.

In the followings, the details of numerical simulation is described at the first including the material properties and how to change them. And the results is shown and discussed.

Method of Numerical Simulation

The transient heat transfer analysis is conducted for the calculated region shown in Fig.1. The scanning length is 5 mm and 3 times scan is simulated. The building conditions are p = 292 W, d = 0.1 mm, v = 610 mm/s, w = 0.15 mm, t = 0.05 mm. The initial temperature of powder and bulk plate was set to 323 K (50° C). The laser power distribution was assumed to Gaussian distribution. On the side and bottom surfaces, it was applied the heat conduction condition of 20 W/(K· m) and the temperature outside of those surfaces was set to 323 K (50° C).

Thermal Properties of Inconel 718 bulk and powder is described as follows. According to the density and specific enthalpy measured experimentally[3], the volumetric specific enthalpy H_v J/m^3 was evaluated as follows,

$$H_v = \begin{cases} -1.1600 \times 10^9 + 2.3613 \times 10^6\ T + 1.3623 \times 10^3\ T^2 & \\ \quad -3.5435 \times 10^{-2}\ T^3 - 1.1272 \times 10^{-5} & 400 < T < 1000\ \text{K} \\ -2.8141 \times 10^9 + 5.4716 \times 10^6 T - 1.4051 \times 10^2 T^2 & 1000 < T < 1528\ \text{K} \\ -4.4581 \times 10^{10} + 3.76667 \times 10^7 T - 3321.76 T^2 & 1528 < T < 1610\ \text{K} \\ -2.5345 \times 10^9 + 6.5088 \times 10^6 T - 1.8983 \times 10^2 T^2 & 1610 < T < 2100\ \text{K} \end{cases}$$

The volumetric specific enthalpy of powder is estimated as $H_V^{Powder} = (1-\epsilon)H_v$, where ϵ is the void ratio of powder. The heat conductivity λ W/(K· m) is

$$\lambda = \begin{cases} 1.1224 + 0.0165T & T < 1528\ \text{K} \\ 4.8985 + 0.0136T & 1600 < T < 2100\ \text{K} \end{cases}$$

The heat conductivity of powder is estimated using Kunii-Smith model[4].

$$\frac{\lambda_{\text{Powder}}(T)}{\lambda_g} = \epsilon\left(1 + \beta\frac{h_{rv}D_p}{k_g}\right) + \frac{\beta(1-\epsilon)}{\dfrac{1}{\dfrac{1}{\phi} + \dfrac{h_{rs}D_p}{\lambda_g}} + \gamma\left(\dfrac{\lambda_g}{\lambda}\right)}$$

Profiles of these thermal properties is shown in Fig.2

To simulate the melting and solidification of powder, finite elements' material properties are changed by their temperature values. For a finite element with powder material property, when its temperature value raise up to the liquidus of Inconel 718, *i.e.* 1610 K, its material property of enthalpy and thermal conductivity changes from the property for powder to for bulk. The enthalpy value is multiplied by $1-\epsilon$. Because the net amount of powder inside an finite element volume is lower by vacant ratio than that of bulk.

Results

Fig.3 shows the temperature distribution on top surface. The melt pool is considered to the region with higher

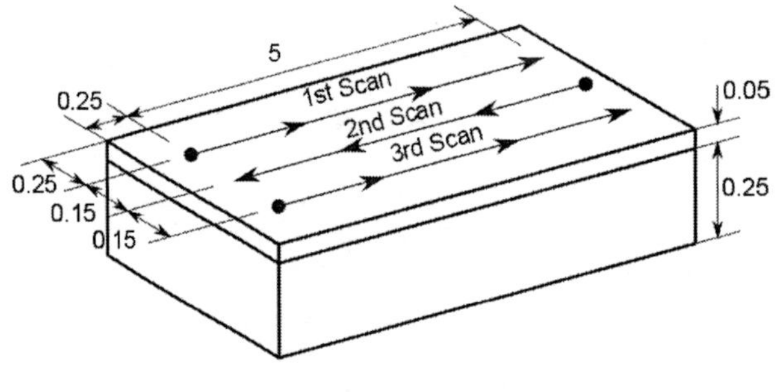

Fig.1 The calculated region.

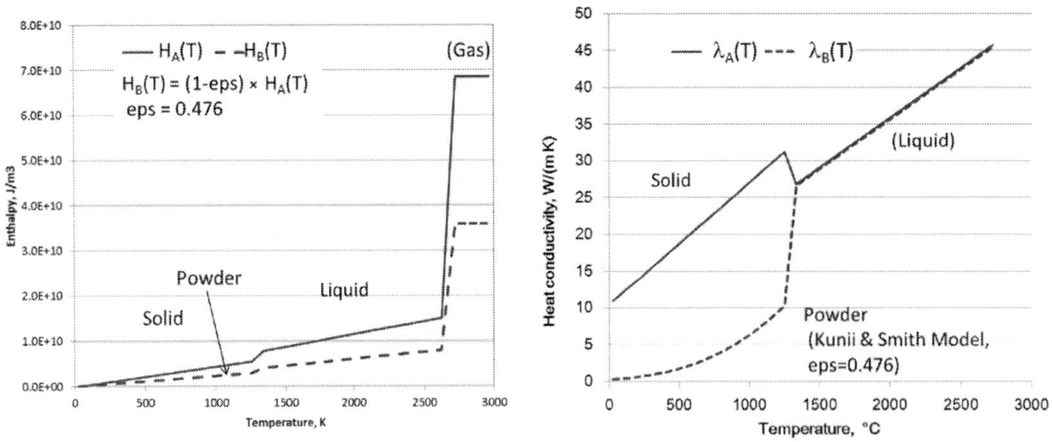

Fig.2 Material properties of Inconel 718 bulk and powder.

temperature than 1360° C or 1633 K. The brightest region in the melt pool was with temperature higher than 3000° C, and the region might be evaporated. But in this calculation the evaporation was not considered. Therefore, the excessive amount of heat might be remained in the calculated region to cause the higher temperature than actual temperature distribution. At 6.15 ms after the onset of calculation, the melt pool shape became stable without big change with time increment. The shape of melt pool was symmetric according to the direction of laser scanning. Because on the first scan, the laser irradiated the surface wholly covered with powder. At 14.3 ms after the onset of calculation, the melt pool shape became stable in the second scan, but it was distorted. The first scan side (upper side in the figure) wider than its opposite side. This is due to the higher thermal conductivity at the first scan side. The first scan side was already solidified and its material property was the bulk one. Its opposite side was still powder region and its thermal conductivity was much smaller than the bulk. This tendency was also observed for the temperature distribution at 22.5 ms after the onset of calculation. The evaporated part which is the region with higher temperature than 3000° C in the melt pool was almost symmetrical, but the melt pool's shape was distorted.

The melted-and-solidified powder region (M/S region) is shown in Fig.4. In the middle part of the calculation region, the M/S region was continuous each other side by side. That means the laser tracks joined side by side with this building conditions. But at the edge of the scanning, or turning parts, M/S region still has the circular shape due to the Gaussian profile of laser. And cleavages were formed between the laser tracks. These part will remain as the voids inside of the build material. To modify these irregular M/S regions, the laser conditions must be changed at the turning points of laser scanning. Or it is necessary to be line-scanned to connect the turning points.

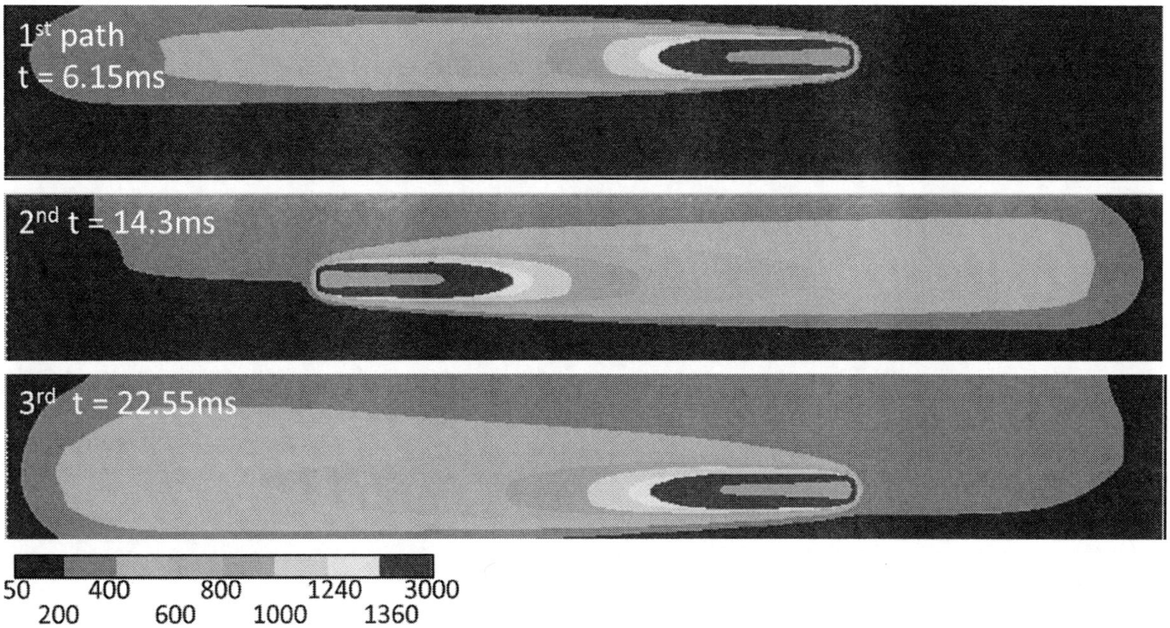

Fig.3 Temperature distribution of the top surface. Temperature in Celsius.

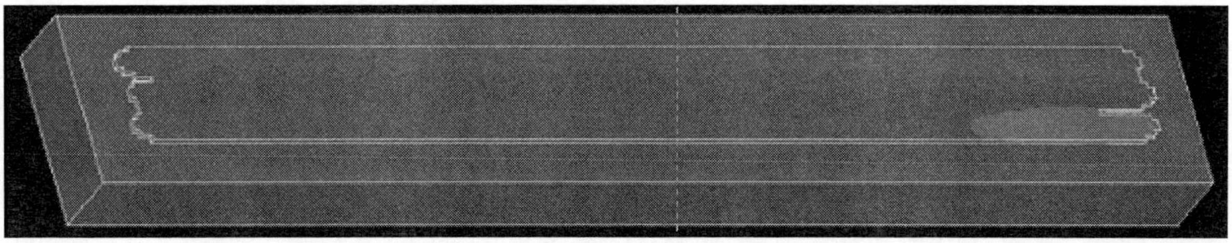

Fig.4 Melted-and-solidified powder region on bulk material.

Conclusions

For the powder bed selective laser melting process of Inconel 719 powder, the numerical transient heat conduction analysis was conducted to estimate the melt pool size. The heat conductivity of powder was estimated using the Kunii-Smith model and the melting and solidification of powder was imitated by changing the material propertied of finite elements. The estimated melt pool size was stabilized at the 3rd path of laser scanning.

Acknowledgments

This research was supported by Technology Research Association for Additive Manufacturing and CASTEM Co. Ltd.

References

[1] M. Matsumoto, M. Shiomi, K. Osakada, F. Abe, Finite element analysis of single layer forming on metallic powder bed in rapid prototyping by selective laser processing, International Journal of Machine Tools and Manufacture, 42 (2002) 61-67.

[2] M.F. Zaeh, G. Branner, Investigations on residual stresses and deformations in selective laser melting, Production Engineering, 4 (2010) 35-45.

[3] G. Pottlacher, H. Hosaeus, E. Kaschnitz, A. Seifter, Thermophysical properties of solid and liquidInconel 718 Alloy, Scandinavian Journal of Metallurgy, 31 (2002) 161-168.

[4] S. Bedarkar, N.N. Viswanathan, N.B. Ballal, Measurement of Thermal Conductivity along the Radial Direction in a Vertical Cylindrical Packed Bed, Journal of Powder Technology, 2015 (2015) 7.

Effects of Parameters on Microstructure and Mechanical Properties of Cu-Cr Alloys Fabricated by Selective Laser Melting

Keiko Kikuchi, Naoyuki Nomura, Takato Kousaka, Akira Kawasaki
Tohoku University, Sendai, Miyagi, Japan
Email kikuchi@material.tohoku.ac.jp

Shinichi Moriya
Japan Aerospace Exploration Agency, Kakuda, Miyagi, Japan

Takayuki Nakamoto, Takahiro Kimura
Technology Research Institute of Osaka Prefecture, Izumi, Osaka, Japan

Abstract
Cu-Cr-Zr alloys were fabricated by selective laser melting (SLM), and the effects of the SLM parameters were investigated in relation to the microstructure and mechanical properties of the alloys. The microstructure was observed by optical microscopy (OM) and scanning electron microscopy (SEM). The crystallographic orientation was examined using X-ray diffractometry (XRD). The mechanical properties were evaluated by tensile testing. By observing the microstructure, the pores were determined to be crescentic and larger than the original powder, suggesting that balling occurred during laser scanning. Interestingly, the tensile strength at a scanning speed of 600 $mm{\cdot}s^{-1}$, was highest, both in Ar or N_2 atmospheres, although no significant difference was observed in the relative density or the oxygen and nitrogen concentrations of the SLM builds. However, the crystallographic orientation varied according to the scanning speed, which suggests that tensile strength was influenced by the crystallographic anisotropy of SLM builds.

Introduction
Cu-Cr-Zr alloys have been used in the regeneratively cooled combustion chamber of a liquid rocket engine due to their high yield strength and high thermal conductivity [1]. The inner wall of the combustion chamber, which is exposed to high-temperature combustion heat, is cooled with liquid fuel flowing through internal cooling channels. However, multi-step and complex processes are required to fabricate these internal cooling channels. In order to simplify the manufacturing process of the combustion chamber, much attention has been paid to additive manufacturing processes, including selective laser melting (SLM) [2-4]. In the SLM process, a laser is scanned over metal powders layer-by-layer, and a near-net-shape of the products is directly fabricated in a single stage from three-dimensionally modeled data. Thus, SLM has the potential to reduce both manufacturing steps and time. However, there are some difficulties in manufacturing products made of copper and its alloys by SLM due to their low laser absorption and high thermal conductivity. In recent years, SLM has utilized higher power lasers, and is more able to produce products made of copper and copper alloys [4]. In this study, Cu-Cr-Zr alloys were fabricated by SLM and the effects of SLM parameters were investigated in relation to the resulting microstructure and mechanical properties.

Experimental procedure
Cu-0.95Cr-0.01Zr (mass%) alloy powders were prepared by gas atomization (Fukuda Metal Foil & Powder Co., LTD., Japan). The particle size distribution of the powder is listed in Table 1. SLM was carried out using EOSINT-M280 with a Yb fiber laser in an Ar or N_2 atmosphere. Cylindrical specimens with a diameter of 6 mm and a height of 40 mm were fabricated by SLM. To examine the effect of the SLM process parameters, the laser scanning speed was varied from 200 to 1000 $mm{\cdot}s^{-1}$. The powder stacking thickness was fixed for each atmosphere, specifically 0.03 mm for Ar and 0.04 mm for N_2. The laser power and hatching distance were fixed at 350 W and 0.1 mm regardless of scanning speed and atmosphere. The input energy, E, ($J{\cdot}mm^{-3}$) was calculated by

Table 1: Particle size distribution of Cu-Cr-Zr powder

	d_5 / μm	d_{50} / μm	d_{90} / μm
Cu-0.95Cr-0.01Zr	6.85	27.06	56.20

the following equation [5];

$$E = P/hvw \quad (1)$$

where P indicates the laser power (W), h, the hatching distance (mm), v, the scan speed ($mm \cdot s^{-1}$), and w, the powder stacking thickness (mm).

A dumbbell specimen (3 mm in diameter and 18 mm in gage length) was fabricated by a machining lathe from a SLMed specimen for tensile testing. The longitudinal direction of the specimens was fixed parallel to the building direction. Uniaxial tensile tests were performed using an Instron universal testing machine at room temperature. The microstructure of SLMed specimens was observed by optical microscopy (OM) and scanning electron microscopy (SEM). The crystallographic orientation was examined using X-ray diffractometry (XRD). The relative density was measured using Archimedes' principle. The oxygen and nitrogen concentrations were evaluated based on the infrared absorption and thermal conductivity, respectively.

Results and discussion

Fig. 1 shows the typical OM image of SLM builds taken from a vertical cross section parallel to the building direction. The pores were crescentic and larger than the original powder, suggesting that balling occurred during laser scanning. The relative density of SLM builds was between 96% and 98% regardless of SLM parameters. In addition, no significant difference was observed in the oxygen and nitrogen concentration of the SLM builds.

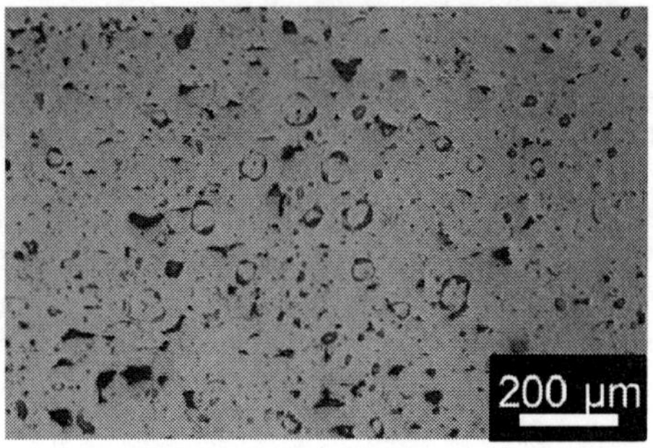

Fig. 1 OM image of SLMed builds at a scan speed of 1000 $mm \cdot s^{-1}$ in an Ar atmosphere.

Fig. 2 shows the 0.2% yield strength and elongation of SLM builds. These mechanical properties at a scanning speed of 600 $mm \cdot s^{-1}$ showed the highest values regardless of the atmosphere. Moreover, SLM builds fabricated in an Ar atmosphere showed relatively higher mechanical properties than those fabricated in a N_2 atmosphere. These results indicated that the mechanical properties of the SLM builds were dependent on the atmosphere and SLM process parameters, although no significant difference was observed in the relative density and oxygen or nitrogen concentration of the SLM builds.

Fig. 3 shows the X-ray profiles of the SLM builds. X-ray diffraction was conducted using a transverse cross section normal to the building direction. Peaks from the Cu phase were dominant for all SLM builds. However, the intensity of each peak changed depending on the scanning speed. *Fig*.4 shows the relationship between the crystal orientation ratio and scanning speed, in which the crystal orientation ratio R_{hkl} was calculated using the following equation:

$$R_{hkl} = I_{hkl} / \sum I_{hkl} \quad (2)$$

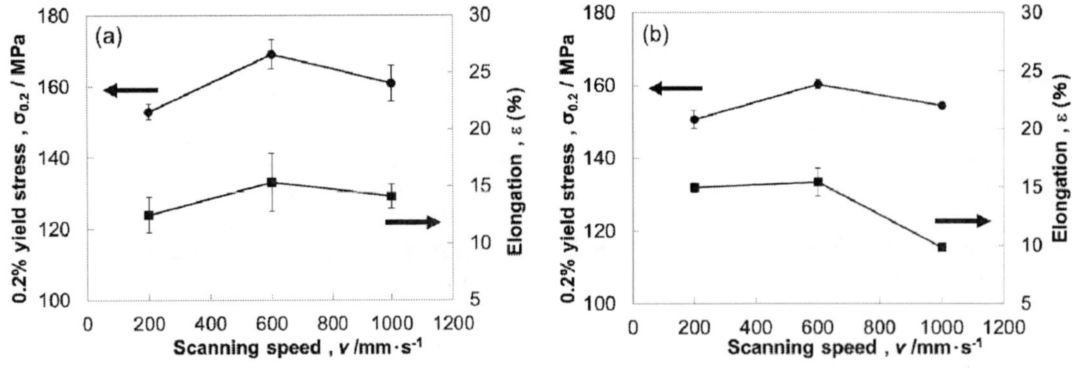

Fig. 2 0.2% yield stress and elongation of SLM builds fabricated in (a) Ar or (b) N_2 atmospheres.

where I_{hkl} indicates the peak intensity of the (h k l) plane. It was found that the crystal orientation ratio of the plane (1 1 1) normal to the building direction was largest at a scanning speed of 600 mm·s^{-1} both in Ar and N_2 atmospheres. Considering the Schmid factor for the slip system of [$\bar{1}$0 1] (1 1 1) in fcc metal, [$\bar{1}$ 1 1] stress axes have smaller values compared with [0 0 1] or [0 1 1] axes. Thus share stress acting on the slip system is smaller when the tensile direction is normal to the (1 1 1) plane, which may be one reason why the mechanical properties are dependent on the scanning speed.

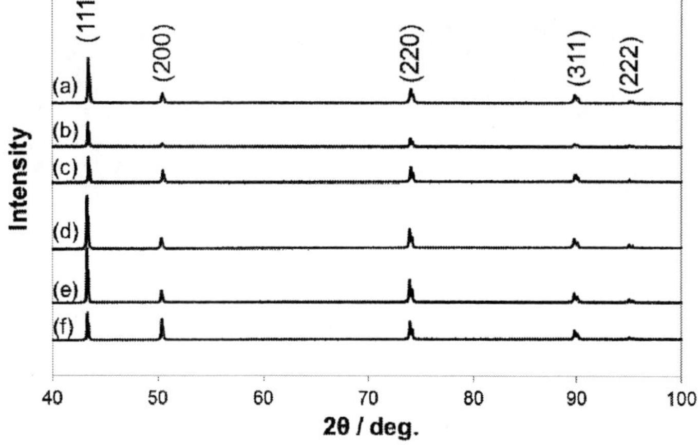

Fig. 3 XRD profiles of SLM builds fabricated with scan speeds of: (a) 200 mm·s^{-1}, (b) 600 mm·s^{-1} and (c) 1000 mm·s^{-1} in Ar atmosphere; (d) 200 mm·s^{-1}, (e) 600 mm·s^{-1} and (f) 1000 mm·s^{-1} in N_2 atmosphere.

The fracture surface observations of the SLM builds are shown in *Fig.* 5, where relatively flat surfaces were observed in the builds SLMed at a scanning speed of 1000 mm·s^{-1}. As indicated in equation (1), SLM builds fabricated at higher scanning speed received lower input energy, which leads to shallower melting depth, affecting the bonding strength between the upper and lower layers. Focusing on the difference in atmosphere, the fracture surface of SLM builds fabricated in N_2 had slightly flat surfaces compared with those fabricated in Ar. The input energy received by each layer of the SLM builds fabricated in N_2 was slightly lower than those in Ar, because the powder stacking thickness for N_2 was larger than that for Ar. Therefore, the melting depth of each layer may be shallower and the bonding strength between the upper and lower layers was affected.

Conclusions

Cu-Cr-Zr alloys were fabricated by SLM and the effects of SLM parameters were investigated in relation to the microstructure and mechanical properties of the alloys. By observing the microstructure, it was found that the pores were crescentic and larger than the original powder, suggesting that balling occurred during laser scanning. A yielding stress of 0.2% and elongation were dependent on the scanning speed and SLMed builds at a scanning speed of 600 mm·s^{-1} showed the highest values in Ar and N_2 atmospheres respectively. No significant difference was observed in the relative density and the oxygen and nitrogen concentrations of the SLM builds. However, the crystallographic orientation varied according to the scanning speed, which suggests the mechanical properties

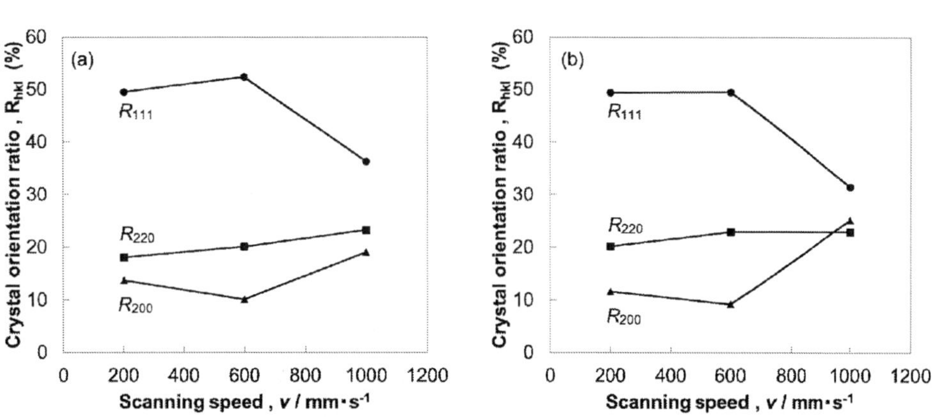

Fig. 4 Relationship between crystal orientation ratio and scanning speed of SLM builds fabricated in (a) Ar or (b) N_2 atmospheres.

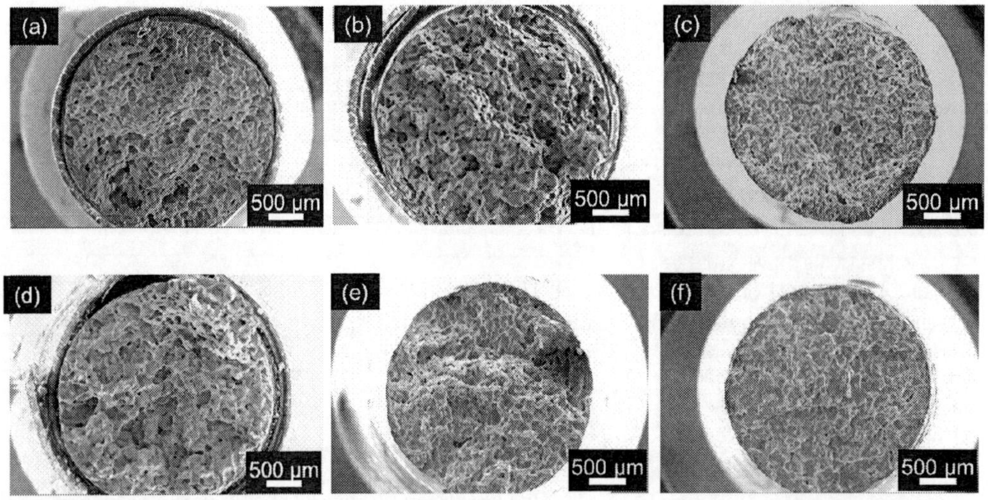

Fig. 5 Fracture surface of SLM builds after tensile testing, with scan speeds of: (a) 200 mm s^{-1}, (b) 600 mm s^{-1} and (c) 1000 mm s^{-1} in an Ar atmosphere; (d) 200 mm s^{-1}, (e) 600 mm s^{-1} and (f) 1000 mm s^{-1} in a N_2 atmosphere.

were influenced by the crystallographic anisotropy of SLM builds. Fracture surface observation revealed that relatively flat surfaces were obtained in builds SLMed at 1000 mm·s^{-1}. In addition, SLM builds fabricated in a N_2 atmosphere had slightly flat surfaces compared with those fabricated in an Ar atmosphere. Low input energy leads to shallow melting depth, affecting the bonding strength between the upper and lower layer.

References

[1] S. Komazaki, *et al*: "Influences of oxidation and hydrogen-charging on mechanical properties of CuCrZr alloy," *Transactions of the Japan Society of Mechanical Engineers, Part A*, Vol. 76, No. 768 (2010), pp. 95-101.

[2] L.E. Murr, *et al*: "Metal Fabrication by Additive Manufacturing Using Laser and Electron Beam Melting Technologies," *Journal of Materials Science and Technology,* Vol. 28, No. 1 (2012), pp. 1-14.

[3] A. Takaichi, *et al*: "Microstructures and mechanical properties of Co–29Cr–6Mo alloy fabricated by selective laser melting process for dental applications," *Journal of the Mechanical Behavior of Biomedical Materials*, Vol. 21 (2013), pp. 67-76.

[4] A. Popovich, *et al*; "Microstructure and mechanical properties of additive manufactured copper alloy," *Materials Letters*, Vol. 179 (2016), pp. 38-41.

[5] A. Shimchi; "Direct laser sintering of metal powders: Mechanism, kinetics and microstructural features", *Materials Science and Engineering A*, Vol. 428, No. 1-2 (2006), pp. 148-158.

Development of a High-speed 3D Printer by Using Laser Metal Deposition Technology

Yasutomo Shiomi, Hiroshi Ohno, Naotada Okada
Corporate Manufacturing Engineering Center, Toshiba Corp., Yokohama-shi, Kanagawa, Japan
yasutomo.shiomi@toshiba.co.jp

Shimpei Fujimaki, Kazuki Wada
Toshiba Machine Co., Ltd., Numazu-shi, Shizuoka, Japan

Abstract

A high performance LMD (Laser Metal Deposition) 3D printer prototype is presented here, demonstrating a building speed over 300 cc/ h, an accuracy of +/- 30 um, and a maximum building size of 300 mm x 300 mm x 160 mm. The prototype consists of a 6 kW fiber laser, a powder focusing nozzle, an inert gas chamber, a metal powder feeding system with inert gas, and a laser polishing system. A 6 kW fiber laser enable about ten times higher building speed than the conventional powder bed fusion. An axial nozzle concentrating the powder stream with a diameter as small as 0.7 mm can improve the accuracy. Inert gas can reduce a content of oxygen in the chamber to less than 50 ppm. Multi-layer parts consisting of multi-materials were built by switching power materials with a metal powder feeding system. The surface of built parts with a roughness, Ra, of 14.1 um was improved to 3.9 um with the laser polishing system. The parts built from Stainless Steel 316L powder satisfied the mechanical properties of JIS G4304.

Introduction

Additive manufacturing (AM) technology by stereolithography was invented by Kodama and et al. in 1980s. In 1997, AM technology for metals was proposed by J.J.Beaman et al. by using SLS(Selective Laser Sintering)[1]. In 2000s and later, products of 3D printer for metals began to be commercially available due to the development of high-power fiber lasers, improvement of powder manufacturing technology, progress in software development such as 3D-CAD. [2].

Advantages of AM are

1. rapid prototyping from a 3D-CAD data,
2. building of a three-dimensional structure which is unable to build by conventional methods.

Taking advantages of these features, the cost reduction of materials can be realized by near-net-shaping with the 3D printer, therefore it has attracted attention from the viewpoint of energy saving .

Additive manufacturing technology has been attracting attention as a breakthrough technology in advanced manufacturing as above. It has, however, technical issues to be solved at the building speed, the building accuracy, and the maximum size of building. Therefore, the Ministry of Economy, Trade and Industry (METI) established a research association, TRAFAM (Technology Research Association for Future Additive Manufacturing), in 2014. TRAFAM's mission is the development of additive manufacturing systems that will meet the world's highest standards in the "3D printer technology development for the next-generation industry". Toshiba and Toshiba machine belong to this association and have developed a 3D printer prototype by using Laser Metal Deposition (LMD). In this presentation, we describe the features of our prototype.

Laser Metal Deposition

The laser metal deposition is a process which uses a laser beam to form a melt pool on a metallic substrate, into which metal powder is introduced through a focusing nozzle with inert carrier gas, as shown Fig.1. The metal powder is deposited through melt and solidification process at the melt pool.

The LMD 3D printer has some advantages, compared with the conventional 3D printer which adopts Powder Bed Fusion. One of advantages is the high potential for improving the building speed depending on the laser power. Second one is possibility to build multi-material with switching the powder source.

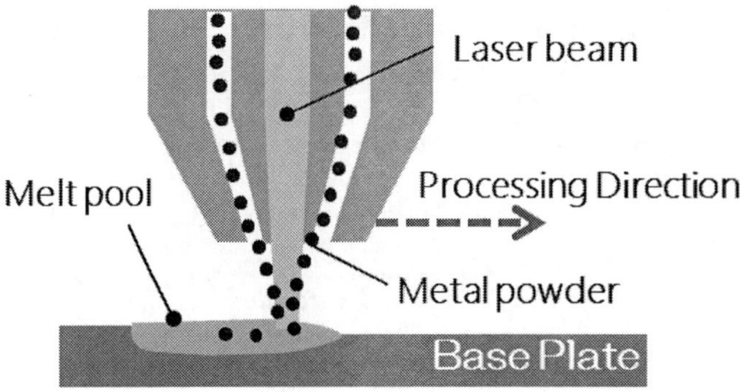

Fig.1 Schematic of LMD.

Feature of LMD 3D printer prototype

Figure 2 shows the outlook of the LMD 3D printer prototype. Maximum building size is 300 mm x 300 mm x 160 mm. The prototype consists of a 6 kW fiber laser, a powder focusing nozzle, an inert gas chamber, a metal powder feeding system with inert gas, and a laser polishing system. A 6 kW fiber laser enable about ten times higher building speed than the conventional powder bed fusion. The laser beam introduced into the building chamber through an optical fiber is focused with a focusing optics on a workpeice at a diameter between 0.2 and 3.0 mm. An axial nozzle concentrating the powder stream with a diameter as small as 0.7 mm can improve the accuracy. Inert gas can reduce a content of oxygen in the chamber to less than 50 ppm. Multi-layer parts consisting of multi-materials can be built by switching power materials with a metal powder feeding system. Metal powders including stainless steel, nickel based alloy, aluminum alloy or copper alloy can be used. Figure 3 shows the schematic of laser polishing. The laser polishing improves the roughness of the building surface by utilizing the surface tension, which is caused by re-melting the irradiated surface with a laser on the surface of the building parts.

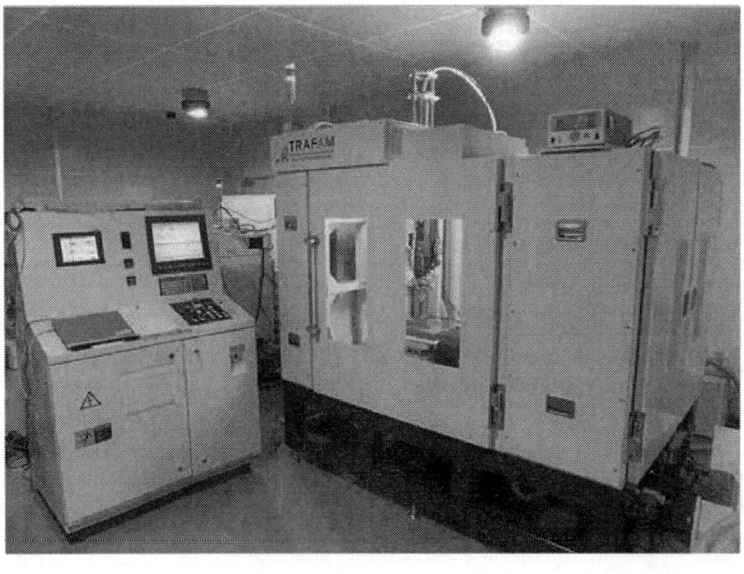

Fig.2 Outlook of LMD 3D printer prototype.

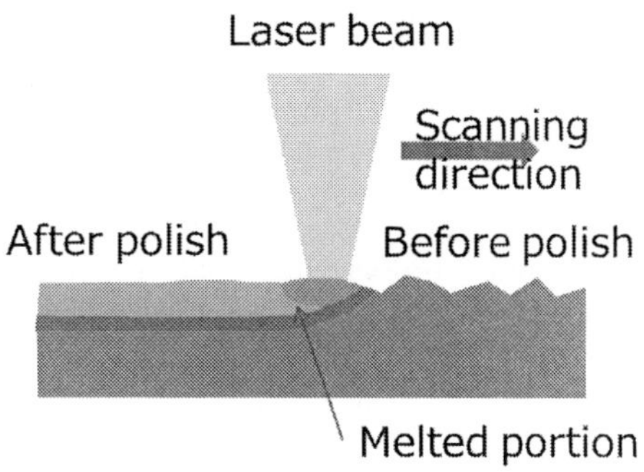

Fig.3 Schematic of Laser polish.

Built parts

Figure 4 shows building processing of a cylinder, which has the diameter of 30mm and the height of 20mm. Powder of SUS316L, a laser output power of 800W and a nitrogen atmosphere having an oxygen concentration less than 100 ppm were used to build this cylinder. At this time, the building speed of 110 cc/h was realized. As building speed is improved depending on the laser power, it was over 300 cc/h for 100 x 100 x 10 mm Inconel 718 workpiece with a laser power over 4 kW. By controlling the laser power, accuracy of the size of column workpiece was +/- 30 μm.

Built parts with the present prototype are shown in Fig. 5. A slight overhung figure is shown in Fig.5(a). The built part composed of SUS316L and Inconel 718 is shown in Fig. 5(b). The powder was switched from SUS316 to Inconel 718 with the powder feeding system after the lower SUS316L part was built. Laser polish of built parts has also been developed. "As built" surface with a roughness, Ra, of 14.1μm was improved to 3.9 μm by re-melting using laser irradiation as shown in Fig. 5 (c).

Fig.4 Building processing of a cylinder.

(a)

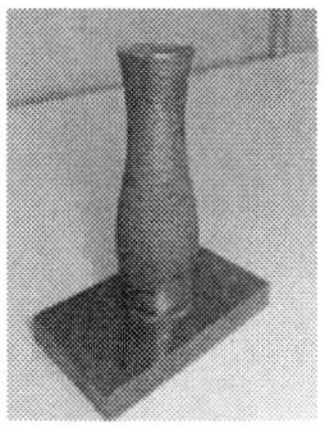

(b)

(c)

Fig.5 Parts built with the prototype. (a) A "cup" figure of SUS316L at a height of 60 mm. (b) A part consists of SUS 316L and Inconel 718. (c) Laser polished workpiece. Roughness of upper polished surface is 3.9 μm and lower "as built" surface is 14.1 μm.

Conclusions

A LMD 3D printer prototype is presented here, demonstrating a building speed over 300 cc/h, an accuracy of +/- 30 μm, and a maximum building size of 300 mm x 300 mm x 160 mm. Multilayer parts consist of SUS316L and Inconel 718 was built by switching power materials with a metal powder feeding system. Laser polishing system improved the surface roughness of the workpiece from 14.1 μm to 3.9 μm.

Acknowledgments

This work was supported by 3D metal printer development project sponsored by Japan's Ministry of Economy, Trade and Industry (METI) under its program "Technological Development for Next-Generation Industrial 3D Printers and Ultra-High-Precision 3D Shaping Systems".

References

[1] J. J. Beaman, *et al*: A New Direction in Manufacturing, Solid Freeform Fabrication (Springer , 1997), p23.
[2] H. Kyogoku: "Recent Trend in Development of 3D Printer", The Review of Laser Engineering, Vol. 45, No. 11 (2014), pp. 833-837.

Friction Stir Processing of 304L Stainless Steel

Michael Miles, Tracy Nelson, Cameron Gunter, Fenchaio Liu

Brigham Young University, United States

Friction stir processing (FSP) has potential as a weld repair technique for irradiated stainless steel, because it operates at much lower temperatures than fusion welding, and is therefore less likely to cause cracking in the HAZ. Experimental evidence of the crack repair capability of FSP will be presented. Numerical simulation of the FSP process in 304L stainless steel was performed using both a Lagrangian and an Eulerian finite element approach. Model input required flow stresses for the large range of strain rates and temperatures inherent in the FSP process. Temperature predictions in three locations adjacent to the stir zone were accurate to within 4% of experimentally measure values. The relationship between FSP process conditions and recrystallized grain size in the stir zone were investigated.

Development of Friction Stir Spot Welding of Advanced High Strength Steel for Automotive

Masahiro Miyake, Ryoji Ohashi, Yoshitaka Kobayashi, Takuya Fukuda
Kawasaki Heavy Industries, Ltd., 1-1, Kawasaki-cho, Akashi, Hyogo, 673-8666 JAPAN
miyake_masahiro@khi.co.jp

Yoshiharu Utsumi
Sumitomo Electric Industries, Ltd., 1-1-1, Koyakita, Itami, Hyogo, 664-0016 JAPAN

Masatoshi Tokunaga
Nippon Steel & Sumitomo Metal Corporation, 20-1, Shintomi, Futtsu, Chiba, 293-8511 JAPAN

Abstract

Introduction

Experimental procedures

Table1

Table 1 Materials used in this study

Material	Characteristic	Nominal tensile strength of base material (MPa)	Thickness (mm)	Size for tensile shear testing (mm)	Size for cross tension testing (mm)
IF steel	Ultra-low carbon steel	270	1.2	30W×100L	50W×150L
DP590	Dual phase steel	590			
DP980		980			
Hot stamped steel	As-quenched steel	1,470			

Fig. 1

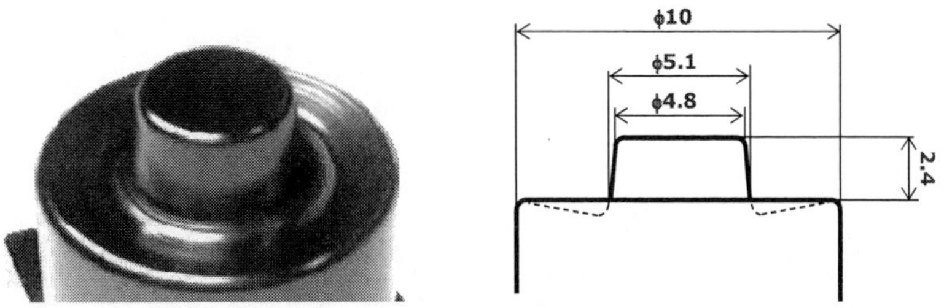

(a) Appearance *(b) Geometry*

Fig. 1 Appearance and geometry of tool tip

Fig. 2

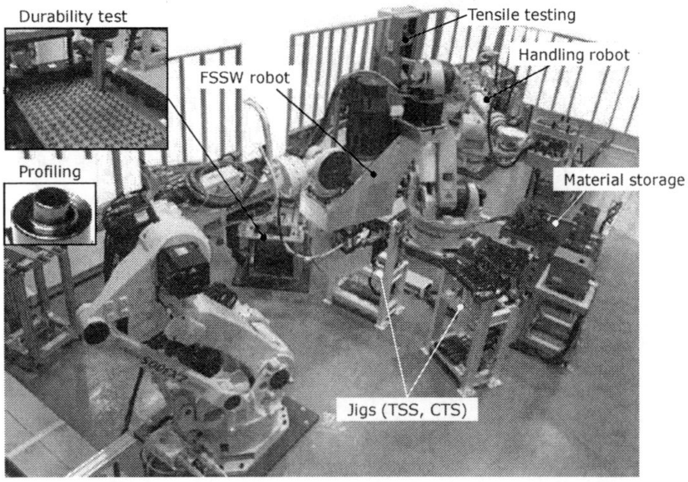

Fig. 2 Evaluation system for FSSW

Results

Fig. 3

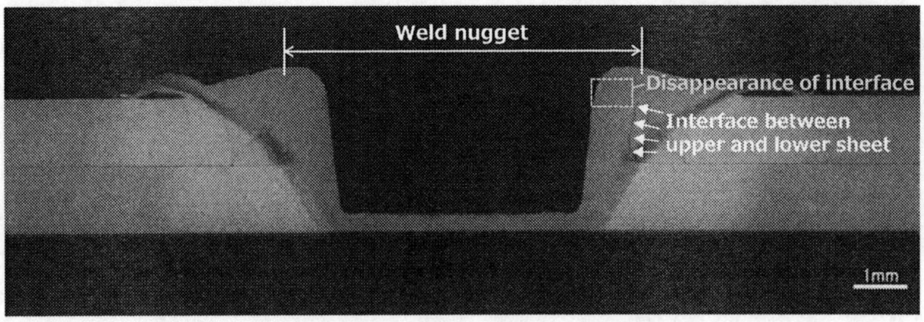

Fig. 3 Typical cross section of the joint (DP980)

Fig. 4

√

√

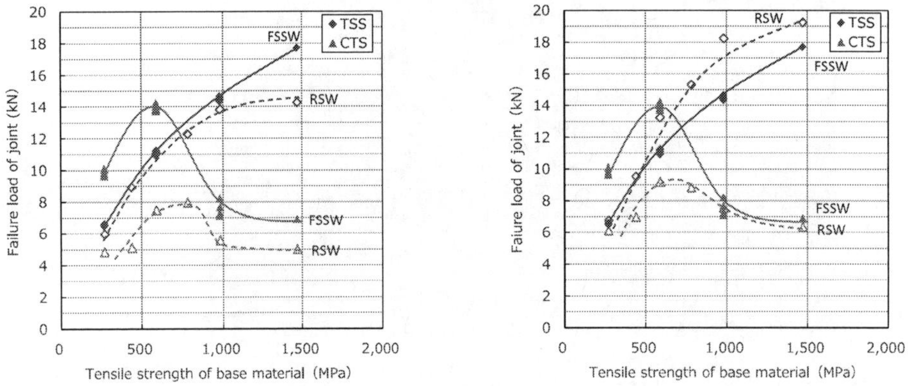

(a)Nugget size of RSW: 4√t *(b) Nugget size of RSW: 5√t*

Fig. 4 Comparisons of the joint strength of FSSW and RSW

Fig. 5

Fig. 6

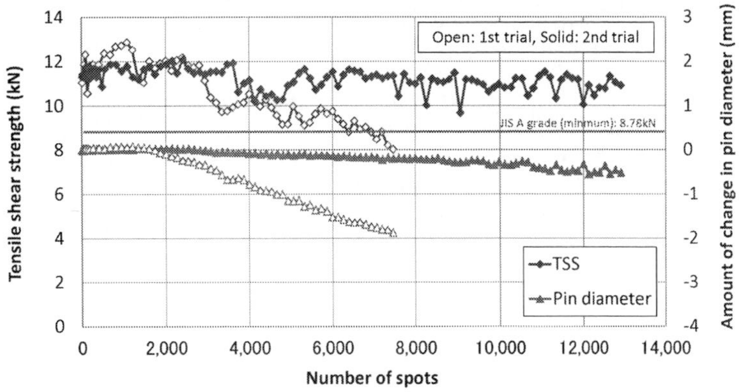

Fig. 5 Tensile shear strength and change in pin diameter during durability test

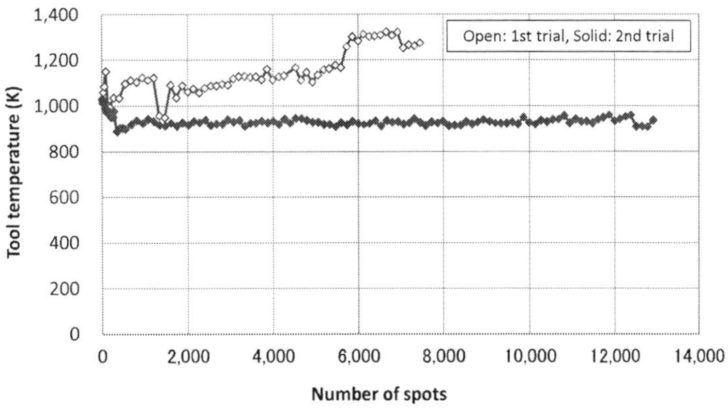

Fig. 6 Tool temperature during durability test

Conclusions

References

[1] M.D. Tumuluru, 'Resistance Spot Welding of Coated High-Strength Dual-Phase Steels', Welding Journal, 2006, 31-37.

[2] S. Ferrasse, P. Verrier, F. Meesemaecker, 'Resistance Spot Weldability of High Strength Steels for Use in Car Industry', Welding in the World, 41(1998) 177-195.

[3] A. Yoshitake, K. Yasuda, 'Vision of Application Technologies for High Strength Steel Sheets Supporting Automobile Weight Reduction', JFE technical report 10 (Dec. 2007), 1-7.

[4] H. Fujimoto, T. Okada, Welding Technology, Mar. 2011, 48-53. (in Japanese)

Development of advanced W-based alloy tool for FSW of stainless steel

A. Tsuji, A. Ikegaya, T. Takida
Allied Material, 2,Iwasekoshi-machi,Toyama-shi,Toyama 931-8543, Japan
Email: ayuri-tsuji@allied-material.co.jp, Phone: +81-76-437-1951, Fax: +81-76-437-1955

Y.S. Sato, H. Kokawa, S. Susukida
Department of Materials Processing, Graduate School of Engineering, Tohoku University, 6-6-02,Aramaki-aza-Aoba, Aoba-ku, Sendai 980-8579, Japan
Email: ytksato@material.tohoku.ac.jp, Phone&Fax: +81-22-795-7352

H. Moriguchi, A. Shibata
Nippon ITF, 575,Kuze Tonoshiro-cho, Minami-ku, Kyoto 601-8205, Japan
Email: moriguchi@nippon-itf.co.jp, Phone: +81-75-931-6040, Fax: +81-75-931-6166

Abstract

In this study, two W-based alloys containing the different hard-particles were designed as cost-effective tool-materials for friction stir welding of steel, and the wear behaviors of the tools, made of the W-based alloys, with 1.7mm long probe were examined during friction stir welding of 304 stainless steel. Two W-based alloys themselves showed the sufficient strength, ductility and thermal conductivity at high temperatures. Among them, the tools with the higher high-temperature strength exhibited more effective suppression of the wear during FSW, which was hardly worn even after the weld length of 20m.

Introduction

FSW is a solid-state joining process with many advantages compared to fusion welding processes. This process has been widely used in constructions of low-melting-temperature materials, such as aluminum alloys. However, application of FSW of steels is still limited because of a lack of the cost-effective tools.

Gan et al. [1] reported that the tool material for steels should have the strength over 400 MPa at 1000°C to prevent the deformation during FSW, suggesting that metallic-based alloys with high melting-temperatures, such as Mo and W alloys, are preferred as the tool materials. Therefore, development of the FSW tools made of new Mo-based alloys containing hard particles was attempted in a preliminary study [2]. The material factors governing the tool damage using several Mo-based alloys containing various hard particles were examined, which clarified that both the high ductility and thermal-conductivity at high temperatures were necessary to suppress the tool damage in addition to the high high-temperature strength.

The Mo-based alloy tool exhibited the good performance during FSW of 304 stainless steel, implying that the tool performance is enhanced by replacing Mo with W because W has the superior physical and mechanical properties than Mo. In this study, two W-based alloys containing the different hard-particles were newly developed for FSW of steels based on the preliminary study. FSW trials were carried out on 304 stainless steel with these tools, and the tool wear behaviors were examined.

Experimental Procedures

Two W-based alloys containing the different hard-particles were newly designed and produced by sintering. Mechanical properties and thermal conductivity of the alloys themselves were examined at high temperatures. 0.2 % proof stress and fracture displacement were obtained at 1200°C through three-point bending test. The thermal conductivity was examined by laser flash method.

FSW tools with 1.7mm-long probe were made of the developed W-based alloys. Some of the tools were coated with a ceramic by physical vapor deposition to examine effect of ceramic coating on tool wear. Appearance and dimension of the tool used is shown in Fig. 1. FSW trials were conducted on a 4mm-thick 304 austenite stainless steel with the W-based alloy tools. The welding parameters consisted of a tool rotational speed of 200 rpm and a travel speed of 1 mm/s. Argon gas shielding was employed to prevent the surface oxidization. A weld length of an FSW pass was 450 mm, which was repeated to increase the total weld length. The tool shape was captured after each FSW pass using a digital microscope to quantify degree of the tool wear during FSW. The tool wear was evaluated with a decrease in cross-sectional area of the tool. The FSW trials were stopped when the decreases in cross-sectional area exceeded 2.0 mm^2. The weld length at the stoppage of the FSW trial was defined as the tool life in this study.

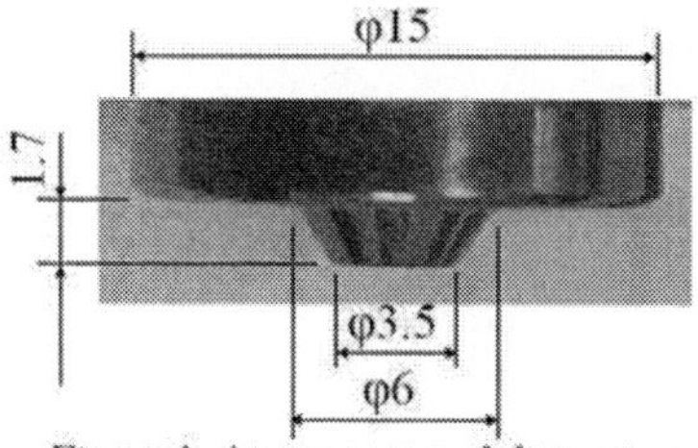

Figure 1. Appearance and dimension (mm) of W-based alloy tool used in this study.

Results and Discussion

Properties of the W-based alloys

Properties of the W-based alloys are shown in Table 1. For comparison, properties of the Mo-based alloy showing the best tool performance in the preliminary study [1], designated as Mo alloy 3, are also given in this table. Two W-based alloys designed in this study showed the higher strength, displacement and thermal conductivity at high temperatures than the Mo-based alloy. In W-based alloys, W alloy 1 exhibited the less strength and higher thermal conductivity than W alloy 2.

Table 1. Properties of Mo-based and W-based alloys.

	Alloy ID		
	Mo alloy 3	W alloy 1	W alloy 2
0.2 % proof stress at 1200°C (MPa)	491	588	950
Displacement at fracture at 1200 °C (mm)	3.6	>6.0	>6.0
Thermal conductivity at 1000 °C (W/m K)	64	100.4	69.3

FSW tests

Effect of weld length on decrease in cross-sectional area of Mo- and W-based alloy tools is shown in Fig. 2. The decrease in cross-sectional area increased with the weld length on all tool materials. This means that all tool materials were worn during FSW, but degrees of tool wear were much different. It is clearly shown that the non-coated tool materials experienced the severer tool wear than the coated ones, meaning that the ceramic coating is very effective to suppress the wear of the tool material. In the coated W-based alloy tools, W alloy 2 exhibited the more excellent wear resistance than W alloy 1. W alloy 1 was hardly worn up to the weld length of 4 m, beyond which its wear resistance was reduced. The precise examination on the tool damage revealed that the tool made of W alloy 1 was slightly deformed at the probe portion at the weld length of 0.9 m, resulting in peeling of the ceramic coating from the surface at the weld length of 4 m. On the other hand, the coated tool made of W alloy 2 was hardly deformed and underwent peeling of the ceramic coating during FSW, resulting from the superior high-temperature strengths. In the coated tool made of W alloy 2, consequently, the decrease in cross-sectional area was less than 1 mm^2, and any cracks and drastic deformation were not found even at the weld length of 20 m. The workpiece used in this study was 304 stainless steel, usually causing severer tool wear than carbon steels, so that the coated tool made of W alloy 2 would show the much excellent wear-resistance on carbon steels. This study successfully developed a new tool exhibiting the excellent performance during FSW of 304 stainless steel using the W-based alloy containing hard particles with a ceramic coating.

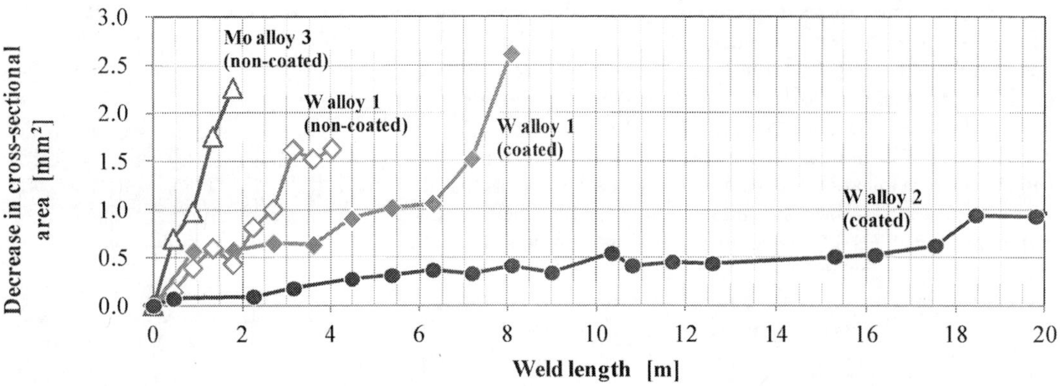

Figure 2. Relationship between weld length and decreased area on cross section.

Conclusions

In this study, the wear behaviors of the newly developed W-based alloy tools with a ceramic coating were examined during FSW of 304 austenite stainless steel. The FSW tool made of the W-based alloy having higher high-temperature strength exhibited the excellent performance on 304 stainless steel. It hardly experienced macroscopic wear, deformation and cracking even at the weld length of 20 m.

References

[1] W. Gan, *et al.*: Sci. Technol. Weld. Joining, Vol. 12 (2007), pp. 610-613.
[2] Y.S. Sato, *et al.*: *Preprints of the National Meeting of JWS*, Vol. 96 (2015), pp. 60-61.

Underwater friction stir welding of medium carbon steel

Tomoko Miyamori[a], Yutaka S. Sato[a], Hiroyuki Kokawa[a]
[a]Department of Materials Processing, Graduate School of Engineering, Tohoku University, Sendai, Japan
Contact e-mail: tomoko.miyamori.r7@dc.tohoku.ac.jp, TEL: +81-22-795-7353

Abstract
In this study, underwater and conventional friction stir welding was applied to 0.45wt% medium carbon steel at various welding parameters, and feasibility of underwater friction stir welding to steels, and the microstructure and hardness profile of the welds were examined. Underwater friction stir welding could produce defect-free welds, but it exhibited the narrower process window, and the higher torque and Z-axial load than conventional friction stir welding. The stir zone was mainly composed of ferrite and pearlite after conventional process, while the underwater operation produced the microstructure consisting of bainite with ferrite and pearlite in the stir zone, causing the higher hardness after underwater process. Formation of the hard microstructure could be explained by the cooling rate and the continuous cooling transformation diagram of the steel.

Introduction
To secure the safety during decommissioning of nuclear power plants, especially of the Fukushima nuclear power plant, the temporary repairing of the defective steel structures is required. In some cases, the defective steel structures are placed in water or in explosive gas atmosphere, so that the underwater repairing process in solid state would be preferred.

Friction stir welding (FSW) is a solid-state welding process using frictional heat and material flow arising from the rotation of the inconsumable tool [1]. FSW is often used as a process to eliminate porosity and/cavities in the cast materials [2]. Some previous studies [3-5] reported that FSW can produce the sound welds in aluminum alloys in the underwater operation. Zhang et al. [3] examined mechanical properties and microstructure of aluminum alloy 2216-T6 produced by underwater FSW, showing that reduction of the hardness in the heat affected zone was suppressed through rapid cooling of underwater FSW. These studies suggest that underwater FSW could be used as a repairing process of the defective steel structures, but underwater FSW of steels has been hardly studied.

In this study, underwater and conventional FSW of medium carbon steel was carried out, and then the microstructure and hardness profile of the welds were examined. Moreover, the microstructural evolution of the stir zone was discussed with thermal history during FSW.

Experimental procedures
Underwater and conventional FSW was performed on 0.45wt% medium carbon steel plate, 5.5mm in thickness. All FSW trials were conducted on the bead-on-plate configuration with a welding tool made of a composite consisting of cBN and W-Re alloy. Appearance and dimensions of the tool used is shown in Fig. 1. Tilting of the tool axis was not employed in this study. The travel speed was constant at 1.0 mm/min, and the rotational speed varied between 150 and 300 rpm. During underwater FSW, the steel plate was placed in an acrylic box filled with flowing water (292 K). Machine outputs, such as torque and Z-axial load, were recorded during FSW. Moreover, thermal history of the weld was measured by R-type thermocouple embedded on the steel plate.

Vickers hardness profile of the weld across the stir zone was measured on the cross-section perpendicular to the welding direction, and the microstructure was characterized by scanning electron microscopy (SEM).

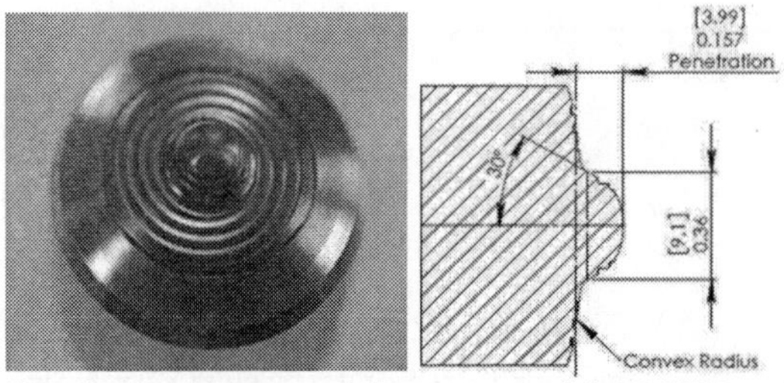

Fig. 1: Appearance and dimensions of the PCBN tool

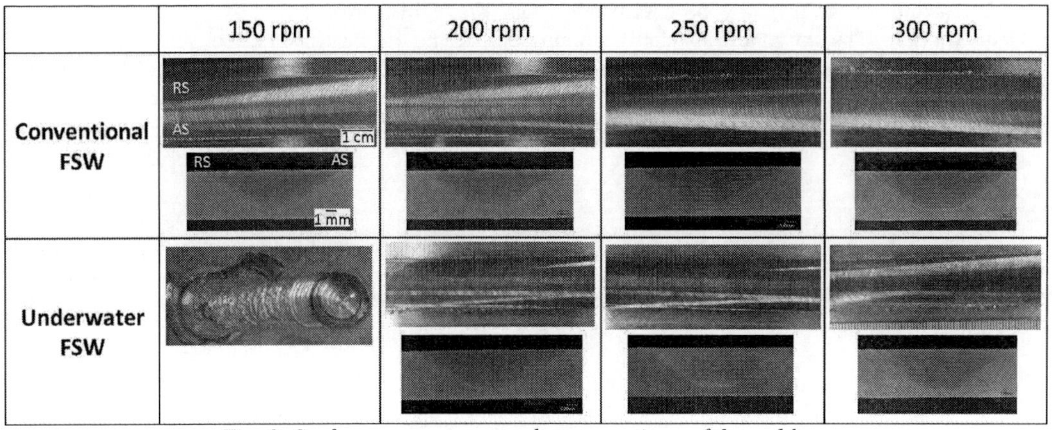

Fig. 2: Surface appearances and cross sections of the welds

Results and discussion

Surface appearances and cross sections of the underwater and conventional friction stir welds are shown in Fig. 2. Underwater FSW resulted in the slightly rougher appearance than the conventional one. Underwater FSW could produce defect-free welds at the most welding parameters used in this study as well as the conventional one, although the probe was broken at the rotational speed of 150 rpm during underwater FSW. This result means that underwater FSW exhibits slightly narrower process window than conventional one. Vickers hardness profiles of the welds across the stir zones are presented in Fig. 3. All stir zones are hardened, and underwater FSW leads to higher hardness values in the stir zone than conventional one, suggesting that cooling was faster after underwater FSW.

Torque and Z-axial load during FSW are shown in Fig. 4. They decrease with increasing rotational speed. It is interesting that underwater FSW resulted in the higher torque and Z-axial load than conventional FSW. The energy consumption per unit length during FSW, which can be calculated using the torque, rotational speed and travel speed [6], was compared between underwater and conventional FSW (Fig. 5). Effect of rotational speed on energy consumption was negligible, but the energy consumption of underwater FSW was larger than that of conventional FSW, implying that the water cooling reduces temperature of the stirred material around the welding tool during FSW even at the same welding parameters.

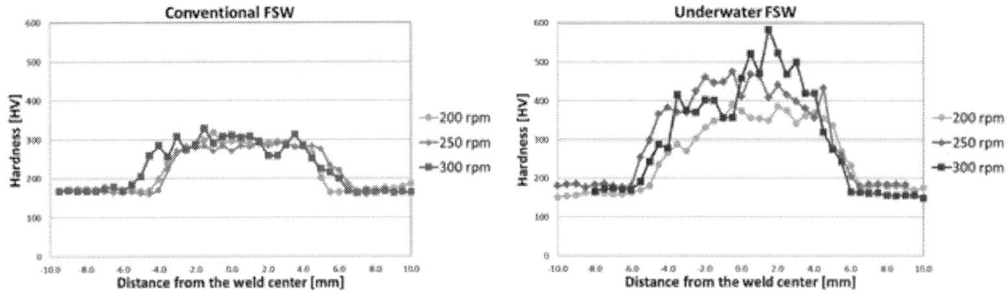

Fig.3: Hardness distribution of the welds

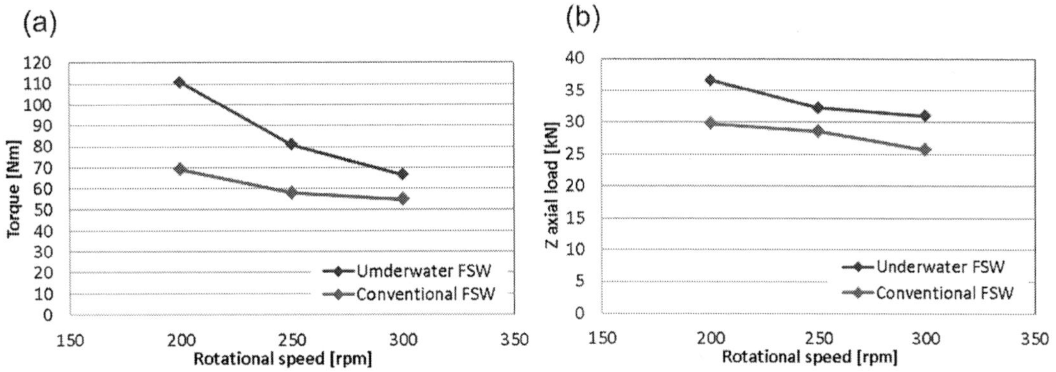

Fig.4: Torque and Z-axial load during welding

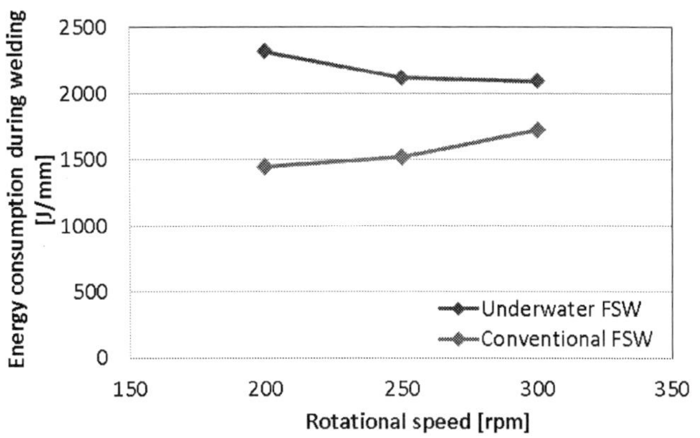

Fig.5: Energy consumption during welding

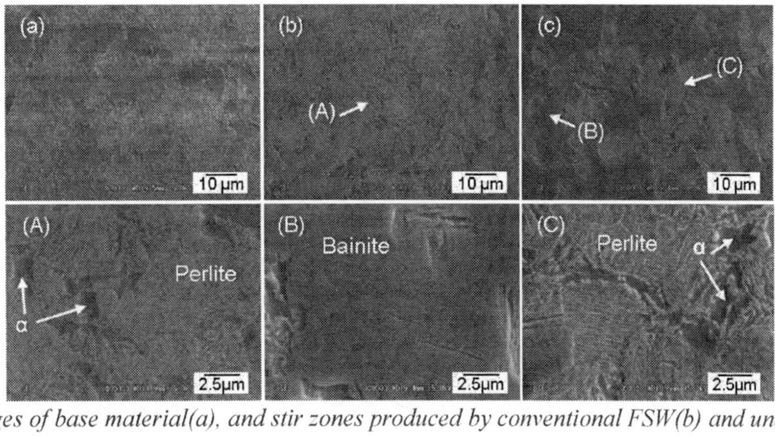

Fig.6: SEM images of base material(a), and stir zones produced by conventional FSW(b) and underwater FSW(c) (250 rpm). Higher magnified images of regions A – C in (b) and (c) are shown in (A) – (C), respectively.

Microstructures of the base material, and the stir zones produced at 250 rpm are shown in Fig.6. The stir zone was composed of ferrite and pearlite after conventional FSW, while underwater FSW produced the microstructure consisting of ferrite, pearlite and bainite. This difference can be explained by difference in cooling rate during FSW and continuous cooling rate (CCT) diagram of 0.45%C steel. Thermal histories during underwater and conventional FSW are given in Fig. 7. From Fig. 7, the cooling rates were estimated to be 312 and 290 K/s in the underwater and conventional FSW, respectively. According to CCT diagram of 0.45%C steel [7], the bainite formation requires the cooling rate higher than 296 K/s, suggesting that the bainite can be formed only during cooling of underwater FSW. The formation of bainite causes the higher hardness in the stir zones produced by underwater FSW.

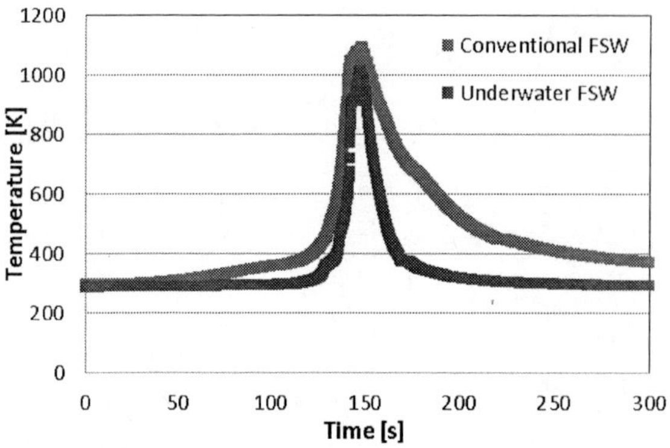

Fig. 7: Thermal histories during the underwater and conventional FSW at 250 rpm

Conclusions

Underwater friction stir welding was applied to 0.45wt% medium carbon steel and the defect-free welds were successfully produced. The underwater FSW consumed higher energy than the conventional one, and resulted in enough higher cooling rate for the bainite formation, causing the higher hardness in the stir zone.

Acknowledgments

The authors are grateful to Mr. A. Honda for technical assistance. This study was partially supported by the Center of World Intelligence Project for Nuclear S&T and Human Resource Development entrusted by the Ministry of Education, Culture, Sports, Science and Technology of Japan.

References

[1] W.M. Thomas, E.D. Nicholas, J.C. Needham, M.G. Murch, P. Templesmith, C.J. Dawes, International Patent Application No. PCT/GB92/02203.

[2] M.L. Santella, *et al*: "Effects of friction stir processing on mechanical properties of the cast aluminum alloys A319 and A356," *Scripta Materialia*, Vol. 53 (2005), pp. 201-206.

[3] H.J. Zhang, *et al*: "Effect of water cooling on the performances of friction stir welding heat-affected zone," *Journal of Materials Engineering and Performance*, Vol. 21 (2012), pp. 1182-1187.

[4] H.J. Zhang, *et al*: "Microstructure and mechanical properties as a function of rotation speed in underwater friction stir welded aluminum alloy joints," *Materials and Design*, Vol. 32 (2011), pp. 4402-4407.

[5] P. Upadhyay, and A.P. Reynolds: "Effects of thermal boundary conditions in friction stir welded AA7050-T7 sheets," *Materials Science & Engineering A*, Vol. 527 (2010), pp. 1537-1543.

[6] J.W. Pew, *et al*: "torque based weld power model for friction stir welding," *Science and Technology of welding and Joining*, Vol. 12, No.4, (2007), pp. 341-347.

[7] G. Totten, *et al*: Handbook of Residual Stress and Deformation of Steel, AMS international (Ohio, 2002), pp. 249.

Friction stir welding of 9%Ni steel pipe

J. Rodriguez.[1], V.F. Pereira[1], E.B. Fonseca[1], A.P. Tschiptschin[1,2], D. Garcia[3], R. Kalume[4], L. Lima[4]

1 Brazilian Nanotechnology National Laboratory (LNNano), Campinas, SP, Brazil.

2 Metallurgical and MaterialsEngineering Dept., University of São Paulo, SP, SP, Brazil.

3 Petrobras, Rio de Janeiro, RJ, Brazil.

4 Vallourec, Belo Horizonte, MG, Brazil.

**corresponding author: johnnatan.fernandez@lnano.cnpem.br*

Abstract

Microstructural characterization and mechanical behavior of 9%Ni steel pipes processed by friction stir welding were studied. Two welding configurations, thickness 8 mm and 12 mm, were used. 12 mm-thick pipe was used for Crack Tip Opening Displacement (CTOD) tests and the remaining tests were carried out using 8 mm-thick pipes. Butt joints were welded at a constant rotation speed of 300 rpm, travel speed between 50-65 mm.min^{-1}, force control of 32 kN and tool offset in Y-axis of -9 mm. Defect free butt joints were produced with full penetration (for 8 mm-thick pipe) and good surface appearance. The microstructure of the affected regions by the process were characterized as martensite with prior-austenite grain (PAG) size variations. The results showed good mechanical behavior for cryogenic applications.

Introduction

Friction stir welding (FSW) is a solid-state joining process developed at The Welding Institute (TWI) in 1991 (1-4). Initially, aluminum and magnesium alloys were welded. Then, this process advanced to weld high melting point materials. Nowadays, one of the main interest of this technique is welding pipes for the oil and gas industry (5, 6). During the process, heat generated by friction and deformation softens the material being welded. Severe plastic deformation and flow of the plasticized metal occurs as the tool travels along the welding direction (2).

Regarding the base material, the microstructure of the 9%Ni steel is composed of martensite along with small quantities of retained austenite at the grain boundaries. The fracture toughness at cryogenic temperatures is increased by the presence of retained austenite (7). Thus, the aim of this work is to explore the application of FSW on steel pipes for cryogenic applications.

Experimental Procedure

Pipes of ASTM A333 Gr8 steel 8 and 12 mm-thick were butt friction stir welded using a W-Re PCBN (polycrystalline cubic boron nitride) tool. 12 mm thick pipes were used to perform the CTOD tests; the remaining tests and the metallographic analyses were carried out in 8 mm-thick pipes. Tool rotation speed of 300 rpm, travel speed between 50-65 mm.min^{-1} and force control of 32 kN were used during the welding. In addition, tool offset in Y-axis of -9 mm was adopted to obtain sound welds.

Microstructural characterization was carried out using optical microscopy (OM) and scanning electron microscopy (SEM). Metallographic samples were prepared using standard metallographic grinding and polishing techniques and etched with Nital 2%. Mechanical properties were assessed by CTOD, Charpy impact and tensile tests.

Results and Discussion

The development of welding parameters allowed choosing the right combination to obtain good surface characteristics and stable heat input during the process. *Fig. 1* shows the surface appearance of the final joints of 9%Ni steel 8 mm-thick pipes. The welds showed good surface appearance without defects. The main characteristic to process this joint was the use of a travel speed changing between 50-65 mm.min^{-1}, which allowed a good material flow and stable head input.

The mechanical behavior of the welded joint was assessed by tensile and bend tests. The tensile test performed perpendicular to the welding direction exhibited good results and the fracture occurred at the base metal far from the welded joint. Root and face bend tests were employed to verify full penetration of the tool and weld integrity, respectively. The bend tests showed a consolidated joint and good integrity of the weld. The absorbed energy of the stir zone (SZ) and heat-affected zone (HAZ) was evaluated and compared to the base metal (BM), as shown in *Table 1*. Even though the SZ and HAZ values are below the absorved energy of the BM, the minimum results obtained in the processed regions were 56 J and 83 J respectively, which are suitable for the applications considered. Concerning the fracture mechanic toughness tests of the HAZ at -90 °C, good CTOD values were observed and these results were above the minimum accepted values according to previous studies (8, 9, 10, 11). These results were also considered suitable for the applications under consideration.

Figure 1. Surface appearance of the friction stir welded pipe of 9%Ni steel. Joint produced using rotation speed of 300 rpm, travel speed between 50-65 mm.min^{-1} and a tool offset in Y-axis of -9 mm.

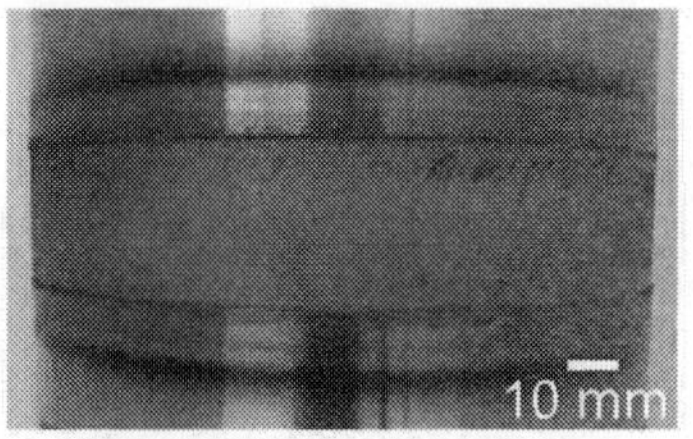

Table 1. Results of sub-size Charpy impact at 0°C test for the affected regions during the FSW process.

Zone	Absorbed Energy (J)
SZ	56
HAZ	83
BM	102

Fig. 2 shows the microstructural characterization of the friction stir welded pipe of 9%Ni steel. The (BM) was composed by tempered martensite with retained austenite at the grain boundaries and acicular bainite, as shown in *Fig. 2b* and *Fig. 3a*. In general, all the affected regions by the process were characterized as martensite with prior-austenite grain (PAG) size variations. The inter-critical heat-affected zone (ICHAZ) showed a partial tempered structure with fine laths of martensite due to the peak temperature reached during the process (500 °C to 700 °C), as shown in *Fig. 2c*. The fine-grained heat-affected zone (FGHAZ) reached temperatures above Ac3 and was composed of a martensitic structure with a minor PAG compared to the BM (*Fig. 2d* and *Fig 3b*). The highest peak temperatures of the HAZ were reached by the coarse-grained heat-affected zone (CGHAZ) and the microstructure was characterized as coarse martensite. In addition, granular and coalesced bainite were observed (*Fig. 2e* and *Fig 3c*). The SZ also exhibited coarse martensite, granular and coalresced bainite similar to the CGHAZ, but with a larger PAG (*Fig. 2f* and *Fig 3d*). Moreover, the peak temperature reached during the process was approximately 1200 °C.

Conclusions

Successful circumferential joints in 8 and 12 mm-thick 9%Ni steel pipes were obtained with full penetration without any defect and good surface appearance. An increase in the previous austenite grain size and formation of bainite and martensite were observed. Fracture toughness of the coarse-grained heat affected-zone was evaluated at -90 °C and the results showed good mechanical behavior for cryogenic applications.

Figure 2. Microstructural characterization of the friction stir welded pipe of 9%Ni steel. Joint produced using rotation speed of 300 rpm, travel speed between 50-65 mm.min-1 and a tool offset in Y-axis of -9 mm. Optical analyses.

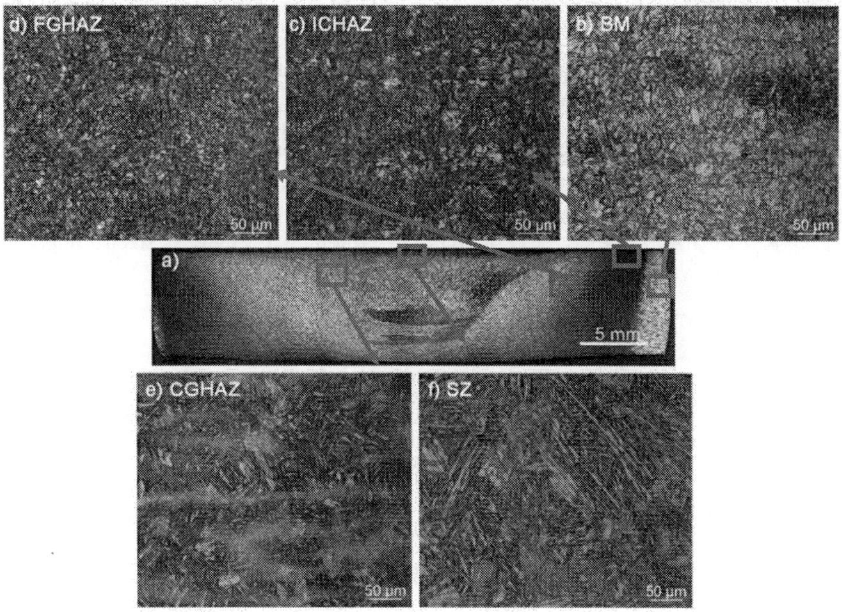

Figure 3. Microstructural characterization of the friction stir welded pipe of 9%Ni steel. Joint produced using rotation speed of 300 rpm, travel speed between 50-65 mm.min-1 and a tool offset in Y-axis of -9 mm. SEM analyses.

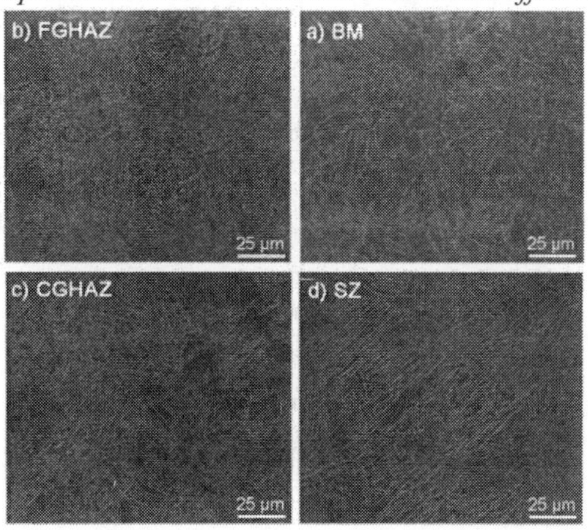

Acknowledgments

The authors gratefully acknowledge Petrobras for financial support, the Brazilian Nanotechnology National Laboratory (LNNano-CNPEM) and the Electron Microscopy Laboratory (LME-LNNano) for technical support.

References

[1] R. S. Mishra, Z.Y. Ma: 'Friction stir welding and processing', Mater. Sci. Eng. R, 2005, 50, 1–78.
[2] R. Nandan, *et al*: Recent advances in friction-stir welding – Process, weldment structure and properties. Prog. Mater Sci., 2008, 53, 980–1023.
[3] K. Colligan: Material Flow Behavior during Friction Stir Welding of Aluminum. Weld. J. July 1999, 229-237.
[4] Z.Y. Ma: 'Friction Stir Processing Technology: A Review', Metall. Mater. Trans. A., 2008, 39A 642-658.

[5] J.A.D. Ávila, *et al*: Fracture toughness assessment at different temperatures and regions within a friction stirred API 5L X80 steel welded plates, Eng. Fract. Mech. 147 (2015) 176–186. doi:10.1016/j.engfracmech.2015.08.006.
[6] M. Abbasi, *et al*: Microstructure-Fracture Toughness Studies in Friction Stir Processed API X80 Pipeline Steel, in: Trends Weld. Res., ASM International, Chicago, 2012: pp. 387–391.
[7] K. Zhan, *et al:* Effect of heating rate before tempering on reversed austenite in Fe-9Ni-C alloys.
[8] D.P. Fairchild, *et al:* A Study Concerning Intercritical HAZ Microstructure and Toughness in HSLA Steels. Welding Journal, v. Supplement, n. December, p. 321s–330s, 1991.
[9] D.P. Fairchild, *et al:* Research Concerning the Friction Stir Welding of Linepipe Steels. Trends in Welding Research, 8th International Conference. v. 8, p.371–380, 2009.
[10] D.P. Fairchild, *et al:* Fractographic Investigation of Cleavage Initiation in Steel Friction Stir Welds. Trends in Welding Research, 9th International Conference. p.193–200, 2012.
[11] A. Kumar, *et al:* Research Progress on Friction Stir Welding of Pipeline Steels. International Pipeline Conference. v. IPC2010-31, p.1–9, 2010.

Arc Behaviour in Multi-Electrode Arc Welding Processes

Shujun Chen, Liang Zhang, Yaxiu Song, Guangqiang Men

Beijing University of Technology, China

In traditional GMAW, the wire is the cathode and the work-piece the anode, the deposition rate and heat input are coupled and unable to be freely optimized. The multi-electrodes coupled arc welding processes are difference from the conventional arc welding processes, which are not based on the fundamental principle, but in a new mode, multi-electrodes or pairs of electrodes are discharged in a common space, the welding pool is just only one polarity of many electrodes, the arcs are coupled by each other. The multi-electrodes coupled arc welding processes can freely provide the needed deposition and heat input as desired. In this paper, the research is focus on the arc behavior, as stable arc behavior can get stable metal transfer and good welds. In the experiments of DE-GMAW, the bypass arc should be established between tungsten and wire, but it was shown that the GTA is bent and deviated, especially when the metal transfer in the form of short circuiting transfer, this phenomenon is obvious, GTA even established from tungsten to work-piece directly. In the experiments of arcing-wire GTAW, the bypass arc can be established between tungsten and wire stably, the length of GTA and the GMA current have no effect on the stability of arcing-wire GTAW. Despite decreasing the length of GTA, GMA has no changes, just the metal transfer changed. In cross arc welding experiments, the main arc is established between tungsten and work-piece, the main arc swung between the two wires. It is deflected by negative of inter-wire arc. The change is that the wire is heated by two arcs. The mass transfer is generated, when the wire is negative. The changes of arc behavior is the result of the electromagnetic force, which can change metal transfer in the multi-electrodes coupled arc welding processes.

Influence of doped metal oxide on tungsten-based electrode evaporation in multiphase AC arc

Taro Hashizume, Seiya Nagao, Manabu Tanaka, and Takayuki Watanabe
Department of Chemical Engineering, Kyushu University, Fukuoka, Japan
Email watanabe@chem-eng.kyushu-u.ac.jp, phone +81-92-802-2745

Abstract

Multiphase AC arc has various advantages such as high energy efficiency, large plasma volume and low gas velocity compared with the conventional thermal plasmas. However, electrode erosion is one of the most important issues to be solved because it determines the electrode lifetime and the purity of the products. Therefore, the purpose of the present study is to investigate the evaporation mechanism from tungsten electrode with doped oxide in the multiphase AC arc.

Metal vapor from tungsten and doped oxide was visualized by high-speed camera and appropriate band-pass filter system. Tungsten evaporated just after the evaporation of doped oxide. Therefore, lower boiling point of doped oxide led to quicker evaporation of tungsten during anodic period at AC cycle. These results suggest that evaporation of doped oxide influences the tungsten electrode evaporation. Addition of the vapor from doped oxide into the arc leads to the arc constriction due to its higher electrical conductivity. Therefore, heat flux from the arc to the electrode become larger after the evaporation of doped oxide, resulting in tungsten evaporation.

Introduction

The multiphase AC arc is generated among multi-electrodes by phase-shifted AC power supplies. It has high energy efficiency because of the generation without converting AC to DC. The multiphase AC arc can be generated without the intermittent discharge because the arc always exists among 12 electrodes. It has various advantages such as high energy efficiency, large plasma volume and low gas velocity compared with the conventional thermal plasmas. The multiphase AC arc is expected to be applied to massive powder processing such as nanomaterial fabrication processes and innovative in-flight glass melting technology [1]. However, fundamental phenomena in the multiphase AC arc have rarely been reported because it is still a new type of the thermal plasma generating system. In particular, electrode erosion is one of the most important issues to be solved in multiphase AC arc because it determines the electrode lifetime and the purity of the products [2].

Tungsten electrode with doped oxide is generally used because of their suitable characteristics for stable arc operation, although the influence of such doped oxide on electrode erosion has not been clarified. In the multiphase AC arc, the erosion of electrode due to evaporation and droplet ejection was observed [3, 4]. In previous work, influence of doped oxide on the droplet ejection has been clarified. Doped oxide in the tungsten-based electrode decreased the electrode erosion rate due to droplet ejection. On the other hand, influence of doped oxide on the tungsten evaporation has not been understood yet. The purpose of this work is to investigate the evaporation mechanism from tungsten electrode with doped oxide in the multiphase AC arc. Metal vapor from tungsten and doped oxide were visualized by high-speed camera system.

Experimental setup

Figure 1 shows a schematic image of the experimental setup. It consisted of 12 electrodes, arc chamber, and AC power supply at 60Hz. The electrodes were divided into two layers, upper six and lower six electrodes. The electrodes were symmetrically arranged by the angle of 30 degrees. To prevent the electrodes from oxidation, argon was injected around the electrode as shield gas at 2-3,5 L/min. The applied voltage between each electrode and the neutral point of the coil of the transformer can be

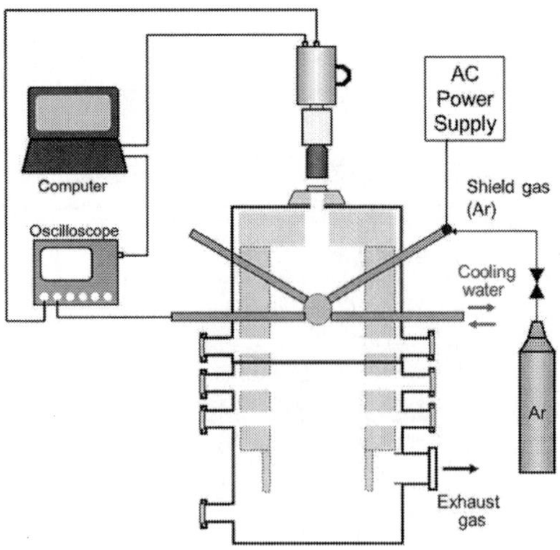

Fig. 1 Schematic image of multiphase AC arc generator.

calculated by the following equation:

$$V_i = V_m \sin\left[\omega t - \frac{2\pi(i-1)}{12}\right] \quad (i=1, 2, \ldots, 12) \qquad (1)$$

where V_i indicates the applied non-load voltage for each electrode number i and V_m indicates the amplitude of the non-load voltage (about 220 V, AC 60Hz). The arc current was 120 A for each electrode.

The electrodes were made of tungsten (98wt%) and doped oxide (2.0wt%) with diameter of 6.0 mm. In this work, three types of electrode, 2.0wt%ThO_2-W, 2.0wt%CeO_2-W and 2.0wt%La_2O_3-W were compared in order to investigate the influence of the doped oxide on the evaporation of tungsten electrode.

Evaporation of tungsten electrode was visualized by the high-speed camera system (FASTCAM SA-5, Photron Ltd., Japan). One of the electrodes was observed by high-speed camera installed on the top of the arc generator. Conventional observation of electrode during arc discharge was prevented by the strong emission of the arc. Therefore, the band-pass filter (Andover Optical Inc., USA) with 393 nm was combined with the high-speed camera system to separate the emission of tungsten vapor from the emission of the arc as shown in *Fig. 2*.

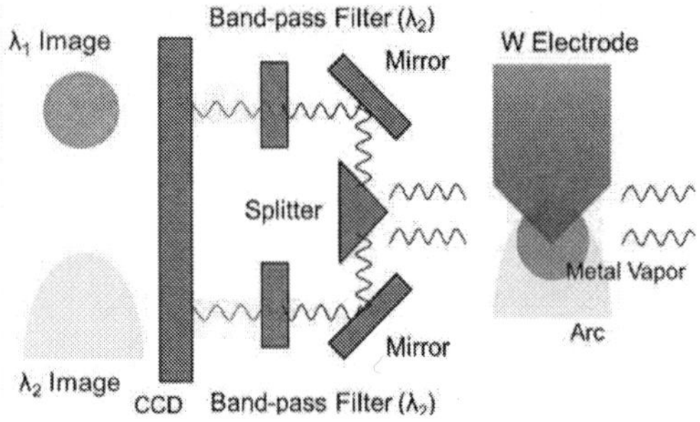

Fig. 2 Schematic diagram of high-speed camera system with band-pass filters.

High-speed visualization of tungsten vapor

Figure 3 shows representatives of high-speed snapshots of the tungsten vapor of ThO_2-W (a), CeO_2-W (b) and La_2O_3-W (c) from 3.0 to 6.5 ms during an Ac cycle. The indicated time in *Figs. 3* (a), (b), (c) correspond to the waveform of arc current in *Fig. 3* (d). In the multiphase AC arc, the evaporation of the tungsten electrode mainly observed at the anodic period. Vapor addition of metal into the arc leads to arc constriction due to its higher electrical conductivity. Therefore, heat flux from the arc to the electrode is enhanced, resulting in severer evaporation of tungsten.

According to high-speed observation, luminance area of the tungsten vapor and timing when tungsten started to evaporate depended on kinds of doped oxide. Evaporation timing in La_2O_3-W was earlier than that in ThO_2-W and CeO_2-W. Tungsten evaporation for La_2O_3-W started before the peak top of the arc current at the anodic period. Tungsten vapor for La_2O_3-W constricted at 4.0 ms as shown in *Fig. 3* (c). In contrast, the tungsten evaporation for ThO_2-W and CeO_2-W started after the peak top of arc current. The tungsten vapor for ThO_2-W constricted at 5.5 ms, and the vapor for CeO_2-W constricted at 4.5 ms as shown in *Fig. 3* (a), (b). This result suggests that the tungsten evaporation is strongly influenced by doped oxide, and electrode erosion rate by the evaporation depends on kinds of doped oxide.

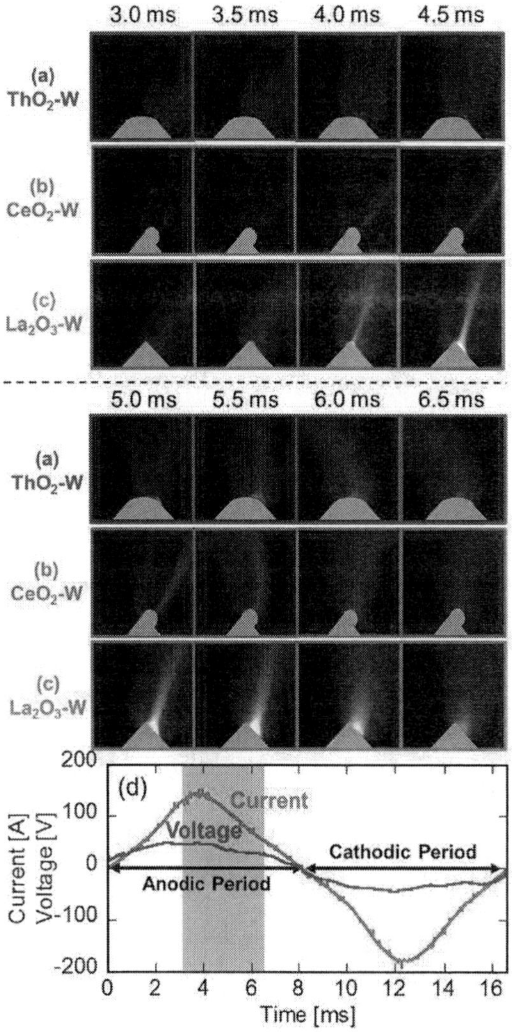

Fig. 3 High-speed snapshots of tungsten vapor of ThO_2-W (a), CeO_2-W (b), La_2O_3-W (c), and synchronized current waveform (d).

Figure 4 shows time variation of area of the tungsten

vapor during an AC cycle with different kinds of doped oxide. Here, the area of tungsten vapor was averaged for more than 5 AC periods. Results indicated that the area of tungsten vapor for La_2O_3-W was about four times larger than that for ThO_2-W and CeO_2-W. Moreover, the area of tungsten vapor for CeO_2-W was larger than that for ThO_2-W. The tungsten vapor and the evaporation timing at the anodic period depend on kinds of doped oxide. The reason for this result will be discussed in following paragraphs.

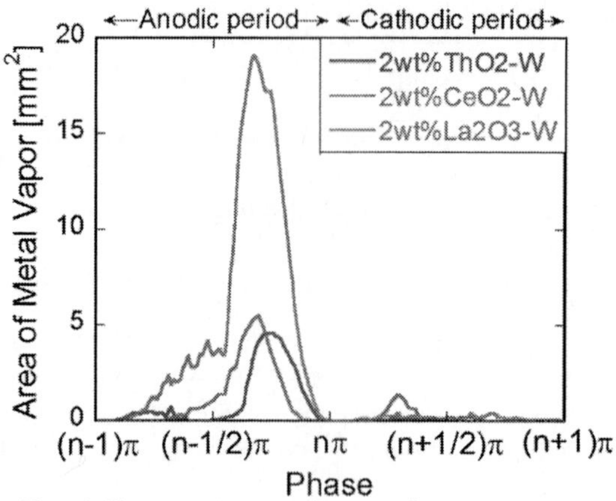

Fig. 4 Time variation of area of tungsten vapor during an AC cycle with different types of doped oxide.

Figure 5 shows the possible mechanism of evaporation from tungsten electrode with doped oxide in multiphase AC arc. First, the doped oxide evaporates before the tungsten evaporation at the anodic period because the boiling point of doped oxide is lower than that of tungsten. Then, addition of the vapor from doped oxide into the arc leads to arc constriction due to its higher electrical conductivity. Consequently, heat flux from the arc to the electrode is enhanced after the evaporation of doped oxide, resulting in tungsten evaporation.

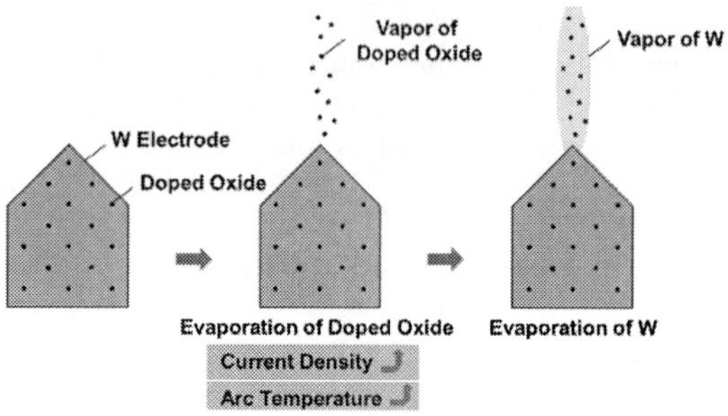

Fig. 5 Schematic diagram of mechanism of evaporation of tungsten electrode with doped oxide.

Evaporation of doped oxide triggers the tungsten evaporation. This suggests that timing of the tungsten evaporation depends on boiling point and vapor pressure of doped oxide. Doped oxide of La_2O_3 has lower melting (2,588 K) and boiling point (4,473 K) than that of ThO_2 (m.p.: 3,445 K, b.p.: 4,673 K). This suggests that La_2O_3 is relatively easier to evaporate. La_2O_3 starts to evaporate at low temperature of the electrode tip. Therefore, tungsten starts to evaporate before the peak top of arc current. Increase of the arc current before the peak top leads to higher electrode temperature, resulting in severer evaporation of tungsten. In contrast, ThO_2 and CeO_2 start to evaporate at high electrode temperature near the peak top of the arc current. Decrease of the arc current after the peak top leads to lower electrode temperature. Therefore, the tungsten evaporation at the peak top of the arc current for La_2O_3-W is relatively severe as shown in *Fig. 4*.

High-speed visualization of metal vapor from doped oxide

Tungsten evaporation with different kinds of doped oxide suggests evaporation of doped oxide triggers the arc constriction, resulting in the severer evaporation of tungsten due to higher heat flux from the arc to the electrode. Therefore, metal vapor from doped oxide was observed by high-speed camera system in order to investigate mechanism of the evaporation of tungsten-based electrode in the multiphase AC arc. In present study, La vapor from La_2O_3-W electrode were visualized by a high-speed camera combined an appropriate band-pass filter.

Spectroscopic measurements (iHR550, Horiba Jobin Yvon) were conducted to determine the appropriate wavelengths for high-speed observation of La vapor. *Figure 6* shows the representative emission spectra at a distance of 2 mm from the electrode tip for La_2O_3-W. Line emissions from La vapor, as well as that from the tungsten vapor can be observed. The band-pass filter with 577 nm was combined with the high-speed camera system to separate the emission of La vapor from the emission of tungsten vapor.

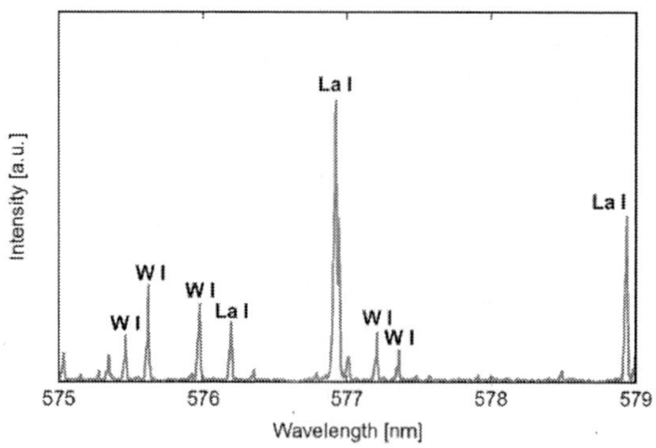

Fig. 6 The emission spectrum of the multiphase arc in the electrode region for La_2O_3-W.

Figure 7 shows representatives of high-speed snapshots of tungsten vapor (a) and La vapor (b) of La_2O_3-W from 0.5 to 4.0 ms during an AC cycle. The indicated time in *Fig. 7* (a), (b) corresponds to the waveform of arc current in *Fig. 7* (c). Evaporation of La started at 0.5 ms, and the La vapor constricted at 1.0 ms as shown in *Fig. 7* (b). On the other hand, tungsten evaporation and the constriction of tungsten vapor started at 1.5 ms as shown in *Fig. 7* (a). Tungsten vapor constricted at the point of the constriction of La vapor. These results suggest that the constriction of La vapor leads to higher heat flux from the arc, resulting in tungsten evaporation.

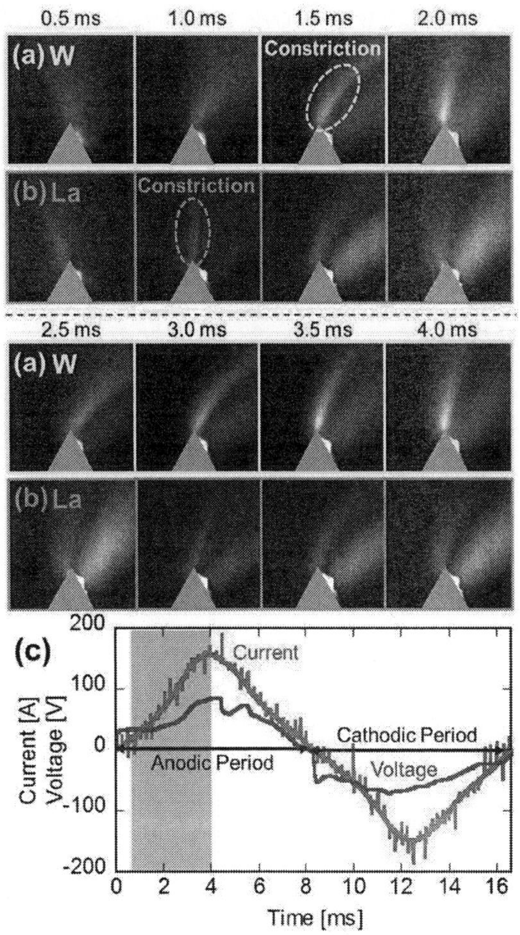

Fig. 7 High-speed snapshots of tungsten vapor (a), La vapor (b) of La_2O_3-W, and synchronized current/voltage waveforms (d).

Conclusions

The dynamic behavior of metal vapor was successfully visualized by the high-speed camera observation combined with the appropriate band-pass filter system. Three types of electrodes, 2.0wt%ThO_2-W, 2.0wt%CeO_2-W and 2.0wt%La_2O_3-W were compared to investigate the influence of the doped oxide on the evaporation of tungsten electrode. Obtained results revealed that the evaporation of doped oxide had an important role on the tungsten evaporation.

The mechanism of evaporation of tungsten-based electrode with doped oxide was investigated by observation of the tungsten vapor and the vapor from doped oxide. Doped oxide evaporated just before the tungsten evaporation. Addition of the vapor from doped oxide into the arc leads to the arc constriction, resulting in higher heat flux from the arc to the electrode. Therefore, evaporation of doped oxide triggers tungsten evaporation in the multiphase AC arc. Tungsten-based electrode with doped oxide which has higher boiling point is suitable to decrease the erosion rate due to evaporation in the multiphase AC arc. Understanding the erosion mechanism in more details enable to realize the practical use of the multi-phase AC arc in various applications.

References

[1] Y. Yao, *et al*: "An innovative energy-saving in-flight melting technology and its application to glass production," *Sci. Technol. Adv. Mat*, Vol. 9 (2008), 025013.

[2] M. Tanaka, *et al*: "Investigation of Electrode Erosion Mechanism of Multi-Phase AC Arc by High-Speed Video Camera," *J. Phys. Conf. Ser.*, Vol. 441 (2013), 012015.

[3] T. Hashizume, *et al*: "Investigation of droplet ejection mechanism from electrode in multi-phase AC arc," *Quart. J. Jpn. Weld. Soc.*, Vol. 33 (2015), 44s.

[4] T. Watanabe, *et al*: "Investigation of Electrode Phenomena in an Innovative Thermal Plasma Process for Glass Melting," *Plasma Chem. Plasma Process.*, Vol. 34 (2014), p.443-456.

Single pass full penetration joining for heavy plate steel using high current GMA process

Hayato Baba, Tetsuo Era, Tomoyuki Ueyama, Hideo Shiozaki
DAIHEN Corporation, KOBE, HYOGO, JAPAN
Email: hayato.baba@daihen.co.jp, Phone: +81-78-275-2019 (representative)

Manabu Tanaka
Joining and Welding Research Institute Osaka University, Suita, Osaka, JAPAN
Email: tanaka@jwri.osaka-u.ac.jp, Phone: +81-6-6879-8640

Abstract
At a range of high current in gas metal arc welding (GMAW) using 100% CO_2 as a shielding gas and a carbon steel solid wire, as an arc voltage decreases, an arc length gets shorter and a surface of molten pool right under the arc becomes dented. And the arc is finally generated at a lower position under the surface of the molten pool and in a space formed by a wall of molten pool. In this metal transfer mode, a spatter generation level is extremely small even at a high current range. This arc phenomenon is referred to as "buried arc" and arc stabilization technique at high current buried arc more than 450A is described in this report.

The arc phenomenon taken by high brightness X-Ray observation system and high speed video camera were observed using JIS SS400 (ASTM A36, EN S275) as a base metal, JIS YGW12 (AWS ER 70S-6, EN G 42 2 C G3Si1) as a carbon steel solid wire of 1.2 mm, 100% CO_2 as the shielding gas and travel speed of 30 cm/min. The high current buried arc stabilized by external characteristic control of digital inverter controlled welding power source (maximum current 1000A, maximum wire feed rate 100 m/min) showed three kinds of metal transfer mode including rotating transfer mode at the current range more than 450A. Consequently, a new current waveform control combined an optimal gradient of the external characteristic and a low frequency voltage oscillation, which realizes more stable high current buried arc, has been developed in order to improve a weld bead formation using these particular metal transfer modes and achieve one side piercing welding of plate with 25mm thickness.

Introduction
When focused on heavy plate joining, a lot of welding processes, which are Submerged Arc Welding, Friction Stir Welding, Electroslag Welding and so on, have been developed and these joining processes are being applied to practical production. These welding processes are very efficient but limited on a usage, for instance, a welding position, a welding direction, a jig system, and so on. On the other hand Gas Metal Arc Welding (GMAW) is relatively flexible in the usage, however, multi-layer welding is required for heavy plate since rated output of GMAW is limited up to 500A in general. Therefore authors considered that a high efficiency welding process based on GMAW is expected to develop. In recent years it has been reported that a maximum penetration of 10 mm per one path of welding was achieved[1-3]. Then authors have developed a new welding system with a new current waveform control which realizes one pass welding of heavy plate more than 10 mm with stable buried arc. In this paper, a principle of this new welding process and its application are presented.

Observation of buried arc phenomenon
In order to observe arc phenomena of buried arc at the high current range more than 500A, a new welding power source (Max. output of 1000A) and a new wire feed system (Max. wire feed rate of 100 m/min) with a wire buffer includes a sub feeder inside have been developed. A high speed video camera (HSV) and a high brightness X-ray transmission system (HBX-ray) were used as observation systems. Table 1 shows common welding condition through the observation of buried arc.

Table 1 Common welding condition through observation of buried arc

Base metal	**JIS SS400 (ASTM A36, EN S275)**	**Welding wire**	**JIS YGW12 (AWS ER 70S-6, EN G 42 2 C G3Si1)**
CTWD	**25 mm**	**Shielding gas**	**100% CO_2**
Welding speed	**30 cm/min**	**Wire diameter**	**1.2 mm**

As the first step of observation of buried arc, a gradient of external characteristic of -1V/100A, which is a general value of CV controlled welding power source, was selected to carry out the bead on plate welding. Fig.1 shows the observation result of buried arc taken by HSV at a wire feed rate of 40 m/min, a welding current of 570A and an arc voltage of 48V and each frame represents the arc phenomenon with a time interval of 10 ms. As shown in Fig. 1 the arc cannot be observed except at the frame of 0 ms since it is generated under the surface of molten pool, which is the buried arc that authors refer in this paper. And the arc phenomenon was very unstable even though the buried arc was realized. A size of opening of the molten pool was periodically fluctuated with less spattering and the opening of molten pool was often closed and both of the welding wire and the molten pool were short-circuited. Then a large amount of spatter was generated due to releasing of short-circuiting and the bead appearance also became irregular.

Time	0ms	10ms	20ms	30ms	40ms
Captured image of HSV	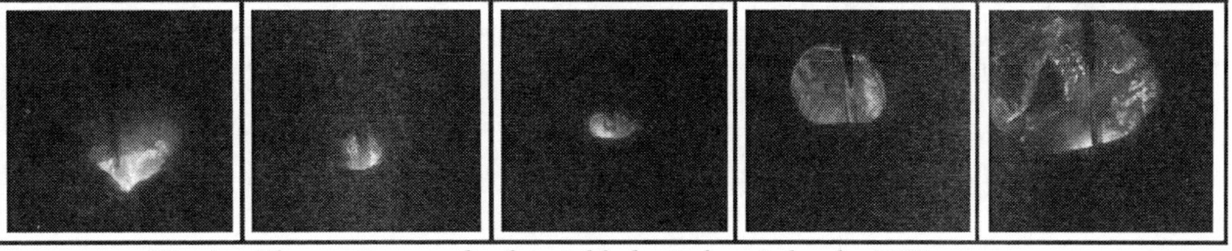				

Fig. 1 Observation result of unstable buried arc taken by HSV

In order to observe the phenomenon in the molten pool, HBX-ray was used since it is very hard to observe the buried arc by HSV. Fig. 2 shows the observation result of buried arc taken by HBX-ray. Each dashed line in frame (a) and (b) shows the welding wire and the white part around the wire tip shows the buried space in the molten pool. The frame (a) shows that the wire tip is lower than the surface of base metal and the buried space is relatively narrow in comparison with the frame (b). The frame (b) shows that the wire tip is higher than the surface of base metal though the buried space was formed. And the buried space is relatively bigger than that of frame (a). The phenomena of frame (a) and (b) were irregularly repeated and it is considered that the weld bead becomes irregular when the position of wire tip and the buried space are greatly fluctuated. Consequently, authors assumed that the small opening of molten pool shows the small size of buried space, and the large one shows the large size of buried space, however, it was still very hard to connect the position of wire tip in Fig. 1 and that in Fig. 2.

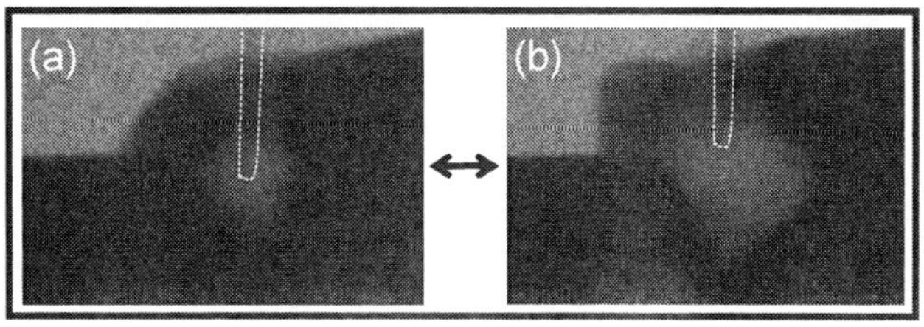

Fig. 2 Observation result of unstable buried arc taken by HBX-ray

Three metal transfer modes of buried arc at high current range

In order to clear the arc phenomenon on each current range, the relatively stable arc phenomena of buried arc were observed using HSV again. Fig. 3 shows the observation results of a wire feed rate of 30 m/min, 50 m/min, and 60 m/min respectively.

In the case of a wire feed rate of 30 m/min, the drop transfer mode was observed. And in the case of a wire feed rate of 60 m/min, the rotating transfer mode[4] was observed.

Welding condition	0ms	0.4ms	0.8ms	Transfer mode
Wire feed rate:30m/min Output current:450A Output voltage:39V	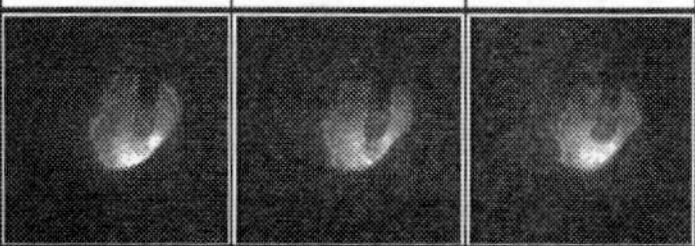			Drop
Wire feed rate:50m/min Output current:620A Output voltage:55V	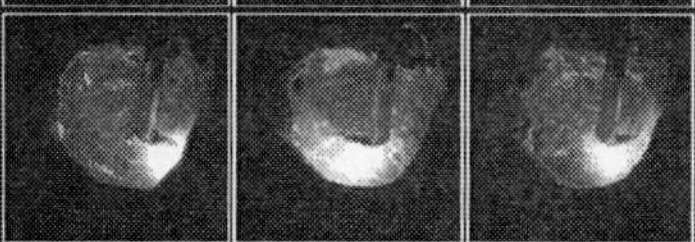			Pendulum
Wire feed rate:60m/min Output current:700A Output voltage:60V	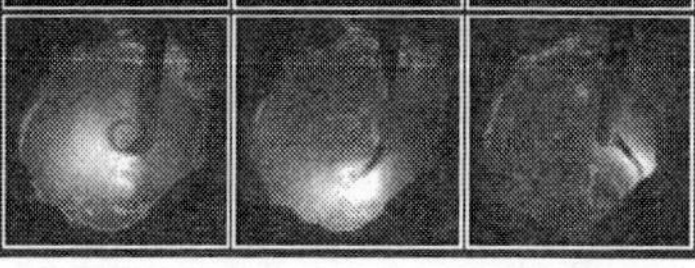			Rotating

Fig. 3 Metal transfer modes of buried arc at high current range

Additionally, in the case of a wire feed rate of 50 m/min, one kind of intermediate transfer was observed as follows. The column of liquid metal (CLM) is formed at the wire tip like the rotating transfer mode; however it swung in the same plane like a pendulum, which was rotating with the high speed. It is noted that this transfer mode is very similar to swinging spray transfer on MAG Welding[5]. Therefor the item of this transfer mode in Fig. 3 was described as "Pendulum". The spatter generation level and the size of opening were intermediate between the drop transfer mode and the rotating transfer mode.

Low frequency modulated voltage control

To stabilize high current buried arc, the low frequency modulated voltage control based on the external characteristic control has been developed. The current waveform can be controlled by the modulated voltage setting with a constant frequency to combine the drop transfer mode with the "Pendulum" or rotating transfer mode. Fig. 4 shows a typical waveform sample of the welding current and the arc voltage of this control scheme.

To investigate the effect of low frequency modulated voltage control, the arc phenomena taken by HSV were observed using a wire feed rate of 40 m/min, a welding current of 570A, a voltage setting of 48±10V, and its frequency of 100 Hz. The observation results are shown in Fig. 4 (A) to (E). The frame (A) shows the drop transfer mode at a low current range of 300A during the low voltage setting. As time advances the control phase shifts to the high voltage section. The frame (B) shows that the position of wire tip is getting high according to increasing of the wire melt rate, and the CLM is formed at the wire tip and the arc widely spreads, and the arc pressure pushes the bottom of buried space. The frame (C) shows the beginning of rotating with the unstable CLM. The frame (D) shows the final stage of high current section, which means at the end of high voltage setting. The small droplet is absorbed into the molten pool and the rotated arc irradiates the wall of buried space. Therefore it is considered that this arc pressure supports the wall of buried space and the opening of buried space can be prevented from becoming small and closed. The frame (E) shows the transient state from the high voltage setting to the low one. The size of arc is getting small as the current decreases, and the position of wire tip is getting low.

Consequently these arc phenomena were repeated stably and periodically and the stable buried space with deep penetration was always maintained without a large amount of scattered spatter. And as shown in Fig. 5, a bead appearance was improved by this low frequency modulated voltage control.

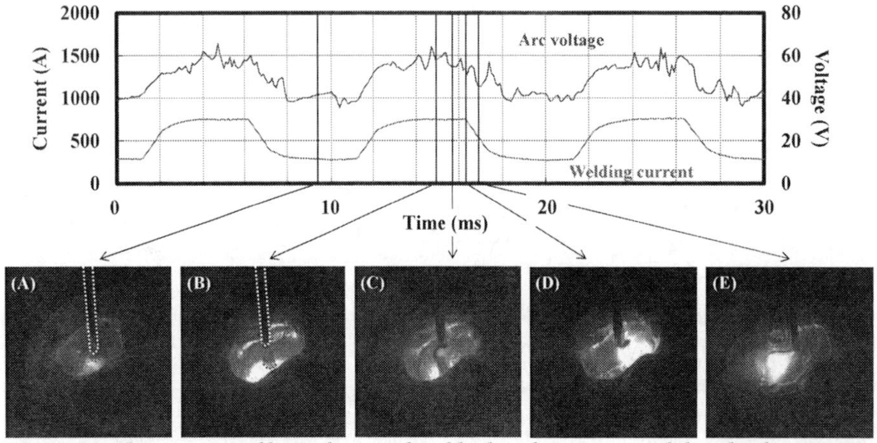

Fig. 4 Arc phenomenon of buried arc realized by low frequency modulated voltage control

Welding condition	Voltage control	Bead appearance (30mm)
Wire feed rate : 40m/min Ave. current : 570A Ave. voltage : 48V External characteristic : -10V/100A		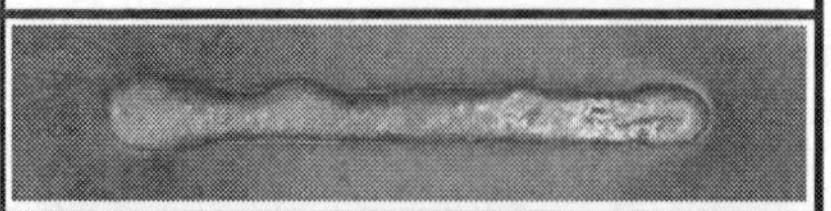
	±10V 100Hz	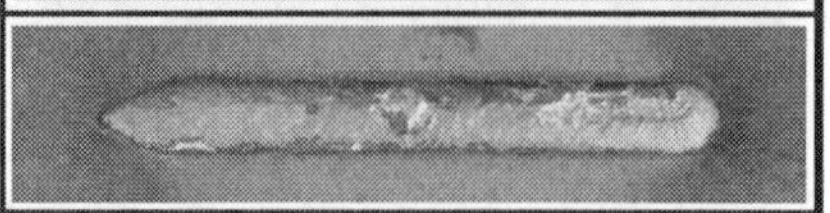

Fig. 5 Improvement effect of bead appearance realized by low frequency modulated voltage control

Single pass full penetration joining of heavy plate

Stabilized high current buried arc was applied to the welding of plate with 25 mm thickness. In order to control the amount of weld reinforcement and to prevent overlap, Y-groove whose angle is 90 degrees and root face is 16 mm was formed. Table 2 shows the welding condition. This experiment was carried out with two kinds of the gradient of external characteristic, which were -10V/100A and -20V/100A. The latter was better, that is, the gradient of external characteristic with smaller change of welding current against the arc voltage error is more efficient for stabilization of buried arc in groove weld, compared with bead on plate weld.

Fig. 6 shows the bead appearance and the cross section of weld bead with the gradient of external characteristic of -20V/100A. As shown in Fig. 6, single pass full penetration joining of plate with 25 mm thickness realized with a satisfactory bead appearance and no weld defects.

Table 2 Welding condition through welding of plate with 25 mm thickness

Base metal	**JIS SS400 (ASTM A36, EN S275)**	**Welding wire**	**JIS YGW11 (AWS ER 70S-G, EN G 42 3 C G0)**
CTWD	**25 mm**	**Shielding gas**	**100% CO_2**
Welding speed	**30 cm/min**	**Wire diameter**	**1.4 mm**

Welding condition	Bead appearance	Cross section
Ave. current : 720A Ave. voltage : 47V External characteristic : -20V/100A Voltage control : ±10V, 100Hz	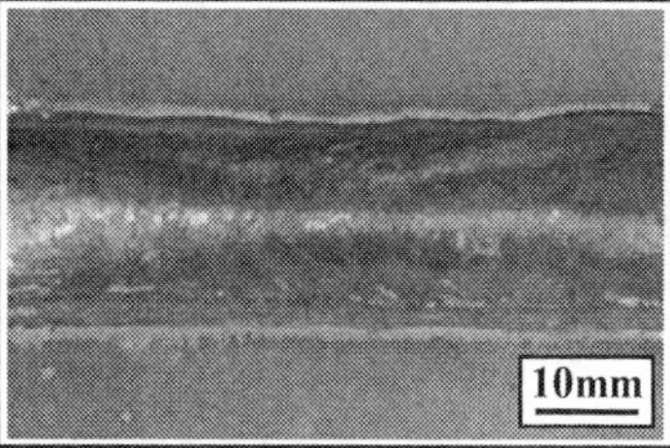	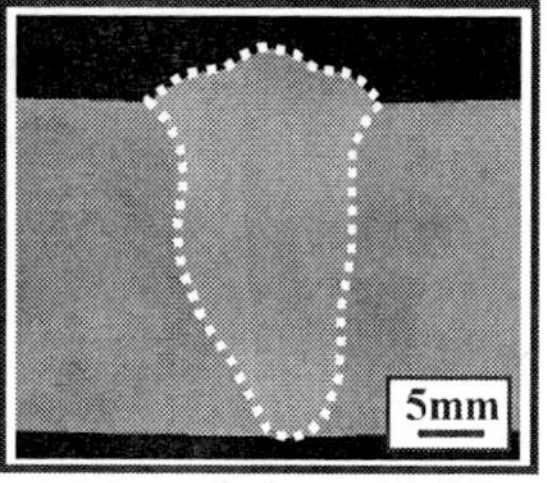

Fig. 6 Bead appearance and cross section in single pass full penetration joining of plate with 25 mm thickness

Conclusions

The given knowledge in this paper can be summarized as follows,

1) Three kinds of metal transfer at a current range of 450A or more are observed. In particular, one of these transfer modes is distinctive: that the droplet at the wire tip consists of the column of liquid metal and it swings in the same plane with rotating at the high frequency.

2) High current buried arc is stabilized with newly developed low frequency modulated voltage control. In this control, it is considered that the wall of buried space is supported periodically by arc pressure in "pendulum" or rotating transfer mode and the opening of buried space can be prevented from becoming small and closed.

3) Stabilized high current buried arc realizes single pass full penetration joining of plate with 25 mm thickness. The gradient of external characteristic with smaller change of welding current against the arc voltage error is more efficient for stabilization of buried arc in groove weld, compared with bead on plate weld.

References

[1] Noriyuki Muraoka, *et al*: "Effect of Power Output Modulation on Weld Profile in CO_2 Welding," Preprints of the National Meeting of JWS, Vol. 2015s (2015), pp. 134-135 (in Japanese).
[2] Hirotaka Adachi, *et al*: "Metal Transfer Phenomena in High-current MAG Welding," Preprints of the National Meeting of JWS, Vol. 2006f (2006), pp. 112-113 (in Japanese).
[3] Masatoshi Murayama, *et al*: "Narrow Gap Gas Metal Arc (GMA) Welding Technologies," JFE TECHNICAL REPORT, No. 20 (Mar. 2015), pp. 147-153.
[4] Katsuyuki Tsuji, *et al*: "Robot Welding Process for medium and Heavy Plate," KOBELCO TECHNOLOGY REVIEW, No. 32 (DEC. 2013), pp. 9-15.
[5] Masao Ushio, *et al*: "Effects of Shielding Gas Composition on Metal Transfer Phenomena in High Current GMA Welding," Trans. JWRI, Vol. 2, No. 1 (1993), pp. 7-12.

Analysis of Heat Distribution on Hollow Cathode Centered Negative Pressure Arc

Fan Jiang [1], Shujun Chen [1], Zhaoyang Yan [1], Ruiying Zhang [1], Yuming Zhang [2]

[1] *Beijing University of Technology, China*
[2] *Beijing University of Technology / University of Kentucky, China / USA*

As a new type of welding heat source, hollow cathode centered negative pressure arc (HCCNPA) is an arc discharge being ignited/maintained at a hollow cathode whose cavity is negatively pressurized. With these changes, the brightness of the cathode tip significantly increased, which considered as the effect of inner arc plasma on the hollow cathode. To better understand of this, the measurements of energy distribution between anode and cathode in HCCNPA were performed respectively by calorimetry calculation. Experiments were conducted to evaluate the influence of welding parameters on this heat distribution of HCCNPA, and the results were compared with the conventional GTAW process. It was found that the application of negative pressure in the hollow cathode greatly modifies the energy distribution between anode and cathode. Though the energy demand for maintaining a stable arc increases with applying negative pressure to hollow cathode, which reflect as the increase of arc voltage, the total energy generation by anode and cathode become lower. In particular, the energy in cathode increases and which in anode decreases as the negative pressure applied, and this differences increase as the current and arc length increase. Otherwise, the application of negative pressure in the hollow make the result arc constrained, so the HCCNPA has a more evenly distributed energy density in the center and a higher energy gradient at the boundary in comparison with the conventional GTA, which would enable the development of high quality applications.

GMA-Tandem-Arc-Splitting for Reducing the Dilution Rate in Cladding Processes

Johannes Schäfer, Uwe Reisgen, Konrad Willms
ISF-Welding and Joining Institute of RWTH Aachen University, Aachen, Germany
schaefer@isf.rwth-aachen.de, +49 241/80-96259

Abstract
The introduced Tandem-Arc-Splitting (**TAS**) welding-process describes a double-electrode GMA welding-process by using two power sources in a special electrical circuitry, which is resulting in the splitting of two arcs out of one welding wire. In contrast to the GMA-Tandem-Welding, where an interaction of both used arcs is given, the interaction of both arcs in this modification is prevented by the orientation of one arc used as a transferred arc to the base-material and one further arc between the two utilized wires of contrary polarity. While the process can be set in different geometrical orientations and angles between the electrodes and the base-material, different geometrical formed welding seams can be produced by the influence of the adjustable energy input in both arcs. Especially this advantage enables a purposeful regulation and control of the arcs with a resulting influence on the dilution rate by a simultaneous high deposition rate for cladding processes.

Introduction
Several welding processes are used for the cladding of metallic materials. Especially when thick metallic layers are required, the cladding in its various processes and variants is a technologically and economically required process step [1]. Applications for such produced layers, which are typically utilized for upgrading the surfaces on inexpensive steels with supporting functions, date back to various fields and industries.

Essential criteria for the quality of a surfaced or cladded layer with dissimilar filler material are the obtained mechanical properties and the effective thickness. This behavior is standing in a direct connection with the dilution rate of the base-material or rather the filler material (corresponding to the penetration). On the one hand, the dilution rate should be as small as possible in order to keep the dilution of the filler metal with the base-material on a low level, but on the other hand, it should be large enough to rule out welding errors like incomplete fusion so that a sufficient adhesion (bond) to the base-material is given.

This dilemma is at least only partially solved for arc processes with consumable electrodes since energy and additional material input are coupled directly with one another. Especially when thick surface layers are required, a high deposition must be provided which leads to an increased energy input into in the base-material and thus to a negative effect on the dilution. Nevertheless, the GMA-Welding with its different versions and variations is the most important method used for surfacing and coatings.

Process Description Tandem-Arc-Splitting
Double electrode processes are known for different arc-processes, while the GMA-Tandem-Welding is the most established and implemented process. It is characterized by the use of two power sources with two welding wire electrodes. While both wires are electropositive and the related arcs are placed close together for the use in a common weld pool, an interaction takes place due to the arc blow. In order to prevent this interaction, a phase shift of 180° is used for pulsed arcs while just thinner electrodes can be implemented due to the interaction.

Irrespective of this welding process, a special electric circuitry enables the forming of two arcs without an interaction and without a necessary phase shift (Fig. 1). Thereby, a transferred arc is oriented from a positive wire electrode on the negative base-material and a non-transferred arc is burning between the positive electrode and a second electrode with a negative potential. The idea of this circuitry was first mentioned in [2] and compared to other double electrode GMA processes. Further experiments with that circuitry were carried out by [3] for reducing the heat input in the base-material by using the electro-gas welding-process.

Therefore, actual research about this circuitry is carried out by the authors in order to transfer and qualify a welding process for cladding and surfacing applications. By pursuing this ideas further, the name "**T**andem-**A**rc-**S**plitting (**TAS**)" is used since two power sources are implemented and a splitting of the arc is given in one welding wire.

Thereby, each arc of the TAS process can be adjusted by voltage and current respectively welding wire speed so that the deposition rates of both and the common dilution rate can be adjusted. In particular, the advantage for the use for cladding processes and applications is given by a low possible energy input by the transferred arc, so that this arc is just used for preheating the base-material and delivering a small part of the deposition rate. The additional non-transferred arc between both electrodes is put on a higher electrical capacity for producing a higher deposition rate. Thereby, the used voltage and current is selected and adjusted to each other in both circuits with a resulting high deposition rate and an adjustable dilution rate for the desired application in which the base-material should be influenced as less as possible.

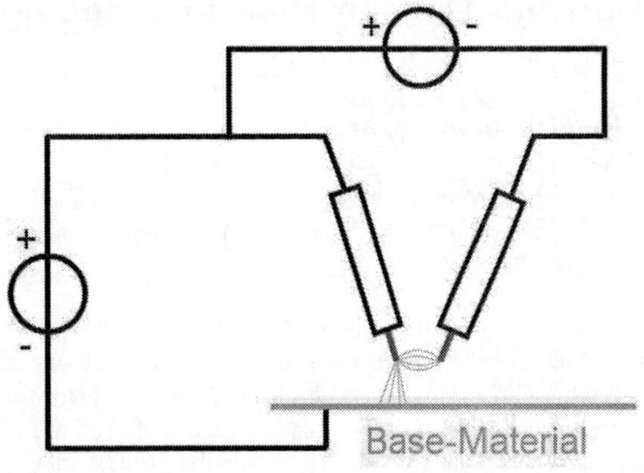

Figure 1: Electrical circuitry of Tandem-Arc-Splitting

Arc and Droplet Behavior

The TAS process enables a wide geometrical range by different positions of both welding wire electrodes. It is, on the one hand possible to vary the angle between the wires and the two angles between the electrodes and the base-material. On the other hand, the orientation of both electrodes in welding direction or perpendicular to the welding direction is possible. From these two orientation possibilites, three weldings with different boundary conditions are illustrated in Table 1 as examples.

Table 1: Experimental choice of boundary conditions (voltage and current measured as mean values; index 1 = transferred arc; index 2 = non-transferred arc)

No.	wire type	wire polarity with diameter	shielding gas	weld. speed [m/min]	U_1 [V]	I_1 [A]	U_2 [V]	I_2 [A]
1	solid wire ER70S-6	+ 1,2 mm - 1,2 mm	82% Ar 18% CO_2	0,5	29,2	42	23,2	226
2	solid wire ER70S-6	+ 1,6 mm - 1,6 mm	82% Ar 18% CO_2	0,5	29,1	144	27,45	181
3	cored wire ISO 14343-A:G 17	+ 1,2 mm - 1,6 mm	97,5% Ar 2,5% CO_2	0,5	27,5	87	27,2	190

The use of solid wire illustrates two possible geometrical orientations of the two used torches/wires perpendicular to the welding direction (Example No. 1 and Fig. 2) and in welding direction (Example No. 2 and Fig. 3). Both examples also clarify the usage of different welding wire diameters, where a combination of 1,2 mm (positive polarity) and 1,6 mm (negative polarity) is also possible. Disregarding the wire diameters, the angles between the wires and the base-material are changeable within a range in which the electrode to electrode angle stays in a range of between 40-100° and the angle to the base-material in both ways is just limited by the peripheral devices. More precisely, peripheral devices are given by the already welded seam in which the electrode with the negative polarity in the position of Fig. 3 will dip into the liquid weld metal with its underside, when this electrode is not oriented before the positive electrode.

For the droplet detachment and forming by using solid wires, a different behavior can be observed. Of the shown example No. 1 (Fig. 2), a continuous chain of small droplets is formed by the transferred arc to the base-material with a current of just 42 A while bigger droplets are formed by a higher current choice of 226 A in the circuitry of the non-transferred arc. Due to the arc pressure of the transferred arc which is oriented around the axis of the electrode with the positive polarity, the free falling droplets formed of the electrode with the negative polarity are oriented in parallel to the droplet chain of the transferred arc. This causes the laterally shifted formation of the already welded seam oriented in the background of Fig. 2.

Disregarding these observations, those of example No. 2 (Fig. 3) are contrary in a geometrical change of the wires. While big droplets are formed of the wire with the positive polarity and thrown out sideways into the liquid weld pool, the melting of the wire with the negative polarity is in a continuous flow oriented in the front of the weld pool. The selected parameters in the two electrical circuits are chosen as well as a split, in which a higher current is adjusted for the non-transferred arc.

The last introduced example No. 3 (Fig. 4) which was made with cored wires differs completely in welding behavior and droplet detachment. The current for the transferred arc has to be increased compared to solid wire examples with the adjustment of the non-transferred arc, which is used with a lower current (compared to solid wire of the same diameter) for preventing a burn off of the wire-containing elements. The corresponding droplet detachment results in nearly equal diameters of both droplets, which differs from the previously mentioned results of solid wires.

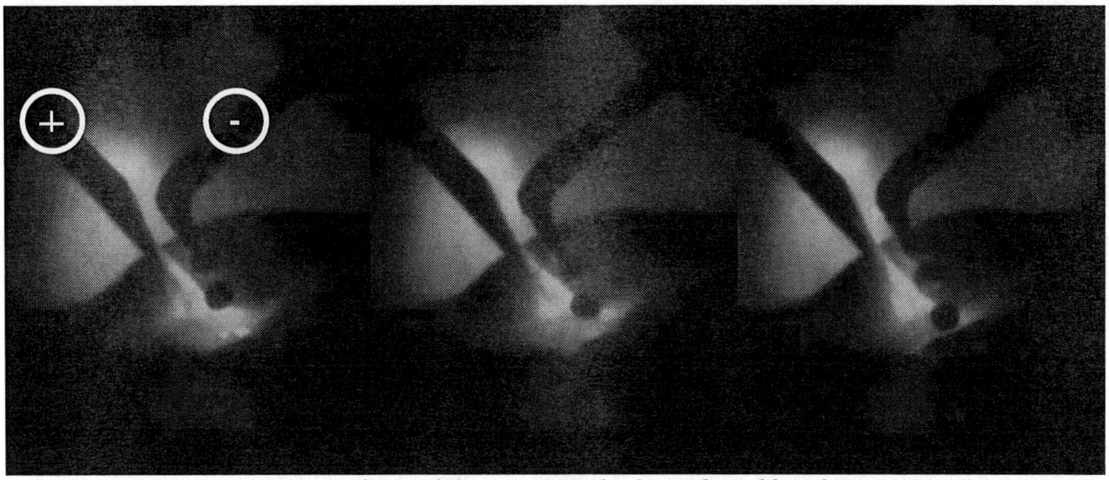

Figure 2: HS-Section; orientation of two solid wires perpendicular to the welding direction (No. 1)

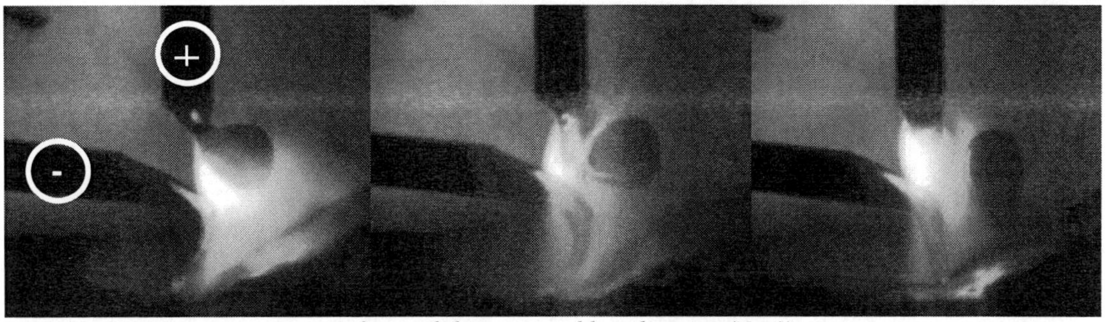

Figure 3: HS-Section; orientation of two solid wires in welding direction (No. 2)

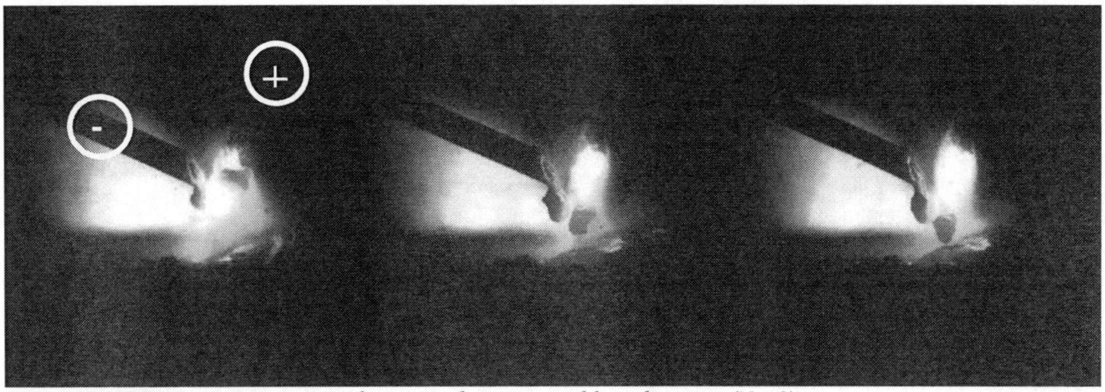

Figure 4: HS-Section; orientation of two cored wires in welding direction (No. 3)

Weld Seam Forming

Besides a steady stable and smooth welding process, the welds are always consistent without any porosity and a constant penetration (Fig. 5, Fig. 6). In particular, the main advantage of a penetration regulation is given by keeping high deposition rates > 10 kg/h. Thereby, the dilution rate is out of changeable penetration adjustable and reducible to nearly 1% and depending on respective applications increasable. The excess weld metal is as well adjustable by the settings of both arcs and the corresponding welding speed. The given examples with a dilution rate of 1% (Fig. 5 left) is resulting in a width of 12 mm and a height of 5 mm, while the dilution rate under the same geometrical setting is changeable to 16% (Fig. 5 right) with a height of 5,5 mm and a width of 10 mm. By using oscillating techniques (Fig. 6) which are often used for cladding and surfacing applications, the width is increasable up to 18 mm by keeping a height of 4,5 mm (welding speed 0,4 m/min, frequency 1,2 Hz). The resulting penetration is also adjustable and reducible (Fig. 6 right) as it is the case for seams without oscillation.

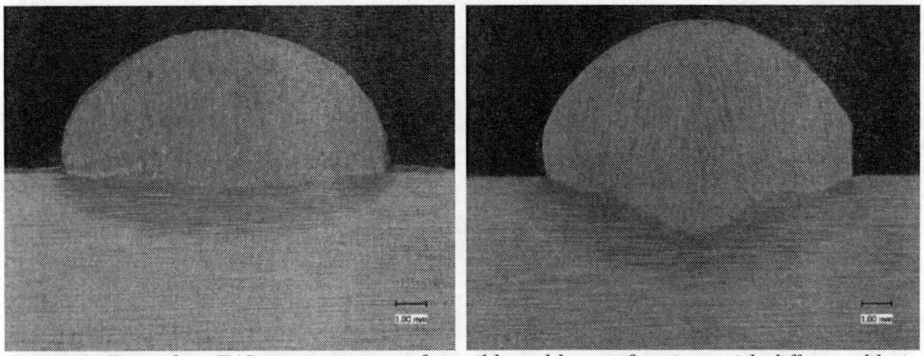

Figure 5: Exemplary TAS macrosections of possible weld seam formings with different dilution rates

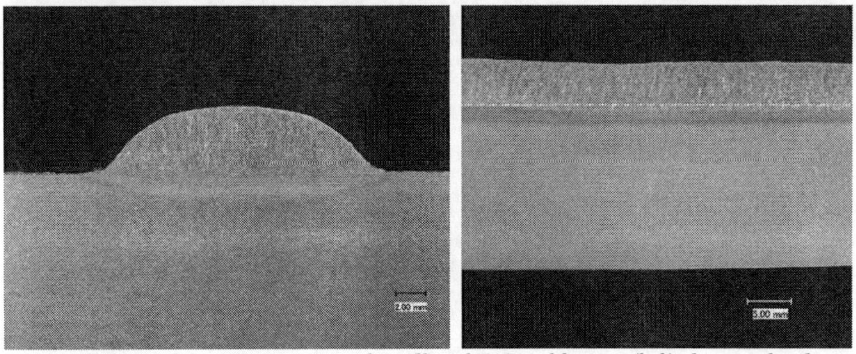

Figure 6: Exemplary macrosection of oscillated TAS weld seam (left); longitudinal penetration (right)

Conclusions

The TAS welding-process provides an opportunity for cladding and even surfacing applications by using solid and cored wires in which, for the latter, the possibility for a selective chemical weld adjustment by the wire composition is given. By the use of two arcs which are not influencing each other as it is the case for GMA-Tandem-Welding, a higher deposition rate can be achieved by resigning pulsed arcs with a phase shift and using higher welding wire feed speed. While the droplet detachment for solid wire is changeable due to the geometrical orientation and the electric setting, the droplet detachment for cored wires is as well different with the forming of nearly equal droplet diameters out of each electrode. The resulting weld seams are formed consistently without any porosity and high excess weld metal for wear resistant layers, so that the use of the TAS circuitry provides advantages compared to conventional GMA-Tandem-Welding.

Acknowledgment

This research work is supported by "Industrielle Gemeinschaftsforschung IGF/AiF" with the reference number 17.885 N. The authors would like to express their thanks for the support.

References

[1] A. Gebert, *et al*: Moderne Beschichtungsverfahren, WILEY-VCH (Weinheim, 2005), pp. 288-298.

[2] K.H. Li, *et al*: "Consumable Double-Electrode GMAW-Part1: the Process", *Welding Journal*, Vol. 87, (2008), pp. 11s-17s.

[3] U. Reisgen, *et al*: "Reduzierung des Wärmeeintrags beim Elektrogasschweißen durch modifizierte Zweidrahttechnik", *Schweissen und Schneiden*, Vol. 64, (2012), pp. 582-587.

Microstructural aspects of hydrogen embrittlement for low carbon FCA steel welds applied in cathodic protection and saltwater environment

Hanji Park, Namhyun Kang
Department of Materials Science and Engineering, Pusan National University, Busan, Korea
hjpark89@pusan.ac.kr, nhkang@pusan.ac.kr

Stephen Liu
Colorado School of Mines, Colorado, USA
sliu@mines.edu

Abstract
Needs of offshore platform have been grown consistently with the increase of worldwide energy demand. Because the small defects may lead to a serious accident in offshore plants, it is necessary to produce and maintain the high quality of materials and weldments. In addition, because of harsh installation environment, particularly saltwater environment that has great risk of corrosion, the need of the cathodic protection method has increased. However such cathodic protection method electrolyzes saltwater into hydrogen atoms at the interface between solution and the structure surface. And some of the hydrogen atoms are absorbed into steel. Hydrogen inside the material causes hydrogen embrittlement (HE), which is sensitive particularly in weldment and/or heat affected zone due to hardened microstructure and residual stress. It is caused by the interaction of environment in which the hydrogen is caused, vulnerable microstructure, and tensile residual stress. Atomic hydrogen is diffused to stress concentration regions occurred at the tip of defects, therefore aggravating the crack propagation to induce brittle fracture. For the reason, it is necessary to understand the hydrogen embrittlement of the welds caused by salt water environments electrolysis. Some studies on HE sensitivity of weld metal or heat-affected zone have been conducted. However, there were only few studies on which factors are critical to HE for the actual welded specimen composed of base metal, heat affected zone (HAZ), and the weldment. This study investigates the role of weld and HAZ microstructures on HE and critical trapping sites of diffusible hydrogen for low carbon FCA steel welds. HE fractography and fracture behavior are analysed to investigate the role of weld and HAZ microstructures on HE. Finally, the optimized condition of cathodic protection will be proposed to inhibit HE for low carbon FCA steel welds in saltwater environment.

Introduction

Offshore platform steels need the cathodic protection or sacrificial anode method to inhibit corrosion. Many studies have investigated the proper cathodic potential for general steel materials, e.g., -0.70V ~ -1.05V vs. Ag/AgCl for low carbon steels in saltwater [1-3]. However, the preferential corrosion behavior of carbon steel weldments has been investigated, and is dependent on a number of factors such as slightly different composition or stress concentration at the weld toe, or contained impurities that make them anodic to nearby base metal [4]. For the reason, the optimum cathodic potential for weldments should be different with that for base metal.

Hydrogen embrittlement (HE) is known to be caused by the interaction of environment in which the hydrogen is caused, vulnerable microstructure, and tensile residual stress. Atomic hydrogen inside the material diffused to the stress concentration regions occurred at the tip of defects and let the crack propagate to induce brittle fracture. So, the hydrogen embrittlement is more sensitive particularly in weldment or heat affected zone (HAZ) due to the hardened microstructure and residual stress.

Many studies have investigated HE properties of carbon steel, weldments and HAZ, respectively. However, there were only few studies on the critical factors of HE for the actual welded specimen consisting of base metal, HAZ, and the weldment. Therefore, the HE properties were investigated in a low carbon steel flux cored arc (FCA) welds. Cathodic protection conditions were applied using in-situ hydrogen charging slow strain rate test (SSRT) on actual welded specimens to evaluate the effects of welded microstructures on HE of the steel welds.

Experimental Procedure

Specimens used in the study were EH36 steel used commonly for offshore structures. The chemical compositions of the steel in mass% were 0.08 C, 0.31 Si, 0.15 Mn, <0.03 Cr, Ni, Nb, Ti, and balanced Fe. Welded specimens were produced by 3mm thickness butt plate, one-pass welding using FCA welding process with AWS A5.29 E81T1-k2c filler metal. Two types of heat input (6kJ/cm and 11.5kJ/cm) was designed. Microstructures and fractography were observed by optical microscopy (OM), and scanning electron microscopy with a field emission gun (FE-SEM). Polarization test was conducted for 0.01V/s scan rate in artificial seawater (ASTM D 1141-98). Working electrode, counter electrode, and reference electrode were specimen, platinum plate, and Ag/AgCl, respectively. Tensile specimens for slow strain rate tests (SSRT) were produced following ASTM E8 subsize specimen without notch. After 24hr pre-charging of hydrogen, in-situ SSRT was conducted with hydrogen charging applying constant cathodic potential -0.75V and -1V vs. Ag/AgCl in artificial seawater (ASTM D 1141-98). The crosshead speed was 0.0015 mm/min, corresponding to a nominal strain rate of 10^{-6}/s.

Results and discussion

Welded specimens have the typical weld microstructure of low carbon steel. Fig. 1a shows the equilibrium phase diagram for the steel, and the numbers ① to ⑤ indicate peak temperatures of each welding positions. Fig. 1b shows microstructures for the each positions in weldments. Position ① indicates the base metal (BM) consisting of polygonal ferrite (PF) and pearlite (P). Position ② is intercritical HAZ (ICHAZ) of PF and degenerated pearlite (DP). Position ③ shows fine grained HAZ (FGHAZ) constituting recrystallized PF and DP. Position ④ indicates coarse grained HAZ (CGHAZ) occupied with bainitic ferrite (BF). Position ⑤ is weldments (WM) having grain boundary ferrite (GBF) and acicular ferrite (AF).

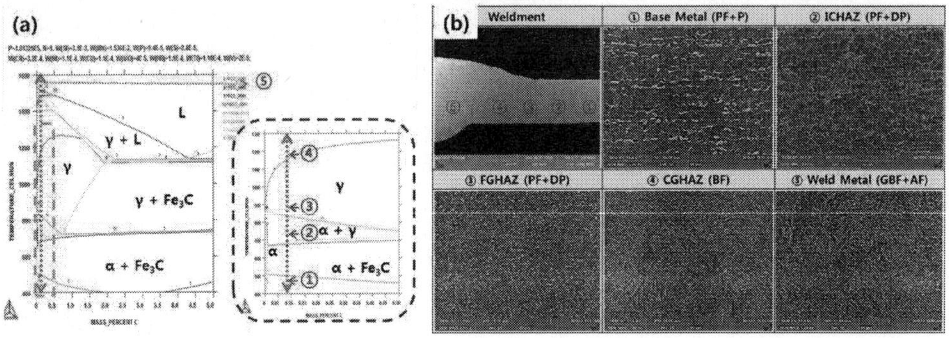

Fig. 1 (a) Phase diagram for the steel. (b) SEM microstructures (x1,000) of each positions in weldments.

Fig. 2 shows polarization curves for BM and each positions of HAZ. The optimized cathodic protection potential of low carbon steel was -0.7~-0.8V vs. Ag/AgCl as reported in earlier study [3]. Weldment showed this potential range of -1V vs. Ag/AgCl. Therefore, BM and weldments indicated the optimized cathodic protection potential is different. Because the current density in cathodic region may be thought as hydrogen evolution rate at interface between specimen and solution, hydrogen evolution at BM and weldments is different for the applied cathodic protection potential. For example, the potential approximately at -0.8V has faster hydrogen evolution rate in weldments than that in BM, and that approximately at -1.2V shows faster hydrogen evolution rate in BM interface than that in weldment interface. So to consider only about effect of microstructural factor, applied cathodic potential is set as -1.0V vs. Ag/AgCl that has approximately same current density for each weld location.

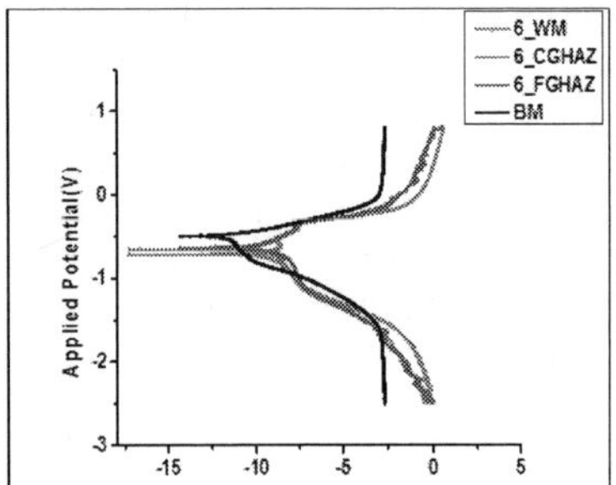

Fig. 2 Polarization curves for base metal and each weld location.

Table 1 shows a variation of SSRT properties before and after charging hydrogen. The hydrogen embrittlement index, E_I is the variation rate of reduction of area using the following equation:

$$E_I = \frac{RoA_{in\ air} - RoA_H}{RoA_{in\ air}} \times 100\%$$

The variation of yield strength and tensile strength are very small. On the other hand, the changes of El and E_I are important to evaluate hydrogen embrittlement sensitivity. Therefore, these changes mean that welded specimens show very high sensitivity particularly when -1V vs. Ag/AgCl is applied. Hydrogen embrittlement resistivity decreased with welds microstructure and with large cathodic potential.

Table 1. The ratio of tensile properties changed with hydrogen charging.

		ΔY.S(%)	ΔT.S(%)	ΔEl(%)	E_I(%)
-1V vs. Ag/AgCl	BM	0.9	1.0	-33.5	46.8
	6kJ/cm	6.0	3.7	-51.4	83.3
	11.5kJ/cm	-3.1	4.8	-48.8	83.1
-0.75V vs. Ag/AgCl	11.5kJ/cm	-11.2	-2.1	-40.3	65.2

Fig. 3 shows different fracture location comparing without hydrogen. Without hydrogen, fracture occurred at ① BM on both BM and welded specimens. With hydrogen, fracture occurred at ② ICHAZ on only welded specimens applied -1V vs. Ag/AgCl. This means the welds microstructure of fracture occurred has large effect in hydrogen embrittlement. And (i-j) show along the pearlite and degenerated pearlite secondary crack occurred and then propagated across applied stress direction.

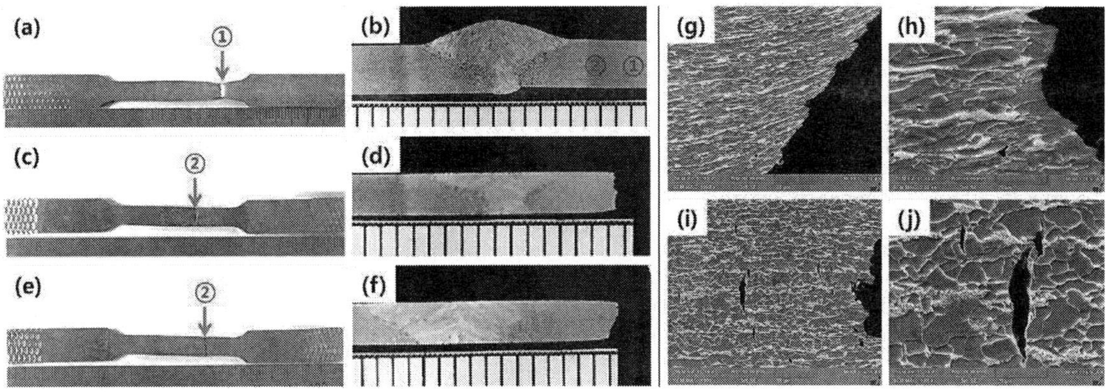

Fig. 3 (a), (c), (e) Fracture location for 6kJ/cm_air, 6kJ/cm_H, 11.5kJ/cm_H, respectively; (b) weldment for 6kJ/cm, the numbers indicate the fracture locations applied to Figs. 3a, 3c, 3e; (d), (f) cross section of fracture surface for 6kJ/cm_H, 11.5kJ/cm_H, respectively; (g), (h) SEM images of cross sections for BM_H, x1,000 and x 3,000 magnification; (i), (j) SEM images of cross sections for 6kJ/cm_air, x 1,000 and x 3,000 magnification.

Fig. 4 shows fracture surface before and after hydrogen charging. Figs. 4a-4c are low magnification fractographs without hydrogen charging, Figs. 4d-4f are low magnification and Figs. 4g-4i are high magnification fractographs with hydrogen charging. Fractographs without hydrogen show lots of necking occurred and consist of dimple fracture only. However, fractographs with hydrogen exhibits no necking and show quasi-cleavage (QC) fracture morphology. Specifically, welded specimens didn't show any dimples and the appearances of QC was more abundant than BM.

HE fractography and fracture behavior were analyzed. Fracture locations were to base metal in air and -0.75V vs. Ag/AgCl cathodic potential. Fracture locations were changed from base metal to ICHAZ by applying -1V vs. Ag/AgCl cathodic potential. At the same time, fracture modes also changed from dimple fracture to QC fracture. This was because the brittle granular cementite in ICHAZ acted more sensitively on HE when a cathodic potential under the specific potential applied.

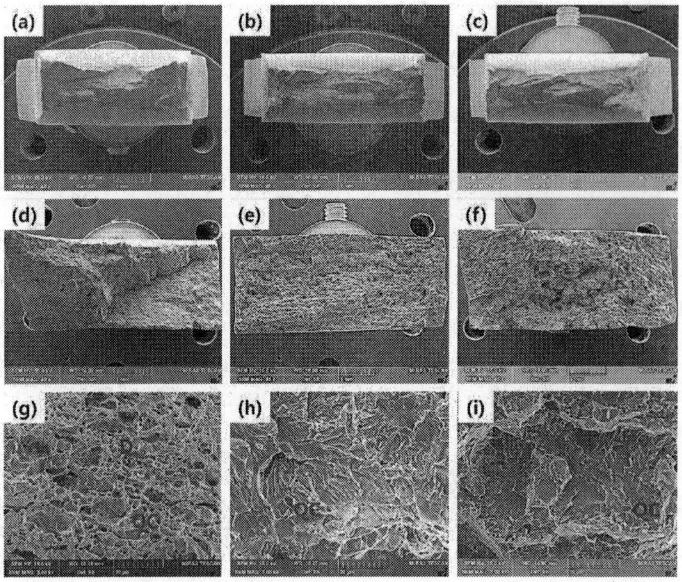

Fig. 4 SEM images for fracture surfaces: (a), (b), (c) low magnification fracture surfaces for BM_air, 6kJ/cm_air, 11.5kJ/cm_air; (d), (e), (f) low magnification fracture surfaces for BM_H, 6kJ/cm_H, 11.5kJ/cm_H; (g), (h), (i) high magnification (x 3,000) fracture surfaces for BM_H, 6kJ/cm_H, 11.5kJ/cm_H, respectively.

Conclusions

The effect of HAZ microstructures on HE were investigated in a low carbon FCA steel welds by cathodic protection. In-situ hydrogen charging slow strain rate tests (SSRT) were conducted on smooth whole weldment specimens to evaluate the effects of welded microstructures on HE of the steel welds.

1) Hydrogen embrittlement sensitivity increased with the existence of weldment.
2) Hydrogen charging induced fracture at ICHAZ. The granular cementite of degenerated pearlite in ICHAZ may act as crack initiation site, and then propagated into ferrite grain.
3) As applied potential was -0.75V vs. Ag/AgCl, the fracture occurred at BM and E_I value was much lower than E_I value as applied potential was -1V vs. Ag/AgCl. Therefore, optimum cathodic protection conditions should be considered not only in corrosion resistance but also in HE resistance.

Acknowledgements

This work was supported by the National Research Foundation of Korea (NRF) grant funded by the Korea government(MSIP) through GCRC-SOP (No. 2011-0030013).

References

[1] BS EN 12495-2000 cathodic protection for fixed steel offshore structures, pp. 9.
[2] Recommended Practice DNV-RP-B401, October 2010, CATHODIC PROTECTION DESIGN, pp. 11.
[3] D.P Hong: "study on cathodic protection of stainless steel" (2002), pp. 46-59.
[4] J.R.Davis: Corrosion of Weldments, ASM International (2006), pp. 19-23.

Evaluation of Pitting Corrosion Resistance in the Weld Joint of a Duplex Stainless Steel by Micro Electrochemical Cell Technique

Masato Sasa, Fu-Gao Wei, Kun Wang
Nippon Yakin Kogyo Co., Ltd., Technical Research Center, Kawasaki, Kanagawa, JAPAN
Email:masato.sasa@nyk.jp, telephone: +81-44-271-3361.

Abstract
The local pitting corrosion resistance in the weld joint of a duplex stainless steel UNS32205 was studied by means of micro electrochemical cell with a heating control unit. Measurement was performed in a 20% NaCl aqueous solution at 65±1℃. The result showed that pitting potential decreases from base metal towards the fusion line and HAZ demonstrated the lowest pitting potential. Pit was found to initiate with α phase. Decrease of pitting potential in HAZ can be explained.

Introduction
Micro electrochemical cell technique has been applied to evaluate the corrosion resistance of individual point in a heterogeneous material such as weld joint [1]. However, in the case of high corrosion resistant stainless steel it is difficult to get pitting potential at room temperature since pitting doesn't occur during anodic polarization. This study aims to evaluate pitting potential in the weld joint of a duplex stainless at a high temperature.

Experimental Procedure
The chemical compositions of the duplex stainless steel UNS 32205 and wire are shown in Table 1. Weldment was prepared using automatic GMAW procedure with Ar - 5% CO_2 gas. Fig. 1 illustrates schematically the micro electrochemical measurement system [2]. This system was equipped with a heating control unit that enables specimen and aqueous solution to be heated to high temperature. Measurement was performed in a 20% NaCl aqueous solution at 65±1℃. The surface was polished with 1μm diamond paste. The micro capillary used in this study was able to measure the pitting potential of an area of about 600μm diameter.

Table 1. Chemical compositions of the duplex stainless steel for study (wt%).

	C	Si	Mn	P	S	Ni	Cr	Mo	N	PRE
UNS 32205 (Base material)	0.014	0.30	0.97	0.028	0.000	5.67	22.41	3.16	0.17	35.6
ER2209 (Wire)	0.014	0.41	1.69	0.017	0.001	8.72	22.59	3.15	0.17	35.7

PRE=Cr+3.3×Mo+16×N

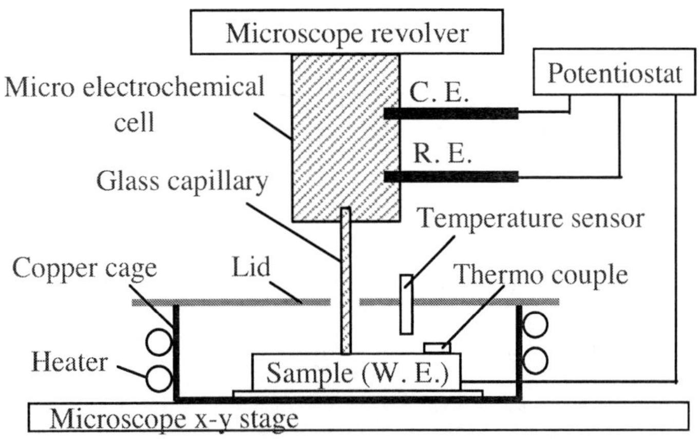

Fig. 1. Schematic illustration of the micro electrochemical cell system with a temperature controler.

Result and discussion

The change in pitting potential from the fusion line to the base metal is shown in Fig. 2. It was found that pitting potential decreases from base metal towards fusion line and the lowest pitting potential is located at HAZ. Fig. 3 shows a pit 1.5 mm away from the fusion line after test. Pitting occurred at the interface between inclusion and α phase.

Thermo-calc calculation of the change in PRE value with increasing temperature revealed that the PRE value of α phase is lower than that of γ phase in the temperature range of HAZ and the PRE value of α phase decreases with increasing temperature. Decrease of pitting potential in HAZ can be explained by the decrease of PRE value in α phase.

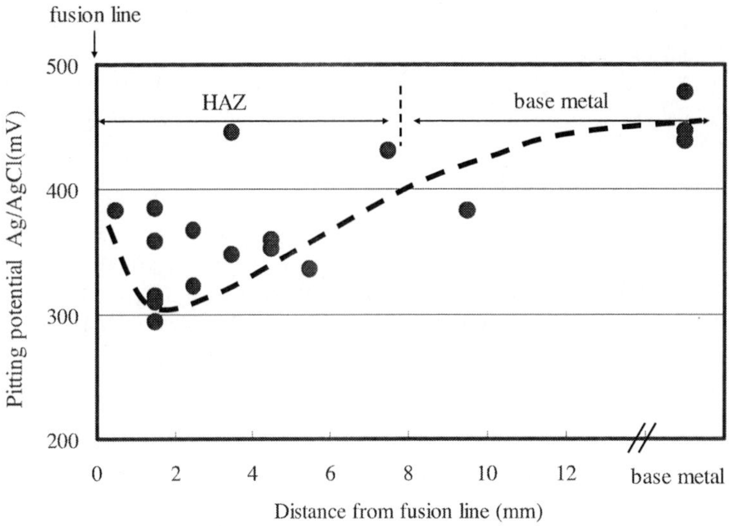

Fig. 2. The change in pitting potential from the fusion line to the base metal.

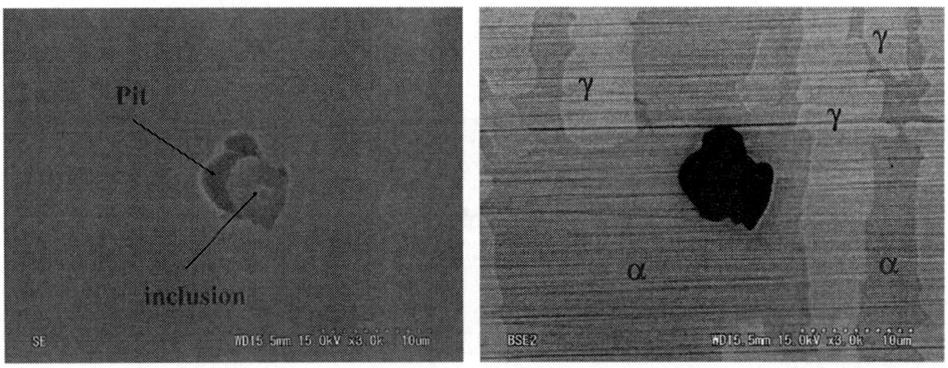

Fig. 3. SEM images of a pit 1.5 mm away from the fusion line after pitting potential measurement.
Left: secondary electron image, Right: backscattered electron image

Conclusions

The change in pitting potential from the fusion line to the base metal was shown. Pitting potential decreased from base metal towards the fusion line and HAZ showed a lowest pitting potential. Decrease in the PRE value of α phase is considered to be responsible for the change in pitting potential. It is clarified that pitting occurred at the interface between inclusion and α phase

References

[1] C.Garcia, *et al*: "Pitting corrosion of welded joints of austenitic stainless steels studied by using an electrochemical minicell," Corrosion science 50 (2008), pp. 1184-1194.

[2] T.Takei: "Pitting corrosion resistance at local region of a high corrosion resistance alloy by using the microelectrochemical system with temperature control function," The 172nd ISIJ Meeting (2016)

Precision Microanalysis of B toward actualization of B doped Al-Si-Mn weathering steel

MEGURO Susumu, KIMURA Takashi, NAKAMURA Terumi, QIU Hai, KAWADA Satoshi
National Institute for Materials Science, Tsukuba-shi, Ibaraki, Japan
Email: MEGURO.Susumu@nims.go.jp

TSUZAKI Kaneaki
Kyushu University, Fukuoka-shi, Fukuoka, Japan

Abstract

Al-Si-Mn weathering steel is a promising material for long-life social infrastructure. This steel consists of ubiquitous elements, while it shows excellent weathering ability. Toughness at HAZ bond area in weld joint of the steel changes poor after welding, but doping of B at ppm-level to the steel solves the problem. In this study, we examined effective range of B concentration to improve toughness at HAZ bond area in the steel. Charpy impact absorbed energy increases with B concentration, but the effect saturates over several tens of ppm. This result shows precision quantitative microanalysis of B is indispensable. Various wet chemical microanalyses are prescribed for B in steels. But dissolution of specimen during the processes is very complicated and looses information on distribution of B at microstructure in the steel. So we applied laser ablation ICP mass spectrometry (LA-ICP-MS) equipment to precision quantitative microanalysis of B without dissolution of the steel specimen. The values of B ion intensity by LA-ICP-MS are correlated closely with B concentration by wet chemical microanalysis. This excellent correlation suggests LA-ICP-MS is a promising method for quantitative microanalysis of B in Al-Si-Mn weathering steel.

Introduction

Al-Si-Mn weathering steel is a promising material for long-life social infrastructure. Conventional weathering steels contain high-priced Ni, Cr and rare elements as major raw materials. So National Institute for Materials Science (NIMS) has developed a new steel to show excellent weathering ability without using these high-priced and rare elements [1]. This Al-Si-Mn weathering steel consists of ubiquitous elements, while it shows weathering ability equal to 1% Ni weathering steel. When welding of this steel, precipitation of ferrite at HAZ bond area in weld joint reduces toughness. But the authors found that doping of B at ppm-level to the steel can improve the toughness. Charpy impact absorbed energy increases with B concentration, but the effect saturates over several tens of ppm [1]. This result shows precision quantitative microanalysis of B is indispensable from the view of quality control of the steel product.

On the other hand, it is becoming clear that doping of B in steel prevents phase transformation at grain boundaries from the result of long years of research [2][3][4]. But some details of this behavior are not yet clear. So precision surface microanalysis is necessary to investigate the relationship between B content and microstructure. ICP atomic emission spectrometry (ICP-AES) and spectrophotometric curcumin method are generally used to microanalysis of B in steels, but these wet chemical analyses include complicated dissolution process of a specimen that loose information on distribution of elements. In this study, we applied laser ablation ICP mass spectrometry (LA-ICP-MS) to precision quantitative microanalysis of B on a surface of the steel. We manufactured B doped Al-Si-Mn weathering steels and showed the relationship between B content and Charpy impact absorbed energy of the steel. Moreover, we applied LA-ICP-MS to quantitative microanalysis of B in the steel and tried to generate calibration curve of B for the steel.

Experimental procedure

Table 1 shows composition of Al-Si-Mn weathering steel. Composition of B is varied from 0 to 58 ppm. Ingots of the steels are manufactured by vacuum melting. Upset forging and solution treatment are carried out to the ingots for forming fine-grained structure and dissolution of precipitations in the ingots. Specimens are formed by groove rolling after these mechanical process and thermal treatment.

Simulated HAZ structure for measurement of Charpy absorbed energy was formed by simulated thermal cycle. The maximum temperature of the process was 1623 K. Charpy impact test was carried out at room temperature. Wet chemical analysis certified by JIS (Japanese Industrial Standards) was carried out to determine accurate B content in the specimens. Spectrophotometric curcumin method is applied to specimens under 20 ppm B doping.

ICP-AES is applied to the rest.

Laser ablation ICP mass spectrometry (LA-ICP-MS) consists of femtosecond pulsed UV Laser system and quadrupole mass spectrometer. This laser system has galvano mirror that enables ablation at any position on a surface of a specimen. Ablation of the pulse laser vaporize a surface in very short period without selective vaporization. This vaporized gas is introduced into quadrupole mass spectrometer with carrier gas. In this study, laser ablation area is about 0.5 mm diameter with spiral scanning of laser and we measured the ion intensity of ^{11}B in the specimens. Moreover, ICP-MS and GD-MS were carried out as comparative experiments by other microanalysis using mass spectrometry.

Table 1: Chemical composition of Al-Si-Mn weathering steel (mass % and mass ppm).

Fe	C	Al	Si	Mn	B
bal.	0.16 %	0.6 %	0.6 %	1.5 %	0-58 ppm

Results and discussion
Charpy impact test

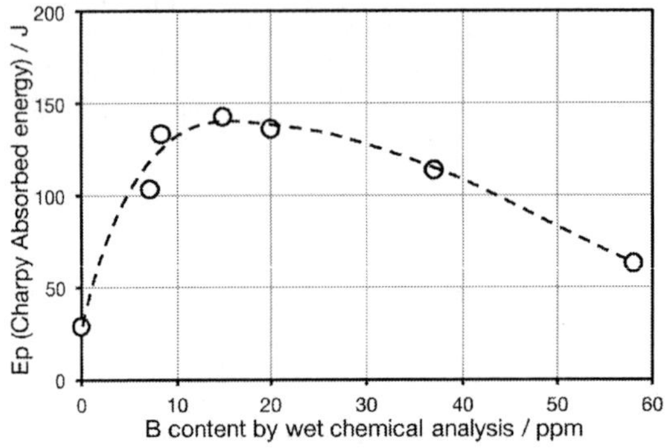

Fig. 1: Relationship between B content and Charpy absorbed energy at simulated HAZ bond area in Al-Si-Mn weathering steel.

Fig. 1 shows the relationship between B content and Charpy absorbed energy at simulated HAZ bond area in Al-Si-Mn weathering steel. Wet chemical analysis is applied to determine B content. Absorbed energy increases with B content and reaches maximum value at about 8 ppm. On the other hand, absorbed energy decreases when B content exceeds 20 ppm.

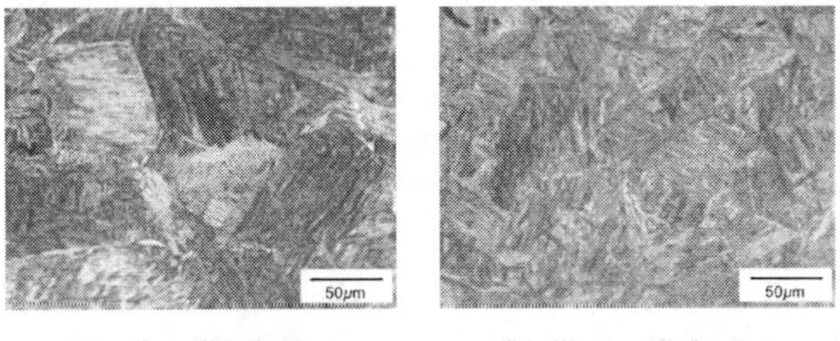

Fig. 2: Optical microscope image of cross setction of HAZ bond area in Al-Si-Mn weathering steel. (a) Without B doping, (b) 15 ppm B doping.

Fig. 2 shows the microstructure of the cross section at simulated HAZ bond area in the specimens. Ferrite structures precipitate around boundaries of former austenite grains during cooling process of welding and reduce toughness of the specimen without B doping. On the other hand, the precipitation is prevented in the specimen with 15 ppm B doping. These results show doping of B at a certain range of ppm-level prevents precipitation of ferrite structure and improves the toughness at HAZ bond area in weld joint. Therefore, precision quantitative microanalysis of B is necessary for control of B content within the most effective range to improve the toughness.

Precision Microanalysis using laser ablation ICP mass spectrometry

Doping of B to Al-Si-Mn weathering steel improve the toughness at HAZ bond area as shown in Fig.1. But wet chemical analysis is very complicated. So we tried to apply LA-ICP-MS to microanalysis of B as a rapid and precision method without dissolution of specimens.
Fig. 3 shows the relationship between B content and normalized ion intensity by LA-ICP-MS. Laser was scanned on the surface of the cross section at simulated HAZ bond area. Correlation coefficient of approximation straight line for the values is 0.999. This result shows B content and ion intensity by LA-ICP-MS have excellent linear relationship even at the range under 10 ppm.

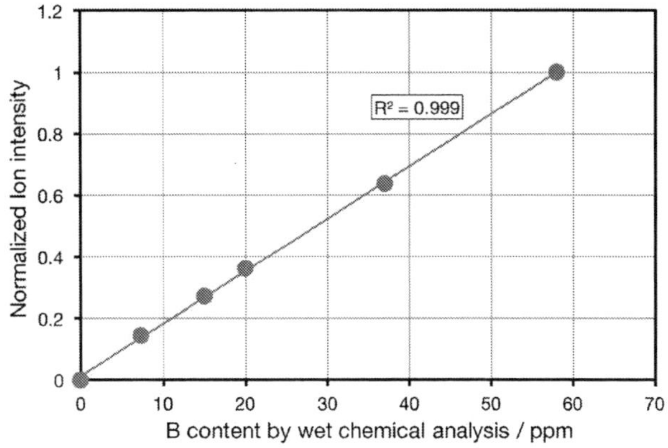

Fig. 3: Relationship between B content and normalized ion intensity by laser ablation ICP-MS.

The ablation area of 0.5 mm diameter is large enough to include many crystal grains on the specimen surface as shown in Fig. 2. So the values of ion intensity are expected to reflect the average B content in the specimens respectively. These results show the specimens manufactured in this study are applicable as standard materials for quantitative microanalysis using LA-ICP-MS. And the approximation straight line can be accepted as the calibration curve of B in Al-Si-Mn weathering steel.

Comparison with other instrumental microanalysis

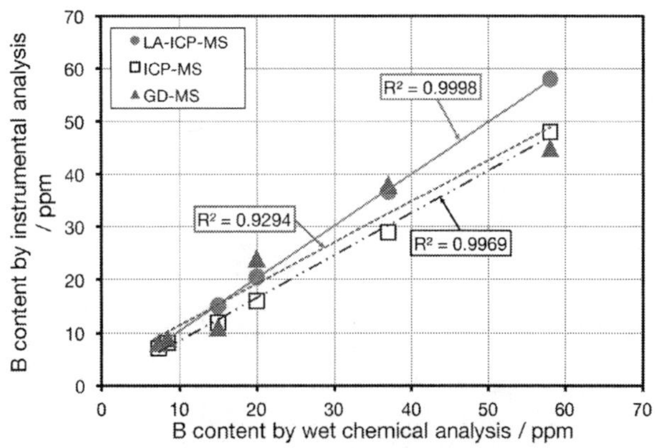

Fig. 4: Comparison of B content by various instrumental analyses.

The result of former paragraph shows LA-ICP-MS is applicable to precision microanalysis of B in Al-Si-Mn weathering steel. In this study, we determined B content using the values of ion intensity and calibration curve by LA-ICP-MS. Moreover, we compared the values of B content with some other analyses using mass spectrometry. ICP-MS is widely used to microanalysis of various elements, but dissolution of a specimen is necessary. GD-MS is a sensitive microanalysis using glow discharge to vaporize a surface of a specimen. Fig. 4 shows the comparison of B content by various instrumental analyses. The vertical axis shows the values of B content calculated from calibration curve and ion intensity respectively. Correlation coefficient between B content by wet chemical analysis and instrumental analysis of LA-ICP-MS, ICP-MS and GD-MS are 0.9998, 0.9969and 0.9294 respectively. These results show LA-ICP-MS has almost equal precision to wet chemical analysis. Therefore, LA-ICP-MS is a promising precision quantitative surface microanalysis of B at ppm-level in steels with rapid processes.

Conclusions

1. Doping of B improves toughness at simulated HAZ bond area of weld joint in Al-Si-Mn weathering steel. Precipitation of ferrite is prevented at the microstructure in the B doped specimen. The effective range of B content is limited under several tens of ppm. This result shows precision microanalysis for control of B content is necessary.
2. Ion intensity measured by LA-ICP-MS and B content by wet chemical analysis have excellent linear relationship at ppm-level. This result shows the precision of LA-ICP-MS is almost equal to wet chemical analysis.
3. LA-ICP-MS is a promising precision quantitative surface microanalysis of B at ppm-level in Al-Si-Mn weathering steel. Calibration curve of B in Al-Si-Mn weathering steel is generated using materials manufactured in this study as standard materials.

Acknowledgments

This study was supported by CREST, Japan Science and Technology Agency.

References

[1] T. Nakamura, *et al*: Japanese Patent Application No. 2012-234603
[2] R. A. Grange and T. M. Garvey: "Factors Affecting the Hardenability of Boron-Treated Steels," *Transactions of the A. S. M.*, Vol.37 (1946), pp. 136-191.
[3] R. A. Grange and J. B. Mitchell: "On the Hardenability Effect of Boron in Steel," *Transactions of the A. S. M.*, Vol. 53 (1961), pp. 157-185.
[4] M. Ueno and K. Itoh: "Distribution of Boron at Austenite Grain-Boundaries and Bainitic Transformation in Low-Carbon Steels," *Transactions of the Iron and Steel Institute of Japan*, Vol. 13, No. 3(1973), pp. 210-217.

Elucidation of metallurgical factor for MIC (Microbiologically Influenced Corrosion) of stainless steel welds using a newly developed in-situ simultaneous Metal/Microbial observation technique

Yasuyuki Miyano
Akita University, 1-1, Tegatagakuen-machi, Akita, Japan
y.miyano@gipc.akita-u.ac.jp

Tomohiro Inaba
National Institute of Advanced Industrial Science and Technology, 16-1, Onogawa, Tsukuba,Japan

Hiroki Watanabe, Tatunori Kiyokawa, Nobuhiko Nomura
University of Tsukuba, 1-1-1, Tennoudai, Tsukuba, Japan

Abstract
The feasibility of a newly developed biofilm observation technique, COCRM (continuous optimizing confocal reflection microscopy) as an investigation technique for simultaneous metal/microbial observation was examined. According to COCRM, images of biofilm formation on the surface of a test specimen and the morphology of a metal surface were clearly observed simultaneously. Moreover, continuous 15-h fixed-point in situ simultaneous metal/microbial observation was performed, then bacterial behavior on the metal surface and the change in the metal surface depending on time were clearly observed. From the results, the feasibility and possibility of COCRM for microbiologically influenced corrosion research was confirmed.

Introduction
In daily life it is common for microbes to adhere onto material surfaces, eventually forming a biofilm. Such interactions between microbes and material sur faces leads to such industrial problems as pollution of pipelines, storage tanks, and ship bottoms, and more serious corrosion degradation for chemical plants and power stations. This is known as microbiologically influenced corrosion (MIC), the induction of corrosion degradation by direct or indirect action of microbes [1]. In certain suspected cases of MIC, a localized corrosion rate as fast as a few millimeters per year was reported for corrosion resistant materials [2].

Stainless steel is a representative corrosion resistant material, but it is not free of MIC [3]. For stainless steels, suspected MIC incidents have been reported to occur mainly for their welds and heat affected zones (HAZ) [4]. Stainless steel is inherently a highly heterogeneous material consisting of many alloying elements and its microstructures are affected widely by the heat cycle, including welding. Thus, some local regions over a stainless steel surface can be subjected to preferential MIC degradation. In fact, a number of research works [5–11] have been published on MIC.

In aforementioned works [5–11], adhesion of microbes to a metal surface is considered as the MIC initiation stage. On this basis, a microscopic investigation has been carried out to elucidate the influence of metallurgical factors on microbe adhesion and subsequent biofilm formation [6–10]. However, microscopic observability is quite different between microbes and a metal surface. Due to this, there are a number of technical limitations on microscopic visualization of microbes over a metal surface.

Thus, to elucidate microbe adhesion behavior over a metal surface with microstructural heterogeneity, the dependences of microbe adhesion and biofilm formation behaviors on microscopic metallurgical factors over the metal surface must be reviewed quantitatively taking into account kinetic factors. In this regard, development of visualization techniques for the MIC process from microbe adhesion over a metal surface through biofilm formation to eventual MIC is considered important. For this, visualization of microbes over a metal surface must be performed in situ without disrupting the biological activity of microbes without staining in an atmospheric environment or setting in vacuum chamber to emit an electron beam.

One promising method relevant for this purpose was recently developed by biology researchers: continuous optimizing confocal reflection microscopy (COCRM) based on reflection microscopy using CSLM (confocal scanning laser microscope) [12]. In conventional CSLM observation, microbes are commonly inspected after staining with fluorescent dye. However, COCRM inspection can also be performed for microbes without fluorescent dye. COCRM has been used for in situ inspection of biofilm formation by microbes over resin surfaces [12]. The authors expected COCRM should be useful for in situ visualization of the behavior of microbes over a metal surface in MIC initiation process.

In this work, COCRM inspection of a stainless steel surface exposed to natural sea water was per formed in search of possibility of in situ inspection of process of microbe adhesion to the metal surface, subsequent formation of biofilm and resultant metal surface degradation by COCRM. In addition, in situ COCRM inspection of an as welded stainless steel specimen exposed to natural sea water was performed. Based on detailed inspection of COCRM surface images obtained by continuous 15-h fixed-point in situ simultaneous metal/microbial observation, the applicability of COCRM for in situ visualization of the MIC initiation process was examined.

Experimental

Test Materials: The specimens were SUS316 and SUS303 sheets 297 × 210 × 6 mm in size. Over the specimen sheet surface, a weld line was prepared along the long edge (297 mm) of the test piece by gas shield tungsten arc welding (GTAW) by the bead on plate technique using YS308L weld metal wire (Ø1.2 mm). To ensure a sufficient amount of weld metal, a double path weld was made. From the specimen sheet with the weld line, a test piece 30 × 6 × 2 mm in size was cut from the base metal part or weld metal part. The surface of the test piece was mirror finished by 1 μm diamond paste.

Test Solution: The test solution was natural sea water sampled at Futtsu Bay at the Nippon Steel & Sumitomo Metal Plant (Chiba Prefecture, Japan). In the as received natural sea water, the microbe concentration was measured as 10^2 cell/cm^2 using an optical microscope and hematometer. This level of microbe concentration in the as received natural sea water was considered too low for the planned experiment. Thus, the as received natural sea water was filtered through a membrane filter with a pore diameter of Ø0.22 μm to collect microbes over the filter; the collected microbes were suspended in sea water containing no microbes to enrich the microbe concentration to 10^8 cell/cm^2.

Exposure Test: To undertake in situ COCRM inspection of microbes on the metal surface of the specimen with sufficient feed of the test solution, a titer plate (6 well joined, bottom surface area 9.4 cm^2, capacity 16 ml) was employed as the exposure test cell. One test piece was placed on a well segment filled with 10 ml of the test solution. The temperature of the test environment was set at 24°C, and the exposure test duration was up to 36 h.

Microscopic Inspection: For in situ microscopic inspection of microbes over the metal surface, an LSM5 Pascal confocal laser microscope (Carl Zeiss) was used. The light source was an Ar gas laser with a wavelength of 488 nm. A reflective microscope image was obtained. The reflected light signal passing through the beam splitter (NFT490) was amplified by a photomultiplier. As reviewed below, when a synthetic reflective microscope image/fluorescent microscope image was obtained, the reflected light and excited fluorescent light were separated using a beam splitter (NFT490) and emission filter (LP505) to obtain amplified signals separately through individual photo multipliers.

Results and Discussion

Figures 1a and 1b show COCRM images of the SUS303 welds test piece after 36 h exposure in natural sea water with enriched microbes. Figure 1a shows the surface area with assumed microbe adhesion; polishing scratches are detectable. As shown in Fig. 1, COCRM was confirmed useful for inspection of the surface morphology of the metal surface exposed in test solution containing microbes. The constructed 3D image of Fig. 1a is shown in Fig. 1b. The colony of microbes at the center of this image in Fig. 1b looks like a mushroom [13], implying that this must be a biofilm that has grown to certain extent. Detailed inspection of Fig. 1a shows the presence of thin deposit layer outside the central mushroomlike biofilm. To identify the microbial cell in the mushroomlike deposit, the test piece was subjected to processing with SYTO9 fluorescent dye. Figure 1c shows a synthetic image from COCRM and fluorescent observations. This image ensures that the mushroomlike deposit was definitely a biofilm. The areas not stained green inside the biofilm and the surrounding of the mushroom deposit have been assumed as extracellular polysaccharide (EPS) secreted by the microbes forming the biofilm. The images in Fig. 1 clearly suggest that COCRM imaging is useful for in situ inspection of microbes over a metal surface when supplemented by fluorescent imaging.

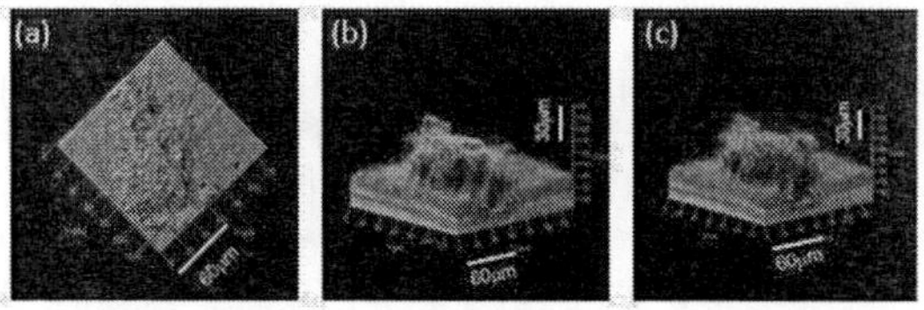

Fig. 1 Fig.1 COCRM(Continuous optimizing confocal reflection microscopy) images of 303 welds surface after 36 hours of exposure tests to the natural sea water. (a) top view surface image, (b) three-dimensional topographical image, and (c) three-dimensional topographical image synthesized with CSLM epifluorescence image.

COCRM images were taken for SUS316 and SUS303 at the base metal part and weld metal part after 12 h exposure after rotary shaking conditions. For clear visualization of the locations of microbe adhesion over the metal surface, staining with SYTO9 was performed for these specimens in exposure state to obtain fluorescent images for synthesis with the COCRM image.

The synthesized images are shown in Fig. 2. For each test condition, such synthetic images were taken for more than ten fields of view; the ones exhibited in Fig. 2 are those that showed typical features in the respective test conditions. It is clearly seen in Fig. 2 that area ratio of microbe deposition was far larger for SUS303 welds than for the other three specimens showing the order SUS303 welds > SUS303 base metal > SUS316 welds> SUS316 base metal. The detected order of area ratios for microbe adhesion suggests a trend of promoted microbe adhesion around welded metal rich in P and S. Walsh et al. [14] reported that sulfide inclusions in metallic alloy promote adhesion of microbes.

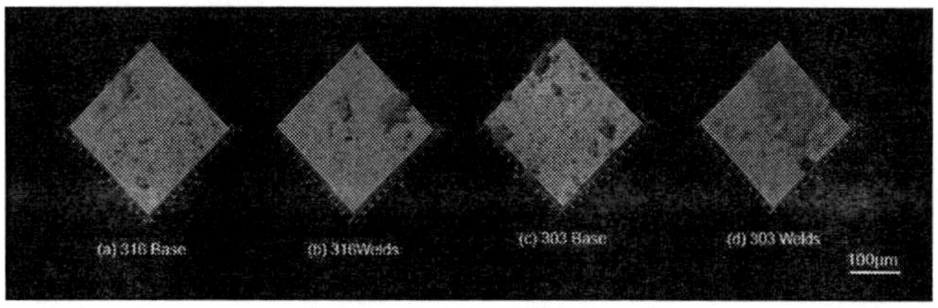

Fig. 2 COCRM top view surface images synthesized with CSLM epifluorescence images of each tested specimens surfaces after 12 hours exposure test. (a) 316 base metal, (b) 316 welds, (c) 303 base metal, and (d) 316 welds.

The COCRM imaging results ensured that a considerable extent of microbe adhesion progressed over the stainless steel specimens in 12 h exposure under the chosen test conditions. After 12 h exposure, red rust became visible to the naked eye for SUS303 with a corrosion resistance inferior to SUS316 and the trend of the red rusting was severer for the weld specimen than for the base metal for a given steel type. Thus, it was decided that more detailed in situ COCRM inspection should be performed for the SUS303 welds specimen during the exposure test to natural sea water with an enriched microbe concentration.

For this, the SUS303 welds test piece was set on a titer plate and continuous COCRM inspection at an arbitrary chosen fixed viewpoint was performed in a 15 min sampling interval during 15 h exposure. The exposure test was performed in an air conditioned laboratory space at a temperature of 24°C. To carry out COCRM inspection at a fixed point, the immersion exposure was performed under stagnant conditions.

The results of COCRM inspection obtained during 15 h exposure are shown in Fig. 3. At the onset of exposure, it appears that the surface view was quite affected by light reflection showing the contrast of the polished surface pattern (Fig. 3a). In Fig. 3a, polishing scratches and pits formed by indentation of the polishing paste particles are visible. It was expected that adhesion of microbes onto the metal surface must start immediately after exposure; however, in the few minutes it took to obtain the COCRM image shown in Fig. 3a, no clear evidence of microbes adhesion was detectable. Figure 3b shows a COCRM image obtained after 4 h; Fig. 3c, after 8 h; Fig. 3d, after 12 h; and Fig. 3e, after 15 h. It is seen in this sequence of COCRM images that the surface area with dark contrast tended to increase with increasing exposure duration. It is also seen in Fig. 3 that some zone with dark contrast can return to brighter contrast later. Some areas became more eminent from the initial metal surface. The changing COCRM image contrast must be due to adhesion and detachment of microbes, while the increased surface height must represent biofilm formation. To draw reliable conclusions on these aspects, further experimental evidence must be obtained. During naked eye inspection of the SUS303 welds specimen surface during exposure, red rust was detectable over the test piece surface and surroundings. Therefore, we cannot rule out the possibility that localized corrosion dissolution took place at some spots in the view area of fixed point COCRM inspection. These problems must be solved by further improving the in situ COCRM imaging technique. Figure 3f is an image of the 15 h exposed SUS303 welds surface after SYTO9 staining. No pre or post staining processing (such as preliminary cleaning) was applied; therefore, the microbe adhesion state in Fig. 3f was considered comparable to that in Fig. 3e. The green zones in Fig. 3f refer to areas with adhered microbes. Compared with the exposure test results obtained under rotary shaking condition (Figs. 1 and 2), the results in Fig. 3 obtained under stagnant conditions show a smaller extent of microbe adhesion. After 15 h exposure, more than 30% of the SUS303 welds surface was deposited with microbes (Fig. 3f). Around the area with adhered microbes, some modification of the surface was also detectable. These results encourage the acceptance of COCRM as an in situ imaging technique to characterize the adhesion behavior of microbes over a metal surface. To draw a convincing conclusion from COCRM imaging, supporting experimental results must be obtained by potential monitoring at a specific spot and by quantitative evaluation of the corrosion damage.

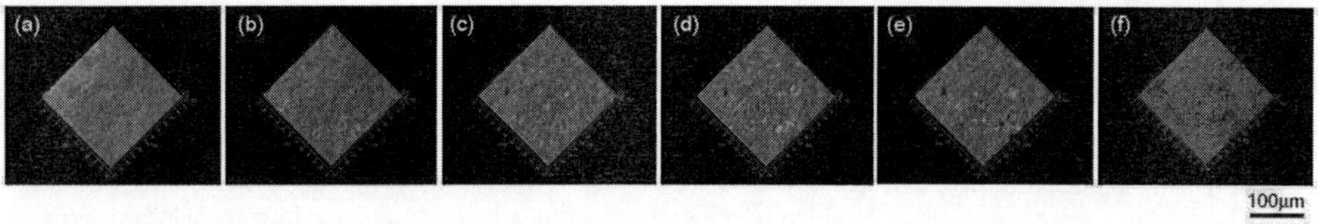

Fig.3 Continuous fixed-point observation images of 303 welds surface acquired by COCRM at regulated intervals during 15 hours exposure test. (a)0 hour (Start period), (b)4 hour, (c)8 hour, (d)12 hour, (e)15 hour and (e)303 base metal, and (d)COCRM image synthesized with epifluorecent of 303 welds surface after 15 hour.

Conclusions

(1) In situ COCRM visualization of the adhesion state of microbes over a stainless steel surface in seawater, verified that an image was obtainable that represents both the microbe adhesion state and the corrosion state of the steel surface morphology. By synthesizing COCRM image with supplemental fluorescent image of microbes, exact positional information of microbes was obtained.

(2) A COCRM fixed-point inspection was performed for the stainless steel welds surface exposed in natural sea water for 15-h to examine the aging transfiguration of microbe adhesion behavior over a given spot of the stainless steel welds surface in a sea water environment. Information representing both the microbe adhesion state and the corrosion state of the steel surface morphology was obtained. As such, the potential of COCRM as a novel in situ characterization technique was demonstrated.

Acknowledgments

This study was financially supported by The Japan Welding Engineering Society and MEXT (Ministry of Education, Culture, Sports, Science and Technology of Japan) through the Kaken hi Basic Research Fund C 16K06792.

References

[1] H.C. Flemming: *Appl. Microbiol. Biotechnol.*, vol. 59, No. 6 (2002), pp. 629-640.
[2] S.A.Wade, *et al.*: *Corr. Mater.*, vol. 36, No.5(2011), pp. 68-79.
[3] S.W. Borenstein: *Microbiorogically Influenced Corrosion Handbook*, (Woodhead publishing)1994, pp. 50-60.
[4] Y. Miyano: *J. Jpn. Wel. Soc.*, vol. 77, No.7, (2008), pp. 650-657.
[5] D.D. Mara, *et al*: *Br. Corr J.*, vol. 7, no. 3, (1972), p. 139.
[6] G.G. Geesey *et al.*: *Corr. Sci.*, vol. 38, No. 1 (1996), pp. 73-95.
[7] J.R. Ibars, *et al.*: *Int. Biodete. Biodeg.*, vol. 29, No. 4 (1992), p. 343-355.
[8] M.A. Javed, *et al.*: *J. Bioad. Biofi. Res.*, vol. 29, No.8 (2013), p. 939-952.
[9] K.R. Sreekumari, *et al.*: *J. Bioad. Biofi. Res.*, vol. 17, (2001), p. 303.
[10] K.R. Sreekumari, *et al.*: *Trans. JWRI.*, vol. 29, No.1 (2000), p. 67-71.
[11] M.V. Biezma, *et al.*: *Int. J. Hydro. En.*, vol. 26, (2001), p. 515-520.
[12] Y. Yawata, *et al.*: *J. Biosci. Bioeng.*, vol. 110, No.3(2010), pp. 377-380.
[13] P. Stoodley, *et al.*: *Ann. Rev. Microbiol.*, vol. 56, (2002), pp. 187-209.
[14] D. Walsh, *et al.*: *J.,Corrosion 94*, no. 612, NACE International, Houston, 1994.

Corrosion & Cracking of HSLA A543 Steel Weld in High Purified Boiler Water Service: Surface Mobility Mechanism

Ali H. AlShawaf
Saudi Basic Industrial Corporation
SABIC
Al-Jubail Industrial city, KSA
shawafaa@sabic.com

Stephen Liu
Colorado School of Mines
Center of Welding, Joining & Coating
Research (CWJCR)
Golden, CO. USA
Sliu@mines.edu

David Olson
Colorado School of Mines
Center of Welding, Joining & Coating
Research (CWJCR)
Golden, CO. USA
Sliu@mines.edu

Abstract

Multiple through-thickness cracks emanating from the circumferential weld on the shell-side in a pressure vessel were observed after a few months of service. The pressure vessel was fabricated using a Q&T HSLA steel. The failure has been diagnosed as Environmentally-Assisted Cracking (EAC), which is a general term for brittle mechanical failures that result from a combination between residual tensile stress, susceptible microstructure and corrosive environment. Slow Strain Rate Testing (SSRT) in a high temperature-high pressure autoclave was carried out using a Constant Extension Rate Test (CERT) machine. Stress corrosion testing of the steel was carried out from 25 to 175 °C at a constant strain rate (10^{-4} s^{-1}) in two alkaline environments. After the fracture of the specimen, Scanning Electron Microscopy (SEM) was used to study the crack morphology as a function of the applied strain. To characterize the stress corrosion cracking (SCC) susceptibility, the plastic strain-to-failure ratio (E_PR) was calculated from data extracted from the stress-strain curves and the reduction-in-area ratio (RAR) estimated from the fracture surface of the broken specimens. Results of SSRT indicate that SCC susceptibility increased with increasing temperature and the percentage of salt in the solution for the welded steel samples while base metal steel samples exhibited lower susceptibility to SCC. Also, advanced corrosion study using Electrochemical Impedance Spectroscopy (EIS) was performed at several conditions to reveal that the A543 steel was prone to form pits in the HAZ. The corrosion study was validated by fitting the EIS results with the Gamry Echem Analyst software. A Constant Phase Elements (CPE) with diffusion model was suggested which confirms the pitting activity on the steel surface. The anodic pits on the HAZ surface led to stress corrosion cracking due to the capture of vacancies by the stressed area at the tip of the pit and the micro-crack. The surface mobility mechanism was controlled by the rate of movement of the vacancies along the surface of the pit and also by the environment.

Introduction

High strength low alloy (HSLA) steels are used widely in the fabrication of heavy structures such as pressure vessels in industries. These steels typically contain less than 0.25wt. pct. carbon and less than 5wt. pct. combined alloying content, and are well recognized for their atmospheric corrosion resistance, high yield and tensile strengths, and good ductility. The quenching and tempering process is to produce microstructures that contain desired amounts of martensite, bainite, and in some compositions, ferrite, for specific mechanical performance. Q&T HSLA steels are often selected because of their high strength and also because of the weight savings, which is a critical factor when considering the manufacture, transport and erection of large vessels [1]. However, the increase in carbon, chromium, nickel, and molybdenum percentage has great influence on the carbon equivalent value of the alloy, welding and corrosion behavior. The carbon equivalent of ASTM A543 steel is 0.93 (minimum chemical composition) and 1.16 (maximum chemical composition). Both carbon equivalent values are located in the third zone of the Graville weldability chart, which indicates that the steel is difficult to weld and that proper pre- and post-weld heat treatment would be required [2]. The ASME Section VIII Div. 2 code, however, does not mandate or prohibit the requirement of pre- and post-weld heat treatment for this type of steel, so the fabrication specification did not indicate any heat treatment requirement. Recently, some cracks were discovered in the closing V-groove welds in the coolers (heat exchangers) [3,4,5]. For safe practice of welding in these coolers, the effects of the environment on the base and welded steel were investigated and reported in this paper. The base metal was welded and samples were extracted from the weld region. The samples were exposed to similar conditions as the base metal.

The double layer theory proposed by Helmholtz, Guoy-Chapman, Frumkin and Grahame led later to the development of the equivalent circuit modeling with the representation of impedance data by Randles and Warburg [4]. Electrochemical Impedance Spectroscopy (EIS) involves the use of a potentiostat in applying a potential (voltage) relative to a reference electrode and measuring the current flowing from the working electrode (specimen) to the counter electrode as shown in Figure 1. Electrical impedance (Z) is related to the ability of a circuit element (the sample) to resist the flow of electrical current [6]. The impedance can be defined by Ohm's law as follow:

$$Z = \frac{E_t}{I_t} = \frac{E_0 \sin(\omega t)}{I_0 \sin(\omega t + \phi)} = Z_0 \frac{\sin(\omega t)}{\sin(\omega t + \phi)} \quad (1)$$

where (I) is the current, (E) is the voltage, ϕ is phase shift, ω is radial frequency ($\omega = 2\Pi f$), and t is the time. Generally speaking, impedance analysis is based on detection of the electrical parameters that develop in the corrosion system in response to an external AC voltage applied to the surface of the sample, i.e. by measuring the corresponding AC current in the cell. Electrochemical Impedance Spectroscopy (EIS) has been successfully used to study the corrosion systems because of the advantage of using small amplitude signals without disturbing the properties of the sample [6,7].

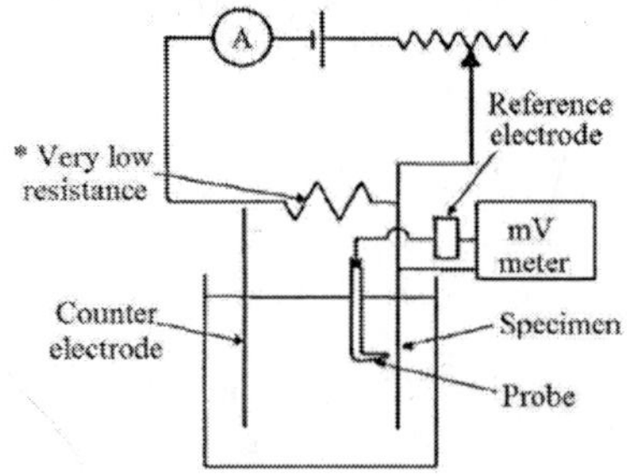

Figure 1. Schematic of EIS electrical circuit [6].

Experimental Procedure

A Gamry Reference 600™ Potentiostat/Galvanostat/ZRA was used to conduct the corrosion experiments and the Gamry Echem Analyst software was used to analyze the electrical signals from the circuit. Electrochemical experiments were conducted to measure the corrosion rate using the linear polarization resistance (LPR) technique and to determine the evolution of corrosion and interfacial processes using electrochemical impedance spectroscopy (EIS). Experiments were carried out in the 1-liter round bottom flask corrosion cell. The flask with multiple necks permits the insertion of electrodes, gas inlet and outlet tubes, and thermometer. The reference electrode (RE) was located close to the working electrode using a bridge tube. One liter of 0.1 wt. pct. $NaNO_2$ dissolved in deionized (DI) water was used as the corrosive medium. The electrochemical test specimens were made of the Q&T HSLA steel with four different conditions. In addition to the base steel, the Q&T HSLA steel was welded with a Low Transformation Temperature Welding (LTTW) wire, MG-S80 in a single pass, and three-pass weld. There were also two samples of the three pass welds, prepared with different interpass temperature, (150°C) and (25°C). The LPR technique was applied to measure the solution resistance and corrosion rate. The working electrode was polarized at ±5 mV and the scan rate was 0.125 mV s^{-1}. To measure the rate of corrosion, Tafel graphs were constructed. The EIS measurements were conducted using 5 mV rms AC voltages at a frequency of 100,000 Hz [7].

Discussion & Results

One can start analyzing the Nyquist plot which consists of a series of points, each of which represent the magnitude and direction of the impedance vector of a particular frequency. The test results for all different conditions including the base metal gave a 45° straight line on these Nyquist plots even after leaving the sample immersed in the solution and tested again after 24 hours as shown in Figure 2. The straight line is explained by Warburg and De Levie [8] that there is surface mass transport taking place on the sample because of pore formation. Figures 2a and b show the result of Q&T HSLA steel welded with MG-S80 wire at two different interpass temperatures. Figure 2c shows the test result of the MG-S80 weld with interpass temperature equal to room temperature. There are two curves in each figure. One curve shows the one hour test and the second curve is after 24 hours test result. Figure 2d shows the different type of pores that could result after the breakage of the protective oxide layer. For example, curve 1 in Figure 2d shows a continuous deep pit while curve 5 shows a sharp bottom pit. Figure 2e and f show the test result for the base metal with no weld and the base Q&T HSLA steel welded with LTTW wire. So, all Nyquist plot results for the base Q&T HSLA steel and welded samples show a 45o straight line.

Bode curves show the phases and the impedance modulus as a function of the frequency of the excitation signal in a corrosion cell. Figure 3a shows by the two arrows the detection of two different signal peaks indicating that there are two activities happening on the sample surface. The first activity is the breakage of the protective oxide layer and the second, the formation of the pit. Figure 3a also shows that conducting the test for one hour (red curve) or 24 hour (black curve) did not change the essential mechanism of the two step process, protective oxide layer breakdown and pit formation, for the MG-S80-A543 weld. Looking at the surface of the welded samples after completing the test, many pores were detected, especially at or near the HAZ. In the case of the LTTW-A543 weld, the one hour test showed a single peak for the formation of a protective film at a higher frequency. This observation is logical since the chemical composition of LTTW is close to that of a stainless steel. Twenty four hour exposure, however, revealed that the LTTW-A543 weld was also susceptible to pit formation with the appearance of the second peak. The MG-S80 weld curves at lower frequencies is an indication of the formation of non-uniform compact protective oxide film in the corrosion solution. The higher chromium and nickel composition of the LTTW weld would produce a more uniform and protective oxide layer shifting the phase maximum to higher frequencies [9,10].

The Nyquist plot for the Q&T HSLA base steel, as shown in Figure 2e, is somehow an intermediate case between the MG-S80, Figure 2c, and the LTTW wire, Figure 2e. In essence, an intermediate thickness protective oxide layer would be formed [11]. The impedance modulus curve show higher impedance in low frequencies when compared to the base metal HSLA steel and the base metal HSLA steel show higher impedance when compared to the LTTW.

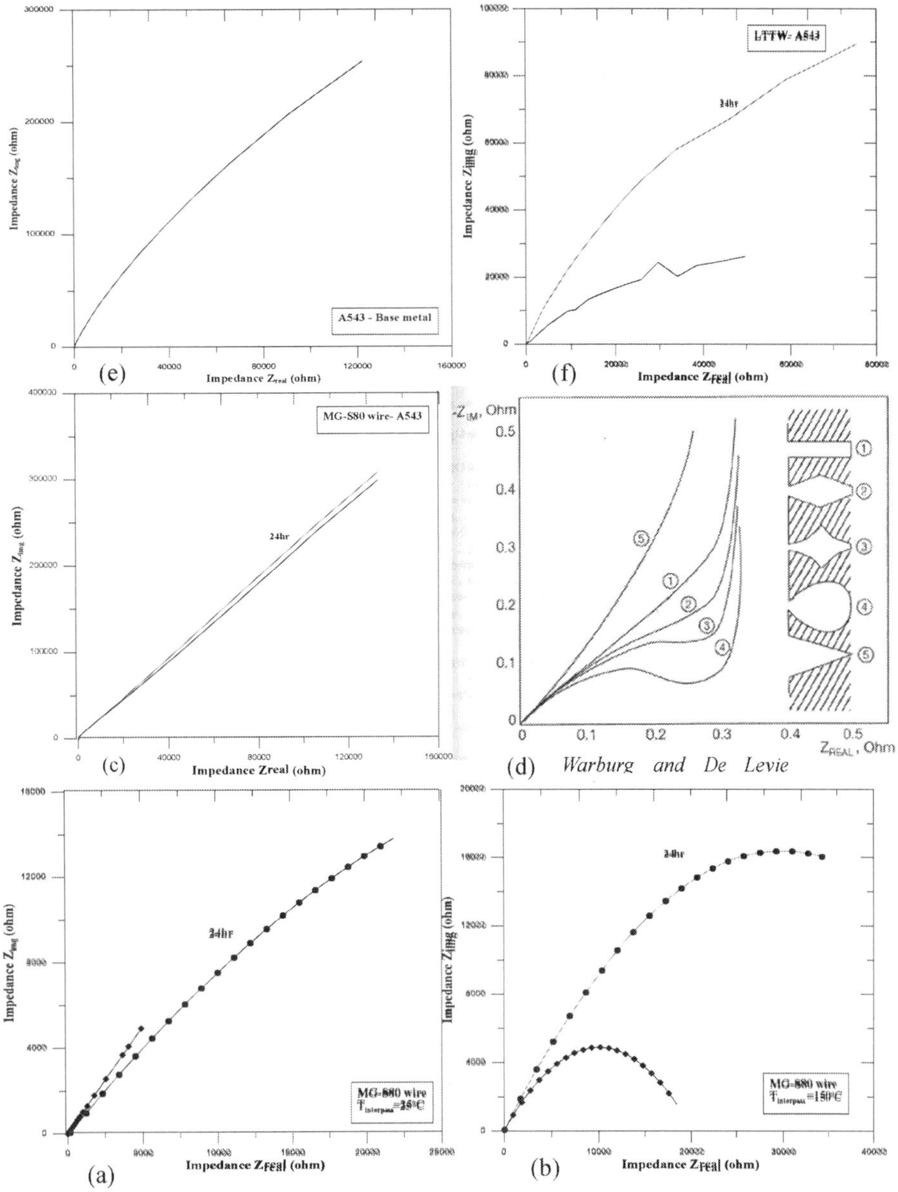

Figure 2.Nyquist curves showing the 45° straight line in a test sample base metal Q&T HSLA steel welded with (a&b) MG-S80 wire at two different interpass temperature, (c)MG-S80 wire at interpass temperature equals to room temperature, (d) Warburg and De Levie model, (e) no weld, (f) LTTW wire.

The Gamry Electrochemical Analyst software (version 5) was used to explain the electrochemical activity between the sample and the corrosive medium. The fitting is done through the representation and series and parallel combination of basic electronic circuit components such as resistors, capacitors, and inductors, as well as other components that represent electrochemical processes, e.g., Constant Phase Element and Warburg (diffusional) Impedance. The fitted model suggested the occurrence of two activities on the sample metal surface as shown by the two dotted lines in Figures 3b [12,13]. The model will be further elaborated later in a later section of this paper.

Corrosion rate was also measured using the Potentiodynamic test as shown in Figure 4. On the left hand side of the figure is a schematic representation of potential-current density diagram of the anodic and cathodic reactions of the corrosion system. It shows the passivation potential and transpassive region where pitting begins. The experimental potential-current density graph on the right shows that the lowest corrosion rate was observed for the base metal HSLA steel welded with MG-S80 wire. However, the corrosion rate increased for the multipass welds. Pore formation was observed for all samples except when welded with the LTTW wire. The potentiodynamic graph also shows that passivation film formed in the LTTW weld. The behavior is different from that showed by the MG-S80 weld where pits formed and broke the passive film. The following paragraph will explain how the oxide layer break down followed by pit formation.

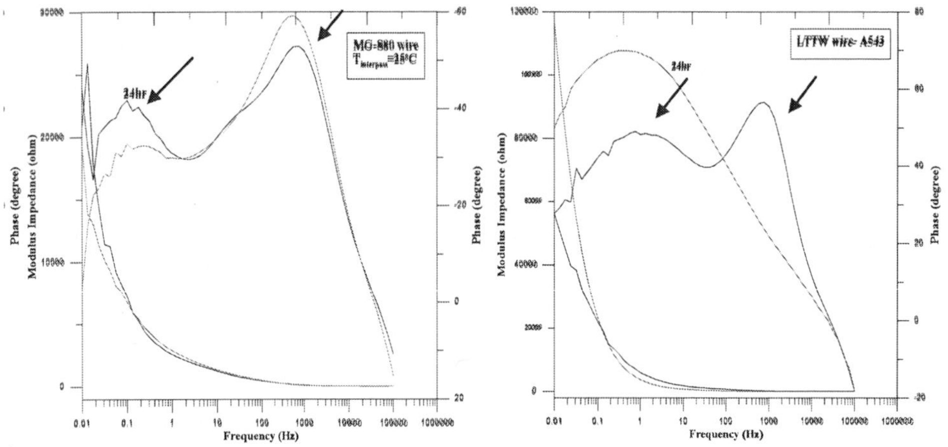

(a)

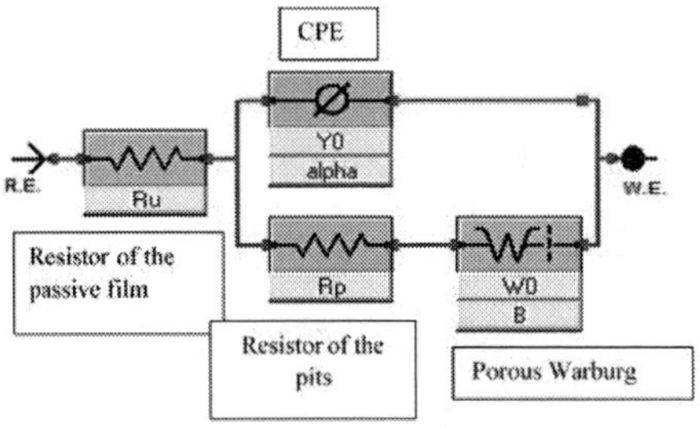

(b)

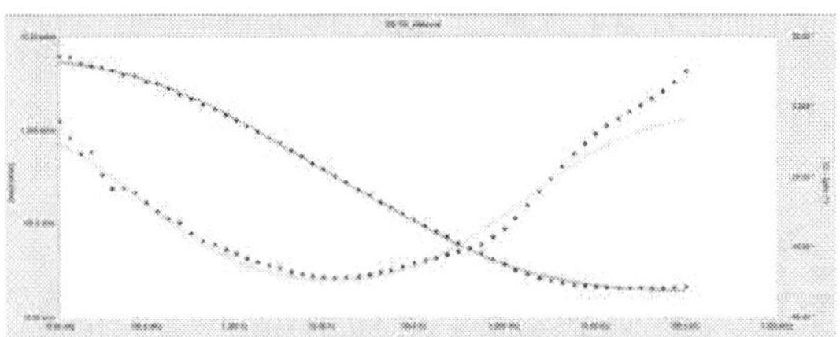

Figure 3. (a)Modulus impedance and phase as function of frequency curves showing the two peaks indicated by the arrows, (b) Electrochemical Analyst software suggested model explaining the electrochemical activity.

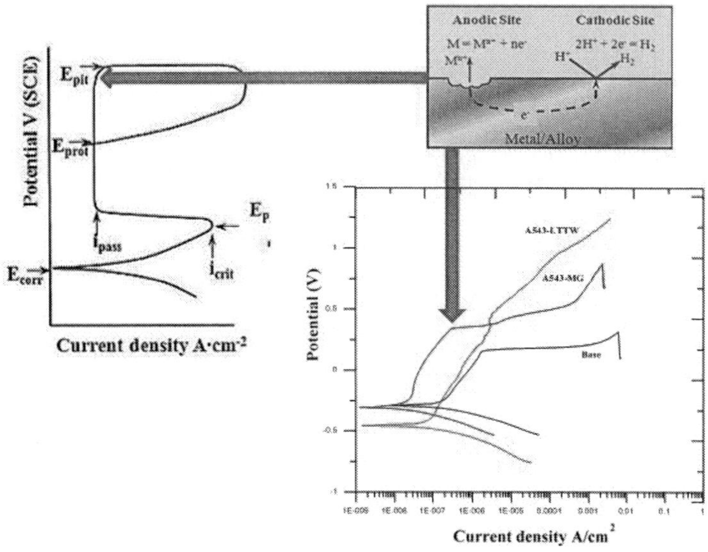

Figure 4. Potentiodynamic results showing the corrosion rates and pits formation.

The surface activities or surface mobility mechanism causes anodic pits to form on the metal surface that is controlled by the rate of movement of the vacancies along the surface of the pit and by the environment [14,15]. The surface mobility mechanism assumes that the stress concentration is established because of the oxidation process at the tip of a pit which generates a localized higher vacancy concentration. Vacancy generation at the stressed micro-crack tip is responsible for the crack propagation process. Figure 5 depicts a schematic drawing of the surface mobility process. The reaction of iron with steam, which produces oxide films, injects vacancies in the surface. At high temperature, there is a clear difference between the growth mechanism of oxides films like FeO, Fe_2O_4 and Fe_2O_3. The growth of FeO and Fe_2O_4 films take place by outward movement of iron ions, while the growth of Fe_2O_3 films is inward, by the movement of oxygen anions or oxygen ion vacancies which keeps the growth of the oxide passive film. This anodic reaction increases the number of pits formed on the metal surface. The cathodic reaction completes the electrochemical reaction in the micro-crack tip. Figures 5a and b show the anodic reaction of the pits. This mechanism described well the observations conducted in this work. Figure 5c shows the pits that formed during the electrochemical impedance test. A cross section of a sample taken from the pressure vessel in the plant confirmed the pits formation on the metal surface as shown in Figure 5c. It is also clearly shown in Figure 5c that the pits shape agree very well with Warburg and De Levie model number 5 [16].

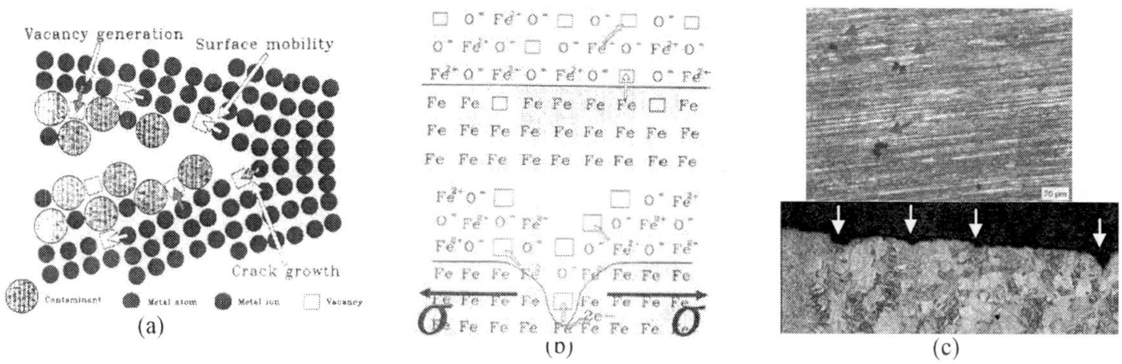

Figure 5. (a)Surface mobility mechanism in the steam environment, (b) Anodic pits formation in steam environment, showing Evans diagram of the electrochemical reaction happening most likely in the micro-crack tip, (c)Pits formation during the electrochemical impedance test (top), pits formation in a sample taken from the plant (bottom) which agree very well with Warburg & De Levie model-curve 5.

Conclusions

The corrosion susceptibility of ASTM A543 steel welds has been investigated in this work. Some conclusions were drawn applying the surface mobility mechanism using EIS.

- Bode curves obtained from electrochemical testing determined that the corrosion behavior can be described by the occurrence of two metal surface activities (surface mobility mechanism). The first one is the formation of a passive film followed by the development of pits that breaks the passivation process.
- The pitting that formed during the electrochemical test in the sodium nitrite medium can become stress concentration points that can potentially develop into cracks in the weld regions and cause a catastrophic failure.

Acknowledgments

The authors would like to acknowledge the Center for Welding, Joining and Coatings Research of the Colorado School of Mines, for the help and support in conducting this research. The authors would also like to thank SABIC management for the financial support.

References

[1] ASTM A543/A543M-09 "Standard Specification for Pressure Vessel Plates, Alloy Steel, Quenched and Tempered Nickel-Chromium-Molybdenum".
[2] B.A. Graville, "Cold Cracking in Welds in HSLA Steels", Welding of HSLA (Microallayed) Structural Steels, Proc. Int. Conf. American Society for Metals, Nov. 1976.
[3] Denny A. Jones, *Principles and Prevention of Corrosion*, 2nd ed., (Prentice Hall, Upper Saddle River, NJ. 1996).
[4] V.S. Sastri, Edward Ghali & Mimoun Elboujdaini, "Corrosion Prevention & Protection, Practical solution", John Wiley & Sons Ltd. 2007.
[5] EIS website "http://www.gamry.com/application-notes/EIS/basics-of-electrochemical-impedance-spectroscopy/ ".
[6] ESI group simulation https://www.esi-group.com.
[7] ASTM G 1–2003, "Standard practice for preparing, cleaning and evaluating corrosion test specimens", Philadelphia, PA, 2003.
[8] D. A. Jones, "Principles and Prevention of Corrosion" second edition, Prentice-Hall, Inc. Simon & Schuster, Upper Saddle River, NJ. USA 1996.
[9] J.L. Crolet, N. Thevenot, S. Nesˇic, "Role of Conductive Corrosion Products on the Protectiveness of Corrosion Layers", Paper No. 96004, (Houston, TX: NACE International, 1996).
[10] J. Chen, B. Young, and B. Uy."Behavior of High Strength Structural Steel at Elevated Temperatures". Journal of Structural Engineering, American Society of Civil Engineering. (2006).
[11] M.B. Kermani, L.M. Smith, "Corrosion Control in Oil and Gas Production - Design Considerations". European Federation of Corrosion Publication no. 23, (Institute of Materials, London, UK, 1997).
[12] Srdjan Nesic, "Key issues related to modeling of internal corrosion of oil and gas pipelines – A review", Corrosion Science 49 (2007) p. 4308–4338.
[13] R.W. Staehle, "Stress Corrosion Cracking and Hydrogen Embrittlement of Iron Base Alloys", (Houston, TX: NACE-5, 1997). p. 193.
[14] A. AlShawaf, S. Liu, A. Elramady."Stress Corrosion Cracking Assessment of a Quench and Tempered HSLA Steel in Two Alkaline Environments", presented and published in the proceedings of the NACE International-Corrosion 2014 Conference, San Antonio, TX, March 9-13, 2014.
[15] J. R. Galvele, "A stress Corrosion Cracking Mechanism Based on Surface Mobility", Corrosion Science, Vol. 27, 1987, pp. 1-33.
[16] V. F. Lvovich, "Impedance Spectroscopy: Application to Electrochemical and Dielectric Phenomena", John Wiley & Sons Inc., Canada 2012.

Welding Residual Stress Modeling and Experimental Validation

Dieter Siegele, Majid Farajian
Fraunhofer Institute for Mechanics of Materials IWM, Freiburg, Germany
dieter.siegele@iwm.fraunhofer.de, +49-761-5142116

Kimiya Hemmesi
Karlsruhe Institute of Technology, Computational Materials Science, Karlsruhe, Germany
kimiya.hemmesi@kit.edu, +49-761-5142268

Abstract
Residual stresses in welded structures are of most importance for their fatigue behavior and in order to provide a reliable lifetime estimation based on analytical or numerical methods the level and distribution of residual stresses from the welding process have to be known. Residual stresses can be measured by non-destructive methods using X-ray or neutron diffraction techniques for example. On the other side numerical methods are powerful tools to calculate welding residual stresses and to provide them in the whole volume of a structure for the following fatigue analysis. The calculated residual stresses depend on the quality of the modeling of the welding process, the material data and the material model used within the finite element software.

In this paper, the welding residual stresses are calculated for a welded pipe as well as for a welded plate using different material models with respect to their hardening behavior. The stress results were compared with measurements using neutron diffraction, hole drilling and contour method. As a result, in summary a good agreement between simulation and measurement is achieved if the calculated temperature field matches with temperature measured during the welding process. Concerning the material model, pure isotropic hardening leads to over conservative predictions of residual stress, particularly in the region where reversed yielding occurs close to the weld. The most accurate predictions of stress are achieved using mixed isotropic-kinematic hardening models. Hereby also the fitting strategies of the model parameter to the cyclic material behavior are of importance.

Introduction
Distortions and residual stresses of welded structures may influence the production and the functionality of welded structures [1]. Therefore, the knowledge of welding residual stresses as well as the distortion due to the welding process is an important task in order to be able to assess the lifetime of a component and to optimize the welding or the manufacturing process. Experimentally, the residual stresses can be measured with different methods. The X-ray diffraction method allows for the determination of surface residual stresses that are of special interest in combination with fatigue assessment since in most cases the damage starts at the surface where stress concentrations from notches, geometry changes or surface roughness are additional factors determining the components lifetime. An alternative non-destructive method is the neutron diffraction technique where residual stresses can be measured in varying depths. But this method is very cost intensive and time consuming and cannot be applied for large components. In contrast to experimental procedures, the simulation of the welding process using finite elements is able to determine the complete residual stress field with all stress components, to separate the influence of each welding parameter, and to provide a detailed understanding of the various effects on distortions and residual stresses while welding. The three-dimensional residual stress field is especially of importance if the fatigue life including the crack propagation in the residual stress field of a welded component have to be assessed [2][3]. Modern finite element codes as SYSWELD not only allows for calculation of deformations and stresses due to the welding process but also take into account the change of microstructure due to different heating and cooling rates.

But in addition to the thermo-mechanical and mechanical material properties also the material plasticity models influence the residual stresses and should be qualified for instance by low cycle fatigue tests in combination with accompanying simulation. In general, an isotropic hardening material model results in higher residual stresses than kinematic hardening models. For austenitic steels, it was shown in [4] that isotropic hardening models estimate the residual stresses close to the measured data. If these results could be transferred to other materials and especially to ferritic steels is still an open question and will be investigated in this paper. For two practical applications a girth weld in a pipe and a bead on plate weld the welding residual stresses are numerically determined and compared with measurements. For the weld bead on a plate the quality of different material models with respect to its hardening rules are studied.

Residual Stress Analysis of a Welded Pipe

As a first example, a girth weld in a pipe is investigated. The pipe consists of a ferritic-pearlitic steel S355J2H which had a single pass bead on plate circumferentially. The welding process used, was TIG (Tungsten Inert Gas) welding without consumable. The experimental setup and the welding parameters are described in [5]. The simulations are performed using the finite element software SYSWELD considering phase transformation. In figure 1 the finite element model which was used for heat source calibration, the cross section of the weld and the comparison of the measured and calculated temperature fields are presented.

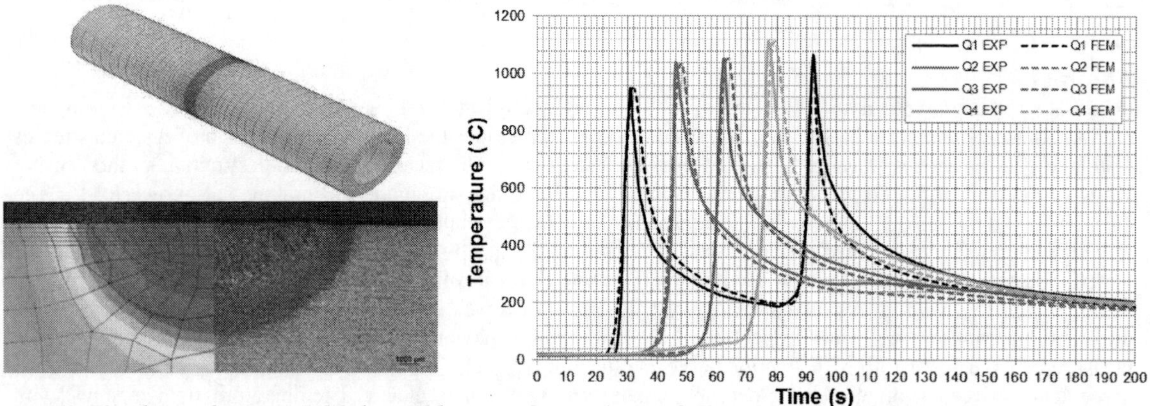

Fig. 1: The finite element model for welding simulation (upper left), temperature field in the cross section compared with the macrograph (lower left) and temperature – time curves at different points along the circumference compared with measurements (right).

An equivalent heat source moving in the welding direction has been used to model the complex heat transfer, melting and fluid dynamics processes in the torch and the melt pool, the geometry and intensity of which has been calibrated by comparing calculated temperature distributions with thermocouple measurements during welding and with melt zone geometries derived from metallographic sections of real welds. Very good agreement between measured and calculated temperature profiles concerning peak temperature and cooling rate are received for the heat source model with gaussian distribution of energy density over the depth. Figure 1 (right) shows the temperature vs. time curves measured with thermocouples close to the weld at different positions of the circumference in comparison with FE results after fitting the heat source model.

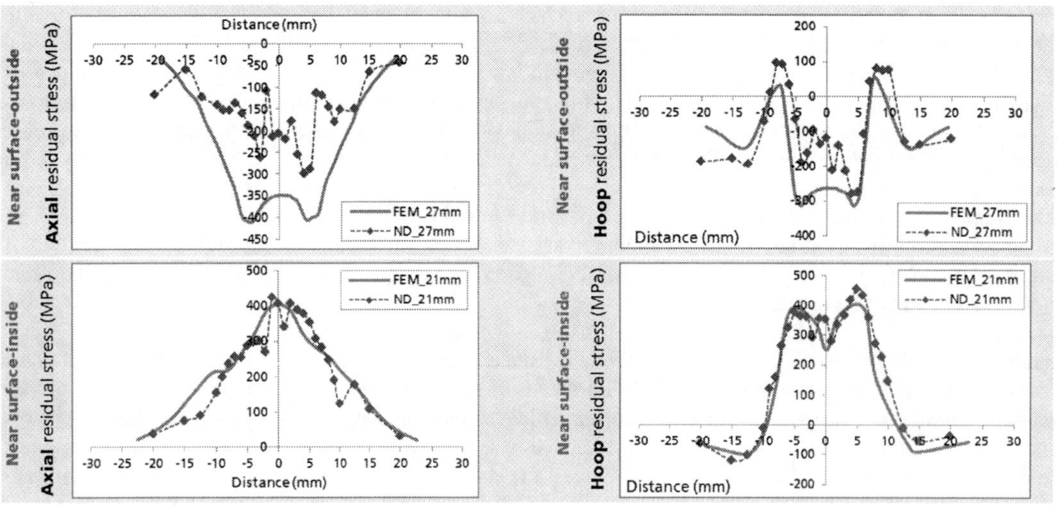

Fig. 2: Calculated residual stresses (red lines) close to surface inside and outside of the pipe compared with neutron diffraction measurements (blue rhombs).

The agreement of the peak temperature and the slope between 800°C and 500°C is a requirement for the correct simulation of the microstructure and the residual stresses. The good agreement between the macro section of the real weld and the calculated isothermal contours of the melting zone and the heat affected zone (left part of figure 1) also confirms the good quality of the simulation.

After the heat source calibration the residual stress field was calculated in the pipe. In figure 2 the comparison of the calculated residual stress in axial and hoop direction with the neutron diffraction measurements is shown. As observed, a very good agreement between calculation and measured residual stresses is obtained for both directions specifically for near surface inside the pipe. For the outer surface the calculated residual stresses in the axial direction are overestimated. In general, it is observed here that approaching the outer surface and thus getting closer to the weld bead makes the welding simulation more complicated and requires better material models which describe the material behavior in these region during welding. The material hardening model used was the standard isotropic hardening model from SYSWELD which showed good results for the ferritic-pearlitic steel in this case.

Residual Stress Analysis of a Bead on Plate

The second example under investigation is a simple bead on plate weld in a European round robin project NeT [6]. The material is an austenitic steel AISI 316L which was solution heat-treated prior to welding to eliminate fabrication residual stresses. A single TIG weld bead was then deposited along the center-line of the plate with 0.8 mm AISI 316L filler material. The details about the experimental setup are described in [6].

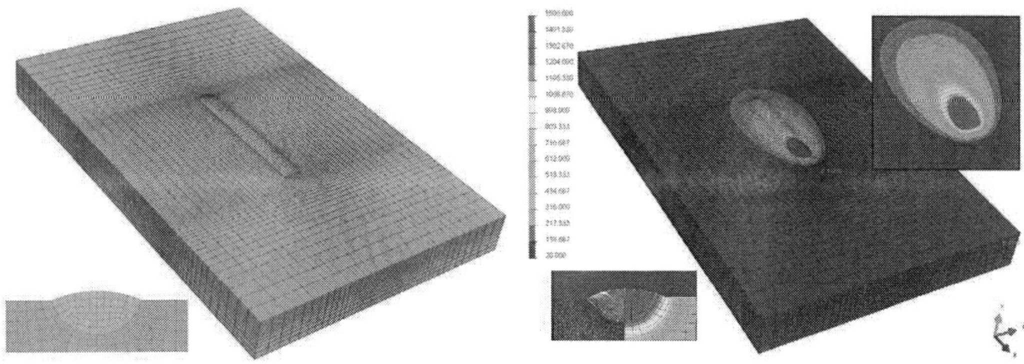

Fig. 3: Finite element model and temperature field during welding of the bead on plate.

Figure 3 shows the plate with the meshing and the temperature distribution during welding as well as the cross section with the temperature contours. As for the first example, the heat source of the model was adjusted to temperature measurements during welding as well as to the molten zone of the macrograph.

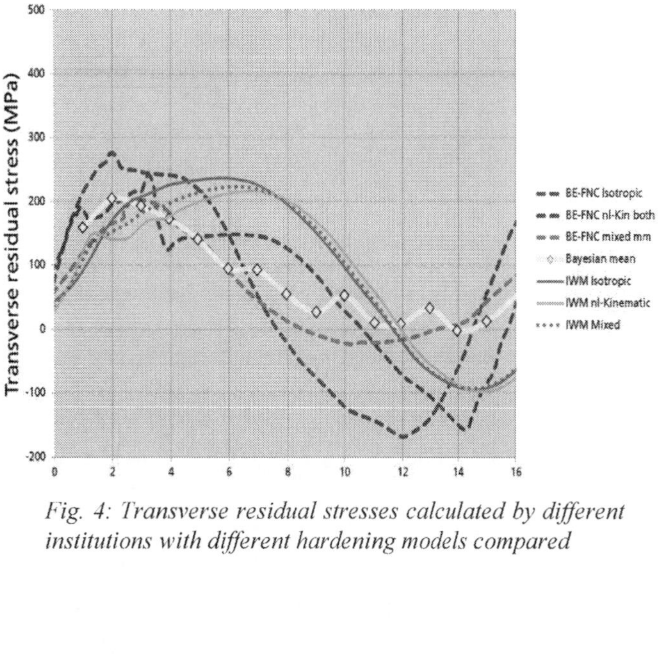

Fig. 4: Transverse residual stresses calculated by different institutions with different hardening models compared

The mechanical analyses of the welding process simulation were performed using different material models available in SYSWELD, i.e. a von Mises plasticity model with pure isotropic hardening, a model with kinematic hardening as well as a model with mixed hardening. Figure 4 depicts the calculated transverse residual stresses in the middle of the weld in this study (IWM) in comparison with the numerical and experimental results in the NeT project (BE FNC). The plotted experimental residual stress depth profile in yellow is the Bayesian mean values obtained from neutron diffraction, hole drilling and contour methods [6].

The differences in both calculations from IWM investigations and the NeT project are because of the differences in material model definitions. In the NeT project the kinematic parameters were fitted to the saturated cyclic response and the cyclic data were again used to derive the isotropic parameters. In the IWM investigation in hand, both isotropic and kinematic parameters are determined based on the stress-strain data from the monotonic tensile tests of the material. The mixed hardening model in IWM is defined considering a 50% fraction for isotropic hardening. In comparison to the experimental results, the calculated residuals stress depth profiles by means of the IMW material models are in reasonable agreement. The best results are however achieved by using the mixed hardening material model based on saturated cyclic response in the Net project (BE-FNC mixed). For the IWM calculations the difference in the residual stress depth profiles obtained from the three used hardening models is not significant. This could be according to the low plastic ranges in the transverse direction.

Conclusions

Numerical simulation of welded specimens showed good agreement with the experimental results for both the steel pipe made of S355J2H and the austenitic steel plate AISI 316L. Once an accurate thermal solution is obtained, then the most important mechanical simulation variable is the material hardening model. Pure isotropic hardening leads to overestimation of the residual stresses, in the region where reversed yielding occurs close to the weld bead. This was found at least for the S355J2H material. For the austenitic plate in this investigations, the calculations with all the three material hardening models i.e. isotropic, kinematic and mix hardening lead to insignificant differences and could reasonably describe the residual stress depth profile.

The most accurate predictions of the residual stresses are however achieved using mixed isotropic-kinematic hardening models using the parameter obtained based on saturated cyclic stress-strain tests. The predicted residual stresses are sensitive to the parameter-fitting strategies adopted for mixed hardening models. The predicted stresses are strongly affected by whether the model is optimized to match the monotonic or the cyclic response.

Acknowledgments

The paper was supported by the German Research Foundation (Deutsche Forschungsgemeinschaft (DFG)) as part of the project "Numerical description of the behavior of welding residual stress field under multiaxial mechanical loading". The authors would like to thank the DFG for its support.

References

[1] D. Radaj, Welding residual stresses and distortion, revised edition of the original German edition entitled: Eigenspannungen und Verzug beim Schweißen – Rechen und Messverfahren, Fachbuchreihe Schweißtechnik Band 143, DVS-Verlag, Düsseldorf 2002.

[2] I. Varfolomeyev, D. Siegele, D. Beukelmann, Analysis of residual stresses and assessment of postulated cracks in a core shroud weld, ASME Pressure Vessels&Piping Conference, PVP, Vancouver, 4-8 August 2002.

[3] I. Varfolomeev, S. Moroz, D. Siegele, J. Baumgartner, Fatigue Life and Crack Growth Assessment for Welded Longitudinal Stiffeners, IIW Document XIII-2489-13, Annual Assembly International Institute of Welding, Essen, Germany, 2013.

[4] H. Wohlfahrt, et al, Residual Stress Calculations and Measurements - Review and Assessment of the IIW Round Robin Results, Welding in the World, September 2012, Volume 56, Issue 9, pp 120-140.

[5] K. Hemmesi, M. Farajian, D. Siegele, Numerical and experimental description of the welding residual stress field in tubular joints for fatigue assessment , Weld World (2016) 60, 741–748

[6] M. C. Smith et al., A review of the NeT Task Group 1 residual stress measurement and analysis round robin on a single weld bead on plate specimen, International Journal of Pressure Vessels and Piping, 120-121, 2014, pp. 93-140.

Opportunities of advanced viscoplastic material modelling for the numerical prediction of thermal and residual stresses in high-strength low-alloy steel welds

N. Häberle, A. Pittner, R. Falkenberg, M. Rethmeier
Bundesanstalt für Materialforschung und -prüfung, Unter den Eichen 87, 12205 Berlin, Germany
nicolas.haeberle@bam.de, +49 30 8104-4645

Abstract

The accurate numerical prediction of thermal stresses, residual stresses, and distortions in high-strength low-alloy steel welds is challenging due to the steels' complex hardening, viscoplastic, and phase transformation behaviour. This work demonstrates the opportunities of applying an advanced viscoplastic material model for the numerical simulation of a single-pass gas-metal arc (GMA) weld using steel of grade S960QL. A very good agreement was found between numerically predicted and experimentally measured residual stresses. The opportunities of advanced viscoplastic material modelling in computational welding mechanics (CWM) are further demonstrated by comparing the simulation results with those obtained by applying a classic rate-independent isotropic hardening approach. The numerical simulations did show that simple isotropic hardening formulations result in inaccurate predictions of residual stresses for non-austenitized base material within the heat affected zone (HAZ).

Introduction

Welding induced residual stresses and distortions are of practical interest due to undesirable effects on the load bearing capacity of structures and potentially increased manufacturing costs due to reduced dimensional accuracy. The evolution of welding induced residual stresses and distortions is governed by the thermomechanical deformation behaviour of the welded material and the thermomechanical loading conditions during welding. Numerical simulations are applied to facilitate the prediction and to enhance the understanding of welding induced thermal stresses, residual stresses, and distortions [1]. One important aspect in CWM is the correct description of a material's thermomechanical deformation behaviour subjected to cyclic thermomechanical loads [2]. For low carbon steel, the employed hardening and viscoplastic formulation was shown to markedly contribute to numerically predicted thermal and residual stresses [3,4]. Furthermore, contributions of phase transformations were shown to be of great importance in the numerical prediction of thermal and residual stresses for welding-like cyclic thermomechanical loads [4-7]. The cyclic hardening behaviour of grade S960QL steel does involve kinematic hardening and strain range memory effects [8] at ambient temperatures. Viscoplastic contributions on hardening do exist at elevated temperatures – in particular for austenitized material [9]. Phase transformations do notably affect thermal expansions and result in transformation-induced plastic strains [10]. In CWM, the phenomenological Leblond [11] model is widely employed to describe the phase transformation kinetics of low carbon steel [2]. The effect of transformation-induced plastic strains is described by rate equations [10,12]. In the absence of phase transformations, the deformation behaviour of metals subjected to cyclic thermomechanical loads can be accurately described by viscoplastic Chaboche material models [13-15]. Coret [16] did introduce the meso framework to couple the phase transformation kinetics model, the rate equation for transformation-induced plastic strains, and the material models used to describe the thermomechanical deformation behaviour of each metallurgical phase. In this work, the meso approach was used to describe the multiphasic deformation behaviour of grade S960QL steel. A viscoplastic Chaboche material model was used to accurately describe the thermomechanical deformation behaviour of base and austenitized material. The multiphasic viscoplastic material model was applied in the numerical simulation of an experimental single-pass GMA weld. The predicted residual stresses were evaluated by comparison to experimental measurements and results obtained from a numerical simulation utilizing a more simple material modelling approach typically employed in CWM.

Methodology

An advanced viscoplastic material model and a classic rate-independent J2 plasticity material model were applied in the numerical simulation of a GMA weld. The advanced viscoplastic material model does utilize the meso approach of Coret [16] to describe the multiphasic thermomechanical deformation behaviour of steel S960QL. Here, the steel was treated as two phase material of base and austenitic phases. Phase transformation kinetics were described by the phenomenological Leblond model [11]. Thermal expansions were described by a linear mixture rule for thermal expansions of base and austenitized material [2]. Transformation-induced plastic strains were accounted for by a typical rate equation [12,16]. The TRIP formulation was equally employed for the formation and the segregation of the austenitic phase. The thermomechanical deformation behaviour of base and austenitized material was described by a viscoplastic Chaboche material model. For the most part, the employed Chaboche model is identical to the formulation proposed in [14] but an extended isotropic hardening formalism in analogy to [15] was employed. Hardening of each metallurgical phase thus was described by internal variables and corresponding evolution equations for nonlinear kinematic hardening and isotropic softening with strain range memory effects. The contribution of viscoplastic effects was accounted for by the employed overstress model and static recovery in kinematic hardening state variables [14]. Furthermore, the evolution equations of kinematic and isotropic hardening state variables were extended to allow for annealing due to phase transformations analogous to the formalism presented in [4]. The advanced viscoplastic material model was implemented in the ABAQUS 6.12.3 FEM software package utilizing the UMAT subroutine. The material model's parameters were identified based on in-house experimental data on the cyclic hardening and stress relaxation behaviour of steel S960QL and further experimental data reported in [10]. With the classic material model, a rate-independent isotropic hardening formulation was employed. A linear mixture rule based on experimental hot tension test data [9] was applied to describe the material's strength in dependence of the metallurgical state. The total strain rate was additively decomposed in a thermoelastic, plastic, and transformation-induced plastic part to account for the multiphasic deformation behaviour [2]. The formulations of the thermoelastic, phase kinetic, and TRIP models were identical to those used with the advanced viscoplastic model. The classic material model was implemented based on UEXPAN, UHARD and USDFLD subroutines.

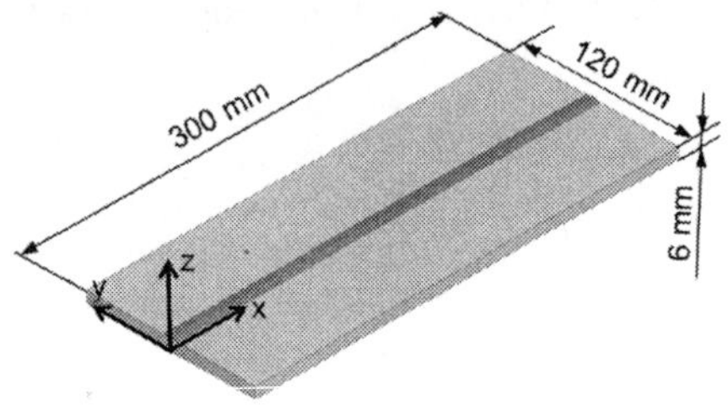

Fig. 1: Geometry and coordinate system of the single-pass GMA weld.

Both material models were applied for the numerical simulation of the experimental GMA weld reported in [10]. In the experiment, two 6 mm thick S960QL plates were joined in one pass using a matching filler wire. Free shrinkage of the plate was allowed during and after welding. The geometric dimensions of the experimental weld are shown in figure 1. In the numerical simulations a half model was employed and symmetry with respect to the xz-plane was assumed. Boundary conditions were applied corresponding to the experimental set-up. The temperature field and metallurgical state were obtained by solving the thermometallurgical problem. Volumetric heat sources were applied to model heat input using the UDFLUX subroutine. The intensity and distribution of the heat input was adjusted to obtain a close agreement to experimental thermocouple measurements and contour of the weld's fusion zone. The deposition of weld filler material was modelled utilizing the chewing gum method [1].

Results and Discussion

The parameters of the advanced viscoplastic material model were successfully identified. The mixed nonlinear kinematic / isotropic softening with strain range memory effects hardening formalism did result in a very accurate reproducibility of the cyclic uniaxial hardening behaviour of steel S960QL at ambient temperatures, figure 2 (a), and elevated temperatures, figure 2 (b). The combination of overstress model and static recovery in kinematic hardening state variables did result in a very accurate description of the strain rate dependent hardening, figure 2 (b), and stress relaxation behaviour, figure 2 (c). Utilizing the meso approach, the load dependent evolution of plastic strains during the gamma-alpha transformation was also properly described, figure 2 (d).

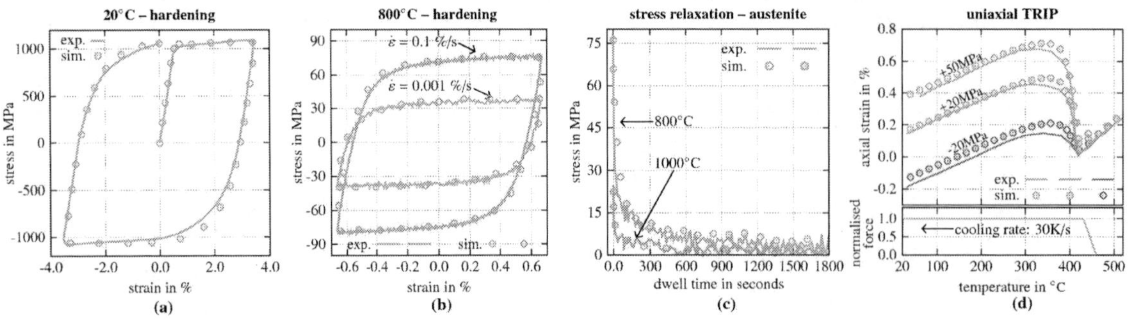

Fig. 2: S960QL thermomechanical deformation behaviour, experiment vs. numerical simulation utilizing the advanced viscoplastic material model. Cyclic hardening at 20°C (a) [6], stationary cyclic hardening at 800°C at two different strain rates (b), stress relaxation (c), and uniaxial TRIP behaviour (d) [9].

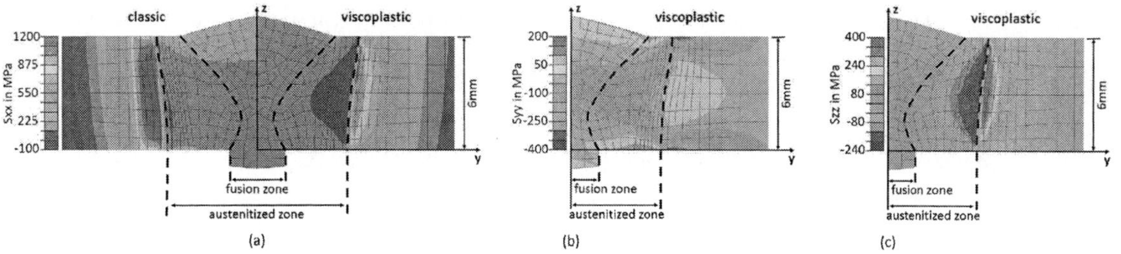

Fig. 3: Numerically simulated longitudinal (a), transversal (b), and through thickness (c) residual stresses at x=150mm.

Longitudinal, transversal, and trough thickness residual stresses obtained from the numerical simulations of the experimental weld are shown in figure 3. Numerically predicted longitudinal residual stresses (Sxx) did depend on the employed material modelling approach, figure 3 (a). For both modelling approaches, longitudinal residual stresses of high magnitude did result in non-austenitized base material in the HAZ. However, the peak magnitude of residual stresses predicted with the classic rate-independent isotropic hardening material model was higher. Material in the non-austenitized HAZ was subject to cyclic thermal and mechanical loads during the simulated weld. The employed hardening formalism was found to significantly contribute to predicted longitudinal residual stresses due to cyclic plastic loading of material in the non-austenitized HAZ. The employed hardening and viscoplastic formalism did also affect the numerically predicted thermal stresses in austenitized material. However, contributions of thermal expansions and transformation-induced plastic strains during the segregation of the austenitic phase were found more significant with regards to resulting longitudinal residual stresses. The observation is in agreement with [3, 4].

For transversal residual stresses (Syy), figure 3 (b), virtually identical numerical predictions of residual stresses did result for both material models. The employed hardening model did not affect the predicted residual stresses in non-austenitized material due to the absence of significant cyclic plastic loading. Also, the contribution of viscoplastic effects was negligible because of a low level of stress and limited duration of high temperature exposure. Predicted transversal residual stresses in the austenitized zone were not subject to contributions of the employed hardening and viscoplastic approach for the same reasons outlined for longitudinal residual stresses.

Numerically predicted through thickness residual stresses (Szz) were also virtually identical for both employed material models and are shown in figure 3 (c). Residual stresses of notable magnitude were confined to the transition band of the non-austenitized and austenitized HAZ. Tensile through thickness residual stresses did result in non-austenitized base material, compressive through thickness residual stresses did result in material temporarily austenitized during welding. The through thickness residual stress distribution was found to be a result of the thermal expansion of austenitized material during its segregation. The hardening and viscoplastic modelling approach did not affect the predicted through thickness residual stresses due to a lack of cyclic plastic loading and a limited time span of high temperature exposure needed for notable stress relaxation to occur.

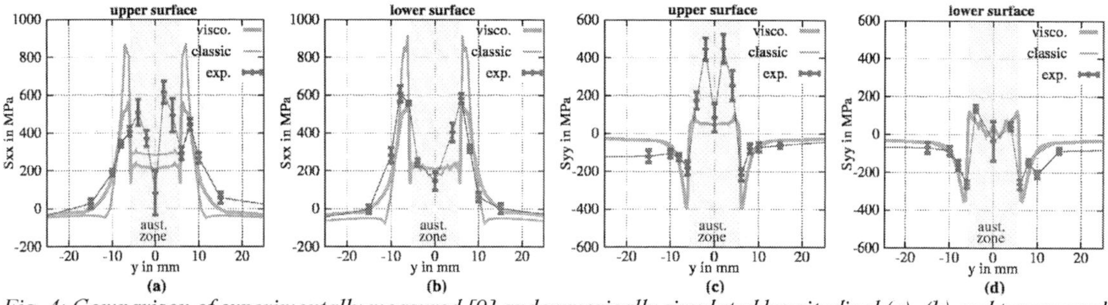

Fig. 4: Comparison of experimentally measured [9] and numerically simulated longitudinal (a), (b) and transversal (c), (d) residual stresses at x=150mm.

Utilizing the advanced viscoplastic material model a very good agreement was obtained between numerically simulated and experimentally measured [9] longitudinal, transversal, and through thickness residual stresses. Only transversal residual stresses on the upper surface of the plate were partially mispredicted. Figure 4 shows selected comparisons of numerically predicted and experimentally measured residual stresses for both applied material models. Longitudinal residual stresses (Sxx) in non-austenitized material were only correctly predicted utilizing the advanced viscoplastic material model, figure 4 (a) and (b). The correct description of steel S960QL hardening behaviour in cyclic plastic loading thus was found necessary for an accurate prediction of longitudinal residual stresses in the non-austenitized HAZ. Transversal residual stresses (Syy) were partially mispredicted in both numerical simulations, figure 4 (c), (d). Experimentally measured tensile transversal residual stresses in the austenitized zone on the upper surface of the plate were underestimated. The simulation results suggest that high tensile thermal stresses in austenitized material can hardly result in high tensile residual stresses due to the material's thermal expansion and the TRIP effect during segregation of the austenitic phase. We thus suspect that high tensile residual stresses in the austenitized HAZ most likely develop during and after the segregation of the austenitic phase. Spatially varying phase transformation kinetics due to austenite grain size effects were neglected in this work but are known to affect residual stresses and are a possible explanation [7]. For the investigated weld, the experimentally observed tensile transversal residual stresses could be explained by correlating a delayed austenite segregation with the region of increased hardness which did occur in the inner and bottom part of the austenitized HAZ – see hardness mapping in [10].

Conclusion

The numerical prediction of welding induced residual stresses for a steel S960QL weld was improved by the application of an advanced viscoplastic material model. Longitudinal residual stresses in the non-austenitized HAZ were only correctly predicted by properly accounting for the steel's hardening behaviour for cyclic plastic loads. Transversal residual stresses were partially underestimated and the neglect of austenite grain size effects in the material modelling approaches was identified as likely cause.

Literature

[1] L.E. Lindgren: "Finite element modeling and simulation of welding part 1: Increased complexity", *Journal of Thermal Stresses*, Vol. 24, No. 2 (2001), pp. 141-192.

[2] L.E. Lindgren: "Finite element modeling and simulation of welding. Part 2: improved material modeling", *Journal of Thermal Stresses*, Vol. 24, No. 3 (2001), pp. 195-231.

[3] Y. Vincent, *et al.*: "Viscoplastic behavior of steels during phase transformations", *Comptes Rendus Mecanique*, Vol. 42, No. 14 (2005), pp. 4077-4098.

[4] Y. Vincent, *et al.*: "Thermo-mechanical consequences of phase transformations in the heat-affected zone using a cyclic uniaxial test", *Int. Journal of Solids and Structures*, Vol. 42, No. 14 (2005), pp. 4077-4098.

[5] A.S. Oddy, *et al.*: "Transformation effects in the 3D finite element analysis of welds", *Proceedings of the Second International Conference on Trends in Welding Research*, Chicago, IL, 1989, pp. 15-19.

[6] Y. Li, *et al.*: "Accurate Prediction of Welding Stress Evolution by Considering Improved Phase Transformation Model", *Materials Transactions*, Vol. 56, No. 5 (2015), pp. 715-719.

[7] C. Heinze, *et al.*: "Dependency of martensite start temperature on prior austenite grain size and its influence on welding-induced residual stresses", *Computational Materials Science,* Vol. 69 (2013), pp. 251-260.

[8] L. Fu, *et al.*: "A modified Yoshida-Uemori constitutive model and its application to cold-bending in Weldox960", *Advanced Materials Research*, Vol. 834 (2014), pp. 407-415.

[9] S. Neubert, *et al.*: "Simulationsgestützte Einflussanalyse der Festphasenumwandlung auf die Schweißeigenspannungen bei Verwendung martensitischer Zusatzwerkstoffe", DFG SCHW 1505/4-1 (2014).

[10] S. Neubert, *et al.*: "Numerical sensitivity analysis of TRIP-parameter K on weld residual stresses and weld distortion ", accepted for publication in *Proceedings of the 11th International Seminar on Numerical Analysis of Weldability*, Graz, Austria, September 2015.

[11] J.B. Leblond and J. Devaux: "A new kinetic model for anisothermal metallurgical transformations in steels including effect of austenite grain size", *Acta Metallurgica*, Vol. 32, No. 1 (1984), pp. 137-146.

[12] F.D. Fischer, *et al.*: "Transformation-induced plasticity (TRIP)", *Applied Mechanics Reviews*, Vol. 49, No. 6 (1996), pp. 317-364.

[13] J.L. Chaboche: "A review of some plasticity and viscoplasticity constitutive theories", *International Journal of Plasticity*, Vol. 24, No. 10 (2008), pp. 1-22.

[14] C. Sommitsch, *et al.*: "Modelling of creep-fatigue in containers during aluminium and copper extrusion", *Computational Materials Science*, Vol. 39, No. 1 (2007), pp. 55-64.

[15] J.L. Chaboche, *et al.*: "Viscoplastic constitutive equations of combustion chamber materials including cyclic hardening and dynamic strain aging", *International Journal of Plasticity*, Vol. 46 (2013), pp. 1-22.

[16] M. Coret: "Étude expérimentale et simulation de la plasticité de transformation et du comportement multiphase de l'acier de cuve 16MND5 sous chargement multiaxial anisotherme", PhD thesis, 2001.

Prediction of Residual Stress Distributions in External Attachment Welding Applications to Evaluate the Need of PWHT

*Rashed Alhajri and Stephen Liu**
Colorado School of Mines, Center for Welding, Joining and Coatings Research, Golden, Colorado, U.S.A.
**Email (sliu@mines.edu), Phone (+1-303-273-3796), Fax (+1-303-384-2189)*

Abstract
There has been clear uncertainty whether post-weld heat treatment (PWHT) should be required for external attachment (non-pressure retaining) welds in petrochemical industry applications where stress corrosion cracking (SCC) is a possibility. An industrial criterion established by NACE SP0472-2010, paragraph 3.5.1, indicates that PWHT is not required if no tensile residual stresses extend through the entire thickness. To investigate this problem, commercial FEA software, SYSWELD, was utilized to predict the extent and level of residual stresses of such welds through the thickness of the pressure vessel shell. The study used the GMAW process to deposit single bead-on-plate welds with ER70S-6 wire on ASTM-516 grade 70 PVQ steel plates of 6.3, 12.7 and 19.0mm thicknesses. Microstructural analysis through the use of metallography and micro-hardness tracing of weldments were performed as a mean to qualify the modeling phase quantification results. Following the through-thickness direction, the peak of tensile residual stresses was invariably observed in the Sub- and Inter-critical HAZ regions. In addition, through-thickness longitudinal tensile stresses existed in 6.3 and 12.7 mm thick plates whereas both tensile and compressive longitudinal stresses were observed in 19.0 mm thick plates. Maximum longitudinal tensile stresses at 85-104% of the base metal yield strength existed in the bottom surface of the 6.3 mm thick plates whereas maximum longitudinal tensile stresses reached only 8-12% of the yield strength in the 19.0 mm thick plates. The conclusion indicates that 6.3 mm thickness applications demand PWHT whereas welding 12.7 and 19.0mm thick plates are possibly safe depending on the amount of SCC threshold stress and loading conditions.

Introduction
Fusion welded joints are extensively used in structural fabrication including petrochemical plants, shipping, offshore structures, and steel bridges. However, welding residual stresses which have been investigated since 1950s along with distortion have imposed inherent challenges in almost all fusion welding processes [1-3]. Such residual stresses can be as high as the local yield strength in weld metal and surrounding HAZ and are well identified to promote service failures including brittle fracture, fatigue and stress corrosion cracking (SCC). Cheng et al. identified that such stresses can be detrimental to structural fatigue life though increasing minimum to maximum internal stress ratio experienced in fatigue cycles. Such increase in stress ratio decreases the threshold stresses below which no fatigue cracking can initiate and therefore results in inferior fatigue strength [4]. Furthermore, many researchers identified SCC threshold stresses above which SCC can occur [5, 6]. Jones et al. estimated that such stresses are usually around 80% of yield strength and can be in other cases as high as 40-50% [7]. Therefore, the presence of residual stresses can affect the safe operation of a pressure vessel in terms of SCC.

Among common post-weld treatment methods, including shot peening, hammering, laser remelting, ultrasonic impact treatment, and high frequency impact treatment, the petrochemical industry has almost always utilized post-weld heat treatment (PWHT) to relieve high residual stresses of weldments used in SCC environments. Many petrochemical equipment construction codes, e.g. ASME BPVC and API 650, mandate and provide detailed PWHT procedures for welds directly exposed to SCC environments. Such PWHT requirement criterion is well realized for through-thickness (pressure-retaining) welds, such as longitudinal and circumferential pressure vessel welds.

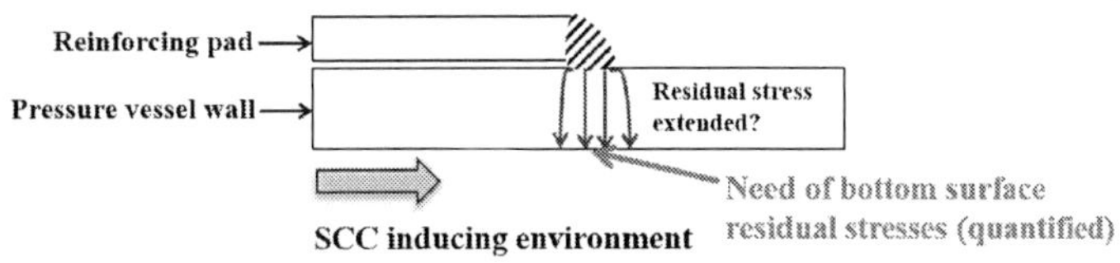

Figure 1: Schematic illustration of external attachment weldment in pressure vessels

There exist, however, other weldments in which the weld metal and HAZ which typically experience highest residual stresses are not directly exposed to the SCC environment. Figure 1 provides an example with schematic illustration of an external attachment weld joining a reinforcing pad or sleeve to a pressure vessel wall. In such application, the threat of extending the weld metal and HAZ tensile residual stresses to the bottom surfaces of the pressure vessel shell targeted for SCC environment exposure has imposed clear uncertainty whether such weldment types should receive a PWHT. According to NACE SP0472, paragraph 3.5.1, PWHT in such weldment types is not required if no tensile residual stresses extend through the entire wall thickness [8]. To resolve such a problem, this research is devoted to determine the residual stress extension behavior from weld metal to the bottom surfaces and quantify the residual stress levels in the bottom surfaces exposed to SCC environments.

Experimental Procedures

To simulate such external attachment welds, single pass bead-on-plate weldments were applied on (178x254mm) ASTM-516 grade 70 C-Mn pressure vessel quality steel plates using the gas metal arc welding (GMAW) process. A conventional GMAW C-Mn steel filler wire, ER70S-6 (1.2mm diameter), and shielding gas of 75%Ar-25%CO_2 were used. Such shielding gas was selected to promote short-circuiting transfer mode and low heat input values. The study was targeted to study low heat input welding conditions due to the fact that if high residual stresses are extended to the bottom surface, PWHT would be required with higher heat input values. To study the effect of heat input, five sets of welding parameters within a heat input range of 327 to 509J/mm were followed. In addition, three base metal thicknesses, 6.3, 12.7 and 19.0mm were included in this study to evaluate the effect of pressure vessel wall thickness on bottom surface residual stress levels.

Meeting AWS A9.5 Standard, Guide for Verification and Validation in Computation Weld Mechanics, five cross section weld macrographs were taken to obtain average weld metal geometry and HAZ depth [9]. Such geometry, from each weld set, was used as input to the model meshing of the finite-element analysis (FEA). In addition, the weld metal and CGHAZ microstructure of each weld set for 6.3 and 12.7mm thickness was characterized and quantified following ASTM E562-11 [10]. The phase quantification was targeted for FEA phase amount computation validation. Micro-hardness line tracing was also performed to correlate the hardness results with the phases identified in weld metal and HAZ.

Modeling Procedures

To predict the residual stress distributions and extent in the bottom surfaces exposed to SCC environments, FEA using SYSWELD^{TM} software by ESI-group was performed. The FEA uses sequentially-coupled-analysis feature where the thermo-metallurgical module is performed first to compute and obtain transient temperature and metallurgical phase results. The mechanical module is performed thereafter utilizing thermo-metallurgical module historical results for transient residual stress and distortion computations. For low distortion applications, such feature has provided satisfactory results and significant time saving as opposed to the simultaneous-coupling feature. The material database used in the FEA included property values quantified as a function of temperature for each metallurgical phase complemented with the utilization of continuous cooling transformation and austenitization diagrams. Some of the properties in the database of A516 grade 70 steel are shown in figure 2.

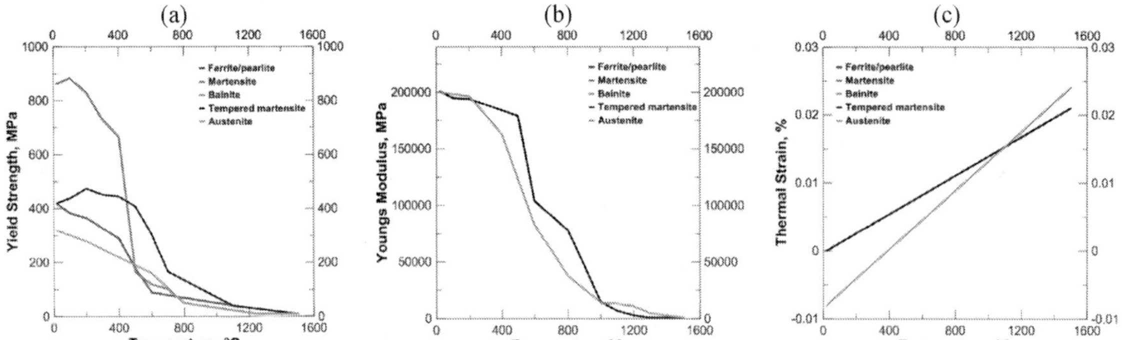

Figure 2: Base metal yield strength (a), Young's modulus (b) and thermal strain (c) included in the A516 grade 70 material database

The meshing methodology is illustrated in figure 3. The base metal and HAZ were mostly composed of quadrilateral 3D elements whereas triangular and quadrilateral 3D elements were created in the weld metal region. In the weld metal vicinity, a through-thickness region with fine-scale 3D elements was created in order to study the residual stress extension behavior from the weld top to the bottom surface. The front view plane meshing represents a separate layer where 60 identical layers were created to complete the total weldment. The interlayer distance was set minimum at the welding path start and end locations and reached a maximum in the welding path center where a thermal steady state condition is reached. Goldak's double-ellipsoid heat source model was used and iteratively calibrated until the average weld metal geometry (from experimentation) was matched in each welding set.

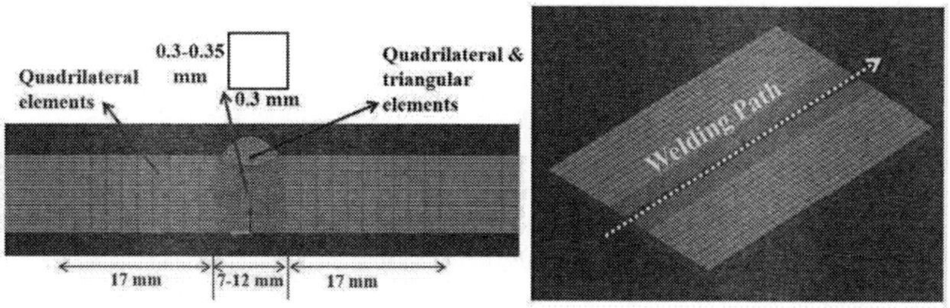

Figure 3: Illustration of FEA meshing in front view plane and isometric view.

Results and Discussion

Figure 4a represents an illustration of 3D elements selected from weld top surface to bottom surface to generate through-thickness paths for the purpose of studying through-thickness residual stress distributions.

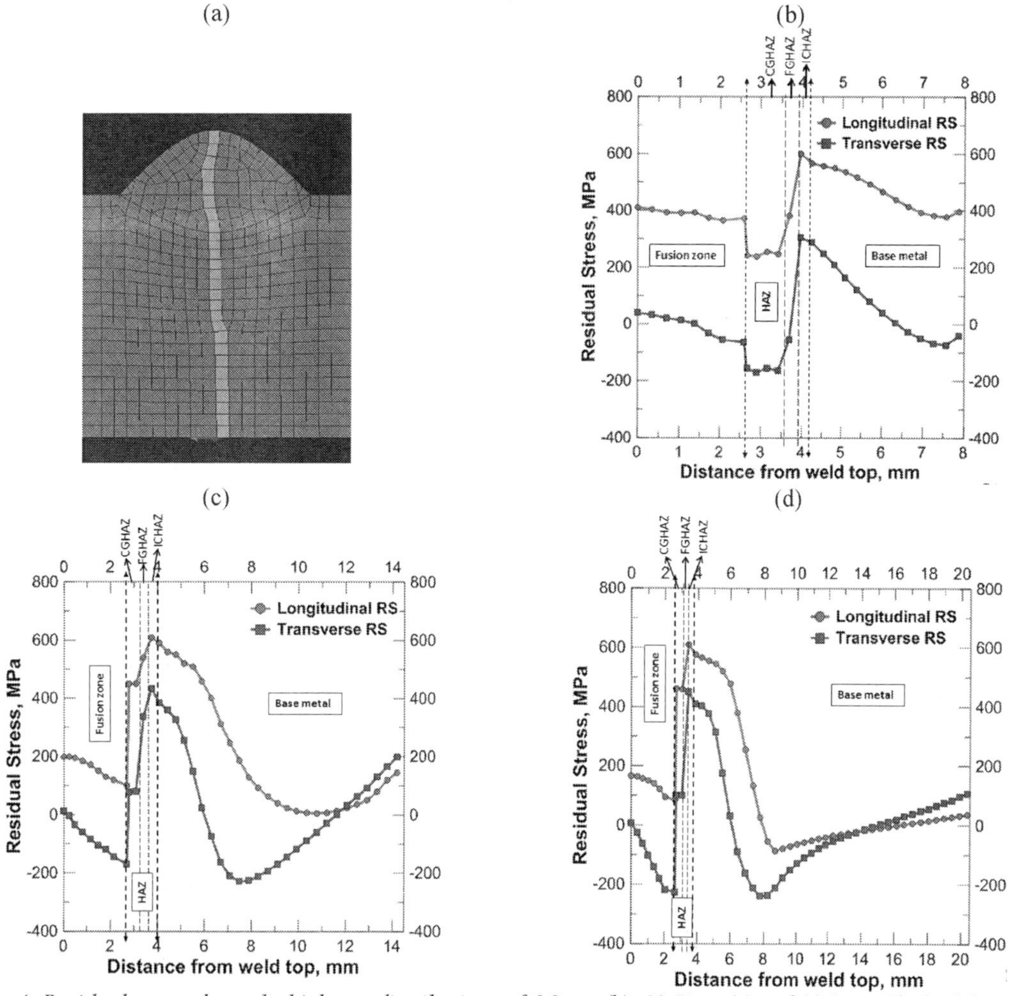

Figure 4: Residual stress through-thickness distributions of 6.3mm (b), 12.7mm (c) and 19.0mm (d) thick base metals following 3D element through-thickness path illustrated in (a).

Following such path, figures 4b, c and d represent through-thickness residual stress distributions for 327J/mm heat input and the three base metal thicknesses considered in the study. It can be observed in all distributions that the distribution point distance exceeds the base metal thickness since the weld metal reinforcement is included. The HAZ is classified into coarse-grained (CGHAZ), fine-grained (FGHAZ) and inter-critical (ICHAZ) regions using the peak temperature reached in the HAZ elements. In 6.3mm thickness, it can be seen that the longitudinal residual stresses are tensile from weld top to bottom surfaces. This observation is consistent with all higher heat input values in 6.3mm base metal thickness. In addition, it is observed that the weld metal and HAZ longitudinal residual stresses are higher than the transverse residual stresses which are in agreement with literature. In contrast to longitudinal residual stresses, the transverse residual stresses were in compressive state in the lower region of the weld metal, CGHAZ and at depth values greater than around 6.5mm from weld top surface. Invariably observed in all thicknesses and welding sets, the lowest residual stresses in the HAZ are consistently located in CGHAZ due to the appreciable amount of martensite present, reaching ~100%. Further increasing distance from weld top, the stresses, in all models, are observed to rise in the FGHAZ and reach their peak in the ICHAZ and sub-critical HAZ regions. This behavior is attributed to the high thermal strains reached in ICHAZ and sub-critical HAZ which are not relieved by appreciable transformation-induced strains as in FGHAZ and CGHAZ. At distance values greater than sub-critical HAZ, the residual stresses gradually decreased until the bottom surface region is reached. It is believed that the eventual rise of stresses prior to reaching the bottom surface is due to mechanical equilibrium. This result can be well explained by the peak temperature distribution in the through-thickness direction where the peak temperature is lower with increasing distance from fusion line. Such distribution leads to higher thermal strains and stresses for base metal regions located at distances closer to HAZ boundaries and lower for regions at further distances from HAZ boundaries. Therefore, such rise of residual stresses prior to reaching bottom surface is not caused by the local thermal strain of the bottom surface but due to a need to mechanically equilibrate the high residual stresses experienced in the weld metal, HAZ, and base metal that experienced higher peak temperatures and located close to the HAZ boundary. With increasing thickness, the distance between the peak residual stresses, at HAZ boundary, and bottom surfaces is increased leading to lower bottom surface peak residual stress values at 12.7 and 19.0mm thick base metals.

To validate the FEA results, the FEA estimated phase amounts, in 12.7mm thickness, are compared with metallographic and hardness results. The weld metal and CGHAZ microstructures, for 327J/mm heat input and 12.7mm base metal thickness, are shown in figure 5, as an example. Lath martensite is dominantly present in weld metal with 386 HVN local micro-hardness. Such hardness is quite typical for weld metal lath martensite and close to that found by Glover et al. [11] for welds generated under close conditions to this study. In addition, the CGHAZ region closest to the fusion line is dominantly martensitic with a minimum micro-hardness of 460 HVN. These values matched well with the FEA computed results of 100% martensite in the CGHAZ of the same heat input and base metal thickness.

(a) (b)

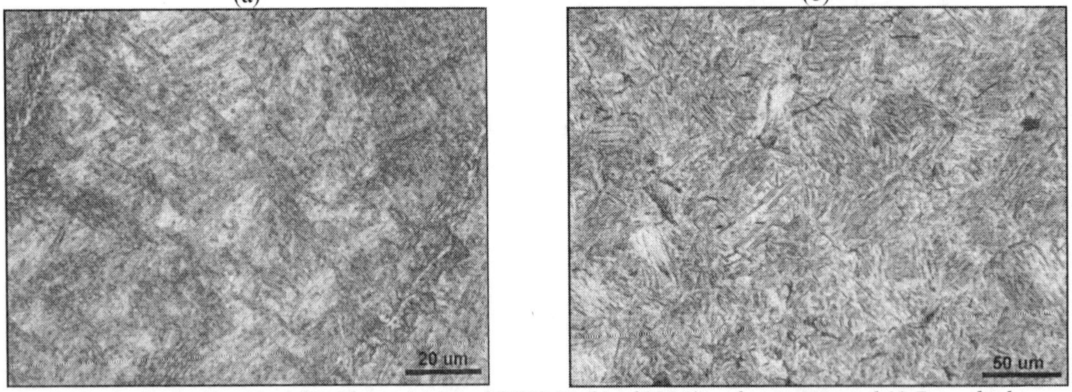

Figure 5: Optical micrographs of weld metal (a) and CGHAZ (b) of 327J/mm heat input and 12.7mm thickness.

Figure 6 represents FEA and metallography weld metal phase amounts in 12.7mm thick base metal varied as a function of heat input. The error bars in the metallography curve indicate the 95% confidence interval of point counting, after ASTM E562. Similar trends are observed by metallography and FEA results which are in agreement with a maximum deviation of 6% observed at the heat input of 509J/mm. Such close trends and phase quantification validate FEA phase computations. In addition, figure 7 illustrates the effect of heat input and base metal thickness on the residual stresses reached in the bottom surface. Such residual stress values represent those observed in the steady state region of the weldment and are normalized by base metal yield strength for relative comparison.

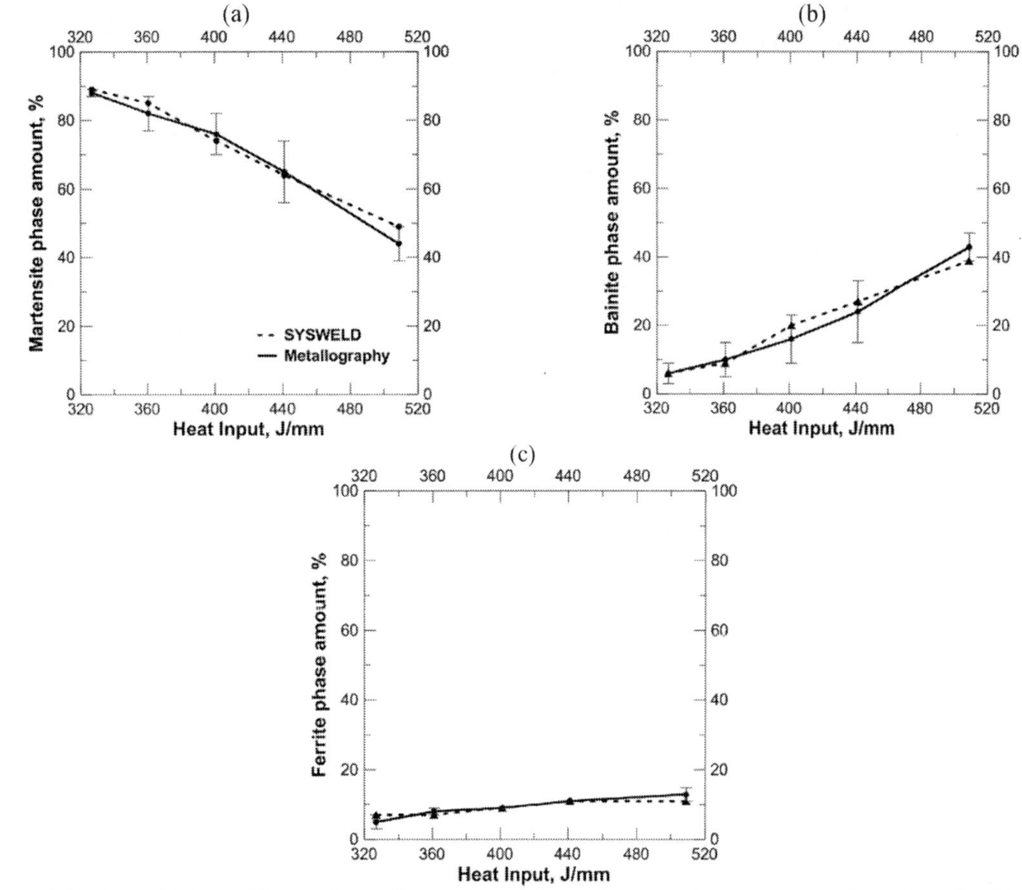

Figure 6: 12.7mm thickness FEA and metallography weld metal martensite (a), bainite (b) and ferrite (c) phase amounts varied by heat input.

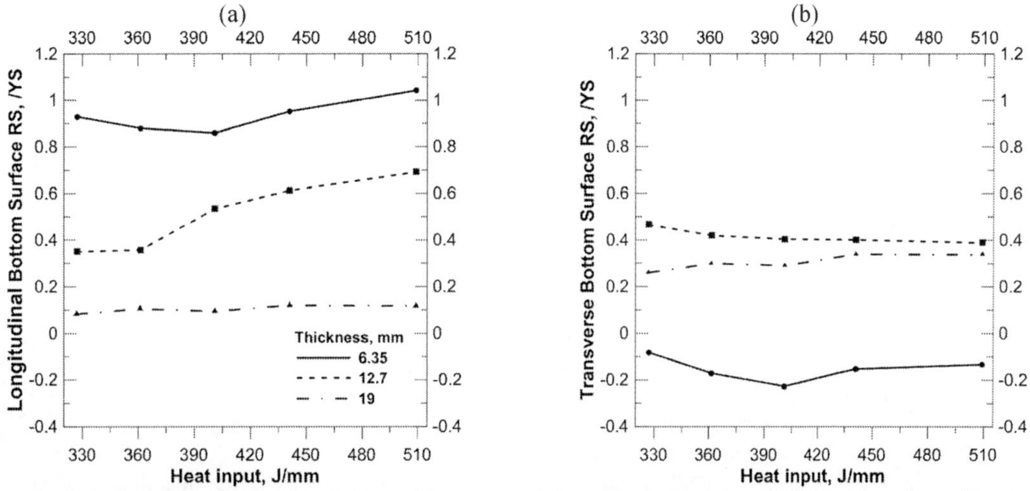

Figure 7: Bottom surface longitudinal (a) and transverse (b) residual stresses varied as a function of heat input and base metal thickness.

It can be seen that, in 6.3mm thick base metal, the bottom surface peak residual stresses exceed 80% of base metal yield strength for all welding sets. The corresponding stresses in 12.7mm thickness reach a range of 47-69% whereas only 25-35% of base metal yield strength was reached in 19.0mm thick base metal. From figure 7 and under the heat input range studied, PWHT would be required for any 6.3mm base metal thickness application whereas for 12.7 and 19.0mm thicknesses, PWHT can be waived depending on SCC environment and loading conditions.

Conclusions

It is obvious from the research results that the base metal thickness represents the most significant effect on bottom surface residual stresses reached. This conclusion is attributed to the distance between the HAZ boundary where peak residual stresses exist and the bottom surface potentially exposed to SCC environment. Invariably observed, the lowest HAZ residual stresses exist in CGHAZ region whereas the peak residual stresses are generated in the ICHAZ and subcritical HAZ regions.

The bottom surface residual stress results, illustrated in figure 7, should be used as an input for PWHT requirement determination along with loading condition and SCC threshold stress considerations. Since the bottom surface residual stresses in 6.3mm thick base metal can exceed 80% of base metal yield strength, PWHT is strongly recommended to avoid in-service SCC. Such high residual stress values observed exceed the 80% yield strength SCC threshold stress suggested by Jones et al. The bottom surface peak residual stresses in 19.0mm thickness application are significantly lower, only reaching a range of 25-35% of yield strength, which for most SCC environments is significantly safer and PWHT requirement can be optional. The peak residual stresses in 12.7mm thickness application reached between 47 and 69% of the yield strength. In such thickness applications, the PWHT requirement can be specified as a function of the specific SCC environment threshold stress. If the threshold stress is exceeded, PWHT is required. It should be emphasized that, for high risk applications, the utilization of the specific SCC threshold stress is essential for rigorous PWHT requirement determination.

Acknowledgments

The authors would like to acknowledge the financial support received from Saudi Aramco and the technical assistance received from ESI group. In addition, the authors would like to thank Dr. Zhenzhen Yu and all research personnel of the Colorado School of Mines - Center for Welding, Joining and Coatings Research.

References

[1] G. Totten, M. Howes and T. Inoue: Handbook of Residual Stress and Deformation of Steel, ASM International, Matal Park, (Ohio, 2002), pp. 391-397.

[2] R. Leggatt: "Residual stresses in welded structures", *International Journal of Pressure Vessel and Piping*, Vol. 85, (2008), pp. 144-151.

[3] R. Colegrove *et al*: "The welding process impact on residual stress and distortion", *Science and Technology of Welding and Joining*, Vol. 14, No. 8 (2009), pp. 717-725.

[4] X. Cheng *et al*: "Residual stress modification by post-weld treatment and its beneficial effect on fatigue strength of welded structures", *International Journal of Fatigue*, Vol. 25, (2003), pp. 1259-1269.

[5] R. Ibrahim, R. Rihan and R. Singh Raman: "Validity of a new fracture mechanics technique for the determination of the threshold stress intensity factor for stress corrosion cracking (K_{ISCC}) and crack growth rate of engineering materials", *Engineering Fracture Mechanics*, Vol. 75, (2008), pp. 1623-1634.

[6] S. Pal, R. Ibrahim and R. Singh Rama: "Threshold stress intensity factor and crack growth rate for stress corrosion cracking of simulated heat affected zone in caustic solution", *Engineering Fracture Mechanics*, Vol. 78, (2011), pp. 13-26.

[7] R. Jones and R. Ricker: Stress-Corrosion Cracking: Material Performance and Evaluation, ASM International, Metal Park, (Ohio, 1992), pp. 1-38.

[8] NACE SP0472-2010: Methods and controls to prevent in-service environmental cracking of carbon steel weldments in corrosive petroleum refining environments, NACE International, (Texas, 2010).

[9] AWS A9.5: Guide for verification and validation in computation weld mechanics, American Welding Society, (Florida, 2013).

[10] ASTM E562-11: Standard test method for determining volume fraction by systematic manual point count, ASTM International, (Pennsylvania, 2011).

[11] A. Glover, J. McGrath, M. Tinkler and G. Weatherly: "The influence of cooling rate and composition on weld metal microstructures in a C/Mn and a HSLA steel", *Welding Journal*, Vol. 56, (1977), pp. 267-273.

Numerical Study on Influence of Specimen Preparation Procedure upon Fatigue Strength of Weld Joints

Hidekazu Murakawa

Joining and Welding Research Institute, Osaka University, Japan

Fatigue strength of weld joint specimen is influenced by the procedure how the specimen is prepared. For example, when small specimens with same size are cut out from long weld specimen, the residual stress and distortion may be different depending on the location from where it is cut. In case of specimen containing several welding lines, the sequence of welding influences the residual stress and distortion. In this report, the influence of specimen preparation procedure is investigated using finite element method and simple method to estimate fatigue life.

Author Index

Author index (A-D)

A

B

C

D

Author index (D-H)

Author index (H-K)

I

J

K

Author index (K-L)

L

Author index (L-M)

Author index (N-P)

N

O

P

Author index (P-S)

Q

R

S

Author index (S-T)

T

*1 Kyushu University *2 Osaka University

Author index (U-Y)

U

V

W

X

Y

Author index (Y-Z)

Z

American Welding Society
8669 NW 36 Street, #130
Miami, Florida 33166-6672
USA

CURRAN ASSOCIATES INC.
proceedings.com

ISBN 978-1-5108-4403-2
9 781510 844032

10th International Conference on Trends in Welding Research and 9th International Welding Symposium of Japan Welding Society (9WS)

Tokyo, Japan
11 – 14 October 2016

Volume 2 of 2

10th International Conference on Trends in Welding Research and 9th International Welding Symposium of Japan Welding Society (9WS)

Tokyo, Japan
11 – 14 October 2016

Volume 2 of 2

ISBN: 978-1-5108-4403-2

Printed from e-media with permission by:

Curran Associates, Inc.
57 Morehouse Lane
Red Hook, NY 12571

Some format issues inherent in the e-media version may also appear in this print version.

Printed by Curran Associates, Inc. (2017)

For permission requests, please contact American Welding Society
at the address below.

American Welding Society
8669 NW 36 Street, #130
Miami, Florida 33166-6672
USA

Phone: 800-443-9353
305-443-9353

www.aws.org

Additional copies of this publication are available from:

Curran Associates, Inc.
57 Morehouse Lane
Red Hook, NY 12571 USA
Phone: 845-758-0400
Fax: 845-758-2633
Email: curran@proceedings.com
Web: www.proceedings.com

Proceedings

10th International Conference on

Trends in Welding Research

&

9th International Welding Symposium of Japan Welding Society (9WS)

October 11 - 14, 2016

Hitotsubashi Hall,
National Center of Sciences Building
Tokyo, Japan

Organizing Society
American Welding Society (AWS)
Japan Welding Society (JWS)

AWS JWS

http://trends2016.org

Program

October 11th (Tuesday)

Program - October 11th (Tuesday)

Room 2

[S3] Laser welding 1
Session chair: Hiroto Yamaoka

Room 3

[S4] Solid-state bonding 1
Session chair: Kazuhiro Ito

Room 4

[S5] Residual stress 1
Session chair: Zhili Feng

Program - October 11th (Tuesday)

Main Hall

14:00-14:40 **Keynote Lecture 2**
Session chair: Toshihiko Koseki

15:00-17:30 **9WS Special Session: Activities of Japan Welding Society, Trends of Welding Research Activities in Japan, Trends of Welding Engineering Activities in Japan**
Fumiyoshi Minami, Manabu Tanaka, Kenji Shinozaki, Satoru Asai, Kenji Oi, Mitsuyoshi Nakatani

Room 1

[S6] Microstructure of steel weld
Session chair: Norbert Enzinger

Room 2

[S7] Laser welding 2
Session chair: Satoshi Igi

Room 3

[S8] Solid-state bonding 2

Session chair: Yansong Zhang

Room 4

[S9] Residual stress 2

Session chair: Michael Andreassen

October 12th (Wednesday)

9:00-9:30 **Registration**

Main Hall

Program - October 12th (Wednesday)

Main Hall

Room 1

Room 2

Room 3

[S18] Dissimilar and carbon materials joining

Session chair: Shoichi Nambu

Room 4

[S19] FSW modeling

Session chair: Satish Kailas

October 13th (Thursday)

9:00-9:30 **Registration**

Main Hall

9:30-10:10 **Keynote Lecture 5**
Session chair: Stan A. David

[S20] Arc welding 4
Session chair: Yoshinori Hirata

Room 1

[S21] Welding of stailess steel
Session chair: Kazuyoshi Saida

Room 2

[S22] Computational fracture modeling
Session chair: Mitsuru Ohata

Room 3

[S23] Additive manufacturing 1

Session chair: Soshu Kirihara

Room 4

[S24] Steel FSW

Session chair: Hidetoshi Fujii

Program - October 13th (Thursday)

Main Hall

[S25] Arc welding 5

Session chair: Anthony Murphy

Room 1

[S26] Welding of stainless steel & Corrosion

Session chair: Kazuhiro Ogawa

Room 2

[S27] Computational welding mechanics

Session chair: Junya Inoue

Room 3

[S28] Additive manufacturing 2

Session chair: Wei Zhang

Room 4

[S29] Dissimilar FSW & FSSW

Session chair: Pedro Vilaca

Program - October 13th (Thursday)

Main Hall

[S30] Arc welding 6
Session chair: Hiroyuki Shimizu

Room 1

[S31] Welding for automotive
Session chair: Yongbing Li

Room 2

[S32] Simulation for weld microstructure
Session chair: Dieter Siegele

Program - October 13th (Thursday)

Room 3

[S33] Surface modification

Session chair: Toshitaka Ikeshoji

Room 4

[S34] Monitoring and inspection

Session chair: Tetsuo Era

October 14th (Friday)

9:00-9:30 **Registration**

Main Hall

9:30-10:10 **Keynote Lecture 6**
Session chair: Toshihiko Koseki

[S35] Simulation for welding process 1
Session chair: Fumikazu Miyasaka

Room 1

[S36] Cold cracking in weld
Session chair: Tadashi Kasuya

Room 2

[S37] Automation
Session chair: Satoru Asai

Room 3

[S38] Nano/micro joining

Session chair: Tomokazu Sano

Room 4

[S39] FSW metallurgy in Al alloys

Session chair: Egoiz Aldanondo

Program - October 14th (Friday)

Main Hall

[S40] Simulation for welding process 2
Session chair: Masaya Shigeta

Room 1

[S41] Hot cracking in weld 1
Session chair: Hiroaki Mori

Room 2

[S42] Welding comsumable
Session chair: Hiroyuki Shimizu

Room 3

[S43] Brazing and soldering

Session chair: Jaepil Jung

Room 4

[S44] FSW metallurgy

Session chair: Yutaka Sato

Program - October 14th (Friday)

Main Hall

[S45] Simulation for welding process 3

Session chair: Shinichi Tashiro

Room 1

[S46] Hot cracking in weld 2

Session chair: Kenji Shinozaki

Room 2

[S47] Pipeline

Session chair: Satoshi Igi

Room 3

[S48] Linear friction welding

Session chair: Masakatsu Maeda

Room 4

[S49] FSW metallurgy in steels

Session chair: Jorge dos Santos

Program - October 12th (Wednesday)

Poster

16:10-18:00

Fundamental and applications of welding and joining process

Weld properties and performance

Program - October 12th (Wednesday)

Modeling and simulation of weld behaviors

Weldability of materials

Microstructure and metallurgical behaviors in welds

Others

An Efficient Numerical Modelling Approach to Predict Residual Stresses in Ti-6Al-4V Linear Friction Welds

C. Buhr, P. A. Colegrove, A.R. McAndrew
Cranfield University, Cranfield, Bedfordshire, UK
c.buhr@cranfield.ac.uk

Abstract

There is increasing industrial interest in linear friction welding (LFW), due to its ability to manufacture high-value aerospace preforms, such as bladed disks. Unfortunately welding can produce significant residual stresses which can negatively impact performance. Computational models can provide an insight into how these residual stresses are formed, however conventional models that include the mechanical processing are extremely computationally expensive. Therefore this work presents an innovative and computationally efficient finite element model for predicting the residual stress formation in Ti-6Al-4V LFW. The model assumes that the residual stresses are predominantly due to the thermal history – not the dynamic oscillations. This assumption is proved by comparing models with and without the mechanical movement and the associated shear stresses and strains. The paper demonstrates the ability of the innovative modelling approach to accurately predict the thermal histories and residual stresses associated with the process. Thermal results are confirmed with thermocouple measurements. The effect of the applied pressure on the development of residual stresses was studied. The modelling predictions suggest that the higher the applied pressure, the lower the peak of tensile residual stress. These findings have significant, practical benefits and may aid further industrialisation of LFW.

1. Introduction

Linear friction welding (LFW) is an established solid state welding process in the aerospace industry to manufacture preforms, such as bladed disks. The titanium alloy Ti-6Al-4V has been found particularly suitable for bonding using LFW. The bond is formed by linearly oscillating one workpiece against another under pressure. A consequence of weld formation is the generation of residual stress due to the non-uniform heating of the weld region. A tensile band of residual stresses is generated at the weld interface, balanced by compressive stresses further away before eventually becoming nil [1]. Measuring residual stresses is difficult and expensive so numerical modelling can be a useful tool for understanding how the stresses are generated. Modelling of residual stress generation has been investigated for an extensive range of welding processes such as arc or friction stir welding, however there are relatively few publications on LFW which are limited to fully coupled models that include the thermomechanical processing that occurs during welding [2], [3]. In this work an innovative modelling approach has been developed to predict residual stresses without modelling the dynamic oscillatory motion.

2. Methodology

2.1 Modelling approach

A numerical model has been developed using the commercial finite element software ABAQUS. The dimensions of the workpiece of interest are presented in Figure 1 a) and one of the workpieces was oscillated in the x direction during the experiment. Dynamic oscillations are not simulated in the model. The temperature of the weld interface is reported to exceed the β transus temperature [4]. Hence, at high temperatures the stresses are expected to have a low magnitude and it is not likely they will affect the final residual stress field. Furthermore, Turner et al. [2] stated that the final residual stress field arises primarily as a consequence of the cooling. The analysis was carried out on an eighth of the structure, as shown Figure 1 b). This is because the deformation and heat flows are approximately symmetric around planes (P1), (P2) and (P3) [5], see Figure 1 a). The mesh is composed of hexahedral elements which grow in length from 0.35 mm at the interface and its vicinity to 1 mm further away. The mesh size was designed to capture the steep thermal and stress gradients expected near the weld interface. Temperature dependent material properties were used in the sequentially coupled thermal-stress analysis performed. Thermal conductivity and specific heat from Boivineau et al. [6] were used while Young's modulus and Poisson's ratio were obtained from Turner et al. [2]. Thermal expansion coefficient values were sourced by the material properties handbook [7] and flow-stress curves under and equal to 700°C were provided by Chen et al. [8] and curves over 800°C by Turner et al. [9]. Above a certain temperature, the Ti-6Al-4V material loses its strength to become highly viscous. During the LFW process, it is known that dynamic recrystallisation occurs at the weld interface and its vicinity [10]. Therefore to account for the loss of hardening memory, an annealing temperature of 800°C was defined in ABAQUS. At this temperature, flow stress curves exhibit a perfectly plastic behavior [9] and any work hardening is removed. Two sets of experimental conditions were used in this work and are presented in Table 1. The second welding condition has a much higher applied pressure than the first.

Thermal and mechanical models are computed separately. The thermal history obtained from the thermal model is mapped on the mechanical model where stresses are calculated over the heating phases and the cooling phase. To account for the heat generation occurring during the different phases, the average heat flux per phase was calculated by post-processing the LFW machine data, using the approach reported in McAndrew et al [4]. These heat fluxes were applied as uniform surface heat fluxes at the weld interface. The axial shortening that occurred during the process is accounted for by removing row of elements at a rate that matches the actual rate of axial shortening or burn-off. A time-step of 0.005 s is used during the heating phases; a time-step of 0.1 s is used for the first 15 seconds of cooling and a time-step of 1 s is used to complete the remaining 185 s of the cooling stage. The applied pressure is maintained during all the heating phases and during the first 15 seconds of cooling. Clamping of the workpieces was included and was modelled with a rigid body and a hard contact interaction defined between the clamp and the workpiece.

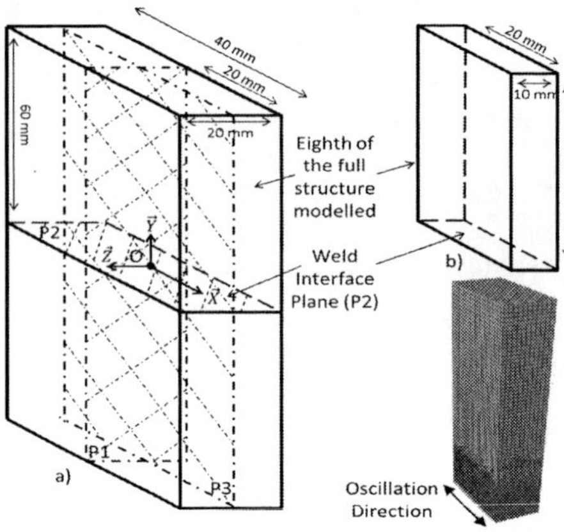

Fig. 1 a) Schematic diagram of the full LFW structure with its symmetrical planes P1, P2 and P3 and b) eighth of the full structure modelled.

Table 1: Welding parameters

	Weld 1	Weld 2
Frequency (Hz)	50	50
Amplitude (mm)	2.7	2.7
Applied pressure (N/mm2)	40	90

2.2 Modelling of the shearing

The authors conducted a study to prove that modelling the shearing process is not necessary to predict residual stresses. However for brevity, only the approach is described in this paper. A model that included the oscillating movement of the workpieces and the associated shear stresses was compared with one where this was ignored. The model that included the oscillatory movement required the full workpiece geometry. Rather than including all the oscillations of the process, a single oscillation was modelled at the end of phase 3. The two pieces composing the full welded structure were considered as a single body upon which a thermal profile was mapped. This approach was successfully used by Turner et al. [9] and McAndrew et al. [5] to model the dynamic oscillatory motion during phase 3. Two amplitudes were compared 0.1 mm and 2.7 mm. Simulations demonstrated that there was little difference in the results from the two models with both a shear stress of around 7 MPa at the weld interface being produced in each case. This stress is significantly lower than the final weld residual stress of 925 MPa, so there was negligible difference between the models that included the oscillatory motion and those that did not.

2.3 Experiments

Linear friction welds were made at TWI using the Thompson E20 machine. In order to record thermal histories, k-type thermocouples were used. Five holes were made through the workpiece at specific depths using EDM. Thermocouples were placed within the holes and an epoxy resin was used to fix them in place. Holes were drilled at 0.3mm, 2.7mm, 4.2mm and 5.2mm from the interface for both welds. A comparison between thermal histories recorded and simulated was made.

3. Results

3.1 Comparison of experimentally recorded thermal histories against simulation

Comparisons of thermal histories between experiments and simulations, for the two sets of experimental conditions presented Table 1, are displayed Figure 2. A reasonable match between experiments and simulations is observed. Differences between the predictions of the model and the temperature recorded arise from the positioning of the thermocouples and the accuracy of the material properties. The temperature gradients generated during the LFW process are very high, so a small error in a thermocouple position results in a large difference in temperature. In addition, the position of the thermocouple may be affected by the oscillatory motion of the process and contact resistance at the material/thermocouple interface can affect the temperature measurements. Finally, obtaining accurate material properties is difficult and Boivineau et al. [6], who is the main source of the thermal properties, estimated that the uncertainty of the specific heat was 3% and the thermal conductivity was 15%.

There is an oscillation of the temperature predicted 0.3 mm from the weld interface in Figure 2 which is due to the approach used to account for the axial shortening with the step-wise removal of the elements and is therefore a modelling artefact. However, it shows that this step-wise removal approach of material manages to capture the quasi-static state of the weld interface temperature. Also if the interface temperature for weld 1 stays almost perfectly constant (see Figure 2 a)), the interface temperature slightly decreases for weld 2 during phase 3, as shown Figure 2 b). A higher pressure was used for weld 2 which resulted in a greater rate of material expulsion in the experiments and consequently in the model. Therefore, the temperature cannot build up at the interface, resulting in a drop of temperature during phase 3 for weld 2.

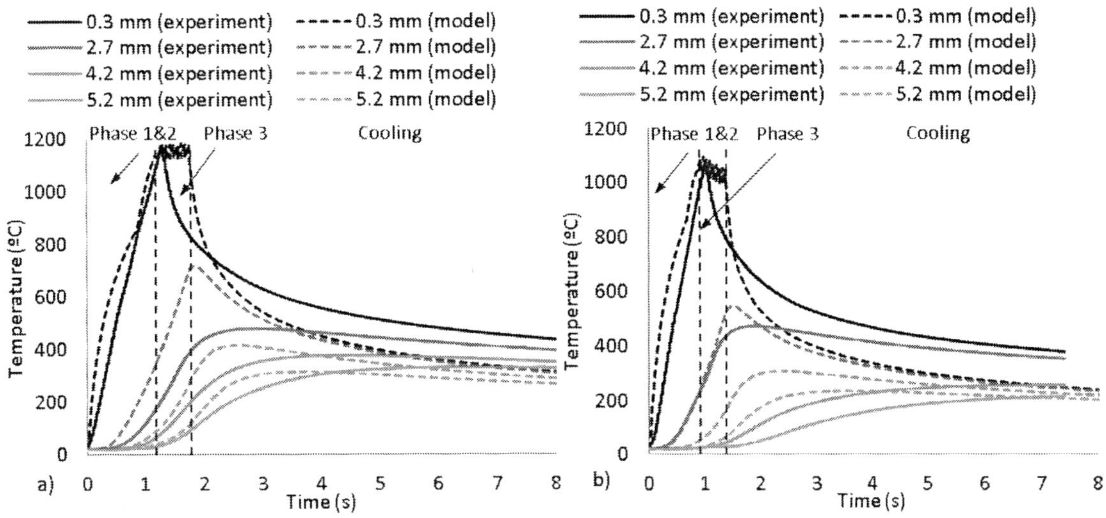

Fig. 2 Thermal histories obtained from experiments and FEA simulations for two sets of experimental conditions: a) weld 1 and b) weld 2.

3.2 Effect of applied pressure on the residual stress field

The residual stress distributions across the interface, in the three directions of space for both welds are displayed Figure 3 a). The highest magnitudes of residual stresses are observed in the directions of the weld interface σ_x and σ_z while the stresses in the direction of the applied pressure σ_y are noticeably lower. The maximum values are 925 MPa for weld 1 and 864 MPa for weld 2. Residual stress magnitudes and distributions predicted are in a good agreement with the literature [1], [2], exhibiting a band of tensile residual stress at the weld interface, which becomes compressive further away and then zero at the edge of the part.

Applying a higher applied pressure resulted in a decrease of the magnitude of the residual stress and narrowed the band of tensile residual stress. Looking at Figure 3 b), it can be seen that using a higher applied pressure causes a lower temperature at the end of phase 3 as a consequence of the higher burn-off rate. Therefore, during cooling, weld 2 will experience less thermal contraction than weld 1, resulting in lowered tensile residual stresses. These findings show that similar results to the fully-coupled numerical model developed by Turner et al. [2] are achieved using this simple modelling approach.

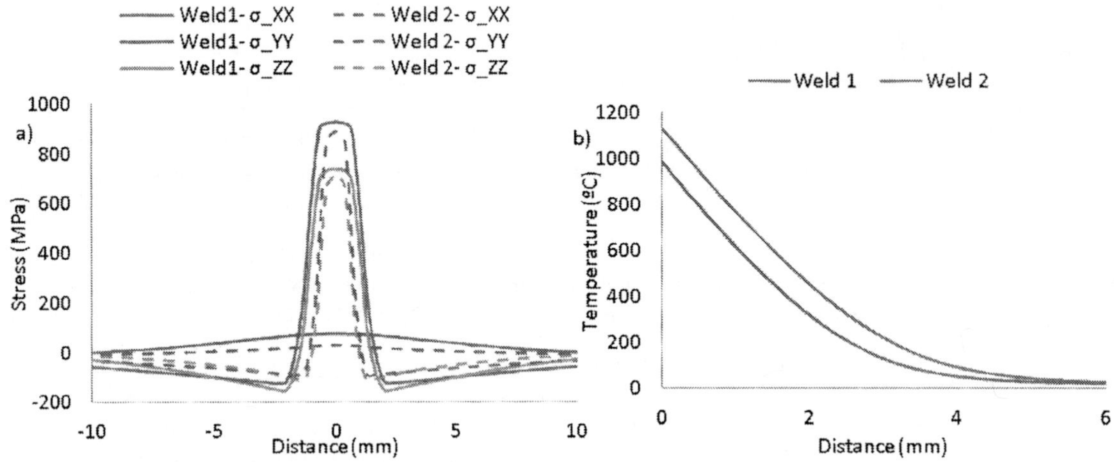

Fig. 3 a) Residual stress distributions in the three direction of space for weld 1 and 2, plotted across the weld interface and b) temperature profiles at the end of the heating stages.

4. Conclusions

This work presented a new modelling approach, bypassing the simulation of the dynamic oscillations of the LFW process, to predict residual stresses. Thermal histories predicted by the numerical models were validated against thermocouples for two sets of experimental conditions and were shown to be in a reasonable agreement. The dynamic oscillations were shown to be not necessary to simulate residual stress. Finally, an increase in the applied pressure reduced the magnitude of the residual stresses by lowering the temperature at the weld interface and its vicinity.

Acknowledgments

The authors wish to thank Honeywell Aerospace for its financial support of this project.

References

[1] P. Frankel, *et al*: "Comparison of residual stresses in Ti–6Al–4V and Ti–6Al–2Sn–4Zr–2Mo linear friction welds," *Mater. Sci. Technol.*, Vol. 25, No. 5 (2009), pp. 640–650.

[2] R. Turner, *et al*: "The Magnitude and Origin of Residual Stress in Ti-6Al-4V Linear Friction Welds: An Investigation by Validated Numerical Modeling," *Metall. Mater. Trans. B*, Vol. 43, No. 1 (2011), pp. 186–197.

[3] X. Song, *et al*: "Residual stresses in Linear Friction Welding of aluminium alloys," *Mater. Des.*, Vol. 50 (2013), pp. 360–369.

[4] A. R. McAndrew, *et al*: "Energy and Force Analysis of Ti-6Al-4V Linear Friction Welds for Computational Modeling Input and Validation Data," *Metall. Mater. Trans. A*, Vol. 45, No. 13 (2014), pp. 6118–6128.

[5] A. R. McAndrew, *et al*: "Modelling the influence of the process inputs on the removal of surface impurities from Ti-6Al-4V linear friction welds," *Mater. Des.*, Vol. 66, No. Part A (2014), pp. 183–195.

[6] M. Boivineau, *et al*: "Thermophysical Properties of Solid and Liquid Ti-6Al-4V (TA6V) Alloy," *Int. J. Thermophys.*, Vol. 27, No. 2 (2006), pp. 507–529.

[7] E. . Boyer, *et al*: Materials Properties Handbook: Titanium Alloys, ASM International (1994),pp. 493–521.

[8] G. Chen, *et al*: "Temperature dependent work hardening in Ti-6Al-4V alloy over large temperature and strain rate ranges: Experiments and constitutive modeling," *Mater. Des.*, Vol. 83, pp. 598–610, 2015.

[9] R. Turner, *et al*: "Linear friction welding of Ti–6Al–4V: Modelling and validation," *Acta Mater.*, Vol. 59, No. 10 (2011), pp. 3792–3803.

[10] A. R. McAndrew, *et al*: "Modelling of the workpiece geometry effects on Ti-6Al-4V linear friction welds," *Mater. Des.*, Vol. 87 (2015), pp. 1087–1099.

System Architectures for Large Scale Wire + Arc Additive Manufacture

P. A. Colegrove, A. R. McAndrew, J. Ding, F. Martina, P. Kurzynski, and S. Williams
Welding Engineering and Laser Processing Centre, Cranfield University, Building 46, Cranfield MK43 0AL, UK.
p.colegrove@cranfield.ac.uk

Abstract

The paper provides an overview of the ground breaking work on Wire + Arc Additive Manufacture (WAAM) undertaken at Cranfield University. Processes based on the common arc welding processes of TIG, MIG and plasma welding have advantages for manufacturing large-scale components. Examples of the kinds of parts that can be produced with these processes are provided as well as the different deposition strategies that can be used. Of particular interest is the improvement in properties that can be obtained with interpass rolling which results in a dramatic reduction in the prior β grain size in titanium parts. There is an improvement in the overall strength as well as producing properties that are isotropic. Both robotic and gantry based systems can be used to manipulate the torch and Cranfield University has examples of both. The High Value Engineering (HiVE) Centre is a gantry system based on a large friction stir welding machine, which provides a large platform for manufacturing parts that incorporate multi-dimensional interpass rolling. We also have a dual robot cell that is able to produce parts that are up to 10 m in length. The capabilities of each of these machines will be discussed and evaluated. In addition, other manufacturing processes such as rolling, machining, NDT and metrology may be incorporated so that quality assured parts can be manufactured.

Introduction

Additive manufacturing techniques currently have a high profile within industry and the wider public because of the perception that almost anything can be manufactured by such processes. The reality is that there are many technical challenges associated with even relatively simple parts that need to be overcome. Most metal-based parts are manufactured with systems based on lasers and powders where the deposition rate is limited. At Cranfield University we are developing Wire + Arc Additive Manufacturing systems where standard plasma, TIG or MIG based welding processes are used for high deposition rate manufacturing of components[1]. With these systems the size of the part is only limited by the size of the manipulator. WAAM is most suited to manufacturing components with a small batch size and the main business drivers are reductions in material cost through reductions in buy to fly ratio and reduced lead times[1].

Part building

Some examples of the components that have been manufactured by WAAM are shown in Fig. 1. The first component (Fig. 1a) shows a wing for a wind tunnel model that was manufactured in steel with Fronius Cold Metal Transfer (CMT) – a low heat input MIG-based process. This shows how varying wall thicknesses may be deposited with a uniform layer height provided the correct process parameters are used; and the ability to incorporate a hollow section within a component for weight saving or the provision of services. The thick wall is produced by oscillating the torch along a path similar to that shown in Fig. 2(a). The second part (Fig. 1b) shows a titanium landing gear rib that was produced for Bombardier. This was deposited with the plasma process within an argon filled tent which minimized surface oxidation of the part. The walls in this part are manufactured by two strategies as described in Bandari et al.[2]: the internal walls are deposited with the oscillating strategy while the exterior enclosed section is deposited with parallel walls – see Fig. 2(b). The part was almost symmetric in the amount of material deposited on either side of the baseplate so that distortion was almost eliminated. One of the challenges with manufacturing such components is obtaining the required geometry at intersections – an enlarged photo of one is shown in the figure. If no allowances are made to the deposition parameters, the deposit will narrow at the intersection due to the greater heat sink in this region which results in an increased layer height. The final component (Fig. 1c) is manufactured in a similar way to the second component, however the wall thickness is lower, so only single pass deposits are required in most areas. In addition, the part is single sided, however to eliminate distortion two identical parts were built on either side of the baseplate. After heat treatment to relieve stresses the parts are split using electro discharge machining.

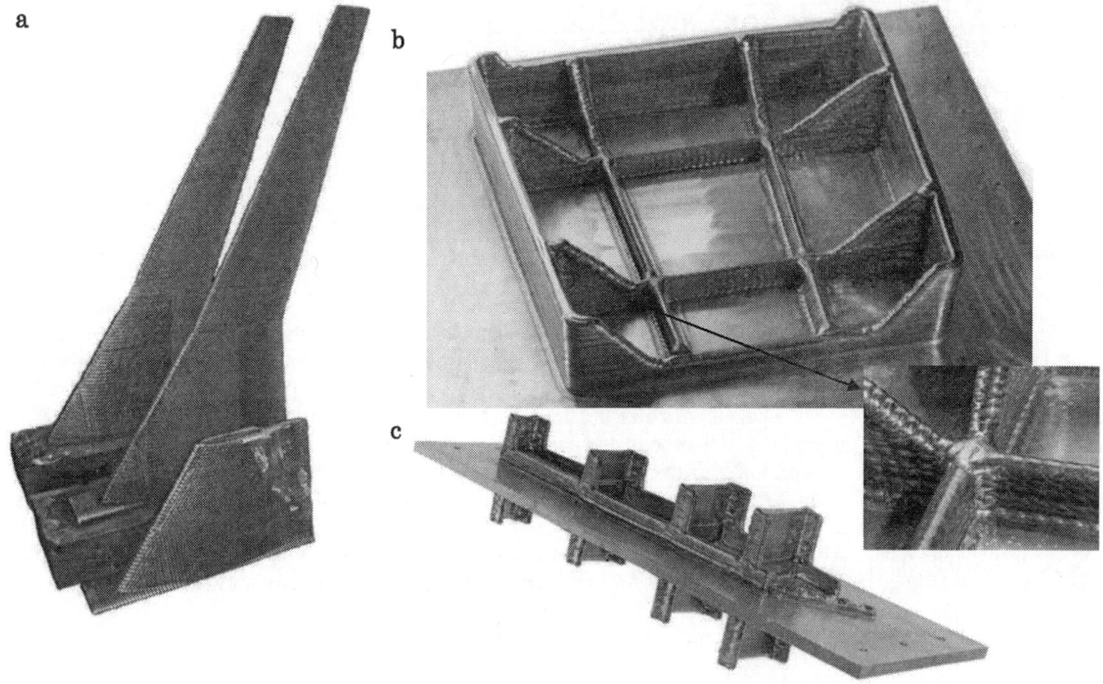

Fig. 1 (a) Steel wing manufactured for a wind tunnel model; (b) titanium landing gear rib for Bombardier and (c) thin-walled titanium component for Fokker.

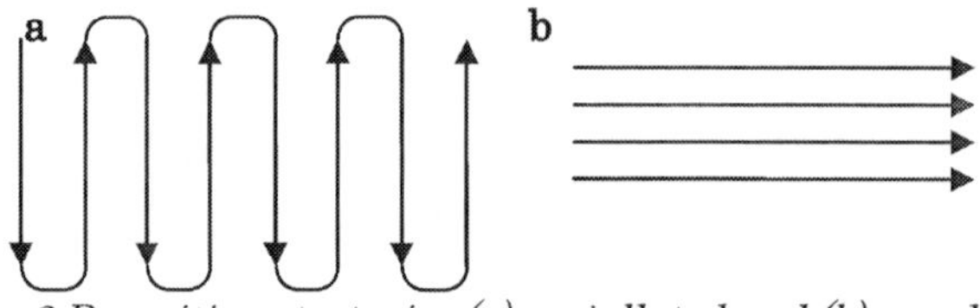

Fig. 2 Deposition strategies (a) osciallated and (b) parallel.

Although a previous publication[1] stated that the preferred deposition method was MIG based systems because of their flexibility, recent work has demonstrated the significant advantages of plasma-based deposition techniques. These have the advantage of allowing independent control of the wirefeed speed and current, which is particularly important around intersecting features.

Grain refinement through interpass rolling

Interpass rolling was originally implemented for residual stress and distortion control, but has subsequently proven to be an extremely effective method for refining the prior β grains in Ti-6Al-4V which under normal circumstances grow epitaxially from the substrate as shown in Fig. 3(a). The improvement in the prior β grain size has been reported in Martina et al. [3]and Donoghue et al. [4]and is shown in Fig. 3(b,c). The benefits to the material properties are evident in Fig. 3(d): there is a significant improvement in the overall strength and the properties are almost isotropic.

For thin walls (e.g. 6 mm) the microstructural refinement can be achieved with either a flat or a profiled roller that conforms to the shape of the deposit. Until relatively recently, achieving the microstructural refinement in thick >10 mm walls was a challenge because of the difficulty in inducing strain into the material. i.e. loading goes from plane stress to plane strain. Therefore alternate roller designs were developed that concentrated the deformation along bands in the deposit. An example of a roller design that was particularly effective is shown in Fig. 4(a), and the corresponding refinement in the microstructure is shown in Fig. 4(c). An unrolled control sample is shown in Fig. 4(b) for comparison.

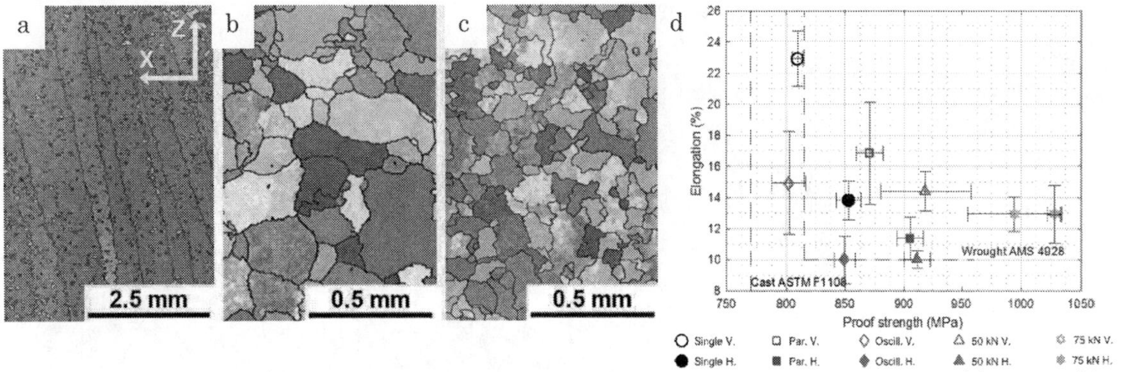

Fig. 3 Prior β grains in (a) as-deposited Ti-6Al-4V and after rolling with a rolling load of (b) 50 and (c) 75 kN. (d) The effect on the material properties, for single, parallel (Par.) and oscillated (Oscill.) walls in horizontal (H.) and vertical (V.) directions.

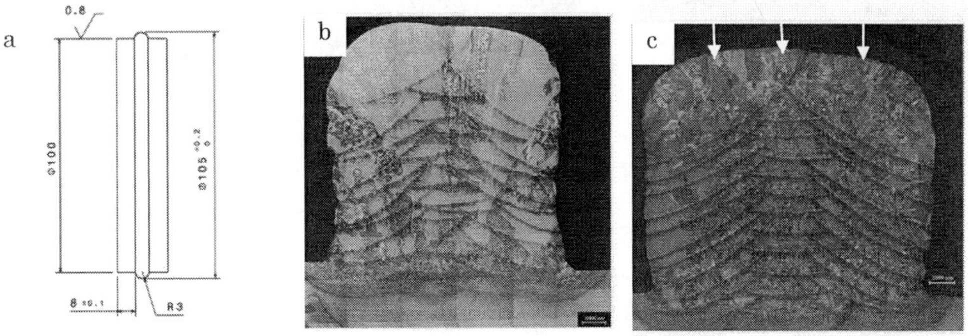

Fig. 4 (a) Roller design used for thick walls and (b) control sample without rolling and (c) microstructural refinement with a rolling load of 90 kN. Note the arrows indicate the tracks where rolling was applied.

Equipment

There are several options when it comes to the equipment that can be used for WAAM. Most of the parts manufactured at Cranfield University have been produced with robotic systems because they are relatively inexpensive and flexible and generally the accuracy is adequate for the kind of parts that we produce. We have recently acquired a dual robot system (Fig. 5(a)) that has two 10 m tracks for each robot – hence we have the capability of producing 10 m long parts. Depositing with two robots will increase productivity which will be important for large parts. We are also developing a tool changing system to incorporate in-process machining, NDT and metrology.

The alternative to such as system is a gantry-based system such as the High Value Engineering (HiVE) Centre which is shown in Fig. 5(b) and is based on what was originally a Friction Stir Welding machine. Gantry-based systems have higher levels of accuracy and most importantly can apply the high rolling loads necessary for interpass rolling of titanium structures. They are however much more costly and are therefore only suited to high-end aerospace applications. An enlarged view of the deposition and rolling equipment is shown in Fig. 4(c). The roller is placed in the spindle of the FSW machine which can rotate to facilitate rolling of 2D parts. Feeding accuracy of the wire is improved by rotating the wire feeder and spool around the torch – without this small deviations in the position of the wire occur as it rotates around the torch due to the camber of the wire. Also shown in this figure is a local shielding device (provided by Air Products/PWP) which eliminates the need for either a tent or chamber. This is particularly useful for large working envelope systems such as the ones described where the amount of argon required to fill an enclosure that encompasses the device becomes prohibitively large. The device shown uses flow straighteners to produce a laminar flow that avoids entrainment of the surrounding air and the details are provided in Ding et al. [5]. This local shielding device is more effective than conventional ones used for welding in that a suitable local environment is obtained at significant distances below the device as required for additive manufacture [4]. Based on surface appearance of the parts produced with this device, it is superior to those produced with either a tent or chamber. This is likely to be due to the argon purity being higher in the local device than that often used with a tent (500 ppm oxygen). The optical appearance of a wall manufactured with this device is compared with a part manufactured in a tent in Fig. 5(d).

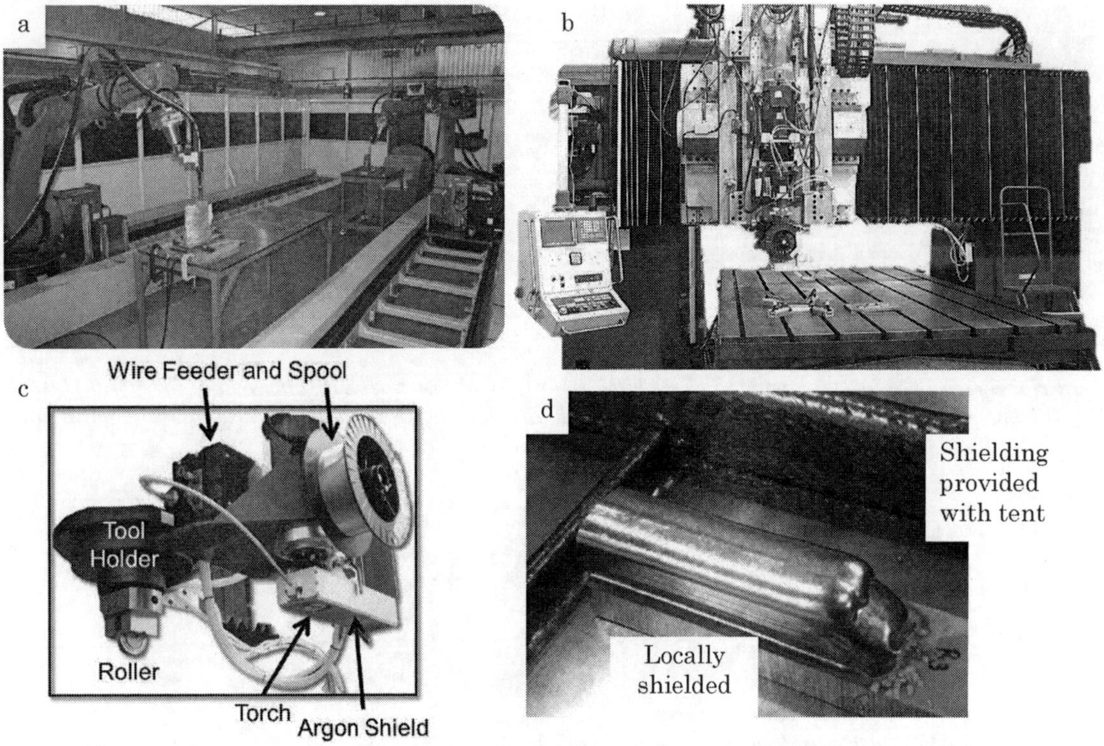

Fig. 5 (a) Dual robot system for WAAM – robots are on 10 m long tracks; (b) gantry based HiVE machine for aerospace applications; (c) depositon and rolling head on the HiVE machine and (d) comparison of parts shielded locally vs. in a tent.

Conclusions

In this paper we have:

- Discussed the advantages of WAAM for manufacturing large-scale parts.
- Described some of the techniques that can be used for depositing thin and thick walls as well as the problems that can occur when making intersections.
- Described how rolling can be used to significantly reduce the prior β grain size in titanium alloys which results in significant improvements to the mechanical properties.
- Compared the benefits of robotic and gantry based equipment for deposition. Robotic systems are adequate for most applications, however a gantry based system is necessary where interpass rolling is used.

Acknowledgments

The authors gratefully acknowledge the financial support of Airbus, Global Robots, Aircraft Research Association, Bombardier, GKN , Air Products and Innovate UK through the RAWFEED project (Project Reference: 101663). In addition the support of Andrew Henstridge and Stephen Porter from Airbus is gratefully acknowledged.

References

[1] S. W. Williams, F. Martina, A. C. Addison, J. Ding, G. Pardal, and P. Colegrove, "Wire + Arc Additive Manufacturing," *Mater. Sci. Technol.*, vol. 32, no. 7, pp. 641–647, 2016.

[2] F. M. Yashwanth K. Bandari, Stewart W. Williams, Jialuo Ding, "Additive manufacture of large structures: robotic or CNC systems?," in *Solid Freeform Fabrication*, 2015, pp. 17–25.

[3] F. Martina, P. A. Colegrove, S. W. Williams, and J. Meyer, "Microstructure of Interpass Rolled Wire + Arc Additive Manufacturing Ti-6Al-4V Components," *Metall. Mater. Trans. A*, vol. 46, no. 12, pp. 6103–6118, Oct. 2015.

[4] J. Donoghue, A. A. Antonysamy, F. Martina, P. A. Colegrove, S. W. Williams, and P. B. Prangnell, "The effectiveness of combining rolling deformation with wire-arc additive manufacture on β-Grain refinement and texture modification in Ti-6Al-4V," *Mater. Charact.*, Feb. 2016.

[5] J. Ding, P. Colegrove, F. Martina, S. Williams, R. Wiktorowicz, and M. R. Palt, "Development of a laminar flow local shielding device for wire+arc additive manufacture," *J. Mater. Process. Technol.*, vol. 226, pp. 99–105, Dec. 2015.

Effect of process parameters on defects in Ti-6Al-4V specimen built with EBM

Masato Yamada, Yuuki Kozue, Masashi Mouri, Koji Nezaki
IHI Corporation, Yokohama, Kanagawa prefecture, Japan
masato_yamada@ihi.co.jp

Abstract

Titanium alloy workpieces built with additive manufacturing (AM) are going to be put to industrial use [1]. However, AM process still contains uncertainties on some of its principles in the build conditions provided by EBM-manufacturer are only applicable to a certain range of workpiece dimensions and shapes. Therefore, it is difficult to ensure the build workpiece without any defects: voids or lack of fusion that are harmful to mechanical properties may exist. This study targets Electron Beam Melting (EBM) using Ti-6Al-4V powder. Purpose of this study is to categorize defects in the build specimen, based on the situation of their occurrence. As a result of this parametric study for each considered situation, the mechanism required for adjusting the process was clarified.

1. Introduction

AM is fabrication method specific to adding material to required positon. Compared to ordinary fabrication method such as machining or welding, it doesn't have a limitation of tools so that it can produce a workpiece of arbitrary shape. Therefore, AM enables free production, in accordance with 3D-CAD design. However, AM fabrication still contains several unknown phenomena that there isn't any proper build condition applicable for any geometries or largeness of workpieces. Therefore, it still requires adjustments on build condition.

2. Experimental procedure

EBM machine (Q20, produced by Arcam AB) was applied for test build in this study. Its process shall be as follows. First, put a start plate, which workpiece shall be built on, into chamber of the machine. Second, metal powder shall be sweeped by rake, a jig with brush, as shown in Fig.1. Rake is plunged into pile of metal powder to draw it and make powder layer of specific thickness onto the start plate. Third, electron beam scans on the layer to preheat and melt to form cross section of required workpiece. And then, repetition of powder sweeping and melting makes completed workpiece. If workpiece has negative surface, where upper side is larger than bottom one, structure to support upper side shall be built under the surface to prevent distortion. This structure is called support. Especially for EBM, wafer support shall be applied for easy removal after build. Square bars were built as specimen, with Ti-6Al-4V powder provided by Arcam AB (See Fig.2). EBM process is made up of several processes and each process has a large number of parameters which can be modified. In this study, condition for melting powder, especially beam current and its scan speed were chosen as parameter to verify heat input during melting. Typical condition is shown in Table 1.

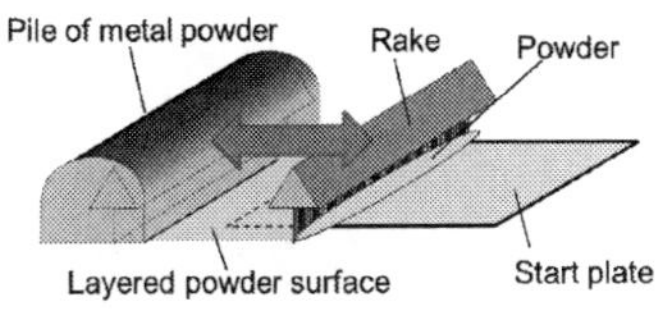

Fig.1: Schematic illustration of powder spreading

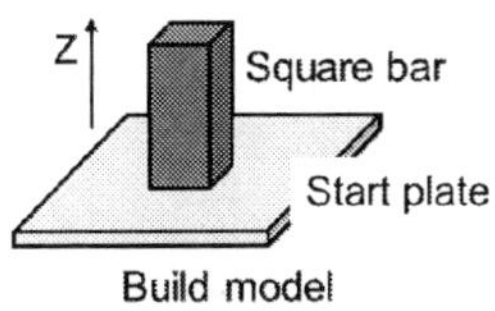

Fig.2: Schematic illustration of specimen

Table 1: Typical build condition

Beam current	5 mA to 28 mA
Voltage	60 kV
Scan speed	150 mm/s to 2000 mm/s
Layer thickness	90 um/Layer

3. Categorize of defects in workpiece

Prior to this study, typical defects observed in workpiece were categorized, as shown in Fig.3. First, top of workpiece was sometimes unfilled and voids were observed with visual inspection. Second, continuous defects along build direction (i.e. vertical direction) were sometimes found in radiographic test. Third, in case that cross section of workpiece was small or narrow locally, powder was not melted completely and remained as defects. In this study, suppositions for occurrence were set up and confirmed for these three types of defects.

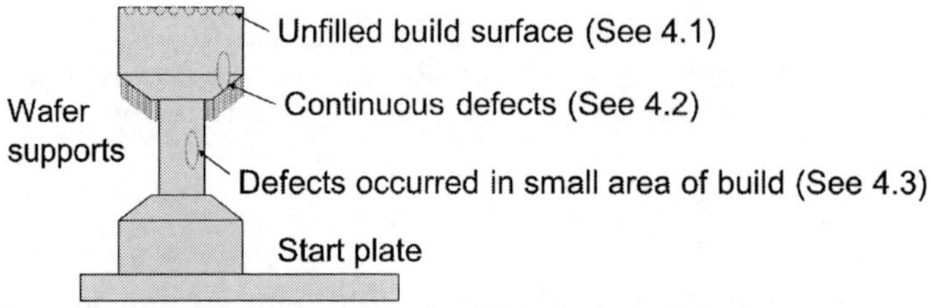

Fig.3: Typical example of defects found in EBM built workpieces.

4. Results and discussions

4.1 Unfilled build surface

Fig.4 shows typical appearance of top of workpiece after melting of specific layer. Z value shows build height completed. Voids opened on surface should be appeared as high-brightness image so that numbers of voids seemed to be increased as build height is increased. Following supposition was set up to explain this phenomenon. As height of blocks built is increased, heat capacity of them will be also increased. In addition, EBM doesn't apply any heat source other than electron beam so that all heat input is given from the top layer. Therefore, there will be temperature gradient from build surface to start plate, as shown in Fig.5. In that case, as block height increased, more heat would be conducted to block which was already built and then less heat can be applied for melting powder on surface. Then, it is estimated that this loss of heat input causes generation of voids on surface. To confirm this supposition, following build test was performed. Shape and height of blocks were set unified and heat input was varied. Specifically, beam current and voltage were kept constant and only scan speed was changed as parameter. As shown in Fig.6, there was a threshold on heat input whether voids were occurred or not. From this result, it is assumed that voids are occurred if reduced heat input goes below this threshold. For future testing, correlation between build height and threshold of heat input shall be clarified.

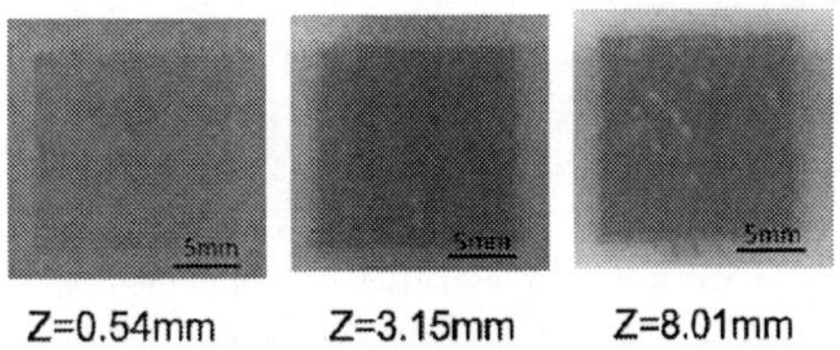

Fig.4: Image of surface on EBM built specimens

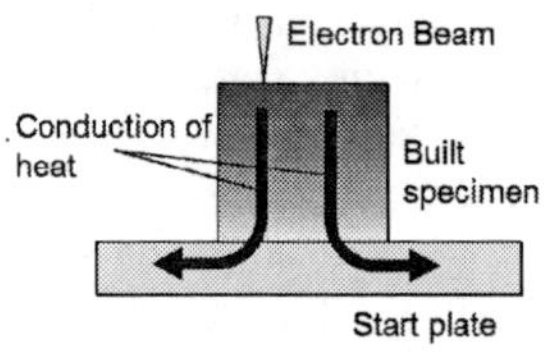

Fig.5: Schematic illustration of heat conduction

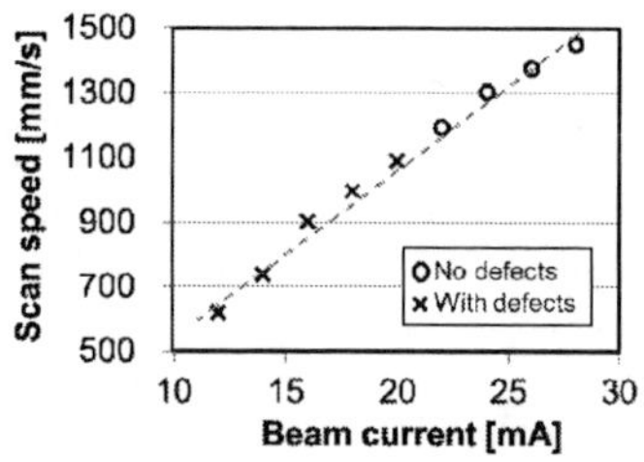

Fig.6: Effect of scan speed and heat input on defects

4.2. Continuous defects along Z-axis of build

Fig.7 shows radiographic test (RT) result of square bar. Specimen built on wafer support had continuous defects inside, along build direction. There are two suppositions for these defects as follows;

1) Melt down of build surface

First supposition is that melt down of build surface makes these defects. Fig.8 shows typical example of a beam shaped specimen built without wafer support. If build is started without any support (such as start plate or wafer support), level of negative surface fell under that of designed. It is estimated that negative surface could not support its weight until solidification and then fell into bottom. On the other hand, wafer supports are required to be removed after build so that they are designed as large-meshed structure, as shown in Fig.9. In this case, it is assumed that negative surface may melted down as well in small area surrounded by wafer support. To confirm this, first layer built on wafer support was observed, as shown in Fig.10. As a result, hollows were found on surface so that this assumption seemed to be reasonable. If powder was swept onto surface with hollows, these hollows would be filled with powder particle so that thickness of powder layer would be partially large, as shown in Fig.11. In that case, following phenomena is estimated: During melting next layer, heat input from electron beam would not be enough to penetrate partially thick layer of powder and melt entire hollows. And then, molten metal over the hollows would not support its weight and it would be fallen to bottom as well. As a result, hollows would be appeared repeatedly and becomes continuous defects. In this case, if molten metal had larger viscosity or solidified more rapidly, molten metal would not be fallen to bottom and defects would be prevented even if there were any space without support underneath. Then, build tests were performed with lower heat input for first few layers on the wafer support. As a parameter, beam current and voltage were kept constant and only scan speed was varied. Varying shall be in accordance with a function of thickness built on wafer support, as shown in Fig.12. Fig.13 shows typical appearance of top of workpiece after melting of layer 1mm above of wafer support. In case that scan speed was accelerated for layers near wafer supports and lowered rapidly, amount of voids were confirmed to be small. As a result, scan speed compensated dependent on build thick restrains occurrence of these defects.

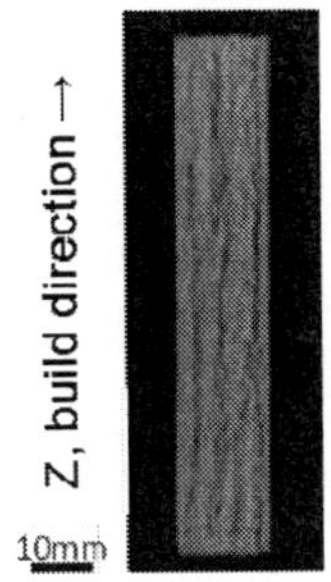

Fig.7: Image of RT result

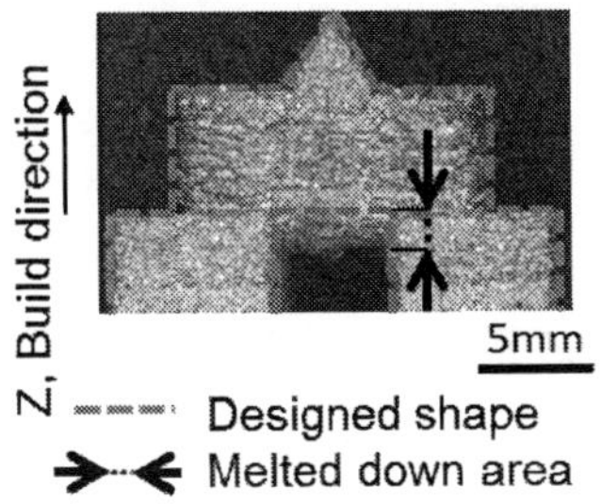

Fig.8: Melt down on beam shaped specimen

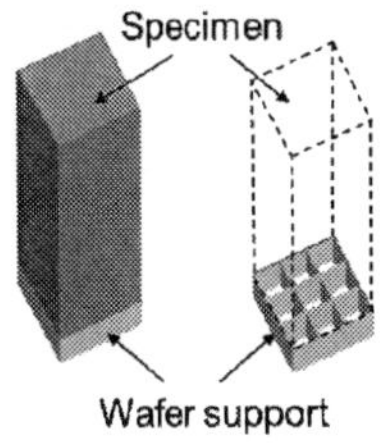

Fig.9: Example of wafer support

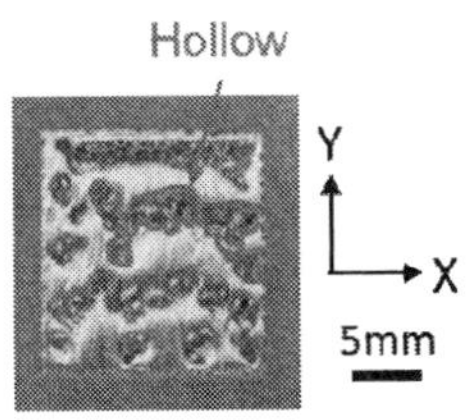

Fig.10: 1-layered thin plate specimen

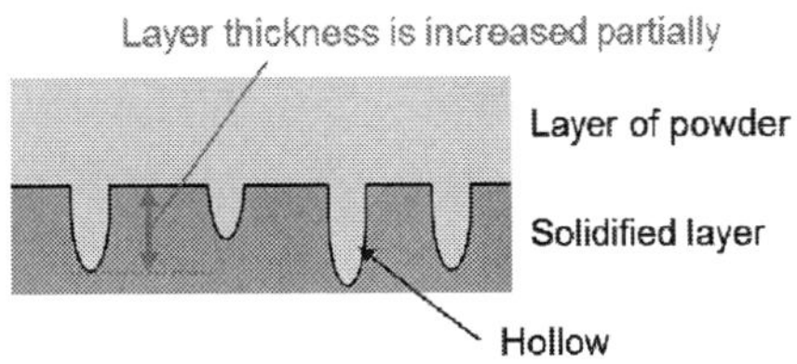

Fig.11: Schematic illustration of partially thick layer

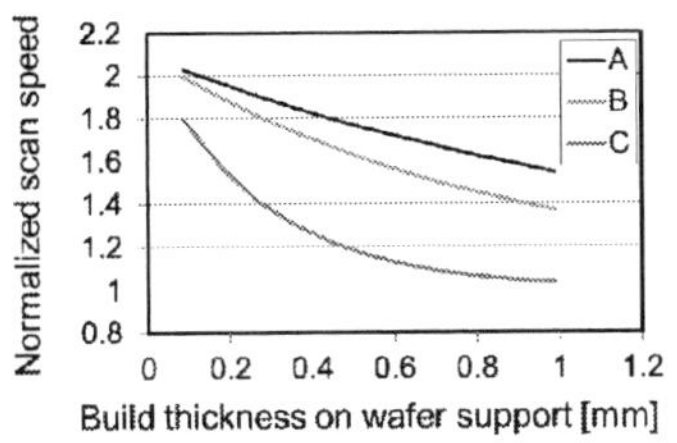

Fig.12: Speed compensation rate applied

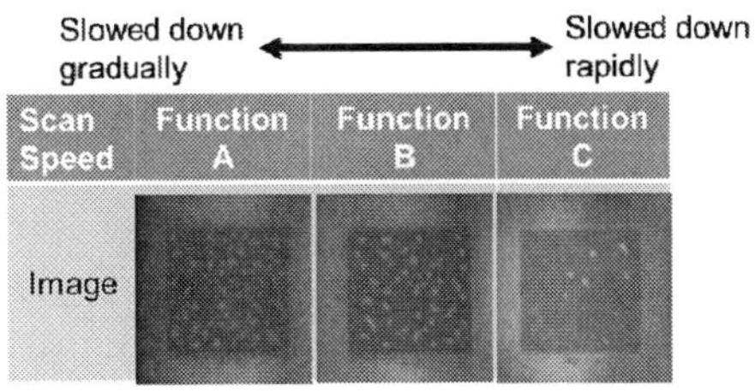

Fig.13: Appearance of top of workpiece

2) Balling phenomena

Balling phenomena is one of problem widely reported [2] on Selective Laser Melting, AM with laser beam heating, as shown in Fig.14. In this situation, line of molten metal is divided by surface tension and it is solidified spherically. If sphere of solidified metal is on the build surface, swiping powder over the sphere would not provide into gap between sphere and surface, as shown in Fig.15. In addition, that gap would be difficult to be melted by electron beam. Thus, it is assumed that defects like voids would be occurred, and then when powder of next layer fills these voids, they may be origin of continuous defects as well as melt down of bead surface.

In this case, it is estimated that if molten pool size during build would be enough large, it cannot be divided by surface tension. Though molten pool size can be verified by either of beam current or scan speed, if heat input per area is varied, it may cause defects in another way such as melt down of surface. Therefore, both of beam current and scan speed is varied in conjunction to keep heat input per area equivalent. Fig.16 shows correspondence relation between parameters and defects. As a result, defects were found in low scan speed and low beam current conditions. Therefore, it is assumed that high scan speed makes molten pools of adjacent scan lines to be united and then it may be enlarged compared to that of one scan line.

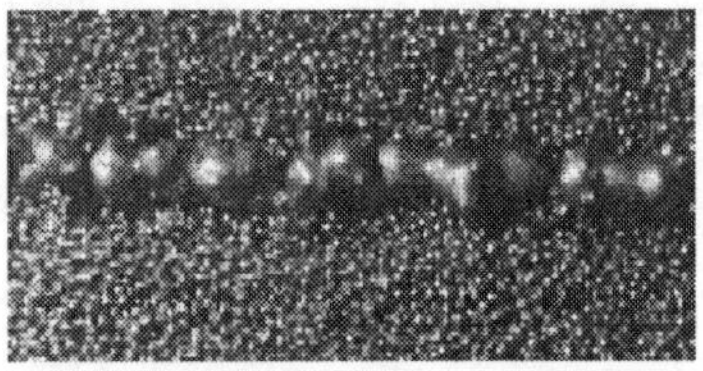

Fig.14: Balling phenomena on SLM (Quoted from Nikolay et al. [2])

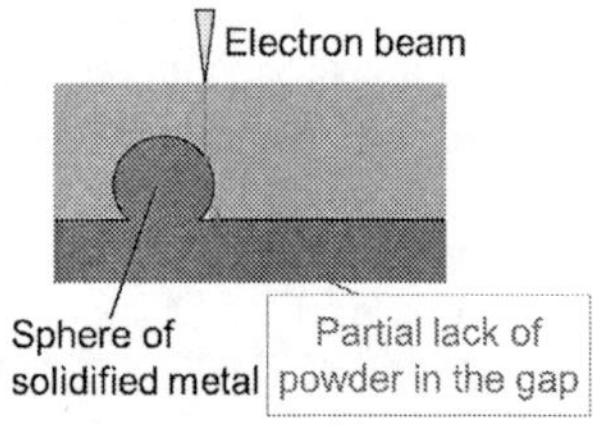

Fig.15: Phenomena around sphere of solidified metal applied.

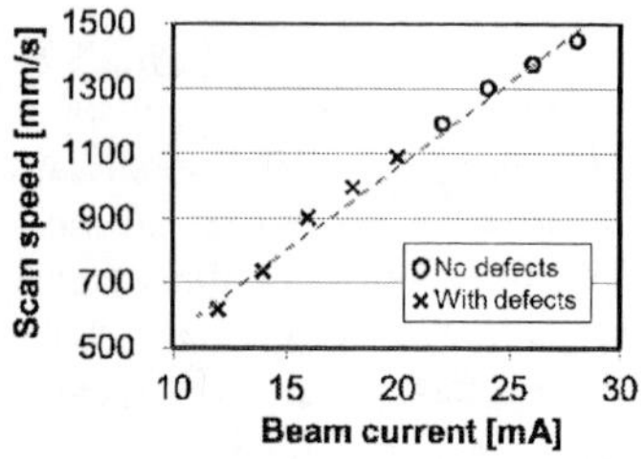

Fig.16: Correspondence relation between defects and parameters

As mentioned above, two different factors are estimated for occurrence of continuous defects. Though both of them are affected by scan speed, which factor is more dominant is not yet clarified so that additional testing shall be required.

4.3. Defects occurred in small area of build

Fig.17 shows an example of cross section of workpiece. There were large defects in narrow area of build. It was estimated that lack of heat input causes generation of these defects. Therefore, it is assumed increasing heat input will lower the viscosity of molten metal and make it spread well on build surface.

To confirm this, block specimens of 40mm and 3mm width were built. Beam current and voltage were kept constant and only scan speed was varied to apply multiple heat input per area. As shown in Fig.18, voids were found on surface of narrow specimen (3mm width), in contrast with wide specimen (40mm width). However, condition with more heat input, voids did not found. Therefore, heat input per area have to be determined dependent on width of build.

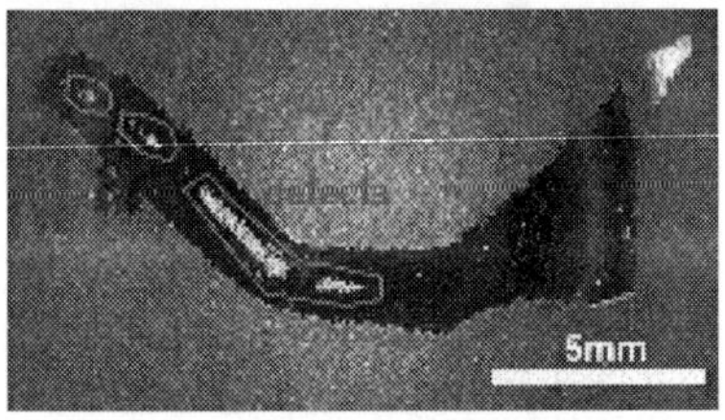

Fig.17: Example of defects in narrow area of build

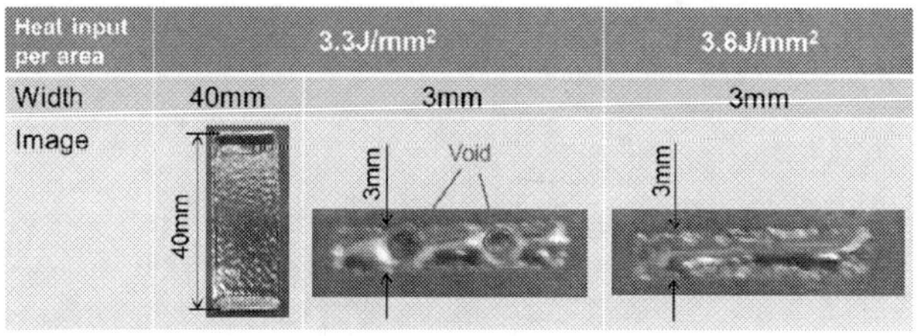

Fig.18: Correspondence relation between heat input, specimen width and defects

5. Conclusions

Typical defects occurred in EBM were categorized dependent on their features. Build surface tended to be unfilled as build process advanced. Conduction of heat to solidified specimen and start plate seemed to cause lack of heat on build surface so that limitation was required for minimum heat input per area. Radiographic test result showed continuous defects along build direction (Z-axis) and they should be restrained by scan speed. In addition, defects occurred in small area of build and they shall be prevented by raising heat input per area. Therefore, these parameters shall be modified and adjusted to ensure acceptable quality of the build.

References

[1] Raphael Salapete: "Comparison between EBM and LBM process for liquid propulsion engineparts," *2nd international symposium: Material Science and Technology of Additive Manufacturing*, Bremen, Germany, Feb 2016, p.21.

[2] Nikolay, K *et al*: "Balling processes during selective laser treatment of powders," *Rapid Prototyping Journal*, Vol. 10, Iss 2, pp. 78-87.

Microstructure Control of Alloy 718 Additively Manufactured by Selective Electron-Beam Melting

Yuichiro Koizumi [1], Shi-Hai Sun [2], Xiao Ding [1], Kenta Yamanaka [1], Kenta Aoyagi [1], Akihiko Chiba [1]

[1] *Institute for Materials Research, Tohoku University, Japan*

[2] *Institute for Materials Research, Tohoku University (Present: Osaka University), Japan*

The control of microstructures in additively manufactured parts is attracting growing interest. In the present study, Alloy 718 rods with cylindrical axes variously deviating from the build directions (BD) were fabricated by selective electron-beam melting: a powder bed fusion type additive manufacturing technology in which 3 dimensional objects are fabricated in layer-by-layer manner by selectively fusing metal powder particles. The microstructures of the rods were investigated with special focus on the effect of building condition and BD on their microstructures. Columnar grain structures or mixtures of columnar and equi-axed grains were obtained depending on the building conditions: preheating temperature, layer thickness, and interval between electron beam scanning lines). Fully oriented columnar structure was obtained under a condition with relatively higher energy input per volume. The <100> direction of the grains tended to be oriented not only in BD but also in the two horizontal EB scanning directions (SDs), which were perpendicular to each other. As a result, choosing their cylindrical axes could control the crystal orientation of the rods. For instance, rods oriented to <100>, <110>, <111> in their cylindrical axes could be fabricated by choosing their cylindrical axes parallel to BD, the face diagonal, and the space diagonal of the cubic space defined by the BD and the two SDs, respectively. To understand the relationship among building condition and microstructure, we investigated the effects of power and scanning speed of electron beam on the crystal growth in single beads formed by melt-run experiments on Alloy 718 plates under various conditions as well. It was revealed the epitaxial growth of the columnar grains occur much more readily than predicted from solidification map. The reason for the discrepancy will be discussed in terms of the influences of the smallness of melt pools, which are smaller than 1 mm.

Stereolithographic Additive Manufacturing of Geometrically Modulated Metal and Ceramic Components

Soshu Kirihara

Osaka University, Japan

Stereolithographic additive manufacturing (STL-AM) of lamination shaping can fabricate solid components according to topological geometry. Two dimensional (2D) cross sectional patterns were created through photo polymerization by ultraviolet laser drawing on spread resin paste including ceramic nanoparticles, and three dimensional (3D) composite models were sterically printed by layer lamination through chemical bonding. An automatic collimeter was equipped with the laser scanner to adjust beam diameter. Fine or coarse beams could realize high resolution or wide area drawings, respectively. As row material of the 3D printing, nanometer sized metal or ceramic particles were dispersed into photo sensitive liquid resins from 40 to 60 % in volume fraction. The resin paste was spread on a glass substrate at 10 μm in layer thickness by a mechanically moved knife edge. An ultraviolet laser beam of 355 nm in wavelength was adjusted from 10 to 300 μm in variable diameter and scanned on the pasted resin surface. Irradiation power was changed automatically from 10 to 200 mW for enough solidification depth for 2D layer bonding. The created 3D composite precursor was dewaxed and sintered in the vacuum and air atmosphere to obtain full metal and ceramic components. Through the computer aided design, manufacture and evaluation (CAD/CAM/CAE), geometrically modulated periodic and self-similar patterns with graded and fluctuated structures were processed. Porous electrode of yttria stabilized zirconia for solid oxide fuel cell, alumina photonic crystals to control electromagnetic waves and artificial bones of calcium phosphate scaffolds were created successfully.

Thermal management in manufacture of thin-walled components produced by arc-based additive manufacturing

Jayprakash P Sharma[1], Suresh Goka[1], Suryakumar Simhambhatla[1], Kazuyuki Kohama[2], Abhay Sharma[1], Kazuhiro Ito[2]

[1] Indian Institute of Technology Hyderabad, Telangana, INDIA

[2] Joining and Welding Research Institute,Osaka University

me12p1003@iith.ac.in, me14resch01002@iith.ac.in, ssurya@iith.ac.in, kohama@jwri.osaka-u.ac.jp, abhay@iith.ac.in, ito@jwri.osaka-u.ac.jp

Abstract

Significant progress has been made in last decade in area of arc-based additive manufacturing wherein complex geometries could be successfully manufactured. Control of inter-layer property variation and surface modulations are current challenges that the present investigation addresses. The challenges are more pertinent for thin wall components wherein surface area to volume area ratio is higher. Thin walled components over 100 layers are produced using cold metal transfer welding under different thermal management schemes such as natural cooling, forced cooling, liquid bed cooling, and their combinations. A consistent inter-pass temperature is maintained and thermal gradient are recorded. The samples are evaluated for surface topology, buy-to fly ration and achievable thickness. It is observed that cooling method significantly affect the surface topography. Use of a liquid bed provides a control on thermal gradients which also leads to a smoother outer surface and minimum post-process of material skinning. The result of present investigation has potential to be coupled with inter-layer heat management through control of process parameters.

Introduction

Additive Manufacturing (AM) is a technique of realizing desired shapes directly from their digital data. In AM process, the component grows from bottom to top till completion and reduces material wastage, tooling and lead time by slicing the 3D geometric model into 2D layers. Based on the energy source used, AM methods for creating metallic components can be classified as laser, electron beam and arc based processes. Of these, the arc based processes has been of particular interest for many researchers due to its high deposition rates, better energy and material utilization and low costs [1]. The current work uses a Gas Metal Arc Welding (GMAW) based weld-deposition system integrated with a CNC setup for producing thin walled metallic components (i.e., objects with single bead-on-bead) [2]. As the parts are made in a layer-by-layer manner, different layers are subjected to varying levels heating cycles and hence thermal management of the same assumes importance. The current work investigates the influence of various control parameters on the process for better thermal management.

Experimental setup

The experimental setup for realizing the thin-walled component consists of a Cold Metal Transfer (CMT) unit for weld-deposition and a CNC vertical milling center as positional interpolator as shown in Fig. 1. A synergic CMT is used as a weld-deposition unit for the present investigation as it offers several benefits in metallic AM over conventional arc welding process such as better arc stability, good quality material deposition, less distortion, low thermal heat input, and nearly zero spatter. The material deposition process in CMT is based on dip-transfer welding. The droplet detachment in CMT is purely due to the to and fro (push pull) motion of the wire and this happens at a frequency of 70-80 Hz. A 0.8 mm ER70S-6 (copper coated mild steel) filler wire is used for all experiments and an 82:18% mixture of $Ar:CO_2$ is used as the shielding gas. Thin walled features of length 180 mm each were created for the studies with a welding torch speed of 400mm/min. The following is the composition of filler wire used: C (0.075%), Si (0.67 %), Mn (1.22 %), S (0.013%) and Ph (0.01 %). The deposition process was performed on a mild steel substrate of dimensions 300 mm X 150 mm X 25 mm. During the deposition process, the substrate was kept stationary and the weld-deposition torch was moving across the substrate. For positioning of the weld-deposition torch at the desired position with the required speed, it was mounted onto a CNC vertical milling centre, as shown in Fig. 1.

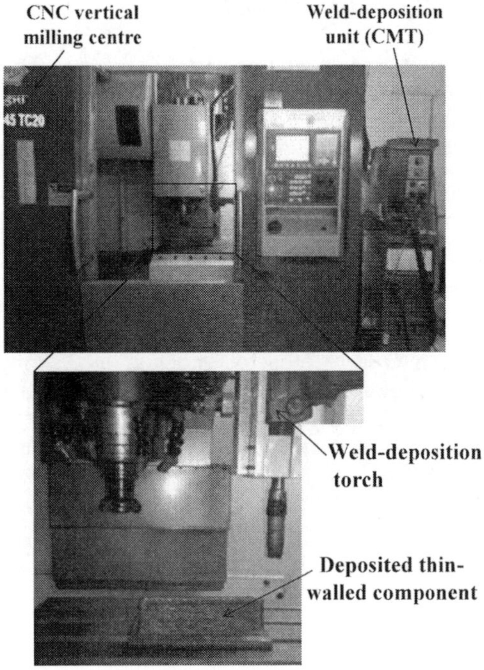

Figure 1: Experimental setup at IIT Hyderabad.

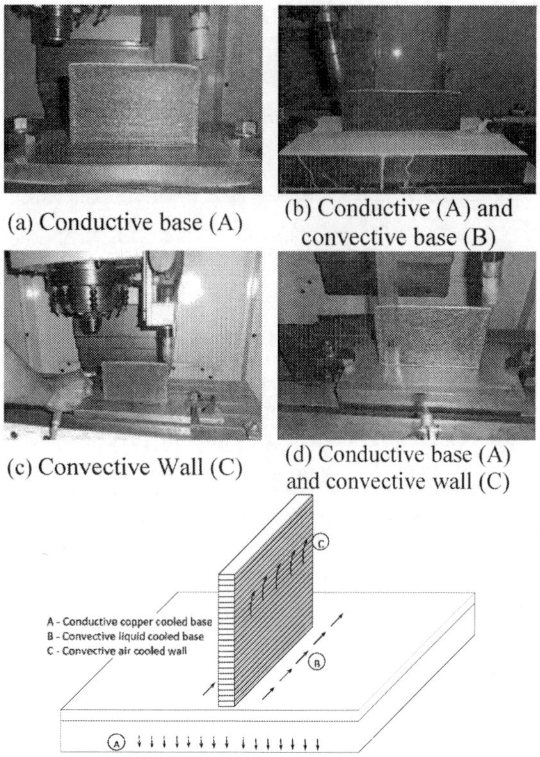

Figure 2: Various cooling arrangements.

Methodology

Control of inter-layer property variation and surface modulations are some of the important challenges in the weld-deposition process. To address the same, the following thermal management schemes have been investigated (Figure 2).

a) Conductive base (A) : Natural cooling with Copper back plate
b) Conductive (A) and convective base (B): Liquid cooling at base with Copper back plate
c) Convective Wall (C): Inter-pass cooling using forced air
d) Conductive base (A) and convective wall (C) : Inter-pass cooling using forced air with Copper back plate

The process is carried out in a layer-by-layer manner while ensuring consistent inter-pass temperature. After a given layer is deposited, the top surface of the layer is allowed to cool down till it reaches 93^{0}C (pre heating temperature of low carbon steel for avoiding thermal cracks) to deposit the next layer. The surface temperature is measured at the midpoint of straight thin wall. This process will continue till the height of the wall reaches to 120mm. In order to avoid crater defects and negate any effect of directionality, the deposition was carried out in alternate directions across the layers [3]. This means that the present (nth) layer deposition direction is quite opposite to the previous (n-1th) layer deposition direction. After straight thin-walled components with all cooling arrangements were made in the above mentioned procedure, Coordinate Measuring Machine (CMM) was used to observe its surface modulations (surface flatness). After the examination of the geometrical features, the thin wall was separated from the substrate using wire cut Electron Discharge Machining (EDM) process and then face milling operation was performed on the both sides of the metallic thin wall to achieve a flat thin wall. The minimum possible material was skinned to obtain parallel faces. Efficacy of different cooling methods is compered by calculating buy-to-fly ratio, vis-a-vis, ratio of weight of raw material used (i.e. deposited material) to weight of final product (i.e. plate).

Results

Figure 3(a) depicts geometrical accuracy for various schemes of cooling. The surface flatness remains same for all cases except convective wall (i.e. air cooled). Forced air facilitates faster cooling of molten material which leads to higher toe angle and produces surface modulation. However, forced air cooling with Copper backing does not results in surface modulations similar to that occurs with use of only forced air. The Copper backing provides directional heat flow towards the base. Thus, faster solidification that occurs in previous case does not occur. The effect of flatness can be seen on plate thickness in Fig. 3(b). The wall is considerably thin with forced air cooling as more material is removed in skinning. The wall thickness is minimum in case of conductive base with convective base (i.e. Copper backing with liquid cooling). However, the reason of the same is not the surface modulations as discussed earlier. The bead height in this case is considerably higher than other cases, as shown in Fig. 3(c). Liquid cooling during welding allows continuous cooling. The new bead rests on comparatively cooler surface and thus solidifies very fast without lateral movement of molten material. This facilitates thinner plate with considerable flat surface.

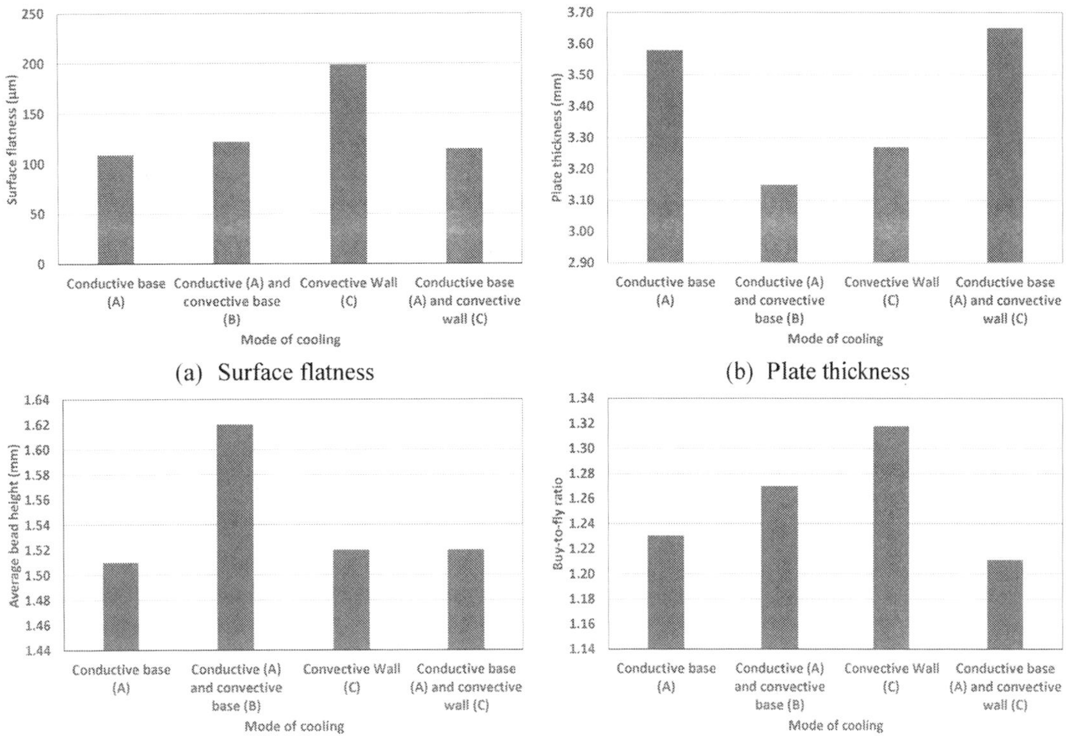

(a) Surface flatness (b) Plate thickness

(c) Average bead height (d) But-to-fly ratio

Figure 3: Effects of cooling methods

The effect of surface modulation can also be seen on buy-to-fly ratio in Fig.3(d). Convective wall cooling results in highest buy-to-fly ratio among all cases. This is because of excessive skinning due to surface modulations. Conductive base with air cooling results in maximum thickness and minimum buy-to-fly ratio which are desirable attributes of AM.

Conclusions

The present investigation reveals important observations on cooling methodologies in arc-based additive manufacturing. The cooling method can influence weld bead shape that eventually affects surface flatness. In turn, buy-to-fly ratio can also bealtered. Forced air cooling leads to rough surface with higher buy-to-fly ratio. Copper base with liquid cooling provides a desirable outcome such as thicker plate with lower buy-to-fly ratio. The results of this investigation can be used to develop a new process variant wherein process parameters and cooling methodology can be coupled to achieve better arc-based additive manufactured products.

References:

[1] P. Kazanas, P. Deherkar, P. Almeida, H. Lockett, and S. Williams, "Fabrication of geometrical features using wire and arc additive manufacture," *Proc. Inst. Mech. Eng. Part B J. Eng. Manuf.*, vol. 226, no. 6, pp. 1042–1051, Jun. 2012.

[2] J. S. Panchagnula and S. Simhambhatla, "Inclined slicing and weld-deposition for additive manufacturing of metallic objects with large overhangs using higher order kinematics," Virtual Phys. Prototyp., vol. 0, no. 0, pp. 1–10, Apr. 2016.

[3] J. Xiong, Z. Yin, and W. Zhang, "Forming appearance control of arc striking and extinguishing area in multi-layer single-pass GMAW-based additive manufacturing," Int. J. Adv. Manuf. Technol., pp. 1–8, Feb. 2016

[4] Y. Ma, D. Cuiuri, N. Hoye, H. Li, and Z. Pan, "The effect of location on the microstructure and mechanical properties of titanium aluminides produced by additive layer manufacturing using in-situ alloying and gas tungsten arc welding," Mater. Sci. Eng. A, vol. 631, pp. 230–240, Apr. 2015.

Solid-State Joining of Dissimilar Materials with Friction Stir Scribe

Yuri Hovanski, Piyush Upadhyay, Leonard S. Fifield, and Saumyadeep Jana
Pacific Northwest National Laboratory, Richland, WA, USA
Yuri.Hovanski@pnnl.gov, 509-375-3940
Piyush.Upadhyay@pnnl.gov, 509-375-6591
Leo.fifield@pnnl.gov, 509-375-6424
Saumyadeep.Jana@pnnl.gov, 509-372-6979

Abstract
The complexities of joining materials with vastly different melting points hampers our ability to effectively realize the benefits of multi-material design. This is further complicated by the introduction of new microstructure-sensitive metals and reinforced composites, which only increases our need to be able to join very application-specific materials to those with dramatically different properties. This is especially challenging in cost sensitive, high volume markets that have largely relied on low-cost welding solutions. These markets have increasingly been trending toward non-welding solutions such as fasteners and adhesives, but this has come at a price. Friction stir scribe technology was engineered to meet the demands of joining materials with drastically different properties and melting regimes. The process enables joining of metals like magnesium and aluminum to metals with much higher melting temperatures like titanium and steel. Furthermore, it creates opportunities for linear joining of polymeric materials to metals. Details will be presented on how this process enables the joining of dissimilar metals and polymer/metal combinations.

Introduction
Invented and developed at the Pacific Northwest National Laboratory, friction stir scribe (FSS) technology has enabled joining of materials with vastly different melting points by producing in situ, rivet-like fastening at the material interface with the use of a modified friction stir welding (FSW) pin [1]. As a derivative of traditional FSW, FSS takes advantage of the hydrostatic pressure condition within the weld nugget to merge two familiar and independent processes, FSW and machining, into a simultaneous event. The FSS tool consists of the typical FSW tool with one modification: at the tip of the tool, a scribe cutter is attached offset from the axis of rotation. A schematic of a typical FSS tool plunged into a material stack to be joined is shown in Figure 1.

Figure 1. Schematic illustration of the FSS tool as it plunges into the dissimilar material lap joint (left). Typical FSS tool (center and right).

Layers to be joined are assembled and appropriately clamped in lap configuration such that the lower melting point material is on the top. While the conventional components of the FSW tool (shoulder and pin) plastically deform the top sheet, the scribe cutter cuts into the higher melting point material underneath. As the tool rotates and traverses forward, the in situ machining of the bottom sheet creates a mechanical interlock in the dissimilar material interface. The depth of cut in the bottom layer is determined by the length of the scribe cutter. A predetermined gap between the tip of the FSW pin and the interface is maintained so that only the scribe cutter is in contact with the bottom sheet while the FSW tool interacts with the top sheet. This aspect—unique to FSS—intrinsically prevents plastic deformation of the bottom layer (higher melting point material). This avoids the propensity of melting and void formation related to joining materials with significantly different melting points using conventional FSW. Herein we present a brief overview of the technique, including tooling and other critical requirements of the process. Progress made and challenges associated with development of joining dissimilar material combinations, including polymer to aluminum, aluminum to steel, and magnesium to steel, are also reported.

Body of paper

Various dissimilar metallic combinations have been investigated. As with traditional FSW, variations in material thickness, alloy, or surface conditions can lead to changes in process parameters and tool design in order to most effectively produce the desired joint characteristics. The combination of AZ31B magnesium alloy and zinc-coated mild steel was successfully joined using the FSS technique (Figure 2). The resulting joint performed significantly better than joints made without any scribe in FSW with an otherwise similar tool. The statistical variation in the tensile strength was significantly lower in the scribe joint (<5% overall). Strength of close to 90% of the base metal steel was regularly achieved. Additional information on this particular set of dissimilar-metal joinings can be found in Jana et al. [2-4]. Combinations of Mg and steel were investigated with coatings and interlayers, including e-coated Mg joined to zinc-coated steels (body steel and high-strength low-alloy steel (HSLA)), coated Mg joined to steel with wet Betamate adhesive (subsequently cured after the FSS process; see Figure 2), and variations of the same.

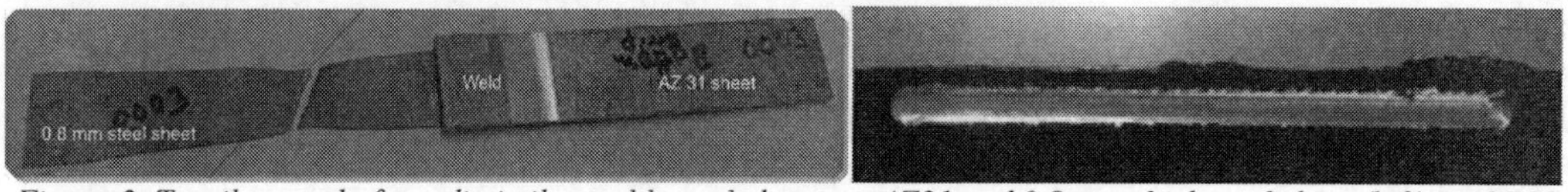

Figure 2. Tensile sample from dissimilar welds made between AZ31 and 0.8 mm thick steel sheet (left), and FSS joint between e-coated AZ31b magnesium and a zinc-coated HSLA steel with a wet adhesive interlayer (right).

In joints between mild steel and aluminum like the one shown in Figure 3, the FSS tool was machined out of H13 tool steel and had a scrolled shoulder and a tapered, tri-flat pin design. The pin tool had 12.7 mm shoulder diameter and the pin length was 0.9 mm. A tungsten carbide scribe (1.9 mm diameter) was inserted at 1.2 mm offset distance from the axis of rotation (Figure 1, right). A hole was drilled into the pin at the appropriate location and the scribe cutter was press-fit into the hole. Care was taken to insure that the bottom of the scribe was completely seated on the base of the hole to avoid any readjustment in the scribe height during joining. At the outset, the scribe height was chosen such that the scribe would engage 20% of the bottom sheet material.

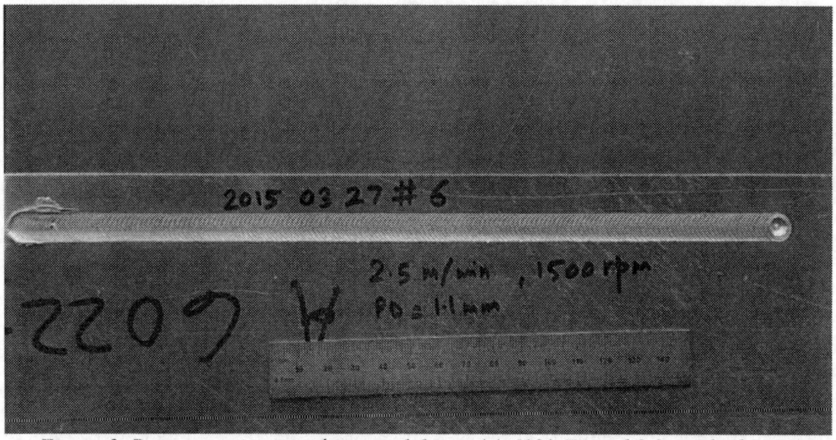

Figure 3. Representative joint between 1.0 mm AA-6022-T4 and 0.8 mm body steel

Several control parameters such as welding speed, rotational speed, and plunge depth were varied, resulting in a wide array of welded coupons. Samples that performed reasonably well in manual peel tests were subsequently tensile tested. Figure 4 shows load vs. elongation obtained from a representative joint sample extracted from an aluminum-steel joint. The welds were made at the welding speed of 750 mm/sec and 1950 rpm and at the plunge depth of 0.99 mm. For reference, the load vs. extension curve for base metal mild steel (0.8 mm) and that for AA6022 are also shown. For all the samples tested, the peak load sustained by the joint was very similar to that of the parent mild steel sheets. The necking and eventual failure took place away from the weld region, as shown on the lower right-hand corner of the figure 4.

There are two aspects of FSS joining that need to be fulfilled in order to obtain a defect-free joint that can perform well in tensile tests. While the FSW on the top sheet must be defect free, it is also important to have a complete hook feature at the bottom sheet to form an in situ joint without causing voids in the plastically deformed aluminum sheet. This requirement demands a very small window of plunge depth and scribe length. The hook feature morphology seen in Figure 5 for three different locations shows very consistent hook formation. Although there are inclusions of some stray steel chips into the aluminum nugget, this does not lead to void formation and eventual fracture initiation sites.

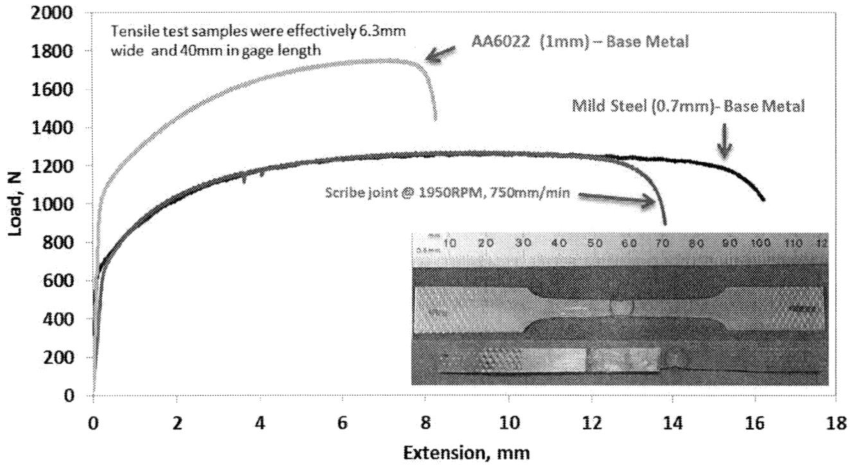

Figure 4. Load vs. extension of a representative FSS joint sample. An E8 sub-size tensile sample for mild steel and AA6022-T4 is shown for side-by-side comparison. The locations of final fractures are marked with red circles.

A wide range of aluminum/steel combinations have been investigated, including wrought, cast, and extruded aluminum products being joined to steel alloys ranging in strength from 280 MPa to 1500 MPa in thicknesses from 0.7 mm to 75 mm. Patterson et al. provide a more detailed look at the joining of aluminum to steel with FSS [5].

Figure 5 shows a representative FSS joint made between 3 mm thick carbon-fiber-filled polyamide and 1 mm thick AA6022 (left). For the case of a polymer to metal joint, in order to avoid excessive crown heating and to insure through-thickness uniformity, a stationary-shoulder tool setup was employed (Figure 5, right). A more detailed discussion of the tools, parameters used, and preliminary joint properties pertaining to polymer/aluminum joints can be found in Upadhyay et al. [6].

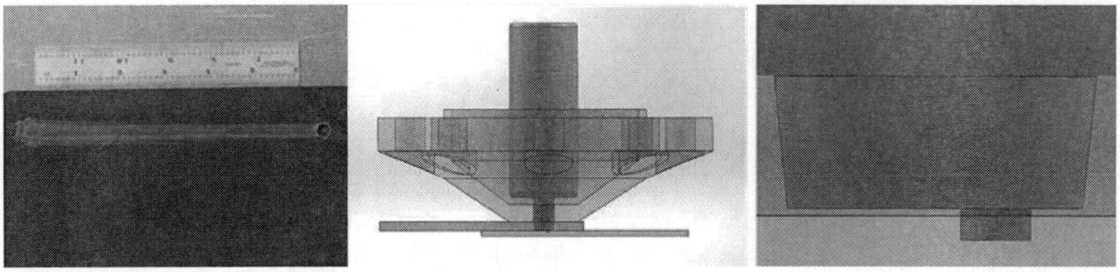

Figure 5. Representative FSS welded joint made between carbon fiber reinforced nylon (LCF50-PA66) and aluminum (left) and illustrative representation of the FSS process with integrated stationary shoulder (center and right).

Conclusions

The FSS technique has been successfully used to join dissimilar materials with vastly different material properties in lap geometries. The continuous, rivet-like in situ mechanical interlocking feature produced during the process was found to provide promising joint strength. While fairly consistent along the length of the joint, several joining and tooling parameters can influence the effectiveness of the necessary hook morphology. It is anticipated that this capability of readily joining dissimilar materials will open new avenues in the use of hybrid structures.

Acknowledgments

Pacific Northwest National Laboratory is operated by Battelle Memorial Institute for the U.S. Department of Energy under contract DE-AC05-76RL01830. This work was sponsored by Dr. Will Joost and Sarah Ollila in association with the U.S. Department of Energy Office of Vehicle Technologies as part of the Lightweight Materials program.

References

[1] Y. Hovanski, G.J. Grant, S. Jana, K.F. Mattlin, U.S. Patent Office, 2013.
[2] S. Jana, Y. Hovanski, S.P. Pilli, D.P. Field, H. Yu, T.Y. Pan, M.L. Santella, Magnesium Technology, (2012) 11-16.
[3] S. Jana, Y. Hovanski, G.J. Grant, Metall. Mater. Trans. A, 41A (2010) 3173-3182.
[4] S. Jana, Y. Hovanski, Sci. Technol. Weld. Joining, 17 (2012) 141-145.
[5] E.E. Patterson, Y. Hovanski, D.P. Field, Metall. Mater. Trans. A, 47A (2016) 2815-2829.
[6] P. Upadhyay, Y. Hovanski, L.S. Fifield, K.L. Simmons, Friction Stir Lap Welding of Aluminum - Polymer Using Scribe Technology, Friction Stir Welding and Processing VIII, John Wiley & Sons, Inc., 2015, pp. 153-161.

Joining Aluminum Alloys to High Strength Steels by Friction Spot Welding

Uceu F.H. Suhuddin, Vanessa F. Fischer, Jorge F. dos Santos

Helmholtz-Zentrum Geesthacht, Germany

In order to reduce car body weight, many car producers try to increase the content of lightweight materials such as Al alloys and high strength steel in their products without sacrificing safety. Consequently joining dissimilar materials of Al to high strength steel is inevitable. Due to large difference in mechanical and physical properties of both materials, welding them has become more challenging for conventional welding process over the years. One of the promising solutions in welding hybrid aluminum-steel combination is by solid-state friction-based joining process. In current presentation, the weldability of dissimilar joints of Al to high strength steel of dual phase steel (about 600 MPa ultimate tensile strength) and hot stamped boron steel (1500 MPa ultimate tensile strength) in lap joint configuration by friction spot welding (FSpW) also known as refill friction stir spot welding were studied. The FSpW process is an energy efficient process, requires no special sample preparation prior welding and produce joints having superior mechanical properties. The results shows the mechanical properties of the joint with regards to lap shear strength exceed the minimum requirement according to American Welding Society standard for resistance spot welding for aerospace applications, AWS D17.2. The detail process and the testing results of the joined materials are reported.

Dissimilar Joining between Al and Mg alloys with FSW

Masatoshi Enomoto JAPAN LIGHT METAL WELDING ASSOCIATION
E-mail: Masatoshi_enomoto@mbe.nifty.com Tel:+81-3-3863-5545
Shigeki Kakiuchi Toyama Industrial Technical Center
E-mail: kakiuchi@itc.pref.toyama.jp Tel:+81-766-21-2121
Masafumi Araki Yamamoto kinzoku Seisakusyo Co.Ltd
E-mail: araki@yama-kin.co.jp Tel:+81-86-239-8585
Prof.Kazuhiro Nakata Osaka University

Abstract

Joining of dissimilar metals with FSW is not established yet due to be formed Intermetallic compounds(IMCs) at interface between base metal and stir zone. Formation and thickness of IMCs are depend ant to heat input and cooling rate during joining. Though is hard to predict the heat distribution during joining, in these days temperature rise of tool or probe during joining can be measured by wireless technology and then basic concept of joining dissimilar metals with FSW and A6005C Al alloy and uninflammable Mg alloy can be joined for 3mm thick test coupons.

Introduction

As is well known, it is hard to weld dissimilar metals due to formation of intermetallic compounds at the interface between base metal and welded metal. Melting point of AL and Mg is very close and these metals form an eutectic structure. So it is afraid that hot crack occurs during arc welding. Friction stir welding (FSW) can be adopted to weld these metals because this process is controllable the temperature rise of base metals during joining.

Uninflammable Mg alloys are invented in Japan. Increasing Ca content in Mg is contributed to increase ignition temperature of Mg and about 1%Ca content, ignition temperature exceeds melting point of Al. So this metal is considered to apply the components of vehicle or rolling stock in next generation. Trial of dissimilar joining between Al alloy and uninflammable Mg alloy is described in this paper.

Experimental Procedure

A6005C and AZX611 are used for joining. Chemical composition of A6005C is Si 0.64wt%,Fe 0.24wt% Cu 0.23wt%, Ti 0.02wt%, Mn0.01wt%, Mg 0.50wt% Al bal. Mechanical properties are: Tensile strength 294Mpa, 0.2%offset 262MPa and elongation 11%.Chemical composition of AZX611 is Al 5.9wt%, Zn 0.61wt%, Mn 0.22wt%,Si 0.03wt%, Ca 1.1wt% and Mg Bal. Mechanical properties of AZX611 are: tensile strength 278MPa, 0.2%offset 197MPa and elongation 16%. Before FSW, Al-Mg equilibrium diagram is investigated and it is shown in

Fig.1. Very low eutectic temperature(450°Cin Al rich side, 437°Cin Mg rich side) is shown in this figure. So, joining is tried under eutectic temperature by measuring the temperature in tool and test coupon.

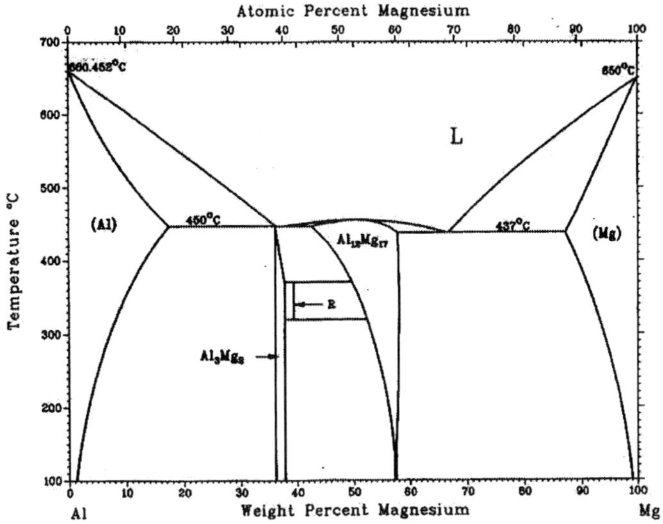

Fig.1 Al-Mg two phase equilibrium diagram

To detect the temperature profile in tool, K type thermocouples are inserted in 3points in tool and one point in AZX611. They are shown in Fig.2.

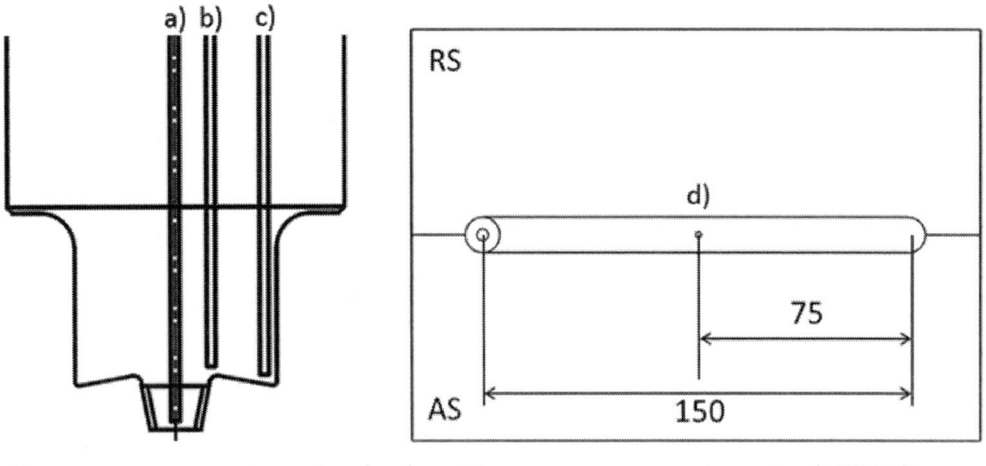

Temperature measuring points(tool) Temperature measuring point (AZX611)

Fig.2 Temperature measuring points of tool and AZX611

Dimension of tool is as follows. Shoulder diameter is 12mm ϕ ,probe diameter is 3mm ϕ ,plunge depth of shoulder is 0.1mm and plunging speed of tool is 30mm/min. Material thickness is 3mm. A6005C is located Retreating side and AZX611 is located Advancing side. Rotation speed and travelling speed of tool are change to investigate the temperature profile of tool and materials. Tool offset is not considered.

Results and Discussion

One example of temperature profile in tool is shown in Fig.3. Top of the probe is shown maximum temperature of tool while those at shoulder are lower than the former and these temperature have a little discrepancy. Difference between the temperature of probe tip and shoulder is as follows. One is the mass difference between probe and shoulder and the other is the reflection of heat at the bottom of test coupon.

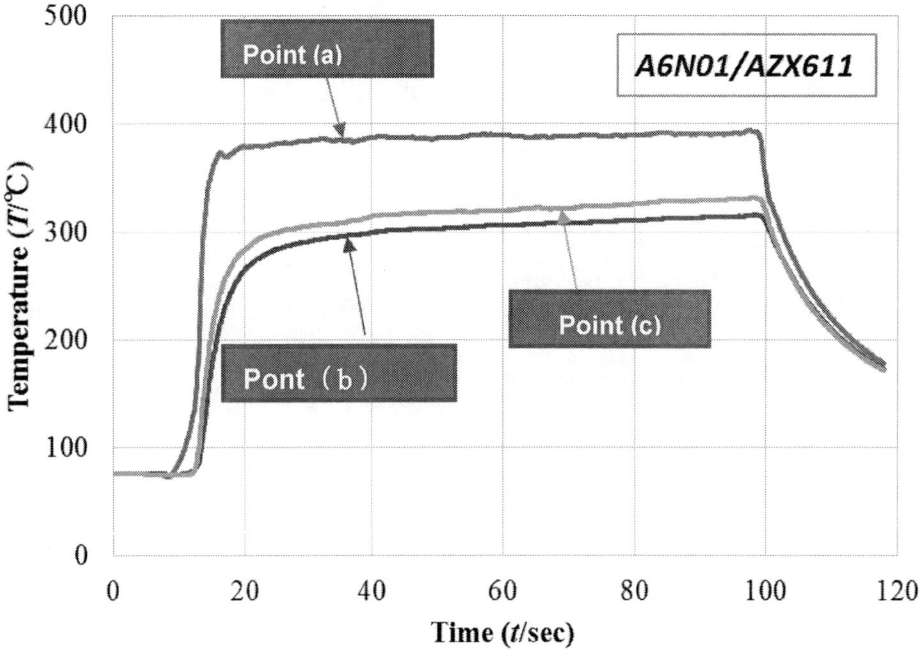

Fig.3 Temperature profile of probe and shoulder during FSW

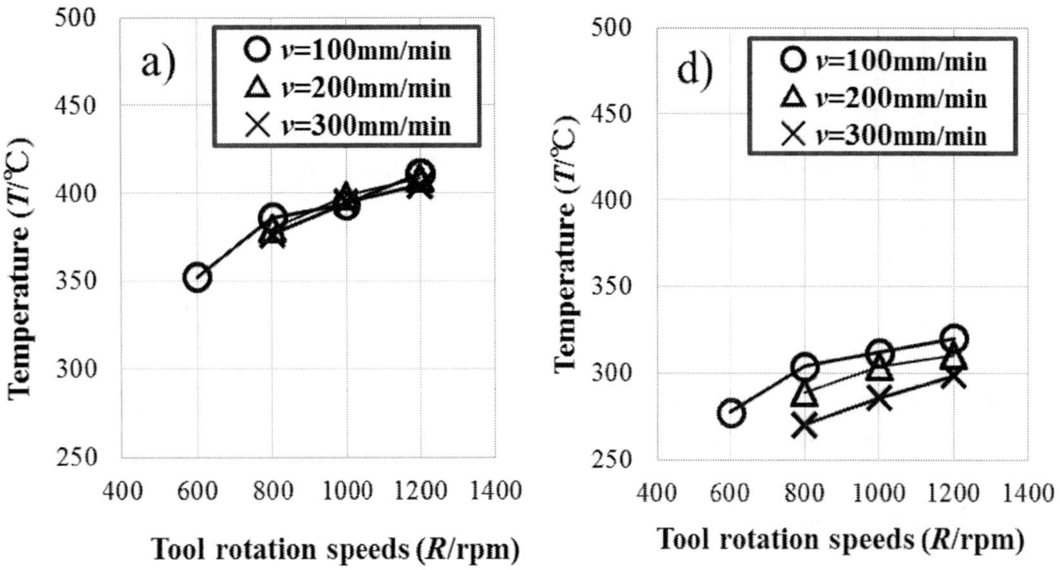

Fig.4 Maximum temperature of tool and temperature rise at the bottom of test coupon

Effect of FSW parameters on the maximum temperature of tool and test coupon are shown in Fig.4 Temperature of tool is not affected by travelling speed and only affected by tool rotational speed. One reason is suggested by the formula of heat input described by P.FERRO et all[1]. This formula does not include travelling speed of tool. On the contrary, temperature of test coupon is affected by tool rotation speed and travelling speed. Temperature of test coupon does not exceed the eutectic temperature within the test parameters. After FSW, transverse section of weld is investigated by microscope and TEM. The results are shown in Fig.5

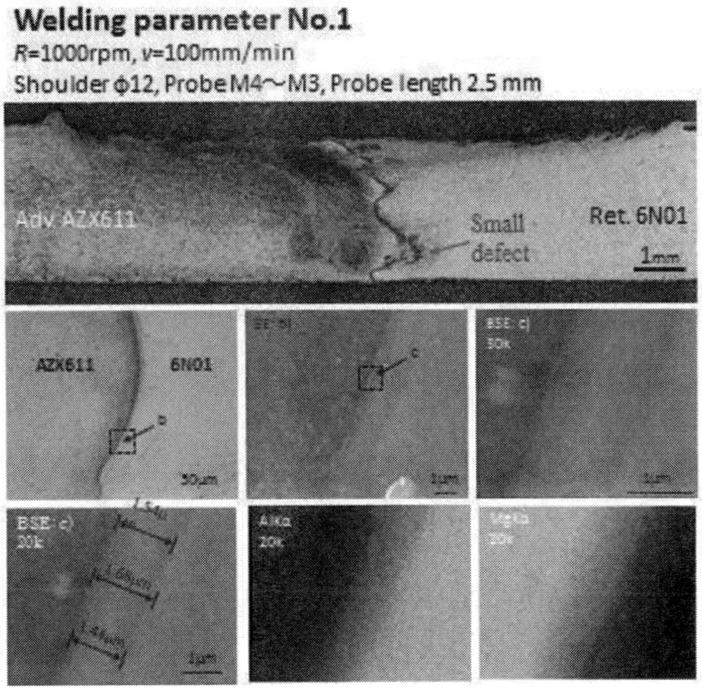

Fig.5 Cross sectional view by microscope and TEM analysis

Though small imperfection is observed by microscope due to probe length is unmatched for thickness of test coupon, thickness of IMC is very thin and it is expected to keep sufficient strength after FSW.

Conclusion

Dissimilar welding with FSW is expected to join using controlling the tool temperature profile.It is important to investigate the equilibrium diagram of two metals to join before starting the FSW.

We are very appreciate for METI (Ministry of Economy and Trade Industry) for financial support.

Reference

[1] P.FERRO and BONOLLO "A semi analytical Thermal Model for Friction Stir Welding" METALLUGICAL AND MATERIALS TRANSACTIONS A Vol.41A FEBRUARY (2010) pp440-pp449

Microstructural Characterization and Mechanical Properties of Refill Friction Stir Spot Welded Aluminum Alloy AA6181-Ti6Al4V Dissimilar Joint

Athos H. Plaine, Uceu F.H. Suhuddin, Jorge F. dos Santos
Helmholtz-Zentrum Geesthacht, Institute of Materials Research, Materials Mechanics, Solid-State Joining Processes, Geesthacht, Germany.
athosplaine@hotmail.com, uceu.suhuddin@hzg.de, jorge.dos.santos@hzg.de; Tel.: +49 4152 87 2070.
Nelson G. Alcântara
Federal University of São Carlos, Materials Engineering Department, Sao Carlos, Brazil.
nelsong@ufscar.br; Tel.: +55 16 3351 8800.

Abstract
Due to their properties, demand of multi-materials structures made of titanium and aluminum has increased to reduce weight and cost (Al alloys), and improve strength and corrosion resistance (Ti alloys). However, joining Al alloys to Ti alloys is challenging due to their chemical incompatibility and also different physical characteristics. Therefore, it is an emergent problem to be solved for future industrial application. In the present study, refill friction stir spot welding or also known as friction spot welding of dissimilar AA6181 Al alloy to Ti-6Al-4V was studied. Sound joints with maximum lap shear strength of 6.4 kN were achieved. It was noticed that rotational speed significantly influences the mechanical performance of the joints. Microstructure analyses and local properties characterization by nanoindentation reveal that medium tool rotational speed produces thin and strong layer of Ti(Al,Si)3 intermetallic at the interface.

Introduction
Multi-materials structures between aluminum and titanium alloys are a technologically attractive field of research and their increasing use in the transportation sector for weight and emission reduction is part of the motivation of this study [1]. Aluminum alloys are desirable due to its low density and cost. Titanium alloys are also promising materials because of their high specific strength and corrosion resistance. Several attempts to join these materials using different technologies have been conducted [2-6]. Mainly owing to the formation of brittle intermetallic phases at the interface, the joints still either limited in performance or too expensive. Therefore, a successful technology for joining of Ti and Al is highly required that could reduce or eliminate the formation of potentially brittle intermetallic compounds.

Refill friction stir spot welding (refill FSSW) or also known as friction spot welding (FSpW) is a solid state process and a potential candidate for joining dissimilar materials, especially for structural transport applications, due to its advantageous of lower processing time and temperature over conventional fusion welding, and geometry compatibility to replace rivets and fasteners. In the process, a non-consumable rotating tool consisting of three independently moving parts (pin, sleeve and clamping ring) is used to join two or more similar/dissimilar sheets in lap configuration.

The objective of this study is to investigate the effects of different welding parameters (tool rotational speed and dwell time) on the lap shear strength (LSS) of AA6181-Ti6Al4V dissimilar joints by using a 3^2 factorial design of experiments and perform the metallurgical characterization of the joint interface.

Experimental results and discussions
Table 1 lists the DOE test matrix along with the experimental mean values of the LSS for the AA6181/Ti6Al4V single overlap joints. The joints exhibited LSS varying from 4769 ± 73 N (condition 9) to 6449 ± 554 N (condition 5). The average LSS of all welds exceed the minimum required value of 3400N, according to AWS D17.2/D17.2M [7], for a 1.6 mm nominal thickness of Al alloys (the weakest material in the joint) having ultimate tensile strength from 240 to 386 MPa. These outstanding results are likely associated with the reduction of the brittle intermetallic compound (IMC) formation at the interface due to the low temperature cycles involved in friction welding processes. Tanaka et al. [8] established that joint strength increased exponentially with a decrease in IMC thickness during friction stir welding. Watanabe et al. [9] confirmed this trend by showing that the presence of IMC leads to a decrease in the mechanical properties of the weld.

Table 1. Summary of the 3^k-full factorial design conditions and the experimental values of LSS.

Condition	RS (rpm)	DT (s)	LSS (N)			Average LSS (N)
			1	2	3	
1	2000	1	5434	5160	5061	5218 ± 193
2	2000	3	4896	5325	5326	5182 ± 248
3	2000	5	5831	5209	5506	5515 ± 311
4	2500	1	5516	5737	5827	5693 ± 160
5	2500	3	6121	7089	6138	6449 ± 554
6	2500	5	5821	5596	5863	5760 ± 144
7	3000	1	5813	5399	5611	5608 ± 207
8	3000	3	4873	4681	5175	4910 ± 249
9	3000	5	4830	4789	4688	4769 ± 73

An analysis of variance (ANOVA) was carried out to quantify the effect of each refill FSSW process parameter and also their interaction on the LSS of the joints. The contribution percentage was calculated by the ratio of the sum of squares (SS) of the selected factor and the total sum of squares. Table 2 shows the ANOVA performed with the acquired data for an interval of confidence of 95%. The factors are physically significant when its percentage of contribution is smaller than the associated error. Among the parameters, RS showed to be the most affecting parameter on the LSS of the joints (49.6%), followed by the interaction RS*DT (30.7%). DT itself seems to have no significant influence on the response LSS for the selected range of welding parameters. The total contribution rate of the refill FSSW parameters was 82.4%.

Table 2. Analysis of variance of LSS values.

Source	SS	df	MS	F-value	Contrib. [%]
RS (rpm)	3729508	2	1864754	25.47	49.66
DT (s)	157706	2	78853	1.08	2.10
RS (rpm) DT (s)	2304202	4	576050	7.87	30.68
Error	1318005	18	73222		17.56
Total SS	7509421	26			

Fig. 1(a) shows the typical cross-section of a representative dissimilar weld. No defects or obvious reductions of thickness were observed in the weld. A higher-magnification micrograph acquired from the center of the joint (Fig. 1(b)) reveals the presence of a continuous layer of approximately 0.8 μm in thickness. Although the exposure time of the interface was relatively short, it was apparently sufficient to induce the interdiffusion of Al and Ti atoms at the interface. The Fig. 1 (c) shows the corresponding EDS concentration profiles across diffusion zones, and it was found that a thin layer of $TiAl_3$ with about 10 at.% of [Si] was formed in the interface. The $TiAl_3$ with [Si] is denoted $Ti(Al, Si)_3$ because Si prefers to substitute on Al sites.

The $TiAl_3$ phase was also reported in other studies involving friction welding techniques [5, 6]. Kim et al. [5] reported that the thickness of the $TiAl_3$ IMC layer produced at the interface is the dominant factor determining the joint mechanical characteristic in friction welds between Al and Ti. According to the results, it was found that the critical thickness of the IMC which directly influenced on the tensile strength and the bending ductility was about 5 μm. Wu et al. [6] confirmed this trend by showing that high strength friction stir welds of AA6061 and Ti6Al4V were achieved due to the formation of a thin IMC layer of only 100 nm at the joint interface. Similarly, the IMC layers formed during refill FSSW in this study were remarkably thin (not exceeding the critical value of 5 μm), resulting in the high failure loads values observed for the refill FSSW joints.

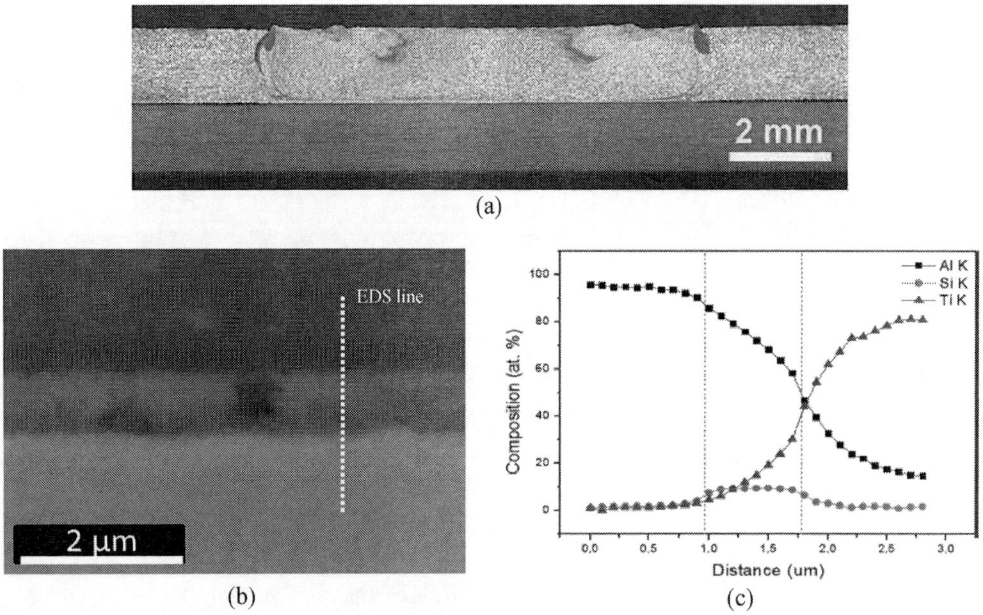

Fig. 1. Typical cross-section of a representative dissimilar weld (a), IMC at the interface (b) and EDS line scan.

Nanoindentation measurements were performed across the joint interface. The results were compared to pure $TiAl_3$ synthesized by reaction annealing of multi-laminated Ti-(SiCP/Al) composites [10]. The average nanohardness and modulus are listed in Table 3. The solid solution of Si in the $TiAl_3$ layer leads to the increase of elastic modulus and hardness compared to pure $TiAl_3$. Fig. 2 presents a typical load–displacement curve for the reaction layer. In comparison to the $TiAl_3$ curve, less pop-ins are found for the curve of the $Ti(Al, Si)_3$. The pop-ins are often associated with the onset of dislocation plasticity. It indicates that the plasticity of $Ti(Al, Si)_3$ is reduced compared to pure $TiAl_3$.

Table 3. Nanohardness and elastic modulus of each phase.

System	Nanohardness (GPa)	Elastic modulus (GPa)
$Ti(Al,Si)_3$ interface	11.20±0.8	288±13
Pure $TiAl_3$	9.20±0.5	220±10
Al base material	0.85±0.2	60±10
Ti base material	1.90±0.3	110±10

These results are theoretically explained in the literature by the first-principles methods from the electronic structure of a supercell model of $Ti(Al, Si)_3$ [10], in which a new strong hybridization occurring between Si and Al and a strong covalent bond between Al and Si lead to the increased modulus of the$TiAl_3$ layer. The model also indicates that the solid solution of Si reduces the plastic properties of $TiAl_3$.

Conclusions

The individual effect of refill FSSW process parameters on the lap shear strength of aluminum AA6181/Ti6Al4V joints was investigated using statistical analysis. Produced joints showed good mechanical performance with lap shear strength varying from 4769 N to 6449 N. Tool rotational speed was the parameter with the largest influence on the lap shear strength of the joints (49.6%), followed by the interaction of rotational speed*dwell time (30.7%). The joint interface was mainly composed of a thin layer consisting of $Ti(Al, Si)_3$, in which the solid solution of Si increases the elastic modulus and the hardness of the $TiAl_3$ intermetallic.

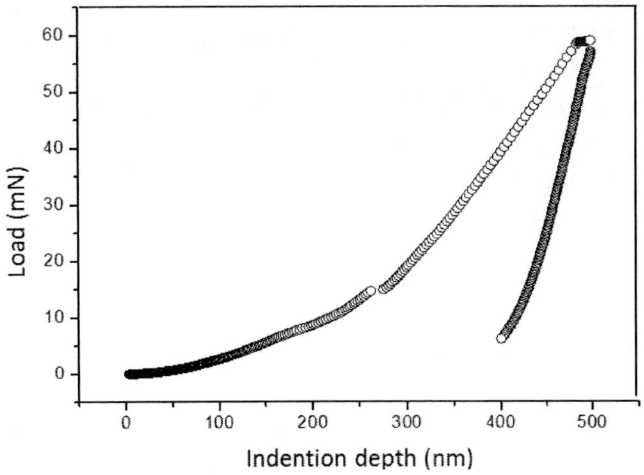

Fig. 2. Typical load-displacement curves for the $Ti(Al,Si)_3$ *interface.*

Acknowledgement

The authors would like to acknowledge Prof. Dr. Erica Lilleodden and Julian Correa from Department of Experimental Materials Mechanics, Institute of Materials Research, Helmholtz Zentrum Geesthacht for the technical support in nanoindentation tests.

References

1. Lehmhus, D.B., M. ; Herrmann, A. ; Kayvantash, K., Structural Materials and Processes in Transportation. John Wiley & Sons. 2013: Wiley-VCH Verlag GmbH & Co KGaA: pp. 345-349.
2. Chang, S.Y., et al., Brazing of 6061 aluminum alloy/Ti–6Al–4V using Al–Si–Cu–Ge filler metals. Journal of Materials Processing Technology, 2012. 212(1): pp. 8-14.
3. Vaidya, W.V., et al., Improving interfacial properties of a laser beam welded dissimilar joint of aluminium AA6056 and titanium Ti6Al4V for aeronautical applications. Journal of Materials Science, 2010. 45(22): pp. 6242-6254.
4. Zhang, C.Q., et al., Microstructural characterization and mechanical properties of high power ultrasonic spot welded aluminum alloy AA6111–TiAl6V4 dissimilar joints. Materials Characterization, 2014. 97: pp. 83-91.
5. Kim, Y.C. and A. Fuji, Factors dominating joint characteristics in Ti – Al friction welds. Science and Technology of Welding and Joining, 2002. 7(3): pp. 149-154.
6. Wu, A., et al., Interface and properties of the friction stir welded joints of titanium alloy Ti6Al4V with aluminum alloy 6061. Materials & Design, 2015. 71: pp. 85-92.
7. International, S., Welding, Resistance: Spot and Seam. 2005, SAE International. pp. 47.
8. Tanaka, T., T. Morishige, and T. Hirata, Comprehensive analysis of joint strength for dissimilar friction stir welds of mild steel to aluminum alloys. Scripta Materialia, 2009. 61(7): pp. 756-759.
9. Kumai, S.W., M., Microstructure and mechanical properties of friction stir welded aluminum alloy / steel lap joint. Journal of Japan Institute of Light Metals, 2007. 57: pp. 529-535.
10. Pang, J.C., et al., Effect of solid solution of Si on mechanical properties of TiAl3 based on the multi-laminated Ti-(SiCP/Al) composite system. Materials Science and Engineering: A, 2013. 579(0): pp. 57-63.

The inhibition of abnormal grain growth in weld nugget via in-situ intermetallic-particle formation during friction stir welding of AA6061

Cunli Liu[1], Muyang Deng[1], Ji Chen[1], Lijie Guo[2], Aidang Shan[1], Ke Chen[1,*]

1. School of Materials Science and Engineering, Shanghai Jiao Tong University, Shanghai 200240, China;
2. Shanghai Aerospace Equipment Manufacturing General Factory, Shanghai 200245, China;
*Corresponding author: chenke83@sjtu.edu.cn

Abstract: Abnormal grain growth (AGG) has been widely observed in many friction stir welded (FSW) joints during post welding heat treatment. Various ways have been reported in previous studies to control it for better mechanical properties, but most of them fail to eliminate this problem completely. In the current research, a new method has been invented and adopted in common FSW and proved to be very effective in limiting AGG. Both macroscopic observation and microscopic characterization were conducted to explain why it succeeded in dealing with AGG. A model was raised to demonstrate the in-situ reaction that took place in multi-pass FSW.
Key words: Abnormal grain growth, friction stir welding, in-situ reaction

1. Introduction

Friction stir welding (FSW) was invented at The Welding Institute (TWI) of UK in 1991[1] and has become one of the most promising methods of joining in the past two decades. It is a solid-state joining which can be applied to various materials. The nature of FSW is very simple: A non-consumable tool is rotated and inserted in to the abutting edges of two plates. The heat generated by the friction between the tool and the workpieces softens the material in both sides and drives it to flow around the tool, during the process of which two plates are joined. [2] The method was brought to the notice of the public right after its invention and then widely studied and used by groups of researchers and engineers because of its energy efficiency, environment friendliness, and versatility. In addition, refined grains can be created in stirred zone through dynamic recrystallization due to the special thermal and mechanical conditions in the process of FSW. Despite its many advantages, a specific problem is aroused when it comes to its application at elevated temperature, under which the welded material experiences pre-treatment[3] or solution treatment[4] in the practice. As has been reported in many previous works, grain coarsening occurs in various zones during post-welding heat-treatment. [5,7,10] Some of these grains can grow up to as large as several millimeters by devouring their neighboring grains.[9] This phenomenon is known as abnormal grain growth (AGG). Efforts have been made for years to find a solution to eliminating AGG. [6,8,11] However, these methods failed to completely eliminate AGG.

According to Humphreys and his fellow colleagues[12], AGG occurs only when the combination of several parameters including the homogeneity of grain structure, the average grain size, the average size and the number of 2nd phase particles, is within the range of certain value. Humphreys also concluded two basic methods to eliminate AGG. One of them is to moderately increase the initial average grain size of material prior to heat treatment. The other is to raise the amount of second-phase particles in the matrix or to reduce the size of them, or even both if possible. In the case of FSW, initial grain size can be controlled by adjusting the welding parameters (rotation speed and welding speed). It is generally accepted that larger grains can be achieved by increasing the rotational speed while decreasing welding speed so that the heat input into the workpieces is higher. However, AGG is prohibited at the cost of mechanical property because larger grain size means lower strength and hardness in most cases. So what can be envisioned is that the input or production of large amount of nano-size second-phase particles is the more effective way to prohibit AGG.

In the present work, post-weld grain size in the stirred zone is controlled by optimizing welding parameters. And through the introduction of ball-milled mixed powder of aluminum and titanium and the later in-situ reaction between the powder and the matrix, nano-size Al_3Ti particles were produced and widely spread in the stirred zone. During multi-pass experiment, and with the combination of clockwise and counter-clockwise rotation of the tool, AGG is negligible throughout the stirred zone even after heat-treatment. A model of the mechanism of the in-situ reaction is also discussed based on the findings.

2. Experiment

6061 aluminum alloy of T6 condition was used in the present study. Its chemical composition is presented in Table 1.

Table.1 6061 aluminum alloy T6

Element	Al	Cu	Fe	Mg	Mn	Si	Cr
%	bal	0.25	0.49	0.97	0.10	0.70	0.15

The raw material was cut into plates, the size which being 3mm in thickness, 40mm in length and 10mm in width. The two plates were butt-welded by a specially designed welding tool. The tool has a shoulder of 18mm in diameter and a cylindrical pin of 6mm in diameter and 2.7mm in height with spiral feature on its side. The plates on both sides have step-like configuration (as is presented in Fig.1) in order that mixed powder can be placed and kept in the (shaded region in Fig.1). The pure aluminum powder and pure titanium powder had been mixed at a weight ratio of 20:1 and then ball-milled into flake-shape particles (Fig.2). Side surfaces of both plates were grounded and then washed by acetone, enabling pure aluminum matrix to react with the powder during FSW. Welding parameters were settled at 1500rpm and 50mm/min based on previous trial set of experiments. Cross section samples were collected to study the macrostructure and microstructure of stirred zone. Selected samples were heat-treated at 560 °C for 1 hour and then quenched in deionized water. Heat treated cross-sections were then mounted and polished to a 1-μm finish and chemically etched for optical microscopy. The etchant used is the modified version of Keller's reagent, which is: 95ml water, 5ml HCl, 2.5ml HNO_3, 1ml HF. Samples for SEM were mechanical polished to a 0.25μm finish.

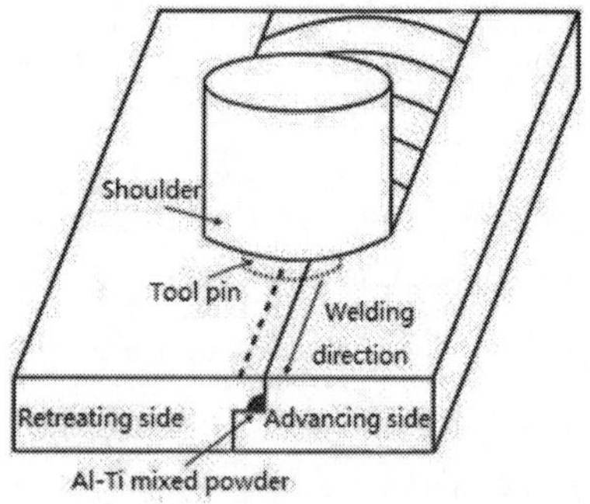

Fig.1 Demonstration of welding method

Fig.2 Ball-milled Al-Ti mixed powder

3. Results

Two groups of welding were carried out to study the effect of Al-Ti mixed powder on AGG, with the former one commonly FSWed and the latter one done in the new method presented in Fig.1. Each one of them included three samples, which were FSWed by 1, 2 and 4 passes correspondingly and then experienced the 560 °C heat treatment for 1 hour and quenched. As is showed in Fig.3(a), the clear majority of the stirred zone is covered by AGG (the area enclosed by the white dash-line), which shrinks a little bit in the 2-pass (Fig.3(b)) and 4-pass samples (Fig.3(c)). It could be primarily concluded that an increase in passes lead to higher thermal stability in the FSW nugget. But on the whole, AGG region still could be widely observed in all the three samples of the first group.

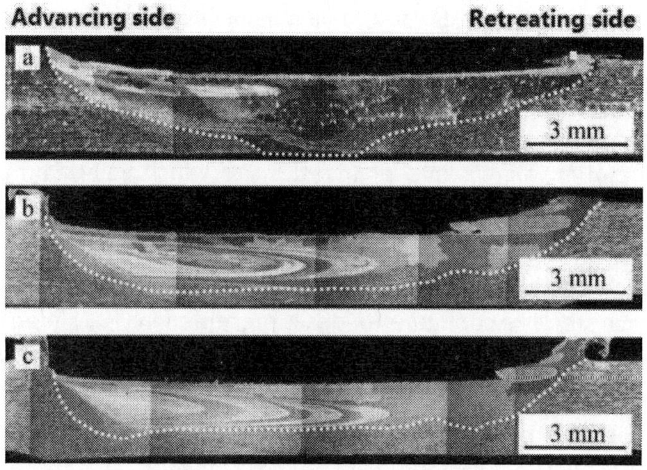

Fig.3 1/2/4 passes without Ti

In the second group, however, only about half of the stirred zone is covered by AGG even in the 1-pass sample (Fig.4(a)), suggesting the added mixed powder have some restrictive effect on the AGG. With more passes conducted, the total area of AGG is gradually reduced, as has been seen in the previous study without Al-Ti mixed powder. In the 4-pass sample, AGG is further prohibited and appears only in the narrow zone on the retreating side.

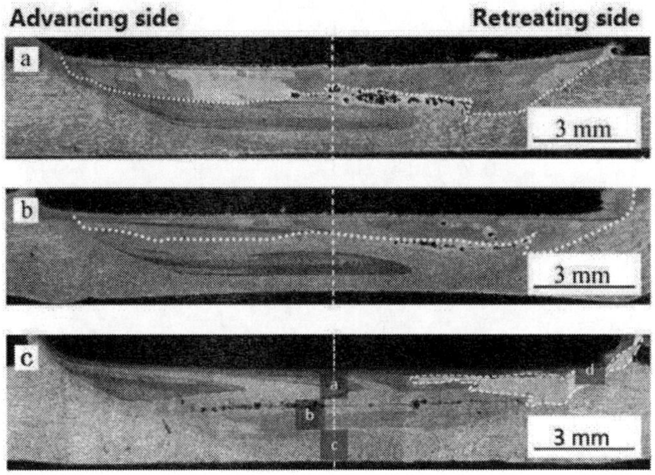

Fig.4 1/2/4 passes with Ti

SEM was carried out to verify the previously discussed results. In the cross-section of the as-welded 2-pass FSWed samples without the involvement of Al-Ti mixed powder (Fig.5), only refined grains with sizes no more than 20μm could be seen all the way from the top to the bottom through the centerline. Grains in the top area are smaller than those in the middle and bottom areas. A layer of ultra-refined grains can also be seen on the very top. In the post-heat-treatment sample, however, AGG could be founded in both the top and middle parts but not in the bottom area, which was in accordance with the former result we have seen in the same sample showed by Fig3.(b). It is also worth mentioning that there are also some small grains remained in both the top and middle parts, suggesting the inhomogeneity of local grain structure.

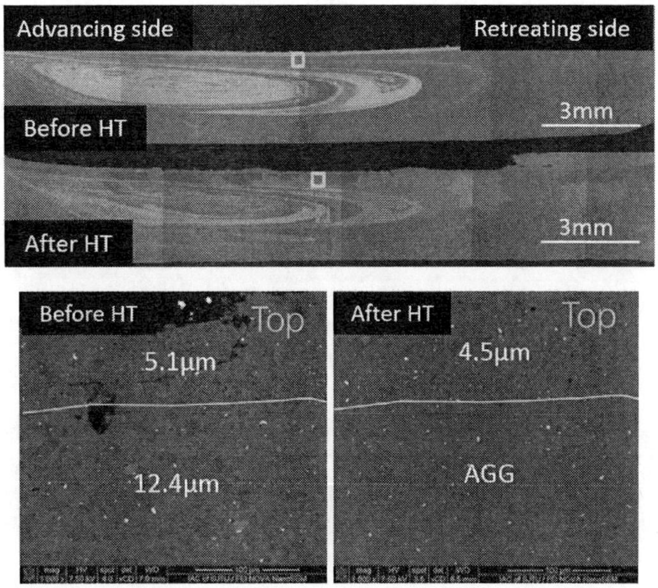

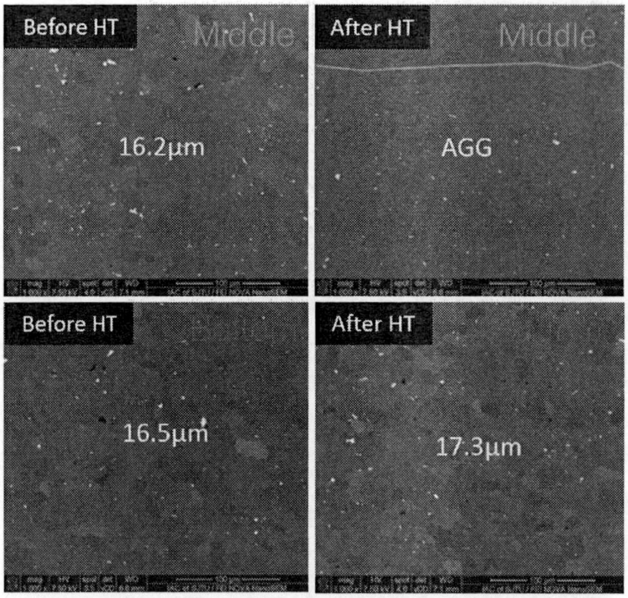

Fig.5 Grain size in various areas in Fig.3(b)

Further SEM observation was also made in the 4-pass samples FSWed with the addition of Al-Ti mixed powder. Four regions (Fig.4(c)) were selected to quantify the grain size. From the top area both in the advancing side and in the centerline to the middle of stirred zone, the average grain size was less than 10μm and no AGG was observed. On the other hand, there is a clear dividing line between refined grains and AGG zone. Again, the grain structure characterized by SEM complied with that observed in OM.

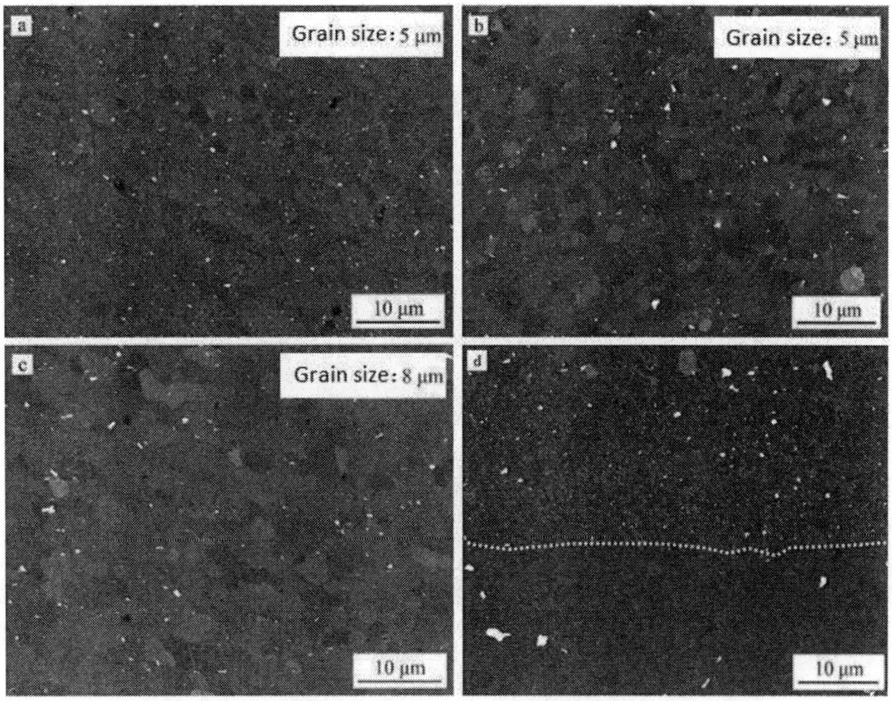

Fig.6 Particular areas in Fig.4(c)

Further 5-pass FSW experiment using the same method was carried out to ensure that the similar effect on limiting AGG is repeatable and to make advance hopefully according to the fact that increased FSW passes result in smaller AGG area. Cross-section revealed that the AGG was greatly suppressed to a narrow corridor on top of the retreating side (Fig.7). In addition, the unreacted Ti powder was distributed evenly along a horizontal line in the heart of the stirred zone.

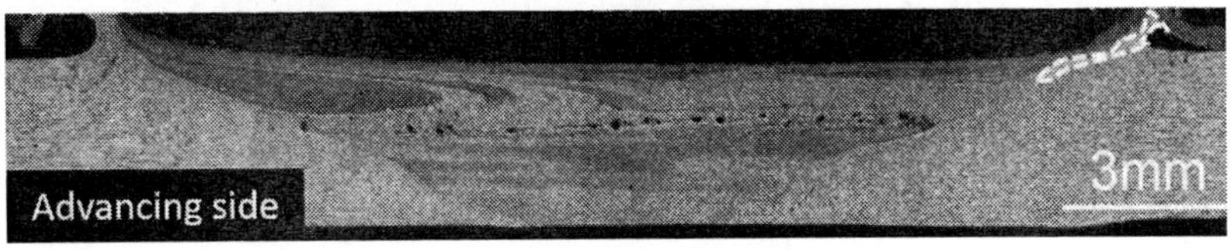

Fig.7 5-pass FSWed welding joint

4. Discussion

4.1 The inhibition of AGG by introducing Ti

According to Humphreys [12], three kinds of grain growth can be roughly seen in materials, which is normal grain growth, abnormal grain growth and no grain growth. Basically, the way that the grains in the material grow depend on two factors: the homogeneity of grain size (X: vertical axis) and the grain structure (Z: horizontal axis). AGG occurs when the point (X,Z) locates at a particular zone. So the key here is to modify the value of X or Z or both in order that the new point (X',Z') locates at "No grain growth" zone. Since the homogeneity of grain size differs little from material to material, increasing the initial grain size and introducing more 2nd-phase particles of smaller size are two methods generally adopted to control AGG. The former one, though good enough to solve the issue of AGG, is a hazard for mechanical property. A comparison between Fig.3 and Fig.4 indicates that the added Al-Ti powder makes all the difference by lowering the area of AGG conspicuously. It is imaginable that the Ti powder experienced mechanical and thermal cycles during FSW and reacted rapidly with matrix as well as the added Al powder. And because the mixed powder has been formerly ball-milled into flake (Fig.2), more flaws and larger surface area are created to allow for the following easier and stronger in-situ reaction, the reactants and product of which are then pressed and cut into smaller pieces by the fierce deformation and material flow. These have maintained in the matrix and served as 2nd-phase particles to stop the grain boundary mobility.

4.2 Multi-pass FSW enhances the effect of Ti

From the result of the experiment without and with the addition of mixed powder (Fig.3, Fig.4), we can draw the conclusion that AGG regions shrink gradually with more FSW passes conducted. It is also worth mentioning that compared with 1-pass and 2-pass samples ((Fig.4(a)and(b)), noticeable mixed powder was spread and located more widely and evenly in 4-pass and 5-pass samples((Fig.4(c)and Fig.7). This phenomenon can be explained by the reasonable assumption that an extra pass of FSW creates another thermal and mechanical cycle, with the in-situ reaction being carried out more completely. Also, the range of area to which the reactants spread to becomes wider when more passes are conducted. The combined effect of these two leads to the shrinkage of AGG area.

4.3 In-situ reaction in FSW

Further SEM (Fig.8) is done in the partially-reacted area to figure out the mechanism of Al-Ti reaction. EDS result (Table.2) confirms that the elongated white particles are pure titanium and that the light grey particles are Al_3Ti.

Table 2 EDS result of Fig.8 (Atom percentage)

	Al(%)	Ti(%)	Phase
1	100.0	N/A	Al
2	72.9	27.1	Al3Ti
3	71.0	29.0	Al3Ti

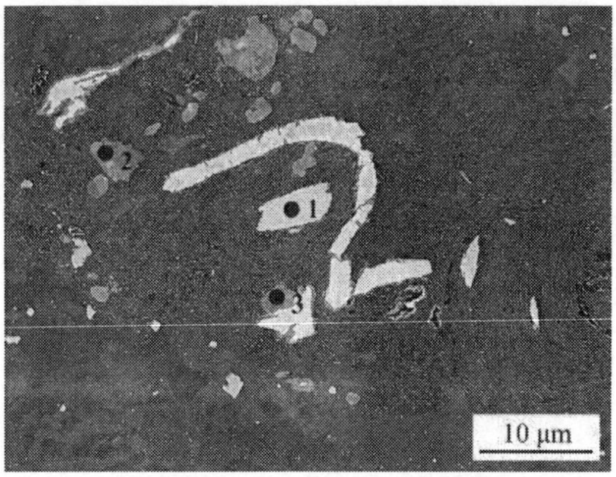

Fig.8 SEM result of partially-reacted area

It is clear that the Ti particle has been severely deformed either during the previous ball-milling treatment or friction stir welding. However, it can be observed from Fig.1 that almost all the powder has been pressed and extruded into flakes, while the Ti powder in Fig.8 has been cut into three sections but still loosely connected to each other. Thus it can be extrapolated that the drastic deformation in FSW suppressed and sheared the Ti powder and created even more flaws and surface area for in-situ reaction to take place. Under the elevated temperature and high stress, aluminum atoms diffuse into titanium particle much more rapidly than they do in room temperature and normal strass. In a few seconds when the FSW tool moves through, a layer of Al_3Ti intermetallic compound is formed. With more passes conducted, the layer becomes thicker until Ti particle is totally swallowed by aluminum matrix. Later, some of the completely-reacted Ti powder (Al_3Ti) remained as a integrate particle in the matrix, while others are tore apart by the deformation and the severe material flow in all directions. The latter one exists in the matrix either in the form of cluster of dissociated Al_3Ti particles or smaller Al_3Ti particles dispersed in the Al matrix, which may well become the pinning force against AGG. An equiaxed Al_3Ti particle was found and confirmed through TEM, proving the existence of new nano-sized second-phase particles.

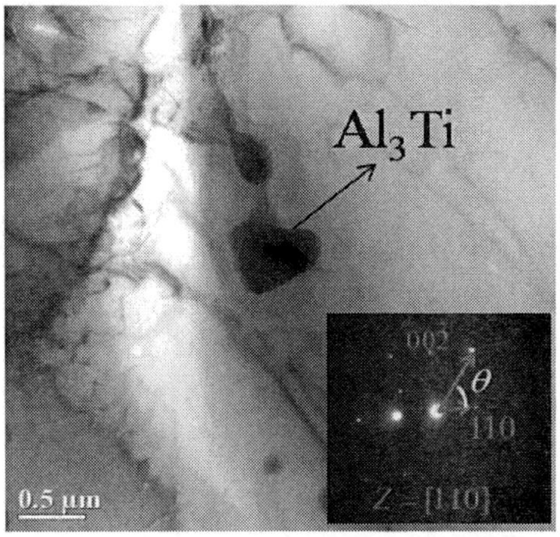

Fig.9 TEM of equiaxed Al3Ti particle

5. Conclusion

(1). The refined grains in the stirred zone of 6061 Al FSWed welding joint are inclined to grow into coarse grains up to several millimeters. The area of AGG shrank by more than 50% when Ti-Al mixed powder was introduced into an internal space before FSW. Multi-pass FSW helps improve the thermal stability of as-welded grain structure either with or without the addition of mixed powder. AGG phenomenon almost disappeared in the 5-pass FSW joint, where the added Ti was relatively evenly distributed.

(2). A new phase was founded and confirmed to be Al_3Ti in the stirred zone. A model of Al-Ti in-situ reaction has been raised based on the SEM findings. TEM result proves the existence of nano-sized Al_3Ti dispersed particles in the Al matrix.

Reference

[1] W.M. Thomas, E.D. Nicholas, J.C. Needham, M.G. Murch, P. Templesmith, C.J. Dawes, G.B. Patent Application No.9125978.8 (December 1991).

[2] Mishra R.S., Ma Z.Y. Friction stir welding and processing[J]. Materials Science and Engineering, 2005, (50):1 - 78

[3] Charit I, Mishra R S. High strain rate superplasticity in a commercial 2024 Al alloy via friction stir processing[J]. Materials and Engineering, 2003, (A359):290-296

[4] Chen Y.C., Feng J.C., Liu H.J, Stability of the grain structure in 2219-O aluminum alloy friction stir welds during solution treatment[J]. Materials Characterization, 2007, (58):174 - 178

[5] Ke Chen, Wei Gan, K.Okamoto, Kwansoo Chung, R.H. Wagoner. The Mechanism of Grain Coarsening in Friction-Stir-Welded AA5083 after Heat Treatment. Metallurgical and Materials Transactions A[J]. 2011, Volume 42(2), :488-507

[6] M.A. García-Bernala, R.S. Mishra, R. Verma, D. Hernández-Silva. Inhibition of abnormal grain growth during hot deformation behavior of friction stir processed 5083 Al alloys. Materials Science & Engineering A 636 (2015) 326 - 330

[7] I. Charit, R.S. Mishra. Abnormal grain growth in friction stir processed alloys. Scripta Materialia 58 (2008) 367 - 371

[8] S. Jana, R.S. Mishra, J.A. Baumann, G. Grant.Effect of process parameters on abnormal grain growth during friction stir processing of a cast Al alloy. Materials Science and Engineering A528 (2010) 189 - 199

[9] J. Ciulik, E.M. Taleff. Dynamic abnormal grain growth: A new method to producesingle crystals. Scripta Materialia 61 (2009) 895 - 898

[10] Kh. A.A. Hassana, A.F. Norman, D.A. Price, P.B. Prangnell. Stability of nugget zone grain structures in high strength Al-alloy friction stir welds during solution treatment. Acta Materialia 51 (2003) 1923 - 1936

[11] Moataz M. Attallaha, Hanadi G. Salem. Friction stir welding parameters: a tool for controlling abnormal grain growth during subsequent heat treatment. Materials Science and Engineering A 391 (2005) 51 - 59

[12] Humphreys, F.J., A unified theory of recovery, recrystallization and grain growth, based on the stability and growth of cellular microstructure. 2. The effect of second-phase particles. Acts materials, 1997,(45): 5031-5039

Melting Efficiency in Anti-Phase Synchronized Twin-wire Gas Metal Arc Welding

Syed Quadir Moinuddin, Abhay Sharma
Indian Institute of Technology Hyderabad, Sangareddy, Hyderabad, India, 502 285
me12p1007@iith.ac.in, abhay@iith.ac.in, +9140 2301 6091

Abstract
Melting of electrode is a fundamental phenomenon in Gas Metal Arc Welding (GMAW) that influences arc stability and weld bead geometry. In this study effect of process parameters on melting efficiencies has been studied in anti-phase synchronized GMAW process wherein lead and trail wires are powered in phased manner such that one arc acts at a time, thus, mutual disturbance can be avoided. Bead-on-plate weld experiments were carried out with an anti-phase of 1 ms. A mathematical relation was developed to calculate melting efficiencies It is observed that effective utilization of heat takes place when dissimilarity is maintained between lead and trail variables because of improved arc stability. In turn, melting efficiency and weld properties also improve. The heat affected zone hardness is analyzed to understand the aforementioned effect.

Introduction
Gas metal arc welding (GMAW) process is one of the widely used arc welding processes in depositing metal for various applications in industries. Chandel [1] investigated GMAW melting efficiency which is defined as amount of energy utilized to form fusion area (A_t) from the total energy input. Similarly, electrode melting efficiency is defined as the amount of energy used to form deposit area (A_d) from total energy input. Welding parameters such as welding current, voltage, electrode diameter, electrode polarity, and electrode extension have a significant influence on the melting efficiency. Dupont and Marder [2] evaluated effect of welding parameters and process types on melting efficiencies. Niles and Jackson [3] showed that plate melting efficiency increases with increase in current level and travel speed in gas tungsten arc welding (GTAW) process. Lee et.al [4] studied effect of process parameters on heat affected zone (HAZ), weld bead properties in submerged arc welding (SAW) and related them with melting efficiency. Fuerschbach and Knorovsky [5] compared melting efficiency in GTAW process and Plasma arc welding (PAW). The melting efficiency is studied by varying the travel speed using laser welding process [6]. In the aforementioned investigations, the researchers have primarily investigated melting efficiency in single wire processes. Tusek and Suban [7] calculated melting efficiency for twin-wires in SAW and GMAW. The twin-wire process came into existence for its high deposition at higher speeds which in turn improving the productivity in industries. Sharma et. al [8,9] investigated arc behavior in twin-wire SAW and observed that dissimilar wire diameters stabilize arc and reduce flux consumption. Moinuddin and Sharma [10, 11] investigated effect of process parameters in twin-wire antiphase synchronized GMAW process on arc stability and weld bead properties. It was found that though arcs at lead and trail wires are phased in time, larger lead current results in stable arc. The arc stability has a close relation with melting efficiency. Effect of arc stability on melting efficiency in anti-phased twin-wire GMAW has not been revealed in literature. Thus, an in-depth investigation is essential to understand the effect of process parameter on melting efficiencies and weld bead properties in twin-wire GMAW process. The present work aims to study the effect of changes in process parameters at lead and trail wires (i.e. current values and wire diameters) on melting efficiencies and their relation with weld bead properties in twin-wire antiphase synchronized GMAW process.

Experimental work
Bead-on-plate (BOP) welds on 10mm thick mild steel plate (Grade A mild steel) with ER70S-6 as filler wire were produced under different welding conditions as shown in Table 1. Fronius make welding power sources were integrated to a six-axis robot (Kuka make). The welds were obtained at constant travel speed at 0.9 m/min and contact tube-to-work-piece-distance (CTWD) at 20 mm. The pulse-synergic mode with an anti-phase synchronization of 1 ms was used. The data acquisition system, Arc client 3000p, attached with non-contact Hall Effect sensor was used to acquire real-time electrical signals which were used in calculating actual heat input and subsequently melting efficiency was obtained. The shielding gas used was a mixture of 82% Argon and 18% CO_2 with constant gas flow rate (GFR) at 16 lpm. Weld macrographs were obtained after polishing and etching with 4% Nital solution (HNO_3 + Ethanol). The melting efficiency calculation method is described in following section.

Melting efficiency calculation

The melting efficiency depends on amount of heat input generated through welding parameters (i.e. current, voltage, and travels speed). The mathematical relation in calculating melting efficiency of twin-wire process is an extension of single wire formulation [1, 3].

Let

HI	=	Heat input in KJ/mm,
V(t)	=	Instantaneous voltage in V
I (t)	=	instantaneous current in A
S	=	Travel speed in mm/sec
A_d	=	Deposited area in mm^2
A_p	=	Penetration area in mm^2
A_t	=	Total fusion area in mm^2
WFR	=	Wire feed rate in m/min
E	=	Melting efficiency

Suffix L and T represents lead and trail wires, respectively.

Table 1 Welding conditions

S.No	Wire Diameters		Currents	
	Lead $Ø_L$ (mm)	Trail $Ø_T$ (mm)	Lead I_L(A)	Trail I_T(A)
1	0.8	0.8	220	220
2	1.2	1.2	220	220
3	0.8	1.2	220	220
4	1.2	0.8	220	220
5	0.8	1.2	180	220
6	0.8	1.2	220	180
7	1.2	0.8	180	220
8	1.2	0.8	220	180

The heat produced at lead and trail wires can be calculated using instantaneous voltage and current as shown in Eq. 1 and 2.

$$HI_L = \frac{\int V_L(t) I_L(t) dt}{S} \quad (1)$$

$$HI_T = \frac{\int V_T(t) I_T(t) dt}{S} \quad (2)$$

Total heat input

$$HI_{total} = HI_L + HI_T \quad (3)$$

One joule of heat melts 0.0854 mm^3 of steel, thus, melting efficiency can be calculated as follows:

Lead electrode melting efficiency

$$E_L(\%) = \frac{A_d * 100}{0.0854 * (HI_L)} * \frac{WFR_L}{WFR_L + WFR_T} \quad (4)$$

Trail electrode melting efficiency

$$E_T(\%) = \frac{A_d * 100}{0.0854 * (HI_T)} * \frac{WFR_T}{WFR_L + WFR_T} \quad (5)$$

Total electrode melting efficiency

$$E_{total}(\%) = \frac{A_d * 100}{0.0854 * (HI_{total})} \quad (6)$$

The following section discusses results on heat consumption at lead and trail wires, melting efficiencies, weld bead properties, and mechanical properties.

Results and discussion

The process parameters determine heat input and control stability of arc welding process. From Fig. 1, it can be observed that for same welding conditions (e.g. both lead and trail wires 0.8 or 1.2 mm diameter and at 220 A current) heat consumption are different at lead and trail wires. More heat is consumed at lead than trail wire. This occurs because lead wire acts on fresh material whereas trail electrode on the molten material [12]. However, for certain welding conditions wherein lead wire has lesser current density, heat consumption at trail wire is more than lead wire. This is due to improper combinations of electrode diameter and current value which were unable to generate sufficient heat to melt fresh material and perform limited pre-heating action by lead wire. These implications directly reflected on melting efficiencies as shown in Fig. 2. When electrode diameter was increased from 0.8 to 1.2 mm, melting efficiency decreased irrespective of wire position (lead/trail) due to current density effect. The current density is inversely proportionally to electrode diameter and thereby influences melting efficiency. This can be clearly understood and visualized in certain case in Fig.2 wherein currents at lead and trail wires were kept constant and the electrode diameters were altered. It can be interpreted that more heat is drawn by lead wire which in turn melts more amount of wire.

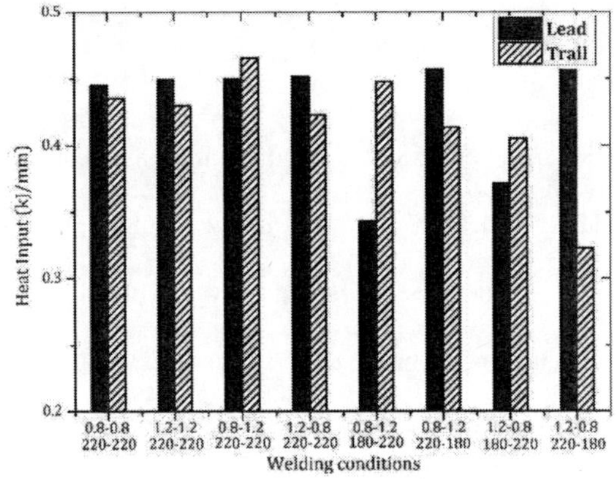

Fig. 1 Heat input at different wire diameters in (mm) and welding currents in (A)

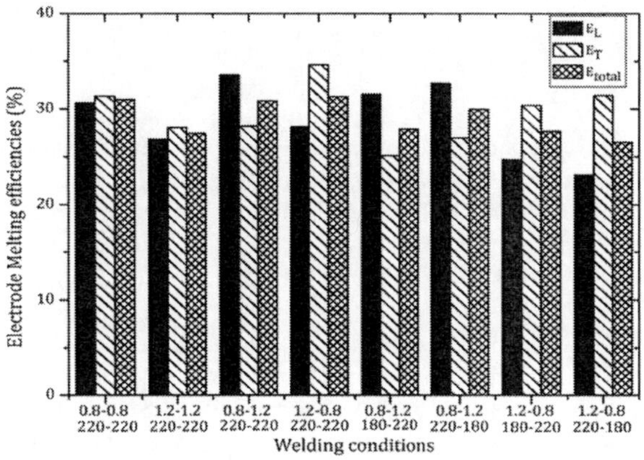

Fig. 2 Melting efficiencies at wire diameters in (mm) and welding currents in (A)

The melting efficiency increases with heat input which also increases reinforcement height and decreases penetration depth as shown in Fig. 3. This gives an understanding of heat input distribution between electrode and base metal. It is also observed that reversing electrode and current position increases electrode melting efficiency and softens (HAZ) due to formation of coarse pearlite and ferrite microstructure as shown in Fig. 4 which in turn reduces hardness. The present investigation suggests that lead current should be greater than trail current which is in good agreement with the literature and also wire diameter at lead should be smaller than trail. Further it increases electrode melting due to increase in current density and enhances melting efficiency along with weld bead properties.

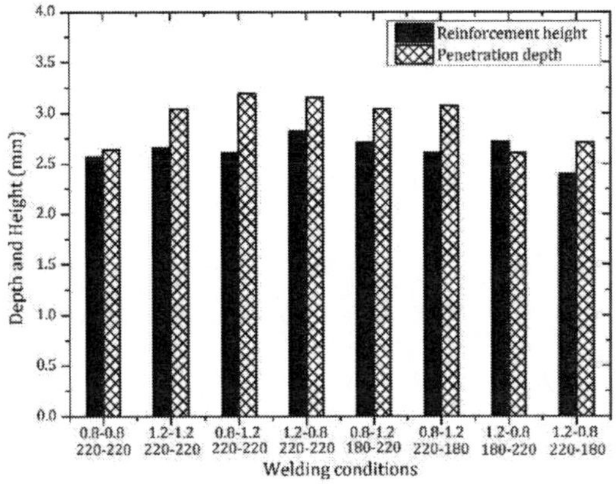

Fig. 3 Reinforcement height and penetration depth at different wire diameters in (mm) and welding currents in (A)

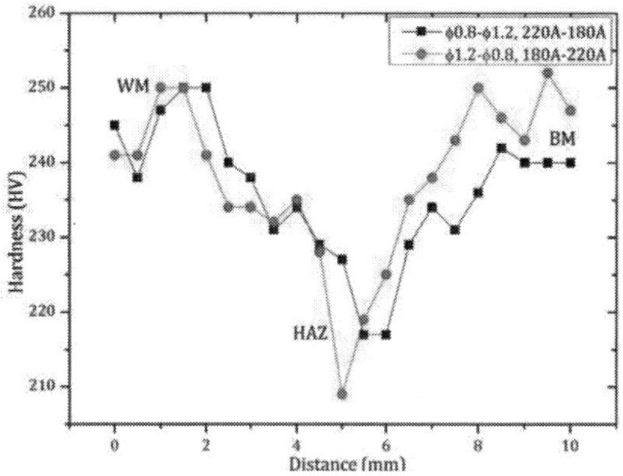

Fig. 4 Weld Hardness

Conclusions

The conclusions drawn from present investigation are as follows:

1. The process parameters at each wire in twin wire anti-phase synchronized GMAW influence heat consumption at either of wires because of mutual interaction. Under ideal conditions, lead electrode draws more heat due to interaction with fresh material.
2. The melting efficiency decreases with increase in electrode wire diameter due to reduced current density irrespective of lead or trail position.
3. Heat input is directly proportional to electrode melting efficiency which also leads to increase in reinforcement height and decrease in penetration depth.
4. Enhancement in melting efficiencies and weld bead properties are achieved when lead diameter is kept smaller and lead current remains greater than trail current.

5. Interchanging process parameters between lead and trail wires influences melting efficiency and hardness of weld metal.

References

[1] R.S. Chandel: "Wire melting and plate melting efficiencies of submerged arc welding and gas metal arc welding," *Materials Science and Technology*, Vol. 6 (1990), pp. 772-777.

[2] J.N. Dupont, A.R. Marder: "Thermal efficiency of arc welding processes," *Welding Journal*, Vol. 74 No.12, (1995), pp. 406-s - 416-s.

[3] R.W. Niles, C. E. Jackson: "Weld thermal efficiency of the GTAW process," *Welding Journal*, Vol. 54 No.12 (1975), 25-s - 32-s.

[4] C.S. Lee, R.S. Chandel, H.P. Seow: "Effect of Welding Parameters on the Size of Heat Affected Zone of Submerged Arc Welding," *Materials and Manufacturing Processes*, Vol. 15, No. 6 (2000), pp. 649-666.

[5] P.W. Fuerschbach, G.A. Knorovsky: "A Study of Melting Efficiency in Plasma Arc and Gas Tungsten Arc Welding," *Welding Journal*, Vol. 70 No. 11 (1991), pp. 287-s - 298-s.

[6] P.W. Fuerschbach, G.R. Eisler: "Effect of Very High Travel Speeds on Melting Efficiency in Laser Beam Welding," *Journal of Materials and Manufacturing,* Vol. 108, (1999), pp. 1-7.

[7] J. Tusek, M. Suban: "High-productivity multiple-wire submerged-arc welding and cladding with metal-powder addition." *Journal of Materials Processing Technology*, Vol. 133, (2003), pp. 207-213.

[8] A. Sharma, N. Arora, B.K. Mishra : "Mathematical modeling of flux consumption during twin-wire welding, *The International Journal of Advanced Manufacturing Technology*, 2008 Vol. 38 No. 11-12, pp.1114-1124.

[9] A. Sharma, N. Arora, S.R. Gupta: "Investigation into arc behavior during twin-wire submerged arc welding," *Materials and Manufacturing Process*, Vol. 25, No. 8 (2010), pp. 873-879.

[10]S.Q. Moinuddin, A. Sharma: "Arc stability and its impact on weld properties and microstructure in anti-phase synchronized synergic-pulsed twin-wire gas metal arc welding," *Materials and Design*, Vol. 67, (2015), pp. 293–302.

[11]S.Q. Moinuddin, A. Kapil, K. Kohama, A. Sharma, K. Ito, M. Tanaka: "On process-structure-property interconnection in anti-phase synchronised twin-wire GMAW of low carbon steel," *Science and Technology of Welding & Joining*, Vol. 21, No. 6 (2016), pp. 452-459.

[12]T. Ueyama, T. Ohnawa, K. Yamazaki, M. Tanaka, M. Ushio, K. Nakata: "High speed welding of steel sheets by the tandem pulsed gas metal arc welding system." *Transactions of JWRI*, Vol. 34, No. 1 (2005), pp. 11-18.

Melting behaviour of wire electrodes acting as cathodes or anodes in GMAW

Gerd Huismann
HSU, Hamburg, Germany
Gerd.Huismann@HSU-HH.de, +49 40 6541-2750

Abstract
MAG welding is normally done with using the wire electrode as anode. This gives a combination of filler metal and heat flow to the workpiece, which forms satisfying beads. A reduction in heat flow to the workpiece can be received by using the wire as cathode. Normally this mode gives too low heat input. AC-MAG gives a combination of both modes and allows a tuning of the heat input into the workpiece. The cathode can get totally different specific heat input. Cathode fall voltage as low as 5V are shown for thermoinic cathodes. 10 to 20V are reported for non thermoinic cathodes. This effect gives a wide range of specific heat input into the cathode. Additionally electron emission gets influence from the oxide formation resulting from the active shielding gas components. Therefore, the heat input on cathodes with the MAG process is influenced by a large variety of factors. The effect of polarity change on the melting behaviour was investigated with respect to the AC-MAG process. A special process, which allows full time negative welding, has been used for this investigation. The effect of current-time functions, gas composition and wire diameter was shown. Using wire electrodes as cathode can double the melting rate.

Introduction
Most of the welding applications are done with consumable wire electrodes. The melting velocity is a main influence factor in the welding performance. It was shown that the specific melting rate [g/As] has been quite unchanged under different welding conditions, when using the wire electrode as anode [4]. The conditions are becoming more complicated, if the wire is used as cathode. The cathode gets more power from the arc than the anode in GMAW. AC-MIG welding gives both modes and the possibility to reduce the process power for a given wire feed rate significantly. Therefore, less heat is flowing into the workpiece, if it is becoming the anode. Further reduction in heat input into the workpiece is received, when the wire electrode is set full time as cathode. This mode is not often used in industry, because the workpiece needs more heat input with normal plate thickness. But some applications on electrode negative [1] and some new patents [2,3] are seen now. Tests on AC-MAG showed a non consistant change of the melting rate. Therefore, an investigation under full time negative polarity for the wire electrode was set up to receive information on the melting behaviour.

Experimental setup
A CSC welding system with bidirectional wire feeder was used for obtaining stable processes under negative polarity. The process sequence is divided into short circuiting and arcing. After detection of the short circuit a current of 60A was set and braking of the wire feed unit was initiated. After stopping of the downward wire movement the feeding direction was set opposite. The current was increased to 150A for all tests 3ms after short circuiting detection. After disconnection of the wire electrode from the weld pool a new process current was set. The upward movement of the wire electrode was stopped. Another current value could be set including pulse phases now. For receiving higher melting rates the current could be set to higher values at the beginning of the arc phase. Different wire sizes were used: 1.0mm, 1.2mm, 1.6mm, and 2.5mm. Welding wire electrodes iron based (G3Si1) were investigated. The distance contact tip work piece was set to 19mm. A variety of shielding gases were used: Ar (4.6), He (4.6), Ar +18%CO2, 100% CO2, Air.

$I_{arc}=\frac{1}{t}\int_0^t i(t) * arcon * dt$, mean arc current with arcon=1 during arc present

Resistance heating was calculated by measuring the mean resistance during short circuiting and using the mean square current. The enthalpies were calculated from mean power flowing into the wire electrode, mean wire feed speed, cross section of the wires, and the density (7.8g/cm³). The estimation of the heat flow from the arc into the wire electrode was done by a specific power of 5.5W/A at the anode and 11W/A at the cathode.

Results
1.6mm wire electrode
The mean current is the most used index for process power in welding. Two different types of behaviour become visible in *Fig. 1*. Both types are nearly proportional to the welding current. With wire electrode positive low

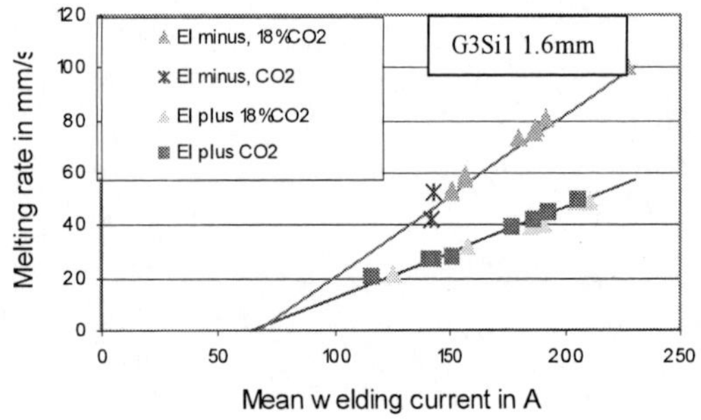

Fig. 1 Melting rate function of the mean welding current

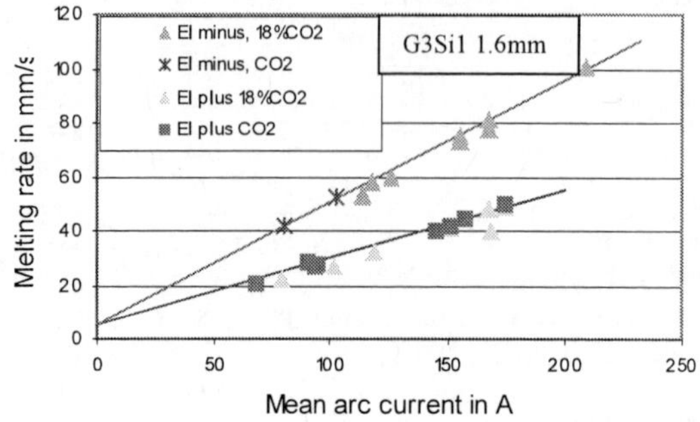

Fig. 2 Melting rate as a function of the arc current, Iarc

melting rates are present. Using negative polarity the melting rate becomes nearly double. The influence of different process gases is small (compare *Fig. 7*, *Fig. 8*). Linear extrapolation of the two lines gives about 70A at zero melting rate. The different process currents were set by using different constant current values and using current pulses. Therefore, an increasing amount of resistance heating was present in the experiments with rising mean current. A linear behaviour as seen in correlation to the mean process current is present with the variable mean arc current (*Fig. 2*). The real difference is visible when extrapolating the lines to zero melting rate. Here both lines are directing nearly to the origin as normally expected. Therefore, this variable can be used to steer the melting velocity on bidirectional wire feeding systems.

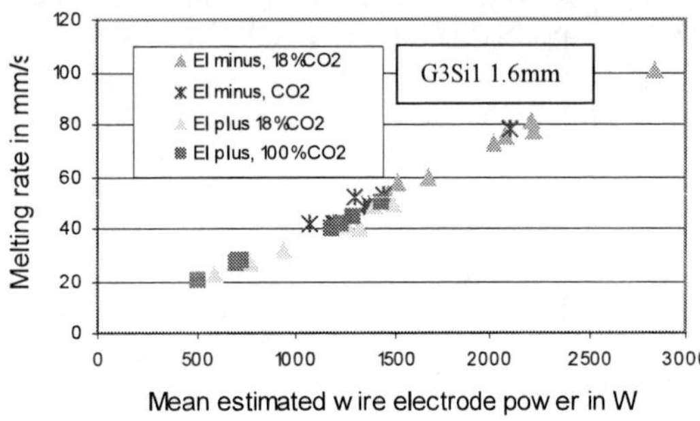

Fig. 3 Melting rate as a function of the mean estimated electrode power

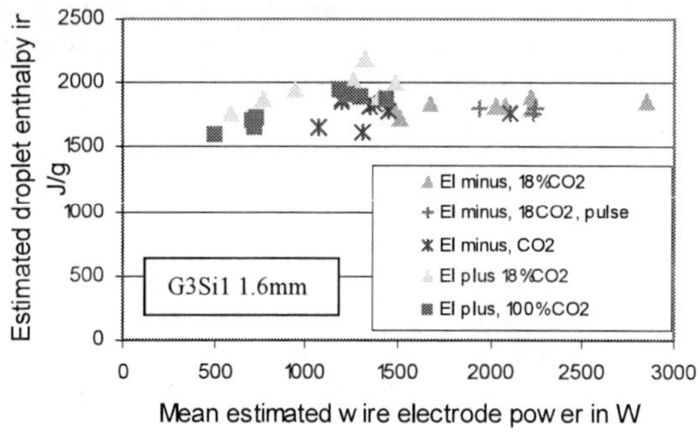

Fig. 4 Droplet enthalpy as a function of the estimated electrode power

Wire electrode positive gives more Joule heating (*Fig. 6*) than electrode negative (*Fig. 8*). The overall Joule heating was much smaller than the arc power. The mean short circuiting resistance was about 9mOhm. The power input from the arc into the wire electrode, P_{fall}, was estimated by using a specific power of 5.5W/A on the anode and 11W/A on the cathode. The melting rates in correlation to the power fed into the wire electrode by arc and

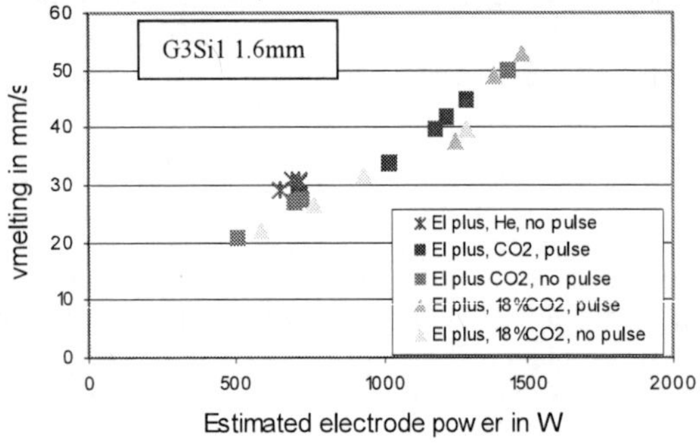

Fig. 5 Melting rate function of the estimated electrode power for electrode positive, comparison of process gases 100% CO2, argon with 18% CO2 and He

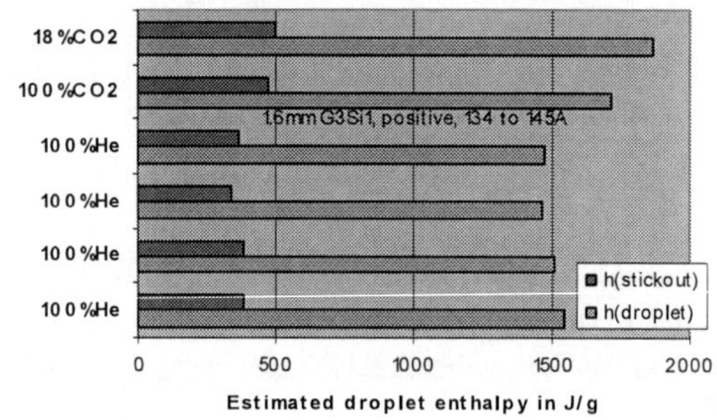

Fig. 6 Calculated enthalpy with different process gases, electrode positive

resistance heating, $P_{droplet} = P_{fall} + P_{resist}$, are shown in *Fig. 3*. The melting rates on anode and cathode conditions are forming lines, which are laying close together (*Fig. 3*). The influence of the different gases is small. 100% CO2 shielding gas gives only a little bit higher melting rate compared to Ar with 18% CO2. The estimated enthalpy $h_{droplet} = (P_{fall} + P_{resist}) / \dot{m}$ stays at about 1800J/g for electrode negative and about 1900J/g for electrode positive (*Fig. 4*).

Different process gases, electrode positive polarity

The plasma shape and properties are changed in a large amount by oxygen and oxygen containing components. The melting rate with pure CO2 compared to Argon with 18% CO2 was for nearly all tests slightly higher (*Fig. 5*). About 10% higher melting rates were obtained under pure CO2. Helium gives an additional 10% extra melting rate. Helium gave the lowest calculated droplet enthalpy (*Fig. 6*).

Different process gases, electrode negative polarity

The tendency on negative polarity (*Fig. 7*) is similar to the behaviour seen on positive polarity (*Fig. 5*). The bandwidth of the results is smaller. The increase of the melting rates on CO2 and He is smaller.

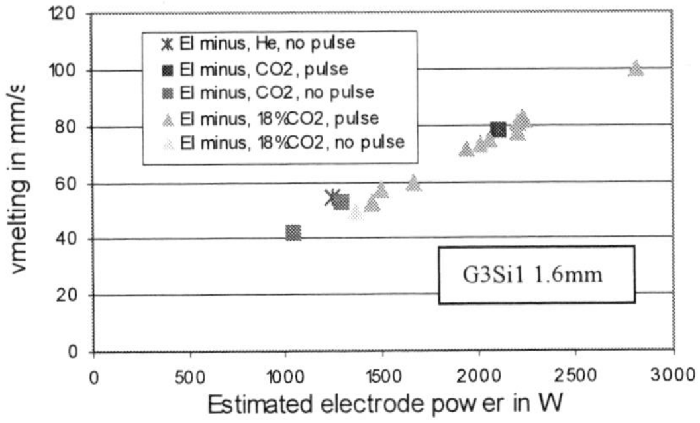

Fig. 7 Melting rate function of the estimated electrode power for electrode negative, comparison of process gases 100% CO2 and argon with 18% CO2 and He

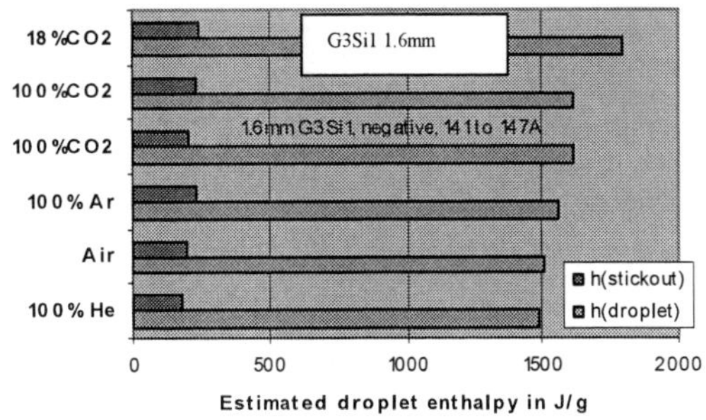

Fig. 8 Calculated enthalpy with different process gases, electrode negative

Different process gases were used on similar process data. *Fig. 8* shows the estimated enthalpy for the different process gases on a melting rate of about 55mm/s. The enthalpy shows low variations. It shows no evident correlation to the ionisation energy. Here Helium should have a much higher cathode heating, if the ionisation potential gives high impact on the cathode fall region. But pure Argon gives a similar melting rate. *Fig. 8* shows for Ar + 18% CO2 the highest estimated droplet enthalpy. This was the process gas, which gave the most stabile process behaviour (of the tested ones). Helium gave the highest problems in running a continuous process.

Wire diameter, electrode negative

The results showed nearly constant droplet enthalpy with electrode negative on 1.6mm wires (*Fig. 4*). Therefore, those parameters were checked on other electrode diameters. The calculated droplet enthalpy was slightly higher on the thinnest wire electrode (1mm wire electrode).

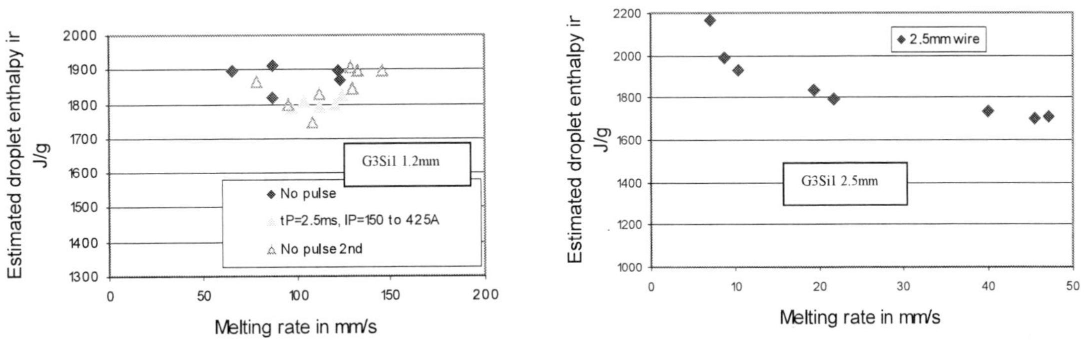

Fig. 9 Estimated droplet enthalpy as function of the melting rate Ar +18%CO2, left 1.2mm electrode, right 2.5mm

Fig. 9 (left) shows a mean droplet enthalpy of 1850J/g for the 1.2mm diameter electrode, which is the same rate as seen in *Fig. 4* for the 1.6mm wire electrode.

Switching to 2.5mm electrode diameter gives a falling tendency over the tested range (*Fig. 9* (right)). The lower melting rates are very slow. Here heat diffusion gets a higher impact. Therefore, the values below 19mm/s were excluded for receiving the mean value.

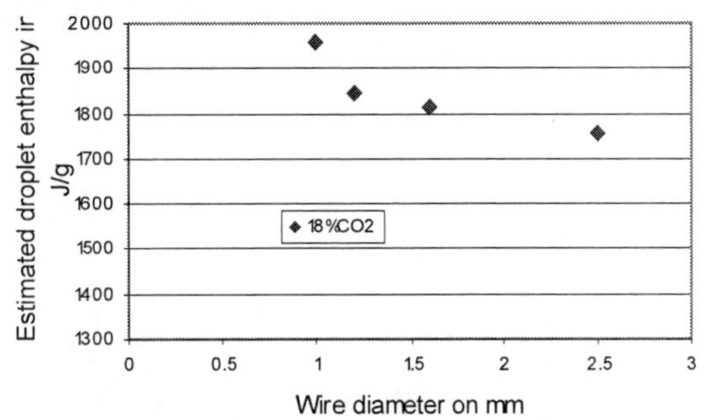

Fig. 10 Estimated droplet enthalpy as function wire electrode diameter

The change of the mean estimated droplet enthalpy was quite small over the wide range of wire diameters (Fig. 10).

Effect of long pulses with high current level

The test results with pulse heating are indicated in the figures above. The results on the parameters shown here gave no evidence on a different behaviour with and without pulse heating for normal pulse durations.

However, long pulse times in combination with high currents showed higher estimated droplet enthalpy (*Fig. 11*). The videos showed heavy metal evaporation on the droplets increasing with pulse time.

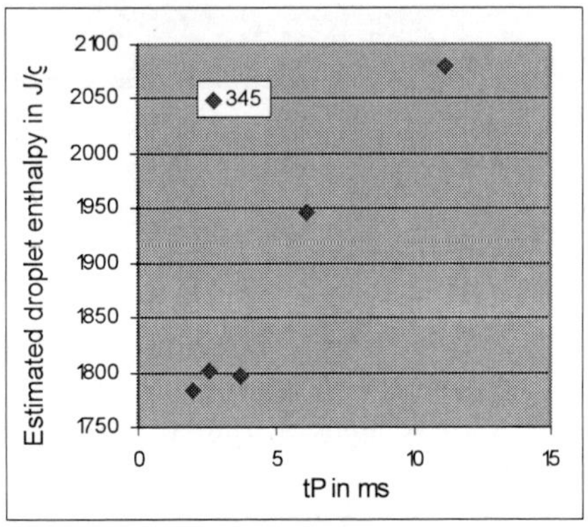

Fig. 11 Influence of pulse time on estimated droplet enthalpy, Ar 18%CO2, 1.2mm wire, Ip=345A

Discussions

The calculated droplet enthalpies showed quite similar values on positive and negative wire electrodes over the wide range of test parameters. The specific cathode heat flow was estimated with 11W/A and 5.5W/A on the anode. Similar results results were shown in [4]. Park [5] showed a doubling of the melting coefficient with 100% EN ratio. With the Daihen AC-MIG power source DW 300+ a two step negative phase is used with the Ar +18%CO2 shielding gas. The second step is combined with high currents. With increasing EN ratio this current and its time was set higher. Such high negative pulse actions can change the melting rate significantly. A similar effect is seen in *Fig. 11* with the rising pulse times. Here more heat flow became necessary to get a specific melting rate. The additional power may flow into metal evaporation as seen on the HS videos.

The minimum droplet enthalpy is equivalent to 1300J/g, which means just melting temperature. Such a droplet cannot melt additional wire material. Therefore, the droplet temperature has to be much higher for being able to melt material from the solid wire. If significant wire melting is present due to heat transfer from the droplet, the droplet temperature has to be significantly higher than the melting temperature. The mean calculated droplet enthalpy was 1900J/g, which is equivalent to 2300°C mean droplet temperature. With short circuiting transfer significant heating of the wire has to be done by the droplet due to the shape of the plasma. It seems technically correct, that the droplet enthalpy should be equal in wire electrode negative and positive mode. Determination of droplet temperatures on the AC-MAG process will be done in the next step.

Using that conclusion gives us evidence that the droplet receives more heat in AC-MAG welding with the Daihen DW300+ power source (Ar +18% CO2) than with DC-MAG welding. Such effects were seen with high metal vapour flow from the droplet during the high current negative phases. Opposite results were shown from AC-MIG welding of aluminium by [6, 7]. Tong [7] claimed lower metal evaporation with AC-MIG. But a large amount of droplet explosions during AC-MAG was seen on all videos with the AC-MAG process on steel wires. That process gives lots of fume and a large amount of fine spatter, which is combined with droplet explosions. It may be different in AC-MIG aluminium, because here less EN ratio has to be used for reducing the input into the workpiece. This effect is based on the low resistivity in aluminium materials (Ieff/Im becomes lower with AC) [6]. The results shown here will be used to correlate melting velocities on AC-MAG with droplet enthalpies.

Conclusions

The use of full time negative current on the wire electrode gave the chance of measuring the melting behaviour without side effects from a positive current phase. That type of process is difficult to handle especially with inert gases. Pre-tests with a conventional power source showed high process instabilities and a very narrow range of parameter combinations for receiving something like a bead. That problem was overcome by the use of the CSC-process. It gives evidence for a double specific heat flow to the cathode compared to the anode with GMAW. The influence of shielding gas composition was smaller than expected. The next step is to get more information on metal evaporation during long time high current phases. That effect gives problems in using high EN ratios in AC-MAG welding. But high EN ratios are necessary for reducing the heat input significantly in AC-MAG welding.

Acknowledegements

The IGF-project "Development of an AC-GMAW process for welding high-strength fine grained steels" (funding number 18.458 B), coordinated by the Research Association on Welding And Allied Processes of the DVS, is funded by the Federal Ministry of Economic Affairs and Energy (BMWi) on basis of a decision by the German Bundestag as part of the AiF Industrial Collective Research (IGF) program.

References

[1] T. Era, et al.: "Controlled bridge transfer (CBT) gas metal arc process for steel sheets joining", *Welding International*, Vol. 27, No. 4, (2013), pp. 268–273,

[2] B. D. Marschke: ELECTRODE NEGATIVE PULSE WELDING SYSTEM AND METHOD, Patent application number: 20140263243, Publication date: 2014-09-18

[3] T. E. Holverson: METHOD AND APPARATUS FOR WELDING, Patent N0.: US 7,335,852 B2, *Feb. 26, 2008

[4] G. Huismann: "Heat transfer to the wire electrode in AC-MIG welding", IIW annual meeting, Istanbul, 2010, paper XII-1998-10

[5] H. J. Park et al.: "The arc phenomenon by the characteristic of EN ratio in AC pulse GMAW", *Int J Adv Manuf Technol* (2013) 66:867–875

[6] Gerd Huismann: Heat transfer by droplets with the GMAW process, Part I aluminium, IIW annual meeting, Istanbul, 2010, SG-212-1185-10

[7] H. Tong et al., "Mechanism of welding fume suppression during AC-pulsed MIG welding of Al-Mg alloys",*Welding International*, 2005. 19(7): p. 509-523.

Study of metal transfer using synchronous wire feeding and current waveform control

By Kohei ONO, Tetsuo ERA, Hisao MIYAHARA, Shugo HIROTA
DAIHEN Corporation, KOBE, HYOGO, JAPAN
Email: k-ono@daihen.co.jp, Phone: +81-78-275-2019 (representative)

Abstract

A gas metal arc welding process to change the wire feed direction according to arc phenomena has been proposed and applied to the field of ultra-thin sheet metal joining with the Gas Metal Arc Welding process, due to ultra-low spatter generation and low heat input into the base metal in a current range of the short-circuit transfer mode. When such a welding process can realize a stable metal transfer in a current range of the globular transfer mode with carbon dioxide shielding gas, it can contribute the further improvement of productivity. Under these circumstances, the authors have developed a synchronous control of wire feeding and current waveform. Utilizing a Buffer Position Control reduces the interference of the wire feed direction between the push feeder and the pull feeder and can compensate the constant mean wire feed rate, and the Pulsed Dip Transfer Process which forms the pulsating output of the welding current during arc duration. This stabilizes the metal transfer with a constant voltage characteristic and improves the heat input into the base metal. In this paper, the principle of these controls and the effect of high speed welding with ultra-low spatter generation of sheet metal joining will be presented.

Introduction

The stabilization of metal transfer by synchronous wire feeding can be traced back for its history in Japan to "the mechanical-short-circuit-transfer arc welding process" [1] reported in 1973. In this welding process, the wire feed is mechanically reciprocated to maintain a repetitive shorting-to-arcing cycle, which was developed with the hint of the bridging transfer by the surface tension of a molten droplet. However, no marketable wire feed motor could be developed at that time, and thus the mechanical-short-circuit-transfer arc welding process could not be used for a practical application. Then, later than 2000, the concept of this process was put into practice with the CSC (Controlled Short Circuit) process [2] and the CMT (Cold Metal Transfer) process. [3] In the domestic market, the Active TAWARS process [4] and the EAGL (Enhanced Arc Robot Welding for Low Spatter) process were introduced in 2010 or later.

In the metal transfer control by synchronous wire feeding, the wire is typically fed to the normal direction for an arcing period and retracted for a short circuit period; consequently, the spatter generation accompanied by re-arcing is believed to be prevented almost completely. Nevertheless, it becomes extremely difficult to stabilize the metal transfer in globular metal transfer mode by the constant voltage control which is premised on the wire feed rate kept constant. On the other hand the penetration with the welding process is shallow though using a CO_2 shielding gas. When the wire is fed to the normal direction for the re-arcing period, not only the extended wire but also part of the wire retracted once into the contact tube are reheated by the arc and Joule heats. Due to the reheating, the melting rate of the wire becomes higher than in general GMA welding, thereby reducing relatively the heat input in the base metal, unless the compensated control is applied for an appropriate wire melting rate. This process can be previously considered to be excellent in low heat-input performance.

In the following paragraphs, discussed are the characteristics and performances of the newly developed welding process that solves the problems with the conventional process mentioned above. The new process is applicable to the ultrathin steel sheet welding in the low current range and to the high speed welding.

Development of new welding system

In order to configure the new welding process, the following systems have been developed: (1) the high-transient-response AC servo motor that can switch the wire direction according to the shorting and arcing status at a high wire-feed rate in conjunction with the servo driver to control the motor; and (2) the control for maintaining the average wire feed rate at a constant level together with the current waveform control for low spatter performance and deep penetration. The outline of the welding system configuration developed is shown in Fig. 1.

In the figure, the connection of cables such as the power cables for the welding power supply is omitted. The pull feeder that is built in the torch mounts the newly developed high-transient-response and low-inertia AC servo motor. Because the feeder is characterized by small size and light weight, it can also be mounted on a 4-kg transportable manipulator. The push feeder that is installed in the wire feed section is controlled basically to feed a welding wire in the normal direction only. In the welding current range where a globular transfer is observed, the average wire feed rate increases; in such a case, a big load is liable to be applied on the wire in the wire feed passage due to the different directions of the wire fed by the push and pull feeders together with the disturbance on the wire feed system. To ensure the proper wire feed control in such a high burden condition, the wire buffer mechanism is installed between the push feeder and the pull feeder to execute the interference control, thereby preventing troubles such as wire-buckling in the wire feed system. With this specific system configuration, designated mean wire feed rate has been achieved, even in a high current range where the average wire feed rate exceeds approx. 6 m/min; as a result, stable droplet transfer has been created in a wide current range from 50 to around 300A by synchronized with the following welding current waveform control.

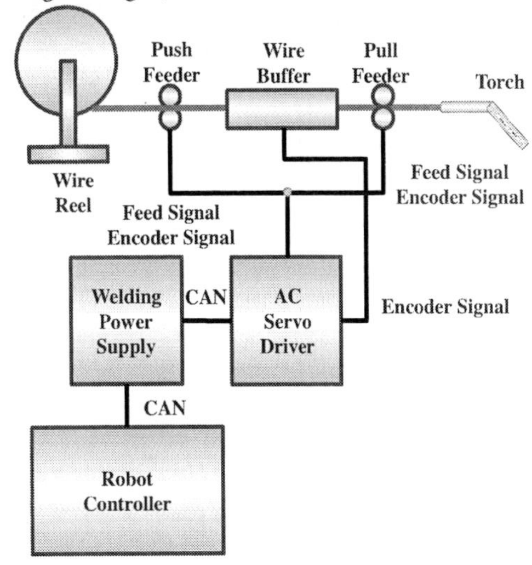

Fig.1 System configuration sample

Development of new current waveform control technology (Pulsed Dip Transfer Process)

In order to realize ultra-low spatter welding with the CO_2 or MAG gas shielding and a mild steel solid wire of 1.2-mm diameter, the current waveform control technology of Pulsed Dip Transfer Process (PDT process) has been developed. The low-spatter high-speed welding of sheet metals is the intended application of this welding system with the current-waveform and wire-feed controls for short-circuit metal transfer. In the PDT process, the principle is the same for both CO_2 and MAG welding; hence, CO_2 welding is taken for the example to show the current waveform, wire feed waveform, and typical metal transfer in Fig. 2. The metal transfer was observed by high-speed video camera (HSV) in the following welding conditions: a welding current of 300A, an arc voltage of 23.0V, and a welding speed of 60 cm/min. The outlines of the metal transfer process will be discussed below.

On detecting a short circuit of droplet, the wire is retracted and then, according to the retraction status of the wire, the optimized short-circuit current is supplied to promote re-arcing. Frame (a) shows the short-circuit phase. Immediately after re-arcing, the wire is still retracted until the wire feed rate becomes 0 cm/min., and then the wire is inched to the normal direction for the re-arcing period. During this period, the welding current is reduced to prevent the molten droplet on the wire tip from becoming excessively large. As shown in Frame (b) for this phase, no spatter generation is observed. When the wire movement is switched to the inching, the high pulsed current is supplied to help the arc force push down the molten pool for deep penetration; simultaneously, the approximately constant voltage control stabilizes the droplet transfer during the pulse current period. This phase is shown in Frame (c). In the latter half of the arcing period, the arc current is limited to the base current to promote the recurrence of short circuit as shown in Fame (d). As mentioned above, the PDT process executes the approximately constant voltage control for the pulsed current period to make the average wire feed rate constant regardless of the transitional switching in the wire feed direction, which is the indispensable function to stabilize the metal transfer. The stabilization of metal transfer is realized by this

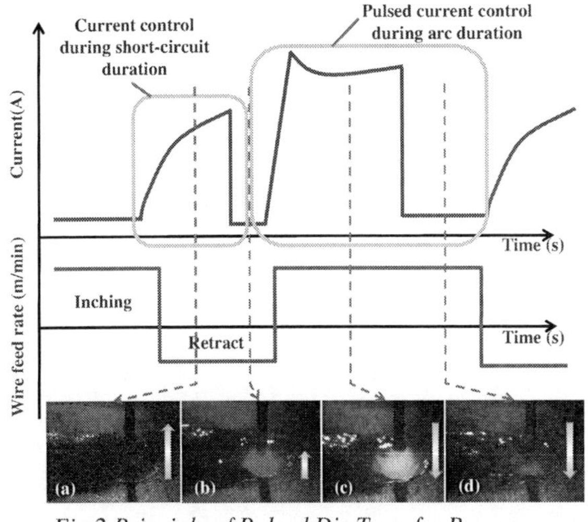

Fig.2 Principle of Pulsed Dip Transfer Process on CO_2 shielded arc welding

function in conjunction with the synergy effect of the wire-feed-rate control by the pull feeder described later.

Buffer Position Control

With the PDT process, stable metal transfer can be obtained at an average wire feed rate of approx. 10 m/min. maximum, and the shorting-to-arcing cycle can be increased up to approx. 100 Hz. For switching the wire feed direction according to the shorting-to-arcing cycle of 10 ms, the peak wire feed rate after the switching is needed to be at least several times the mean wire feed rate; hence, the wire feed system needs, other than the transient response, the interference control to reduce its burden and thereby contribute greatly to the stabilization of metal transfer. Fig. 3 shows schematically the original mechanism of Buffer Position Control (BP control) with S-shaped buffer. According to the wire direction fed by the pull feeder, the buffer interferes with the direction, while the rotary encoder that is built in the buffer can detect the angle of rotation as positional information in real time. Base on the positional information, the mean wire feed rate by the push feeder and the designated wire feed rate by the pull feeder are optimized to accord with each other, and in turn the interference is controlled to be minimal. In addition, since the mean wire feed rate by the pull feeder is compensated, the metal transfer can be stabilized by the approximately constant voltage control for the pulse current period in the PDT process. On the other hand, if the process detects an error that disables the BP control, it can also output such an abnormality in the wire feed system to stop the welding operation, thereby avoiding any situations difficult to recover from such as wire-buckling.

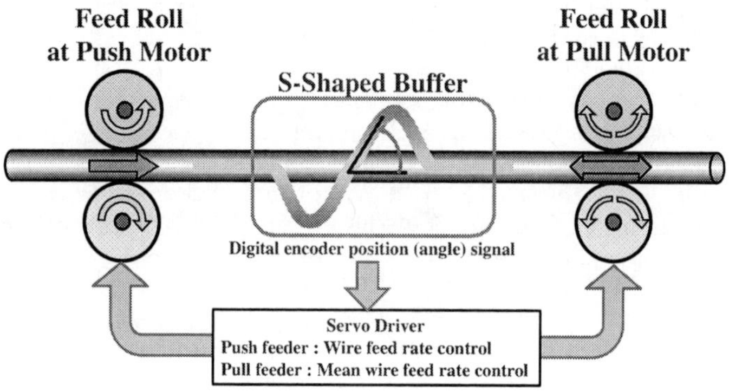

Fig.3 Schematic of buffer position control

Low-Spatter Performance of Developed System

Fig. 4 shows the amounts of spatter for one minute in CO_2/MAG welding with a 1.2-mm dia. mild steel solid wire of YGW12 to compare the developed system with the conventional low spatter welding.

The welding speeds were set at 100 cm/min. at 200A or higher, 75 cm/min. at 150A, and 50 cm/min. at 100A. In MAG welding, a shielding gas mixture of 80%Ar+20%CO_2 was used. When the amounts of spatter with the developed system are compared to those with the conventional process, as recognized in general, MAG welding generates lower amounts of spatter than does CO_2 welding for both welding processes. The amount of spatter was less than 0.1 g/min. for the measurements of developed process. Clearly, with the developed system, the amounts of spatter were less at most 98% than with the conventional process.

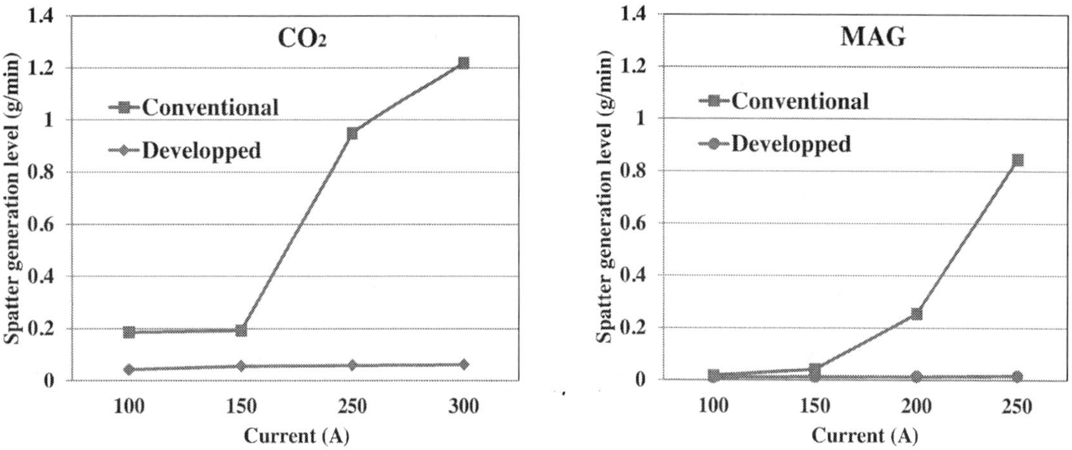

Fig.4 Comparison of spatter generation level in CO_2/MAG welding

Welding Test Results

With a shielding gas of 100%CO_2 and a mild steel solid wire of 1.2-mm diameter, gas shielded metal arc welding was carried out on SPCC-grade steel sheets having several different thicknesses and joint configurations, the results of which are discussed in the following. Fig. 5 shows the results of lap fillet welding and T-joint fillet welding on 3.2-mm thick steel sheets. In every test result, the weld bead exhibits sufficient penetration, no undercut, and good shape with uniform toes. Fig. 6 shows the results of welding 0.6-mm thick steel sheets with butt joint and 0.6-mm gapped lap joint. In the general low-heat-input GMA welding of ultrathin steel sheets, excessive melt-through is apt to occur, and the bead shape is prone to become convex. In contrast to this, in the present welding results, every weld bead exhibited the good fusion at bead toes, sufficient flank angles, and no excessive melt-through.

	Bead appearance	Cross section
3.2mm Lap joint 300A 22.5V 100cm/min	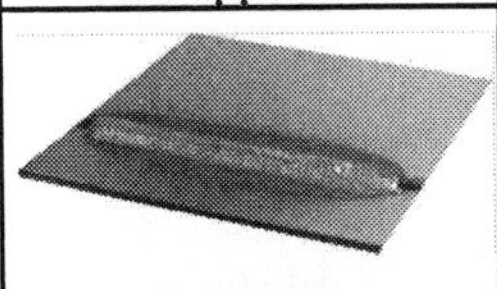	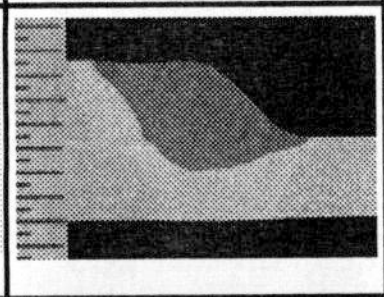
3.2mm T-fillet joint 250A 23.0V 50cm/min	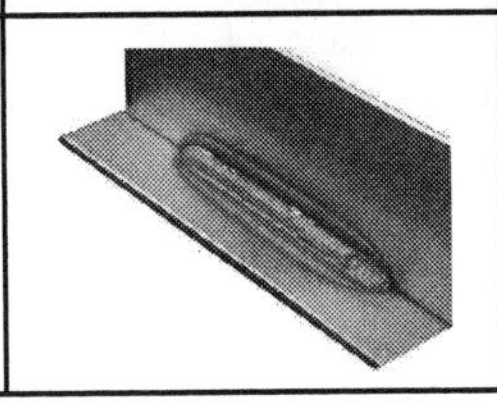	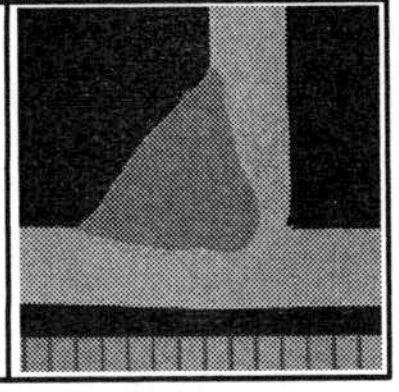

Fig. 5 Bead appearance and cross section of 3.2 mmt

	Bead appearance	Cross section
0.6 mm Butt joint 50A 18.0V 70 cm/min	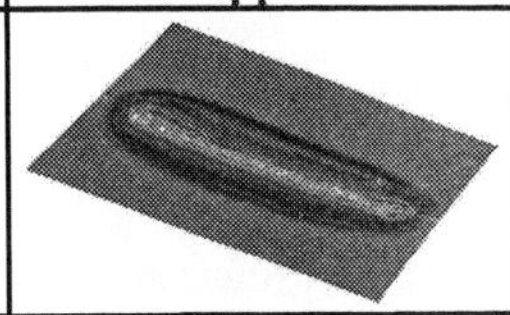	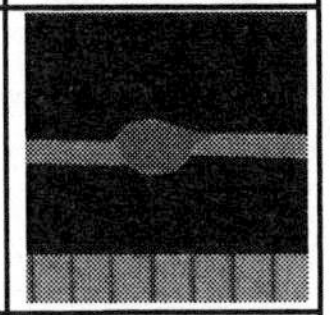
0.6 mm Lap joint 0.6mm gap 50A 18.0V 70cm/min	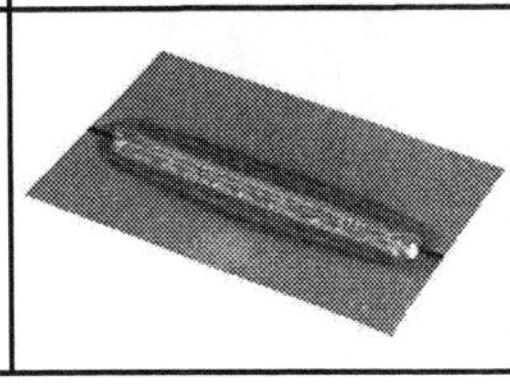	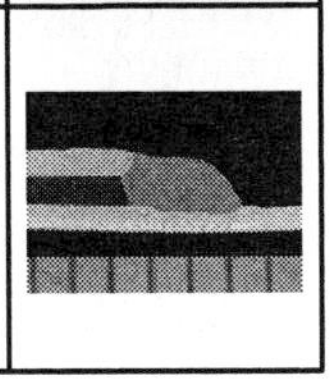

Fig. 6 Bead appearance and cross section of 0.6 mmt

Conclusions

1) The new metal transfer control using synchronous control of wire feeding and current waveform has been developed for practical uses.

2) Buffer Position Control can make the mean wire feeding rate constant, which can help the arc length regulation in the globular transfer mode using CV characteristic of welding power sources and prevent from troubles on the wire feed system.

3) Pulsed Dip Transfer Process can realize an ultra-low spatter with sufficient penetration and the high travel speed.

4) As the welding test results, the developed welding system offers ultra-low spatter performance in the use of a shielding gas of CO_2 or MAG gas, a 1.2-mm dia. solid wire of mild steel, and a wide range of welding currents from 50 to 300A. With a low current of 50A, a wide weld bead could be obtained without excessive melt-through on ultrathin steel sheets. Even with a high current of 250-300A, a smooth and glossy bead with sufficient penetration could be created.

References

[1] H. Yamamoto et al. : Study on short-circuit time ratio control in arc welding (4th report), Development of mechanical short-circuit transfer arc welding process, Preprints of the National Meeting of JWS, Vol.12 (1973) pp.294-295 (in Japanese)

[2] G. Huismann : Proc International Conference on 'Advances in welding technology 2000' ('Gas metal arc welding for the 21st century'), Orland, FL, USA, December 2000, The American Welding Society, p. 165-172

[3] K. Himmelbauer : The CMT-Process – A revolution in welding technology, IIW Doc. XII-1875-05 (2005)

[4] J. Fujiwarea : The new welding robot fused with welding power source which realizes super low spatter [Active TAWERS], Welding technology, Vol.59 (2011) No.2, pp.51-56 (in Japanese)

Study of the electrode metal transfer in GMAW by combining of two modeling approaches

Uwe Reisgen, Oleg Mokrov
ISF of the RWTH Aachen University, Aachen, Germany
office@isf.rwth-aachen.de, +49 241 80-93-870

Igor Krivtsun, Oleksii Semenov
The E.O.Paton Electric Welding Institute, Kyiv, Ukraine
krivtsun@paton.kiev.ua, +38 044 206-18-24

Abstract
The current work presents two mathematical models for describing of electrode wire melting, metal transfer and arc plasma in GMAW. The proposed models takes into account interaction of thermal, hydrodynamics, capillary and electromagnetic processes in the system 'electrode extension - droplet - arc plasma' by self-consistent way. The governing equations of the model are solved by means of FEM. For computer realization of the model the specialized software has been developed. Extensive numerical analysis of the physical processes inside the electrode metal during GMAW has been performed. The simulation results of the droplet shape dynamics along with temperature and velocity fields inside the melt for globular and spray transfer modes are presented.

Introduction
The productivity and quality of GMAW process to great extent depend on electrode melting rate and metal transfer mode. Considerable attention has been devoted separately to theoretical and experimental investigation of heat processes in electrode metal [1-3] and metal transfer [4-6].
The main purpose of the investigation of heat transfer processes in electrode is to define the influence of technological parameters of the welding process upon the electrode melting rate. It was found that electrode melting rate is defined by welding current, electrode diameter, shielding gas and dimensions of electrode extension [2]. It turned out that the wire melts due to Joule heating taking place in electrode and heat input from arc plasma. The relative importance of these two heat sources in electrode heating depends on electrode extension and diameter, as well as welding current [3]. The simplified model of heat transfer in droplet was reported in paper [7]. In this study the main reason of the convective motion in melt is supposed to be thermocapillary Marangoni effect. Indeed, this phenomenon can be ascribed in part to Lorentz force which arose due to interaction of welding current with its own magnetic field.

The most simple models for metal transfer description are the Static Force Balance Theory (SFBT) [4], and Pinch Instability Theory (PIT) [5]. SFBT describes droplet formation well in globular mode whereas PIT can be used for description of spray mode. In paper [6] the dynamic model of droplet formation based on slender jet approximation was developed. By including non-stationary terms this model can quite successfully describe droplet formation up to pinch off but heat transfer processes are beyond the scope of this model. A unified models of heat, mass and charge transfer in GMAW have been presented in [8], [9]. Such a models use Volume of Fluid (VOF) method for free surface tracking and allows to reveal physical phenomena taking place in molten metal and in nearby plasma but heat transfer in electrode extension is not considered in these studies. For a detailed review of existing models of metal transfer the reader is referred to paper [10].

In current paper we present the self-consistent mathematical model of GMAW electrode melting and metal transfer which allows heat, mass and charge transfer processes taking place in the system "electrode extension – droplet". The present paper contains a detailed numerical analysis of the processes in question for the case of globular and spay modes. Such important characteristics of the GMAW process as droplet radius, its velocity and temperature, electrode melting rate can be numerically analyzed by means of the model developed.

Mathematical formulation and numerical procedure
In the developed model further assumptions have been made:
(a) The physical processes taking place in the system considered are supposed to be axisymmetric
(b) The liquid metal is a Newtonian fluid
(c) Distribution of the heat flux along the metal surface is prescribed as described later.

(d) The droplet is surrounded by arc plasma with homogeneous properties
The unified mathematical model includes models of heat and mass transfer in electrode metal and model of electromagnetic processes in system „electrode metal-arc plasma".

The electrode extension length is significantly greater than droplet diameter, therefore the temperature field inside the wire is far from melting front and can be described as a simplified 1D approach. In current study the heat transfer processes in metal described by the 2D model (inside the droplet and in small region above liquid front) along with 1D model (inside solid phase far from liquid front). Such approach results in great computational savings. Heat absorption due to melting is taken into account by means of effective heat capacity method [11].

In such approach equations with corresponding boundary conditions were used to cover the following processes: heat transfer, hydrodynamic and electromagnetic. On the boundary between 1D and 2D domains conjunctions conditions which specify temperature and heat flux continuity were imposed. The model of hydrodynamic processes is based on incompressible Navier-Stokes equations. Viscosity in solid phase was specified as 10^6 times greater than viscosity in liquid phase. Such approach allows suppressing metal motion inside the region occupied by solid metal and is referred to as effective viscosity method [12]. Electromagnetic processes are described by current continuity equation, Ohm's law and Maxwell's equation. Unlike the approach used in [9] in present paper the current density distribution along droplet surface is unspecified but is calculated automatically. Following [3] we suppose that heat flux from plasma to the droplet is equal to welding current multiplied by voltage equivalent of heat. Furthermore the voltage equivalent of heat does not changes with current. In addition we use linear dependence to specify distribution of heat flux on anode surface [9]. Governing equations of the model are solved by means of Galerkin's finite element method. Triangular elements are used for solution of 2D problems whereas heat transfer equation for the area AH is approximated on uniform mesh (*Fig. 1*). The calculation stops when electrode extension is no longer changes that corresponds to achievement of thermal equilibrium in electrode.

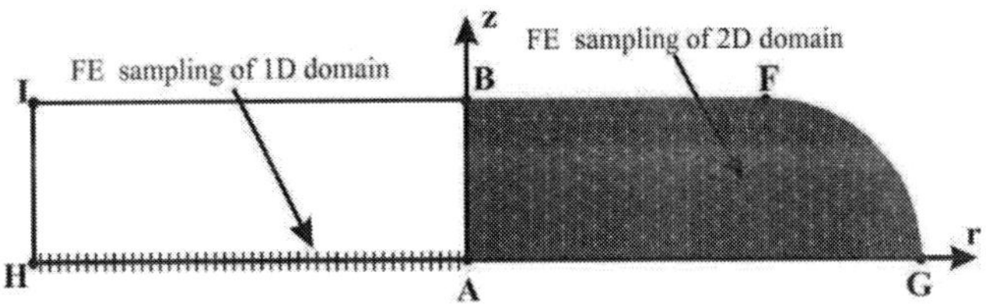

Figure 1. Finite element mesh in metal

Results and discussion

Our calculations were performed for mild steel as base metal and argon as shielding gas. Two different modes of metal transfer have been considered: globular and spray. Input parameters for modelling are listed in *Table 1* [1]. Physical properties of the material in use taken from [13] and [14]. The voltage equivalent of heat was taken as 6 V [3]. Electrical conductivity of plasma was set as 8745 $(\Omega\cdot m)^{-1}$ that corresponds to the temperature of 18000 K [15]. Ambient temperature was set to 300 K.

Table 1: Input parameters for modelling.

Metal transfer mode	Wire radius	Electrode extension	Welding current	Wire feed rate
Spray	0.8 mm	18 mm	352 A	85 mm/s
Globular	0.8 mm	50 mm	138 A	30 mm/s

Calculations show that in spray transfer mode the fluid flow inside the melt is unidirectional whereas in globular transfer the flow pattern has a vortex structure. Such a convective melt motion inside the droplet is ascribed to influence of electromagnetic forces. Typical value of the metal velocity is about several dozens centimeters per second. In globular mode the secondary vortex appears in the upper part of the neck prior to detachment (*Fig. 2*). This leads to the flattening of the melting front.

Temperature distribution inside the solid phase of the electrode is close to linear (*Fig. 3*). Heat flux undergoes discontinuity on the melting front due to latent heat absorption. In the vicinity of droplet neck the local temperature maximum is observed owing to substantial joule heating in this region. It should be mentioned that calculated metal temperature on the droplet tip is greater than boiling temperature. Therefore, it is important to take into account metal vaporization from droplet surface.

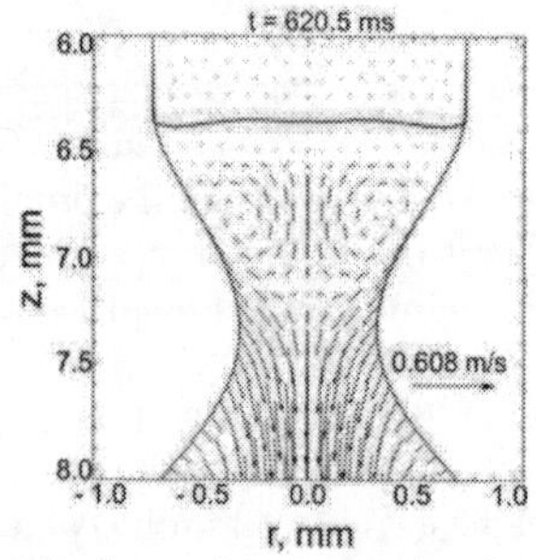

Figure 2. Velocity field inside the neck prior to droplet detachment (I_{arc} = 138 A)

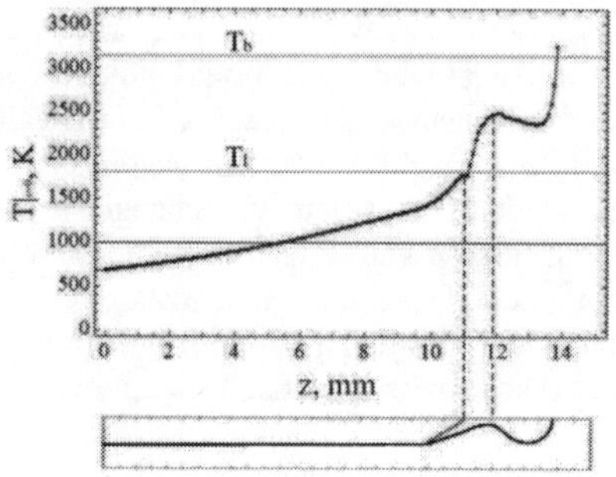

Figure 3. Temperature distribution along electrode axis prior to droplet detachment in 2D domain (I_{arc} = 352 A)

When droplet neck reaches certain dimensions the region on the droplet surface through which the current flows in the metal from plasma appears (*Fig. 4b*). This is because of increase of the droplet neck resistance which in its turn depends on neck radius.

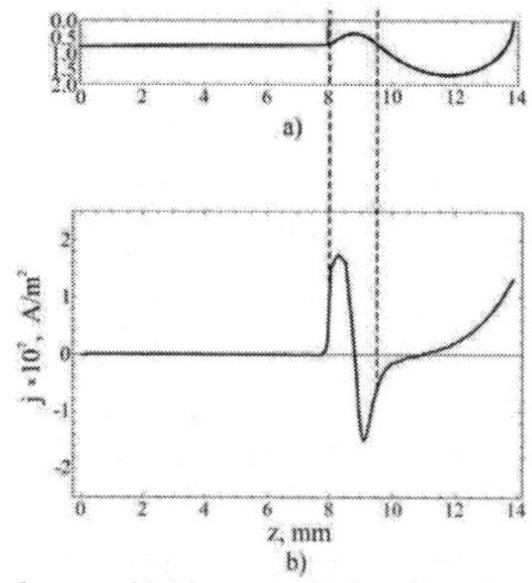

Figure 4. a) droplet shape before detachment, b) current density distribution along droplet surface (I_{arc} = 138 A)

Characteristics of the detached droplet

The model developed allows calculating such characteristics of the detached droplet as its radius, mean temperature, centre of gravity velocity and frequency of droplet detachment. All aforementioned parameters for considered metal transfer modes are presented in *Table 2*. It should be mentioned that the largest droplet temperature corresponds to the case of globular transfer mode.

Table 2: Calculated characteristics of the droplet.

Metal transfer mode	R_d	V_d	T_d	τ_d
Spray	0.73 mm	0.854 m/s	2911 K	13 ms
Globular	1.91 mm	0.128 m/s	3058 K	600 ms

The minimum discrepancy between calculated and experimental values of electrode extension is observed when voltage drop in contact area (V_c) is supposed to be equal to 0.49 V. At a given V_c calculated values of electrode extension for globular and spray transfer modes are 49.4 mm and 20.8 mm respectively, that rather well corresponds to experimental data (*Table 1*).

Conclusions

A mathematical model of wire melting and electrode metal transfer in GMAW have been developed. The model takes into account following phenomena: heat transfer, hydrodynamic and electromagnetic. These processes taking place in the system "electrode extension-droplet" by self-consistent way. The model predict metal transfer characteristics and analyze heating and melting of electrode. On the basis of the model a software has been developed by using self-written code, allowing for simulation of the aforementioned processes. Numerical analysis of the heat and mass transfer processes for the case of globular and spray transfer modes have been completed. In current paper the Marangoni effect and vapour recoil pressure are not analyzed. For more accurate simulation of heat processes in electrode the value of contact voltage should be specified.

Acknowledgments

This work was carried out with the financial support of the German Research Foundation DFG RE-2755/36-1 "Development of a physical-mathematical model of the gas metal arc welding process for pulsed mode by using multicomponent filler materials and molecular gases". The authors wish to express their thanks for this funding.

References

[1] J.H. Waszink: "Heat generation and heat flow in the filler metal in GMA welding," *Welding Journal*, Vol. 61, No. 8 (1982), pp. 269-282.

[2] Y.S. Kim, D.M. McEligot, T.W. Eagar: "Analysis of electrode heat transfer in gas metal arc welding," *Welding Journal*, Vol. 70, No. 1 (1991), pp. 20-31.

[3] V.A. Nemchinsky: "Heat transfer in an electrode during arc welding with a consumable electrode", *Journal of Physics D: Applied Physics*, Vol. 31, No. 6 (1998), pp. 730-736.

[4] J.C. Amson: "Lorentz force in the molten tip of an arc electrode," *British Journal of Applied Physics*, Vol. 16 (1965), pp. 169-1179.

[5] C.J. Allum: "Metal transfer in arc welding as a varicose instability. II. Development of model for arc welding," *Journal of Physics D: Applied Physics*, Vol. 18, No. 7 (1985), pp. 1447-1468.

[6] O. Semenov, V. Demchenko, I. Krivtsun et al: "A dynamic model of droplet formation in GMA welding," *Modelling and Simulation in Materials Science and Engineering*, Vol. 20, No. 4 (2012), p. 045003.

[7] V.A. Nemchinsky: "Heat transfer in a liquid droplet hanging at the tip of an electrode during arc welding," *Journal of Physics D: Applied Physics*, Vol. 30, No. 7 (1997), pp. 1120-1124.

[8] J. Haidar, J.J. Lowke: "Predictions of metal droplet formation in arc welding," *Journal of Physics D: Applied Physics*, Vol. 29, No. 12 (1996), pp. 2951-2960.

[9] F. Wang, W.K. Hou, S.J. Hu et al: "Modelling and analysis of metal transfer in gas metal arc welding," *Journal of Physics D: Applied Physics*, Vol. 36, No. 9 (2003), pp. 1143-1152.

[10]A.P. Semyonov: "Methods of mathematical modelling of the processes of electrode metal drop formation and transfer in consumable electrode welding (Review)," *The Paton Welding Journal*, No. 10 (2014), pp. 2-10.

[11]A.A. Samarskii, P.N. Vabishchevich, O.P. Iliev, A.G. Churbanov: "Numerical simulation of convection/diffusion phase change problems—a review," *International journal of heat and mass transfer*, Vol. 36, No. 17 (1993), pp. 4095-4106.

[12]S. Kou, D.K. Sun: "Fluid flow and weld penetration in stationary arc welds," *Metallurgical Transactions A*, Vol. 16, No. 2 (1985), pp. 203-213.

[13]Hu2007

[14]J.H. Waszink, G. Van den Heuvel: "Measurements and calculations of the resistance of the wire extension in arc welding," *Proc. Int. Conf. Arc Physics and Weld Pool Behaviour*, London, (1979).

[15]R.S. Devoto: "Transport coefficients of ionized argon," *Phys. Fluids.*, Vol. 16 (1973), pp. 616-623.

Influence of pulse profile on droplet transfer dynamics and its implication on gas metal arc weld bead penetration and cooling rate

E.B.F. Santos, Abdelbaset R. H. Midawi, A.P. Gerlich
Department of Mechanical and Mechatronics Engineering, University of Waterloo, Waterloo, ON, Canada

R. Pistor
Liburdi Engineering Limited, Dundas, ON, Canada

Abstract

Four pulsed current profiles were compared during gas metal arc welding, which were distinct in terms of the peak to background current ratio and ramp rate from peak to background current. The influence of pulse profile was investigated at four different wire feed speeds ranging from 150 to 300 in/min. Duty cycle and travel speed were varied such that the average current and voltage were approximately the same with all profiles for the same wire feeding speed, and that a one drop-per-pulse (ODPP) transfer condition dominated. High speed videography at 5000 frames per second, with synchronized voltage and current measurements at a 20kHz sampling rate were used to observe and characterize the metal transfer for the different profiles. The results revealed that the greater the difference between peak and background current, the velocity of the droplet transferred to the weld pool is up to 200% faster. At the same time, the instantaneous average current and arc power required for stable metal transfer is reduced compared to pulse profiles with lower peak to background current ratios. In addition, with a longer transition time from high to low current (as long tailed square wave form), the droplet detachment occurs during the current transition region at higher current values. In comparison, for a more ideal square waveform with a shorter transition time from peak to background current, the droplet detachment takes place during the background current, which promotes droplet transfer at lower arc current. The average power required to transfer the same amount of metal when using the nearly square wave form is also lower than for a long tailed profile. Implications of these findings suggest that pulse parameters can be optimized to achieve high penetration with low heat input.

Introduction

Spray is the common metal transfer modes in gas metal arc welding (GMAW), since it produces a highly regular and stable transfer with no significant amounts of spatter [1]. Therefore, it is usually used to produce beads with a good surface finish. However, in order to achieve spray transfer, high values of current and voltage are required; which leads to high heat input values that can be an issue when welding heat sensitive materials like pipeline steels [2]. Additionally, due to the high amount of heat, a large weld pool size is produced which often limits the welding operation to the flat position. One way to overcome this situation and achieve spray transfer at low heat input (allowing welding in all positions) is to use pulsed GMAW (GMAW-P) [3]. Due to the control of current wave forms in a constant pulse frequency the spray transfer mechanism involving one droplet per pulse (ODPP) can be achieved at average currents well below the spray transition current (which is usually a minimum current required for spray transfer in constant voltage GMAW). Since the invention of GMAW-P in the mid-1960s [4–7], advances in electronics have allowed the application of pulses with various shapes such as square, trapezoidal and exponential decay wave forms [3]. Each of these profiles are claimed to have their own advantages to specific weld characteristics and applications. The energy transferred to the droplet and base metal have been reported to be proportional to the area under the current profile, and therefore the welding heat input can be controlled by changing this area.

Besides the metallurgical implications of lower heat input during pulsed GMAW, there might also be implications on the geometric features of the weld the bead, such as bead penetration. Murray and Scotti [8], found the main factor influencing the penetration during GMAW could be defined by two dimensionless terms, mass and heat transfer coefficients, and showed that despite the role of mass transfer, the heat transfer has a major influence on total bead penetration. Therefore, it is expected that beads performed by pulsed GMAW would present low penetration values, due to the lower heat input when compared to conventional constant voltage GMAW. However it has also been reported that the bead penetration is somewhat related to the momentum transferred by the impinging droplet on the welding pool and therefore the mass and velocity of the droplet might have a contribution to the overall bead penetration [9,10].

The present work aims to investigate the influence of pulse wave profile on droplet formation mechanism and transfer dynamics, namely droplet size and velocity, and the possible implications that those might have on the bead penetration and heat input. The ultimate goal of this research is to understand the effect of different pulse profiles and pulse parameters on the droplet velocity in order to determine how parameters affect the deposit of material with the lowest heat input and achieve the highest penetration.

Experimental Procedure

The four pulse profiles were investigated in this work are schematically shown in *Fig. 1*. Two profiles present an exponential wave form, P1 and P3, and two have a nearly square wave form, P2 and P4. Bead on plates were performed at four different wire feed speeds 150, 200, 250 and 300 in/min for each profile form. The shielding gas was 85%Ar-15%CO_2, and an ER70S consumable of 1.13 mm diameter was used for all tests. A high speed camera, operating at 5000 frames per second, with synchronized voltage and current measurements at a 20kHz sampling rate.

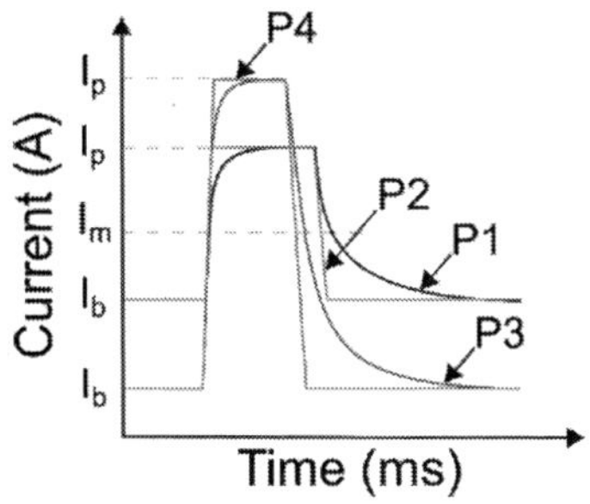

Fig. 1: Schematic of the four different profiles investigated.

The welding parameters were chosen to maximize metal transfer in the one droplet per pulse regime (ODPP). For each wire feeding speed, an effort was made to keep the voltage for the different profiles as equal as possible, i.e. resulting in the same arc length. The welding parameters for each wire feeding speed for all the profiles are shown in *Tab. 1*.

Table 1: Welding parameters showing the average and the confidence interval according to the t-student test.

Wire Feeding Speed	Profile	I_p	I_b	DC	f	Instantaneous Average Current	Instantaneous Average Voltage	Instantaneous Average Power	Travel Speed	Nominal Heat Input
in/min		A	A	%	Hz	A	V	W	in/min	kJ/mm
150	P1	375	50	27.0±0.1	140	137.9±0.5	21.5±0.1	3308±20	12	0.651
	P2	375	50	19.9±0.6	140	114.8±1.9	21.5±0.2	2840±44		0.559
	P3	510	30	16.6±0.2	110	109.6±0.9	20.2±0.0	2842±12	10	0.671
	P4	510	30	12.5±0.3	110	90.0±1.6	20.0±0.2	2338±47		0.552
200	P1	380	75	30.3±0.6	170	167.4±1.8	22.6±0.1	4110±32	10	0.809
	P2	380	75	25.4±0.8	170	152.4±2.4	22.5±0.8	3788±70		0.746
	P3	520	40	23.9±0.7	150	154.6±3.3	21.1±0.1	4143±87	10	0.815
	P4	520	40	17.4±0.4	150	123.3±2.1	22.0±0.1	3396±56		0.669
250	P1	400	100	39.3±0.6	220	217.8±1.7	24.1±0.0	5514±33	14	0.930
	P2	400	100	27.7±1.0	220	183.0±3.0	24.9±0.6	4860±178		0.820
	P3	520	45	27.5±0.4	180	175.8±1.7	23.9±0.0	4874±41	12	0.959
	P4	520	45	21.7±0.6	180	148.3±3.0	23.3±0.1	4232±84		0.833
300	P1	400	115	43.0±0.7	240	237.5±2.1	25.6±0.0	6362±57	14	1.074
	P2	400	115	34.1±1.4	240	2112.2±3.9	26.2±0.4	5924±190		1.000
	P3	530	55	35.6±0.8	220	224.3±4.0	24.9±0.0	6344±92	12	1.249
	P4	530	55	25.8±0.4	220	177.5±2.1	24.4±0.1	5188±52		1.021

In order to remove the effect of oxides presents on the plate surfaces, they were sand blasted prior to welding. This was done to avoid any interference of from surface active elements on the weld pool fluid flow and consequently penetration. Lastly, it should be mentioned that for each condition 5 beads were performed, one trial and 4 replicates, and that the margin of error presented are all based on the t-distribution test.

Results and discussion

Figure 2 shows the results for droplet diameter, droplet velocity, and penetration at different wire feed speed. It can be noted that for certain wire feeding speeds, droplet size is larger for profiles 3 and 4, which is consistent with the fact that the pulse frequency of those profiles are lower, refer to *Tab. 1*. Interestingly, it is also possible to note in *Fig. 2* that for the same wire feeding speed, the droplet speed is larger for profiles 3 and 4. Two main factors are responsible for this behaviour. First, the higher peak current, which increase the acceleration of the droplet due to the electromagnetic force, and second the lower background current, which in turn decreases the arc pressure prior and after the droplet detachment, when compared to profiles 1 and 2. It should be pointed out that the arc pressure is considered to be a retaining force during the transfer of a pendant drop.

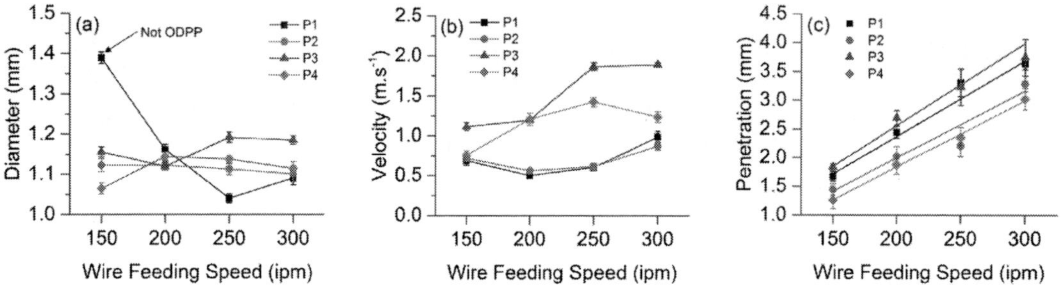

Figure 2: Variation of droplet (a) diameter and (b) velocity; and (c) bead penetration for the different welding profiles at various wire feeding speeds.

In *Fig. 2(c)*, one can note that the penetration of the bead performed using profiles 2 and 4 are the smallest of all the profiles for all parameters investigated in this work. Based on the findings and mass coefficient defined by Murray et al. [8], the lowest penetration of profiles 2 and 4 are attributed to the lowest heat transfer coefficients produced by these profiles. This can be further verified by plotting penetration against the mass transfer coefficient defined by Murray et al. [8]. This is shown in *Fig. 3(a)* where it is possible to confirm that for the same mass transfer coefficient (MTC), profiles 1 and 3 have higher penetration depth and therefore suggesting that the heat transfer coefficient is larger for these profiles.

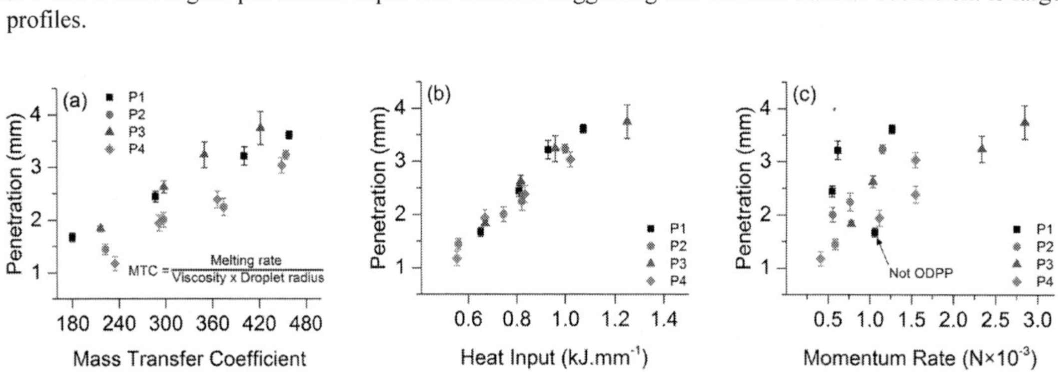

Figure 3: Variation of bead penetration as a function of (a) mass transfer coefficient; (b) heat input and (c) effective momentum.

The hypothesis of profiles 2 and 4 present lower penetration due to a lower heat transfer factor is supported by the lowest instantaneous average power of these profiles, and possible implications of the different droplet transfer mechanism observed through the high speed videos observation for profiles 2 and 4, such as the droplet detachment occuring at background current, which can potentially decrease the amount of heat carried to the weld pool by the droplet. The hypothesis that the lower penetration is due to the difference in the heat transfer to the weld pool is further supported by the correlation presented in *Fig. 3(b)* where it can be seen that the penetration is proportional to the nominal heat input, calculated based on the welding parameters shown in *Tab. 1*.

The implications of the combination of droplet mass and velocity on penetration observed here can be quantified by calculating the momentum rate of the droplets, which has been reported to have a direct correlation to penetration [9,10]. In *Fig. 3(c)* the penetration is plotted against the momentum rate and it can be noted that, despite the fact that between momentum rate of 0.5×10^{-3} N to 1.5×10^{-3} N. Profile 1 exhibits the highest penetration, again suggesting that heat transfer is responsible for this observation.

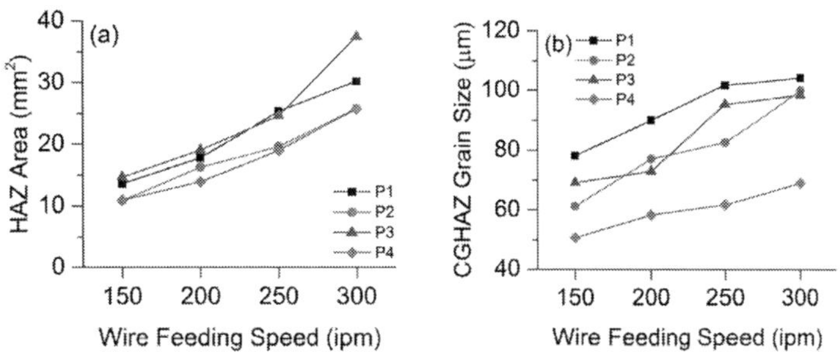

Figure 4: Variation of (a) HAZ are and (b) apearent prior austenit grain size for the different welding profiles at various wire feeding speeds.

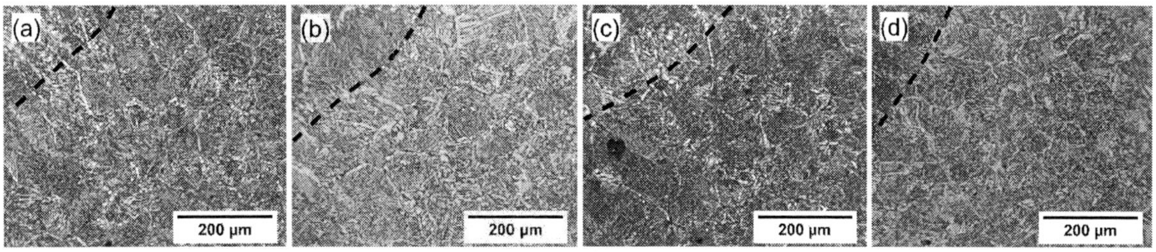

Figure 5: Microstructure of the CGHAZ for the different welding profiles for the wire feeding speed of 300 ipm: (a) profile 1, (b) profile 2, (c) profile 3 and (d) profile 4.

To confirm the hypothesis of lower heat input using profiles 2 and 4, the heat affected zone (HAZ) and the appearent prior austenite grian size in the coarse grained heat affected zone (CGHAZ) at the fusion line was measured, and the results are presented in *Fig. 4.* It was possible to verify that profiles 2 and 4 has the lowest HAZ area and smallest CGHAZ grain size, as expected to be the the case for lower heat input conditions. On *Fig. 5* the microstructure of the CGHAZ for the four different profiles for the wire feeding speed of 300 ipm is also shown for comparison.

Conclusions

Based on the results observed, it is concluded that pulse profiles with nearly an ideal square shaped wave form required the lowest instantaneous average power to transfer the same amount of welding metal. It was noted that the higher peak and lower background current, the faster tend to be the velocity of the droplet being transferred. It was also possible to verify that the bead penetration is proportional to the heat input and momentum rate due to the droplet speed and mass.

Therefore, is it concluded that the beads performed using profile 3 (long tailed square wave with reduced background current) presented the highest penetration due to the highest momentum rate and heat transfer terms.

Acknowledgments

The authors would like to acknowledge Liburdi automation Ltd. for their assistance and technical support. The authors would also like to acknowledge TransCanada Pipeline and NSERC for funding this project.

References

[1] A. Scotti, V. Ponomarev, W. Lucas, A scientific application oriented classification for metal transfer modes in GMA welding, J. Mater. Process. Technol. 212 (2012) 1406–1413. doi:10.1016/j.jmatprotec.2012.01.021.

[2] I.K. Nuruddin, Effect of welding thermal cycles on the heat affected zone microstructure and toughness of multi-pass welded pipeline steels, 2012. http://dspace.lib.cranfield.ac.uk/handle/1826/7755.

[3] P. Praveen, P.K.D.V. Yarlagadda, M.J. Kang, Advancements in pulse gas metal arc welding, J. Mater. Process. Technol. 164-165 (2005) 1113–1119. doi:10.1016/j.jmatprotec.2005.02.100.

[4] J.C. Needhan, Control of transfer in aluminium consumable electrode welding, Phys. Weld. Arc, Inst. Welding, London. (1962) 114–124.

[5] J.C. Needham, Pulsed Current for Gas Shielded Arc Welding, IEEE Trans. Ind. Gen. Appl. IGA-2 (1966) 225–233. doi:10.1109/TIGA.1966.4180658.

[6] E.P. Vilkas, Square-Wave Welding, IEEE Trans. Ind. Gen. Appl. IGA-3 (1967) 525–530. doi:10.1109/TIGA.1967.4180827.

[7] E.H. Daggett, A power supply for pulsed power welding, IEEE Spectr. 5 (1968) 67–70. doi:10.1109/MSPEC.1968.5215635.

[8] P.E. Murray, A. Scotti, Depth of penetration in gas metal arc welding, Sci. Technol. Weld. Join. 4 (1999) 112–117. doi:10.1179/136217199101537644.

[9] W.G. Essers, R. Walter, Heat transfer and penetration mechanism in GMAW and Plasma-GMA welding, Weld. J. 60 (1981) 37s–42s.

[10] A. Scotti, C.E.A.L. Rodrigues, Determination of momentum as a mean of quantifying the mechanical energy delivered by droplets during MIG/MAG welding, Eur. Phys. J. Appl. Phys. 45 (2009) 11201. doi:10.1051/epjap:2008196.

Resistance Spot Welding of New Generation of High Strength Automotive Steels

M. Tumuluru, United States Steel Corporation, Munhall, PA, USA
mtumuluru@uss.com, +1- 412-433-7261

Abstract

A new generation of high strength steels termed as Generation 3 (GEN3) steels has recently been introduced for automotive applications, specifically for body-in-white applications. These steels have tensile strengths in excess of 1000 MPa coupled with higher elongation compared to the Generation 1 steels, such as the dual phase and Transformation-Induced Plasticity (TRIP) steels that are currently in use. Goals in the development of the new generation of steels include achieving improved formability over the Generation 1 steels and making them more affordable than the Generation 2 steels, such as the Twinning-Induced Plasticity (TWIP) steels. Studies were undertaken to characterize the weldability of one such recently developed 980 MPa GEN3 steel using the resistance welding process. In this study the 980 GEN3 steel was tested both in the bare (uncoated) and electrogalvanized conditions. Welding lobes (current ranges at three weld times) were determined to assess the range of currents over which welds of acceptable quality could be made. Weld mechanical properties were evaluated through tensile testing and microhardness determination. Microstructural characterization was conducted to identify any potential problems with welding of the steel. The study showed that the steel possessed excellent current ranges and tensile properties, suggesting that it has robust welding behavior that makes it suitable for industrial use. In this presentation an introduction to some of the new grades of automotive steels is provided. A brief discussion on various generations of automotive steels is made. Characteristics that distinguish GEN3 steels from previous generations of steels are provided.

Introduction

Since 2000 several high strength steels with a minimum tensile strength in excess of 500 MPa have been successfully developed and introduced into body-in-white applications in the automotive industry (Ref. 1, 2). These steel grades contain mostly ferrite based microstructures and include dual-phase and Transformation-induced plasticity (TRIP) steel grades. They are generally known as Generation 1 (Gen1) Advanced High strength Steels. Generation 2 (or Gen2) steel that include such grades as Twinning-Induced Plasticity (TWIP) steels have higher strength and ductility combination. However, their use has been somewhat limited due to a lack of global availability and high cost. Efforts have been underway for some time to develop Generation 3 (or Gen3) steels that are more affordable and possess improve ductility over Gen1 at equivalent strength levels.

Several automotive customers have expressed a keen interest in using Generation 3 (Gen3) advanced high strength (AHSS) steels in body-in-white (BIW) applications. Gen3 steels offer the advantages that they have high formability due to much better tensile elongation than conventional (or Generation 1) AHSS grades and are far less expensive than Generation 2 (such as Twinning-Induced Plasticity steel) or aluminum alloys. Despite their high strength and elongation properties, published literature indicated that TWIP steels have proven to be more difficult to weld than conventional AHSS (Ref. 3- 6). Similarly, gas metal arc welding of Fe-Mn (TWIP) steels have shown difficulties in welding (unacceptable seam profiles, poor wetting characteristics) with conventional arc welding methods, suggesting that obtaining defect-free welds requires the use of advanced joining techniques, such as cold metal transfer or pulsed arc methods (Ref. 7). In the case of boron added hot stamped grades welding issues have been noted for zinc coated steels (Ref. 8). Specifically, the weld lobes were found to be smaller than those obtained with steels coated with aluminum-silicon type coatings.

For use in BIW applications, weldability of a given grade of steel is perhaps the most important attribute for its use. Hence, any steel grade intended for BIW use must possess good weldability to allow fabrication of components in the automotive assembly plants. Any concerns or problems with weldability of a given grade will very likely prevent its use in BIW applications. Therefore, prior to testing any new grade of steel for potential use, automotive companies require steel companies to demonstrate to them the grade of steel can be easily weldable in their plants. To satisfy this need, an investigation was undertaken to examine the weldability of commercially produced 980 Gen3. The thickness of the sheet selected was 1.2 mm. The steel was tested in the as-cold rolled condition (bare or uncoated) as well in electrogalanized (EG) condition with a plating weight of 60 g/m^2 per side.

Procedures

The mechanical properties of the steel used are shown in Tables I. The carbon equivalent for the steel was calculated using two different equations, including the equations used by some automotive companies and the International Institute of Welding (IIW) (Ref. 9) and is provided in Table I. Carbon equivalent provides an indication of the weldability of a given grade of steel.

Table I. Typical Mechanical Properties of the 980 GEN3 Steel

Detail	980 GEN3 Steel
Yield Strength, MPa	607
Ultimate Tensile Strength, MPa	999
Total Elongation, percent	23.4
Hole Expansion ratio (HER)	38.4
Carbon Equivalent [1]	0.38
Carbon Equivalent [2]	0.62

1. *Based on Ceq = C+ Mn/20 + Si/3 0 +Cr/20 + Mo/15 + 2 P+ 4S*
2. *Based on Ceq = C + Mn/6 + Si/30 + (Cr+Mo+V+Nb)/5 + (Cu+Ni)/15 [International Institute of Welding (IIW) formula]*

During preparation of welds the welding current, weld time, and voltage were monitored on the secondary side of the machine using a Miyachi MG3 weld checker. Two different tests were conducted to evaluate the weldability of the steel. These were the determination of the weld lobes (welding current ranges at three different weld times) as well as weld shear and cross tension strengths. The weld shear and cross tension tests were done both on the minimum weld size and the expulsion weld size to assess the weld strength as a function of the welding current, hence the weld size. All of the weld testing was conducted with a pedestal-type Taylor Winfield welding machine using alternating current. The welding conditions chosen in this investigation are typical for the thickness of the material being welded, and are considered to be within the capability of the welding equipment currently used in automotive assembly plants.

Prior to determining the current range, the electrode tips were conditioned by making 50 welds on flat panels. Current ranges were then determined by first determining the lowest welding current that produced the minimum acceptable weld size. The current was then gradually increased in 200 Amp increments until weld metal expulsion resulted. At both the minimum weld size and the expulsion weld size, peel test coupons, three shear tension, and three cross tension test samples were prepared. After peeling open the peel test coupons, weld sizes were measured and recorded. Weld size measurements were also made from the shear and cross tension test samples after the completion of the tensile tests.

Weld cross tension and shear tension tests were completed per the procedure described in Ref. 10.

Microhardness traverses were determined at room temperature using a Vickers hardness tester. A force of 1000 g (9.8N) was used for the microhardness measurements. The hardness indentations were spaced 0.2 mm apart. The microhardness traverses were completed in such a way as to cover as much area in the weld and heat affected zones as possible to determine the hardness uniformity in the weld areas. Weld microstructures were examined using both an optical microscope and a scanning electron microscope (SEM) equipped with an energy dispersive x-ray spectrometer (EDS).

Results and Discussion

The result of the weld lobe determination for the steel are shown in Figure 1, where the plot shows the welding current range obtained at three weld times expressed in cycles (at 60 hz A.C.).

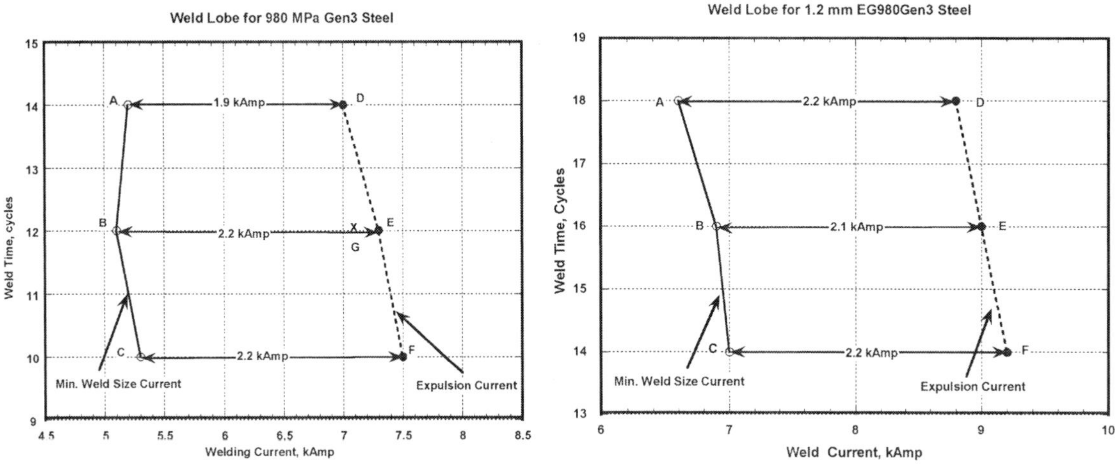

Figure 1. Weld lobes for bare and electrogalvanized steels.

The useful current ranges (the range from the current that provides the minimum weld size to the one that causes expulsion of the weld metal) for the cold rolled Gen3 steel were determined to be around 2.0 kAmp, which is considered excellent from an automotive assembly welding point of view. Similar current ranges were also obtained for the EG steel (Fig. 1). These results clearly demonstrate that the steel possess robust welding characteristics and should be readily weldable under a variety of shop-floor conditions.

The results of the weld shear and cross tension tests are shown in Table II. The shear and cross tension strengths indicate that as the weld size was varied from the minimum weld size to the expulsion weld size through increased weld current the weld strength also increased due to the dependence of weld strength on the weld size. Further, the strength values obtained are consistent with the base material strength. It is important to note that the cross tension strength of the welds remained fairly high to obtain a high ratio between cross tension and shear tension strength ratios. The reason for this trend could be understood from the weld microhardness traverse performed on the steel, which is shown in Fig 2.

2.Table II. Results of Weld Shear and Cross Tension Strengths

Point	Ave. Weld Size, mm	Ave. Shear Tension Strength, Newton	Fracture Mode	Ave. Cross Tension Strength, Newton	Fracture Mode
A	5.9	19154	FB	8563	FB
B	5.7	17964	IF	5909	PIF
C	5.6	16827	IF	7968	FB
D	7	22706	FB	8668	FB
E	6.8	22125	FB	9267	FB
F	6.5	21087	FB	9235	FB

FB – Full Button Pullout
IF – Interfacial Fracture
Average CTS/STS Ratio ~ 0.4
Averages based on three tests

Weld cross sectional view for a typical weld made at 6√t, where t is the thickness of the base material, is shown in Figure 2. As can be seen in this photograph the weld was well formed and exhibited no visible imperfections. The microhardness plot shown at the bottom of the photograph shows the weld area hardness to be uniform. No significant softening in the heat affected zone was observed.

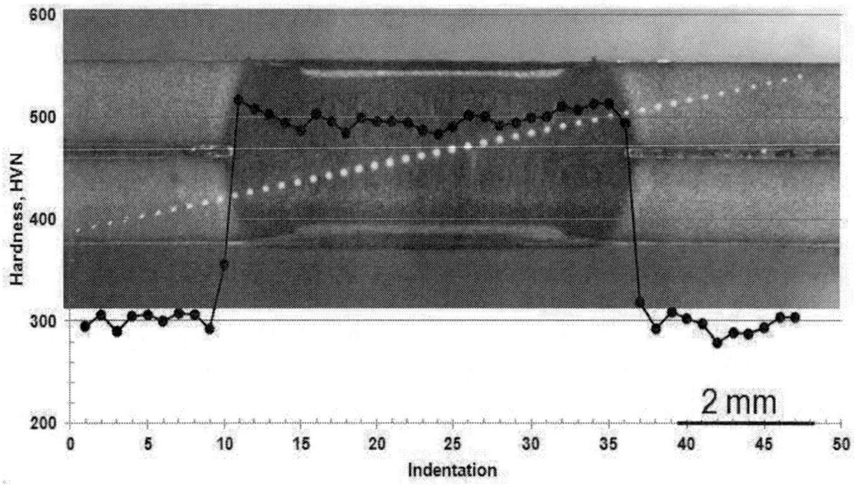

Figure 2. Weld cross section showing hardness indentations and hardness plot across the weld.

Weld microstructural examination indicated that there were no weld imperfections such as cracks or voids in the cold rolled steel.

Conclusions

The following conclusions can be drawn from this study:

1. The 980 MPa Gen3 steel possessed wide welding current ranges to provide welds of acceptable quality.
2. The steel showed good shear and cross tension strength consistent with its base material strength.
3. The weld displayed uniform hardness in the weld zone. No noticeable softening in the far heat affected zone was noted.
4. The welds examined showed no imperfections,

References

1. M. Tumuluru, "Resistance spot welding coated high strength dual phase steels", *Welding Journal*, American Welding Society, Miami, FL, Vol. 85, No. 8 2006.
2. M. Tumuluru, "Some considerations in the resistance spot welding of dual phase and TRIP steels" Proceedings of the International seminar on "*Advances in Resistance Welding*", Toronto, Sept 24-26, 2008.
3. L.MujicaL, S. Weber, G. Hunold, W. Theisen . "Development and characterization of novel corrosion-resistant TWIP steels." Steel Res. Int. 2011; 82:26.
4. D. C. Saha, S. Han, K. G. Chin, I. Choi, Y. D. Park. "Weldability evaluation and microstructure analysis of resistance-spot-welded high-Mn steel in automotive application." Steel Res Int 2012; 83:352.
5. D. C. Saha., Y. Cho Y, Y. D. Park. "Metallographic and fracture characteristics of resistance spot welded TWIP steels." Science and Technology Welding and Joining 2013; 18:711.
6. S. A. Harm and W. Bleck, "Mechanical property and microstructure of resistance spot welded twinning induced plasticity-dual phase steels joint." Science and Technology of Welding and Joining 2013; 18:485.
7. D. Keil, M. Zinke, and H. Pries, "Weldability of novel high Fe-Mn High Strength Steels for automotive applications." IIW Document II-1738-10 (II-A-217-10). International Institute of Welding, Paris.
8. C. Ji, Y. Kim, Y. Park, "Effects of types of coating on weld growth mechanism of hot stamped boron steels in resistance spot welding". IIW Document 1665-13, www.iiw.org
9. ASM Handbook Vol. 6, "Welding Brazing and Soldering", ASM International, www.asminternational.org.
10. American Welding Society Specification D8.9M-2012, American Welding Society, Miami, FL.

Challenges of joining lightweight materials for dissimilar joints

Peter Mayr, S. Habisch, A. Haelsig, W. Georgi
Institute of Joining and Assembly, TU Chemnitz, Germany

The application of fibre-reinforced plastics, aluminium and magnesium in modern constructions became more and more state of art. One of the biggest challenges is to join these material combinations without losing their favourable properties. A quickly formed and stable oxide layer on the surface, the formation of brittle intermetallic compounds along the interface of the joints and the recrystallization of fine-grained aluminium and magnesium alloys at elevated temperatures are the biggest challenges to overcome for metallic dissimilar welds. To solve this challenge, a glove box directly coupled to a diffusion bonding chamber was used to pre-treat surfaces mechanically or chemically and transfer the samples into the joining chamber under continuous inert atmosphere protection. The results have shown, that a systematically reduction of the joining temperature reduces the intermetallic Al-Mg compound in width as well as reduces the hardness by 20% and limits the grain growth in both materials. The joint strength was still limited by the Al-Mg intermetallics along the interface, so the role of different interlayer materials were examined. Another approaches are thermal and mechanical joining strategies for joining fibre-reinforced polyamide to steel or aluminium alloys without additional adhesives. High strength joints are realised by the combination of different heat inputs generated by friction or conduction, and pressure in the joining zone and an optional additional mechanical joining process. In order to improve joint strength, the influence of several surface pre-treatments has been investigated. Highest joint strength was achieved when using the adhesive promoter or the application of pin structures on the metal side. Furthermore, force- and form-locked joining technologies like clinching in combination with ultrasonic heat input or self-piercing blind riveting were developed. Mechanical joints of metal-plastics reference samples doubled strength levels up to 25 MPa in shear mode compared to conventional bonds.

Efficiency comarison in 1kHz and 10kHz DC welding system

Boguslaw Grzesik [1], Mariusz Adam Stepien [1], Zygmunt Mikno [2]

[1] *Silesian University of Technology, Poland*
[2] *Welding Institute, Poland*

The paper is focused on the comparison of measurement results of power losses and efficiency in a structure of DC resistance spot welding systems operating at different frequencies of inverter, 1 kHz and 10 kHz. The system includes inverter, transformer and high current rectifier. The frequency of inverter influences on operation of the inverter and following transformer. Increased frequency enables increasing of power density (reduction of mass and volume at the same rated power), but it can results with higher power losses in the system. The analysis will appoint the most important sources of power losses and their dependence on frequency. The measurements are carried out keeping the same output rectifier. The result of the described analysis carried out in the paper is the answer how increasing of frequency reduces power losses and increases energy density in the system.

Experimental Study of Lap Shear Strength and Fracture Mechanism of Al-Steel Resistance Spot Welds

Nannan Chen[1], Hui-Ping Wang[2], Min Wang[3], Blair E Carlson[4] and David R. Sigler[5]
1, School of Material Science Engineering of Shanghai Jiao Tong University, 800 Dongchuan Road, Shanghai, 200240, China. cnhfut@126.com, +86 13721057259
2, General Motors Global R&D Center, 30500 Mound Rd, Warren, MI 48092, USA. hui-ping.wang@gm.com, +1 2482758209
3, School of Material Science Engineering of Shanghai Jiao Tong University, 800 Dongchuan Road, Shanghai, 200240, China. wang-ellen@sjtu.edu.cn, +86 21 34202943
4, General Motors Global R&D Center, 30500 Mound Rd, Warren, MI 48092, USA. blair.carlson@gm.com, +1 5868647698
5, General Motors Global R&D Center, 30500 Mound Rd, Warren, MI 48092, USA. david.r.sigler@gm.com, +1 5868644022

Abstract
A resistance spot weld between an aluminum sheet and a steel sheet is composed of an aluminum nugget in the Al sheet growing from the Al-steel faying interface, a thin intermetallic compound (IMC) layer generated due to inter-diffusion between two materials, and sometimes a separate steel nugget inside the steel sheet. The strength of the weld depends highly on the properties of the IMC layer, aluminum fusion zone, aluminum heat affected zone and base metals. In this paper, we experimentally study the lap shear strength of resistance spot welds between 1.2mm AA6022 sheet and 2mm zinc-coated low carbon steel, and their fracture modes. The effects of factors, such as IMC thickness profile, aluminum nugget dimension and level of steel bulging at the faying interface, on the fracture mechanism are investigated. It is observed that there are three major fracture modes: semi-brittle fracture in the IMC layer, ductile thickness fracture in the aluminum nugget, and ductile fracture in the aluminum base metal or heat affected zone which leads to the engineering preferred button pullout. Furthermore, it is found that the IMC thickness is the dominant factor of the strength and fracture mode. The button pullout fracture or thickness fracture is related to thin IMC thickness to be within 3μm, while increasing the Al nugget size or steel bulge height helps to increase lap shear strength and promote button pullout fracture. On the other hand, thick IMC layer with thickness beyond 3μm tends to cause interfacial fracture in the IMC layer, even with large Al nugget size and severe steel bulging.

Introduction
Automotive manufacturers have put great effort in reducing weight of vehicles in order to improve the fuel efficiency and overall performance. The adoption of lightweight materials such as aluminum alloys to replace heavy steel parts in the body structure of vehicles has been a common approach. As a result, there is a need to join aluminum components to steel components. However, very different physical properties between aluminum alloys and steels and incompatibility in one solid phase of these two materials lead to great challenges in joining the two materials[1-4]. Common issues in resistance spot welding aluminum to steels are resistance heat imbalance, growth of Al-steel intermetallic compounds (IMC), expulsion of aluminum material, separate melting of two materials and defect accumulation at the Al-steel interface. The resulting welds often fail along the interface or in the nugget, not in the preferred button pullout mode. Understanding the factors that influence strength and fracture mode of Al-steel welds is the purpose of this study. In the following, we first generate Al-steel resistance spot welds by various welding schedules. Then, the aluminum nugget diameter, sheet thinning, weld hardness, IMC thickness profile and level of steel bulging into aluminum are measured for the generated welds. Lap shear tests are carried out next, and their corresponding lap shear strength and failure modes are recorded. The relationship between the measured factors, weld strength and fracture path are examined.

Resistance spot welding of Al sheet to Zinc-coated steel
Resistance spot welding was carried out with the aluminum sheet contacting the positive electrode and steel contacting the negative electrode. Figure 1 shows the stackup and electrode being used. All coupons were pre-cleaned by an ultrasonic cleaner before welding to avoid the test variation caused by sheet surface contaminations. Welding was carried out with an electrode force of 750 lb. Six types of samples are generated with different nugget size, IMC thickness, steel bulging and Al thinning etc. All the generated welds show common features of Al-steel resistance spot welds: fusion in the aluminum sheet starting next to the steel, IMC growth at the interface, indentations on both Al and steel by the electrodes, steel bulging into the aluminum sheet along with thinning of the Al sheet. No porosity defects in the nugget or at the interface were observed.

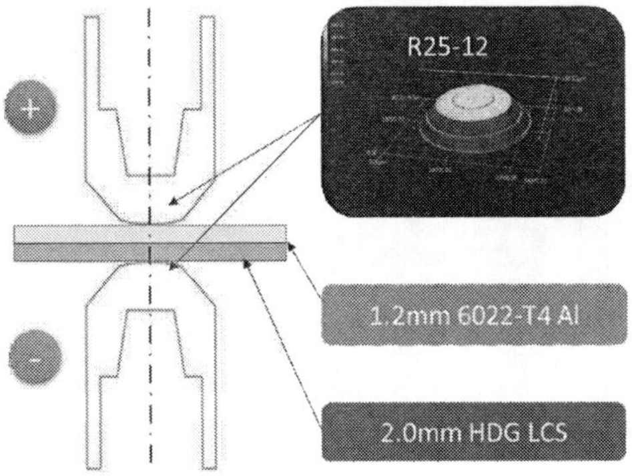

Figure 1. Illustration of Resistance spot welding Test Setup

Analysis of weld strength and fracture modes

Figure 2 gives the measured IMC thickness distributions. The IMC thickness for Samples #1, #2 and #3 were fairly low with the greatest thickness being ~3μm, observed in the center of the weld. Samples #4, #5 and #6 have overall higher IMC thicknesses than Samples #1, #2 and #3. Sample #4 has an IMC thickness slightly greater than 4μm in the weld center. Also, the IMC thickness profile for Sample #5 and Sample #6 have an interesting M shape. Using numerical simulation of Al-steel resistance spot welding process, Wan et al.[5] explained that the drop of IMC growth in the center was caused by thinning of aluminum in the center which caused increased cooling by the electrode.

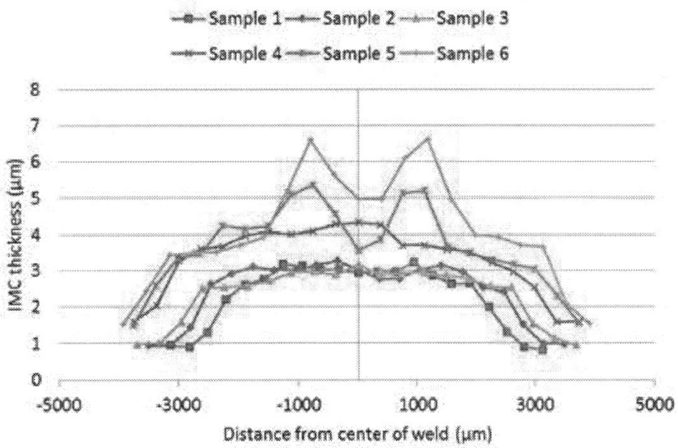

Figure 2. The IMC thickness distributions of the Al-steel welds

Figure 3 plots the lap shear strength and aluminum nugget diameter for Samples #1 through #6. For joints made using similar materials, the fusion zone or weld nugget size is typically the dominant factor for lap shear strength with strength increasing along with the fusion zone size[6]. However, the Al-steel welds' lap shear strength does not always increase with aluminum nugget diameter as shown by Figure 3. For Samples #1, #2 and #3, lap shear strength increases with the final nugget diameter. For Samples #1, #4, #5 and #6, the lap shear strength ends up decreasing as Al nugget diameter increases. Figure 4 plots the lap shear stress and average IMC thickness for Samples #1 through #6. The lap shear stress is calculated by dividing the lap shear strength by the fusion zone area at the interface, $\pi D^2/4$, where D is the aluminum nugget diameter. Figure 4 shows that for IMC thickness of 2 to 3 microns the lap shear stress is constant. Also, as the average IMC thickness increases to be above 3 μm, the lap shear stress drops and continues to decrease as the average IMC thickness increases. To better understand the results the joint fracture modes must be considered.

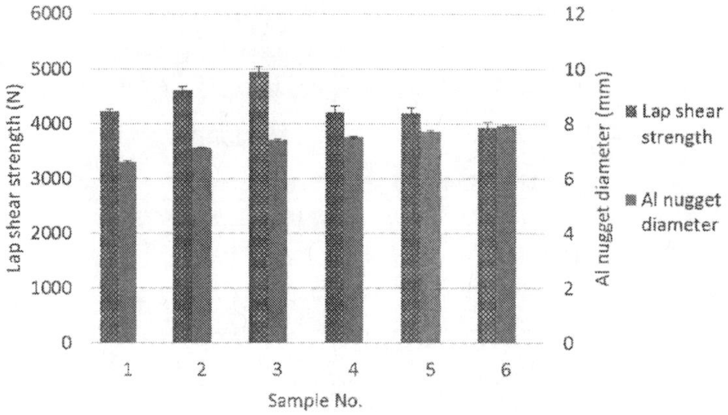

Figure 3. Comparison between Lap shear strength and Al nugget diameter

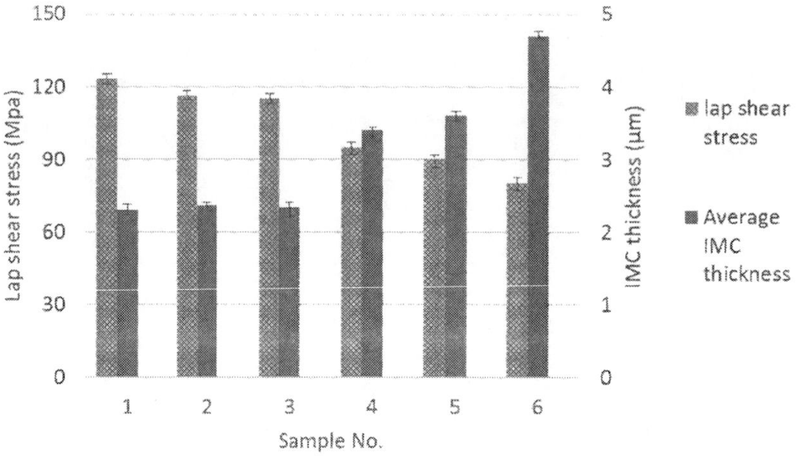

Figure 4. Comparison between averaged Lap shear stress and average IMC thickness

Figure 5 summarizes the three common fracture modes observed in the Al-steel lap shear test: fracture in the aluminum sheet heat affected zone or base metal with a pullout button, fracture in the aluminum nugget with partial or no button, and fracture in the intermetallic compound layer with no button. The 1st fracture mode is the same as the button fracture mode observed in welds made with similar materials. The crack starts due to strain localization in the heat affected zone or from the notch tip of the weld. The 2nd fracture mode is similar to the interfacial fracture mode, also referred to as thickness fracture in similar material welds, where semi-brittle fracture occurs in the weld nugget starting at the faying surface. The last fracture mode is unique to Al-steel welds due to existence of the IMC layer between the aluminum and steel sheets.

It was observed that when the average IMC thickness is thin, less than 3μm, the fracture mode is either thickness fracture or button fracture. Interfacial fracture mode was not observed for these samples. As the aluminum nugget grows, the fracture mode transitions to button fracture mode with increasing lap shear strength. We also observed that as the average IMC thickness increases beyond 3μm for Samples #4, #5 and #6, the joints all fail in an interfacial fracture mode, i.e., within the IMC layer. This fracture mode reduces lap shear strength even though the aluminum nugget size and level of steel bulging increased.

These observations confirm that the IMC thickness is the dominant factor influencing the weld lap shear strength and fracture mode. To generate a strong and ductile Al-steel joint, it is critical to maintain a thin IMC layer at the faying interface, preferably to be within 3μm.

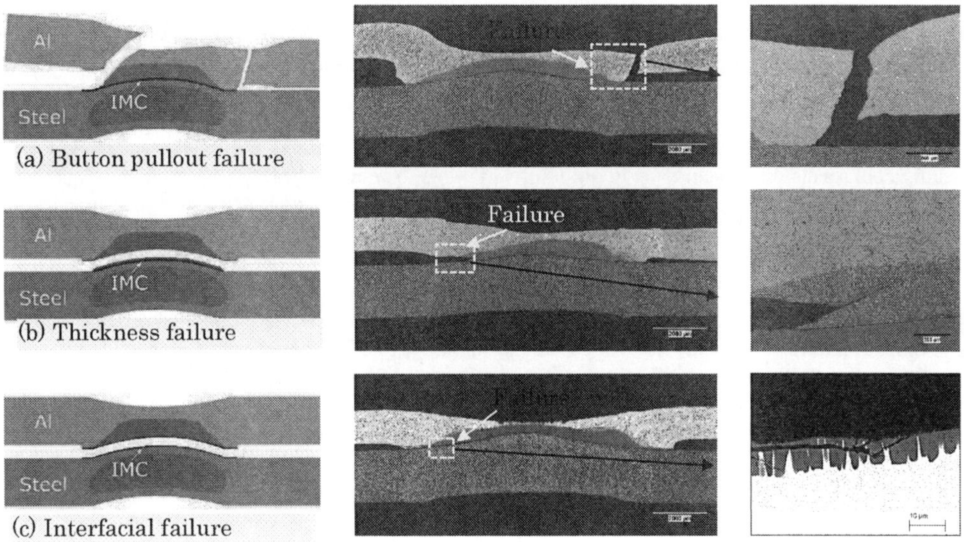

Figure 5. Illustration of observed three typical Al-steel RSW fracture modes

Conclusions

In this study, we experimentally investigated the lap shear strength and fracture modes of resistance spot welds between 1.2mm AA6022 sheet and 2mm zinc-coated low carbon steel made using different weld schedules. It was observed that there are three major fracture modes for Al-steel welds: 1) semi-brittle fracture in the IMC layer, 2) ductile thickness fracture in the aluminum nugget, and 3) ductile fracture in the aluminum base metal or heat affected zone which leads to the engineering preferred button pullout.

Furthermore, it was found that the IMC layer thickness is the dominant factor for determining the weld's strength and fracture mode. The button pullout fracture or thickness fracture occurs with thin IMC layers of less than 3μm. Increasing the Al nugget size or level of steel bulging into the aluminum sheet further helps to increase lap shear strength and promote button pullout failure. On the other hand, thick IMC layers where thickness exceeds 3μm tends to cause interfacial fracture of the IMC layer, even with large aluminum nugget sizes and severe steel bulging. The steel bulging into the aluminum sheet does not play a role in fracturing thick IMC layers. To generate strong and ductile Al-steel joints, it is recommended that a thin IMC layer be produced at the faying interface, preferably less than 3μm, while growing the aluminum nugget diameter.

References

[1] I.H. Hwang, T. Watanabe, Y. Doi. Dissimilar Metal Welding of Steel to Al-Mg Alloy by Spot Resistance Welding, Advanced Materials Research Vols. 15-17 (2007) pp 381-386

[2] Kenji Miyamoto, Shigeyuki Nakagawa etc. Dissimilar Joining of Aluminum Alloy and Steel by Resistance Spot Welding, SAE 2009-01-0034, Int. J. Mater. Manuf. | Volume 2 | Issue 1, P58-P67

[3] Seyedeh Nooshin Mortazavi, Pirooz Marashi, Majid Pouranvari, Maryam Masoumi, Investigation on Joint Strength of Dissimilar Resistance Spot welds of Aluminum Alloy and Low Carbon Steel, Advanced Materials Research Vols. 264-265 (2011) pp 384-389

[4] J. Wang, H.P. Wang, F. Lu, B.E. Carlson, D.R. Sigler. Analysis of Al-steel resistance spot welding process by developing a fully coupled multi-physics simulation model. International Journal of Heat and Mass Transfer, 89(2015), pp. 1061-1072

[5] Z. Wan, H.P. Wang, M. Wang, B.E. Carlson and D.R Sigler, Numerical simulation of resistance spot welding of Al to Zinc-coated steel with improved representation of contact interactions. International Journal of Mass and Heat Transfer, 2016, to appear.

[6] M. Pouranvaria, S.P.H. Marashi, Failure mode transition in AHSS resistance spot welds. Part I. Controlling factors, Materials Science and Engineering A 528 (2011) 8337– 8343

Evaluation of hot stamped component assembled by 2D brazing

Tasuku Zeniya, Hitomi Nishibata, Shinji Kodama, Masanori Yasuyama
Nippon Steel & Sumitomo Metal Corporation, Amagasaki, Japan
zeniya.s9r.tasuku@jp.nssmc.com, +81-6-7670-8022,

Abstract

High strength steel sheets have been used in automotive bodies to improve safety and achieve a light body. Press hardened steel (PHS) has been increasingly applied to structural channels such as the center pillar to ensure the safety of the occupants' cabin in the event of side impact. Although these components are usually reinforced partly by attaching reinforcement with spot welding, it is not sufficient to enhance bending performance. Therefore, to bond the interface between the main component and reinforcement fully, a new brazing method has been developed. It is called 2D brazing. 2D brazing is furnace brazing using the hot stamping process and it enables large area brazing at a time. This study highlights the optimization of 2D brazing conditions, and evaluation of the bending performance of components. The optimized condition of 2D brazing using brass (Cu-35Zn) was 1000 degrees C for 10 min for a bonding pressure of 1000Pa. Note that the quenching temperature should be taken into consideration in order to prevent liquid metal embrittlement. A comparison of the joint strengths 2D brazing and spot welding showed that tensile shear strength and cross tension strength in 2D brazing is 2.2 and 2.7 times higher than those in spot welding, respectively. Bending performance of hat channel specimens was evaluated by the three-point bending test. Compared with the hat channels assembled by conventional spot welding or spot patchwork-tailored welded blanks (PW-TWB), the maximum load of the hat channel assembled by 2D brazing was approximately two times higher. Under the condition that total thickness of steel sheet is the same, the hat channel assembled by 2D brazing achieved the same level to the maximum load as that of the hat channels made of a single sheet. This suggests that the 2D brazing hat channel performs similarly to the single sheet hat channel.

1. Introduction

High strength steel sheets have recently been used in automotive bodies to improve safety and achieve a light body [1-3]. Press hardened steel (PHS) has been increasingly applied to structural channels such as the center pillar to ensure the survival space of occupants in a crash [4]. In many cases, these channels are reinforced partly through attaching reinforcement by spot welding in order to suppress the bending deformation. In this conventional method, main channel and reinforcement are formed, respectively. After that, they are assembled together by welding. Therefore, there is a gap of a few millimeters between the main channel and reinforcement. In such cases, it is difficult to obtain the effective reaction force in a crash because the two steel sheets deform respectively.

Therefore, we developed a fully bonding method between the main channel and reinforcement. In this method, PHS sheets are bonded by the two dimensional (2D) brazing method in the hot stamping (HS) process. Here, we report the 2D brazing process, joint strength, and bending performance of the hat channel assembled by this method.

2. 2D brazing method utilizing HS process

3.

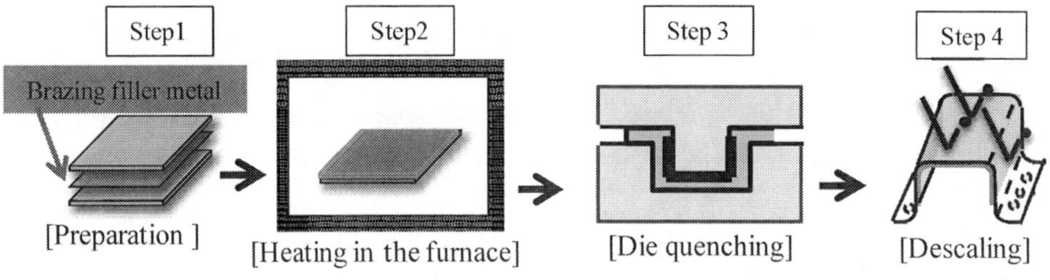

Fig. 1 Outline of 2D brazing method utilizing the HS process.

Figure 1 illustrates the outline of the 2D brazing method utilizing the HS process.

Step1…Brazing filler metal is arranged between two steel blanks.

Step2…Test specimens are heated in the furnace. In this step, PHS steel is austenitized and brazing filler metal is melted.

Step3…Test specimens are simultaneously formed and quenched. To prevent liquid metal embrittlement (LME) at the bonding interface, the brazing filler metal needs to solidify before bending.

Step4…The scale on the hat channel surface is removed by shot blasting.

3. Materials and experimental procedures

3.1 Materials

The base metal used here is a PHS steel sheet that has 1.5GPa tensile strength after quenching. The chemical compositions of base metals are shown in Table 1.

Table 1 Chemical compositions of base metal

Materials	C	Si	Mn
PHS steel	0.21	0.23	1.26

For quenching in the HS process, steel sheets should be transformed into the austenite phase. Accordingly, in the 2D brazing method, steel sheets are brazed using a brazing filler metal that can be used in a furnace at the austenite transformation or a higher temperature. In addition, in consideration of LME, the brazing filler metal should have a solidus temperature close to the furnace temperature. Therefore, the brazing materials used in this study are foil brass and bronze, as shown in Table 2.

Table 2 Chemical compositions and properties of brazing filler metals

Materials	Chemical compositions (mass%)	Liquidus temperature (°C)	Solidus temperature (°C)	Use of application
Brass	Cu: 64%–68% Zn: Bal.	930	903	Evaluation of joint strength and performance of hat channels
Bronze	Cu: 80% Sn: 20%	925	770	Evaluation of joint strength and LME crack

3.2 Evaluation of LME crack

Bending high temperature, LME might occur depending on the brazing filler metal. Therefore, the high-temperature bending test was performed to clarify the relationship between LME cracking and solidus temperature of the brazing filler metal.

The specimen dimensions were 30 $mm^W \times 60\ mm^L \times 1.4\ mm^t$. Specimens with a brazing material inserted into the entire area between the two steel sheets were heated at 1000°C for 5 min. The bending temperatures were 800 or 750°C, bending radius was 5mm, and bending angle was 90°. Subsequently, the appearances of the test pieces were observed, and the interface between the two sheets was checked for LME cracking.

3.4 Evaluation method of joint strength

The load applied to the brazed interface of actual channels is assumed to be the force resulting from the force in the shearing and peeling-off directions. Accordingly, joint strength was examined using tensile shear strength (TSS) and cross tension strength (CTS) tests. Figure 2 illustrated the shape of TSS and CTS specimens. The dimensions of the TSS specimen were 25 $mm^W \times 90\ mm^L \times 1.8\ mm^t$, and the lapped area of 25 $mm^W \times 15\ mm^L$ was brazed entirely; the dimensions of the CTS specimen were 50 $mm^W \times 150\ mm^L \times 1.8\ mm^t$, and the lapped area of 50 $mm^W \times 50\ mm^L$ was brazed entirely.

In addition, using JIS Z 3136 as a reference, the joint strength of spot welding was examined in both TSS and CTS tests to compare the joint strength of 2D brazing. The dimensions of the TSS spot welding joint were 30 $mm^W \times$ 100 $mm^L \times 1.8\ mm^t$, and the lapped area was 30 $mm^W \times 30\ mm^L$. The size of the CTS spot welding joint was the same as that of the 2D brazed joint. The center of the lapped area was welded once. The tension rate of the TSS and CTS tests was 10 mm/min.

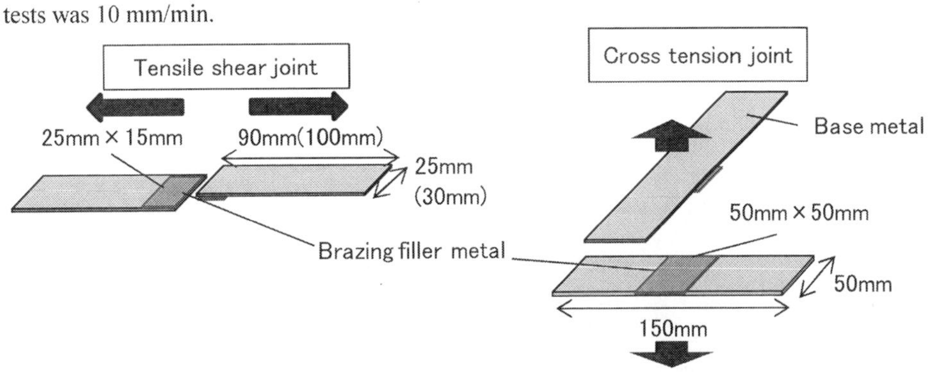

Fig. 2 Shapes of TSS and CTS specimens.

3.5 Three-point bending test of hat channel

The hat channel was fabricated by assuming an automotive channel. The main channel and reinforcement of the hat channel area of 170 mm × 595 mm at 1000°C for 10 min were entirely brazed. The brazing filler metal inserted was 100 μm thick, and the quenching temperature was 800°C. As shown in Figure 3, the hat channel was evaluated by the bending load obtained from the three-point bending test. The supporting points of the hat channels were formed with a radius of 30 mm and 500 mm interval. The punch radius was 150 mm, and the three-point bending test was conducted with a maximum crushing stroke of 50 mm and crushing speed of 15 mm/min.

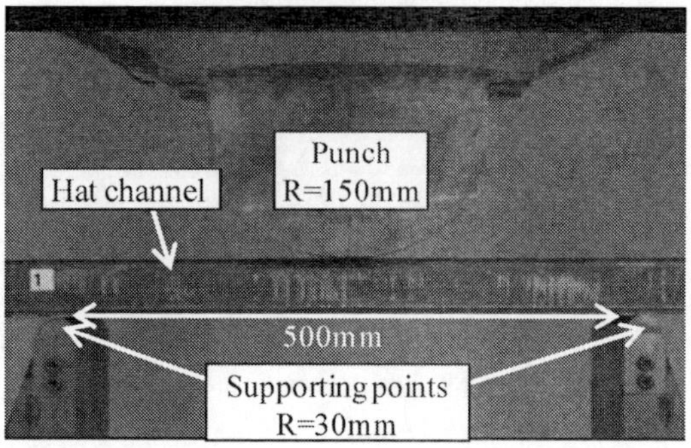

Fig. 3 Schematic of three-point bending test.

4. Results and discussions

4.1 Effects of bending temperature on LME cracking

Figure 4 shows the bending interface at 800°C and 750°C, which are higher and lower, respectively, than the solidus temperature of brazing filler metal. LME cracking occurs under bending at the solidus or a higher temperature, whereas it does not occur at temperatures corresponding to the solidus or a lower temperature. This indicates that specimens should be subjected to bending lower than the solidus temperature of the brazing filler metals to prevent LME cracking. At 800°C bending of brass, it was confirmed that the LME cracking does not occur.

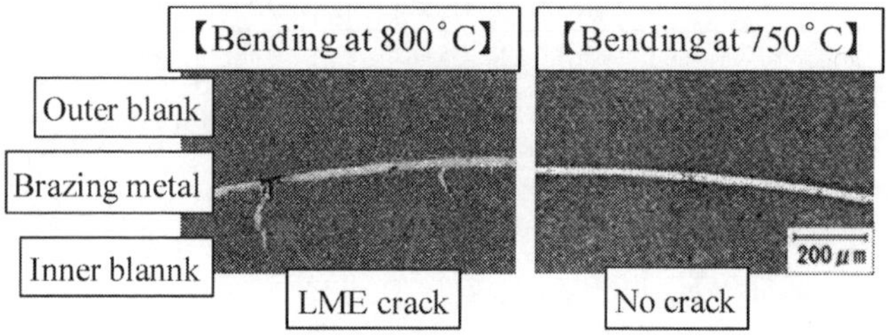

Fig. 4 Interface of 2D brazed specimens bending at 800°C and 750°C.

4.2 Evaluation of joint strength

Figure 5 compares the joint strengths by brass brazing, bronze brazing and spot welding. TSS in brass brazing is 1.3 and 2.2 times higher than that in bronze brazing and spot welding. CTS in brass brazing is 5.7 and 2.7 times higher than that in bronze brazing and spot welding. These results indicate that the channel assembled by the 2D brazing method is expected to show good performance. Also, brass is more suitable for the assembling channel in 2D brazing than bronze.

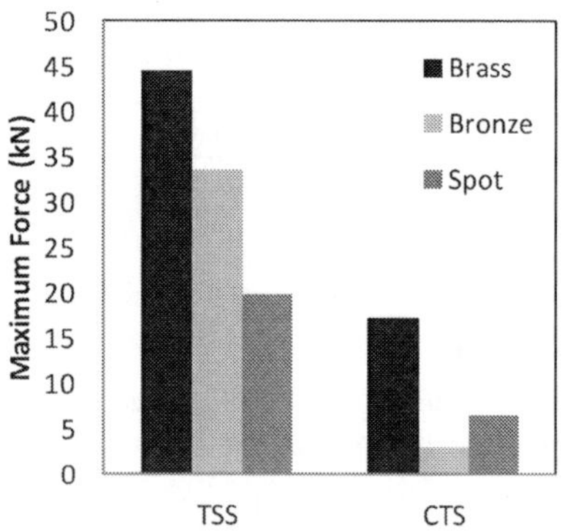

Fig. 5 Comparison of joint strengths.

4.3 Bending performance of 2D brazing channel in three-point bending.

Hat channels were evaluated through a three-point bending test.
(a) Hat channel assembled by conventional spot welding.
(b) Hat channel assembled by 2D brazing using brass.
(c) Hat channel assembled by spot PW-TWB.
(d) Hat channel made of a steel sheet 2.6mm thick without bonding.
Conventional hat channel (a), 2D brazing hat channel (b) and spot PW-TWB hat (d) are made of a 1.2-mm main channel and 1.4-mm reinforce channel, respectively.

Figure 6 presents the bending loads in three-point bending for each channel. Compared with the conventional hat channel (a), the maximum bending load of 2D brazing hat channel (b) is two times higher (50 kN → 100 kN). In addition, the maximum load of 2D brazing hat channel (c) is the same as that of a single sheet hat (c). This result suggests that the 2D brazed hat channel performs similarly to the single sheet hat channel.

Compared with the SP PW-TWB hat channel (d), the maximum bending load of 2D brazing hat channel (b) is 1.6 times higher (61 kN → 100 kN). The decisive difference between 2D brazing hat (c) and SP PW-TWB hat (d) is whether the main channel and reinforcement are integrated. In the case of 2D brazing hat (c), reinforcement is integrated to the main channel. Therefore, 2D brazing hat (c) has a stiffness of the single sheet nearly 2.6 mm thick. On the other hand, in the case of SP PW-TWB hat (d), since a large area is not welded, hat has stiffness of 1.2mm thick and 1.4mm thick.

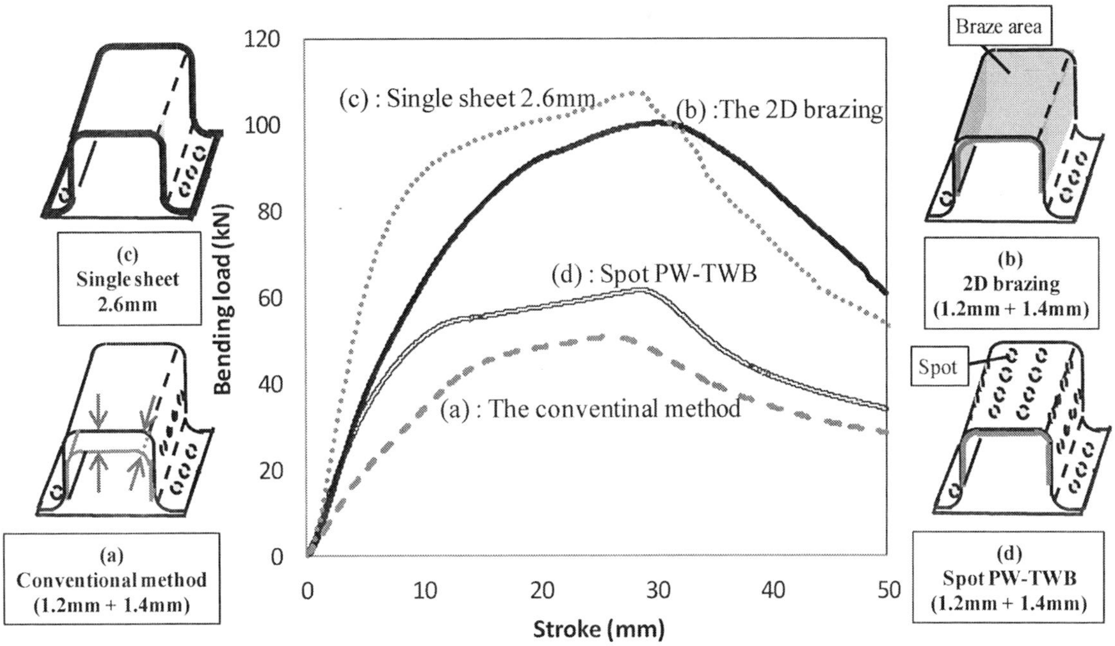

Fig. 6 Comparison of bending loads in a three-point bending test.

5. Conclusions

To develop the 2D brazing method, joint strength and bending performance of the hat channel assembled by 2D brazing were investigated. The obtained results are summarized as follows.

(1) Joint strength such as TSS and CTS of 2D brazing is higher than that of conventional spot welding.
(2) Compared with the hat channel assembled by conventional spot welding, the maximum bending load of the hat channel assembled by 2D brazing is two times higher.
(3) The maximum load of the hat channel by 2D brazing (1.2 mm + 1.4 mm) achieves the same level as the maximum load of the hat channel made of a 2.6-mm thick single sheet. This result suggests that the 2D brazing hat channel performs similarly to a hat channel made of a single sheet.
(4) Compared with the SP PW-TWB hat channel, the maximum bending load of the 2D brazing hat channel is 1.6 times higher. The decisive difference between the 2D brazing hat and SP PW-TWB hat is whether the main channel and reinforcement are integrated.

References

[1] K. Steinhoff et al.: Press Hardening-from Galvanized UHSS to Body-in White Application: Galvatech 2011conference, Genova, Italy, (2011), pp. 319–326.

[2] T. Sugiyama : Recent Progress and Future on High Strength Steel Sheets, JOURNAL OF THE JAPAN SOCIETY FOR TECHNOLOGY OF PLASTICITY, 46(534), p.552-555

[3] Y. Okuri : Journal of the Adhesion Society of Japan 39(12), 455-460

[4] M. Enoki et al. : Material Technology in the Last Ten Years, Journal of Society of Automotive Engineers of Japan, Vol.64 (2010) No.1, p.79-84

Numerical simulation of Pearlite formation during welding of rails

Weingrill Leonhard, Nasiri Mohammad Bagher, Enzinger Norbert
Institute of Material Science and Welding, Graz University of Technology, Graz. Austria
leonhard.weingrill@tugraz.at, +43 (316) 873 - 1671

Abstract
In this work, a thermo-metallurgical coupled simulation model of welding of R350HT pearlitic rail steel was realized to estimate the mechanical properties in the HAZ. The applied approach was to estimate the hardness of pearlitic phase fractions by calculating the interlamellar spacing. This was realized in a methodology of multiple calculation steps. Therein, MATLAB®-routines are used to calculate parameters for a metallurgical model to be used in SYSWELD®, based on a simulated TTT-diagram from JMatPro® and an experimentally determined welding- CCT-diagram of R350HT rail steel. This so established metallurgical model was verified by a simulation of dilatometry tests in SYSWELD® and subsequent quantitative comparison to experimental results. In a final step three more MATLAB®-routines are used to calculate phase fractions, interlamellar spacing and hardness for a practical example. Therefore, measured temperature curves from instrumented flash-butt welding (FBW) experiments of standard rails were used as input. The results of the simulation are compared to real welds' phase fractions and hardness values in the HAZ and showed very good agreement.

Introduction
Due to various reasons – such as increased traffic, higher market shares of high speed trains, a general increase in traveling speeds and axle loads, but also increased safety regulations – the requirements for railways have increased in recent decades [1] und thus have tightened the cost situation. Consequently, the intention of rail operators to extent maintenance intervals and overall lifetime of tracks in order to cut down operation cost is more vital as ever. In answer to this development manufacturers seek to provide rails with improved properties against wear and rolling contact fatigue (RCF) [2], [3]. However, when welding modern rail steel grades undesirable microstructures are formed within the HAZ, which deteriorates the designed properties locally [4], [5]. As a countermeasure in today's common rail welding processes thermal cycles are adjusted to reduce the dimension of this so called soft zone. However with further increasing mechanical properties of newly development rail steel grades also requirement for correspondingly optimized welding processes is given.

Interlamellar spacing is an informative aspect in order to characterize the mechanical properties of pearlitic steels such as strength, hardness and thus wear-resistance [6]–[8]. The pearlitic phase transformation is influenced beside others by the thermal history of the material [9]. Thus, by adapting the thermal cycle during welding also the properties of the weld joint can be controlled.

Table 1: Chemical composition and Brinell hardness at running surface of investigated pearlitic rail steel [10].

steel grade acc. to [10]	main alloying elements in weight -%						HBW
	C	Si	Mn	P max.	S max.	Cr	on running surface
R350HT	0,72-0,80	0,15-0,58	0,70-1,20	0,020	0,025	≤0,15	350- 390

Work content
The objective of this work was to have a tool to estimate the influence of microstructure changes in the HAZ of R350HT rail steel – s. *Table 1* – on the mechanical properties as a function of the thermal history due to welding. Therefore a simulation model was developed which consists of numerical calculations, but also uses experimental methods for data gathering and validation. It calculates pearlitic phase fractions and the interlamellar spacing, as well as the hardness.

Approach
The thermo-metallurgical numerical simulation was realized by using MATLAB® routines in combination with SYSWELD®. The general approach is schematically depicted in a flow chart in *Fig. 1* and calculation steps are correspondingly described in the following passages.

In the first preparatory step (i) required material and process data are gathered. This is done with the help of:

- Simulation of a TTT-diagram of R350HT rail steel in JMatPro®, based on measured chemical composition and austenitization conditions comparable to welding.
- A welding-CCT-diagram of R350HT rail steel, obtained by dilatometry experiments.
- Instrumented flash-butt welding experiments to capture T(t)-curves (T_{ij}) in the HAZ at the rail head at different distances (points of measurement P_{ij}) from the weld centerline (WCL).

These so retrieved data are than preprocessed to the corresponding format and provided for further deployment.

In the subsequent calculation step (ii) MATLAB®-routine (a) is used in order to determine parameters of the metallurgical model of the type SYSWELD® uses, s. [11]. Johnson-Mehl-Avrami-Kolmogorov (JMAK) type kinetics model for diffusion controlled but non-isothermal transformations, as also described in [12], is used. For martensitic transformations Koistinen-Marburger type model is used [13]. For both, the routine initially calculates parameters based on the data from the simulated TTT-diagram and experimental welding CCT-diagram. Subsequently an iterative algorithm, based on additivity rule, as also described in [14], calculates the final phase fractions P_f of Martensite, Bainite and Pearlite for each cooling rate $\dot{T}_i$. In order to optimize the results the parameter $F(\dot{T}_i)$ is determined by comparing numerical and experimental results and recomputing P_f at every repetition cycle until sufficient agreement is reached. Finally, all determined parameters are put together to establish the metallurgical database file for SYSWELD®.

In the next step (iii) the accuracy of the model is verified. Therefore, the beforehand conducted dilatometry experiments are simulated using the thermo-metallurgical coupled calculation option of SYSWELD®. Simplified sample geometry and artificially adapted cooling parameters are used. The cooling curves from the experimental welding CCT diagram are accurately reproduced. Phase fractions of Martensite, Bainite and Pearlite are quantitatively compared to the experimental results in order to verify the established metallurgical model.

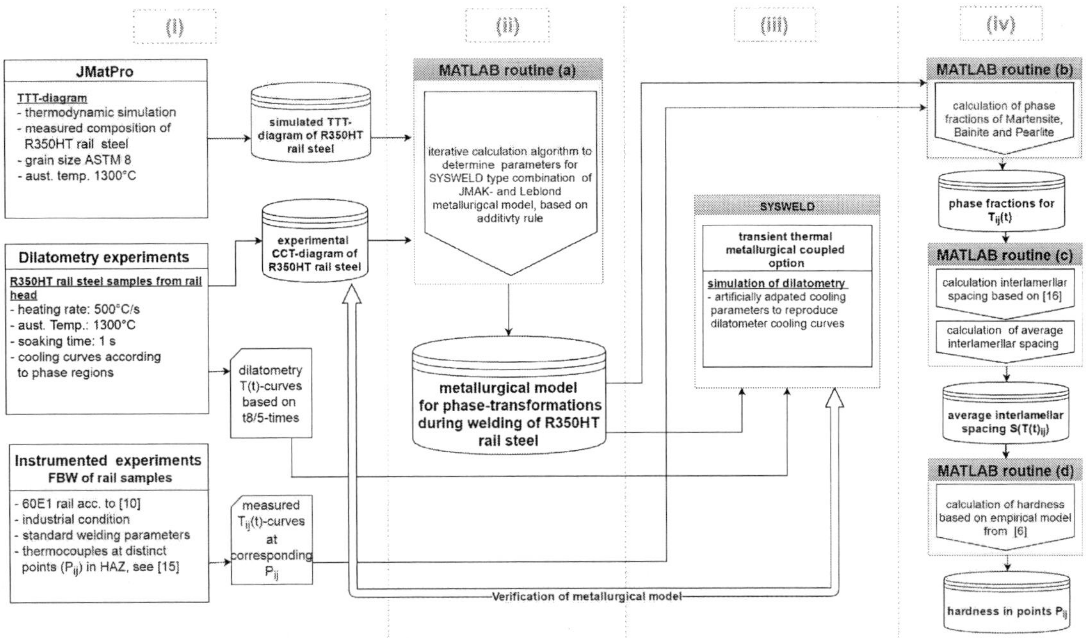

Fig. 1 Schematic flow chart of the used calculation approach.

In step (iv) the calculation is finalized by using three more MATLAB® routines. First, routine (b) is fed with two T(t)-curves measured in the HAZ during welding, gathered during experiments in a recently published work [15]. Based on these so defined thermal cycles the metallurgical model established in step (ii) is used to calculate the phase fractions of Martensite, Bainite and Pearlite in distinct points. Subsequently, MATLAB®-routine (c) calculates the interlamellar spacing S of the computed pearlitic phase fraction and corresponding $T_{i}(t)$-curve, respectively the corresponding cooling rate. The calculation of S is based on an empirical expression proposed in [16] and depicted in *Equation 1*. Thereafter the interlamellar spacing in eutectoid steel can be predicted for isothermal transformation as a function of specific alloying elements' content, the temperature T, at which pearlite formation happens, and the eutectoid temperature T_E. In order to take into consideration the non-isothermal formation of the Pearlite during welding, a weighted average of the interlamellar spacing is calculated subsequently for each cooling rate $\bar{S}(\dot{T}_i)$ based on *Equation 2* subsequently. It integrates the interlamellar spacing S(T) over the derivatives of formed phase fractions dP at the corresponding calculation step for a given cooling rate and finally sets it in relation to the final phase fraction P_f. Finally MATLAB®-routine (d) calculates the hardness based on an empirically determined equation proposed in literature [6], s. *Equation 3*.

Equation 1 [16]:

$$\log(S) = -2.2358 + 0.09863 \times Mn - 0.05427 \times Cr + 0.03367 \times Ni - \log\left(\frac{T_E - T}{T_E}\right)$$

Equation 2:

$$\bar{S} = \frac{1}{P_f}\int_0^{P_f} S(T)\,dP$$

Equation 3 [6]:

$$HV10 = 150 + 2{,}15\,\bar{S}^{-1/2}$$

Results

First of all the results of step (iii) from the dilatometry simulation in SYSWELD® are compared in *Fig. 2* to experimental results from the welding CCT-diagram of R350HT. It can be derived that the simulation shows very good agreement to the experimental data for all three relevant phases and in all investigated cooling rate ranges. The calculation of formed phase fractions based on the new metallurigcal model of this work and two experimental T(t)-curves in step (iv) show fully pearlitic transformation in the HAZ. This is in very good agreement with metallography data provided by rail steel manufactrurers [17]. *Table 2* shows the final result for computed interlamellar spacing at two distinct points in the HAZ – represented by their distances from WCL– together with computed hardness. Accordingly, the interlamellar spacing inside the HAZ of FBW rail decreases with increasing distance from the WCL resulting in an increased hardness. This can be understood as a result of the higher cooling rate. For comparison reasons the computed values are converted and compared to values from a literature source [4]. The computed hardness values are in very good agreement to experimental results. Thus, the model is able to very accurately calculate the hardness based on interlamellar spacing as a function of the cooling rate.

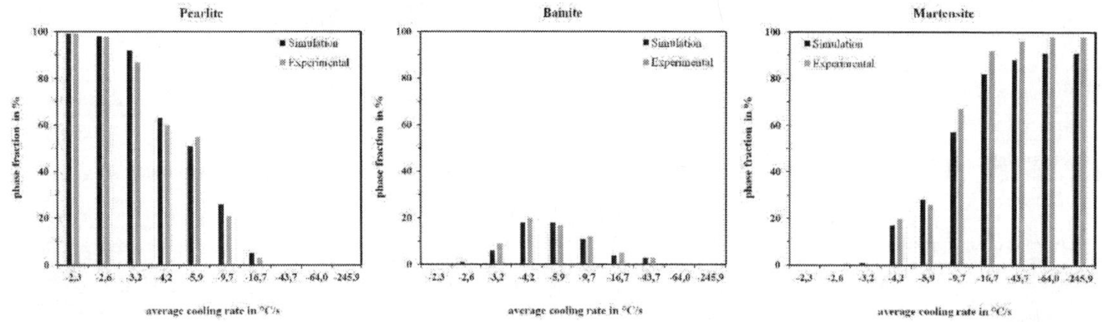

Fig. 2. Calculated vs.measured phase fractions as a functions of cooling rates for R350HT rail steel.

Table 2: Calculated interlamellar spacing and corresponding calculated hardness for 2 exemplary thermal cycles in the HAZ for R350HT 60E1 rail and comparison to experimental values.

distance from weld centerline (WCL) in mm	average cooling rate from 800°C to 500°C	peak temperature in °C	calculated average interlamellar spacing $\bar{S}(\dot{T})$ in μm	calculated hardness HV10	HV10 converted to HRC acc. to [18]	experimental HRC of FBW of rails acc. to [4]
1	-0,9321	1355	0,1078	357	36,6	38,5
10	-1,1565	854	0,0809	388	39,5	39

Conclusions

A methodology of combining MATLAB® routines, SYSWELD® and JMatPro®, as well as data from experimental work was developed to implement a multistep thermo-metallurgical simulation of phase transformations in the HAZ during welding of pearlitic rail steel R350HT. With the help of the determined metallurgical model phase fractions and interlamellar spacing was calculated as well as the hardness was calculated for different distances from the WCL in the HAZ at the rail head for a practical example. The comparison of calculated and experimental results showed that very good agreement was obtained. Thus, it was shown that the proposed approach to estimate the hardness in the HAZ by calculating the interlamellar spacing as a function of thermal history in a numerical welding simulation is a very practicable method.

Outlook

In a next step the model will be further extended in order to estimate the softening in the annealed zone of the weld. Therefore additional metallurgical phenomena which are believed to also have influence on the softening in the HAZ of pearlitic rail steel will be studied. Furthermore, additional T(t)-curves at the corresponding locations in the HAZ and proper material parameters are needed. Based on this, an extended approach will be developed and implemented to further enhance the here presented simulation model.

Acknowledgments

K-Project Network of Excellence for Metal JOINing is fostered in the frame of COMET - Competence Centers for Excellent Technologies by BMWFW, BMVIT, FFG, Land Oberösterreich, Land Steiermark, Land Tirol and SFG. The programme COMET is handled by FFG.

References

[1] European Commission, "Fourth report on monitoring development of the rail market," pp. 1–43, 2014.
[2] L. I. Xiao-fei, P. Langenberg, S. Münstermann, and W. Bleck, "Recent Developments of Modern Rail Steels," *HSLA Steels 2005*, vol. 2, 2005.
[3] R. Kuziak and T. Zygmunt, "A new method of rail head hardening of standard-gauge rails for improved wear and damage resistance," *Steel Res. Int.*, vol. 84, no. 1, pp. 13–19, 2013.
[4] P. Micenko and H. Li, "Double Dip Hardness Profiles in Rail Weld Heat-affected Zone — Literature and Research Review Report," Brisbane, Australia, 2013.
[5] P. Mutton, J. Cookson, C. Qiu, and D. Welsby, "Microstructural characterisation of rolling contact fatigue damage in flashbutt welds," *Wear*, pp. 1–10, 2016.
[6] A. J. Perez-Unzueta and J. H. Beynon, "Microstructure and wear resistance of pearlitic rail steels.pdf," *Wear*, vol. 162–164, pp. 173–82, 1993.
[7] B. L. Bramfitt and B. S. Corporation, "Structure / Property Relationships in Irons and Steels," *Met. Handb. Desk Ed. Second Ed.*, pp. 153–173, 1998.
[8] J. Toribio, B. Gonzalez, J. C. Matos, and F. J. Ayaso, "Role of the microstructure on the mechanical properties of fully Pearlitic eutectoid steels," *Frat. ed Integrita Strutt.*, vol. 30, pp. 424–430, 2014.
[9] K. M. Wu and H. K. D. H. Bhadeshia, "Extremely fine pearlite by continuous cooling transformation," *Scr. Mater.*, vol. 67, no. 1, pp. 53–56, 2012.
[10] "ÖNORM EN 13674-1: 2011 Bahnanwendungen - Oberbau -Schienen Teil 1: Vignolschienen ab 46kg/m;" Standards Institute Austria, Vienna, Austria, 2011.
[11] "SYSWELD 2016 Reference Manual." ESI Group, Paris, France, 2016.
[12] P. Krüger, "On the relation between non-isothermal and isothermal Kolmogorov-Johnson-Mehl-Avrami crystallization kinetics," *J. Phys. Chem. Solids*, vol. 54, no. 11, pp. 1549–1555, 1993.
[13] D. P. Koistinen and R. B. Marburger, "A general equation prescribing the extent of the austenite-martensite transformation in pure iron-carbon and plain carbon steels," *Acta Metall.*, vol. 7, 1959.
[14] J. B. Leblond and J. Devaux, "A new kinetic model for anisothermal metallurgical transformations in steels including effect of austenite grain size," *Acta Metall.*, vol. 32, no. 1, pp. 137–146, 1984.
[15] L. Weingrill, J. Krutzler, and N. Enzinger, "Temperature field evolution during flash-butt welding of railway rails." THERMEC 2016, Graz, Austria., 2016.
[16] M. Takahashi, "Reaustenitisation from bainite in steels." Ph.D. thesis, University of Cambridge, 1992.
[17] "Interne Doku Abbrennstumpfschweißung R350HT Probe 6, Schweißstoß 624." voestalpine Schienen GmbH, Leoben/Donawitz AUSTRIA, 2010.
[18] "ÖNORM EN ISO 18265:2003 : Metallische Werkstoff - Umwertung von Härtewerten." Standards Institute Austria, Vienna, Austria, 2004.

TOWARDS A THROUGH-PROCESS MODEL FOR WELDING OF 9CR STEEL

Padraig Mac Ardghail[1, 2], Richard A. Barrett[1, 2], Noel Harrison[1, 2], Sean B. Leen[1, 2]
[1]Mechanical Engineering, College of Engineering and Informatics, NUI Galway, Ireland
[2]Ryan Institute for Environmental, Marine and Energy Research, NUI Galway, Ireland
p.macardghail1@nuigalway.ie, 00353 91 495955

Abstract

A through-process modelling methodology has been developed to predict the effect of the gas tungsten arc welding (GTAW) process on tensile stress-strain response within the as-welded inter-critical heat-affected zone (ICHAZ) of a 9Cr steel. A physically-based, unified viscoplasticity model, with dislocation density evolution and plasticity damage, was implemented and validated across a range of temperatures for tempered and untempered P92 martensitic steel phases. A dual-phase model was developed, using a rule of mixtures approach, to predict the variation in temperature-dependent tensile behaviour within the ICHAZ of a GTAW butt-welded thick-walled P92 pipe, based on the predicted phase distributions. The model was calibrated against experimental test data for both tempered and untempered martensite across a range of temperatures.

Introduction

Efforts to improve thermal power plant efficiency, in order to reduce fuel wage and CO_2 emissions, require higher operating temperatures and steam pressures. This increases the risk of creep failure in power plant, particularly at welded connections. At the same time, increased use of renewable energy sources has led to conventional power plant operating under intermittent or load-following conditions. This leads to fluctuations in heating and pressure load cycles, with associated increased risk of thermo-mechanical fatigue, including creep-fatigue and high temperature fatigue.

The creep failure (e.g Type IV cracking) region in welded connections is typically located in the heat-affected zone (HAZ) [1, 2]. In order to understand and predict the thermo-mechanical behaviour of welded power plant connections, it is considered necessary to characterise the constitutive behaviour of the different weld zones; the coarse- and fine-grain HAZ (CGHAZ, FGHAZ) and the ICHAZ. The ICHAZ is of particular interest to this research, as it is reported that Type IV cracking occurs at its interface with the FGHAZ [1, 2]. Of particular interest, also, is the inhomogeneity of response associated with the ICHAZ region. Here, the parent material partially transforms into the austenite phase during welding, resulting in a heterogeneous material region upon cooling. In order to characterise this evolution of solid-state phases and associated mechanical response within the HAZ, the complex thermal history is required. Hence, in this study, a thermal model of the GTAW process is first developed, allowing the associated evolution of solid-state phases to be predicted.

Methodology

A transient thermal model for multi-pass GTAW has been implemented using the finite element (FE) code Abaqus, following the detailed work of Yaghi et al. [3]. The 2D axisymmetric model simulates a 36-pass GTAW process of a P92 steel pipe with 40 mm wall thickness. The model includes temperature-dependent heat-transfer effects, with temperature input defined by the welding torch settings, via a volumetric heat flux. Figure 1 shows sample temperature contour distributions from the present implementation, for different numbers of completed passes. The temperature history for the entire welding process was sampled at discrete points, e.g. A, B and C in Figure 1, and the temperature-time results were processed via a phase-transformation module based on the continuous-cooling-transformation (CCT) diagram for P92, to predict the phase mixture at those points. Phase change was assumed to occur instantaneously and in proportion to the temperature between the start and finish temperatures such that the transformed volume fraction for a certain solid state phase, *i*, V_i, is defined as:

$$V_i = 1 - \frac{T_{fi} - T}{T_{fi} - T_{si}} \tag{1}$$

where T_{fi} is the temperature at which transformation phase *i* finishes (e.g A_{c3} for austenite or M_f for martensite), T_{si} is the temperature at which transformation to phase *i* starts (e.g A_{c1} or M_s) and T is the current temperature between T_{fi} and T_{si}. The transformation temperatures used in this research are as follow: A_{c3} is 924 °C, A_{c1} is 830 °C, M_s is 400 °C [4], M_f is assumed to be 200°C lower than M_s.

The ICHAZ is idealised, here, as consisting of tempered and untempered martensite (TMR and UMR, respectively). Homogeneous regions of the welded joint are the parent metal (PM), consisting of as-received TMR, and the fine- and coarse-grained HAZ (FGHAZ, CGHAZ), both consisting of UMR. It is assumed that as-received TMR transforms to austenite, on heating (due to the welding process), and cools to harder, UMR. These two phases were modelled separately and validated against experimental data from Yaghi et al. [4]. The constitutive model utilises a unified viscoplastic flow rule to determine viscoplastic strain rate, as follows:

$$\dot{\varepsilon}_p = \alpha \sinh\left[\beta\left(\frac{\sigma}{1-D} - R - k\right)\right] \tag{2}$$

where D is the damage parameter, R is the isotropic hardening variable, k is yield stress and α and β are rate-controlling constants for the viscoplastic strain-rate. The model is physically-based, relying on normalised dislocation density, similar to work by Li et al. [5], as follows:

$$\bar{\rho} = 1 - \frac{\rho_i}{\rho} \tag{3}$$

where ρ_i refers to initial dislocation density for the material and ρ refers to the dislocation density in the deformed material. The normalised dislocation density evolves from 0 to 1 under loading conditions, according to:

$$\dot{\bar{\rho}} = c_1(1-\bar{\rho})|\dot{\varepsilon}_p| - c_2\bar{\rho} \tag{4}$$

where the first term represents accumulation of dislocations, due to viscoplastic strain, and the second term represents dislocation annihilation, due to thermal recovery of dislocation density, at higher temperatures. A physically-based, Taylor hardening parameter is adopted here, based on the predicted evolution of dislocation density, as follows:

$$\dot{R} = \frac{B\dot{\bar{\rho}}}{2\sqrt{\bar{\rho}}} \tag{5}$$

where isotropic hardening coefficient B is described as [6]:

$$B = \frac{aMG\sqrt{b}}{\sqrt{KL}} \tag{6}$$

In the latter equation, a (different from α in eq. 1) is a constant, taken here as 3.3, M is the Taylor factor, G is shear modulus, b is the magnitude of the Burgers vector, L is the dislocation mean free path and K is a constant associated with the rate of recovery. The nucleation of damage is proportional to dislocation density evolution, in order to represent the initiation of micro-voids caused by dislocation motion, as follows:

$$\dot{\omega}_1 = a_1\dot{\bar{\rho}} \tag{7}$$

where a_1 is a fitted parameter. A second damage evolution term is implemented, to represent viscoplasticity-induced void growth, as follows:

$$\dot{\omega}_2 = a_2 D(1-D)^{\psi_1}|\dot{\varepsilon}_p| + a_3 D(1-D)^{\psi_2}|\dot{\varepsilon}_p| \tag{8}$$

where both terms represent damage growth. The use of two damage variables increases flexibility with respect to capturing the detailed failure behaviour of the martensitic phases, as shown below. The total damage is thus:

$$\dot{D} = \dot{\omega}_1 + \dot{\omega}_2 \tag{9}$$

The elastic rate response is:

$$\dot{\sigma} = E(\dot{\varepsilon} - \dot{\varepsilon}_p) \tag{10}$$

The elastic constants (E, k, ν, G) were determined from the elastic portion of the tensile response. Poisson's ratio, ν, is assumed to be 0.3 at all temperatures. The isotropic hardening constants a, M, b, K, L are physically based and temperature independent, e.g. see [6]. The viscoplasticity constants α and β were identified from the uniform elongation portion of the stress-strain responses. The remaining constants were found by an iterative process of fitting. The constant, c_1, was chosen to fit the stress-strain curve with the most rapid hardening and held constant for all other curves. The resultant hardening could be suppressed at lower temperatures with the constant a_1 or with c_2 at higher temperatures. For all temperatures, a_2 and ψ_1 were fitted first, to capture softening behaviour, with a_3 and ψ_2 fitted last. The complete set of constants is contained in Table 1. An extra check was required against high positive values of ψ_1 and ψ_2, to prevent a non-physical response with respect to the cavitation damage rate, Eq. 8. In order to predict the dual-phase response, a rule of mixtures approach is applied to the individual constitutive parameters, as follows, for the case of α, for example:

$$\alpha = \alpha_{UMR}V_{UMR} + \alpha_{TMR}V_{TMR} \tag{11}$$

Results & Discussion

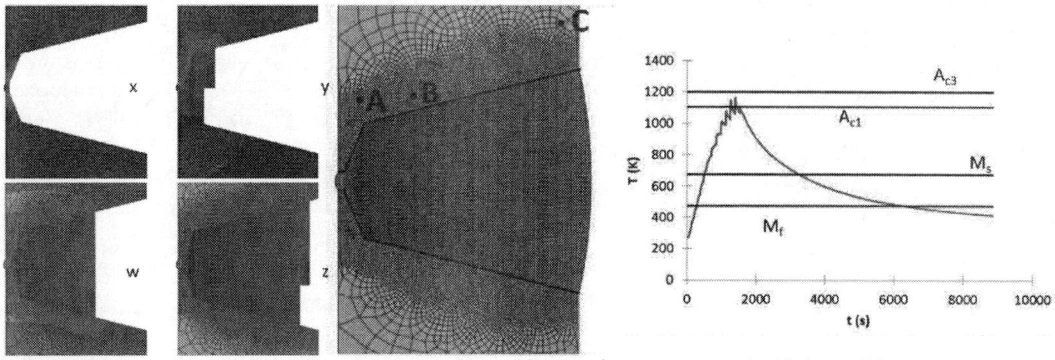

Fig. 1: Predicted temperature contour plots at 1, 5, 18 and 29 weld passes (x, y, w, z, respectively),the model geometry showing weld fusion line and sample locations A, B and C in the ICHAZ, and the temperature-time history of sample point C from the 36-pass welding simulation.

The predicted responses for points A and B are very similar to that of point C (see Figure 1), except with slightly lower maximum temperatures. All three points are predicted to reach maximum temperatures between A_{c1} and A_{c3} and, hence, partially transform into austenite. The predicted volume fractions of UMR are 0.11, 0.47 and 0.58 (for points A, B and C, respectively); the balance is TMR. Figure 2 shows the validation of predicted tempered and untempered martensite responses against the test data for these individual phases, at three different temperatures, 20°C, 200°C and 650°C, as well the corresponding predicted dual-phase response for the point C at each temperature. Figure 3 shows the corresponding predicted responses for points A, B and C, within the ICHAZ, at the same three temperatures.

Table 1: Identified constitutive parameters for (a) tempered and (b) untempered martensitic P92.

(a)

T(°C)	E(GPa)	k(MPa)	c_1	c_2	a_1	a_2	a_3	Ψ_1	Ψ_2	α (1/s)	β (1/MPa)
20	215	600	0.4	0	0.3	35	0.0005	2.5	-15.5	0.00027	0.24
200	207	550	0.4	0	0.3	52	0.0012	4	-15.5	0.00038	0.216
400	198	490	0.4	0	0.3	55	0.0012	4.2	-15.5	0.00056	0.192
550	156	406	0.4	0.07	0.3	55	0.1	8	-8	0.00074	0.176
650	114	300	0.4	0.7	0.3	150	0.1	24	-4.9	0.0009	0.166
850	40.2	83	0.4	0.7	0.3	150	0.1	50	-4.9	0.00133	0.148

(b)

T(°C)	E(GPa)	k(MPa)	c_1	c_2	a_1	a_2	a_3	Ψ_1	Ψ_2	α (1/s)	β (1/MPa)
20	215	1170	9	0	0.6	0.5	0.0005	-3	-15.5	0.00027	0.24
200	207	1080	9	0	0.3	1.2	0.03	-3	-15.5	0.00038	0.216
400	198	954	9	0	0.2	7	0.03	0	-15	0.00056	0.192
550	156	900	9	0	0.1	80	0.03	6.2	-11.7	0.00074	0.176
650	114	475	9	2	0.1	200	0.03	15	-5	0.0009	0.166
850	40.2	85	9	10	0.1	200	0.03	75	-5	0.00133	0.148

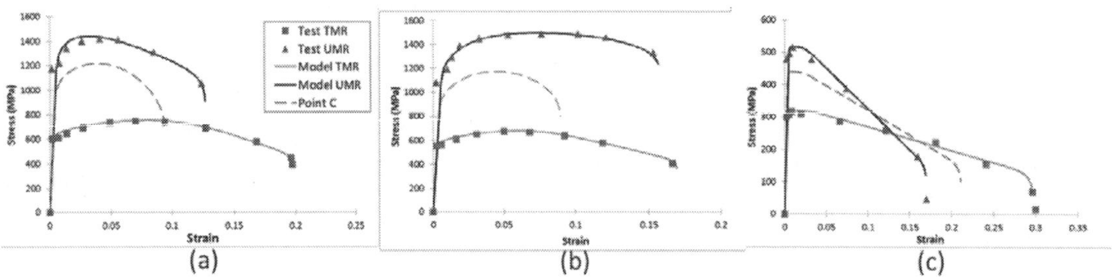

Fig. 2: Experimental and predicted responses of TMR, UMR and dual-phase mixed ICHAZ (point C) at a) 20°C, b) 200°C and c) 650°C.

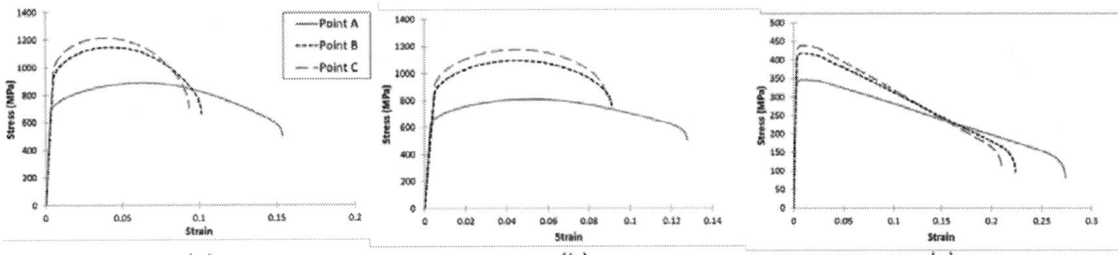

(a) (b) (c)

Fig. 3: Comparison of predicted dual-phase ICHAZ tensile responses at points A, B and C for a) 20°C, b) 200°C and c) 650°C.

In Figure 1, the welding simulation model demonstrates high temperature weld beads transferring heat to the rest of the geometry. Point C shows increasing temperature as more weld beads are laid, exceeding A_{c1} during the 28th and 31st weld passes. The tensile responses shown in Figures 2 and 3 reveal that the material model fits well with test data across a range of temperatures. The dual-phase results show considerable reduction in ductility in the ICHAZ with increasing UMR fraction, particularly at lower temperatures. Depending on the phase mixture and temperature, the ICHAZ is predicted to be the least ductile region weld region and therefore a source of failure. This is consistent with observed Type IV cracking, hence giving some qualitative validation. The use of a direct rule-of-mixtures' approach is considered acceptable under iso-strain conditions, which is assumed before yield, immediately after yield and after hardening completes. According to eq. 10, however, the plastic strain rate must numerically exceed the total strain rate to allow the material to soften. At this point the plastic strain rates for both materials are different and the rule-of-mixtures may be inappropriate. For the present study, the rule-of-mixtures' approach represents a simple solution to the complicated issue of heterogeneous material and it provides interesting insight for the ICHAZ. Over-tempered martensite (OTMR), created in the ICHAZ when TMR is heated without transforming, is not considered in this work. It is assumed the OTMR will be a softer, more ductile material than TMR, as evident from results presented in [7], where the effects of tempering temperature on P92 were investigated, but the through-process methodology would remain the same and the results would be expected to follow similar trends.

This work presents the results of the through-process methodology at sample points within a simulated welded joint. The successful application of this methodology to the entire weld model will permit the simulation of the mechanical response of the entire heterogeneous weld joint by predicting a TMR-UMR distribution within the joint, with an associated distribution of mechanical properties and response.

Conclusion

A through-process modelling methodology for the characterisation of the as-welded ICHAZ region of P92 steel, directly linking a welding process to mechanical response via solid-state phase transformation, has been implemented at sample points within a welding simulation. The mechanical model has been calibrated against published test data and predicts reduced or minimum ductility in the ICHAZ of a welded joint, depending on the amount of solid-state transformation that occurs during the welding process.

References

[1] J. A. Francis, et al: "Type IV cracking in ferritic power plant steels", *Mat Sci Tech*, Vol. 22, No. 12 (2006) pp. 1387 – 1395

[2] J. S. Lee, K. Maruyama, "Mechanism of microstructural deterioration preceding type IV failure in weldment of Mod.9Cr-1Mo steel", *Met Mater Int*, Vol. 21, No. 4 (2015) pp. 639 - 645

[3] A. H. Yaghi et al. "Residual stress simulation in welded sections of P91 pipes", *J Mat Proc Tech*, Vol. 167 (2005) pp. 480 - 487

[4] A. H. Yaghi et al. "Finite element simulation of residual stresses induced by the dissimilar welding of a P92 steel pipe with weld metal IN625", *Int J Press Vessel Piping*, Vol. 111 – 112 (2013) pp. 173 – 186

[5] H. Li et al. "Modelling mechanical recovery of a linepipe steel in annealing process", *Int J Plasticity*, Vol. 25 (2009) pp. 1049 - 1065

[6] X. Wei et al. "Micromechanical modelling of bending under tension forming behaviour of dual phase steel 600", *Comp Mat Sci*, Vol., 108 (2015) pp. 72 – 79

[7] D. R. Barbadikar et al. "Effect of normalizing and tempering temperatures on microstructure and mechanical properties of P92 steel", *Int J Press Vessel Piping*, Vol. 132 – 133 (2015) pp. 97 - 105

Multi-phase-field simulation of austenite-to-ferrite transformation in heat affected zone of Fe-C-Mn alloy

Akinori Yamanaka, Masahito Segawa
Tokyo University of Agriculture and Technology, Koganei-shi, Tokyo, Japan
a-yamana@cc.tuat.ac.jp

Munekazu Ohno
Hokkaido University, Sapporo-shi, Hokkaido, Japan
mohno@eng.kokudai.ac.jp

Toshiyuki Koyama
Nagoya University, Nagoya-shi, Aichi, Japan
koyama@numse.nagoya-u.ac.jp

Jyunya Inoue
The University of Tokyo, Bunkyo-ku, Tokyo, Japan
inoue@material.t.u-tokyo.ac.jp

Abstract
The sequential simulation framework for simulating microstructure evolutions in the heat affected zone (HAZ) of steels using the phase-field and the multi-phase-field methods. This paper presents the sequential simulation of the formation of the grain-boundary allotriomorph ferrite (α_g) and the morphological change of α_g to the Widmanstätten ferrite plates in a Fe-C alloy that is often observed in the HAZ of steels.

Introduction
Mechanical properties of steels are strongly affected by their underlying microstructures. In the heat affected zone (HAZ) of steels, various microstructures are formed by grain growth, solid-solid phase transformations and precipitation. In order to improve strength and toughness of welded steels, it is essential to control size, distribution and morphology of microstructures in the HAZ. However, understanding the microstructure formations in the HAZ is very difficult only by experimental techniques, because the microstructure evolutions depends on various factors, such as chemical composition of the steel and temperature.

Recently, the phase-field (PF) and the multi-phase-field (MPF) methods have been widely used for simulating the microstructure evolutions in steels, such as solidification [1], grain growth [2], the austenite-to-ferrite ($\gamma \rightarrow \alpha$) transformation [3], the pearlite transformation [4], and the martensitic transformation [5]. The purpose of this study is to develop a sequential simulation framework for simulating the microstructure evolutions in the HAZ of steels using the PF and MPF models. This paper presents the simulation results of the formation of the grain-boundary allotriomorph ferrite (α_g) and the morphological change of α_g to the Widmanstätten ferrite (α_w) plates in a Fe-C binary alloy.

Multi-phase-field model
The formation of α_g during continuous cooling in a Fe-C alloy is simulated using the MPF model [3,6]. Using the MPF model, the total Gibbs free energy of a system is defined as

$$G = \int_V \left\{ \sum_{i=1}^{N} \sum_{j=1}^{N} \left(-\frac{a_{ij}^2}{2} \nabla\phi_i \cdot \nabla\phi_j + W_{ij}\phi_i\phi_j \right) + g_{ch} \right\} dV \tag{1}$$

where N describes the total number of crystal grains in the system. ϕ_i is the phase-field variables which describe local volume fraction of ith grain. ϕ_i takes 1 inside ith grain and 0 in other grains. In a grain boundary region, ϕ_i changes from 0 to 1 smoothly. a_{ij} and W_{ij} are the gradient energy coefficient and the height of the potential energy, respectively. These parameters are given as a function of the interfacial energy between ith and jth grain, σ_{ij}, and the thickness of the grain boundary, δ [3]. g_{ch} is the chemical free energy density of the system, which is given by

$$g_{ch} = \sum_{i=1}^{N} g_i\left(C, T\right)\phi_i \tag{2}$$

where g_i is the chemical free energy density of ith grain and is a function of the carbon concentration, C, and temperature, T. In this study, g_i is evaluated by the Gibbs free energy function of a Fe-C alloy assessed by

Gustafson [7].

The time evolution of ϕ_i, which corresponds to the migration of the α/γ interface, is given by

$$\frac{\partial \phi_i}{\partial t} = -\sum_{j=1}^{N} \frac{2M_{ij}}{n} \left[\sum_{k}^{n} \left\{ (W_{ik} - W_{jk})\phi_k + \frac{1}{2}\left(a_{ik}^2 - a_{jk}^2\right)\nabla^2\phi_k \right\} + \frac{8}{\pi}\sqrt{\phi_i\phi_j}\Delta g_{ch} \right] \quad (3)$$

where M_{ij} is the mobility of the phase-field variables. n is the number of the grains which consist of a grain boundary. Δg_{ch} is the chemical driving force of the $\gamma \rightarrow \alpha$ transformation, which is given as

$$\Delta g_{ch} = g_i(c_i) - g_j(c_j) + (c_i - c_j)\mu \quad (4)$$

where c_i is the carbon concentration in ith grain and μ is the diffusion potntial. Noted that in order to calculate the carbon concentration in each grain, c_i, we employ the Kim-Kim-Suzuki model [8], which is described as follow:

$$\mu = \frac{\partial g_i}{\partial c_1} = \frac{\partial g_i}{\partial c_2} = \frac{\partial g_i}{\partial c_3} = \ldots = \frac{\partial g_i}{\partial c_N} \quad (5)$$

The diffusion of the carbon atom during the $\gamma \rightarrow \alpha$ transformation is described by the following equation.

$$\frac{\partial C}{\partial t} = \nabla \cdot \left(\sum_{i=1}^{N} \phi_i D_i \nabla c_i \right) \quad (6)$$

Here, D_i is the diffusion coefficient of the carbon atom in ith grain. In the numerical simulation, Eq. (3) and Eq. (6) are solved by the finite difference (FD) method.

Phase-field model

The morphological change of α_g to α_w is simulated by the PF model [9]. Using the PF model, the total free energy of the system is described by

$$G = \int_V \left\{ g_{ch}^s + \frac{\epsilon(\theta)^2}{2}|\nabla\phi|^2 \right\} dV \quad (7)$$

where ϕ is the phase-field variable which is defined as 1 in the α phase and 0 in the γ phase. The chemical free energy density of the system, g_{ch}^s, is given by

$$g_{ch}^s = p(\phi)g^\alpha + \{1 - (\phi)\}g^\gamma + W_s q(\phi) \quad (8)$$

where g^α and g^γ are the Gibbs free energy density of the α and the γ phases, respectively. These are evaluated by the same way for above-mentioned MPF model. $p(\phi)$ is an interpolation function. $q(\phi)$ is the double-well potential function and W_s describes the height of the double-well potential. In this study, we assume that α_w is formed by strong anisotropy of the interfacial energy between the α and the γ phases, σ. Therefore, the anisotropy of the gradient energy coefficient, $\varepsilon(\theta)$, is described by the following equation.

$$\epsilon(\theta) = \sqrt{\frac{3\sigma\delta}{b}}\eta(\theta) \quad (9)$$

Here, b is a constant that determines the interfacial thickness. $\eta(\theta)$ is the anisotropic function [9] and θ represents the preferential growth direction of α_w.

The time evolution of ϕ in a two-dimensional space is given by the following equation.

$$\frac{\partial \phi}{\partial t} = M_\phi \left\{ \nabla \cdot (\epsilon^2 \nabla\phi) - \frac{\partial g_{ch}^s}{\partial \phi} - \frac{\partial}{\partial x}\left(\epsilon \frac{\partial \epsilon}{\partial \theta}\frac{\partial \phi}{\partial y}\right) + \frac{\partial}{\partial y}\left(\epsilon \frac{\partial \epsilon}{\partial \theta}\frac{\partial \phi}{\partial x}\right) \right\} \quad (10)$$

Here M_ϕ is the mobility of the phase field variable and given as,

$$M_\phi = \frac{b}{3\delta}M_{\alpha/\gamma} \quad (11)$$

where $M_{\alpha/\gamma}$ is the interfacial mobiliy of the α/γ interface.

On the other hand, the time evolution of the carbon concentration during the α_w formation is given as,

$$\frac{\partial u_C}{\partial t} = \nabla \cdot \left\{ M_C \left(\frac{\partial^2 g_{ch}^s}{\partial u_C^2}\nabla u_C + \frac{\partial^2 g_{ch}^s}{\partial u_C \partial \phi}\nabla\phi \right) \right\} \quad (12)$$

where, u_C is the u-fraction of the carbon atom [9], which is defined as,

$$u_C = \frac{x_C}{1 - x_C} \quad (13)$$

where x_C is the molar fraction of the carbon atom. In the PF simulation, Eq. (10) and Eq. (12) are solved using the finite element (FE) method and the adaptive mesh refinement (AMR) algorithm.

Simulation results

In this study, firstly, the MPF simulation of the formation of α_g in a Fe-0.1mass.%C alloy during a continuous cooling process is performed. Subsequently, the mophological change of α_g to α_w is simulated by the PF model using the distributions of the phase-field variables and the carbon concentration obtained by the MPF simulation.

Formation of grain-boundary allotriomorph ferrite

The size of computational domain used for the MPF simulation is 20 × 20 μm^2. The number of FD meshes is 200 × 200 and the spacing of FD mesh is $\Delta x = \Delta y = 0.1$ μm. Periodic boundary condition is used for all direction. The interfacial energy is assume to be isotropic and constant at $\sigma_{ij} = 0.5$ J/m^2. The interfacial thickness is $\delta = 7\Delta x = 0.7$ μm. The initial and final temperature are set to be 1023 K and 1011 K, respectively. The diffusion coefficient of the carbon atom in the α and the γ phases are refered to the literature [10]. For the initial condition, four hexagonal γ grains are created by the grain growth simulation. Because the MPF model can not describe the nucleation of α grain, eight α grains are nucleated at the triple point of the γ grain boundary.

Fig. 1 shows the evolution of carbon concentration during α_g formation. As temperature is decreased, α grains grow into the γ phase with carbon diffusion from the γ phase to the α phase. It is seen that the center γ grain is surrounded by α_g at 1013 K.

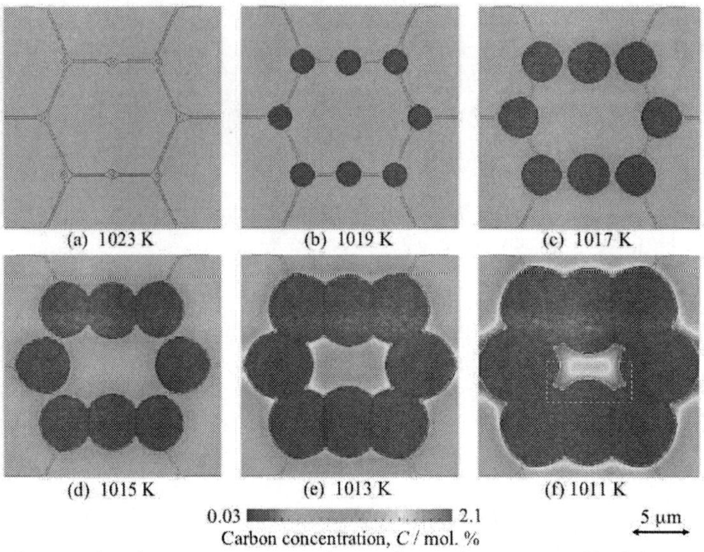

Fig. 1 Evolution of carbon concentration during the α_g formation in Fe-0.1mass%C alloy.

In the next subsection, we present the PF simulation result of the morphological change of α_g to α_w in the rectangle region indicated by dotted line in *Fig. 1* (f). The size of the rectangle region is 8 × 3 μm^2. The distributions of the phase-field variables, in other words, the distribution of the α and γ phases, and the carbon concentration are directly mapped to the FE mesh used in the PF simulation.

Widmanstätten ferrite formation

In the PF simulation, the square FE mesh with four nodes is used. The minimum size of the FE mesh is $\Delta x = \Delta y = 3.3 \times 10^{-3}$ μm. The interfacial thickensss is $\delta = 6\Delta x = 19.8$ nm. The interfacial energy is set to be $\sigma = 0.6$ J/m^2. Using the strength of interfacial anisotropy used in the previous study [9], the maximum interfacial energy is 0.83 J/m^2 at the tip of a α_w plate. While, the minimum interfacial energy is 0.24 J/m^2 at the lateral side of a α_w plate.

Fig. 2 shows the evolution of the carbon concentration during the α_w formation. In the figure, black solid line indicates the position where $\phi = 0.5$. At the early stage (0 ~ 100 μs), the α/γ interface is flucutated due to the interfacial anisotropy. Furthermore, many tips of α_w plate are formed at 100 μs. Subsequently, it is found that the competitive growth and the coarsening of the α_w plates proceed with time. It is worth mentioning that the growth rate of the α_w plates that is initially formed at the convex part of the α_g is high. This is because the carbon supersaturation (or the driving force of the phase transformation) becomes high in the central region of γ grain.

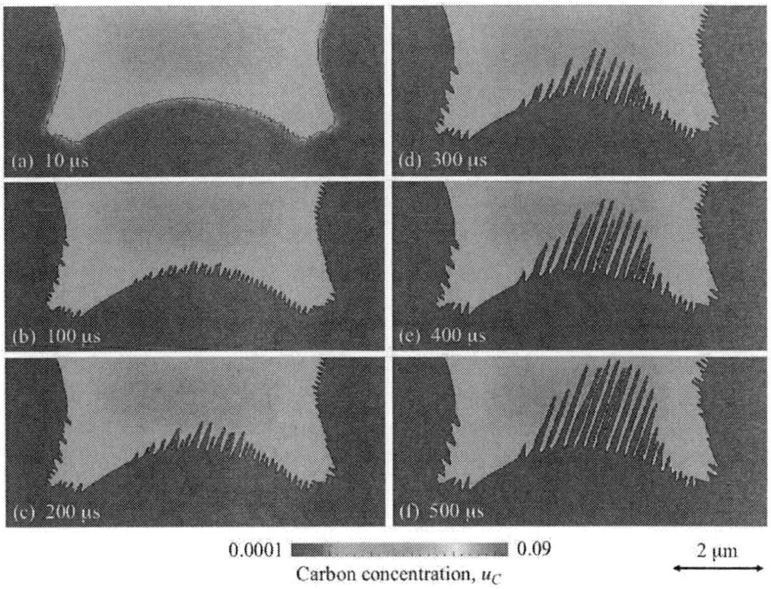

Fig. 2 Evolution of carbon concentration during the α_w formation in the rectangle area indicated in Fig. 1 (f).

Conclusions

The formation of α_g and the morphological change of α_g to α_w plates in a Fe-0.1mass%C alloy were simulated using the MPF and the PF models. The sequential simulation was realized by mapping the distributions of the phase-field variables and the carbon concentration calculated by the MPF simulation to the computational domain used in the PF simulation. Furthermore, according to the results, it was essential to use the Gibbs free energy function of a Fe-C alloy assessed by the CALPHAD method in order to simulate realistic microstructure formations. In future work, the formations of α_g and α_w plates in a Fe-C-Mn alloy will be simulated using our simulation methodology.

Acknowledgments

This work was supported by Council for Science, Technology and Innovation (CSTI), Cross-ministerial Strategic Innovation Promotion Program (SIP), "Structural Materials for Innovation, Materials Integration" (Funding agency: JST).

References

[1] M. Ode, T. Suzuki, S. G. Kim, W. T. Kim: "Phase-field model for solidification of Fe-C alloys," *Sci. Tech. Adv. Mater.*, Vol. 1, (2000), pp. 43-49.
[2] M. Ohno, T. Yamaguchi, D. Sato, K. Matsuura: "Existence or nonexistence of thermal pinning effect in gain growth under temperature gradient," *Comp. Mater. Sci.*, Vol. 69, (2013), pp. 7-13.
[3] A. Yamanaka, T. Takaki, Y. Tomita: "Multi-phase-field modeling of diffusive solid phase transition in carbon steel during continuous cooling transformation," *J. Crystal Growth*, Vol. 310, (2008), pp. 1337-1342.
[4] K. Nakajima, M. Apel, I. Steinbach: "The role of carbon diffusion in ferrite on the kinetics of cooperative growth of pearlite: A multi-phase-field study," *Acta Mater.* Vol. 54, (2006), pp. 3665-3672.
[5] Y. Tsukada, Y. Kojima, T. Koyama, Y. Murata: "Phase-field simulation of habit plane formation during martensitic transformation in low-carbon steels," *ISIJ Int.* Vol. 55, (2015), pp. 2455-2462.
[6] I. Steinbach, F. Pezzolla: "A generalized field method for multiphase transformations using interface fields," *Physica D*, Vol. 134, (1999), pp. 385-393.
[7] P. Gustafson: "Thermodynamic evaluation of the Fe-C system," *Scand. J. Metall.*, Vol. 14, (1985), pp. 259-267.
[8] S. G. Kim, W. T. Kim, T. Suzuki, "Phase-field model for binary alloys," *Phys. Rev. E*, Vol. 60, (1999), 7186.
[9] A. Yamanaka, T. Takaki, Y. Tomita: "Phase-field simulation of austenite to ferrite transformation and Widmanstätten ferrite formation in Fe-C alloy," *Mater. Trans.*, Vol. 47, (2006), pp. 2725-2731.
[10] Japan Institute of Metals: Metal data book, Maruzen Publishing Co., Ltd., (Tokyo, 2004), pp. 21-22.

Austenite grain growth simulation in weld heat affected zone of Ti bearing steel

Naoto Fujiyama, Toshinobu Nishibata, Hiroyuki Hirata
Nippon Steel & Sumitomo Metal Corporation, Futtsu, Japan
fujiyamama.2yc.naoto@jp.nssmc.com

Akira Seki, Kazuhiro Ogawa
Nippon Steel & Sumikin Technology Corporation, Amagasaki, Japan

Abstract

This study proposes a new calculation model for predicting the prior-austenite grain size in the weld heat affected zone (HAZ) of Ti bearing steel. The model considers the solute-drag effect of multi-element systems based on the theory of grain boundary movement by elemental transition. The pinning effect of TiN is also taken into account. Additionally, for modeling, the grain boundary segregation of each element and the weakening of the pinning effect due to the dissolution of TiN during the thermal cycle are taken over. The prior-austenite size in HAZ of 0.05C-0.14Si-1.6Mn-0.012Ti was calculated using the proposed model during thermal cycle heated at a temperature ranging from 1473 to 1673 K. The calculation was also conducted using the conventional theoretical model. Furthermore, the grain growth behavior was experimentally examined in the simulated HAZ test specimen. As a result, the theoretical grain growth estimated by the proposed model was consistent with the experimental grain growth. In particular, at a thermal cycle of over 1623 K, which dissolves TiN, the proposed model properly describes the actual grain growth behavior in HAZ of low alloy steel bearing Ti compared with the conventional model. Additionally, the calculation considering temperature gradient in the area from the fusion line to 0.5 mm HAZ was performed. This calculation method was found to provide a good analysis of the austenite grain growth behavior.

1. Introduction

High heat input welding is often employed for low alloy steel structures in industrial fields such as ship building. In high heat input welding, the heat affected zone (HAZ) adjacent to the fusion line is heated to nearly the melting point of steel by arc energy. Therefore, the austenite grain size in the area coarsens, and the toughness of the HAZ is generally lower than that of the base metal. A fine HAZ microstructure is effective for obtaining high HAZ toughness, and some technical proposals such as the pinning effect or intragranular ferrite transformation have been proposed in this regard [1-5]. However, the optimum condition of inclusions, including their size and distribution, for improving HAZ toughness is not yet clear. As, it is difficult to clarify each factor's influence on HAZ toughness through experiments, it is important to develop numerical simulations for austenite grain growth. In previous research on austenite grain growth simulation, Burke proposed the following grain growth equation [6].

$$R^n - R_0^n = kt \qquad (1)$$

R is the radius of the grain (m), t is time (s), k is the kinetic constant ($m^2 \cdot s^{-1}$), and n is selected from 2 to 4 based on the rate-limiting steps. In addition, Hillert proposed the following kinetic equation of grain growth [7].

$$\frac{dR}{dt} = M^{gb} \times \frac{2\sigma V}{R} \qquad (2)$$

M^{gb} is grain boundary mobility ($m \cdot s^{-1} \cdot J^{-1} \cdot mol$), σ is grain boundary energy ($J \cdot m^{-2}$), and V is the molar volume ($m^3 \cdot mol^{-1}$). This equation represents the mobility and capillary effect, and it can be replaced with Burke's equation. The solute-drag effect is also useful for restraining austenite grain growth in low alloy steel. This effect has a lower grain kinetic constant resulting from the segregation of impurities, and it has been researched by Lücke [8], Cahn [9], Hillert [10], and Purdy [11]. Although many researchers have discussed the solute-drag effect with mathematical models and experiments, most have been binary or ternary systems. Furthermore, these predict only the solute-drag effect, and there are few researches on how much influence this effect has on austenite grain growth. In addition to the solute-drag effect, the pinning effect is a widely used technique for improving HAZ toughness in low alloy steel. The pinning effect suppresses austenite grain growth using inclusions and TiN in particular [12,13]. However, the pinning effect is not obtained by predicting the effect on austenite grain growth in combination with the solute-drag effect. Therefore, we proposed a method for predicting the austenite grain growth behavior of Ti bearing steel that considers both the solute-drag effect for multi-element systems and pinning effects under welding thermal cycles of HAZ.

2. Experimental Procedure

A Ti bearing steel plate (0.05C-0.14Si-1.6Mn-0.012Ti-0.0035N), which was prepared in an actual production process, was used as the test material. The simulating HAZ test specimens were machined from the test material and heated in the thermal cycle shown in Fig.1. Furthermore, the submerged arc welded joint was produced with a heat input of 5.0 kJ/mm. The microstructures of the simulated HAZs and HAZ in submerged welded joint were observed by optical microscope.

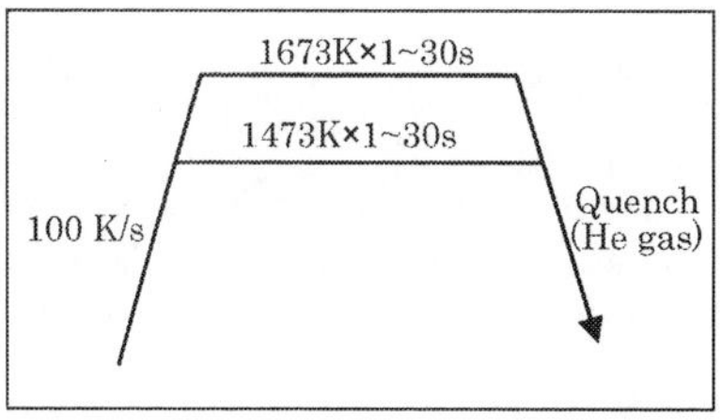

Fig. 1 Heat condition simulated HAZ

3. Calculation method of austenite grain growth

Grain boundary mobility M^{gb} considering the solute-drag effect for multi-element systems was proposed [14].

$$M^{gb} = \frac{1}{4}\frac{1}{\left(1-\sum_{i\neq Fe} X_i\right)+\sum_{i\neq Fe}\frac{(X_i)^2}{x_i}} M^{gb}_{Fe} \qquad (3)$$

X_i is the austenite grain boundary concentration (mol), x_i is the molar concentration of the solute (mol), and M^{gb}_{Fe} is the grain boundary mobility of pure iron ($m\cdot s^{-1}\cdot J^{-1}\cdot mol$). From Eq. (3), the austenite grain boundary concentration needs to be obtained. Figure 2 shows a schematic diagram of free energy curves. We employed Hillert's theory [15] under the assumption that the austenite grain boundary phase is a liquid phase. Thus, it is possible to estimate the equilibrium concentration of element i at the austenite grain boundaries from the intersection point between the grain boundary energy of the liquid phase and the slope of a line through μ^{gb}_{Fe} and μ^{gb}_{i}. The calculations were performed using Thermo-Calc, and the TCFE6 database was used. Figure 3 shows the calculation results of grain boundary mobility at each temperature of Ti bearing steel. Grain boundary mobility increases with increasing temperature. This means that the grain tends to grow at higher temperatures. Austenite grain growth behavior at isothermal process was calculated by Eq. (4).

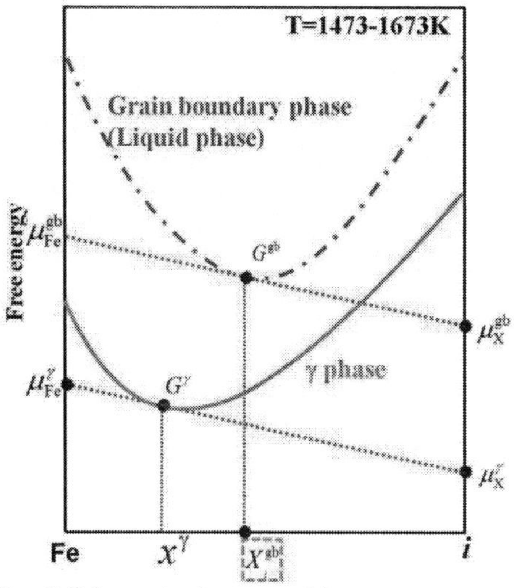

Fig. 2 Schematic diagram of free energy curves

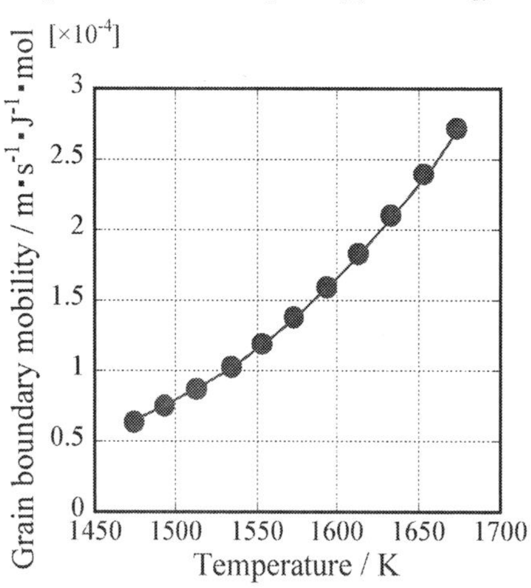

Fig. 3 Temperature dependency of grain boundary mobility

$$\frac{dR}{dt} = \frac{1}{4} M^{gb} \times \left(\frac{\sigma V}{R} - \Delta G_{pin}\right) \qquad (4)$$

ΔG_{pin} is the pinning energy ($J\cdot mol^{-1}$), which can be represented by Eq. (5) [16, 17].

$$\Delta G_{pin} = \frac{3}{4}\frac{\sigma V f^{2/3}}{r} \qquad (5)$$

f is the volume fraction of TiN, r is radius of TiN (m). In Eq. (4), TiN dissolution was considered because it dissolves over 1623 K and that reduces the pinning effect. This is represented by Eq. (6) [18].

$$r^2 - r_0^2 = -2\left(\frac{C^{\gamma/\theta}_{Ti(N)} - C^{0}_{Ti(N)}}{C^{\theta}_{Ti(N)} - C^{\gamma/\theta}_{Ti(N)}}\right) D^{\gamma}_{Ti(N)} \cdot t \qquad (6)$$

R_0 is the initiation radius of TiN (m), $C^{\gamma/\theta}_{Ti(N)}$ is the γ/θ interface concentration of Ti(N), $C^{0}_{Ti(N)}$ is the balk concentration of Ti(N), $C^{\theta}_{Ti(N)}$ is the precipitate phase (θ) concentration of Ti(N), $D^{\gamma}_{Ti(N)}$ is the γ diffusion coefficient of Ti(N)($m^2\cdot s^{-1}$). The solubility product constant of TiN is represented by the following equation.

$$\log\left[C^{\gamma/\theta}_{Ti}\right]\left[C^{\gamma/\theta}_{N}\right] = -\frac{16750}{T} + 5.89 \qquad (7)$$

The calculation results of Eq. (6) at Ti and N are equal, so Eq. (6) can replace Eq. (8). From Eq. (7) and (8), the TiN dissolution can be calculated.

$$\left(\frac{C^{\gamma/\theta}_{\mathrm{Ti}}-C^{0}_{\mathrm{Ti}}}{C^{\theta}_{\mathrm{Ti}}-C^{\gamma/\theta}_{\mathrm{Ti}}}\right)D^{\gamma}_{\mathrm{Ti}}=\left(\frac{C^{\gamma/\theta}_{\mathrm{N}}-C^{0}_{\mathrm{N}}}{C^{\theta}_{\mathrm{N}}-C^{\gamma/\theta}_{\mathrm{N}}}\right)D^{\gamma}_{\mathrm{N}} \quad (8)$$

Austenite grain growth simulation considering the temperature gradient of the welding thermal cycle at HAZ was conducted on the basis of the phase-field method using the phase-field equation [19]:

$$\frac{\partial\phi_i}{\partial t}=M^{gb}\sum_{j=1}^{n}\frac{1}{\nu}\left\{\sum_{k=1}^{n}\left[\left(\sigma_{jk}-\sigma_{ik}\right)\left(\frac{\pi^2}{\delta^2}\phi_k+\nabla^2\phi_k\right)\right]+\frac{2\pi}{\delta}\sqrt{\phi_i\phi_j}\,\Delta G_{ij}\right\} \quad (9)$$

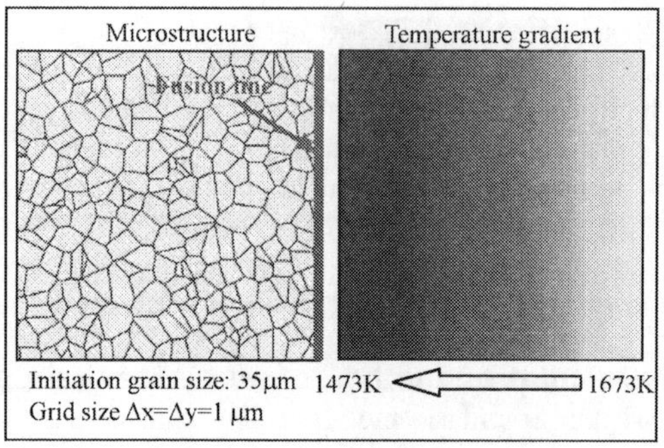

Fig. 4 Schematic of austenite grain growth calculation condition in MICRESS®

where $\phi_{(i,\ j,\ k)}$ is the phase field parameter, $\nu = 1\sim3$ depending on the balk and region of the interface, σ_{jk} is the interface energy between grain j and k, σ_{ik} is the interface energy between grain i and k, δ is the interface width, and ΔG_{ij} is the driving force of grain growth between grain i and j. In this method, each grain is distinguished by an ID number used to track the growth and disappearance of each grain. Eq. (14) includes the Gibbs-Tomson effect, which is the driving force of grain growth. Each ID has a phase-field parameter and the same mobility. The calculation was performed using MICRESS® [20]. Figure 4 shows a schematic of the calculation conditions. The calculated area is 500 × 500 μm with an interface width of 6 μm. The temperature was calculated with a gradient so as to be lower with distance from the fusion line. The grain boundary mobility was input using the value shown in Fig.3. In this report, it is assumed that the grain boundary energy is constant (σ=0.8(J·m^{-2})). The pinning force κ to input in MICRESS® needs to be input from Eq. (10).

$$\kappa=\frac{\Delta G_{pin}}{2\sigma V} \quad (10)$$

4. Results of austenite grain growth simulation at HAZ

Figure 5 shows a comparison of the theoretical and the experimental austenite grain growth behaviors at isothermal heating at (a) 1473 K and (b) 1673 K. The calculations were conducted under the following conditions: (I) without considering the solute-drag effect and pinning effect, (II) considering only the solute-drag effect and (III) considering the solute-drag effect and pinning effect. For calculation, the volume fraction of TiN was estimated using the thermodynamics calculation software, Thermo-Calc. The radius of TiN was supposed as 50 nm. Calculated in conditions (I) and (II), the estimated austenite grain size was much larger compared with the prior austenite grain size experimentally-measured. Meanwhile, the theoretical grain growth behavior calculated under condition (III), which considers both the solute-drag effect and pinning effect of TiN, was consistent with the prior austenite grain size experimentally-measured. Figure 6 shows the results of the austenite grain growth simulation considering the temperature gradient simulated heat input of 5.0 kJ/mm. In the case of calculation without consideration of the solute-drag and pinning effects, the calculated austenite grain size was much larger than the prior austenite grain size in an actual submerged welded joint. On the contrary, the grain size distribution, which was estimated with consideration of both the solute-drag and pinning effects of TiN, had excellent correspondence with that in an actual welded joint.

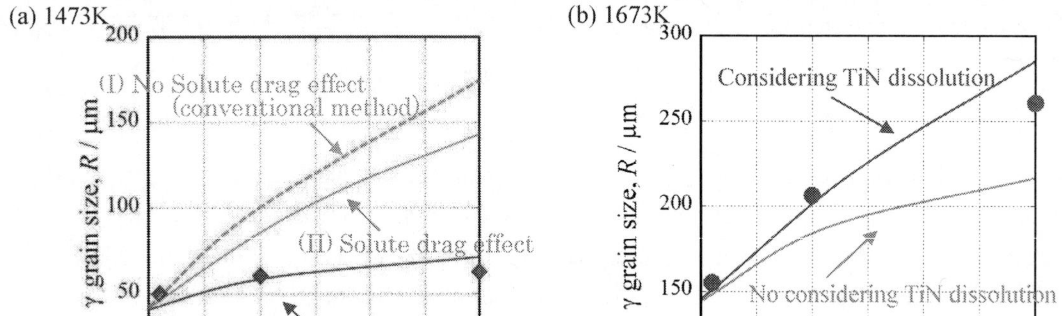

Fig.5 Result of austenite grain growth simulation considering the solute-drag effect and pinning effect in an isothermal process

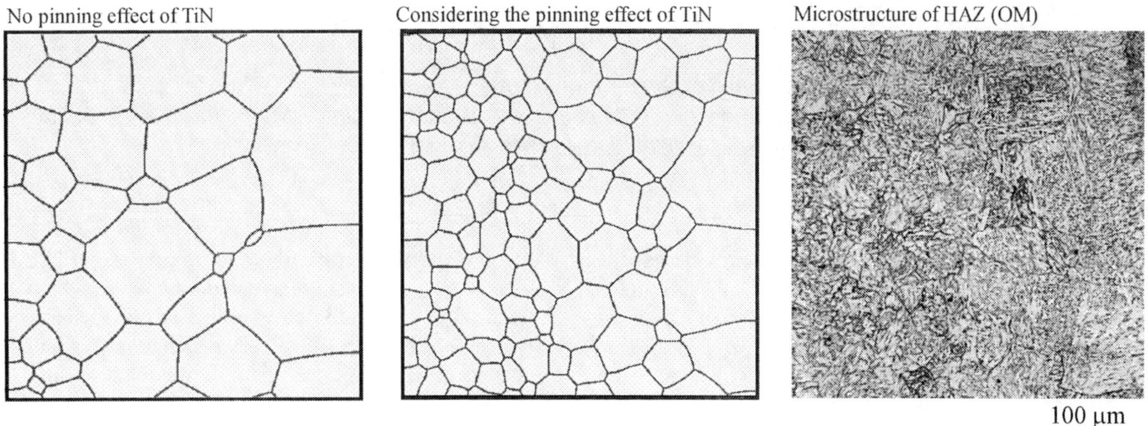

Fig. 6 Result of austenite grain growth similation at HAZ (5.0kJ/mm)

5. Conclusion

1) A new calculation model for predicting austenite grain growth behavior at HAZ is proposed. This calculation model considers the solute-drag for multi-element systems effect and pinning effect of TiN.
2) The theoretical grain growth estimated by the proposed model was consistent with the experimental grain growth in an isothermal process. Furthermore, calculation considering the temperature gradient described the distribution of the prior austenite grain size in HAZ of a submerged arc welded joint.

References

[1] S. Kanazawa, et al. : Tetsu-to-hagané, 61 (1975), 2590.
[2] A. Kojima, et al. : *CAMP-ISIJ*, 16(2003), 360.
[3] T. Funakoshi, et al. : *Tetsu-to-Hagané*, 63(1977), 303
[4] K. Yamamoto, et al. : Residual and Unspecidied Elements in Steels. ASTEM STP1042(1988), 266.
[5] Y. Tomita, et al. : ISIJ International, 34(1994), 829.
[6] J. E. Burke and D. Turnbull : Recrystallization and Grain Growth, Progress in Physics 3, Ed. B. Chalmers, Pergamon, London (1952), 220.
[7] M. Hillert : *Acta Metall.*, 13 (1965), 227.
[8] K. Lücke and K. Detert : *Acta Met.*, 5 (1957), 628.
[9] J. W. Cahn : *Acta Met.*, 10 (1962), 789.
[10] M. Hillert and B. Sundman : *Acta Met.*, 24 (1976), 731.
[11] G. R. Purdy and Y. J. M. Brechet : *Acta Met.*, 43 (1995), 3763.
[12] Nakanishi, et al. : Journal of Japan Welding Society (1983), 117.
[13] Mukae, et al. : Quarterly journal of Japan Welding Society, 3 (1985), 567.
[14] Fujiyama, et al. : Preprints of the National Meeting of the Japan Welding Society 89 (2011), 212.
[15] M. Hillert : Lectures on the theory of Phase Transformations, Ed. H.I. Aaronson, The Metallurgical Society of AIME, Pennsylvania (1975), 36.
[16] Zener, C. quoted by Smith, S. : Trans. AIME. 15 (1948), 175.
[17] Nishizawa, T. et al. : Mat Trans. JIM. 38 (1997), 950.
[18] Fujiyama, et al. : Preprints of the National Meeting of the Japan Welding Society 93 (2013), 176.
[19] J. Eiken, et al. : Phys. Rev. E. 73 (2006), 066122-1-9.
[20] MICRESS® : http://www.micress.de(accessed 2015-06-17)

Two kinds of variant of friction stir welding (FSW) process: friction stir brazing (FSB) and friction stir spot brazing (FSSB)

Gui-Feng ZHANG, Jitao KUANG, Hao LIU*
State Key Laboratory for Mechanical Behaviour of Materials, Xi'an Jiaotong University, Xi'an 710049, China
gfzhang@mail.xjtu.edu.cn, +86 2982668807

Abstract
Friction stir lap welding (FSLW) cannot provide excellent results in dispersing oxide film and mixing extent like friction stir butt welding. Additionally, FSLW suffers from hook formation and small welded area for one pass. To expulse oxide film with liquid phase and then extend bonded area and to eliminate hook formation, two kinds of variant of FSW process of friction stir brazing (FSB) and friction stir spot brazing (FSSB), in which a pinless tool and braze are used, were developed by the present author. For instance, FSB of Al/steel dissimilar assembly using Zn foil braze and FSSB of steel/Ti dissimilar assembly using Cu foil braze were attempted. For FSB of Al/Steel using a special pinless tool with large shoulder of 65 mm diameter, Zn braze foil was melted and expulsed, bonding was established by intimate contact at central region and intermetallic compound layer of 16 μm at outer region. For FSSB of steel/Ti assembly, intimate contact and quite thin eutectic reaction layer of 2 μm were achieved at steel/Cu and Cu/Ti interface, respectively.

1. Introduction
Friction stir welding (FSW) has been applied to surface modification. For example, friction stir process (FSP) was developed by Mishra as a novel surface modifying technique to fabricate surface composite on a monolithic substrate [1,2]. Similar to friction surfacing (FS) process, which is a solid state coating process based on the plastic deformation of a metallic consumable rod [3], friction stir lap welding (FSLW) should be a promising technique for surface modification by cladding a dissimilar metal sheet in solid state on metallic substrate.

However, FSLW cannot provide excellent results in dispersing oxide film and mixing extent like friction stir butt welding. Additionally, FSLW suffers from hook formation and small welded area for one pass. To overcome the problems in FSLW associated with plastic flow (e.g. pin wear, narrow bonded area, hook defect) by metallurgical reaction instead of simple plastic flow, a novel process named friction stir brazing (FSB) or friction stir soldering (FSS) was developed [4-6] and patented in China [7] by the present authors for dissimilar joining. In FSB, as shown in Fig. 1, a pinless tool is used to remove both keyhole and wear of pin end, and a brazing filler metal is used (i) to expulse oxide film fragments out of interface with liquid phase with the aid of mechanical actions of the rotating shoulder, and (ii) to increase bonded area and productivity for one pass.

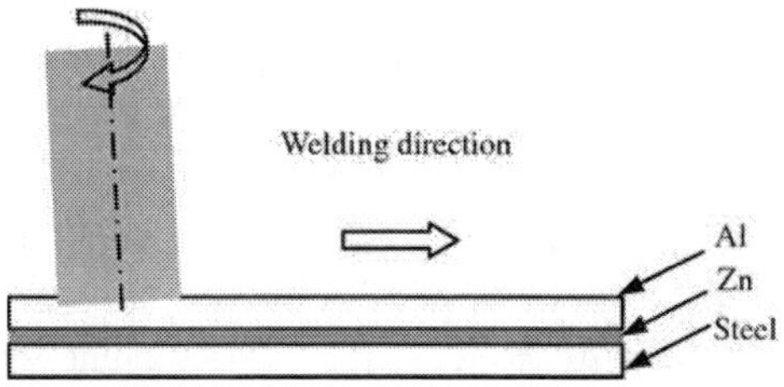

Fig. 1 Schematic diagram of friction stir brazing (FSB) technique [4]

Also, a simplified FSB without tool traverse motion was developed by the present author, which is called friction stir spot brazing (FSSB) [8]. The FSSB process is similar to friction stir spot welding (FSSW), except that a pinless tool and a braze foil are used. As a result of the use of pinless tool and braze foil, keyhole and hook defects can be eliminated.

As a novel surface modifying technique, FSB has been used for producing clad composite. In our previous work on feasibility study, small pinless tool with a shoulder diameter of 20 mm was used [4]. Up to now, the feasibility of FSB of Al/x and Cu/x (x = Al, Cu, Fe, Ti, Mg, SS) assembly for small tool have been proved by the current authors.

This work aims to increase production efficiency in manufacturing clad composite using the FSB process for surface modification by using large pinless tool.

2. Experimental materials and methods

The materials used were 2 mm and 3mm thick 1060 Al sheet (as top clad) and 2 mm thick mild steel sheet (as substrate). The Al and steel sheets were cut into the size of 120 mm (W) × 100 mm (L) and 60 (W) × 100 mm (L), respectively. The Al and mild steel components were assembled in lap configuration as top clad and bottom substrate, respectively. A commercial Zn foil of 0.03 mm thickness was inserted into the combination as brazing filler metal. The used pinless tool made of a high-speed tool steel of W18Cr4V had a large diameter of 65 mm. Considering the large diameter of tool, a slow rotating speed of 950 rpm was selected. Traverse speed of 95 mm/min, tilted angle of 1.3° and plunge depth of 1 mm (caused by tool large diameter) were adopted. No shielding gas or flux was used during the FSB process.

Friction stir spot brazing (FSSB) was performed on the assembly of steel/Ti (top/bottom) sheet of 2 mm in thickness, in which a pinless tool made of cemented carbide shoulder of 13 mm diameter (embedded in a steel holder) was used.

After FSB or FSSB, joint appearance and microstructure were observed to investigate their feasibility of the new processes.

3. Results and discussion

3.1 FSB of Al/steel for preparing clad composite using large tool (Ø 65 mm)

Fig. 2 shows the FSB joint appearance produced at 950 rpm and 95 min/min using the large pinless tool with a diameter of 65 mm. The surface of weld bead on the Al sheet was smooth and no obvious fragment and flashing was found, which showed perfect weld forming ability. Especially, no crack and adhesion of Al to shoulder was observed. While pushing deformation at advanced side (AS) and several droplets expulsed from interface were observed, showing that the rotating tool has considerable mechanical effects on oxide film disruption by pushed and forged deformation. Compared with general furnace brazing process, the mechanical effect of the tool on oxide film disruption was a key beneficial factor to ensure wetting at interface and to achieve a reliable joint.

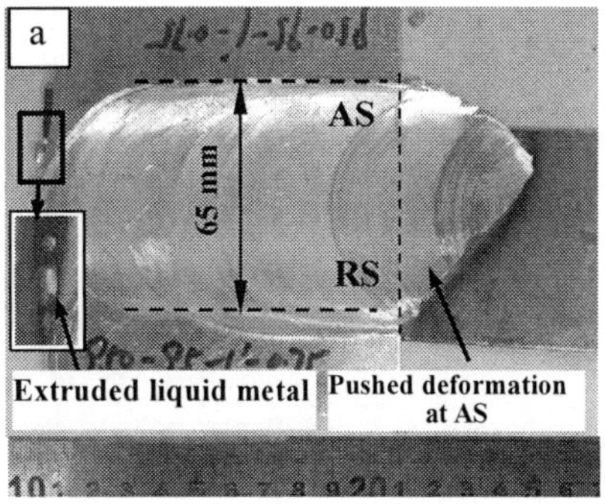

Fig. 2 Smooth FSB joint appearance produced with the large tool of a diameter of 65 mm showing the absence of crack and adhesion and the presence of pushed deformation at AS and droplet expulsed from interface

Fig. 3 shows the microstructure of the FSB joint using the large pinless tool with a diameter 65 mm (950 rpm × 95 min/min). From the center to the edges, the interface can be classified into four regions according to the microstructure feature: direct bonded region (5 mm in the central region), direct bonding and separated IMC particle co-existing region (4 + 4 mm), continuous IMC layer region (16 + 13 mm) and separated region (7 + 11 mm). It can be seen that so wide region (up to 42 mm) was bonded, showing increased production efficiency for one pass. In the bonded regions, no crack and evident Zn were observed, the average thickness of the continuous IMC layer was ~ 16 μm. The absence of evident Zn at central region demonstrated that the frictional heat was sufficient for melting Zn foil braze and the pressure was sufficient to expulse molten Zn. The results would be beneficial to cleaning interface. Additionally, it should be emphasized that a number of disrupted steel particle were observed inside the separated region (see Fig. 3 e~j), indicating that the FSB process was able to disrupt oxide film on mild steel surface by strong pushing and torquing actions of the moving and rotating tool.

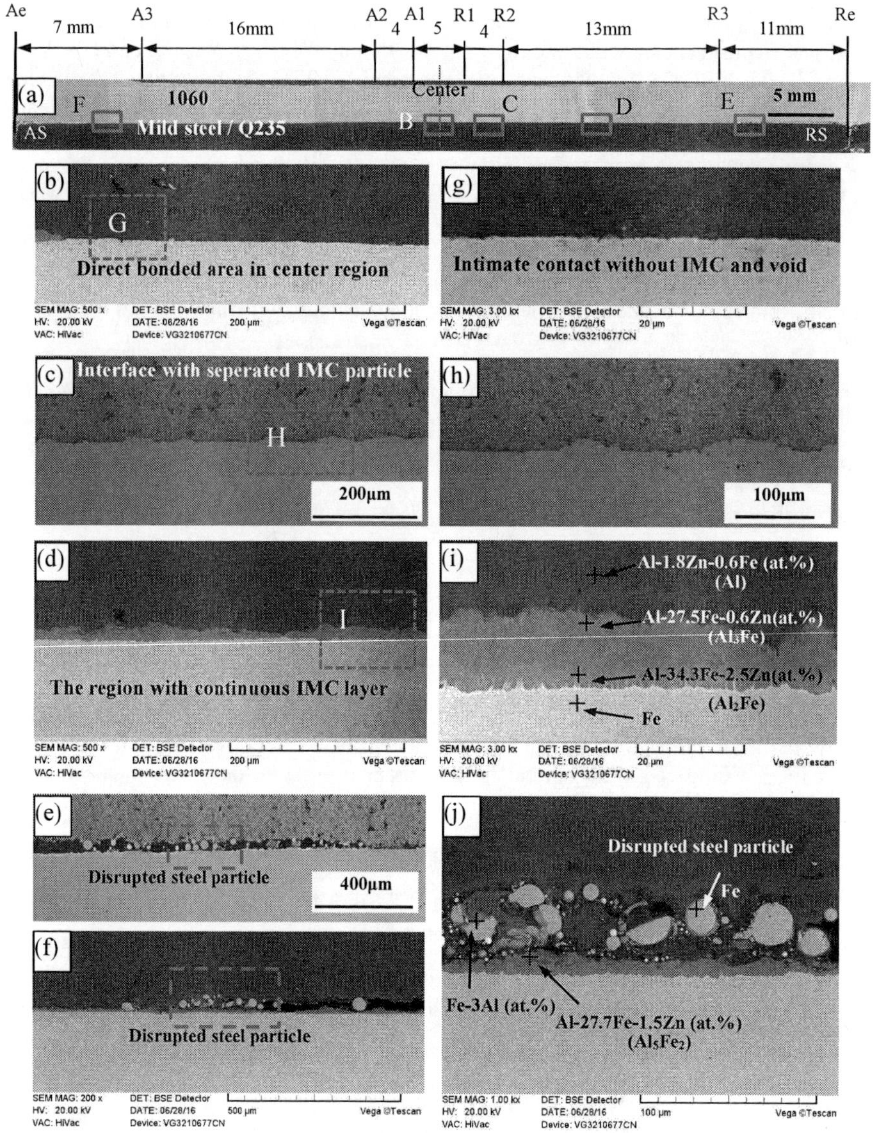

Fig. 3 Microstructure of the FSB joint using the large pinless tool with a diameter of 65 mm (950 rpm × 95 min/min) showing wide bonded region up to 42 mm with different microstructure features

3.2 Friction stir spot brazing (FSSB) of steel/Ti assembly

Fig. 4 shows FSSB steel/Ti joint appearance (Fig. 4a) and microstructure. At Cu/Ti interface, Cu foil was partially melted by eutectic reaction, and the liquid eutectic was partially expulsed (especially at centric position and in periphery ring region), while steel/Cu interface has become dense contact by deformation in solid state (Fig. 4c). The presence of retained solid Cu foil adjacent to steel would weaken the steel/Cu interfacial bonding, but was beneficial to reduce the thermal stress. The joint exhibited tensile shear strength of ~ 77 MPa.

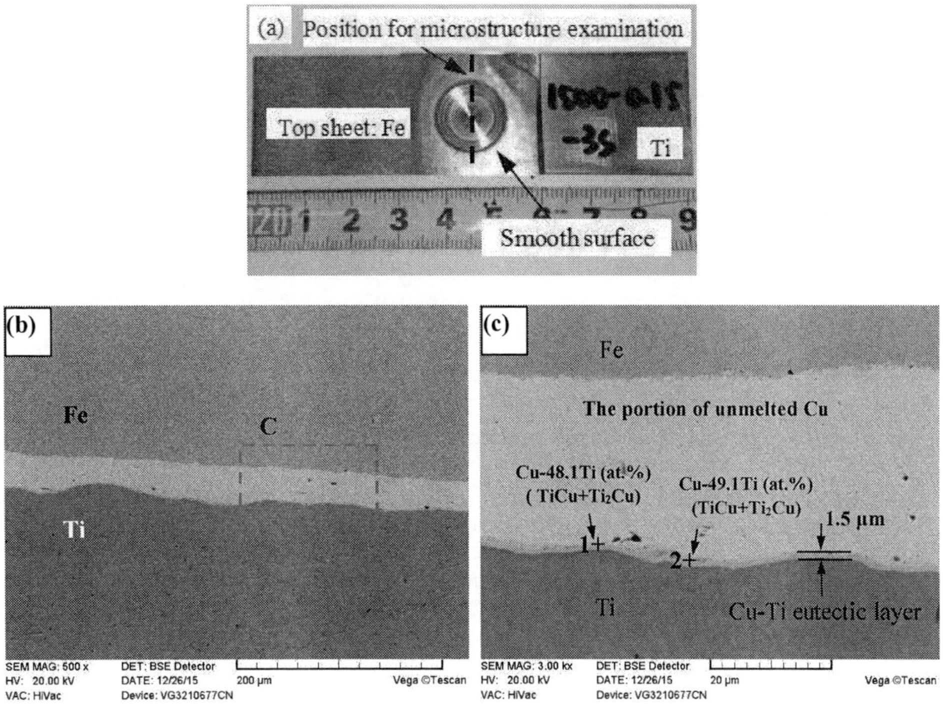

Fig. 4 FSSB Joint appearance and microstructure showing smooth surface and intimate contact at interface (1500rpm×0.15mm×3s)

4. Conclusions

FSB has potentials in producing dissimilar joint (for spot and line joining) and clad composite because of its unique advantages of increased production efficiency for one pass (especially when using large tool) and improved ability to remove oxide film by strong pushing and torquing actions of the moving and rotating tool to disrupt bottom component surface and expulsion oxide fragment with liquid phase.

References

[1] R.S. Mishra, Z.Y. Ma, I. Charit: " Friction stir processing: a novel technique for fabrication of surface composite," *Mater. Sci. Eng. A*, Vol. 341, (2003): pp. 307-310.

[2] Z.Y. Ma: "Friction stir processing technology: a review," *Metall. Mater. Trans. A*, Vol. 39, No. 3 (2008), pp. 642–658.

[3] J. Gandra, H. Krohn, R.M. Miranda, P. Vilac, L. Quintino, J.F. dos Santos: "Friction surfacing-A review," *J. Mater. Process. Technol*, Vol. 214, (2014), pp. 1062-1093.

[4] G.F. Zhang, W. Su, J. Zhang: "Friction stir brazing: a novel process for fabricating Al/steel layered composite and for dissimilar joining of Al to steel," *Metall. Mater. Trans. A*, Vol. 42,(2011) , pp. 2850–2861.

[5] G.F. Zhang, K. Zhang, L.J. Zhang, J.X. Zhang: " Approach to disrupting thick intermetallic compound interfacial layer in friction stir brazing (FSB) of Al/Cu plates," *Sci. Technol. Weld. J.* Vol. 19 No 7 (2014) pp. 554–559.

[6] G.F. Zhang, K. Zhang, Y. Guo, J.X. Zhang: "A comparative study of friction stir brazing and furnace brazing of dissimilar metal Al and Cu plates," *Metall. Microstruct. Anal*, Vol. 3, (2014), pp. 272–280.

[7] G.F. Zhang,W. Su, J.X.Zhang: "A method to fabricate layered composite by friction stir brazing," China Patent ZL 2009 1 021918.3, July 4, 2012.

[8] G.F. Zhang, L. Zhang, C. Kang, J. Zhang: "Development of friction stir spot brazing (FSSB)," *Mater. Des*, Vol. 94,(2016), pp. 502–514.

Surface Modification of Medium-carbon Steel by Friction Stir Powder Processing with Cr Particles

SUN Jiawen, LIU Huihong, MORISADA Yoshiaki, FUJII Hidetoshi
Joining and Welding Research Institute of Osaka University, Osaka, Japan
Email: j-sun@jwri.osaka-u.ac.jp, Phone & FAX: +81-6-6879-8663

Abstract
In this study, the Friction Stir Powder Process (FSPP) was applied to the surface modification of medium-carbon steel with chromium (Cr) powder dispersion. Instead of adopting the conventional Friction Stir Processing (FSP) that uses hard ceramic particles as the reinforcement, we focused on the interaction between the Cr particles and the matrix to develop a novel surface modification method. Cr was chosen as the reinforcement for two reasons. First, Cr diffuses more rapidly in ferrite and austenite than most alloying elements. Therefore, a certain amount of Cr may diffuse into the matrix during the FSPP, even though the processing time is short and the processing temperature is low. Second, Cr has many positive influences on steel, such as improving the toughness of the FSW-ed carbon steel. After the processing, the nano-sized Cr carbides were uniformly dispersed, and the spherical cementites precipitated in the Cr-rich region of the stir zone. It is considered that the diffused Cr promoted the precipitation of Cr carbides as a strong carbide former, and restrained the coarsening of the cementite. Additionally, due to the pinning effect of these small carbides, grain growth of the matrix was restricted in the region where the carbide precipitation occurred. The mechanical properties of the modified region were improved by the FSPP with Cr particles due to the precipitation of nano-sized carbides, the formation of bainites, and the grain refinement. Especially, under the processing condition of 300 rpm tool rotational speed and 50 mm/min tool travel speed, the ultimate tensile strength was enhanced without reducing the elongation by the FSPP using Cr powder.

Introduction
Friction Stir Powder Process (FSPP) is a processing technique that directly originated from Friction Stir Welding (FSW) technology, which has been widely considered as a promising surface modification technique. Generally speaking, FSPP is performed on a groove of the base material, where the powdered reinforcement material was filled. In other words, the reinforcement powder serves as both a filler material and reinforcement source. Advantages like flexibility and economy make the FSPP suitable for not only complicated structures, but also partial surface modification where unnecessary processing efforts need to be avoided for cost reduction or productive convenience. More importantly, the microstructural evolution and mechanical property improvement brought by FSPP are tremendous, i.e. effects like grain refinement and hardness improvement have been proved in many successful applications of producing metal matrix composites by FSPP.

However, it is worth noting that most previous studies concentrated on using hard ceramic particles like SiC, C60, etc., as reinforcement particles, which is a methodology relying on the original hardness of the reinforcement particles although there are derivative effects like grain reduction because of introducing foreign particles. In this study, attention will be paid on the interaction between the reinforcement particles and the matrix to develop a new method for fabricating metal matrix composites by FSPP. Chromium (Cr) powder was chosen for the FSPP for two reasons. First, Cr diffuses more rapidly in ferrite than most alloying elements [1]. It is reasonable to expect that a large amount of diffused Cr could be found in the matrix after the FSPP. Secondly, Cr shows many positive influences on steel. According to a previous study on the effect of Cr on the mechanical property of the FSW joint, the carbon steel S20C joint experienced a significant improvement in toughness with more than 1.5% Cr addition [2]. Therefore, it is expected that a remarkable improvement in toughness could be achieved by directly alloying Cr into carbon steel by FSPP.

Experimental Procedure
Medium-carbon steel S45C was used as the base material. A $1mm^D \times 1mm^W$ (D: depth, W: width) groove was trenched on the surface of the base material plate and filled with Cr powder (mean diameter: 10 μm, 99% purity). Two welding passes were consecutively conducted for improving the dispersion of the Cr powder. In the 1st pass, a flat tool (without probe) made of WCo was used to cover the groove and retain the Cr powder within the groove, while in the 2nd pass, a normal tool (with 1.5mm probe) made by W-Co was used to crush and disperse the Cr

powders in order to facilitate the diffusion of Cr into the matrix. During processing, the probe was inserted into the groove filled with the Cr powder. A constant tool rotating rate of 300 rpm was adopted and the travel speed was varied from 50 to 250 mm/min. The tool tilt angle of 3° was used. The detailed processing conditions are shown in table 1.

Table 1. Processing Conditions for FSPP using Cr powders.

Normal Force (KG)	Rotation Speed (rpm)	Travel Speed (mm/min)	Tilt Angle (Degree)
1750~1200	300	50~250	3

The microstructure was examined by optical microscopy (OM), X-ray Diffraction (XRD), Energy-dispersive X-ray spectroscopy (EDS), Scanning Electron Microscopy (SEM) and Electron backscatter diffraction (EBSD). Transverse sections of the FSPP-ed samples were cut, then mechanically polished for the OM observations. The distribution of the Cr particles and carbides was observed by SEM. The Cr concentration was determined by EDS. XRD was used to identify the type of Cr carbides. The grain sizes of the FSPP-ed samples were evaluated by EBSD. The mechanical property was evaluated by a tensile test.

Results and Discussion

Distribution of Cr Particles

To analyze the distribution of the Cr particles, we observed the stir zone in the compositional mode (BSE) of the SEM and identified the dark black part as Cr particles as shown in Fig. 1. The non-defect stir zone can be obtained under all three process parameters and no discernible defects and porosities were observed. As the travel speed decreased, the Cr particles were much more evenly distributed. Compared to the transverse speed of 250 mm/min, the transverse speeds of 100 mm/min and 50 mm/min resulted in a relatively uniform Cr particle distribution on the surface of the stir zone.

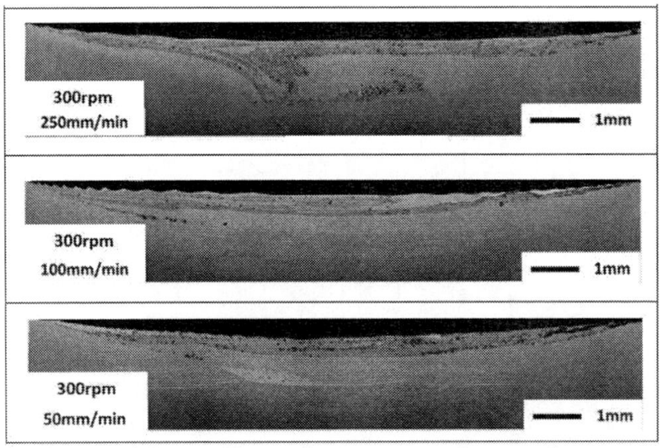

Fig.1 Distribution of Cr particles in the stir zone under varied transverse speeds.

Microstructure Development

As shown in Fig.1, the Cr particles were well dispersed on the surface of the stir zone. In these Cr-rich regions, two main microstructural features were observed; one is the precipitation of nano-sized carbides, and the other is grain refinement. As shown in Fig.2, evenly-distributed nano-sized carbides precipitated in the Cr-rich region of the stir zone. This is because Cr is a strong carbide former and the enthalpies of formation of the Cr carbides are relatively low. [3] The carbides with an average size of around 200 nm were identified as the complex cubic carbide $Cr_{23}C_6$ using XRD. Furthermore, Cr can retard the coarsening of cementite [3] and increase its tendency to spheroidize [4] so that some of the cementites become smaller due to the effect of Cr. Therefore, the addition of Cr has an effect on facilitating the formation of nano-sized carbides in the matrix, which is expected to improve the toughness of the modified layer.

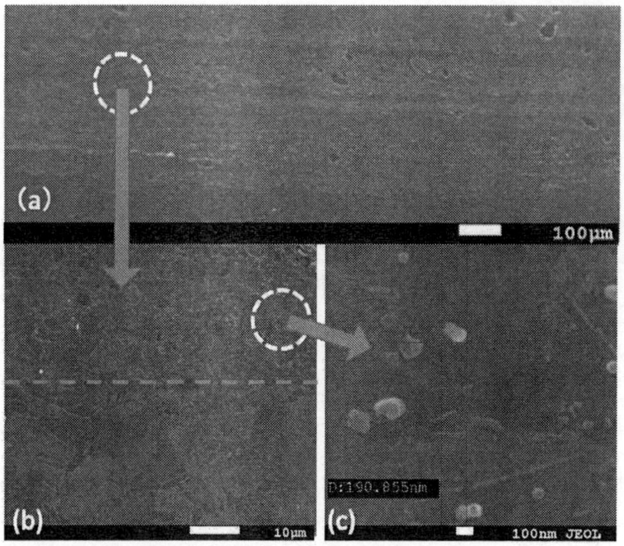

Fig.2. SEM images of nano-sized carbides precipitated in the Cr-rich region.

Additionally, Fig.3(a)(b) showed that there was an significant difference in the microstructure between the Cr-rich region and the Cr-lean region. The microstructure of the Cr-rich region (1.36 mass%) consisted of a ferrite and perlite matrix with uniformly precipitated Cr carbides, while the Cr-lean region (0.4 mass%) consisted of bainite and perlite without carbide precipitation. Furthermore, the EBSD mapping result (Fig.3(c)) showed that the grains in the Cr-rich region were clearly refined compared with those in the Cr-lean region. This grain refinement was attributed to the dynamic recrystallization caused by the FSSP and uniformly precipitated nano-sized carbides which can inhibit the grain growth due to the pinning effect [5].

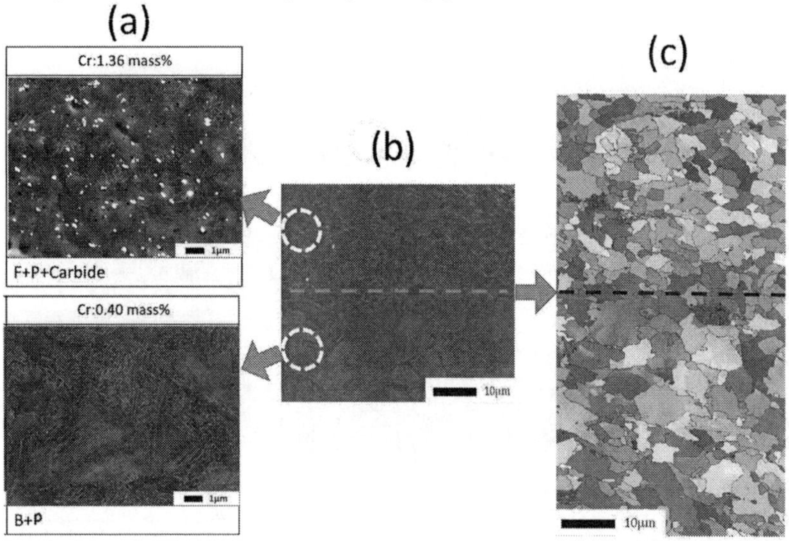

Fig. 3. SEM and EBSD images of the Cr-rich region and the Cr-lean region in the stir zone.

In order to clarify the effect of the small Cr carbides on grain refinement, the grain size distributions under all three processing conditions with and without Cr addition were analyzed and compared. It was clearly shown that all three samples experienced grain refinement after the addition of Cr. Furthermore, without Cr addition, the grain size significantly increased as the temperature increased. However, with the addition of Cr, the grain size increased relatively slightly, which indicated that the grain growth was restrained due to the presence of Cr, even with an increasing temperature.

Table 2. Average grain size under varied processing parameters with and without Cr addition.

Process Parameter (Constant 300rpm)	Average Grain Size	
	Without Cr	With Cr
250mm/min	2.18μm	1.92μm
100mm/min	2.91μm	2.24μm
50mm/min	3.60μm	2.85μm

Mechanical Properties

A tensile test was performed to evaluate the toughness of the modified layer. For the samples with the travel speed of 50 mm/min., it was shown that after the Cr addition, the ultimate tensile strength was improved and the elongation remained at the same level. Furthermore, for the samples with Cr addition, but under different travel speeds of 50 mm/min and 100 mm/min., the ultimate tensile strength was similar for the samples, but the elongation was reduced at the higher travel speed. It is possible that some small defects, which formed due to unevenly distributed Cr powders at the high travel speed, deteriorated the tensile property.

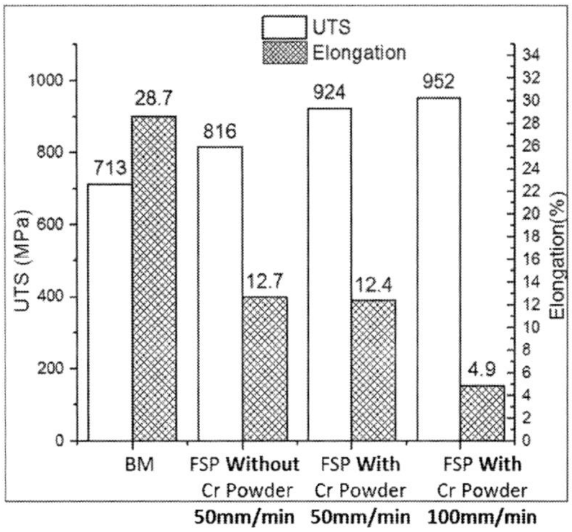

Fig. 4. Tensile properties of the base material and Cr reinforced modified layers under varied process parameters.

Conclusions

A lower travel speed can lead to a more uniform dispersion of the Cr powder, thereby facilitating a better quality being achieved in the modified layer. After adding the Cr powdes into the matrix, a large amount of nano-sized carbides precipitated in the Cr-rich region which led to grain refinement after the process. Overall, it is possible to improve the ultimate tensile strength with little cost in the elongation by adding Cr powders by the FSPP.

References

[1] H.K.D.H. Bhadeshia, *et al*: Steels: Microstructures and Properties, Elsevier Ltd (Burlington, 2006), pp. 200-201.

[2] Shuo. Wu: "Rare metal reduction of high tensile strength steel by friction stir welding," *the National Meeting of JWS*, Nara, Japan, September 2012

[3] G. Zhang: "Effects of Mn, Si and Cr addition on the dissolution and coarsening of pearlitic cementite during intercritical austenitization in Fe-1mass%C alloy," *Materials Characterization*, Vol. 81, (2013), pp. 56-67.

[4] H.K.D.H. Bhadeshia, *et al*: Steels: Microstructures and Properties, Elsevier Ltd (Burlington, 2006), pp. 82

[5] Li, Shu-Xin, et al. "Evaluation of the effect of grain size on chromium carbide precipitation and intergranular corrosion of 316L stainless steel." Corrosion Science 66 (2013): 211-216.

Splat Formation Issue in Thermal Spray Process

Masahiro Fukumoto, Daiki Suzuki and Masayoshi Jinbo
Toyohashi University of Technology, Toyohashi, Aichi, 441-8580, Japan
fukumoto@tut.jp

Abstract

Transition in a flattening behavior of the thermally sprayed metallic materials has been recognized in our previous systematic investigations. Based on the results, a hypothesis, ultra-rapid cooled chill structure preferentially formed at the bottom part of the splat may play an essential role for the generation of the disk splat, has been proposed. Universality of this hypothesis beyond material difference was verified experimentally by using several kinds of ceramic materials, having different thermal properties. To perform this, Al_2O_3, Y_2O_3 and YSZ powder materials were plasma sprayed onto AISI304 stainless steel substrate, and fraction change of the disk splat with substrate temperature increase was investigated followed by the precise observation on the cross-section microstructure of the splats. The results showed that unique amorphous and chill structure were observed in Al_2O_3 and Y_2O_3 splat at their bottom part, respectively, indicating that similar disk formation mechanism may act in these materials. On the other hand, only normal columnar structure was recognized in YSZ splat. As a result, our hypothesis was verified partially regardless of the materials difference.

Introduction

Objective of our series of research [1] on the flattening behavior of a single splat thermally sprayed is to establish the controllability of the thermal spray process. Since an individual particle thermally sprayed is a fundamental element for the coating formation, the flattening behavior of the sprayed particle onto the flat substrate surface has been precisely investigated. The results showed that a transitional phenomenon was recognized in the flattening behavior both with the substrate temperature and ambient pressure changes. That is, the flattening of the splat changes transitionally from a distorted shape with splash to a disk-shaped splat at certain critical temperature and pressure values, respectively. A transition temperature, T_t and transition pressure, P_t were defined and introduced by the author [2], respectively, for those critical conditions. Correspondingly, it was confirmed that the coating adhesive property on the blasted substrate surface changes transitionally on those critical temperature and pressure ranges, respectively. Based on the results, three dimensional transition curvatures by combining both transition curves was proposed for each sprayed material, as a controlling principle for the practical usage in the thermal spray industries [3].

From an academic viewpoint, on the other hand, restraint mechanism for the splashing flow have to be explained based on the fluid dynamics, namely, from the splashing parameter or so-called Sommerfeld Number, *K*. In the observations on the metallic materials, it was indicated that ultra-rapid solidified chill structure preferentially formed at the bottom part of the splat plays an essential role for the disk splat formation [4]. To verify the universality of this hypothesis beyond material difference between metal and ceramic, three kinds of ceramic powder particles with different thermal conductivity, Al_2O_3, Y_2O_3 and YSZ, were thermally sprayed onto the metallic substrate. The effect of solidification behavior at bottom part of the splat on the flattening was investigated precisely in the present study.

Table 1 Metallic and ceramic powder feedstocks

	Cu	Al_2O_3	Y_2O_3	YSZ
Heat conductivity [W·m^{-1}·K^{-1}]	401	30.0	14.5	3.0
Melting point [K]	1,358	2,345	2,647	2,715
Specific heat [kJ/kg·K]	0.39	0.80	0.47	0.46
Specific gravity	8.9	3.9	5.0	6.0

Experimental Procedures

Commercially available metallic and ceramic powders with diameter of several tens micrometers were used as feedstock powders as shown in Table 1. AISI304 stainless steel plates with dimensions of 25 mm × 25 mm × 6 mm finally polished with 0.3 μm

Al_2O_3 buff were used as substrates. Sprayed particles were collected on the substrate surface using both atmospheric and low pressure plasma spray equipment. The splat shape was classified and the number of both disk splat and splash splat were counted on every trial. The fraction of the disk splat was defined as the ratio of the number of disk splat to the total splat. The cross section microstructure of the splat was precisely observed by means of a scanning ion microscope, SIM and transmission electron microscope, TEM after cutting the splat by focused ion beam microscope, FIB.

Results and Discussion

Transition Mechanism in Metallic Materials

In general, it is well known that the break-up phenomenon of the liquid film generated by the collision of the liquid particle onto the flat solid surface can be evaluated by the splashing parameter, K, in the fluid dynamics field [5]. K is defined as $We^{0.5}Re^{0.25}$, and it is based on the in-flight kinetic information of the liquid particle. Here, $We = \rho d v^2/\gamma$, and $Re = \rho d v/\eta$: where ρ is density, d is diameter, v is velocity, γ is surface tension and η is viscosity, respectively. Here, K has an empirically decided critical value $Kc = 57.7$, and if the K value of the particle exceeds Kc, the liquid film shows the break-up after the collision onto the solid surface. To verify the actual K value of the regular thermally sprayed particles, in-flight measurement of the sprayed particles was conducted by using the DPV-2000 system. By introducing the measured temperature into each physical constant (i.e., density, surface tension, and viscosity), the actual K value was calculated for a few kinds of sprayed metallic materials. As a results, K values were remarkably larger compared to Kc regardless of the sprayed material [3]. It means that the regular thermally sprayed particles have enough higher driving force for the splashing. The results indicate that the essential point of the transition issue in this case lies in not why and how the splashing occurs on the cold substrate surface, but why and how the disk splat appears on the hot substrate surface or in the reduced pressure condition instead of the splash splat.

To verify the transition mechanism in metallic materials, solidification microstructure on the bottom surface of the splat was precisely observed. On the bottom surface of the splat, after peeled off from the substrate surface, obtained at heated substrate or reduced pressure conditions indicated that ring shaped dense zone with fine microstructure was observed near peripheral regions. This fine microstructure near peripheral region indicates that rapid solidification might be occurred at this region. To confirm the solidification inside of the splat directly and to elucidate the effect of rapid solidification on the flattening, cross section microstructures of the splats were precisely observed by SIM. The results were summarized and shown in ***Fig. 1***.

In the splash-typed splat collected on the room temperature substrate under atmospheric pressure (***Fig. 1(a)***), coarser columnar grains perpendicular to the substrate surface are observed in whole part of the splat. On the other hand, in the disk-typed splat collected on the 773K heated substrate, finer columnar grains inside of the splat are observed in whole part of the splat (***Fig. 1(b)***). Finer structure indicates that faster solidification occurred in the disk splat given on the heated substrate. However, since it is well known that the columnar solidification structure itself is generally formed after finishing the flowing of the liquid, it is found that the solidification inside of the splat never affect the flattening of

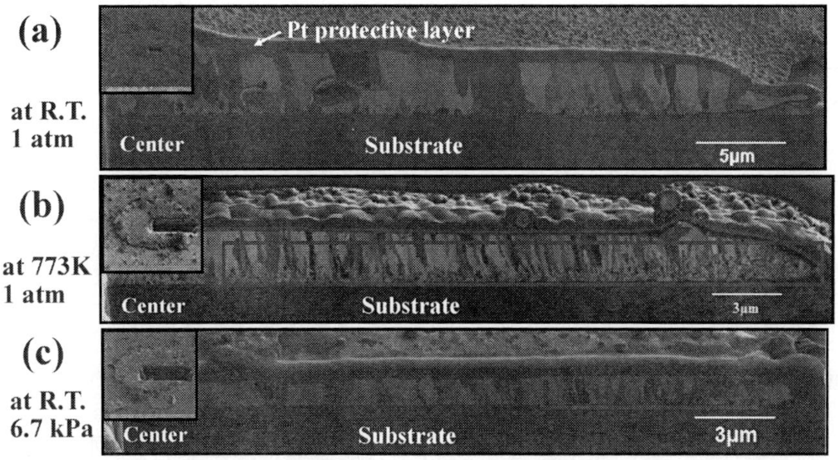

Figure 1. Cross section microstructures of plasma sprayed Cu splats obtained on AISI304 stainless steel substrate at a) room temperature substrate under atmospheric pressure, b) 773K heated substrate under atmospheric pressure, c) room temperature substrate under 6.7 kPa pressure conditions.

the splat. Hence, difference in the mean cooling rate, and solidification behavior inside of the splat cannot become a reason for the difference of the splat shape.

More precise observation was conducted on the cross section microstructures of the splats, and a layer composed with equiaxial fine grain, so-called chill crystal structure which usually caused by ultra-rapid cooling, was found at the bottom part of the splat, indicated as blue rectangular shown in ***Fig. 1(b) and (c)***. The essential point is that chill crystal layer appeared on the heated substrate is much thicker than that on the unheated substrate. Moreover, the thickness of this ultra-rapid solidified chill layer becomes thicker at near the periphery region of the splat compared with the central region. This indicates that ultra-rapid cooling and resultant ultra-rapid increase of the viscosity and/or solidification occurred near the peripheral region and this ultra-rapid material change might act as the restrain for the splashing. This is exactly a new finding found in the relating research field.

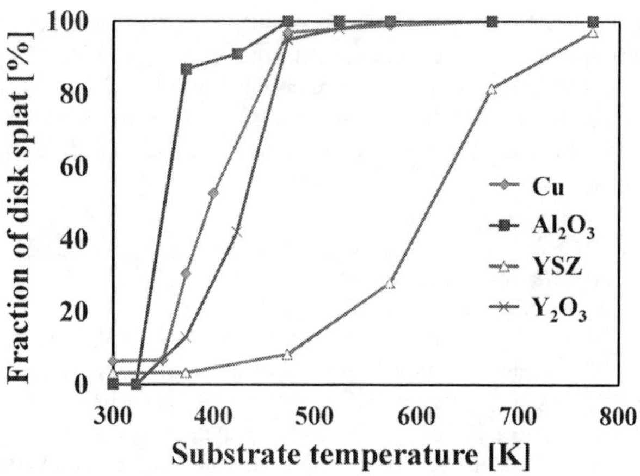

Figure 2. Change of fraction of disk splat with substrate temperature increase for each material.

Transition Mechanism in Ceramic Materials

Universality of the new finding, namely ultra-rapid solidified chill structure in the splat has a restraint role for the splashing, obtained in the metallic materials was verified in the ceramic materials, since the similar transition behavior is recognized also in the ceramic splats. The change of the fraction of disk splat with substrate temperature increase for each sprayed ceramic material was summarized in ***Fig. 2***, including the data on Cu particle. It was found from the figure that the ceramic material with higher thermal conductivity has a tendency having a lower transition temperature. The result shows that the ceramic material having a higher heat conductivity tends to show a transitional behavior more easier, indicating that a certain solidification property of the material may affect the flattening. To verify this, the cross-section microstructure of the disk splats for each ceramic material collected on the heated substrate surface was observed. The results were summarized and shown in ***Fig. 3***.

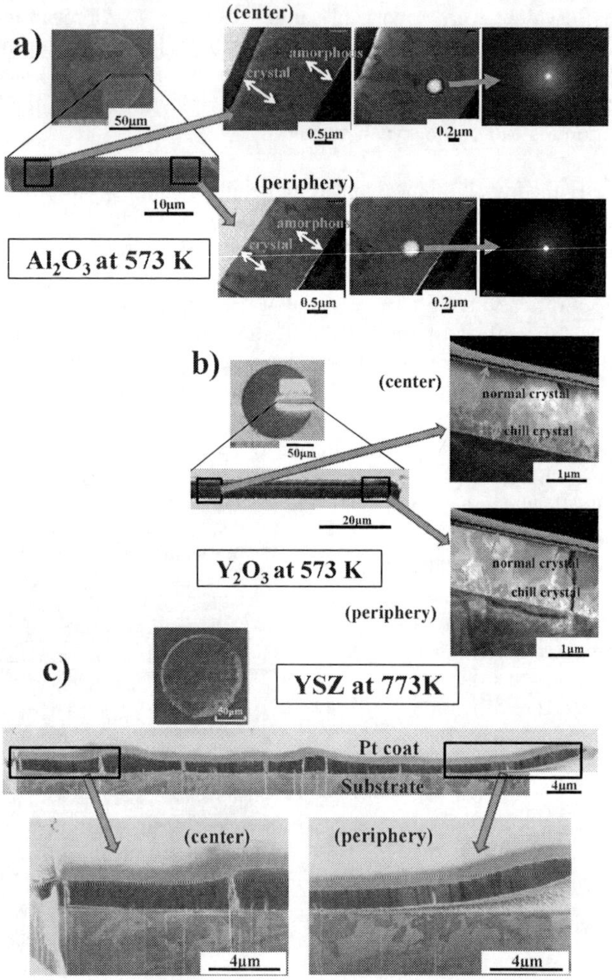

Figure 3. Cross section microstructures of plasma sprayed ceramic splats obtained on heated AISI304 substrate surface. a) Al_2O_3, b) Y_2O_3, c) YSZ.

In Al_2O_3 splat shown in ***Fig. 3(a)***, having the highest heat conductivity, an amorphous layer, mostly half thickness of the splat, was clearly observed at bottom part of the splat entirely whole part in the radius direction. In case of Y_2O_3 shown in ***Fig. 3(b)***, on the other hand, having a middle value of the heat conductivity, chill crystal layer similar to the metallic materials at bottom part of the splat was clearly observed entirely whole part in the radius direction. In these two materials, ultra-rapid solidification at bottom part of the splat may cause the disk splat, namely, the result supports the above hypothesis given in the metallic materials. However, in ZrO_2 splat shown in ***Fig. 3(c)***, only the columnar structure similar to the metallic material was observed without any typical ultra-rapid solidified microstructure at bottom part of the splat, even in the disk splat. In this case, solidification cannot be used for the explanation in transition behavior. Some other mechanism like rapid viscosity increase due to the ultra-rapid cooling has to be adopted for the explanation. Moreover, the intrinsic meaning of the relationship between thermal conductivity of each ceramic material at room temperature range and the solidification tendency in the splat as shown in ***Fig. 3*** is not clear at this moment. More precise, systematic observation has to be done in the future study.

Conclusions

The hypothesis, namely, ultra-rapid solidification at bottom part of the splat may act as the restrain for the splashing, was verified through the observation on the flattening behavior of several kind of the ceramic materials. Since an exception was observed in case of ZrO_2 splat, however, more precise, systematic verification has to be conducted in the future study.

References

[1] P. Fauchais, M. Fukumoto, A. Vardelle and M. Vardelle: "Knowledge concerning splat formation: An invited review", *J. of Thermal Spray Technology,* **Vol. 13, No. 3** (2004) pp.337-360

[2] M. Fukumoto, S. Katoh and I. Okane: "Splat Behavior of Plasma Sprayed Particles on Flat Substrate Surface", *Proceedings of 14th International Thermal Spray Conference*, Osaka, May 1995, pp. 353-358.

[3] M. Fukumoto, M. Shiiba, H. Kaji and T. Yasui: "Three-dimensional transition map of flattening behavior in the thermal spray process", *Pure Appl. Chem.*, **Vol. 77, No. 2** (2005), pp. 429-442.

[4] M. Fukumoto, D. Suzuki and N. Maeda: "Flattening mechanism in thermally sprayed ceramic particles", *Proceedings of 7th Asian Thermal Spray Conference*, Xi'an, November 2015, CD.

[5] Chr. Mundo, M. Sommerfeld and C. Tropea: "Droplet-wall collisions: Experimental studies of the deformation and breakup process", *Int. J. Multiphase Flow*, **Vol. 21, No.2** (1995), pp. 151-173.

Acknowledgments

This research was supported partly by the Grant-in-Aid for Scientific Research of the Ministry of Education, Science, Culture and Sports in Japan (#25249095).

Surface Modification of Steel by YAG Laser Irradiation

Mingzhong Tang, Tomiko Yamaguchi
Kyushu Institute of Technology, Kitakyushu, Fukuoka, Japan
Email: p347831m@mail.kyutech.jp phone: 070-1947-5678

Kazumasa Nishio
Nishinippon Institute of Technology, Kitakyushu, Fukuoka, Japan

Abstract
YAG laser surface modification of S45C steel in surrounding of N_2 or Ar gas was studied. The microstructure, phases and properties of the surface were investigated by X-ray diffraction (XRD), field emission -electron probe micro analysis (FE-EPMA), optical microscope (OM), Nano-indentation hardness test and vickers-hardness (HV). The results showed that N was identified in any of the samples on the laser irradiation portion surface. With the increase of the irradiation times, the surface irradiation layer, which contains a small amount of nitrides, became thicker and thicker. In addition, the variation of Nano-indentation hardness from surface to substrate showed that surface Nano-indentation hardness was greatly improved. After quenching, laser irradiation on pre-coated laser non dross 371 in the nitrogen atmosphere, and laser cladding in the argon atmosphere, S45C steel's surface hardness was greatly improved, which became 837HV, 955HV and 985HV, respectively,

1. Introduction

Currently, to improve the surface hardness and wear resistance of steels, many surface modification methods have been carried out, such as PVD, CVD, carburizing, nitriding, and so on. Among them, the nitriding methods include salt bath nitriding, gas nitriding and plasma nitriding. However, these conventional methods are usually used to treat all the surface, they cannot selectively process a specific site of materials.

A continuous wave (CW) Nd:YAG laser has various advantages, such as a small spot size, a higher energy density and a higher energy absorption rate of the workpiece. Particularly, YAG laser surface nitriding, with the high directionality, the capability for rapid processing, net-shape manufacturing, and low heat effect on the substrate, has aroused extensive study. In addition, laser cladding, as a new surface strengthening and repair technology, has advantages such as limited heat affect zone, small stress deformation, low dilution ratio and better metallurgical bonding with the substrate.

Therefore, in this study, the methods of forming a surface modification nitride layer on S45C steel by YAG laser irradiation have been investigated.

2. Experimental procedures

S45C steel was used as the substrate material with a dimension of 90×30×6 mm. The surface of samples was polished using emery paper (# 1200) to remove the oxide layer and then cleaned with acetone. On one hand, in order to improve the absorption of the laser irradiation beam, the samples were pre-coated with a laser non dross 371 (Ishihara Chemicals Co., Ltd. Japan) on the surface, its chemical composition as shown in Table 1. On the other hand, about 200μm thicknesses mixed powders (SHA: BN=9.5: 0.5) were pre-placed onto the substrate surface, then dried at 60 ℃ for 4 hours before laser cladding. The chemical composition of SHA is listed in Table 2, and its particle diameter is -45μm. A YAG laser processor (MW2000, Sumitomo Heavy Industry, Japan) in continuous wave mode was used for the laser irradiation. Experimental conditions of laser irradiation is shown in Table 3. What's more, laser irradiation was done in a nitrogen and argon atmosphere. Schematic diagram of laser irradiation is shown in Fig.1.

Table 1: Chemical composition of Laser Non-doross 371(mass %)

C	CH_2CH_3OH	$CH_3CH_2CH_2CH_3$
5-10	35-45	20-30

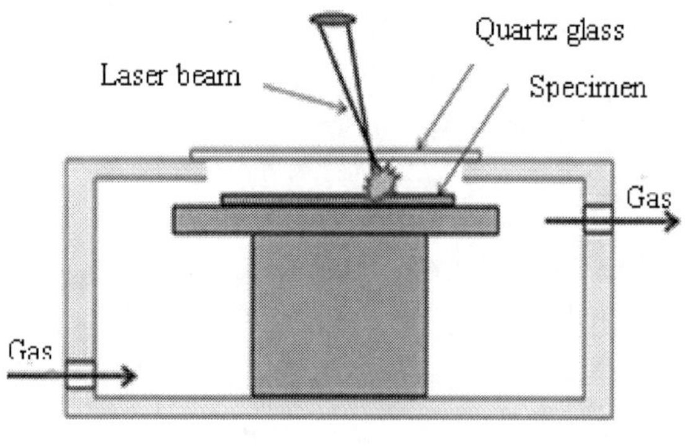

Fig. 1: Schematic diagram of laser irradiation

Table 2: Chemical compositions of SHA (mass %)

C	Si	Cr	Mo	Ni	Fe
4.99	0.99	9.84	4.92	4.83	Bal.

Table 3: Experimental conditions of laser irradiation

Samples	Power (kW)	Focal distance (mm)	Traveling speed (mm/min)	Gas flow speed (L/min)	Types of gases	Surface preprocess	Irradiation times
S45C	1.5	45	500	20	N2	Laser non dross 371	5, 10, 15
	1.5	15	100, 300, 500	20	Ar	Fe-BN	1

The microstructure was observed using an optical microscopy (OM, Nikon-L150). The surface chemical composition was analyzed using X-ray diffraction (XRD, JDX-3500 K). The elemental distribution was determined by an electron probe micro-analyzer (EPMA, JXA-8900R). The hardness distribution along the laser irradiation layer depth direction was measured by a Nano-indentation hardness testing device (ENT-110a) with a Berkovich indenter, using a load of 0.49 N at a step interval of 20 msec. The hardness distribution along the laser cladding layer depth direction was measured by a Vickers-hardness tester (HV, MHT-1), with a load of 4.9 N, a holding time of 15 s, and an interval of 0.1 mm. Finally, in order to testify if there is an obvious improvement after the laser irradiation on pre-coated laser non dross 371 samples in the nitrogen atmosphere, S45C steel's surface hardness was tested by a Vickers-hardness tester after quenching and laser irradiation.

3. Results and discussion

3.1 XRD and EPMA analysis

The surface XRD patterns of laser irradiation layer is shown in Fig. 2. The FE-EPMA mapping analysis results of laser irradiation 10 times' layer are shown in Fig. 3. The XRD and FE-EPMA results indicate that N is identified in any samples on the laser irradiation portion surface. The pre-coated laser non dross 371 enhanced the absorption of the laser beam, the sufficiently absorbed energy made nitrogen diffuse into the steel. According to nitrogen content and Fe-N phase diagram, it can be inferred that solid solution α-phase of nitrogen is formed on the laser irradiation portion surface.

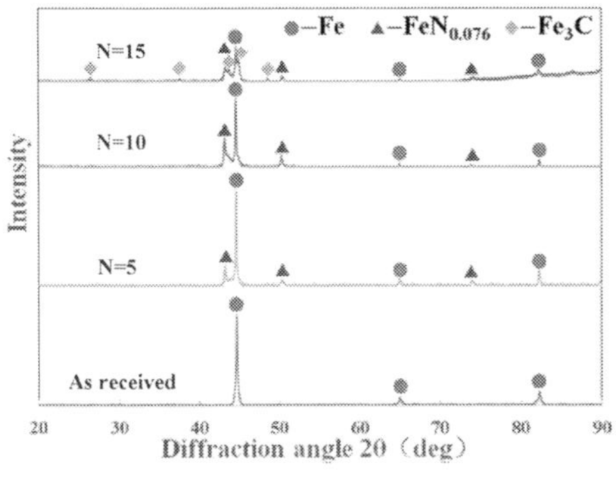

Fig. 2: XRD results of samples as received and different irradiation times

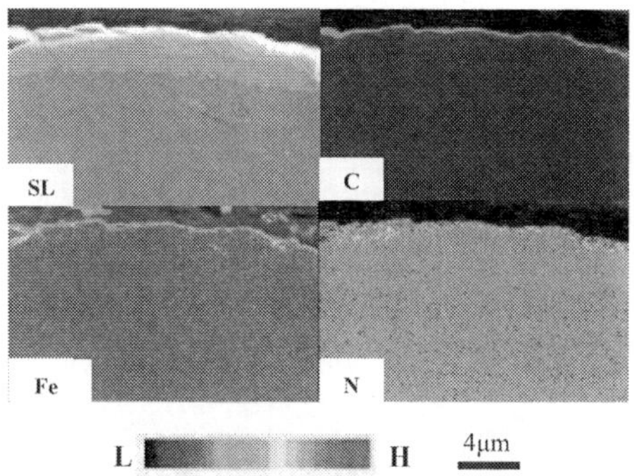

Fig. 3: FE-EPMA analysis results of laser irradiation 10 times' layer

Fig. 4 shows the surface XRD patterns of laser cladding layer. The reason of identified Fe, N or B compounds on laser cladding surface is also due to the high laser energy, which separated BN and recombined the Fe, N or B compounds.

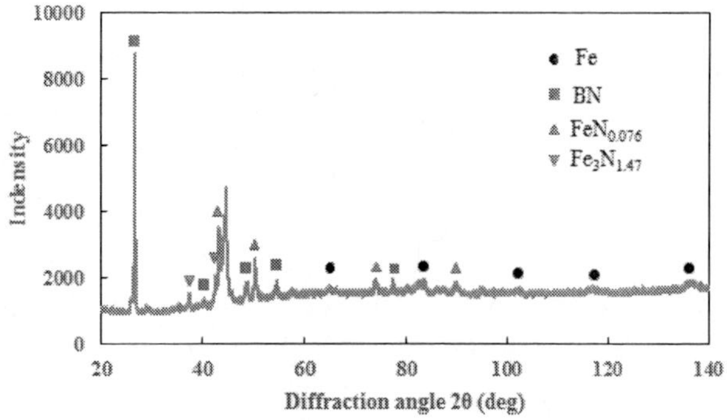

Fig. 4: XRD results of laser cladding layer samples

3.2 Microstructure

Fig. 5 shows that microstructures of samples as received and different laser irradiation times on the pre-coated laser non dross 371 samples. It can be observed that with the increase of the irradiation times, the surface irradiation layer, which contains a small amount of nitrides, becomes thicker and thicker. When the laser irradiation times are 5, 10, 15 times, the thickness of surface modification layer is about 20, 30, 40μm, respectively. When the laser irradiation times are 10 times, it can be observed that there are almost no cracks and pores in the surface modification layer. When the laser irradiation times are 15 times, a few of cracks are formed on the surface. It might be that in the rapid heating and cooling conditions (10^5~10^6℃/s), stress relief makes the surface form microcracks. Then, with the increase of laser irradiation times, microcracks become bigger.

In addition, below the surface modification layer, needle-plate-like martensite can be observed obviously. The rapid heating conditions (10^5~10^6℃/s), significant of overheating degree cause the large phase transformation driving force ($\triangle G^{\alpha\to\gamma}$), it increases the number of crystal grain, simultaneously, for the sake of the rapid cooling after heating, the fine martensite is too late to grow up, which has a good influence on the high hardness and excellent wear resistance.

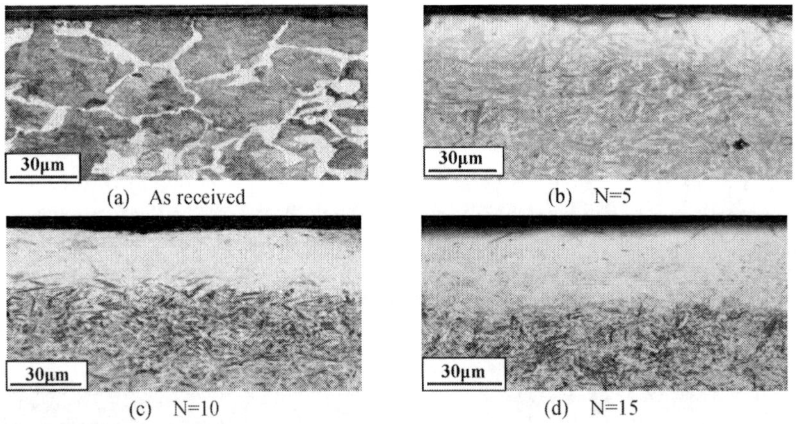

Fig. 5: Microstructures of samples as received and different laser irradiation

Fig. 6: Microstructure of Fe-BN sample by laser irradiation

Fig. 6 shows the cross sectional morphology of laser cladding sample observed by the optical microscope. The cross section of laser cladding sample could be divided into three parts, including the laser cladding layer, the heat affected zone (HAZ) and the non-heat affected zone (Non-HAZ). Among them, the laser cladding layer's thickness is about 400μm, and the heat affected zone's (HAZ) thickness is about 600μm.

3.3 Hardness

Fig. 7 shows cross sectional Nano-indentation hardness of the pre-coated laser non dross 371 samples as received and different laser irradiation times. With the increase of sample depth, the hardness becomes lower and lower. The variation of Nano-indentation hardness from surface to substrate indicates that surface Nano-indentation hardness was greatly improved.

Fig. 8 shows cross sectional Vickers-hardness of laser cladding samples as received and different laser irradiation speeds (100 mm/min, 300 mm/min and 500 mm/min). As shown in Fig. 8, in any of the torch moving speed, the surface hardness of the samples is almost the same, which is about 985HV. When the torch moving speed becomes higher, the hard layer becomes thinner.

It can be observed that the hardness, below the surface modification layer, is also improved by about three times. Because there are fine crystal grains and needle-plate-like martensite in this area, which has a positive impact on high hardness of the samples.

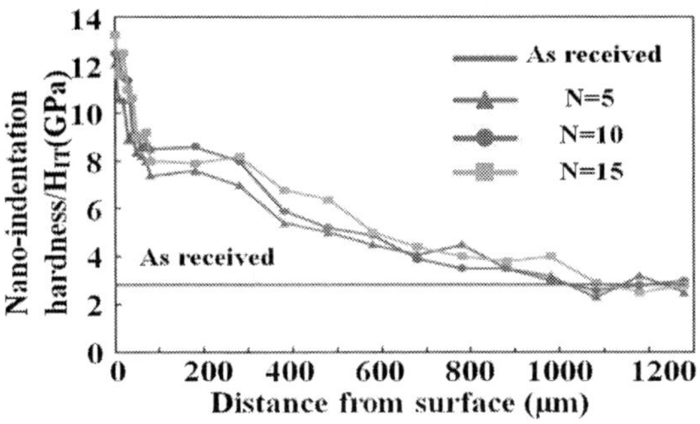

Fig. 7: Cross sectional hardness of Laser non dross 371 samples as received and different laser irradiation times

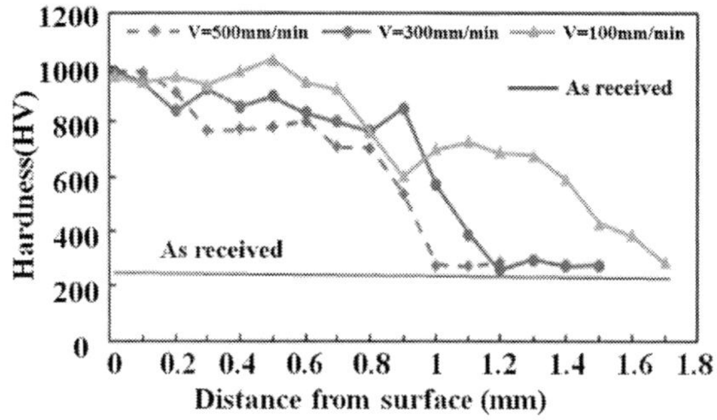

Fig. 8: Cross sectional hardness of Fe-BN samples as received and different laser irradiation speeds

Additionally, after quenching and laser irradiation on pre-coated laser non dross 371 in the nitrogen atmosphere, S45C steel's surface hardness becomes 837HV and 955HV, respectively, which means that there is an improvement after the laser irradiation on pre-coated laser non dross 371 in the nitrogen atmosphere more than quenching. It can be inferred that the improved hardness is not only due to martensite, but also a small amount of nitrides.

4. Conclusions

(1) N is identified in any of the samples on the laser irradiation portion surface, because of the high laser energy.

(2) With the increase of the irradiation times, the surface irradiation layer becomes thicker and thicker. Due to the rapid heating and cooling conditions (10^5~10^6℃/s), fine crystal grains and needle-plate-like martensite can be observed.

(3) After quenching, laser irradiation on pre-coated laser non dross 371 in the nitrogen atmosphere, and laser cladding in the argon atmosphere, S45C steel's surface hardness was greatly improved. Among them, laser cladding samples' hardness is the highest, which is 985HV.

References

[1] D.Höche, J.Kaspar, P.Schaaf: "Laser Surface Engineering," Woodhead Publishing (UK, 2015), pp. 33-58.

[2] Hui Zhang, Yong Zou, Zengda Zou, Dongting Wu: "Microstructure and properties of Fe-based composite coating by laser cladding Fe–Ti–V–Cr–C–CeO2 powder," Optics & Laser Technology, Vol. 65, (2015), pp. 119-125.

[3] P.P. Psyllaki, A.Griniari, D.I.Pantelis: "Parametric study on laser nitriding of 1.5919 steel," Journal of Materials Processing Technology, Vol. 195, (2004), pp. 299-304.

[4] WU Gan, WANG Rui, YANG Jang-ping: "Study of laser nitriding on the GCR15 steel surface," Physics Procedia, Vol. 18, (2011), pp. 285-290.

[5] Satyajit Chatterjee, P. Ganeshc, R. Palai: "Effect of h-BN addition on the properties of nanostructured Al2O3–TiB2–TiN based coatings developed by combined SHS and laser surface alloying," Surface & Coatings Technology, Vol. 204, (2010), pp. 1702–1709.

Effect of additive materials compositions on laser weak-magnetic treatment

Shinya Okamoto, Xudong Zhang
Hitachi, Ltd., Hitachi-shi, Ibaraki-ken, Japan
shinya.okamoto.zw@hitachi.com

Nobuaki Kobayashi, Takahiro Saito, Akihiro Yamazaki
Hitachi Automotive Systems, Ltd., Isesaki-shi, Gunma-ken, Japan

Abstract
Ferromagnetic stainless steels, including AISI 430 stainless steel, are widely used for electromagnetic-controlled components due to their strong magnetic properties. In some conditions, partial weak magnetic property is required in order to reduce the leakage of magnetic flux and improve the responsiveness of sliding part. It is well known that the chemical composition, especially Ni and Cr, has significant influence on the microstructure of stainless steel, which will influence the magnetic property. Therefore, it is possible that partial weak-magnetic or non-magnetic property can be realized by adjusting the composition at a local area. In this paper, a weak magnetic treatment process was developed by using a laser local modifying process. A 2 kW fiber laser was utilized to conduct treatment procedure. AISI 430 stainless steel pipes were used as test pieces. Four kinds of stainless wires, pure Ni, Ni-Cr alloy, AISI 316 and AISI 310 stainless steels, were used as the additive materials. The laser power and the wire feed rate were changed in the experiment. After the laser treatment, the chemical compositions and microstructures of the laser treatment range were observed. The saturated magnetizations were also measured for the evaluation of magnetic property of laser treated range. The results showed the microstructure and saturated magnetizations varied with different wire compositions and laser treatment conditions. In specific, at appropriate treatment conditions, the microstructure of laser treated range could be changed from a full ferritic microstructure to a ferritic-austenitic mixed or a full austenitic microstructure. Furthermore, the saturated magnetization of laser treatment range was reduced to 0.07 T from the original 1.6 T for the AISI 430 stainless steel pipe. The mechanism of laser weak-magnetic treatment was discussed.

Introduction
AISI 430 stainless steel is being widely used for architectural material and automotive material as a kind of ferromagnetic stainless steels. The main composition of AISI 430 stainless steel is that about 17 %Cr is included. It combines high stress corrosion resistance [1-2]. Many electromagnetic-controlled components, which are usually made with AISI 430 stainless steel, are required to get more rapid responsiveness. In this case, partial weak magnetic property is required in order to reduce the leakage of magnetic flux and improve the responsiveness of the components.

In order to obtain a partial weak or non-magnetic property for a ferromagnetic material, three methods can be considered. One is to insert a non-magnetic material, such as AISI 304 austenitic stainless steel, between two pieces of AISI 430 stainless steels. Another is composed of two steps. The first step is to convert a non-magnetic material, such as AISI 304 austenitic stainless steel, from non-magnetic into ferromagnetic by cold-working. The second step is to reconvert a part of the converted ferromagnetic material by partial heating. The other is to make laser treatment with adding some non-magnetic materials to AISI 430 stainless steel. In order to obtain a partial weak magnetic range inside a full ferromagnetic material, the ferrite content of the laser treated range should be reduced. According to the Schaeffler diagram of stainless steels, the ferrite content varies with the chemical compositions, in specific, the ratio of Cr equivalent and Ni equivalent has significant effect on the ferrite contents [3]. Although nickel and nitrogen is a kind of strong austenitic elements, the nitrogen content solved in the melt metal of ferritic stainless steel is very low [4]. Therefore, Ni is considered as the main additive element to realize weak magnetic treatment. In this paper, four kinds of materials including Ni are selected as the additive materials, pure Ni, Ni-Cr alloy, AISI 316L and AISI 310S stainless steels.

For weak magnetic treatment, a laser beam was used to form melting pool and then add the additive material into the melt pool. High heating and cooling rate, which can obtain low distortion, is the reason for the using laser beam as the heat source. In this paper, a 2 kW fiber laser was utilized to conduct treatment procedure. AISI 430 stainless steel pipes were used as the test pieces and four kinds of metal wires, pure Ni, Ni-Cr alloy, AISI 316L and AISI 310S stainless steels, were used as the additive materials. The chemical compositions and microstructures of the laser treatment range were analyzed. The saturated magnetizations were also measured for the evaluation of magnetic property of laser treated range. The mechanism of laser weak-magnetic treatment was discussed.

Experimental

AISI 430 stainless steel pipe, whose sizes were about 6 mm in the diameter and 70 mm in length, were used as test pieces. Table 1 shows the chemical composition of AISI 430 stainless steel pipe (Base metal) and four kinds of additive wires. For pure Ni, Ni-Cr alloy and AISI 316L stainless steel wires, the diameter was 0.6 mm in diameter. AISI 310S stainless wire was 0.8 mm in diameter. The laser utilized for conducting experiment was a fiber laser, which had a maximum laser power of 2 kW. The beam parameter product (BPP) was 4.5 mm*mrad and the core diameter of fiber was 0.1 mm. A laser irradiation head contained a collimation lens and a set of lens which transformed a circular beam profile into a line-shaped beam profile. The collimation lens had a focal length of 125 mm. The size of the line-shaped beam had a 5.0 mm in width. Before conducting the investigation of effect of wire composition on weak-magnetic treatment, a preliminary test was conducted by using the Ni wire. The effect of process parameters, including laser power, laser beam moving speed, wire feed speed, wire feed angle and position, on the process stability was preliminarily investigated with plate material. The image observed by a high speed CMOS camera shows that a stable weld pool with a width of about 2.0 mm (less than the laser beam width (5.0 mm)) was formed and the wire tip was fed in the center of weld pool from the front of laser beam. In this experiment, nitrogen gas was used as the shielding gas to protect the melt pool from oxidization. After the appropriate process condition was established, four kinds of wires were separately used to perform weak-magnetic treatment. All the process conditions were the same for the four kinds of wires, in spite of AISI 310S stainless steel having its diameter of 0.8 mm. The processes were conducted by the conditions that laser power was 0.8 ~ 2 kW and wire feeding speed was 700 mm/min. The laser beam was moved in a speed of 1500 mm/min and the wire was fed from the front of the laser beam. The angle between the wire and laser beam was 60 degrees. After weak-magnetic treatment, all of the treated samples were cut to observe the geometry and microstructure of the treatment range. For the observation of the treatment range, the etching of base metal (AISI 430) by nitro-hydrochloric acid and the electrolytic etching of melted metal by 10 % chromic oxide aqueous solution were performed. The microstructures were observed by using an optical microscope. The dilution rate of the laser-melted metal was calculated based on the ratio of additive wire and the melted base metal volumes. Then the chemical composition of the laser-melted metal was calculated on the basis of the dilution rate and the chemical compositions of base metal and additive wires. In addition, the Cr equivalent and the Ni equivalent of the laser-melted metal was calculated on the basis of the chemical composition of it. In order to evaluate the weak-magnetic treatment results, a vibration sample magnetometer was used for measuring of the saturated magnetizations. The laser-melted metal was cut to a piece of about 1.5 mm in width. The saturated magnetization was measured with vibrating the piece in applied magnetic field. Magnetic flux density in the case that the intensity of the magnetic field was about 1.6×10^6 A/m was defined as the saturated magnetization of the laser-melted metal. The saturated magnetization of it was used to be discussed.

Table 1: Chemical compositions of base metal and additive wires

Unit: mass%

Elements	C	Si	Mn	P	S	Ni	Cr	Mo	Cu	Ti	Fe
AISI430 stainless steel pipe (Base metal)	0.01	0.44	0.25	0.029	0.001	0.19	16.49	-	-	0.20	Bal.
Pure Ni wire	0.03	0.10	0.20	-	0.002	99.40	-	-	<0.10	-	0.20
Ni-Cr alloy wire	0.03	1.25	<0.01	-	-	78.4	19.79	-	-	-	0.01
AISI 316L wire	0.01	0.49	0.85	0.025	0.000	12.13	17.50	2.06	-	-	Bal.
AISI 310S wire	0.04	0.64	1.29	0.022	0.000	19.31	25.62	-	-	-	Bal.

Results and Discussion

Chemical composition of melted metal

The dilution rate for the four kinds of wires was considered almost same as 60 ~ 65 % from the cross-section geometry of laser melted area. Then the chemical composition of the melted metal was calculated based on the dilution rate. The Cr equivalent and Ni equivalent were calculated by using the chemical composition of the melted metal. Based on the Cr equivalent and Ni equivalent, in addition to the each calculated dilution rate, the location of each additive wire was plotted in Schaeffler diagram, as shown in Fig. 1. It shows that for pure-Ni wire and Ni-Cr wire, the melted metal has fully austenitic microstructure. For AISI 316L and AISI 310S stainless steel wire, however, a mixed microstructure which consists of austenitic, martensitic and ferritic should be formed in the melted metal.

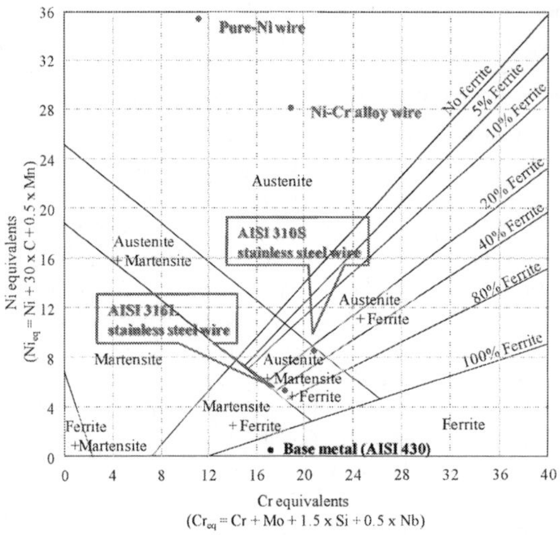

Fig. 1: Location of each wire at Schaeffler diagram

Microstructures of laser melted metal

Fig. 2 shows the microstructures of laser melted metal in the case that each wire was used as the additive material. The photo (a) is the case of pure-Ni wire, the photo (b) is the case of Ni-Cr alloy wire, the photo (c) is the case of the AISI 310S stainless steel wire, and the photo (d) is the case of the AISI 316L stainless steel wire. The photo (a) indicates that almost 100 % austenitic dendrite formed in the whole melted metal. The average arm size of the dendrite column is about 3 ~ 6 μm, which is much smaller than the grain size of the base metal (30 ~ 100 μm). The photo (b) indicates that almost 100 % austenitic dendrite formed in the whole melted metal, in spite of partial residual no-melted wire. This residual no-melted wire will be disappeared if the process parameters are changed. In this case, the microstructure is predicted to be same full austenite. Moreover, the average arm size of the dendrite column is also about 3 ~ 6 μm, which is almost the same as that of pure-Ni wire. The photo (c) indicates that the laser melted metal had a microstructure that γ (austenite) and α (ferrite + martensite) is mixed. The average grain size of the mixed microstructure was about 40 ~ 120 μm, which is the same order as that of base metal. The photo (d) indicates that a high proportion of α (ferrite + martensite) was formed in the microstructure of melted metal. The average arm size of the dendrite column is about 20 ~ 100 μm, which is much larger than those of pure-Ni wire and Ni-Cr alloy wire.

Saturated magnetization

Fig. 3 shows the effect of additive wire compositions on saturated magnetization as laser power increases. The saturated magnetization of the laser-treated range formed by adding pure-Ni wire or Ni-Cr alloy wire could be considerably reduced as compared with that of the base metal. It is due to the almost full austenitizing of the microstructure after laser melt treatment with the additive wire as mentioned above. The saturated magnetization decreased with increasing laser power, and when laser power was higher than 1200 W, the saturated magnetization was obtained less than 0.1 T. Therefore, this result means that ferromagnetic material was transformed into non-magnetic material by the laser melt treatment. The reason why the saturated magnetization got higher at lower laser power is considered that residual no-melt wire and residual ferrite still existed in the microstructure, and eventually, a homogeneous full austenitic microstructure did not formed. On the contrary, for AISI 316L stainless steel wire adding, the saturated magnetization was almost not reduced under all the laser power conditions. This result indicates that the melted metal has a full α (ferrite + martensite) microstructure. By contrast, for AISI 310S stainless steel wire adding, the decrease rate of the saturated magnetization remained about 40 % of the base metal at appropriate process condition and was larger than the case of pure-Ni wire adding or Ni-Cr alloy wire adding.

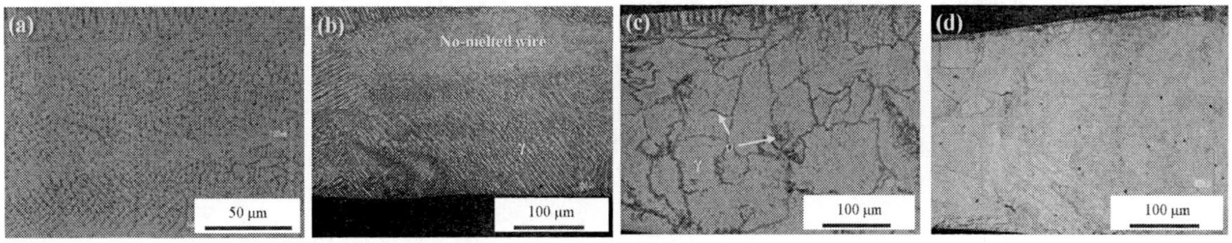

Fig. 2: Microstructures of laser melted for each wire
(a): pure-Ni wire, (b): Ni-Cr alloy wire, (c): AISI 310S stainless steel wire, (d): AISI 316L stainless steel wire

This result indicates that γ (austenite) formed during laser weak-magnetic treatment is contained in the mixed microstructure. The reason why the saturated magnetization increases with laser power is considered that the solidification speed decreases as laser power increases. As a result, it is suggested that more ferrite was preliminarily solidified.

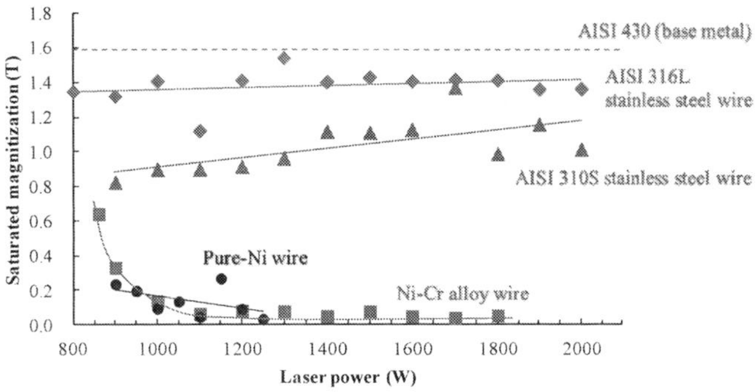

Fig. 3: Effect of wire compositions on saturated magnetization

Conclusions

In order to partially reduce the saturated magnetization of AISI 430 ferromagnetic stainless steel pipes, a laser weak-magnetic treatment process by adding Ni-contained wire was investigated. The chemical compositions, microstructures and saturated magnetizations of the laser-melted metal were analyzed. The following results were obtained.

1) The microstructure of the melted metal was fully austenitic columnar dendrite when pure-Ni or Ni-Cr alloy wire was used. For AISI 310S stainless steel wire, the melted metal was consisted of the microstructure which γ (austenitic) and α (ferrite + martensite) were mixed. For AISI 316L stainless steel wire, it was consisted of a high proportion of α (ferrite + martensite).

2) By adding pure-Ni or Ni-Cr alloy wires, the saturated magnetization was significantly reduced. In specific, when the laser power was higher than 1200 W, the saturated magnetization was obtained less than 0.1 T. The ferromagnetic material was transformed into non-magnetic material by the laser melt treatment.

3) For AISI 316L stainless steel wire adding, the saturated magnetization was almost not reduced under all the laser power conditions.

4) For AISI 310S stainless steel wire adding, the decrease rate of the saturated magnetization remained about 40 % of the base metal and was larger than the case of pure-Ni wire adding or Ni-Cr alloy wire adding.

References

[1] W.S. Brandão, V.T.L. Buono, P.V. Marques, P.J. Modenesi, Welding Int., 6 (1992) 713.
[2] D. O'Sullivan, M. Cotterell, S. Cassidy, D.A. Tanner, I. Mészáros, J. Magnetism and Magnetic Materials, 271 (2004) 381-389.
[3] A. L. Schaeffler, Metal Progress, 56-11 (1949) 680.
[4] W. Dong, H. Kokawa, Y.Sato and S. Tsukamoto, Metall. Mater. Trans. B, 34-1 (2003) 75-82.

Tracking of the welding line in lap welding using pattern matching

Satoshi Yamane, Hiroyuki Ohyama
Saitama University, Saitama, Saitama, Japan
yamane@mail.saitama-u.ac.jp

Abstract
In order to trace the welding line in the lap welding, weld pool in the pulsed MAG welding was taken by using the CMOS camera. The weld pool was observed in the front and rear to the welding direction. The features of the weld pool were investigated to recognize the welding line. There are some spatters in lap welding under the shielding gas of Ar 80% and CO_2 20%. Since arc affected the brightness of the weld pool, the timing of the shutter in CMOS camera was synchronized with the current waveform. The welding line was detected by using pattern matching method as image processing and the digital controller was designed to trace the welding line.

Introduction
Recently, it is required to reduce the production cost in the joining of materials in a thin board. The automation and the high efficiency arc welding have attracted in the attention of many researchers[1)], since the skill workers are reduced year by year. The mechanical strength of the welded joint in the lap welding depends on the welding quality and the tracking of the welding line. If the groove welding was carried out, it was easy to trace the welding line by an arc sensor due to the weaving[2)]. Since torch is not weaved in the lap welding, it is difficult to apply the arc sensor to the tracking of the welding line. Welders watched the molten region by eyes with the face shield and operated the welding torch so as to trace the welding line. Kuo and Wu took the welding line with a CCD camera and an assist lamp and proposed the image processing method to detect the welding line[3)].

The authors tried to use a CMOS camera instead of human eyes during the GMAW welding. As the arc light is too strong, the taking of the weld pool is difficult. In pipe GTAW, Tsai et al[4)] took the weld pool on back side of base metals without the arc light and adjusted the welding current to control a quality of the welding. It is required to take clear image of the clear weld pool. The welding current was synchronized with the timing of the shutter of the CMOS camera. The authors proposed the visual welding system to take a clear weld pool image[5)], in which the welding current reduced to 30A at the opening of the shutter. To apply this system conventional welding power sources was difficult. There are spatters in GMAW. It is one of the reasons in image processing error. In this research, the welding current is not reduced for the opening of the shutter. Since the pulsed welding current was used, the shutter opens at the base current duration. Since the image depends on the exposure time of the CMOS camera, the relationship between the exposure time and the image was investigated. It is important to get the features of the taken image to detect the welding line. During the welding, the weld pool was observed from the front and the rear of the welding torch. From experiments, it is shown that it is easy to detect the welding line in the rear image. The taken image of the weld pool was processed to detect the welding line by using pattern matching. After that, the digital controller was designed and the experiment was carried out to verify the validity of the image processing method and the digital controller.

System configuration
Robotic welding system is shown in Fig.1. There is three kinds of personal computers, PCs. Two are for the real time robot control and the welding power source control. Welding conditions were stored in the PC controlling the robot. This PC send the welding conditions to the PC controlling the power source. The trigger signal to open the shutter of the CMOS camera, which is Basler acA1300-60gmNIR camera, was synchronized with the current waveform. Its signal was generated by PC for the power source control and was sent to the CMOS camera. The camera sent the taken image to third PC with gigabit Ethernet as digital value. Its PC processed its image, detected the welding line and sent to the PC controlling the robot with RS-232C. The CMOS camera was fixed on the torch axis as shown in Fig.2. In this welding, the torch is inclined to the lap area.

Waveform of welding current and the timing of trigger signal
The pulsed MAG welding was applied in the lap welding as shown in Fig.3. The peak current and the base current are 450 A and 60 A, respectively. The frequency of the pulsed current is 60 Hz. The weld pool is taken with the

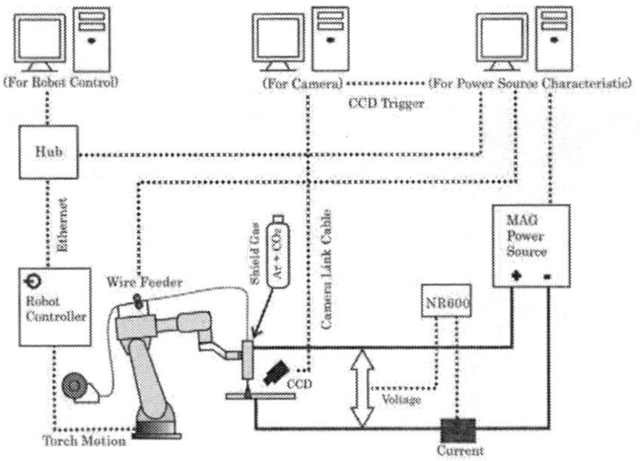

Fig.1 Robotic welding system.

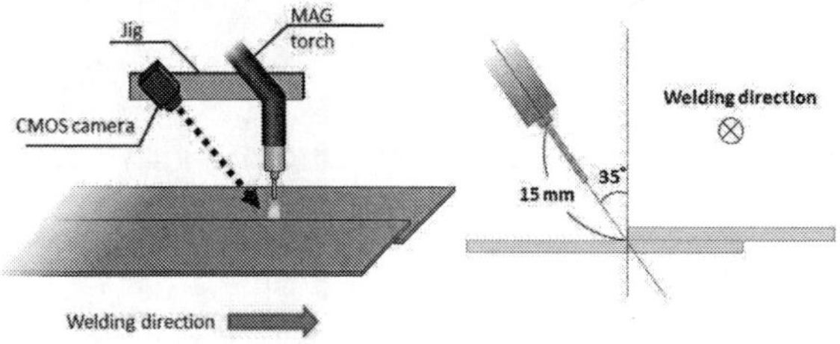

Fig.2 Relationship between CMOS camera and welding position.

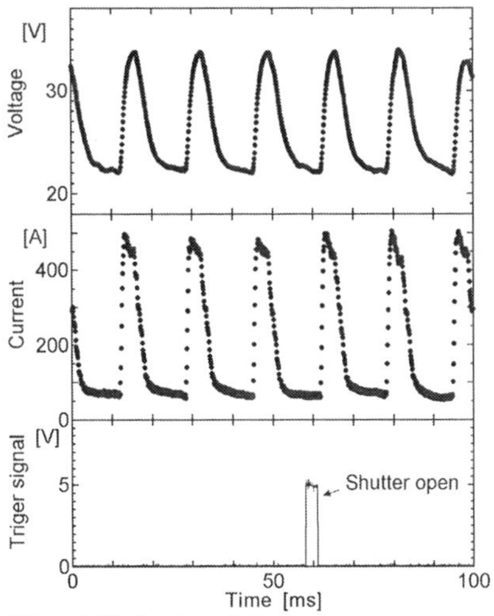

Fig. 3 Relationship between the current and trigger of camera.

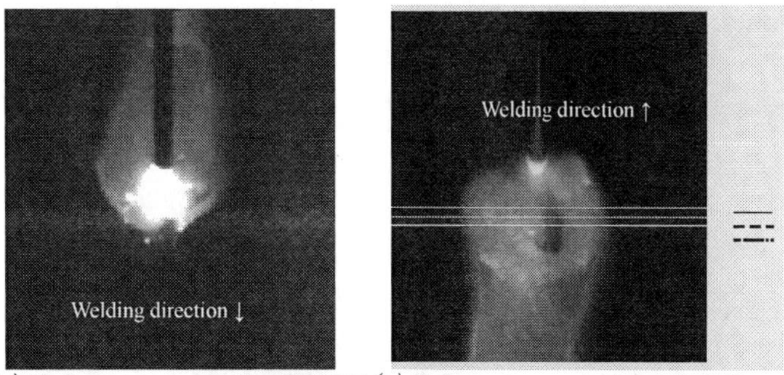

(a) Taken image with CMOS camera at the front position. *(b) Taken image with CMOS camera at the rear position.*

Fig. 4 Typical weld pool image of pulsed MAG welding.

CMOS camera, of which the shutter is opened by generating the trigger signal in the base current duration to reduce the affection of the arc light. Moreover, the interference filter of 950 nm was used to reduce the arc light[6,7], too. The typical weld pool images of the weld pool from the front and the rear were shown in Fig.4. When the weld pool was taken from the front, it is difficult to recognize the welding line, as shown in Fig.4 (a).

Image processing of the weld pool

It is important to recognize the welding line corresponding to the lap area. The authors tried to process the weld pool image by using HALCON[8] due to ability of maintenance. The electrode wire position in the image was found during the set-up of the system. As the welding line under the arc cannot be found, first, the arc region was found in the weld pool image. Although the brightness near the welding line becomes dark than weld pool, it is higher than outside of the weld pool. As shown in Fig.4(b), it is difficult to detect the welding line area. The template matching method was applied to detect the welding line area. The typical image of the welding line was selected as the template image. Pattern matching was based on regular cross-correlation by using

$$Rcc = \frac{\sum_{i=0}^{M}\sum_{j=0}^{N} I(i,j)T(i,j)}{\sqrt{\sum_{i=0}^{M}\sum_{j=0}^{N} I(i,j)^2 \times \sum_{i=0}^{M}\sum_{j=0}^{N} T(i,j)^2}} \tag{1}$$

where $I(i,j)$ is the brightness of the taken image at i row and j column, and $T(i,j)$ is the brightness of the template image at ith row and jth column.

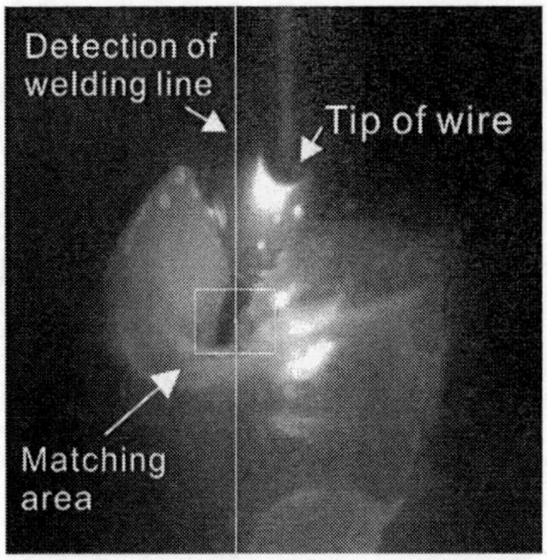

Fig.5 Pattern matching of the welding line.

When *Rcc* is close to 1, the image is similar to the welding line area. The detection was carried out by using value of *Rcc*. The result of the pattern matching was shown in Fig.5.

Design of the digital controller

The characteristic of the digital controller in the control block diagram as shown in Fig.6 is

$$D[z] = \frac{1}{G[z]} \frac{C[z]}{R[z] - z^{-1}C[z]} \tag{2}$$

where G[z], R[z] and C[z] is dynamic behavior of the robot, the reference and the desired response, respectively. The discrete time system to calculate the manipulator value Δu[n] was found from Eq. (2).

$$\Delta u[n] = d_0 e[n] + d_1 e[n-1] + d_2 e[n-2] + \cdots + d_{12} e[n-12] \;\; - f_1 \Delta u[n-1] - f_2 \Delta u[n-2] - \cdots - f_{12} \Delta u[n-12] \tag{3}$$

The manipulating value is calculated by the error and the manipulating variables, i.e., Eq. (3) corresponds to the weighted mean calculation. Even if the deviation $e[n]$ has some noise, its affection is small in Eq. (3). Tracking experiment at the welding current of 180 A, the travelling speed of 1 mm/s and the wire feed rate of 98 mm/s was carried out under when the start point of the welding is on the welding line and the end point of the welding was shifted from the welding line by 3mm. The suitable position of the torch is 0.5 mm away from the lap. The experiment result and the bead appearance are shown in Fig.7 and Fig.8, respectively. In the case, the detection is almost kept constant without the teaching point to the robot, as shown in Fig.7. The variation of the detection is below 15 pixels. Since the image of the molten area was processed, the image was influenced by reflection of the arc in the molten area. As the digital controller has robustness, the influence was small to trace the welding line. Figure 8 shows the image and the bead appearance. The weld pool image was almost kept constant. A good tracking result was obtained and the validity of the tracking system was verified.

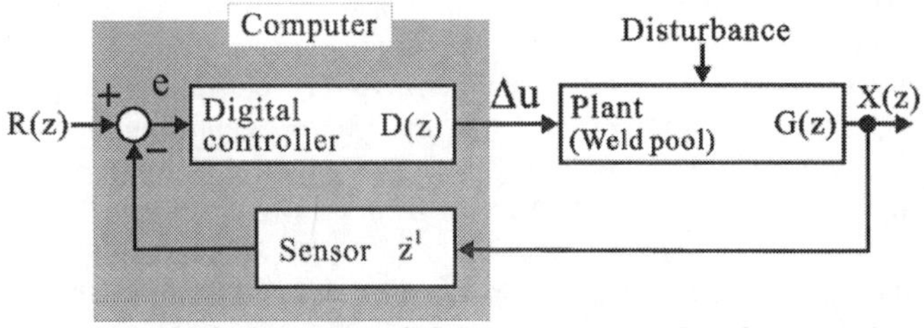

Fig.6 Block diagram of the penetration depth control.

Conclusions

It is important to trace the welding line. But it is difficult to apply the arc sensor in lap welding. The weld pool image was observed with the CMOS camera. The interference filter of 950nm was used and the shutter of the CMOS camera was synchronized with the base current duration to avoid the affection of the arc light. The welding line was found by using image processing. When the exposure time is long, the image becomes stable in

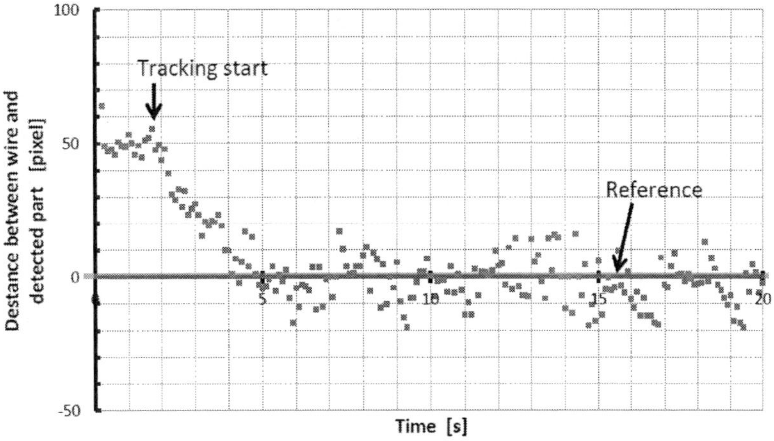

Fig. 7 Tracking result.

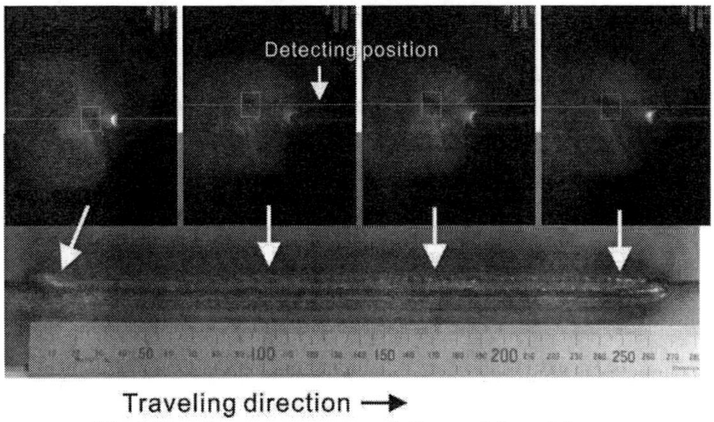

Fig.8 Bead appearance and the weld pool image.

the lap welding and it is easy to process the image. The deviation from the welding line can be detected by using the image processing. The tracking of the welding line was carried out by using the digital controller so that the deviation becomes small. The validity of the image processing by the template matching and the tracking system was verified by carrying out the welding experiment.

References

[1] T.Ueyama, T.Ohanawa, M.Tanaka, K.Nakata, STWJ, 6(2005) 750-759

[2] S.Yamane, H.Yuzawa, Y.Kaneko, H.Yamamoto, M.Hirakawa, K.Oshima,: Image Processing and Control of Weld pool in Switch Back Welding without Backing Plate, Quar. J. JWS, 23(2005),65-70 (in Japanese)

[3] Hsing-Chia Kuo, Li-Jen Wu: An image tracking system for welded seams using fuzzy logic, J. of Mat. Proc. Tech., 120 (2002), 169–185

[4] Chiung-Hsin Tsai, Kuang-Hua Hou, Han-Tung Chuang: Fuzzy control of pulsed GTA welds by using real-time root bead image feedback, J. of Mat. Proc. Tech., 176 (2006), 158-167

[5] K.Oshima, et al : Digital Control of Torch Position and Weld Pool in MIG Welding Using Image Processing Device, Trans. of IEEE IAS, 28(1992), 607-612

[6] W Lucas, L Mcllroy,J.S Smith : Review of Industrial Sensors for Control of Arc Welding, Proc. of IIW Commission XII, Doc.XII-1566-99(1999), 45-61

[7] H.Wada, Y.Manabe, S.Inoue, H.Yoritaka : Group-control System of Narrow-Gap MIG Welding, Sensors and Control Systems in Arc Welding(1990), II-61-II-64(in Japanese)

[8] MVTec Software GmbH: HALCON 12 User's Manuals (2014)

Digital Radiographic Inspection of Weld Quality in Joined Thick Shipbuiding Steel by the Computed Radiography

Armando H. Shinohara, Emilly C.P. Silva, Ayako Ono
Federal University of Pernambuco (UFPE), Recife, Pernambuco, Brazil
shinohara@ufpe.br, 81-21268230

Estefen Farid Segen
Federal University of Rio de Janeiro (UFRJ), Rio de Janeiro, Rio de Janeiro, Brazil
segen@lts.coppe.ufrj.br, 21-25627791

Abstract

Traditionally, the radiographic inspection of weld quality of thick steel plate in shipyards is conducted using the gamma-ray source and X-ray film. In that procedure, the radiographic inspection is conducted at night due to the severe requirement of radiation protection. In order to overcome not only the limitation of conduction the radiography inspection during the daylight with gamma-ray sources, but also, aiming to improve the quality of radiographic imaging using a high sensitivity detector for inspection of discontinuities and failures in weld, a nondestructive testing laboratory was established to conduct research works on the development of new technologies for industrial application of radiography and also the human resource development in the weld quality and integrity inspection with establishment of national laboratory in welding and joining at UFPE – Federal University of Pernambuco.

As a result, a high sensitivity computed radiographic system was developed. Testing was conducted to inspect MIG welded steel plates with 19 mm in thickness in the laboratory by using a light battery powered and portable X-ray source of 120 kV with sub-milimeter focal spot made by ICM and high resolution of imaging plates (blue type IP) from Perkin-Elmer. The IP reader was conducted using the Perkin-Elmer Cyclone with 16 bit resolution. As a result, in general, use of iridium gamma-ray source of high activity is required for inspection of thick steel plate, but good quality radiographic images have been obtained by the system tested and to enhance details of region of interest was possible with digital radiographic images by dedicated softwares.

Introduction

Recently, several new shipyards are in operation in Brazil to build oil and gas tankers, drillships vessels for domestic exploration of new oil and gas reservoir in the pre-salt region, platforms. The welding is the one of the main techniques employed to join the steel plates to assembly the ships and platforms. One of major difficulties in the shipbuilding is the welding deformation resulting after welding of plates [1] and all efforts to mitigate it is great interest. Furthermore, the evaluation of quality of weld is of vital importance, especially for offshore structures due to the long period of operation in the ocean. So, radiographic and ultrasonic for weld inspection techniques are the two most common nondestructive testing (NDT) used to detect discontinuities within the internal structure of welds. Particularly, radiographic technique is considered the most reliable method for inspection.

The different welding processes can produce high residual stresses accompanied by the effects of deformations in welded plates. Since the welding residual stresses are present in all welded steel plate, the weld quality has no direct influence on the occurrence of this phenomenon, but a poor weld quality may aggravate the weakness of the joining of ship plates, generating, for example, stress concentrations at levels that compromise the structure.

The new nondestructive weld testing technology in industry is the computed radiography [2]. It is a technique of great interest because it is 100 to 1000 times more sensitive than X-ray film and allows disclosure of internal defects such as cracks, slag inclusion, blistering, corrosion, weld quality and integrity by checking welds in images, which are not accessible with other techniques. It is also interesting from radiation protection point of view due to the considerable reduction of exposure time.

At the present work, in order to show potential application of the computed radiography in the shipyard for thick welede steel plates, a high sensitive and compact computed radiography system was tested using small and portable X-ray source of 120 kV, imaging plate and shipbuilding steel plates of 19 mm in thickness.

Materials and Method

Shipbuilding steel plates of 19 mm in thickness were welded by the MIG dual wire at the laboratory of Federal

University of Rio de Janeiro, as shown in figure 1 and 2.
The computed radiography was consisted of battery powered X-ray source weighting 7 kg with maximum out power of 120 kV and 1 mA. The focal size was 0.8 mm X 0.25 mm and model portable CP120B made by ICM, Belgium. The detector was IP - Imaging Plate was from Perkin-Elmer using two types: SR (Super Resolution) and SS (Super Sensitive) with 12.5 cm wide and 43 cm long; Laser scanner with a resolution of 16-bit, model Perkin-Elmer Cyclone controlled by the computer was used for for IP reader. Digital imaging processing was conducted by using dedicated software Isee! (BAM).

Figure 1.Original naval steel plate sample stuck in form holder. Later, it was cut and three parts of this plate were sent to UFPE for inspection of weld.

Figure 2. Specimen number 3 was inspected by the computed radiographic system.

The process of preparing radiographic imaging by the computed radiography Imaging Plates (IP) is similar to the process of developing common radiographs, as show in figure 3. It consists of an X-ray source, a sample to be imaged and an IP, replacing the ordinary films. The IP keeps track according to the intensity of radiation with which it is exposed. This information is then scanned and transformed into a digital image. This image can then be processed, expanded and manipulated for better viewing of the details of the object imaged.

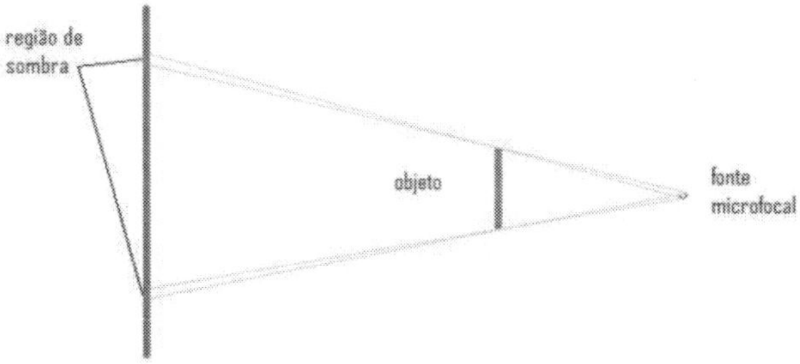

Figure 3. View of the concept of the art "Projection Magnification" with a submilimiter and microfocal source.

Compared with other image sensors of radiation, the main advantages of Imaging Plates are:
1) Ultra-high sensitivity: 100 to 1000 of times more sensitive than conventional film, and casually even thousands of times, depending on the sample.
2) Extended Dynamic Range.
3) Excellent linearity: The photo emission is proportional to exposure throughout the range.
4) Increased spatial resolution: When compared with other electronic systems, a higher pixel density can be obtained to meet the system's purpose though less freely than in films.
5) The electrical digital signals are available directly from the scanner. The computing and processing or combination with other electronic systems is easy.
6) Due to the use of a full detector, the IP produces less error detection, even with high flux densities, which often happens with pulse type detectors such as proportional counting tube and scintillation counters.
7) The Imaging Plate can be erased and reused. It replaces the conventional radiographic films, viewing the image radiographic imaging with high sensitivity through the digital process also used in conventional radiographic films and also makes it possible to quantify the position and intensity of the radiography details.

Results and Discussion

The images that follow show how the variation of parameters in obtaining the radiographs can leverage a compact system. The X-ray source used has a manufacturer's specification for use in steel plates with thicknesses less than 15 mm. With the computed radiography technique it was possible to obtain radiographic images of good quality plates naval thicknesses of 19mm, beating expectations.

Figure 4 – Radiographic image with 30s, 40cm and IP SS

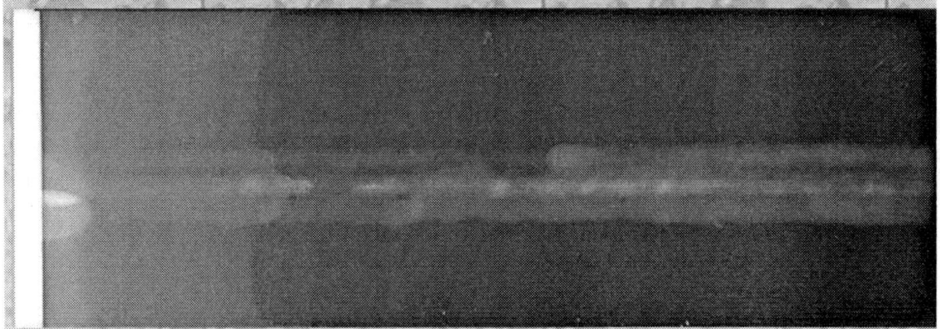

Figure 5 – Radiographic image with 160s, 40cm and IP SR

Figure 6 – Radiographic image with 60s, 25cm and IP SS

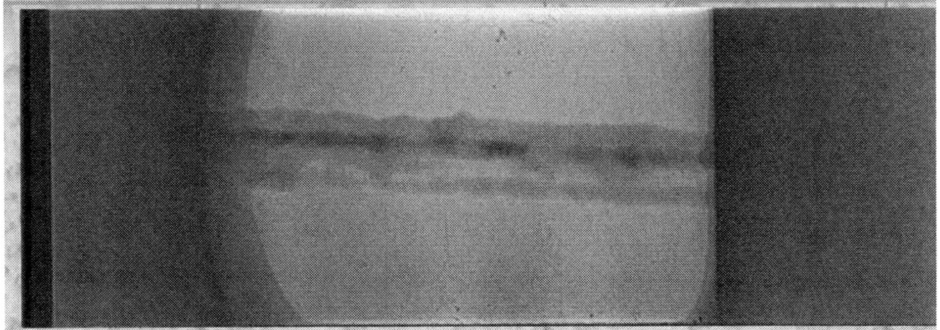

Figure 7– Radiographic image with 60s, 15cm and IP SS

Conclusions

Relatively simple and lightweight Programs as the iSee!, Matlab and others, for example, Origin, can serve to locate and identify defects in weld seams, as well as quantify them as long as precautions are taken. It proved that you can analyze and measure radiographic images to determine the integrity of the weld beads with relatively simple tools and methods.

In general, the examined welded naval plate is of good quality. For the application in the marine industry it is within the tolerances set by the standards commonly used, such as ASME VIII, API 1104 or AWS.

The result caused considerable optimism was the fact of having succeeded in obtaining good quality radiographs in naval plates with a thickness of 19 mm which should be added a further increase of welding, resulting in a welding thickness over 20 mm using an X-ray source of relatively low power.

Notable was the fact, when starting to get X-ray radiographies in the laboratory using the Imaging Plate, the information regarding the x-ray-source indicated by the manufacturer the use for plates of steel to a maximum of 15 mm thickness when the present naval plate examined presented thickness of 19 mm plus a few millimeters in the actual weld. This was only possible by using the Imaging Plate, instead of traditional radiographic film because the Imaging Plate shows sensitivity 100 times higher than traditional film.

Acknowledgments

The authors acknowledge FINEP.

References

[1] Masubuchi, K., Analysis of welded structures: residual stresses, distortion, and their consequences, Pergamon Press, 1980.

[2] Halmshaw, R., Industrial radiology: Theory and practice, 2nd edition, Chapman & Hall, 1995.

Development of a Blowhole Detection System for Robotic MAG Welding

Yosuke SASAKI, Yukio HIOKI, Yingda DAI, Siyang TU
Hitachi Zosen Corporation, Osaka, Osaka, Japan
sasaki_yo@hitachizosen.co.jp, +81-6-6551-9239

Mitsuyoshi NAKATANI
Osaka University (Hitachi Zosen Corporation), Ibaraki, Osaka, Japan
nakatani_m@hitachizosen.co.jp, +81-6-6879-8644

Abstract
Multi-layered pulse MAG welding by robots is used for large structures. The sensing technology for the detection of defects is crucial towards improving the productivity and quality of robotic welding. The authors have developed a system that detects welding defects (blowholes) by utilizing welding current and voltage waveform data. By processing the measurement data, it is possible to detect welding defects with a probability of 95% or better. The developed defect detection system has been introduced in the company's robotic welding systems.

1. Introduction

Automatic multi-layered MAG welding is used for large structures [1]. To perform proper welding with lower spatter, pulse MAG welding may be used [2]. Because multi-layered welds are formed successively in automatic robotic welding, blowholes caused by poor shielding or defects due to contamination during the welding process will lead to a significant loss of time, since inspections and repairs have to be made after the completion of all welding operations. Therefore, the reduction of repair time was targeted in developing a system that can detect blowholes during the layered welding process.

2. Blowhole Detection Method

2.1. Pulse MAG Welding

Figure 1 shows waveforms of I-V characteristics of a pulse MAG welding power supply unit for proper welding condition. In pulse MAG welding, the peak and base values appear with a certain cyclicity in the current and voltage waveforms. The peaks and bases in these waveforms are synchronized with each other and, if the welding was performed properly, the peak and base current and voltage values appear periodically. For this reason, if the current and/or voltage deviate from the proper pattern due to a disturbance in the welding for some reason, as shown in Fig. 2, it is possible to detect them as abnormal values.

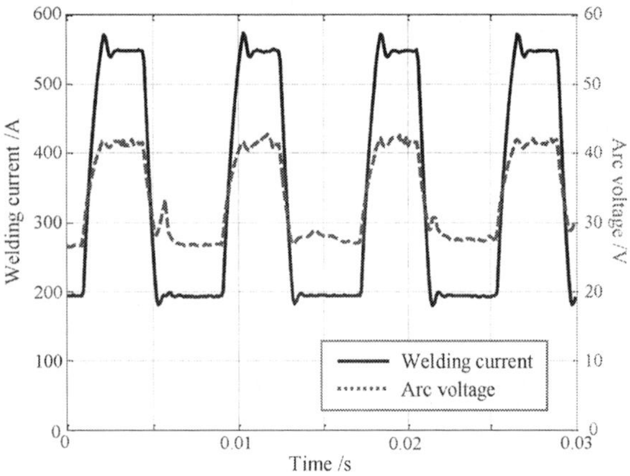

Fig. 1 Current/Voltage waveforms in pulse MAG welding (Proper values)

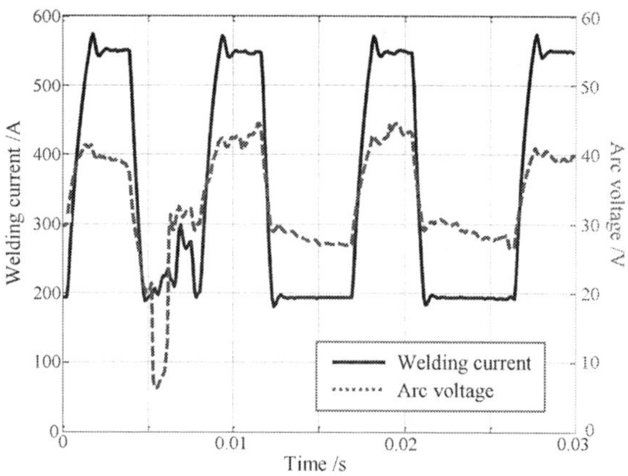

Fig. 2 Current/Voltage waveforms in pulse MAG welding (Abnormal values)

2.2. Blowhole Detection Algorithm

2.2.1. Area Division of Current/Voltage Distribution Charts

Since the current and voltage change synchronously on a pulse MAG power supply unit, the frequency distribution of the current and voltage was checked first. Figure 3 shows the distribution of the current, voltage and frequency on the X, Y and Z axes, respectively. It can be seen that the peak and base values of the current and voltage are concentrated at the pulses in proper welding. Here, it is possible to draw a straight line through the most frequently observed values for the peaks and bases of the current and voltage, respectively. Furthermore, by setting certain distances in the current and voltage axes from the most frequently observed peak and base values, the current and voltage distribution charts can be divided into six areas as shown in Fig. 4. The four values of I1, I2, I3 and I4 are set on the current axis, as well as the four values of V1, V2, V3 and V4 on the voltage axis. The chart consists of six areas: the base area of the pulse waveform *PB*, the rising/falling area *rf*, the peak area *PP*, the short-circuit area on the base side *sB*, the short-circuit area in rising/falling *srf*, and the short-circuit area on the peak side *sP*.

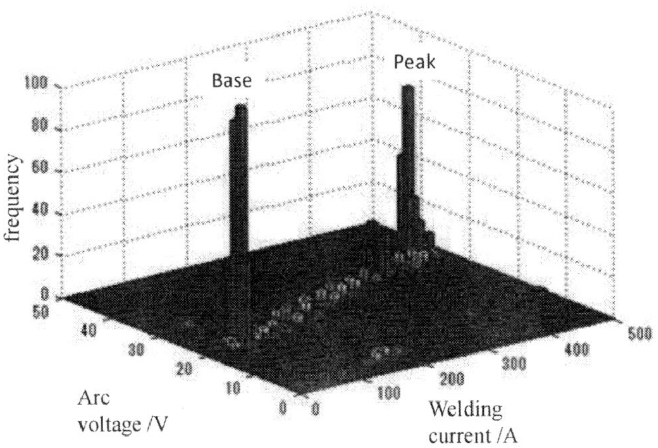

Fig. 3 Current/Voltage frequency distribution chart

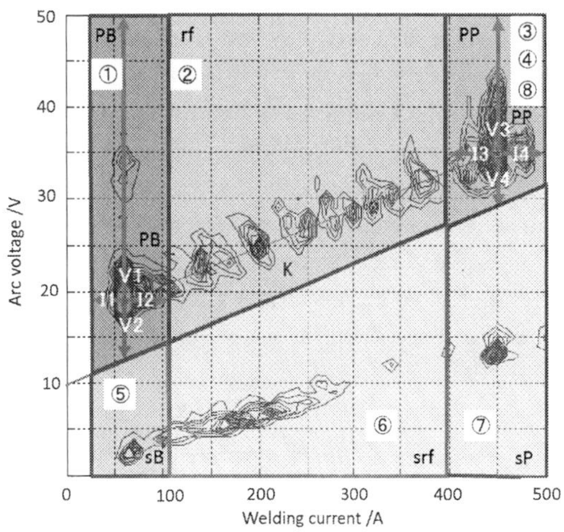

Fig. 4 Current/Voltage distribution chart divided into six areas

2.2.2. Calculation of Characteristic Quantities

In development work, the following 9 characteristic quantities were selected for the purpose of blowhole detection. Each characteristic quantity was calculated to determine if it exceeded the reference value and the presence of blowholes was determined by evaluating them as a whole.

① Distribution in the base area
② Distribution in the rising/falling area
③ Distribution in the peak area
④ Frequency in the peak area
⑤ Frequency in the short-circuit area on the base side
⑥ Frequency in the short-circuit area in rising/falling
⑦ Frequency in the short-circuit area on the peak side
⑧ The highest frequency in the peak area
⑨ Standard deviation of cyclic frequency for the welding current waveform

3. Blowhole Detection System

3.1. System Configuration

Figure 5 shows the configuration of the blowhole detection system. The rate of measurement is 10 kHz and the system supports single and tandem welding. If a judgment was made that there is a defect, signals to stop the robot and indicate a warning will be sent to the control panel of the welding robot. This system makes it possible to detect blowholes early in the welding process and reduce the time required for repair.

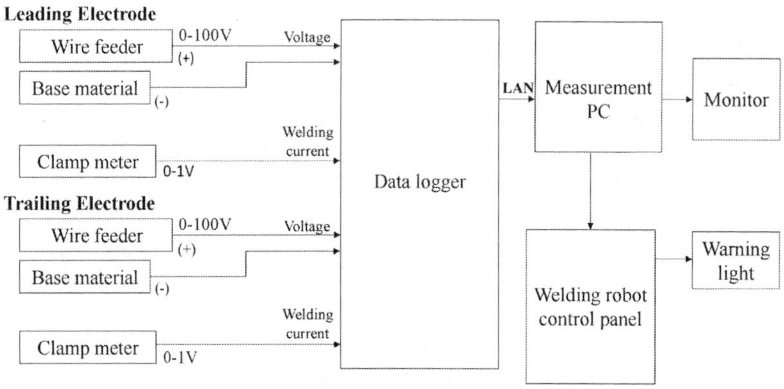

Fig. 5 Defect detection system configuration

3.2. Blowhole Detection Process

The process for detecting blowholes consists generally of the three processes of preparation, calculation of characteristic quantities and judgment of blowhole presence. In the preparation process, current and voltage data obtained from the data logger are converted into CSV format. After eliminating data in the periods 5 seconds after the start and 10 seconds before the end of welding, data are grouped at intervals of one second. Next, the characteristic quantities are calculated for the obtained data every second. After generating frequency distribution charts for the current and voltage, the charts are divided into six areas using straight lines that pass through the highest peak and base values and settings. The selected 9 characteristic quantities are calculated for the six areas and recorded in a data file on the computer. These processes are repeated for the entire length of the weld beads. Finally, the presence of blowholes is determined. The judgment is made by checking if each characteristic quantity calculated every second exceeds the threshold. If there were a certain number of characteristic quantities exceeding the thresholds and such a condition lasted more than 2 seconds, it is concluded that there are blowholes. Figure 7 shows an example of blowhole detection using the characteristic quantities obtained from the current/voltage frequency distribution charts. It is also possible to add a weight to each characteristic quantity used for judgment. Since the thresholds for characteristic quantities need to be adjusted to the current, they are modified automatically according to the cyclic frequency of welding current waveform. Should a conclusion be made that there are blowholes, the welding robot stops before the next path. At the same time, the warning lamp up to inform the operator that blowholes have been detected. The points at which blowholes are considered present are displayed on the monitor screen in such a way that the actual points on the weld bead are identified.

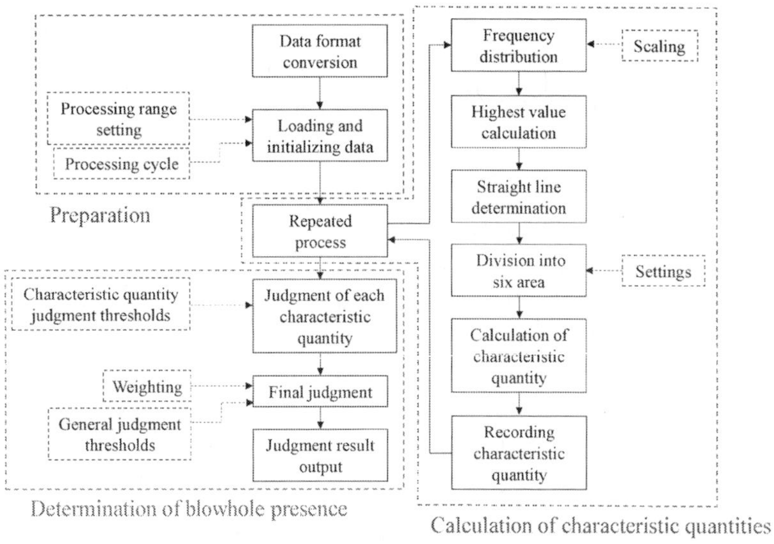

Fig. 6 Blowhole detection process

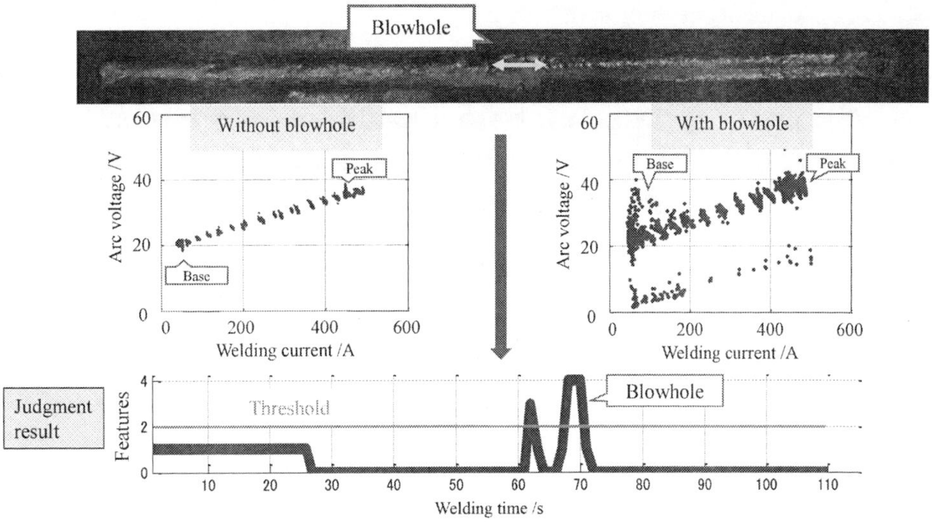

Fig. 7 Blowhole detection using characteristic quantities

3.3. Blowhole Detection Experiment

The developed blowhole detection system was applied to a case, in which blowholes were intentionally generated by a cross wind at some point of the welding process. An example of obtained the current/voltage frequency distribution chart is shown in Fig. 8. It can be seen that the frequency distribution varies significantly with and without the presence of cross-wind. Based on these distribution charts, characteristic quantities were calculated and the presence of blowholes was determined. As a result, the points of the blowholes obtained by the developed system agreed well with the actual locations of the blowholes and, thus, it has been verified that the developed detection method makes it possible to detect blowholes during welding.

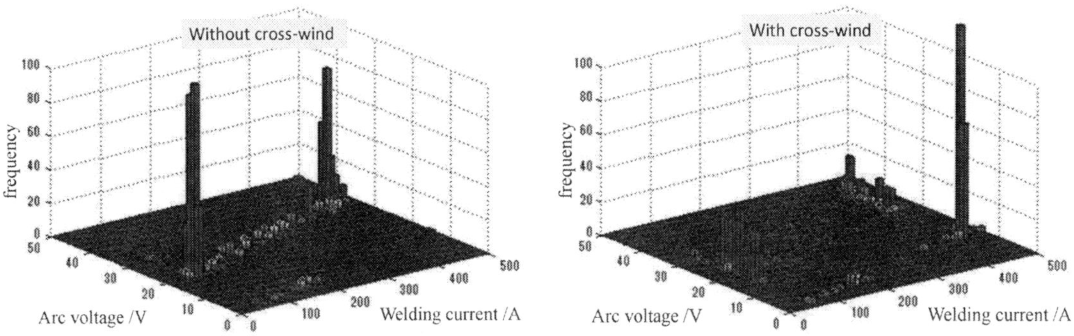

Fig. 8 Effects of cross-wind on the current/voltage frequency distribution

4. Conclusions

A system has been developed to detect blowholes in robotic MAG layered welding using current and voltage waveforms during the welding process. It has been found that the presence of blowholes can be determined by dividing the current/voltage distribution chart of pulse MAG welding into six areas and calculating characteristic quantities in each area. In an experiment to generate blowholes with a cross-wind, a defect detecting rate of 95% or better was achieved.

References

[1] K.Tsuji, *et al*: "Robot Welding Process for Medium and Heavy Plate" *Kobe Steel Engineering Reports,* Vol.63, No.1 (2013), pp.48-53

[2]A.Matsunawa, *et al*: "Pulse Arc Welding" Journal of the Japan Welding Society, Vol.53, No.6 (1984), pp.20-58

"A holistic approach for steel health monitoring and annihilation/control concerning welding in shipyards, shipping & offshore platform operations"

S. Papadopoulou, E. Hristoforou
National Technical University of Athens. Zografou Campus, Athens 15780, Greece.
eh@metal.ntua.gr, T: +30 210 7722178

M. Zavali, K. Kalkanis, S. Mores
Nomasico LTD. 46 Stasinou Street, 1st Floor, Office 28 CY-2002, Nicosia
s.mores@nomasico.com, T: +30 210 9717386, F: +30 210 9652423

Abstract
In this paper, we present a technique to monitor stress distribution in steels and their welds, by using the correlation between magnetostriction and stresses, accompanied by a novel method to provide stress rehabilitation in the desired stress window by applying an electromagnetic push-pull hammer.

Introduction
The implementation of holistic approaches that integrate condition monitoring systems with appropriate data analytics and simulation techniques under a single umbrella dedicated to advanced asset management, with regard to structural integrity monitoring bear significant benefits in risk minimization and safety improvement, as especially the waterborne industry still faces quite a lot of incidents attributable to loss of structural stability leading to accelerated fatigue and failure in vessels and offshore constructions [1, 2]. Nomasico LTD aims at applying a holistic methodology for the real-time monitoring of the structural integrity and annihilation/control of critical steel structures, in steel production, manufacturing and use. Particular attention is given to steel welds in shipyards, onboard vessels, offshore platforms and other maritime applications. Furthermore, the same approach can be used in automotive and rail industry lines, in manufacturing-repairing-conditioning of cylinder liners, weld seams, axle arms, camshafts and wheel flanges, including their machinery equipment. The approach combines two novel technologies in the fields of structural integrity assessment and treatment. An innovative steel health monitoring method which integrates computer tools based on magnetic methods that implement advanced data analytics and simulation models so as to derive early warnings and prognosis on potential upcoming failures [3]. The gathered information is utilized to increase knowledge and insight of when and how fatigue develops in such structures and to early detect areas requiring attention. Apart from that, another technique providing contactless stress annihilation and control with a new pending patent based on electromagnetic forming [4] completes this methodology. The presented methodology achieves lifecycle risk reduction, intrinsic risk mitigation, resilience, integrity, and improved survivability of critical steel structures and machinery under normal and extreme operating conditions, always taking into consideration cost-based factors.

Steel health monitoring
The present sensor is related to a method and apparatus for monitoring the distribution of the hydraulic stresses (type stresses I) and the residual stresses (type stresses II and III) in ferromagnetic steels within the elastic region and plastic deformation region, determining the deforming step and use them for deletion and / or control of the stresses in the steel. The measuring sensor set up (Figure 1) has the ability to measure magnetostriction, function $\lambda(H)$. Magnetostriction is directly related to the algebraic sum of the residual (type stresses I and II) and hydraulic stresses (type stresses I), including possible extension to the plastic deformation region. We have realized such measurement in rod shape steels, with the future target to visualize it in planar and large radius cylindrical shapes.

As illustrated in Figure 1, the excitation/receiving coil (3a) and (3b) and the bias coils (2a) and (2b) are disposed at the large steel rod axis (5).

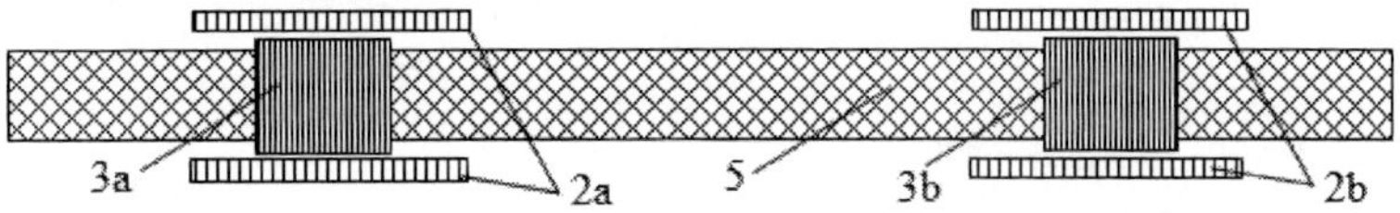

Fig. 1: Sensor arrangement in under test steel.

One of the two excitation/receiving coils (3a) is acting as the excitation coil, forcing a low frequency magnetic field which penetrates through the under measurement steel (5), creating elastic waves due to the movement of the 90° domain walls. These elastic waves are basically transmitted longitudinally in the direction of the long axis of the examined steel (5), having many forms of elastic waves (modes). The longitudinal form of the elastic wave, being the fastest form of elastic wave, reaches first and quite isolated from the rest of the elastic waves in the steel region (5) beneath the other excitation/receiving coil (3b) which receives the mentioned elastic wave as an output voltage, due to the inverse magnetostrictive effect and thus functions as the receiving coil. Thereafter, the supply of adjustable intensity bias current (H) in the biasing coil (2a) located around the excitation/receiving coil (3a) and thus the biasing field in the magnetizable area of the examined steel (5), results in the modulation of the output voltage (V_o) of the excitation/receiving coil (3b), the maximum value of which is proportional to the mean value of the magnetostriction (λ) of the volume of the examined steel (5) located beneath the excitation/receiving coil (3a), according to the formula:

$$V_o = C_1 \frac{d\lambda}{dH} \qquad (1)$$

where C_1 is a constant, determined from the experimental data [5]. This way and moving the coil device (2a)–(3a) and (2b)–(3b) along the steel rod, the measurement of the distribution monitoring of the magnetostriction function is performed. Bearing in mind that the magnetostriction sign changes monotonically with respect to strain from the elastic till the plastic deformation region, then the algebraic sum of the residual stresses (type II and III stresses) and the hydraulic stresses (type I stresses) is obtained.

Electro Magnetic Impact Treatment

A novel integrated technique Electro Magnetic Impact Treatment (EMIT) has been generated [4] for performing non-contact mechanical treatment of conductive and non-conductive surfaces, which method may be universally applied to perform mechanical treatment operations of all kinds including impact treatment, cyclic deformation, surface polishing, oxide removal, electromagnetic forming and electromagnetic welding, as well as to perform measurement of the stress tensor distribution. Furthermore, EMIT portrays the exclusivity of delivering both tensile and/or compressive forces and the ability of adjusting the direction and the amplitude of the applied force vector. Overall, EMIT delivers the above-mentioned treatments in a faster, better and cheaper way than any of the devices and methods currently available in the industry. For this particular scientific venture, this technique is an integral part of this holistic approach, by implementation of Mechanical treatment for tailoring the mechanical properties of a material subject to treatment with a scope of obtaining an approved profile of stress tensor distribution. After having a discrete knowledge of the stress tensor cross the area to be processed (SteHeMon), the exact required forces are applied. The component is afterwards reassessed by the SteHeMon module. Thus a closed loop system is created which leads to certification of critical steel structures.

The specified application of EMIT, an equivalent electromagnetic hammer device, is utilized to provide mechanical treatment via a system of conductors supplied with pulsed currents. In Figure 2a a primary conductor (2) is arranged to transmit pulsed current through the predetermined path (3) within the material subject to treatment (1) and a secondary conductor (4) positioned over it. A layer of insulating material (5) is lined intermediately between conductors. A simultaneous application of the pulsed currents in the same direction results in applying a tensile force and exerting a pull effect onto the material subject to treatment and application of pulsed currents in opposite directions results in applying a compressive force and exerting a push effect onto the material subject to treatment. Furthermore, introduction of auxiliary conductors (7, 8) bilaterally with respect to the initial conductor configuration provide a capacity of controlling the angle in which the aforementioned tensile or compressive forces are being applied onto the surface of the material subject to treatment. The resultant force can be controlled to be directed at any angle from -90^0 to $+90^0$. In cases of localized damage and correspondingly focused treatment (9), the secondary conductor is of a V–shape (10) as shown in Figure 2b.

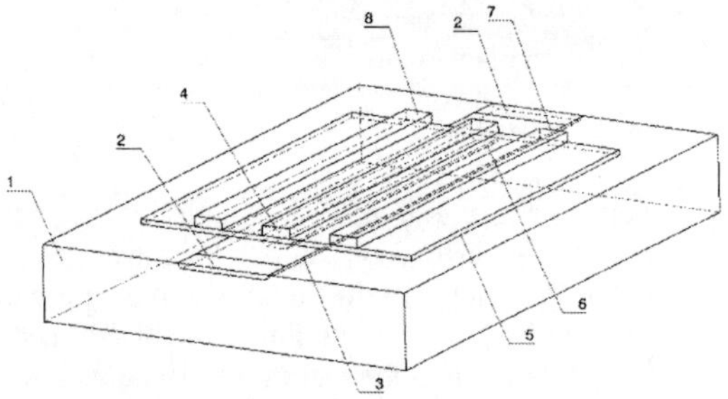

Fig. 2a: EMIT configuration enabling control of force direction.

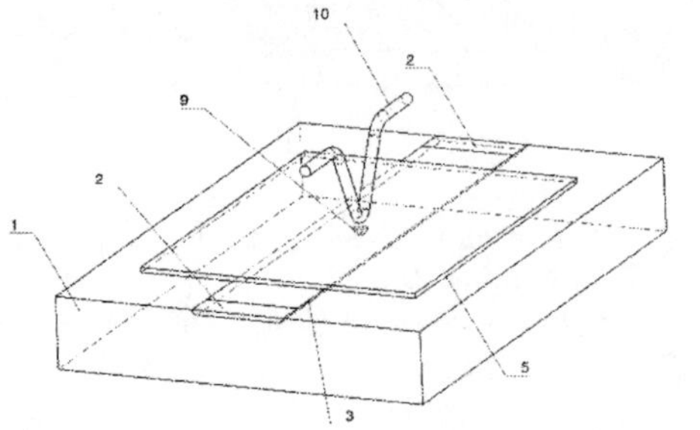

Fig. 2b: EMIT configuration enabling localized treatment.

Another application of such impact treatment refers to the treatment of steady-state conductive cylinders (Figure 3a) and tubes (Figure 3b) utilizing the same principle. In Figure 3a the cylinder subject to treatment (13) is covered by a thin insulating film (5) which in turn is covered by a conductive tube (14). According to this set-up, conductive elements with cylindrical shape can be processed by impact or cyclic deformation treatment by passing a first pulsed current through the whole cylinder (13) and a second pulsed current through the surrounding conductive tube (14) that is set on top of the thin insulating film (5) that covers the cylinder subject to treatment (13). In this way, tensile or compressive forces are applied throughout the infinitesimal volume (15) (skin effect) of the cylinder subject to treatment (13).

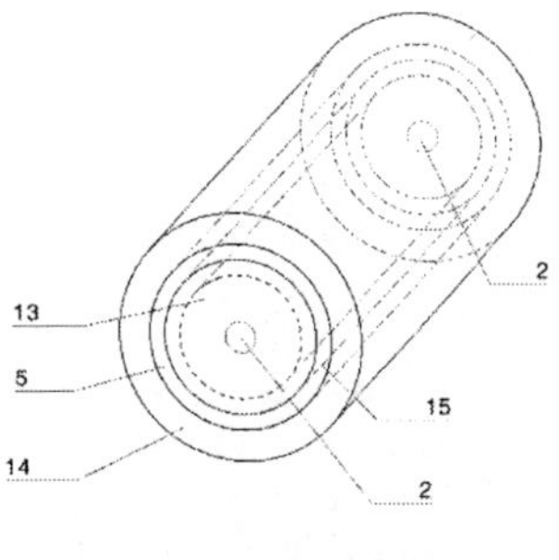

Fig. 3a: EMIT configuration for the treatment of steady-state conductive cylinders.

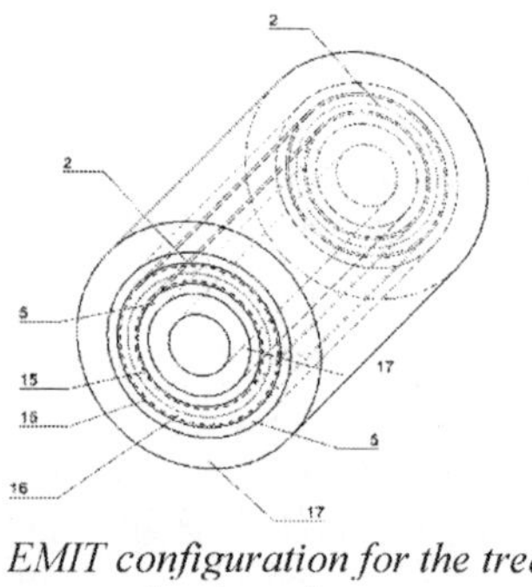

Fig. 3b: EMIT configuration for the treatment of steady-state conductive tubes.

In Figure 3b the tube subject to treatment (16) is interiorly and exteriorly covered by a first and a second thin insulating film (5). A first conductive tube (17) is provided interiorly to the first insulating film (5) and a second conductive tube (17) is provided exteriorly to the second insulating film (5) respectively. According to this set-up, both the interior and the exterior circumference of the conductive tube subject to treatment (16) can be processed by impact or cyclic deformation treatment. To provide this effect a first pulsed current is passed through the conductor (2) of the entire tube subject to treatment (16) and a second pulsed current passes through the first conductive tube (17) that is positioned in the interior of the tube subject to treatment (16) and through the second conductive tube (17) being positioned in the exterior of the tube subject to treatment (16). In this way,

tensile or compressive forces are applied throughout the infinitesimal volumes (15) of the tube subject to treatment (16).
Scanning Electron Microscopy (SEM) images of various specimens have shown that even with the minimum of a single pulse by the electromagnetic hammer device (Figure 4) a profound effect on the surface is present.

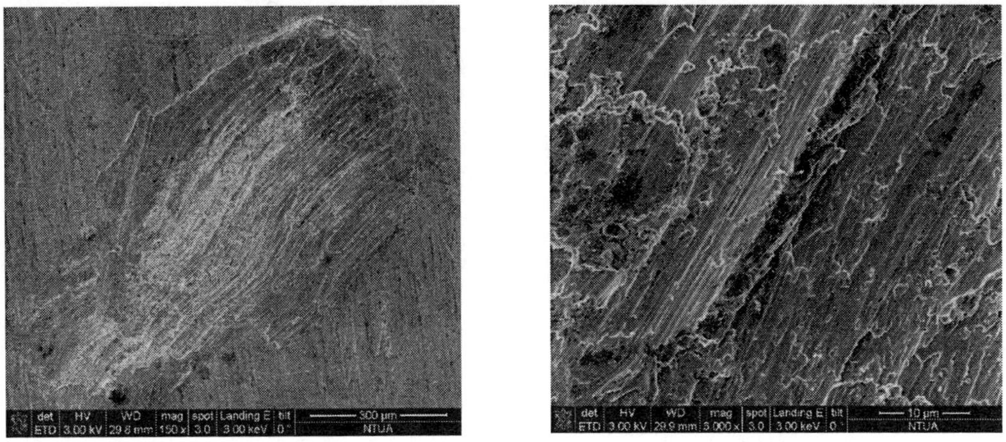

Fig. 4: SEM for Aluminium specimens' surface treated by a single pulse. The process parameters were: Voltage: 2000V, Current: 150kA. Pulse: 1. Insulating film: Mica (phyllosilicate mineral).

Conclusions

Our presented holistic method allowing for the monitoring of plastic deformation and residual stress distribution in steels is the key-point and the feedback element for steel health monitoring, which is calibrated by reference methods in the surface or the bulk of steels. After monitoring the residual stress distribution on the surface of the examined steel, the necessary following action refers to the stress gradient rehabilitation using EMIT. Components and assemblies considered perfect by the conventional inspection methods, will by this methodology be subjected to stress rehabilitation so as to obtain optimal stress gradient levels to allow for long-life operation. The trans-disciplinary dependence is clear between magnetic and structural properties of steel as well as, most importantly, between technical considerations and economic factors. Conclusively, our monitoring and rehabilitation methods are the tools for the new operation approach regarding steel structures in shipping industry, including offshore applications, especially in extreme environments and conditions. Furthermore, the combination of the stress monitoring and the stress annihilation/control systems will provide a new design concept for the manufacturing of ships and offshore platforms, particularly in welding process, aiming at establishing a new method for optimization of steel structure manufacturing.

References

[1] Robert M. Conachey, Randal L. Montgomery: "Application of Reliability-centered Maintenance Techniques to the Marine Industry". Originally presented at Institute of Marine Engineers, India, November 2002.
[2] Inho Kim: "Ten years after the enactment of the Oil Pollution Act of 1990: a success or a failure". College of Law, Ewha Womans University, 11-1 Daehyun-dong, Seodaemun-gu, Seoul, 120-750, Republic of Korea
Available online 21 February 2002.
[3] P. Vourna, A. Ktena, A. Mpalliou, A. G. Mamalis, E. Hristoforou: "Steel Health Monitoring Using Magnetic Techniques", Materials Science Forum, Vol. 792, pp. 139-143, 2014.
[4] PCT Submitted (I.P.O. P.N.: 3427-13/06/2016) and pending patent: "Electromagnetic hammer device for the mechanical treatment of materials and method of use thereof". Inventor: Stylianos Mores.
[5] E. Hristoforou and P. Dimitropoulos: "A new method for M-H and λ-H determination using the magnetostrictive delay line technique". Journal of Magnetism and Magnetic Materials, 304, pp 164-167, 2006.

Gap Bridging by Advanced Adaptive Weaving controlled GMA Welding of High Strength Steel Crane Structures

Sven-F. Goecke
Technical University of Applied Sciences Brandenburg, Brandenburg an der Havel, Germany
goecke@th-brandenburg.de

Abstract
In mobile cranes under limited gross vehicle weight, the aim is an increase of the performance characteristics by a considerable reduction the total mass of the vehicle undercarriage. Therefore, a challenging lightweight design was realised, not only by introducing high strength fine-grained steels mild steels but also by a totally newly developed integrated lightweight monocoque design. Furthermore, pre-manufacturing tolerances and weld distortion are the motivators to implement sensors in GMAW in order to create a robust robot welding process, adapting the real situation in a similar manner like professional manual welders are qualified to, but with much higher travel speeds and at minimum heat input. To compensate varying gaps, causing weld defects, two approaches were carried out, an in-process measuring of the joint edges position including the gap geometry and second, a temperature field measuring on the weld and work piece surface. In single-pass welding of 2.5 mm HSS corner welds with varying gaps up to 3.0 mm without backing, two sensors were implemented, a high-power Laser camera and a two-colour pyrometer camera. Thus, the hitherto manual welding was completely substituted by a fully mechanised robot welding under rough industrial condition like large tolerances. The approach was to toughen up a commercial GMA welding robot by integration of real-time control to provide a mechanised zero-defect high strength steel welding production of lightweight cranes. The resulting real-time compensation of a wide range of typical industrial joint tolerances enables a zero-defect GMAW robot welding of HS steel in lightweight mobile cranes for the first time.

Introduction
High strength steels are increasingly used not only in automotive but also in mobile crane structures. According to national road traffic regulations, the gross vehicle weight is limited to e.g. 3.5 t for trailers or 7.5 t for trucks. Since the mobile truck and trailer crane manufacturing industry underlies a hard competition in Western Europe, the development of modern crane structures is conducted by a minimising of the total mass accompanied by a concurrent maximising of strength and stiffness, condensed in a multiparametric lightweight design of the crane undercarriage and the revolving superstructure. The challenge of this approach is, to enhance the cranes performances such as the crane hook height, the lifting capacity, the work height and radius without weight increase in the overall vehicle mass. On the other hand, with reduced sheet thicknesses, large pre-manufacturing and positioning tolerances as well as welding distortion comes in the fore and cause varying gaps, leading to weld defects right up to rejects on the other hand. First attempts on adaptive in-line controlling the welding parameters in dependence of the actual gap between overlapped ultra-high strength Boron Manganese steels were made in automotive application, [1]. A determination of the temperature field was investigated on the melt pool in GMA welding, [2, 3]. On basis of an in-situ quotient pyrometer measurement during the running arc welding process [3] realised its application for a heat input control.

Approach
With the introduction of higher strength steel associated with thinner sheets, the increased tendency of pre-manufacturing tolerances and weld distortion were the motivators to implement optical sensors in GMAW in order to create a robust robot welding process, adapting the real situation in a similar manner like professional manual welders are qualified to, but with much higher travel speeds and at minimum heat input.

To create a robust welding in automated rough industrial environments, two approaches were carried out in this work, an in-process measuring of the joint edges position including the actual gap geometry between the work pieces and second, a temperature field measuring on the weld and the work piece surface.

Experimental Set-up
This application is based on a single pass welding of 2.5 mm S500MC and S700MC HS steel corner weld pieces

with varying gaps from zero up to 3.0 mm without any backing. Two sensors were thus implemented, a single-strip-Laser high power triangulation camera and additionally, a 2-colour quotient pyrometer camera. For robot welding based on the integration of a Laser camera system, a real-time control has been developed by adapting the welding parameters wire feed rate WFR, welding travel speed TS and weaving amplitude WA and frequency WF to the current joint situation.

Fig. 1 shows the welding set up exemplary for corner joints with a wedge-like raising gap from 0.0 … 3.0 mm. The precursory installed Laser triangulation camera Coletrax SH 5-HR detects the current gap. Besides the seam tracking well known in many industrial applications, the Laser camera was applied to the measure the actual gap in order to adapt the welding process parameters all as a function of it, Fig. 2 right.

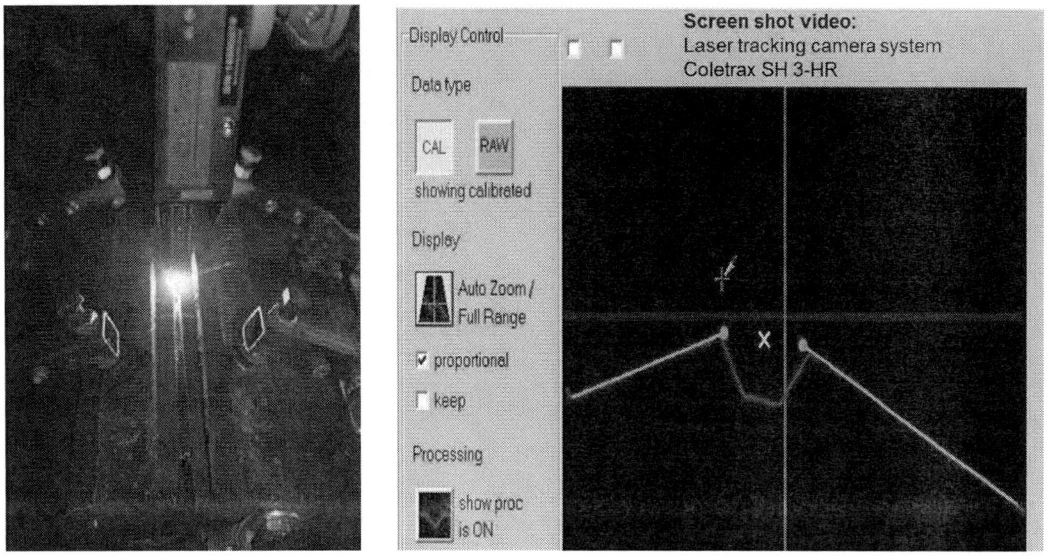

Fig. 1: Experimental welding set-up (left) for the gap variation at corner joints with integrated Laser-camera, screen shot of the thus identified edges and the gap measrued by means of image processing (right)

Results and discussion

By an extensive experimental-analytical analysis on ensuring appropriate welding results, the process characteristics were identified to finally achieve a 4-parameter-algorithm as a function of the in-line during the welding procedure measured actual gap. For robot welding of complex mobile crane frame structures, a real-time control has been developed by means of adapting the welding process parameters to the actual joint situation in order to preserve the maximal joint strength under rough industrial environments.

Furthermore, a 2-colour HDRC pyro camera IMS Chips PyroCam was simultaneously implemented for measuring the trailing transient temperature field on the joint and the melt pool and also in order to obtain the real cooling rates in the HAZ during the running welding procedure. This PyroCam enables an absolute in-situ temperature measuring without any knowledge of the non-accessible specific emission coefficients [4], Fig. 2.

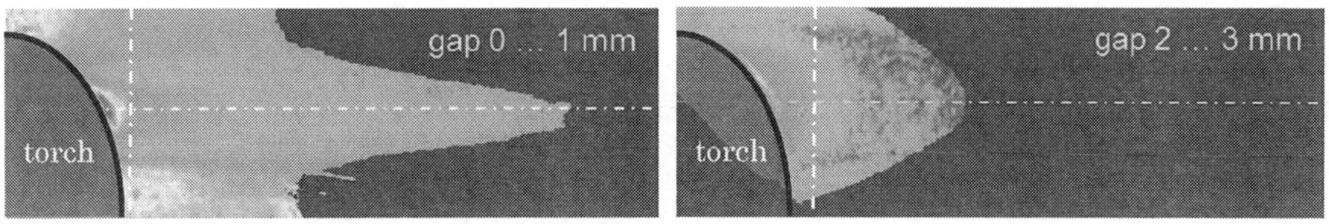

Fig. 2: Temperature profiles along the weld as function of the gap width, the doted lines mark the profiles along (horizontally) resp. across (vertically) the weld. Welding travel direction to the left.

From the temperature fields in Fig. 2 specific line profiles have been extracted along and across the weld, Fig. 3. These profiles depict relatively longer temperature fields resp. weld pool length paired with a narrower weld width in case of small gaps and all reverse for larger gaps. An extended cooling rate goes along with smaller gaps. This indicates the influence of the real-time controlled welding parameters, e.g. the welding travel speed TS that is lowered markedly in order to bridge the larger gaps.

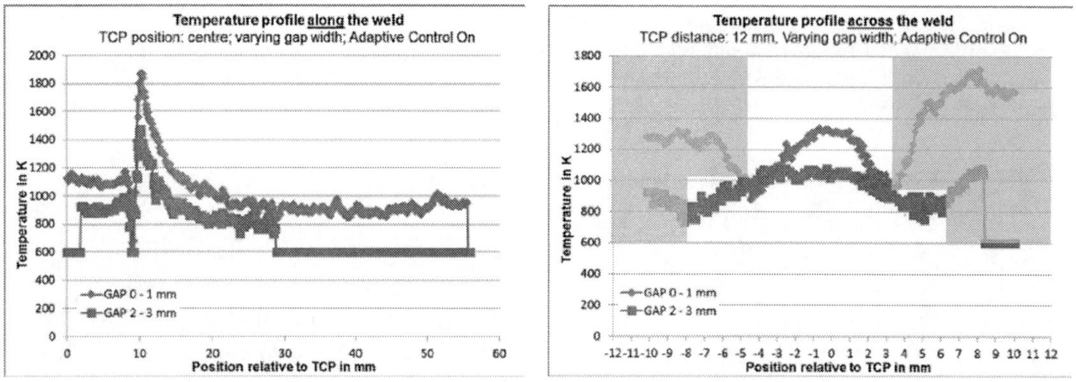

Fig. 3: Temperature profiles along (left) and across (right) the weld as function of the gap width (outer overexposed areas have been greyed)

In Fig. 4 the effect of the gap-adapted welding parameter on the corner joint hardnes profiles is depicted. It can be deduced that, a tendency to wider and a bit lower hardness profile appears with larger gaps, although an absolute deviation of less than 20 HV1. The reason is the higher heat input caused by the markedly lowered travel speed.

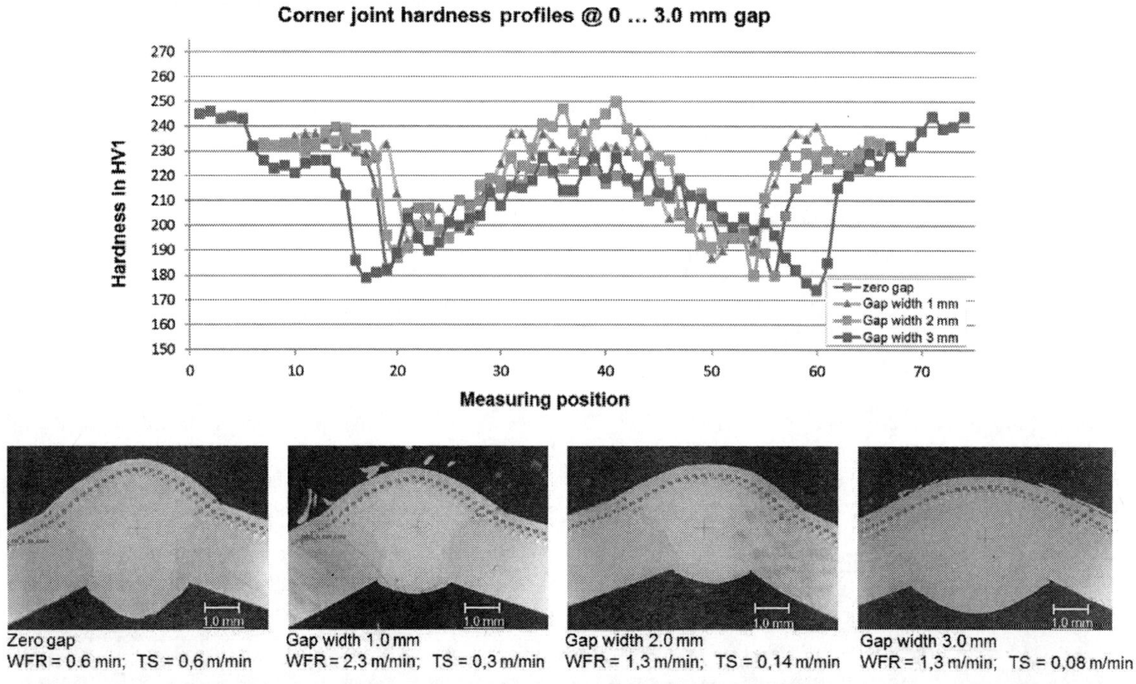

Fig. 4: Hardness profiles as function of the gap width and corresponding cross sections with hardness identions

A real-time control was designed and realised by means of a multi-parameter-algorithm as a function of the current gap measured in-line precursory to the arc and in-situ during the welding procedure. The controlled parameters were (TS, WFR, WA, WF) = f (gap). For this purpose, two alternative principles have been realised, an easy to handle look-up table and furthermore an exponential function set extracted from a regression analysis.

Besides the single-pass welding of 2.5 mm HSS corner welds, also butt welds and flange welds with varying gaps up to 3.0 mm were realised without any backing to prove this system.

Conclusions

This approach of two a stepped in-situ sensed and feed-back controlled GMA welding process was settled for an gap-adapted optimisation of the multiple parameters during the welding procedure of crane components under the presence of large and varying gaps in order to avoid any reject. Concluding, this work represents a successful approach in realising an innovative new integrated monocoque lightweight design for mobile truck and trailer cranes that will be welded robustly with an adaptively controlled welding robot.

The final overall result is a new crane generation with up to 20 % enhanced work height and radius at constant total crane vehicle weight, that is now manufactured in a newly developed fully mechanised robot production line. The resulting real-time adaptive feed-back controlled compensation of a wide range of typical industrial joint tolerances enables a zero-defect GMA robot welding of HS steel in lightweight mobile cranes under rough industrial environments for the first time.

Acknowledgments

The author like to thank the German Federal Ministry of Economic Affairs and Energy (BMWE) within the scope of the "The Central Innovation Programme for SMEs - ZIM" (BMWE KF2337707DF4) for supporting the carried out research project "ZD-LeiK".
Further, the author gratefully acknowledge BIAS - Bremen Institute for Applied Beam Technology GmbH for supplying a test version of the evaluation and control software developed by them for the IMS Chips PyroCam.
We also thank Linde Gas for supplying gas and for sheet material.

References

[1] S.-F. Goecke and M. Ebert-Spiegel. Towards a more robust automated MAG welding. Proc. *IEEE CASE 2015, The eleventh annual IEEE Int. Conf. on Automation Science and Engineering Gothenborg*, August 24 to 28, 2015

[2] R Kozakov, G Gött, H Schöpp, D Uhrlandt, M Schnick, M Häßler, U Füssel, S Rose. Spatial structure of the arc in a pulsed GMAW process. *Journal of Physics D: Applied Physics*, vol. 46, no. 22 (2013), pp. 224001

[3] S.-F Goecke, Y. Ogino, Y. Hirata, 2015. In-situ temperature determination of MAG welding and its application for heat input control. *IIW-Doc.XII-2254-15. 68th IIW Annual Assembly*. Helsinki, Finland, 28th June - 3rd July 2015, p. 35-36

[4] F.X. Hutter, D. Brosch, H.-G. Graf, W. Klingler, M. Strobel, J. N. Burghartz, "A 0.25 µm Logarithmic CMOS Imager for Emissivity-Compensated Thermography", *Proc. IEEE ISSCC 2009, Sensors and Mems*, February 11, 2009, pp. 354-355

October 14th
(Friday)

Grain boundary engineering to suppress weld degradations

Hiroyuki Kokawa
Tohoku University, Department of Materials Processing, Sendai, Japan
Email: kokawa@tohoku.ac.jp

Abstract
Grain boundary engineering (GBE) primarily intends to prevent the initiation and propagation of intergranular degradation along random boundaries by frequent introduction of coincidence site lattice (CSL) boundaries into the grain boundary networks in polycrystalline materials. A high frequency of CSL boundaries by GBE thermomechanical processing results in disruption of random boundary connectivity and leads to high resistance to grain boundary degradations. The resulting austenitic stainless steels indicated remarkably high resistance to intergranular corrosion even to weld-decay and knife-line attack during welding. GBE could have a high potential to suppress many of grain boundary-related degradations of welded polycrystalline materials.

Introduction
Grain boundary phenomena strongly depend on the grain boundary structure and character, and low energy grain boundaries such as CSL boundaries have strong resistance to intergranular degradations, as contrasted with high-energy random boundaries [1,2]. The concept of 'grain boundary design and control' [3] has been developed and refined as grain boundary engineering [4]. Grain boundary engineered materials (GBEMs) which are characterized by high frequencies of CSL boundaries are resistant to intergranular degradations of materials. The author's group has achieved very high frequencies of CSL boundaries in austenitic stainless steels by a single-step slight strain plus anneal GBE processing [5,6]. The GBEMs showed excellent intergranular corrosion resistance indicating significantly smaller corrosion rates than those of the as-received base materials. Also the GBEMs indicated strong resistance to intergranular degradations during welding such as weld-decay [7] and knife-line attack [8] in austenitic stainless steels. This paper introduces GBE of austenitic stainless steels by the single-step thermomechanical processing and discusses the potential of GBE to suppress intergranular degradations during welding by reviewing our GBE studies [5-11].

Thermomechanical GBE processing
As an example, the optimization during the single-step thermomechanical processing with 304 austenitic stainless steel is shown as follows [5]. The 304 base material (BM) was pre-strained and annealed. The pre-strain was given by cold-rolling, resulting in a 0-60% reduction of specimen thickness. The ranges of annealing temperatures and times were 1200-1600 K and 0.5-72 h, respectively. The maximum CSL frequency of 87% was obtained at 5% pre-strain and annealing at 1200 K for 72 h for the 304 steel. The optimized 304 steel was called as 304 GBEM. The same optimization sequences for many other commercial austenitic stainless steels, such as 304L, 316, 316L, 321, 347 steels, etc. [6,8,9] resulted in high CSL frequencies of over 80% during at small pre-strain plus annealing during the single-step thermomechanical processing. The grain boundary character distributions (GBCDs) of BM and GBEM for 304 steel are shown in grain boundary maps by electron backscatter diffraction (EBSD) technique in Fig. 1. In the maps, black and gray lines indicate random and CSL boundaries, respectively. The corrosive random boundary network is well fragmented by CSL boundaries in 304 GBEM. Fragmentation and disconnection of random boundary network were also observed in GBCDs of GBEMs with the other examined commercial austenitic stainless steels [8].

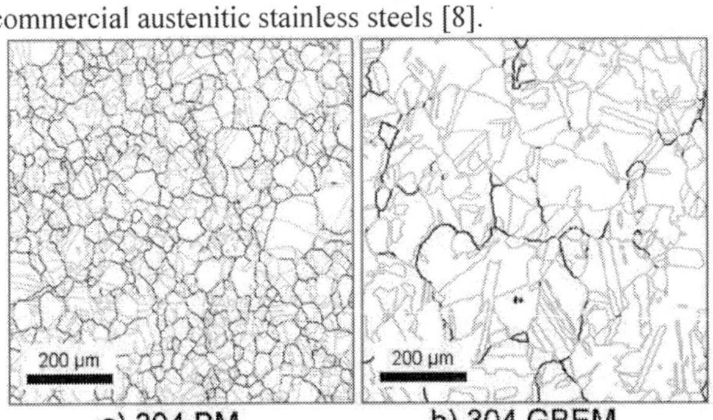

Fig. 1 GBCDs of 304BM and 304GBEM by EBSD [5].

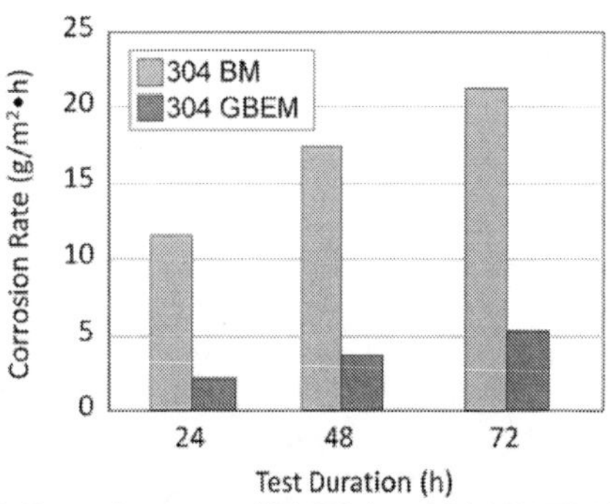

Fig. 2 Corrosion rate of 304BM and 304GBEM [5].

Suppression of intergranular corrosion by GBE of 304 steel

Figure 2 shows the corrosion rate of the 304 BM and 304 GBEM during ferric sulfate-sulfuric acid test after sensitization. The intergranular corrosion propagates three-dimensionally along grain boundaries and causes mass-loss due to grain dropping from the surface of material during the ferric sulfate-sulfuric acid test. The corrosion (mass-loss) rate of GBEM was less than one-fourth that of BM for every test duration in the 304 steel, as shown in Fig. 2 [5]. The uniform distribution of highly dense CSL boundaries in GBEM suppressed the deep propagation of intergranular corrosion from the specimen surface due to extensive disruption of the random boundary network which transmits intergranular degradation, as schematically illustrated in Fig. 3.

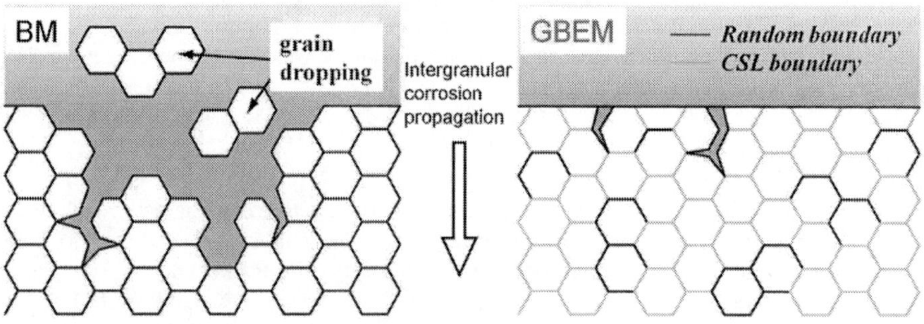

Fig. 3 Suppression mechanism of intergranular corrosion from the surface by GBE [8].

Suppression of weld-decay by GBE of 304 steel

Figure 4 shows the optical micrographs in the weld HAZs for the 304 BM (a) and the 304 GBEM (b) which were obtained from the areas shown in the schematic illustration of the weld cross-section perpendicular to the welding direction. Typical weld-decay region is detected in the weld HAZ of 304 BM, i.e., characteristic deep grooving along grain boundaries due to sensitization is clearly seen between 4 mm and 12 mm areas away from the weld fusion interface (Fig. 4(a)). In contrast, the 304 GBEM did not show any clear grain-boundary grooving in the weld HAZ (Fig. 4(b)). Further examinations confirmed that the suppression effect by GBE on intergranular corrosion and weld-decay of 304 steel was maintained even during post-GBE strain-sensitization and –welding [10,11].

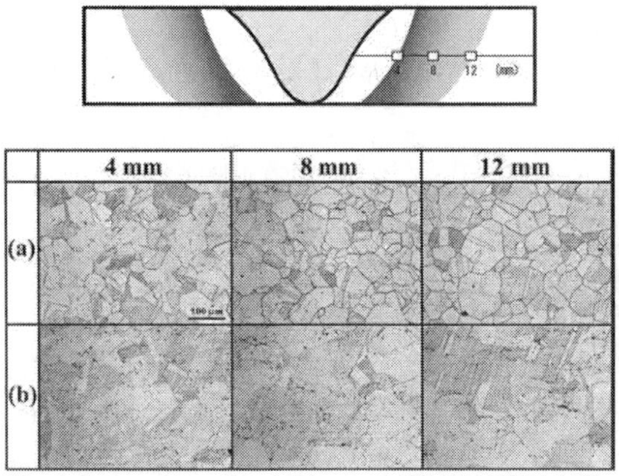

Fig. 4 Suppression of weld-decay, (a)BM (b)GBEM [7].

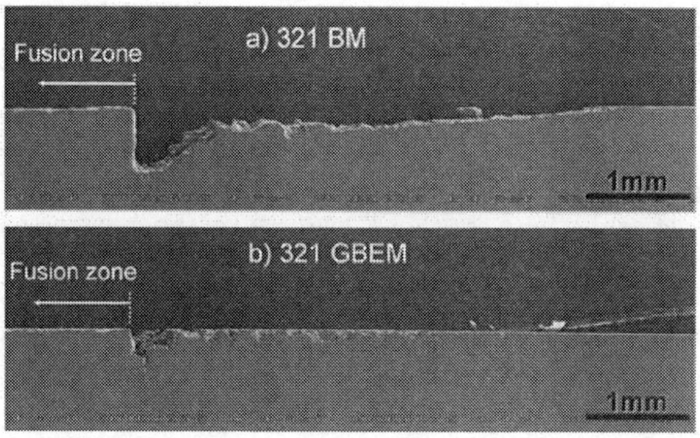

Fig. 5 Suppression of knife-line attack in 321 weld HAZ [8].

Suppression of knife-line attack by GBE of 321 steel

A titanium stabilized 321 austenitic stainless steel with a commercial chemical composition was solution-heat treated and termed the starting base material (321 BM) with a low CSL frequency of 53%. The 321 BM was thermomechanically processed by the single-step pre-strain plus annealing. The pre-strain was given by cold-rolling, resulting in a 0-5% reduction of specimen thickness. The ranges of annealing temperatures and time were 1220-1300 K for 72 h. The thermomechanically processed 321 steel specimens were annealed at 1173 K for 1 h to stabilize carbon as TiC by titanium so as to prevent chromium carbide precipitation. The specimen which achieved an optimum GBCD with the highest CSL frequency of 89% was termed as 321 GBEM [9]. The stabilized 321 BM and 321 GBEM blocks were alternately aligned together and welded at one same pass by bead-on-plate GTA welding. Figure 5 shows the optical micrographs of the cross-sectional views perpendicular to the welding directions in the HAZs for 321 BM (a) and 321 GBEM (b) after sensitization and subsequent ferric sulfate-sulfuric acid test. Deep knife-line attack corrosion is detected in the weld HAZ of 321 BM, while it is significantly suppressed in that of the 321 GBEM. The 321 GBEM could maintain the high frequency of CSL boundaries over 80 % in the whole weld HAZ. This fact approves that the optimized GBCD of the 321 GBEM possesses a high thermal-stability during welding and shows the high resistance to the knife-line attack corrosion of stabilized austenitic stainless steels [8,9], as same as the case of weld-decay in non-sensitized austenitic stainless steels [5,6].

Conclusions

The twin-induced GBE utilizing the optimized single-step thermomechanical processing with small pre-strain and subsequent annealing, introduced a very high frequency of CSL boundaries into austenitic stainless steels. The resulting steels indicated remarkably high resistance to intergranular corrosion even to weld-decay and knife-line attack during welding. GBE has a high potential to suppress variety of grain boundary-related degradations of welded austenitic stainless steels.

Acknowledgments

The author wishes to thank to Professor Z.J. Wang, Professor Y.S. Sato, Professor H. T. Fujii, Dr. M. Shimada, Dr. W.Z. Jin, Dr. M. Michiuchi, Mr. T. Inoguchi, Mr. K. Sakai, Mr. M. Miyagi, Mr. T. Yokoyama, Mr. S. Sato, Mr. T. Oyamada, Mr. K. Kurihara, Mr. G. Yamada, Ms. M. Ikeshoji, Mr. S. Tokita, Mr. Y. Suzuki, Mr. S. Kodama, and Mr. A. Honda for their useful discussions and technical assistance.

References

[1] P.H. Pumphrey: "Special high angle grain boundaries" in Grain boundary structure and properties, Academic Press (London, 1976), pp. 139-200.

[2] H. Kokawa, M. Shimada, Y.S. Sato: "Grain-boundary structure and precipitation in sensitized austenitic stainless steel," *JOM*, Vol. 52, No. 7 (2000), pp. 34-37.

[3] T. Watanabe: "An approach to grain-boundary design for strong and ductile polycrystals," *Res Mechanica*, Vol. 11, No. 1 (1984), pp. 47-84.

[4] G. Palumbo, E.M. Lehockey, P. Lin: "Applications for grain boundary engineered materials," *JOM*, Vol. 50, No. 2 (1998), pp. 40-43.

[5] M. Shimada, H. Kokawa, Z.J. Wang, Y.S. Sato, I. Karibe: "Optimization of grain boundary character distribution for intergranular corrosion resistant 304 stainless steel by twin-induced grain boundary engineering," *Acta Materialia*, Vol. 50, No. 9 (2002), pp. 2331-2341.

[6] M. Michiuchi, H. Kokawa, Z.J. Wang, Y.S. Sato, K. Sakai: "Twin-induced grain boundary engineering for 316 austenitic stainless steel," *Acta Materialia*, Vol. 54, No. 19 (2006), pp. 5179-5184.

[7] H. Kokawa, M. Shimada, M. Michiuchi, Z.J. Wang, Y.S. Sato: "Arrest of weld-decay in 304 austenitic stainless steel by twin-induced grain boundary engineering," *Acta Materialia*, Vol. 55, No. 16 (2007), pp. 5401-5407.

[8] H. Kokawa: "Potential of grain boundary engineering to suppress welding degradations of austenitic stainless steels," *Science and Technology of Welding and Joining*, Vol. 16, No. 4 (2011), pp. 357-362.

[9] K. Kurihara, H. Kokawa, S. Sato, Y.S. Sato, H.T. Fujii: "Grain boundary engineering of titanium-stabilized 321 austenitic stainless steel," *Journal of Materials Science*, Vol. 46, No. 12 (2011), pp. 4270-4275.

[10]G. Yamada, H. Kokawa, Y. Yasuda, S. Tokita, T. Yokoyama, Y.S. Sato, H.T. Fujii, S. Tsurekawa: "Effect of post-GBE strain-sensitisation on corrosion resistance of grain boundary engineered 304 austenitic stainless steel," *Philosophical Magazine*, Vol. 93, No. 10-12 (2013), pp. 1443-1453.

[11]S. Tokita, M. Ikeshoji, G. Yamada, K. Kurihara, H. Kokawa, Y.S. Sato, H.T. Fujii: "Effect of post-GBE strain on weld decay resistance of grain boundary engineered 304 austenitic stainless steel," *Science and Technology of Welding and Joining*, Vol. 20, No. 8 (2015), pp. 702-707.

Computational Simulation and Optimization of Steel Weldment for Thermal Load Reduction

Charles R. Fisher, Matthew F. Sinfield

Naval Surface Warfare Center - Carderock Division, USA

The integrated computational materials engineering (ICME) paradigm has shown expanded use in manufacturing industries by linking software tools to improve efficiency and decrease costs. For welding, computational weld mechanics (CWM) and design of experiments (DOE) show promise to enable reduction in production costs before even striking an arc. This project used SYSWELD and MiniTab software to systematically vary welding joint details and heat inputs to study variable sensitivity to local heat transfer from welding. A full-factorial design investigated how heat input, plate thickness, backing bar width, and backing bar thickness could be modified to optimize both peak temperature and time at elevated temperature of interior surfaces. Results enabled reduction of physical testing of the various weld configurations from sixteen specimens to only two specimens, saving on both material and production costs.

Comparison of forming, welding and heat treatment simulations in LS-DYNA and MSC Marc

Joachim Steffenburg-Nordenström
GKN Aerospace Sweden, Trollhättan, Sweden
joachim.steffenburg-nordenstrom@gknaerospace.com, +46 520 9400

Lluís Pérez Caro, Eva-Lis Odenberger
IUC, Olofström, Sweden
lluis.perez@iuc-olofstrom.se, eva-lis.odenberger@iuc-olofstrom.se, +46 454 97500
Mats Oldenburg
Luleå University of technology, Luleå, Sweden
mats.oldenburg@ltu.se, +46 920 491752
Lars-Erik Svensson
University West, Trollhättan, Sweden
lars-erik.svensson@hv.se, +46 520 223000

Abstract
The manufacturing of components for aero engine structures from a flat sheet to the final shape usually requires several steps that may introduce residual stresses and shape distortions in the part. Depending on the magnitude, sign and distribution with respect to the stresses induced by the service load, the remaining stresses may affect the service life of a component, especially when submitted to cyclic loading. Nowadays, several types of software that have the ability to predict the residual stresses and the final shape of a component subjected to various process steps are available. However, literature shows a lack of comparison studies among different software tools for multi-step simulations of a manufacturing process. In this study, the manufacturing process chain of an aerospace component including forming, welding and heat treatment in the nickel-based superalloy 718 is modelled and simulated using the two finite element software codes LS-DYNA and MSC.Marc. The results from the displacement of the blank in the punch stroke direction, the equivalent plastic strain and the von Mises stress are compared between both FE codes. The displacement of the blank after forming is slightly higher in LS-DYNA compared to MSC.Marc, as well as the equivalent plastic strain and the von Mises stress values. This tendency is also observed after trimming and welding. It can also be noted that the distribution of both strains and stresses on the trimmed and welded parts varies between the two compared codes, presumably due to the choice of different solver options, explicit and implicit.

Introduction
Residual stresses are present in many fabricated structures due to local plastic deformation from both thermal and mechanical operations during the manufacturing process. The remaining stresses will affect the service life of a component especially when submitted to cyclic loading. The residual stresses could either be beneficial or detrimental depending on magnitude, sign and distribution with respect to the stresses induced by the service load. Mostly, there are several manufacturing steps carried out sequentially and therefore it is important how the remaining stresses evolve throughout the manufacturing routine. There are several FE software tools available that can predict the material behavior during the manufacturing process chain i.e. ABAQUS and MORFEO, to predict weld residual stresses by simulations. ABAQUS is a multi-purpose FE tool while MORFEO is more dedicated to the welding process [1]. A comparison of the FE software tools ABAQUS and DEFORM for simulation of the face turning of an Inconel 718 disk with experimental validation can be found in [2]. Moreover, three-dimensional welding simulations were carried out with the FE software tool MSC.Marc and two-dimensional welding simulations in MSC.Marc and ANSYS [3]. Literature available shows a lack of comparison studies among different software tools for multi-step simulations of a manufacturing process.
In this paper, a comparison between the finite element software LS-DYNA and MSC.Marc is done in a simulation of a manufacturing process chain comprising sheet metal forming with springback and trimming, welding and stress relief heat treatment.

Method

The prerequisites are that both software tools use an equivalent FE discretisation (mesh) and geometry for the blank and the forming tools. Moreover, the mesh was kept constant during all simulation steps (no refinement of the mesh) and the same material data is used for the nickel-based aerospace material alloy 718. Both FE tools use the Goldak double ellipsoid formulation for the shape of the heat source [4], while the stress relaxation in the stress relief heat treatment the creep law is driven by *Eq. 1*, as implemented in both MSC.Marc and LS-DYNA [5, 6], where A, m and n are material constants which have to be determined by material testing.

$$\dot{\varepsilon}_{creep} = A\sigma^m T^n \qquad \textit{Eq. 1}$$

The setup and meshing of the blank, punch and die is done in LS-PrePost v4.3 (hereafter denoted as LS-PrePost), while the simulations are carried out in LS-DYNA v971 (hereafter denoted as LS-DYNA) [6]. The LS-PrePost is used for pre- and post-processing of calculations in the finite element software LS-DYNA. Both tools are developed by the Livermore Software Technology Corporation (LSTC). The second FE tool used is MSC.Marc (hereafter denoted as Marc) and its pre- and post-processor MSC.Mentat [5] (hereafter denoted Mentat). The mesh was done in LS-PrePost and thereafter transferred to Marc. The mesh consists of eight-node hexahedral solid elements, where the initial mesh had 37,200 elements and final mesh had 6,188 elements. The analysis was carried out using Marc and using the assumptions for all steps regarding large deformations, strains and plastic incompressibility. In Marc the staggered approach was used to solve the coupled thermo mechanical problem and constant temperature was used in each element in the mechanical step [7]. LS-DYNA uses explicit solver for the forming simulation and implicit solver for the subsequent steps while Marc uses the implicit method throughout the simulation chain. The results from the last time step for each simulation are stored and imported as the initial conditions to the next simulation.

Forming

The forming simulation consists of three parts, the blank, and the die and punch tools. On the one hand, the die and punch are in the LS-DYNA case made of shell elements acting as rigid bodies; on the other hand, the forming tooling in the Marc case is made of surfaces which are acting as rigid bodies, while in both cases the blank is made of four solid elements through the thickness, having a nominal thickness of 2.54 mm. The draw beads are represented by constraining the nodes on the edges during the operation. The punch stroke length in the forming operation was about 88 mm and gave in Marc at most a deformation at the sheet of 77 mm which became 75 mm after tool release, *Fig. 1*, where also the equivalent plastic strain and the von Mises stress after forming can be seen. The deformation obtained in LS-DYNA after tool release was 78 mm, see *Fig. 2*. The trimming operation in Marc gave a small decrease in stress due to elastic springback as shown in *Fig. 3*. Similar results have been obtained after trimming in LS-DYNA, see *Fig. 4*. The major part of the plastic strains and the stresses after forming is within the part of the blank that trimmed away which is showed in the comparison between *Fig. 1* and Fig. *3* of the results from Marc and similar for LS-DYNA in *Fig. 2* and *Fig. 4*. Furthermore, the trimming operation gave a small reduction in stress due to elastic springback as shown when comparing *Fig. 1C* with *Fig. 3B* and *Fig. 2c* with *Fig. 4b*.

Fig. 1: Displacement in the punch stroke direction [mm] (A), equivalent plastic strain (B) and von Mises stress [MPa] (C) after the forming and before the trimming operation in Marc.

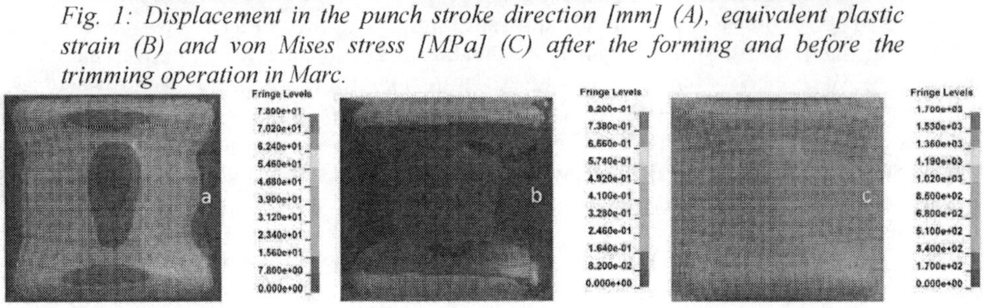

Fig. 2: Displacement in the punch stroke direction [mm] (a), equivalent plastic strain (b) and von Mises stress [MPa] (c) after the forming and before the trimming operation in LS-DYNA.

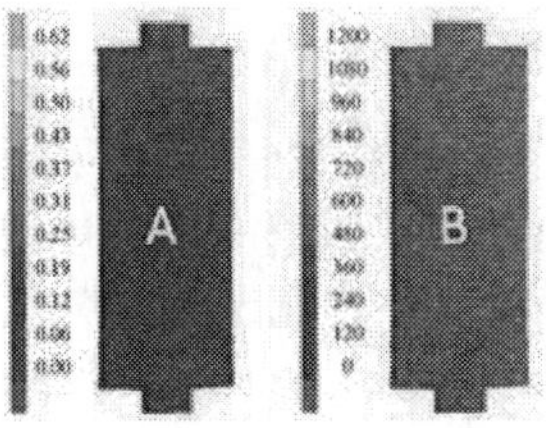

Fig. 4: The equivalent plastic strain (A) and the von Mises stress [MPa] (B) after the trimming operation in Marc.

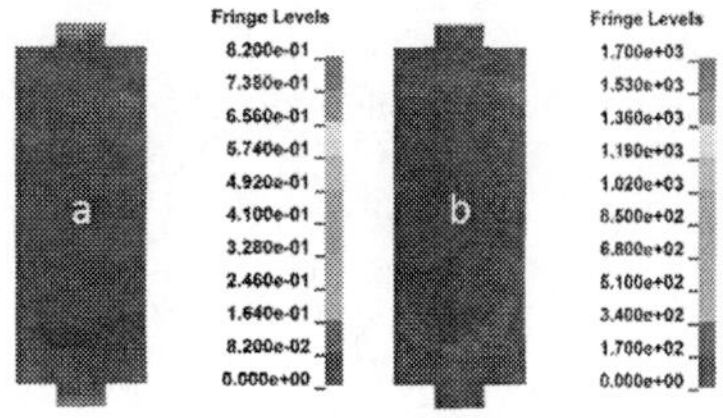

Fig. 3: The equivalent plastic strain (a) and the von Mises stress [MPa] (b) after the trimming operation in LS-DYNA.

Welding

The welding simulation started from the formed blank after springback and trimming to the wanted geometry. The weld simulated was a bead on plate weld, meaning a weld on one piece without an existing gap. The parameters used for the weld correspond to a TIG (Tungsten Inert Gas) weld, with the welding source travelling at 2.8 mm/s. The welding simulation is done in LS-DYNA using the material model *MAT_CWM with ghost element functionality [8]. In *Fig. 5* and *Fig. 6* the simulated plastic strain and von Mises stress after release from fixture are shown done in Marc and LS-DYNA respectively. The accumulated equivalent plastic strain history is reset to the materials initial state where the weld melt pool has passed as is shown *Fig. 5A* and *Fig. 6a* in Marc and LS-DYNA respectively.

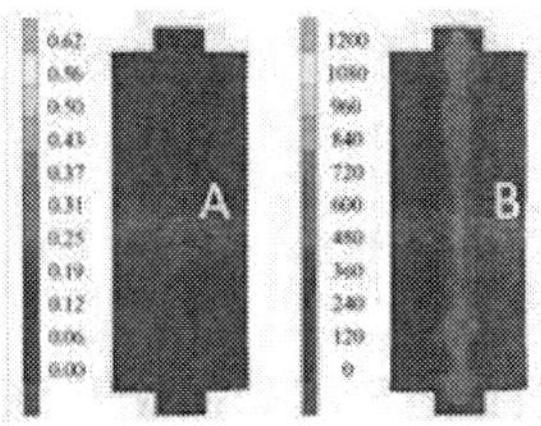

Fig. 5: The equivalent plastic strain (A) and the von Mises stress [MPa] (B) after the welding operation in Marc.

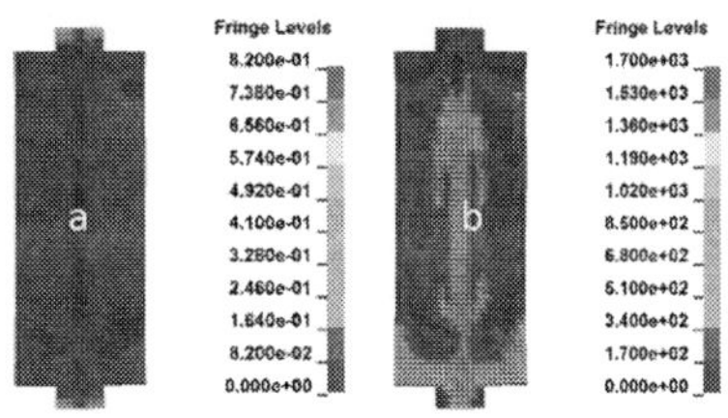

Fig. 6: The equivalent plastic strain (a) and the von Mises stress [MPa] (b) after the welding operation in LS-DYNA.

Stress relief heat treatment

The stress relief heat treatment consists of heating the part to an elevated temperature, holding it for a specific amount of time and cooling it down to room temperature. The stress relaxation phenomenon is described mathematically in *Eq. 1* and by test in [9-11]. The creep model used in LS-DYNA is found in the *MAT_188 keyword. As a future work, a stress relief heat treatment simulation will be carried out to capture the stress relaxation and change of the geometry due to shape distortions.

Conclusions

The results implies that both software tools are capable of simulating a manufacturing process chain which here comprises sheet metal forming, welding and stress relief heat treatment. However, there are discrepancies in the result levels between the software tools. The equivalent strain level is about 30% higher in LS-DYNA than for the result in Marc. Also the corresponding stress level is about 40% higher in LS-DYNA than for Marc. This discrepancy might be due to the different solver option (implicit and implicit) used in the forming simulation for each software tool. Thus, the allover behavior in strains and stresses ar similar for the results in MSC.Marc and LS-Dyna.

Acknowledgments

The work presented in this paper has been carried out, with financial support from the Swedish KK-foundation, within the research school SiCoMaP (Simulation and Control of Material affecting Processes) at University West Trollhättan Sweden and also the support from GKN Aerospace Sweden AB, VINNOVA – Swedish Governmental Agency for Innovation Systems NFFP6 program for SME, Swedish Armed Forces and Swedish Defence Materiel Administration is greatly appreciated. Grant No. 2013-01173.

References

[1] C. Acevedo, *et al.*: "Residual Stresses in As-Welded Joints: Finite Element Modeling and Neutron Difraction Stress Measurements," *Key Engineering Materials*, Vol. 488, No. (2012), pp. 335-338.
[2] T. Ozel, *et al.*: "3D finite element modelling of chip formation process for machining Inconel 718: comparison of FE software predictions," *Machining Science and Technology*, Vol. 15, No. 1 (2011), pp. 21-46.
[3] Z. Barsoum and A. Lundbäck: "Simplified FE welding simulation of fillet welds–3D effects on the formation residual stresses," *Engineering Failure Analysis*, Vol. 16, No. 7 (2009), pp. 2281-2289.
[4] J. Goldak, *et al.*: "A new finite element model for welding heat sources," *Metallurgical transactions B*, Vol. 15, No. 2 (1984), pp. 299-305.
[5] "*MSC Marc Mentat Manual*," (2013, 20150610), Available: http://simcompanion.mscsoftware.com/infocenter/index?page=content&channel=DOCUMENTATION&cat=MARC_DOCUMENTATION.
[6] Livermore Software Technology Corporation: "*LSTC Product Manuals*," (2016, 20160613), Available: http://www.lstc.com/download/manuals.
[7] L. Lindgren: Computational Welding Mechanics. Thermomechanical and Microstructural Simulations, Woodhead Publishing (Cambridge, 2007).
[8] M. Schill and E.-L. Odenberger: "Simulation of Residual Deformation from a Forming and Welding Process using LS-DYNA," *13th International LS-DYNA Users Conference*, 2014.
[9] J. Ritter and R. McPherson: "Anisothermal stress relaxation in a carbon-manganese steel," *J Iron Steel Inst*, Vol. 208, No. 10 (1970), pp. 935-941.
[10] C. Leymonie: "Utilization of Anisothermal Stress Relaxation Tests," *International Conference on Engineering Aspects of Creep*, 1980, pp. 115-120.
[11] J. Steffenburg-Nordenström: "The evolution of residual stresses in a stress relief heat treatment of test specimen of alloy 718," *Proceedings of the 11th International Symposium on Friction Stir Welding*, 2015.

Cross-Wire Projection Welding of Aluminium Alloys in relation to Pneumatic and Servomechanical Electrode Force Systems

Zygmunt Mikno
Welding Institute, Gliwice, Poland
Email: Zygmunt.Mikno@is.gliwice.pl, phone: +48 32 33 58 360

Boguslaw Grzesik, Mariusz Stepien
The Silesian University of Technology, Gliwice, Poland
Email: mariusz.stepien@polsl.pl, boguslaw.grzesik@polsl.pl, phone: +48 32 237 12 47

Abstract

The article presents the analysis of the cross-wire projection welding of aluminium wire in relation to conventional, i.e. pneumatic, and servomechanical electrode force systems. The analysis was performed using FEM calculations, which were next subjected to experimental verification. The calculations and experiments were performed using wire made of Al 5182 (diameter of 4.0 mm) as well as various force values (pneumatic system) and various values of electrode travel (servomechanical system). The FEM calculations were used to analyse a number of parameters, including the penetration of the wires into each other, waveforms instantaneous power as well as the diameter and volume of the weld nugget. The experiments involved the recording of welding current and voltage as well as electrode force and travel. The tests also involved destructive (peeling test) and metallographic tests.

The tests were focused on process optimisation in relation to force control and, as a result, to the control of welding power space distribution enabling the concentration of energy in the central zone of the weld. When the pneumatic force system was used, the above named zone was characterised by an incomplete fusion instead of interpenetration. The practical implementation required the use of the servomechanical electrode force system.

Introduction

Force is one of the three main technological parameters of a welding process. The other two parameters are the value of current and the time of current flow. During cross wire projection welding (especially for aluminium alloys) in a conventional application i.e. with a pneumatic force system, it is very difficult to make weld with full/completely weld nugget. Pneumatic force system is characterised by high inertia and cannot be applied for fast force changes. For this reason, the value of pre-set force is usually constant. When the force is too high, this causes high deformation of welded elements (wires). If the force is too low, this could lead to the formation imperfections of a projection joint (high temperature in contacts, expulsion). In pneumatic systems, force applied during welding results from a specific force pre-set in a pneumatic cylinder. The travel of electrodes is a consequence of the action of this force and of the changeable mechanical resistance of materials being welded.

An alternative solution requires another method for exerting force on materials being welded. [1][2][3][4][5][6]. In a solution, which is analysed in this paper, with servomechanical force system, the travel of electrodes is pre-set and resultant force depends on the travel of electrodes and the resistance of deformation of contact area being heated. Available reference publications do not contain information concerning such a method of electrode movement control as the method presented in this article [7][8][9].

Publication [10] presents an entirely different approach in which the electrode travel rate is reduced. This solution is new and its advantages are of important meaning. Using this technology very desirable lower deformation (interpenetration) of wires can be obtained and more energy can be produced in the contact area of wires. This new solution consist in electrode travel control, which significantly changes the previous approach to the resistance welding process (especially to projection cross wire welding) and significantly affects the development of the entire research area (pressure welding).

FEM Calculations

The research-related FEM calculations were conducted using both the SORPAS software model 2.5D and 3D [11]. The software application enables carrying out associated analyses, including coupled electro-thermo-mechanical analyses. This software features a module including the effect of a new force solution and a precise electrode movement solution, i.e. by the servomechanical system. 3D model was used due to its ability to better demonstrate the process and higher accuracy of calculations. However for the reason of the lacking of the necessary options (detection of expulsion, weld force run in cease of electrode travel control) in 3D model, for optimisation of the process 2.5D model was used. In the final phase new idea of optimisation was tested using 3D model. Cross-wire projection welded joint is characterised by a thin layer of melted material being welded. The

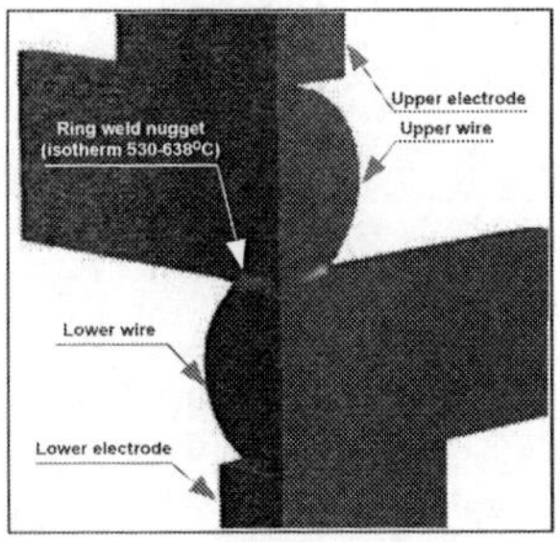

weld is being formed at temperature lower than the melting point. This results from the specificity of the process, i.e. material melting and squeezing of plasticised material outside a weld as a result of electrode force action. A weld takes a shape like in Fig. 1 (ring weld, lack of a weld in central part).

Fig. 1. Temperature distribution in welding area (model 3D).

In calculations and experimental tests wire of aluminium series Al. 5182 (solidus - 577°C, liquidus - 638°C) was used. For better demonstration and exact analysis of a process while performing FEM calculations, the welding area has been marked with isotherm of the temperature 530°C. It is a temperature below aluminium melting point, however for such a temperature it is possible to obtain a correctly welded joint [11]. This area hereinafter is referred to as a *weld.*

Calculation Model

The FEM calculations model of the resistance cross-wire welding of aluminium wire is shown in Fig. *2*. The following model assumptions were made: copper electrodes (A2/2), material being welded – aluminium wires of AA5182 grade and diameter ϕ=4 mm, numerical model 2.5D (1600 elements) and 3D (nodes: 13600, elements: 11700). The analysis was conducted for the pneumatic and servomechanical force systems. Assumed criteria for calculations were the following:

1) maximum temperature at the point between the electrode and a wire being welded is 500 ^{O}C (no. 494 Fig. *2*),
2) depth of wires interpenetration max. 20% (Δl=1.6 mm),
3) lack of expulsion,
4) lack of clear deformation (sagging) of wires,
5) maximum time of the main welding current flow 60 ms.

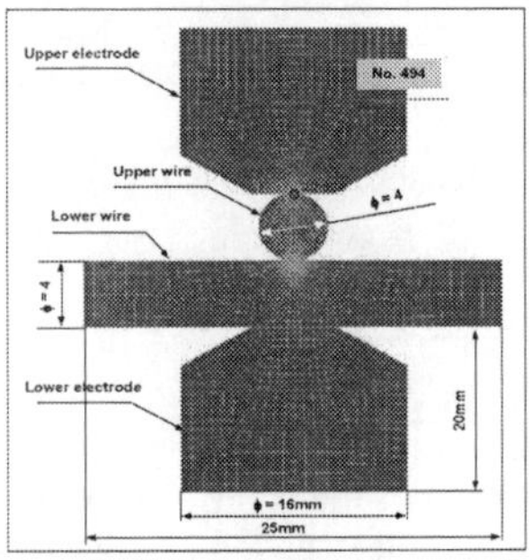

Fig. 2. FEM calculation model of the process of aluminium wires welding (5182 cross-wire projection welding).

Process Parameters

Numerical calculations were conducted for the pneumatic force system in a wide range of welding current values, i.e. 8-12 kA (8/10/12), electrode force 1.5-0.5 kN (1.5/1.0/0.5 kN) and welding time maximum 60 ms (additional up-slope 3 ms) for each parameters combination [13]. Other parameters of resistance welding technology: initial electrode force (squeeze) time– 20 ms, final electrode force (hold) time – 100 ms, DC inverter welding machine (1kHz). Optimisation of the process was conducted in the case of servomechanical force system for 8kA, which is the lowest analysed current value in pneumatic system and additionally for the lower value, i.e. 6kA. Numerical calculations were carried out until the moment of occurrence or exceeding of the one of assumed criterions.

Results and Analysis of Numerical Calculations

The results of the numerical calculations are presented in Table 1 as well as Figs. *3*, *4* and *5*.

Table 1: Welding parameters, characteristic parameters of the process – FEM calculations (model 2.5D)

no.	Denotation of a FEM calculation file	Current	Time (slop+main time)	Electrode force	Interpenetration Δl	Weld diameter	Weld volume	Energy	T_{e-m}no. 494	Notes
		kA	ms	kN	mm		mm^3	kJ	^{O}C	
	1	2	3	4	5	6	7	8	9	10
Pneumatic force system										
1	01a1	8.0	3+60	1.5	3.0	1.4	2.1	234	351	too high interpenetration and wires sagging
2	01a3	10.0	3+60	1.5	4.1	4.8	23.0	322	448	
3	01a5	12.0	3+60	1.5	3.6	5.6	48.9	328	460	
4	1b1b	8.0	3+37	1.0	1.65	4.5	24.0	188	383	
5	1b2b	10.0	3+17	1.0	1.0	3.0	12.1	143	388	OK, ϕ_{weld} > 3.0mm
6	1b3b	12.0	3+12	1.0	1.1	3.0	14.2	145	468	
7	1c1b	8.0	3+27	0.5	0.4	2.8	13.2	167	500	Temperature T_{e-m} > 500°C, too small weld diameter
8	1c2b	10.0	3+9	0.5	0.1	1.9	6.0	94	526	
9	1c3b	12.0	3+4	0.5	0.1	1.8	6.1	66	528	
10	01b0	6.0	3+60	1.0	1.3	1.5	3.7	183	310	nominal weld diameter < 3mm
Servo-mechanical force system										
11	11a1	8.0	3+32	servo1	1.0	3.0	16.6	184	435	OK, ϕ_{weld} = 3.0mm
12	21a2	6.0	3+57	servo4	1.2	3.0	12.7	193	395	

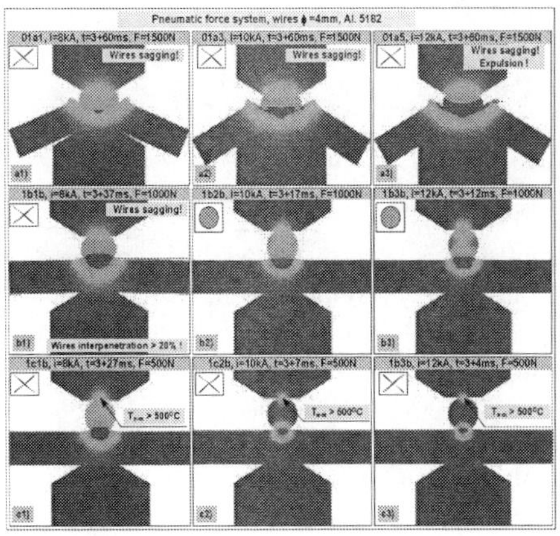

Fig. 3. The temperature distribution in the welding area for aluminium wires (5182) in the case of pneumatic force system in the end of welding process

Analysis of the results of numerical calculations (for pneumatic force system) is as follows:
Electrode force 1.5 kN is too high and causes too big interpenetration of wires (Table 1 – line 1, 2 and 3, column 5) together with very big sagging (deformation) of wires: Fig. 3a1/a2/a3. For aluminium wires (5182) and pneumatic force system lower electrode force should be used.
For electrode force 1.0 kN it is possible to obtain the assumed weld diameter (3 mm), but only for higher values of welding current being within the range of 10.0 kA - 12.0 kA (Table 1 – line 5 and 6, column 6, Fig. *3*b2/b3). Interpenetration of wires is restricted within the assumed range, less than 20% of total diameter of wires (Table 1 – line 5 and 6, column 5). Wire sagging is not observed (Fig. 3b2/b3). Current of 8 kA is however too low to provide correct (intensive) heating of welding area. Finally a weld has a diameter of 4.5 mm but the interpenetration of wires is too high (Table 1, line 4, column 5). Too high sagging of wires can be also observed.
For electrode force 5.0 kN the criterion of the appropriate weld diameter is not fulfilled (Table 1, line 7, 8 and 9, column 6 and Fig. *3*c1/c2/c3). For lower electrode force the critical become the parameter of the exceeding of the temperature in the contact point between electrode and material being welded over the assumed value of 500^{0}C and the phenomenon of expulsion. Due to exceeding of this parameter and expulsion the time of welding was reduced, which caused too small diameter of a weld.

Optimisation of the process

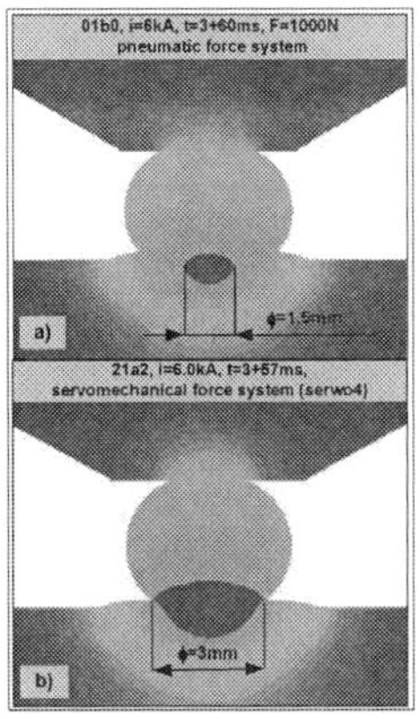

Optimisation of the welding process was conducted for lower values of welding current, i.e. 8 and 6 kA. Value of 6kA is definitely the lowest value of welding current for which the joint can be obtained using pneumatic force system. For comparison, the calculations were done for pneumatic system for electrode force of 1.0 kN being recognised as the most favourable and next the process was subjected to optimisation using the same reduced welding current value, i.e. 6kA, but with the appropriate force profile using servomechanical system.
The comparative results in the form of the images of temperature distribution in a welding area is shown in Fig. *4*, while Table 1 contains the results in the numerical form (line 11 and 12). For the applied servomechanical force system in the option of controlling the electrode travel during current flow, the assumed diameter of 3.0 mm obtained.

Fig. 4. Temperature distribution in a welding area for welding current of 6.0 kA:
a) pneumatic system, b) servomechanical system.

The manner of the process optimisation is shown in Fig. *5*. The initial parameter is electrode travel (the depth of wires interpenetration). For pneumatic system the weld diameter obtained is too small and amounts to 1.5 mm (Table 1, line 10, column 5). Too low value of current causes only the plasticisation of material which in effect results with too large contact area. For low welding current (6kA) the current density is too small and prevents melting of material being welded and only its heating and plasticisation are possible.
Final value of wire interpenetration in this case (pneumatic system) amounts to 1.2 mm (Fig. *5*d, course 1) and from the viewpoint of formation of a weld, this value and the course of a curve of wire interpenetration itself are undesirable. While using servomechanical force system there is a tendency to the obtainment of lower final value of wire interpenetration than that for pneumatic system, for instance 0.8 mm (Fig. *5*d, course 3). In consequence the power course (instantaneous) of higher value in the whole range is acquired (Fig. *5*c, course 3) if compared to pneumatic system (Fig. *5*c, course 1). Such a power course enables the obtainment of nominal weld diameter (3 mm) in the time of 63 ms (Fig. *5*b, course 3 and table 1 line 12, column 6). Final effect of the control in the case of servomechanical system is the acquiring the desirable course of electrode force. This is just the electrode force that influences the value of resistance, especially in contact areas (particularly in the configuration: welded wire – welded wire) and this in turn influences the adequate space distribution of welding power. Next it translates into the adequate temperature distribution in welding area and in effect into material melting and formation of a weld of adequately larger dimension.

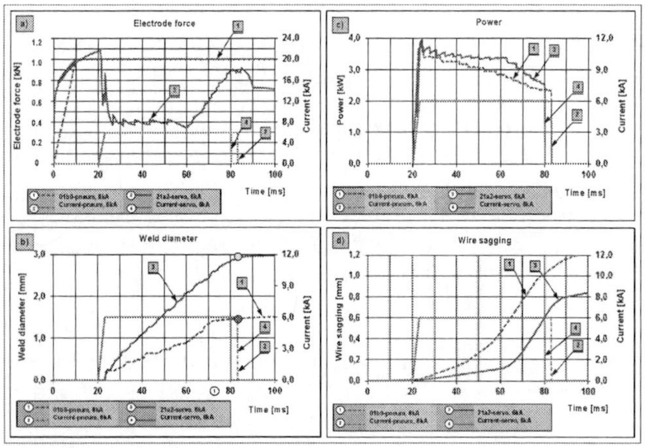

Fig. 5. Results of FEM calculations: a) electrode force, b) weld diameter, c) instantaneous power, d) depth of wires interpenetration (sagging):
1, 2) pneumatic electrode force (i=6.0kA, F=1.0kN),
3, 4) servomechanical electrode force (i=6.0kA, serwo4),

To sum up this part of the analysis concerning the value of welding current 6 kA (being recognised as too low for the obtainment the correct joint in the pneumatic force system) it should be specified that in the result of the servomechanical system usage and adequate control (of force and/or travel) it is possible to improve the welding process and to achieve the assumed weld diameter (3 mm) together with the fulfilment of other requirements (quality criteria).

Experimental Verification

The numerical calculations have been verified experimentally. The examples of welded joints produced using pneumatic force system are shown in Fig. *6*a, and servomechanical system in Fig. *6*b.

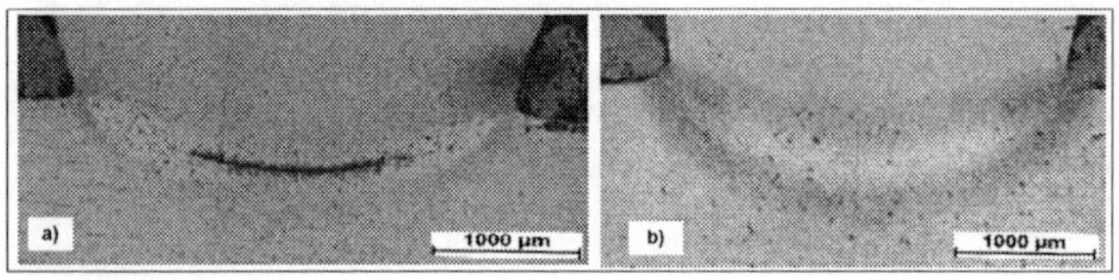

Fig. 6. The result of metallographic examination:
a) pneumatic system (i=8kA, t=40ms, F=1.0kN),
b) servomechanical system (i=6.8kA, t=63ms, F=serwo 4)

Conclusions

The selection of the optimal parameters for cross-wire projection welding especially of soft materials such as aluminium alloys and pneumatic electrode force system is very difficult. The limitations result from the lack of fast controlling of force during the welding process. Importantly, the parameter being set is force (of the electrode) and the resulting parameter is travel (of electrodes) which is not monitored and controlled.

Application of servomechanical electrode force system is the way to improve the welding process performance (extending the parameter window). In the mode of operation with electrode travel it is possible to set the favourable trajectory of electrode movement, which enables the desirable distribution of current density and then more favourable space welding power distribution. As a result this translates into the enlargement of welding area and the obtainment of a weld of larger dimensions.

What is the most important the *full weld* is being obtained, i.e. welded materials (wires) are joined (they are welded together) in the whole contact area (without a gap, so called ring weld).

Acknowledgments

This work was supported by the Polish National Centre for Research and Development (TANGO1/267374/NCBR/2015).

References

[1] X. Zhang, G. Chen, Y. Zhang, and H. Lai. 2009. *Improvement of resistance spot weldability for dual-phase (DP600) steels using servo gun*, Journal of Materials Processing Technology.
[2] H. Tang W. Hou, and S. Hu. 2002. *Forging force in resistance spot welding*. Proceedings of the Institution of Mechanical Engineers, Part B: Journal of Engineering Manufacture, Volume 216, Number 7.
[3] J. Gould. *Joining Aluminium Sheet in the Automotive Industry-A 30 Year History. 2012.* Welding Journal (Welidng Res.) vol91. January,pp.23-34
[4] J. S. Agapiou, T. A. Perry, 2013. *Resistance mash welding for joining of copper conductors for electric motors*. Journal of Manufacturing Processes, Volume 15, Issue 4, October, pp. 549–557.
[5] X. Q. Zhang, G. Chen. 2008. *On-line evaluation of electrode wear by servo gun*, Int J Adv Manuf Technol, Vol 36.
[6] S. A. Slavick. 1999.*Using Serwoguns for Automated Resistance Welding*, Welding Journal vol. 78, no7.
[7] Z. Mikno, Z. Bartnik, A. Ambroziak, and A. Pietras. 2012. Patent P. 401723 *Method for Projection Resistance Welding of Steel Plates with Embossed Projections*.
[8] Z. Mikno, B. Grzesik, M. Stępień. 2015. Patent P. 412615. *Cross-wire resistance welding method especially of aluminium wires*.
[9] Z. Mikno, Z. Bartnik. *Projection welding with pneumatic and servomechanical electrode pressure system in FEM calculation*. The 7th International Seminar on Advances in Resistance Welding 12-14.09.2012, Busan, Korea.
[10] Z. Mikno. 2016. *Projection Welding with Pneumatic and Servomechanical Electrode Operating Force Systems*. Welding Journal (Welidng Research) 2016 vol 95. August, pp. 1-13 (forthcoming publication).
[11] Swantec Inc. SORPAS model 2.5D Software Version 11.2 and model 3D Version 4.0x64 http://swantec.com/
[12] W. Xu, J. Liu, G. Luan, Ch. Dong. *Temperature evolution, microstructure and mechanical properties of friction stir welded thick 2219-O aluminum alloy joints*. Material and Design 2009.
[13] H. Papkala. *Zgrzewanie oporowe metali*. *Wydawnictwo KaBe* Krosno 2003.

Temperature Histories of Structural Steel Laser and Hybrid Laser-GMA Welds Calculated Using Volumetric Constraints

S.G. Lambrakos,[1] L. Massa,[2] A. Shabaev,[3] and L. Huang[2]
[1]Center for Computational Materials, Code 6390, Materials Science and Technology Division, Naval Research Laboratory, Washington, DC, [2]Hunter College, New York, NY, [3]George Mason University, Virginia

Abstract
Inverse thermal analyses of structural steel deep-penetration welds are presented. These analyses employ a methodology that is in terms of numerical-analytical basis functions and volumetric constraint conditions for inverse thermal analysis of steady state energy deposition in plate structures. These analyses provide parametric representations of weld temperature histories that can be adopted as input data to various types of computational procedures, such as those for prediction of solid-state phase transformations and mechanical response. In addition, these temperature histories can be used to construct parametric-function representations of thermal histories for welds corresponding to other process parameters or welding processes whose process conditions are within similar regimes. The present study applies an inverse thermal analysis procedure that uses volumetric constraint conditions whose two-dimensional projections are mappings onto transverse cross sections of experimentally measured solidification boundaries.

Introduction
Inverse thermal analyses of structural steel deep-penetration welds are presented, which provide parameterization of temperature histories for prediction of properties within the Heat Affected Zone (HAZ) of welds for the regime considered. The present study applies an inverse thermal analysis procedure that uses volumetric constraint conditions whose two-dimensional projections are mappings onto transverse cross sections of experimentally measured solidification boundary, or any type of specified boundary within the temperature field, including thermocouple measurements. For the present study, which considers deep-penetration welds of structural steel, experimentally observed solidification boundaries are adopted as temperature-field constraint conditions. The inverse analyses presented in this study provide parametric representations of weld temperature histories that can be adopted as input data to computational procedures, such as those that predict solid-state phase transformations and mechanical response. In addition, these temperature-history parametric representations permit construction of a model-parameter space that can be adopted for inverse analysis of welds corresponding to other process parameters or welding processes whose process conditions are within similar regimes. The construction of temperature fields according to spatially and temporally distributed constraint conditions using linear combinations of basis functions, whose fundamental trend features are consistent with those of the underlying process, represents a convenient approach to inverse analysis of energy deposition processes.

The formal structure of the inverse analysis approach applied in this study is that of parametric model representation of the temperature field in terms of numerical-analytical basis functions. The foundation of this approach was introduced by [1], and provides a general reduction of model complexity for purposes of weld thermal analysis. Reduction of model complexity is achieved by adopting numerical-analytical models for either the heat source or temperature field (or both) in combination with numerical methods. Similarly, the present study contributes to a parameter space relating shapes of weld solidification and transformation boundaries, or shapes of any type of specified boundary within the temperature, to weld temperature histories, as well as to weld process parameters. A specific aspect of the inverse analysis methodology applied here, which permits convenient adjustment of parameters, is construction of discrete volumetric source distributions. The mathematical foundation of the inverse analysis methodology for calculation of temperature histories using linear combinations of numerical-analytical basis functions is that of least-squares parameter optimization [2]. As discussed previously, the inverse modeling approach compensates for lack of information concerning material properties [2-7].

Methodology for Inverse Thermal Analysis
For processes involving heat deposition within structures characterized by a finite thickness, a consistent parametric representation of the temperature field is given by

$$T(\hat{x},t) = T_A + \sum_{k=1}^{N_k}\sum_{n=1}^{N_t} C(\hat{x}_k)G(\hat{x},\hat{x}_k,\kappa,n\Delta t) \quad \text{(Eq 1)}$$

where

$$G(\hat{x},\hat{x}_k,t,\kappa) = \frac{1}{t}\exp\left[-\frac{(x-x_k-V_k t)^2+(y-y_k)^2}{4\kappa t}\right] \times\left\{1+2\sum_{m=1}^{\infty}\exp\left[-\frac{\kappa m^2\pi^2 t}{l^2}\right]\cos\left[\frac{m\pi z}{l}\right]\cos\left[\frac{m\pi z_k}{l}\right]\right\}, \quad \text{(Eq 2)}$$

and

$$T(\hat{x}_n^c,t_n^c) = T_n^c \quad \text{(Eq 3)}$$

where $T_k(\hat{x},\hat{x}_k,\kappa,t)$ represent an effectively complete set of basis functions for representation of the temperature field within a bounded region. The quantity T_A is the ambient temperature of the workpiece and the locations $\hat{x}_n^c$ and temperature values T_n^c specify constraint conditions on the temperature field. The functions $T_k(\hat{x},\hat{x}_k,\kappa,t)$ represent an optimal basis set of functions for given sets of boundary conditions and material properties. The quantities $\hat{x}_k = (x_k, y_k, z_k)$, $k=1,...,N_k$, are the locations of the elemental source or boundary elements. The constraint conditions are imposed on the temperature field spanning the bounded spatial domain of the workpiece by minimization of the value of the objective functions defined by

$$Z_T = \sum_{n=1}^{N} w_n\left(T(\hat{x}_n^c,t_n^c)-T_n^c\right)^2 \quad \text{(Eq 4)}$$

where T_n^c is the target temperature for position $\hat{x}_n^c = (x_n^c, y_n^c, z_n^c)$.

The input of information into the inverse model defined by Eqs. (1)-(4) is effected by: the assignment of individual constraint values to the quantities T_n^c; the form of the basis functions adopted for parametric representation; and specifying the shapes of inner and outer boundaries, which bound the temperature field within a specified region of the workpiece. The constraint conditions and basis functions, i.e., $T(\hat{x}_n^c,t_n^c) = T_n^c$ and $T_k(\hat{x},\hat{x}_k,\kappa,t)$, respectively, provide for the inclusion of information that can be obtained from both laboratory and numerical experiments The procedure for inverse analysis defined by Eqs. (1)-(4) entails adjustment of the parameters $C(\hat{x}_k)$, $\hat{x}_k$ and Δt defined over the entire spatial region of the workpiece.

The inverse analysis methodology defined by Eqs.(1)-(4) is based on a parametric numerical-analytical model, which combines numerical integration with optimization of linear combinations of numerical-analytical basis functions, which include fundamental solutions to the heat conduction equation [8] and their Fourier-series representation. In particular, Eq.(1) defines a discrete numerical integration over time, where the time step Δt is specified according to the average energy deposited during the time Δt, for transition of the temperature field to steady state. It should be noted that the formulation of the inverse analysis methodology defined by Eq.(1)-(4) is equipped with a mathematical structure that satisfies all boundary conditions associated with welding of plate structures (see [2] for further discussion).

Inverse Thermal Analysis of Structural Steel Laser and Hybrid Laser-GMA Welds

In this section results of inverse thermal analyses of structural steel deep-penetration welds are described, which correspond to different weld process conditions and associated process-control parameters. The significance of the inverse-problem approach for these analyses is that the nature of the coupling of the energy source to the workpiece, which is a function of beam power and process control parameters, is in principle difficult to specify relative to analysis based on the direct-problem approach.

The structural steel deep-penetration welds, whose inverse analysis is presented here, consist of laser and laser-GMA hybrid welds [9]. The analyses presented here entail calculation of the steady state temperature field for a specified range of sizes and shapes of inner surface boundaries S_i defined by the experimentally observed estimate of the solidification boundary. The shapes of these boundaries are determined experimentally by analysis of transverse weld cross sections showing microstructure revealing estimated solidification boundaries. For calculations adopting the solidification boundaries as constraints, the parameter values assumed are $\kappa = 5.88 \times 10^{-6}\ m^2s^{-1}$, $T_M = 1503.0\ ^oC$. As discussed previously [2], reasonable estimates of κ and T_M are sufficient for inverse analysis. This assumption is sufficient, within reasonable estimates, in that the set of parameters $C(\hat{x}_k)$, $k=1,...,N_k$, and κ are not uniquely determined by inverse analysis. Thus, changing estimated values of κ would require different values of $C(\hat{x}_k)$ in order to satisfy specified constraint conditions associated with T_M. With respect to inverse analysis, the interpretation of κ as both an estimated material property and adjustable parameter is emphasized within the following.

The goal of the present analysis is determination of a set of parameters that can serve as initial estimates for adjustment with respect to deep penetration welds of steels, whose process parameters are within similar regimes. Parameter adjustment with respect to other welds, which assume the results of this study as initial estimates, would adopt κ and T_M as adjustable parameters, as well as the discrete source function $C(\hat{x}_k)$. Values of the workpiece thickness l and welding speed V defined in Eq. (2), and the upstream boundary constraints on the temperature field, $T_c = T_M$ for (y_c,z_c) defined in Eq. (3) for the solidification boundaries, are given in the figures below. Shown in Figs. 1(a) and 2(a) 4 are experimentally measured transverse weld cross sections of solidification boundaries [9], which are adopted as constraint conditions, and planar slices of the steady state temperature field that have been calculated according to these constraint conditions. Referring to the planar slices of the calculated temperature fields shown in these figures, it can be seen that all boundary conditions are satisfied, namely the condition $T = T_M$ at the solidification boundary and $\nabla T \cdot \hat{n} = 0$ at surface boundaries, where $\hat{n}$ is normal to the surface.

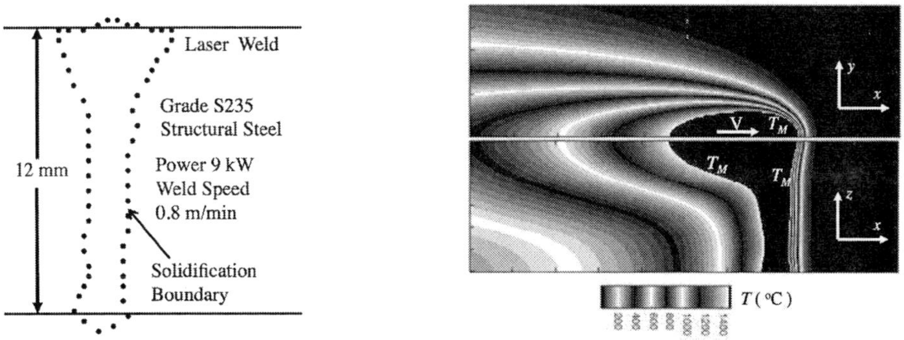

Figure 1. (a) Experimentally measured transverse weld cross sections of solidification boundary for steel laser weld, where V=1.333 cm/s. (b) Two-dimensional slices, at half workpiece top surface and longitudinal cross section at symmetry plane, of three-dimensional temperature field (°C) calculated using cross section information shown in Fig. 1 for solidification boundary.

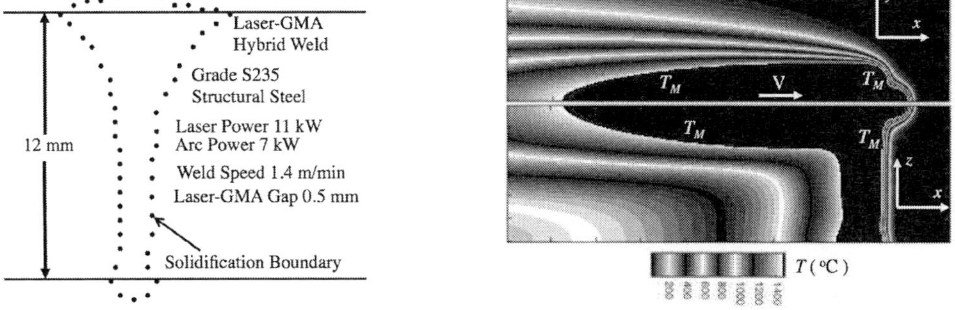

Figure 2. (a) Experimentally measured transverse weld cross sections of solidification boundary for steel laser-GMA hybrid weld, where V=2.333 cm/s. (b) Two-dimensional slices, at half workpiece top surface and longitudinal cross section at symmetry plane, of three-dimensional temperature field (°C) calculated using cross section information shown in Fig. 3 for solidification boundary.

Conclusions

The inverse analysis procedure entails calculating a three-dimensional solidification boundary using experimentally measured constraint conditions, and the temperature field consistent with the isothermal surface associated with this boundary. Shown in Figs. 1(b) and 2(b) are two-dimensional slices of the calculated three-dimensional temperature field obtained using the cross-section constraint conditions shown in Figs. 1(a) and 2(a), respectively, for the measured solidification boundary, which are parallel to the relative motion of laser or laser-GMA source and workpiece. Referring to the calculated temperature fields shown above, it is important to note that the constraint conditions on the calculated three-dimensional solidification boundaries are that the projections of all their two-dimenional transverse slices, as a function of time, are consistent with the experimentally measured transverse cross sections of the solidification boundaries, respectively. The final stage of the analysis entails calculation of temperature histories as a function of transverse position within the cross section of the weld. This is accomplished by constructing a steady state three-dimensional temperature field that is consistent with experimentally measured constraint conditions for the solidification boundary.

Acknowledgments

This work was supported by a Naval Research Laboratory (NRL) internal core program.

References

[1] D. Rosenthal, "The theory of moving sources of heat and its application to metal treatments," Trans ASME, Vol. 68 (1946), pp. 849-866.

[2] S.G. Lambrakos, "Inverse Thermal Analysis of Stainless Steel Deep-Penetration Welds Using Volumetric Constraints," Journal of Materials Engineering and Performance, published online 2014, DOI: 10.1007/s11665-014-1023-7, Volume 23(6), June 2014, pp. 2219-2232.

[3] M.N. Ozisik and H.R.B. Orlande: Inverse Heat Transfer, Fundamentals and Applications, Taylor and Francis, New York, 2000. K.

[4] Kurpisz and A.J. Nowak: Inverse Thermal Problems, Computational Mechanics Publications, Boston, USA, 1995.

[5] O.M. Alifanov, Inverse Heat Transfer Problems. Springer, Berlin, 1994.

[6] J.V. Beck, B. Blackwell, C.R. St. Clair, Inverse Heat Conduction: Ill-Posed Problems, Wiley Interscience, New York, 1985.

[7] J.V. Beck, "Inverse Problems in Heat Transfer," Mathematics of Heat Transfer, G.E. Tupholme and A.S. Wood eds., Clarendon Press, (1998), pp. 13-24.

[8] H. S. Carslaw and J. C. Jaegar: Conduction of Heat in Solids, Clarendon Press, Oxford, 2nd ed, 374, 1959.

[9] J. K. Kristensen, "Laser and Hybrid Laser-GMA Welding of Structural Steels, A Challenge to Research and Industry for Two Decades," Trends in Welding Research, Proceedings of the 8th International Conference, Editors: S.A. David, T. DebRoy, J.N. DuPont, T. Koseki, H.B. Smartt, ASM International, 645-641, 2009.

Microstructural Approach on Cold Cracking for High Strength FCA Steel Welds of YS 600 MPa Grade

Namhyun Kang, Myungjin Lee, Kyungmox Cho
Pusan National University, Busan, Korea
nhkang@pusan.ac.kr, mj.lee@pusan.ac.kr

Stephen Liu
Colorado School of Mines, Colorado, USA
sliu@mines.edu

Abstract

Cold cracking is observed at HAZ and weld metal at temperatures below 200°C. It is caused by the interaction of susceptible microstructures, diffusible hydrogen, and tensile residual stress in welds. To minimize cold cracking, preheating and/or post-heating processes need to be applied for welding. Steel plates for the ocean plants are normally welded by using flux-cored arc (FCA) to maximize weld productivity. However, FCA-weld metals are more sensitive to cold cracking compared to solid wire because of their large moisture adsorption. The increased use of high strength steels with yield strengths (YS) greater than 600 MPa has also changed the weld microstructure from a mixture of proeutectoid ferrite and acicular ferrite to bainite and martensite. However, there were only few studies on the effect of preheating temperature on microstructure and cold cracking properties in high strength steel welds. This study investigated the role of weld metal composition and preheating temperature on the cold cracking of welds deposited with YS 600 MPa and low-hydrogen FCA weld consumables. Cold cracking fractography and crack propagation behavior were analyzed to investigate the relationship between cold cracking and microstructure in the high strength steel welds.

Introduction

As the scale of offshore facilities becomes increasingly massive, the strength of steels used in these ocean plants must also increase. To acquire high strength, steel for the ocean plants is prepared by applying various alloying elements and heat treatment processes [1, 2]. As a part of this process, the carbon equivalent (Ceq) content of high-strength steels is increased, which promotes cold cracking after welding.

There are various microstructures in the weld metal, with grain boundary ferrite and acicular ferrite being the dominant microstructures for weldments having a yield strength of less than 600 MPa. In ferritic steel welds, an acicular ferrite microstructure offers the greatest resistance to cold cracking while improving strength and toughness. Acicular ferrite grows through heterogeneous nucleation on non-metallic inclusions. However, inclusions also contribute adversely to the mechanical properties of the weld by nucleating brittle fractures and creating voids that may lead to ductile fracture. The relationship between ferrite and cold crack sensitivity has been previously studied [3, 4]; however, there are few studies on the effect of martensite on the cold cracking of weld metals, which is known to be significantly sensitive to cold cracking.

The objective of the study is to investigate, with the Y-groove test, cold crack behavior in low-hydrogen FCA-weld metals having a yield strength of 600 MPa. Weld metal composition and preheating temperature were varied to study the effect of microstructure and hardness on the cold crack ratio. Cold cracking fractography and crack propagation behavior were analyzed to investigate the relationship between cold cracking and inclusions in the high-strength steel welds.

Experimental Procedure

Table 1 shows the chemical composition and diffusible hydrogen content of two types of low hydrogen FCA-weld metals having a yield strength of 600 MPa. The weld metals A and B had various Ceq values from 0.41 to 0.48, and the same level of HD.

A cold crack experiment was conducted using the Y-groove test according to the ISO standard [5]. Thick EH36 plates of 50 mm were used as the base metal for the Y-groove test. The welding voltage and currents to produce the Y-groove specimens were 27 V and 240 A, respectively. The welding speed was 50 cm/min and the heat input for welding was 0.7 kJ/mm. One hundred percent CO_2 was used as a shielding gas. Various preheating temperatures were employed during the Y-groove test: room temperature, 50, 100 and 150°C.

The Y-groove weld specimens were etched with 2 % natal solution, and the microstructure was analyzed by optical microscopy and scanning electron microscopy. The cold crack cross-sections were examined using scanning electron microscopy and electron back-scatter diffraction (EBSD). Because of the fine grain size in the weldments, the microstructure volume fractions were measured using EBSD IQ maps at ×3000 magnification and characterized by image analysis (iSolution DT). In order to find the influence of segregation on cold cracking, electron probe micro-analysis (EPMA) was conducted in the cold crack cross-section.

Table 1. Chemical composition and diffusible hydrogen content of base metal and weld wire (wt.%)

No	C	Si	Mn	Ni	Ti	Al+Nb+V+B	Ceq	HD (ml/100g)
EH36	0.08	0.29	1.43	-	0.014	0.048	0.32	-
A	0.062	0.22	1.19	2.06	0.041	0.033	0.41	4.15
B	0.059	0.42	1.44	2.55	0.058	0.032	0.48	4.39

Results and Discussion

Figures 1a and 1b show the cold cracking in the Y-groove weld for specimens A and B, respectively. All cold cracking occurred in the weld, and no cracks were observed in the HAZ or base metal. Cracks initiated at the weld root and the primary cracks propagated almost vertically towards the weld surface. The primary crack initiated at the weld root propagated in the near vertical direction to the weld surface and the secondary cracks propagated parallel to the columnar dendrites growing from the fusion line. Figure 1c shows the primary crack propagating along the martensite boundary and flat fracture surface inside the crack path. Figure 1d shows the EBSD IPF and IQ micrograph of the primary and secondary cracks. Primary cracks and some secondary cracks depicted a linear and straight path of crack propagation along prior austenite grain boundaries, therefore indicating intergranular fracture (IF) path.

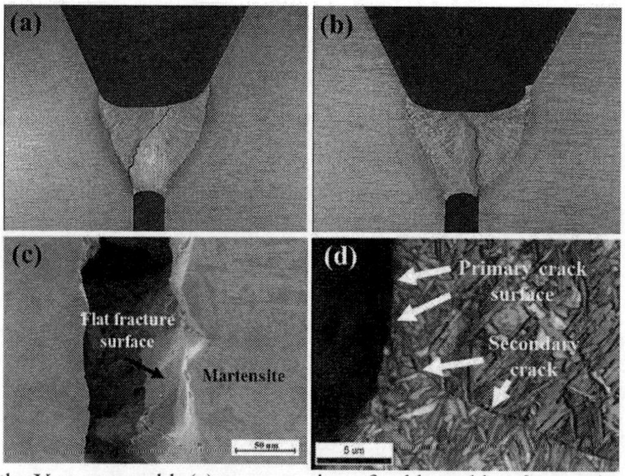

Figure 1. Cold cracking in the Y-groove weld: (a) cross-section of cold cracking for specimen A, (b) cross-section of cold cracking for specimen B, (c) primary crack propagating along martensite boundary, and (d) electron backscatter diffraction (EBSD) inverse pole figure and image quality (IPF+IQ) map indicating an intergranular crack path along prior austenite grain boundaries.

Cold cracking fracture is known to exhibit several morphologies: dimple fracture, quasi-cleavage fracture (QCF), cleavage fracture and intergranular fracture (IF). The cold crack can be characterised as IF and QCF. The IF with flat and shiny fracture surfaces was nucleated and propagated along the prior austenite grain boundary. A QCF indicated rough fracture surfaces. The etched IF shows the fine lath morphology that is characteristic of bainite and martensite. IF nucleated at and propagated through the prior austenite boundaries. No acicular ferrite microstructure nucleated directly on prior austenite grain boundaries. The etched QCF fracture indicates that there were inclusions of various sizes. Therefore, inclusions are the important microstructural constituent associated with QCF generation.

With no preheating, microstructure of specimen A consisted of acicular ferrite of 45.7% and martensite of 54.3% with no preheating, and that of specimen B was acicular ferrite of 28.6% and martensite 71.4%. Larger amount of martensite for specimen B was due to the larger Ceq (0.48), promoting martensitic transformation. As the preheat temperature increased from room temperature to 150°C, the volume fraction of martensite decreased and that of acicular ferrite increased continuously for the specimens A and B. The specimen B indicated the average Vickers hardness larger than the specimen A and the hardness of the specimens A and B decreased as the preheat temperature increased from room temperature to 150°C. Therefore, microstructural behaviour of acicular ferrite and martensite fractions was consistent with the hardness behaviour as a function of the preheat temperature.

Table 2 shows the cold crack ratio (CCR) as a function of preheat temperature. With no preheating, the specimen A having a low Ceq (0.41) showed CCR of 39.5% and the specimen B of 48%. As the preheat temperature increased to 50°C, the specimen A had no cold crack and the specimen B indicated CCR of 41.7%. Above the preheat temperature of 100°C, no cold cracks were observed in the Y-groove test. Therefore, the CCR decreased with martensite fraction and hardness as the preheat temperature increased.

Morphology of cold cracking fracture consisted of the mostly same fraction of IF and QCF for the specimen A with no preheating. The specimen B had a dominant morphology of IF with no preheating and increased fraction of QCF with a decreased IF as the preheat temperature increased to 50°C. The reason for the specimen B having a larger fraction of IF was due to the larger fraction of martensite contributing to nucleation and propagation through the prior austenite boundaries. Furthermore, the reason for the specimen B decreasing IF fraction and increasing QCF fraction with increasing the preheat temperature was due to the decreased cooling rate and the increased fraction of acicular ferrite with increasing the preheat temperature.

Table 2. Cold crack ratio as a function of preheat temperature

Preheat Temp.	Specimen A	Specimen B
RT	39.5%	48%
50°C	0%	41.7%
100°C	0%	0%
150°C	0%	0%

Inclusions are the important microstructural constituent associated with QCF generation. Small inclusions with diameter below 0.4 μm had a probability less than 0.2 of nucleating cold cracks from inclusion. The probability of crack nucleation increased with increasing inclusion size. Inclusions larger than 1 μm diameter increased the crack nucleation probability to greater than 0.6. Most inclusions in the Y-groove welds were found to be smaller than 1 μm, and the average size of the inclusion was approximately 480 nm. However, if the inclusions were larger than 2 μm, the probability of crack nucleating from inclusion approached to 1. Therefore, if the size of inclusion is increased, the cleavage fracture stress is reduced because of the increasing surface energy of inclusion.

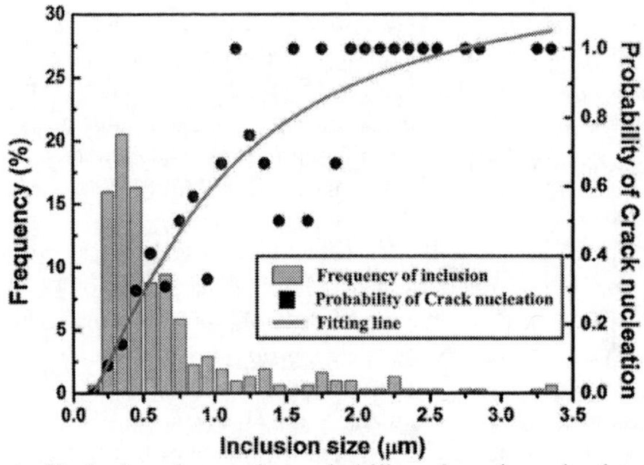

Figure 2. Effect of inclusion size on the probability of crack nucleating from inclusion

Conclusions

The study investigated the effect of Ceq and preheat temperature on cold crack ratio (CCR). As Ceq decreased and preheat temperature increased, the fraction of acicular ferrite and quasi-cleavage fracture (QCF) increased and the fraction of martensite and intergranular fracture (IF) decreased. To increase the cold crack resistivity of high strength steel welds, it is necessary to decrease martensite fraction and to increased acicular ferrite fraction. The role of inclusion contributing to cold cracking in high-strength steel welds showed similar results as previous studies on low-strength steel welds. Inclusions in low-strength steel welds increased the number of cold cracks when the crack encountered inclusions, and crack was initiated at and propagated through inclusions. In this study, inclusions in high strength steel welds were observed in QCF surface and secondary crack path. Micro-cracks were initiated from inclusions and propagated to other inclusions. Specifically, inclusion larger than 2 μm served to increase the sensitivity of cold cracking in high-strength steel welds. However, inclusions smaller than 2 μm, occupying the welds predominantly, increased the resistance of cold cracking by contributing to the nucleation of acicular ferrite. Therefore, in high-strength steel welds, the size of inclusions should be controlled below 2 μm to increase the nucleation of acicular ferrite, therefore increasing the resistance of cold cracking. Specifically, inclusions between 0.4 and 1 μm diameter were essential for acicular ferrite nucleation and significantly decreased the probability of crack nucleation in high strength steel welds.

Acknowledgements
This work was supported by a grant from National Research Foundation of Korea (NRF) funded by Ministry of Education (grant no. NRF-2013R1A1A2013125), and the National Research Foundation of Korea (NRF) grant funded by the Ministry of Science, ICT and Future Planning through GCRC-SOP (grant no. 2011-0030013).

References

[1] T. Kitada, K. Fukuda: Nippon Kokan Technical Report (1986), p.47
[2] J. W. Fisher, R. J. Dexter: "High-performance steels for Americal's bridges" Weld J 73(1), pp.35-43
[3] J. S. Seo, H. J. Kim, H. S. Ryoo: "Effect of grain boundary ferrite on susceptibility to cold cracking in high-strength weld metal" Met Mater Int 14(4), pp.515-522
[4] J. H. Kim, J. S. Kim, H. S. Ryoo, K. H. Kim, M. Y. Huh: "Effect of weld metal microstructures on cold crack susceptibility of FCAW weld metal" Met Mater Int 14(2), pp.239-245
[5] ISO 17642-2 (2005) "Destructive tests on welds in metallic materials - cold cracking tests for weldments - arc welding processes part 2. Self-restraint tests"

Weldability of Q&T HSLA A543 Steel: Microstructural Simulation and Modified Implant Testing

Ali H. AlShawaf
Saudi Basic Industrial Corporation SABIC
Al-Jubail Industrial city, KSA
shawafaa@sabic.com

Stephen Liu
Colorado School of Mines
Center of Welding, Joining & Coating Research (CWJCR)
Golden, CO. USA
Sliu@mines.edu

Abstract
This paper reports on the investigation of weldability of modern quenched and tempered (Q & T) high strength low alloy (HSLA) steel. During service, multiple through-thickness cracks emanating from the circumferential weld on the shell-side of a pressure vessel were observed after a short time of service in the heat exchangers. A standardized weldability test known as the Implant testing was used to study the susceptibility of the Q&T HSLA steel in question to hydrogen cracking. The charged hydrogen content for each weld was recorded against the applied load and test time during Implant testing. The lack of understanding in detail of the interaction between hydrogen and each HAZ subzone of these modern steels in the Implant testing led to the development of additional tests to obtain more data about their weldability. The most crack susceptible HAZ subzones identified by the Implant testing were simulated using two techniques: standard heat treating and Gleeble® thermomechanical simulation. The simulated HAZ subzone specimens were pre-charged with hydrogen to different concentration levels to mimic prior exposure of the material to high humidity environment during welding. Fractographical and microstructural characterization of the HAZ subzones were conducted using SEM (Scanning Electron Microscopy). A modified Implant test using a tensile testing machine was also conducted for each simulated HAZ subzone sample to observe the effects of hydrogen on the cracking behavior on that specific HAZ subzone (microstructure) and to collect all the tensile properties during the test. It was concluded that different from many other structural steels, the region with the highest susceptibility to hydrogen cracking was not located only in the CGHAZ. Instead, the lower intercritical and subcritical HAZ specimens exhibited lower load to fracture in the modified Implant testing as well as shorter time to fracture than the CGHAZ. It was also concluded that hydrogen cracks were intergranular that propagated to the base steel following the retained austenite constituents.

Introduction
In the fabrication of a pipeline or a chemical reactor, circumferential or girth welds are often required to join sections of tubular structures together. In the presence of moisture during welding fabrication, hydrogen pick up is expected which together with thermal stresses, can lead to hydrogen cracking in the weld zone. Hydrogen cracking is also referred to as delayed or cold cracking, because cracking is often delayed for minutes, hours, or even longer after solidification has completed. Requisites for cold cracking include tensile stress (external loading and/or residual stresses from welding), crack-susceptible microstructure (martensite), and the presence of hydrogen. Cold cracking can occur in the weld fusion zone in the form of microfissuring, or in the HAZ as underbead cracking [1, 2]. There has recently been a trend towards manufacturing large pressure vessels using ASTM A543 steel. This type of Q&T HSLA steel is selected because of its high strength and potential for weight savings, which is a critical factor when manufacturing, transporting and erecting such large vessels [3]. However, the increase in carbon, chromium, nickel, and molybdenum content has great influence on the carbon equivalent of the alloy. The carbon equivalent of the target steel is between 0.93 wt. pct. and 1.16 wt. pct. (for the minimum and maximum nominal chemical composition). Both carbon equivalents are located in the third zone of the Graville chart, which indicates that the steel is difficult to weld and that proper heat treatment would be required [4]. The ASME Section VIII Div. 2 code, however, does not mandate or prohibit the requirement of pre- and post-weld heat treatment for this type of steel, so the fabrication specification of one such reactor did not include any heat treatment. Recently, some cracks were discovered in the closing V-groove welds in the coolers.

Experimental Procedure
Each sub-zone in the HAZ is characterized by a different microstructure and hence different mechanical properties. The presence of a susceptible microstructure in the HAZ is one of the factors besides stress and diffusible hydrogen responsible for hydrogen-induced cracking (HIC). So, during welding, the chemical composition along with the cooling time/rate is two main factors that can develop different sub-zone microstructures in the HAZ. Figure 1 demonstrates the replication of HAZ sub-zone microstructures, from martensite to bainite and then ferrite, using a Gleeble® thermomechanical test system based on computation simulation using a commercial software named SYSWELD. After producing the HAZ sub-zone using the peak

temperature and cooling rate data obtained from SYSWELD simulation, tensile specimens were machined according to ASTM E8/E8M-13a specification. A series of tests were performed with hydrogen introduced through cathodic charging to the Q&T HSLA A543 steel. An extensometer was used to measure the elongation of the specimens during the modified Implant test at a displacement rate of 0.2 in/min. The elongation and strength of the hydrogen-charged samples were compared with the uncharged ones for all the different heat treated samples [5].

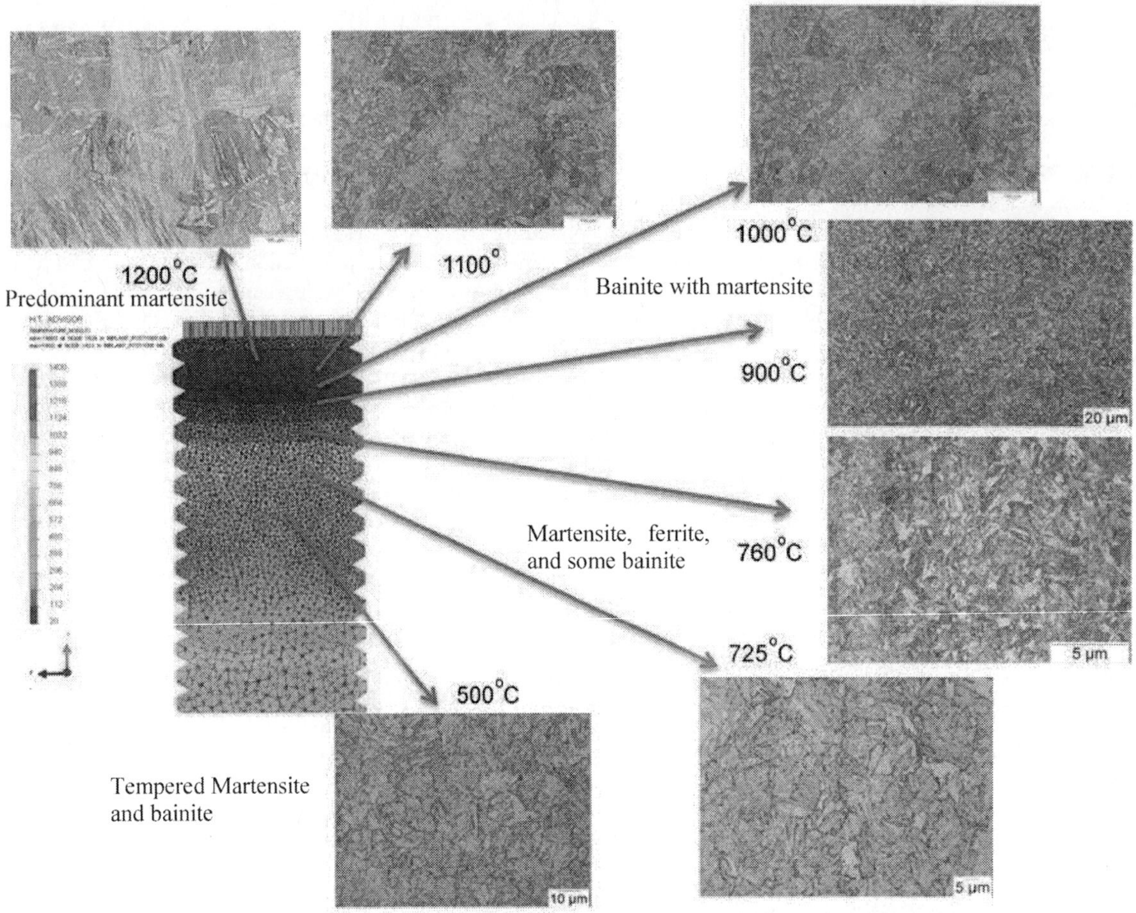

Figure 1.The replicated HAZ sub-zones using Gleeble® thermomechanical test system. The temperatures represented peak temperatures during the thermal cycles.

Discussion & Results

To investigate hydrogen effects on mechanical strength of the tested steel, tensile specimens were prepared from the two different heat treatments, in standard heat treating furnace and in Gleeble® thermomechanical test system, respectively. The fracture surface was examined and analyzed based on surface features like intergranular cracking, micro-voids, and mixed zone. It is, however, important to understand the interaction between the hydrogen and the steel before analyzing the fracture surface. Examples of hydrogen trapping sites in steels are for example, impurities (inclusions), structural defects, and microstructural constituents. Binding between hydrogen and the trapping sites usually happen due to physical trapping, electric field, stress field, temperature gradient, or chemical gradient. The trapping sites can generally be identified as two types, mobile and stationary trapping sites. Examples of the mobile trapping sites are dislocations and stacking faults. Grain boundaries, carbide particles, inclusions, and individual solute atoms are examples of the stationary trapping sites.

Generally speaking, the two heat treatments resulted in similar fracture behavior, i.e. similar interaction between hydrogen and the different simulated microstructures. Freshly broken specimens were examined using Scanning Electron Microscopy. Ductile fracture with cup and cone features was observed in all the uncharged simulated

HAZ heat treated samples. At high magnification, the fracture is characterized by the presence of a great number of microvoids (dimples). The influence of hydrogen on ductile fractures can be attributed to an accumulation of hydrogen in the trap sites, which lowers the mechanical strength [19,20]. High temperature simulated microstructures like CGHAZ (peak temperatures of 1200°C, 1100°C) are susceptible to intergranular hydrogen cracking as shown in Figure 2.

After one hour of cathodic charging, around 2 ppm hydrogen was introduced into the specimen. Samples that were treated to 1100 and 900°C showed fracture composed of two parts: a brittle area near the sample surface followed by a ductile part in the center and in the remaining area of the surface. It seems that hydrogen is migrating from the outer surface to the center of the specimen. The almost 100% intergranular fracture is a strong indication of hydrogen embrittlement. The susceptibility to fracture also increases as the hydrogen concentration increases in the CGHAZ. Increasing hydrogen concentration also promotes multiple crack initiation and branching into secondary. Similar behavior was noticed in the simulated FGHAZ (1000°C, 900°C) microstructures but with much less intergranular cracking.

Most fractures on samples treated to lower temperatures like 760, 725, and 500°C still show the ductile fracture mode with dimples and micro-voids at higher magnification. Intergranular brittle fracture is observed only along some of the surface area of the CGHAZ and FGHAZ specimens. After about nine hours of charging, more hydrogen (around 5 ppm) entered the specimens. The brittle fracture grew toward the central area for the CGHAZ as shown in Figure 2. Due to the loss of ductility and plasticity, the intergranular brittle fracture area increased in comparison to specimens after one hour of charging [6-18].

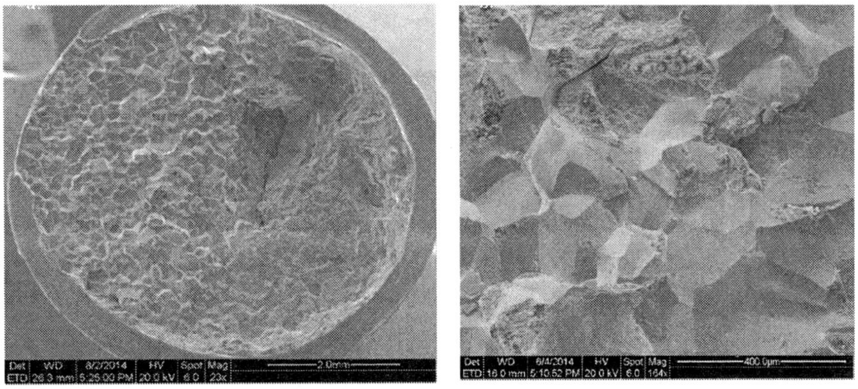

Figure 2. SEM images showing the fracture surface of CGHAZ sub-zone with 5ppm H_2 charging.

Tensile testing of hydrogen charged specimens showed an expected trend, Figure 3. Materials that experienced higher simulated peak temperatures saw dramatic changes in their yield and tensile strength, as well as elongation. For the 2 ppm H_2 samples, the tensile strength increased with increasing peak temperature up to 900°C and then decreased. At the same time, the ductility of the sample decreased. In fact, the drop in tensile strength for the samples treated to higher temperatures is likely a result of great drops in ductility rather than strength impairment. The simulated CGHAZ specimen with 1200°C peak temperature had the lower ductility. This behavior is even clearer with the 5 ppm H_2 specimens. The FGHAZ samples had the largest ductility while the CGHAZ samples had decreasing ductility and lower nominal tensile strengths with peak temperature.

Although the overall hydrogen level in the specimens was about 5 ppm, it appeared that the hydrogen concentration was much higher near the outer diameter surface of the test samples. These samples also exhibited typical brittle fracture mode with some secondary cracking. In the central region of the sample, a mixed fracture mode was observed due most likely to the lower hydrogen concentration. So, the specimens were mostly brittle with intergranular cracking in the high temperature simulated CGHAZ and intergranular cracking with secondary cracking in the FGHAZ simulated samples.

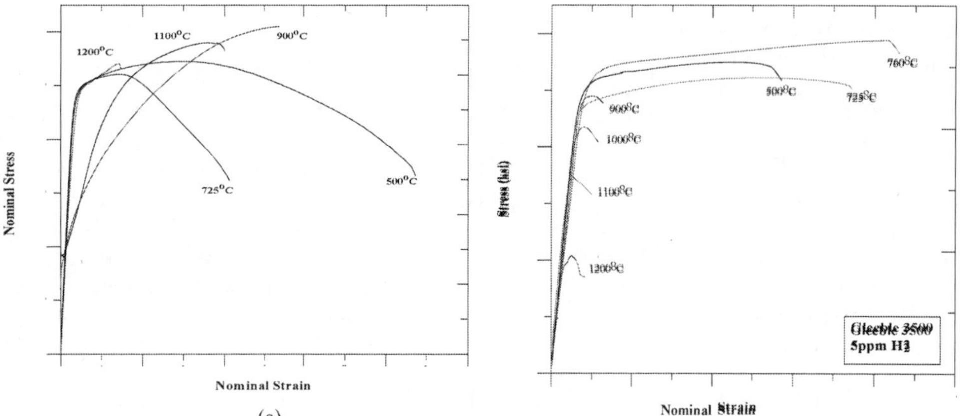

(a) (b)

Figure 3. Mechanical properties of Q&T HSLA steel in (a) 2ppm H_2 charging and (b) 5ppm H_2 Charging.

Figure 4 shows the time to fracture versus the tested load for the simulated HAZ subzones samples. On this diagram, higher load to fracture and longer time to fracture indicate greater resistance to weld cracking. In general, when a material exhibits a plot low on the graph, its resistance to weld cracking would also be low. The lower critical load is defined as the threshold load below which no fracture would occur even after 24 hours. After testing of the simulated samples using the modified Implant setup, it was observed that the lower inter-critical and sub-critical HAZ exhibited lower load to fracture during testing as well as shorter time to fracture than the CGHAZ.

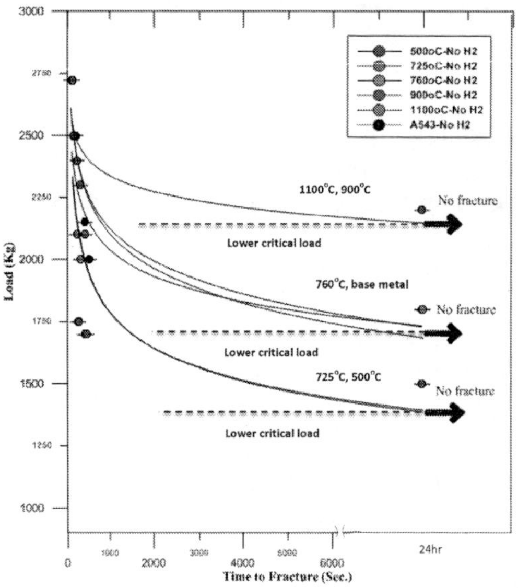

Figure 4. Tensile test versus time-to-fracture results using constant load for the simulated samples.

The fractured samples were examined. A sample cross section of the fracture surface was shown in Figure 5 to illustrate the microstructural features after hydrogen charging. Figures 5a and 5b show hydrogen cracking in the as-etched 1200°C and 1100°C simulated sample, respectively. The crack is propagating into the microstructure in an intergranular manner. The same behavior is observed in the FGHAZ simulated samples as shown in Figure 5c. While a small crack is noticed in the upper inter-critical zone simulated sample (Figure 5d), no cracks are observed in the lower inter-critical and subcritical tempering zones as shown in Figure 5e [21].

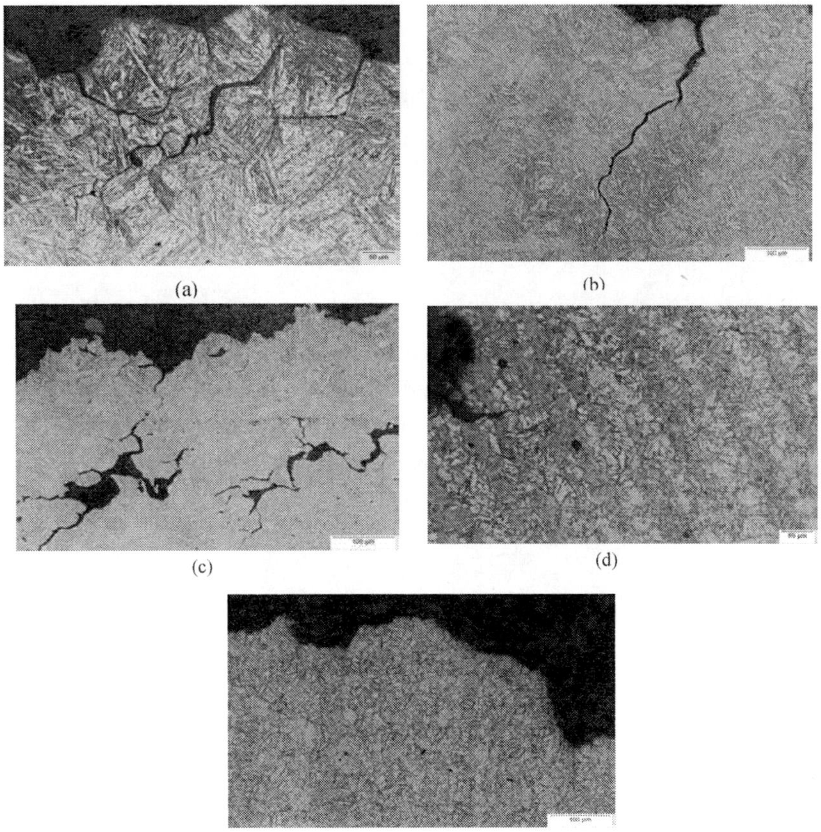

(e)

Figure 5. As-etched microstructures of samples heat treated to a) 1200°C, b) 1100°C CGHAZ, c) 1000°C FGHAZ, d) 760°C inter-critical zone, and e) 500°C peak temperature simulating the different sub zones of weld HAZ.

Color etching (color metallography) is widely used for visualizing different phases in the metals. It is an excellent technique to recognize phases such as martensite, austenite and bainite. The etchant solution produces thin chemical films on the surface of the specimens that appear in different colors under polarized light. The etchant used in this work was the LePera solution. When etched with the LePera solution, ferrite would appear as blue or dark blue, bainite and martensite would be brown and retained austenite would be white [22]. Figure 5 shows examples of the CGHAZ and FGHAZ simulated samples after etch. It can be observed that the cracks follow the white phase which is believed to be the retained austenite of the Martensite-Austenite (MA) constituents. Most of these constituents are located near the grain boundaries (traced in heavy lines in Figure 5). There are many known hydrogen trapping sites in steels and grain boundary is one such site. So, the cracks are propagating into the microstructure and following either the grain boundary and/or the MA constituent which is also a hydrogen trap site. Hence, hydrogen cracks were intergranular and propagating through the steel following the retained austenite constituents or grain boundaries, eventually leading to failure of the Q&T HSLA specimens. This is very likely one of the causes of failure embedded into the system early on during the construction phase of the pressure vessel.

Conclusions

This is part of a research work intended to determine ranges of safe operation of the coolers. The weldability of this steel has been investigated in this paper. Conclusions that can be drawn on the steel susceptibility to weldability cracking are in the following.

1. The influence of hydrogen extends beyond CGHAZ. It affects the entire heat affected zone in these high strength steels. The lower intercritical and subcritical HAZ exhibited lower load to fracture, as well as shorter time to fracture during modified implant tensile testing than the CGHAZ.

2. Hydrogen cracks were found to follow the MA constituents leading to failure of the modified implant test specimens. The presence of MA constituents provide easy path for crack propagation in the simulated coarse grained heat affected zone.

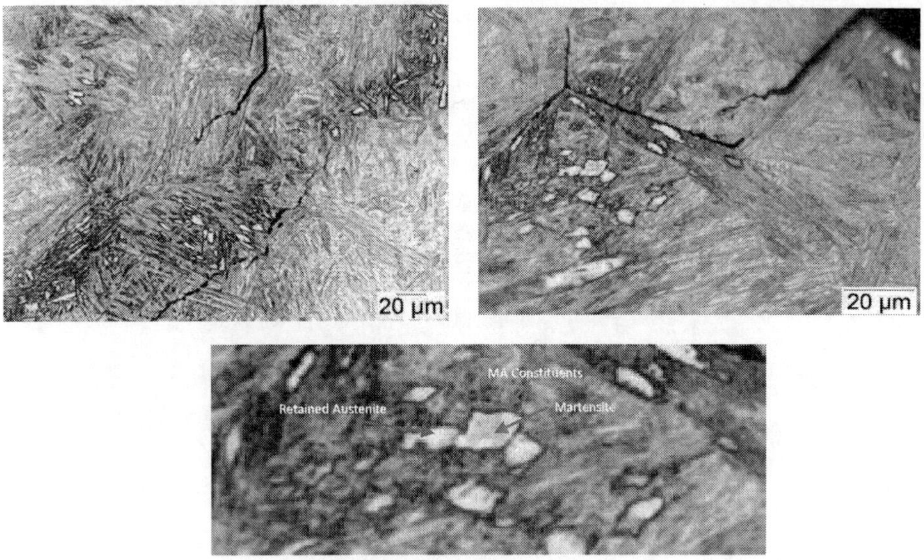

Figure 5. LePera etchant for the CGHAZ and FGHAZ samples, ferrite is blue, bainite and martensite are brown, and the retained austenite white.

Acknowledgments

The authors would like to acknowledge the Center for Welding, Joining and Coatings Research of the Colorado School of Mines, for the help and support in conducting this research. The authors would also like to thank SABIC management for the financial support.

References

[1] J. Cwiek, "Hydrogen Degradation of High-Strength Steels", Journal of Achievements in Materials and Manufacturing Engineering, vol. 37, 2009.

[2] J. Woodtli and R. Kieselbach, " Damage Due to Hydrogen Embrittlement and Stress Corrosion Cracking", Engineering Failure Analysis, vol. 7, pp. 427-450, 2000.

[3] ASTM A543/A543M-09 "Standard Specification for Pressure Vessel Plates, Alloy Steel, Quenched and Tempered Nickel-Chromium-Molybdenum".

[4] B.A. Graville, "Cold Cracking in Welds in HSLA Steels", Welding of HSLA (Microalloyed) Structural Steels, Proc. Int. Conf. American Society for Metals, Nov. 1976.

[5] ASTM E 8-2004, "Standard test method for tension testing of metallic materials", Philadelphia PA, 2004.

[6] M. L. Martin, I. M. Robertson, and P. Sofronis, "Interpreting Hydrogen-induced Fracture Surfaces in Terms of Deformation Processes: A new Approach", Acta Materialia, vol. 59, pp. 3680-3687, 2011.

[7] B.D. Craig "Hydrogen Damage in Corrosion: Fundamentals, Testing and Protection" ASM Handbook, ASM International, vol. 13A, pp. 367-380, 2003.

[8] Failures Related to Welding, Failure Analysis and Prevention, Failure Analysis and Prevention: ASM Handbook, ASM International, vol. 11, Materials Park, OH, pp. 156–191, 2002.

[9] G. Pressouyre and I. Bernstein, "An Example of the Effect of Hydrogen Trapping on Hydrogen Embrittlement", Metallurgical and Materials Transactions, vol. A12, pp. 835-844, 1981.

[10] Y.D. Park, I.S. Maroef, A. Landau, and D. L. Olson, "Retained Austenite as a Hydrogen Trap in Steel Welds", Welding Journal, pp. 27-35, 2002.

[11]D. L. Olson, "High Strength Steel Weldment Reliability: Weld Metal Hydrogen Trapping", Center for Welding and Joining Research, CSM, Golden CO. USA, 1998.

[12]M. I. Luppo and J. Ovejero-Garcia, "The Influence of Microstructure on the Trapping and Diffusion of Hydrogen in a Low Carbon Steel", Corrosion Science, vol. 32, pp. 1125-1136, 1991.

[13]G.P. Tiwari, A. Bose, J.K. Chakravartty, S.L. Wadekar, M.K. Totlani, R. N. Arya, and R. K.Fotedar, "A Study of Internal Hydrogen Embrittlement of Steels", Materials Science and Engineering, vol. A286, pp. 269–281, 2000.
[14]Failures Related to Welding, *Failure Analysis and Prevention*, Failure Analysis and Prevention: *ASM Handbook*, ASM International, vol. 11, Materials Park, OH, pp. 156–191, 2002.
[15]S. A. Gedeon and T. W. Eagar, "Assessing Hydrogen-Assisted Cracking Fracture Modes in High-Strength Steel Weldments", Welding Research Supplement 214s, 1990.
[16]G. Magudeeswaran, V. Balasubramanian, G. Madhusudhan Reddy. "Hydrogen Induced Cold Cracking Studies on Armour Grade High Strength Quenched-and-tempered Steel Weldments", International Journal of Hydrogen Energy 33, pp 1897-1908, 2008.
[17]W. Kim, S. Koh, B. Yang, and K. Kim." Effect of Environmental and Metallurgical Factors on Hydrogen Induced Cracking of HSLA Steels" Corrosion Science, 2008.
[18]H. K. Birnbaum and P. Sofronis, "Hydrogen-Enhanced Localized Plasticity- a Mechanism for Hydrogen-related Fracture", Materials Science and Engineering, vol. A176, pp. 191-202, 1994.
[19]D.M. Perez Escobar, L. Duprez, K. Verbeken, C. Miñambres, M. Verhaege, "Identification of the damage in high strength steels after electrochemical hydrogen charging," *Metallurgists 49th Conference Proceedings*, 2010.
[20]A.K. Sinha, Defects and Distortion in Heat-Treated Parts, *Heat Treating*, Vol 4, *ASM Handbook*, ASM International, 1991, p 601–619.
[21] I. Maroef, D. L. Olson, M. Eberhart, and G. R. Edwards, " Hydrogen Trapping in Ferritic Steel Weld Metals", International Materials Reviews, vol. 47, 2002.
[22] H. Zakerinia, A. Kermanpur, and A. Najafizadeh, "Color Metallography: a Suitable Method for Characterization of Martensite and Bainite in Multiphase Steels", International Journal of ISSI, vol. 6, pp. 14-18, 2009.

Interpretation of diffusible hydrogen with respect to grain size using in situ slow strain rate test and thermal desorption spectrometry

Cheolho Park, Namhyun Kang
Pusan National University, Busan, Korea
Chulho84@naver.com, nhkang@pusan.ac.kr

Stephen Liu
Colorado School of Mines, Colorado, USA
sliu@mines.edu

Abstract

Many studies have investigated the role of grain size in the occurrence of hydrogen embrittlement (HE). However, previous studies on the effects of grain size on HE provided inconsistent results. Some researchers reported that there are no appreciable differences between coarse grains and fine grains but the others reported that HE resistivity increased with grain size refinement. The previous studies dealt with different alloy compositions corresponding to microstructures such as the phase type and grain size. This study investigated the effects of solely the grain size on HE resistivity by in situ slow strain rate tests (SSRTs) for offshore platform steels. Hydrogen was electrochemically charged into specimens and the hydrogen content that evolved from a diffusible hydrogen-trap site was measured by thermal desorption spectrometry (TDS). A simplified microstructure with polygonal ferrite and pearlite as the primary and secondary phases, respectively, formed under various solution heat treatment times. The fine-grained (14 μm) steel showed superior HE resistivity despite exhibiting higher total hydrogen content than the coarse-grained (35 μm) steel. The ferrite grain boundary was the main trapping site of diffusible hydrogen. Quasi-cleavage fracture resulted in a significant decrease of displacement as grain size increased. During in situ SSRT with hydrogen charging, the coarse-grained steel underwent small plastic deformation and the coarse grain suffered from significant HE. Decreasing grain size resulted in a larger grain boundary area, thereby distributing more of the diffusible hydrogen trapped in the grain boundary and reducing the hydrogen trapped in the unit grain boundary near the notched area. The hydrogen content in a unit boundary was defined as the hydrogen released from the grain boundary divided by the grain boundary density. Grain size refinement was effective in increasing the HE resistivity.

Introduction

Hydrogen embrittlement (HE) is known to be related with the factors such as microstructure, grain size, strain rate, residual stress, and diffusible hydrogen. And, the HE sensitivity increases with increasing strength level due to the formation of low temperature transformation microstructures such as bainite and martensite. Grain size refinement reduces the stress concentration and mitigates impurity segregation at grain boundaries. Specifically, grain boundaries block dislocation movement, and therefore, grain size refinement plays an important role in preventing dislocation-induced hydrogen transportation.

Many studies have investigated the role of grain size in the HE occurrence. However, previous studies on the effects of grain size on HE provided inconsistent results. Some researchers reported that there are no appreciable differences between coarse grains and fine grains but the others reported that HE resistivity increased with grain size refinement. The previous studies dealt with different alloy compositions corresponding to microstructures differing in aspects such as the phase type and grain size. Therefore, by means of an in situ slow strain rate test (SSRT), the present study investigated the effects of solely the grain size on the HE resistivity while keeping the chemical composition, phase type, and strain rate constant.

Experimental Procedure

Offshore platform steels were produced by TMCP with a plate thickness 60 mm. The chemical composition of the alloy is as follows: C 0.14 wt.%, Si 0.35 wt.%, Mn 1.5 wt.%, Cr 0.25 wt.%, Mo 0.12 wt.%, Ni 0.7 wt.%, and balanced Fe. In order to change solely the grain size, the steels were reheated to 1423 K and subjected to various solution heat treatments.

Microstructures were observed by optical microscopy (OM), scanning electron microscopy with a field emission gun (FE-SEM), and electron backscattered diffraction (EBSD). The specimens were mechanically polished and etched with an aqueous solution of 2% HNO_3.

Specimens for the in situ SSRT were prepared in the rolling direction. A notch was introduced at the centre of the gauge section for all specimens. The cleaned specimen was pre-charged for 12 h at a current density of 0.5 mA/cm^2. The solution used for hydrogen charging was an aqueous mixture of 3% NaCl and 0.3% NH_4SCN. The specimens were immediately set into an SSRT cell, and the in situ SSRT was performed at the same current density and solution as that in the pre-charging experiment. The crosshead speed was 0.005 mm/min, corresponding to a nominal strain rate of 10^{-6}/s.

The hydrogen content in the specimen was measured immediately after the fracture of the in situ SSRT specimens. TDS with gas chromatography was performed at constant heating rates of 200 K/h. Hydrogen desorption curves were measured at least 3 times for each grain size, and the diffusible hydrogen contents were averaged from all measurements.

Results and discussion

As-rolled specimen fabricated by thermal-mechanical controlled process (TMCP) consisted of fine, elongated ferrite grain as the primary phase, with various kinds of secondary phases such as finely dispersed carbides and bainite. The base material was occupied with the primary phase (ferrite) and the complex secondary phases formed in the as-rolled specimen. Therefore, various times of solution heat treatment resulted in the formation of a simplified microstructure with polygonal ferrite and pearlite. Fig. 1 shows the SEM-EBSD images of specimens after solution heat treatment at 1423 K with various solution heat treatment time. The average grain sizes of the specimens were 14 μm, 26 μm, and 35 μm for solution heat treatment times of 1 h, 6 h, and 40 h, respectively.

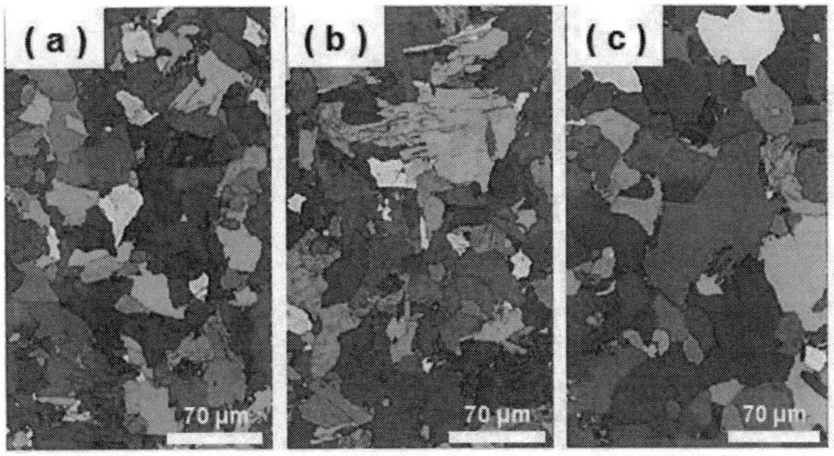

Fig. 1 SEM-EBSD microstructure of specimen after solution heat treatment at 1423 K for (a) 1 h, (b) 6 h, and (c) 40 h.

Tensile stress–displacement was measured for uncharged and cathodically charged specimens having various grain sizes. An increase of grain size resulted in a decrease in the yield strength (YS) and ultimate tensile strength (UTS) regardless of hydrogen charging. With hydrogen charging, the total displacement and UTS suffered a significant loss as the grain size (G.S) increased. The YS also decreased with hydrogen charging, but its decrease with increasing grain size was insignificant as the decrease of the UTS and displacement. Table 1 shows the strength and displacement measured by the in situ SSRT for various grain sizes and hydrogen charging conditions. Regardless of hydrogen charging, the YS for each grain size remained mostly constant. However, the hydrogen charging condition decreased the UTS and displacement significantly for each grain size. Hydrogen charging caused a decrease in the YS/TS/fracture strength relative to the respective strengths before hydrogen charging. Specifically, the displacement showed a significant decrease with increasing grain size after hydrogen charging.

Table 1. Tensile properties of specimens with various grain sizes.

G.S (μm)	Atmosphere	YS (MPa)	UTS (MPa)	Displacement(mm)	Fracture stess (MPa)
14	Air	625	758	2.51	459
14	H_2	623	688	2.10	452
26	Air	579	736	2.52	448
26	H_2	568	623	1.94	417
35	Air	549	707	2.70	426
35	H_2	538	568	1.62	308

Fig. 2 shows the detailed effect of hydrogen charging on the fractographic transition investigated as a function of the grain size. Prior to hydrogen charging, ductile fractured surfaces with micro-void coalescence and a dimpled feature were observed regardless of the grain size. However, the in situ SSRT specimens consisted of quasi-cleavage (QC), intergranular (IG), and dimpled fracture regions for various grain sizes as indicated in Figs. 2a–2c. The red boundaries in Figs. 2a–2c represent the borders between the QC and dimpled fracture surfaces. The red arrows in Figs. 2a–2c indicate the mean depth of the QC fracture initiated from the notch edges of the specimens. The fractions of QC fracture were ~30%, ~48%, and ~68% for the grain sizes of 14 μm, 26 μm, and 35 μm, respectively. Fig. 2d shows a significant increase in the fraction of the QC fracture and a decrease in that of the dimpled fracture with an increase of the grain size. As can be found from the total displacement loss in Table 1, this result implies that the coarse-grained specimen is more vulnerable to HE than is the fine-grained specimen.

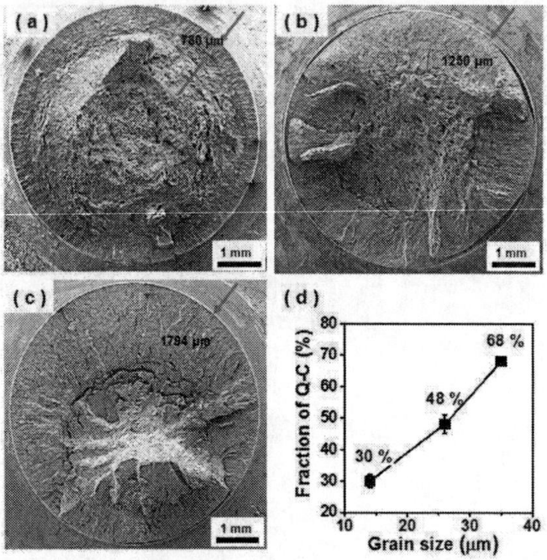

Fig. 2 Fractography of specimens after in situ SSRT with hydrogen charging as a function of grain size: (a) 14 μm, (b) 26 μm, and (c) 35 μm. (d) fraction of quasi-cleavage fracture as a function of grain size.

Fig. 3a shows the hydrogen desorption rate as a function of temperature and grain size. The hydrogen contents corresponding to peak area integration were 2.2 $\pm$ 0.36 wppm, 1.9 $\pm$ 0.34 wppm, and 1.6 $\pm$ 0.25 wppm for the grain sizes of 14 μm, 26 μm, and 35 μm, respectively. The hydrogen desorption curves showed a bimodal pattern: the first peak was at approximately 330 K and the second peak was at approximately 390 K. The first peak at 330 K is attributed to the hydrogen desorbed from normal lattice sites and the second peak is attributed to the grain boundaries. Therefore, the measured curves were deconvoluted into two Gaussian curves after fixing the peak locations to 330 K and 390 K, as indicated in Fig. 3b. Despite exhibiting the largest amount of total hydrogen desorbed from the grain boundary, the fine grain size gave the best SSRT result, i.e. the smallest total displacement loss, as indicated in Table 1.

To explain the characteristics of the hydrogen amount and tensile properties during the in situ SSRT, the parameter of hydrogen content in a unit boundary is defined as hydrogen in unit grain boundary of ferrite, i.e. the hydrogen released from the grain boundary divided by the grain boundary density: where the grain boundary density was calculated by taking the ratio of the total grain boundary length to the total measurement area. The density values were measured by EBSD. Clearly, the grain boundary density decreased as the grain size increased. The hydrogen amounts desorbed in the unit grain boundary were 3.5 ± 0.18 wppm-μm, 4.0 ± 0.17 wppm-μm, and 7.3 ± 0.12 wppm-μm for the grain sizes of 14 μm, 26 μm, and 35 μm, respectively.

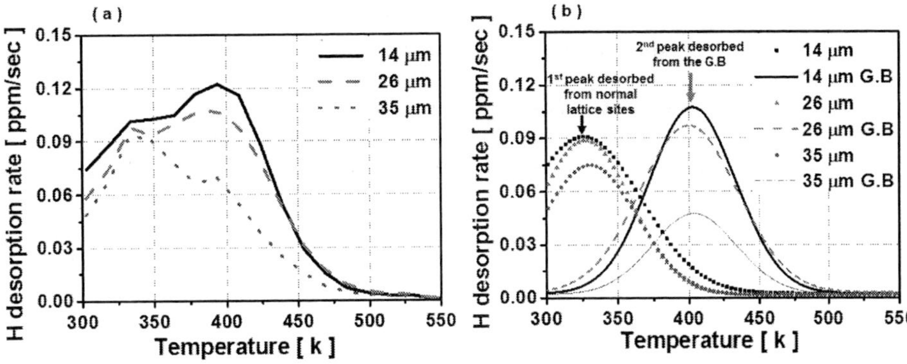

Fig. 3 Rate of hydrogen desorption from specimen as a function of grain size: (a) curves obtained after in situ SSRT and (b) peaks deconvoluted from thermal desorption curves shown in (a).

Finally, the fine-grained steel exhibited a lower hydrogen content in the unit grain boundary and superior tensile properties against HE in comparison to the large-grained steel, although the fine-grained steel had a high total hydrogen content measured by the TDS. A decrease in the grain size resulted in an increase in the grain boundary area, thereby distributing more of the diffusible hydrogen trapped in the grain boundary and reducing the amount of hydrogen trapped in the unit grain boundary. The pre-notched area was the location of initiation of tensile fracture during the in situ SSRT. For this reason, grain size refinement was effective in reducing the hydrogen content in the unit grain boundary near the notched area and in increasing the HE resistivity.

Conclusions

1) Hydrogen charging decreased the tensile/fracture strengths relative to the respective strengths before hydrogen charging. The fine-grained (14 μm) steel showed superior HE resistivity, although it contained a higher total hydrogen content than the coarse-grained (35 μm) steel.

2) The fractions of QC fracture were approximately 30%, 48%, and 68% for the grain sizes of 14 μm, 26 μm, and 35 μm, respectively. QC fracture resulted in a significant decrease of displacement with increasing grain size.

3) The fine-grained steel exhibited the lowest hydrogen content in a unit grain boundary. For this reason, the fine-grained steel exhibited superior tensile properties for the in situ SSRT despite having the largest amount of total hydrogen and hydrogen desorbed from the ferrite grain boundary. Grain size refinement was effective in increasing the HE resistivity.

Acknowledgements

The authors are thankful for the financial support of the Basic Science Research Program through the National Research Foundation of Korea (NRF) grant funded by the Ministry of Education (No. 2013R1A1A2013125).

References

[1] Y.Matsumoto: "Reduction of Delayed Fracture Susceptibility of Tempered Martensitic Steel through Increased Si Content and Surface Softening", *ISIJ int*, Vol. 53, (2013), pp. 714-722.
[2] K. Takasawa: "Effects of grain size and dislocation density on the susceptibility to high-pressure hydrogen environment embrittlement of high-strength low-alloy steels", *Int. J. Hydrogen Energy*, Vol. 37 (2012), pp. 2669–2675.
[3] M. Wang: "Effect of hydrogen on the fracture behavior of high strength steel during slow strain rate test", *Corros. Sci*, Vol. 49, (2007), pp. 4081–4097.
[4] K. Takasawa: "Effects of Grain Size on Hydrogen Environment Embrittlement of High Strength Low Alloy Steel in 45 MPa Gaseous Hydrogen", *Mater. Trans*, Vol. 51, (2010), PP. 347–353.

Comprehensive analysis of cold crack testing for high strength FCA welds using Y- and y-groove

Hyunbin Nam, Namhyun Kang
Pusan National University, Pusan, Korea
hbnam3206@gmail.com, nhkang@pusan.ac.kr

Jaeseok Yoo, Kwanghee Yun, Kihyung Han
Daewoo Shipbuilding & Marine Engineering Co. Ltd, Geoje, Korea
yjs0303@DSME.co.kr, gjan@DSME.co.kr, khhan@DSME.co.kr

Abstract

In modern times, energy resources in polar region and abyss receives more attention due to the increased possibility of economic production of crude oil and natural gas resources. As the offshore plant and vessels installed in polar region or abyss are operated in extreme environments, the high strength and low temperature toughness properties of a structural steel is required, and the high strength weld consumables of low hydrogen type need to be applied for the requirement of structural integrity. However, when we add alloying elements to meet the high strength and low temperature toughness of steels and weld consumables, the weldability is normally degraded due to increase of the carbon equivalent of the welds. Heat-affected zone (HAZ) was the critical flaw for cold cracking until we used the steels to medium strength, but the location of cold cracking was incrementally shifted to the deposited metal with increasing the use of high strength steels. Therefore, previous studies of cold crack testing consists of two geometry, i.e. Y-groove and y-groove. Many studies were conducted to examine the effect of preheating temperature and diffusible hydrogen content using various FCA consumables on the cold crack. However, there is few studies to collaborate the effect of strength grade, diffusible hydrogen content, base metal thickness, preheating temperature, and groove geometry on the cold crack. The main objective of the study is to understand how the groove geometry and base metal thickness influence the cold cracks for each strength grade and diffusible hydrogen content. We used a various strength grade of welding consumable from YS375 to 500 MPa and base metal thickness of 20mm and 40mm. Comprehensive analysis of hardness distribution, microstructure, and diffusible hydrogen content was also performed in the weld metals to understand cold crack ratio and cold crack propagation behaviour.

Introduction

Continuously, energy consumption is increasing and it is not possible to satisfy the demand of energy because of resources depletion. Therefore, submarine resources and natural gas need to be excavated in the polar region or abyss in the shallow sea. It is essential to request the high strength of the offshore plant and vessels. For welding the structural steel of medium strength, cold cracking was frequently observed in the heat affected zone (HAZ). However, for the high strength steel, cold cracking location has been shifted to the weld metal. Cold cracking occurs by complex factors such as the amount of diffusible hydrogen, susceptible microstructure, and residual tensile stress. Research on cold cracking occurred in HAZ has been evaluated by the y-groove test. Cold cracking was improved with the control of the diffusible hydrogen, microstructure, and tensile residual stress. On the other hand, cold cracking in the weld metal increased for high strength steels and it was also evaluated by the Y-groove as well as the y-groove. In order to apply the high-strength steels and weld metal, the studies on weld/flux materials, preheat temperature, and the application of weld material in accordance with strength grade of the base material reduced the cold cracking successfully. However, there is no fundamental and systematic study of the diffusible hydrogen, the base material thickness, and groove shape in accordance with welding metal of each strength grade on the cold cracking. Therefore, the study investigated the comprehensive analysis of the cold cracking with respect to the diffusible hydrogen content, groove shape, and thickness of the base material in the weld metal of each strength grade (YS375/500 MPa grade).

Experimental Procedure

The base metals used in the Y/y-groove tests were YS235/500 MPa-grade steel plates of thickness 20 mm and 40 mm. This research applied four types of YS375/500 MPa grade FCA welding consumables. Table 1 shows the chemical compositions and diffusible hydrogen content of the base and weld metals. For nomenclature of welding and base materials, R stands for rutile type flux, B for basic type flux, S for seamless, and F for folded type consumables.

Table 1. Chemical compositions and diffusible hydrogen of base metal and weld metal.

No		C	Si	Mn	Ni	Cr	Cu+Mo+B	Ceq	Pcm	HD(ml/100g)
Welding Material	WM375-R	0.05	0.73	1.40	0.02	0.04	0.03	0.30	0.15	3.64
	WM375-B	0.07	0.64	1.41	0.01	0.02	0.03	0.31	0.16	0.82
	WM500-S	0.06	0.32	1.37	1.37	0.01	0.27	0.40	0.19	1.49
	WM500-F	0.06	0.37	1.37	2.07	0.01	0.02	0.44	0.20	2.39
Base Metal	BM235	0.16	0.28	0.68	0.01	0.02	0.01	0.28	0.20	-
	BM500	0.06	0.20	1.54	0.19	0.22	0.35	0.42	0.19	-

The cold cracking evaluation was carried out with reference to the ISO17642-2 standard. During welding, the welding heat input was 8 KJ/mm. In order to determine cold cracking ratio (CCR), the weld zone of 5 equal parts were cut to observe 5 cross-sections through a stereoscopic microscopy and light optical microscopy (LOM) since 48 hours past to welding. CCR was calculated by the following equation: $Cs = Hc / H X 100$ where Cs is cold cracking ratio in a cross section (%), Hc is the root cracking height (mm), and H is the minimum thickness of the test beads (mm). The Y/y-groove weld specimens were etched with 2 % nital solution, and the microstructure was analysed by LOM and scanning electron microscopy (SEM). The volume fraction of microstructure was analyzed from the microstructural photographs taken at x500 magnification by applying a point counting method. Hardness properties of the weld metal was measured by micro-vickers hardness, a load of 0.5 kg for 10 seconds. Hardness was taken at 0.5mm below from the surface of the specimen and 0.5 mm above from the root section, and it was measured at 0.5mm interval.

Results and discussion

1. Effect of alloying constituent, diffusible hydrogen content, hardness, and microstructure on cold cracking

Figures 1 show the CCR of the weld metal as a function of the alloying constituent, respectively. As the strength of the weld metal increased, Ceq and CCR increased. For the same strength grade, CCR increased with increasing diffusible hydrogen content in weld metals. Basic type flux and seamless type FCA consumables had less amount of diffusible hydrogen than rutile and folded type FCA consumables. This result was consistent with the previous result of cold cracking [1,2]. Therefore, the CCR values of weld metals were related to Ceq and diffusible hydrogen content.

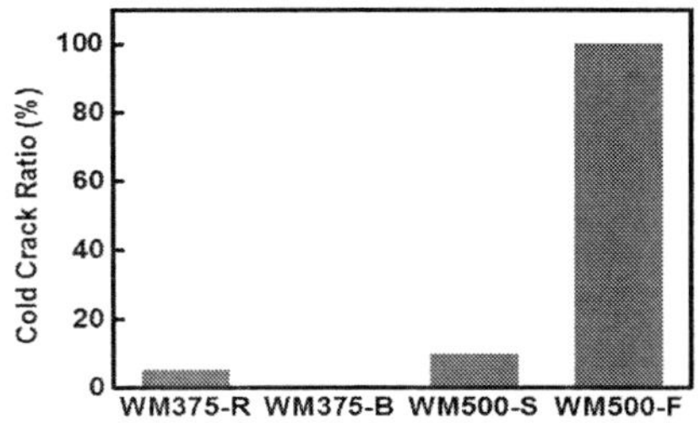

Figure 1. Cold crack ration (CCR) for each weld metal.

Table 2 shows the volume fractions of the microstructure and average hardness values as a function of the alloying constituents of each weld metal. With increasing weld metal strength grade and Ceq (Ceq: $0.30 < 0.31 < 0.40 < 0.44$), the type of the main microstructure varied [AF → bainite and martensite] and the fraction of main microstructure and the average hardness of the weld metal increased. Therefore, the increased cold cracking susceptibility was promoted by the bainite and martensitic transformation [3].

In addition, the Ceq values of the weld metal for the same strength grade (WM375-R/B), microstructural fraction of AF, and the average hardness were essentially the same. However, the CCR of WM375-B was significantly lower than that of WM375-R. For the reason, we made a conclusion of the major factor on CCR: Ceq and diffusible hydrogen content were the primary factors on CCR, and hardness and microstructural fraction were the secondary factors on CCR [4,5].

Table 2. The volume fractions of microstructures and hardness values as a function of the alloying constituents.

Specimen	GBF&WF	AF	B&M	Hardness (Hv)
WM375-R	20%	80%	-	284
WM375-B	32%	68%	-	276
WM500-S	-	20%	80%	318
WM500-F	-	10%	90%	364

2. Effect of the groove shape on cold cracking

Figure 2a shows the CCR of each strength grade weld metal as a function of the groove shape. Normally, the Y-groove is used to evaluate the cold-cracking resistance of the weld metal, whereas the y-groove is used to evaluate the cold-cracking resistance of the HAZ. However, regardless of the groove shape, cold cracking of the weld metal occurred in the weld metal for all specimens. The low-strength-grade weld metal, regardless of the groove shape, showed a low and similar CCR. However, the specimens of high-strength grade increased the CCR for the y-groove as compared with the Y-groove.

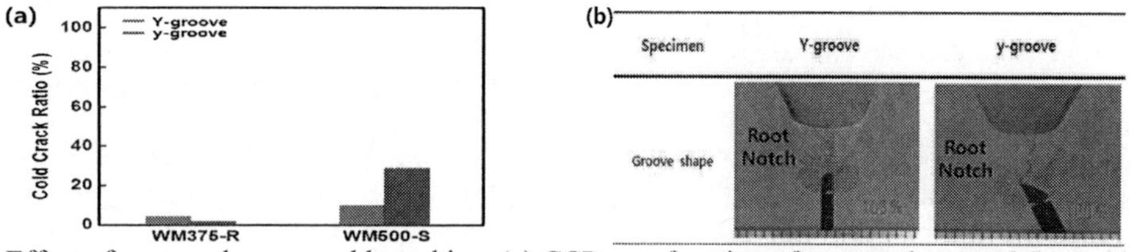

Figure 2. Effect of groove shape on cold cracking: (a) CCR as a function of groove shape and (b) Cold-cracking starting point and propagation behavior.

Figure 2b shows the cold-cracking starting point and propagation behavior as a function of the groove shape. Lack of weld penetration was formed at the root of the specimen, in which the stress concentration could be maximized during cooling. The Y-groove specimen indicated large curvature radius of the root notch, therefore reducing the stress concentration. However, the y-groove specimen showed sharpness of the root notch, inducing the stress concentration more serious than the Y-groove specimen.

Table 3 shows the volume fraction of the various microstructures and average hardness values as a function of the groove shape of each weld metal. Regardless of the strength grade of the weld metal, the fraction of microstructure and the average hardness were mostly same with respect to the groove shape. The simulation of residual stress-strain distribution is underway to understand CCR behavior of the Y-groove and y-groove for each weld metal.

Table 3. The volume fractions of microstructures and hardness values as a function of the groove shape.

Specimen		GBF&WF	AF	B&M	Hardness (Hv)
WM375-R	Y-groove	20%	80%	-	284
	y-groove	25%	75%	-	272
WM500-S	Y-groove	-	20%	80%	318
	y-groove	-	17%	83%	324

3. Effect of the base metal thickness on cold cracking

Figure 4 shows the CCR of each strength grade of weld metal as a function of the thickness of the base metal. Regardless of strength-grade of the weld metal, 40-mm thick base metals showed a increased CCR as compared with 20-mm thick base metals. As the thickness of the base metal increased (20 mm → 40 mm), the cooling rate and restraint stress of the weld metal was maximized. Therefore, the CCR increased.

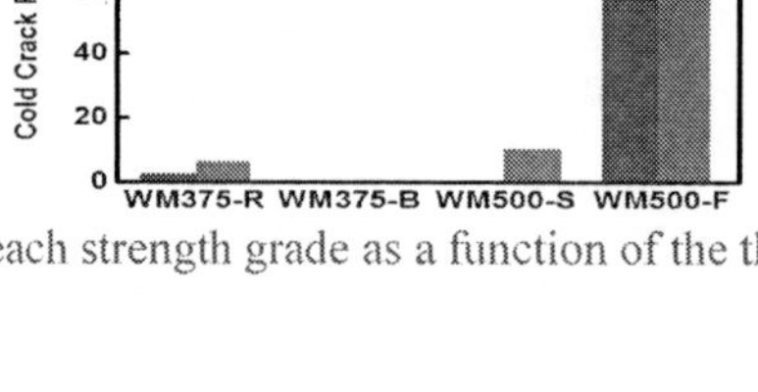

Figure 4. The CCR of each strength grade as a function of the thickness of the base metal.

Table 4 shows the fraction of the microstructures and average hardness values associated with differences in the thickness of the base metal. As the strength grade of weld metal increased, the type of the main microstructure changed from AF to B and M. Furthermore, the thickness of the base metal increased the fraction of the main microstructure. Normally, the cooling rate, which depends on the thickness of the steel plate, makes a notable difference on the fraction of microstructure and hardness [6]. Therefore, the simulation of residual stress-strain distribution and cooling rate associated with continuous cooling transformation (CCT) curves is underway to understand CCR behavior of the base metal thickness for each weld metal.

Table 4. The volume fractions of microstructures and hardness values the thickness function of the base metal

Specimen		GBF&WF	AF	B&M	hardness values(Hv)
WM375-R	20mm	29%	71%	-	257
	40mm	20%	80%	-	284
WM375-B	20mm	43%	57%	-	259
	40mm	32%	68%	-	276
WM500-S	20mm	-	30%	70%	303
	40mm	-	20%	80%	318
WM500-F	20mm	-	15%	85%	362
	40mm	-	10%	90%	364

Conclusions

The study investigated the comprehensive analysis of the cold cracking with respect to the diffusible hydrogen content, groove shape, and thickness of the base material in the weld metal of each strength grade (YS375/500 MPa grade) and made a conclusion as follows:

1) With increasing strength grade of the weld metal, the fraction of martensite vulnerable to cold cracking increased. In addition, for two weld metals with the same strength grade, the specimen with the smaller diffusible hydrogen content showed a lower cold-cracking ratio. Therefore, it was determined that cold cracking was reduced by the combined effect of diffusible hydrogen content and carbon equivalent.

2) The low-strength-grade weld metal exhibited similar cold-cracking ratios, regardless of groove shape. However, the high-strength-grade y-groove weld metals exhibited slightly increased cold-cracking ratios. The y-groove stress concentration and the restraining stress exerted on the root notch shape should be the main reason to increase the cold cracking ratio (CCR).

3) The CCR of each strength grade of weld metal was investigated as a function of the base metal thickness. The cold cracking ratio increased as the thickness of the base material increased. With increasing thickness of the base metal, the cooling rate and restraint stress of the weld metal should be maximized.

References

[1] McParlan, M., and Graville, B. A., "Hydrogen Cracking in Weld Metals," Welding Journal, 55-5, 95s-102s, (1986).

[2] Hart, P. H. M., "Resistance to Hydrogen Cracking in Steel Weld Metal," Welding Journal, 65-1, 14s-22s, (1986).

[3] Lee. M. J., and Kang. N. H., "The effects of microstructure on cold crack in high-strength weld metals," J KWJS Vol. 32, No. 1, pp. 22–27 (2014)

[4] Sevensson, L. E., and Gretoft, B., "Microstructure and Impact Toughness of C-Mn Weld Metals." Welding Journal, Vol. 75, No. 8, pp.251s-260s (1990).

[5] Evans, G, M., "Microstructure and Properties of Ferritic Steel Welds Containing Ti and B," Welding Journal, Vol. 75, No. 8, pp.251s-260s (1990).

[6] Adams, C. M., "Cooling Rates and Peak Temperatures in Fusion Welding," Welding Journal Research Supplement, pp. 210~215, (1958)

Research and development on the welding process for reducing diffusible hydrogen content

Naoki MUKAI, Tokuji MARUYAMA, and Reiichi SUZUKI
Technical Center, Welding Business, Kobe Steel, LTD. , Kanagawa, Japan
Email : mukai.naoki@kobelco.com, phone : +81-466-20-3381, fax : +81-466-20-3289

Abstract

Cold cracking sensibility is severe problem in welding fabrication of large-scale steel structures. It is well-known that diffusible hydrogen strongly influences this cracking. However, it isn't understood well what kinds of factors in welding consumables affect diffusible hydrogen content. This study shows the effect of factors related to welding wire on diffusible hydrogen content in GMAW and FCAW. Furthermore, it shows the effectiveness of "low-hydrogen welding process [1-2]" on each factor. The factors related to welding wire on diffusible hydrogen content are divided into the following classes. (a)Surface lubricant, (b)Initial moisture in flux and (c)Moisture absorbed after production. Some welding wires were prepared and measured the diffusible hydrogen content in the weld metal with the conventional and the low-hydrogen welding process. As results, the low-hydrogen welding process removed the influence of factor (a) and (c). About factor (b), there is a slight effect of the low-hydrogen welding process in seamless flux cored wire tests. However, a great effect of decreasing diffusible hydrogen content is obtained from in seamed wire tests. There is a possibility that welding operator or supervisor will be able to control the diffusible hydrogen content in weld metal by applying this welding process and solid or seamed flux cored wire.

Introduction

In response to the industry request for higher performance efficiency, the enlargement of steel structures has rapidly been progressed. For scaling up the size of a structure, the components are designed with thicker/higher-strength materials in order to increase their strength and rigidity. However, associated with the use of thicker/higher-strength steel materials, the cold crack sensitivity increases when welding is applied. Meanwhile, cold cracking is affected greatly by diffusible hydrogen [3].

In the cases of gas metal arc welding (GMAW) with a solid wire and flux cored arc welding (FCAW) with a flux-cored wire (FCW), The diffusible hydrogen content may vary markedly depending on the design of a welding consumable; e.g. solid wires result in lower values. By contrast, FCWs result in higher values than with solid wires, and the seamed FCW in particular has been considered more likely to increase the diffusible hydrogen than with the seamless FCW because the former tends to absorb moisture through the seam in the air with the time elapsed after being unpacked. However, this accepted understanding has not been verified by statistics or mechanical descriptions at an academic place.

Investigation of diffusible-hydrogen content with commercially available welding wires

The diffusible hydrogen content in the weld metal was researched with FCWs (*Table 1*) available in several countries in order to review such a general view that the diffusible hydrogen content with a FCW varies depending on FCW's structure (seamless or seamed). In the results shown in *Fig.1*, no clear relation was found between seamless and seamed FCWs in terms of the diffusible hydrogen content, unlike the accepted understanding about the nature of FCW's structure. In contrast, the diffusible hydrogen content has a good relationship with the strength class of the weld metal (*Fig.2*), which suggests that the diffusible hydrogen content is controlled basically by the design of a welding wire.

Table 1 Tested flux-cored wires

Wire	Seam type	Flux type	Strength class	Shielding gas	Manufacturer
A	Seamless	Metal	760MPa (110ksi)	Ar-20%CO_2	H
B		Rutile			
C			620MPa (90ksi)		I
D			490MPa (70ksi)	CO_2	J
E	Seamed		780MPa	Ar-20%CO_2	K
F			550MPa (80ksi)	CO_2	L
G			490MPa		K

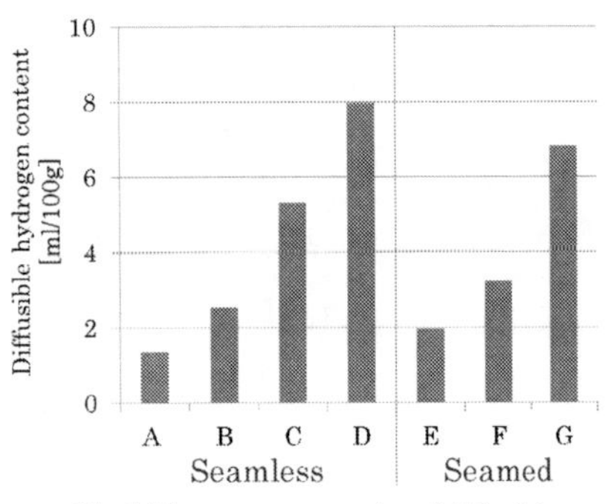

Fig.1 Measurement results of diffusible-hydrogen content for various FCWs

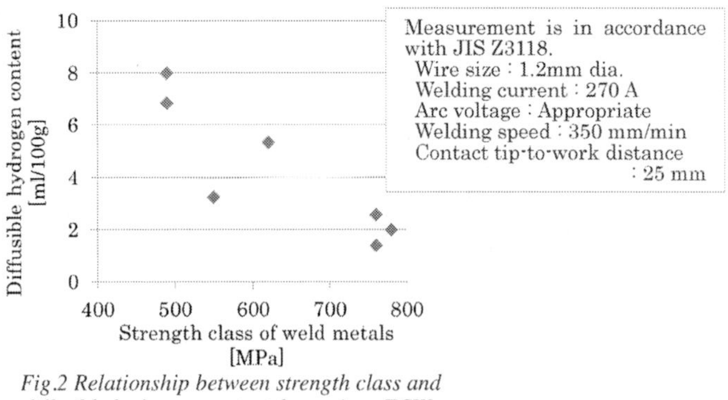

Fig.2 Relationship between strength class and diffusible hydrogen content for various FCWs

Purposes for this study

It is hard say that the factors affecting the diffusible hydrogen content of the weld metal made by FCAW is well understood. Therefore, in the present study, these factors were classified and investigated its influence individually. Meanwhile, various methods of reducing the diffusible hydrogen in a weld metal have been researched over the past [4-6]. Moreover, a new diffusible-hydrogen-reducible welding process (the low-hydrogen welding process) with a special welding torch is researched in recent years [1-2]. This study also intended to clarify the effectiveness of the low-hydrogen welding process in relation to the individual factors affecting diffusible hydrogen content.

Diffusible-hydrogen-reducible welding process with special welding torch

For gas-shielded arc welding, proposed is an innovative welding process that can reduce the diffusible hydrogen content in the weld metal by using a special torch that sucks the gas near the welding wire [1-2]. The low-hydrogen welding process uses the double-nozzle torch system that sucks a part of the shielding gas through the suction nozzle installed between the contact tip and the shield nozzle as shown in *Fig. 3*.
The wire in the contact tip-to-work distance is heated by the Joule and arc heats generated by the welding current, and such a thermal energy may release the hydrogen source from the wire. In the low-hydrogen welding process, the gas near the wire, which contains much of the hydrogen source released from the wire, is discharged outside the welding atmosphere through the suction nozzle, and thus the hydrogen quantity led to the arc atmosphere is reduced, thereby reducing the hydrogen to be absorbed in the weld metal.

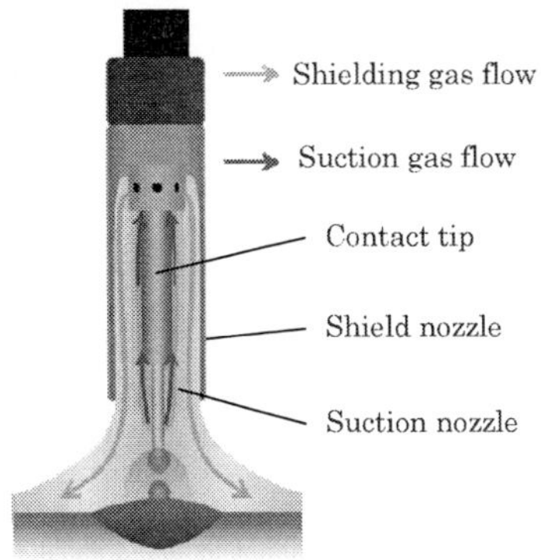

Fig.3 Welding-torch structure used in low-hydrogen welding process

Classification of welding-wire-related factors affecting the diffusible hydrogen content

The diffusible hydrogen sources contained in the welding wires can largely be classified into the following three factors (*Table 2*): (a) surface lubricant, (b) initial moisture of flux, and (c) moisture absorbed after production. The applicability of the factors (a)~(c) was considered variable depending on the wire structure. Other than these hydrogen sources, an organic matter contained as a flux impurity and absorbed moisture in the rust on the wire surface can also be considered to be hydrogen sources. However, the amount of organic impurities can be considered little under the usual production control for the wire, and the absorbed moisture in the rust is an abnormal factor; hence, such hydrogen sources were not taken into account in this study.

Table 2 Welding-wire-related factors affecting the diffusible hydrogen in GMAW and FCAW

Factors of hydrogen source	Solid wire		Seamless flux cored wire		Seamed flux cored wire	
(a) Surface lubricant	○	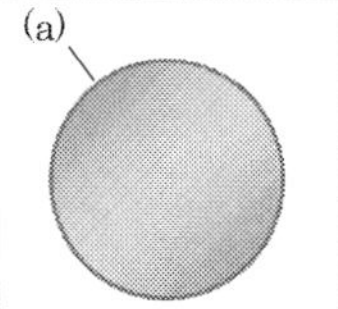	○	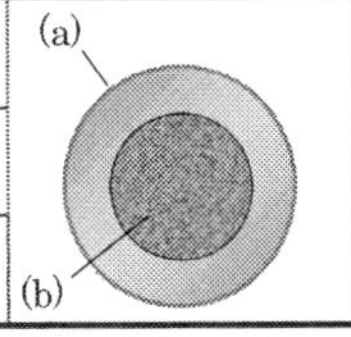	○	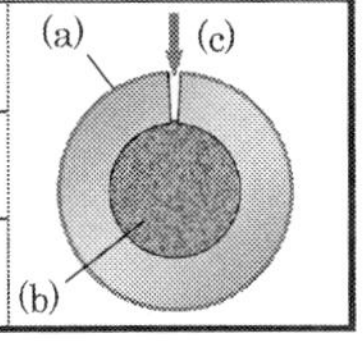
(b) Initial moisture of flux	—		○		○	
(c) Moisture absorbed after production	—		—		○	

Note ○: Applicable, —: Negligible

Effectiveness of low-hydrogen welding process on diffusible hydrogen in relation to welding-wire-related factors

In the research paper [3-4] of the low-hydrogen welding process (refer to LHTP: low-hydrogen torch process), reported are (1) the assumed process and mechanism by which the hydrogen source of the welding wire is absorbed in the weld metal, (2) the design concept of the special torch based on the said assumption, and (3) the effect of the said torch on reducing the diffusible hydrogen content. However, it was unclear how the LHTP affects the factors of several hydrogen sources of the welding wire. To clarify this, LHTP was tested on its effectiveness in relation to the each welding-wire-related factor (shown in *Table 2*) that affects the diffusible hydrogen.

Effectiveness of LHTP on surface lubricant

The effect of "(a) surface lubricant" alone on the diffusible hydrogen content in the weld metal was investigated with varied amounts of the lubricant adhered on a solid wire. For solid wires, the effects of the factors of (b) and (c) were considered negligible. The testing wires prepared are shown in *Table 3*. The trial wires were examined with conventional welding process (CWP) and LHTP to measure the diffusible hydrogen content in the weld metal. The suction gas flow rate in the LHTP was set at 5 l/min. In *Fig.4*, the measurement results are plotted.

Table 3 Trial solid wires for testing the effect of surface lubricant

Sample No.	Surface oil amount [ppm]	Applicable standard	Wire diameter [mm]
a1	0	JIS Z 3312 YGW11 (AWS A5.18 ER70S-G)	1.2
a2	60		
a3	110		
a4	540		

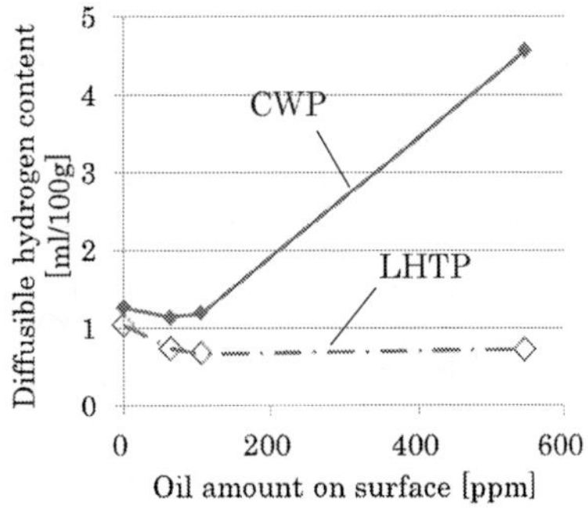

Measurement is in accordance with JIS Z3118 except the contact tip-to-work distance.

Shielding gas : CO_2
Welding current : 300 A
Arc voltage : 32 V
Welding speed : 350 mm/min
Contact tip-to-work distance : 25 mm
Shielding gas flow rate : 25 l/min
Suction gas flow rate : 5 l/min (LHTP)

Fig.4 Diffusible hydrogen content by CWP and LHTP as a function of the oil amount on surface of solid wire

In the case of CWP, the diffusible hydrogen content of the high oil wire (a4) increased by approx. 3 ml/100g as compared with the low oil wires. By contrast, LHTP is so effective to reduce the diffusible hydrogen and thus can eliminate the influence of the factor "(a) surface lubricant". For the oil amount on surface of 110 ppm or less (shown in *Table 3*, a1- a3), the data plots for LHTP show a descending tendency on the right side, but they should probably be the similar diffusible hydrogen measurements within a certain measurement error.

Effectiveness of LHTP on initial moisture of flux

The effect of "(b) initial moisture of flux" alone on the diffusible hydrogen content in the weld metal was investigated with FCWs according to the experimental procedure stated below.

The seamless and seamed FCWs were prepared with the constant raw material chemical composition and the two levels of moisture content adjusted by the preliminary process. The factor (a) was considered negligible for both types of FCWs because the amount of surface lubricant was kept at the low level shown in *Fig.4*, and the factor (c) for seamless FCWs was deemed to be ignorable in this experiment. The seamed FCWs were stored in a low-humidity atmosphere after production so that the effect of the factor (c) could be minimized. The FCWs prepared are shown in *Table 4*. The trial wires were examined with CWP and LHTP to measure the diffusible hydrogen content in the weld metal. In *Fig.5*, the results are plotted.

Table 4 Trial FCWs for testing the effect of initial moisture of flux

Sample No.	Seam type	Moisture content of flux [ppm]	Applicable standard	Wire diameter [mm]
b1	Seamless	350	JIS Z 3313 T49J0T1-1 (AWS A5.20 E71T-1C)	1.2
b2		1210		
b3	Seamed	350		
b4		1210		

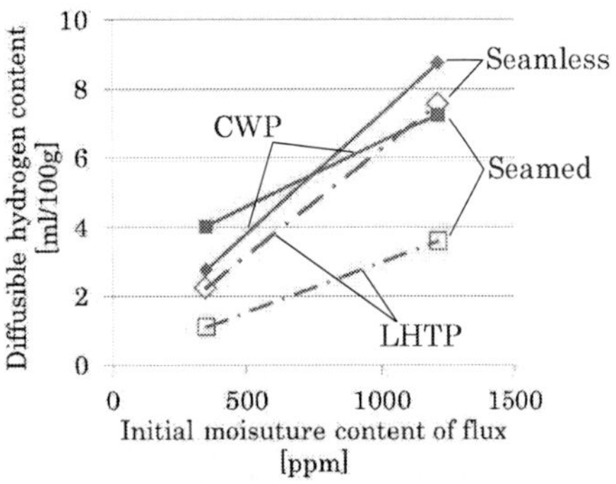

Measurement is in accordance with JIS Z3118.

Shielding gas : CO_2
Welding current : 270 A
Arc voltage : 32 V
Welding speed : 350 mm/min
Contact tip-to-work distance : 25 mm
Shielding gas flow rate : 25 l/min
Suction gas flow rate : 5 l/min (LHTP)

Fig.5 Diffusible hydrogen content by CWP and LHTP in relation to initial moisture content of flux and wire structure (seamless/seamed FCW)

In the case of CWP, as the initial moisture content of flux increased, so did the amount of diffusible hydrogen content by the FCWs (both seamed and seamless).

When comparing the CWP and LHTP with seamless FCWs, the reduction effect of LHTP was as small as 0.5-1 ml/100g. That is, the effect of LHTP on the diffusible hydrogen reduction for the factor "(b) initial moisture content of flux" can be said to be slight.

By contrast, in the use of seamed FCWs, the reduction effect of LHTP was as big as 3-4 ml/100g. In other words, it has been verified that LHTP enables to eliminate a large amount of influence of the initial moisture. The difference in the behavior of diffusible hydrogen reduction between seamless and seamed FCWs can be understood with the following mechanism. When the wire was heated by the Joule heat in the contact tip-to-work distance, the gaseous hydrogen source generated in the flux could not be released from the seamless FCW with no releasing path. In contrast, the seamed FCW released such a hydrogen source gas easily through its seam, and the shielding gas with a large quantity of hydrogen source gas was effectively ejected though the suction nozzle.

Effectiveness of LHTP on moisture absorbed after production

The effect of "(c) moisture absorbed after production" alone on the diffusible hydrogen content in the weld metal was investigated with seamed FCWs. The sample wires were dried for 3 hours in an isothermal oven kept at 80°C and then exposed for varied periods of time in an constant temperature/humidity chamber kept at 30°C×80%RH for moisture absorption; subsequently, they were subjected to the measurement of the diffusible hydrogen content. The amount of surface lubricant was kept at the low level, and the sample wires for drying/moisture absorption were cut from the continuous wires beforehand; therefore, the factor (a) and the factor (b) of original level were considered negligible in this test. The seamed FCWs prepared are shown in *Table 5*.

The trial wires were examined with CWP and LHTP to measure the diffusible hydrogen content in the weld metal. In *Fig.6*, the results are plotted.

With CWP, the diffusible hydrogen content tended to increase with an increase in the moisture absorption time; by contrast, LHTP exhibited no such a tendency and resulted in almost the constant diffusible hydrogen content as the moisture absorption time elapsed. The moisture absorbed after production probably consisted much of easy-to-remove adhesion water, and hence LHTP probably worked effectively.

Table 5 Trial seamed FCWs for testing the effect of moisture absorbed after production

Sample No.	Applicable standard	Strength class	Moisture absorption time [hr]	Wire diameter [mm]
c1-1	JIS Z 3313 T49J0T1-1 (AWS A5.20 E71T-1C)	490 MPa	0	1.2
c1-2			24	
c1-3			288	
c2-1	JIS Z 3313 T556T1-1 (AWS A5.29 E81T1-K2C)	550 MPa	0	
c2-2			24	
c2-3			72	
c2-4			232	

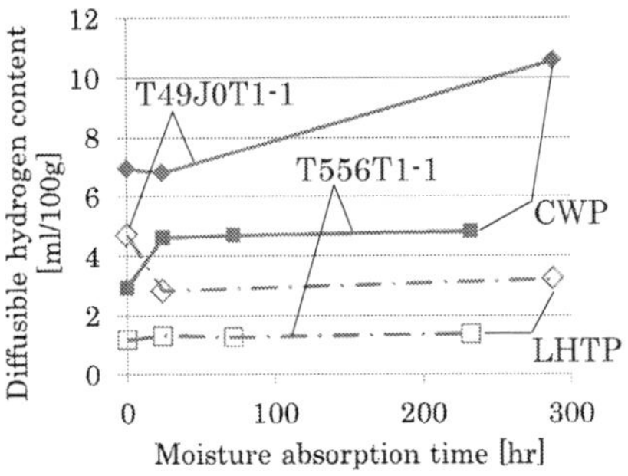

Measurement is in accordance with JIS Z3118.

Shielding gas : CO_2
Welding current : 270 A
Arc voltage : 32 V
Welding speed : 350 mm/min
Contact tip-to-work distance : 25 mm
Shielding gas flow rate : 25 l/min
Suction gas flow rate : 5 l/min (LHTP)

Fig.6 Diffusible hydrogen content by CWP and LHTP with two types of seamed FCWs as a function of moisture absorption time

As regards the factors "(a) surface lubricant" and "(c) absorbed moisture after production," LHTP was so effective that influences of both of the factors could be decreased totally. For the factor "(b) initial moisture of flux," LHTP was not so effective with seamless FCWs to reduce the diffusible hydrogen but was sufficiently effective with seamed FCWs to reduce it. The above-mentioned results are summarized in *Table 6*.

Table 6 Summary of effectiveness of LHTP

Factors of hydrogen source	Solid wire	Seamless flux cored wire	Seamed flux cored wire
(a) Surface Lubricant	Total decrease	Total decrease	Total decrease
(b) Initial moisture of flux	-	Slight decrease	Great decrease
(c) Absorbed moisture after production	-	-	Total decrease

As discussed above, many of the diffusible hydrogen sources of the welding wires can be decreased by the LHTP with a seamed FCW. When using a seamed FCW in combination with LHTP, the diffusible hydrogen content can likely be controlled by the users to a certain degree.

Conclusions

1) A difference could be seen between the accepted and actual understanding, related to diffusible hydrogen content.
2) Welding-wire-related factors, affecting diffusible hydrogen content, were classified into the three factors.
3) The effectiveness of the low-hydrogen torch process on each diffusible hydrogen factor was examined, and results are summarized in the following.
3-1) The low-hydrogen torch process affects surface lubricant and moisture absorption after production significantly.
3-2) The effectiveness on the initial moisture of flux depended on wire structure. With seamed FCW, the low-hydrogen torch process can greatly decrease, while with seamless FCW, there was only a slight decrease.

References

[1] Kawabe, et al.: Development of low hydrogen welding process for gas shielded arc welding, Preprints for the National Meeting of The Japan Welding Society, Vol.96(2015), p172-173.
[2] Kawabe et al: Development of low hydrogen welding process for gas shielded arc welding, IIW Doc.XII-2236-15(2015).
[3] Kasuya: Cold cracking assessment of steels by experiments and numerical simulations, Journal of The Japan Welding Society, Vol.84(2015), No.1, p45-50.
[4] Godai et al.: The hydrogen sources in covered arc welding, Journal of The Japan Welding Society, Vol.45(1976), No.7, p14-20.
[5] Kasuya, et al.: Annealing of flux cored wires and its effect on diffusible hydrogen content, Preprints for the National Meeting of The Japan Welding Society, Vol.84(2009), p311-312.
[6] Fiore et al.: A new method for reducing diffusible hydrogen in weld metal, IIW Doc.XII-2103-12(2012).

Study on Human Welder Interactive Behavior with Weld Pool in GTAW

*Gang ZHANG[1, 2], Yu SHI[*1], YuMing ZHANG[2], YuFen GU[1], Ding FAN[1]*
1. State Key Laboratory of Advanced Processing and Recycling Non-ferrous Metals, Lanzhou University of Technology, Lanzhou, GanSu, P.R.China
gang.zhang@uky.edu, shiyu@lut.cn, guyf@lut.cn, fand@lut.cn

2. Electrical and Computer Engineering Department and Institute of Sustainable Manufacturing, University of Kentucky, Lexington, KY, USA
yuming.zhang@uky.edu

Abstract
Human welder adjusts welding parameters including the torch attitude, traveling speed and position to control weld penetration and avoid weld defects, through real time observing the varying weld pool surface, for precision joining using GTAW. However, welder behavior or adjustment appears to be complex. To understand this complex behavior, a new experimental system is established to synchronously measure the torch attitude and the 3D weld pool surface in this paper. The system uses a wire inertial measuring unit (IMU) to sense torch attitude and a laser vision based method to measure the 3D weld pool surface. An image processing algorithm has been developed to extract the characteristic parameters of the weld pool surface from the laser stripes reflected from the pool surface which is specular or mirror like. Then series of experiments are conducted by different level welder. Experiments verified that the ability of welder to regulate or control the torch attitude represents the operating skills and welding experience of welder. The varying weld pool surface reflects the effects of torch adjustment and provides feedback for welder to make a reasonable, smart adjustment or decision. While novice welders tend to regulate the welding torch more irregularly/randomly reflecting less interactive consciousness and producing poorer weld beads, skilled welders are capable of better regulating the torch to keep the optimal weld metal fluid flow based on closer interaction with the weld pool surface assuring production of quality weld beads.

Introduction
Gas tungsten arc welding (GTAW) operated by human welder is still applied in high standards required components or structures welding-pressure vessels, gas transportation pipelines, etc., due to the rich welding experience and skills of human welder enabling high quality welds under variable welding conditions. Unfortunately, inconsistent concentration, fatigue, and stress do build up such that human welder's capabilities may degrade during daily operations, which seriously affects the production and weld quality. However, robotic welding may solve the human welder's problems caused body limitation. Unfortunately, current welding robot has low intelligent skills likewise the accurate decision and timely adjustment of skilled welder. Thus, understanding and learning skilled welder's responses and operation is critical for developing intelligent welding robot.

For learning welder's behavior, University of Kentucky Prof. Zhang has conducted lots of work. They have modeled the response of skilled human welder to weld pool surface, and the human welder's behavior is analyzed from adjusting welding current point view [1-5]. However, the variable torch information enables to better represent the welder's complex behavior. Particularly, torch attitude strongly affects the fluid flow status of weld pool surface which may cause the formation of weld defect. In addition, understanding the complex mechanism that torch attitude changes the fluid flow status can assist to deeply illustrate the occurred mechanism and avoid it. With author's limited ability, there are little works. Parvez et al. [6] have established a three-dimensional numerical model to simulate the effect of torch angles on arc properties and weld pool shape in stationary GTAW. Hashimoto et al. [7] have utilized four cameras to record skilled welder's movement and orientation of the electrode tip TIG welding. Because the designed training system and image processing algorithm are over complex, and large training sample data is needed, this method cannot attractive more attention of researchers. Zhang et al. [8] has used a miniature wireless inertial measurement unit to measure the three-dimensional welding torch orientation in manual arc welding process. However, gyroscope zero drift cannot be well resolved.

This paper focuses on simultaneously measuring the torch attitude and the weld pool geometry. The measured data are utilized to qualitatively investigate the welder's responses on the weld metal fluid flow status and the inverse actions on welding torch. A new image processing algorithm is developed to extract the characterized parameters of weld pool. The prepared experiments with different welders are conducted, and the results are presented and analyzed. The principle diagram of the proposed method is shown in Figure 1. In this procedure, three characterized parameters, width, length and curvature diameter are utilized to describe the fluid flow status. For measuring the welding torch attitude, a tilt sensor includes a tri-axial accelerator and a tri-axial gyroscope is utilized to acquire the acceleration and angle acceleration of welding torch. Using Kaman-based filter fuses the integrated angle acceleration and to reduce the zero offset error generated by gyroscope. Finally, the welding torch attitude is achieved with acceptable accuracy. The framework is shown in Figure 2.

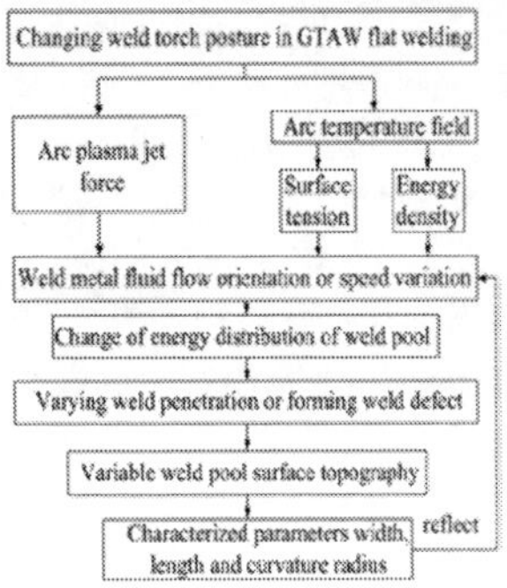

Fig. 1 Measurement of weld pool

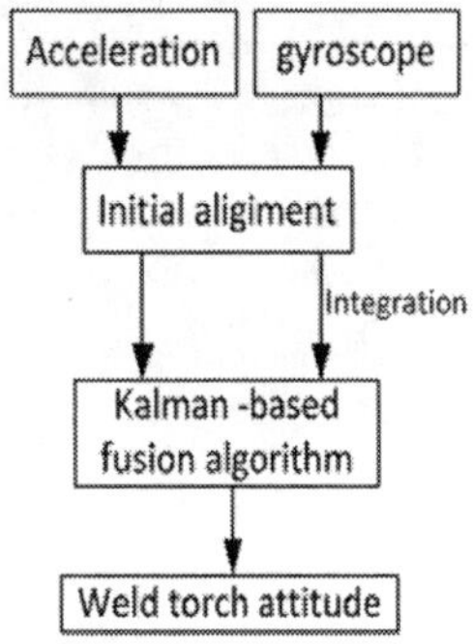

Fig. 2 Measurement of welding torch attitude

Experimental system

In experimental system shown in Figure 3, a welder holds the cardon fixed torch and timely adjusts the torch attitude by observing the geometry, fluid flow of the weld pool or other information. A 50-mW laser generator at a wavelength of 660nm with a variable focus was used to project 5-paralle line pattern structured light on the weld pool surface, and the laser stripes reflected by the weld pool surface were intercepted by an imaging plane and recorded by a high speed camera. The imaging plane was placed with a distance about 70mm from the torch. During welding process, the imaging plane, the laser generator, and the camera are stationary while the work-piece is moving, and its speed can be adjusted by a computer. Sensor consists of a type MPU6050 module including a high precision tri-axial accelerometer and six-in-one serial conversion module, and a captured software. Serial communication program developed based on LABVIEW platform was utilized to real-time record the angle information of measurement module.

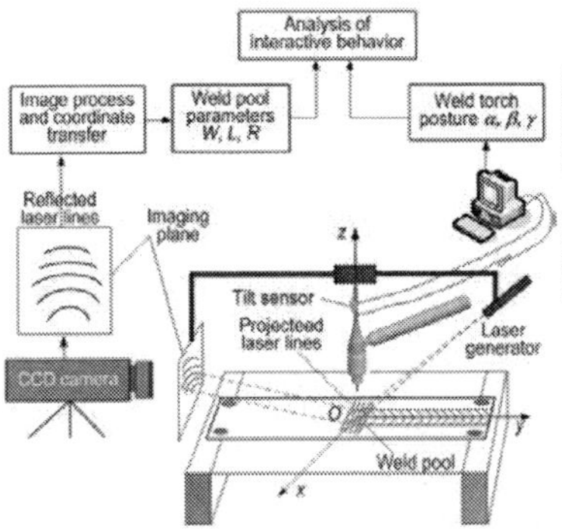

Fig. 3 Experimental system

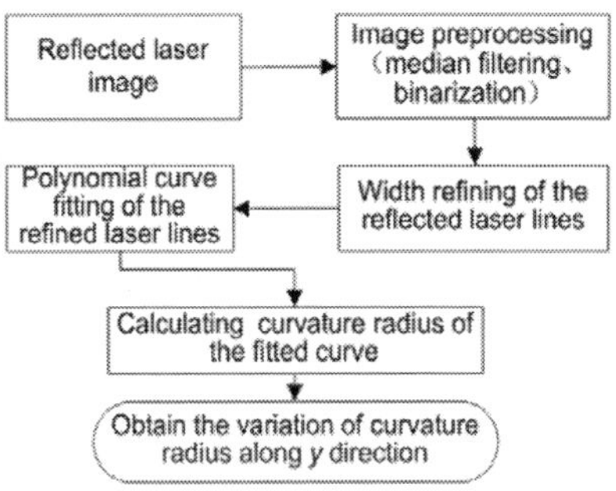

Fig. 4 calculating curvature diameter

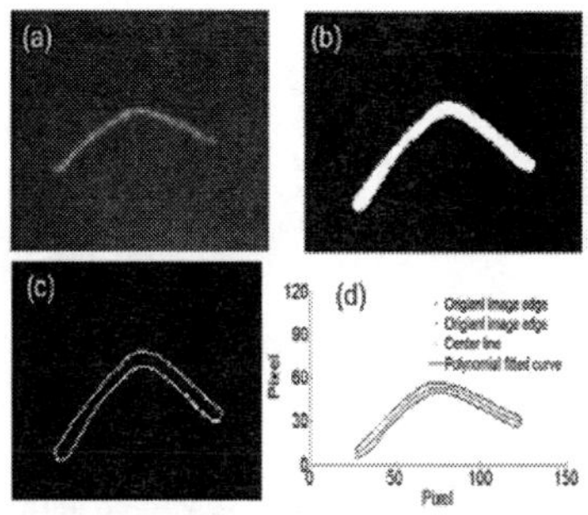

Fig. 5 Image processin results

Image processing algorithm

In our previous study [9], it has been demonstrated that the curvature and distortion of the reflected laser stripe has a close relationship with the 3D weld pool surface and the weld metal fluid flow. In this paper, a new image processing algorithm is developed and its calculation framework is shown in Figure 4. Figure 5 depicts the image processing results. Comparing with the original image of reflected laser stripes, fit curve's shape is very similar with it, which can reflect the original geometric information. Finally, fit curve's curvature diameter was calculated by curvature diameter solution formula. Other reflected laser stripes in the image can be refined, fit and its curvature diameter is calculated by using the above same algorithm.

Results and Discussion

A set of experiments were conducted on type 304 stainless steel plates with size 100×50×3mm. Four sets of them were selected and the data are processed and analyzed. Argon was considered as shielding gas at a flow rate of 10L/min. The sampling frequency both of CCD camera and sensor sets100Hz. To analyze the capability of human welders adjusting the torch and keep a particular weld metal fluid flow status, reflected laser stripes videos were synchronously captured in experiments and automatically framed. Ten reflected laser images were selected from experimental group *A* and group *E* at interval 0.3s, which were shown in Figure 6.

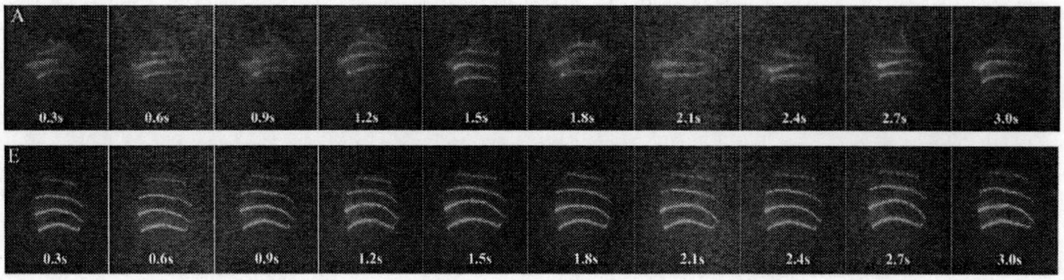

Fig. 6 *Reflected laser images selected at interval 0.3s (A-novice (No. A), E-skilled (No. E))*

In novice welder's welding process, as Figure 6A shows that a large change of the laser stripe's shape and irregular variation is occurred. The central laser stripe has an obviously heavy bending at both horizontal and vertical direction. The reason is assumed that the intense flow of weld metal fluid caused by variable welding torch leads it. For a skilled human welder as shown Figure 6E, the change of curvature of each laser stripe is gentler, and the bending of four stripes successively decreases from bottom to top in a vertical direction

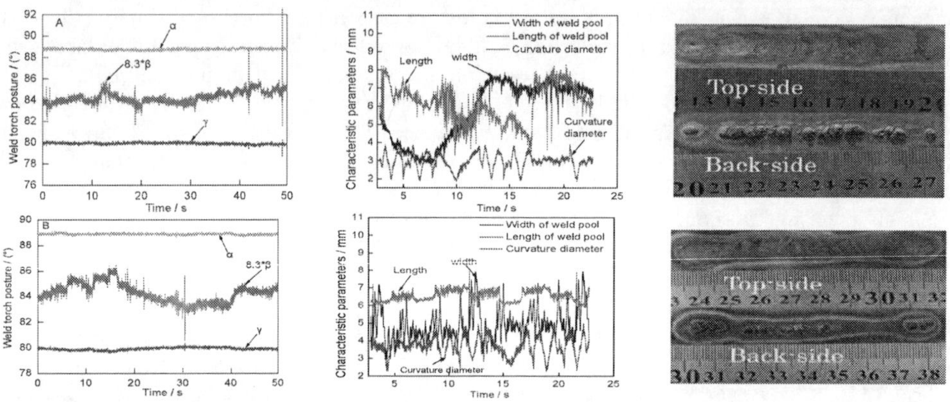

Fig. *7 Welding torch angle and weld metal flow parameters(novice welder, A-welding current 65A, welding speed 1.33mm/s², arc length 3~5mm;B- welding current 72A, welding speed 1.20mm/s², arc length 3~5mm)*

Figure 7A and 7B show that the novice welder's torch angle and the corresponding characteristic parameters changes irregularly and randomly, and the inconsistent weld penetration, uneven width, height and curving weld bead are obtained. From the variable torch attitude's curve, it is also achieved that a large fluctuation is occurred in angle β, but angle α and angleγ is changed in a small range of 88.5° and 80°, which shows the singular angle adjustment of the novice welder. When the welding time arrives to the 11th second, the novice welder observes a collapse of weld pool surface (the increase of diameter of curvature) and tries to decrease angle β to make the arc centre and high temperature weld metal move ahead of weld pool for decreasing the heat-input of arc to weld pool. Although the adjustment of only angle β can successfully lows the heat-input to avoid the excessive weld joint penetration, the strong fluid flow of weld metal caused by arc force and surface tension makes more melted metal of weld pool move to the front of weld pool so that the moved weld metal cannot be moved back when the weld pool starts solidification, and then a hollow and spindly weld bead is formed, which is similar with humping weld. From the dynamical operation of novice welder reflected by the captured welding torch angle, it would be achieved further that the novice welder lacks of welding experience and has not a good welding skill, and the manual GTAW process is a comprehensive process based on welder's experience and welding skills.

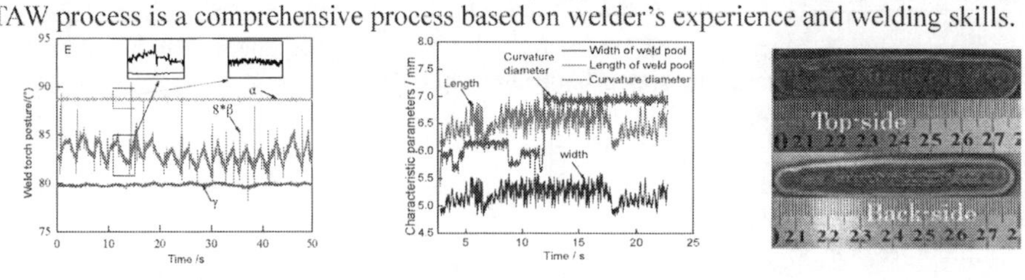

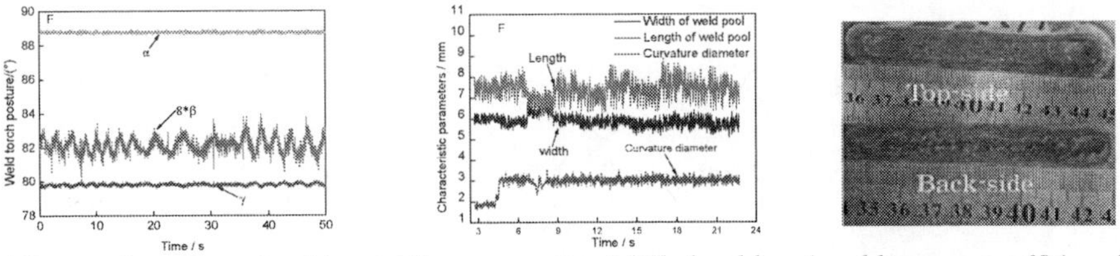

Fig. 8 *Welding torch angle and weld metal flow parameters(skilled welder, A-welding current 65A, welding speed 1.33mm/s², arc length 3~5mm;B- welding current 72A, welding speed 1.20mm/s², arc length 3~5mm)*

Figure 8E and 8F show that the skilled welder can adjust welding torch attitude regularly to keep weld metal fluid flow in stable status. At the 9th second, he/she observes a collapse of the weld pool surface, and then he/she regularly adjusts angle β, angle α and γ in small area to keep a gentle weld metal fluid flow under the impact of the welding arc force, producing a sound weld bead, which shows his/her superior ability to control welding torch attitude and abundant welding experience. Besides, at the 9~12s time in Figure 8E, the curvature diameter is decreased due to the cave formation of weld pool surface caused by more heat-input to work-piece. When the skilled welder observes this case, he/she knows that the complete weld joint penetration has been achieved and an excessive weld joint penetration or welding hole will be generated if some actions are not adopted to maintain the heat balance of weld metal fluid flow, therefore, he/she tries to increase the angle β and γ to move the high temperature weld metal to tail of weld pool so that lows the heat-input of weld pool centre, because the weld pool centre is thought the easiest instability area of the weld pool. When welding time arrives 12.5s, the curvature diameter is increased and maintained a small fluctuated stability about 4mm by the regular adjustment of welding torch attitude. Finally, a consistent and complete weld joint penetration weld bead is achieved.

Conclusions

(1) An experimental system is established to successfully measure the torch attitude and weld pool. This system mainly includes a 5-line pattern laser vision system and a dynamic tilt sensed system. Based on it, original reliable data reflected the human welder's skills and experience and weld pool fluid flow status are obtained.
(2) A close correlation is existed in the dynamic welding torch attitude and weld pool parameters with human welder's experience and skills.
(3) A novice welder has poor ability to adjust welding torch attitude to keep weld metal fluid flow in a particular status by observing the weld pool surface pattern in the welding process, and a bad weld bead appearance is achieved. A skilled welder can regularly adjust welding torch attitude to keep weld metal fluid flow in a particular status by observing weld pool surface pattern, which presents his good welding skills and abundant welding experience, and leads to a sound weld shape.

Acknowledgments

This work is funded by the 973 Program Prophase Research Project of China (#2014CB660810), National Natural Science Foundation of China (No.61365011), Young Creative Talent.

References

[1] W.J. Zhang, *et al*: "Characterization of three-dimensional weld pool surface in GTAW," *Welding Journal*, Vol. 91, No. 7 (2012), pp. 195s-203s.
[2] W.J. Zhang, *et al*: "Modeling of human welder response to 3D weld pool surface: part1: principles," *Welding Journal*, Vol. 91, No. 11 (2012), pp. 310s-318s.
[3] W.J. Zhang, *et al*: "Modeling of human welder response to 3D weld pool surface: part2: results and analysis," *Welding Journal*, Vol. 91, No. 12 (2012), pp. 329s-353s.
[4] Y.K. Liu, *et al*: "Skilled human welder intelligence modeling and control: part 1-Modeling," *Welding Journal*, Vol. 93, No. 2 (2014), pp. 46s-52s.
[5] Y.K. Liu, *et al*: "Skilled human welder intelligence modeling and control: part 2-Application," *Welding Journal*, Vol. 93, No. 5 (2014), pp. 162s-170s.
[6] S. Parvez *et al*: "Effect of torch angle on arc properties and weld pool shape in stationary GTAW," *Journal of Engineering Mechanics*, Vol. 139, No.9 (2013), pp. 1268–1277.
[7] N. HASHIMOTO: "Difference of improving welder's skill through training progression," *Bulletin of Hiroshima Institute of Technology Research*, No. 49(2015), pp. 75-81.
[8] W.J. Zhang, *et al*: "Measurement of three dimensional welding torch orientation for manual arc welding process," *Measurement Science and Technology*, Vol. 25, No. 3 (2014), pp. 035010-031527.
[9] G. Zhang, *et al*: "Research on the correlation between the status of three-dimensional weld pool surface and weld penetration in TIG welding," *Acta Metallurgica Sinica*, Vol. 50, No. 8 (2014), pp. 995-1002.

Advanced Control of Human Welder Movement

Shujun Chen [1], Ning Huang [2], Yukang Liu [3], YuMing Zhang [3]

[1] *Beijing University of Technology, China*
[2] *Beijing University of Technology, USA*
[3] *University of Kentucky, USA*

This paper summarizes our recent efforts in modeling and control of human welder arm motion. Novice welders typically have challenges to maintain the welding speed at the desired level. This directly affects the heat input and weld penetration. To combat this issue especially facing the increasing shortage in the supply of skilled welders, the authors proposed to use a machine to instruct the novice welder to adjust the travel speed. To this end, an actuator will be needed to interface between the machine and human such that the command from the machine can instruct the welder. The machine will also measure the actual welding speed, compare it with the desired speed, and determine how the instruction applied by the actuator to the human welder should be changed using a feedback control algorithm. To design the feedback control algorithm, the behavior of the human welder in responding to the applied instruction needs to be modeled. As such, we have proposed/experimented to use vibration and visual arrow as the instruction/actuator to instruct the welder to adjust the speed, modeled how the welder speed responds to the instruction, designed an advanced feedback control algorithm to determine how to adjust the instruction, tested the results in welder arm speed control, and performed welding experiments using the machine assisted welder arm movement control. It was found that the human response is an uncertain dynamic process and the proposed interval model control and adaptive neurofuzzy control both worked well in controlling this uncertain dynamic process. With the machine assisted control, the novice welder can maintain the travel speed around the desired level to make quality welds.

Qualification of Trackless Autonomous Crawling All-Position Arc Welding Robot for Welding of Large Engineering Structures

Wei Zhou
School of Mechanical and Aerospace Engineering, Nanyang Technological University, 50 Nanyang Avenue, Singapore 639798
Email wzhou@cantab.net, phone +65 67904700

Xiaoming Jiang, Xuefeng Zhou, Xiaoguang Liu, Taobo Cheng
Guangdong Institute of Intelligent Manufacturing, No. 13 Building, No. 100 Xianliezhong Road, Guangdong 510070, P. R. China

Abstract
Considerable effort has been made to develop trackless autonomous welding robots with good adsorption ability and capability of huge payload and crawling on curved surfaces in all positions. Such robots are expected to have great commercial prospects; however, unfortunately, so far they have been rarely used in industry. This is mainly due to the lack of work to qualify the new robotic welding technology. A roadmap is drawn for the purpose of systematic qualification of the welding technology.

Introduction
Manual welding is highly flexible and widely used, but it is associated with the problems of low productivity, inconsistent quality, and increasingly high labour cost. In contrast, robotic and automated welding processes lead to improved quality as well as higher productivity. Sophisticated robots have been increasingly used to carry out welding in many different industrial sectors [1]. However, for welding of very large engineering structures, commonly used robotic arms are not suitable.

NOMAD project was carried out in Europe a decade ago to develop an autonomous, mobile welding robot capable of fabricating large-scale customised structures [2]. The project consortium consists of a balanced mix of prestigious suppliers, technology enablers and end users in many different countries across Europe, including Caterpillar Belgium S.A., Robosoft SA (France), TWI Ltd (UK), Fraunhofer IFF (Germany), Reis Robotics (Germany), ESAB AB (Sweden), and Nusteel Structures Ltd (UK). The concerted effort for many years made it possible to weld parts placed arbitrarily within a workcell area using a 6-axis robot, which is moved around by an autonomous robot transport vehicle (RTV). The RTV developed in this project can move on ground only (i.e., without wall-climbing capability); therefore, the robotic system is still not capable of welding very large engineering structures.

For welding of very large engineering structures (e.g., a nuclear pressure vessel of 12 m in height or hulls of large ship), fixed-track mechanized welding machines are usually used [3]. Pre-installation of the tracks can be done reasonably efficiently for straight seams on plane surfaces but becomes very difficult and time-consuming for non-straight seams or curved surfaces, such as those shown in *Fig. 1*. As a result, manual welding is frequently carried out in these cases. For manual welding of large structures such as ship hulls (as shown in *Fig. 2*), large-scale scaffolding has to be erected, and this incurs additional cost and further reduces productivity.

It makes much economic sense to weld the large structures using a trackless autonomous welding robot with wall-climbing capability (as illustrated *Fig. 3*). The paper briefly reviews such technology and highlights the importance of qualifying the technology according to well-established standards.

Fig. 1 Example of large curved surface with non-straight welds.

(a) (b)

Fig. 2 (a) Large-scale scaffolding erected for manual welding of ship structures;
(b) Close-up look showing long welds, which are suitable for automated robotic welding.

Fig. 3 Illustration showing welding of a large structure with curved weld using trackless autonomous welding robots with wall-climbing capability.

Development of Trackless Autonomous Welding Robots

Control and seam tracking are certainly important to development of trackless autonomous welding robots, but such technologies are not reviewed in the paper, since they are readily available in the age of AGV (automated guided vehicle) and driverless cars. The real technological challenge is to design a proper adsorption mechanism, which must yield the sufficient loading capacity and in the meantime maintain the good surface adaptability.

The surface adsorption can be achieved by using vacuum cups. Such an adsorption mechanism can be conveniently used for surface cleaning [4] but are not suitable for welding, where environmental temperature is high and movement of welding torch must be very steady.

Wall-climbing mobile robots for welding usually gain desirable level of surface adsorption through using permanent magnets [5-6]. If a wheel-type design is used (as illustrated in *Fig. 4(a)*), the robot possesses good surface adaptability but suffers poor surface adsorption because the wheels provide limited contact area. Use of permanent magnet caterpillar belts (as shown in *Fig. 4(b)*) improves the loading capacity significantly but at the expense of surface adaptability. It can be envisaged to design hybrid wheel-belt type configuration to gain the optimum combination of load-carrying capacity and surface adaptability, as illustrated in *Fig. 5*.

Technology Qualification

The currently available trackless autonomous welding robots may be already good enough for real engineering applications even without further improvement. For example, the mobile robot developed by Pan et al. [5] has reasonably good surface adaptability and a pay load as high as 120 kg. However, in reality, such robots are rarely used in industry.

The major reason lies in lack of research work to qualify the robotic welding technology. Stacy and Canfield [6] did outstanding work to qualify their mobile welding robot according to AWS B2.1:2005 [7] and AWS D1.1/D1.1M:2002 [8]; however, unfortunately, they chose to qualify procedures for welding in vertical position only. Such welding can be much more economically carried out without using the expensive mobile robot.

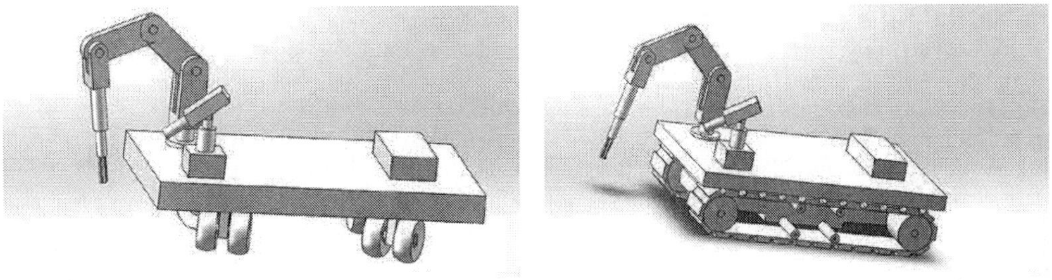

(a) (b)

Fig. 4 (a) Wheel-type design with good adaptability but poor surface adsorption; and (b) caterpillar belt-type design with good surface adsorption but poor adaptability.

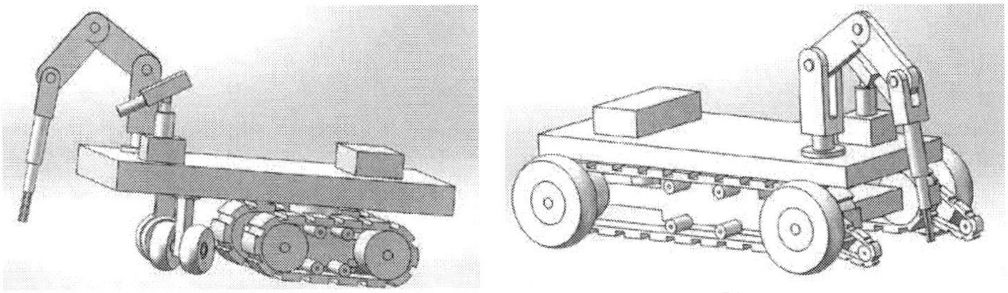

(a) (b)

Fig. 5 Wheel-belt hybrid design for possible optimization of both surface adsorption and adaptability.

It is important to define a systematic roadmap for qualification of the robotic welding technology, which covers both welding personnel qualification and welding procedures qualification.

Welding personnel qualification

Trackless autonomous welding robots are very smart and adaptable, but it must be noted that the welding robots cannot possibly climb onto the structures and carry out welding entirely by themselves. They have to be operated by welding personnel, and the welding personnel must be qualified to operate the robots correctly.

The AWS D16 Committee on Robotic and Automatic Welding was organized in 1985 to provide centralized source for the exchange of technical information between manufacturers, installers, and operators of robotic and automated welding equipment. One of the standards produced by the committee, AWS D16.4M/D16.4:2015, provides specifications for the qualification of robotic arc welding personnel [9].

The AWS standard defines three levels of qualification: Level 1 – Certified Robotic and Automated Welding Operator (CRAW-O); Level 2 – Certified Robotic and Automated Welding Technician (CRAW-T); and Level 3 – Certified Robotic and Automated Welding Engineer (CRAW-E). It is alarming to take note that most robotic welding researchers (e.g., PhD students) lack practical welding experience or skills of visual weld inspection and thus may fail to qualify at any of the three qualification levels according to AWS D16.4M/D16.4:2015. The lack of personnel qualification would severely limit the researchers' ability to test and qualify the robotic welding technology.

ISO 14732:2013 specifies requirements for qualification of welding operators and also weld setters for mechanized and automatic welding [10]. The standard itself does not contain the term "robotic welding" but is applicable to robotic welding since it is a type of "fully automated welding" as defined in the standard.

The ISO standard deals with personnel qualification in a more comprehensive way by covering the following five major parts: methods of qualification, essential variables and range of qualification, period of validity, certificate, and documentation.

ISO 14732:2013 provides four methods to qualify welding operators or weld setters: qualification based on a welding procedure test in accordance with the relevant part of ISO 15614 [11]; qualification based on a pre-production welding test in accordance with ISO 15613 [12]; qualification based on a test piece in accordance with the relevant part of ISO 9606 [13]; and qualification based on a production test or production sample test.

Welding procedure qualification

Welding procedure specification (WPS) can be tested according to AWS standard D1.1/D1.1M [8] or preferably according to the more recently developed international standards ISO 15614-1 [11], ISO 15613 [12], and ISO 15609-1 [14]. What is most important is to qualify welding procedures in horizontal position and in curved surfaces as well (as shown in *Fig. 3*) rather than just qualify welding the straight vertical position.

The mobile robots have to climb on surface of large structures; therefore, the welding procedure testing would consume lots of structural steels. It is suggested that considerable saving can be achieved by machining a "window" in the large structures and using inserted plates inside the window for the subsequent non-destructive and mechanical testing.

Conclusions

Trackless autonomous welding robots with good adsorption ability and capability of huge payload have been developed but rarely used in industry. The problem is attributed to the lack of work to qualify the mobile robotic welding technology. A roadmap is drawn for the purpose of systematic qualification of the welding technology. Especially, it is important to qualify welding procedures for the horizontal position and in the curved surfaces. An effective method for saving testing materials is proposed.

Acknowledgments

The authors would like to acknowledge financial support by the high-level leading talent introduction program of GDAS (project number 2016GDASRC-0106) and by Nanyang Technological University and to thank Mr. L. C. Cao for help to draw *Figs. 3*-5.

References

[1] S. B. Chen: "On the Key Intelligentized Technologies of Welding Robot," *Lecture Notes in Control and Information Sciences*, Vol. 362 (2007), pp. 105–116.

[2] S. Mulligan, G. Melton, A. Lylynoja, and K. Herman: "Autonomous Welding of Large Steel Fabrications," *Industrial Robot: An International Journal*, Vol. 32, No. 4 (2005), pp. 346 – 349.

[3] D. H. Lee: "Robotics and Computer-Integrated Manufacturing," *Robotics and Computer-Integrated Manufacturing*, Vol. 30, No. 5 (2014), pp. 442 -450.

[4] N. Alexandresu, C. Udrea, D. Duminica, and T. C. Apostolescu: "Research on the Vacuum Attachment System of a Cleaning Robot with Vertical Displacement" *Proc. 2nd Inter. Conf. on Advanced Material Research*, Chengdu, China, Jan. 2012, pp. 1673-1677.

[5] J. L. Pan, B. Y. Yan, L. S. Gao, H. Zhang, and Q. Y. Liu: U.S. Patent 7,498,542 B2.

[6] J. Stacy and S. Canfield: "Theoretical and Empirical Verification of a Mobile Robotic Welding Platform," *Welding Journal*, Vol. 91, No. 12 (2013), pp. 338s-345s.

[7] AWS B2.1:2005, *Specification for Welding Procedure and Performance Qualification*. Miami, Fla.: American Welding Society.

[8] AWS D1.1/D1.1M:2002, *Structural Welding Code — Steel*. Miami, Fla.: American Welding Society.

[9] AWS D16.4M/D16.4:2015, *Specification for the Qualification of Robotic Arc Welding Personnel*, American Welding Society, 2015.

[10] ISO 14732:2013, *Welding Personnel — Qualification Testing of Welding Operators and Weld Setters for Mechanized and Automatic Welding of Metallic Materials.*

[11] ISO 15614-1, *Specification and Qualification of Welding Procedures for Metallic Materials — Welding Procedure Test — Part 1: Arc and Gas Welding of Steels and Arc Welding of Nickel and Nickel Alloys.*

[12] ISO 15613, *Specification and Qualification of Welding Procedures for Metallic Materials — Qualification based on Pre-production Welding Test.*

[13] ISO 9606-1, *Qualification Testing of Welders — Fusion Welding — Part 1: Steels.*

[14] ISO 15609-1, *Specification and Qualification of Welding Procedures for Metallic Materials — Welding Procedure Specification — Part 1: Arc Welding.*

Automatic Narrow Gap GTAW System For On-site Welding of Thermal Power Generation Piping

S. Nakamura*, T. Ogawa*, T. Kato*, M. Otsuki* and S. Asai**

*Toshiba Corporation, Yokohama-city, Japan, **Osaka university, Suita-city, Japan

shinichiro.nakamura@toshiba.co.jp, +81-45-510-5381

Abstract

Recently, there is a tendency that 9Cr-1Mo steel has been more likely to be applied to thermal power generation thick-walled piping due to the rise of operating steam temperature. In order to reduce construction period of thermal power plant, high efficient welding method with satisfying required quality, especially automatic narrow gap GTAW of multi-handling, has been expected to be used in on-site welding of 9Cr-1Mo steel thick-wall piping. However, there were several obstacles for applying this method. The machine has to be operated at outside and narrow working area. This means that not only the risk of blow hole induced by wind would become drastically high, but also narrow working area restricted the operating space of machine. Also, low fitting accuracy of on-site groove joint caused the bad penetration bead. Therefore, automatic narrow gap GTAW system for on-site welding of thermal power generation piping was developed, including jig for wind protection, suitable device configuration for on-site welding, and welding head design permitted to improve the robustness of automatic GTAW. Also, the automatic weld line tracking by image-processing was developed for multi-handling, which is the method that operator can control two weld machines at the same time. Developed system was applied in construction of combined thermal power plant, and contributed to reduce construction period with showing good weld quality.

Introduction

In construction of combined thermal power plant, many fixed pipes of 9Cr-1Mo steel between steam turbine and Heat Recovery Steam Generator (HRSG) have to be joined by welding. As the number of skilled welders has been decreasing, high efficient and high quality welding method are required in on-site welding. Especially, automatic narrow gap GTAW with multi-handling are considered being the most suitable method for solving its problem, but in on-site welding, there are several obstacles for automatic GTAW. As shown in *Fig.1*, operators have to control automatic GTAW at outside working area, with caring the risk of blow hole induced by wind. And, its working area is so narrow that the possibility of interference between machine and jigs are high (*Fig. 2*). Also, *Fig.3* shows that it was difficult to fit narrow groove joint without any gap and misalignment as turbine piping line were generally consisted of many weld joints. Then, the risk of bad penetration shape in first layer welding became drastically high. For above problems, the suitable automatic systems and equipment have been studied.

Fig.1 Outside working area

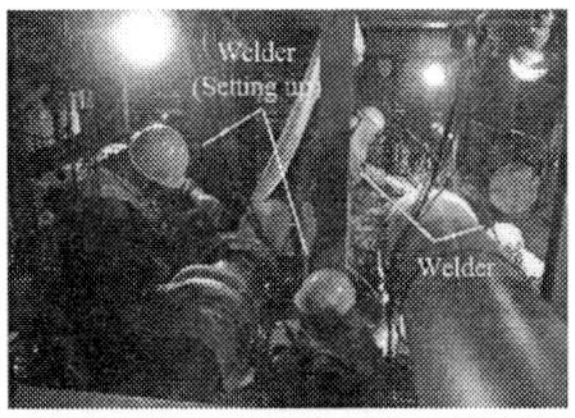

Fig.2 Narrow working environment

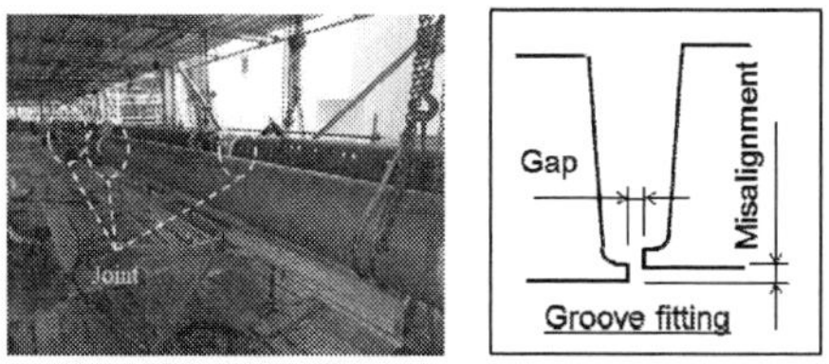

Fig.3 Fitting accuracy for Narrow groove

System configuration

Fig.4 shows the developed configuration of On-site automatic GTAW system. Considering workability of narrow space, the devices mainly installed in working space are only welding head, controller and welding monitor. Welding head runs through the rail setting up in fixed pipe. Operator controls welding head with checking welding condition by monitor. Controller and monitor are designed to be wireless in order to improve the usability in operating two machines installed in distant area at the same time. Welding power source and hot wire power source, which are almost no necessary to be moved after installation, are placed at distant from working area. Fig.5 shows the schematic view of welding head. This device can weld pipe ranged from 250A to 800A, and its height is less than 250mm regardless of the pipe diameter to avoid crashing to jigs.

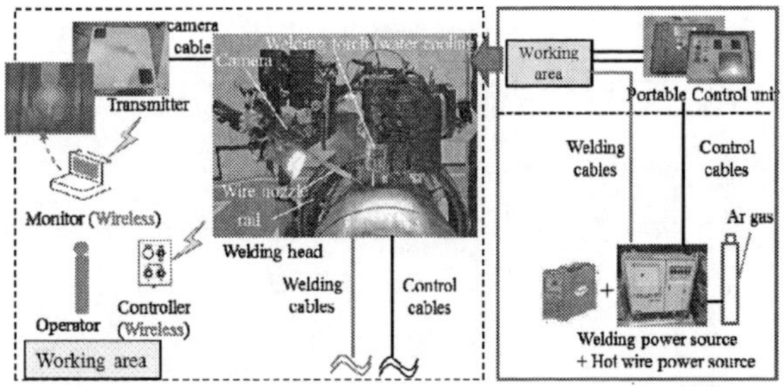

Fig.4 Configuration of On-site automatic GTAW device

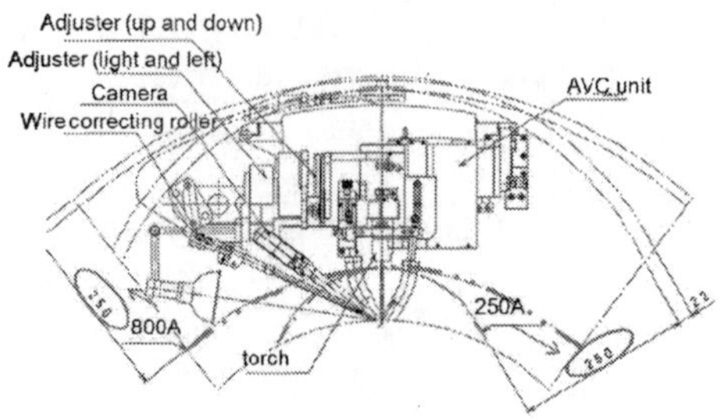

Fig.5 Schematic view of Welding head

Original Narrow Gus nozzle and robustness of first layer welding

As the on-site fitting accuracy of narrow groove was considered to be low, there is a need to improve the robustness of first layer welding for automatic GTAW. Also, the suitable shape of Gus nozzle for narrow groove has to be developed at first, since groove width is about 10mm in thick wall-piping. Thus, narrow Gus nozzle was designed as shown in *Fig.5*. Considering usability in on-site working, it was designed to be put in commercial welding torch and insulator with using the Gus lens which its shape was modified. In order to optimize Flow rate and tungsten extension for this nozzle, Shlieren method testing was conducted (*Fig.6*). In addition, weld-ability was confirmed with using each combination of flow rate and extension, as shown in *Fig.7*. In this test, following welding parameters were applied. Namely, current was 100 (A), voltage was 10 (V) and speed was 60 (cm/min). From *Fig.6* and *7*, it can be found that acceptable range of flow Gus rate was largest when tungsten extension was 20mm. Also, good weld-ability was confirmed regardless of extension when Flow Gus rate was 15(L/min). Thus, the most suitable condition was decided that flow rate was 15 (L/min) and extension was 20 (mm).

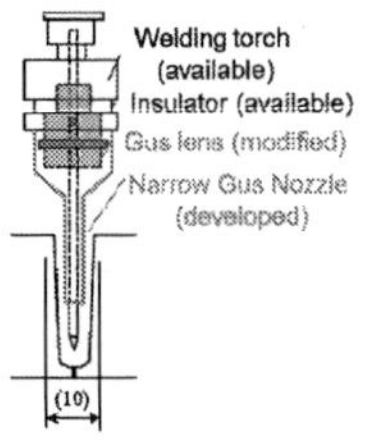

Fig.5 Narrow nozzle

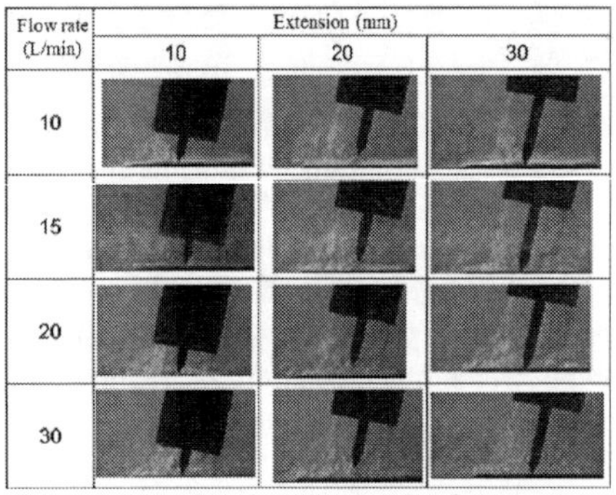

Fig. 6 Shlieren method testing

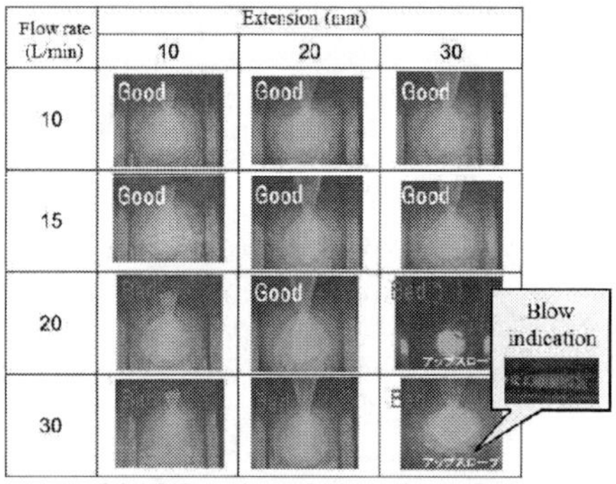

Fig.7 Weld-ability for Flow rate and Extension

Table 1

Fig.8

Fig.8

Fig.9

Table 1 Welding parameters of Automatic GTAW

ITEM	Parameter
Current (Pules)	140~210 (A)
Voltage	8~10 (V)
Speed	50~90 (mm/min)
Wire feed rate	100~150 (mm/min)

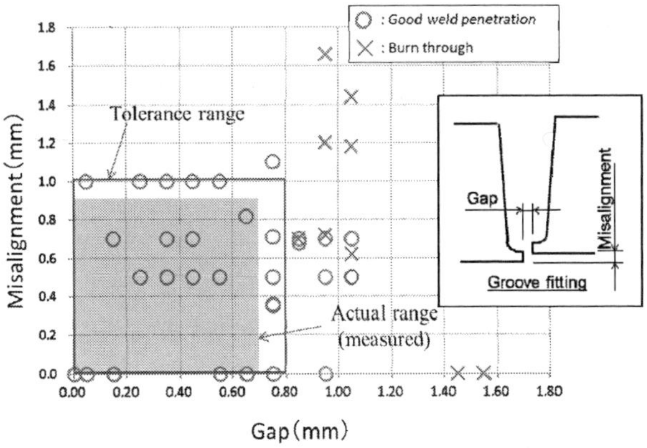

Fig. 8 penetration shape against gap and misalignment

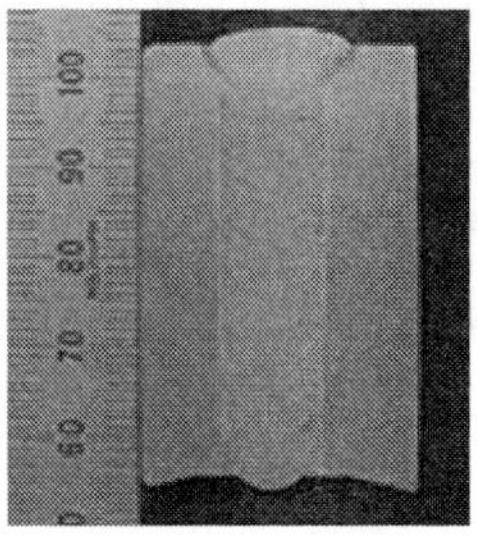

Fig.9 Macro of 9Cr-1Mo steel

On-site wind protection equipment

Fig.1

Fig.10

Fig.11

Fig.12

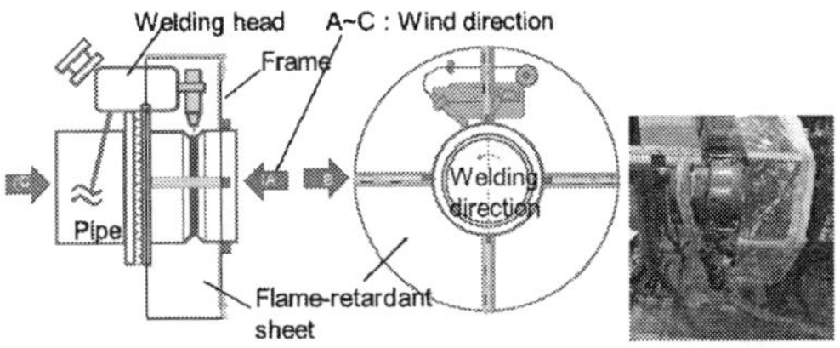

Fig.10 Concept of local wind protection

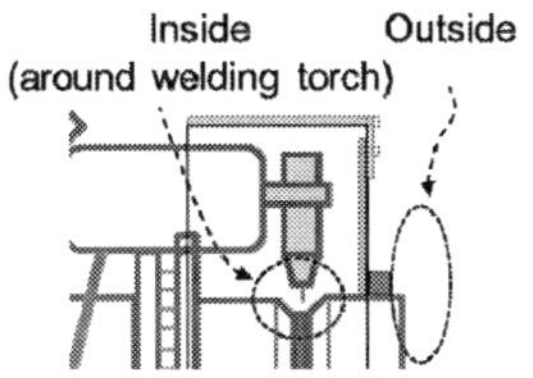

Wind direction	Wind Speed (m/s) Outside	Inside	Weld quality
A	8.5	0.0	○
B	4.0	1.5	×(Blow hole)
B	6.5	2.5	×(Blow hole)
C	5.0	1.4	×(Blow hole)
C	7.0	1.8	×(Blow hole)
C	8.2	2.5	×(Blow hole)

Fig.11 weld-ability for wind direction and speed

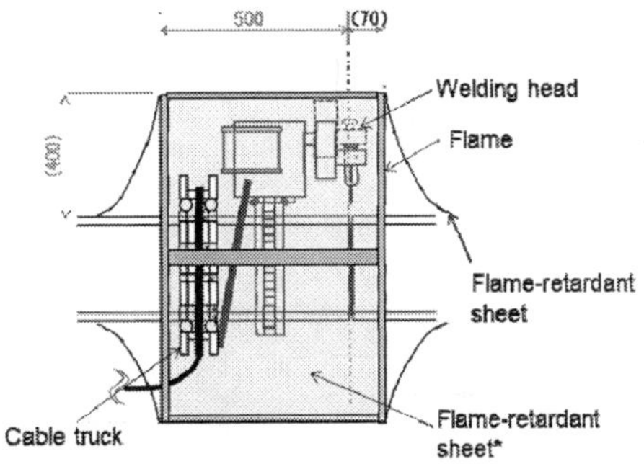

Fig.12 Suitable concept of local wind protection

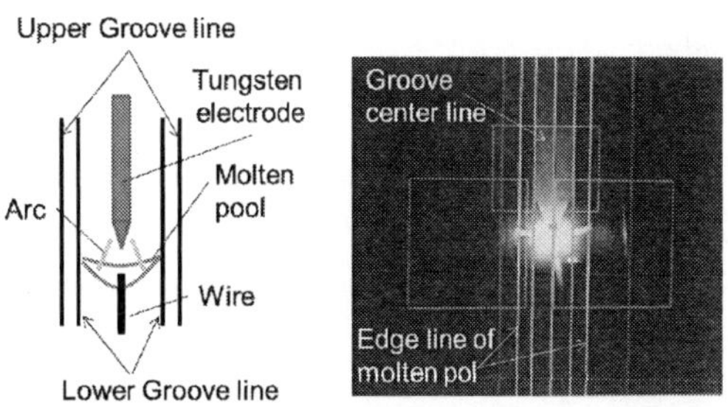

Fig.13 Automatic weld-line tracking system

Automatic weld-line tracking with image-processing

In automatic welding, operators previously check the location of tungsten electrode at all the time since it would strongly affect the weld quality, and they have to fix the position by controller when it was not located at the center of groove. Thus, automatic weld-line tracking system with image-processing was developed. The outline of this system was revealed in *Fig.13*. Welding condition reflected in monitor was captured into Image-processing unit. Center of tungsten electrode and edge line of molten pool was recognized by luminance difference induced by welding, and then it can be possible to derive the gap between tungsten center line and groove center line. If the gap was confirmed in this system, tungsten automatically moved suitable position. Thanks to this system, operators have almost no need to control the device in welding and multi-handling became easy for them.

Conclusions

Fig.14

-
-
-
-

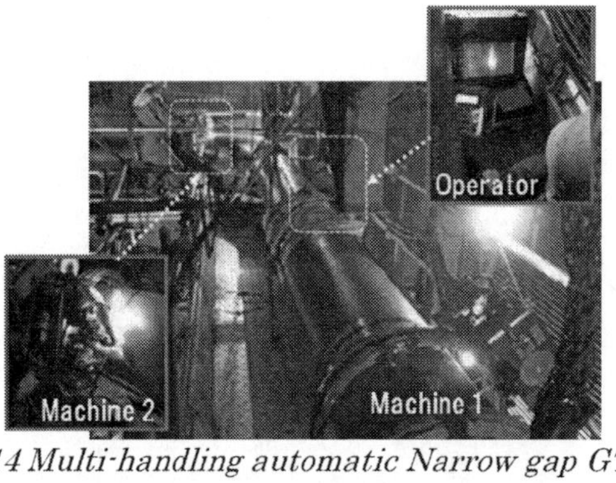

Fig.14 Multi-handling automatic Narrow gap GTAW

References

[1] S. Asai: "Trend of Automatic Welding System and In-process Quality Control System," Journal of Japan Welding Society 81(1), 34-44, 2012-01

[2] T. Kato: "Trend of welding technology for Plant Piping," Journal of Japan Welding Engineering Society 63(8), 40-45, 2015-08

[3] T. Ogawa: "Development of Visual Sensing System for Hot Wire Orbital Narrow GTAW," Proceedings of the 7th International Symposium of Japan Welding Society, Vol.1, .217-222

[4] K. Kamo: "Development of Automatic GTAW Technology Using Visual Sensor in Narrow Gap All Position," Proceedings of IIW, Doc. XII, 1760-03, 2003

Highly Efficient, Less-Spatter Welding Process for Thick Steel Plates

Yuji Taniguchi and Yoshihide Inoue
KOBE STEEL, LTD., Fujisawa, Kanagawa, 251-8551 Japan
taniguchi.yuji1@kobelco.com, +81-466-20-3037

Abstract
Robotization of the welding process for thick steel plates has been promoted to enjoy its high performance and quality. For these robot systems, there are increasing demands for better performances like higher efficiency, less cycle time, and less spatter. To respond to these needs, various new welding processes have been developed recently. In this paper, the new welding processes such as "Tandem Arc Welding," "High-Current MAG Welding," and "CO_2 Pulse Arc Welding (REGARC™)" are described.

The "Tandem Arc Welding" process offers higher metal deposition by using the two torches aligned in tandem. Due to its high deposition rate, this process is used in some robot systems, but still there are some difficulties to use this process in a robot system. That is, with two torches in one process, the robot welding operator must adjust the electrode's aiming position for the two torches together to each welding line. "Dual Arc Sensor" is the newly developed robotic function for this process to adjust these two torches automatically to a particular welding line.

The new process using a single torch has also been developed, which is called "High-Current MAG Welding." All the welding wire, welding torch, and power source have specially been developed for this process. By using this process, the welding deposition rate can be increased to 300 g/min or higher.

For the merits of lower welding cost and deeper penetration, CO_2 shielding gas is used in many robotic systems. In the CO_2 welding process, however, it is known that spatter increases than in MAG welding, and thus the need of removal of adhered spatters on the weldment leads to higher costs and lower efficiency in production. These problems have been desired to solve. To answer these needs, the authors have developed the new CO_2 pulse welding process called "REGARC™." With the special welding wire and welding current waveform, the process can reduce spatters even in CO_2 globular transfer welding. This paper also describes how the waveform is controlled and how less the spatter and fume are in the new process.

Introduction

In the mid-and-thick plate welding industries where higher welding techniques are required than in the thin plate welding fields such as the car industry, semiautomatic welding was applied at a high usage ratio. In recent years, however, the use of the arc welding robot has been expanded to improve the welding efficiency, compensate for the shortage of skilled workers, and to stabilize the weld qualities. Of these improvements, higher efficiency, higher welding speeds, lower spatter, and cost savings have always been demanded, and hence development of new welding processes has been desired.

In this paper, three high-value-added welding processes will be introduced and explained: firstly, the high-deposition, high-speed welding process with two electrodes, "Tandem Arc Welding," which reduces spatter to a great extent; secondly, the "High-Current MAG Welding" process which offers the high-current, high-deposition performance with one electrode; finally, the "REGARC™" process which offers extra-low spatter in CO_2 arc welding.

(1) Tandem Arc Welding

In tandem arc welding with two wires, if either wire deviates from the welding line, welding defects such as undercut and insufficient penetration will occur. Therefore, when tandem arc welding was carried out by robot, there was such a problem that the teaching work for robot positioning had to be conducted precisely so that both two electrodes aimed at the welding line. In addition, in the case of the welding of mid-and-thick plate work, even when the teaching work was carried out precisely, the electrode aiming position was prone to deviate due to the adverse effects of the insufficient dimensional accuracy and thermal distortion of the work. To solve this problem, the arc sensor function has been used. With the arc sensor for the conventional system for tandem arc welding, the deviated tracking position of the leading electrode was sensed by a change in the welding current of the leading electrode, and then the positions of both the leading and trailing electrodes were corrected together in the right/left and up/down directions — Fig. 1-1. In the actual fabrication site, however, when the aiming position of the trailing electrode was mistaken to store accurately in the memory during the teaching work, or the trailing electrode wire on arcing is curved irregularly, the deviation from the welding line could become different between the leading and trailing electrodes, and thus the deviated aiming of the trailing electrode sometimes caused weld imperfections. In order to cope with the deviated aiming of the trailing electrode, the new arc sensor that can sense the deviated tracking position of the trailing electrode has been developed.

(a) Development of dual arc sensor

In tandem arc welding, the trailing electrode generates an arc on the weld pool made by the leading electrode, and thus there is almost no change in the electrode-to-base metal distance during the weaving period of time; as a result, the deviated tracking position of the trailing electrode could not be sensed correctly by the similar control method for the leading electrode. In order to overcome this problem, the new control method to be used exclusively for the trailing electrode has been developed. For correcting the deviated tracking position of the trailing electrode, the newly developed method adjusts it by rotating the robot with respect to the axis of the leading electrode. With this method, the deviated aiming of the trailing electrode can be corrected without affecting the leading electrode. By using the new arc sensor for the trailing electrode together with the conventional arc sensor for the leading electrode, the deviated aiming of both the leading and trailing electrodes has become possible to control — Fig.1-2.

(b) Performance of dual arc sensor

In order to verify the performance of the dual arc sensor, tandem arc welding was carried out with the trailing electrode the tracking position of which was deviated by 5 mm. Fig. 1-3 shows the bead appearance made with the conventional arc sensor, while Fig. 1-4 shows that made with the dual arc sensor. From the bead appearance shown in Fig. 1-4, it is obvious that the dual arc sensor is good for correcting the deviated tracking position of the trailing electrode, providing the good result without weld imperfection. With this new sensing function, tandem arc welding in the actual fabrication sites will be able to provide higher-quality welds.

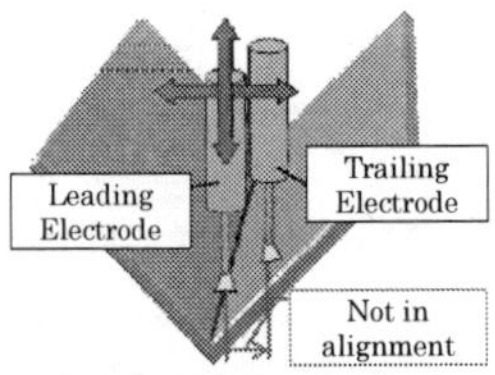

Fig. 1-1 Correction directions by conventional arc sensor

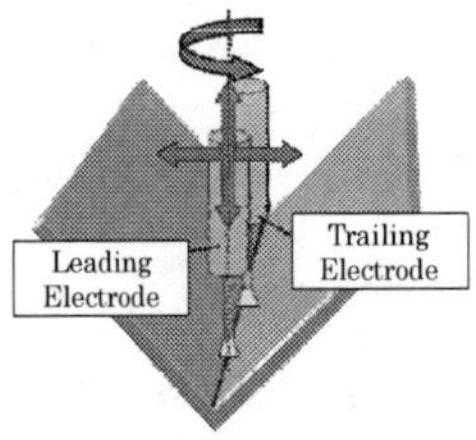

Fig. 1-2 Correction directions by dual arc sensor

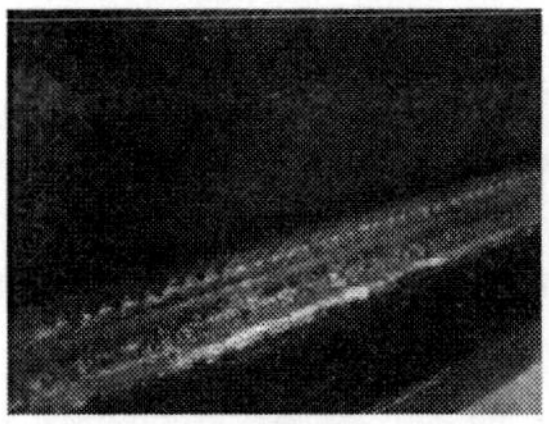

Fig. 1-3 Bead appearance by conventional arc sensor

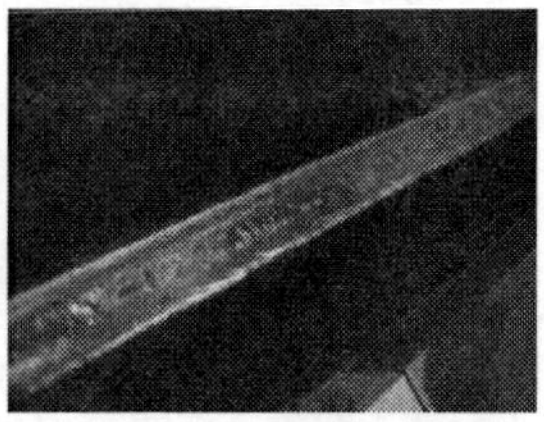

Fig. 1-4 Bead appearance by dual arc sensor

(2) High-current MAG welding process

In the conventional high-current GMA welding with carbon-steel solid wires, the spatter caused by the rotating droplet transfer has been a big problem. This phenomenon occurs in such a way that the wire extension part is fused by the electric resistance heat before reaching the arcing point, thereby losing its rigidity and thus scattering the molten droplets while being rotated by the electromagnetic force. To solve this problem, the cored flux of a flux-cored wire (FCW) was taken into account for the functional possibility of preventing the rotation of wire extension part, and then its slag formation and resistance to nitrogen were optimized to develop the dedicated FCW for high-current MAG welding process.

(a) Development of the electric current/voltage waveform control

The newly developed FCW can prevent the rotation of the wire extension part to create a stable spray transfer. However, the arc voltage (arc length) was varied irregularly when using it with a usual constant-voltage power source. In order to improve this phenomenon, it was examined to stabilize the molten state of the wire extension part by using the welding current/arc voltage waveform control. Consequently, the arc length has been stabilized by using the dedicated waveform control with periodical low-current periods that reduce the arc and Joule heats, thereby decreasing the heat input to the wire.

(b) Performance of the high-current MAG welding process

The amount of spatters in conventional MAG welding with solid wire and that in high-current MAG welding are shown in Fig. 2-1, and the nitrogen content of each weld metal is shown in Fig. 2-2. With the high-current MAG welding process, the stable spray transfer can be maintained, and thus low-spatter performance can be obtained consistently even in the high current range. Thanks to the spray transfer, the air inclusion can be reduced, and in turn the nitrogen content of weld metal can be kept low. Furthermore, the chemical composition of the dedicated FCW was designed with an emphasis on nitrogen resistant, and thus porosity occurrence can be prevented even at a high melting rate of 300 g/min. The welding conditions for the flat fillet weld with a leg length of 15 mm by the conventional MAG and high-current MAG welding processes are shown in Table 2-1, the cross-section macrograph of the weld is exhibited in Fig. 2-3, and the bead appearance of the weld is presented in Fig. 2-4. In conventional MAG welding, such a large-leg fillet weld requires two passes; by contrast, the high-current MAG welding process can finish it with one pass, and in turn the welding efficiency can be increased by 2.4 times. In addition, it is obvious that the fine-looking bead appearance with almost no spatter adhered can be obtained even with such a high welding current as 600A.

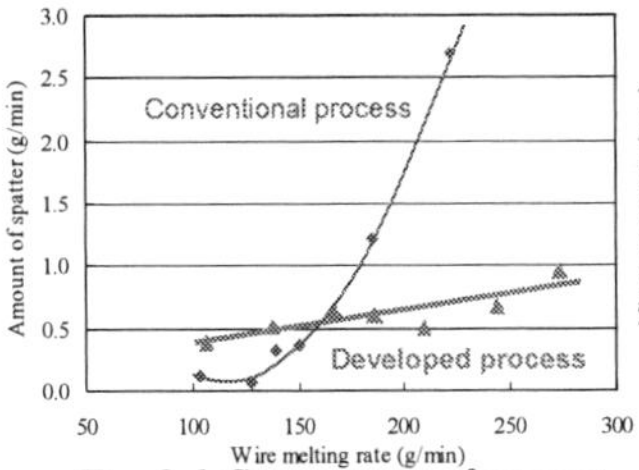

Fig. 2-1 Comparison of amount of spatters between conv. and developed processes

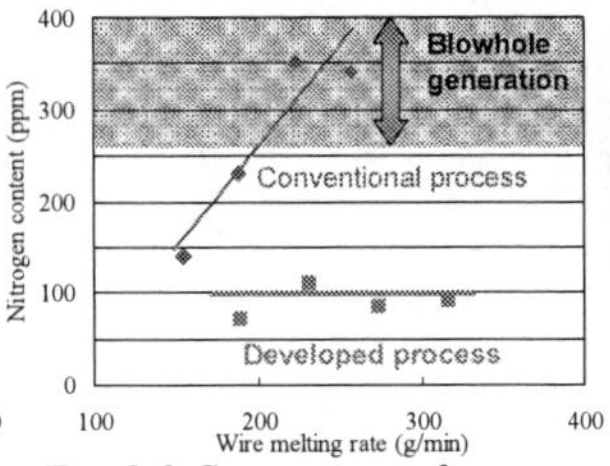

Fig. 2-2 Comparison of nitrogen content of weld metal between conv. and developed processes

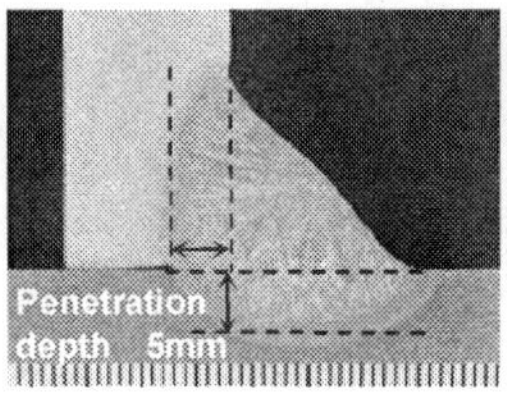

Fig. 2-3 Macrostructure of fillet weld's cross section

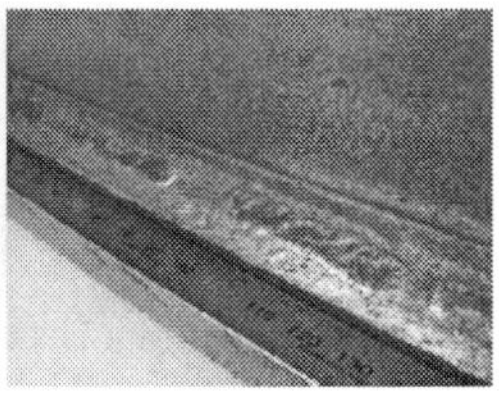
Fig. 2-4 Bead appearance of fillet weld

Table 2-1 Welding conditions for conventional and developed processes

	Conventional process*		Developed process
Layers	1	2	1
Welding current(A)	420	420	600
Deposition rate(g/min)	125	125	268
Welding Speed(cm/min)	34.0	23.0	33.0
Ratio of weld efficiency	1		2.4

*Solid, 1.4mm dia. ,MAG, Ext 25mm

(3) REGARC™

In conventional CO_2 arc welding, with an increase in the welding current, the droplet transfer mode becomes globular in which the molten droplet formed on the wire tip becomes big, generating many spatters. In order to reduce spatters in the globular transfer range, the following three measures are important: (1) Suppress an excessive growth of the molten droplet to prevent the short circuiting between a molten droplet and the weld pool; (2) Reduce the arc's reactive force to prevent an excessive effect of pushing a molten droplet upwards; and (3) Reduce the welding current immediately after a molten droplet was detached. The REGARC™ process has solved these issues by controlling the output waveform for reducing spatters.

(a) Development of REGARC™

The output waveform of the REGARC™ process is shown in Fig. 3-1. In this process, the pulsed waveform for forming a molten droplet and that for detaching the droplet are alternately output, and one droplet is transferred to the weld pool at every cycle of these two kinds of pulsed currents. For forming a molten droplet, the pulsed waveform set according to the wire feed rate is used. With this waveform, the energy for fusing the wire becomes constant, and thus the size of a molten droplet becomes consistent, thereby preventing an excessive growth of the droplet to reduce the spatter caused by short-circuiting. The droplet formation pulse with the optimized peak current and time prevents the arc from exerting an excessive reaction force, thereby reducing the spatter expelled by pushing a molten droplet upwards. For detaching a molten droplet, the droplet detaching pulse with the peak current higher than that for the droplet formation pulse is used to increase the electromagnetic pinch force that is one of the forces to create the droplet's squeezed part or the starting point for detaching the droplet. Immediately after the droplet was detached, the pulse waveform is shifted promptly to the base period to generate a low current arc at the wire tip from which the droplet was just detached. This technique extremely reduces the chance of expelling the molten part of a wire tip — Fig.3-2.

(b) Performance of REGARC™

The spatter generation rates in the flat fillet welding by the conventional and REGARC™ processes are shown in Fig. 3-3. With the REGARC™ process, the spatter generation rate is reduced greatly, and the adhered spatters feature a small particle size and extremely low quantity. The wire melting rate as a function of the mean welding current is shown in Fig. 3-4. When the mean welding current is the same, the REGARC™ process can increase the melting rate, thereby shortening the arc time. When the conventional process and the REGARC™ process are compared on the mean welding current at the same wire melting rate, the latter process offers a lower current by approx. 10% than the former one. Therefore, the heat input can be reduced, by which the mechanical properties of weld metal may be improved, and the distortion by weld strain may be decreased. Finally, the fume emission rates for each unit weight of wire are compared in Fig. 3-5. In the REGARC™ process, the fume emission rate decreases clearly by half as compared with the conventional one. In the REGARC™ process, the generation of metal vapor from the droplet's surface is reduced because short-circuiting and re-arcing hardly takes place, and the time for heating the molten droplet is shortened due to a short cycle of metal transfer. Therefore, the REGARC™ process can improve the welding operation environment.

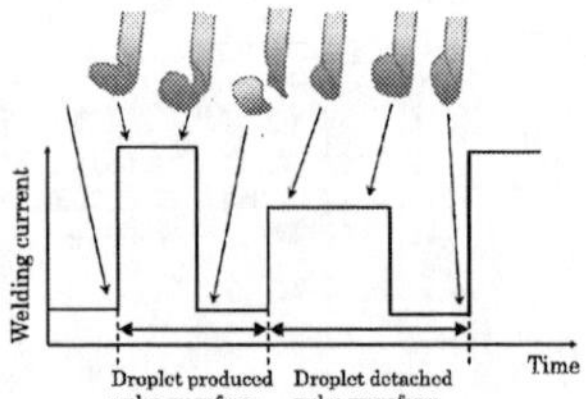

Fig. 3-1 An output waveform for REGARC™ processes

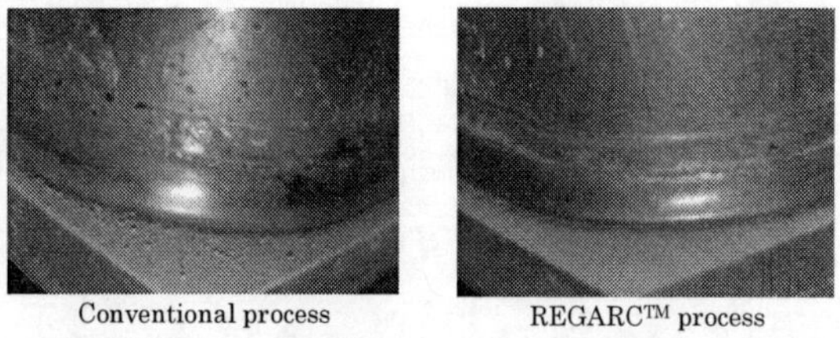

Fig. 3-2 Comparison of spatter adhered to diaphragm and structural hollow section between conv. And REGARC™ process

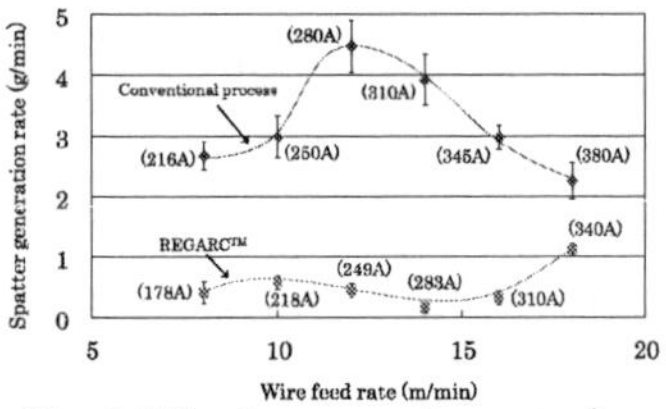

Fig. 3-3 Spatter generation rates by conv. and REGARC™ processes

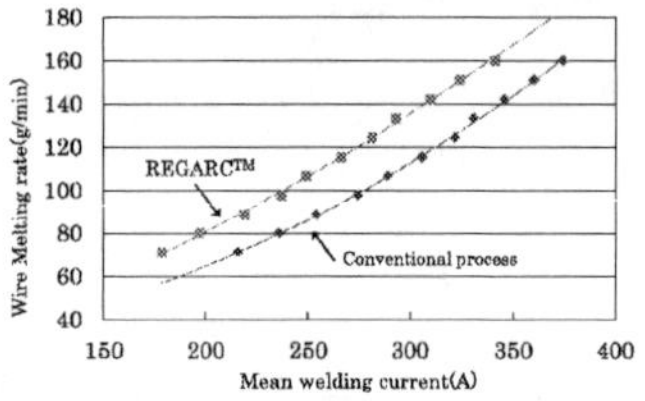

Fig. 3-4 Wire melting rate vs. mean welding current in conv. and REGARC™ processes

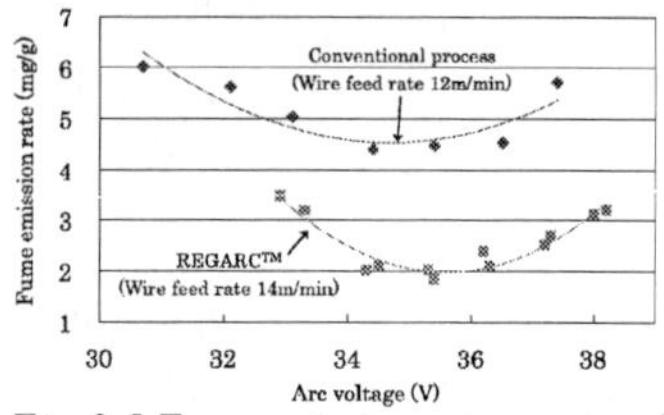

Fig. 3-5 Fume emission rate vs. arc voltage in conv. and REGARC™ processes

Conclusions

In this paper, three high-value-added welding processes for mid-and-thick plates applications have been introduced and explained. The tandem-arc welding process is good for maximizing the welding speed, and the high-current MAG welding process excels in decreasing the number of weld passes and in increasing the weld penetration; hence, these processes can be employed for various applications according to the individual requirements. In addition, the REGARC™ process has realized the ultra-low spatter performance in CO_2 arc welding which had been considered difficult. These welding processes are strongly expected to exert their capabilities in increasing the welding work efficiency, improving the welding environment, and thus in cost savings for the users. The authors are determined to provide users with new solutions through multiple approaches with the combination of welding consumables, welding processes, and welding robot systems for the future.

References

[1] M. Shigeyoshi: Dual arc sensor for Tandem arc welding,
KOBE STEEL TECHNICAL REPORTS/Vol. 48 2008-12
[2] K. Tsuji, et al: Robot Welding Process for Medium and Heavy Plate,
KOBE STEEL ENGINEERING REPORTS/Vol. 63 No.1 (Apr.2013)
[3] M. Yokota, et al: Robotic Welding System for Architectural Steel Frames Equipped with REGARC™ Process,
KOBE STEEL ENGINEERING REPORTS/Vol. 63 No.1 (Apr.2013)

Effect of Nanoparticle Additions on Microstructure, Solderability and Mechanical Properties of Sn-Ag-Cu solder

Dohyun Jung [1], Ashutosh Sharma [2], Donguk Lim [1], Songhee Yim [3], Jonghyuk Yoon [3], Bumgyu Baek [3], Jaepil Jung [1]

[1] *University of Seoul, South Korea*
[2] *University of Seoul, India*
[3] *KD One, Co., Ltd, South Korea*

In this study, the addition of nano-sized AlN and graphene particles has been investigated in Sn-3.0wt%Ag-0.5wt%Cu (SAC305) lead-free solder alloy. The various weight percentages of nano-particle have been successfully reinforced into SAC305 solder matrix by mechanical mixing and melting route. The influences of nano-sized particles on microstructure, mechanical properties and solderability such as spreadability and wettability have been carried out. The structural and morphological features of the nanocomposite solder are characterized by scanning electron microscope (SEM), energy dispersive spectroscopy (EDS), and transmission electron microscope (TEM). The experimental results show that good combination of solderability and mechanical properties is obtained by the addition of AlN in the solder matrix. The reinforced composite solder with optimum AlN nanoparticles shows 11.71% improvement in zero-cross time (ZCT), and 4.05% increase in the spreadability. In addition, the results of microstructural analysis of composite solder indicate that the nanocomposite solder, especially reinforced with AlN nanoparticles, exhibits better microstructure, improved tensile strength and elongation properties as compared to monolithic SAC305 solder. In case of the addition of graphene nanoplatelets into solder, the composite solder showed significantly improved spreading ratio around 80% compared to 70% of monolithic SAC305 solder alloy.

Effect of Metal Coating on Sn Whisker Growth at Welded Zone

Junpei Sasaki, Shun Watanabe, Tomomi Sakakida and Osamu Kamiya
Akita University, 1-1 Gakuenmachi, Tegata, Akita 010-8502, Japan
Email: lkamiya@gipc.akita-u.ac.jp, phone: +81188892730

Abstract
From the early 1980s, the problem of Sn whiskers in welded electronic components was suppressed by adding Pb as an alloying element to Sn solder. However, since the beginning of the 21st century, Pb has been phased out from solder because of its toxicity, and Sn whiskers have become a problem again. Sn whisker growth can cause an electrical short resulting in system failure. In this study, the effects of metal coating on Sn whisker growth in welded electronic parts were observed. In particular, we observed that Sn electroplating prevented Sn whisker growth in the welded zone. After producing a lead wire for an aluminum electrolytic capacitor by arc stud welding, the weld zone was coated by Sn electroplating. The samples were subjected to an ambient storage test at room temperature. Only one whisker was observed, even after 4000 h ambient storage, demonstrating that Sn electroplating strongly suppressed whisker growth. The whisker suppression mechanism was different from the effect of pressing such as in conformal coating. We concluded that the whisker suppression was caused by atomic diffusion of Sn during electroplating and the Sn plating coating absorbed the residual stress of the interior Sn.

1. Introduction
In the 1940s in the United States, short circuits occurred frequently in telephone exchanges. The cause was identified as single crystals of Sn, known as Sn whiskers. Sn whiskers mainly grow from the surface of Sn plating with diameters of 1–3 μm and reach a maximum length of several millimeters. This problem was resolved by using a Sn-Pb alloy solder, and the mechanism of Sn whisker generation was not studied. However, in 2006, "the Restriction of Hazardous Substances Directive" (RoSE) directive issued by the European Union has meant that harmful substances, such as Pb, can no longer be used in electronic and electrical equipment. Since Pb-free electrical and electronic equipment has been introduced, the problem of Sn whiskers has reemerged, even though Pb-free solder forms reliable joints. Therefore, it is important to elucidate the formation mechanism of Sn whiskers. Sn whiskers grow in devices after plating before they are shipped, and thus manufacturers must suppress the growth of Sn whiskers. Currently, suppression methods include using a Sn-Bi alloy, heat treatment by annealing, and pressing a conformal coating onto the welded zone. In this study, metal coatings were applied to a welded zone and the occurrence of Sn whiskers and their growth were investigated.

2. Experimental
A wire lead for an aluminum electrolytic capacitor was used in this experiment. The wire, which was made from a CP wire (Sn-plated, copper-covered wire), and an Al tab were welded together by arc stud welding. The weld zone was plated with Sn and Zn by electrolysis. Each plating samples had matte plating. Whisker growth was not suppressed by bath plating. 3 types of samples were prepared, Sn plating time was 5 and 10 min, Zn plating time was 10 min. All samples were stored at room temperature in an ambient storage test and the whiskers were observed by scanning electron microscopy (SEM; JSM-5800, JEOL), the whisker organization was determined electron backscatter diffraction (EBSD; XL30S-FEG, FEI), and focused ion beam (FIB; FIB NOVA200, FEI) was used to observe the whisker cross sections.

3. Results
Figure 1 (a) shows an SEM image of a welded zone after ambient storage for 1440 h, and no whiskers are observed. Figures 1 (b)–(d) show SEM images of samples that were plated after welding. No whiskers are visible in Figures 1 (b) (c). Therefore, plating the welded zone suppressed whisker growth. SEM images of samples with different Sn plating thicknesses are shown in Figure 2. Whiskers grew where the plating was less than 5 μm thick (Figure 2 (a)). When the plating layer was thicker than 10 μm (Figure 2 (b)), only one whisker was observed after 4000 h storage. Therefore, whisker growth decreased as the thickness of the plating layer increased.

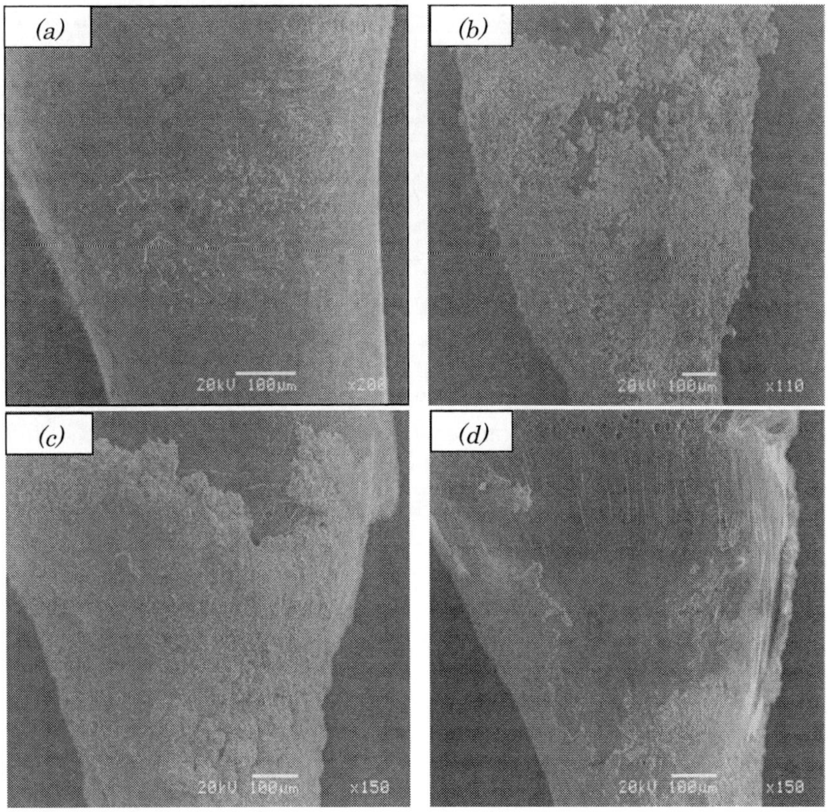

Fig. 1 Whisker suppression by plating the welded zone. SEM images of the welded zones (a) without plating, and with (b) Zn plating, (c) Sn plating, and (d) incomplete Sn plating.

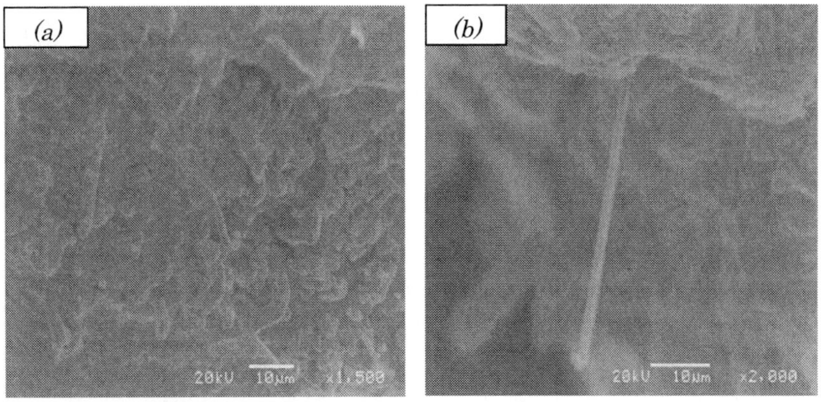

Fig. 2 SEM images of whiskers in samples with Sn plating thicknesses of (a) 5 and (b) 10 μm.

4. Discussion

4.1 Grain size and plating thickness

SEM images of the surface and a cross section of the plating are shown in Figure 3. The grain size in the Sn plating was 3–5 μm. When the plating time was less than 5 min, plating thickness becomes on the size of the less than grain diameter. Therefore, the whisker suppressing effect was reduced. However, when the plating time was more than 10 min, plating thickness becomes on the size of more than grain diameter, and polycrystalline in the thickness direction. Therefore, strong whisker suppressing effect was obtained. These suppressing effects were discussed as follows.

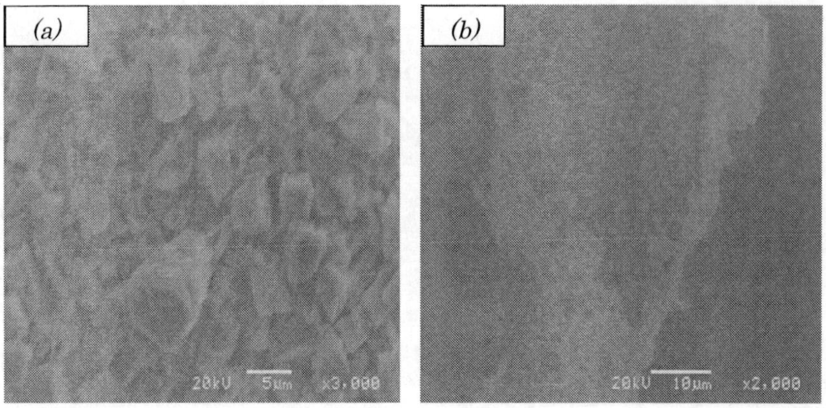

Fig. 3 SEM images of plating. (a) Plating surface and (b) plating section.

4.2 Relationship between whisker generation mechanism and plating layer

Figure 4 shows a cross-sectional FIB image and EBSD image of a whisker [1]. The whisker grew in the [001] direction from where the Sn appeared on the surface in the network-like mixed region of Sn and Al. The growth direction of the Sn network formed by rapid solidification after welding was also in the [001] direction, making it was easy for a Sn whisker to grow vertically from surface. The atomic diffusion of Sn occurred along the (100) [001] slip system in the network. The relationship between the whisker generation mechanism and plating layer is shown in Figure 5. Whiskers grew when the plating layer was a single crystal in the thickness direction.

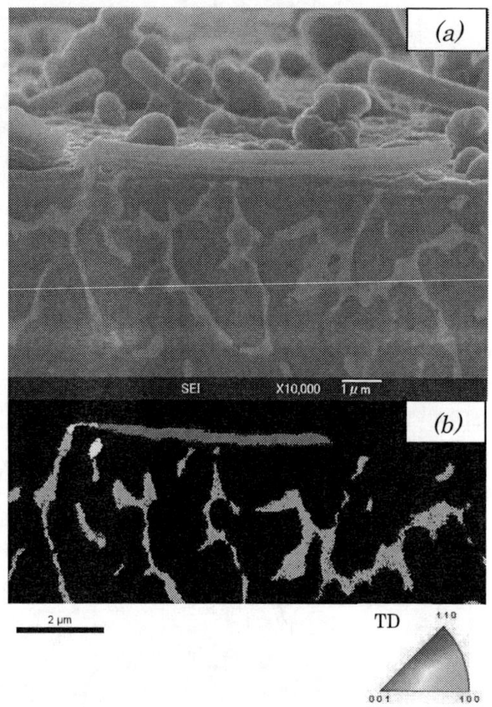

Fig. 4 (a) Cross-sectional FIB image and (b) EBSD orientation of a whisker.

4.3 Growth speed of a Sn whisker expressed by a diffusion equation

The chemical potential energy in the Sn in the welded zone is the weld residual stress during the solidification process. An atom moves along the grain boundary and the interface, driven by the chemical potential gradient. When the normal stress, σ, acts on a grain boundary, the chemical potential, μ, in the grain boundary is expressed as[2]

$$\mu = \mu_0 - \sigma\Omega \qquad (1)$$

where μ_0 is the chemical potential of the bulk and Ω is the atomic volume. Atomic diffusive flux J_B, which spreads along the grain boundary, is expressed as[2]

$$J_B = -\frac{n_L D_b}{kT} grad\mu \qquad (2)$$

where D_b is the grain boundary diffusion coefficient, k is Boltzmann's constant, T is absolute temperature, gradμ is the chemical potential gradient, and n_L is the number of lattice sites per unit volume. Equation (3) can be derived from equations (1) and (2).

$$J_B = \frac{n_L D_b \Omega}{kT} grad\sigma \qquad (3)$$

Here, gradσ is the stress gradient. An atom moves from the high stress region to the low stress region along the grain boundary. gradσ and whisker growth speed calculated by equation (3) are shown in Table 1. For example, when the stress gradient is 300 N/mm^3, whisker growth speed is 3.59 μm over 1000 s, and this can be explained the phenomenon of after welding. As the stress gradient is decreased by atomic diffusion over time, the growth speed decreases. It is important whether this stress can be eased in internal plating to the stress gradient that occurs in the welded zone. Generally, the whiskers grew about 10–50 μm 24 h after welding. The stress gradient was calculated from this growth speed as 9.67–48.3 N/mm^3. In contrast, the whisker length for the electrolysis plated sample was a maximum of 40 μm 1 month after welding (Figure 2), and the stress gradient was calculated as 1.29 N/mm^3. The stress gradient was decreased by electroplating. because the supply of Sn atoms to the surface was decreased owing to the slow atomic diffusion. Weld residual stress arising during solidification was the main driving force behind whisker growth. However, stress occurs in the long term because of the thermal expansion difference between the metal, and the increase in volume caused by oxidation and the generation of intermetallic compounds. The weld residual stress is large compared with the long-term stress, and is at most about 500 MPa according to shrink-fit theory[1]. Therefore, it is expected that atomic diffusion is caused easily. For the Zn plated sample, the Sn atoms that diffused because of the stress gradient generated an alloy of Sn and Zn in a boundary region between the plating layer and the welded surface, which suppressed whisker formation. Therefore, the increase in the thickness of the plating layer suppressed whisker formation.

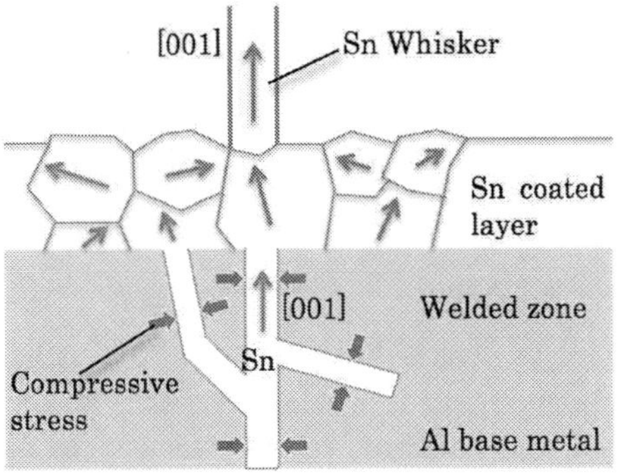

Fig. 5 Schematic of Sn whisker formation and the Sn coated layer on a welded metal zone.

Table 1 Stress slope (gradσ) and whisker growth speed.

$grad\sigma$ [N/mm^3]	J_B [1/m^2·s]	Whisker growth speed [μm/s]
1000	3.56×10^8	1.20×10^{-2}
500	1.78×10^8	5.98×10^{-3}
300	1.07×10^8	3.59×10^{-3}
100	3.56×10^7	1.20×10^{-3}
50	1.78×10^7	5.98×10^{-4}

5. Conclusions

We electroplated a welded wire for an aluminum electrolytic capacitor and we investigated the growth of Sn whiskers. A large number of whiskers grew on normal welded wires, whereas electroplating suppressed whisker growth. Increasing the plating thickness increased the suppression of whisker growth. We propose the following mechanism to explain our results.

1) A whisker grows in the [001] direction. A single Sn crystal in the plating layer in the thickness direction will form a whisker. Whisker growth is suppressed when there are many crystals with different crystal directions.
2) A Sn atom moves along the grain boundary, driven by the stress gradient. The diffusion equation indicates that the stress gradient is decreased by electroplating, and the diffusion of Sn atoms occurs less readily suppressing whisker growth.

References

[1] T. Kubouchi and O. Kamiya: Study of Sn Whisker Growth Mechanism in Welding Lead, Journal of smart Processing Society, Vol. 3, No. 2, (2013) pp. 135-141.

[2] C Herring: Diffusional Viscosity of a Polycrystalline Solid, Journal of Applied Physics 21, 437 (1950), pp. 437-445.

Quantitative assessment of thermally induced stresses for assemblies with TLP bonded Si chips

A. Lis[1,2], A. Hirose[2], C. Leinenbach[1]

[1]Empa, Swiss Federal Laboratories for Materials Science and Technology, Ueberlandstrasse 129, 8600 Duebendorf, Switzerland

[2] Osaka University, Division of Materials and Manufacturing Science, Graduate School of Engineering, 2-1 Yamada-oka, Suita, Osaka 5650871, Japan

adrian.lis@mapse.eng.osaka-u.ac.jp, hirose@mapse.eng.osaka-u.ac.jp, christian.leinenbach@empa.ch

S. Kicin[3], F. Brem[3]

[3]ABB Corporate Research, Segelhofstrasse 1K, 5405 Baden-Daettwil, Switzerland

slavo.kicin@ch.abb.com, franziska.brem@ch.abb.com

Abstract

(Micro-)electronics industries are currently facing demanding challenges due to severe restrictions on widely used Pb based solders and upcoming trends for miniaturization and 3D integrated circuits. Low temperature processes up to 300 °C and low volume interconnects with a few microns of thickness are of particular interest. Transient liquid phase (TLP) bonding has been identified as a promising joining technique to overcome these issues.

The present study aims at assessing thermally induced stresses in TLP bonded Si chips to metallic substrate materials after joining and during service. Although thermal stresses strongly impact the life-time of such modules, the experimental determination is very difficult in thin interlayer arrangements. In this context, a novel procedure was developed which combines experimental and numerical methods. Shear tests were conducted on TLP bonded Cu-Cu components with Ag-Sn-Ag, Ni-Sn-Ni and Ag-Sn-Ni interlayers and low thermal expansion mismatch. The obtained information on shear strength and failure mode and position was used to adjust a numerical shear test model with finite element methods (FEM) which provided interlayer dependent failure criteria. In the following, a numerical parameter study on TLP bonded Si chips allowed for quantitative assessment of thermally induced stresses including failure prediction and design optimization while the effect of the interlayer system, process and service conditions as well as metallization layers was investigated. The method serves as a guideline to choose the right TLP system for defined requirements and can be expanded to other joining techniques.

1. Introduction

Transient liquid phase (TLP) bonding is considered as a promising low temperature joining technique for packaging applications in semiconductor high power modules. This joining technique uses thin interlayer arrangements when isothermal solidification of a liquid low melting point material is attained by dissoluting higher melting point interlayers and forming intermetallic compounds (IMC) [1]. After bonding the remelting temperature of the joint is raised. As a consequence, TLP joints are expected to provide an enhanced resistance against elevated process temperature compared to solutions with Pb or Sn based solders.

[2] investigated the impact of process parameters on TLP bonded Si/SiC dies with Ag-Sn interlayers. The bonding pressure was varied between 0 – 15 MPa while the shear strength of around 30 MPa only remarkably degraded with no applied pressure. Contamination-free surfaces, good surface finish and protecting process atmosphere act beneficially on the performance of the TLP bonds [2,3]. [4] identified an ideal process window for TLP bonding at 280 – 300 °C regarding process efficiency and void appearance. The bending strength of samples with a continuous IMC layer sandwiched between Cu metallization layers increased from 290 to 400 MPa after a heat treatment due to partial transformation into Cu-Sn solid solution [5]. An FE assisted study in [6] derived the following trends: Increasing the IMC thickness from 5 to 40 μm resulted in a linear increase of residual stresses in the joint by 30%. Attaching a fillet at the joint edge increased the residual stresses by 20%. Geometric modifications were suggested when slits in the metallic substrates reduced the residual stress within the joint by 20%. [7] calculated residual

stresses in TLP bonded Si chips considering interlayer sequences of Cu_6Sn_5 – $(Ag,Cu)_3Sn$ – Cu_3Sn (case A) as well as $(Ag,Cu)_3Sn$ – Cu_3Sn (case B). For case A, residual stresses in the $(Ag,Cu)_3Sn$ layer were by a factor of 2 lower compared to case B. [15] suggested a hybrid packaging solution including TLP bonding and soldering when residual stress peaks could be reduced by 20%. [9,10] pointed out that highest stresses arise in the IMC layer of TLP bonded packages. On the other hand, highest plastic strains in soldered packages appear within the solder. It was concluded that TLP bonded packages will be less susceptible to failure caused by creep and low cycle fatigue.

The present study is motivated by the fact that thermal stresses play an important role on the reliability and performance of TLP bonded chips while the experimental assessment is very difficult. Therefore, a method was developed which combines experimental results, i.e. shear strength and fractographic analyses, as well as FE simulations. Derived failure criteria are used to compare and assess thermal (residual) stresses in TLP bonded Si chips for different IMC-interlayer systems, metallization layers, process and service conditions.

2. Experimental and Numerical

Cu plates with the dimensions 110 x 60 x 4 mm were electroplated with 14 μm Ag and with 17 μm Ni, respectively. Subsequently, the plates were cut into small samples of 5 x 5 x 4 mm by Electrical Discharge Machining (EDM). Sn foil with a thickness of 8 μm and a purity of 99.75% was applied (Goodfellow Cambridge Ltd). Substrates and foil were assembled according to *Fig. I a)* with Ag-Sn-Ag (samples A), Ni-Sn-Ni (samples B) and Ag-Sn-Ni (samples C) interlayer combinations. All joining experiments were conducted in a tube furnace system under Argon 6.0 shielding gas atmosphere while applying a bonding pressure of 0.3 MPa. After 60 min of holding time at 300 °C slow furnace cooling was applied to minimize thermal gradients within the samples. After testing the shear strength of the samples fractographic characterization was conducted. The experimental shear test conditions were modeled with FEM (Abaqus 6.14). *Fig. I b)* shows the 2D FE model where the upper Cu substrate is loaded with the exp failure load. The TLP joint was modeled in good approximation to the microstructural observations in [4] when interlayer arrangement for sample A consisted of planar $Ag-Ag_3Sn-Ag$ with a thickness of 10 μm each. The TLP joint of sample B was equally modeled with an interlayer sequence of $Ni-Ni_3Sn_4-Ni$. For sample C, two planar IMC layers of Ag_3Sn and Ni_3Sn_4 with a thickness of 5 μm were sandwiched between the base metallization of Ag and Ni. The model was meshed with 4-nodal plane strain elements (CPE4R). The averaged element size in the Cu substrates was set to 0.005 mm whereas the TLP interlayers were meshed with 0.0025 mm. The IMC layers were modeled as linear-elastic whereas bilinear elastic-plastic behavior was considered for the substrates and base metallizations. Thermo-mechanical FE calculations were performed on a model representing a TLP bonded Si chip in a high power module as shown in *Fig. II*. The FE model was meshed with 3-nodal coupled temperature-displacement elements (CPE3T) under plane strain conditions. The TLP joint with its interlayers (each 8 μm thick) was meshed with an element edge size of 0.001 mm, neighboring layers with 0.002 mm at the interface towards the TLP joint and increasing mesh size up to 0.01 mm with increasing distance from the joint. TLP joints consisting of Ag(M1)- Ag_3Sn(IMC)-Ag(M2), Ni(M1) Ni_3Sn_4(IMC)- Ni(M2) and Ni(M1)- Ni_3Sn_4/Ag_3Sn(IMC)- Ag(M2) layers are placed between the Si chip and the metallized AlN.

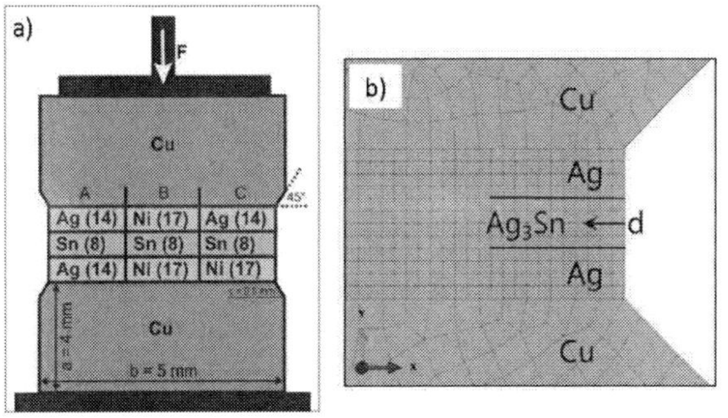

Fig. I: a) schematic TLP interlayers arrangements (all values in bracketsare in microns, b) FE model representing the samples from a)

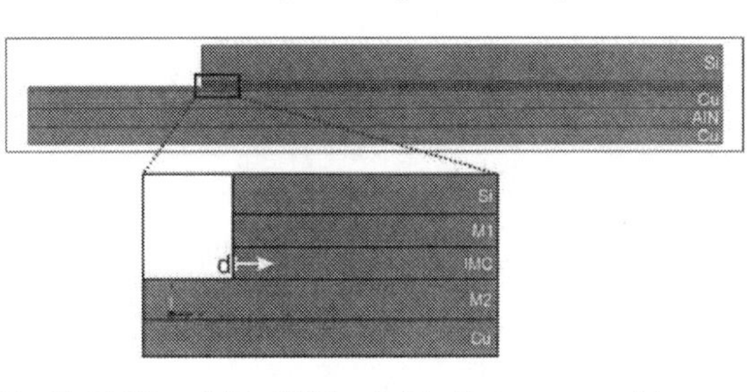

Fig. II: 2D FE model for TLP bonded Si chip in power module

3. Results and Discussion

Mechanical shear strength was measured highest for samples A (Ag-Ag_3Sn-Ag) with 60.4 MPa, followed by samples C (Ag-Ag_3Sn-Ni_3Sn_4-Ni) with 40.7 MPa and lowest for samples B (Ni-Ni_3Sn_4-Ni) with 27.4 MPa [4]. Cohesive crack propagation occurred in all cases within the IMC layer where bonding defects like pores weaken the structural integrity. The calculated shear stress within the FE model is visualized in *Fig. III a)* confirming a highly inhomogeneous loading on the TLP bonded samples during shear testing. This result motivates a local examination of the stress state in the TLP joint. Since failure occurs in the IMC layers, the calculated shear stress was plotted within the IMC layer along the local parameter *d* (cf. *Fig. I b)*). When the experimental failure load is applied, the local critical shear stress in the IMC layer can be derived as a function of the TLP interlayer system. For sample A, a maximal shear load $\tau_{12,max}$ = -71 MPa appears within the Ag_3Sn IMC layer immediately before failure as shown in *Fig. III b)*. Sample B fails with a local shear stress $\tau_{12,max}$ = -62 MPa in the Ni_3Sn_4 IMC layer. The failure criterion for sample C was extracted from the peak shear stress of $\tau_{12,max}$ = -70 MPa. Choosing a critical shear stress as failure criterion seems to be suitable for two reasons: 1) Despite the inhomogeneous stress distribution during shear testing, shear stress is the main loading mode leading to the joint fracture. 2) TLP joints between a chip and a metallized surface with large CTE mismatch are highly loaded under shear stresses after bonding and in service. The obtained interlayer dependent local failure criteria can be normalized to $\tau_{12,norm} = \tau_{12}/\tau_{12,max}$. A value for $\tau_{12,norm}$ larger than 1 indicates failure.

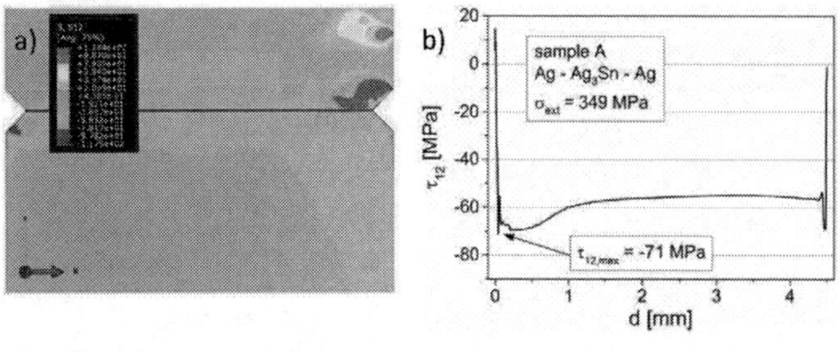

Fig. III: a) Shear stress distribution on sample A before failure, b) local shear stress before failure along local coordinate d in sample A

The calculation of thermally induced residual stresses along the local coordinate d is plotted in *Fig. IV a)* for a TLP interlayer system consisting of (Ag)-Ag_3Sn-Ag sandwiched between Cu and Si chip. For the case with M1 = 0 µm, highest values of $\tau_{12,norm}$ are calculated in near-edge regions of the IMC layer with 0.55 and 0.66 after cooling down from 250 and 300 °C, respectively, as plotted in *Fig. IV a)*. The implementation of M1 = 8 µm reduces the absolute peak values to 0.34 and 0.39. The lowest peak of $\tau_{12,norm}$ = 0.21 is achieved with M1 = 2 µm (for T_h = 250 °C) indicating that thinner M1 layers are beneficial in terms of residual stresses. For TLP bonded Si chips with (Ni)-Ni_3Sn_4-Ni interlayers, the calculations in *Fig. IV b)* predict failure without metallization M1 and T_h = 300 °C with $\tau_{12,norm}$ = 1.11 (> 1) (which was experimentally verified) while at lower holding temperature of 250 °C and M1 = 0 µm, the normalized peak shear stress of 0.96 is slightly below the failure criterion. The addition of M1 = 8 µm leads to significantly reduced and uncritical values of $\tau_{12,norm}$ with 0.49 and 0.56 after a process temperature of 250 and 300 °C, respectively. TLP chip bonding with dissimilar metallization on chip (Ni) and substrate (Ag) will lead to an interlayer arrangement of (Ni)-Ni_3Sn_4-Ag_3Sn-Ag in the TLP joint. Highest normalized stress of 0.98 evolves along the local parameter d without metallization M1 for a joining temperature of 300 °C (intact joint was experimentally verified). With M1 = 8 µm, $\tau_{12,norm}$ strongly decreases to 0.33 and 0.4 when holding temperatures of 250 and 300 °C, respectively, are implemented to the FE model.

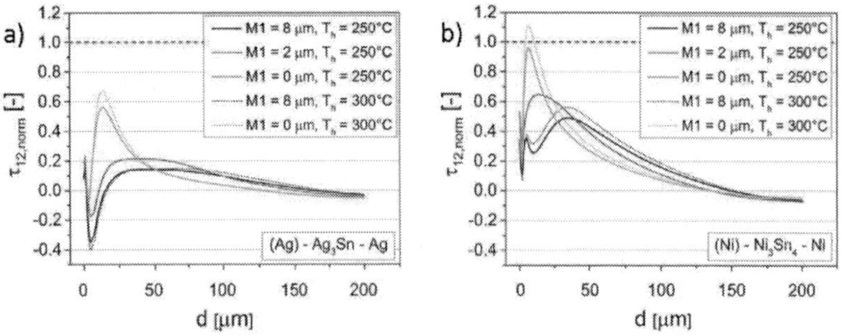

Fig. IV: Normalized shear stress along d with a) (Ag)-Ag_3Sn-Ag and b) (Ni)-Ni_3Sn_4-Ni interlayers

FE calculations on residual stresses after TLP bonding revealed pronounced differences as a function of the applied interlayers. The binary Ag-Sn system is favored due to the lowest normalized shear stresses in the critical Ag_3Sn

IMC layer after TLP bonding. Stress values are far below the critical value of $\tau_{12,norm} = 1$ even when the IMC layer is in close contact to the chip (M1 = 0 μm). Increased stiffness combined with lower strength of the Ni_3Sn_4 IMC results in significantly higher normalized shear stress when the binary Ni-Sn interlayer system is applied for TLP chip bonding under identical process conditions. Failure prediction for M1 = 0 μm and T_h = 300 °C was experimentally confirmed. The combined interlayer system with Ni-Sn-Ag layers is an option when as-delivered components may be TLP bonded without additional coatings. Calculations on residual stresses do not exceed the failure criterion in any of the considered configurations. However, the stress evolution is strongly governed by the Ni_3Sn_4 IMC layer in the TLP joint where failure is expected to occur. The calculations for all three interlayers indicate that a residual metallization layer on the chip side (M1) after TLP bonding significantly reduces residual thermal stresses. Consequently, structural optimization can be achieved by having a sufficiently thick metallization layer on the chip which is not fully consumed by the TLP reaction (M1 + Sn → IMC). It was found that the M1 metallization deforms plastically and protects the critical IMC layer from shear stress peaks which act on the IMC layer in near-edge areas when M1 = 0 μm. Highest CTE mismatch is shifted from the chip-IMC interface to the chip-M1 interface.

4. Conclusions

The combination of experimental and numerical methods enabled a detailed thermo-mechanical characterization of TLP bonded Si chips in power modules. The developed procedure allowed for predictive stress assessment and recommendations for design optimization considering realistic process (and service conditions). FE calculations on residual stresses pointed out that the metallization layer M1 between chip and IMC layer is very beneficial to reduce critical residual stresses. Regarding semiconductor chip packaging and the thermo-mechanical performance, the Ag-Sn-Ag TLP system fulfills the requirements in power modules best. Although providing the cheapest TLP solution, the use of Ni-Sn-Ni layers contains drawbacks for advanced semiconductor modules. The combined Ag-Sn-Ni system can be seen as a compromise between the both binary systems and might be an interesting candidate to join as-delivered substrates with asymmetric coatings.

5. References

[1] G.O. Cook, et al: Overview of transient liquid phase and partial transient liquid phase bonding, J Mater Sci, Vol. 46, (2011), pp. 5305-5323.

[2] A.A. Bajwa, et al: Foil Based Transient Liquid Phase Bonding as a Die-Attachment Method for High Temperature Devices, VDE Verlag, CIPS, Vol. 141, (2014), pp. 359-364.

[3] S.H. Eo, et al: Fluxless eutectic bonding of GaAs-on-Si by using Ag/Sn solder, Electron Mater Lett, Vol. 9, (2013), pp. 787-791.

[4] A. Lis, et al: Influence of process and service conditions on TLP bonded components with (Ag,Ni-)Sn interlayer combinations, J Electron Mater, Vol. 44, (2015), pp. 4576-4588.

[5] N.S. Bosco, et al: Strength of joints produced by transient liquid phase bonding in the Cu-Sn system, Acta Mater, Vol. 53, (2005), pp. 2019-2027.

[6] M.N. Matsushima, et al: Effects of material property and structural design on the stress reduction of the joint in electronics devices, Mater Sci Forum, Vol. 783-786, (2014), pp. 2765-2770.

[7] T. Takahashi, et al: Development of Ag_3Sn Intermetallic Compound Joint for Power Semiconductor Devices, Electrochem Solid St, Vol. 12, (2009), pp. H263-H265.

[8] D.Q. Yu, Development of reliable low temperature wafer level hermetic bonding using composite seal joint, Microelectron Reliab, Vol. 52, (2012), pp. 589-594.

[9] R. Dudek, et al: Reliability investigations for high temperature interconnects, Solder Surf Mt Tech, Vol. 26, (2014), pp. 27-36.

[10] R. Dudek, et al: Studies on the reliability of power packages based on strength and fracture criteria, Microelectron Reliab, Vol. 52, (2012), pp. 1255-1265.

TLP Diffusion Bonding of Pure Aluminum A1070 with Copper Foil as an Insert Metal

Haoming GUO, Tomiko YAMAGUCHI,
Kyusyu Institute of Technology, Kitakyusyu, Fukuoka, Japan
p347829k@mail.kyutech.jp, +81-93-884-3357

Kazumasa NISHIO
Nishinippion Institute of Technology, Kanda, Fukuoka, Japan

Abstract

During the diffusion bonding process of aluminum alloys, the existence of aluminum oxide interferes with direct contact between atoms form parent materials and leads to bad influence on the bond strength. To overcome this problem, TLP (Transit Liquid Phase) diffusion bonding, which is different form ordinary solid state diffusion bonding, was applied. In the experiments, industrial pure aluminum A1070 was used as bonding material and pure copper foils with thickness of 7 and 50μm were used as an insert metal. Bonding temperature was set around 560 degrees (Celsius) lasting for 5 minutes using high frequency induction heating under bonding pressure of 2MPa. Additionally, temperature gradient around bonding zone was achieved by adjusting the position of heating coil. As a result of SEM observation, a significant difference in the form of bonding surface was confirmed. In the experiments without temperature gradient, the bonding surface turned out to be uniform while bonding surface showed a discontinuous form when temperature gradient was introduced, which was considered as a sign of destroyed aluminum oxide. Also, the EDS line analysis results revealed the diffusion of atoms after the bonding process and oxidation reaction around the bonding surface.

1. Introduction

Aluminum alloys with properties such as low specific gravity, high specific strength and high corrosion resistance, are often used in industrial products. However, now by the current welding methods of the mainstream of aluminum, since the low oxidation resistance, melting point and hardness of aluminum, the overlarge and deformation by heat input are likely to cause problems during welding process. To solve this problem, investigations of diffusion bonding of aluminum as well as FSW have become popular in recent years.

On the other hand, diffusion bonding of aluminum is different from the iron or titanium type, which means in the bonding process, a very stable oxide layer of the junction interface becomes the inclusions and voids in the interface joined, leading to a decrease of bonding strength[1].According to the previous research of diffusion bonding of aluminum, solid-state bonding of aluminum by easy means (low degree of vacuum, low bonding pressure, short time) has not obtained a bonded joint of the reasonable strength [2]. To resolve this problem caused by the bad influence of aluminum oxide during solid-state bonding process, TLP (Transit Liquid Phrase liquid phase diffusion) junction was cited as a countermeasure. In the liquid phase diffusion bonding, after temporarily melt the insert material is to join by isothermal solidification utilizing the diffusion of atoms. Compared with traditional solid-phase diffusion bonding, in the fluidity of the liquid phase the insert metal can easily ensure the adhesion of the bonding interface at low pressures, which is possible to suppress the generation of voids. Then, the insert material in the liquid phase is likely to destroy oxidization film by creating the intermetallic compound and obtain stronger bonding joints. In this study, pure copper foil was subjected to TLP diffusion bonding of aluminum as an insert metal and the influence of heating condition was discussed.

2. Experimental procedures

Industrial pure aluminum A1070 round rod with a diameter of 18mm and length of 40mm was used as the bonding material. Table 1 shows the chemical compositions of A1070. As an insert metal, we used pure copper foil having a thickness 7 and 18μm. Fig. 1 shows a schematic view of a vacuum diffusion

Table 1: Chemical compositions of A1070 (mass%)

Si	Fe	Cu	Mn	Mg	Cr	Zn	Ga, V, Ni	Ti	Al
0.02	0.25	0.04	0.03	0.03	0.01	0.04	0.05	0.03	Bal.

bonding. For uniform heating, the heating coil of high frequency induction heating was disposed to the center of the bonding materials, while in order to create the temperature gradient across the bonding interface, the heating coils was set to a position below the junction interface, and ceramic jig was also adopted to keep the bonding material form cooling.

As pre-bonding procedure, the surface of the aluminum was prepared by mechanical polishing until ♯1200 and stored with acetone. Insert metal foil was cut into round plate with diameter of 19 mm and washed with 10% dilute sulfuric acid before degreased. Bonding was conducted in the temperature of 540°C~560°C, vacuum degree of 2×10^{-1}Pa with the loading pressure of about 2MPa. Heat-up time was set as 3min and bonding time was set as 5min.

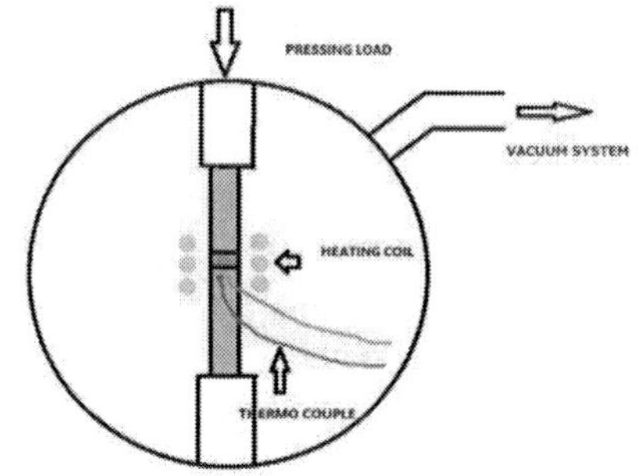

Fig.1: Schematic diagram of diffusion bonder

3 Results and discussion

3.1 Mircostrucure of bonding surface

As shown in Fig. 2, due to the presence or absence of temperature gradients, the shape s of the bonding interface were found to vary greatly. When there was no temperature gradient, change in shape of the insert metal was hardly observed and showed nearly uniform bonding lines in these joint interface. On the other hand, with the introduction of temperature gradient, copper foil varies significantly, resulting in discontinuous bonding interface. Also, on the bonding interface, both well bonded region and voids were confirmed. Furthermore, when the insert foil with the thickness of 7μm was used, significant change in the shape of bonding line was not observed.

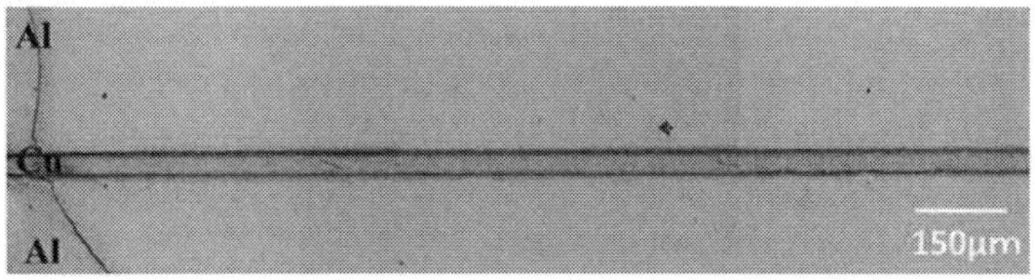

a): specimen without temperature gradient

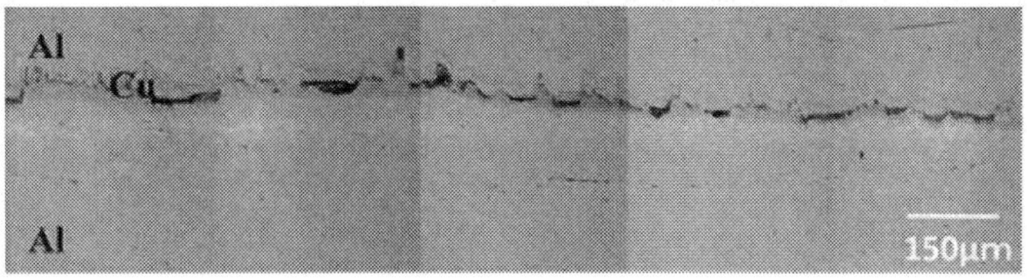

b): specimen with temperature gradient

Fig.2: Microstructure of bonding surface

3.2 Three-point bending test

To evaluate the bonding strength, three-point bending test was adopted. The distance between support spans was 40 mm and the moving speed of the pin was 1 mm/min. As a result, the bonding material with a temperature gradient, achieved a load of up to 3850N, which was about twice compared to the result when no temperature gradient was involved. However, the average strength of the sample with temperature gradient (1848N), remained the same level of the sample without temperature gradient (1353N). As a discussion on these results, the distribution of voids in the bonding interface was thought to be have a considerable impact on the results of the three-point bending test.

3.3 EDS linear analysis

The result of specimen without temperature gradient was shown in Fig. 3 a), the concentration of aluminum can be divided into four stages and with the change of distance from copper-aluminum interface, the different degrees of diffusion reactions were confirmed. On the other hand, by the introduction of temperature gradient, the peak of the copper across bonding interface was unclear, proving that the insert material has sufficiently diffused into the bonding material. Moreover, partial oxidation

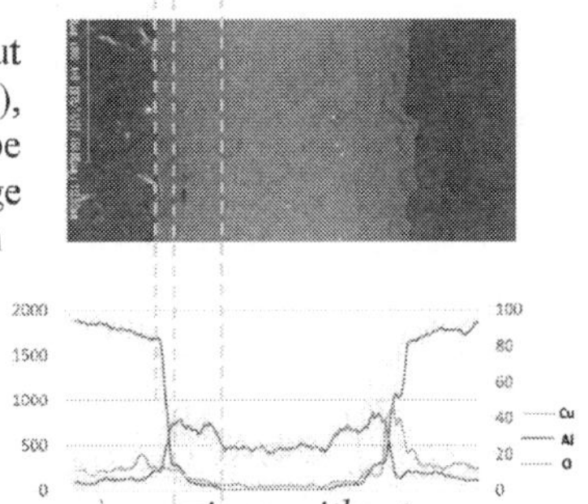

a): specimen without temperature gradient

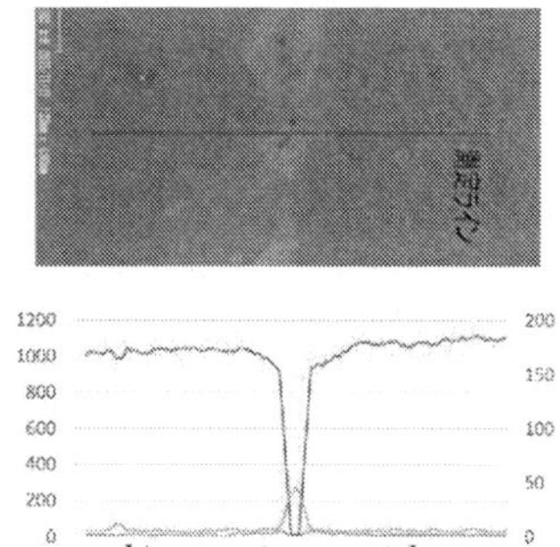
b): specimen with temperature gradient

Fig.3: Results of EDS linear analysis

reaction has also occurred according to the peak of oxygen in the diagram.

4 Conclusions

1) In the TLP diffusion bonding of aluminum, depending on the presence or absence of heating temperature gradient near the bonding interface, the shape of the interface has changed greatly. Wherein, the well-bonding region was observed. In addition, the influence of the thickness of the insert material was little.

2) By introducing temperature gradient, the average bonding strength did not changed substantially, while the maximum joint strength was improved about twice.

References

[1] Osamu Ohashi, *et al*: Welding appendix, Maruzen (1998)
[2] Kenji Amiwaki: "Diffusion bonding of aluminum with silver foil", master thesis of Kyushu Institute of Technology, 1995

Expansion of Hardness Prediction System for Temper Bead Welding of A533B Steel to Low-Alloy Steels

Lina Yu[a], Kazuyoshi Saida[a], Masahito Mochizuki[a], Masashi Kameyama[b], Naoki Chigusa[c] and Kazutoshi Nishimoto[a]

[a] *Osaka University, Osaka, Japan*

E-mail: yulina@mapse.osaka-u.ac.jp , Tel.: +81 6 6879 7543; fax: +81 6 6879 7543

[b] *Japan Nuclear Safety Institute, Tokyo, Japan*

E-mail: kameyama.masashi@genanshin.jp, Tel.: +81 3 5404 3874; fax: +81 3 5404 3882

[c] *The Kansai Electric Power Co., Inc., Fukui, Japan*

E-mail: chigusa@wano-tc.or.jp, Tel.: +81 770 32 3670; fax: +81 770 32 3659

Abstract

Based on the experimentally obtained hardness database, the neural network-based hardness prediction system of heat affect zone (HAZ) in temper bead welding for A533B steel was constructed. In the actual industry, not only A533B steel but also the other low-alloy steels were needed to temper bead welding. In order to construct the neural network-based hardness prediction system of A533B steel, about 500 hardness database sets were required, which were subjected to the simulated thermal cycles in HAZ of temper bead welding. For the hardness prediction of the other low-alloy steels, it would take a lot of time and cost if using the same method. Because the hardness after tempering had a linear relationship with LMP (Larson-Miller parameter) for the low-alloy steels without secondary hardening, the expansion of the previously constructed hardness prediction system of A533B steel to the other low-alloy steels without secondary hardening was proposed from the view of engineering. Through the new extended method, only about 100 hardness database sets of the other steels were needed to obtain. Using the expanded hardness prediction system of A533B steel, the hardness distribution in HAZ of temper bead welding for three kinds of steels (A508, 1.5%Ni and SM490 steels) was calculated based on the thermal cycles computed by finite element method. The experimental results revealed that the predicted hardness were in good accordance with the measured ones. It followed that the currently proposed expansion method was effective for estimating the tempering effect during temper bead welding for the low-alloy steels without secondary hardening.

Keywords: Hardness prediction; Temper bead welding; Low-alloy steel; Neural network; Heat conduction analysis

1. Introduction

Low alloy steels have been widely applied for the production of pipes, automotive and aerospace bodies, railway lines and offshore and onshore structural engineering plates. A low-alloy steel ASTM A533B possessing superior low-temperature toughness and weldability is typically used as the material for pressurized water reactor (PWR) vessels in nuclear power plants. Based on the experimentally obtained hardness database, the neural network-based hardness prediction system of heat affect zone (HAZ) in temper bead welding for A533B steel has been constructed by the authors [1]. However in actual industry, not only A533B steel but also the other low-alloy steels are required to temper bead welding. In order to construct the neural network-based hardness prediction system of A533B steel, about 500 hardness database sets were needed, which took a lot of time and cost. Therefore, the expansion of the previously constructed hardness prediction system of A533B steel to the other low-alloy steels without secondary hardening was proposed from the view of engineering.

2. Materials and experimental procedures

The chemical compositions of A533B, A508, 1.5%Ni and SM490 low-alloy steels are shown in Table 1. Specimens of low-alloy steels (5×5×5 mm) were heated by a high frequency induction heating device to synthesize the as-welded and temper-processed HAZ. The multi-pass welded samples of A508, 1.5%Ni and SM490 steels were produced by GTA welding with the welding conditions shown in Tables 2 and 3. The temper bead welding was performed using the consistent layer technique [2]. The cross-sectional surfaces after 1-layer and 6-layer welding were cut from the multi-pass welded samples. The Vickers hardness was measured in the cross-section of the specimens after polishing and etching with 3% nital solution. The Vickers hardness measurement was performed at a load of 9.8N, and the mean value was taken after excluding the maximum and minimum values from each multiple measurement. The thermal cycles in temper bead welding were calculated using FEM software developed by the authors specifically for welding simulation [2].

Table 1 Chemical compositions of the low-alloy steels

Steel	C	Mn	P	S	Si	Ni	Cr	Mo	Cu	Al	Fe
A533B	0.12	1.43	0.006	0.002	0.26	0.53	0.01	0.51	0.02	0.003	Bal.
A508	0.18	1.48	0.005	0.002	0.18	0.76	0.12	0.5	0.2	0.003	Bal.
1.5%Ni	0.08	1.01	0.011	0.003	0.25	1.093	0.16	0.07	0.05	-	Bal.
SM490	0.17	1.37	0.018	0.018	0.33	0.07	0.06	0.008	0.01	-	Bal.

Table 2 Welding conditions of multi-pass welding of A508 and SM490 steels

Layer	Current [A]	Voltage [V]	Wire sending speed[mm/s]	Welding speed [mm/s]	Heat Input [kJ/mm]	Number of passes
1st	250	15.3	50	1.67	2.29	3
2nd	120	12.5	25	5	0.3	5
3rd	120	12.5	25	5	0.3	4
4th	200	14.5	42	1.67	1.74	3
5th	200	14.5	42	1.67	1.74	3
6th	200	14.5	42	1.67	1.74	3

Table 3 Welding conditions of multi-pass welding of 1.5%Ni steel

Layer	Current [A]	Voltage [V]	Wire sending speed[mm/s]	Welding speed [mm/s]	Heat Input [kJ/mm]	Number of passes
1st	250	17	33.3	3.33	1.7	11
2nd	140	14.5	16.7	1.67	3.3	25
3rd	140	14.5	16.7	1.67	3.3	26
4th	200	16.5	23.3	2.33	1.7	14
5th	200	16.5	23.3	2.33	1.7	14
6th	200	16.5	23.3	2.33	1.7	13

3. Results and discussions

Because the hardness after tempering has a linear relationship with LMP (Larson-Miller parameter) for the low-alloy steels without secondary hardening [1], the expansion of the previously constructed hardness prediction system of A533B steel to the other low-alloy steels without secondary hardening was proposed. The basic concept of expanding the hardness prediction system is expressed as formulas (1) and (2):

$$\mathrm{QR} = (Hv_\alpha - Hv0_\alpha)/(\mathrm{Hv}M_\alpha - \mathrm{Hv0}_\alpha) \quad (1)$$

$$Hv_\beta = \left(HvM_\beta - Hv0_\beta\right) \times QR + Hv0_\beta \quad (2)$$

where, α indicates A533B steel and β indicates the other steels; HvM is the maximum hardness, and Hv0 is the minimum hardness. Through this method, the QR prediction system can be constructed by neural network [3], as shown in Fig. 1.

Using the QR prediction system, the hardness of A508, 1.5%Ni and SM490 steels after thermal cycle was calculated.

As an example, the comparison between the original hardness and the calculated hardness after 1-cycle+temper thermal cycle are shown in Fig. 2. The correlation coefficient for the A508, 1.5%Ni and SM490 steels were 0.93, 0.92 and 0.90, respectively. The high correlation coefficients suggested that the hardness of the other low-alloy steels without secondary hardening could be predicted by using the expanded hardness database of A533B steel.

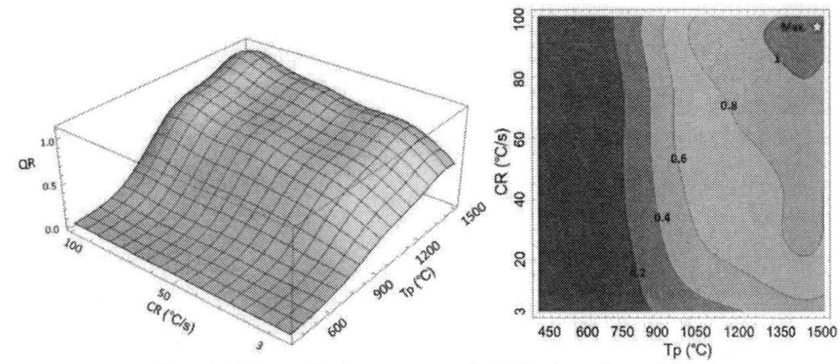

Fig. 1 NN-prediction system of QR in 1-cycle welding

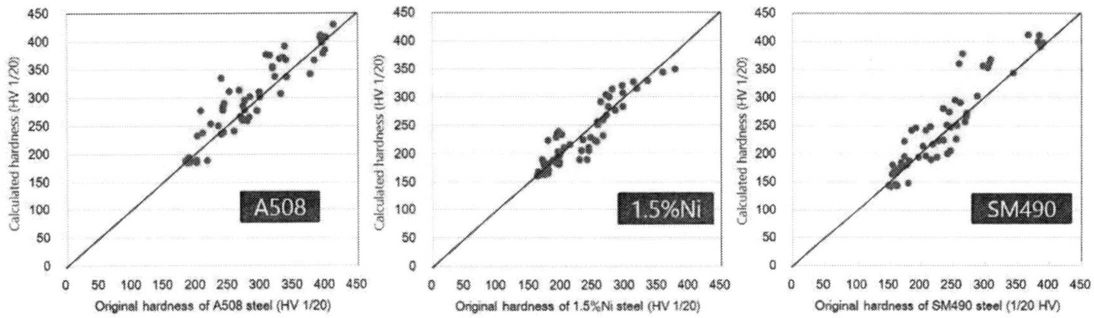

Fig. 2 Comparison between the original and the calculated hardness of A508, 1.5%Ni and SM490 steels after 1-cycle+temper thermal cycle

The temperature distribution in welds during temper bead welding was calculated using three-dimensional finite element analysis code, developed by the authors specifically for welding simulation [2]. The mesh model was the same size with the experimental welding sample, and the welding conditions were the same as the experimental conditions. The simulated temperature after multi-pass welding are shown in Fig. 3. Based on the calculated thermal history of every grid node, the hardness at the grid node was calculated. The two-dimensional hardness distribution in the HAZ of A508, 1.5%Ni and SM490 steels are shown as color contour maps in Fig. 4. Besides the red weld metal (WM) and grey base metal (BM) regions, the hardness in HAZ is shown with rainbow colors depending on the different hardness levels.

Furthermore, comparison of the predicted and the measured hardness after multi-pass welding of A508, 1.5%Ni and SM490 steels are shown in Fig. 5. The hardness profiles were measured along the dotted line, 1.0 mm from the specimen surface. The blue points indicated the experimentally measured hardness, and the red points indicated the calculated hardness using our proposed method, based on the FEM simulated thermal history. The predicted hardness was in good accordance with the experimentally measured hardness for all steels. It followed that the proposed hardness prediction method was useful and effective for estimating the tempering effect in temper bead welding of the low-alloy steels without secondary hardening.

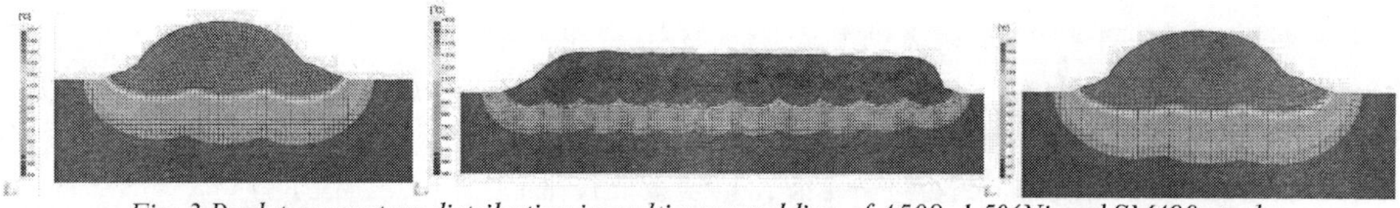

Fig. 3 Peak temperature distribution in multi-pass welding of A508, 1.5%Ni and SM490 steels

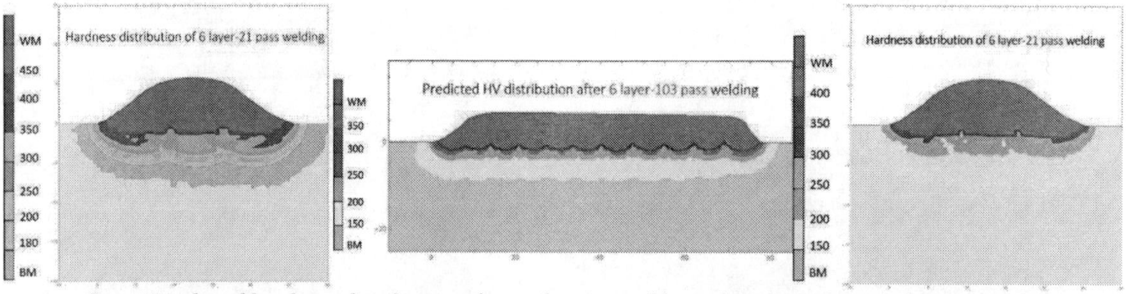

Fig. 4 Predicted hardness distribution after multi-pass welding of A508, 1.5%Ni and SM490 steels

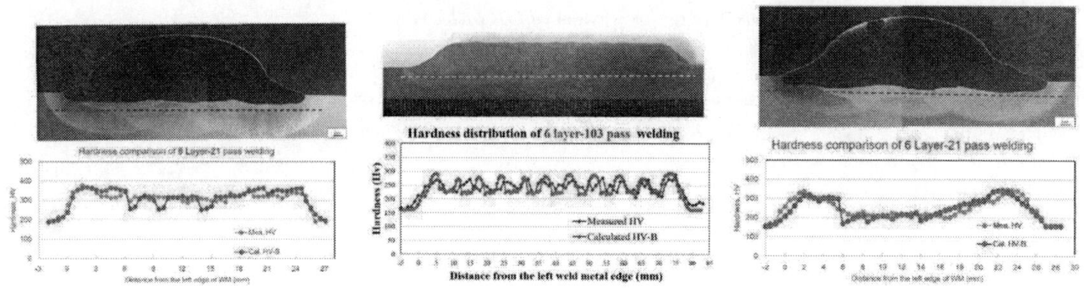

Fig. 5 Comparison of the predicted and the measured hardness after multi-pass welding of A508, 1.5%Ni and SM490 steels

Conclusions

Using the expanded hardness prediction system of A533B steel, the hardness distribution in HAZ of temper bead welding for A508, 1.5%Ni and SM490 steels were calculated based on the thermal cycles computed by finite element method. The experimental results revealed that the predicted hardness were in good accordance with the measured ones. It followed that the currently proposed expansion method was effective for estimating the tempering effect during temper bead welding for the low-alloy steels without secondary hardening. Through this method, the hardness could be successfully predicted before the actual temper bead welding, thus the appropriate welding conditions would be able to optimize before the actual welding, which was very useful for the repair welding for large structures.

References

[1] L. Yu, et al: "Neural Network Prediction of Hardness in HAZ of Temper Bead Welding using the Proposed Thermal Cycle Tempering Parameter (TCTP)", *ISIJ International*, Vol. 51, No.9 (2011), pp.1506-1515.

[2] L. Yu, et al: "Neural Network-Based Toughness Prediction in HAZ of Low-Alloy Steel Produced by Temper Bead Welding Repair Technology", *Science and Technology of Welding and Joining*, Vol.18, No.2 (2013), pp. 120-134.

[3] D. Casasent, et al: "Radial Basis Function Neural Networks for Nonlinear Fisher Discrimination and Neyman-Pearson Classification", *Neural Networks - 2003 Special issue: Advances in neural networks research - IJCNN'03*, Vol.16, No.5-6 (2003), pp. 529-535.

A study of deformation and microstructural evolution in friction stir welding of Al and Ti

Amlan Kar[1], Devinder Yadav[1], Satyam Suwas[2], Satish V. Kailas[1*]
[1]Department of Mechanical Engineering, Indian Institute of Science, Bangalore, India.
[2]Department of Materials Engineering, Indian Institute of Science, Bangalore, India.
** satvk@mecheng.iisc.ernet.in*

Abstract

Friction stir welding (FSW) is a novel technology used to join similar and dissimilar materials. Welding of Al to Ti is a challenge due to wide differences of their physical and thermal properties, especially their melting points. In the present experiment, friction stir welding of commercial purity Al and Ti is carried out. X-ray tomography result shows that Ti particles are inhomogeneously distributed in the weld and their distribution depends on particle's morphology. Differences in deformation behavior and microstructural evolution are investigated using electron back scatter diffraction (EBSD). Ti particles form by the development of adiabatic shear band (ASB) in Ti interface. Severe deformation of materials at high strain rates by the FSW tool leads to formation adiabatic shear band in Ti. Microstructure of Al in the nugget zone is dynamic recrystallized and a substantial grain refinement is observed. Microstructural evolution in Al is gradual than Ti due to the differences in their microstructural response. Hardness data of the weld indicates a composite property of weld nugget and it is found that the microstructure of the weld nugget is stable at higher temperature. Microstructural stability depends on the nature of mechanical mixing and particle distribution. A proper control of Ti particles distribution by optimizing the process parameter can give a good weld with failure occurring in the parent Al material. It is possible to use this method to produce composites with unique properties.

Introduction

Welding of Al to Ti is always a challenge due to wide differences in their physical and thermal properties, especially the melting points. Welding of Al to Ti, by fusion process leads to the formation of brittle intermetallic phases and cavities at the joint. Solid state welding methods, such as pressure welding [1], diffusion bonding and friction welding have been used to join these two dissimilar materials and control the formation of brittle phases. Van Loo et al., [2] reported that the compound layer of Al_3Ti was the primary intermetallic phases that formed in the Al/Ti diffusion bonding. Therefore, for dissimilar metals, solid state joining processes such as friction stir welding is a possible attractive option to restrict the formation of cavities and intermetallics, to improve the mechanical properties of the weld. During FSW, the tool (consisting of pin and shoulder) is made to rotate at high speed and a downward force is applied to it so that the pin goes into the plates and the shoulder just touches them. The tool is then traversed along the joint line at a particular traverse speed. It should be noted that the deformation behaviour of Al and Ti is different both at ambient and at elevated temperatures. Hence, defects can form due to improper material flow if the process parameters are not optimised or controlled. It was shown by Kumar et al., [3] that the material flow is not symmetric during FSW and hence the tool is usually given an offset towards the softer metal for better material mixing. In dissimilar metal welding, material flow and elemental distribution strongly influences the weld properties [3]. In order to understand the material mixing and overcome the shortcomings of dissimilar metal welding, a fundamental study on FSW of Al and Ti was undertaken in the present work and the nugget zone microstructure was characterised in detail.

Experimental procedure

Four millimetre thick plates of commercially pure Al (cp-Al) (0.002% Cu, 0.003% Mg, 0.017% Si, 0.012% Fe and balance Al in wt %) and commercially pure Ti (cp-Ti) (0.3% Fe, 0.080% C and balance Ti in wt %) sheet were taken as the starting materials. Cp-Ti and cp-Al were fixed on the advancing and retreating side of the tool, respectively. FSW was carried out along the joint on a custom-built 5-axis machine (BiSS-IISc, Bangalore, India) using a cylindrical shaped welding tool made of WC-8%Co alloy with shoulder diameter, pin diameter and pin length of 20 mm, 4 mm and 3.65 mm, respectively. A tool rotation speed of 800 rpm, traverse speed of 20 mm/min, plunge depth of 3.5 mm and a tool offset of 2 mm from the weld line on the Al side were used. The parameters were optimised after several trials. X-ray micro-computed tomography (X-ray micro-CT) was carried out in the nugget zone of the weld. For EBSD, metallographically polished samples were electropolished and observed in FEI SEM equipped with TSL-OIM data acquisition software. Vickers micro hardness was measured across the weld nugget using a load of 200 g and dwell time of 10 sec.

Results and discussion

FSW lead to fracture of the Ti metal and generation of particles with a wide size distribution. These particles were mixed in Al in the weld nugget zone. The detailed 3D distribution of Ti particle in the nugget of the weld is shown in Fig 1. The particles were recognized by selecting an interactive value belonging to Ti and the corresponding images were reconstructed. The particles (shown in blue colour) were separated based on their size and the volume range corresponding to each size is shown if Fig. 1(a) through (f). Very fine particles (volume less than 10^3 µm^3) could not be captured due to instrument limitation. A careful observation shows that the fine particles are uniformly distributed in the nugget ((a), (b) and (c)) unlike the coarse particles and flakes ((d), (e) and (f). This can be attributed to the lesser drag force the smaller particles experience during the material flow. The coarse Ti particles and flakes appear to align along certain path. Their alignment and arrangement represent the trajectory of the material flow due to the drag force that develops at the interface of the particle and the semisolid Al, as shown by red line in Fig. 1(d).

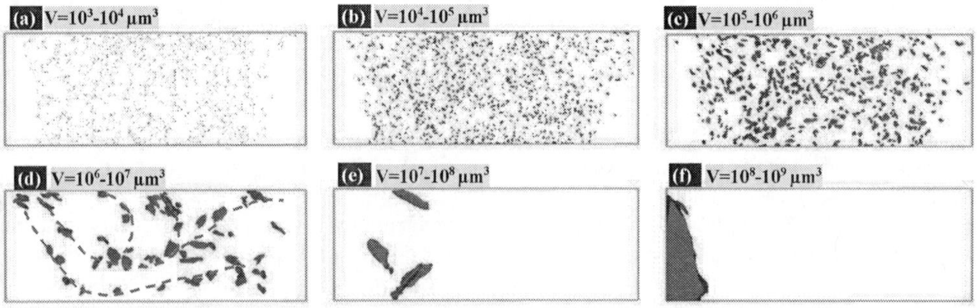

Figure 1(a) to (f): Filtered X-ray tomography images corresponding to different volume range (10^3 to 10^9 µm^3) of Ti particles Seperated distribution reveal homogeneous and heterogeneous distribution for finer and coarser particles, respectively depending on volume of particles.

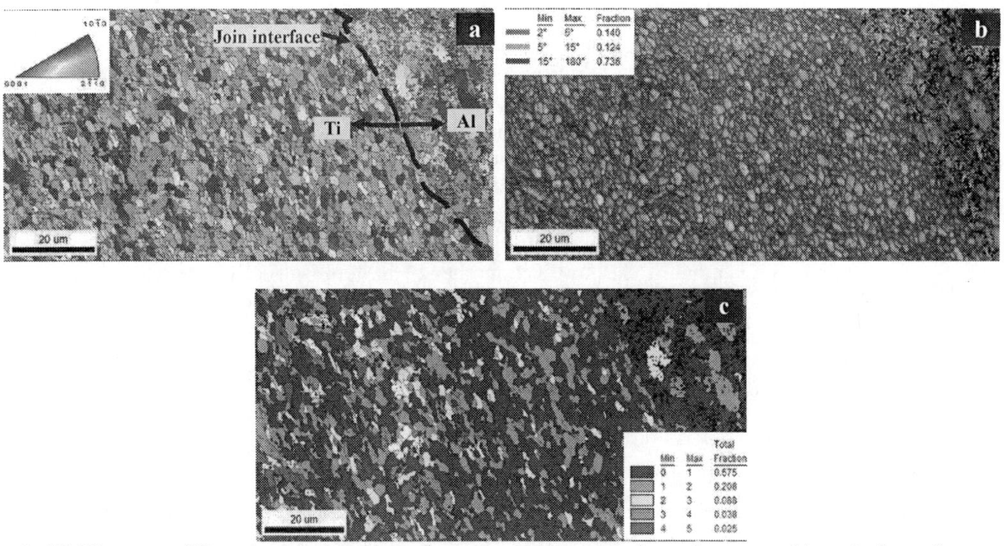

Figure 2: EBSD maps of Ti at the join interface showing a)IPF+grain boundary map, b) grain boundary map and (c) grain orientation spread map.

To understand the possible mechanism of Ti particle formation, EBSD scans at the interface on the Ti side were carried out. The inverse pole figure (IPF) map (Fig 2(a)) shows that Ti underwent significant grain refinement during FSW. The average grain size was 5 µm compared to 40 µm of the base Ti. Further, the microstructure was characterized by a high fraction (> 70%) of high angle grain boundaries (> 15° misorientation) (Fig. 2(b)) and the grain orientation spread (GOS) map in Fig. 2(c) shows that majority of the grains (78%) have an orientation spread of less than 2 degrees indicating them to be strain free. There were more strain free grains near the interface than away from it. In addition, no twins (which can occur in Ti up to 400°C) were observed in the microstructure. These fine grained structure and lack of twining points to occurrence of dynamic recrystallization (DRX) in Ti, which usually occurs at much higher temperatures than the temperatures experienced during the weld, which was around 400 °C [4]. The temperature of the weld is dictated by the melting point of Al. We believe that the combination of high strain rate and lower temperature during FSW causes adiabatic shear banding (ASB) in Ti [5]. These bands are strain hardened and fracture during the material flow thereby generating particles of different sizes. However, within ASB, the temperature can be significantly higher than that of the bulk temperature leading to significant grain refinement and recrystallization of Ti[5, 6].

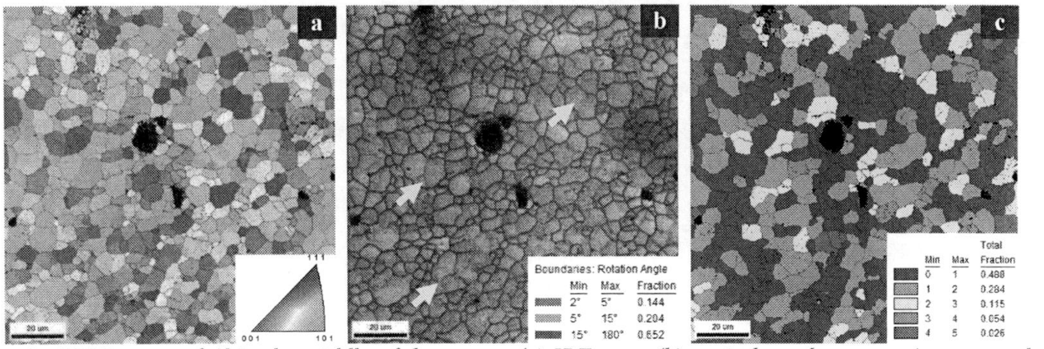

Figure 3: EBSD maps of Al in the middle of the nugget (a) IPF map, (b) grain boundary map. Arrows mark the boundaries with mixed character and (c) GOS map.

The nugget zone contains more Al due to the tool offset. EBSD scans were also carried out on the Al side of the weld (in the middle of the nugget). A confidence index cut off of 0.1 was used to remove the Ti particles (appear in black colour in the images). The IPF map in Fig. 3(a) shows equiaxed fine grains with average grain size being 7 µm compared to 40 µm of the base Al. More than 65% of the grain boundaries were high angle (Fig. 3(b)). When carefully looking at the grain boundary map, it was found that some of the boundaries exhibited mixed character, i.e., part of the grain boundary have high angle character (>15°) and part have low angle character (5-15°) (indicated by arrows). This points to gradual transformation of grain boundaries from low angle to high angle which is a characteristic feature of continuous dynamic recrystallization (CDRX), where there is no nucleation step [7]. In Al dynamic recovery readily occurs during high temperature deformation due to its high stacking fault energy (~ 188 mJ/mm^2) which leads to arrangement of dislocations into subgrain boundaries during high temperature deformation. More than 76% of the grains had GOS of less than 2 degree.

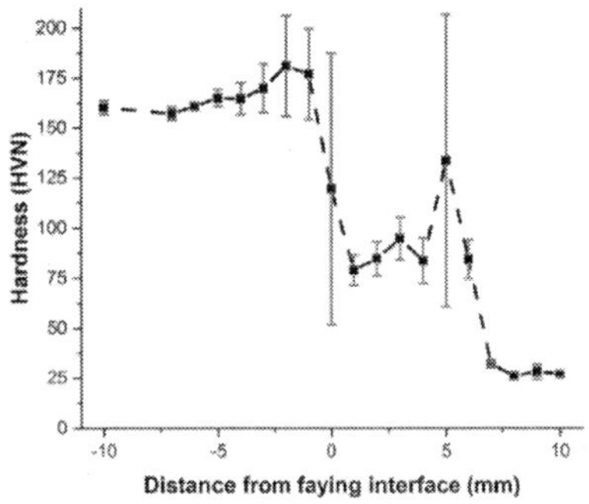

Figure 4: Hardness variation across the weld corresponding to different tool offsets. Variation in hardness at the weld nugget is partly due to indentation on the particle and variation in composition of particles.

The hardness across the weld cross-section is shown in Fig. 4. A rather large scatter of hardness is observed in the nugget zone, which is attributed to the presence of Ti particles and possible intermetallic phase. Hardness in the weld nugget is higher than that of base Al and therefore, it is expected that failure will possibly occur in the parent Al material instead of weld nugget during tensile testing.

Thermal stability of microstructure was studied by heating the sample to 500°C for 60 minutes. It is worth mentioning here that this temperature and time is higher than that occurring during welding. Abnormal grain growth was observed in the regions where the Ti particles were absent. The image quality map and the IPF map of the heat treated samples are shown in Fig. 5(a) and 5(b), respectively. The regions with higher density of Ti particles showed no sign of any grain growth indicating pinning of the grain boundaries by the fine Ti particles. This suggests that this process could be used to produce Al-Ti particulate composites with unique properties.

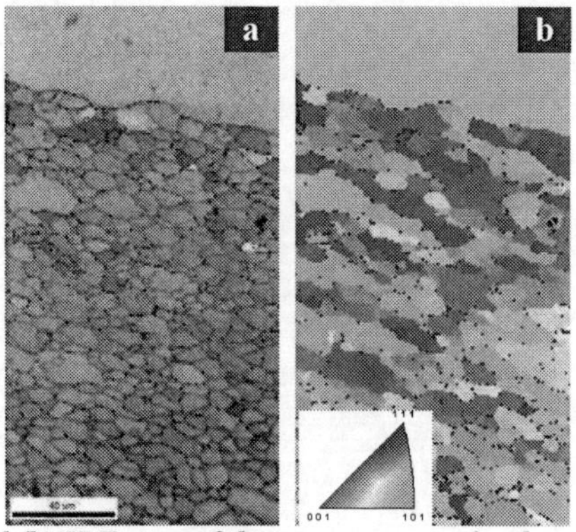

Figure 5: (a) image quality map and (b) IPF map of the nugget zone after heat treatment (500 °C for 60 min).

Conclusions

In the present work, friction stir welding of commercially pure Al to commercially pure Ti was successfully carried out. The microstructure in the nugget was characterized in detail. Following conclusions can be drawn from the present study.

1. Morphology of the Ti particles governs their distribution in the weld nugget. Fine particles were homogeneously distributed compared to bigger particles and flakes due to the drag force. The particle distribution shows the material flow pattern during the process.
2. Deformation during FSW depends on the material properties. Deformation and fragmentation of Ti occurs by the formation of adiabatic shear bands whereas Al underwent dynamic recrystallization. Fine grains were observed both in Ti and Al.
3. Distribution of fine Ti particles improved the hardness across the nugget zone and also improved the thermal stability of the microstructure against abnormal grain growth due to pinning of grain boundaries.
4. FSW can possibly be used to produce Al-Ti particulates composites with unique mechanical properties.

Acknowledgments

The authors thank Defence Research & Development Organization (DRDO), India for funding. Thanks are also due to Mr. Vishnu Vijayan for his assistance in FSW experiment.

References

1. V.D. Saprygin, Pressure welding of aluminium–steel and titanium–aluminium transition pieces for low-temperature service. Weld Prod, 1975. **22**: p. 29–3.
2. van Loo, F.J.J. and Rieck, G.D., Diffusion in the titanium-aluminium system—I. Interdiffusion between solid Al and Ti or Ti-Al alloys. Acta Metallurgica, 1973. **21**(1): p. 61-71.
3. Ouyang, J.H. and Kovacevic, R., Material flow and microstructure in the friction stir butt welds of the same and dissimilar aluminum alloys. Journal of Materials Engineering and Performance, 2002. **11**(1): p. 51-63.
4. Prasad, Y.V.R.K. and Seshacharyulu, T., Processing maps for hot working of titanium alloys. Materials Science and Engineering: A, 1998. **243**(1–2): p. 82-88.
5. Kailas, S.V., A study of the strain rate microstructural response and wear of metals. Journal of Materials Engineering and Performance, 2003. **12**(6): p. 629-637.
6. Yang, Y. and Wang, B.F., Dynamic recrystallization in adiabatic shear band in α-titanium. Materials Letters, 2006. **60**(17–18): p. 2198-2202.
7. Yadav, D. and Bauri, R., Microstructural characterisation of friction stir processed aluminium. Materials Science and Technology, 2011. **27**(7): p. 1163-1169.

Microstructure and Mechanical Properties of 1050 Aluminum Joints Formed by Underwater Friction Stir Welding

Masakatsu Maeda, Masaru Terai, Kazuyoshi Katoh

College of Industrial Technology, Nihon University, Japan

Appropriate temperature control during friction stir welding (FSW) guarantees the performance of the joints. Several methods to control the temperature during FSW has been demonstrated up to now. The most frequently used method is to control the heat input by adjusting the revolutionary pitch and by applying assist-heating source. However, there are few reports which attempted to control the heat evacuation. To control both incoming and outgoing heat flux must be essential for achieving appropriate temperature control. In the present study, work-hardened 1050 aluminum thin plates were welded by FSW to investigate the temperature history during the process, the resultant microstructure, and the mechanical properties of the welds.

Effect of cooling rate on microstructure of friction-stir welded aluminum alloy 1100

S. Mironov, D. Yi, Y.S. Sato, H. Kokawa
Department of Materials Processing, Graduate School of Engineering, Tohoku University, Sendai, Japan
smironov@material.tohoku.ac.jp, ydl11@mails.tsinghua.edu.cn, ytksato@material.tohoku.ac.jp, kokawa@tohoku.ac.jp

Abstract
In this study, the microstructural changes occurring during cooling of friction-stir welded aluminum alloy 1100 were evaluated. Below 0.6Tm (Tm being the melting point) the stir zone material was found to experience no significant changes during cooling. At higher welding temperatures, however, notable grain growth and material softening occurred.

Introduction
It is well known that friction-stir welding (FSW) typically leads to drastic microstructural changes [1]. It is worth noting that some portion of the changes may occur *after* welding during cooling to ambient temperature. The importance of this effect is obvious for materials exhibiting phase transformations. In single-phase materials, however, the cooling step may also play a role. To suppress the cooling effect and produce a fine-grained structure in the stir zone, the cooling rate is sometimes enhanced by the pouring of coolant mixtures [e.g. 2] or the application of submerged FSW [e.g. 3]. Despite the presumable importance of the cooling effect, it has not been studied systematically so far. The present work was undertaken in an attempt to fill this gap in our knowledge.

Material and Experimental Procedures
The commercial purity aluminum alloy (AA) 1100 was used as a program material in this study. To provide a diversity of temperatures and cooling rates, FSW was performed at tool rotational speeds ranging from 600 to 3000 rpm and tool travel speeds ranging from 300 to 600 mm/min. The welding temperatures and cooling rates were measured by K-type thermocouples placed beneath the welded sheets. Microstructure observations were performed by electron-backscatter diffraction (EBSD) and focused in the central part of the stir zone. A 15° criterion was employed to differentiate low-angle boundaries (LABs) versus high-angle boundaries (HABs).

Results and Discussion
Relationship between welding variables and cooling rate
The effect of tool rotational speed and tool travel speed on the cooling rate is shown in Fig. 1a. It is clear that the cooling rate was primarily dependent on the tool travel speed, whereas the influence of the tool rotational speed was relatively small. As expected, the cooling rate typically decreased with decreasing tool travel speed.

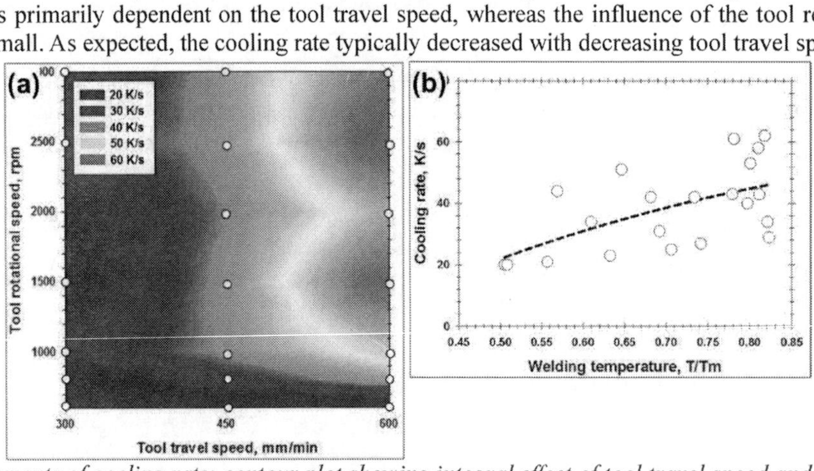

***Fig. 1**. Measurements of cooling rate: contour plot showing integral effect of tool travel speed and tool rotational speed (a), and relationship between the welding temperature and the cooling rate (b). In (a), the open circles indicate the welding conditions at which measurements were actually done.*

At the tool rotation speeds below 800 rpm, however, this relationship was found to break down. This interesting effect may be explained as follows. Since the cooling process is driven by a temperature difference, the cooling rate should naturally decrease with the welding temperature, as shown in Fig. 1b. At very low welding temperatures, therefore, the cooling rate should be relatively small, irrespective of the tool travel speed.

Microstructure morphology and grain size

The effect of the welding temperature and cooling rate on microstructure morphology and mean grain size is presented in Figs. 2 & 3. The microstructures produced at relatively low welding temperatures were fine-grained and completely recrystallized (Fig. 2a). As expected, the increase in welding temperature led to grain coarsening (Fig. 3). Moreover, it also gave rise to the appearance of partially recrystallized microstructures consisting of an irregular mixture of LABs and HABs (Figs. 2b & c). As suggested in the previous work [4], the observed incompleteness of the recrystallization process may originate from rapid annihilation of dislocations at high temperatures which inhibits the evolution of deformation-induced boundaries. This effect became particularly pronounced at ~0.7Tm (Fig. 2b), providing an abrupt increase in mean HAB spacing (Fig. 3b). As a result, the microstructures produced at higher welding temperatures were relatively coarse-grained (Fig. 3) and contained a high LAB fraction (Fig. 2c).

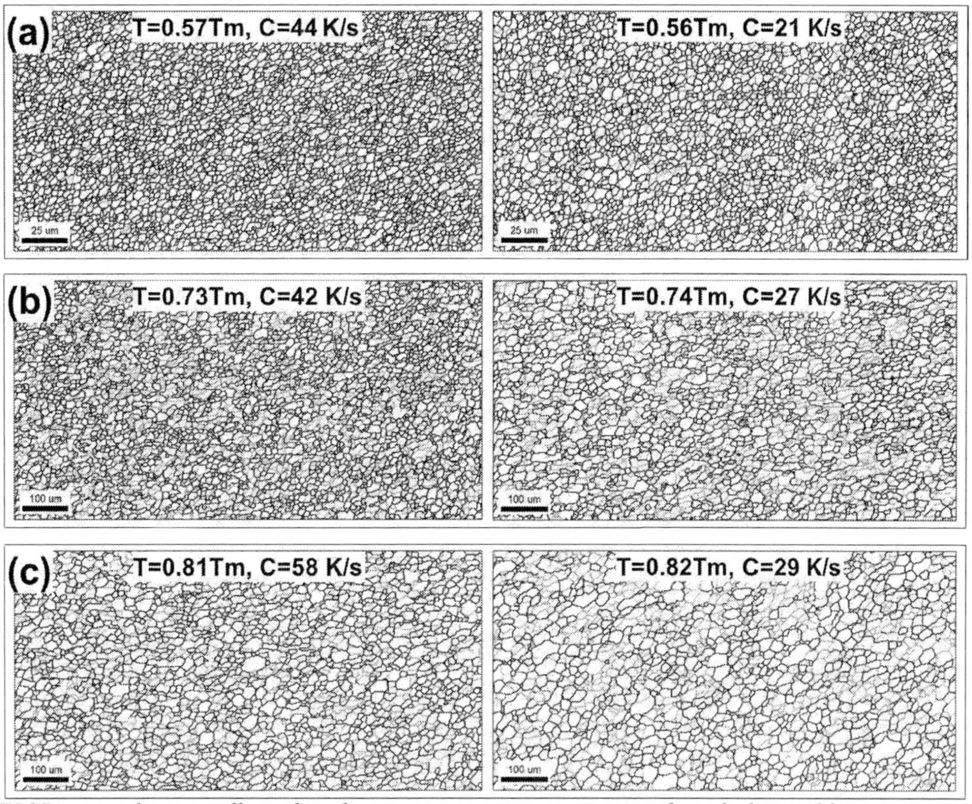

Fig. 2. EBSD maps showing effect of cooling rate on microstructure at relatively low welding temperature (a), moderate welding temperature (b) and high welding temperature (c). In the maps, LABs and HABs are depicted as red and black lines, respectively. Note: T and C are welding temperature and cooling rate, respectively.

The influence of the cooling rate on the microstructure was even more complex. At welding temperatures below ~0.65Tm, the cooling rate had no significant effect on the microstructure morphology (Fig. 2a) and mean grain size (Fig. 3a). This perhaps evidences that the material experienced no substantial changes during the weld cooling cycle. Considering the relatively short exposure of the material above ~0.5Tm at these welding conditions, this observation seems to be reasonable.

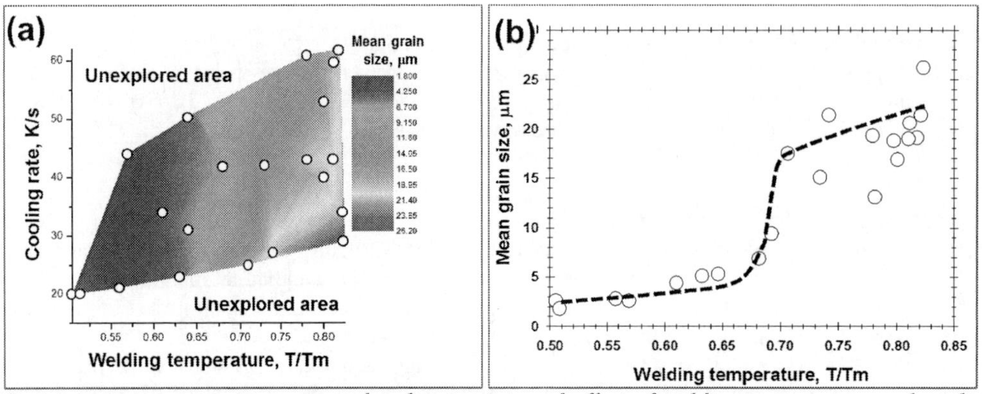

Fig. 3. Grain size measurements: contour plot showing integral effect of welding temperature and cooling rate (a), and scatter plot illustrating the individual influence of the welding temperature (b). In (a), open circles show experimental conditions at which measurements were actually done.

In the range of 0.7~0.8Tm, in contrast, the reduction of the cooling rate was found to lead to significant microstructure coarsening (Figs. 2b & 3a). The maximal effect was predicted at ~0.78Tm, where the grain size difference was nearly twofold (Fig. 3a). The high sensitivity of the microstructure to the cooling conditions presumably indicates the occurrence of substantial grain growth during the cooling stage. Above ~0.8Tm, however, the cooling effect was fond to decrease (Figs. 2c & 3a). This surprising result is thought to be attributable to the relatively coarse-grained microstructures produced in this temperature range (>15 µm in size) and the respective reduction of driving force for further grain growth.

Misorientation distribution

Attempting to provide additional insight into the microstructural changes occurring during cooling, misorientation-angle distributions were derived from EBSD maps and analyzed in present section. The influence of the welding temperature and cooling rate on the HAB fraction is summarized in Fig. 4. The effect of the welding temperature was relatively simple: the HAB percentage gradually decreased from ~85% to ~60% in the range of 0.50-0.75Tm, and then tended to become saturated (or even slightly increased) at higher temperatures (Fig. 4b).

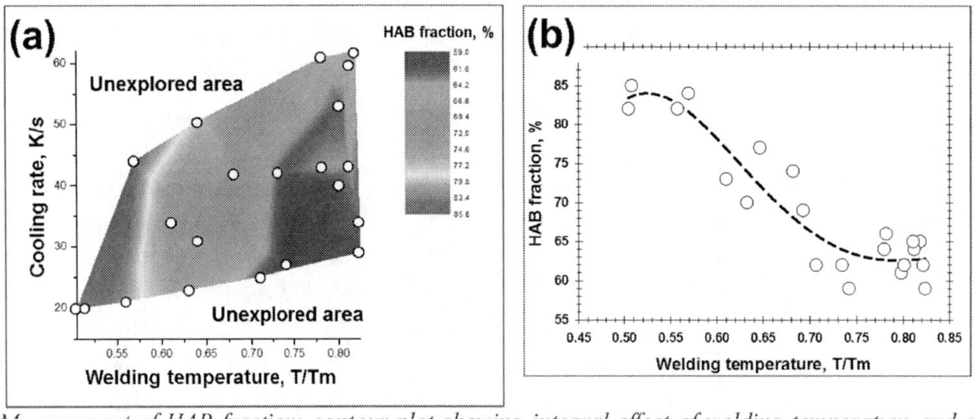

Fig. 4. Measurement of HAB fraction: contour plot showing integral effect of welding temperature and cooling rate (a), scatter plot illustrating the individual influence of the welding temperature (b). In (a), open circles show experimental conditions at which measurements were actually done.

The relationship between the HAB proportion and the cooling rate was more complex. Below ~0.55Tm, the HAB fraction was nearly independent from the cooling rate (Fig. 4a). This probably means that the HAB percentage was not measurably altered during the cooling step, a result which agrees well with microstructure- and texture stability, as discussed in the previous section. In the range of ~0.65-0.75Tm, the HAB portion somewhat decreased with decreasing cooling rate (Fig. 4a). This observation may indicate the reduction of the HAB fraction during the FSW cooling cycle. This effect correlates well with the grain coarsening during cooling under these welding conditions (Fig. 3a). Above ~0.8Tm, the HAB fraction weakly depended on the cooling rate (Fig. 4a), thus probably indicating only small changes of the HAB fraction during FSW cooling cycle.

Microhardness

The influence of the welding temperature and cooling rate on the mean microhardness in the stir zone is shown in Figs. 5a. Below ~0.65Tm, the cooling rate had no significant effect on the microhardness (Fig. 5a). This agrees well with microstructure stability during cooling at low FSW temperatures, as discussed in the previous sections. At higher welding temperatures, however, the microhardness was found to decrease with the cooling rate, but this effect did not exceed ~10% (Fig. 5a). This perhaps indicates the occurrence of subtle material softening during the FSW cooling cycle and is in agreement with the grain coarsening discussed above.

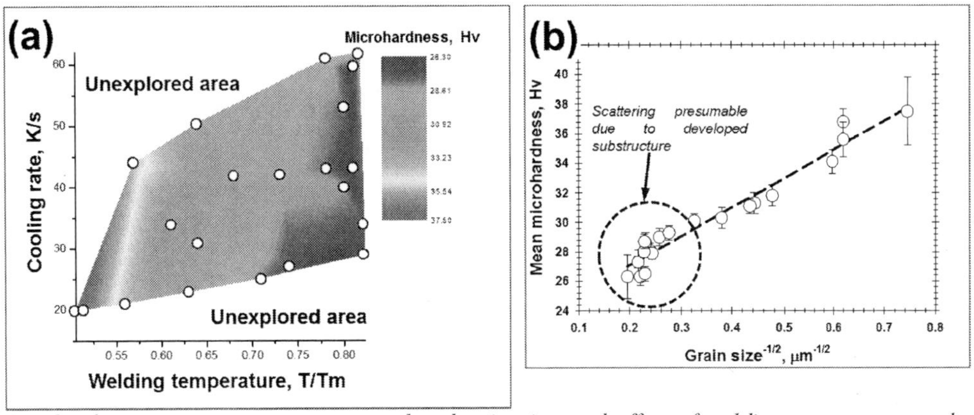

Fig 5. Microhardness measurements: contour plot showing integral effect of welding temperature and cooling rate (a), scatter plot illustrating the Hall-Petch relationship (b). In (a), open circles show experimental conditions at which measurements were actually done. In (b), error bars show standard deviation.

To examine the possible relationship between the stir zone microstructure and microhardness, the experimental data are arranged as a Hall-Petch plot in Fig. 5b. The relatively good fit between the mean grain size and microhardness indicates that the strength of the stir zone material was primarily governed by the grain-boundary hardening. On the other hand, the notable scattering in the coarse-grained range (selected area in Fig. 5b) was presumably due to the development of an LAB substructure in this material condition, as shown in Figs. 2b & c.

Conclusions

In this work, microstructural changes occurring during the FSW cooling cycle of aluminum alloy 1100 were evaluated. At welding temperatures below ~0.6Tm, the stir zone material experienced no significant changes during the FSW cooling stage. This microstructure stability was attributed to the relatively short duration of the thermal cycle at these welding conditions. At higher FSW temperatures, however, the substantial microstructure coarsening was deduced to occur during the FSW cooling cycle. The cooling effect was predicted to be most pronounced in the range of the welding temperatures of 0.7-0.8Tm. Above 0.8Tm, the cooling effect was found to decrease due to the formation of a relatively coarse-grained structure in the stir zone and the respective reduction of driving force for the subsequent grain growth.

References

[1] R.S. Mishra, Z.Y. Ma: "Friction stir welding and processing", *Mater. Sci. Eng. R*, Vol. 50 (2005), pp. 1–78.
[2] N. Xu, R. Ueji, H. Fujii: "Enhanced mechanical properties of 70/30 brass joint by rapid cooling friction stir welding", *Mater. Sci. Eng. A*, Vol. 610 (2014), pp. 132–138
[3] D.C. Hofmann, K.S. Vecchio: "Submerged friction stir processing (SFSP): An improved method for creating ultra-fine-grained bulk materials", *Mater. Sci. Eng. A*, Vol. 402 (2005), pp. 234–241.
[4] S. Mironov, K. Inagaki, Y.S. Sato, H. Kokawa: "Effect of welding temperature on microstructure of friction-stir welded aluminum alloy 1050", *Metal. Mater. Trans. A*, Vol. 46 (2015), pp. 783-790.

Microstructure evolution and tensile fracture behavior of friction stir welded Al-Li alloy

Dingrui Ni, Yao Tao, Bolv Xiao, Zongyi Ma

Institute of Metal Research, Chinese Academy of Sciences, China

As-rolled 2060-T8 sheets were selected and subjected to friction stir welding (FSW), with the aim to study the effect of welding parameters on the microstructure evolution and mechanical properties of the FSW joint. The main results are as following. The FSW joint showed good tensile properties, and joint efficiency was obtained as high as 85%. With the increase of the traverse speed, the hardness of the lowest hardness zone (LHZ) of the joint increased, and the UTS and elongation (El) increased, too. With the increase of the rotation rate, the UTS and El increased due to the increased hardness of the nugget zone (NZ) and LHZ. The latter is inconsistent with the results for normal precipitation strengthened Al alloys, in which the LHZ is hardly changed with the rotation rate, and the reason should be attributed to dissolution of T phase and subsequent natural aging process. The fracture location could occur at NZ, heat affected zone (HAZ), and thermo-mechanical affected zone (TMAZ) respectively, according to different parameters; however, for the joint under a higher rotation rate and welding speed, a strange phenomenon occurred: it failed at TMAZ which is not correspond to the LHZ. The reason is that the fracture is determined by the joint action of the hardness, texture, and precipitates.

High Speed, Solid-State Welding of Dissimilar Alloy Aluminum Tailor-Welded Blanks

Yuri Hovanski, Piyush Upadhyay, and Ayoub Soulami
Pacific Northwest National Laboratory, Richland, WA, USA
Yuri.Hovanski@pnnl.gov, 509-375-3940
Piyush.Upadhyay@pnnl.gov, 509-375-6591

John Carsley and Blair Carlson
General Motors, Warren, MI, USA
John.Carsley@gm.com, 586-986-2928
Blair.Carlson@gm.com, 586-864-7698

Mark Eisenmenger, Tom Luzanski, and Dustin Marshall
TWB Company, Monroe, MI, USA
Mark.Eisenmenger@twbcompany.com, 734-289-6471
Tom.Luzanski@twbcompany.com, 734-289-6468
Dustin.Marshall@twbcompany.com, 734-289-6537

Brandon Landino and Glenn Jarvis
Alcoa, Farmington, MI, USA
Brandon.Landino@alcoa.com, 248-489-4379
Glenn.Jarvis@alcoa.com, 212-836-2600

Abstract
While several welding methods are currently available for the production of aluminum tailor-welded blanks, each with their own advantages and challenges, only friction stir welding enables both linear and curvilinear joint configurations in similar and dissimilar thickness and alloy combinations. An evaluation of high speed friction stir welding was performed using dissimilar alloy combinations with dissimilar thickness sections. Various combinations of precipitation strengthened and work hardenable alloys were welded together to evaluate the feasibility of joining alloys with different high temperature yield strengths, strengthening mechanisms, and tempers. The effect of welding speed on the precipitation strengthened alloys is especially of interest, as high welding speeds demonstrated shifts in softening mechanisms. Evaluation of combinations of AA7075, AA6022, and AA5182 in similar and dissimilar thickness combinations are presented for both linear and curvilinear applications.

Introduction

The production of welded blanks for high volume automotive stampings is a core technology enabling the efficient use of materials throughout the body-in-white [1, 2]. This approach of welding dissimilar thicknesses and alloys into a common sheet or coil aligns manufacturers with the type of lean engineering tools necessary to manage costs while simultaneously lowering weight and reducing material waste (Figure 1). Recently, high speed friction stir welding (FSW) has been demonstrated as an effective technology to enable high volume welding of aluminum tailor-welded blanks [3-5]. Eliminating the more stringent cleaning requirements for laser welding of aluminum blanks, FSW provides a more robust solution that enables linear and curvilinear joining solutions (Figure 2).

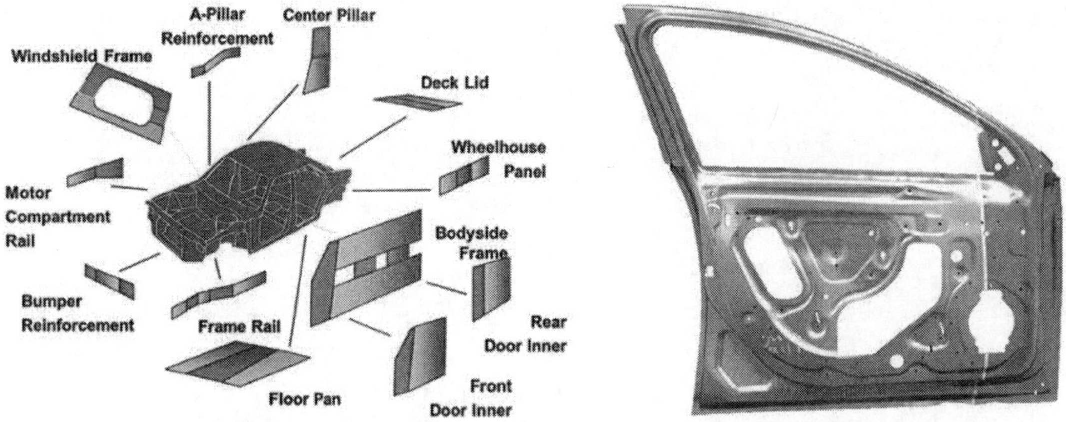

Figure 1. Illustration of possible welded blank solutions on the automotive body-in-white (left); door inner panel produced from dissimilar thickness aluminum tailor-welded blanks (right).

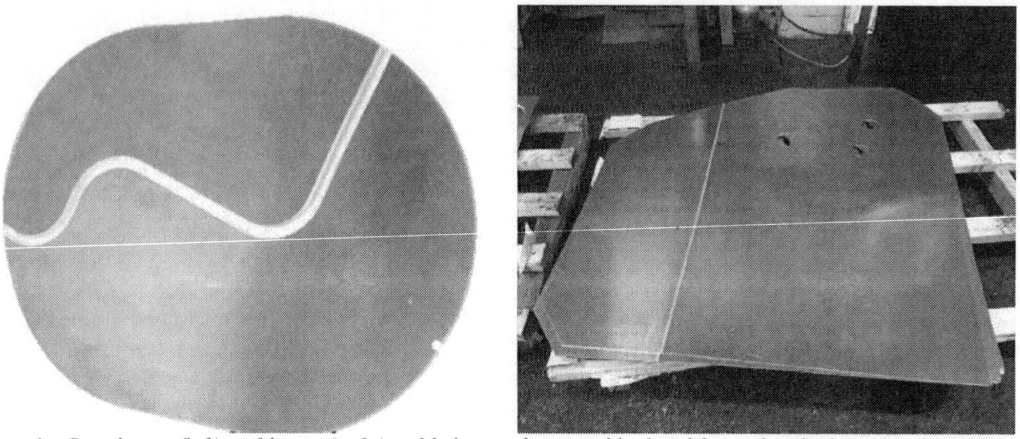

Figure 2. Curvilinear (left) and linear (right) welded joints between blanks of dissimilar thicknesses of AA5182-O.

Body of paper

With a successful demonstration completed of both linear and curvilinear high speed FSW of aluminum blanks in dissimilar thickness sections, a detailed investigation into expanding the technology for joining dissimilar alloy combinations at high volume production velocities was undertaken. While preliminary investigations demonstrated the feasibility of joining dissimilar aluminum alloys via FSW [6, 7], a systematic investigation of high speed FSW of dissimilar alloys has not been reported.

With the goal of joining various precipitation strengthened alloys such as AA6011, AA6022, and AA7075 to a work hardenable alloy such as AA5182, a comparison between the high temperature yield stresses of each alloy was prepared. Figure 3 shows a comparison between the flow stresses of AA6011 and AA5182 across the temperature range of 350°C–450°C. Weld temperatures were measured for various alloys across a range of linear velocity from 1 m/min to 6 m/min, in order to determine a welding window that provided a comparable process window in which the flow stresses of the dissimilar alloys were not divergent.

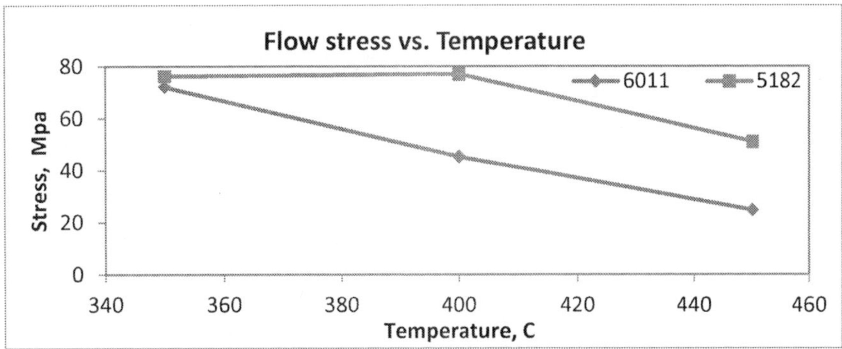

Figure 3. Chart showing the diverging behavior of the flow stresses of AA6011 and AA5182 across a 100°C temperature range.

Weld temperatures, measured in the tool via thermocouple, are plotted in Figure 4 to show the effect of welding velocity on temperature. Welding speeds near 4 m/min showed peak temperatures dropped to nearly 350°C, which, based on data similar to that shown for AA6011 and AA5182 in Figure 3, may be more amenable to the simultaneous local extrusion present in FSW. Combining thermal telemetry of the weld as a function of welding speed with elevated-temperature material properties data allows weld process windows to be identified where flow stress deviations will not encumber the welding process.

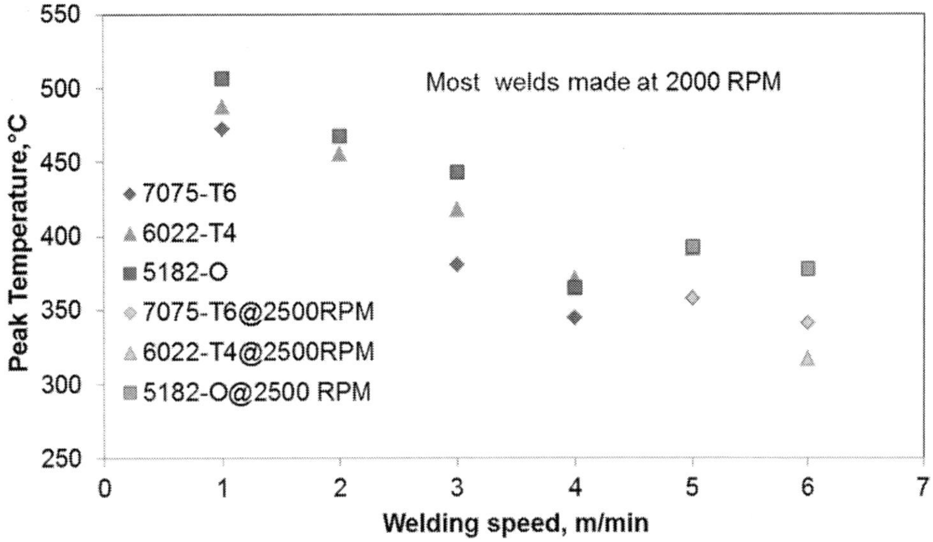

Figure 4. Peak temperatures of welds in various aluminum alloys across a range of welding speeds from 1 m/min to 6 m/min.

Using information like that shown in Figures 3 and 4 enabled deterministic process development that led to successful welding of 2 mm AA7075-T6 to 1.1 mm AA5182-O, as shown in Figure 5.

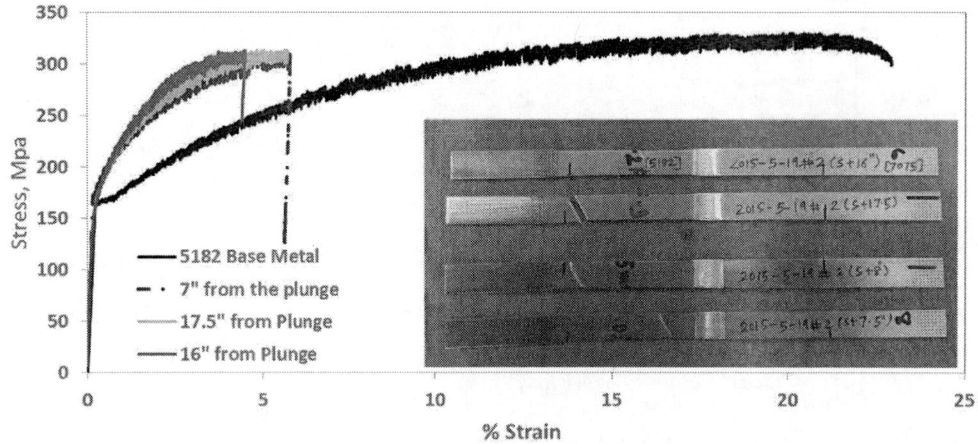

Figure 5. Tensile results from welds of dissimilar blanks: 2.0 mm AA7075-T6 joined to 1.1 mm AA5182-O.

Conclusions

Combining elevated-temperature materials properties with thermal telemetry obtained from high speed FSW of various aluminum alloys provided information that allowed for deterministic selection of weld process windows. An example of AA7075-T6 joined to AA5182-O is shown, which was welded at 2.6 m/min.

Acknowledgments

Pacific Northwest National Laboratory is operated by Battelle Memorial Institute for the U.S. Department of Energy under contract DE-AC05-76RL01830. This work was sponsored by Dr. Will Joost and Sarah Ollila in association with the U.S. Department of Energy Office of Vehicle Technologies as part of the Lightweight Materials Program.

References

[1] G.W.E.S.M. Specification, GMW16772, 2012.

[2] H. Mohrbacher, in: J.R. Duflou, Geiger, M., Kals, H.J., Shirvani, B., Singh, U.P. (Ed.) 9th International Conference on Sheet Metal, Leuven, Belgium, 2001, pp. 305-312.

[3] Y. Hovanski, P. Upadhyay, J. Carsley, T. Luzanski, B. Carlson, M. Eisenmenger, A. Soulami, D. Marshall, B. Landino, S. Hartfield-Wunsch, JOM, 67 (2015) 1045-1053.

[4] Y. Hovanski, P. Upadhyay, S. Pilli, B. Carlson, J. Carsley, S. Hartfield-Wunsch, M. Eisenmenger, Light Metals 2014, (2014) 265-270.

[5] Y. Hovanski, Carsley, J., Carlson, B., Hartfield-Wunsch, S., SAE Int. J. Mater. Manf., 7 (2014) 537-544.

[6] G. Grant, R. Davies, E. Stephens, S. Wazny, L. Kaunitz, D. Waldron, 2005 SAE World Congress, April 11, 2005 - April 14, 2005, SAE International, Detroit, MI, United States, 2005.

[7] A. Von Strombeck, Sheikhi, S., Dos Santos, J.F., 4th International Friction Stir Welding Symposium, TWI, Park City, UT, USA, 2003.

Development of numerical simulation model of weld pool phenomena in flux cored arc welding

Yosuke OGINO, Yoshinori HIRATA, Satoru ASAI
Graduate school of engineering, Osaka University, Suita, Osaka, Japan
ogino@mapse.eng.osaka-u.ac.jp

Abstract
Arc welding process is indispensable technology in various fields of industry. In order to develop a high quality and high productivity welding process, it is important to control the welding process. There are many reports about controlling the welding process from view points of the welding power source, the welding materials and the shielding gas. In flux cored arc welding (FCAW), a flux cored wire (FCW) is used. The FCW is composed of a metal sheath and a central flux core. There are various types of the flux such as a deoxidizing agent, a stabilizer of the arc plasma, a slag-forming agent and so on. When the slag-based FCW in which includes a lot of the slag-forming agent is used, a large amount of slag generates and the molten slag and the molten metal coexist in the weld pool. Since the slag covers the weld metal surface, the appearance of weld metal after the welding process is beautiful and the FCW can be applied to all-position welding. In this study, a numerical simulation model of weld pool phenomena in FCAW is constructed. The model includes fluid flow in the weld pool and surface deformation of the weld pool. Influence of the material properties such as the density and the surface tension of the slag are numerically investigated by using the model. In addition, the numerical model is applied to the vertical upward welding, and the numerical result shows that the slag-based FCW can prevent the molten metal from dripping off from the weld pool.

Introduction
Arc welding process is an indispensable technology in various fields of industry. High-productivity and high-quality welding process has been strongly demanded for many years. In order to develop demanded welding processes, it is very important to predict and control the process. There are many reports about controlling the process from view points of the welding power source, the welding material and the welding shield gas.
In flux cored arc welding (FCAW), a flux cored wire (FCW) is used, and a large variety of FCW are developed. The FCW is composed of a metal sheath and a central flux core. There are various types of flux such as a deoxidizing agent, a stabilizer of the arc plasma, a slag-forming agent and so on. For example, a slag-based FCW in which includes much of slag forming agent can be applied to all-position welding. Fig. 1 shows the bead shape of vertical upward welding using a solid wire (SW) and a slag-based FCW. When the SW is used, molten metal drops off the molten pool. On the other hand, in case of the slag-based FCW, molten metal does not drop off and a sound bead can be formed. In this case, a large amount of slag generates during the welding process, and molten metal and molten slag coexist in the molten pool. At the rear of the molten pool, the slag covers the molten pool surface.

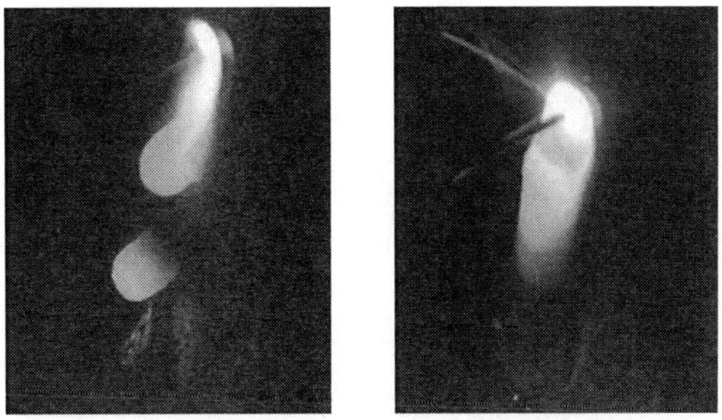

(a) SW *(b) Slag-based FCW*

Fig. 1 Weld bead shape of vertical upward welding

It is also important to understand the welding process in order to control the process. However, the welding process is not sufficiently understood since the phenomena are very complicated. In order to visualize and understand the welding phenomena, numerical simulation is one of the effective means. A lot of numerical models about the arc welding process are reported, and they are helpful tools for understanding the arc welding phenomena [1-3].
In this study, a numerical model of weld pool phenomena including both of the molten metal and molten slag is constructed. We apply this model to vertical upward welding, and numerically investigate about difference of the phenomena when the SW and the slag-based FCW are used.

Simulation model

In this study, a 3D weld pool model including convection in the molten pool and the surface deformation is constructed. The molten metal and the molten slag are treated as non-mixed fluid, and diffusion of the slag material is ignored. The flow field is calculated by follow equations:

$$\nabla \cdot \vec{v} = 0 \tag{1}$$

$$\frac{\partial \vec{v}}{\partial t} + \nabla \cdot (\vec{v}\vec{v}) = -\frac{1}{\rho}\nabla P + \frac{1}{\rho}\nabla \cdot \boldsymbol{\tau} + \vec{g} + \vec{F}_{ex} \tag{2}$$

where $\vec{v}$ is the velocity [m/s], t is the time [s], ρ is the density [kg/m^3], P is the pressure [Pa], $\boldsymbol{\tau}$ is the viscous stress tensor [Pa], $\vec{g}$ is the gravitational acceleration [m/s^2], $\vec{F}_{ex}$ is the driving forces [N/m^3]. In this model, the surface tension, the arc pressure and the Marangoni force are considered as the driving force. The temperature distribution is calculated by follow equation:

$$\frac{\partial H}{\partial t} + \nabla \cdot \vec{v}H = -\frac{1}{\rho}\nabla \cdot (-\kappa \nabla T) + \frac{1}{\rho}W \tag{3}$$

where H is the enthalpy [J/kg], κ is the thermal conductivity [W/mK], T is the temperature [K], W is the heat input from the heat source [W/m^3]. VOF method is used to track the free-surface of the weld pool [4]. In VOF method, the shape of fluid is described by the fraction of volume occupation in each calculation cell. In this model, both of the fraction of volume occupation of the molten metal and the molten slag are calculated, as shown in Fig. 2.

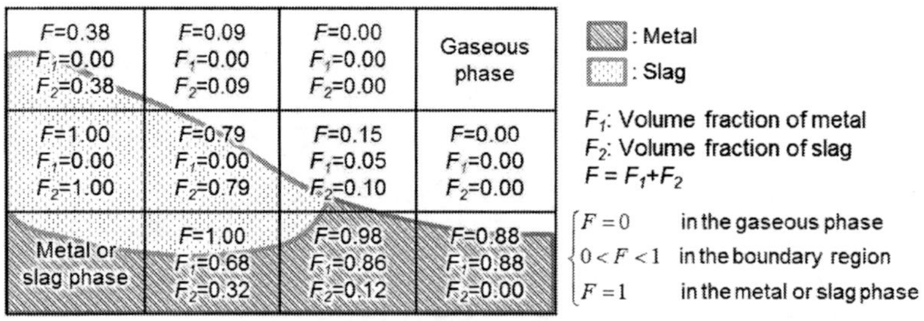

Fig. 2 VOF method

In this model, the heat input and the pressure from the arc plasma are approximated by Gaussian distribution as shown in Fig. 3. The droplet is approximated by sphere shape. The shape of Gaussian distribution and the temperature and velocity of the droplet are input parameters of this model. The slag is provided from the droplet into the molten pool.

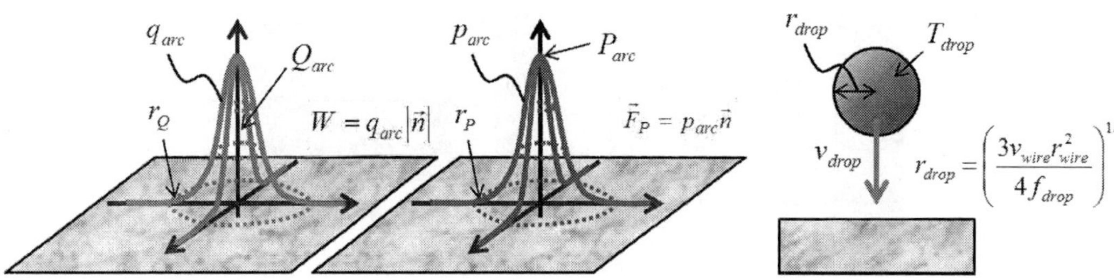

(a) Heat input and pressure from arc plasma (Gaussian distribution) *(b) Droplet*

Fig. 3 Heat source model used in this model

The material properties used in this study are shown in Table 1. The metal is mild steel. The slag is lower density, higher melting point, and lower surface tension compared with the metal. The other properties are same as the metal. The density and melting point of the slag are set in reference to the value of TiO2. In this model, the temperature dependence of the surface tension is ignored, but the values of surface tension are different between the metal and the slag, the Marangoni force is generated by the concentration gradient of the slag. In addition, the interfacial tension between the metal and the slag is also considered, and the value is 1.2 N/m.

Table 1 Material properties used in the model

	SW	FCW	
		Metal	Slag
Density [kg/m3]	7200		4000
Viscosity [Pas]	6×10^{-3}		6×10^{-3}
Thermal conductivity [W/mK]	26~52		26~52
Specific heat [J/kgK]	695~800		695~800
Surface tension [N/m]	1.5		1.2
Solidus temperature [K]	1750		1850
Liquidus temperature [K]	1800		1900
Latent heat of fusion [J/kg]	2.47×10^{5}		2.47×10^{5}

Simulation results and discussion

Next, simulation results are shown. Table 2 shows the heat source parameters used in the simulation.

Table 2 Heat source parameters used in the simulation

Wire feeding rate [cm/min]	800
Welding speed [cm/min]	30
Heat input [W]	3500
Heat input radius [mm]	7.0
Peak of pressure [Pa]	1000
Radius of pressure [mm]	2.0
Frequency of metal transfer [Hz]	40
Temperature of droplet [K]	2700
Velocity of droplet [m/s]	0.2

When the slag-based FCW is used, 50% of volume of the droplet is the slag. Fig. 4 shows the simulation results of

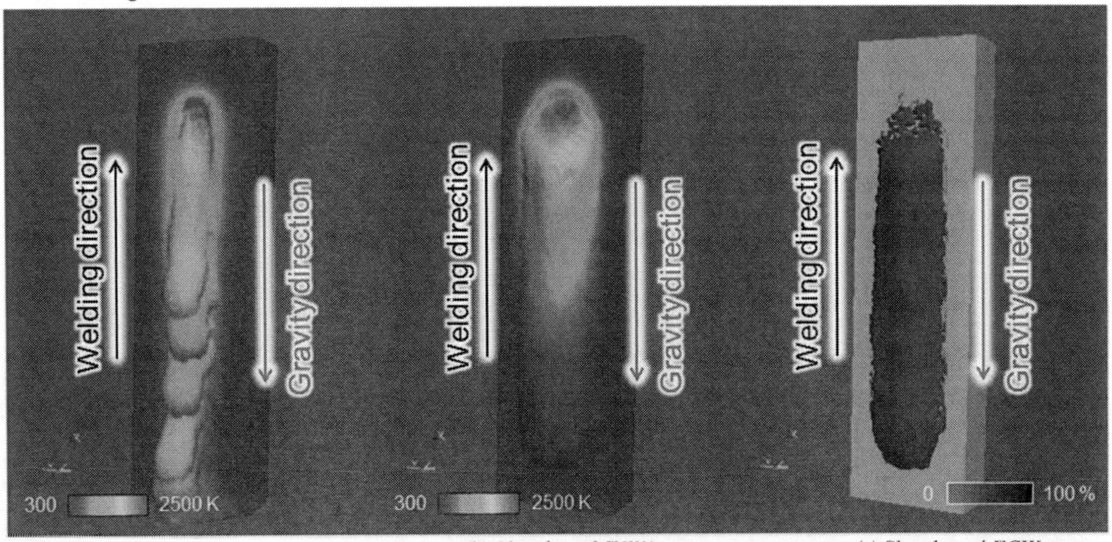

(a) SW (temperature distribution) *(b) Slag-based FCW (temperature distribution)* *(c) Slag-based FCW (slag concentration distribution)*

Fig. 4 Simulation results of vertical upward welding

vertical upward welding. When the SW is used, the molten metal drops off the molten pool as shown in Fig. 4(a). On the other hand, as shown in Fig. 4(b), in case of the slag-based FCW, the molten metal does not drop off the molten pool and a sound bead is obtained. The slag covers the molten pool and the weld bead as shown in Fig. 4(c). Fig. 5 shows the longitudinal cross-section of the weld pool. In this simulation, the melting point of the slag is higher than that of the metal, and the slag solidified before the metal solidified at the rear of the molten pool. The solidified slag holds the molten metal and prevents from dropping off the molten pool.

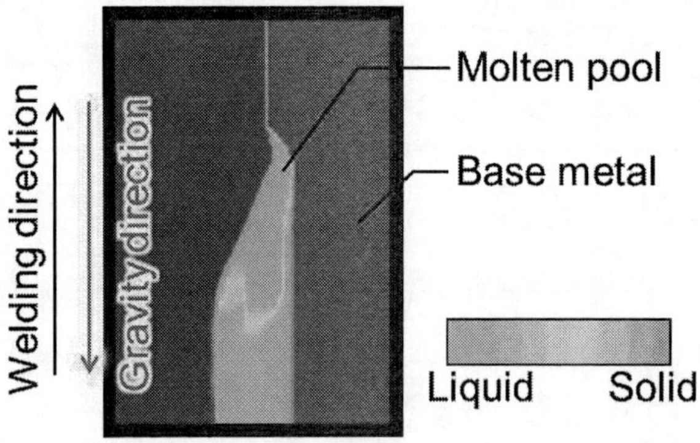

Fig. 5 Longitudinal cross-section of weld pool in case of using the slag-based FCW

Conclusions

In this study, a numerical simulation model for FCAW in which the molten metal and the molten slag coexist was constructed, and the model was applied to the vertical upward welding. When the slag-based FCW is used, the slag holds the molten metal and prevents the molten metal from dropping off the molten pool. In the future, if precise properties of the molten metal and the molten slag are obtained, the phenomena of FCAW will be understood more deeply.

References

[1] M. Schnick, U. Fuessel, M. Hertel, M.Haessler, A. Spille-Kohoff and A. B. Murphy: "Modelling of gas-metal arc welding taking into account metal vapour", Journal of Physics D: Applied Physics, Vol. 43 (2010), pp.434008

[2] M. Hertel, A. Spille-Kohoff, U. Fuessel and M. Schnick: "Numerical simulation of droplet detachment in pulsed gas-metal arc welding including the influence of metal vapour", Journal of Physics D: Applied Physics, Vol. 46 (2013), pp. 224003

[3] Y. Ogino, Y. Hirata and A. B. Murphy: "Numerical simulation of GMAW process using Ar and an Ar-CO2 gas mixture", Welding in the World, Vol. 60 (2016), pp. 345-353

[4] C. W. Hirt and B. D. Nichols: "Volume of Fluid (VOF) method for the dynamics of free boundaries", Journal of Computational Physics, Vol. 39 (1981), pp. 201-225

Numerical Simulation of Three Dimensional TIG Welding Using MPS Calculation

Shun SASO, Masashi MOURI
IHI Corporation, 1, Shin-Nakahara, Isogo, Yokohama 135-8501, JAPAN
shun_sasou@ihi.co.jp, +81-45-759-2812

Abstract
Recent developments allow the calculation of welding arc parameters using several ways. From this we can obtain information on the arc properties such as its current, its temperature, the arc pressure, etc. This study shows how we were able to calculate both the welding arc and the molten pool conditions by combining the two aforementioned numerical simulation methods. Indeed, we applied the Particle Method for the simulation of the molten pool in order to predict its shape, and used the grid method for the simulation of the arc. The Moving Particle Semi-Implicit (MPS) method was applied.

In this study, a three dimensional particle simulation model of the molten pool was developed. The convection in the molten pool and the phase change could be analyzed using three dimensional particle models. It should be noted that a difference in welding conditions causes a different convection and therefore alters the penetration shape, but that the developed simulation was able to predict this change.

1. Introduction

Generally, the shape of welding penetration affects the strength of welding joint. In order to provide a trusted product, the technique of predicting welding penetration is required in production department. However, the welding process is very complex and is difficult to model accurately. For this reason, the penetration shape is difficult to predict.

Recent years, it has been actively developed to predict the welding by numerical calculation. The welding phenomena are usually calculated by Finite Volume Method (FVM) or Finite Element Method (FEM)[1]. We can get some information of the arc, for example, current, arc pressure, temperature, etc...

Another available option is the particle method, an emerging numerical analysis technology. This method is interesting when calculating large deformations and is able to account for material phase changes. Some particle methods such as Smoothed Particle Hydrodynamics (SPH) and Moving Particle Semi-implicit (MPS) are currently available. SPH can be used for various problems, and it has been applied to the welding phenomena analysis by Shigeta[2]. The MPS method developed by Koshizuka[3]-[4] is easy to use.

We have investigated the merits of this particle method ever since its early stages and have applied it to analyzing welding conditions[5]. To determine welding pool phenomena, we developed an MPS method that incorporates a thermal conduction model, heat transfer and radiation model, melting and solidification model, and surface tension model. However, determining both the welding arc and welding pool phenomena using the MPS method is difficult because it is extremely time consuming and has high calculation cost. As such, welding arc analysis is performed using the grid-based Finite Volume Method (FVM). To reduce computational cost, welding arc data are then used by the MPS method as boundary conditions or source calculations.

2. A Model of welding

2.1. Moving Particle Semi-implicit (MPS) model

In this study, we used the MPS developed by Koshizuka to perform an incompressible flow analysis. We also used the source code of program provided by the "MPS Code User Group." The momentum equation of an incompressible fluid and the energy conservation equation express in Lagrange form are given as follows;

$$\frac{d\mathbf{u}}{dt} = -\frac{1}{\rho}\nabla P + \nabla\cdot(\nu\nabla\mathbf{u}) + g + \frac{\mathbf{F}_s}{\rho V} + \frac{\mathbf{F}_{em}}{\rho V}. \tag{1}$$

$$\rho\frac{Dh}{Dt} = \nabla\cdot(\kappa\nabla T) + Q_{in} + Q_{out}. \tag{2}$$

$$h = C_p T. \tag{3}$$

In Equation (1), u, ρ, P, ν, g, V, F_s, and F_{em} are the velocity, density, pressure, kinematic viscosity, acceleration due to gravity, volume, surface tension, and electromagnetic force, respectively. In Equation (2) and (3), h, ρ, T, κ, Q_{in}, Q_{out} and C_p are the enthalpy, density, temperature, thermal conductivity, heat input, radiation and specific heat, respectively. Equation (1) and (2) include both gradient and Laplacian differential operators. These operators are discretized by the particle interaction model of a common MPS method.

This model considers the interaction between the particles. The model of the gradient differential operator and Laplacian differential operator are given as follows (4)-(5);

$$\nabla \phi = \frac{d}{n^0} \sum_{j \neq i} \left[\frac{\phi_j - \phi_i}{|\mathbf{r}_j - \mathbf{r}_i|} \frac{(\mathbf{r}_j - \mathbf{r}_i)}{|\mathbf{r}_j - \mathbf{r}_i|} w(|\mathbf{r}_j - \mathbf{r}_i|) \right]. \tag{4}$$

$$\nabla^2 \phi = \frac{2d}{\lambda n^0} \sum_{j \neq i} \left[(\phi_j - \phi_i) w(|\mathbf{r}_j - \mathbf{r}_i|) \right]. \tag{5}$$

Here, ϕ, d, n^0, λ, $\mathbf{r}$ and w are a physical quantity, the number of dimensions, particle number density at the initial condition of a disposed particle, normalization parameter, the position vector of the particle, and the weight function, respectively. Index i and j refer to the attention particle and the particles surrounding attention particle respectively. The weight function w varies with the distance between the attention particle i and surrounding particles j. The effect of the surrounding particles is modified by the weight function w when their interaction with the calculation particle is determined. Moreover, parameter λ , the weight function w and number density n_i of attention particle i is defined as follows (6)-(8);

$$\lambda = \frac{\sum_{j \neq i} |\mathbf{r}_j - \mathbf{r}_i|^2 w(|\mathbf{r}_j - \mathbf{r}_i|)}{\sum_{j \neq i} w(|\mathbf{r}_j - \mathbf{r}_i|)}. \tag{6}$$

$$w(|\mathbf{r}_j - \mathbf{r}_i|) = \begin{cases} \left(\dfrac{r_e}{|\mathbf{r}_j - \mathbf{r}_i|} \right) - 1 & (|\mathbf{r}_j - \mathbf{r}_i| < r_e) \\ 0 & (r_e < |\mathbf{r}_j - \mathbf{r}_i|) \end{cases}, \tag{7}$$

$$n_i = \sum_{j \neq i} w(|\mathbf{r}_j - \mathbf{r}_i|). \tag{8}$$

Here, $r_{e,}$ r_i and r_j are the effective radius, the position vector of particle i, and particle j, respectively. The effect of surrounding particles in the effective radius of particle i is considered by weight function and equation (4) and (5).

The momentum equation and energy equation are formulated as follows. First, the gravity and surface tension terms are calculated by an explicit method and temporarily particle position is determined by reference [6]. Second, the pressure is calculated from temporarily position of the particles. Third, the viscosity term is determined by an implicit method. Forth, the particles are moved to new position based on the new step velocity. Finally, the temperature is calculated. In that way, the flow of the molten metal is calculated.

2.2 Additional model for welding

In order to express a behavior of molten metal, it is necessary to calculate the high-viscosity fluid flow. However, it is difficult to calculate high-viscosity fluid flow by explicit method because the time step become smaller and the calculation become unstable condition. Then, in order to use large time step, the viscosity term is calculated using implicit method.

Additionally, the dynamic melting temperature is applied in order to keep the stable calculation. Dynamic melting temperature is the temperature that the yield stress of the materials becomes almost zero. If the temperature of solid particles become more than dynamic melting temperature, the solid particle is changed a fluid particle which has quiet high viscosity. In this way, the positions of solid particles are modified in order to keep distance between surround particles. The melt limit coefficient is also applied. The melt limit coefficient is parameter that prevent particles behavior until fluid particles is appeared enough for stable calculation in order to avoid the unstable condition in early stage of melting calculation.

In order to reduce the calculate cost and consider the influence of welding arc into the molten pool calculation, some physical parameters are calculated by grid method. After the arc calculation, the parameter of the arc is assigned as a boundary condition of the molten pool calculation. In the same way, after the molten pool calculation, the temperature distribution of molten pool is assigned as a boundary condition of the arc calculation. Fig.1 shows procedure of calculation. In this method, the arc is calculated at once to 500 steps of molten pool calculation.

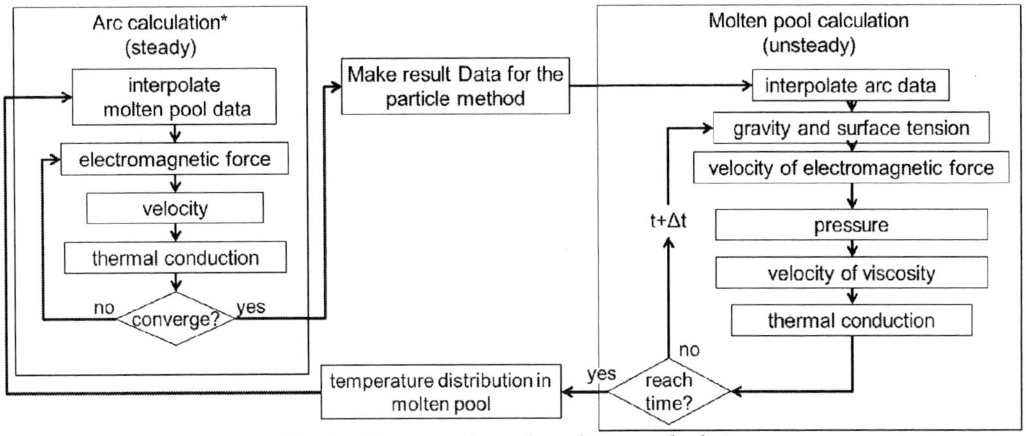

Fig. 1 The procedure of combining calculation

3. Welding calculation

3.1. The conditions of calculation

The molten pool was calculated by three dimensional MPS program. Fig. 2 shows the three dimensional model of the particles from the small iron plate. The center of the welding arc was indicated by red arrow. Table 1 lists the parameters used in the calculation. Calculated welding arc data (heat input, pressure, velocity, current, magnetic flux) were input to match the center of particle model. The temperature of the out sides of boundary was constant at 300 K. The boundary condition of temperature was constant at 300K.

Table. 1 The condition of molten pool calculation and welding arc.

Molten pool	Material	Pure iron	
	Thermal conductivity [W/mK]	Solid	83.5~34.6
		Liquid	40.9~45.2
	Specific heat [J/kgK]	Solid	450
		Liquid	795
	Kinematic viscosity [m2/s]	3.5×10-7~7.8×10-7	
	Latent heat [kJ/kg]	277.5	
	Surface tension Coefficient	$-4.9\times10^{-4}T+2.63$	
	Particle size [mm]	0.15	
	Number of particle	266833	
Arc	Current [A]	150	
	Flow rate [l/min]	10.0	
	Arc length [mm]	5	
	Gas	Argon	

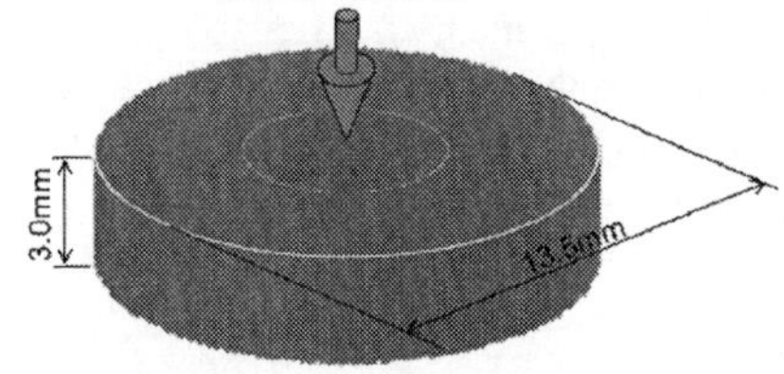

Fig. 2 The region of calculation

3.2. Result of calculation

Fig.3 shows a result of molten pool calculation. The color shows temperature distribution. A indented surface was generated by arc forceand convections in the molten pool. After that the molten pool was spread. The depth of molten pool was shallow and the fluid particles move apart direction by 0.1 seconds. After 0.1 seconds, the depth of molten pool was about 0.5 mm. Then, the flow which goes into the outside direction and the flow of depth direction were caused at the center of molten pool. These flows in molten pool transported the heat energy from the welding arc and grew the molten pool.

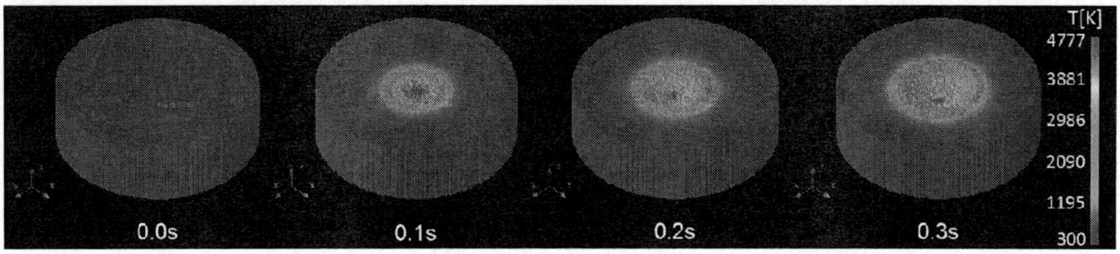

Fig.3 The result of molten pool calculation in birds-eye view. The color shows temperature.

4. Conclusion

We developed the three dimensional particle simulation model of the molten pool. This model took the arc data into consideration by combining a grid method and a particle method. Additionally, we modified the temperature calculation and added the some ideas. In this way, the behavior of molten metal became stable at the early stage of a melting.

The convection caused by some forces in molten pool was able to calculate by combining the grid method and particle method. The tendency of expansion molten pool was expressed. It is assumed that the molten pool calculation is useful to predict a tendency of molten pool behavior.

5. Referance

[1] Tanaka M, Ushio , Lowke JJ,(2005) Numerical Analysis for Weld Formation Using a Free-Burning Helium Arc at Atomospheric Pressure, *JSME International Jounal Series B*, Vol.48, No3

[2] Ito M, Izawa S, Fukunishi Y, Shigeta M, (2014) Numerical Simulation of a Weld Pool Formation in a TIG Welding Using an Incompressible SPH method, *Jounal of the Japan Welding Society*, Vol.32, No4 (Japanese)

[3] Koshizuka, Oka Y, (1996) Moving particle semi-implicit method for fragmentation of incompressible fluid. *Nuclear Science and Engineering* , 123: 421-434

[4] Kondo M, Koshizuka S, (2011) Improvement of stability in moving particle semi-implicit method, *Int. J. Numer. Methods Fluids*, 65 : 638–654

[5] Saso S, Mouri M, Tanaka M, Koshizuka S, (2016) Numerical analysis of two-dimensional welding process using particle method, *Welding in the World*

[6] Kondo, M., Suzuki, K., Koshizuka, S., Takimoto, M. , Surface Tension Model Using Inter-Particle Force in Particle Method 2007 Proceedings of the 5th Joint ASME/JSME Fluids Engineering Summer Conference, FEDSM 2007 SYMPOSIA (PART A): 93-98

Numerical analysis of fluid flow during swing arc narrow gap GMA welding

Guoxiang Xu, Qingnan Cao, Jie Zhu, Jiayou Wang
Jiangsu University of Science and Technology, Zhenjiang, Jiangsu, China
Email: xugxiang@163.com, phone: +86-511-84401184

Abstract

As a new type of narrow-gap arc welding, swing arc narrow gap GMA welding (swing arc NG-GMAW) has the advantages including high welding quality, compact torch, improved arc directivity and parameter controllability, good adaptability as well as low cost, thus having great potential of wide application in large-scale and thick-wall welding structures. But, this novel narrow gap welding involves more welding parameters, leading to difficulty of process optimization. Therefore, it is necessary to investigate the heat and mass transfer phenomena deeply for improving reliability of welding quality. To date, however, most studies are focused on the welding procedure and equipment and there is little research involving internal mechanism due to its complexity, thereby limiting its expanding. In this paper, based on the theories of heat transfer and fluid dynamics, a three-dimensional numerical model of fluid flow in swing arc NG-GMAW is developed with considering the effect of moving feature of arc heat source and joint geometric shape. Arc heat input is modeled as an double-ellipsoid heat source. The droplet heat transfer is assumed as the process of liquid metal with high temperature flowing into the weld pool from the certain area above it. The temperature field and fluid flow under different welding conditions are calculated using Fluent software. The characteristics of velocity field and temperature profile are analyzed and the effect of different welding parameters on the weld pool dynamic behavior is discussed. The results show that the built model can describe the fluid flow pattern accurately and reasonably, which lays foundation for deeply understanding the mechanism and optimizing the welding parameters in swing arc NG-GMAW.

Introduction

Large scale and thick wall weld structure are widely used in shipbuilding, pipeline for transferring oil and natural gas and other manufacturing fields. Thus, narrow gap welding process suitable for the joining of thick plate has received increasing attention. As a new type of narrow gap arc welding technology, swing arc NG-GMAW [1]overcomes the shortcomings of the traditional narrow gap arc welding and has the features including the high deposition rate, compact torch, improved parameter controllability, and so on, thus having significant patent of wide application in industries. But, this novel technology involves more welding parameters. Although the increased number of process parameters allows a flexible adjustment of the narrow gap welding process, it places higher demands on the process optimization, thereby limiting its expanding in industries. To improve the reliability of welding quality, it is necessary to deeply investigate the heat and mass transfer phenomena in swing arc narrow gap through mathematical modeling.

To date, extensive simulation efforts have been made to investigate the fluid flow in GMAW. Wu[2] proposed a three dimensional model to calculate the fluid flow in GMA bead-on-plate welding. Tsai[3] developed a two dimensional unified model to study the interaction among the droplet, arc and molten pool in bead-on-plate welding. Debroy[4] numerically analyzed the influence of fluid flow on weld shape and size in fillet weld of GMAW through a model similar to that of Wu. Farson[5] studied the hump formation in high speed GMAW from the view point of fluid flow pattern using a three dimensional mathematical model based on the commercial software. However, for swing arc narrow gap GMA welding, most studies are focused on the welding process and design of equipment and there are few works involving the mechanism for this type of welding process. Wang[6] experimentally studied the effect of different welding parameters on weld formation in narrow gap welding. Lan[7] numerically studied the temperature field using the finite element model without considering the flow field in weld pool. According the above review, it is conclude that no much information on heat and mass transfer is available for swing arc NG-GMAW.

In this paper, considering the welding process feature, a three dimensional numerical analysis model for fluid flow in swing arc narrow gap GMA welding is established and the calculation is conducted using FLUENT software. Base on this model, the fluid flow feature in swing arc NG-GMAW is investigated

Experimental

Fig.1 presents the schematic of swing arc NG-GMAW. During welding, the micro-bent conductive rod is rotated around its central axis and GMAW torch moves in the welding direction at a given speed. Fig.2 provides the moving route of arc center. In this study, the base metal is mild steel with a 16 mm thickness and the filler metal is solid welding wire with a 1.2 mm diameter. I type groove configuration is used, which has a gap of 12 mm. Welding current is 300 A and welding speed was set at 3.38 mm/s. The shielding gas is Ar-20%CO_2 and the flow rate is 40 L/min. The swing frequency of arc takes 0 Hz and 1.5 Hz, respectively. For the latter case, the duration time of arc at the sidewall is kept at 100ms. Other parameters are 68 ° swing angle and 8° bending angle of conductive rod. To save the computing time, only one weld pass was produced.

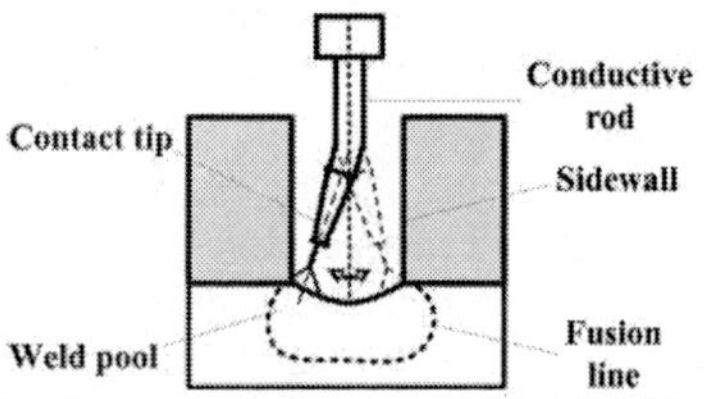

Fig.1 Schemtic of swing arc NG-GMAW

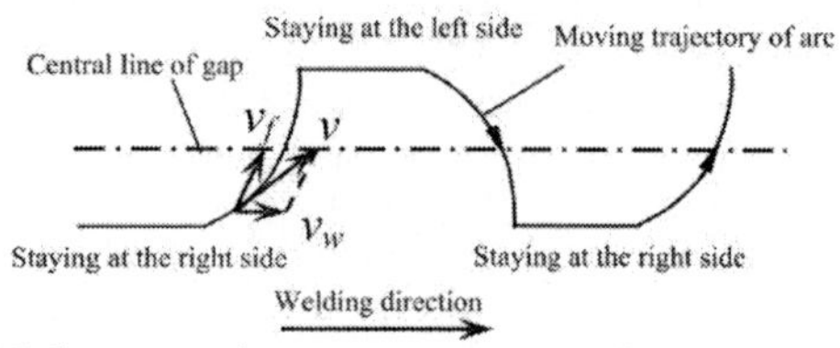

Fig.2 Schematic of moving trajectory of arc center

Mathematical model

In the calculation, the arc heat input is modeled as one double ellipsoid heat source model, which takes into account the moving trajectory of arc heat source in swing arc NG-GMAW. The droplet transfer is regarded as the process of high temperature liquid metal flowing into weld pool from the certain region above it, which also considers the travelling feature of electrode tip. The temperature of droplet is assumed to be 2400 K. The detailed description of arc heat source and droplet transfer model can be available in Ref. [8]. Weld pool free surface is traced by means of the volume of fluid (VOF) method. Though FLUENT software, the temperature and velocity profiles in swing arc NG-GMAW were calculated.

Results and Discussion

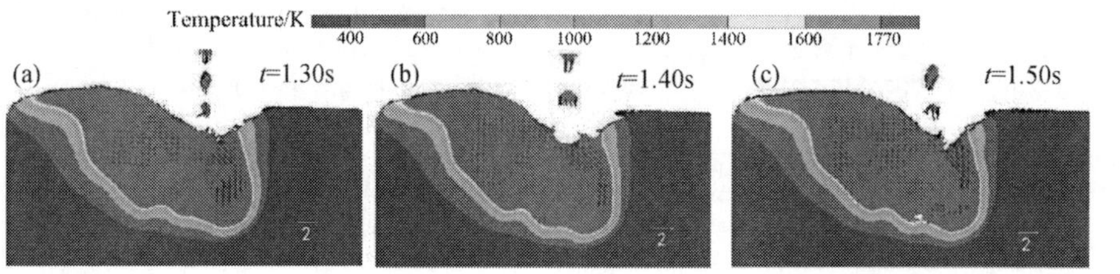

Fig.3 Evolution of temperature and velocity at longitudinal section in non-swing arc welding process

Fig.3 provides the temperature and velocity fields at longitudinal section of weld pool for the 0 Hz swing frequency. It is seen that, owing to large arc pressure and droplet impingement force in the case of high welding current, a heavy distortion of weld pool free surface is generated underneath the arc. At this location, the liquid metal quickly flows downward and then flows backward by way of the bottom of weld pool. A clockwise vortex is produced in the rear part of weld pool.

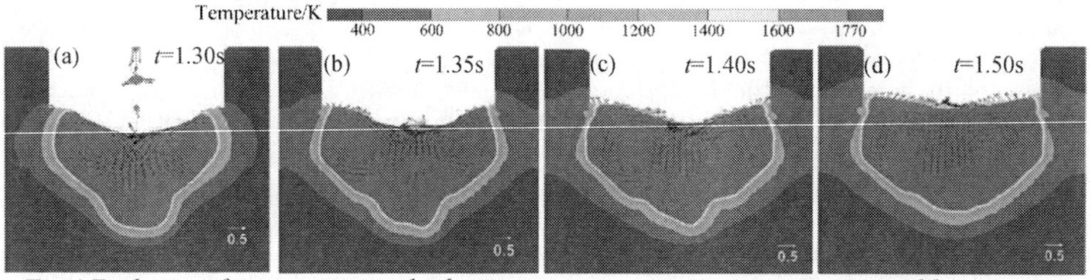

Fig.4 Evolution of temperature and velocity at cross section in non-swing arc welding process

Fig.4 shows the temperature and velocity fields at the cross section in non-swing process for different times. It is observed that, in this condition, the groove sidewall cannot be heated by the arc directly and its penetration depth is small. At t=1.3s, under the combined actions of arc pressure, droplet impingement force as well as electromagnetic force, the molten metal underneath the arc flows toward the bottom edge of weld pool at a high velocity, leading to the formation of finger penetration at the bottom plate. The liquid metal flowing downward is redirected towards the sidewall near the bottom of weld pool and then flows upward when encountering the groove sidewall. A vortex appears at either side of weld pool symmetrical axis. This flow pattern greatly contributes to the increase of weld penetration at the bottom plate since it carries heat and momentum to the bottom of weld pool. Due to existence of the groove sidewall, the lateral spreading of filler metal is limited greatly. Therefore, there exists a thick liquid layer underneath the arc during welding. The momentum of molten metal caused by arc pressure and filler metal droplets is expected to be absorbed and dissipated to large extent, leading to that the flow velocity of liquid metal becomes low when it approaches the sidewall. Thus, the fluid flow pattern in non-swing process has a very limited effect on the enhancement of sidewall penetration depth. With the arc heat source moving, the depression of weld pool surface decreases gradually due to reduction in effect of arc pressure and droplet and weld pool also starts to solidify. In addition, it should be noted that, because of low temperature, the molten metal solidifies more rapidly near the groove sidewall than that near the center of weld pool, which effectively suppresses the backfilling of the concave region near weld pool center. Hence, the final weld bead top surface is not plat but concave.

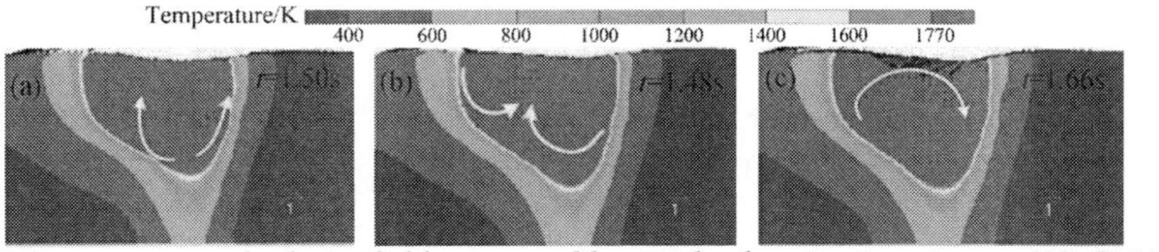

Fig.5 Temperature and velocity fields at central longitudinal section in swing arc NG-GMAW

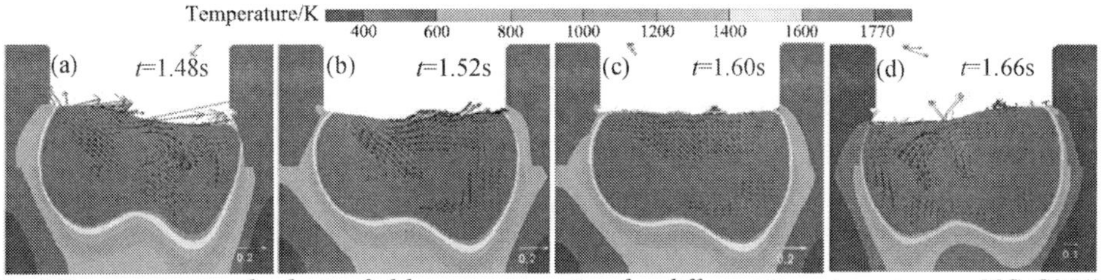

Fig.6 Temperature and velocity fields at cross section for different times in swing arc NG-GMAW

Fig.5 and Fig.6 present the temperature and velocity fields at the longitudinal section and transverse cross section (where heat source center is located) for different times in swing arc NG-GMAW, respectively. It is seen that, the weld pool dynamic behavior in swing arc NG-GMAW is obviously different to those in non-swing arc welding process. In this case, the electric arc can act on the groove sidewall directly, permitting more arc heat to transfer into the sidewall. Consequently, the penetration depth of the groove sidewall is increased largely and twin peaks appear at weld pool bottom. Besides, owing to swing of the welding arc, the influence of the arc pressure and droplet on weld pool free surface profile is weakened, especially for the latter. Thus, weld pool surface deformation is relatively low. Meanwhile, the flow mode of liquid metal is changing with time during welding.

At t=1.48s, the arc is in the stage of staying at the right groove sidewall. Due to the effect of arc force, a deformation of weld pool free surface also emerges in this zone and the liquid metal at upper part of weld pool flows toward the right groove sidewall with high velocity, as seen in Fig.6 (a). This strong toward-sidewall flow of high temperature molten metal helps to further increase the penetration depth of the sidewall. After hitting the right edge of weld pool, the molten metal with high momentum is deflected and simultaneously divided into two sub-streams. Small part of liquid metal flows up ward and the rest flows downward along the weld pool right edge and then is deflected toward the left groove sidewall near the bottom of the weld pool. The latter benefits large weld penetration depth at the bottom plate near the groove sidewall, thus inducing the appearance of twin peaks of weld bead bottom. When the arc leaves its right position and swings toward the left side of gap, the general fluid flow pattern in weld pool changes very limitedly but the flow velocity of liquid metal diminishes. Besides, the size of the weld pool close to the right sidewall tends to further increase in the swing process of arc, as illustrated in Fig.6 (b). In Fig.6(c), it is seen that, while the arc just moves to the left staying position of the groove at t=1.6s, the flow velocity of liquid metal is low. However, its basic flow pattern is still similar to that at t=1.48s and the weld pool dimensions at the right side are also still lager than the ones at the right side, demonstrating that the welding arc has minor effect on the fluid flow mode and temperature field during the swing process of arc. This is probably related to the swing velocity of arc. In the condition of this study, the swing velocity has a large value (it takes only 0.1s of arc to swing from one groove sidewall to the other), leading to that there is no enough time for the arc to show its thermal and force action. But, owing to low welding speed, after staying near the groove sidewall for a certain time, the influence of arc starts to emerge. From Fig.6 (d), it is revealed that, when time reaches 1.66s, under the actions of the arc pressure, a weld pool deformation occurs near the left sidewall and molten metal also flows toward the left groove sidewall at a high velocity. Table 1 compares the calculated size of weld bead with the experimental data, which have a general agreement, demonstrating the accuracy of the developed model. But, there is still a discrepancy in weld bead surface curvature, which will be modified in the future work.

Table 1 Comparison of calculated weld bead size with the experimental data

Swing frequency/ Hz	Sidewall penetration/mm		Thickness of weld bead/mm		Bead surface depression/mm	
	Measured	Calculated	Measured	Calculated	Measured	Calculated
0	1.01	0.75	10.4	10.8	1.7	1.1
2.5	2.0	1.8	8.5	8.6	2.05	1.2

Conclusions

(1) A three dimensional transient numerical model is developed to investigate the fluid flow in swing arc narrow gap GMA welding. This model considers the moving characteristic of the welding arc, joint geometric feature as well as the coupling of droplet and weld pool. The temperature and velocity fields under different conditions are analyzed. The calculated results agree generally with the experimental data.

(2) For non-swing process, there exists strong downward fluid flow in and near the symmetrical axis of weld pool and a vortex appears at either side of weld pool, which almost has no effect on the enhancement of sidewall weld depth. For swing arc NGW, the high temperature molten metal impinges on the sidewall with high momentum, and the fluid flow plays a positive effect on increase of penetration depth of the groove sidewall.

Acknowledgments

The present work is financially supported by the National Natural Science Foundation of China (Grant No. 51475218) and the Specialized Research Fund for the Doctoral Program of Higher Education (Grant No. 20133220110001).

References

[1] J.Y. Wang, J. Zhu, P. Fu, *et al*: "A swing arc system for narrow gap GMA welding," *ISIJ International*, Vol. 52, No.1 (2012), pp.110-14.

[2] J.S. Sun and C.S. Wu: "Modeling the weld pool behaviors in GMA welding," *International Journal for the Joining of Materials*, Vo. 11, No. 4(1999), Vol.11(4), pp. 158-166.

[3] J. Hu, H.L. Tsai: "Heat and mass transfer in gas metal arc welding. Part II: The metal," *International Journal of Heat and Mass Transfer*, Vol. 50, No. 5(2007), pp. 833-846.

[4] A. Kumar and T .Debroy: "Heat transfer and fluid flow during gas-metal-arc fillet welding for various joint configurations and welding positions," *Metallurgical and Materials Transactions A*, Vol. 28, No. 3, pp.506-519.

[5] M.H. Cho, D.F. Farson: "Simulation study of a hybrid process for the prevention of weld bead hump formation," *Welding Journal*, Vol. 86, No. 9 (2007), pp. 253-262.

[6] J. Y. Wang, J. Zhu, C. Zhang, G. X. Xu *et al*: "Effect of arc swing parameters on narrow gap vertical GMA weld formation," *ISIJ International*, Vol. 56, No.5 (2016), pp. 844-850.

[7] L. Lan, H.J. Zhang, A.J. Chen, et al: "Numerical simulation on dynamics process and thermal physical properties of narrow gap MAG vertical welding," *Transactions of the China Welding Institution*, Vol. 36, No. 7(2016), pp. 77-82

[8] G.X. Xu, Q.N. Cao, Q.X. Hu, *et al*: "Modelling of bead hump formation in high speed gas metal arc welding," *Science and Technology of Welding and Joining*, 2016. (Online, DIO: 10.1080/13621718.2016.1146427)

Numerical Simulation of Fume Formation Process in GMA Welding

Shinichi Tashiro, Manabu Tanaka
Joining and Welding Research Institute, Osaka University, Ibaraki, Osaka, Japan
tashiro@jwri.osaka-u.ac.jp

Anthony B. Murphy
CSIRO Manufacturing, PO Box 218, Lindfield NSW 2070, Australia

Abstract
In order to clarify fume formation mechanism in arc welding, quantitative investigation based on understanding of interaction among the electrode, arc and weld pool is indispensable. A fume formation model consisting of homogeneous nucleation model, heterogeneous condensation model and coagulation model has been developed and coupled with a GMA welding model. A series of processes such as evaporation of metal vapor from molten metal, transport of the metal vapor in the arc, fume formation from the metal vapor, transport of the fume to the surrounding was totally investigated by employing this simulation model. This study aims to visualize the fume formation process and clarify the fume formation mechanism theoretically through numerical analysis. Furthermore, the reliability of the simulation model is also evaluated through comparison of the simulation result with experimental result. As a result, it was clarified that fume was formed mainly in the upstream region of the arc near the droplet and also in the downstream region of the arc near the base metal. However, large part of the former was transported into the arc and evaporated and returned to the metal vapor again. Fume formation rate peaked near the end of peak current due to rapid decrease in arc temperature near the outer edge of the arc column. Averaged total fume formation rate was nearly agreed with experimental results.

Introduction
In arc welding process, reduction and harmless of the fume based on the understanding of the fume formation process are one of the key issues.

Tashiro et al have developed the unique numerical simulation model which enables to analyze sizes and shapes of the fume by giving the concentration of metal vapor and cooling rate of the arc plasma obtained by a GMA welding model as calculation conditions to clarify the fume formation process. Consequently, it was found that the fume with a spherical shape whose size exceeded 1μm was produced in the dense metal vapor region around the droplet. On the other hand, the fume with a chain-like shape whose size was several tens of nanometer was produced near the weld pool surface [1].

Furthermore, Tashiro et al improve the above model to consider influences of particle charging and non-equilibrium of the arc. The large amount of the fume is produced in low temperature region around 2000K in the downstream of the plasma jet. It is pointed out that the plasma in this region deviates from the LTE condition and tends to be in non-equilibrium because of very low electron density. As a result, it was clarified that aggregation rate of the fume decreased especially where the plasma was strongly in non-equilibrium and the electron temperature was high because of the increase in repulsive electromagnetic force caused by the particle charging. It led to the production of the fume with smaller size [2].

On the other hand, distributions of temperature, flow velocity, concentration of metal vapor in GMA welding accompanying metal transfer largely change depending on time. Therefore, in order to clarify fume formation mechanism in GMA welding, quantitative investigation based on understanding of interaction among the electrode, arc and weld pool is indispensable. A fume formation model consisting of homogeneous nucleation model, heterogeneous condensation model and coagulation model has been developed and coupled with a GMA welding model. A series of processes such as evaporation of metal vapor from molten metal, transport of the metal vapor in the arc, fume formation from the metal vapor, transport of the fume to the surrounding are totally investigated by employing this simulation model. This study aims to visualize the fume formation process and clarify the fume formation mechanism theoretically through numerical analysis. Furthermore, the reliability of the simulation model is also evaluated through comparison of the simulation result with experimental results.

Simulation model

Since it is well known that large amount of the fume is produced especially in CO_2 arc welding, the most important target of this study is of course CO_2 arc welding. However, as the first step of the study, the fume formation process in argon MIG welding with pure iron wire is discussed in this presentation.

Fig.1 shows an axisymmetric two-dimensional simulation domain (r,z) with a radius of 15mm and a height of 15mm consisting of an arc region and a wire region assuming stationary welding for simplicity. Regions for a contact tip, a shielding gas inlet and a nozzle are defined on the top boundary. The outer diameter of the contact tip and the inner diameter of the shielding gas inlet are 2.4mm and 14mm, respectively. On the axis, the wire region with the diameter of 1.0mm and the initial extension of 10mm corresponding to arc length of 5mm is defined. The side boundary is pressure outlet. The bottom boundary corresponds to the surface of the base metal. The region inside the base metal are not calculated. Non-uniform mesh is used. The maximum and the minimum sizes are 0.2mm and 0.05mm, respectively. The wire feed speed is 5m/min. The argon shielding gas is introduced from the top boundary at the flow rate of 10L/min. Fig.1 also shows the pulsed current waveform. It has the primary peak current of 450A, the secondary peak current of 150A and the base current of 30A. The frequency of the pulsed current is 100Hz.

In this model, distributions of flow velocity, temperature, metal vapor concentration and current density are obtained mainly by solving conservation equations of mass, momentum, energy, mass of metal vapor and current. The magnetic field is obtained from vector potential. The plasma is assumed to satisfy the LTE condition. The thermodynamic and transport properties of the arc under the LTE condition are calculated as functions of temperature and metal vapor concentration [3]. Diffusion coefficient of metal vapor is taken from Ref. [4]. Only iron is assumed for metal vapor composition. The thermodynamic and transport properties of mild steel are used for the wire [5]. In this model, Volume of Fluid (VOF) method is used for tracking the free surface.

The fume formation model considers homogeneous nucleation, heterogeneous condensation and aggregation. In this study, moment method is employed for reducing calculation cost [6-9]. In this model, it is assumed that the fume is spherical, the temperature of the fume is the same as the plasma and the fume size follows unimodal lognormal distribution.

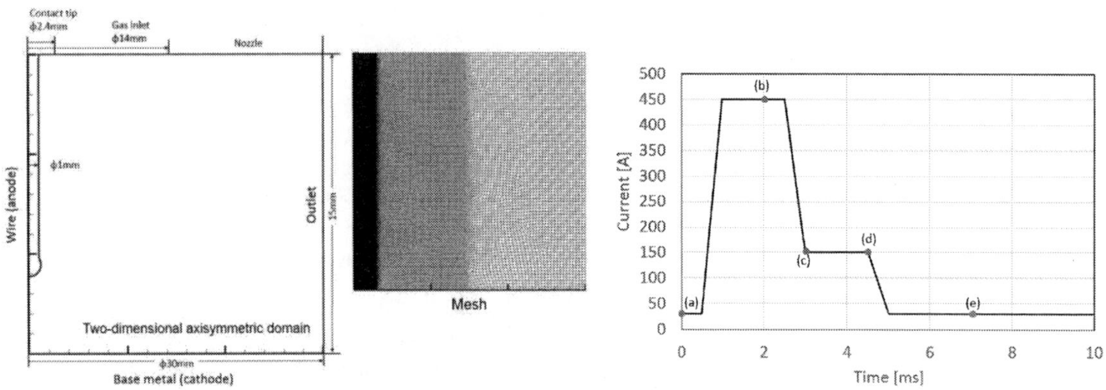

(a) Simulation domain (b) Pulsed current waveform

Fig. 1: Simulation domain and current waveform

Results and discussion

Fig. 2 shows distributions of temperature, mole fraction of metal vapor, fume formation rate and fume mass density at five moments (a)~(e) presented in Fig. 1. Fig. 3 shows time variation in the total fume formation rate obtained by integrating the fume formation rate over the entire simulation domain in Fig. 3 (iii).

The arc temperature reached approximately 22,000K at the maximum during the primary peak current of 450A as shown in (b) of Fig. 2 (i). The temperature inside the droplet at the wire tip was raised to 2,500K on average due to the heating by the electron condensation and the thermal conduction from the arc plasma. Consequently, large amount of metal vapor was evaporated from the droplet surface and most of the metal vapor was transported to the base metal by the plasma jet. The metal vapor transported to the low temperature arc plasma around 2,000K produced the fume through the nucleation. Then, the fume was transported to the surrounding due to the convection and the diffusion. As a result, it was clarified that the fume was formed mainly in the upstream region of the arc near the droplet and also in the downstream region of the arc near the base metal. However, large part of the former was transported into the arc and evaporated and returned to the metal vapor again. Fume formation rate peaked near the end of the secondary peak current as shown in (d) of Fig. 2 (iii) due to rapid decrease in arc temperature near the outer edge of the arc column.

The total fume formation rate changed between 2mg/s~24mg/s depending on the instantaneous current. The averaged total fume formation rate was 7.31mg/s which nearly agreed with experimental results of 1mg~10mg [10].

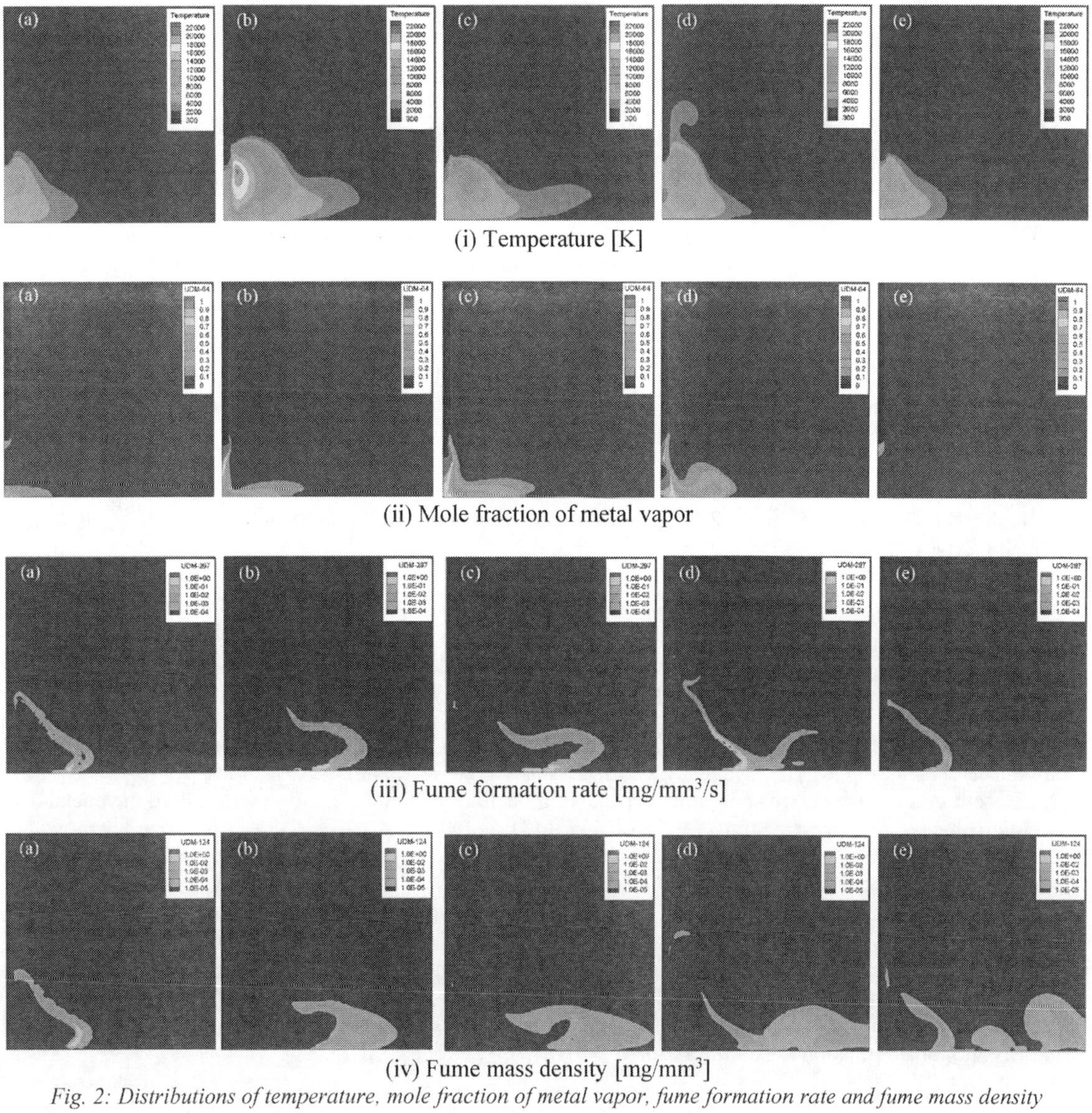

(i) Temperature [K]

(ii) Mole fraction of metal vapor

(iii) Fume formation rate [mg/mm^3/s]

(iv) Fume mass density [mg/mm^3]

Fig. 2: Distributions of temperature, mole fraction of metal vapor, fume formation rate and fume mass density

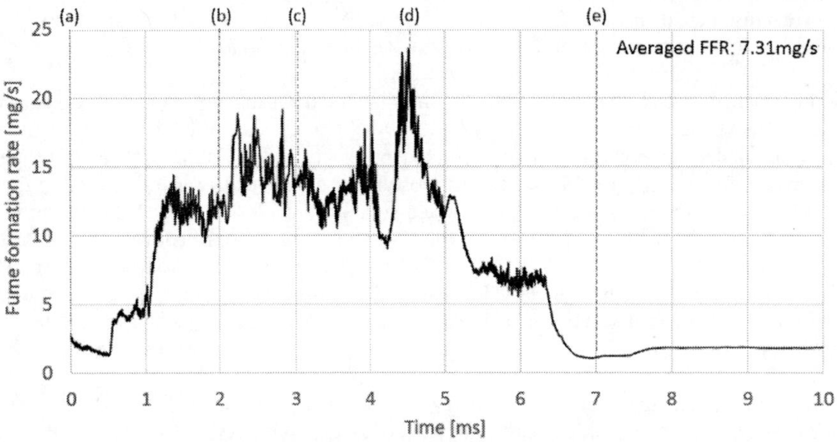

Fig. 3: Time variation in total fume formation rate

Conclusions

Since it is well known that large amount of the fume is produced especially in CO_2 arc welding, the most important target of this study is CO_2 arc welding. However, as the first step of the study, the fume formation process in argon MIG welding with pure iron wire was investigated in this presentation. As a result, it was clarified that fume was formed mainly in the upstream region of the arc near the droplet and also in the downstream region of the arc near the base metal. However, large part of the former was transported into the arc and evaporated and returned to the metal vapor again. Fume formation rate peaked near the end of the secondary peak current due to rapid decrease in arc temperature near the outer edge of the arc column. The averaged total fume formation rate was nearly agreed with experimental results.

Acknowledgments

This work was supported by JSPS KAKENHI Grant Number JP25420779.

References

[1] S. Tashiro et al: "Numerical analysis of fume formation mechanism in arc welding", *J. Phys. D: Appl. Phys.*, Vol. 43 (2010), 434012.

[2] S. Tashiro et al: "Numerical analysis of the influence of particle charging on the fume formation process in arc welding", *J. Phys. D: Appl. Phys.*, Vol. 46 (2013), 224007.

[3] M Tanaka et al: "Time-dependent calculations of molten pool formation and thermal plasma with metal vapour in gas tungsten arc welding", *J. Phys. D: Appl. Phys.*, Vol. 43 (2010), 434009.

[4] M. Schnick et al: "Numerical investigations of arc behaviour in gas metal arc welding using ANSYS CFX", *Front. Mater. Sci.*, Vol. 5, 2 (2011), 98.

[5] M Hertel el al: "Numerical simulation of droplet detachment in pulsed gas–metal arc welding including the influence of metal vapour", *J. Phys. D: Appl. Phys.*, Vol. 46 (2013), 224003.

[6] M. Boselli et al: "Two-dimensional time-dependent modelling of fume formation in a pulsed gas metal arc welding process", *J. Phys. D: Appl. Phys.*, Vol. 46 (2013), 224006.

[7] S. E. Pratsinis et al.: "Simultaneous nucleation, condensation, and coagulation in aerosol reactors", *Journal of Colloid and Interface Science*, Vol. 124 (1988), 416.

[8] G. M. Phanse et al.: "Theory for Aerosol Generation in Laminar Flow Condensers", *Aerosol Science Technology*, Vol. 11(1989) 100.

[9] F. Aristizabal et al.: "Modeling of the production of ultra fine Aluminium particles in rapid quenching turbulent flow", *Aerosol Sciense*, Vol. 37 (2006) 162.

[10]I. Pires et al: "Analysis of the influence of shielding gas mixtures on the gas metal arc welding metal transfer modes and fume formation rat", *Materials and Design*, Vol. 28 (2007), 1623.

Development of TIG Orbital Welding System using Welding Pool Model

Seokhyoung Lee, ***Yong-Baek Kim***
Hyundai Heavy Industries, Ulsan, South Korea
Email: {s.lee, ybkim0913}@hhi.co.kr

Abstract
The applicable mathematical model of the weld pool oscillation was investigated experimentally using arc voltage in obital (pipe) TIG welding. The main point of this work is to compare mathematical characteristics between relationships of pool shape and oscillation signals with its frequencies. To figure out the characteristics of pool oscillation signals in a full penetration weld pool under orbital welding conditions, an orbital automatic welding system based on real-time data aquisition was specially developed for applying a mathematical weld pool model to the experimental condition ns of an orbital (pipe) welding. The developed system includes real-time remote controller for changing weld conditions, real-time DAQ boards, signal/data/video image processing units of the molten pool surface continuously.

From the experimental results, pool oscillation through arc voltage showed normal coherence. In traveling full penetration condition, several oscillation modes existed in the molten pool which showed multiple oscillation frequencies according to welding position. However, there still remains some difficulties for generalization of weld pool model due to different and changeable welding conditions with respect to orbital (pipe) weld positions.

Introduction
For the joining of metals in industry environments, an arc welding process is widely employed. The main reason of this is that the heat to melt the workpiece or the metals to be joined can be generated and controlled easily in a relatively inexpensive way. However as is some difficulties in any welding processes, in arc welding process, variation of weld penetration during arc welding is one of the most important problems that influence weld quality.

In order to reduce the rate of defective welds, lots of approaches have been investigated in a view of the automatic welding systems. A typical method of them is known as a seam tracking. This seam tracking has been also one of main issues with respect to welding automation. Several attempts have been successful for reducing the rate of defective welds, because a seam tracking method is able to measure the weld pass in a real time. However we still cannot say that it is enough to manage the state of weld defects on a molten pool at any time. So if the state of molten pool can be managed in inexpensive way, the automatic welding systems can be developed which is able to guarantee high weld quality. As an alternative method of them, a weld pool modeling approach is considered to manage or control the weld conditions in this paper. These kinds of approaches have been also investigated with theoretical and experimental results [1, 2].

This paper presents the applicable mathematical model of the weld pool oscillation which is experimentally using arc voltage in TIG welding. The main point of this work is to compare mathematical characteristics between relationships of pool shape and oscillation signals with its frequencies. To figure out the characteristics of pool oscillation signals in a full penetration weld pool under orbital welding conditions, an automatic welding system based on real-time data acquisition was developed for applying a mathematical weld pool model to the experimental conditions of an orbital (pipe) welding, and the comparison between a mathematical calculation and an experimental analysis is also presented.

Welding Pool Impedance Model

In automatic welding process, it is important to examine the physical state which is related with the weld geometry. A lot of weld monitoring techniques using measurable parameters have been studied. In this paper, a method for sensing extent of full penetration is to use a mathematical model of a weld pool at full penetration which is considered as a classical model. The brief derivation of the model is as follows [2].

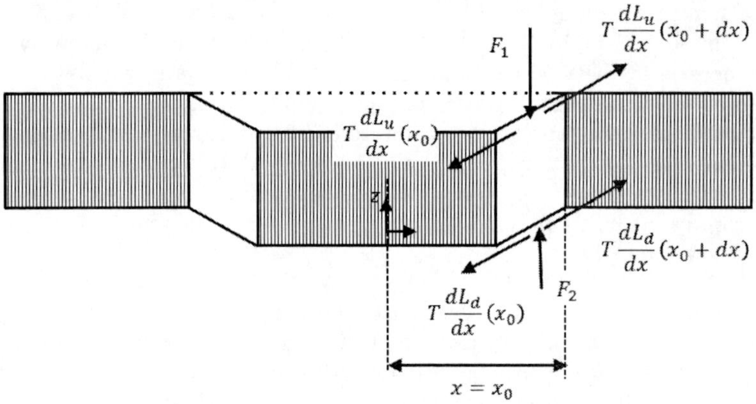

Figure 1: Related forces on a full penetration model

The welding pool dynamics can be obtained by the combination of the gravity, surface tension and outside forces (Figure 1). The free body equation of the motion is

$$T\frac{1}{x}\frac{d}{dx}\left(x\frac{d}{dx}(L_u+L_d)-\frac{g^2}{T}(L_u-L_d)\right)=\frac{1}{2}\rho(L_u-L_d)\frac{d^2(L_u+L_d)}{dt^2}+F_1-F_2, \tag{1}$$

where T and ρ are the constant surface tension and density, L_u and L_d are the upper and lower surface shape, and F_1, F_2 and g are the outside above, below forces and gravity constant [2].

Here, there are three main assumptions that

1. The molten metal is incompressible and inviscid,
2. The molten metal is homogeneous,
3. Fundamental (natural) frequency will dominate.

Based on the assumptions and equation (1), the natural frequency f_0 of the welding pool surface motion is obtained by

$$f_0=\frac{2.4}{x_0}\sqrt{\frac{2\,T}{\rho\,h}}, \tag{2}$$

where h is a material thickness of a workpiece, and surface tension T is normally considered as 1 N/m [2].

Of course, there are many studies about a mathematical model of the welding pool with natural frequency and its dominated parameters. In this work, however, only a classical equation (2) was considered for investigating welding pool shapes.

Welding Experiment

To figure out the characteristics of welding pool oscillation signals in a full penetration weld pool. An automatic welding system was developed, which can control a motion of a torch part based on a welding pool surface frequencies. The system is normally able to be applied to pipe welding, and includes real-time remote controller for changing weld conditions, real-time DAQ boards, signal/data/video image processing units of the molten pool surface continuously.

At first, the developed system was applied to a TIG welding with two flat work pieces. The main issue of this experiment is how well a motion of a torch part can be controlled for satisfying a full penetration welding condition with equation (2).

Figure 2: Result of welding with a flat work piece (x_0 = 2.5mm)

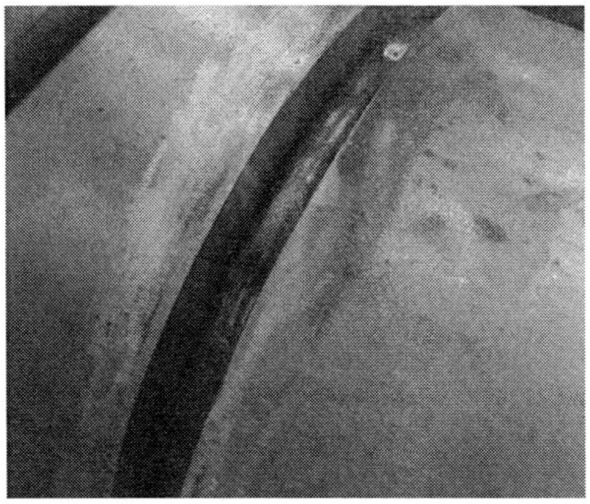

Figure 3: Result of orbital welding

Under the condition of the thickness $h = 3$mm and $\rho = 2.7 \times 10^{-6}$ in Figure 2, the natural frequency f_0 is calculated as about 477Hz. However, according to the analysis of the signals from DAQ boards, a dominant frequency was calculated as about 520Hz sometimes, but normally calculated as about 220Hz ~ 700Hz. Furthermore at the second experiment, the dominant frequency was calculated as about 350Hz ~ 1.2kHz.

Of course, there would be some strong welding arc noises, measurement errors, measuring time delays, and so on. Sometimes dominant frequency was calculated as the similar to natural frequency. But, that was not a normal phenomenon.

Conclusions

From the experimental results, pool oscillation through arc voltage showed normal coherence. In traveling full penetration condition, several oscillation modes existed in the molten pool which showed multiple oscillation frequencies according to welding position. However, there still remain some difficulties for generalization of weld pool model due to different and changeable welding conditions with respect to orbital (pipe) weld positions.

References

[1] C.D. Yoo, R.W. Richardson: "An Experimental Study on Sensitivity and Signal Characteristics of Weld Pool Oscillation", *Transactions of the Japan Welding Society*, Vol. 24, No. 2 (1933), pp.54-62.

[2] M. Jacksenhouse, D.E. Hardt: "Weld Pool Impedance Identification for Size Measurement and Control", Transaction of the ASME, Vol. 105 (1983), pp.179-184.

Prediction of solidification cracking in multi-pass weld metal of modified 9Cr1Mo steel during hot-wire laser welding with a narrow gap groove

Ichihu Ryu, Masatomo Todo, Motomichi Yamamoto, Kenji Shinozaki

Hiroshima University, Japan

The aim of this study is to predict solidification cracking in multi-pass weld metal of modified 9Cr1Mo steel during hot-wire laser welding with a narrow gap groove. ASTM A213 Grade T91 steel (Thickness: 30mm) was used as a base metal. A filler wire according to AWS A 5.28: ER90S-G was used. The multi-pass hot-wire laser welding with a laser power of 5.5 and 6kW were applied for a narrow groove with a gap width of 4 and 8mm. The bead heights of each pass were varied from 2 to 6 mm. Solidification cracks occurred in weld metal more than second pass of some joints. Three-dimensional finite element method was carried out to estimate an equivalent plastic strain history at the temperature range from liquidus to solidus in a weld metal of each pass during solidification considering residual stress of each pass. Equivalent plastic strain at high temperature increased with an increase in weld pass. It seemed that solidification crack was easy to occur dependent of increment of weld pass. Moreover, according to the principle of prediction of solidification cracking by the competition between a ductility curve and a high temperature strain calculated by FEM, the occurrence of solidification cracking can be predicted by comparing the strain rate with the ductility curves of weld metal. Therefore, several narrow gap weld joints with multi-pass weld metal were tried to predict solidification cracking. It was found that the prediction approach of solidification cracking was practically succeeded.

Evaluation of solidification cracking susceptibility of stainless steels using extra high-purity steels for ferrite-free welding

Tomo Ogura*, Toshimitsu Masaki and Kazuyoshi Saida
Division of Materials & Manufacturing Science, Graduate School of Engineering, Osaka University, Osaka, Japan
** tomo.ogura@mapse.eng.osaka-u.ac.jp*

Kazutoshi Nishimoto
Department of the Application of Nuclear Technology, Fukui University of Technology, Fukui, Japan

Kiyoshi Kiuchi
Nuclear Science and Energy Directorate, Japan Atomic Energy Agency, Ibaragi, Japan

Junpei Nakayama
Natural Resources & Engineering Business, Kobe Steel, Ltd., Hyogo, Japan

Abstract
Solidification cracking susceptibility of extra high-purity type 304 and 316 stainless steels was evaluated by the transverse-Varestraint test using arc and laser welding, in order to clarify the feasibility of ferrite-free welding. The solidification brittle temperature ranges (BTRs) of 304EHP and 316EHP steels were slightly increased with a decrease in the amount of δ-ferrite in the weld metal, while they were considerably lower than those of commercial-purity steels with FA solidification mode during arc as well as laser welding. It follows that the ferrite-free welding of austenitic stainless steels could be accomplished by applying the extra high-purity steels. Theoretical approach to the BTR was made to elucidate the depression mechanism of solidification cracking susceptibility in extra high-purity steels. From the fact that solidification segregation of minor and impurity elements such as C, P and S was considerably depressed in the extra high-purity weld metal, the solidification cracking susceptibility would be reduced even when the amount of ferrite was extremely small or free, due to the inhibition of solidification segregation. Numerical analysis of the BTR suggested that the ferrite-free welding of type 304 and 316 stainless steels could be achieved at the compositional range of P+S+0.1C<80ppm in the weld metal.

Introduction
Type 304 and 316 austenitic stainless steels have applied to components of chemical and power plants due to good corrosion resistant and high formability. These alloys are designed to solidify to give primary ferrite and secondary austenite (FA mode solidification) to minimize the occurrence of hot cracks. However, high welding speed decreases the amount of δ-ferrite and increases hot cracking susceptibility [1]. In addition, precipitated σ phase in δ-ferrite causes embrittlement at 700-800°C [2]. Therefore, change in microstructures and corresponding mechanical properties of the alloys during long-time operation of plants at high temperature should be considered. Recently, extra high purity (EHP) stainless steels, whose total contents of minor and impurity elements, such as C, P, S, N and O, were reduced to <100 ppm, have been developed as anti-intergranular corrosion alloys to replace existing ULC grade stainless steels [3,4]. The authors have investigated the weldability (especially hot cracking susceptibility) of EHP type 310 stainless steel and 25%Cr-35%Ni stainless steel for fusion welding and reported that this steel was barely susceptible to hot cracking despite its fully austenitic structure and that refining to extra high purity grade (purification) would be one of the most innovative countermeasures for solving the hot cracking issue in welding of fully austenitic alloys. These results indicates that widely-used austenitic stainless steels such as type 304 and 316 have the possibility to weld ferrite-free using EHP technique. The objective of this study is to evaluate solidification cracking susceptibility of EHP type 304 and 316 stainless steels in order to clarify the feasibility of ferrite-free welding.

Experimental procedure
Specimens
Chemical compositions of stainless steels used in this study is shown in table 1. EHP type stainless steels with several Cr_{eq}/Ni_{eq} ratio was used to vary solidification mode. For the comparison, commercial purity type of

304ULC and 316L stainless steels were also used.

Table 1 Chemical compositions of stainless steels used (mass%)

	Steel	C	Si	Mn	P	S	Ni	Cr	Mo	Al	N	O	Fe
Type304EHP	304EHP-A	0.0012	0.01	0.01	0.0005	0.0005	10.48	19.94	-	-	-	-	Bal.
	304EHP-B	0.0011	<0.02	<0.01	0.0017	<0.0005	12.30	18.00	-	0.010	0.0019	0.0068	Bal.
	304EHP-C	0.0020	<0.02	<0.01	0.0013	<0.0005	13.10	17.70	-	0.010	0.0011	0.0055	Bal.
	304EHP-D	0.0024	<0.01	<0.01	0.0009	<0.0005	15.03	18.06	-	0.008	0.0007	0.0027	Bal.
	304EHP-E	0.0048	<0.01	<0.01	0.0009	<0.0005	11.04	18.04	-	0.010	0.0013	0.0029	Bal.
	304EHP-F	0.0040	<0.01	<0.01	0.0006	<0.0005	10.08	18.03	-	0.010	0.0014	0.0025	Bal.
Type 316EHP	316EHP-A	0.0021	0.01	0.01	0.0005	0.0005	14.55	17.77	2.43	-	-	-	Bal.
	316EHP-B	0.0012	<0.01	<0.01	0.0009	<0.0005	12.60	16.20	2.30	0.020	0.0019	0.0068	Bal.
	316EHP-C	0.0022	<0.01	<0.01	0.0008	<0.0005	13.60	15.90	2.31	0.020	0.0011	0.0055	Bal.
	316EHP-D	0.0018	<0.01	<0.01	0.0008	<0.0005	16.05	16.08	2.05	0.022	0.0007	0.0015	Bal.
	316EHP-E	0.0028	<0.01	<0.01	0.0006	<0.0005	15.06	16.09	2.04	0.020	0.0011	0.0019	Bal.
	316EHP-F	0.0026	<0.01	<0.01	0.0008	<0.0005	10.11	16.13	2.05	0.019	0.0009	0.0016	Bal.
Commercial purity	304ULC	0.0129	0.39	1.47	0.0138	0.0011	10.85	18.93	-	-	-	-	Bal.
	316L-A	0.0130	0.59	0.88	0.0200	0.0073	12.25	16.67	2.14	-	-	-	Bal.
	316L-B	0.0100	0.60	0.80	0.0140	0.0100	12.10	17.80	2.00	-	-	-	Bal.

Experimental procedure

Solidification cracking susceptibility of EHP type 304 and 316 stainless steels was evaluated by the transverse-Varestraint test using arc and laser welding. Table 2 shows conditions of transverse-Varestraint test in this study. The detail of laser transverse-Varestraint test has reported to ref.[9]. Laser power was adjusted to obtain a half–penetration bead at each welding speed for the transverse welding. During the welding, argon was adopted as a shield gas to protect the bead surface from oxidation. The augmented strain was varied between 0.25 and 2.44 %. A length of the transverse weld bead was adjusted to 30 mm in every welding speed. The crack surface occurring in the Varestraint test was observed using a scanning electron microscope (SEM).

Table 2 Conditions of transverse-Varestraint test

GTAW transverse-Varestraint test	
Arc current	120 A
Arc voltage	14 V
Welding speed	1.67 mm/s
Arc distance	2 mm
Shield gas flow rate	Ar: 15 L/mim
Augmented strain	0.25-2.44%

Results and discussion

3.1 Hot cracking susceptibility evaluated by GATW transverse Varestraint test

Table 3 shows amount of ferrite measured by ferrite scope and solidification mode and Cr_{eq}/Ni_{eq} in weld metals. Solidification mode was identified by SEM observation of weld metals in steels. Cr_{eq}/Ni_{eq} was measured by Delong formula. Solidification mode of the investigated steels was FA or AF/A and the amount of ferrite was 0-10%. In this study, the cracks after GATW transverse-Varestraint test were identified as solidification cracking in all samples using SEM. Fig. 1shows relationship between BTR and amount of ferrite in samples. The BTRs of 304EHP and 316EHP steels were slightly increased with a decrease in the amount of δ-ferrite in the weld metal, while they were considerably lower than those of commercial-purity steels with FA solidification mode during arc welding.

Table 3 Amount of ferrite measured by ferrite scope, solidification mode and Cr_{eq}/Ni_{eq} in weld metals

Steel	Amount of ferrite (%)	Solidification mode	Cr_{eq}/Ni_{eq}
304EHP-A	5.6	FA	1.90
304EHP-B	0.25	AF	1.46
304EHP-C	0.11	AF	1.35
304EHP-D	1.4	AF	1.20
304EHP-E	4.6	FA	1.61
304EHP-F	9.5	FA	1.76
304ULC	5.4	FA	1.68
316EHP-A	4.5	AF	1.38
316EHP-B	1.9	AF	1.46
316EHP-C	1.6	AF	1.34
316EHP-D	0	AF/A	1.05
316EHP-E	0.1	AF	1.12
316EHP-F	9.3	FA	1.62
316L-A	0.9	AF	1.37
316L-B	5.2	FA	1.54

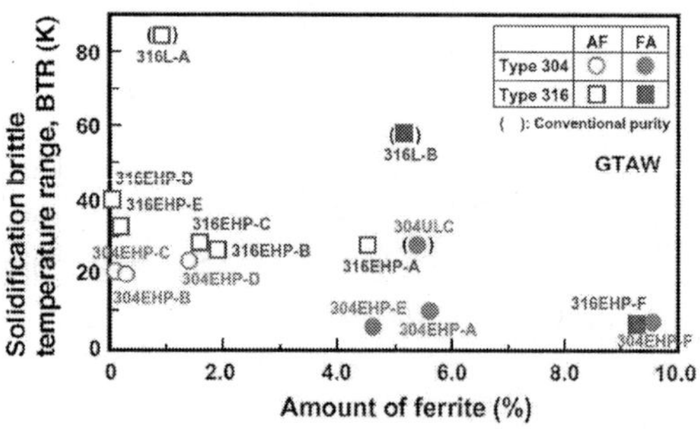

Fig. 1 Relationship between BTR and amount of ferrite

3.2 Hot cracking susceptibility evaluated by LBW transverse Varestraint test

Microstructural analysis of 304EHP and 316EHP stainless steel welds using laser welding was performed and the effect of welding speed on the amount of ferrite was investigated. Fig. 2 shows the microstructures of weld metals of 316EHP-F with several welding speed. The amount of ferrite was decreased with an increase of welding speed. Hot cracking susceptibility was evaluated by LBW transverse Varestraint test. In this study, the cracks after LBW transverse-Varestraint test were also identified as solidification cracking in all samples. Fig. 3 shows relationship between BTR and welding speed. The results clearly shows that the increase of BTR was very slightly (<10K) with an increase of welding speed. The BTR of 304EHP and 316EHP stainless steels were considerably lower than those of commercial-purity steels with FA solidification mode during arc as well as laser welding. It follows that the ferrite-free welding of austenitic stainless steels could be accomplished by applying the extra high-purity steels.

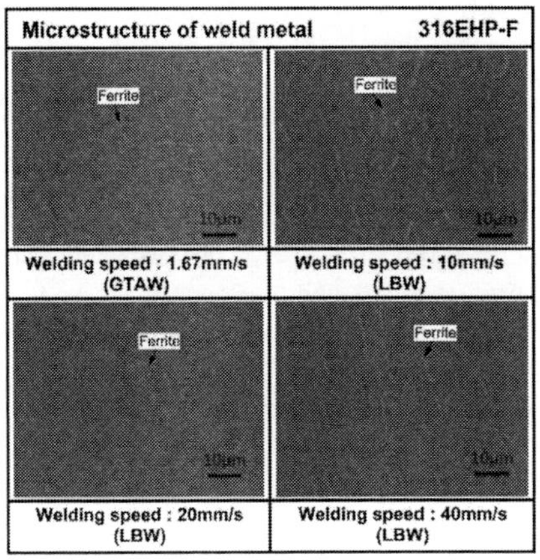

Fig. 2 Change in microstructure of LB weld metal of 316EHP-F with welding speed

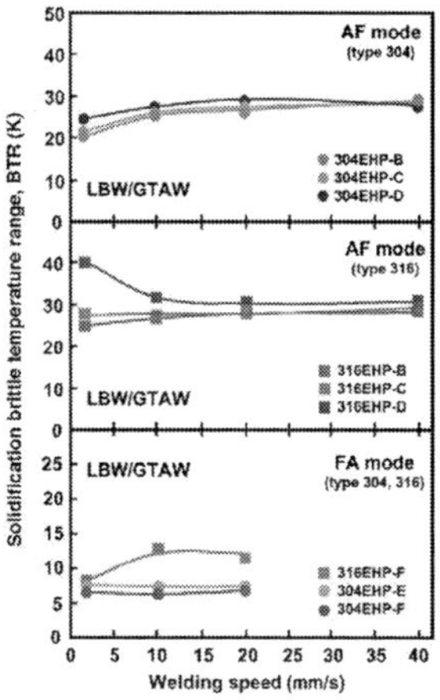

Fig. 3 Relationship between BTR and welding speed

3.3 Theoretical approach to BTR

In order to clarify the mechanism of the improved solidification cracking susceptibility of the extra high-purity steels, the solidification segregation behavior of impurity elements was theoretically analyzed. The calculation procedures for the solidification segregation are similar to those previously reported [7, 8, 10]. The good positive reaction between the measured BTR and calculated solid/liquid coexistence temperature range was confirmed. Fig. 4 shows relationship between segregated concentration (P and S) and amount of ferrite. With an increase of the amount of ferrite, the segregation of P and S was suppressed in EHP stainless steels, however, it is clear that the amount of segregated concentrations in EHP steels with AF solidification mode was significantly lower than that in stainless steels with FA solidification mode. Therefore, it is considered that BTR shrinkage by the reduction of the solid/liquid coexistence temperature range of the weld metal was due to the inhibited solidification segregation of solute elements (mainly P and S) and the solidification cracking susceptibility would be reduced even when the amount of ferrite was extremely small or free, due to the inhibition of solidification segregation.

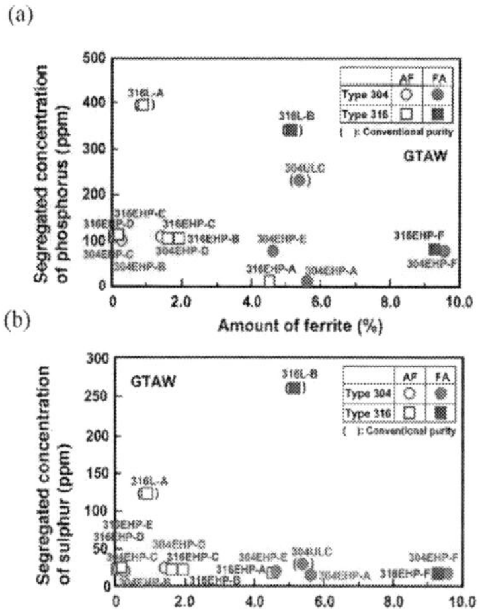

Fig. 4 Relationship between segregated concentration and amount of ferrite.(a) Phosphorus and (b) Sulphur

3.4 Numeral analysis of feasible compositional range for Ferrite-free welding

Numeral analysis was carried out to estimate the compositional range for ferrite-free welding in stainless steels. In this study, BTR was calculated in type 316 stainless steel with A solidification mode with verifying concentrations of C, P and S. The compositional ranges were C:20-400ppm, P:10-400ppm and S:10-300ppm. The welding speed was 1.67mm/s. Fig. 5 shows contour maps of BTR of synthetic type 316 stainless steel with A mode solidification. BTR was decreased with the increase of the amount of P and S. This tendency was clearer with the increase of C. P, S, and C was formulated as P+S+0.1C<80ppm when BTR was estimated below 40K, which was limit of solidification cracking. Therefore, numerical analysis of the BTR suggested that the ferrite-free welding of type 304 and 316 stainless steels could be achieved at the compositional range of P+S+0.1C<80ppm in the weld metal

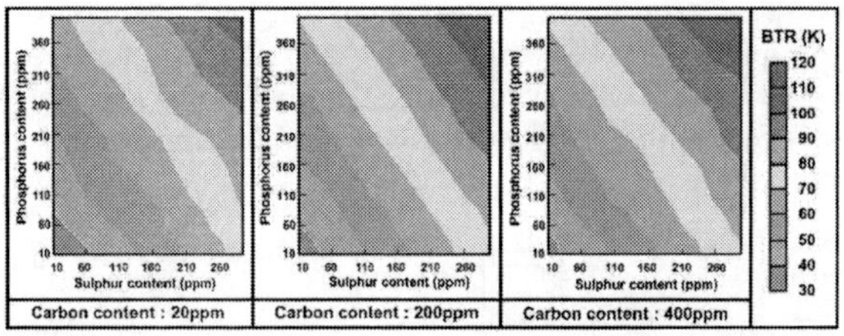

Fig. 5 Contour maps of BTR of synthetic type 316 stainless steel with A mode solidification

Conclusions

The transverse-Varestraint test showed that BTR of 304EHP and 316EHP stainless steel welds with AF solidification modes were relatively lower than those of conventional 304ULC and 316L stainless steel welds with FA solidification mode. A numerical simulation of the solid/liquid coexistence temperature range using a solidification segregation model explained the mechanism of the BTR shrinkage by the reduction of the solid/liquid coexistence temperature range of the weld metal due to the inhibited solidification segregation of solute elements (mainly P and S). Numerical analysis of the BTR suggested that the ferrite-free welding of type 304 and 316 stainless steels could be achieved at the compositional range of P+S+0.1C<80ppm in the weld metal.

References

[1] Y.Nakao, K.Nishimoto and W.P.Zhang: Effect of Rapid Solidification by Laser Surface Melting on Solidification Modes, -Study on Laser Surface Modedication of Stainless Steels (Report 1)-, Quar. J. JWS, 7-3 (1989), p.414-421 (in Japanese).

[2] E.J.Chun, H.Baba, K.Nishimoto and K.Saida: Precipitation of Sigma and Chi Phases in δ-ferrite of Type 316FR Weld Metals, Mater. Characterization, 86 (2013), p.152-166.

[3] J.Nakayama, T.Noura, H.Yamada and K.Yamamoto: Extra High Purity Alloy EHP for Next Generation Nuclear Systems, Kobe Steel Eng. Rept., 59-2 (2009), p.94-97 (in Japanese).

[4] K.Kiuchi and K.Ioka: Development of extra high purity stainless steels for next generation nuclear systems Genshiryoku-eye, Vol.55 No.5 (2009), p.54-58 (in Japanese).

[5] K.Nishimoto, K.Saida, Y.Okabe, K.Hata, K.Kiuchi and J.Nakayama: Hot Cracking Behaviour and Susceptibility of Extra High-Purity Type 310 Stainless Steels, Sci. & Technol. Welding & Joining, 15-1 (2010), p.87-96.

[6] K.Saida, H.Matsushita, K.Nishimoto, K.Kiuchi and J.Nakayama: Quantitative Influence of Minor and Impurity Elements on Hot Cracking Susceptibility of Extra High-Purity Type 310 Stainless Steel, Quar. J. JWS, 31-1 (2013), p. 56-65 (in Japanese).

[7] K.Saida, H.Matsushita, K.Nishimoto, K.Kiuchi and J.Nakayama: Quantitative Influence of Minor and Impurity Elements on Solidification Cracking Susceptibility of Extra High-Purity Type 310 Stainless Steel, Sci. & Technol. Welding & Joining, 18-7 (2013), p.616-623.

[8] K.Saida, Y.Nishijima, K.Nishimoto, K.Kiuchi and J.Nakayama: Contribution of Phosphorus and Sulfur on Hot Cracking Susceptibility of Extra High-Purity 25Cr-35Ni Stainless Steels, W. J., 92-11 (2013), p.322s-331s.

[9] E.J.Chun, H.Baba, K.Nishimoto and K.Saida: Development of Laser Beam Welding Varestraint Test for Assessment of Solidification Cracking Susceptibility in Laser Welds, Metals and Materials International, 21-3 (2015), p.543-553.

[10]E.J.Chun, H.Baba, K.Nishimoto and K.Saida: Theoretical Analysis for Solidification Cracking Susceptibility in Type 316FR Stainless Steel Laser Welds, Quar. J. JWS, 33-2 (2015), p.20s-24s.Please be reminded that the full paper must be with us no later than June 30, 2016. The USB flash drive including all papers can be circulated to all participants at the registration desk of the conference.

Effect of Chemical Composition on Solidification Cracking Susceptibility in Austenitic Weld Metals

Kota Kadoi, Kenji Shinozaki
Hiroshima university, Higashi-Hiroshima, Hiroshima, Japan
kadoi@hiroshima-u.ac.jp, +81-82-424-757

Abstract

The aim of this work is to investigate the influence of chemical compositions in austenitic metals such as chromium equivalent, nickel equivalent and so on, on the solidification cracking susceptibility. In order to change the chemical composition of specimen systematically within austenite single phase region in Schaeffler diagram, the specimens were fabricated using several base materials and filler wires of Ni-based alloy and stainless steel. Distribution of the solidification cracking susceptibility in Schaeffler diagram could be evaluated quantitatively by measuring brittle temperature range (BTR) and crack length. The influence of niobium content on the cracking susceptibility is quite significant due to low partition coefficient. Thus, the susceptibility distribution in Schaeffler diagram could be identified by classifying according to niobium content. The distribution showed different tendency depending on the content.

Introduction

Austenitic metals such as stainless steels and Ni-based alloy have great features such as high corrosion and heat resistance and so on. However, these materials are high susceptible to weld hot cracking such as solidification cracking [1,2]. Recently, multi-material structures of mechanical components are increasing for high functionality and streamlining design. Thus, the cracking susceptibility must increase with increasing in dissimilar welding using the austenitic metals. In order to achieve the requirement of the multi-material structure, welding of an austenite single phase material is highly demanded even though the materials have high hot cracking susceptibility. Therefore, it is necessary to prevent from the crack initiation by understanding the influence of chemical composition change in the region during dissimilar welding on the susceptibility.

Schaeffler diagram which consists of chromium equivalent Cr_{eq} (= %Cr + %Mo + 1.5×%Si + 0.5×%Nb) and nickel equivalent Ni_{eq} (= %Ni + 30×%C + 0.5×%Mn), is often used to predict containing phases during dissimilar welding of austenitic metals [3]. It is known that affecting element on solidification and the cracking susceptibility are nickel, chromium, phosphorous, sulfur, silicon, niobium and so on [4]. The formation of austenite phase as primary phase requires $Cr_{eq} / Ni_{eq} < 1.48$ [5]. On the other hand, phosphorous and sulfur as an impurity element, and silicon, niobium, titanium etc as alloying element are known as elements to enhance the susceptibility. However, it is not clear about the affection of these elements with changing chemical compositions in an austenite single phase region in Schaeffler diagram. The aim of this work was to investigate the influence of chemical compositions, especially niobium content, Cr_{eq} and Ni_{eq}, on solidification cracking susceptibility in single austenite phase region systematically and quantitatively.

Experimental procedure

Various base materials and filler wires were used in order to control chemical composition in evaluated area, namely weld metal, especially changing Cr_{eq}, Ni_{eq} and niobium content. Type 304 and Type 347 stainless steels (110 × 145 × 5 mm^3) were used as base materials and were machined to make U-type groove of several widths from 2.0 to 6.0 mm in the center of the width direction. Eight type of filler wires of stainless steels and Ni-based alloy were also used as filler wires for specimen fabrication. Table 1 shows chemical compositions of the base materials and the filler wires.

A hot-wire laser welding process was employed for specimen fabrication owing to its advantages of controllable deposit ratio and reduced dilution of the base material [6,7]. The chemical composition in the specimen was controlled by selecting various combinations of the base material, the groove shape, the filler wire and the process parameters. Fabricated specimens were cut into size of 110 × 70 × 5 mm^3 for evaluation of solidification cracking susceptibility and temperature measurement. The rest of the specimen after cutting were used for measurement of the dilution ratio of the base materials by cross section microstructure. Figure 1 shows distribution of the chemical compositions of specimens in Shaeffler diagram. It is confirmed that the specimens which have chemical compositions with wide distribution in austenite single phase region in Shaeffler diagram.

Laser Trans-Varestraint test [8] was carried out to evaluate solidification cracking susceptibility. Fiber laser was used as a heat source. Laser spot size was 0.4 mm (just focus) and laser power was 1.0 kW. The laser head was tilted 25°. Welding speed was 0.2 m/min. Applied augment strain was 4.3 % which was enough for saturated to open the cracks [7]. Total crack length (TCL) and maximum crack lengths (MCL) were measured using scanning electron microscopy. Temperature profiles were measured by direct insertion of an optical-fiber radiation thermometer with a high response rate into the rear of the molten pool during welding. Brittle temperature range (BTR) was measured by conversion from the maximum crack length into the temperature range using the temperature history.

Table 1 Chemical compositions of base materials and filler wires. (mass%)

		Fe	Cr	Mo	Si	Nb	Ni	C	Mn	P	S	Co	Ti	Cu	Al	W	V
B.M.	SUS304L	Bal.	18.14	-	0.69	-	10.06	0.006	1.42	0.034	0.002	0.18	-	-	-	-	-
	SUS347	Bal.	19.18	0.01	0.27	0.58	10.34	0.017	1.66	0.020	0.002	-	-	0.03	0.004	-	-
Wire	YS310S	Bal.	26.18	0.08	0.48	-	21.1	0.050	1.59	0.022	0.001	-	-	0.10	-	-	-
	Inver	Bal.	0.03	<0.01	0.15	-	36.0	0.020	0.35	0.005	0.001	-	-	0.02	0.003	-	-
	ERNi-1	0.015	-	-	0.14	-	Bal.	0.007	0.34	0.003	0.0006	-	2.37	-	0.57	-	-
	ERNiCr-3	0.97	20.45	-	0.01	2.84	Bal.	0.005	2.98	0.002	0.002	-	0.54	0.01	-	-	-
	ERNiCrFe-7	9.99	29.23	0.04	0.13	0.01	Bal.	0.025	0.24	0.002	0.001	0.004	0.55	0.02	0.65	-	-
	ERNiCrMo-10	4.60	21.2	13.4	0.04	-	Bal.	0.003	0.15	0.010	-	1.0	-	0.05	-	3.0	0.05
	Alloy 600	6.40	14.7	-	0.24	-	Bal.	0.040	0.25	0.006	0.001	-	-	0.01	-	-	-
	ERNiCrFe-13	8.79	29.5	3.51	0.11	2.51	Bal.	0.023	0.31	0.004	<0.0005	<0.01	0.18	0.05	0.13	-	-

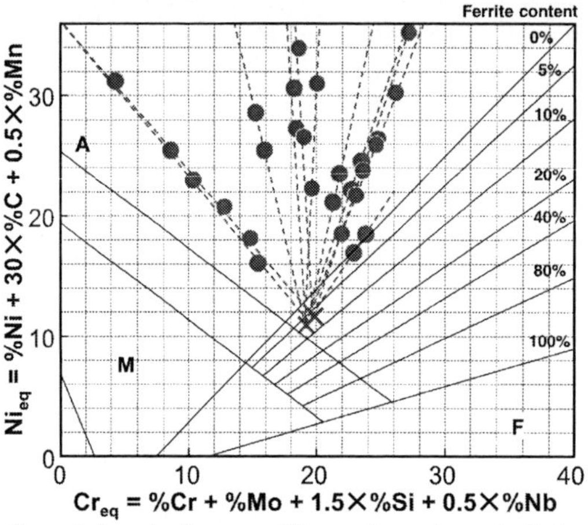

Fig. 1 Distribution of chemical compositions of specimen in Schaeffler diagram.

Results

Figure 2 shows solidification cracking distributions after Trans-Varestraint test (base material: Type 304L, filler wire: Alloy600, Cr_{eq}=18.3, Ni_{eq}=30.6). All of cracks were distinguished solidification crack from ductility-dip crack and solidification cracks were measured the initiation position and the length. The longest solidification crack initiates from the trailing edge of the rear center of the molten pool backside. The number of the crack was 21, TCL was 4.9 mm and MCL was 0.57 mm. Beside, BTR was 129 °C measured using the temperature history of the specimen. Identical evaluations were carried out using all the other specimens. Measured MCLs and BTRs were placed on each chemical composition in Schaeffler diagram after the evaluations. However, there were not clear tendency of the distributions of MCL and BTR with changing Ni_{eq} and Cr_{eq} equivalent. Thus, these results were distinguished by average niobium content $Nb_{ave.}$ of the combination of used base material and filler wire since niobium is knows as enhancing factor on solidification cracking susceptibility due to a low partition coefficient as follows; $Nb_{ave.}$ 0%: Type 304L + Nb free wire, $Nb_{ave.}$ 0.41%: Type 347 + Nb free wire and $Nb_{ave.}$ 0.84%: Type 304L + Nb [2].

Figure 3 shows distribution of MCLs and BTRs of each $Nb_{ave.}$ in Shaeffler diagram. The distributions of MCL and BTR have a similar tendency in the diagram. Both values of MCL and BTR increase with increasing in $Nb_{ave.}$. Besides, it can be confirmed that the distribution depends on $Nb_{ave.}$ value. In the case of $Nb_{ave.}$ of 0 %, high value of MCL and BTR are within $22 \leqq Ni_{eq} \leqq 30$ and $16 \leqq Cr_{eq} \leqq 22$ and those values reduce toward marginal lines of single austenitic single phase region. In the case of $Nb_{ave.}$ 0.41 %, the high values are within $19 \leqq Ni_{eq} \leqq 27$ and $19 \leqq Cr_{eq} \leqq 22$ and the values decrease with decreasing in Cr_{eq}. On the other hand, in the case of $Nb_{ave.}$ 0.84 %, there are no clear tendency of the distributions and the values are much higher than those of the other $Nb_{ave.}$ in spite of Ni_{eq} and Cr_{eq}. Theses tendency was also found in TCL distribution in Schaeffler diagram. Thus, niobium affects strongly on SC susceptibility and the influence depends on Ni_{eq} and Cr_{eq}.. This should mean partition behavior of niobium alter with Ni_{eq} and Cr_{eq}.. Therefore, it is necessary to pay attention in the case of practical welding such as first layer during cladding using materials including niobium as an alloying element.

It is known that solidification cracking occurs within temperature range of mush zone. Thus, temperature difference ΔT between liquidus temperature T_L and solidus temperature T_S was calculated in each chemical composition of the specimen using Thermo-Calc (equilibrium calculation, database: SSOL5). In spite of niobium content, similar tendency of the distributions in Schaeffler diagram between BTR and ΔT was found. Figure 4 shows relationship between BTR and ΔT. There is linear relationship between each value and the correlation coefficient is 0.87 while absolute values are different. Therefore, the tendency of BTR distribution must be predicted by ΔT calculation. Moreover, the calculation is one of useful candidate to predict the susceptibility.

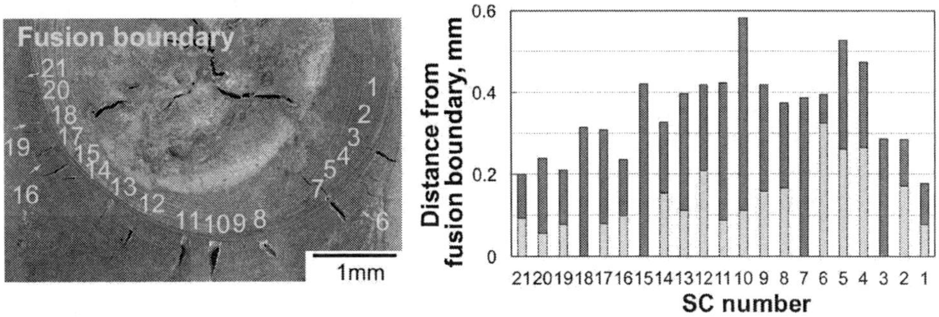

Fig. 2 Distribution of solidification crack. (base material: Type 304L, filler wire: Alloy600, Cr_{eq}=18.3, Ni_{eq}=30.6)

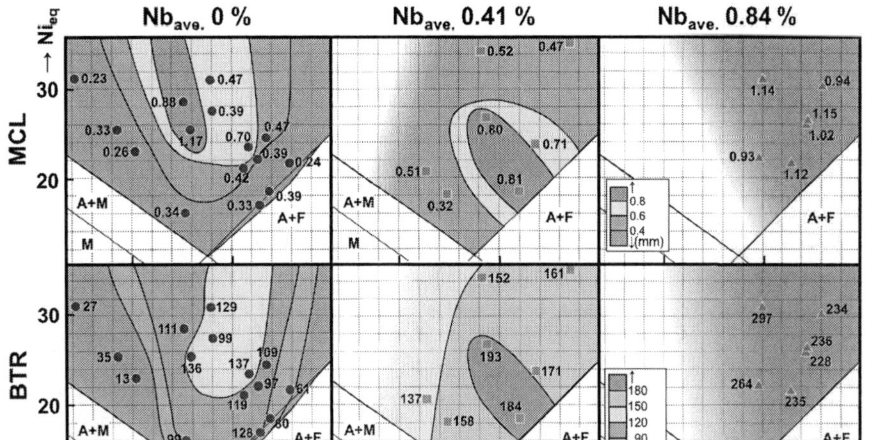

Fig. 3 Distribution of maximum crack length and BTR in Schaeffler diagram.

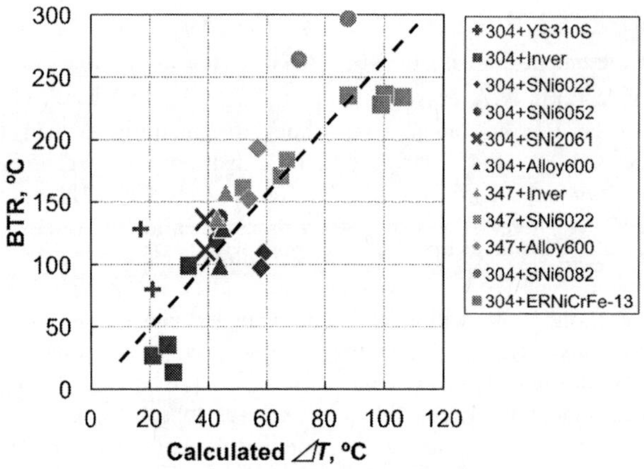

Fig. 4 Relationship between BTR and ΔT.

Conclusions

Solidification cracking susceptibility such as BTR of austenite single phase metals could be evaluated quantitatively and systematically by Trans-Varestraint test using the specimens controlled the chemical composition by Hot-wire laser welding. The distributions of MCL and BTR in austenite single phase region in Schaeffler diagram could reveal high susceptible region to solidification cracking. Beside, containing niobium increased the susceptibility and different distribution of MCL and BTR were confirmed in each niobium content. BTR has good agreement of the tendency with calculated ΔT. Therefore, the thermodynamics calculation is a one of useful candidate to predict the susceptibility.

References

[1] S. Kou: Welding Metallurgy, John Wiley & Sons (New Jersey, 2003)

[2] J.C. Lippold and D.J. Kotecki: Welding Metallurgy and Weldability of Stainless Steels, John Wiley & Sons (New Jersey, 2005).

[3] A. L. Shaeffler: "Constitution diagram for stainless steel weld metal", *Metal Progress*, Vol. 56, No. 11 (1949), pp. 680 - 680B.

[4] T. Ogawa: "*Prevention of Weld Crack; Hot Cracking*", Journal of Japan Wekding Society, Vol. 60, No.3 (1991), pp. 188-193.

[5] V Kujanpaa, N Suutala, T Takalo, T Moisio: "*Correlation between solidification cracking and microstructure in austenitic and austenitic-ferritic stainless steel welds*", Welding Research International, Vol. 9, No. 2 (1979), pp. 55-75.

[6] R. Phaoniam, K. Shinozaki, M. Yamamoto, K. Kadoi, S. Tsuchiya, A. Nishijima, "*Development of a highly efficient hot-wire laser hybrid process for narrow-gap welding - welding phenomena and their adequate conditions*", Welding in the World, Vol. 57, (2013), pp. 607-613.

[7] K. Kadoi, K. Shinozaki, M. Yamamoto, K. Owaki, K. Inose and D. Takayanagi: "*Development of High-efficiency / High-quality Hot-wire Laser Fillet Welding Process*", Quarterly Journal of Japan Welding Society, Vol. 29, No. 3 (2011), pp. 62s-65s.

[8] D. Wang, S. Sakoda, K. Kadoi, K. Shinozaki and M. Yamamoto: "*Investigation of Evaluation Method for Hot Cracking Susceptibility of 310S Stainless Steel during Laser Welding using Trans-Varestraint Test*", Quarterly Journal of Japan Welding Society, Vol. 33, No. 2 (2015), pp. 39s-43s.

Evaluation of Solidification Cracking Susceptibility in Laser Welds of SS310S using Laser Trans-Varestraint Test with In-situ Observation and Two-color Thermometry

Daisuke ABE, Yu MURAKAMI, Fumio MATSUZAKA, Takaaki MATSUOKA, Hiroto YAMAOKA
IHI Corporation, Yokohama, JAPAN
daisuke_abe@ihi.co.jp, +81-45-759-2812

Abstract
This study was performed to evaluate more accurately the solidification cracking susceptibility in laser welds of austenitic stainless steel by in-situ observation and two-color thermometry of Trans-Varestraint test. Fiber laser is applied to Trans-Varestraint test as a power source instead of TIG arc in order to evaluate the welds of laser welding. High-speed camera and optical system for two-color thermometry was combined to laser head to observing crack initiation and measuring the temperature distribution from straight above. The evaluation area near by laser welds was covered and decompressed by the original local chamber to reduce plume which may negatively affect to observation. Using this developed method, crack occurring on laser welds of SS310S during Trans-Varestraint test was observed in the 1000fps and measured temperature distribution from every video image. BTR (brittleness temperature range) obtained by temperatures at both tips of solidification crack was approximately 108K.

Introduction
It is well known that laser welding has a lot of advantages compared with the conventional arc welding process. It includes high efficiency and lower strain and so on. From these advantages, laser welding is expected to applying actual fabrication to improve productivity. However it is reported that laser welding in some conditions causes higher solidification cracking susceptibility. [1] Because of this reason, evaluating cracking susceptibility in laser welding is needed before actual fabrication. Trans-Varestraint test is one of the evaluation tests of solidification cracking susceptibility. In this test TIG arc is used as a heat source. However it is considered that the temperature distribution and/or solidification behavior in laser welding is different from TIG arc welding. So that the conventional Trans-Varestraint test using TIG arc process seems not to accurate to evaluate solidification cracking susceptibility in laser welding.

In this study, fiber laser is applied to Trans-Varestraint test as a heat source. To measure the temperature at both tips of occurring solidification cracking in rapidly cooling in laser welding, in-situ observation by a high-speed camera and two-color thermometry are also applied. In addition, to reduce plume and fume occurred in laser welding which interrupt observation, the evaluated region in Trans-Varestraint test is decompressed by special case box like a vacuum chamber. Using developed facilities for Laser Trans-Varestraint test, the crack initiation in this test is observed and temperatures at cracks are measured directly. After then BTR (brittleness temperature range) is obtained as a solidification cracking susceptibility.

Procedure
Schematic illustrations of Laser Trans-Varestraint test is shown in *Fig.1.* A high-speed camera and original optical system for two-color thermometry is combined to a laser head. Optical path to high-speed camera and it to laser beam are arranged coaxially. The special case box is set between laser head and a specimen. Inside of the case box is decompressed using ejection effect of venture tube. It helps more accurate observation by high-speed camera to reduce plume and fume occurred by laser welding.[2] The laser head and optical system are covered by a chamber filled by N_2. It helps weld metal to prevent oxidation after bending.

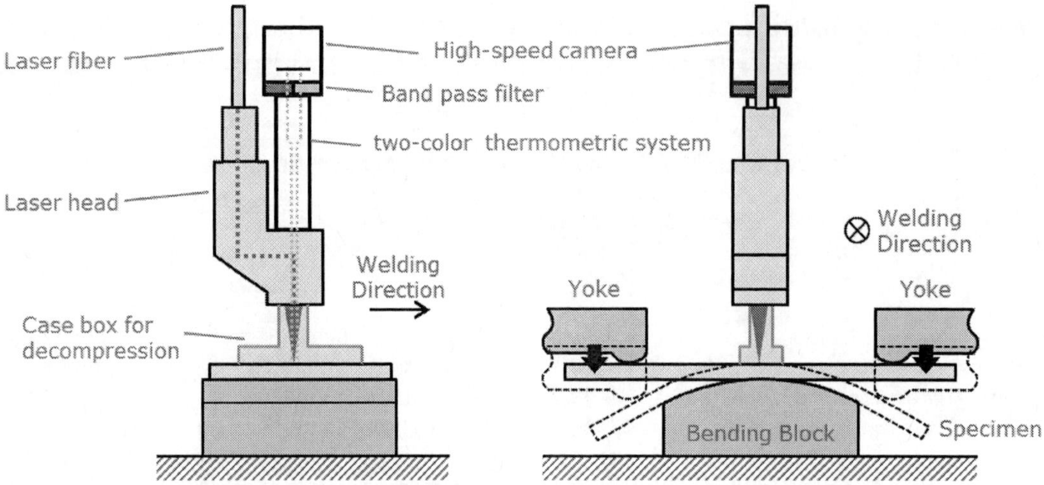

Fig.1 Schematic illustrations of Laser Trans-Varestraint test machine developed

Two-color thermometry

The brightness of emitted light is expressed as a function of wavelength (λ), temperature (T), emissivity (ε) shown in (1). But it is difficult to measure the temperature of the target in high accurate since the emissivity seems to change by kind of material and/or surface conditions. On the other hand, using luminance ratio of two wavelengths like expression (2), the temperature shall be measured not using the emissivity. Two-color thermometry is the temperature measurement method using this expression. From this reason, this method needs two images which wavelength is different of each other.

$$I = \varepsilon \cdot f(\lambda, T) \qquad \cdot\cdot\cdot (1)$$

$$\frac{I_1}{I_2} = \frac{\varepsilon \cdot f(\lambda_1, T)}{\varepsilon \cdot f(\lambda_2, T)} \qquad \cdot\cdot\cdot (2)$$

In this study, a high-speed camera and the original optical system for two-color thermometry are used for in-situ observation and measuring the temperature in Laser Trans-Varestraint test. The appearance of this optical system is shown in *Fig.2*. By this original optical system, the observed image is divided to two images by a prism, and then these two images are filtered in two band pass filters (850nm to 950nm), and finally they achieve to one image pickup device of a high-speed camera. The sample image is shown in *Fig.3*.

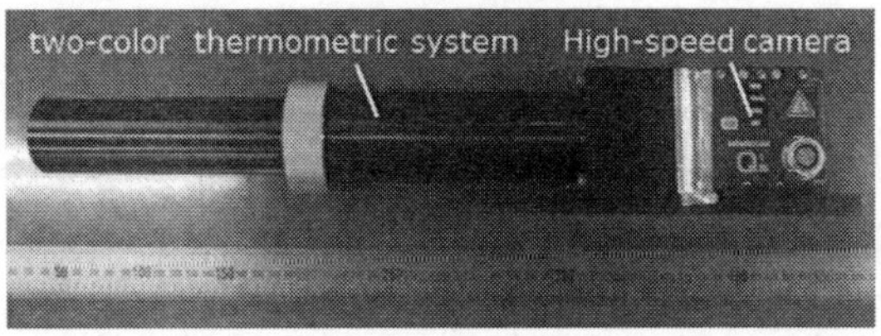

Fig.2 High-speed camera and two-color thermometric system used

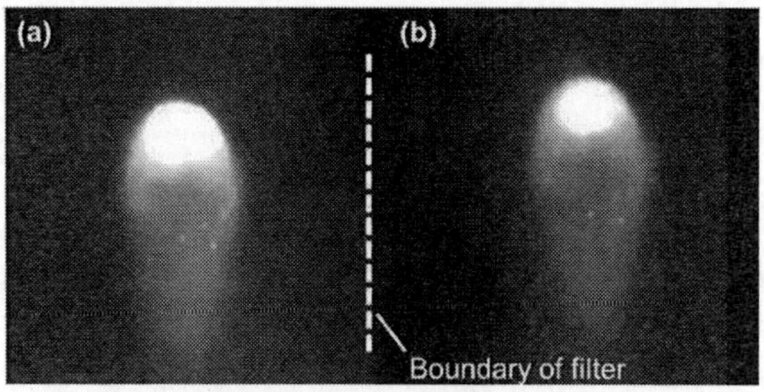

Fig.3 Sample image through two-color thermometric system (a) through 850nm filter, (b) through 950nm filter (one image of high-speed camera)

Materials

In this study, SS310S is used for testing material. The compositions of it are shown in *Table 1*.

Table1 Chemical compositions of material used

Material	C	Si	Mn	P	S	Cu	Ni	Cr	Mo	Nb	Co	Ti	Fe
SS310S	0.05	0.68	0.82	0.024	0.000	—	19.22	24.86	—	—	0.13		Rem

Procedure of Trans-Varestraint Test and Observation

Testing parameters of Laser Trans-Varestraint Test are shown in *Table 2*. Laser power and velocity is selected to put molten pool in the image pickup field of a high-speed camera. Augmented strain is determined enough to occur solidification cracking. Bend speed is maximum of the machine. Delay time from bend start to laser suspend is 0.2mm so that molten pool is not solidified during bending. Degree of vacuum in case box is approximately 0.13atm during laser welding. Oxygen concentration inside the chamber set is controlled less than 0.5% by N_2 shielding to prevent oxidation. Observation parameters are shown in Table3. Shutter speed is determined in which cracks are identified by high-speed camera image in maximum aperture value. Flame rate is highest in this shutter speed. Band pass filters are selected 850nm and 950nm.

Table2 Testing and observation parameter of Laser Trans-Varestraint test

Testing Parameters		Observation Parameters	
Laser Power	1.5 kW	High-Speed Camera	nac Q1m
Velocity	16.6 mm/s	Shutter Speed	1000 flames/s
Bend Speed	400 mm/s	Flame Rate	1/1000 sec
Augmented Strain	4.8%	Filters	850nm, 950nm
Delay Time of Laser Stop	0.2 s		
Degree of Vacuum	0.13 atm		
Shielding	N_2		

Solidification Cracking Phenomena in Laser Trans-Varestraint Test

SEM images of the appearance of specimen after Laser Tarns-Varestraint test is shown in Fig.4. Some of cracks are located at behind of instantaneous solid liquid interface. These cracks have angle of approximately 40 ~ 70° with respect to welding direction. The fracture surface of the crack which is indicated "A" in Fig.4 (a) shows dendrite in Fig.4 (b) so that it is defined as a solidification cracking.

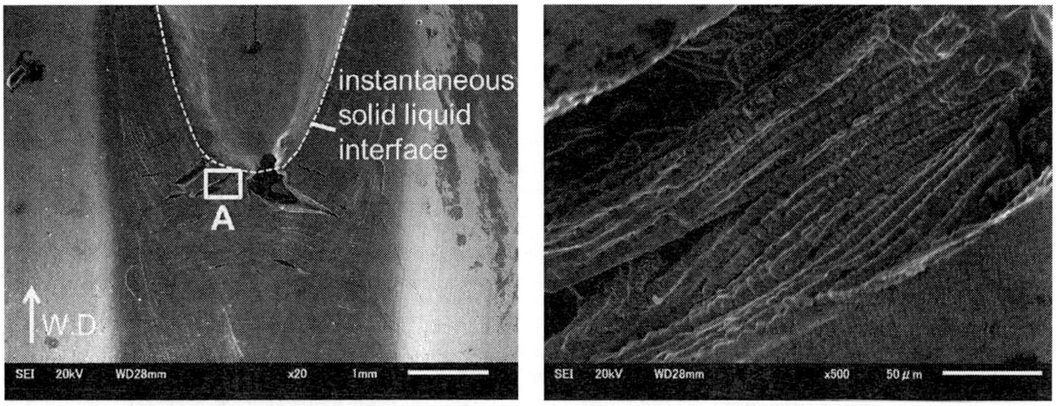

(a) Appearance of bead *(b) Enlarged A*

Fig.4 SEM observations of the surface of specimen after Laser Trans-Varestraint test

High-Speed Camera Image and Temperature Distribution during Cracking

Some images of high-speed camera at crack initiation to 10ms during bending in Laser Trans-Varestraint test are shown in *Fig.5*. These images are filtered 850nm filter. Some of cracks are observed clearly in these high-speed camera images. Cracks are enlarged in 10ms and the lengths of cracks at 10ms after initiation is achieved the lengths at after testing. Temperature distributions calculated by Fig.5 images are shown in Fig.6. It is cleared that these temperature map indicates the temperature not only at molten pool but also solid region around molten pool including cracks. The temperatures of both tips of the longest crack which is indicated "A" in Fig.4 at 10ms after crack initiation are measured. The temperature at solid liquid interface is 1374 deg.C, and the temperature at crack tip furthest from solid liquid interface is 1266 deg.C. The temperature range between further tip of the longest crack in Trans-Varestraint test and solid liquid interface indicates BTR as a solidification susceptibility. Then the BTR of laser welding in SS310S is defined 108K.

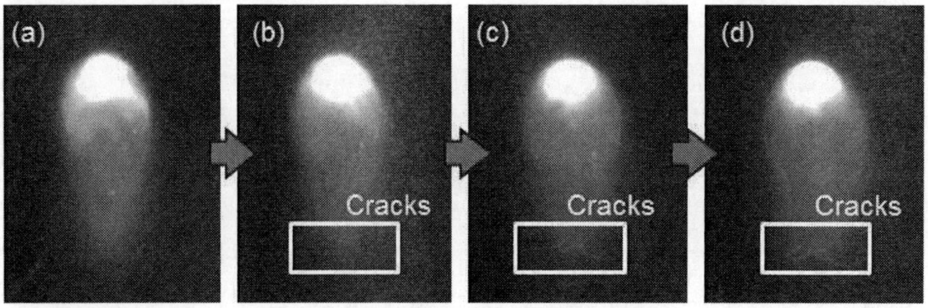

Fig.5 High-speed camera images in Laser Trans-Varestraint test (a) at crack initiation, (b) 3ms after initiation, (c) 5ms after initiation, (d) 10ms after initiation

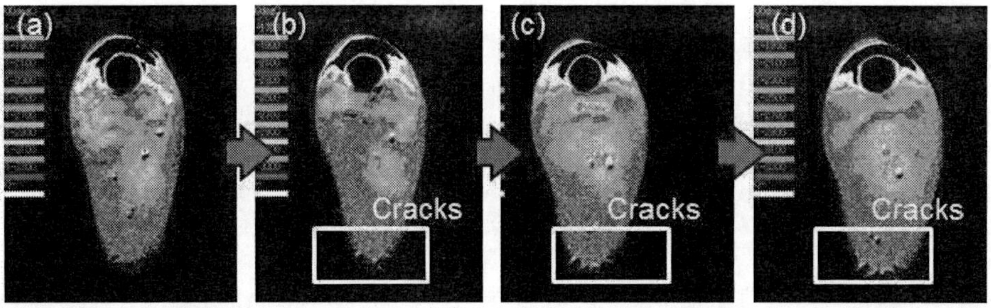

Fig.6 Temperature distributions calculated by two-color thermometry in Laser Trans-Varestraint test (a) at crack initiation, (b) 3ms after initiation, (c) 5ms after initiation, (d) 10ms after initiation

Conclusions

In this study, Laser Trans-Varestraint test was conducted to evaluate more accurately the solidification cracking susceptibility in laser welds of austenitic stainless steel. Not only applying fiber laser as a power source instead of TIG arc, but also high-speed camera observation and temperature measuring by two-color thermometry in order to measure BTR directly in rapid cooling of laser welding. As the result of this study, BTR of laser welding of SS310S was obtained as 108K.

References

[1] E.J.Chun, H.Baba, K.Terashima, K.Nishimoto and K.Saida: "Numerical Estimation of Solidification Cracking Susceptibiilty in Laser Welding of Type 316FR Stainless Steel", Proceedings of International Conference on Joining Materials.

[2] Y.Abe, M.Mizutani, Y.Kawahito and S.Katayama: "High-Power Laser Weldability under Low-Vacuum", Preprints of the national meeting of J.W.S, Vol.87, pp-, (2010)

Effect of resistance spot welding parameters on liquid metal embrittlement of Zn coated high Mn TWIP steels

Siva Prasad Murugan [1], Jin-Hee Bae [1], Md Anwarul Haque [2], Yeong-Do Park [2]

[1] *Department of Advanced materials and Industrial Management Engineering, Dong-Eui University, Busan, Republic of Korea*

[2] *Department of Advanced materials Engineering, Dong-Eui University, Busan, Republic of Korea*

Liquid metal embrittlement (LME) is characterized by extremely rapid penetration of a liquid metal into a susceptible material preferentially along its grain boundaries and simultaneous crack propagation. Critical tensile stress (or strain), critical temperature, threshold strain rate, susceptible microstructure along with the presence of liquid metal are the driving forces for the occurrence of LME. Among them, temperature, stress and strain rate are strongly dependent on the welding parameters such as weld current, weld time and hold time. The aim of this study is to investigate the influence of various welding parameters on the liquid metal embrittlement of Zn coated high Mn TWIP steel. Resistance spot welding of TWIP steels were carried out with different weld current, weld time and hold time. In-situ observations were made for the initiation and propagation of LME assisted cracks with a high speed imaging technique. The SORPAS simulations were also conducted in order to obtain the temperature history and stress distribution on welding. The results reveal that the maximum LME crack is initiated in the super critical LME area in the gap between the electrode and workpiece at the periphery of the electrode during the weld time and it continues to propagate even after the completion of weld time. It extends the study towards a new finding that, the hold time has significant influence on LME.

The effect of alloying elements of gas metal arc welding (GMAW) wire on weld pool flow and slag formation location in cold metal transfer (CMT)

Md. R. U. Ahsan[1,3], *Muralimohan. Cheepu*[2], *Yeong-Do Park** [2,3]

[1]*Department of Mechanical Engineering, International University of Business, Agriculture and Technology, Dhaka 1230, Bangladesh.*
r.ahsan06me@gmail.com
[2]*Department of Advanced Materials and Industrial Management Engineering, Dong-Eui University, Busan 47340, Republic of Korea.*
muralicheepu@gmail.com
[3]*Department of Advanced Materials Engineering, Dong-Eui University, Busan 47340, Republic of Korea.*
**ypark@deu.ac.kr, +82-51-890-290.*

Abstract

The surface tension driven flow or Marangoni flow in welding plays an important role in governing weld bead shape hence affecting the weld joint quality. The surface tension of molten metal usually has a negative temperature coefficient causing the weld pool to flow from the center towards the toe. This temperature coefficient of the surface tension can be altered to be a positive one in the presence of surface-active elements like sulfur (S), oxygen (O), selenium (Se) and tellurium (Te). The amount of deoxidizing elements present in the consumables governs the amount of oxygen present in the weld metal. The presence of a lower amount of deoxidizers results in higher concentration of oxygen in the weld metal. The presence of adequate amount of oxygen can result in a positive temperature coefficient of surface tension gradient in the weld pool. In such situation, the weld pool flows from the toe towards the direction of the center. As a result, of pyrometallurgical reactions in the arc and the weld pool various oxides of the alloying elements are former which are referred as slag. The slags float on the weld pool surface and accumulate following the weld pool flow pattern. As a result, slags form either along the center of the weld bead or the toe depending on the weld pool flow pattern. The slags need to be removed as they degrade the weld bead appearance and paint adhesiveness. Due to easy detachability, slag formation near the center of the weld bead is desired. From in-situ high-speed videography of weld pool, weld metal chemical composition analysis, the effect of consumables alloying elements on weld pool flow pattern and slag formation location are disclosed, which can facilitate the selection of the welding consumables for better productivity in CMT-GMAW.

Introduction

The weld pool convection or weld pool flow pattern is an important factor in governing the weld geometry hence affecting the weld quality [1]. The weld pool flow pattern is affected by buoyancy force, electromagnetic force, plasma arc pressure and surface tension driven force or Marangoni force [2]. Existing studies show that the surface tension driven force or Marangoni force governs the weld pool convection [3], [4]. The weld pool surface shows a temperature gradient as the weld pool direct below the arc has a higher temperature than the locations near the edge [3]. A surface tension gradient is also created across the weld pool surface along with the temperature gradient as the surface tension has temperature dependent property [5]. In most cases, molten metals show a negative surface tension gradient [6]. However, this surface tension gradient can be changed to be a positive one by adding some minor elements like Sulphur, Oxygen, Selenium and Tellurium [7–9]. Heiple *et. al.* disclosed the mechanisms involved in altering the surface tension gradient by addition of the surface active elements [7–9].

Altering the surface tension gradient changes, the weld pool flow pattern. Most of the researchers have been studied on the effect of surface active elements in GTAW [2], [7–11] and Laser welding [5], [12–14]. However, the gas metal arc welding (GMAW) process incorporates the impingement of droplets in to the weld pool involves mixing of mass and momentum along with thermal energy [15]. This makes the GMAW process more complicated than GTAW and Laser welding for studying the surface tension driven flow [15]. Hence, there are very few experimental or theoretical studies available on the weld pool flow and the effect of surface active elements in GMAW. Among the surface active elements, oxygen can be added into the weld metal during welding through the shielding gas. However, alloying elements like silicon and manganese, which are present in the base metal and the wire, have a high affinity to react with oxygen and form silicon oxide and manganese oxide. These oxides accumulate on the surface of the weld pool and forms slag [16]. As these slags float on the weld pool surface, the weld pool flow pattern can govern the slag formation location. Besides, the slag formation location can also give an indication of the weld pool flow pattern.

In this work, two ER70S grade wires are used with the different amount of deoxidizing elements silicon and manganese to study particularly the effect of oxygen content in weld metal in the weld pool flow pattern and the slag formation location. The slag formation and accumulation location are studied using high-speed videography of the weld pool and observing the post weld bead appearance. This study leads to an understanding of the weld pool flow pattern in GMAW depending on the amount of surface active element present in the weld metal.

Experimental

The welds were made on a hot rolled steel plates of thickness 2.3 mm and strength of 440 MPa. The sheets were welded with two 1.2 mm diameter ER70S grade wires named as wire 1 and wire 2, which fall in ER70S-G and ER70S-6 category respectively according to AWS specification. The chemical composition of the base metal is (wt.%) Fe-0.092C-0.011Si-1.3Mn-0.015P-0.005S-0.008Ni-0.023Cr-0.018Cu-0.001Ti. The steel was welded using a Fronious transpuls synergic® 3200 machine and Yaskawa mottoman® robot with an Argon-CO_2 mixture (Ar:CO_2=9:1) as the shielding gas. A constant welding current of 200 Amp and the welding speed of 80 cpm were used throughout the study. The contact tip to work piece distance (CTWD) was maintained at 15 mm.

Results and Discussion

Figure 1 shows the weld bead images for the weldments produced with the mentioned two wires. As observed in Fig. 1(a), the weldment with wire 1 (High Si, Mn wire) has larger slags in the form of islands. The slag islands are entirely located at the toe sections of the weld bead on both sides. In the case of wire 2 (see Fig.1(b)) (Low Si, Mn wire), the slag islands are situated on both sides of the center line for the weldment. There are no visible slag islands at the toe sections for wire 2. The location of the slag islands carries a great significance in terms of slag detachability. The presence of slag on the weld bead surface degrades the appearance along with the paint or primer coating adhesiveness on the bead when there is such post weld painting [17]. Hence, easy detachability of slags is desired. Slag islands located at the center are easy to brush off.

As the liquid slags float on the surface of molten metal, the weld pool flow pattern governs the slag accumulation location. To understand the weld pool flow pattern, *in-situ* high-speed videography was performed for both wire using a frame rate of 5000 frames per second. The high-speed videography data are discussed in the following section.

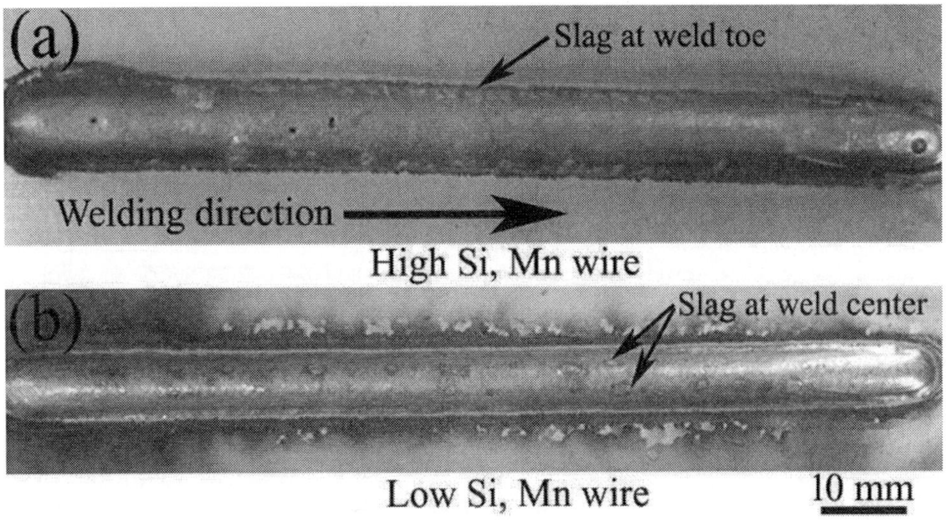

Fig. 1: Weld bead surface images showing the slag formation location for (a) wire 1 and (b) wire 2.

Study of weld pool flow

Frames from high-speed videography for both wires are presented in the Fig. 2 along with schematic representations of the weld pool flow patterns. The high-speed video frames for wire 1 show that the slag particles come from below the arc. The tiny slag particles then flow towards the edge of the weld zone. The tiny slag particles accumulate at the weld toe and grow in size. With the further accumulation of slag particles over time forms large slag islands at the toe, which is also observed in the weld bead image in Fig. 1(a). In the case of the wire 2 (see Fig. 2), the weld pool flow pattern observed through the high-speed camera shows an entirely different behavior. The tiny slag particles come from the arcing region and they move behind the arc following the fusion zone boundary. However, instead of accumulating at the toe, the slag particles follow a converging flow pattern. The converging currents from both ends collide at the center and deflect the flow direction. Thus, two opposite directional rotating region emerged in the weld pool. The slag particles follow this flow pattern and result in slag islands on both sides of the weld pool as observed in the weld bead image presented in the Fig. 1(b)

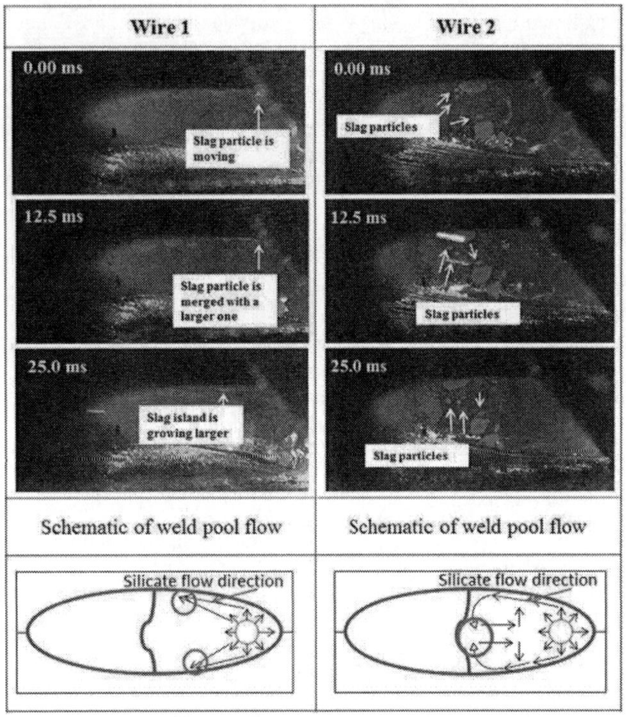

Fig. 2: High-speed movie frames and schematic showing the weld pool flow pattern and the slag formation location for wire 1 and wire 2.

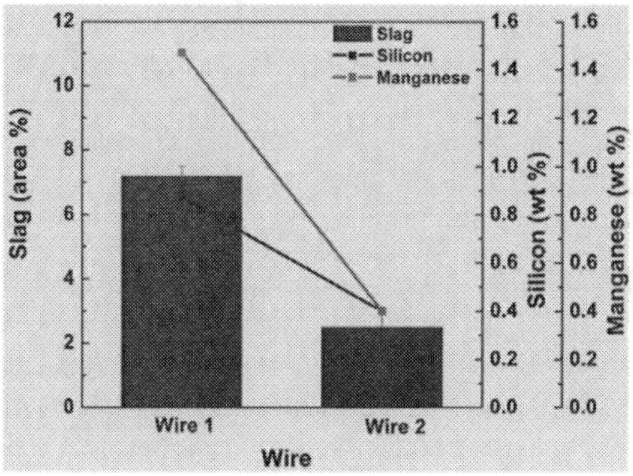

Fig. 3: Quantitative analysis data on slag formation for different wire.

The widely different weld pool flow pattern and slag accumulation location are attributed to the chemical composition of the weld metal. To achieve a better understanding, chemical composition of the weld metals was analyzed through Inductively coupled plasma (ICP) mass spectrometry and dedicated analyzer for oxygen and sulfur. The chemical composition of weld metal for wire 1 is (wt.%) 0.002S-0.002O-0Se-0Te and wire 2 is (wt.%) 0.003S-0.012O-0Se-0Te.

The chemical composition analysis data show the weld metals do not contain any Se and Te as none of the base metal and welding wires contained these two elements. There is a slight difference in Sulfur content for two weldments. However, the oxygen content in the weld metal for wire 2 is significantly higher than the wire 1. Previous researches showed that an oxygen content higher than 0.006 wt.% is required to achieve a positive surface tension gradient [7], [8]. In the case of wire 2, the oxygen content is significantly higher than the threshold value of 0.006 wt.%. The presence of higher amount of oxygen in the weldment produced with wire 2 is attributed to the presence of low amount of deoxidizer elements such as silicon and manganese. Due to the lower amount of deoxidizers present in the system, more oxygen was available for dissolving into the weld metal and altering the surface tension gradient to be a positive one. The lower amount of deoxidizer elements also leads to the formation of less slag as seen in the quantitative analysis data presented in the Fig. 3.

Conclusion

In the present work, the surface tension driven flow or Marangoni flow is investigated for cold metal transfer (CMT) Gas metal Arc Welding (GMAW). The effect of deoxidizing elements present in the welding wire on weld pool flow has also been disclosed. The weld pool flow pattern is governed by the surface tension gradient which has a negative temperature gradient in usual case. Hence, weld pool usually flows into outward direction. However, the presence of surface active elements like Oxygen over the threshold value (>0.006 wt.%) in the weld metal can cause a positive temperature gradient. The amount of deoxidizers present in the welding consumable can affect the amount of oxygen present in the weld metal. Due to high amount of silicon and manganese in wire 1, the weld metal contains a much lower amount of oxygen than the threshold. Hence, the weld pool flows from the center to outward direction and the slags accumulate at the toe. However, in the weldment with wire 1, oxygen content in weld metal is significantly higher than the threshold. This is attributed to the low amount of silicon and manganese in the consumables, which allowed a higher content of oxygen in the weld metal. Due to the higher oxygen content in weld metal, weld pool flows from the toe towards the center. As a result, slag was formed along the center of the weld pool, which is more desired due to easy detachability. Hence, a wire with a lower amount of deoxidizers more desired to avoid additional delay involved in cleaning up the weld bead.

References

[1] S. Lu, H. Fujii, and K. Nogi: "Marangoni convection and weld shape variations in He-CO2 shielded gas tungsten arc welding on SUS304 stainless steel," *J. Mater. Sci.*, Vol. 43, No. 13 (2008), pp. 4583–4591.

[2] Y. Wang and H. L. Tsai: "Effects of surface active elements on weld pool fluid flow and weld penetration in gas metal arc welding," *Metall. Mater. Trans. B*, Vol. 32, No. 3 (2001), pp. 501–515.

[3] P. Sahoo, T. Debroy, and M. J. McNallan: "Surface tension of binary metal-surface active solute systems under conditions relevant to welding metallurgy," *Metall. Trans. B*, Vol. 19, No. 2 (1988), pp. 483–491.

[4] M. J. Mcnallan and T. Debroy: "Effect of Temperature and in Fe-Ni-Cr Alloys Containing Sulfur,"*Metall. Trans. B*,Vol. 22, No. 4 (1991) pp. 557-560.

[5] S. Kou, C. Limmaneevichitr, and P. S. Wei: "Oscillatory Marangoni flow: A fundamental study by conduction-mode laser spot welding," *Weld. J.*, Vol. 90, No. 12 (2011), pp. 229–240.

[6] M. Hasegawa, M. Watabe, and W. H. Young: "Theory of the surface tension of liquid metals," *J. Phys. F Met. Phys.*, Vol. 11, No. 8 (2000), pp. 173–177.

[7] C. Heiple and J. Roper: "Effect of selenium on GTAW fusion zone geometry," *Weld. J.*, (1981), pp. 143–145.

[8] C. R. Heiple and J. R. Roper: "Mechanism for Minor Element Effect on {GTA} Fusion Zone Geometry," *Weld. J.*, Vol. 61, (1982)pp. 97–102.

[9] C. Heiple, J. Roper, R. Stagner, and R. Aden: "Surface active element effects on the shape of GTA, laser and electron beam welds," *Weld. J.*, (1983) pp. 72–77.

[10] C. R. Heiple and P. Burgardt: "Effects of SO_2 Shielding Gas Additions on GTA Weld Shape," *Weld. J.*, (1985) pp. 159–162.

[11] P. F. Mendez, and T. W. Eagar: "Penetration and Defect Formation in High-Current Arc Welding," *Weld. J.*, (2003) pp. 296–306.

[12] B. Ribic, S. Tsukamoto, R. Rai, and T. DebRoy: "Role of surface-active elements during keyhole-mode laser welding," *J. Phys. D. Appl. Phys.*, Vol. 44, No. 48 (2011), pp. 485–203.

[13] C. Limmaneevichitr and S. Kou, "Experiments to simulate effect of Marangoni convection on weld pool shape," *Weld. J.*, Vol. 79, (2000)pp. 231–237.

[14] C. Limmaneevichitr and S. Kou: "Visualization of Marangoni convection in simulated weld pools containing a surface-active agent," *Weld. J.*, vol. 79, No. 11 (2000), pp. 324–330.

[15] Y. Wang and H. L. Tsai: "Impingement of filler droplets and weld pool dynamics during gas metal arc welding process," *Int. J. Heat Mass Transf.*, Vol. 44, No. 11 (2001), pp. 2067–2080.

[16] S. Liu: "Pyrometallurgical Studies of Molten Metal Droplets for the Characterization of Gas Metal Arc Welding," Proc 9th*Trends in Welding Research Conf.*, Chicago, Illinois, June 2012, pp. 353–361.

[17] Y. Umehara, R. Suzuki and T. Nakano: "Development of the innovative GMA wire improving the flow direction of molten pool" *Quart. J. Japan Weld. Soc.*, Vol. 27, NO. 2 (2009), pp. 163–168.

Development of welding materials for pressure vessels made of 9Cr-1Mo-V steel

Kazuya Ikai[*1], Tetsunao Ikeda[*1], Ken Yamashita[*1], Tomoaki Nakanishi[*2], Shigenobu Nanba[*3]
*1 Kobe Steel, Ltd., Welding Business, Fujisawa, Japan
*2 Kobe Steel, Ltd., Industrial Machinery Division, Takasago, Japan
*3 Kobe Steel, Ltd., Materials Research Laboratory, Kobe, Japan

Abstract
The efficiency of hydro processing and heavy crude oil is demanded to cope with foreseen depletion of conventional fossil fuels. 2.25Cr-1Mo-V steel has been widely used for heavy-wall pressure vessels since the 1990's, and ASME Code, Section VIII, Div.2 limits its design temperature up to 482°C. If the process temperature can be increased around 500°C, the process efficiency will be improved significantly. Heat-resistant 9Cr-1Mo-V ferritic steel is excellent in high-temperature mechanical properties and has been used for tubing and piping materials in power boiler plants. For manufacturing the heavy wall pressure vessels made of 9Cr-1Mo-V steel, the welding materials for submerged arc welding of shell rings and for internal overlay by strip welding have been developed taking into account the effect of long-time post weld heat treatment (PWHT) on mechanical properties.

1. Introduction

The steel materials for oil-refining pressure vessels contain such alloying elements as Cr and Mo so that the vessels can work for a long time in the high-temperature high-pressure environment. Since the first half of the 1960's, 2.25Cr-1Mo steel has mainly been used, and since the 1990's, 2.25Cr-1Mo-V steel has been practiced to respond to the increasing requirements for the improved process efficiency by means of higher-temperature and higher-pressure operations. 2.25Cr-1Mo-V steel has excellent mechanical properties, but its design temperature is limited to 482°C in accordance with ASME Code, Section VIII, Div.2. In recent years, development of the upgrading of heavy oil or coal-to-liquid process that can operate at high temperature of 500°C or higher was promoted actively, and thus the process came into practice for a limited application. However, the pressure vessel for such a high-temperature application is made of the 2.25Cr-1Mo-V steel wall whose internal surface is covered with insulation material called refractory, the structure of which causes problems in the maintenance and in-service inspections. In order to overcome such problems, it is desired to develop the single wall pressure vessel made of heat-resistant ferritic steel and suitable welding materials that are applicable for the said process.

For a candidate of heat-resistant ferritic steel that can be used at 500°C or higher, 9Cr-1Mo-V steel was selected. 9Cr-1Mo-V steel has already been proven in many applications for the main steam pipes and reheater tubes in thermal power equipment. Figure 1 shows the maximum allowable stress values for several kinds of Cr-Mo steel in accordance with ASME Code, Section II, Part D. 9Cr-1Mo-V steel can be used with a higher allowable stress in comparison with 2.25Cr-1Mo steel and 2.25Cr-1Mo-V steel at the same temperature and thus is beneficial to reducing the weight of a pressure vessel.

Oil-refining pressure vessels are the structures fabricated by welding the components of forged shell rings, hemispherical heads and nozzles. The pressure vessel has dimensions of 40~70 m in the total length and few meters in the girth diameter for a large one. The shell ring can be as thick as 200~300 mm, and hence narrow-groove weld joints are used to minimize the mass of weld. For such welding joints, submerged arc welding (SAW) is used in order to increase the welding efficiency. The inner surface of the pressure vessel is overlaid with austenitic stainless steel by strip welding in order to protect from corrosion. In this report, introduced are the newly developed welding materials for SAW and internal strip overlay for the high-temperature process.

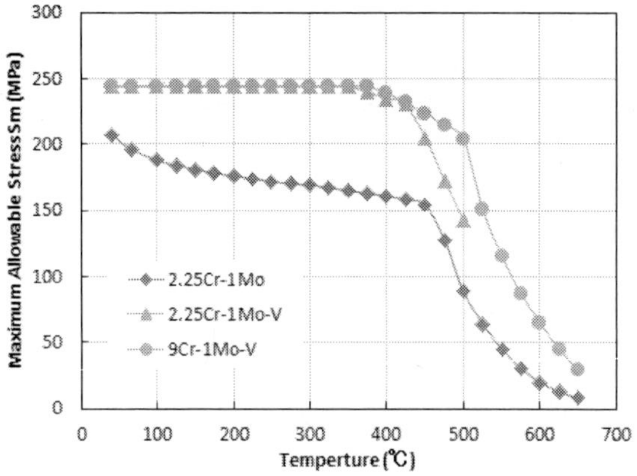

Fig.1 Maximum Allowable Stress of Cr -Mo steel

2. SAW materials for 9Cr-1Mo-V steel

SAW was conducted with a combination of 9Cr-1Mo-V steel wire and low-hydrogen bonded flux, and then the deposited metal was evaluated on various performances. Table 1 shows the welding conditions used in this test. The testing base metal was prepared with 2.25Cr-1Mo steel plates whose beveled surfaces and backing plate were cladded by buttering in advance by using the welding wire having the same chemistry as that of the tested welding wire. Table 2 shows the chemical compositions of deposited metals. The chemical analysis samples were taken from the 8-mm dia. area at the center of deposited metal. Welding material which is conventionally used in power generation industry containing 8.5Cr-0.9Mo-1.5Mn-0.6Ni-0.07C-0.2V-0.03Nb in deposited metal was employed as a comparative standard. The developed welding material is designed with lower amounts of C, Mn+Ni, and Cr together with higher amounts of N and Nb when compared with the conventional ones. The design concept for the chemical composition of the developed material is described in the following.

The circumferential welding of shell ring forgings requires a highly efficient welding procedure with a narrow groove. The design concept of low C is to minimize the hot-crack susceptibility of weld metal. In order to compensate for the decrease in high-temperature tensile strength associated with lower C, the amounts of intergranular carbides and nitrides are increased with higher amounts of N and Nb to improve the performance of weld metal. Since decrease in C and Mn+Ni content promotes the precipitation of soft δ ferrite in the solidification process, ferrite phase former element Cr is reduced. Figure 2 shows the uniform martensitic structure of as casted zone in the final welding pass by optical microscope observation.

Table 1 Welding condition

Wire diameter	4.0 φ
Flux	Bonded Type
Polarity	AC-AC Tandem
Current	L 410A /T 400A
Voltage	L 30V /T 31V
Welding speed	55cpm
L-T distance	20 mm

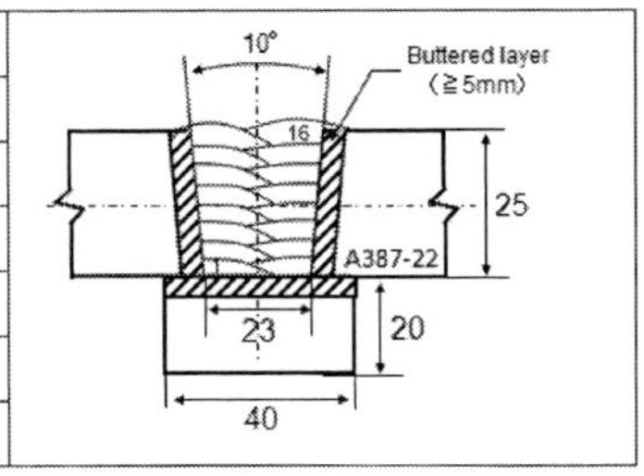

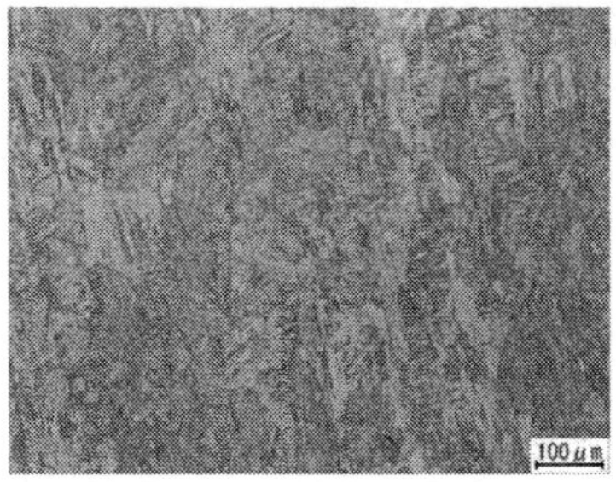

Fig.2 Optical micrograph of deposited metal

Table 2 Chemical composition of deposited metal (mass%)

	C	Si	Mn	P	S	Ni	Cr	Mo	V	Nb	N	Mn+Ni
Conventional	0.07	0.16	1.52	0.008	0.003	0.58	8.46	0.89	0.21	0.03	0.02	2.10
Developed	0.05	0.13	0.85	0.007	0.004	0.77	8.26	0.93	0.23	0.05	0.05	1.62

Oil-refining pressure vessels are fabricated by the welding and PWHT processes repeated several times, and thus the PWHT time can vary for the individual weld joints. Therefore, the welded joints are required to meet the specified performances in the most stringent PWHT condition for the pertinent performance. For instance, the tensile strength of weld metal is required to meet the specified values after 3~4 times PWHT (refer to Max. PWHT); by contrast, the impact value of weld metal is required to meet the specified values after 1-time PWHT (refer to Min. PWHT). The

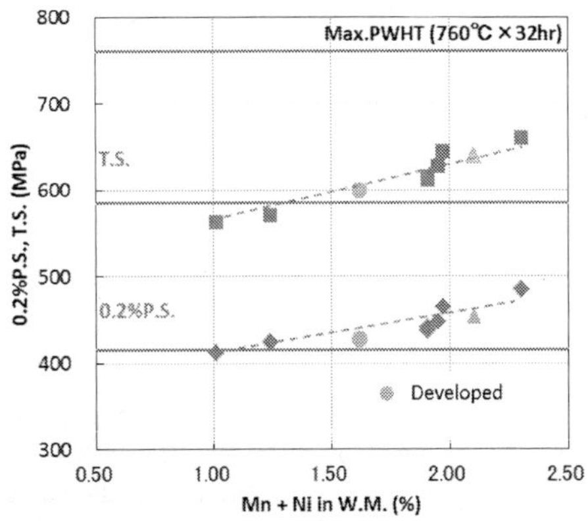

Fig.3 Relationship between Mn + Ni and 0.2%PS, TS at room temperature. (PWHT: 760°C x 32hr)

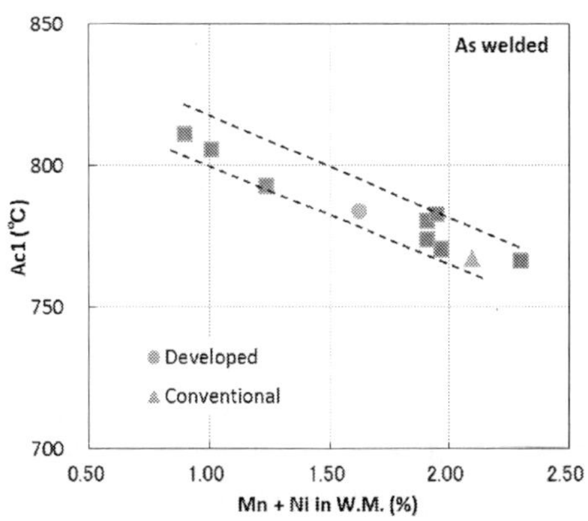

Fig.4 Relationship between Mn+Ni and Ac_1 T.T. of deposited metal

amount of Mn+Ni of the deposited metal made with the developed welding materials was determined based on the room-temperature strength after Max. PWHT and Ac_1 transformation temperature (Ac_1 T.T.). Figure 3 shows the room-temperature tensile properties of deposited metal as a function of its Mn+Ni content. With an increase in the amount of Mn+Ni, the 0.2% proof strength and tensile strength exhibit higher values. The figure clarifies that the Mn+Ni content of deposited metal should be 1.25% or higher to satisfy the targeted values.

Figure 4 shows the Ac_1 T.T. of deposited metal as a function of its Mn+Ni content. For determining the Ac_1 T.T. of deposited metal, the cylindrical test specimen was heated at a constant rate by the high-frequency induction method in the high-vacuum chamber, and then a change in specimen's diameter was read by the contact-free LED, which was used for the calculation of Ac_1 T.T. As the amount of Mn+Ni increases, Ac_1 T.T. decreases. When PWHT temperature exceeds Ac_1 T.T., the deposited metal may transform to martensite after the reverse transformation. The Ac_1 T.T. of deposited metal with developed material is around 780°C and it is above applicable PWHT temperature 730-760°C. In the fabrication of oil-refining pressure vessels, the impact value is required as an important performance. Figure 5 shows the relationship between Charpy impact energy and test temperature after Min. PWHT. The test specimens were taken from the center of plate thickness and the notch was located at the inter-pass region. The impact values of the deposited metal made with the developed welding material were higher than those obtained with the conventional one, and the mean value of Charpy impact absorbed energy was verified to satisfy the specified value (55J min.) in accordance with ASME Sec VIII, Div. 2.

The developed welding material was successfully applied to circumferential welding of full-scale forged shell ring implemented as a mock-up test (Figure 6).

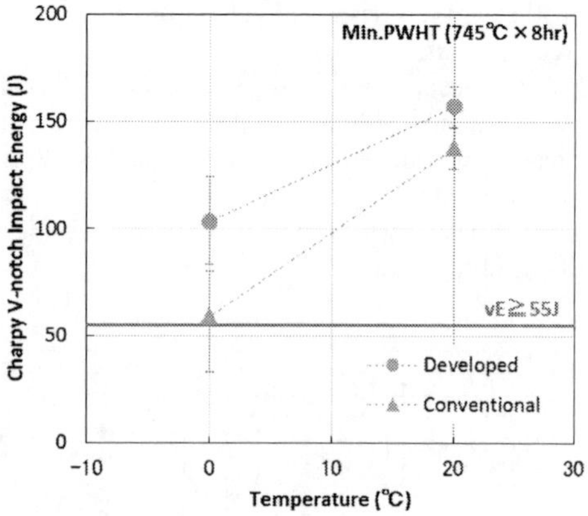

Fig.5 Test result of Charpy V-notch Impact Test

Fig.6 Circumferential welding of full-scale forged shell ring

3. Stainless steel weld overlay

Inside of the pressure vessel is weld overlaid with stainless steel and is also subject to PWHT. The effect of PWHT temperature on sigma phase embrittlement was studied by using type 347 weld metal deposited with 309LNb strip electrode in single layer overlay by electro slag welding (ESW). Figure 7 shows constituents in extracted residue of 347 overlay weld metals by XRD analysis after PWHT at as-weld condition (AW), 705, 750, 775, 800 and 825°C. Sigma phase increases at PWHT 750-800°C than as-weld condition and causes failure in side bend test due to the loss of ductility. The influence of higher PWHT and process temperature on sensitization of weld overlay is also studied. Figure 8 shows reactivation ratios by EPR (Electrochemical Potentiokinetic Reactivation) test method. Reactivation ratio of 347 specimen subject to PWHT at 750°C followed by aging at 500°C is lower than that of the one subject to PWHT at 705°C followed by aging at 450°C. It represents PWHT and process temperature for 9Cr-1Mo-V pressure vessels is favorable in terms of sensitization resistance.

309LNb type strip electrode which provides overlay weld metal with sufficient resistance to sigma phase embrittlement and sensitization was developed by optimizing chemical composition (Table 3). Mn, Cr and Nb are reduced to suppress sigma phase formation. Table 4 shows side bend tests which strain at the bent surface is altered from 11 to 28% demonstrating weld metal sustains sufficient ductility after PWHT at 775°C for 32 hours with developed material. EPR test result also verified sensitization resistance after exposure at operation temperature 500 °C is comparable to conventional weld metal after exposure at 450 °C.

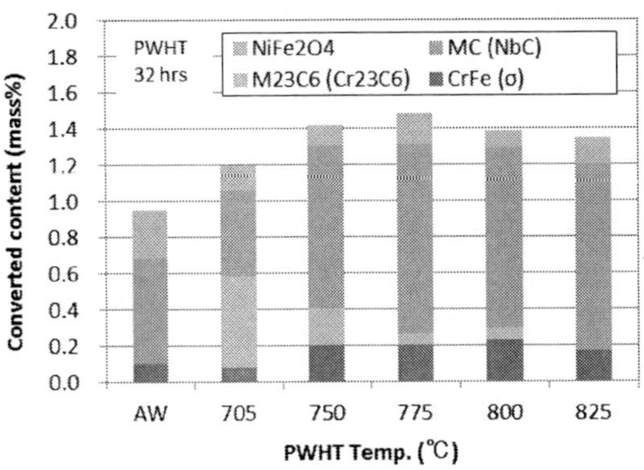

Fig.7 XRD analysis of extracted residue

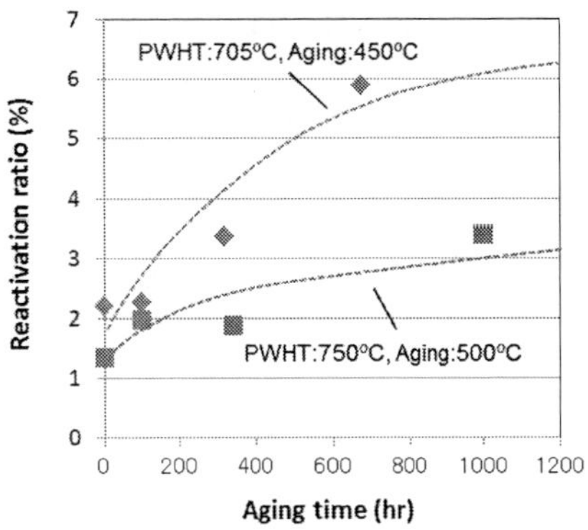

Fig.8 Reactivation ratio by EPR test

Table 3 Chemical composition of overlay weld metal (mass%)

	C	Si	Mn	P	S	Ni	Cr	Mo	Nb	FN*
Conventional	0.03	0.49	1.53	0.017	0.010	10.5	19.3	0.13	0.46	7
Developed	0.03	0.40	0.61	0.004	0.012	9.4	18.2	0.19	0.31	5

* Ferrite Number by WRC1992

Table 4 Bending strain and results in side bend tests

	T(°C)	Bending strain				
		11%	17%	20%	23%	28%
Conventional	705	A	A	A	A	A
Conventional	775	A	A	B	B	C
Developed	775	A	A	A	A	A

A: No crack in all two specimens, B: Crack in one of two specimens C: Cracks in all two specimens

4. Conclusions

The welding materials for SAW and internal stainless steel overlay have been developed for the fabrication of the pressure vessel for the 500°C or higher temperature process. Consequently, the following conclusions have been obtained.

○ SAW materials for 9Cr-1Mo-V steel

・The chemical composition of deposited metal is designed with low C to prevent hot cracking together with high amounts of N and Nb to ensure high-temperature strength. Mn+Ni content of deposited metal is targeted to 1.6% in order to balance the tensile strength and Ac_1 T.T.

・The microstructure of deposited metal is martensite single phase. The impact value of deposited metal is improved as compared with the conventional one.

○ Stainless steel weld overlay

・347 stainless steel overlay weld metal deposited with developed 309LNb type electrode contains less Cr, Mn and Nb than conventional one. It is effective to prevent sigma phase embrittlement and demonstrates comparable sensitization resistance to conventional weld metal after its specific PWHT and aging condition.

The Effects of Weld Filler Material and Cooling Time $t_{8/5}$ on the Dynamic Impact Behaviour of High-Strength Laser-Hybrid Welds

Thora E. Falkenreck[1,2], Maximilian Klein
[1]Rheinmetall MAN Military Vehicles, Kassel, Germany
Thora.Falkenreck@rheinmetall.com, Maximilian.Klein@rheinmetall.com

Thomas Böllinghaus, Marco Lammers
[2]BAM Bundesanstalt für Materialforschung und Prüfung, Berlin, Germany
Thomas.Boellinghaus@bam.de, Thora.Falkenreck@bam.de, Marco.Lammers@bam.de

Abstract
The use of high-strength steels is wide spread in vehicle and crane manufacturing due to light weight reasons. These steels are used for impact of crash loaded components and therefore it is important to investigate high-strength welds at dynamic impact. Investigations of welds at high strain rates have been rarely conducted in the past.
To determine the dynamic impact behaviour of hybrid laser-arc welds, the Split Hopkinson Pressure Bar (SHPB) technique was used. The base material was a quenched and tempered fine-grained structural steel with yield strength of 1100 MPa. The influence of two variables, cooling time $t_{8/5}$ and strength of filler material, on the impact behaviour was studied. A matching and an under matching filler material were used. The cooling time $t_{8/5}$ was varied to influence the microstructure in the heat affected zone (HAZ) and to analyse the effect on hardening and softening. Hardness measurements and optical microscopy was used to analyse the weld microstructure before and after the SHPB test. The investigations showed a correlation between the overall hardness of the weld, influenced by filler material and cooling time, on the maximum stress level during dynamic impact.

Introduction
Welding is one of the most widely used joining technique for metals in industry. Weldments are often required to operate at dynamic loads rather than under static conditions and therefore must also be studied under impact forces. Dynamic impact properties are only well documented for base materials and in fewer cases for filler materials [1–6]. There is a lack of information regarding the impact behaviour of the HAZ of weldments.
The effect of under matching filler material, specifically current filler materials at high strain rates up to 10^3 s^{-1}, has not been deeply investigated [5]. The present study uses the SHPB to investigate the dynamic impact behaviour of weldments welded with different cooling time $t_{8/5}$ and filler materials.

Experimental
Materials
A quenched and tempered high-strength structural steel ($R_{p0.2}$= 1100 MPa) with a sheet thickness of 8 mm was used. For laser-hybrid welding, two different filler materials were chosen to determine the influence of under matching as compared to nearly matching material on dynamic impact behaviour. The under matching filler material "U" was a ferritic solid wire with a yield strength of 620 MPa. As nearly matching material, a medium alloy solid wire electrode referred to as filler material "M" with yield strength of 930 MPa, was chosen. The chemical compositions are summarized in Table 1.

Table 1 : Chemical compositions of materials used, in wt.-%, Fe balanced

	C	Si	Mn	P	S	Cr	Mo	Ni	V	CET[b]
Base material[a]	0.13	0.27	0.92	0.014	0.002	0.62	0.6	1.98	0.006	0.36
Filler material "U" [c]	0.08	0.6	1.8	≤0.015	≤0.018	-	0.4	1	-	0.325
Filler material "M" [d]	0.12	0.8	1.9	-	-	0.45	0.55	2.35	-	0.45

[a] Spark emission spectroscopy; [b] ISO/TR 17671-2 [7], [c] manufacturer's data, EN 12534: G62 4 M Mn3Ni1Mo [8], [d] manufacturer's data, EN 12534: G89 5 M Mn4Ni2,5CrMo [9]

Specimen dimensions and welding experiments
The investigated weld geometry was a 400 mm long butt joint, consisting of a single-layer hybrid laser-arc weld. The metal sheet was prepared for welding with an opening angle of 6 degrees. The butt joint was chosen because of the homogenous characteristic of the HAZ compared to e.g. fillet welds which allowed taking reproducible SHPB specimens out of the weld.
The GMAW process was carried out at a current of 375 ± 10 V and a voltage of 30.2 ± 0.5 A with a 92 % Ar and 8 % CO_2 shielding gas. The Ytterbium laser produced a laser power of 6.2 kW at a welding speed of 1.6 m/min.

The angle between GMA-Torch and laser beam axis was 30°. The hybrid laser-arc welding (HLAW) process was performed in leading arc configuration. An automated temperature measurement apparatus was invented by the Federal Institute of Materials Research and Testing (BAM) to measure the $t_{8/5}$ inside the weld pool. Additionally, the cooling time $t_{8/5}$ was monitored by a two-colour pyrometer with a measuring range of 350 °C to 1300 °C. The cooling rate measured inside the weld pool was slightly higher than measured on the weld surface by the pyrometer. This can be traced back to the influence of radiation from the laser on the pyrometer. Preheating at 120 °C prolonged the cooling time $t_{8/5}$ from 4 s to 6.5 s on average.

Split Hopkinson Pressure Bar Test

The high strain rate experiments were conducted by using a Split Hopkinson Pressure Bar (SHPB) test. The classic studies of Hopkinson [10], Davies [11] and Kolsky [12] are the foundation for the state-of-the-art experimental methods described in [13].
Figure 1 shows the SHPB specimens and their location inside the completed weld. The cylindrical specimens (Ø5 mm, height 4 mm) were taken adjacent to weld centre line to consist of a combination of the HAZ and filler material. The specimens from the upper side of the weld had a higher percentage of filler material than the specimens taken from the lower side of weld. In the discussion, the specimens will be named either "UP" or "LOW" depending on their position inside the weld.
The specimen surfaces were carefully finished to ensure close contact during impact deformation. A strain rate of $3 \cdot 10^3\ s^{-1}$ was achieved with a gun pressure of 170 kPa and a launch velocity of 12.5 m/s.

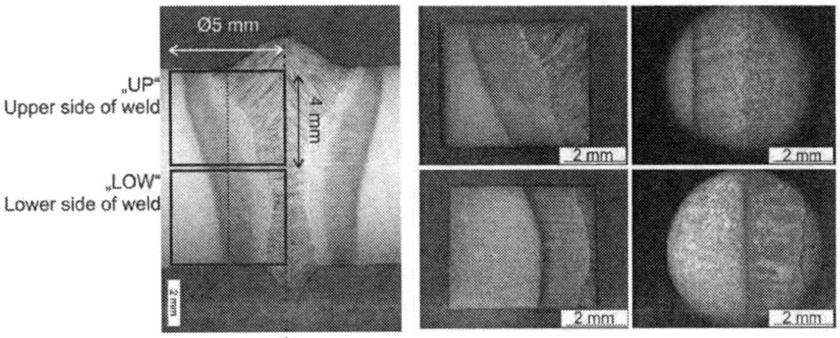

Figure 1: Schematic representation of SHPB specimen location in welded plates; cross-section and longitudinal polish

Results

Hardness measurements

To characterise the laser-hybrid weld microstructure heterogeneity, hardness mapping of the entire weld was performed on the cross section of the welds, as depicted in Figure 2. All specimens showed a non-homogeneous HAZ with hardening and softening zones. Furthermore, the filler material hardness was affected by preheating.
The maximum hardness in the HAZ of the non-preheated welds was significantly higher as compared to preheated welds. Nevertheless, the preheating did not affect the minimum hardness in the softening zone – it only affected the width of the softened zone which was much more pronounced in preheated welds. A softened region is characterised by a hardness drop and a change in microstructure [14]. Compared to the average base material hardness of approximately 470 HV 0.2, the welded joints showed a hardness drop (ΔHV) of more than 130 HV 0.2 in the softening zone.

Stress-Strain Characteristics

The stress-strain curves of specimens from the upper side of the weld, measured by the SHPB test, are shown in Figure 3. The effect of friction was not considered. Figure 3 a compares the stress-strain curves of the base material with the results of different filler materials strength levels. The maximum stress withstood by the base material was higher than the maximum stress of welded material. The welds with under matching filler material ($R_{P0.2}$ = 620 MPa) showed slightly lower maximum stress than the welds with nearly matching ($R_{P0.2}$ = 930 MPa) filler material. Figure 3 b compares preheated and non-preheated specimens and shows that preheating causes a lower maximum stress level at dynamic impact.

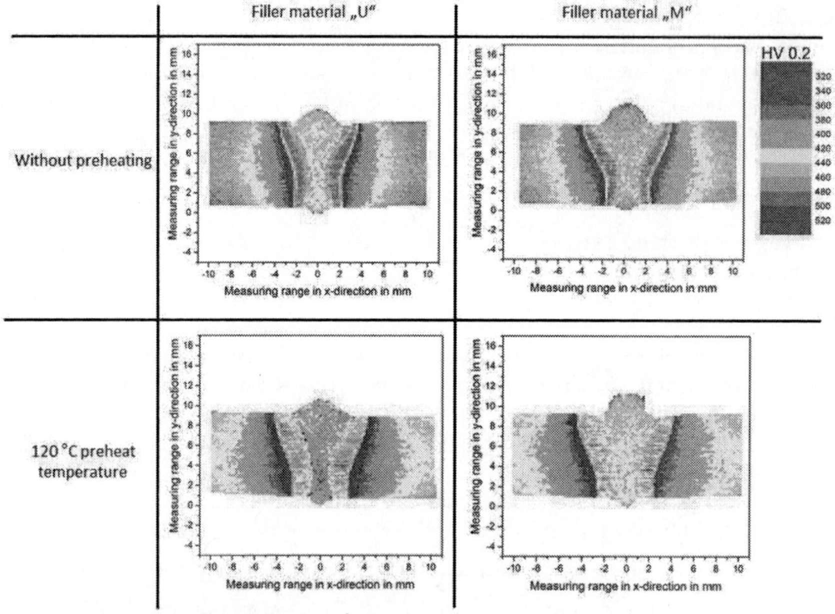

Figure 2: Hardness HV 0.2 over entire weld

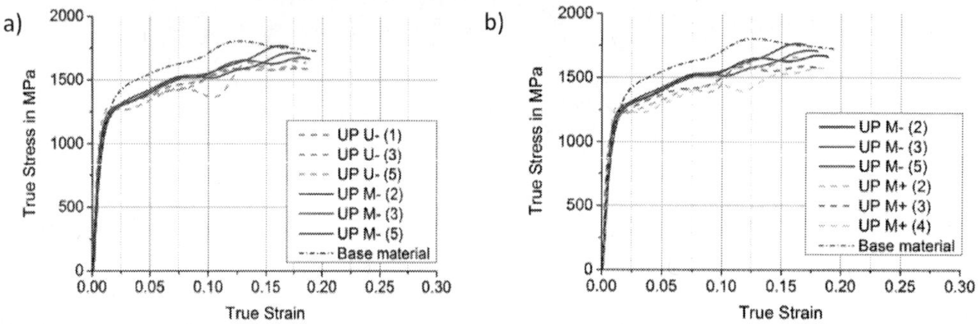

Figure 3: True Stress vs. True Strain for specimens from upper side of weld, a) influence of filler material strength on dynamic behaviour, b) influence of preheating temperature on dynamic behaviour; (U: under matching filler material, M: nearly matching filler material, -: without preheating, +: 120 °C preheating temperature)

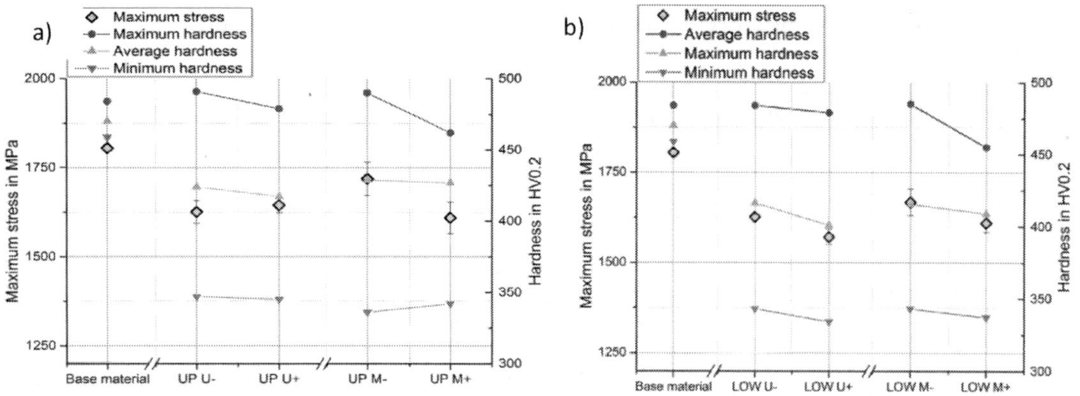

Figure 4: Maximum stress during SHPB experiment and hardness in specimens vs. material, position, filler and preheating condition, a) upper side of the weld, b) lower side of weld

To investigate the effect of preheating and under matching filler material on the maximum stress, the maximum stress was correlated with the maximum, minimum and average hardness per specimen (Figure 4). A correlation between maximum stress and average hardness inside the specimen was detected. This investigation coincides

with studies of Weerasooriya and Moy [15].
The impact caused plastic deformation of the specimens and the average size of the specimens after impact was Ø 5.6 mm and height 3.3 mm. The microstructures were investigated by optical microscopy before and after impact. Figure 5 compares the microstructure of the coarse grained HAZ (CGHAZ) and intercritical HAZ (ICHAZ). The CGHAZ adjacent to the fusion line suffered temperatures up to the A_{c3} temperature. It experienced fast grain growth due to dissolution of carbides and nitrides and consisted mostly of martensitic phase. The peak temperature range for the ICHAZ is between A_{c1} and A_{c3} temperature and caused partial austenitisation resulting in softening which was confirmed by hardness measurements of the welds. The investigation of the microstructure by light microscopy did not show any significant differences between unimpacted specimens and specimens after the SHPB test. Twinning was not detected. Further analyses by Electron Backscatter Diffraction are necessary.

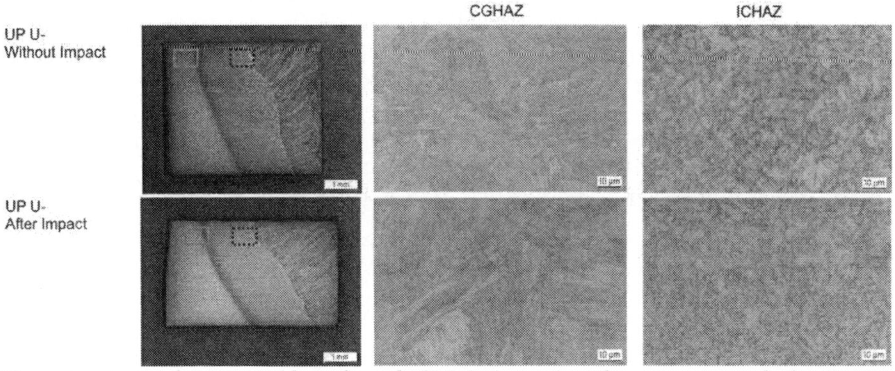

Figure 5: Microstructure of CGHAZ (dotted) and ICHAZ (green) without impact and after SHPB experiment

Conclusions

This study has reported upon the impact compression behaviour of high-strength laser-hybrid welds at a strain rate of $3 \cdot 10^3$ s^{-1}. During welding, the filler material strength and preheating temperature was varied. Preheating at 120 °C caused less hardening in the HAZ and a more distinct softening zone than in non-preheated welds. Cylindrical specimens for the SHPB test were taken adjacent to weld centre line and consisted of a combination of the HAZ and filler material. The investigation of "true stress/true strain" curves confirmed that the base material withstood a higher maximum stress than any welded material. Furthermore, weldments that consisted of nearly matching filler material or without preheating achieved higher stress levels than weldments with preheating or having under matching filler material. A correlation was detected by comparing the maximum stress level and the average hardness level of the respective specimens. Investigations of the microstructure using light microscopy of untested and tested specimens did not show significant differences; no twinning was detected.
Further studies will investigate the dynamic behaviour of physically simulated single weld microstructure (i.e. by Gleeble^{TM}, dilatometer etc.) which will allow a homogenous microstructure inside SHPB specimens.

References

[1] B. Burgan: Elevated temperature and high strain rate properties of offshore steels. Ascot, Steel Construction Institute, 2001

[2] G.E. Linnert: Welding Metallurgy, Carbon and Alloy Steels, Volume 2. American Welding Society, 1967, 674 pp.

[3] B. Cary, Howard; S.C. Helzer: Modern Welding Technology. Pearson/Prentice Hall, 2004, 715 pp.

[4] B. Wang; S. Yi: Dynamic plastic behavior of 63 wt % Sn 37 wt % Pb eutectic solder under high strain rates. In: Journal of Materials Science Letters, Vol. 21, 2002, pp. 697–698

[5] B. Wang; G. Lu: Dynamic Strength of Steel Welds under High Strain Rate Loading. In: Advanced Materials Research, Vol. 9, 2005, pp. 87–92

[6] W. Lee; C. Lin; C. Liu; C. Cheng: The Effects of Strain Rate and Welding Current Mode on the Dynamic Impact Behavior of Plasma-Arc-Welded 304L Stainless Steel Weldments. In: Metallurgical and Materials Transactions A, Vol. 35, Nr. May, 2004

[7] ISO/TR 17671-2: Welding - Recommendations for welding of metallic materials - Part 2, 2002

[8] Oerlikon: CARBOFIL NiMo1 GMAW solid wire, 2016

[9] Boehler Welding: Union X 96 GMAW solid wire, 2005

[10] B. Hopkinson: A Method of Measuring the Pressure Produced in the Detonation of High Explosives or by the Impact of Bullets. In: Philosophical Transactions of the Royal Society of London, Vol. 213, 1914

[11] R.M. Davies: A Critical Study of the Hopkinson Pressure Bar. In: Philosophical Transactions of the Royal Society of London A, Vol. 240, Nr. 821, 1948, pp. 437

[12] H. Kolsky: An Investigation of the Mechanical Properties of Materials at very High Rates of Loading. In: Vol. 676, 1949

[13] W. Chen; B. Song: Split Hopkinson (Kolsky) Bar- Design, Testing and Applications. Springer, 2011, 388 pp.

[14] H.G. Pisarski; R.E. Dolby: The significance of softened HAZs in high strength structural steels. In: Welding in the World, Vol. 47, Nr. 5/6, 2003, pp. 32–40

[15] T. Weerasooriya; P. Moy: Effect of Strain-Rate on the Deformation Behavior of Rolled-Homogeneous- Armor (RHA) Steel at Different Hardnesses. In: Proceedings of the 2004 SEM International Congress, 2004

Experimental Study on fatigue strength for U-rib welded joint of Metal Deck Bridge using Laser Arc Hybrid welding with LTT filler wire

Koutarou INOSE [1], Junko KANBAYASHI [1], Daisuke ABE [1], Naoyuki MATSUMOTO [1], Daiki OKITA [1], Seiji OKADA [2]

[1] *IHI Corporation, JAPAN*
[2] *IHI Infrastructure System, JAPAN*

Fatigue strength of Metal Deck Bridge has been improved by using optimum controlled Laser arc hybrid welding (LA-Hybrid welding) with Low Transformation Temperature filler wire (LTT filler wire). Metal deck consists of stiffened deck plate by U-shaped closed section rib welded by one side welding. In the case of conventional arc welding, appropriate penetration of one side welded joint require skillful welding work. Lack of penetration is cause of fatigue crack initiation. Experimental assessment method for metal deck welded joint has been established to estimate fatigue strength of conventional arc welded joint. SN curve of fatigue strength for conventional arc welded metal deck joint has been derived. Fatigue life can be calculated using local stress rage at 5mm distance from root of the welded joint and number of the cycle. Fatigue life has been estimated using the SN curve with 72 hours stress range record measured on existing heavy traffic bridge. Roughly, the calculated fatigue life can be account for damage of the real bridge. For improving fatigue strength, LA-Hybrid welding including its adequate welding condition using LTT welding filler wire has been investigated. LTT filler wire is made of 3%Ni, 12%Cr, and the others. Martensitic transformation temperature of all deposit metal is 312 degrees Centigrade. Fatigue life of welded joint by deep penetrated LA-Hybrid welding with LTT filler wire is at least as 2 times long compare with conventional arc welded joint

Three Axis Residual Stress State at Stress Anomaly Point of Brazed Ceramic/Metal

Shun-Ichiro Tanaka
NICHe, Tohoku University, Sendai, Japan
sitanaka@tagen.tohoku.ac.jp

Abstract

Measured residual stress tensor distributions around interfaces of Si_3N_4/Cu/Steel and Si_3N_4/Cu brazed by Ag-Cu-Ti using micro-area X-ray 2D method were shown. Around a stress singularity point of the interface, normal stress values perpendicular to the interface reached only 130 MPa in tensile, on the other hand maximum principal stress reached 300 MPa by the influence of shear stress. Stress values in depth direction reached as high as 100 MPa in tensile.

Introduction

Experimental efforts have been made in author's laboratory since 1982 to measure the actual stress/strain in a micro-area around ceramic/metal interfaces [1]. The measurement methods being developed and applied are X-ray micro-$\sin^2\psi$ method, micro-area X-ray two dimensional (2D) method, surface acoustic wave velocity, micro-Raman scattering, micro-FTIR and convergent beam electron diffraction (CBED) in TEM. These methods provide tools to evaluate the actual stress/strain quantitatively in a micro-area for all material. The diameter of measured area ranged from sub-millimeters to nanometers.

Why we need to measure micro-area stress/strain? First it affects various properties of materials and the joint, second for quality assurance using actual stress/strain, and third to check theory or computer simulation. Using X-ray diffraction it is possible to evaluate stresses in each phase composing joint or composite and to predict a crack initiation point without loading. Stress measuring method using X-ray diffraction is applicable to polycrystal materials and $\sin^2\psi$ method has been widely utilized as a conventional method [2]. However, it is limited to obtain two axial stress obtain based on plain stress hypothesis. X-ray stress 'two dimensional (2D) method' can be applied to obtain three axis stress states utilizing integration points along Debye ring detected by two dimensional detector and stress in depth direction is succeeded to measure lead to obtain stress tensor in the confined region [3].

In this paper I will introduce recent outputs of stress tensor measurement around brazed interface, especially at the stress anomaly points of the joint.

Experimental

Joined specimens were Si_3N_4/t0.2Cu/stainless steel and Si_3N_4/Cu brazed by Ti-Ag-Cu filler at 1143K in vacuum. Residual stress tensor distributions were obtained around interfaces using micro-area X-ray 2D method which provides three dimensional tensor of normal stresses, σii, shear stresses, σij, where i, j =1, 2 & 3 denoted in the specimen coordinate, principal stresses σI-III and Euler angles. X-ray condition for Si_3N_4 was Cr-Kα 38 kV, 16 mA at (212) $2\theta_0$=131.55 deg and irradiated area in Si_3N_4 was 0.1 mm in diameter with collimator, whereas diffraction plane in Cu, Ag and γ phase in stainless steel were (220) at $2\theta_0$=127.2 deg, (311) at $2\theta_0$=136.8 deg and (220) at $2\theta_0$=128.5 deg, respectively, and irradiated area was 1.0 mm in diameter for all metals. Obtained principal stresses and Euler angles were drawn as 'Stress Ellipsoid' to help understanding of the feature.

Measured residual stress distribution at an anomaly point and interface

Residual stresses in Si_3N_4 at a stress singularity point of brazed specimen

Around a singularity point of the interface nearby edges of Si_3N_4 part normal stress values perpendicular to the interface reached only 130 MPa in tensile and normal stress values in depth direction reached 100 MPa in tensile. On the other hand maximum principal stress reached 300 MPa by the influence of shear stresses, as shown in Fig. 1, which value is comparable to the tensile strength of Si_3N_4. It is important to evaluate 3 axis stress values at stress anomaly points.

Residual stresses in metal parts of brazed Si_3N_4/metal

Residual stress tensor distributions around brazed interface in Cu, Ag and steel were also obtained in detail compared with those in Si_3N_4. Although stress values showed large tensile at anomaly points, stresses in metal parts showed compressive stress compensating tensile stress in Si_3N_4. Equivalent stress such as von Mises stress showed minimum inside anomaly points and maximum at the center of the joint.

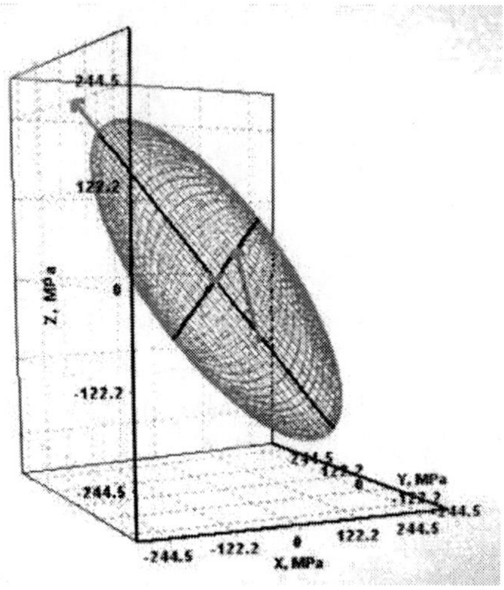

Fig. 1 Stress ellipsoid at a stress anomaly point of brazed Si_3N_4/Cu/SUS specimen

Conclusions

1. Stresses should be measured by 3 dimensional analysis, such as X-ray 2D method, around stress anomaly points of joint where three axis tensile stress states remained.

2. Around a singularity point of in Si_3N_4 part normal stress values perpendicular to the interface reached only 130 MPa in tensile and normal stress values in depth direction reached 100 MPa in tensile. Maximum principal stress reached 300 MPa comparable to the strength of Si_3N_4.

3. In case of Si_3N_4/Cu/steel brazed joint, axial normal stress values in Cu, Ag, stainless steel are compressive except stress singular points.

4. Von Mises stress showed minimum inside anomaly points and maximum at the center of the joint.

References

[1] Shun-Ichiro Tanaka: "Measurement of Stress/Strain Distributions in a Micro-Area around a Heterogeneous Interface", *Materials Science Research International, Special Technical Publication-1* (2001) pp. 312-315
[2] Shun-Ichiro Tanaka and Yumiko Takahashi: "Effects of X-Ray Beam Collimation on the Measurement of Residual Stress Distribution in a Si_3N_4/Steel Joint", *ISIJ International*, **30** (1990) 1086-1091
[3] Bob B. He: Two-Dimensional X-Ray Diffraction, John Wiley & Sons (2009)

Bonding between Cr-alloyed SiAlON ceramic and heat-resistant alloys

Toko Tokunaga [1], Masayuki Koshiyama [2], Munekazu Ohno [1], Kiyotaka Matsuura [1]

[1] *Faculty of Engineering, Hokkaido University, Japan*

[2] *Graduate School of Engineering, Hokkaido University (Currently Mitsubishi Materials Corporation), Japan*

Recently, it has been strongly required to improve the heat efficiency of the thermal power generation system to save fuel resourccs and prevent the global warming. In order to improve the heat efficiency, advanced-ultra supercritical technique has been developed to increase the steam temperature up to around 700 ℃. In order to realize such a high steam temperature in the system, appropriate heat-resistant alloys are essential. Although as promising heat-resistant alloys such as Ni-based superalloy and SUS304 stainless steel have been proposed until now, there still remain some concerns such as high-temperature corrosion by flaming gas and erosion by combustion ash. Therefore, in the present study, SiAlON ceramic coating on INCONEL X-750 and SUS304 is proposed, because SiAlON has excellent heat, wear and corrosion resistances. Furthermore, 10 mass% Cr addition to the SiAlON ceramic is also attempted to improve the oxidation resistance. In the present study, brazing was applied as the bonding method, and Cu and Ag were used as the filler to minimize the stress generation due to the difference in thermal shrinkage between the ceramic and alloy. Y-α-SiAlON powder was sintered by the spark plasma sintering method and disk-like samples were produced. The sample was bonded with the alloys having the similar disk shape at 1150 ℃ with the Cu filler, while at 1050 ℃ with the Ag filler. The holding time at the bonding temperatures was varied from 10 to 60 minutes, and the effect of the holding time on the bondability was investigated. It has been found that although SiAlON/INCONEL bonding was feasible only with Ag filler, SiAlON/SUS304 bonding was feasible with both Ag and with Cu fillers. As for the Cr-alloyed SiAlON, same as the case of SiAlON without Cr, the SiAlON/INCONEL bonding with Ag filler and the SiAlON/SUS304 bonding with both fillers were feasible.

Vacuum free self fusion solidification bonding of aluminum by periodic motion

Hiroshi Kawakami, Kenta Matsuda, Seiya Kodama, Toshiki Nagao, Jippei Suzuki
Mie university, Tsu, Mie, Japan
kawakami@mach.mie-u.ac.jp, +81-59-231-9474

Abstract

Introduction

Al oxide film behaviors as a barrier of solid state diffusion at the bonding interface. Al/Cu interface can be melted with increase of the solute concentration in each metal by the reaction diffusion over eutectic temperature[1-3]. When the solute concentration of the liquid decreases to that of solid by the progress of the diffusion, the liquid solidified. In self fusion solidification bonding process, fusion and solidification of the bond occurs under isothermal condition. Periodic motion of the bonding interface breaks Al oxide film by the impact force effectively and obtains the contact at Al/Cu interface[3]. Increase of impact number may promote the fracture of Al oxide film and the reaction diffusion between the Al/Cu interfaces. It is cleared that number of impact between each metal surface increases with increase of the frequency of the periodic motion. Effect of the frequency of periodic motion of the bonding interface on this bonding process was investigated in this study.

Experimental procedure

The rods of pure aluminum and the disk of pure copper with the diameter of 6mm were used in the bonding experiment. The bonding specimen was heated in the air. The bonding temperature was 893K. The temperature of bonding experiment was decided between melting points of each metal and eutectic temperature of Al-Cu system. Al base metal and Cu insert metal were occurred isothermal fusion and solidification by reaction diffusion at the bonding temperature. Periodic motion of the upper bonding interface was occurred by the programmed linear actuator. The maximum bonding pressure of each cycle of periodic motion was measured and controlled by the lower load cell. Typical bonding pressure histories in each periodic motion are shown in Fig.1. Set point of maximum bonding pressure in each frequency was 1MPa.

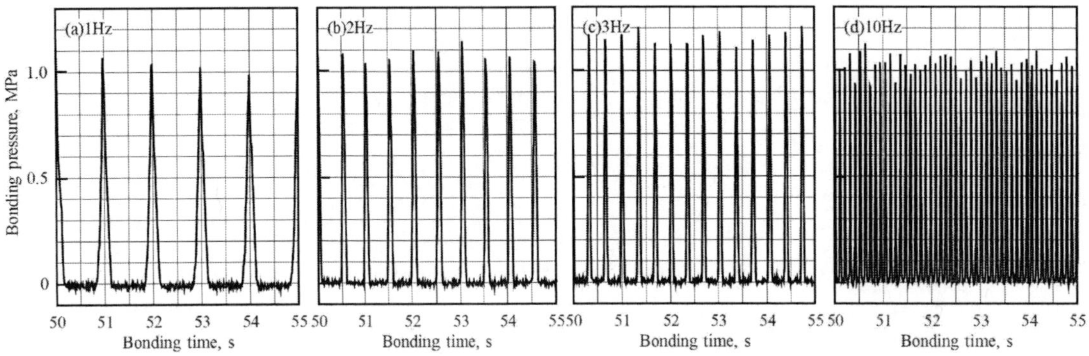

Fig.1 : Examples of waveforms of periodic motion at four frequency bonding conditions.

Experimental result

Fig.2 shows the relationship between the frequency of periodic motion and the joint strength. The effect of frequency on the initial bonding time was confirmed at low frequency bonding conditions below 3Hz. The initial bonding time depends on the number of impact of bonding interfaces. However, time saving effect on the initial stage of self fusion solidification bonding process by increase of frequency disappeared at high frequency bonding conditions like 10Hz. Effect of insert metal thickness on the change of the joint strength was confirmed by this experiment results. Fig.3 shows the relationship between the frequency of periodic motion and the bonding area ratio. The effect of frequency on the fracture of Al oxide film was confirmed at low frequency bonding conditions below 3Hz. The initial bonding time depends on the number of impact of the bonding interfaces. However, time saving effect on the Al oxide film fracture in self fusion solidification bonding process by increase of frequency disappeared at high frequency bonding conditions like 10Hz. The effect of insert metal thickness on the bonding area was confirmed as well as the change of the joint strength in Fig.2.

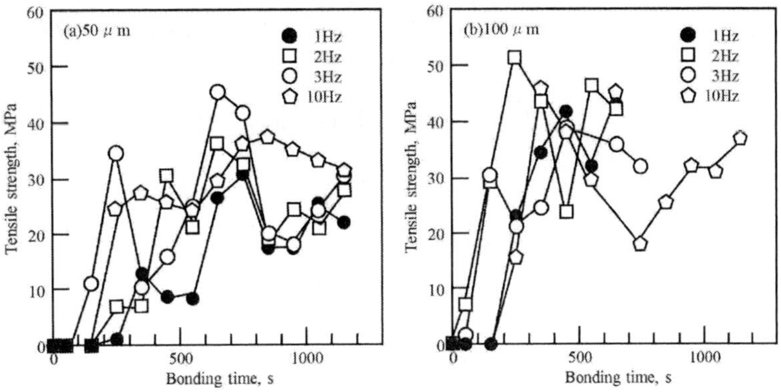

Fig.2 : Change of joint strength at four frequency bonding conditions.

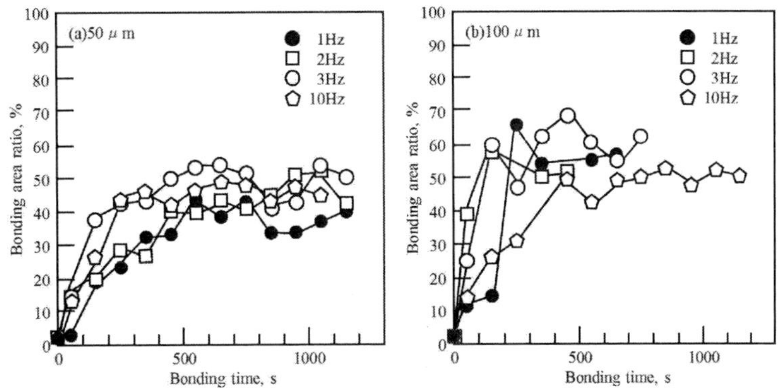

Fig.3 : Change of bonding area ratio at four frequency bonding conditions.

Conclusions

Effect of frequency on the change of the joint strength and the bonding area in Al self fusion solidification bonding process were confirmed at low frequency bonding conditions below 3Hz. At high frequency bonding condition like 10Hz, this effect disappeared.

Acknowledgments

This work is supported by JSPS KAKENHI Grant Number 15K06464.

References

[1] H. Kawakami, J. Suzuki, M. Fujiwara, J. Nakajima: Effect of bonding condition on Al/Cu dissimilar bonding with liquefaction by reaction diffusion in air, Quarterly Journal of the JWS, 25-1(2007), pp.24-30. (in Japanese)

[2] H. Kawakami, J. Suzuki, J. Nakajima: Bonding process of Al/Cu dissimilar bonding with liquefaction in air, Quarterly Journal of the JWS, 25-1(2007), pp.51-58. (in Japanese)

[3] H. Kawakami, T. Nagao, K. Kimura J. Suzuki :Study of displacement-load control for vacuum free self-induced fusion solidification bonding of aluminum with copper insert metal, Journal of Smart Processing, 4-2(2015), pp.83-88.

Margin and file formats
Margin requirements are listed in Table 1. Microsoft Word 2000 or higher version is the preferred file format. You may use other programs, but you must follow the formatting specifications below, and the manuscript must still be submitted as a word processing document. Do not use Corel Ventura or Corel Draw to format your manuscript. Mac users may convert this file to the Mac version of Word 2000 or higher, and save it with a ".doc" extension in the filename. The file size must NOT exceed 10MB.

Table 1: Table titles appear above the table. There is one line space between the table title and the actual table. One line separates the final line of the table and continuation of the narrative.

Page Setup for Manuscripts	
Top	25mm
Bottom	25mm
Left	20mm
Right	20mm

Paragraphs
Do not indent paragraphs; use a blank line to separate them. To separate the sections, use two blank lines. Use only one space between sentences.

Tables and figures
Arrange figures and tables to allow text to flow around them. If they are created in a spreadsheet or graphics program, simplify them for readability. All images contained in the manuscript must be embedded in the file and of a minimum 300 dpi print resolution. Photographs (color and B/W) and scanned images are acceptable when embedded in the file. Acceptable bitmap graphic files are EPS, TIFF, and JPEG.

Before each table, type *the table number and caption in italics*. The table number (using the format, *Table X*) must be referenced in the text. All tables must be center-aligned.

After each figure, type *the figure number and the caption in italics*. The figure number (using the format, *Fig. X*) must be referenced in the text. All figures must be center-aligned.

Effect of additional element on interfacial strength at Al/Fe dissimilar metal weld by TIG arc brazing

Hiroki Furuya[a], Yoshikatsu Sato[a], Yutaka S. Sato[a], Hiroyuki Kokawa[a], Yujiro Tatsumi[b]
[a]Department of Materials Processing, Graduate School of Engineering, Tohoku University, Sendai, Japan
Contact e-mail: hiroki.furuya.p7@dc.tohoku.ac.jp, TEL: +81-22-795-7353
[b]NIPPON STEEL & SUMITOMO METAL CORPORATION, Chiba, Japan

Abstract
Dissimilar welding of aluminum alloys containing Ni, Cr, Ti, Mn, Zn or Si individually, to interstitial free steel was conducted by tungsten inert gas arc brazing, and then effect of alloying element on joint strength associated with the interfacial microstructure was examined. Joint strength of the weld interface was improved by addition of alloying element to aluminum at appropriate range of each content, and the Ni addition exhibited the largest improvement. The Ni addition to aluminum refined the grain structure of the η-Fe_2Al_5 layer, increasing joint strength of the Al/Fe dissimilar interface.

Introduction
Dissimilar welding of aluminum alloy to steel is expected as a weight saving technology in automotive industry to improve driving performance and energy efficiency [1]. However, dissimilar welding of aluminum alloy to steel is often difficult because Al-Fe intermetallic compound (IMC) layer is easily formed at the dissimilar interface. Formation of Al-Fe IMC layer, having low toughness, results in a significant reduction of joint strength of the weld [2].

IMC formation at Al/Fe dissimilar weld interface is often affected by alloying element in aluminum. Some studies have reported that addition of Si to aluminum drastically reduced the thickness of IMC layer during a reaction between liquid aluminum and solid steel [3-4]. Furthermore, Song et al. [5] reported that the reduction of IMC thickness by Si addition improved joint strength of the Al/Fe dissimilar weld interface. Moreover, the IMC formation was suppressed by addition of the other alloying elements, such as Ti, Ge, Cu and Sr [6-7]. These studies suggest that the addition of alloying element to aluminum has a potential to improve joint strength of the Al/Fe dissimilar weld, but effect of alloying element, except for Si, on joint strength of the weld has not been fully understood.

This study conducted dissimilar welding of aluminum containing alloying element to steel by tungsten inert gas (TIG) arc brazing. Joint strength of the dissimilar welds, and effect of alloying element on IMC formation at the dissimilar weld interface were examined. Moreover, the relationship between the joint strength and resulting microstructure was studied.

Experimental procedures
Interstitial free steel was used in this study, whose chemical composition was given in *Table 1*. Dimension of the steel included 100 mm in length, 60 mm wide, and 2 mm in thickness. To examine effect of alloying element in aluminum on IMC formation at the dissimilar weld interface, aluminum alloys containing Ni, Cr, Ti, Mn, Zn or Si individually, were prepared by casting. The cast aluminum alloys were rolled up to 2 mm in thickness, and then cut to 100 mm in length and 50 mm in wide.

Dissimilar lap welding of aluminum alloy to steel was conducted by TIG arc brazing. Schematic illustration of the Al/Fe dissimilar brazing used in this study is shown in *Fig. 1*. Prior to welding, material surfaces were polished with 400 grits emery paper to remove the surface oxides. A flux, TOYO METAL TF-170, was put on only the steel surface to improve Al/Fe dissimilar weldability. The sheets were tilted 10 degrees from the horizontal, as shown in *Fig. 1*. TIG arc brazing was carried out at the following parameters: welding current of 140 A, arc voltage of 9 V, welding speed of 200 mm/min, electrode diameter of φ2.4 mm and arc length of 1 mm. Ar gas shielding was employed during TIG arc brazing.

Table 1: Chemical composition of interstitial free steel used in this study

Element	C	Si	Mn	P	S	Fe
wt.%	0.002	<0.00	0.39	0.017	0.007	Bal.

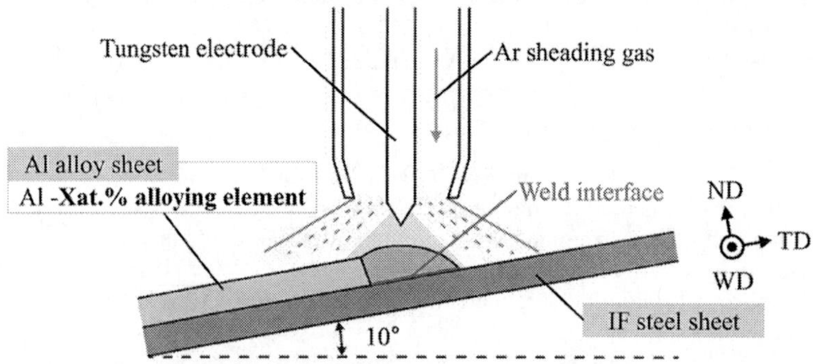

Fig. 1: Schematic illustration of the Al/Fe dissimilar lap TIG arc brazing

Joint strength of the Al/Fe dissimilar weld interface was examined by tensile shear strength testing. The tensile specimens were cut perpendicular to the welding direction. A cross-head speed of 0.3 mm/s was employed. The fractured specimens and cross-section of the Al/Fe dissimilar welds were examined by optical microscopy (OM), scanning electron microscopy (SEM), X-ray diffraction (XRD) analysis, electron probe micro-analysis (EPMA) and electron backscatter diffraction (EBSD) technique.

Results and discussion

Joint strength of the welds

All tensile specimens of the Al/Fe dissimilar welds were broken at the weld interface regardless of the alloying elements. Effect of alloying element on tensile shear strength, i.e., interfacial strength of the dissimilar welds, is shown in *Fig. 2*. All alloying elements improved tensile shear strength of the Al/Fe dissimilar welds at appropriate range of each content. Among these tensile specimens, the dissimilar weld containing 3.5 at.% Ni showed the highest tensile shear strength.

XRD spectra obtained from fracture surfaces of the tensile specimens without alloying element and with 3.5 at.% Ni addition are given in *Fig. 3*, respectively. A strong peak of η-Fe_2Al_5 was detected at both sides of the fracture surfaces regardless of the Ni addition. This result shows that the fracture occurred at the η-Fe_2Al_5 layer in the Al/Fe dissimilar welds.

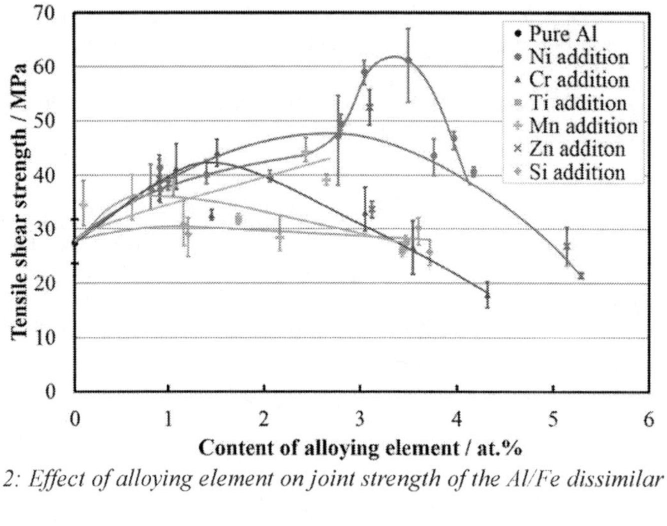

Fig. 2: Effect of alloying element on joint strength of the Al/Fe dissimilar weld

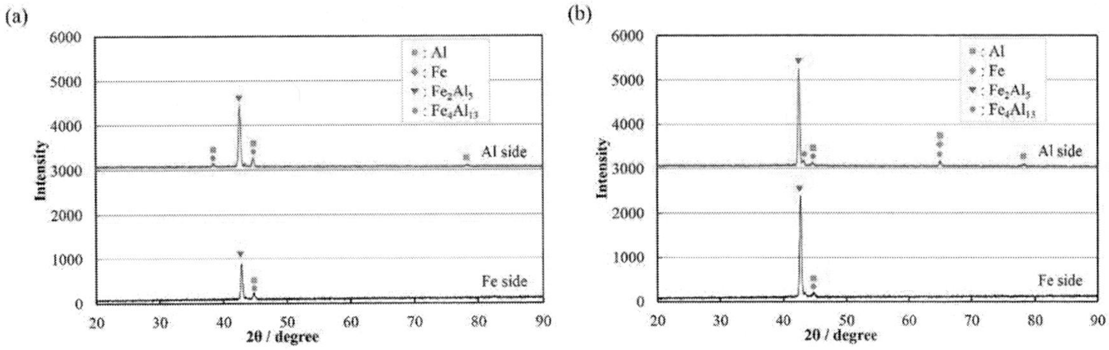

Fig. 3: XRD spectra obtained from fracture surfaces of the Al/Fe dissimilar weld (a) without alloying element and (b) with 3.5 at.% Ni addition

Microstructural characteristics of dissimilar interface

Backscatter electron (BSE) image, and profiles of Fe, Al and Ni contents across the dissimilar interface without alloying element and with 3.5 at.% Ni addition were shown in *Figs. 4* and *5*, respectively. The IMC layers of these dissimilar interfaces have a tongue-like structure mainly consisting of 27 at.% Fe and 73 at.% Al, i.e., η-Fe_2Al_5, being independent of the Ni addition. It should be noted that Ni was detected at the interface between aluminum alloy and η-Fe_2Al_5 layer. The weld with Ni addition had the thicker η-Fe_2Al_5 layer, although it exhibited the higher tensile shear strength, as shown in *Fig. 2*.

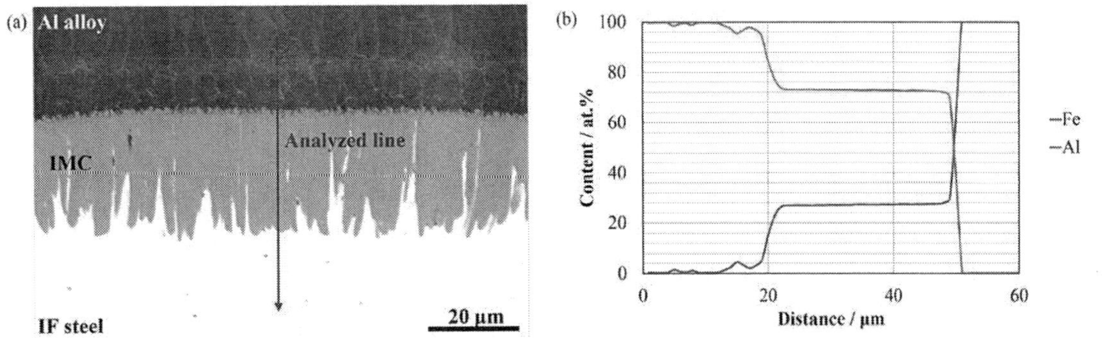

Fig. 4: (a) BSE image and (b) profiles of chemical compositions by EPMA at the Al/Fe dissimilar weld interface produced without alloying element

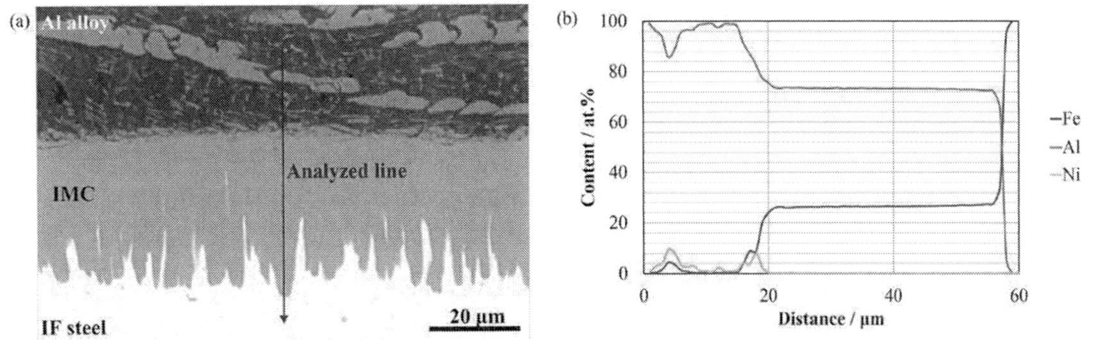

Fig. 5: (a) BSE image and (b) profiles of chemical compositions by EPMA at the Al/Fe dissimilar weld interface produced with 3.5 at.% Ni addition

To clarify a reason for the improvement of tensile shear strength at the Al/Fe dissimilar weld by Ni addition, the grain structure of η-Fe_2Al_5 was precisely examined by EBSD technique. This analysis revealed that the Ni addition resulted in formation of fine grain structure of η-Fe_2Al_5. Grain width of η-Fe_2Al_5, defined as grain boundary spacing parallel to the Al/Fe dissimilar weld interface, was used to quantify the size of the η-Fe_2Al_5 structure because the crack propagated parallel to the η-Fe_2Al_5 layer during tensile testing. The relationship between tensile shear strength and grain width of η-Fe_2Al_5 at the Al/Fe dissimilar weld interface is presented in *Fig. 6*. The tensile shear strength increased with decreasing grain width of η-Fe_2Al_5, suggesting that the grain refinement of η-Fe_2Al_5 by Ni addition would contribute to the improvement of the Al/Fe dissimilar weld strength. Moreover, this study implies that the grain width of η-Fe_2Al_5 is one of the dominant factors governing tensile shear strength of the Al/Fe dissimilar weld.

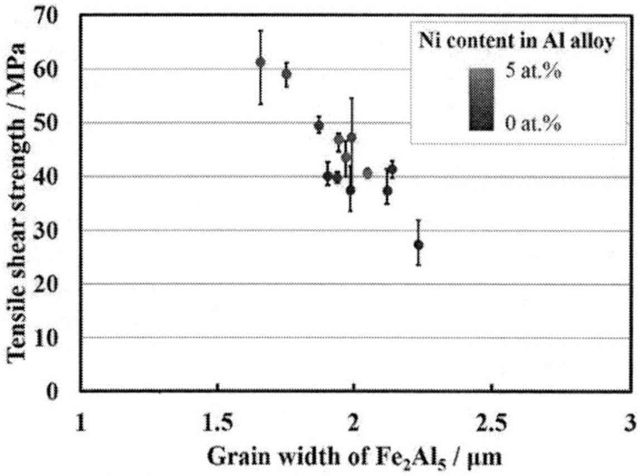

Fig. 6: Relationship between tensile shear strength and grain width of η-Fe_2Al_5 at the Al/Fe dissimilar weld interface

Conclusions

Addition of alloying element, such as Ni, Cr, Ti, Mn, Zn, and Si, to aluminum could increase tensile shear strength of the Al/Fe dissimilar weld interface. The dissimilar weld at Ni content of 3.5 at.% exhibited the highest strength among all welds. The Ni addition hardly changed the chemical composition of the fractured η-Fe_2Al_5 layer, but refined the grain structure of η-Fe_2Al_5. Reduction of the grain width of η-Fe_2Al_5 by Ni addition improved tensile shear strength of the Al/Fe dissimilar weld.

Acknowledgments

The authors are grateful to Mr. A. Honda and Dr. K. Kobayashi for technical assistance and acknowledge Prof. N. Hirata for preparation of aluminum alloys used in this study.

References

[1] W.S. Miller, *et al*: "Recent development in aluminium alloys for the automotive industry," *Materials Science and Engineering A*, Vol. 280, (2000), pp. 37-49.

[2] M. Yasuyama, *et al*: "Spot welding of Aluminum and Steel Sheet with insert of Aluminum Clad Steel Sheet," *Quarterly Journal of the Japan Welding Society*, Vol. 14, No. 2, (1996), pp. 314-320.

[3] T. Heumann, *et al*: "Uber die Kinetik der Reaktion von festem und flûssigem Aluminium mit Eisen," *Zeitschrift fur Metallkunde*, Vol. 50, (1959), pp. 617-625.

[4] W.J. Cheng and C.J. Wang: "Effect of silicon on the formation of intermetallic phases in aluminide coating on mild steel," *Intermetallics*, Vol. 19, (2011), pp. 1455-1460.

[5] J.L. Song, *et al*: "Effects of Si additions on intermetallic compound layer of aluminum–steel TIG welding–brazing joint," *Journal of Alloys and Compounds*, Vol. 488, (2009), pp. 217–222.

[6] M.V. Akdeniz and A.O. Mekhrabov: "The effect of substitutional impurities on the evolution of Fe-Al diffusion layer," *Acta Materialia*, Vol. 46, No. 4, (1998), pp. 1185-1192.

[7] K.A. Nazari and S.G. Shabestari, "Effect of micro alloying elements on the interfacial reactions between molten aluminum alloy and tool steel," *Journal of Alloys and Compounds*, Vol. 478, (2009), pp. 523–530.

Formation mechanism and elimination of pores for Ag-Sn low temperature TLP-bonded joint

Huakai Shao[1,*], Aiping Wu[1,2,3], Yudian Bao[1]
1. Department of Mechanical Engineering, Tsinghua University, Beijing 100084
2. State Key Laboratory of Tribology, Tsinghua University, Beijing 100084
3. Key Laboratory for Advanced Materials Processing Technology, Ministry of Education, Tsinghua University, Beijing 100084
Email: shk13@mails.tsinghua.edu.cn, Tel.: +86-10-62773859

Abstract

This study investigated the mechanism of pore formation during Ag-Sn low temperature TLP bonding, and two novel methods were proposed to prevent pores from the joint accordingly. Experimental results showed that formation of pores was attributed to volume shrinkage during Sn-islands conversion into intermetallic compounds (IMCs), and depended on the bonding temperature, pressure and thickness of the interlayer. It was found that replacement of one Ag substrate by pure Cu rod to form dual layers of dissimilar-based IMCs could reduce porosity of the TLP-bonded joint due to interface migration. Furthermore, thin layer of Cu film coated on one Ag substrate to form Cu_6Sn_5 phases that occupied micro-pores in the IMCs, also eliminated pores between two neighboring Ag_3Sn scallops. The novel methods were also able to avoid the defect for other systems during low temperature TLP bonding and improve the mechanical properties.

Key word: TLP bonding; pore formation; intermetallic compounds; interface migration; grain boundary

1. Introduction

Low temperature transient liquid phase (TLP) bonding is a potential high temperature (HT) packaging technology in application to interconnection for wide band gab (WBG) semiconductors such as SiC and GaN, which are expected to be used in automotive, aerospace, deep-well drilling and energy production industries [1, 2]. This method is capable of ensuring devices operated at higher temperatures without failing, for which joint is fully composed of intermetallic compounds (IMCs) at a low temperature (<300 °C) by diffusion mechanism. However, some defects particularly pores are found in the TLP-bonded joints [3-5]. As a result, the residual voids may degrade joint reliability such as thermal stability, mechanical properties, thermal and electrical conductivities.

Three types of voids, Kirkendall void, processing pore and contraction pore, can be categorized according to the formation characteristics during isothermal solidification. Kirkendall voids are introduced by unbalanced diffusion for different elements at the interface, and are regarded as familiar defects in Cu-Sn system, especially distributed along the Cu/Cu_3Sn interface [3, 6]. Processing pore are found in the centerline of the IMC layer, and proposed to be formed prior to the occurrence of the liquid phase, due to solid state reactions during heating stage consume most of the coated interlayer and no enough liquid is available to successfully accomplish TLP bonding at targeted temperature [4, 7]. Contraction pore is the defect that easily exists in Ag-Sn [8], Ni-Sn [9] TLP system, and generally locates in the middle of the IMC layer, as is similar to the processing pore. The root cause of this type of pores formation is volume shrinkage during isolated liquid phase islands conversion into IMCs.

Although many researchers have paid attention to the characteristics of the shrinkage pores, there are no effective measures to eliminate them from TLP-bonded joint. In the present work, contraction pore for Ag-Sn TLP bonding was experimentally verified, and the formation mechanism was also discussed deeply. Furthermore, two novel methods, dual dissimilar-based (e.g. Cu_6Sn_5/Ag_3Sn or Cu_3Sn/Ag_3Sn) IMC-layers formation and second dissimilar-based IMC-particles introducing, were proposed to effectively eliminate pores form the joint.

2. Experimental procedure

Polished pure Cu (99.9%) rods with dimensions of Φ10 mm×5.5 mm and Φ6 mm × 5.5 mm were electroplated with 25 μm-thick Ag layer as substrates, and commercially available pure Sn (99.9%) foil of thickness 10 μm was adopted as interlayer. To remove contaminates and oxides, the Ag-plated substrates were cleaned in 5 vol. % nitric acid solution with an ultrasonic bath for 5 min, followed by rinsing with deionized water and anhydrous alcohol, and drying at last. The substrates were coated with small flux and assembled with Sn foil into sandwich structure, such that simulated Ag/Sn/Ag system low temperature TLP bonding. However, it was found that many pores were formed at the centerline of the IMCs joint in our previous work.

Two novel methods were proposed to eliminate pores in the TLP-bonded joint. On the one hand, the bottom Ag-plated substrate was substituted for pure Cu rod to form dual layers of dissimilar-based IMCs (*Fig. 1a*), which was denoted as "Cu/Sn/Ag" in the next section. On the other hand, a thin layer (1.5μm) of Cu film was electroplated on the upper Ag-plated substrate to introduce Cu-Sn IMCs to fill the pores located at the Ag_3Sn grain boundaries (*Fig. 1b*), which was regarded as "Ag(Cu)/Sn/Ag" in the following. After cleaning as mentioned above, sandwich structures were prepared and placed into resistance furnace (KSL-1100X).

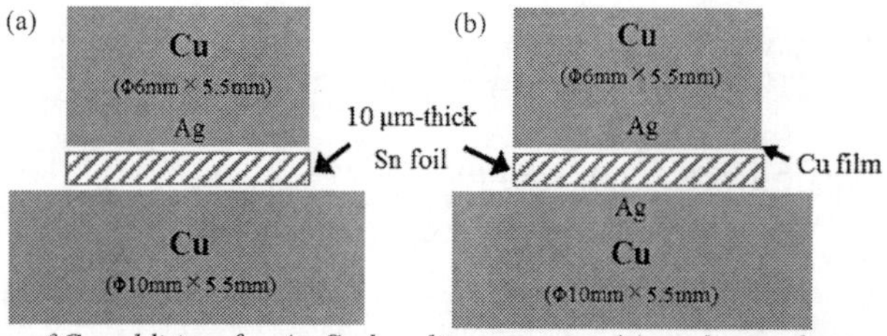

Fig. 1. Schematic diagram of Cu addition for Ag-Sn bonding process: (a) replacing bottom Ag-plated substrate with pure Cu rod; (b) coating thin Cu film on upper Ag-plated substrate

The joining experiments were carried out under ambient air condition, and bonding pressure of 0.3 MPa was applied by stainless steel blocks. The bonding temperature was changed among 250 °C, 300 °C and 350 °C, and bonding time was determined by thickness of the interlayer and varied from 60 min to 420min. After bonding process, the samples were prepared to metallographic specimens. Microstructure was investigated by optical microscope, scanning electron microscopy (SEM) and energy dispersive X-ray spectroscopy (EDS). Shear tests were conducted on Gleeble-1500D thermal simulated test machine with shear rate of 2 mm/min, to determine the mechanical behaviors of the joint without defects. At least three shearing specimens were employed to estimate averaged shear strength under one condition.

3. Results and discussion

3.1 Mechanism of pore formation

In order to clearly observe appearance of the pores, 25 μm-thick Sn foil was used to join Ag substrates under vacuum condition, and the microstructure was showed in *Fig. 2*. It was found that some big-sized pores were formed when two different sides of Ag_3Sn scallops contacted with each other though amount of Sn still remained (*Fig. 2a*), and their sizes significantly increased after Sn-islands were totally consumed (*Fig. 2b*). We suggested that formation of the pores was attributed to the volume shrinkage of residual Sn-islands during isothermal solidification, based on the change of the pore size. It was calculated that the conversion process of Sn into Ag_3Sn leaded to volume contraction of approximately 9% in theory [8]. According to the mass conservation, the size of shrinkage pore was determined as follow

$$3Ag + Sn \rightarrow Ag_3Sn \quad (1)$$

$$\Delta V = \frac{m_{Sn}}{\rho_{Sn}} + \frac{3(1-\eta)m_{Sn}}{\eta\rho_{Ag}} - \frac{m_{Sn}}{\eta\rho_{Ag_3Sn}} = \alpha m_{Sn} \quad (2)$$

where m_{Sn} is the mass of the Sn-island, ρ_{Sn}、ρ_{Ag} and ρ_{Ag3Sn} are the densities of Sn, Ag and Ag_3Sn respectively, η is the molar mass fraction of Sn in Ag_3Sn, and α was calculated as 0.537. From Eq. 2, the probability of contraction porosity was evaluated by the remained Sn-islands separated by two different sides of Ag_3Sn scallops.

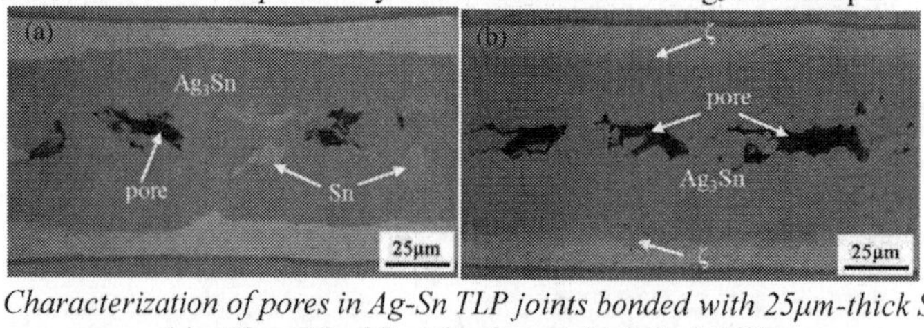

Fig. 2. Characterization of pores in Ag-Sn TLP joints bonded with 25μm-thick Sn foil: (a) with residual Sn-islands and (b) full of IMCs

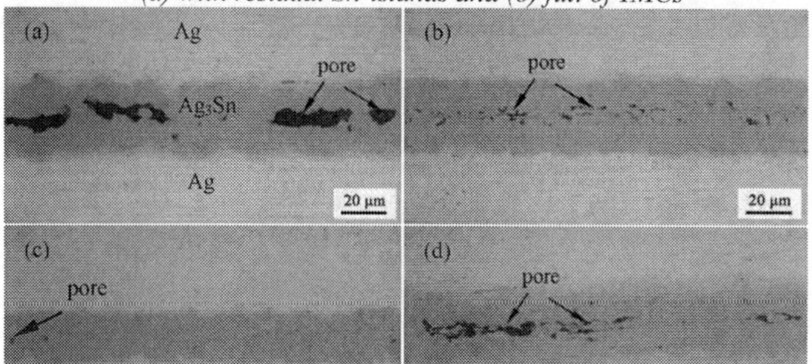

Fig. 3. Effect of bonding process on pores formation for the Ag_3Sn joints: (a) 0.2MPa-300℃-90min; (b) 0.3MPa-300 ℃-90min; (c) 0.3MPa-250℃-240min and (d) 0.3MPa-350℃-60min

Fig. 3 showed the cross-sectional microstructure of Ag_3Sn joints (majorly be composed of Ag_3Sn IMCs) bonded under different conditions. As can be found, the amount and size of the pore were significantly affected by TLP process of temperature, pressure and thickness of the interlayer. In detail, the probability of pores formation increased with increasing temperature (*Figs. 3b, 3c* and *3d*), thicker interlayer of Sn foil (*Figs. 2b-25 μm, 3a-10 μm*) or decreasing process pressure (*Figs. 3a, 3b*). Lis et.al [10] used the interfacial roughness to describe the scalloped shape of Ag_3Sn, as well as the amount of the remaining Sn-islands to some extent. They found that the value was strongly depended on temperature, and increased with higher temperature. Furthermore, the previous study indicated that some other shapes of Ag_3Sn like column were more easily to generate at higher temperature [11]. These phenomena explained effect of bonding temperature on pores formation.

3.2 Defect elimination by Cu/Sn/Ag TLP bonding

Fig. 4 showed the cross-sectional microstructure of the Cu/Sn/Ag TLP joint bonded at different temperatures. At 250 °C, it could be noted that a few pores still remained along Cu_6Sn_5/Ag_3Sn interface after bonding for a short time (*Fig. 4a*), in particular close to the side of Ag_3Sn layer such as Ag_3Sn grain boundaries. However, such pores were totally disappeared with prolonged dwell time to 420 min though similar microstructure of $Cu_3Sn/Cu_6Sn_5/Ag_3Sn$ layered structure existed in the joint (*Fig. 4*b). Wave-shaped interface was presented between Cu_6Sn_5 and Ag_3Sn layers, and turned more planar with long-term heat treatment before full transformation of the Cu_6Sn_5 IMCs.

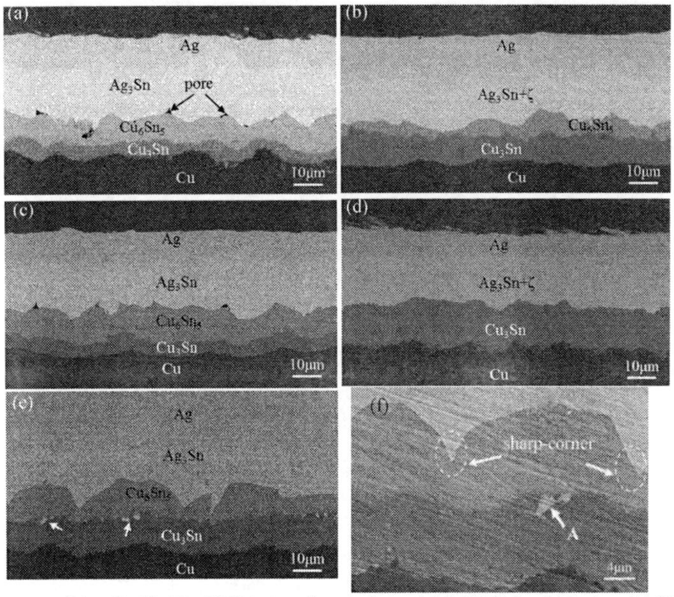

Fig. 4. Transverse section of the Cu/Sn/Ag TLP joints bonded at: (a) 250°C for 240min; (b) 250°C for 420min; (c) 300°C for 90min; (d) 300°C for 420min; (e) 350°C for 60min; and (f) magnification for arrow marks

For the joint bonded at 300 °C, a significant reduction of defect size and amount was observed after bonding for 90 min in comparison to the Ag/Sn/Ag TLP joints (*Fig. 4c*), and such pores almost could not be found with increasing dwell time to 420 min any more (*Fig. 4d*), accompanied by Cu_6Sn_5 transformed into Cu_3Sn completely. Interestingly, the wave-shaped interface between Cu_6Sn_5 and Ag_3Sn layers converted into more planar interface of Cu_3Sn/Ag_3Sn, which revealed that interface migration took place during scalloped Cu_6Sn_5 changed to columnar Cu_3Sn dendrite grains.When bonded at 350 °C, no evident pores retained between Cu_6Sn_5 and Ag_3Sn layers longer, but the waved shape of Cu_6Sn_5/Ag_3Sn interface was approximate to be a zigzag form (*Fig. 4e*). Furthermore, many bright micron-particles, were determined to be Ag_3Sn by EDS analysis, embedded in the Cu_6Sn_5/Cu_3Sn interface (*Fig. 4f*).

It was suggested that pores elimination in this method associated with interface migration between Cu-Sn IMCs and Ag-Sn IMCs during solid-state reaction. The Gibbs free energy of the interface can be given by:

$$G = \sigma A \quad (1)$$

where σ is the surface energy and equals to the surface tension in numerical value; A is the area of the interface. From this equation, when the value of A reaches to the minimum, the Gibbs free energy of this system will be obtained with the minimum. Therefore, wave-shaped interface obviously turned into planar interface to make the lowest energy with bonding time. In this work, it was proposed that the driving force of interface migration was the interface curvature according to the report of Upmanyu et.al [12]. However, no inter-diffusion of Cu, Sn or Ag atoms occurred across the interface due to the difference of crystal structures for Cu-Sn IMCs and Ag-Sn IMCs.

3.3 Defect elimination by Ag(Cu)/Sn/Ag TLP bonding

Fig. 5 showed the cross-sectional microstructure of the Ag(Cu)/Sn/Ag TLP joint bonded at different temperature. For the joint obtained at 300 °C (*Fig. 5*a), many spherical or irregular-shaped particles, were determined to be mixed structures of Cu_6Sn_5 and Cu_3Sn IMCs by EDS analysis, dispersedly embedded in the middle of the Ag_3Sn layer and filled the original sites of the possible pores. The IMC-particle was clearly exhibited in *Fig. 5b*, and principally composed of big-sized Cu_6Sn_5 core (light area) partially covering with thin-layered Cu_3Sn shell (dark area). There were similar microstructure for the joints bonded at 250 °C (*Fig. 5c*) and 350 °C (*Fig. 5d*), however, only Cu_6Sn_5 was examined for 250 °C while a thicker Cu_3Sn shell was detected for 350 °C. Furthermore, the size of Cu-Sn IMCs particles increased with increasing bonding temperature but the distributions were more dispersive at the same time, which resulted from different diffusivities of the atoms at different temperatures in reaction.

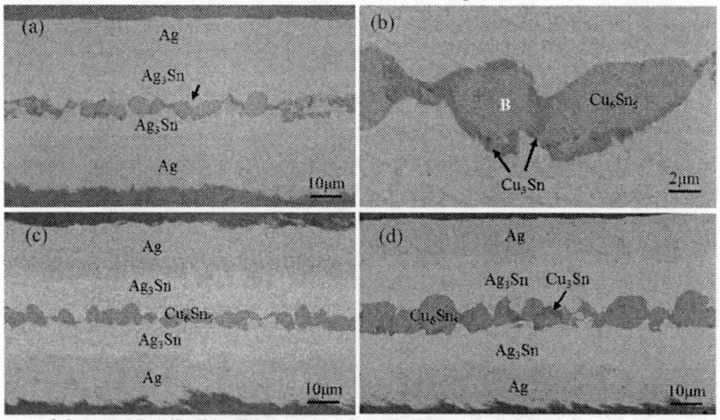

Fig. 5. Transverse section of the Ag(Cu)/Sn/Ag TLP joints bonded at: (a) 300°C for 90min; (b) magnification for arrow mark; (c) 250°C for 240min; (d) 350 °C for 60 min

The Cu film addition could not change the scalloped shape of Ag_3Sn grains and interfacial roughness of Ag_3Sn/Sn interface, though a second IMCs particle phase might impact growth kinetics of the IMC layer [13]. We considered that Cu_6Sn_5 particles was introduced to fill the pores and reduce the contraction porosity, which was the principal mechanism of pores elimination for this method.

3.4 Mechanical behaviors of TLP joint

The results of shearing tests were showed in *Fig. 6*. Shear strength of the Cu/Sn/Ag TLP joints was generally higher than that of the Ag/Sn/Ag TLP joints apart from the process condition of 250 °C, since some pores still existed in the IMCs after short-term homogenization at lower temperature. However, shear strength of the Cu/Sn/Ag TLP joint bonded at 250 °C increased with prolonging hold time and exceeded that of the Ag/Sn/Ag TLP joint, which was in agreement with the change of the microstructure. Furthermore, shear strength of the Cu/Sn/Ag TLP joint obtained at 300 °C also increased with prolonging dwell time to 420 min, due to further suppression of the pores after continuing homogenization and combination between Cu_3Sn layer and Ag_3Sn layer. This indicated that adhesive strength of Cu_3Sn/Ag_3Sn interface was higher than that of the Cu_6Sn_5/Ag_3Sn interface.

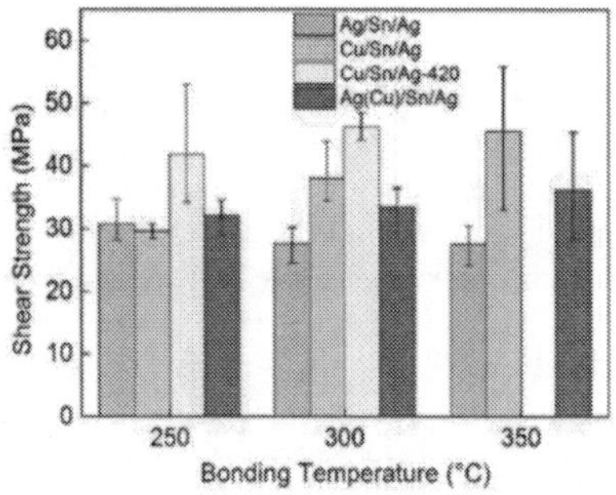

Fig. 6. Shear strength of the different-typed TLP joints depended on bonding temperature. Wherein, Ag/Sn/Ag, Cu/Sn/Ag and Ag(Cu)Sn/Ag represents shear strength of the corresponding TLP joints, and Cu/Sn/Ag-420 represents the bonding time is prolonged to 420 min.

Shear strength of the Ag(Cu)/Sn/Ag TLP joints was also slightly higher than that of the Ag/Sn/Ag TLP joints, but obviously lower than that of Cu/Sn/Ag TLP joints. This revealed that effects of replacement of Ag substrate with Cu on pores suppression was more prominent than that of coating thin Cu film on Ag substrate. On the other hand, it was found that shear strength of Ag/Sn/Ag TLP joint decreased with increasing temperature, whereas both Cu/Sn/Ag and Ag(Cu)/Sn/Ag TLP joints had contrary results. That was to say, the effect of interface migration induced by Sn diffusions was more dominant than the change of Sn/Ag_3Sn interfacial morphology resulted from temperature on the pore formation.

Conclusions

Pores formation for Ag/Sn/Ag TLP bonding was experimentally verified to volume shrinkage of isolated Sn-islands conversion into Ag_3Sn IMCs during isothermal solidification, and strongly depended on the temperature, interlayer thickness and process pressure. Replacement of one Ag substrate by Cu rod to form dual dissimilar-based IMCs layers could effectively eliminate shrinkage pores due to interface migration occurred in homogenization. Coating thin Cu film on Ag substrate to introduce second dissimilar-based IMCs phase of Cu_6Sn_5 particles was also available to expel shrinkage pores thoroughly, since these particles filled the micro-pores sufficiently. The two novel methods could significantly improve mechanical performance of the TLP joint.

Acknowledgments

This research is financially supported by the National Science Foundation of China under Grant No. 51375260, which entitled "Technology and Mechanism of Low Temperature Transient Liquid Phase Bonding".

References

[1] V.R. Manikam, *et al*: "Die Attach Materials For High Temperature Applications: A Review," *IEEE-Trans-CPMT*, Vol. 1, No. 4 (2011), pp. 457-478.

[2] P.G. Neudeck, *et al*: "High-temperature electronics-a role for wide bandgap semiconductors?" *P-IEEE*, Vol. 90, No. 6 (2002), pp. 1065-1076.

[3] H. Liu, *et al*: "Aasmundtveit, N. Hoivik, Intermetallic Compound Formation Mechanisms for Cu-Sn Solid－Liquid Interdiffusion Bonding," *J-Electron-Mater*, Vol. 41, No. 9 (2012), pp. 2453-2462.

[4] P.J. Wang, *et al*: "Silver Microstructure Control for Fluxless Bonding Success Using Ag-In System," *IEEE Trans-CPT*, Vol. 33, No. 2 (2010), pp. 462-469.

[5] K. Chu, *et al*: "A comparative study of Cn/Sn/Cu and Ni/Sn/Ni solder joints for low temperature stable transient liquid phase bonding," *Scripta-Mater*, Vol. 109 (2015), pp. 113-117.

[6] C. Yu, *et al*: "Fine grained Cu film promoting Kirkendall voiding at Cu3Sn/Cu interface," *J-Alloy-Compd*, Vol. 660 (2016), pp. 80-84.

[7] N.S. Bosco, *et al*: "Critical interlayer thickness for transient liquid phase bonding in the Cu－Sn system," *Acta-Mater*, Vol. 52, No. 10 (2004), pp. 2965-2972.

[8] A. Lis, *et al*: "Effect of Process and Service Conditions on TLP-Bonded Components with (Ag,Ni－)Sn Interlayer Combinations," *J-Electron-Mate*r, Vol. 44, No. 11 (2015), pp. 4576-4588.

[9] K. Chu, *et al*: "A comparative study of Cn/Sn/Cu and Ni/Sn/Ni solder joints for low temperature stable transient liquid phase bonding," *Scripta-Mater*, Vol. 109 (2015), pp. 113-117.

[10] A. Lis, *et al*: "Early stage growth characteristics of Ag3Sn intermetallic compounds during solid－solid and solid－liquid reactions in the Ag－Sn interlayer system: Experiments and simulations," *J-Alloy-Compd*, Vol. 617 (2014), pp. 763-773.

[11] H. Shao, *et al*: "Mechanism of Ag3Sn grain growth in Ag-Sn transient liquid phase soldering," *T-Nonferr-Metal-Soc*, (accepted).

[12] M. Upmanyu, *et al*: "Atomistic simulation of curvature driven grain boundary migration", *Interface-Sci*, Vol. 6, No. 1-2 (1998), pp. 41-58.

[13] H.Y. Chuang, *et al*: "Elimination of voids in reactions between Ni and Sn: A novel effect of silver," *Scripta-Mater*, Vol. 66, No. 3-4 (2012), pp. 171-174.

MICROSTRUCTURAL FEATURES IN FRICTION STIR WELDED LAP JOINTS

Egoitz Aldanondo, Ekaitz Arruti, Julen Sarasa, Alberto Echeverria
IK4 LORTEK, Ordizia, Spain
Egoitz@lortek.es, +34 943 882303

Abstract
Friction Stir Welding of aluminium alloys in lap joint configuration has shown a great potential for lightweight structure manufacturing but it is very important to consider specific issues relative to lap joint formation that are different from the more commonly used butt joint configuration. Typical features such as the hook formation at the advancing side or the cold lap defect or joint uplifting at the retreating side are the most characteristic defects in the lap joint configuration. The formation of such defects is mostly influenced by the FSW tool design and welding parameters. It is very important to identify appropriate processing conditions that avoid or minimize these defects in order to create welds with good properties that can be applied in final products.

The effect of several FSW variables on the formation of lap joint features is presented in this work. FSW joints in lap joint configuration were performed using an AA7075-T6 aluminum alloy and different welding conditions including tool designs and main welding parameters. The microstructural characteristics of lap joints were analysed by metallographic examination and the formation of the characteristic features was evaluated. Thus the influences of FSW variables such as welding tool design, rotational speed and welding speed on the formation of hook features or joint uplifting was established.

Introduction
Although the majority of applications based on FSW technology manufacturing have been focused in butt joint configurations, the lap joint configuration is highly interesting too representing a number of applications or potential developments [1, 2]. Therefore many research works have been conducted with the main objective of understanding the joint formation mechanisms, microstructural features and the joint strength of FSW lap joints [3-6]. According to those works, there are a number of factors that influence the joint properties such as tool design, welding parameters or number of passes. Depending on those factors typical characteristics of FSW lap joints have been defined as hook features, cold lap defects or the joint uplifting. These characteristics are the main factors that determine the joint strength and microstructural properties, and are largely influenced by the tool design used to perform FSW lap joints as shown in Figure 1. Therefore it is very important to optimize the tool design, as well as other welding parameters, in order to produce FSW lap joints without unwanted features that could reduce the joint strength.

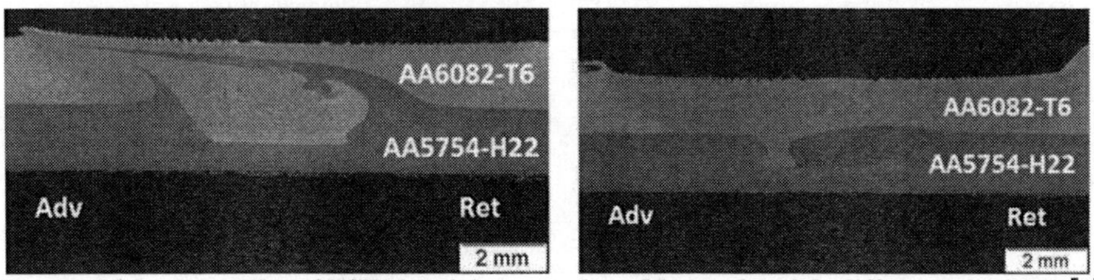

Figure 1: Macrographs of FSW lap joints produced by tools with different designs [4].

Experimental details
FSW lap joints were performed between a 2 mm thick AA7075-T6 aluminium alloy on top of a 3 mm sheet of the same alloy. FSW tests were performed in a I-STIR PDS4 machine using two different tool designs and two different welding conditions. Thus a cylindrical threaded tool and a tool containing three flats were used both having a shoulder diameter of 10 mm and a pin length of 2,5 mm. Different welding conditions were represented from a heat input point of view, using 1200 rpm and 152 mm/min as high heat input conditions; while 800 rpm and 254 mm/min were used as low heat input welding conditions. FSW tests were conducted initially in position control in order to identify the appropriate axial or Z force and torque for each welding condition and FSW tool as

shown in Figure 2. It can be observed that high heat input conditions require lower Z force and torque values than lower heat input conditions for both FSW tool designs used for the tests. Final FSW samples used for the metallographic examination were produced in force controlled FSW tests applying the appropriate Z force for each tool and welding condition.

For metallographic examination cross section specimens were obtained from the FSW lap joints and carefully prepared following standard metallographic procedures comprising cross-sectioning, mounting, polishing and etching. Keller's reagent was used as etchant and the specimens were submerged for 5 seconds, rinsed in water and dried in a warm airflow. An Olympus GX51 light optical microscope was used to investigate the microstructural properties and defects of the FSW lap joints.

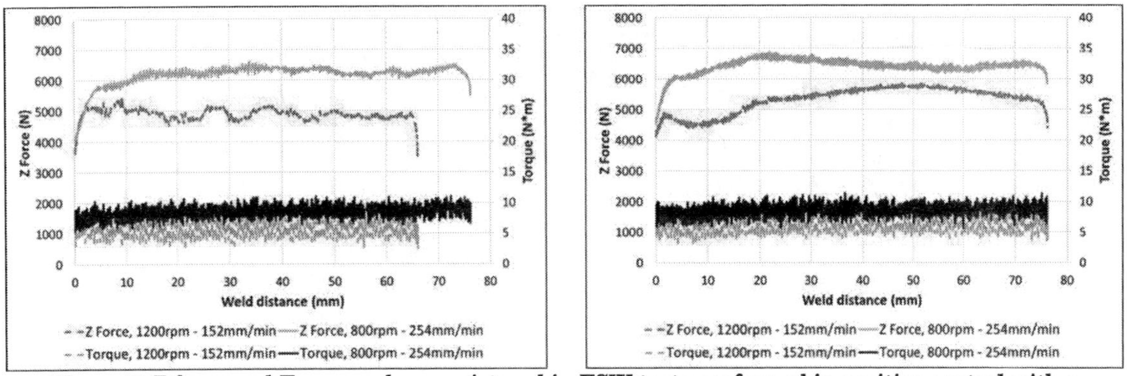

Figure 2: Z force and Torque values registered in FSW tests performed in position control with a cylindrical threaded tool (left) and a 3 flat tool (right).

Results and discussion

Cross sections of FSW lap joints produced by the two different tools and the two different welding conditions are shown in Figure 3. Obvious volumetric defect free joints were obtained except for the cylindrical threaded tool operated at high heat input welding conditions (1200 rpm, 152 mm/min). A continuous wormhole was identified in that sample at the advancing side suggesting an insufficient material flow and consolidation at these welding conditions. Larger stirred zones and TMAZ extensions were observed in the joints produced by the 3 flat tool in comparison with the joints produced by the cylindrical threaded tool at equivalent welding conditions (images "c" and "d" vs "a" and "b" at Figure 3). The same effect was seen by using low heat input conditions vs the high heat input conditions, achieving larger stirred zones and TMAZ extensions at low heat input conditions (images "a" and "c" vs "b" and "d" at Figure 3). This effect could be attributed to the larger sticking effect between the tool and the material at low heat input conditions resulting in larger material flow regions and larger stirred zones and TMAZ extensions.

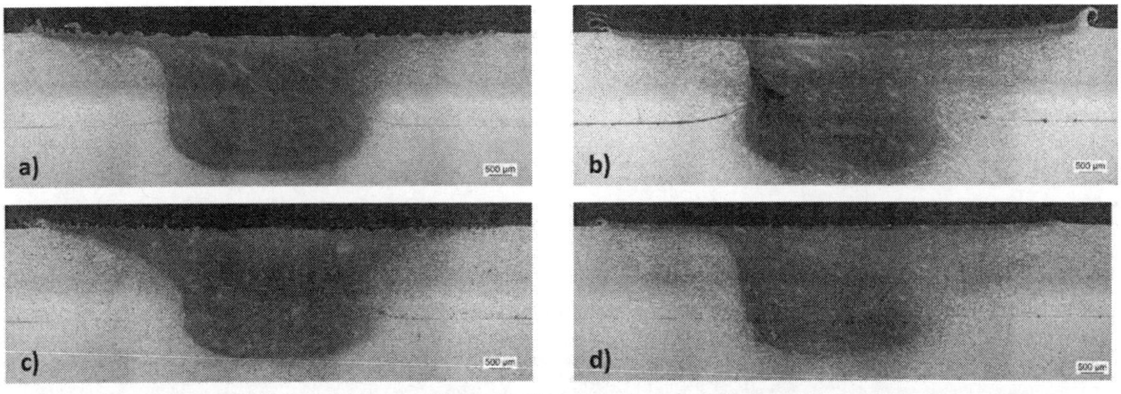

Figure 3: Cross-sections of FSW lap joints performed with a) Cylindrical threaded tool at 800rpm, 254mm/min and 6,4kN; b) Cylindrical threaded tool at 1200rpm, 152mm/min and 5,1kN; c) 3 flat tool at 800rpm, 254mm/min and 6,6kN; d) 3 flat tool at 1200rpm, 152mm/min and 5,1kN.

Apart from volumetric defects at the joint made with the cylindrical threaded tool operated at high heat input conditions, other microstructural features such as hooking and joint line uplifting were observed in the joints as shown in Figure 4, which are characteristic of FSW lap joints. These features can act as top sheet thinning or crack initiation originators so they can be considered as weld defects that are responsible to reduce the joint strength. Taking into account the importance of these features in joint properties, the hook size at the advancing side (H1) and the joint line uplifting (H2) at the retreating side were estimated by optical microscopy and an image processing software, as shown in the detail images at Figure 4.

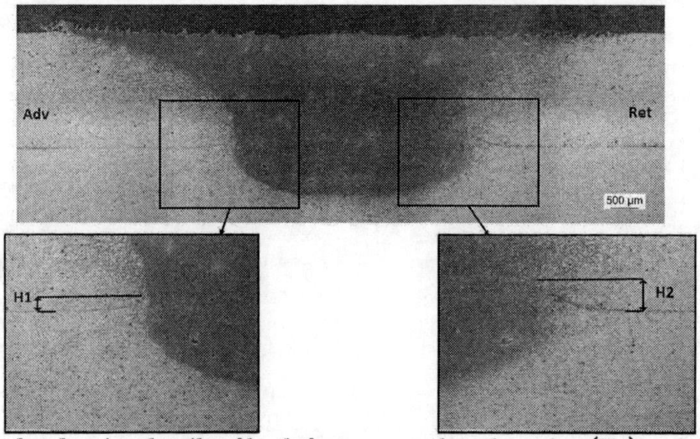

Figure 4: Micrographs showing details of hook features at the advancing (H1) and the retreating (H2) sides of a FSW weld performed at 800rpm and 254mm/min with a 3 flat tool.

The results of hook size measurements at the advancing side (H1) and the retreating side (H2), which were performed in the metallographic specimens obtained from the FSW lap joints produced with the two different tools and two different welding conditions, are plotted in Figure 5. Smaller hook sizes were generally observed in the joints performed with the 3 flat tool in comparison with ones produced with the more conventional cylindrical threaded design. This effect was magnified at high heat input welding conditions (1200 rpm – 152 mm/min) due to the extensive vertical flow of plasticized material at the TMAZ proximities, which is induced in a larger extent by the cylindrical threaded tool in comparison with the 3 flat tool. Therefore larger hook sizes in the range of ~0,4-0,5 mm were identified in joints produced by the cylindrical threaded tool at high heat input conditions in both advancing (H1) and retreating sides (H2). However, a different trend in hook feature formation was identified at the joints performed by the 3 flat tool. In these cases, smaller hook sizes were identified at the advancing side (H1) as well as at the retreating side (H2) of the joints produced using high heat input conditions. From these results it can be concluded that the vertical flow of plasticized material at the TMAZ proximities is minimal in the joints produced by the 3 flat tool, meaning that this tool design is more appropriate for FSW lap joints than the cylindrical threaded tool design.

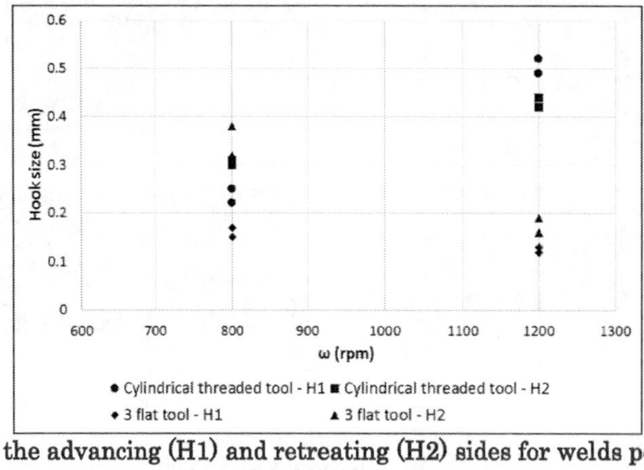

Figure 5: Hook size at the advancing (H1) and retreating (H2) sides for welds performed with different tools and welding parameters.

Therefore, the smallest hook sizes were observed at the joints produced by the 3 flat tool operated at high heat input conditions, being these sizes in the range of ~0,1-0,2 mm. These joints would be expected to offer highest joint strengths although other effects that could reduce the joint properties such as the overaging of the AA7075-T6 alloy at the HAZ should be taken into account.

Conclusions

Microstructural features of AA7075-T6 alloy FSW lap joints produced by two different tool designs and two different welding parameter sets were investigated. The following conclusions can be presented:

- A continuous wormhole defect was produced by the cylindrical threaded tool operated at high heat input welding conditions.
- Larger stirred zones and TMAZ extensions were observed generally using low heat input conditions and the 3 flat tool design to perform the FSW lap joints.
- Using a cylindrical threaded tool design and high heat input conditions, largest hook sizes in the range of ~0,4-0,5 mm were observed.
- Using a 3 flat tool design and high heat input conditions, smallest hook sizes in the range of ~0,1-0,2 mm were observed.
- Different flow patterns were produced by the cylindrical threaded and the 3 flat tool. The 3 flat tool showed the capabilities to produce reduced hook size features and to be operated in a wider parameter window, being a more promising tool design for FSW lap joints.

References

[1] M.J. Booker, *et al*: "Applying Friction Stir Welding to the Arianne 5 Main Motor Thrust Frame" *Proc 2nd International Symposium on Friction Stir Welding*, Gothenburg, Sweden, 2000.

[2] R. Talwar, *et al*: "Friction Stir Welding of Airframe Structures" *Proc 2nd International Symposium on Friction Stir Welding*, Gothenburg, Sweden, 2000.

[3] L. Cederquist, A.P. Reynolds: "Factors affecting the properties of friction stir welded aluminium lap joints", *Weld. J. Res. Suppl.*, 80 (2001), pp. 281.

[4] E. Aldanondo, *et al*: "Mechanical and microstructural properties of lap joints" *Proc 142nd TMS Annual Meeting & Exhibition – Friction Stir Welding and Processing VII*, San Antonio, USA, March 2013, pp. 195-205.

[5] M.K. Kulekci, *et al*: "Effects of tool rotation and pin diameter on fatigue properties of friction stir welded lap joints", *Int. J. Adv. Manuf. Technol.*, 36 (2008), pp. 877-882.

[6] S.B. Jung, *et al*: "Lap joint properties of FSWed dissimilar formed AA5052 and AA6061 Al alloys with different thicknesses", *J. Mat. Sci.*., 43 (2008), pp. 3296.

Grain Refinement via Friction Stir Processing with 10-nm-size Zirconia Particles

Y. Kimoto, T. Nagaoka, H. Watanabe, M. Fukusumi, T. Takeuchi
Osaka Municipal Technical Research Institute, Joto-ku, Osaka, Japan
Email:y-kimoto@omtri.or.jp, phone: +81-6-6963-8157

Y. Morisada, H. Fujii
Joining and Welding Research Institute, Osaka University, Ibaraki, Osaka, Japan
Email:morisada@jwri.osaka-u.ac.jp, phone: +81-6-6879-8663

Abstract
Friction stir processing (FSP) with 10-nm-size zirconia nanoparticles has been examined for the grain refinement of a coarse-grained cast pure magnesium plates. A multi-pass FSP has been performed over the slotted hole filled with the zirconia nanoparticles. After 3-pass FSP, the nanoparticles have been widely spread among the stir zone (SZ), and the grain size has reduced from 740 μm to 0.57 μm. It has been proved that FSP with 10-nm-size zirconia particles can produce ultra-fine-grained (UFG) metal matrix composites (MMCs) from coarse-grained cast magnesium.

Introduction
FSP is an application of the friction stir welding to the surface modification of metal and alloy plates [1]. It can refine the grains in the SZ through the dynamic recrystallization due to the severe strain, heating and rapid cooling [2]. FSP with particles has also been studied as a simple way to fabricate metal MMCs in the SZ [3, 4]. The grains in the SZ can be further refined by the pinning effect of the additive particles against the grain growth [5]. However, it is difficult to form homogeneous composite in a single-pass FSP, and the multi-pass FSP along the same line is effective to improve the homogeneity. Our recent literature survey showed that 3- or 4-pass FSP with carbide or oxide particles smaller than 100 nm can produce UFG MMCs with the grain size less than 1 μm [6]. In this study, we have examined 1- to 3-pass FSP with 10-nm-size zirconia nanoparticles for coarse-grained cast pure magnesium, and explore the possibility of the production of UFG MMCs from coarse-grained cast metal via FSP with nanoparticles.

Experimental Procedures

FSP conditions
FSP was performed for cast pure Mg plates with 25 mm thickness. The tool had a columnar shoulder with a threaded probe. The shoulder diameter is 25.4 mm, and the diameter and length of the threaded probe are M12 and 10 mm, respectively. Zirconia nanoparticles with primary particle diameter of 10 nm (Kanto Denka Kogyo Co., Ltd.) were filled in the slotted hole with 4-mm diameter. The volume fraction of the particles in the composite was estimated to be 1.7 vol.%. The center axis of the slotted hole was positioned at 5 mm depth from surface so as to be nearly coincident with the center of the SZ as illustrated in *Fig. 1*. The applied load was in the range from 550 to 575 kgf. The tool rotating rate, travelling speed and tilt angle were 1500 rpm, 50 mm/min and 3°, respectively. The multi-pass FSP with 100 % overlap was employed to homogenize the particle distribution among the SZ. The no. of passes was from one to three.

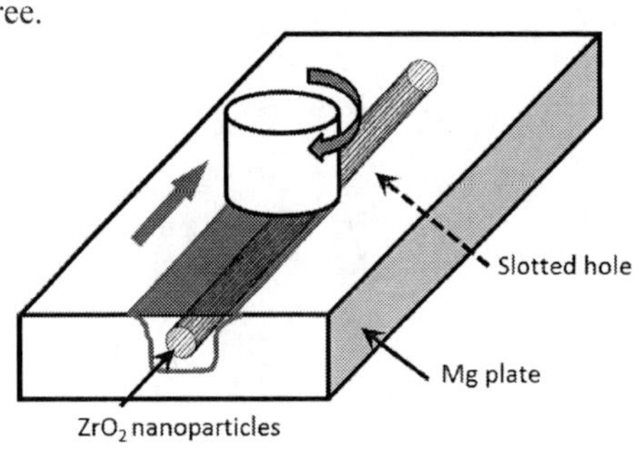

Fig. 1: Schematic illustraion of FSP with nanoparticles.

Characterization of the FSPed samples

The cross section of FSPed samples were polished and etched, and the macrostructures were observed by optical microscope (OM) Nikon AZ100. The microstructures were observed by a scanning electron microscope (SEM) JEOL JSM-6610 LA and a laser microscope (LM) KEYENCE VK-8500. 2-dimensional Vickers microhardness mappings were conducted for the cross section using programmable microhardness tester (Future-Tech FM-300/ASV-90) with a load of 100 g. The average grain sizes before and after FSP were measured by applying the linear intercept method to the OM and LM images, respectively.

Results and Discussions

Macrostructures and microhardness mappings

Macrostructures and microhardness mappings of the cross sections of FSPed samples are shown in *Fig. 2*. The particles were widely spread among the SZ after 2-pass FSP. The homogeneities of the particles distribution and microhardness mapping have been enhanced with the no. of passes. The microhardnesses near the center of the SZ after 2-pass and 3-pass FSP samples were 51 HV and 55 HV, respectively, corresponding to 1.5 and 1.6 times as hard as the base metal (34 HV). An inner defect is significant in the 3-pass sample. The higher heat input condition (e.g. the higher applied load) will eliminate the inner defect.

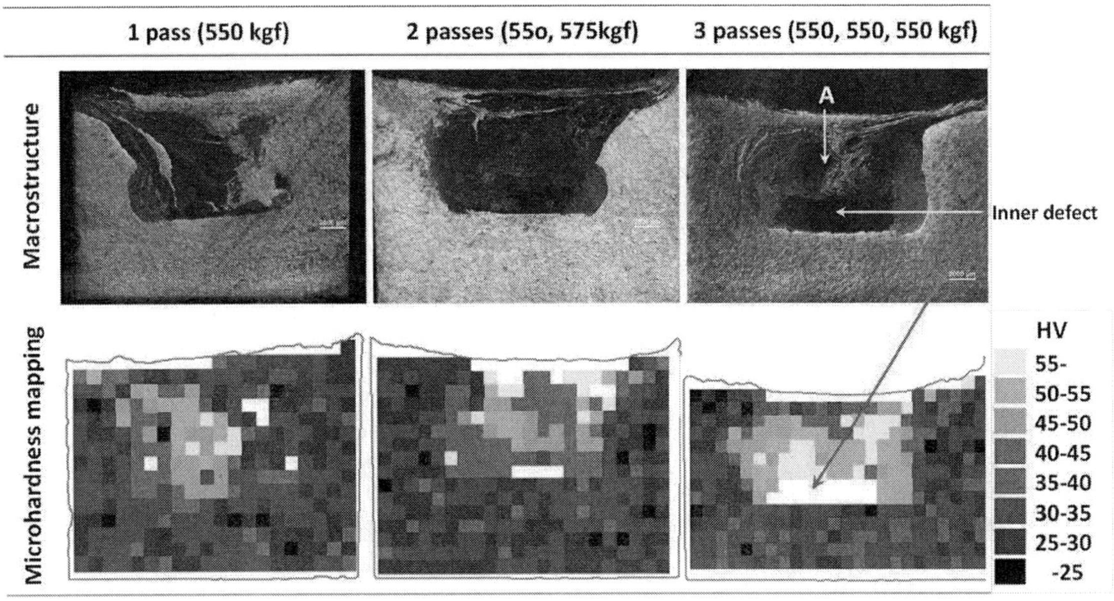

Fig. 2: Macrostructures and microhardness mappings of the cross sections of FSPed samples.

Microstructures

Fig. 3 shows the SEM and LM images of the microstructure of the 3-pass FSPed sample around the point A in *Fig. 2*. The grain size has been 0.56 μm, achieving UFG structure (<1 μm). The agglomerated zirconia particles can be seen in the SEM image, and the cluster sizes are 3 μm or less. Assuming the homogeneous distribution of the single particles, the final grain diameter D_f after balancing between the driving force of grain growth and pinning force can be theoretically estimated by $D_f=4r/(3f)$, where r is the particle size and f is the volume fraction of the particle [7]. Substituting r = 0.010 (mm) and f = 0.017 gives D_f = 0.77 (μm). Although the nanoparticles have not been singly dispersed in *Fig. 3 (b)*, D_f is roughly comparable to the experimental value (0.57 μm). The achieved grain size (0.57 μm) has been less than one thousandth of that of the base metal (740 μm). It has been demonstrated that FSP with 10-nm-size zirconia particles can produce UFG MMCs even from a coarse-grained cast metal.

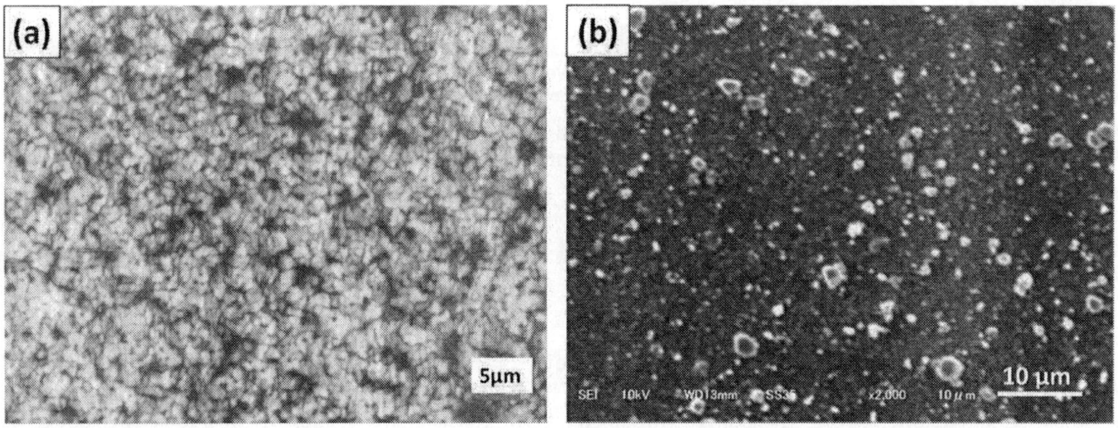

Fig. 3: (a) LM and (b) SEM images of the microstructure of the 3-pass FSPed sample around the point A in Fig. 2.

Conclusions

A multi-pass FSP with 10-nm-size zirconia particles has been performed for a coarse-grained cast pure magnesium plates. After 3-pass FSP, the nanoparticles have been widely spread among the SZ, and the grain size has reduced from 740 μm to 0.57 μm. It has been proved that FSP with 10-nm-size zirconia particles can produce ultra-fine-grained (UFG) metal matrix composites (MMCs) from coarse-grained cast magnesium.

Acknowledgments

This study was supported by Japan Science and Technology Agency (JST) under Collaborative Research Based on Industrial Demand "Heterogeneous Structure Control: Towards Innovative Development of Metallic Structural Materials" and The Amada Foundation, grant no.AF-2014019.

References

[1] R.S. Mishra and Z.Y. Ma: "Friction stir welding and processing", *Mater. Sci. Eng.* R 50 (2005), pp. 1-78.
[2] C. I. Chang, *et al.*: "Relationship between grain size and Zener–Holloman parameter during friction stir processing in AZ31 Mg alloys", *Scr. Mater.* 51 (2004), pp. 509-514.
[3] R.S. Mishra and Z.Y. Ma: "Friction stir processing: a novel technique for fabrication of surface composite", *Mater. Sci. Eng. A* 341 (2003), pp. 307-310.
[4] H.S. Arora, et al.: "Composite fabrication using friction stir processing—a review" *Int. J. Adv. Manuf. Technol.* 61 (2012), pp. 1043-1055.
[5] Y. Morisada, *et al.*: "MWCNTs/AZ31 surface composites fabricated by friction stir processing", *Mater. Sci. Eng. A* 419 (2006), pp. 344-348.
[6] Y. Kimoto, *et al.*: "Empirical process design for fabricating ultra-fine grained metal matrix composites via friction stir processing", *Proc 4th Int Conf on Scientific and Technical Advances on Friction Stir Welding & Processing*, San Sebastian, Spain, October 2015, pp. 60-64.
[7] J. F. Guo, *et al.*: "Effects of nano-Al_2O_3 particle addition on grain structure evolution and mechanical behaviour of friction-stir-processed Al", *Mater. Sci. Eng. A*602 (2014), pp. 143-149.

Role of tool profile in grain structure formation and texture development during friction stir welding of thick welds

Murshid Imam[a1], Yufeng Sun[a], Hidetoshi Fujii[a], Nishu MA[b], Seiichiro Tsutsumi[a], Hidekazu Murakawa[a]

[a] *Joining and Welding Research Institute, Osaka University, Ibaraki 5670047, Japan*

[b] *JSOL Corporation, Engineering Technology Division, Nishi-ku, Osaka, 550-0001*

Abstract

The present study successfully obtained defect-free welds of 20 mm thick 5083-O aluminium alloy. It was shown that the material behavior at the thread and flat faces of the tool are different and can be linked with the microstructure developed after completion of the weld.

Key words: Friction stir welding, Probe profile, Stop action weld, Microtextures, Grain structure

1. Introduction

For the FSW of thick aluminium welds, both the tool material selection and tool geometrical features have a large impact on the process forces, plastic flow, heat generation, tool life and achievable welding speeds. A recently developed Triflat^{TM} design FSW tool has shown advantages in material mixing along the thickness direction, withstanding higher welding forces and obtaining defect-free welds in thicker plates [1]. This type of tool has a scrolled shoulder and probe in the shape of a truncated cone with threads and three flats [1-3]. A detailed study on the individual effect of the threads and flats of such a tool on the development of microstructures has not yet been fully clarified. Therefore, in this study, a 20 mm thick 5083-O aluminium alloy plate was welded using this tool. To develop a better understanding of the microstructure formed after the completion of the weld, a stop action weld was prepared and sectioned into thin slices with the probe in place. Analysis was made on both the plan-view and cross-section of the obtained weld. An attempt was made to understand the influence of the tool geometrical features on the microstructure development after completion of the weld. This study has important implications and perhaps the first report providing the details regarding the changes in the present tool design based on a scientific principle.

2. Experimental procedure

The experiment was conducted using a 20 mm thick 5083-O aluminium alloy plate with the process parameters shown in Table 1. The chemical composition of the plate is listed in Table 2. A common Triflat^{TM} design FSW tool was used to fabricate the joint. A 'stop action' weld sample was obtained to study the individual effects of the threads and flats of the FSW tool used in the present case. Microstructural studies focused on the slices prepared with the probe in place from both the transverse and plan cross-sections. The thin slices were examined metallographically by optical microscope (OM). More details about the microstructures were also obtained from the electron backscatter diffraction (EBSD) analysis. For the EBSD, the JSM-7001FA field emission type scanning electron microscope (FE-SEM) operating at 25 kV was used.

Table 1: Process parameters used in this work.

Tool pin profile	Rotational speed (rpm)	Welding speed (mm/min)	Pin height (mm)	Axial load (kN)	Tool tilt angle (°)
Threaded Triflat ^{TM} design	500	50	19.9	28	3

Table 2: Chemical compositions of 5083-O aluminium alloy.

Element	Si	Fe	Cu	Mn	Mg	Cr	Zn	Ti	Al
Weight %	0.13	0.26	0.04	0.65	4.30	0.11	0.01	0.02	Bal

3. Results and discussion

An overview of the material flow and grain structure formation around tool probe in the sectioned 'stop action' weld, schematics of the shear stresses around the probe and at the tool shoulder/workpiece interface and the different tool positioning during one complete tool rotation are shown in Fig. 1. It can be clearly seen that the thermo-mechanically affected zone (TMAZ) is wider on the retreating side than on the advancing side because all the material displaced by the tool probe is extruded around this side driven by the clockwise rotation of the tool [4]. The presence of voids indicate that the material in this region (back side of advancing edge) is mechanically detached from the probe where the tool imparts no force on the material (zero shear stress, $\tau = 0$). In addition, the material entering the curved and flat portions has different velocities and changes with the tool rotation during the welding. For the partial sliding/sticking condition under the actual situation, the material velocity is shown in Fig. 1 c. To develop a better understanding of the effect of the thread and flat faces on the microstructure development, the sample was also prepared from the weld cross-section with the probe in place which is shown in Fig. 2. It can be clearly seen that the thread and flat of the FSW tool give rise to different material flows. During the FSW, the plasticized material around the probe experienced two kinds of forces, i.e., the force F_n acting normal to the tool probe surface due to the tool rotation and the other is the traction force F_t which was parallel to the tool probe surface as a consequence of the friction between the surface and the plasticized material [5]. The original position of the threads and flats for the stop action weld and the orientation relationship of the forces exerted on the plasticized material are also schematically shown in Fig. 2. The combined effect of both these forces determines the direction of material flow along the resultant force (F) direction. At the thread portion of the tool probe, the material experiences a significant downward vertical movement within the rotational zone. This material is highly deformed and sloughs off behind the probe in an arc shape as can be clearly seen from this figure. On the other hand, the material from the front side entering on the retreating side is opposite to the welding direction, therefore, the traction force acts in the same direction. Since the thread portion of the tool probe is on the front when the flat is on the retreating side, it is believed that the material on the front experiences the same downward vertical movement before it encounters the flat face. Consequently, the material flow is affected due to change in both the pressure and velocity fields at the flat face. Furthermore, the crystallographic texture in the form of 111 pole figures at the thread/workpiece and flat/workpiece interfaces were also studied which is also shown in Fig. 2. A simple schematic showing the formation of regular 111 tetrahedrons and shear stress during the FSW is shown in Fig. 3. The cube texture component present in the base material is changes to the shear texture components at the tool/workpice interface. The presence of the shear texture components provides an indication of the relative strain introduced by the thread and flat of the tool probe [4].

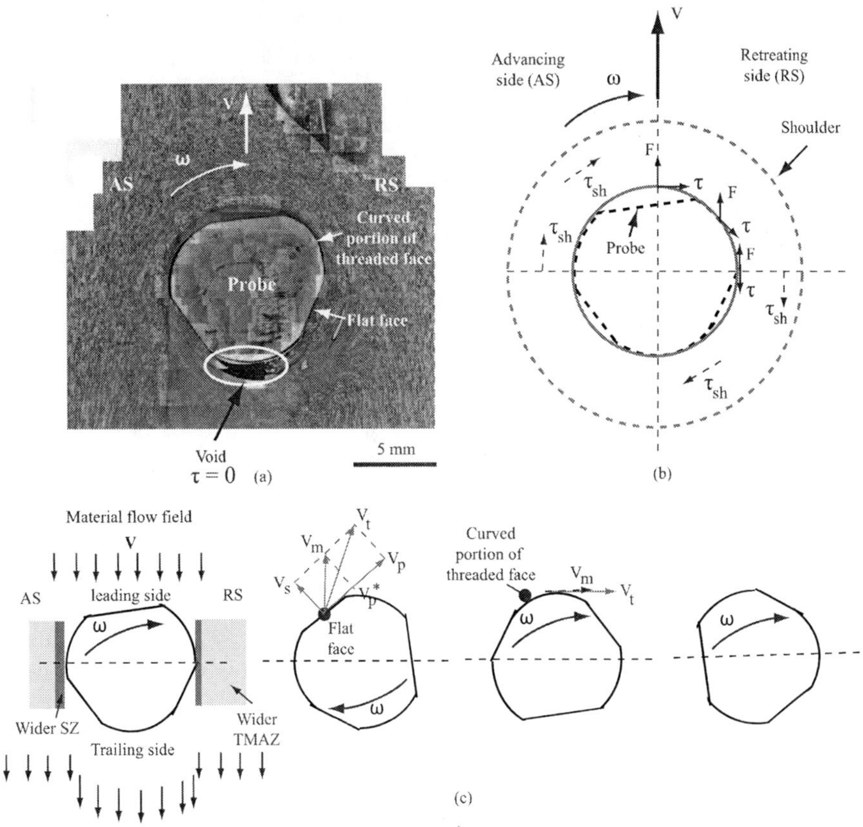

Figure 1: (a) Overview of the 'stop action' weld obtained from stiching together orientation contrast optical images, (b) schematics showing the distribution of shear stresses under tool shoulder and around probe when the tool rotation is clockwise, and (c) different condition of tool positioning during one complete rotation of tool. Note that here τ_{sh} and τ are the shear stresses under tool shoulder and around probe. Also note that the V_t and V_m are the tool velocity and material velocity and V_{sh} and V_{sh} are the tool velocities in the parallel and perpendicular directions, respectively.

4. Conclusions

A better understanding of the effect of the tool geometrical features on the development of the weld final microstructure is presented. The grain structure and microtextures formed at the thread and flat faces of the tool suggest that the thermal and material flow behaviors are of great importance for the optimal tool/probe design, operating parameters and suggest new improvements to the process.

Acknowledgements

The authors would like to thanks the staffs of the welding workshop of the Joining and Welding Research Institute of Osaka University for their cooperation and support.

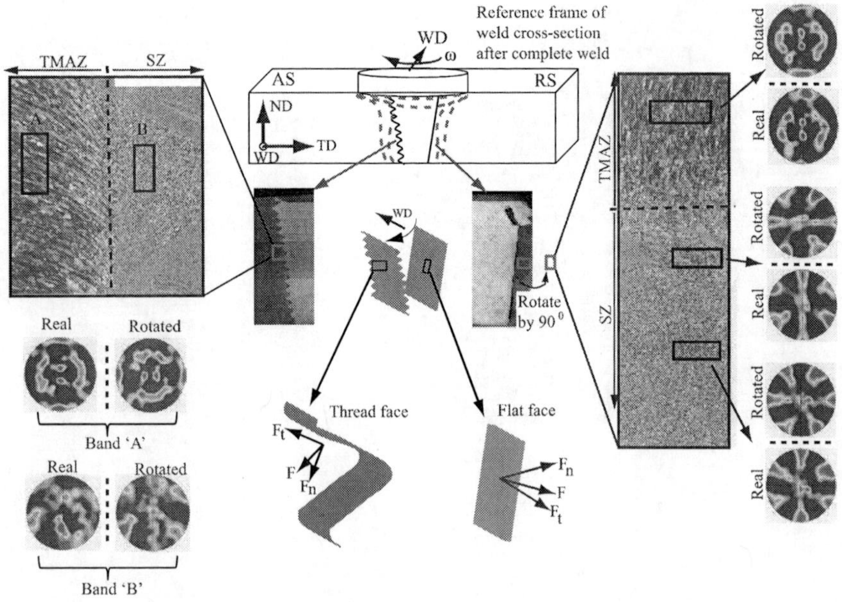

Figure 2: Weld cross-section optical macrograph and EBSD orientation color maps showing grain structure and microstextures in SZ and TMAZ when the material encounter at the thread and flat surface of tool probe on the advancing side (AS) and retreating side (RS).

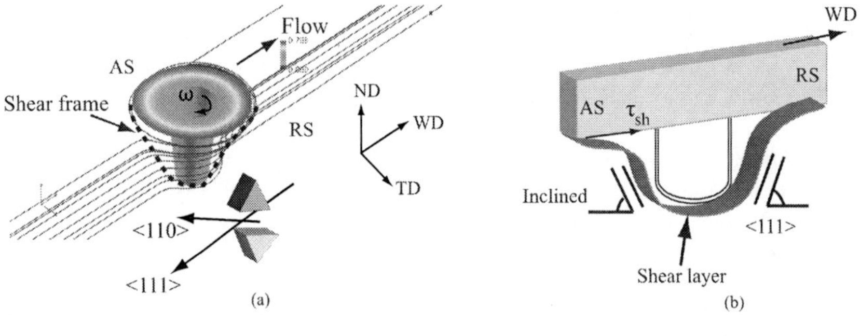

Figure 3: Schematics showing the (a) 111 tetrahedrons formed and (b) the effect of shear stress arising from the tool shoulder on the shear plane inclination at the weld center during the welding.

References

1. J.G. Perrett, J. Martin, P.L. Threadgil, M.M.Z. Ahmed, Proc. 6th World Congress, Al. 2000, Florence, Italy, 2007. pp. 13 - 17.
2. P.Upadhyay, A. Reynolds, Met. Mater. Trans., A, 45A (2014) 2091-2100.
3. R. Nandan, T. DebRoy , H.K.D.H. Bhadeshia, Prog. Mater. Sci., 2008, 53 (6): 980-1023.
4. P. B. Prangnell, C.P. Heason, Act. Mater. 53 (2005) 3179-3192.
5. P. A. Colegrove and H.R. Shercliff, Sci. Tech. Weld. Join., 9 (4) (2004) 352-361.

Effect of welding parameters and B_4C contents on the microstructure and mechanical properties of friction stir weldedB_4C/6061Al joints

Li Yuze, Wang Quanzhao

Institute of Metal Research, Chinese Academy of Sciences, China

Annealed 2.8mm thick B_4C/6061Al composite sheets with various B_4C contents(15,20,25,30%) were friction stir welded (FSW)at a tool rotation rate of 1000rpm and traverse speeds of 50 and 150mm/min. Microstructure characterization of the composites was investigated using optical microscopy (OM) and scanning electron microscopy (SEM). The nugget zone (NZ) was divided into top and bottom parts: NZ-I and NZ-II. The size of NZ-I increased with the increase of the B_4C content, and decreased with the increase of the welding speed. "S" line and a B_4C depleted region were formed at the top and the bottom of the NZ, respectively. FSW resulted in obvious homogenization and fragmentation of B_4C, thereby remarkably increased the hardness of NZ.The tensile strength of all the joints were close to or even up to that of the base material with the fracture occurring at the base material.

Effect of microstructure on corrosion behavior of friction stir welded Mg-Al-Zn alloys

Qu Liu[1,2], *Qingyu Shi*[1,2]
1. Department of Mechanical Engineering, Tsinghua University, Beijing 100084
2. The State Laboratory of Tribology, Tsinghua University, Beijing 100084
shqy@mail.tsinghua.edu.cn

Abstract: The microstructure and corrosion behavior of friction stir welded(FSW) AZ91D magnesium alloy was examined. Prominent refinement and homogenization of second phase β-$Al_{12}Mg_{17}$ morphology were acquired via severe plastic deformation and high temperature in FSW. The grain size in nugget zone(NZ) was refined to 3.95±2.7μm as compared to cast base materials. Meanwhile, plastic deformation flow of β-$Al_{12}Mg_{17}$ phase was observed in thermal-mechanic affected zone(TMAZ). Potentiodynamic polarization curve, Electrochemical impedance Spectroscopy (EIS) and immersion test were taken to evaluate the corrosion property of different zones in the FSW joint. The NZ in FSW joint revealed highest corrosion potential E_{corr} VS SSCE and less value of current density i_{corr} in comparison to as-cast metal as well as TMAZ in FSW joint. The TMAZ in FSW joint was observed more negative corrosion potential and lower current density in comparison to the as-cast alloy. The charge-transfer resistance and polarization resistance of different samples were observed in the following comparative order: NZ>TMAZ>as-cast. The corrosion rate of as-cast alloy was nearly two times higher than the FSW joint in constant immersion test The improvement of corrosion resistance in NZ was mainly attributed to grain refinement, significant breakup and re-dissolution of the coarse β-$Al_{12}Mg_{17}$ phase as a result of FSW.

Keywords: Mg-Al-Zn alloys; friction stir weld; corrosion behavior

Introduction

Owing to their lower density and high strength-to-weight ratio, magnesium alloys had tremendous applications as structural materials in portable electronic devices, mobile communication and automotive[1]. Lower mechanical performance at elevated temperature and poor corrosion resistance is the important factors that limit the utilization of magnesium alloys. Adding of other alloying elements had been attempted to improve the mechanical properties and corrosion resistance of magnesium alloys. Mg-Al-Zn alloys with Al and Zn alloying elements had been identified as an excellent magnesium alloy which widely used to produce die-cast components. However, microstructural variation and compositional heterogeneities easily appear in the welding process which will affect the corrosion behavior of the joints. The decrease of corrosion resistance of joints will seriously affect the reliability of Mg alloy structure in service environment. Hence, the corrosion mechanism and relationship with microstructure morphology has attracted various researchers to focus on[2-3].

The conventional fusion welding of magnesium alloys have lower corrosion resistance compared to as-received alloy owing to the coarse precipitates formed during solidification creates more micro-galvanic sites in the joint[4]. Friction stir welding (FSW) is a solid-state joining process invented by The Welding Institute (TWI) in 1991[5]. In FSW procedure, the rotating non-consumable tool develops frictional heat during tool plunge into the plates. The heat softens the material and causes plasticized material flow with the rotating tool. Then, the joint was formed with appropriate axis force. This process avoids melting defects in welding zone due to the temperature consistently below the melting point of the materials. Furthermore, the microstructure modification and grain refinement occurred during severe plastic deformation processes[5]. Therefore, FSW process is an effective technique for joining magnesium alloy and has significant influence on the corrosion resistance of magnesium joints. However, the effect of FSW process on the corrosion behavior of magnesium alloys has diverse results. Different materials and microstructure have diverse corrosion resistance in the FSWed joints. The objectives of the present study were to investigate the effect of microstructure variation in FSWed AZ91D alloy joint and the relationship between the microstructure with the corrosion behavior .

Experimental

The as-received material is commercial AZ91D cast alloy with chemical composition of 9.3Al-0.67Zn-0.25Mn-0.027Si-0.005Cu-0.003Fe-0.001Ni (wt%). The plates with 5mm thick were machined from the billets and subjected to FSW at a tool rotation of 650rpm and a traverse speed of 30mm min^{-1}. A tool with a shoulder diameter of 16mm and a threaded conical pin with 6mm on bottom, 4mm on tip and 4.7mm in length was used. The plunge depth along Z axis is 0.3mm to ensure no porosity generated in the weld nugget. Microstructural observation and analysis was carried out by optical microscope (OM). The samples were prepared by mechanical polishing and etching using a solution of 1g oxalic acid, 1ml nitric acid, 1ml acetic acid and 150ml distilled water.

Corrosion behavior of the different samples were studied using potentiodynamic polarization and electrochemical impendence spectroscopy (EIS) and constant immersion test in 3.5wt% NaCl solution at room temperature. The samples were ground down to 2000 grit and then polished up to 0.5μm surface by conventional metallographic steps. The polished specimen surface was cleaned with acetone and preserved in the drying vessel before subjecting to corrosion tests. Electrochemical studies were performed on Zahner's electrochemical workstation using a three-electrode condition consisting of saturated silver chloride electrode(SSCE) as reference electrode, platinum mesh as counter electrode and as-cast/FSWed samples as working electrode in a 3.5wt% NaCl solution. Prior to polarization scan, the samples were given an incubation period of 30 minutes to stabilize the initial condition. Potential was scanned from -1400mV vs. SSCE to -800mV vs. SSCE at a scan rate of 0.5mV/s. EIS measurements were carried out over a frequency ranging from 100kHz to 10mHz using 5mV amplitude sinusoidal voltage. Constant immersion tests were carried out in the 3.5wt% NaCl solution for 1week. During immersion test, each sample was cleaned by immersing in a mixture of 200g/L chromium trioxide and 10g/L silver nitrate solution for 5 minute to dissolve corrosion productions. Then, the samples were cleaned by acetone and measured the weight loss.

Results and discussion

The microstructure of as-received AZ91D magnesium alloy (seen in Figure 1a) was characterized by α-Mg matrix with coarse intermetallic β-$Mg_{17}Al_{12}$ network distributed at the grain boundaries, which is a typical cast microstructure of AZ91D. After FSW procedure, the microstructure in the nugget zone of joint (seen in Figure 1b) was characterized by fine equiaxed grains (3.95±2.7μm) and little fine second-phase particles distributed along material plastic flow line. The coarse second phase β-$Mg_{17}Al_{12}$ network were broken down and disappeared in the nugget zone after FSW. The microstructure in the thermal and mechanical affect zone(TMAZ) (seen in Figure 1c)was different. The β-$Mg_{17}Al_{12}$ phase mainly be distorted, refined and arranged along the direction of material flow. It could be indicated that prominent grain refinement of matrix and scattered of second phase β-$Al_{12}Mg_{17}$ were occurred via severe plastic deformation and high temperature in FSW.

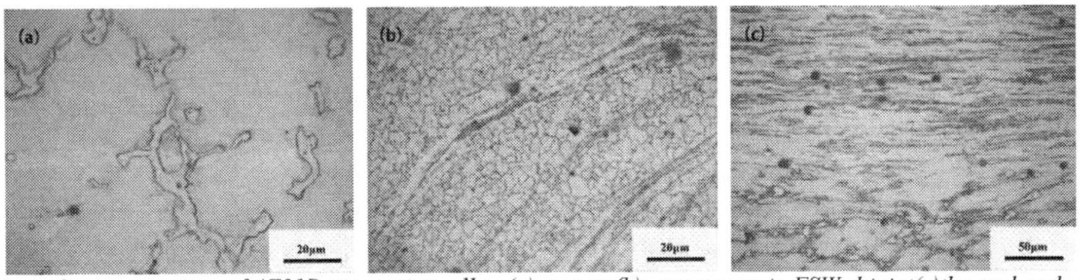

Figure 1 Typical microstructure of AZ91D magnesium alloy (a) as-cast(b)nugget zone in FSWed joint(c)thermal and mechanical affected zone in FSWed joint

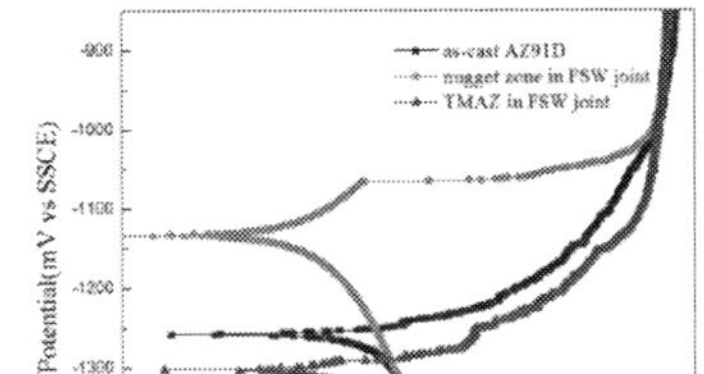

Figure 2 Potentiodynamic polarization curves of as-cast as well as FSWed joint

The potentiodynamic polarization curves of as-cast and as-FSWed AZ91D magnesium alloys in 3.5wt% NaCl solution at room temperature were shown in Fig 3. It is evident that the polarization curve of nugget zone in FSW joint shifts toward a more noble corrosion potential than as-cast and TMAZ in FSW joint samples, implying the reduction of corrosion tendency. There is no stable passive region in the curves of as-cast and TMAZ in FSW joint of AZ91D magnesium alloy. However, a small passive region in the anodic branch of nugget zone in FSW joint was observed. It is indicated that the nugget zone in FSW joint is appeal to generate more compact and firm passivation film than as-cast alloy. This is mainly attributed to the fine grains and fine second phase in the nugget zone. Electrochemical parameters obtained from these curves were summarized in the Table 1. The corrosion potential (E_{corr}) of the as-cast, nugget zone of FSWed joint and TMAZ of FSWed joint is -1258mV, -1133mV and -1303mV vs. SSCE, respectively. Furthermore, corrosion current density (i_{corr}) of these alloys is 3.16×10^{-2}mA/cm^2, 5.02×10^{-3}mA/cm^2 and 1.94×10^{-2}mA/cm^2, respectively. It is apparent the nugget zone of FSW joint had the highest corrosion potential and lowest current density. It is indicated that the corrosion resistance of nugget zone in FSW joint is significantly higher than other samples. The TMAZ in FSW joint had more negative corrosion potential and lower current density in comparison to the as-cast alloy, implying the TMAZ was slightly less susceptible to corrosion attack than as-cast alloy.

Table 1 Electrochemical parameters obtained from potentiodynamic polarization curves

Sample conditions	Corrosion potential E_{corr}(mV vs. SSCE)	Current density i_{corr}(mA/cm^2)	Cathodic slope(mV/decade)
As-cast	-1258	3.16×10^{-2}	-170
NZ of FSWed joint	-1133	5.02×10^{-3}	-141
TMAZ of FSWed joint	-1303	1.94×10^{-2}	-197

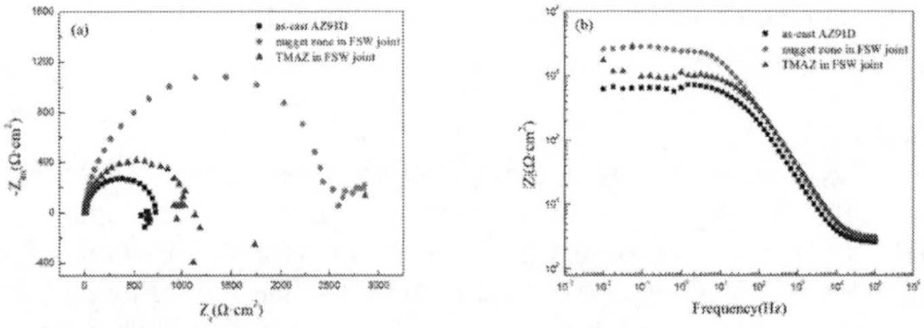

Figure 3 Electrochemical impedance spectroscopy results of as-cast and FSWed joint (a)Nyquist plot(b)Bode plot.

Electrochemical impendance Spectroscopy is universally applied to compare the impedance of different materials in the corrosion environment. The EIS test results of as-cast as well as FSWed AZ91D alloys were represented in the Nyquist and Bode plots, as shown in Fig 3. Based on the Nyquist plot, a large capacitive loop for both three kinds of samples was observed. The diameter of the capacitive loop indicates charge-transfer resistance[6]. It is apparent that the nugget zone in FSW joint had the largest diameter of capacitive loop. The diameter of capacitive loop in TMAZ zone of FSW joint was slightly larger than as-cast alloy. So the comparative charge-transfer resistance of different samples was observed in the following order: NZ>TMAZ>as-cast. Based on the Bode plot, the similar solution resistance in the high frequency range was about 5Ω for as-cast as well as FSW AZ91D alloys. However, the polarization resistance in the low frequency was found to be about 3000Ω, 1000Ω and 650Ω for NZ in FSW joint, TMAZ in FSW joint and as-cast alloy, respectively. The polarization resistance of different samples followed the same comparative order with the charge-transfer resistance. Therefore, it is indicated the impedance of AZ91D magnesium alloy was improved via FSW. This is in agreement with the results of the potentiodynamic polarization experiments.

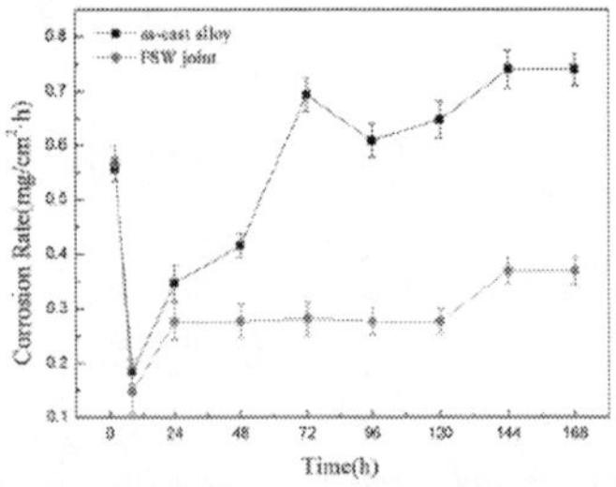

Figure 4 Corrosion rate of as-cast as well as FSW AZ91D magnesium alloys obtained from immersion corrosion tests.

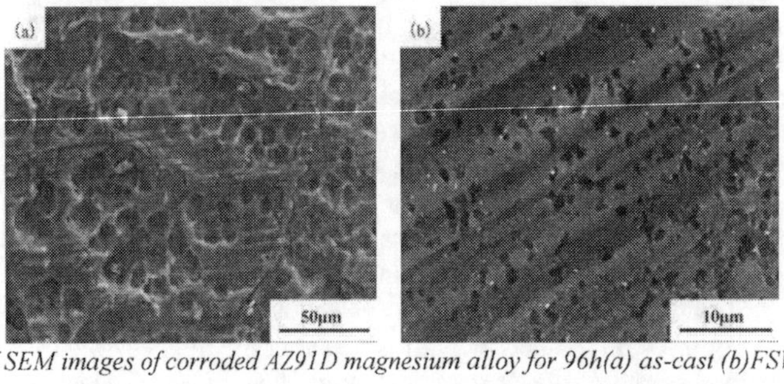

Figure 5 SEM images of corroded AZ91D magnesium alloy for 96h(a) as-cast (b)FSWed joint

Constant immersion test is a direct method to estimate the corrosion resistance of the materials. The weight loss experienced during constant immersion test in the 3.5wt% NaCl solution represents the corrosion rate of materials. Fig 4 showed the variation of corrosion rate with respect to time of as-cast as well as FSWed AZ91D magnesium alloy. It is observed that during the initial hours of immersion, the corrosion rate was higher in as-cast as well as FSW joint samples. After eight hours, the corrosion rate of both samples decreased to a low level, the extent of which is comparatively lower in the case of FSW joint sample. This is mainly attributed to the formation of protective passivation film. However, the Pilling Bedworth ratio for Mg is less than 1, so the oxide film formed is not able to provide efficient corrosion protection because of the compressive stresses that generate as a result of oxide/metal mismatch[7]. Therefore, the corrosion rate increased in both samples after 24 hours immersion time. The corrosion rate of FSW joint sample showed the tendency to become nearly uniform with respect to time. However, the corrosion rate of as-cast alloy continuously increased with respect to time. Furthermore, the value of corrosion rate of as-cast alloy was nearly two times higher than the FSW joint after 72 immersion hours. It is indicated that the FSW joint had better corrosion resistance than as-cast alloy. The SEM images of corroded samples immersed in the 3.5wt% NaCl solution for 96h were shown in Fig 5. It is revealed that the α-Mg which located surrounding the bigger network β-$Mg_{17}Al_{12}$ phase in the as-cast alloy corroded vigorously and rapidly propagated to greater depths. The surface of the FSW joint showed comparatively slight pitting corrosion due to the fine grain and second particles uniformly distribute in the matrix. The size of corrosion pit was remarkably smaller in the FSW joint's surface. It represented the corrosion extent and mass loss of the FSW joint is less than the as-cast AZ91D alloy.

The corrosion resistance of AZ91D magnesium alloy was significantly improved after FSW procedure. This is mainly attributed to significant breakup and re-dissolution of the coarse eutectic β-$Mg_{17}Al_{12}$ phase and remarkable grain refinement. First, the fine second phase acted as cathodic sites in the microgalvanic cells as well as fine grains mades the general corrosion occurred in magnesium matrix. And the fine second phase worked as a barrier to inhibit the extension of corrosion. Second, the re-dissolution of Al element into the magnesium matrix via FSW procedure made the passivation formed on the α-Mg matrix more compact to breakdown. In all, Combining prominent grain and second phase refinement with re-dissolution of β-$Al_{12}Mg_{17}$ phase significantly improved the corrosion resistance of the AZ91D magnesium alloy.

Conclusions

(1) The corrosion resistance of the FSWed AZ91D magnesium joint was significantly improved by as-cast microstructure modification. This is mainly attributed to remarkable grain refinement, significant breakup and re-dissolution of the coarse β-$Al_{12}Mg_{17}$ phase via severe plastic deformation and high temperature in FSW procedure.

(2)The corrosion resistance varied across the different FSW zones. The NZ in FSW joint revealed highest corrosion potential and lowest corrosion current density in comparison to as-cast metal as well as TMAZ in FSW joint. The TMAZ was found to slightly less susceptible to corrosion attack than as-cast alloy. The corrosion resistance a different zones in samples were observed in the following comparative order: NZ>TMAZ>as-cast.

Acknowledgments

The authors gratefully acknowledge the support of State Laboratory of Tribology in Tsinghua University.

References

[1] B.L. Mordike, T. Ebert: "Magnesium: properties-applications-potential," Mater. Sci. Eng. A 302(2001),pp. 37-45.
[2] L.M. Liu, R.Z. Xu: "Investigation of the corrosion behavior of laser-TIG hybrid welded Mg alloys" Corros. Sci.52(2010),pp. 3078-3085.
[3] G.B. Hamu, et al: "The relation between microstructure and corrosion behavior of Mg-Y-RE-Zr alloys" J. Alloy. Compd. 431(2007),pp. 269-276.
[4] G. Ben-Hamu, et al: "The relation between microstructure and corrosion behavior of GTA welded AZ31B magnesium sheet" Mater. Sci. Eng. A 452-453(2007),pp. 210-218.
[5] R.S. Mishra, Z.Y. Ma: "Friction stir welding and processing" Mater. Sci. Eng. R. 50(2005),pp. 1-78.
[6] Z. Pu et, al: "Grain refined and basal textured surface produced by burnishing for improved corrosion performance of AZ31B Mg alloy" Corros. Sci. 57(2012),pp. 192-201.
[7] K.D. Ralston, N. Birbilis. "Effect of Grain Size on Corrosion: A Review" Corros. Sci. 66(2010),pp. 501-513.

Imaging and simulation of shielding gas flows during GMAW

I. Bitharas[1], N.A. McPherson[2], W. McGhie[3], D. Roy[3], A. J. Moore[1]*

[1] *Institute of Photonics and Quantum Sciences, Heriot-Watt University, Edinburgh, UK*
**Email: ib92@hw.ac.uk*

[2] *Department of Mechanical and Aerospace Engineering, University of Strathclyde, Glasgow, UK*

[3] *BAE Systems Maritime – Naval Ships, Glasgow, Scotland, UK*

Abstract

Schlieren imaging was utilised to image density gradients that arise from the inert Ar gas flows during GMAW. In addition, 2D and 3D numerical magneto-hydrodynamic (MHD) analyses of the equivalent steady-state flows were used to predict air contamination and density gradients under the same conditions. X-ray radiography and micro/macrographs of welded DH36 mild steel samples determined porosity levels and weld suitability. Optimal flow conditions were investigated for a number of cases, relevant to automated and semi-automated processes.

Introduction

Large steel deposition volumes made using gas metal arc welding (GMAW), such as during shipbuilding or wire-arc additive manufacture (WAAM), require the volumetric flow rate of the shielding gas to be high enough to achieve the required weld quality. Naturally, the long term operation of an automatic or semi-automatic production unit can be optimised to minimise gas wastage. The objective of this study is therefore to establish the minimum flow conditions with which to adequately shield an acceptable weld but also to determine the sensitivity of such limits. Moreover, how heavily coverage quality depends on common process parameters was investigated. To that end, the transparent Ar flow was visualised through schlieren imaging. Additionally, a detailed computational model was employed to aid in the interpretation of the images as well as in a predictive manner.

In some early schlieren visualisation work, Gibson and Moet [1] studied the effects that nozzle standoff, diameter and flow rate have on the effectiveness of the gas shield. They showed that a smaller standoff distance improves the shielding quality and that while nozzle size and cross-drafts affected coverage, no simple relation between them was evident. Okada *et al.* produced laser shadowgraphs of gas tungsten arc welding (GTAW) [2]; they found that the gas flow was heavily affected by variations in arc current, indicating that the most significant contribution to the overall observed turbulence levels was that due to Lorentz forces.

Simulating the Ar plasma spray process, the k-ε turbulence model was previously used to describe air entrainment within a high velocity Ar jet in [3], [4]. Compared with a series of experimental measurements by Fincke *et al.* [5], the k-ε model adequately predicted the general flow features, provided initial conditions for the turbulent kinetic energy were well posed. Of particular relevance to this study, the combination of modelling with schlieren photography has previously been used to examine the gas distribution during arc welding [6], [7]. The imaged flow features were shown to be in agreement with numerical predictions. In [7], torch and geometry features were numerically and experimentally examined for GMAW and GTAW, determining which ones affected coverage the most. They showed that an increase in standoff distance of the order of 10 mm significantly affected coverage and had to be compensated. A limit of 50 ppm air over the weld was taken as the maximum allowed for a good quality weld, especially when joining sensitive metals such as aluminium.

Experimental setup

Fig. 1 shows the schlieren imaging system layout and placement relative to the welding torch [10]. In this setup, the workpiece was moved at a constant speed while the torch and the optics were kept stationary to facilitate recording. It was assumed that any effects on the flow were negligible; this can be justified when considering the relative magnitude of the torch travel speed (5 mm/s) and the average shielding gas speed (~5 m/s).

Broadband white light from the filament of a 200 W tungsten-halogen lamp was collected through a fast condenser lens (L1) and focused onto the source slit. The shape and size of the source slit determines the sensitivity and measurement range of the schlieren system [8]. We used a 2 mm × 6 mm rectangular source slit with the long edge vertical, as recommended by Siewert *et al.* for welding applications [9]. The source slit was positioned at the focus of M1, to produce an image of the source at the focus of M2.

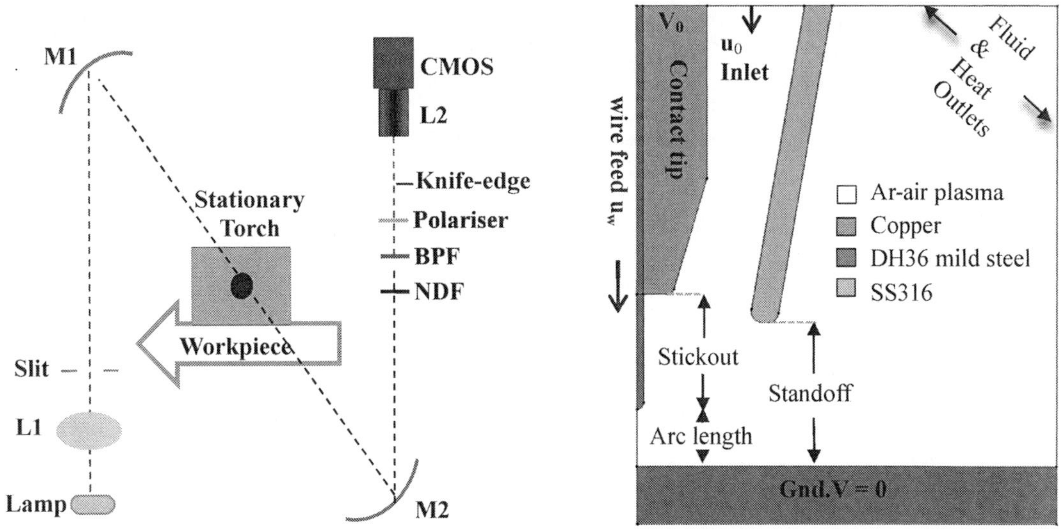

Fig. 1 (Left) Schematic of Z-type schlieren and welding setup, comprised of a tungsten filament lamp, parabolic mirrors, a combination of filters and a CMOS sensor. (Right) 2D axisymmetric geometry, with modelling outline.

A 633 nm ±1 nm full width at half maximum band- pass filter (BPF) was used to eliminate as much arc light as possible [2], [10]. A polariser was used to remove glare from the workpiece, and along with a neutral density filter (NDF), the overall light intensity was brought to the appropriate level. Placement of a vertical knife's edge (matched with half of the source slit image) at the focus of M2 resulted in a final ~60% drop in intensity whilst highlighting positive and negative refractive index/density gradients in the x-direction.

The flow visualisations were intended to be characteristic of a steady-state process, representative of long weld seams in an enclosed workshop environment. Therefore, the recorded images contained gradient information that were averaged over relatively long exposure times. The image sequences were thus recorded at 150 frames per second (fps) with a resolution of 1280 × 1024 pixels using a Mono CMOS Flea 3 camera, fitted with a variable focus macro lens (L2).

Magneto-hydrodynamic (MHD) flow modelling

Simulations of the Ar/air mixture flow, including the complex interactions between contact tip, electrodes and nozzle were carried out using COMSOL multiphysics. 2D axisymmetric (Fig. 2) and 3D welding cases were analysed, depending on the potential for simplification presented by a given geometry.

The Ar-air plasma was included under the local thermodynamic equilibrium (LTE) principle, therefore the characteristic thermophysical properties namely density, heat capacity, viscosity, thermal and electrical conductivity were expressed as a function of temperature for the range of 300 – 30,000 K as well as the mass fraction of air. Consequently, the plasma was treated as a multicomponent gas comprised of Ar and air with a distributed mass fraction. Such data were generated by Murphy through the Gibbs free energy minimisation method [11], but including some recently updated collision integrals for the Ar and N species, as presented in [12].

By assuming that the highest degree of entrainment occurs during the highest current and voltage period (Lorentz body forces are maximum), the dynamics of such variation were omitted from the simulation. Instead, steady-state flow conditions were utilized, making use of the governing equations in a time independent formulation. The well-established MHD equations were solved numerically, using the finite element method.

State of the art treatment of the plasma's thermodynamics, fluid mechanics but also turbulent species mixing and de-mixing was included in the simulation. To allow more space for results, precise PDE formulations are omitted from this work but will be reported elsewhere [13]. Typical solution times ranged from 15 min for a single 2D case with 10^4 elements, up to 6.5 hours for a half-symmetry 3D cases and $2 \cdot 10^5$ domain elements.

Results

In order to establish how effectively the process was shielded, the model and associated image sequences were used to investigate a wide range of cases. Indeed, in order to determine the most efficient use of gas, the process variables examined included torch positioning and design, arc waveform and settings, shielding gas mixture (such as Ar vs Ar +20% CO_2) or the type of wire used (solid vs flux cored).

In Fig. 2, the temperature iso-surfaces show the characteristic bell shape of the arc, reaching maximum temperatures of over 17 kK for the case of a torch under a 20° angle (pulling). The grayscale plot shows the resulting steady state temperature distribution in the metal. The omission from the calculations of Fe vapour or Marangoni convection, both cooling effects, suggests that the total heat transfer to the workpiece is exaggerated. In order to determine the shield's effectiveness, the oxygen concentration over the average length of molten material was examined numerically.

The arrows are proportional to the local velocity, but truncated to 8 m/s in order to exclude the influence of the ~200 m/s plasma jet. The direction of the gas shield flow vectors under the nozzle show how much the fluid is accelerated downwards and towards the wire tip. This 'pinch' effect can be explained by the magnitude of the Lorentz body forces: the flow tends towards the inter-electrode arc region of higher current and magnetic field densities. By increasing Ar inlet flowrate, the downwards momentum of the shield increases in comparison to the pull by the arc. This in turn leads to decreased overall mass fractions over the weld hence more effective coverage, as expected.

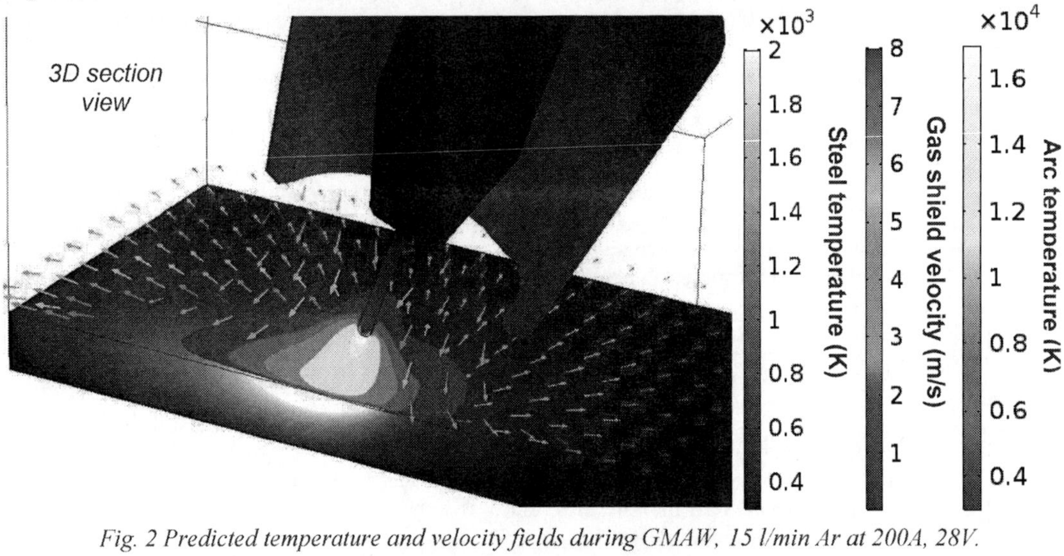

Fig. 2 Predicted temperature and velocity fields during GMAW, 15 l/min Ar at 200A, 28V.

Complementing the numerical results, the refractive index gradients in the horizontal direction $\partial n/\partial x$ were imaged under the same conditions, shown in Fig. 3. Over the arc, a dark, bell-shaped gradient is visible in both images. By comparison with Fig. 2, it is clear that this feature is formed due to the thermal gradient of the arc but also due to the separation between 'pinched' and bulk flow.

Below the nozzle sidewalls, a long vertical and curved gradient indicative of the bulk gas flow is visible, Fig. 3. The relative blur of the features for the low flowrate (Fig 3. – left) suggests an overall greater degree of intermixing and therefore an increased air mass fraction is expected under the nozzle. Additionally, a trailing or leading vortex was observed in the image sequences. The distance between the gas shield gradient and the steady-state vortex was also found to be characteristic total coverage area.

To validate the combined results, the corresponding bead-on-plate welds were tested through x-ray radiography. For the case reported in this work, clusters of porosity were present in the DH36 samples at 6 l/min, while at 9 l/min or greater, the welds were within standards.

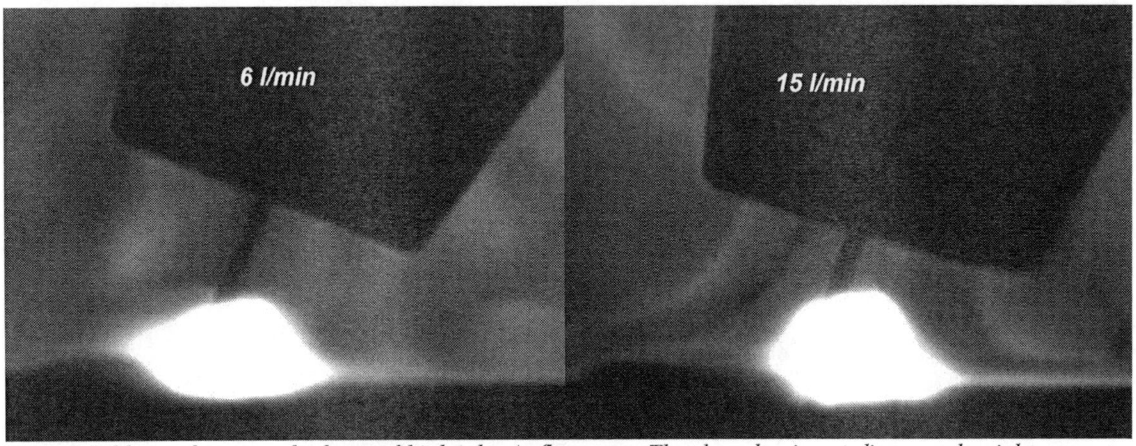

Fig. 3 Schlieren frames under low and high inlet Ar flow rates. The clear density gradients on the right suggest less Ar-air mixing, a lesser degree of air entrainment in the Ar jet and therefore better coverage.

Conclusions

Detailed optical flow diagnostics and computational analyses allowed characterisation of the Ar gas flow through which a weld was shielded. The MHD model's results were in good agreement with schlieren images captured under shielding gas flowrates of 6 – 15 l/min. Through this approach, investigation of the effects of changes in torch design but also general machine operation through selection of the appropriate parameters is facilitated.

References

[1] G. J. Gibson, "Gas flow requirements for inert-gas shielded arc welding," *Weld. J.*, vol. 32, no. 4, p. 128s–208s, 1953.

[2] T. Okada, H. Yamamoto, and S. Harada, "Observation of the shielding gas flow pattern during arcing by the use of a laser light source," *Arc Phys. Weld Pool Behav.*, vol. 1, pp. 203–213, 1980.

[3] R. L. Williamson, J. R. Fincke, D. M. Crawford, S. C. Snyder, W. D. Swank, and D. C. Haggard, "Entrainment in high-velocity, high-temperature plasma jets. Part II: computational results and comparison to experiment," *Int. J. Heat Mass Transf.*, vol. 46, no. 22, pp. 4215–4228, Oct. 2003.

[4] K. Cheng and X. Chen, "Prediction of the entrainment of ambient air into a turbulent argon plasma jet using a turbulence-enhanced combined-diffusion-coefficient method," *Int. J. Heat Mass Transf.*, vol. 47, no. 23, pp. 5139–5148, Nov. 2004.

[5] J. R. Fincke, D. M. Crawford, S. C. Snyder, W. D. Swank, D. C. Haggard, and R. L. Williamson, "Entrainment in high-velocity, high-temperature plasma jets. Part I: experimental results," *Int. J. Heat Mass Transf.*, vol. 46, no. 22, pp. 4201–4213, Oct. 2003.

[6] M. Schnick, M. Dreher, I. J. Zschetzsche, U. Füssel, and A. Spille-Kohoff, "Visualization and optimization of shielding gas flows in arc welding," *Weld. World*, vol. 56, no. 1–2, pp. 54–61, 2012.

[7] M. Dreher, U. Füssel, S. Rose, M. Häßler, M. Hertel, and M. Schnick, "Methods and results concerning the shielding gas flow in GMAW," *Weld. World*, Mar. 2013.

[8] G. S. Settles, *Schlieren and Shadowgraph Techniques: Visualizing Phenomena in Transparent Media.* Springer, 2001.

[9] E. Siewert, G. Wilhelm, M. Häßler, J. Schein, T. Hanson, M. Schnick, and U. Füssel, "Visualization of Gas Flows in Welding Arcs by the Schlieren Measuring Technique," *Weld. J.*, vol. 94, p. 1s–5s, Jan. 2014.

[10] I. Bitharas, S. W. Campbell, A. M. Galloway, N. A. McPherson, and A. J. Moore, "Visualisation of alternating shielding gas flow in GTAW," *Mater. Des.*, vol. 91, pp. 424–431, Feb. 2016.

[11] A. B. Murphy, "Transport coefficients of air, argon-air, nitrogen-air, and oxygen-air plasmas," *Plasma Chem. Plasma Process.*, vol. 15, no. 2, pp. 279–307, Jun. 1995.

[12] A. B. Murphy, "Transport coefficients of plasmas in mixtures of nitrogen and hydrogen," *Chem. Phys.*, vol. 398, pp. 64–72, Apr. 2012.

[13] I. Bitharas, N.A. McPherson, W. McGhie, D. Roy and A.J. Moore, "Visualisation and optimisation of gas coverage during MIG welding", In preparation (2016)

Three-dimensional Numerical Analysis of Heat Transfer and Fluid Flow in Keyhole Laser Welds Based on Thermal-hydraulic

Qiaofeng ZHOU [1], Katsuma HORIO [1], Fumikazu MIYASAKA [1], Hiroaki MORI [1], Masami MIZUTANI [2], Yousuke KAWAHITO [2], Seiji KATAYAMA [2]

[1] *Osaka University, Graduate School of Engineering, Japan*
[2] *Osaka University, Joining and Welding Research Institute, Japan*

Laser welding is one of useful methods for micro-joining in manufacturing processes of MEMS and/or other micro-size devices required high production quality, automation potential and flexibility. But, one of the significant obstacles to overcome for practical applications of laser welding is prevention of weld defects such as porosity, hot cracking, and so on. The related studies can be divided into experimental and numerical approaches. However, at present, it's very difficult even to observe micro-bubbles and cracking behavior during laser micro welding process owing to shortage of spatial resolution of high speed cameras for micro-scale in-situ observation. That is, the mechanisms of these micro-size defects have not been clarified yet. Therefore, a numerical approach is favorable for comprehending how the driving forces affect the flow characteristics of a molten pool. In this study in order to clarify the mechanism of porosity in weld metal on laser spot welding process and suggest proper welding conditions without any defects, in-situ observation by X-ray transmission real-time imaging system and calculation of micro-bubbles' behavior using the numerical model based on the fundamental fluid dynamic theories was conducted. On the basis of numerical analysis, A three-dimensional bubble model is used to simulate pore generation with a keyhole collapse, keyhole and micro-bubbles's behavior could be qualitatively expressed and/or replicated. From the observation, keyhole behavior was able to be clearly observed and two kinds of major dynamic flows were recognized in molten pool during laser welding. Micro-bubbles formed at the bottom of keyhole were maintained nearby the bottom due to the bottom side flow in molten aluminium. Therefore, these results suggest that one of the causes of porosity in laser welding of aluminium is micro-bubbles maintained at the bottom due to molten metal flow.

Numerical simulation of temperature field and flow field of PAKW process in quasi-steady state

Bin Xu, Shujun Chen, Fan Jiang
Engineering research center of advanced manufacturing technology for automotive components, ministry of education, Beijing, China
sjchen@bjut.edu.cn, (010)67391620

Abstract

The plasma arc keyhole welding (PAKW) has been widely applied in engineering project as a high energy beam welding process. But high current and strong arc force may lead to arc instability and produce the welding defects such as undercut and welding bubble. Furthermore, it was found that the flow field of liquid metal and the stability of keyhole had a critical influence on keeping weld seam uniform in quasi-steady state. Therefore, in this work, a three dimensional transient thermal mechanical coupled model of the plasma arc was established to describe the flow field, temperature field and keyhole behavior of PAKW. A combined volumetric heat source model of double ellipsoid and conical body was employed. The energy of the conical source changed in a parabolic way as the depth of the keyhole increased. A keyhole inside solid agitated (KISA) calculation method was proposed to maintain stability of the keyhole in the quasi-steady state. To improve the computational efficiency, the calculated region was limited within the action region of the heat source. The time to reach quasi-steady state from arc igniting was calculated for PAKW of stainless steel plates with thickness of 6 mm. The offsets between top side and bottom side of the keyhole center at different welding times were analyzed. The thermal fluid flow on longitudinal section and different horizontal cross section was numerical simulated, and its effect on stability of keyhole was considered. The fusion line and keyhole image were measured. The calculated results of size of fusion zone and keyhole were in agreement with the experimental results.

1. Introduction

Plasma arc keyhole welding (PAKW) is a kind of high efficient welding method. The workpiece can melt very quickly and a keyhole establish through the whole workpiece. The molten pool and keyhole extend along the welding direction keeping stable[1]. This process involves very complicated fluid dynamics and heat transfer processes coupling keyhole and molten pool[2]. In order to obtain the information of this process some experimental researching methods were proposed. Mainly, they are high-speed camera shooting to capture the images of molten pool and keyhole[3-6], as well as the metallographic way to analyze the evolution of keyhole and molten line[7,8]. However, with these methods the PAKW process was only analyzed indirectly. The evolution situation inside the molten pool and keyhole cannot be accessed. In recent years, with the development of computer technology and numerical simulation theory, using computer simulation technology to study welding process has become an effective method.

To the unique heating form of PAKW process, it is necessary for simulating the complex process accurately to establish a reasonable heat source model. Considering convection and radiation loss, researchers [9, 10] establish a two-dimensional model to analyze the temperature distribution and fluid flow. But this method has a certain gap with the actual welding process. In order to get more accurate calculated results, various forms of three dimensional volumetric heat source model has been proposed [11]. Especially, the cone heat source model[12], the transient plasma arc welding model(TPAW) and double ellipsoidal combined cylinder volumetric heat source[14,15] accord to the actual PAKW process further. Welding arc is the heat source as well as the force source for the welding process. So in addition to the thermal effect, the other important role is the plasma arc force. Huo [16 to 18] analyzed the penetrating process by establishing calculated model of keyhole based on mechanical equilibrium conditions. But this model did not couple the thermal effect. To establish PAKW thermal-mechanical coupled model, researchers[19-21] simulated the position welding coupling heating and force process based on fluid dynamics. Wu [22 to 24]and Li [25] established a model of moving ignition process to analyze the mobile keyhole and molten pool. But the calculated penetrating time had a big difference with experimental results. Li [26] simulated a more accurate penetrating process by establishing a thermal mechanical coupled model which changed as the size of keyhole increasing. Through the coupled calculation of plasma arc and molten pool, Jian [27] obtained the characteristic of PAKW including arc and molten pool in the same calculated process. Wang [28,29] established the calculated model of variable polarity plasma arc welding. The effect of droplet to keyhole and molten pool was considered. But the thermal mechanical interaction was not considered. In this work, a thermal mechanical coupled model is established to describe the quasi-steady process of PAKW. This model is used to simulate the evolution of keyhole and molten pool of the quasi-steady process.

2. Mathematical model

It is critical to establish a reasonable coupled model for the accurate numerical simulation of quasi-steady state process [30]. Therefore, A combined volumetric heat source model of double ellipsoid and conical body, as well as a double ellipsoid plasma arc force model, are proposed. And the energy of the conical source and arc pressure changed in a parabolic way as the depth of the keyhole increased. The coupled model is described as follows.
The heat flux density in front half of the ellipsoid at point (x, y, z) with x>0, z≤c;

$$q_{\mathrm{f}} = \frac{12\sqrt{3}Qm_1}{\pi\sqrt{\pi}b(a_{\mathrm{f}}+a_{\mathrm{r}})c}\exp\left(\frac{-3x^2}{a_{\mathrm{f}}^2}-\frac{3y^2}{b^2}-\frac{3z^2}{c^2}\right) \tag{1}$$

The heat flux density in rear half of the ellipsoid at point (x, y, z) with x>0, z≤c;

$$q_{\mathrm{r}} = \frac{12\sqrt{3}Qm_1}{\pi\sqrt{\pi}b(a_{\mathrm{f}}+a_{\mathrm{r}})c}\exp\left(\frac{-3x^2}{a_{\mathrm{r}}^2}-\frac{3y^2}{b^2}-\frac{3z^2}{c^2}\right) \tag{2}$$

The heat flux density of cone heat source at point (x, y, z);

$$q(r,z) = \frac{9Qm_2e^3}{\pi H_2(e^3-1)(r_{\mathrm{e}}^2+r_{\mathrm{e}}r_{\mathrm{i}}+r_{\mathrm{i}}^2)}\exp\left(-\frac{3r^2}{r_0^2}\right) \tag{3}$$

The arc pressure in the front half of the ellipse at point (x, y, z):

$$P_{\mathrm{a1}} = \frac{C_{arc}\mu_0 I^2}{4\pi^2}\left[\left(\frac{1}{r_2^2}-\frac{1}{r_1^2}\right)+C_{\mathrm{j}}\frac{1}{2\sigma_{\mathrm{j}}^2}\exp\left(-\frac{3(x-u_0t)^2}{A_1^2}-\frac{3y^2}{B^2}\right)\right](x\geq u_0t) \tag{4}$$

The arc pressure in the rear half of the ellipse at point (x, y, z):

$$P_{\mathrm{a2}} = \frac{C_{\mathrm{arc}}\mu_0 I^2}{4\pi^2}\left[\left(\frac{1}{r_2^2}-\frac{1}{r_1^2}\right)+C_{\mathrm{j}}\frac{1}{2\sigma_{\mathrm{j}}^2}\exp\left(-\frac{3(x-u_0t)^2}{A_2^2}-\frac{3y^2}{B^2}\right)\right](x< u_0t) \tag{5}$$

m1, m2 are the distribution coefficient of energy. Q is the effective heat power of plasma arc. a_f, a_r, b, c, H_2, r_e, r_i are shape parameters of the heat source. C_{arc} is corrected coefficient of arc pressure. μ_0 is magnetic permeability. *I* is the welding current. u_0 is the welding speed. *t* is the welding time; r is the distance between the point (x, y, z) with the center of cone heat source. r_0 is the radius at different z coordinates. r_1 is the plasma arc radius at the surface of the workpiece. r_2 is the plasma arc radius at the nozzle; σ_j is the distribution parameters of current density; A_1, A_2, B are shape parameters of the double ellipse. C_j is adjustment coefficient of plasma arc pressure. e is the natural constant. i and j are sign which has no specific meaning.

It is found that the penetrating rate reduces as the depth of the keyhole increasing[7]. Therefore, the heat source model of double ellipsoid and cone as well as the arc force of double ellipse are corrected. The energy of cone heat source increase as a parabolic way and the plasma arc force reduce also in a parabolic way.
The distribution of coefficient of energy is corrected as follows:

$$m_1 = 1 - m_2 \tag{6}$$

$$m_2 = \begin{cases} 0 & h_{\mathrm{kh}} < c \\ k_1\left(\frac{h_{\mathrm{kh}}}{L}\right)^2 + k_2\left(\frac{h_{\mathrm{kh}}}{L}\right) + k_3 & c \leq h_{\mathrm{kh}} \leq L \end{cases} \tag{7}$$

h_{kh} is the depth of the keyhole. L is the thickness of the workpiece. k_1, k_2, k_3 are the adjustment parameter, where $k_1<0$. The energy of the plasma arc is distributed to double ellipsoid heat source and cone heat source. So m_1 plus m_2 is 1. the energy of cone heat source changes with the keyhole depth increasing according to equation (7). In this study, (h_{kh}, m_2) = (1.5, 0), (h_{kh}, m_2) = (6, 0.25) and (h_{kh}, C_{arc}) = (0, 1) are adopted. The relationship between adjustment parameters is as follows:

$$k_2 = \frac{1}{3} - \frac{5}{4}k_1 \tag{8}$$

$$k_3 = \frac{1}{4}k_1 - \frac{1}{12} \tag{9}$$

The plasma arc force reducing pattern with the depth of keyhole increasing is defined as follows.

$$C_{arc} = 1 - \varphi(\frac{h_{kh}}{L})^2 \quad (10)$$

φ is the adjustment parameters ($0 < \varphi < 1$)。

In the quasi-steady state process, the plasma stream like a solid bar with high temperature inserting inside the molten pool agitates the liquid metal. In order to simulate this process, the plasma arc force is applied on the boundary of the keyhole and the gas inside the keyhole, which is named "keyhole inside solid agitated (KISA) calculation method".

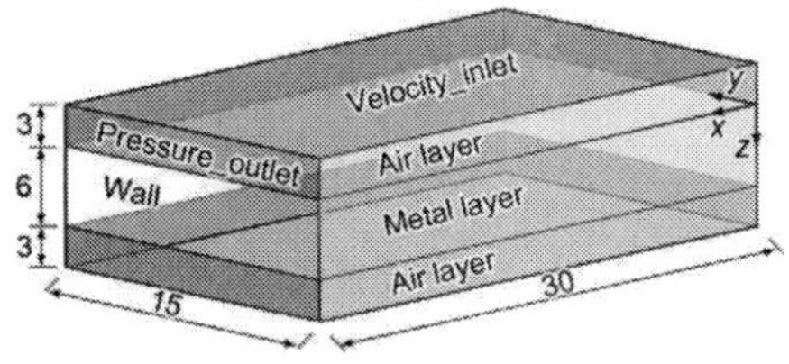

Fig.1 Computational domains and boundary condition (unit: mm)

Fig.1 shows the calculated domain established with a preprocessing software GAMBIT. The domain contains three layers, which are air-metal-air. The thickness of air layer is 3 mm while that of metal layer is 6 mm. The width and length of the domain respectively are 15 mm and 30 mm. Minimum discrete grid boundary is set 0.3 mm. The boundary conditions are distinguished by different colors

3. Analysis and discussion

The welding material is 304 stainless steel of 6 mm thick. The welding speed is 2 mm/s. Welding current is 140 A. The quasi-steady state welding process of PAKW was simulated with FLUENT software, tracking keyhole boundary with VOF method. The features of keyhole and molten pool were analyzed through the simulated results. Fig.2 shows the calculated results of molten pool, keyhole and flow field at different welding time. The keyhole become stable after 3 s from ignition. And its shape is like a bugle which become narrow gradually from top to bottom. The front wall of the keyhole inclines along the thickness direction because energy decrease from top to bottom. Comparing the keyhole shape of different welding time, it is can be seen that the shapes are stable relatively, which demonstrated quasi-steady state has been achieved.

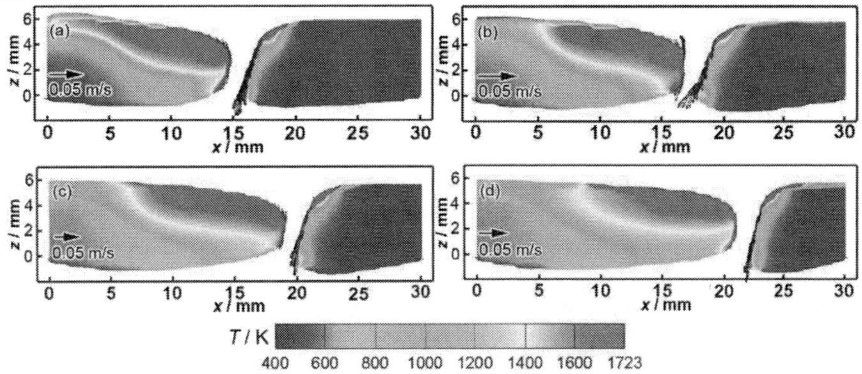

Fig.2 Temperature(T) and flow field distributions at longitudinal section of keyhole plasma arc molten pool for 3.0 s (a), 4.0 s (b), 5.0 s (c) and 6.0 s (d)

After reaching quasi-steady state, fluid flow of liquid metal in molten pool will reach a relative stable state. Liquid metal flow from front to rear and from top to bottom of molten pool. As is shown in Fig.2, liquid metal in the molten pool of front edge flow to the bottom of the workpiece while that on the rear edge flow to different direction. One part flow to back of the molten pool along the upward direction. And the other part flow to the bottom along downward direction. The maximum flow rate appears on the front boundary. In order to analyze the fluid flow inside the molten pool, the trace line that shown in Fig.3 is obtained. When the welding time is 3 s the keyhole is just established. So the flow field inside the molten pool is not stable completely, which performed as that a few counterclockwise vortex and clockwise vortex appear in the back of the molten pool at the same time. When the welding time reach 4 s counterclockwise vortex form stably. The reason why this phenomenon appear is that the liquid metal flow from high temperature zone to low temperature zone. Then the fluid contact to the solid metal and there will be a returning flow, which lead to the stable counterclockwise vortex.

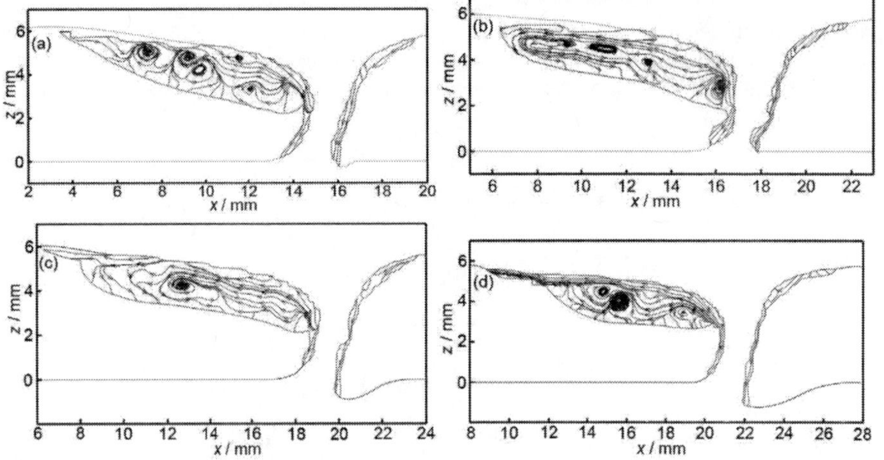

Fig.3 Flow trace variations in molten pool at welding time 3.0 s (a), 4.0 s (b), 5.0 s (c) and 6.0 s (d)

In the process of PAKW, the melting rate is different at variable thickness due to the energy decreasing of plasma arc along the depth direction of keyhole, which lead the keyhole incline. The offset between the center of the bottom keyhole and the center of heat source along the welding direction is defined "the bottom keyhole offset". The measure method of bottom keyhole offset is shown in Fig. 4 (a). The lines of a, b, c, d represent the four auxiliary dotted line respectively. Line a is the vertical tangent line of the rear wall of the keyhole and the tangency point is B while those of front wall are line c and point C. Line d, intersect at A point with the top surface of the workpiece, is the vertical line of welding direction through the center of heat source. Among them, the combination of every two lines represent the distance between the two lines. For instance, ac represents the distance between line a and line c. Line b made by ab=bc intersect at D point with the bottom surface of the workpiece. And D point is defined as the center of bottom keyhole. Therefore, bd, the distance between A and D along the welding direction, is the offset of bottom keyhole center to the center of the heat source.

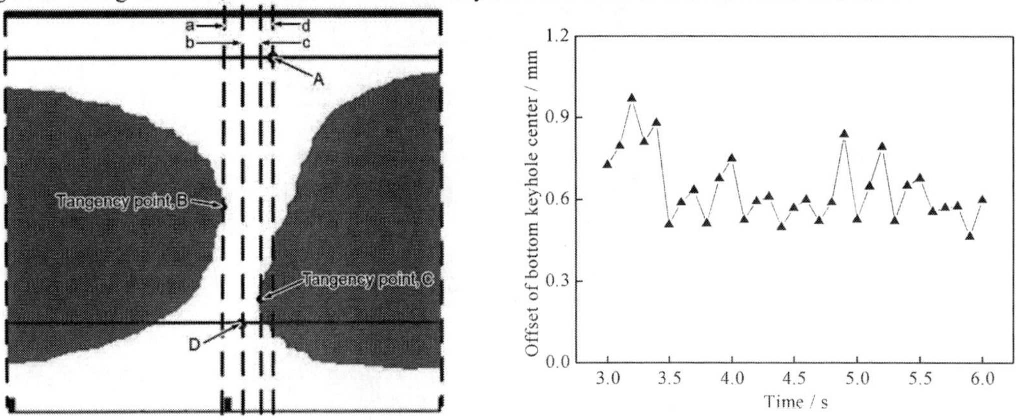

(a) The measure method of bottom keyhole offset *(b) Keyhole offset at different welding times*

Fig.4 the bottom keyhole offset

The offsets of keyhole at different welding time are measured with the above mentioned method. As is shown in Fig.4 (b), the offsets fluctuate at different welding time. The maximum 0.97 mm appears at the instable period of keyhole just establishing. At other time, the offsets fluctuate around 0.6 mm and there is a decreasing tendency overall as the time goes on, which demonstrate that the welding process reach stable period gradually.

In order to verify the accuracy of calculated model, the PAKW experiments were carried out with the same parameters of the calculated process. The image of the keyhole on the bottom side of the workpiece was obtained using the high speed camera. Fig. 5 shows the image and schematic boundary of keyhole. The keyhole is charactered by the length and width. The width of weld seam on bottom surface of workpiece is determined by the width of keyhole. If the width of the keyhole is too big, the defect of undercut on the bottom side will appear. On the contrary, if the width of keyhole is too small, the keyhole is easy to close, causing the bottom seam interruption. The measured and calculated width of keyhole are compared to verify the accuracy of the calculated model. As is shown in Fig. 6, the experimental and calculated results are in good agreement with each other, which verify the accuracy of the calculated model furthermore.

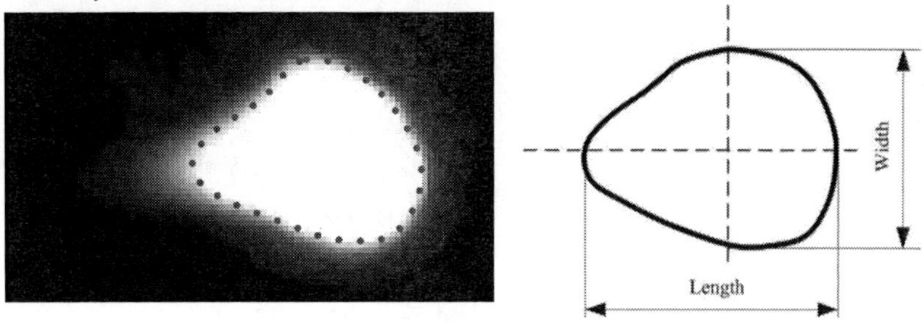

Fig.5 Schematic boundary of measured keyhole

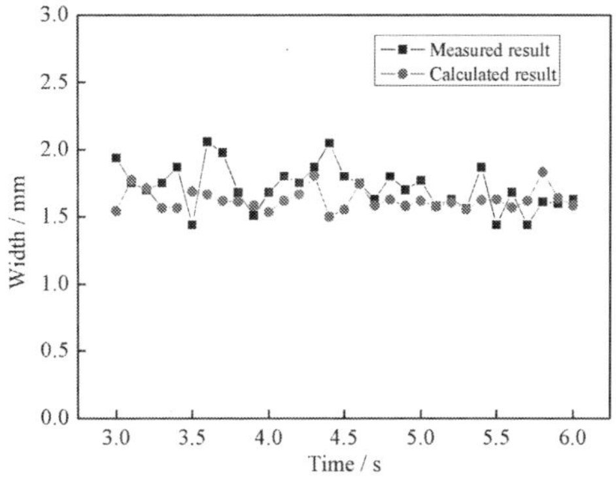

Fig.6 Comparison between measured and simulated keyhole widths

4. Conclusions

(1) A optimized mathematical model of PAKW was established and the welding process of quasi-steady state was achieved to simulate. The evolution of molten pool, keyhole and flow field are analyzed, which show welding process reach quasi-steady state after 3 s from ignition.

(2) The molten pool near the front wall of keyhole is thin. The average thickness is 0.6 mm. there is stable counterclockwise vortex in the back molten pool. And the range of keyhole offset is within 1 mm. All the above demonstrate the welding process is stable in quasi-steady state of PAKW.

(3) The calculated model is verified by PAKW experiments. The experimental and simulated size of the keyhole are in good agreement with each other.

Acknowledgments

Supported by National Natural Science Foundation of China (No. 51205176)

References

[1] Jenney C L, O'Brien A. Welding handbook vol2, 9th, Miami: American Welding Society, 2001: 303
[2] Wu C S. Thermal Press and Molten pool Behaviour. Beijing: China Machine Press, 2007: 8
[3] Wu C S, Gao J Q, Liu X F, Zhao Y H. Proc Inst Mech Eng, 2003; 217: 879
[4] Liu Z M. PhD Dissertation, Shandong University, Jinan, 2013
[5] Liu Z M, Wu C S, Gao J Q. Int J Therm Sci, 2013; 63: 38
[6] Wu C S, Liu Z M, Chen J. J Mech Eng, 2011; 47(6): 45
[7] Hu Q X, Wu C S, Zhang Y M. China Weld, 2007; 16(1): 6
[8] Pei L C. Master Thesis, Harbin Institute of Technology, 2006
[9] Dong H G, Gao H M, Wu L. Trans China Weld Inst, 2002; 23(4): 24
[10] Hsu Y F, Rubinsky B. Int J Heat Mass Transfer, 1988; 31(7): 1409
[11] Hu Q X. PhD Dissertation, Shandong University, Jinan, 2007
[12] Wang H G, Wu C S, Zhang M X. Trans China Weld Inst, 2005; 26(7): 49
[13] Wu C S, Wang H G, Zhang M X. Acta Metall Sin, 2006; 42: 311
[14] Wu C S, Hu Q X, Gao J Q. Comput Mater Sci, 2009; 46: 167
[15] Hu Q X, Wu C S, Zhang Y M. China Weld, 2007; 16(2): 55
[16] Huo Y S, Wu C S, Chen M A. Acta Metall Sin, 2011; 47: 706
[17] Huo Y S, Wu C S, China Weld, 2009; 18(2): 17
[18] Wu C S, Huo Y S. J Manuf Process, 2013; 15: 593
[19] Fan H G, Kovacevic R. J Phys, 1999; 32D: 2902
[20] Li T Q, Wu C S, Feng Y H, Zheng L C. Int J Heat Fluid Flow, 2012; 34: 117
[21] Wang X J, Wu C S, Chen M A. Acta Metall Sin, 2010; 46: 984
[22] Wu C S, Zhang T, Feng Y H. Int J Heat Fluid Flow, 2013; 40: 186
[23] Zhang T, Wu C S, Chen M A. Acta Metall Sin, 2012; 48: 1025
[24] Zhang T, Wu C S, Trans China Weld Inst, 2011; 32(7): 87
[25] Li Y, Feng Y H, Zhang X X, Wu C S. Acta Metall Sin, 2013; 49: 804
[26] Li T Q. PhD Dissertation, Shandong University, Jinan, 2014
[27] Jian X X, Wu C S. Int J Heat Mass Transfer, 2015; 84: 839
[28] Wang H X, Wei Y H, Yang C L. Comput Mater Sci, 2007; 38: 571
[29] Wang H X, Wei Y H, Yang C L. Comput Mater Sci, 2007; 40: 213
[30] Zhang T. PhD Dissertation, Shandong University, Jinan, 2011

Deformation analysis of friction welding of heat-resistant steels

Shigekazu Miyashita, Masashi Takahashi
Toshiba Corporation Energy Systems & Solutions Company, Yokohama, Kanagawa, Japan
shigekazu.miyashita@toshiba.co.jp, +81-(0)45-510-6682

Masaaki Kimura
University of Hyogo, Himeji, Hyogo, Japan,

Abstract
Friction welding is widely used in various industrial fields as a solid-state welding process. A consistent and stable method for predicting suitable process parameters by numerical simulation is required in order for friction welding processes to be applied in the assembly of new materials in industrial applications. The main purpose of the present study is to develop a simulation method for predicting deformation behavior at the weld interfaces and the shape of the weld flashes in joints. Coupled thermo-mechanical FEM analysis was performed with a 2D axial symmetry model. In the simulation, the states of each element on the weld interface were identified as "wear state" or "seizure state" according to the temperature and stress conditions at each analysis step. The friction heat values of the weld interface were determined by process parameters, such as friction speed and friction pressure, and coefficients of friction, which correspond to the states of the elements of the weld interface. The simulation was carried out for friction welding of heat-resistant steel pipe with a 13.5 mm outer diameter and an 8.0 mm inner diameter, made with a friction speed of 27.5 s^{-1} and a friction pressure of 30 MPa. The shapes of the weld flashes at each friction times in the present simulation were similar to the experimental results. The simulation method showed the possibility of predicting the deformation behavior for practical use.

Introduction
The friction welding process is one of the most effective and popular solid-state welding methods. The advantages of the friction welding process are its short process time, high quality of joints, good process robustness, and so on. This unique joining process has been successfully applied in various industrial fields. However, the complex process itself is not yet fully understood. Various studies have been conducted for gaining an adequate understanding of the joining phenomena involved in the friction welding process. Kimura et al. investigated the joining mechanism and the changes of the weld interface states during the friction welding process for similar joints in low carbon steel [1]. Their results revealed that the change of the weld interface into seizure state from wear state during the early stage of the friction process and 100% joint efficiency of steel joints (joints had the same tensile strength as that of the base metal) could be achieved using the friction process up to the initial peak torque without increasing forge pressure. Based on these experimental findings, a method for analyzing the friction torque and interfacial temperature during the early stage of the process by considering the changes of the weld interface state was also proposed [2]. Isshiki et al. reported a temperature estimation method on the basis of a simple model of friction heat generation [3-4].

Such a numerical analysis approach is a powerful tool for not only investigating the joining mechanism of friction welding but also optimizing the process parameters, compared with experimental approaches, which are seriously limited due to cost and time restrictions. In particular, prediction of the plastic deformation behavior of the weld interface is one of the most interesting subjects, together with the friction torque and interfacial temperature. Predicting the deformation behavior by a rapid and convenient simulation method would be quite beneficial for further development of the friction welding process in engineering applications. Hence, in this study, it is aimed to develop a simple and convenient simulation method for predicting the weld interface deformation. The following describes the simulation model and the analysis results, obtained using commercial FEM software. Experimental results are also presented. The industrial applicability of this simulation method is argued by comparing the simulation and experimental results.

Calculation Model
Heat generation model
The friction force in the general Coulomb friction phenomenon is believed to originate in microscopic adhesion at the interface. In friction welding, it was observed that the friction phenomenon changes from microscopic adhesion into macroscopic seizure as a result of plastic flow. Therefore, it is quite reasonable that the friction heat value changes according to the surface state, namely, the wear state or the seizure state.

In friction welding, friction heat is generated on the weld interface by conversion of mechanical energy. The friction heat per unit time, q_e , and which is generated on an element e of the micro radius area in Fig.1, is expressed by

$$q_e = F_e \cdot 2\pi N r_e \qquad (1)$$

where F_e is the friction force, N is the friction speed, and r_e is the radius of element e. In this case, it is assumed that the conversion efficiency to the frictional heat is 1. When the friction pressure is assumed to be uniform in the entire friction area, the surface pressure applied in each element is equal to the friction pressure P_f. From the definition of friction coefficient, the friction force of the wear state is obtained as given by equation (2) below. In case of the seizure state, the frictional force is the shear fracture force of the entire area of element e, and it is expressed by equation (3):

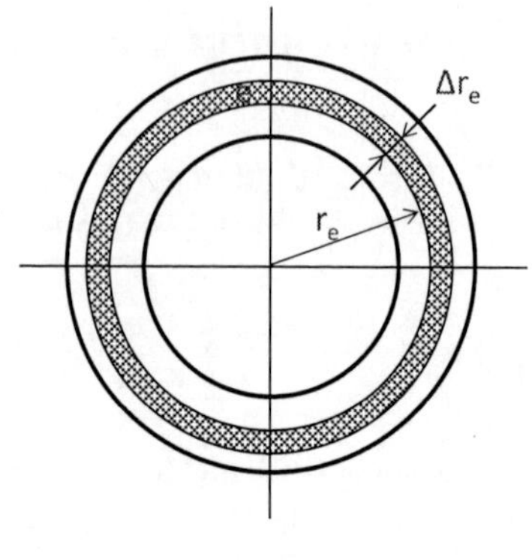

Fig. 1 Schematic model of weld interface.

$$F_e = \mu P_f A_e \qquad (2)$$

$$F_e = S_e A_e \qquad (3)$$

where μ is friction coefficient, A_e is the area of element e, and S_e is the shear strength of element e. The friction heat of the wear state can be calculated from μ and the process parameters from equations (1) and (2). Similarly, the friction heat of the seizure state can be obtained from S_e and the process parameters from equations (1) and (3). To discriminate the seizure phenomenon, seizure temperature was proposed by Kimura et al. [2]. The seizure temperature, which depends on the contact pressure and temperature at the weld interface, is precisely determined from experimental data. In this study, it is attempted to estimate the seizure behavior analytically from the view point of engineering convenience. Although the details of the seizure phenomenon in friction welding remain unclear, a study about the friction behavior of a cutting tool concluded that yielding of the material surface caused adhesion to a cutting tool [5]. Following the fundamental idea about seizure, yielding of the weld interface is used as the criteria for seizure in this study. Thus, the heat flux in a different radius area of the weld interface can be calculated from the process parameters, the temperature dependency of the material properties, and the stress condition of the weld interface.

Calculation procedure

In this study, MARC 2013.1 software was used in the coupled deformation and heat flow analysis during friction welding. The subject was similar combination of friction welded joints for seamless carbon steel pipes (JIS STPT410-S-C, STPT410) with an outer diameter of 13.5 mm and an inner diameter of 8.0 mm. The simulation was performed at a friction speed of 27.5 s^{-1} and a friction pressure 30 MPa. In this simulation, upsetting was not applied because the early stage of the process was mainly focused on. Fig. 2 shows a two-dimensional axisymmetric model used in this simulation. The model was 30 mm long and was created using 2552 quad elements. The friction pressure was applied to one end face in the longitudinal direction, and the above-mentioned friction heat was applied to the elements on the weld interface. If a stress condition of an element at a certain radius satisfied the von Mises yield criterion, the state of the element was assumed to change to the seizure state from the wear state. In accordance with adhesive friction theory, the temperature dependency of μ can be approximated by those of hardness and shear strength of base metal [6]. The temperature dependency of S_e was assumed to be equal to that of tensile strength. The detailed mechanical and thermal properties of the material used were those reported for low carbon steel [7-9]. In this simulation, no friction heat was generated on non-contact elements on the weld interface. Sticking to each weld interface was not considered. Even if the new surface became a contact state, the friction heat was only given to the weld interface defined initially. The suitability of the simulation was evaluated by comparing the results with those from experiments performed under the same conditions.

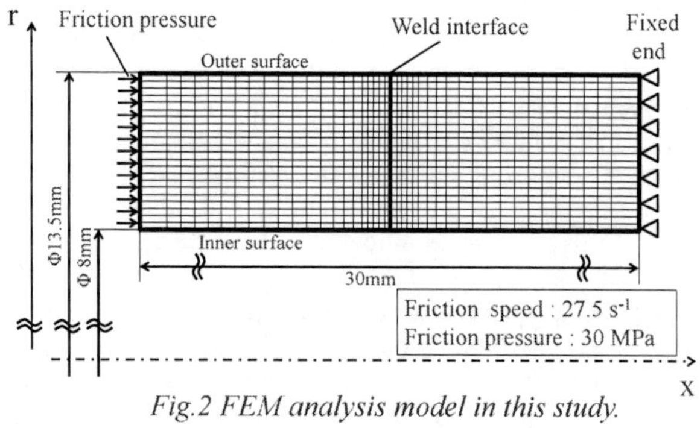

Fig.2 FEM analysis model in this study.

Results and Discussion

The accuracy of the calculated friction torque is an important factor for achieving adequate deformation simulation because the friction torque directly corresponds to the friction heat. Fig. 3 shows the calculated friction torque profile, which was obtained from the total friction heat on the entire weld interface, compared with the measured one. Both of the friction torques gradually rose to an initial peak torque with friction time. Although the calculated elapsed time for the initial peak was approximately 0.15 s shorter than the actual value, the initial peak torque value and its increasing rate were quite similar. Focusing on the early stage, within about 0.1 s, the measured torque started from zero and increased with the friction time, whereas the calculated value was about 10 Nm from the start. The experimental results implied that there was no actual contact area of the weld interface at the moment of contact. On the contrary, the contact immediately and homogeneously occurred on the entire weld interface in the simulation. Hence, it is possible that the difference in elapsed time for initial peak is caused by the difference of the initial contact state between the actual phenomenon and the simulation.

Fig.4 shows the calculation results of the weld interface temperatures at the inner surface, the middle of the wall thickness, and the outer surface. The maximum temperatures of each part were almost the same; however, the heating rates were obviously different. Although the relative speed of the inner surface position was the slowest, the heating rate was much higher than other parts.

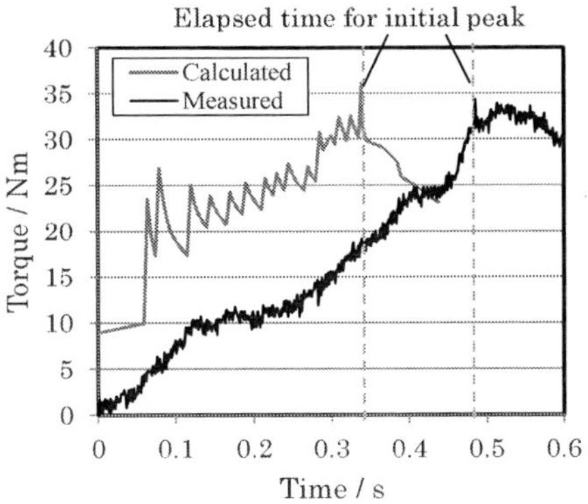

Fig. 3 Calculated friction torque profile compared with measured profile.

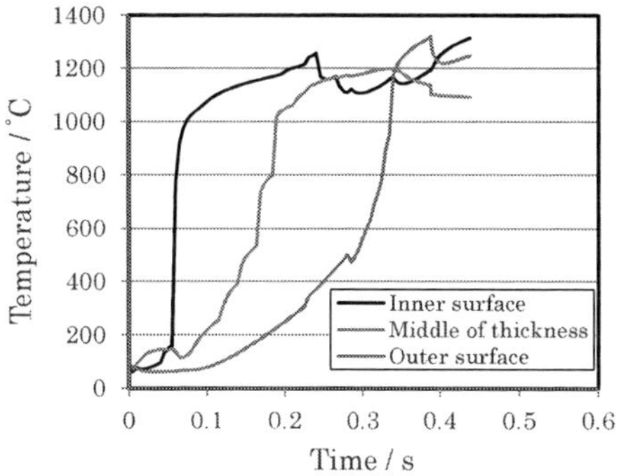

Fig. 4 Calculation results of the weld interface temperatures.

The difference in heating rates is explained by changing the contact conditions during the process. Fig. 5 shows the simulation results for deformation near the weld interface at a friction time of 0.05 s. The degree of deformation in the figure is magnified by 100 times. It can be seen that the outer surface part did not maintain the contact state by the radial thermal expansion of the local longitudinal area. In the simulation, contact of the weld interface gradually occurred from the inner side to the outer side. As a result, the inner surface part was heated up earlier than the outer part. However, there is no experimental evidence about such a partial separation, and further investigation will be required.

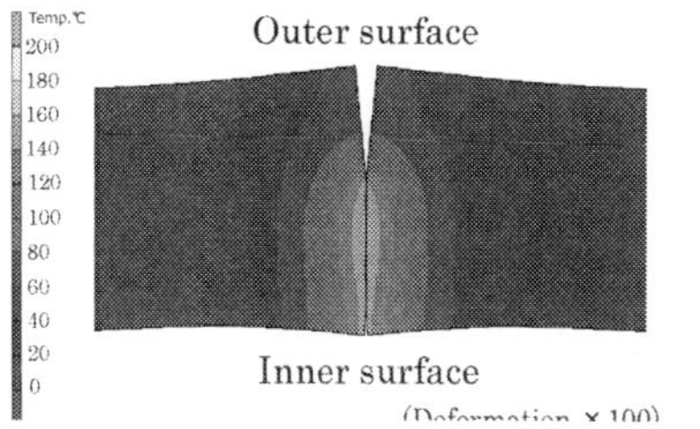

Fig. 5 Contact condition of the weld interface of 0.05 s. (Deformation ×100)

Figure 6 shows the calculated deformation and temperature field with various friction times. Before the initial peak torque (about 0.35 s), the temperature of the weld interface was not so high. Accordingly, the amount of flashes was little. At the moment of the initial peak torque, obvious flashes were generated on the inner surface. After the initial peak torque, remarkable flashes were generated on both the inner and outer surfaces. Experimental results for a carbon steel rod showed that the flashes began to occur just before the initial peak torque and rapidly increased immediately after the initial peak torque [1]. The simulation results showed the same tendency as these results.

Figure 7 shows a comparison of the cross-sectional appearances between the simulation and experiments. Because the elapsed times for the initial peak in both cases were slightly different, the friction times were normalized by the elapsed time for the initial peak. In both cases, the amount of flashes was small before the initial peak and remarkably increased after the initial peak. Although the sizes of the flashes on the simulation were slightly smaller, the shapes of the flashes were similar to the experimental results. The analysis cannot take into account sticking to each element on the weld interface and plastic flow induced by the rotation. These assumptions might be one of the causes for the slight difference in deformation behavior between the simulation and actual phenomenon. Although there is room for further consideration about the simulation model, this simulation method can predict the deformation behavior for practical use.

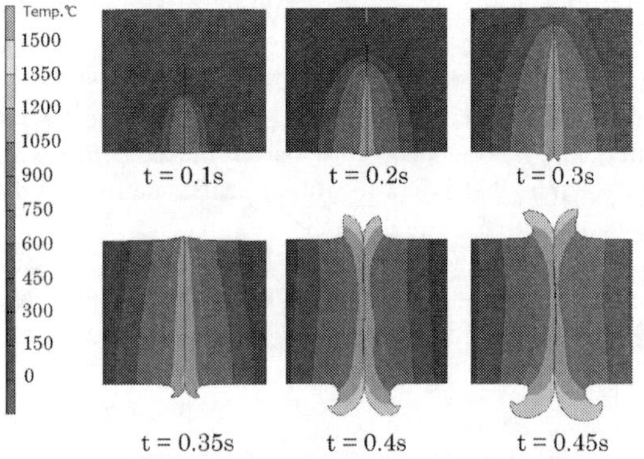

Fig. 6 Calculated deformation and temperature field with various friction times.

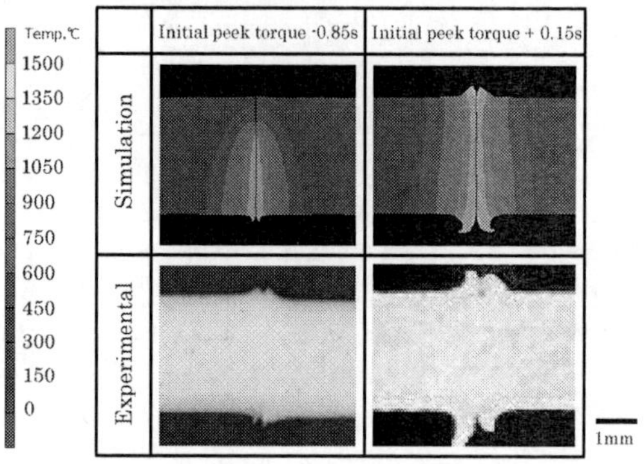

Fig. 7 Comparison of the cross-sectional appearances between the simulation and experiment.

Conclusions

In order to predict the deformation behavior of the weld interface during the friction welding process, we proposed a simple and convenient thermo-mechanical FEM simulation. The following results were obtained.

(1) The calculated initial peak torque value and increasing rate were similar to the measured results, although the calculated elapsed time for the initial peak torque was slightly shorter than the measured value.

(2) The simulation results showed that the heating rate of the inner surface part was higher than that of the outer part due to the non-contact state of the outer weld interface during the friction process.

(3) The calculated flash shapes before and after the initial peak were similar to the experimental results. The simulation method shows the possibility of predicting the deformation behavior for practical use.

Note; Product names mentioned herein may be trademarks of their respective companies.

References

[1] M.Kimura et al.: "Observation of Joining Phenomena in Friction Stage and Improving Friction welding Method," JSME International Journal Series A, Vol.46, No.3 (2004), pp.384-390

[2] M.Kimura et al.: "Analysis Method of Friction Torque and Weld Interface Temperature during Friction Process of Steel Friction Welding," Journal of Solid Mechanics and Materials Engineering, Vol.4, No.3 (2010), pp.401-413

[3] Y.Isshiki et al.: "Numerical Analysis of Temperature Distribution in Friction Welding of Carbon Steel," Journal of High Temperature Society of Japan, Vol.31, No.4 (2005), pp.225-231 (in Japanese)

[4] Y.Isshiki et al.: "Numerical Analysis of Temperature Distribution in Friction Welding of Two Similar Materials of S25C or SUS304," Journal of High Temperature Society of Japan, Vol.32, No.4 (2006), pp.236-243 (in Japanese)

[5] T.Hatta et al.: "Elucidation of Friction Behavior on Rake Surface of Cutting Tool using Newly-Formed Surface Friction Test, " DENKI-SEIKO (ELECTRIC FURNACE STEEL), Vol.82, No.2 (2011), pp.109-115 (in Japanese)

[6] S.Miyashita et al.: "Friction welding simulation of hollow tubular carbon steel," Proceedings of the 2014 National Symposium on Welding Mechanics and Design, Osaka, Japan, pp.369-372 (in Japanese)

[7] The Japan Society of Thermophysical Properties (ed.): Thermophysical Properties Handbook, Yokendo, Tokyo, (2008) p.210 (in Japanese)

[8] T.Tominaga: "High Temperature hardness test" Journal of the Japan Society for Precision Engineering, Vol. 55, No. 8 (1989), pp.1337-1341

[9] S.Kato et al.: "High-temperature strength tests of a material of liner plate," JAEA-Research, PNC-TN9450 97-011, (1996) pp.15,59-60

Challenges in development and importance of heat source models for simulation of fusion welding processes

Yadaiah Nirsanametla [1], Sohini Chowdhury [1], Swarup Bag [2], Md S Mujaheed Khan [1], Nabam Teyi [1], Manapuram Muralidhar [1]

[1] *Department of Mechanical Engineering, North Eastern Regional Institute of Science and Technology (NERIST), Nirjuli, Itanagar 791109, Arunachal Pradesh, India*

[2] *Department of Mechanical Engineering, Indian Institute of Technology Guwahati, Guwahati 781039, Assam, India*

Fusion welding process is complex in nature as it involves several interactive physical phenomena such as localized heating, melting, evaporation, solidification etc. Real-time measurement of the transient growth of temperature and material flow field during fusion welding is extremely difficult and there is uncertainty to measure the residual stress of weld joint accurately. Alternatively, computational model of increasing complexity based on scientific principle is an effective route to analyze the differential influence of process parameters. The conduction mode heat transfer analysis is often preferred due to reduced amount of computational time, ease of modeling and ease of coupling with stress analysis model. To approximate the influence of convective heat transport within the weld pool, the volumetric heat is often used in conduction heat transfer model. However, the major difficulty of conduction heat transfer analysis is the selection of suitable heat source model and a prior definition of model parameters which are limited by the definition of weld dimensions only from experimental measurement. The present work dealt with an examination of the different heat source models, namely, semi-ellipsoid, double ellipsoidal and egg configuration heat source model for simulation of linear fusion welding processes. This work also includes the challenges in development and importance of heat source models; and merits and demerits of different heat source models. It is observed that the egg configuration heat source model provides an asymmetric power density distribution with less number of model parameters and conserves the continuity in profile during linear welding simulation. Using different volumetric heat source models, a 3D finite element based transient heat transfer analysis have been performed to examine thermal behavior. The numerically estimated weld bead shapes and dimensions are compared with experimentally measured results. The results showed a fair agreement between the computed and experimental results.

Mechanistic study on hot cracking of 30%Cr-Ni based alloys

Toshio Yonezawa [1], Yuichi Miyahara [2], Mikko J Peltonen [3], Teemu Sarikka [3], Anssi Brederholm [3], Hannu Hanninenn [3]

[1] *Tohoku University, Japan*
[2] *Central Research Institute of Electric Power Industry, Japan*
[3] *Aalto University, Finland*

Hot cracking susceptibility of 18 laboratory heats of 30% Cr-Ni based alloys was studied by Longi-Varestraint test, On-heating and On-cooling hot ductility test, five passes four-layered beads build-up test, Gleeble test, etc. Hot cracking susceptibility based on the Longi-Varestraint and On-heating and On-cooling hot ductility test did not correlate with that of five passes four-layered beads build-up (FPFLB) test. Hot cracking susceptibility based on the Longi-Varestraint and On-heating and On-cooling hot ductility test seems to be appropriate for evaluation of solidification cracking and liquation cracking susceptibility, but not so suitable for evaluation of ductility dip cracking (DDC). The authors studied the thermal cycling test under tensile loading of the round-notched specimen, using Gleeble test machine, based on the recent hypothesis of the strain ratcheting mechanism for the DDC. The Gleeble test specimens were machined from the large two-pass weld beads. Using the Gleeble tests, the DDC was reproduced and seems to be most sensitive in the temperature range of 1000 to 1100oC for the 30% Cr-Ni based alloys. This tendency is different from the earlier observations and hypotheses of DDC. In the paper, the authors discuss the DDC mechanism and propose the new quantitative test technique for evaluating the DDC susceptibility of the 30% Cr-Ni based alloys.

Effect of Thermal Cycles on Hot Crack Susceptibility in Multi-pass Welds of Reduced Activation Ferritic/Martensitic Steel F82H

Hiroaki Mori [1], Takanori Hirose [2], Taichiro Kato [2], Hiroyasu Tanigawa [2]

[1] *Osaka University, Japan*
[2] *Japan Atomic Energy Agency, Japan*

The reduced activation ferritic/martensitic steel F82H has been developed as a structural material of the blanket module, which is the device set on the inner wall of fusion reactors to transmit kinetic energy of neutrons to thermal energy. In order to construct the modules, it is necessary to establish the welding methods and conditions for each joint shapes and thickness. In the case of welding for thick plates of the steel, multi-pass welding has to be adopted to the joints. Hot cracking in welds reheated following weld passes is one of concerned issues in multi-pass welding. Especially, hot crack susceptibilities in welds repeatedly reheated during multi-pass welding should be examined as an important evaluation matter as well as other weld defects. Therefore, the crack susceptibility in multi-pass welds of the steel, the longitudinal Varestraint test with multi-pass was conducted for weld metal of the steel. Ductility-dip cracks were observed in the first bead reheated by following weld passes in the Varestraint tests with double or triple weld beads. The crack susceptibility was raised with the number of re-heating. Such a tendency was remarkably recognized in the weld metal of the steel with higher Ta content. In weld metal heat-affected by following weld passes, the hardness in grains was seemed to be slightly hardened by increase in reheating times. From microstructure observation for reheated weld metal and X-ray diffraction analysis for extraction residue taken from the weld metal, a lot of microscopic precipitation were found in grains and these were identified as TaN. Based on these results, it is suggested that one of the causes to make hot crack susceptibility in the reheated weld metal increased is hardness in grains raised by the microscopic precipitation mainly consisting with TaN owing to reheating during multi-pass welding thermal cycles.

Effect of Welding Speed and Sulfur Segregation on Hot Cracking Susceptibility of Carbon Steels

Toru Amaya
Caterpillar Japan Ltd., Akashi, Japan
amaya_toru@cat.com

Toshio Yonezawa, Kazuhiro Ogawa
Tohoku University, Sendai, Japan
t-yonezawa@fri.niche.tohoku.ac.jp, kogawa@rift.mech.tohoku.ac.jp

Hannu Hänninen, Mikko Peltonen
Aalto University School of Engineering, Espoo, Finland
hannu.e.hanninen@aalto.fi, mikko.j.peltonen@aalto.fi

Abstract
It is well known that the hot cracking susceptibility of carbon steels increases with the increasing sulfur content of the steels. Recently, it was reported that the hot cracking of weld metal of carbon steels depends on the sulfur and/or phosphorus segregation along the dendrite boundaries during peritectic reaction due to the difference of the solubility of sulfur and/or phosphorus between the delta-ferrite and the austenite. Authors have applied the low carbon and low sulfur filler metal for carbon steel welding, without hot cracking. However, hot cracking was experienced when the welding conditions were changed. Authors have evaluated how the hot cracking susceptibility can be changed by the welding condition. In order to clarify the cause of hot cracking of low carbon and sulfur steels, the effect of the welding speed on the hot cracking susceptibility and the identification of sulfur and/or phosphorus segregation on the hot cracking were studied. The hot cracking susceptibility of carbon steels increases with the increasing welding speed. Carbon content of the weld metal increases with the increasing welding speed. Sulfur content of the weld metal does not change with the varying welding speed. In addition, the sulfur segregates along the dendrite boundaries on the hot crack. It is concluded that the increased carbon content of the weld metal with the increasing welding speed results in an increasing amount of austenite phase in the solidified weld metal and sulfur segregates along the dendrite boundaries during peritectic reaction.

Introduction
Susceptibility of hot cracking of carbon steels increases with the increasing sulfur content of the steels [1],[2]. However, hot cracking has been detected in a welded structure made of carbon steels even though sulfur content is low. Solute sulfur content in delta-ferrite is larger than that in austenite [2]. Therefore, sulfur segregates along the dendrite boundaries when the amount of the delta-ferrite transforming into the austenite increases based on the carbon content of carbon steels in the peritectic reaction (Delta-ferrite+Liquid → Austenite). As a result, hot cracking susceptibility increases due to sulfur segregation along the dendrite boundaries even though sulfur content of carbon steels is low [3]. In addition, it is generally agreed that the hot cracking susceptibility of carbon steels depends on the welding conditions. However, only few reports are available on the effect of welding conditions on the hot cracking susceptibility of the carbon steels [4] and the identification of sulfur segregation even though sulfur content in the weld metal is low. Therefore, in order to identify the cause of hot cracking of low carbon and low sulfur steels, the effect of the welding conditions on the hot cracking susceptibility and the identification of sulfur and/or phosphorus segregation on the fracture surface of a hot crack using Auger Electron Spectroscopy were studied.

Experimental Procedures
Sample, fixture design and welding conditions
The hot cracking susceptibility was evaluated with two kinds of materials, a low carbon steel and a medium carbon steel. Conventional filler metal was used for the carbon steel. The chemical compositions of the steel plates and filler metal are shown in Table 1. Low carbon and medium carbon steels were cut into samples with dimensions of 300 mm length x 50 mm width each. Thicknesses of three test plates were, 20 mm, 40 mm and 80 mm. The bevel with 6 mm width x 12 mm depth for the low carbon steel and 3 mm width x 12 mm depth for the medium carbon steel were milled on one side of the sample. The other three sides of the samples were processed to 10 mm thickness and 15 mm width as shown in Fig. 1. The samples were combined butt joint welds between low carbon and medium carbon steels by low carbon filler metal. The samples were restrained between the bed plate and restraining plate in the fixture. All bolts were fixed to ensure consistent restraint conditions. The weld beads were applied in 2 passes on the groove of butt joint to study susceptibility for liquation cracking of 1st and

Table 1: Chemical composition of the low carbon steel, the medium carbon steel and the filler metal.

	Chemical composition (wt.-%)					
	C	Si	Mn	P	S	Ti +Zr
Low carbon steel	0.12	0.14	1.20	0.019	0.002	---
Medium carbon steel	0.34	0.24	0.82	0.013	0.003	---
Filler metal	0.06	0.61	1.16	0.012	0.006	0.10

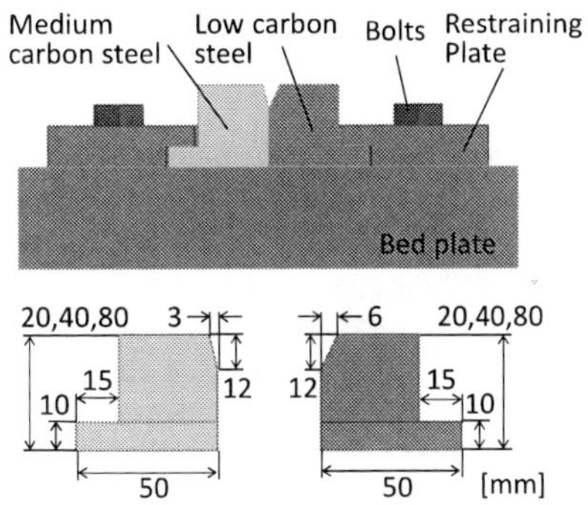

Fig. 1: Samples and fixture design.

Table 2: Welding Conditions.

	Welding conditions of 1st and 2nd pass		
	Current (A)	Voltage (V)	Welding Speed (cm/min)
Case 1	300	28.3	20
Case 2			27
Case 3			34
Case 4			49
Case 5			64
Case 6			80

2nd passes. Welding bead location was 1 mm to the medium carbon steel from the center of 1st pass and 5 mm to the medium carbon steel from the center of 2nd pass. Welded bead length was 50 mm. Six welding speeds were used to vary the heat input as shown Table 2. Diameter 1.4 mm filler metal wire was used. The shielding gas was Ar-20%CO_2 using a MAG semi-automatic welding machine.

Chemical composition and hot cracking observation in the weld metal

The welded sample was cut at the center of welded path length in two pieces. Quantitative analysis of chemical composition of weld metals was conducted using Glow Discharge Optical Emission Spectrometry for the first sample. The analyzed location was center of the weld metal. Hot cracking was observed on the section using optical microscopy of the second sample.

Identification of elemental distribution on the fracture surface of hot crack

The distribution of elements on the hot crack fracture surface was analyzed using Auger Electron Spectroscopy (AES). The samples were prepared to include hot cracks in the analyzing location and samples were fractured using Slow Strain Rate Test at 0.001 mm/min in the ultrahigh vacuum. The intrinsic fracture surface of a hot crack was studied by the elemental mapping to identify the distribution of impurity elements.

Results and Discussion

Carbon and sulfur contents in the weld metal depending on the welding speed

Carbon and sulfur contents in the weld metal were analyzed to determine the influence of the welding speed. Fig. 2 and Fig. 3 show carbon and sulfur contents in the weld metal of 1st and 2nd passes vs. the welding speed,

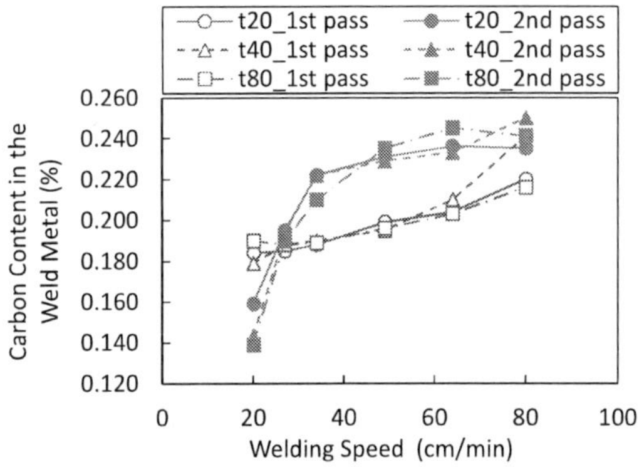

Fig. 2: Carbon content in the weld metal vs. welding speed.

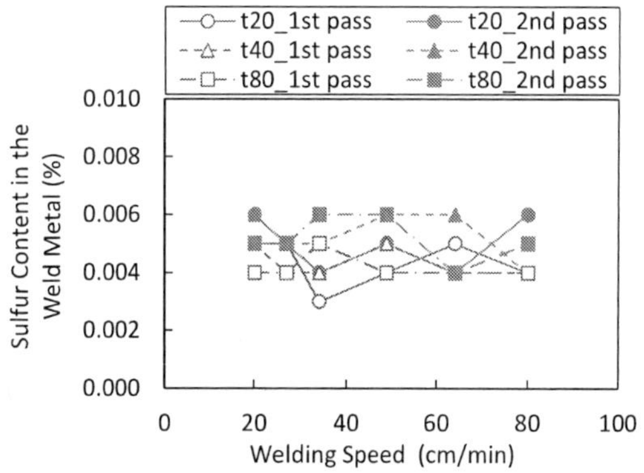

Fig. 3: Sulfur content in the weld metal vs. welding speed.

respectively. The carbon content in the weld metal of 1st pass increases gradually from 0.18 to 0.22 wt.-% with the increasing welding speed. However, the sulfur content in the weld metal did not differ by the welding speed. It is estimated that the low carbon content of the weld metal results from the higher filler metal ratio in the weld pool due to the low welding speed. On the other hand, small variation of sulfur content in the weld metal due to the welding speed is due to small difference between the samples and the filler metal. The carbon content in the weld metal of 2nd pass increases gradually from 0.14 to 0.24 wt.-% with the increasing welding speed, as in the 1st pass. Carbon content increases significantly depending on the size of weld pool of 1st pass with welding speed from 20 cm/min to 34 cm/min. Sulfur content in the weld metal did not differ by the varying welding speed, as in the 1st pass.

Hot cracking susceptibility and carbon content in the weld metal

Fig. 4 shows the liquation cracking in the 1st and 2nd passes of the Case 4 with 80 mm plate thickness. The liquation crack, which is surrounded with a red rectangle, was observed in the 2nd pass. The total crack length of liquation cracks was determined (in this case, all hot cracking was considered as liquation cracking, as described above) to compare the correlation between the hot cracking susceptibility and carbon content in the weld metal. As a result, the total liquation crack length in the 1st and 2nd passes depends on the carbon content of the weld metal as shown in Fig. 5. In addition, it can be seen that the total liquation crack length in the 1st and 2nd passes increases above approximately 0.19 wt.-%C. Fig. 6 shows the total liquation crack length in the heat affected zone (HAZ) of 1st pass after welding 2nd pass vs. the carbon content of the weld metal. Maximum total liquation crack length in the HAZ was approximately 0.3 mm for 0.18 to 0.19 wt.-%C in the weld metal. Total liquation crack length in the HAZ is significantly smaller than that in 1st and 2nd passes. Liquation cracking in the HAZ was observed when liquation cracking in the 1st pass was not present. The carbon content in the weld metal that showed liquation cracking in the HAZ is lower than 0.19 wt.-%C. Therefore, it is considered that liquation cracking was not present in the 1st pass. However, liquation cracking in the HAZ is possible due to the segregation of sulfur based on the welding heat input by the 2nd pass.

The tendency to a smaller total liquation crack length with lower carbon content is considered with Fe-C phase diagram [5]. The liquation structure is divided in two regions of the diagram:

(1) Region I (C:0.10~0.18 wt.-%): Delta-ferrite + Liquid → Delta-ferrite + Austenite.

(2) Region II (C:0.18~0.51 wt.-%): Delta-ferrite + Liquid → Austenite + Liquid.

The delta-ferrite exists in the liquation structure after the peritectic reaction in region I. The delta-ferrite transforms into the austenite in region II. The austenite content gradually increases from 0.10 wt.-%C. Solubility of sulfur in delta-ferrite is larger than that in austenite [2]. Sulfur segregates on the dendrite boundaries with increasing amount of austenite phase due to increasing carbon content. The increasing hot cracking susceptibility results from sulfur segregation due to the increasing amount of austenite phase in the solidified weld metal over 0.18 wt.-%C. In contrast, the hot cracking susceptibility decreases below 0.18 wt.-%C because of the presence of delta-ferrite.

It is considered that the increased carbon content of the weld metal with the increasing welding speed results in an increasing amount of austenite phase in the solidified weld metal and sulfur segregates along the dendrite boundaries during peritectic reaction.

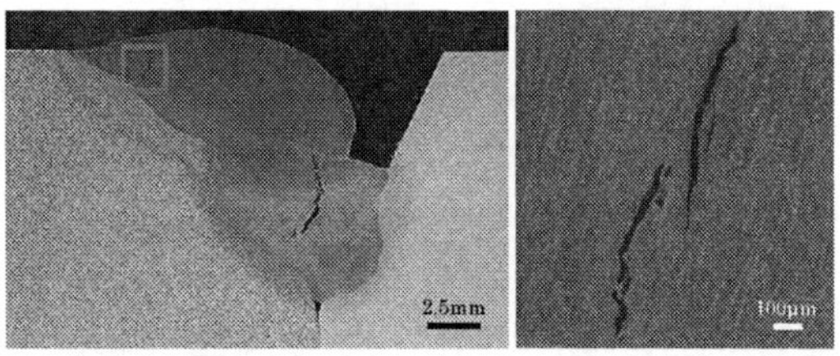

Fig. 4: Liquation cracking in the 1st and 2nd passes.

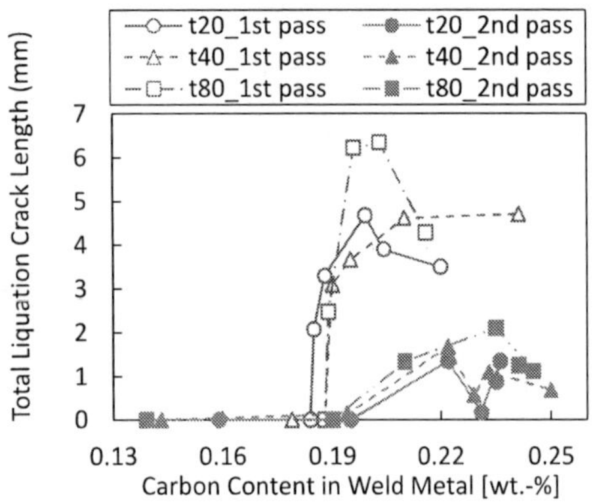

Fig. 5: Total liquation crack length vs. carbon content in the 1st and 2nd passes.

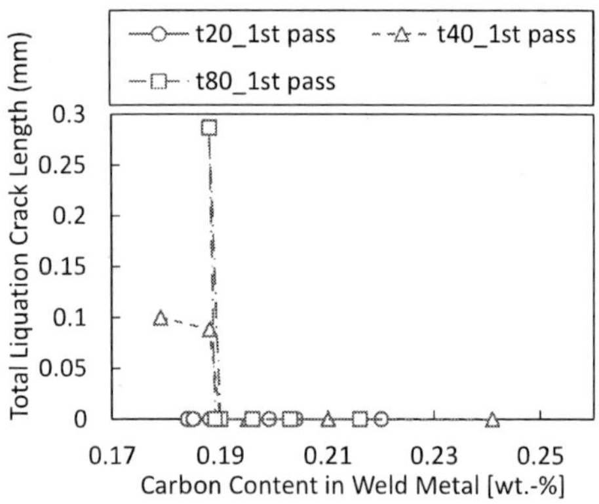

Fig. 6: Total liquation crack length vs. carbon content in the heat affected zone.

Distribution of elements in hot cracking

In the previous section, it was considered that sulfur segregates along the dendrite boundaries during peritectic reaction. To evaluate this further, the AES elemental mapping was conducted to identify the distribution of elements on the fracture surface of a hot crack. Fig. 7 shows the fracture surface of Slow Strain Rate Test and AES elemental mapping results. The liquation cracks were observed in two regions on the fracture surface. One liquation crack, which is surrounded with a red rectangle, has an intrinsic fracture surface without oxide layer. On the other hand, the liquation crack in the opposite side was exposed to atmospheric air and an oxide layer was observed. Therefore, the dendrite boundary of the liquation crack with intrinsic fracture surface was analyzed by the elemental mapping.

Iron was distributed in most of the dendrite boundary. The dendrite boundary, which does not show iron contained manganese and sulfur as MnS and sulfur segregation. In addition, manganese was present only as MnS phase inclusions. In contrast, sulfur was distributed on most of the dendrite boundary as FeS phase. An area of ductile fracture, which is surrounded with a blue rectangle, was observed. Iron was present at the ductile fracture area and manganese and sulfur were not present there. Phosphorous was not observed on the dendrite boundaries. FeS, which is low melting compound, has formed on the dendrite boundaries of the liquation cracks during peritectic reaction even though sulfur content of the weld metal is low. Therefore, sulfur segregation is the main cause of hot cracking.

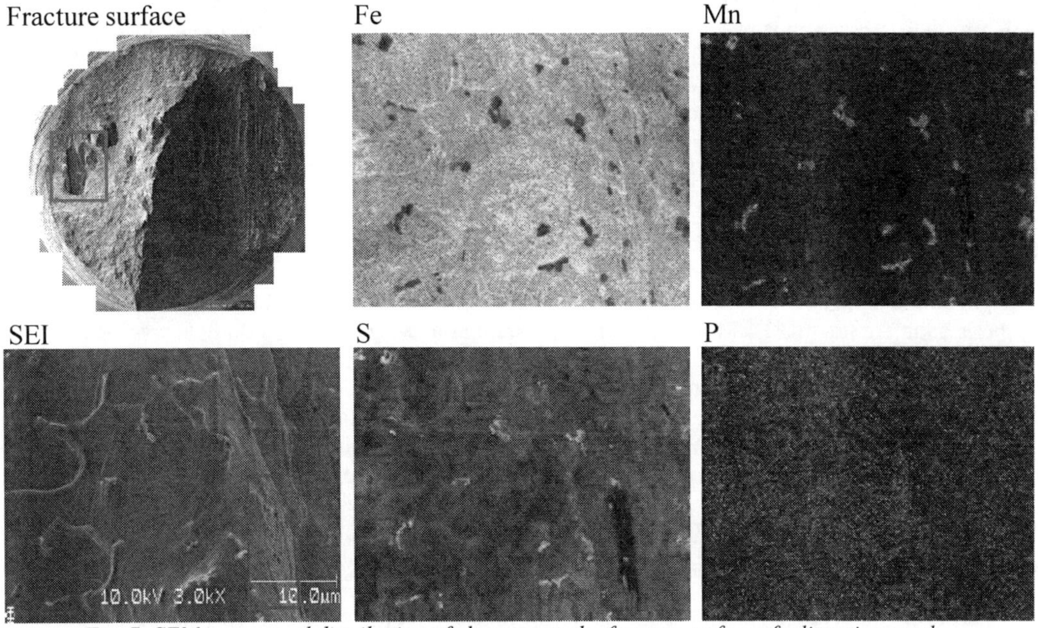

Fig. 7: SEM images and distribution of elements on the fracture surface of a liquation crack.

Conclusions

Carbon content of the weld metal increases with the increasing welding speed. Amount of filler metal in the weld pool decreases with the increasing welding speed. It is concluded that the increased carbon content of the weld metal with the increasing welding speed results in an increasing amount of austenite phase in the solidified weld metal and sulfur segregation along the dendrite boundaries during peritectic reaction increases the hot cracking susceptibility of carbon steel.

References

[1] T. Senda, F. Matsuda, H. Nakagawa, S. Nishino, "Sulphide and Its Effect on Solidification Cracking in Weld Metal of Steel (Report 5)", J. Jpn. Weld. Soc., vol. 43, pp. 57-65, 1974

[2] I. Masumoto, H. Ozaki, "Effect of Carbon Content on the Hot Cracking in Solidified Steel", J. Jpn. Weld. Soc., vol. 42, pp. 674-684, 1973

[3] K. Tamaki, H. Kawakami, J. Suzuki, "Effect of Carbon Content and Peritectic Reaction on the Hot Cracking of Weld Metal of High Carbon Steel", Q. J. Jpn. Weld. Soc., vol. 20, pp. 266-275, 2002

[4] S. Ohshita, N. Yurioka, N. Mori, T. Kimura, "Prevention of Solidification Cracking in Very Low Carbon Steel Welds", Welding Journal, vol. 62, pp. 129s-136s, 1983

[5] O.A. Bannykh, K. Enan, S. Nagasaki, "Iron Alloy Phase Diagram Book", Agne Gijutsu Center, pp. 12-17, 2001

GMA Welding of Al-Mg-Si alloys using novel similar filler wires fostering grain refinement in the weld metal

A. Pittner, Th. Michael, R. Haecker, M. Rethmeier
Federal Institute for Materials Research and Testing, Berlin, Germany
andreas.pittner@bam.de, +493081043696
R. Lahnsteiner
MIG Weld International Inc., Landau, Germany
robert.lahnsteiner@migweld.de, +499951601230

Abstract

The application of welded aluminum constructions has increased over the last decades due to the high demand for lightweight design. In this context aluminum alloys have a significant importance especially the heat treatable Al-Mg-Si alloys. With regards to welding the heat input causes the dissolving of the precipitates yielding a degradation of the strength properties. Furthermore, there is the risk of solidification cracking during welding for Al-Mg-Si alloys, especially AW 6082-T6. As a consequence Al-Mg-Si alloys are welded using a dissimilar filler material based on 5xxxx aluminum alloys that yields a reduction of resulting strength properties in comparison to the base material. Therefore, the paper presents the application of a novel massive filler wire for Al-Mg-Si alloys. The wire consists of AW 6082 with additionally added AlTi5B1 master alloy fostering the grain refinement of the weld metal. Exemplarily for the base material AW 6082-T6 GMA welding experiments are performed using the novel filler wire 6082TiB. It could be shown that solidification cracks could totally be avoided. The tensile strength of the welded joints are evaluated and compared with weldments using a conventional 5183 filler wire. Artificial aging is performed after welding in order to improve the strength properties by solution hardening showing the potential of these alloys.

Introduction

Lightweight design plays an important role in order to reduce the raw material consumption. Concerning this issue aluminum is an interesting material because of its high strength-to-weight ratio [1]. In industry fusion welding is still one of the most relevant joining methods for the fabrication of complex aluminum components. However, the integrity of the welding joint is governed by the heat effects of the welding process. With respect to the alloy system Al-Mg-Si no similar filler wires are available mainly caused by the high risk for solidification cracking [2]. Therefore, gas metal arc (GMA) welding of Al-Mg-Si alloys is done using dissimilar fillers wires as 4043 (AlSi5) or 5xxx (AlMg). As a consequence, the weld metal often does not achieve the strength properties of the base material. The joint efficiency (tensile strength joint / tensile strength base material) obtained by GMA welding of Al-Mg-Si alloys using 4043 filler wire may decrease down to 50 % [3]. Furthermore, the precipitation hardened Al-Mg-Si alloys are very heat sensitive, especially concerning a degradation of the strength properties in the heat affected zone (HAZ). Fine dispersed Mg_2Si-precipitates (β''-phase) obtained by a proper heat treatment are the prerequisite for the high yield and tensile strength values of Al-Mg-Si alloys [4]. The degradation is caused by dissolution of these Mg_2Si-precipitates as an effect of the heat input of the welding process. In addition, the precipitates may grow or transform from coherent precipitates to incoherent forms that is referred to as overaging. The latter effect occurs in regions of lower peak temperatures having a higher distance to the fusion line than the dissolution area [5]. Besides reducing the heat input per unit length of the welding process itself [6] post weld heat treatment (PWHT) is generally required to restore the material properties of the heat affected zone. If PWHT is neglected, the resulting joint efficiency is poor (<50%) for the as welded condition as shown in [7] for 6082-T6 base material welded with 4043 filler material. The necessary heat treatment for restoring the strength properties comprises solution treatment, quenching and artificial aging fostering the precipitation. The solution treatment is done at temperature ranges between 520 °C and 540 °C [5] and a holding time between 0,5-2h. After quenching in water, artificial aging is performed within a temperature range of 160°C – 175°C [8, 9]. Scott [10] analyzed the resulting strength and toughness properties during GMA welding of 6082 and 5083 base material applying 4043 and 5556 filler. In the as welded condition the 5083-5556 joint achieves higher tensile strength values (about 300 MPa) in comparison to the 6082-4043 (220 MPa) or 6082-5556 (233 MPa) joint. If the corresponding 6082-4043/5556 joints were solution treated at 530 °C for 0,5 h, then quenched in water and artificial aged at 175°C for 8 h, the tensile strength values could be increased remarkably. In case of 6082 base material welded with 5556 the tensile strength could be increased up to 285 MPa with a failure position in the weld metal. The highest increase could be achieved for the joint 6082-4043 yielding a tensile strength of 295 MPa which is in the range of the 6082-T6 base material. However, the toughness properties of the weld metal are reduced in case of

4043 filler wire, if full heat treatment is applied. The effect that the weld metal obtained by fusion welding of Al-Mg-Si alloys with 4043 (AlSi5) filler is sensitive to heat treatment was also observed by [11]. Due to the dilution of base material with filler wire that can reach values up to 50 % in case of V-groove GMA welds, the resulting weld metal contains manganese enabling the precipitation of Mg_2Si-particles. In [11] it was also shown, that post weld heat treatment significantly improves the mechanical properties of the joint. Furthermore, a fine grained weld metal additionally yields to better mechanical performance. Consequently, this paper presents the application of novel massive filler wires for welding Al-Mg-Si-alloys. The filler wires are based on the 6082 alloy with additional AlTi5B1 fostering the grain refinement in the weld metal which is necessary to reduce the susceptibility for solidification cracking [12, 13]. In previous work the authors [14-17] realized grain refinement in the weld metal by inoculation during GTA remelting of AlTi5B1-alloyed inserts for 6082 base material. The results obtained are now transferred to GMA welding of 6082 base material with the similar filler wire 6082TiB.

Methodology

GMA welding experiments have been conducted using the base material 6082-T6 and two filler wires. Besides the commercially available filler wire S Al 5183 after ISO 18273 the newly developed filler wire 6082TiB was used for the investigations. The chemical composition of the materials analyzed is given in Tab. 1. Concerning the in-house measured element concentrations the base material AW 6082-T6 agrees well with the standard DIN EN 573-3. The same is true for the filler wire S Al 5183 that is used for welding both the 5083 and 6082 aluminum alloy. With regards to the mechanical properties the Al-Mg-Si alloy is in the condition T6 (solution treated and artificially aged) after the standard DIN EN 485-2 and DIN EN 515. Additionally, the new filler wire based on the alloy 6082 is listed which is named 6082-TiB. The prefix S Al is avoided since the wire is not included in the corresponding standard yet. It is worth noticing that the chemical element composition agrees with the alloy AW 6082 except the content of boron and titanium. The titanium and boron content of the wires is realized by adding the master alloy AlTi5B1 into the melt. The titanium and boron is now available as intermetallic phases like Al_3Ti as well as TiB_2 that forms particles of around 1 micron in size. Especially the TiB_2 particles are very active heterogeneous solidification nuclei of aluminum grains since they do not dissolve in the melt. The authors also observed that TiB_2-particles are covered by a thin layer of Al_3Ti [14]. Finally, during fabrication of the wire the manganese content was slightly increased compared to the base material 6082 to account for element losses caused by the welding process.

Tab. 1: Chemical composition of base material and filler wires (in-house spectroscopic measurements)

	Si	Fe	Cu	Mn	Mg	Cr	Zn	Ti	B	Al
					Base material					
AW 6082-T6	0,9	0,48	0,07	0,45	0,7	0,02	0,07	0,03	-	balanced
					Filler wire					
S Al 5183	0,04	0,11	0,005	0.86	5,12	0,11	0,005	0,15	-	balanced
Filler 6082-TiB	1,1	0,35	0,003	0,65	**1,4**	0,1	0,007	**0,3**	**0,062**	balanced

The aluminum base material plates of thickness 5 mm had a V-groove joint preparation with 50° opening angle. A pulsed GMA welding process with pure argon as shielding gas was used to perform the welding experiments. The corresponding welding wires had a diameter of 1.6 mm.

For the welding of 6082 base material plates with the commercial filler wire S Al 5183 the heat input per unit length was 0.37 kJ/mm. The wire feed rate was set to 4.7 m/min considering a troch travelling speed of 0.45 m/min. The welding experiments for the S Al 5183 wires were done based on the Mg-characteristic of the welding power source.

However, for the filler wire 6082-TiB a new characteristic of the welding power source had to be developed based on a Mg- and Si-characteristic. As a result the heat input per unit length was 0.475 kJ/mm for a wire feed rate of 4.2 m/min. The welding speed was set accordingly to 0.45 m/min.

Post weld heat treatment was performed in order to restore the mechanical properties of the base material and for the weld metal in case of 6082-TiB filler material. After in-house experiments for the base material 6082 it was found that solution treatment (ST) at 540°C for 2 hours, subsequent quenching in water and artificial aging (AA) at 165°C for 14 hours yield to mechanical strength properties that correspond to the condition 6082-T6 after the standard DIN EN 485. The same heat treatment conditions were applied for the welding joints. For the material combination 6082-6082TiB the influence of a full heat treatment (ST+AA) as well as solely artificial aging (AA) on the tensile strength was analyzed. For the weld joints 6082-5183 only artificial aging was considered and compared with the as welded condition. The transverse tensile testing specimens had a thickness that corresponds to the welded plates. The specimens were fabricated according to the standard DIN EN ISO 4136 with removed weld reinforcements. The tensile tests were conducted with a traverse speed of 0.02 mm/s.

Results and Discussion

The metallurgical effect of the new filler wire 6082TiB on the grain size of the weld metal is shown in Fig. 1. As it can be seen a significant grain refinement is obtained, if a TiB alloyed filler wire is used instead of a dissimilar AlMg-wire. An important result is that the similar filler 6082TiB can be welded without the risk for solidification cracking. A slightly higher likelihood for pore formation in the weld metal is observed for the 6082TiB wire in comparison to the S Al 5183 filler wire. However, the pore sizes are within a tolerable range in accordance to the quality level B of EN ISO 10042.

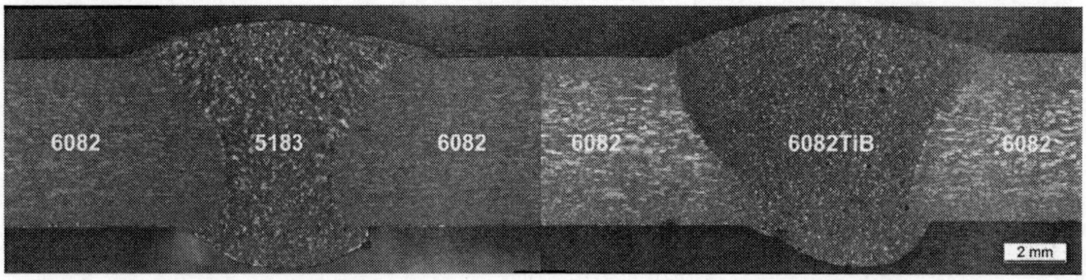

Fig. 1: Grain size distribution- left: weld joint 6082-5183; right: weld joint 6082-6082TiB

The effect of post-weld heat treatment on the mechanical properties is illustrated in Fig. 2. As seen in the left figure the heat treatment has a significant influence on the hardness evolution for the 6082-6082TiB weld. For the as welded condition a loss of hardness in the weld metal and heat affected zone (HAZ) occurs. If artificial aging is applied the hardness values can be increased in both the weld metal and HAZ. However, the hardness values for the HAZ as well as the weld metal can be improved up to the range of the base material, if full heat treatment is applied (ST+AA). Consequently, the weld metal hardness is strongly affected by solution treatment which means that the heat input of the welding process does not cause a solution treatment of the weld metal itself. This is in agreement to what is observed by Scott [10] using 5556 filler wire for welding 6082. In his analyses, a full heat treatment of the weld joint also caused the highest tensile strength values in a range of 285-300 MPa.

In Fig. 2 right the tensile strength of different welding experiments is shown. In this context, the combination 6082-5183 corresponds to the state of the art welding technology and achieves a tensile strength value of 248 MPa for the as welded condition. Interestingly, artificial aging also improves the strength value of this material combination up to 284 MPa which matches the values of [10]. Due to dilution of filler wire with base metal enough manganese and silicon particles are available for the intermetallic precipitates Mg_2Si. This effect was also observed by Ambriz [11] for the filler AlSi5 used for welding of 6082. The resulting weld metal strength was increased by heat treatment but the toughness properties decreased significantly. It is important to note that the most interesting results occur if the 6082 base material is welded with the new 6082TiB wire and heat treated subsequently. If only artificial aging is performed the maximum strength is in the range of the 5183 filler wire having a value of 271 MPa. Nevertheless, the strength can be increased remarkably up to a joint efficiency of 95 %, if full heat treatment is performed resulting in a tensile strength value of 328 MPa.

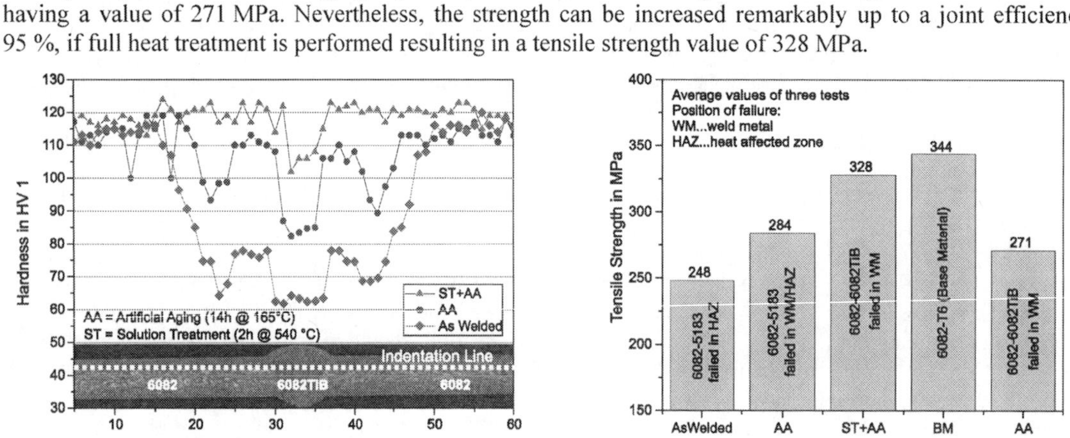

Fig. 2: Effect of heat treatment on mechanical properties – left: Hardness values for 6082-6082TiB welds; right: Tensile strength in dependence on heat treatment conditions and material combinations

Conclusion

A new massive filler wire containing titanium and boron was developed and applied for GMA welding of the Al-Mg-Si alloy 6082. In comparison to the filler wire S Al 5183 generally used for these alloys the TiB-alloyed filler material fosters the grain refinement in the weld metal. Hence, sound welds were realized since the susceptibility for solidification cracking has reduced significantly. Moreover, post weld heat treatment enabled to improve the tensile strength of the weld up to 328 MPa that corresponds to a joint efficiency of 95 %.

Acknowledgments

The research work was funded in the framework of the Central Innovation Program for small and medium enterprises on behalf of the German Federal Ministry of Economic Affairs and Energy (BMWi) under the grant number KF 2201061SU3.

References

[1] W. S. Miller, L. Zhuang, J. Bottema, A. J. Wittebrood, P. De Smet, A. Haszler, A. Vieregge, "Recent development in aluminium alloys for the automotive industry", Materials Science and Engineering A, Vol. 280, p. 37-49, 2000

[2] C. Huang, S. Kou, "Liquation Cracking in Full-Penetration Al-Mg-Si Welds", Welding Research Supplement, Vol. 83, p. 111-s-122-s, 2004

[3] A. K. Lakshminarayanan, V. Balasubramanian, K. Elangovan, "Effect of welding processes on tensile properties of AA6061 aluminium alloy joints", The International Journal of Advanced Manufacturing Technology, Vol. 40, p. 286-296, 2007

[4] C. D. Marioara, S. J. Andersen, J. Jansen, H. W. Zandbergen, "Atomic Model for GP-Zones in a 6082 Al-Mg-Si System", Acta Materialia, Vol. 49, p. 321-328, 2001

[5] S. Kou, "Welding Metallurgy", 2nd Edition, New Jersey, John Wiley & Sons Inc., 2002.

[6] R. Ahmad, M. A. Bakar, "Effect of a post-weld heat treatment on the mechanical and microstructure properties of AA6061 joints welded by the gas metal arc welding cold metal transfer method", Materials & Design, Vol. 32, p. 5120-5126, 2011

[7] S. Missori, A. Sili, "Mechanical behaviour of 6082-T6 aluminium alloy welds", Metallurgical Science and Technology, Vol. 18, p. 12-18, 2000

[8] K. Elangovan, V. Balasubramanian, "Influences of post-weld heat treatment on tensile properties of friction stir-welded AA6061 aluminum alloy joints", Materials Characterization, Vol. 59, p. 1168-1177, 2008

[9] T. Saito, C. D. Marioara, J. Røyset, K. Marthinsen, R. Holmestad, "The effects of quench rate and pre-deformation on precipitation hardening in Al–Mg–Si alloys with different Cu amounts", Materials Science and Engineering: A, Vol. 609, p. 72-79, 2014

[10] M. H. Scott, M. F. Gittos, "Tensile and Toughness Properties of Arc-Welded 5083 and 6082 Aluminum Alloys", Welding Research Supplement, Vol. 62, p. 243-s-252-s, 1983

[11] R. R. Ambriz, G. Barrera, R. García, V. H. López, "The microstructure and mechanical strength of Al-6061-T6 GMA welds obtained with the modified indirect electric arc joint", Materials & Design, Vol. 31, p. 2978-2986, 2010

[12] T. Yuan, S. Kou, Z. Luo, "Grain refining by ultrasonic stirring of the weld pool", Acta Materialia, Vol. 106, p. 144-154, 2016

[13] S. Kou, Y. Le, "Nucleation Mechanisms and Grain Refining of Weld Metal", Welding Research Supplement, Vol. 65, p. 305-s-313-s, 1986

[14] P. Schempp, C. E. Cross, A. Pittner, G. Oder, R. S. Neumann, H. Rooch, I. Dörfel, W. Österle, M. Rethmeier, "Solidification of GTA aluminium weld metal: Part I - Grain morphology dependent upon alloy composition and grain refiner content", Welding Research Supplement, Vol. 93, p. 53-s - 59-s, 2014

[15] P. Schempp, C. E. Cross, A. Pittner, M. Rethmeier, "Solidification of GTA aluminium weld metal: Part 2 - Thermal conditions and model for columnar-to-equiaxed transition", Welding Research Supplement, Vol. 93, p. 69-s - 77-s, 2014

[16] P. Schempp, C. E. Cross, A. Pittner, M. Rethmeier, "Influence of solute content and solidification parameters on grain refinement of aluminum weld metal", Metallurgical and materials transactions A, Vol. 44A, p. 3198-3210, 2013

[17] P. Schempp, A. Pittner, M. Rethmeier, Z. Tang, T. Seefeld, C. E. Cross, "Influence of alloy and solidification parameters on grain refinement in aluminium weld metal due to inoculation", in Trends in welding research - Proceedings of the 9th international conference, ASM international, pp. 98-107, 2013, Chicago, Illinois, USA, 2012/06/04-2012/06/08

Hot cracking sensitivity of 2A14 high strength aluminum alloy in fiber laser welding

Wei Wang [1], Jian Huang [1], Bang Yao Zhao [2], Pei Pei Hu [2]

[1] *Shanghai Jiao Tong University, Shanghai Key Laboratory of Materials Laser Processing and Modification, China*
[2] *Shanghai Spaceflight Precision Machinery Research Institute, China*

It is a known fact that 2A14 aluminum alloy is susceptible to crack in the weld metal when welded with fusion processes. Hot cracking sensitivity of 2A14 aluminum alloy in fiber laser welding is first evaluated experimentally. The cracking behaviors of the specimens are evaluated by fishbone cracking test. The fracture morphologies of the weld are observed by optical microscope (OM) and scanning electron microscope (SEM). The results show that the crack formed in fiber laser welding with filler wire of 2A14 aluminum alloy is solidification crack. There is an obviously laser power threshold to form thermal cracks in different welding speeds. Hot cracking tendency is obvious when the laser power is below the threshold. There is no hot cracking when the laser power is greater than or equal to the threshold. The hot cracking rate of the welding speed 3.0m/min is slightly higher than the welding speed 4.5m/min. The microstructure of the weld metal consists of &alpha(Al), Al2Cu and Mg2Si.

Pipeline Design to Withstand Strong Ground Motion or Large Permanent Ground Deformations

Nobuhisa Suzuki, Takekazu Arakawa and Teruo Yamaura
JFE Steel Corporation, Tokyo, Japan
bell-wood@muh.biglobe.ne.jp, t-arakawa@jfe-steel.co.jp and t-yamaura@jfe-steel.co.jp

Abstract

This paper describes pipeline design methods to withstand strong ground motion or slope movement for which strain based design method in the elastic deformation range or stress based design method in the plastic deformation range shall be employed. The strain based design method shall be applied to displacement controlled deformations and the nonlinear stress based design method shall be used for load controlled deformations. Dimensionless parameters are defined to discriminate between the displacement and load controlled deformations.

Introduction

Strain based design shall be employed in the cases pipeline deforms in displacement controlled condition where soil springs behave elastically. However, when the ground deformation increases and soil springs turn to behave in the plastic deformation range, pipeline shall be subjected to uniform external loads in the longitudinal or transverse direction [1-3]. In these cases, the load controlled deformation shall be induced in pipeline and the stress based design in the plastic range shall be employed to investigate deformation stability instead of the strain based design [3]. Dimensionless parameters are defined to distinguish the displacement and load controlled deformations [3], which parameters consist of pipe dimensions, soil spring properties, a max ground displacement, and a breadth or length of ground deformation. Furthermore design formulas to calculate strain demands, stress demands, and moment demands to withstand strong ground motion or large ground deformation are presented [3].

The ground deformations and the deformations induced in pipeline

The ground deformations imposed by seismic waves, the longitudinal and transverse ground movements that may occur in an unstable slope are presented in Figs. 1 to 3, respectively. In the figures, the left side figure illustrates the ground movement and the right side figures present the ground displacements and a pipeline. Figure 1 explains that the seismic waves propagate along a pipeline which may be subjected to axial compression at the origin of the coordinate, y=0, and axial tension at y=L/2 and –L/2 [1]. Figure 2 explains the longitudinal slope movement which generates longitudinal forces on a pipeline, due to which forces axial compression and tension are induced in a pipeline at 0<x<L/2 and -L/2<x<0, respectively [1]. Figure 3 shows that bending deformation is generated along a pipeline due to the transverse slope movement where a max bending deformation is generated at y=0, the center of slope movement [1].

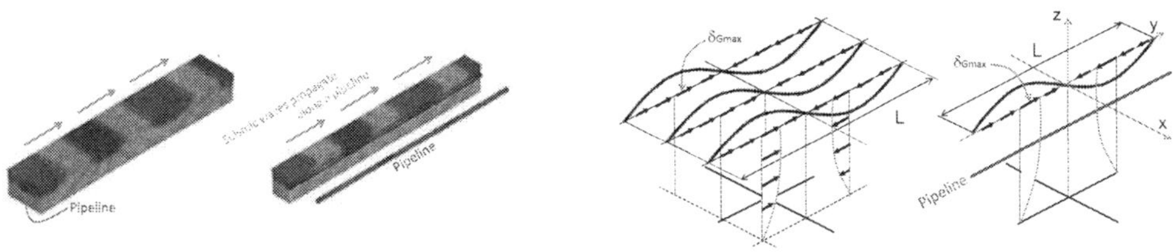

Fig. 1 Longitudinal deformation induced in the ground and a pipeline by seismic waves [3]

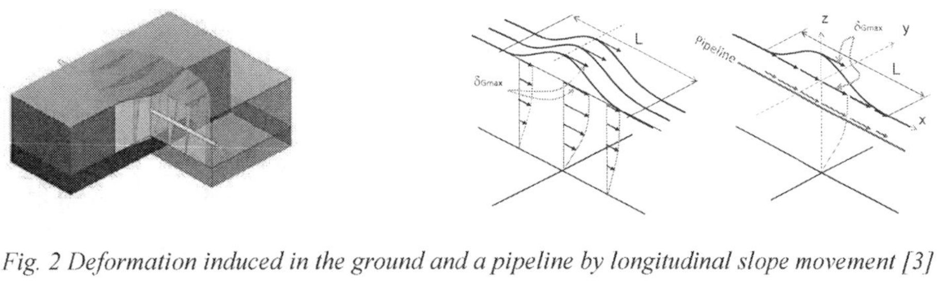

Fig. 2 Deformation induced in the ground and a pipeline by longitudinal slope movement [3]

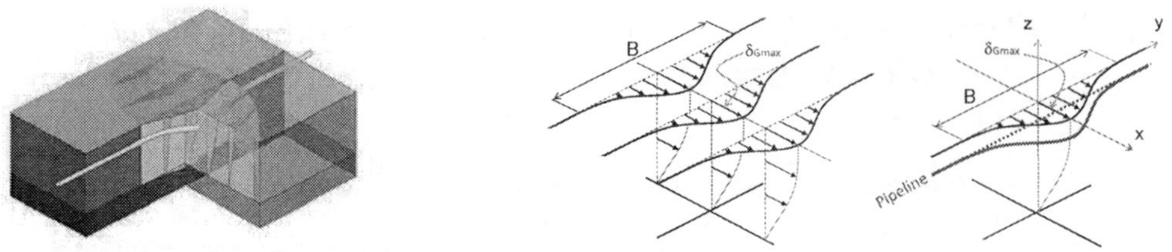

Fig. 3 Deformation induced in the ground and a pipeline by transverse slope movement [3]

Pipe-soil interaction and soil spring properties

The ground deformations will be transmitted to pipeline through soil springs as schematically illustrated in Fig. 4. The soil spring properties are shown in Fig. 5 where the properties are represented with an elastic-perfect plastic bilinear function both in the longitudinal and the transverse directions [1, 3]. The deformation induced in the pipeline shall be displacement controlled as long as the soil springs behave elastically however that may turn to be load controlled after some soil springs begin to yield [3].

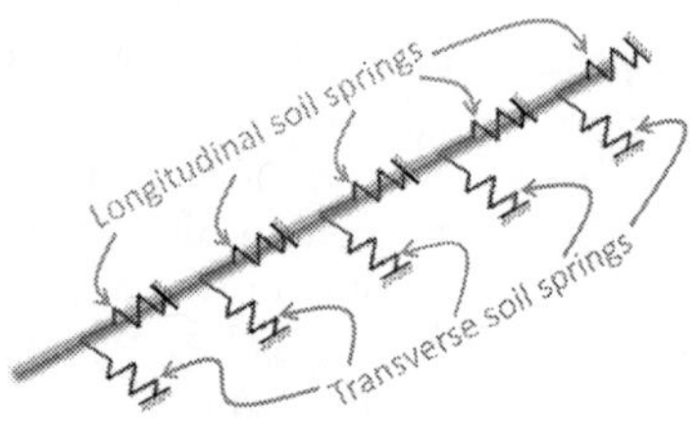

Fig. 4 Pipe-soil interaction
(Pipeline constrained by soil springs)

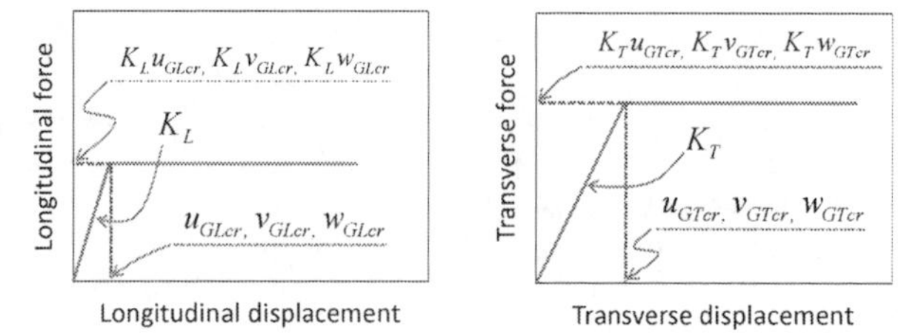

Fig. 5 Soil spring properties
(Left: longitudinal spring, Right: transverse spring)

Dimensionless parameters to distinguish deformation modes of pipeline

The dimensionless parameters αL and βB are listed in Table 1, which were obtained by elastic solutions of a pipeline subjected to the ground deformations [3]. Pipeline may deform in the displacement controlled condition in the cases αL or βB satisfies the corresponding equation presented in the left side column. More when a max ground displacement increases, pipeline tends to present the load controlled deformation where the dimensionless parameter satisfies the equation presented in the right side column [3].

Table 1 Dimensionless parameter to distinguish the deformation modes of pipeline [3]

Ground Deformation	Displacement Controlled Condition	Load Controlled Condition
Seismic waves	$2(\delta_{Gmax}/v_{GLcr}-1)^{1/2}\pi \leq \alpha L$	$\alpha L < 2(\delta_{Gmax}/v_{GLcr}-1)^{1/2}\pi$
Longitudinal slope movement	$3.46(\delta_{Gmax}/u_{GLcr}-1.20)^{3/5} \leq \alpha L$	$\alpha L < (\delta_{Gmax}/u_{GLcr}-1.20)^{3/5}$
Transverse slope movement	$(2\delta_{Gmax}/u_{GTcr})^{1/4}\pi \leq \beta B$	$\beta B < (2\delta_{Gmax}/u_{GTcr})^{1/4}\pi$

The displacement and load controlled deformations of pipeline are discriminated in Fig. 6 using the dimensionless parameter where pipeline deforms due to longitudinal slope movement. In addition, examples are represented in Fig. 7 with respect to two pipelines with diameters of 12 and 24-in and a D/t of 50 which may be subjected to longitudinal slope movement with a slide length up to 200 m and a max ground displacement up to 20 cm. The longitudinal soil spring properties were assumed as $k_L=10\ N/cm^3$, $K_L=\pi D k_L$, and $u_{Lcr}=1.0\ cm$.

Strain demand and load demand

Strain demand in compression or tension for pipeline, which is the max longitudinal pipe strain induced by the strong ground motion or the large permanent ground deformations, shall be calculated with the corresponding formula represented in the left side column in Table 2 [3]. Axial load demand in compression or tension and bending moment demand, which are a max axial compressive or tensile force and a max bending moment, can be calculated by the corresponding formula presented in the right side column in Table 2 [3].

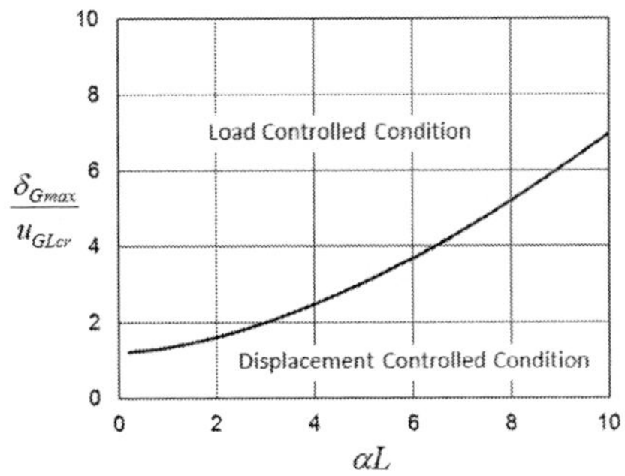

Fig. 6 Discrimination of the deformation modes induced by longitudinal slope movement

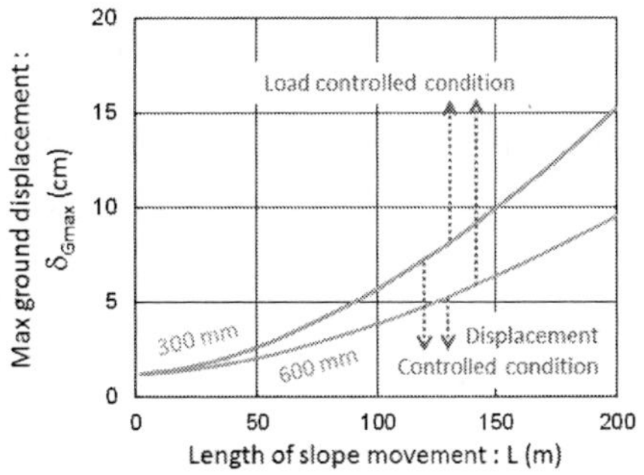

Fig. 7 Deformation modes of 12 and 24-in pipelines induced by longitudinal slope movement

Table 2 Strain demand in displacement controlled condition and load demand in load controlled condition [3]

Ground Deformation	Displacement Controlled Condition	Load Controlled Condition
	Strain Based Design Strain Demand in Compression & Tension	Load Based Design Load Demand in Comp., Tension & Moment
Seismic waves	$\varepsilon_{PC\,max} = F_{Cseis}\frac{2\pi}{L}\delta_{G\,max}$	$N_{PC\,max} = K_L\left\{\frac{F_{Cseis}F_{FRseis}}{\alpha^2}\varepsilon_{GC\,max} + \left(\frac{L}{4} - S_{elas}\right)v_{GLcr}\right\}$
Longitudinal slope movement	$\varepsilon_{PC\,max} = F_{SCslope}\frac{\pi}{L}\delta_{G\,max}$	$N_{PC\,max} = K_L\left[\{U_R(L_0) - U_R(S_0)\} + u_{GLcr}S_0\right]$
Transverse slope movement	$\varepsilon_{PC\,max} = \frac{\pi^2 D}{B^2}\delta_{G\,max}$	$M_{max} = max\left[\, M(0),\ M(B_0/2),\ M_{outside}\,\right]$

Max strain induced in the 12" & 24" pipelines due to longitudinal slope movement

Max longitudinal strain of the 12 and 24-in pipelines caused by the longitudinal slope movement are plotted in Figs. 8 and 9, respectively. The slide length of slope movement varies from 50 to 200 m and the max ground displacement of 20 cm was assumed. The max longitudinal pipe strain for each case was calculated by the equation presented in the left side column in Table 2. Observing the plots, the max longitudinal pipe strain and the corresponding critical ground displacement tend to increase with increasing max ground displacement. The max pipe strain induced in the 24-in pipeline is small compared to the corresponding data of the 12-in pipeline.

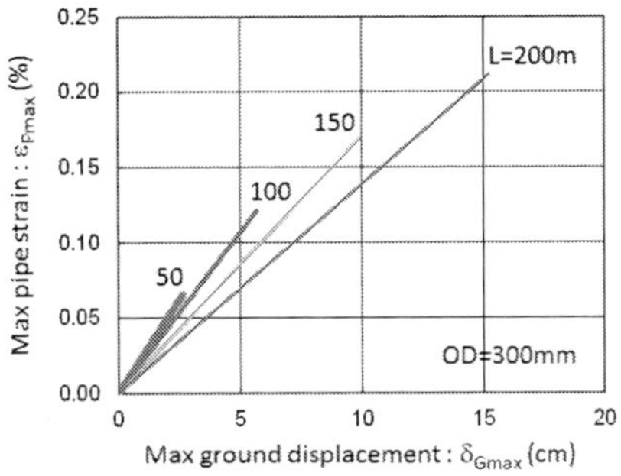

Fig. 8 Max longitudinal strain induced in a 12-in pipeline in displacement controlled condition

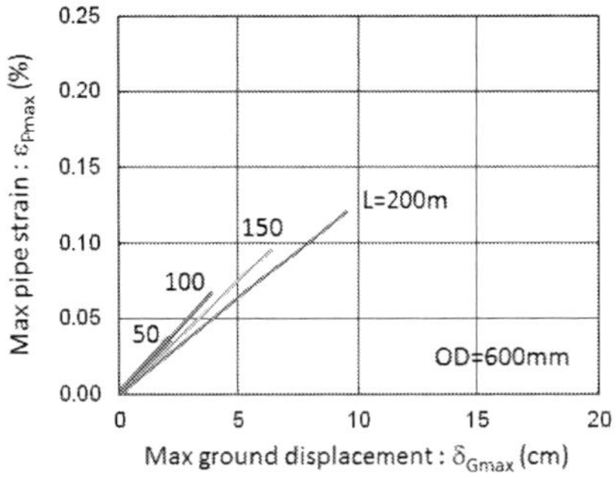

Fig. 9 Max longitudinal strain induced in a 24-in pipeline in displacement controlled condition

The strain capacity in compression of Round-House type pipe and Luders-Elongation type pipe with a D/t of 50 is estimated as from 0.80 to 0.85% and from 0.10 to 0.20%, respectively [3]. As mentioned above, the max pipe strain induced in the 24-in pipeline shall be small compared to the strain capacity in compression of the pipe therefore it is recognized that the pipeline integrity can be ensured in the case of displacement controlled condition. The max pipe strain of the 12-in pipeline caused by the 200 m long slope movement can develop to 0.21% however the LE type pipeline may buckle at the max ground displacement of about 15 cm when the displacement controlled deformation terminates. The max pipe strain with a slide length shorter than 150 m is smaller than 0.18% hence the LE type pipeline will be able to ensure the integrity.

Max stress induced in the 12 and 24-in pipelines due to longitudinal slope movement

Variations of the max longitudinal stress of the 12 and 24-in pipelines imposed by longitudinal slope movement are plotted in Figs. 10 and 11, respectively. The max ground displacement up to 100 cm is plotted in the figures. The axial stress demand, which is obtained dividing the load demands presented in Table 2 by the section area, is plotted in the figures. Observing the curves, the max longitudinal pipe stress increases with increasing max ground displacement at the beginning of deformation in load controlled condition. Furthermore the longitudinal pipe stress tends to be constant with increasing max ground displacement where the yield zone of longitudinal soil springs spreads over almost the entire length of pipeline.

The max longitudinal pipe stress generated by the longitudinal slope movement with a long slide zone tends to be high compared to that generated by that with a short slide length. It is apparent that the max pipe stress induced in the 12-in pipeline is larger than that induced in the 24-in pipeline. Figure 11 explains that the 24-in pipeline will be able to withstand the longitudinal slope movement however Fig. 10 implies the 12-in pipeline may buckle in the cases the sliding length becomes 150 and 200 m.

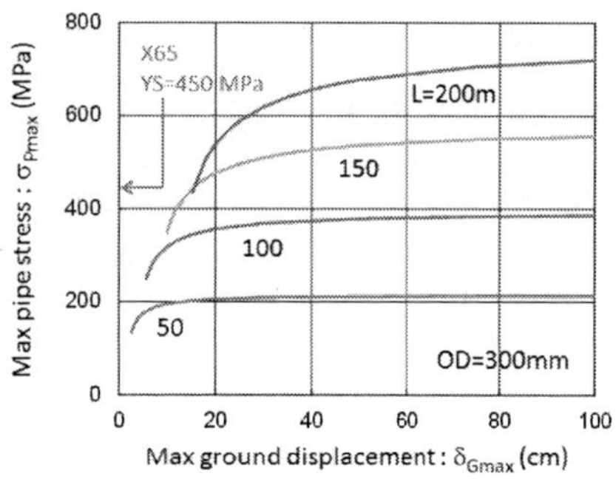

Fig. 10 Max longitudinal stress induced in a 12-in pipeline during load controlled condition

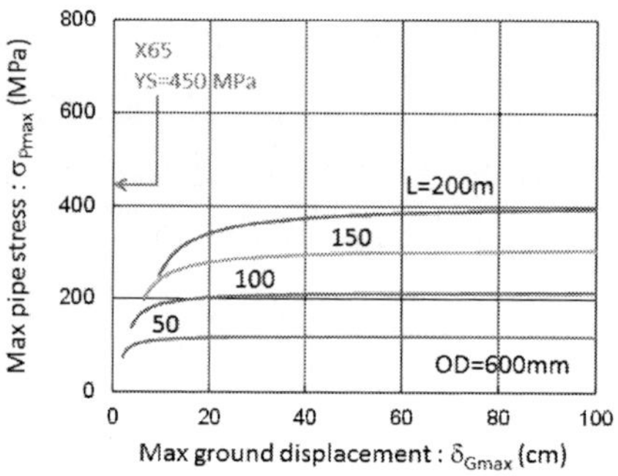

Fig. 11 Max longitudinal stress induced in a 12-in pipeline during load controlled condition

Conclusions

The dimensionless parameters are proposed to simply discriminate between the displacement and load controlled deformations. Design formulas to calculate the strain demand and the load demand are also proposed. The strain based pipeline design shall be consistent with displacement controlled deformation. And the stress based pipeline design beyond the elastic range shall be employed when the deformation of pipeline is load controlled, where the deformation stability of pipeline shall be investigated referring to axial load capacity or moment capacity.

Nomenclature

The constants and variables used in this paper are as follows; A : cross sectional area of a pipe, B : Breadth of ground deformation, D : Diameter of a pipe, E : Young's modulus, I : moment of inertia of a pipe, L : wavelength or length of ground deformation, α : $(K_L/EA)^{1/2}$, β : $(K_T/4EI)^{1/4}$, δ_{Gmax} : max ground displacement, ε_{PCmax} : max pipe strain (L. direction), N_{max} : max compressive axial force (L. direction), and M_{max} : max bending moment.

Acknowledgments

The authors wish to appreciate JFE Steel Corporation for the permission of publication of this paper and the contribution by Izumi Kubo in preparing this paper was greatly appreciated.

References

[1] Suzuki, N. (1995): Seismic Design of Pipelines Considering Liquefaction-Induced Permanent Ground Deformation, Ph. D. Thesis, Tohoku University.

[2] Suzuki, N, Arakawa, T. and Arabey, A. (2014): Strain-Based Pipeline Design in Harsh Environments Using Large Diameter High Strain Line Pipes, Proc. of the 10th IPC, IPC2014-33421.

[3] Suzuki, N, Arakawa, T. and Arabey, A. (2016): Pipeline Integrity in Seismic and Extremely Cold Regions, Proc. of the 11th IPC, IPC2016-64301.

The Tolerability of Welding Defects in Pipelines.

Hamad Almostaneer [1], Mansour Al-Elyani [2], Ahmad Shahreer [2]

[1] *SABIC - MCE - MCC - Materials, Corrosion, and Static Equipment Domain, SAUDI ARABIA*
[2] *SABIC - T& I - Regional Analytical Lab, SAUDI ARABIA*

The problem is on a new pipeline that have been newly installed. It is about 1.2 km pipeline constructed of seam welded ASTM A312 type 316/316L stainless steel. It is longitudinally seam-welded and the circumferential welds to connect the pipeline segments are field welds. During radiography testing of the circumferential welds, defects (probable slag inclusions or porosity) were observed on the longitudinal seam welds. It was found on about twenty (20) joints sides (longitudinal) out of 100 joints (on the short distances close to the circumferential weld joints covered by the radiography). Analysis were conducted on the longitudinal seam-weld to study the manufacturer welding integrity. Nonetheless, based on the lab testing and failure investigation, the materials were found to meet the provided operating parameters. The welding defects are found to be tolerable in such condition.

APPLICATION OF COLD-WIRE TANDEM SUBMERGED ARC WELDING FOR LOW CARBON MICROALLOYED LINEPIPE STEELS

Mohsen Mohammadijoo, Hani Henein, Douglas G. Ivey
Advanced Materials Processing Laboratory, University of Alberta, Edmonton, AB, Canada T6G 1H9
mo1@ualberta.ca, hhenein@ualberta.ca, divey@ualberta.ca

Stephen Kenny
Research and Development Division, Evraz Inc. NA, Regina, SK, Canada
Stephen.kenny@evrazna.com

Abstract
To improve productivity of submerged arc welding (SAW) for steel pipeline manufacturing, many fabricators have resorted to increasing the number of electrodes with arcs, i.e., tandem/twin submerged arc welding (TSAW), which may result in an increase in the overall heat input due to the increase in welding current and voltage needed for a higher deposition rate. In the present work, cold-wire tandem submerged arc welding (CWTSAW) has been developed to improve the productivity of pipeline manufacturing by increasing the welding travel speed and deposition rate, while retaining adequate joint geometry and properties without increasing the welding heat input. In addition to increasing productivity, incorporating a cold wire in the CWTSAW process moderates the heat input by consuming the energy introduced to the weld puddle, which affects the properties and geometry of the heat-affected zone (HAZ). The weld joints of an X70 microalloyed steel were made at various travel speeds and cold-wire feed rates. The HAZ toughness was investigated using Charpy impact testing. Scanning electron microscopy (SEM) and tint etching optical microscopy (TEOM) were utilized to characterize the microstructure alterations in the HAZ. Analysis of the microstructure and mechanical properties indicated a reduction in the prior austenite grain (PAG) size and martensite-austenite (M-A) constituent fraction, resulting in an improvement in the HAZ toughness, when the cold wire was fed at 25.4 cm/min. An increase in the welding travel speed and deposition rate by 26% and 22%, respectively, was achieved by the addition of the cold-wire at 58 cm/min with no detrimental effect on the microstructure and toughness of the HAZ.

Introduction
Submerged arc welding (SAW) has been preferred over other welding processes in pipe production due to its inherent properties, such as high deposition rate, deep penetration and capability of welding thick sections [1]. To improve welding productivity in a global economy, many fabricators have resorted to some techniques to increase the SAW deposition rate. The commonly used techniques to improve the welding deposition rate are increasing the electrode stick-out, twin-wire SAW, tandem SAW (TSAW), applying a metal-cored wire, varying electrode polarity and addition of a cold-wire to a single electrode SAW or in a twin-wire SAW. Certain applications, such as pressure vessels, girth welds in pipes or spiral pipe mills can benefit from the techniques employed to improve the deposition rate and, consequently, the productivity of the SAW process [2]. TSAW is a common process employed by industry to improve the welding deposition rate. However, some of these techniques present challenges to the welding process, such as an increase in the overall welding heat input, reduction in the weld penetration depth and an adverse effect on the geometry and properties of the weld metal and the heat-affected zone (HAZ). The addition of a cold wire to the one electrode SAW process was initially proposed by Mruczek et al. in 2005 [3]. In the current research work, tandem submerged arc welding with an additional cold-wire (CWTSAW) has been developed to improve the productivity in terms of deposition rate and welding travel speed, while retaining adequate weld geometry and properties. A cold wire addition also improves the macrostructure, mechanical properties and microstructure of the HAZ [4]. These improvements, as a result of cold wire addition, are attributed to changes in the actual heat introduced to the weldment, the peak temperature and the cooling rate in the HAZ, when cold wire is added to the TSAW process. The developed CWTSAW process involves two electrodes with arcs and one electrically cold electrode, which is fed at a lagging position close to the trail electrode. Since an appropriate understanding of the welding conditions is always essential in the development of a welding process to guarantee requisite weld geometry, appearance and mechanical properties, the CWTSAW process was initially optimized and investigated in terms of the welding parameters (i.e., heat input, voltage, travel speed, and cold wire feed speed, position and angle) and this work has been presented in an earlier publication [5]. The present CWTSAW process setup has been developed based on the optimized welding parameters. The cold wire provides a higher deposition rate and better productivity for the welding process without increasing heat input compared with the TSAW process. Furthermore, incorporating a cold wire in TSAW moderates the heat input by consuming the energy of the trail electrode as the wire melts into the weld puddle. Accordingly, better quality welds are expected with deep penetration at lower heat inputs per mass of deposited material and with a substantial reduction in arcing time, causing the formation of a smaller and shorter weld pool (compared with TSAW without a cold wire). As such, CWTSAW is a promising technique for spiral welds commonly used in the pipeline industry. In addition, in this

research work, the effect of cold wire addition on welding travel speed is studied. The microstructure and toughness in the HAZ of a typical microalloyed steel, and their evolution by varying the cold wire feed rate and travel speed in the CWTSAW process, are also studied. Microstructural characterization is carried out using tint etching optical microscopy (TEOM) [4] and scanning electron microscopy (SEM). Charpy V-notch impact testing is performed to investigate the property changes in the HAZ of samples prepared by CWTSAW.

CWTSAW Process Development

In TSAW, the lead electrode is positioned to generate sufficient penetration depth of the weld and the trail electrode(s) provides filling of the bevel, resulting in adequate joining of the metal pieces of interest. Accordingly, in the CWTSAW process, DCEP (direct current electrode positive) and ACSQ (square wave alternating current) polarities were selected for the lead and trail electrode, respectively, since DCEP on the lead electrode causes deeper penetration and ACSQ on the trail electrode provides increased deposition rate relative to DCEP polarity. The cold wire was fed at a lagging position close to the trail electrode and consumes some of the heat of the trail electrode. Our previous work [5] indicated that the addition of a cold wire at a lagging position with an angle of 63° resulted in an overall improvement in the weld geometry, dilution and micro-hardness. In the current work, to assess the influence of cold wire addition on the HAZ properties and welding productivity, the heat input and the cold wire feed speed were varied to investigate the properties, deposition rate and travel speed of welding. As such, seven weld samples were prepared by CWTSAW of 13.4 mm thick plates of X70 microalloyed steel. A 90° V-shaped bevel with 4.5 ± 0.5 mm depth was machined in the steel plates prior to welding. The compositions of the microalloyed steel and the consumable electrode are indicated in *Table 1*. The 4 mm diameter wires for the electrodes and cold wire were selected based on EN756/EN14295 (BA-S2Mo, Bavaria, Germany) and AWS A5.17/A5.23 (EA2). The consumable flux was chosen according to EN 760 (BF6.5, Bavaria, Germany).

***Table 1**. X70 microalloyed steel and electrode compositions (wt%).*

X70 composition (wt%)									
C	P	S	Mn	Si	N	V+Mo+Nb+Ti	Cu+Ni+Cr+Sn+Al+Ca	Pcm	
0.043	0.013	0.001	1.73	0.28	0.008	0.21	0.57	0.175	
Electrode and cold-wire composition (wt%)									
Symbol	**C**	**P**	**S**	**Mn**	**Si**	**Mo**	**Ni**	**Cr**	**Cu**
BA-S2Mo	0.10	0.007	0.01	1.04	0.1	0.56	0.02	0.03	0.03

The CWTSAW process set-up and the fixed welding parameters are illustrated in *Fig. 1*. The welding conditions to fabricate the microalloyed steel joint are presented in *Table 2*. Sample T1 was welded by TSAW and the rest of the welds were fabricated by the CWTSAW process. Accordingly, the CWTSAW samples (i.e., C1, C2, C3, C4, C5 and C6) were compared with the TSAW sample (T1). Relatively high current, voltage and cold wire feed speed were employed to evaluate the extreme value of the travel speed that can be achieved by cold wire addition, while maintaining adequate HAZ and weld geometry, appearance and properties. Visual inspection was performed on the prepared welds prior to any further investigation to ensure that the welds were free from the macro-level defects, such as surface porosity, hot-cracking, undercutting and burn-through.

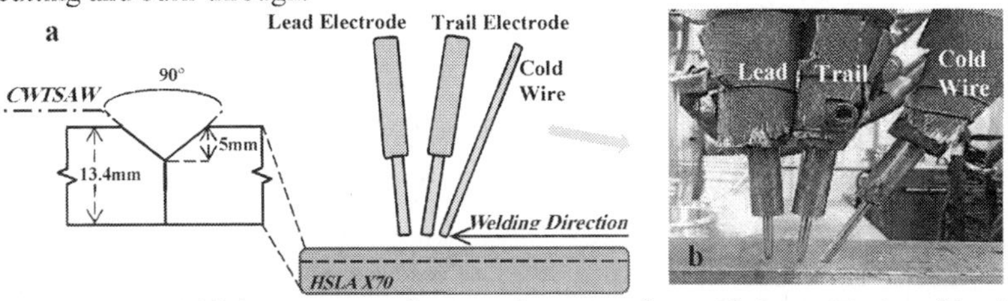

***Figure 1**. CWTSAW process setup: (a) Schematic view of joint configuration along with the positioning of the electrodes and cold wire and (b) welding setup designed at Evraz Inc. NA.*

Results and Discussion

The specimens welded by CWTSAW and TSAW were tested by Charpy impact testing at three different temperatures. *Fig. 2* depicts the CVN results for the HAZ of the prepared weld samples. The average of at least four Charpy results for each weld at each temperature is presented. As shown, the impact toughness of the C1 sample was increased (cold wire feed rate was 25.4 cm/min) at the same welding heat input compared with the T1 sample, due to the lower fraction of finely distributed martensite-austenite (M-A) constituents formed in the coarse grained heat-affected zone (CGHAZ). However, the Charpy impact toughness values for the HAZ of the C2 weld were reduced, relative to the T1 sample, when the cold wire was fed at 63.5 cm/min and welding heat input was reduced. This reduction in toughness is attributed to the formation of slender shaped M-A constituents, which formed mostly along the prior-austenite grain (PAG) boundaries as a result of the faster cooling rate in the CGHAZ of the C2 sample. The HAZ toughness was slightly increased for the C3 and C4 welds (higher cold wire feed rates) relative to the T1 sample. A significant reduction in the toughness of the C5 and C6 weld samples, relative to the T1 sample, was observed. The macro and microstructural analysis indicate the formation of the weld defects along the fusion line, which led to the brittle behavior of the C5 and C6 samples. A small amount of the granular flux used to shield the arcs was entrapped in the weld metal along the fusion line. According to the Halmoy equation [6], there is a direct relationship between the arc force and the current. The arc current of the C5 and C6 samples was not sufficient to produce enough arc force to remove all the shielding flux while welding. However, our observations indicate that increasing the arc current resulted in a negative effect on the weld properties due to an increase in the overall welding heat input as well as some welding technical challenges such as arc blowing, arc sparking and burn-through. The variations in the toughness of the HAZ of the T1 and C1-C4 weld samples are a result of microstructural alterations in the HAZ for the different welding conditions. The microstructural alterations in the HAZ, as a result of cold wire addition, are attributed to changes in the actual heat introduced to the weldment and, consequently, the cooling rate in the CGHAZ, when cold wire is added to the TSAW process. The results of our previous studies [4] have indicated that the microstructure and mechanical properties of the HAZ are altered by a variation in the cold wire addition rate. Mruczek et al. [3] and Ramakrishnan et al. [7] have proposed that addition of a cold wire to SAW reduces the actual welding heat input since the cold wire consumes the heat of the electrode. In addition, the HAZ cooling rate increases as the actual welding heat input decreases [8]. Accordingly, addition of the cold wire in the conventional TSAW process affects the microstructure and mechanical properties of the HAZ due to reduction in the actual heat introduced to the weldment.

***Table 2.** Welding process parameters.*

Process Parameter		*Weld sample ID*						
	Unit	*T1*	*C1*	*C2*	*C3*	*C4*	*C5*	*C6*
Current-Lead Electrode	A	1040	1060	1070	1200	1250	1300	1350
Current-Trail Electrode	A	830	830	840	950	1000	1050	1050
Voltage-Lead Electrode	V	30	31	31	32	33	34	36
Voltage-Trail Electrode	V	34	34	34	36	37	38	39
Travel Speed	cm/min	160	167	177	190	203	216	229
Welding Heat Input	kJ/cm	22.2	22.1	20.5	23.0	23.2	23.5	23.6
Cold Wire Feed Speed	cm/min	NA	25.4	63.5	51	58	66	71

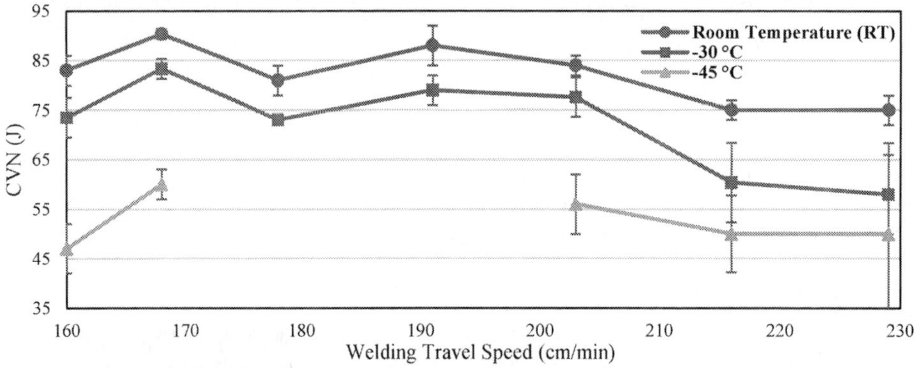

***Figure 2.** Charpy impact energy of the HAZ for steels welded at different travel speeds and cold wire feed rates.*

The type and fraction of microstructural features, along with the PAG formed in the HAZ, are altered due to the heat input, high peak temperature and thermal cycles during welding. The PAG size influences the phase transformation temperature and kinetics during cooling. Qiao et al. [1] and Shome [8] reported that the size of PAGs in the HAZ depends on the local thermal cycle and the PAG size increases by increasing the welding heat input. Refining the PAG size influences the transformation products, particularly the M-A constituents, within the HAZ, which affect the toughness. In addition, Bhadeshia [9] and Yan et al. [10] have suggested that, in addition to the PAG size, cooling rate affects the morphology of the M-A constituents, which also affect the toughness of the welded steel. Kim et al. [11] have reported that M-A islands are the main metallurgical factor, which contribute to local embrittlement of microstructures of welded microalloyed steels. They also stated that the Charpy impact toughness of the CGHAZ of the high strength low-carbon steels is a function of the fraction, morphology and carbon content of the M-A islands. The work done by Lan et al. [12] indicated that the relatively fast cooling rate in the HAZ of a low carbon steel led to the formation of slender M-A constituents with higher carbon levels and segregated silicon in the M-A islands, which resulted in an increase in the martensite hardness in the HAZ. As such, there is a concurrent effect of PAG size and cooling rate on the characteristics of the transformation products, in particular the M-A constituents, in the HAZ of microalloyed steels. To evaluate the effect of microstructure alterations on the toughness, four weld samples, i.e., T1, C1, C2 and C3, were analyzed using TEOM [4,5] and SEM. *Fig. 3* depicts optical micrographs and SEM images from the CGHAZ of the weld samples, revealing the microstructural features, particularly M-A constituents shown as shiny white islands. The PAG size in the CGHAZ for T1, C1, C2 and C3 samples was measured as 68 μm, 55 μm, 42 μm and 54 μm, respectively. The PAG size in the CGHAZ of C1, C2 and C3 was decreased compared with the T1 sample due to the addition of the cold wire and lower actual welding heat input. The M-A fraction in the CGHAZ of T1, C1 and C2 samples was 4.1%, 3.0% and 2.4%, respectively, indicating a reduction in the fraction of M-A as a consequence of PAG size reduction by cold wire addition [5,10]. The M-A fraction in the CGHAZ of C3 was 2.9%, which were finely distributed within the PAGs. The M-A fraction in the C1 and C2 samples was decreased due to the reduction in PAG size. However, the formation of slender shaped M-A constituents in the C2 samples, as a result of the relatively faster cooling rate, resulted in a reduction in the HAZ toughness compared with the C1 weld sample. The reduction in the PAG size in the CGHAZ is attributed to a reduction in the actual heat introduced to the weldment, the reduction in the retention time in the austenitization temperature range (1100-1400°C) and an increase in the cooling rate by adding the cold wire and lowering the welding heat input.

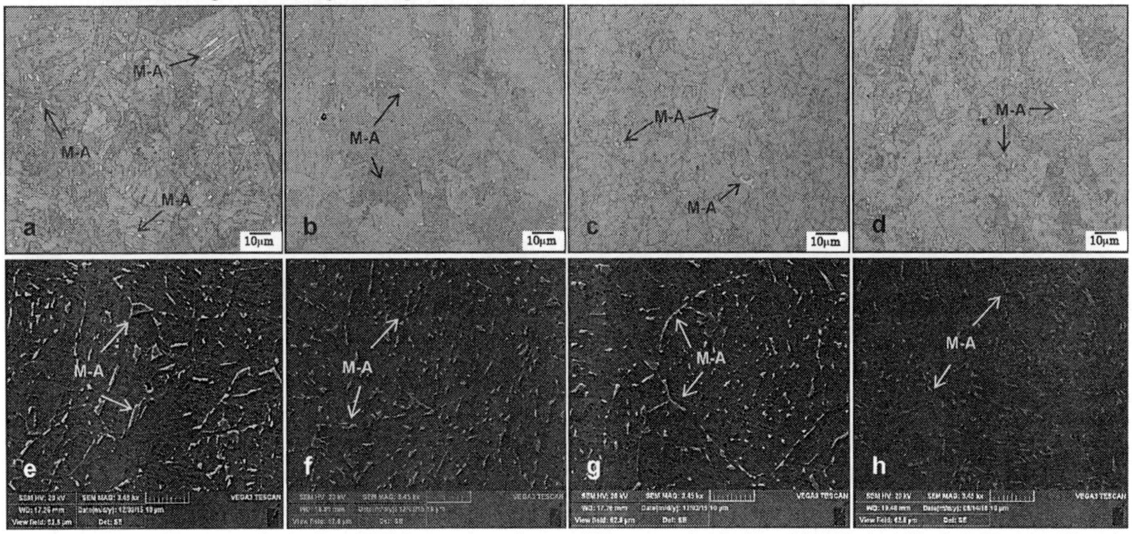

***Figure 3**. Optical micrographs (a-d) and SEM secondary electron (SE) images (e-h) of the CGHAZ. (a,e) T1, (b,f) C1, (c,g) C2 and (d,h) C3 samples.*

Conclusions

Cold-wire tandem submerged arc welding (CWTSAW) has been developed to improve the productivity and properties of welded microalloyed pipeline steels. Cold wire addition at 25.4 cm/min resulted in an overall improvement in the heat affected zone (HAZ) properties and microstructure. An increase in the welding travel speed and deposition rate by 26% and 22%, respectively, was achieved by the addition of the cold wire at 58 cm/min with no detrimental effect on the weld geometry, microstructure and toughness of the HAZ.

Acknowledgments

The authors would like to thank the Natural Sciences and Engineering Research Council (NSERC) of Canada and Evraz Inc. NA for financial support and providing equipment and technical assistance to conduct the welding runs.

References

[1] G.Y. Qiao, *et al*: "Effects of Contents of Nb and C on Hot Deformation Behaviors of High Nb X80 Pipeline Steels," *Trans Nonferrous Met Soc China*, Vol. 19 (2009), pp. 1395-1399.
[2] S. Massey: "Increasing Productivity with Submerged Arc Welding," EWI Publication (2012).
[3] M.F. Mruczek: "Cold wire Feed Submerged Arc Welding," Technical Report, Advanced Technology Institute, SC (2006).
[4] M. Mohammadijoo, *et al*: "Effect of Cold-wire Addition in the TSAW Process on Microstructure and Mechanical properties of the HAZ of X70 Microalloyed Steel," *Proc 11 International Pipeline Conference*, ASME Publication, Calgary, AB, Sept. 2016.
[5] M. Mohammadijoo, *et al:* "Influence of Cold-wire Tandem Submerged Arc Welding Parameters on Weld Geometry and Microhardness of Microalloyed Pipeline Steels," *Int J Adv Manuf Tech,* In Press (2016).
[6] E. Halmoy: "The Pressure of the Arc Acting on the Weld Pool," EWI Publication (Cambridge, 1979).
[7] M. Ramakrishnan, *et al*: "Application of Submerged Arc Welding Technology with Cold Wire addition for Drum Shell Long Seam Butt Welds of Pressure Vessel Components," *Int J Adv Manuf Tech,* Vol. 65 (2013), pp. 945-956.
[8] M. Shome: "Effect of Heat-Input on Austenite Grain Size in the Heat-Affected Zone of HSLA-100 Steel," *Mater Sci Eng. A.*, Vol. 445-446 (2007), pp. 454-460.
[9] H.K.D.H. Bhadeshia: "About Calculating the Characteristics of the Martensite Austenite Constituent," *Proc CBMM and The Minerals, Metals and Materials Society*, USA, 2013, pp. 99-106.
[10] P. Yan, *et al*: "Austenite-ferrite transformation in enhanced niobium steel," *Mater Sci Technol*, Vol. 31 (2015), pp. 1066-1076.
[11] B.C. Kim, *et al*: "Microstructure and Local Brittle Zone Phenomena in High-strength Low-alloy Steel Welds," *Metall Mater Trans A,* Vol. 22A (1991), pp. 139-149.
[12] L. Lan, *et al*: "Analysis of Martensite-Austenite Constituent and its Effect on Toughness in Submerged Arc Welded Joint of Low Carbon Bainitic Steel," *J Mater Sci,* Vol. 47 (2012), pp. 4732-4742.

Linear Friction Welding of Ti 6Al 4V for Aerostructure Applications

Bertrand Flipo, Kathryn Beamish
TWI Ltd, Cambridge, United Kingdom
bertrand.flipo@twi.co.uk, kathryn.beamish@twi.co.uk +441223899000

Bryan Humphreys, Martin Wood
CAV Advanced Technologies Ltd, Consett, United Kingdom
b.humphreys@cav-at.com, m.wood@cav-at.com, +441207593540

Alan Shilton
Ten Solutions Ltd, Wednesbury, United Kingdom
alanshilton@tensolutions.co.uk, +447525029611

Abstract

The increasing use of composite materials observed in new aircraft builds leads to a significant demand for titanium alloy structural parts. The rising material costs and popularity, together with their restrictions in supply and processing, is driving the aerospace industry to make increasingly efficient use of available material.

Linear Friction Welding (LFW), a rapid, high integrity, solid-state forging process, has the potential to decrease the buy-to-fly ratio, production time and time to market of aerostructure components. It has already proven its dependability for the production of some of the latest generation aero engines.

The LFW process is not yet used for aerostructures. This is primarily due to the LFW process not being widely known, and not having been developed or proven for aerostructure applications; but also due to a lack of performance data on aerostructures manufactured by LFW being currently available to the aerostructure supply chain and design community.

To address this issue, a large series of Ti-6Al-4V weldments was produced and assessed via metallographic examinations and mechanical testing. The matrix of experiments was able to capture the LFW process window of this titanium alloy, and to measure the impact of the parametric conditions. Metallographic examination revealed a high integrity weld free from contaminants and oxides at the weld interface.

The metallurgy of the welds revealed a characteristic fine-grained equiaxed microstructure. As-welded joints were tested under tensile and alternating fatigue conditions to provide an extended set of joint performance data. Finally, a techno-economic assessment was conducted which demonstrated the viability and potential of the LFW process for a representative structural airframe component.

Manufacturing of a Dissimilar Joint Between Ti6Al4V and IN718 Using the Linear Friction Welding Technology

Priti Wanjara [1], Kosuke Watanabe [2], Qi Yang [1], Koji Nezaki [2], Javad Gholipour [1]

[1] *National Research Council Canada, Canada*
[2] *IHI Corporation, Japan*

With applications increasingly targeting multi-functional requirements, there is an avid interest to develop joining technologies for dissimilar materials, such as titanium alloys to nickel-based superalloys. For aero-engine structures and components, Ti-6%Al-4%V (Ti6Al4V) and INCONEL718 (IN718) remain the "workhorse" alloys of the industry due to their excellent thermal and mechanical properties. A joint between these workhorse alloys targets to combine the high strength-to-weight ratio of Ti6Al4V with the high temperature oxidation resistance of IN718 into one multi-functional assembly. However, due to differences in their physical and chemical properties there are considerable challenges associated with developing a welding process that will render a functional joint between these two alloys without cracking due to the formation of brittle Tix-Niy intermetallic phases at the joint interface. This work presents a detailed analysis of a joint between Ti6Al4V and IN718, in the solution treated and aged (STA) condition, that was manufactured by linear friction welding (LFW), which is an emerging solid-state joining technology that has been demonstrated to effect bonding through localized heating and a rapid thermal cycle, two important processing aspects for minimizing the formation of brittle intermetallic phases at the interface. Specifically, the paper overviews the characteristics of the interface through microscopic examination, elemental analysis using energy dispersive x-ray spectroscopy (EDS) and microhardness measurements across the Ti6Al4V/IN718-STA joints in the as-welded and stress relieved conditions.

The effect of the welding parameter to mechanical property and welding process of Linear Friction Welded Dissimilar Titanium Joint

Takahiko Shinohara, Koji Nezaki
IHI Corporation, 1 Shin-Nakahara-cho, Isogo-ku,
Yokohama-shi, Kanagawa, 235-8501, Japan
takahiko_shinohara@ihi.co.jp, +81-45-759-2812

Abstract

In recent aircraft jet engines, the application of blisks is increasing to reduce the weight and improve the performance. Blisks are normally manufactured by milling process from large forging materials. But with this method, large amount of material is wasted as cutting chips. Linear Friction Welding, so called LFW, is the technique of manufacturing Blisks which is expected to reduce the material waste by joining the blade to the disks. LFW is also able to manufacture dissimilar material joined Blisks. Ti-6Al-4V (Ti64) is the typical material which has been used for fan and compressor blades. It has good fatigue property and toughness, which is important property because fan and compressor blades are subject to foreign object damage. Beta processed titanium alloys such as Ti-5Al-2Sn-2Zr-4Mo-4Cr (Ti17) have a high static and fatigue strength, and it is often used for fan and compressor disks. The combination of these materials may allow further weight reduction of Blisks.
In Ti64 LFWed joint, thin and sharp flash normally expelled out from the both sides of the joint. However, in case of Ti17/Ti64 dissimilar joint which was welded in the same welding parameter of homogeneous joint of Ti alloy, the expulsion of Ti17 was little and most of flash was formed from Ti64. This joint fractured at the weld interface and elongation and fatigue strength of the joint were low in comparison with those of Ti64.
This study found that increasing the axial shortening of the joint improved the amount of flash of Ti17 and this joint exhibited excellent mechanical property and fatigue property, and the difference of flash amount between Ti17 and Ti64 is not only material strength but also travelling of the center of material flow from interface of Ti17/Ti64 to Ti64 side.

Introduction

LFW is a solid state joining process to join the works using frictional heat generated by linear reciprocated motion. Vairis and Frost reported that joining process of LFW can divided into four phases: initial phase (Stage I), in which the stationary workpiece starts contacting the sliding workpiece and frictional heat is generated; transition phase (Stage II), in which the materials start to soften and form flash at the interface of the two pieces (burn-off begins); equilibrium phase (Stage III), in which the materials continue to soften and the burn-off rate becomes constant; and Stage IV in which oscillation motion ends and work pieces cool down [1-2]. Figure 1 (a) shows Ti64/Ti64 joint. Around the welded area, thin sharp flashes was extruded from both side, left side (oscillation side) and right side (forge side) equality.

In this study, the authors joined two different alloy. Ti17 and Ti64, in the same welding parameter using for

(a) (b)

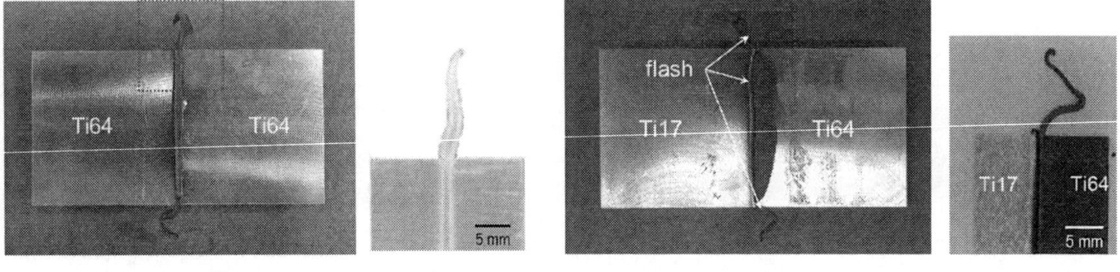

Fig.1 LFW joint and cross section of (a) Ti64/Ti64, (b) Ti17/Ti64

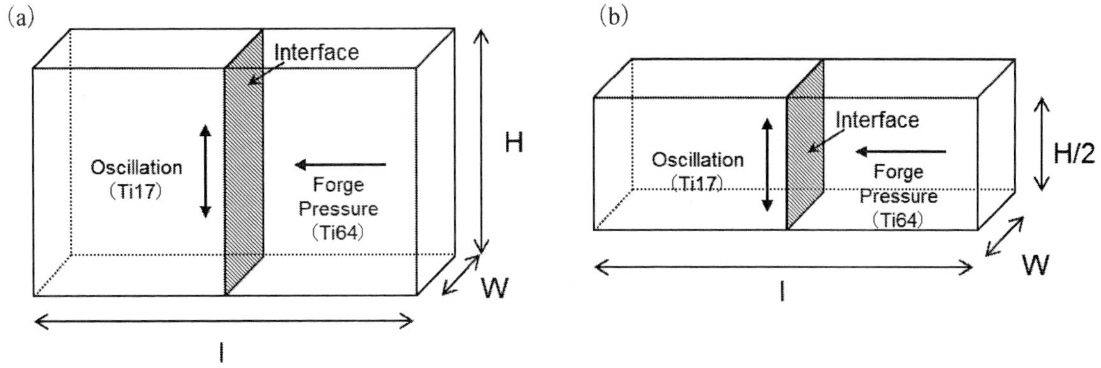

Fig.2 Schematic of workpiece geometry for (a) microstructure observation and tensile test, (b) observation of time series variations of joining process

joining of homogeneous joint of Ti alloy. There was a difference of amount of deformation between Ti17 and Ti64. In figure 1 (b), the flash mainly expelled out from Ti64 side, and flash of Ti17 was very small. In this case, there is a risk that contaminations and oxides remain on the weld interface and bad influence on mechanical property. Thus, this research aims are that investigating the relationship between welding parameter, amount of flash of Ti17 and mechanical property, and understanding of the deformation mechanisms of Ti17/Ti64 dissimilar joint.

Experimental procedure

In this study, Ti64 (AMS4928) and Ti17 (AMS4995) used for workpieces. The workpiece shape is rectangle and figure 2 shows the workpiece geometry. As shown in Figure 2 (a) , the oscillation direction is parallel to the height (H) direction of the workpiece. Before joining, weld surface of the workpiece was cleaned using ethanol. Ti17 and Ti64 joined using the LFW machine developed internally. During the joining process, a constant pressure is applied in the perpendicular direction to the weld interface. When the Burn-off distance reaches a predetermined value, oscillation and welding end.

The main welding parameters of LFW are friction force, frequency, amplitude of oscillation and Burn-off distance. In this study, on the basis of the welding parameter using for joining Ti64/Ti64 joint, frequency and amplitude were constant, and friction force and burn-off distance used several levels. Friction force were 1F, 1.3F, 1.8F. F is the friction force used for homogeneous joint of Ti alloy. And burn-off distance were 1L, 1.6L, 2.6L, 3.6L. L is the burn-off distance used for homogeneous joint of Ti alloy. And also, to observe the time series variation of joining process, particularly the width of the plastic flow area, the joints which stopped the joining on the way made using the half size specimens showing figure 2 (b).

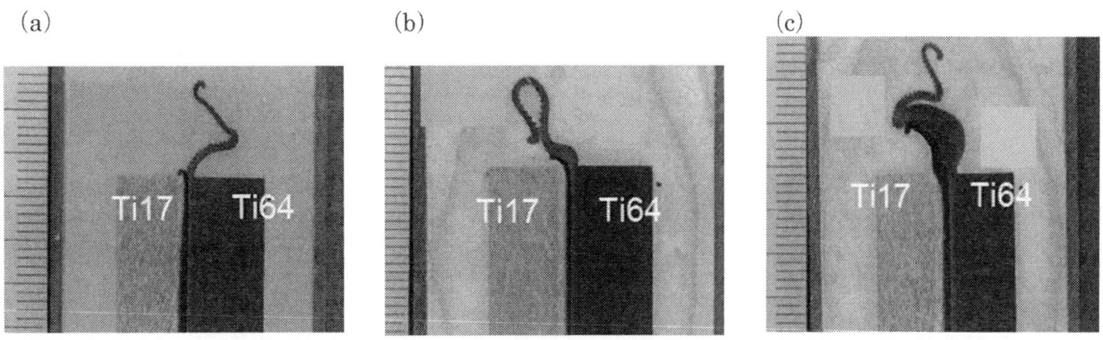

Fig.3 Cross section of the Ti17/Ti64 dissimilar joint (a) 1F, 1L (b) 1.8F, 1L (c) 1F, 3L

Results and discussions

Figure 3 shows the cross section of the joint, (a) is 1F, 1L, (b) is 1.8F, 1L and (c) is 1F, 3L. Each joint, the flash mainly expelled out from Ti64 side, and the amount of flash of Ti17 is small. One factor of this result is that the material strength of Ti64 is lower than Ti17. However, the tensile strength of Ti64 and Ti17 are 971 MPa and 1125 MPa respectively, and the difference of the strength in these material is not so large. It indicates that another factor is thought to be caused the difference of the amount of the flash in addition to the material strength. Figure 4 shows time series variations of the width of the plastic flow area (the distance from weld center to outside of TMAZ) of Ti64 in Ti64/Ti64 joint, Ti64 in Ti17/Ti64 dissimilar joint and Ti17 in Ti17/Ti64 dissimilar join. In Ti64/Ti64 joint, after starting friction, plastic flow area increased and showed a peak 0.84mm.

While in Ti17/Ti64 dissimilar joint, it found that Ti64 started plastic flow earlier than Ti17, and similar to the result of the Ti64 joint, plastic flow area of Ti64 increased and showed a peak 1.16 mm. Comparing the peak value of the width of the plastic flow area of Ti64, that in Ti17/Ti64 joint is wider than that in Ti64/Ti64 joint. It indicates that the area that had reached high temperature travelled from weld interface to low strength side, Ti64. Sato et al. reported the travelling phenomena of rotational plane during rotational frictional welding (RFW) for stainless steel and rolled steel. [3]

Not only in RFW but also in LFW, as shown in Figure 5, in case of joining two workpieces whose strength are different each other by using LFW, relatively lower strength workpiece deformed earlier and the heat generated by plastic deformation accelerated the temperature rising in it. As a result, the peak of temperature distribution and the center of plastic flow travelled from weld interface to lower strength side (Ti64 side), and Ti64 deformed intensively. It is suggested that this is one of the reason why the difference of the amount of flash between Ti17 and Ti64 occurred.

Figure 6 shows the relationship between the welding parameter and burn-off ratio of Ti17 (burn-off distance of Ti17/ total burn-off distance of the joint.). The joint welded in base welding parameter (1F, 1L) had 7% of little burn-off ratio of Ti17.

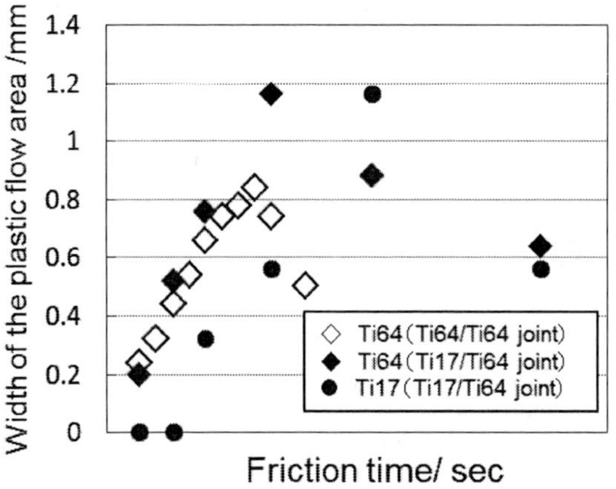

Fig.4 The width between weld center and TMAZ of Ti64 in Ti64 joint

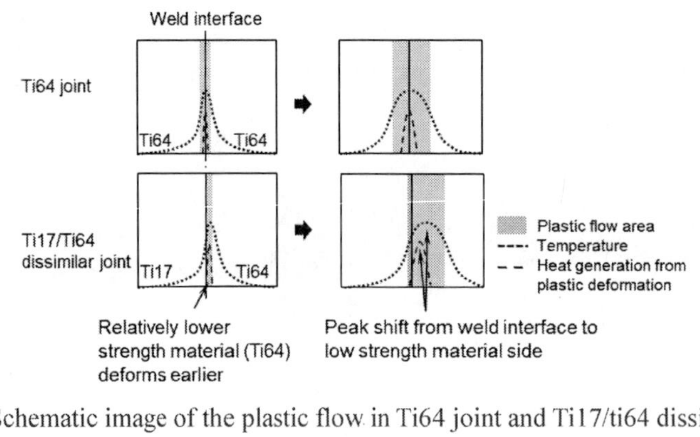

Fig.5 Schematic image of the plastic flow in Ti64 joint and Ti17/ti64 dissimilar joint

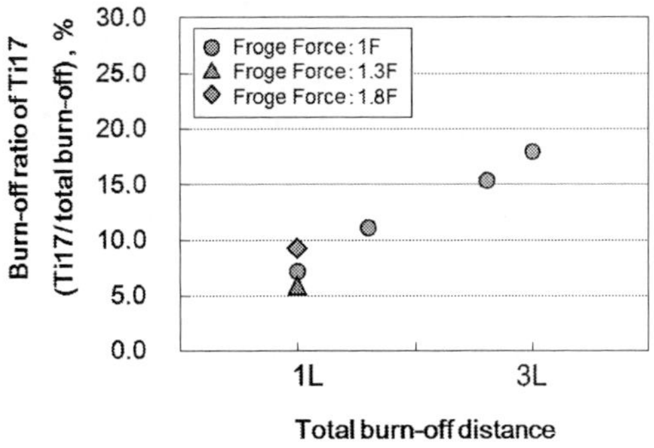

Fig.6 Burn-off ratio of Ti17 in total burn-off

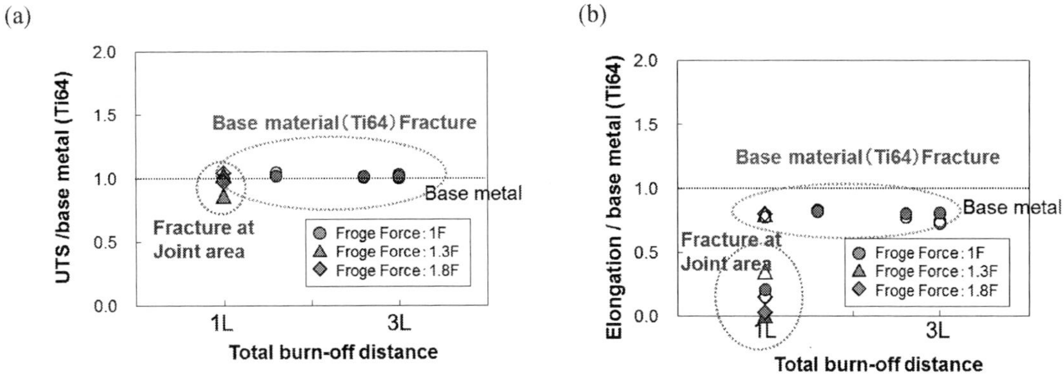

Fig.7 Effects of total burn-off on (a) tensile strength, (b) elongation

In case of increasing frictional force, the amount of flash and burn-off ratio of Ti17 remained small. Increasing the frictional force was not effective to improve the burn-off ratio of Ti17. While, the joint increasing the burn-off distance of the joint (1F, 3L) expelled out large amount of flash of Ti17, and improved the burn-off ratio of Ti17. It is suggested that the total heat input was increased and Ti17 was softened sufficiently to be expelled out, because increasing total burn-off distance of the joint had extended the welding time.

Figure 7 indicates the result of tensile test. UTS and elongation of the joint whose total burn-off was more than 1L were equivalent to those of base metal (Ti64) and fractured at base metal (Ti64 side). While some specimens cut up from the joint whose total burn-off length is 1L fractured at Ti17/Ti64 interface and their elongation were small regardless of the frictional force. This result indicates it is very important for improving the mechanical property of Ti17/Ti64 dissimilar joint to force out Ti17 sufficiently.

Conclusions

- In case of joining Ti17 and Ti64 using LFW, there was large difference of the amount of flash between Ti17 and Ti64. The reason is not only the material strength but also travelling phenomena of the center of material flow from interface of Ti17/Ti64 to lower material strength, Ti64, side.
- Increasing the total burn-off distance was effective to obtain large burn-off distance of Ti17.
- By using sufficient burn-off distance, tensile property of Ti17/Ti64 LFW joint were equivalent of base metal of Ti64.

References

[1] A. Vairis, M. Frost, 1998, High frequency linear friction welding of a titanium alloy, Wear, 217 (1998), pp. 117–131.

[2] A. Vairis, M. Frost, 2000, Modeling the linear friction welding of titanium blocks, Mater Sci Eng A, 292 (2000), pp. 8–17.

[3] K. Fukakusa,T. Satoh, 1984, Travelling Phenomena of Rotational Plane during Friction Welding -Experimental Results and Travelling Mechanisms-, Q. J. Jpn. Weld. Soc., 50 (1981) pp.953-958

High Temperature Fatigue of Dissimilar Metal Joints of IN718:IN713LC by Linear Friction Welding

Ruoru Ye, Hangyue Li, Tim Doel, Paul Bowen

University of Birmingham, UK

Linear friction welding was adopted to join IN718 to IN713LC in the current study. The fatigue performance of the joints has been investigated at a temperature of 630℃ in air using notched test-pieces (K_t=2.29) on which the weld interface were intended to locate at the middle of the notches. All tests were carried out under constant amplitude loading with a load ratio of 0.1. The results obtained from the weld test-pieces were compared to those of the parent materials. For stress levels tested, the fatigue life of the weld test-pieces was found to be slightly shorter than those of parent IN718, but longer than those of parent IN713LC. Interestingly, the failure location move from IN713LC to IN718 with increase of the stress level. Despite being located within the notch region, the weld zone (WZ) forced the failure to occur outside when tested in high cycle fatigue regime. Often the fatigue crack was initiated at internal casting pores in the IN713LC base material (BM) or IN713LC heat affected zone (HAZ). However, failure occurred in the notch region in the IN718 WZ or IN718 HAZ when tested in low cycle fatigue regime. Detailed fractographic analysis was performed throughout to provide insight of the failure mechanisms.

Application of Linear Friction Welding to a Thin Carbon Steel Plate

Ryosuke KUROIWA[a], Yasuhiro AOKI[a], Hidetoshi FUJII[a], Gen MURAYAMA[b] and Masanori YASUYAMA[b]

[a] Joining and Welding Research Institute, Osaka University, Ibaraki, Osaka, Japan
Email: fujii@jwri.osaka-u.ac.jp , Phone&FAX +81-6-6879-8663
[b] Nippon Steel & Sumitomo Metal Corporation, Futtu, Chiba, Japan

Introduction

Medium and high carbon steels can provide high strength and high ductility at a low cost by increasing the content of carbon and microstructure control. Therefore, the medium and high carbon steels are the promising materials for applications like transportation equipment which requires a high strength and weight reduction.

For the fusion welding of the medium and high carbon steels, problems such as hot cracking, porosity, martensitic transformation, etc. [1, 2], occur during the solidification and phase transformation, which decrease the strength of the joint. Therefore, it is important to establish a welding technique which can be performed at a low cost without reducing the strength of the joint. Recently, many studies have been conducted on obtaining a better joint by a solid-state joining process.

Linear Friction Welding (LFW) is a solid-state joining process, in which a joint is obtained through the relative motion of two components under forge force. LFW has two major advantages. First, it has a high energy efficiency because a joint is obtained using just frictional heat. Second, LFW can be applied to non-axial symmetric objects, while the conventional Rotary Friction Welding is still limited to cylindrical objects. However, this process was originally established for the fabrication of the bladed disk (blisk) assembly in aeroengines [3, 4] and the research for this LFW has been mostly limited to the joining of titanium alloys [5, 6] and nickel alloys [7, 8] which are used for high-added value aeroengine applications. In a previous study, medium carbon steel of 5mm thickness was successfully welded by linear friction welding. In order to confirm the new potential of this LFW process, the applicability to transportation equipment, thinner medium carbon steel was welded with different parameters to determine the optimum processing conditions in this study. However, LFW can solve the solidification problem of a steel joint, but it cannot avoid martensitic transformation during the conventional LFW process. This paper describes the difference between the 5mm thickness and 2.6mm thickness.

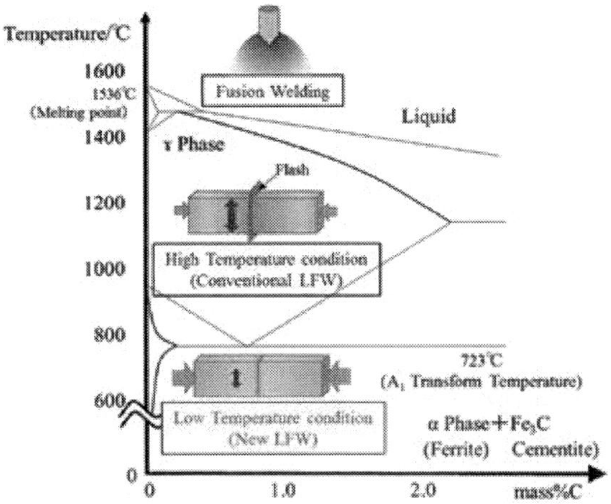

Fig.1 The relationship between Fe-C diagram and weld temperature

Experimental procedure

In this study, 0.45% carbon steels (JIS-S45C; 0.48wt%C -0.77wt%Mn -0.23wt%Si -0.08wt%Cr –bal. Fe) with 5mm×25mm×62mm size (5mm thickness) and 2.6mm×50mm×62mm size (2.6mm thickness) were used as the base metals. The microstructure of theses base metals contained ferrite and pearlite as shown in Fig. 2.

Generally, the LFW process consists of four stages. The first stage is the contact stage such that the two parts (reciprocating part and stationary part) contact each other under a load without any linear reciprocating motion (oscillation). The second stage is the conditioning stage when the stationary part is forced against the reciprocating part, while applying a reciprocating motion in a linear direction. The third stage is the burn-off stage when the weld interface is expelled as flash. The final stage is the forge stage involving rapid cessation of the oscillation motion and forge-force. For each stage, the frequency, amplitude and loading force are independently set. The loading force was constant during the conditioning stage, burn-off stage and forge stage in this study. The welding temperature was measured by an infrared camera (30 f/s) at 15-50 Hz and the same forge force (200 MPa) in this study. Microstructure observations and hardness testing were conducted. The joints were cut to observe the microstructure on the cross-section as shown in Fig.3. The cross-sections were mechanically polished for the metallographic analysis. An optical microscope (OM) and scanning electron microscope (SEM) were used to characterize the microstructure. The hardness was investigated by the micro Vickers hardness test. The micro Vickers hardness of the cross-sections was measured at the applied load of 100gf with the holding time of 15sec.

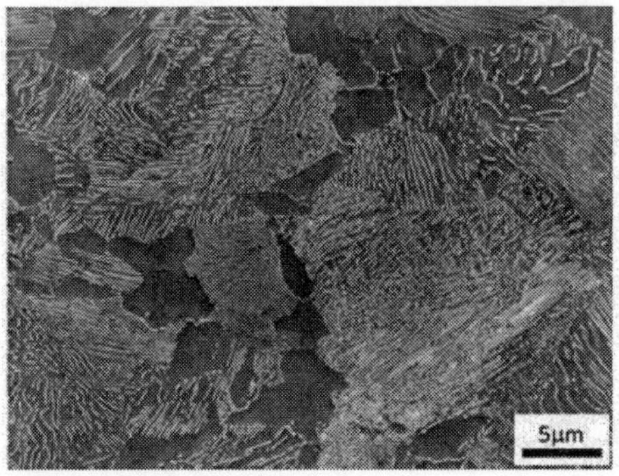

Fig.2 SEM microstructure images of the base metal

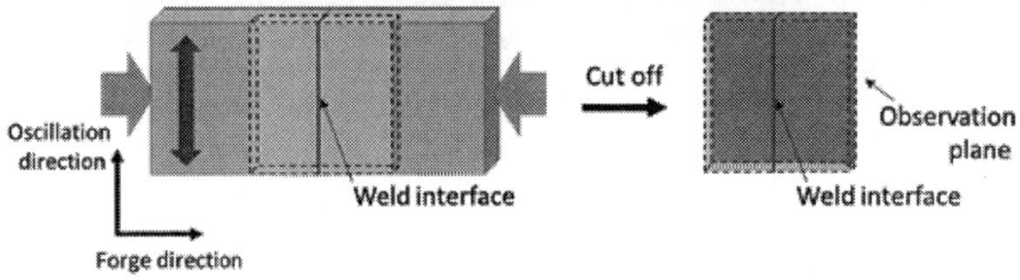

Fig.3 Schematic illustration of sampling method

Results and discussion

Temperature measurements

Fig.4 shows the effect of the frequency on the maximum temperature of the joint surface. For all of the measurements, the maximum temperatures were observed at the joint interfaces. According to Fig.4, the surface temperature decreased with a reduction in the frequency regardless of the specimen thickness. This indicated that the friction heat input becomes lower with a decrease in the sliding velocity. Moreover, the surface temperature decreased with a reduction in the specimen thickness. This indicated that the process window changes due to a reduction of the specimen thickness.

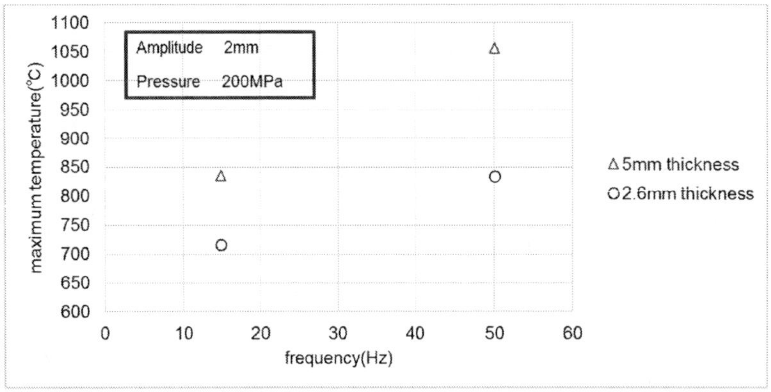

Fig.4 Effect of applied pressure on maximum temperature of the joint surface

Microstructure of the welds in the thin plate

Optical micrographs of the cross-section of the joints are shown in Fig.5. The macrostructure of the joint at 50Hz was quite different from that of the joint at 15Hz. A white etched region was observed around the center of the joint at 50Hz. On the other hand, most of the area at 15Hz was mixed in the darkly etched and white etched region as shown in Fig.5 (c). These results imply that the microstructures of the joints at 50Hz and 15Hz are different.

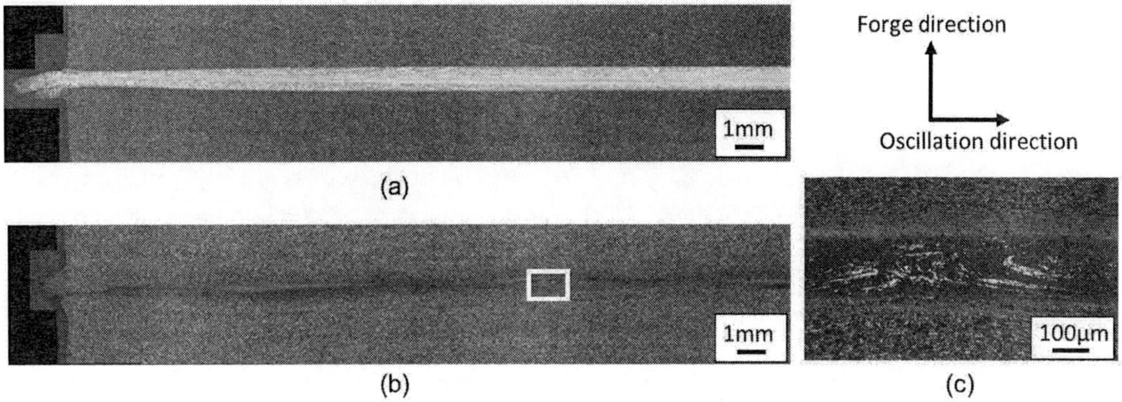

Fig.5 Optical microstructure images of cross-section of LFWed joints:
(a) 50Hz, (b) 15Hz and (c) white square region in (b)

SEM microstructure images at 50Hz and 15Hz are shown in Fig.6. Regarding the joint at 50Hz, martensite was observed at the joint center. These regions where a martensite structure is observed correspond to the white etched region shown in Fig.5. This means that the area around the joint interface was transformed into austenite. Regarding the joint at 15Hz, most of the area showed fine ferrite and spheroidal cementite, and martensite was observed in the white etched region. The spheroidized cementite is usually found in carbon steel after plastic deformation and annealing of the ferrite and pearlite microstructure [9]. This indicates that the joint at 15Hz was obtained around the A_1 temperature. This means that it must be joined by slightly lowering thewelding temperature. Namely, it can be said that it is possible to weld below the A_1 point by lowering the frequency or amplitude which affect the heat input.

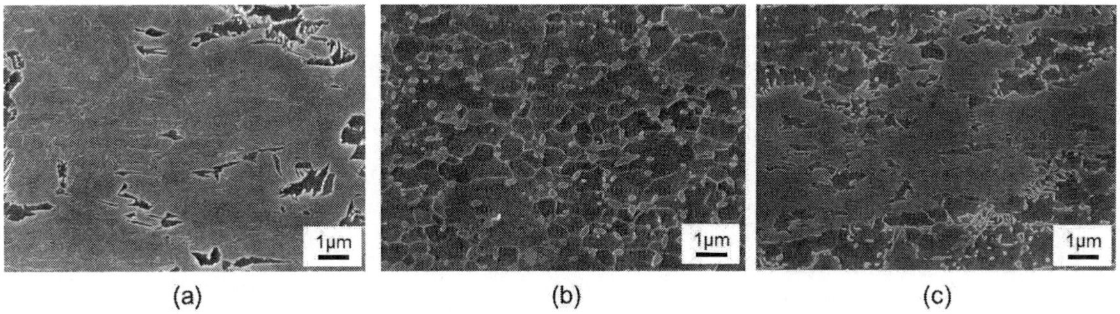

Fig.6 SEM microstructure images of LFWed joint:
(a)center of the joint at 50Hz, (b)center of the joint at 15Hz and (c)white etched region at 15Hz

Martensite microhardness of the welds at high frequency

Fig. 7 shows the Vickers hardness distributions of both the 2.6mm thickness and 5mm thickness joints at 50Hz along the oscillation direction (OD). The 2.6mm thick joint has a high hardness due to martensite at all the locations. The hardening region about 800Hv was detected as shown in Fig.7 (a). This value is approximately the hardness of the as-quenched martensite in the 0.45%C steel [10]. The 2.6mm thick joint has a higher hardness than the 5mm thick joint. Thus, the hardness increased with a reduction in the specimen thickness. This indicates that the cooling rate of the 2.6mm thick joint is higher than that of the 5mm thick joint.

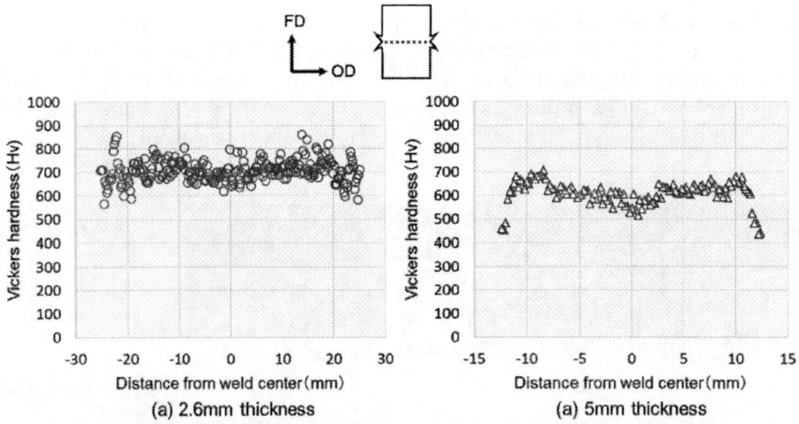

Fig. 7 Vickers hardness distributions along the oscillation direction: (a) 2.6mm thickness and (b) 5mm thickness

Conclusions

Medium carbon steel of 2.6mm thickness could be successfully welded by linear friction welding.
(1) The surface temperature decreased with a decrease in the frequency. This indicated that the friction heat input becomes lower with a decrease the sliding velocity.
(2) The maximum temperature of the joint surface decreased with a reduction in the specimen thickness.
(3) The thick plate joint welded at 50Hz has a high hardness due to the martensite formation at all the locations and the hardness increased with a reduction in the specimen thickness. This indicates that the cooling rate of the 2.6mm thick joint is higher than that of the 5mm thick joint.

Acknowledgments

This study was partly supported by the New Energy and Industrial Technology Development Organization (NEDO) under the "Innovation Structural Materials Project (Future Pioneering Projects)".

References

[1] A. Matsunawa, J. D. Kim, N. Seto, M. Mizutani and S. Katayama "Dynamics of keyhole and molten pool in laser welding" *Journal of Laser Applications*, **10**, 247-254 (1998).
[2] J. G. Garland and P. R. Kirkwood "Towards improved submerged arc weld metal" *Metal Construction*, **7**, 275 (1975).
[3] I. Bhamji, M. Preuss, P. L. Threadgill and A. C. Addison "Solid state joining of metals by linear friction welding: a literature review" *Materials Science and Technology*, **27**, 2-12 (2011).
[4] A. Chamanfar, M. Jahazi and J. Cormier "A Review on Inertia and Linear Friction Welding of Ni-Based Supwralloys" *Metallugical and Materials Transactions A*, **46A**, 1639-1669 (2015).
[5] P. Wanjara and M. Jahazi "Linear friction welding of Ti–6Al–4V: processing, microstructure, and mechanical-property inter-relationships" *Metallugical and Materials Transactions A*, **36A**, 2149–2164 (2005).
[6] E. Dalgaard, P. Wanjara, J. Gholipour, X. Cao and J. J. Jonas "Linear friction welding of a near-β titanium alloy" *Acta Materialia*, **60**, 770-780 (2012).
[7] M. Karadge, M. Preuss, P. J. Withers and S. Bray: 'Importance of crystal orientation in linear friction joining of single crystal to polycrystalline nickel-based superalloys', *Mater. Sci. Eng. A*, **A491**, 446–453 (2008).
[8] T. J. Ma, X. Chen, W. Y. Li, X. W. Yang, Y. Zhang and S. Q. Yang "Microstructure and mechanical property of linear friction welded nickel-based superalloy joint" *Materials and Design*, **89**, 85-93 (2016).
[9] R. Kuroiwa: "Linear Friction Welding of Medium Carbon Steel" *Proc 11th International Symposium on Friction Stir Welding*, Cambridge, Cambridgeshire, May 2016.
[10] G. Krauss "Martensite in steel: strength and structure", Materials Science and Engineering A, 273-275, 40-57 (1999).

Increase in Fatigue Strength and Charpy Absorbed Energies of TIG-Welded SS400 Steel Plates by Friction-Stir-Processing Grain-Refinement

Kazuhiro Ito, Tatsuya Okuda, Hiroki Izumi, Makoto Takahashi, Kazuyuki Kohama, Hajime Yamamoto, Hidetoshi Fujii
Joining and Welding Research Institute, Osaka University, Ibaraki, Osaka, 567-0047, Japan
ito@jwri.osaka-u.ac.jp, +81-6-6879-8659

Abstract
To improve fatigue strength of high-tensile-steel welds, friction-stir-processing (FSP), which is based on the principle of friction stir welding, is an effective grain-refinement technique for the topmost layer of the welds. In this study, its effects on fatigue strength/life of the tungsten inert gas (TIG)-welded SS400 plates and Charpy absorbed energy of the FSP-modified microstructure were investigated. FSP produced microstructure consisting of ultrafine eqiaxial ferrite grains beneath the TIG-welded surface. A cross-sectional TEM image indicates that cementite phases were segregated at their grain boundary, which formed about 2 nm-thick thin layers in a certain FSP condition. The microstructure increased three-point-bending fatigue strength/life in comparison with those of TIG-welded specimens. Their Charpy absorbed energies exhibited about 110 J in the temperature range between -110°C and 40°C, which is higher than the values of TIG-welded specimens. Grain refinement provided by FSP with lower heat input could further decrease brittle-ductile transition temperature. FSP modification of a topmost area of the welds is useful to increase fracture toughness as well as fatigue strength/life of the welds.

Introduction
Hear-affected zone (HAZ) ordinary degrades fatigue strength and toughness of welded joints. Fatigue strength of steels increases with increasing their yield stress, but that of their welds does not [1]. To improve fatigue strength of their welds, various surface-treatment techniques such as shot peening [2], hammer peening [3, 4], ultrasonic peening [5, 6] and laser peening [7] have been studied and some of them were conducted on the welds. The techniques provide compressive residual stress beneath the surface of the welds due to compressive plastic deformation at the surface. However, it may not contribute to the improvement of their toughness. On the other hand, friction-stir-processing (FSP), which is based on the principle of friction-stir-welding, is an effective surface-microstructural modification technique to increase fatigue strength/life of weld joints, as are re-melting, hammering, and blasting for aluminum alloys [8, 9]. FSP can also be one of techniques to increase fatigue strength/life of steel welds [10]. In this study, its effects on fatigue strength/life of the tungsten inert gas (TIG)-welded SS400 plates and Charpy absorbed energy of the FSP-modified microstructure were investigated.

Effects of FSP on Fatigue Strength
Fatigue strength of welds with FSP modified surfaces
FSP was performed on TIG-welded beads with small tungsten carbide tools containing a 12-mm-diameter shoulder and a 0.8-mm-long and 4-mm-diameter probe. FSP tools were operated in Ar atmosphere in applied-stress control with counter-clockwise rotation at 400 rpm and 140 mm/min with 14.7 kN (hereafter FSP(H)) and at 200 rpm and 400 mm/min with 24.5 kN (hereafter FSP(L)) in welding direction. The heat input produced by FSP(L) was lower than that by FSP(H). Figs. 1(a)-1(f) show surface and cross-sectional optical images of TIG-welded SS400 steel plates without FSP, with FSP(H), and with FSP(L). The weld metal depth and width were about 2.5 mm and 6 mm, respectively (Fig. 1(b)). HAZ was about 8 mm wide and reached the bottom of the steel plates. The FSP region was about 1 mm deep and 9 mm wide, with which the weld metal and HAZ were covered (Fig. 1(d) and 1(f)). Microstructural modification by FSP was limited to the topmost of the welds.

Fig. 1(g) shows results of three-point bending fatigue test at room temperature (RT) for the weld specimens with and without FSP. The tests were performed with a sinusoidal waveform at 20 Hz as a function of maximum applied stress (σ_{max}) between 600 and 800 MPa with a stress ratio of 0.1. The number of cycles to failure is plotted as a function of applied stress amplitude ((σ_{max} - σ_{min})/2). The number of cycles to failure of TIG specimens exhibited log-linear increase with decreasing stress amplitude (broken line). TIG+FSP(H) specimens also exhibited log-linear increase of the number of cycles to failure in the stress amplitude range between 270 and 315

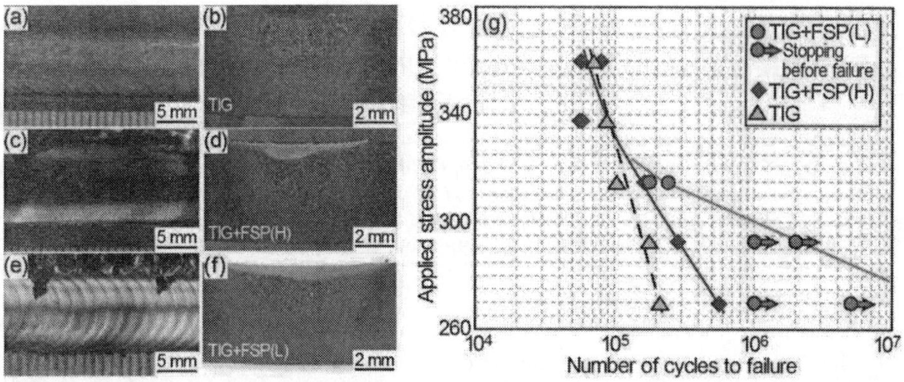

Fig. 1 Surface and cross-sectional optical images of TIG-welded SS400 steel plates (a),(b) without FSP, (c),(d) with FSP(H), and (e),(f) with FSP(L). (g) S-N curves (stress amplitude v.s. number of cycles to failure) of three-point bending fatigue at RT for the TIG-welded steel plates with and without FSP.

MPa (♦, solid blue line), and its increase rate was higher than that of the TIG specimens. TIG+FSP(L) specimens exhibited similar feature and its increase rate was much higher. The TIG+FSP(H) and TIG+FSP(L) specimens exhibited 5.6×10^5 and more than 5.0×10^6 cycles to failure at an applied stress amplitude of 270 MPa, respectively. Those are about 170% and more than 2000% increase in comparison with the number of cycles to failure for the TIG specimens. The bending displacements of the TIG+FSP(H) and TIG+FSP(L) specimens were remarkably smaller than those of TIG specimens at all cycle numbers and at both the minimum and maximum applied stresses. Similarly, the displacement of the TIG+FSP(L) specimens at the maximum applied stress was obviously smaller than that of the TIG+FSP(H) specimens.

FSP modified microstructure related fatigue of welds

Microstructures of the specimens before the three-point bending fatigue tests were observed by scanning electron microscopy (SEM) equipped with an electron backscattering diffraction (EBSD) pattern detector, and transmission electron microscopy (TEM). Figs. 2(a)-2(c) show portions of cross-sectional SEM/EBSD images beneath the surface of the TIG, TIG+FSP(H), and TIG+FSP(L) specimens. FSP(H) and FSP(L) provided microstructure consisting of ultrafine equiaxial ferrite grains with average grain sizes of about 2.6 μm (Fig. 2(b)) and 1.9 μm (Fig. 2(c)), respectively. The microstructure modified by FSP was observed in the region between the surface and about 1 mm below the surface. The as-TIG-welded coarse grains were observed in the region below the FSP modified microstructure, which is similar to the SEM/EBSD image obtained beneath the surface of the TIG specimen (Fig. 2(a)). Increasing three-point bending fatigue resistance of the TIG+FSP(H) and TIG+FSP(L) specimens can be explained mainly by increasing stiffness of a surface portion due to microstructure consisting of ultrafine grains produced by FSP. Decreasing FSP heat input promoted further refinement of surface microstructure, leading significant increase of their fatigue strength/life.

Fig. 2(d) shows a portion of transverse-sectional TEM bright-field image beneath the surface of the TIG+FSP(L) specimen. Ferrite-grain boundaries were not shown in sharp outline or contrast. The selected area diffraction (SAD) image obtained from the grain boundary indicated presence of a single-crystalline cementite phase (Fig.

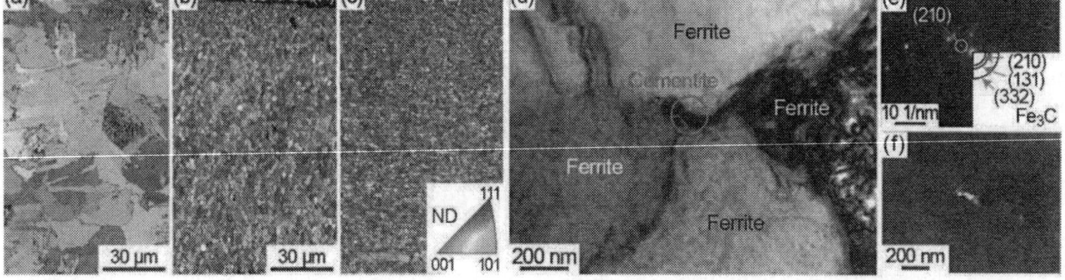

Fig. 2 Cross-sectional SEM-EBSD images of (a) TIG, (b) TIG+FSP(H), and (c) TIG+FSP(L) specimens. (d) A TEM bright-field image in a stir zone of TIG+FSP(L). (e) A SAD image obtained from the ferrite-grain boundary (a red circle in (d)). (f) A dark-field image taken using the diffraction spot of (210) (a white circle in (e)).

2(d)). It existed at the grain boundary in a leaf shape with thickness of several ten nm as shown in Fig. 2(f). Similar TEM analysis provided that ferrite-grain boundaries were fully covered with several cementite leafs. The microstructure consisting of ultrafine ferrite grains covered with a cementite layer increased fatigue strength/life. FSP(H) provided similar microstructure consisting of ultrafine ferrite grains and cementite phases at ferrite-grain boundaries. Its layer thickness became relatively thicker with increasing heat input. In addition, some of them segregated in grain-boundary triple-points into mass. Grain refinement as well as such the distribution of a cementite layer contributes increasing fatigue strength/life of TIG welds with FSP, and its difference between with FSP(L) and with FSP(H) can be explained by difference of ferrite-grain sizes and cementite distribution.

Effects of FSP on Charpy Absorbed Energy

The microstructure consisting of ultrafine ferrite grains covered with a cementite layer is not on the Fe-C binary phase diagram and related diagrams. This can be a FSP-induced new microstructure, which is not simply explained by dynamic recrystallization. Noted that cementite is generally believed to be a brittle phase, and its presence at the grain boundary is expected to degrade ferrite-grain-boundary toughness. Even though the microstructure formed at weld surfaces increased its stiffness, it is never welcome for the microstructure to degrade its toughness. Thus, to clarify its toughness, the Charpy impact tests using specimens with the FSP modified microstructure were conducted.

Charpy absorbed energy of FSP-modified welds

The 2.5 mm-thick sub-size Charpy-impact-testing specimens were prepared perpendicular to the weld beads with and without FSP, and their bar-type samples had a size of $10 \times 55 \times 2.5$ mm^3, which is based on JIS Z2242. The V notch with 2 mm in depth was produced in the center of the FSP-modified-microstructure region with a size of $10 \times$ about 12×2.5 mm^2. The Charpy impact tests were conducted at 196°C and in the temperature range between -110°C and 40°C. Fig. 3(a) shows temperature dependence of Charpy absorbed energy of TIG, TIG+FSP(H), and TIG+FSP(L) specimens. The plotted values of Charpy absorbed energy were simply four times as much as the values obtained using the 2.5 mm-thick sub-size specimens. The values of the TIG-welded specimens with FSP were much larger than those of the TIG specimens in all the temperature range. The TIG+FSP(H) and TIG+FSP(L) specimens exhibited about 120 J and 110 J, respectively, in the temperature range between -110°C and 40°C, in comparison with about 80 J in the temperature range between -40°C and 40°C for the TIG specimens. An inset in Fig. 3(a) shows an optical image of the TIG+FSP(L) specimen after a Charpy impact test at -110°C. It shows M-shaped fracture surface at the end portion of the V notch at a red square, indicating that FSP(L)-modified regions were ductile enough for significant plastic deformation and TIG-welded regions remained partly at the center groove were brittle. The value at -196°C of the TIG+FSP(L) specimens exhibited about 50 J, while that of the TIG and TIG+FSP(H) specimens exhibited about 0 J. This indicates that a part of the FSP(L)-modified microstructure still remained ductile at -196°C. Thus, FSP can be concluded to be a useful technique to increase both fatigue strength/life and fracture toughness of the welds.

Fracture surface of FSP-modified welds

Figs. 3(b)-3(c) show SEM images of a fracture surface after Charpy impact tests at -110°C for the TIG+FSP(L) and TIG+FSP(H) specimens, respectively. The M-shaped fracture surfaces were observed at the end portion of the V notch for the TIG+FSP(L) specimen, while uniform fracture surface for the TIG+FSP(H) specimen. This is

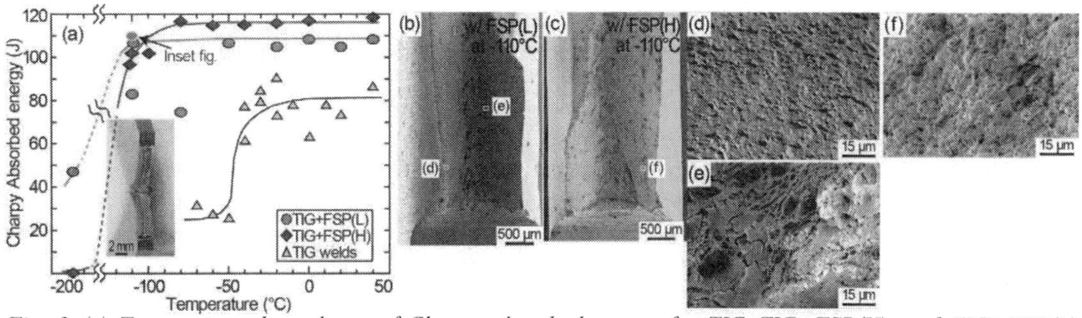

Fig. 3 (a) Temperature dependence of Charpy absorbed energy for TIG, TIG+FSP(H), and TIG+FSP(L) specimens and an inset image shows an optical image of the TIG+FSP(L) specimen fractured after a Charpy impact test at -110°C. SEM images of a fracture surface after Charpy impact tests at -110°C for (b), (d), (c) TIG+FSP(L) and (c), (f) TIG+FSP(H) specimen.

because FSP(L) remained TIG-welds at the specimen center, but FSP(H) provided modified regions at any thickness. In enlarged SEM images of square areas in Figs. 3(b) and 3(c), dimple fracture was observed in Figs 3(d) and 3(f), while brittle fracture was observed beside the dimple fracture in Fig. 3(e). The dimple fracture indicates that FSP(L) and FSP(H) modified microstructure remains ductile at -110°C, in contrast the remained TIG-welded microstructure at the center groove in the TIG+FSP(L) specimen becomes brittle at -110°C. The dimple sizes in the TIG+ FSP(L) specimens tended to be smaller than those in the TIG+FSP(H) specimen due to average-grain-size difference. Although smaller dimples are expected to increase their Charpy absorbed energy, the Charpy absorbed energy of the TIG+FSP(L) specimens was lower than those of the TIG+FSP(H) specimens above -100°C. This is attributed the remained TIG-welded microstructure at the specimen center in the TIG+FSP(L) specimens. A part of fracture surface of the TIG+FSP(L) specimen after Charpy impact tests at -196°C shows dimple fracture similar to the case at -110°C. This indicates that FSP(L) modified microstructure remains ductile at -196°C. In contrast, uniform brittle fracture surface with river patterns was observed in the TIG+FSP(H) specimen after Charpy impact tests at -196°C. Further grain refinement provided by FSP(L) with lower heat input than FSP(H) can decrease brittle-ductile transition temperature. A layer of cementite phase at the ultrafine ferrite-grain boundaries did not degrade the Charpy absorbed energy. This suggests that FSP-modification for a topmost area of the welds is useful to increase fracture toughness as well as fatigue strength/life of the welds.

Conclusions

Surface-microstructural modification by FSP is effective and practical for increasing fatigue strength and Charpy absorbed energy of the TIG-welded SS400 plates. The findings in this study are three-fold:

1. FSP(L) and FSP(H) produced microstructure beneath the TIG-welded surface, consisting of ultrafine ferrite grains with average grain sizes of about 1.9 μm and 2.6 μm respectively. Decreasing heat input in FSP provided further grain refinement, and the ferrite grains were surrounded with layers of cementite;
2. TIG+FSP(L) and TIG+FSP(H) specimens exhibited more than 5.0×10^6 cycles and 5.6×10^5 cycles to failure at an applied stress amplitude of 270 MPa, respectively. Those are more than 2000% and about 170% increase in comparison with the number of cycles to failure for the TIG specimens;
3. Charpy absorbed energies for the TIG+FSP(L) and TIG+FSP(H) specimens exhibited about 110 J in the temperature range between -110°C and 40°C, in comparison with about 80 J in the temperature range between -40°C and 40°C for the TIG specimens. The value at -196°C of the TIG+FSP(L) specimen exhibited about 50 J, in contrast that of the TIG+FSP(H) specimen exhibited about 0 J.

References

[1] K. Anami, *et al*: "Fatigue Strength of Welded Joints Made of High-Strength Steels," *Progr. Struct. Eng. Mater.*, Vol. 3, No. 1 (2001), pp.86-94.

[2] N.A.C. Lah, *et al*: "The Effect of Controlled Shot Peening on Fusion Welded Joints," *Mater. Design*, Vol. 31, No. 1 (2010), pp.312-324.

[3] C.M. Branco, *et al*: "Fatigue Behaviour of Welded Joints with Cracks, Repaired by Hammer Peening", *Fatigue Fract. Engng. Mater. Struct.*, Vol. 27, No. 9 (2004), pp.785-798.

[4] M. Tai, *et al*: " Fatigue Strength Improvement by Hammer Peening Treatment—Verification from Plastic Deformation, Residual Stress, and Fatigue Crack Propagation Rate", *Weld. World*, Vol. 58, No. 3 (2014), pp.307-318.

[5] S. Roy, *et al*: "Fatigue Resistance of Welded Details Enhanced by Ultrasonic Impact Treatment (UIT)," *Int. J. Fatigue*, Vol. 25, No. 9-11 (2003), pp.1239-1247.

[6] W. Ting, *et al*: "Discussion on Fatigue Design of Welded Joints Enhanced by Ultrasonic Peening Treatment (UPT)," *Int. J. Fatigue*, Vol. 31, No. 4 (2009), pp.644-650.

[7] Y. Sakino, *et al*: "Major Factor Causing Improvements in Fatigue Strength of Butt Welded Steel Joints after Laser Peening without Coating," *Sci. Technol. Weld. Joi.*, Vol. 17, No. 5 (2012), pp.402-407.

[8] C.B. Fuller, *et al*: "The Effect of Friction Stir Processing on 5083-H321/5356 Al Arc Welds: Microstructural and Mechanical Analysis," *Metall. Mater. Trans.*, Vol. A37, No. 12 (2006), pp.3605-3615.

[9] J. da Silva, *et al*: "Fatigue Behavior of AA6082-T6 Welded Butt Joints Improved by Friction Stir Processing," *Mater. Design*, Vol. 51 (2013), pp. 315-322.

[10] K. Ito, *et al*: "Increase of Bending Fatigue Resistance for Tungsten Inert Gas Welded SS400 Steel Plates Using Friction Stir Processing," *Mater. Design*, Vol. 61 (2014), pp. 275-280.

Effect of Alloying Elements on Microstructure of Friction Stir Welded Thick Structural Steel Plate

Muneo Matsushita, Rinsei Ikeda

JFE Steel Corporation, Japan

Friction stir welding (FSW) was performed with single-sided one-pass butt welding using 12 mm thick structural steel plates of 400-500 N mm-2 grade of tensile strength. The tested plates includes the plain carbon-manganese steel and the micro-alloyed thermo-mechanically processed steel. The microstructures and mechanical properties such as Vickers hardness, tensile strength, and Charpy impact toughness of the weld joints were investigated. The investigation exhibited the inhomogeneous feature of microstructure and variation of toughness within the stir zone (SZ). In the case of plain carbon-manganese steel, Charpy impact values at -40 Celsius in the advancing side (AS) of SZ obviously tended to be higher, showing more than 50% of shear area (SA), than in the retreating side (RS), with less than 50% of SA. It was verified that this discrepancy of toughness within the SZ was attributed to the microstructural difference. In fact, the higher fraction of coarse ferrite structure was observed in the RS than in the AS. On the other hand, the weld of micro-alloyed steel with titanium addition exhibited the high values in Charpy impact test at -40 Celsius, more than 50% of SA, in both the AS and RS of SZ and its microstructure mainly consisted of fine ferrite structure throughout the SZ. It has been reported that the effect of grain refinement of ferrite structure in the heat affected zone of steel weld joints by means of the addition of precipitation forming elements such as titanium. In this study, the effect of those elements on the formation of ferrite microstructure in the SZ is investigated concerning the unique thermo-mechanical history of friction stir welding, unlike the conventional welding processes.

Properties of Multilayer Friction Stir Welding Joint in SUS316L Clad Steel

Yuta Honma*, Rinzo Kayano*, Yutaka S. Sato**, Hiroyuki Kokawa**
**The Japan Steel Works Ltd, Hokkaido Japan*
E-mail yuta_honma@jsw.co.jp, Tel +81-143-22-0392 fax +81-143-22-4180
***Tohoku University, Miyagi Japan*

Abstract
In clad pipe that has been used for natural gas transport, good toughness at low temperatures is required for safety reasons. The good toughness is needed not only in the backing steel but also in the welded parts. The advantages of friction stir welded (FSW) joints are formation of the fine grains and very little dilution owing to the solid-state process with intensive plastic deformation. However, applying FSW to thick plate might require two-passes welding from both the top and bottom surfaces because of limitations of the tool shape and the machine capacity. In general, there are concerns about decreasing the mechanical properties of the joints due to the thermal effect of a second pass. Therefore, various properties of multilayer FSW joints in SUS316L clad steel were examined in this study. Our results showed that the corrosion resistance of FSW joints in the SUS316L steel decreased, but that the toughness of API-X65 steel part of the FSW joints was better than that of the arc-welded joints.

Introduction
As the global economy expands, demand for natural gas is expected to keep increasing. Transporting natural gas from wellheads to pre-treatment equipment through pipelines requires high corrosion resistance in pipelines because natural gas includes corrosive gases, such as methane, sulphur compounds, chlorine compounds and carbon dioxide. Therefore, the clad steel consisting of the backing steel and cladding material has been used for the line pipe because it provides the sufficient strength of the backing steel and the superior corrosion resistance of cladding material. In the clad steel, API 5L grade pipeline steel has been widely used as the backing steel, and SUS 316L steel or a Ni based alloy has been used as the cladding material in a strongly corrosive environment. Recently, there is a demand for the backing steel to increase the strength while also to reduce the weight and cost of pipe lines for long distance transportation. Moreover, the good toughness is also required in the clad steel for safety reasons. These properties are needed not only in the base metal of the backing steel but also in the weld parts, especially in heat-affected zone (HAZ). Methods to improve the toughness in the HAZ have been examined in the previous studies[(1),(2)], which suggest presence of fine precipitates, such as TiN, to suppress the grain coarsening in the HAZ. However, it is difficult to control the precipitation phenomena during the manufacturing process, and also the precipitates might be dissolved by welding. Thus, it is still a great challenge to ensure stable toughness in the HAZ.

Friction stir welding (FSW) is a solid-state joining process invented by The Welding Institute (TWI) in the UK in 1991. It is well known that the stir zone (SZ) consists of fine equiaxed recrystallized grains. For various reasons, FSW is expected to provide a joint having good toughness. However, the application of FSW to thick steel plates is restricted because of limitation of the tool shape and the machine capacity. In the joints of clad steel, moreover, mixing of the backing steel and cladding material needs to be essentially prevented. In general, there are concerns about decreasing the mechanical properties of the weld joints owing to thermal effect of the second pass. However, there have been few reports on mechanical properties of the multilayer FSW joints of the clad steel. In this study, the mechanical properties and corrosion resistance of the multilayer FSW joints of SUS316L clad steel were examined.

Experimental procedure
The workpieces used in this study were a 7.0 mm-thick API 5L-X65 Grade pipeline steel and a clad plate consisting of API 5L-X65 pipeline steel with SUS316L cladding. The pipeline steel without cladding material was used to examine mechanical properties of the backing steel in the FSW joints, while the clad plate was used to evaluate the corrosion resistance of the cladding material in the FSW joints.

These steel plates were friction-stir-welded using a scrolled shoulder step spiral probe tool made of W-Re/cBN composite material in the double-sided bead-on-plate configuration. The probe length of the tool was 4 mm. The rotation speed varied from 100 rpm to 300 rpm at a constant travel speed of 60 mm/min, as shown in Table 1. FSW was performed from the backing steel first, and then from the cladding material.

Optical metallography was conducted on a cross section to observe the defects and the microstructures of the welded joints. Transverse tensile testing of the joint was carried out at room temperature. The specimens, whose dimension was 5×1×0.34 mm^t, were cut from the welded joints so as that the border of the SZ was located at the center of the gauge part of the tensile specimen, as shown Fig.1. The constant crosshead speed of 0.5 mm/min was used until fracture occurred. A Charpy impact test was conducted using a sub-sized specimen (5×10×50mm, 2 mm V notch). A schematic illustration of the specimen locations is shown in Fig. 2. The notch locations were the center of the SZ and a region showing OB + 1 mm. OB was defined as the overlapping boundary of multilayer passes in this study. In order to obtain a better understanding of effect of FSW on toughness, Charpy impact test was also conducted using a joint produced by submerged arc welding (SAW). Vickers hardness measurements were carried out at a load of 19.6 Nf. Hardness profiles were measured at the center of the SZ in the thickness direction. Moreover, scanning electron microscopy (SEM) and electron backscatter diffraction (EBSD) method were also employed. A specimen for SEM was cut perpendicular to the welding direction and etched in a 2 vol% nitric acid + 98 vol% ethanol solution.

Table 1. Welding parameters used

Tool rotational speed (rpm)	Travel speed (mm/min)	Tool tilt angle (°)
100~300	60	0

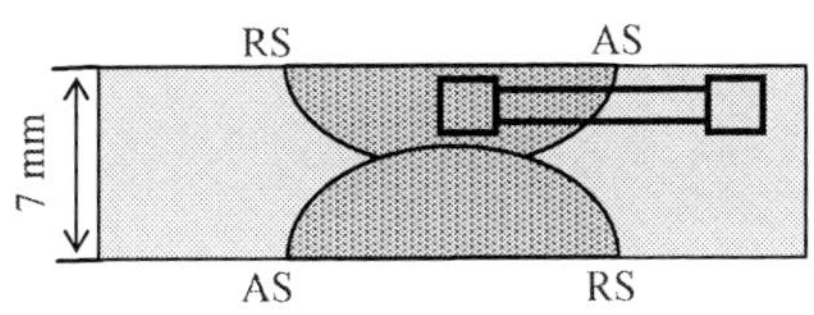

Fig. 1 Schematic illustration of tensile specimen location

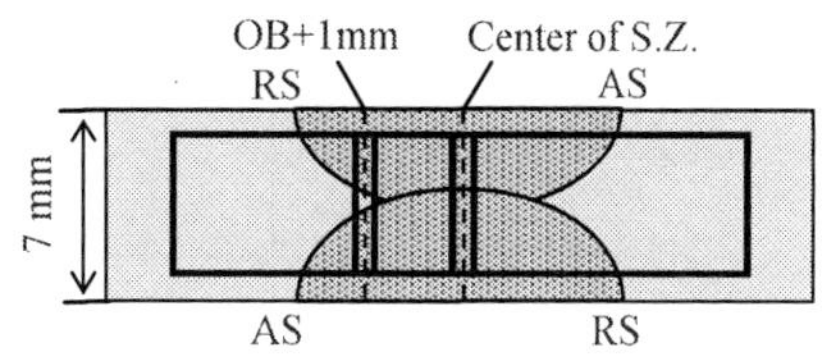

Fig. 2 Notch location of sub sized Charpy impact specimen

Pitting testing based on ASTM G 48 Method A and intergranular corrosion testing based on ASTM G28 Method A were conducted to examine the corrosion resistance of the cladding material in the welded joint. The pitting test solution used was a 6 % $FeCl_3$ solution. The test temperature was 5 °C maintained to an accuracy of ±1 °C and the immersion time was 72 h. The intergranular corrosion testing was performed for 120 h in a boiled 50 % H_2SO_4 + $Fe_2(SO_4)_3$ solution. The intergranular corrosion resistance was evaluated by the corrosion rate calculated from the weight reduction during the immersion. The sampling locations were at the center of the SZ, and the borders of the SZ at the advancing side (AS) and the retreating side (RS).

Result and Discussion

Mechanical properties of Multilayer FSW Joint in API 5L-X65 Grade steel

The cross-sectional macrostructures produced at rotational speeds of 100 rpm and 200 rpm are shown in Fig. 3. At these parameters, FSW successfully produced the defect-free multilayer joints. Generally, the heat input increases with increasing rotation speed, resulting in increasing total width of the thermo-mechanically affected zone (TMAZ) and HAZ. The results of the tensile testing of the welded joint are shown in Table 2. Both specimens were fractured in the base metal region because FSW could produce an overmatched joint. The results of the Charpy impact testing of the specimens with a notch at the center of the SZ are shown in Fig. 4. The results of the base metal are also shown in this figure for comparison. The fracture appearance transition temperature (FATT) decreased with decreasing rotation speed, and the toughness of the welded joint produced at a rotational speed of 100 rpm was superior to that of the base metal. These results suggest that FSW resulted in good tensile and impact properties even at a rotational speed of 200 rpm, and that the thermal effect of the second pass on the toughness is negligible in the FSW joint.

Effect of rotational speed on hardness and grain structure of the SZ was examined. The hardness distributions at the center of the SZ in the thickness direction are shown in Fig.5. The hardness increases with increasing rotational speed. SEM images of the SZs are shown in Fig. 6. The number of precipitates in the SZ increased with increasing rotational speed. The stir zone had the smaller grain structure than the base metal by intensive plastic deformation of FSW. The average grain sizes of the stir zones were 4.6 μm and 7.1 μm at the rotational speeds of

100 rpm and 200 rpm, respectively. The grain size of the stir zone increases with increasing rotation speed, being in agreement with the total width of the TMAZ and HAZ, as previously stated. It is generally known that the grain coarsening decreases toughness. These results imply that the reduction of toughness in the SZ produced at rotational speed 200 rpm arises from the hardness increasing and grain coarsening.

In order to examine effect of FSW on toughness of the TMAZ and HAZ, a Charpy impact test with a notch at the OB + 1 mm region was conducted. The results are shown in Fig. 7. The toughness of the HAZ of the FSW joint produced at a rotational speed of 200 rpm was roughly the same as that of the SAW joint, whereas the toughness was clearly improved in the FSW joint produced at the rotational speed of 100 rpm. This result shows that FSW can produce the TMAZ and HAZ with the good toughness.

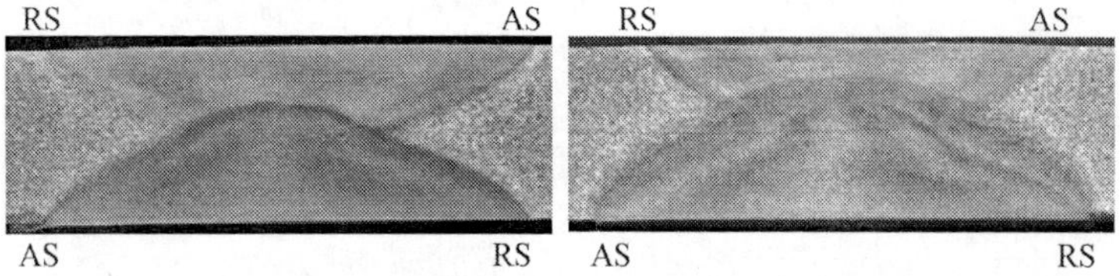

(a) Rotational speed: 100rpm *(b) Rotational speed: 200rpm*

Fig.3 Macroscopic overviews of the welded joints

Table 2 Results of tensile testing of the welded joint

T.P.	Rotationnal speed (rpm)	T.S. (MPa)	R.A. (%)	Rupture Position
1	100	590	67	BM
2	200	587	61	BM
BM	-	587	70	-

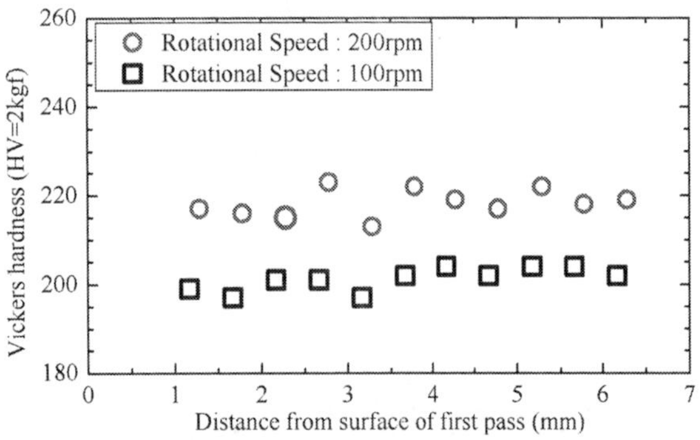

Fig. 5 Vickers hardness distributions of center line of SZ

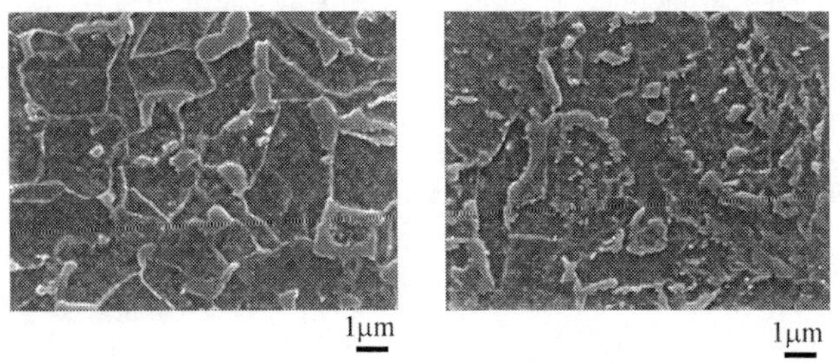

(a) Rotational speed: 100rpm *(b) Rotational speed: 200rpm*

Fig.6 SEM images of center of SZ

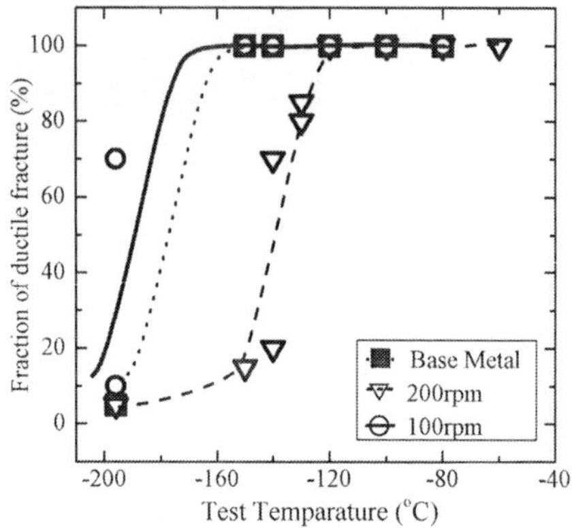

Fig. 4 Effect of rotation speed on results of Charpy impact testing (Notch location : center of SZ)

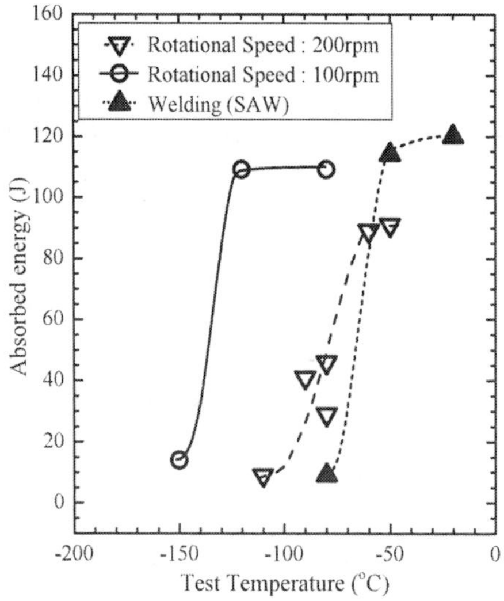

Fig. 7 Charpy impact test results of HAZ (Notch location : OB+1mm)

Properties of FSW Joint in SUS316L clad steel

Cross-sectional overviews of the welded joint produced at a rotational speed of 150 rpm and 300 rpm are shown in Fig. 8. Even on the clad steel, the good joints without defects were obtained. The backing steel was slightly mixed with the cladding material by stirring. EBSD analysis revealed that fine grains were produced both in SUS316L cladding as well as in API 5L-X65 backing steel by FSW, and that the grain size increased with increasing rotation speed.

The results of the pitting test and intergranular corrosion test using specimens sampled from the center of the SZ are shown in Table 3. Even though there are no pits in the base metal of SUS316L steel, pitting occurred in the SZ at 5 °C. The intergranular corrosion testing showed that the corrosion rate of the SZ was faster than that of the base metal of SUS316L steel. Furthermore, the corrosion rate increased with increasing rotation speed. The relationship between the corrosion rate and sampling position in the specimen produced at a rotational speed of 150 rpm is shown in Fig. 8. The corrosion rate is found to be higher at the AS than at the SZ and RS. This result shows that the welded joint exhibits the inferior corrosion resistance to the base metal of SUS316L steel. The reason for the decreased corrosion resistance of the welded joints might be related to changes in the precipitation behavior, such as Cr carbonitride, due to the high temperatures and high strains during FSW. The Precise examination to clarify the reason for the reduction of the corrosion resistance is needed in the future. Based on these results, this study suggests that FSW should be done at low heat-input parameters to effectively suppress the reduction of the good corrosions resistance of SUS316L cladding material in the welded joint.

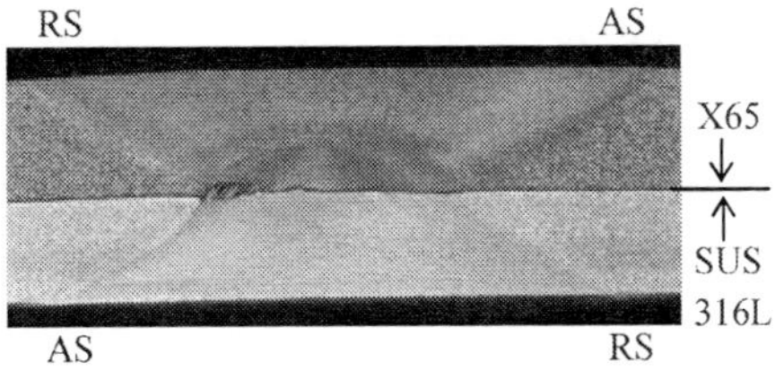

(a) Rotation speed: 150rpm

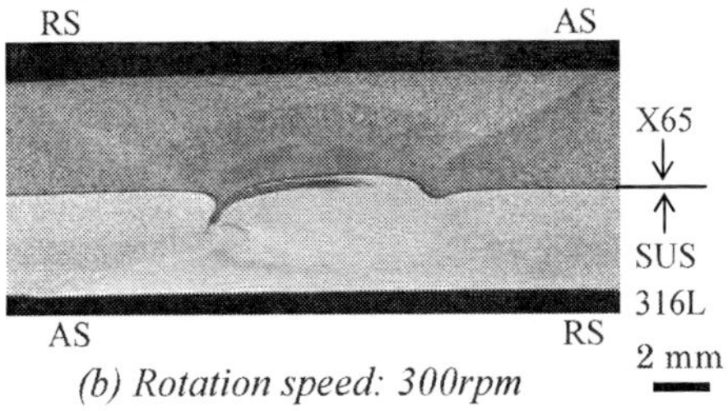

(b) Rotation speed: 300rpm

Fig. 7 Macro observation of cross section

Table 3 Corrosion test results of SZ

	Pitting test (Test temperature: 5 °C) Pitting	Intergranular corrosion test (mm/year)
SUS 316L	O (None)	0.89
300 rpm	×	1.58
150 rpm	×	1.13

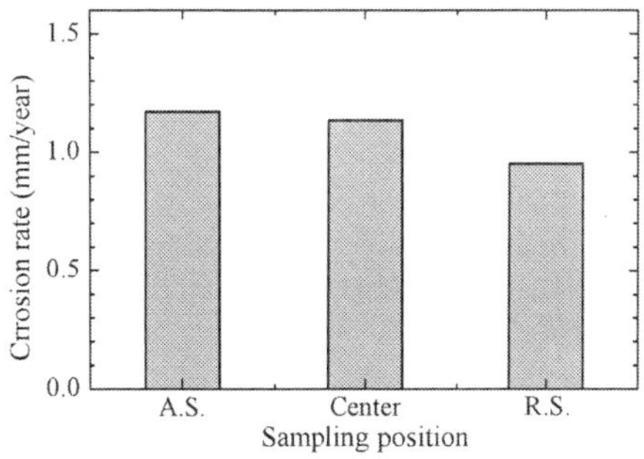

Fig. 8 Relationship between corrosion rate and sampling position

Conclusions

In this study, various properties of the multilayer FSW joints in SUS316L clad steel were examined. FSW successfully produced the defect-free multilayer joints in the clad steel consisting of API 5L-X65 pipeline steel with SUS316L cladding. FSW resulted in good tensile and impact properties, and that the thermal effect of the second pass on the toughness was negligible in the FSW joint. Moreover, the SUS316L cladding material of the FSW joint exhibited the lower corrosion resistance than the base material of SUS316L steel. However, FSW could produce the welded joint with sufficient corrosion resistance in SUS316L cladding material in the FSW joint at low heat-input parameters.

References

[1] T. Taira et al., HSLA Steels Technology and Applications, ASM, Metals Park (OH), 1983, p. 723-731

[2] S. Sato, et al.: "Development of High Toughness X70 Grade Clad Steel Pipe for Natural Gas Transportation Pipelines" The Japan Steel Works Technology Review, No.60 (2009), pp.41-47 (In Japanese)

Increase in Fatigue Strength of Butt-Welded High-Tensile Steel Joints by Friction Stir Processing

Hajime Yamamoto, Kazuhiro Ito, Makoto Takahashi, Kazuyuki Kohama, Hidetoshi Fujii
Joining and Welding Research Institute, Osaka University, Ibaraki, Osaka, Japan
h.yamamoto@jwri.osaka-u.ac.jp, +81-6-6879-8664

Abstract

Fatigue strength of the high-tensile steels increases with increasing their yield stress, but that of their welds does not. It is caused mainly by tensile residual stress and microstructure-related issues in heat-affected zone (HAZ) due to fusion welding. Friction stir processing (FSP), which is based on the principle of friction stir welding, is an effective surface-microstructural modification technique to increase fatigue strength of the welds. In this study, FSP was conducted on HAZ surfaces of the butt-welded joints for the 490 MPa and 780 MPa grade high-tensile steels (hereafter HT490 and HT780, respectively) to investigate whether or not FSP is an effective technique to increase their tensile fatigue strength. FSP for the butt-welded HT490 and HT780 joints was able to increase their fatigue strength. Failure occurred at a base metal and FSP-modified region in the butt-welded HT490 and HT780 joints with FSP, respectively, while at HAZ in the both grade as-welded steel joints. Thus, FSP can be concluded to be an effective technique to increase their tensile-fatigue strength. The increase can be attributed to grain refinements. The effect of FSP on increase in the tensile fatigue strength is higher in the butt-welded joints of HT490 than that of HT780. This can be explained by increase of high angle boundary and a hard cementite layer wrapping an ultrafine ferrite grain observed in the butt-welded joints of HT490 with FSP.

Introduction

For energy efficiency improvement in a means of transportation, it is essential to reduce the weight and to optimize the design of each system component with guaranteed necessary strength. Therefore, yield stress and tensile strength of the developed high-tensile steels go on increasing and are expected to continue to increase. Fatigue strength of the high-tensile steels increases with increasing their yield stress, but that of their welds does not [1]. It is caused mainly by tensile residual stress and microstructure-related issues in heat-affected zone (HAZ) due to fusion welding. To improve fatigue strength of their welds, surface-treatment techniques such as shot peening [2], hammer peening [3], ultrasonic peening [4-6] and laser peening [7] have been studied and some of them are conventionally conducted on the welds. The techniques provide compressive residual stress beneath the surface of the welds. However, they may not contribute to the improvement of microstructures in HAZ. On the other hand, friction stir processing (FSP), which is based on the principle of friction stir welding, is an effective surface-microstructural modification technique to increase fatigue strength of the welds, as are re-melting, hammering, and blasting for aluminum alloys [8, 9]. We also suggested that FSP can be one of techniques to increase fatigue strength of the tungsten inert gas (TIG) welded SS400 mild steel plates [10]. It was explained mainly by increasing stiffness of a surface portion due to microstructure consisting of ultrafine grains produced by FSP. In this study, FSP was conducted on HAZ surfaces of the but-welded joints for the 490 MPa and 780 MPa grade high-tensile steels (hereafter HT490 and HT780, respectively) to investigate whether or not FSP is an effective technique to increase their tensile fatigue strength.

Experimental procedures

The double-sided metal active gas (MAG) butt-welding was conducted in CO_2 shield gas on 4.5 mm-thick HT490 and HT780 plates by using MX-Z200 and DW-A80L filer metals, respectively. MAG welding was operated at 330 A with a head travel speed of 900 mm/min. After removing excess weld metals by grinder, the double-sided TIG welding was conducted on the weld beads, operating at 180 A with a head travel speed of 50 mm/min to provide tensile residual stress around the welds. FSP was performed on each fusion line in the same direction to modify microstructures of HAZ surface with a small tungsten carbide tool containing a 12 mm-diameter shoulder and a 0.8 mm-long and 4 mm-diameter probe in Ar atmosphere. The tool was operated in applied stress control with counter-clockwise rotation at 400 rpm and 140 mm/min with 14.7 kN in the welding direction. Specimen-surfaces after FSP were smoothed by manual polishing. Tensile fatigue tests were performed at room temperature (RT) with a sinusoidal waveform at 20 Hz as a function of maximum applied stress (σ_{max}) with a stress ratio of 0.1. Vickers hardness tests were performed beneath the specimen-surfaces with the applied load of 9.8 N and the loading time of 35 s. Microstructures of the butt-welded specimens with and without FSP were observed by optical microscopy, field emission scanning electron microscopy (FE-SEM) equipped with an electron backscattering diffraction (EBSD) pattern detector, and transmission electron microscopy (TEM).

Results and discussion

Results of tensile fatigue test

Fig. 1 shows results of tensile fatigue test for the but-welded HT490 and HT780 specimens with and without FSP. The number of cycles to failure is plotted as a function of applied stress amplitude ((σ_{max} - σ_{min})/2). In the both grade welded specimens with (●, solid line) and without FSP (▲, broken line), the number of cycles to failure exhibited log-linear increase with decreasing stress amplitude and fatigue failure did not occur below a certain stress amplitude. The stress amplitudes of specimens with FSP are higher than those without FSP at any number of cycles to failure. The stress amplitude at fatigue limit for the butt-welded HT490 specimens with and without FSP is estimated to be about 180 MPa and 145 MPa, respectively. Similarly, those for the butt-welded HT780 specimens with and without FSP are 195 MPa and 180 MPa, respectively. The increases due to FSP are about 35 MPa and 15 MPa for the butt-welded HT490 and HT780 specimens, respectively. Failure occurred at a base metal and FSP-modified region, respectively, in the butt-welded HT490 and HT 780 specimens with FSP, in contrast at HAZ in the both grade as-welded specimens. Thus, FSP can be concluded to be an effective technique to increase their tensile fatigue strength, while there is difference in their increase between the butt-welded HT490 and HT780 specimens with FSP. The former increase was larger than the latter increase.

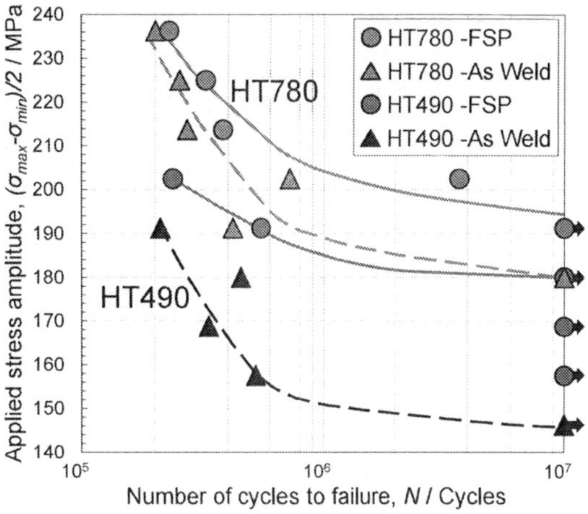

Fig. 1: S-N curves (stress amplitude v.s. number of cycles to failure) obtained in tensile fatigue tests at RT for the butt-welded HT490 and HT780 specimens with and without FSP.

Results of Vickers hardness test

Fig. 2 shows Vickers hardness distribution beneath the surface of the butt-welded *(a)* HT490 and *(b)* HT780 specimens with and without FSP. Hardness of HAZ near a base metal was lower than that of a base metal in the both grade as-welded specimens. Hardness of the weld metal for HT490 was higher than that of a base metal, but that for HT780 was higher near HAZ and lower in the center. On the other hand, hardness of the FSP-modified HAZ exhibited higher than that of HAZ without FSP. The hardness increase due to FSP for the butt-welded HT 490 and HT780 specimens are estimated to be about 55HV and 150HV, respectively.

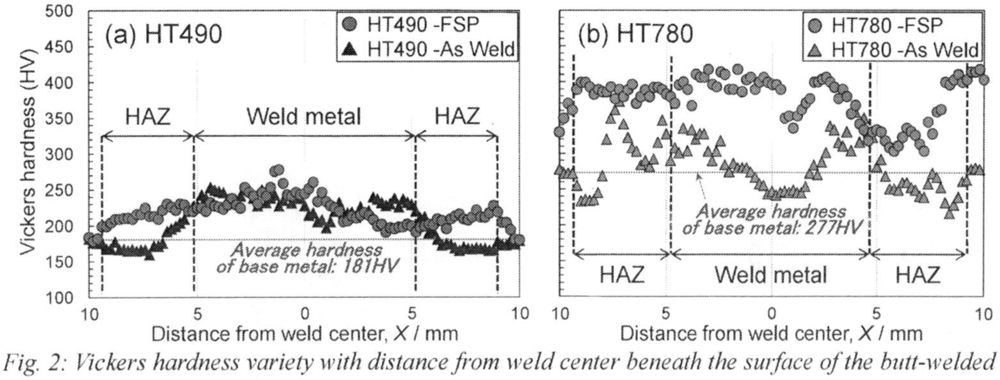

Fig. 2: Vickers hardness variety with distance from weld center beneath the surface of the butt-welded (a) HT490 and (b) HT780 specimens with and without FSP.

Microstructural evaluation

Fig. 3 shows cross-sectional FE-SEM/EBSD images beneath the coarse grain HAZ (CGHAZ) surface of the butt-welded HT490 and HT780 specimens with and without FSP. Upper bainite formation was observed in the butt-welded HT490 specimens (*Fig. 3(a)*). In contrast, FSP produced microstructures consisting of ultrafine equiaxial grains with average grain sizes of about 8.18 μm (*Fig. 3(c)*). Similarly, FSP produced microstructures consisting of ultrafine equiaxial grains with average grain sizes of about 4.00 μm (*Fig. 3(g)*), in contrast to lath martensite formation observed in the butt-welded HT780 specimens (*Fig. 3(e)*). The microstructures modified by FSP were observed in the region between the surface and about 1 mm below the surface. Thus, increase in HAZ hardness and tensile fatigue strength due to FSP for the both grade butt-welded steel joints can be attributed to grain refinement strengthening. Misorientation-axis distribution plots for *Fig. 3(a), 3(c), 3(e) and 3(g)* were shown in *Fig. 3(b), 3(d), 3(f) and 3(h)*, respectively. Yellow color as depicted "1" in the indicator indicates uniform existence probability, and warm and cold colors indicate increasing and decreasing existence probability from "1", respectively. Although there were lacking in uniformity in the both grade as-welded specimens, FSP decreased the lacking in uniformity in HAZ of the both grade welded specimens, and totally random crystal orientation was observed in HAZ of the butt-welded HT780 specimens after FSP. In addition, FSP dramatically increased total length of high angle boundaries (HAB) in the butt-welded HT490 specimens, while increase of total length of low angle boundaries (LAB) was obvious in the butt-welded HT780 specimens. HAB is effective for the resistance to fatigue crack propagation.

Fig. 4 shows transverse-sectional TEM bright-field images beneath the surface of the butt-welded *(a)* HT490 and *(d)* HT780 specimens with FSP. A small grain with a dark contrast was observed at the ferrite grain boundary in the butt-welded HT490 specimens with FSP (*Fig. 4(a)*). The selected area diffraction (SAD) image obtained from in a white circle in *Fig. 4(a)* indicated a single-crystalline diffraction pattern, which corresponds to a cementite phase (*Fig. 4(c)*). A dark-field image taken using the diffraction spot in a white circle in *Fig. 4(c)* is shown in *Fig. 4(b)*, where the grains with a white contrast indicated a cementite phase with the same crystal orientation (*Fig. 4(c)*). Similar observation found that ferrite-grain boundary was covered with several single-layer cementite grains. The cementite grains in a layer of about 200 nm (*Fig. 4(b)*) are thicker than those of about 20 nm obtained in FSP with the smallest heat input conducted in TIG-welded SS400 steels [11]. This suggests that increasing heat input in FSP promotes agglomeration of cementite along the ferrite grain boundary. Noted that increasing fatigue strength of the butt-welded HT490 joints can be attributed to a hard cementite layer wrapping an ultrafine ferrite grain as well as grain refinement. In contrast, the precipitation such as a carbide was not observed in FSP-modified microstructure of the butt-welded HT780 specimens. *Figs. 4(e) and 4(f)* show presence of a retained austenite phase, which has no specific orientation relationship with a martensite matrix. It indicated that a phase transformation did not occur in FSP-modified region due to small FSP heat input. Thus, the difference in FSP-modified microstructure between the both grade butt-welded steel joints would result in the difference of their fatigue strength increase.

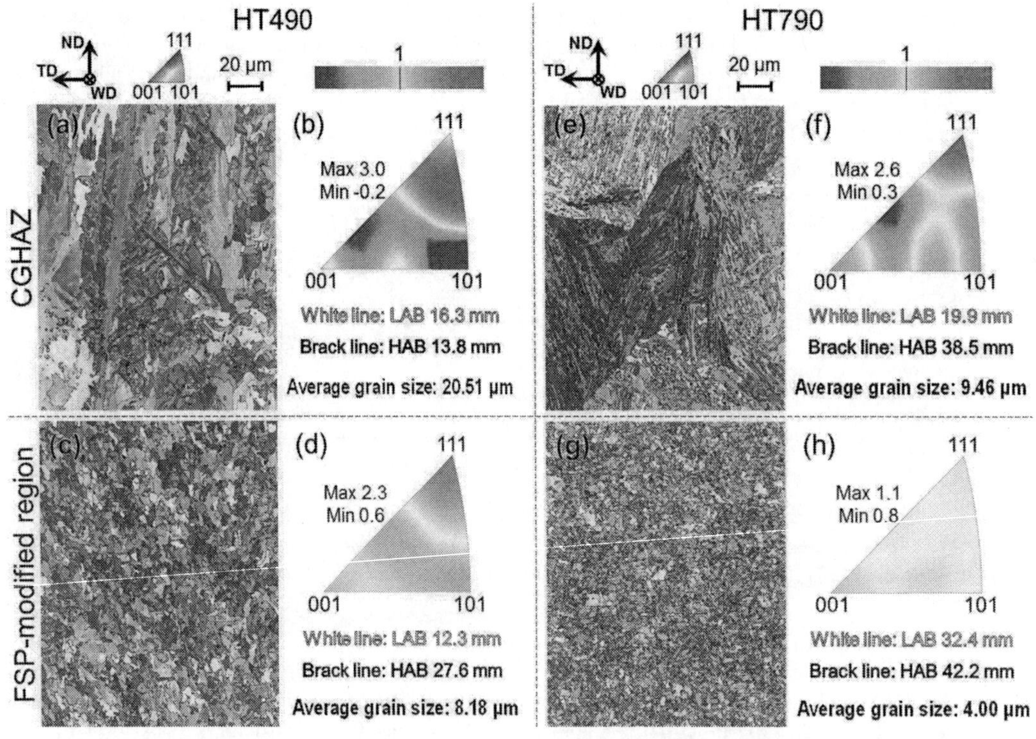

Fig. 3: Cross-sectional FE-SEM/EBSD images and their misorientation-axis distribution plots beneath CGHAZ-surface of the butt-welded HT490 and HT780 specimens with and without FSP.

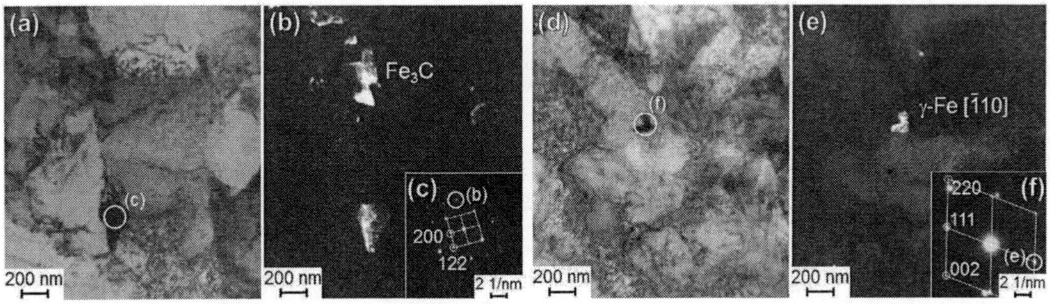

Fig. 4: Transverse-sectional TEM bright-field images beneath the surface of the butt-welded (a) HT490 and (d) HT780 specimens with FSP, and (c), (f) SAD images obtained in a white circle in (a) and (d), respectively. (b), (e) Dark-field images taken using the diffraction spot in a white circle in (c) and (f), respectively.

Conclusions

In this study, FSP was conducted on HAZ surfaces of the but-welded HT490 and HT780 joints. FSP could increase tensile fatigue strength of the both grade butt-welded steel joints. The important findings are as follows:

1) The stress amplitude at fatigue limit for the butt-welded HT490 and HT780 specimens with FSP exhibited about 180 and 195 MPa, respectively, which are about 35 MPa and 15 MPa increase in comparison with those without FSP.
2) FSP produced microstructures consisting of ultrafine equiaxial grains with average grain sizes of about 8.18 μm and 4.00 μm beneath the surface of the butt-welded HT490 and HT780 specimens, respectively. FSP increased HAZ hardness in the both grade butt-welded steel joints due to grain refinement strengthening.
3) FSP dramatically increased HAB in the butt-welded HT490 joints, while LAB increase was obvious in the butt-welded HT780 joints. In addition, FSP decreased the lacking in uniformity in HAZ of the both grade welded specimens, and totally random crystal orientation was observed in HAZ of HT780 after FSP.
4) The microstructure consisting of ultrafine ferrite grains covered with cementite grains in a layer of about 200 nm produced by FSP can contribute to increasing fatigue strength of the butt-welded HT490 joints. In contrast, the precipitations were not observed in FSP-modified microstructure of the butt-welded HT780 specimens.

References

[1] K. Anami, *et al*: "Fatigue Strength of Welded Joints Made of High-Strength Steels," *Progress in Structural Engineering and Materials*, Vol. 3, No. 1 (2001), pp. 86-94.
[2] N.A.C. Lah, *et al*: "The Effect of Controlled Shot Peening on Fusion Welded Joints," *Materials and Design*, Vol. 31, No. 1 (2010), pp.312-324.
[3] C.M. Branco, *et al*: "Fatigue Behavior of Welded Joints with Cracks, Repaired by Hammer Peening," *Fatigue and Fracture of Engineering Materials and Structures*, Vol. 27, No. 9 (2004), pp.785-798.
[4] S. Roy, *et al*: "Fatigue Resistance of Welded Details Enhanced by Ultrasonic Impact Treatment (UIT)," *International Journal of Fatigue*, Vol. 25, No. 9-11 (2003), pp.1239-1247.
[5] W. Ting, *et al*: "Discussion on Fatigue Design of Welded Joints Enhanced by Ultrasonic Peening Treatment (UPT)," *International Journal of Fatigue*, Vol. 31, No. 4 (2009), pp.644-650.
[6] T. Deguchi, *et al*: "Fatigue Strength Improvement for Ship Structures by Ultrasonic Peening," *Journal of Marine Science and Technology*, Vol. 17, No. 3 (2012), pp.360-369.
[7] Y. Sakino, *et al*: "Major Factor Causing Improvements in Fatigue Strength of Butt Welded Steel Joints After Laser Peening without Coating," *Science and Technology of Welding and Joining*, Vol. 17, No. 5 (2012), pp.402-407.
[8] C.B. Fuller, *et al*: "The Effect of Friction Stir Processing on 5083-H321/5356 Al Arc Welds: Microstructural and Mechanical Analysis," *Metallurgical and Materials Transactions A*, Vol. 37A, No. 12 (2006), pp.3605-3615.
[9] J. da Silva, *et al*: "Fatigue Behavior of AA6082-T6 Welded Butt Joints Improved by Friction Stir Processing," *Materials and Design*, Vol. 51 (2013), pp. 315-322.
[10] K. Ito, *et al*: "Increase of Bending Fatigue Resistance for Tungsten Inert Gas Welded SS400 Steel Plates Using Friction Stir Processing," *Materials and Design*, Vol. 61 (2014), pp. 275-280.
[11] K. Ito, *et al*: "Increase in Fatigue Strength and Charpy Absorbed Energies of TIG-Welded SS400 Steel Plates by Friction-Stir-Processing Grain-Refinement," *Proc 10th International Conference on Trends in Welding Research*, Tokyo, October 2016, submitted.

Microstructure and Mechanical Properties of Friction Stir Processed Ferritic Steel

Lauri Luumi, Yutaka S. Sato, Hiroyuki Kokawa
Tohoku University, Sendai, Japan
luumi.lauri.tapio.s5@dc.tohoku.ac.jp, ytksato@material.tohoku.ac.jp, kokawa@tohoku.ac.jp

Hiroki Komuro, Yasuyuki Fujiya
Mitsubishi Heavy Industries, Nagoya, Japan
hiroki_komuro@mhi.co.jp, yasuyuki_fujiya@mhi.co.jp

Abstract
Friction stir processing (FSP) was applied to a cast steel with an inhomogeneous microstructure consisting of the banded structure. The objective of this study was to examine effect of FSP on the microstructure and mechanical properties of the cast steel. The banded structure of the base material was deformed, and then experienced phase transformation during FSP. FSP left the inhomogeneous microstructure containing regions with high fraction of martensite in the stir zone, resulting in an inhomogeneous hardness distribution. Degree of inhomogeneity of the microstructure decreased with increasing rotational speed.

Introduction
As a solid-state process, friction stir processing (FSP) has been suggested as a possible method for removing cast defects to overcome some of the difficulties associated with fusion welding [1]. The low-alloyed cast steel used in this study often includes cast defects and had a highly inhomogeneous microstructure. Effect of FSP on microstructure, mechanical properties and inhomogeneity on such steel has hardly been reported in the literature.

FSP usually produces the hard microstructure with low toughness in ferritic steels because the peak temperature exceeds the A_{c3}-point of the steel, resulting from formation of the martensitic microstructure during cooling cycle. However, Chung et al. [2] have shown that reduction of toughness can be prevented when the peak temperature is below the A_{c1}-point. Moreover, it is also reported that high toughness can be kept in the stir zone if the peak temperature during welding is in the intercritical temperature range which produces a fine-grained dual-phase microstructure of ferrite and martensite [3]. Some studies suggest that a dual-phase microstructure with a high martensite content provides optimal toughness and strength [4, 5]. These studies suggest that the peak temperature during FSP should be lower than the A_{c3}-point to obtain the good strength and toughness in the stir zone of ferritic steels.

On the other hand, the low-temperature FSP would reduce degree of intermixing and diffusion in the stir zone, suggesting that the inhomogeneous microstructure in the stir zone of the cast steel might be hardly homogenized. In this study, a low-alloyed cast steel with an inhomogeneous microstructure attempted to be FSPed at the low temperature range to suppress the reduction of the toughness in the stir zone. Effect of FSP on microstructure and hardness of the cast steel was examined.

Experimental methods
The base material (BM) was a low-alloyed cast steel with 0.2 wt% carbon. It contained the other alloying elements, such as Mn, Ni and Mo, resulting in a carbon equivalent of 0.62. The FSW tool used in this study was made of polycrystalline cubic boron nitride (PCBN) and had a conical 2 mm probe. During FSP, torque was measured. FSP was performed at rotation speeds of 150, 200 and 400 rpm, and travel speeds of 60 and 120 mm/min. Plunge depth of the tool was kept constant at 2 mm.

The microstructure was examined on the cross section perpendicular to the FSP direction by scanning electron microscopy (SEM) after etching with 3% Nital. Phase fraction was quantified by point counting. Grain size was measured with Electron Back-Scatter Diffraction (EBSD). The mechanical properties were examined by Vickers hardness testing. Hardness profiles of the stir zones were measured along a line 1 mm below the top surface with 500 μm intervals on the cross section at a load of 500 gf.

Results

The recorded torque ranged from 46 Nm to 71 Nm, and it decreased linearly with increasing rotational speed because the higher rotation speed results in the higher peak temperature. The higher travel speed causes the higher torque because of the lower heat-input, but effect of rotation speed on torque was considerably larger.

The base material used in this study had the banded microstructure consisting of light and dark bands parallel to the transverse direction of the plate, as shown in *Fig 1*. SEM images of these bands are shown in *Fig. 2*. Their hardness values are added on the pictures, showing that the light bands are harder than the dark bands. Both the bands consisted of ferrite and bainite, but the bainite fraction was different. The light bands have the higher fraction of bainite, causing the higher hardness than the dark bands.

Fig. 1: Macroscopic overview of the base material.

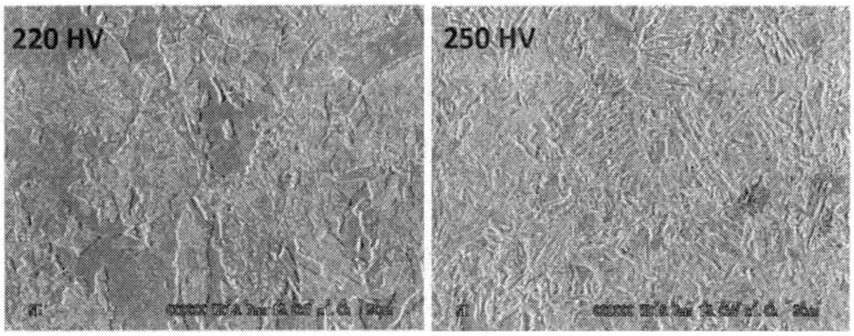

Fig. 2: SEM images and hardness values of dark and light bands of the base material.

Macroscopic overview of the stir zone produced at 120 mm/min and 150 rpm is shown in *Fig. 3*. The banded structures with the different contrasts are distinctly observed even in the stir zone. It seems that the banded structure in the base material is retained in the stir zone after FSP. The most stir zones had such banded structures, and the contrasts of the banded structure became weaker in the stir zones produced at the higher rotational speed.

Hardness testing showed that the stir zones were significantly hardened and that hardness distributions of the stir zones were largely scattered. The dark bands were harder in the stir zone. Difference in hardness between light and dark bands tended to decrease with increasing rotation speed. SEM images of the light and dark bands in the stir zone are given in *Fig. 4*. They have the fine-grained microstructure consisting of ferrite and martensite, but the martensite fraction was higher in the darker bands. The grain size and martensite fraction increased with increasing rotation speed. Formation of fine-grained microstructure including martensite caused the significant increase in hardness in the stir zone.

Fig. 3: Macroscopic overview of the stir zone produced at 120 mm/min and 150 rpm.

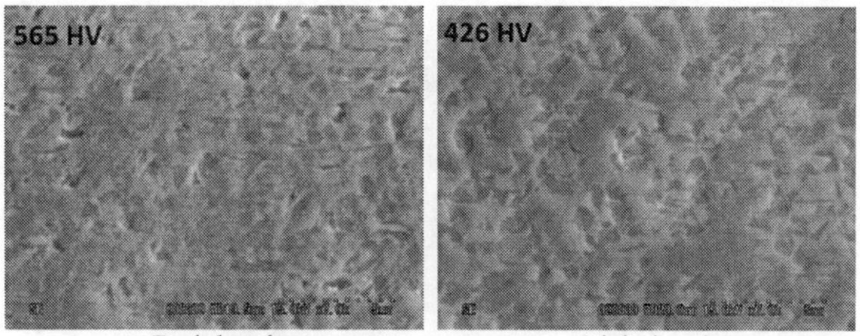

Fig. 4: SEM images of the dark and light bands of the stir zone produced at 120 mm/min and 150 rpm.

Conclusions

A low-alloyed cast steel with an inhomogeneous microstructure consisting of two bands was FSPed at various parameters, and effect of FSP on the microstructure and mechanical properties was investigated. The banded structure was deformed, and then experienced phase transformation in the stir zone during FSP. The stir zone also had the banded structure as well as the base material. Degree of inhomogeneity decreased with increasing rotational speed.

References

[1] P. Salami, *et al*: "Friction stir welding/processing as a repair welding," M.K.B. Givi, P. Asadi (Eds.), Advances in Friction-Stir Welding and Processing, Woodhead Publishing, Cambridge (2014), pp. 427–457.

[2] Y.D. Chung, *et al*: "Friction stir welding of high carbon steel with excellent toughness and ductility," *Scripta Materialia*, Vol. 63 (2010), pp. 223-226.

[3] P. Xue, *et al*: "Achieving ultrafine dual-phase structure with superior mechanical property in friction stir processed plain low carbon steel," *Materials Science & Engineering A*, Vol. 575 (2013), pp. 30–34.

[4] M.A. Maleque, *et al:* "The effect of intercritical heat treatment on the mechanical properties of AISI 3115 steel," *Journal of Materials Processing Technology*, Vol. 153-154 (2004), pp. 482-487.

[5] J.Y. Koo and G. Thomas: "Metallurgical factors controlling impact properties of two phase steels," *Scripta Metallurgica*, Vol. 13 (1979), pp. 1141-1145.

Poster

Laser tailor welded blanks of hot press forming steel with Al-Si coating

Cheolhee Kim, Minjung Kang

KITECH, South Korea

Al-Si coating to prevent surface oxidation during the hot press forming process can be diluted into the weld metal during laser welding for tailor welded blanks. Even after the successive hot press forming process, The diluted Al-Si coating could not be homogeneously diffused into the weld metal and softened microstructure used to be observed in the Al-rich zone. The removal of Al-Si coating prior to laser welding was recommended in order to achieve sound tailor welded blanks made of Al-Si coated hot press forming steel. To avoid expensive coating removal processes, three alternative methods were implemented in this study. Firstly, the effects of heating duration and temperature were examined and nearly full martensitic morphology was achieved by increasing the heating duration and temperature. Secondly, preheating process was applied before laser welding. By the preheating process using gas tungsten arc, Al segregation within welds decreased and homeogenous hardness was achieved in welds. Finally, so called undermatching was attempted using a relatively low strength steel, Ductibor™ which is specially developed for hot press forming. The tensile test specimens were made with these methods and the soundness of welds was confirmed by fracture at the base metal during the tensile test.

Laser-MAG Hybrid Welding of SUS301L-MT Stainless Steel T-joint

Xiaoyi Yang, Hui Chen, Zongtao Zhu, Shuang He
Southwest Jiaotong University, Chengdu, China
yangxiaoyi@my.swjtu.edu.cn, +86-028-87600726

Abstract

The SUS301L-MT stainless steel T-joint welded by laser-MAG hybrid welding (LMHW) that used in next generation city rail train is focused in this study. The angel of laser incidence was changed compared with original hybrid welding to improve weld beam forming and decrease deformation. Optical microscope (OM), scanning electron microscope (SEM), and hardness were carried out to investigate the microstructure and properties of LMHW SUS301L-MT stainless steel T-joint.

The results indicate that the angular deformation of optimized LMHW T-joint is about 3°. The microstructure of weld metal (WM) is columnar austenite dendrite mixed with little δ-ferrite. Heat affected zone (HAZ) and base metal (BM) are mainly rolling banded structure contained original austenite grains. Hardness test shows that the minimum value appears at the weld seam and the width of heat affected zone is almost 3 mm.

Introduction

Stainless steel is wildly applied in rail train manufacture and petrochemical industry because of its environment-friendly and corrosion-resistant [1]. The conventional arc welding is sensitive to the welding deformation and Cr-rich carbides deposition along the grain boundaries, which weakens the properties of the joints [2]. Laser-arc hybrid welding combined with the advantages of laser welding and arc welding is a promising welding technique, which is attractive in recent years [3].

Literature [4] and [5] introduced an optimized laser hybrid welding method for T-joint that could avoid the disadvantages of traditional laser hybrid welding. However, the SUS301L-MT stainless steel T-joint welded by laser-MAG hybrid welding (LMHW) have been reported seldom. Especially, the deformation and microstructure of LMHW T-joint. In this study, 6 mm SUS301L-MT stainless steel T-joint were welded by LMHW. Then, characterization of macrostructure, microstructure, and microhardness distribution had been investigated.

Materials and methods

Welding process

The LMHW process is shown in *Fig. 1*. The base metal used in this study was 6 mm×100 mm×150 mm SUS301L-MT austenitic stainless steel, and the horizontal plate was prepared in 40° groove and 3mm roof face. The filler metal was ER308L of 1.2 mm diameter. The chemical compositions of the base metal and the filler wire used are given in *Table 1*. Hybrid welding were performed by IPG YLS-4000 fiber laser and KEMPPI KempArc-450 pulse welding machine. Shielding gas for welding area was 95% Ar+5% CO_2 with a constant flow rate of 35 l/min. The welding parameters used are shown in *Table 2*. Laser leading model was chosen in this study, and the distance between filler wire and laser focal point (D_{la}) was 3mm. The angles of laser to vertical direction α_1, α_2 are both 10°, and the angles of MAG torch to horizontal direction θ_1, θ_2 were 50°, 45° respectively.

Table 1: The chemical compositions of SUS301L-MT austenitic stainless steel and ER308L filler wire.

Materials	C	Si	Mn	P	S	Cr	Ni	Cu	N
SUS301L-MT	0.02	0.39	1.12	0.034	0.00	17.24	7.10	0.0460	0.10
ER308L	0.02	0.42	1.65	—	—	20.1	10.3	—	—

Table 2: The parameters of LMHW.

Laser power /kW	Welding speed /m·min^{-1}	Wire feeding speed /m·min^{-1}	Current /A	Voltage /V	D_{la} /mm	Flow rate /L·min^{-1}
4.0	1.2	11.0	263	30.3	3	35

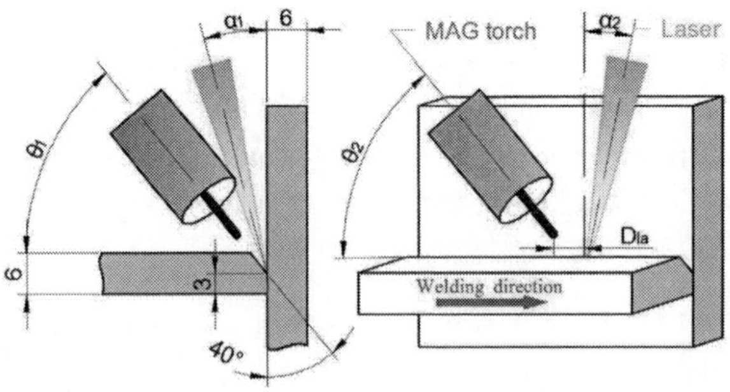

Fig. 1: Schematic of the LMHW process.

Weld formation and microstructure characterization

After welding, the LMHW T-joint was tested with protractor to measure the angular deformation. The metallographic specimens were polished with silicone paper and etched with $FeCl_3$-HCl solution (3 mg $FeCl_3$+50 ml HCl+100 ml H_2O). The macroscopic morphology of T-joint was obtained by Carl Zeiss Stemi 2000-C stereomicroscope and scanning electron microscopy (SEM) was performed by HITACHI SU8010. The microhardness test was performed by HVS-30 vivtorinox hardness tester along vertical and horizontal direction, with 10 kg force load and 10 s dwell time. The position of microhardness was in the middle of plate.

Results and discussion

Weld formation

The surface of weld (*Fig. 2(a)*) appears clean and white, except a small quantity of spatters. Surface defects, e.g. pores and cracks, are not found by visual inspection. The roof of weld (*Fig. 2(b)*) is continuous and uniform, which indicates the LMHW process is stable and enable to achieve favorable penetration. *Fig. 2(c)* presents typical cross-section of LMHW T-joint, which is a wine-glass shape. The morphology of base metal (BM) is typical banded structure which is formed during rolling process. The columnar grains of weld metal (WM) are quite obvious and their growing direction is almost vertical to fusion line between WM and heat affected zone (HAZ). The upper part of weld is arc dominating zone, and it's much larger than laser dominating zone that exists at the lower part of weld. However, there is an angle deformation of 3° to vertical direction inevitably (as shown in *Fig. 2(c)*), which is caused by welding heat input.

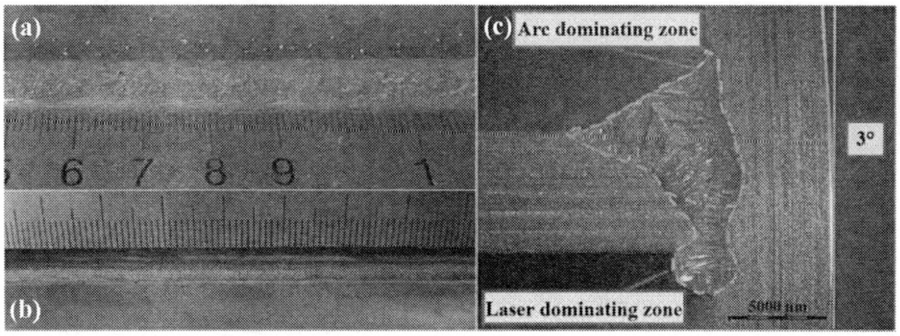

Fig. 2: The macroscopic morphology and weld transverse cross-section of LMHW T-joint: (a) the surface of weld, (b) the roof of weld, and (c) the transverse cross-section of LMHW T-joint.

Microstructure

Fig. 3(a) ~ (e) is the morphology of LMHW T-joint. *Fig. 3(b)* shows the microstructure of WM, which consisted of austenite with a little δ-ferrite. In addition, the left side of orange line is vermicular primary austenite mixed with netlike δ-ferrite, which indicates the solidification mode of weld metal is A-F mode. On the contrary, the solidification mode of right side is F-A mode and a microstructure composed of lath primary ferrite and austenite is formed. Thus, the microstructure of WM presents a degree of inhomogeneity. According to the literature [6], local variations in solidification mode is affected by incomplete mixing, it can also causes practical problems in e.g. risk for hot cracking or differences in corrosion behavior. It can be seen from *Fig. 3(c)* that the FZ presents the characteristic of associated crystallization. The WM near FZ is columnar austenite dendrite and netlike δ-ferrite. The microstructure of HAZ and BM is almost the same, which is rolling banded structure and retain original austenite grains. But there are some δ-ferrite separated out along rolling strip in HAZ, because of welding heat, shown in *Fig. 3(d)*. And a small amount of lath martensite is formed in BM during rolling process, shown in *Fig. 3(e)*.

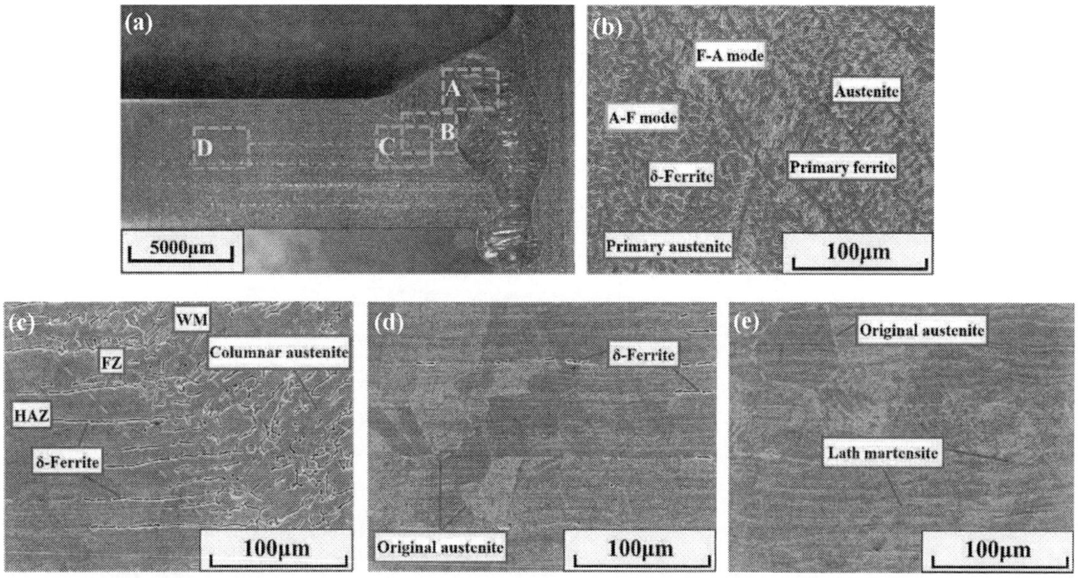

Fig. 3: The microstructure of LMHW T-joint: (a) the transverse cross-section of weld, (b) weld metal (zone A), (c) fusion zone (zone B), (d) heat affect zone (zone C), and (e) base metal (zone D).

Microhardness

The microhardness maps are shown in *Fig. 4*. In the horizontal direction of LMHT T-joint, the minimum microhardness value appears in the weld metal, which is 167.7~181.35 HV10. The average width of HAZ is about 3 mm, where the microhardness increases with the distance from the fusion zone. The maximum microhardness value of LMHT T-joint is observed in BM, ranging from 228.2~ 263.6 HV10. As illustrated in *Fig. 4(b)*, there is a relatively wide HAZ in the vertical direction, which is approximately 9 mm. The reasons for this phenomenon are likely to be welding heat effect and structure of joint.

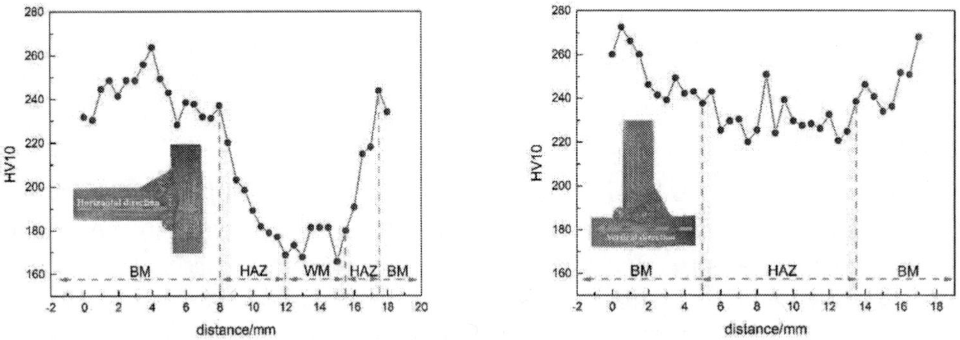

Fig. 4: The microhardness maps of LMHW T-joint: (a) the microhardness distribution along horizontal direction, and(b) the microhardness distribution along vertical direction.

Conclusions

The following conclusions were drawn from above experimental results and discussion.

Laser-MAG hybrid welding can obtain favorable SUS301L-MT stainless steel T-joint, and the angle deformation to vertical direction is 3 degrees. Austenite and a little δ-ferrite is formed in weld metal, where the incomplete mixing is observed. The width of HAZ is approximately 3 mm in horizontal direction.

References

[1] J. Yan, *et al*: "Study on microstructure and mechanical properties of 304 stainless steel joints by TIG, laser and laser-TIG hybrid welding," *Opt Lasers Eng*, Vol. 48, No. 4 (2010), pp. 512-517.

[2] V. Muthupandi, *et al*: "Effect of weld metal chemistry and heat input on the structure and properties of duplex stainless steel welds," *Mater Sci Eng A*, Vol. 358, No. 1-2 (2003), pp. 9-16.

[3] B. Ribic, *et al*: "Problems and issues in laser-arc hybrid welding," *Int Mater Rev*, Vol. 54, No. 4 (2009), pp. 223-244.

[4] M. Wahba, *et al*: "Hybrid welding with fiber laser and CO2 gas shielded arc," *J Mater Process Technol*, Vol. 221, 2015, pp. 146-153.

[5] M. Mazar Atabaki, *et al*: "Hybrid laser/arc welding of advanced high strength steel to aluminum alloy by using structural transition insert," *Mater des*, Vol. 75, 2015, pp. 120-135.

[6] Veli Kujanpää: "Thick-section laser and hybrid welding of austenitic stainless steels," *8th International Conference on Photonic Technologies LANE 2014*, Fürth, Germany, Sept 2014, pp. 630-636.

Characteristics of residual magnetism in square narrow gap arc welding groove

Jiayou Wang [1], Yuan Tong [1], Ping Huang [2], Jie Zhu [1], Guoxiang Xu [1]

[1] *Jiangsu University of Science and Technology, China*
[2] *Xinyu University, China*

Narrow gap arc welding is an attracting process of high efficiency and high quality, and is thus paid more attention in thick-wall manufacturing industry. However, high strength alloy steel is readily magnetized particularly due to machining, storage and transportation, which leads to the occurrence of serious residual magnetism in narrow gap welding groove. This residual magnetism will deflect the arc by Lorentz force, and finally affect the stability of welding process due to arc blow. So, it is necessary to investigate the characteristics of the residual magnetism, to understand deeply the phenomena and to reduce appropriately this harm. The present work uses a two-dimensional Gauss meter to detect the intensity of residual magnetism vectors that point to the length and width directions of square narrow gap welding groove for 700PMa grade high strength steel, and then physically models the residual magnetism. Since the magnitude of residual magnetic vector is much greater across groove width than across groove length, two-dimensional surface and curve cloud maps and three-dimensional surface cloud maps were drawn particularly for the width-directional magnetic vector on Matlab software platform to express visually. Experimental results show that the residual magnetism along groove length direction is strong near both ends and weak in the middle, and is of opposite polarity at both sides of groove middle for the opposite groove surfaces of unlike magnetic poles, while the magnetic field of residual magnetism is almost well-distributed along groove width direction. When adding metallic subplate on the bottom of groove or assembling two groove surfaces in manner of same magnetic poles, furthermore, the residual magnetism obviously weakens. It is thus demonstrated that the same-magnetic-pole assembling of groove and the utilization of bottom subplate are the simple and practicable ways to reduce the residual magnetism in narrow gap arc welding.

Resistance Spot welding of Mg alloy with Mg-Zn-Y alloy insert

Chihiro Iwamoto [1], Yoshihito Kawamura [2], Mitsuru Hano [2]

[1] *Ibaraki University, JAPAN*
[2] *Kumamoto University, JAPAN*

Mg alloy has great potential for reducing weight application because of its high specific strength. To extend the application of the Mg alloy, development of the welding technique is indispensable. In the light of the reliability and productivity, resistance spot welding is an attractive method. However, due to the high thermal and electrical conductivity of the commercial Mg alloy, extremely high currents need to join the Mg alloys leading to promote electrode degradation. In this study, we proposed a new technique of the Mg alloy resistance spot welding; the Mg alloy spot welding with Mg-Zn-Y plate insert. In contrast to the commercial Mg alloy plates, the Mg-Zn-Y plates were found to need less welding currents to obtain joints. The Mg-Zn-Y insert plate was put between two 1mm thick AZ31B plates, and the sandwiched sample was joined using stationary spot welding machine. The insert plates were Mg96Zn2Y2 and Mg85Zn6Y9 alloy with 0.3mm in thickness. Fracture load of joints showed that the joints with the Mg85Zn6Y9 insert exhibit relatively high fracture load at low welding current compared to that without insert alloy. At higher welding current, however, fracture load of all kinds of joints tended to converge. Microstructure analysis of the nugget of the joint with insert alloy showed that AZ31B plate and insert alloy melt and mixed to produce nugget. In the nugget obtained by the low welding current with the Mg85Zn6Y9 insert, many Mg compounds were observed at the grain boundary. In contrast, in the nugget of the joint with the higher welding current, the Mg compounds were only produced sparsely along the grain boundary. The high density of the reaction product was thought to relate to the mechanical properties of the joints with the Mg85Zn6Y9 insert.

Analysis of resistance welding process and expulsion of liquid metal from a weld nugget

Zygmunt Mikno, Adam Pilarczyk, Kazimierz Czylok
INSTYTUT SPAWALNICTWA (INSTITUTE OF WELDING), Gliwice, Poland
Email: zygmunt.mikno@is.gliwice.pl, adam.pilarczyk@is.gliwice.pl, phone: +48-32 33 58 360

Abstract

The article presents the process of resistance welding in relation to the expulsion of liquid metal from the weld nugget. The tests involved the synchronic recording of welding process parameters such as welding current and voltage as well as electrode force and travel. The phenomenon of expulsion was filmed using a high speed camera. The tests aimed to determine the most effective parameter as regards the detection of expulsion as well as the accurate determination of the moment of expulsion in relation to the above named parameter. During the tests it appeared that the most favourable parameter was the force of electrodes. The tests required the precise synchronisation of the recording of process parameters with the recording of images (using the camera). The uncertainty of expulsion time determination was estimated at 0.1 ms.

Presently, research-related tests are focused on the possibility of eliminating expulsion by stopping (blocking) the flow of welding current. The tests are performed using an inverter welding machine having an internal transformation frequency of 10 kHz.

Introduction

The phenomenon of expulsion in resistance welding is disadvantageous in nature as regards the aesthetics of joints and welded components and their strength. Due to the short duration (current flow) and large dynamics of resistance welding processes, the phenomenon of expulsion is difficult to eliminate [1] [2].

The phenomenon of expulsion was analysed at Instytut Spawalnictwa in Gliwice, Poland, using a high-speed camera, paying special attention to the monitoring of parameters being characteristic of resistance welding, i.e. welding current and voltage, electrode travel and force [3].

The research was aimed at the determination of possibilities of detecting the expulsion phenomenon on the basis of characteristic parameters as well as the accurate determination of the moment of expulsion occurrence.

Testing station

The testing station is presented in Fig. 1, whereas components of the measuring system are shown in Fig. 2.

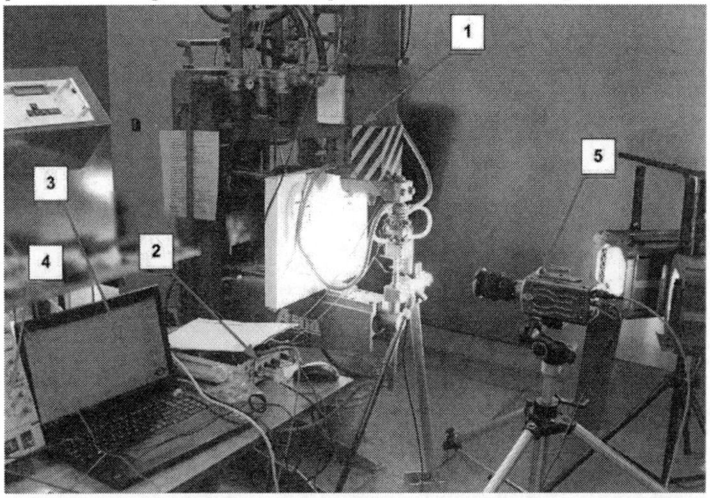

Fig. 1. Testing station: 1) DC resistance welding machine, 2) LogWeld4 system, 3) PC for graphic presentation of waveforms 4) oscilloscope, 5) high-speed camera.

Fig. 2. Components of the measuring system: 1) high-speed camera, 2) resistance welding machine, 3) head with the sensing element for welding force measurement, 4) displacement sensor, 5) cables for measuring welding voltage.

Resistance welding parameters were recorded using a LogWeld4 measuring system, designed and made at Instytut Spawalnictwa, and provided with a synchronisation system featuring a Phantom high-speed camera made by Vision Research company. Images were recorded at a speed of 10 000 fps.

Special emphasis was laid on the developing of fast and not causing delays, electronic module synchronising welding current and voltage as well as electrode travel and force being recorded by LogWeld4 device featuring high-speed camera. Measured maximum delay during synchronisation did not exceed several tenths of ns (nanoseconds).

While recording using high-speed camera, lightning of the recorded area is of a great importance, due to the fact that the quicker recording of the image the shorter time of exposure [4] [5]. One can say that it cannot be too much light, especially when during expulsion strong flash occurs. Therefore two reflectors equipped with Fresnel lenses of luminous flux over 50 000 lm each, were installed in the distance of 50 cm from the recorded area.

The synchronisation of both recording processes was an important issue. The precise determination of the initial moment of resistance welding process recording and of the beginning of filming also enabled the precise determining of the moment of expulsion formation.

The uncertainty of the determination of the beginning of recording of parameters for LogWeld4 measuring device and camera was also determined. The determined uncertainty is +/- 100 μs (microseconds).

Testing results

The test results are shown in Figs. 3, 4 and 5.

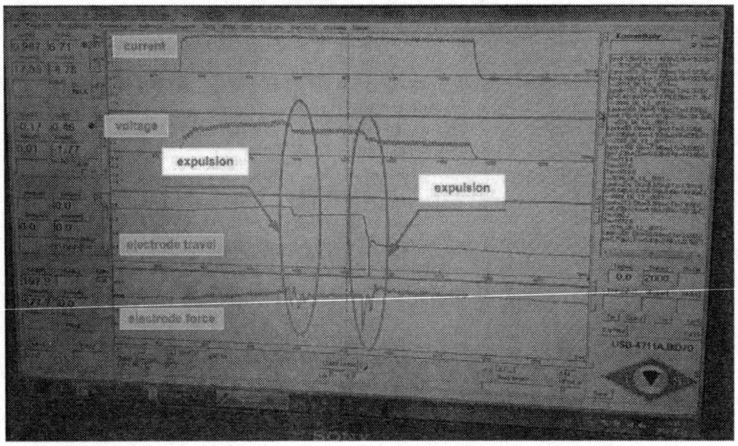

Fig. 3. LogWeld4 system with recorded resistance welding parameters (welding current and voltage, electrode travel and force).

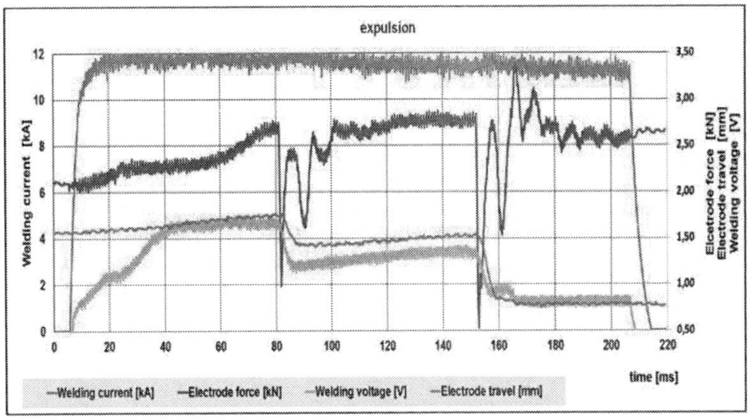

Fig. 4. Consolidated diagram of resistance welding parameters restricted within the range of 0 to 200 ms, recorded using the LogWeld4 system.

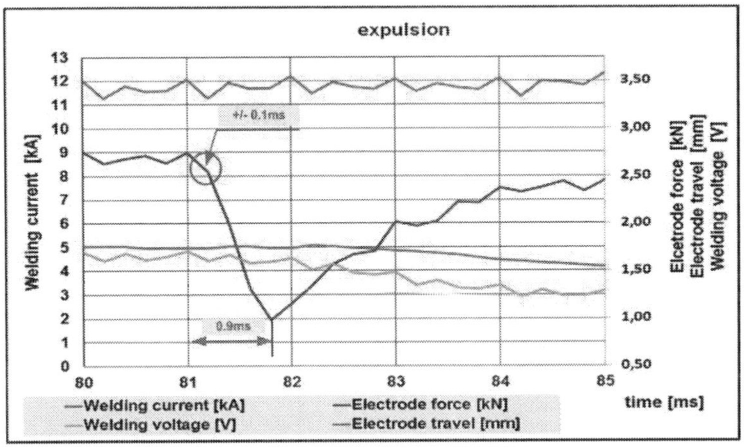

Fig. 5. Detailed analysis of friction force during expulsion. Uncertainty of expulsion time determination is +/- 0.1 ms maximum.

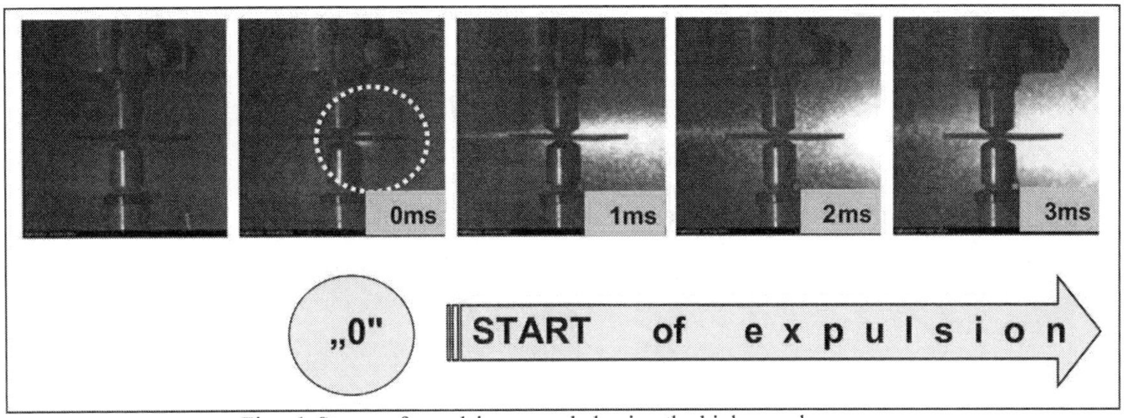

Fig. 6. Stages of expulsion recorded using the high-speed camera.

The selected frames of the film are presented in Fig. 6 – the image of expulsion course and its intensity.

The phenomenon of expulsion can be seen most clearly in measurement curves of voltage, electrode displacement and force. The signal of electrode force is characterised by the highest dynamics. The other two signals (voltage and electrode travel) have an approximate delay of 5 ms.

Image recording at a speed of 10 000 fps enables the observation and analysis of the process frame by frame at every 100 μs.

On the basis of delay time measurements related to recordings performed by both measuring systems, i.e. the recording of welding process parameters and filming, the uncertainty of expulsion time determination was defined at a level of tenths of a millisecond (+/- 0.1 ms). The beginning and duration of expulsion are shown in Fig. 5.

Conclusion

The conducted research revealed that recorded resistance welding process parameters, such as welding current and voltage as well as electrode displacement and force provide information on liquid metal expulsion from a weld nugget. On the basis of the conducted analysis it was ascertained that the most effective way of detecting and further monitoring of expulsion phenomenon, involves the use of the welding force signal. During expulsion, this signal is characterised by the most dynamic changes. The amplitude of those changes and the area restricted by the course of the registered signal determine the intensity of expulsion, which will be the subject of more detailed research in the nearest future.

The solution related to the detection and elimination of expulsion has been submitted to the Patent Office (P.416596).

Acknowledgments

Research was financed by NCBiR within the confines of PBS3/B4/12/2015 project.

References

[1] *H. Zhang, J. Senkara, Resistance Spot Welding, Fundamentals and Applications, 2nd Edition, CRC Press Taylor & Francis Group, 2011.*

[2] *J. Senkara, H. Zhang, Mechanism of Expulsion of Liquid Metal During Resistance Spot Welding, Welding Technology Review, no. 5-7, 2004.*

[3] Research project no. PBS3/B4/12/2015: *Development of innovative high-efficient compact resistance welding machine of higher frequency*. Realisation 2015-2017.

[4] A. Pilarczyk, Lecture and publication: *Visual recording of the welding arc with simultaneous recording of current-voltage parameters of MIG/MAG welding process*, XVI. Scientific Welding Conference 25-27.05.2010, Międzyzdroje, Poland.

[5] Research Project no. Da-105/2, *Research into the possibility of visual recording of welding arc synchronised with the recording of welding process parameters*. Realisation 2009-2010.

parameter study of droplet transfer in GMAW

Jochen Schein, Matthias Bredack, Erwan Siewert

Universitaet der Bundeswehr, Germany

The droplet transfer in GMA Welding has a large influence on the weld quality. The kinetic energy as well as the droplet temperture are responsible to a large extend for the heat distribution in the workpiece. Thus it is desirable to understand all physical mechanisms involved in particel production. Involved simulations have been developed which produce very good results, but also require significant computing power and time. In order to have a controlled process where online measurements are used to adjust the droplet parameters faster methods are required. In order to determine the important parameters that affect droplet size and temperature a simple model of the heat transfer into the droplet based on the child-langmuir law has been developed and will be compared to experimental results. Droplet size and temperature were measured using optical methods. The model attempts to predict the arc attachment at the electrode to determine the resulting theoretical droplet parameters. A comparison between the theoretical and experimental results will be presented.

SiC/SiC Joining Utilizing Paste-Like Si Powder in a Graphite Pipe Heated Electrically

Kazuyuki Kohama, Kazuhiro Ito, Shunichi Terada, Soshu Kirihara
Joining and Welding Research Institute, Osaka University, Ibaraki, Osaka, Japan
Email: kohama@jwri.osaka-u.ac.jp, Phone/Fax: +81-6-6879-8664

Abstract

Diffusion bonding of SiC-sintered materials was conducted using paste-like Si powder as the insert material to reduce bonding temperature to less than 1400°C while maintaining joint high-temperature reliability (e.g. oxidation resistance). The SiC/SiC joints were successfully fabricated at 1300°C for 60 min in a graphite pipe heated electrically due to formation of the porous insert layer consisting of sintered Si powder (sintered Si layer). The shear fracture stress of the joints increased as the densified sintered Si layer was obtained. It was achieved by reducing the Si particle median diameter and increasing the particle volume fraction in the paste-like Si powder, as long as its fluidity and dispersed state still remained. The maximum shear fracture stress of ~60 MPa was obtained. Such high fracture stress was explained by increase in strength of the sintered Si layer itself and bonding area at the Si-layer/SiC interface.

Introduction

SiC is a ceramic material with superior physical properties such as light weight, high-temperature strength and oxidation resistance [1, 2]. In recent years, fiber-reinforcement technologies for SiC have been extensively studied to increase its fracture toughness, and it is recognized as one of the alternative materials for Ni-based superalloys used in high-temperature applications such as aircraft engines [3, 4]. As is the case with many ceramics, utilization of SiC products in high-temperature applications requires joining methods to produce reliable SiC/SiC joints. Metal insert layers are widely used for joining SiC materials by diffusion bonding [4 – 9], but they degrade the joint high-temperature reliability (e.g. oxidation resistance). Although direct diffusion bonding has been studied as a possible method to produce refractory SiC/SiC joints [10 – 12], its practical utilization is limited because the self-diffusion process needs to be conducted at an extremely high bonding temperature of no less than 1800°C.

In this study, paste-like Si powder was proposed as the insert material for SiC/SiC joining. It has the potential to be sintered at the joint interface below 1400°C while the sintered Si powder exhibits its inherent melting point of over 1400°C. In addition, Si promotes formation of Si-oxide scale which protects the joints against high-temperature oxidation. Based on powder metallurgy, the sintering behavior in the paste-like Si powder is considered to depend on the powder properties such as median diameter (D_{Si}) and volume fraction (V_{Si}) of particle in the paste-like Si powder. Therefore, several types of paste-like Si powder with different D_{Si} and V_{Si} were used, and their effects on the SiC/SiC joint strength related with the interfacial microstructure was investigated to discuss an optimized fabrication process to obtain high-strength SiC/SiC joints.

Formation of SiC/SiC Joint Using Paste-Like Si Powder

SiC-sintered materials with dimensions of 4 × 4 × 6 mm^3 were used. The bonding surface (4 × 4 mm^2) was polished using diamond slurry, and ultrasonically cleaned with acetone. To prepare paste-like Si powder, it was dispersed in poly-ethylene glycol solvents. D_{Si} and V_{Si} varied in the range of 0.1–16 μm and 40–65%, respectively. The paste-like Si powder was inserted between two SiC samples, then diffusion bonding was conducted using a spark plasma sintering (SPS) system [11 – 12]. Fig. 1(a) shows a schematic illustration of the sample position arranged in the SPS chamber. The process can be conducted with a rapid heating rate in reducing atmosphere in low vacuum of about 20 Pa due to the graphite fixtures Joule-heated by pulsed direct current. A compressive stress was not applied initially to the samples but increased gradually with increasing temperature because coefficient of thermal expansion of the metal-carbide spacers is much larger than those of SiC and graphite. The bonding temperature and holding time was 1300°C and 60 min, respectively. Fig. 1(b) shows a typical photograph of the SiC/SiC joint. Excess Si powder flowed out from the joint interface and formed a sintered Si compact on outer surface of the samples, suggesting the bonding mechanism based on sintering of the Si powder. No defects such as cracks or oxidation were visible in the SiC/SiC joints. All other joints were similar in appearance, and it is thought that sample oxidation was suppressed because of the reducing atmosphere generated by the heated graphite fixtures.

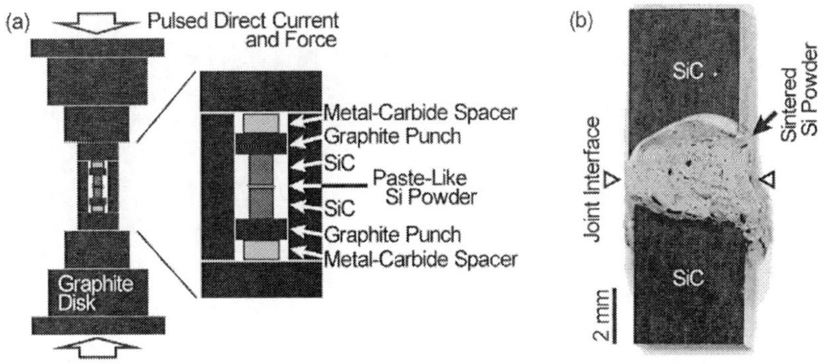

Fig. 1 (a) A schematic illustration of SiC sample position in SPS chamber. (b) Appearance of the SiC/SiC joint bonded at 1300°C for 60 min using paste-like Si powder with D_{Si} of 5 μm and V_{Si} of 56%.

Correlation between Joint Strength and Interfacial Microstructure Depended on Median Diameter of Particles in Paste-Like Si Powder

In order to clarify the microstructure around the joint interface depended on D_{Si}, scanning electron microscopy (SEM) studies were carried out for the joints bonded at 1300°C and 60 min using the paste-like Si powder with D_{Si} of 5, 13 and 16 μm. V_{Si} values used for these powder were 56, 60 and 60%, respectively. The selectable range of V_{Si} values was limited by the D_{Si} values. To enhance the sintering, V_{Si} was increased as much as possible for each powder within the range that its fluidity still remained. Note that the joint was not formed when D_{Si} was 0.1 μm, since the powder could not be applied uniformly onto the bonding surface due to its agglomeration. Fig. 2 shows cross-sectional SEM backscattered electron (BSE) images around the joint interface. Porous insert layers consisting of sintered Si powder (hereafter referred to as the "sintered Si layers") were formed. Their thickness and porosity as well as the pore size increased with increasing D_{Si}. In particular, most of the pores seemed to be formed in contact with SiC when D_{Si} was 5 μm, while those were dispersively distributed inside the sintered Si layers when D_{Si} was larger. These results indicate that the densified sintered Si layers can be obtained by refinement of Si powder while suppressing its agglomeration, which is presumably derived from increase in surface energy and size uniformity of Si particles.

To evaluate the joint strength related with the interfacial microstructure, shear tests at room temperature were performed. Fig. 3(a) shows the average shear fracture stress (σ_f) represented as a function of D_{Si}. The error bars indicate the maximum and minimum values. The joint strength increased as the D_{Si} values increased from 0.1 to 5 μm, and further D_{Si} increase decreased σ_f. The maximum σ_f value of about 60 MPa was obtained. Fig. 4(b) shows a typical plan-view SEM-BSE image of the sintered Si layer that appeared on the fracture surface for the joint bonded using the paste-like Si powders with D_{Si} of 5 μm. The sintered Si layer covered a large part of the fracture surface and its morphology was relatively flat and smooth. The opposite-side (conjugate) fracture surface mainly consisted of SiC (image not shown). These indicate that the fracture dominantly occurred at interface between the sintered Si layer and SiC sample when D_{Si} was 5 μm. In contrast, brittle fracture surface with river pattern was visible for the joint bonded using the paste-like Si powder with D_{Si} of 16 μm (Fig. 3(c)), indicating the fracture inside the sintered Si layer. Considering these results, the joint strength increased as the densified sintered Si layer was obtained, which can be explained by increase in strength of the sintered Si layer itself.

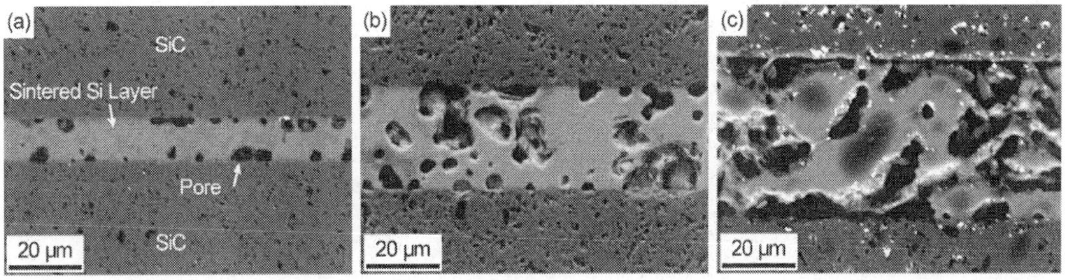

Fig. 2 Cross-sectional SEM-BSE images around the joint interface after bonding at 1300°C for 60 min using paste-like Si powder with D_{Si} and V_{Si} of (a) 5 μm and 56%, (b) 13 μm and 60%, and (c) 16 μm and 60%.

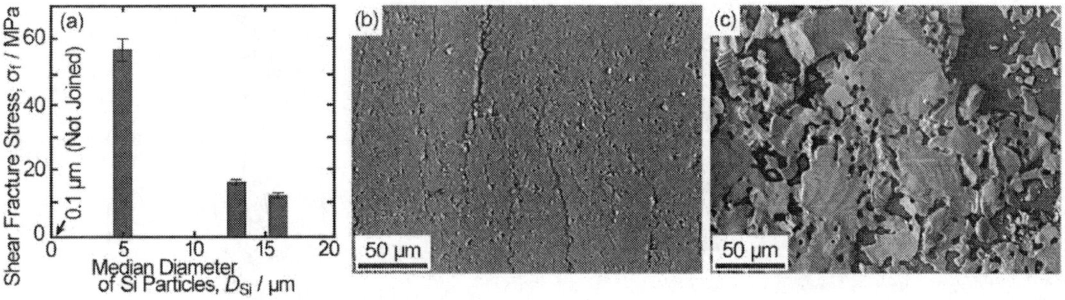

Fig. 3 (a)Average σ_f represented as a function of D_{Si}. Plan-view SEM-BSE images of the sintered Si layers that appeared on fracture surface for the joints bonded at 1300°C for 60 min using paste-like Si powder with D_{Si} and V_{Si} of (b) 5 μm and 56%, (c) 16 μm and 60%.

Effects of Densification Behavior in the Sintered Si Layers on Joint Strength in Relation with Volume Fraction of Particles in Paste-Like Si Powder

Figs. 4(a) – (c) show cross-sectional optical microscope images around the joint interface for the joints bonded at 1300°C for 60 min using the paste-like Si powder with V_{Si} of 56%, 50% and 40%, respectively. D_{Si} was fixed at 5 μm. When V_{Si} was larger than 56%, it was impossible to apply the paste-like Si powder onto the bonding interface because its fluidity significantly decreased. In these images, the sintered Si layers with similar thickness of about 10 μm were obtained regardless of the V_{Si} values, whereas their porosity significantly increased with decreasing V_{Si}. Fig. 4(d) summarizes σ_f for these joints. It increased as the V_{Si} values increased up to 56%. As with the previous paragraph, these results suggest that the joint strength increased as the densified sintered Si layers was obtained. It could be achieved by increasing volume fraction of particles in the paste-like Si powder as much as possible while maintaining its fluidity.

To understand how the joint strength was determined, the fracture surface was systematically investigated. Figs. 5(a) and 5(b) show plan-view SEM-BSE images of the fracture surfaces for the joints bonded using the paste-like Si powder with V_{Si} of 56% and 40%, respectively. Similar to Fig. 3(b), the densified sintered Si layers that have flat and smooth surface were observed with a continuous and bright contrast for both samples. The opposite-side (conjugate) fracture surface mainly consisted of SiC (images not shown). Meanwhile, the area in which the densified sintered Si layer was formed (A_{Si}) seemed to increase significantly as V_{Si} increased. These suggest that fracture at the Si-layer/SiC interface was dominant regardless of the V_{Si} values when D_{Si} was 5 μm, while the joint strength increased with increasing the Si-layer/SiC interface area. To verify this, a ratio of A_{Si} to the total cross-sectional area of the joints is defined as the bonding area fraction (x_{Si}), and relationship among σ_f, x_{Si} and V_{Si} are shown in Fig. 5(c). The graph reveals that σ_f increased as x_{Si} increased in relation with V_{Si} increase. It leads to a principle that the joint strength increases with increasing the bonding area at the Si-layer/SiC interface, which can be achieved by densification of the sintered Si layers.

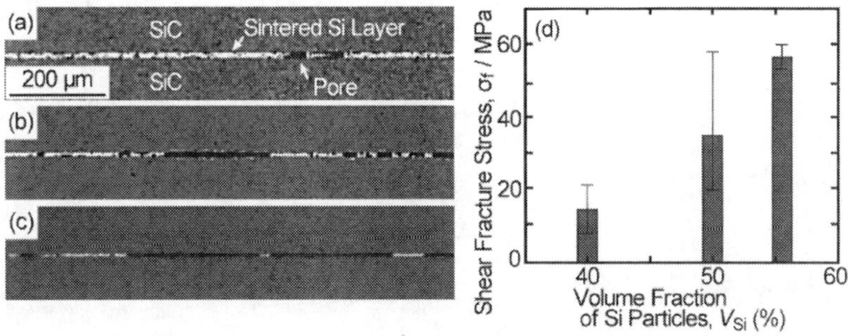

Fig. 4 Cross-sectional optical microscope images around the joint interface after bonding at 1300°C for 60 min using paste-like Si powder with V_{Si} of (a) 56%, (b) 50% and (c) 40%. (d) Dependence of σ_f on V_{Si}.

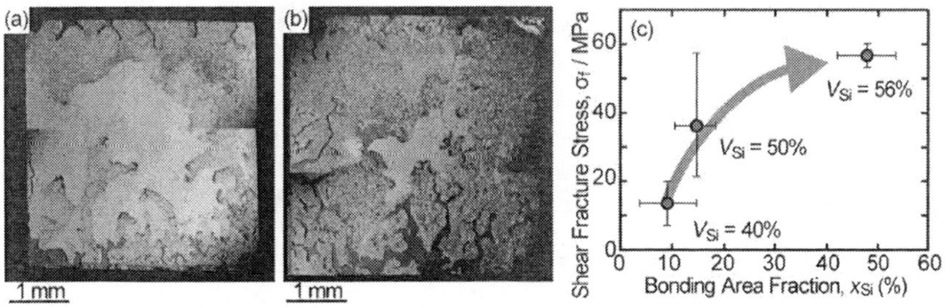

Fig. 5 Plan-view SEM-BSE images of the fracture surfaces for the joints bonded at 1300°C for 60 min using the paste-like Si powder with V_{Si} of (a)56% and (b) 40%. (c) Plots of σ_f versus x_{Si} in relation with V_{Si}.

Conclusions

The paste-like silicon powder was used as the insert material for SiC/SiC joining to reduce the bonding temperature to less than 1400°C while maintaining the joint high-temperature reliability. The median diameter (D_{Si}) and volume fraction (V_{Si}) of particles in the paste-like Si powder varied in the range of 0.1–16 μm and 40–65%, respectively. The SiC-sintered materials were successfully bonded at 1300°C for 60 min due to formation of the sintered Si layers at the joint interface. The shear fracture stress of the joints (σ_f) increased as the D_{Si} and V_{Si} values increased up to around 5 μm and 56%, respectively. Further D_{Si} and V_{Si} increase decreased σ_f. The maximum σ_f value of about 60 MPa was obtained. Based on SEM observation, the joint strength increased as the densified sintered Si powder was obtained, which can be achieved by decreasing D_{Si} and increasing V_{Si} within the range that the powder can be applied uniformly onto the bonding surface. High σ_f was concluded to be increased by increase in strength of the sintered Si layer itself and bonding area at the Si-layer/SiC interface.

Acknowledgments

The authors would like to thank Masayoshi Kamai, Joining and Welding Research Institute, Osaka University, for his help with the SPS work. This work was supported financially by JSPS KAKENHI [16K18246].

References

[1] R. L. Fleischer: "High-Temperature, High-Strength Materials — An Overview," *JOM*, vol. 37, no. 12, (1985) pp. 16–20.
[2] J. Roy, *et al*: "Oxidation behaviour of silicon carbide - A review," *Rev. Adv. Mater. Sci.*, vol. 38, no. 1, (2014) pp. 29–39.
[3] L. Giancarli, *et al*: "Design requirements for SiC/SiC composites structural material in fusion power reactor blankets," *Fusion Eng. Des.*, vol. 41, no. 1–4, (1998) pp. 165–171.
[4] C. Jiménez, *et al*: "Joining of ceramic matrix composites to high temperature ceramics for thermal protection systems," *J. Eur. Ceram. Soc.*, vol. 36, (2015) pp. 443–449.
[5] B. V. Cockeram: "Flexural Strength and Shear Strength of Silicon Carbide to Silicon Carbide Joints Fabricated by a Molybdenum Diffusion Bonding Technique," *J. Am. Ceram. Soc.*, vol. 88, no. 7, (2005) pp. 1892–1899.
[6] G. W. Liu, *et al*: "SiC/SiC and SiC/Kovar joining by Ni–Si and Mo interlayers," *J. Mater. Sci.*, vol. 45, no. 16, (2010) pp. 4299–4307.
[7] M. C. Halbig, *et al*: "Characterization of silicon carbide joints fabricated using SiC particulate-reinforced Ag–Cu–Ti alloys," *Ceram. Int.*, vol. 39, no. 4, (2013) pp. 4151–4162.
[8] Y. Katoh, *et al*: "Radiation-tolerant joining technologies for silicon carbide ceramics and composites," *J. Nucl. Mater.*, vol. 448, no. 1–3, (2014) pp. 497–511.
[9] M. C. Halbig, *et al*: "Diffusion bonding of SiC fiber-bonded ceramics using Ti/Mo and Ti/Cu interlayers," *Ceram. Int.*, vol. 41, no. 2, (2015) pp. 2140–2149.
[10] R. Aroshas, *et al*: "Silicon Carbide Diffusion Bonding by Spark Plasma Sintering," *Mater. Manuf. Process.*, vol. 30, no. 1, (2015) pp. 122–126.
[11] A. Stern, *et al*: "Silicon Carbide Bonding for High Temperatures Resistant Joints," *Ann. "DUNAREA JOS" Univ. GALATI Fascicle XII Weld. Equip. Technol.*, vol. 23, (2012) pp. 5–44.
[12] A. Allemand and F. Audubert: U.S. Patent 8956478 B2.

Tracking Using Pattern Matching of Keyhole in Visual Robotic Plasma Welding

Satoshi Yamane, Weixi Wang, Takaaki Hirano
Graduate School of Science and Engineering, Saitama University, Saitama, Japan
Email: yamane@mail.saitama-u.ac.jp

Kazumichi Hosoya, Toru Nakajima, Hikaru Yamamoto
Hitachi Construction Machinery Co., Ltd, Ibaraki, Japan

Abstract

Compared to conventional welding technique, plasma arc welding (PAW) is known without spatter and fume. Owing to the characteristic of high welding quality, PAW becomes more and more attractive in welding industry. In order to achieve stable welding for a long time, it is essential to employ an automatic control during welding process. Welding quality in PAW is considered depending on the performance of tracking results and maintenance of keyhole that an image processing method for automatic tracking of weld line has been developed under monitoring of keyhole. This method can help welding engineers to make a visible automatic control instead of human sight to teach plasma torch on condition that plasma electrode hides in the torch. As a consequence, successful tracking has been achieved in welding results that has well improved the efficiency of welding work, and effectively ensure the welding quality.

Introduction

In the so-called keyhole welding, welding quality in PAW is considered depending on the performance of tracking results of weld line and maintenance of keyhole. On the other hand, energy efficiency of thick material welding can be improved under proper welding conditions and exact control of the torch, since high-density current has been employed in PAW [1]. In order to achieve stable welding for a long time, it is essential to employ an automatic control during welding process. Thus, process control can greatly improve the quality and efficiency of welding, which is divisible into two broad strategies: weld formation control and seam tracking [2]. Since it is necessary to pay close attention and make real-time control during welding process, visual sensor is considered to be more appropriate to make an exact observation of welding condition [3]. Visual sensor extracts the torch position from image characteristics of weld zone, enabling automatic tracking of weld line in the welding process. By using visual sensor the research so far carried out by authors can be divided into two parts. One is according to boundary detection of gouging domain of keyhole photographed by CCD camera in front of plasma torch, both profile and height control of plasma torch has automatically proceeded in the welding [4]. The other is aiming at a possible gap and triangle groove ahead of weld zone photographed behind plasma torch, and in this way successful tracking of weld line has been achieved [5].

Prior to tracking experiments, numerous fundamental experiments have been conducted for acquisition of the characteristic quantity of weld line from weld zone. Because electrode hides inside the plasma torch, it cannot be taught accurately by remote controller within a reasonable time frame. To reduce the time of torch guidance, an image processing method for automatic tracking of weld line has been developed under monitoring of keyhole. This method can help welding engineers to make a visible automatic control instead of human sight by an image processing procedure implemented in Halcon software.

Robotic welding system

Schematics of PAW system are shown in *Fig. 1a*. Main components include a plasma welding robot with its movement controller, an image-processing PC, and a CMOS camera equipped with an infrared (IR) 950-nm filter, which intercepts all of the visible light down to near-infrared region (NIR). Since it is considered that CMOS camera has a higher sensitivity in the NIR, it has been utilized in this research rather than CCD camera used before. By using GIGE, communication speed of CMOS camera becomes considerably faster without occupying PC memory, and its shutter timing has been controlled by a 16-bit digital controller (dsPIC). Welding robot communicates with its movement controller through an Ethernet, and welding conditions are conveyed to the dsPIC through an RS-232C, which greatly reduces electric noise. Heat of plasma arc is generated between the tungsten electrode and base metal, and ejects through a small bore nozzle. In PAW, power source for main arc and high-frequency power supply for pilot arc are assembled together. Once orifice gas is supplied, a voltage is applied between the electrode and tip, and the ionizing gas in pilot arc generates the plasma. *Figure 1b* shows the state of main electric current flowing through the base metal. Plasma gas and small-diameter nozzle ensure a high current density in the arc.

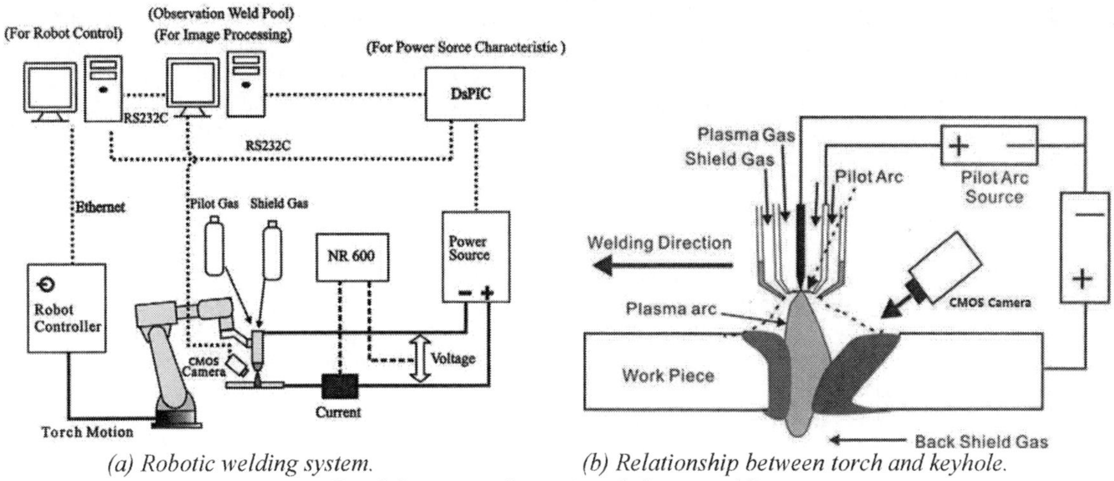

(a) Robotic welding system. *(b) Relationship between torch and keyhole.*

Fig. 1 System configuration of plasma welding.

In the welding system, CMOS camera has been set in the rear of plasma torch. Vertical center line of photographed image is merged with the torch center (electrode position) before the welding experiment. Moreover, the two instruments' positional relation has been fixed and they operate together during the welding process. In this way, the image photographed by CMOS camera directly reflect real-time welding condition and torch position. On the other hand, through welding characteristic quantities from image, it is available to indirectly detect the location of weld line to make an automatic tracking of weld line possible. Eventually, key point becomes how to acquire the characteristic quantities hidden in the image of weld zone

Experimental Procedure

To search for the characteristic quantity of weld line from keyhole area, a series of fundamental experiments have been carried out without tracking process. Above all, welding experiment is conducted by following the weld line. Typical image of weld zone photographed by CMOS camera is in *Fig. 2a*. Since CMOS camera equipped with an infrared (IR) 950-nm filter, the image is photographed clearly regardless of high light from plasma arc. It is possible to make an observation of keyhole right under the cap of plasma torch. Since heat is spreading around through the plasma arc in the center, molten metal in the weld pool with higher temperature appears to be brighter in the groove. On the other hand, cooling and solidification is from the surrounding. Solidified metal in the rear boundary is with temperature dropping, and appears to be black confirmed in *Fig. 2a*. It is noticeable of the boundary line of weld pool and due to the symmetry, boundary center of keyhole and center of image almost coincide with each other.

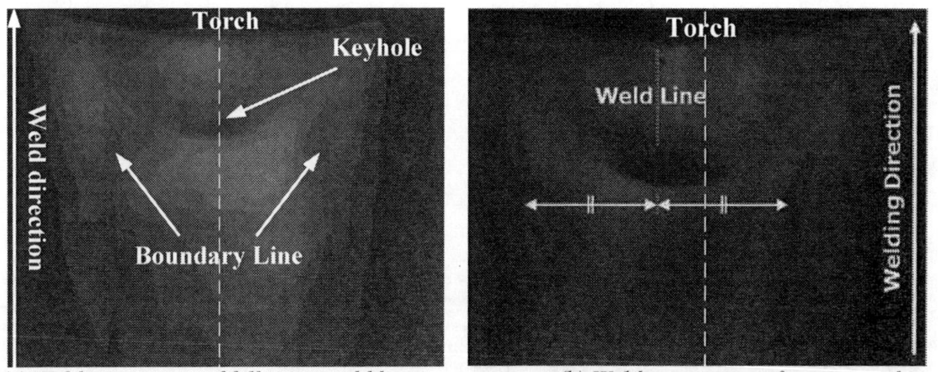

(a) Weld zone image of following weld line *(b) Weld zone image of wrong teaching*

Fig. 2 Typical image of no-tracking

To make a distinct comparison, the image with a deviation from the endpoint has been photographed in *Fig. 2b*. Extremely small weld seam has been deliberately left as white imaginary line to determine the position of weld line and related characteristic quantity. Yellow imaginary line in the center of image is electrode position and gradually deviating from the weld line. Based on the analysis of lots of fundamental experiments, it is determined that as long as the center line of image is still in the V-type groove, keyhole will pass through the weld line. Moreover, the center of keyhole will deviate along with that of image to some extent, and its area will grow bigger. It's worth noting that the center of boundary lines behind keyhole almost coincides with the weld line in *Fig. 2b*. And according to a large number of deviation experiments, it is found out not to be an individual phenomenon, which certainly matches the pervious result when following the weld line.

Since it will take processing time even if keyhole does not exist, it is needed to identify whether there is a keyhole or not at the beginning. Specific way is to draw a centerline in vertical direction across the image. And if the average brightness is less than 10, it is considered that keyhole doesn't exist to remove the image of welding start and end. On the other side, it regards the 100th pixel from the bottom up in the centerline as average brightness to make a binarization processing for the whole centerline. When continuous pixels darker than average brightness from the bottom up are more than 300, it is considered that keyhole hasn't formed yet to remove the image before beginning of steady welding state. Since keyhole's shape is maintained in a stable welding, template matching has been used for the detection of keyhole. This is a method through calculating the similar degree between a template image and target image to detect an object among the images. Above all, it is to appoint a small rectangle containing keyhole center in the steady state as a template, and a relatively larger template containing the both boundary lines.

Provided that there is no reference point matching with the target image, due to the angle of CMOS camera or some other reasons leading to a particular shape of keyhole temporarily, another different model is available to be used at the same time. And if it still doesn't work, the program will come back to the acquisition for next image. Since it is so significant to identify the existence of keyhole, the model rectangle is to be denoted with the change of its size and angle. It is capable of calculating the transformation matrix through the difference of size and angle between the model and target image. And due to the transformation matrix, it is able to carry on affine transformation and combined transformation of parallel translation and linear movement to denote the change of both rectangles. Since the right part of image will be brighter than the left due to a deviation of arc, brightness distribution is not the same. Two rectangular domains of 100 pixel height are appointed in both sides below the center of the small rectangle to each side of big rectangle. It is to make a detection area for finding out the black boundary lines in *Fig. 3a*. There is often more than one black line in the domain which needs to prevent the false detection of boundary line. Since edge detection is on the basis of looking for the area with big change of brightness, it is available to eliminate the noise with topical change of brightness by the filter using Gaussian.

Inflection point, which is the cross point between brightness curve and horizontal axis of pixel will be used for extraction of the edge. If length of detected edge is shorter than 150, it will be regarded as the noise and start for a loop. In addition, average brightness of each edge will be calculated. And if minimum value of edges is bigger than the average brightness of rectangular domain, which the edges seem brighter, it is considered lowering smoothing quantity to acquire the dark edge easily. And if there is no other edge meeting the above condition with length from 100 to 2000, it will return to the loop beginning of image acquisition.

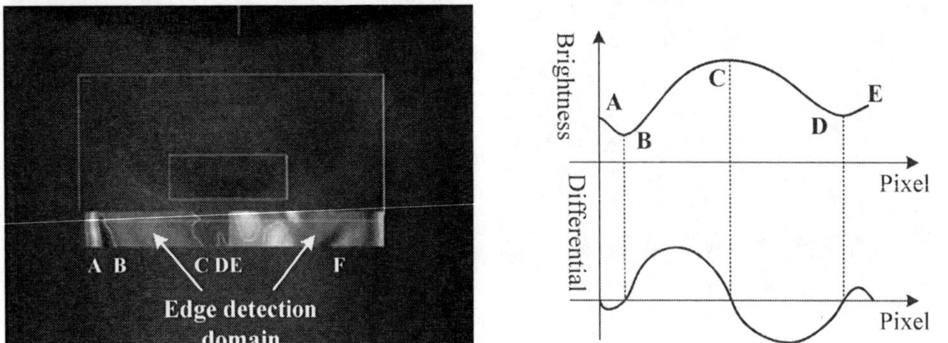

(a) Edge detection in both sides *(b) Determination of boundary line*

Fig. 3 Boundary acquisition by using differential filter of Gaussian function

In the edge detection domain, there are more than one remaining edges whose differential are equal to zero, such as edge A, B, C, D and E, but only edge B with smallest average brightness will be regarded as the boundary line in *Fig. 3b*. It is going to calculate the central abscissa of both ends of edge B to determine the boundary in one side. Determination of boundary in other side will proceed in the same way. In this case, there is black area such as F in the figure. Since luminance change here is not acute, F is not considered as edge by the differential filter. Furthermore, weld line is located in the center of connecting line by two boundaries. Integer part of x coordinate of weld line will be written in the serial communication.

Result and Discussion

Welding object is a 9-mm-thick plate of SS400 mild steel with V-type groove. Plasma gas and shield gas are pure argon. In the experiments, the standoff was 6mm, and start and end points were the groove center and a 3-mm displacement from groove center, respectively. Welding current was constant current of 235A and travelling speed was 15cm/min. Tracking result of image processing program is presented in *Fig. 4*. This time, automatic tracking of PAW was monitored. In processing program, deviations et[i], is the difference between reference positions Rt and abscissa value of central point. Simultaneously, command pulse of the position is sent to the robot. After numerous fundamental experiments, desired positions Rt of welding torch was determined as 640 pixels (Ref 0 mm) in vertical direction. A deviation of 1 mm corresponds to approximately 70 pixels in the target image. Welding robot was started after an initial 4-s stop period for the generation of plasma arc, and automatic tracking began at about 10 s. It can also confirm that detection points of the torch distance from weld line were approaching the reference line of 0 after tracking started.

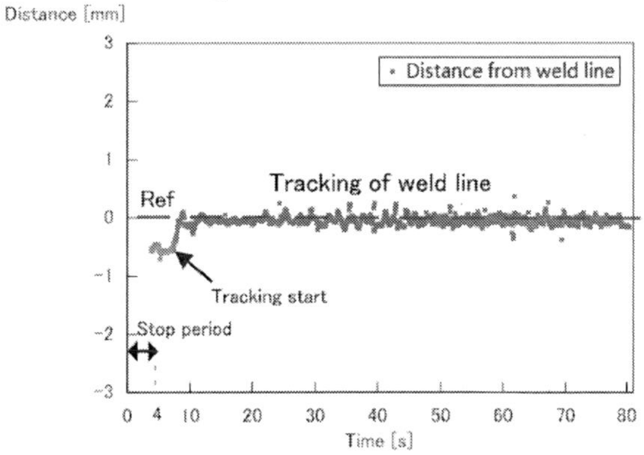

Fig. 4 Tracking result of weld line

Conclusions

PAW improves the efficiency and working environment of manufacturing, but it is difficult to control by human vision and manipulation. In automated welding of PAW, welding conditions especially keyhole status should be accurately observed in real-time. Since welding quality in PAW is considered depending on the performance of tracking results and maintenance of keyhole, we integrated automatic profile control of weld line into an image processing method under monitoring of keyhole. Welding results has proved that an efficient, high-quality, and stable welding process has been achieved in this research.

References

[1] Her-Yueh Huang: "Research on the activating flux gas tungsten arc welding and plasma arc welding for stainless steel," *Metals and Materials International*, Vol. 16, No. 5 (2010), pp. 819-825.

[2] C.S. Wu, L. Wang, W.J. Ren, X.Y. Zhang: "Plasma arc welding: Process, sensing, control and modeling," *Journal of Manufacturing Processes*, Vol. 16, No. 1 (2014), pp. 74-85.

[3] Hongyuan Shen, Tao Lin, Shanben Chen, Laiping Li: "Real-Time Seam Tracking Technology of Welding Robot with Visual Sensing," *Journal of Intelligent & Robotic Systems*, Vol. 59, No. 3 (2010), pp. 283-289.

[4] Weixi Wang, Satoshi Yamane, Tomoyuki Koike, Jiro Touma, Kazumichi Hosoya, Toru Nakajima, Hikaru Yamamoto: "Image processing method for automatic tracking of the weld line in plasma robotic welding," *The International Journal of Advanced Manufacturing Technology*, DOI: 10.1007/s00170-015-8311-8.

[5] Weixi Wang, Satoshi Yamane, Tomoyuki Koike, Jiro Touma, Kazumichi Hosoya: "Tracking and height control in plasma robotic welding using digital CCD camera," *The International Journal of Advanced Manufacturing Technology*, DOI: 10.1007/s00170-016-8437-3

Effect of Co Addition on Interface Reaction between Sn-Ag-Cu Solder and Cu Substrate

Jianxin Wang *, Taikun Fan, Yun Zhou, Qing Zhu
Jiangsu Provincial Key Laboratory of Advanced Welding Technology, Jiangsu University of Science and Technology, Zhenjiang, Jiangsu, China
** Corresponding author: wangjx_just@126.com, +86-0511-8440-1184*

Abstract
Sn-Ag-Cu (SAC) alloys are considered as the most promising Pb-free solders in electronic industry. The solidification microstructure and interface reaction behaviors of SAC alloys are therefore of fundamental importance for the service reliability of electronic devices. This is particularly true for these SAC alloys with low silver contents, partially because the coarsened interfacial intermetallics (IMCs) of these low-Ag SAC alloys with higher surface tension than conventional near-eutectic SAC alloys. As a result, it is desirable to refine the grain size of interfacial IMCs between low-Ag alloys and common substrates such as Cu. In this work, the effects of addition of trace amount of Co on the interface reaction between both conventional SAC305 and low-Ag SAC107 alloys on Cu substrate have been studied by reflowing experiments at temperature close to 260 °C. In addition, effects of Co additions on the solid state growth of interfacial IMCs have been studied at 150 °C after ultra-long annealing treatment for 384h, 768, and 1536h. Both top-view and cross-section micro-graphs have been obtained using electron microscopes. It has been found that addition of trace amount of Co can significantly refine the interfacial Cu_6Sn_5 IMCs grains after reflow process and impede the growth of Cu_3Sn after annealing treatment. This attributed to the replacement of Cu atoms by Co atoms in Cu_6Sn_5 crystals, which in turn depresses the diffusion of Cu and impedes the transformation from Cu_6Sn_5 to Cu_3Sn during aging.

Keywords

lead-free solder, Sn-Ag-Cu, intermetallic compound, microstructure

Introduction
Sn-Ag-Cu (SAC) alloys are considered as the most promising Pb-free solders in electronic industry. In order to further reduce the cost of SAC solder, new SAC alloys with lower Ag content have been widely studied [1], and alloying elements, such as Ni, Bi, Co, are selected as additions into these alloys [2-4]. In particular, the addition of trace amount of Co can effectively reduce the undercooling of Sn-Ag-Cu solder alloys, improve the mechanical properties, retard the formation of Cu3Sn IMCs, and suppress the void formation during interfacial reactions with Cu substrate [5-7].

In this work, the effects of addition of trace amount of Co on the interface reaction between both conventional SAC305 and low-Ag SAC107 alloys on Cu substrate have been studied by reflowing experiments at temperature close to 260°C. In addition, effects of Co additions on the solid state growth of interfacial IMCs have been studied at 150°C after ultra-long annealing treatment for 384h, 768, and 1536h. Both top-view and cross-section micro-graphs have been obtained using electron microscopes.

Experimental
Sn-3.0Ag-0.5Cu-xCo (x=0, and 0.2 wt.%) and Sn-1.0Ag-0.7Cu-xCo (x=0, and 0.07 wt.%) alloys were prepared, and then reflowed on Cu substrate using a RMA flux. The maximum soldering temperatures for Sn-3.0Ag-0.5Cu-xCo and Sn-1.0Ag-0.7Cu-xCo alloys were 260°C and 265°C, respectively, and the samples were cooled in the air after reflowing for 30 seconds. To reveal the effects of Co additions on the interface morphology of IMCs, microstructure observation was performed using an optical microscopy (VHX-900) and electron microscopy (JSM 6480).

Results and discussion
The cross-sections of IMC layers at solder/ Cu interface in the as-soldered samples are shown in Fig. 1. The

morphologies of Cu_6Sn_5 are scallop-like in the as-soldered Sn-3.0Ag-0.5Cu and Sn-1.0Ag-0.7Cu samples (Fig. 1(a, c)), and thicker interfacial IMCs can be observed in the other two samples.

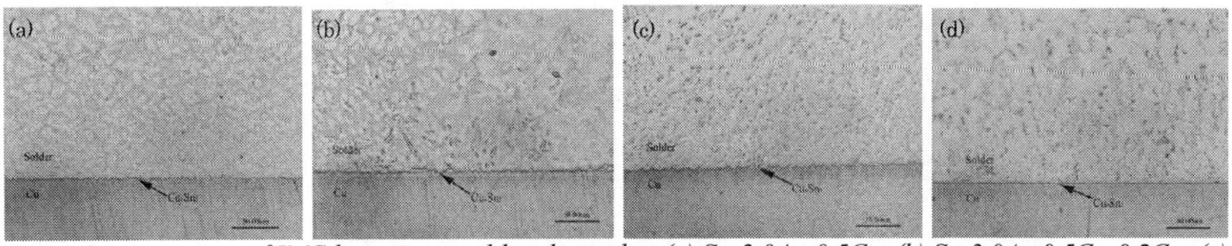

Fig. 1 Cross-sections of IMC layer in as-soldered samples. (a) Sn-3.0Ag-0.5Cu, (b) Sn-3.0Ag-0.5Cu-0.2Co, (c) Sn-1.0Ag-0.7Cu, (d) Sn-1.0Ag-0.7Cu-0.07Co.

In the top-views as shown in Fig. 2, it is found that the morphology of interfacial IMC is round shape in all the as-soldered samples after deep etching, while the sizes of interfacial IMCs grains with Co addition are much smaller than those of without Co addition. In other words, the addition of trace amount of Co can significantly refine the interfacial Cu_6Sn_5 IMCs grains after reflow process.

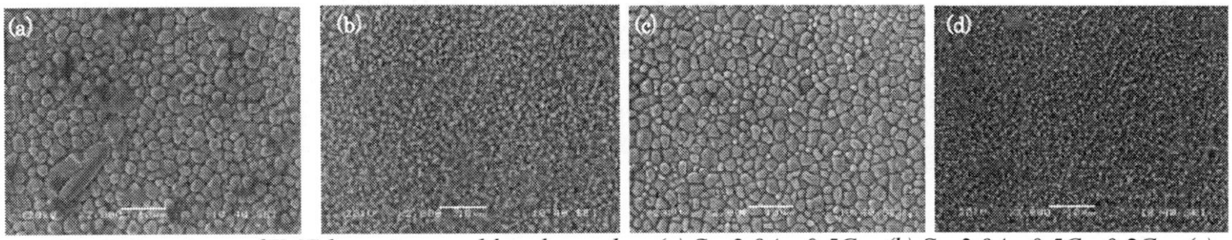

Fig. 2 Top-views of IMC layer in as-soldered samples. (a) Sn-3.0Ag-0.5Cu, (b) Sn-3.0Ag-0.5Cu-0.2Co, (c) Sn-1.0Ag-0.7Cu, (d) Sn-1.0Ag-0.7Cu-0.07Co.

Table 1 shows the EDX results of interfacial IMC layers in as-soldered Sn-3.0Ag-0.5Cu-0.2Co and Sn-1.0Ag-0.7Cu-0.07Co samples in Fig. 2, 3, in order to determine the composition of the interfacial IMC layer between the Sn-Ag-Cu-Co solder and Cu substrate. The intermetallic phase is likely to be expressed as $(Cu, Co)_6Sn_5$ phase, which is considered to be formed by substituting Cu atoms in binary compounds by Co atoms, consistent with previous result reported by Nishikawa et al. [7].

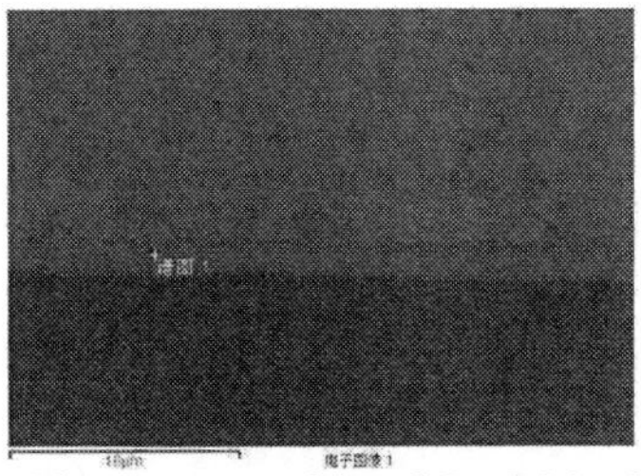

Fig. 3 EDX test of IMC layer in as-soldered Sn-1.0Ag-0.7Cu-0.07Co sample.

Table 1. EDX results of interfacial IMC layers in as-soldered samples.

Region	Solder	Element (at.%)		
		Sn	Cu	Co
Cross-section in Fig. 3	SAC107-0.07Co	46.68	50.45	2.87
Top-view in Fig. 2	SAC305-0.2Co	44.75	47.13	8.12
Top-view in Fig. 2	SAC107-0.07Co	43.06	53.90	3.04

The morphologies of IMCs layer at the interface change obviously after isothermal aging, and severe coarsening and planarization of IMC grains can be found. The morphologies of Cu_6Sn_5 in Sn-3.0Ag-0.5Cu and

Sn-1.0Ag-0.7Cu samples are planarized to be a thick layer after isothermal aging, while the interfacial IMC are much thicker with addition of trace amount of Co, as the cross-sections of the interface aged at 423 K for 384h in Fig. 4. In the top-views of the IMC layers as shown in Fig. 5, the morphologies of interfacial IMC are changed from roundshape into polyhedron-shape after isothermal aging. The grain sizes of interfacial IMC with Co addition are still smaller, and the grain refinement effect is obtained with 0.2% Co addition.

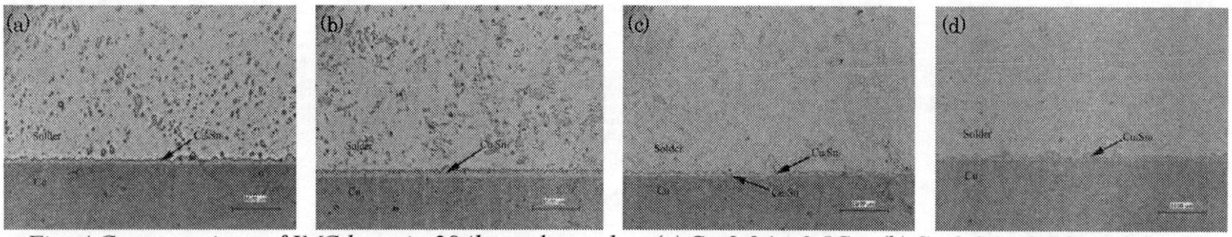

Fig. 4 Cross-sections of IMC layer in 384h aged samples. (a) Sn-3.0Ag-0.5Cu, (b) Sn-3.0Ag-0.5Cu-0.2Co, (c) Sn-1.0Ag-0.7Cu, (d) Sn-1.0Ag-0.7Cu-0.07Co.

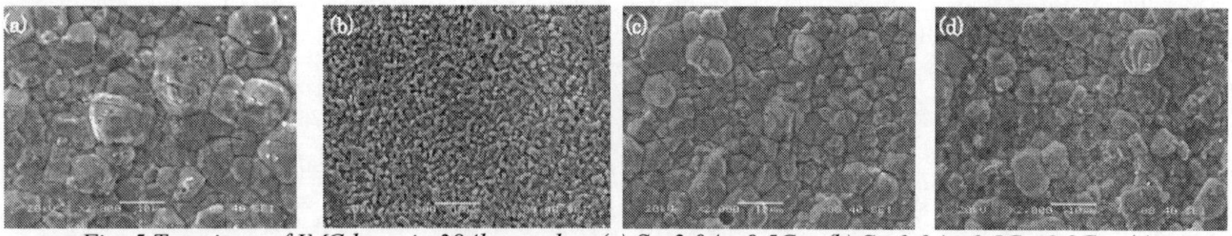

Fig. 5 Top-views of IMC layer in 384h samples. (a) Sn-3.0Ag-0.5Cu, (b) Sn-3.0Ag-0.5Cu-0.2Co, (c) Sn-1.0Ag-0.7Cu, (d) Sn-1.0Ag-0.7Cu-0.07Co.

After aged for 768h and 1536h, the Cu_3Sn layer can be obtained obviously at the cross-sections of the interface between SAC alloys and Cu substrates as shown in Fig. 6, 8, which decreases the mechanical strength of solder joint due to its brittleness nature and different CTE [8]. It is found that addition of trace amount of Co can impede the growth of Cu_3Sn after aging treatment, attributed to the replacement of Cu atoms by Co atoms in Cu_6Sn_5 crystals, which in turn depresses the diffusion of Cu and inhibits the transformation from Cu_6Sn_5 to Cu_3Sn during thermal aging. At the same time, the grain refinement effect of Co addition becomes weaker as the aging time increasing, and the sizes of interfacial IMC grains in all samples become the same after 768h aging, according to the top-views in Fig. 7, 9.

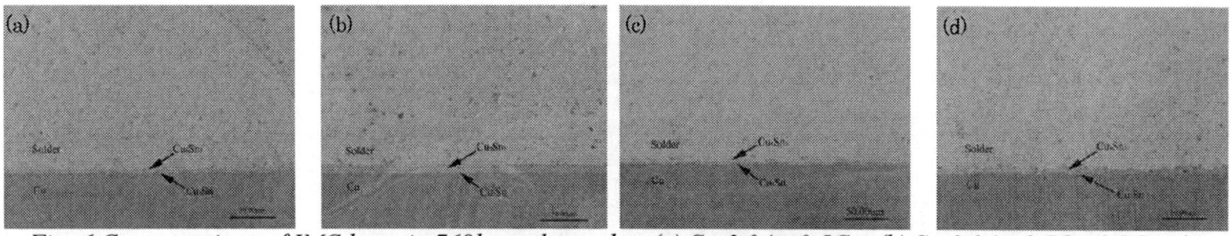

Fig. 6 Cross-sections of IMC layer in 768h aged samples. (a) Sn-3.0Ag-0.5Cu, (b) Sn-3.0Ag-0.5Cu-0.2Co, (c) Sn-1.0Ag-0.7Cu, (d) Sn-1.0Ag-0.7Cu-0.07Co.

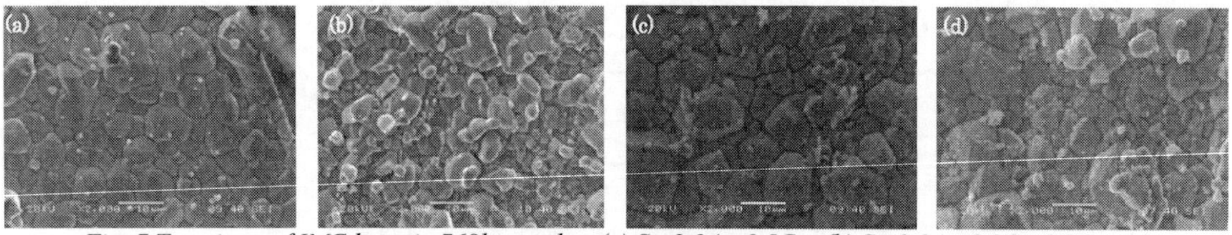

Fig. 7 Top-views of IMC layer in 768h samples. (a) Sn-3.0Ag-0.5Cu, (b) Sn-3.0Ag-0.5Cu-0.2Co, (c) Sn-1.0Ag-0.7Cu, (d) Sn-1.0Ag-0.7Cu-0.07Co.

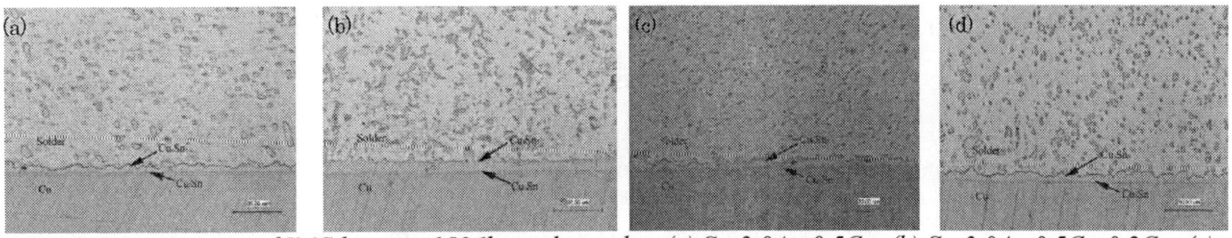

Fig. 8 Cross-sections of IMC layer in 1536h aged samples. (a) Sn-3.0Ag-0.5Cu, (b) Sn-3.0Ag-0.5Cu-0.2Co, (c) Sn-1.0Ag-0.7Cu, (d) Sn-1.0Ag-0.7Cu-0.07Co.

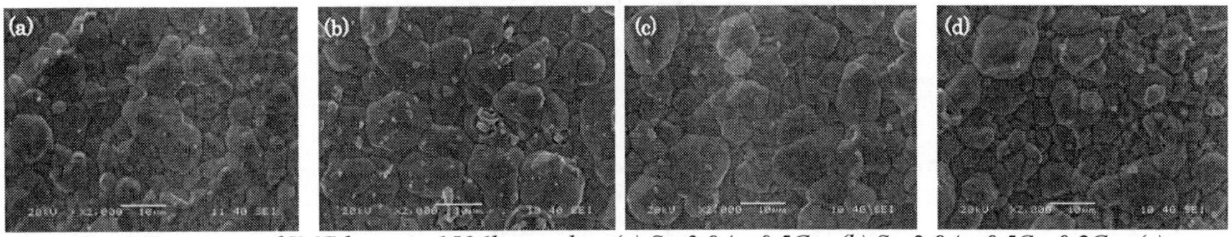

Fig. 9 Top-views of IMC layer in 1536h samples. (a) Sn-3.0Ag-0.5Cu, (b) Sn-3.0Ag-0.5Cu-0.2Co, (c) Sn-1.0Ag-0.7Cu, (d) Sn-1.0Ag-0.7Cu-0.07Co.

Conclusions

(1) Trace amount of Co can significantly refine the interfacial Cu_6Sn_5 IMCs grains after soldering.
(2) Co can inhibit the growth of Cu_3Sn after annealing treatment.
(3) Cu atoms are replaced by Co atoms in Cu_6Sn_5 crystals, depressesing the diffusion of Cu and impeding the transformation from Cu_6Sn_5 to Cu_3Sn during aging.

Acknowledgments

The present work was carried out with the support of National Natural Science Foundation of China (51305173), China Postdoctoral Science Foundation (General Financial Grant 2014M550289, and Special Financial Grant 2015T80548), Natural Science Fund for Universities in Jiangsu Province (12KJB460003), and a Project Funded by the Priority Academic Program Development of Jiangsu Higher Education Institutions.

References

[1] X. Chen, J. Zhou, F. Xue, J. Bai, and Y. Yao: "Microstructures and Mechanical Properties of Sn-0.1 Ag-0.7 Cu-(Co, Ni, and Nd) Lead-Free Solders." J Electron. Mater. Vol. 44, No. 2 (2015), pp. 725-732.
[2] H. Nishikawa, A. Komatsu, and T. Takemoto: "Effect of Ni or Co addition to Sn-Ag solder on microstructure and joint strength at interface," Mater. Trans. Vol. 49, No. 7 (2008), pp. 1518-1523.
[3] I.E. Anderson: "Development of Sn-Ag-Cu and Sn-Ag-Cu-X alloys for Pb-free electronic solder applications," J Mater. Sci-Mater. El. Vol. 18, No. 1-3 (2007), pp. 55-76.
[4] F. Gao, T. Takemoto, and H. Nishikawa: "Effects of Co and Ni addition on reactive diffusion between Sn-3.5Ag solder and Cu during soldering and annealing," Mater. Sci. Eng. A, Vol. 420, No. 1-2 (2006), pp. 39-46.
[5] A. Haseeb, and T.S. Leng: "Effects of Co nanoparticle addition to Sn-3.8Ag-0.7 Cu solder on interfacial structure after reflow and ageing," Intermetallics. Vol. 19, No. 5 (2011), pp. 707-712.
[6] D.H. Kim, M.G. Cho, S.K. Seo, and H.M. Lee: "Effects of Co addition on bulk properties of Sn-3.5Ag solder and interfacial reactions with Ni-P UBM," J Electron. Mater. Vol. 38, No. 1 (2009), pp. 39-45.
[7] H. Nishikawa, A. Komatsu, and T. Takemoto: "Morphology and pull strength of Sn-Ag (-Co) solder joint with copper pad," J Electron. Mater. Vol. 36, No. 9 (2007), pp. 1137-1143.
[8] D.K. Mu, S.D. McDonald, J.Read, H. Huang and K. Nogita: "Critical properties of Cu_6Sn_5 in electronic devices: Recent progress and a review," Curr. Opin. Solid. St. M. Vol. 20, No. 2 (2015), pp. 55-76.

Variation of Arc Deflection Length Affected by Frequency Increment of Transverse AC Magnetic Field

Yoko Ishii *, Toru Shiino*, Yoshifumi Maeda*, Shinji Yamamoto*, Toru Iwao*
**Tokyo City University, Setagaya, Tokyo 158-8557, Japan*
g1315007@tcu.ac.jp

Shinichi Tashiro, Masaya Shigeta**, Manabu Tanaka****
***Joining and Welding Research Institute, Osaka University*
11-1,Mihogaoka, Ibaraki,Osaka 567-0047, Japan

Abstract

Gas metal arc welding using shielding gas is often used in the welding industry. However, the arc deflection affected by lateral gas is problem because of inappropriate heat transfer. Shielding gas is used in order to prevent the instability affected by the arc deflection. However, shielding gas causes turbulence, then weld defect occurs. Thus, the magnetic field is applied to the arc in order to stabilize the arc using low amount of shielding gas. The arc is deflected by electromagnetic force caused by transverse magnetic field, and moves from right and left side alternatively under AC magnetic field. However, the arc does not enable to follow with increasing the frequency. The migration length of the electron in the arc under transverse AC magnetic field is shorter than that under DC magnetic field, especially in the case of high frequency. Thus, the arc is stabilized against the disturbance. In this paper, the variation of the arc deflection length affected by frequency increment of transverse AC magnetic field was elucidated. The frequency of transverse AC magnetic field plays an important role for the arc deflection.

Introduction

Gas metal arc welding using shielding gas is often used in the welding industry and obtains high quality welding. However, the arc deflection affected by lateral gas is problem because of inappropriate heat transfer. Shielding gas is used in order to prevent the instability affected by the arc deflection. However, shielding gas causes turbulence, then weld defect occurs. Thus, the magnetic field is applied to the arc in order to stabilize the arc using low amount of shielding gas. The arc is deflected by electromagnetic force caused by transverse magnetic field [1], and moves from right and left side alternatively under AC magnetic field. When the AC magnetic frequency increases, the arc deflection amplitude decreases, and the arc deflection length affected by lateral gas decreases in the case of high magnetic frequency, 500-1200 Hz. This characteristic has elucidated in previous research [2]. We assume that the deflection length affected by transverse AC magnetic field and lateral gas decreases because the arc cannot follow with the AC magnetic field in the case of high frequency. In this research, the variation of the arc deflection length affected by frequency increment of transverse AC magnetic field in the case of low frequency is elucidated. The arc deflection is expected to follow with AC magnetic field completely at low frequency region. The arc deflection amplitude corresponds the deflection length affected by transverse static magnetic field as shown in *Fig. 1*. When the arc affected by transverse AC magnetic field and lateral gas alternatively move, the arc motion took turns the case *(a)* and *(b)* as shown in *Fig. 2*. The case *(a)* is the case of the electromagnetic force with drag force direction, and the case *(b)* is the case of the electromagnetic force against drag force direction. When the arc deflection follows AC magnetic field completely, it is expected that the arc deflection amplitude in the case *(b)* corresponds the sum of the deflection length affected by lateral gas and the deflection length affected by transverse static magnetic field as shown in *Fig. 3*.

Experimental method

Fig. 4 shows the experimental setup. The arc is generated in argon atmosphere chamber between the tungsten cathode and anode made by water-cooled copper. The transverse AC magnetic field is generated using the Helmholtz coil. Lateral gas is argon and flows to the arc. The arc mode is observed by high speed video camera (HSVC). The electrode distance, the peak of AC magnetic flux density and lateral gas flow density are constant. *Table 1* shows the experimental condition.

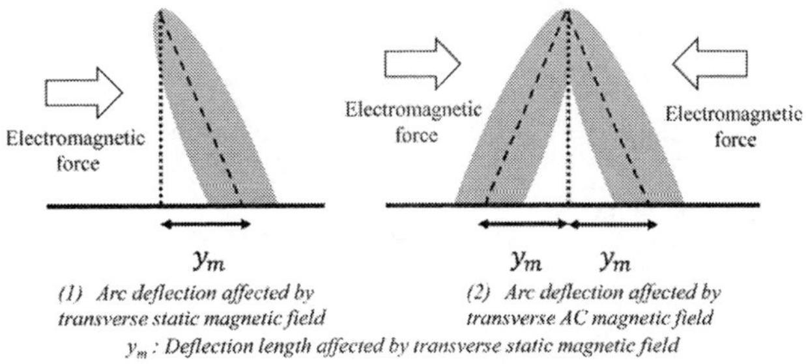

Figure 1: Expectation model of arc under transverse AC magnetic field

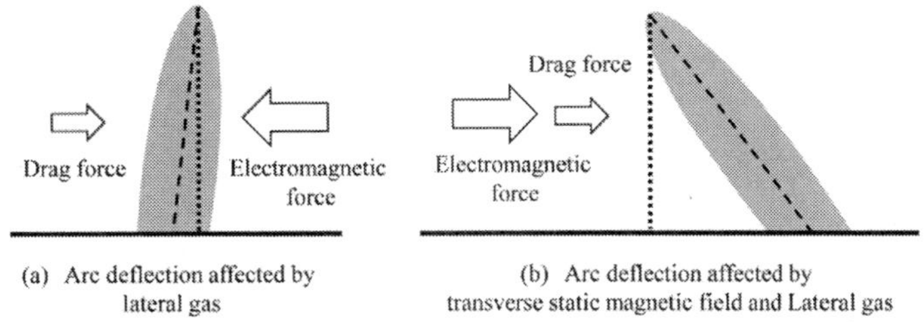

Figure 2: Arc motion affcted by transverse AC magnetic field and lateral gas

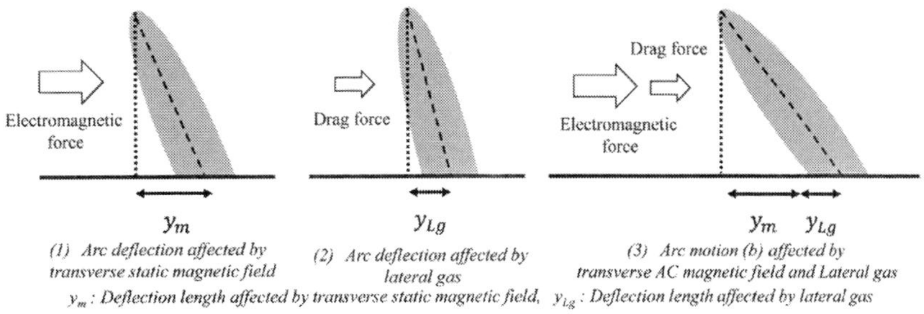

Figure 3: Expectation model of arc under transverse AC magnetic field and lateral gas

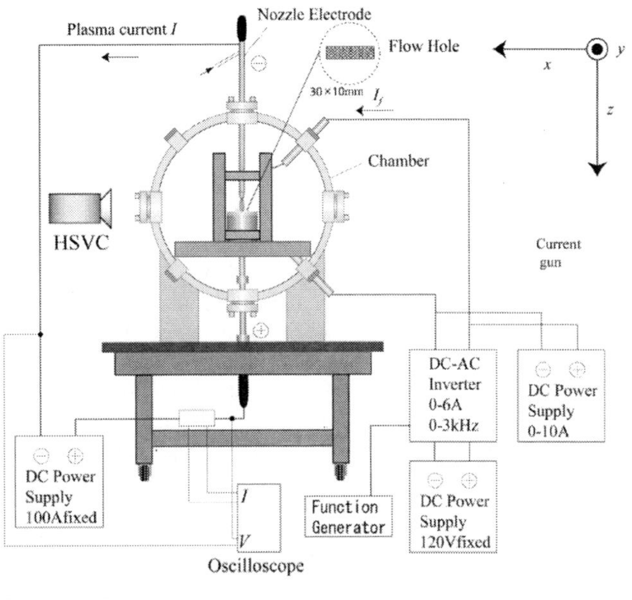

Figure 4: Experimental setup

Table 1: Experimental condition

Experimental condition	
Atmosphire gas	Ar
Pressure	0,1 MPa
Electrode distance	10 mm
Arc current	100 A
Cathode material	W
Anode material	Cu
Shield gas flow rate	3 L/min
Lateral gas flow rate	1.67 m/s
Field current	4 $A_{p\text{-}p}$
Field frequency	10~150 Hz

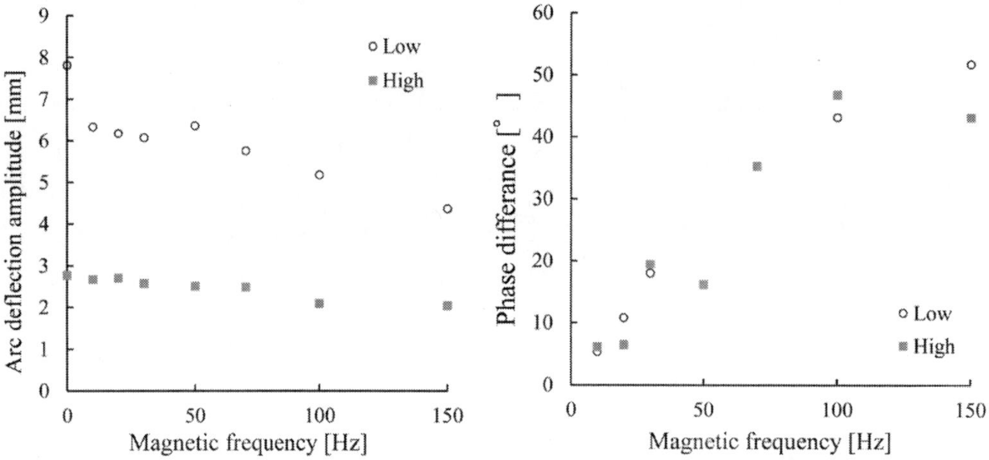

(a) Arc deflection amplitude

(b) Phase difference between arc deflection and AC magnetic field

Figure 5: Arc deflection affected by transverse AC magnetic at 10~150 Hz of magnetic frequency.

Result

Fig. 5 shows the arc deflection affected by transverse AC magnetic field. The arc deflection is measured at the high brightness area and the low brightness area observed by HSVC. We assume that the movements of two area have difference because of the difference of effect of electromagnetic force depending on the amount of electrons and ions. *Fig. 5 (a)* shows the arc deflection amplitude. The arc deflection amplitude decreases with increasing the magnetic frequency. The deflection length of 0 Hz is the deflection length affected by transverse static magnetic field. In the case of the low brightness area, the frequency that the deflection length corresponds the deflection length of 0 Hz does not exist. In the high brightness area, the magnetic frequency does not decreases largely. *Fig. 5 (b)* shows the phase difference between the arc deflection and the AC magnetic field. The phase difference increases with increasing the magnetic frequency. The arc deflection does not follow with the AC magnetic field completely, and the phase difference still exists at 10-150 Hz of the magnetic frequency. *Fig. 5* suggests that the phase difference still exists at 10 Hz or less of frequency.

Fig. 6 shows the arc deflection affected by transverse AC magnetic field and lateral gas. *Fig. 6 (b)* shows the arc deflection length. The arc deflection length means that the deflection of AC motion to axial direction in the case of the arc affected by transverse AC magnetic field and lateral gas. The deflection length at 0 Hz is the deflection length affected by lateral gas. The axial deflection affected by transverse AC magnetic field and lateral gas is larger than that affected by lateral gas. At 150 Hz or less of the magnetic frequency, the arc deflection length does not decrease with increasing the frequency in the case of the low brightness area. The characteristic of the deflection length decrement with increasing the magnetic frequency at 150-1200 Hz is not observed at 10-150 Hz. In the case of the high brightness area, the deflection length decreases with increasing the magnetic frequency at 10-70 Hz, and is not change at 70 Hz or more.

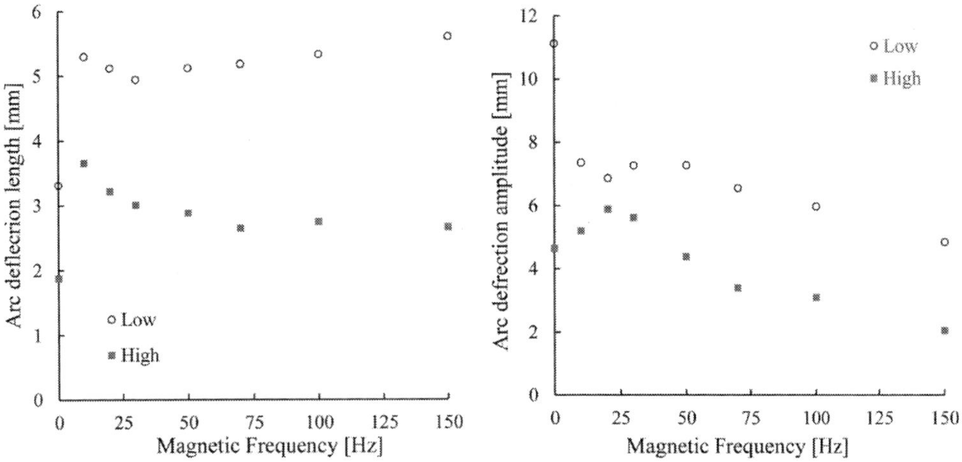

(a) Arc deflection length *(b) Arc deflection amplitude*

Figure 6: Arc deflection amplitude affected by AC magnetic field at 10~150 Hz of magnetic frequency.

Figure. 6 (b) shows the arc deflection amplitude affected by transverse AC magnetic field and lateral gas. The amplitude of 0 Hz is the sum of the deflection length affected by lateral gas and the deflection length affected by transverse static magnetic field. In the case of the high brightness area, the arc deflection amplitude increase at 0-20 Hz, and decreases at 20-150 Hz with increasing the magnetic frequency. In the case of the low brightness area, the arc deflection amplitude decreases with increasing the magnetic frequency totally, and increases at 20-50 Hz.

Conclusions

In this paper, the variation of arc deflection length affected by frequency increment transverse AC magnetic field in the case of low frequency was elucidated. The main results are shown below.

(1) The arc deflection amplitude affected by transverse AC magnetic field decreases with increasing the magnetic frequency at 150 Hz or less.

(2) The phase difference between the arc deflection and AC magnetic field exists at 150 Hz or less of the magnetic frequency.

(3) The arc deflection length does not decrease with increasing the frequency in the case of the low brightness area. In the case of the high brightness area, the deflection length decreases wish increasing the magnetic frequency at 10-70 Hz, and is not change at 70 Hz or more.

(4) The arc deflection amplitude increases at 0-20 Hz, and decreases at 20-150 Hz with increasing the magnetic frequency in the case of the high brightness area. In the case of the low brightness area, the arc deflection amplitude decreases with increasing the magnetic frequency totally, and increase at 20-50 Hz.

References

[1] T. Shiino, *et al*: "Arc Deflection Affected by Transverse Magnetic Field with Lateral Gas", Joint Technical Meeting on "Electrical Discharges", "Static Apparatus" and "Switching and Protecting Engineering", IEE Japan, Jun 2016, pp. 27-32.

[2] Y. Ishii, *et al*: "Arc Deflection Affected by Transverse AC Magnetic Field with Lateral Gas", Joint Technical Meeting on "Electrical Discharges", "Static Apparatus" and "Switching and Protecting Engineering", IEE Japan, Jun 2016, pp. 65-69.

Analysis of arc acoustic signal in laser enhanced hyperbaric underwater MIG welding

zhu jia lei, Jiao xiang dong

Beijing Institute of Petrochemical Technology, Beijing,CHINA

Hyperbaric underwater welding is the main underwater welding technology for important structures in marine engineering. laser enhanced welding can reduce adverse effects of ambient pressure on metal transfer state and welding process stability. The welding arc acoustic signal is closely related to the state of welding process, and has a good prospect in engineering applications in hyperbaric underwater welding. The arc acoustic signal of laser enhanced MIG welding in atmospheric environment and 0.2MPa and 0.4MPa ambient pressure are acquired, combined with synchronizing electrical signal and high speed image, the influence of ambient pressure and laser enhancement on arc acoustic signal in welding process are analyzed, The results show that: With the increase of ambient pressure, the arc acoustic frequency has certain regularity changes. When the arc is stable, the arc acoustic frequency is fixed at 19kHz,its strength can be used as an important indicator to judge the stability of the arc. Laser induced acoustic signal is a fixed frequency stationary signal. Based on the comprehensive analysis of the arc acoustic signal, the stability of the laser enhanced hyperbaric underwater MIG welding process can be determined.

Flux-free Brazing of Aluminum Alloys under Ultra-low Oxygen Partial Pressure using Zirconia Oxygen Pump

Tomoyuki SHINODA[1], Shumpei OZAWA[1], Kazuhiko KURIBAYASHI[1], Tomoki YAMAYOSHI[2], and Yasunaga ITOH[2]

[1] Department of Advanced Materials Science and Engineering, Chiba Institute of Technology, Narashino, Chiba, Japan

shumpei.ozawa@it-chiba.ac.jp, phone & fax +81-47-478-0131

[2] UACJ corporation, Nagoya, Aichi, Japan

Abstract

Flux-free brazing of aluminum alloys was carried out using the clearance filling test. The influences of additives to the core material and brazing filler alloys, heating time of the sample, and oxygen partial pressure of atmospheric gas were investigated with the aim of improving the flux-free brazability. The flux-free brazability of the aluminum alloys was much improved by adding some magnesium to the sample. When the oxygen partial pressure of atmospheric gas was much reduced to the order of 10^{-30}atm by using zirconia oxygen pump, a good brazability was obtained even after a long heating of the sample.

Introduction

When automotive heat exchangers are manufactured by brazing of aluminum alloys, a fluoride flux is indispensable to clean the brazing surface from its contamination such as oxidization and to prevent oxides from re-forming during the heating process [1]. However, the flux residuals remaining in clearances of heat exchangers would involve the risk of hardware damage of electronic components which have been widely used for recent cars. Thus the development of flux-free brazing of aluminum alloys are strongly required. One way to address it is to find alternatives for preventing the oxidization of the aluminum alloys instead of using the flux. Since the Gibbs energy for formation of magnesium oxide is lower than that of aluminum oxide, some additive of magnesium into the sample may have value for improving the flux-free brazability. A heating time of the sample and an oxygen partial pressure of atmospheric gas would influence the oxidation.

In the present study, flux-free brazing of aluminum alloys were performed using the clearance filling test. The purpose of this study was to investigate the influences of the magnesium addition, heating time, and oxygen partial pressure of atmospheric gas on the flux-free brazability.

Experimental procedure

Figure 1 depicts a schematic diagram of the experimental set-up for the clearance filling test, which consists of an electrical resistance furnace, a quartz glass chamber, a zirconia oxygen pump, and three zirconia oxygen sensors. Since a stabilized zirconia is a solid electrolyte, it can work as an oxygen pump where oxygen ions migrate through the dividing wall of a zirconia tube from the cathode to the anode by an applied voltage according to the Nernst equation [2]. A stabilized zirconia can also work as an oxygen sensor when the voltage induced by the difference of oxygen activities between the process gas and reference gas is monitored at the dividing wall. The oxygen pump and sensors were calibrated by oxidization/reduction reaction of pure metals [3]. For the sample of clearance filling test, a vertical plate of A3003 alloy was positioned on a flat brazing sheet, in which one end of the plate was raised on a resting stainless steel wire to make a clearance between them. The brazing sheet with 0.4 mm in thickness was prepared by cladding the brazing filler materials on one side of the core of aluminum alloys through a hot rolling followed by a cold rolling. The basic compositions of the brazing filler material and core alloy used in this study were Al-10 mass% Si and Al-1.2 mass% Mn, respectively. The cladding ratio was 10 %.

The assembled sample was heated up to 600 °C and then maintained for 6 minutes, at which the heating time of the sample between 350 °C - 577 °C was controlled to be 7 - 22 minutes (the corresponding heating rates were about 33 °C/min to 10°C /min). High purity nitrogen gas with the oxygen partial pressure of 10^{-6} atm was used as the atmosphere. Furthermore, it was reduce to the order of 10^{-30} atm by the oxygen pump operated at 600 °C. Furthermore, the oxygen partial pressure of atmospheric gas was confirmed by oxygen sensors maintained at 600 °C. The brazability of the sample was evaluated from the shape and length of fillet of brazed joint.

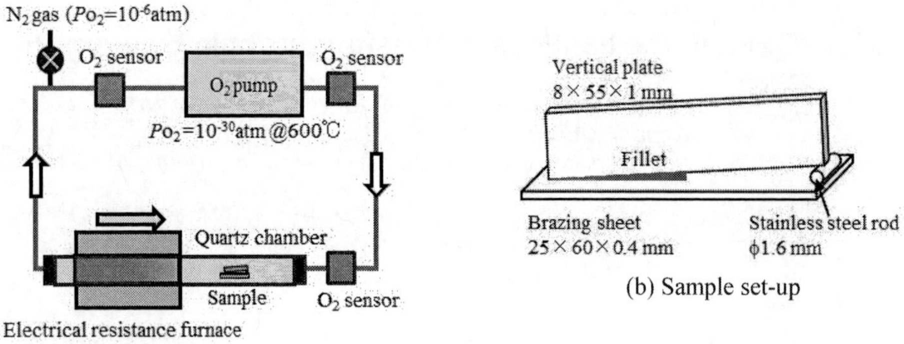

(a) Experimental set-up (b) Sample set-up

Figure 1 Experimental set-up of clearance filling test for flux-free brazing of aluminum alloys

Results and discussion

Figure 2 shows the typical appearances of the samples after experiments under the high purity nitrogen gas with the oxygen partial pressure of 10^{-6} atm, in which the time to heat the sample from 350 °C to 577 °C was 7 minutes. When Al-Si binary alloys were used as a brazing filler material, no fillet of brazed joint was formed at all unless some flux was used as shown in Fig. 2 (a) and (b).

However, comparatively a long fillet more than 17 mm was successfully formed in the sample without using any flux when some magnesium was added into the Al-Si brazing filler alloy or the Al-Mn core alloy in the brazing sheet. In particular, a very long fillet of 32 mm was obtained when 0.6 mass% magnesium was added into the core alloy. The effect of additive magnesium to improve the brazability became higher in the core material of the brazing sheet than in the filler material.

	Flux brazing	Flux-free brazing	
	a 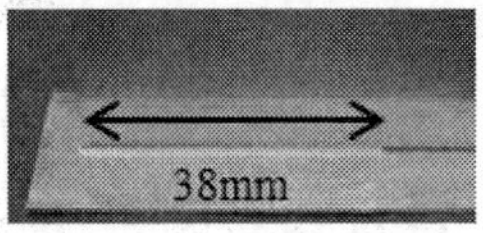	b 	c 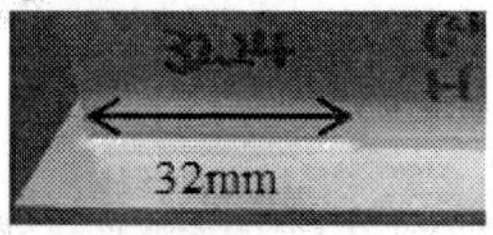
Filler brazing alloy	Al-10%Si	Al-10%Si	Al-10%Si
Core alloy	Al-1.2%Mn	Al-1.2%Mn	Al-1.2%Mn-0.6%Mg

(mass%)

Figure 2 The typical appearances of the samples maintained at 600 °C for 6 minutes under high purity nitrogen gas with the oxygen partial pressure of 10^{-6} atm. The heating time of the samples was controlled at 7 minutes when elevating it from 350 °C to 577 °C.

Although the flux-free brazing of aluminum alloys was achieved by the modifications of composition of brazing sheet as mentioned above, it is reasonable to suppose that the brazability would also be influenced by the heating time and oxygen partial pressure of the atmospheric gas because aluminum alloys are oxidized in a relatively easy manner. The time to heat the sample from 350°C to 577 °C was varied from 7 to 22 minutes. Furthermore, the oxygen partial pressure of atmospheric gas was much reduced to the order of 10^{-30} atm by the oxygen pump. Figure 3 shows the fillet length of the sample as functions of the heating time and oxygen partial pressure, in which the Al-10 mass% Si and Al-1.2 mass% Mn-0.6mass% Mg alloys were used for the filler material and the core of the brazing sheet, respectively. The fillet length of brazed joint becomes shorter with increasing the heating time regardless of the oxygen partial pressure of atmospheric gas.

When the oxygen partial pressure of the atmospheric gas is reduced to the order of 10^{-30} atm, however, the fillet length becomes longer than that under the oxygen partial pressure of 10^{-6} atm. This result confirms that the ultra-

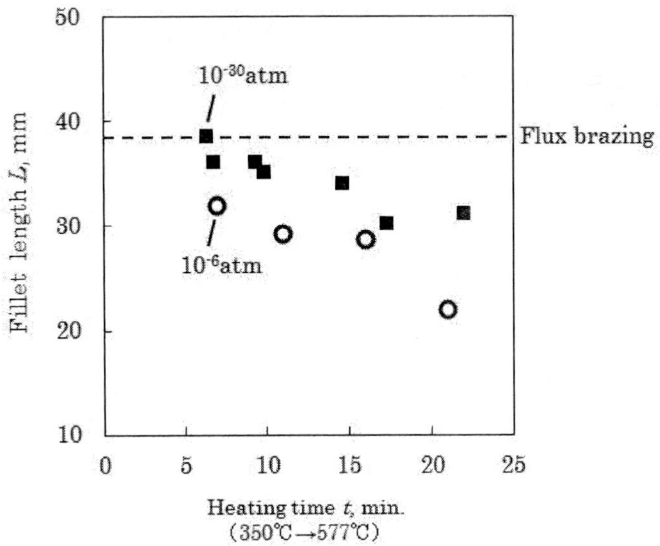

Figure 3 the length of fillet of brazed joint in the clearance filling test as functions of heating time and oxygen partial pressure of atmospheric gas. The Al-10mass%Si filler alloys and Al-1.2mass%Mn-0.6mass%Mg core alloys were used for the brazing sheet in the experiment. The open circle (○) and closed square (■) are corresponding to the results obtained under the oxygen partial pressure of 10^{-6} atm and 10^{-30} atm, respectively.

low oxygen partial pressure of atmospheric gas is one of the ways to improve the flux-free brazing of aluminum alloys. Noteworthy thing is that so excellent brazability was obtained as in the case of the flux brazing when the heating time is comparatively short of 7 minutes under the oxygen partial pressure of the order of 10^{-30} atm. The oxidization of the sample would have been prevented from such a short heating time and an ultra-low oxygen partial pressure of the atmospheric gas.

Conclusions

In order to improve the flux-free brazability of aluminum alloys, the influences of sample composition, the heating time, and oxygen partial pressure were investigated. As a result, it was improved by adding magnesium into the filler brazing material of Al-Si or the core of Al-Mn alloy in the brazing sheet. The effect of additive magnesium became higher in the core than in the filler material. The brazability became better as the heating time of the sample was shorter. Furthermore, ultra-low oxygen partial pressure of 10^{-30} atm for the atmospheric gas improved the flux-free brazability of aluminum alloys.

Acknowledgments

This work was partially supported by JSPS KAKENHI Grant Number 15552844. We would like to thank Mr. Nagasawa (Canon Machinery Inc) for his supports.

References

[1] Y. Itoh: "History of Aluminum Brazing Technology," Sumitomo Light Metal Technical Reports, Vol. 53, No. 1 (2012), pp. 79-91.

[2] W. Nernst: "Uber die elektrolytische Leitung fester Koorper bei sehr hohen Temperaturen, Z. Elektrochem, Vol. 6, No. 2 (1899), pp. 41-43.

[3] S. Ozawa, Y. Kawanobe, K. Kuribayashi, T. Nagasawa, "Influence of Trace Impurities on Oxygen Activity for High purity Nitorogen Gas Processed by Zirconia Oxygen Pump", Int. J. Microgravity Sci. Appl., Vol. 33, No. 2 (2016), 330214.

Effect of shape for weld faying parts on joint strength of steel pipe friction welded joints

Shuhei Iwamoto, Graduate Student, Graduate School of Engineering, University of Hyogo
Masaaki Kimura, Masahiro Kusaka and Koichi Kaizu, University of Hyogo
University of Hyogo, Himeji, Hyogo, Japan
mkimura@eng.u-hyogo.ac.jp
Shigekazu Miyashita and Masashi Takahashi, Toshiba Corporation
Toshiba Corporation Energy Systems & Solutions Company, Yokohama, Kanagawa, Japan

Abstract

This paper describes the effect of shape for weld faying parts on joint strength of steel pipe friction welded joints. The weld faying parts specimens had flat and taper shapes, and the taper shape specimens had a concave slope with about 15.5degrees on the inner part of the weld faying surface. Then, the steel pipe had the outer diameter of 13.5mm and the inner diameter of 8.0mm, and the joints was made with a friction speed of 27.5s^{-1} and a friction load of 2.79kN (friction pressure of 30MPa for the flat shape specimen). When the joint was made with flat shape specimens and a friction time of 0.5s, i.e. the friction torque reached the initial peak, almost all joints had the fracture in the base metal with 100% joint efficiency. However, one of the joints fractured between the weld interface and the base metal, and it had about 75% joint efficiency. Then, all joints fractured at the base metal were obtained at a friction time of 0.7s. Also, the inner flashes almost filled the hole of the steel pipe at this friction time. On the other hand, when the joint was made with taper shape specimens, the friction torque reached the initial peak at a friction time of 0.25s, and all joints had the fracture in the base metal with 100% joint efficiency. It was considered that taper shape was effective for decreasing of flash. Although further investigation will be necessary to clarify the effect of taper shape, it is desirable that the joints are made with taper shape specimens to decrease the flashes.

Introduction

The circular pipe or tube is mainly used for the piping components of transport equipment for liquid or gas. They are widely used as the various components or parts of a plant, pipeline, vacuum vessel, the cases for a variety of sensors or actuators, and so on. To use such structures, it is necessary that they are joined to use the pipe shape. Friction welding is well known amongst solid state joining methods. Many researchers [1] have reported that the results of the pipe joint with good characteristics by friction welding method were demonstrated. On the other hand, it is desirable that the pipe joint did not have the inner flash (burr and collar) for the usage of itself, because it is conceivable that various liquids or cables will be inserted in it. Nevertheless, it will be difficult to achieve the joint that had both the tensile strength of the base metal and no inner flash because it was considered that some flashes should be exhausted from the weld interface for obtaining the weld completely on it. In particular, the joining mechanism of friction welding for pipe shapes has not been fully clarified, so that the friction welding conditions are determined by trial and error. To determine the theoretical friction welding conditions for the ideal pipe joint, of which had the tensile strength of the base metal with no inner flash, is necessary to clarify the joining phenomena during the friction process.

The authors have been carrying out research to clarify the joining mechanism in friction welding for the ideal pipe joint. In the present work, the authors show the effects of the shape for weld faying parts and friction time on the tensile strength of steel pipe friction welded joints. Then, the authors demonstrate the shape of weld faying parts for obtaining the joint, which have the tensile strength of the base metal and the lesser inner flashes.

Experimental procedure

The material used was cold drawn seamless carbon steel pipes for high temperature service (JIS STPT410-S-C, STPT410) with an outer diameter of 13.8mm and an inner diameter of 7.8mm. The ultimate tensile strength of this material was 484MPa. This pipe was machined by a lathe to the pipe shape as shown in Fig.1a, which were the outer diameter of 13.5mm and the inner diameter of 8.0mm. Furthermore, one of the specimens had the flat shape for the weld faying (contacting) surface, and another had the concave slope shape (referred to as taper shape) for it as shown in Fig.1b. Then, the same shape specimens for the weld faying parts were welded.

A direct drive friction welding machine was used for the joining. During the friction welding operations, the friction welding condition was set to the following combinations: a friction speed of 27.5s^{-1} (1650rpm), a friction load of 2.79kN (the value equivalent to a friction pressure of 30MPa for the flat shape specimen), a range of friction times from 0.1 to 0.8s, a forge load of the same value (an identical value) of friction loads, and a forge time of 6.0s. The joining behavior during the friction process was recorded by a digital video camera, and the friction torque was measured with a load-cell. Then, the friction torque was recorded with a personal computer through an A/D converter. Furthermore, the specimens were joined by using an electromagnetic clutch in order to

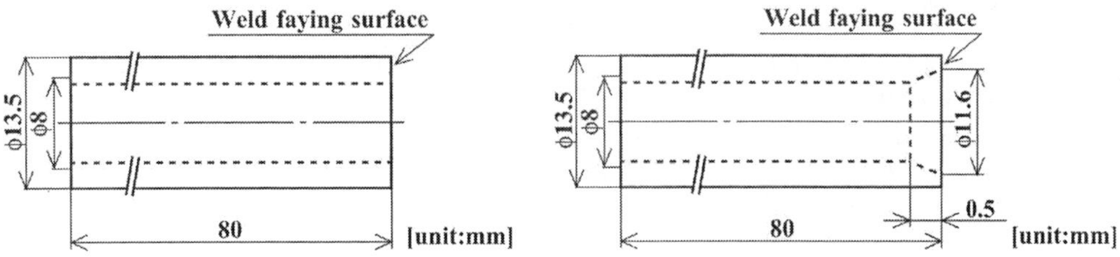

(a) Flat shape specimen *(b) Taper shape specimen*

Fig.1 Shapes and dimensions of friction welding specimens.

prevent deformation during braking time. When the clutch was released, the relative speed between both specimens instantly decreased to zero. The joint tensile test specimen had retained the inner and outer flashes that were expelled from the weld interface, i.e. the flashes were not removed. The joint tensile test was carried out with as-welded condition at room temperature.

Results and discussion

Relationship between joining behavior and friction torque

Figure 2 shows the relationship between the joining behavior and the friction torque curve during the friction process of joint with flat shape specimens. Photos ① to ④ in Fig.2a correspond to the friction torque of ① to ④ in Fig.2b, respectively. Photo ① shows the state at the faying surfaces as they contacted each other, and the friction torque was increased. Then they sparkled from the weld interface as Photo ②. When the friction torque reached to the initial peak of ③, the initial peak torque was approximately 50Nm, and the elapsed time for the initial peak was about 0.5s. Thereafter the friction torque was maintained nearly constant from ③, and the flashes were increased with increasing friction time as shown in photos 4). Then, the friction torque was decreased.

Figure 3 shows the relationship between the joining behavior and the friction torque curve during the friction process of joint with taper shape specimens. Photo ① shows the state at the faying surfaces as they contacted each other, and the friction torque was increased. However, the friction torque of the joint with the taper shapes was rapidly increased in comparison with that of the flat shapes. When the friction torque reached to the initial peak of ③, the initial peak torque was about 50Nm, and the elapsed time for the initial peak was about 0.25s. Thereafter, the joining behavior and friction torque were similar to that of flat shape one. It was clarified that the elapsed time for the initial peak differed with the shape for weld faying parts although the numerical value of initial peak torque was same.

Joint efficiency

Figure 4 shows the relationship between the friction time and the joint efficiency of the joints with flat shape

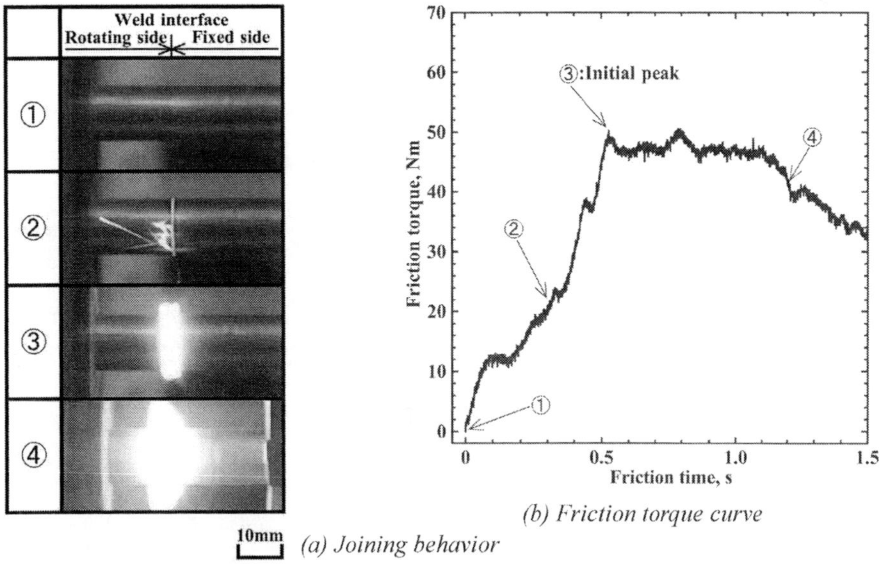

(b) Friction torque curve

(a) Joining behavior

Fig.2 Joining behavior and friction torque curve during friction process; flat shape specimen.

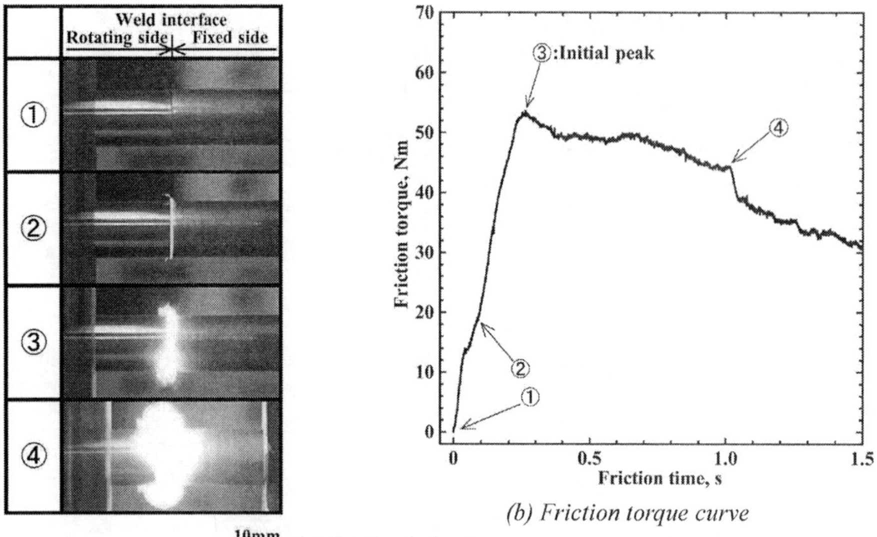

(a) Joining behavior

(b) Friction torque curve

Fig.3 Joining behavior and friction torque curve during friction process; taper shape specimen.

specimens, plotted alongside the friction torque curve. The joint efficiency was defined as the ratio of joint tensile strength to the ultimate tensile strength of the base metal. Figure 5 shows the appearances of joint tensile tested specimens for joints with flat shape one. When joints were made at a friction time of 0.3s, (the weld interface had sparkled) the joint efficiency of those was about 20%, and which was showed in Fig.4. Also, almost all joints fractured at the weld interface indicated by solid circular symbols in Fig.5, though one of the joints fractured between the weld interface and the base metal (mixed mode fracture) indicated by solid triangular symbols.

The joint efficiency increased with increasing friction time. One of the joints at a friction time of 0.5s (the initial peak) had the joint efficiency of 75% with the mixed mode fracture, although some joints had the joint efficiency of 100% with the base metal fracture, which was indicated by solid rhombus symbols. When joints were made with a friction time of 0.7s, all joints had the joint efficiency of 100% and the base metal fracture. Thereafter, the joint efficiency was kept about 100% and all joints had the fracture on the base metal. Therefore, it was clarified that the joint, which had the joint efficiency of 100% with the fracture on the base metal, was obtained when it was made with long friction time such as 0.7s or longer.

Figure 6 shows the relationship between the friction time and the joint efficiency of joints with taper shape specimens, plotted alongside the friction torque curve. Figure 7 shows the appearances of joint tensile tested

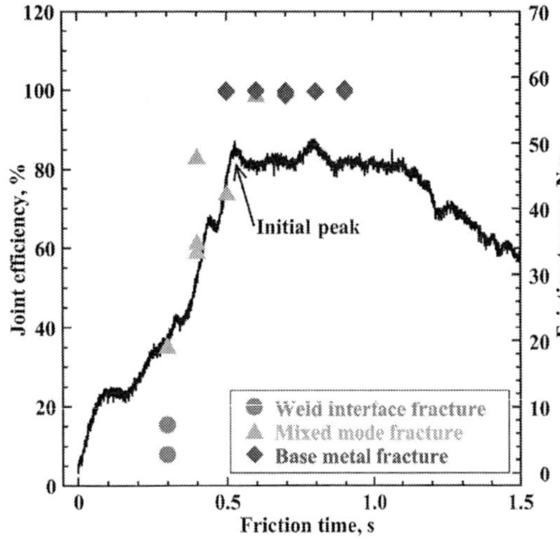

Fig.4 Relationship between joint efficiency and friction time, in relation to friction torque; flat shape specimen.

Fig.5 Examples of joint tensile tested specimens; flat shape specimen.

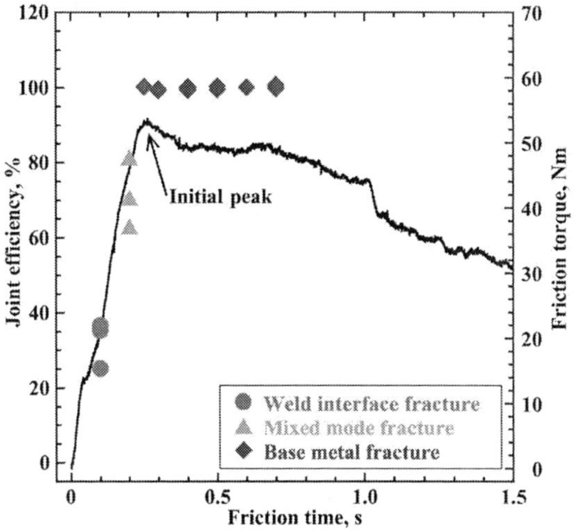

Fig.6 Relationship between joint efficiency and friction time, in relation to friction torque; taper shape specimen.

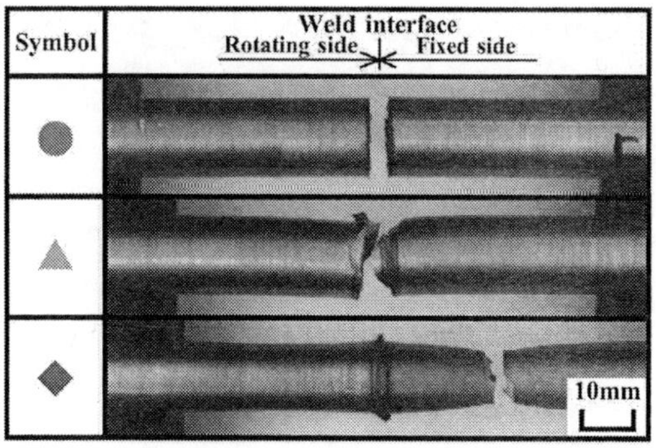

Fig.7 Examples of joint tensile tested specimens; taper shape specimen.

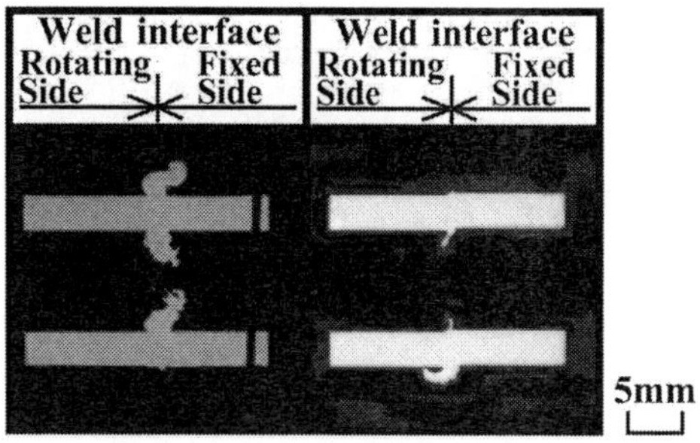

(a) Flat shape (b) Taper shape

Fig.8 Cross sections of joins.

specimens for joints with taper shape specimens. When joints were made at a friction time of 0.1s (the weld interface had sparkled) the joint efficiency of those was about 32%, and which was showed in Fig.6. The joint efficiency increased with increasing friction time, and then the fractured mode of joints changed to the mixed mode fracture (solid triangular symbols) from the weld interface fracture (solid circular symbols). Then, the joint efficiency achieved 100% when joints were made at a friction time of 0.25s, i.e. the friction torque just reached the initial peak. Also, all joints fractured in the base metal, which was indicated by solid rhombus symbols in Fig.7. Thereafter, the joint efficiency was kept about 100% and all joints had the fracture on the base metal. Therefore, it was clarified that the joint, which had the joint efficiency of 100% with the fracture on the base metal, was obtained when the friction torque reached the initial peak.

Figure 8 shows examples of the appearances of the cross-section for joints with the fracture on the base metal at various friction times. When the joint with flat shape specimens was made with a friction time of 0.7s, the inner flashes almost filled the hole of the steel pipe as shown in Fig.8a. On the other hand, when the joint with taper shape specimens was made with a friction time of 0.25s (the initial peak), it had little inner flashes as shown in Fig.8b. Therefore, it was considered that the taper shape was effective for decreasing the flash, although the further investigation will be necessary to clarify the effect of the shape for weld faying parts.

Conclusions

This paper described the effect of shape for weld faying parts on joint strength of steel pipe friction welded joints. The prepared specimens have two types of the flat and taper shapes, and those were welded with the combination of the same shapes by a direct drive friction welding machine that has an electromagnetic clutch. The following conclusions are provided.

(1) When the joint was made with flat shape specimens, the initial peak was approximately 50Nm and the elapsed time for the initial peak was about 0.5s. The joint, which had the joint efficiency of 100% with the fracture on the base metal, was obtained when it was made with long friction time such as 0.7s or longer.

(2) When the joint was made with taper shape specimens, those values were approximately 50Nm and about 0.25s, respectively. The joint, which had the joint efficiency of 100% with the fracture on the base metal, was obtained when the friction torque reached the initial peak, i.e. a friction time of 0.25s.

(3) It was clarified that the inner flash of the joint with the fracture on the base metal for taper shape specimens was smaller than that of flat shape one.

Reference

[1] For example, M.V.Kumar, V.Balasubramanian: "Microstructure and tensile properties of friction welded SUS 304HCu austenitic stainless steel tubes", International Journal of Pressure Vessels and Piping, Vol. 113(2014), pp. 24-31.

Observation of Welding Phenomenon on Horizontal Steel Stud of 19mm Diameter

Yoshimoto Nakaseko, Jippei Suzuki, Hiroshi Kawakami and Hitoshi Ozaki
Graduate School of Engineering, Mie University, tsu-city, mie, Japan
Email: 412da02@m.mie-u.ac.jp phone: +81-59-231-9373

Introduction

In previous study, we have found two types of lack of fusion. One is insufficient melting of interface by low heat input for welding and the other is burn through by excessive heat input. However a good weld can be expected with the appropriate heat input between both these, a width of the optimum range is narrow [1][2]. If the area of weld puddle is larger than a cross-section of the stud, and if the molten metal can be prevented from falling from weld puddle, a sound weld can be obtained under the wider range of welding condition. Therefore, the weld defects were investigated for two kinds of horizontal stud welds, which were welded by using the ferrules having different gap between ferrule and stud. The changing the set direction of the ferrule was actually adopted in order to change the gap.

Experimental methods

A steel stud of 19mm diameter was made from carbon steels for machine structural use (JIS S10C), and a base metal plate was rolled steel for general structure(JIS SS400) (100×100×16 mm). Figure1(a) shows the setup of the commercial ferrule for flat usage in horizontal position. In this case, the ferrule has a large gap to make a sufficiently big collar. In this study, experiments were carried out with setting a ferrule in opposite direction to get a small gap, as shown in Fig.1(b), and the influence of the gap volume was studied. Hereafter, the experiment with large gap is called the normal direction and that with small gap is called the opposite direction.

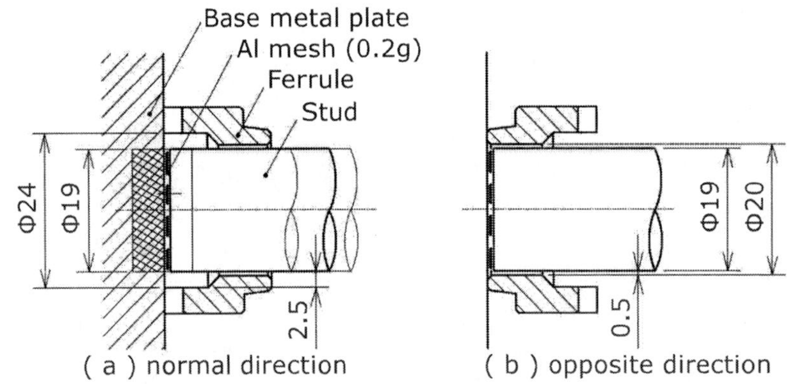

Fig.1 Motion of stud during welding process

The aluminum net(0.2g) was cut out at the same size as the end face of stud. And it was set up between stud and base metal plate instead of the aluminum ball(0.2g) embedded at the end face of commercial stud to ignite an arc. At the start of welding, the distance was 0.5mm between end face of stud and base metal plate. Weldings were carried out by using the power supply of direct current in the straight polarity(stud negative). The stud was separated 3.5 mm from base metal plate at the ignition of welding arc. After a given time, stud was forced into weld puddle, and the arc disappeared, and a welding was completed. The displacement of stud from the initial position to the final position is called the plunge distance. The set value of the plunge distance was 6mm. The actual plunging movement was stopped on the way to a given final position, because the solid metal collided each other.

As described above, there are two types of lack of fusion in the weld. One is the insufficient melting of the interface between stud and base metal plate due to the shortage of heat input, and this defect is shortly written unmelted zone. And the other is the burn through of molten metal caused by the elimination of excessive molten metal from the weld puddle. Both weld defects were judged by visual examination of the weld and those areas were measured with X-ray inspection.

Results

Figure2(a) shows an example of burn through in a weld observed by X-ray inspection. So a porosity was found in weld as shown in Fig.2 (b), a porosity was also examined.

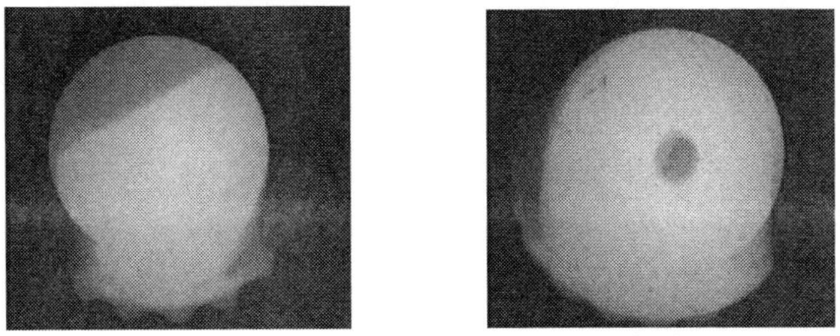

(a) (b)

Fig.2 Examples of welds including the defect of (a) burn through and (b) porosity observed by X-ray inspection

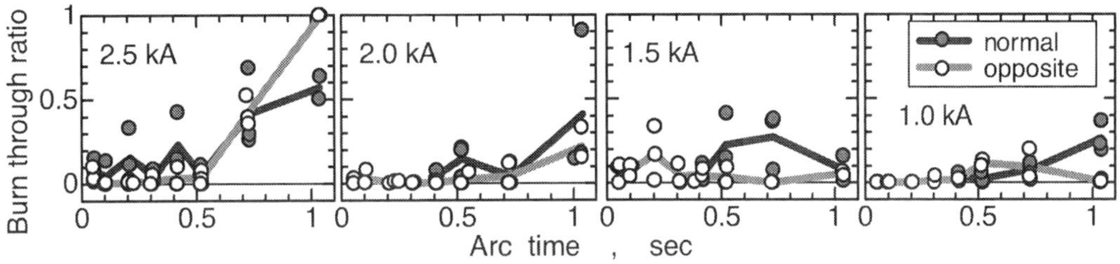

Fig.3 Influence of arc time on burn throufh ratio

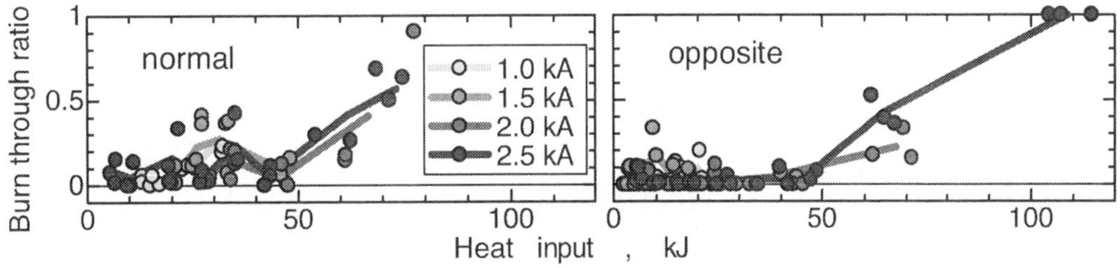

Fig.4 Influence of heat input on burn through ratio

The ratio of the projected area of burn through to the cross-sectional area of the stud is shortened to burn through ratio. Figure3 shows the burn through ratio with increasing arc time. When an arc time exceed a certain value, burn through ratio increases suddenly. This tendency becomes remarkable with an increase of welding current. The burn through ratio in normal direction is higher than that in opposite direction. As we consider that the burn through is caused by excessive heat input, the experimental results in Fig.3 are plotted against the heat input in Fig.4. In the case of the opposite direction with the heat input of 30~40kJ, the burn through ratio was low.

As the other type of lack of fusion, Fig.5 indicates the influence of arc time on unmelted ratio. A insufficient melting occurs frequently at short arc time, and the ratio is decreased with the increase in arc time. Arc time at when the unmelted zone does not occur is shifted to the short arc time with the increase of welding current. We also thought occurrence of unmelted zone is affected by heat input, we show in Fig.6 the influence of heat input on unmelted ratio.

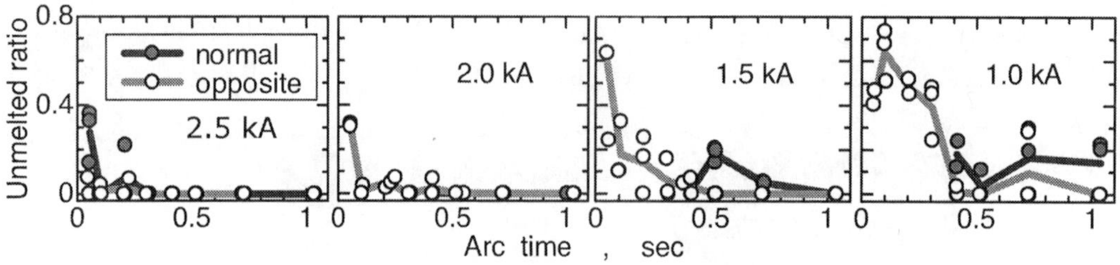

Fig.5 Influence of arc time on unmelted ratio

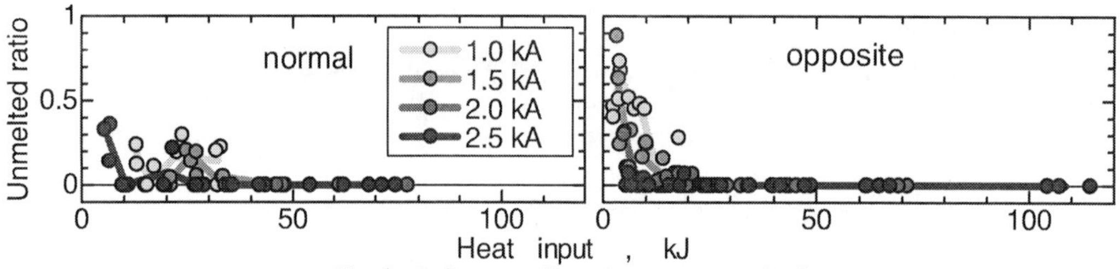

Fig.6 Influence of heat input on unmelted ratio

Unmelted ratio is decreased with the increase of heat input, and in the case of the opposite direction, the circumference of stud was melted by heat input smaller than that of the normal direction.

From the results of measurement of porosity, the porosity ratio becomes high in low welding current and in long period of arc time at the normal direction, but the porosity ratio itself is low. The influence of arc time on porosity ratio is indicated in Fig.7.

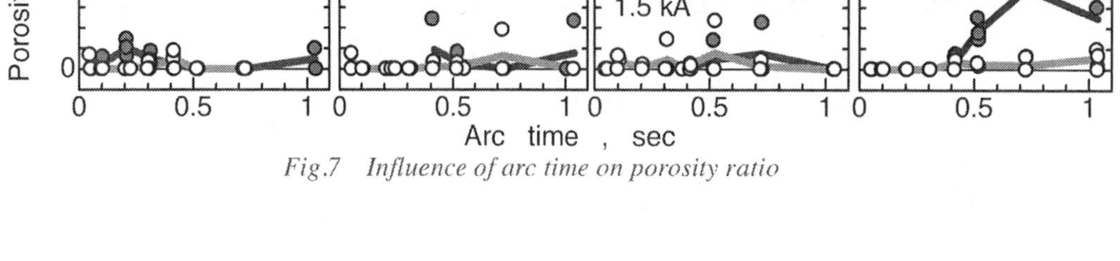

Fig.7 Influence of arc time on porosity ratio

The defect ratio is the index representing the influence of weld defect on the reduction of the cross-sectional area of the welds. The total defect ratio is the sum of the ratio of burn through, unmelted zone and porosity. The influence of heat input on total defect ratio, is shown in Fig.8.

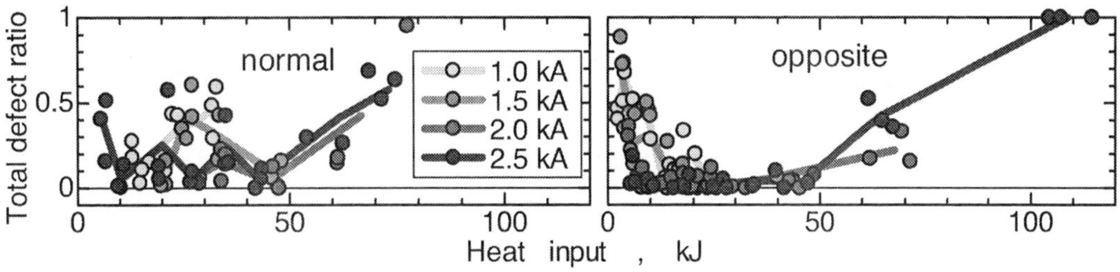

Fig.8 Influence of heat input on total defect ratio

The total defect ratio at the normal direction is high over the whole period of arc time in each welding current insofar as the experimental results, and the change of ferrule direction reduces the welding defect at about 30kJ of heat input.

Conclusions

The following conclusions may be drawn from this study:

1. Narrow gap between the stud and the ferrule suppress the weld defect when the heat input is about 30kJ for the stud of 19mm in diameter.
2. Narrow gap decreases the occurrence of porosity in low welding current and long arc time.

References

[1] NAKASEKO, *et al* : PRE PRINT OF THE NATIONAL MEETING OF J.W.S., Vol.93, p.146-147

[2] NAKASEKO, *et al* : PRE PRINT OF THE NATIONAL MEETING OF J.W.S., Vol.94, p.24-25

Strain Measurement in Arc Welding using Digital Image Correlation

Zongyao Chen[1,2], Jian Chen[1], Zhili Feng[1,2]
1. Oak Ridge National Laboratory, Oak Ridge, TN, USA
2. University of Tennessee, Knoxville, TN, USA
Email:chenz@ornl.gov,chenj2@ornl.gov,fengz@ornl.gov 18655745974

Abstract

A newly developed experimental system using Digital image correlation (DIC) technic was applied to the strain measurement during tungsten inert gas (TIG) welding. When the measured distance is close to the fusion line during welding, the thermal disturbance caused by the heated air flow will generate large noise in the calculated strain. In this paper, the influence of the thermal disturbance on the DIC strain calculation was investigated and the strain noise was effectively reduced via data filtering. Finally, a welding experiment was conducted using the proposed method to measure the strain near the fusion zone on a DP600 plate.

Keywords: Digital image correlation, arc welding, thermal disturbance

Introduction

The strain analysis is essential to understand the formulation of residual stress and distortion during arc welding. The conventional strain gauge method is not suitable to the in-situ measurement due to the elevated temperatures. Recently, digital image correlation (DIC), as a non-contacted method to obtain the full field strain [1], has been attempted for strain measurement during welding. Due to the intense arc light and high temperatures, most of the in-situ DIC measurements were conducted on the backside or far away from the arc. Strycker et al. [2] first applied DIC on the tube during arc welding. Coules et al. [3] applied DIC to measure the strain at 20mm and 30mm away from the weld line.

Recently, a novel high-temperature DIC technique has been developed to measure the strain evolution near the fusion line during welding, thanks to the development of high-temperature speckle patterns and a special illumination method [4]. Further investigations found that relatively large noise existed in the measured strain data, especially in the locations close to the fusion line. This was mainly due to the thermal disturbance (the turbulence of heated air) causing shimmering and distortion of the DIC images [5].

In this research, experiment was conducted to investigate the influence of thermal disturbance on DIC measurement as a function of temperature. The Savizky-Golay smooth filter was applied on the calculated strain to effectively reduce the noise of the strain measurement. Finally, bead-on-plate welding experiments were performed on a carbon steel DP600 plate to verify the feasibility of the proposed method.

Experiments

Thermal disturbance from heated air is a major source of noise for DIC measurement during arc welding. In order to evaluate the influence of the thermal disturbance, an experiment was conducted to quantify the noise as a function of temperature. The high-temperature-resisting speckle patters were painted on the sample surface. The sample was placed on a heating plate and continuously heated from 27° C to 300° C in 10 min. Temperature was measured by thermal couples. Meanwhile, the image sequence was recorded at 15 Hz for DIC strain measurement.

The schematic of the welding experiment setup for strain measurement is shown in Fig. 1. The work piece was clamped firmly on a workbench. Thermocouples were welded on the plate to measure the temperature. Special treatment on surface speckle patterns was conducted to improve its resistance to high temperature. A special illumination and optical filtering system was applied to eliminate the interference of arc light so that high contrast images of the speckle patterns can be captured by the digital camera during the welding process.

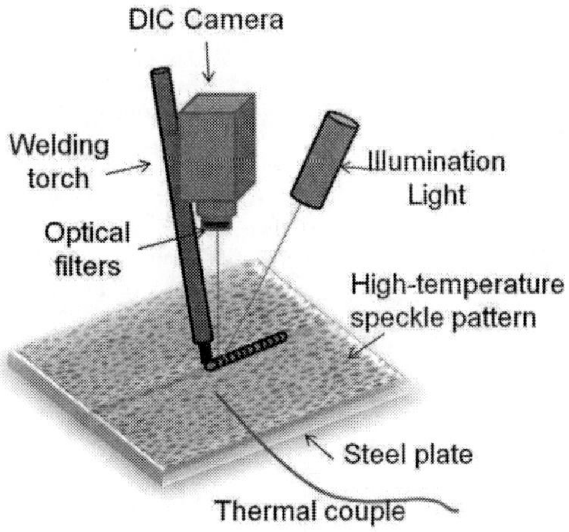

Fig. 1 Schematic of Experiment setup during arc welding

Bead-on-plate TIG welding experiments were performed on a 6mm-thick-stainless steel plate with the welding parameters listed in Table 1. Argon shielding gas at 10CFH is used to protect the surface from oxidation. During the welding, the digital camera and the illumination light were fixed on a tripod, while the welding torch was moving towards left. The digital camera was perpendicular to the work piece, and the welding torch was tilted as shown in Fig.1.

Table 1: Welding parameters

Current	160A
Voltage	12V
Welding speed	10 inch·min^{-1}
Wire feeding speed	20inch·min^{-1}
Wire diameter	0.045 inch

Result and Discussion

1) Study of Noise from Thermal Disturbance

In the experiments where the sample was heated by the heating plate, three surface strain components (ε_{xx}, ε_{yy} and ε_{xy}) as a function of temperature were plot in Fig. 2. The subscripts x and y represent two orthogonal directions on DIC images. Overall, two normal strain components increased with temperature due to the material's thermal expansion. The noise level of both normal and shear strain components also increased with temperature. At room temperature, the noise is less than $\pm 10^{-3}$, while it reached the magnitude of $\pm 10^{-3}$at about 300° C. Savizky-Golay filter was applied to reduce the noise as shown in Fig. 2. The filtered strain curves are also shown in the figures.

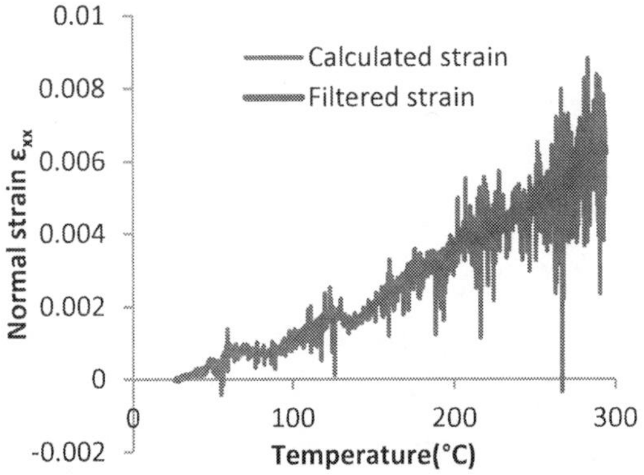

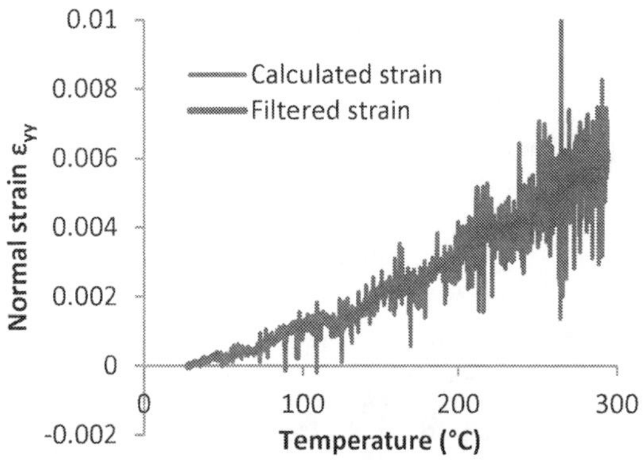

Fig. 2 (a)Normal strain ε_{xx} in horizontal direction *(b)Normal strain ε_{yy} in vertical direction*

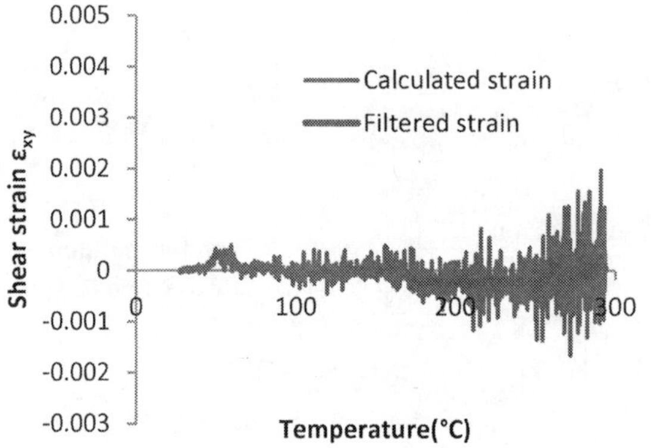

(c) Shear strain ε_{xy} with increasing temperature

2) ***Strain Measurement during Arc Welding***

In Fig. 3, an in-situ DIC measurement was applied for TIG welding on DP600 steel plate. The influence of the arc light was eliminated by the band pass filter. The red circle indicates the welding pool position which was blocked by the welding torch. The color map indicates the distribution of the transverse strain (perpendicular to the welding direction) on the selected frame.

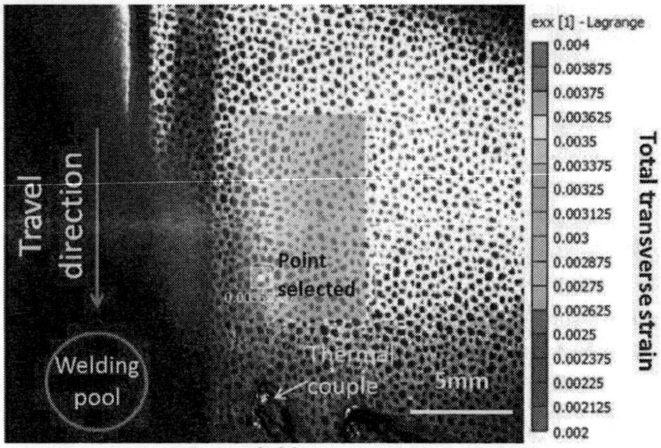

Fig. 3 Strain field measurement using DIC

The time history of strain at the selected point (6.5mm apart from the fusion line) in Fig. 3 is shown in Fig. 4(a). Savizky-Golay filter was applied to effectively reduce the noise. Results show that, at the selected location, the transverse strain quickly increased to a peak value of 0.0032 within approximately 5s and slowly reduced to approximately zero after 100s. It is worth to mention that measured strain during welding includes not only the thermal expansion, but also the material's elastic and plastic deformation. Fig. 4(b) shows the temperature curve measured by the thermal couple at another location (indicated in the figure). The distance from the fusion line was also 6.5mm.

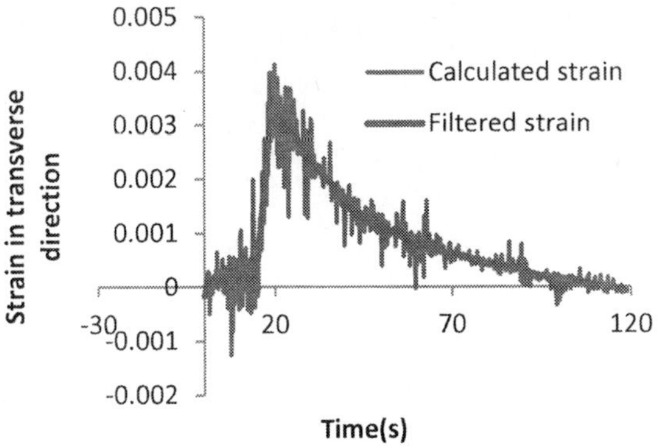

Fig.4 (a) DIC strain data during arc welding

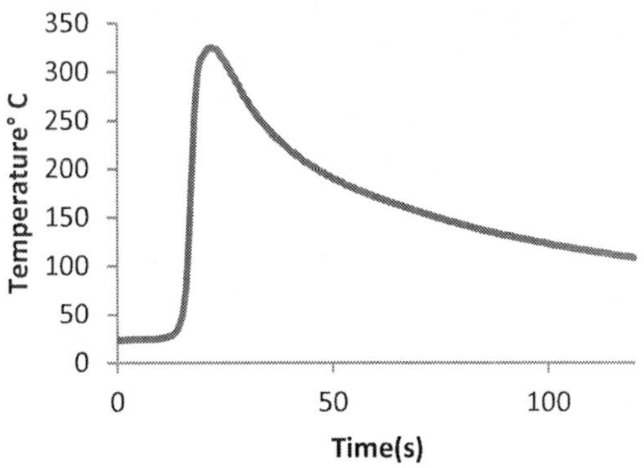

Fig.4(b) Measured temperature from thermal couple

Conclusions

The thermal disturbance during arc welding introduces large noise on the strain measurement with DIC. The strain noise caused by the thermal disturbance is related with the temperature. And this noise can be removed by applying Savizky-Golay filter. By using the newly developed inspection experiment system, in situ strain measurement on DP600 can be conducted at the very close distance to the welding edge.

Acknowledgments

Research supported by US Department of Energy, Office of Nuclear Energy. This research was sponsored by the US Department of Energy, Office of Nuclear Energy, for Nuclear Energy Enabling Technologies Crosscutting Technology Development Effort, under a prime contract with Oak Ridge National Laboratory (ORNL). ORNL is managed by UT-Battelle, LLC for the U.S. Department of Energy under Contract DE-AC05-00OR22725.

References

[1] M. A. Sutton, et al: Image correlation for shape, motion and deformation measurements: basic concepts, theory and applications: Springer Science & Business Media, 2009.
[2] M. De Strycker, P. Lava, W. Van Paepegem, L. Schueremans, and D. Debruyne, "Measuring welding deformations with the digital image correlation technique," Welding Journal, vol. 90, pp. 107-112, 2011.
[3] H. Coules, P. Colegrove, L. Cozzolino, and S. Wen, "Experimental measurement of biaxial thermal stress fields caused by arc welding," Journal of Materials Processing Technology, vol. 212, pp. 962-968, 2012.
[4] J. Chen, X. Yu, R. G. Miller, and Z. Feng, "In situ strain and temperature measurement and modelling during arc welding," Science and Technology of Welding and Joining, vol. 20, pp. 181-188.
[5] Z.Chen, J.Chen, Z.Feng,Y.Zhang, "In situ strain monitoring in gas tungsten arc welding process"(Accepted)

Fatigue property of A7N01P-T4 Al alloy plates joined by hybrid laser-MIG welding

Qiuying Wang, Hui Chen, Xiaoyi Yang, Zongtao Zhu, Lidong Xu

Southwest Jiaotong University, China

High strength Al alloy (e.g. A7N01) is widely used on high-speed train body. For Al alloy high-speed train body, fatigue cracking one of the main failure modes. Welded joints are highly risky positions of fatigue failure. Hybrid laser-MIG welding method is considered as one of the methods which can replace MIG welding method in the manufacturing of high-speed train body. Research on fatigue property of hybrid laser-MIG welded joint is very necessary. A7N01P-T4 plates were joined by hybrid laser-MIG welding. Fatigue strength of hybrid laser-MIG welded joint was tested. Fractures were examined by optical microscopy (OM) and scanning electron microscopy (SEM). Microstructures of A7N01P-T4 welded joints were observed by OM, SEM and transmission electron microscopy (TEM). After natural aging for one month, fatigue strength of A7N01P-T4 hybrid welded joint could reach 141MPa, which was 25% higher than that in as-welded state and 41% higher than that of MIG welded joint. All the fracture positions of the fatigue specimens are at weld zone. Fatigue initiations were porosities or sites at the surface or near surface. Fatigue crack growth rate of weld zone of hybrid welded joint was also studied, and compared with that of MIG welded joint. And the relationship between mechanical properties and microstructure was discussed.

Bonding strength of friction stir welded joint of WC-12Co and medium carbon steel

Toru Nagaoka, Yoshihisa Kimoto, Takashi Takeuchi, Masao Fukusumi
Osaka Municipal Technical Research Institute, Osaka, Japan
Email: nagaoka@omtri.or.jp, TEL:+81-6-6963-8157

Yoshiaki Morisada, Hidetoshi Fujii
Joining and Welding Research Institute, Osaka University, Osaka, Japan

Abstract
Friction stir welding of WC-12 mass% Co and 0.44 mass% C steel was investigated. A shear test was carried out to clarify the bonding strength. Sound joints were obtained after friction stir welding at a forging load of 9.8 kN. The obtained joint showed higher strength than a brazed joint. Fracture occurred in the cemented carbide during the shear test. No compounds were formed on the bonding interface during the welding process.

Introduction
Hard metal comprises of WC particles and metallic binder, resulting in high hardness and high toughness. Therefore, they are widely used for cutting tools, dies, molds, and so on [1-3]. Due to their poor machinability and high cost, they are often bonded to metallic materials by brazing [4,5]. However, the brazed joints have some problems such as low bonding strength and corrosion due to a flux residue. On the other hand, friction stir welding (FSW) does not need any insert materials and any fluxes [6]. It could be a better process to reduce the joining cost and produce high strength joints. In this study, FSW of WC-12 mass% Co and 0.44 mass% C steel was conducted. A shear test was carried out to clarify the bonding strength.

Experimental procedures
A WC-12 mass% Co plate and a 0.44 mass% C steel plate were bonded by friction stir welding. The chemical composition of the steel was given in Table 1. The size of the hard metal plate and the medium carbon steel plate were 15 mm × 45 mm × 1.0 mm and 15 mm × 45 mm × 2.4 mm, respectively. The hard metal plate was overlapped by the steel plate. FSW was carried out as shown in Fig. 1. A tool with a probe was used. The diameter of the shoulder was 12 mm. The length of the probe was 2.4 mm, which was same as the thickness of the steel plate. The diameter of the probe tip was 4.9 mm. The rotating speed and the traveling speed were 800 rpm and 50 mm/min, respectively. The tool tilt angle was 3°. The forging load was varied from 4.9 kN to 14.7 kN. The processing area was shielded by Ar gas (20 L/min).

Table 1: Chemical composition of medium carbon steel in this study (mass%).

C	Si	Mn	P	S	Fe
0.44	0.23	0.69	0.010	0.005	Balance

An optical microscope (Nikon, EPIPHOT-TME) and a scanning electron microscope (Jeol, JSM-6610LA) were used for observing microstructures. Microhardness of the cross-section was measured by using a micro-Vickers hardness tester (Future-tech, FM-300) at a load of 2.94 N. A bonding tester (Rhesca, PTR-1102) was used for measuring shear strength. The test piece was cut off from the processing centerline. The size of the projection portion comprising of WC-12Co was 1 mm × 1 mm × 1 mm. The tool speed in the shear test was 1 mm/min.

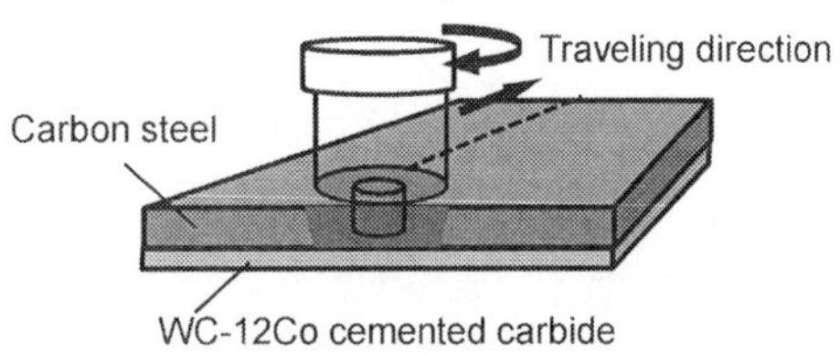

Fig. 1 Schematic illustration of friction stir welding of WC-12Co and carbon steel.

Results and discussion

FSW of WC-12Co to carbon steel was successfully carried out. Fig. 2 shows the lap joint obtained by FSW at a load of 9.8 kN. Friction stir weld was observed on the steel side. No cracks were observed on the hard metal side. When the forging load was 4.9 kN or 7.4 kN, the probe could not be plunged into the steel plate. On the other hand, the hard metal plate was broken during FSW at a forging load of 12.3 kN or 14.7 kN.

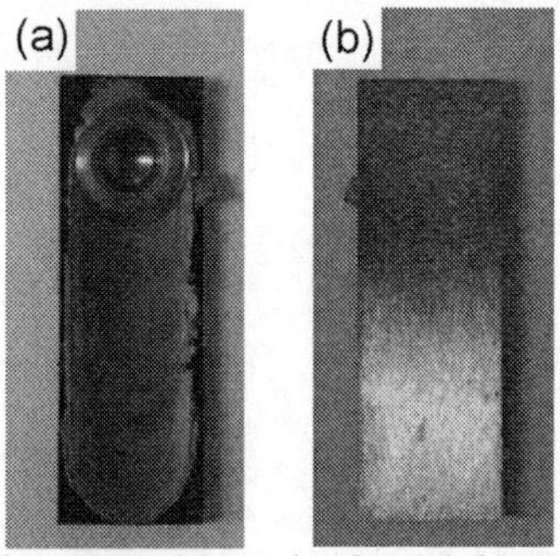

Fig. 2 WC-12Co/carbon steel lap joint after FSW: (a) steel side and (b) WC-12Co side.

Fig. 3 shows the cross-section of the joint welded at a load of 9.8 kN. Although a tunnel hole defect was observed on the advancing side, the stir zone in the steel plate reached to the surface of the hard metal plate. The weld width on the interface was 5.1 mm, which was broader than the probe tip diameter of 4.9 mm. Material flow around the probe could contribute to broadening of the weld area.

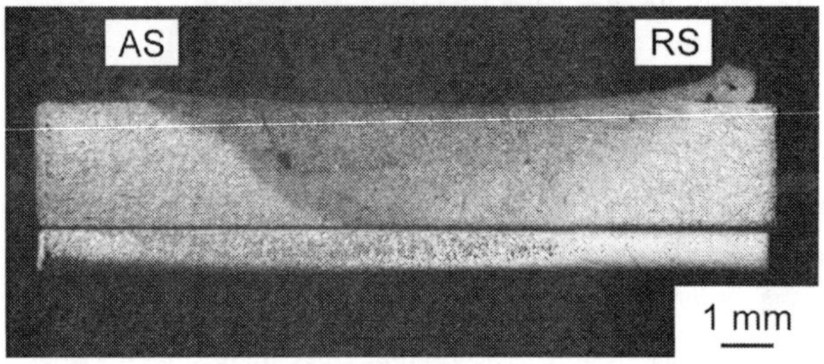

Fig. 3 Cross-section of WC-12Co/carbon steel joint after FSW.

Hardness distribution of the welded joint is shown in Fig. 4. The hardness of WC-12Co after FSW was equivalent to that before FSW (1590–1760 HV). The hardness of carbon steel was harder than that before FSW (200 HV). The stirring effect refined the microstructure in the stir zone, which led to higher hardness.

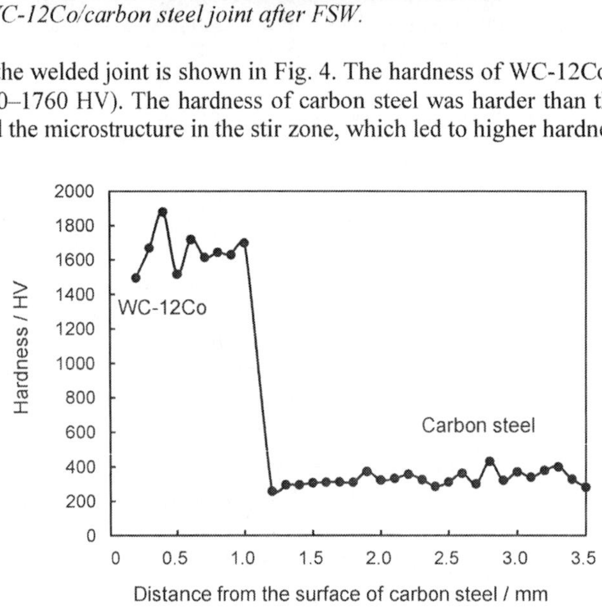

Fig. 4 Hardness profile across bonding interface of WC-12Co/carbon steel joint.

Fig. 5 shows elemental EDS line profile across the bonding interface. No intermetallic compounds were observed on the interface. Based on the result of elemental analysis, W was diffused to the steel side during FSW. It is considered that some WC particles near the bonding interface were dissolved into the carbon steel.

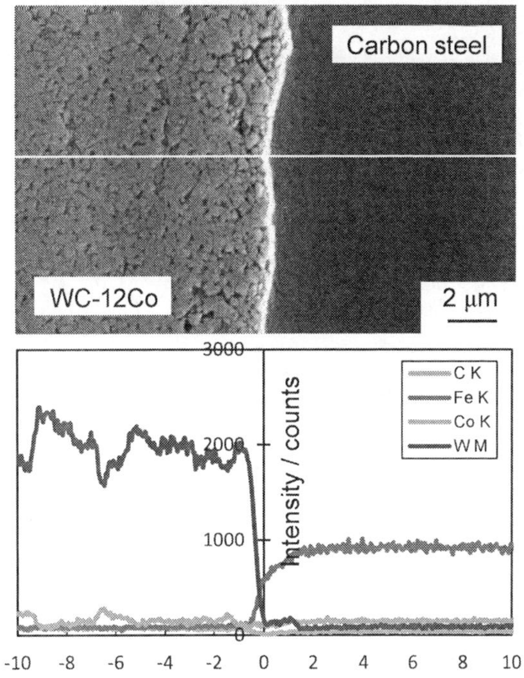

Fig. 5 Elemental EDS line profile across bonding interface of WC-12Co/carbon steel joint.

As a result of shear test, shear strength of the welded joint was 495–736 MPa, which is higher than that of a brazed joint of WC-Co and carbon steel (346–441 MPa) [4]. It is because FSW can produce a joint without inserting any soft materials. Fig. 6 shows fracture surface of the welded joint after shear test. Fracture occurred in the hard metal, which means that the interface was sufficiently bonded by FSW.

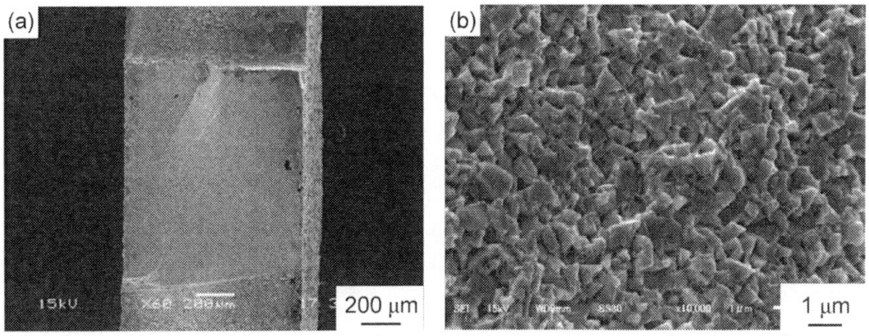

Fig. 6 Fracture surface of WC-12Co/carbon steel joint after shear test: (a) Low magnification and (b) High magnification.

Conclusions

FSW of a WC-12Co plate to a medium carbon steel plate was carried out. Obtained results are summarized as follows:

(1) Hard metal and carbon steel were successfully bonded by FSW at a load of 9.8 kN. When the forging load was lower, the probe of the rotating tool could not be plunged into the steel plate. On the other hand, the hard metal plate was broken during FSW at a higher load.

(2) As a result of shear test, shear strength of the friction stir welded joint was higher than that of a brazed joint. Fracture occurred in the hard metal of the welded joint.
(3) Although diffusion of W into the steel was observed near the bonding interface, generation and growth of intermetallic compounds were suppressed during FSW.

References

[1] S. Mi, T.H. Courtney: "Synthesis of WC and WC-Co cermets by mechanical alloying and subsequent hot isostatic pressing", *Scripta Materialia*, Vol. 38 (1998), pp. 171–176.

[2] G.S. Upadhyaya: "Materials science of cemented carbides - an overview", *Materials & Design*, Vol. 22 (2001), pp. 483–489.

[3] H.O. Andren: "Microstructures of cemented carbides", *Materials & Design*, Vol. 22 (2001), pp. 491–498.

[4] H. Ohmura, K. Kawashiri, T. Yoshida: "Cemented tungsten carbide/carbon steel joint brazed with copper", *Quarterly Journal of The Japan Welding Society*, Vol. 6, No. 4 (1988), pp. 499–504.

[5] S. Yaoita, T. Watanabe, T. Sasaki: "Brazing of Cemented Carbide Using Ag-based Filler Metals with a Low Melting Point", *Quarterly Journal of The Japan Welding Society*, Vol. 29, No. 3 (2011), pp. 204-209.

[6] W.M. Thomas, E.D. Nicholas, J.C. Needham, M.G. Murch, P. Templesmith, C.J. Dawes: International Patent Application PCT/GB92/02203 and GB Patent Application G.B. Patent Application 9125978.8, 1991.

Effect of process parameters on tensile strength in laser-MIG hybrid welding for aluminum alloy

Yong Chen, Hui Chen*, Tao Yang, Minhao Zhu
School of Materials Science and Engineering, Southwest Jiaotong University, Chengdu, China.
E-mail: yognchen@my.swjtu.edu.cn, phone: +86 028 87600726

Abstract
Hybrid laser arc welding technology attracts a considerable attention recently due to its advantages. This technology has been widely applied to weld aluminum alloy. This paper aims at revealing the correlation of hybrid laser arc welding process parameters and tensile strength of the aluminum alloy. A central composite design was used to perform experiments for obtaining input-output data. The results shown that the confidence level of the developed model was over ninety-five percent. Influences of the single factor and interactive effect on tensile strength are analyzed. Laser power is the most significant factor on tensile strength. The interaction effect of laser power and welding speed on tensile strength is significant. Using the response surfaces plots, the optimal hybrid laser arc welding parameters for maximum tensile strength are calculated.

Introduction
Aluminum alloys are widely employed in many fields, such as aerospace, shipbuilding, and automotive, for its low density, superior strength-to-weight ratio, corrosion resistance, good formability and recyclability [1, 2]. Laser-MIG hybrid welding, combining the merits of conventional MIG welding and laser welding, is designed to overcome the problems for welding aluminum alloy. In recent years, laser-MIG hybrid welding has been demonstrated to be a powerful tool for welding aluminum alloy due to high weld speed and penetration necessary, the gap bridging ability, and lower heat input [3, 4].

As we know, tensile strength (TS) is the most important mechanical property of weld joints. It is also widely used to judge the welding quality. However, softening of the HAZ in the base metal has greatly limited the application of aluminum alloys [5, 6]. In laser-MIG hybrid welding process, the mechanical properties of weld joint are directly influenced by the welding current, laser power, welding speed, and so on. A great deal of previous research has been done to study the correlation between parameters and weld bead geometry and formation [4, 7]. But, the key problems in laser-MIG hybrid welding for aluminum alloy are how to obtain expected tensile strength and how the process parameters affect the tensile strength. Consequently, it is essential to reveal the relationship between process parameters and the tensile strength, and optimize the tensile strength.

Nevertheless, to our knowledge little systematic investigation has been reported to disclose the relationship between welding process parameters and tensile strength, as well as the effect of process parameters on tensile-strength of laser-MIG hybrid welded joints for aluminum alloy. This paper aims at establishing a mathematical model of laser-MIG hybrid welding parameters on tensile strength, for understanding the multiple process parameter interactions and achieving optimum welding parameters for aluminum alloy.

Experimental materials and equipment
The workpiece is Al-Mg-Si aluminum alloy with dimensions of 300 mm×150 mm×3.6 mm. Butt joint with no clearance is used as the joint form for the laser-MIG hybrid welding. Filling metal with a diameter of 1.2 mm was adopted. The main chemical compositions of the base metal and filler metal are shown in *Table 1*. Argon gas (99.999%) was supplied at a constant flow rate of 25 L/min as shielding gas.

Table 1: Chemical compositions (wt. %) of aluminum alloy

element	Si	Cu	Mn	Mg	Cr	Zn	Ti	Fe	Al
base metal	0.5-0.9	0.3	0.5	0.4-0.7	0.3	0.2	0.1	0.35	Bal.
filler metal	≤0.25	≤0.10	0.05-0.20	4.5-5.5	0.05-0.20	≤0.10	0.06~0.20	0.4	Bal.

The Laser-MIG welding system was composed of an IPG YLS-4000 laser and a Kemppi pulse MIG welding power supply. The incident laser beam was inclined by 10° in order to avoid any damage to the laser equipment from the laser reflection, and the angle of the welding torch to the workpiece was 67°, as detailed in Fig. 1.

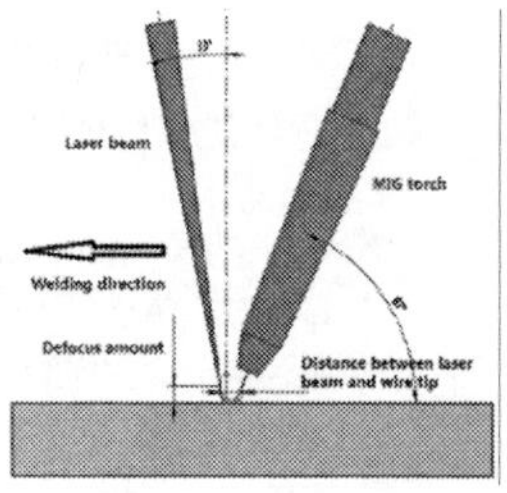

Fig. 1: Geometrical parameters of the hybrid welding

As we all know, there are many process parameters determine the tensile strength of laser-MIG hybrid welded joints for aluminum alloy. Optimization through response surface method has been applied Statistical software Design-Expert 8.05 was used to develop the mathematical model and to analyze and optimize the laser-MIG hybrid welding parameters. The main process parameters and its range were identified from preliminary experiments, their limits and units are shown in *Table 2*. The experiment was designed based on a central composite rotatable six-factor five-level factorial design with 1/2 replication.

Table 2: The main process parameters of laser-MIG hybrid welding

Parameters	MIG voltage (V)	MIG current (A)	Welding speed (m/min)	laser power (kW)	distance of laser and wire (mm)	Defocus amount(mm)
Symbol	U_M	I_M	v	P	d	f
+2.378	23.00	165.00	1.60	3.50	4.00	4.00
+1	22.13	154.86	1.43	3.07	3.13	1.68
0	21.50	147.50	1.30	2.75	2.50	0.00
-1	20.87	140.14	1.17	2.43	1.87	-1.68
-2.378	20.00	103.00	1.00	2.00	1.00	-4.00

The coded values of these factors were designated as -2.378, -1, 0, + 1, +2.378, respectively. Actual and coded values of various factors used in the study are shown in Table 2. The relation between the actual and coded value of a factor is shown as follows:

$$Z_i = \frac{2[2Z - (Z_{max} + Z_{min})]}{Z_{max} - Z_{min}} \qquad (1)$$

Where Z is the value of a variable rang between Z_{min} and Z_{max}, Z_i is the corresponding coded value, Z_{max} and Z_{min} are the upper and low limit of the variable, respectively. A universal tester machine is used for qualification of their relative tensile strength with a speed of 3 mm/min. The average of at least three results of tensile strength are calculated and presented in *Table 3*.

Table 3: Laser-MIG hybrid welding design matrix and measured responses

No.	Z_1	Z_2	Z_3	Z_4	Z_5	Z_6	TS(MPa)	No.	Z_1	Z_2	Z_3	Z_4	Z_5	Z_6	TS(MPa)
1	-1	-1	-1	-1	-1	-1	218.0	27	-1	1	-1	1	1	1	219.3
2	1	-1	-1	-1	-1	1	215.0	28	1	1	-1	1	1	-1	224.0
3	-1	1	-1	-1	-1	1	225.7	29	-1	-1	1	1	1	1	217.0
4	1	1	-1	-1	-1	-1	226.0	30	1	-1	1	1	1	-1	225.7
5	-1	-1	1	-1	-1	1	201.0	31	-1	1	1	1	1	-1	230.7
6	1	-1	1	-1	-1	-1	205.0	32	1	1	1	1	1	1	223.0
7	-1	1	1	-1	-1	-1	213.0	33	-2.38	0	0	0	0	0	226.0
8	1	1	1	-1	-1	1	208.0	34	2.38	0	0	0	0	0	224.0
9	-1	-1	-1	1	-1	1	223.0	35	0	-2.38	0	0	0	0	208.0
10	1	-1	-1	1	-1	-1	229.3	36	0	2.38	0	0	0	0	228.0
11	-1	1	-1	1	-1	-1	230.0	37	0	0	-2.38	0	0	0	233.7
12	1	1	-1	1	-1	1	222.7	38	0	0	2.38	0	0	0	214.0
13	-1	-1	1	1	-1	-1	229.0	39	0	0	0	-2.38	0	0	202.0
14	1	-1	1	1	-1	1	213.0	40	0	0	0	2.38	0	0	231.0
15	-1	1	1	1	-1	1	219.0	41	0	0	0	0	-2.38	0	218.0

No.	Z_1	Z_2	Z_3	Z_4	Z_5	Z_6	TS(MPa)	No.	Z_1	Z_2	Z_3	Z_4	Z_5	Z_6	TS(MPa)
16	1	1	1	1	-1	-1	225.0	42	0	0	0	0	2.38	0	203.0
17	-1	-1	-1	-1	1	1	202.0	43	0	0	0	0	0	-2.38	221.0
18	1	-1	-1	-1	1	-1	213.0	44	0	0	0	0	0	2.38	200.0
19	-1	1	-1	-1	1	-1	222.0	45	0	0	0	0	0	0	229.0
20	1	1	-1	-1	1	1	216.0	46	0	0	0	0	0	0	231.7
21	-1	-1	1	-1	1	-1	207.0	47	0	0	0	0	0	0	230.3
22	1	-1	1	-1	1	1	197.3	48	0	0	0	0	0	0	226.0
23	-1	1	1	-1	1	1	206.0	49	0	0	0	0	0	0	230.7
24	1	1	1	-1	1	-1	213.0	50	0	0	0	0	0	0	228.7
25	-1	-1	-1	1	1	-1	228.0	51	0	0	0	0	0	0	232.3
26	1	-1	-1	1	1	1	210.0	52	0	0	0	0	0	0	228.7

Results and discussion

To estimate the effect of process variables on the tensile strength, a second-order polynomial is used for the mathematical model. According to the experimental design, when the number of variables is k, the quadratic regression equation is generally expressed as:

$$y = b_0 + \sum_{i=1}^{k} b_i x_i + \sum_{i=1}^{k-1} \sum_{j=i+1}^{k} b_{ij} x_i x_j + \sum_{i=1}^{k} b_{ii} x_i^2 + \varepsilon_R \tag{2}$$

Where y is the response value of the tensile strength, x_i and x_j are the variables, b_0, b_i, b_{ij} and b_{ii} are the regression coefficients, and ε_R is the experimental error. The coefficient was calculated using the design-expert software. F-value was 32.02, implying that the model is significant. The "Lack of Fit F-value" of 1.66 implies that the "Lack of Fit" is not significantly. The value of the determination coefficient (adjusted R^2) was calculated to be 0.9730 for the response which indicated that the model can predict 97.22% of the experimental data. The final developed empirical relationships are given below:

$$\begin{aligned} y = & -1713.04 + 11.72 I_M - 85.73 v + 226.21 P - 28.74 d - 12.11 f - 0.71 I_M P + 0.08 I_M f \\ & + 65.60 vP + 24.39 vd - 1.69 Pf - 0.73 df - 1.82 U_M^2 - 0.04 I_M^2 - 22.40 P^2 - 8.27 d^2 - 1.16 f^2 \end{aligned} \tag{3}$$

The impact of single factor on the tensile strength can be confirmed by the single-effect regression coefficient. Within the set range of factors, laser power is the most significant process parameter on tensile strength.

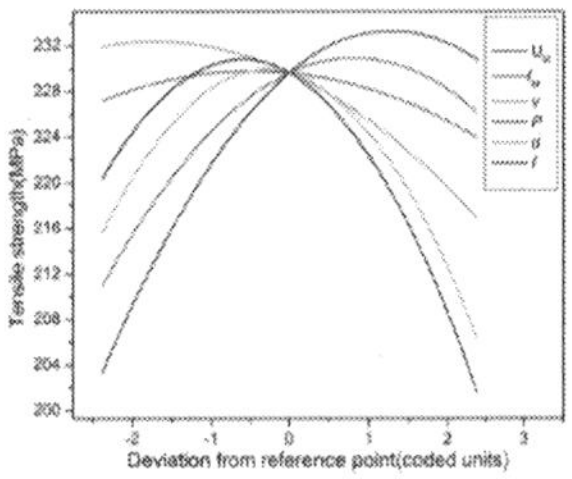

Fig. 2: Effect of process parameters on tensile strength of the welding joint

The effect of process parameters on the tensile strength is demonstrated in *Fig. 2*. It can be seen that with the increase of process parameters the tensile strength firstly increases, and then decreases. Laser beam as a kind of high power density heat source, laser power has the greatest effect on the hybrid welding heat input. With increase of laser power, welding heat input increases and the laser keyhole effect enhances, which will increase the weld penetration. The result showed that hybrid welding weld penetration is mainly determined by the laser power. In laser-MIG hybrid welding process, the spot size and the power density on the sheet surface are determined by defocus amount, which also plays an important role in the weld quality. Therefore, when the negative defocus was used, the laser energy can deliver deeper to the work piece and results in the decreases of weld penetration. Consequently, the joint strength is improved. And, the effect of welding speed on the tensile strength can be attributed to the influence of the welding heat input. The study shows that, for the laser-MIG hybrid welding, the distance of laser and wire mainly affects the synergic effect between the two heat sources. At the optimal distance of laser and wire of 2.3 mm, the synergic effect between laser and arc occurs, a deeper weld penetration and larger tensile strength is obtained. The results demonstrates that, welding parameters mainly affect the heat input and the energy distribution, which will influence the weld penetration and softening zone, eventually determine the strength of the welding joint.

As showed in *Fig. 3*, the interaction between the laser power and welding speed on tensile strength is significant. This is because the line energy is mainly decided by the laser power and varies with the welding speed. From the plot, it is clear that, the tensile strength first increases and then decreases by increasing laser power at a lower welding speed. While at a higher welding speed increasing laser power causes the increases of the joint strength. At a lower welding speed, the increased laser power will increase the line energy. But, as the heat input is higher than a certain number, further increase of the laser power will lead to the sagging of weld zone and the reduction of the tensile strength. Nevertheless, the energy density and the keyhole depth will be increased, leading to a deeper weld penetration if laser power is increased with a higher welding speed.

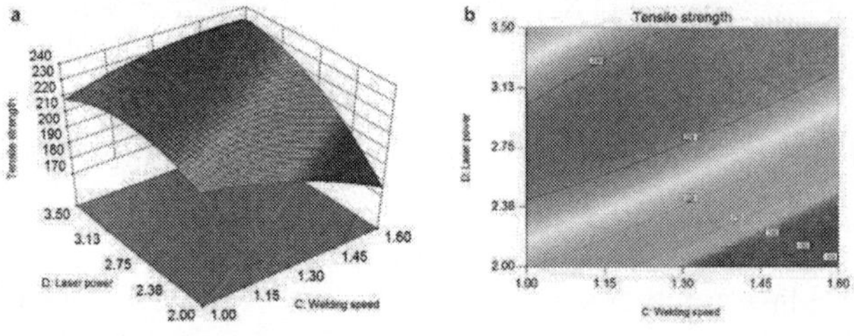

Fig. 3: Response surface (a) and contours plot (b) plot showing the effect of interaction between parameters laser power and welding speed on the tensile strength

By analyzing the response surfaces and contour plots, the maximum achievable tensile strength of the laser-MIG hybrid joint of aluminum is found to be 234.0 MPa. It is could be attained under the corresponding parameters, U_M = 20.98 V, I_M = 144.97 A, v = 1.34 m/min, P = 2.58 kW, d = 2.58 mm, f = -0.98 mm.

Conclusions

In this work, the following conclusions can be drawn based on the range of values of parameters considered.
1. The confidence level in the developed models was over than 95%. The experiment results indicate that the optimized model has high precision.
2. It can be analyzed from the results that the laser power has a significant effect on tensile strength. Laser power and welding speed, MIG current and laser power have stronger interaction effects on tensile strength.
3. The maximum tensile strength that could be attained in the laser-MIG hybrid welded joints of aluminum alloy was 234.00 MPa, as MIG voltage was 20.98 V, MIG current was 144.97 A, welding speed was 1.34 m/min, laser power was 2.58 kW, distance between laser beam and wire tip is 2.58 mm, defocus amount is -0.98 mm.

Acknowledgments

The authors gratefully acknowledge the financial support provided by the National Natural Science Foundation of China under grant no. 51474178.

References

[1] W.S Miller, *et al*: "Recent development in aluminium alloys for the automotive industry," Mat Sci Eng A, 280 (2000), pp. 37-49.
[2] G Casalino, *et al*: "Study on arc and laser powers in the hybrid welding of AA5754 Al-alloy," Mater Design, 61 (2014), pp. 191-198.
[3] SH Yan, *et al*: "Hybrid laser-Metal Inert Gas welding of Al-Mg-Si alloy joints: Microstructure and mechanical properties," Mater Design, 611 (2014), pp. 60-167.
[4] A.M. Mazar, *et al*: "Experimental and numerical investigations of hybrid laser arc welding of aluminum alloys in the thick T-joint configuration," Opt Laser Technol, 59 (2014), pp. 68-92.
[5] K.P Vijaya, *et al*: "Microstructure, mechanical and corrosion behavior of high strength AA7075 aluminium alloy friction stir welds-Effect of post weld heat treatment," Def Technol, (2015), pp. 1-8.
[6] A Karaaslan, *et al*: "Effect of aging temperature and of retrogression treatment time on the microstructure and mechanical properties of alloy AA7075," Met Sci Heat Treat, 49 (2007), pp. 9-10.
[7] H Shi, *et al*: "Applying statistical models optimize the process of multi-pass narrow-gap laser welding with filler wire," Int J Adv Manuf Technol, 75 (2014), pp. 279-291.

Effects of friction time and forge pressure on tensile strength of 5083 aluminum friction stud welded joints

Yukio Sano, Graduate Student, Graduate School of Engineering, University of Hyogo
Masaaki Kimura, Masahiro Kusaka and Koichi Kaizu, University of Hyogo
University of Hyogo, Himeji, Hyogo, Japan
mkimura@eng.u-hyogo.ac.jp

Abstract

This paper describes the effects of friction time and forge pressure on tensile strength of 5083 aluminum alloy (A5083) friction stud welded joints. In this research, the work side specimen had 32mm diameter and the stud one had 12.8mm diameter, and then those were friction welded. When the joint was made with a friction pressure 30MPa and a forge pressure 30MPa, the upsetting (deformation) occurred at only the stud side. The joint at a friction time of 1.4s (close to the initial peak torque) had approximately 29% joint efficiency. However, the joint efficiency was maintained approximately 35% in spite of increasing friction time, because the joint had the not-joined region at the weld interface. The joint efficiency increased with increasing forge pressure. Then, when the joint was made at a friction time of 1.4s and a forge pressure of 210MPa, it had approximately 75% joint efficiency. However, the joint efficiency slightly decreased with increasing forge pressure, and it was approximately 62% at a forge pressure of 360MPa. On the other hand, the joint, which was made at a friction time of 5.0s and a forge pressure of 210MPa, had approximately 63% joint efficiency. Almost all joints fractured at the initial weld interface, although a few joints fractured between at the weld interface and the base metal. To obtain high joint efficiency, the joint should be made with suitable friction time near 1.4s and forge pressure such as 210MPa.

Introduction

The joint between plate and bar is used with various industrial fields such as some construction parts and electric parts. The joint of that combination often is called as the stud joint. However, the stud joint has some problems for actual welding construction, because the joint properties will be dependent on operator skill or working conditions. Friction welding is well known amongst solid state joining methods. This method has several advantages over fusion welding, e.g., high energy efficiency, narrower heat affected zone (HAZ), and low welding cost. Hence, this friction welding method is suitable for minimizing heat input in the welding of aluminum (Al) and its alloys. In particular, friction welding is able to easily produce joints with high reliability, therefore it is widely used in the automobile industry and applied to fabricate important parts such as drive shafts, and engine valves. Accordingly, it is considered that the stud joint will be made by friction welding method. Some researchers [1] have reported that the results of the stud joint with good characteristics by friction welding method were demonstrated. However, the joining phenomena were differed with between the combination of the same weld faying surface diameter and the combination of the different weld faying surface diameter, i.e. the joining mechanism of the stud joint made by friction welding (friction stud welding) has not been fully clarified. To determine the theoretical friction welding conditions is necessary to clarify the joining phenomena of friction stud welding.

The authors have been carrying out research to clarify the joining mechanism of friction stud welding. In the present work, we show the effects of friction time and forge pressure on the tensile strength of the joint of Al-Mg alloy.

Experimental procedure

The materials used were Al-Mg alloy (JIS A5083-O, A5083) in the form of rods with diameters of 35mm and 20mm. The ultimate tensile strength of 35mm rod was 364MPa and that of 20mm rod was 349MPa, respectively. About the diameter of the weld faying (contacting) surface, 35mm rod was machined to 32mm and 20mm rod was machined to 12.8mm. In this connection, when the friction welding with the ratio of the large diameter to the small diameter was 1.75 or over, this specimen combination was able to consider the friction stud welding, because the large side specimen did not have heavily deformation [2]. All weld faying surfaces of specimens were polished with a surface grinding machine before joining in order to eliminate the effect of surface roughness on the mechanical properties of joints.

A direct drive friction welding machine was used for the joining. During the friction welding operations, the friction welding condition was set to the following combinations: a friction speed of $27.5s^{-1}$ (1650rpm), a friction pressure of 30MPa, a range of friction times from 0.8 to 5.0s, a range of forge pressures from 30 to 360MPa, and a forge time of 6.0s. The joining behavior during the friction process was recorded by a digital video camera, and the friction torque was measured with a load-cell. Then, the friction torque was recorded with a personal computer through an A/D converter. Furthermore, the specimens were joined by using an electromagnetic clutch in order to prevent deformation during braking time. When the clutch was released, the relative speed between both specimens instantly decreased to zero. The joint tensile test specimens were machined to 12.8mm in diameter and 66mm in parallel length. The joint tensile test was carried out with as-welded condition at room temperature.

Results and discussion

Relationship between friction time and joint efficiency

Figure 1 shows the relationship between the friction time and the joint efficiency, plotted alongside the friction torque curve. The joints were produced at a forge pressure of 30MPa. The joint efficiency was defined as the ratio of joint tensile strength to the ultimate tensile strength of the base metal. Figure 2 shows examples of the appearances of the tensile tested specimens. The friction torque rapidly increased when both specimens contacted each other. After that, it reached to the initial peak at a friction time of about 1.1s. The joints before the initial peak fractured at the initial weld interface, which were indicated by solid circular symbols. The joint efficiency increased with increasing friction time. The joints had a joint efficiency of 29% at a friction time of 1.4s, i.e. after the initial peak torque. These joints had the fracture between the initial weld interface and base metal side (mixed mode fracture), which were indicated by solid triangular symbols. Then, the joint efficiency maintained approximately 35%. The joints at a friction time of 5.0s fractured from the final weld interface, which were indicated by solid square symbols.

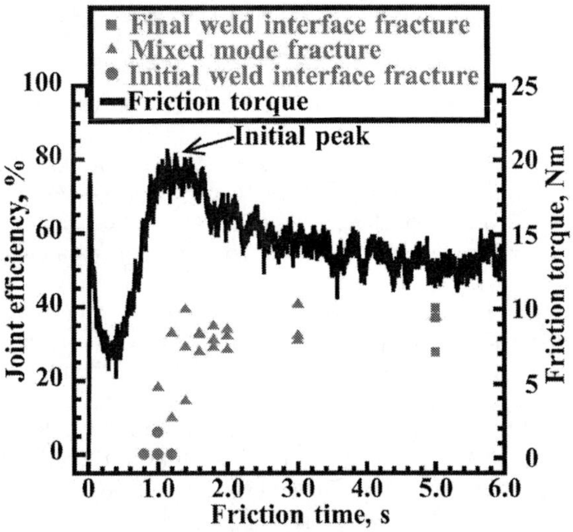

Fig.1 Relationship between joint efficiency and friction time of joints, in relation to friction torque.

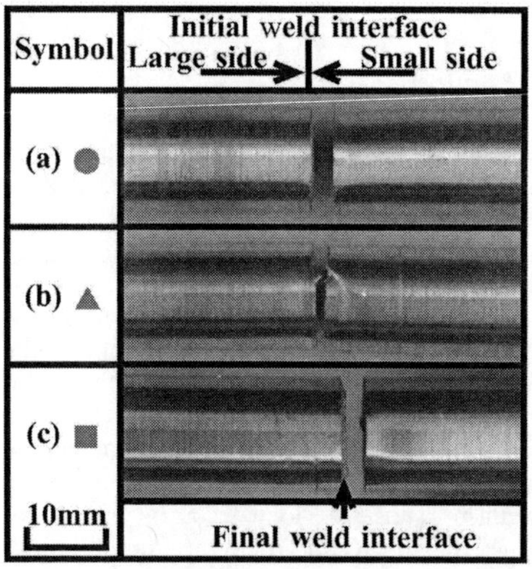

Fig.2 Examples of joint tensile tested specimens.

Relationship between forge pressure and joint efficiency

To improve joint efficiency, the joint was made with high forge pressure. Figure 3 shows the relationship between the forge pressure and joint efficiency. In this case, the joint was made with friction times of 1.4 and 5.0s. Joint efficiency of joints at both friction times increased with increasing forge pressure. The maximum joint efficiency was obtained at a forge pressure of 210MPa, that at a friction time of 1.4s was approximately 75% and at 5.0s was approximately 63%, respectively. However, almost all joints fractured at the initial weld interface as shown in Fig.2a. Thereafter, the joint efficiency slightly decreased.

Figure 4 shows examples of the fractured surfaces of joints. The half radius portion (area A) was rough, and the fractured surface showed two type of dark and bright contrast such as area A and B. Figure 5 shows the optical overviews at the fractured surface, and those shows detail of areas A and B on the fractured surfaces in Fig.4. A

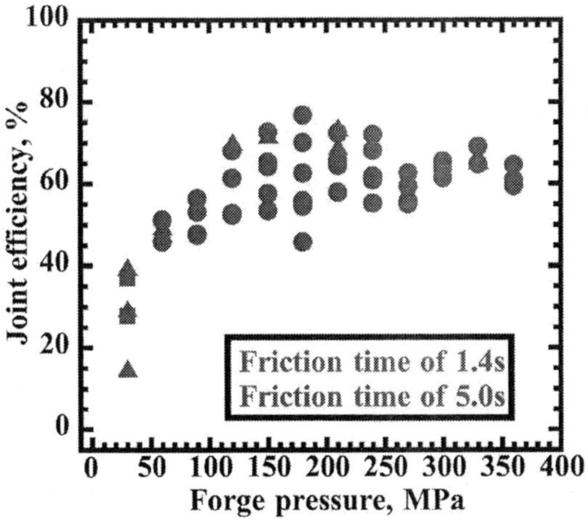

Fig.3 Relationship between forge pressure and joint efficiency of joints at various times.

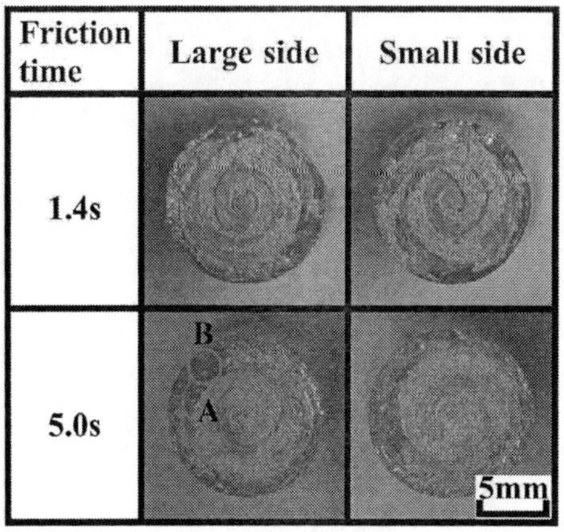

Fig.4 Appearances of fractured surfaces of joint tensile tested specimens.

surface of area A resembled the fractured surface of the base metal. That is, the area A had a dimpled pattern, and the area B had the wear state pattern.

Figure 6 shows examples of the cross section of joints at a forge pressure of 210MPa. Flush (burr or collar) exhausted from the small side specimen, and it increased with increasing friction time. In particular, the weld interface was not kept to the first contacted position (initial contact position) of both weld faying surfaces, and that traveled to the longitudinal direction of the small side specimen. That is, the traveling phenomena of the weld interface were observed in those joints. In the case of the combination of different diameters, the large side is difficult to heat in comparison with the small side, and the generated friction heat is transferred to the regions of the large side specimen that are not in contact. Thus, the temperature of the weld interface at the initial position is

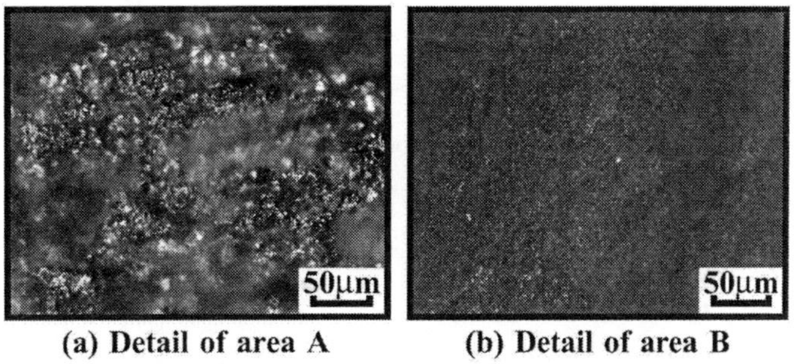

Fig.5 Micrographs of fractured surfaces of joint tensile tested specimens.

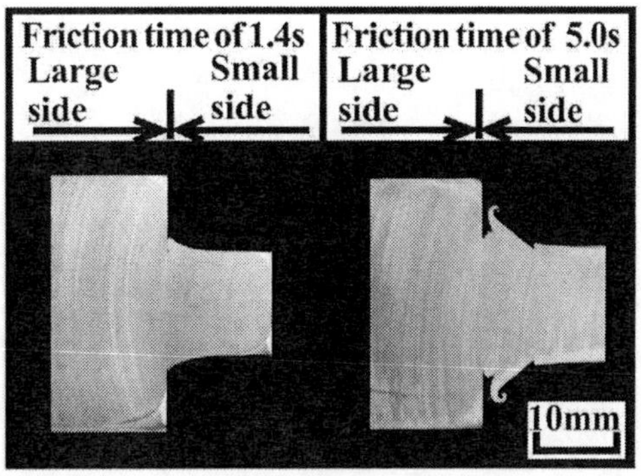

Fig.6 Cross-sectional appearances of weld interface of joints at various friction times.

decreased and the welding interface is transferred to the high temperature of the small side specimen. On the other hand, the traveled distance of the joint at a friction time of 1.4s was smaller than that of the joint at a friction time of 5.0s (see Fig.6). Hence, it was considered that the joint efficiency of the joint at a friction time of 1.4s was higher than that of a friction time of 5.0s. Therefore, to obtain high joint efficiency, the joint should be made with suitable friction time near 1.4s and forge pressure such as 210MPa before the weld interface traveled, although further investigation is necessary to increasing joint efficiency.

Conclusions

This paper described the effects of friction time and forge pressure at friction pressure of 30MPa on tensile strength of 5083 aluminum alloy (A5083) friction stud welded joints. The following results were obtained.

(1) The joint, which was made with a friction time of 1.4s and a forge pressure of 30MPa, had a joint efficiency of approximately 29%. This joint had the fracture between the initial weld interface and the base metal side (mixed mode fracture).

(2) The joint efficiency increased with increasing forge pressure. Then the joint with a forge pressure of 210MPa had a joint efficiency of approximately 75% at a friction time of 1.4s.

(3) The traveling phenomena of the weld interface were observed in those joints. Therefore, to obtain high joint efficiency, the joint should be made with suitable friction time near 1.4s and forge pressure such as 210MPa before the weld interface traveled.

References

[1] H.Ochi, K.Morikawa, T.Moritani, Y.Issiki, G.Kawai: "Strength of 5083 aluminum alloy stud joints", Journal of the Japanese Society for Strength and Fracture of Materials, Vol. 47, No. 1 (2013), pp. 1-6. [in Japanese]

[2] K.Okita, M.Aritoshi, W.Kishimoto, S.Yamada, H.Hira: "Study on the Friction Welding of Different Diameter Bars (Report 1) -Effect of Relative Difference of Bar Diameter on the Friction Welding Phenomenon-", Journal of the Japan Welding Society, Vol. 50, No. 2 (1981), pp. 189-195. [in Japanese]

The effect of electromagnetic stirring on the microstructure and mechanical properties of 309L stainless steel welds

Chao-Yuan, Kung[1*], Ge-Ping Yu[1], Jia-Hong Huang[1], Sheng-Long Jeng[2]
1. National Tsing Hua University, Hsinchu, Taiwan, R.O.C.
2. Institute of Nuclear Energy Research, Taoyuan,, Taiwan, R.O.C.
*Corresponding author. Jordan10041@hotmail.com

Abstract

This study is to investigate the effect of electromagnetic stirring on the microstructure and mechanical properties of 309L welds. 304 stainless steel plates and 309L filler wire were used for the base metal and the weld metal, respectively. Multi-pass welding was performed with a mechanical gas tungsten arc welding process with/without electromagnetic stirring (EMS). The results revealed that EMS could change the grain orientation and fine the crystallite size of 309L welds, while the composition of weldment didn't vary significantly along the cross section. As a result, the yield strength and tensile strength of 309L weldment with EMS both were improved with the same elongation.

Introduction

Electromagnetic stirring (EMS) is usually equipped with Gas Tungsten Arc Welding (GTAW) and Plasma Arc Welding (PAW) to refine the weld microstructure and change grain boundary properties [1-9]. The external electromagnetic field enhances the Lorentz force that causes fluid flow and stirring effect. The formation of fine equiaxed grains in fusion zones helps reducing solidification cracking and improving the mechanical properties of the weld metal [4, 7, 10].The base metal and filler metals used in this study all belong to 300 series stainless steel. 309L/308L filler metal is common used for the buffer layers of weld overlays[11]. However, hot cracking and DDC can be found in austenitic stainless steel welds as well[12].
The present research planned to use electromagnetic stirring during the welding, for the purpose to evaluate the effect of electromagnetic stirring on the mechanical properties and microstructure of 309L stainless steel welds.

Materials and experimental procedure

The base metal and backing plate used in this experiment was 304 stainless steel of 15 mm thick, and dimensions were 400mm (l) x 120mm (w). The filler metal was 309L, and the composition of filler metal is as commercially used. Before welding, the base metal and backing fixed on the working table and cleaned with alcohol. Fig. 1(a) shows the weldment design. An mechanical welding machine equipped with electromagnetic stirring apparatus and wire feeder was used for this experiment. The gas tungsten arc welding (GTAW) was used for this study. The welding system is schematically shown in Fig. 2. The welding current and welding speed were controlled to an average level of 120A and 100 mm/min, respectively. The electromagnetic stirring frequency was set at 3Hz. Weldments have about 100 passes as the grooves were fulfilled.
After welding, the weldments were checked by radiographic test (RT) to make sure there was no welding defect, and began subsequent sample machining. The specimens for microstructural observations were cut from the direction parallel to the welding direction (Longitudinal section), and prepared following a standard mechanical grinding and polishing process, followed by electrolytic etching using reagent of 10g $H_2C_2O_4$ filled by H_2O to 100ml under a potential of 6V DC for about 30 s. These specimens were analyzed by laser scanning confocal microscopy (LSCM) and JEOL scanning electron microscope (SEM) equipped with an electron backscatter diffraction (EBSD) and energy dispersive spectrometers (EDS) instrument. The hardness measurements of the weldments were taken using a Vickers hardness tester under a load of 300g. Tensile test were performed with a tensile testing machine at room temperature. Fig. 1(b) shows the dimensions of a tensile test specimen. The welding zone made up the gauge length zone of tensile specimens. The strain rate was set at 3×10^{-4} .The fracture sections were observed with SEM after tensile test.

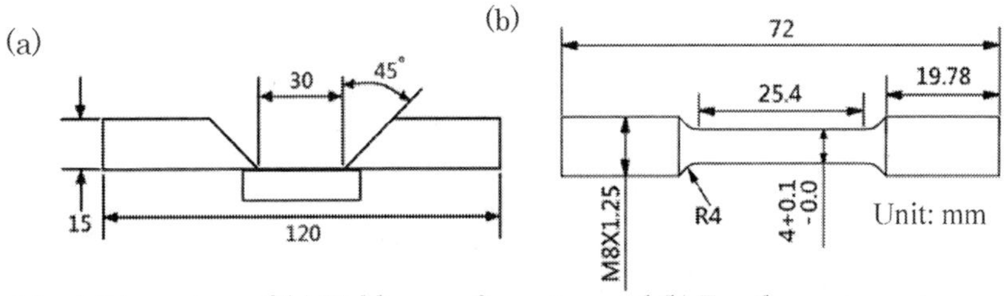

Fig. 1 Dimensions of (a) Welding configuration and (b) Tensile specimen size

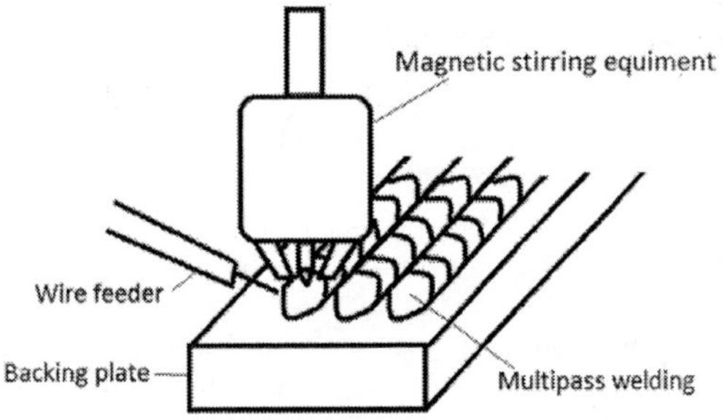

Fig.2 Schematic diagram of welding system with electromagnetic stirring.

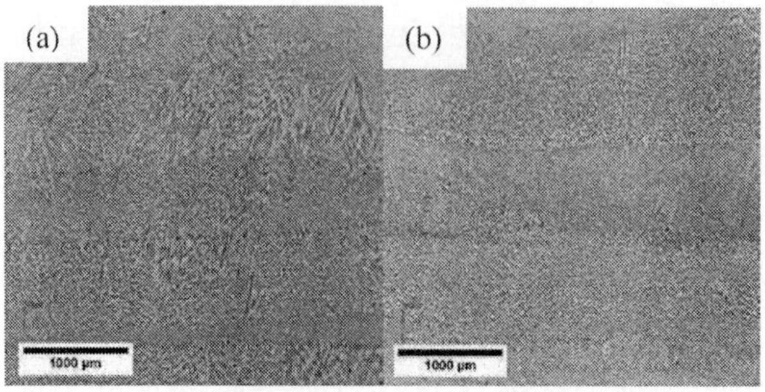

Fig. 3 laser scanning confocal microscope results from 309L weldments (a) without electromagnetic stirring (b) with electromagnetic stirring.

Results and Discussion

The laser scanning confocal microscopy micrograph of longitudinal sections of fusion zone in the unstirred and electromagnetically stirred welds of 309L are shown in Fig. 3. The micrograph shows that dendrite structure was changed while applied electromagnetic stirring (EMS). The 309L weldment has section of column dendrite structure without electromagnetic stirring (NEMS), and reduced with electromagnetic stirring. We can see the trace of electromagnetic stirring. N. Suutala et al [13-16] classified stainless steel welds of the AISI/AWS 300 series into three types A, B and C through welds microstructure and morphology of the delta ferrite. In this study, the Cr_{eq}/Ni_{eq} ratio of 309L is approximately 1.53, where Cr_{eq} and Ni_{eq} are the chromium and nickel equivalents on the Schaeffler diagram, correspond to type B whether with or without electromagnetic stirring. In addition, we can know that the weldment is primary ferrite structure, that is, liquid first solidify as ferrite then transform to austenite through solid state transformation[13, 16, 17]. It is obvious that dendrites become smaller in the weldment with electromagnetic stirring than without electromagnetic stirring.

According to Fukuhisa Matsuda et al[9] and Takehio Watanabe et al[5], electromagnetic stirring can make a compulsive fluid flow that can cause dendritic fragmentation. When dendrite arms were broken by the fluid flow and carried away, it can serve as nuclei of new grains, and come out to be refinement of grains.

From results of energy dispersive spectrometer we can know that EMS does not change the composition distribution of 309L weldments, there is no composition difference between weldments with EMS and without EMS, it may attribute to the mechanism of the transformation mentioned above, because EMS can only affect the weldment when it is liquid, the solid state transformation take place when the weldments solidified and result in unchanged compositional distribution.

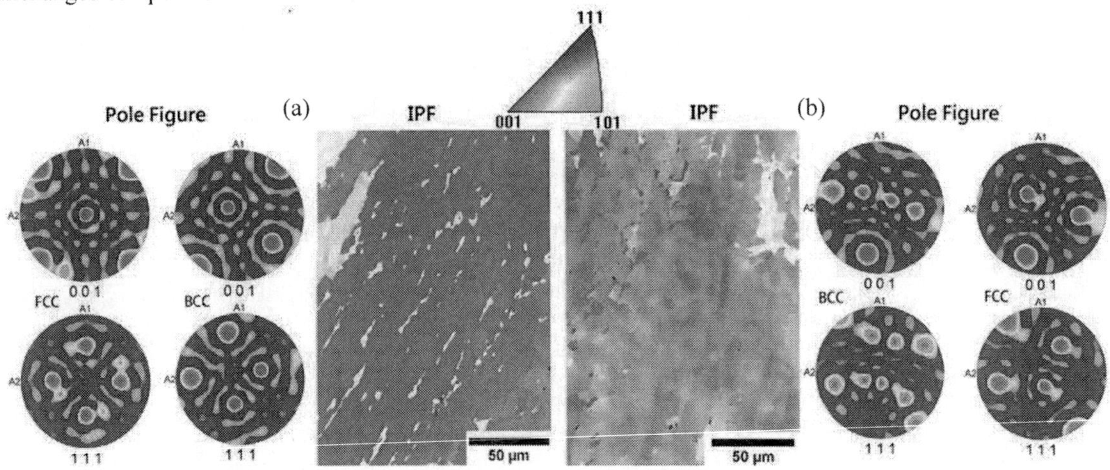

Fig.4 Inverse pole figure [Normal Direction] and pole figure ([001] and [111]) of 309L weldments (a) without electromagnetic stirring (b) with electromagnetic stirring.

Previous research shows that the orientation of austenite tends to have [100] texture and the texture of ferrite is correlated to austenite [13-15]. The EBSD results of present work correspond to previous study while without EMS, as shown in Fig. 4, however, the texture changed when EMS was applied during welding. The orientations of austenite and ferrite switch to near [111] when EMS was applied. Also the inverse pole figure (IPF) shows the orientation difference

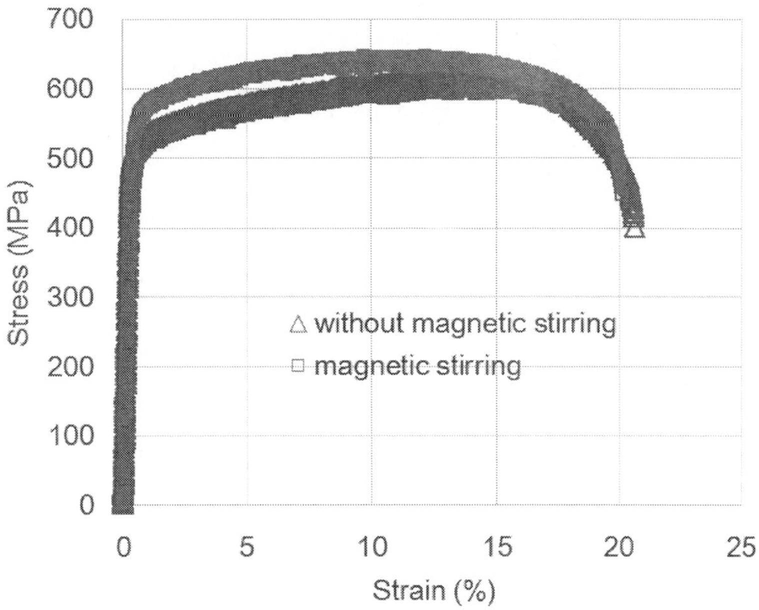

Fig. 5 Tensile stress-strain curves for 309L weldments with/without electromagnetic stirring.

of 309L weldments with/without EMS. The statistical data of grain sizes from ten points of 309L weldment with/without EMS reveals that the grains are refined, which is correspond to the previous studies [1, 3, 4, 6, 8, 9, 18-20]. Fig. 5 shows the stress and strain curve of tensile test of 309L weldments with/without electromagnetic stirring. The figure shows that the yield strength and tensile strength of specimen with EMS are better than the specimen without EMS, and the elongation is almost the same, moreover, through tensile data of average value of three specimens of 309L weldments with/without EMS, the result shows that the yield strength and tensile strength are enhanced about 5.2% and 3.3% for weldment with EMS compared to the weldment without EMS, respectively. The result of EBSD shows that the size of grain with EMS reduced about 13.5% compared to the grain without EMS, but the yield strength and tensile strength do not enhance as well. According to the Hall-petch equation: $\sigma_{ys}=\sigma_i+k_yd^{-1/2}$, where σ_{ys} is yield strength, σ_i is lattice resistance, k_y is locking parameter and d is grain size, respectively. 309L is austenitic stainless steel, the primary structure is face-centered cubic (FCC), because of the considerable slip system of FCC that allows dislocation to slip much easily, the locking parameter is relatively small, as a result, the finer grain size does not have obvious effect on the strength of 309L weldments[21].

Conclusions

1. Applying electromagnetic stirring during welding process can refine dendritic structure of 309L multipass weldment.
2. Electromagnetic stirring have no effect on composition distribution of 309L welds.
3. Electromagnetic stirring changes the grain orientation of 309L welds from [100] to [111].
4. The yield strength and tensile strength of 309L weldment with electromagnetic stirring both were improved with the same elongation.

Acknowledgments

This research was funded by the National Science Council of the Republic of China under the contracts 104-NU -E-007 -003 -NU. The authors also appreciate the instrumental and tensile test support of Institute of Nuclear Energy Research of the Republic of China.

References

[1] Lim, Y.C., et al., Effect of magnetic stirring on grain structure refinement: Part 1 – Autogenous nickel alloy welds. Science and Technology of Welding and Joining, 2010. 15(7): p. 583-589.

[2] Lim, Y.C., et al., Effect of magnetic stirring on grain structure refinement Part 2 – Nickel alloy weld overlays. Science and Technology of Welding and Joining, 2010. 15(5): p. 400-406.

[3] Mousavi, M.G., et al., Grain refinement due to grain detachment in electromagnetically stirred AA7020 welds. Science and Technology of Welding and Joining, 2003. 8(4): p. 309-312.

[4] Ram, G.D.J., R. Murugesan, and S. Sundaresan, Fusion zone grain refinement in aluminum alloy welds through magnetic arc oscillation and its effect on tensile behavior. Journal of Materials Engineering and Performance 1999. 8(5): p. 513-520.

[5] Watanabe, T., H. Nakamura, and K. Ei, Solidification Control of Austenitic Stainless Steel Weld Metal by Electromagnetic Stirring. Transactions of the Japan Welding Society, 1990. 21: p. 37-43.

[6] Malinowski-Brodnicka, M., G.D. Ouden, and W.J.P. Vink, Effect of Electromagnetic Stirring on GTA Welds in Austenitic Stainless Steel. Welding Journal, 1990: p. 52-59.

[7] Kou, S. and Y. Le, Grain structure and solidification cracking in oscillated arc welds of 5052 aluminum alloy. Metallurgical Transactions A, 1985. 16(7): p. 1345-1352.

[8] Pearce, B.P. and H.W. Kerr, Grain refinement in magnetically stirred GTA welds of aluminum alloys. Metallurgical Transactions B 1981. 12(3): p. 479-486.

[9] Matsuda, F., K. Nakata, and N. Sano, Effect of Electromagnetic Stirring on Weld Solidification Structure of Austenitic Stainless Steels. transactions of JWRI, 1986. 15(12): p. 327-338.

[10] S. Kou and Y.LE, Nucleation mechanism and grain refining of weld metal. Welding Journal, 1986: p. 305-313.

[11] McCracken, S.L. and R.E. Smith, Behavior and Hot Cracking Susceptibility of Filler Metal 52 M (ERNiCrFe-7A) Overlays on Cast Austenitic Stainless Steel Base Materials. 2011: p. 333-352.

[12] Lippold, J.C. and D.J. Kotecki, Welding Metallurgy and Weldability of Stainless Steels. 2005: John Wiley & Sons, Inc.

[13] Suutala, N., T. Takalo, and T. Moisio, Ferritic-austenitic solidification mode in austenitic stainless steel welds. Metallurgical Transactions A, 1980. 11(5): p. 717-725.

[14] Takalo, T., N. Suutala, and T. Moisio, Austenitic solidification mode in austenitic stainless steel welds. Metallurgical Transactions A, 1979. 10(8): p. 1173-1181.

[15] Suutala, N., T. Takalo, and T. Moisio, Single-phase ferritic solidification mode in austenitic-ferritic stainless steel welds. Metallurgical Transactions A, 1979. 10(8): p. 1183-1190.

[16] Suutala, N., T. Takalo, and T. Moisio, The relationship between solidification and microstructure in austenitic and austenitic-ferritic stainless steel welds. Metallurgical Transactions A, 1979. 10(4): p. 512-514.

[17] Brooks, J.A. and A.W. Thompson, Microstructural Development and Solidification Cracking Susceptibility of Austenitic Stainless-Steel Welds. International Materials Reviews, 1991. 36: p. 16-44.

[18] Sivaprasad, K. and S.G.S. Raman, Influence of magnetic arc oscillation and current pulsing on fatigue behavior of alloy 718 TIG weldments. Materials Science and Engineering: A, 2007. 448(1-2): p. 120-127.

[19] Vollertsen, F. and C. Thomy, Magnetic stirring during laser welding of aluminum. Journal of Laser Applications, 2006. 18(1): p. 28.

[20] Campanella, T., C. Charbon, and M. Rappaz, Grain refinement induced by electromagnetic stirring a dendrite fragmentation criterion. Metallurgical and Materials Transactions A 2004. 35(10): p. 3201-3210

[21] Li, W.J., et al., The effects of rolling and sensitization treatments on the stress corrosion cracking of 304L stainless steel in salt-spray environment. Corrosion Science, 2013. 68(0): p. 25-33.

Investigation of Main Factor on Enhancing Fatigue Performance of Welded Structural Steel Plates due to Laser Peening

Yoshihiro SAKINO [1], Yuji SANO [2]

[1] *Kindai University Faculty of Engineering, JAPAN*
[2] *Japan Science and Technology Agency, JAPAN*

Laser peening, used in this study, is an innovative surface enhancement technology for introducing compressive residual stress to metallic materials without any surface preparation such as coating. Laser peening is effective for preventing the initiation and propagation of stress corrosion cracking, and therefore, it has been used as a preventive maintenance technology of reactor components of Japanese nuclear power plants. Recent studies have revealed that laser peening dramatically improves the fatigue performance of some metallic materials. However, there are only a few open literatures on the application of laser peening to structural steels. The authors studied irradiation conditions for laser peening of structural steels by measuring the residual stress and performing hardness tests. Then, we also studied changes in the surface residual stress, the depth distribution of residual stress, the hardness distribution, and the surface roughness in four types of structural steel with different strengths. Moreover, it is clarified that the fatigue performance of welded steel plates were improved due to laser peening for various stress ranges. In this study, butt-welded specimens of structural steel were used to study changes in the residual stress and the hardness near the toe of the welded zone caused by laser peening. Then surface roughness, residual stress and hardness were measured, and fatigue test were conducted after applying stress relief annealing to the specimens treated by laser peening. The results were compared with those of the unpeened specimens to confirm the main factor determining the enhancement of the fatigue performance caused by laser peening. It is concluded that the generation of compressive residual stress due to laser peening is the main factor enhancing the fatigue performance, because the reduction in compressive residual stress due to stress relief annealing decreased the fatigue performance to the same level as that of butt-welded joints without laser peening.

Effects of Heat Input and Groove Angles on Distortion and Residual Stress on Submerged Arc Welded Ship Steel Plates ASTM A131 Grade A

Tanawat Sirisatien [1], Sasawat Mahabunphachai [2], Pornsak Thasanaraphan [3], Kittichai Sojiphan [1]

[1] *Department of Welding Engineering Technology, College of Industrial Technology, King Mongkut's University of Technology North Bangkok, Thailand*

[2] *National Metal and Materials Technology Center, National Science and Technology Development Agency, Thailand*

[3] *Thailand Institute of Scientific and Technological Research, Thailand*

Submerged arc welding process is one of welding processes widely used in ship building industry. One of the typical problems found is distortion induced by residual stress. This major problem is crucial for rectification and it causes many subsequently severe losses in terms of quality and time for the repair of steel plates using welding and joining processes. In this research, the relationship and the effects of residual stress caused during submerged arc welding using varying heat inputs and different groove angles was studied. The residual stress was determined experimentally using X-ray diffraction method. The welding parameters including current voltage, welding travel speed, electrode stick-out, and weld gap were carefully controlled during welding experiments. The ship steel plate used for this experiment is naval steel plate ASTM A131 grade A which is commonly used in ship building industry. One side and one pass welding technique have been used together with ceramics backing. Temperature distribution was compared to experimental results of temperature measured during welding using thermocouple type K. Also, the distortion measured experimentally was compared for different groove angles. It was found that increasing heat input leads to higher residual stress and distortion of welded steel plates. The results from this study can be used to predict the behavior of distortion induced by residual stress in ship steel plate welding and allow engineers to select proper process parameters and joint design to avoid or minimize the welding distortion.

Solid State (Friction) Welding: Influences of In-situ Pre-Heat

D. Adams, Z. Roelofs, M. Skinner

Manufacturing Technology, Inc., USA

Friction welding dates back to the early 1950's when companies commercialize rotary fiction welding for joining a wide range of metallic products. Friction welding is a solid-state welding process that generates heat through mechanical friction between workpieces in relative motion to one another. The mechanism for such heating is known to transition from sliding friction to one based on continual plastic deformation. The transition between these heating regimes is characterized by a short duration spike in the shear loads necessary to maintain the desired relative motion. Because no melting occurs, the benefits associated with friction welding are significant and have been well documented.

These instantaneous high shear loads govern designs for current welding system solutions. This work investigates the use of in-situ heating for mitigating such shear loads. A hybrid system has been developed that incorporates in-situ pre-heat and conventional friction welding. Studies have been conducted to define relationships between processing conditions and resulting shear load variations for some representative material systems (steels, Al-alloys).

An evaluation was performed on the mechanical properties and metallurgical changes in the microstructure. The transverse sectional view of the welds in the as-welded condition along with micro structural details in the center and fusion boundary region of welds are presented.

The results were compared to samples produced by a conventional friction welding process (no pre-heat). The metallographic changes and mechanical properties were similar in comparison. The pre-heating had a significant impact on the process parameters and resulting forces.

Numerical Simulation of Localized Distribution of Stress and Diffusible Hydrogen Related to Cold Cracking in High-Strength-Steel Weld Metal

Yoshiki Mikami, Nao Kawabe, Masahito Mochizuki
Osaka University, 2-1 Yamadaoka, Suita, Osaka 565-0871, Japan
mikami@mapse.eng.osaka-u.ac.jp, mmochi@mapse.eng.osaka-u.ac.jp

Nobuyuki Ishikawa
JFE Steel Corporation, 1-1 Minamiwatarida-cho, Kawasaki-ku, Kawasaki 210-0855, Japan
nob-ishikawa@jfe-steel.co.jp

Abstract
In this study, an evaluation procedure for cold cracking in high-strength-steel weld metal based on local stress and locally accumulated hydrogen concentration was developed. A critical condition for cold cracking was determined considering the stress concentration and the accumulation of diffusible hydrogen. The histories of local stress and locally accumulated hydrogen concentration at the root of the weld metal of the cracking test specimen were calculated. The local stress and hydrogen concentration were compared with the determined critical condition, and the occurrence of cold cracking was estimated. The estimated results were in good agreement with the experimental results.

Introduction
Cold cracking is normally investigated by performing slit-type weld cracking tests. These tests provide appropriate conditions to avoid cold cracking in terms of factors such as the chemical composition, hydrogen concentration, welding conditions, plate thickness, and groove geometry. However, cold cracking is observed at the weld root where stress concentration and diffusible hydrogen accumulation occur. Therefore, the evaluation of cold cracking should be based on more microscopic criteria. In this study, local stress and locally accumulated hydrogen concentration were numerically simulated to evaluate the cold cracking of high-strength-steel weld metal. First, a critical condition for the high-strength-steel weld metal was determined by performing a series of slow strain rate tensile (SSRT) tests. Finite element simulations of stress distribution and hydrogen concentration in the SSRT test specimens were performed to determine the local stress and locally accumulated hydrogen concentration at fracture. Second, finite element simulations of the weld stress and hydrogen diffusion in the Y-slit weld cracking test specimen were performed. Based on the calculated histories of local stress and hydrogen concentration at the weld root and the determined critical condition for cold cracking, the occurrence of cold cracking was discussed.

Numerical formulation of hydrogen diffusion
Hydrogen diffusion was numerically simulated based on the following formulation [1] by the finite element method. Assuming that the driving force of mass diffusion is the gradient of the chemical potential μ, the concentration flux $\boldsymbol{J}$ is described by Eq. (1):

$$\boldsymbol{J} = -\frac{\boldsymbol{D}c}{RT}\frac{\partial \mu}{\partial \boldsymbol{x}} \tag{1}$$

where $\boldsymbol{D}$ is the diffusion coefficient, c is the concentration, R is the gas constant, T is the temperature, and $\boldsymbol{x}$ is the position vector. The chemical potential is defined by Eq. (2):

$$\mu = \mu^0 + RT\ln\phi + p\overline{V}_{\mathrm{H}} \tag{2}$$

where μ^0 is the standard chemical potential, ϕ is the normalized concentration (= c/s, where s is the solubility), p is the hydrostatic pressure, and $\overline{V_{\mathrm{H}}}$ is the partial molar volume of hydrogen in the solid solution. Equations (1) and (2) yield Eq. (3):

$$\boldsymbol{J} = -s\boldsymbol{D}\left(\frac{\partial \phi}{\partial \boldsymbol{x}} + \phi\ln\phi\frac{\partial \ln T}{\partial \boldsymbol{x}} + \frac{\phi\overline{V_{\mathrm{H}}}}{RT}\frac{\partial p}{\partial \boldsymbol{x}}\right) \tag{3}$$

The terms on the right side of Eq. (3) represent the driving force of diffusion due to the gradient of concentration, temperature, and hydrostatic stress, respectively. The numerical simulation in this study was performed using the finite element software Abaqus, in which the above formulation was incorporated.

Determination of critical condition for cold cracking

SSRT test

The critical condition for cold cracking was determined by performing the SSRT test on hydrogen-charged specimens. The specimen material was HT980-class high-strength steel quenched at 1200°C. Round-bar tensile specimens with a circumferential notch (Fig. 1) were used. To change the local stress distribution at the notch root, the notch root radius was selected to be 0.25 and 0.1 mm. Diffusible hydrogen was electrochemically charged into the specimens. The concentration of the charged hydrogen was measured by carrying out thermal desorption analysis (TDA) and was found to range from 0 (uncharged) to 1.75 ppm. The specimen surface was galvanized soon after the hydrogen charging to avoid discharging. The crosshead displacement rate in the SSRT test was 5 μm/min, and the test was continued until fracture.

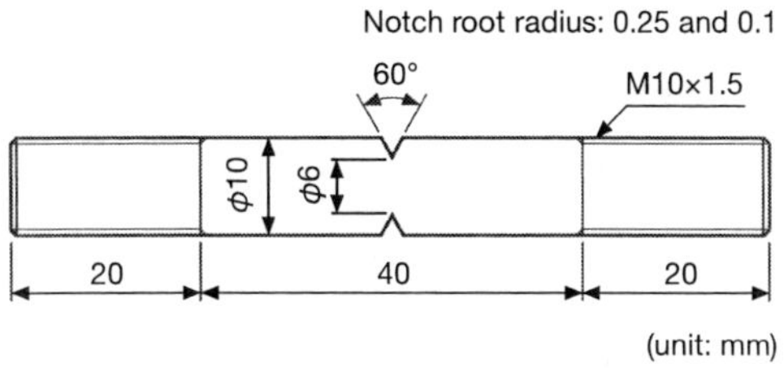

Fig. 1: Specimen dimensions.

Numerical simulation of SSRT test

The SSRT test was modeled by carrying out elastic–plastic analysis and hydrogen diffusion analysis. The specimen was modeled using an axisymmetric model. A displacement boundary condition was applied to the edge of the model to reproduce the displacement rate obtained in the experiment. The calculated hydrostatic pressure was imported into the hydrogen diffusion analysis. The hydrogen diffusion coefficient and solubility used were 3.4×10^{-5} mm^2/s and 0.0708 ppm·mm/N$^{0.5}$, respectively. The hydrogen concentration measured by TDA was assumed to distribute uniformly in the specimen under the initial condition. The discharging of hydrogen was not considered because the specimen surface was galvanized.

Critical condition based on local stress and localized hydrogen concentration

Figure 2 shows the distribution of the maximum principal stress and hydrogen concentration along the line from the notch root to the center of the specimen under the fracture load. The critical condition for cracking in the SSRT test is usually represented by the macroscopically applied stress at fracture and the initially charged average hydrogen concentration. However, because the stress concentration and hydrogen accumulation occur as shown in Fig. 2, it is more appropriate to use the local stress and localized concentration to determine the critical condition for cold cracking. The local critical condition for cracking was determined by this approach for each specimen in the SSRT test with different notch root radii and different initially charged hydrogen concentrations. The determined critical condition is shown in Fig. 3.

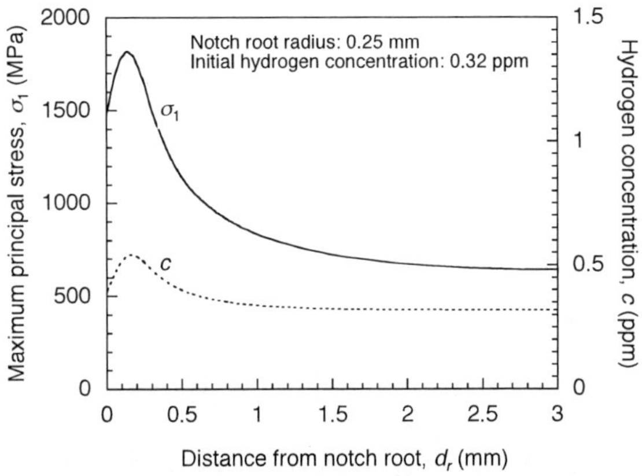

Fig. 2: Distribution of stress and hydrogen concentration in the specimen.

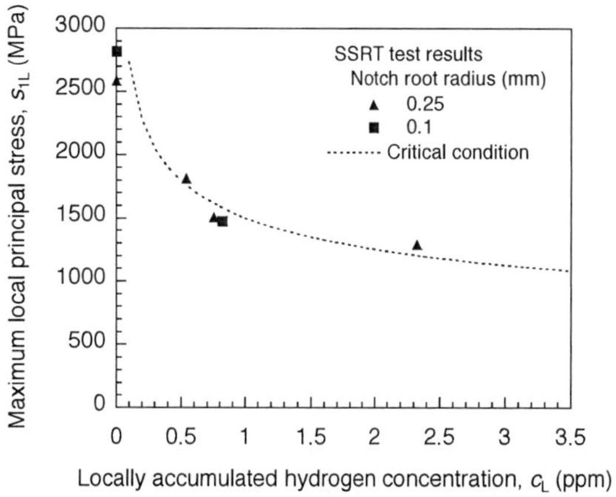

Fig. 3: Critical condition based on local stress and locally accumulated hydrogen concentration.

Evaluation of cold cracking in Y-slit test

The determined critical condition was applied to the Y-slit weld cracking test, and the occurrence of cold cracking was estimated.

Y-slit weld cracking test

The Y-slit weld cracking test was performed on the high-strength-steel weld metal. The detailed procedure and results are presented in the literature [2]. A 980 MPa-class high-strength steel plate having a thickness of 25 mm was welded with a corresponding wire by metal inert gas welding. The welding current, arc voltage, and welding speed were set to 280 A, 30 V, and 5 mm/s, respectively. During welding, Ar+5%CO_2 was used as the standard shielding gas, and H_2 gas was mixed to vary the diffusible hydrogen concentration contained in the weld metal. The percentage of the mixed H_2 gas was varied as 0%, 0.04%, 0.08%, 0.125%, and 0.25%, and the corresponding diffusible hydrogen concentration in the weld metal was found to be 0.8, 2.3, 3.0, 3.8, and 5.8 ppm, respectively. The cross section of the weld metal was observed 48 h after the welding. Five cross sections along the welding direction were observed per specimen. No crack was found in the weld metal with a hydrogen concentration of 0.8 ppm; however, cracks were found in the weld metal with hydrogen concentrations greater than or equal to 2.3 ppm. The crack penetrates through the weld metal in the direction of the throat thickness.

Simulation of Y-slit test

The three-dimensional thermoelastoplastic weld residual-stress simulation of the Y-slit weld cracking test specimen was performed by a finite element simulation that considered the effect of phase transformation [3]. A moving welding heat source was used, and the heat input condition was determined based on the experiment. The residual stress distribution in the weld metal was validated by comparing it with the neutron diffraction measurement results. The calculated histories of the temperature and stress were imported into the hydrogen diffusion simulation. The initial hydrogen concentration in the weld metal was set to be in the range 0.8–5.8 ppm based on the experimental results. The hydrogen concentration at the surface of the specimen was fixed to 0 ppm as a boundary condition.

Behavior of stress and hydrogen concentration in weld metal

Figure 4 shows the history of the temperature and maximum principal stress at the root of the weld metal. The horizontal axis represents the elapsed time since the deposition of the weld metal and is plotted on a logarithmic scale. The four time points (t_1, t_2, t_3, and t_4) presented in Fig. 4 represent the times at which the values of the stress and hydrogen concentration are used in the later discussion on cold cracking. Time t_1 represents the instant just after the weld metal is deposited. At time t_4, the weld is almost cooled and the stress is almost saturated. Times t_2 and t_3 are located between times t_1 and t_4, wherein the stress is in the process of increasing. Figure 5 shows the histories of the hydrogen concentration at the root of the weld metal for varied initial hydrogen concentrations. The hydrogen concentration initially set in the weld metal starts to decrease just after the deposition of the weld metal. The decrease in the hydrogen concentration continues until the end of the calculation. The behavior of the hydrogen concentration is affected by the boundary condition at the surface; however, this aspect is not discussed here because of space limitations.

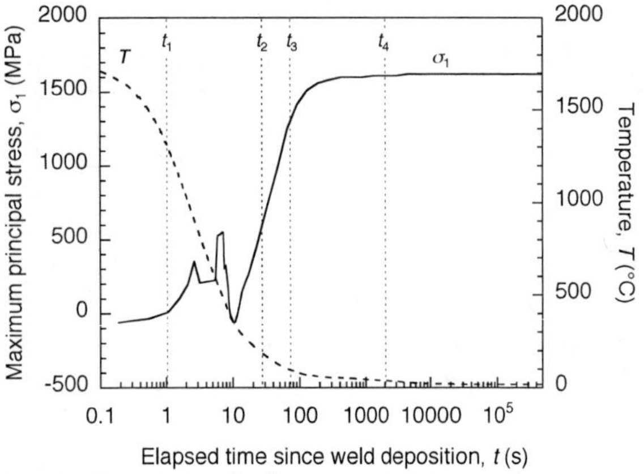

Fig. 4: Histories of the temperature and maximum principal stress at the root of the weld metal.

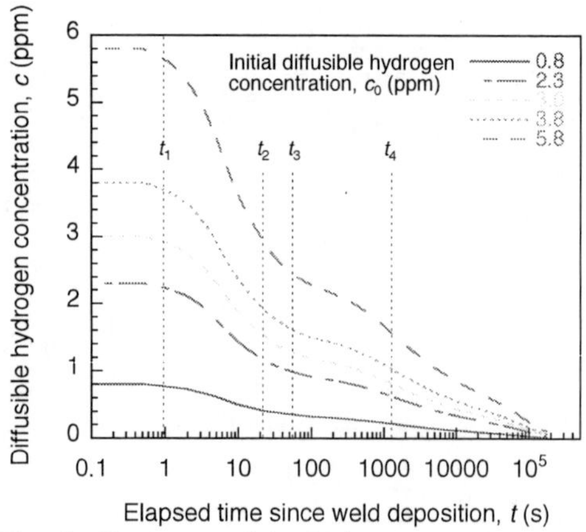

Fig. 5: Histories of the hydrogen concentration at the root of the weld metal for different initial hydrogen concentrations.

Evaluation of cold cracking

The histories of the stress and hydrogen concentration at the weld root are obtained by numerical simulation, as shown in Figs. 4 and 5. The evolution of the stress and hydrogen concentration at times t_1, t_2, t_3, and t_4 was plotted over the critical condition shown in Fig. 3, and the result is shown in Fig. 6. At time t_1, which is the instant just after the deposition of the weld metal, the hydrogen concentration does not decrease from the initial value. The hydrogen concentration at this time is the highest in its entire history; however, the stress is quite low. Therefore, the point determined by the stress and hydrogen concentration is located considerably lower than the curve representing the critical condition. At time t_2, the stress is in the process of increasing; however, the value is still lower than the value for the critical condition. In addition, the hydrogen concentration decreases clearly compared to that at time t_1 because the temperature around the weld is high and the diffusion is sufficiently fast. At instants between times t_2 and t_3, the stress continues to increase; however, the hydrogen concentration remains almost constant at its value at time t_2 or slightly decreases. This causes the point determined by the stress and hydrogen concentration to move upward and surpass the curve representing the critical condition in the case where the initial hydrogen concentration is high. Because the weld has already cooled to room temperature at time t_3, the increase in the stress from times t_3 to t_4 is modest. When the initial hydrogen concentration is high, the stress and hydrogen concentration could possibly increase beyond the critical condition at time t_4. The simulation results indicate that in the cases where the initial hydrogen concentration is greater than 3.0 ppm, the point determined by the stress and hydrogen concentration exceeds the curve representing the critical condition, i.e., cold cracking can occur. In the experiment, cold cracking occurred in the case where the initial hydrogen was 2.3 ppm. The simulation value for the initial hydrogen concentration of 2.3 ppm did not exceed the value for the critical condition, but the point at time t_4 was located very close to the curve. Hence, the simulation can be considered to provide a fair estimation.

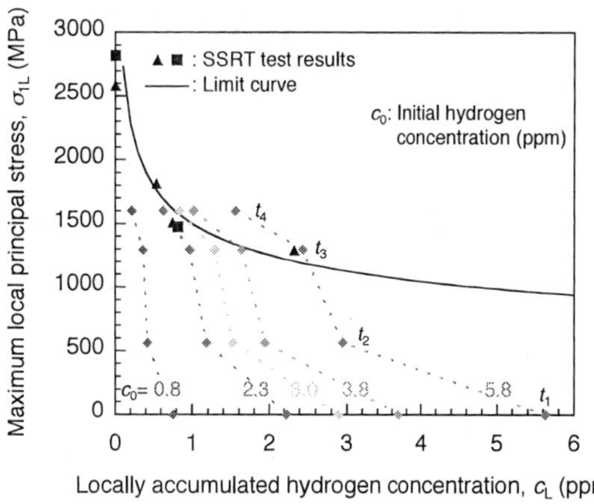

Fig. 6: Histories of the stress and hydrogen concentration at the root of the weld metal in the Y-slit weld cracking test in relation to the critical condition for cold cracking determined by the SSRT test.

Conclusions

In this study, cold cracking of high-strength-steel weld metal was evaluated in terms of the local stress and locally accumulated hydrogen concentration. The following results were obtained:

(1) The critical condition for cold cracking based on the local stress and locally accumulated hydrogen concentration was determined by the SSRT test and corresponding numerical simulation.

(2) The occurrence of cold cracking at the weld root of the Y-slit weld cracking test specimen was evaluated by comparison with the critical condition determined by the SSRT test, and the obtained results were in good agreement with the experimental results.

References

[1] P. Sofronis, *et al*: "Numerical Analysis of Hydrogen Transport near a Blunting Crack Tip", *Journal of Mechanics and Physics of Solids*, Vol. 37, No. 3 (1989), pp. 317–350.

[2] N. Ishikawa, *et al*: "Critical Condition for Hydrogen Induced Cold Cracking of 980MPa Class Weld Metal", *Quarterly Journal of the Japan Welding Society*, Vol. 29, No. 3 (2011), pp. 218–224. (in Japanese)

[3] Y. Mikami, *et al*: "Angular Distortion of Fillet Welded T Joint Using Low Transformation Temperature Welding Wire", *Science and Technology of Welding and Joining*, Vol. 14, No. 2 (2009), pp. 97–105.

Effects of Heat Affected Zone Softening Extent on Strength of Advanced High Strength Steels Resistance Spot Weld

Hassan Rezayat, S. S. Babu
University of Tennessee, Knoxville, Tennessee, USA
hrezayat@utk.edu, sbabu@utk.edu

Hassan Ghassemi-Armaki
ArcelorMittal Global R&D, East Chicago, Indiana, USA
hassan.ghassemi@arcelormittal.com

Abstract
Spot welds in advanced high-strength steels (AHSS) exhibit heat affected zone (HAZ) softening due to tempering of pre-existing martensite into a mixture of ferrite and cementite. Despite the high strength level for the base metal, the HAZ softening may lead to inferior joint strength during tensile-shear and cross-tension testing. In this work, we investigated the effects of the HAZ softening on the global loading response for AHSS steels with three different volume fraction of martensite. Microhardness mapping was used as a measure of martensite tempering and extent of softening. Based on the data, the softening was identified in sub-critical heat affected zone. The hardness drop with magnitude of 6%, 18%, and 42% was observed in steels with 16%, 52% and 100% of martensite, respectively. In order to model the welded joint loading response using finite element methods (FEM), there is a need to represent the softening in terms of stress-strain relationships. Local stress-strain curves for different weld zones were obtained by scaling the base metal constitutive properties with local hardness ratio. Finite element simulations of tensile shear tests showed that HAZ softening can impact the maximum tensile shear load of specimens more significantly when the base metal tensile strength is above 1000 MPa. The paper will discuss the validity of the above finite element approach for describing experimental results and future directions.

Introduction
Advanced High Strength Steels (AHSS) are among the new generation of steels that are being developed to satisfy recent global standards for vehicle safety, fuel economy and emission requirements. With precisely selected chemical composition and controlled manufacturing process, AHSS possess a combination of strength, ductility, toughness, and fatigue properties due to their complex microstructure consisting ferrite, martensite, bainite, and austenite [1, 2]. DP steels consist of a soft ferrite matrix and hard second phase martensite islands. Increasing martensite volume fraction (MVF) increases the strength of DP steels. For martensitic steels, the entire austenite formed during annealing or hot-rolling transforms almost entirely to martensite during quenching process on the run-out table. Dual-phase (DP) and Martensitic (MS) steels are AHSS grades that are being used widely in automotive industry for fabricating auto parts, mainly using resistance spot welding (RSW) [1, 3]. Despite having desirable as-received mechanical properties, existence of metastable phases, particularly martensite, can make AHSS sensitive to higher ranges of temperature that occur during RSW. Previous researches [4, 5] have shown that in a process, known as martensite tempering, in temperatures between 100 °C and Ac1 line of steel, i.e. in SCHAZ, carbon atoms in carbon-supersaturated martensite microstructure get activated, start diffusing into interlath spacing and dislocations and form clusters, and finally form cementite, e.g. Fe_3C, and coarsen. The final product of this process is tempered martensite consisting ferrite matrix and cementite precipitates, with lower strength, higher ductility, and lower hardness. This hardness drop is mainly due to martensite tempering, and change in ferrite hardness has been shown to be slight [6]. It has been shown that time and temperature can affect the extent of softening in AHSS containing martensite. In these studies [4-7], it was shown that increasing the temperature up to Ac1 line or increasing tempering time can increase the extent of softening. Strength and failure behavior of spot welded joints under certain loading modes have been investigated in several studies and full and partial interfacial failures (FIF and PIF), round button (RB), and round button in SCHAZ (RBH) failure modes were observed [8-10]. Computational modeling has also been used to understand mechanical behavior of quasi-static tensile shear and cross tension specimens with particular attention to local deformation behavior and weld geometry [11-13].

In this work, we initially measure the hardness of spot welds made from AHSS with different volume fraction of martensite. Hardness maps and microstructural analysis are then used to build the geometry of spot weld for FE simulations. Spot weld zones are then partitioned into cells with constitutive behavior obtained by scaling base metal true stress-strain curve with average hardness ratio of that cell. Results from running FE simulations are finally compared with those from tensile shear testing for same steels.

Experimental Approach

Dual-phase and MS AHSS sheet steels with thickness of 1.6 mm and MVF of 16%, 52%, and 100% were chosen to study the extent of softening in SCHAZ and its effects on global loading response in tensile shear specimens. Steel sheets from similar grades were spot welded together with welding parameters resulting nugget diameters of 6 and 8 mm. The welding procedure, specimens' geometry, and testing conditions were chosen in accordance with AWD8.9[14]. Some of these weld specimens were cross-sectioned in rolling direction and prepared for microstructure analysis and hardness measurements, according to ASTM E3[15]. Microhardness measurements were performed on cross section of spot welds with 300 gr load and 15 seconds dwell time. Local hardness drop, if any, was calculated as the $(H_{BM}-H_{Local})/H_{BM}$ where H_{BM} and H_{Local} are BM and local hardness. Hardness drop was used later to correlate the extent of softening to steel's chemistry and initial microstructure. The correlations between extent of softening with steel's carbon content and martensite carbon content (MCC), were then investigated to find which the better representation of softening is. Microhardness maps were also used along with SEM observation to recognize the weld zones and extract weld zones geometry for FE modeling of tensile-shear testing. BM stress-strain relationship was obtained from uniaxial tensile testing. Local hardness ratio was used to scale the BM stress-strain curve to predict the local constitutive behavior, a required input for computational modeling of tensile-shear testing.

Computational Approach

Based on symmetry and to minimize the computational time, half model of tensile shear specimen was modeled in Abaqus commercial FE package. For geometry and boundary conditions for this quasi-static testing, AWS D8.9 standard was used. Using hardness maps and microstructural analysis, spot weld in this three-dimensional model was partitioned into five cells for SCHAZ to allow for assigning loading response of material in form of true stress-strain curves by scaling the BM curves with cell's averaged hardness ratio. For fusion zone (FZ), upper-critical (UCHAZ), and inter-critical (ICHAZ) one cell was created. For DP590, since ICHAZ was wider with a sharp hardness change, two cells were used to allow a smoother transition in material properties. It was assumed that this method of partitioning allows for more realistic results for global loading response in comparison with using a single partition for HAZ regions. Typical meshing based on tetrahedron elements with appropriate refinement was performed to ensure solution converges. Figure 2 shows a view of this mesh along with hardness map, partitions, and hardness ratio. In this work, FE models with softening and without softening in SCHAZ were used to understand how HAZ softening impact spot weld strength.

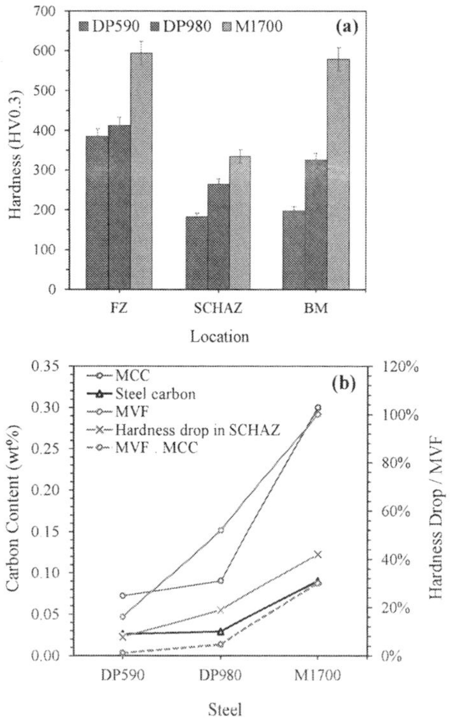

Fig. 1: Hardness magnitude in three regions of spot weld for steels used in this study (a), Correlation between hardness drop and MVF and carbon content (b).

Results and Discussion

Softening Extent in SCHAZ of Spot Welds

The average hardness for FZ, SCHAZ and BM are brought in Figure 1 (a) for comparison. From the results, it was observed that BM hardness increases proportionally as the MFV increases. It was also observed that when MFV in BM decreases from 100% to 52% and 16%, for M1700, DP980, and DP590, respectively, the hardness ratio of FZ increases, significantly. This indicated the effect of base metal MFV on the hardness of the FZ. Comparing results for the SCHAZ, as shown in Figure 1 (b), shows that as the MVF increases the hardness drop in this region increases. For DP590, DP980, and M1700 with 16%, 52%, and

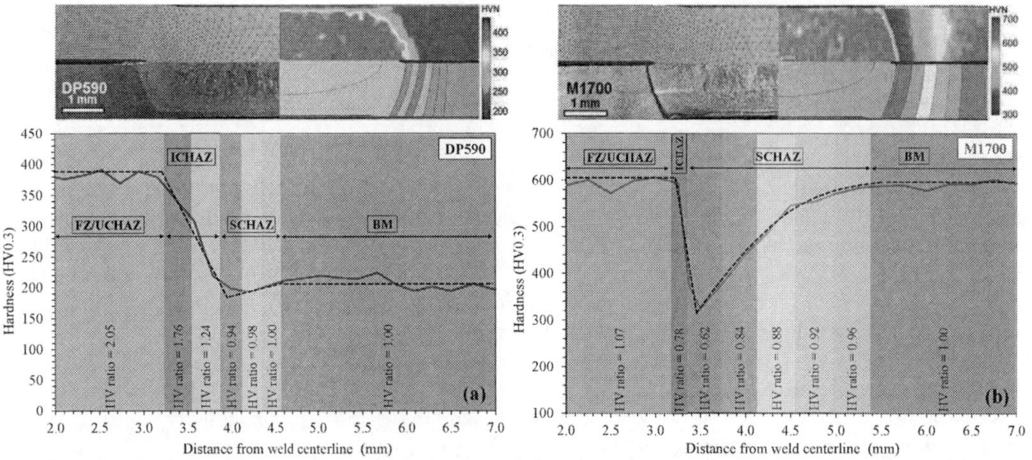

Fig. 2: Hardness maps, generated partitions with local constitutive behavior, and mesh used for computational modeling of spot weld for DP590 (a) and M1700 (b).

100% martensite in initial BM microstructure, the hardness drop was measured to be 6%, 18%, and 42%, respectively. This indicated that a direct correlation can be made between the MVF and hardness drop as illustrated in Figure 1 (b). For DP590 the softening is very low due to the small amount of martensite phase in microstructure. The severe softening in M1700 is evident and a result of higher MVF in this fully martensitic steel. Trends of plots in Figure 1 (b) show that softening extent can be predicted more accurately by using both MVF and MCC. Here and for example, we have multiplied MVF by MCC and added results to Figure 1 (b). This new plot suggests that a better correlation can be made with combining both these two parameters for prediction of softening extent. However, more investigations are required to understand this relationship with more detail. Results from micro-hardness mapping were also used to build the spot weld geometry in computational models. These results and partitions created for spot weld zones along with the mesh on the cross-section for DP590 and M1700 spot welds are presented in Figure 2. Local constitutive behavior for each region of weld was obtained by scaling BM true stress-true strain curve by the average hardness ratio of that region.

Effects of SCHAZ Softening on Tensile Shear Global Response

Finite element simulations were run for tensile shear models built for DP590, DP980, and M1700 spot welds with nugget diameter (ND) of 6 mm for models with and without softening in HAZ and results are presented in Figure 3 (a). One observation from these results is that the maximum load at failure point increases as the MVF increased in both experimental and computational work. From this result, it is observed that the max load at failure has 5%, 17%, and 34% error, for DP590, DP980, and M1700, respectively. This increase in error can be explained by the extent of softening meaning that as the extent of softening increases, the error in results from computational modeling increases as well. This hardness ratio was used to scale the BM properties and also the weld zones were partitioned into a discrete number of zone. In addition to these, hardness ratio was averaged through each partition. In contrast with FE model, real weld has material changes smoothly through and between zones. Therefore these errors can be acceptable with simplifications that have been used in this study. Another observation from Figure 3 (a) can be the difference between load from weld with softening and that from weld without softening. As it can be seen as the extent of softening

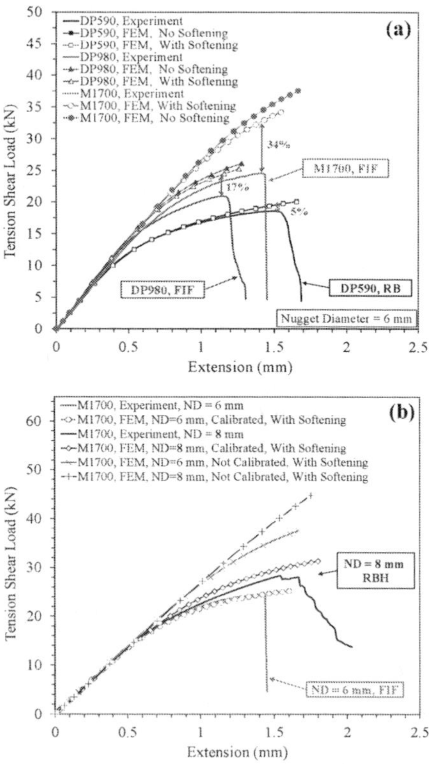

Fig. 3: Results from FE simulation and experiments for DP590, DP980, and M1700 ND of 6 mm (a) and for M1700 with ND of 6 and 8 mm (b).

increases, this difference increases as well. This implies that when the base metal has higher MVF, the impact of the maximum load for the welded joint reduces more when compared with simulations without SCHAZ softening. This of course is valid for this certain ND. From FE simulations, the difference between load at experiments failure for model with, and the one without HAZ softening is 0%, 3%, and 7% for steels with 6%, 18%, and 42% hardness drop in SCAHZ. For AHSS with BM strength above 1000 MPa this difference is considerable and accurate measurements of local constitutive properties especially for SCHAZ become more important. In another attempt to study how the softened-SCHAZ and ND can affect the weld strength, the model for M1700 spot weld was calibrated by using hardness ratios less than measurements, for SCHAZ cells and modified stress-strain curves for these cells were assigned to the FE model and simulations were run for welds with softening in SCHAZ. The results from recent simulations are plotted in Figure 3 (b) along with those from experiments. FE simulations show that after calibrating the model for M1700 spot weld with ND of 6 mm, model is able to predict the load-extension curve for the weld with 8 mm ND better but not overlapping with experiment. This shows that if the representative local properties are available, the model can predict the loading response for both nugget diameters and failure modes more accurately.

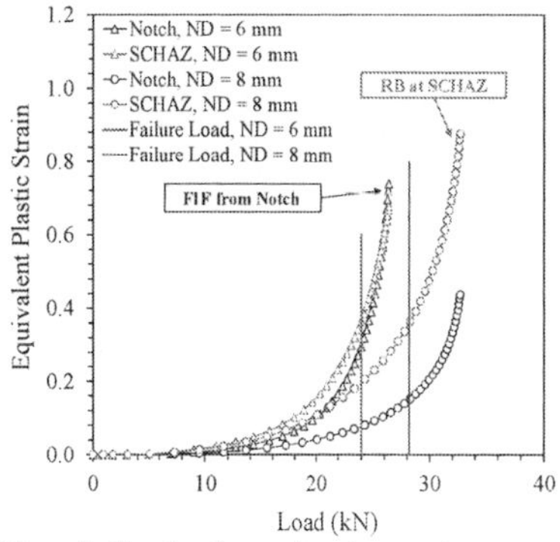

Fig. 4: Equivalent plastic strain at notch and SCHAZ for M1700 tensile shear specimen with nugget diameters of 6 and 8 mm.

To understand the role of ND and softened-SCHAZ on failure mode, equivalent plastic strain (PEEQ) at notch and SCHAZ inner edge for M1700 spot welds with 6 and 8 mm ND are plotted in Figure 4 (a). Results from experiments showed that failure mode for weld with ND of 6 and 8 mm are FIF and RBH, respectively. Comparing PEEQ at notch and SCHAZ in Figure 4 (a) shows that PEEQ at SCHAZ for ND of 6 mm is similar to notch. Considering PEEQ as a parameter that can be used to evaluate the failure mode, this indicates that for the 6 mm, material in notch will reach the failure criterion that can cause crack initiation, growth and finally failure in FIF mode. Using the same approach for weld with 8 mm ND, it can be seen that PEEQ in SCHAZ is significantly higher than that in notch indicating that material in this region meets the failure plastic strain before notch leading to necking in sheet thickness and finally failure in RBH mode. Another observation from FE simulations was made based on the cross-section inplane shear stress in the center of FZ for welds with 6 and 8 mm ND. Results are presented in Figure 5. As it can be seen, for weld with ND of 6 mm, shear stress increases significantly as the extension increases while for that with ND of 8 mm this curve is smoother with lower magnitude. It was also observed that for weld with ND of 6 mm that failed with FIF mode, the inplane shear stress is significantly higher than that for ND of 8 mm where failure happens at SCHAZ.

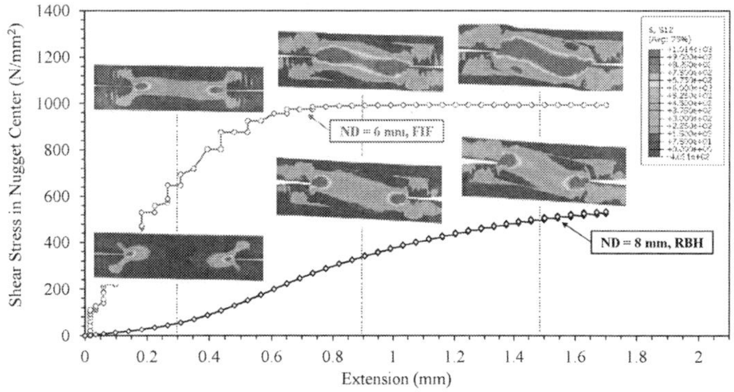

Fig. 5: Shear stress in the center of nugget for M1700 tensile shear specimen with nugget diameters of 6 and 8 mm.

Conclusions

Based on the data from microhardness measurements, softening was identified in SCHAZ. The hardness drop with magnitude of 6%, 18%, and 42% was observed in steels with 16%, 52% and 100% of martensite, respectively. The hardness ratio evaluated and used to represent as the extent of softening.

Maximum hardness drop was observed in SCHAZ close to the Ac1 line. Hardness drop increases as steel carbon content, MVF, and MCC increase; however, it was observed that a combination of MVF and MPC can be used to better predict the hardness drop or extent of softening in SCHAZ.

FE model for tensile-shear test with partitioned weld zones is able to predict the global deformation behavior of specimens when the extent of softening in SCHAZ is negligible, for instance DP590. In steels with higher extent of softening, for example DP980 and M1700, contribution of SCHAZ and its local constitutive behavior increases in prediction of deformation response.

Acknowledgments
The authors acknowledge the support of this work by the ArcelorMittal R&D. Valuable and helpful discussions with Dr. Shrikant Bhat and Dr. Sriram Sadagopan from ArcelorMittal Global R&D are appreciated.

References
[1] R. Kuziak, R. Kawalla, and S. Waengler, "Advanced high strength steels for automotive industry," *Archives of Civil and Mechanical Engineering*, vol. 8, pp. 103-117, 2008.
[2] C. M. Tamarelli, "The Evolving Use of Advanced High-Strength Steels for Automotive Applications," *Steel Market Development Institute, Michigan*, 2011 2011.
[3] J.-O. Sperle and K. Olsson, "High strength and ultra high strength steels for weight reduction in structural and safety-related applications," 1996, pp. 115-125.
[4] V. H. Baltazar Hernandez, S. S. Nayak, and Y. Zhou, "Tempering of Martensite in Dual-Phase Steels and Its Effects on Softening Behavior," *Metallurgical and Materials Transactions A*, vol. 42, pp. 3115-3129, 2011.
[5] E. Biro, "Heat-Affected Zone Softening Kinetics in Dual-Phase and Martensitic Steels," 2014 2014.
[6] V. H. Baltazar Hernandez, S. K. Panda, Y. Okita, and N. Y. Zhou, "A study on heat affected zone softening in resistance spot welded dual phase steel by nanoindentation," *Journal of Materials Science*, vol. 45, pp. 1638-1647, 2009.
[7] E. Biro, S. Vignier, C. Kaczynski, J. R. McDermid, E. Lucas, J. D. Embury, *et al.*, "Predicting Transient Softening in the Sub-Critical Heat-Affected Zone of Dual-Phase and Martensitic Steel Welds," *ISIJ International*, vol. 53, pp. 110-118, 2013.
[8] Y. J. Chao, "Ultimate Strength and Failure Mechanism of Resistance Spot Weld Subjected to Tensile, Shear, or Combined Tensile/Shear Loads," *Journal of Engineering Materials and Technology*, vol. 125, p. 125, 2003.
[9] M. Pouranvari and S. P. H. Marashi, "Failure mode transition in AHSS resistance spot welds. Part I. Controlling factors," *Materials Science and Engineering: A*, vol. 528, pp. 8337-8343, 2011 2011.
[10] M. Pouranvari and P. Marashi, "Failure behaviour of resistance spot welded low carbon steel in tensile-shear and coach-peel tests: A comparative study," *Metallurgical Journal of Metallurgy*, vol. 15, pp. 149-157, 2009 2009.
[11] Y. P. Yang, S. S. Babu, F. Orth, and W. Peterson, "Integrated computational model to predict mechanical behaviour of spot weld," *Science and Technology of Welding & Joining*, vol. 13, pp. 232-239, 2008 2008.
[12] S. Dancette, D. Fabregue, R. Estevez, V. Massardier, T. Dupuy, and M. Bouzekri, "A finite element model for the prediction of Advanced High Strength Steel spot welds fracture," *Engineering Fracture Mechanics*, vol. 87, pp. 48-61, 2012.
[13] S. Dancette, D. Fabrègue, V. Massardier, J. Merlin, T. Dupuy, and M. Bouzekri, "Experimental and modeling investigation of the failure resistance of Advanced High Strength Steels spot welds," *Engineering Fracture Mechanics*, vol. 78, pp. 2259-2272, 2011.
[14] "AWS D8.9M. Test methods for evaluating the resistance spot welding behavior of automotive sheet steel materials, third ed. American Welding Society (AWS), Miami, Fl, USA, 2012.
[15] "Standard Guide for Preparation of Metallographic Specimens," ed: ASTM International, 2011.

Heat Flux and Electromagnetic Force Affected by Current Frequency in Argon Pulsed Arc Welding Mixed with Oxygen

Yoshifumi Maeda, Tatsuro Tanaka*, Shinji Yamamoto*, Toru Iwao**
**Tokyo City University, Setagaya, Tokyo, 158-8557, Japan*
g1681339@tcu.ac.jp

Shinichi Tashiro**, Masaya Shigeta**, Manabu Tanaka**
***Joining and Welding Research Institute, Osaka University*
11-1, Mihogaoka, Ibaraki, Osaka, 567-0047, Japan

Abstract
TIG (Tungsten Inert Gas) welding is high quality joining technology. However, the TIG welding cannot melt the welding pool deeply. In the case of the shallow welding pool, the accident caused by a welding defect may occur. It is necessary to prevent the accident caused by a welding defect. Current waveform, shielding gas species and the others should be controlled in order to improve the weld shape. Some researchers have researched that the weld depth increases using the dissociation gas, such as oxygen. It is considered that thermal pinch effect occurs because the specific heat and thermal conductivity increase. The temperature distribution and current path become small. Thus, the current density is high, and the temperature derived from Joule heating increases at anode surface. However, the weld depth decreases with increasing the contamination ratio of the dissociation gas excessively. It is necessary to control the temperature distribution affected by the current waveform change in order to increase the current density to the arc center on the anode surface at a low contamination ratio of the dissociation gas. According to other researches, the current is concentrated to the arc center and the convection increases because the temperature transient response is affected by a high current frequency. Therefore, the deep weld depth with the combination between pulsed arc and dissociation gas is expected at a low contamination ratio of the dissociation gas. In this paper, the heat flux and electromagnetic force affected by the current frequency in argon pulsed arc welding mixed with oxygen is elucidated by using EMTF(electromagnetic thermal fluid) simulation. As a result, the temperature near anode with the pulsed current was lower than that near anode with the steady current. The heat transfer of electron inflow caught up the changing current and increased with increasing the frequency because the times of increment of the current increases. The electromagnetic force with 1000 Hz of pulsed current frequency was higher than that with 250 Hz of pulsed current frequency.

Introduction
TIG welding is high quality joining technology. However, it cannot melt the welding pool deeply because the high current cannot use for preventing the erosion of tungsten electrode. The needs for the welding industry is that the welding pool is melted deeply with the low current in order to improve the productivity [1][2]. Thus, the pulsed current is used to improve the welding speed and the quality. This welding can control the heat flux to anode and the electromagnetic force to cause the convection of the weld depth direction affected by the pulsed current parameters. Therefore, the welding pool becomes deep with increasing the frequency because a transient response of the temperature occurs [3][4]. The temperature distribution depended on thermal inertia is changed at various times [5]. In addition, the temperature and the thermal inertia are changed by using the different gases. The weld depth greatly increases using the dissociation gas because of the thermal pinch effect caused by increment of specific heat and thermal conductivity [6][7]. Therefore, the deep weld depth with combination of the pulsed arc and dissociation gas is expected at a low contamination ratio of the dissociation gas. In this paper, the heat flux and electromagnetic force affected by current frequency in argon pulsed arc welding mixed with oxygen is elucidated in order to know the weld depth for preventing a welding defect.

Calculation methods
This calculation is analysed with one model of electrode and arc, simultaneously, under the boundary condition [2]. The calculation condition of the plasma arc is assumed by the local thermal equilibrium, chemical equilibrium, optically thin plasma, incompressibility, laminar flow, flat welding, non-vapor and the 2D cylindrical coordinates. This simulation is analysed by the governing equations and the SIMPLER method [4][8]. The positive sign energy transferred between the arc and anode is obtained from the heat conduction and electron inflow. This added as a generation term of the energy conservation equation on the anode boundary [5][9].

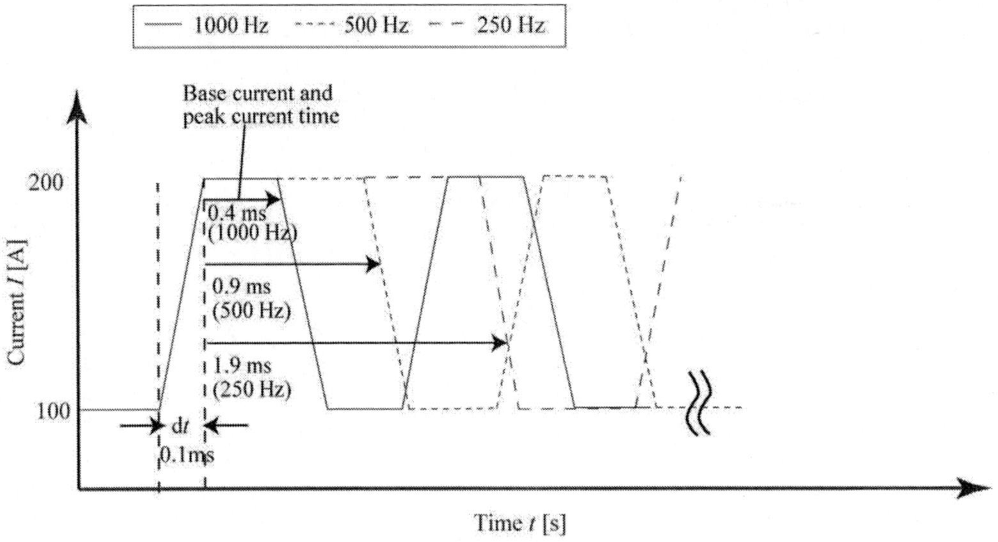

Fig. 1: Current waveform.

The cathode is tungsten, the anode is SUS304, the electrode gap is 5 mm, the ambient gas is Ar(95%)-O_2(5%) and the flow rate is 10 slm. *Fig. 1* shows the current waveform. The base current and peak current are 100 and 200 A, respectively. The frequency is 250, 500, 1000 Hz and a 50 % duty rate. In addition, the initial value was set by the temperature, pressure, flow velocity and electric field to an initial current. The welding pool exists because the anode has already melted.

Results and discussion

Fig. 2 shows the temperature distribution with the steady current and pulsed current at each time. The temperature increased with increasing the current and decreased with decreasing the current at each frequency. At the current increasing time such as (b), the temperature near anode cannot catch up the changing current, and the electrical conductivity depended on the temperature is low. The temperature near anode with the pulsed current was lower than that near anode with the steady current. In comparison with the steady current. A temperature distribution near anode shrank with the pulsed current because of thermal pinch effect derived from the dissociation of oxygen at 4000 K. In addition, the temperature distribution near anode shrank with increasing the frequency because the peak current time is low. The current density near anode is high with increasing the frequency in order to maintain the current continuity in the narrow current radius. This is because that the electrical conductivity to the center depended on the temperature is low. *Fig. 3* shows the time transition of heat transfer within 2 mm of radial distance from the center. The heat transfer of heat conduction cannot catch up the changing current and was constant with increasing the frequency. The heat transfer of electron inflow caught up the changing current and increased with increasing the frequency because the times of current increment increases. Therefore, the heat transfer within 2 mm of the radial distance increased with increasing the frequency. *Fig. 4* shows the radial distribution of the heat flux and electromagnetic force at anode surface. If the electromagnetic force is positive sign, the direction of the electromagnetic force is the weld depth direction. The heat flux of electron inflow at anode surface increased with increasing the frequency at 2 mm from the center. This is because that the current density is high in order to maintain the current continuity. Therefore the heat transfer increased with increasing the frequency. In addition, the electromagnetic force with 1000 Hz of pulsed current frequency was higher than that with 250 Hz of pulsed current frequency. Therefore, the weld depth increment is suggested with increasing the frequency because of the convection to the weld depth direction in the anode.

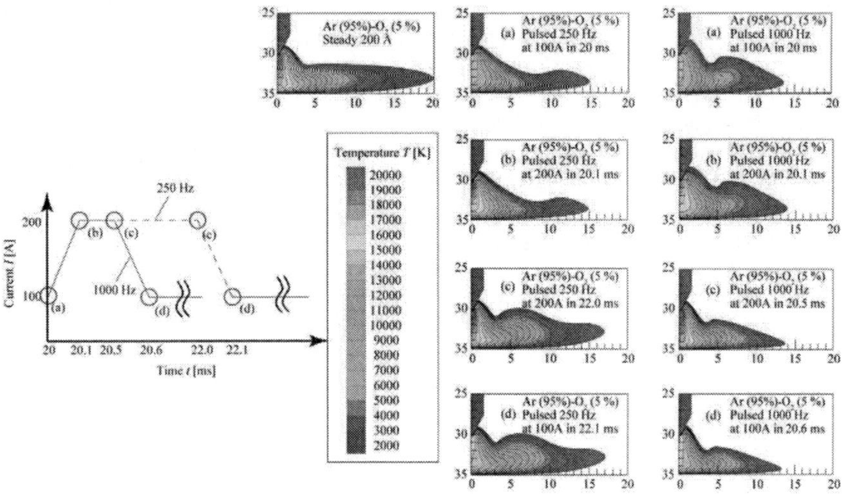

Fig.2: Temperature distribution with steady current and pulsed current at each time.

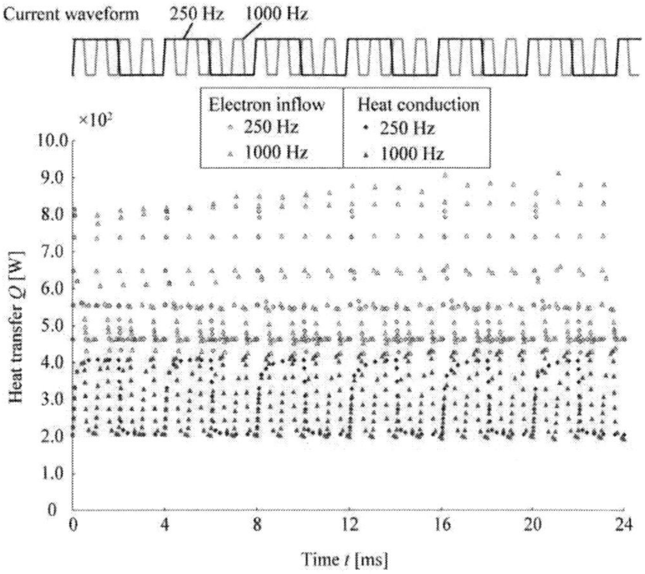

Fig. 3: Time transition of heat transfer within 2 mm of radial distance from the center.

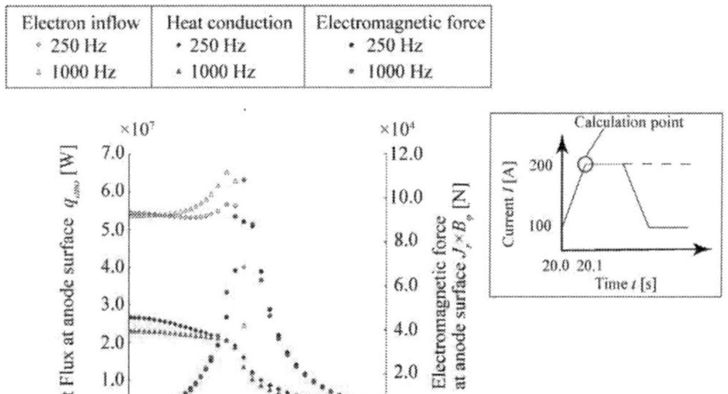

Fig. 4: Radial distribution of heat flux and electromagnetic force at anode surface.

Conclusions

The heat flux and electromagnetic force affected by current frequency in argon pulsed arc welding mixed with oxygen was elucidated in order to know the weld depth for preventing a welding defect. The main results are shown below.

(1) The temperature near anode with the pulsed current was lower than that near anode with the steady current. The temperature distribution near anode shrank with the pulsed current because of thermal pinch effect derived from the dissociation of oxygen at 3000 K. The current density near anode is high with increasing the frequency in order to maintain the current continuity in the narrow current radius. This is because that the electrical conductivity to the center depended on the temperature is low.
(2) The heat transfer of electron inflow caught up the changing current and increased with increasing the frequency because the times of increment of current increases. The heat flux of electron inflow at anode surface increases with increasing the frequency at 2 mm from the center. This is because that the current density is high in order to maintain the current continuity.
(3) The electromagnetic force with 1000 Hz of the pulsed current frequency was higher than that with 250 Hz of the pulsed current frequency at anode surface.

Therefore, it is elucidated that the convection to the weld depth direction in the anode increases with increasing the pulsed current frequency at a low contamination ratio of the dissociation gas.

References

[1] T. Sakai, H. Taki, T. Iwao, S. Tashiro, M. Tanaka and M. Yumoto: "Driving Force Variation in Weld Pool Affected by Current density and Flow Velocity of Gas Tungsten Arc Welding," IEEJ Trans, Vol. 132, No. 5 (2012), pp. 486-492.

[2] S. LU, H. Fujii, K. Nogi and T. Sato: "Effect of Oxygen Content in He-O_2 Shielding Gas on Weld Shape in Ultra-deep Penetration TIG Welding," Quarterly J. Japan Welding Soc, Vol. 25, No. 1 (2007), pp. 196-202

[3] B.J.Qi, M.X.Yang, B.Q.Cong and F.J.Liu: "The effect of arc behavior on weld geometry by high-frequency pulse GTAW process with 0Cr18Ni9Ti stainless steel," International Journal of Advanced Manufacturing Technology, Vol. 66 (2013), pp. 1545-1553.

[4] T. Momii, T. iwao and M. Yumoto: "Contribution for Heat Transfer and Heat Flux to Anode Affected by Rise Current Transition Time in Pulsed Arc," IEEJ Trans, Vol. 133, No. 5 (2013), pp. 409-416.

[5] Y. Mori, T. Iwao, M. Yumoto, S. Tashiro and M. Tanaka: "Effects of Metal Vapor for Temperature Distribution of Transient Arc in Pulsed TIG Welding," Quarterly J. Japan Welding Soc, Vol. 27, No. 2 (2009), pp. 8-12

[6] S. LU, D. LI, H. Fujii and K. Nogi: "Time Dependant Weld Shape in Ar-O2 Shielded Stationary GTA Welding," Journal of Materials Science and Technology, Vol.23, No. 5 (2007), pp. 650-654.

[7] H. Fujii, T. Sato, S. Lu and K. Nogi: "Development of an advanced A-TIG (AA-TIG) welding method by control of Marangoni convection," Journal of Materials Science and Engineering A, Vol.495 (2008), pp. 296-303.

[8] S. V. Patanker: Numerical Heat Transfer and Fluid Flow, Hemisphere Publishing Corp. (1980), pp. 116-139.

[9] M. Tanaka: "An introduction to physical phenomena in arc welding processes," Welding International, Vol.18 (2004), pp.845-851

Linking a CFD and FE analysis for Welding Simulations in Ti-6Al-4V

R.P. Turner*, C. Panwisawas, Y. Sovani, B. Perumal, R.M. Ward, H.C. Basoalto and J.W. Brooks
PRISM[2] Research Group, School of Metallurgy & Materials, University of Birmingham, Birmingham, B15 2TT, UK
Contact: Email: r.p.turner@bham.ac.uk; Tel: (+44) 121 4147565*

Abstract

Finite element (FE) modelling of fusion welding methods has become an established numerical tool used by high-value manufacturing industries and academic communities, largely due to its capabilities to predict residual stress and distortion. However, a major drawback of this type of approach is the requirement to perform a test weld at the relevant process parameters, geometry and material to understand the size and shape of the weld pool formed. With this knowledge a priori the FE model can then be used to best-fit the thermal cycles to the part, and from the thermal field predict the mechanical response to this thermal loading. This well-established method of FE simulation reduces the predictive capabilities of the model. Thus, an improved method of using a different modelling strategy to feed the thermal cycles in to the FE model is desirable. A computational fluid dynamics (CFD) modelling capability has been developed which is able to predict not just weld pool shape, but using real physical phenomena such as surface tension and thermo-capillary forces, buoyancy forces and interfacial phenomena between solid-liquid and liquid-gas phases, can predict thermal fluid flow lines within the molten region, the presence of regions susceptible to porosity and the formation of the keyhole phase, containing metallic vapor. Using simplistic Cartesian co-ordinates the fusion boundary can be extracted from CFD analysis for entry in to an FE model for structural analysis in terms of residual stress and distortions. Therefore, the modelling approach predicts both fluid type and structural type properties of the transient welding operation.

Introduction

The titanium alloy Ti-6Al-4V remains by far the most commonly used titanium alloy in high-technology manufacturing [1], with particularly widespread use in the aerospace and aero-engine industries, owing largely to its excellent strength : weight ratio [2]. In the manufacture of modern gas-turbine jet-engines, titanium alloys are used to manufacture both rotating and stationary components, all of which are subject to very closely controlled manufacturing tolerances. Indeed, given the strict guidelines for the use of newer-generation materials in to a jet-engine component, there remains a distinct advantage in retaining the use of fully certified materials from engines of previous generations, such as Ti-6Al-4V. Components including turbine discs and blades, the compressor drum and the fan remain commonly manufactured using Ti-6Al-4V. The complex geometries of aero-engine components lead to the requirement for welding techniques to join parts structurally. Welding methods used as standard within the aerospace industry include older arc-type welds such as tungsten inert-gas (TIG) welding [1], as well as newer beam-type processes such as laser and electron-beam welding [3]. The underlying physical phenomena associated with beam processes are more complex as the material experiences localized vaporization in these high power-density processes.

Given that Ti-6Al-4V therefore remains an industrially critical alloy for the aero-engine manufacturers, the methods of welding and joining safety-critical components of this material remain of great interest. As such, numerical modelling methods which allow for greater through-process understanding, material flow, process optimization and parametric studies can be used to enhance the welding engineer's understanding of the micro- and macro-mechanical materials behavior that leads localized distortions at the crown and the underbead, and residual stress formation. It is understood that the local deformation at the location of the weld pool, such as the weld crown and underbead, is primarily formed due to thermal fluid-flow relationships of the molten material [4], whereas larger-scale bulk deformation [5] and the associated residual stresses generated are formed due to thermal cycles within the solidified and rapidly cooling material. Therefore, in to model accurately some aspects of the process we require a computational fluid dynamics (CFD) approach to fully capture fluid behavior, whilst to accurately model others needs a solid mechanics finite element (FE)-type approach.

Additionally, typical welding simulation often requires an a-priori understanding of the weld pool size and shape. Traditionally, an experimental weld at the required process parameters has been performed and sectioned for analysis. As such, it becomes of great interest to develop a bespoke modelling framework, whereby a CFD model can predict the weld pool size and shape, as well as localized weld crown deformation due to fluid dynamics, and this CFD model be coupled with a FE model to predict global deformations and residual stresses (see Fig. 1).

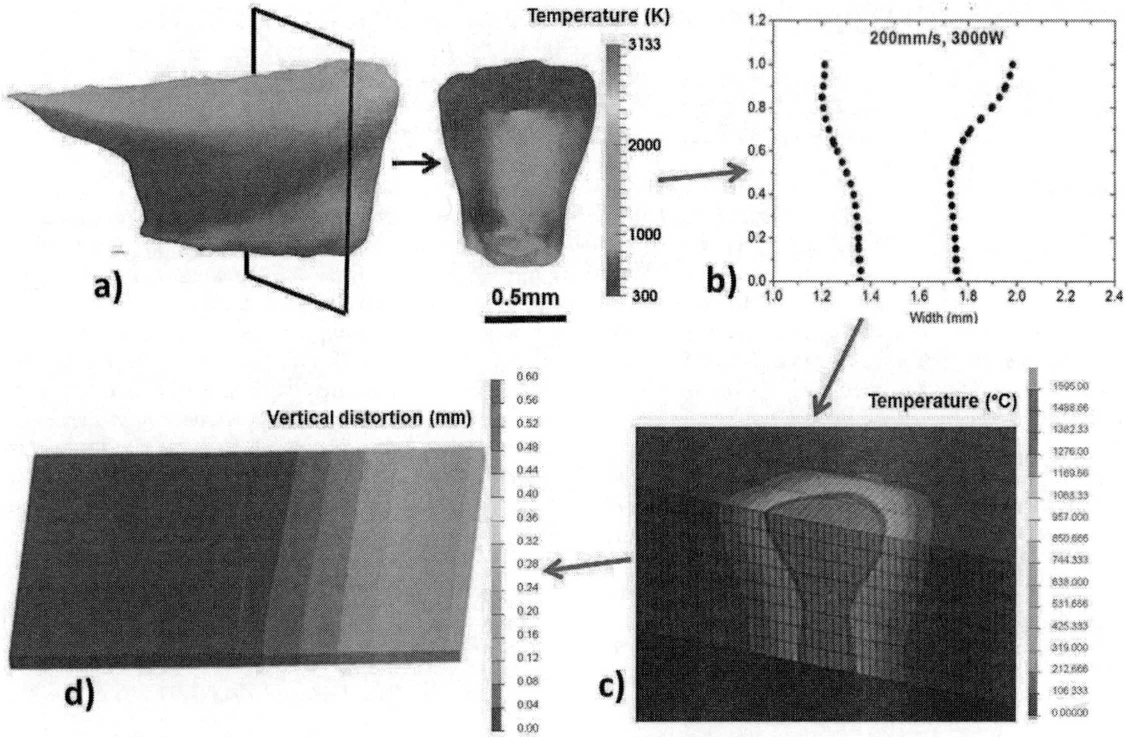

Fig. 1: A graphical representation of the modelling framework developed, linking a CFD model (a) to an FE model (c,d) by extracting the isotherm at the liquidus temperature, thus generating the molten pool shape (b).

Computational fluid dynamics (CFD) modelling

In order to simulate the fluid and vapour-phase interactions of the Ti-6Al-4V material under the presence of an external heating source, the C++ open source Open Field Operation and Manipulation (OpenFOAM) package has been utilised. This has allowed the researchers to develop a model capable of predicting the presence of the so-called "keyhole" in a beam-type weld, where vapour-phase material sits. The vapour pressure causes the keyhole to open. The CFD model uses two incompressible, non-isothermal immiscible fluids using a volume of fluid (VOF) phase-fraction based interface capturing approach. Further details can be found in the literature [6]. The thermal fluid dynamics calculation is computed based upon the evolution of the thermal field coupled with the Navier-Stokes equation for fluid flow [6]. The flow calculation has included vapourisation (recoil pressure), solid-liquid and liquid-vapour phase changes, surface tension, and thermal-capillary effects or Marangoni flow; and crucially an adaptive volumetric heat source model, modified from literature [7] has been used to simulate the energy reflection inside the keyhole. Heat loss due to conduction, convection and radiation has been physically captured at the metallic/gaseous interface. However, despite the simplification of this fluid and thermal dynamics model when compared to others, run-times are still relatively long. CFD models were calculated on a HPC cluster, parallelised over 4 cores, and each model had a run-time of ~4-5 days.

The OpenFOAM CFD model was simulated for long enough to establish a so-called "steady-state" weld shape. The model was constructed to consider the laser welding of a Ti-6Al-4V plate of 1mm thickness. This same plate geometry is therefore considered when results are passed through to a FE model for mechanical analysis. Material properties for the titanium alloy were obtained through materials databases such as JMatPro [8]. An example of the predicted weld pool, viewed from side-on through the plate, can be seen in Fig. 2. The CFD model can be further interrogated to understand; (i) the size and shape of this fusion boundary and (ii) the impact of the thermal fluid flow upon the local area surrounding the molten pool. Indeed, the CFD model is capable of predicting the localised surface deformations that produce the weld crown and the underbead. Although FE software has also been shown capable of predicting these, this is generally due to the soft material and the applied forces causing the deformations. In the case of this CFD model, the correct physical phenomena, namely the Marangoni force, is the driving mechanism for the weld crown and underbead. Notably the CFD model highlights the fact that the weld never truly achieves steady-state because the vapour pressure inside the keyhole is highly unstable, thus the keyhole region is constantly pulsing and momentarily collapsing in on itself.

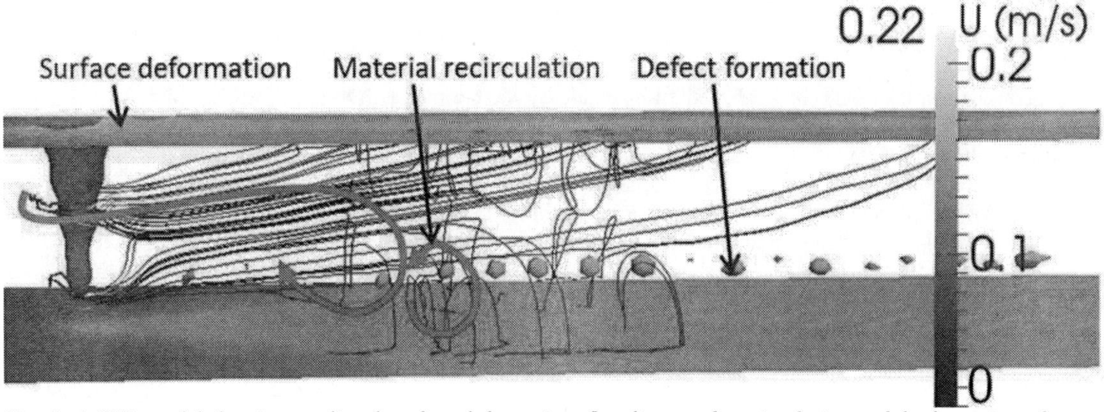

Fig. 2: A CFD model showing predicted surface deformation, flow lines and re-circulation and the formation of porous defects in the weld path.

Additionally, the CFD model can predict the material re-circulation in the fluid region caused by thermal fluid flow, and importantly predictions are made for the formation of sub-surface defects in the material after the molten pool has passed by. The presence of sub-surface porosity defects has been observed experimentally [9], and the CFD model can help to explain their formation. As the fluid flow lines are circulating, the vortices can set-up "dead-zones" where material cannot escape. The vapour-phase keyhole represents the only chance for trapped pores to vacate the material. If the defect makes its way in to the keyhole then the trapped atmospheric gases can exit the weld. However, once the keyhole region of the weld has passed by, the porosity defects become trapped in the structural part as the material around them solidifies.

Coupling the CFD to a FE model

Whilst the CFD model will of course give a complete, full 3D representation of the predicted weld pool shape, some pragmatism must be adopted when attempting to fit this weld pool shape in to the FE model. The FE code Sysweld is used for the solid mechanics analysis of the structural joint. The Sysweld user-interface requires information about the steady-state cross section of the weld pool size and shape. Typically a heat source is fitted using the so-called Goldak function [10]. However, in this work, a script has been written to determine the isotherm at the liquidus temperature within the CFD model, and extract this isotherm as a series of Cartesian co-ordinates [11]. These can then be entered in to the Sysweld database to describe the shape of the molten pool of material as it traverses through the material at a desired welding speed. Hence, once a validated CFD model is developed, the need for FE models to perform experimental welds to understand pool size will be removed. A comparison of the CFD prediction of a weld pool, the FE representation once this shape is entered as a series of Cartesian co-ordinates, and a real weld at this parameter set sectioned for analysis, is shown in Fig. 3.

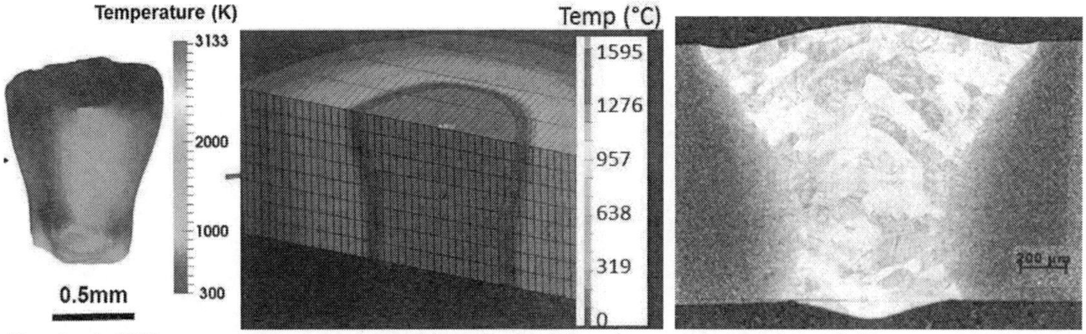

Fig. 3: A CFD representation of the weld pool, the FE software interpretation of this representation, and an experimentally measured weld pool, for the same weld parameter set.

The CFD modelling method shows considerable capabilities capturing the tapering of the cross-sectional width at the 'waist' region of the weld pool, a phenomenon that is observed experimentally in the sectioned weld pool. The FE software makes a reasonable attempt to replicate this tapering at the 'waist', although the amount of taper actually produced is dependent upon ta fine-enough mesh size at this location. The CFD model is reasonable at predict the flaring out at the bottom surface of the pool width. However, the FE approximation struggled with this, and compensated by making the 'waist' region slightly wider and the flaring-out at the base smaller.

Finally, a comparison of the mechanical predictions made by the coupled CFD-FE modelling framework against real weld data is required. In order to assess how accurate the coupled model was, the angular mode of distortion observed from the process of welding was analysed experimentally, and by FE model. Results of the FE modelled predictions for angular distortion, critically compared to experimental welding trials, are shown in Fig. 4. The coupled CFD-FE modelling methodology shows some capability to predict a sensible amount of angular distortion based upon the welding parameters entered, although the model can under- / over-predict by up to 25%.

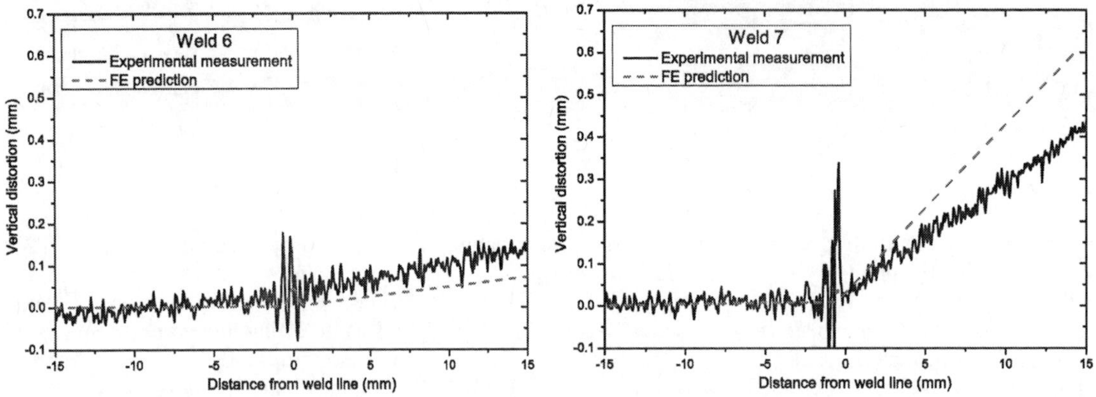

Fig. 4: FE predicted angular distortion compared to experimentally measured distortion, for two welds. Weld 6 considered a 3kW, 100mm/s parameter set, whilst Weld 7 considered a 1.5kW, 200mm/s set.

Conclusions

A modelling framework has been established coupling a CFD and an FE code to allow for thermal fluid flow and solid mechanics predictions of a component as it undergoes a welding operation. The framework developed uses OpenFOAM as the CFD tool and links it with Sysweld, an FE tool for weld simulations. The following conclusions can be drawn:

- CFD methods capture with reasonable accuracy the size and intricate shape of the weld pool.
- Transferring this data to an FE model is possible with the Cartesian co-ordinates of the isotherm, although the weld pool shape accuracy depends upon meshing.
- Mechanical predictions from the FE-part of the coupled model show reasonable predictive capabilities, when critically compared with experiment.

Acknowledgments

The research presented here was carried out as a part of a collaborative project, called CASiM^2, between the University of Birmingham, the MTC and Rolls-Royce plc. The authors wish to acknowledge the financial support of this project provided by the European Regional Development fund (ERDF).

References

[1] N. Cumpsty, Jet Propulsion, Cambridge Univ. Press (UK, 2003).
[2] I.J. Polmear, Light Alloys – 2nd Ed, Edward Arnold (UK, 1989), pp211-271
[3] J.C. Williams and E.A. Starke, *Acta Mater.*, 51; (2003), pp5775-5799.
[4] H. Schultz, Electron Beam Welding, Woodhead Publishing – Elsevier (Holland, 1994), pp65-66
[5] H. J. Stone, S. M. Roberts, R. C. Reed, *Metall Trans A*, 31; 9, (2000), pp2261-73
[6] C. Panwisawas, Y. Sovani, R. Turner, J.W. Brooks, H.C. Basoalto, ***Article in preparation*** (2016).
[7] G.X. Xu, C.S. Wu, G.L. Qin, X.Y. Wang, S.Y. Lin , *Int. J Adv. Manuf. Tech.*, 2011; 57:245–55.
[8] JMatPro, Sente Software Ltd, Occam Rd, Surrey.
[9] J.L. Huang, N. Warnken, J-C, Gebelin, M. Strangwood, R.C. Reed, *Acta Mater*, 60; 6-7, (2012) pp.3215-25
[10] J. Goldak, A. Chakravarti, M. Bibby, *Metall Trans. B*,(1984); 15:299–305.
[11] R. Turner, Y. Sovani, C. Panwisawas, R.M. Ward, H.C. Basoalto, J.W. Brooks, *Met. Trans B*, 2016; 47, 1: 485-94

Magnetic Pulse Welding of Al/Cu Sheets using 6-Turn Flat Coil

Tomokatsu Aizawa
Tokyo Metropolitan College of Technology, Tokyo, Japan
Email: aizawa@mem.iee.or.jp

Abstract
A high pulse current (e.g. peak 150-200 kA) is discharged through a one-turn flat coil which is usually used for magnetic pulse welding of sheet metals. Little attention has been paid to decrease the current. In this study the current was decreased using a 6-turn flat coil and the weldability of dissimilar sheet metals was checked. The 6-turn coil could generate the magnetic flux of the same value by the current of about 1/6 in comparison with the one-turn flat coil. Al/Cu sheets were overlapped with a gap and fixed on the 6-turn coil. When a magnetic flux was suddenly generated around the coil, eddy currents were induced in the sheets. Electromagnetic force acted on the Al sheet nearer the coil. This sheet was collided to the Cu sheet and welded. The peak current required for this welding was 25 kA at the discharge energy of 0.6 kJ.

Introduction
In magnetic pulse welding (MPW), electromagnetic forces are used to weld metal work-pieces. For a long time, the MPW method has been used for welding tubular work-pieces [1,2]. Recently, this method is applied to flat work-pieces [3-5]. The most common configurations consist of a lap joint of similar or dissimilar metals. The mechanism and features of MPW are considerably well known. Its technique is based on discharging a high pulse current through a coil to produce electromagnetic forces. The coil shape is chosen to become strong in the electromagnetic coupling between the coil and the work-pieces. A solenoid coil (usually a multi-turn coil) is used for the tubular work-pieces. On the other hand, a one-turn flat coil is mainly used for the flat work-pieces.

The flat work-pieces are overlapped sheet metals, which are usually fixed on the insulated E-shaped one-turn flat coil [3,4]. The peak discharge current for the welding is, for examples, 150-200 kA. About the improvement of such a high current, MPW of sheet metals has not been studied enough. Little attention has been paid to the MPW technique using a multi-turn flat coil. This flat coil had difficulty in design and manufacture in comparison with the multi-turn solenoid coil. A few examples of the multi-turn flat coil have been reported in Japan [6,7].

The objective of this study is to decrease the high current technically. The E-shaped 6-turn flat coil was adopted to decrease the current and the weldability of dissimilar sheet metals was tested. The current for the welding decreased to about 1/6. The properties of the Al/Cu sheets welded by the 6-turn flat coil were similar to the case of the one-turn flat coil. The collision speed of the Al sheet to the Cu sheet was calculated and the weld interfaces were observed. The deformation of the coil after the welding was also checked.

Experimental procedures
Figure 1 is a sketch of the E-shaped 6-turn flat coil for MPW. The coil comprises six pieces of E-shaped metal plates which are electrically insulated and are laminated. The discharge current flows into the narrow central part of the plate No.1 upward and flows out from the two wide outside parts downward. The divided currents join at the lower parts of the plate No.1. The joined current flows again into the central part of the plate No.2. The discharge current flows through each central part in series and does six turns. Each central part of the plates is just piled with the same width. A high-density magnetic flux is generated around the piled central part as the 6-turn flat coil. Therefore, this coil can generate the magnetic flux of the same value by the current of about 1/6 in comparison with the one-turn flat coil if the coil-design and other conditions are suitable.

The sheet metals (Al/Cu) for the welding are fixed on the 6-turn coil like Fig. 2. The lower Al sheets are insulated in a DC manner from the coil, but there is electromagnetic coupling between them. When the discharge current flows into the coil, the high-density magnetic flux is suddenly generated around the central part. The flux intersects the lower Al sheet and eddy currents are induced electromagnetically in the skin of the Al sheet. The eddy currents near the central part flow in opposite direction to the discharge current of the central part. As a result, the electromagnetic force ***f*** acts on the Al sheet (Fig. 2). The Al sheet is accelerated upward and collides with

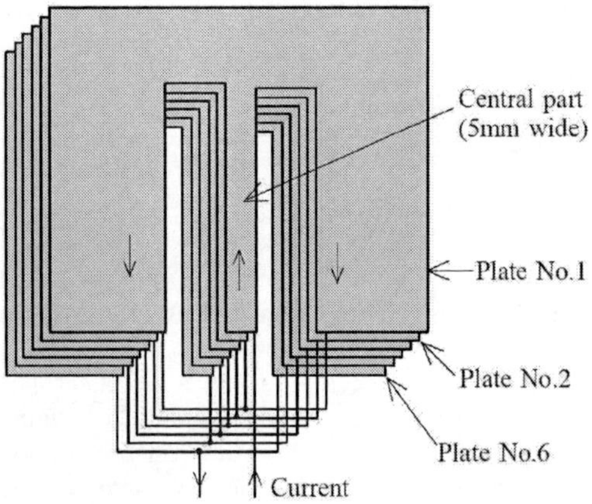

Fig.1 Sketch of 6-turn flat coil.

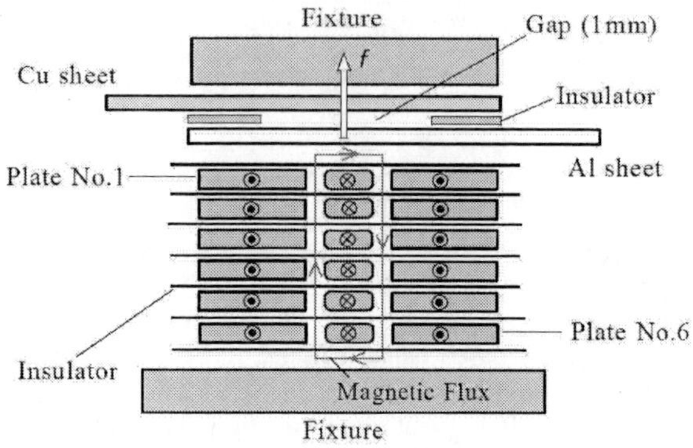

Fig.2 Fixation of coil and Al/Cu sheets.

the upper Cu sheet at very high speed. The Al/Cu sheets will be welded along the central part by the decreased discharge current of about 1/6. The type of the welding is such a seam welding.

The electromagnetic force ***f*** to act on the Al sheet is proportional to the square of the discharge current. The discharge current is proportional to the magnetic flux density. The details of these are shown elsewhere [4].

Experimental apparatus and materials

The discharge system consisted of the capacitor bank, a discharge switch, the 6-turn flat coil and others. The capacitor bank (capacitance C=24 μF) consisted of four capacitors of 6 μF in parallel. The bank was connected to the discharge switch and the coil by low inductance transmission lines. The charged energy E of the bank was 0.6 kJ. When the switch was closed, the discharge current suddenly flowed through the coil and the part of energy E moved to the coil as magnetic energy. The welding process was finished in one shot of the discharge.

The coil was mainly made by six E-shaped plates (0.3 mm thick) of Cu alloy. The narrow central part of each plate was 5 mm wide and 110 mm long. The thickness as the 6-turn flat coil was approximately 3 mm including the insulators and about the same as that of the conventional one turn flat coil. Each insulator was polyimide of about 0.1 mm thick. Figure 3 shows the appearance of the 6-turn coil consisting of the plates which made gaps for photography without the insulators. The inductance of the 6-turn coil was about 1.3 μH and larger enough than the inductance to exist (or remain) in the discharge circuit except the coil. The energy E will move more to the 6-turn coil as magnetic energy in comparison with the one-turn coil of smaller inductance.

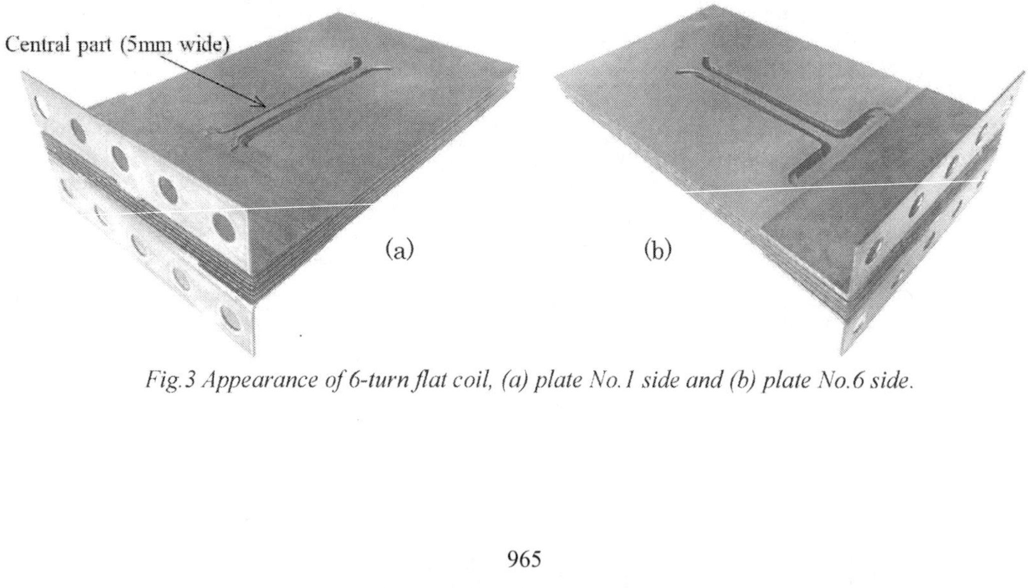

Fig.3 Appearance of 6-turn flat coil, (a) plate No.1 side and (b) plate No.6 side.

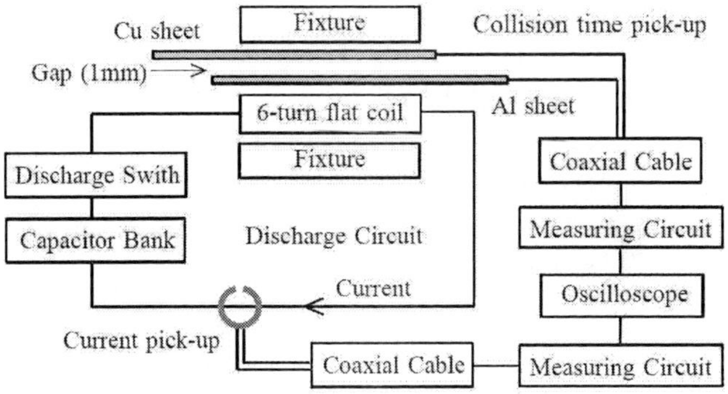

Fig.4 Block diagram of discharge and measuring system.

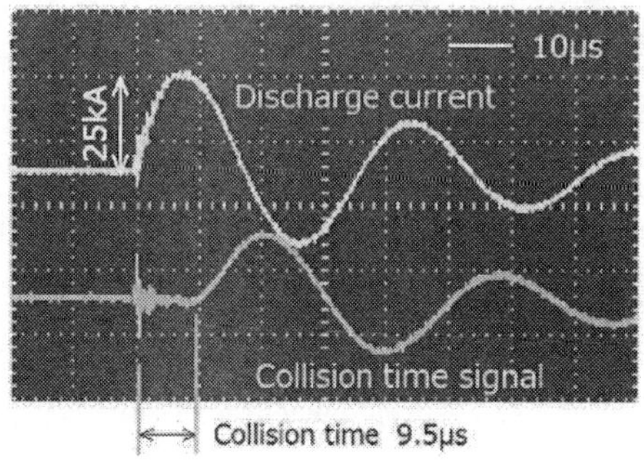

Fig.5 Current and collision time signals.

The Al (1050) and Cu (1220) sheets were 100 mm long, 100 mm wide and 0.5 mm thick. The Al/Cu sheets were overlapped 50 mm in length with the gap of 1.0 mm and fixed on the coil covered by the thin insulator such as Fig.2. The discharge current signal and a collision time signal of the Al/Cu sheets were obtained by using a measuring system as shown in Fig.4. The interfaces of welded Al/Cu sheets were observed by microscopes.

Experimental results

Typical discharge current and collision time signals are shown in Fig. 5. The damping and oscillating current flowed through the coil for duration of about 100 μs. In measurement the oscillating period was 36 μs, and so the skin depth was 0.5 mm for the Al sheet. The peak discharge current was 25 kA at the discharge energy of 0.6 kJ. The sheets collided in around the time for the peak current and after 9.5 μs from the start of the current. The collision speed when the sheets collided was calculated roughly with 210 m/s.

Figure 6 shows a typical appearance of the welded Al/Cu sheets taken from the Al sheet side. A part of the Al sheet deformed toward the Cu sheet along the narrow central part. The weld zones (usually two seams) were in the deformed region which was 100 mm in length to the longitudinal direction and about 10 mm in width direction. The welded sheets were cut and polished for the interfacial observation.

Figure 7 shows a cross sectional appearance of the welded Al/Cu sheets (or weld interface) at the deformed region. Two weld zones marked by arrows (↓) and a middle clearance (non-weld zone) were seen in Fig. 7. This result was similar to the result using the one-turn flat coil [3-5]. Figure 8 is a typical optical micrograph of the interface at the one weld zone and Fig.9 a SEM image of the same interface. Wavy pattern morphology was seen in the weld interfaces as shown in Fig. 9. Such morphology was also seen in the case of the one-turn flat coil [3-5].

Figure 10 shows the appearance of the coil without the insulators after the welding experiments of about 40 times. Each coil plate and each insulator were not damaged by these experiments. The deformation such as wrinkles was

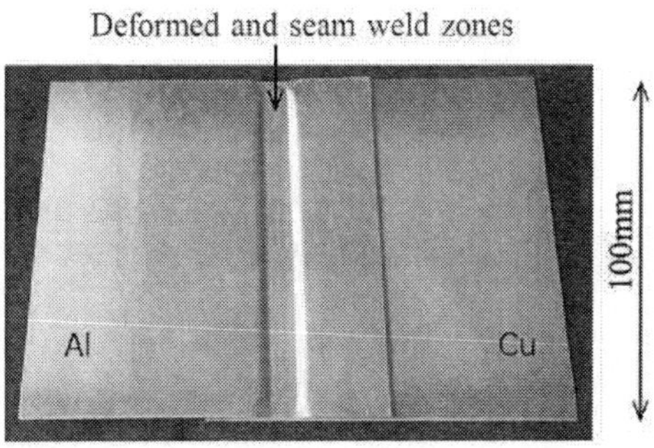

Fig.6 Appearance of welded Al/Cu sheets.

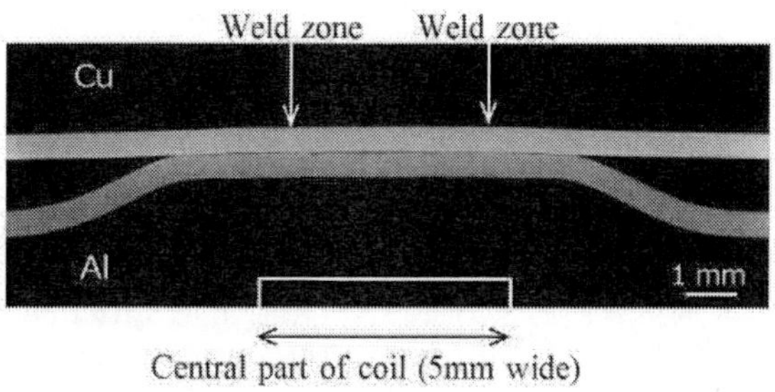

Fig.7 Cross sectional appearance of welded Al/Cu sheets.

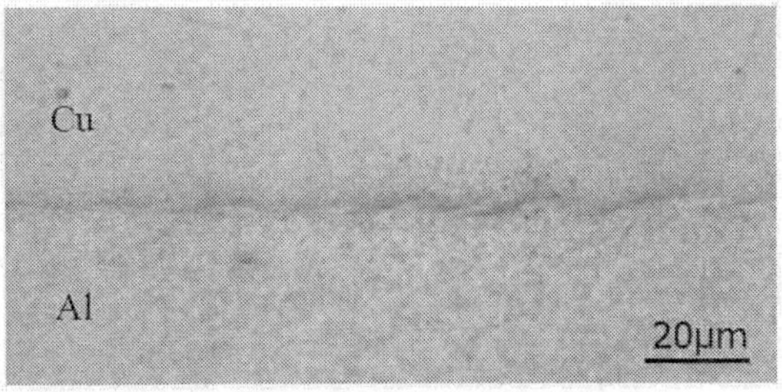

Fig.8 Optical micrograph of weld interface at one weld zone.

Fig.9 SEM image of weld interface at one weld zone.

(a)

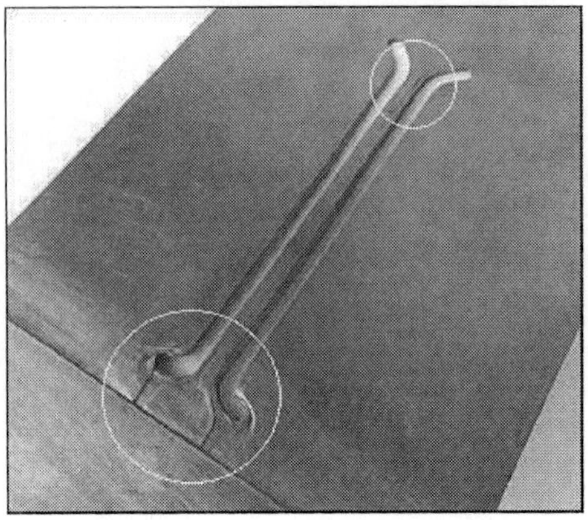

(b)

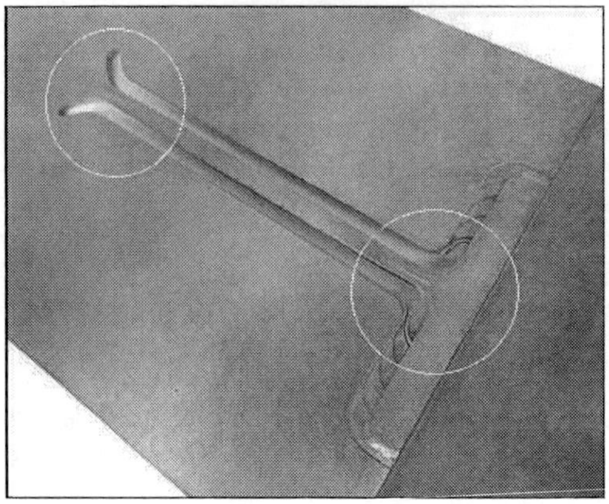

Fig.10 Appearance of 6-turn flat coil after experiments of about 40 times, (a) plate No.1 side and (b) plate No.6 side. Clear deformations occurred in parts of white line circles.

seen partially in each coil plate. The relatively clear deformation occurred in the parts of white line circles (Fig. 10). It was possible to continue the experiments with this coil, but the coil was destroyed before long.

Conclusions

We found that the peak discharge current required for MPW of sheet metals was about 25 kA by using the E-shaped 6-turn flat coil. The welding results with this coil were similar with the case of the one-turn flat coil. The welded Al/Cu interface indicated wave interface morphology like explosive welding. The deformation (or damage) of the coil was also shown. Future work will need to improve the coil life and to increase the coil turns.

References

[1] V. Shribman, A. Stern, Y. Livshitz, O. Gafri: "Magnetic pulse welding produces high-strength aluminum welds," *Welding Journal*, Vol. 81 (2002), pp. 33-37.

[2] M. Kimchi, H. Shao, W. Cheng, P. Krishnaswamy: "Magnetic Pulse Welding Aluminium Tubes to Steel Bars," *Welding in the World*, Vol. 48 (2004), pp. 19-22.

[3] M. Watanabe, S. Kumai, T. Aizawa: "Interfacial micro-structure of magnetic pressure seam welded Al-Fe, Al-Ni and Al-Cu lap joints," *Materials Science Forum*, Vol. 519-521 (2006), pp. 1145-1150.

[4] T. Aizawa, M. Kashani, K. Okagawa: "Application of Magnetic Pulse Welding for Aluminum Alloys and SPCC Steel Sheet Joints," *Welding Journal*, Vol. 86 (2007), pp. 119s-124s.

[5] S. D. Kore, P. P. Date, S. V. Kulkarni: "Effect of process parameters on electromagnetic impact welding of aluminum sheets," *International Journal of Impact Engineering*, Vol.34 (2007), pp. 1327-1341.

[6] T. Aizawa: "Flat multiple-winding coil," Japanese Unexamined Patent No.2007-305555.

[7] T. Aizawa: "Magnetic Pulse Welding of Sheet Metals using Flat Three-Turn Coil," *Int. Workshop on Explosion, Shock-wave and High-velocity Phenomena*, Kumamoto, Japan, Sep 2015, pp. 25-28.

Weldability of TEMPALOY AA-1 and T92 dissimilar joints

Michael URZYNICOK
ZELKOT, Koszęcin, POLAND
mu@zelkot.pl

Robert JACHYM, Krzysztof KWIECIŃSKI, Jacek SŁANIA
INSTITUTE OF WELDING, Gliwice, POLAND

Abstract
Constant development of steels used in power generation industry especially for pressure elements in USC boilers creates new welding challenges. The introduction of new combinations of alloying agents aimed at obtaining the best possible mechanical properties, including creep resistance, affects the weldability of new steel grades. Each of the latter have to deeply tested, particularly as regards cold and hot bending and welding, in order to enable the development of technologies ensuring failure-free production and assembly of boiler elements like headers and superheaters. 9% Cr martensitic steels which are commonly used in production of steam superheaters, are characterized by good creep resistance and, at the same time, low oxidation resistance at a temperature in excess of 600°C. In turn, 12% Cr martensites are characterized by significantly higher oxidation resistance but lower strength at higher temperatures, which translates to their limited application in the production of boilers operating at the highest parameters.

The gap between the above mentioned steep types is perfectly filled by austenitic steels where creep resistance and oxidation resistance are unquestionable. This article presents experience gained during welding dissimilar joints of advanced TEMPALOY AA-1 and T92 grades, with the use of EPRI P87, Inconel 82 and Inconel 617 filler metals.

Introduction
The actual tendency aimed at reducing power generation costs is closely related to increasing the efficiency of power units in conventional power plants. An increase in efficiency can be obtained by increasing steam pressure and temperature. As for today, obtaining higher parameters of steam boilers is possible only by applying modern steels and alloys able to bear increased operational loads and ensure sufficiently high resistance at higher temperatures.

Martensitic steels such as P92 (X10CrWMoVNb9-2), VM12 (X12CrCoWVNb12-2-2) or PB2 (X12CrMoCoVNbNB9-2-1) make it possible to design power plant systems in which the temperature of superheated steam is 625°C. An increase in steam pressure and temperature has a direct effect on the operational conditions of boiler elements like steam superheaters and reheaters. Martensitic steels are characterised by high creep resistance and moderate oxidation resistance at high temperature (P92, PB2) or vice versa (VM12), therefore the development of modern steels is focused on materials with austenitic microstructure [1-4].

Many years of research have led to the development of TEMPALOY AA-1 austenitic steel by TenarisNKKTubes. This article shows possible combinations of welds in real-life situation where a header is made out of P92 with dissimilar connecting tubes of T92 welded to a TEMPALOY AA-1 superheater coils. New EPRI P87 filler metal has been evaluated/tested along with Inconel 82 and 617.

General characteristics of tested steel

TEMPALOY AA-1 is an austenitic steel containing 18% Cr and 10% Ni alloyed with niobium and titanium with addition of 3% Cu, the microstructure of which does not contain delta ferrite. Factors which directly improve creep resistance are niobium and titanium carbides (Nb, Ti)C present along grain boundaries as well as inside the grain.

In operating temperature range 600÷750°C, the microstructure of the steel is strengthened by Cr-rich carbides ($M_{23}C_6$). Addition of copper determines the precipitation of fine and diffuse Cu-rich phase which significantly increases creep resistance.

Hot corrosion and steam oxidation resistance are conditioned by chromium content. The metal loss of TEMPALOY AA-1 material due to hot corrosion is around 25% lower than that of TP321H steel, while the oxidation resistance in the environment of fumes and boiler ash at a temperature of 700°C is similar for both. The performing of shot blasting on the internal surface enables the formation of a stable Cr-rich layer which reduces further and substantially the steam oxidation rate. There was no grain boundary corrosion observed on any of the TEMPALOY AA-1 test specimens.

Martensitic steel 9Cr-1Mo-0.25V-0.06Nb-0.1C designated as P/T91 was developed in the USA in order to meet the needs of nuclear and conventional power engineering. In Europe the steel is designated as X10CrMoVNb9-1.

The modifications of the chemical composition of P91 steel consisting in the reduction of molybdenum content by approx. 0.5% and addition of 1.7% tungsten resulted in the development of steel whose creep resistance is by approx. 30% higher than that of P91 grade.

The steel obtained through the aforesaid modifications was designated as P92 or NF616 or, otherwise as, X10CrWMoVNb9-2. Tungsten, similarly to molybdenum, is an element enhancing solid solution strengthening as it increases recrystallisation temperature thus improving long-term stability of matrix at higher temperature. Through stabilising carbides a microaddition of boron used in P92 steel increases its creep resistance and limits excessive growth of carbides at grain boundaries [1-6].

Advanced structural examinations

Structural analysis was performed by Scanning Electron Microscope (SEM) on Hitachi S-3400N in a non-etched state with Back Scattered Electrons (BSE) technique as well as in etched state with Secondary Electrons (SE) technique. Chemical composition microanalysis was performed by Energy-Dispersive X-Ray Spectroscopy (EDS) on ThermoNoran System Six.

On figures I, II it can be observed that the microstructure in the transition zone between P92 and EPRI P87 filler is cellular converting to mixed cellular-dendritic. No microimperfections were found.

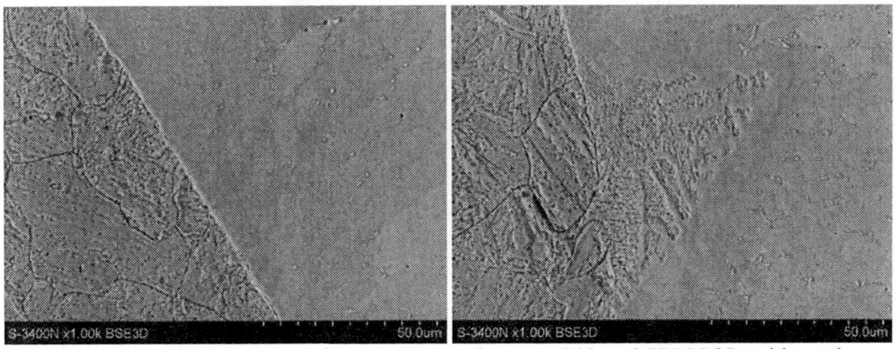

Fig. I, II. Microstructure of transition zone between T92 and EPRI P87 weld metal

On figures III, IV in transition zone between P92 and Inconel 82 filler the microstructure is based on Fe-γ and Ni-γ with no clear segregation of alloying elements into interdendritic area.

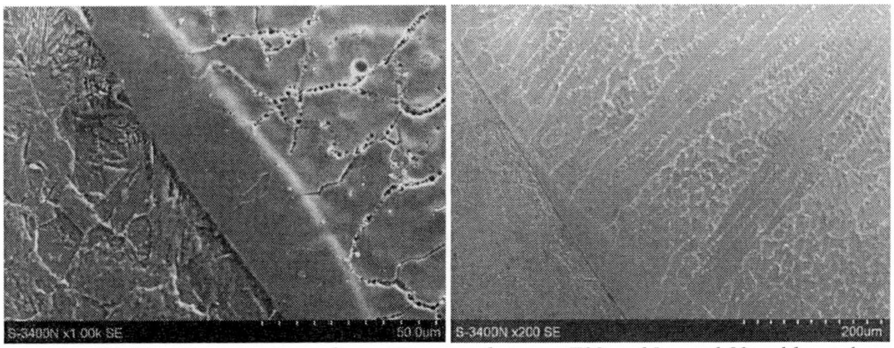

Fig. III, IV. Microstructure of transition zone between T92 and Inconel 82 weld metal

During examinations some microcracks were found as shown of figure V.

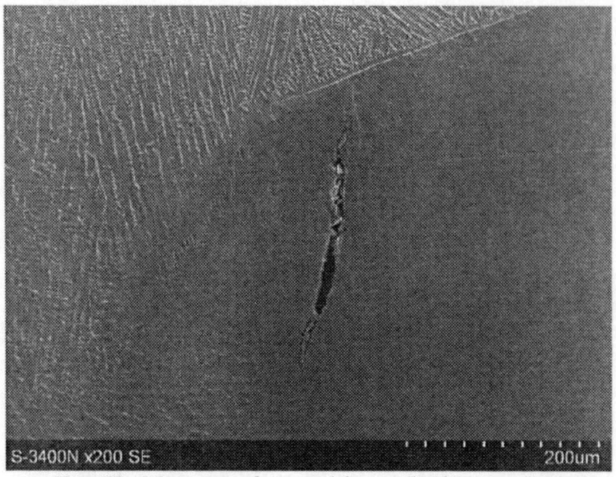

Fig. V. Microcrack in weld metal of Inconel 82

On figures VI, VII in transition zone between P92 and Inconel 617 filler the microstructure is a heterogeneous dendritic with a lot of Mo and Ti carbides. No microimperfections were found.

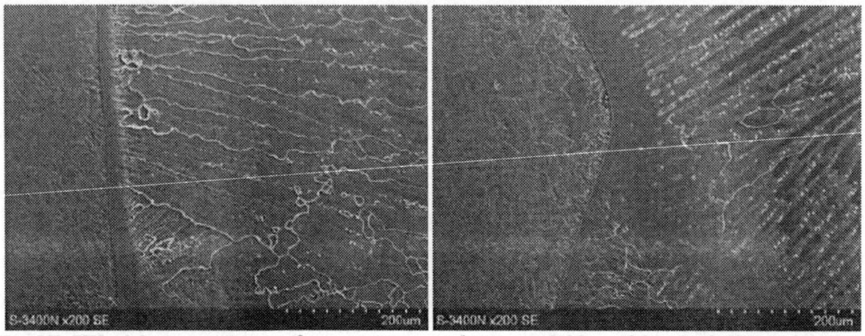

Fig.VI, VII. Microstructure of transition zone between T92 and Inconel 617 weld metal

Microchemical analysis (Figure VIII) clearly shows Fe, Ni and Cr distribution along the transition zone between P92 and weld metal. Cr level in EPRI P87 weld stays at the same level as P92 while in welds performed with Inconel 82 and Inconel 617 Cr content increases significantly. It may be the explanation of different transition zone sizes and its probable decarbonisation.

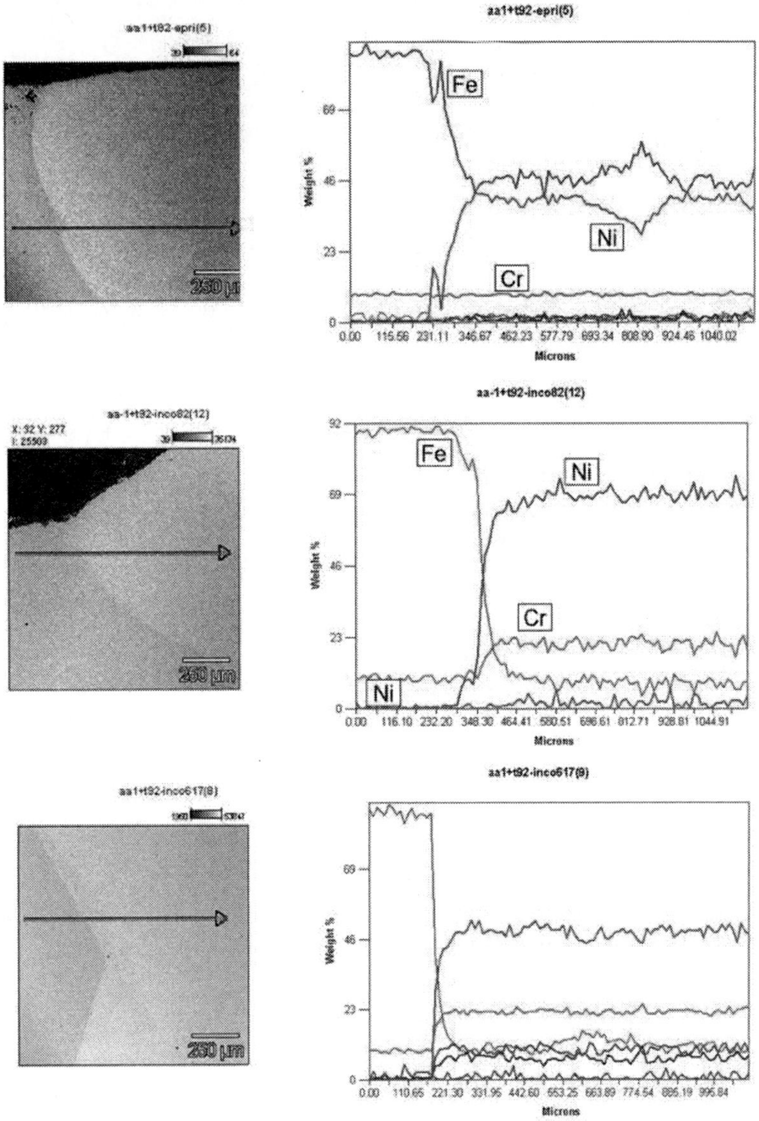

Fig. VIII. Microchemical analysis of transition zone between T92 and different weld metals

Conclusions

On the basis of performed examinations it was possible to come to the following conclusions:

1. Dissimilar joints made of steel grades TEMPALOY AA-1 and T92, GTAW welded in 5G uphill position are characterised by high quality, which is confirmed by the results of destructive and non-destructive tests.
2. The application of filler metal Metrode EPRI P87 made it possible to obtain very good results on a dissimilar welded joint and the investigations confirmed the absence of welding micro-imperfections. The transition zone is very narrow, which favourably affects the operational properties of the joint.
3. It may be concluded based on SEM and EDS examinations supported by micro-hardness tests that the transition zone is in fact carbon depleted zone but to support this theory extensive TEM (Transmission Electron Microscope) examinations shall be done.
4. EPRI P87 is not yet included in the EN code thus its usage may be limited in European projects.

Acknowledgements

The authors would like to thank TenarisDalmine, Italy and TenarisNKKTubes, Japan for providing TEMPALOY AA-1 and T92 tunes, Metrode, UK for providing EPRI P87, Inconel 82 and Inconel 617 rods and ZELKOT, Poland for general idea of research and producing the welded joints, Institute of Welding, Poland for all carried examinations.

References

[1] Caminada S., Cumino G., Cipolla L., Venditti D., Di Gianfrancesco A., Minami Y., Ono T.: Creep properties and microstructural evolution of austenitic TEMPALOY steels. International Journal of Pressure Vessels and Piping 87 (2010)

[2] Urzynicok M., Kwieciński K., Szubryt M., Słania J.: Application of new GMAW welding methods used in prefabrication of P92 (X10CrWMoVNb9-2) pipe butt welds. 9th Liege Conference on Materials for Advanced Power Engineering, September 27th - 29th, Liege, Belgium, 2010

[3] Urzynicok M., Kwiecinski K., Slania J., Szubryt M., Mariani P.: Comparison of SMAW and GMAW Methods used in P92 Butt Welds. 1st Australasian International Welding, Inspection and NDT (iWIN2013) Conference, 10-13 March 2013, Perth, Australia

[4] Urzynicok M., Kwiecinski K., Szubryt M.: Testing butt joints in pipes made of VM12-SHC martensitic steel, Welding International, iFirst article 2012, 1-6

[5] Kwieciński K., Urzynicok M., Łomozik M.: Practical experience with welding new generation steel PB2 grade assigned for power industry. Archives of metallurgy and materials, Vol. 56, Issue 1, 2011

[6] Jachym R., Kwieciński K., Urzynicok M., Łomozik M.: Properties of thickwelded similar and dissimilar joints made of TEMPALOY A-3 steel assigned for power industry. 9th International Conference on Trends in Welding Research, June 4-8 2012, Chicago IL, USA

[7] Schingledecker J.: Update on EPRI P87 Filler Metal. Boiler Reliability Interest Group, Baltimore 22.04.2010

Effect of friction welding conditions on joining phenomena and tensile strength of friction welded joint between OFC and 304 stainless steel

Koki Ohara, Graduate Student, Graduate School of Engineering, University of Hyogo
Masaaki Kimura, Masahiro Kusaka and Koichi Kaizu, University of Hyogo
University of Hyogo, Himeji, Hyogo, Japan
mkimura@eng.u-hyogo.ac.jp

Abstract

This paper describes the effect of friction welding conditions on joining phenomena and tensile strength of friction welded joint between oxygen-free copper (OFC) and austenitic stainless steel (SUS304). When the joint was made with a friction pressure of 30MPa, only the OFC side was deformed, which was exhausted as flash. The temperature of the weld interface increased with increasing friction time, and it achieved approximately 900K at a friction time of 3.2s. The joint efficiency increased with increasing friction time, and it reached approximately 49% at a friction time of 3.6s. However, all joints fractured at the weld interface. To improve the joint efficiency, the joint was made with high forge pressure. When the joint was made with a friction time of 2.8s, it had approximately 88% joint efficiency with a forge pressure of 240MPa. Also this joint fractured at the weld interface. On the other hand, when the joint was made with a friction time of 3.2s and a forge pressure of 210MPa, all joints fractured on the OFC side. In addition, this joint had the approximately 93% joint efficiency and the softened region at the weld interface of the OFC side. In conclusion, to obtain higher joint efficiency and the fracture at the OFC side, the joint should be made with a friction time of 3.2s that the temperature at the weld interface reached to approximately 900K, and with a forge pressure of 210MPa or higher.

Introduction

Copper (Cu) and many of its alloys are well-known materials that have highly attractive characteristics in terms of metallurgical property and workability, e.g. high electrical and thermal conductivity, good corrosion resistance, and good high elasticity. They are widely used for important components in a wide variety of electrical and mechanical parts, such as air conditioner heat exchangers, condensers for ships or plants, and so on. On the other hand, fusion welds between Cu and other metals such as stainless steel and carbon steel have poor mechanical properties due to the brittle intermetallic compound interlayer produced at the joint interface. Moreover, fusion welds between Cu or its alloys and various steels have some problems, e.g. cracking of the joint interface. Therefore, a welding process for dissimilar joints with Cu that will result in less degradation of the mechanical and metallurgical properties of the joint is urgently required. Many researchers [1] have reported that the mechanical and metallurgical properties of friction welded joints between Cu or its alloys and stainless steel show desirable characteristics. However, the joining mechanism of friction welding between dissimilar materials such as Cu or its alloys and stainless steel has not been fully clarified, so that the friction welding conditions for material combinations are determined by trial and error. To determine the theoretical friction welding conditions is necessary to clarify the joining phenomena between dissimilar materials in friction welding.

The authors have been carrying out research to clarify the joining mechanism between dissimilar materials in the friction process. In the present work, we show the effects of friction time and forge pressure on the tensile strength of the joint between oxygen-free copper (OFC) and austenitic stainless steel (SUS304).

Experimental procedure

The materials used were oxygen free Cu (JIS C1020 BD-H, OFC) and AISI 304 austenitic stainless steel (JIS SUS304, SUS304) in the form of a rod with a diameter of 16mm. The ultimate tensile strength of OFC was 321MPa and that of SUS304 was 663MPa, respectively. Both rods were used for this experiment as received condition. Those rods were machined to 12mm in diameter of the weld faying (contacting) surface. In addition, the temperature changes during the friction process at the centerline, half radius and periphery portions of the 1.0mm longitudinal direction from the

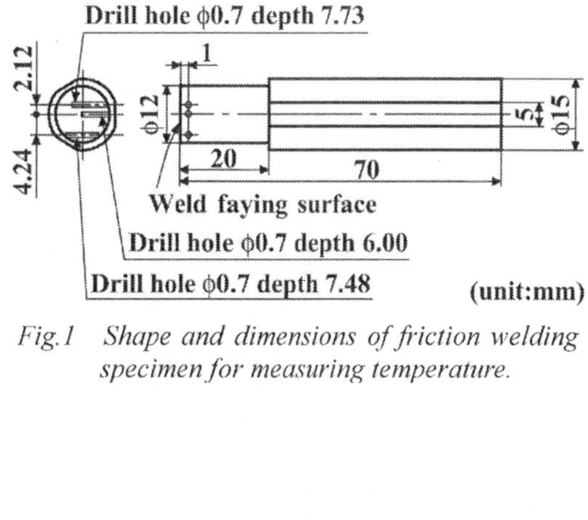

Fig.1 Shape and dimensions of friction welding specimen for measuring temperature.

weld faying surface were measured by using the SUS304 specimen as shown in Fig.1. All weld faying surfaces of specimens were polished with a surface grinding machine before joining in order to eliminate the effect of surface roughness on the mechanical properties of dissimilar material joints [2].

A direct drive friction welding machine was used for the joining. During the friction welding operations, the friction welding condition was set to the following combinations: a friction speed of $27.5s^{-1}$ (1650rpm), a friction pressure of 30MPa, a range of friction times from 2.0 to 4.0s, a range of forge pressures from 30 to 240MPa, and a forge time of 6.0s. The joining behavior during the friction process was recorded by a digital video camera, and the friction torque was measured with a load-cell. The mineral insulated thermocouple with chromel-alumel was inserted into a drill hole of the SUS304 specimen, which was showed in Fig.1. The friction torque and temperature were recorded with a personal computer through an A/D converter. Furthermore, the specimens were joined by using an electromagnetic clutch in order to prevent deformation during braking time. When the clutch was released, the relative speed between both specimens instantly decreased to zero. The joint tensile test specimens were machined to 12mm in diameter and 66mm in parallel length. The joint tensile test was carried out with as-welded condition at room temperature.

Results and discussion

Relationship between joining behavior and friction torque curve

Figure 2 shows the relationship between the joining behavior and the friction torque. Photo (1) to (3) in Fig.2b corresponds to the friction torque of (1) to (3) in Fig.2a, respectively. Photo (1) shows the state at the faying

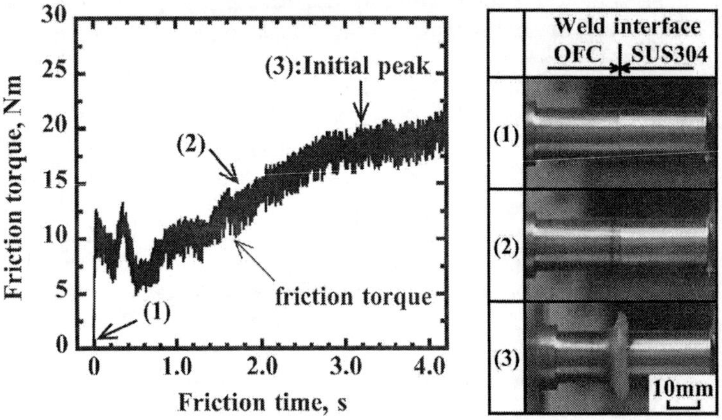

(a) Friction torque curve *(b) Joining behavior*

Fig.2 Joining behavior and friction torque curve during friction process.

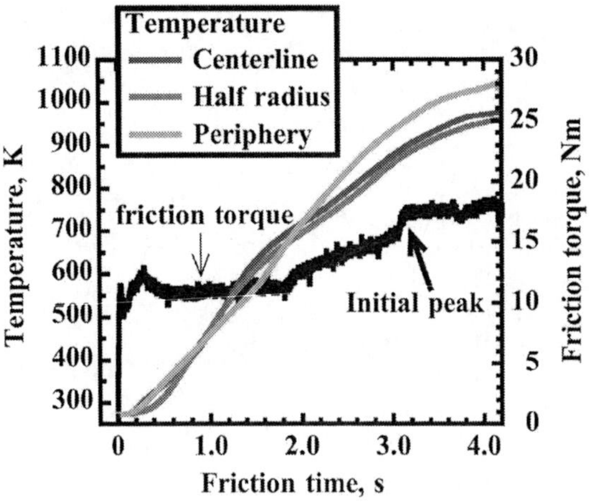

Fig.3 Temperature change and friction torque during friction process.

surfaces as they contacted each other, then the friction torque was increased. The friction torque was kept with approximately 13Nm although that had slightly scattering. The OFC side was slightly upset (deformed) when the color of the outer surface at the adjacent region of the weld interface on the OFC side changed as shown in photo (2). Thereafter, the upsetting and the flash (burr and collar) of OFC increased with increasing friction time, although the SUS304 side was not upset. Then, the friction torque reached to the initial peak of (3) that was approximately 20Nm when the friction time was about 3.2s. However, the initial peak torque was not so clear.

Temperature changes during friction process

Figure 3 shows the relationship between the friction time and the temperature changes in relation to the friction torque during the friction process. Temperatures of each measuring points on the weld interface of the SUS304 side increased with increasing friction time. Those temperatures were almost the same, although the temperature at the periphery portion was slightly high after a friction time of 2.0s. However, the difference in each three temperatures was approximately 50K. Then, when the friction time was 3.2s, i.e. the friction torque just reached to the initial peak, those temperatures achieved approximately 900K or over.

Relationship between friction time and joint efficiency

Figure 4 shows the relationship between the friction time and the joint efficiency of the joint, plotted alongside the friction torque curve. The joint efficiency was defined as the ratio of joint tensile strength to the ultimate tensile strength of the OFC base metals. Figure 5 shows an example of the appearance of the joint tensile tested specimen. In this case, forge pressure was applied at an identical friction pressure, i.e. 30MPa. The joint efficiency at a friction time of 2.0s was approximately 4% because a sufficient quantity of heat for welding could not be produced during this friction time. Then, the joint efficiency increased with increasing friction time. When the joint was made with a friction time of 3.2s (the initial peak), the joint efficiency was approximately 26%. Thereafter, the joint efficiency was approximately 49% at a friction time of 3.6s or longer. All joints fractured at the weld interface, which had an OFC adhering to the weld interface on the SUS304 side, as shown in Fig.5.

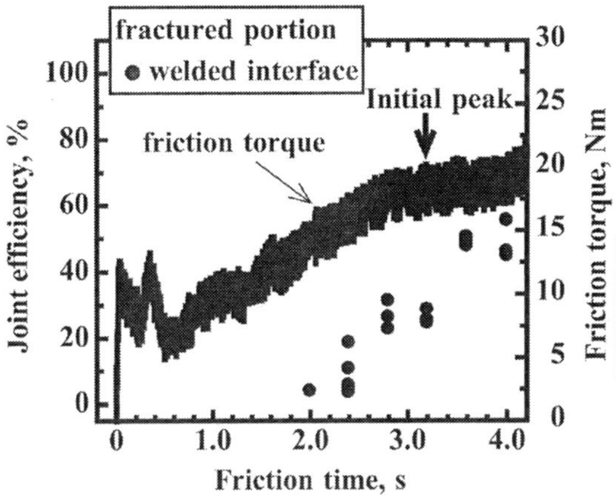

Fig.4 Relationship between friction time and joint efficiency, in relation to friction torque.

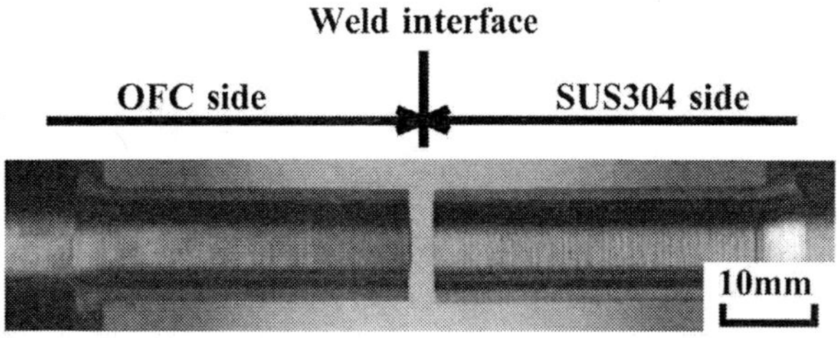

Fig.5 Example of joint tensile tested specimen appearance.

Improving joint efficiency

In an attempt to improve the joint efficiency, the joints were made by adding forge pressure. Figure 6 shows the relationship between the forge pressure and the joint efficiency of joints at a friction time of 3.2s. Figure 7 shows examples of appearances of the joint tensile tested specimens. The joint efficiency of joints with a forge pressure of 30MPa was approximately 26%, and those joints fractured at the weld interface as shown in Fig.5. The joint efficiency increased with increasing forge pressure. Some joints with a forge pressure of 60 to 180MPa fractured between the weld interface and the OFC side (mixed mode fracture) as shown in Fig.7(a), and others fractured from the OFC side as shown in Fig.7(b). Then, when the joint was made with a forge pressure of 210MPa, all joints fractured from the OFC side as shown in Fig.7(b) although it had the joint efficiency of approximately 93%. However, the adjacent region of the weld interface on the OFC side of this joint softened, although the data was not shown here. On the other hand, the joint with short friction time did not have the fracture on the OFC side

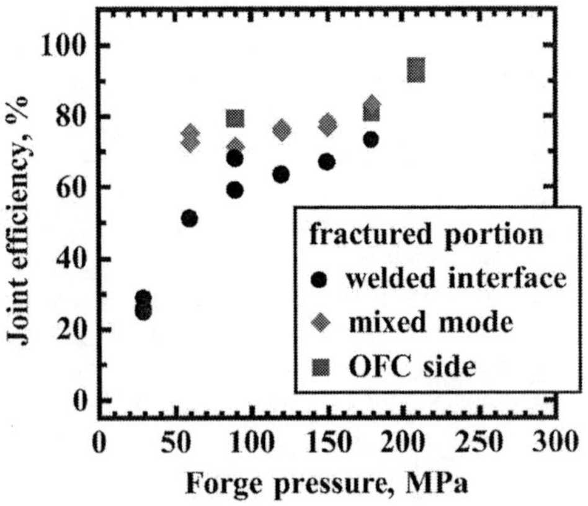

Fig.6 Relationship between forge pressure and joint efficiency of joints at friction time of 3.2s.

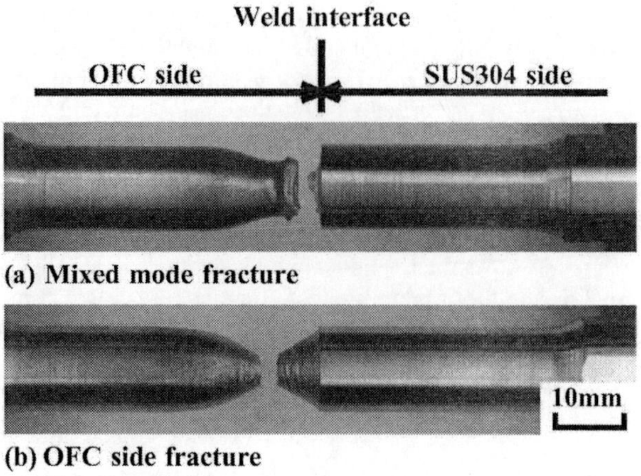

Fig.7 Examples of joint tensile tested specimens appearances.

because the weld interface did not have enough temperature for welding. For instance, when the joint was made with a friction time of 2.8s, it had approximately 88% joint efficiency with a forge pressure of 240MPa. However, the joint fractured at the weld interface although it had high joint efficiency. Therefore, to obtain higher joint efficiency and fracture in the OFC side, the joint should be made with low friction pressure and high forge pressure, and with the friction time at which the friction torque reaches the initial peak.

Conclusions

This report described the effect of friction welding conditions on joining phenomena and tensile strength of friction welded joint between OFC and 304 stainless steel. The following conclusions are provided.

(1) When the joint was made at a friction pressure of 30MPa with a friction speed of $27.5s^{-1}$, the upsetting (deformation) occurred at the OFC side. In addition, the friction torque reached to initial peak that was approximately 20Nm when friction time was about 3.2s.

(2) When the friction time was 3.2s, i.e. the friction torque just reached to the initial peak, the temperatures at each measured portions of the SUS304 side achieved approximately 900K or over.

(3) The joint efficiency increased with increasing friction time, and it reached approximately 49% joint efficiency at a friction time of 3.6s. However, all joints fractured from the weld interface.

(4) The joint efficiency at a friction time of 3.2s increased with increasing forge pressure, and it reached approximately 93% joint efficiency at a forge pressure of 210MPa. This joint fractured at the softened OFC region adjacent to the weld interface.

(5) To obtain higher joint efficiency and the fracture at the OFC side, the joint should be made with a friction time of 3.2s that the temperature at the weld interface reached to approximately 900K, and with a forge pressure of 210MPa or higher.

Acknowledgments

This research was partially supported by the Ministry of Education, Culture Sports, Science and Technology, Japan, Grant-in-Aid for Scientific Research (C) (MEXT KAKENHI), Grant Number 16K06748, 2016.

References

[1] For example, C.Shanjeevi, S.S.Kumar, P.Sathiya: "Evaluation of Mechanical and Metallurgical properties of dissimilar materials by friction welding", Procedia Engineering, Vol. 64 (2013), pp. 1514-1523.

[2] M.Kimura, Y.Saitoh, M.Kusaka, K.Kaizu, and A.Fuji: "Effect of Friction Welding Condition and Weld Faying Surface Properties on Tensile Strength of Friction Welded Joint between Pure Titanium and Pure Copper", Journal of Solid Mechanics and Materials Engineering, Vol. 5, No. 12 (2011), pp. 849-865.

RECYCLING OF SLAG AND ITS EFFECTS ON FLUX CONSUMPTION IN SUBMERGED ARC STAINLESS STEEL CLADDING

Jatinder Garg [1], Kulwant Singh [2]

[1] *Baba Hira Singh Bhattal Institute of Engineering & Technology, Lehragaga (Sangrur), India*
[2] *Sant Longowal Institute of Engineering & Technology, Longowal, India*

Weld cladding finds a wide variety of applications in manufacturing industry. Huge quantity of slag is generated during submerged arc cladding which is discarded as waste. Disposal of this slag poses environmental threats since it is non biodegradable in nature. Minerals and other natural resources which are ingredients of the flux also get wasted. An attempt was made to recycle the slag generated during submerged arc cladding of stainless steel. The recycled slag can be reused as a substitute of fresh flux without affecting the product quality. Since the flux contributes a major portion of the total cost in submerged arc cladding, controlling flux consumption can play a significant role in reducing the process cost. Further it also influences the extent of transfer of various alloying elements between slag and the metal. Therefore it is essential to study the flux consumption behavior of recycled slag and compare it with the fresh flux. In the presented work, the effect of welding parameters and type of flux (fresh/recycled) on the flux consumption has been evaluated. The experiments were conducted using fractional factorial design technique to develop a mathematical model of flux consumption. Results show that the flux consumption increases with increase in open circuit voltage, decreases with increase in welding current and remains unaffected to welding speed. Further it was found that use of recycled slag does not produce any significant effect on the flux consumption. The methodology adopted for recycling of slag has also been discussed briefly in this paper.

Research on the characteristics of arc acoustic sensing in rotating arc narrow gap MAG welding

Wenhang Li, Jielong Wang, Jiayou Wang
Material science and Engineering School, Jiangsu University of Science and Technology, Zhenjiang City, jiangsu Province, China P.R.
Lwh_abc@qq.com, +8651184401184
Wenhang Li
also Material science and Technology School, Nanjing University of Aeronautics and Astronautics, Nanjing City, jiangsu Province, China P.R.

Abstract

Rotating arc narrow gap MAG welding is a high efficient and high quality welding method. It requires real time effective welding quality supervision to overcome the interference. Due to the complexity of the welding process, multi-sensor information fusion has obvious advantages over a single sensor. Besides arc sensing and vision sensing, the arc acoustic sensing obtains rich information but it is seldom studied. In this paper, the arc acoustic source in the welding method was analyzed, and an arc-vision-sound sensing system was set up. Therefore, the time domain and frequency domain characteristics of arc acoustic signal and its cause were studied and analyzed under different welding currents, arc rotating frequencies, gas flow rates and different surface states. The results showed that arc acoustic signal had effective and reliable time domain or frequency domain characteristics under different welding process parameters or conditions. Acoustic sensing was useful for further application of welding quality supervision.

Keywords Rotating arc narrow gap MAG welding . Acoustic sensing . Welding quality supervision

Introduction

Thanks to the arc rotation, rotating arc narrow gap MAG welding can reduce heat input and improve weld formation[1,2]. It usually needs sensing, modeling and control to ensure welding quality and overcome the interference[3-6]. Because of the complexity and the uncertainty factors in welding, a single sensor usually has limitation in information richness and an alternative can be multi-sensor information fusion technology based on accurate and reliable sensing method, which can be used for better welding quality control.

Compared with the arc sensing and visual sensing in narrow gap welding, acoustic sensing[7] obtains equally abundant information but was less studied The experienced welder can monitor the welding quality by arc sound, which inspires research on on acoustic sensing[8]because it is insensitive to welding arc, splash and other instant disturbance. Therefore, it is significant to study the characteristics of arc acoustic sensing in rotating arc narrow gap MAG welding. In this paper, we first set up a system that can obtained the arc, vision and sound information synchronously. Then, by changing the welding process parameters or conditions, the characteristics of acoustic signal sensing for the rotating arc narrow gap MAG welding was studied. The remainder of the paper is organized as follows. In section 2, the rotating arc narrow gap MAG welding and its sensing systems were introduced. In section 3, the experiments were designed and the data were analyzed on the time domain and frequency domain characteristics. At last, some conclusions were drawn.

The rotating arc narrow gap MAG welding and acoustic sensing systems

Fig. 1 showes the rotating arc narrow gap MAG welding and its acoustic sensing systems. For the rotating arc torch, the biased hole on the tip makes the wire end deviate from the tip axis so that the arc rotates during welding. The detail principle of welding torch can be found in Wang's work[1].As shown in *Fig. 1*, the sensing system includes arc sensing part and acoustic sensing part. The acoustic sensing system includes microphone and vibration debug module. The microphone's model is brüel & kjær 4944-A, and it was placed in the symmetric plane of the narrow gap groove to collect the arc acoustic signal. The vibration debug module is brüel & kjær 2690-A for acoustic signal amplification, isolation and denoising. The welding current and voltage was obtained by the Hall sensor. In addition, the power source is Fronius TPS 5000, and the welding parameters were shown in *Table 1*. The shape of work piece was shown in *Fig. 2*.

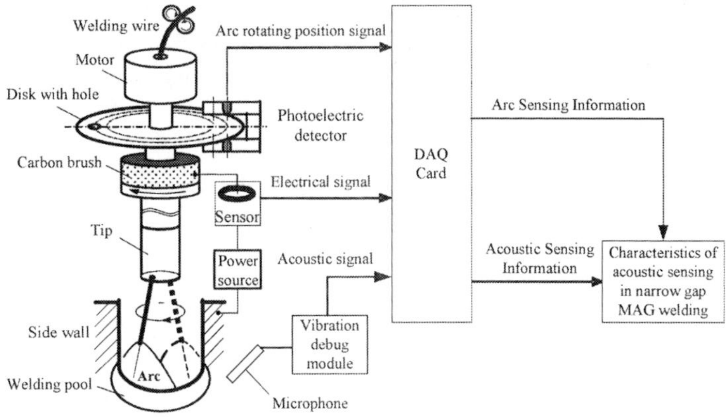

Fig. 1 Structure diagram of rotating arc welding torch and sensing systems

Table 1: Narrow gap welding parameters

Parameter	Value
Torch height H, mm	20
Wire diameter, mm	1.2
Work piece	Q235
Welding voltage U, V	31
Welding speed v, mm·min^{-1}	230

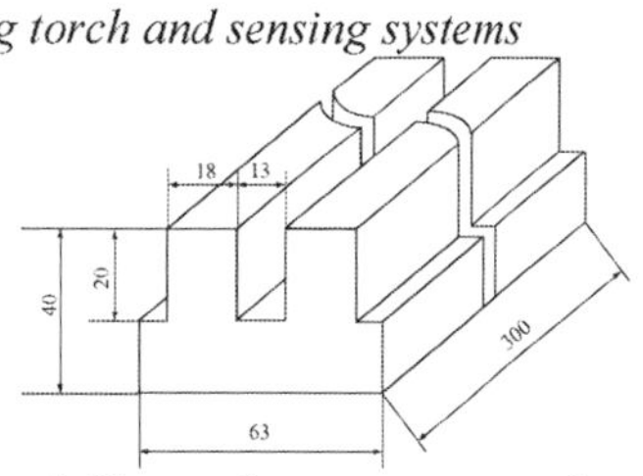

Fig. 2 Shape of narrow gap work piece

Experiments and data analysis

The welding sound is from four sources. Fristly, the variation of arc shape at different rotation position generates vibration and sound. Secondly, the short circuit of welding instantly generates heat and sound. Thirdly, the wire burns or oxide with inadequate or without shield gas, which also generates sound. The last sound source is from the gasified water in expanding rust on the work piece Obviously, the main factors affecting the arc sound is welding current, arc's rotating frequency, shiled gas flow rate and the workpiece state. So, we designed experiments to study the impact by changing these parameters respectively.

Changing the welding current

Usually, the welding current was set to 300A, and the welding current and sound's waveform were shown in *Fig. 3*a. the time domain waveform of the arc acoustic signal was relatively stable and there was almost none strong amplitude. By using Fronius Rob5000 module, the welding current can be adjusted in real time when welding. We set the welding current to change in triangular or trapezoidal wave, and the real welding current and sound signal were shown in *Fig. 3*b. It showed that when the welding current was high enough, there will be short circuit, and a ringing acoustic signal occurred shortly after the short circuit, as shown in *Fig. 3*c. The Power Spectral Density(PSD) of sound with ringing had higher energy in low frequency(<2000Hz) than the one without short circuit or ringing, as shown in *Fig. 3* d, e.

Changing rotating frequency

To study the relationship between the arc rotating frequency and the acoustic signal, the frequencies was set to 50Hz, 35Hz and 20Hz respectively. The result was similar and the acoustic signal of 50Hz was shown in *Fig. 4*. It showed that the vibration of the time domain waveform of the arc sound signal was periodic, and the cycle consistent with arc rotation cycle. The reasons are as follows. The action of arc rotating heating, arc blowing made the weld pool concave down. When the arc rotated to the position near groove sidewall, the arc length was pressed and becomes shorter, so welding current increased, instantaneous short circuit was more likely to happen, and the acoustic signal was periodic.

The state of work piece

To study the impact of rust on the arc acoustic signal, we compared results of rusted and clean work pieces. *Fig. 5* shows that the time domain waveform of arc acoustic signal was extremely unstable and strong irregular peak occurred for rusted work piece. The cause might be that the rust will reduce the electrical conductivity of work piece and arc stability of rotating arc narrow gap MAG welding decrease to cause the instantaneous short circuit (proved by the current waveform) to occur much more easily. In addition, the water or other foreign substances in the rust will be gasified and expand when the arc heat, and it produces sound too.

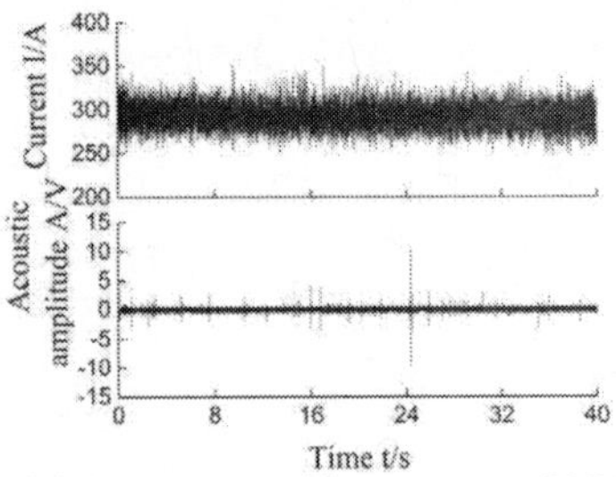

(a) Constant welding current 300A and sound information

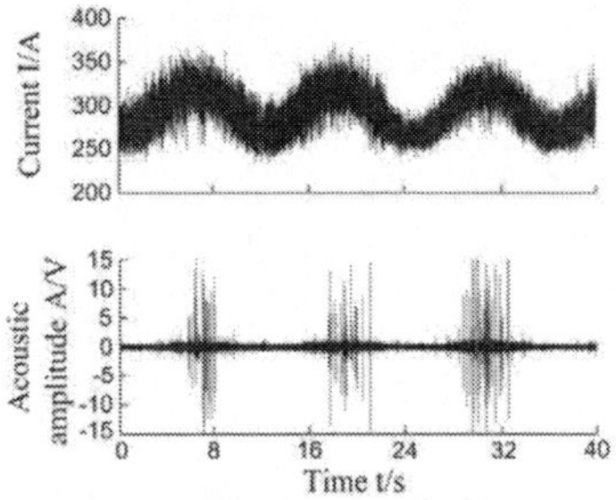

(b) Changing current and according sound informaiton

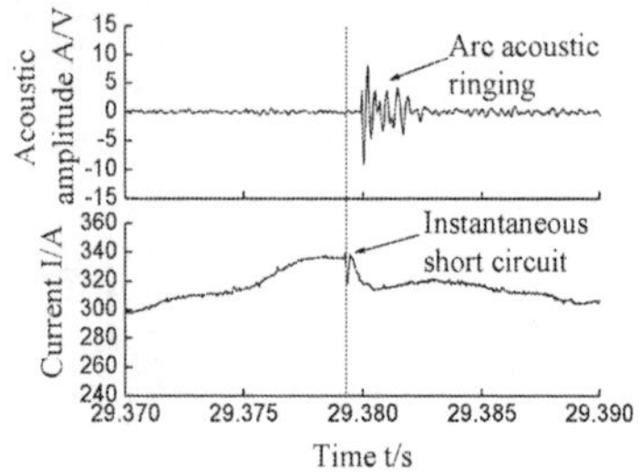

(c) Part signal of (b)

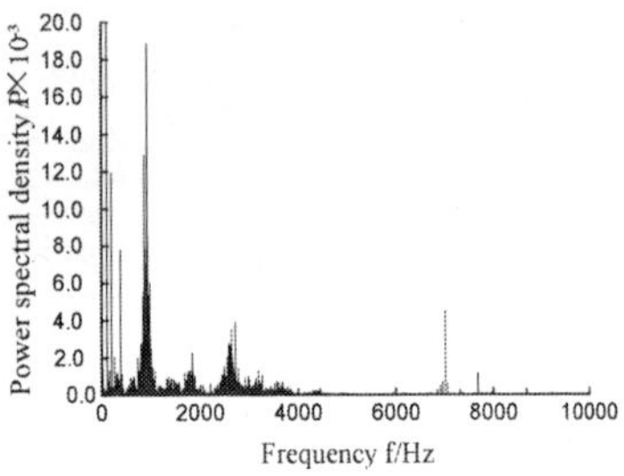

(d) Arc acoustic signal PSD diagram of ringing

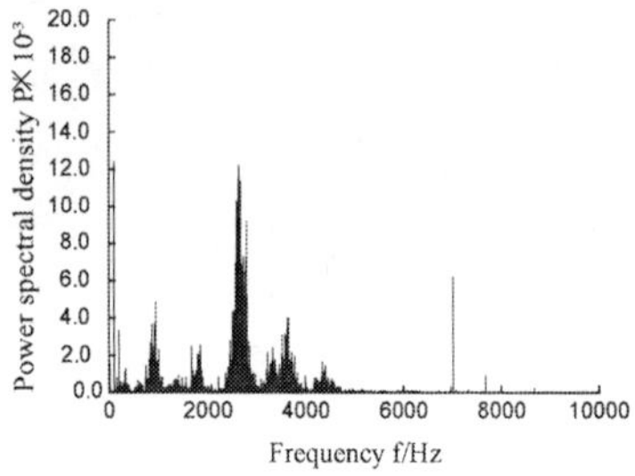

(e) Arc acoustic signal PSD diagram without ringing

Fig. 3 Current's impact on acoustic signal

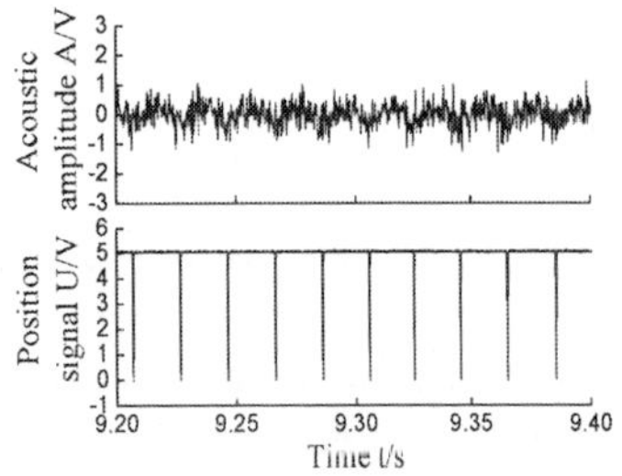

Fig. 4 Arc rotating frequency of 50Hz

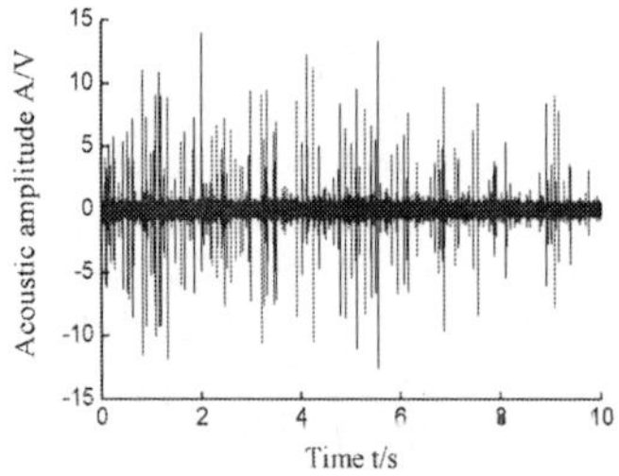

Fig. 5 Work piece with rust on its surface

Shielding gas

In the welding method, the normal shielding gas flow rate was 30L/min. When the shielding gas flow rate was 30L/min, 20L/min and 10L/min respectively, multiple continuous spikes begin to appear and become higher in the frequency range of 300~2500Hz, as shown in *Fig. 6*b. The reasons are as follows. When the gas flow rate decreased, the wire or welding pool was likely to oxide or burn, dramatic changes in energy would generate sound.

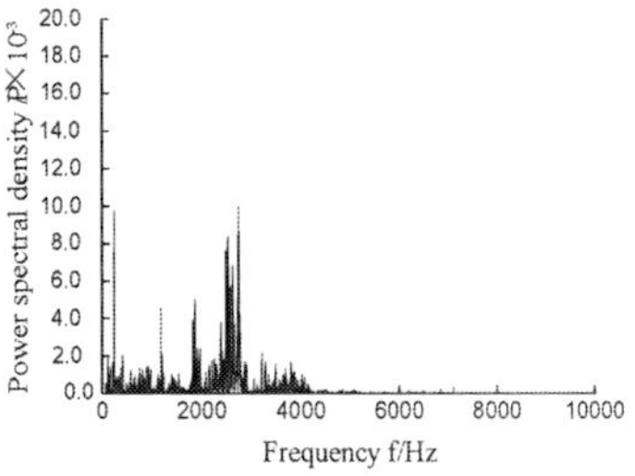

(a) Normal arc acoustic signal power spectral density diagram

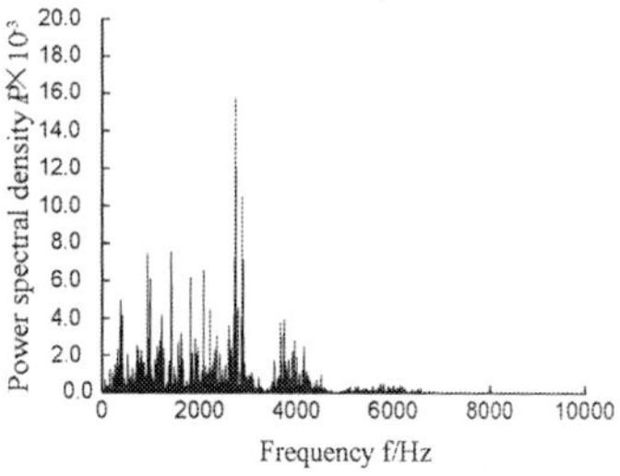

(b) Shielding gas flow rate of 20L/min

Fig. 6 Arc acoustic signal power spectral density diagram under different flow rate of shielding gas

Conclusions

In this paper, by changing the welding current, arc rotating frequency, work piece surface state and shielding gas flow rate, characteristics of arc acoustic sensing signal were studied. Some conclusions were drawn as following:

(1)When the welding current changed to a large value, short-circuit was apt to occur and an according strong ringings appeared in the time domain of arc acoustic signal. The frequency domain characteristic of ringing area was different from that without ringing.

(2)The vibration cycle of arc acoustic signal the time domain waveform was almost the same as the arc rotating cycle. It was because that the welding current became high when the arc rotated near the groove sidewall.

(3)The decrease of the shielding gas flow rate had an intense effect on some frequency band of the arc acoustic signal.

(4)Rust on the work piece surface leads to the irregular occurrence of the strong ringing because of the reduced electrical conductivity and the random distribution of the rust on the work piece surface.

Acknowledgments

The authors would like to thank the financial support from the National Natural Science Foundation of China (No 51475218), Specialized Research Fund for the Doctoral Program of Higher Education (20133220110001), the Qing Lan Project of Jiangsu province, and China Postdoctoral Science Foundation (2015M580423, 2016T90453).

References

[1] Wang, J.Y., et al.: "Novel rotation arc system for narrow gap MAG welding." *Science and Technology of Welding and Joining*, 2007. 12(6): p. 505-507.

[2] Yang, C.L., et al.: "Application of rotating arc system to horizontal narrow gap welding." *Science and Technology of Welding and Joining*, 2009. 14(2): p. 172-177.

[3] Xu, Y., et al.: "The acquisition and processing of real-time information for height tracking of robotic GTAW process by arc sensor." *The International Journal of Advanced Manufacturing Technology*, 2013. 65(5): p. 1031-1043.

[4] Chen, S.B., J. Wu, and Q.Y. Du: "Non-linear modelling and compound intelligent control of pulsed gas tungsten arc welding dynamics." *Proceedings of the Institution of Mechanical Engineers*, Part I: Journal of Systems and Control Engineering, 2011. 225(1): p. 113-124.

[5] Wu, J. and S.B. Chen: "Software System Designs of Real-Time Image Processing of Weld Pool Dynamic Characteristics," *Robotic Welding, Intelligence and Automation*, T.-J. Tarn, S.-B. Chen, and C. Zhou, Editors. 2007, Springer Berlin Heidelberg: Berlin, Heidelberg. p. 303-309.

[6] Li, W., et al.: "Groove sidewall penetration modeling for rotating arc narrow gap MAG welding." *The International Journal of Advanced Manufacturing Technology*, 2015. 78(1): p. 573-581.

[7]Liu, L., et al.: "Feature extraction and dimensionality reduction of arc sound under typical penetration status in metal inert gas welding." *Chinese Journal of Mechanical Engineering*, 2012. 25(2): p. 293-298.

[8] Saad, E., H. Wang, and R. Kovacevic: "Classification of molten pool modes in variable polarity plasma arc welding based on acoustic signature." *Journal of Materials Processing Technology*, 2006. 174(1–3): p. 127-136.

Effects of Argon-Nitrogen Backing Gas on Microstructure and Corrosion Resistance of Duplex Stainless Steel Pipe ASTM A790 Welds by Gas Tungsten Arc Welding

Somsak Pamuk [1], Siriporn Daopiset [2], Kittichai Sojiphan [1]

[1] *Department of Welding Engineering Technology, College of Industrial Technology, King Mongkut's University of Technology North Bangkok, Thailand*

[2] *Department of Production Engineering, Faculty of Engineering, King Mongkut's University of Technology North Bangkok, Thailand*

Nowadays, there are more needs for using duplex stainless steels in offshore and petrochemical industries due to their good mechanical properties and corrosion resistance. The microstructure of duplex stainless steel has two phases including austenite and ferrite. During welding of duplex stainless steels, welding parameters as well as gas mixtures have direct effects on the balance of the phases as well as the precipitation of intermetallic particles such as chromium nitride when mixture of argon and nitrogen gas is used. The current research is aimed to study the effects of increasing nitrogen content into argon gas mixture on the changes in microstructure and corrosion resistance. In this study, five gas mixing ratios of 100% Ar, 5%N2+95%Ar, 10%N2+90%Ar 15%N2+85%Ar and 100%N2 were selected as backing gas for gas tungsten arc welding process of duplex stainless steel ASTM A790 pipe using ER 2209 filler metal. The welding process parameters were controlled the heat input and controlled with the interpass temperature not exceed 150 deg-C. After welding, samples were prepared for microstructure characterization using optical microscope to evaluate phase morphology and volume fractions of ferrite in the weld metal and the heat affected zone compared to the base metal. Finally, Electrochemical test were obtained to determine the effect of adding nitrogen on the pitting corrosion resistance in the weld. The results show some correlations between increasing nitrogen contents to the improvement in microstructure and pitting corrosion resistance in the weld.

Influence of Welding Process on Microstructure and Corrosion Behaviour of Nickel Free High Nitrogen Stainless Steel Welds

Raffi Mohammed [1], G Madhusudhan Reddy [2], K Srinivasa Rao [1]

[1] *Andhra University, India*
[2] *Defence Metallurgical Research Laboratory, India*

Nickel free high nitrogen stainless steels are recommended for manufacturing defence battle tanks by replacing existing armour grade steel. Welding is the major fabricating technique to join the structural components made of high nitrogen steel. Even though wrought high nitrogen steel possesses excellent mechanical properties and corrosion resistance but fusion welding may result in poor mechanical and corrosion resistance due to composition changes and nitrogen loss in the weld metal. An attempt has been made in the present investigation to weld high nitrogen steel of 5mm thick plates using various welding processes i.e., shielded metal arc welding (SMAW), gas tungsten arc welding (GTAW) and electron beam welding (EBW) process. Present work is aimed at studying the microstructural changes and its effects on mechanical properties and corrosion resistance. Microscopic studies were carried out using optical microscopy (OM) and field emission scanning electron microscopy (FESEM). Electron back scattered diffraction (EBSD) method was used to determine the phase analysis, grain size and orientation image mapping. Hardness and tensile studies were done using vickers hardness tester and universal tensile testing machine. Potentio-dynamic polarization testing was carried out to study the pitting corrosion resistance in aerated 3.5% NaCl environment using a GillAC basic electrochemical system. Stress corrosion cracking studies were carried out under constant load 45%MgCl2 in boiling solution. Results of the investigation revealed that the welding process may influence the mechanical properties and corrosion resistance significantly. Relatively narrow width of weld zone, low heat input, faster cooling rates and finer grain size in electron beam welding may be the reason for better combination of properties when compared to the arc welds.

Hydrogen in Welded Microstructures of Creep-resistant T24 Boiler Steel: Effect on Mechanical Properties and Corresponding Hydrogen Diffusion

Michael Rhode, Joerg Steger, Thomas Boellinghaus, Thomas Kannengiesser
BAM Federal Institute for Materials Research and Testing, Berlin, Germany
michael.rhode@bam.de, +49-30-8104-4279

Hans Hoffmeister, Christoph Muenster
Institute for Failure Analysis and Failure Prevention ISSV e.V., c/o Helmut Schmidt University of the Federal Armed Forces Hamburg, Hamburg, Germany
hans.hoffmeister@hsu-hh.de, +49-40-6541-2738

Abstract

Low-alloyed CrMoV steels, such as T24, are widely used for welded components in fossil power stations due to their excellent creep-strength. Spectacular failure cases in the recent years exhibited severe cracking in T24 welds. The results showed that hydrogen-assisted cracking (HAC) occurring up to 200 degree Celsius cannot be excluded. Hence, a basic understanding is necessary on how hydrogen affects the material properties of welded microstructures. In this regard, each weld microstructure (HAZ and weld metal) has influence on the HAC susceptibility and respective hydrogen diffusion. Thus, the present contribution summarizes different results obtained from experiments with grades T24 (CrMoV alloy) and T22 (CrMo) and thermally simulated HAZ. Tensile test were conducted with hydrogen charged specimens and were compared to hydrogen-assisted stress corrosion cracking results obtained from slow strain rate tests (SSRT) up to 200 degree Celsius. Electrochemical permeation and degassing experiments were performed to identify a particular weld microstructure influence on hydrogen diffusion and trapping (especially in the HAZ). The results showed that the T24 base material has improved resistance to hydrogen-assisted degradation/cracking. In contrast, the as-welded HAZ has remarkably increased susceptibility. Tensile tests confirmed that hydrogen concentration of 1 to 2 ppm has remarkable effect on the HAZ ductility. SSRT experiments confirmed this at elevated temperatures for both T24 and T22. Hence, the evaluation of a particular degradation of the mechanical properties should be performed independently for each weld microstructure. In addition, the HAZ showed decreased diffusion coefficients (at room temperature) of approximately one magnitude compared to the base materials. Trapped hydrogen was determined in the T24 at temperatures up to 120 degree Celsius compared to 75 degree Celsius in the T22. This has to be considered in case of changing operational temperatures in terms of the start-up and shutdown processes of boiler components.

Introduction

Safety and reliability of weld joints are important for later operating of welded components in fossil-fired power plants. A particular example is represented by high-strength low-alloyed CrMo steels like T22 or the CrMoV steel T24. These steel grades are particularly used for boiler components in power plants [12]. Spectacular failure cases of boiler components (water wall panels) in the recent years exhibited severe cracking in the tubular T24 TIG welds [3]. Since the first failures occurred, an ongoing discussion started including metallurgical facts, welding technique, and power plant start-up procedures. Anodic stress corrosion cracking (SCC) in high temperature water between 150 °C and 280 °C (accompanied by dissolved oxygen) was identified as reason for component failure [4] despite others [5,6]. For this purpose, hydrogen-assisted SCC in the higher temperature regime of up to 200 °C cannot be excluded [7], particularly not in those cases when newly manufactured boilers underwent numerous cycling through this temperature range. In general, hydrogen might cause degradation of mechanical properties or hydrogen assisted cracking (HAC) in low-alloyed creep-resistant steels like the T24 [8-10]. In this scope, the addition of V to CrMo steels delays/decreases the hydrogen diffusion by fine dispersed nanosized VC-precipitates [10,11]. These particles act as (reversible) hydrogen traps. Nevertheless, trapping increases with decreasing temperature. As a result, a (crack) critical hydrogen concentration can remain in the material in case of changing service temperature, for example due to the intended numerous cycling of fossil-fired power plants in future. Thus, this contribution summarizes experimental results of mechanical properties of T24 and T22 base materials and T24 HAZ and results of hydrogen diffusion and trapping at elevated temperatures.

Experimental

Two different creep-resistant steels were investigated: CrMoV steel T24 and CrMo steel T22. A representative T24 HAZ microstructure was simulated by austenizing the material at 1,150 °C (SSRT) and 1,200 °C (tensile test) followed by quenching in water. *Table 1* shows the chemical composition of the investigated steel grades. For evaluation of the mechanical properties in presence of hydrogen, two different techniques were used: (1) Tensile test of electrochemically hydrogen charged tensile specimens and (2) slow strain rate test (SSRT) at elevated temperatures. The applied experimental techniques and used specimen geometries were shown in detail elsewhere: the tensile test in [12,13] and the SSRT in [7]. The applied strain rate was $8.3*10^{-4}$ 1/s in case of the tensile test and $1.85*10^{-7}$ 1/s to $1.11*10^{-3}$ 1/s in case of the SSRT. The hydrogen diffusion coefficients (at room temperature) were calculated from electrochemical permeation experiments [14]. The used permeation method and the calculation of the hydrogen diffusion coefficient were shown in detail in [15]. In addition, isothermal degassing experiments at elevated temperatures were conducted to identify possible trapped hydrogen amounts at elevated temperatures. Thus, the carrier gas hot extraction (CGHE) technique with quadrupole mass spectrometer (MS) was applied in accordance to [16]. The experiments were conducted for temperature range from 50 °C to 315 °C. In advance, cylindrical specimens with 3 mm diameter and a length of 10 mm were machined and electrochemically hydrogen charged in accordance to [12]. Subsequently, these specimens were degassed at the desired temperature for at least 2 hours to determined the diffusible hydrogen (HD I). Finally, they were heated to 900 °C to activate the remaining "residual" hydrogen (HD II).

Table 1: Chemical composition of steels (mass-%)

Grade	C	Cr	Mo	V	Ti	B	Si	Mn
T24	0.08	2.44	1.00	0.27	0.07	0.005	0.25	0.52
T22	0.13	2.38	1.05	0.02	<0.01	-	0.36	0.56

Results of Mechanical Properties

The next figures present the results from the tensile test of hydrogen charged specimens. The results of the T24 are taken from [12]. Independent of the microstructure, the tensile properties tensile strength " R_m" and yield strength "$R_{p0.2}$" (*Fig. 1*) are less influenced by hydrogen compared to the significantly reduced ductility (fracture strain) in hydrogen charged condition (*Fig. 2*). For example, the fracture strain of the T24 HAZ is reduced from 15 % in hydrogen free condition to 5 % at a hydrogen concentration (HD) of2 ml/100g Fe. Nevertheless, the CrMo steel T22 had the lowest resistance to the particular degradation of the fracture strain. At HD of 2 ml/100g Fe, the remaining ductility was approx. 5 %, which is below the 15 % of the T24 BM, which has been also in tempered condition. Hence, two effects in the examined microstructures can be stated: 1) hydrogen degradation effect was intensified in case of untempered condition (HAZ vs. T24 BM) and 2) CrMoV steel T24 had increased resistance to a certain hydrogen effect compared to the CrMo steel T22. The reason is the addition of V, which increases the number of traps in the microstructure (VC-precipitates, in accordance to [10,11]). The SSRT experiments revealed a similar behavior of the investigated microstructures at elevated temperatures (for example in *Fig. 3*). In all cases, the SSRT ductility was decreased independent of the applied strain rate from $1.85*10^{-7}$ 1/s to $1.11*10^{-3}$ 1/s [7]. In general, the time to fracture (*Fig. 3*) increased with decreasing strain rate accompanied by

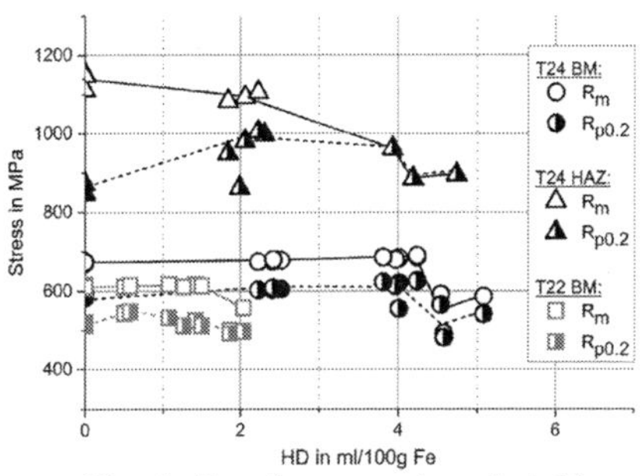

Fig. 1: Tensile strength and yield strength at 25 °C, T24 data from [12]

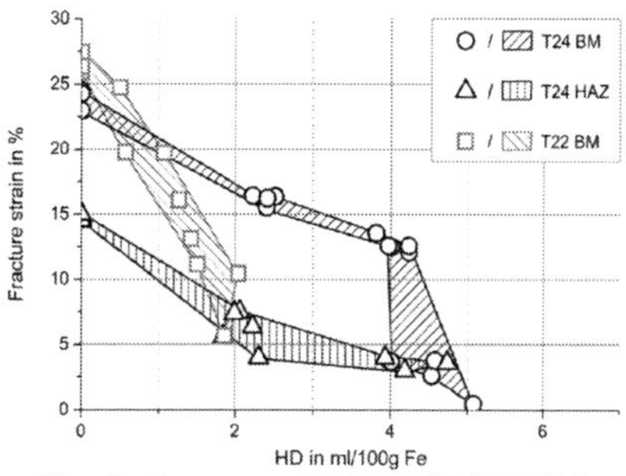

Fig. 2: Fracture strain at 25 °C, T24 data from [12]

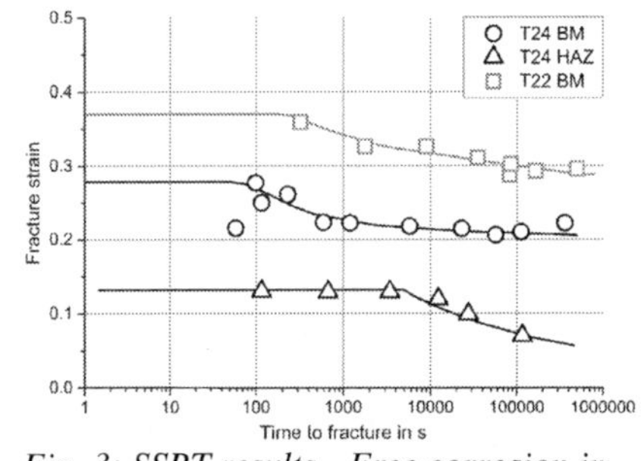

Fig. 3: SSRT results - Free corrosion in distilled water at 200 °C [7]

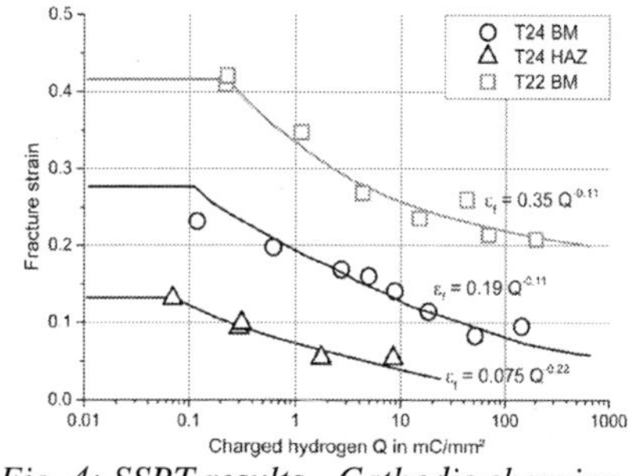

Fig. 4: SSRT results - Cathodic charging in 0.05 % H_2SO_4 at 200 °C [7]

moderately reduced SSRT ductility. Nevertheless, testing in acidic solution (*Fig. 4*) resulted in remarkably decreased SSRT ductility compared to the free corrosion in distilled water for all microstructures. However, remarkable differences are regarded if the SSRT results of the BM grades are compared to those of the tensile test (*Fig. 1* and *Fig. 2*). Thus, the CrMo steel T22 exhibited higher SSRT ductility compared to the CrMoV steel T24 at 200 °C. This is independent of hydrogen absorption in case of free corrosion (*Fig. 3*) or in case of cathodic charging (*Fig. 4*). Nonetheless, the as-quenched T24 HAZ had the lowest tensile test and SSRT ductility of all three microstructures. Hence, the as-quenched HAZ is regarded as the most susceptible microstructure, independent of the applied hydrogen charging condition and temperature. In addition, this was confirmed for the as-quenched T22 HAZ [7]. In general, a post weld heat-treatment (PWHT) should improve the mechanical properties of welded microstructures like the HAZ, i.e. the resistance to certain hydrogen assisted degradation/cracking. This is in accordance to the findings of Parvathavarthini et al. [17]. The authors concluded that tempering or annealing of a microstructure changes the number and density of available hydrogen traps. In addition, these traps decrease the apparent hydrogen diffusion. Hence, a closer view is necessary on possible trapped hydrogen amounts and corresponding hydrogen diffusion coefficients.

Results of Hydrogen Diffusion and Solubility

Table 2 shows the hydrogen diffusion coefficients obtained from the permeation experiments. The diffusion coefficient was lowest in the T24 HAZ. This can be attributed to the untempered microstructure, grain size or dissolved precipitates [17]. In addition, the CrMoV grade T24 has a nearly two times decreased diffusion coefficient at 25 °C compared to the CrMo grade T22. Hence, the T24 must have increased hydrogen solubility due to increased trapping. This was confirmed by the CGHE experiments at elevated temperatures, which is shown in *Fig. 5*. The absorbed hydrogen concentration had a clear dependency on the chemical composition. The CrMoV T24 showed for both microstructures (BM and as-quenched HAZ) a total charged HD of approximately 4.2 ml/100g Fe (HD I + HD II in *Fig. 5*) in contrast to the total charged HD of 2 ml/100g Fe in case of the CrMo T22. The increased hydrogen solubility by trapping at VC-precipitates [10,11] is responsible for increased resistance of the T24 microstructures to certain degradation at room temperature. In addition, the temperature for hydrogen trapping (HD II) in T24 BM and HAZ remarkably increased at temperature below 120 °C. In contrast, this temperature is approximately 75 °C for the T22 BM. Further experiments confirmed a trapped hydrogen amount (HD II) of 1.6 ml/100g Fe even after aging of 24 h at 25 °C in case of the T24 BM. Hence, during start-up of or shut-down, trapped hydrogen has to be considered in components of T24 even at elevated temperatures. This is important if these components undergo a localized anodic dissolution of the material surface that might result in hydrogen generation and diffusion into the bulk material. In addition, each weld microstructure has a unique diffusion behavior (see *Table 2*).

Table 2: Diffusion coefficients at 25 °C

Grade	T24 BM [15]	T24 HAZ	T22 BM
D in 10^{-5} mm²/s	6.7	1.6	10.6

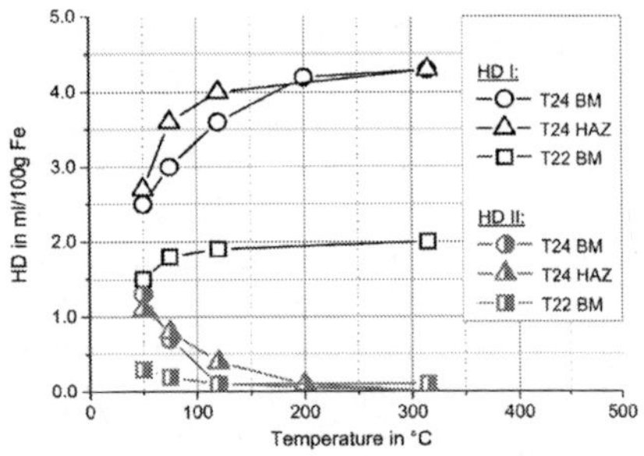

Fig. 5: Temperature dependent diffusible HD I and trapped HD II

Conclusions

This contribution summarizes the experimental results of the hydrogen effect on the mechanical properties, diffusion at room temperature and trapping characteristics for the T24 BM and HAZ as well as T22 BM. The following conclusions can be drawn:

1.) Independent of the used charging condition and temperature, the (as-quenched) HAZ is regarded as most susceptible weld microstructure in terms of hydrogen effect on mechanical properties. Nonetheless, the ductility of all examined microstructures was significantly reduced by hydrogen even in tempered material conditions (BMs). A further PWHT is recommended for the reduction of a degradation of the mechanical properties.

2.) In contrast to the tensile test, SSRT experiments showed that T22 BM is less susceptible to degradation at 200 °C compared to the T24 BM. This can be attributed to the decreased hydrogen concentration gradient in case of the tensile test (due to the previous hydrogen charging compared to the in-situ hydrogen charging during SSRT) combined with the increased trapping efficiency of the T24 that might result in increased sub-surface hydrogen concentration favoring a crack initiation.

3.) The T24 HAZ showed decreased diffusion coefficients at 25 °C compared to the tempered T24 BM. This is attributed to the missing tempering of the HAZ. Remarkable hydrogen trapping occurred in the T24 BM and HAZ at elevated temperatures (approximately 1.6 ml/100g Fe below 125 °C).

4.) The obtained mechanical properties and absorbed hydrogen amounts suggest that start-up and shut down procedures might be optimized (for example to pass quickly the critical temperature range below 200 °C). Especially, the mentioned trapping in the T24 weld microstructures might result in hydrogen super-saturation of component sub-surfaces (for example exposed to corrosion) with increasing risk for HAC initiation.

References

[1] B. Hahn, W. Bendick: „Rohrstaehle fuer moderne Hochleistungskraftwerke", *3R Int* Vol. 47 (2008), pp. 3-12.

[2] A. Dhooge, J. Vekeman: "New generation 21/4Cr steels T/P 23 and T/P 24 weldability and high temperature properties", *Weld World* Vol. 49 (2005), pp. 75-93.

[3] R. Nowack, C. Goette, S. Heckmann: "Quality management at RWE using T24 boiler material as an example", *VGB Powertech Journal* Vol. 11 (2011), pp. 1-5.

[4] J. Boewe, M. Becker: "Modified commissioning procedure for USC boilers using material T24", *Proc. of Power-Gen Europe Conference*, Vienna, Austria, June 2013, Access via: http://pennwell.websds.net/2013/vienna/pge/papers/T4S5O3-paper.pdf .

[5] R.U. Husemann, S. Devrient, R. Kilian: "Cracking mechanism in high temperature water - T24 Root cause analysis program", *Proc. 38th VDI Annual Conference on Failure Analysis*, Düsseldorf, Germany, October 2012, pp. 87-103.

[6] G. Luedenbach: "Stress corrosion cracking of T24 – Project achievements and ongoing investigations", *39th MPA Seminar on Power Plant Monitoring and Availability - Component Calculation and Materials Utilization*, Stuttgart, Germany, October 2013, Access via: http://seminar.mpa.uni-stuttgart.de/s-2013/Download/Stress%20Corrosion%20Cracking%20of%20T24%20-%20Project%20Achievements%20and%20Ongoing%20Investigations.pdf .

[7] H. Hoffmeister, Th. Boellinghaus: "Modeling of combined anodic dissolution/hydrogen-assisted stress corrosion cracking of low-alloyed power plant steels in high-temperature water environments", *Corros Sci* Vol. 70 (2014), pp. 563-578.

[8] M. Garet, A.M. Brass, C. Haut, F. Guttierez-Solana: "Hydrogen trapping on non-metallic inclusions in Cr-Mo low alloyed steels", *Corros Sci* Vol. 40 (1998), pp. 1073-1086.

[9] S.K. Albert, V. Ramasubbu, N. Parvathavarthini, T.P.S. Gill: "Influence of alloying on hydrogen-assisted cracking and diffusible hydrogen content in Cr-Mo steel welds", *Sadhana* Vol. 28 (2003), pp. 383-393.

[10] R.C. Brouwer: "Hydrogen diffusion and solubility in vanadium modified pressure vessel steels", *Scripta Metall Mater* Vol. 27 (1992), pp. 353-358.

[11] A.M. Brass, F. Guillon, S. Vivet: "Quantification of hydrogen diffusion and trapping in 2.25Cr-1Mo and 3Cr-1Mo-V steels with the electrochemical permeation technique and melt extraction", *Metallurgical and Materials Transactions A*, Vol. 35 (2004), pp. 449-1464.

[12] M. Rhode, J. Steger, Th. Boellinghaus, Th. Kannengiesser: "Hydrogen degradation effects on mechanical properties in T24 weld microstructures", *Weld World* Vol. 60 (2016), pp. 201-216.

[13] M. Rhode, J. Steger, E. Steppan, Th. Kannengiesser: "Effect of hydrogen on mechanical properties of heat affected zone of a reactor pressure vessel steel grade", *Weld World* Vol. 60 (2016), pp. 623-638.

[14] M.A.V. Devanathan, Z. Stachurski: "A technique for the evaluation of hydrogen embrittlement characteristics of electroplating bath", *Journal of the Electrochemical Society* Vol. 110 (1963), pp. 886-890.

[15] M. Rhode, Th. Kannengiesser, J. Steger (2014) Approach for calculation of apparent hydrogen diffusion coefficients with permeation experiments in CrMoV steel weld joints", *Proc. 2nd International Conference on Metals and Hydrogen,* Ghent, Belgium, April 2014, pp. 670-674.

[16] S. Salmi, M. Rhode, S. Juettner, M. Zinke: "Hydrogen determination in 22MnB5 steel grade by use of carrier gas hot extraction technique", *Weld World* Vol. 59 (2015), pp. 137-144.

[17] N. Parvathavarthini, S. Saroja, R.K. Dayal, H.S. Khatak: "Studies on hydrogen permeability of 2.25% Cr–1% Mo ferritic steel: correlation with microstructure", *J Nucl Mater* Vol. 288 (2001), pp. 187-196.

Clarification of microstructure evolution of aluminum during friction stir welding using liquid CO_2 rapid cooling

X.C. Liu, H. Fujii
Joining and Welding Research Institute, Osaka University, 11-1 Mihogaoka, Ibaraki, Osaka 567-0047, Japan
Email: liu@jwri.osaka-u.ac.jp, phone, fax: 81-6-6879-8663

Introduction

It is difficult to obtain fine grains at elevated temperature in high stacking fault energy (SFE) metal Al due to easy dynamic recovery (DRV) taking place during straining. As for friction stir welding (FSW), a hot-plastic working technique, to obtain fine grain structure of aluminum and its alloy is a perennial issue. Therefore, it is very importantto fully understand the grain structure development during FSW. Although there have been many reports about the microstructure evolution of aluminum and its alloys during friction stir welding (FSW), the inherent mechanism is still not fully understood [1-3]. This is because the thermal-mechanical coupling during FSW makes it difficult to clarify what happens during the different welding stages. Liquid CO_2 has a strong cooling capacity due to its higher specific heat capacity than other ultralow temperature cooling media. This makes it possible to freeze the microstructure rapidly when combined with the tool's "sudden stop action". If a subsequent short-time annealing is adopted, it will reproduce the similar microstructure during the normal cooling stage. This technique makes it possible to separate the deformation stage from the normal cooling stage [5] and to investigate the microstructure evolution individually. By this process, Xu et al. [4] investigated the microstructure evolution of low SFE metal copper during the deforation stage and cooling stage of the FSW. According to their results, the initial large grains are subdivided into fine grains with a large number of low angle boundaries with the increasing strain and temperature during the deformation stage. During the subsequent annealing stage, static recrystallization occurs with selected grain growth and the disappearance of dislocations. In this study, a similar technique was used for the high SFE metal Al 1050 FSW. The microstructure evolution during the cooling stage is expected to be clarified.

Experiment

Cold-rolling Al 1050 sheets with dimensions of $200_L \times 60_W \times 3_T$ mm were butt welded perpendicular to the rolling direction by RCFSW in this study. A schematic of the RCFSW (Rapidly cooling FSW) using liquid CO2 [5]is shown in Fig. 1, in which flowing liquid CO_2 was simultaneously jetted on the weld bead from the tool rear side during the normal FSW. A tungsten carbide tool containing a concave shoulder and a cylindrical probe of 2.8 mm length with a smooth surface was used. The diameters of the shoulder and the probe were 12 mm and 4 mm, respectively. The tool tilt angle was 3°. The welding speed and rotating speed were 800 rpm and 200 mm/min, respectively. The FSW machine used in this study was in the displacement control mode. The position of the shoulder was an automatic check-out value at a pressure of 50 N. In order to investigate the initial microstructure after stirring and the subsequent microstructure evolution after several annealings with different times and temperatures, the tool "stop action" was adopted to freeze the microstructures around the tool and provide a static presentation of the microstructures that dynamically changed during the normal welding process. The "stop action" means that the tool suddenly lost power and was instantaneously pulled out. In order to detect the possible microstructure evolution during the cooling stage, five annealings with gradually increased temperatures and times were used in this study. They were 250℃+10s, 300℃+10s, 300℃+30s, 300℃+3min and 400℃+3min in chronological order. After every annealing, ice water quenching was used, and subsequently, the microstructure was characterized by an electron back-scattered diffraction (EBSD) technique. For the EBSD measurement, electrolytic polishing using a solution of $HClO_4$:C_2H_5OH=1:4 in volume at 20 V and 0 ℃ was adopted. The EBSD measurements were carried out in a JEM-7001FA field-emission scanning electron microscope (FE-SEM) operating at 15 kV with the TSL OIMTM system. The step size of the EBSD scanning was 0.2 μm. To avoid spurious boundaries caused by orientation noise, a lower limit boundary misorientation cut-off of 2° was employed. A 15° criterion was used to differentiate between the LABs and HABs. The grain size was quantified by determining the area of each grain in the EBSD map and calculating its circle equivalent diameter. The grains were defined as crystallites bordered by a continuous perimeter in the EBSD maps. The measured positions were selected on the trailing side of the keyhole, horizontal cross-section with a distance of 0.5 mm to the top surface of the weld. The distances of 0.1mm, 1mm, 2mm and 4mm to the wall of the keyhole were used to show the difference in the stir zone. Also, the weld zone was characterized by transmission electron microscopy (TEM).

For the TEM sample preparation, thin polished samples were double-jet electro-polished using a solution of $HClO_4$:C_2H_5OH=1:9 in volume at 20 V and 0 °C.

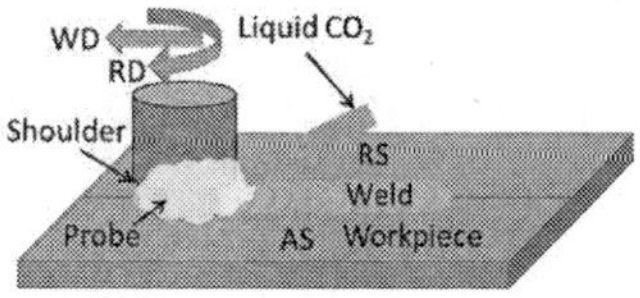

Fig. 1 Schematic of RCFSW using liquid CO_2

Results and discussion

Fig. 2 shows the grain structure changes with respect to the distance to the wall of the keyhole. It can be seen that the grain structure is not uniform from the wall of the keyhole to the weld in the stir zone. Close to the wall of the keyhole, there exists a layer of ultrafine grains (mean grain diameter 1.07 μm) with a high aspect ratio, as shown in Fig. 2a. This structure is something like the lamellar structure occurring during other SPD methods such as accumulative roll bonding ARB and equal-channel angular pressing (ECAP). However, the formation mechanism should be somewhat different due to the significant difference in the processing temperature. In other SPD methods, the processing temperature is usually room temperature, but in FSW, the temperature is certainly high near the probe. This increases the difficulty to form an ultrafine grain layer. Therefore, the possible reason for the generation of this ultrafine grain layer is that the large velocity gradient close to the probe surface causes a very high strain rate. The recrystallized grain size can be estimated by the Zener–Hollomon parameter (Z) and a higher strain rate corresponds to a high Z value, and further corresponds to a smaller grain [6]. From distance of 1 mm to 4 mm to the keyhole, the grain structure has no obvious change, as shown in Fig. 2b-d and Table 1. The grain structure is characterized by approximately equixed grains with a mean grain diameter of 1.57~1.67 μm, which is a little larger than the ultrafine grain layer. In addition, the number fraction of LABs showed a slight increase, which indicates that the recrystallization in the outer part of the stir zone is more incomplete. The possible reason for this phenomenon is that the inner layer and the outer layer of the stir zone undergo different deformation processes caused by material flow.

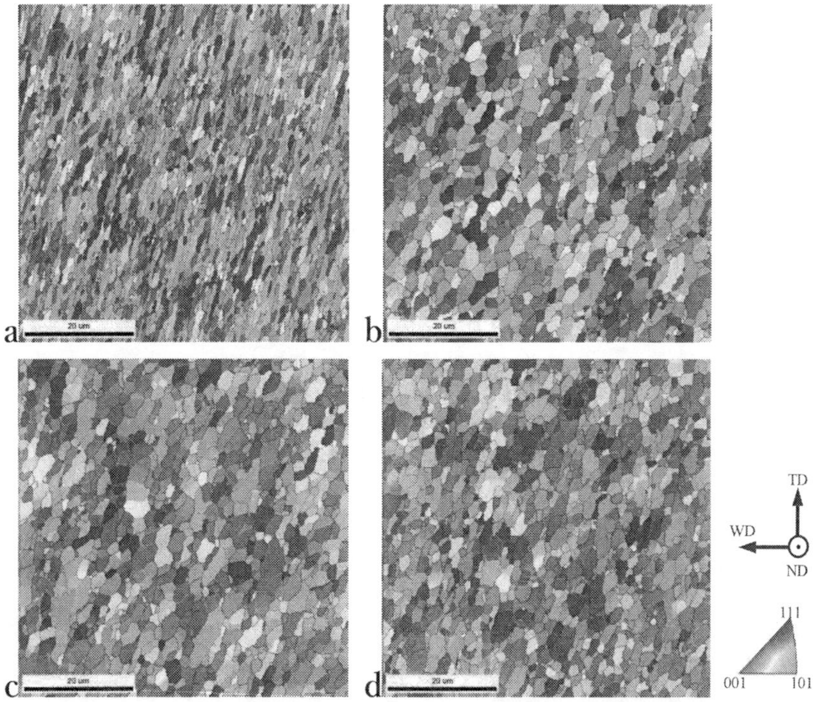

Fig. 2 Inverse pole figure of grain structure with various distances to the wall of the keyhole after welding measured by EBSD: a, 0.1 mm; b, 1 mm; c, 2 mm and d, 4 mm (in which the black line denote the HABs and the white line denote the LABs)

Table 1: Grain size and number fraction of LAB for various distances to the wall of the keyhole after welding

Distance to keyhole (mm)	0.1	1	2	4
Grain size (diameter; μm)	1.07	1.57	1.67	1.57
Number fraction of LABs (%)	18.5	18.6	21.8	23.1

Actually, after the first to fourth annealings, the grain structure nearly has no change compared with the initial state after welding. Accordingly, the fifth annealing at 400℃+3min was employed. 400℃+3min is sufficient for the grain to change. As shown in Fig. 3, it can be seen that the grain slightly grows homogeneously in contrast to the initial state. As for the ultrafine grain layer, the mean grain diameter increased from 1.07 μm to 1.32 μm. However, the grain shape still remains intact, i.e., lamellar structure with HABs. For the grains at the distances of 1 mm ~ 4 mm to the keyhole, the size also increases from ~1.6 μm to ~1.9 μm, but the grain shape remained equixed. At the same time, the number fraction of the LABs slightly decreases in the outer layer, but slightly increases in the inner layer of the stir zone. It is easy to understand the LAB decrease during grain growth. This is because the HAB migration during grain growth absorbs the LABs. However, the increase of the LABs in the ultrafine grain layer should be attributed to the extremely elongated grain shape. It can be seen that the aspect ratio in Fig. 3a is slightly lowerthan that in Fig. 2a. This means that the grain spheroidization happened during the fifth annealing. During the grain spheroidization, some LABs have to be formed to harmonize the HABs migration.

During the whole annealing process, there is no static recrystallization or selected grain growth occuring. This indicates that the stored strain energy in the frozen grain structure is not sufficient to drive grain growth. Fig. 4 shows the microstructure of the weld by TEM. It can be seen that the dislocation density is quite low. Liquid CO_2 is considered as the best cooling medium. It has also been observed that high dislocation density deformation structure exists after rapid cooling during the low SFE pure copper FSW[4]. If we believe the cooling rate is enough, it can be concluded that in high SFE pure aluminium FSW, once the stirring stops, the dislocation density is relatively low due to the facile dynamic recovery taking place. This reduces the drive force for grain growth or static recrystallization during the subsequent cooling stage. In other words, during pure aluminium FSW, the determining factor of the grain size is the deformation temperature and strain rate, while the cooling rate is not very important.

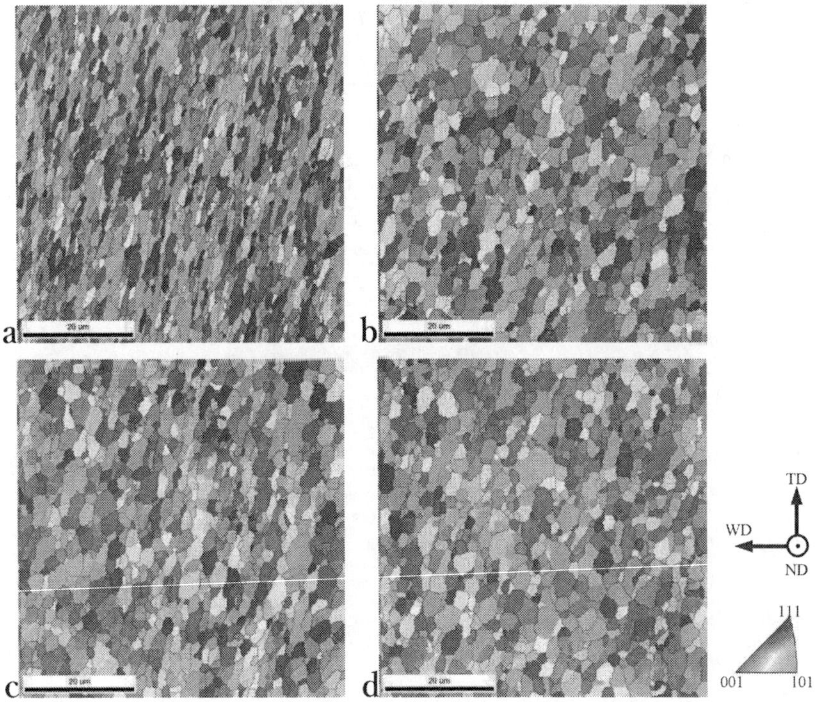

Fig. 3 Inverse pole figure of grain structure after five annealings measured by EBSD: a, 0.1 mm; b, 1 mm; c, 2 mm and d, 4 mm (in which black line denotes the HABs and the white line denotes the LABs)

Table 2: Grain size and number fraction of LAB after five annealings

Distance to keyhole (mm)	0.1	1	2	4
Grain size (diameter; μm)	1.32	1.88	1.88	1.89
Number fraction of LABs (%)	22.9	19.5	19.8	20.3

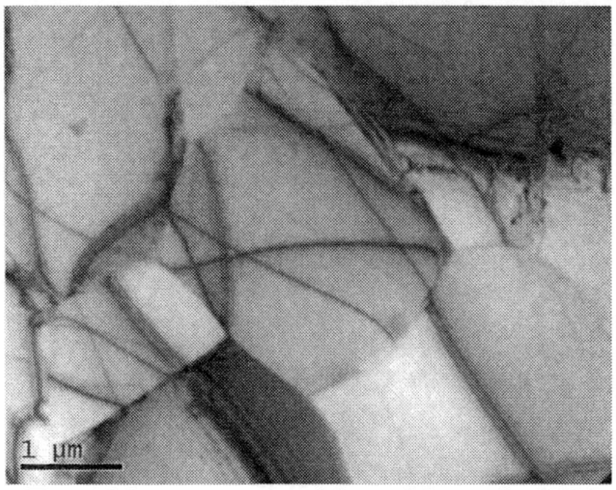

Fig. 4 Microstructure of the weld zone by TEM

Conclusions

Based on the above analysis, it can be concluded that during the high SFE metal aluminum FSW, static recrystallization and obvious grain growth are difficult to occur during the cooling stage due to the relatively low dislocation density once the tool stirring stops. This is because the dynamic recovery easily takes place very quickly.

Acknowledgments

This study was partly supported by the New Energy and Industrial Technology Development Organization (NEDO) under the "Innovation Structural Materials Project (Future Pioneering Projects)" and a Grant- in-Aid for Science Research from the Japan Society for Promotion of Science.

References

[1] J.Q. Su, T.W. Nelson, C. J. Sterling: "Microstructure evolution during FSW/FSP of high strength aluminum alloys,". Mater Sci Eng A, Vol. 405, No. 1 (2005), pp. 277-286.

[2] R.W. Fonda, J.F. Bingert, K.J. Colligan: "Development of grain structure during friction stir welding," Scr Mater, Vol. 51, No. 3 (2004), pp. 243-248.

[3] P.B. Prangnell, C.P. Heason: "Grain structure formation during friction stir welding observed by the 'stop action technique'," Acta Mater, Vol. 53, No. 11 (2005), pp. 3179-3192.

[4] N. Xu, R. Ueji, H. Fujii: "Dynamic and static change of grain size and texture of copper during friction stir welding," J Mater Process Technol, Vol. 232, (2016), pp. 90-99.

[5] H.Fujii, Y.D.Chung and Y.F.Sun, "Friction Stir Welding of AISI 1080 Steel Using Liquid CO_2 for Enhanced Toughness and Ductility", Sci. Tech. Weld. Join., 18 (2013) 500-506.

[6] M. Pietrzyk, L. Cser, J.G. Lenard: Mathematical and physical simulation of the properties of hot rolled products. Elsevier (Oxford, 1999), pp. 69.

Effects of Nitrogen on Strength of Fiber Laser and Electron Beam Weld Metal

Erik A PFeif[1], Stephen Liu[2]

[1] NIST, USA
[2] Colorado School of Mines, USA

The use of nitrogen as a solid solution strengthening element in high nitrogen austenitic stainless steels is well known; furthermore, the vaporization of nitrogen during many welding processes has been investigated and remedidied using nitrogen as a shielding gas. Full penetration, keyhole mode, electron beam and autogenous laser welds in nitrogen and argon on 3 mm thick Nitronic 40 stainless steel heat containing nitrogen contents varying between 0.24 and 0.31 wt. pct. were produced. From these welds: hardness maps were produced, bulk nitrogen contents were analyzed, and microtensiles were produced in the longitudinal direction from the top, middle and bottom 1 mm of the welds. Small differences in the nitrogen contents were found to occur in the metals, while larger differences in mechanical strength were found to occur between electron beam welds and laser beam welds. A consistent nitrogen solid solution strengthening of approximately 513 MPa per wt. pct. nitrogen was found. Both vermicular and lacy ferrite were found to occur in the laser weld metal and which were characterized using optical microscopy and Electron Backscatter Diffraction. Through the use of a multi-scale mechanical property model developed by Hansen it was found that the formation of low angle grain boundaries were the predominant microstructural feature responsible for strengthening within the weld metal and that this contribution must be accounted for and can be applied when predicting yield strength of the weld metal.

Mechanical-Electrical Composite Design and Test Technique of the Insulating Substrate in Power Semiconductors for High-voltage Applications

Yuji Hayase, Keisuke Yamashiro and Tetsumi Takano
Fuji Electric Co.,Ltd., Hino-city, Tokyo JAPAN
hayase-yuuji@fujielectric.com, +81-42-583-6796

Abstract

This paper introduces the approach from the junction shape and bonding process to electrical reliability that has not been discussed so far. Power semiconductors are the fruit of advanced bonding technology and process technology. The insulating substrate bonded with semiconductor is a core component for releasing heat from the semiconductor. Therefore, high reliability will be required in that component. However, the failure mode is a complex of mechanical disruption and electrical breakdown. Accordingly, it is necessary to advance the understanding and design from both sides. Especially, the junction edge fillet shape of the insulating substrate decides the mechanical reliability and electrical reliability trade-offs. We calculate electric field distribution generated at the junction edge is determined by the fillet sharpness and the dielectric constant of surrounding material of junction edge. Partial discharge inception voltage (PDIV) is determined by this electric field distribution. If there partial discharge occurs, it is leading to repeated partial discharges and breakdown. We propose the method of individual dry bonding process of ceramic and electrode with the eaves shape serving as an electric field shield, replacing the current etching process. By the methods described above it is possible to produce a highly functional reliability of the component.

Introduction

The development of the semiconductor power module technology is closely related to the development of joining technology. In recent years, along with the turning point to the SiC/GaN devices from Si devices, higher reliabilities (resistances to higher voltage more than 5 kV, high repetition switching more than 20 kHz and high heating repetition more than 150 K temperature difference) are required in component of these modules [1]. *Fig.1* is shows typical power modules, cross section structure of it and enlarge view of the insulating substrate. This insulated type module usually sealed with silicone gel or other resin. The insulating substrate made by ceramic of Al_2O_3, AlN or Si_3N_4 bonded with semiconductor, is a core component to insulate the circuit from the base plate. And this part also role to releasing heat from the semiconductor. Therefore the complex load is applied to the insulating substrate, especially triple-junction part (metal-substrate-sealing) formed at the frame of the substrate. This triple-junction, it is often become the breaking point at the development test or in the actual aging products.

We scoped the triple-junction because it has the possibility of technological development, not only a material combination of metals, insulating substrate and insulation sealing but also its manufacturing process, subsequent steps, and operating environment. In this paper, we introduce an electric incomplete destruction (known as partial discharge) started from triple-junction of ceramic substrate, and investigated results of how to suppress the destruction of both electrical and mechanical by varied the shape of triple-junction.

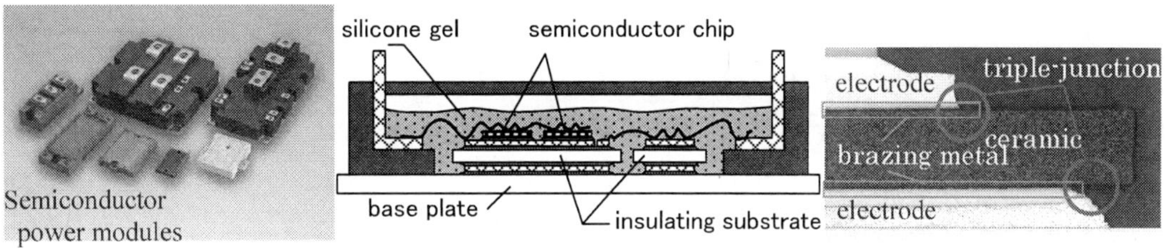

Fig. 1: Typical semiconductor power modules, cross section of it and enlarge view of insulationg substrate.

Visualization of electric problems with partial discharge experiment

At first, we visualized the electric problem at triple-junction with partial discharge experiment. For ceramic insulating substrate, there are dielectric breakdown test and partial discharge test as reliability tests. In the dielectric breakdown test, it is applied to more than twice the rated operation voltage in order to verify that can be operated at the rated voltage after the aging. In the partial discharge test, it is applied 1.5 times the rated voltage to verify the partial discharge occurs or not, then 1.2 times the rated voltage to verify the partial discharge disappears or not [2]. If partial discharge occurs in the power modules, organic insulating materials will rapidly deterioration by the heat, gas generation (including the oxidation), high energy light, and impact of expansion or others. In other words, the occurrence of partial discharge is directly linked to the life time of the product.

Experimental setup

We tried to observe partial discharge propagation process on ceramic insulating substrate under AC stress. AC stress profile for this experiment as shown in *Fig.2* left side. Measurement for partial discharge, apply AC stress to insulating substrate sample through the double low pass filter as shown in *Fig.2* right side. Discharge amount detected by impedance method using CD-6 (tuning type) connected under the capacitance. Both AC voltage and discharge amount were observed by DSO simultaneously. In this experiment, we observed light emitting using optical camera (still camera) in dark room. We kept open the shatter while 20 seconds from 10 pC (pico coulomb) partial discharge was detected.

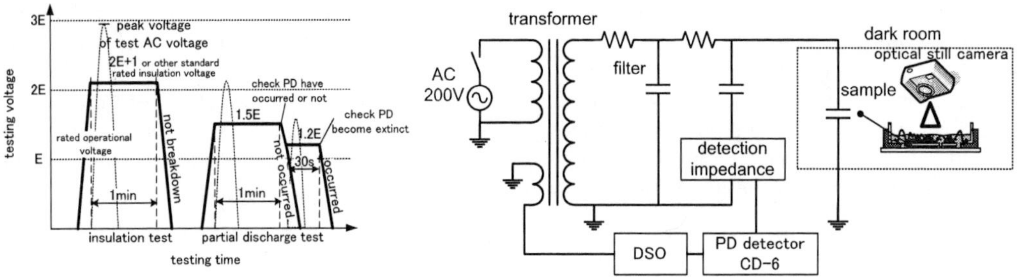

Fig. 2: Applied AC stress profile in dielectric breakdown and partial discharge test and experimental setup.

Experimental results

Fig.3 shows the typical experimental results of partial discharge test. In this figure, focus on the triple junction at the edge of copper electrode in center picture, two longer large light emitting spread over at the ceramic-gel interface to the right side, two shorter light emitting passes run at the interface to the below of picture and weak light emitting along at the triple-junction (edge of copper electrode) were observed. We thought thin light or weak light shows one unit of partial discharge.

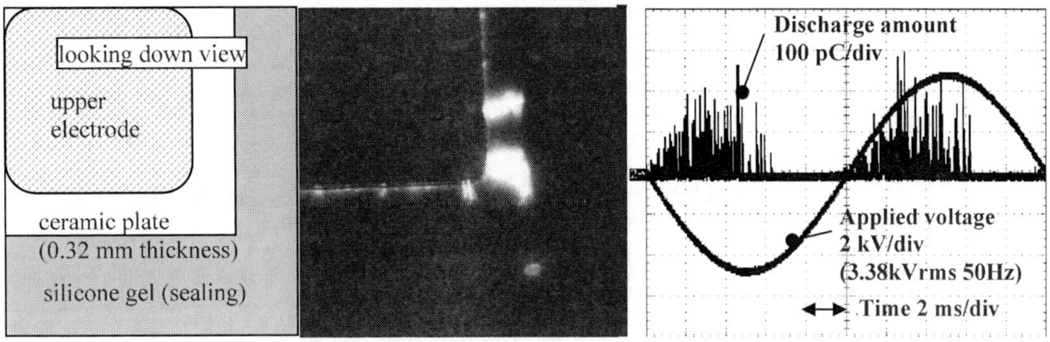

Fig. 3: Typical experimental results of partial discharge test.

In addition, partial discharge signals gradually increased after the initial state; finally the peak value was 300 pC with the intensity distribution of the sawtooth wave shape in phase range of 0° to 110°. The mountain shape distribution of the first half phase is often observed when a number of partial discharge from peeling defects in dielectric material [3]. This is consistent with the result which are optically observed, it can be said that these partial discharge occurred from the triple-junction and spread at the interface between ceramic insulating substrate and insulating sealing gel, then it is composed a new peeling defects. In our reference, this propagation process is named bubble tree [4]. Therefore, deterioration of insulating materials occurs in short time even with a small discharge, it is not acceptable if the detectable discharge would occur.

Development direction of the power module from electric problems

It is necessary to advance the dielectric strength that means improvement of partial discharge inception voltage, if we want to control higher voltage with SiC device. In response to this challenge, power modules have been developed. One direction is the conventional extended method, which enhance the insulating using high insulating gel and high heat resistance gel. The advantage of this approach, it is not necessary to change the manufacturing process and it can be adapted quickly. The disadvantage, it is very difficult to increase the adhesive strength of the sealing gel over a wide range of temperature, and furthermore the dielectric strength of the interface is not determined by performance of one side material alone. The other direction is the resin molding. The advantage to be expected is strong adhesion to the ceramic and high heat resistance, and it will withstand higher voltage than sealing gel. Moreover, it is considered to be using the proven technology of small discrete semiconductors. Unfortunately, power modules are too large to keep the whole temperature in uniform; there is a concern that the heat stress becomes very large by expansion coefficient difference.

However, the power module failure mode is not simple; it seems complex of mechanical disruption and electrical breakdown. Accordingly, it is necessary to advance the understanding and design from both sides. Especially, the metal-ceramic junction edge fillet shape of the insulating substrate decides the mechanical reliability and electrical reliability trade-offs.

Numerical calculation setup and results

In this section, we introduce numerical calculated results of electrical (electric field at the interface) and mechanical (thermal stress in brazing metal) stress around the triple-junction using a same edge fillet (triangle shape) model. Material parameters are shown in *Table 1*, and the model structure is shown in *Fig.4*. Calculated results are also shown in the same figure and results are organized by edge fillet contact angle of brazing metal. In many cases, this fillet shapes are produced by wet etching process, and would not be controlled in comparison with the solder fillet. In fact, there is also a contact metal between ceramic and the brazing metal, but ignored. We used FEM (finite element method) and temperature difference was 153 K (difference of 22°C and 175°C) for thermal stress. On the other, we used BEM (boundary element method) and applied potential difference to 0.635 mm ceramic plate was 7000 V (peak value) for electric field. The BEM is effective in the calculation of the fine interval and boundary surface.

Table 1: Material parameters used in the numerical calculations.

	Linear expansion coefficient (ppm/°C)	Young's modulus (GPa)	relative permittivity
copper electrode	18	110	-
solder/brazing metal	24	22	-
ceramic plate	4.8	320	8.8
base plate	9.4	96	-
silicone gel/oil	-	-	2.6

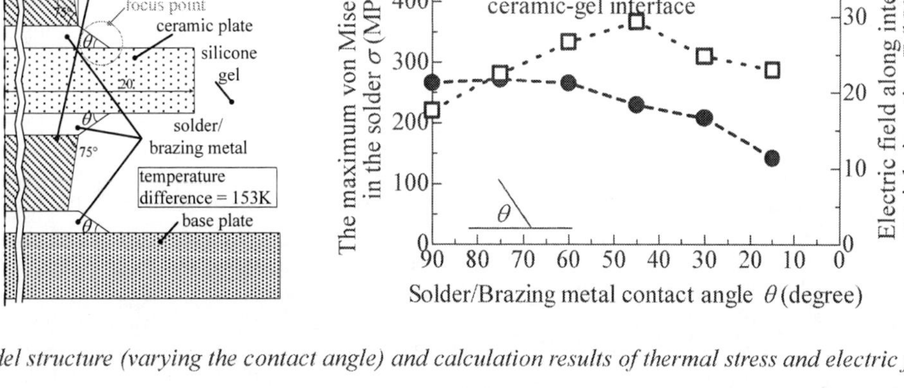

Fig. 4: Model structure (varying the contact angle) and calculation results of thermal stress and electric field.

From the results, the triple-junction is the both stress and electric field concentration point, and further, these are in a trade-offs. The maximum von Mises stresses tend to be relaxed with the decrease of contact angle. Conversely, the maximum electric fields along the interface tend to be increasing with the decrease of contact angle until the 45 degree. In the smaller angle, electric fields along the interface are decreased by divergence of the vector direction.

Consideration for problem-solving of electro-mechanical trade-offs on triple-junction

The triple-junction is in an electro-mechanical trade-offs as previously mentioned. To solve this problem, one way is to reduce as much as possible the contact angle in reference to the above calculation result. However, in the wet etching process where mainstream production method of insulating substrate, it is difficult to keep the low contact angle stable. In addition, an effective implementation area will be reduced by being etched into the low angle.

Another way is to move the concentration point of electrical or mechanical load either. Form an electrical field shield using electrode shape and it can reduce the electric field concentration of triple-junction. In this way, it can keep the low contact angle and low mechanical stress without increasing the electric field [5]. *Fig.5* shows the comparison previous model and proposed concept model with the eaves (*hisashi*) shape electrode for high voltage device. It can cause the construction failure of the gel sealing, but it can be solved by the control of the construction process and the pressure [6][7]. Actually create an insulation substrate with eaves shape model, as a result of the partial discharge test, it was confirmed that the partial-discharge inception voltage was improved.

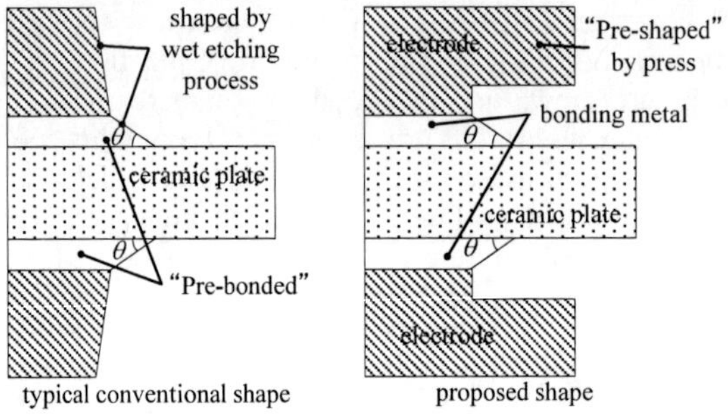

Fig.5: Comparison conventional shape and proposed concept shape with the eaves shape electrode.

Conclusions

We visualized the electric problem in the semiconductor power module with partial discharge experiment. It was observed partial discharge gradually increased started from the triple-junction. We also calculate thermal stress and electric field determined by the fillet sharpness of triple-junction. These stress are in the trade-offs associated with the contact angle. In finally, we propose the method of individual dry bonding process of ceramic and electrode with the eaves shape serving as an electric field shield, replacing the current etching process. By the methods described above it is possible to produce a highly functional reliability of the component.

References

[1] A. A. Abdelmalik, *et al*: "Influence of fast rise voltage and pressure on partial discharges in liquid embedded power electronics", *IEEE Trans-DEI*, Vol. 22, No. 5 (2015), pp. 2770-2779.

[2] IEC61287-1:2014, Railway applications - Power converters installed on board rolling stock

[3] T.M. Do, O. Lesaint, J.-L. Augé: "Streamers and Partial Discharge Mechanisms in Silicone Gel Under Impulse and AC Voltages", *IEEE Trans-DEI*, Vol. 15, No. 6 (2008), pp. 1526-1534.

[4] Y. Hayase, K. Yamashiro and T. Takano: "OBSERVATION OF PARTIAL DISCHARGE PROPAGATION PROCESS IN INSULATION SILICONE GEL", *ISH2013*, Seoul, KOREA, Aug 2013, pp. 938-941.

[5] Y. Hayase, T. Torii: JP 2014-120728 A 2014.6.30

[6] M. Sato, A. Kumada, K. Hidaka, *et al*: "Simplified Void-free Encapsulation Technique for Power Semiconductor Devices using Silicone Gei," *IEEE Trans-Fundamentals and Materials*, Vol. 134, No. 11 (2014), pp. 572-557.

[7] K.Yamashiro, M. Sato, *et al* : JP 2015-41740 A 2015.3.2

Effect of Current Load at Temperature on Microstructure of Copper Joined by Sn-based alloy

U.Harutoshi, N.Moriyama , and Y.Miyazawa
Tokai University 4-4-1 Kitakaname Hirartsuka-shi Kanagawa-ken 259-1292 japan
TEL81-463-58-1211(Ext,4208)ymiyazawa@tokai-u.jp

Abstract
In these days, there are fine joints on electric substrate, because of miniaturization and high integration of electronic devices. The microstructual changing at the fine joints is accompanied by an increase in the current density. On the other hand, microstructure at the joint influenced strongly mechanical and electric properties in general.According to previous works at our laboratory, It became clear that microstructural changing was caused by the electro migration and the heating. It was not possible to clarify the extent of both the effect. In this study, new temperature control system was developed with using heatsink device. Then, the effect of current load at room temperature on the microstructure at the joint was investigated in this study.

1.Introduction (Required)

Until now, the soldering Sn-37Pb has been commonly used. However, Pb is easily eluted with acid rain of about pH5.6, so that cumulative toxicity is high and it is said to be responsible for groundwater contamination. So lead-free solder has been developed. At present, alloys such as Sn-Ag system are used in place of Sn-Pb system. In particular, Sn-Ag-based solder has excellent mechanical properties because of the microstructure. Sn-3.0Ag-0.5Cu is used as a standard lead-free solder. However, it is expensive because Ag is a precious metal, further, it is liquidus temperature is about 218 °C. [1]

In these days, there are fine joints on electric substrate, because of miniaturization and high integration of electronic devices such as LSI. So that current density at the fine joints increased drastically. And the microstructure at the fine joint was changed by high current density. On the other hand, microstructure at the joint influenced strongly mechanical and electric properties in general. According to previous works at our laboratory, it appeared that microstructural changing was caused by electro migration and generated heat by high current density. It was not possible to clarify the extent of both the effect. [2]

In this study, new temperature control system was developed with using heatsink device. And effect of current load at room temperature on the microstructure at the joint was investigated in this study.

2. Experimental method

2.1 Specimen preparation

The specimen preparation shown in **Fig.1**. Tough pitch copper bar was employed as base material, and Sn-based alloy was used as solder. Cross-section of the specimen was 1mm^2, and joint gap was 0.3mm. And Cu bars were joined with Sn-based alloy filler with hot plate system. Joining temperature was 250°C and joining time was 30sec.**Table 1** showed the experimental conditions of this study.

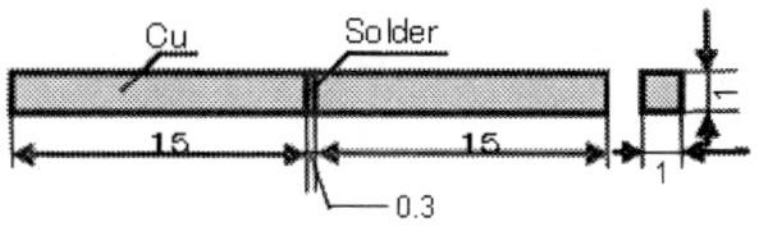

Fig.1 Specimen schematic

Table:1 Experimental conditions			
Condition	Solder composition	Temperature (°C)	Current (A/mm²)
1	Sn	25	50
2	Sn	60	50
3	Sn	100	50
4	Sn	120	50
5	Sn	50	100
6	Sn	100	100
7	Sn0.7Cu	25	50
8	Sn0.7Cu	40	50
9	Sn0.7Cu	70	50
10	Sn0.7Cu	100	50
11	Sn0.7Cu	25	100
12	Sn0.7Cu	100	50
13	Sn0.3Ag0.7Cu	25	50
14	Sn0.3Ag0.7Cu	100	50
15	Sn0.7Cu1.5In	25	50
16	Sn0.7Cu1.5In	60	50
17	Sn0.7Cu1.5In	100	50
18	Sn0.3Ag0.7Cu1.5In	25	50
19	Sn0.3Ag0.7Cu1.5In	100	50
20	Sn0.1Ag0.7Cu1.5In	20	50
21	Sn0.1Ag0.7Cu1.5In	40	50
22	Sn0.1Ag0.7Cu1.5In	60	50
23	Sn0.1Ag0.7Cu1.5In	100	50
24	Sn0.1Ag0.7Cu0.5In	100	50
25	Sn0.1Ag0.7Cu1.0In	50	50
26	Sn0.1Ag0.7Cu1.0In	100	50

2.2 Microstructure observation of the specimen

Microstructure was observed with the microscope. We used ferric chloride aqueous solution as corrosion solution. After metallographic obserbation, we determine the area of IMC with area measourment software. In this research, it was difficult to distinguish η layer and ε layer. Therefore, we defined the area of the IMC layer, the total area of the η layer and the ε layer. The microstructure schematic and formula fo calculating the avarage thickness of the IMC layer shown in **Fig.2**.

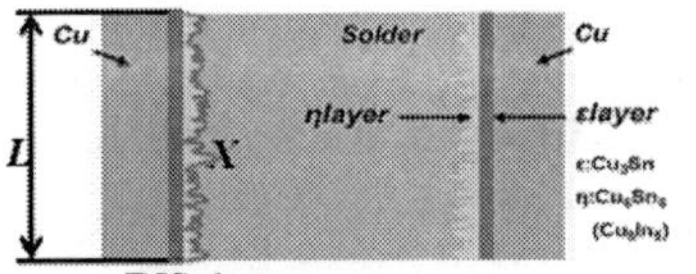

$$\frac{\text{IMC Area}}{L} = \text{IMC thickness}$$

Fig.2 Formula of the thickness of IMC and Microstructure schematic

2.3 Current load experiment

We did current load experiment to study the influence of the current load on the IMC with the constant current power supply and a thermostatic bath. Current was loaded in the thermostatic bath. In the case of a relatively low temperature, room temperature, a temperature-control device was invented and manufactured in this laboratory. In this device, when the temperature rise due to joule heat generated in the current load, it was restrained by placing the specimen on the cold heat sink. In addition, we have established a thermocouple at soldered joint, and measured temperature. **Fig.3** showed this device. During current load, at the predetermined time interval, specimen was taking off from device, and surface was polished to observe microstructure.

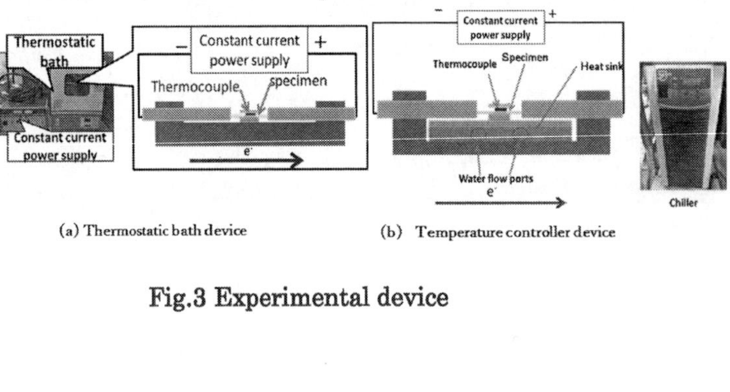

(a) Thermostatic bath device (b) Temperature controller device

Fig.3 Experimental device

3. Results and Discussion

Cross-sectional microstructure of the intermetallic compound layer in that case of using of Sn0.7Cu were shown in **Fig4**. Region surrounded by a yellow line in these figures were IMC. **Fig.5** showed the effect of treatment time and temperature on IMC thickness. **Fig.6** showed the effect of treatment temperature on IMC thickness change rate. **Fig.7** showed the effect of current density and temperature on IMC thickness.

From **Fig.4a**, and **Fig.5** IMC thickness at the anode side was increased with increasing holding time at 100°C. IMC thickness at the cathode side was decrease slightly with increasing holding time. The anode side IMC was changed to layered microstructure from scalloped structure, the thickness of the IMC was increased with increasing load time at 100°C. The current concentrated at the valley of scalloped IMC microstructure, IMC layer was considered to have grown from the valley preferentially. It was considered that the increase of the current density was caused by the defference of electrical resistance between the solder layer and the IMC layer. IMC microstructure at cathode side changed in layers of microstructure from the scalloped structure. It was considered that IMC degraded by current load.

From **Fig.4b**, in the case of a current density of 50A / mm^2 at a holding temperature 25°C, cathode side thickness was significantly decreased. From **Fig.6**, It was decreasing changes of the IMC layer with decreasing temperature. From **Fig.4c**, In the case of a current density of 100A / mm^2 at a holding temperature 25°C, cathode side thickness was decreased. Anode side thickness was increased. From **Fig.7**, the thickness of the IMC layer at cathode side was almost the same, the thickness of the IMC layer at anode side was increased, when current density increase.

Table:2 showed the rate of change of the IMC after the current load.

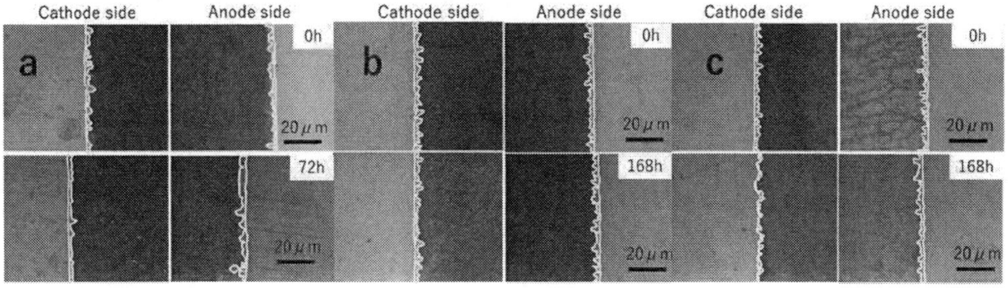

Fig.4 The cross-sectional microstructure of the soldering joint

(a) Sn0.7Cu solder. Holding temperature 100℃ current density 50A/mm^2.

(b) Sn0.7Cu solder. Holding temperature 25℃ current density 50A/mm^2.

(c) Sn0.7Cu solder. Holding temperature 25℃ current density 100A/mm^2.

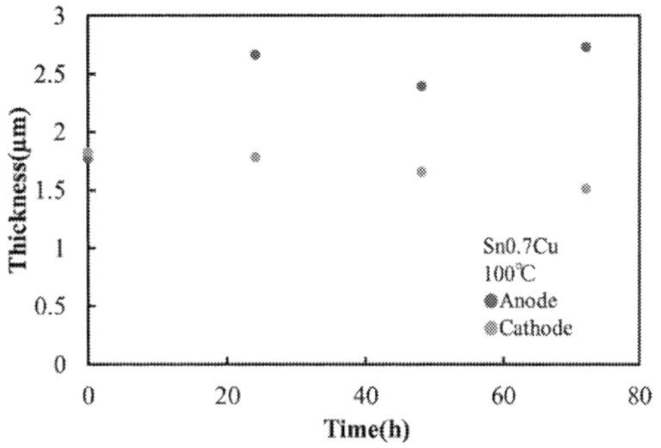

Fig.5 **the effect of treatment time and temperature on IMC thickness.**

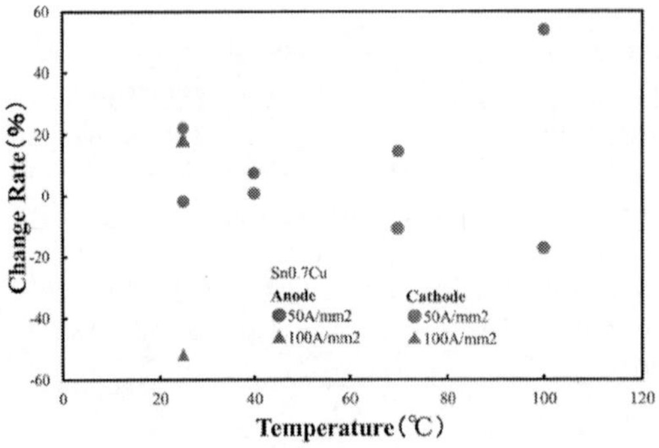

Fig.6 the effect of treatment temperature on IMC thickness change rate.

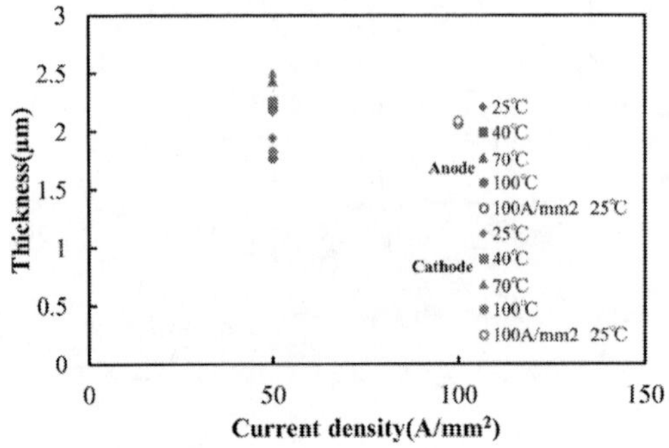

***Fig.7* the effect of current density and temperature on IMC thickness.**

Table:2 The rate of change of the IMC after the current

	100°C		25°C		25°C (336h)	
	Cathode	Anode	Cathode	Anode	Cathode	Anode
Sn	-33%	168%	-4%	24%		
Sn0.7Cu	-17%	54%	-2%	22%		
Sn0.3Ag0.7Cu	-57%	108%	-9%	-14%	-18%	6%
Sn0.7Cu1.5In	-24%	65%	6%	21%	-5%	17%
Sn0.3Ag0.7Cu1.5In	-45%	15%	-11%	12%	-17%	19%
Sn0.1Ag0.7Cu1.5In	-41%	126%				
Sn0.1Ag0.7Cu1.0In	-57%	86%				
Sn0.1Ag0.7Cu0.5In	-68%	29%				

4. Conclusions

(1) IMC layer grow preferentially from scalloped recess.
(2) Over time, each cathode side, reduction of IMC has been confirmed 100 ℃, both around 25 ℃.
(3) Appears difference in the change in thickness of the IMC by the composition of the solder.
(4) When the current density is increased, the change in the IMC is active.
(5) Changes in the cathode side of the IMC was strongly influenced current density than the temperature.

5. References

[1] Katsuaki S:Lead-free soldering technology(2001)
[2] N.Moriyama U.Tokai A doctoral dissertation Effect of current density and temperature on the alloyed layer of the soldered joint(2016)

Development of a Dynamically Evaluated Thermodynamic Property Database for Metallurgical Systems

Erik A Pfeif, Boris Wilthan, Vladimir Diky, Kenneth Kroenlein, Andre Kazakov

NIST, USA

The thermodynamic Research Center at NIST has participated in the collection, curation, and communication of thermochemical and thermodynamic data in the organics realm for the past 75 years. The result of 75 years is the largest collection of experimental thermophysical property data for organics systems including uncertainties on every measurement. Recently NIST has undertaken an effort to expand the thermodynamics database into metallic systems in an effort to aid data driven research, and also provide research tools for identifying potential research avenues, prevent unnecessary duplicative research, and provide a means for validating published results. Approximately every 16 years the total amount of data is doubling; Consequently, the importance of refined data inquiries with evaluated databases are also increasing in the age of Integrated Computational Materials Engineering (ICME) and High throughput computational methods. Furthermore, standards associated with data storage and dissemination are needed to improve the findability, accessibility, interoperability, and reusability of the data. NIST is working with the Materials Genome Initiative to improve data quality, communication, and providing data services to facilitate future alloy development, modelling, and publication quality. NIST is developing an effective XML-based communication standard, similar to ThermoML, an IUPAC accepted XML standard for organic thermodynamic data representation.

Metal Micro-Droplets Deposition Forming Technology Based on Pulsated Orifice Ejection Method (POEM)

Wei Dong, Shiwei Kang, Yuting Wei, Fumin Xu, Pengting Li
Dalian University of Technology, Dalian, Liaoning, China
w-dong@dlut.edu.cn

Akira Kawasaki
Tohoku University, Sendai, Miyagi, Japan

Abstract

Rapid prototyping technology has been developed for decades. It has an important position in manufacturing industry for its unique principle, and it has been widely applied in many fields. In this paper, a novel deposition forming technology is developed based on rapid prototyping technology in combination with mono-sized droplets prepared by Pulsated Orifice Ejection Method (POEM). Mono-sized metal micro-droplets are precisely deposited on three-dimensional platform one by one. The position accuracy of micro droplets in vertical deposition and the influence of velocity of working platform on lines deposition were carried out. The good reproducibility and accuracy of droplets deposition were proved by high speed camera and micro-structure observation. The overlapping with no voids between adjacent droplets was obtained when the speed of working platform reaches around 600μm/s in lines deposition. Based on the above results, several thin wall components with good flatness were successfully obtained.

Keywords: Metal droplets deposition forming; Pulsated Orifice Ejection Method; mono-sized metal droplets

Introduction

Rapid prototyping (RP) has been under development the past thirty years. Without mold or any mechanical equipment, RP can decrease production costs by up to 70% and the time to market by 90% [1]. Prototypes can be constructed rapidly using a layer-by-layer additive method controlled by three dimensional computer-aided model data by RP [2]. In selective laser sintering (SLS), a typical method in RP, prototypes can be produced by metallic, ceramic or thermoplastic powder using a laser to fuse the particles together. However, due to its limitations, including internal defects and mechanical properties, all parts require post-treatment [3]. Another rapid prototyping method is called droplet-based manufacturing. Materials which can be melted and ejected as droplets can be used as raw materials. Orme (1993) proposed a droplet-based method to fabricate parts by droplets deposition [4]. Yao et al, (2004) proposed a method where droplets were produced by a sinusoidal vibration on molten capillary stream, in which the flight trajectory of the droplets could be precisely controlled by charging and deflecting [5]. One common feature of all these systems mentioned above is that the droplets are produced by vibration, but none of these systems can produce uniform droplets, which can affect droplets deposition accuracy. Another droplet-based manufacturing method is under development at Dalian University of Technology [6]. In this reported process, uniform droplets are produced by the pulsated orifice ejection method (POEM) and deposited on a platform to fabricate parts controlled by three dimensional model data and a programmable multi-axes controller (PMAC). 3D metal parts composed of the Sn63-Pb37 alloy have been fabricated using this reported technique.

Experimental Procedure

The schematic diagram of the droplets-deposition apparatus based on POEM is shown in Fig.1 (a). The primary components of the apparatus were droplet generator (shown as Fig.1 (b)) and 3D platform with its controller. The droplet generator consisted of piezoelectric actuator, amplifier, crucible, rod and induction heater. The process was conducted by first heating the raw material in the crucible by the heater. After the material was melted, a select pressure of Ar gas was applied to the molten material using a pressure controller and a supply of Ar. Then a periodic vibration signal was transmitted through the amplifier to the piezoelectric actuator and rod placed in the molten material. The gas pressure and periodic movement of the rod caused droplets to be ejected from an orifice, which was located at the bottom of the crucible. As this occurred, the platform moved along its trajectory controlled by a PC and the PMAC information to ensure that droplets were deposited in accurate positions to fabricate the desired part layer by layer. The Sn63-Pb37 alloy was used in these experiments. A number of parameters were varied including heating temperature of alloy (493K), the orifice diameter (350 μm), pressure difference between the crucible and chamber (0-5.43kPa), the ejecting frequency (1-20Hz) and the deposition distance (40-80mm). All of these parameters can be adjusted to control the droplet size and productivity. The key factor for good deposition accuracy and quality is that the coordination of the droplet ejection frequency and the movement of the three dimensional platform. The movement and velocity of the three dimensional platform can be altered to match the frequency of the droplet ejection, which was all controlled by PC and PMAC in real-time.

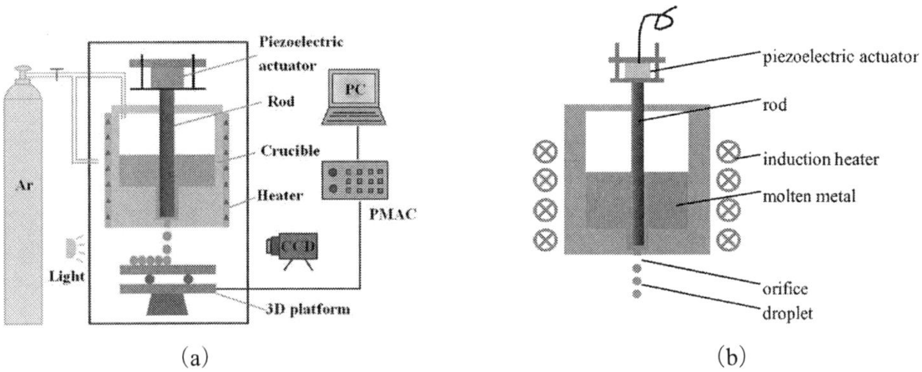

Fig.1 Schematic diagram of (a) droplets-deposition apparatus based on POEM, (b) droplet generator

Results and Discussion

The micrographs of the droplets and the particle size distribution produced by the POEM are shown as Fig. 2 (a). High deposition accuracy is a key factor in forming a good quality part. However, the actual position of the droplet deposit may deviate from the programmed position. This is caused by the deviation of droplet ejection and the accuracy in positioning the 3D platform. The deposition accuracy was determined by depositing a 5-5 dots array on the platform. The distance of each dot in each row and column in the resulting array was 10 mm when the distance between the orifice and platform was altered from 40 to 80 mm. The accuracy of the 3D platform in this experiment was ±2 μm. The deposition distance was altered and the spacing between each dot were measured.

The standard deviation of each row and column was calculated and the results are shown in Fig.2 (b). As shown, for each dot array, row and column with different distance, five data were given to compare. It can be seen from the data that the standard deviation increased with the increase of deposition distance. Most of these data were below 0.1 mm, which was smaller than the diameter of the droplet. Under these conditions, the deposition accuracy can be ensured.

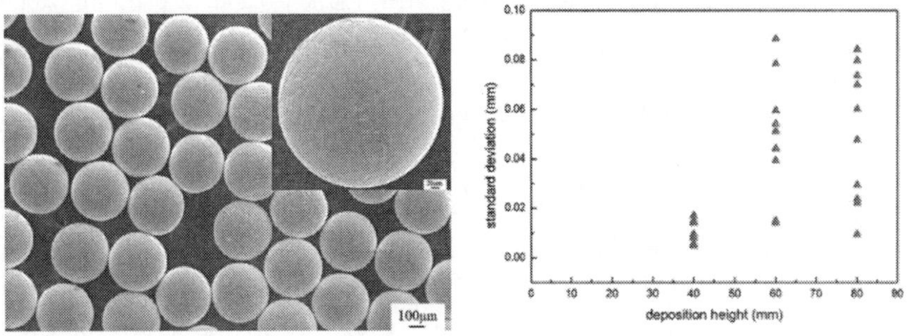

(a) (b)

Fig.2. (a)shape of uniform droplets produced by POEM and (b) the standard deviation of droplet spacing and deposition distance

It is important to synchronize the ejection frequency and the movement of platform. The platform must move appropriately in three dimensions to ensure that the droplets are placed correctly to form the desired part. The movement velocity of platform is determined by the distance of two nearest droplets and the droplet diameter. The ejection frequency and the deposition distance were set at 2 Hz and 80 mm. High-speed photographs of a line deposited with various platform velocities of platform are shown in Fig.3. The platform velocities ranged from 400 μm/s to 900 μm/s. As shown in Fig.3, the platform velocity can affect the combinations of the droplets. Higher velocities, such as 900 μm/s, led to a weak combination of droplets and the distance of the two nearest droplets was larger than the diameter of droplet. Lower velocities, such as 400 μm/s, resulted in overlap of droplets in the line, which adversely affected later deposition. Platform velocities of 500 μm/s and 600 μm/s proved to be good choices that produced good droplet combinations and advancement to the next deposition layer.

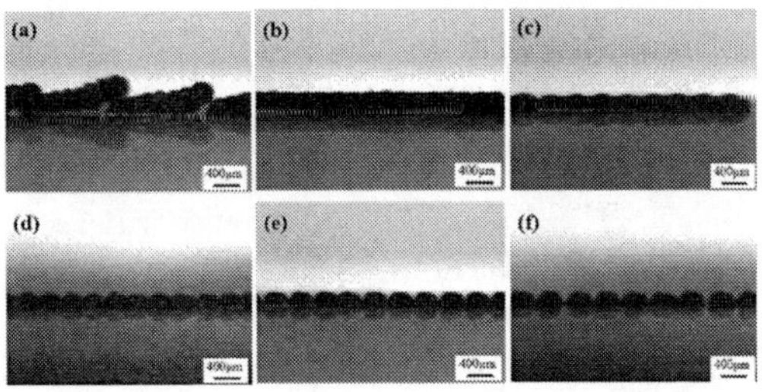

Fig.3. High-speed photographs of a droplet line deposited at different platform velocities

Micrographs of a triangular prism product and microstructure are shown in Fig.4. It can be seen in the images that the structure of the formed part resembled uniform layers of metal droplets. There were no spaces between any of these droplets, which indicated that the droplets combined evenly with the former one. A regular boundary line appeared in the structure, because of the impact and vibration when the droplet deposited on former one. This indicated that the defined experimental parameters were appropriate. However, there were still some issues in the use of this method. The parts fabricated by this method do not reach its full density and the surface quality was not very good. These drawbacks may be solved by increasing the heating temperature and post-treatment.

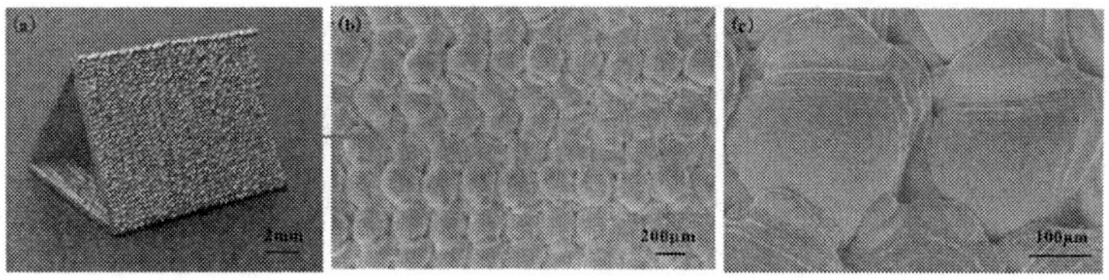

Fig.4. A triangular prism part and its microstructure image

Conclusions

This study was initiated to explore the process parameters of the POEM. The results have produced the following conclusions. (1) Based on POEM, a new droplet deposition manufacturing method was developed. This method can be used to fabricate small machine parts and thin-walled parts. (2) An experiment was initiated in this effort to determine the accuracy of droplet deposition based on the calculated standard deviation of the deposition. The results showed that the standard deviation of the droplet deposition was less than 0.1mm, which was less than the diameter of the droplet confirming the accuracy of the technique. (3) The droplet ejection frequency and movement of the receiving platform were coordinated experimentally. The optimum platform velocity was found to be 500μm/s and 600μm/s in coordination with a droplet ejection frequency of 2Hz. (4) From the microstructure of a typical machine part, it was clear that it was composed of a uniform layered structure. There were no obvious voids and a regular boundary line was present, which suggested acceptable deposition quality.

Acknowledgments

This work was supported by "the Fundamental Research Funds for the Central Universities", DUT14QY45.

References

[1] N.A. Wateraman: "Rapid Product Development in the USA, Europe and Japan," *World Class Design to Manufacture*, Vol. 1, No. 3, pp. 27-36.

[2] D.T. Pham: "A Comparison of Rapid Prototyping Technologies." *International Journal of Machine Tools & Manufacture*, Vol. 38, No. 10-11, pp. 1257-1287.

[3] W.B. Cao: "Freeform Fabrication of Aluminum Parts by Direct Deposition of Molten Aluminum," *Journal of Materials Processing Technology*, Vol. 173, No. 2, pp. 209-212.

[4] M. Orme: "A Novel Technique of Rapid Solidification Net-Form Materials Synthesis," *Journal of Materials Engineering & Performance*, Vol. 2, No. 3, pp. 399-405.

[5] Y.X. Yao: "Rapid prototyping based on uniform droplet spraying," *Journal of materials processing technology*, Vol. 146, No. 3, pp. 389-395.

[6] D. Lu: "Fabrication of BGA Solder Balls by Pulsated Orifice Ejection Method," *Proceedings of 14th International Conference on Electronic Packaging Technology*, Dalian, Liaoning, August 2013, pp. 694-697.

Author Index

Author index (A-D)

A

B

C

D

Author index (D-H)

Author index (H-K)

I

J

K

Author index (K-L)

L

Author index (L-M)

M

Author index (N-P)

N

O

P

Author index (P-S)

Author index (S-T)

T

*[1] Kyushu University *[2] Osaka University

Author index (U-Y)

U

V

W

X

Y

Author index (Y-Z)

Z